Symbols for Amino Acids

Ala	alanine
Arg	arginine
Asn	asparagine
Asp	aspartic acid
Cys	cysteine
Gln	glutamine
Glu	glutamic acid
Gly	glycine
His	histidine
Ile	isoleucine
Leu	leucine
Lys	lysine
Met	methionine
Phe	phenylalanine
Pro	proline
Ser	serine
Thr	threonine
Trp	tryptophan
Tyr	tyrosine
Val	valine

Greek Alphabet

Lower Case	Capital	Name
α	A	alpha
β	B	beta
γ	Γ	gamma
δ	Δ	delta
ϵ	E	epsilon
ζ	Z	zeta
η	H	eta
θ	Θ	theta
ι	I	iota
κ	K	kappa
λ	Λ	lambda
μ	M	mu
ν	N	nu
ξ	Ξ	xi
o	O	omicron
π	Π	pi
ρ	P	rho
σ	Σ	sigma
τ	T	tau
υ	Υ	upsilon
ϕ	Φ	phi
χ	X	chi
ψ	Ψ	psi
ω	Ω	omega

Chemical Abbreviations

Ac	acetyl, $CH_3\overset{O}{\underset{\|}{C}}-$
Boc	t-butoxycarbonyl, $(CH_3)_3CO\overset{O}{\underset{\|}{C}}-$
n-Bu	n-butyl, $CH_3CH_2CH_2CH_2-$
t-Bu	t-butyl, $(CH_3)_3C-$
Cbz	benzyloxycarbonyl, $C_6H_5CH_2O\overset{O}{\underset{\|}{C}}-$
DCC	dicyclohexylcarbodiimide, $C_6H_{11}N{=}C{=}NC_6H_{11}$
DIBAL	diisobutylaluminum hydride, $[(CH_3)_2CHCH_2]_2AlH$
diglyme	bis(2-methoxyethyl) ether, $(CH_3OCH_2CH_2)_2O$
DMF	dimethylformamide, $(CH_3)_2NCHO$
DMSO	dimethyl sulfoxide, $(CH_3)_2SO$
DNP	2,4-dinitrophenyl, $2,4-(O_2N)_2C_6H_3-$
Et	ethyl, CH_3CH_2-
glyme	1,2-dimethoxyethane, $CH_3OCH_2CH_2OCH_3$
HMPT	hexamethylphosphoric triamide, $[(CH_3)_2N]_3PO$
LAH	lithium aluminum hydride, $LiAlH_4$
LDA	lithium diisopropylamide, $LiN[CH(CH_3)_2]_2$
Me	methyl, CH_3-
PPA	polyphosphoric acid
THF	tetrahydrofuran, $\overline{CH_2CH_2CH_2CH_2O}$
TMS	tetramethylsilane, $(CH_3)_4Si$
Ts	p-toluenesulfonyl, p-$CH_3C_6H_4SO_2-$

Equilibria and Free Energy

$$A \rightleftharpoons B \qquad K = \frac{[B]}{[A]}$$

$$\Delta G^\circ = -RT \ln K$$

K	Percent B	Percent A	ΔG°, kcal mole^{-1} (25°)
0.0001	0.01	99.99	+5.46
0.001	0.1	99.9	+4.09
0.01	0.99	99.0	+2.73
0.1	9.1	90.9	+1.36
0.33	25	75	+0.65
1	50	50	0
3	75	25	−0.65
10	90.9	9.1	−1.36
100	99.0	0.99	−2.73
1000	99.9	0.1	−4.09
10000	99.99	0.01	−5.46

Introduction to Organic Chemistry

Introduction to Organic Chemistry

SECOND EDITION

Andrew Streitwieser, Jr.
Clayton H. Heathcock

UNIVERSITY OF CALIFORNIA, BERKELEY

Macmillan Publishing Co., Inc.
NEW YORK
Collier Macmillan Publishers
LONDON

Earlier edition © 1976 by Macmillan Publishing Co., Inc. Selected illustrations have been reprinted from *Orbital and Electron Density Diagrams: An Application of Computer Graphics,* by Andrew Streitwieser, Jr., and Peter H. Owens, copyright © 1973 by Macmillan Publishing Co., Inc.

MACMILLAN PUBLISHING CO., INC.
866 Third Avenue, New York, New York 10022

COLLIER MACMILLAN CANADA, LTD.

Library of Congress Cataloging in Publication Data

Streitwieser, Andrew, (date)
Introduction to organic chemistry.

"The chemical literature": p.
Includes index.
1. Chemistry, Organic.   I. Heathcock, Clayton H., joint author. II. Title.
QD251.2.S76 1981   547   80-10305
ISBN 0-02-418050-5 (Hardcover Edition)
ISBN 0-02-978310-8 (International Edition)

Printing: 3 4 5 6 7 8   Year: 2 3 4 5 6 7

Preface

THE response to the first edition of our *Introduction to Organic Chemistry* has been gratifying. However, organic chemistry, like all science, has continued to develop at a rapid pace, and there are subjects that have matured during the last five years to a point where they deserve coverage. Consequently, we have been encouraged to undertake this revision. In the process, we have been able to smooth out a few rough spots and also to implement several helpful suggestions from users of the first edition.

Those familiar with the first edition will notice a change in the organization in this version. Underlying the first edition was the premise that the important bio-organic topics, carbohydrates and peptides, should be treated as soon as possible. This required deferring most of aromatic chemistry. Now, since virtually all suggestions we received were for unifying the treatment of aromatic chemistry at the expense of deferring the treatment of carbohydrates and peptides, we have put more aromatic chemistry early in this edition. All of the important chemistry of the aromatic ring is now introduced in Chapters 22 and 23, following the general treatment of conjugated systems in Chapter 20. This change allows us to treat the chemistry of aliphatic and aromatic amines in a unified manner in Chapter 24, which leads naturally into Chapter 25, a discussion of other nitrogen-containing functional groups, including the important diazonium compounds.

A second significant change from the first edition is our treatment of alicyclic compounds. We originally deferred this subject until Chapter 23, to get into reactions as soon as possible. However, many users prefer to deal with cyclohexane conformations and the subject of ring strain early, along with the conformations and thermodynamics of acyclic hydrocarbons. Accordingly, we have redistributed the material of the old Chapter 23 as appropriate in the revision.

A third important change in this edition is the placement of organometallic compounds. This topic has been deferred and now appears as Chapter 15, where we can reasonably include the most important organometallic reaction, the Grignard synthesis of alcohols. We have also introduced transition metal organometallic chemistry in this chapter. This is an area of almost explosive current growth, and a well-educated chemistry student of the 1980s must be exposed to it.

Another area in which organic chemistry is undergoing a revolution is spectroscopy. In the last five years, carbon nuclear magnetic resonance (cmr) has become as important to the practicing chemist as proton nuclear magnetic resonance. Consequently, we have expanded our discussion of cmr in Chapter 9 and have included sections on the characteristic cmr properties of alcohols and ethers (Chapter 10), alkenes (Chapter 11), and aromatic compounds (Chapter 22).

Other, less significant, organizational changes have been made in the interest of more even distribution of material. For example, the original Chapter 2, which we perceived to be a review of freshman chemistry, has been split into two chapters. The new Chapter 2 concerns itself with the simple ideas of bonding and orbitals, and *is* a review of topics the student will be familiar with. The new Chapter 3 treats the general topic of organic structures, with emphasis on the concept of molecular shape, and contains an introduction to systematic nomenclature. Chapter 4, which introduces the student to the important concepts of organic reactivity,

v

has been expanded to include a section on acids and bases. Although this subject is covered in detail in freshman chemistry courses, its importance in organic chemistry justifies its review at an early point. The chemistry of sulfur and phosphorus compounds has been excised from the old Chapters 11 and 18 and is now presented in a somewhat expanded form as Chapter 26. The new chapters on alcohols and ethers (Chapter 10) and derivatives of carboxylic acids (Chapter 19) are now shorter, as requested by many of our advisors.

The perspective diagrams of electron density and wave functions, which were unclear to many students, have been replaced with another type of computer rendering. The new diagrams, though more traditional in appearance, are also based on actual calculations and should help students to distinguish between reality and symbols.

We have retained the three-dimensional (stereoscopic) diagrams and have added a few more. We are aware that many readers have had difficulty in obtaining suitable lenses for three-dimensional viewing. Those who have been able to perceive the three-dimensional images have been enthusiastic in their responses to these figures. It turns out that, with practice, it is possible for some people to see a stereo image without special viewers (see D. W. Russel, *Chemistry in Britain*, **13**, 354 [1977]). Nevertheless, it has been our experience that most students can visualize the stereo image only with the aid of a viewer. An inexpensive cardboard viewer is available from the Taylor-Merchant Corporation, 25 West 45 Street, New York, NY 10036. Most bookstores will not stock these viewers unless the Professor in a course requests it. Therefore, instructors should remember to include the stereo viewers on their textbook requisition lists along with such other optional items as molecular model kits.

By popular request we have added exercises within chapters, putting them between sections so as not to inhibit the flow of study. These exercises, usually in the form of drill questions, enable the student to review a section before going on to the next topic. The problems at the end of each chapter also include some drill questions. However, the problems sections also contain a number of thought questions that (a) will assist the student in assimilating new knowledge by using several pieces of information to solve a problem or (b) serve to introduce material that is a modest extension of concepts discussed in the chapter. Within each problems section, the first few questions are usually drill, and the remainder are arranged more or less in order of increasing difficulty. Supplemental problems for each chapter are included in the Student Study Guide, which has been revised and greatly expanded by Professor Paul A. Bartlett. The new guide also includes outlines and highlights for each chapter, study hints for the student, and worked-out answers to the exercises and problems in the text.

In developing the new edition we are indebted to a number of individuals, both instructors and students, for corrections, comments, and suggestions. Especially helpful suggestions came from the following individuals, to whom we are grateful: Professor Joseph Casanova (California State University, Los Angeles), Professor Tom Clark (Humboldt State University, Arcata), Professor Guido Daub (University of New Mexico), Professor Gordon Evans (Tufts University), Professor Ed Harris (California State University, Long Beach), Mr. David Hom (Brooklyn, New York), Professor Andy Kende (University of Rochester), Professor Steve Kent (Rockefeller University), Professor Henry Klostergaard (California State University, Northridge), Professor Heinz Koch (Ithaca College), Professor Nelson Leonard (University of Illinois), Professor Jerrold Lokensgard (Lawrence University, Professor James Moore (University of Delaware), Professor George Olah (University of Southern California), Ms. Ila Ottinger (Cheney, Washington), Professor Warren Smith (University of Alaska), Professor Thomas Tidwell (University of Toronto), Mr. William Tumas (Ithaca, New York), and Professor Frederick Ziegler (Yale University). In addition, we are indebted to David Grier for new figures developed with the use of the QCPE program of Professor William Jorgensen of Purdue University, and to Steve Young and Todd Blumenkopf for assistance in checking some reactions and running a number of spectra for the new edition. Finally, we thank Mary Browne, Marcy Yim, Cheri Hadley, and Wendy Zukas for their secretarial skills.

ANDREW STREITWIESER, JR.
CLAYTON H. HEATHCOCK

Berkeley, CA

Contents

Stereoscopic Representations

A Note to the Student

BEFORE you begin your adventure in organic chemistry it is perhaps appropriate for you to take a few minutes to plan your journey. The first chapter of this book provides a succinct history of the development of chemical science up to the beginnings of organic chemistry in the middle of the last century. Immediately following is a brief review of the important concepts of orbitals and chemical bonds. Although Chapter 2 is a review of topics you learned in your general chemistry course, it is essential that you be familiar with this material before proceeding with your study. Therefore, take an hour or so to go over this chapter and work the problems, even if your instructor does not specifically repeat the review material in lecture.

Chapters 3 and 4 are intended to introduce you to the two important aspects of organic chemistry—structures and reactions. Although some general chemistry courses will have covered the subject matter of these two chapters, many will not. Again, you should be thoroughly acquainted with the material in Chapters 3 and 4 before going further with your studies.

In Chapter 5 you will encounter the simplest organic compounds, those made up solely of carbon and hydrogen. This chapter also introduces two basic principles—thermodynamics and conformations. In Chapter 6 you will find the first detailed study of an organic reaction, free radical halogenation, and you will be able to put into practice the general ideas of reaction mechanisms and thermodynamics presented in Chapters 4 and 5. Chapter 7 will introduce you to a fascinating topic—stereochemistry. This special aspect of molecular structure is of fundamental importance to organic chemistry and to biochemistry. Although you may find thinking in three dimensions difficult at first, practice pays off. Once you can freely visualize organic compounds as three-dimensional objects just like the familiar objects of your everyday life, you will discover that organic chemistry is suddenly "much easier" than you thought.

The displacement reaction mechanism, which is treated in Chapter 8, is one of the fundamental mechanisms of organic chemistry. It is important that you grasp the generality of this mechanism because you will find that the same relationships of structure to reactivity recur over and over again in organic chemistry. By acquiring an early understanding of the principles of the displacement reaction, you can avoid mindlessly memorizing dozens of reactions; you will be able to recognize that many "new" reactions are only different versions of reactions you already understand.

At Chapter 9 you will shift gears and turn from chemical reactivity back to chemical structure. In this chapter you will have your first encounter with spectroscopy, which is the way organic chemists find out how the atoms of a molecule are joined together. In all, four chapters of the book are devoted to various forms of spectroscopy—Chapter 9 (nuclear magnetic resonance spectroscopy), Chapter 14 (infrared spectroscopy), Chapter 17 (mass spectrometry), and Chapter 21 (ultraviolet spectroscopy). Each kind of spectroscopy gives us different pieces of the molecular jigsaw puzzle, but nuclear magnetic resonance spectroscopy is the most important.

As you progress in your study of organic chemistry, you will discover that the science is conveniently organized in terms of *functional groups*—the parts of organic compounds other than the carbon–carbon and carbon–hydrogen single bonds that are common to all organic structures. Several chapters systematically treat the chemistry of various functional group classes of organic compounds— alcohols and ethers (Chapter 10), alkenes (Chapter 11), alkynes (Chapter 12), aldehydes and ketones (Chapter 13), organometallic compounds (Chapter 15), carboxylic acids (Chapter 18), and derivatives of carboxylic acids (Chapter 19). In each of these chapters the topical sequence is similar.

1. The functional group itself—its characteristic geometry and its effect on the geometry of the hydrocarbon part of the molecule containing it.
2. How compounds of the class are named.
3. The common physical properties of the class of compounds under consideration, including characteristic spectral properties.
4. The chemical reactions that are characteristic of the functional group, to which the bulk of the chapter is devoted.

In most organic reactions one functional group is typically transformed into another. Thus, you will find that an organic reaction can usually be thought of both as a characteristic of a given class of compounds and as a characteristic method of preparation of another class of compounds. In this book you will find that reactions are generally introduced as a characteristic property of a class of compounds. However, each functional group chapter also contains a section on preparative methods. In general, the emphasis in these preparation sections is on the practical aspects of the reactions rather than on the mechanistic aspects.

Chapter 16 is rather different from the other chapters in the book in that it is essentially a review of the organic chemistry learned up to that point. In some ways, learning organic chemistry is like learning a language. The simple reactions and mechanistic principles are like the vocabulary of the language. As in learning a language, you must first learn the vocabulary. However, if you only know the words of a language you will not be able to compose a poem, or even rent a hotel room with hot and cold running water. It is necessary also to learn how the words are put together to make sentences—the grammar and syntax of the language. In organic chemistry we learn to put several simple reactions together to achieve an overall transformation that cannot be accomplished by any single reaction. Chapter 16 will give you an opportunity to practice multistep synthesis using the reactions you have learned.

In the first half of the book, you will consider the typical chemical properties of molecules having a single functional group. In Chapter 20 you will discover that compounds having two functional groups can have properties that are very different from those of compounds having only one of the groups. This study of "conjugated systems" is fundamental to the study of ultraviolet spectroscopy (Chapter 21) and aromatic chemistry (Chapters 22 and 23). The remainder of the book contains a good deal of chemistry of such "polyfunctional" compounds. For example, Chapter 27 treats the special chemistry of compounds with two oxygen-containing functional groups. Chapters 28 and 29 cover the chemistry of two important families of polyfunctional "natural products"—carbohydrates and amino acids.

Up to Chapter 24 your study of organic chemistry will have dealt with compounds made up mainly of carbon, hydrogen, oxygen, and the halogens. In Chapters 24–26 you will encounter organic compounds of nitrogen, phosphorus, and sulfur. Of these compounds, the amines (Chapter 24) are the most important, but the other nitrogen functions (Chapter 25) are also important, especially in aromatic chemistry. Although your instructor may choose to omit Chapter 26 (sulfur and phosphorus compounds), if you are bound for a career in medicine or in the health sciences, you will want at least to read through this chapter.

Chapters 30–32 deal with further aspects of aromatic chemistry. The astute student will recognize that there are virtually no new concepts in these three chapters; rather, they serve to add flesh to the bones of the subject. However, it is interesting flesh, and the future chemical engineer or physician will find in these chapters many hints of things to come.

The final two chapters of the book are optional reading. Chapter 33 is an introduction to the literature of chemistry. Although you may not need to use the chemical literature at this point in your career, many of you will need this knowledge later. Chapter 33 will give you a start at the appropriate time. Chapter 34 is a collection of brief essays on topics somewhat beyond the scope of a general introduction. These essays are provided to give the interested student a glimpse of some of the exciting areas of modern research.

It is also appropriate at this point to mention several tools we have provided to assist you in learning organic chemistry. The first is the "indented sections," which are in smaller type and, as a further aid to their recognition, are set apart by brackets at the left and right. These sections, which are found at various points within each chapter, contain several types of information. Some give more detailed information on the topic immediately preceding. Others contain specific reaction conditions for a reaction that has been used for an example. Still others convey information of interest about specific compounds, often inorganic compounds that are employed as reagents in organic chemistry. These indented sections are set apart so that they may be skipped over by the student who is just reviewing the important principles of the chapter. Our rule of thumb at Berkeley is that the material in these sections is for enrichment, and that students are not held responsible for it on examinations. You should ask your instructor about the policy in your course.

A second invaluable tool is the exercises and problems in each chapter. The exercises, which are at the ends of most sections, are cast mainly in the form of "drill" to provide you with immediate practice in using new principles or reactions you have just learned. For many of you, these exercises will seem ridiculously easy, as you will be asked to write out an equation you have just learned. However, they are an important part of the learning process. Everyone has had the experience of "daydreaming" while reading merrily along. It is possible to read several pages and be totally unconscious of what you have read. The exercises force you to pause periodically and check to see that you have really been assimilating what you have been reading. The problems at the end of each chapter also contain some drill questions, but the parts of a single question may draw from many different sections of the chapter. Thus, these questions provide for a second check on your retention of the various reactions and principles you have studied. There are also "thought questions" that ask you to take several reactions or principles and put them together to solve a problem or in some cases to extend your knowledge and discover something for yourself.

To enable you to derive the greatest benefit from the exercises and problems, we have prepared a Student Study Guide. This paperback book contains worked-out answers to all of the exercises and problems as well as a key-word index and study hints for each chapter. In addition, the study guide contains supplemental problems, with answers. It is important that you give any problem a good try before looking up the answer. It is human nature to quit worrying about a problem as soon as the answer is known, and it is also true that we learn more from a problem we have labored over than from one we haven't given much thought to.

One teaching device we have used requires special mention—the three-dimensional stereoscopic projections that are distributed throughout the book (see list, page xvii). These computer-generated images are designed so that you may see the figure in three dimensions. By using a suitable viewer one may cause each eye to focus independently on one of the two images of such a projection, and there is an illusion of depth to the resulting picture one sees. An inexpensive cardboard viewer is available from the Taylor-Merchant Corporation, 25 West 45 Street, New York, NY 10036. Most bookstores will not stock these viewers unless the Professor in a course requests it. If your bookstore does not have a supply of viewers, ask that they be ordered. It is actually possible, albeit a bit more difficult, to see a stereo image without the special viewer. To do this, hold the page about 20 inches from your eyes and focus on a point behind the book in such a way that the two images merge. Generally, the merged image will suddenly seem three-dimensional. One caution: with this method the right eye will sometimes focus on the left image and vice versa. The result is the perception of the mirror image, a concern when precise stereochemistry is important, as in Chapter 7.

With these general suggestions in mind, it remains only for us to wish you luck as you set out upon your journey through organic chemistry. Both of us look back with fond remembrance upon our own discovery of this fascinating subject; we hope that you will find it as rewarding.

ANDREW STREITWIESER, JR.
CLAYTON H. HEATHCOCK

Chapter 1
Introduction

Although chemistry did not emerge as a coherent science until the seventeenth century, its roots extend back into antiquity. Chemical changes were probably first brought about by paleolithic man when he discovered that he could make fire and use it to warm his body and roast his food. Being a curious and a resourceful creature, man observed and exploited other natural phenomena. By neolithic times he had discovered such arts as smelting, glass making, the dyeing of textiles, and the manufacture of beer, wine, butter, and cheese.

Matter and changes of matter were not systematically discussed in a theoretical sense until the period of the Greek philosophers, beginning in about 600 B.C. The popular theory that emerged during this period saw all matter as being made up of the four "elemental" substances: fire, earth, air, and water. For a time, the atomist school, of which Democritus was the chief spokesman, gained popularity. In this theory, all matter was considered to be made up of hypothetical particles called atoms, of which there were assumed to be but a finite number of different kinds. Although the atomists held sway for several centuries, the notion was highly speculative, being based on nothing directly observable. The demise of this theory was foreshadowed when it was rejected by the highly respected Aristotle; its burial was assured with the advent of stoicism and the subsequent rise of the popular religious movements in the Western world. The idea of fundamental particles was not resurrected for almost two millenia.

Around the time of Christ, the Greek philosophers hit upon the idea of changing (or "transmuting") base metals such as lead and iron into gold and silver. Although alchemy was first practiced in a serious sense by the Greeks, it quickly spread to other cultures and continued as a lively discipline throughout the world for over a thousand years. This alchemical period has often been put down as a "dark age" of science. However, one must recognize that there is nothing inherently wrong with the notion that one metal may be transformable into another. Chemistry is, in fact, based upon changes in the state of matter. The alchemists had no way of recognizing the elemental nature of the metals with which they dealt.

Although they were uniformly unsuccessful in their quest for the philosopher's stone, the alchemists contributed a great deal to the technology of handling matter. Not only did they develop numerous processes for the production of relatively pure compounds but they also invented tools and apparatus, many of which persist in similar form to the present day—beakers, flasks, funnels, mortars, crucibles. Perhaps the most important invention of alchemy was the **still.** The important technique of distillation was probably discovered by the early Greek alchemists when they noticed condensate on the lid of a vessel in which some liquid was being heated. It was only a short step from this observation to the realization that this technique could be used to separate volatile substances from nonvolatile animal and vegetable matter. Although the still was quite inefficient in its infancy, its design improved steadily. By 1300 actual fractionation was being practiced, and alcoholic distillates of fairly high alcohol concentration were available. The production of whiskey and brandy became an established industry in short order.

The invention and development of the still by the alchemists had an interesting consequence in another area—medicine. Through the Middle Ages, medicine was practiced as a mystical blend of magic and folklore. It had long been noticed that certain animal and plant substances seemed to possess curative powers. With the advent of the still, it became possible to concentrate the "essence" of various natural materials. The use of various distillates as medical remedies quickly became a widespread practice. For several hundred years, physicians and their associates distilled all manner of natural substances. In the process, a number of relatively pure organic compounds were isolated, such as acetic acid from vinegar and formic acid from ants.

During this pre-1600 period, as the tools for handling matter were being developed and as numerous relatively pure chemical substances were being discovered, there was relatively little serious experimentation and no advance at all in the theory of matter. However, during the seventeenth and eighteenth centuries, chemistry was born as a science in Europe. The first area of serious investigation was gases. Although Boyle, Cavendish, Priestley, and Scheele made important breakthroughs, it was Lavoisier who laid the real foundation for modern chemistry. During this period, there evolved the notion of elements and combining weights. By 1789, Lavoisier had assembled a Table of the Elements, containing 33 substances, most of which appear in the modern periodic table.

In this formative stage in the science of chemistry, the substances derived from the animal and vegetable worlds were largely ignored. These materials were recognized as being different—more complex—than the compounds of the atmosphere or those compounds derived from the mineral kingdom. Lavoisier himself noted that **organic compounds,** as they came to be known, differed from the inorganic compounds in that they all seemed to be composed of carbon and hydrogen and occasionally nitrogen or phosphorus. For a time it was thought that organic compounds did not obey the new law of definite proportions, and people came to believe that a **vital force,** present only in living organisms, was responsible for the production of organic compounds.

The vitalism theory persisted until the middle of the nineteenth century. In 1828 Frederick Wöhler, working in Heidelberg, reported that, upon treating lead cyanate with ammonium hydroxide, he obtained urea. Since urea was a well-known organic compound, having been isolated from human urine by Roulle in about 1780, Wöhler had succeeded in preparing an organic compound in the laboratory for the first time. Although the synthesis of urea was recognized by the leading chemists of the day, the concept of vitalism did not die quickly. It was not until the synthetic work of Kolbe in the 1840s and Berthelot in the 1850s that the demise of vitalism was complete.

At this time, chemists recognized that it was not the vital force which imparted uniqueness to organic chemistry but rather the simple fact that organic compounds are all compounds of carbon. This definition—organic chemistry is the chemistry of carbon compounds—has persisted.

Simultaneously with the discovery of methods for the laboratory preparation of a multitude of organic compounds, analytical methods were also being perfected. With the advent of these methods, particularly the technique of combustion analysis, organic chemistry began to take on new dimensions. For the first time accurate formulas were available for fairly complicated organic compounds. There ensued a confusing period, which lasted from about 1800 until about 1850, during which various theories were advanced in an attempt to explain such complexities as isomerism (the existence of two compounds with the same formula) and substitution (the substitution of one element for another in a complex organic formula).

Organic chemistry began to emerge from this chaotic period in 1852 when Frankland advanced the concept of valence. In 1858, Kekulé and Couper, working independently, introduced a simple, but exceedingly important, concept. Making use of the new structural formulas shown, which had come into vogue

$$\left.\begin{array}{c}H\\H\\H\end{array}\right\}N \qquad \left.\begin{array}{c}H\\H\end{array}\right\}O \qquad \left.\begin{array}{c}C_2H_5\\H\end{array}\right\}O \qquad \left.\begin{array}{c}C_2H_5\\C_2H_5\end{array}\right\}O$$

ammonia   water   alcohol   ether

since 1850, Kekulé and Couper proposed that the carbon atom is always tetravalent and that carbon atoms have the ability to link to each other.

A third event, which ushered organic chemistry into its modern period, was the demonstration by Cannizzaro in 1858 that Avogadro's hypothesis, available since 1811, allowed the determination of accurate molecular weights for organic compounds. With this last piece of the structural puzzle available, it was possible to think in terms of molecular structure and the chemical bond. Kekulé introduced the idea of a bond between atoms and depicted it with his "sausage formulas" in the first edition of his textbook in 1861.

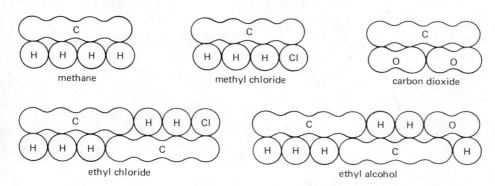

methane   methyl chloride   carbon dioxide

ethyl chloride   ethyl alcohol

In the century since Kekulé, organic chemistry has matured as a scientific discipline in its own right. Well over 95% of all known chemical compounds are compounds of carbon. Over one half of present day chemists classify themselves as organic chemists. The organic chemical industry plays a major role in world economy. Finally, because organic chemicals are literally the "stuff of life," the significant advances in unravelling the nature of life are discoveries in organic chemistry.

Why study organic chemistry? There are different answers to this question depending upon who you are. It may be that you will devote your life to a career in organic chemistry *per se*, although if this is the case, you probably do not know it yet. Or you may plan to specialize in some other area of chemistry and want a knowledge of organic chemistry as an adjunct to your specialty area. You may be a future chemical engineer; if so, organic chemistry will be an important part of your life, since most of the industrial processes you will encounter will be organic reactions. If you are headed for medicine or nursing, simply note that pharmaceutical products amounted to over $8 billion in 1978 and that chemotherapy is one of the major techniques in modern medicine. You may be going into biochemistry, molecular biology, or some other life science to which organic chemistry is essential; biochemistry is simply a study of organic chemistry as it goes on in living organisms, and the molecules of molecular biology are organic molecules.

Even if you do not have any of the foregoing "reasons" to study organic chemistry, there are many other justifications. The previous arguments are vocational motivations. But organic chemistry can also provide a fascinating area of "natural philosophy" for the student who wants only to obtain a broad liberal arts education. If you approach the subject in the proper frame of mind, you will find it to be an extremely stimulating intellectual pursuit. Organic chemistry has a highly logical structure. As you will see, we make much use of symbolic logic, the logical principle of analogy, and deductive reasoning. In fact, it has been intimated that medical school admission boards value organic chemistry courses as much for the test in logical thinking that they provide as for their factual content.

Finally, organic chemistry has a unique content as an *art form*. The building up of complex molecular architecture by appropriate choice of a sequential combination of reactions provides syntheses that are described as "elegant" and "beautiful." The design of an experiment in reaction mechanism can be similarly imaginative. Such elegantly conceived experiments can evoke that delightful feeling of pleasure that one obtains from the appreciation of human creativity—but only in the mind of the knowledgeable spectator. These unique works of art can only be appreciated by those who know some organic reactions and have tried to design some simple syntheses and experiments themselves, such as those suggested in problem sets throughout this textbook. Only one who has played chess can feel that special pleasure of following a game between Grand Masters.

Chapter 2
Electronic Structure and Bonding

2.1
Periodic Table

The periodic table of the elements was developed just over 100 years ago. At that time it was an empirical organization based on the chemical and physical properties of the known elements. The table now embraces over 100 elements. Compounds of carbon, organic compounds, are known that contain virtually all of the elements except the noble gases. However, only a small part of this organization is important in the introductory study of organic chemistry. In the condensed form of the periodic table shown in Table 2.1, the most important elements are emphasized with bold type: **C, H, N, O, S, Mg, Cl Br,** and **I.** Secondary but still important elements are in italics: *Li, B, F,* and *P.*

TABLE 2.1
Abbreviated Periodic Table

first period:	**H**							He
second period:	*Li*	Be	*B*	**C**	**N**	**O**	*F*	Ne
third period:	Na	**Mg**	Al	Si	*P*	**S**	**Cl**	Ar
							Br	
							I	

2.2
Lewis Structures

The "noble" gases, He, Ne, Ar, Kr, Xe, and Ra, are almost inert chemically. Paradoxically, it is this very inert character that dominates much of the chemistry of all the rest of the elements. The noble gases have characteristic numbers of electrons, 2 for helium, 10 for neon $(2 + 8)$, 18 for argon $(2 + 8 + 8)$, and so on. They are described as having "filled shells" or, for neon and argon, as having filled outer **octets.** Other elements can achieve such stable electronic configurations by gaining or losing electrons.

The energy required to lose an electron is known as the **ionization potential,** IP.

$$M \longrightarrow M^+ + e^- \qquad \Delta H° = IP$$

For elements at the far left of the periodic table, loss of an electron produces the electronic configuration of the next lower noble gas. Examples are

$$Li \xrightarrow{-e^-} Li^+$$
3 electrons   2 electrons
(same as helium)

$$Na \xrightarrow{-e^-} Na^+$$
11 electrons   10 electrons
(same as neon)

Such elements have relatively low ionization potentials and are described as being **electropositive.**

The energy liberated when an electron is acquired is called the **electron affinity, EA.**

$$X + e^- \longrightarrow X^- \qquad -\Delta H° = EA$$

The electron affinity of an atom is also the ionization potential of the corresponding anion. Elements at the far right of the periodic table readily acquire electrons to produce the stable electronic configuration of the next higher noble gas. Examples are

$$F + e^- \longrightarrow F^-$$

9 electrons $\qquad$ 10 electrons (same as neon)

$$Cl + e^- \longrightarrow Cl^-$$

17 electrons $\qquad$ 18 electrons (same as argon)

Such elements have relatively high electron affinities and are described as being **electronegative.**

The electrons outside the shell of the next lower noble gas are the **valence electrons** and are the only ones normally included in symbols. The above ionization processes are then symbolized as

$$Li \cdot \longrightarrow Li^+ + e^- \qquad IP = 123.6 \text{ kcal mole}^{-1}$$

$$Na \cdot \longrightarrow Na^+ + e^- \qquad IP = 118.0 \text{ kcal mole}^{-1}$$

$$:\ddot{F}\cdot + e^- \longrightarrow :\ddot{F}:^- \qquad EA = 78.3 \text{ kcal mole}^{-1}$$

$$:\ddot{C}l\cdot + e^- \longrightarrow :\ddot{C}l:^- \qquad EA = 83.3 \text{ kcal mole}^{-1}$$

Electropositive elements such as the alkali metals tend to lose electrons to electronegative elements such as the halogens to form pairs of ions. Such compounds are described as having ionic bonding. Typical examples are lithium chloride ($Li^+ :\ddot{C}l:^-$) and sodium fluoride ($Na^+ :\ddot{F}:^-$).

For elements in the middle of the periodic table, too much energy is required to gain or lose sufficient electrons to form similar octet ions. Compare the energy required to generate a triply positive boron or quadruply positive carbon with the energies required to form Li^+ or Na^+.

$$\cdot \dot{B}\cdot \longrightarrow 3 e^- + B^{3+} \qquad IP = 870.4 \text{ kcal mole}^{-1}$$

$$\cdot \dot{C}\cdot \longrightarrow 4 e^- + C^{4+} \qquad IP = 1480.7 \text{ kcal mole}^{-1}$$

Consequently, such elements tend to acquire their electron octets by *sharing* electrons, as in the following examples.

$$\begin{matrix} & H & & H & & & & H \\ H:\overset{..}{C}:H & & H:\overset{..}{N}: & & H:\overset{..}{O}:H & & H:\overset{..}{C}:\overset{..}{F}: \\ & H & & H & & & & H \end{matrix}$$

methane $\qquad$ ammonia $\qquad$ water $\qquad$ methyl fluoride

Such bonds are described as **covalent bonds.**

The symbols used to describe the foregoing examples are called **Lewis struc-**

tures. Such structures not only provide simple and convenient representations of ions and compounds but are also valuable in providing an accurate accounting for electrons. They form an important basis for predicting relative stabilities. Lewis structures are important in the study of organic chemistry and the student should be able to write them with facility. The following general rules are useful for deriving suitable structures.

1. *All valence electrons are shown.* The total number of such electrons is equal to the sum of the numbers contributed by each atom, with addition or subtraction of another number to account for any ionic charge. Some examples are worked out in Table 2.2.

**TABLE 2.2
Valence Electrons**

Species	Atomic Contributions	−	Cation Charges	+	Anion Charges	=	Total Valence Electrons
CH_4	$4(C) + 4 \times 1\,(H) = 8$	−	0	+	0	=	8
NH_3	$5(N) + 3 \times 1\,(H) = 8$	−	0	+	0	=	8
H_2O	$6(O) + 2 \times 1\,(H) = 8$	−	0	+	0	=	8
H_3O^+	$6(O) + 3 \times 1\,(H) = 9$	−	1	+	0	=	8
HO^-	$6(O) + \quad 1\,(H) = 7$	−	0	+	1	=	8
BF_3	$3(B) + 3 \times 7\,(F) = 24$	−	0	+	0	=	24
NO_2^-	$5(N) + 2 \times 6\,(O) = 17$	−	0	+	1	=	18
CO_3^{2-}	$4(C) + 3 \times 6\,(O) = 22$	−	0	+	2	=	24

2. *Each element should, to the greatest extent possible, have a complete octet.* Exceptions are hydrogen, which has a duet shell, and elements beyond the first row, such as sulfur and phosphorus, which may accommodate more than eight valence electrons ("expand their octets") in certain circumstances.

Correct Structure *Incorrect Structures*

$$:\overset{..}{O}::C::\overset{..}{O}: \qquad :\overset{..}{O}:C::\overset{..}{O}: \quad :\overset{..}{O}:C:\overset{..}{O}: \quad :\overset{..}{O}::C:\overset{..}{O}:$$

$$H:\overset{..}{Cl}: \qquad\qquad\qquad\qquad :H:\overset{..}{Cl}:$$

$$:N:::N: \qquad\qquad\qquad :\overset{..}{N}::N: \quad :\overset{..}{N}:N:$$

3. **Formal charges** *are assigned by dividing each bonding pair of electrons equally between the bonded atoms.* The number of electrons "belonging" in this way to each atom is compared with the neutral atom and appropriate positive or negative charges are assigned. Lone pairs "belong" to a single atom.

$$\begin{array}{c} H \\ H:\overset{\textstyle H}{\underset{\textstyle H}{N^+}}:H \\ \end{array} \qquad \begin{array}{c} H \\ H:\overset{\textstyle H}{\underset{\textstyle H}{C}}:\overset{..}{\underset{..}{O}}:^- \\ \end{array} \qquad \begin{array}{c} :\overset{..}{O}:^- \\ -:\overset{..}{O}:S^{2+}:\overset{..}{O}:^- \\ :\overset{..}{O}:^- \\ \end{array}$$

ammonium ion   methoxide ion   sulfate ion

In NH_4^+, each electron pair is divided between N and H. This gives one electron for each H, the same as a hydrogen atom. N has a total of 4, one less than atomic nitrogen; hence, the formal charge of +1 is associated with N. This procedure assigns the entire positive charge of $(NH_4)^+$ to the nitrogen; in practice, the electrons are spread over the entire molecule. However, this method of assigning formal charges does keep strict account of the total numbers of electrons and charges present and, when used with care, it helps to interpret chemistry. For example, the formal charge of -1 assigned to the oxygen of methoxide ion helps to explain why this ion is a strong base that readily adds a proton to the oxygen.

The example of sulfate ion is more complex. Some students tend to write this ion as $^-:\ddot{O}:\ddot{O}:\ddot{S}:\ddot{O}:\ddot{O}:^-$, an arrangement that has the proper number of valence electrons and a less complex formal structure assignment. However, sulfate ion is known experimentally to have each oxygen bound to sulfur in an equivalent manner.

When a species has an incomplete octet, it is usually unstable or highly reactive. Examples are methyl cation and methyl radical.

$$
\begin{array}{cc}
\text{H} & \text{H} \\
\text{H}:\overset{..}{\underset{+}{\text{C}}}:\text{H} & \text{H}:\overset{..}{\underset{.}{\text{C}}}:\text{H} \\
\text{methyl cation} & \text{methyl radical}
\end{array}
$$

Multiple bonds are handled in a straightforward manner.

$$
\begin{array}{ccc}
\begin{array}{cc}\text{H} & \text{H}\\ :\text{C}::\text{C}: \\ \text{H} & \text{H}\end{array} & \text{H}:\text{C}:::\text{C}:\text{H} & ^-:\text{C}:::\text{N}: \\
\text{ethylene} & \text{acetylene} & \text{cyanide ion}
\end{array}
$$

A further simplifying convention is to replace each electron-pair bond by a line. For convenience, electron pairs are frequently omitted.

$$
\begin{array}{cccc}
\text{H} & \text{H} & \text{O}^- & \text{H} \\
| & | & | & | \\
\text{H}-\overset{+}{\text{N}}-\text{H} & \text{H}-\text{C}-\text{O}^- & ^-\text{O}-\overset{2+}{\text{S}}-\text{O}^- & \text{H}-\overset{+}{\text{C}}-\text{H} \\
| & | & | & \\
\text{H} & \text{H} & \text{O}_- &
\end{array}
$$

$$
\begin{array}{cccc}
\text{H} & & & \\
| & \overset{\text{H}}{\underset{\text{H}}{\diagup}}\text{C}=\text{C}\overset{\text{H}}{\underset{\text{H}}{\diagdown}} & \text{H}-\text{C}\equiv\text{C}-\text{H} & ^-\text{C}\equiv\text{N} \\
\text{H}-\overset{.}{\text{C}}-\text{H} & & &
\end{array}
$$

In these symbolic representations, the lone-pair electrons are understood to be present and their presence is signified by appropriate formal charges. This is another reason for assigning formal charges properly. *The use of such symbols is widespread in organic chemistry, and practice in reading and writing these electronic representations cannot be overemphasized.* The simplified symbols correspond to an earlier notational system proposed by August Kekulé. Accordingly, such symbols, in which each electron-pair bond is represented by a line and the lone-pair electrons are omitted, are frequently called **Kekulé structures.**

The use of a "dative" or "coordinate covalence" bond is sometimes convenient. In this convention, an arrow represents a two-electron bond in which both electrons are considered to "belong" to the donor atom for the bookkeeping purpose of assigning formal charges.

$$
\begin{array}{ccc}
& \text{O} & \text{O} \\
& \diagup\!\!\diagup & \diagup\!\!\diagup \\
\text{H}-\text{O}-^+\text{N} & \quad\text{or}\quad & \text{H}-\text{O}-\text{N} \\
& \diagdown & \diagdown\!\!\!\rightarrow \\
& \text{O}_- & \text{O}
\end{array}
$$

This type of symbolism finds most use in representing ligands in inorganic complexes and will rarely be used in this text.

EXERCISE: Rewrite the following Kekulé structures as Lewis structures, including all of the electrons.

(a) chloride ion, Cl^-

(b) water, $H—O—H$

(c) hydroxide ion, $H—O^-$

(d) hypochlorite ion, $Cl—O^-$

(e) ozone, $^-O—O^+{=}O$

(f) ammonia, $H—\underset{\underset{\displaystyle H}{|}}{N}—H$

(g) cyanogen chloride, $Cl—C{\equiv}N$

(h) nitric oxide, NO

2.3
Geometric Structure

One of the really important achievements of the physics of a half century ago was the determination of crystal structures by x-ray diffraction. Other methods that may be used for the precise determination of molecular structures include electron diffraction and microwave spectroscopy. These experimental approaches have yielded a wealth of detailed structures at the molecular level. For example, H_2O is known to have a structure with a bent $H—O—H$ bond angle of 104.5° and an O—H bond distance of 0.96 Å ($1 Å \equiv 1 \times 10^{-8}$ cm).

It should be emphasized that water is not a rigid molecule with the atoms fixed in this geometry. The atoms are constantly in motion, even at a temperature of absolute zero. This motion is conveniently described in terms of the bending and stretching of bonds. At any instant of time, the actual O—H distance may vary from 0.96 Å by several hundredths of an Ångstrom, but the average distance will be that given. Similarly, bond angles are constantly changing, and the value given is an average value.

An important result has emerged from these many structural studies. Specific bonds retain a remarkably constant geometry from one compound to another. For example, the O—H bond distance is almost always 0.96–0.97 Å.

Compound	O—H Bond Distance, Å
HO—H, water	0.96
HOO—H, hydrogen peroxide	0.97
H_2NO—H, hydroxylamine	0.97
CH_3O—H, methyl alcohol	0.96

In fact, it is this consistency that allows us to treat the O—H bond as an individual unit in different compounds.

Lewis structures can be useful in the interpretation of bond distances. For example, the N–O bond distance is longer in hydroxylammonium ion than in nitronium ion.

$$H_3\overset{+}{N}{-\!\!-\!\!-}OH \qquad O{=}\overset{+}{N}{=}O \ \equiv\ :\overset{..}{O}::\overset{+}{N}::\overset{..}{O}:$$

$$\text{1.45 Å} \qquad\qquad \text{1.15 Å}$$

hydroxylammonium ion   nitronium ion

The hydroxylammonium ion should not be confused with ammonium hydroxide, $NH_4^+ OH^-$. The hydroxylammonium ion is ammonium ion with one N—H bond replaced by an N—OH bond.

In $HONH_3^+$, one electron pair binds the nitrogen to the oxygen, and the compound is said to have a N—O **single bond.** In NO_2^+, the Lewis structure shows that each nitrogen–oxygen bond involves *two* pairs of electrons and is therefore said to be a **double bond.** At the equilibrium bond distance the electrostatic forces between the negative electrons and positive nuclei are balanced. For the N—O single bond, two bonding electrons are involved, and the balance of attraction and repulsion occurs at an internuclear distance of 1.45 Å. For the N=O double bond, more electrons are involved with consequent greater net electrostatic attraction to the nuclei. The increased internuclear repulsion required to reach a balance occurs at the shorter distance of 1.15 Å.

EXERCISE: Considering the Lewis structure you wrote for the exercise on page 9, what N–O bond distance would you anticipate in nitric oxide?

2.4
Resonance Structures

In some cases, it is not possible to describe the electronic structure of a species adequately with a single Lewis structure. An example is nitryl chloride, NO_2Cl.

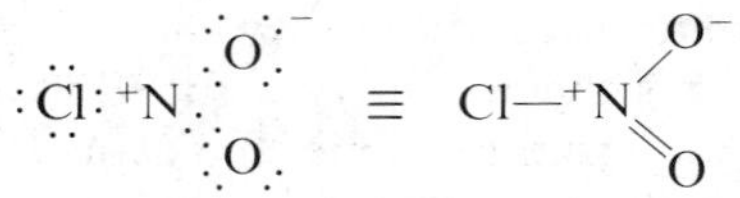

nitryl chloride

The Lewis structure shown has one N—O single bond and one N=O double bond. However, it has been determined experimentally that both N–O bonds are equivalent. Furthermore, the N–O bond distance of 1.21 Å is intermediate between the nitrogen–oxygen single- and double-bond distances described in the previous section. Actually, two alternative structures may be written for nitryl chloride. The two structures differ only in the positions of electrons.

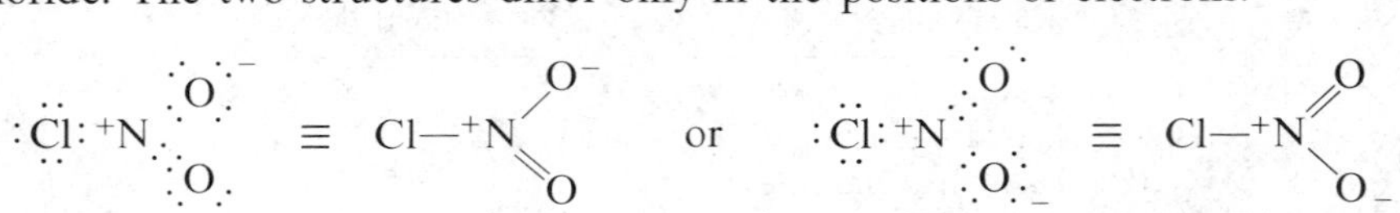

The actual electronic structure of NO_2Cl is a composite or weighted average of the two Lewis structures. The two alternative structures are called **resonance structures,** and the molecule is said to be a **resonance hybrid.**

It is important to recognize that nitryl chloride *has only one geometric structure:* that in which the two N—O bonds are equivalent. It is *not* 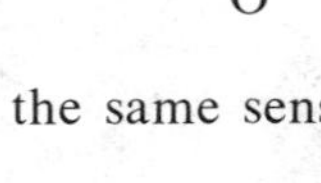 half of the time and 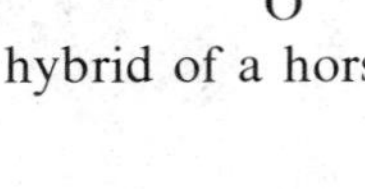 the other half. It is a hybrid in the same sense that a mule is a hybrid of a horse and a donkey. Resonance structures are necessary only because of inadequacies in our simplified system for describing bonding and electron distribution in molecules. When one conventional Lewis structure does not adequately describe what we know to be the actual structure of a species, we use two or more structures (resonance structures) for the species and bear in mind that the species has some characteristics of each structure.

Resonance structures are written with a double-headed arrow, and the resonance hybrid is frequently written with dotted lines to represent partial bonds. Even in such cases, the individual Lewis structures provide an accurate accounting of the electrons and are frequently preferred to dotted-line formulas. In the case of nitryl chloride, the Lewis structures indicate that the N–O bond is halfway between single and double, and we expect an intermediate bond distance. Because each N–O bond is single in one resonance structure and double in the other, the N–O bond in the resonance hybrid is said to have a bond order of $1\frac{1}{2}$.

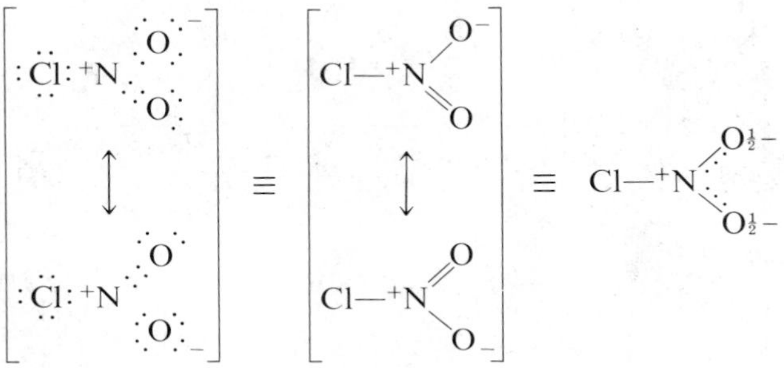

Another species that is not adequately described by a single structure is formate ion, HCO_2^-. As in the case of nitryl chloride, formate ion is a hybrid of two resonance structures.

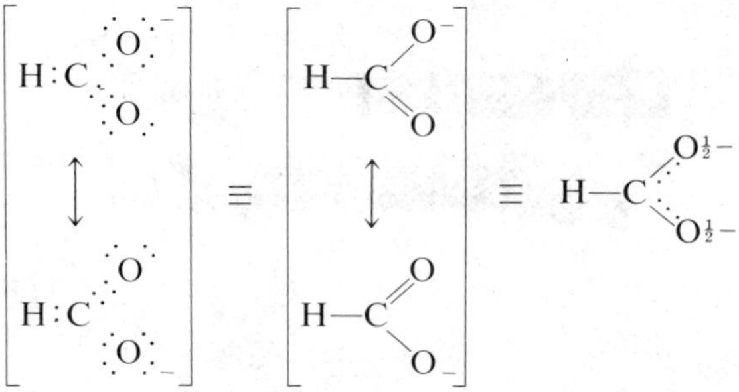

Both of the C–O bonds have a bond order of $1\frac{1}{2}$. Accordingly, the C–O bond distance of 1.26 Å is intermediate between the C=O double-bond distance of 1.20 Å in $H_2C{=}O$ and the C—O single-bond distance of 1.43 Å in $HO{—}CH_3$.

Carbonate ion, CO_3^{2-}, is somewhat more complicated in that three resonance structures are required. The resonance hybrid has three equivalent C—O bonds, each having a bond order of $1\frac{1}{3}$. Because the C—O bonds in carbonate ion (order $1\frac{1}{3}$) have more single-bond character than those in formate ion (order $1\frac{1}{2}$), they are slightly longer (1.28 Å).

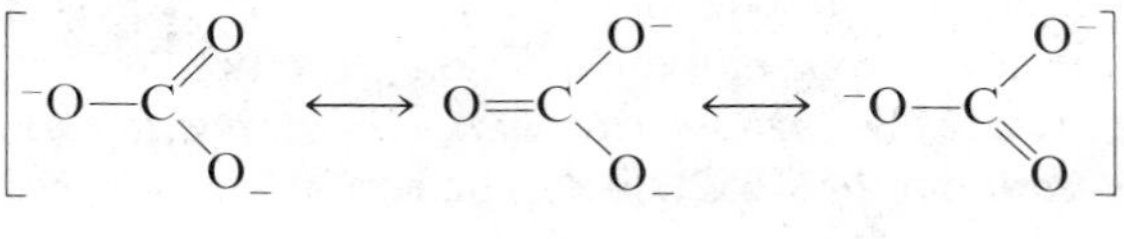

In each of the foregoing examples, the important resonance structures are equivalent. In some cases, a species is best described by two or more resonance structures that are not energetically equivalent. One such species is protonated formaldehyde, $(H_2COH)^+$. Formaldehyde itself may be represented by a Lewis structure in which there are two C—H single bonds and a C=O double bond.

$$\overset{H}{\underset{H}{\cdot}} \cdot C::\ddot{O}: \quad \equiv \quad \overset{H}{\underset{H}{\diagup}} C=O$$

formaldehyde

In protonated formaldehyde, an additional O—H single bond is present. Two Lewis structures may be written for $(H_2COH)^+$.

$$\left[\begin{matrix} \overset{H}{\underset{H}{\cdot}}\cdot C::O^+:H \\ \updownarrow \\ \overset{H}{\underset{H}{\cdot}}\cdot C^+:\ddot{O}:H \end{matrix}\right] \equiv \left[\begin{matrix} \overset{H}{\underset{H}{\diagup}} C=^+OH \\ \updownarrow \\ \overset{H}{\underset{H}{\diagup}} C^+—\ddot{O}H \end{matrix}\right] \equiv \overset{H}{\underset{H}{\diagup}} C^{\delta+}\cdots\cdots^{\delta+}OH$$

In one structure there is a C=O double bond, and the positive charge is assigned to oxygen. This **oxonium ion structure** is analogous to the hydronium ion, H_3O^+.

$$\overset{H}{\underset{H}{\cdot}}\cdot C::O^+:H \qquad\qquad H:\overset{H}{\underset{H}{\ddot{O}}}:{}^+$$

oxonium ion structure   hydronium ion

In the alternative structure there is a C—O single bond, and the positive charge is assigned to carbon. This **carbocation structure** is analogous to the methyl cation, CH_3^+.

$$\overset{H}{\underset{H}{\cdot}}\cdot C^+:\ddot{O}:H \qquad\qquad H:\overset{H}{\underset{H}{C^+}}$$

carbocation structure   methyl cation

Which structure more adequately represents protonated formaldehyde? The C-O bond length in $(H_2COH)^+$ is 1.27 Å, which is much closer to the normal C=O double-bond length of 1.20 Å than to the normal C—O single-bond length of 1.43 Å. On this basis, we conclude that $(H_2COH)^+$ is more nearly described by the oxonium ion structure than by the carbocation structure. However, the C-O bond length *is* significantly longer than a normal double bond, and calculations show that there *is* a substantial partial positive charge on carbon. We shall see in Chapter 13 that much of the chemistry of $(H_2COH)^+$ is best explained by the contribution of the less important carbocation structure.

Again, let us reiterate that *neither oxonium nor carbocation structure provides a totally accurate description of* $(H_2COH)^+$. The actual ion is a resonance hybrid of the two structures. It "looks" more like the oxonium ion structure than like the carbocation structure, and it has some of the characteristics of each. The C–O bond order is something between 1 and 2, but closer to 2. The positive charge is spread over both atoms, but is mostly borne by oxygen. Because oxygen is more electronegative than carbon, the positive charge would rather be on carbon. However, in this structure, carbon does not have an electron octet. In order for carbon to fill its octet, the positive charge must be borne by the more electronegative oxygen. *In cases such as this, the more important resonance structure is generally that one with all octets filled, even if a positive charge is assigned to the more electronegative atom.*

An extreme example of this principle is trifluoromethyl cation, CF_3^+. It has been calculated that the C–F bond length in this ion is 1.27 Å, much less than the normal C—F single bond length of 1.38 Å. Thus, the fluoronium ion structure is a major contributor to the resonance hybrid, even though the positive charge must be assigned to fluorine, the most electronegative of all the elements.

Another interesting example is provided by protonated methyleneimine, $(H_2CNH_2)^+$. The C–N bond length of 1.29 Å is almost exactly the same as the C=N double-bond length in methyleneimine itself (1.27 Å) and is much less than the normal C—N single-bond length of 1.47 Å.

methyleneimine
C=N length = 1.27 Å

protonated methyleneimine
C–N length = 1.29 Å

methylamine
C—N length = 1.47 Å

In this case, the ammonium ion structure dominates the hybrid even more than the oxonium ion structure does in the case of $(H_2COH)^+$ because the difference in electronegativity between carbon and nitrogen is less than that between carbon and oxygen.

In summary, let us set out some empirical rules for assessing the relative importance of the resonance structures of molecules and ions:

1. Resonance structures involve *no* change in the positions of nuclei; only the electron organization is involved.
2. Structures in which all second-period atoms have filled octets are more important than structures with unfilled octets. The contribution of the nonoctet structure increases as the difference in electronegativity between the atoms increases.

More Important   *Less Important*

$$\left[\begin{array}{c} \underset{H}{\overset{H}{>}}C=\overset{+}{\overset{\cdot\cdot}{O}}\!-\!H \end{array} \longleftrightarrow \begin{array}{c} \underset{H}{\overset{H}{>}}\overset{+}{C}\!-\!\overset{\cdot\cdot}{\underset{H}{O}}: \end{array} \right]$$

$$\left[\begin{array}{c} \underset{H}{\overset{H}{>}}C=\overset{+}{\underset{H}{\overset{H}{N}}} \end{array} \longleftrightarrow \begin{array}{c} \underset{H}{\overset{H}{>}}\overset{+}{C}\!-\!\overset{H}{\underset{\cdot\cdot}{\underset{H}{N}}} \end{array} \right]$$

$$\left[\quad H\!-\!C\!\equiv\!\overset{\cdot\cdot}{O}:^{+} \longleftrightarrow \quad H\!-\!\overset{+}{C}=\overset{\cdot\cdot}{O}: \quad \right]$$

3. The more important structures are those involving a minimum of charge separation.

More Important   *Less Important*

$$\left[\begin{array}{c} :\overset{\cdot\cdot}{O}:\\ \| \\ H\!-\!\overset{\cdot\cdot}{O}\!-\!C\!-\!H \end{array} \longleftrightarrow \begin{array}{c} :\overset{\cdot\cdot}{O}:^{-}\\ | \\ H\!-\!\overset{+}{O}\!=\!C\!-\!H \end{array} \right]$$

$$\left[\begin{array}{c} H\!-\!\overset{\cdot\cdot}{\underset{H}{N}}\!-\!C\!\equiv\!N: \end{array} \longleftrightarrow \begin{array}{c} H\!-\!\overset{+}{\underset{H}{N}}\!=\!C\!=\!\overset{\cdot\cdot}{N}:^{-} \end{array} \right]$$

$$\left[\begin{array}{c} \overset{\cdot\cdot}{\underset{\cdot\cdot}{F}}:\\ :F\!-\!B\!\!\overset{\cdots}{\underset{\cdot\cdot}{F}}: \end{array} \longleftrightarrow \begin{array}{c} \overset{\cdot\cdot}{\underset{\cdot\cdot}{F}}:\\ :\overset{+}{F}\!=\!\overset{-}{B}\!\!\overset{\cdots}{\underset{\cdot\cdot}{F}}: \end{array} \right]$$

In cases such as these, however, the less important charge-separated structure still contributes significantly, and we shall find this contribution useful in interpreting some chemical reactions.

In some cases, Lewis structures with complete octets cannot be written without charge separation. In such alternative structures, the more important structure is again that in which the negative charge is borne by the more electronegative element and the positive charge by the more electropositive element.

More Important   *Less Important*

$$\left[\begin{array}{c} H\!-\!C\!=\!\overset{+}{N}\!=\!\overset{-}{\overset{\cdot\cdot}{N}}:\\ | \\ H \end{array} \longleftrightarrow \begin{array}{c} H\!-\!\overset{\overline{\cdot\cdot}}{C}\!-\!\overset{+}{N}\!\equiv\!N:\\ | \\ H \end{array} \right]$$

diazomethane

$$\left[\begin{array}{c} H\\ | \\ H\!-\!C\!-\!C\!\equiv\!\overset{+}{N}\!-\!\overset{\cdot\cdot}{\underset{\cdot\cdot}{O}}:\\ | \\ H \end{array} \longleftrightarrow \begin{array}{c} H\\ | \\ H\!-\!C\!-\!\overset{\overline{\cdot\cdot}}{C}\!=\!\overset{+}{N}\!=\!\overset{\cdot\cdot}{O}:\\ | \\ H \end{array} \right]$$

acetonitrile oxide

Elements beyond the second period form structures with an apparent expansion of their octets. Examples are provided by sulfur hexafluoride, SF_6, and phosphorus pentachloride, PCl_5. Such compounds are probably best considered in terms of "no-bond" contributions. See how the P–Cl bond-length variations can be explained by simple octet resonance structures.

$$\left[\begin{array}{ccc} \underset{\substack{\displaystyle Cl\\}}{\overset{\substack{Cl\\ 2.19\ \text{Å}}}{\underset{\displaystyle Cl}{\overset{\displaystyle Cl}{P}}}}\!\!-\!Cl & \longleftrightarrow & \overset{Cl^-}{\underset{Cl}{\overset{Cl}{P^+}}}\!\!-\!Cl & \longleftrightarrow & \underset{Cl^-}{\overset{Cl}{\underset{Cl}{P_+}}}\!\!-\!Cl \end{array} \right]$$

In the same manner, some compounds of these elements are often written as resonance hybrids with expanded octet resonance structures. An example is sulfuric acid.

$$\left[\begin{array}{ccccc} & \overset{\displaystyle O^-}{} & & & \overset{\displaystyle O}{\parallel} \\ H\!-\!O\!-\!\underset{\displaystyle O^-}{S^{2+}}\!-\!O\!-\!H & \longleftrightarrow & H\!-\!O\!-\!\underset{\displaystyle O^-}{S^+}\!-\!O\!-\!H & \longleftrightarrow \\ & \overset{\displaystyle O^-}{} & & & \overset{\displaystyle O}{\parallel} \\ H\!-\!O\!-\!\underset{\displaystyle O}{S^+}\!-\!O\!-\!H & \longleftrightarrow & H\!-\!O\!-\!\underset{\displaystyle O}{S}\!-\!O\!-\!H \end{array} \right]$$

The normal Lewis octet structure at the far left has a formal charge of +2 on sulfur.
Sulfuric acid is known to be a strong acid, and the high formal charge on sulfur helps
to explain the ease of loss of a proton.

EXERCISE: In this section we have made use of the concept of bond order and the
generalization that the higher the bond order (degree of bonding), the shorter (and
stronger) the bond. An example is the application to nitric acid and nitrate ion. Write
two equivalent resonance structures for nitric acid and three for nitrate ion. How
would you expect the three different N–O bonds in those two compounds to rank in
length?

2.5
Atomic Orbitals

Careful use of Lewis structures and the related straight-line structural shorthand is clearly important in understanding the physical and chemical properties
of molecules. But these structures are themselves only symbolic representations of
electronic structures. In the real world, electrons do not stand still in octets. A
more complete understanding of the chemical bond and the structure of molecules requires a discussion of the modern theory of electronic structure in terms of
wave functions or orbitals. Unfortunately, this theory involves new and unfamil-
iar concepts that do not relate to human experience. Atomic and molecular orbitals are usually covered in depth in courses on physical chemistry, but the qualitative aspects are so important to understanding modern organic chemistry that a
brief survey of some results of quantum mechanics is highly desirable at this
point. In the next few sections, we shall review those aspects of atomic and molecular orbital theory that are particularly important in the study of organic
chemistry.

As mentioned in Section 2.1, the periodic table of the elements was first conceived in a purely empirical fashion. The various known elements were arranged
into groups and rows on the basis of similarities in their chemical and physical
properties. The "periodicity" of the table first became understandable with the
early development of electronic theory. This early theory was based on the Bohr

model, which is often taught by comparing an atom to a miniature solar system in which electrons are pictured as revolving in fixed orbits around a nucleus much as planets revolve about the sun.

With the advent of quantum mechanics about a half century ago, this analogy was shown to be seriously deficient in an extremely important respect. A basic tenet of quantum mechanics is the **Heisenberg uncertainty principle,** which states that it is not possible to determine simultaneously both the precise position and momentum of an electron. In other words, the laws of nature are such that we cannot determine an exact trajectory for an electron. The best we can do is to describe a probability distribution that gives the probability of finding an electron in any region around a nucleus. The mathematical description that leads to this probability distribution has the same form as that which describes a wave. Thus, we may use the mathematics and concepts of wave motion to describe electron distributions. Consequently, it is common to refer to the motion of an electron around a nucleus as a "wave motion" or in terms of a "wave function." This does not mean that the electron actually bobs up and down like a cork in a stormy sea. It is only a convenient language that helps to characterize the mathematical equations that describe the electron probability distribution.

In quantum mechanics, an **atomic orbital** is defined as a one-electron wave function, ψ. For each point in space there is associated a number whose square is proportional to the probability of finding an electron at that point. Such a probability function corresponds to the more useful concept of an **electron density** distribution. The mathematical function that describes an atomic orbital has all of the properties associated with waves; hence, it is called a **wave function.** It has a numerical magnitude (its amplitude), which can be either positive or negative (corresponding to a wave crest or a wave trough, respectively), and nodes. A **node** is the region where a crest and a trough meet. For the three-dimensional waves characteristic of electronic motion, the nodes are two-dimensional surfaces at which $\psi = 0$. Consequently, atomic orbitals may be characterized by their corresponding nodes as given by quantum numbers (Table 2.3).

If one recognizes the relationship between quantum numbers and the number and character of the nodes in a wave, it is clear why quantum numbers are integers; that is, it is meaningless to talk of a fraction of a node. In labeling a particu-

TABLE 2.3
Atomic Quantum Numbers

Quantum Number	Symbol	Possible Values	Relationship to Nodes
principal	n	1, 2, 3, . . .	one more than the total number of nodes[a]
azimuthal or angular momentum	l	0, 1, . . . , $n-1$	number of nonspherical nodes
magnetic	m	$-l, \ldots, 0, \ldots +l$	character (planes or cones) and orientation of nonspherical nodes
spin	—	$-\frac{1}{2}, +\frac{1}{2}$	none

[a] Because atomic orbitals are exponential functions, they have very small values at distances far from the nucleus but never reach zero. The extra node could therefore be taken at infinity, but such a node could never be represented in conventional symbols. If the node "at infinity" is included, the principal quantum number is the same as the total number of nodes.

lar atomic orbital, the principal, azimuthal, and magnetic quantum numbers are specified. The three quantum numbers are expressed in the order nlm, but in a particular manner. The principal quantum number n is given as the appropriate integer. The azimuthal number l is expressed in code, where $0 = s$, $1 = p$, $2 = d$, and $3 = f$. In spatial descriptions m is not given explicitly, but is implied in a subscript code that defines the orientation of the orbital.

Examples.

$1s$ no nodes. This wave function is a spherically symmetric function whose numerical value decreases exponentially from the nucleus.

$2s$ one spherical node

$2p_x$ one node, the yz-plane

$2p_y$ one node, the xz-plane

$2p_z$ one node, the xy-plane

These orbitals are the most important for the organic compounds we will study. They are usually represented symbolically as in Figure 2.1. The plus and minus signs in Figure 2.1 have no relationship to electric charge. They are simply the arithmetic signs associated with the wave function, much as a positive sign for an ocean wave is a wave crest and a negative sign denotes a wave trough. We shall see in the next section that the positive and negative signs determine how two or more wave functions combine when they interact.

In the symbolic representations given in Figure 2.1, the solid line represents the angular part of the wave function and defines a three-dimensional closed surface. A useful approximation is to regard the surface as a locus of points of constant value of ψ such that some given, but arbitrary, proportion of the total electron density is contained within the surface. For example, the value of ψ may be selected so that the resulting surface will enclose 80%, 90%, 95%, and so on of the

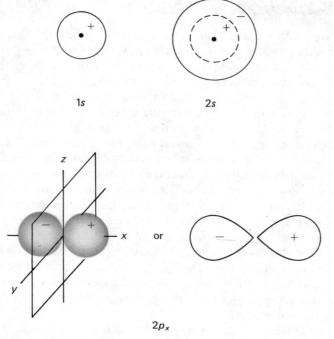

FIGURE 2.1. *Symbolic representation of some atomic orbitals.*

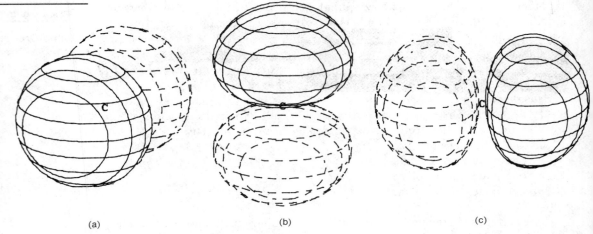

FIGURE 2.2. *Perspective diagrams of* $2p_x$-, $2p_y$-, *and* $2p_z$-*orbitals.*

electron density. The dotted lines in Figure 2.1 represent nodal surfaces. These nodes are a sphere for the 2s-orbital and a plane for the 2p-orbital. The strange shape of the p-orbital is determined by the central attractive force of the nucleus and the constraint of a planar node. The representation of 2p-orbitals in Figure 2.2 gives a better perspective for the three-dimensional shape and shows the three orthogonal directions of p_x-, p_y-, and p_z-orbitals.

2.6
Electronic Structure of Atoms

The Pauli principle applied to atoms states that no two electrons can have identical quantum numbers. Three quantum numbers characterize an atomic orbital. Electrons have a fourth quantum number associated with the characteristics of spin. This quantum number may have a value of either $+\frac{1}{2}$ or $-\frac{1}{2}$. Consequently, each atomic orbital may have associated with it no more than two electrons, and these two electrons must have "opposite spin."

In general, the more nodes a wave function has, the higher is its energy. In atoms that have more than one electron, the energies of atomic orbitals increase in the order $1s < 2s < 2p < 3s < 3p$, and so on (see Figure 2.3).

The first electron is put into the lowest energy atomic orbital, 1s, to produce the hydrogen atom. The helium atom has two electrons, and the second electron can also be put into a 1s-orbital if the second electron has a spin opposite that of the first electron. These two electrons "fill" the 1s-shell, and helium has the filled-shell configuration characteristic of noble gases. The third electron of lithium must be put into a higher energy atomic orbital, 2s. The fourth electron of beryllium can also be put into the 2s-orbital if its spin is opposite that of the third electron. The 2s-orbital is now also filled, and the additional electrons of the first row elements must go into 2p atomic orbitals. The $2p_x$-, $2p_y$-, and $2p_z$-orbitals may each accept two electrons, giving a total of six for the p-set. Consequently, eight electrons fill the $n = 2$ shell and again give the stable filled-shell electronic configuration characteristic of the noble gases.

The process of filling successive atomic orbital levels with electron pairs is used to build up the entire periodic table. The atomic configurations of the first ten

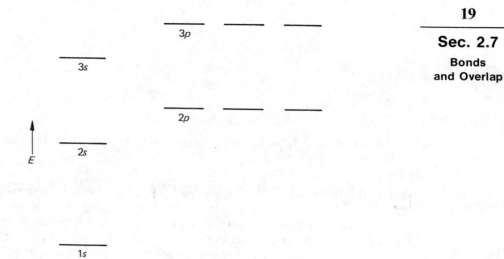

FIGURE 2.3. *Order of energy levels in an atom with more than one electron.*

elements are summarized in Table 2.4. Each filled principal quantum shell corresponds to a stable noble gas. Other elements react in such a way as to achieve the stability associated with filled orbital shells. One way of achieving this higher stability is by combining atomic orbitals into molecular orbitals, as discussed in the next section.

TABLE 2.4
**Electronic Configurations
of Some Elements**

First Period		Second Period	
H	$1s$	Li	$1s^2 2s$
He	$1s^2$	Be	$1s^2 2s^2$
		B	$1s^2 2s^2 2p$
		C	$1s^2 2s^2 2p^2$
		N	$1s^2 2s^2 2p^3$
		O	$1s^2 2s^2 2p^4$
		F	$1s^2 2s^2 2p^5$
		Ne	$1s^2 2s^2 2p^6$

2.7
Bonds and Overlap

One of the useful concepts derived from treating atomic orbitals as wave functions is that two such orbitals may overlap to form a **bond**. The combination of two waves having the same sign is reinforcing (Figure 2.4). This is true for light waves, sound waves, or the waves of an ocean. It is also true for the combination of two electron waves or wave functions having the same sign.

The increased magnitude of the wave function between the atoms corresponds to higher electron density in this region. Electrons are attracted electrostatically to both nuclei, and the net effect of increased electron density between the nuclei

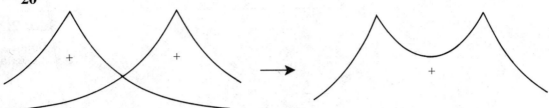

FIGURE 2.4. *Two interacting waves or wave functions of the same sign add or reinforce.*

counterbalances the internuclear repulsion. The result is a **covalent bond.** An example is the combination of two 1s atomic orbitals to give a new wave function. This wave function now encompasses both nuclei and is a **molecular orbital.** Figure 2.5 shows a symbolic representation of a molecular orbital (b) formed by the overlap of two atomic orbitals as in (a). Part (c) of Figure 2.5 is a contour diagram that depicts the value of the wave function in such a covalent bond as a function of distance from the nuclei, which are symbolized by the two heavy dots. The diagram represents a plane passing through the nuclei; each contour line connects points having the same value of the wave function in the same way that a contour line on a map connects points having the same altitude relative to sea level. In the example shown, all of the contours are positive. Figure 2.5d is a perspective view of a contour surface of such an orbital for a given contour value. Each bond that we have heretofore symbolized by a shared electron pair or by a straight line may now be interpreted as a two-center molecular orbital (an orbital encompassing two nuclei). Each such two-center molecular orbital contains two electrons of opposite spin.

When two waves of opposite sign interact, they interfere or cancel each other. It is this characteristic of waves that can produce regions of darkness in the interaction of two light beams or regions of silence from the combination of two sound waves. At the point of interference the wave function has the value of zero; that is, interference of waves creates a node. The same pattern holds for electron waves. The interaction of two orbitals of opposite sign produces a node between the

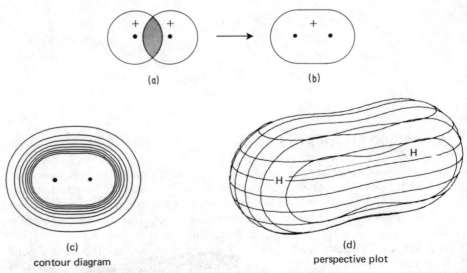

FIGURE 2.5. *The combination of two H 1s orbitals to form* H_2.

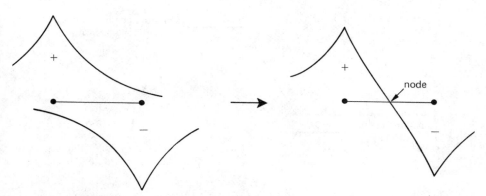

FIGURE 2.6. *Interaction of two waves of opposite sign gives subtraction of wave
functions or* **interference.**

nuclei, as illustrated in Figure 2.6. Because there is no electron density at a node,
the net effect of the reduced electron density between the nuclei in this case does
not compensate for nuclear repulsion, and the net result is a higher energy or
lower stability than that which corresponds to the noninteracting orbitals. Such a
molecular orbital is called **antibonding.** A diagram of such an antibonding molecular orbital is shown in Figure 2.7.

The interaction of two orbitals can be positive or reinforcing to give a bonding
molecular orbital, or the combination can be negative or interfering to give an
antibonding molecular orbital. The bonding combination corresponds to a decrease in energy (greater stability); the antibonding combination corresponds to
an increase in energy (lower stability). Two atomic orbitals give rise to two molecular orbitals. The two paired electrons of opposite spin available for the bond can
be put into the bonding molecular orbital. We will not refer to antibonding molecular orbitals for most of the normal compounds important in organic chemistry,
but we will make use of them in some reactions and in electronic spectroscopy
(Chapter 21).

The energy relationships of two combining orbitals are summarized in Figure 2.8. Note how the energies of the two starting orbitals separate or spread apart
when they interact to form the two molecular orbitals. The amount of the separation depends on the degree to which the orbitals overlap. A slight overlap gives

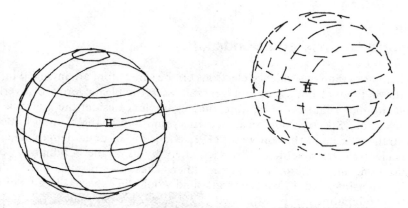

FIGURE 2.7. *A two-center antibonding molecular orbital. The dashed lines indicate
a wave function of negative sign.*

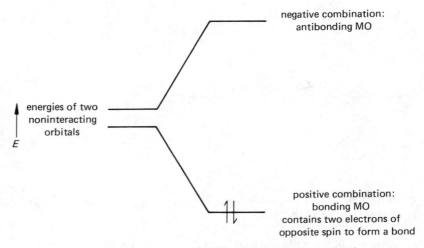

FIGURE 2.8. *Energy relationships of combining orbitals.*

two molecular orbitals that differ little in energy; a large overlap results in strong separation such as that shown in Figure 2.8. For axially symmetric orbitals, such as *p*-orbitals, the greatest overlap occurs when the orbitals are allowed to interact along the nuclear axis, that is, to form straight bonds. We shall see later that if orbitals are so constrained that overlap is not along the internuclear axis, then the resulting "bent bonds" (Figure 2.9b and c) are weaker than equivalent straight bonds (Figure 2.9a).

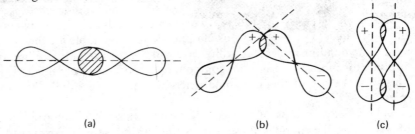

FIGURE 2.9. *Illustrating the overlap of two p-orbitals* (a) *along the internuclear axis and* (b and c) *off the internuclear axis. The molecular orbital resulting from the overlap in* (a) *has lower energy than the other two.*

2.8
Hybrid Orbitals and Bonds

When more than two valence electrons on the same atom are involved in bonding, the individual bonds are not generally describable in terms of overlap of simple atomic orbitals as in the foregoing example. Consider the molecule BeH_2 as an example. Spectroscopic measurements show that the three atoms in BeH_2 lie in a straight line and that the two Be—H bonds are of equal length. These two bonds clearly cannot be described adequately by using the beryllium 2s-orbital for one bond and a 2p-orbital for the other. These two orbitals have different spatial extensions and different energies and would be expected to give different bonds. The bonding can be explained if we construct *two equivalent hybrid orbitals* by combining the 2s- and a 2p-orbital. This is done mathematically by taking the sum and the difference of the two orbitals, as in Figure 2.10. This example

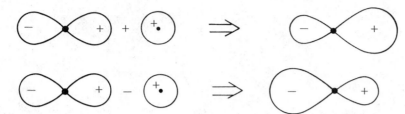

FIGURE 2.10. *Mathematical combination of* s- *and* p-*orbitals to yield two* sp-*hybrid orbitals.*

shows how the mathematical signs of wave functions enter into arithmetic operations.

The two orbitals that result from this operation are designated *sp*-hybrid orbitals because they are each constructed from equal amounts of an *s*- and a *p*-orbital. The *sp*-hybrid orbital is shown in contour form in Figure 2.11a and as the three-dimensional perspective plot in Figure 2.11b. The two hybrid orbitals are each suited for bond formation by overlapping with an H 1*s*-orbital. They are equivalent and are directed opposite each other. Furthermore, the two lobes of a hybrid orbital are unequal in "size"—the larger lobe can overlap well with another orbital. That is, overlap at the large lobe can occur readily in a straight line to produce stronger bonding.

Why does beryllium form bonds in this manner rather than by overlap of the simple atomic orbitals? The answer is simply that stronger bonds and a more stable structure result when the system H–Be–H is linear and the two bonds are of equal length. In this manner *the two electron pairs involved in the bonds are directed as far apart from each other as possible.* This principle is a useful method for predicting the geometry of a molecule in which several groups are bonded to a central atom. In general, the bonding may be described by constructing as many hybrid orbitals from the simple *s*- and *p*-atomic orbitals as are needed to accommodate all of the valence electrons associated with the central atom. In the BeH_2 example, we used one *s*- and one *p*-orbital and constructed two equivalent hybrids. Each such hybrid is described as 50% *s* and 50% *p*. In constructing such combinations, we must again obey the "rule of conservation of orbitals." We must end up with as many orbitals as we started with. The beryllium atom has, of

(a)   (b)

FIGURE 2.11. *Contour and perspective plots of an* sp-*hybrid orbital.*

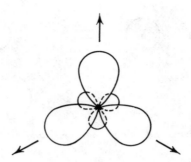

FIGURE 2.12. *Three sp²-hybrids.*

course, two remaining *p*-orbitals that are not occupied by electrons in the molecule BeH_2.

As a further example, consider a species in which three groups are to be bonded to a central atom. From an *s*-orbital and two *p*-orbitals—for example, a p_x- and p_y-orbital—we may construct three equivalent *sp²*-hybrids. Each such hybrid is $\frac{1}{3}s$ and $\frac{2}{3}p$. The three equivalent hybrids lie in the *xy*-plane (the same plane defined by the two *p*-orbitals) and are directed 120° from each other (Figure 2.12).

Methyl cation, CH_3^+, is an example of such a species. It is planar and the three C–H bonds are equal in length. It may be regarded as being derived from overlap of three equivalent carbon *sp²*-orbitals with hydrogen 1*s*-orbitals. Each bond may be represented as C_{sp^2}—H_{1s}. The remaining carbon *p*-orbital is perpendicular to the molecular plane and contains no electrons. In this process of conceptual development, we have used the sequence of combining three atomic orbitals to form three hybrid obitals (Figure 2.12) that are allowed to overlap (Figure 2.13a) to form three two-center molecular orbitals (Figure 2.13b). Each of these molecular orbitals contains two electrons, and the carbon also has two electrons in its 1*s*-orbital that are not normally represented in our simple valence symbols.

Finally, from an *s*-orbital and three *p*-orbitals we may derive four *sp³*-hybrids directed to the corners of a tetrahedron with an interorbital angle of 109.5°, the tetrahedral angle. Each such hybrid orbital is 25% *s* and 75% *p*. A three-dimensional perspective plot of one *sp³*-hybrid orbital is shown in Figure 2.14. Note that the "small lobe" is much larger than in the *sp*-hybrid depicted in Figure 2.11b.

The tetrahedral structure of methane, CH_4, is illustrated in stereo plot in Figure 2.15a or by the perspective model in Figure 2.15b. Each bond between C and H may be described as a C_{sp^3}—H_{1s} bond. Each such bond is derived by the interaction of a C_{sp^3}-hybrid orbital with a hydrogen 1*s*, as in Figure 2.15c, to produce the resulting two-center molecular orbital shown in Figure 2.15d. The

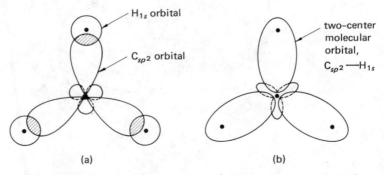

(a)   (b)

FIGURE 2.13. *Development of the electronic structure of* CH_3^+.

FIGURE 2.14. *Three-dimensional perspective view of an sp³-hybrid orbital.*

actual wave function—the mathematical form of the molecular orbitals—for which Figure 2.15d is only a symbolic representation is shown in contour form in Figure 2.15e.

The hybrid orbitals considered thus far are equivalent, but it is not necessary that all orbitals on an atom be equivalent when the molecule lacks symmetry. It is possible to have a hybrid orbital that is, for example, 23% s and 77% p. In NH_3, for example, the H—N—H bond angle of 107.1° does not correspond to any

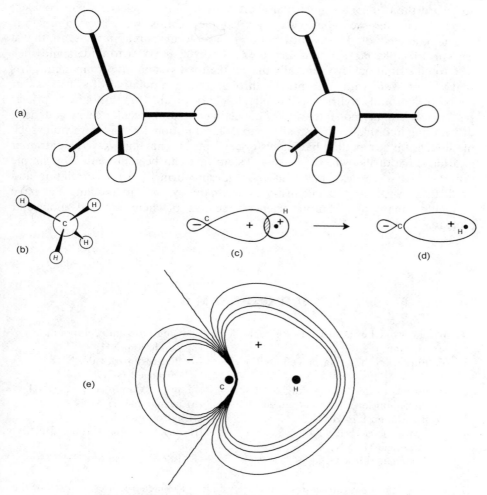

FIGURE 2.15. *Methane, CH_4, and its C_{sp^3}—H_{1s} bond.*

simple hybrid. Recall that sp^3-hybridization corresponds to a bond angle of 109.5°, sp^2-hybridization corresponds to 120°, and that pure p-orbitals form a 90° angle. In addition to its three N—H bonds, ammonia also has a nonbonding pair of electrons on the nitrogen. These electrons are in an orbital which has slightly more s-character than in a simple sp^3-orbital. Consequently, the three hybrid orbitals that overlap with the three hydrogen atoms contain slightly less s-character than in a sp^3-orbital (actually, these orbitals are each approximately 23% s and 77% p). Electrons in s-orbitals have lower energy than electrons in p-orbitals. Therefore, bonds with more s-character tend to be stronger. However, an electron pair in a bond is affected by two nuclei, whereas nonbonding electrons are attracted only by a single nucleus. Hence, s-character is more important for lone-pair electrons than for bonding electron pairs. In dividing the available s-orbital among bonds and lone pairs generally receive a higher proportion.

This type of result is general. In water, for example, the H—O—H bond angle is 104.5° and each O—H bond clearly involves an oxygen hybrid with more p-character than in ammonia. The two oxygen lone pairs require a large fraction of the available oxygen 2s-orbital. In HF, the H—F bond is an almost pure F_{2p}—H_{1s} bond, and the fluorine 2s-orbital is used almost entirely for the three lone pairs. Nevertheless, despite these complexities, it is frequently convenient and sufficient to regard the two-center molecular orbitals that comprise electron-pair covalent bonds to be composed *approximately* of simple hybrids: sp, sp^2, sp^3, and so on.

The total electron density distribution in a molecule is real in the sense that it can, in principle, be seen and measured. However, in order to understand such electron distributions, we generally dissect the total system into component parts that we can work with conceptually through the manipulation of symbols. Our concepts of orbitals, hybrids, and bonds should be regarded in this light. In principle, there are many possible ways of dissecting a total molecular electron density distribution into smaller and smaller parts. Our traditional way is merely one such method, but it is a method having historical roots and having evolved a grammar and language of its own. It is also a system that can be represented by simple symbols and serves as a powerful and widespread method for correlating and predicting a wide range of chemistry. As such, this symbolism and language have permeated many neighboring sciences such as biochemistry and molecular biology.

PROBLEMS

1. Write a valid Lewis structure for each of the following inorganic compounds.
   (a) bisulfate ion, HSO_4^-
   (b) amide ion, NH_2^-
   (c) nitrite ion, NO_2^- (arranged ONO)
   (d) dinitrogen trioxide, N_2O_3 (arranged ONONO)
   (e) nitrous oxide, N_2O (arranged NNO)
   (f) hydroxylamine anion, $(ONH_2)^-$
   (g) nitronium ion, NO_2^+ (arranged ONO)
   (h) cyanamide, H_2NCN
   (i) hydronium ion, H_3O^+
   (j) nitrosoniun ion, NO^+
   (k) hydrazoic acid, HN_3 (arranged HNNN)
   (l) azide ion, N_3^- (arranged NNN)
   (m) carbonate ion, CO_3^{2-}
   (n) isocyanic acid, OCNH

2. Write out the Lewis structures and corresponding Kekulé structures for each of the following organic compounds.

(a) ethyl cation, $CH_3CH_2^+$

(b) ethyl radical, CH_3CH_2

(c) ethyl anion, $CH_3CH_2^-$

(d) methylacetylene, $CH_3C{\equiv}CH$

(e) methyl ethyl ketone, $CH_3COCH_2CH_3$

(f) dimethyl ether, CH_3OCH_3

(g) methylamine, CH_3NH_2

(h) methylammonium cation, $CH_3NH_3^+$

(i) methoxide ion, CH_3O^-

(j) methyloxonium ion, $CH_3OH_2^+$

(k) vinyl chloride, $CH_2{=}CHCl$

(l) formyl cation, HCO^+
(arranged HCO)

3. For each of the following compounds, describe each bond in terms of its component atomic orbitals.

(a) ethane, CH_3CH_3

(b) ethyl anion, $CH_3CH_2^-$

(c) ethyl cation, $CH_3CH_2^+$

(d) methylborane, CH_3BH_2

(e) methylberyllium hydride, CH_3BeH

(f) methanol, CH_3OH

4. Which of the following pairs of Kekulé structures do *not* constitute resonance structures?

(a) $CH_3\overset{\overset{\displaystyle O}{\|}}{C}{-}O^-$ and $CH_3\overset{\overset{\displaystyle O^-}{|}}{C}{=}O$

(b) $CH_3\overset{\overset{\displaystyle O}{\|}}{C}{-}OH$ and $CH_3\overset{\overset{\displaystyle OH}{|}}{C}{=}O$

(c) $CH_3\overset{\overset{\displaystyle O}{\|}}{C}CH_3$ and $CH_3\overset{\overset{\displaystyle OH}{|}}{C}{=}CH_2$

(d) $^+CH_2{-}CH{=}CH_2$ and $CH_2{=}CH{-}\overset{+}{C}H_2$

(e) $CH_2{=}CH{-}\overset{\overset{\displaystyle O}{\|}}{C}H$ and $^+CH_2{-}CH{=}\overset{\overset{\displaystyle O^-}{|}}{C}H$

(f) $CH_3CH{=}CHCH_3$ and $CH_3CH_2CH{=}CH_2$

(g) $CH_2{=}C{=}CH_2$ and $CH_3C{\equiv}CH$

(h) $H{-}C{\equiv}\overset{+}{N}H$ and $H{-}\overset{+}{C}{=}NH$

(i) $CH_3{-}\overset{\overset{\displaystyle O}{\|}}{\underset{+}{N}}{-}O^-$ and $CH_3{-}\overset{\overset{\displaystyle O^-}{|}}{\underset{+}{N}}{=}O$

(j) $CH_3\overset{\overset{\displaystyle OH}{|}}{C}HCH_3$ and $CH_3CH_2\overset{\overset{\displaystyle OH}{|}}{C}H_2$

(k) $CH_3N{=}C{=}O$ and $CH_3O{-}C{\equiv}N$

(l) $CH_3N{=}C{=}O$ and $CH_3\overset{+}{N}{\equiv}C{-}O^-$

(m) $^-O{-}\overset{+}{O}{=}O$ and $O{=}\overset{+}{O}{-}O^-$

(n) $\overset{\displaystyle \cdot}{N}{=}O$ and $^-N{=}O\cdot^+$

5. For each of the following resonance hybrids, rank the contributing structures in order of their relative importance.

(a) $H{-}\overset{\overset{\displaystyle :\ddot{O}:}{\|}}{C}{-}\ddot{N}H_2 \longleftrightarrow H{-}\overset{\overset{\displaystyle :\ddot{O}:^-}{|}}{C}{=}\overset{+}{N}H_2$

(b) $H{-}\overset{\overset{\displaystyle :\ddot{O}:}{\|}}{C}{-}\ddot{C}H_2 \longleftrightarrow H{-}\overset{\overset{\displaystyle :\ddot{O}:^-}{|}}{C}{=}CH_2$

(c) 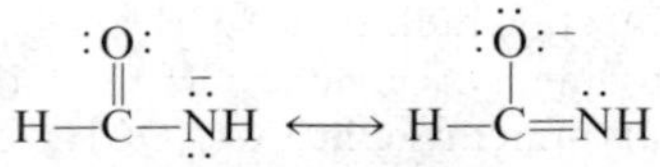 $\overset{:O:}{\overset{\|}{H-C-\overset{..}{N}H}} \longleftrightarrow \overset{:\overset{..}{O}:^-}{H-C=\overset{..}{N}H}$

(d) 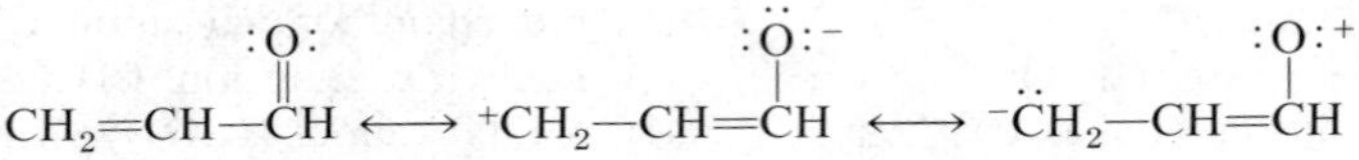 $\overset{:O:}{\overset{\|}{CH_2=CH-CH}} \longleftrightarrow {}^+CH_2-CH=\overset{:\overset{..}{O}:^-}{CH} \longleftrightarrow {}^-\overset{..}{C}H_2-CH=\overset{:O:^+}{CH}$

(e)

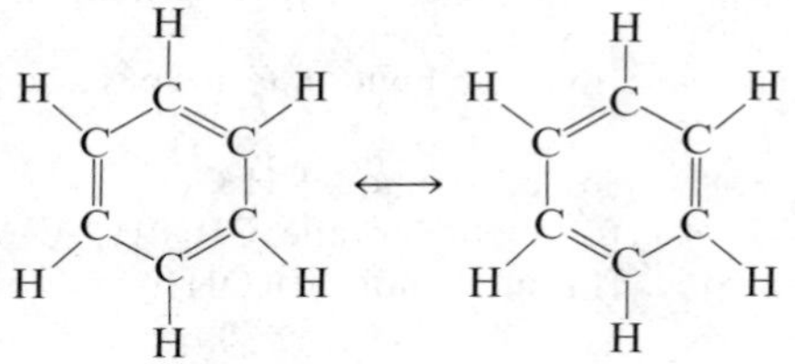

(f) 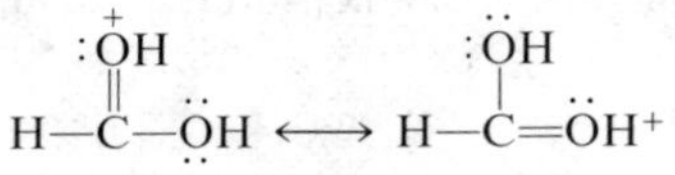 $\overset{:\overset{+}{O}H}{\overset{\|}{H-C-\overset{..}{O}H}} \longleftrightarrow \overset{:\overset{..}{O}H}{H-C=\overset{..}{O}H^+}$

(g) $CH_3-\overset{+}{N}\equiv C-O^- \longleftrightarrow CH_3-{}^-N-C\equiv O^+$

(h) ${}^-O-\overset{+}{O}=O \longleftrightarrow O=O=O$

(i) $\overset{.}{N}=O \longleftrightarrow {}^-N=O\cdot{}^+$

6. Use the information given on page 6 to show that the reaction $\quad Na\cdot + Cl\cdot \longrightarrow$ $Na^+ + Cl^-\quad$ is endothermic in the gas phase. Sodium chloride has m.p. 801°C and boils at 1465°C. The vapor consists of ion pairs, Na^+Cl^-, held together by electrostatic attraction at a bond distance of 2.36 Å. A proton and electron at a distance of 1 Å have an electrostatic attraction of 330 kcal mole^{-1}. What is the electrostatic energy of a proton and electron at the bond distance of sodium chloride? As a model for Na^+Cl^-, is this value enough to make the reaction $\quad Na\cdot + Cl\cdot \longrightarrow Na^+Cl^-\quad$ exothermic?

Chapter 3
Organic Structures

3.1
Introduction

By the middle of the nineteenth century, organic chemistry was being actively explored in Europe. Many new compounds had been isolated from natural sources, and it was found increasingly that one organic compound could be transformed into another. However, the fundamental concept of **structure** had not yet evolved. The atomic theory of matter was well established, and it was clear that, by and large, inorganic compounds could be characterized by simple formulas, such as NH_3, P_2O_5, HNO_3, or H_2SO_4.* The satisfying theory of valency was well on its way to providing organization to the multitude of known chemical substances. However, some vexing problems remained, largely with the "organic" compounds. For example, "marsh gas" (methane) was shown to have the formula CH_4, which agreed with the quadrivalence of carbon in inorganic compounds such as CS_2. However, "olefiant gas" (ethylene) was found to have the formula C_2H_4, in which the apparent valence of carbon was two. To make matters worse, a gaseous hydrocarbon (acetylene) discovered by Edmund Davy in 1836 was found to have the formula C_2H_2 in which carbon had the ridiculous valence of one!

A second difficult problem was presented by the phenomenon of **isomerism** (the existence of two or more compounds having the same molecular formula). Isomerism had first been discovered in the 1820s by Liebig and Wöhler when they found that silver cyanate (AgNCO) and silver fulminate (AgOCN) have the same atomic composition. Gay-Lussac made the then-revolutionary suggestion that the two compounds differed "in the way the elements were combined together." From this point on, it was clear that a formula alone was not adequate to characterize a compound uniquely.

The third important event which paved the way for the invention of chemical structures was the discovery of **substitution reactions** (replacement of an atom or group of atoms in a compound by another atom or group of atoms). Although substitution had been noted in the late eighteenth century, it was first actively investigated by Dumas and Laurent in the 1830s and 1840s. One of the first substitution reactions studied was halogenation (Chapter 6). For example, it was quickly found that chlorine combines with many organic compounds by replacing hydrogen atoms one-for-one, as in the conversion of methane to chloroform.

$$CH_4 + 3Cl_2 \longrightarrow CHCl_3 + 3HCl$$

However, it was also soon found that in some compounds all of the hydrogens did not seem to be equivalent. For example, acetic acid was found to undergo substitution of only three of its four hydrogens, no matter how vigorous the reaction. The product, trichloroacetic acid, was found to have properties which differed only quantitatively from those of acetic acid itself.

---

* Actually, the formulas in use in the early nineteenth century were not standardized, since different chemists were prone to use different atomic weights for carbon and oxygen. For example, Liebig used C = 6, O = 8, while Dumas used C = 6, O = 16. Berzelius used atomic weights which agree with the present-day values, but he also used double volume formulas. In this section, we use only the modern formulas.

$$HC_2H_3O_2 + 3Cl_2 \longrightarrow HC_2Cl_3O_2 + 3HCl$$

Thus, it became increasingly clear that it must matter *how* a given hydrogen is joined to the remainder of the molecule.

> At the time, a considerable controversy raged over the work of Laurent and Dumas. In fact, strange as it may now seem, the very existence of substitution was not even accepted by Berzelius, since it appeared to be leading toward the nullification of the "dualistic theory," of which he was the architect and principal advocate. The controversy was not solely the result of inflexibility on the part of Berzelius, since Dumas was prone to make rather rash extensions of his experiments, such as his claim that even carbon atoms were susceptible to substitution by halogens. The argument eventually led Wöhler, who was something of a practical joker, to perpetrate an amusing hoax. In an article published in the German journal *Liebigs Annalen* under the pen name S. C. H. Windler, Wöhler described work on the reaction of chlorine with manganese acetate, $Mn(C_2H_3O_2)_2$. He reported that the hydrogen was first replaced by chlorine as expected. However, he reported that on longer reaction the oxygen and manganese were also replaced and that finally even the carbon was replaced by chlorine. There was obtained a substance which analyzed for 100% chlorine, but which had all the physical and chemical properties of manganese acetate!

Gradually, it came to be realized that the atoms in a molecule are "hooked together" in a certain way and that this assembly or *structure* is a more accurate way of describing a substance than a simple molecular formula. As shown on page 3, the first attempts to depict structures were fairly clumsy. However, these initial attempts to depict the manner in which a compound's atoms are joined together soon gave way to a standard formalism in which a line was used to symbolize a point of connection, or "bond," between two atoms. With this simple concept came the realization that "valency," which worked so well for inorganic compounds, could be easily extended to organic compounds by the expedient of using "double" and "triple" bonds, as in ethylene and acetylene.

$$
\begin{array}{cc}
H & H \\
\diagdown & \diagup \\
C = C \\
\diagup & \diagdown \\
H & H
\end{array}
\qquad
H-C\equiv C-H
$$

It was much later that the idea of the "electron pair" or covalent bond was introduced, and the lines of the nineteenth century acquired a physical significance as shown in the Lewis "electron dot" structures we reviewed in Chapter 2. In modern orbital terms each of these lines represents a two-center molecular orbital, an orbital "localized" on a pair of atoms and derived from the overlap of two atomic or hybrid orbitals. These **structural formulas** are often further abbreviated for convenience by omitting the lines. The resulting expressions are called **condensed formulas.**

$$
\begin{array}{ccc}
& H & \\
& | & \\
H - & C & - H \equiv CH_4 \\
& | & \\
& H &
\end{array}
\qquad
\begin{array}{ccc}
& H & \\
& | & \\
H - & C & - \ddot{Cl}: \equiv CH_3Cl \\
& | & \\
& H &
\end{array}
$$

methane   methyl chloride

$$
\begin{array}{ccc}
& H & \\
& | & \\
H - & C & - \ddot{O} - H \equiv H - C - O - H \equiv CH_3OH \\
& | & \\
& H &
\end{array}
$$

methyl alcohol

An important characteristic of organic compounds is the ubiquity of C—C bonds. Although some other atoms can bond to themselves to form short or long chains, carbon is unique in the extent and versatility of its catenation (chain formation; L., *catena,* a chain). Such C—C bonds are treated in the same way as others, as shown by the following examples.

$$
\begin{array}{cc}
\underset{\underset{\displaystyle H}{|}}{\overset{\overset{\displaystyle H}{|}}{H-C}}-\underset{\underset{\displaystyle H}{|}}{\overset{\overset{\displaystyle H}{|}}{C}}-H \equiv CH_3CH_3
&
\underset{\underset{\displaystyle H}{|}}{\overset{\overset{\displaystyle H}{|}}{H-C}}-\underset{\underset{\displaystyle H}{|}}{\overset{\overset{\displaystyle H}{|}}{C}}-Cl \equiv CH_3CH_2Cl
\end{array}
$$

ethane   ethyl chloride

$$
\begin{array}{cc}
\underset{\underset{\displaystyle H}{|}}{\overset{\overset{\displaystyle H}{|}}{H-C}}-\underset{\underset{\displaystyle H}{|}}{\overset{\overset{\displaystyle H}{|}}{C}}-O-H \equiv CH_3CH_2OH
&
\underset{\underset{\displaystyle H}{|}}{\overset{\overset{\displaystyle H}{|}}{H-C}}-\underset{\underset{\displaystyle H}{|}}{\overset{\overset{\displaystyle H}{|}}{C}}-\underset{\underset{\displaystyle H}{|}}{\overset{\overset{\displaystyle H}{|}}{C}}-H \equiv CH_3CH_2CH_3
\end{array}
$$

ethyl alcohol   propane

The compounds involve C—H bonds that are all approximately C_{sp^3}—H_{1s}. Correspondingly, all of these C—H bonds are about the same length, 1.10 Å. Similarly, all of the C—C bonds in these compounds are approximately C_{sp^3}—C_{sp^3}, and these bond lengths are all about the same, 1.54 Å.

Compounds with multiple bonds can also be represented by condensed formulas.

$$
\underset{H}{\overset{H}{{\diagdown}}}C=C\underset{H}{\overset{H}{\diagup}} \equiv CH_2{=}CH_2 \qquad H-C{\equiv}C-H \equiv HC{\equiv}CH
$$

ethylene   acetylene

The C—H bonds in ethylene are approximately C_{sp^2}—H_{1s} and are slightly shorter than C_{sp^3}—H_{1s} bonds. Similarly, the C—H bonds in acetylene are approximately C_{sp}—H_{1s} and are shorter still. The C—C double and triple bonds in ethylene and acetylene are also shorter and stronger than single bonds and are discussed in detail in subsequent chapters (Sections 11.1 and 12.1).

Further examples in which several different types of bonds are involved are

$$
\begin{array}{cc}
\underset{\underset{\displaystyle H}{|}}{\overset{\overset{\displaystyle H}{|}}{H-C}}-\underset{}{\overset{\overset{\displaystyle O}{\|}}{C}}-\underset{\underset{\displaystyle H}{|}}{\overset{\overset{\displaystyle H}{|}}{C}}-H \equiv CH_3COCH_3
&
\underset{\underset{\displaystyle Cl}{|}}{\overset{\overset{\displaystyle Cl}{|}}{H-C}}-Cl \equiv CHCl_3
\end{array}
$$

acetone   chloroform

$$
O{=}C{=}O \equiv CO_2 \qquad
\underset{\underset{\displaystyle H}{|}}{\overset{\overset{\displaystyle H}{|}}{H-C}}-\overset{\overset{\displaystyle O}{\|}}{C}-O-H \equiv CH_3COOH
$$

carbon dioxide   acetic acid

In our subsequent discussion of organic structures, we shall make frequent use of these simple bonding concepts and symbols. We shall find them to be common and powerful devices for understanding physical properties and reactions. Organic structures are generally so large and complex that it is essential to have such systematic methods for dissecting the whole molecule into component parts and individual bonds.

EXERCISE: Write structural and condensed formulas for two compounds having the formula C_4H_{10} and three compounds having the formula C_5H_{12}.

3.2
The Shape of Molecules

Molecular shape is a fundamental concept in organic chemistry. Molecules are three-dimensional, and the spatial interactions between different parts of a molecule can be very important in determining a compound's chemical and physical properties. Consequently, the student must cultivate the ability to think of organic molecules as "objects" having a definite shape. Since this is a difficult task for most people, especially with complex molecules, various aids to visualization are employed. One such technique, which we shall use in this book, is *stereoscopic projection* (see the list of stereo drawings on page xvii).

Other tools which are indispensable to visualizing spatial interactions in organic molecules are *molecular models*. Since bonds in organic compounds are formed from hybrid orbitals which approximate simple *sp-*, *sp²-*, and *sp³*-hybrids, and bond angles and distances are relatively constant from one molecule to another, it is possible to use simple objects as models for various atoms. These objects may be joined together in the same manner as "Tinkertoys" to produce models of molecules. Various types of model sets are available for this purpose. Some are expensive precision constructions used primarily for research purposes, but several are relatively inexpensive and are designed for student use in the study of organic chemistry. Some of the sets available are summarized below and illustrated in Figure 3.1.

Dreiding stereomodels are skeletal models constructed from welded stainless steel tubing. The bond lengths and angles are precisely proportional to the average molecular dimensions. They are relatively expensive and are widely used by professional chemists for research purposes.

An inexpensive student set is marketed by Science Related Materials, Inc. These models use small plastic nuclei and plastic tubing. The Prentice-Hall Framework molecular models are similar, but use metal nuclei. The Godfrey stereomodels use larger plastic nuclei and a more flexible type of tubing. The Benjamin–Maruzen models use plastic atoms with holes drilled to accommodate bonds. Models such as these are relatively inexpensive and are recommended as an aid in visualizing three-dimensional aspects of organic structures and reactions.

Corey–Pauling–Koltun (CPK™) molecular models are an example of space-filling models. The models are constructed from an acrylic polyester plastic and are proportional to the covalent and atomic radii of the atoms. They are held together by connectors made of a hard rubber-like elastomer. They are fairly expensive and are mainly used by professional chemists for constructing models where a knowledge of molecular shape and intramolecular interactions is important.

EXERCISE: If you have purchased a molecular model set, construct models of some of the organic molecules previously mentioned in this chapter. If you do not own a model set, you may wish to experiment with making a few simple models using toothpicks for bonds and either corks or gumdrops for nuclei.

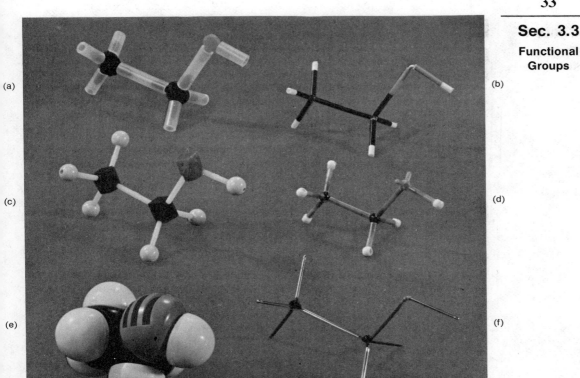

FIGURE 3.1. *Some molecular model representations of ethyl alcohol,* CH₃CH₂OH. *Models used are* (a) *Godfrey,* (b) *Framework,* (c) *Benjamin–Maruzen,* (d) *Science-Related Materials,* (e) *CPK*ᵀᴹ, *and* (f) *Dreiding.*

3.3
Functional Groups

The OH group in water reacts avidly with potassium to form potassium hydroxide and molecular hydrogen.

$$H_2O + K \longrightarrow K^+ OH^- + \tfrac{1}{2} H_2$$

A similar reaction occurs with the OH groups in nitric acid, sulfuric acid, and other acids containing hydroxy groups.

$$O_2NOH + K \longrightarrow K^+ NO_3^- + \tfrac{1}{2} H_2$$

$$HOSO_2OH + 2\,K \longrightarrow 2\,K^+ SO_4^{2-} + H_2$$

It somes as no surprise, therefore, to find the same reaction occurring with methyl alcohol.

$$CH_3OH + K \longrightarrow CH_3O^- K^+ + \tfrac{1}{2}H_2$$

A large number of different alcohols are known. Each consists of an OH group attached to a carbon framework, and all show this same reaction. Because of this constancy in the chemical properties of the OH group, it is unnecessary to study in detail the reactions of each of these many alcohols. Instead, it suffices to study alcohols as a class of organic compounds characterized by the chemical properties of the hydroxy group. This is a fortunate situation, for it gives organic chemistry a logical and systematic structure. There are a number of atoms or groups of atoms that show a relative constancy of properties when attached to different carbon chains. Such groups are called **functional groups.** In our systematic study of organic chemistry we shall examine the chemistry of the important functional groups.

The simplest organic compounds are those that have no functional groups. These compounds consist only of carbon and hydrogen and are molecules in which carbons are joined to each other only by single bonds. These **saturated hydrocarbons** ("saturated" means having no double or triple bonds) may be non-cyclic (the **alkanes**) or cyclic (the **cycloalkanes**). They form the framework to which functional groups may be attached. The symbol R is often used to denote an alkyl group, the simplest being the methyl group, CH_3—. Since the simplest alkane is methane, CH_4, we see that with this symbolism the alkane class may be represented by RH. We shall see in Chapter 6 that alkanes undergo only a limited number of reactions—precisely because they have no functional groups.

Some Alkanes

$$CH_4 \qquad CH_3CH_3 \qquad CH_3CH_2CH_3 \qquad CH_3CH_2CH_2CH_3$$

Some Cycloalkanes

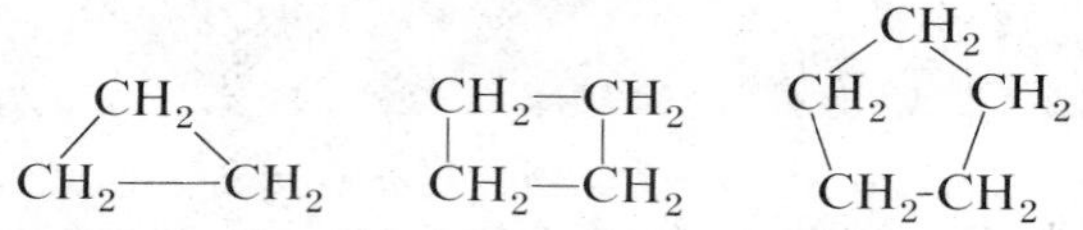

In a similar manner, all hydrocarbons containing one or more C=C double bonds form a logical class, the **alkenes.** The hydrocarbons having a C≡C triple bond form a third structurally similar set, the **alkynes.**

Some Alkenes

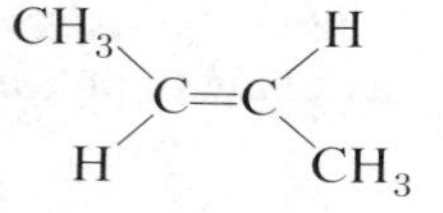

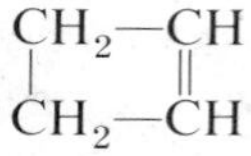

Some Alkynes

$$CH_3-C≡C-H \qquad CH_3CH_2CH_2-C≡C-CH_3 \qquad H-C≡C-H$$

We will find a number of reactions characteristic of C—C multiple bonds that are not shared by single bonds.

Organic compounds that contain C—O single bonds are classed as **alcohols** or **ethers,** depending on whether or not the oxygen is also bonded to a hydrogen.

Some Alcohols

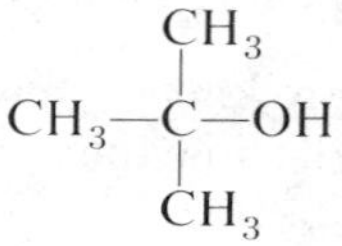

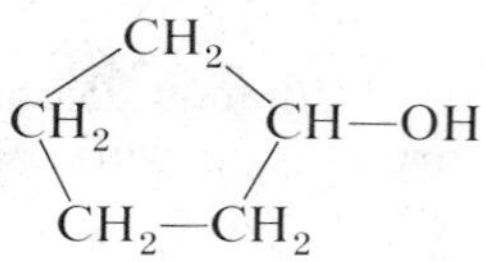

Some Ethers

$$CH_3\!-\!O\!-\!CH_3 \qquad CH_3\!-\!O\!-\!CH\overset{\displaystyle CH_3}{\underset{\displaystyle CH_3}{\Big\langle}}$$

The C=O double bond, the carbonyl group, is found in **aldehydes** and **ketones**.

Combined with an OH group, it becomes a carboxy group, $-\overset{\displaystyle O}{\overset{\|}{C}}\!-\!OH$. Compounds containing this functional group are **carboxylic acids**.

Some Aldehydes

$$H_2C\!=\!O \qquad CH_3\overset{\displaystyle O}{\overset{\|}{C}}H \qquad CH_3CH_2\overset{\displaystyle O}{\overset{\|}{C}}H$$

Some Ketones

$$CH_3\overset{\displaystyle O}{\overset{\|}{C}}CH_3 \qquad CH_3\overset{\displaystyle CH_3}{\underset{}{C}}H\!-\!\overset{\displaystyle O}{\overset{\|}{C}}CH_3$$

Some Carboxylic Acids

$$H\overset{\displaystyle O}{\overset{\|}{C}}OH \qquad CH_3\overset{\displaystyle O}{\overset{\|}{C}}OH$$

Table 3.1 lists a number of the important functional groups. The structures and names of these groups should be committed to memory. They form an essential part of the language of organic chemistry. In our subsequent studies we will develop the chemistry of the individual functional groups in terms of structural and electronic theory, nomenclature (names), physical properties, the preparation from other functional groups, and the reactions that produce other groups.

Interconversions of functional groups constitute a large proportion of organic chemistry. After the individual groups have been studied, the effect of one group on another can be considered, for the organic chemistry of compounds with more than one functional group is not simply the sum of the parts. Groups affect each other, sometimes in complex ways. One of the reasons for studying the theory of organic chemistry is that the mutual interactions of functional groups can be understood.

The aromatic ring in Table 3.1 is written with three C=C double bonds. Nevertheless, we shall see later (Chapters 22 and 23) that compounds containing this ring system differ substantially in their chemistry from the alkenes. Compounds containing this ring system are known collectively as **aromatic compounds**. Compounds with no aromatic ring are known as **aliphatic compounds**.

EXERCISE: Using R = ethyl, write structural and condensed formulas for a compound in each of the classes of organic compounds in Table 3.1.

TABLE 3.1

Class	General Structure	Characteristic Functional Group	Example
alkanes	R—H	none	CH_4
alkenes	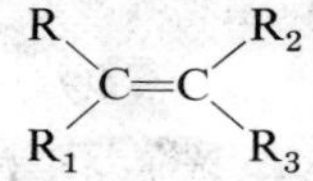 $R\!-\!\underset{R_1}{\overset{R}{C}}\!=\!\underset{R_3}{\overset{R_2}{C}}$	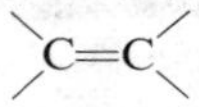 $\diagdown C\!=\!C \diagup$	$CH_3CH\!=\!CH_2$
aromatic ring	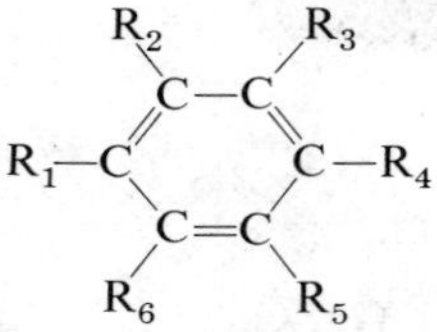 (aromatic ring structure with R_1, R_2, R_3, R_4, R_5, R_6)	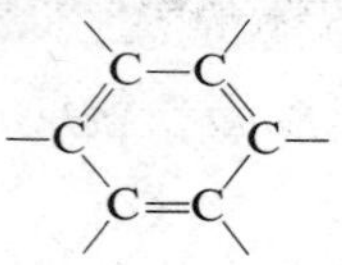 (aromatic ring functional group)	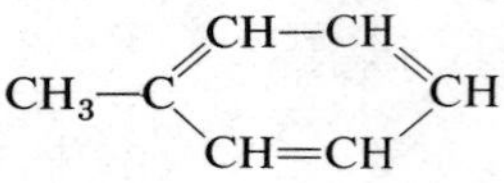 $CH_3\!-\!C$ (aromatic ring)
alkynes	$R\!-\!C\!\equiv\!C\!-\!R'$	$-\!C\!\equiv\!C\!-$	$CH_3\!-\!C\!\equiv\!C\!-\!CH_3$
alkyl halides	RF, RCl, RBr, RI	—F, —Cl, —Br, —I	CH_3Cl
alcohols	R—OH	—OH	CH_3OH
ethers	R—O—R'	—O—	CH_3OCH_3
amines			
primary amines	$R\!-\!NH_2$	$-\!NH_2$	CH_3NH_2
secondary amines	R—NH—R'	$\diagup N\!-\!H$	CH_3NHCH_3
tertiary amines	$R\!-\!\underset{R''}{N}\!-\!R'$	$\diagup N\!-$	$CH_3\underset{CH_3}{N}CH_3$
thiols	R—SH	—SH	CH_3SH
sulfides	R—S—R'	—S—	CH_3SCH_3
disulfides	R—S—S—R'	—S—S—	CH_3SSCH_3
boranes	R_3B	—B—	$(CH_3)_3B$
organometallic	RM, R_2M, R_3M	—M	$CH_3Li, (CH_3)_2Mg, (CH_3)_3Al$
aldehydes	$R\!-\!\overset{O}{\overset{\|}{C}}\!-\!H$	$-\!\overset{O}{\overset{\|}{C}}\!-\!H$	$CH_3\!-\!\overset{O}{\overset{\|}{C}}\!-\!H$
ketones	$R\!-\!\overset{O}{\overset{\|}{C}}\!-\!R'$	$-\!\overset{O}{\overset{\|}{C}}\!-$	$CH_3\!-\!\overset{O}{\overset{\|}{C}}\!-\!CH_3$
imines	$R\!-\!\overset{N\!-\!R'}{\overset{\|}{C}}\!-\!R''$	$-\!\overset{N}{\overset{\|}{C}}\!-$	$CH_3\!-\!\overset{N\!-\!CH_3}{\overset{\|}{C}}\!-\!H$
carboxylic acids	$R\!-\!\overset{O}{\overset{\|}{C}}\!-\!OH$	$-\!\overset{O}{\overset{\|}{C}}\!-\!OH$	$H\!-\!\overset{O}{\overset{\|}{C}}\!-\!OH$

TABLE 3.1 (continued)

Class	General Structure	Characteristic Functional Group	Example
esters	$\overset{\displaystyle O}{\overset{\|}{R-C-OR'}}$	$\overset{\displaystyle O}{\overset{\|}{-C-O-}}$	$\overset{\displaystyle O}{\overset{\|}{CH_3-C-OCH_3}}$
amides	$\overset{\displaystyle O}{\overset{\|}{R-C-NR_2'}}$	$\overset{\displaystyle O}{\overset{\|}{-C-N\backslash}}$	$\overset{\displaystyle O}{\overset{\|}{CH_3-C-NH_2}}$
acyl halides	$\overset{\displaystyle O}{\overset{\|}{R-C-X}}$	$\overset{\displaystyle O}{\overset{\|}{-C-X}}$	$\overset{\displaystyle O}{\overset{\|}{CH_3-C-Cl}}$
nitriles	$R-C\equiv N$	$-C\equiv N$	$CH_3C\equiv N$
nitro compounds	$R-NO_2$	$-NO_2$	CH_3-NO_2
sulfones	$R-SO_2-R'$	$-SO_2-$	$CH_3-SO_2-CH_3$
sulfonic acids	$R-SO_2-OH$	$-SO_2-OH$	CH_3-SO_2-OH

3.4
The Determination of Organic Structures

In previous sections, we have reviewed some basic concepts of electronic structure and bonding and have introduced the subject of organic structures and functional groups. In subsequent chapters, we shall take up the structures and chemical reactions of various classes of organic compounds and examine them in detail. At this point, an additional question must be addressed before we embark upon our systematic study of organic chemistry. How does the chemist know the structure of a compound? The question is an important one, and it is encountered over and over again by researchers in the field. In fact, the rate of development of organic chemistry as a science has been intimately related to our ability to *determine structure*.

Mentally transport yourself back a hundred years—imagine that you are a nineteenth-century scientist and have laboriously purified an organic substance from some source. How do you determine its structure? The substance has various physical properties that can be measured—boiling point, melting point, density, refractive index. These properties constitute the **characterization** of a compound. Just as no two people have identical fingerprints, no two compounds have *all* physical properties in common. It is relatively easy to assemble a catalog of physical properties for the compound and to decide that it is different from other previously isolated substances. Your new compound also undergoes various chemical reactions, and from its behavior with various reagents you may be able to decide what kinds of functional groups are present. For example, suppose that your material reacts with sodium to produce hydrogen gas. Since simple alcohols such as methyl alcohol are known to undergo this reaction, you can make the deduction that your new compound contains the functional group —OH. But still, from all these data, how do you write a molecular structure for the compound? This problem challenged chemists for over a hundred years.

The first major breakthrough came with the development of methods of elemental analysis. The first attempts at elemental analysis were made by Lavoisier in the late eighteenth century in connection with his pioneering work on the reactions of oxygen. Lavoisier examined the combustion products from various compounds and could deduce which elements were present in the substance burned. For example, combustion of methane gives carbon dioxide and water. Hence, methane must be built up from carbon and hydrogen in some way.

$$\text{methane} + O_2 \longrightarrow CO_2 + H_2O$$

Although Lavoisier's method of qualitative analysis sufficed to indicate which elements are present in various compounds, his results complicated rather than simplified organic chemistry. It showed that all known organic compounds were built up from a relatively small number of elements.

The next significant advance came in 1831 when Liebig developed the Lavoisier method into a precise quantitative technique for elemental analysis. For the first time, it was possible to determine accurate empirical formulas for organic compounds. In connection with methods for the determination of molecular weights, it was then possible to determine molecular formulas.

The method of combustion analysis, as developed by Lavoisier and Liebig, is conceptually very simple. A weighed quantity of the sample to be analyzed is burned in the presence of red-hot copper oxide, which is reduced to metallic copper. The sample is swept through the combustion tube with pure oxygen gas, which reoxidizes the copper to copper oxide.

$$C_{10}H_{14} + 27\,CuO \longrightarrow 10\,CO_2 + 7\,H_2O + 27\,Cu$$
naphthalene

$$2\,Cu + O_2 \longrightarrow 2\,CuO$$

The combustion products are swept through a calcium chloride tube, which absorbs the water formed, and then through a tube containing aqueous potassium hydroxide, which absorbs the carbon dioxide produced. The two tubes are weighed before and after combustion to determine the weights of water and carbon dioxide produced. From the weights of the two products, the weight of sample burned, and the atomic weights of carbon and hydrogen, it is possible to compute an empirical formula for the substance burned.

$$\text{weight of H in sample} = \text{weight of } H_2O \times \frac{2.016}{18.016}$$

$$\text{weight of C in sample} = \text{weight of } CO_2 \times \frac{12.01}{44.01}$$

$$\%\ \text{H in sample} = \frac{\text{weight of H}}{\text{weight of sample}} \times 100$$

$$\%\ \text{C in sample} = \frac{\text{weight of C}}{\text{weight of sample}} \times 100$$

If the percentages of carbon and hydrogen do not add up to 100 and no other element has been detected by qualitative tests, the deficiency is taken as the percentage of oxygen.

As an example, consider the analysis of propyl alcohol, C_3H_8O. An ideal analysis on a 0.5000 g sample would give 0.600 g of H_2O and 1.099 g of CO_2. The calculations proceed as follows.

$$\text{weight of H in sample} = 0.600 \text{ g} \times \frac{2.016}{18.016} = 0.067 \text{ g}$$

$$\text{weight of C in sample} = 1.099 \text{ g} \times \frac{12.01}{44.01} = 0.300 \text{ g}$$

$$\% \text{ H in sample} = \frac{0.067}{0.500 \text{ g}} \times 100 = 13.4$$

$$\% \text{ C in sample} = \frac{0.300}{0.500} \times 100 = 60.0$$

The percentages of hydrogen and carbon add up to 73.4%, and the remaining 26.6% is taken as the percentage of oxygen in the sample. In actual practice, the analytical values are usually accurate to $\pm 0.3\%$.

From the elemental analysis of a compound, one may easily calculate its **empirical formula,** which expresses the ratio of the elements present. In the present case, for example, the analysis tells us that 100 g of propyl alcohol contains 60.0 g of carbon, 13.4 g of hydrogen, and 26.6 g of oxygen. Dividing each of these weights by the appropriate atomic weights gives us the number of moles of each element in 100 g of sample.

$$\frac{60.0}{12.01} = 5.00 \text{ moles of carbon}$$

$$\frac{13.4}{1.008} = 13.29 \text{ moles of hydrogen}$$

$$\frac{26.6}{16} = 1.66 \text{ moles of oxygen}$$

This gives us an empirical formula of $C_{5.00}H_{13.29}O_{1.66}$. However, because the atoms in a molecule must be present in whole numbers, the initially derived formula must be normalized. If we divide each of the factors derived above by the smallest, we have

$$C_{5.00/1.66}H_{13.29/1.66}O_{1.66/1.66} = C_{3.01}H_{8.01}O_{1.00}$$

Thus, the empirical formula of propyl alcohol is calculated from its elemental analysis to be C_3H_8O. The **molecular formula** expresses the total number of each atom present and is the same as the empirical formula or some multiple of it. For example, if the molecular formula of propyl alcohol were $C_6H_{16}O_2$, the percentages of carbon, hydrogen and oxygen would be the same. (Actually, because of the rules of valence, $C_6H_{16}O_2$ is an impossible formula, as a little trial and error will readily reveal.)

The Lavoisier–Liebig method of analysis provided a tremendous boost to the development of organic chemistry but required relatively large amounts of sample, on the order of 0.25–0.50 g. In 1911 Pregl introduced a technique of microanalysis that allows combustion analysis to be carried out on 3–4 mg of sample. Elements such as N, S, Cl, Br, I, and P are determined on a micro scale by other analytical methods that we shall not detail. Highly accurate molecular formulas may now be determined on a few micrograms of substance by the technique of high-resolution mass spectrometry (Chapter 17).

From the molecular formula, the next step is to derive a molecular structure. How are the atoms bonded to one another? For our present example of C_3H_8O, which of the following structures corresponds to propyl alcohol?

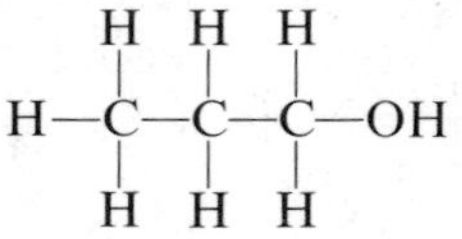

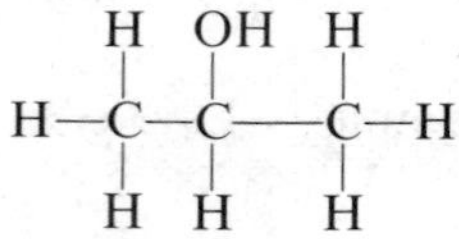

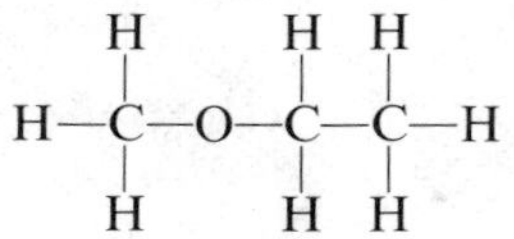

$$
\begin{array}{ccc}
\text{H}\;\;\text{H}\;\;\text{H} & \text{H}\;\;\text{OH}\;\;\text{H} & \text{H}\;\;\;\;\;\text{H}\;\;\text{H} \\
|\;\;\;\;|\;\;\;\;| & |\;\;\;\;|\;\;\;\;| & |\;\;\;\;\;\;\;\;|\;\;\;\;| \\
\text{H--C--C--C--OH} & \text{H--C--C----C--H} & \text{H--C--O--C--C--H} \\
|\;\;\;\;|\;\;\;\;| & |\;\;\;\;|\;\;\;\;| & |\;\;\;\;\;\;\;\;|\;\;\;\;| \\
\text{H}\;\;\text{H}\;\;\text{H} & \text{H}\;\;\text{H}\;\;\text{H} & \text{H}\;\;\;\;\;\text{H}\;\;\text{H}
\end{array}
$$

As indicated previously, some insight may be gained by a consideration of the gross chemical properties of the material. For example, if we know that the compound reacts with sodium to liberate hydrogen, this would indicate the presence of the —OH functional group and would eliminate the third possible structure above. In a similar manner, an examination of *other* chemical properties could lead to the elimination of other candidate structures. For example, we shall see in later chapters that the —OH functional group has slightly different chemical properties when it is bonded to a carbon having two hydrogens than when it is bonded to a carbon having only one hydrogen. Thus, a careful consideration of the properties of a substance could eventually lead to a structure consistent with all the data. The modern chemist relies heavily on spectroscopic methods for the determination of structure. As we shall see in subsequent chapters, *spectroscopy* is the experimental evaluation of the manner in which a substance interacts with electromagnetic radiation. Thus, by examining the various *spectra* of a material, the chemist is evaluating a physical property of the substance. In fact, this particular physical property is a very powerful one, and we shall consider various kinds of spectroscopy in our study of organic chemistry: nuclear magnetic resonance (Chapter 9), infrared spectroscopy (Chapter 14), mass spectrometry (Chapter 17), and ultraviolet spectroscopy (Chapter 21).

EXERCISE: How much CO_2 and H_2O are produced by the combustion of 3.74 mg of $C_6H_{12}O$?

3.5
n-Alkanes, the Simplest Organic Compounds

The straight-chain alkanes constitute a family of hydrocarbons in which a chain of —CH_2— groups is terminated at both ends by a hydrogen. They have the general formula H—$(CH_2)_n$—H or C_nH_{2n+2}. Such a family of compounds, which differ from each other by the number of CH_2 groups in the chain, is called an **homologous series.** The individual members of the family are known as **homologs** of one another. Straight-chain alkanes are called **normal alkanes,** or simply **n-alkanes,** to distinguish them from the branched alkanes, which we shall study later.

Alkanes are sometimes called **saturated** hydrocarbons. This term means that the carbon skeleton is "saturated" with hydrogen. That is, in addition to its bonds to other carbons, each carbon bonds to enough hydrogens to give a maximum covalence of 4. In saturated hydrocarbons, there are only single bonds. Later, we shall study **unsaturated** hydrocarbons, compounds that contain double and triple

C—C bonds. The normal alkanes are named according to the number of carbon atoms in the chain (Table 3.2).

TABLE 3.2

n	Name	Formula
1	methane	CH_4
2	ethane	CH_3CH_3
3	propane	$CH_3CH_2CH_3$
4	butane	$CH_3CH_2CH_2CH_3$
5	pentane	$CH_3(CH_2)_3CH_3$
6	hexane	$CH_3(CH_2)_4CH_3$
7	heptane	$CH_3(CH_2)_5CH_3$
8	octane	$CH_3(CH_2)_6CH_3$
9	nonane	$CH_3(CH_2)_7CH_3$
10	decane	$CH_3(CH_2)_8CH_3$
11	undecane	$CH_3(CH_2)_9CH_3$
12	dodecane	$CH_3(CH_2)_{10}CH_3$
13	tridecane	$CH_3(CH_2)_{11}CH_3$
14	tetradecane	$CH_3(CH_2)_{12}CH_3$
15	pentadecane	$CH_3(CH_2)_{13}CH_3$
20	eicosane	$CH_3(CH_2)_{18}CH_3$
21	heneicosane	$CH_3(CH_2)_{19}CH_3$
22	doeicosane	$CH_3(CH_2)_{20}CH_3$
30	triacontane	$CH_3(CH_2)_{28}CH_3$
40	tetracontane	$CH_3(CH_2)_{38}CH_3$

These names derive from the generic name alkane with the **alk-** stem replaced by a stem characteristic of the number of carbons in the chain. The first four members of this series, methane, ethane, propane, and butane, are names which came into widespread use before any attempts were made to systematize nomenclature. The remaining names derive quite obviously from Greek numbers; compare **pent**agon, **oct**al, **dec**imal, and so on. The student should memorize the names of the *n*-alkanes up through dodecane and know the logical procedure for developing names for larger compounds.

A group is a portion of a molecule in which a collection of atoms is considered together as a unit. For purposes of naming more complicated compounds, it is necessary to have names for such groups. A group name is derived by replacing the **-ane** of the corresponding alkane name by the suffix **-yl.**

Alkane	*Group*	*Sample Molecule*
CH_4	CH_3-	CH_3-OH
meth**ane**	meth**yl** group	methyl alcohol
CH_3CH_3	CH_3CH_2-	CH_3CH_2-Cl
eth**ane**	eth**yl** group	ethyl chloride
$CH_3CH_2CH_3$	$CH_3CH_2CH_2-$	$CH_3CH_2CH_2-Br$
prop**ane**	prop**yl** group	propyl bromide

3.6
Systematic Nomenclature

There is only one compound having each of the formulas CH_4, C_2H_6, and C_3H_8. There are two **isomeric** compounds having the formula C_4H_{10}. **Isomers** are defined as compounds that have identical formulas but differ in the nature or sequence of bonding of their atoms or in the arrangement of their atoms in space. One of the C_4H_{10} isomers is *n*-butane, discussed previously. The other is isobutane.

$$CH_3CH_2CH_2CH_3 \qquad CH_3\overset{\overset{\displaystyle CH_3}{|}}{\underset{\underset{\displaystyle H}{|}}{C}}CH_3$$

n-butane   isobutane

In general, isomers have different physical and chemical properties. Of the two C_4H_{10} compounds, isobutane has the lower melting point and boiling point. The lower boiling point reflects the branched-chain structure of isobutane, which provides less effective contact area for van der Waals attraction.

Interconversion of the two butane isomers requires breaking bonds. Since C—C bonds have bond strengths of about 80 kcal mole^{-1}, these isomers are completely stable under normal conditions. Interconversion requires very high temperatures or special catalysts.

There are three C_5H_{12} isomers.

$$CH_3CH_2CH_2CH_2CH_3 \qquad CH_3CH_2\overset{\overset{\displaystyle CH_3}{|}}{C}HCH_3 \qquad CH_3\overset{\overset{\displaystyle CH_3}{|}}{\underset{\underset{\displaystyle CH_3}{|}}{C}}CH_3$$

pentane   isopentane   neopentane

The **iso-** prefix serves to name one of these isomers, and the new prefix **neo-** provides an additional name. Three of the hexanes can be named using these special prefixes.

$$CH_3CH_2CH_2CH_2CH_2CH_3 \qquad CH_3\overset{\overset{\displaystyle CH_3}{|}}{C}HCH_2CH_2CH_3 \qquad CH_3\overset{\overset{\displaystyle CH_3}{|}}{\underset{\underset{\displaystyle CH_3}{|}}{C}}CH_2CH_3$$

n-hexane   isohexane   neohexane

However, these are the only special prefixes in general use. With more carbons the number of possible isomers increases rapidly. As shown in Table 3.3, there are 5 possible hexanes, 9 heptanes, and 75 decanes, and with larger alkanes the number of possible isomers becomes astronomic. Clearly, in this situation an essential requirement is a systematic nomenclature so that each different compound may be assigned an unambiguous name.

This problem was solved by an international group of chemists who met in Geneva as part of the first meeting of the International Union of Pure and Applied Chemistry. The Geneva rules of 1892 are continuously updated and extended as new kinds of compounds are discovered. They now comprise a consistent and detailed nomenclature known as the IUPAC rules. The IUPAC system of alkane nomenclature is based on the simple fundamental principle of considering

TABLE 3.3
Number of Isomers of C$_n$H$_{2n+2}$

n	Number of Isomers
4	2
5	3
6	5
7	9
8	18
9	35
10	75
11	159
12	355
13	802
14	1,858
15	4,347
16	10,359
17	24,894
18	60,523
19	148,284
20	366,319

all compounds to be **derivatives of the longest single carbon chain** present in the compound. Appendages to this chain are designated by appropriate prefixes. The chain is then numbered from one end to the other. The end chosen as number 1 is that which gives the smaller number at the first point of difference.

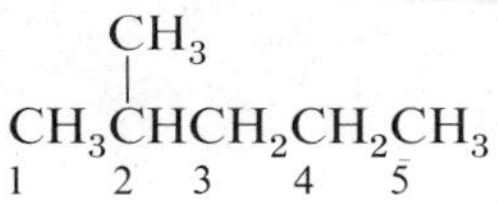

$$\underset{1 \quad 2 \quad 3 \quad 4 \quad 5}{\overset{\overset{\displaystyle CH_3}{|}}{CH_3CHCH_2CH_2CH_3}} \qquad \overset{\overset{\displaystyle CH_3}{|}}{CH_3CH_2CHCH_2CH_2CH_3}$$

2-methylpentane   3-methylhexane
(not 4-methylpentane)   (not 4-methylhexane)

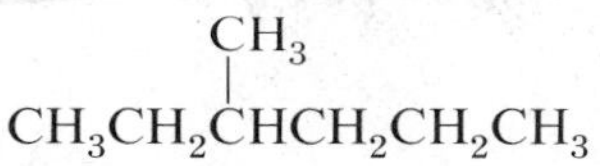

The modifying prefixes di-, tri-, tetra-, penta-, hexa-, and so on are used to indicate multiple identical appendages, but every appendage group still gets its own number.

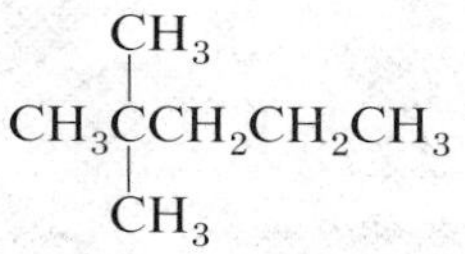

$$\underset{\underset{\displaystyle CH_3}{|}}{\overset{\overset{\displaystyle CH_3}{|}}{CH_3CCH_2CH_2CH_3}} \qquad \underset{\underset{\displaystyle CH_3 \quad CH_3}{| \qquad |}}{\overset{\overset{\displaystyle CH_3 \quad CH_3}{| \qquad |}}{CH_3C-CH_2-CCH_3}}$$

2,2-dimethylpentane   2,2,4,4-tetramethylpentane
(not 2-dimethylpentane)

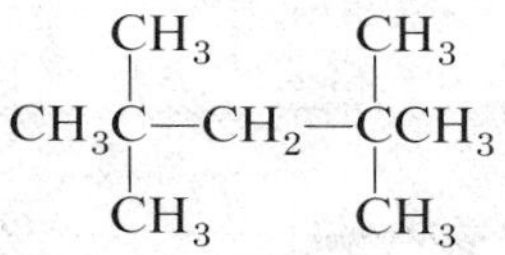

When two or more appendage locants are employed, the longest chain is numbered from the end which produces the lowest series of locants. When comparing one series of locants with another, that series is lower which contains the lower number at the first point of difference.

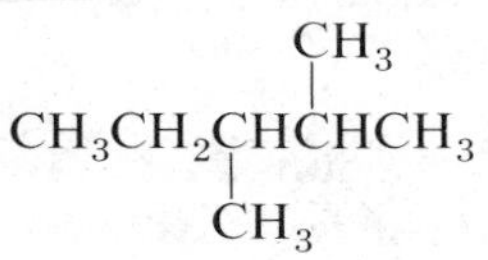

$$\underset{\underset{\displaystyle CH_3}{|}}{\overset{\overset{\displaystyle CH_3}{|}}{CH_3CH_2CHCHCH_3}} \qquad \underset{\underset{\displaystyle CH_3}{|}}{\overset{\overset{\displaystyle CH_3 \qquad CH_3}{| \qquad\qquad |}}{CH_3CHCH_2CH_2CHCHCH_2CH_3}}$$

2,3-dimethylpentane   2,5,6-trimethyloctane
(not 3,4-dimethylpentane) (not 3,4,7-trimethyloctane)

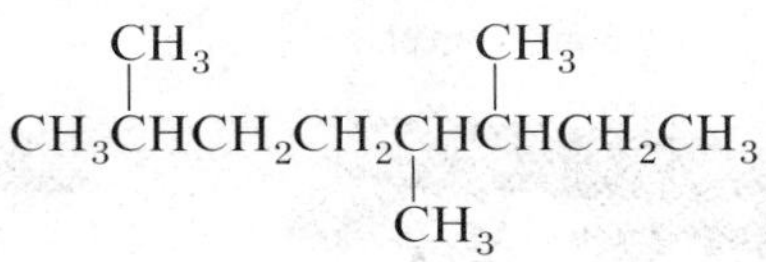

$$CH_3\overset{\underset{\displaystyle CH_3}{|}}{CH}CH_2\overset{\underset{\displaystyle CH_3}{|}}{CH}CH_2CH_2\overset{\underset{\displaystyle CH_3}{|}}{CH}CH_3$$

2,4,7-trimethyloctane
(not 2,5,7-trimethyloctane)

$$CH_3\overset{\underset{\displaystyle CH_3}{|}}{\overset{\displaystyle CH_3}{C}}CH_2\overset{\underset{\displaystyle CH_3}{|}}{CH}CH_3$$

2,2,4-trimethylpentane
(not 2,4,4-trimethylpentane)

Groups derived from the terminal position of a *n*-alkane are named as shown in Section 3.5. Several other common groups have special names that must be memorized by the student.

$$CH_3\overset{\underset{\displaystyle CH_3}{|}}{CH}-$$

isopropyl

$$CH_3\overset{\underset{\displaystyle CH_3}{|}}{CH}CH_2-$$

isobutyl

$$CH_3CH_2\overset{\underset{\displaystyle CH_3}{|}}{CH}-$$

sec-butyl

$$CH_3\overset{\underset{\displaystyle CH_3}{|}}{\overset{\displaystyle CH_3}{C}}-$$

tert-butyl
or *t*-butyl

$$CH_3\overset{\underset{\displaystyle CH_3}{|}}{\overset{\displaystyle CH_3}{C}}CH_2-$$

neopentyl

A more complex appendage group is named as a derivative of the **longest carbon chain in the group** starting from the carbon that is attached to the principal chain. The description of the appendage is distinguished from that of the principal chain by enclosing it in parentheses.

$$
\begin{array}{c}
CH_3 \\
| \\
H-C-CH_3 \\
| \\
H-C-CH_3 \\
| \\
CH_3CH_2CH_2CH_2CHCH_2CH_2CH_2CH_3
\end{array}
$$

5-(1,2-dimethylpropyl)nonane

When two or more appendages of different nature are present, they are cited as prefixes in alphabetical order. Prefixes specifying the number of identical appendages (di-, tri-, tetra-, and so on) and hyphenated prefixes (*tert-* or *t-*, *sec-*) are ignored in alphabetizing except when part of a complex substituent. The prefixes cyclo-, iso-, and neo- count as a part of the group name for purposes of alphabetizing.

$$CH_3CH_2\overset{\underset{\displaystyle CH_2CH_3}{|}}{\overset{\displaystyle CH_3}{C}}CH_2CH_3$$

3-ethyl-3-methylpentane

$$CH_3CH_2CH_2\overset{\underset{\displaystyle CH_2CH_2CH_3}{|}}{\overset{\displaystyle CH_3CHCH_3}{CH}}CHCH_2CH_2CH_3$$

4-isopropyl-5-propyloctane

When chains of equal length are competing for selection as the main chain for purposes of numbering, that chain is selected which has the greatest number of appendages attached to it.

$$CH_3-\underset{\underset{\displaystyle CH_2}{|}}{\overset{\overset{\displaystyle CH_2CH_3}{|}}{C}}-CH_3$$

$$CH_3CH_2CH_2CH_2CH_2\underset{|}{CH}CH_2\underset{\underset{\displaystyle CH_2CH_3}{|}}{CH}CH_2CH_3$$

5-(2-ethylbutyl)-3,3-dimethyldecane
[not 5-(2,2-dimethylbutyl)-3-ethyldecane]

When two or more appendages are in equivalent positions, the lower number is assigned to the one that is cited first in the name (that is, the one that comes first in the alphabetic listing).

$$CH_3CH_2\underset{\underset{\displaystyle CH_3CH_2}{|}}{CH}CH_2\underset{\underset{\displaystyle CH_3}{|}}{CH}CH_2CH_3$$

$$CH_3CH_2CH_2\underset{\underset{\displaystyle CH_2CH_3}{|}}{\overset{\overset{\displaystyle CH_3CHCH_3}{|}}{CH}}CHCH_2CH_2CH_3$$

3-ethyl-5-methylheptane
(not 5-ethyl-3-methylheptane)

4-ethyl-5-isopropyloctane
(not 4-isopropyl-5-ethyloctane)

Note that the direction of numbering of the main chain may *already* have been decided by application of a higher priority rule.

$$CH_3CH_2\underset{\underset{\displaystyle CH_3}{|}}{\overset{\overset{\displaystyle CH_3}{|}}{C}}CH_2CH_2\underset{\underset{\displaystyle CH_2CH_3}{|}}{CH}CH_2CH_3$$

$$CH_3\underset{\underset{\displaystyle CH_3}{|}}{CH}CH_2CH_2CH_2\underset{\underset{\displaystyle CH_2CH_3}{|}}{CH}CH_2CH_3$$

6-ethyl-3,3-dimethyloctane
[3,3,6 is lower than 3,6,6 at
the first point of difference
(underlined)]

6-ethyl-2-methyloctane
(2,6 is lower than 3,7)

The complete IUPAC rules actually allow a choice regarding the order in which the appendage groups may be cited. One may cite the appendages alphabetically, as above, or in order of increasing complexity. In this book, we shall adhere to the alphabetic order in citing appendage prefixes. The alphabetic order is also used by *Chemical Abstracts* for indexing purposes.

The prefix halo- is used as a generic expression for the halogens, which are treated in the same manner as alkyl appendages for purposes of nomenclature. The individual halogen prefixes are fluoro-, chloro-, bromo-, and iodo-.

$$CH_3\underset{\underset{\displaystyle Cl}{|}}{CH}CH_2\underset{\underset{\displaystyle CH_3}{|}}{CH}CH_3$$

$$CH_3CH_2\underset{\underset{\displaystyle CH_2Cl}{|}}{CH}CH_2CH_3$$

2-chloro-4-methylpentane

3-(chloromethyl)pentane

$$ClCH_2CH_2CH_2CH_2Br$$

$$BrCH_2\underset{\underset{\displaystyle CH_3}{|}}{CH}FCHCH_2I$$

$$BrCH_2CH_2CHBrCH_2Br$$

1-bromo-4-chlorobutane

1-bromo-2-fluoro-4-
iodo-3-methylbutane

1,2,4-tribromobutane

As with many other classes of organic compounds, haloalkanes can be named with systematic names, such as the foregoing, or with *common names* (names which evolved before attempts were made to systematize nomenclature). Simple haloalkanes are often named as though they were salts of alkyl groups—as *alkyl halides.*

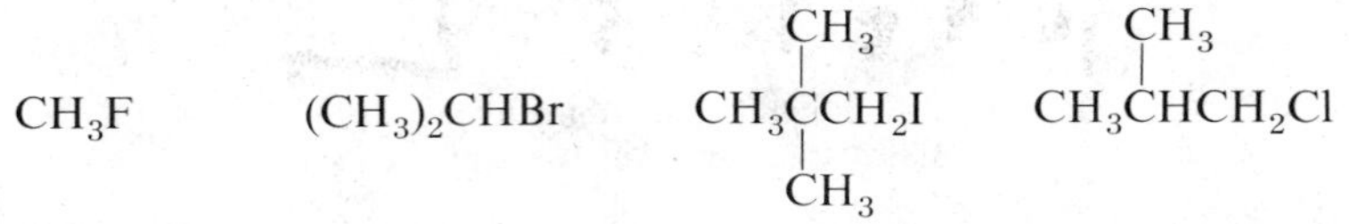

$$CH_3F \qquad (CH_3)_2CHBr \qquad CH_3\overset{\overset{\displaystyle CH_3}{|}}{\underset{\underset{\displaystyle CH_3}{|}}{C}}CH_2I \qquad CH_3\overset{\overset{\displaystyle CH_3}{|}}{CH}CH_2Cl$$

methyl fluoride　　isopropyl bromide　　neopentyl iodide　　isobutyl chloride

Nomenclature is an essential element of organic chemistry for several reasons. First, it is our basic tool for *communicating* about the subject. It is not always convenient to draw a structure on every bottle you want to label or for every compound you want to talk about. Therefore, it is important that you be able to assign a name to every compound, and it is essential that the name you use correspond to *only one compound.* This is the most fundamental rule of nomenclature. If more than one structure can be written which corresponds to a name, or if the name is so ambiguous that no unique structure can be written, then the name is incorrect.

However, another important use of nomenclature is in *searching the chemical literature* for the physical and chemical properties of various compounds. We shall consider the chemical literature in Chapter 33; however, a brief mention at this point is appropriate to the subject of nomenclature. New chemical information is made public for the first time as scientific **papers** which appear in various chemical magazines called **journals.** Examples are the *Journal of the American Chemical Society,* the *Journal of Organic Chemistry,* and *Chemische Berichte.* Hundreds of such journals are published mostly on a monthly or twice-monthly basis. The back issues of these basic journals contain all of the accumulated knowledge of the science. To facilitate the retrieval of information from this mass of data, the American Chemical Society publishes a reference journal known as *Chemical Abstracts.* This journal is published twice monthly and contains short abstracts of all of the chemical papers published in the basic journals. At the end of each year, an extensively cross-referenced index is published. At the end of each 5-year period (10-year period before 1957), a **cumulative index** covering that period is published.

In order to search the chemical literature for a compound *by name* it is necessary to know the correct name of any given structure. For such a name search to be successful, it follows that every structure correspond to *only one name.* The IUPAC system was the first attempt to construct such an unambiguous system of nomenclature. The actual rules are quite extensive and allow for all sorts of special situations. More complete versions of the rules may be found in standard reference works such as the *Chemical Rubber Handbook of Physics and Chemistry.* Unfortunately, the rules as they have been formulated are not totally unambiguous, and all of the special situations which may arise were not anticipated. Moreover, in some situations the rulemakers were permissive rather than compulsory, as in allowing the retention of certain common names and in allowing a choice in deciding which appendage is cited first in a name. Therefore, one finds that the same compound may be named several different ways in different important reference works. In fact, in the most important reference source, *Chemical Abstracts,* one will find that the name used for a compound has often been changed over the years. Thus, in order to use the name of a compound unambiguously in searching the literature, it is necessary to be able to name the compound in all of the ways that a knowledgeable chemist might use. Fortunately, the necessity of using compound names only for searching the literature is partially alleviated by the existence of formula indices. In the future, literature searches will probably be done by computer and use symbols for structures instead of names.

It is essential that the student thoroughly learn the simplified IUPAC system that we have presented here, which may be summarized as follows.

1. Find the longest carbon chain in the compound.
2. Name each appendage group that is attached to this principal chain.
3. Alphabetize the appendage groups.
4. Number the principal chain from one end in such a way that the smaller number is used at the first point of difference.
5. Assign to each appendage group a number signifying its point of attachment to the principal chain.

We shall see that this basic system is used in naming all other classes of organic compounds. For naming substances which have functional groups, the foregoing system is simply modified by the use of appropriate prefixes or suffixes. For example, the characteristic suffix denoting the functional group —OH is *-ol* and that for the group —NH$_2$ is *-amine*. Thus, alcohols and amines are named as follows:

$$\overset{\overset{\textstyle CH_3}{|}}{CH_3CH_2CHCH_2CH_2OH} \qquad CH_3CH_2CH_2CH_2CH_2NH_2$$

3-methyl-1-pentanol   1-pentanamine

Complete details on the modifications which are necessary to adapt the basic system to a given class of compound will be given when we discuss each class.

PROBLEMS

1. From the analytical values for each compound, derive its empirical formula.
   (a) hexanol: 70.4% C, 13.9% H
   (b) benzene: 92.1% C, 7.9% H
   (c) pyrrole: 71.6% C, 7.5% H, 20.9% N
   (d) morphine: 71.6% C, 6.7% H, 4.9% N
   (e) quinine: 74.1% C, 7.5% H, 8.6% N
   (f) DDT: 47.4% C, 2.6% H, 50.0% Cl
   (g) vinyl chloride: 38.3% C, 4.8% H, 56.8% Cl
   (h) thyroxine: 23.4% C, 1.4% H, 65.3% I, 1.8% N

2. In each of the following examples, qualitative analysis shows the presence of no elements other than C, H, and O. Calculate the empirical formula for each case.
   (a) Combustion of 0.0132 g of camphor gave 0.0382 g of CO_2 and 0.0126 g of H_2O.
   (b) Combustion of 1.56 mg of the sex-attractant of the common honeybee (*Apis mellifera*) gave 3.73 mg of CO_2 and 1.22 mg of H_2O.
   (c) Benzo[a]pyrene is a potent carcinogenic compound that has been detected in tobacco smoke. Combustion of 2.16 mg gave 7.50 mg of CO_2 and 0.92 mg of H_2O.

3. Halogen may be determined by burning the sample under conditions such that the halogen is converted into halide ion, which is then determined by titration with standard silver nitrate solution. Carbon and hydrogen are determined by conversion into CO_2 and H_2O as usual. Calculate the empirical formula for the following case: 2.03 mg gave 4.44 mg of CO_2 and 0.91 mg of H_2O; 5.31 mg gave a chloride solution which required 4.80 mL of 0.0110 N $AgNO_3$.

4. Write condensed formulas and IUPAC names for the eight isomers having the molecular formula $C_5H_{11}Br$.

5. Write condensed formulas for seven isomers having the molecular formula $C_4H_{10}O$.

6. The empirical formula C_4H_8 may correspond to many molecular formulas: C_8H_{16}, $C_{12}H_{24}$, etc. Explain why the empirical formula C_5H_{12} can only correspond to one molecular formula.

7. Convince yourself by trial and error that the molecular weight of an organic compound containing only carbon, hydrogen, and oxygen will always be an even number and that the molecular weight for compounds having an odd number of nitrogens will be an odd number.

8. Each of the following molecules contains one principal functional group. Locate and name the group, and classify the molecule for each case.

(a)
(b)
(c)

(d)
(e)
(f)

(g)
(h)
(i)

(j)
(k)
(l)

(m)
(n) $CH_3CH_2CH_2CH_2Li$

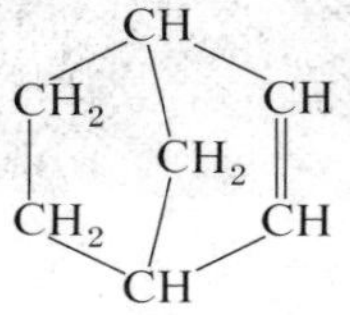

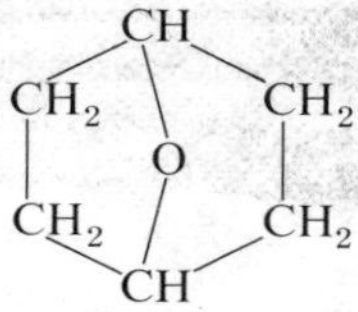

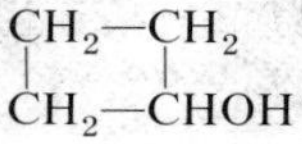

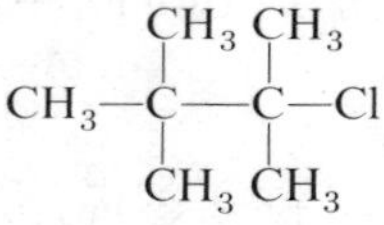

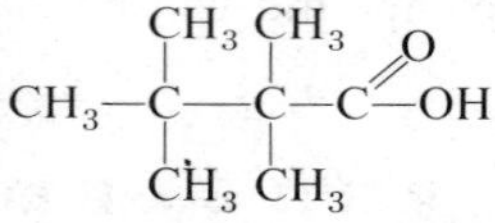

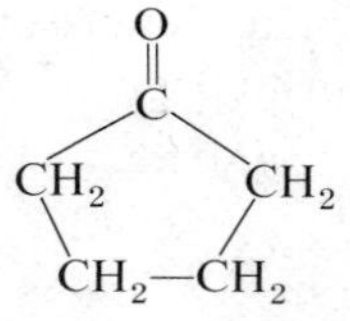

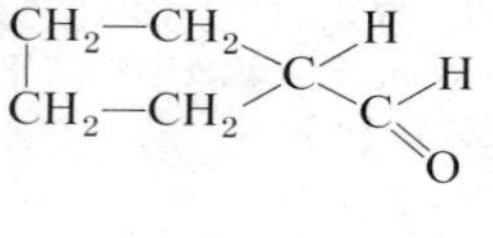

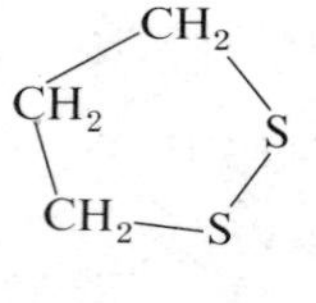

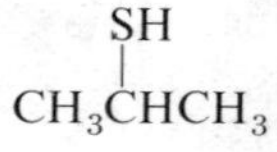

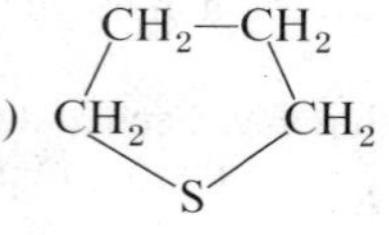

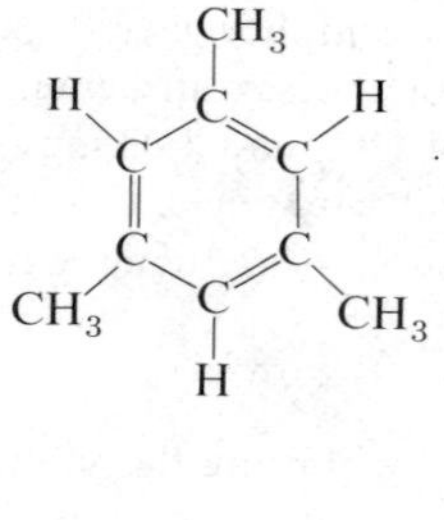

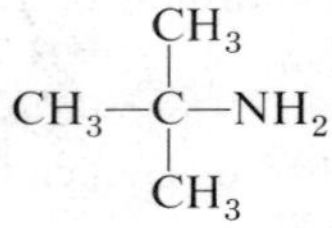

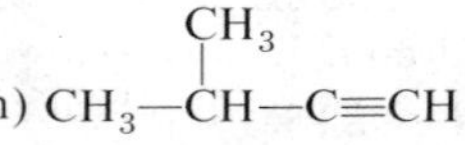

9. Each of the following structures is a significant organic molecule. Identify and name every functional group in each structure.

(a)

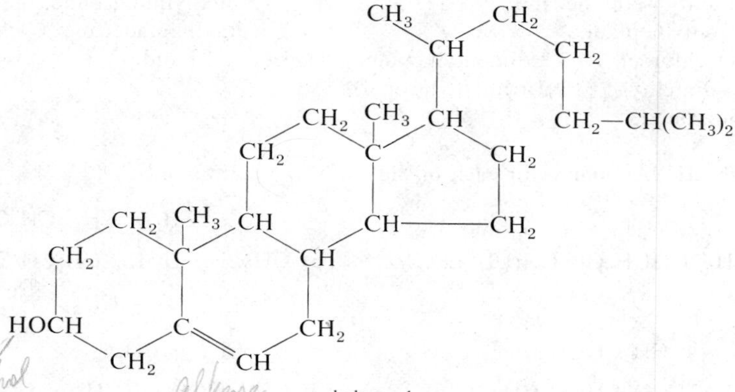

CH_3 CH_2 CH CH_2 CH_2 CH_3 CH $CH_2-CH(CH_3)_2$ CH_2 CH_2 C CH_2 CH_2 CH_3 CH CH CH_2 CH_2 C CH *alcohol* $HOCH$ C CH_2 CH_2 CH *alkene*

cholesterol

(b) *alcohole* *alkene*

$HO-C$ C O *keytone* O *ester* $HO-C$ $CHCHOHCH_2OH$

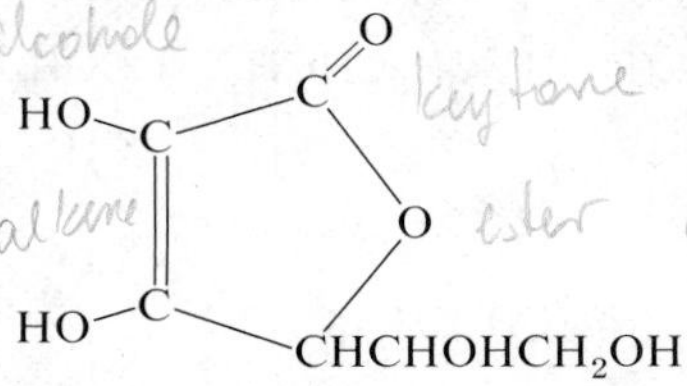

vitamin C

(c) *ethers* *aromatic ring* *primary amide*

$(C_2H_5)_2NCH_2CH_2OC-C$ $CH-CH$ $C-NH_2$ $CH=CH$ *tertiar amide* *keytone*

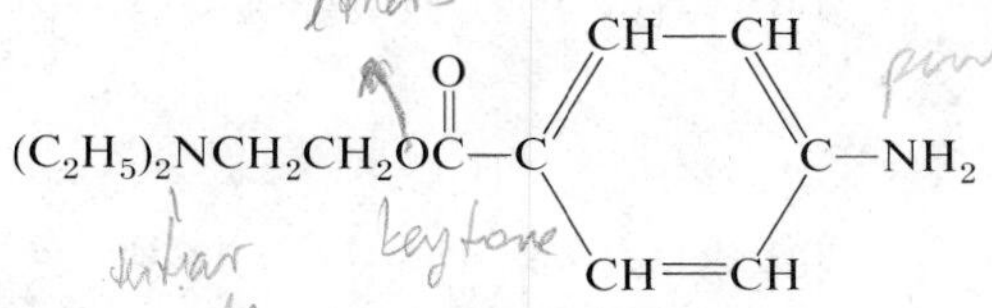

novocaine

(d) *sulfide* *acid*

CH_3 CH_3-C $CHC-OH$ S *tertiary amine* N CH $C=O$ *keytone* CH $NHCOCH_2O-C$ $CH-CH$ *aromatic compd.* CH *ester* $CH=CH$ *(2nd amine)*

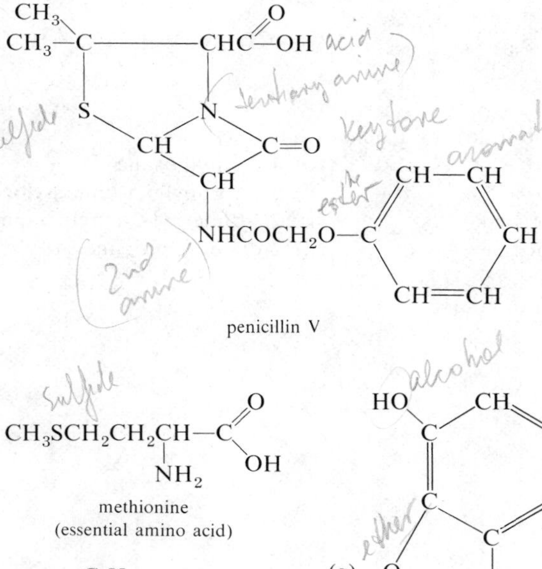

penicillin V

(e) *aromatic compd.*

CH_3 C CH_2 O C CH_3-C-CH_3 CH_2 CH_2 C H

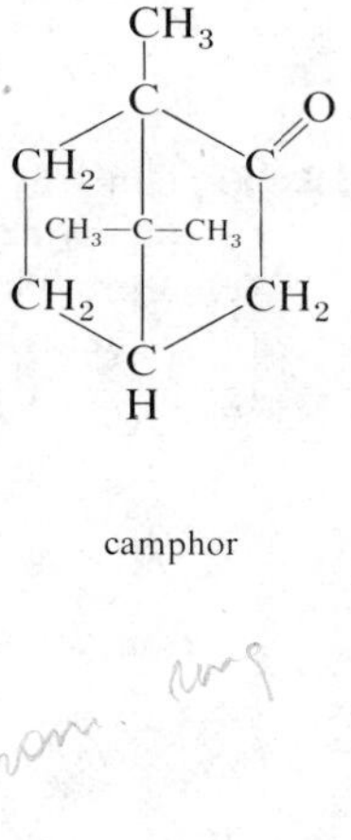

camphor

(f) *sulfide*

$CH_3SCH_2CH_2CH-C$ O NH_2 OH

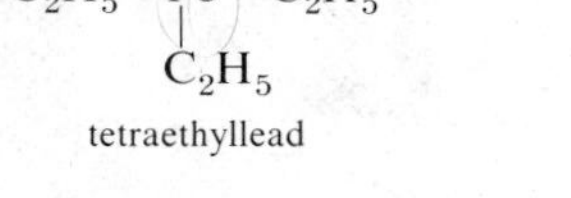

methionine
(essential amino acid)

(g) *alcohol* *arom ring*

$HO-C$ $CH-CH$ C *ether* C O CH_2 C CH CH NCH_3 *3rd. amine* $CH-CH$ CH_2 CH_2 *alcohol* $HOCH$ CH *alkene*

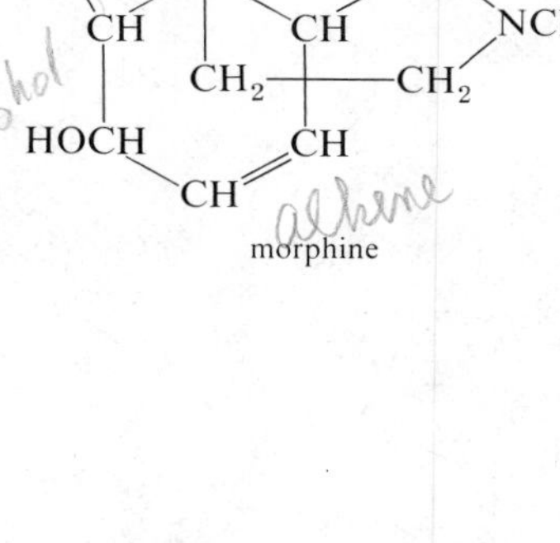

morphine

(h) C_2H_5 $C_2H_5-Pb-C_2H_5$ C_2H_5

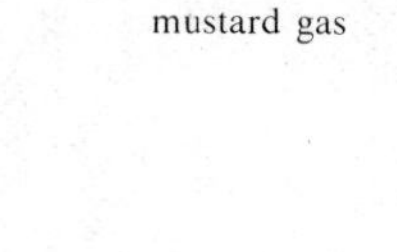

tetraethyllead

(i) $ClCH_2CH_2SCH_2CH_2Cl$

mustard gas

10. Write the structural formula of each of the following compounds.
 (a) neopentane
 (b) isobutane
 (c) *t*-butyl bromide
 (d) 3,4,5-trimethyl-4-propyloctane
 (e) 3-fluoro-4-ethylhexane
 (f) 6-(3-methylbutyl)undecane
 (g) 4-*t*-butylheptane
 (h) 2-methylheptadecane
 (i) 4-(1-chloroethyl)-3,3-dimethylheptane
 (j) isobutyl iodide
 (k) 6,6-dimethyl-5-(1,2,2-trimethylpropyl)-
  dodecane

11. Give the IUPAC name for each of the following compounds.

(a) $(CH_3)_2CHCH_2CH_2CH(CH_3)_2$

(b) 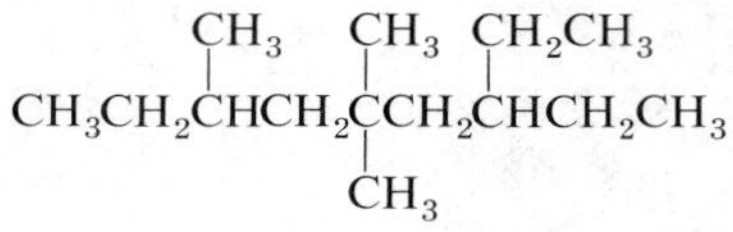
$$\underset{\qquad\qquad\quad\underset{\displaystyle CH_3}{|}}{CH_3CH_2\overset{\displaystyle\overset{CH_3}{|}}{C}HCH_2\overset{\displaystyle\overset{CH_3}{|}}{C}CH_2\overset{\displaystyle\overset{CH_2CH_3}{|}}{C}HCH_2CH_3}$$

(c)
$$CH_3CH_2CH_2\overset{\displaystyle\overset{CH_3CHCH_2CH_3}{|}}{C}HCH_2CH_3$$

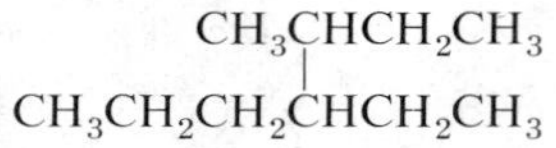

(d) 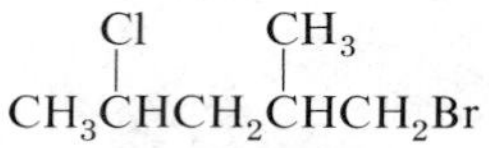
$$CH_3\overset{\displaystyle\overset{Cl}{|}}{C}HCH_2\overset{\displaystyle\overset{CH_3}{|}}{C}HCH_2Br$$

(e) 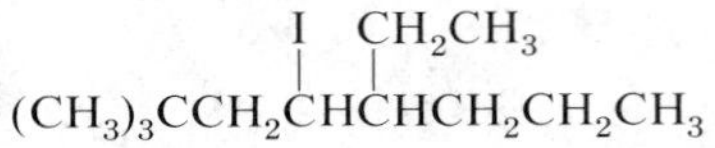
$$(CH_3)_3CCH_2\overset{\displaystyle\overset{I}{|}}{C}H\overset{\displaystyle\overset{CH_2CH_3}{|}}{C}HCH_2CH_2CH_3$$

(f) 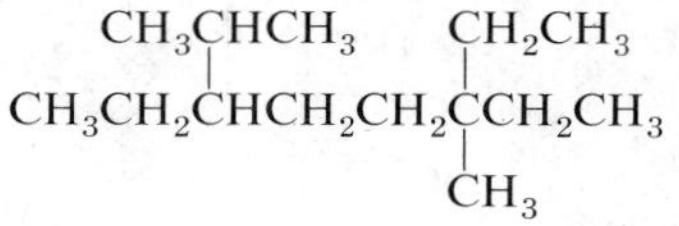
$$CH_3CH_2\overset{\displaystyle\overset{CH_3CHCH_3}{|}}{C}HCH_2CH_2\overset{\displaystyle\overset{CH_2CH_3}{|}}{\underset{\displaystyle\underset{CH_3}{|}}{C}}CH_2CH_3$$

(g) 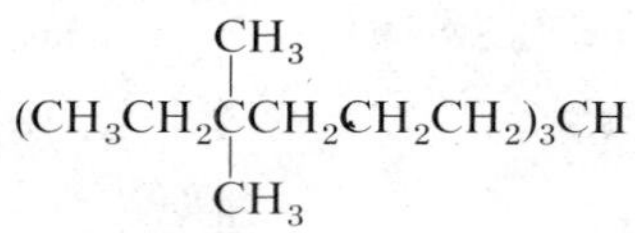
$$(CH_3CH_2\overset{\displaystyle\overset{CH_3}{|}}{\underset{\displaystyle\underset{CH_3}{|}}{C}}CH_2CH_2CH_2)_3CH$$

(h) $(CH_3CH_2)_4C$

(i) 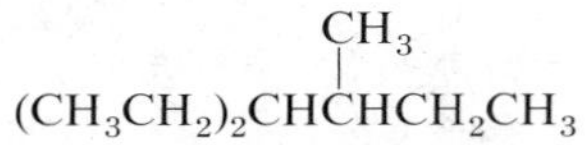
$$(CH_3CH_2)_2CH\overset{\displaystyle\overset{CH_3}{|}}{C}HCH_2CH_3$$

(j) 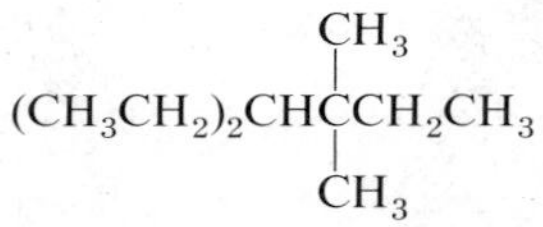
$$(CH_3CH_2)_2CH\overset{\displaystyle\overset{CH_3}{|}}{\underset{\displaystyle\underset{CH_3}{|}}{C}}CH_2CH_3$$

12. Explain why each of the following names is incorrect.
 (a) methylheptane
 (b) 4-methylhexane
 (c) 3-propylhexane
 (d) 3-isopropyl-5,5-dimethyloctane
 (e) 3-methyl-4-chlorohexane
 (f) 2,2-dimethyl-3-ethylpentane
 (g) 3,5,6,7-tetramethylnonane
 (h) 2-dimethylpropane

Chapter 4
Organic Reactions

4.1
Introduction

The two principal components of organic chemistry are **structure** and **reactions.** Each of these components has experimental and theoretical aspects, and they are interrelated. Structures in terms of bond angles and bond distances are available from the interpretation of experimentally obtained rotational spectra and x-ray or electron diffraction patterns. In Chapter 2 we reviewed the symbolism used to represent such structures—structural formulas and Lewis structures—and their modern significance in terms of atomic, hybrid, and molecular orbitals.

In this chapter we introduce some concepts concerning reactions. Many reactions are known in organic chemistry that allow us to convert one structure to another. In this connection we must distinguish between **equilibrium** and **rate.** Equilibrium refers to the relative amounts of reactants and products expected by thermodynamics, *if a suitable pathway exists between them.* A simple example is glucose in the presence of oxygen. These two reagents can exist together for indefinite periods without change, but if the sugar is ignited it will burn to produce the equilibrium products CO_2 and H_2O. Alternatively, the same result is accomplished in living organisms by a series of catalysts (enzymes) that allow this oxidation to occur in a sequence of controlled steps.

Reactants can reach equilibrium at a variety of rates ranging from immeasurably slow to exceedingly fast. The rate at which equilibrium is reached depends on the reaction and on the structures of the reactants. Consequently, we will be much concerned with the effect of structural change on reactivity. We will also find that many reactions are characteristic of individual functional groups and form much of the chemistry of functional groups. Finally, we will also be concerned with the pathway, or *mechanism,* by which reactants find their way to products.

4.2
An Example of an Organic Reaction: Equilibria

Although methyl chloride is a gas at room temperature, it is sufficiently soluble in water to give a solution of about 0.1 M concentration. If the solution also contains hydroxide ion, reaction occurs to form methyl alcohol and chloride ion.

$$HO^- + CH_3Cl \rightleftharpoons CH_3OH + Cl^- \qquad (4\text{-}1)$$

$$\underset{\text{methyl chloride}}{CH_3Cl} \qquad \underset{\text{methyl alcohol}}{CH_3OH}$$

At equilibrium all four compounds are present, but the equilibrium constant K is such an exceedingly large number that the amount of methyl chloride present in the equilibrium mixture is vanishingly small.

$$K = \frac{[CH_3OH][Cl^-]}{[CH_3Cl][OH^-]} = 10^{16}$$

If the reaction started with 0.1 M CH_3Cl and 0.2 M NaOH, at the end of the reaction we would have a solution of 0.1 M CH_3OH, 0.1 M NaCl, 0.1 M NaOH, and 10^{-17} M CH_3Cl. That is, 1 mL of such a solution would contain only a few thousand molecules of CH_3Cl.

Such a reaction is said to **go to completion.** In practice a reaction may be considered to go to completion if the final equilibrium mixture contains less than about 0.1% of reactant.

The reaction of methyl chloride and hydroxide ion may also be characterized by the **Gibbs standard free energy change $\Delta G°$** at equilibrium.

$$\Delta G° = -RT \ln K$$

$\Delta G°$ for reaction (4-1) is -22 kcal mole^{-1}, a rather large value. We may speak of this reaction as having a large **driving force.** That is, the driving force for a reaction is a qualitative description of an equilibrium property and is related to the overall free energy change. For comparison, $\Delta G°$ for a reaction that proceeds to 99.9% at equilibrium is 4.1 kcal mole^{-1} at 25° C.*

There is an important difference between ΔG and $\Delta G°$. ΔG is the free energy of a given system. $\Delta G°$ is the free energy of that system with the components in their standard states. For gases, the standard state is generally the corresponding ideal gas at 1 atmosphere pressure. For solutions, the standard state is normally chosen to be the ideal 1 M solution. The standard free energy of a system $\Delta G°$ is defined as the free energy ΔG of an ideal solution in which each reactant and product is present in a concentration of 1 M. For such a system, $\Delta G = \Delta G°$. When reaction has reached equilibrium, the free energy of the system $\Delta G = 0$. The concentrations of the components at this point are given as

$$\Delta G = \Delta G° + RT \ln K$$

where

$$K = \frac{[\text{products}]_{eq}}{[\text{reactants}]_{eq}}$$

Hence, the standard free energy is given by

$$\Delta G° = -RT \ln K$$

Note also that the units of K are determined by the standard states.

The Gibbs standard free energy may be dissected into **enthalpy $\Delta H°$** and **entropy $\Delta S°$** components.

$$\Delta G° = \Delta H° - T \Delta S°$$

Enthalpy is the heat of reaction and is generally associated with bonding. If stronger bonds are formed in a reaction, $\Delta H°$ is negative and the reaction is **exothermic.** A reaction with positive $\Delta H°$ is **endothermic.** Entropy is best thought of as freedom of motion. The more a molecule or portion of a molecule is restricted in motion, the more negative is the entropy. Both the formation of stronger bonds and greater freedom of motion can contribute to a favorable driving force for reaction (negative $\Delta G°$).

For reaction (4-1), $\Delta H° = -18$ kcal mole^{-1} and $\Delta S° = +13$ eu (ΔS is usually specified in entropy units, eu, which have the units cal deg^{-1}). The driving force in this case comes mostly from bond energy changes; a C—O bond is stronger than a

---

* In general, temperatures will be given in degrees centigrade (Celsius) and the symbol C will be omitted, except where confusion might arise.

C—Cl bond. The formation of stronger bonds is usually an important component of the driving force of a reaction.

In the vapor phase, where intermolecular interactions are negligible, the strength of the internal bonds in a molecule is especially important in determining its stability. In solutions, however, one must also consider the intermolecular interactions with solvent molecules **(solvation).** Solvent interactions that involve varying degrees of ionic and covalent bonding are particularly important for ions. They provide the main driving force for breaking up the stable crystal lattices when ionic substances dissolve. Although solvation of an ion provides bonding stabilization which is reflected in $\Delta H°$, it is partially offset by a decrease in entropy $\Delta S°$. The crowding of several solvent molecules around an ion restricts the freedom of motion of these molecules. In the present case the entropy of reaction $\Delta S°$ is positive because chloride ion is less strongly solvated than hydroxide ion. That is, the solvent molecules are less restricted after reaction than before. Since $\Delta S°$ is positive, the quantity $(-T\Delta S°)$ contributes a negative value to $\Delta G°$ and provides an additional driving force for reaction to occur.

EXERCISE: Using the basic thermodynamic relationships given in this section, calculate the equilibrium constant K for a reaction that has $\Delta H° = -10$ kcal mole^{-1} and $\Delta S° = -22$ eu (a) if the reaction is carried out at 27°C (300° Kelvin) and (b) if the reaction is carried out at 227°C.

4.3
Reaction Kinetics

Because it has such a large driving force, it seems remarkable that the reaction of methyl chloride with hydroxide ion is relatively slow. For example, a 0.05 M solution of methyl chloride in 0.1 M aqueous sodium hydroxide will have reacted only to the extent of about 10% after 2 days at room temperature. It is an important principle in all reactions that favorable thermodynamics is not enough; a suitable reaction pathway is essential.

Reactions generally involve an **energy barrier** that must be surmounted in going from reactants to products. This barrier is called the **activation energy** or the **enthalpy of activation** and is symbolized by $\Delta H^{\ddagger}$. In the reaction of methyl chloride with hydroxide ion, $\Delta H^{\ddagger}$ is about 25 kcal mole^{-1}. This appears to be a rather formidable hurdle when one realizes that the average kinetic energy of molecules at room temperature is only about 0.6 kcal mole^{-1}. However, this latter number is only an average. Molecules are continually colliding with each other at rapid rates and exchanging kinetic energy. At any given instant some molecules have less than this average energy, some have more, and a few even have very large energies—like 25 kcal mole^{-1}.

The relative number of molecules with any given energy is given by the Boltzmann distribution function, shown schematically in Figure 4.1. Most of the molecules have an energy close to the average energy represented by the large hump. Only the minute fraction of molecules in the far end of the asymptotic tail have sufficient energy to overcome the barrier to reaction.

At a higher temperature, the average kinetic energy of the molecules is greater, and the entire distribution function is shifted to higher energies, as shown by the dashed curve in Figure 4.1. The fraction of molecules with kinetic energy sufficient for reaction is larger, and the rate of reaction is correspondingly larger. For example, the reaction of methyl chloride with hydroxide ion is 25 times faster at

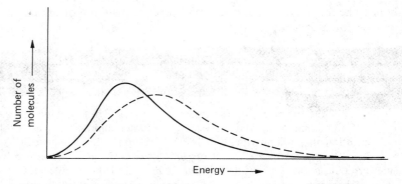

FIGURE 4.1. *A Boltzmann distribution function. Dotted line shows a higher temperature.*

50° than it is at 25°. A useful rule of thumb for many organic reactions is that a 10° change in temperature causes a two- to- threefold change in rate of reaction.

The rate of a chemical reaction depends not only on the fraction of molecules that have sufficient energy for reaction but also on their concentration because this determines the probability of an encounter that could lead to reaction. Reaction rates are directly proportional to the concentrations of the reactants, and the proportionality constant is called a **rate constant** k. The reaction of methyl chloride with hydroxide ion is an example of a **second-order reaction,** since the rate depends on two concentrations.

$$\text{rate} = k[CH_3Cl][OH^-] \tag{4-2}$$

"Rate" involves a change in concentration of something per unit time, usually expressed as moles per liter per second, $M\ sec^{-1}$. In equation (4-2), therefore, k must have units of $M^{-1}\ sec^{-1}$.

$$(M^{-1}\ sec^{-1})(M)(M) = M\ sec^{-1}$$

The "something" whose concentration is changing is either a reactant or a product.

$$\text{rate} = -\frac{\Delta[CH_3Cl]}{\Delta t} = -\frac{\Delta[OH^-]}{\Delta t} = \frac{\Delta[CH_3OH]}{\Delta t} = \frac{\Delta[Cl^-]}{\Delta t} \tag{4-3}$$

The minus signs for the reactants indicate that their concentrations decrease with increasing time. All of the changes shown are equal by stoichiometry.

$$\left[\text{In the language of calculus, this rate equation becomes} \right.$$
$$\left. -\frac{d[CH_3Cl]}{dt} = -\frac{d[OH^-]}{dt} = \frac{d[CH_3OH]}{dt} = \frac{d[Cl^-]}{dt} = k[CH_3Cl][OH^-] \right]$$

The actual value of k at 25° is $6 \times 10^{-6}\ M^{-1}\ sec^{-1}$.

The reaction of methyl chloride with hydroxide ion may be compared with its reaction with water. The reaction with water is an example of a **first-order reaction.** The reaction involves water molecules, but, because water is the solvent, it is present in large excess. Therefore the concentration of water remains effectively the same, even after all of the methyl chloride has reacted. Since the concentration of water appears not to change during the reaction, it does not appear in the kinetic expression; thus the rate of reaction depends only on the concentration of methyl chloride.

$$\text{rate} = k[CH_3Cl] \tag{4-4}$$

Because only one concentration is involved, equation (4-4) is the equation of a first-order reaction. The rate is a change in concentration per unit time, for example, moles per liter per second, $M \text{ sec}^{-1}$. Therefore, k has the units of sec^{-1}. For methyl chloride at $25°$, $k = 3 \times 10^{-10} \text{ sec}^{-1}$.

The reaction of methyl chloride with water is experimentally a first-order reaction because the concentration of one reactant (water) does not change significantly during the reaction. A second-order reaction becomes effectively a first-order reaction if the concentration of one component is much greater than the other. For example, in the reaction of a solution of $0.01\ M\ CH_3Cl$ with $1\ M\ NaOH$ the concentration of hydroxide ion changes from $1\ M$ to $0.99\ M$ during the reaction. That is, its concentration remains essentially constant and the reaction of methyl chloride under these conditions appears to be a first-order reaction with a rate constant of $6 \times 10^{-6} \text{ sec}^{-1}$. Such a reaction is also called a **pseudo first-order reaction.**

The student should be careful not to confuse "rate" and "rate constant." The rate constant k for a reaction is simply a numerical measure of how fast a reaction can occur if the reactants are brought together. It relates the actual concentrations of the reactants to the rate of reaction, which is the "through-put" or "flux" of the reaction.

EXERCISE: Using the rate constant $k = 6 \times 10^{-6}\ M^{-1}\ \text{sec}^{-1}$ for the reaction of methyl chloride with hydroxide ion, calculate the initial rate of reaction for each of the following initial reactant concentrations.
(a) $0.1\ M\ CH_3Cl$ and $1.0\ M\ OH^-$ (b) $0.1\ M\ CH_3Cl$ and $0.1\ M\ OH^-$
(c) $0.01\ M\ CH_3Cl$ and $0.01\ M\ OH^-$

4.4
Reaction Profiles and Mechanism

In the reaction of methyl chloride and hydroxide ion atoms must move around and bonds must change in order that the products methyl alcohol and chloride ion may be produced. One of the important concepts in organic chemistry involves the consideration of the structure of the system as reaction proceeds. Each configuration of the atoms during the process of changing from reactants to products has an associated energy. Since reaction generally involves bringing the reactants close together and breaking bonds, these structures generally have higher energy than the isolated reactants. That is, as the reactants approach each other and start to undergo the molecular changes that will eventually result in products, the potential energy of the reacting system increases. As the reaction encounter continues, the potential energy continues to increase until the system reaches a structure of **maximum energy.** Thereafter the changes that result in the final products continue, but the structures represent lower and lower energy until the products are fully formed.

The difference in the energy of the isolated reactants and the maximum energy structure which the system passes through on the path to products is the **activation energy** of the reaction. This maximum energy corresponds to a definite structure called the **transition state.** The measure of the progress of reaction from reactants to products is the **reaction coordinate.** This coordinate is usually not specified in detail because the qualitative concept is usually sufficient, but in our reaction, for example, it could be represented by the C—O bond length or C—Cl bond length as the reaction progresses, or by the net electronic charge on chlorine. Whatever measure is used, the general reaction profile is given by Figure 4.2. In this figure, the energy shown is the potential energy. This quantity is related to but is not identical with $\Delta H°$. Similarly, the difference in potential energy between reactants and products contributes to $\Delta G°$ for the reaction. The magnitude of this difference determines the *position* of equilibrium. The magnitude of the activation energy determines the *rate* at which equilibrium is established.

The energy quantities involved in reactions are given more precise definitions in the **theory of absolute rates.** In this theory the transition state is characterized by thermodynamic properties: free energy, enthalpy, and entropy. The rate constant for reaction is related to the Gibbs free energy difference between the transition state, sometimes called an **activated complex,** and the reactant state by the equation

$$k = \nu^{\ddagger} e^{-\Delta G^{\ddagger}/RT} = \nu^{\ddagger} e^{-\Delta H^{\ddagger}/RT} e^{\Delta S^{\ddagger}/R}$$

The proportionality constant, $\nu^{\ddagger}$, is a kind of frequency. Its magnitude is 6.2×10^{12} sec^{-1} at 25°, a magnitude comparable to ordinary vibration frequencies. In fact, the reaction process can be described as one of the modes of vibration of the activated complex. For this reason, the activated complex or transition state is not a normal molecule and is only a transient phase in the course of reaction.

The structure of the transition state is an important feature of a reaction. If we can estimate its energy, we can predict the reaction rate, at least roughly. For example, a transition state in which several bonds are broken is likely to correspond to high energy and a slow reaction. Furthermore, and most important, from the structure of the transition state we can often evaluate how a given change in structure will change the rate. Unfortunately, we cannot directly observe a transition state—we cannot take its spectrum or determine its structure by x-ray diffraction. Instead, we must infer its structure indirectly.

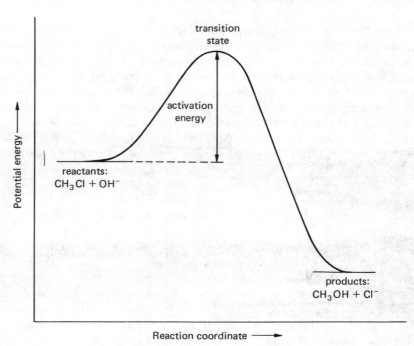

FIGURE 4.2. *A reaction profile for the reaction* $CH_3Cl + OH^- \longrightarrow CH_3OH + Cl^-$.

The reaction of hydroxide ion with methyl chloride is an example of an S_N2 reaction, which we will study in Chapter 9. The structure of the transition state will be developed at that time. We will find, for example, that the S_N2 reaction probably involves a single step with one transition state, as suggested in Figure 4.2. Many other reactions, however, involve more than one step. A reaction profile such as that in Figure 4.3 is not uncommon. Such a reaction involves one or more intermediates, and each intermediate is flanked by transition states. Reaction intermediates correspond to energy minima on the reaction coordinate diagram. They may be sufficiently stable that they can be isolated and stored in bottles, or they may have such fleeting existence that their presence must be inferred from subtle observations of cleverly designed experiments.

The **reaction mechanism** is a sequential account of each transition state and intermediate in a total reaction. The overall rate of reaction is determined approximately by the transition state of highest energy in the sequence, so that this structure has particular importance. The step involving this transition state is often called the **rate-determining step.**

The reaction profile shown in Figure 4.3 corresponds to the equations

$$\text{reactants} \underset{k_{-1}}{\overset{k_1}{\rightleftharpoons}} \text{intermediate} \underset{k_{-2}}{\overset{k_2}{\rightleftharpoons}} \text{products}$$

The relative energies in this figure correspond to rate constants having the relationship

$$k_{-1} > k_1 > k_2 > k_{-2}$$

Remember that the lower the energy barrier, the larger the rate constant. Because

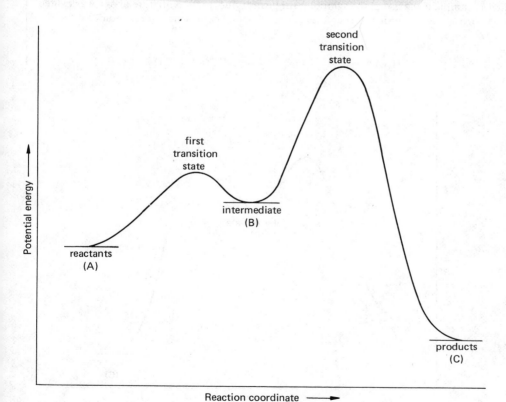

FIGURE 4.3. *Profile of a more complex reaction.*

it is the highest point between reactants and products, the second transition state in Figure 4.3 corresponds to the rate-determining step.

Again, it must be emphasized that rate constant and rate are not the same. For example, in Figure 4.3 the first step has a greater rate constant than the second step (i.e., $k_1 > k_2$). It is also clear from the graph that since B is less stable than A, the equilibrium between A and B is such that the concentration of B at any given time is much less than the concentration of A, or $[A] > [B]$. Thus, the first step in the reaction has a faster rate than the second ($k_1[A] > k_2[B]$).

As a further example of this point, consider the reaction coordinate diagram for a two-step reaction shown in Figure 4.4. Again, the highest energy transition state occurs in the second step, which is therefore the rate-determining step. In this example $k_2 > k_1$. However, since $[A] \gg [B]$, the rate of the first step is still faster than the rate of the second.

An example of a reaction with intermediates is the hydrolysis of ethyl acetate with aqueous sodium hydroxide.

$$CH_3\overset{\overset{\displaystyle O}{\|}}{C}OCH_2CH_3 + OH^- \xrightarrow[H_2O]{25°} CH_3CO_2^- + CH_3CH_2OH$$

ethyl   acetate   ethyl
acetate   ion   alcohol

> This example illustrates the way in which organic reactions are typically written. The arrow shows the direction of the reaction and implies that the equilibrium lies far to the right. Reaction conditions such as solvent, temperature, and any catalysts used are written with the arrow as shown. Abbreviations are often used in this formulation. An example is the use of the symbol Δ for heat. If our reaction mixture above was heated or refluxed in order to speed reaction, we could represent the reaction as
>
> $$CH_3\overset{\overset{\displaystyle O}{\|}}{C}OCH_2CH_3 + OH^- \xrightarrow[H_2O]{\Delta} CH_3CO_2^- + CH_3CH_2OH$$

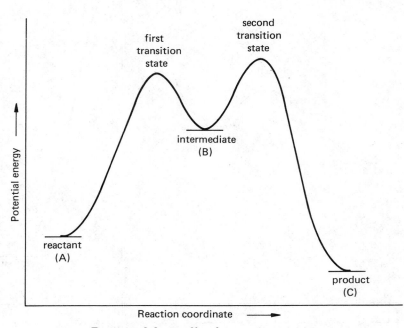

FIGURE 4.4. *Profile of a two-step reaction.*

The rate expression for this reaction is

$$\text{rate} = 0.1 \, [\text{OH}^-][\text{CH}_3\text{COOCH}_2\text{CH}_3]$$

The second-order rate constant $0.1 \, M^{-1} \, \text{sec}^{-1}$ is relatively large and corresponds to a rather fast reaction. As we shall learn in Chapter 19, the mechanism of this reaction appears to be

$$\underset{\overset{\displaystyle \|}{\text{O}}}{\text{CH}_3\text{C}}\text{OCH}_2\text{CH}_3 + \text{OH}^- \rightleftharpoons \underset{\overset{\displaystyle |}{\text{OH}}}{\overset{\overset{\displaystyle \text{O}^-}{|}}{\text{CH}_3\text{C}}}-\text{OCH}_2\text{CH}_3$$

$$\underset{\overset{\displaystyle |}{\text{OH}}}{\overset{\overset{\displaystyle \text{O}^-}{|}}{\text{CH}_3\text{C}}}-\text{OCH}_2\text{CH}_3 \longrightarrow \text{CH}_3\text{COOH} + \text{CH}_3\text{CH}_2\text{O}^-$$

The reaction is effectively irreversible because the strong base $\text{CH}_3\text{CH}_2\text{O}^-$ reacts immediately with acetic acid to produce ethyl alcohol and acetate ion.

$$\underset{\text{acetic acid}}{\text{CH}_3\text{COOH}} + \text{CH}_3\text{CH}_2\text{O}^- \xrightarrow{\text{fast}} \text{CH}_3\text{CO}_2^- + \text{CH}_3\text{CH}_2\text{OH}$$

The reaction profile for this reaction is shown in Figure 4.5.

Reaction profiles such as those in Figures 4.2–4.5 can be useful because they illustrate a rather complex pattern in a diagrammatic way that is easy to visualize. But such diagrams have only a qualitative significance. The reason is that the rate constant for a given reaction step depends on the Gibbs free energy difference between the reactants and transition state, $\Delta G^{\ddagger}$, a thermodynamic quantity that has significance only for a statistical collection of species. It is not defined for the distorted structures that result as the reactant molecule is changing to the product molecule. The potential energy is defined for individual molecules and distorted structures, but is only one

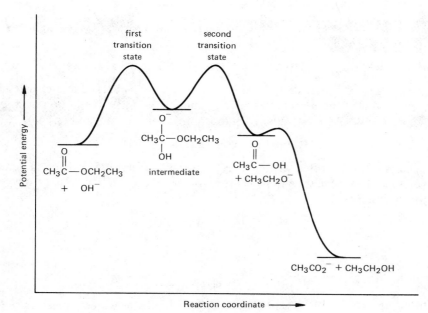

FIGURE 4.5. *Reaction profile for hydrolysis of an ester.*

component of the free energy. Nevertheless, it is generally (but not always) true that a larger potential energy increase between reactants and transition state is associated with a larger $\Delta G^{\ddagger}$ and a smaller rate constant. Moreover, for these same reasons the term "rate-determining step" has only approximate significance. In multistep reactions the actual rate of reaction may be a complex function of time and reactant concentrations that does not lend itself to a simple pictorial representation.

The elucidation of reaction mechanisms is a fascinating branch of organic chemistry. In our study of organic chemistry, we will deal frequently with reaction mechanisms because they help enormously in our classification and understanding of the vast array of organic reactions known. In some important cases, such as the hydrolysis of ethyl acetate, we will also study some of the experimental evidence from which the reaction mechanisms and transition state structures have been deduced.

EXERCISE: Construct a reaction coordinate diagram for a reaction $A \rightleftharpoons B \rightleftharpoons C$ in which the relative stabilities of the three species are $C > A > B$ and for which the relative order of the four rate constants is $k_2 > k_{-1} > k_1 \gg k_{-2}$. Which is the rate-determining step in your diagram? What is the relative order of the four activation energies, E_{act}? Of the two forward steps, which has the faster rate? Is there any way you can adjust the relative levels within the foregoing specified conditions so that the rate of the other step becomes faster?

4.5
Acidity and Basicity

One of the most important reactions in chemistry is that associated with acidity and basicity. **Acidity** refers to the loss of a proton.

$$HY \rightleftharpoons H^+ + Y^-$$

Such reactions can be measured for the gas phase and are invariably highly *endothermic*. For example,

$$HCl(g) \rightleftharpoons H^+(g) + Cl^-(g) \qquad \Delta H° = 328.8 \text{ kcal mole}^{-1}$$

This reaction is extremely endothermic because a bond is broken and charges are separated; both of these processes require energy. In aqueous solution acidity is defined in terms of a dissociation equilibrium involving solvated species.

$$HCl(aq) \rightleftharpoons H^+(aq) + Cl^-(aq)$$

The species $H^+(aq)$ is often written as H_3O^+ for convenience. However, in actuality even more extensive solvation occurs involving hydrogen bonds to the oxygens of other water molecules.

$$H_2O\text{----}H \diagdown \underset{\underset{\underset{OH_2}{|}}{\overset{+}{O}}}{} \diagup H\text{----}OH_2$$

Similarly, the Cl^- ion is solvated by hydrogen bonds to water molecules.

Hydrogen bonds will be discussed in greater detail in Section 10.3. They result from the electrostatic attraction between lone-pair electrons, such as those on oxygen and chlorine, and hydrogens that are bound to an electronegative element. Such hydrogens are good "receptors" for these electrostatic bonds because of the partial positive charge they bear. These **solvation bonds** are sufficiently strong in the aggregate to compensate for the energy that must be supplied to break the H—Cl bond and the electrostatic energy required to separate negative from positive charges.

Acidity measurements have been made in many different solvents. Nevertheless, water is by far the most common solvent and is assumed to be the reference solvent if none is specified. The acidity of an acid HA in water is defined as the equilibrium constant for the reaction

$$HA(aq) \; \overset{K}{\rightleftharpoons} \; H^+(aq) + A^-(aq)$$

$$K = \frac{[H^+][A^-]}{[HA]}$$

Note that the concentration of water does not appear in the expression and that K normally has units of M or moles liter^{-1}.

Acidity equilibrium constants vary over a wide range. Acids with $K > 1$ are referred to as strong acids; acids with $K < 10^{-4}$ are weak acids, and many compounds are very weak acids for which K is exceedingly small. Methane, for example, is a very weak acid with $K \approx 10^{-50}$. This number is so small (it corresponds to approximately one pair of dissociated ions per universe of solution) that it is known only approximately and must be measured indirectly.

Acidity equilibrium constants are usually expressed as an exponent of 10 in order to accommodate this large range of possible values. The pK is defined as the negative exponent of ten, or as

$$pK = -\log K$$

The pK values for some common acids are summarized in Table 4.1. A more extensive list is given in Appendix IV. The term "acidity" is also used in a qualitative sense to refer to acidic character relative to water. Solutions of acids that are substantially more acidic than water have significant hydrogen ion concentrations. That is, their aqueous solutions have pH values less than 7. Recall that the pH is defined as the negative logarithm of the hydrogen-ion concentration.

$$pH = -\log [H^+]$$

Neutral water has $[H^+] = 1.0 \times 10^{-7} \, M$ or pH $= 7.0$. Solutions with pH < 7 are "acidic" and have a distinctive sharp taste at the tip of the tonque. The strong acids HI, HBr, HCl, HNO_3, and H_2SO_4 are commonly referred to as "mineral acids." Their solutions have the low pH values characteristic of solutions that are "acidic." Acetic acid is a weaker acid, but its aqueous solutions also have pH values that are definitely "acidic." Alcohols have about the same acidity as water itself and are therefore not "acidic" in this sense.

TABLE 4.1
Acidities of Common Acids at 25°

Name	Formula	pK
acetic acid	CH_3COOH	4.76
ammonium ion	NH_4^+	9.24
hydriodic acid	HI	-5.2
hydrobromic acid	HBr	-4.7
hydrochloric acid	HCl	-2.2
hydrocyanic acid	HCN	9.22
hydrofluoric acid	HF	3.18
hydrogen selenide	H_2Se	3.71
hydrogen sulfide	H_2S	6.97
methyl alcohol	CH_3OH	15.5
nitric acid	HNO_3	-1.3
nitrous acid	HNO_2	3.23
phosphoric acid	H_3PO_4	2.15 (7.20, 12.38)[a]
phenol	C_6H_5OH	10.00
sulfuric acid	H_2SO_4	-5.2 (1.99)[a]
water	H_2O	15.74

[a] Values in parentheses are the constants for dissociation of the second and third protons from phosphoric acid and the second proton from sulfuric acid.

Table 4.1 gives second and third dissociation constants where appropriate. Thus, HSO_4^- with $pK = 1.99$ is about as acidic as the first dissociation of H_3PO_4, $pK = 2.15$. The acidity of $H_2PO_4^-$, $pK = 7.20$, is comparable to that of H_2S, $pK = 6.95$, and both pKs are comparable to the pH of neutral water, 7. The significance of this point is apparent from the following analysis. When an acid is exactly 50% dissociated, the remaining acid concentration [HA] is equal to the concentration of the conjugate anion [A⁻]. For such a case, the hydrogen-ion concentration of the solution is numerically equal to the acidity constant, or pH = pK.

$$K = \frac{[A^-][H^+]}{[HA]}$$

$$pK = -\log K = -\log [H^+] = pH$$

For aqueous solutions pK values in the range of about 2–12 are known fairly accurately. The acidity constants of stronger acids ($pK < 2$) are known with less precision because such acids are extensively dissociated in aqueous solution.

We have discussed acidity in terms of the process of acid HA giving up a proton to the solvent water. But the process is an equilibrium, the reverse of which involves the reaction of the conjugate anion A⁻ to accept a proton from the solvent. We could speak equally of the acidity of HA or of the **basicity** of A⁻.

$$HA \rightleftharpoons H^+ + A^-$$

conjugate   conjugate
acid   base

The stronger HA is as an acid, the weaker A⁻ is as a base. We could define basicity in terms of removing a proton from water.

$$A^- + H_2O \overset{K_b}{\rightleftharpoons} HA + OH^-$$

$$K_b = \frac{[HA][OH^-]}{[A^-]}$$

The negative logarithm of K_b can be defined as pK_b.

$$pK_b = -\log K_b$$

To distinguish the acid and base equilibria, use is often made of the symbol K_a. The constants K_a and K_b are not independent; they are related through the auto-protolysis constant of water K_w in the following way.

$$K_w = [H^+][OH^-] = 1.0 \times 10^{-14} \, M^2 \quad (\text{at } 25°)$$

$$K_a = \frac{[H^+][A^-]}{[HA]} = \frac{K_w}{[H^+][OH^-]} \cdot \frac{[H^+][A^-]}{[HA]} = \frac{K_w}{K_b}$$

$$pK_a + pK_b = 14$$

Because of these relationships it is not necessary to consider acidity and basicity separately. We can refer to basicity in terms of the acidity of the conjugate acid; that is, K refers generally to K_a and pK refers generally to pK_a. Remember that a strong base has a weak conjugate acid and a weak base has a strong conjugate acid. The pKs of acids weaker than water are difficult to measure and are almost always determined indirectly. Their conjugate bases are strong bases whose reactions with water lie far to the right.

$$\underset{\substack{\text{strong} \\ \text{base}}}{A^-} + H_2O \rightleftharpoons OH^- + \underset{\substack{\text{weak} \\ \text{acid}}}{HA}$$

The relationship between structure and acidity is extremely important in chemistry. It is no less important in organic chemistry, and we shall refer frequently throughout our study to the acidities of different functional groups and the manner in which these acidities change as the structure is perturbed. We shall find that the mechanisms of many organic reactions involve intermediates functioning in one way or another as acids or bases. For comparisons of the acidities of two or more compounds, some generalizations are useful. Removing a proton from a molecule involves breaking a bond to hydrogen and putting a negative charge on the remaining system. Thus, **weaker bonds generally lead to greater acidity.** As we move from element to element down the periodic table, bond strengths diminish because valence atomic orbitals become larger and more diffuse and their overlap to a hydrogen $1s$-orbital is less effective. This factor can have a dominating effect on acidity. Compare the acidity series $H_2O < H_2S < H_2Se$ and $HF < HCl < HBr < HI$.

Stabilization of the negative charge on an anion also generally results in greater acidity of the acid from which it is derived. Along any given row of the periodic table electronegativity increases as we move to the right. This is the dominating factor in such acidity comparisons as $HF > H_2O$ and $HCl > H_2S$. Stabilization of the anion can also result from more extensive distribution of charge. As we shall discuss in more detail in Section 18.4, one reason for carboxylic acids being more acidic than alcohols is that each carboxylate oxygen bears only one-half of the negative charge.

$$ROH \rightleftharpoons H^+ + RO^-$$

$$\underset{}{\overset{O}{\underset{\|}{RC}}}\!-\!OH \rightleftharpoons H^+ + \left[\overset{O}{\underset{\|}{RC}}\!-\!O^- \longleftrightarrow RC\!=\!\overset{O^-}{} \right]$$

In Table 4.1 we see that phenol is a weaker acid than acetic acid but stronger than methyl alcohol. Phenol has an OH group attached to a benzene ring, and in Chapter 30 we will learn that one reason for this acidity order is that part of the negative charge in the phenoxide anion is distributed to the benzene ring.

The presence of a formal positive charge results in an increase in acidity because of the electrostatic attraction between positive and negative charges. Some examples are

$$\underset{\text{sulfuric acid}}{HO\!-\!\overset{O^-}{\underset{O\_}{S^{2+}}}\!-\!OH} > \underset{\text{sulfurous acid}}{HO\!-\!\overset{O^-}{\underset{\cdot\cdot}{S^+}}\!-\!OH}$$

$$\underset{\text{nitric acid}}{O\!=\!\overset{O^-}{\underset{\cdot\cdot}{N^+}}\!-\!OH} > \underset{\text{nitrous acid}}{O\!=\!\underset{\cdot\cdot}{N}\!-\!OH}$$

A simple corollary is that a nearby negative charge will reduce acidity because of charge repulsion in the resulting anion. Thus the second dissociation constants are generally smaller than the first. Examples of such acidity orders are

$$HO\!-\!\overset{O^-}{\underset{O\_}{S^{2+}}}\!-\!OH > HO\!-\!\overset{O^-}{\underset{O\_}{S^{2+}}}\!-\!O^-$$

$$\overset{OH}{\underset{OH}{^-O\!-\!P^+}}\!-\!OH > \overset{O^-}{\underset{OH}{^-O\!-\!P^+}}\!-\!OH > \overset{O^-}{\underset{O\_}{^-O\!-\!P^+}}\!-\!OH$$

Other effects on acidity will be discussed in detail later in this text. Examples are the role of s-character in the C—H bonds of certain carbon acids and the effect of nearby polar groups ("inductive effects").

To summarize, acidity and basicity are important chemical properties. Many organic reactions involve proton-transfer equilibria, and we need to appreciate the effect of structural change on such equilibria. Moreover, some organic compounds are related chemically to corresponding inorganic acids. For example, methanesulfonic acid, CH_3SO_3H, is related to sulfuric acid, $HOSO_3H$, and is also a strong acid. Ethylamine, $CH_3CH_2NH_2$, is related to ammonia, NH_3, and has comparable base strength. Some important generalizations are

1. Acidity for Y—H increases as Y is further down a given column of the periodic table.
2. Acidity for Y—H increases as Y is toward the right along a given row of the periodic table.
3. Nearby positive charges increase acidity; negative charges decrease acidity.

EXERCISE: Calculate the pH of a solution of 1 mole of each of the following acids in 1 L of water.
(a) HI   (b) HCl   (c) HF   (d) acetic acid   (e) H_2S

PROBLEMS

1. In this chapter we discuss the hydrolysis of methyl chloride *in aqueous solution.* Consider the same reaction in the gas phase at 25°

$$CH_3Cl + H_2O \rightleftharpoons CH_3OH + HCl$$

(a) $\Delta H° = 7.3$ kcal mole^{-1}; $\Delta S° = 0.3$ eu. Calculate $\Delta G°$.
(b) Calculate the equilibrium constant.
(c) Can this reaction be said to "go to completion" in the direction shown?

2. Consider the equilibrium between butane and ethane plus ethylene.

$$C_4H_{10} \rightleftharpoons C_2H_6 + C_2H_4$$

(a) At 25° $\Delta H° = 22.2$ kcal mole^{-1} and $\Delta S° = 33.5$ eu. What is $\Delta G°$? On which side does the equilibrium lie?
(b) Calculate $\Delta G°$ at 800°K (527°C) and determine the position of equilibrium.
(c) How does the relative effect of $\Delta H°$ and $\Delta S°$ change with temperature? (Actually, $\Delta H°$ and $\Delta S°$ change somewhat with temperature, but the effect is not large enough to change the qualitative result.)

3. (a) At room temperature what change in free energy in units of kcal mole^{-1} will change an equilibrium constant by a factor of 10? By a factor of 100? This energy quantity is a handy number to remember.
(b) These numbers can be converted to equivalent ΔH and ΔS values. Consider ΔG for the factor of 10 change in equilibrium constant. What is the equivalent value for ΔH in kcal mole^{-1} if $\Delta S = 0$; what is the equivalent value for ΔS in eu (cal deg^{-1}) if $\Delta H = 0$?

4. During the course of reaction, the concentration of reactants decreases; hence, the rate of reaction is reduced.
(a) In the example of 0.05 M methyl chloride and 0.10 M OH$^-$ discussed on page 53, what is the rate of reaction at the start of the reaction, using the rate constant given on page 54?
(b) Using this rate, determine the time required for 10% reaction. What are the concentrations of reactants after 10% reaction? What is the rate of reaction at this point? Using this rate, determine how long it takes for the second 10% of reaction to occur.
(c) Repeat the calculation to estimate the time for 50% completion of the reaction.

5. The equilibrium reaction in the gas phase of ethylene and HCl to give ethyl chloride, $C_2H_4 + HCl = C_2H_5Cl$, has a favorable enthalpy, $\Delta H° = -15.5$ kcal mole^{-1}, but an unfavorable entropy, $\Delta S° = -31.3$ eu.
(a) Why is the entropy negative?
(b) What is $\Delta G°$ at room temperature (25°)?
(c) If the reaction mixture started with 1 atm pressure each of HCl and C_2H_4, what pressure of each is left at equilibrium?
(d) For the system to be at equilibrium with all three components present in equal amounts, what total pressure is required?
  Incidentally, in this system a mixture of pure, dry HCl and C_2H_4 will not react at room temperature. Establishment of the equilibrium requires a suitable catalyst.

6. Consider the following reaction sequence in which B is an intermediate. Sketch energy profiles for each of the possible relationships among rate constants shown.

$$A \underset{k_2}{\overset{k_1}{\rightleftharpoons}} B$$

$$B \xrightarrow{k_3} C$$

The back-reaction from C is negligible.
(a) k_1 and k_2 large; k_3 small.
(b) k_1 large; k_2 and k_3 large, but $k_2 > k_3$.
(c) k_1 and k_3 large; k_2 small.
(d) k_1 small; k_2 and k_3 large, but $k_3 > k_2$.
Identify the rate-determining transition state for each of the four cases.

7. Consider the hypothetical two-step reaction

$$A \underset{k_2}{\overset{k_1}{\rightleftharpoons}} B \underset{k_4}{\overset{k_3}{\rightleftharpoons}} C$$

that is described by the following energy profile.

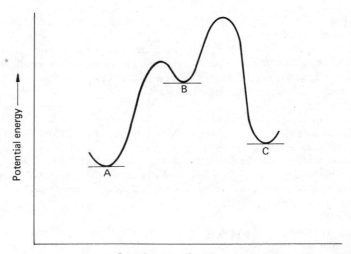

(a) Is the overall reaction (A $\longrightarrow$ C) exothermic or endothermic?
(b) Label the transition states. Which transition state is rate-determining?
(c) What is the correct order of magnitude of rate constants?
 (i) $k_1 > k_2 > k_3 > k_4$ (ii) $k_2 > k_3 > k_1 > k_4$
 (iii) $k_4 > k_1 > k_3 > k_2$ (iv) $k_3 > k_2 > k_4 > k_1$
(d) Which is the thermodynamically most stable compound?
(e) Which is the thermodynamically least stable compound?

8. *t*-Butyl chloride reacts with hydroxide ion according to the following equation.

t-butyl chloride *t*-butyl alcohol

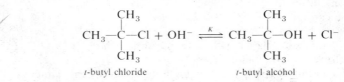

We will learn in Chapter 9 that the reaction is believed to proceed by the following mechanism.

(1) $CH_3-\underset{\underset{CH_3}{|}}{\overset{\overset{CH_3}{|}}{C}}-Cl \underset{k_2}{\overset{k_1}{\rightleftharpoons}} CH_3-\underset{\underset{CH_3}{|}}{\overset{\overset{CH_3}{|}}{C}}+ \ + Cl^-$

(2) $CH_3-\underset{\underset{CH_3}{|}}{\overset{\overset{CH_3}{|}}{C}}+ \ + OH^- \underset{k_4}{\overset{k_3}{\rightleftharpoons}} CH_3-\underset{\underset{CH_3}{|}}{\overset{\overset{CH_3}{|}}{C}}-OH$

The order of rate constants is $k_3 > k_2 > k_1 \gg k_4$.
(a) Construct a reaction coordinate diagram for the reaction.
(b) Is the first step exothermic or endothermic?
(c) Is the overall reaction exothermic or endothermic?
(d) Does the first or second step govern the rate of disappearance of t-butyl chloride?

9. Ammonia is a weak acid with a pK estimated as 33. The conjugate anion NH_2^-, amide ion, is available as alkali metal salts such as sodium amide, $NaNH_2$. Calculate the pH of a solution prepared by adding 0.1 mole of sodium amide to 1 L of water. Does this pH differ appreciably from that of a solution prepared from 0.1 mole of NaOH in 1 L of water? How does the pK of phosphine, PH_3, compare with that of ammonia?

10. Hydride ion, H^-, is known in the form of salts such as sodium hydride, NaH. When sodium hydride is added to water, it is converted completely into hydrogen. What does this say about H_2 as an acid relative to water?

11. Explain the following acidity orders.
(a) $H_2SO_4 > HSO_4^-$
(b) Nitric acid ($HONO_2$) > nitrous acid (HONO)
(c) $H_2Te > H_2Se$
(d) Nitrous acid > hydroxylamine (H_2NOH)
(e) $H_2S > PH_3$
(f) Perchloric acid ($HOClO_3$) > hypochlorous acid (HOCl)

(g) Oxalic acid ($HO-\overset{\overset{O}{||}}{C}-\overset{\overset{O}{||}}{C}-OH$) > oxalate ion ($HO-\overset{\overset{O}{||}}{C}-CO_2^-$)
(h) A sulfinic acid (RSO_2H) > a sulfenic acid (RSOH)

12. Consider the reaction of A + B as a second-order reaction for which rate = $k[A][B]$. If A and B start off with equal concentrations, how has the rate changed at 50% reaction?

13. (a) The reaction of methyl chloride with water was described as a first-order reaction because the concentration of water does not change during the reaction. If the reaction with water is exactly analogous to the reaction with hydroxide ion, we should write the kinetic equation as

$$rate = k_2[CH_3Cl][H_2O]$$

What is the value of $[H_2O]$ in this expression?
(b) Because $[H_2O]$ remains constant, this case is an example of pseudo first-order kinetics. For the expression

$$rate = k_1[CH_3Cl]$$

we found that $k_1 = 3 \times 10^{-10}$ sec^{-1}. Using the value of $[H_2O]$ found above, derive k_2. How does the value of k_2 for the reaction of methyl chloride with water compare with that for reaction with hydroxide ion?

(c) For a first-order reaction, the time for half of the remaining reactant to react—the half-life—is given by

$$t_{1/2} = 0.693/k$$

From the value of k_1, calculate the half-life in years of an aqueous solution of methyl chloride.

14. Problem 4 was solved in an approximate manner. Using the methods of differential and integral calculus, derive the exact answers.

Chapter 5
Alkanes

5.1
n-Alkanes: Physical Properties

Table 5.1 lists the boiling points, melting points, and densities of some *n*-alkanes. These properties vary in a regular manner. The alkanes from methane through butane are gases at room temperature, pentane boils just above room temperature, and the remaining alkanes show regular increases in boiling point with each additional methylene unit. This regularity of physical properties stems from a regularity of structure. In all of the alkanes the bonds to carbon are nearly tetrahedral and the C—H bond lengths are all essentially constant at 1.095 ± 0.01 Å. Similarly, the C—C bonds are uniformly 1.54 ± 0.01 Å in length.

The boiling point of a substance is defined as the temperature at which its vapor pressure is equal to the external pressure, usually 760 torr.

[Torr is a unit of pressure; 1 torr is equal to 1 mm of mercury at 25°. At 25°, standard atmospheric pressure is 760 torr.]

The vapor pressure of a compound is inversely related to the energy that causes the molecules to attract one another. If the intermolecular attractive force is weak, little energy must be supplied in order for vaporization to occur and the compound has a high vapor pressure. If the intermolecular attractive force is large,

TABLE 5.1
Physical Properties of *n*-Alkanes

Hydrocarbon	Boiling Point, °C	Melting Point, °C	Density[a] d^{20}
methane	−161.7	−182.5	
ethane	−88.6	−183.3	
propane	−42.1	−187.7	0.5005
butane	−0.5	−138.3	0.5787
pentane	36.1	−129.8	0.5572
hexane	68.7	−95.3	0.6603
heptane	98.4	−90.6	0.6837
octane	125.7	−56.8	0.7026
nonane	150.8	−53.5	0.7177
decane	174.0	−29.7	0.7299
undecane	195.8	−25.6	0.7402
dodecane	216.3	−9.6	0.7487
tridecane	235.4	−5.5	0.7564
tetradecane	253.7	5.9	0.7628
pentadecane	270.6	10	0.7685
eicosane	343	36.8	0.7886
triacontane	449.7	65.8	0.8097
polyethylene			0.965

[a] Note that densities vary with temperature; d^{20} refers to the density in grams per milliliter at 20°C.

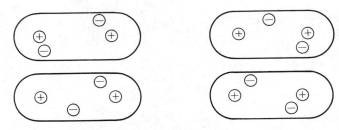

FIGURE 5.1. *Van der Waals attraction. Electronic motion is such as to produce net electrostatic attraction at every instant.*

more energy must be supplied to cause vaporization and the compound has a low vapor pressure. Interactions between neutral molecules generally result from **van der Waals forces,** dipole–dipole electrostatic attraction, and hydrogen bonding. For hydrocarbons, only the van der Waals interaction is important. This force of attraction results from an electron correlation effect also called the **London force** or **dispersion force.**

Although we normally think of atoms and molecules in terms of smeared-out electron-density distributions, it should be emphasized that this is a time-average picture. At any given instant electrons will be positioned as far from each other as possible although these positions are different from one instant to the next. Consider the simplified models shown in Figure 5.1. The system of charges on the left has a small net attraction that binds the two molecules together. In the system on the right, the electrons have all moved but there is still net attraction. The motion of the electrons is mutually *correlated* to produce net attraction at all times. This attractive force is sensitive to distance and varies as $1/r^6$. It is significant only for molecules close to each other—but not too close. As molecules get too close, the electron charge clouds overlap appreciably and electron repulsion dominates.

Van der Waals attraction depends on the approximate "area" of contact of two molecules—the greater this area, the greater is the attractive force. Because of the tetrahedral nature of carbon, alkane chains tend to have a zigzag geometry. For example, one of the geometric arrangements adopted by butane is shown in stereo-plot form in Figure 5.2. This type of zigzag arrangement of butane and

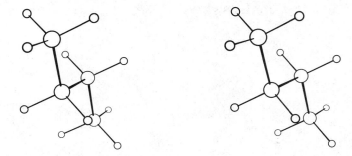

FIGURE 5.2. *Stereo representation of one conformation of butane.*

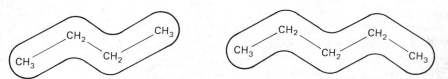

FIGURE 5.3. *Zigzag geometry of alkanes.*

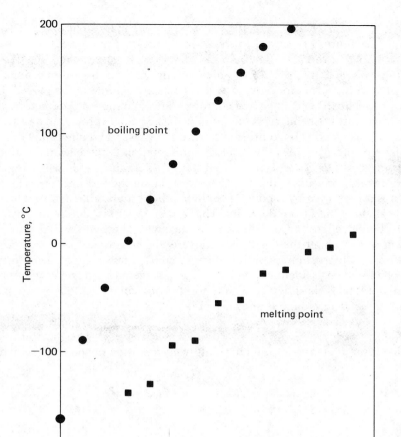

FIGURE 5.4. *Boiling points and melting points of* n*-alkanes.*

pentane is symbolized in Figure 5.3. Each additional methylene unit provides an additional area of contact that increases the total attractive force and gives rise to a greater boiling point. The energy of attraction per methylene group is approximately 1–1.5 kcal mole^{-1}.

Van der Waals forces are even greater in solids, and there is a progressive change in melting point with increasing chain length. Because of different packing requirements in the crystal for odd and even zigzag chains, there is an alternation of melting points with increasing chain length (Figure 5.4).

5.2
n-Alkanes; Barriers to Rotation

In 1937 the entropy of ethane was found experimentally to be somewhat lower than that calculated for a molecule in which the methyl groups can rotate freely about the central C—C bond. This reduced entropy was interpreted to mean that the methyl groups have reduced freedom of motion. Subsequent experiments of several kinds have confirmed that the **eclipsed** structure for ethane is 3 kcal mole^{-1} higher in energy than the more stable **staggered** structure. These two different structures are shown in stereo-plot form in Figure 5.5.

Structures that differ only by rotation about one or more bonds are defined as **conformations** of a compound. In order to represent the three-dimensional character of such conformations, two useful systems are commonly employed. In Figure 5.6 the eclipsed and staggered conformations of ethane are depicted as "sawhorse" structures. In this representation a dashed bond projects away from the viewer, a heavy wedge bond projects toward the viewer, and a normal bond lies in the plane of the page. Another useful representation is the Newman projection.

Newman projections for eclipsed and staggered ethane are shown in Figure 5.7. In a Newman projection the C—C bond is being viewed end on. The nearer carbon is represented by a point. The three other groups attached to that carbon radiate as three lines from the point. The farther carbon is represented by a circle with its bonds radiating from the edge of the circle. These projections show that a rotation of 60° about the C—C axis converts the staggered form to the eclipsed structure. As the rotation is continued another 60°, a new staggered conformation is reached, which is identical with the first staggered conformation. A plot of potential energy versus degree of rotation for one complete 360° rotation about the C—C bond in ethane is shown in Figure 5.8.

Thus, in rotating about the C—C bond, there is a 3.0 kcal mole^{-1} energy barrier in passing from one staggered conformation to another. The instability of the eclipsed form of ethane appears to result from repulsion of some of the hydrogen

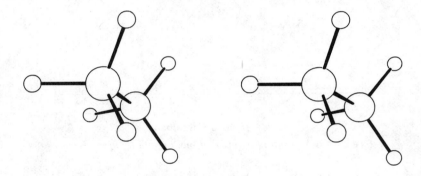

(a) eclipsed structure of ethane

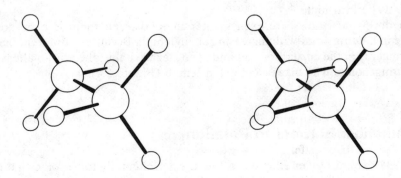

(b) staggered structure of ethane

FIGURE 5.5. *Stereo plots illustrating the eclipsed and staggered structures of ethane.*

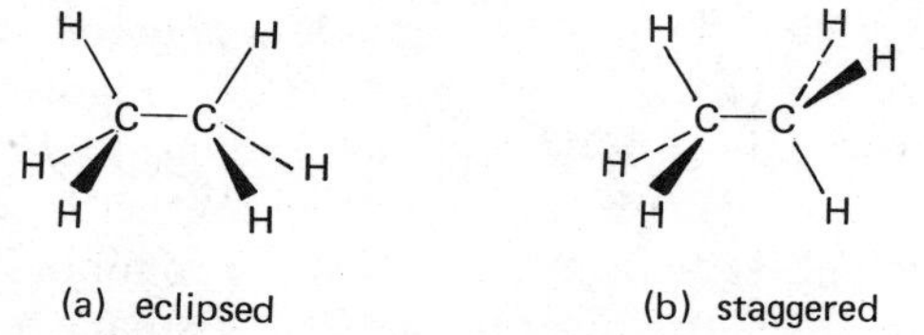

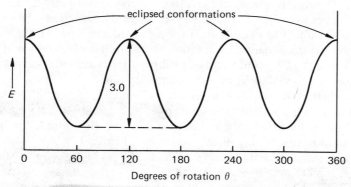

FIGURE 5.6. *Sawhorse structures illustrating the eclipsed and staggered conformations of ethane.*

(a) eclipsed   (b) staggered

FIGURE 5.7. *Newman projections illustrating the eclipsed and staggered conformations of ethane.*

eclipsed conformations

3.0

E

0   60   120   180   240   300   360

Degrees of rotation θ

FIGURE 5.8. *Potential energy of ethane as a function of degree of rotation about the C—C bond.*

orbitals. The hydrogen orbitals on one methyl group are rather far from those on the other. But they are closer in the eclipsed conformation than in the staggered one. The internuclear H–H distance in staggered ethane is 2.55 Å, whereas it is only 2.29 Å in the eclipsed form. Orbital overlap between hydrogens on adjacent atoms is antibonding or repulsive. At these distances the magnitude of this repulsion is small, only about 1 kcal mole^{-1} per pair of hydrogens, but this small energy effect has staggering structural consequences. As we shall see, staggered conformations are general in alkane chains and cycloalkane rings.

In propane the barrier to rotation now involves a C—CH$_3$ bond and is slightly higher in energy at 3.4 kcal mole^{-1}. The most stable conformation is illustrated in Figure 5.9. In this conformation the bonds are completely staggered.

A potential energy plot for rotation about the C$_2$—C$_3$ bond in butane is shown in Figure 5.10. The conformations at the various unique maxima and minima are shown in Figure 5.11, both as Newman projections and as stereo plots. Note that there are two different kinds of staggered conformation: **anti,** in which the two methyl groups attached to carbons 2 and 3 are farthest apart, and **gauche,** in which these two methyl groups are adjacent. These two kinds of staggered conformation have different energies. The *anti* conformation is more stable than the

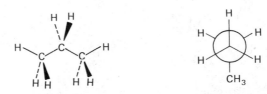

FIGURE 5.9. *Most stable conformation of propane.*

gauche by 0.9 kcal mole^{-1}. At room temperature butane is a mixture of 72% *anti* and 28% *gauche* conformations.

If these two structures could be isolated, they would have different physical properties such as density, spectra, and melting points. However, the energy barrier separating them is rather small, only 3.8 kcal mole^{-1}. A barrier of such magnitude is far too small to permit isolation of the separate *anti* and *gauche* conformations at normal temperatures. In order to separate these two species, one would have to slow the conversion by working at very low temperatures, below approximately −230°.

Also note that there are two distinct eclipsed conformations. One of these maxima is passed in rotation from the *anti* to a *gauche* conformation. In this conformation, there are two CH$_3$–H and one H–H eclipsed interactions. This conformation is 3.8 kcal mole^{-1} less stable than the *anti* conformation. The other eclipsed conformation, which is passed in rotation from one *gauche* conformation to the other, has one CH$_3$–CH$_3$ and two H–H eclipsed interactions. Its energy is about 4.5 kcal mole^{-1} above that of the *anti* conformation.

Finally, note that the *gauche* conformations labeled B and E, although energetically equivalent, are not really the same. They are actually mirror images of one another. They are the same only in the sense that your right and left hands are the same. We shall return to this phenomenon in Chapter 7.

The same principles apply to larger alkanes. In general, the most stable structure is the completely staggered one with all alkyl groups having an *anti* relationship to one another. However, keep in mind that *gauche* conformations are only slightly less stable, and there will always be a sizable fraction of the molecules with these conformations. The two conformations of pentane given here are shown in stereo-plot form in Figure 5.12.

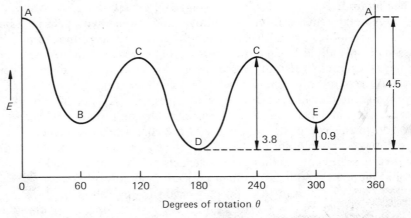

Degrees of rotation θ

FIGURE 5.10. *Potential energy of butane as function of degree of rotation about the C$_2$—C$_3$ bond.*

(a) eclipsed (A)

(b) *gauche* (B)

(c) eclipsed (C)

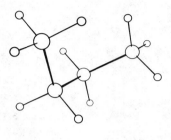

(d) *anti* (D)

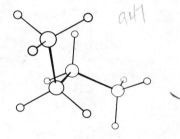

(e) *gauche* (E)

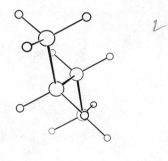

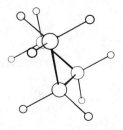

FIGURE 5.11. *Conformations of butane.*

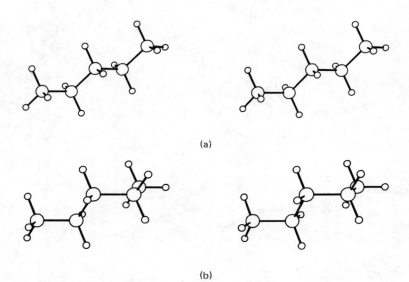

FIGURE 5.12. *Stereo representations of two pentane conformations:* (a) anti-anti; (b) anti-gauche.

When writing the structures of such compounds, it is usually inconvenient to depict the full geometry as is done in the preceeding structures. It is important to recognize that a given structure may be written in many ways. For example, the various structures that follow all represent pentane.

$$CH_3-CH_2-CH_2-CH_2-CH_3 \qquad CH_3-CH_2-CH_2 \atop CH_2-CH_3 \qquad CH_3-CH_2 \atop CH_2 \atop CH_2-CH_3$$

$$CH_3 \diagdown {CH_2} \diagdown {CH_2} \diagdown CH_3 \qquad CH_3 \diagdown {CH_2} \diagdown CH_2 \atop CH_2 \diagdown CH_3$$

EXERCISE: Using molecular models, compare the *gauche* and *anti* conformations of butane. In particular, estimate the distance between different pairs of hydrogens. Which pair of hydrogens is closest together? Compare their separation with that of eclipsed hydrogens in ethane. Does this comparison suggest a principal reason for the higher energy of the *gauche* conformation?

5.3
Branched-Chain Alkanes

The branched-chain alkanes also exist as mixtures of rapidly interconverting staggered conformations. For example, the two staggered conformations of iso-pentane for rotation about the C_2-C_3 bond, $(CH_3)_2CH-CH_2CH_3$, are shown in Figure 5.13. The rotational barrier separating these two conformation is about 5 kcal mole^{-1}. At room temperature isopentane exists as a mixture of the two

FIGURE 5.13. *Staggered conformations of isopentane.*

conformations, 90% of the one with only one *gauche* interaction and 10% of the one with two *gauche* interactions.

According to the definition of isomers given in Section 5.2, different conformations of a compound are isomers since they have the same formula but differ in the arrangement of the atoms in space. However, it is convenient to distinguish such isomers, which rapidly interconvert at ordinary temperatures, from other kinds of isomers that interconvert at high temperatures or not at all. Consequently, we refer to these easily interconvertible spatial isomers as **conformational isomers,** as **conformations,** or as **conformers,** to distinguish them from **structural isomers,** such as butane and isobutane.

Branched-chain hydrocarbons are more compact than their straight-chain isomers. For this reason, branched hydrocarbons tend to have lower boiling points and higher densities than their straight-chain isomers. Physical properties for some branched hydrocarbons are summarized in Table 5.2.

Isolated branches interfere with the regular packing of linear alkanes in the crystal and cause a reduced melting point. That is, branched isomers have a smaller area of contact and van der Waals attraction is reduced. Consequently, melting points and boiling points are lower. However, when a molecule has sufficient symmetry, it forms a crystal lattice *more* easily and therefore has a higher melting point but a relatively low boiling point. For an example of such a case, compare the melting points of pentane, isopentane, and neopentane. An extreme example is given by 2,2,3,3-tetramethylbutane, which boils only a few degrees above its melting point. Hydrocarbons having a high degree of symmetry or "ball-like" character tend to sublime rather than boil. On heating they pass directly from the solid to vapor state without passing through the intermediate liquid state.

This result can be cast into entropy concepts in a straighforward way. In the crystal, molecules are locked in and have greatly restricted movement. In the liquid, molecules have enhanced freedom of movement. Consequently, the entropy of melting is a positive quantity whose magnitude is a measure of this increased freedom of move-

TABLE 5.2
Physical Properties of Some Branched Alkanes

	Boiling Point, °C	Melting Point, °C	Density d^{20}
isobutane	−11.7	−159.4	0.5572
isopentane	29.9	−159.9	0.6196
neopentane	9.4	−16.8	0.5904
2-methylpentane	60.3	−153.6	0.6532
3-methylpentane	63.3		0.6644
2,2-dimethylbutane	49.7	−100.0	0.6492
2,3-dimethylbutane	58.0	−128.4	0.6616
2,2,3,3-tetramethylbutane	106.3	100.6	0.6568

ment. The entropy of melting of pentane is $+14$ eu, whereas that of neopentane is only $+3$ eu. Both molecules have increased freedom of translational motion in the liquid hydrocarbon. In addition, *n*-pentane has a floppy chain with many rotational degrees of freedom so that the liquid hydrocarbon is a mixture of many staggered conformational isomers having relatively high entropy. Rotation about the C—C bonds in neopentane, however, always gives back the same structure, and it has a lower entropy. Note that isopentane has an intermediate value of the entropy of melting of $+11$ eu.

EXERCISE: Write Newman projections for two different staggered conformations of 2,3-dimethylbutane for rotation about the C_2—C_3 bond.

5.4
Cycloalkanes

Carbon chains can also form rings. Because there are no ends to the carbon chain in a cyclic alkane, the general formula is $(CH_2)_n$ or C_nH_{2n}. Like straight-chain alkanes, they are saturated hydrocarbons. They are named according to the number of carbons in the ring with the prefix **cyclo-**.

cyclopropane	cyclobutane	cyclopentane
C_3H_6	C_4H_8	C_5H_{10}

The physical properties of some cycloalkanes are summarized in Table 5.3. Note that their symmetry and more restricted rotations result in higher melting points and boiling points than the comparable *n*-alkanes.

Because of symmetry, there is only one monosubstituted cycloalkane, and a number to designate the position of the appendage is not necessary.

methylcyclohexane   ethylcyclopentane

TABLE 5.3
Physical Properties of Some Cycloalkanes

	Boiling Point, °C	Melting Point, °C	Density d^{20}
cyclopropane	-32.7	-127.6	
cyclobutane	12.5	-50.0	
cyclopentane	49.3	-93.9	0.7457
cyclohexane	80.7	6.6	0.7786
cycloheptane	118.5	-12.0	0.8098
cyclooctane	150.0	14.3	0.8349

When there is more than one substituent, numbers are required. One substituent is always given the number 1, and the other is given the next lowest possible number.

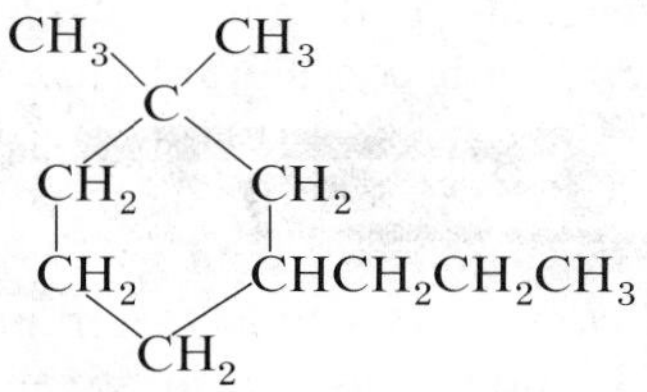

$$\begin{array}{c} \text{CH}_3 \quad \text{CH}_3 \\ \text{C} \\ \text{CH}_2 \quad \text{CH}_2 \\ \text{CH}_2 \quad \text{CHCH}_2\text{CH}_2\text{CH}_3 \\ \text{CH}_2 \end{array}$$

1,1-dimethyl-3-propylcyclohexane

In more complex compounds, the cycloalkyl radical may be named as a prefix.

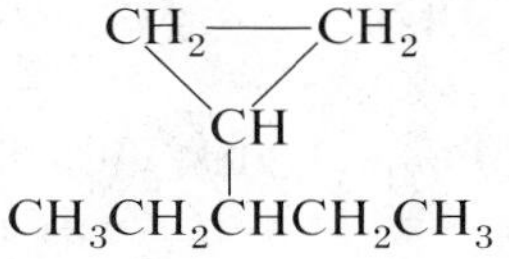

$$\begin{array}{cc} \text{CH}_2\!-\!\text{CH}_2 & \begin{array}{c}\text{CH}_2\\ \text{CH}_2 \quad \text{CH}_2\end{array} \\ \text{CH} & \text{CH} \\ \text{CH}_3\text{CH}_2\text{CHCH}_2\text{CH}_3 & \text{CH}_3\text{CHCHCH}_3 \\ & \text{CH}_3 \end{array}$$

3-cyclopropylpentane 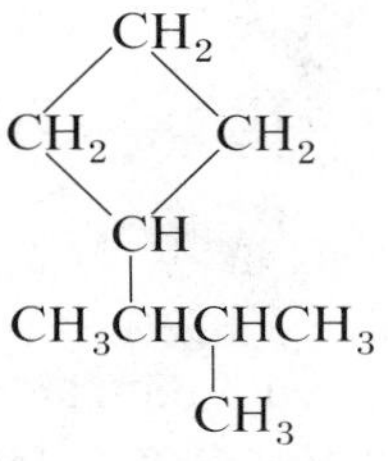   2-cyclobutyl-3-
                          methylbutane

Cycloalkanes are often symbolized by simple geometric figures in which a carbon atom with its appropriate number of attached hydrogens is understood to be present at each apex. The foregoing compounds may be rewritten in this shorthand notation as

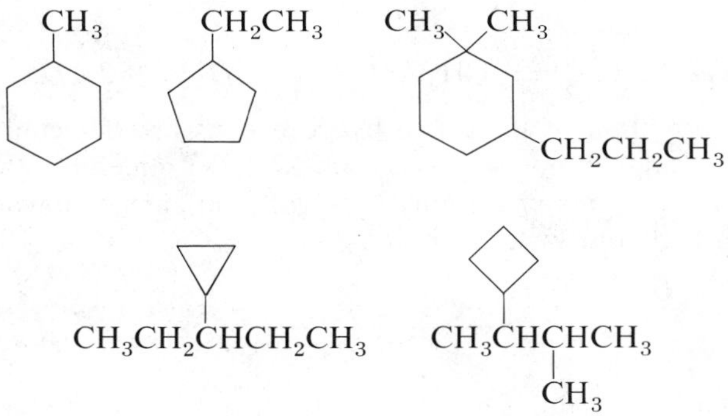

The alkanes and cycloalkanes are the parent structures in the general class of aliphatic compounds. Most of the chemistry of cycloalkanes is similar to that of the alkanes. There are some differences in conformations and in stability, which will be discussed in Sections 5.6 and 5.7.

EXERCISE: Write structures, using simple geometric structures as required, for the following cycloalkanes.
(a) 1,1,2-trimethylcyclopentane
(b) *n*-propylcycloheptane
(c) 1,1,4,4-tetramethylcyclooctane

5.5
Heats of Formation

The **heat of formation** of a compound from its elements in their standard states is a thermodynamic property with considerable use in organic chemistry. This quantity, symbolized ΔH_f°, is defined as the enthalpy of the reaction of elements in their standard states to form the compound. The standard state of each element is generally the most stable state of that element at 25° and one atmosphere pressure. The standard state of carbon is taken as the graphite form, whereas those of hydrogen and oxygen are H_2 and O_2 gases, respectively. By definition, ΔH_f° for an element in its standard state is zero. The standard heat of formation of butane is -30.36 ± 0.16 kcal mole^{-1} and that of isobutane is -32.41 ± 0.13 kcal mole^{-1}.

$$4\,C\,(graphite) + 5\,H_2(g) = n\text{-}C_4H_{10}(g) \qquad \Delta H^\circ \equiv \Delta H_f^\circ(n\text{-}C_4H_{10})$$
$$= -30.36 \text{ kcal mole}^{-1}$$

$$4\,C\,(graphite) + 5\,H_2(g) = (CH_3)_3CH(g) \qquad \Delta H^\circ \equiv \Delta H_f^\circ(i\text{-}C_4H_{10})$$
$$= -32.41 \text{ kcal mole}^{-1}$$

We shall see in Chapter 6 how these hypothetical enthalpies of reaction are determined. For now, suffice it to say that they *can* be determined, although indirect methods are required. The ΔH_f° of a compound may be either negative, as in the two foregoing examples, or positive. A negative ΔH_f° means that heat would be liberated if the compound could be prepared directly by combination of its elements. That is, *n*-butane and isobutane are both *more stable* (have lower enthalpy) than four carbon atoms and five hydrogen molecules in their standard states. The heats of formation of the butane isomers reveal that isobutane is more stable than *n*-butane by 2.05 kcal mole^{-1}. Thus, in the following hypothetical equilibrium, isobutane would predominate.

$$CH_3CH_2CH_2CH_3 \overset{K}{\rightleftharpoons} \overset{\displaystyle CH_3}{\underset{\displaystyle}{CH_3CHCH_3}} \qquad \Delta H^\circ = -2.05 \text{ kcal mole}^{-1}$$

The heats of formation of these two hydrocarbons are depicted graphically in Figure 5.14. In using energy diagrams such as these, remember that *down* represents less energy and greater stability ("downhill in energy"), whereas *up* represents higher energy and lower stability.

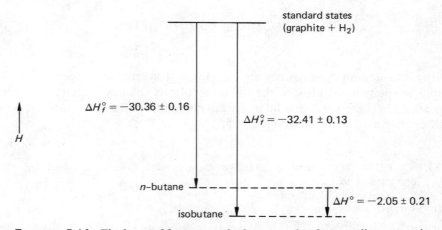

FIGURE 5.14. *The heats of formation of* n-*butane and isobutane, illustrating the use of* ΔH_f° *values to compute* ΔH° *for a simple reaction.*

TABLE 5.4
Some Heats of Formation

Compound	Heat of Formation at 25° $\Delta H_f°$, kcal mole^{-1}
CH_4	-17.9
CH_3CH_3	-20.2
$CH_3CH_2CH_3$	-24.8
$CH_3CH_2CH_2CH_3$	-30.4
$(CH_3)_3CH$	-32.4
$CH_3CH_2CH_2CH_2CH_3$	-35.1
$(CH_3)_2CHCH_2CH_3$	-36.9
$(CH_3)_4C$	-40.3
CO	-26.4
CO_2	-94.1
$H_2O(g)$	-57.8
$H_2O(l)$	-68.3
H_2	0
O_2	0
C (graphite)	0

Some values for heats of formation are listed in Table 5.4. A more complete list is given in Appendix I. These $\Delta H_f°$ values are useful for estimating possible reactions, providing that a pathway or reaction mechanism is possible. For example, the hydrogenation of butane to ethane is exothermic by 10 kcal mole^{-1}.

$$C_4H_{10} + H_2 = 2\,C_2H_6$$
$$\Delta H_f°: \quad -30.4 + 0 \quad 2 \times -20.2 \quad \Delta H° = -10.0 \text{ kcal mole}^{-1}$$

If a suitable catalyst or reaction pathway could be found, this reaction would proceed toward the right. However, no such catalyst or pathway is known at ordinary temperatures. The reaction remains hypothetical even though, if realized, it would be exothermic. This example illustrates the difference between thermodynamics and kinetics. A given reaction may have favorable thermodynamics but will occur only if a pathway with a sufficiently low activation barrier can be found. Because of the importance of pathways our studies of organic reactions will also often include discussions of reaction mechanism. The importance of enzymes in biochemical reactions is that they provide such pathways for reaction.

The hydrogenation of ethylene to ethane is also highly exothermic.

$$C_2H_4 + H_2 = C_2H_6$$
$$12.5 + 0 \quad -20.2 \quad \Delta H° = -32.7 \text{ kcal mole}^{-1}$$

In this case a number of catalysts are known that provide a reaction pathway, and this reaction is an important general reaction of alkenes (Section 11.4).

One important limitation on the use of heats of formation is that equilibria are determined by free energy rather than by enthalpy alone.

$$\Delta G° = -RT \ln K = \Delta H° - T\Delta S°$$

That is, an entropy change plays a large role in determining an equilibrium constant. For example, in the equilibrium discussed above between *n*-butane and isobutane, $\Delta H°$ is -2.05 kcal mole^{-1}, but $\Delta G°$ is only -0.89 kcal mole^{-1}, corresponding to an equilibrium constant at $25°$ of 4.5. Since entropy is a measure of freedom of motion, the largest entropy changes result from a difference in numbers of molecules on the two sides of an equilibrium. The magnitude of this effect depends on physical state (gas, liquid, and so on), molecular weight, and temperature. For a gas at ordinary temperature and pressure a difference of one molecule on the two sides of an equilibrium (for example, A = B + C) corresponds to about 30–40 eu, which is equivalent to 9–12 kcal mole^{-1} in enthalpy at room temperature. At higher temperatures any entropy change has a still greater effect. For example, at $25°$ the conversion of n-butane to one molecule of ethane and one of ethylene is highly endothermic.

$$C_4H_{10} = C_2H_6 + C_2H_4$$
$$\Delta H_f°: \quad -30.4 \quad -20.2 \quad 12.5$$

$$\Delta H° = \Delta H_f°(\text{products} = -20.2 + 12.5) - \Delta H_f°(\text{reactants} = -30.4)$$
$$= +22.6 \text{ kcal mole}^{-1}$$

Even though this reaction involves one molecule going to two, the resulting $\Delta S°$ of 33 eu still leaves a positive free energy change at room temperature: $\Delta G° = +12.7$ kcal mole^{-1}. At $500°$, although the equilibrium is still highly endothermic in enthalpy, the positive entropy change gives a $\Delta G°$ of -3.8 kcal mole^{-1}. The equilibrium now favors the products. As we shall see in Section 6.2, this reaction is involved in the refining of petroleum ("cracking"). However, the thermodynamics of the molecules demands that the reaction be carried out at high temperature, as the foregoing simple calculations show.

EXERCISE: Calculate $\Delta H°$ for the isomerization reactions (a) n-pentane $\rightleftharpoons$ isopentane and (b) isopentane $\rightleftharpoons$ neopentane. Note that each new branch provides 2–3 kcal mole^{-1} in stabilization.

5.6
Cycloalkanes; Ring Strain

Ring strain is an energy effect that can be seen clearly in the heats of formation of the cycloalkanes. In alkanes each CH_2 group contributes about -5 kcal mole^{-1} to $\Delta H_f°$ of a molecule; that is, the heats of formation of compounds differing by only one CH_2 differ by a regular increment of about 5 kcal mole^{-1}.

		$\Delta H_f°$, kcal mole^{-1}
$4 C + 5 H_2 \longrightarrow n\text{-}C_4H_{10}$		-30.4
$5 C + 6 H_2 \longrightarrow n\text{-}C_5H_{12}$		-35.1
$6 C + 7 H_2 \longrightarrow n\text{-}C_6H_{14}$		-39.9

Since cycloalkanes have the empirical formula $(CH_2)_n$, one can obtain the $\Delta H_f°$ for each CH_2 group by simply dividing $\Delta H_f°$ for the molecule by n. The heats of formation for a number of cycloalkanes are tabulated in Table 5.5. Examination of the table shows that most of these cycloalkanes have less negative values of $\Delta H_f°/n$ than the alkane value of about -5 kcal mole^{-1}. That is, many cycloalkanes have a higher energy content per CH_2 group than a typical acyclic alkane. This excess energy is called ring strain. The total excess energy of a cycloalkane is

TABLE 5.5
ΔH_f° of Cycloalkanes, $(CH_2)_n$

n	Cycloalkane	ΔH_f°, kcal mole^{-1}	$\Delta H_f^\circ/n$, kcal mole^{-1} per CH_2 group	Total Strain Energy, kcal mole^{-1}
2	ethylene	+12.5	+6.2	22
3	cyclopropane	+12.7	+4.2	27
4	cyclobutane	+6.8	+1.7	26
5	cyclopentane	−18.4	−3.7	6
6	cyclohexane	−29.5	−4.9	(0)
7	cycloheptane	−28.2	−4.0	6
8	cyclooctane	−29.7	−3.7	10
9	cyclononane	−31.7	−3.5	13
10	cyclodecane	−36.9	−3.7	12
11	cycloundecane	−42.9	−3.9	11
12	cyclododecane	−55.0	−4.6	4
13	cyclotridecane	−58.9	−4.5	5
14	cyclotetradecane	−57.1	−4.1	12
15	cyclopentadecane	−72.0	−4.8	2
16	cyclohexadecane	−76.9	−4.8	2

simply the excess energy per CH_2 multiplied by the number of CH_2 groups in the particular cycloalkane.

Cyclohexane shows essentially no ring strain; its CH_2 groups have essentially the same ΔH_f° as those of normal alkanes. For the purpose of computing the ring strain of a particular cycloalkane, cyclohexane is considered to be strain-free; it is the standard for comparison. For cyclohexane $\Delta H_f^\circ = -29.5$ kcal mole^{-1} and $\Delta H_f^\circ/n = -29.5/6 = -4.92$ kcal mole^{-1}. This value is taken as ΔH_f° for a "strainless" CH_2 group. For example, ΔH_f° for a hypothetical "strainless" cyclopentane would be -4.92 kcal mole$^{-1} \times 5 = -24.6$ kcal mole^{-1}. Hence, the strain energy of cyclopentane $= (-18.4) - (-24.6) = +6.2$ kcal mole^{-1}. Cyclopentane is 6 kcal mole^{-1} less stable than it would be if each CH_2 group were in some hypothetical strain-free state.

The inherent ring strain of a given molecule results from three distinct conditions: bond strain ("bent bonds"), eclipsing of adjacent pairs of C—H bonds, or transannular nonbonded interaction (the "bumping together" of two hydrogens that are bonded to atoms "across the ring" from one another.) We shall examine these different kinds of ring-strain effects in Section 5.7 as we consider the conformations of the various cycloalkanes.

5.7
Cycloalkanes: Conformation

A. *Cyclopropane and Cyclobutane*

A bond is strongest when it is formed by the overlap of two atomic orbitals along the internuclear bond axis. The strength of the bond is reduced if overlap of the constituent orbitals is not along the bond axis. The structure of cyclopropane

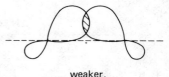

stronger,
more efficient overlap

weaker,
less efficient overlap

is shown in Figure 5.15. For purely geometric reasons the internuclear C—C—C angle in cyclopropane is 60°. The natural bond angle for C_{sp^3}-orbitals overlapping linearly would be 109.5°. For hybrid orbitals having more p-character the natural angle is smaller, but even with pure p-orbitals the natural bond angle cannot be less than 90°. In practice, the C—C bond orbitals in small rings do have more p-character than sp^3 (between sp^4 and sp^5), and the resulting orbitals form **bent bonds** (Figure 5.16). As a result the C—C bonds in cyclopropane are weaker than those in normal alkanes. This reduced bond strength shows up in the chemistry to be discussed subsequently and also as a ring strain in the ΔH_f°. The ring strain in cyclopropane results primarily from bent bonds.

To compensate partially for ring strain, extra s-character is used for the C—H bonds. These bonds are somewhat stronger than alkyl C—H bonds, and the H—C—H bond angle is greater than tetrahedral. Another factor that contributes to the ring strain in cyclopropane is the eclipsing of the C—H bonds. Recall that the eclipsed conformation of ethane is 3.0 kcal mole^{-1} less stable than the staggered conformation; each pair of eclipsed hydrogens raises the energy by 1.0 kcal mole^{-1} (Section 5.2). In cyclopropane there are six pairs of eclipsed hydrogens, which could contribute a maximum of 6 kcal mole^{-1} to the energy of the molecule. However, the eclipsed hydrogens are farther apart in cyclopropane than they are in ethane because of the small C—C—C angle in the former. Therefore, the

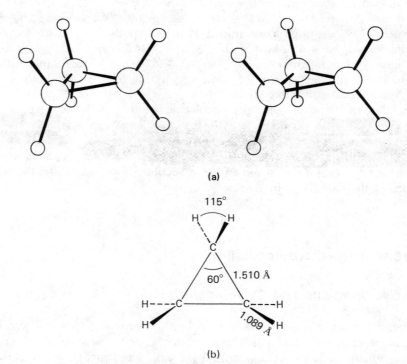

(a)

(b)

FIGURE 5.15. *Cyclopropane:* (a) *stereo representation;* (b) *geometric structure.*

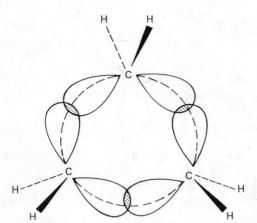

FIGURE 5.16. *Orbital structure of cyclopropane ring showing bent-bond strain.*

actual magnitude of the eclipsing interaction is somewhat less than the maximum
of 6 kcal mole^{-1}.

In cyclobutane the internuclear angles of 90° are not so small as in cyclopropane. The C—C bonds are not so bent, and there is less strain per bond. However,
there are four strained bonds rather then three, and there are eight pairs of
eclipsed hydrogens rather than six. Also the eclipsing in a planar cyclobutane
would be more important than in cyclopropane because the hydrogens are closer.
The result is that the total ring strain in the two compounds is about the same.

Since three points define a plane, the carbon framework of cyclopropane must
have a planar structure. However, cyclobutane can exist in a nonplanar conformation. Spectroscopic studies show that cyclobutane and many of its derivatives
do have nonplanar structures in which one methylene group is bent at an angle of
about 25° from the plane of the other three ring carbons. In this structure, shown
in Figure 5.17, some increase in bond-angle strain is compensated by the reduction in the eclipsed-hydrogen interactions.

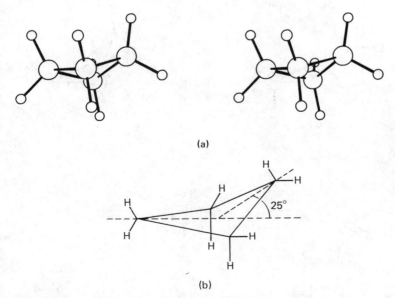

FIGURE 5.17. *Bent cyclobutane:* (a) *stereo representation;* (b) *illustrating the
angle of bend.*

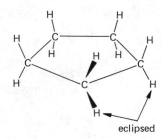

FIGURE 5.18. *Planar structure of cyclopentane showing eclipsed hydrogen pairs.*

B. *Cyclopentane*

A planar pentagonal ring structure for cyclopentane would have C—C—C bond angles of 108°, a value so close to the normal tetrahedral angle of 109.5° that no important strain effect would be expected. However, all of the hydrogens are completely eclipsed in such a structure (Figure 5.18), and it would have about 10 kcal mole^{-1} of strain energy.

The molecule finds it energetically worthwhile to twist somewhat from a planar conformation. The actual structure has the "envelope" shape shown in Figure 5.19. The additional bond-angle strain involved in this structure is more than compensated by the reduction in eclipsed hydrogens. The out-of-plane methylene group is approximately staggered with respect to its neighbors.

As also shown in Figure 5.19, the **envelope** structure of cyclopentane is dynamic. By twisting about the various C—C bonds, successive conformations are reached in which four carbons are in a plane and the fifth is out of plane. The concerted up and down movement of all of these carbons produces a series of structures that appear as if the molecule were rotated through 360° in 72° steps and constitutes a form of molecular motion known as **pseudorotation.**

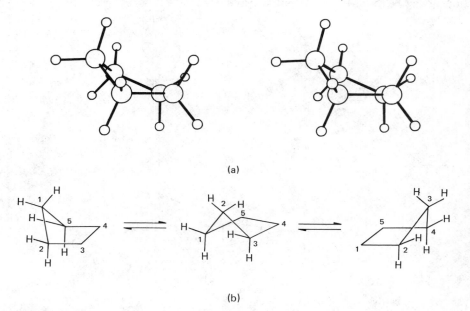

(a)

(b)

FIGURE 5.19. *Envelope structure of cyclopentane:* (a) *stereo representation;* (b) *pseudorotation.*

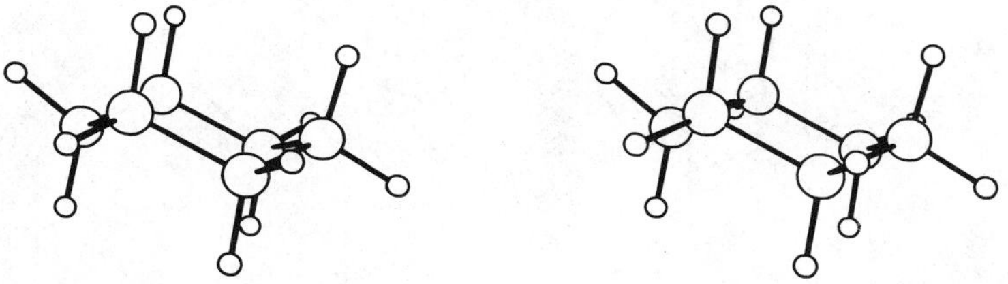

(a)

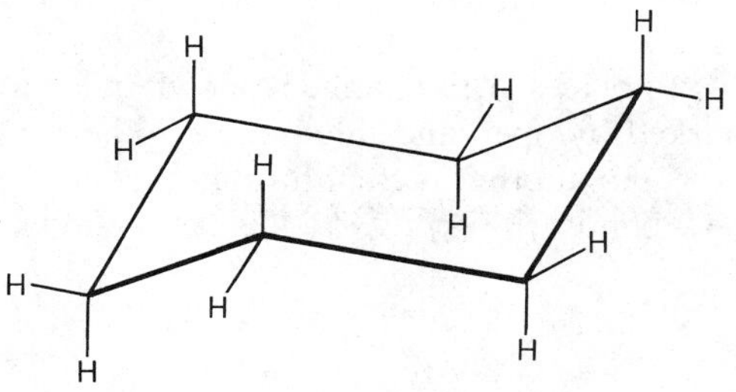

(b)

FIGURE 5.20. *Chair conformation of cyclohexane:* (a) *stereo representation;* (b) *conventional perspective drawing.*

C. *Cyclohexane*

Cyclohexane is the most important of the carbocycles; its structural unit is widespread in compounds of natural origin. Its importance no doubt stems from the fact that it can adopt a conformation that is essentially strain-free. This structure, shown in Figure 5.20, is known as a **chair conformation.** In this structure the bond angles are all close to tetrahedral, and all pairs of hydrogens are completely staggered with respect to each other. The latter point can easily be seen by looking down each C—C bond in turn to produce the Newman projections shown in Figure 5.21. Cyclohexane has neither bond-angle strain nor eclipsed-hydrogen strain.

The chair conformation has two distinct types of hydrogens. These different hydrogens correspond to two sets of exocyclic bonds, the **axial** and **equatorial** bonds shown in Figure 5.22.

The chair conformation of cyclohexane is so important that the student should learn to draw it legibly. Notice should be taken of the sets of parallel lines in the structure shown in Figure 5.23. The molecular axis shown in Figure 5.23 is a threefold axis; rotation by 120° about this axis leaves the molecule unchanged.

Cyclohexane is also a dynamic structure. A concerted rotation about the C—C bonds changes one chair conformation to another in which the axial and equato-

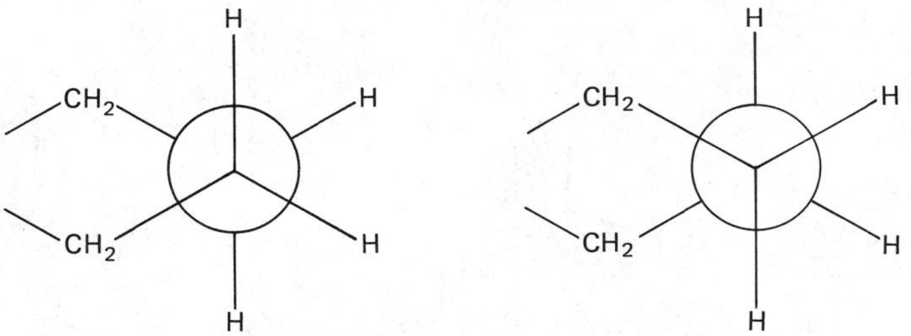

FIGURE 5.21. *Newman projections of* C—C *bonds in cyclohexane.*

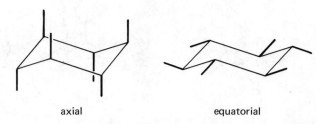

axial   equatorial

FIGURE 5.22. *Cyclohexane bonds.*

rial bonds have changed places. This change is shown in Figure 5.24, in which two sets of bonds are marked by open and filled circles. The conversion of cyclohexane from one chair form to another is a conformational change that involves only rotation about C—C bonds. The process has an energy barrier of 10.8 kcal mole^{-1}.

D. *Larger-Ring Cycloalkanes*

The larger cycloalkanes are less important, and we will not dwell on them. In general, the medium-ring cycloalkanes C_7–C_{12} have conformations in which some form of hydrogen repulsions is inescapable.

The most stable conformation of cycloheptane appears to be that represented in Figure 5.25. This conformation is a type of twisted-chair structure in which the hydrogens are all at least partially staggered with each other, but not completely, the partial eclipsing gives rise to a strain energy.

Cyclooctane apparently is a mixture of several conformations, which differ little from each other in energy. For all of the structures partial eclipsing of hydrogens is unavoidable, and substantial strain energy results. The most stable conformation appears to be the boat-chair structure represented in Figure 5.26.

When the carbocyclic ring is sufficiently large, the constraint of ring formation is no longer significant. This point is reached by about C_{15}. Segments of such rings behave much as long linear alkanes; in such large rings there are generally a number of possible conformations in which the hydrogens are sufficiently separated and staggered from each other.

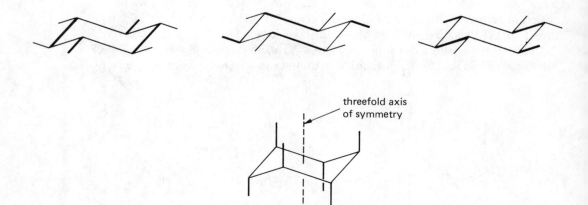

threefold axis of symmetry

FIGURE 5.23. *Construction of chair conformations.*

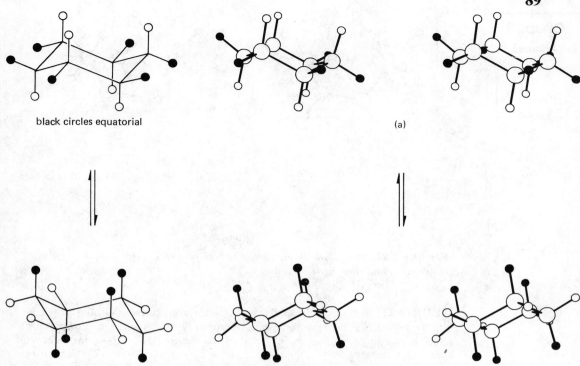

black circles equatorial

(a)

black circles axial

(b)

FIGURE 5.24. *Two chair conformations of cyclohexane:* (a) *black balls equatorial;* (b) *black balls axial. Left: normal projection; right: stereo.*

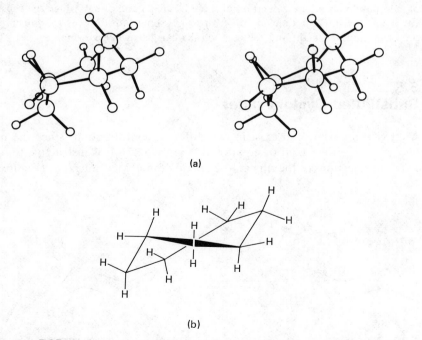

(a)

(b)

FIGURE 5.25. *Stable conformation of cycloheptane:* (a) *stereo representation;* (b) *conventional perspective drawing.*

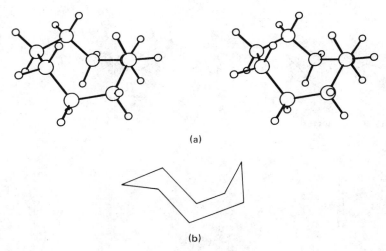

(a)

(b)

FIGURE 5.26. *Boat-chair conformation of cyclooctane:* (a) *stereo representation;* (b) *conventional line representation.*

EXERCISE: (a) Using a molecular model set, construct a model of cyclohexane. Place the model in a chair conformation (reference to the stereoscopic projection of cyclohexane in Figure 5.24 may be useful in this regard), and identify the axial and equatorial hydrogens. Affix a gumdrop to (or otherwise mark) the end of one of the C—H bonds. Experiment with "flipping" the model from one chair conformation to the other, noting that the gumdrop or mark is in an axial position in one conformation and in an equatorial position in the other.

(b) Practice drawing a chair conformation of cyclohexane, paying careful attention to the sets of parallel lines as shown in Figure 5.23. Compare the drawing you have made to the molecular model you constructed in part (a). Draw a chair conformation of chlorocyclohexane in which the substituent is axial and one in which the substituent is equatorial. Do not include the hydrogens. Show your drawings to your professor or teaching assistant and see if he or she can tell which is which.

5.8
Substituted Cyclohexanes

Methylcyclohexane can exist in two different conformations since the methyl group may be either axial or equatorial (Figure 5.27). It is instructive to look at Newman projections of the ring C_1—C_2 bond of methylcyclohexane (Figure 5.28).

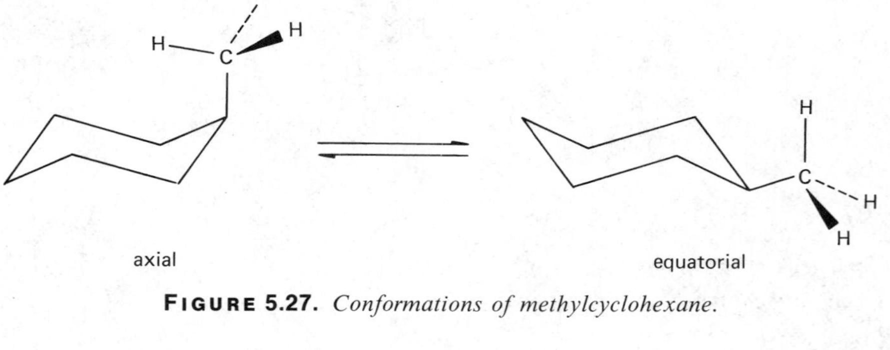

axial

equatorial

FIGURE 5.27. *Conformations of methylcyclohexane.*

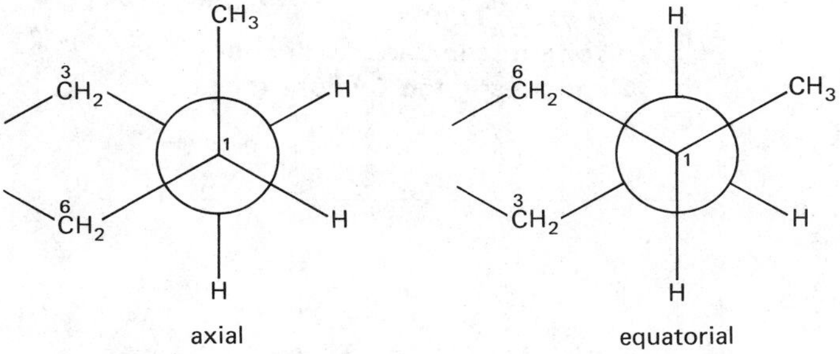

FIGURE 5.28. *Newman projections of methylcyclohexane.*

In the equatorial conformation the methyl group is *anti* to the C-3 CH_2 group of the ring, whereas in the axial conformation these groups have a *gauche* relationship. The interaction of a methyl hydrogen with the axial hydrogen of the C-3 CH_2 group is much like the interaction of the corresponding hydrogens in a *gauche* conformation of butane (Section 5.2). Recall that this interaction in butane causes an enthalpy increase of 0.9 kcal mole^{-1}. There are two such interactions in axial methylcyclohexane—between the methyl group and the C-3 CH_2 and the C-5 CH_2, as shown in Figure 5.27. Correspondingly, the axial conformation of methylcyclohexane is expected to be about 1.8 kcal mole^{-1} less stable than the equatorial conformation. This difference in energy for the two conformations may be approximated as $\Delta G°$ and transformed into an equilibrium constant for the equilibrium

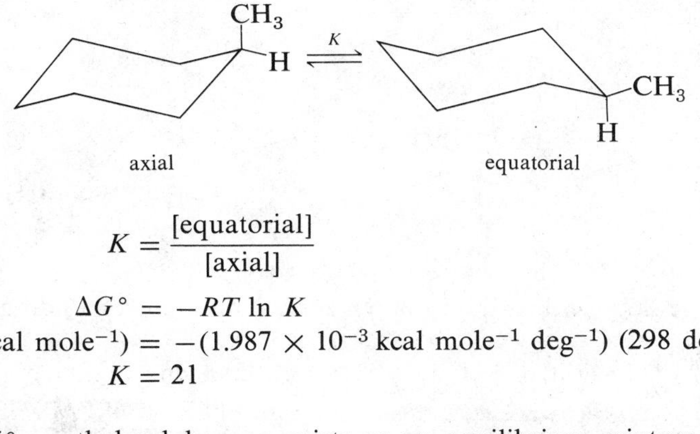

$$K = \frac{[\text{equatorial}]}{[\text{axial}]}$$

$$\Delta G° = -RT \ln K$$

$$(-1.8 \text{ kcal mole}^{-1}) = -(1.987 \times 10^{-3} \text{ kcal mole}^{-1} \text{ deg}^{-1})(298 \text{ deg}) \ln K$$

$$K = 21$$

Thus, at 25°, methylcyclohexane exists as an equilibrium mixture of the two conformations, with *95% of the molecules having the equatorial-methyl structure and 5% having the axial-methyl structure.* Because of interaction of axial groups with the other axial hydrogens on the same side of the ring, *axial conformations of substituted cyclohexanes are generally less stable than the corresponding equatorial conformations.* Actual energy differences for various substituents, expressed as $\Delta G°$ values, are summarized in Table 5.6.

Bulky groups such as isopropyl and *t*-butyl have such strong interactions in the axial position that the proportion of axial conformation in the equilibrium mixture is small. For example, the $\Delta G°$ of 3.1 kcal mole^{-1} for the group C_6H_5 (phenyl) corresponds to an equilibrium constant at 25° of 189; for phenylcyclohexane 995 out of every 100 molecules have the C_6H_5 group equatorial.

Like cyclohexane, 1,1-dimethylcyclohexane can exist in two chair conformations. In each conformation, one methyl group is axial and one is equatorial.

TABLE 5.6
Conformational Energies for
Monosubstituted Cyclohexanes

Group	$-\Delta G°$ (axial $\rightleftharpoons$ equatorial), kcal mole^{-1} (25°)
F	0.25
Cl	0.5
Br	0.5
I	0.45
OH	1.0
OCH_3	0.55
$OCOCH_3$	0.71
CH_3	1.7
CH_2CH_3	1.8
$C\equiv CH$	0.41
$CH(CH_3)_2$	2.1
$C(CH_3)_3$	$\sim$5–6
C_6H_5	3.1
COOR	1.3
COOH	1.4
CN	0.2
$OSO_2C_6H_4CH_3$	0.52

Of course, the $\Delta G°$ for this equilibrium is zero, and the corresponding equilibrium constant is 1.

For other disubstituted cyclohexanes, more than one isomer is possible. For example, in 1,4-dimethylcyclohexane the two methyl groups may both project *upward* from the general plane of the ring. This isomer is called *cis*-1,4-dimethylcyclohexane (L., *cis,* on this side). In flat projections of substituted ring compounds, substituents that project upward from the general ring plane are indicated by heavy bonds, and groups that project downward are indicated by dashed bonds. Hydrogens are usually omitted for convenience.

cis-1,4-dimethylcyclohexane

In the other 1,4-dimethylcyclohexane isomer, one methyl group projects upward from the ring and the other downward. Since the substituents are on opposite sides of the ring, this isomer is called *trans* (L., *trans,* across).

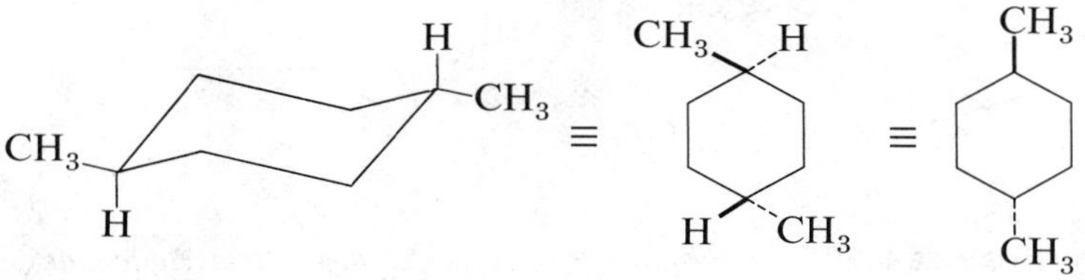

trans-1,4-dimethylcyclohexane

The lower energy of equatorial substituents compared to axial is also seen in the disubstituted compounds. For example, *trans*-1,4-dimethylcyclohexane ($\Delta H_f^\circ = -44.1$ kcal mole^{-1}) is more stable than the *cis* isomer ($\Delta H_f^\circ = -42.2$ kcal mole^{-1}) by 1.9 kcal mole^{-1}. In the *trans* isomer both methyl groups can be accommodated in equatorial positions, whereas in the *cis* hydrocarbon one methyl must be axial.

Both *cis*- and *trans*-1,4-dimethylcyclohexane can exist in two chair conformations. For the *cis* isomer, the two conformations are of equal energy, since each has one axial substituent and one equatorial substituent.

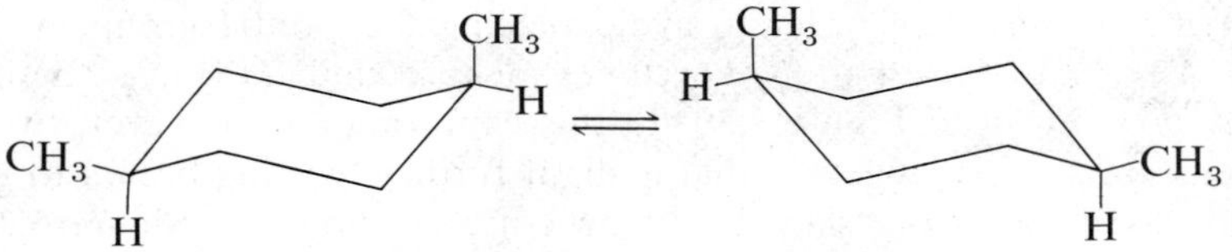

cis-1,4-dimethylcyclohexane, $\Delta G^\circ = 0$

For the *trans* isomer, one conformation has both substituents axial and the other has both substituents equatorial. The diequatorial conformation predominates greatly at equilibrium ($\Delta G^\circ \approx 3.4$ kcal mole^{-1}).

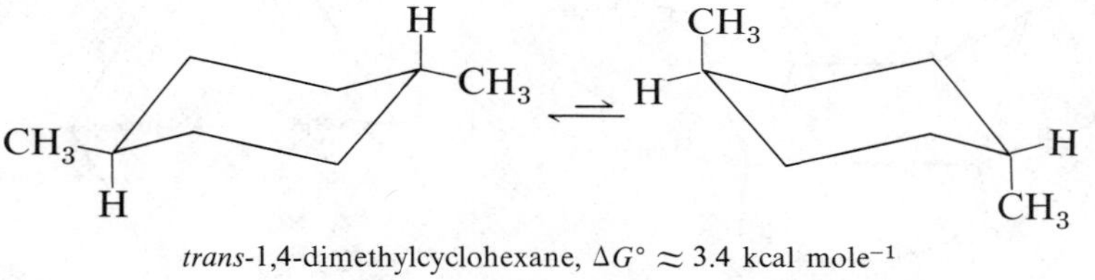

trans-1,4-dimethylcyclohexane, $\Delta G^\circ \approx 3.4$ kcal mole^{-1}

For 1,3-dimethylcyclohexane the *cis* isomer is more stable than the *trans*. In *cis*-1,3-dimethylcyclohexane both methyls can be equatorial, whereas one methyl must be axial in the *trans* isomer.

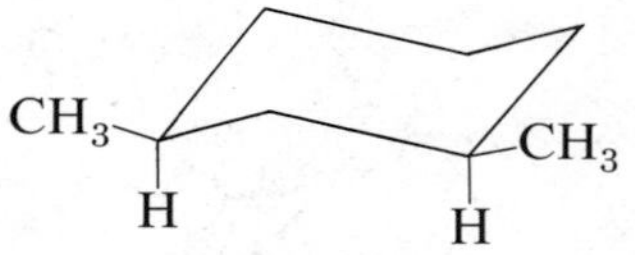

cis-1,3-dimethylcyclohexane
both methyls equatorial

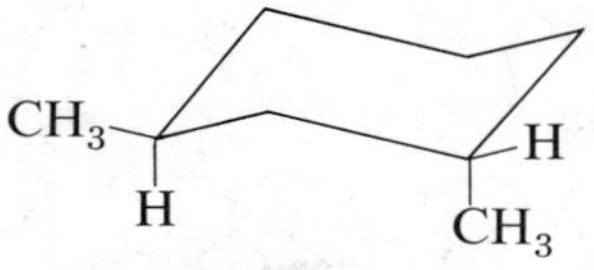

trans-1,3-dimethylcyclohexane
one methyl equatorial

The *t*-butyl group is so bulky that it effectively demands an equatorial position. Indeed, an axial *t*-butyl group represents so strained a structure that the ΔG°

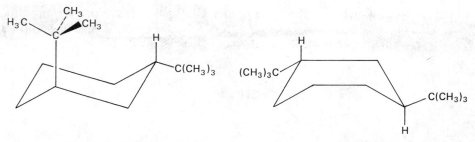

FIGURE 5.29. *Conformations of* trans-*1,3-di-t-butylcyclohexane.*

value in Table 5.6 for the difference between axial and equatorial *t*-butyl groups is only a rough estimate. In *cis*-1-*t*-butyl-4-methylcyclohexane, for example, the conformation with axial methyl and equatorial *t*-butyl groups dominates completely.

When excessive strain is involved, a distortion of the cyclohexane ring occurs. For example, phenyl and *t*-butyl are both rather bulky groups. A crystal sructure analysis of a compound that has a *cis*-1-*t*-butyl-4-phenylcyclohexane structure shows that the ring has been stretched out somewhat but still has essentially a chair conformation with axial phenyl and equatorial *t*-butyl groups.

In *trans*-1,3-di-*t*-butylcyclohexane a chair-cyclohexane ring would require one *t*-butyl group to be axial as in Figure 5.29. Actually, in this compound the cyclohexane ring is twisted in order to avoid placing the *t*-butyl group in an axial position. This new conformation of cyclohexane is related to the hypothetical **boat** conformation shown in Figure 5.30. In this conformation, however, two of the hydrogens are so close together that a slight further twisting occurs to give the so-called "twist-form" or "skew-boat" structure as shown in Figure 5.31. This skew-boat form occurs in several compounds containing bulky groups, but is not an important conformation for cyclohexane itself. In the skew-boat conformtion

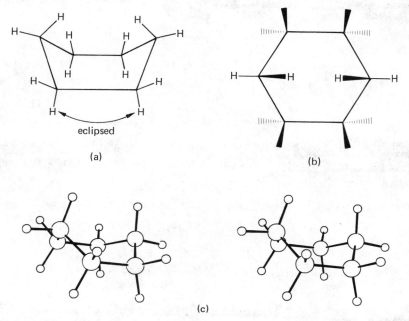

eclipsed

(a)

(b)

(c)

FIGURE 5.30. *Boat conformation of cyclohexane:* (a) *side view;* (b) *top view;* (c) *stereo view.*

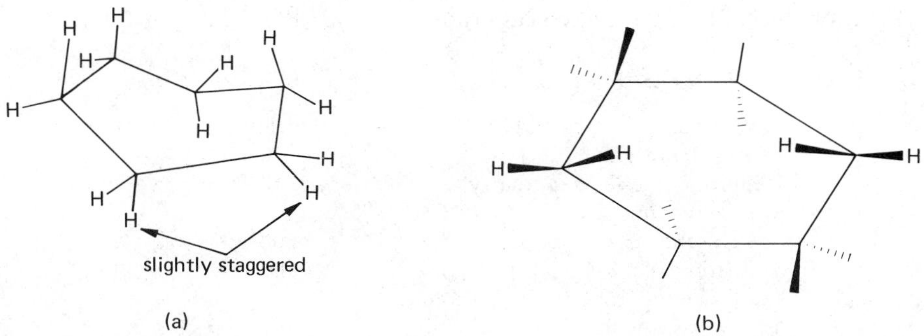

FIGURE 5.31. *Skew-boat conformation of cyclohexane:* (a) *side view;* (b) *top view.*

several hydrogens are partially eclipsed. The structure has a strain energy of about 5 kcal mole^{-1} relative to chair cyclohexane.

> **EXERCISE:** Make perspective drawings of the two chair conformations of *cis*-1,2-dimethylcyclohexane. What can you say about the relative stabilities of the two conformations? Work this exercise as well for the *trans* isomer.

5.9
Occurrence of Alkanes

Alkanes are widespread natural products on earth. They are primarily the product of living processes. Methane is produced by the anaerobic bacterial decomposition of vegetable matter under water. Because it was first isolated in marshes, it was long called "marsh gas." It is also an important constituent of the gas produced in some sewage-disposal processes. Methane also occurs in the atmosphere of coal mines, where it is called "fire damp" because of the explosive nature of methane–air mixtures.

Natural gas is a mixture of gaseous hydrocarbons and consists primarily of methane and ethane, along with small amounts of propane. Natural-gas production in the United States in 1978 was 22 trillion (22×10^{12}) cubic feet (ft^3), corresponding to about 10^{12} lb or 400 million tons of methane! The smaller alkanes are also byproducts of petroleum refining operations. For example, propane is the major constituent of liquefied petroleum gas (LPG), a domestic fuel used in mobile homes, among other places.

Petroleum itself is a complex mixture of hydrocarbons, mostly alkanes and cycloalkanes. It is the end result of the decomposition of animal and vegetable matter which has been buried in the earth's crust for long periods of time. The hydrocarbon mixture collects as a viscous black liquid in underground pockets, whence it is obtained by drilling wells. The resulting **crude oil** is refined by distillation into useful fuels and lubricants. Crude oil has a very broad boiling range. The more volatile constituents are propane, which is used as LPG, and butane, which is used as a chemical raw material. **Light petroleum ether** consists of pentanes and hexanes and boils from 30 to 60°. **Ligroin** is a mixture of heptanes and boils from 60 to 90°. These relatively volatile mixtures are often used as solvents, both in industry and in chemical laboratories. The most important petroleum distillates are kerosene and gasoline.

Fractional distillation of a typical crude oil yields the following fractions.

	Boiling Range, °C
natural gas (C_1 to C_4)	below 20
petroleum ether (C_5 to C_6)	30–60
ligroin or light naphtha (C_7)	60–90
straight-run gasoline (C_6 to C_{12})	85–200
kerosene (C_{12} to C_{15})	200–300
heating fuel oils (C_{15} to C_{18})	300–400
lubricating oil, greases, paraffin wax, asphalt (C_{16} to C_{24})	over 400

In 1978 total world production of crude oil was 19.7 billion barrels (bbl). About 50% of this total was converted into gasoline, 30% into heating fuels, and the remainder mostly into kerosene and jet fuel.

One of the major problems facing mankind is energy. As consumption increases, the supply of nonrenewable fossil fuels obviously decreases. Actually, the fossil-fuel "crisis" has been a bit exaggerated. Between 1976 and 1979 worldwide crude oil production amounted to 62 billion bbl. However, new discoveries added 45 billion bbl to the "proven reserves" (unmined oil which is economically recoverable using only existing technology), so that proven reserves decreased only 17 billion bbl, from 659 to 642 billion bbl. Furthermore, proven reserves of natural gas increased in the same period from 2,232 to 2,502 trillion ft³; the increase of 270 trillion ft³ is equivalent to about 20 billion bbl of crude oil. Hence, the total proven reserves of crude oil and natural gas actually *increased* slightly. However, even though discoveries of new deposits have kept pace with usage so far, it is clear that one day the known reserves will begin to decline and we must eventually turn to alternative sources of fuel.

Although other sources of energy will undoubtedly replace the fossil fuels as energy sources, there will still be a need for the fossil fuels as a source of **carbon.** At present, petroleum and coal hydrocarbons are the basic raw materials of much chemical industry. As the reserves become depleted, it is essential that we develop new sources of carbon raw materials to augment and eventually replace petroleum and natural gas. One possible source is **shale oil,** petroleum that is not collected in pockets from which it is easily retrieved but is interspersed throughout a porous rock formation. There are enormous reserves of shale oil, particularly in the Western Rockies of the United States, and active research is directed toward improving the economics of its recovery.

An obvious source of additional petroleum is the vegetable matter from which it derives in the first place. However, the natural production of petroleum by the decomposition of vegetation requires eons of time. Some current research is directed toward developing ways to speed up this process, since vegetation may be grown relatively quickly and is therefore replaceable. An interesting recent development in this area uses animal wastes as a starting material for the production of petroleum. The total amount of animal waste produced in the United States alone is astronomical; it is estimated that in 1979 more than 2.5 billion tons of chicken, pig, and cattle manure was produced. A process has been developed in which this manure is heated with carbon monoxide at 1200 psi and 380°. The product is a crude oil with properties similar to those of oil obtained from natural sources, and it is produced in high yield (about 3 bbl of oil from a ton of manure). The process is not a solution to the energy crisis because considerable energy is actually expended in manufacturing the product, but it may provide a future source for petroleum products when sufficient inexpensive nuclear or solar power is available. It also disposes of the animal wastes, which are becoming major environmental pollutants.

Hydrocarbons also result from some inorganic reactions. Examples are the production of methane by the hydrolysis of beryllium carbide or aluminum carbide.

$$Be_2C + 4 H_2O \longrightarrow CH_4 + 2 Be(OH)_2$$

$$Al_4C_3 + 12 H_2O \longrightarrow 3 CH_4 + 4 Al(OH)_3$$

Methane is also an important constituent of the atmospheres of the outer planets Jupiter and Saturn. It has even been suggested that in some distant time these planets may supply the earth's hydrocarbon needs.

Methane and ethane are odorless, but many of the higher hydrocarbons have distinctive odors. Despite the absence of typical functional groups, hydrocarbons can effect the changes at olfactory centers that we sense as odor. At least one alkane functions as a **pheromone,** a chemical used for communication in nature. For example, 2-methylheptadecane is the sex-attractant for tiger moths.

PROBLEMS

1. Give the IUPAC name for each of the following compounds.

(a) [structure: cyclohexane with CH₂CH₃ and CH₃ substituents]

(b) [structure: cyclohexane with CH₃ and CH(CH₃)₂ substituents]

(c) [structure: decalin with C(CH₃)₃ substituent]

(d) [structure: cyclopropane with two CH₃ substituents]

(e) [structure: cyclopentane with —CH₂CHCH₃ and CH₃]

(f) [structure: cyclobutane with CH₂CH₂CHCH₂CH₃ and CH₃]

(g) [structure: cyclohexane with CH₃ and Br]

(h) [structure: cyclopentane with I and —CH₂CH₃]

2. Write out the structures of the nine possible heptanes and assign IUPAC names. Which structures correspond to the *common* names isoheptane and neoheptane? The b.p. of heptane is 101°. By referring to the b.p.s of the isomeric hexanes in Table 5.2, estimate the b.p.s of the heptane isomers. Check your answers by looking up these compounds in the *Handbook of Chemistry and Physics* or *Lange's Handbook of Chemistry*. Not all of these hydrocarbons are listed in these handbooks. Browse through your library and see if you can find their properties in other reference works.

3. Write Newman projections showing the possible staggered conformations about the C-2—C-3 bond of pentane. Which ones correspond to the two stereo projections shown in Figure 5.12?

4. Write Newman projections and "sawhorse" structures for the three possible staggered conformations of 2,3-dimethylbutane. Note that two of these conformations are equivalent. The two different types of conformation differ in enthalpy by 0.9 kcal mole^{-1}. Which has the lower energy? Assume that $\Delta G° = \Delta H°$ and calculate the equilibrium composition at room temperature and compare your answer with that given for butane on page 74.

5. Examine a molecular model of adamantane. Give a rough estimate of its b.p. Would you expect the m.p. to be far below the b.p.? Look up its m.p. and b.p. in a handbook.

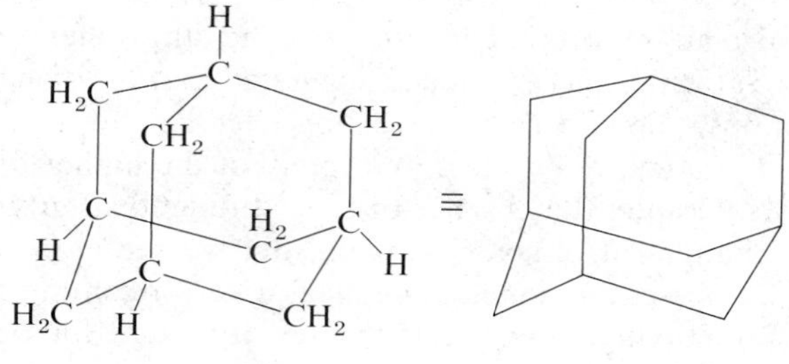

6. With a set of molecular models find each of the four staggered conformations of pentane. Sketch each of these structures using dashed bonds and wedges as appropriate. Try to rank these conformations in order of increased energy (remember that a *gauche* conformation is less stable than *anti*).

7. Construct potential energy diagrams for rotation about the C-2—C-3 bonds in isopentane, 2,3-dimethylbutane, and 2,2,3,3-tetramethylbutane. For each unique energy maximum or minimum, illustrate the structure with a Newman projection.

8. Using the heats of formation in Appendix I, calculate $\Delta H°$ for each of the following reactions.

(a) $CH_2{=}CH_2$ (ethylene) $+ H_2 \longrightarrow CH_3CH_3$
(b) $CH_2{=}CH_2 + HCl \longrightarrow CH_3CH_2Cl$
(c) $CH_2{=}CH_2 + H_2O \longrightarrow CH_3CH_2OH$ (ethanol)

What do your calculations tell you about the equilibrium constant for each reaction? What do they tell you about the rates of the three reactions?

9. Ethylcyclohexane and cyclooctane are isomers. Using the heats of formation of the two cycloalkanes (Appendix I) and assuming $\Delta H° = \Delta G°$, calculate the equilibrium constant for the equilibrium

$$\text{cyclooctane} \overset{K}{\rightleftharpoons} \text{ethylcyclohexane}$$

The actual equilibrium constant at 25° is 6.7×10^8. Explain.

10. From the data in Table 5.6, calculate the percentage of molecules having the substituent in the equatorial position for each of the following compounds.

(a)

(b)

(c)

(d)

11. (a) Make a model of cyclopentane in the envelope conformation. How many eclipsed interactions are there?

(b) Make a model of methylcyclohexane. Examine the model with the methyl group equatorial and compare it with a model of butane. How many "*gauche*" interactions are there in the equatorial methylcyclohexane structure that involve the methyl group? Repeat the comparison for axial methylcyclohexane.

12. (a) Write flat projection structures for each of the seven dimethylcyclohexanes. Use heavy bonds to indicate substituents that project above the ring and dashed bonds for substituents that project below.

(b) Sketch the two chair forms for 1,1-dimethylcyclohexane. What is the free energy difference between the two conformations?

(c) Answer part (b) for *cis*-1,2-dimethylcyclohexane, *trans*-1,3-dimethylcyclohexane, and *cis*-1,4-dimethylcyclohexane.

(d) Answer part (b) for *trans*-1,2-dimethylcyclohexane and *trans*-1,4-dimethylcyclohexane.

(e) Sketch the two chair forms for *cis*-1,3-dimethylcyclohexane. What interaction is present in one chair conformation of this isomer that does not occur in any of the other five isomers?

(f) For the seven dimethylcyclohexanes in parts (a)–(e), locate the *gauche* butane interactions in each of the chair conformations.

13. Draw a Newman projection for the C-1—C-2 bond of *cis*-1,2-dimethylcyclohexane.

14. We found in Section 5.5 that at 25° the equilibrium butane $\rightleftharpoons$ isobutane has $\Delta H° = -2.05$ kcal mole^{-1}; however, $\Delta G°$ is only -0.89 kcal mole^{-1}. Calculate the entropy change for the reaction and explain the direction of the effect. Calculate the equilibrium constant at 25°.

15. The free energy of formation $\Delta G_f°$ is defined in a manner analogous to the enthalpy of formation $\Delta H_f°$. Some values of $\Delta G_f°$ (25°, g) are pentane, -2.00; isopentane, -3.54; neopentane, -3.64 kcal mole^{-1}. Calculate the composition of the equilibrium mixture at 25°.

16. Careful inspection of the heats of formation in Table 5.4 will show regular increments per CH_2 group in a homologous series. In fact, $\Delta H_f°$ increments can be associated with the groups

$$CH_3, \quad \diagdown{CH_2}\diagup, \quad -\overset{|}{\underset{|}{CH}} \text{ and } -\overset{|}{\underset{|}{C}}-$$

Determine average values for these groups from Table 5.4, or, if you have access to a small computer, calculate the values that give the best least squares fit to the experimental data. Use your results to estimate $\Delta H_f°$ for hexane, 2-methylpentane, 3-methylpentane, 2,2-dimethylbutane, and 2,3-dimethylbutane. Compare with the experimental values in Appendix I.

You can see how far your "group equivalents" will go by comparing your calculated value with the experimental $\Delta H_f°$ for nonane of -54.7 kcal mole^{-1}. You should agree to several tenths of a kilocalorie per mole. However, compare your calculated value for 2,2,4,4-tetramethylpentane with the experimental $\Delta H_f°$ of -57.8 kcal mole^{-1}. Why is there a discrepancy? (*Hint:* Look at a molecular model. Will steric interferences or *strain* increase or decrease $\Delta H_f°$?)

17. Isobutane is thermodynamically more stable than butane. Which has the lower boiling point? Is there any relationship between thermodynamic stability and boiling point? Would you expect such a relationship between thermodynamic stability and melting point?

Chapter 6
Reactions of Alkanes

6.1
Bond-Dissociation Energies

Heat is kinetic energy. When a substance is heated, this kinetic energy increases the motion of atoms and molecules. When methane is heated, much of the added energy goes into translational motion, and the molecules move about faster relative to one another. However, some of the energy absorbed appears as increased vibrational and rotational motion. Figure 6.1 is a schematic diagram of the potential energy of the C—H bond as a function of the bond distance. It shows the vibrational quantum states for bond stretching. At room temperature, only the lowest quantum state is significantly populated.

> Remember that even at absolute zero the atoms are still vibrating. If the atoms were at rest, we would know both their position and momentum exactly—in violation of the Heisenberg uncertainty principle. This lowest vibrational quantum state has an energy ε_0 above the potential minimum. The quantity ε_0 is called the zero-point energy of the vibration.

As heat is applied, higher vibrational states are increasingly populated. In the higher vibrational quantum states, the average C—H bond distance during a vibration is greater. When sufficient energy D is absorbed, the bond breaks. The distance r_{C-H} increases to infinity and a hydrogen atom and a methyl radical result.

$$CH_4 \longrightarrow CH_3{\cdot} + H{\cdot}$$

The value of D is about 102 kcal mole^{-1}. It is generally more convenient to refer to the enthalpy of reaction at 25°, $\Delta H°$. This quantity ($\Delta H°$ for dissociation of a

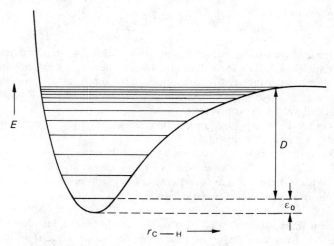

FIGURE 6.1. *Schematic diagram of potential function for a* C—H *bond in methane. Horizontal lines represent the various vibrational energy levels.*

bond at 25°) is given the special symbol $DH°$. For methane $DH°$ is 104 kcal mole^{-1}.

> The value of $DH°$ usually differs from D by a small amount. D is an energy quantity at 0° K, where only the lowest energy state is populated. At higher temperatures, various rotational and vibrational states are populated and contribute to the experimental bond dissociation energy $DH°$. The value of D is estimated by extrapolation to 0° K.

A $DH°$ of 104 kcal mole^{-1} represents a rather strong bond. Temperatures of the order of 1000° are required for dissociation of methane to occur at an appreciable rate. A C—H bond in ethane has a slightly lower $DH°$ (98 kcal mole^{-1}), but $DH°$ for the C—C bond is only 88 kcal mole^{-1}. Consequently, when ethane is heated, C—C fission occurs more rapidly than does C—H bond fission.

$$CH_3CH_3 \longrightarrow 2\ CH_3\cdot$$

This reaction occurs at about 700°.

In general, pyrolysis of a compound results in fission of the weakest bond. The products are **free radicals.**

> Free radicals contain an odd number of electrons. The Lewis structure for methyl radical is
>
> $$\begin{array}{c} H \\ H:C\cdot \\ H \end{array}$$
>
> Alkyl radicals can exist only at low concentrations at ordinary temperatures. Nevertheless, many such radicals have been "seen" by various spectroscopic methods. For example, methyl radical has been shown by spectroscopic measurements to be essentially flat—all four atoms lie in the same plane. Free radicals are important intermediates in many organic reactions, and we shall encounter them in this context later in this chapter.

The bond-dissociation energies, $DH°$, of several hydrocarbons are listed in Table 6.1. A more extensive table for a variety of compounds is given in Appendix II.

Note that $DH°$ depends on the character of the radical products. The $DH°$ for dissociation of a terminal C—H of an alkane is always about 98 kcal mole^{-1}. The

TABLE 6.1
Bond-Dissociation Energies for Some Alkanes[a]

Compound	$DH°$, kcal mole^{-1}	Compound	$DH°$, kcal mole^{-1}
CH_3—H	104	CH_3—CH_3	88
C_2H_5—H	98	C_2H_5—CH_3	85
$CH_3CH_2CH_2$—H	98	C_3H_7—CH_3	85
$(CH_3)_2CHCH_2$—H	98	C_2H_5—C_2H_5	82
$(CH_3)_2CH$—H	94.5	$(CH_3)_2CH$—CH_3	84
$(CH_3)_3C$—H	91	$(CH_3)_3C$—CH_3	80

[a] The bond dissociated is shown as a bond.

product of this bond cleavage is called a **primary** alkyl radical. It has the structure $RCH_2\cdot$, where R is any alkyl group.

$$CH_3CH_3 \longrightarrow CH_3CH_2\cdot + H\cdot \qquad \Delta H° = 98 \text{ kcal mole}^{-1}$$

When an interior C—H bond of a linear alkane is broken, the product $R_2CH\cdot$ is a **secondary** alkyl radical. Such bonds have a lower $DH°$ of 94–95 kcal mole^{-1}.

$$CH_3CH_2CH_3 \longrightarrow CH_3\overset{\cdot}{C}HCH_3 + H\cdot \qquad \Delta H° = 94.5 \text{ kcal mole}^{-1}$$

A C—H bond at a branch point is the weakest type of C—H bond. Such bonds have $DH°$ of about 91 kcal mole^{-1}. The product is a **tertiary** alkyl radical, $R_3C\cdot$.

$$\underset{\displaystyle CH_3\overset{\displaystyle\overset{CH_3}{|}}{C}HCH_3}{} \longrightarrow \underset{\displaystyle CH_3\overset{\displaystyle\overset{CH_3}{|}}{\underset{\cdot}{C}}CH_3}{} + H\cdot \qquad \Delta H° = 91 \text{ kcal mole}^{-1}$$

The relative stability of alkyl radicals depends on the number of alkyl groups attached to the radical carbon; alkyl radicals have the order of stability

$$\text{tertiary} > \text{secondary} > \text{primary} > \text{methyl}$$

The same principle applies to C—C bonds. The strength of this bond also depends on the relative stabilities of the radical products.

$$CH_3CH_3 \longrightarrow 2\ CH_3\cdot \qquad \Delta H° = 88 \text{ kcal mole}^{-1}$$

$$\underset{\displaystyle CH_3\overset{\displaystyle\overset{CH_3}{|}}{\underset{\underset{\displaystyle CH_3}{|}}{C}}CH_3}{} \longrightarrow \underset{\displaystyle CH_3\overset{\displaystyle\overset{CH_3}{|}}{\underset{\underset{\displaystyle CH_3}{|}}{C}}\cdot}{} + CH_3\cdot \qquad \Delta H° = 80 \text{ kcal mole}^{-1}$$

Additional C—C bond-dissociation energies are also tabulated in Table 6.1.

Consider fission of the two types of C—H bonds in isobutane (Table 6.1). In order to break one of the terminal C—H bonds, 98 kcal mole^{-1} of energy must be absorbed. In order to break the C—H bond at the branch point, only 91 kcal mole^{-1} is required.

$$\underset{\displaystyle CH_3\overset{\displaystyle\overset{CH_3}{|}}{C}HCH_3}{} \left\langle \begin{array}{l} \longrightarrow CH_3\overset{\displaystyle\overset{CH_3}{|}}{C}HCH_2\cdot + H\cdot \qquad \Delta H° = 98 \text{ kcal mole}^{-1} \\[2em] \longrightarrow CH_3\overset{\displaystyle\overset{CH_3}{|}}{\underset{\underset{\displaystyle CH_3}{|}}{C}}\cdot + H\cdot \qquad \Delta H° = 91 \text{ kcal mole}^{-1} \end{array} \right.$$

We start at the same point for the two reactions and, because one of the products is the same in each case (H·), the difference in $\Delta H°$s for these reactions is a direct measure of the difference in stability of the two alkyl radicals. *The t-butyl radical is more stable than the isobutyl radical by 7 kcal mole^{-1}* (Figure 6.2).

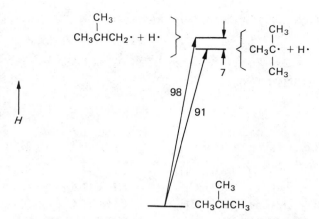

FIGURE 6.2. *The DH°s for the two C—H bonds in isobutane.*

These results can also be expressed in terms of heats of formation. The heat of formation of a radical is derived from the ΔH_f° of the reactants and the DH° of the bond-breaking reaction; for example,

$$H_2 \longrightarrow 2H\cdot \qquad \Delta H^\circ = DH^\circ = 104 \text{ kcal mole}^{-1}$$

$$\Delta H_f^\circ(H\cdot) = [DH^\circ + H_f(H_2)] \div 2 = 52 \text{ kcal mole}^{-1}$$

$$\Delta H_f^\circ[(CH_3)_2CHCH_2\cdot] = \Delta H_f^\circ[(CH_3)_3CH] + DH^\circ[(CH_3)_2CHCH_2\cdot] - \Delta H_f^\circ(H\cdot)$$

$$= -32.4 + 98 - 52 = 14 \text{ kcal mole}^{-1}$$

Heats of formation of some radicals are summarized in Table 6.2. A more complete list is given in Appendix I.

TABLE 6.2
Heats of Formation of Some Radicals

Compound	ΔH_f°, kcal mole^{-1} at 25°	Compound	ΔH_f°, kcal mole^{-1} at 25°
H·	52	$(CH_3)_3C\cdot$	7
$CH_3\cdot$	34	$CH_3CH_2CH_2CH_2\cdot$	16
$C_2H_5\cdot$	26	$(CH_3)_2CHCH_2\cdot$	13
$CH_3CH_2CH_2\cdot$	21	$CH_3CH_2\dot{C}HCH_3$	12
$(CH_3)_2CH\cdot$	17.5		

EXERCISE: For each of the reactions discussed in this section, calculate the bond-dissociation energies using the ΔH_f° values in Table 6.2 and Appendix I.

6.2
Pyrolysis of Alkanes: Cracking

When a molecule is broken up by heat, the process is called **pyrolysis** (Gk., *pyros*, fire; *lysis*, a loosening). When alkanes are pyrolyzed, the C—C bonds cleave to produce smaller alkyl radicals. With higher alkanes, the cleavage occurs randomly along the chain.

$$CH_3CH_2CH_2CH_2CH_3 \begin{cases} \longrightarrow CH_3\cdot + CH_3CH_2CH_2CH_2\cdot \\ \\ \longrightarrow CH_3CH_2\cdot + CH_3CH_2CH_2\cdot \end{cases}$$

One possible reaction of these radicals is recombination to form an alkane. A mixture of different alkanes is produced.

$$CH_3\cdot + CH_3\cdot \longrightarrow CH_3CH_3$$
$$CH_3\cdot + CH_3CH_2\cdot \longrightarrow CH_3CH_2CH_3$$
$$CH_3CH_2\cdot + CH_3CH_2CH_2CH_2\cdot \longrightarrow CH_3CH_2CH_2CH_2CH_2CH_3$$

Another reaction that occurs is **disproportionation.** In this process one radical transfers a hydrogen atom to another radical to produce an alkane plus an alkene.

$$CH_3CH_2\cdot + CH_3CH_2CH_2\cdot \longrightarrow CH_3CH_3 + CH_3CH{=}CH_2$$

The net result of pyrolysis is the conversion of a large alkane to a mixture of smaller alkanes and alkenes. This reaction is *not* a useful one in the organic laboratory where the aim is generally to produce a *single* pure compound in high yield. However, thermal cracking of hydrocarbons has been an important industrial process. As it comes from the ground, crude oil varies widely in composition depending on its source. For example, fractional distillation of a typical light oil affords 35% gasoline, 17% kerosene, and only a trace of asphalt, the balance being mainly high-boiling heating and lubricating oils. On the other hand, a typical "heavy oil" affords only 11% gasoline, 10% kerosene, and 50% asphalt. In order to reduce the average molecular weight of heavy oils and increase the production of the desirable more volatile fractions, various *cracking* techniques have long been employed. The oldest such method was the thermal-cracking method. However, thermal cracking has all but disappeared in recent years, although the process (which is actually more complex than indicated in the simple equation on this page) is still used to some extent for the production of low molecular weight alkenes. Modern cracking methods employ various catalysts, mainly composed of alumina and silica, which accomplish degradation of the large hydrocarbons into smaller ones at lower temperatures. Catalytic cracking probably involves cationic rather than free-radical intermediates.

6.3
Halogenation of Alkanes

A. *Chlorination*

When a mixture of methane and chlorine is heated to about 120° or irradiated with light of a suitable wavelength, a highly exothermic reaction occurs.

$$CH_4 + Cl_2 \xrightarrow[\text{or light}]{\text{heat}} CH_3Cl + HCl \qquad \Delta H° = -24.7 \text{ kcal mole}^{-1}$$

methyl
chloride

The reaction is a significant industrial process for preparing methyl chloride. It has limited usefulness as a laboratory preparation because the reaction does not stop with the introduction of a single chlorine. As the concentration of methyl chloride builds up, this compound can be chlorinated in competition with methane.

$$CH_3Cl + Cl_2 \longrightarrow CH_2Cl_2 + HCl$$
methylene
chloride

$$CH_2Cl_2 + Cl_2 \longrightarrow CHCl_3 + HCl$$
chloroform

$$CHCl_3 + Cl_2 \longrightarrow CCl_4 + HCl$$
carbon
tetrachloride

The actual product of the reaction of methane and chlorine is a mixture of methyl chloride (b.p. 23.8°), methylene chloride (CH_2Cl_2, b.p. 40.2°), chloroform ($CHCl_3$, b.p. 51.2°), and carbon tetrachloride (CCl_4, b.p. 76.8°). The composition of the mixture depends on the relative amounts of starting material used and the reaction conditions. In this case the products can readily be separated by fractional distillation because of the difference in boiling points.

A good deal of experimental evidence is in accord with the mechanism presented below for the chlorination of methane. The reaction begins with the **homolysis** of a chlorine molecule to two chlorine atoms (equation 6.1).

> When a covalent bond breaks in such a way that each fragment retains one electron of the bond, the process is called **homolytic cleavage** or **homolysis**.
>
> $$A:B \longrightarrow A\cdot + B\cdot$$
>
> When one fragment retains both electrons, the process is termed **heterolytic cleavage** or **heterolysis**.
>
> $$A:B \longrightarrow A + B:$$

Since molecular chlorine has a rather low bond-dissociation energy ($DH° = 58$ kcal mole^{-1}), chlorine atoms may be produced by light of relatively long wavelength or by heating to moderate temperatures.

$$Cl_2 \xrightarrow[\text{or } h\nu]{\Delta} 2\,Cl\cdot \tag{6-1}$$

Once chlorine atoms are present in small amount, a **chain reaction** commences. A chlorine atom reacts with a methane molecule to give a methyl radical and HCl (equation 6.2). The methyl radical then reacts with a chlorine molecule to give methyl chloride and a chlorine atom (equation 6-3).

$$Cl\cdot + CH_4 \rightleftharpoons CH_3\cdot + HCl \qquad \Delta H° = +1\ \text{kcal mole}^{-1} \tag{6-2}$$

$$CH_3\cdot + Cl_2 \longrightarrow CH_3Cl + Cl\cdot \qquad \Delta H° = -25.7\ \text{kcal mole}^{-1} \tag{6-3}$$

The chlorine atom produced in equation (6-3) can react with another methane molecule to continue the chain. Reaction (6-1) is called the **initiation** step, and reactions (6-2) and (6-3) are called the **propagation** steps.

In principle, only one chlorine molecule need homolyze in order to convert many moles of methane and chlorine to methyl chloride and HCl. In practice, the chain process only goes through, on the average, about 10,000 cycles before it is **terminated.** Termination occurs whenever two radicals happen to collide, for example, equations (6-4) and (6-5).

$$CH_3\cdot + Cl\cdot \longrightarrow CH_3Cl \qquad (6-4)$$

$$CH_3\cdot + CH_3\cdot \longrightarrow CH_3CH_3 \qquad (6-5)$$

Another possible termination step involves the collision of two chlorine atoms— reverse of the initiation step (equation 6-1). However, when two chlorine atoms collide to form Cl_2, the resulting molecule has as vibrational energy all of the kinetic energy of translation of the two atoms. This energy is always in excess of the bond energy, and the two atoms simply separate again. Only if collision occurs in the presence of a third body or on the wall of the reaction vessel to remove some of this energy does the chlorine molecule formed stay intact.

$$Cl\cdot + Cl\cdot + M \longrightarrow Cl_2 + M$$

With polyatomic molecules the translational energy of the reactants of equations (6-4) and (6-5) can be transferred into other bond vibrations, and reaction occurs directly. That is, the C—H bonds in these reactions serve as the third body.

Other reactions that may (and probably do) occur are unproductive and do not terminate the chain reaction.

$$CH_4 + CH_3\cdot \longrightarrow CH_3\cdot + CH_4$$

$$Cl_2 + Cl\cdot \longrightarrow Cl\cdot + Cl_2$$

Let us look at each of the foregoing propagation steps in some detail. Reaction (6-2) is slightly endothermic and reversible, but it has a low activation energy of only about 4 kcal mole^{-1}. The reaction may be considered in further detail in terms of attack by Cl· on hydrogen.

$$Cl\cdot + H{-}CH_3 \longrightarrow [Cl\cdots H\cdots CH_3]^{\ddagger} \longrightarrow Cl{-}H + \cdot CH_3$$

The H—Cl and H—CH$_3$ bonds have similar strength [the $DH°$s are 103 and 104 kcal mole^{-1}, respectively (Appendix II)]. As the Cl—H bond forms and becomes stronger, the H—C bond becomes weaker and breaks. The product methyl radical appears to be planar (Figure 6.3). Methyl radical can be described to a good approximation in terms of three $C_{sp^2}{-}H_{1s}$ bonds with the odd electron contained in the remaining C_{2p}-orbital. At the transition state the methyl group has started to flatten out from its original tetrahedral structure.

For the reverse process, HCl + CH$_3\cdot \longrightarrow$ CH$_4$ + Cl·, the same mechanism applies in reverse. The carbon radical attacks the hydrogen of HCl at the rear of the H—Cl bond and a C—H bond begins to form. As the forming C—H bond distance decreases and the bond strength increases, the remaining C—H bonds begin to bend back toward their tetrahedral geometry in CH$_4$. At the same time, the H—Cl bond distance increases.

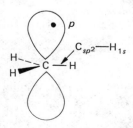

FIGURE 6.3. *Methyl radical.*

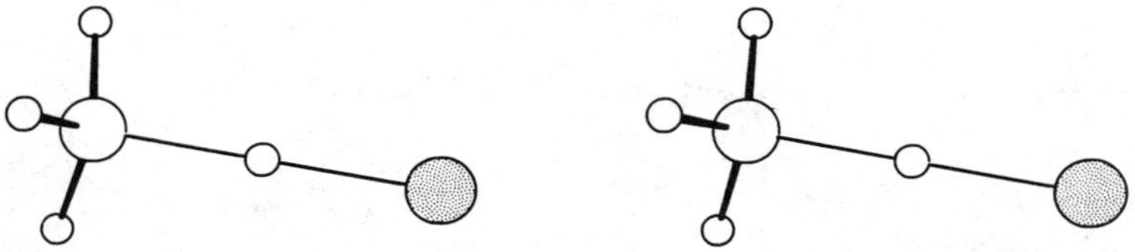

FIGURE 6.4. *Stereo representation of the transition state for the reaction*
$CH_4 + Cl\cdot \rightleftharpoons CH_3\cdot + HCl$.

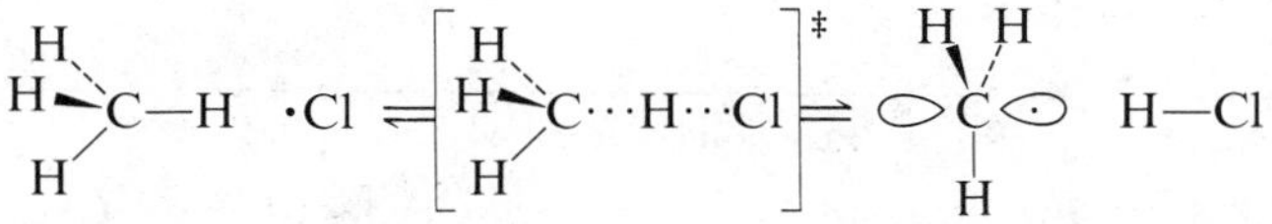

A stereo representation of the transition state is shown in Figure 6.4.

> The structure of the transition state is the same for both directions by the **principle of microscopic reversibility.** That is, the reverse reaction from products to reactants must have the same reaction mechanism as the forward reaction. If it did not, we could, in principle, set up a perpetual motion machine in violation of the second law of thermodynamics.

An equivalent description may be given in orbital terms. As the chlorine orbital containing one electron overlaps with the hydrogen $1s$-orbital, electron repulsion causes a decrease in the overlap of the H_{1s}-orbital with the C_{sp^3}-orbital, and the C—H bond begins to lengthen and become weaker. As this C—H bond gets weaker, it has less demand for s-orbital character, and the carbon s-orbital is used more for bonding to the other C—H bonds. Rehybridization occurs progressively from sp^3 toward sp^2. The carbon begins to flatten out, and the remaining C—H bonds become somewhat shorter and stronger. The structure of the transition state is depicted in terms of component atomic orbitals in Figure 6.5. A reaction coordinate diagram for reaction (6-2) is shown in Figure 6.6.

Reaction (6-3) has only a small activation energy of about 1 kcal mole^{-1}. This reaction is rapid and highly exothermic. The reverse reaction is highly endothermic and has a correspondingly high activation energy of $25.7 + 1 = 26.7$ kcal mole^{-1}. Consequently, the overall forward reaction is effectively irreversible. A reaction coordinate diagram for this step is shown in Figure 6.7. The transition state for this reaction is one in which the C—Cl bond is partly formed and the Cl—Cl bond is partly broken.

$$[CH_3\cdots\cdots Cl\cdots Cl]^\ddagger$$

A stereo representation is given in Figure 6.8.

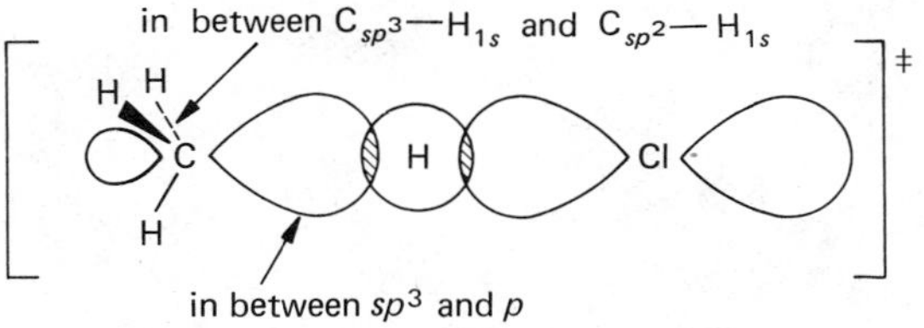

FIGURE 6.5. *Orbital description of transition state for the reaction* $CH_4 +$ $Cl\cdot \rightleftharpoons CH_3\cdot + HCl$.

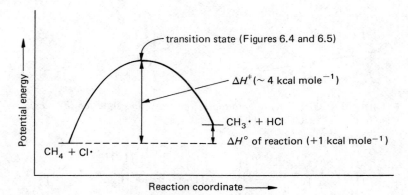

FIGURE 6.6. *Reaction profile for the reaction* $CH_4 + Cl\cdot \rightleftharpoons CH_3\cdot + HCl$.

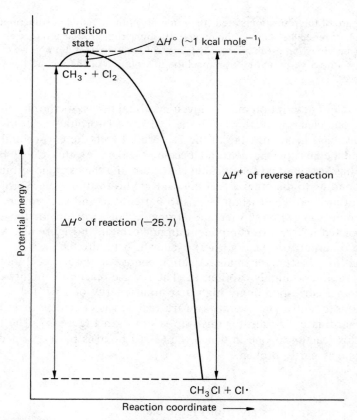

FIGURE 6.7. *Reaction profile for the reaction* $CH_3\cdot + Cl_2 \rightleftharpoons CH_3Cl + Cl\cdot$.

FIGURE 6.8. *Stereo representation of the transition state for the reaction* $CH_3\cdot + Cl_2 \rightleftharpoons CH_3Cl + Cl\cdot$.

The overall $\Delta H°$ of the net chlorination reaction may be obtained by summing equations (6-2) and (6-3).

$$\Delta H°, \ kcal \ mole^{-1}$$

$$(6\text{-}2): \quad CH_4 + \cancel{Cl}\cdot \longrightarrow \cancel{CH_3}\cdot + HCl \qquad\qquad +1$$
$$(6\text{-}3): \quad \cancel{CH_3}\cdot + Cl_2 \longrightarrow CH_3Cl + \cancel{Cl}\cdot \qquad\qquad -25.7$$
$$\overline{CH_4 + Cl_2 \longrightarrow CH_3Cl + HCl \qquad\qquad -24.7}$$

> Note that $\Delta H°$ for the initiation step is *not* added to the $\Delta H°$ values for the propagation steps in deriving $\Delta H°$ for the overall reaction. If one does this, one is actually calculating $\Delta H°$ for *another reaction*.
>
> $$\Delta H°, \ kcal \ mole^{-1}$$
>
> $$Cl_2 \longrightarrow 2\ Cl\cdot \qquad\qquad +58$$
> $$CH_4 + \cancel{Cl}\cdot \longrightarrow \cancel{CH_3}\cdot + HCl \qquad\qquad +1$$
> $$\cancel{CH_3}\cdot + Cl_2 \longrightarrow CH_3Cl + \cancel{Cl}\cdot \qquad\qquad -25.7$$
> $$\overline{Cl_2 + CH_4 + Cl_2 \longrightarrow CH_3Cl + HCl + 2\ Cl\cdot \qquad\qquad +32.3}$$
>
> This equation is just the sum of the overall chlorination reaction and the chlorine homolysis.
>
> This is often a point of confusion because the student reasons that heat had to be put in to initiate the reaction. However, the question is not how much heat is applied, but what is $\Delta H°$, the *heat of the reaction*?

Chlorination of higher alkanes is similar to chlorination of methane except that the product mixtures are more complex. Ethane gives not only ethyl chloride, but also 1,1-dichloroethane and 1,2-dichloroethane.

$$CH_3CH_3 + Cl_2 \longrightarrow CH_3CH_2Cl + HCl$$

$$CH_3CH_2Cl + Cl_2 \longrightarrow \quad CH_3CHCl_2 \quad + \quad ClCH_2CH_2Cl + HCl$$
$$\text{1,1-dichloroethane} \qquad \text{1,2-dichloroethane}$$

With propane, two monochloro products may be formed. Both *n*-propyl chloride and isopropyl chloride are formed, but not in equal amounts.

$$CH_3CH_2CH_3 + Cl_2 \longrightarrow CH_3CH_2CH_2Cl + CH_3CHClCH_3$$
$$\text{(43\%)} \qquad\qquad \text{(57\%)}$$
$$\textit{n}\text{-propyl chloride} \quad \text{isopropyl chloride}$$

In carbon tetrachloride solution at 25°, the two isomers are produced in the relative amounts 43:57. Further reaction gives a mixture of the four possible dichloropropanes.

Let us examine the monochlorination of propane in greater detail. Recall that $DH°$ for the secondary hydrogen in propane is about 3.5 kcal mole^{-1} lower than $DH°$ for the primary hydrogen (Table 6.1). We might anticipate, then, that the secondary hydrogen would be removed by a chlorine atom more easily than a primary hydrogen. However, there are six primary hydrogens that may be replaced, whereas there are only two secondary hydrogens. The **relative reactivity** per hydrogen is then

$$\frac{\text{secondary}}{\text{primary}} = \frac{57/2}{43/6} = \frac{4}{1}$$

A similar trend is noticed in the monochlorination of isobutane, which gives 36% *t*-butyl chloride and 64% isobutyl chloride.

$$(CH_3)_3CH + Cl_2 \longrightarrow (CH_3)_3CCl + (CH_3)_2CHCH_2Cl$$

(36%)   (64%)

t-butyl chloride   isobutyl chloride

The relative reactivity of tertiary and primary hydrogens on a per-hydrogen basis is

$$\frac{\text{tertiary}}{\text{primary}} = \frac{36/1}{64/9} = \frac{5.1}{1}$$

Thus, the relative rates of reaction of different hydrogens with Cl· are just as we expect on the basis of $DH°$ for the various hydrogens

tertiary > secondary > primary

However, the degree of preference is relatively low. That is, there is less difference between the activation energies for the various reactions than there is between the heats of reaction (Figure 6.9).

For example, in chlorination of propane, the Cl· can abstract a hydrogen from the methyl group or from the methylene group. In the former case $\Delta H°$ is -5 kcal mole^{-1}, and in the latter it is -9 kcal mole^{-1}. Thus, the difference in heats of reaction $\Delta\Delta H° = 4$ kcal mole^{-1}. However, $\Delta H^{\ddagger}$ for abstraction of a CH_2 hydrogen is 3 kcal mole^{-1}, whereas $\Delta H^{\ddagger}$ for abstraction of a hydrogen from a CH_3 group is 2 kcal mole^{-1}; the difference in activation energies $\Delta\Delta H^{\ddagger} = 1$ kcal mole^{-1}. This result becomes reasonable when one realizes that in the transition state the free radical is not yet fully formed. Whatever it is that causes a secondary free radical to be more stable than a primary free radical will also affect the two transition states. However, that effect will be muted in the transition state to the extent that carbon has not achieved complete free radical character.

With even more complicated alkanes, chlorination mixtures are hopelessly complex. Hence, chlorination of alkanes is *not a good general reaction for preparing alkyl chlorides*. There is one type of compound for which chlorination has practical utility in laboratory preparations. When all hydrogens are equivalent, there is only one possible monochloro product. In such cases the desired product can generally be separated from hydrocarbon and di- and higher chlorinated

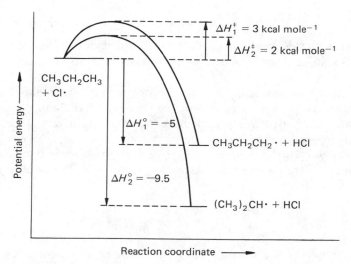

FIGURE 6.9. *Reaction profiles for the reaction of* Cl· *with* C_3H_8.

species by fractional distillation. Two examples are the chlorination of cyclohexane and neopentane.

$$\text{(cyclohexane)} + Cl_2 \longrightarrow \text{(cyclohexyl chloride, Cl)} + HCl$$

cyclohexyl chloride

$$CH_3-\overset{\overset{\displaystyle CH_3}{|}}{\underset{\underset{\displaystyle CH_3}{|}}{C}}-CH_3 + Cl_2 \longrightarrow CH_3-\overset{\overset{\displaystyle CH_3}{|}}{\underset{\underset{\displaystyle CH_3}{|}}{C}}-CH_2Cl + HCl$$

neopentyl chloride

Since handling gaseous chlorine in the laboratory is inconvenient, such chlorinations are often done with sulfuryl chloride, SO_2Cl_2, instead.

[Sulfuryl chloride is a colorless liquid, b.p. 69°, produced by reaction of Cl_2 and SO_2. It fumes in moist air because it reacts rapidly with water according to the reaction

$$SO_2Cl_2 + 2\,H_2O \longrightarrow 2\,HCl + H_2SO_4$$]

When sulfuryl chloride is used as a chlorinating agent, a special **initiator** must be used to provide the free radicals that start the chain reaction. **Peroxides** are often used for this purpose because the O—O bond is weak and readily broken at relatively low temperatures (see Appendix II).

$$ROOR \longrightarrow 2\,RO\cdot$$

The chlorination of cyclohexane by sulfuryl chloride provides a typical example.

$$\text{(cyclohexane)} + SO_2Cl_2 \longrightarrow \text{(chlorocyclohexane, Cl)} + HCl + SO_2$$

[A mixture of 1.8 mole of cyclohexane, 0.6 mole of sulfuryl chloride, and 0.001 mole of benzoyl peroxide, $(C_6H_5COO)_2$, is refluxed for 1.5 hr. Fractional distillation gives 89% of chlorocyclohexane, b.p. 143°, and 11% of a mixture of dichlorocyclohexanes.]

EXERCISE: Write equations showing the initiation, propagation, and termination steps for the chlorination of ethane. Compute the $\Delta H°$ for each step and the overall $\Delta H°$ of reaction.

B. *Halogenation with Other Halogens*

The mechanism discussed in the preceding section for chlorination may also be applied to the other halogens, but the actual reactions show important differences. The overall enthalpies of halogenation of methane by various halogens are summarized in Table 6.3.

The reaction with fluorine is so highly exothermic that controlled fluorination is

TABLE 6.3
$CH_4 + X_2 = CH_3X + HX$

X	$\Delta H°$, kcal mole^{-1}
F	-102.8
Cl	-24.7
Br	-7.3
I	$+12.7$

difficult to accomplish. The energy liberated is sufficient to break most bonds. The HF bond is so strong ($DH° = 136$ kcal mole^{-1}) that the following reaction is endothermic by only 6 kcal mole^{-1}.

$$CH_4 + F_2 \longrightarrow CH_3\cdot + F\cdot + HF \qquad \Delta H° = +6 \text{ kcal mole}^{-1}$$

Consequently, when methane and fluorine are mixed, a few radicals form spontaneously and initiate chain reactions. The heat of reaction, which is liberated, causes a rapid rise in temperature, and more bonds break to form radicals, which initiate more chain reactions. A radical chain reaction that is highly exothermic and produces radicals faster than they are destroyed results in an explosion. Organofluorine compounds are important because they frequently have unique and desirable properties. However, they are generally *not* made by direct fluorination, and this reaction is *not* a general laboratory preparation.

Iodination is at the opposite extreme. As shown in Table 6.2, the reaction of methane with iodine is endothermic. In fact, methyl iodide reacts with HI to generate CH_4 and I_2. Iodine atoms are relatively unreactive. For example, reaction with methane is so endothermic that no significant reaction occurs at ordinary temperatures.

$$CH_4 + I\cdot \longrightarrow HI + CH_3\cdot \qquad \Delta H° = +33 \text{ kcal mole}^{-1}$$

Any iodine atoms produced ultimately dimerize to reform I_2.

The bromination of methane is less exothermic than is chlorination. Of the two chain propagation steps only one is relatively exothermic.

$$CH_4 + Br\cdot \longrightarrow CH_3\cdot + HBr \qquad \Delta H° = +16.5 \text{ kcal mole}^{-1}$$
$$CH_3\cdot + Br_2 \longrightarrow CH_3Br + Br\cdot \qquad \Delta H° = -23.8 \text{ kcal mole}^{-1}$$

Consequently, bromination is much slower than chlorination. It is instructive to examine the bromination of methane from a mechanistic standpoint. The two propagation steps are plotted in reaction coordinate form in Figures 6.10 and 6.11.

In its reactions with other alkanes, bromine is a much *more selective* reagent than chlorine. For example, bromination of propane at 330° in the vapor phase gives 92% isopropyl bromide and only 8% *n*-propyl bromide.

$$CH_3CH_2CH_3 + Br_2 \longrightarrow CH_3CH_2CH_2Br + (CH_3)_2CHBr$$
$$\qquad\qquad\qquad\qquad\qquad (8\%) \qquad\qquad (92\%)$$
$$\qquad\qquad\qquad\qquad \text{\textit{n}-propyl bromide} \quad \text{isopropyl bromide}$$

The hydrogen abstraction steps for formation of the two isomers are

$$CH_3CH_2CH_3 + Br\cdot \longrightarrow CH_3CH_2CH_2\cdot + HBr \qquad \Delta H° = +10.5 \text{ kcal mole}^{-1}$$
$$CH_3CH_2CH_3 + Br\cdot \longrightarrow (CH_3)_2CH\cdot + HBr \qquad \Delta H° = +6.0 \text{ kcal mole}^{-1}$$

The two reactions are plotted in reaction coordinate form in Figure 6.12.

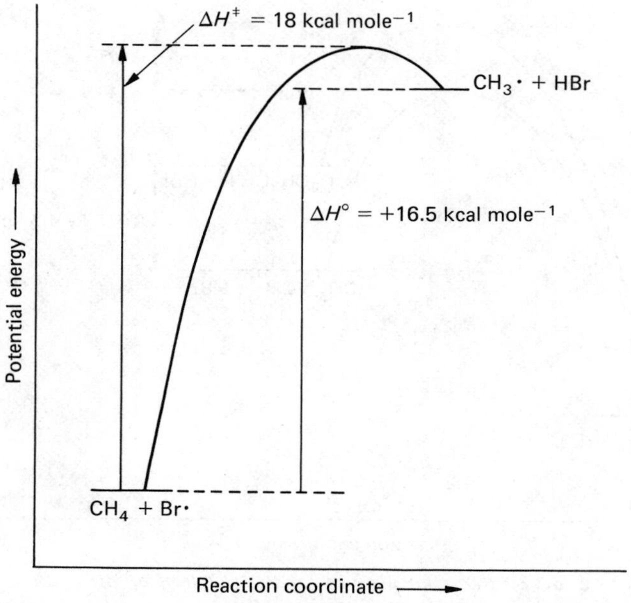

$\Delta H^{\ddagger} = 18$ kcal mole^{-1}

$CH_3\cdot + HBr$

$\Delta H^{\circ} = +16.5$ kcal mole^{-1}

$CH_4 + Br\cdot$

Potential energy

Reaction coordinate

FIGURE 6.10. *Reaction profile for the reaction* $CH_4 + Br\cdot \rightleftharpoons CH_3\cdot + HBr$.

The rates of reaction of a bromine atom with the two types of hydrogen in propane are given by

$$\text{rate } (1^{\circ}) = k_{1^{\circ}}[CH_3CH_2CH_3][Br\cdot]$$
$$\text{rate } (2^{\circ}) = k_{2^{\circ}}[CH_3CH_2CH_3][Br\cdot]$$

The ratio of the products formed is simply the ratio of the two rate constants.

$$\frac{\text{rate } (2^{\circ})}{\text{rate } (1^{\circ})} = \frac{k_{2^{\circ}}}{k_{1^{\circ}}}$$

For two similar reactions such as these, the ratio of the rate constants is related in

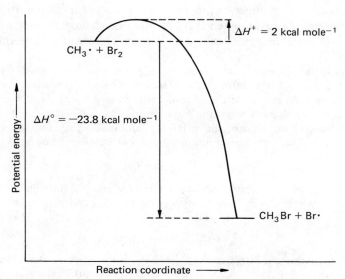

$\Delta H^{\ddagger} = 2$ kcal mole^{-1}

$CH_3\cdot + Br_2$

$\Delta H^{\circ} = -23.8$ kcal mole^{-1}

$CH_3Br + Br\cdot$

Potential energy

Reaction coordinate

FIGURE 6.11. *Reaction profile for the reaction* $CH_3\cdot + Br_2 \rightleftharpoons CH_3Br + Br\cdot$

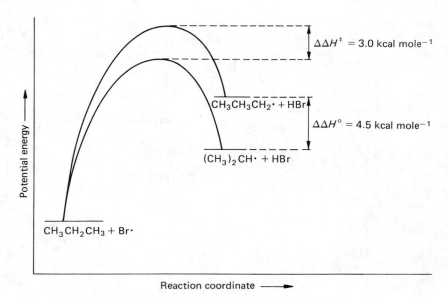

FIGURE 6.12. *Reaction profiles for the reaction of* Br· *with* C_3H_8.

an exponential manner to the two activation energies. The reaction with the larger activation energy has the smaller rate constant. In the chlorination of propane $\Delta\Delta H^{\ddagger}$ is only 1 kcal mole^{-1}, and consequently chlorination is relatively nonselective. For bromination $\Delta\Delta H^{\ddagger}$ is 3 kcal mole^{-1}, and hence bromination gives a greater *ratio* of secondary to primary products.

The selectivity of bromine relative to chlorine is even more apparent when there are tertiary hydrogens in the alkane. For example, 2,2,3-trimethylbutane undergoes bromination to give more than 96% of the tertiary bromide, even though the alkane has only one tertiary hydrogen and fifteen primary hydrogens.

$$(CH_3)_3CCH(CH_3)_2 + Br_2 \xrightarrow[CCl_4]{h\nu} (CH_3)_3CCBr(CH_3)_2$$
$$(>96\%)$$

Thus, bromination is a somewhat more useful process for preparative purposes than chlorination. However, when there is only one tertiary hydrogen and many secondary hydrogens in a molecule, complex mixtures will still be produced.

It is interesting to speculate why bromine atoms are so much more discriminating than chlorine atoms. A straightforward explanation is based on the concept of "early" and "late" transition states. Chlorination is believed to proceed through an "early" transition state. By an early transition state, we mean that C—H bond cleavage has not proceeded very far. Consequently, H—Cl bond making has not happened to a very great extent. Such an early transition state may be symbolized as

$$[R\text{---}H\text{--------}Cl]^{\ddagger}$$

In contrast to chlorination, bromination is believed to proceed through a "late" transition state in which C—H bond breaking and H—Br bond making are well advanced. Such a late transition is symbolized as

$$[R\text{--------}H\text{---}Br]^{\ddagger}$$

If the transition state is early and the C—H bond has not been stretched very much, the carbon is still essentially tetrahedral. Its geometry resembles the starting alkane. Conversely, if the transition state is late and C—H bond breaking has proceeded to a greater extent, then the geometry of carbon is more nearly planar. It resembles the product free radical.

In a reaction proceeding through an early transition state, the stability of the product has little influence on rate because the structure and energy of the transition state are similar to the reactants. Thus, in chlorination of isobutane, the two possible transition states are of similar energy and the two competing reactions proceed at comparable rates.

However, if the transition state is late, the stability of the product is reflected in the transition state. In the bromination of isobutane, one transition state looks like a primary free radical and the other resembles a tertiary free radical. These two transition states differ in energy considerably. Consequently, the reaction leading to a tertiary free radical is much more rapid.

The foregoing discussion is an application of a fundamental principle called the Hammond Postulate. As initially formulated by Hammond, the postulate states that "If two states, as for example, a transition state and an unstable intermediate, occur consecutively during a reaction process and have nearly the same energy content, their interconversion will involve only a small reorganization of the molecular structure." That is, along a given reaction path, states which differ little in energy will also differ little in geometry. It is a corollary of the postulate that processes having very low activation energies must proceed through transition states differing little in structure from the reactants themselves, since a low energy of activation energy implies that the energies of the starting and transition states are similar, as in Figure 6.11. It also follows that the higher the activation energy for a process, the less the transition state for that process will structurally resemble the reactants. In the case of the halogenation reactions under consideration, we saw that abstraction of a hydrogen by $Br\cdot$ is more endothermic by 15.5 kcal mole^{-1} than abstraction by $Cl\cdot$. Thus, the transition state for abstraction by $Cl\cdot$ should be an "early" one, while that for abstraction by $Br\cdot$ should be a "late" one.

EXERCISE: Using the data in Appendices I and II, calculate the heats of reaction for

$$(CH_3)_3CH + X_2 \longrightarrow (CH_3)_3CX + HX$$

for $X_2 = F_2$, Cl_2, Br_2, and I_2.

6.4
Combustion of Alkanes

In terms of the mass of material involved, combustion of alkanes is one of the most important organic reactions. All burning of natural gas, gasoline, and fuel oil involves mostly the combustion of alkanes. However, this combustion is an atypical organic reaction in two respects. First, mixtures of alkanes are normally the "reactants" in this reaction. Second, the desired product of the reaction is not the chemical products but the heat of reaction. Indeed, the chemical products are frequently undesirable, and their sheer mass creates significant problems of disposal. The equation for complete combustion of an alkane is simple.

$$C_nH_{2n+2} + \left(\frac{3n+1}{2}\right)O_2 \longrightarrow n\,CO_2 + (n+1)\,H_2O$$

However, many combustion processes, such as the burning of gasoline in an internal combustion engine, do not result in complete combustion. In an automobile, 1 gal of gasoline produces more than 1 lb of carbon monoxide. There are many other products resulting from incomplete combustion. Among these other products are aldehydes (RCHO), compounds that contribute significantly to the smog problem.

The mechanism by which alkanes react with oxygen is an exceedingly complex one and has not been worked out in detail. There are many partially oxidized intermediates. Radical chain steps are certainly involved. An especially important reaction is the combination of alkyl radicals with oxygen to give **alkylperoxy** radicals, which abstract hydrogen from an alkane to give intermediate alkyl hydroperoxides.

$$R\cdot + O_2 \longrightarrow ROO\cdot$$
$$ROO\cdot + RH \longrightarrow ROOH + R\cdot$$

Alkyl hydroperoxides contain a weak O—O bond ($DH° \approx 44$ kcal mole^{-1}), which breaks readily at elevated temperatures to produce more radicals.

$$ROOH \longrightarrow RO\cdot + HO\cdot$$

Thus combustion is another example of a radical-multiplying reaction, which leads to explosions under proper conditions.

When such an explosion occurs in the reaction chamber of an internal combustion engine, the piston is driven forward with a violent rather than a gentle stroke. Such premature explosions cause the phenomenon known as "knocking". The tendency of a fuel to knock depends markedly on the nature of the hydrocarbons used. In general, branching of an alkane chain tends to inhibit knocking. The knocking characteristic of a fuel is expressed quantitatively by an "octane number." On this arbitrary scale, n-heptane is given a value of 0 and 2,2,4-trimethylpentane ("iso-octane") is assigned the value of 100. An octane number of 86, typical of a medium-grade "standard" or "regular" gasoline, has a knocking characteristic equivalent to that of a mixture of 86% 2,2,4-trimethylpentane and 14% n-heptane. The octane rating may be upgraded by the addition of small amounts of tetraethyllead ($(C_2H_5)_4Pb$), which is called an "antiknock" agent. Its function is to control the concentration of free radicals and prevent the premature explosions which are characteristic of knocking. In recent years, the proliferation of "catalytic converters" for smog control has reduced the use of leaded fuel, since the lead compounds produced in the combustion process inactivate the catalysts. Thus, for environmental protection reasons, there has been a shift away from the use of lead compounds for increasing the octane rating of gasoline. However, the production of unleaded gasoline of comparable octane rating results in less efficient use of our limited petroleum reserves, since more processing is required, with attendant losses. For this reason, there is currently an active program under way to find other compounds that will improve the knocking characteristics of gasoline without damaging smog-control devices. Two leading candidates are t-butyl alcohol and methyl t-butyl ether (Chapter 10).

The **heat of combustion** is defined as the enthalpy of the complete oxidation. The heat of combustion of a pure alkane can be measured experimentally with high precision ($\pm 0.02\%$) and constitutes an important thermochemical quantity. For example, the heat of combustion of n-butane is $\Delta H° = -634.82 \pm 0.15$ kcal mole^{-1}, whereas that of isobutane is $\Delta H° = -632.77 \pm 0.11$ kcal mole^{-1}. The general equation for combustion of these two isomers is the same.

$$C_4H_{10} + 6\tfrac{1}{2} O_2 \longrightarrow 4\,CO_2 + 5\,H_2O$$

Heats of combustion are often expressed in terms of $H_2O(l)$ as one product. To avoid confusion with other energy terms in this text we have given values for the heats of

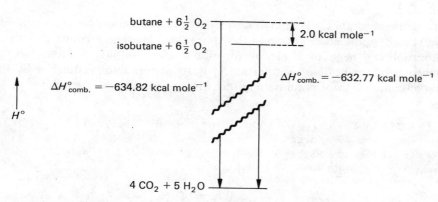

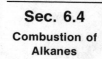

FIGURE 6.13. *Illustrating the heats of combustion of* n-*butane and isobutane.*

$\lfloor$ combustion with $H_2O(g)$ as one product. This difference is the heat of vaporization of $\rfloor$
$\lfloor$ water, 10.52 kcal mole^{-1}. $\rfloor$

A direct comparison of these two heats of combustion shows that the branched hydrocarbon is 2.0 kcal mole^{-1} more stable than the straight-chain hydrocarbon at room temperature (Figure 6.13). The products, carbon dioxide and water, are more stable than the reactants. Because the products have a lower energy content, energy is released as heat—the heat of combustion. The *less* stable the reactants, the *more* heat is evolved. Since *n*-butane has a heat of combustion of higher magnitude than isobutane, *n*-butane must have a higher energy content and is less stable thermodynamically than isobutane.

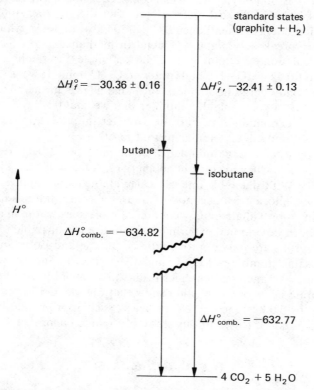

FIGURE 6.14. *The relationship between the heats of formation and combustion of butane and isobutane.*

It is from accurate heats of combustions that heats of formation (Section 5.5) have been determined. Figure 6.14 illustrates the relationship between heats of combustion and heats of formation for *n*-butane and isobutane. Note that the conversion of the heats of combustion to heats of formation requires only the heats of combustion of graphite and of hydrogen.

> **EXERCISE:** The heat of combustion of graphite (per carbon) and hydrogen (H_2) are −94.05 and −57.80 kcal mole^{-1}, respectively. From the heats of combustion of *n*-butane and isobutane shown in Figure 6.14, calculate ΔH_f° for both hydrocarbons and compare with the values in Figure 6.14.

6.5
Average Bond Energies

Appendix I includes heats of formation for a number of free atoms. With these values we can calculate **heats of atomization,** the enthalpy required to dissociate a compound into all of its constituent atoms. For example, the heat of atomization of methane is 397 kcal mole^{-1}.

$$CH_4 = \quad C \quad + 4 H$$
$$\Delta H_f^\circ: \ -17.9 \quad 170.9 + (4 \times 52.1) \quad \quad \Delta H^\circ = 397.2 \text{ kcal mole}^{-1}$$

(Note that ΔH_f° of atomic carbon is much higher than ΔH_f° of C(graphite), carbon bound as graphite, which is defined as the standard state.) This reaction requires breaking four C—H bonds. Hence, we can consider each bond to have an **average bond energy** of 397/4 = 99 kcal mole^{-1}. Note that this number differs from the bond dissociation energy of methane ($DH^\circ = 104$ kcal mole^{-1}), which is the energy required to break only *one* C—H bond in methane.

A similar calculation for ethane gives 674.6 kcal mole^{-1} as the heat of atomization required to break six C—H bonds and one C—C bond. If we *assume* that the average C—H bond energy in ethane is the same as it is in methane, we obtain 675 − (6 × 99) = 81 kcal mole^{-1}, a number that we could call the average bond energy of the C—C bond in ethane. If the same technique is applied to propane, we find a C—C bond energy similar to that in ethane.

In practice, data for a large number of compounds have been used to derive best overall values for such average bond energies. A table of such values is given in Appendix III. With this table one can calculate heats of atomization that are accurate to a few kilocalories per mole. The use of such a table is important for determining the approximate energy content of molecules whose heats of formation have not been determined experimentally or are too unstable to be isolated. Note that the results are only approximations. Butane and isobutane, for example, have the same numbers of C—C and C—H bonds. Such an approximate calculation using average bond energies results in identcial heats of atomization; however, accurate heats of combustion show that butane and isobutane differ in energy content by 2.0 kcal mole^{-1}. Nevertheless, such approximate values will be found to have important uses in our study of organic chemistry.

PROBLEMS

1. (a) What products are expected from thermal-cracking of pentane?
(b) Write reaction mechanisms leading to each product.

(c) From heats of formation calculate the enthalpy of each of the net reactions involved.

2. Using the appropriate $DH°$ values from Appendix II, calculate $\Delta H°$ for each of the reactions shown.
(a) $Br_2 \longrightarrow 2\ Br\cdot$
(b) $CH_3CH_3 + Br\cdot \longrightarrow CH_3CH_2\cdot + HBr$
(c) $CH_3CH_2\cdot + Br_2 \longrightarrow CH_3CH_2Br + Br\cdot$
What is the overall $\Delta H°$ for bromination of ethane given as the sum of reactions (b) and (c)?
(d) $CH_3CH_3 + Br_2 \longrightarrow CH_3CH_2Br + HBr$
How does this value compare with that obtained using heats of formation?

3. In the course of the bromination of ethane (problem 2), both bromine atoms and ethyl radicals will be present but not in equal amounts. Which is present in larger quantity? Explain.

4. The reaction of the unusual hydrocarbon spiropentane with chlorine and light is one of the best ways of preparing chlorospiropentane.

spiropentane   chlorospiropentane

(a) Explain why chlorination is such a useful preparative method in this case.
(b) Write the reaction mechanism.

5. For each of the following compounds, write the structures of all of the possible monochlorination products and predict the relative amounts in which they will be produced.
(a) butane   (b) 2-methylbutane
(c) 2,2,4-trimethylpentane   (d) 2,2,3-trimethylbutane
(e) pentane

6. Answer problem 5 for bromination, using the relative reactivities of C—H bonds toward bromine atoms at 40°: prim., 1; sec., 220; tert., 19000.

7. In the chlorination of ethane, the observed reaction is

$$CH_3CH_3 + Cl_2 \longrightarrow CH_3CH_2Cl + HCl$$

An alternative reaction that might have occurred is

$$CH_3CH_3 + Cl_2 \longrightarrow 2\ CH_3Cl$$

(a) Calculate $\Delta H°$ for each reaction.
(b) Propose a radical chain mechanism by which the alternative reaction might occur. Calculate $\Delta H°$ for each of the propagation steps.
(c) Suggest a reason why the alternative reaction does not occur.

8. From Appendix I calculate the heat of atomization of *n*-butane. Compare this value with the approximate one obtained from the use of average bond energies (Appendix III).

9. From the heats of formation given in Appendix I calculate the heat of combustion of cyclopropane and cyclohexane. For combustion of an equal *weight* under the same conditions, which is the better fuel?

10. Chlorine fluoride, ClF, is a colorless gas (b.p. $-101°$) which behaves chemically as a reactive halogen. In principle, it can react with methane in two alternative ways

$$CH_4 + ClF \begin{array}{c} \nearrow CH_3F + HCl \\ \searrow CH_3Cl + HF \end{array}$$

(a) From $\Delta H_f° (ClF) = -12.2$ kcal mole^{-1} and other data in Appendix I, calculate $\Delta H°$ for both reactions.

(b) $\Delta S°$ for both reactions is expected to be approximately the same; hence, which set of products is expected to predominate at equilibrium?

(c) The reaction mechanism involves radical chain reactions similar to the reactions with chlorine except that the reaction of methyl radical with ClF can take two possible courses.

$$CH_3\cdot + ClF \begin{array}{c} \nearrow CH_3Cl + F\cdot \\ \searrow CH_3F + Cl\cdot \end{array}$$

What is the difference in $\Delta H°$ for these two reactions? To the extent that the difference in activation energies reflects the difference in $\Delta H°$, which reaction is expected to be the faster?

11. Nitromethane, CH_3NO_2, is prepared by reaction of methane with nitric acid in the gas phase at temperatures over 400°. Appendix I includes values for some nitrogen compounds. Calculate $\Delta H°$ (298°K) for the equilibrium

$$CH_4 + HNO_3 \Longrightarrow CH_3NO_2 + H_2O$$

It may seem strange that such an exothermic reaction requires such a high temperature. The actual reaction steps are believed to be

$$CH_4 + \cdot NO_2 \Longrightarrow CH_3\cdot + HNO_2$$
$$HNO_2 + HNO_3 \Longrightarrow 2 NO_2\cdot + H_2O$$
$$CH_3\cdot + NO_2\cdot \Longrightarrow CH_3NO_2$$

Calculate $\Delta H°$ for each step. Which step is expected to be the slow step that requires the high temperature? The reaction is initiated by traces of oxygen or radicals to produce some $NO_2\cdot$ radicals which start the reaction. Note that this reaction sequence involves a radical chain propagation. Nitrogen dioxide is a rather stable radical, and its concentration in the reaction mixture is relatively high. It reacts rapidly with the methyl radicals and keeps these radicals at a very low concentration so that alternative free radical chain reactions are kept to minor importance; that is, nitrogen dioxide *scavenges* the methyl radicals. List several possible reactions of methyl radicals with nitric acid to produce methyl alcohol or nitromethane directly and show that these reactions are exothermic.

Nitrogen dioxide is a resonance-stabilized radical in which the odd electron can be placed on both oxygens and nitrogen. Write Lewis structures to demonstrate this point. In view of these structures it may seem surprising that methyl radical reacts with the nitrogen of NO_2. Actually, reaction at oxygen also occurs to give methyl nitrite, CH_3ONO. Compare $\Delta H°$ for this reaction with that for production of nitromethane:

$$CH_3NO_2 \longleftarrow CH_3\cdot + NO_2\cdot \longrightarrow CH_3ONO$$

Methyl nitrite is unstable under the reaction conditions and gives other products. The entire reaction is complex, and the discussion has treated only the most important of the many reactions that actually occur in this system.

12. In the chlorination of methane, the propagation steps (equations 6-2 and 6-3) do not have the same activation energies, as shown by Figures 6.6 and 6.7. However, by varying the relative concentrations of the methane and chlorine, either step can be made to be the slow step. Explain. Which step is rate-determining when $[CH_4] = [Cl_2]$? Under these conditons of relative concentrations, which free radical will be present in higher concentration, $CH_3\cdot$ or $Cl\cdot$? Show that the nature of the principal termination step depends on which propagation step is faster.

13. The description of rotational barriers in the alkanes (Section 5.2) was actually over-simplified, since, like all motion, the torsional motion around a C—C bond is quantized and a molecule will exist in one or more torsional-rotational states. Construct a more accurate version of Figure 5.10 to show this quantization. Assume that the zero-point energy level is about 0.5 kcal mole^{-1} above the potential minimum and that the torsional quantum levels are separated by about 1.0 kcal mole^{-1} increments. Which quantum level corresponds to continuous, uninhibited rotation about the bond?

Chapter 7
Stereoisomerism

7.1
Chirality and Enantiomers

The two objects depicted in Figure 7.1 appear to be identical in all respects. For every edge, face, or angle on one there is a corresponding edge, face, or angle on the other. And yet the two objects are not superimposable upon each other and are therefore *different objects*. They are related to one another as an object is related to its mirror image.

Another pair of familiar objects related to each other in this way are your right and left hands. They are (to a first approximation) identical in all respects. Yet your right hand will fit into a right glove and not into a left glove. The general property of "handedness" is called **chirality.** An object that is not superimposable upon its mirror image is **chiral.** If an object and its mirror image can be made to coincide in space, then they are said to be **achiral.**

Careful inspection of 2-iodobutane reveals that it is a chiral compound. There are actually two isomeric 2-iodobutanes, which are nonsuperimposable mirror images (Figure 7.2). Two compounds that differ in handedness in this way are called **enantiomers** and are said to have an **enantiomeric** relationship to each other. In order to convert one of the enantiomers into the other, it is necessary to break and re-form bonds. Such a process requires substantial energy; consequently, there is a rather large energy barrier to interconversion of this type of enantiomers. One may have a flask which contains only one of the enantiomers, and it will be stable indefinitely under normal conditions. This type of isomerism is called **stereoisomerism. Stereoisomers** are compounds that have the same sequence of covalent bonds and differ in the relative disposition of their atoms in space.

2-Iodobutane owes its chirality to C-2, which has *four different groups* attached to it (C_2H_5, CH_3, I, H). Such a carbon is said to be **asymmetric.** Notice that 1,1-dichloroethane, which contains three different groups on C-1 (C_2H_5, Cl, H), is achiral; it is superimposable upon its mirror image (Figure 7.3). When a compound has one asymmetric carbon atom, the molecule is always chiral. However,

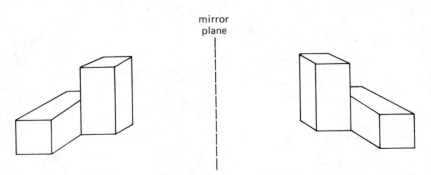

mirror
plane

FIGURE 7.1. *Two nonsuperimposable mirror-image objects.*

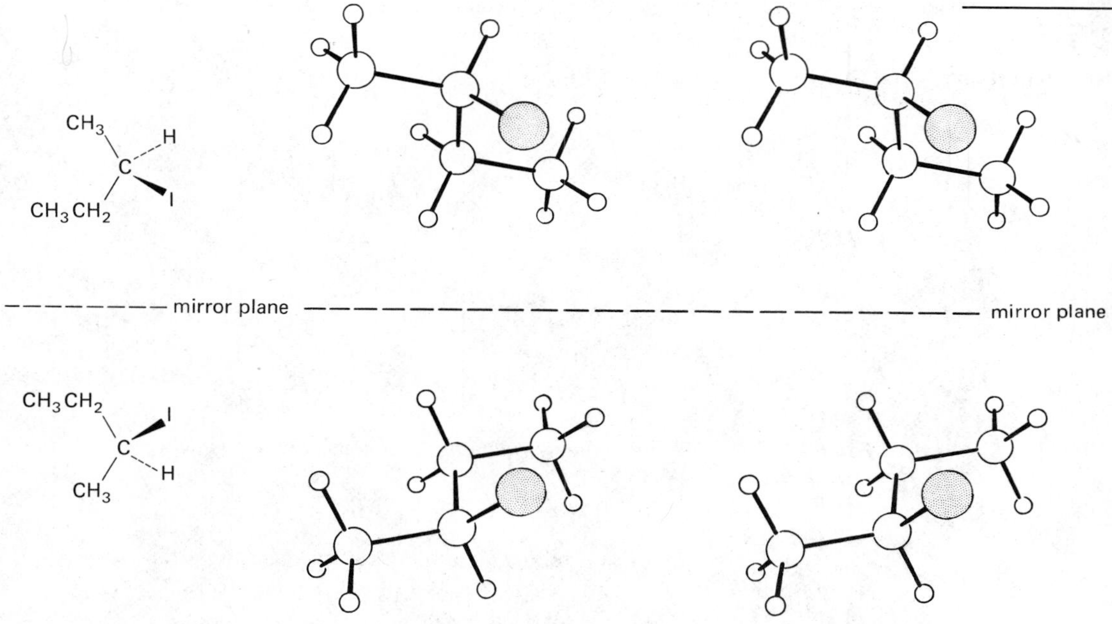

FIGURE 7.2. *The mirror-image relationship of the two 2-iodobutanes.*

an asymmetric atom is not a necessary condition for chirality, as we shall soon see. Also, as we shall see in Section 7.5, a molecule may still be achiral if it contains more than one asymmetric atom.

Two of the conformational isomers of butane (Section 5.2) also have an enantiomeric relationship to each other (Figure 7.4). In this case, however, the two enantiomers may interconvert simply by rotation about the central C—C bond.

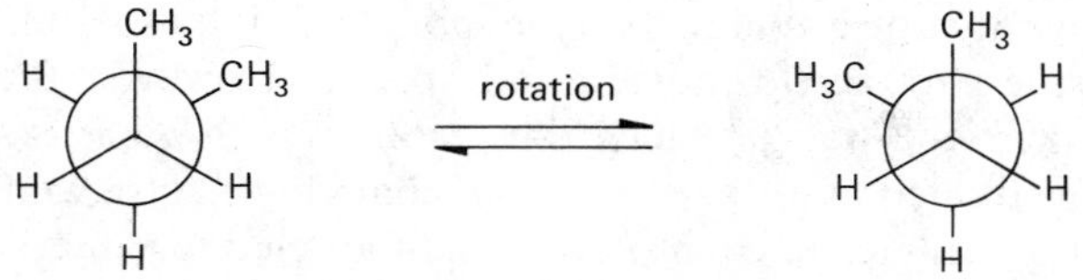

Since rotational barriers are generally quite small, enantiomers such as these interconvert rapidly at room temperature. The individual enantiomers could be obtained in a pure state only by working at exceedingly low temperatures, on the order of $-230°$ (page 74).

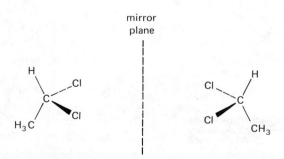

FIGURE 7.3. *The achirality of 1,1-dichloroethane.*

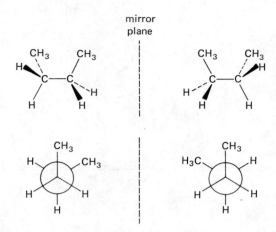

FIGURE 7.4. *The enantiomeric relationship between the two* gauche *conformations of butane.*

EXERCISE: Which of the following familiar objects are chiral?
(a) a football   (b) an egg   (c) a corkscrew
(d) a golf club   (e) a crescent wrench   (f) a person
(g) a catcher's mitt   (h) a spiral staircase   (i) a pencil
(j) a slide rule   (k) a pair of scissors   (l) a screw-cap bottle top
(m) a Greek vase   (n) a portrait

7.2
Physical Properties of Enantiomers: Optical Activity

Most of the physical properties of the two enantiomeric 2-iodobutanes are identical. They have identical melting points, boiling points, solubilities in common solvents, densities, refractive indices, and spectra. However, they differ in one important respect, the way in which they interact with **polarized light.**

Light may be treated as a wave motion of changing electric and magnetic fields which are at right angles to each other. When an electron interacts with light, it oscillates at the frequency of the light in the direction of the electric field and in phase with it. In normal light, the electric field vectors of the light waves are oriented in all possible planes. **Plane-polarized light** is light in which the electric field vectors of all the light waves lie in the same plane, the **plane of polarization.**

normal light   plane-polarized light

Plane-polarized light may be produced by passing normal light through a polarizer, such as a polaroid lens or a device known as a **Nicol prism.**

In a molecule, an electron is not free to oscillate equally in all directions; that is, its polarizability is **anisotropic,** which means different in different directions. When electrons in molecules oscillate in response to plane-polarized light, they generally tend, because of their anisotropic polarizability, to oscillate out of the

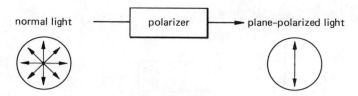

plane of polarization. Because of its interaction with the oscillating electrons, the light has its electric and magnetic fields changed. Thus, when plane-polarized light interacts with a molecule, the plane of polarization rotates slightly. In a large collection of achiral molecules, however, for any orientation of a molecule that changes the plane of polarization of the light, there is apt to be another molecule with a mirror-image orientation which has the opposite effect. Consequently, when a beam of plane-polarized light is passed through such a compound, it emerges with the plane of polarization unchanged.

For molecules such as one of the enantiomers of 2-iodobutane, however, no such mirror-image orientations exist, and the plane of polarization of the light is usually measurably altered in its passage through the sample. Such compounds are said to be **optically active.** If a compound causes the plane of polarization to rotate in a clockwise (positive) direction on facing the beam, it is called **dextrorotatory.** If it causes the plane to rotate in a counterclockwise (negative) direction, it is called **levorotatory.** The amount by which the plane is rotated is expressed as the angle of rotation α and by the appropriate sign which shows whether rotation is in the dextro ($+$) or levo ($-$) sense.

Rotations are measured with a device called a polarimeter. Since the degree of rotation depends on wavelength, monochromatic light (light having a single wavelength) is necessary. Common polarimeters use the sodium D line (5890 Å). The monochromatic light is first passed through the polarizer (usually a Nicol prism) from which it emerges polarized in one plane. The plane-polarized light is then passed through a tube that contains the sample, either as a liquid or dissolved in some achiral solvent. It emerges from the sample with the plane of polarization rotated in either the plus or minus direction by some amount. The light beam then passes through a second Nicol prism, which is mounted on a circular marked dial (the analyzer). The analyzer is rotated by an amount sufficient to allow the light beam to pass through at maximum intensity. Readings are compared with and without the sample tube to obtain the rotation value. Precision polarimeters using the sodium yellow line (D line) or the mercury green line are generally precise to about $\pm0.01°$. Modern spectropolarimeters use photocells in place of visual observation and can give even more precise data over a wide spectral region. A schematic representation of a polarimeter is shown in Figure 7.5.

The student may easily experience the phenomenon of optical rotation by performing a simple experiment. Take two pairs of Polaroid sunglasses and line them up, one in front of the other. Look through one lens of each pair of glasses at a bright light. Now

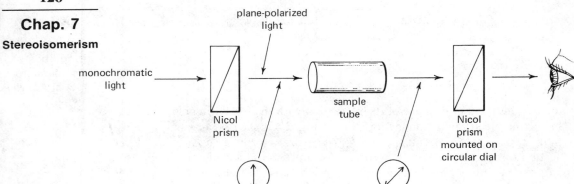

FIGURE 7.5. *Polarimeter schematic.*

rotate one of the lenses. When the glasses are parallel, the maximum amount of light is transmitted. When they are oriented at right angles to each other, no light is transmitted. What you have constructed is a simple polarimeter. The first pair of glasses corresponds to the polarizer and the second to the analyzer. Now dissolve several tablespoons of table sugar (sucrose, an optically active compound) in a small glass of water and place the glass between the two sunglasses. Again rotate one pair of glasses and note that the orientation for maximum and minimum transmission of light is now different. It is easier to observe the change at the point of minimum transmission.

The observed angle of rotation α is proportional to the number of optically active molecules in the path of the light beam. Therefore, α is proportional to the length of the sample tube and to the concentration of the solution being observed. The **specific rotation** $[\alpha]$ is obtained by dividing α by the concentration (expressed in g mL^{-1} solution) and by the length of the cell (expressed in decimeters). The wavelength of light used is given as a subscript, and the temperature at which the measurement was made is given as a superscript.

$$[\alpha]_D^t = \frac{\alpha}{l \cdot c} \quad \text{(for solutions)}$$

The decimeter is used as the unit of length simply because a 1 dc (10 cm) tube is a common length for measurements of rotation. For a pure liquid the definition of c (g mL^{-1}) is simply the density of the compound d.

$$[\alpha]_D^t = \frac{\alpha}{l \cdot d} \quad \text{(for liquids)}$$

When the temperature is not given, the rotation is assumed to be that at room temperature.

Actually, it is not possible to determine whether the rotation is $(+)$ or $(-)$ from a single measurement. Is a reading of 60° to be interpreted as $+60°$ or $-300°$? The sign may be determined by measuring the rotation at different sample concentrations. For example, if a 1 M sample gives a reading of 60 on the polarimeter, this may be either $+60°$ or $-300°$. For a 1.1 M sample, the values would be either $+66°$ or $-330°$, which are easily distinguished.

As was mentioned earlier, enantiomers differ from one another in the manner in which they interact with plane-polarized light. In fact, two enantiomers cause the plane of polarization to rotate by exactly the same amount, but in *opposite*

127

Sec. 7.3

Nomenclature of
Enantiomers:
The R,S
Convention

directions. For example, one of the two enantiomeric 2-iodobutanes has $[\alpha]_D^{24°} = +$ 15.9°, and the other has $[\alpha]_D^{24°} = -15.9°$. This knowledge still does not tell us which enantiomer is which. *There is no simple relationship between the sign of α and the absolute stereostructure of a molecule.*

Absolute stereostructure can be determined by x-ray diffraction using a technique known as **anomalous dispersion.** Although the technique is too sophisticated to discuss here, suffice it to say that absolute stereostructures for some optically active compounds have been established in this way. Once the absolute stereostructures for a few optically active compounds are known, other molecular configurations may be determined by correlating them chemically with the compounds of known structure. We shall show how this is done in later sections. By these methods, the structures of (+)- and (−)-2-iodobutane are known to be

(+)-2-iodobutane
$[\alpha]_D^{24°} = +15.9°$

(−)-2-iodobutane
$[\alpha]_D^{24°} = -15.9°$

EXERCISE: The specific rotation of sucrose is +66°. Assuming that 5 tablespoons weighs 60 g and that a small glass of 5 cm diameter holds 300 mL of solution, calculate how much rotation should have been observed in the experiment on page 125.

7.3
Nomenclature of Enantiomers: The R,S Convention

Suppose we have one bottle containing only one of the two enantiomeric 2-iodobutanes and another bottle containing only the other enantiomer. What labels do we attach to the two bottles? We cannot simply label each bottle "2-iodobutane" because they contain different compounds. We can label the bottles "(+)-2-iodobutane" and "(−)-2-iodobutane." By this we mean: "This bottle contains the 2-iodobutane that rotates the plane of polarized light in the dextro sense." And "This bottle contains the 2-iodobutane that rotates the plane of polarized light in the levo sense." Since it has also been determined which absolute stereostructure corresponds to (+)-2-iodobutane, these labels are sufficient to define unambiguously which compounds are in the bottles.

However, if a chemist were to encounter a bottle labeled (+)-2-iodobutane, chances are that he would not know which of the two absolute configurations correspond to dextrorotation. For this reason, it is highly desirable to have a system whereby the **absolute configuration** may be specified in the name of the compound. The system of nomenclature that has been adopted for this purpose by the IUPAC is called the R,S convention, or the "sequence rule."

The application of the sequence rule to naming enantiomers which owe their chirality to one or more asymmetric atoms is quite straightforward and involves the following simple steps.

1. Identify the four different substituents attached to the asymmetric atom. Assign to each of the four substituents a priority *a, b, c,* or *d,* using the sequence rule, such that $a > b > c > d$.
2. Orient the molecule in space so that one may look down the bond from the asymmetric atom to the substituent with lowest priority, *d.* When one looks

along that bond, one will see the asymmetric atom with the three attached substituents *a, b,* and *c* radiating from it like the spokes of a wheel. Trace a path from *a* to *b* to *c*. If the path describes a *clockwise* motion, then the asymmetric atom is called (R) (L., *rectus,* right). If the path describes a *counterclockwise* motion, then the asymmetric atom is called (S) (L., *sinister,* left).

Stereo representations of R and S structures are shown in Figure 7.6.

The sequence rule is the method whereby the four substituents are assigned priorities *a, b, c,* and *d* so that the symbols (R) and (S) may be assigned. There are a number of parts to the sequence rule, but we need only consider four aspects of it.

1. For the four atoms directly attached to the asymmetric atom, **higher atomic number precedes lower.** In some cases, this will be sufficient to rank the four substituent groups. For example, in 1-bromo-1-chloroethane, the four atoms involved are Br, Cl, C, and H.

$$c \; H_3C \diagdown \atop a \; Br \diagup \!\!\! C \!\!\! \diagdown Cl \; b \atop \diagup H \; d$$

(S)-1-bromo-1-chloroethane

2. In cases where two of the attached atoms are isotopes of each other, **higher atomic mass precedes lower.** In 1-deuterio-1-fluoroethane, the four groups are therefore ranked $F > C > D > H$.

$$a \; F \diagdown \atop c \; D \diagup \!\!\! C \!\!\! \diagdown CH_3 \; b \atop \diagup H \; d$$

(R)-1-deuterio-1-fluoroethane

3. For many chiral compounds, two of the atoms directly attached to the asymmetric carbon will be the same. In this case, work outward concurrently along the two chains atom by atom until a point of difference is reached. **The priorities are then assigned at that first point of difference,** using the considerations of atomic number and atomic mass. In 2-iodobutane, the iodine is assigned *a* and the hydrogen is assigned *d*. The two remaining groups are —CH_2—CH_3 and —CH_2—H. The first point of difference is at the two carbons attached to the asymmetric atom. The group —CH_2—CH_3 takes priority over —CH_2—H because carbon has a higher atomic number than

(a)

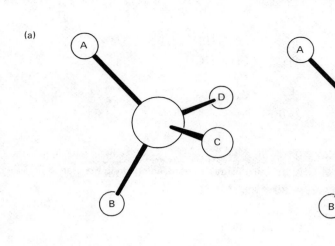

(b)

FIGURE 7.6. *Stereo diagrams of an asymmetric carbon, illustrating the arrangement of* a, b, c, *and* d *priority groups for assignment of configurations as* (a) S *and* (b) R.

hydrogen. Thus, we may assign (R) and (S) configurations to the two enantiomers as

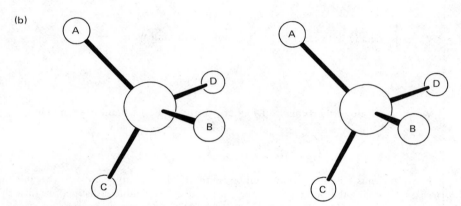

For 4-chloro-2-methyloctane, the four atoms attached to the asymmetric carbon are Cl (*a*), H (*d*), C, and C. In order to rank the isobutyl and butyl groups, we work along the chains until we reach the second carbon from the asymmetric carbon before we reach a point of difference. The group

$$-CH_2CH \begin{array}{c} CH_3 \\ \\ CH_3 \end{array}$$

takes priority over $-CH_2CH_2-CH_2CH_3$ because at the first point of difference the carbon in isobutyl has *two other carbons* attached to it, whereas the analogous carbon in *n*-butyl has only *one other attached carbon*.

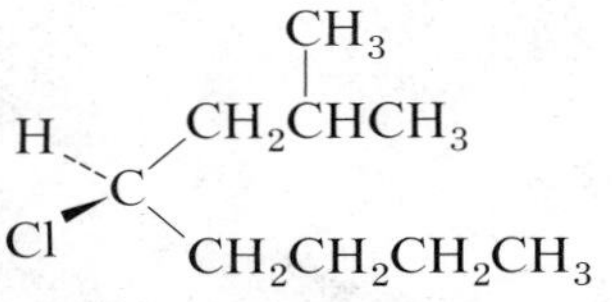

$$\begin{array}{c} CH_3 \\ | \\ CH_2CHCH_3 \end{array}$$

$$\underset{Cl}{\overset{H}{\underset{\diagdown}{\cdots}}} C \underset{CH_2CH_2CH_2CH_3}{\diagup}$$

(R)-4-chloro-2-methyloctane

In some cases, one must make a choice at a branch point as to which branch to follow. The rule here is that, if possible, one decides by **proceeding along the branch of higher priority.** Consider the following example.

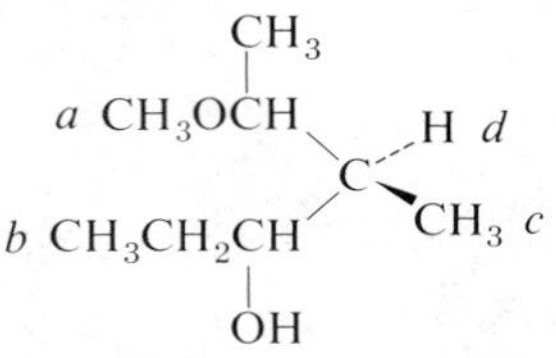

$$\begin{array}{c} CH_3 \\ | \\ a\ CH_3OCH \\ \\ b\ CH_3CH_2CH \\ | \\ OH \end{array} \overset{H\ d}{\underset{CH_3\ c}{\overset{\diagup}{C}}}$$

We must decide between two groups, both beginning with

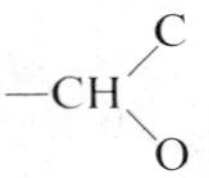

$$-CH \overset{C}{\underset{O}{\diagup}}$$

We proceed along the branch of higher priority in each chain, oxygen. Thus,

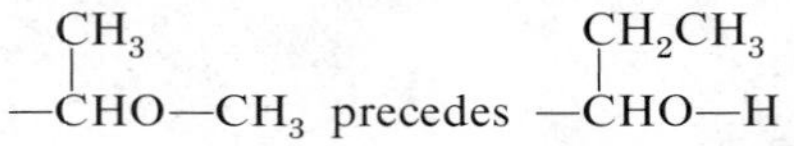

$$\begin{array}{cc} CH_3 & CH_2CH_3 \\ | & | \\ -CHO-CH_3 \text{ precedes } -CHO-H \end{array}$$

because C has a higher atomic number than H. The example shown is (S). In some cases, following the branch of higher priority does not lead to a distinction. In such a case, assignment must be made by following the secondary branches.

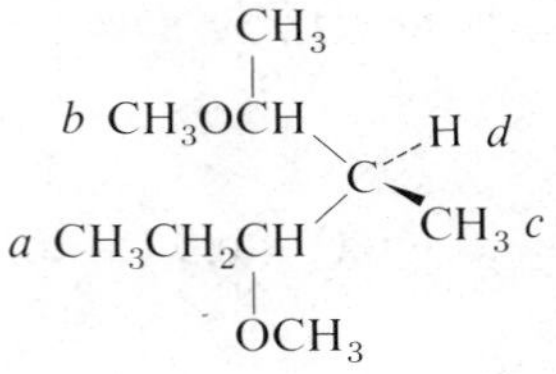

$$\begin{array}{c} CH_3 \\ | \\ b\ CH_3OCH \\ \\ a\ CH_3CH_2CH \\ | \\ OCH_3 \end{array} \overset{H\ d}{\underset{CH_3\ c}{\overset{\diagup}{C}}}$$

(R)

4. **Double and triple bonds are treated by assuming that each such bonded atom is duplicated or triplicated.**

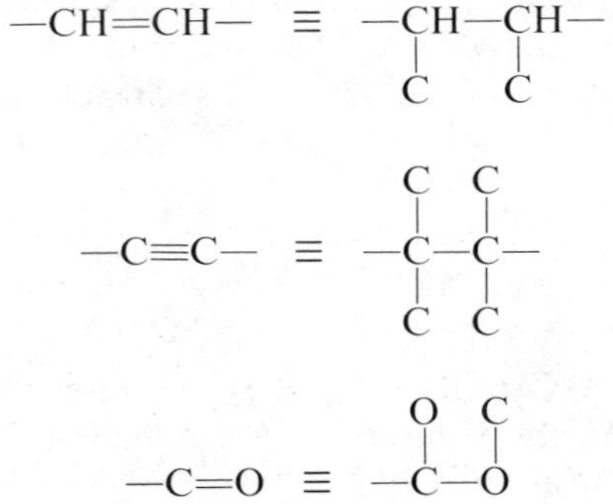

$$-CH=CH- \quad \equiv \quad \begin{array}{cc} -CH-CH- \\ | \quad | \\ C \quad C \end{array}$$

$$-C{\equiv}C- \quad \equiv \quad \begin{array}{cc} C \quad C \\ | \quad | \\ -C-C- \\ | \quad | \\ C \quad C \end{array}$$

$$-C=O \quad \equiv \quad \begin{array}{cc} O \quad C \\ | \quad | \\ -C-O \end{array}$$

$$-C\equiv N \equiv -\overset{\displaystyle N \quad C}{\underset{\displaystyle N \quad C}{C-N}}$$

Several examples of compounds containing multiple bonds follow.

a CH_2=CH, b CH_3CH_2 —C— H d, CH_3 c
(S)

a HC (O), c CH_3CH_2 —C— H d, $C\equiv N$ b
(R)

b CH_2=CH, c $(CH_3)_2CH$ —C— H d, $C\equiv CH$ a
(S)

b CH_3CH=CH, c CH_3CH_2CH —C— H d, $C\equiv CH$ a, CH_3
(S)

It is important to remember that the R and S prefixes discussed in this section are *nomenclature devices*. They specify the absolute configuration of individual molecules. The terms $(+)$ and $(-)$ refer to the experimental property of optical activity.

EXERCISE: Draw perspective structures (using wedges and dashed lines) for the (R)-enantiomers of (a) 2-bromo-2-chlorobutane, (b) 3-methylhexane, and (c) 2,3-dimethylhexane.

7.4
Racemic Mixtures

An equimolar mixture of two enantiomers is called a **racemic mixture** or a **racemate.** Since a racemic mixture contains equal numbers of dextrorotating and levorotating molecules, the net optical rotation is zero. A racemic mixture is often specified by prefixing the name of the compound with the symbol $(\pm)$, for example, $(\pm)$-2-iodobutane.

The physical properties of a racemic mixture are not necessarily the same as those of the pure enantiomers. A sample composed solely of right-handed molecules will experience different inermolecular interactions than will a sample composed of equal numbers of right- and left-handed molecules. (In order to verify this in a simple way, use your right hand to shake hands with another person. The interaction is clearly different depending on whether the other person extends his right or his left hand.)

A racemic mixture may crystallize in several ways. In some cases, separate crystals of the $(+)$ and $(-)$ forms result. In this case, the crystalline racemate is a mechanical mixture of two different crystalline compounds. The melting-point diagram for such a mixture is like that for any other mixture of two compounds (Figure 7.7). The eutectic point in such a case is always at the 50:50 point. Addition of a little of either pure enantiomer will cause the melting point of the mixture to increase. The racemate may also crystallize as a **racemic compound.** In

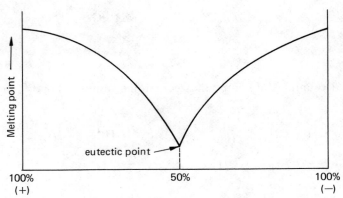

FIGURE 7.7. *Melting point diagram for a racemic mixture.*

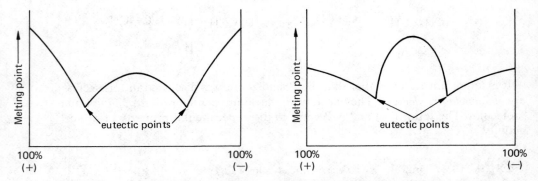

FIGURE 7.8. *Representative melting point diagrams for racemic compounds.*

this case only one type of crystal is formed, and it contains equal numbers of (+) and (−) molecules. The racemic compound acts as though it were a separate compound; its melting point is a peak on the phase diagram. However, the racemic compound may melt either higher or lower than the pure enantiomers. Addition of a small amount of either pure enantiomer causes a melting-point depression (Figure 7.8).

Because of these differential intermolecular interactions, racemates frequently differ from the pure enantiomers in other physical properties. Differences have been observed in density, refractive index, and in various spectra.

The process whereby a pure enantiomer is converted into a racemic mixture is called **racemization.** Racemization may be accomplished in a trivial sense by simply mixing equal amounts of two pure enantiomers. Racemization may also result from chemical interconversion; we shall see many examples of this in future chapters. We have already encountered one racemization process in Section 7.1, the interconversion of the two enantiomeric *gauche* forms of butane by rotation about the central C—C bond.

7.5
Fischer Projections

For acyclic compounds having only one asymmetric atom, the projection representations used so far in this chapter are useful and relatively unambiguous in meaning. However, for compounds with more than one asymmetric atom, such representations become awkward. An alternative projection system, which is widely used, is called the Fischer system. In a **Fischer projection** an asymmetric atom is represented by the intersection point of the two lines of a cross. The horizontal lines extending to the left and right of this point represent bonds extending forward from the plane of the paper. The two vertical lines extending to the top and bottom represent bonds extending back away from the plane of the paper. A Fischer projection for (R)-2-iodobutane is compared to a wedge-and-dotted-line structure as follows.

$$
\begin{array}{ccc}
& CH_3 & CH_3 \\
I \!-\!\!\!-\! H & \equiv & I \!\leftarrow\! C \!\rightarrow\! H \\
& CH_2CH_3 & CH_2CH_3
\end{array}
$$

It is important to remember that Fischer projections are two-dimensional representations of three-dimensional objects. For purposes of visualizing whether or not two structures are identical, these projections can be manipulated only in certain ways. In order to change one Fischer projection to another correct projection for the same enantiomer, one may interchange any *two* pairs of substituents. If only one pair of groups is interchanged, a projection for the enantiomer is generated.

$$
\begin{array}{ccccccc}
& Br & & CH_3 & & H & & H \\
CH_3\!-\!\!-\!H & \equiv & Br\!-\!\!-\!Cl & \equiv & Cl\!-\!\!-\!Br & \neq & Cl\!-\!\!-\!CH_3 \\
& Cl & & H & & CH_3 & & Br
\end{array}
$$

$$
\underbrace{\qquad\qquad\qquad\qquad}_{(S)} \qquad \underbrace{\qquad}_{(R)}
$$

A Fischer projection may be rotated in the plane of the paper by 180° but *not by 90°*.

$$
\begin{array}{ccc}
CH_3 & & H \\
Br\!-\!\!-\!Cl & \equiv & Cl\!-\!\!-\!Br \\
H & & CH_3
\end{array}
$$

$$
\begin{array}{ccc}
CH_3 & & Br \\
Br\!-\!\!-\!Cl & \neq & H\!-\!\!-\!CH_3 \\
H & & Cl
\end{array}
$$

The complete set of Fischer projections for (R)-1-bromo-1-chloroethane is shown in Figure 7.9. If in doubt about the identity of two Fischer projections, use wedge-and-dotted-line projections or molecular models to settle the matter. In order to ascertain whether a Fischer projection portrays the (R) or (S) absolute configuration, the following method is useful. First, change the projection to one in which the lowest priority group is at the top and the highest priority group is on the bottom. If the motion from priority group *b* to priority group *c* is to the

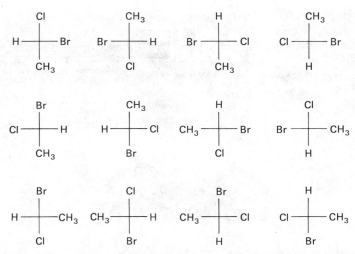

FIGURE 7.9. *Fischer projections for (R)-1-bromo-1-chloroethane. These structures are all equivalent.*

right, the projection represents the (R) configuration; if to the left, the projection represents the (S) configuration.

(S)   (R)

7.6
Compounds Containing More Than One Asymmetric Atom: Diastereomers

If a molecule has more than one asymmetric atom, the number of possible stereoisomers is correspondingly larger. Consider 2-chloro-3-iodobutane as an example. There are four isomers, which are depicted in Figure 7.10. Of the four stereoisomeric 2-chloro-3-iodobutanes, two pairs bear an enantiomeric relationship to one another. The (2R,3R) and (2S,3S) compounds are one enantiomeric pair, and the (2R,3S) and (2S,3R) compounds are another enantiomeric pair. As with the other enantiomeric pairs previously discussed, the (2R,3R) and (2S,3S) compounds have identical boiling points, melting points, densities, solubilities, and spectra. They cause the plane of polarized light to rotate to the same degree but in opposite directions; one is dextrorotatory, and the other is levorotatory. A similar correspondence in physical properties is observed for the (2R,3S) and (2S,3R) compounds.

Fischer projections for the four 2-chloro-3-iodobutanes are shown in Fig-

135

Sec. 7.6

Compounds
Containing
More Than One
Asymmetric
Atom;
Diastereomers

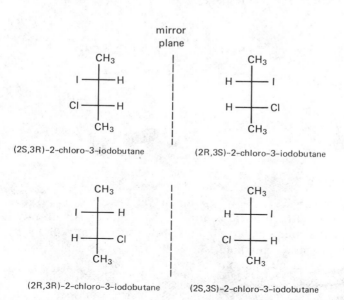

FIGURE 7.10. *Stereoisomers of 2-chloro-3-iodobutane.*

ure 7.11. For compounds such as these, in which there is more than one asymmetric carbon, the Fischer projection is written in such a way that the *main chain bonds extend from top to bottom and the appendage bonds extend to the right and left.*

Compounds that are stereoisomers of one another, but are not enantiomers, are called **diastereomers** and are said to have a **diastereomeric relationship.** The stereoisomeric relationships for a compound having two unlike asymmetric atoms are summarized in schematic form in Figure 7.12.

In general, the maximum number of possible stereoisomers for a compound having n asymmetric atoms is given by 2^n. Thus, for a compound with one asymmetric atom, there are $2^1 = 2$ stereoisomers. For a compound with two asymmetric atoms, there may be $2^2 = 4$ stereoisomers. In some cases, there are fewer than

mirror
plane

CH₃ | I—H | Cl—H | CH₃
(2S,3R)-2-chloro-3-iodobutane

CH₃ | H—I | H—Cl | CH₃
(2R,3S)-2-chloro-3-iodobutane

CH₃ | I—H | H—Cl | CH₃
(2R,3R)-2-chloro-3-iodobutane

CH₃ | H—I | Cl—H | CH₃
(2S,3S)-2-chloro-3-iodobutane

FIGURE 7.11. *Fischer projections of the 2-chloro-3-iodobutane isomers.*

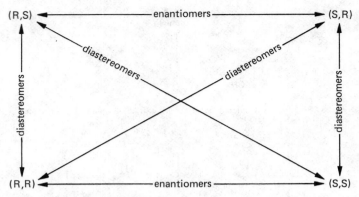

FIGURE 7.12. *Stereoisomeric relationships for a compound having two unlike asymmetric atoms.*

the maximum number of possible stereoisomers. As an example, consider 2,3-dichlorobutane. The (2R,3R) and (2S,3S) compounds are enantiomers of one another.

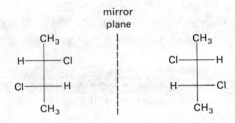

However, careful inspection reveals that the (2R,3S) and (2S,3R) compounds are actually the same compound (mentally perform a 180° rotation of the entire molecule in the plane of the page).

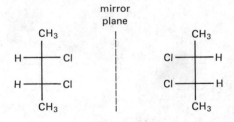

Since this isomer of 2,3-dichlorobutane is achiral, it is not optically active. Such a compound, which has asymmetric atoms yet is achiral, is called a **meso** compound. It is important not to confuse meso compounds with racemic mixtures, which are actually equimolar mixtures of two enantiomers. Both show no optical activity, but a meso compound is a single achiral substance, whereas a racemic mixture is a 50 mole % mixture of two chiral substances.

Meso compounds may be recognized by looking for a plane or a center of symmetry *within a molecule* which has asymmetric atoms. When such an element of symmetry exists, the maximum number of possible stereoisomers is less than

2^n. In one of the eclipsed conformations of meso-2,3-dichlorobutane the plane of symmetry is clearly obvious.

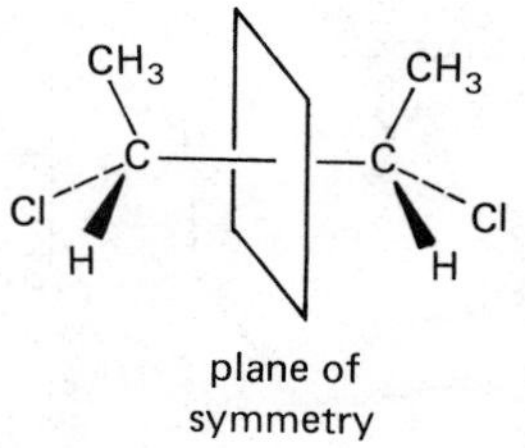

plane of
symmetry

A center of symmetry may be seen in one of the staggered conformations.

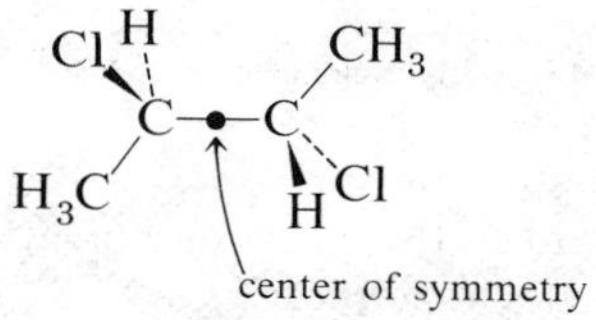

center of symmetry

An object has a center of symmetry (or is centrosymmetric) when the identical environment is encountered at the same distance in both directions along any line through a given point.

EXERCISE: Draw Fischer projections for all stereoisomers of 2,4-dichloropentane and 2-bromo-4-chloropentane. Identify all pairs of enantiomers and all pairs of diastereomers in each case. Assign (R) and (S) notations to the asymmetric carbons in each compound. Identify any meso compounds.

7.7
Stereoisomeric Relationships in Cyclic Compounds

In Section 5.8 we pointed out that there are actually two isomers of 1,4-dimethylcyclohexane (page 93). We may now recognize cis- and trans-1,4-dimethylcyclohexane as a pair of diastereomers. Both compounds are achiral, as seen from the symmetry plane passing through C-1 and C-4 and the two methyl groups. Note that even in the "flat" projection structure of a cyclohexane derivative the symmetry planes are clearly evident.

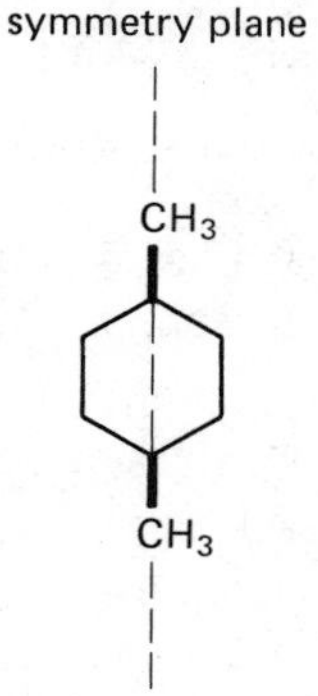

cis-1,4–dimethylcyclohexane 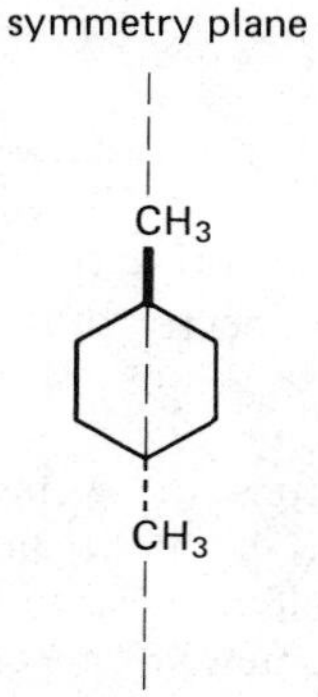   trans-1,4–dimethylcyclohexane

In the chair perspective structure, the symmetric planes may also be seen, but with somewhat more difficulty.

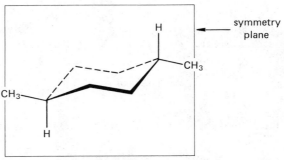

trans-1,4–dimethylcyclohexane

The 1,3-dimethylcyclohexane stereoisomers comprise an interesting case. A plane of symmetry is apparent in the *cis* isomer. Since this compound (in contrast to the 1,4-dimethyl isomer) has asymmetric carbons, *cis*-1,3-dimethylcyclohexane is a meso compound.

However, no symmetry plane or center of symmetry exists in the *trans* isomer. Thus, there are two enantiomeric *trans*-1,3-dimethylcyclohexanes.

(1R,3R)–1,3-dimethylcyclohexane   (1S,3S)–1,3-dimethylcyclohexane

Note that with the absolute configuration for each asymmetric carbon specified it is not necessary to specify that either is a *trans* isomer.

EXERCISE: Construct a molecular model of *trans*-1,4-dimethylcyclohexane. Carefully examine the model in both chair conformations and note the symmetry planes. Do the same exercise with a model of the *cis* isomer. Next, construct models of the two enantiomeric forms of *trans*-1,3-dimethylcyclohexane. Convince yourself that there is no conformation in which the two isomers are superimposable.

7.8
Chemical Reactions and Stereoisomerism

When a chemical reaction involves only achiral reactants, solvents, and reagents, the products of the reaction must be achiral or racemic mixtures. As an example, consider the monochlorination of butane. After the monochlorobutane fraction has been isolated, it is found to be a mixture of 1-chlorobutane and 2-chlorobutane. The 1-chlorobutane is, of course, achiral. The 2-chlorobutane formed in the reaction is a racemic mixture; it is an equimolar mixture of (R)-2-chlorobutane and (S)-2-chlorobutane.

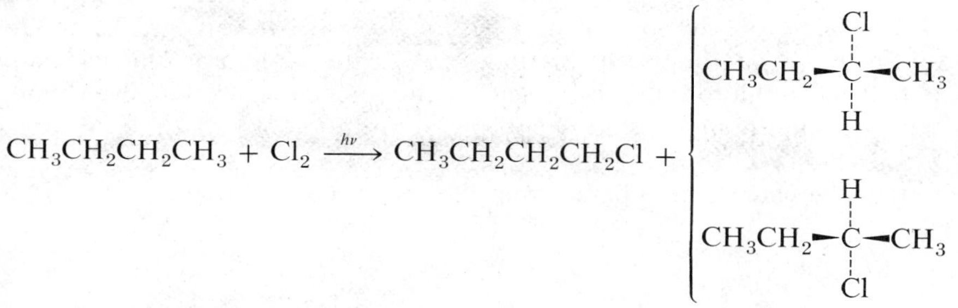

Moreover, recall that the reactive intermediate in the reaction leading to 2-chlorobutane is the *sec*-butyl free radical, which is approximately planar. Being planar, it is achiral and may react with Cl_2 on either side of the molecule. Reaction on one side yields (R)-2-chlorobutane, and reaction on the other side yields (S)-2-chlorobutane. Since reaction is equally probable on the two faces, a racemic mixture results. Consequently, any reaction that involves an achiral intermediate will give racemic products.

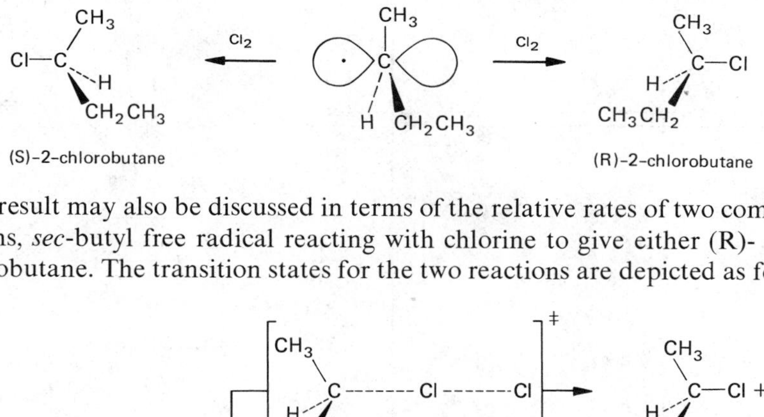

This result may also be discussed in terms of the relative rates of two competing reactions, *sec*-butyl free radical reacting with chlorine to give either (R)- or (S)-2-chlorobutane. The transition states for the two reactions are depicted as follows.

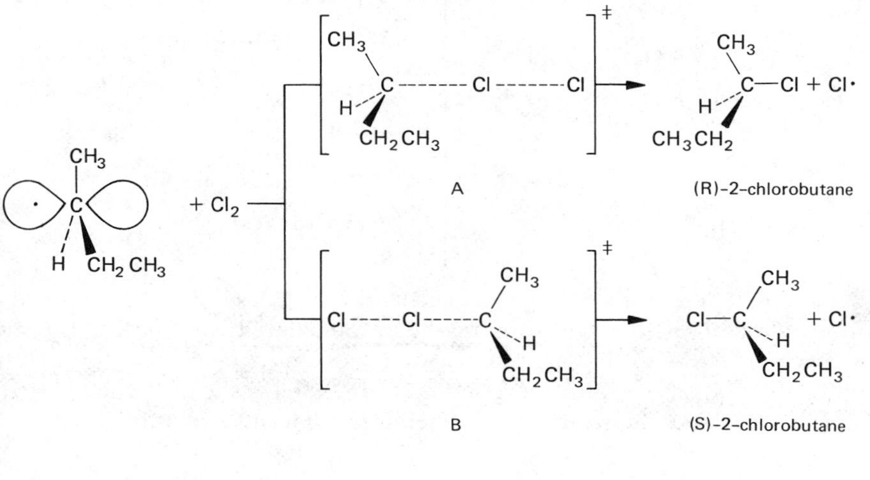

Notice that transition state A leading to the (R) enantiomer and transition state B leading to the (S) enantiomer are themselves enantiomeric. Because they are enantiomeric, they have identical physical properties, including bond angles, bond lengths, *and free energies of formation*. Since the two competing reactions begin at the same place and pass through transition states of equal energy, they have identical activation energies, and a 50:50 mixture of (R)- and (S)-2-chlorobutane results (Figure 7.13).

Let us now examine the vapor phase chlorination of a chiral compound, (S)-2-fluorobutane. The monochlorination fraction of the reaction product contains 1-chloro-2-fluorobutane, 2-chloro-2-fluorobutane, 2-chloro-3-fluorobutane, and 1-chloro-3-fluorobutane. Substitution of one of the C-1 hydrogens by chlorine yields the 1-chloro-2-fluoro isomer. Since no bond to the asymmetric atom is broken in the formation of this product, this 1-chloro-2-fluoro isomer is chiral. (Note that the 1-chloro-2-fluoro isomer has the (2R) configuration, even though the starting 2-fluorobutane is 2S.) Similarly, substitution of a C-4 hydrogen gives the 1-chloro-3-fluoro isomer, which is also chiral and has the configuration (3S). (Note that the chain is numbered from the other end in this isomer.)

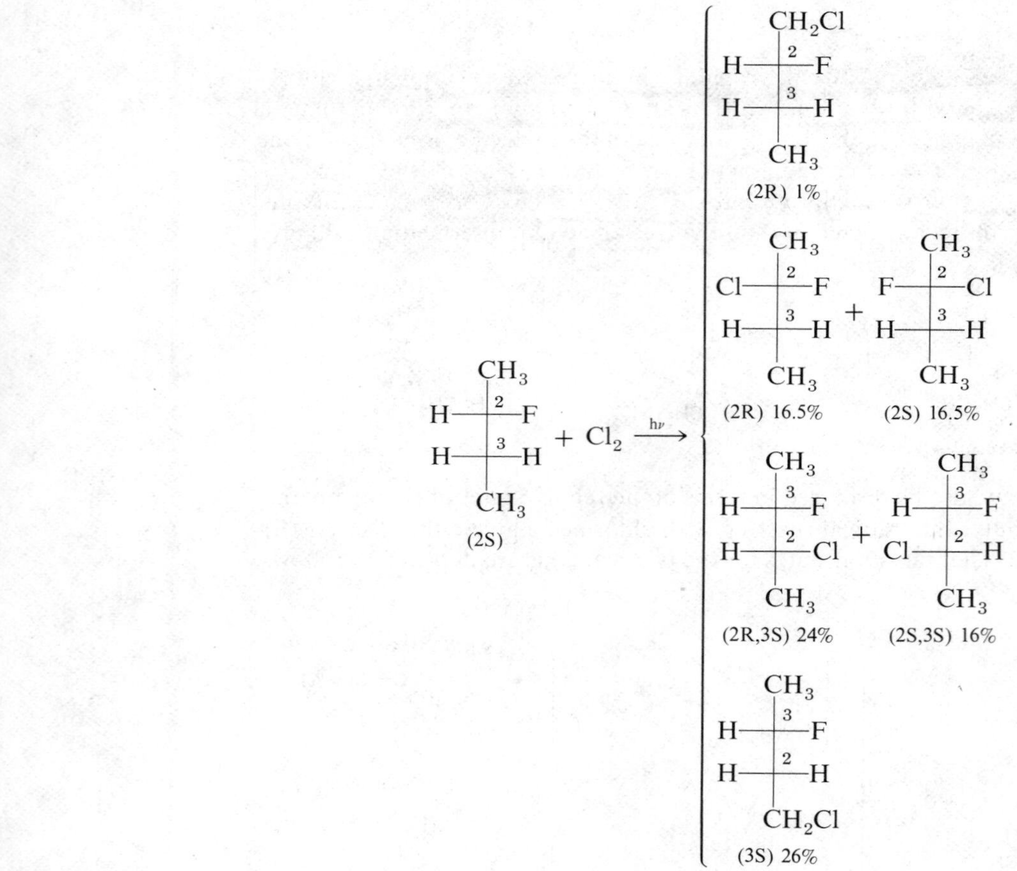

Substitution at C-2 involves forming the intermediate free radical at this position. Since the free radical is achiral, reaction with chlorine proceeds through two enantiomeric transition states and gives equal amounts of the (2R) and (2S) products. The 2-chloro-2-fluoro product is therefore a racemic mixture.

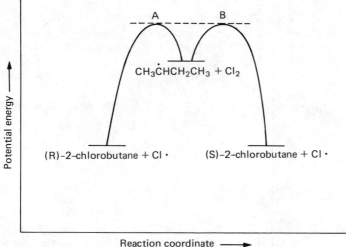

FIGURE 7.13. *An achiral intermediate gives enantiomeric transition states with equal activation energies.*

Substitution of a hydrogen at C-3 generates a new asymmetric atom, but no bond to the existing asymmetric atom is broken in the process. Therefore the absolute configuration at C-3 in the 2-chloro-3-fluoro products must be the same as it is in the starting 2-fluorobutane. However, the diastereomeric 2-chloro-3-fluorobutanes *need not be formed in equal amounts.* In order to understand this, consider the reaction of the intermediate free radical with chlorine. Again, the free radical carbon is approximately planar and has the odd electron in a *p*-orbital. However, in this case the species is chiral because of the asymmetric carbon at C-3.

This free radical may react with Cl$_2$ from either side, giving either transition state C or D. Transition state C yields the (2R,3S) diastereomer, and transition state D yields the (2S,3S) diastereomer.

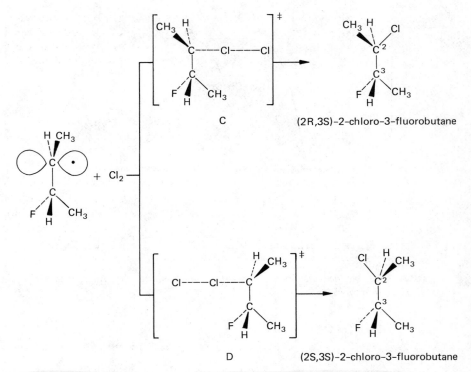

C

(2R,3S)–2–chloro–3–fluorobutane

D

(2S,3S)–2–chloro–3–fluorobutane

In this case the two transition states C and D *are not enantiomeric* but are diastereomeric. Since they are not enantiomeric, they will have different physical properties, including different free energies of formation. Since the two competing reactions start at the same place and pass through transition states of different energies, the two activation energies are different. Therefore one diastereomer will be formed in greater amount than the other (Figure 7.14).

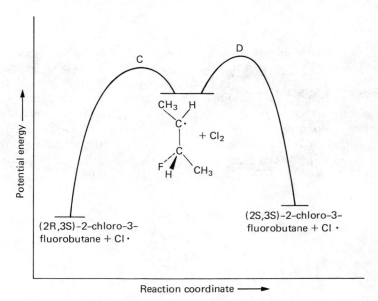

FIGURE 7.14. *A chiral intermediate yields diasteriomeric products, generally in unequal amounts.*

The reaction is actually a good deal more complicated than it appears in the fore-going simplified discussion. There are several transition states that can lead to either diastereomer. Nevertheless, because of the asymmetric atom already present in the molecule, the free radical is chiral and it will usually give rise to unequal amounts of the two diastereomers.

In the reaction leading to the 2-chloro-3-fluoro products, asymmetry is generated at C-2 in a preferred sense, due to the asymmetric atom already present in the compound. This phenomenon is called **asymmetric induction.**

Now let us consider briefly one further aspect of chemical reactivity and stereoisomerism, the relative reactivity of enantiomers. *Enantiomers show equal reactivities toward achiral reagents.* Thus, (R)-2-fluorobutane and (S)-2-fluorobutane will undergo chlorination at exactly the same rate. In order to see this clearly, consider the transition states for abstraction of, for example, the C-2 hydrogen by a chlorine atom from the two enantiomers. The two transition states are enantiomeric and therefore have equal energies.

$$
\left[\text{Cl} \text{------} \text{H} \text{---} \text{C} \underset{\substack{\text{CH}_3}}{\overset{\substack{F \\ \text{C}_2\text{H}_5}}{}} \right]^{\ddagger}
\qquad
\left[\underset{\substack{\text{C}_2\text{H}_5 \;\;\; \text{CH}_3}}{\overset{\substack{F}}{}} \text{C} \text{---} \text{H} \text{------} \text{Cl} \right]^{\ddagger}
$$

Since the reactants have equal energies and the transition states have equal energies, the two reactions must proceed at the same rate.

When two enantiomers of a chiral compound react with another chiral compound, the two enantiomers generally do not react at equal rates because, in this case, the two transition states are diastereomeric rather than enantiomeric. Since diastereomers are not necessarily equal in energy, two diastereomeric transition states are generally unequal in energy, and one of the enantiomers will react faster with the chiral reagent than the other. To visualize this relationship, consider the interaction of your right and left hands with an achiral object such as a baseball and a chiral object such as a right glove. You can grasp the achiral object equally well with either hand, but your right hand interacts much more easily than your left with the right glove.

10,14

2-6

P R O B L E M S

1. Calculate $[\alpha]_D$ for each of the following compounds.
   (a) A 1 M solution of 2-chloropentane in chloroform in a 10 cm cell gives an observed α of $+3.64°$.
   (b) A solution containing 0.96 g of 2-bromooctane in 10 mL of ether gives an observed α of $-1.80°$ in a 5 cm cell.

2. How many stereoisomers may exist for each of the following compounds?

$$\text{(a)} \quad \underset{\substack{| \quad\quad |}}{\text{CH}_3\text{CH}\!-\!\text{CHCH}_3} \overset{\text{OH} \quad\text{OH}}{}$$

$$\text{(b)} \quad \underset{\substack{| \quad\quad |}}{\text{CH}_3\text{CH}_2\text{CH}\!-\!\text{CHCH}_3} \overset{\text{OH} \quad\text{OH}}{}$$

$$\text{(c)} \quad \underset{\substack{| \quad\quad |}}{\text{CH}_3\text{CH}\!-\!\text{CHCH}_3} \overset{\text{Cl} \quad\text{OH}}{}$$

$$\text{(d)} \quad \underset{\substack{| \quad\quad | \quad\quad |}}{\text{CH}_3\text{CH}\!-\!\text{CH}\!-\!\text{CHC}_2\text{H}_5} \overset{\text{Cl} \quad\text{Cl} \quad\text{Cl}}{}$$

$$\overset{\text{OH}}{\underset{}{}} \quad \overset{\text{Cl}}{\underset{}{}} \quad \overset{\text{Cl}}{\underset{}{}}$$

(e) 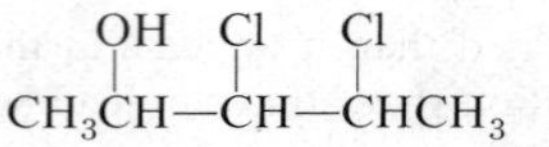 $CH_3CH{-}CH{-}CHCH_3$

(f)

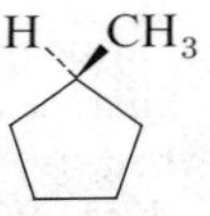

(g) 1,2-dimethylcyclopropane

(h) 1,1-dimethylcyclopropane

$$\overset{\text{Cl}}{\underset{\text{Cl}}{}}$$

(i) 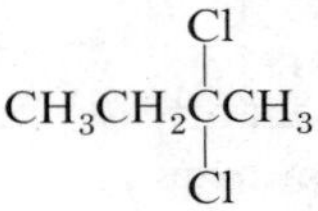 $CH_3CH_2CCH_3$

3. For parts (a)–(e) in problem 2, write Fischer projections for the different stereoisomers. Show which pairs of stereoisomers are enantiomeric and which pairs are diasteromeric. Assign (R) or (S) to each asymmetric atom.

4. Write each of the following compounds in Fischer projection and assign (R) or (S) to each asymmetric atom.

(a)

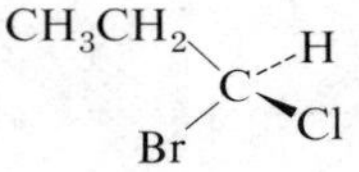

(b)

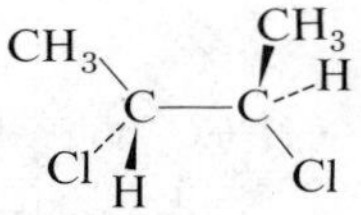

(c)

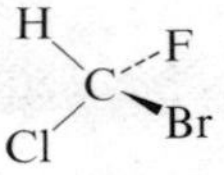

(d)

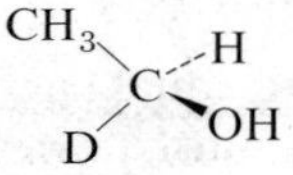

(e)

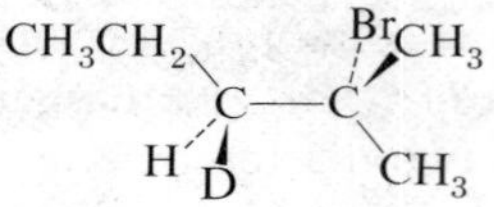

(f)

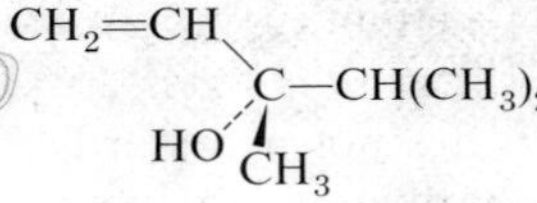

(g)

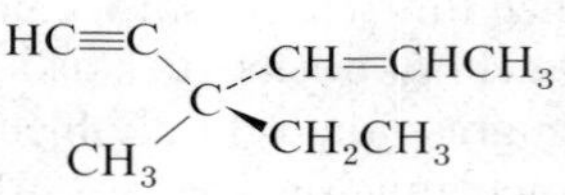

(h)

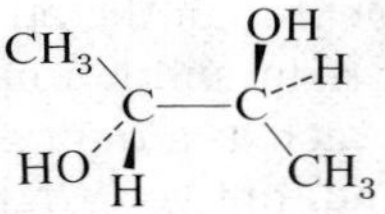

5. Given the Fischer projection

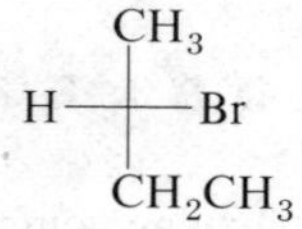

$$\begin{array}{c} CH_3 \\ H{-}\!\!\!-\!\!\!-Br \\ CH_2CH_3 \end{array}$$

is this structure (R) or (S)? Determine whether each of the following structural symbols is equivalent to the above Fischer projection or to its enantiomer.

(a)

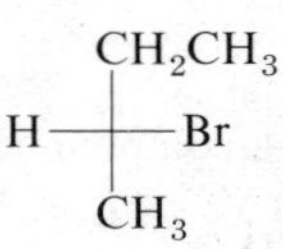

(b)

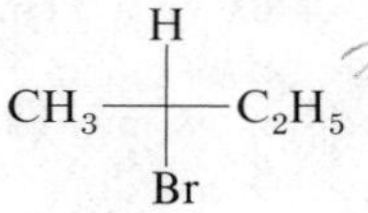

(c)

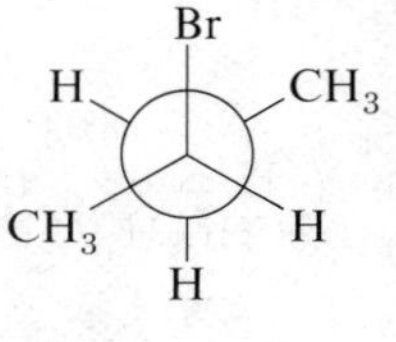

(d)

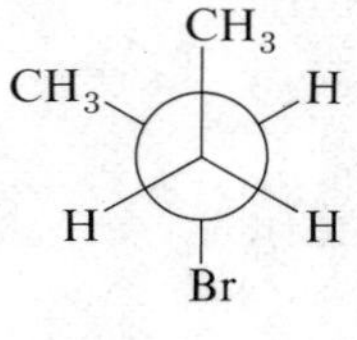

(e) [structure: H, H, CH₃, CH₃, C, C, H, Br] (f) [structure: CH₃, CH₃, C, C, H, H, Br, H]

6. How many stereoisomers exist for 1,3-dichloro-2,4-dimethylcyclobutane? Write all the structures. Which are chiral and which are achiral? Identify all planes and centers of symmetry in the various structures.

7. Consider the chlorination of 3-methylpentane.
(a) Write all of the different monochloro products that may be obtained.
(b) Which pairs of isomers in part (a) are enantiomers?
(c) Point out all diastereomeric relationships.
(d) Which isomers are achiral?

8. Answer problem 7 for methylcyclopentane.

9. Write the 2-iodobutane enantiomers of Figure 7.2 in Fischer projection and assign (R) and (S) appropriately.

10. Assign (R) or (S) to each asymmetric atom in the following compounds:

(a) [Fischer projection: CH₃ / H—Br / H—Cl / CH₃] (b) [Fischer projection: Cl / H—Br / H—Br / Cl] (c) [Fischer projection: CH₃ / H—OH / H—OH / H—OH / H—OH / CH₃]

(d) [Fischer projection: CH₃ S / H—OH / H—Cl S / H—OH R / CH₂CH₃] (e) [Fischer projection: CH₃ / H—OH S / Cl—H R S / CH₃] (f) [structure: H, C=O R / H—OH / CH₂OH]

11. (S)-1-chloro-2-methylbutane has been shown to have (+) rotation. Among the products of light-initiated chlorination are (−)-1,4-dichloro-2-methylbutane and (±)-1,2-dichloro-2-methylbutane.
(a) Write out the absolute configuration of the (−)-1,4-dichloro-2-methylbutane produced by the reaction and assign the proper (R) or (S) label. What relationship does this example show between sign of rotation and configuration?
(b) What does the fact that the 1,2-dichloro-2-methylbutane produced is totally racemic indicate about the reaction mechanism and the nature of the intermediates?

12. Draw a chair perspective structure for *cis*-1,2-dimethylcyclohexane. Is the molecule in the conformation you have drawn chiral or achiral? Now draw the structure for the *other* chair conformation. What relationship do the two conformations have?

13. Consider the chlorination of optically active (2R,3R)-2-chloro-3-methylpentane. If the dichloro fraction is analyzed by gas liquid phase chromatography (glpc), how many peaks may be seen? Suppose you isolate each fraction by preparative glpc. How many of the isolated fractions show optical activity?

14. How many stereoisomers exist for 3-bromo-2,4-dichloropentane? Write Fischer projections for each isomer. Attempt to assign (R) and (S) configurational labels to each asymmetric carbon in each isomer. What special problems arise with two of these compounds?

Chapter 8
Alkyl Halides; Nucleophilic Substitution and Elimination

8.1
Structure

Alkyl halides are the first group of organic compounds we will study in which there is a functional group. The halogen group can be converted to many other functional groups, and some of these reactions are among the most important in organic chemistry.

The carbon atoms in alkyl halides are essentially tetrahedral. The C—X bond may be regarded to a good approximation as resulting from overlap of a C_{sp^3}-orbital with a hybrid orbital from the halogen. Molecular orbital calculations suggest that the hybrid halogen orbital is mostly p, with only a small amount of s-character. In methyl fluoride, for example, the hybrid orbital from fluoride in the C—F bond is calculated to be about 15% s and 85% p. The reason for the relatively small amount of s-character is that the halogen has three lone pairs and most of the X_{2s} atomic orbital is used to bind these lone-pair electrons. Only a small amount of s-orbital is available for the orbital bonded to carbon. Note that the hybridization of the halogen must be computed; it is not amenable to experimental tests with currently available methods.

The carbon-halogen bond lengths of the methyl halides are shown in Table 8.1. The size of the halogen atoms increases as we go down the periodic table. The fluorine atom is somewhat larger than hydrogen, but smaller than carbon: compare the C—F bond distance of 1.385 Å with C—C, 1.54 Å, and C—H, 1.10 Å. The higher halogens are all substantially larger than carbon.

The *van der Waals radius* of a group is the effective size of the group. As two molecules approach each other, the van der Waals attractive force (Section 5.1) increases to a maximum, then decreases and becomes repulsive (Figure 8.1). The van der Waals radius is defined as one-half the distance between two equivalent atoms at the point of the energy minimum. It is an equilibrium bond distance and is usually evaluated from the structures of molecular crystals. Van der Waals radii for several atoms and groups are summarized in Table 8.2. The van der Waals radius of bromine (1.95 Å) is about the same as that for a methyl group (2.0 Å).

Although the C—X bonds in alkyl halides are covalent, they have a polar

TABLE 8.1
Bond Lengths of Methyl Halides

Compound	r_{C-X}, Å
CH_3F	1.385
CH_3Cl	1.784
CH_3Br	1.929
CH_3I	2.139

148

Chap. 8

**Alkyl Halides;
Nucleophilic
Substitution and
Elimination**

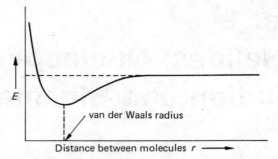

E

van der Waals radius

Distance between molecules r →

FIGURE 8.1. *Van der Waals forces.*

TABLE 8.2
Van der Waals Radii, Å

H	N	O	F
1.2	1.5	1.4	1.35
CH_2	P	S	Cl
2.0	1.9	1.85	1.8
CH_3			Br
2.0			1.95
			I
			2.15

character because halogens are more electronegative than carbon. That is, the "center of gravity" of the electron density does not coincide with the center of nuclear positive charge. This imbalance results in a **dipole moment** μ, which is expressed as the product of the charge q and the distance of separation d: $\mu = qd$. The distance involved has direction; hence, dipole moments are vectors. In the case of methyl halides this vector is directed along the C—X bond and is usually symbolized as

$$\overset{\longrightarrow}{CH_3\text{—}X}$$

The direction of the arrow is from positive to negative charge. The magnitudes of the dipole-moment vectors for methyl halides are summarized in Table 8.3. Since the charge involved is on the order of 10^{-10} esu and the distance is on the order of 10^{-8} cm, dipole moments are on the order of 10^{-18} esu cm. This unit is named the

TABLE 8.3
**Dipole Moments of Methyl
Halides (Vapor Phase)**

Compound	μ, D
CH_3F	1.82
CH_3Cl	1.94
CH_3Br	1.79
CH_3I	1.64

Debye, abbreviated D, after the late Professor Peter Debye who discovered this molecular property. Thus, if a positive charge and a negative charge of 10^{-10} esu are separated by 10^{-8} cm, the system has a dipole moment of 1 Debye or 1 D.

8.2
Physical Properties

The lower molecular weight *n*-alkyl halides are gases at room temperature. Starting with *n*-butyl fluoride, *n*-propyl chloride, ethyl bromide, and methyl iodide, the alkyl halides are liquids at room temperature. This result comes mostly from the increasing effective "size" of the halogens as we proceed down the periodic table. We saw in the previous section how this changing size is reflected in increasing C—X bond distances along the series from fluorine to iodine. Increasing size carries with it an increase in the effective "area of contact" at the van der Waals radius that produces van der Waals attraction.

However, size does more than increase this area of contact. We learned on page 70 that van der Waals attraction results from the mutual correlation of electronic motions. The movement of one electron describes a changing electric field. The ability of a second electron to respond to such a changing field is measured by its **polarizability.** The smaller and "tighter" the atom, the lower the polarizability of its electrons and the lower the van der Waals attraction for a given area of contact. Consequently, along the series F, Cl, Br, I the polarizability increases. Furthermore, lone-pair electrons are generally held more loosely than bonding electrons and can be more polarizable. Although bromine has a van der Waals radius similar to that of a methyl group, it has much higher polarizability, and an alkyl bromide, RBr, has a much higher boiling point than that of the corresponding RCH_3. We have emphasized the role of van der Waals attractions in boiling points, but it should be mentioned that molecular weight also plays a role because of the effect of mass on kinetic energy. Along a given homologous series, however, the van der Waals force depends on the overall size of the molecule which, in turn, parallels the molecular weight.

The tightly held electrons and consequent low polarizability of fluorine results in the unique and distinctive properties of fluorocarbons, compounds composed entirely of carbon and fluorine. The boiling points of fluorocarbons are much closer to those of related hydrocarbons than might have been expected from the difference in molecular weights or size; for example, C_2H_6 has b.p. $-89°$; C_2F_6 has b.p. $-79°$.

Increasing the alkyl chain also causes a normal progressive increase in boiling point. As with alkanes themselves, branched systems have lower boiling points than isomeric linear systems. Some boiling point data are summarized in Table 8.4 and in Figure 8.2.

Alkyl halides are insoluble in water but soluble in most organic solvents. They vary greatly in stability. Monofluoroalkanes are difficult to keep pure; on distillation they tend to lose HF to form alkenes. Chlorides are relatively stable and generally can be purified by distillation. However, higher molecular weight tertiary alkyl chlorides tend to lose HCl on heating and must be handled more carefully. Indeed, this property holds for most tertiary alkyl halides. Note in Table 8.4 that *t*-butyl iodide decomposes on attempted distillation at atmospheric pressure.

Chloroform slowly decomposes on exposure to light. This tendency is dimin-

150

Chap. 8

**Alkyl Halides;
Nucleophilic
Substitution and
Elimination**

TABLE 8.4
Boiling Points of Alkyl Halides (RX)

R	X =	H	F	Cl	Br	I
CH$_3$—		−161.7	−78.4	−24.2	3.6	42.4
CH$_3$CH$_2$—		−88.6	−37.7	12.3	38.4	72.3
CH$_3$(CH$_2$)$_2$—		−42.1	−2.5	46.6	71.0	102.5
CH$_3$(CH$_2$)$_3$—		−0.5	32.5	78.4	101.6	130.5
CH$_3$(CH$_2$)$_4$—		36.1	62.8	107.8	129.6	157.
CH$_3$(CH$_2$)$_5$—		69.0	91.5	134.5	155.3	181.3
CH$_3$(CH$_2$)$_6$—		98.4	117.9	159.	178.9	204.
CH$_3$(CH$_2$)$_7$—		125.7	142.	182.	200.3	225.5
(CH$_3$)$_2$CH—		−42.1	−9.4	34.8	59.4	89.5
(CH$_3$)$_2$CHCH$_2$—		−11.7		68.8		
CH$_3$CH$_2$CH— with CH$_3$		−0.5		68.3	91.2	120.
(CH$_3$)$_3$C—		−11.8		50.7	73.1	dec.

ished by the presence of small amounts of alcohol. Commercially available chloroform has about 0.5% alcohol added as a stabilizer. Alkyl bromides and iodides are also light-sensitive. Upon exposure to light they slowly liberate the free halogen and turn brown or violet, respectively. Thus, these halides are generally stored in opaque vessels or brown bottles and should generally be redistilled before use.

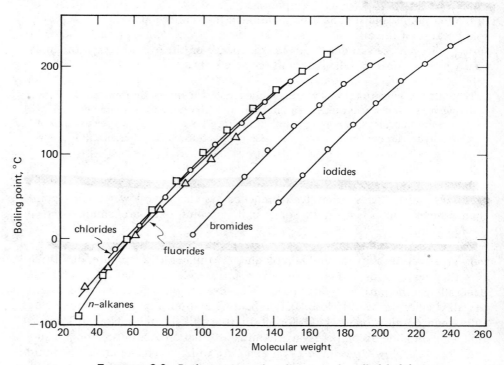

FIGURE 8.2. *Boiling points of* n-*alkanes and* n-*alkyl halides.*

EXERCISE: (a) Which has the higher melting point, *n*-butyl bromide or *t*-butyl bromide? Explain. Compare your answer with the melting points found in a handbook.

(b) From the generalizations and data provided in this section and in Chapter 5, estimate the boiling points of $CH_3CH_2CH_2CH(CH_3)CH_2CH_2Cl$ and $CH_3(CH_2)_3$ $CH(C_2H_5)CH_2Cl$. Look up these boiling points in a handbook.

8.3
Conformations

Barriers to rotation about C—C bonds bearing halogens are comparable to those in hydrocarbons, and these compounds also prefer staggered conformations. Some rotation barriers are summarized in Table 8.5. Note that there is no simple relationship between the barrier and the size of the halogen. One reason is that as the size of a halogen increases, its bond length to carbon also increases. The same principle operates in the halogenated cyclohexanes; recall that the halogens show a relatively small and almost constant preference for the equatorial conformation (Table 5.6). In alkyl halides where rotation involves eclipsing C—H with C—H and C—H with C—X, the barriers are about 3.2–3.7 kcal mole^{-1}. Even for hexa-fluoroethane, where rotation involves eclipsing three pairs of C—F bonds, the barrier is only 3.9 kcal mole^{-1}. However, rotation of one C—Cl bond past another is more difficult; the barrier in hexachloroethane is 10.8 kcal mole^{-1}.

1,2-Dichloroethane, like butane, exists in two conformations, *gauche* and *anti*.

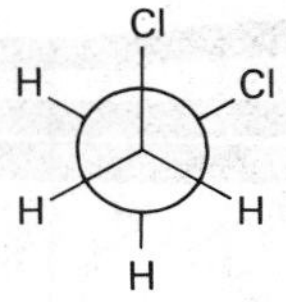

gauche	*anti*

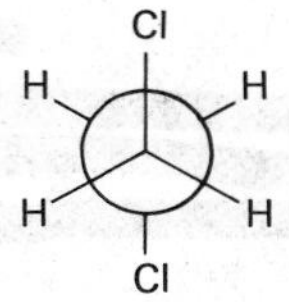

One conformation is converted to another by rotating a C—Cl bond past a C—H bond. It is not necessary to rotate C—Cl past C—Cl. Accordingly, the barrier between these conformations is only about 3.2 kcal mole^{-1}, not much different than for rotation in ethyl chloride.

TABLE 8.5
Barriers to Rotation in Alkyl Halides

Compound	Rotation Barrier, kcal mole^{-1}
$CH_3—CH_2F$	3.3
$CH_3—CHF_2$	3.2
$CH_3—CF_3$	3.25
$CF_3—CF_3$	3.9
$CH_3—CH_2Cl$	3.7
$CH_3—CHCl_2$	3.5
$CCl_3—CCl_3$	10.8
$CH_3—CH_2Br$	3.7
$CH_3—CH_2I$	3.2

152

Chap. 8

**Alkyl Halides;
Nucleophilic
Substitution and
Elimination**

We saw in Chapter 5 that the *gauche* and *anti* conformations of butane differ in energy by about 0.9 kcal mole^{-1}. We might expect, therefore, that the two analogous conformations of 1,2-dichloroethane will also have different energies. In fact, in the vapor phase, the anti conformation is more stable by 1.2 kcal mole^{-1}. Remarkably, however, the energy difference in the pure liquid is about zero!

How do we account for this interesting observation? One explanation involves two opposing factors, dipole repulsion and van der Waals attraction. Each C—Cl bond has an associated dipole moment. The electrostatic repulsion for two dipoles oriented in the *anti* conformation is lower than that for two dipoles oriented in the *gauche* conformation.

anti conformation   *gauche* conformation

The other factor involved is van der Waals attraction. Two chlorines separated by little more than the sum of their van der Waals radii attract each other in exactly the same manner as two neighboring alkanes (Section 5.2). Such van der Waals attraction is especially important for the large halogen atoms because the lone-pair electrons are spread through a relatively large volume and respond easily to changing neighboring charge fields. That is, such electrons have relatively high polarizability.

The net result for the *gauche* and *anti* conformations is a balance. Van der Waals attraction favors the *gauche* conformation, but dipole–dipole repulsion favors the *anti* conformation. In the vapor phase the dipole effect dominates, and the *anti* structure is more stable. In the liquid phase there are many other molecules close by that reduce the importance of the intramolecular dipole factor. The two effects now just cancel.

8.4
Some Uses of Halogenated Hydrocarbons

The simple alkyl halides and polyhaloalkanes are readily available and are used extensively as solvents. Chlorides are most important because of the low cost of chlorine relative to bromine and iodine. In fact, chlorine is one of the basic raw materials of the chemical industry. In 1979 the United States produced 12,110,000 tons of chlorine, most of which was used to produce chlorinated hydrocarbons.

The polychloromethanes are produced industrially by the chlorination of methane. Carbon tetrachloride has been used extensively in drycleaning establishments. However, it must be handled with care because it is an accumulative poison that causes severe liver damage. Consequently, its use in drycleaning has declined. Chloroform was once used as an anesthetic, but its use for this purpose has now been abandoned because it is toxic and a suspected carcinogen. More recently, the mixed halogenated compound $CF_3CHClBr$, "Halothane," has found important use as an inhalative anesthetic because it is effective and relatively nontoxic.

Several theories have been proposed for the action of anesthetics, but the detailed mode of action is not yet known. Anesthetics can be chemically inert; for example, xenon has anesthetic action. Different compounds vary in the concentration required;

even nitrogen under pressure functions as an anesthetic. Nitrogen narcosis is a danger to deep divers. It is remarkable that the effective concentration of an anesthetic is species-independent. The same partial pressure of anesthetic functions as well in man as in a goldfish. The application of anesthetics needs to be carefully monitored. Lethal concentrations are typically only about double the useful anesthetic concentration.

A number of partially fluorinated alkanes are widely marketed for use as cooling fluids in refrigeration systems and as aerosol propellants. These compounds are often known by their trade names.

Compound	Trade Name	Systematic Name
$CFCl_3$	Freon 11	trichlorofluoromethane
CF_2Cl_2	Freon 12	dichlorodifluoromethane
CF_3Cl	Freon 13	chlorotrifluoromethane
CF_4	Freon 14	tetrafluoromethane

Ethyl chloride is used as a local anesthetic. It is a gas at ambient temperature (b.p. 12°) and is kept in pressurized containers. When it is sprayed onto the skin, rapid vaporization occurs. The heat required to cause vaporization is drawn from the local surroundings, in this case the skin, and the resultant cooling deadens the nerve endings.

Another significant use of chlorinated hydrocarbons is as pesticides, a general term that includes fungicides, herbicides, insecticides, fumigants, and rodenticides. There are three main types of pesticides in use: carbamates, organophosphorus compounds, and chlorinated hydrocarbons. The use of such compounds for the control of disease-bearing pests has increased sharply during the past three decades.

The most well-known pesticide is DDT, which has been used extensively since 1939.

1,1,1-trichloro-2,2-bis(p-chlorophenyl)ethane
DDT

DDT is effective against many organisms, but its most spectacular success has been in control of the *Anopheles* mosquito, which transmits malaria. Malaria has been a scourge of mankind for centuries. According to the World Health Organization, malaria is still the chief cause of human death in the world, aside from natural causes. The disease acquired its name in ancient Rome (L. *mala*, bad; *aria*, air), where it was believed to be a result of the bad air in the city. It is actually caused by a parasite of the *Plasmodium* family which infects and ruptures erythrocytes in the blood stream. The organism has a complex life cycle requiring both vertebrate and invertebrate hosts. Humans are infected by sporozoites of the organism which are injected into the bloodstream by the bite of an infected mosquito.

Although malaria may be treated, the most effective method of controlling it is to eliminate the insect vector which is essential for its transmission. DDT is especially effective for this purpose, and malaria has been essentially eliminated from

154

Chap. 8

**Alkyl Halides;
Nucleophilic
Substitution and
Elimination**

large areas of the world through its use. It has been estimated that because of the efficacy of DDT in checking malaria and other mosquito-borne diseases (yellow fever, encephalitis), more than 75 million human deaths have been averted. A striking example is Sri Lanka (the island of Ceylon). In 1934–35, there were 1.5 million cases of malaria resulting in 80,000 deaths. After an intensive mosquito-abatement program using DDT, malaria effectively disappeared and there were only 17 cases reported in 1963. When the use of DDT was discontinued in Sri Lanka, malaria rebounded, and there were over 600,000 cases reported in 1968 and the first quarter of 1969.

In spite of its obvious value in combatting diseases such as malaria, DDT has been abused. It is a "hard" insecticide, in that its residues accumulate in the environment. Although it is not especially toxic to mammals (the fatal human dose is 500 mg kg^{-1} of body weight, about 35 g for a 150 lb person), it is concentrated by lower organisms such as plankton and accumulates in the fatty tissues of fish and birds. The toxicity of DDT was first noted in 1949 by the Fish and Wildlife Service, but indiscriminate use as an agricultural pesticide for the control of crop-destroying pests continued to grow. In 1962, following the publication of *Silent Spring* by the late biologist Rachel Carson, an intensive campaign against the use of pesticides such as DDT commenced. In 1972 its use as an agricultural pesticide in the United States was banned by the Environmental Protection Agency. Active research is being directed toward developing new types of pesticides that are species-specific and biodegradable and will not accumulate in the environment.

Alkyl halides are important as reagents. Many reactions are known for transforming the halogens to other functional groups. Some of these reactions will be discussed in the remainder of this chapter. For industrial reactions, chlorides are used almost exclusively because of the high cost of bromine and iodine. For laboratory uses, where cost is not as great a consideration, bromides are used preferentially because alkyl bromides are generally more reactive than alkyl chlorides. Methyl iodide is a commonly used laboratory reagent because it is the only methyl halide which is liquid at room temperature. The preparation of alkyl halides from alcohols and alkenes will be discussed in Chapters 10 and 11, respectively.

8.5
The Displacement Reaction

The replacement of the halogen in an alkyl halide by another group is one of the most important reactions in organic chemistry. In Section 4.2, we took a brief look at one such reaction, the reaction of methyl chloride with hydroxide ion.

$$HO^- + CH_3Cl \xrightarrow{H_2O} CH_3OH + Cl^-$$

Another example is the reaction of ethyl bromide with potassium iodide in acetone solution.

$$CH_3CH_2Br + K^+I^- \xrightarrow{acetone} CH_3CH_2I + KBr\downarrow$$

Although this is an equilibrium process, the reaction proceeds virtually to completion because potassium iodide is soluble in acetone and potassium bromide is not.

Like the reaction discussed in Section 4.2, the reaction of ethyl bromide with iodide ion is relatively slow. In order for complete reaction to occur, it is necessary to heat the mixture for several hours. The rate of the reaction may be determined by following the rate of disappearance of reactants or the rate of appearance of products. It is proportional to the *product* of the concentrations of the two reactants.

$$\text{rate} = -\frac{d[C_2H_5Br]}{dt} = -\frac{d[I^-]}{dt} = \frac{d[C_2H_5I]}{dt} = \frac{d[KBr]}{dt} = k[C_2H_5Br][I^-]$$

> This equation is expressed in the symbolism of calculus. The expression
>
> $$\frac{d[C_2H_5Br]}{dt}$$
>
> means simply the rate with which the concentration of C_2H_5Br changes with time. The negative sign indicates that the concentration of C_2H_5Br decreases as time increases. Note that in this case [KBr] refers to amount rather than concentration because of its low solubility in acetone.

The concentrations of C_2H_5Br, I^-, C_2H_5I, and KBr may be determined at different times during the reaction by chemical or spectroscopic analysis. As the reaction proceeds, the concentrations of the reactants become reduced and the rate of reaction decreases. For example, at 50°, with the two reactants each present in an initial concentration of 0.1 M, the reaction is 50% complete in 7 min but only 95% complete after 2 hr. Furthermore, the reaction has an activation energy, $\Delta H^{\ddagger}$, of 19 kcal mole^{-1}; it is 10 times slower at 25° than it is at 50°. This activation energy is considerably higher than those of the free radical reactions we studied in Chapter 6.

When the rate of a chemical reaction depends on the concentration of two species, as in this case, it is said to display **second-order kinetics.** This suggests a **bimolecular** mechanism, one in which one molecule of each reactant collide and react. The relatively high activation energy shows that only a minute fraction of such collisions actually result in reaction—those involving reactant molecules with sufficient kinetic energy. We might imagine that a straightforward mechanism would be one in which the attacking group, I^- in this case, simply displaces the *leaving group,* Br^-, from its bond to carbon.

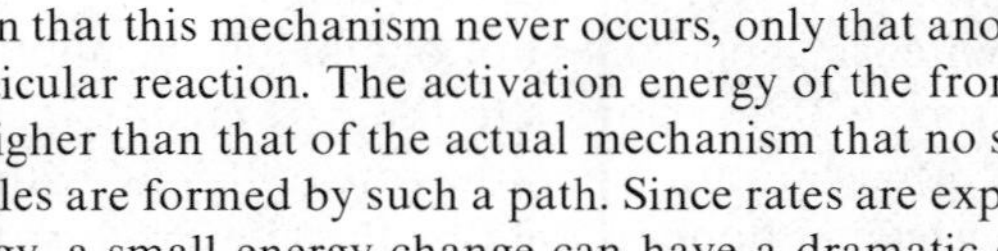

However, a large mass of evidence has been accumulated which shows that this **front-side** attack is *not* the mechanism of this reaction.

> This does not mean that this mechanism never occurs, only that another mechanism is better for this particular reaction. The activation energy of the frontal-attack mechanism is so much higher than that of the actual mechanism that no significant number of product molecules are formed by such a path. Since rates are exponential functions of activation energy, a small energy change can have a dramatic effect on rate. For example, if two reactions differ in activation energy by 10 kcal mole^{-1}, the ratio of their reaction rates is 10,000,000:1. Thus, it is only necessary that we consider the most probable reaction mechanisms for a given reaction—those with the lowest activation energies.

156

Chap. 8

Alkyl Halides;
Nucleophilic
Substitution and
Elimination

Instead of such a simple frontal attack, the reaction actually proceeds by a more complex mechanism that involves attack at the *rear of the C—Br bond.*

$$:I:^- \quad \overset{H}{\underset{CH_3}{\overset{|}{\underset{|}{C}}}}-\overset{..}{B}r: \longrightarrow I-\overset{H}{\underset{CH_3}{\overset{|}{\underset{|}{C}}}}H \quad + \quad :\overset{..}{B}r:^-$$

The foregoing equation illustrates the use of a symbolism for *flow of electrons* in the course of a reaction which finds much use in organic chemistry. An electron pair is thought of as originating at the end of an arrow and flowing in the direction of the arrow. In this case, a pair of electrons belonging to iodide ion flows toward the bromine-bearing carbon and eventually forms a new carbon-iodine bond. Simultaneously, the pair of electrons comprising the carbon-bromine bond flows away from carbon and eventually ends up as a fourth lone pair on the product bromide ion.

8.6
Stereochemistry of the Displacement Reaction

It has long been known that when an optically active alkyl halide is exposed to halide ion in solution, the optical activity gradually diminishes to zero. This is an example of racemization; the optically active halide is converted to an equimolar mixture of (+) and (−) enantiomers. The rate of racemization is dependent on the concentration of both the alkyl halide and the added halide. The rate constant for the racemization process is called k_α. The conditions under which such racemization occurs are the same as those that lead to substitutions such as those discussed in Section 8.1.

The fact that optical activity is lost shows unequivocally that the frontal-attack mechanism cannot be the sole mechanism for such substitution reactions because that mechanism leads to retention of absolute configuration at the asymmetric atom. For example, if the frontal-attack mechanism were the only mechanism for substitution in (R)-2-iodobutane, then a system containing this optically active halide would always contain only alkyl halides of the (R) configuration.

Frontal attack mechanism

$$\underset{C_2H_5}{\overset{CH_3}{C}}\overset{H}{\underset{I}{\diagdown}} + X^- \rightleftharpoons \underset{C_2H_5}{\overset{CH_3}{C}}\overset{H}{\underset{X}{\diagdown}} + I^-$$

$$\text{(R)} \qquad\qquad \text{(R)}$$

There must be a mechanism for substitution that allows racemization at C-2. To determine how often each replacement of one halide by another is accompanied by a change from the (+) enantiomer to the (−) enantiomer, it is necessary to know both the rate of racemization and the rate of the substitution reaction itself.

This experiment was first done in 1935 by the late Professor E. D. Hughes and his colleagues at the University of London. The experiment was an extremely elegant one, which used sodium iodide containing a radioactive isotope of iodine, [128]I, as the attacking group and optically active 2-iodooctane containing normal [127]I as the alkyl halide.

157

Sec. 8.6

**Stereochemistry
of the
Displacement
Reaction**

$$\left[\begin{array}{l} ^{128}\text{I is prepared by exposing normal } ^{127}\text{I to neutrons. The radioisotope decays with a} \\ \text{half-life of 25 min.} \end{array} \right]$$

At equilibrium, the radioactive iodine is distributed equally between 2-iodo-octane and iodide ion.

$$^{128}\text{I}^- + \text{C}_6\text{H}_{13}\overset{\text{I}}{\underset{|}{\text{C}}}\text{HCH}_3 \rightleftharpoons \text{I}^- + \text{C}_6\text{H}_{13}\overset{^{128}\text{I}}{\underset{|}{\text{C}}}\text{HCH}_3$$

The reaction was found to have a rate constant, at 30°, $k_{exch} = (1.36 \pm 0.11) \times 10^{-3}$ M^{-1} sec^{-1}. Under identical conditions the rate constant for racemization of optically active 2-iodooctane was found to be $k_\alpha = (2.62 \pm 0.03) \times 10^{-3}$ M^{-1} sec^{-1}. Within experimental error *the rate constant for racemization is exactly twice that for exchange.* Consequently, each act of replacement of one iodide by the other is accompanied by a change from one enantiomer to another. In this way, the rotation of one molecule of product cancels that of one reactant molecule. Two such molecules constitute a R,S-pair and are equivalent to two molecules of race-mic material. Looking at it in a slightly different way, *all* of the beginning mole-cules must undergo reaction with $^{128}\text{I}^-$ before incorporation of the radioisotope is complete, whereas only half of the beginning molecules must undergo inversion of stereochemistry before racemization is complete. Thus, if inversion of stereo-chemistry occurs every time substitution occurs, k_α must be exactly twice k_{exch}, as is found.

$$\text{I}^- + \underset{\text{(S)}}{\overset{\text{C}_6\text{H}_{13}}{\underset{\text{CH}_3}{\overset{|}{\text{H} \diagup \text{C} - \text{I}}}}} \rightleftharpoons \underset{\text{(R)}}{\overset{\text{C}_6\text{H}_{13}}{\underset{\text{CH}_3}{\overset{|}{\text{I} - \text{C} \diagdown \text{H}}}}} + \text{I}^-$$

This observation has been combined with many other studies to infer that in a bimolecular displacement reaction the attacking group attacks at the carbon atom at the rear of the bond to the leaving group. During reaction, the carbon forms a progressively stronger bond to the attacking group, while the bond to the leaving group is being weakened. During this change, the other three bonds to the central carbon progressively flatten out and end up on the other side of the carbon in a manner similar to the spokes of an umbrella inverting in a windstorm.

$$\text{I}^- + \overset{\text{C}_6\text{H}_{13}}{\underset{\text{H} \quad \text{CH}_3}{\text{C} - \text{I}}} \longrightarrow \left[\overset{\text{C}_6\text{H}_{13}}{\underset{\text{H} \quad \text{CH}_3}{\overset{\delta^-}{\text{I}} \cdots \text{C} \cdots \overset{\delta^-}{\text{I}}}} \right]^{\ddagger} \longrightarrow \overset{\text{C}_6\text{H}_{13}}{\underset{\text{CH}_3}{\text{I} - \text{C} \diagdown \text{H}}} + \text{I}^-$$

The reaction is formulated as above, with the transition-state geometry in brack-ets. The dotted lines indicate a partially formed or partially broken bond. The symbols $\delta-$ indicate that the negative charge is spread over both iodines in the transition state.

During the course of the reaction, the reacting system has greater potential energy than either the reactants or the products. The two weak bonds to the entering and leaving groups are weaker than the single bond in either the reactant or the product. Hence, energy is required in order for reaction to occur. The necessary potential energy is supplied by the conversion of kinetic energy. But only the minute fraction of reactants that have sufficient kinetic energy can react. Furthermore, even if the colliding reactants have sufficient kinetic energy, they

158

Chap. 8

**Alkyl Halides;
Nucleophilic
Substitution and
Elimination**

must have the proper orientation or they will simply bounce apart. Recall that the point of highest energy is called the *transition state*. It is important to remember that the transition state is a point of *maximum* energy. It is not a discrete molecule that can be isolated and studied. In fact the whole act of displacement occurs in the space of about 10^{-12} sec, the period of a single vibration, so the system has the transition-state geometry for only a fleeting moment.

The above discussion is further complicated by **solvation energy.** The transition state is, in effect, a larger ion with a more diffuse charge and therefore a lower solvation energy. That is, solvent dipoles are less oriented by the diffuse charge and the transition state is less stabilized by solvation forces. This loss of solvation energy is an important component in the activation energy required to achieve reaction. Solvation effects are discussed in greater detail in Section 8.9.

The geometry of the transition state appears to be that in which the incoming and leaving groups are both weakly bonded to carbon in a linear fashion and in which the three remaining bonds to carbon lie in a plane perpendicular to the two weak bonds. The reaction mechanism for reaction of an entering group Y^- and a leaving group X^- is shown in Figure 8.3, where the structure of the reacting system at several points along the reaction coordinate is illustrated. At point (b) the C—X bond has started to lengthen and the central carbon has started to flatten out. At the transition state point (c), the central carbon is approximately flat and both bonds to the leaving and entering groups are long. Point (d) occurs on the final road to products (e); the central carbon has bent, the C—Y bond is approaching normal length, and the leaving group X is receding. The structures at points (a) through (e) are represented in stereo form in Figure 8.4.

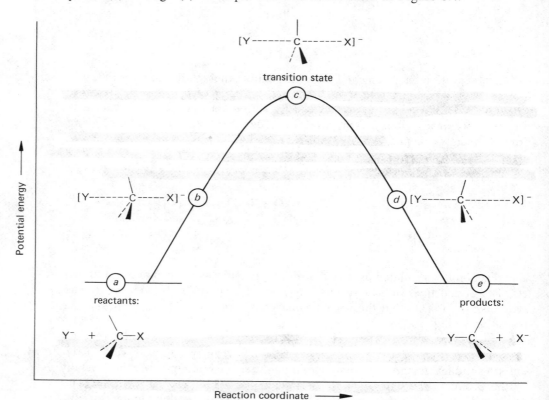

FIGURE 8.3. *Reaction mechanism profile for a displacement reaction by Y^- on RX.*

159

Sec. 8.6

**Stereochemistry
of the
Displacement
Reaction**

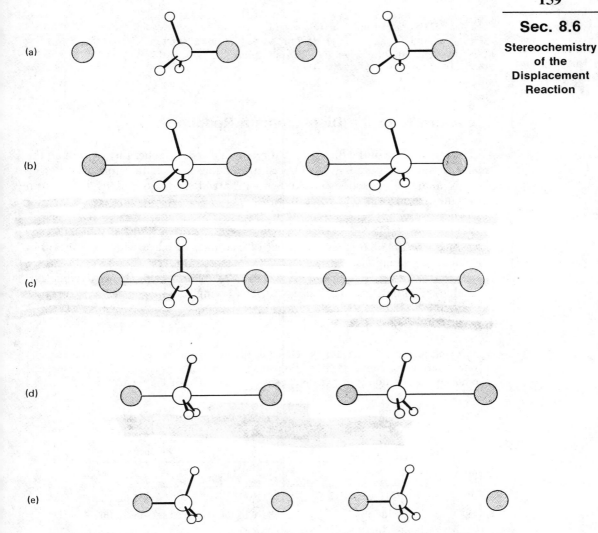

FIGURE 8.4. *The structure of the reaction system at points* (a)–(e) *in Figure 8.3.*

In orbital terms, both the reactant and the product are tetrahedral. The C—X bond in each case is C_{sp^3}—X. In the transition state, the weak bonds to X and Y may be considered to derive from overlap of a halogen orbital with the two lobes of a p-orbital on the central carbon. The other three bonds to this carbon are formed from sp^2-hybrid orbitals, as shown in Figure 8.5.

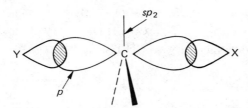

FIGURE 8.5. *Orbital formulation of the transition state of a displacement reaction.*

160

Chap. 8

Alkyl Halides;
Nucleophilic
Substitution and
Elimination

EXERCISE: Consider the reaction of *cis*-1-iodo-4-methylcyclohexane with iodide ion in acetone. What is the product of this displacement reaction? What is the composition of the mixture when equilibrium is reached (note Table 5.6)?

8.7
Generality of the Displacement Reaction

The importance of the displacement reaction lies in its generality. Although the reaction was introduced in the preceding sections with halide ions as entering groups, analogous reactions are known with a wide range of anions and neutral molecules. The only requirement is that the attacking group be a base. The base can be the anion conjugate base of a protic acid, or it can be a neutral molecule having an available lone pair of electrons. Such bases are often referred to as Lewis bases. Table 8.6 gives a number of examples of bases which are useful in displacement reactions. When used in displacement reactions, such bases are often referred to as **nucleophilic reagents** or simply **nucleophiles** (L., *nucleus,* kernel; Gr., *philos,* loving; hence "nucleus loving"). The reaction is then also called a **nucleophilic displacement reaction,** or as an S_N2 reaction for **substitution, nucleophilic, bimolecular.**

> The **molecularity** of a reaction is defined as the number of molecules involved in the rate-determining transition state. It is sometimes, but not always, equal to the **kinetic order** or the reaction. For example, consider a reaction of the type

TABLE 8.6
Some Displacement Reactions with Ethyl Bromide

Attacking Reagent		Product	
Formula	Name	Formula	Name
HO^-	hydroxide ion	C_2H_5OH	ethyl alcohol
$C_2H_5O^-$	ethoxide ion	$CH_3CH_2OCH_2CH_3$	diethyl ether
HS^-	hydrosulfide ion	CH_3CH_2SH	ethanethiol
SCN^-	thiocyanate ion	CH_3CH_2SCN	ethyl thiocyanate
CN^-	cyanide ion	CH_3CH_2CN	ethyl cyanide, propionitrile
N_3^-	azide ion	$CH_3CH_2N_3$	ethyl azide
NH_3	ammonia	$CH_3CH_2NH_3^+\ Br^-$	ethylammonium bromide
H_2O	water	$CH_3CH_2OH_2^+\ Br^-$	ethyloxonium bromide
$CH_3CO_2^-$	acetate ion	$CH_3CO_2C_2H_5$	ethyl acetate
NO_3^-	nitrate ion	$CH_3CH_2ONO_2$	ethyl nitrate
$P(CH_3)_3$	trimethylphosphine	$C_2H_5P(CH_3)_3^+\ Br^-$	ethyltrimethyl-phosphonium bromide
$N(C_2H_5)_3$	triethylamine	$(C_2H_5)_4N^+\ Br^-$	tetraethylammonium bromide
$S(C_2H_5)_2$	diethyl sulfide	$(C_2H_5)_3S^+\ Br^-$	triethylsulfonium bromide

$$A + B \underset{\underset{\text{fast}}{k_{-1}}}{\overset{k_1}{\rightleftharpoons}} C \xrightarrow[\text{slow}]{k_2} D$$

The rate of appearance of the product D is given by the rate law

$$\text{rate of formation of D} = \frac{d[D]}{dt} = \frac{k_1 k_2}{k_{-1}}[A][B] = k'[A][B]$$

which shows second-order kinetics. However, in the slow step (k_2), the rate-determining step, only one molecule, species C, is involved. Hence, the reaction is unimolecular.

Table 8.6 shows that the mechanism label S_N2 covers a wide variety of specific reactions. All of these reactions are bimolecular and occur with inversion of configuration at the reacting carbon. In such reactions we need to be concerned not only with the position of equilibrium but with rate. The reaction can range from incredibly fast even at low temperatures to painfully slow even at higher temperatures depending on the nature of the alkyl group, the nucleophile, the solvent, and the leaving group. We shall discuss each of these variables in turn. The important principles that enter into evaluating the effects of these variables recur frequently in organic chemistry and therefore warrant careful study at this time.

EXERCISE: Give the structure and name of the product of the displacement reaction of each of the nucleophiles (attacking groups) in Table 8.6 with methyl iodide.

8.8
Effect of Substrate Structure on Displacement Reactions

A large variety of alkyl halides undergo substitution by the S_N2 mechanism. The ease of reaction depends markedly upon the structure of the alkyl group to which the halogen is attached. Reactivities vary widely and in a consistent manner. Branching of the chain at the carbon where substitution occurs (the α-carbon) has a significant effect on the rate of reaction. Relative rates of S_N2 reactions for methyl, ethyl, isopropyl, and t-butyl halides are approximately as shown in Table 8.7.

These effects on reaction rate are interpreted with the concept of **steric hindrance** to attack of the attacking nucleophile. The rear of a methyl group is relatively exposed to such attack. As the hydrogens of the methyl group are replaced by methyl groups, the area in the rear of the leaving group becomes more encumbered. It becomes more difficult for the attacking group to approach closely

TABLE 8.7
Effect of Branching at the α-Carbon on the Rate of S_N2 Reactions

Alkyl Halide	Relative Rate
CH_3-	30
CH_3CH_2-	1
$(CH_3)_2CH-$	0.02
$(CH_3)_3C-$	~0

162

Chap. 8

**Alkyl Halides;
Nucleophilic
Substitution and
Elimination**

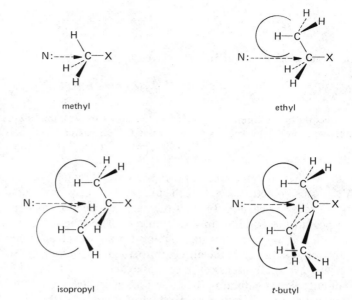

FIGURE 8.6. *Effect of α-branching on* S_N2 *reactions.*

enough to the rear of the C—X bond for reaction to occur, and the rate of reaction diminishes (Figure 8.6).

A similar effect may be seen in branching at the β-carbon. Some typical relative rates are shown in Table 8.8. This reduction in rate is also attributable to steric hindrance. In one conformation, the rear of a *n*-propyl carbon is seriously blocked (Figure 8.7a), but in two other conformations the situation is no worse than for ethyl (Figure 8.7b). Consequently, *n*-propyl halides undergo S_N2 displacement only slightly less readily than do ethyl halides.

For the isobutyl group, it is possible to rotate both of the β-methyl groups out of the way of the attacking group, but the resulting conformation is highly congested and has relatively high energy (Figure 8.8). Accordingly, isobutyl halides are much less reactive than either ethyl or *n*-propyl compounds.

Neopentyl halides are particularly interesting because there is no conformation in which a blocking methyl group can be avoided (Figure 8.9). Neopentyl halides are essentially unreactive in S_N2 reactions except under very drastic conditions.

Substitution of sites more remote than the β-carbon have little or no effect on the ease of S_N2 reactions. For example, *n*-butyl and *n*-pentyl halides react at essentially the same rate as *n*-propyl halides.

The type of steric interaction we have discussed here forces groups to bend

TABLE 8.8
**Effect of Branching at the β-Carbon
on the Rate of S_N2 Reactions**

Alkyl Halide	Relative Rate
CH_3CH_2—	1
$CH_3CH_2CH_2$—	0.4
$(CH_3)_2CHCH_2$—	0.03
$(CH_3)_3CCH_2$—	0.00001

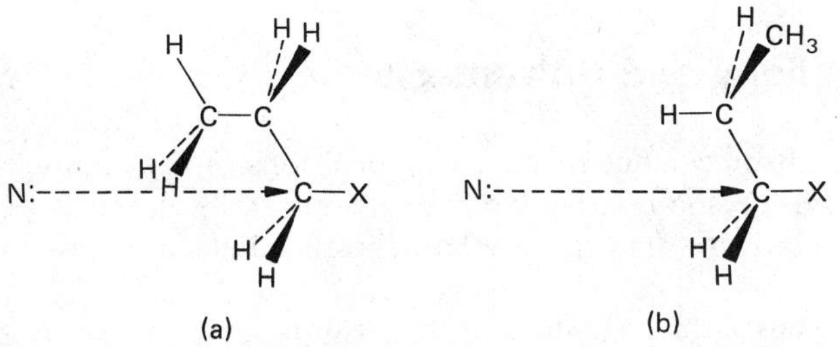

163

Sec. 8.8

**Effect of
Substrate
Structure on
Displacement
Reactions**

FIGURE 8.7. S_N2 *attack at two conformations of* n-*propyl compounds.*

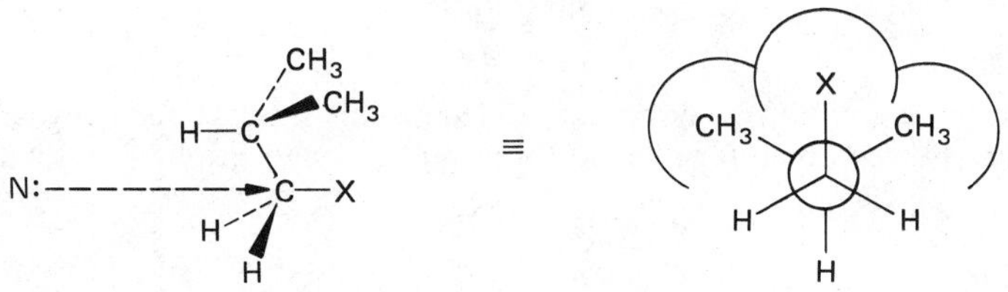

FIGURE 8.8. S_N2 *reaction at isobutyl systems.*

away from each other. Such deformation often forces orbitals to overlap in a noncolinear fashion. Recall that the resulting bent bonds are weaker than the corresponding straight bonds (Section 5.7.A).

In summary, the effect of the structure of the alkyl group on the rate of S_N2 reaction is apparent in two ways.

1. **Branching at the α-carbon hinders reaction: rate order is methyl > primary > secondary >> tertiary.**
2. **Branching at the β-carbon hinders reaction: neopentyl compounds are particularly slow.**

Displacements that proceed by the S_N2 mechanism are most successful with primary compounds having no branches at the β-carbon. Yields are poor to fair with secondary halides and with primary halides having branches at C-2. Neopentyl systems undergo the reaction only under very drastic conditions, and tertiary halides do not react by this mechanism at all. When the rate of the S_N2 reaction is slowed down by these structural effects, alternative side reactions begin to compete. With tertiary halides, and to an important degree with secondary and highly branched primary halides, the side reactions tend to dominate. These side reactions are discussed in Sections 8.11 and 8.12.

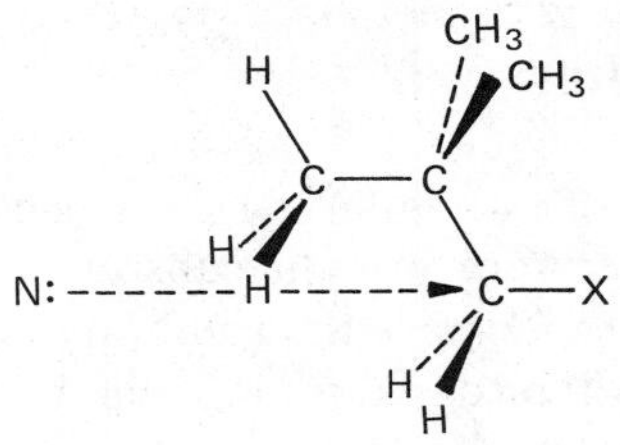

FIGURE 8.9. S_N2 *reaction at neopentyl compounds.*

164

Chap. 8

Alkyl Halides;
Nucleophilic
Substitution and
Elimination

8.9
Nucleophilicity and Solvent Effects

Nucleophilic displacement reactions involve ions either as nucleophiles or as products. For this reason relatively polar solvents are required. A *polar* solvent is one having a relatively high dipole moment and dielectric constant.

> The dielectric constant of a substance is the factor by which an electrostatic interaction in a vacuum is reduced by a medium of the substance. Coulomb's law for the electrostatic interaction of two charges, q_1 and q_2, separated by a distance r in a medium of dielectric constant D is given as
>
> $$E = \frac{q_1 q_2}{Dr}$$
>
> As the dielectric constant is reduced, the electrostatic interactions between ions increases and they combine together to form ion pairs and insoluble crystals.
>
> $$M^+ + X^- \rightleftharpoons M^+ \ X^- \rightleftharpoons (M^+X^-)$$
> $$\text{ion pair} \qquad \text{crystal}$$
>
> Ion pairs are relatively unreactive in S_N2 reactions.

Examples of solvents important in displacement reactions are summarized in Table 8.9. It is convenient to consider them in two categories: *hydroxylic* (water, alcohols) and *polar aprotic* (acetonitrile, dimethylformamide, dimethyl sulfoxide, acetone).

TABLE 8.9

	Boiling Point °C	Dipole Moment μ, D (g)	Dielectric Constant ϵ (25°)
water, H_2O	100	1.85	78.5
methanol, CH_3OH	65	1.70	32.6
ethanol, CH_3CH_2OH	78.5	1.69	24.3
acetonitrile, CH_3CN	81.6	3.92	36.2
dimethylformamide, DMF, $HCON(CH_3)_2$	153	3.82	36.7
dimethyl sulfoxide, DMSO, CH_3SOCH_3	189	3.96	49
acetone, CH_3COCH_3	56.5	2.88	20.7
hexamethylphosphoric triamide, HMPT, $[(CH_3)_2N]_3PO$	232	4.30[a]	30

[a] In solution.

A. *Hydroxylic Solvents*

Ethanol and methanol are particularly useful because they are inexpensive, relatively inert, and dissolve many organic substrates and inorganic salts. Sometimes some water is added to increase the solubility of the inorganic salt used as the displacing agent. Some typical examples follow.

$$CH_3O^- + CH_3CH_2CH_2CH_2Br \xrightarrow{CH_3OH} CH_3CH_2CH_2CH_2OCH_3 + Br^-$$

> *n*-Butyl bromide is refluxed with sodium methoxide in methanol for $\frac{1}{2}$ hr. Water is added, and the organic layer is separated, dried, and distilled to give methyl *n*-butyl ether.

$$\underset{\text{Na}^+\text{SCN}^-}{} + \underset{\underset{\text{CH}_3\text{CHCH}_3}{|}}{\overset{\text{Br}}{\overset{|}{}}} \xrightarrow[\text{H}_2\text{O}]{\text{C}_2\text{H}_5\text{OH}} \underset{\underset{\text{CH}_3\text{CHCH}_3}{|}}{\overset{\text{SCN}}{\overset{|}{}}} + \text{Na}^+\text{Br}^-$$

> Isopropyl bromide and sodium thiocyanate (NaSCN) are refluxed in 90% aqueous ethanol for 6 hr. The precipitated sodium bromide is filtered. The filtrate is diluted with water and extracted with ether. Distillation gives isopropyl thiocyanate, $(\text{CH}_3)_2\text{CHSCN}$, in 76–79% yield.

To begin our discussion of the reactivity of various nucleophiles in the S_N2 process, let us first note that there is a formal relationship between the reaction of an anion with a proton (its basicity) and the reaction of the same anion with an alkyl halide such as methyl iodide (its nucleophilicity).

Basicity

$$\text{B:}^- + \text{H}^+ \overset{K}{\rightleftharpoons} \text{B—H}$$

Nucleophilicity

$$\text{B:}^- + \text{CH}_3\text{I} \xrightarrow{k_2} [\text{B}^{\delta-} \text{---CH}_3 \text{---I}^{\delta-}]^{\ddagger} \longrightarrow \text{CH}_3\text{B} + \text{I}^-$$

In both cases the lone pair of B:^- is forming a covalent bond and B:^- is losing some of its negative charge.

Recall (Section 4.5) that a quantitative measure of the *basicity* of a base is the acidity or pK_a of its conjugate acid. A weak conjugate acid (more positive pK_a) corresponds to a strong base. Alternatively, a strong base tends to give up its charge and to bond to H^+. We might anticipate that a strong base might also tend to give up its charge and to bond to carbon; that is, a stronger base may be expected to react faster with methyl iodide. Consequently we might expect to find some correlation between the energy involved in the protonation of a base and the activation energy for its reaction in a displacement reaction. Instead of energies or enthalpies, many correlations of this type have been found using the Gibbs free energies ΔG. The pK of an acid is proportional to the standard free energy of the acid-base equilibrium.

$$\Delta G° = 2.303 \, RT \, \mathrm{p}K$$

> This equation follows by combining the relation between an equilibrium constant and the standard free energy with the definition of pK
>
> $$\Delta G° = -RT \ln K$$

Recall from Chapter 4 that rate constants are also related to free energies of activation $\Delta G^{\ddagger}$

$$k = \text{constant} \times e^{-\Delta G^{\ddagger}/RT}$$

$$-\log k = \frac{\Delta G^{\ddagger}}{2.303 \, RT} - \log (\text{constant})$$

To test the existence of a correlation between basicity and reactivity in displacement reactions, we first need to determine the second-order rate constants for reaction of a series of bases with some common substrate under the same experimental conditions, that is, for the same solvent and temperature. Next, we plot the

166

Chap. 8

**Alkyl Halides;
Nucleophilic
Substitution and
Elimination**

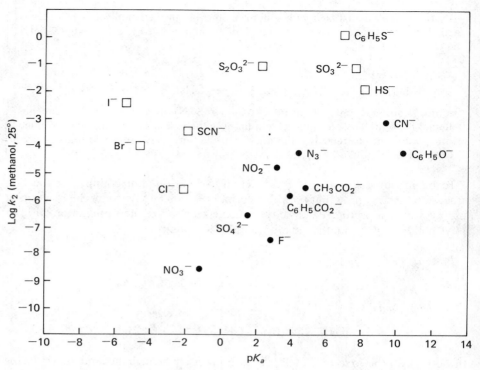

FIGURE 8.10. *Comparison of nucleophilicities and basicities of various anions.*

logarithms of these rate constants against the corresponding pK_a values. For a perfect correlation every point will fall exactly on a straight line. Figure 8.10 shows a plot of this type for the reactions of a series of bases with methyl iodide in methyl alcohol solution.

At first sight we see points scattering uniformly all over with no apparent correlation at all. On closer inspection, however, a pattern does emerge. Anions in which reaction occurs on an atom in the second period of the periodic table (C, N, O, F) do give a roughly linear correlation. These points are filled circles in Figure 8.10. When the attacking atom is the same, the correlation is quite good. Note, for example, that both basicity and nucleophilicity show the same order for $C_6H_5O^- > CH_3CO_2^- > C_6H_5CO_2^- > SO_4^{2-} > NO_3^-$.

We next note that reactions with atoms in higher rows of the periodic table (S, Cl, Br, I; open squares in Figure 8.10) all lie above their second period counterparts. *Third- and fourth-period elements are invariably more nucleophilic than second period elements of comparable basicity.* One reason for this generalization appears to be that the larger elements have relatively diffuse lone pairs, which are more polarizable. These more diffuse lone pairs tend to bond more strongly to the more diffuse p-orbital of the S_N2 transition state than to the small, tight $1s$-orbital of a hydrogen. Moreover, this difference results in an important difference in solvation. Anions in hydroxylic solvents are solvated by hydrogen bonds. Such hydrogen bonding is stronger and more important the smaller the anion and more concentrated the negative charge. In the transition state for an S_N2 reaction the negative charge is less concentrated and solvation is less important. The more concentrated lone pairs of second-period systems thus might give up more solvation stabilization in an S_N2 reaction and react more slowly than comparable larger nucleophiles.

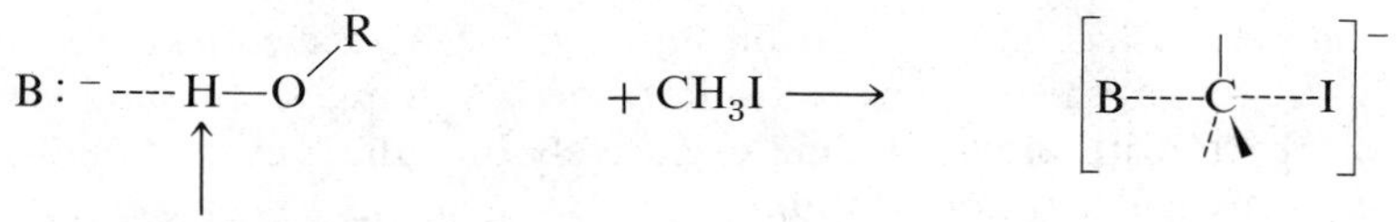

$$B: ^- ---- H—O^{\diagup R} \qquad + CH_3I \longrightarrow \qquad \left[B ---- \underset{\diagup \diagdown}{C} ----I \right]^-$$

hydrogen-bond solvation
more important for smaller atoms (C, N, O, F)

much larger ion
hydrogen-bond solvation less important

Examples of the net results of the polarizability and solvation effects of position in the periodic table of nucleophiles are shown in the following comparisons of reactivity toward methyl iodide in methanol. In each case the nucleophilic atom is in the same *column* of the periodic table and the more nucleophilic system is also the *less* basic.

$$C_6H_5S^- > C_6H_5O^-$$

$$Cl^- > F^-$$

$$(CH_3)_2S > CH_3OH$$

$$(C_2H_5)_3P > (C_2H_5)_3N$$

Note also that the generalization applies as well to *neutral* nucleophiles.

At this point we can now interpret an interesting experimental result. Sulfate ion is straightforward in its reaction with either H^+ or CH_3I in an S_N2 reaction. Both types of reaction occur at an oxygen

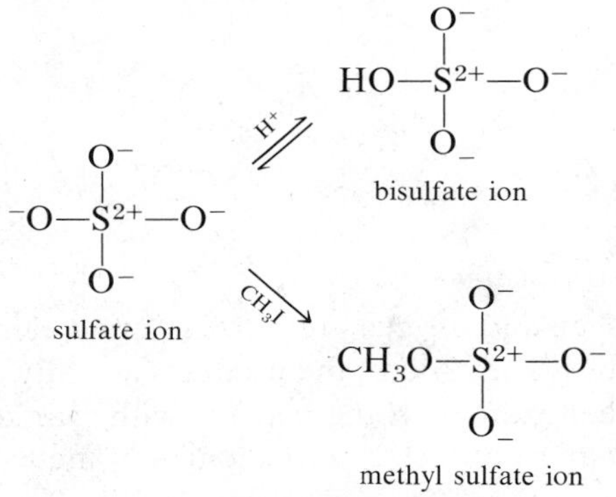

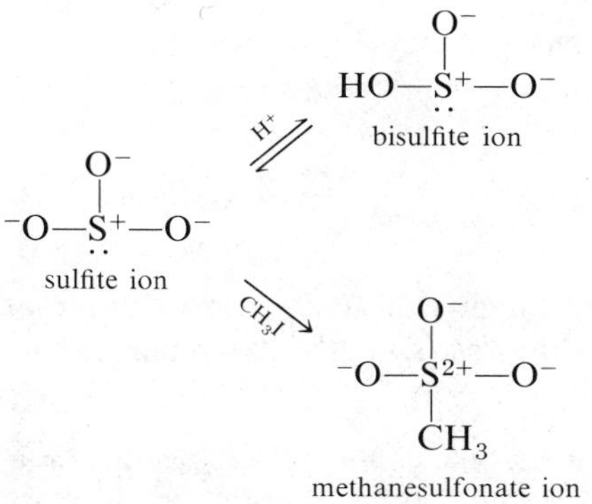

The oxygen in sulfite ion is the more basic atom and prefers to attack H^+, but the lone pair on sulfur is more nucleophilic and has preference in the S_N2 transition

168

Chap. 8

**Alkyl Halides;
Nucleophilic
Substitution and
Elimination**

state. Sulfate ion has no lone pair on sulfur, and both reactions have no alternative but to occur at the oxygen. Thiosulfate ion is a simple sulfur analog of sulfate. This ion reacts with methyl iodide exclusively on sulfur, even though there are three oxygens and only one sulfur.

$$^-S—\underset{\underset{O_-}{|}}{\overset{\overset{O^-}{|}}{S^{2+}}}—O^- + CH_3I \longrightarrow CH_3S—\underset{\underset{O_-}{|}}{\overset{\overset{O^-}{|}}{S^{2+}}}—O^-$$

thiosulfate ion   methylthiosulfate ion

Finally, there are some nucleophiles that show measurable nucleophilic properties at two different atoms. Nitrite ion is an example. The ion undergoes protonation exclusively on oxygen to give nitrous acid.

$$\left[{}^-\ddot{O}—\ddot{N}{=}O \longleftrightarrow O{=}\ddot{N}—\ddot{O}^- \right] \xrightarrow{H^+} HO—\ddot{N}{=}O$$

nitrite ion   nitrous acid

However, the reaction of nitrite ion with methyl iodide gives both methyl nitrite and nitromethane.

$$^-O—\ddot{N}{=}O + CH_3I \longrightarrow$$

$$CH_3O—\ddot{N}{=}O$$
methyl nitrite

$$\left[CH_3—\overset{+}{N}\overset{\overset{O}{\parallel}}{\underset{\underset{O^-}{}}{}} \longleftrightarrow CH_3—\overset{+}{N}\underset{\underset{O}{}}{\overset{\overset{O^-}{}}{}} \right]$$
nitromethane

In this case both nitrogen and oxygen are second-period elements and have comparable nucleophilicities. The ratio of the products actually depends on the reaction conditions. Another example is the reaction with cyanide ion. In addition to methyl cyanide, the major product, small amounts of methyl isocyanide are also produced.

$$^-{:}C{\equiv}N{:} + CH_3I \longrightarrow$$

$$CH_3—C{\equiv}N{:} \quad \text{(major)}$$
methyl cyanide

$$CH_3—\overset{+}{N}{\equiv}\overset{-}{C}{:} \quad \text{(minor)}$$
methyl isocyanide

Anions such as these, which can react at two different positions, are called **ambident** (L., *ambo,* both; *dentis,* tooth), "two-fanged" nucleophiles.

EXERCISE: In Figure 8.10 a number of reactions are in the neighborhood of log $k_2 = -4$; that is, $k_2 = 1 \times 10^{-4} M^{-1}$ sec^{-1}. Consider the reaction of a solution 0.1 M in methyl iodide and 1 M in a nucleophile having this rate constant. How long would it take to achieve 99.9% reaction (10 half-lives)?

B. *Polar Aprotic Solvents*

Polar aprotic solvents are defined as solvents having relatively high dipole moments and dielectric constants but without acidic hydrogens such as those on hydroxy groups. Such solvents can dissolve salts but, because of the absence of hydrogens suitable for hydrogen bonding, do not solvate anions particularly well. As a result, displacement reactions in polar aprotic solvents are frequently much faster than they are in hydroxylic solvents. For example, the reaction of methyl bromide with iodide ion is about 500 times faster in acetone than in methyl alcohol. An even more striking example is the reaction of chloride ion with methyl iodide, which is a *million times* faster in dimethylformamide than it is in methyl alcohol.

Because of the differences in solvation, pK values for aqueous solutions are not generally useful in evaluating reactivities in polar aprotic solvents. The relative reactivities of different nucleophilic groups can change dramatically. For example, the order of reactivity of halide ions in S_N2 reactions in water or in alcohols is

$$I^- > Br^- > Cl^-$$

In acetone the reactivities tend to be closer together, and in dimethylformamide the order is even reversed! This result emphasizes that nucleophilicity is not a simple and invarient property but depends on the specific reagents, conditions, and solvents. We make frequent use of qualitative rather than quantitative generalizations, and it is still useful to know that iodide ion is *frequently* more reactive than chloride ion, and that third-period elements are *usually* more reactive in displacement reactions than second-period elements. Even though anions are generally more reactive in polar aprotic solvents than in aqueous alcohols, the aqueous solvents are still frequently used in practice because they are inexpensive, convenient, and, for many reactions, serve perfectly adequately.

8.10
Leaving Groups

Alkyl chlorides, bromides, and iodides all react satisfactorily by the S_N2 mechanism. The ease of reaction is dependent on the nature of the **leaving group,** alkyl iodides reacting most rapidly and alkyl chlorides most slowly. Alkyl fluorides are essentially unreactive by the S_N2 mechanism. Since chlorine is much cheaper than bromine, alkyl chlorides are the least expensive alkyl halides. However, for laboratory uses where only small amounts of material are involved, alkyl bromides are commonly used because they are 50–100 times more reactive than the corresponding chlorides. Iodides are somewhat more reactive than bromides but are quite a bit more expensive, and this slightly increased reactivity does not justify their additional cost. In industrial processes, where massive amounts of materials are involved and cost is a prime consideration, alkyl chlorides are used almost exclusively.

The S_N2 reaction is not restricted to alkyl halides. Any group that is the conjugate base of a strong acid can act as a leaving group. An example is bisulfate ion, HSO_4^-, which is the conjugate base of sulfuric acid, pK_a -5. Dimethyl sulfate is an inexpensive commercial compound, which reacts readily by the S_N2 mechanism. The leaving group is the methyl sulfate ion, which is similar in its base strength to bisulfate ion.

$$CH_3CH_2CH_2O^- + CH_3O-\overset{\overset{O^-}{|}}{\underset{\underset{O^-}{|}}{S^{2+}}}-OCH_3 \longrightarrow CH_3CH_2CH_2OCH_3 + {}^-O-\overset{\overset{O^-}{|}}{\underset{\underset{O^-}{|}}{S^{2+}}}-OCH_3$$

methyl sulfate ion

The chief disadvantage of dimethyl sulfate is its toxicity. It is water-soluble and reacts readily with the nucleophilic groups in body tissues and fluids. Although dimethyl sulfate is the only sulfate in common use, alkyl sulfonates are often employed. Sulfonic acids, RSO_2OH, are similar to sulfuric acid in acidity, and the sulfonate ion, RSO_3^-, is an excellent leaving group. Alkyl benzenesulfonates, alkyl p-toluenesulfonates, and alkyl methanesulfonates are extremely useful substrates for S_N2 reactions. These compounds are readily prepared from alcohols as described in Sections 10.7.D and 27.5.

methyl benzenesulfonate   ethyl p-toluenesulfonate   methyl methanesulfonate

Alkyl nitrates undergo reaction by the S_N2 mechanism because nitric acid is a strong acid and nitrate ion is a weak base. However, alkyl nitrates are more prone to side reactions than the corresponding halides and, therefore, the yield of substitution product is lower. Consequently, nitrates are rarely used.

The facility with which a group can function as a leaving group in an S_N2 reaction is related to its **basicity.** If a group is a weak base (that is, the conjugate base of a strong acid), it will generally be a "good" leaving group. This is readily understood on recalling that the leaving group L gains electron density in going from the reactant to transition state.

$$N:^- + \ \overset{}{\underset{}{C}}-L \longrightarrow \left[\overset{\delta-}{N}\cdots\overset{|}{\underset{|}{C}}\cdots\overset{\delta-}{L}\right]^{\ddagger} \longrightarrow N-C\diagup + :L^-$$

The more this electron density or negative charge is stabilized, the lower is the energy of the transition state and the faster is the rate of reaction. The degree to which a group can accommodate a negative charge is also related to its affinity for a proton, its basicity. The acids HCl, HBr, HI, and H_2SO_4 are all strong acids because the anions Cl^-, Br^-, I^-, and HSO_4^- are stable anions. These anions are also good leaving groups in S_N2 reactions.

HCN is a weak acid ($pK_a = 10$) and the displacement of cyanide is never observed.

$$N:^- + RCN \ \overset{}{\longrightarrow}\!\!\!/\!\!\!\longrightarrow R-N + CN^-$$

Hydrazoic acid (HN_3) and acetic acid (CH_3CO_2H) are also weak acids (pK_as of 5.8 and 4.8, respectively). Correspondingly, azide ion and acetate ion are extremely poor leaving groups.

$$N:^- + R-N_3 \ \overset{}{\longrightarrow}\!\!\!/\!\!\!\longrightarrow RN + N_3^-$$

$$N:^- + R-O\overset{\overset{O}{\|}}{C}CH_3 \ \overset{}{\longrightarrow}\!\!\!/\!\!\!\longrightarrow RN + CH_3CO_2^-$$

The reason that alkyl fluorides are ineffective substrates in the S_N2 reaction is related to the relatively low acidity of HF ($pK_a = 3$).

By comparing S_N2 reactivity with relative acidity, we can understand the operation of **acid catalysis** in certain displacement processes. Alcohols do not undergo S_N2 reactions because hydroxide ion is too basic (the pK_a of its conjugate acid, H_2O, is 15.7).

$$N\!:^- + R\!-\!OH \xrightarrow{\;\;/\!/\;\;} R\!-\!N + OH^-$$

However, in the presence of a strong mineral acid such as HCl, HBr, or H_2SO_4, the alcohol oxygen is protonated. An S_N2 reaction can now occur because the leaving group is water, which is a much weaker base than OH^- (the conjugate acid of water, H_3O^+, has $pK_a = -1.7$).

$$R\!-\!OH + H^+ \rightleftharpoons R\!-\!\overset{+}{O}H_2 + N\!:^- \longrightarrow R\!-\!N + H_2O$$

Note that the same principles of electron density and relative basicity are involved in this reaction, even though the leaving group is not an anion. This is a useful and important reaction of alcohols and it will be developed more fully in Section 10.7.

EXERCISE: Write out the complete displacement reactions of trimethylamine, $(CH_3)_3N$, with (a) $CH_3OSO_2CH_3$ and (b) $(CH_3)_3S^+$.

8.11
E2 Elimination

One of the side reactions that occurs in varying degree in displacement reactions is the elimination of the elements of HX to produce an alkene.

$$B\!:^- \quad H\!-\!\overset{|}{C}\!-\!\overset{|}{C}\!-\!X \longrightarrow BH + \,\diagup\!\!\!C\!=\!C\!\diagdown + X^-$$

Under appropriate conditions, this reaction can be the principal reaction and becomes a method for preparing alkenes. Accordingly, it is discussed in more detail in Section 11.5.A. For the present, it suffices to know that this reaction occurs by attack of a base on a hydrogen with concomitant formation of a C=C double bond and breaking of the C—X bond to form halide ion.

Mechanistically, the reaction is classified as **bimolecular elimination,** or **E2.** Since attack on a proton is involved, it is the *basicity,* rather than the nucleophilicity, of the Lewis base that is important. Strongly basic species such as alkoxide or hydroxide ions favor elimination; highly nucleophilic species such as second- and third-row elements favor substitution. For example, the S_N2 and E2 reactions of 3-bromo-2-methylbutane with chloride ion in acetone occur at about equal rates.

$$\underset{\underset{Br}{|}}{\overset{\overset{CH_3}{|}}{CH_3CHCHCH_3}} + Cl^- \xrightarrow{acetone} \underset{\underset{Cl}{|}}{\overset{\overset{CH_3}{|}}{CH_3CHCHCH_3}} + \overset{\overset{CH_3}{|}}{CH_3C}\!=\!CHCH_3 + \overset{\overset{CH_3}{|}}{CH_3CH}CH\!=\!CH_2$$

$$(\sim 50\%) \qquad\qquad\qquad\qquad (\sim 50\%)$$

With acetate ion, however, elimination is about 8 times faster than substitution.

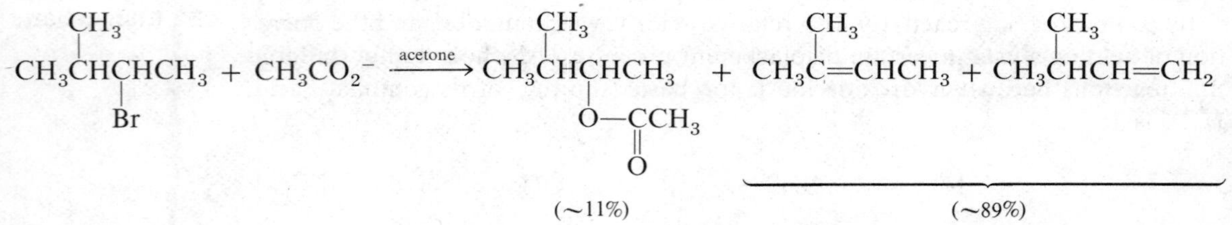

$$\underset{\underset{Br}{|}}{CH_3CHCHCH_3} + CH_3CO_2^- \xrightarrow{acetone} \underset{\underset{O-CCH_3}{\underset{\parallel}{O}}}{CH_3CHCHCH_3} + CH_3C{=}CHCH_3 + CH_3CHCH{=}CH_2$$

with CH₃ groups on the appropriate carbons:

(~11%)   (~89%)

The structure of the substrate compound is also important in determining the substitution/elimination ratio. Straight-chain primary compounds show little tendency toward elimination, mainly since the S_N2 reaction is rapid. For example, ethyl bromide reacts with N_3^-, Cl^-, or $CH_3CO_2^-$ in acetone to give only substitution products. Even the strong base sodium ethoxide in ethanol gives virtually none of the elimination product. However, with more highly branched compounds, the S_N2 reaction is slower, and attack at hydrogen can compete more favorably. Consequently, larger amounts of the elimination product are obtained. Some data are presented in Tables 8.10 and 8.11 for the reactions of various alkyl halides with acetate ion in acetone and ethoxide ion in ethanol.

Note the resulting generalizations that elimination by-products are quite minor with simple primary halides, but with branching at either the α- or β-carbon, the alkene elimination products become increasingly important. The behavior of tertiary halides is more complex. Since they undergo S_N2 reactions so very slowly, one would expect that tertiary halides would give complete elimination, even with

TABLE 8.10

$$R{-}Br + CH_3CO_2^- \xrightarrow{acetone} R{-}\overset{\overset{O}{\parallel}}{O}CCH_3 + Br^-$$

RBr	Percent Substitution	Percent Elimination
CH₃CH₂Br	100	0
(CH₃)₂CHBr	100	0
(CH₃)₃CBr	0	100
(CH₃)₂CHCHBrCH₃	11	89

TABLE 8.11

$$R{-}Br + CH_3CH_2O^- \xrightarrow{CH_3CH_2OH} ROCH_2CH_3 + Br^-$$

RBr	Percent Substitution	Percent Elimination
CH₃CH₂Br	99	1
CH₃CH₂CH₂Br	91	9
CH₃CH₂CH₂CH₂Br	90	10
(CH₃)₂CHCH₂Br	40	60
(CH₃)₂CHBr	20	80
CH₃CH₂CHBrCH₂CH₃	12	88

weak bases. However, tertiary halides can undergo substitution by another mechanism (next section). Consequently, the elmination/substitution ratio for tertiary halides is highly dependent on reaction conditions. In general, they give mainly the elimination products, especially under conditions that favor the bimolecular mechanism (high concentrations of strong base).

8.12
$S_N 1$ Reactions: Carbocations

Ethyl bromide reacts very rapidly with ethoxide ion in refluxing ethanol (78°); reaction is complete after a few minutes. If ethyl bromide is refluxed in ethanol not containing any added sodium ethoxide, $S_N 2$ displacement still occurs, but the reaction is exceedingly slow. After refluxing for 4 days, the reaction is only 50% complete. This reactivity difference is due to the fact that the negatively charged ethoxide ion is much more nucleophilic than the neutral ethanol molecule. On an equal concentration basis, ethoxide ion is more than 10,000 times more reactive than ethanol itself. In general, anions are much more basic and nucleophilic than their conjugate acids:

$$CH_3O^- > CH_3OH$$

$$CH_3S^- > CH_3SH$$

$$CH_3CO_2^- > CH_3COOH$$

$$SO_3{}^{2-} > HSO_3{}^-$$

We saw earlier that tertiary alkyl halides do not react by the $S_N 2$ mechanism. Yet t-butyl bromide reacts quite rapidly in pure ethanol; in refluxing ethanol the half-life for reaction is only a few minutes! Various observations show that this reaction does not proceed by an $S_N 2$ displacement even though the principal product is ethyl t-butyl ether, $(CH_3)_3COCH_2CH_3$. For one thing, the addition of ethoxide ion to an ethanol solution of ethyl bromide causes a large increase in the rate of reaction. The rate law for the reaction of ethyl bromide is

$$\text{rate} = k_1[C_2H_5Br] + k_2[C_2H_5Br][C_2H_5O^-]$$

The first term represents the reaction of ethyl bromide with neutral ethanol; because ethanol is the solvent and its concentration remains virtually unchanged, its concentration does not appear in the rate equation. The second term represents the reaction of ethyl bromide with ethoxide ion. For ethyl bromide, as we saw above, the reaction with ethoxide is much faster than the reaction with ethanol, that is, $k_2 \gg k_1$. Since k_1 is so small relative to k_2, the rate expression is approximately

$$\text{rate} = k_2[C_2H_5Br][C_2H_5O^-]$$

The rate of reaction of t-butyl bromide (t-BuBr) is given by a similar equation.

$$\text{rate} = k_1[t\text{-BuBr}] + k_2[t\text{-BuBr}][C_2H_5O^-]$$

Here, however, the second term is unimportant; $k_1 \gg k_2$. Addition of sodium ethoxide has no effect on the rate of reaction. Therefore, in this case the rate is effectively

$$\text{rate} = k_1[t\text{-BuBr}]$$

[This is true only for small concentrations of sodium ethoxide. As we saw earlier (Section 8.11), high concentrations of strong base lead to bimolecular elimination.]

This change in kinetic behavior is consistent with a change in mechanism. With the tertiary alkyl halide, the rear of the molecule is effectively blocked and the S_N2 mechanism cannot operate. However, a competing mechanism is possible. A great deal of experimental work over the past several decades has established that this mechanism involves two steps. In the first step, ionization of the C—Br bond occurs and an intermediate carbocation is produced. The carbocation then reacts rapidly with solvent or whatever nucleophiles are around.

$$
(1) \quad CH_3\underset{\underset{CH_3}{|}}{\overset{\overset{CH_3}{|}}{C}}Br \underset{slow}{\rightleftharpoons} CH_3\underset{\underset{CH_3}{|}}{\overset{\overset{CH_3}{|}}{C^+}} + Br^-
$$

$$
(2) \quad CH_3\underset{\underset{CH_3}{|}}{\overset{\overset{CH_3}{|}}{C^+}} + C_2H_5OH \xrightarrow{fast} CH_3\underset{\underset{CH_3}{|}}{\overset{\overset{CH_3}{|}}{C}}\overset{+}{\underset{H}{O}}CH_2CH_3 \longrightarrow CH_3\underset{\underset{CH_3}{|}}{\overset{\overset{CH_3}{|}}{C}}OCH_2CH_3 + H^+
$$

When a compound reacts with the solvent, as is the case here, the process is called a **solvolysis reaction.** This solvolysis reaction is classified mechanistically as **unimolecular nucleophilic substitution** or **S_N1** because only one species is involved in the rate-limiting step. A reaction coordinate diagram for the reaction is shown in Figure 8.11. In the intermediate carbocation, the central carbon has only a sextet of electrons (compare with the Lewis structure of methyl cation, Section 2.2). In orbital terms, the ion is best described in terms of a central sp^2-hybridized carbon with an empty p-orbital (Figure 8.12).

[The terms *carbocation* and *carbanion* are generic names for organic cations R$^+$ and anions R$^-$, respectively. For many years the term *carbonium ion* was used as a generic name for organic cations having an electron-deficient carbon with a sextet of elec-]

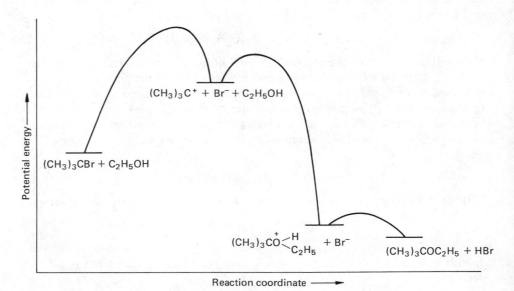

FIGURE 8.11. *Reaction coordinate diagram for the solvolysis of* t-*butyl bromide.*

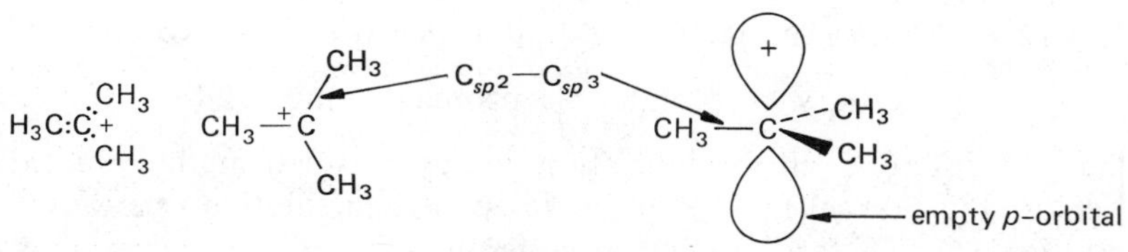

FIGURE 8.12. *The* t-*butyl cation.*

trons. The term carbonium ion has been considered to be nonsystematic because comparable oxonium ions (for example, H_3O^+) and ammonium ions (NH_4^+) have electron octets; the term *carbenium ion* has been suggested instead, but is not yet in common use. To avoid confusion, we will frequently use the term carbocation as a generic name for organic cations with positive formal charge on carbon, but specific ions will be referred to as alkyl cations (methyl cation, *t*-butyl cation, etc.).

Note that alkyl cations are planar; the central carbon and its three attached atoms all lie in one plane. The structure is favored because it places the three groups as far apart from one another as possible and allows the carbocation to achieve the strongest bonds and lowest energy. Alternatively, electrons in a 2s-orbital are more stable than electrons in a 2p-orbital. By using the s-orbital to form three sp^2-hybrid orbitals, it is involved most effectively in bonding; the remaining 2p-orbital is left vacant.

It is important to point out that cations are highly reactive reaction intermediates. They have only a short, but finite, lifetime in solution. Under normal conditions they cannot be observed directly in the reaction mixture because they react almost as soon as they are produced. However, carbocations vary widely in stability with structure. The enthalpies for ionization of various alkyl halides *in the gas phase* are summarized in Table 8.12. Note that ethyl cation is more stable than methyl cation, but *n*-propyl and *n*-butyl cations are of similar stability. These systems may be described as *primary carbocations,* RCH_2^+. Isopropyl cation is an example of a *secondary carbocation,* R_2CH^+, and *t*-butyl cation is a *tertiary carbocation.*

CH_3^+ methyl cation

$CH_3CH_2^+$ ethyl cation, a *primary carbocation*

$(CH_3)_2CH^+$ isopropyl cation, a *secondary carbocation*

$(CH_3)_3C^+$ *t*-butyl cation, a *tertiary carbocation*

TABLE 8.12
Enthalpy for Ionization of
Alkyl Chlorides in the Gas Phase
$$RCl = R^+ + Cl^-$$

R	$\Delta H°$, kcal mole^{-1}
CH_3	228
CH_3CH_2	191
$CH_3CH_2CH_2$	193
$CH_3CH_2CH_2CH_2$	193
$(CH_3)_2CH$	167
$(CH_3)_3C$	151

176

Chap. 8

**Alkyl Halides;
Nucleophilic
Substitution and
Elimination**

Table 8.12 shows that the relative stabilities of various cations are

$$\text{tertiary} > \text{secondary} > \text{primary} > \text{methyl}$$

Note that the difference in the ionization energy between methyl chloride and *t*-butyl chloride is 50 kcal mole^{-1} in the gas phase. In solution, the ions are solvated and ionization is facilitated. Consequently, the $\Delta H°$ for ionization is much lower than given in Table 8.12. Thus, the energy required to ionize *t*-butyl chloride in hydroxylic solvents is low enough that reaction proceeds at normal rates. Tertiary alkyl cations are rather common intermediates in organic reactions. Secondary alkyl cations are considerably less stable and much more difficult to produce, but they do occur as intermediates in some reactions. Primary alkyl cations are so much less stable that they virtually *never occur as reaction intermediates in solution*. These generalizations may be summarized as follows.

Alkyl Group	Occurrence of Carbocation Intermediates in Solution	Occurrence of S_N2 Displacement Reactions
tertiary	common	never
secondary	sometimes	sometimes
primary	rare	common
methyl	never	common

The order of stabilities of carbocations is in large part to be attributed to the greater polarizability of alkyl groups compared to hydrogen. Another factor is an interesting aspect of overlapping orbitals. Consider the methyl cation, CH_3^+. The C—H bonds in methyl cation *lie in the nodal plane of the vacant $2p_z$-orbital and hence cannot overlap with it*. In the ethyl cation, $CH_3CH_2^+$, there can be some overlap between this empty orbital and one of the bonds of the methyl group (Figure 8.13). This type of overlap is readily shown by quantum mechanical calculations. It has the effect of stabilizing the ion because electron density from an adjacent bond can "spill over" into the empty orbital. This results in spreading the positive charge over a larger volume. We shall see frequently that ions with concentrated charge are less stable and more reactive than ions in which the charge is spread over a greater volume. As more alkyl groups are attached to the cationic carbon, it becomes even more stable.

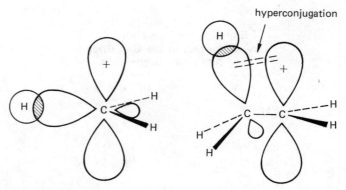

FIGURE 8.13. *Overlap of a C—H bond orbital with the empty p-orbital of a carbocation.*

The interaction of a bond orbital with a *p*-orbital as shown in Figure 8.13 is referred to as *hyperconjugation*. Its relationship to *conjugation* will become apparent when we get to Chapter 20.

The relative stabilities of carbocations can be understood on the basis of simple electrostatics. Electrons are attracted to nuclei and are repelled from other electrons. The electrons in a bond repel each other, but they are prevented from getting too far apart because of their attraction to the two nuclei in the bonded atoms. When there is an adjacent atom with a vacant orbital available for overlap and a positive charge, the original bonding pair of electrons can reduce their mutual repulsion by getting farther apart; they can do this and still maintain the stability of being associated with a positive nucleus. Electron repulsion is decreased in such a carbocation and it is convenient to describe the result in terms of a spreading out of positive charge.

The first step of the S_N1 reaction is an ionization process, which involves separation of charge.

$$R—X \longrightarrow R^+ \ X^-$$

Accordingly, the reaction is highly sensitive to solvent effects. The reaction is rapid in aqueous and hydroxylic solvents and slow in nonpolar solvents.

The solvolysis reaction of tertiary alkyl halides is only in part an S_N1 reaction. The intermediate carbocation can react rapidly in several alternative ways. Two of these reaction paths are

1. Reaction with any nucleophiles present (the S_N1 process)
2. Elimination of a proton (the E1 process).

These reactions are illustrated by the behavior of *t*-butyl bromide in ethanol at 25°. The solvolysis product, ethyl *t*-butyl ether, is produced along with a significant amount of the elimination product, an alkene.

$$(CH_3)_3CBr \xrightarrow[C_2H_5OH]{slow} (CH_3)_3C^+ \begin{cases} \xrightarrow{fast} (CH_3)_3COCH_2CH_3 \\ \quad\quad (81\%) \\ \\ \xrightarrow{fast} (CH_3)_2C{=}CH_2 \\ \quad\quad (19\%) \end{cases}$$

In a mixed solvent, the carbocation can react with both components in addition to eliminating a proton. For example, in a mixture of 80% ethanol and 20% water, *t*-butyl bromide gives three products.

$$(CH_3)_3CBr \xrightarrow[\substack{H_2O \\ 25°}]{C_2H_5OH} (CH_3)_3C^+ \begin{cases} \xrightarrow{C_2H_5OH} (CH_3)_3COCH_2CH_3 \\ \quad\quad (29\%) \\ \\ \xrightarrow{H_2O} (CH_3)_3COH \\ \quad\quad (58\%) \\ \\ \xrightarrow{-H^+} (CH_3)_2C{=}CH_2 \\ \quad\quad (13\%) \end{cases}$$

178

Chap. 8

Alkyl Halides;
Nucleophilic
Substitution and
Elimination

Because product mixtures are so frequently obtained, solvolysis reactions are generally *not important* synthetic methods. Such reactions have been studied in great detail over the past several decades, but primarily for the purpose of studying the properties and relative stabilities of carbocations.

8.13
Summation: Elimination Versus Substitution; Unimolecular Versus Bimolecular

As we have seen in the preceding sections, alkyl halides may react with bases or nucleophiles in a variety of ways. We must recognize competitive elimination and substitution and must also consider whether the mechanism is unimolecular or bimolecular. How does one predict which of the four mechanisms S_N1, S_N2, E1, or E2 will operate? Can the chemist control the reaction so as to produce a given product? There are no simple answers to these questions, but we may set out some broad generalizations.

1. Primary alkyl halides that are not branched at C-2 always react by the bimolecular mechanisms S_N2 or E2. With "good" nucleophiles, such as I^-, Br^-, Cl^-, HO^-, RO^-, R_3N, and R_2S, the S_N2 mechanism dominates and high yields of the substitution products are obtained. Elimination may be achieved only by using bulky bases. We shall discuss the latter point in Section 11.5.A when we consider the preparation of alkenes.

Primary alkyl halides that are branched at C-2 give more elimination because the S_N2 process is retarded. With good nucleophiles, substitution is still the principal product. With poor nucleophiles which are strong bases, elimination may dominate. Neopentyl halides undergo substitution only very slowly and cannot undergo elimination because there are no hydrogens on C-2.

2. In the absence of a strong base, tertiary halides react only by the unimolecular mechanisms S_N1 and E1. Mixtures of substitution and elimination products result, and such reactions are usually not synthetically useful. However, the relative amounts of substitution and elimination products are strongly dependent on reaction conditions. In the presence of a strong base, elimination increases.

$$(CH_3)_3CBr + C_2H_5O^-Na^+ \xrightarrow[25°]{C_2H_5OH} (CH_3)_3COCH_2CH_3 + (CH_3)_2C{=}CH_2$$
$$\text{(2 }M\text{)} \qquad\qquad\qquad (7\%) \qquad\qquad (93\%)$$

In this reaction the substitution product and part of the elimination product are produced by the unimolecular pathway, just as above. However, the main product is now the elimination product, most of which is produced by the E2 mechanism. Since the rate of the E2 process increases as we increase the base concentration, whereas the rates of the S_N1 and E1 processes do not, tertiary halides give more elimination at high base concentrations. Such behavior is typical for tertiary alkyl halides. Another example is 2-bromo-2-methylbutane, which gives the corresponding ether by the S_N1 mechanism and a mixture of two elimination products.

$$\underset{\underset{CH_3}{|}}{\overset{\overset{CH_3}{|}}{CH_3CH_2CBr}} + C_2H_5O^- \xrightarrow[25°]{C_2H_5OH} \underset{\underset{CH_3}{|}}{\overset{\overset{CH_3}{|}}{CH_3CH_2COCH_2CH_3}} + \underset{}{\overset{\overset{CH_3}{|}}{CH_3CH{=}CCH_3}} + \underset{}{\overset{\overset{CH_3}{|}}{CH_3CH_2C{=}CH_2}}$$

$$\underbrace{\qquad\qquad\qquad\qquad}_{\substack{\text{substitution}\\ S_N1}} \qquad \underbrace{\qquad\qquad\qquad\qquad\qquad}_{\substack{\text{elimination}\\ \text{E1 and E2}}}$$

TABLE 8.13
Reaction of 2-Bromo-2-methylbutane
with NaOC$_2$H$_5$ in C$_2$H$_5$OH

[C$_2$H$_5$O$^-$], M	Percent Substitution	Percent Elimination
0	64	36
0.02	54	46
0.08	44	56
1.00	2	98

Data for the product composition as a function of base concentration are given in Table 8.13. Thus, in many cases *elimination* of a tertiary alkyl halide is a useful preparative process.

3. Secondary alkyl halides present the most complex behavior. With a good nucleophile, the S$_N$2 mechanism is favored, and good yields of the substitution product may be obtained. With high concentrations of a strong base, the E2 mechanism is favored, and good yields of the elimination products may be obtained.

8.14
Ring Systems

Cyclic systems provide some special features in substitution reactions of alkyl halides. For example, the reaction of amines with alkyl halides is a typical S$_N$2 reaction (Section 24.5).

$$RCH_2NH_2 + R'CH_2Br \longrightarrow RCH_2\overset{+}{N}H_2CH_2R' + Br^-$$

When both functional groups are present in the same molecule, the reaction is an **intramolecular** S$_N$2 reaction, which creates a ring. Ring formation necessarily has been initiated at the transition state; hence, the relative energies of transition states depend on the ring size. Relative rates for reactions of ω-bromoalkylamines, Br(CH$_2$)$_{n-1}$NH$_2$, are given in Table 8.14. 1-Amino-4-bromobutane, which gives a

TABLE 8.14
Relative Rates of Cyclization
of Aminoalkyl Halides

$$Br(CH_2)_{n-1}NH_2 \longrightarrow (CH_2)_{n-1}NH_2^+ \; Br^-$$

n (Ring Size)	Relative Rate
3	0.1
4	0.002
5	100
6	1.7
7	0.03
10	10^{-8}
12	10^{-5}
14	3×10^{-4}
15	3×10^{-4}
17	6×10^{-4}

180

Chap. 8

Alkyl Halides;
Nucleophilic
Substitution and
Elimination

TABLE 8.15
Relative Rates of Reaction of
Alkyl Bromides with Lithium
Iodide in Acetone

Alkyl Group	Relative Rate
Isopropyl	1.0
Cyclopropyl	<0.0001[a]
Cyclobutyl	0.008
Cyclopentyl	1.6
Cyclohexyl	0.01
Cycloheptyl	1.0
Cyclooctyl	0.2

[a] Approximate upper limit; no reaction
was detected.

five-membered ring, is the most reactive compound, followed by 1-amino-5-bromopentane, which gives a six-membered ring. We see that the stability of the ring is not the only factor—the probability that the ends can get together is also important. The three-membered ring, for example, has high energy, but the functional groups are so close together that ring formation is relatively probable. The pattern shown in Table 8.14 is common. The general order of ring formation in intramolecular S_N2 reactions is $5 > 6 > 3$ with other rings being formed much more slowly.

Large rings are relatively unstrained, but the groups are so far apart that their probability of getting together for reaction is low. Indeed, it becomes more probable for reaction of one amino group to occur with another bromide. That is, **intermolecular** S_N2 reaction is an important side reaction for such cases unless conditions of high dilution are used. At ordinary concentrations the reaction product is a polymer chain.

Displacement reactions on cyclic compounds show large effects of ring size. Table 8.15 summarizes some relative rates of reaction of alkyl bromides with lithium iodide in acetone.

Halocyclopropanes do not undergo S_N2 reactions. At the transition state of an S_N2 reaction, the central carbon has sp^2-hybridization in which the normal bond angle is 120°. The imposition of such an increased bond angle on a cyclopropyl ring would result in additional bond angle strain (Figure 8.14). The same effect is apparent in the slow reactions of cyclobutyl systems, but since the bond angle strain is lower, the effect is not as great.

Cyclopentyl compounds undergo S_N2 reactions at rates comparable to open-chain systems. The reactions usually go in good yield.

Cyclohexyl compounds react rather slowly in S_N2 reactions. Ring strain does

FIGURE 8.14. *Transition state for S_N2 reaction on a cyclopropyl halide.*

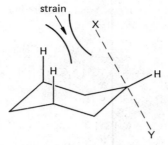

FIGURE 8.15. *Transition state for* S_N2 *reaction with cyclohexyl compounds.*

not appear to be an important factor for five- and six-membered rings, but a novel kind of strain involving axial hydrogens does appear to be significant for cyclohexyl systems (Figure 8.15). Because the displacement reaction has a reduced rate, E2 reactions are frequently important side reactions with cyclohexyl compounds and often dominate. For larger rings, the reactions are roughly comparable to open-chain systems.

PROBLEMS

1. Write Newman projections for the conformations of 1,1,2-trichloroethane. Two of these are the same, and the third is different. The two types of conformation differ in energy by 2.6 kcal mole^{-1} in the vapor phase. Which is the more stable? This energy difference reduces to 0.2 kcal mole^{-1} in the liquid. Explain why. Interconversion of the two similar conformations requires about 2 kcal mole^{-1}, but conversion of either to the third structure requires about 5 kcal mole^{-1}. Explain why these two rotation barriers differ.

2. (R)-2-Butanol, labeled with ^{18}O, is subjected to the following sequence of reactions.

$$\underset{\text{CH}_3\text{CH}_2\overset{\displaystyle ^{18}\text{OH}}{\underset{|}{\text{C}}}\text{HCH}_3}{} \xrightarrow{\text{CH}_3\text{SO}_2\text{Cl}} \underset{\text{CH}_3\text{CH}_2\overset{\displaystyle ^{18}\text{OSO}_2\text{CH}_3}{\underset{|}{\text{C}}}\text{HCH}_3}{} \xrightarrow[\substack{\text{H}_2\text{O} \\ \text{dioxane}}]{\text{OH}^-} \underset{\text{CH}_3\text{CH}_2\overset{\displaystyle \text{OH}}{\underset{|}{\text{C}}}\text{HCH}_3}{} + \text{CH}_3\text{SO}_2{-}^{18}\text{O}^-$$

What is the absolute configuration of the product?

3. 2-Bromo-, 2-chloro- and 2-iodo-2-methylbutanes react at different rates with pure methyl alcohol but produce the same mixture of 2-methoxy-2-methylbutane, 2-methyl-1-butene, and 2-methyl-2-butene as products. Explain these results briefly in terms of the reaction mechanism.

4. Explain each of the following observations.
(a) (S)-3-Bromo-3-methylhexane reacts in aqueous acetone to give racemic 3-methyl-3-hexanol.

$$\underset{\overset{\displaystyle |}{\underset{\displaystyle \text{CH}_3}{\text{CH}_3\text{CH}_2\text{CH}_2\overset{\displaystyle \text{Br}}{\text{C}}\text{CH}_2\text{CH}_3}}}{} \xrightarrow[\text{acetone}]{\text{H}_2\text{O}} (\pm)\underset{\overset{\displaystyle |}{\underset{\displaystyle \text{CH}_3}{\text{CH}_3\text{CH}_2\text{CH}_2\overset{\displaystyle \text{OH}}{\text{C}}\text{CH}_2\text{CH}_3}}}{}$$

(b) Reaction of *trans*-1-chloro-4-methylcyclohexane with potassium iodide in acetone gives a mixture of 32% *cis*-1-iodo-4-methylcyclohexane and 68% of the *trans* isomer.

5. For each of the following pairs of reactions, predict which one is faster and explain why.

(a) $(CH_3)_2CHCH_2Cl + N_3^- \xrightarrow{C_2H_5OH} (CH_3)_2CHCH_2N_3 + Cl^-$

$(CH_3)_2CHCH_2I + N_3^- \xrightarrow{C_2H_5OH} (CH_3)_2CHCH_2N_3 + I^-$

(b) $(CH_3)_3CBr \xrightarrow[\Delta]{H_2O} (CH_3)_3COH + HBr$

$(CH_3)_2CHBr \xrightarrow[\Delta]{H_2O} (CH_3)_2CHOH + HBr$

(c) $CH_3CH_2\overset{\overset{\displaystyle CH_3}{|}}{C}HCH_2Br + CN^- \longrightarrow CH_3CH_2\overset{\overset{\displaystyle CH_3}{|}}{C}HCH_2CN + Br^-$

$CH_3CH_2CH_2CH_2Br + CN^- \longrightarrow CH_3CH_2CH_2CH_2CN + Br^-$

(d) $CH_3CH_2Br + SH^- \xrightarrow[\text{(solvent)}]{CH_3OH} CH_3CH_2SH + Br^-$

$CH_3CH_2Br + SH^- \xrightarrow[\text{(solvent)}]{(CH_3)_2NCHO} CH_3CH_2SH + Br^-$

(e) $(CH_3)_2CHBr + NH_3 \xrightarrow{CH_3OH} (CH_3)_2CH\overset{+}{N}H_3\ Br^-$

$CH_3CH_2CH_2Br + NH_3 \xrightarrow{CH_3OH} CH_3CH_2CH_2\overset{+}{N}H_3\ Br^-$

(f) $CH_3I + Na^+OH^- \xrightarrow{H_2O} CH_3OH + Na^+I^-$

$CH_3I + Na^+SH^- \xrightarrow{H_2O} CH_3SH + Na^+I^-$

(g) $CH_3Br + (CH_3)_3N \longrightarrow (CH_3)_4N^+\ Br^-$

$CH_3Br + (CH_3)_3P \longrightarrow (CH_3)_4P^+\ Br^-$

(h) $SCN^- + CH_3CH_2Br \xrightarrow{aq.\ C_2H_5OH} CH_3CH_2SCN + Br^-$

$SCN^- + CH_3CH_2Br \xrightarrow{aq.\ C_2H_5OH} CH_3CH_2NCS + Br^-$

(i) $N_3^- + (CH_3)_2CHCH_2Br \xrightarrow{C_2H_5OH} (CH_3)_2CHCH_2N_3 + Br^-$

$C_6H_5S^- + (CH_3)_2CHCH_2Br \xrightarrow{C_2H_5OH} (CH_3)_2CHCH_2SC_6H_5 + Br^-$

[*Note:* $pK_a(HN_3) \approx pK_a(C_6H_5SH)$.]

(j) $CH_3CO_2^- + $ [Cl–cyclobutane] $\longrightarrow$ [CH₃COO–cyclobutane] $+ Cl^-$

$CH_3CO_2^- + $ [Cl–cyclopentane] $\longrightarrow$ [CH₃COO–cyclopentane] $+ Cl^-$

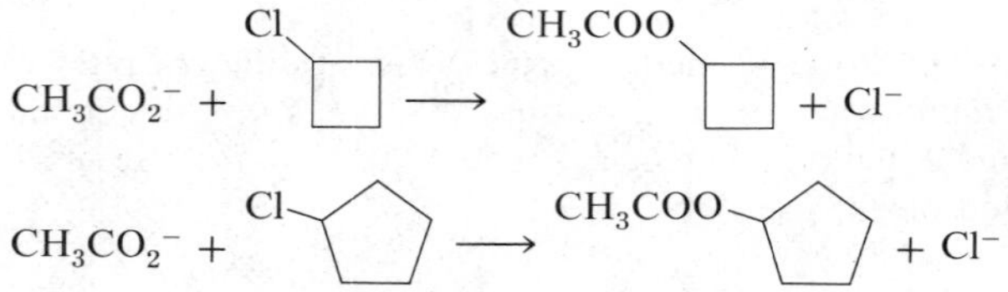

(k) $^-OCH_2CH_2Cl \longrightarrow \overset{\overset{\displaystyle O}{\diagup\!\diagdown}}{CH_2\!-\!CH_2} + Cl^-$

$^-OCH_2CH_2CH_2Cl \longrightarrow \begin{matrix} O\!-\!\!-\!CH_2 \\ | \qquad\quad | \\ CH_2\!-\!\!-\!CH_2 \end{matrix} + Cl^-$

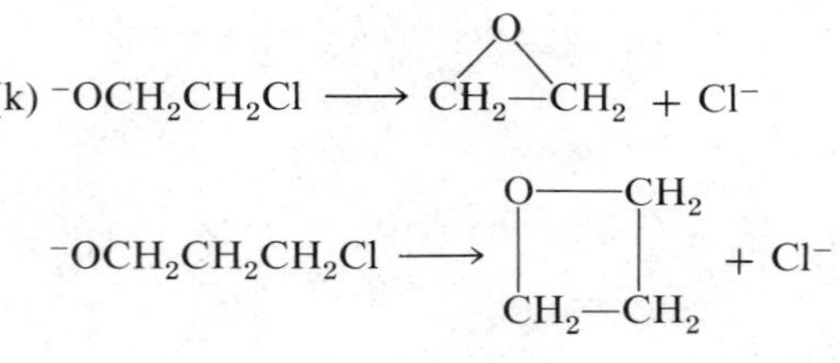

6. Of the following nucleophilic substitution reactions, which ones will probably occur and which will probably not occur or be very slow. Explain.

(a) $CH_3CN + I^- \longrightarrow CH_3I + CN^-$

(b) $CH_3F + Cl^- \longrightarrow CH_3Cl + F^-$

(c) $(CH_3)_3COH + NH_2^- \longrightarrow (CH_3)_3CNH_2 + OH^-$

(d) $CH_3OSO_2OCH_3 + Cl^- \longrightarrow CH_3Cl + CH_3OSO_3^-$

(e) $CH_3NH_2 + I^- \longrightarrow CH_3I + NH_2^-$

(f) $CH_3CH_2I + N_3^- \longrightarrow CH_3CH_2N_3 + I^-$

(g) $CH_3CH_2OH + F^- \longrightarrow CH_3CH_2F + OH^-$

(h) cyclopropyl bromide $+ N(CH_3)_3 \longrightarrow C_3H_5N(CH_3)_3^+ \ Br^-$

7. Give a specific example of two related reactions having different rates for which each of the following is the principal reason for the relative reactivities.

(a) The less basic leaving group is more reactive.

(b) Sulfur is more polarizable than nitrogen.

(c) Tertiary carbocations are more stable than secondary carbocations

(d) Steric hindrance.

(e) Protic solvents can form hydrogen bonds.

(f) E2 elimination is favored by less polarizable bases.

8. Of the following statements, which are true for nucleophilic substitutions occurring by the S_N2 mechanism?

(a) Tertiary alkyl halides react faster than secondary.

(b) The absolute configuration of the product is opposite to that of the reactant when an optically active substrate is used.

(c) The reaction shows first-order kinetics.

(d) The rate of reaction depends markedly on the nucleophilicity of the attacking nucleophile.

(e) The probable mechanism involves only one step.

(f) Carbocations are intermediates.

(g) The rate of reaction is proportional to the concentration of the attacking nucleophile.

(h) The rate of reaction depends on the nature of the leaving group.

9. Answer problem 8 for nucleophilic substitutions occurring by the S_N1 mechanism.

10. The reaction of methyl bromide with methylamine to give dimethylammonium bromide is a typical S_N2 reaction that shows second-order kinetics.

$$CH_3Br + CH_3NH_2 \longrightarrow (CH_3)_2NH_2^+ \ Br^-$$

However, the analogous cyclization of 4-bromobutylamine shows first-order kinetics. Explain.

The foregoing **intramolecular** displacement reaction is a useful method for making **cyclic amines.** However, a competing side reaction is the intermolecular displacement

$$Br(CH_2)_4NH_2 + Br(CH_2)_4NH_2 \longrightarrow Br(CH_2)_4\overset{\overset{\displaystyle H}{|}}{\underset{\underset{\displaystyle H}{|}}{N^+}}(CH_2)_4NH_2 \ Br^-$$

Suggest a way in which this side reaction may be minimized.

184

Chap. 8

**Alkyl Halides;
Nucleophilic
Substitution and
Elimination**

11. Consider the reaction of isopropyl iodide with various nucleophiles. For each pair, predict which will give the larger substitution/elimination ratio.
(a) SCN^- or OCN^- (b) I^- or Cl^-
(c) $N(CH_3)_3$ or $P(CH_3)_3$ (d) CH_3S^- or CH_3O^-

12. HCN has $pK_a = 9.21$; acetic acid has $pK_a = 4.76$.
(a) What is the difference in the standard free energies ($\Delta\Delta G°$) for these two acid-base equilibria?
(b) What is the equilibrium constant and $\Delta G°$ for the reaction

$$HCN + CH_3CO_2^- \longrightarrow CN^- + CH_3COOH$$

(c) The second-order rate constants, k_2, for reaction with methyl iodide in methyl alcohol at 25° for cyanide ion and acetate ion are, respectively, 6.5×10^{-4} and 2.7×10^{-6} M^{-1} sec^{-1}. What is the relative rate, $k_2(CN^-)/k_2(CH_3CO_2^-)$? To what value of $\Delta\Delta G^{\ddagger}$ does this relative rate correspond?

13. (a) In Figure 8.10 the nucleophiles with second-period elements form an approximately linear relation. What is the slope of this correlation line?* What does the slope suggest about the amount of charge remaining on the attacking group?
(b) Methyl iodide is 40 times more reactive than methyl chloride toward thiosulfate ion in water solution. HI is about 3.0 pK_a units more acidic than HCl. What do these data imply concerning the amount of negative charge on the leaving group in the transition state of this S_N2 reaction?

14. How might cyclic ether A be prepared from 1,5-dibromopentane? What dibromide would be required for the synthesis of cyclic ether B by this method? What special reaction conditions are necessary for the preparation of cyclic ether B?

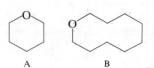

A   B

15. Optically active 3-bromo-2-butanol is treated with KOH in methyl alcohol to obtain an optically inactive product having the formula C_4H_8O. What is the structure of this material?

16. When a solution of *cis*-1-*t*-butyl-4-chlorocyclohexane in ethanol is refluxed for several hours, the major product is found to be *trans*-1-*t*-butoxy-4-ethoxycyclohexane. However, if the solution is also made 2.0 M in sodium ethoxide, the major product after the same treatment is found to be 4-*t*-butylcyclohexene. Explain.

17. Show how the following compounds may be prepared from alkyl halide.
(a) $CH_3CH_2CH_2SH$ (b) $(CH_3)_2CHCH_2CH_2CN$
(c) $CH_3CH_2CH_2OCH_3$ (d) $CH_3CH_2CH_2OH$
(e) CH_3ONO_2 (f) $CH_3CH_2CH_2CH_2N_3$

18. Suggest an explanation for each of the following observations.
(a) Compound C reacts faster by the S_N2 mechanism than the compound D.

*If you have access to a small computer or programmable calculator, calculate the least squares slope and the standard deviation and correlation coefficient of the slope.

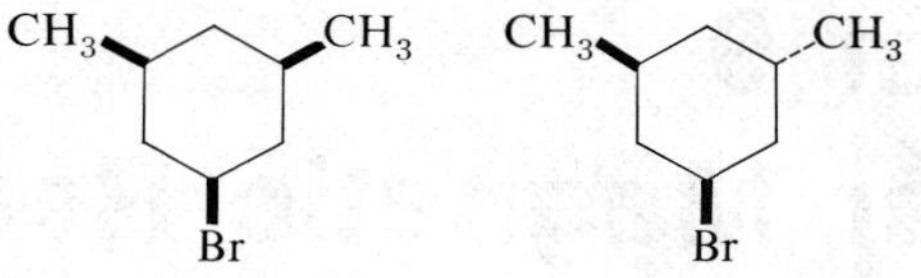

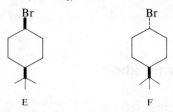

C   D

(b) Compound E reacts faster by the S_N1 mechanism than compound F.

Br   Br

E   F

(c) *t*-Butyl chloride reacts in methanol solution at 25° eight times faster than it does in ethanol solution.

Chapter 9
Nuclear Magnetic Resonance Spectroscopy

9.1
Structure Determination

Structure determination is one of the fundamental operations in chemistry. How does the chemist determine the structure of a compound? Imagine that we carry out a reaction between propane (C_3H_8) and chlorine, both of which are gases at room temperature. After the reaction is completed, we obtain a liquid product. We distill this liquid and obtain two main fractions, one boiling at 36° and one at 47°. These two liquids are obviously reaction products because they have different physical properties (b.p.) from the reactants. What are they?

As a first step, we might perform an elemental analysis (Section 3.4) and determine their empirical formulas. When we do this, we find that they both have the formula C_3H_7Cl. We conclude that a reaction has occurred in which a hydrogen has been replaced by a chlorine and that two isomeric products have been produced in the reaction. Since there are only two types of hydrogen in propane, we can write structures for the two products. One is 1-chloropropane and the other is 2-chloropropane. The reaction is therefore

$$CH_3CH_2CH_3 + Cl_2 \longrightarrow CH_3CH_2CH_2Cl + CH_3\overset{\displaystyle Cl}{\underset{\displaystyle |}{C}}HCH_3$$

But which is which? Is the product that boils at 36° 1-chloropropane or 2-chloropropane?

One way to answer this question is to look up the boiling points of 1-chloropropane and 2-chloropropane in a handbook. But suppose for a moment that the two compounds have never been prepared before and their boiling points are not known. Another way to answer our question would be to convert the two isomers into compounds that *are* known. For example, suppose we have samples of 1-propanol and 2-propanol and that we know which is which. We take our two C_3H_7Cl isomers and treat each with aqueous KOH. The isomer that boils at 47° gives 1-propanol and the isomer that boils at 36° gives 2-propanol. (In addition, both isomers give some propene.) If we assume that the nucleophilic substitution occurs without rearrangement, then we may assign structures to the two isomers on the basis of their conversion to products of known structure.

$$CH_3CH_2CH_2Cl + OH^- \longrightarrow CH_3CH_2CH_2OH + CH_3CH{=}CH_2$$

1-chloropropane   1-propanol
b.p. 47°

$$CH_3\overset{\displaystyle Cl}{\underset{\displaystyle |}{C}}HCH_3 + OH^- \longrightarrow CH_3\overset{\displaystyle OH}{\underset{\displaystyle |}{C}}HCH_3 + CH_3CH{=}CH_2$$

2-chloropropane   2-propanol
b.p. 36°

A more direct method of structure determination involves a consideration of the physical properties of the compound of unknown structure. The most useful properties for this purpose are spectra. Spectroscopy is a powerful tool for structure determination. There are many different types of spectroscopy. In the following section we shall have a brief introduction to spectroscopy generally, and then we shall take up one specific type of spectroscopy, nuclear magnetic resonance spectroscopy, in detail.

9.2
Introduction to Spectroscopy

Molecules are associated with several different types of **motion.** The entire molecule rotates, the bonds vibrate, and even the electrons move—albeit so rapidly that we generally deal only with electron density distributions. Each of these kinds of motion is **quantized.** That is, the molecule can exist only in distinct states that correspond to discrete energy contents. Each state is characterized by one or more quantum numbers. The energy difference between two such states, ΔE, is related to a light frequency ν by Planck's constant h (Figure 9.1).

$$\Delta E = h\nu \qquad (9\text{-}1)$$

Spectroscopy is an experimental process in which the energy differences between allowed states of a system are measured by determining the frequencies of the corresponding light absorbed.

The energy difference between the different quantum states depends on the type of motion involved. Thus, the wavelength of light required to bring about a transition is different for the different types of motion. That is, each type of motion corresponds to the absorption of light in a different region of the electromagnetic spectrum. Because the wavelengths required are so vastly different, different instrumentation is required for each spectral region. For example, the energy differences between molecular rotational states are rather small, on the order of 1 cal mole^{-1}. Light having this energy has a wavelength of about 3 cm and is called microwave radiation. The energy spacings of molecular rotational states depend on bond distances and angles and the atomic masses of the bonded atoms (moments of inertia). Hence, **microwave spectroscopy** is a powerful tool for precise structure determination. However, the technique must be applied in the vapor phase and it is restricted to rather simple molecules. Although it is an important technique in the hands of a specialist, it is not commonly used by organic chemists.

Energy differences between different states of bond vibration are of the order 1–10 kcal mole^{-1} and correspond to light having wavelengths of 30–3 μ (1 $\mu \equiv 10^{-3}$ mm). This is the **infrared** region of the spectrum. Infrared spectrometers are relatively inexpensive and easy to use, and infrared spectroscopy is an important technique in organic chemistry. It is used mainly to determine which

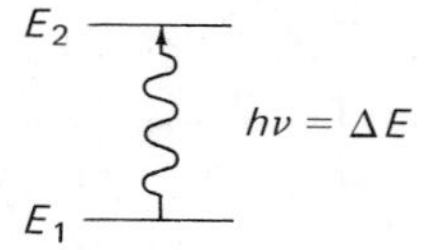

$$E_2 \underline{\qquad} \qquad h\nu = \Delta E \qquad E_1 \underline{\qquad}$$

FIGURE 9.1. *Light of frequency ν corresponds to an energy difference ΔE between states corresponding to energies E_1 and E_2.*

188

Chap. 9

**Nuclear
Magnetic
Resonance
Spectroscopy**

functional groups are present in a compound. We will study its use in more detail in Chapter 14.

Different electronic states of organic compounds correspond to energies in the visible (4000–7500 Å; 1 Å $\equiv 10^{-8}$ cm; 70–40 kcal mole^{-1}) and ultraviolet (1000–4000 Å; 300–70 kcal mole^{-1}) regions of the electromagnetic spectrum. Spectrometers for this region are also common, and ultraviolet-visible spectroscopy is an important technique in organic chemistry, especially for conjugated systems. Such compounds will be discussed in detail later, and our study of this spectroscopy will be deferred to Chapter 21.

9.3
Nuclear Magnetic Resonance

Nuclear magnetic resonance (nmr) spectroscopy has only been important in organic chemistry since the mid-1950s, yet in this relatively brief time it has taken its place as one of our most important spectroscopic tools. In nmr spectroscopy, a solution of the sample is inserted in the instrument—actually the sample tube is fitted precisely between the poles of a powerful magnet—and the spectrum is recorded as a curve. A typical example is the nmr spectrum of 1-chloropropane shown in Figure 9.2. Some appreciation of the usefulness of this technique can be sensed by comparing the spectrum of the isomeric compound, 2-chloropropane, shown in Figure 9.3. In this chapter we will develop the rules for interpreting such spectra. We will find it rather simple, for example, to *deduce* the structures of 1-chloropropane and 2-chloropropane from their nmr spectra. It is possible to treat the nmr spectrometer as a "magic box" and simply memorize a few rules that suffice for deducing the structure of a compound from its spectrum. In this chapter we will also go into some of the theoretical background of nmr spectroscopy to show why the rules take the form they do.

Nuclear magnetic resonance spectroscopy differs from the spectroscopic techniques discussed in the previous section in that the differences in energy in the

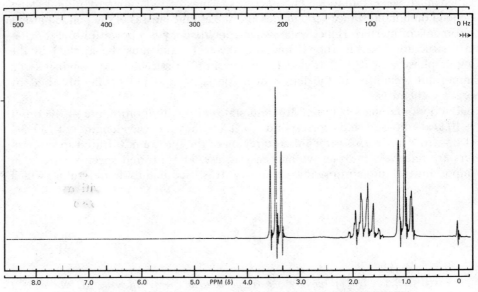

FIGURE 9.2. *Nmr spectrum of 1-chloropropane,* $CH_3CH_2CH_2Cl$.

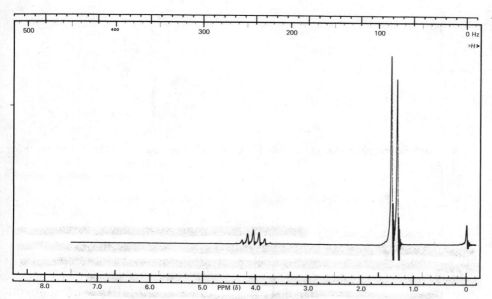

FIGURE 9.3. *Nmr spectrum of 2-chloropropane,* $CH_3CHClCH_3$.

states being examined are created by a magnetic field. That is, the molecules are placed in a powerful magnetic field to create different energy states which are then detected by absorption of light of the appropriate energy. In the absence of the magnetic field, these different states all have nearly the same energy.

The motion involved in nmr spectroscopy is that of **nuclear spin.** The nucleus of many atoms behaves as though it were spinning on an axis. Since they are positively charged, such nuclei follow the physical laws of spinning charged particles. A moving charge, positive or negative, is associated with a magnetic field. Consequently, the spinning nuclei behave as though they were tiny bar magnets; that is, such nuclei have **magnetic moments.** These magnetic moments are oriented in random fashion in field-free space, but they have the important quantization property that in a magnetic field only certain discrete orientations are allowed. For some important nuclei, ^1H (but not ^2H), ^{13}C (but not ^{12}C), and ^{19}F (the only common fluorine isotope), the nuclear spin can have only two alternative values associated with the quantum numbers, $+\frac{1}{2} (= \alpha)$ and $-\frac{1}{2} (= \beta)$. When these nuclei are placed in a magnetic field, their magnetic moments tend either to align with the field (corresponding to α-spin) or against the field (corresponding to β-spin) (Figure 9.4).

In an applied field, a magnetic moment tends to align with the field (for example, a compass needle in the earth's magnetic field). A magnet aligned against the magnetic field is in a higher energy state than one aligned with the field. For ^1H, ^{13}C, and ^{19}F and β-spin state with the magnetic moment aligned against the field corresponds to a higher energy state than the α-spin state. If the system is irradiated with light of the proper frequency or wavelength, a nucleus with α-spin can absorb a light quantum and be converted to the higher-energy β-spin state, a process colloquially described as "flipping the spin" (Figure 9.5).

To recapitulate, the nuclei of ^1H, ^{13}C, and ^{19}F have spinning nuclei with spins of $\pm\frac{1}{2}$. Because of the restrictions imposed by quantization, only two orientations are permitted for these nuclei in a magnetic field: **α-spin ($+\frac{1}{2}$), nuclear magnetic moment aligned with the applied magnetic field, lower energy; β-spin ($-\frac{1}{2}$), nuclear magnetic moment aligned against the applied magnetic field, higher energy.**

190

Chap. 9

**Nuclear
Magnetic
Resonance
Spectroscopy**

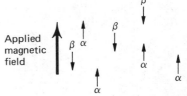

nuclear magnetic moments with no magnetic field nuclear magnetic moments in a magnetic field
Nuclei with α-spin are aligned with the field;
those with β-spin are aligned against the field.

FIGURE 9.4. *Orientation of nuclear magnetic moments.*

Many nuclei have no spin. All even-even nuclei (those having an even number of protons and an even number of neutrons) are in this class. In this important class, which includes ^{12}C and ^{16}O, individual pairs of protons and neutrons have opposed spins so that the net spin of the nucleus as a whole is zero. Other nuclei have three or more possible spin states in a magnetic field. ^{14}N is an important example that has three such states. We will not consider such cases, but will restrict our attention primarily to those nuclei that have spin of $\pm\frac{1}{2}$: ^{1}H, ^{13}C, ^{19}F, and so on. The energy difference between the two states is given by

$$\nu = \frac{\gamma H}{2\pi} \qquad (9\text{-}2)$$

in which **H** is the magnetic field strength *at the nucleus* and γ is the magnetogyric ratio of the nucleus. This quantity is the ratio of the angular momentum (from the rotating nuclear mass) and the magnetic moment (from the rotating nuclear charge) and is characteristic and different for each nucleus. That is, the energy difference between the α- and β-spin states in a magnetic field is proportional to the strength of the magnetic field with a proportionality constant that is characteristic of the nucleus (Figure 9.6).

For the proton γ has the value 2.6753 × 10⁴ radians sec⁻¹ gauss⁻¹. When **H** is given in units of gauss, the frequency ν is given in units of cycles per second or Hertz, Hz. This energy unit is used commonly in nmr, and it may be converted to the more familiar units of cal mole⁻¹ by the conversion

$$E \text{ (cal mole}^{-1}) = 9.54 \times 10^{-11} \nu \text{ (Hz)} \qquad (9\text{-}3)$$

According to equation (9-2), the energy differences involved are proportional to the magnetic field and are exceedingly small. For example, for an isolated hydrogen atom in a magnetic field of 14,092 gauss, the energy difference between α- and β-spin states is given by $\Delta E = (26,753)(14,092)/2\pi = 60 \times 10^6$ Hz = 60 MHz (megaHertz). From equation (9-3) this energy value is equivalent to only 0.0057 cal mole⁻¹ (not kcal!). The frequency of 60 MHz corresponds to a wavelength of 500 cm and is in the radio region of the electromagnetic spectrum. A field of 14,000 gauss is a rather strong magnetic field but one that is readily

magnetic
field ↑ ↑ $\xrightarrow{h\nu}$ ↑ ↓
α β

FIGURE 9.5. *Absorption of light of proper frequency changes the nuclear spin state.*

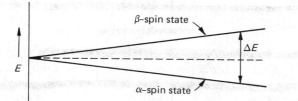

FIGURE 9.6. *The energy difference ΔE between the α- and β-spin states is a function of the magnetic field at the nucleus.*

accessible with modern technology. The 60 MHz nmr spectrometer is now relatively common, and commercial instruments are also available with larger field strengths in which the proton "flip" corresponds to 100 MHz. With superconducting magnets, 220 MHz, 360 MHz, and even 500 MHz instruments are now available.

9.4
Chemical Shift

If nmr spectroscopy related only to free protons floating in a magnetic field, we would hardly expect to find the thousands of nmr spectrometers now spread in laboratories throughout the world. It is when we look at magnetic phenomena in bonds to protons in molecules that we find why nmr is such an invaluable asset to the organic chemist. A proton in a molecule is surrounded by a cloud of electronic charge. In a magnetic field electrons move in such a way that their motion induces a magnetic field characterized by a magnetic moment that is **opposed** to the applied field. Consequently, the net magnetic field at the hydrogen atom is slightly less than the applied field (Figure 9.7).

A frequent source of confusion is the difference between electron flow and electrical current. By a convention established before the discovery of electrons, current flows from anode (+) to cathode (−), exactly the opposite of the actual movement of electrons. The figures in this book (e.g., Figure 9.7) represent the actual flow of electrons and are the reverse of the direction of a positive electrical current.

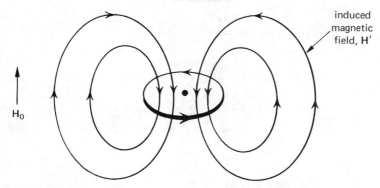

FIGURE 9.7. *An external magnetic field induces an electron flow in an electron cloud that, in turn, induces a magnetic field. At the nucleus the induced field opposes the external field.*

192

Chap. 9

**Nuclear
Magnetic
Resonance
Spectroscopy**

The magnetic field **H** experienced by the nucleus is therefore

$$\mathbf{H} = \mathbf{H}_0 - \mathbf{H}' \qquad (9\text{-}4)$$

where $\mathbf{H}_0$ is the applied field and $\mathbf{H}'$ is the induced field. Because the nucleus experiences a smaller magnetic field than that applied externally, it is said to be **shielded.** This particular type of shielding is called **diamagnetic shielding.** If we are irradiating the proton with radiowaves of exactly 60 MHz frequency, the change of $\alpha \rightarrow \beta$ energy states of "spin flipping" requires a field of 14,092 gauss *at the proton.* However, since the nucleus is shielded by electrons, *the applied field has to be made somewhat higher than 14,092 gauss* in order for the field at the nucleus to have the **resonance** value of 14,092 gauss, the field strength that corresponds to the radio frequency required to produce spin-flipping. Protons in different electronic environments experience different amounts of shielding, and the resonance absorption of light energy will occur at different values for the applied field or irradiating light frequency. These changes are referred to as **chemical shifts.**

A nuclear magnetic resonance spectrometer is arranged schematically as shown in Figure 9.8. A liquid sample or solution contained in a narrow glass tube is put between the poles of the powerful magnet. The magnetic field creates the two energy states for various hydrogen nuclei in the sample. The sample is irradiated with radio waves from a simple coil. In one mode of operation we fix the radio frequency at, say, 60 MHz. We then vary the magnetic field, and as the field at each kind of proton reaches the **resonance** value, energy is absorbed from the radio waves as the nuclear spins "flip," and this absorption is measured and recorded on a graph.

For example, 1,2,2-trichloropropane, $CH_3CCl_2CH_2Cl$, gives the spectrum shown in Figure 9.9. This spectrum consists of two sharp peaks corresponding to the methylene group and the methyl group. The CH_2 group is attached to chlorine, an electronegative element, which withdraws electrons from carbon and hydrogen. Since there is less electron density around the methylene protons, the diamagnetic shielding of these protons is less than it is for the methyl protons. The induced magnetic field is therefore lower at CH_2 than at CH_3, and the applied field must be increased less in order to achieve resonance. Consequently, the methylene protons appear to the left or **downfield** compared to the methyl protons. Note that the difference is exceedingly minute—about 0.004 gauss compared to a total applied field of about 14,000 gauss.

Alternatively, we can keep the magnetic field constant and vary the frequency

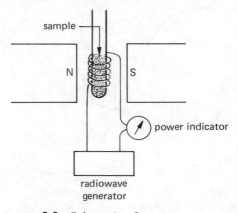

sample

N   S

power indicator

radiowave
generator

FIGURE 9.8. *Schematic of an nmr spectrometer.*

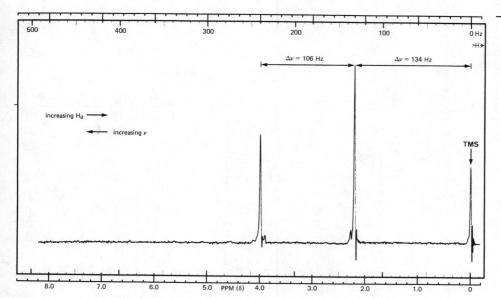

FIGURE 9.9. *Nmr spectrum of* $CH_3CCl_2CH_2Cl$ *at 60 MHz.*

of the radio electromagnetic irradiation. The lower the electronic shielding of the nucleus, the higher the effective magnetic field at a proton and the higher the frequency required to reach the resonance condition. If we plot the frequency increasing from right to left, the resulting spectrum looks exactly like Figure 9.9. The methylene hydrogens now appear at higher frequency than the methyl hydrogens, the frequency difference being 106 Hz. Frequency differences can be measured more precisely than differences in magnetic field strength. Consequently, the difference in peaks is always given in frequency units, regardless of the specific mode of operation of the nmr spectrometer. The kind of language in common use is illustrated by the statement that in Figure 9.9 the methylene group appears downfield with a frequency difference of 106 Hz. Note again that this is a small difference between large numbers; if the methyl group resonates at 60,000,000 Hz, the methylene is at 60,000,106 Hz! Since these *absolute numbers* are difficult to reproduce, in practice we compare differences relative to a standard.

The standard compound used for most proton nmr spectra is tetramethylsilane, $(CH_3)_4Si$, commonly abbreviated **TMS**, a volatile liquid, b.p. 26.5°. It is inert to most reagents and soluble in most organic liquids. A small amount has been added to our sample of 1,2,2-trichloropropane, and it gives rise to the peak at the far right in Figure 9.9. All of the hydrogens in TMS are equivalent and give rise to the single sharp line 134 Hz upfield from the methyl of the trichloropropane. Furthermore, silicon is electropositive relative to carbon and tends to donate electron density to the methyl groups, thereby increasing their shielding. The relatively high shielding of the protons in TMS causes it to resonate upfield from most other protons commonly encountered in organic compounds.

When the spectrum is run on a spectrometer operating at 23,487 gauss, the resonance frequency of hydrogen is 100 MHz. The large magnetic field induces a large electron current, which causes a larger diamagnetic shielding at the nucleus. The difference in diamagnetic shielding is proportionally larger and the peaks spread apart, as shown in Figure 9.10. The chemical shift of the methyl group is now 223 Hz downfield from TMS instead of 134 Hz. Because different nmr instru-

194

Chap. 9

**Nuclear
Magnetic
Resonance
Spectroscopy**

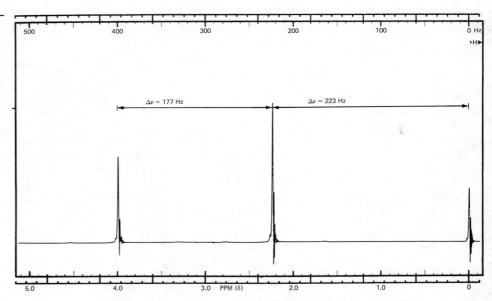

FIGURE 9.10. *Nmr spectrum of* $CH_3CCl_2CH_2Cl$ *at 100 MHz.*

ments are in common use, it is convenient to define a unitless measure that is independent of field strength. The unit used is δ. It is simply the ratio of the chemical shift of the resonance in question, in Hertz, to the total light frequency used. Since the resulting number is small, it is multiplied by 10^6. Thus, δ has the units of parts per million (ppm) and represents a chemical shift downfield (higher frequency) from TMS.

$$\delta_i = \frac{\nu_i - \nu_{\text{TMS}}}{\nu_0} \times 10^6 \text{ ppm} \tag{9-5}$$

In equation (9-5) δ_i is the chemical shift of proton i, ν_i is the resonance frequency of that proton, ν_{TMS} is the resonance frequency of TMS, and ν_0 is the operating frequency of the instrument. Thus, for $CH_3CCl_2CH_2Cl$, $\delta(CH_3) = 134/60 = 2.23$ ppm, and $\delta(CH_2) = 240/60 = 4.00$ ppm. If a resonance is upfield from TMS, its δ value has a negative sign.

In the preceding discussion, it seemed quite natural to expect both hydrogens in the methylene group to absorb in the same place because they appear to be equivalent. However, if we examine the structure of 1,2,2-trichloropropane in more detail, this equivalence is not so apparent. The compound actually exists as an equilibrium mixture of three conformations, symbolized by the following Newman projections.

In structure A both hydrogens, H_a and H_b, are clearly equivalent, but this is not the case in B and C. In B, for example, H_b is flanked by two chlorines and would be expected to be deshielded relative to H_a. Why, then, do we not see two or more peaks for these hydrogens?

The answer comes from the Heisenberg uncertainty principle of quantum mechanics. One expression of this principle is

$$\Delta E \, \Delta t \approx h/2\pi \qquad \text{or} \qquad \Delta \nu \, \Delta t \approx 1/2\pi$$
$$(\text{since } \Delta E = h\Delta \nu)$$

where $\Delta \nu$ and Δt are the uncertainties in energy and time in units of Hertz and seconds, respectively. That is, we cannot know precisely both the energy and the lifetime of a given state. The longer-lived the state, the more precisely can its energy content be evaluated. In our nmr case above, suppose that $\delta(H_a)$ and $\delta(H_b)$ differ by 1 ppm. This amount in a 60 MHz instrument corresponds to an energy difference of 60 Hz or 6×10^{-9} cal mole^{-1}, an exceedingly small energy quantity. In order to measure this small difference for H_a and H_b as separate states, they would have to have lifetimes in each conformation of at least

$$\Delta t \approx 1/(2\pi \, \Delta E) = 1/(2\pi \cdot 60) = 0.0027 \text{ sec}$$

But with an energy barrier of only 3–4 kcal mole^{-1} between one conformer and another, the average lifetime of a given conformation is only about 10^{-10} to 10^{-11} sec! (See Section 5.3.) In other words, the lifetime of a given methylene hydrogen in the magnetic environment of a given conformation is too short to permit us to distinguish it from the other methylene hydrogen. The "state" measured in a nmr spectrometer is a weighted average of all of the rotational conformations. The energy differences measured in nmr are so small that one frequently refers to the "nmr time scale," a time period ranging from milliseconds to seconds.

To summarize, a consequence of the Heisenberg uncertainty principle and the small energy changes characteristic of nmr spectroscopy is that two hydrogen states that are interconvertible but which have separate lifetimes of more than about 1 sec can be seen as two sharp peaks whose separation can be measured accurately. If the lifetimes are less than about 1 msec, they can be seen only as a combined single sharp peak; that is, on the "nmr time scale" the two hydrogens are **magnetically equivalent.** If the lifetimes are in an intermediate region, a broad peak results. An interesting example of this phenomenon is the chair–chair interconversion of cyclohexane discussed in Section 9.9.

The foregoing discussion also shows why the nmr sample tube is spun rapidly between the magnet faces. It is difficult to prevent slight changes in a magnetic field at different places. In a tube placed between the pole faces of even high-quality magnets, different protons would experience slightly different fields at different points. The result would be a rather broad nmr signal. Rapid spinning of the tube causes all of the protons to experience the same average field *on the nmr time scale.*

9.5
Relative Peak Areas

We saw in the previous section that we can obtain a valuable piece of information from an nmr spectrum—the number of magnetically different hydrogens in the compound. The amount of energy absorbed at each resonance frequency is proportional to the number of nuclei that are absorbing energy at that frequency. By measuring the **areas** of each of the resonance lines, we may determine the relative number of each different kind of hydrogen. In practice, this is accomplished with an electronic integrator. After the nmr spectrum has been recorded, the instrument is switched to an "integrator mode" of operation and the spectrum is recorded again. The recorder output in this mode of operation is illustrated in Figure 9.11. The integral line for each of the two peaks in the spectrum of 1,2,2-trichloropropane is shown superimposed on the appropriate peak. The ratio of the

196

Chap. 9

**Nuclear
Magnetic
Resonance
Spectroscopy**

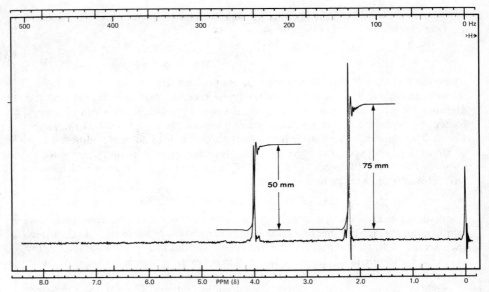

FIGURE 9.11. *Integrated intensities superimposed on the nmr spectrum are proportional to the relative numbers of hydrogens.*

heights of the two integral lines is equal to the ratio of the number of protons giving rise to the two peaks, in this case 3/2. In the remaining sample spectra in this book, we will usually not show the integral lines, but will simply indicate the relative areas of peaks where necessary.

We should note that the nmr experiment does not measure all of the protons but just those that have α-spin and absorb energy in "flipping" to β-spin. The difference in the population of α- and β-spins is rather small. We learned above that the energy difference between the proton α- and β-spin states in the magnetic field of a 60-MHz nmr instrument is only 0.006 cal mole^{-1}. At equilibrium, the population difference is given by the Boltzmann distribution as

$$\frac{N_\alpha}{N_\beta} = e^{0.006/RT} = 1.00001$$

When we now turn on the applied radiofrequency field, we excite the slight excess of α- nuclei to β. If there were no other mechanism for converting β back to α, we would quickly have exactly equal populations in both spin states and could no longer observe any absorption of energy; there would be no spectrum. With a sufficiently strong radio frequency field this can generally be done and the system is then said to be **saturated.** However, in normal operation, hydrogens in the β-spin state continually **relax** back to the α-state because of local fields, and equilibrium imbalance is maintained. These local fields are associated with other spinning nuclei. That is, even in the absence of the applied radiofrequency field, individual protons convert from one state to another quite readily because, in moving about the liquid, they experience the magnetic fields of other nearby nuclear magnetic moments. Occasionally, such moving and changing fields happen to have the resonance value, and energy interchange can occur resulting in spin flipping. In the nmr experiment the net result of all this activity is the conversion of our measuring radio waves into heat within the sample. In normal operation the distribution of α- and β-spins remains close to the equilibrium value and the amount of energy absorbed is proportional to the number of protons.

EXERCISE: What are the relative peak areas for the different hydrogens in methyl acetate (CH_3COOCH_3), 1,1,4,4-tetramethylcyclohexane, and 1,1,3-trichloro-2,2-dimethylpropane?

9.6
Spin-Spin Splitting

We have seen that the nmr spectrum of 1,2,2-trichloropropane shows two sharp peaks that are easy to interpret. From the fact that there are two peaks, we deduce that the compound has two types of hydrogen that are magnetically nonequivalent, and, from the relative areas of the two peaks, we conclude that they are present in a ratio of 3:2. Let us now consider the nmr spectrum of a related trihaloalkane, 1,1,2-tribromo-3,3-dimethylbutane, $(CH_3)_3CCHBrCHBr_2$ (Figure 9.12).

We recognize the small peak at $\delta = 0.0$ as that of TMS added as a standard to define the zero on our scale. The large peak at $\delta = 1.2$ ppm comes from the nine equivalent methyl protons. The other two protons are responsible for the downfield resonances; the downfield shifts are explained by their proximity to the electronegative bromines. But these resonances are now represented by a pair of peaks. Each resonance appears as a **doublet**. This "splitting" of peaks is common in nmr spectra—it is an additional complication, which requires study, but it is also a powerful tool for the determination of molecular structure. The phenomenon has its origin in the magnetic field associated with each individual spinning proton. These small magnetic fields affect the total magnetic field experienced by another proton. For convenience we will label these hydrogens as H_a and H_b.

$$\underset{\underset{\displaystyle H_a}{|}}{Br-\overset{\overset{\displaystyle Br}{|}}{C}}-\underset{\underset{\displaystyle H_b}{|}}{\overset{\overset{\displaystyle Br}{|}}{C}}-C(CH_3)_3$$

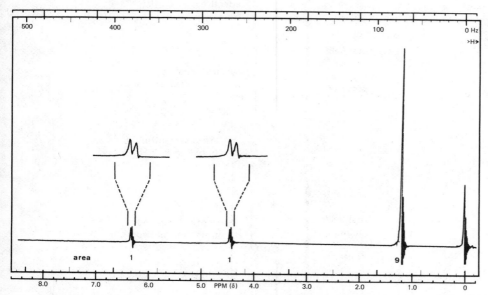

FIGURE 9.12. *Nmr spectrum of* $(CH_3)_3CCHBrCHBr_2$. *The scale expansion indicated shows the two doublets more clearly.*

198

Chap. 9

**Nuclear
Magnetic
Resonance
Spectroscopy**

In the applied magnetic field of the nmr spectrometer we would expect H_a normally to show up as a single peak. However, the magnetic field associated with the spin of the nearby proton, H_b, contributes to the effective field experienced by H_a. This is not a through-space effect of the magnetic field associated with the spinning nucleus H_b but results instead from interaction between each H nucleus and electron spins. That is, the spin of H_b is relayed to H_a by way of shared electrons. For most cases, if H_b has α-spin, the effect is as if the total magnetic field at H_a were slightly greater than that provided by the nmr instrument's applied field alone. Consequently, *less* applied field is required to achieve resonance than in the absence of H_b and we find a slight downfield shift (Figure 9.13). But only half of the H_b nuclei have α-spin. The rest have β-spin, in which the opposite effect results. For these molecules the effect is as if the effective magnetic field at H_a were slightly *weaker* than that given by the applied field alone. The nmr spectrometer must then provide slightly *more* magnetic field in order to achieve the "spin-flipping" resonance condition with H_a. Now the result is an upfield shift (Figure 9.13). H_a and H_b are said to be "coupled." The coupling is an energy term or frequency, but it is convenient to treat it for the present purposes in terms of the equivalent magnetic field.

Let us recapitulate the conditions of the experiment. We start with a low magnetic field in the nmr instrument and irradiate our sample with a radio signal of an accurately constant frequency (energy). As we slowly increase the magnetic field, we reach a point where the effective magnetic field at the half of the H_a protons which are in molecules where the H_b protons have α-spin now matches the energy of the irradiating radio waves. H_a protons of α-spin absorb radio photons and "flip" to β-spins. Motion in the liquid sample provides a mechanism for the β-spins to change to α with the excess energy given up as heat. The absorption of radio waves is recorded by the nmr instrument as a "peak." As the applied magnetic field is increased still more, the resonance condition is destroyed

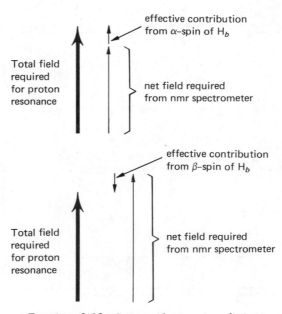

FIGURE 9.13. *Source of spin-spin splitting.*

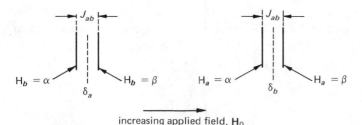

increasing applied field, H_0

FIGURE 9.14. J_{ab} *causes equal spin-spin splitting on both* H_a *and* H_b.

and the recorder pen returns to the base line (usually with the oscillations known as "ringing," a normal phenomenon that we will not detail). At a still higher applied field, we reach a point where the other half of the H_a protons absorb radio energy. These H_a protons are in molecules where H_b has β-spin whose coupling effect subtracts from the applied field, and a stronger field must therefore be applied to achieve resonance.

Note that only one kind of hydrogen is in resonance at any given point in the spectrum. Both lines in the low-field doublet in Figure 9.13 correspond to transitions of H_a. At these field strengths H_b, with its different chemical shift, is not in resonance even though its presence is "felt" by H_a, and it produces two resonance positions instead of one. As the field strength is increased still further, the H_b nuclei eventually come into resonance. However, now we must reckon with the effect of α- or β-spin of H_a on the effective magnetic field experienced by H_b.

The effect is an exact reciprocity—the effect of H_a on H_b is exactly the same as the effect of H_b on H_a. Consequently, the splitting of the H_a peaks has the same magnitude as that of the H_b peaks. The spacing between the peaks is conventionally labeled J and is given in units of cycles per second or Hertz. J is the **coupling constant** between two protons. For our case $J_{ab} = 1.6$ Hz. These relationships are illustrated in Figure 9.14. Since J arises from the spin of the proton, its magnitude is not dependent on the applied magnetic field. That is, the same J value applies for spectra determined at 60 MHz, 100 MHz, 250 MHz, and so on.

[The discussion above was given for J_{ab} having a positive value. Negative J values are
known, but the effect of the sign of J is seen only in more complex cases.]

In the foregoing example, we note that there is no coupling to the methyl protons. Because the coupling effects of proton spin are relayed via shared electrons, the effect is attenuated rapidly with the number of bonds and is usually quite small if more than two atoms intervene between the protons. Thus, the methyl protons do indeed couple to the other two protons in our example, but the magnitude of each such J is so small as to be unobservable in normal spectrometers. As a further example, in the spectrum of 1,2,2-trichloropropane, which was discussed in Section 9.4, the protons on C-1 and C-3 do not noticeably split each other.

$$\underset{\text{C--C}}{\overset{H_a\ H_b}{|\ \ |}} \quad \text{spin-spin splitting normally observed}$$

$$\underset{\text{C--C--C}}{\overset{H_a\qquad H_b}{|\qquad\ |}} \quad \text{spin-spin splitting normally not observed}$$

200

Chap. 9

Nuclear
Magnetic
Resonance
Spectroscopy

Now let us consider a slightly more complex spectrum, that of 1,1,2-trichloro-ethane (Figure 9.15). In this compound, there are two types of hydrogen.

$$\begin{array}{ccc} & Cl & H_b \\ & | & | \\ Cl - & C - & C - Cl \\ & | & | \\ & H_a & H_b \end{array}$$

The spectrum shows two resonances, a triplet centered at $\delta = 5.8$ ppm and a doublet centered at $\delta = 3.9$ ppm. The triplet is associated with H_a, which is more deshielded because it is bonded to a carbon that also has two chlorines. The two equivalent H_b hydrogens are less deshielded because their carbon has only one attached chlorine. In any given molecule, the two H_b nuclei may have their spins as $\alpha\alpha$, $\alpha\beta$, $\beta\alpha$, or $\beta\beta$. When both H_b nuclei have α-spin, the effect of coupling is to augment the applied field and less field is required for H_a to be in resonance. When the H_b spins are $\alpha\beta$ or $\beta\alpha$, there is no effect on the field experienced by H_a because the opposed spins of the two H_b nuclei cancel. When both H_b nuclei have β-spin, the magnetic field associated with their coupling subtracts from H_0 and a greater applied field is necessary to achieve resonance at H_a. Thus, the H_a resonance appears as three lines with relative intensities of $1:2:1$.

At the resonance frequency of the two H_b hydrogens we find two lines because H_a can have either α- or β-spin. In this case the coupling constant J has the value 7 Hz (Figure 9.16). The chemical shift of the H_a triplet corresponds to the center line. The chemical shift of H_b is the midpoint of the doublet. The combined area of the triplet is one, and the combined area of the doublet is two. *The chemical shift of a given proton depends on itself, but the nature of its splitting depends on its proton neighbors.* Consequently, the splitting phenomenon in nmr spectroscopy is an extremely valuable tool for determining structure.

Extension to different numbers of equivalent neighboring hydrogens is straight-forward. Three hydrogens, as in a methyl group, can have the possible spin states $\alpha\alpha\alpha$; $\alpha\alpha\beta$, $\alpha\beta\alpha$, $\beta\alpha\alpha$; $\alpha\beta\beta$, $\beta\alpha\beta$, $\beta\beta\alpha$, and $\beta\beta\beta$. An adjacent proton would there-

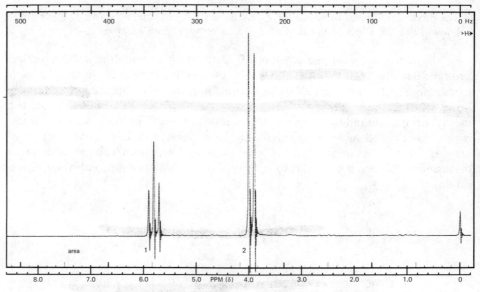

FIGURE 9.15. *Nmr spectrum of* $CHCl_2CH_2Cl$.

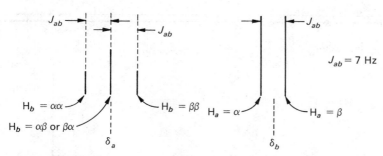

FIGURE 9.16. *Spin-spin splitting analysis of 1,1,2-trichloroethane.*

fore give four peaks with area ratios of approximately $1:3:3:1$. These numbers are simply the binomial coefficients and are summarized in Table 9.1. In general, n neighboring equivalent hydrogens cause splitting into $n+1$ peaks.

These simple prescriptions apply for cases where splitting is small compared to the difference in chemical shift between the neighboring hydrogens; that is, $J \ll \Delta\nu$. As $\Delta\nu$ is reduced (as the peaks for two nonequivalent hydrogens approach each other), the inner peaks increase in intensity and the outer ones diminish. In practice, such perturbations are almost always apparent. This effect may be seen in the spectrum in Figure 9.15.

Ethyl chloride and ethyl iodide present an interesting comparison (Figures 9.17 and 9.18). In each case the three hydrogens of the methyl group give a quartet for the methylene hydrogens, and the two methylene hydrogens in turn produce a triplet with equal splitting for the methyl hydrogens. The area ratio of all four methylene peaks to the three methyl peaks is still $2:3$, the ratio of the total number of hydrogens involved. For ethyl chloride, the highly electronegative chlorine produces a large downfield shift for the methylene hydrogens. That is, $\Delta\nu$ between CH_2 and CH_3 is rather large compared to J, and the peak intensities differ little from the simple ratios expected. For ethyl iodide, however, the less electronegative iodine has a smaller effect on δ and $\Delta\nu$ is smaller. Note that the asymmetry in both groups of peaks is now greater.

If $\Delta\nu$ is too small compared to J, the simple rules do not apply at all. Such spectra are quite complex and require a detailed analysis beyond the scope of this book. One especially simple case, however, is the extreme one for which $\Delta\nu = 0$. Such hydrogens are *magnetically equivalent and do not split each other*. In the cases discussed previously, for example, a methylene group was treated as a unit—because the two hydrogens are magnetically equivalent, they have no effect on each other. This effect is a direct and exact outcome of the quantum mechanics of magnetic resonance and the detailed reason is not important for our purposes.

TABLE 9.1

Number of Equivalent Adjacent Hydrogens	Total Number of Peaks	Area Ratios
0	1	1
1	2	$1:1$
2	3	$1:2:1$
3	4	$1:3:3:1$
4	5	$1:4:6:4:1$
5	6	$1:5:10:10:5:1$
6	7	$1:6:15:20:15:6:1$

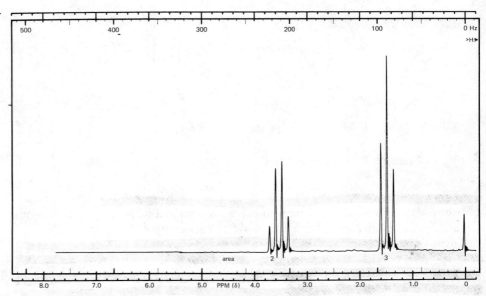

FIGURE 9.17. *Nmr spectrum of ethyl chloride,* CH_3CH_2Cl.

The following simple explanation may be helpful. Because the nmr experiment does not distinguish magnetically equivalent hydrogens, different spin properties cannot be assigned to individual protons. For example, in a methylene group in which the protons have α- and β-spin, we cannot assign one spin to one proton and the other spin to the remaining proton. Instead, the $\alpha\beta$-spin property belongs to the methylene protons *as a unit*. Since we cannot assign individual spins to individual equivalent protons, it follows that we should not observe the splitting or J-coupling normally associated with such assignments.

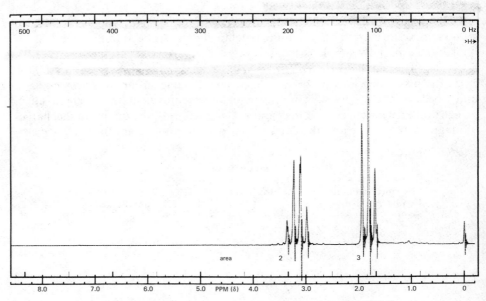

FIGURE 9.18. *Nmr spectrum of ethyl iodide,* CH_3CH_2I.

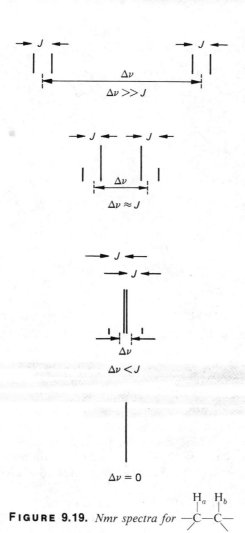

FIGURE 9.19. *Nmr spectra for* $-\overset{\overset{\displaystyle H_a}{|}}{\underset{\diagdown}{C}}-\overset{\overset{\displaystyle H_b}{|}}{\underset{\diagdown}{C}}-$

The effect of the relative magnitudes of J and $\Delta \nu$ is illustrated in Figure 9.19. The first spectrum, for $\Delta \nu \gg J$, is that given by the "first-order" analysis that we outlined above. As $\Delta \nu$ and J become of comparable magnitude, the inner peaks increase in intensity and the outer ones fade until, in the limit where $\Delta \nu = 0$, the two inner peaks have merged and the outer peaks have vanished. Note that our example of ethyl iodide (Figure 9.18) shows the beginnings of breakdown of the simple rules in the slight additional splitting of the peaks. This splitting results entirely from $\Delta \nu$ being insufficiently large compared to J; first-order analysis is only a first approximation.

Some simple generalizations are important in our use of nmr. J-values are generally significant for hydrogens on adjacent carbons and are usually of the order of magnitude of 4–10 Hz. The presence nearby of electronegative elements can give abnormally low coupling constants. Note that the J-value of 1.6 Hz observed in Figure 9.12 is low because of the three bromines.

Another important factor is conformation. The J-value between adjacent axial hydrogens in cyclohexanes is 10–13 Hz, whereas J between axial and equatorial or between two equatorial hydrogens is 3–5 Hz.

204

Chap. 9

**Nuclear
Magnetic
Resonance
Spectroscopy**

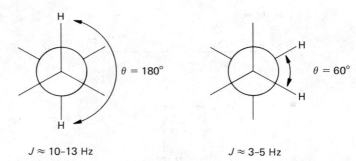

$J \approx 10$–13 Hz $\qquad\qquad\qquad\qquad J \approx 3$–5 Hz

FIGURE 9.20. J_{HH} *and conformation. The angle θ is the dihedral angle between the hydrogens.*

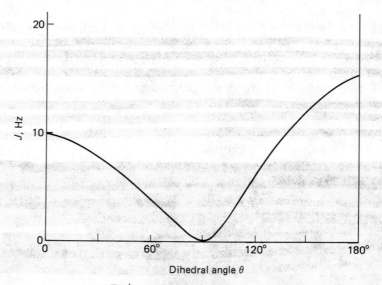

$J_{aa'} = 11.1$ Hz $\qquad\qquad\qquad J_{ee'} = 2.7$ Hz
$J_{ae} = 4.3$ Hz $\qquad\qquad\qquad J_{ae} = 3.0$ Hz

The dihedral angle between two axial hydrogens is 180°, whereas the axial-equatorial and equatorial-equatorial dihedral angle is 60° (Figure 9.20).

In many cyclohexane derivatives the nmr spectra are sufficiently complex that this distinction between J-values is not useful. The relation between J and the dihedral angle between hydrogens has been established theoretically and may be depicted in a familiar graphic form known as the Karplus curve (Figure 9.21). The coupling constant between two hydrogens reaches a minimum when the dihedral angle between them is 90°. The curve can be expressed analytically by two equations.

$$J_{HH'} = 10 \cos^2 \theta \qquad (0 \leq \theta \leq 90°)$$
$$J_{HH'} = 16 \cos^2 \theta \qquad (90 \leq \theta \leq 180°)$$

FIGURE 9.21. *Karplus curve.*

9.7
More Complex Splitting

Splitting becomes more complex if coupling occurs to more than one type of hydrogen. Consider the case of H_a coupled to two different hydrogens, H_b and H_c, with $J_{ab} > J_{ac}$.

$$\begin{array}{ccc} H_b & H_a & H_c \\ | & | & | \\ C & \!\!-C\!\!- & C \end{array}$$

H_a is split into a doublet by H_b, and each of the lines of the doublet is split into a further doublet to give a total of four lines, as shown in Figure 9.22. The four lines will have approximately the same intensity. The four lines correspond to transitions of H_a when the H_b and H_c nuclei have the following spin states.

Line	Spin of H_b	Spin of H_c
1	α	α
2	α	β
3	β	α
4	β	β

In this example, $J_{ab} > J_{ac}$. This means that the effect of nucleus H_b on H_a is greater than the effect of nucleus H_c on H_a. Thus, line 2 appears at lower field than the resonance position of H_a in the absence of H_b and H_c because the α- and

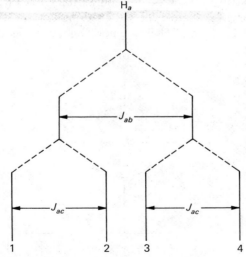

FIGURE 9.22. *Effect of two J couplings, $J_{ab} > J_{ac}$.*

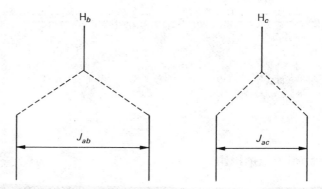

FIGURE 9.23. *The* H_b *and* H_c *resonances of the system* H_b—H_a—H_c *when* $J_{ab} > J_{ac}$.

β-spins of the two nuclei do not cancel. Similarly, line 3, which corresponds to a transition of H_a when H_b is β and H_c is α, is at slightly higher field than the resonance position of H_a in the absence of the other nuclei. The remainder of the spectrum will show H_b and H_c each as doublets due to their respective couplings to H_a (Figure 9.23).

Note what happens if $J_{ab} = J_{ac}$. This will occur if H_b and H_c are magnetically equivalent (that is, if they have the same chemical shift) or if the two Js accidentally have the same value. In such a case the two inner lines of the quartet occur at the same point and appear as a single line of double the intensity. The net result is a triplet with intensity ratios of 1:2:1 (Figure 9.24).

The spectra of 1,3-dichloropropane (Figure 9.25) and 1-bromo-3-chloropropane (Figure 9.26) are interesting examples. In 1,3-dichloropropane, the four CH_2Cl protons are magnetically equivalent. Each CH_2Cl group is adjacent to the middle CH_2 and therefore appears as a 1:2:1 triplet centered at $\delta = 3.66$ ppm. The center CH_2 appears at $\delta = 2.1$ ppm. This resonance is a quintet with an approximate intensity ratio of 1:4:6:4:1 due to splitting by the four magnetically equivalent CH_2Cl hydrogens (Table 9.1).

In 1-bromo-3-chloropropane the CH_2Cl hydrogens and the CH_2Br hydrogens are not magnetically equivalent, and therefore they do not have the same chemical shift. Each is adjacent to two hydrogens (the center CH_2 group) and each appears as a 1:2:1 triplet. The chemical shifts for CH_2Cl ($\delta = 3.66$ ppm) and CH_2Br ($\delta = 3.54$ ppm) are almost the same, so the two triplets **overlap** one another. Even though CH_2Cl and CH_2Br are not magnetically equivalent, the two Js

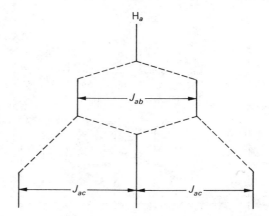

FIGURE 9.24. *Effect of equal J values,* $J_{ab} = J_{ac}$.

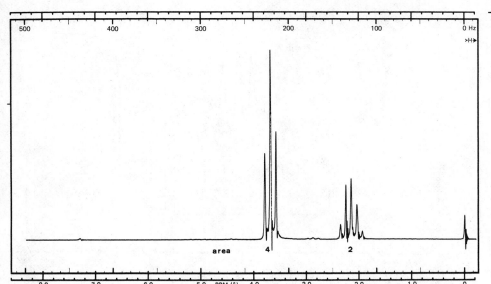

FIGURE 9.25. *Nmr spectrum of 1,3-dichloropropane,* $ClCH_2CH_2CH_2Cl$.

are accidentally equal. Therefore, the center CH_2 still appears as a $1:4:6:4:1$ quintet with $\delta = 2.15$ ppm.

Figure 9.27 shows the spectrum of 1,1,2-trichloropropane, a compound in which there are two different coupling constants. There are three different types of hydrogen, which we may label H_a, H_b, and H_c.

$$
\begin{array}{ccc}
 & Cl & Cl \\
 & | & | \\
Cl- & C- & C-CH_{3(c)} \\
 & | & | \\
 & H_a & H_b
\end{array}
$$

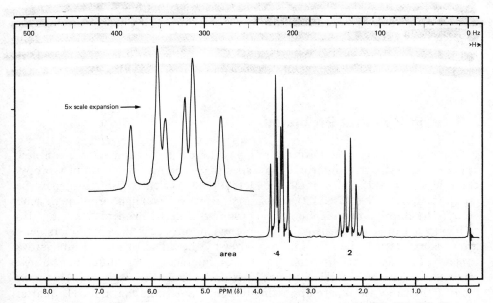

FIGURE 9.26. *Nmr spectrum of 1-bromo-3-chloropropane,* $BrCH_2CH_2CH_2Cl$.

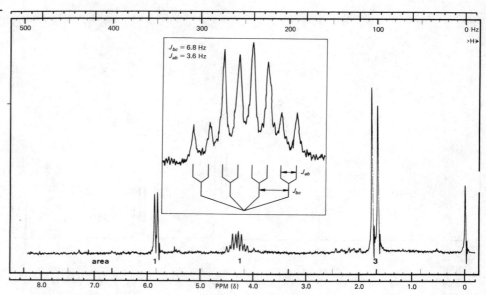

FIGURE 9.27. *Nmr spectrum of 1,1,2-trichloropropane,* $CH_3CHClCHCl_2$. *The center band is expanded in the insert.*

The H_a hydrogen is most deshielded and appears as a low-field resonance with relative area of unity. It is a doublet due to coupling with H_b, and the separation between the two lines, J_{ab}, is 3.6 Hz. The three equivalent CH_3 hydrogens, H_c, are least deshielded and appear as a high-field resonance of area 3. They are also coupled to H_b and appear as a doublet. In this case the separation between the lines, J_{bc}, is 6.8 Hz. The two coupling constants in this case are unequal. The resonance for H_b is in between those of H_a and H_c, and it has a relative area of unity. Because of the two unequal Js, it appears as a "doubled quartet" and may be analyzed as shown on the insert in Figure 9.27. The chemical shift for H_b is $\delta = 4.3$ ppm, the midpoint of the multiplet.

Finally, remember that J is independent of applied field, whereas the normal shielding by electrons $\Delta\nu$ results from an induced field and is proportional to the applied field. One of the incentives for seeking larger magnetic fields for nmr instruments is the spreading of $\Delta\nu$ relative to J. At a sufficiently high field strength, all spectra reduce to the simple "first-order" type ($\Delta\nu \gg J$) we have treated.

9.8
Solving Spectral Problems

Going back to our spectra of the propyl chlorides in Section 9.2, we can now apply our knowledge to interpret those spectra. In the nmr spectrum of *n*-propyl chloride (Figure 9.2), for example, the methyl group is clearly distinguished as the group furthest upfield ($\delta = 1.2$), split into a triplet by its neighboring methylene group. The chlorine-bearing methylene group is furthest downfield ($\delta = 3.6$), also split into a triplet by its neighboring methylene group. The center methylene group is expected to be split into a quartet by the adjacent methyl and into a triplet by the adjacent methylene. If these two interactions had different J values, we would indeed see a total of twelve lines under sufficient resolution. However, because the two J values are approximately the same (note that the CH_3 quartet and the CH_2Cl triplet have approximately equal splittings), the splitting in the middle CH_2 group is that expected for five magnetically equivalent hydrogens, namely six peaks.

The spectrum of isopropyl chloride in Figure 9.3 is simpler. Both methyl groups are equivalent and appear as a doublet caused by the C-2 hydrogen and having a total area six times that of the downfield resonance of the single C-2 hydrogen. The downfield position of $\delta = 4.05$ ppm results from deshielding of the neighboring electronegative chlorine. This peak is split into seven peaks by the six adjacent methyl hydrogens and appears as a multiplet.

The chemical shift of a given proton depends on its immediate neighborhood. Hence, hydrogens in different functional groups tend to have characteristic δ values, and the appearance of peaks having such δ values can be diagnostic of the presence of such functional groups. The relative area indicates the number of hydrogens of a given type, and the splitting provides information as to neighboring hydrogens. Extensive tables of δ values associated with different kinds of hydrogens have been compiled; an example is given in Appendix V. As we discuss different functional groups, we will also discuss any characteristic features of their spectra.

At the present time, we have discussed alkanes and halogenated alkanes in some detail. The appropriate nmr characteristics of these compounds are summarized in Table 9.2. Several important generalizations should be learned; these suffice for solving many problems without resorting to detailed tables:

1. Alkyl hydrogens have $\delta \approx 1$ ppm.
2. δ (tertiary) $> \delta$ (secondary) $> \delta$ (primary).
3. A halogen atom on the same carbon causes a downfield shift by 2–3 ppm.
4. A halogen removed by one carbon still has an effect of about 0.5 ppm.
5. J for neighboring hydrogens on an alkane chain is usually in the range of 4–10 Hz.

TABLE 9.2
Some Nmr Characteristics

Type of Hydrogen	δ, ppm (measured downfield from TMS)
RCH_3	0.9
R_2CH_2	1.25
R_3CH	1.5
RCH_2I	3.15
R_2CHI	4.2
RCH_2Br	3.3
R_2CHBr	4.1
RCH_2Cl	3.4
R_2CHCl	4.0
$-CH_2-\overset{\textstyle |}{\underset{\textstyle |}{C}}-X$	~1.7

An example of the way in which data are frequently presented and a structural problem is solved is shown in the Example.

Example. A compound, $C_4H_7Cl_3$, has the following spectrum.

δ, ppm: 0.9 (t, 3H); 1.7 (m, 2H); 4.3 (m, 1H); 5.8 (d, 1H).

210

Chap. 9

**Nuclear
Magnetic
Resonance
Spectroscopy**

In this shorthand the δ value in ppm is given for the center of a group of peaks. The number of peaks in the group is indicated by the code: s = singlet, d = doublet, t = triplet. Quartet and quintet are obvious, but it may not always be possible to resolve all of these peaks, and such multiple peak groups are frequently recorded as m = multiplet. Finally, the number of hydrogens as determined from the area of each group of peaks is indicated.

To solve the problem shown, we generate hypotheses of structural units from the information given about δ values and put the units together with the help of the splitting information. The peaks at $\delta = 0.9$ ppm clearly correspond to a methyl group; the number of hydrogens indicated fits this hypothesis. The group at $\delta = 4.3$ ppm corresponds roughly to H—C—Cl but is somewhat shifted downfield, and $\delta = 5.8$ ppm is so far downfield it must correspond to —CHCl$_2$. We are left with a —CH$_2$— group to assign to $\delta = 1.7$ ppm. Our structural units are

$$CH_3—$$
$$—CH_2—$$
$$—CHCl—$$
$$CHCl_2—$$

Since the methyl group is a triplet, it must be attached to the —CH$_2$— group. Since —CHCl$_2$ is a doublet, it must be attached to —CHCl—. The entire structure then becomes

$$CH_3—CH_2—CHCl—CHCl_2$$

The CH$_2$ and CHCl protons give rise to complex multiplets because of the unequal coupling constants to their adjacent neighbors (see Figure 9.25).

9.9
Nmr Spectroscopy of Other Nuclei

Up until now, we have discussed solely proton nmr (pmr) because this is the technique most commonly used by organic chemists. However, nmr experiments may be done with any element whose nuclei have a net magnetic spin. A few examples are given in Table 9.3.

^{19}F nmr and ^{13}C nmr (cmr) are used extensively. Cmr, in particular, is of great use in organic chemistry. Cmr spectrometers are not inexpensive (>\$80,000), but many are available and the method is being used more and more. Within the next few years, cmr spectroscopy will probably become as indispensable to the practicing organic chemist as proton nmr is today.

Cmr is a perfect complement to pmr. While pmr allows us to "see" the protons attached to the carbon framework of an organic compound, cmr allows us to see

TABLE 9.3
The Magnetic Properties of Some Nuclei

Isotope	Natural Abundance, %	Spin States	Resonance Frequency at 14,092 Gauss, MHz
^{1}H	99.88	$\pm\frac{1}{2}$	60
^{13}C	1.1	$\pm\frac{1}{2}$	15.1
^{19}F	100	$\pm\frac{1}{2}$	56.4
^{31}P	100	$\pm\frac{1}{2}$	24.4

the carbons themselves. However, there is one important difference between the two techniques. In pmr, we are observing the most abundant isotope, ^{1}H. For carbon, the most abundant isotope, ^{12}C, has an even-even nucleus that has no net nuclear spin or magnetic moment. Therefore, we must observe the isotope ^{13}C, which has a natural abundance of only about 1%. This low abundance is both a blessing and a curse. A simplifying feature is that since the natural abundance of ^{13}C is so low, the chance that we will find two ^{13}C nuclei adjacent to each other in the same molecule is very small ($10^{-2} \times 10^{-2} = 10^{-4}$). Therefore, one does not observe spin–spin splitting between the carbon nuclei. However, the low abundance means that the spectrometer must be much more sensitive than is sufficient for pmr.

The problem of sensitivity has been overcome in several ways. One obvious way is to use much larger samples. However, in most cases this method is impractical. Another method for enhancing the weak ^{13}C signals is to run the spectrum over and over again and store the individual spectra in a computer. The computer averages the spectra and plots out the accumulated spectrum in a conventional manner. In this way, the many small signals eventually add up to give signals of sufficient intensity that they may be distinguished from the normal electronic "noise." Since the electronic noise is random, it eventually averages out to a relatively low value. In practice, many scans of the spectrum are required to achieve the required enhancement in the signal-to-noise ratio (100–10,000). In a normal mode of operation a scan requires on the order of 5–10 minutes. Therefore, this technique would be very time-consuming (8 hr to 5 weeks for a single spectrum). By using a pulse technique in connection with Fourier transform (FT) mathematical analysis, this barrier is overcome. We shall not review the principles of the method here but simply mention that it enables a single scan to be accomplished in about 1 sec; it is thus possible to obtain cmr spectra with good signal-to-noise ratios in reasonable periods of time.

Figure 9.28 shows the cmr spectrum of 1,2,2-trichloropropane. The spectrum was determined on a Varian CFT-20 spectrometer operating at a field strength of 18,665 gauss, which corresponds to a frequency of 20 MHz for ^{13}C. The spectrum was determined by a method called "proton off-resonance decoupling." In this mode of operation one observes only one-bond couplings, that is, ^{13}C—H, and these are reduced in magnitude by a considerable factor. Two-bond couplings, ^{13}C—C—H, are not observed.

In the spectrum in Figure 9.28 note that we see three signals corresponding to the three carbon atoms. The small multiplet at the right is our standard, TMS, which appears as a quartet due to coupling to the three hydrogens joined to each methyl carbon. In cmr, as in pmr, the TMS resonances appear at high field because of the electropositive silicon. The three carbon atoms in trichloropropane appear as a singlet, triplet, and quartet due to spin–spin splitting by their attached hydrogens (zero, two, and three, respectively). As in pmr, the electronegative chlorines result in C-1 and C-2 resonating downfield from C-3. Note that the chemical shifts in cmr are much greater than in pmr.

The spectrum in Figure 9.29 is also of 1,2,2-trichloropropane, but in this case the spectrum was measured while simultaneously applying a strong radiofrequency field of 80 MHz. At the field strength of 18,665 gauss protons resonate at this frequency. Since the hydrogen nuclei are being constantly excited, they do not spend sufficient time in either the α- or β-spin state to couple with the ^{13}C nuclei. That is, on the nmr time scale each hydrogen is in an average or effectively constant state, and the result is that no coupling is observed. This process is called **decoupling,** and the spectrum in Figure 9.29 is said to be **proton-decoupled.** Each

212

Chap. 9

**Nuclear
Magnetic
Resonance
Spectroscopy**

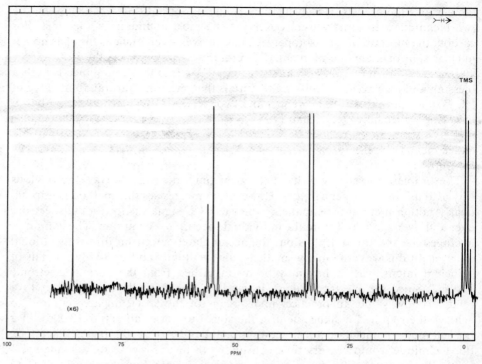

FIGURE 9.28. *Cmr spectrum of 1,2,2-trichloropropane.*

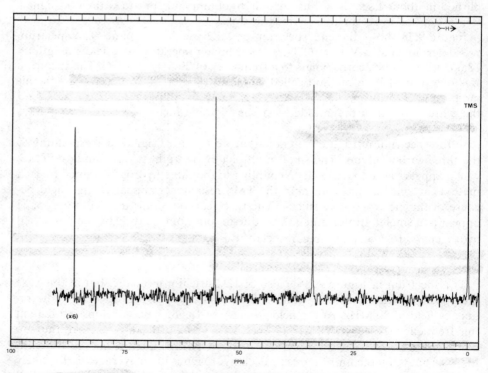

FIGURE 9.29. *Proton-decoupled cmr spectrum of 1,2,2-trichloropropane.*

carbon nucleus now appears as a sharp singlet and the entire spectrum is greatly simplified.

In Figures 9.28 and 9.29 the signal due to C-2, which appears 88.3 ppm downfield from TMS, has been amplified by a factor of 6. In the Fourier transform method the peak areas are not always proportional to the number of atoms involved. Some carbon nuclei, particularly those with no attached hydrogens, relax from an excited spin state to the ground state rather slowly compared to the pulse time. Hence, such carbons tend to be magnetically partially saturated.

For structure work, it is convenient to obtain both types of spectra. For complex molecules the proton-decoupled spectrum often allows one to "see" each carbon resonance and to measure its chemical shift accurately. The proton-coupled spectrum then allows the analyst to determine the number of hydrogens attached to each carbon. By using these data together with the pmr spectrum, even complex structures may be solved.

Because of their simplicity, proton-decoupled cmr spectra are particularly suited for detecting symmetry in fairly complicated molecules. For example, Figures 9.30 and 9.31 are the proton-decoupled spectra of two diastereomers of 1,3,5-trimethylcyclohexane. Note that one of the isomers shows only three resonances. This must be the isomer in which all methyl groups are *cis*. Since all of the methyl groups occupy equatorial positions, there is only one kind of methyl resonance. Likewise, there are only one kind of methylene and only one kind of methine. In the other diastereomer the two equatorial methyls are equivalent, but the axial one gives rise to a different resonance. Similarly, this isomer has two different kinds of CH$_2$ resonance and two different kinds of CH resonance. Note that the six resonances of this molecule occur in three sets with the relative intensity being 2:1 in each set.

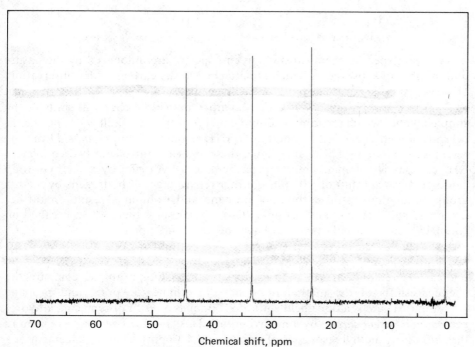

FIGURE 9.30. *Cmr spectrum of all-*cis *1,3,5-trimethylcyclohexane.*

214

Chap. 9

Nuclear
Magnetic
Resonance
Spectroscopy

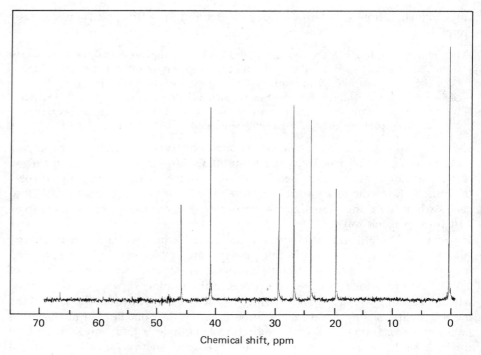

FIGURE 9.31. *Cmr spectrum of 1-cis-3,trans-5-trimethylcyclohexane.*

all-*cis* 1,3,5-trimethylcyclohexane   1,*cis*-3,*trans*-5-trimethylcyclohexane

As in pmr spectra, chemical shifts in cmr spectra are influenced by the nature and number of substituents which are attached to the carbon under observation. However, simple alkyl substitution has a much greater effect on cmr chemical shifts than it does on pmr shifts. For example, notice the chemical shifts of the simple alkanes which are tabulated in the first four lines of Table 9.4. The chemical shift of methane is −2.1 ppm (that is, CH_4 resonates 2.1 ppm *upfield* from the reference compound TMS). In ethane, the chemical shift of the two equivalent CH_3 carbons is 5.9 ppm. Thus, replacement of a hydrogen by a methyl group causes a *downfield* shift of 8.0 ppm. Further replacement of hydrogens by methyl group, as in the central carbons of propane and isobutane, results in further downfield shifts of 10.2 and 9.1 ppm. This regular substituent effect is called an "α-effect" and is usually taken to have the value +9 ppm.

Another difference between cmr and pmr chemical shifts is the rather large effect which results from substitution at the adjacent carbon and at the position one carbon removed from that under observation. For example, compare the chemical shifts of the methyl resonances in ethane (5.9 ppm) and propane (15.6 ppm). There is a downfield shift of 9.7 ppm as a result of replacing a hydrogen *on an adjacent carbon* by a methyl group. This "β-effect" may also be seen in the shift of the methyl resonance of isobutane (24.3 ppm). Thus, replacement of a second hydrogen by a methyl group results in a further downfield shift of 8.7 ppm. The β-effect is usually taken to have the value +9.5 ppm.

TABLE 9.4
Cmr Chemical Shifts for Alkanes

	C-1	C-2	C-3	C-4	C-5	C-6	Other
methane	−2.1						
ethane	5.9	5.9					
propane	15.6	16.1	15.6				
isobutane	24.3	25.2	24.3				
neopentane	31.5	27.9	31.5				
n-butane	13.2	25.0	25.0	13.2			
2-methylbutane	22.0	29.9	31.8	11.5			
2,3-dimethylbutane	19.3	34.1	34.1	19.3			
2,2,3-trimethylbutane	27.2	32.9	38.1	15.9			
n-pentane	13.7	22.6	34.5	22.6	13.7		
2-methylpentane	22.5	27.8	41.8	20.7	14.1		
3-methylpentane	11.3	29.3	36.7	29.3	11.3		18.6[a]
3,3-dimethylpentane	6.8	25.1	36.1	25.1	6.8		4.4[a]
n-hexane	13.9	22.9	32.0	32.0	22.9	13.9	

[a] C-3 methyl carbon(s).

Finally, examination of the chemical shifts tabulated in Table 9.4 reveals that there is a γ-effect which results in an *upfield* shift of 2–3 ppm. It may be seen clearly in the C-1 resonance of propane and butane and the C-4 resonance of isopentane. The γ-effect is usually taken to have the value −2.5 ppm.

Let us see how these empirical substituent effects might be useful. Suppose we want to predict the chemical shifts of the various carbon resonances in the cmr spectrum of 2-methylhexane. We may do this by adding the appropriate substituent corrections (α-, β-, and γ-effects) to the known chemical shifts of the various carbons of *n*-hexane (Table 9.4). The calculation is illustrated below.

Estimation of CMR Chemical Shifts for 2-Methylhexane

Carbon	Chemical Shift in *n*-Hexane	Correction	Estimated Chemical Shift in 2-Methylhexane	Actual Chemical Shift in 2-Methylhexane
1	13.9	β, +9.5	23.4	22.4
2	22.9	α, +9.0	31.9	28.1
3	32.0	β, +9.5	41.5	38.9
4	32.0	γ, −2.5	29.5	29.7
5	22.9	none	22.9	23.0
6	13.9	none	13.9	13.6

Although such calculations are admittedly crude, they are useful in assigning observed resonances to the proper carbons, and in some cases they may be useful in assigning a structure to a compound from its cmr spectrum.

Halogen substituents also show regular substituent effects (Table 9.5). Note that

TABLE 9.5
Halogen Substituent Effects

	Cl	Br	I
α-effect	+31	+20	−6
β-effect	+11	+11	+11
γ-effect	−5	−3	−1

the magnitude of the α-effect correlates with electronegativity of the substituent in that the chlorine produces the greatest downfield shift. It is interesting that the α-effect of iodine is actually negative. That is, replacement of a hydrogen by iodine actually results in an *upfield* shift of the carbon resonance. The β-effects of the three halogens are the same and not very different from the β-effect of a methyl group. As with methyl substituents, the halogens cause a small upfield shift at the γ-position, and the effect again parallels electronegativity, with chlorine producing the greatest effect and iodine the smallest.

EXERCISE: Using the data in Table 9.4, along with the appropriate substituent effects, estimate the chemical shifts of all resonances for each of the following compounds.

(a) 2,5-dimethylhexane   (b) 1-chloropentane
(c) 1-chloro-4-iodobutane   (d) 1-bromo-4-methylpentane
(e) 2,4-dimethylpentane

For each compound, sketch the expected appearance of the proton-coupled spectrum.

9.10
Dynamical Systems

Reactions that have half-lives of the order of minutes or hours can generally be determined without difficulty. Faster reactions, with half-lives of the order of seconds, can frequently be measured with careful work. Nmr spectroscopy provides an excellent extension for determining reactions having half-lives in the nmr time scale of about 10^{-3} to 10^0 second. The chair–chair interconversion of cyclohexane provides an example of the type of techniques used. At room temperature cyclohexane gives a sharp singlet. The rate of interconversion is so fast that the nmr measures only the average state of the axial and equatorial hydrogens. At sufficiently low temperature, $< -70°$, however, the rate of interconversion is so slow that the molecular state measured by nmr is a single conformation.

In one chair conformation, all equatorial hydrogens are equivalent but different from the axial hydrogens. The two sets of hydrogens have different chemical shifts and give rise to two broad bands separated by $\delta = 0.5$ ppm with $\delta_{\text{equatorial}} > \delta_{\text{axial}}$. The bands are broad because of J-splittings between the two sets of protons. The nmr spectrum of cyclohexane-d_{11} is a simpler case to interpret because the J-coupling between the proton nucleus and a deuteron is sufficiently small, and each cyclohexane now has only a single proton. The nmr spectrum is reproduced in Figure 9.32 as a function of temperature. At the lowest temperature $(-89°)$ half of the deuteriated cyclohexane molecules have their lone proton in an axial position and the other half have the proton equatorial. Interconversion of the two isomers is slow, and since the chemical shifts differ, we see two sharp singlets. At the highest temperature $(-49°)$ the ring interconversions are rapid, and the nmr spectrometer "sees" only a time-average position, a singlet with δ midway between δ_{axial} and $\delta_{\text{equatorial}}$. At intermediate temperatures, the rate of interconversion of the conformations is comparable to the frequency difference between the states, and a broad signal results. The results can be analyzed completely to give rate constants as a function of temperature and an enthalpy of activation, $\Delta H^{\ddagger}$, of 10.8 kcal mole^{-1}. This value is relatively high compared to other conforma-

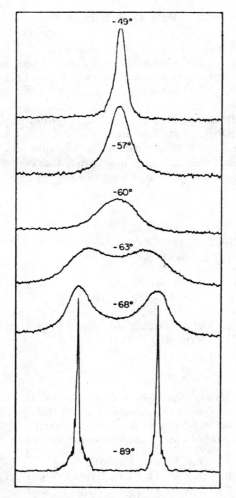

FIGURE 9.32. *Nmr of cyclohexane-*d_{11} *at different temperatures.* [*Reproduced with permission from F. A. Bovey,* Nuclear Magnetic Resonance Spectroscopy, *Academic Press, New York, 1969.*]

tional interchanges we have studied. The transition state involves a partially planar cyclohexane that now has both bond angle strain and eclipsed hydrogen strain.

EXERCISE: 1,3,5-Trioxane exists in a chair conformation similar to cyclohexane.

$$
\begin{array}{c}
O \\
CH_2 \quad CH_2 \\
O \qquad O \\
CH_2 \\
\end{array}
$$

1,3,5-trioxane

What is the expected appearance of its nmr spectrum at room temperature and at $-100°$.

218

Chap. 9

Nuclear
Magnetic
Resonance
Spectroscopy

PROBLEMS

1. Fill in the blank spaces in the following statement:

In the nmr spectrum of ethyl bromide, the methyl hydrogens have $\delta = 1.7$ ppm, the methylene hydrogens have $\delta = 3.3$ ppm, and $J = 7$ Hz. The number of peaks given by the methyl hydrogens is _____ with the approximate area ratio: _____ . These peaks are separated by _____ Hz. The number of peaks given by the methylene hydrogens is _____ with the approximate area ratio: _____ . These peaks are separated by _____ Hz. The total area of the methyl peaks compared to the methylene peaks is in the ratio: _____ . Of these two groups of peaks, the _____ peaks are further downfield. The chemical shift difference between these peaks of 1.6 ppm corresponds in a 60 MHz instrument to _____ Hz.

2. The nmr spectra for some isomers of $C_5H_{10}Br_2$ are summarized as follows. Deduce the structure corresponding to each spectrum.
(a) δ, 1.0 (s, 6H); 3.4 (s, 4H).*
(b) δ, 1.0 (t, 6H); 2.4 (quart, 4H).
(c) δ, 0.9 (d, 6H); 1.5 (m, 1H); 1.85 (t, 2H); 5.3 (t, 1H).
(d) δ, 1.0 (s, 9H); 5.3 (s, 1H).
(e) δ, 1.0 (d, 6H); 1.75 (m, 1H); 3.95 (d, 2H); 4.7 (quart, 1H).
(f) δ, 1.3 (m, 2H); 1.85 (m, 4H); 3.35 (t, 4H).

3. Free radical chlorination of propane using 1 mole of C_3H_8 and 2 moles of Cl_2 gives a complex mixture of chlorination products. By careful fractional distillation of the product mixture, one may isolate four dichloropropanes, A, B, C, and D. From the nmr spectra of the four isomers, deduce their structures.

Compound A: (b.p. 69°) δ 2.4 (s, 6H)
Compound B: (b.p. 88°) δ 1.2 (t, 3H), 1.9 (quint, 2H), 5.8 (t, 1H)
Compound C: (b.p. 96°) δ 1.4 (d, 3H), 3.8 (d, 2H), 4.3 (sext, 1H)
Compound D: (b.p. 120°) δ 2.2 (quint, 2H), 3.7 (t, 4H)

4. There are nine possible isomers (not counting stereoisomers) of $C_4H_8Br_2$. Two of them have the following nmr spectra. Deduce the structures of each and indicate the logic used in your assignment.
(a) δ 1.7 (d, 6H), 4.4 (quart, 2H)
(b) δ 1.7 (d, 3H), 2.3 (quart, 2H), 3.5 (t, 2H), 4.2 (m, 1H)

5. Sketch the expected nmr spectra of the following compounds. Be sure to represent the expected δ for each group of peaks, the relative areas and the splittings.
(a) $CH_3CBr_2CH_3$ (b) CH_3CH_2Br
(c) $CH_3CHBrCHBrCH_3$ (d) $CH_3CBr_2CH_2CH_3$
(e) $CH_3CHClCH_2CH_2Cl$

6. Deduce the structure corresponding to each of the following nmr spectra.

(a) C₂H₃Cl₃

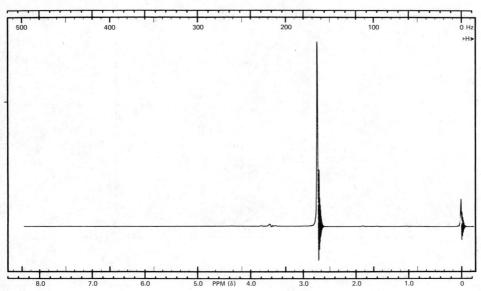

(b) C₂H₃Br₃

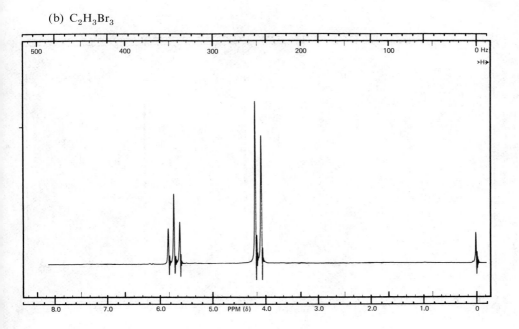

220

Chap. 9

**Nuclear
Magnetic
Resonance
Spectroscopy**

(c) C_4H_9Br

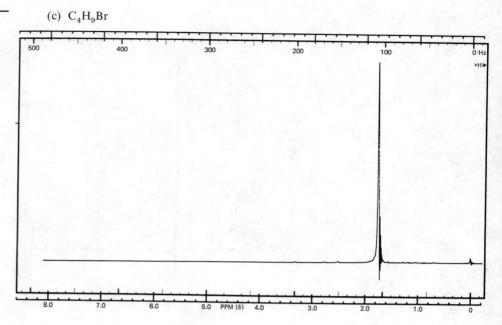

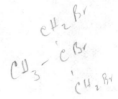

(d) $C_4H_7Br_3$

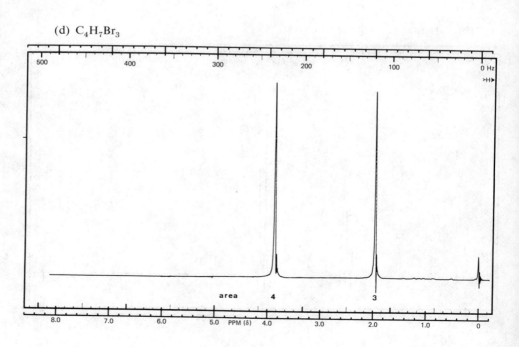

area   4   3

(e) $C_2H_4Br_2$

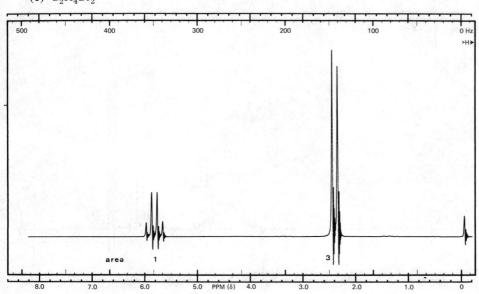

(f) C_3H_7Br

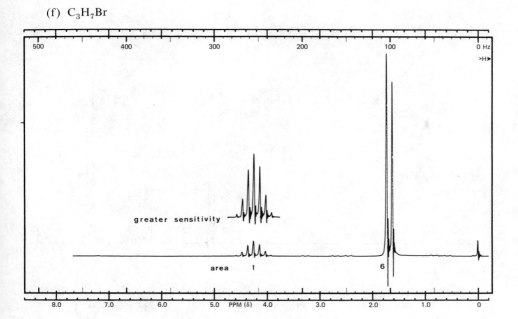

222

Chap. 9

**Nuclear
Magnetic
Resonance
Spectroscopy**

(g) $C_4H_8Br_2$

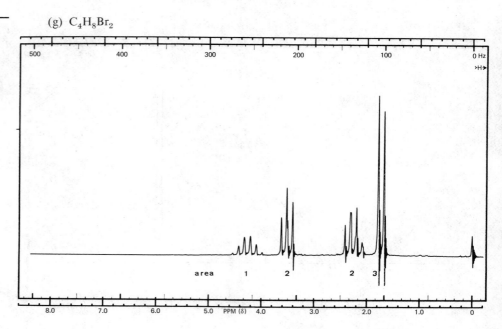

(h) C_4H_9Cl

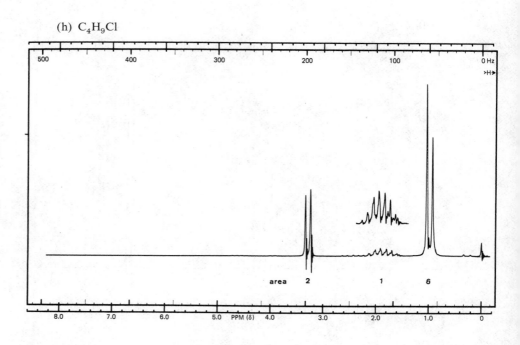

(i) C₄H₉Cl

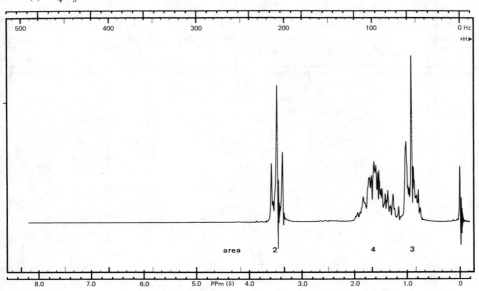

(j) C₄H₉Cl

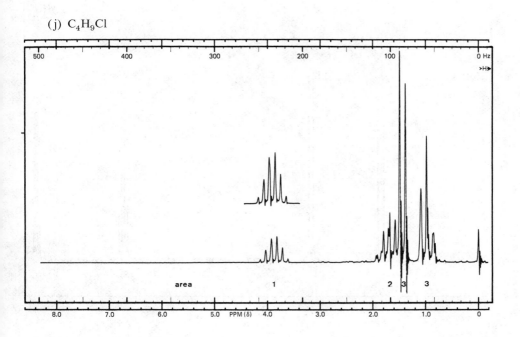

224

Chap. 9
**Nuclear
Magnetic
Resonance
Spectroscopy**

7. While in the process of writing this chapter, the authors ordered a sample of 1,2,2-
trichloropropane from a chemical supplier in order to obtain its nmr spectra. The
proton-coupled and -decoupled cmr spectra of the commercial sample were determined first and are reproduced below. The bottle was obviously mislabeled. What is
the actual structure of this compound?

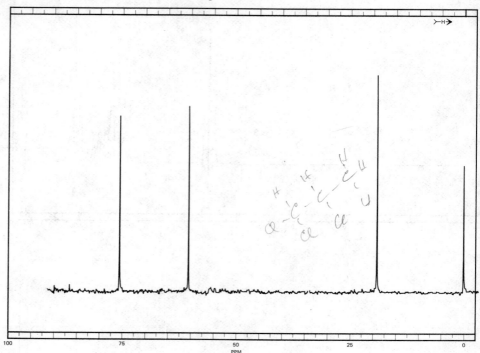

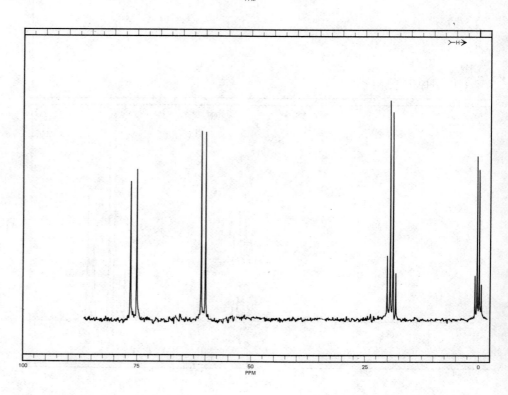

8. The nmr spectrum of chloroform shows a single intense peak at $\delta = 7.27$ ppm. Careful examination, however, shows a small peak 104.5 Hz above and below the main peak. These peaks are associated with $^{13}CHCl_3$. Explain. What sort of cmr spectrum would you expect for $^{13}CHCl_3$?

9. Deduce the structure of each of the following compounds from the nmr and proton-decoupled cmr spectra.

(a) $C_4H_8Cl_2$

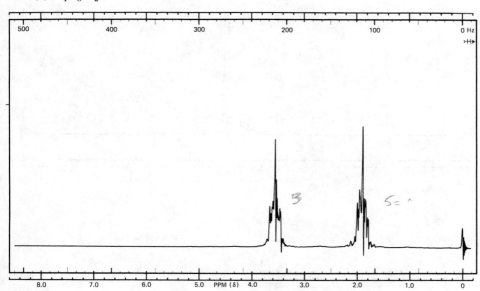

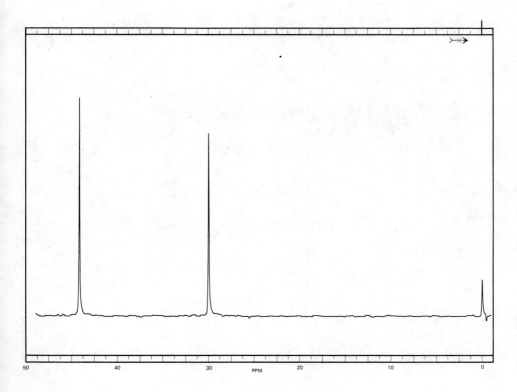

226

Chap. 9

**Nuclear
Magnetic
Resonance
Spectroscopy**

(b) $C_3H_5Cl_3$

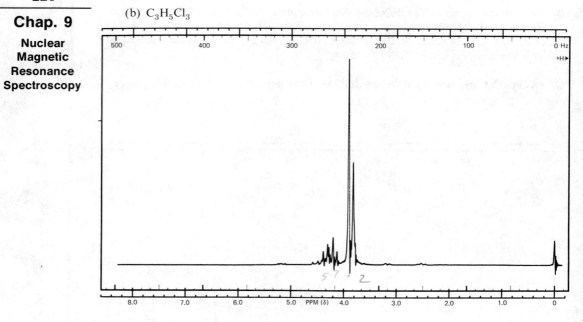

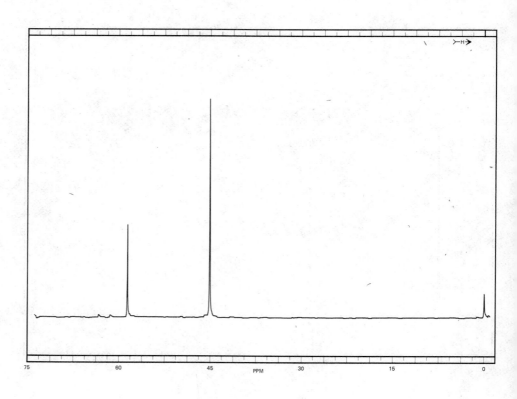

10. Deduce the structure of the following compound (C_4H_9Br) from its proton-decoupled cmr spectrum. What will its nmr spectrum look like?

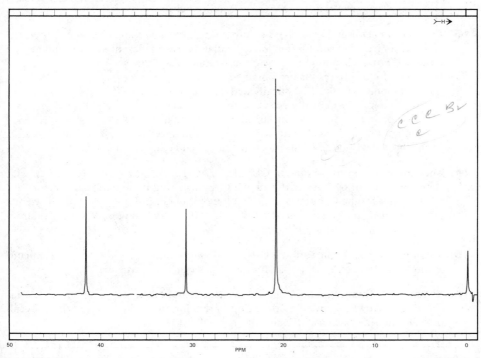

11. There are four compounds with the formula C_4H_9Cl. The proton-decoupled cmr spectrum of one of the isomers is shown below. Which isomers are eliminated by the spectrum? Which isomers might give such a spectrum? Describe how the proton-coupled cmr spectrum can be used to decide which C_4H_9Cl isomer the compound is.

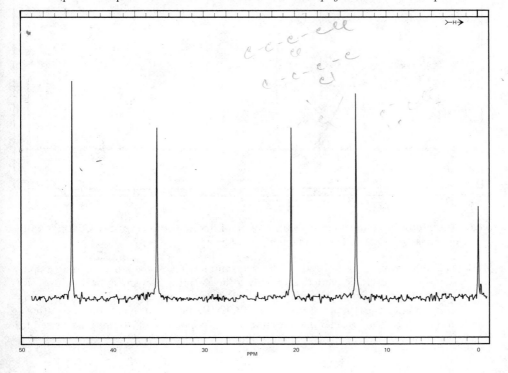

228

Chap. 9

**Nuclear
Magnetic
Resonance
Spectroscopy**

12. Explain why the cmr spectra of *cis*-1,2-dimethylcyclohexane shows only one methyl resonance, even though one methyl is axial and the other is equatorial. How would you expect the spectrum to change on cooling below $-70°$.

13. Free radical chlorination of (R)-2-chlorobutane gives a mixture of isomeric products which is subjected to careful fractional distillation. Five different dichlorobutanes are obtained. For each compound, the optical rotation and the cmr spectrum are measured. The following results are obtained.

  Compounds E and F: optically active; cmr, 4 resonances
  Compound G: optically active; cmr, 2 resonances
  Compound H: optically inactive; cmr, 4 resonances
  Compound I: optically inactive; cmr, 2 resonances

From these data, make whatever structural assignments are possible.

14. The cmr spectrum of *n*-butanol, $CH_3CH_2CH_2CH_2OH$, shows four resonances with chemical shifts of 61.7, 35.3, 19.4, and 13.9 ppm. Using these data, calculate α-, β-, and γ-effects for the OH group. Chlorination of *n*-butanol gives several products, one of which shows the cmr resonances 62.0, 29.7, 30.0, and 45.4 ppm. What is the structure of this chlorination product?

15. A compound having the formula $C_5H_{10}Br_2$ has a proton-coupled cmr spectrum consisting of a doublet, a triplet, and a quartet. What is its structure?

16. The nmr spectrum of difluoromethane measured as a dilute solution in carbon tetrachloride is shown below. Provide an interpretation of the spectrum. (*Hint:* Recall that the fluorine nucleus, ^{19}F, also has spin of $\pm\frac{1}{2}$.)

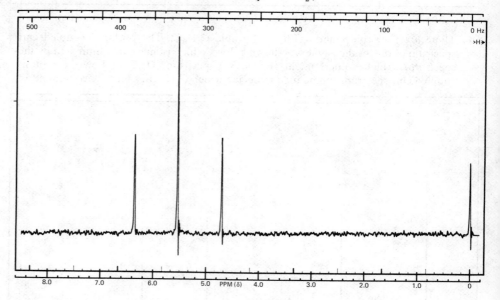

17. The proton-decoupled cmr spectrum of ethyl fluoride consists of two doublets with chemical shifts of 79.3 and 14.6 ppm. Explain.

18. The radio waves used to irradiate the nmr sample are absorbed in converting α-spin states to β and are converted ultimately to heat. To see how much heat is involved calculate approximately the temperature increase produced in 1 mL of an nmr solution containing 0.01 mole of protons in a 60 MHz nmr instrument. For the purpose of this calculation consider that the entire excess population of α-spins is converted to β and that the heat capacity of the solution is 1 cal deg^{-1} mL^{-1}.

Chapter 10
Alcohols and Ethers

10.1
Introduction: Structure

Alcohols are compounds in which an alkyl group replaces one of the hydrogens of water. They are organic compounds that contain the functional group —OH: As we shall see, this functional group dominates the chemistry of alcohols. **Ethers** are analogs of water in which both hydrogens are replaced by alkyl groups.

$$CH_3—OH \qquad CH_3—O—CH_3$$

an alcohol   an ether

(methyl alcohol)   (dimethyl ether)

The geometry of methyl alcohol, as determined by microwave spectroscopy, is shown as follows.

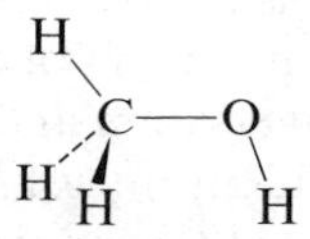

Bond Lengths, Å		Bond Angles, deg	
C—H	1.10	H—C—H	109
O—H	0.96	H—C—O	110
C—O	1.43	C—O—H	108.9

The hybridization of carbon is approximately sp^3. The hybridization of oxygen may also be described as approximately sp^3. Oxygen makes one bond to carbon and one to hydrogen. The O—H bond distance is precisely the same as the O—H bond distance in water. Conformationally, the molecule exists in a structure that has the O—H bond staggered between two C—H bonds and the barrier to rotation about the C—O bond is 1.1 kcal mole^{-1}. It is frequently useful to consider the oxygen lone-pair electrons to occupy orbitals that are each approximately sp^3; such lone-pair orbitals are each staggered between two adjacent C—H bonds.

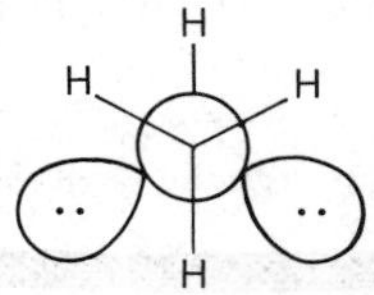

In dimethyl ether the methyl groups are bent apart to minimize nonbonded interactions between the hydrogens marked *a* in the following structural formula. There are two O—C—H and two H—C—H angles, but all of the angles are close to the tetrahedral value of 109.5°.

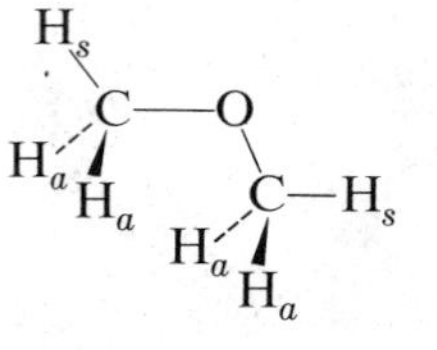

Bond Distances, Å		Bond Angles, deg	
C—H	1.10	C—O—C	111.7
C—O	1.41	O—C—H_a	110.8
		O—C—H_s	107.2
		H_a—C—H_a	108.7
		H_a—C—H_s	109.5

The simple alcohols are important industrial materials. They are also used extensively as laboratory reagents and as solvents. The most important representative of the class is undoubtedly ethyl alcohol. Dilute solutions of this compound containing flavorsome impurities have been known since the dawn of civilization. Indeed, it has been suggested that the discovery of alcohol fermentation and the physiological effects of its product provided a major incentive for agriculture and the start of civilization! Alcohols can be prepared readily from many other classes of compounds and can, in turn, be transformed into many others. For this reason, alcohols play a key role as synthetic intermediates.

10.2
Nomenclature of Alcohols

Like most other classes of organic compounds, alcohols can be named in several ways. Common names are useful only for the simpler members of a class. However, common names are widely used in colloquial conversation and in the scientific literature. In order to communicate freely, the student must know common names. Since the systematic IUPAC names are often used for indexing the scientific literature, the student must be thoroughly familiar with systematic names in order to retrieve data from the literature.

A. *The Alkyl Alcohol System*

The name of an alcohol is derived in one system of common nomenclature by combining the name of the alkyl group with the word alcohol. The names are written as two words.

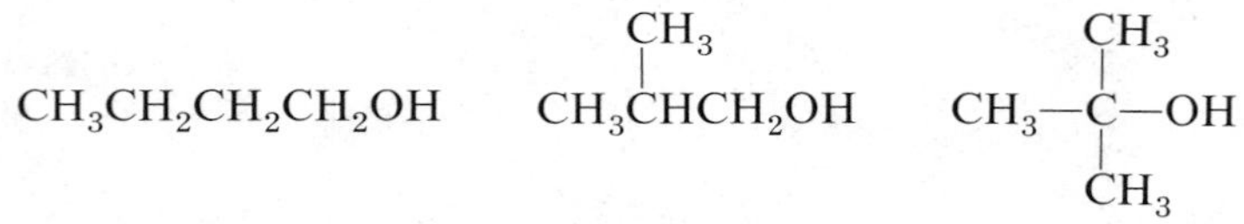

$$CH_3CH_2CH_2CH_2OH \qquad CH_3CHCH_2OH \qquad CH_3—\overset{CH_3}{\underset{CH_3}{C}}—OH$$

n-butyl alcohol   isobutyl alcohol   *t*-butyl alcohol

In this common system, the position of an additional substituent is indicated by letters of the Greek alphabet rather than by numbers.

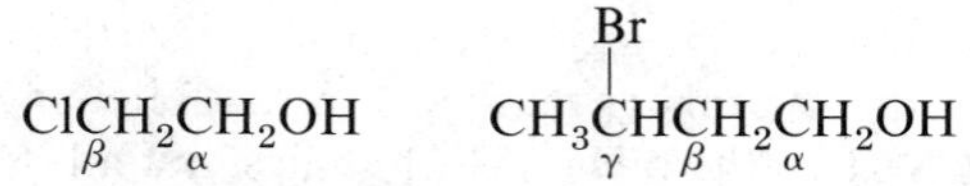

$$ClCH_2CH_2OH \qquad CH_3CHCH_2CH_2OH$$
$$_{\beta}_{\alpha} \qquad _{\gamma}_{\beta}_{\alpha}$$

β-chloroethyl alcohol   γ-bromobutyl alcohol

This use of the Greek alphabet is widespread in organic chemistry, and it is important to learn the first few letters, at least through delta (the entire Greek alphabet is given inside the front cover of this book). Many of the letters, small and capital, have evolved standard meanings in the mathematical and physical sciences (for example, the number π). In organic chemistry, the lowercase letters are used more frequently than the capital letters.

The last letter of the Greek alphabet is omega, ω. Correspondingly, this letter is used to refer to difunctional compounds when the secondary substituent is on the end carbon of the chain.

$$Br(CH_2)_n OH$$
ω-bromo alcohols

Any simple group that has a common name may be used in the alkyl alcohol system, with one important exception. The grouping C_6H_5—has the special name **phenyl,** but the compound C_6H_5OH is **phenol** not phenyl alcohol.

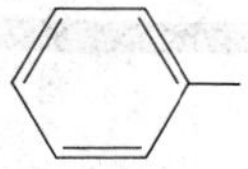

phenyl group

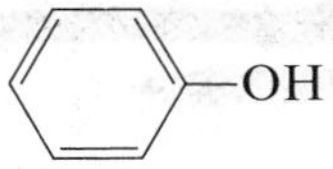

phenol
(not phenyl alcohol)

Substituted phenols are named as derivatives of the parent compound phenol.

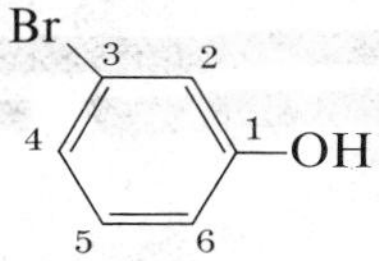

3-bromophenol

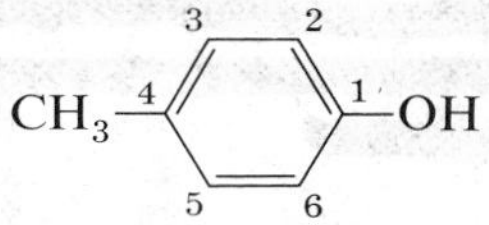

4-methylphenol

The reason for this difference is historical and arose from the fact that phenol and its derivatives have many chemical properties that are very different from those of alkyl alcohols. In this text, phenols are considered as a separate class of compounds (Chapter 30).

However, phenyl-substituted alkyl alcohols are normal alcohols and often have common names. Examples are

$C_6H_5CH_2OH$
benzyl alcohol

$C_6H_5\overset{\underset{|}{OH}}{C}HCH_3$
α-phenylethyl alcohol

$C_6H_5CH_2CH_2OH$
β-phenylethyl alcohol

B. The Carbinol System

In the carbinol system the simplest alcohol, CH_3OH, is called **carbinol.** More complex alcohols are named as alkyl-substituted carbinols. The names are written as one word.

$CH_3\overset{\underset{|}{OH}}{C}HCH_2CH_3$
ethylmethylcarbinol

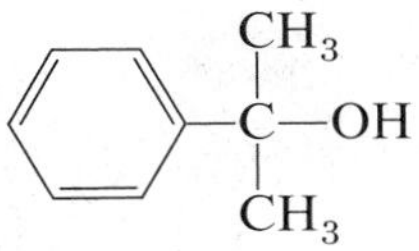

$(CH_3CH_2)_3COH$
triethylcarbinol

dimethylphenylcarbinol

The number of carbons attached to the carbinol carbon distinguishes primary, secondary, and tertiary carbinols. As in the case of the alkyl halides, this classification is useful because the different types of alcohols show important differences in reactivity under given conditions. The carbinol system of nomenclature has been falling into disuse in recent years and is no longer recommended. However, it is found extensively in the older organic chemical literature.

C. The IUPAC System

In the IUPAC system of nomenclature alcohols are named by replacing the **-e** of the corresponding alkane name by the suffix **-ol,** that is, as **alkanols.**

$$CH_3OH \qquad CH_3CH_2OH$$

methanol   ethanol

The **alkan-** stem corresponds to the longest carbon chain in the molecule *which contains the —OH group*. The chain is numbered so that the OH group gets the smaller of two possible numbers.

$$CH_3CH_2CH_2CH_2OH \qquad CH_3\overset{\text{OH}}{C}HCH_2CH_3$$

1-butanol   2-butanol
or butan-1-ol   or butan-2-ol

Substituents are appended as prefixes and are numbered according to the numbering system established by the position of the OH group. Names are written as one word with no spaces.

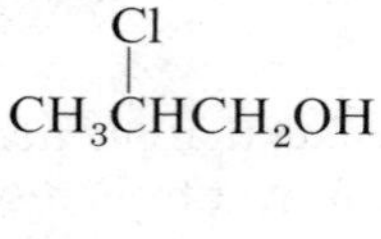

$$CH_3\overset{\text{Cl}}{C}HCH_2OH$$

2-chloro-1-propanol

$$CH_3\overset{\text{CH}_3}{C}H\overset{\text{OH}}{C}HCHCH_3$$

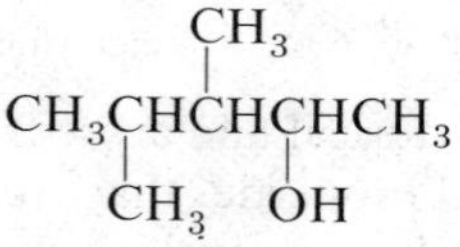

3,4-dimethyl-2-pentanol
(not 2,3-dimethyl-4-pentanol)

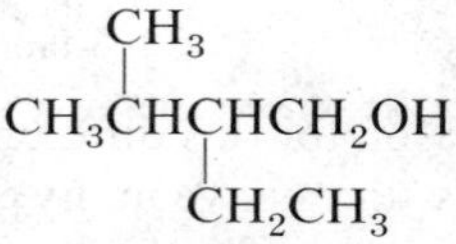

$$CH_3\overset{\text{CH}_3}{C}H\overset{}{C}HCH_2OH \\ \overset{}{C}H_2CH_3$$

2-ethyl-3-methyl-1-butanol
(not 2-isopropyl-1-butanol;
choose chain with
most substituents)

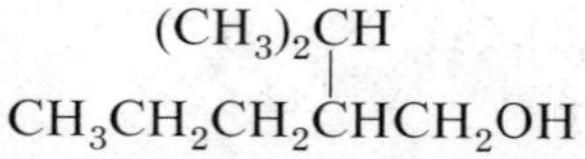

$$CH_3CH_2CH_2\overset{(CH_3)_2CH}{C}HCH_2OH$$

2-isopropyl-1-pentanol
(not 3-methyl-2-propyl-1-
butanol; pentanol has
longer chain)

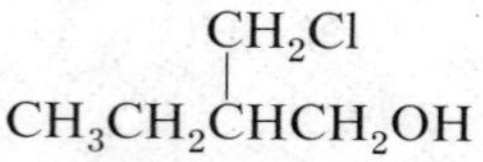

$$CH_3CH_2\overset{CH_2Cl}{C}HCH_2OH$$

2-(chloromethyl)-1-butanol

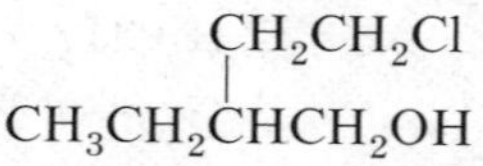

$$CH_3CH_2\overset{CH_2CH_2Cl}{C}HCH_2OH$$

4-chloro-2-ethyl-1-butanol

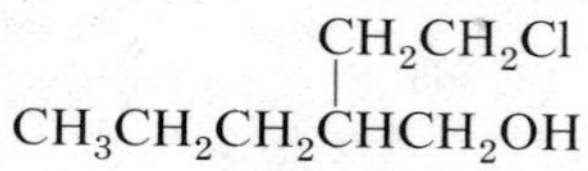

$$CH_3CH_2CH_2\overset{CH_2CH_2Cl}{C}HCH_2OH$$

2-(2-chloroethyl)-1-pentanol
(not β-chloroethyl; do not
mix common and IUPAC systems)

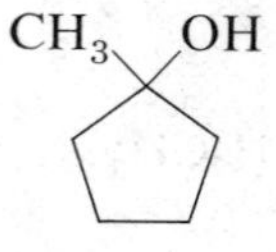

1-methylcyclopentanol

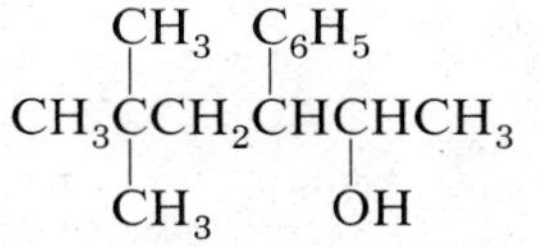

$$CH_3 \quad C_6H_5$$
$$CH_3CCH_2CHCHCH_3$$
$$CH_3 \quad OH$$

5,5-dimethyl-3-phenyl-2-hexanol

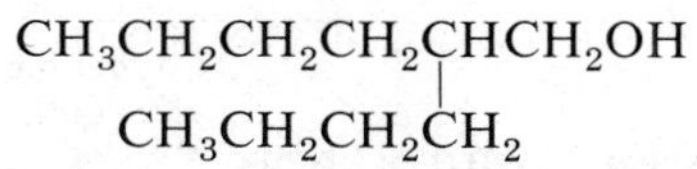

$$CH_3CH_2CH_2CH_2CHCH_2OH$$
$$CH_3CH_2CH_2CH_2$$

2-butyl-1-hexanol

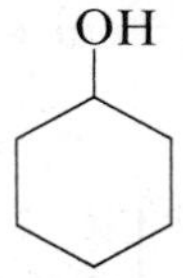

cyclohexanol

The general rule in the IUPAC system is that a functional group named as a suffix becomes a parent system that dominates the numbering scheme. Prefix groups are substituents or appendages to the parent.

Some kinds of alcohols are too difficult or cumbersome to name as alkanols. For such compounds it is preferable to use the appropriate hydroxyalkyl name as a prefix.

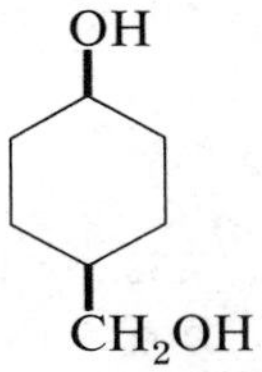

cis-4-(hydroxymethyl)cyclohexanol

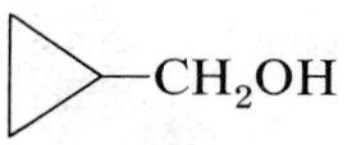

(hydroxymethyl)cyclopropane

EXERCISE: Name each of the following compounds by the carbinol and IUPAC systems.

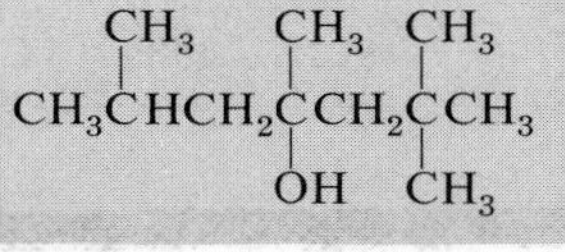

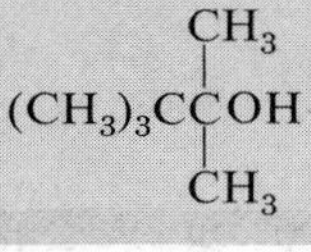

$$CH_3 \quad CH_3 \quad CH_3$$
$$CH_3CHCH_2CCH_2CCH_3$$
$$OH \quad CH_3$$

$$CH_3$$
$$(CH_3)_3CCOH$$
$$CH_3$$

10.3
Physical Properties of Alcohols

The lower alcohols are liquids with characteristic odors and sharp tastes. One striking feature is their relatively high boiling points (Table 10.1). The OH group is roughly equivalent to a methyl group in size and polarizability, but alcohols have much higher boiling points than the corresponding hydrocarbons; for example, compare ethanol (mol. wt. 46, b.p. 78.5°) and propane (mol. wt. 44, b.p. −42°). A plot of boiling point versus molecular weight for straight-chain alcohols and alkanes is shown in Figure 10.1.

The abnormally high boiling points of alcohols are the result of a special type of dipolar association in the liquid phase. Both the C—O and the O—H bonds are polar because of the different electronegativities of carbon, oxygen, and hydrogen. These polar bonds contribute to the substantial dipole moments. However, the dipole moments of alcohols are no greater than those of corresponding chlorides.

$$CH_3OH \qquad CH_3CH_2OH$$
$$\mu = 1.71 \text{ D} \qquad \mu = 1.70 \text{ D}$$

$$CH_3Cl \qquad CH_3CH_2Cl$$
$$\mu = 1.94 \text{ D} \qquad \mu = 2.04 \text{ D}$$

We expect electrostatic interaction between dipoles of the type ⟶ ⟶, but

TABLE 10.1
Physical Properties of Alcohols

Compound	Common Name	IUPAC Name	Melting Point, °C	Boiling Point, °C	Density d^{20}	Solubility, g/100 mL H_2O
CH_3OH	methyl alcohol	methanol	−97.8	65.0	0.7914	∞
CH_3CH_2OH	ethyl alcohol	ethanol	−114.7	78.5	0.7893	∞
$CH_3CH_2CH_2OH$	n-propyl alcohol	1-propanol	−126.5	97.4	0.8035	∞
$CH_3CHOHCH_3$	isopropyl alcohol	2-propanol	−89.5	82.4	0.7855	∞
$CH_3CH_2CH_2CH_2OH$	n-butyl alcohol	1-butanol	−89.5	117.3	0.8098	8.0
$CH_3CH_2CHOHCH_3$	sec-butyl alcohol	2-butanol	−114.7	99.5	0.8063	12.5
$(CH_3)_2CHCH_2OH$	isobutyl alcohol	2-methyl-1-propanol		107.9	0.8021	11.1
$(CH_3)_3COH$	tert-butyl alcohol	2-methyl-2-propanol	25.5	82.2	0.7887	∞
$CH_3(CH_2)_4OH$	n-pentyl alcohol	1-pentanol	−79	138	0.8144	2.2
$C_2H_5(CH_3)_2COH$	tert-pentyl alcohol	2-methyl-2-butanol	−8.4	102	0.8059	∞
$CH_3CH_2CH_2CHOHCH_3$	—	2-pentanol		119.3	0.809	4.9
$CH_3CH_2CHOHCH_2CH_3$	—	3-pentanol		115.6	0.815	5.6
$(CH_3)_3CCH_2OH$	neopentyl alcohol	2,2-dimethyl-1-propanol	53	114	0.812	∞
$CH_3(CH_2)_5OH$	n-hexyl alcohol	1-hexanol	−46.7	158	0.8136	0.7

alcohols boil much higher than the corresponding chlorides. In fact, alkyl chlorides differ very little in boiling point from alkanes of corresponding molecular weight (Figure 8.2). It would seem, then, that dipolar attraction is not the cause of the elevated boiling points of alcohols. Or is it? The magnitudes of the individual dipole moments are not the only important factor. How closely the negative and positive ends of the dipoles can approach one another is also important.

By Coulomb's law two opposite charges attract each other with an energy proportional to $1/r$ where r is the distance between the charges. The electrostatic energy of two dipoles depends on $1/r^3$ and therefore falls off sharply with distance. In alkyl halides the negative end of the dipoles is out at the lone-pair electrons, but the positive end is in the C—X bond close to carbon. Because of the van der Waals size of carbon, the positive and negative ends of adjacent dipoles cannot get close together and the electrostatic energy of dipole–dipole attraction is relatively small.

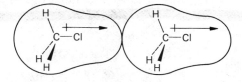

Consequently, such dipole association does not have much of an effect on the energy required to separate alkyl halide molecules.

For alcohols the negative end of the dipole is out at the oxygen lone pairs, and

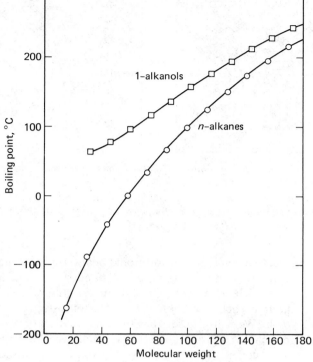

FIGURE 10.1. *Boiling points of 1-alkanols as a function of molecular weight.*

the positive end is close to the small hydrogen. For hydrogen atoms bonded to electronegative elements dipole–dipole interaction is uniquely important and is called a **hydrogen bond.** This proximity of approach is shown by bond-distance data. The O—H bond length in alcohols is 0.96 Å. The hydrogen-bonded H···O distance is 2.07 Å, about twice as large. In fact, this distance is sufficiently small that some hydrogen bonds may have a significant amount of covalent or shared-electron character. In condensed phases, alcohols are associated via a chain of hydrogen bonds.

A three-dimensional view of liquid methanol showing the hydrogen-bonding network is given in Figure 10.2. The figure is of a random cube of the liquid, showing only the C—O—H parts of the various molecules. Some of the oxygen atoms are blackened in the figure. The six at the top of the cube are the oxygens of a "chain" of methanol molecules; the molecules are held in the chain by hydrogen bonds. At the bottom of the cube may be seen a cyclic tetramer of methanol molecules.

The O—H bond in alcohols is stronger than most C—H bonds. It has a bond dissociation energy of 103 kcal mole^{-1}. The hydrogen bond is far weaker—only

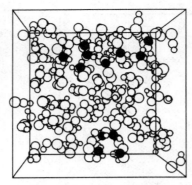 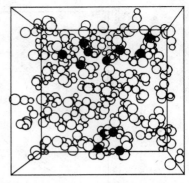

FIGURE 10.2. *Three-dimensional view (stereo plot) of a random cube of liquid methanol, showing hydrogen-bonded networks. The large circles represent methyl groups, the small circles hydrogen, and the intermediate circles oxygen. Darkened oxygens describe a chain and a cycle of molecules. [Courtesy of W. L. Jorgensen, Department of Chemistry, Purdue University.]*

about 5 kcal mole^{-1}. Nevertheless, this additional heat term in the heat of vaporization results in relatively high boiling points. A bond strength of 5 kcal mole^{-1} does not sound like much. However, when there are many such bonds, as in polyhydroxy compounds such as carbohydrates, the total strength is sufficient to hold up tall redwoods. We will also see that the combination of small individual strength and large combined group strength is exactly the kind of bonding required for the genetic code. Life as we know it is impossible without hydrogen bonds.

Since alcohols and water both contain the OH group, they have many properties in common. We should emphasize that water is a remarkable substance. Its boiling point of 100° is exceedingly high for a compound having a molecular weight of only 18. The extensive hydrogen-bond networks in liquid water make it a highly polar liquid having a dielectric constant of 78.5 at 25°. In aqueous solution the interaction between ions is relatively small; hence water is a good solvent for ionic compounds. The lower alcohols also have relatively high dielectric constants (Table 10.2). As the carbon chain gets longer, the importance of the OH group is reduced and the dielectric constant approaches the alkane value of about 2.

Methanol and ethanol are reasonably good solvents for salt-like compounds. Because they are also good solvents for organic compounds, they are used frequently for organic reactions such as S_N2 displacement reactions.

TABLE 10.2
Dielectric Constants of Alcohols

Compound	Dielectric Constant
H_2O	78.5
CH_3OH	32.6
CH_3CH_2OH	24.3
$CH_3(CH_2)_3OH$	17.1
$CH_3(CH_2)_4OH$	13.9
$CH_3(CH_2)_{11}OH$	6.5

237

Sec. 10.4

**Acidity
of Alcohols:
Inductive
Effects**

The OH group of alcohols can participate in the hydrogen-bond network of water. The lower alcohols are completely soluble in water. As the hydrocarbon chain gets larger, the compound begins to look more like an alkane, and more of the hydrogen bonds in water must be broken to make room for the hydrocarbon chain. Since the hydrogen bonds that are lost are not completely compensated by bonding to the alcohol OH, solubility decreases as the hydrocarbon chain gets larger. A rough point of division is four carbons to one oxygen. Above this ratio, alcohols tend to have little solubility in water. This guideline is only approximate because the shape of the hydrocarbon portion is also important. *t*-Butyl alcohol is much more soluble than *n*-butyl alcohol because the *t*-butyl group is more compact and requires less room or broken water hydrogen bonds in an aqueous solution. A similar phenomenon is seen with the branched pentyl alcohols.

10.4
Acidity of Alcohols: Inductive Effects

One of the important properties of water is its self-ionization.

$$H_2O + H_2O \underset{}{\overset{K_w}{\rightleftharpoons}} H_3O^+ + OH^-$$

In pure water, the concentrations of H_3O^+ and OH^- are very low, only 10^{-7} mole L^{-1}. The ion product or self-dissociation constant K_w is defined as

$$K_w = [H_3O^+][OH^-] = 1.0 \times 10^{-14} \text{ mole}^2 \text{ L}^{-2} \text{ (or } M^2)$$

> Remember that this is not a normal equilibrium constant, which includes the concentrations of reactants and products. For water, the concentration is 55.5 moles L^{-1}. The equilibrium constant for dissociation is therefore
>
> $$K = \frac{(10^{-7})(10^{-7})}{(55.5)(55.5)} = 3.25 \times 10^{-18}$$
>
> Note that K is unitless. The relationship between K and K_w is
>
> $$K_w = K \times (55.5 \ M)^2$$

Alcohols also undergo self-dissociation, but to a much smaller extent than water.

$$2\ CH_3OH \rightleftharpoons CH_3OH_2^+ + CH_3O^-$$

For methanol the ion product $K_{CH_3OH} = 1.2 \times 10^{-17} \ M^2$, and for higher alcohols the value is even smaller.

$$K_{CH_3OH} = [CH_3OH_2^+][CH_3O^-] = 1.2 \times 10^{-17} \ M^2$$

The reduced value comes in large part from the lower dielectric constant of alcohols—it takes greater energy to separate charges. But the relative acidities and basicities of alcohols are also important.

The acidity of an alcohol in water is defined in the usual way.

$$ROH + H_2O \underset{}{\overset{K_a}{\rightleftharpoons}} H_3O^+ + RO^-$$

The acid dissociation constant K_a is defined as

$$K_a = \frac{[H_3O^+][RO^-]}{[ROH]}$$

Since these equilibria refer to dilute water solutions, the concentration of water is generally omitted in the expression for an equilibrium constant and K_a has units of mole L^{-1} or molarity M. Recall that the acid dissociation constant is generally such a small number that it is usually more convenient to refer to the negative logarithm or pK_a.

$$pK_a = -\log K_a$$

Values of pK_a for some alcohols are listed in Table 10.3 and are compared to some common inorganic acids. Note that the K_a for water is obtained by dividing K_w by the concentration of water, 55.5 moles L^{-1}. This change is necessary to put all of the ionizations on the same scale and in the same units. Recall that the ion product of water K_w has units of moles^2 L^{-2} or M^2, whereas K_a values are given in units of moles L^{-1} or M.

Methanol and ethanol are about as acidic as water itself. The higher alcohols are less acidic. Water and the alcohols are generally much less acidic than other compounds commonly regarded as acids. Strong acids, such as HI, HBr, HCl, H$_2$SO$_4$, have negative pK_a values. Such compounds are completely dissociated in water and are 10^{15} to 10^{20} more acidic than alcohols. Typical "weak acids" such as acetic acid (to be discussed in Chapter 18), HF, H$_2$S, and HOCl are still 10^7 to 10^{10} stronger than alcohols. For a review of acidity, see Section 4.5.

The character of alcohols as weak acids emerges primarily in reactions with strong base. Alcohols, like water, react with alkali metals to liberate hydrogen and form the corresponding metal alkoxide.

$$CH_3OH + Na \longrightarrow CH_3O^- + Na^+ + \tfrac{1}{2}H_2$$

The reaction tends to be less vigorous than that with water. In fact, isopropyl alcohol is often used to decompose scraps of sodium in the laboratory because its reaction is relatively slow and moderate. When sodium reacts with water, the reaction is so rapid that the heat produced cannot be dissipated quickly enough; the evolved hydrogen catches fire, and an explosion results. Tertiary alcohols react so sluggishly with sodium that potassium must often be used to convert such an alcohol to the alkoxide.

Potassium has a relatively low m.p., 64°. In laboratory use it is often converted to a finely divided state in order to render it more reactive. Solid pieces of potassium are added to benzene (b.p. 80°), and the mixture is heated to reflux. The potassium melts, and the mixture is allowed to cool with vigorous stirring. The potassium solidifies as small particles of "potassium sand." The alcohol is added to this mixture with stirring

TABLE 10.3
pK_a Values for Alcohols and Some Acids

Compound	pK_a	Compound	pK_a
H$_2$O	15.7	HCl	−2.2
CH$_3$OH	15.5	H$_2$SO$_4$	−5
C$_2$H$_5$OH	15.9	H$_3$PO$_4$	2.15
(CH$_3$)$_3$COH	~18	HF	3.18
ClCH$_2$CH$_2$OH	14.3	H$_2$S	6.97
CF$_3$CH$_2$OH	12.4	HOCl	7.53
C$_6$H$_5$OH	10.0	H$_2$O$_2$	11.64
CH$_3$COOH	4.8		

239

Sec. 10.4

**Acidity
of Alcohols:
Inductive
Effects**

and generally reacts readily because of the large surface area of the potassium sand. This procedure is especially useful with tertiary alcohols..

Another reagent commonly used instead of sodium itself is sodium hydride, NaH. This compound is a nonvolatile, insoluble salt, Na^+H^-, and reacts readily with acidic hydrogens.

$$CH_3OH + Na^+H^- \longrightarrow CH_3O^- + Na^+ + H_2$$

The reaction may be regarded as a combination of hydride ion with a proton.

$$H:^- + H^+ \longrightarrow H_2$$

The sodium salts of primary alcohols are common reagents in organic chemistry.

$$Na^+ \ ^-OCH_3 \qquad Na^+ \ ^-OCH_2CH_3$$

sodium methoxide   sodium ethoxide

Because sodium and sodium hydride react so sluggishly with tertiary alcohols, the corresponding potassium salts are more commonly used as reagents.

$$K^+ \ ^-O-\overset{\overset{\displaystyle CH_3}{|}}{\underset{\underset{\displaystyle CH_3}{|}}{C}}-CH_3$$

potassium *t*-butoxide

Note in Table 10.3 that 2-chloroethanol is significantly more acidic than ethanol. This difference in acidity is best understood in terms of the electrostatic interaction of the C—Cl dipole with the negative charge of the alkoxide ion.

$$\overset{Cl}{\underset{}{\diagdown}}CH_2-CH_2 \diagdown O^-$$

The positive end of the dipole is closer to the negative charge than is the negative end of the dipole. Electrostatic attraction exceeds repulsion, and the result is a net stabilization of the anion. Stabilization of the anion increases its ease of formation and the conjugate acid, 2-chloroethanol (or β-chloroethyl alcohol), is more acidic than the unsubstituted alcohol (Figure 10.3).

$$\Delta H^\circ_{ion.} (ClCH_2CH_2OH) < \Delta H^\circ_{ion.} (CH_3CH_2OH)$$

FIGURE 10.3. *The effect of a dipolar substituent on the ionization energy of an alcohol. The stabilizing effect of the substituent is greater in the anion (charge-dipole interaction) than in the alcohol (dipole–dipole interaction).*

This effect is generally called an **inductive field effect** or, more simply, an **inductive effect.** The magnitude of the effect falls off as the distance between the dipolar group and the charged group is increased. The effect increases with the number of dipolar groups. Note the relatively large effect of the three fluorines in β,β,β-trifluoroethyl alcohol. Halogen groups are said to be **electron-attracting** and to stabilize anions. The effect is present in inorganic systems as well: HOCl is a stronger acid than HOH. Conversely, alkyl groups are generally considered to be somewhat **electron-donating** and, therefore, to weaken acids. We will make use of inductive effects frequently in our subsequent discussions of the effects of structure on reactivity.

Alcohols, like water, are not only acids, but bases. Hydrogen chloride passed into an alcohol protonates the oxygen just as in water.

$$H_2O + HCl \longrightarrow H_3O^+ + Cl^-$$

$$\underset{\text{alkyloxonium chloride}}{ROH + HCl \rightleftharpoons ROH_2^+ Cl^- \rightleftharpoons ROH_2^+ + Cl^-}$$

The initially formed species in such a protonation is an **ion pair** (two ions in close juxtaposition). In water and the lower alcohols, the dielectric constant is sufficiently high that the initially formed ion pairs can largely dissociate to free ions. As the dielectric constant becomes smaller, however, too much work is required to separate the ion pairs, and the oxonium chloride remains largely associated.

EXERCISE: Compare relative acidities of the following pairs of compounds.

(a) $ClCH_2CH_2CH_2OH$ and $CH_3CHClCH_2OH$

(b) $CH_3CH_2CH_2CH_2OH$ and $CH_3OCH_2CH_2OH$

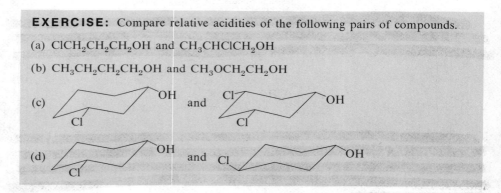

10.5
Nuclear Magnetic Resonance

Primary and secondary alcohols have hydrogen at a carbon bearing an electronegative oxygen. We expect to find resonance of such protons downfield from normal alkane protons, as in the case of alkyl halides. The downfield shift is 2.3–2.5 ppm from the corresponding alkane position. The actual positions depend on the number of protons on the carbinol carbon.

	$CH_3{-}R$	$R'CH_2{-}R$	$R'R''CH{-}R$
δ, ppm:	0.9	1.25	1.5
	$CH_3{-}OH$	$R'CH_2{-}OH$	$R'R''CH{-}OH$
δ, ppm:	3.4	3.6	3.85

Protons β to oxygen are shifted slightly downfield, about 0.1–0.3 ppm.

The hydroxy proton itself shows more complex behavior. In rigorously purified alcohols, the hydroxy proton shows normal splitting by adjacent carbinol protons.

In this case the proton exchange caused by autoprotolysis is sufficiently slow that a given proton is associated with a given oxygen on the nmr time scale. However, the protons are still hydrogen-bonded, and this leads to deshielding. The spectrum of pure ethyl alcohol in Figure 10.4 is illustrative. Note that the H—C—C—H and the H—C—O—H couplings have different magnitudes. The CH₂ resonance appears as a complex multiplet rather than a simple quartet.

Traces of acid or base cause the resonance of the hydroxy proton to collapse to a sharp singlet (see Figure 10.5). In such cases proton exchange is rapid on the nmr time scale, and the "state" observed is that of a proton in a weighted average of a number of environments. No spin-spin splitting is observed for such a proton.

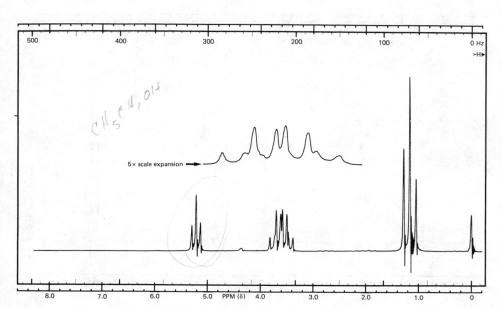

FIGURE 10.4. *Nmr spectrum of pure ethyl alcohol.*

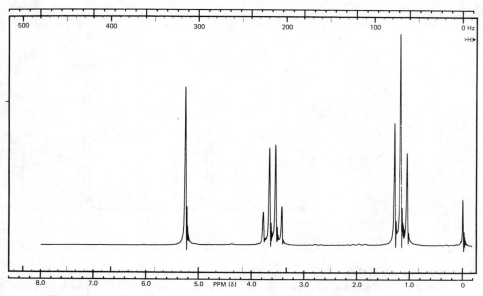

FIGURE 10.5. *Nmr spectrum of ethyl alcohol containing 1% formic acid.*

When an alcohol is diluted by an inert solvent, its hydroxy proton resonance shifts to higher field because hydrogen bonding becomes less important (compare Figure 10.4 with Figure 10.6). In very dilute solutions the hydroxy proton may resonate as high as 0.5 ppm. Often these extreme cases are not observed in the nmr spectrum of an alcohol. Usually the hydroxy proton, because of a combination of hydrogen bonding and some exchange, is observed as a broad featureless peak at a position varying from 2 to 4.5 ppm.

The exact appearance and position depend on the solvent, purity, temperature, and structure. One simple diagnosis for an OH group in the nmr is the addition of D_2O to the nmr solution. Rapid exchange replaces the OH groups by OD

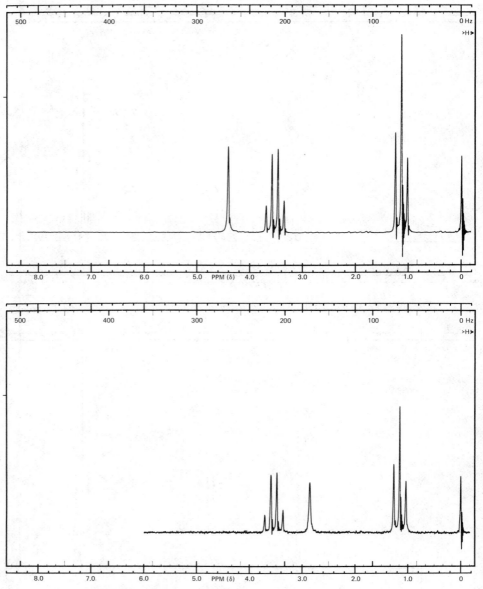

FIGURE 10.6. *Nmr spectra of slightly impure ethyl alcohol in carbon tetrachloride. The ethyl alcohol concentration in the top spectrum is 1.0* M; *in the bottom 0.25* M.

$$ROH + D_2O \rightleftharpoons ROD + DOH$$

and the nmr signal for OH vanishes or becomes less intense.

Table 10.4 summarizes the cmr chemical shifts for some simple alcohols. By comparing these data to those given for alkanes in Table 9.4, it may be seen that the hydroxy group causes a large downfield shift of the carbon to which it is attached. This OH α-effect is generally taken to be +48 ppm. The β- and γ-effects are about the same as for a methyl group, +9.5 ppm and −2.5 ppm, respectively.

TABLE 10.4
Cmr Chemical Shifts for Some Alcohols

	C-1	C-2	C-3	C-4	C-5
methanol	49.3				
ethanol	57.3	17.9			
1-propanol	63.9	26.1	10.3		
2-propanol	25.4	63.7	25.4		
1-butanol	61.7	35.3	19.4	13.9	
2-butanol	22.9	69.0	32.3	10.2	
2-methyl-1-propanol	69.2	31.1	19.2		
2-methyl-2-propanol	31.6	68.7	31.6		
1-pentanol	62.1	32.8	28.5	22.9	14.1
2-pentanol	23.6	67.3	41.9	19.4	14.3
3-pentanol	10.1	30.0	74.1	30.0	10.1

EXERCISE: (a) The CH_2 resonance of pure ethyl alcohol shown in Figure 10.3 is a double quartet with coupling constants of J = 5 and 7 Hz. Using a piece of graph paper, plot the expected appearance of such a double quartet. Remember that a simple doublet has relative intensities of 1:1 and a simple quartet has relative intensities of 1:3:3:1. What appearance would the CH_2 resonance have if the two Js were equal?

(b) Sketch the expected appearance of the nmr spectrum of highly purified isopropyl alcohol.

10.6
Preparation of Alcohols

Alcohols can be obtained from many other classes of compounds. Preparations from alkyl halides and from hydrocarbons will be discussed in this section. The following important ways of preparing alcohols will be discussed later as reactions of the appropriate functional groups.

1. Reduction of aldehydes and ketones (Section 13.8).

2. Addition of organometallics to aldehydes and ketones (Section 15.6).

3. Reduction of carboxylic acids (Section 18.7).

$$-\overset{\overset{\textstyle O}{\|}}{C}-OH \xrightarrow{[H]} -CH_2OH$$

4. Reduction of esters (Section 19.9).

$$-\overset{\overset{\textstyle O}{\|}}{C}-OR \xrightarrow{[H]} -CH_2OH + ROH$$

5. Addition of organometallics to esters (Section 19.6).

$$-\overset{\overset{\textstyle O}{\|}}{C}-OR + R'M \longrightarrow \xrightarrow{H_2O} \overset{\overset{\textstyle R'}{|}}{\underset{\underset{\textstyle R'}{|}}{C}}-OH + ROH$$

6. Additions to alkenes (Section 11.6).

$$\overset{\diagdown}{\diagup}C=C\overset{\diagup}{\diagdown} \longrightarrow H-\overset{|}{\underset{|}{C}}-\overset{|}{\underset{|}{C}}-OH$$

A. *Preparation from Alkyl Halides*

Hydrolysis of alkyl halides in aqueous solvents may occur by either the S_N1 or S_N2 mechanism. With some halides, elimination is a major side reaction (Chapter 8). The hydrolysis of most primary halides occurs by the S_N2 path and is sufficiently clean that this reaction is a good preparative method.

$$CH_3CH_2CH_2CH_2Cl + OH^- \xrightarrow[100°]{H_2O} CH_3CH_2CH_2CH_2OH + Cl^-$$

The reaction can be carried out in refluxing aqueous sodium hydroxide, especially with lower halides. Although alkyl halides are only slightly soluble in water, a two-phase reaction takes place. If the alcohol is water-soluble, the end of the hydrolysis is marked by a homogeneous solution. Alternatively, the reaction can be carried out in a mixture of water and some inert organic solvent such as dioxane.

$$\underset{\underset{\textstyle dioxane}{}}{\overset{\overset{\textstyle O}{\diagup\diagdown}}{\underset{\underset{\textstyle O}{\diagdown\diagup}}{\overset{CH_2 \quad CH_2}{\underset{CH_2 \quad CH_2}{}}}}$$

dioxane

> Dioxane is a colorless liquid, b.p. 100°, completely miscible with water. It is relatively inert to many reagents and is frequently used in mixtures with water to increase the solubility of organic compounds such as alkyl halides.

For secondary alkyl halides and for primary halides with a β-branch, elimination is an important side reaction and may be the principal reaction.

$$(CH_3)_2CHCH_2Br \xrightarrow{OH^-} (CH_3)_2C=CH_2$$

An alternative procedure that avoids the use of strong base makes use of acetate ion $(CH_3CO_2^-)$ as the nucleophilic reagent. Since acetate is much less basic than hydroxide (the pK_a of acetic acid, CH_3COOH, is 4.8, whereas that of water is 15.7), the E2 mechanism is suppressed, and alkene formation is minimized.

$$CH_3CH_2CH_2CH_2Br + CH_3\overset{\overset{O}{\|}}{C}O^- \ K^+ \xrightarrow[100°]{DMF} CH_3CH_2CH_2CH_2O\overset{\overset{O}{\|}}{C}CH_3 + K^+Br^-$$

$$(95\text{–}98\%)$$

The product, an ester, can be readily hydrolyzed to the desired alcohol (Section 19.6).

$$CH_3CH_2CH_2CH_2O\overset{\overset{O}{\|}}{C}CH_3 \xrightarrow[H_2O]{OH^-} CH_3CH_2CH_2CH_2OH + CH_3CO_2^-$$

Since displacement by acetate ion proceeds by the S_N2 mechanism, the alcohol product has an absolute configuration opposite that of the starting halide.

$$CH_3CH_2CH_2\overset{\overset{\displaystyle Cl}{|}}{\underset{\underset{\displaystyle D}{|}}{C}}H + CH_3CO_2^- \longrightarrow$$

(R)-1-chloro-1-deuteriobutane
(R)-1-chlorobutane-*1-d*

$$CH_3CH_2CH_2\overset{\overset{\displaystyle H}{|}}{\underset{\underset{\displaystyle D}{|}}{C}}O\overset{\overset{O}{\|}}{C}CH_3 \xrightarrow[H_2O]{OH^-} CH_3CH_2CH_2\overset{\overset{\displaystyle H}{|}}{\underset{\underset{\displaystyle D}{|}}{C}}OH + CH_3CO_2^-$$

(S)-butyl-*1-d* acetate   (S)-butanol-*1-d*

Implicit in the foregoing statement is the assumption that the newly formed C—O bond is not broken in the ester hydrolysis step. As we shall see in Section 19.6, the normal mechanism for this reaction involves cleavage of the other C—O bond, as indicated below.

This bond cleaves.
$$R—O \overset{\overset{O}{\|}}{\diagdown} C—R'$$

That is, ester hydrolysis *does not proceed by the S_N2 mechanism,* even when the ester group is attached to a primary carbon.

The nomenclature used for isotopically labeled compounds is implied in this example. The use of *-d* for deuterium and *-t* for tritium is common, although often the isotope is specified by the atomic symbol and a prefixed superscript giving the atomic mass of the isotope, for example, (R)-1-chlorobutane-1-^2H. Finally, deuterium may be specified by a prefix as in (R)-1-chloro-1-deuteriobutane.

Tertiary halides also undergo hydrolysis, but this reaction occurs by the S_N1 rather than the S_N2 mechanism. The reaction is best carried out by shaking the halide with aqueous sodium carbonate. The carbonate neutralizes the acid formed by hydrolysis

$$RX + H_2O \longrightarrow ROH + HX$$

but avoids the high concentration of hydroxide ion that encourages E2 elimination. Elimination by the E1 path is more difficult to avoid. However, this side reaction may be minimized by using highly aqueous solvents and by operating at low temperature. Nevertheless, hydrolysis of tertiary halides goes by way of carbocation intermediates, and such intermediates have several modes of reaction available besides reaction with water. Rearrangements can occur as discussed in Section 10.7.B.

The conversions of alkyl halides to alcohols discussed in this section are, by and large, *not important synthetic laboratory processes*. This is not due solely to deficiencies in the methods (although there obviously are some) but also to the practical fact that the halides are commonly obtained from alcohols in the first place. Hydrolysis of sulfonate esters is also a perfectly good reaction, but they are invariably prepared from alcohols. Hydrolysis of halides is an important industrial reaction for those halides obtained commercially by direct halogenation of hydrocarbons. More important laboratory syntheses of alcohols will be discussed in subsequent chapters as reactions of other functional groups.

B. *Preparation from Hydrocarbons*

Hydrocarbons which have particularly weak C—H bonds (that is, those which give stable free radicals upon loss of a hydrogen atom) react with oxygen to form hydroperoxides. For example, isobutane is converted into *t*-butyl hydroperoxide in this manner.

$$(CH_3)_3CH + O_2 \longrightarrow (CH_3)_3COOH$$
$$\text{\textit{t}-butyl hydroperoxide}$$

The reaction, which is called "autoxidation," requires the presence of a free radical initiator and proceeds through a free radical chain mechanism.

(1) $(CH_3)_3CH + rad\cdot \longrightarrow (CH_3)_3C\cdot + radH$ initiation

(2) $(CH_3)_3C\cdot + O_2 \longrightarrow (CH_3)_3COO\cdot$

(3) $(CH_3)_3COO\cdot + (CH_3)_3CH \longrightarrow (CH_3)_3COOH + (CH_3)_3C\cdot$ $\Big\}$ propagation

The bond-dissociation energy of the ROO—H bond is 90 kcal mole^{-1}, which is comparable to that of a tertiary R—H bond but weaker than primary and secondary C—H bonds. Consequently, the reaction is highly selective and is important only for tertiary hydrogens in alkanes. Because of the ready decomposition of alkyl hydroperoxides, the oxidation is generally carried only to low conversion. The reaction is a significant industrial process but is not a common organic laboratory reaction.

Alkyl hydroperoxides contain a weak O—O bond and decompose at temperatures over 100°, sometimes with explosive violence. They must be handled with extreme care. They may be reduced by catalytic hydrogenation to the corresponding alcohol.

$$ROOH \xrightarrow{\text{H}_2/\text{Pt or H}_2/\text{Pd}} ROH + H_2O$$

C. *Special Preparations*

Methanol at one time was prepared commercially by the dry distillation of wood and once had the commercial name of **wood alcohol.** It is now prepared on a large scale by catalytic hydrogenation of carbon monoxide. Methanol is toxic;

ingestion of small amounts causes nausea and blindness, and death can result from ingestion of 100 mL or less. It is an important industrial solvent and reagent and is also used to denature ethanol.

Ethanol is prepared for consumption in beverages by fermentation of sugars, but industrial alcohol is prepared by other routes such as the hydration of ethylene (Section 11.6.B). Ethanol is toxic in large quantities but is a normal intermediate in metabolism and, unlike methanol, is metabolized by normal enzymatic body processes. It is an important solvent and reagent in industrial processes and in such use is often "denatured" by addition of toxic and unappetizing diluents. Alcohol for consumption is heavily taxed, but denatured alcohol is not (for example, the 1979 price of 95% ethanol was $2.82 per gal; the federal excise tax was $4.00 per gal).

t-Butyl alcohol has a much higher octane number than straight-run gasoline (108 versus 70). For some time, it has been used for increasing the octane rating in the production of unleaded gasoline. It has the advantage of being a "clean" octane booster, in contrast to tetraethyllead, which is converted to potentially dangerous lead oxides upon combustion. Unfortunately, the supply of *t*-butyl alcohol is sufficient to treat only a minute fraction of the gasoline produced in the United States. It is manufactured in a process involving the oxidation of propylene by *t*-butyl hydroperoxide, which is in turn prepared by the oxidation of isobutane, as discussed in the previous section.

$$(CH_3)_3COOH + (CH_3)_2C{=}CH_2 \longrightarrow (CH_3)_3COH + (CH_3)_2\overset{O}{\overset{\triangle}{C{-}CH_2}}$$

The oxidation of alkenes by peroxides will be discussed in Section 11.5.E.

10.7
Reactions of Alcohols

The reactions of alcohols generally involve breaking one or more of three types of bonds in the carbinol structure and may be summarized as

    acidity, formation of esters

    displacement, carbocations, elimination

    oxidation

A. *Acidity: Alkoxide Ions*

The alkali alkoxides, produced by reaction of alcohols with alkali metals, are important reagents as bases in nonaqueous media and as nucleophilic reagents. An example of the latter use is

$$(C_2H_5)_3COH + K \longrightarrow \tfrac{1}{2} H_2 + (C_2H_5)_3CO^- \ K^+$$

$$(C_2H_5)_3CO^- \ K^+ + CH_3I \longrightarrow (C_2H_5)_3COCH_3 + KI$$

This is a typical S_N2 reaction and works best with primary halides having no β-substituents. In such cases the reaction is a good method for preparing ethers (Section 10.10.A). Other kinds of halides give more or less elimination.

$$(C_2H_5)_3CCl + CH_3O^- K^+ \longrightarrow (C_2H_5)_2C{=}CHCH_3 + KCl + CH_3OH$$
100% elimination

In cases where the amount of elimination is high, the reaction is an important route to alkenes (Section 11.5). Potassium *t*-butoxide and potassium *t*-pentoxide are frequently used as reagents for dehydrohalogenation because of their high basicity and because they are moderately soluble in nonpolar organic solvents such as benzene (C_6H_6).

$$K^+ \ ^-O{-}\overset{\displaystyle CH_3}{\underset{\displaystyle CH_3}{\overset{|}{\underset{|}{C}}}}{-}CH_2CH_3$$

potassium t-pentoxide

B. Alkyloxonium Salts

Sodium bromide is slightly soluble in ethanol. Such a solution can be refluxed indefinitely with no reaction; ethyl bromide and hydroxide ion are *not* formed. Recall that the *reverse* reaction, hydrolysis of ethyl bromide, can be carried out readily.

$$C_2H_5OH + Br^- \overset{K}{\rightleftharpoons} C_2H_5Br + OH^-$$

For this system, the thermodynamics is such that equilibrium lies far to the left (Section 4.2)

$$K = \frac{[C_2H_5Br][OH^-]}{[C_2H_5OH][Br^-]} = \ \sim 10^{-19}$$

and the rate of reaction of ethanol with bromide ion is very slow. However, if some sulfuric acid is added or if a mixture of ethanol and hydrogen bromide is refluxed, a reaction does occur.

$$C_2H_5OH + HBr \longrightarrow C_2H_5Br + H_2O$$

Why this dramatic difference? The displacement reaction in this case actually occurs on the intermediate **alkyloxonium salt** which is formed when the strong acid protonates the ethanol.

$$C_2H_5OH + HBr \rightleftharpoons C_2H_5OH_2^+ + Br^-$$

$$Br^- + \overset{\displaystyle}{\underset{\displaystyle CH_3}{\overset{|}{CH_2}}}{-}\overset{+}{O}H_2 \longrightarrow Br{-}\overset{\displaystyle}{\underset{\displaystyle CH_3}{\overset{|}{CH_2}}} + H_2O$$

Protonation converts the substrate from one with a very poor leaving group (hydroxide ion) to one with a better leaving group (water). The reaction is therefore only another example of an S_N2 reaction. For primary alcohols, the reaction is carried out by refluxing the alcohol with a mixture of concentrated sulfuric acid and either sodium bromide or hydrobromic acid.

> A mixture of 71 mL of 48% hydrobromic acid, 30.5 mL of concentrated sulfuric acid, and 37 g of *n*-butyl alcohol is refluxed for 2 hr. The product is separated, washed, and distilled to yield 50 g (95%) of *n*-butyl bromide.

For preparation of the corresponding chlorides, more vigorous conditions are required because chloride ion is a poorer nucleophile than bromide. A mixture of concentrated hydrochloric acid and zinc chloride, the so-called Lucas reagent, is frequently used. Zinc chloride is a powerful Lewis acid which serves the same purpose as does a proton in coordinating with the hydroxy oxygen.

$$Cl^- \quad \overset{\frown}{\underset{R}{C}H_2}\!\!-\!\!\overset{+}{O}\!\overset{ZnCl_2}{\underset{H}{\diagup}} \longrightarrow RCH_2Cl + HOZnCl + Cl^-$$

Secondary alcohols react more readily under these conditions than primary alcohols, and tertiary alcohols react the most rapidly of all. Tertiary alcohols are converted to alkyl chlorides by simply shaking with concentrated hydrochloric acid in the cold. The reaction clearly follows a rate trend that is characteristic of a carbocation process (tertiary > secondary > primary).

$$ROH_2^+ \rightleftharpoons R^+ + H_2O$$
$$R^+ + Cl^- \longrightarrow RCl$$

Similarly, cold concentrated hydrobromic acid converts tertiary alcohols to the bromides. One convenient procedure is to pass the hydrogen halide gas into the alcohol at 0°; reaction is complete within minutes.

One important drawback in carbocation reactions is the alternative reaction pathways available. We have already discussed one such reaction, E1 elimination. The electron-deficient carbocation center tends to attract electron density from adjacent bonds, and these bonds become weaker. One result is the ready loss of a proton to a basic solvent molecule. In some systems, another important side reaction can occur—rearrangement. The hydrogen attached by the weakened bond *and its bonding electrons can move to the cationic center,* thus generating a new carbocation.

Note that in this process the positive charge moves to the carbon to which the hydrogen was originally attached. Such rearrangements are especially important when the new carbocation is more stable than the old, but the reaction can occur even when both carbocations have comparable stability. Such reactions are common for secondary and tertiary carbocations but almost never involve primary carbocations. For example, treatment of 3-methyl-2-butanol with HBr gives solely the rearranged product 2-bromo-2-methylbutane.

$$CH_3\overset{\underset{\displaystyle CH_3}{|}}{\underset{\displaystyle |}{C}}\text{--}CHOHCH_3 \underset{}{\overset{H^+}{\rightleftharpoons}} CH_3\overset{\underset{\displaystyle CH_3}{|}}{\underset{\displaystyle |}{C}}\text{--}CHCH_3 \underset{}{\overset{-H_2O}{\rightleftharpoons}}$$

$$CH_3\overset{\underset{\displaystyle CH_3}{|}}{\underset{\displaystyle |}{\overset{+}{C}}}\text{--}\overset{H}{|}CHCH_3 \rightleftharpoons CH_3\overset{+}{\underset{\underset{\displaystyle CH_3}{|}}{C}}\text{--}\overset{H}{|}CHCH_3 \overset{HBr}{\longrightarrow} (CH_3)_2\overset{\underset{\displaystyle Br}{|}}{C}CH_2CH_3$$

secondary carbocation   tertiary carbocation   actual product
                                  (more stable)

Upon heating with hydrobromic acid, both 2-pentanol and 3-pentanol give a mixture of 2- and 3-bromopentane.

$$CH_3CH_2CH_2CHOHCH_3 \overset{HBr}{\longrightarrow} CH_3CH_2CH_2CHBrCH_3 + CH_3CH_2CHBrCH_2CH_3$$
$$(86\%) \qquad\qquad\qquad (14\%)$$

$$CH_3CH_2CHOHCH_2CH_3 \overset{HBr}{\longrightarrow} CH_3CH_2CH_2CHBrCH_3 + CH_3CH_2CHBrCH_2CH_3$$
$$(20\%) \qquad\qquad\qquad (80\%)$$

The mechanism whereby these two products are formed is

$$CH_3CH_2CH_2\underset{\underset{\displaystyle {}^+OH_2}{|}}{CHCH_3} \rightleftharpoons CH_3CH_2CH_2\overset{+}{C}HCH_3 \overset{Br^-}{\longrightarrow} CH_3CH_2CH_2CHBrCH_3$$

$$\Updownarrow$$

$$CH_3CH_2\underset{\underset{\displaystyle {}^+OH_2}{|}}{CHCH_2CH_3} \rightleftharpoons CH_3CH_2\overset{+}{C}HCH_2CH_3 \overset{Br^-}{\longrightarrow} CH_3CH_2CHBrCH_2CH_3$$

Note that equilibration of the carbocations is not complete. The rate of reaction of the carbocation with bromide ion is comparable to the rate of rearrangement. It is important to recognize when this reaction gives rearrangement products or mixtures.

1-Pentanol gives a good yield of 1-bromopentane with HBr; this reaction occurs by the S_N2 mechanism, and carbocations are not involved.

$$CH_3(CH_2)_2CH_2CH_2\overset{+}{O}H_2 + Br^- \longrightarrow CH_3(CH_2)_2CH_2CH_2Br + H_2O$$
$$\Big\downarrow\!\!\!/$$
$$CH_3(CH_2)_2\overset{+}{C}HCH_3 + H_2O$$

Even isobutyl alcohol on heating with HBr and H_2SO_4 gives mainly isobutyl bromide. The formation of a relatively stable tertiary carbocation in this case does result in the formation of some *t*-butyl bromide.

$$(CH_3)_2CHCH_2Br \overset{Br^-}{\longleftarrow} CH_3\overset{\underset{\displaystyle CH_3}{|}}{\underset{\displaystyle |}{C}}\text{--}\overset{H}{|}CH_2\text{--}\overset{+}{O}H_2 \longrightarrow CH_3\overset{\underset{\displaystyle CH_3}{|}}{\underset{\displaystyle |}{\overset{+}{C}}}\text{--}\overset{H}{|}CH_2 \overset{Br^-}{\longrightarrow} (CH_3)_3CBr$$
$$(80\%) \qquad\qquad\qquad\qquad\qquad\qquad\qquad\qquad\qquad (20\%)$$

Primary alcohols having a quaternary carbon (a carbon attached to four other carbons) next to the carbinol carbon react with complete rearrangement.

$$CH_3-\underset{\underset{CH_3}{|}}{\overset{\overset{CH_3}{|}}{C}}-CH_2-OH_2^+ \xrightarrow{-H_2O} CH_3-\underset{\underset{CH_3}{|}}{\overset{\overset{CH_3}{|}}{\overset{+}{C}}}-CH_2 \xrightarrow{Br^-} (CH_3)_2CBrCH_2CH_3$$

neopentyloxonium ion   tertiary carbocation   *t*-pentyl bromide

Recall that S_N2 reactions at neopentyl-type carbons are very slow (Section 8.8). The alternative reaction of rearrangement thus is able to compete and becomes the dominating reaction. Note also that alkyl groups can migrate as well as hydrogen.

In the transition state for the rearrangement of neopentyloxonium ion, the C—O bond has lengthened and weakened and the methyl group has started to migrate in order to give the transition state some of the character of a tertiary carbocation. Many experiments have shown that the migrating alkyl group attacks the carbon at the rear of the leaving group, much as in S_N2 displacement. The structure of this transition state is given approximately in the stereo plot of Figure 10.7.

Cycloalkylcarbinols frequently undergo rearrangement with resultant **ring expansion** under acidic conditions. Ring expansion is particularly prone to occur when there is a decrease in ring strain.

$$\underset{}{\square}\overset{\overset{\displaystyle H \quad CH_3}{|}}{\underset{\underset{CH_3}{|}}{C}}-OH + HCl \longrightarrow \text{(cyclopentane ring with Cl, CH_3, CH_3)}$$

This reaction involves formation of an intermediate carbocation, which undergoes rearrangement of one of the ring bonds to give a cyclopentyl cation.

$$\underset{}{\square}\overset{\overset{\displaystyle H \quad CH_3}{|}}{\underset{+}{C}}-CH_3 \longrightarrow \text{(cyclopentyl cation with H, CH_3, CH_3)}$$

The driving force for this rearrangement is relief of ring strain—the cyclobutyl system has about 26 kcal mole^{-1} of strain, while the cyclopentyl cation has only about 6 kcal mole^{-1} (see Table 5.5). Note that in this case the ring expansion occurs even though it involves rearrangement of a tertiary to a secondary carbocation.

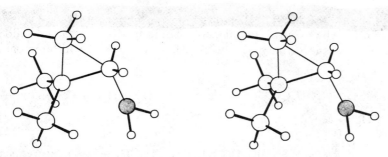

FIGURE 10.7. *Stereo representation of the transition state for rearrangement of neopentyloxonium cation.*

When a primary alcohol is treated with sulfuric acid alone, the product is the **alkylsulfuric acid.** The reaction is an equilibrium process—the product alkyl-sulfuric acid is readily hydrolyzed by excess water. If excess (xs) sulfuric acid is used at $0°$, the reaction proceeds to completion and the alkylsulfuric acid may be isolated.

$$C_2H_5OH + xs\ H_2SO_4 \xrightarrow{0°} C_2H_5OSO_2OH + H_2O$$

However, if ethanol is *heated* with concentrated sulfuric acid, ethyl ether is produced in high yield.

$$C_2H_5OH + H_2SO_4 \xrightarrow{140°} C_2H_5OC_2H_5 + H_2O$$
$$(95\%)$$

The detailed mechanism of the reaction under these conditions is not known. Bisulfate ion is a poor nucleophile, and one possibility is that the initial alkyl-oxonium ion is attacked by another alcohol molecule instead.

$$RCH_2\overset{H}{\underset{R}{O}} \quad CH_2\overset{+}{-}OH_2 \longrightarrow RCH_2\overset{H}{\underset{+}{-}O-}CH_2R + H_2O \rightleftharpoons (RCH_2)_2O + H_3O^+$$
dialkyl ether

Alternatively, the alkylsulfuric acid may be an intermediate.

$$RCH_2\overset{H}{\underset{R}{O}} \quad CH_2\overset{}{-}OSO_3H \longrightarrow RCH_2\overset{H}{\underset{+}{-}O-}CH_2R + HSO_4^-$$

In the case of primary alcohols, the reaction is an acceptable way of preparing symmetrical ethers.

> *n*-Butyl alcohol and concentrated sulfuric acid are refluxed with provision to remove water as it is formed, either with a suitable trap or by a fractionating column. The reaction mixture is maintained at 130–140°. The reaction mixture is allowed to cool and, after washing and drying, is distilled to give di-*n*-butyl ether.

At still higher temperature, elimination occurs to give the alkene.

$$C_2H_5OH \xrightarrow[170°]{H_2SO_4} CH_2=CH_2 + H_2O$$

Secondary and tertiary alcohols give only elimination without forming an ether.

> 2-(or 3-)Pentanol is heated with 50% sulfuric acid with distillation of product until the reaction mixture reaches 120°. The distillate is washed, dried, and distilled to give 2-pentene.

Tertiary alcohols eliminate water readily on heating with even traces of acid. If elimination is to be avoided, the alcohols should be distilled at low temperature (vacuum distillation) or in apparatus that has been rinsed with ammonia.

$$CH_3CH_2\underset{\underset{CH_3}{|}}{\overset{\overset{OH}{|}}{C}}CH_2CH_2CH_2CH_3 + H^+ \rightleftharpoons CH_3CH_2\underset{\underset{CH_3}{|}}{\overset{\overset{+OH_2}{|}}{C}}CH_2CH_2CH_2CH_3 \xrightarrow[-H_2O]{\Delta}$$

$$CH_3CH_2\overset{+}{\underset{\underset{CH_3}{|}}{C}}CH_2CH_2CH_2CH_3 \xrightarrow{-H^+} \text{mixture of alkenes}$$

EXERCISE: Write equations illustrating the following situations. Do not use the same examples used in the text.
(a) Conversion of a primary alcohol to a primary alkyl halide.
(b) Conversion of a secondary alcohol to a tertiary alkyl halide.
(c) A reaction which involves equilibrating isomeric secondary carbocations.
(d) A reaction involving ring expansion.

C. *Formation of Organic Esters*

Alcohols react with carboxylic acids to give esters.

$$CH_3OH + CH_3COOH \overset{H^+}{\rightleftharpoons} CH_3O\overset{\overset{\displaystyle O}{\|}}{C}CH_3 + H_2O$$

The reaction is catalyzed by strong acids and is discussed in detail in Section 18.7.C. Reaction with acyl halides is also an important way of preparing esters.

$$CH_3OH + CH_3\overset{\overset{\displaystyle O}{\|}}{C}Cl \longrightarrow CH_3\overset{\overset{\displaystyle O}{\|}}{C}OCH_3 + HCl$$

This reaction is discussed in Section 20.6.

D. *Formation of Inorganic Esters and Conversion to Alkyl Halides*

Various inorganic halides may be regarded as mixed anhydrides of some inorganic acid and HCl or HBr. Important examples are

1. Thionyl chloride, $SOCl_2$, a colorless liquid with b.p. 79°, is the mixed anhydride of sulfurous acid and HCl. It is corrosive and attacks rubber. It reacts rapidly with water to give sulfur dioxide and HCl.

$$SOCl_2 + H_2O \longrightarrow 2 HCl + SO_2$$

2. Phosphorus tribromide, PBr_3, is the mixed anhydride of phosphorous acid and HBr. It is a dense, colorless liquid with b.p. 173° and is prepared by the direct reaction of phosphorus with bromine. It reacts with water to give phosphorous acid and HBr.

$$PBr_3 + 3 H_2O \longrightarrow 3 HBr + H_3PO_3$$

3. Phosphorus pentachloride, PCl_5, is the mixed anhydride of phosphoric acid and HCl. It is a yellowish white solid with m.p. 162°. Upon hydrolysis, it yields phosphoric acid and HCl.

$$PCl_5 + 4 H_2O \longrightarrow 5 HCl + H_3PO_4$$

These compounds react readily with alcohols to form products that are esters of inorganic acids. Since the inorganic acids are strong acids, their anions are good leaving groups for subsequent S_N1 and S_N2 reactions. An example is the reaction of 1-butanol with thionyl chloride. The intermediate chlorosulfite ester may be isolated if desired. However, reaction with the chloride ion produced in the reaction occurs simply upon warming the alcohol with $SOCl_2$. The products are 1-chlorobutane, SO_2, and HCl.

$$CH_3(CH_2)_2CH_2OH + SOCl_2 \longrightarrow CH_3(CH_2)_2CH_2OSOCl + HCl$$

n-butyl chlorosulfite

$$Cl^- \overset{\frown}{} CH_2\!-\!O\!-\!\overset{\overset{\textstyle O}{\|}}{S}\!-\!\overset{\frown}{Cl} \longrightarrow CH_3(CH_2)_2CH_2Cl + SO_2 + Cl^-$$

(CH$_2$)$_2$CH$_3$

A tertiary amine, R$_3$N, is often used to catalyze the reaction by forming chloride ion from the HCl produced.

$$R_3N + HCl \rightleftharpoons R_3\overset{+}{N}H + Cl^-$$

Similarly, reaction of an alcohol with PBr$_3$ produces first a dibromophosphite ester which immediately reacts further.

$$(CH_3)_2CHCH_2OH + PBr_3 \longrightarrow (CH_3)_2CHCH_2OPBr_2 + HBr$$

isobutyl dibromophosphite

$$Br^- \overset{\frown}{} CH_2 \overset{\frown}{} OPBr_2 \longrightarrow (CH_3)_2CHCH_2Br + Br_2PO^-$$

CH(CH$_3$)$_2$

The dibromophosphite ion produced reacts with more alcohol so that the net reaction is

$$3\,(CH_3)_2CHCH_2OH + PBr_3 \longrightarrow 3\,(CH_3)_2CHCH_2Br + H_3PO_3$$

[Isobutyl alcohol is maintained at 0° with PBr$_3$ for 4 hr. The product is washed, dried, and distilled to give 60% of isobutyl bromide.]

This reaction also works well with many secondary alcohols.

Reaction of an alcohol with PCl$_5$ can be carried out at a 1:1 molar ratio to produce the alkyl chloride.

$$ROH + PCl_5 \longrightarrow RCl + HCl + POCl_3$$

Phosphorus oxychloride is a liquid, b.p. 105°, and will also react further with alcohols to form alkyl chlorides and phosphoric acid.

The above reagents are commercially available and are common laboratory chemicals. Phosphorus triiodide is a red solid that decomposes on heating. It is usually prepared *in situ* by heating red phosphorus, iodine, and the appropriate alcohol.

$$3\,CH_3OH + \tfrac{3}{2}I_2 + P \longrightarrow 3\,CH_3I + H_3PO_3$$

[Iodine is added over a period of several hours to a refluxing mixture of red phosphorus and methanol. Methyl iodide is distilled, washed, dried, and redistilled: yield 94%.]

Acid chlorides of sulfonic acids are also readily available. Common examples are benzenesulfonyl chloride, C$_6$H$_5$SO$_2$Cl, and *p*-toluenesulfonyl chloride, CH$_3$C$_6$H$_4$SO$_2$Cl (Section 26.5). These compounds react with primary and secondary alcohols, generally in the presence of a tertiary amine such as pyridine, to produce sulfonate esters.

$$ClCH_2CH_2CH_2CH_2OH + \left\langle\!\!\bigcirc\!\!\right\rangle\!\!-SO_2Cl + C_5H_5N \longrightarrow$$

benzenesulfonyl
chloride   pyridine

$$ClCH_2CH_2CH_2CH_2OSO_2\!\!-\!\!\left\langle\!\!\bigcirc\!\!\right\rangle + C_5H_5\overset{+}{N}H + Cl^-$$

4-chlorobutyl benzenesulfonate

Although these esters are not esters of inorganic acids, they behave in much the same manner as the chlorosulfite and dibromophosphite esters just discussed. Since the sulfonate anion is the conjugate base of a strong acid (notice the similarity of benzenesulfonic acid to sulfuric acid), it is a weak base and a good leaving group in displacement reactions (Section 8.10). Such esters react with halide ion in inert solvent to give alkyl halides and the sulfonate ion.

$$Na^+I^- + ClCH_2CH_2CH_2CH_2\!\!-\!\!OSO_2\!\!-\!\!\left\langle\!\!\bigcirc\!\!\right\rangle \xrightarrow{\text{acetone}}$$

$$ClCH_2CH_2CH_2CH_2I + \left\langle\!\!\bigcirc\!\!\right\rangle\!\!-SO_3Na\downarrow$$

Sodium and potassium iodides are somewhat soluble in hot acetone, but the benzenesulfonate salts are not and precipitate. Note that the benzenesulfonate ion is displaced *selectively* in this reaction, even though displacement of chloride ion might also have occurred. The benzenesulfonate ion is about 100 times more reactive as a leaving group than chloride.

Other common reagents of this type are lithium chloride in dimethylformamide or ethanol and sodium bromide in dimethylformamide or dimethyl sulfoxide.

$$CH_3CH_2CHCH_2CH_3 \xrightarrow[\substack{CH_3SOCH_3 \\ 25°}]{NaBr} CH_3CH_2CHBrCH_2CH_3$$

$$\mid$$
$$OSO_2\!\!-\!\!\left\langle\!\!\bigcirc\!\!\right\rangle\!\!CH_3$$

1-ethylpropyl p-toluenesulfonate   (85%)
                                3-bromopentane

Note that all of the preceding reactions apply to primary and secondary alcohols. With tertiary alcohols, the S_N2 displacement reaction is so slow that side reactions (mainly elimination) dominate.

We thus have a number of reactions available for accomplishing the important conversion ROH $\longrightarrow$ RX, and we have various complications to watch out for. The best overall methods may be summarized as follows.

1. **Primary alcohols with no β-branching**
   chloride: $SOCl_2$ + pyridine, C_5H_5N (generally better than PCl_5 or $ZnCl_2$–HCl)
   bromide: PBr_3 or HBr–H_2SO_4
   iodide: P + I_2

2. **Primary alcohols with β-branching**
   chloride: $SOCl_2$ + pyridine
   bromide: PBr_3
3. **Secondary alcohols**
   chloride: $SOCl_2$ + pyridine
   bromide: PBr_3 (low temperature, less than 0°C)
   (The two-step sequence alcohol ⟶ sulfonate ester ⟶ halide gives a product of higher purity.)
4. **Tertiary alcohols**
   chloride: HCl at 0°
   bromide: HBr at 0°
   (Rearrangement in particularly sensitive cases cannot generally be avoided.)

EXERCISE: Show how each of the following alkyl halides may be prepared in good yield from an alcohol.
(a) $(CH_3CH_2)_3CCl$ (b) $CH_3CH_2CH_2CH_2CH_2CH_2CH_2Cl$
(c) $(CH_3CH_2)_2CHCH_2Br$ (d) $(CH_3CH_2CH_2)_2CHBr$
Which method would you use to convert (R)-2-octanol into (S)-2-chlorooctane if you wanted to maximize the optical purity of the product halide?

E. *Oxidation of Alcohols*

Primary and secondary alcohols can be oxidized to carbonyl compounds.

$$RCH_2OH \xrightarrow[-2H]{[O]} R-\overset{\displaystyle O}{\overset{\|}{C}}-H$$

an aldehyde

$$R-CHOH-R' \xrightarrow[-2H]{[O]} R-\overset{\displaystyle O}{\overset{\|}{C}}-R'$$

a ketone

Many procedures are available for accomplishing these transformations, but the most common general oxidizing agent is some form of chromium(VI), which becomes reduced to chromium(III).

Chromium trioxide, CrO_3, also known as chromic anhydride, forms red, deliquescent crystals. It is very soluble in water and in sulfuric acid.

Sodium or potassium dichromate forms orange aqueous solutions that convert to the yellow chromate salt under basic conditions.

$$2\,CrO_4^{2-} + 2\,H^+ \rightleftharpoons Cr_2O_7^{2-} + H_2O$$
yellow   orange

Primary alcohols give aldehydes on warming with sodium dichromate and aqueous sulfuric acid. However, aldehydes are also readily oxidized under these conditions to give carboxylic acids (Section 13.8.A). This method is only successful for aldehydes of sufficiently low molecular weight that they may be distilled from solution as formed. In this way, *n*-butyl alcohol gives butyraldehyde in 50% yield.

$$CH_3CH_2CH_2CH_2OH + H_2Cr_2O_7 \longrightarrow CH_3CH_2CH_2CHO$$

(50%)

butyraldehyde

Only aldehydes that boil significantly below 100° can be conveniently prepared in this manner. Since this effectively limits the method to the production of a few simple aldehydes, it is not an important synthetic method. Other special oxidants have been developed that help to circumvent this problem. They will be discussed in Section 13.8.A.

Balancing Oxidation-Reduction Reactions. Organic redox reactions may be balanced by any method that works. The method of half-cells often taught in beginning chemistry courses for inorganic redox reactions applies just as well to organic reactions.

$$3 \times [RCH_2OH \rightleftharpoons RCHO + 2\,H^+ + 2\,e^-]$$
$$\underline{[6\,e^- + Cr_2O_7{}^{2-} + 14\,H^+ \rightleftharpoons 2\,Cr^{3+} + 7\,H_2O]}$$
$$3\,RCH_2OH + Cr_2O_7{}^{2-} + 8\,H^+ \rightleftharpoons 3\,RCHO + 2\,Cr^{3+} + 7\,H_2O$$

Alternatively, one may use a method based on hypothetical oxygen equivalents, [O].

$$3 \times [RCH_2OH + [O] \rightleftharpoons RCHO + H_2O]$$
$$\underline{[Cr_2O_7{}^{2-} + 8\,H^+ \rightleftharpoons 2\,Cr^{3+} + 3\,[O] + 4\,H_2O]}$$
$$3\,RCH_2OH + Cr_2O_7{}^{2-} + 8\,H^+ \rightleftharpoons 3\,RCHO + 2\,Cr^{3+} + 7\,H_2O$$

In this method, each half-reaction is balanced for charges before the oxidation equivalents are added for chemical balance. One or both half-reactions must be multiplied by appropriate factors such that the [O]s cancel out. Both methods, of course, give the same total balanced equation.

Since ketones are more stable to general oxidation conditions that aldehydes, chromic acid oxidations are more important for secondary alcohols. In one common procedure a 20% excess of sodium dichromate is added to an aqueous mixture of the alcohol and a stoichiometric amount of acid.

4-ethylcyclohexanol   4-ethylcyclohexanone (90%)

An especially convenient oxidizing agent is Jones reagent, a solution of chromic acid in dilute sulfuric acid. The secondary alcohol in acetone solution is "titrated" with the reagent with stirring at 15–20°. Oxidation is rapid and efficient. The green chromium salts separate from the reaction mixture as a heavy sludge; the supernatant liquid consists mainly of an acetone solution of the product ketone.

cyclooctanol   cyclooctanone (95%)

Chromium(VI) oxidations are known to proceed by way of a *chromate ester* of the alcohol. If the alcohol has one or more hydrogens attached to the carbinol position, a base-catalyzed elimination occurs, yielding the aldehyde or ketone and a chromium(IV) species. The overall effect of these two consecutive reactions is oxidation of the alcohol and reduction of the chromium.

$$(1) \quad R_2CHOH + H_2CrO_4 \rightleftharpoons R_2CH-O-\overset{\overset{\displaystyle O}{\|}}{\underset{\underset{\displaystyle O}{\|}}{Cr}}-OH + H_2O$$

$$(2) \quad B: \underset{\curvearrowright}{R_2C} \overset{\curvearrowleft}{-O} \overset{\overset{\displaystyle O}{\|}}{\underset{\underset{\displaystyle O}{\|}}{Cr}}-OH \longrightarrow R_2C{=}O + HCrO_3^- + BH^+$$

[The chromium(IV) produced in the elimination undergoes rapid reaction with a molecule containing chromium(VI) to produce two chromium(V) species. These chromium(V)-containing molecules function further as two-electron oxidants to produce more aldehyde or ketone and two chromium(III) species, the ultimate form of the chromium in such reactions.]

Under conditions such as these, tertiary alcohols do not generally react, although under proper conditions the chromate ester can be isolated.

$$2\,(CH_3)_3COH + H_2CrO_4 \rightleftharpoons (CH_3)_3CO-\overset{\overset{\displaystyle O}{\|}}{\underset{\underset{\displaystyle O}{\|}}{Cr}}-OC(CH_3)_3 + 2\,H_2O$$

Since there is no carbinol proton to eliminate in the case of a tertiary alcohol, such esters are stable. If the chromate ester is treated with excess water, simple hydrolysis occurs with regeneration of the tertiary alcohol and chromic acid.

More vigorous oxidizing conditions result in cleavage of C—C bonds. Aqueous nitric acid is such a reagent. Oxidation all the way to carboxylic acids is the normal result. Such oxidations appear to proceed by way of the intermediate ketone, which undergoes further oxidation (Section 13.8.A).

$$\underset{\text{cyclohexanol}}{\overset{\text{H} \quad \text{OH}}{\bigcirc}} \quad \xrightarrow[\substack{55-60°}]{\substack{50\% \text{ HNO}_3 \\ V_2O_5}} \quad \underset{\substack{\text{(60\%)} \\ \text{adipic acid}}}{\overset{\text{COOH}}{\diagdown\diagup}\text{COOH}}$$

Nitric acid may also be used as an oxidant for primary alcohols; again the product is a carboxylic acid.

$$\underset{\text{3-chloropropanol}}{ClCH_2CH_2CH_2OH} \quad \xrightarrow[25-30°]{71\% \text{ HNO}_3} \quad \underset{\substack{\text{(78\%)} \\ \text{3-chloropropanoic acid}}}{ClCH_2CH_2COOH}$$

Instead of oxidation, direct dehydrogenation can be accomplished with various catalysts and conditions. The reaction is of industrial interest but is not much used in the laboratory because of the specialized equipment and conditions required.

Catalysts include copper metal, copper chromite, or copper–chromium oxides prepared in special ways. Examples of dehydrogenation are

$$CH_3CH_2CH_2CH_2OH \xrightarrow[300-345°]{\text{Cu–Cr oxides on Celite}} CH_3CH_2CH_2CHO$$

n-butyl alcohol   (62%)

butyraldehyde

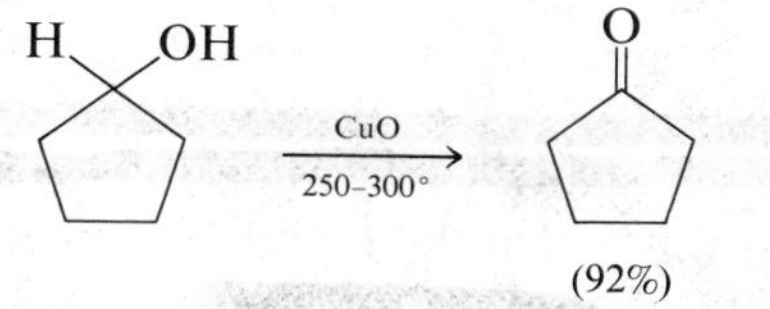

cyclopentanol   cyclopentanone

(92%)

EXERCISE: Write complete and balanced equations for the following chromium(VI) oxidations.
(a) $(CH_3CH_2CH_2)_2CHOH + CrO_3 = (CH_3CH_2CH_2)_2C{=}O + Cr^{3+}$
(b) $CH_3CH_2CH_2CH_2CH_2OH + H_2CrO_4 = CH_3CH_2CH_2CH_2COOH + Cr^{3+}$
(c) $CH_3OH + CuO = CH_2{=}O + Cu^0$

10.8
Nomenclature of Ethers

The common names of ethers are derived by naming the two alkyl groups and adding the word ether.

$CH_3OCH_2CH_3$ $CH_3OC(CH_3)_3$

ethyl methyl ether   *t*-butyl methyl ether

In symmetrical ethers the prefix di- is used. Although the prefix is often omitted, it should be included to avoid confusion.

$CH_3CH_2OCH_2CH_3$ $(CH_3)_2CHOCH(CH_3)_2$

diethyl ether   diisopropyl ether

(or ethyl ether)   (or isopropyl ether)

In the IUPAC system ethers are named as alkoxyalkanes. The larger alkyl group is chosen as the stem.

$$\overset{\displaystyle CH_3}{\underset{\displaystyle |}{CH_3CH_2OCHCH_2CH_2CH_3}}$$

2-ethoxypentane

$$CH_3\overset{\displaystyle CH_3}{\underset{\displaystyle \underset{\displaystyle CH_3}{|}}{\overset{\displaystyle |}{C}}}CH_2OCH_3$$

1-methoxy-2,2-dimethylpropane

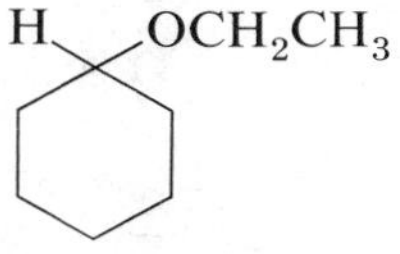

ethoxycyclohexane $CH_3OCH_2CH_2OCH_3$

1,2-dimethoxyethane

10.9
Physical Properties of Ethers

The physical properties of some ethers are listed in Table 10.5. Note that dimethyl ether is a gas at room temperature and that diethyl ether has a boiling point only about 10°C above normal room temperature. Diethyl ether is an important solvent and has a characteristic odor. It was once used as an anesthetic, but it has been largely replaced for this purpose by other compounds.

The rule of thumb that compounds having no more than four carbons per oxygen are water-soluble holds for ethers as well as for alcohols. Dimethyl ether is completely miscible with water. The solubility of diethyl ether in water is about 10 g per 100 g of H_2O at 25°. Tetrahydrofuran (b.p. 67°) is another important solvent. This cyclic ether, commonly abbreviated THF, has essentially the same molecular weight as diethyl ether, but it is much more soluble in water.

$$\begin{matrix} CH_2-CH_2 \\ | \quad\quad | \\ CH_2 \quad CH_2 \\ \backslash \; / \\ O \end{matrix} \equiv \bigcirc_{O}$$

tetrahydrofuran (THF)
(b.p. 67°; miscible with H_2O in all proportions at 25°)

$$CH_3 \overset{CH_2}{\diagup} O \overset{CH_2}{\diagdown} CH_3$$

diethyl ether
(b.p. 34.5; 10 g dissolves in 100 g H_2O at 25°)

Because of its cyclic structure, THF has lone-pair electrons that are more accessible for hydrogen bonding than those of diethyl ether. The "floppy" ethyl groups in the acyclic compound interfere with hydrogen bonding and cause the water solubility of diethyl ether to be lower. Also note that the cyclic compound has a significantly higher boiling point; its more compact structure allows for more efficient van der Waals attraction between molecules.

As we shall see, ethers are fairly inert to many reagents. Because of their unreactivity, they are not generally important as chemical reagents. However, their general lack of reactivity, combined with their favorable solvent properties, makes ethers useful solvents for many other reactions. Several ethers that are

TABLE 10.5
Physical Properties of Ethers

Compound	Name	Melting Point, °C	Boiling Point, °C
CH_3OCH_3	dimethyl ether	−138.5	−23
$CH_3OCH_2CH_3$	ethyl methyl ether		10.8
$(CH_3CH_2)_2O$	diethyl ether	−116.62	34.5
$CH_3CH_2OCH_2CH_2CH_3$	ethyl propyl ether	−79	63.6
$(CH_3CH_2CH_2)_2O$	dipropyl ether	−122	91
$\begin{matrix} CH_3 \\ | \\ (CH_3CH)_2O \end{matrix}$	diisopropyl ether	−86	68
$(CH_3CH_2CH_2CH_2)_2O$	dibutyl ether	−95	142

important as solvents are dioxane (page 244), 1,2-dimethoxyethane (glyme), and
bis-β-methoxyethyl ether (diglyme).

$$CH_3OCH_2CH_2OCH_3 \qquad CH_3OCH_2CH_2OCH_2CH_2OCH_3$$

1,2-dimethoxyethane   bis-β-methoxyethyl ether
(glyme, DME)   (diglyme)

Ethers have nmr spectra similar to those of alcohols except for the absence of an
OH signal. Hydrogens on the same carbon as the ether oxygen usually absorb in
the range $\delta = 3.3$–3.9 ppm. Hydrogens at carbons β to the ether oxygen are only
slightly affected (Figure 10.8).

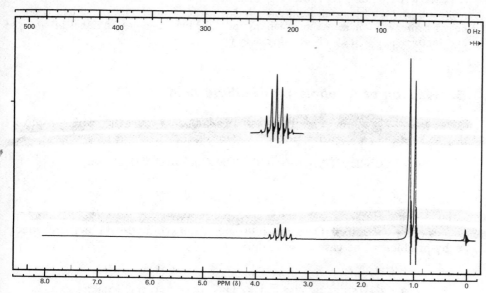

FIGURE 10.8. *Nmr spectrum of diisopropyl ether,* $(CH_3)_2CHOCH(CH_3)_2$.

10.10
Preparation of Ethers

Ethers may be prepared by the Williamson ether synthesis or by the reaction of
alcohols with sulfuric acid.

A. *Williamson Ether Synthesis*

The method of A. W. Williamson, one of the pioneers in organic chemistry, is
simply an S_N2 displacement of a primary alkyl halide or sulfonate ester by an
alkoxide ion. The alkoxide may be derived from a primary, secondary, or tertiary
alcohol, but the substrate must be primary and have no β-branches.

Other halides and sulfonates give too much elimination for the reaction to be of
preparative value.

$$CH_3O^- \, Na^+ \; + \; \overset{\displaystyle OSO_2C_6H_5}{\underset{\displaystyle H}{\bigcirc}} \quad \xrightarrow{\text{DMSO}} \quad \bigcirc \quad + \quad \overset{\displaystyle OCH_3}{\underset{\displaystyle H}{\bigcirc}}$$

(85%)   (5%)

cyclohexene   methoxycyclohexane

Since the reaction is a classic example of an S_N2 reaction, one must keep in mind the principles of that mechanism (Chapter 8) when planning an ether synthesis by this route.

> **EXERCISE:** Methyl neopentyl ether can be prepared readily from sodium neopentoxide and methyl benzenesulfonate but not from sodium methoxide and neopentyl benzenesulfonate. Write equations for these two reactions and explain why one combination works and the other does not.

B. Reaction of Alcohols with Sulfuric Acid

The alcohol–sulfuric acid reaction, which was discussed in Section 10.7, is most often used for the conversion of simple primary alcohols into symmetrical ethers.

$$ClCH_2CH_2OH \xrightarrow{H_2SO_4} ClCH_2CH_2OCH_2CH_2Cl$$

(75%)

bis-β-chloroethyl ether

Secondary and tertiary alcohols undergo predominant dehydration when subjected to these conditions. Occasionally some of the symmetrical ether is formed as a by-product in the case of secondary alcohols

$$\underset{\displaystyle}{CH_3\overset{\displaystyle CH_3}{\underset{|}{C}}HOH} \xrightarrow{H_2SO_4} CH_3CH{=}CH_2 + CH_3\overset{\displaystyle CH_3}{\underset{|}{C}}H{-}O{-}\overset{\displaystyle CH_3}{\underset{|}{C}}HCH_3$$

major product   by-product

The method is generally useless for the preparation of unsymmetrical ethers because complex mixtures are formed.

$$ROH + R'OH \xrightarrow{H^+} ROR + ROR' + R'OR'$$

An exception is the case where one alcohol is tertiary and the other alcohol is primary or secondary. Since tertiary carbocations form under very mild conditions, this method is generally a satisfactory synthetic method.

$$CH_3\overset{\displaystyle CH_3}{\underset{\displaystyle \underset{|}{CH_3}}{\underset{|}{C}}}OH + CH_3\overset{\displaystyle CH_3}{\underset{|}{C}}HOH \xrightarrow[H_2O]{NaHSO_4} CH_3\overset{\displaystyle CH_3}{\underset{\displaystyle \underset{|}{CH_3}}{\underset{|}{C}}}{-}O{-}\overset{\displaystyle CH_3}{\underset{|}{C}}HCH_3$$

(82%)

> t-Butyl alcohol is added to a mixture of 15% aqueous sodium bisulfate and isopropyl alcohol at room temperature. The mixture is neutralized with NaOH and distilled to yield t-butyl isopropyl ether (82%).

The tertiary cation may also be produced by protonation of an alkene in the presence of a primary or secondary alcohol (Section 11.6.B).

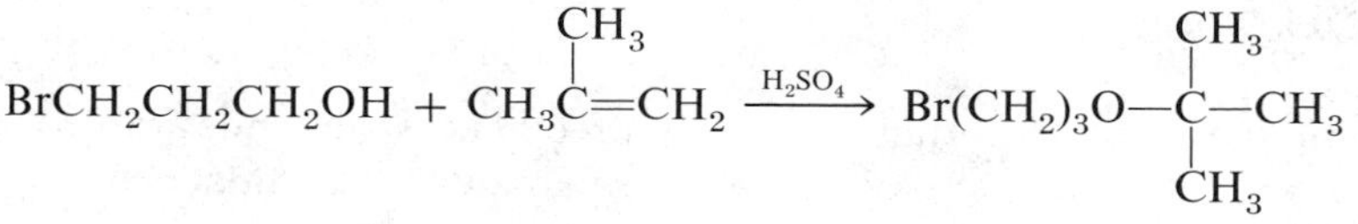

$$BrCH_2CH_2CH_2OH + CH_3\overset{\overset{\displaystyle CH_3}{|}}{C}=CH_2 \xrightarrow{H_2SO_4} Br(CH_2)_3O-\overset{\overset{\displaystyle CH_3}{|}}{\underset{\underset{\displaystyle CH_3}{|}}{C}}-CH_3$$

(80%)

EXERCISE: Write a mechanism, showing each step, for the last reaction.

10.11
Reactions of Ethers

A. *Reaction with Acids*

Ethers are relatively inert to most reagents. They are stable to base, to catalytic hydrogenation, and to most other reducing agents. They are stable to dilute acid but do react with hot concentrated acids. Strong HBr or HI causes cleavage.

$$CH_3CH_2OCH_2CH_3 + 2\,HBr \xrightarrow{\Delta} 2\,CH_3CH_2Br + H_2O$$

The mechanism of the reaction involves an S_N2 displacement by bromide ion on the protonated ether. The alcohol produced reacts further with HBr to yield more alkyl bromide.

(1) $CH_3CH_2OCH_2CH_3 + HBr \rightleftharpoons CH_3CH_2\overset{\overset{\displaystyle H}{|}}{\underset{+}{O}}CH_2CH_3 + Br^-$

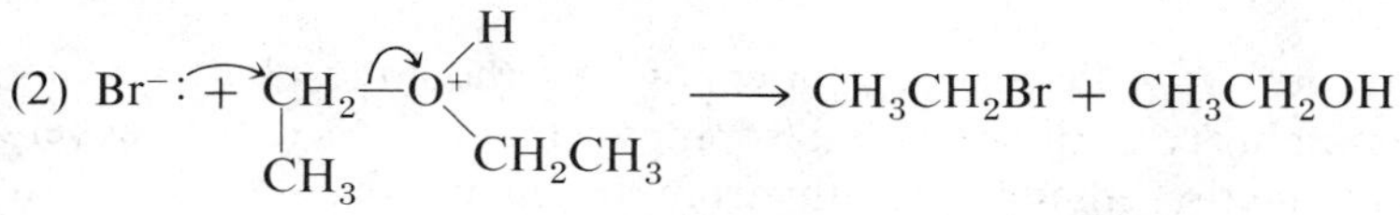

(2) $Br^-: + \overset{}{C}H_2\overset{\frown}{-}\overset{\overset{\displaystyle H}{}}{\underset{\underset{\displaystyle CH_2CH_3}{}}{O^+}} \longrightarrow CH_3CH_2Br + CH_3CH_2OH$
$|$
CH_3

(3) $CH_3CH_2OH + HBr \rightleftharpoons CH_3CH_2\overset{+}{O}H_2 + Br^-$

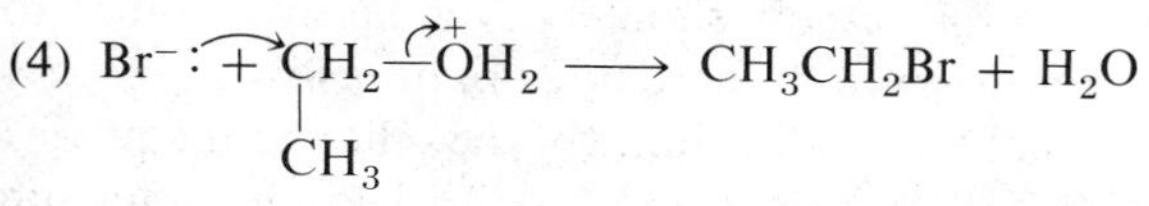

(4) $Br^-: + \underset{\underset{\displaystyle CH_3}{|}}{C}H_2\overset{\frown}{-}\overset{+}{O}H_2 \longrightarrow CH_3CH_2Br + H_2O$

With secondary and tertiary ethers, carbocations are involved and the reactions tend to be much more complex. Heating such ethers with strong acid generally leads to elimination as the major reaction.

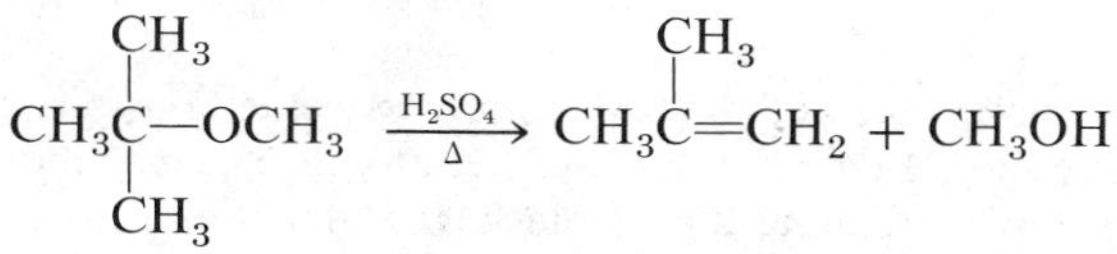

$$CH_3\underset{\underset{\displaystyle CH_3}{|}}{\overset{\overset{\displaystyle CH_3}{|}}{C}}-OCH_3 \xrightarrow[\Delta]{H_2SO_4} CH_3\overset{\overset{\displaystyle CH_3}{|}}{C}=CH_2 + CH_3OH$$

t-butyl methyl ether　　　　　isobutylene

The reaction is not generally useful for preparations unless one of the alkyl groups of the ether is a small tertiary group. In such a case the alkene formed upon elimination volatilizes as it is produced.

B. *Oxidation*

One of the most important reactions of ethers is an undesirable one—the reaction with atmospheric oxygen to form peroxides **(autoxidation).**

$$CH_3CH\underset{\underset{CH_3}{|}}{-}O\underset{\underset{CH_3}{|}}{-}CHCH_3 + O_2 \longrightarrow CH_3\underset{\underset{CH_3}{|}}{\overset{\overset{OOH}{|}}{C}}-O\underset{\underset{CH_3}{|}}{-}CHCH_3$$

a hydroperoxide

Autoxidation occurs by a free radical mechanism.

$$(1)\ RO\underset{|}{\overset{\overset{H}{|}}{-C-}} + R'\cdot \longrightarrow RO\overset{.}{-\underset{|}{C}-} + R'H$$

$$(2)\ RO\overset{.}{-\underset{|}{C}-} + O_2 \longrightarrow RO\underset{|}{\overset{\overset{OO\cdot}{|}}{-C-}}$$

$$(3)\ RO\underset{|}{\overset{\overset{OO\cdot}{|}}{-C-}} + RO\underset{|}{\overset{\overset{H}{|}}{-C-}} \longrightarrow RO\underset{|}{\overset{\overset{OOH}{|}}{-C-}} + RO\overset{.}{-\underset{|}{C}-}$$

or

$$RO\underset{|}{\overset{\overset{OO\cdot}{|}}{-C-}} + RO\overset{.}{-\underset{|}{C}-} \longrightarrow RO\underset{|}{-C}-OO\underset{|}{-C}-OR$$

a peroxide

Ethers of almost any type that have been exposed to the atmosphere for any length of time invariably contain peroxides. Isopropyl ether is especially treacherous in this regard, but ethyl ether and tetrahydrofuran are also dangerous. Peroxides and hydroperoxides are hazardous because they decompose violently at elevated temperatures, and serious explosions may result. When an ether that contains peroxides is distilled, the less volatile peroxides concentrate in the residue. At the end of the distillation, the temperature increases, and the residual peroxides may explode. For this reason, ethers should never be evaporated to dryness, unless care has been taken to exclude peroxides rigorously.

A simple test for peroxides is to shake a small volume of the ether with aqueous KI solution. If peroxides are present, they oxidize I^- to I_2. The characteristic purple-to-brown color of I_2 is diagnostic of the presence of peroxides. Contaminated ether may be purified by shaking with aqueous ferrous sulfate to reduce the peroxides.

Several years ago one of the authors had an impressive demonstration of the violence of a peroxide explosion. In a laboratory adjacent to his office, a laboratory technician was engaged in purifying about 2 L of old THF, later found to contain substantial amounts of peroxides. The material exploded, virtually demolishing the laboratory, and moving the wall several inches toward the author's desk. Several large bookshelves were knocked over and emptied their contents onto his desk and chair. Only by good fortune was no one injured.

A. *Epoxides: Oxiranes*

Heterocyclic compounds are cyclic structures in which one or more ring atoms are hetero atoms, a hetero atom being an element other than carbon. The three-membered ring containing oxygen is the first class of heterocyclic compounds to be considered. Because they are readily prepared from alkenes (Section 11.6.B.3), they are commonly named as olefin oxides. Hence, the parent member is generally called ethylene oxide.

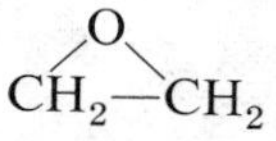

$$\overset{\displaystyle O}{\overset{\displaystyle \triangle}{CH_2-CH_2}}$$
ethylene oxide

This type of compound is frequently called an epoxide, although the formal IUPAC nomenclature is oxirane. Substituents on the oxirane ring require a numbering system. The general rule for heterocyclic rings is that the hetero atom gets the number 1.

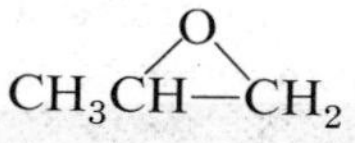

$$CH_3\overset{\displaystyle O}{\overset{\displaystyle \triangle}{CH-CH_2}} \qquad (CH_3)_2\overset{\displaystyle O}{\overset{\displaystyle \triangle}{C\quad CH_2}}$$
propylene oxide   isobutylene oxide
2-methyloxirane   2,2-dimethyloxirane

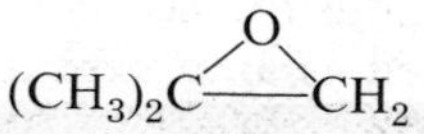

Ethylene oxide is a significant commercial item and is prepared industrially by the catalyzed air oxidation of ethylene.

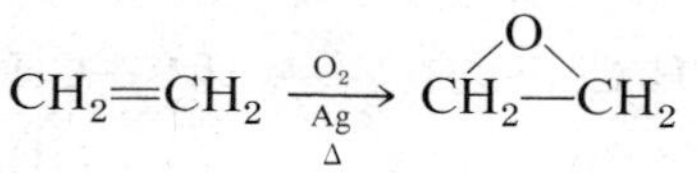

$$CH_2{=}CH_2 \xrightarrow[\Delta]{\underset{Ag}{O_2}} \overset{\displaystyle O}{\overset{\displaystyle \triangle}{CH_2-CH_2}}$$

Several general laboratory preparations are available. One reaction is an internal S_N2 displacement reaction starting with a β-halo alcohol.

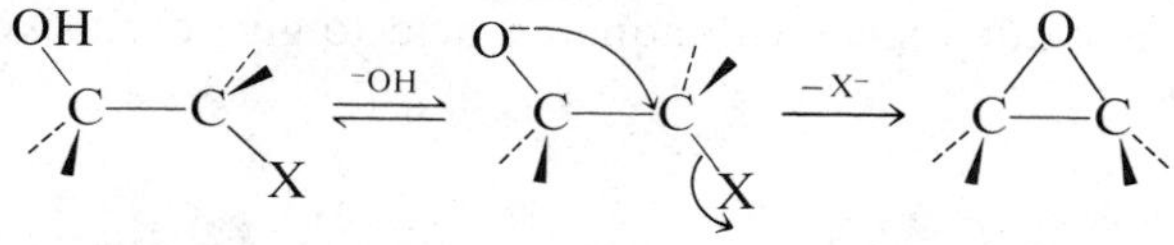

The internal S_N2 reaction is an intramolecular backside reaction and requires that the reacting groups have a *trans* or *anti* relationship. As we shall see in Section 11.6.B.2, the required stereoisomer is readily available from an alkene.

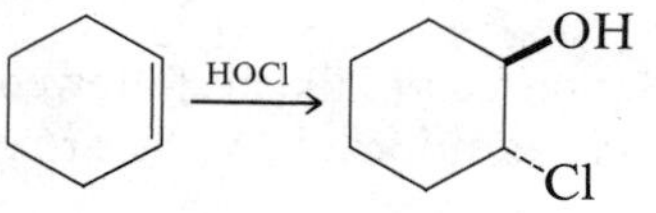

(70–73%)

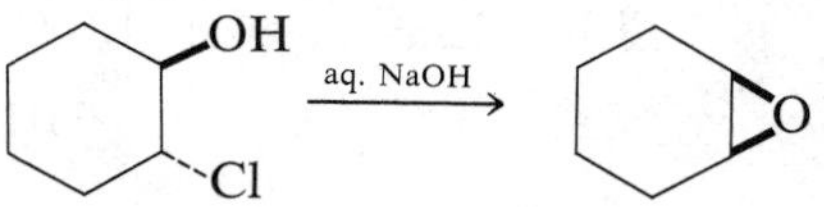

(70–73%)
cyclohexene oxide

Like other ethers, oxiranes undergo C—O bond cleavage under acidic conditions. However, because of the large ring strain associated with the three-membered ring (about 25 kcal mole^{-1}, see Section 5.7), they are much more reactive than normal ethers.

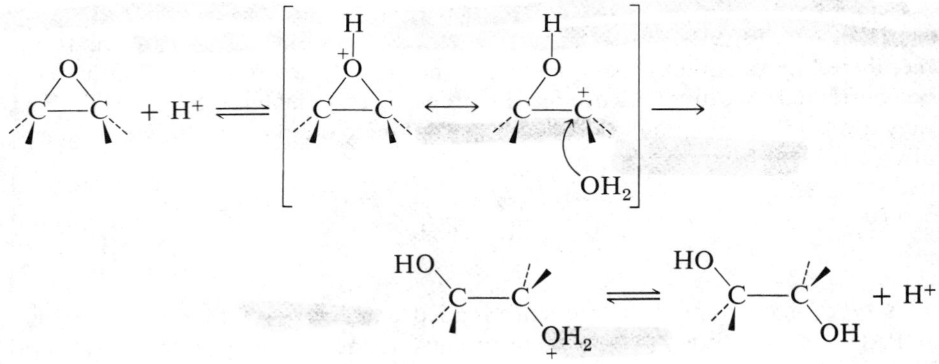

The product is a 1,2-diol (Section 28.3). As a class, 1,2-diols are called "glycols." They are important industrial compounds. The simplest 1,2-diol, "ethylene glycol," is commonly used as antifreeze for automobile radiators.

As shown by the foregoing mechanism, ring opening of an epoxide is another example of the S$_N$2 mechanism (Chapter 8) in which the attacking nucleophile is H$_2$O and the leaving group is ROH.

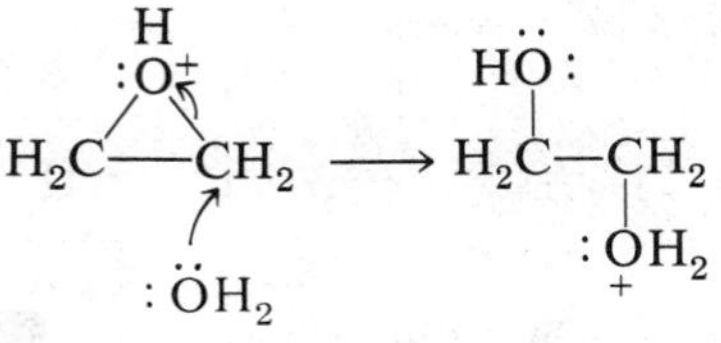

The stereochemical outcome of the reaction is *inversion* at the reaction center. Thus cyclohexene oxide reacts with aqueous acid to give exclusively *trans*-cyclohexane-1,2-diol.

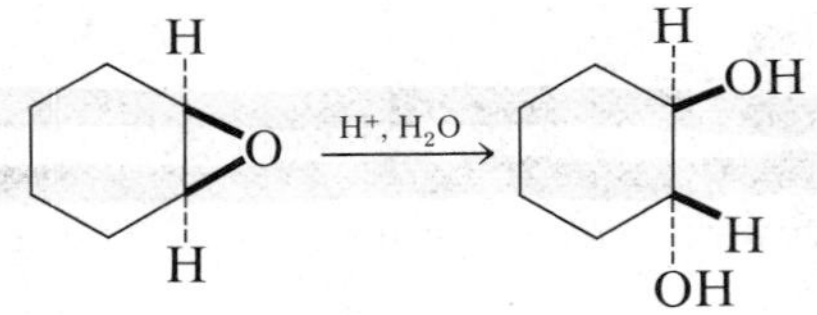

The two-step process of epoxidation of an alkene followed by hydrolysis of the resulting epoxide amounts to *anti* addition of two hydroxy groups to a double bond.

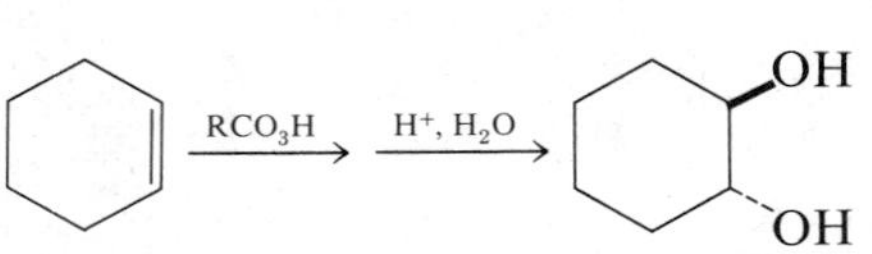

The oxirane ring is so prone to undergo ring-opening reactions that it even reacts with aqueous base.

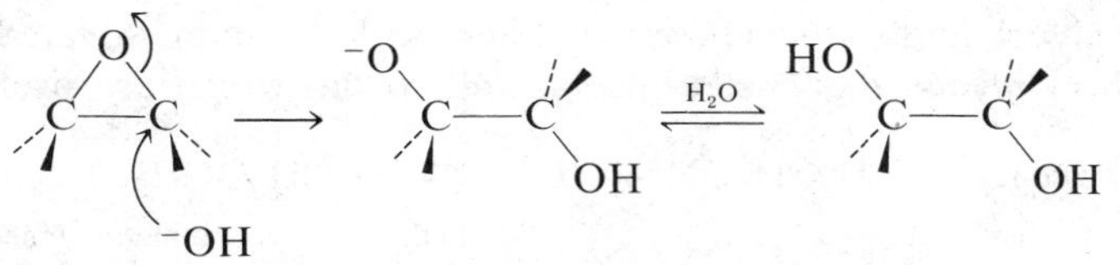

Base-catalyzed opening of an epoxide is an S_N2 displacement reaction with an alkoxide ion as the leaving group and has no counterpart with normal ethers. It occurs with epoxides only because the relief of ring strain provides a potent driving force for reaction.

The two ring-opening reactions have different orientational preferences. The acid-catalyzed reaction is essentially a carbocation reaction, and reaction tends to occur at that ring carbon which corresponds to the more stable carbocation; that is, reaction with solvent occurs at the more highly substituted ring carbon. Like other S_N2 reactions, the base-catalyzed reaction is subject to steric influences. Reaction occurs at the less hindered carbon. The difference is exemplified by reaction of isobutylene oxide in methanol under acidic and basic conditions.

Acidic conditions

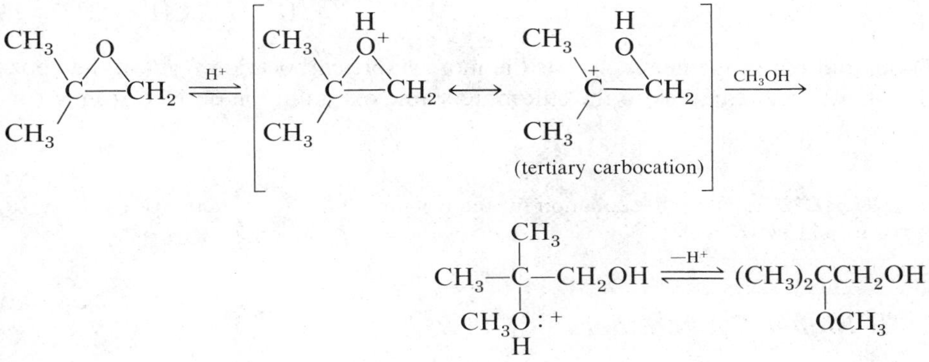

(tertiary carbocation)

$CH_3-\underset{CH_3\overset{+}{O}:}{\underset{H}{\overset{CH_3}{C}}}-CH_2OH \rightleftharpoons^{-H^+} (CH_3)_2CCH_2OH$ with OCH_3

Basic conditions

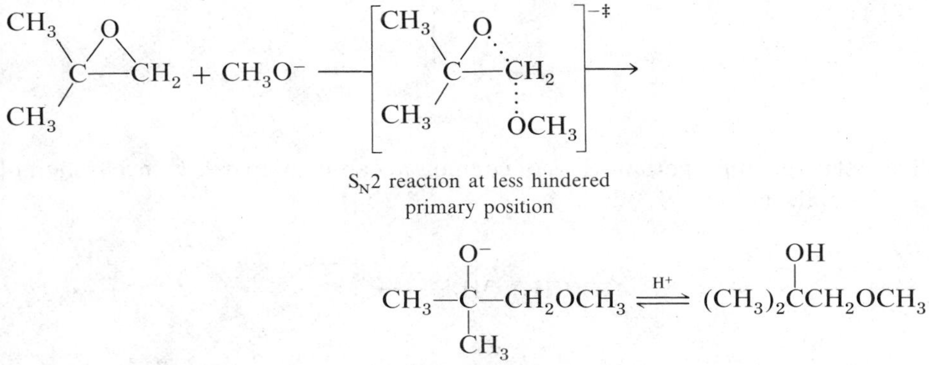

S_N2 reaction at less hindered primary position

$CH_3-\underset{CH_3}{\overset{O^-}{C}}-CH_2OCH_3 \rightleftharpoons^{H^+} (CH_3)_2\overset{OH}{C}CH_2OCH_3$

Under the proper conditions the first product from the reaction of ethylene oxide with hydroxide ion can react with more ethylene oxide.

$HO^- \xrightarrow{\triangle} HOCH_2CH_2O^- \xrightarrow{\triangle} HOCH_2CH_2OCH_2CH_2O^- \longrightarrow$ etc.

The final products are polyether alcohols, which are called diethylene glycol, triethylene glycol, and so on. These polyether alcohols are methylated to produce the polyethers **glyme, diglyme, triglyme,** and so on, which are useful solvents.

$$HOCH_2CH_2OCH_2CH_2OH \longrightarrow CH_3OCH_2CH_2OCH_2CH_2OCH_3$$

diethylene glycol   diethylene glycol dimethyl ether
(diglyme)

In the absence of a reactive nucleophilic reagent, a protonated epoxide rearranges to the more stable carbonyl compound. The rearrangement reaction is best carried out with a Lewis acid.

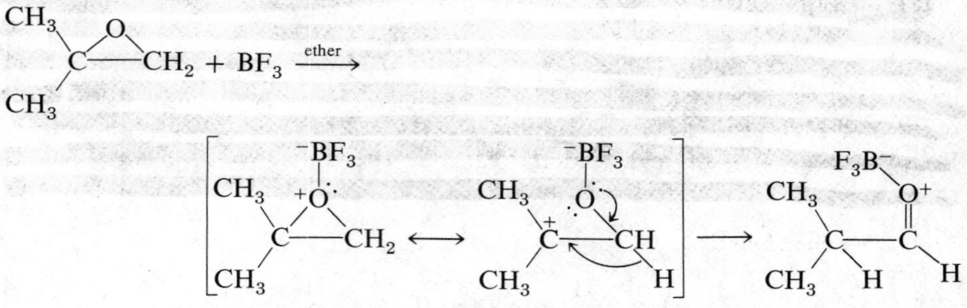

Note that rearrangement converts the more stable carbocation form of the epoxide–Lewis acid complex to the still more stable oxonium ion derived from a carbonyl group.

EXERCISE: Write the equation for the reaction of *trans*-2-butene oxide with sulfuric acid in methanol. What is the stereostructure of the product?

B. *Higher Cyclic Ethers*

Oxetanes are four-membered ring ethers.

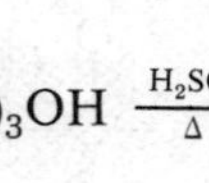

oxetane

This relatively unimportant class of compound can be prepared by **cyclization** of a 3-halo alcohol or a 1,3-diol.

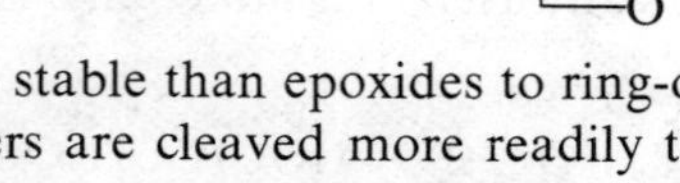

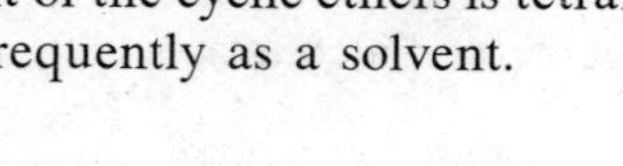

The oxetane ring is more stable than epoxides to ring-opening reactions, but the four-membered-ring ethers are cleaved more readily than open-chain ethers.

One of the most important of the cyclic ethers is tetrahydrofuran, THF, a material we have encountered frequently as a solvent.

tetrahydrofuran
THF

Like many other ethers, tetrahydrofuran reacts with oxygen to form dangerous peroxides. The five-membered ring is stable to ring opening, but ring opening can be accomplished under conditions that cause ether-cleavage reactions of open-chain ethers.

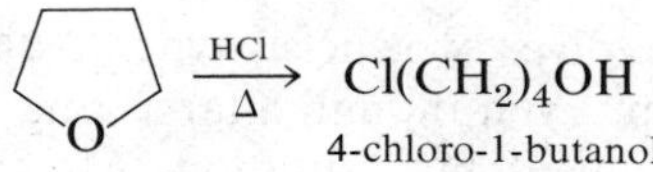

$$\xrightarrow[\Delta]{\text{HCl}} \quad Cl(CH_2)_4OH$$

4-chloro-1-butanol

A group of large-ring polyethers that have attracted a good deal of recent attention is the **crown ethers.** The compounds are cyclic polymers of ethylene glycol, $(-OCH_2CH_2)_n$, and are named in the form x-crown-y, where x is the total number of atoms in the ring and y is the number of oxygens. An example is 18-crown-6, the cyclic hexamer of ethylene glycol.

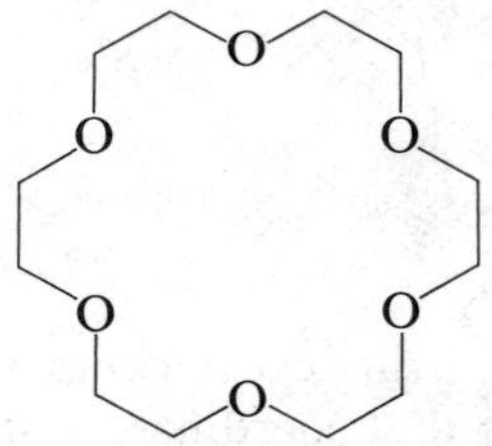

18-crown-6

The crown ethers are important for their ability to solvate cations strongly. The six oxygens in 18-crown-6 are ideally situated to solvate a potassium cation, just as water molecules would normally do. In the resulting complex the cation is solvated by the polar oxygens, but the exterior has hydrocarbon properties. As a result, the complexed ion is soluble in nonpolar organic solvents. For example, the complex of 18-crown-6 and potassium permanganate is soluble in benzene.

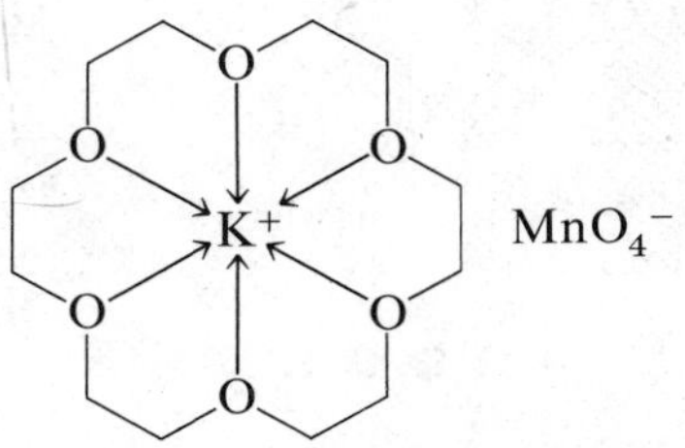

MnO_4^-

18-crown-6–KMnO$_4$ complex
soluble in benzene

18-Crown-6 may be prepared by treating a mixture of triethylene glycol and the corresponding dichloride with aqueous KOH.

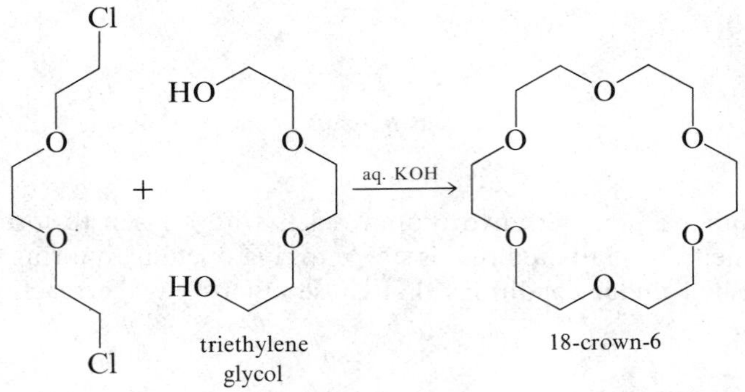

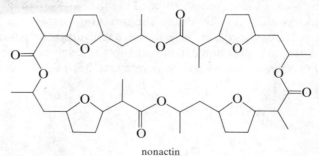

triethylene
glycol

18-crown-6

The reaction mechanism involves two successive S_N2 displacements with chloride ion being the leaving group. Even though a large ring is being formed, the reaction need not be carried out at high dilution. After the initial alkylation the potassium cation apparently acts as a "template" to bring the two reacting ends of the long chain close together for rapid reaction.

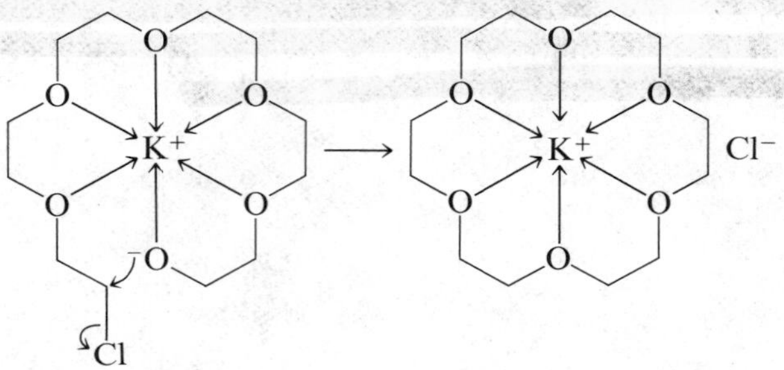

A number of naturally occurring cyclic compounds are now known with oxygens or nitrogens in the ring that coordinate with metal cations. Some of these compounds are involved in the transport of ions across biological membranes. An example is nonactin, an antibiotic that functions by transporting sodium ions into bacteria until the resulting osmotic pressure causes rupture of the cell wall.

nonactin

PROBLEMS

1. Give the structure corresponding to each of the following names.

(a) isobutyl methyl ether

(b) neopentyl alcohol

(c) 3-ethoxy-2-methylhexane

(d) 4-methyl-2-pentanol

(e) triphenylmethanol

(f) tetrahydrofuran

(g) dioxane

(h) 4-*t*-butyl-3-methoxyheptane

(i) 2,3-dimethoxybutane   (j) α,β-diphenylethyl alcohol
(k) 18-crown-6   (l) 15-crown-5

2. Give the IUPAC name corresponding to each of the following structures.

(a) 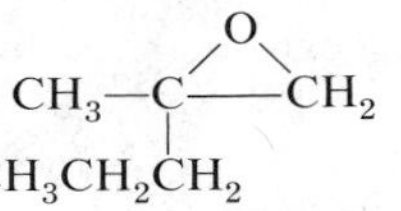
$$CH_3-C\overset{O}{\underset{|}{\triangle}}CH_2$$
$$CH_3CH_2CH_2$$

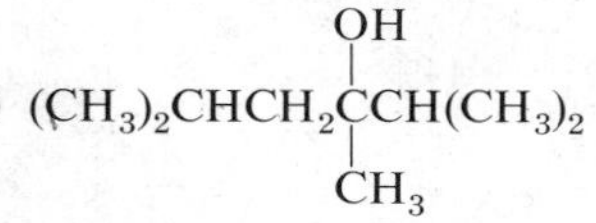

(b)
$$\overset{OH}{\underset{|}{(CH_3)_2CHCH_2CCH(CH_3)_2}}$$
$$CH_3$$

(c)
$$\overset{CH_2OH}{\underset{|}{CH_3CH_2CHCHCH_3}}$$
$$CH_2CH_3$$

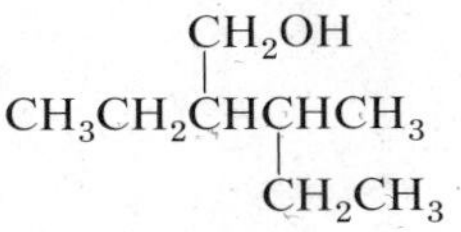

(d) 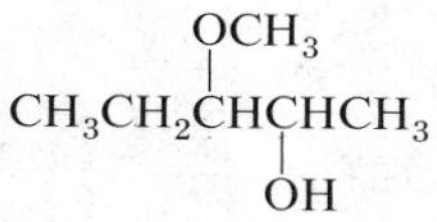
$$\overset{OCH_3}{\underset{|}{CH_3CH_2CHCHCH_3}}$$
$$OH$$

(e) 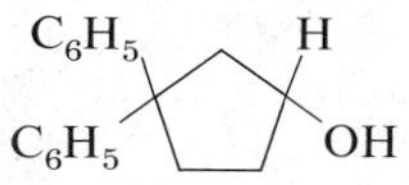
$$C_6H_5 \quad H$$
$$C_6H_5 \quad OH$$

(f) 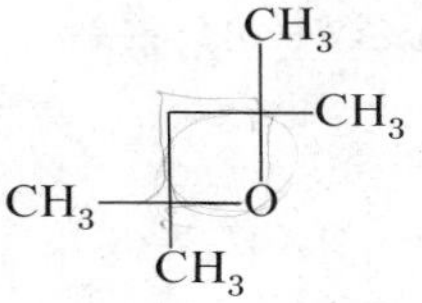
$$CH_3 \quad CH_3$$
$$CH_3 \quad O$$
$$CH_3$$

(g) 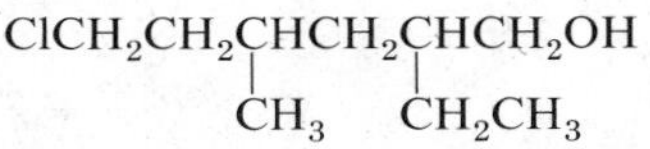
$$ClCH_2CH_2CHCH_2CHCH_2OH$$
$$CH_3 \quad CH_2CH_3$$

3. How many isomeric ethers correspond to the molecular formula $C_5H_{12}O$? Give common and IUPAC names for each structure and sketch the expected nmr spectrum of each. Which of these ethers is capable of optical activity? Write out the structures of the two mirror images and show that they are not superimposable.

4. There are 17 isomeric alcohols of the formula $C_6H_{13}OH$. Write out the structure and give the IUPAC name of each one. Identify the primary, secondary, and tertiary alcohols.

5. A naive graduate student attempted the preparation of $CH_3CH_2CDBrCH_3$ from $CH_3CH_2CDOHCH_3$ by heating the deuterio alcohol with HBr and H_2SO_4. He obtained a product having the correct boiling point, but a careful examination of the spectral properties by his research director showed that the product was a mixture of $CH_3CHDCHBrCH_3$ and $CH_3CH_2CDBrCH_3$. What happened?

6. (a) In a popular handbook the compound $CH_3CH_2C(CH_3)_2OCH_3$ is listed as ether, *sec*-butyl methyl, 2-methyl-. Does this name accord with any approved nomenclature you have studied? Give correct common and IUPAC names.
(b) Note the resemblance in appearance of this ether to 3,3-dimethylpentane. Compare the boiling points of both compounds as given in a handbook.
(c) 2,2-Dimethyl-3-pentanol and *t*-butyl isopropyl ether are isomers that are not listed in common handbooks. How would you expect their boiling points to compare? See if you can find these compounds and their boiling points in the important compendium *Beilstein,* found in almost all chemistry libraries. (*Beilstein,* a series of volumes written in German, is discussed in Chapter 33.)

7. Give the principal product(s) from each of the following reactions.

(a) $CH_3CH_2CH_2OH \xrightarrow[130°]{H_2SO_4}$

(b) $CH_3CH_2CH_2O^- + (CH_3)_3CCl$

(c) $(CH_3)_3CO^- + CH_3CH_2CH_2Br$

(d) $CH_3CH_2CH_2OH + K_2Cr_2O_7 + H_2SO_4 \xrightarrow{50°}$

(e)

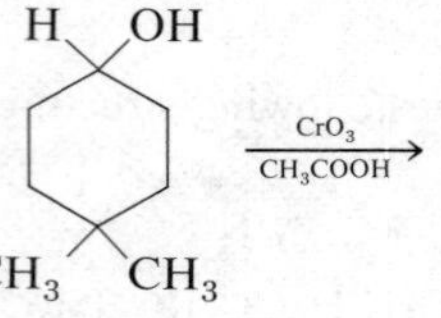

$\xrightarrow[CH_3COOH]{CrO_3}$

(f) 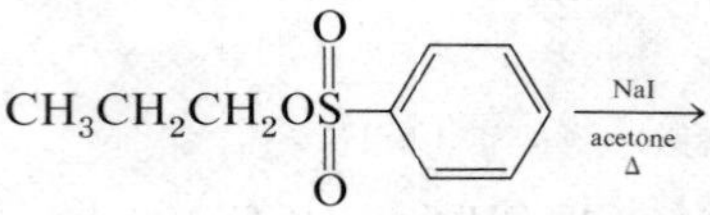 $CH_3CH_2CH_2O\overset{\displaystyle O}{\underset{\displaystyle O}{S}}$— $\xrightarrow[\substack{acetone \\ \Delta}]{NaI}$

(g) $CH_3CH_2CH_2OH \xrightarrow[\substack{H_2SO_4 \\ \Delta}]{HBr}$

(h) $(CH_3CH_2)_3CCH_2OH \xrightarrow[\substack{H_2SO_4 \\ \Delta}]{HBr}$

(i) $(S)\text{-}CH_3\overset{\displaystyle OH}{\underset{\displaystyle |}{C}}HCH_2CH_3 + SOCl_2 \xrightarrow{pyridine}$

(j) $(CH_3CH_2)_2CHOH + KH \longrightarrow$

(k) $(CH_3CH_2CH_2)_2O + HI \text{ (excess)} \xrightarrow{\Delta}$

(l) $CH_3CH_2\overset{\displaystyle |}{\underset{\displaystyle OCH_3}{C}}HCH_2CH_3 + HBr \text{ (excess)} \xrightarrow{\Delta}$

8. Explain why 2-cyclopropyl-2-propanol reacts with HCl to give 2-chloro-2-cyclopropylpropane instead of 1-chloro-2,2-dimethylcyclobutane.

9. Give the reagents and conditions for the best conversions of alcohol to alkyl halide as shown.

(a) $CH_3CH_2CH_2CH_2OH \longrightarrow CH_3CH_2CH_2CH_2Cl$

(b) $CH_3CH_2\overset{\displaystyle CH_3}{\underset{\displaystyle |}{C}}HCH_2OH \longrightarrow CH_3CH_2\overset{\displaystyle CH_3}{\underset{\displaystyle |}{C}}HCH_2Cl$

(c) $CH_3CH_2\overset{\displaystyle CH_3}{\underset{\displaystyle |}{C}}HCH_2OH \longrightarrow CH_3CH_2\overset{\displaystyle CH_3}{\underset{\displaystyle |}{\underset{\displaystyle Cl}{C}}}—CH_3$

(d) $CH_3CH_2CH_2CHOHCH_3 \longrightarrow CH_3CH_2CH_2CHICH_3$

(e)

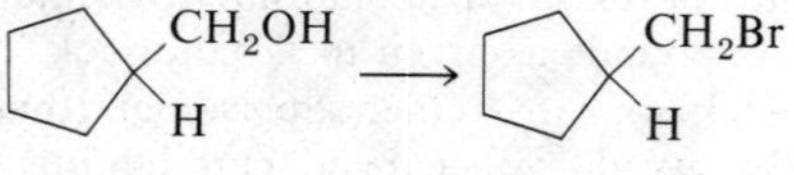

(f)

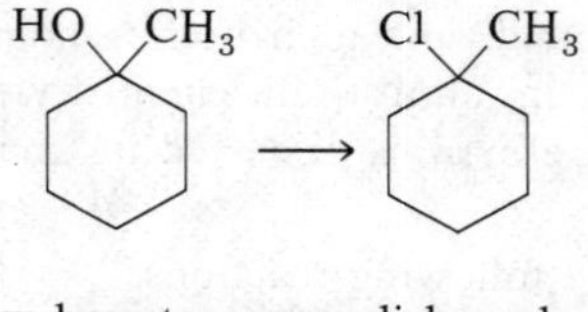

10. Show how to accomplish each of the following conversions in a practical manner.
(a) $(CH_3)_3CCH_2OH \longrightarrow CH_3CH_2CCl(CH_3)_2$
(b) $(CH_3)_2CHCHOHCH_3 \longrightarrow (CH_3)_2CHCHBrCH_3$
(c) $(CH_3CH_2)_3COH \longrightarrow (CH_3CH_2)_2C=CHCH_3$
(d) $CH_3CH_2CH_2CH_2OH \longrightarrow CH_3CH_2CH_2CH_2CN$

(e) $(CH_3)_3CCl + CH_3I \longrightarrow (CH_3)_3COCH_3$

(f) $(CH_3)_2CHCH_2OH \longrightarrow (CH_3)_2CHCH_2SCH_3$

11. Write balanced equations for the following reactions.

(a) 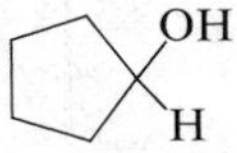 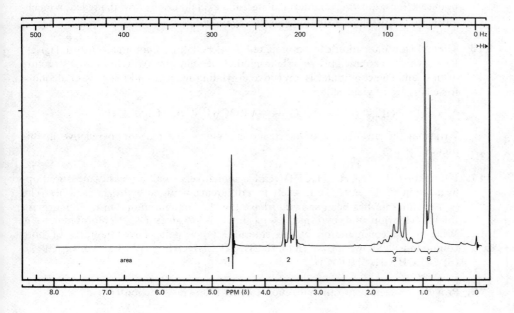 + $HNO_3 \longrightarrow HOOC(CH_2)_3COOH + H_2O + NO_2$

(b) $ClCH_2CH_2CH_2OH + HNO_3 \longrightarrow ClCH_2CH_2COOH + NO_2 + H_2O$

(c) $CH_3CHOHCH_3 + CrO_3 + H^+ \longrightarrow CH_3COCH_3 + H_2O + Cr^{3+}$

(d) 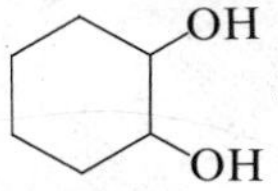 + $K_2Cr_2O_7 + H^+ \longrightarrow HOOC(CH_2)_4COOH + Cr^{3+} + K^+ + H_2O$

12. (a) Although isobutyl alcohol reacts with HBr and H_2SO_4 to give isobutyl bromide, with little rearrangement, 3-methyl-2-butanol reacts on heating with conc. HBr to give only the rearranged product, 2-bromo-2-methylbutane. Explain this difference using the reaction mechanisms involved.

(b) Unsymmetrical ethers are generally not prepared by heating two alcohols with sulfuric acid. Why not? Yet, when *t*-butyl alcohol is heated in methanol containing sulfuric acid, a good yield of *t*-butyl methyl ether results. Explain this result by means of the reaction mechanism.

(c) Prolonged reaction of ethyl ether with HI gives ethyl iodide. Write out the reaction mechanism.

13. Fusel oil is a toxic by-product of carbohydrate fermentation that consists mostly of two five-carbon alcohols. It concentrates in the higher-boiling residues of distillation of ethanol in fermentation and, if such distillation is not monitored, will distill to give the product that special toxic "kick" associated with bootleg liquor. From the nmr spectra given here, determine the structures of the two principal constituents of fusel oil. One of these compounds is obtained in optically active form by fermentation and has the trivial name "active amyl alcohol." Which structure belongs to it?

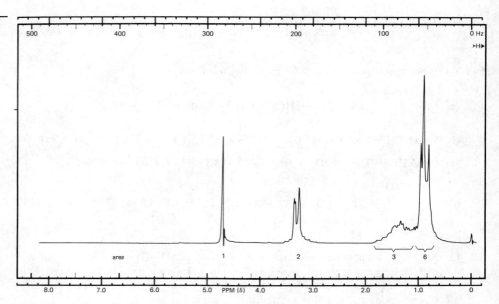

14. The thermodynamics of reactions in a solution are often quite different from those in the gas phase because of the importance of solvation energies. However, thermodynamic data for solvents other than water are sparse, whereas many heats of formation are now available for the gas phase. If one considers reactions in which the number of ions or ion pairs remains the same, the relative gas phase values can be instructive. Compare $\Delta H°$ for the following two reactions in the gas phase.

$$CH_3CH_2OH + Br^- \rightleftharpoons CH_3CH_2Br + OH^-$$
$$CH_3CH_2OH + HBr \rightleftharpoons CH_3CH_2Br + H_2O$$

In addition to the values in Appendix I, the following $\Delta H_f°$ are required: Br^-, -50.8 kcal mole^{-1}; OH^-, -32.9 kcal mole^{-1}.

15. *n*-Butyl ethyl ether is cleaved by hot conc. HBr to give both ethyl bromide and *n*-butyl bromide. *t*-Butyl ethyl ether, however, is cleaved readily by cold conc. HBr to give primarily *t*-butyl bromide and ethyl alcohol. Write the reaction mechanisms of both reactions, showing all intermediates, and explain briefly how the reaction mechanisms relate to these experimental observations.

16. Triethyloxonium cation can be prepared as a crystalline fluoborate salt, $(C_2H_5)_3O^+$ BF_4^-, which is appreciably soluble in methylene chloride. Write the Lewis structure of this salt. The compound is a reactive ethylating reagent and reacts with alcohols, for example, to yield ethers.

$$ROH + (CH_3CH_2)_3O^+ \longrightarrow ROC_2H_5 + (C_2H_5)_2O + H^+$$

Write out the mechanism of the reaction. Why is the reagent so reactive in this process?

17. 1-Butanol-*1-d*, $CH_3CH_2CH_2CHDOH$, has a relatively small but easily measured optical activity, $[\alpha]_D = 0.5°$, due solely to a difference between hydrogen isotopes. The $(-)$ enantiomer has been shown to have the (R) configuration. On treatment with thionyl chloride and pyridine, $(-)$-1-butanol-*1-d* gives $(+)$-1-chlorobutane-*1-d*. What is the configuration of the chloride? Draw perspective diagrams of both compounds. Show how (R)-$CH_3CH_2CH_2CHDOH$ may be converted to (S)-$CH_3CH_2CH_2CHDOCH_3$.

18. Provide a mechanism rationalization for the following reaction course.

19. Identify each of the following compounds from the data presented. Cmr data are given as ppm downfield from TMS, with the multiplicity of the proton-coupled resonance given in parentheses (s = singlet, d = doublet, t = triplet, q = quartet).
(a) $C_5H_{11}ClO$; 62.2(t), 45.4(t), 32.9(t), 32.1(t), 23.7(t).
(b) $C_8H_{18}O$; 71.2(t), 33.1(t), 20.3(t), 14.6(q).
(c) $C_6H_{14}O$; 75.1(d), 35.3(s), 25.8(q), 18.2(q); the resonance at 25.8 ppm is much more intense than the others.
(d) $C_6H_{14}O$; 65.5(d), 49.2(t), 25.1(d), 24.3(q), 23.5(q), 22.7(q).
(e) $C_6H_{12}O_2$; 71.1(d), 33.9(t).

20. Optically active (2R,3S)-3-chloro-2-butanol is allowed to react with sodium hydroxide in ethanol to give an optically active oxirane, which is treated with potassium hydroxide in water to obtain 2,3-butanediol. What is the stereostructure of the diol? What can you say about its optical rotation?

21. Optically active 5-chloro-2-hexanol is allowed to react with KOH in methanol. The product, $C_6H_{12}O$, is found to have $[\alpha]_D = 0$. What can you say about the stereochemistry of the reactant?

22. 18-Crown-6 is a useful catalyst for some reactions of potassium salts. For example, KF reacts with 1-bromooctane to give 1-fluorooctane much faster in the presence of 10 mole % 18-crown-6 than in its absence. However, it is ineffective as a catalyst if NaF is used. Explain.

23. Treatment of 1,2,2-trimethylcycloheptanol with sulfuric acid gave a mixture of 1,7,7-trimethylcycloheptene, 1-t-butylcyclohexene, and 1-isopropenyl-1-methylcyclohexane. Write a mechanism that accounts for the formation of these products.

24. By using a collection of point charges, show that the electrostatic energy between a charge and a dipole varies as $1/r^2$ and that between two dipoles varies as $1/r^3$, where r is the distance to the center of the dipole. The dipole can be treated as two point charges close together relative to r.

25. (a) One useful measure of the energy of hydrogen bonding derives from heats of vaporization, ΔH_v. The following table gives the vapor pressure of ethanol at different temperatures.

Temperature, °C	Vapor Pressure, torr
−31.3	1
−2.3	10
+19.0	40
34.9	100
64.5	400
78.4	760

From

$$\frac{d\ln P}{dT} = -\frac{\Delta H_v}{RT^2} \quad \text{or} \quad \frac{d\ln P}{d(1/T)} = -\frac{\Delta H_v}{R}$$

plot ln P versus $1/T$ and calculate ΔH_v for ethanol. For comparison, ΔH_v of propane is 4.49 kcal mole^{-1}.

(b) This difference can be compared to an electrostatic model. The dipole moment of ethanol is 1.7 D. Consider the approximation that this dipole results from partial charges at oxygen and hydrogen. The O—H bond distance is 0.96 Å. What fraction of positive and negative electronic charges separated by 0.96×10^{-8} cm correspond to a dipole moment of 1.7×10^{-18} esu cm? The charge on an electron is 4.8×10^{-10} esu. Calculate the electrostatic energy of attraction of a pattern of such charges arranged as

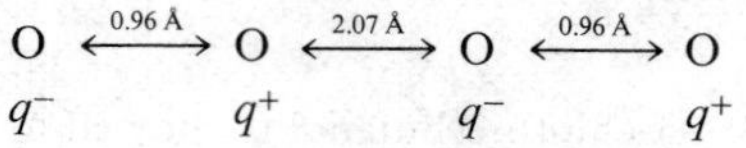

$$\text{O} \xleftrightarrow{\text{0.96 Å}} \text{O} \xleftrightarrow{\text{2.07 Å}} \text{O} \xleftrightarrow{\text{0.96 Å}} \text{O}$$
$$q^- \qquad q^+ \qquad q^- \qquad q^+$$

An electron and a proton 1 Å apart have an electrostatic energy of attraction of 332 kcal mole^{-1}. This electrostatic energy calculation is only a crude model for a hydrogen bond

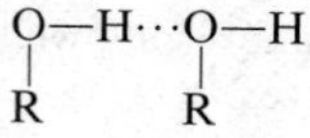

$$\begin{array}{ccc} \text{O}-\text{H}\cdots\text{O}-\text{H} \\ | & | \\ \text{R} & \text{R} \end{array}$$

but it does give a rough idea of the magnitude of the energy quantities involved.

26. Propose a mechanism for the following reaction.

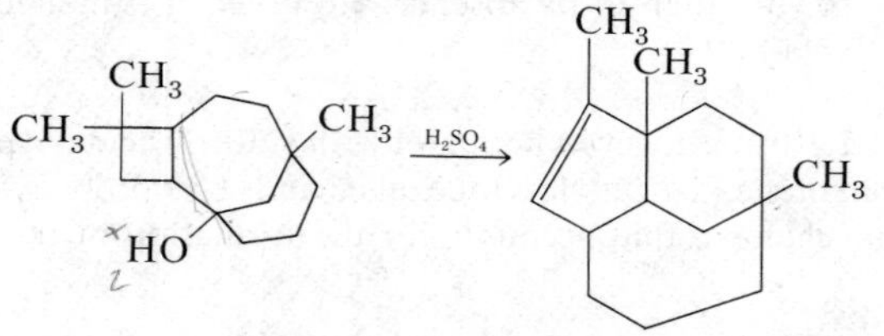

Chapter 11
Alkenes

Alkenes are hydrocarbons with a C=C double bond. The double bond is a stronger bond than a single bond, yet paradoxically the C=C double bond is much more reactive than a C—C single bond. Unlike alkanes, which generally show rather nonspecific reactions, the double bond is the site of many specific reactions and is a functional group.

11.1
Electronic Structure

The geometric structure of ethylene, the simplest alkene, is well known from spectroscopic and diffraction experiments and is shown in Figure 11.1. The entire molecule is planar, as shown in the stereo plot in Figure 11.2.

In the Lewis structure of ethylene the double bond is characterized as a region with two pairs of electrons.

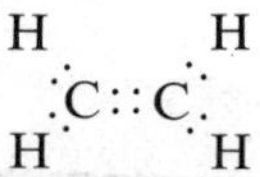

$$\begin{array}{ccc} H & & H \\ & \ddot{C}::\ddot{C} & \\ H & & H \end{array}$$

In orbital descriptions, we need one orbital for each pair of electrons. Hence, we need two orbitals between the carbons to accommodate the electrons. Many possible schemes can be devised to arrange such orbitals, but only two are important.

A. Bent-Bond Model

In one scheme we take two equivalent hybrid orbitals on each carbon and allow them to overlap in a nonlinear fashion as in Figure 11.3. For this kind of "bent-bond" overlap, better overlap results from hybrids that have greater p-character. To a useful approximation, each of the hybrid orbitals making up the two bent bonds has sp^5-hybridization. That is, each of the orbitals is made up from $\frac{1}{6}$ of an

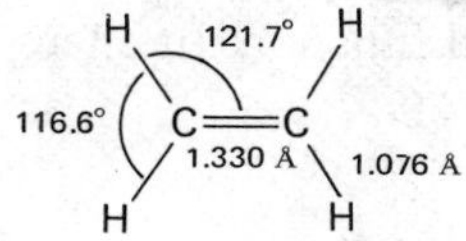

FIGURE 11.1. Structure of ethylene.

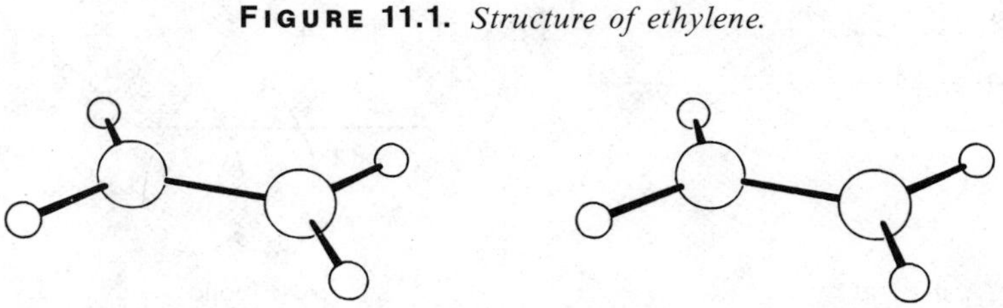

FIGURE 11.2. Stereo representation of ethylene.

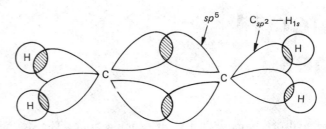

FIGURE 11.3. *Bent-bond orbital model of ethylene.*

s-orbital and $\frac{5}{6}$ of a p-orbital. The two hybrid orbitals together use up $\frac{1}{3}$ of an s-orbital and $1\frac{2}{3}$ p-orbitals. This leaves just enough for each carbon to have two additional sp^2-hybrid orbitals for bonding to the hydrogens.

As shown in Figure 11.2, all six atoms lie in the same plane. Bonds with more s-character tend to be shorter and stronger. The C—H bonds are 1.076 Å long and are slightly shorter than the bonds in methane, 1.085 Å. The H—C—H bond angle is much wider than the tetrahedral angle of 109.5°, but is not quite as large as the angle expected for two sp^2-hybrids, 120°. The orbital picture given is an approximate one. The C=C double bond distance of 1.33 Å is much shorter than the normal C—C single bond of 1.54 Å. The extra electron density between the carbon nuclei provides additional attraction to the nuclei to help overcome the added nuclear repulsion of the shorter internuclear distance.

B. π-bond Model

A different orbital model of ethylene starts with the two sp^2-hybrids from each carbon to the hydrogens. A third sp^2-hybrid on each carbon is used to form a C_{sp^2}—C_{sp^2} single bond. This leaves a p-orbital "left over" on each carbon. This p-orbital lies perpendicular to the plane of the six atoms. The two p-orbitals are parallel to each other and have regions of overlap above and below the molecular plane. This type of bond, in which there are two bonding regions above and below a nodal plane, is called a π-bond. This notation is used in order to distinguish it from the type of bond formed by overlap of two carbon sp^2-orbitals. Such a bond has no node and is called a σ-bond (Figure 11.4). This orbital picture of ethylene is shown in Figure 11.5.

Superficially, this model appears to be completely different from the bent-bond model illustrated in Figure 11.3, but this difference is totally in the inadequacies of the kinds of symbolic representations used in these figures. *The two orbital models are actually exactly equivalent!* They are simply two different visualizations of the same mathematical function. In effect, we have taken the same total

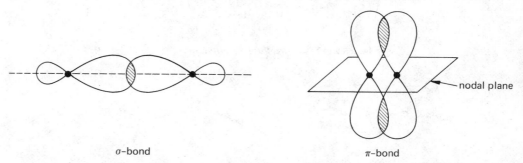

FIGURE 11.4. *σ- and π-bonds.*

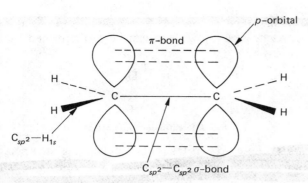

FIGURE 11.5. $\sigma-\pi$-bond model of ethylene.

electron density distribution between the carbons and have split it in two different ways. The total electron density is the same for both. The $\sigma-\pi$ description is a common and useful one.

An interesting view of the electron density in the ethylene double bond is seen in Figure 11.6. Each contour diagram represents the electron density of a slice cut through the midpoint of the C—C bond (see plane in Figure 11.6a). Recall that each line in such a contour diagram represents a constant definite value of the electron density. For example, the innermost oval in Figure 11.6b represents a rather high value of the electron density, and the outermost oval has a rather low value. The σ-electron density is almost cylindrically symmetric but not quite. The outermost contours are oval rather than circular because of electron repulsion by the π-electrons. The π-electron density in Figure 11.6d is much less than the σ-density, and the total electron density in Figure 11.6b has a smooth oval character. This total electron distribution does not have the appearance one might expect from the simple kinds of representations in Figures 11.3 and 11.5. One important consequence of the noncylindrically symmetric electron density about the

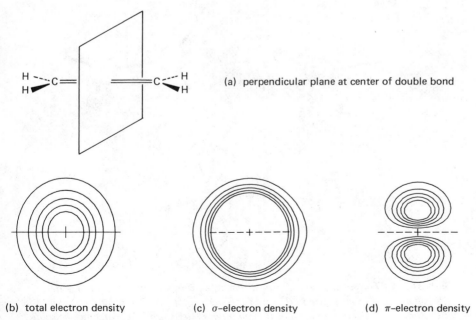

(a) perpendicular plane at center of double bond

(b) total electron density   (c) σ-electron density   (d) π-electron density

FIGURE 11.6. *Electron-density distribution in center of ethylene double bond.*

C—C bond axis is that there is a barrier to rotation about this axis. For example, two dideuterioethylenes are known and can be distinguished by their different spectroscopic properties.

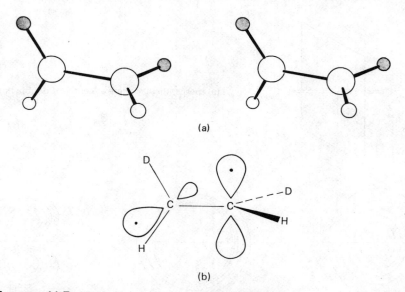

two different dideuterioethylenes

Interconversion of these isomers takes place only at high temperatures (~500°) and has an activation energy of 65 kcal mole⁻¹. The transition state for the reaction has a half-twisted structure in which the *p*-orbitals have zero overlap. This structure is represented in Figure 11.7.

The two forms of dideuterioethylene represent another case of stereoisomerism (Chapter 7). Recall that stereoisomers are compounds having the same sequence of covalently bonded atoms but differing in the orientation of the atoms in space. The two dideuterioethylene stereoisomers may be interconverted by rotation about a bond, just as the *anti* and *gauche* stereoisomers of butane.

However, in one case the barrier to interconversion (the rotational barrier) is 65 kcal mole⁻¹, and in the other it is only 3.3 kcal mole⁻¹. The dideuterioethylene stereoisomers may be obtained separately, and each isomer is perfectly stable at

(a)

(b)

FIGURE 11.7. *The half-twisted transition state for interconversions of dideuterio-ethylenes. (a) The structure is shown in the stereo plot. (b) The orbital representation shows one electron in each of the noninteracting p-orbitals, which are at right angles to each other.*

normal temperatures. On the other hand, the *anti* and *gauche* butane stereoisomers interconvert easily at temperatures above about −230°. This difference in the ease of interconversion has resulted in the two types of stereoisomers having different names. Stereoisomers that can be easily interconverted by rotation about a bond are called **conformational isomers.** Stereoisomers that are not easily interconverted are called **configurational isomers.** The dideuterioethylenes are two such configurational isomers.

11.2
Nomenclature of Alkenes

Historically, hydrocarbons with a double bond were known as **olefins.** This rather strange class name comes from the Latin words *oleum,* an oil, and *ficare,* to make. It arose because derivatives of such compounds often had an oily appearance.

As with other classes of organic compounds, two systems of nomenclature are used: common and systematic. In the common system, which is only used for fairly simple compounds, the final **-ane** of the alkane name is replaced by **-ylene.**

$$CH_2{=}CH_2 \qquad CH_3CH{=}CH_2 \qquad CH_3\overset{\overset{\displaystyle CH_3}{|}}{C}{=}CH_2$$

ethylene   propylene   isobutylene

A few simple molecules are named as derivatives of ethylene.

tetramethylethylene   trichloroethylene

Configurational isomers are distinguished by the use of the prefixes *cis-* (L., on this side) and *trans-* (L., across), as with disubstituted cycloalkanes (pp. 92–93).

cis-dibromoethylene   *trans*-dibromoethylene

Some monosubstituted ethylenes are named as radical combinations in which the $CH_2{=}CH—$ group is called **vinyl.**

$$CH_2{=}CHCl \qquad\qquad CH_2{=}CHOCH_3$$

vinyl chloride   methyl vinyl ether

A few special trivial names are in common use.

$$C_6H_5CH{=}CH_2$$

styrene   *trans*-stilbene

In the IUPAC system alkenes are named as derivative of a parent alkane. The **alk-** stem specifies the number of carbons in the chain and the **ene** suffix specifies a double bond. A number is used to indicate the position of the double bond along the chain. Since the double bond joins one carbon to a carbon with the next higher number, only one number need be given. Finally, a prefix *cis-* or *trans-* is included where necessary.

$$CH_2{=}CH_2 \qquad CH_3CH{=}CH_2 \qquad CH_3CH_2CH{=}CH_2$$

     ethene   propene   1-butene

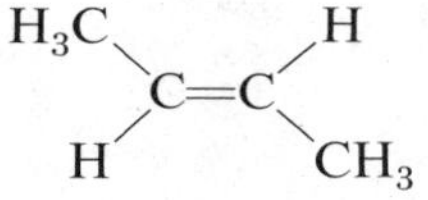

    trans-2-butene 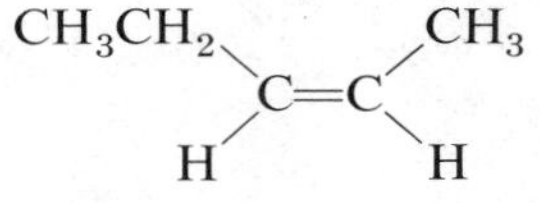   *cis*-2-pentene 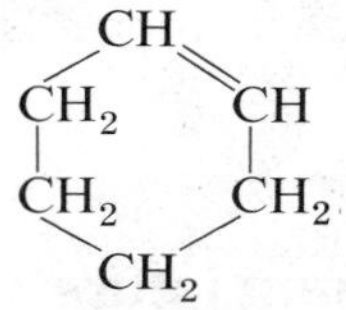   cyclohexene
                                  (no number necessary)

Substituent groups are included as prefixes with appropriate numbers to specify position. Since the **-ene** stem is a suffix, it dominates the numbering. That is, the parent alkene chain is named first, including *cis-* or *trans-* where necessary, and then the substituents are appended as prefixes.

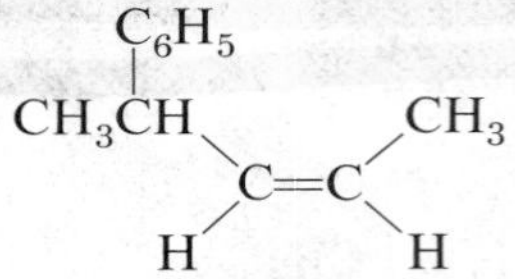

    4-phenyl-*cis*-2-pentene   3-chloro-*trans*-3-hexene
    (not 2-phenyl-3-pentene)

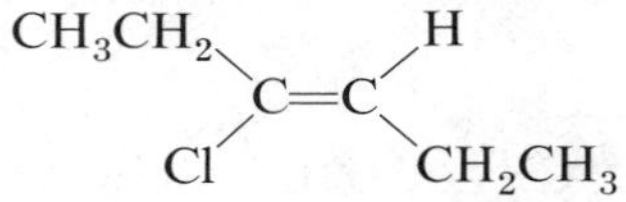

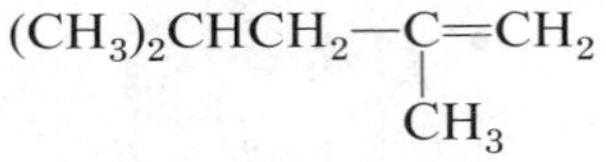

    2,4-dimethyl-1-pentene 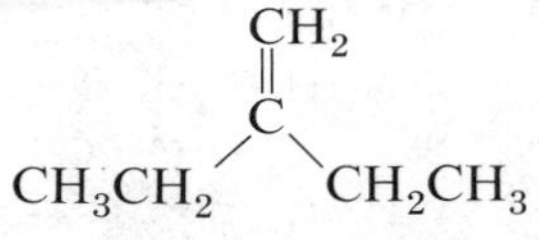   2-ethyl-1-butene
                               (parent is longest chain that
                                 includes the double bond)

The *cis-trans* system for naming configurational isomers frequently leads to confusion. The Chemical Abstracts Service has proposed an unambiguous system that has been adopted by the IUPAC. In this system, the two groups attached to each end of the double bond are assigned priority numbers as is done in naming enantiomers by the (R-S) system (Chapter 7). When the two groups of higher-priority number are on the **same side** of the molecule, the compound is the **Z** isomer (Ger., *zusammen*, together). When the two groups of highest priority are on **opposite sides** of the molecule, the compound is the **E** form (Ger., *entgegen*, opposite).

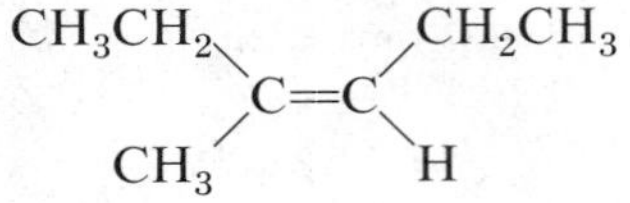

    (Z)-3-methyl-3-hexene 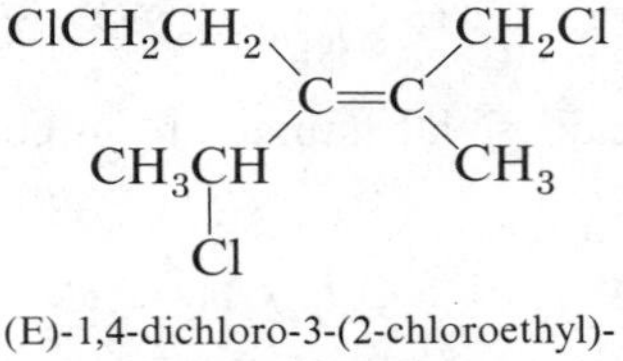   (E)-1,4-dichloro-3-(2-chloroethyl)-
                                    2-methyl-2-pentene

In normal use, several common compounds are almost invariably called by their common or trivial names as given above, but more complex compounds are named by the IUPAC system.

In compounds having both a double bond and a hydroxy group, two suffixes are used. The -ol suffix has priority, and the longest chain *containing both the double bond and the OH group* is numbered in such a way as to give the OH group the smaller number.

$$CH_2=CHCH_2CH_2OH$$

but-3-en-1-ol
(not but-1-en-4-ol)

$$\begin{array}{c} CH_2=CH \\ | \\ CH_3CH_2CHCH_2OH \end{array}$$

2-ethylbut-3-en-1-ol

EXERCISE: Give the correct IUPAC name for each of the following compounds.

(a)
$$\begin{array}{cc} CH_3 & CH_3 \\ C=C \\ H & CH_2CH_3 \end{array}$$

(b)
$$\begin{array}{cc} BrCH_2 & CH_2C(CH_3)_3 \\ C=C \\ CH_3 & CH_3 \end{array}$$

(c)
$$\begin{array}{c} Cl \\ CH_3CH_2CHCH_2 H \\ C=C \\ H CH_2CH_3 \end{array}$$

(d)
$$\begin{array}{c} CH_2CH_2Br \\ CH_2=C \\ CH_2CH_2CH(CH_3)_2 \end{array}$$

11.3
Physical Properties of Alkenes

Physical properties of some alkenes are summarized in Table 11.1. These properties are similar to those of the corresponding alkanes, as shown by the boiling-point plot in Figure 11.8. The lower members are gases at room temperature. Starting with the five-carbon compounds, the alkenes are volatile liquids. Isomeric alkenes have similar boiling points, and mixtures can be separated only by careful fractional distillation with efficient columns. 1-Alkenes tend to boil a few degrees lower than internal olefins and can be separated by such careful fractionation.

A. *Dipole Moments*

In an alkyl-substituted double bond the carbon orbitals making up the $=C-C$ bond have different amounts of *s*-character. Such a bond may be approximated as $C_{sp^2}-C_{sp^3}$. The resulting change in electron density distribution gives such bonds an effective dipole moment in the direction

$$\overset{\longleftrightarrow}{C}-C=C$$

These dipole moments are small for hydrocarbons, but still permit a distinction between *cis* and *trans* isomers. For example, *cis*-2-butene has a small dipole moment, whereas *trans*-2-butene has a resultant dipole moment of zero because of its symmetry.

vector sum

$$\begin{array}{cc} H_3C & CH_3 \\ C=C \\ H & H \end{array}$$
$$\mu = 0.33 \text{ D}$$

$$\begin{array}{cc} H_3C & H \\ C=C \\ H & CH_3 \end{array}$$
$$\mu = 0 \text{ D}$$

TABLE 11.1
Physical Properties of Alkenes

Name	Structure	Boiling Point, °C	Density d^{20}
ethylene	$CH_2{=}CH_2$	−103.7	
propene	$CH_3CH{=}CH_2$	−47.4	0.5193
1-butene	$CH_3CH_2CH{=}CH_2$	−6.3	0.5951
cis-2-butene	$CH_3\overset{H}{C}{=}\overset{H}{C}CH_3$	3.7	0.6213
trans-2-butene	$CH_3\overset{H}{C}{=}\underset{H}{C}CH_3$	0.9	0.6042
2-methyl-1-propene	$(CH_3)_2C{=}CH_2$	−6.9	0.5942
1-pentene	$CH_3CH_2CH_2CH{=}CH_2$	30.0	0.6405
cis-2-pentene	$CH_3CH_2\overset{H}{C}{=}\overset{H}{C}CH_3$	36.9	0.6556
trans-2-pentene	$CH_3CH_2\overset{H}{C}{=}\underset{H}{C}CH_3$	36.4	0.6482
2-methyl-2-butene	$(CH_3)_2C{=}CHCH_3$	38.6	0.6623

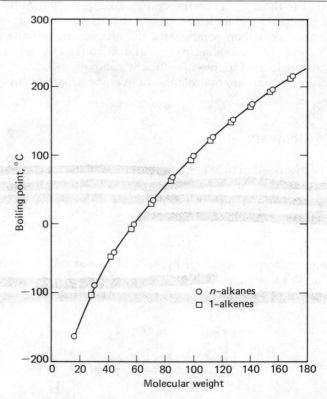

FIGURE 11.8. *Boiling point relationships of alkenes and alkanes.*

With substituents such as halogens the dipole moment differences are greater.

$$\mu = 2.95 \text{ D} \qquad \mu = 0 \text{ D}$$

B. *Nmr Spectra*

Hydrogens attached to a double bond ("vinyl hydrogens") resonate at about $\delta = 5$ ppm. The actual values vary from about $\delta = 4.7$ ppm when the vinyl hydrogens are at the end of a chain to about $\delta = 5.3$ ppm when the vinyl hydrogen is at some other position along the chain.

$$\text{CH}_3\text{CH}_2\text{CH}{=}\text{CH}_2 \qquad \text{CH}_3\text{CH}{=}\text{CHCH}_3$$
$$\delta = 5.3 \text{ ppm} \quad \delta = 4.7 \text{ ppm} \qquad \delta = 5.3 \text{ ppm}$$

These values may appear to be rather far downfield for a C—H function. It is true that the increased *s*-character of the carbon orbital makes the carbon effectively more electronegative, but the observed change is too large to be a simple electronegativity effect.

The effect has its origin in the induced motion of bond electrons, just as discussed earlier in the diamagnetic shielding (Section 9.4) by electrons around the hydrogen nucleus. In a double bond the π-electrons are more polarizable than σ-electrons and are freer to move in response to a magnetic field. Such an electron cloud is said to be magnetically **anisotropic.** Thus when the molecule is oriented so that the plane of the double bond is perpendicular to the applied magnetic field, the π-electrons tend to circulate about the direction of the applied field. As shown in Figure 11.9, the circular motion of the π-electrons produces an induced magnetic field opposed to the applied field *at the middle of the double bond.* Out by the vinyl hydrogens, the magnetic lines of force are in the same direction as the

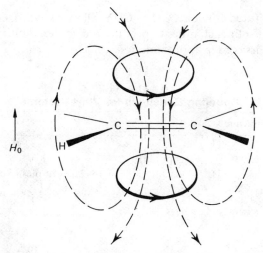

FIGURE 11.9. *Induced motion of π-electrons of a double bond in a magnetic field; the induced field has the same direction as the applied field at a vinyl proton.*

applied field. Hence, a lower applied field is necessary for the total field at the hydrogen nucleus to have the resonance value. For double bonds with orientations other than that shown, the effect will be smaller and the actual effect will be the average for all orientations because the tumbling and rotation of molecules in the normal nmr technique is fast on the nmr time scale.

It should be emphasized that vinyl hydrogens are still subject to the normal diamagnetic shielding effects of nearby electron clouds. The net effect is still a chemical shift far upfield from a bare proton. However, the effect of the **π-electron circulation** partially opposes the normal effect of local electrons in such a way that resonance occurs downfield from that observed for a saturated C—H by a significant amount. The resulting chemical shift of vinyl hydrogens provides an important analytical method for establishing both their presence and number.

> The applied magnetic field also induces electron currents in C—C and C—H single bonds, and it is these induced currents rather than electronegativity effects that give rise to the characteristic pattern of chemical shifts for differently substituted alkyl hydrogens.
>
> $$\delta(\geq\!CH) > \delta(-CH_2-) > \delta(-CH_3)$$

But nmr can tell even more about the structure of double bonds in alkenes. The magnitude of the coupling constant J between vinyl hydrogens varies with structure, as shown in Table 11.2. For two hydrogens that are attached to the same carbon (**geminal** hydrogens) the coupling constant is relatively small. In simple alkenes, J_{ab} is about 2 Hz and becomes smaller in alkenes with electron-withdrawing substituents.

The magnitude of J differs for *cis* and *trans* hydrogens. Although the ranges of both sets overlap, for a pair of isomeric *cis* and *trans* alkenes J_{trans} is invariably greater than J_{cis}. It may seem strange that hydrogens separated by the greater distance should have the greater effect on each other, but magnetic effects are frequently rather subtle. The coupling effect of hydrogen nuclei is transmitted through bonding electrons—not through space. Orientation of the bonds is as important as distance. The difference between J_{cis} and J_{trans} is an important tool for distinguishing *cis* and *trans* alkenes. Of course, this technique is useful only when the two hydrogens are not equivalent because equivalent hydrogens do not split each other.

In a monosubstituted ethylene, CH_2=CHY, all three hydrogens are nonequivalent. They all each split each other to produce a complex multiplet that cannot be analyzed by our first-order approximation.

TABLE 11.2
Coupling Constants for Vinyl Protons

Structure		J_{ab}, Hz
C=C with H_a and H_b	(geminal)	0–3 (~2 in simple alkenes)
H_a / C=C \ H_b	(*cis*)	5–14
H_a / C=C \ H_b	(*trans*)	11–19

Finally, a C—H group attached to a double bond is shifted downfield by about 0.8 ppm. The important generalizations to remember about nmr of alkenes are that $\delta(=C-H) \approx 5$ ppm, $CH_2=CH-$ is generally a complex multiplet, and $J_{trans} > J_{cis}$. These generalizations are exemplified in the nmr examples shown in Figures 11.10–11.12. Note in Figures 11.10 and 11.11 that the two vinyl hydrogens have different chemical shifts because of the different inductive effects of Cl and CN.

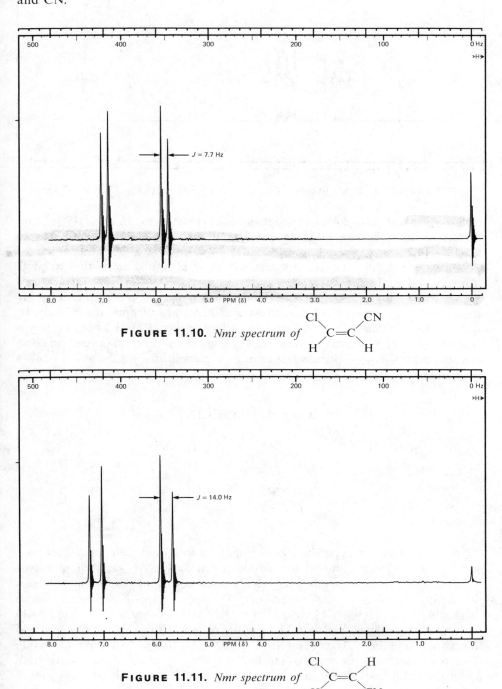

FIGURE 11.10. *Nmr spectrum of*

FIGURE 11.11. *Nmr spectrum of*

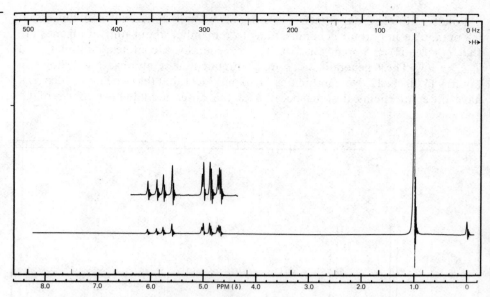

FIGURE 11.12. *Nmr spectrum of 3,3-dimethyl-1-butene,* $(CH_3)_3CCH{=}CH_2$.

Cmr chemical shifts for some simple alkenes are collected in Table 11.3. Comparison of this table with Table 9.4 shows that sp^2-hybridized carbons resonate at 90–120 ppm downfield from sp^3-hybridized carbons having the same degree of substitution. Substituent effects are about the same as for alkanes, with one significant exception. The γ-effect is highly dependent on the steric relationship between the carbon being observed and the substituent. For example, the CH_2 resonances in *cis*- and *trans*-3-hexene occur at 20.6 and 25.8 ppm, respectively. In the 3-methyl analogs, the corresponding carbon resonates at 21.3 ppm in the *cis* isomer and at 21.5 ppm in the *trans* isomer. Thus, the γ-effect on the carbon *cis* to the substituent is -4.3 ppm, while that on a carbon *trans* to the substituent is negligible.

In fact, this sterically dependent γ-effect may be seen in all pairs of *cis* and *trans* isomers. For example, the methyl carbon resonances in *cis*- and *trans*-2-butene occur at 11.4 and 16.8 ppm, respectively. In essence, each methyl in the *cis* isomer is exerting a γ-effect of -5.4 on the other. The appearance of alkene cmr spectra is shown in Figures 11.13 (1-octene) and 11.14 (*trans*-4-octene). In Figure 11.13 only seven resonances are seen because the chemical shifts of the C-4 and C-5 resonances are both 31.1 ppm. In Figure 11.14 (page 290) note that only four resonances are observed owing to the symmetry of the molecule. In both spectra a small 1:1:1 triplet may be seen at about 75 ppm. This triplet is caused by $CDCl_3$, which is a common solvent for cmr measurements.

TABLE 11.3
Cmr Chemical Shifts for Alkenes

	C-1	C-2	C-3	C-4	C-5	C-6
propene	115.4	135.7	18.7			
1-butene	112.8	140.2	23.8	9.3		
cis-2-butene	11.4	124.2	124.2	11.4		
trans-2-butene	16.8	125.4	125.4	16.8		
1-pentene	113.5	137.6				
cis-2-pentene	12.0	122.8	132.4	20.3	13.8	
trans-2-pentene	17.3	123.6	133.2	25.8	13.6	
1-hexene	113.5	137.8				
cis-2-hexene	12.3	123.7	130.6	29.3	23.0	13.5
trans-2-hexene	17.5	124.7	131.5	35.1	23.1	13.4
cis-3-hexene	14.3	20.6	131.0	131.0	20.6	14.3
trans-3-hexene	13.9	25.8	131.2	131.2	25.8	13.9

EXERCISE: For each of the following compounds, sketch the expected nmr spectrum.

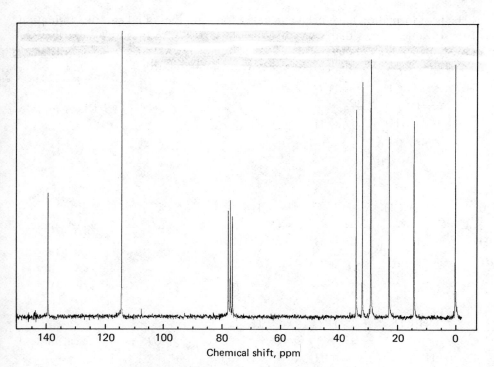

FIGURE 11.13. *Cmr spectrum of 1-octene.*

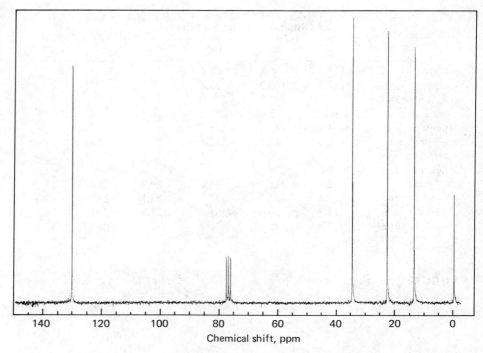

FIGURE 11.14. *Cmr spectrum of* trans-*4-octene.*

11.4
Relative Stabilities of Alkenes: Heats of Formation

Heats of formation have been evaluated for a number of alkenes. Examination of these values shows that *trans* alkenes are generally more stable than the isomeric *cis* alkenes by about 1 kcal mole⁻¹. (Remember that a *more negative* heat of formation ΔH_f° corresponds to a *more stable* compound.)

$$H_3C \underset{H}{\overset{CH_3}{>}}C=C\underset{H}{\overset{}{<}} \rightleftharpoons H_3C\underset{H}{\overset{H}{>}}C=C\underset{CH_3}{\overset{}{<}} \qquad \Delta H^\circ = -1.1 \text{ kcal mole}^{-1}$$

$\Delta H_f^\circ: \qquad -1.9 \qquad\qquad\qquad -3.0$

$$H_3C\underset{H}{\overset{CH_2CH_3}{>}}C=C\underset{H}{\overset{}{<}} \rightleftharpoons H_3C\underset{H}{\overset{H}{>}}C=C\underset{CH_2CH_3}{\overset{}{<}} \qquad \Delta H^\circ = -0.9 \text{ kcal mole}^{-1}$$

$\Delta H_f^\circ: \qquad -7.0 \qquad\qquad\qquad -7.9$

$$H_3C\underset{H}{\overset{CH(CH_3)_2}{>}}C=C\underset{H}{\overset{}{<}} \rightleftharpoons H_3C\underset{H}{\overset{H}{>}}C=C\underset{CH(CH_3)_2}{\overset{}{<}} \qquad \Delta H^\circ = -1.0 \text{ kcal mole}^{-1}$$

$\Delta H_f^\circ: \qquad -13.7 \qquad\qquad\qquad -14.7$

The distance between the adjacent methyl groups in *cis*-2-butene is about 3 Å. Since the sum of the van der Waals radii for two methyl groups is 4 Å, the hydrogens in these two groups are sufficiently close that there is a net repulsion not present in the *trans* compound. This effect of repulsion for sterically congested systems is called **steric hindrance** (Figure 11.15).

Monosubstituted ethylenes are 2–3 kcal mole^{-1} less stable than disubstituted ethylenes. The examples in the following table may be compared with the corresponding isomers listed above.

Compound	ΔH_f°, kcal mole^{-1}
$CH_3CH_2CH{=}CH_2$	-0.2
$CH_3CH_2CH_2CH{=}CH_2$	-5.3
$(CH_3)_2CHCH_2CH{=}CH_2$	-12.3

The stabilizing effect of substituents on the double bond continues with additional substituents, although the incremental effect is reduced because of *cis* interactions. For example, 2-methyl-2-butene, with $\Delta H_f^\circ = -10.1$ kcal mole^{-1}, is the most stable five-carbon alkene. Similarly, 2,3-dimethyl-2-butene is the most stable six-carbon alkene.

These results are most simply interpreted on the basis of relative bond strengths. A C_{sp^2}—H bond is a stronger bond than a C_{sp^3}—H bond. If this were the only important factor, the least substituted alkenes would be the most stable. However, a C_{sp^2}—C_{sp^3} bond is also stronger than a C_{sp^3}—C_{sp^3} bond, and it seems that putting more s-character in a C—C bond has a greater effect than in a C—H bond. This hybridization effect may also be observed in the bond lengths because bond lengths are inversely related to bond strengths. We noted earlier that the C—H bond in ethylene is about 0.01 Å shorter than the C—H bond in ethane. The C—C bond in propene (1.505 Å) is 0.03 Å shorter than the C—C bond in propane. The difference $r_{C_{sp^3}-C_{sp^3}} - r_{C_{sp^2}-C_{sp^3}}$ is greater than $r_{C_{sp^3}-H} - r_{C_{sp^2}-H}$.

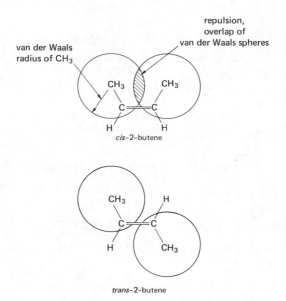

FIGURE 11.15. *Steric hindrance in* cis-*2-butene, relative to* trans-*2-butene.*

The difference in stability of various alkene isomers is only a few kilocalories per mole, but this makes an important difference in equilibria. From the thermodynamic equation

$$\Delta G^\circ = -RT \ln K$$

a free energy difference of 1.4 kcal mole^{-1} at room temperature corresponds to an equilibrium constant of 10 (Table 11.4). Consequently, in an equilibrium mixture of alkenes the more highly substituted isomers predominate. For example, the equilibrium composition of the butenes is

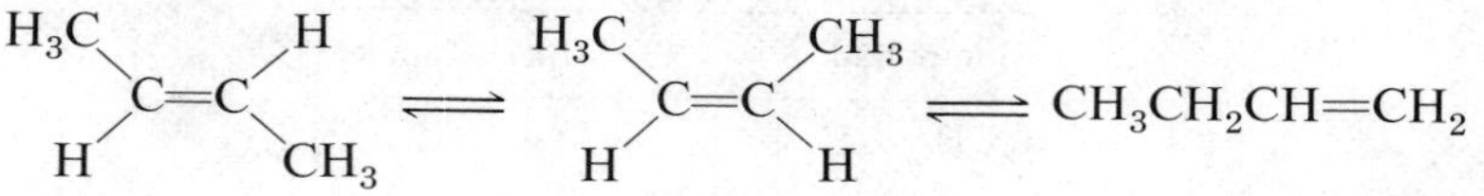

We will find that such equilibria can be established and that the relative stabilities of alkene isomers are important in some synthetic methods.

In principle, the double bond in a cyclic alkene can be either *cis* or *trans*. In a *cis* cycloalkene two carbons that are attached *cis* to one another on the double bond are also part of the ring. Several examples are shown below.

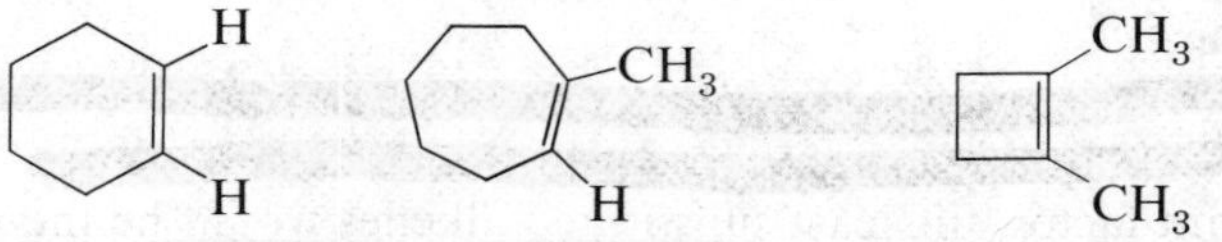

cyclohexene   1-methylcycloheptene   1,2-dimethylcyclobutene

In a *trans* cycloalkene the two ring carbons are *trans* with respect to the double bond. In practice, the *trans* isomers are too strained to exist at room temperature for three- through seven-membered rings. The smallest cycloalkene for which an isolable *trans* isomer is known is cyclooctene. Even in this case there is significant strain—*cis*-cyclooctene is 9.1 kcal mole^{-1} more stable than the *trans* isomer.

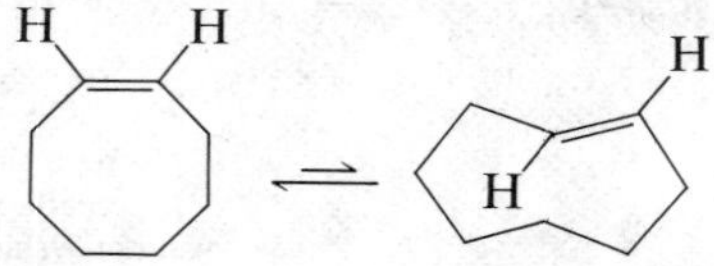

cis-cyclooctene   trans-cyclooctene

TABLE 11.4

Equilibrium Concentrations as a Function of ΔG° at 25°C

A $\rightleftharpoons$ B

ΔG°, kcal mole^{-1}	Percent A	Percent B
-5	0.02	99.98
-2	3.3	96.7
-1	15.6	84.4
-0.5	30.1	69.9
0	50.0	50.0
$+0.5$	69.9	30.1
$+1$	84.4	15.6
$+2$	96.7	3.3
$+5$	99.98	0.02

EXERCISE: Using the fundamental thermodynamic relationship $\Delta G° = -RT \ln K$, verify the equilibrium concentrations given in Table 11.4. Calculate the equilibrium concentration at 25° for *cis*- and *trans*-cyclooctene. Using a molecular model set, construct models of the stereoisomeric cyclooctenes. Note that it is much more difficult to construct a model of the *trans* isomer than of the *cis*. Demonstrate that the *trans*-cyclooctene is chiral, whereas the *cis* isomer is not.

11.5
Preparation of Alkenes

The important preparations of alkenes discussed thus far have all been elimination reactions: E1 and E2 eliminations of alkyl halides and dehydration of alcohols. Other important procedures for introducing double bonds will be discussed in subsequent chapters.

A. *E2 Bimolecular Elimination of Alkyl Halides*

This reaction was discussed previously (Section 8.11), but primarily as a side reaction in S_N2 displacement reactions. E2 elimination can often be made the principal reaction by using a strong base in a nonpolar solvent. One common reagent used for this purpose is potassium hydroxide in refluxing ethanol. This solution is really a solution primarily of potassium ethoxide in ethanol because of the equilibrium

$$OH^- + C_2H_5OH \rightleftharpoons C_2H_5O^- + H_2O$$

This method gives satisfactory results for secondary and tertiary halides, but not for most primary halides. For primary halides, especially with no β-branches, the S_N2 reaction is so facile that it dominates, and ethers are the principal products.

$$n\text{-}C_5H_{11}Br \xrightarrow[\substack{\text{EtOH} \\ 55°C}]{\text{EtONa}} \underset{(12\%)}{CH_3CH_2CHCH{=}CH_2} + \underset{(88\%)}{n\text{-}C_5H_{11}OC_2H_5}$$

Bimolecular elimination almost invariably gives a mixture of the possible alkene products. The mixture usually reflects the thermodynamic stabilities of the isomeric alkenes; the most stable isomers tend to predominate.

$$CH_3CH_2CH_2CHBrCH_3 \xrightarrow[\Delta]{\substack{\text{EtOK} \\ \text{EtOH}}} \underset{\substack{(41\% \; trans) \\ (14\% \; cis)}}{CH_3CH_2CH{=}CHCH_3} +$$

$$\underset{(25\%)}{CH_3CH_2CH_2CH{=}CH_2} + \underset{(20\%)}{CH_3CH_2CH_2CH(OEt)CH_3}$$

In this example note that the *trans* isomer is produced in greater amount than is the *cis* and that the combined 2-pentenes are produced to a greater extent than is 1-pentene. However, the more basic and bulkier reagent potassium *t*-butoxide in the less polar solvent *t*-butyl alcohol tends to give more of the terminal olefin.

$$CH_3CH_2CBr(CH_3)_2 \xrightarrow[\text{EtOH}]{\text{EtOK}} CH_3CH=C(CH_3)_2 + CH_3CH_2\overset{\overset{\displaystyle CH_3}{|}}{C}=CH_2$$

(71%)   (29%)

$$\xrightarrow[\text{HOC(CH}_3)_3]{\text{KOC(CH}_3)_3} \qquad (28\%) \qquad\qquad (72\%)$$

$$\xrightarrow[\text{HOC(C}_2\text{H}_5)_3]{\text{KOC(C}_2\text{H}_5)_3} \qquad (11\%) \qquad\qquad (89\%)$$

Potassium *t*-butoxide gives good yields of E2 product even with straight-chain primary halides.

$$n\text{-}C_{18}H_{37}Br \xrightarrow[\substack{\text{HOC(CH}_3)_3 \\ 40°}]{\text{KOC(CH}_3)_3} n\text{-}C_{16}H_{33}CH=CH_2 + n\text{-}C_{18}H_{37}OC(CH_3)_3$$

(85%)   (12%)

These various effects of structure are best rationalized in the context of reaction mechanism. A correct mechanism needs to explain these facts as well as several other generalizations that can be made about such eliminations.

1. The rate of E2 reactions depends on the concentration of both the alkyl halide *and* the base.
2. The rate of reaction depends on the nature of the leaving group. In general, bromides react faster than chlorides.
3. The reaction has a high primary hydrogen isotope effect; that is, C—D bonds are broken more slowly than C—H bonds.
4. The reaction is **stereospecific**. The leaving hydrogen must generally be con-formationally **anti** to the leaving halide.

The first generalization has an obvious consequence on any proposed mecha-nism. Both the base and the alkyl halide must be involved in the rate-determining step. In fact, it is this observation that gives the mechanism its name—E2 or "elimination, bimolecular." [In another type of elimination reaction, the rate of reaction depends only on the concentration of the alkyl halide. This type of elimi-nation is distinguished by the name E1 or "elimination, unimolecular." E1 elimi-nation is one of the possible modes of reaction of carbocation intermediates (Sec-tions 8.12 and 10.7.B).]

The second generalization also has an obvious consequence: The bond to the leaving halide must be partially broken in the transition state. Bonds that are broken more easily lead to a lower energy transition state and a faster reaction rate.

The third generalization establishes that the bond to the leaving hydrogen is also partially broken in the transition state. To a first approximation, it effectively takes more energy to break a C—D bond than it does to break a C—H bond.

To be more precise, isotope effects originate in the nature of vibrational energy levels for bonds. These quantum states were discussed previously in connection with bond dissociation energies of alkanes (Section 6.1), and we will encounter them again in studying infrared spectra (Section 14.2). For the present purpose, consider the two C—H bond motions in Figure 11.16. In the strong C—H bond the potential energy is very sensitive to the value of the C—H bond distance. Thus, in order to accommodate the Heisenberg uncertainty principle, the lowest vibrational energy state is relatively high above the potential minimum. That is, the zero point energy, ε_0, is relatively large. For the C—H bond in alkanes ε_0 is about 4 kcal mole^{-1}. For a weaker bond, as shown in Figure 11.16b, a given uncertainty in the position of the hydrogen corre-sponds to a smaller change in potential energy. Hence, ε_0 is smaller.

A deuterium atom is twice as heavy as hydrogen because its nucleus has a neutron as well as a proton. Because of its greater mass, the same momentum corresponds to

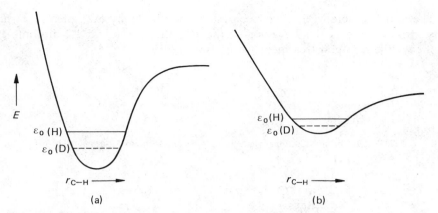

FIGURE 11.16. *Potential energies for stretching motion of* (a) *a strong and* (b) *a weak* C—H *bond.*

a slower velocity. A given uncertainty in momentum corresponds to a smaller uncertainty in position compared to the hydrogen case. Accordingly, the zero point energy for a C—D bond is less than that for C—H, as shown in Figure 11.16.

In a reaction in which a C—H bond is broken, the bond is weaker in the transition state. In the change from reactant to transition state, the C—H bond has lost some zero point energy. In the corresponding case of a C—D bond the loss in zero point energy is lower because the heavier isotope had less zero point energy to begin with. As a result, as shown in Figure 11.17, the activation energy for breaking a C—D bond is greater than that for a C—H bond. The difference in reaction rates can be substantial. For hydrogen isotopes, the effect on ε_0 is approximately the square root of the ratio of masses. If $\varepsilon_0(H)$ is 4 kcal mole^{-1}, the corresponding $\varepsilon_0(D)$ is 2.8 kcal mole^{-1}. If all of this zero point energy were lost in the transition state, the difference in activation energies would correspond to a reaction rate difference of a factor of about 9. In practice, primary isotope effects for E2 reactions have been observed commonly in the range of 4–8.

$$CD_3CHBrCD_3 \xrightarrow{\text{EtO}^-} CD_2{=}CHCD_3$$

$$\frac{k_{\text{CH}_3\text{CHBrCH}_3}}{k_{\text{CD}_3\text{CHBrCD}_3}} = 6.7$$

If no bond to hydrogen is broken at the transition state, isotope effects are generally rather small. For example, in S_N2 reactions deuterium compounds react at virtually the same rates as the corresponding hydrogen compounds.

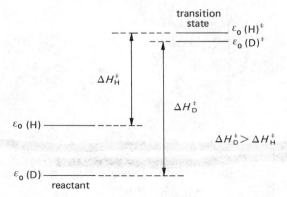

FIGURE 11.17. *The effect on the activation energy of loss of zero point energy in a reaction.*

The fourth generalization has been established by many examples. A particularly striking case is the behavior of the two diastereomeric 1,2-dibromo-1,2-diphenylethanes upon treatment with base.

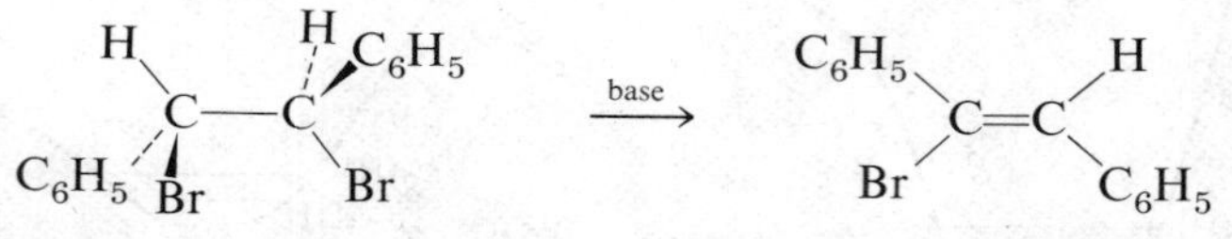

(R,R)-1,2-dibromo-1,2-diphenylethane   (Z)-1-bromo-1,2-diphenylethene

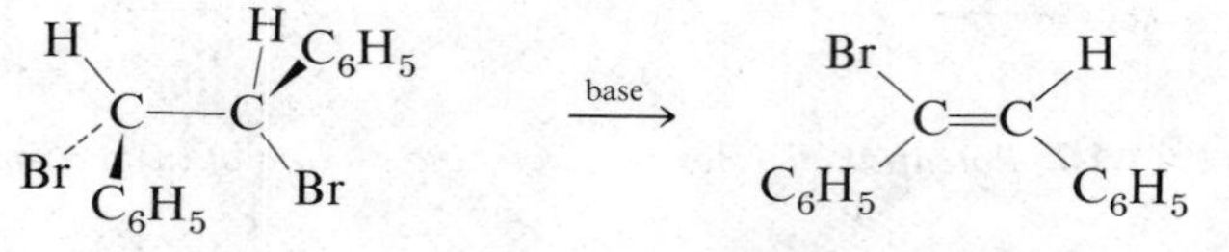

meso-1,2-dibromo-1,2-diphenylethane   (E)-1-bromo-1,2-diphenylethene

The two isomers react by the E2 mechanism to give totally different products. From a knowledge of the configurations of the two reactant dibromides and the product that each gives, it may be deduced that each isomer reacts in a conformation that has H *anti* to Br. The staggered conformations of the two compounds are shown in Newman-projection form in Figure 11.18. Note that removal of H and Br from opposite sides of the molecule results in conversion of the (R,R) stereoisomer into the alkene having the two phenyl groups *trans*.

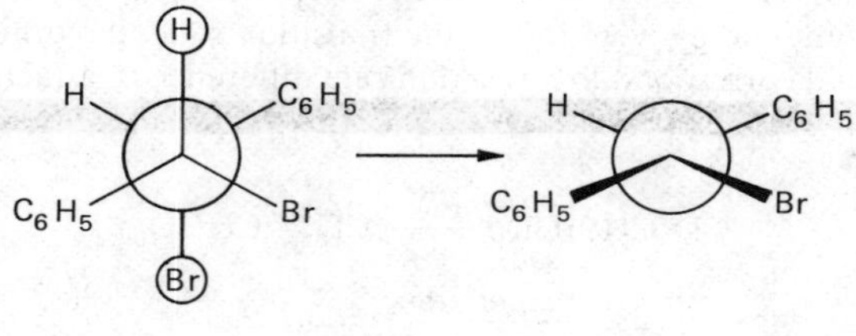

Similarly, *anti* elimination in the *meso* stereoisomer gives the alkene having the phenyl groups *cis*.

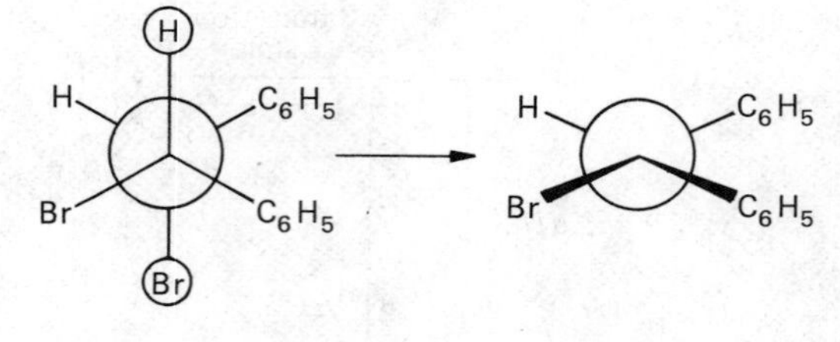

This example demonstrates that *anti* elimination can give either a *cis* or a *trans* alkene, depending on the structure of the starting halide.

The strict requirement for elimination may also be seen in elimination reactions of cyclic alkyl halides. For example, *trans*-1-bromo-4-*t*-butylcyclohexane reacts rapidly with sodium ethoxide in ethanol to give 4-*t*-butylcyclohexene. However, the *cis* isomer undergoes elimination extremely slowly.

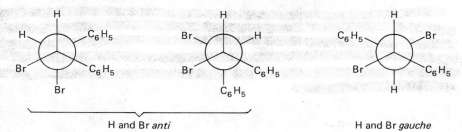

H and Br *anti*   H and Br *gauche*

(a) staggered conformations of (R,R)-1,2-dibromo- 1,2-diphenylethane

H and Br *anti*   H and Br *gauche*

(b) staggered conformations of *meso*-1,2 dibromo-1,2-diphenylethane

FIGURE 11.18. *Newman projections illustrating the staggered conformations of the stereoisomeric 1,2-dibromo-1,2-diphenylethanes.*

In this example the *t*-butyl group is so sterically demanding that it effectively controls the conformation of the two molecules. *cis*-1-Bromo-4-*t*-butylcyclohexane exists almost entirely in the conformation shown. To determine the relative energy of the axial-*t*-butyl, equatorial-bromo conformation, the energies in Table 5.6 may be used.

relative energy: 0 + 0.5 = 0.5 kcal mole^{-1} ~5 + 0 = ~5 kcal mole^{-1}

$\Delta G° \approx 4.5$ kcal mole^{-1}

The corresponding equilibrium constant, $K \approx 10^{-3}$; that is, *cis*-1-bromo-4-*t*-butylcyclohexane exists at least 99.9% in the axial-bromo conformation.

Applying the same approach to the *trans* isomer the conformation with axial bromo is only about 0.01% of the total.

relative energy: 0 + 0 = 0   ~5 + 0.5 ≈ 5.5 kcal mole^{-1}

$\Delta G° \approx 5.5$ kcal mole^{-1}

Thus, only in the *cis* isomer are there any significant number of molecules in which the bromine has an *anti* relationship to a hydrogen.

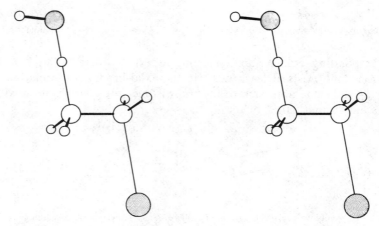

hydrogens *anti* to Br   no hydrogens *anti* to Br

The mechanism that results from putting these facts together is one in which the attacking base removes a proton concurrently with loss of halide ion from the other side of the molecule. As the C—H bond lengthens and weakens, the C—X bond also lengthens and weakens. The remaining four groups on the two carbons start to move into coplanarity, and a C=C double bond starts to form. We may represent this reaction as

reactants   transition state   products

Recall that each curved arrow represents the movement of one pair of electrons. The proton has been displaced in a typical acid-base proton-transfer reaction, and the leaving halide has been displaced by a pair of electrons at the rear of the C—X bond. The transition state for the E2 reaction is represented in stereo form in Figure 11.19.

In orbital terms the bonds on the alkyl halide all involve C_{sp^3} orbitals. In the product alkene some orbitals are C_{sp^2} and others are C_p. At the transition state the orbitals have an intermediate character, as illustrated in Figure 11.20. Because π-bonding is significant in the transition state, those effects that stabilize double bonds, such as added substituents, also stabilize the transition state. However, inasmuch as π-overlap is only partial, these stabilizing effects are not as important as they are in the product olefins.

It is clear that for π-bonding to be significant in the transition state, the C—H

FIGURE 11.19. *Stereo representation of the transition state of an E2 reaction.*

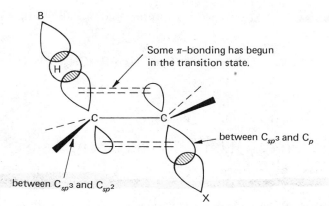

FIGURE 11.20. *Orbital representations of E2 transition state.*

and C—X bonds to be broken must lie in the same plane, but we may well ask why this requires *anti* elimination. The answer is simply that in the corresponding *syn* elimination groups are eclipsed to each other and represent a structure of higher energy.

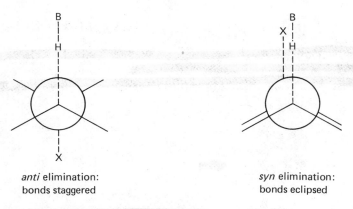

anti elimination:
bonds staggered

syn elimination:
bonds eclipsed

Finally, primary hydrogens are less sterically hindered and more open to attack than secondary or tertiary hydrogens. Refer back to the reaction of 2-bromo-pentane with potassium ethoxide in ethanol (page 293). 1-Pentene is formed in 25% yield although at equilibrium it would constitute only about 3% of a mixture of pentene isomers. With the larger bases this disparate amount of the 1-alkene is even greater (compare the examples on page 294).

EXERCISE: Explain why base-catalyzed elimination of *cis*-1-bromo-2-methylcyclo-hexane gives both 1-methylcyclohexene and 3-methylcyclohexene, but *trans*-1-bromo-2-methylcyclohexane gives only 3-methylcyclohexane. If the answer is not obvious after inspection of appropriate perspective drawings, construct models of the two isomers using molecular models. Arrange each model in a conformation having the bromide axial, and then identify all hydrogens that are *anti* to the bromine.

B. *Alcohol Eliminations*

As we mentioned in Section 10.7, alcohols undergo dehydration upon heating with strong acids. In the case of ethyl alcohol, the reaction requires concentrated sulfuric acid at 170°. At lower temperatures (140°) diethyl ether is the major reaction product.

$$2\ CH_3CH_2OH \xrightarrow[140°]{H_2SO_4} CH_3CH_2OCH_2CH_3 + H_2O$$

$$CH_3CH_2OH \xrightarrow[170°]{H_2SO_4} CH_2{=}CH_2 + H_2O$$

n-Propyl alcohol may be similarly dehydrated to propene.

$$CH_3CH_2CH_2OH \xrightarrow[170°]{H_2SO_4} CH_3CH{=}CH_2 + H_2O$$

With primary alcohols larger than propyl, mixtures of alkenes result.

$$CH_3(CH_2)_6CH_2OH \xrightarrow[\Delta]{H_3PO_4} \underbrace{CH_3(CH_2)_4CH{=}CHCH_3 + CH_3(CH_2)_3CH{=}CHCH_2CH_3}_{\text{major products}}$$

Essentially no 1-octene is produced in this reaction. The problem with the larger primary alcohols is isomerization of the alkene that is the initial product.

$$CH_3(CH_2)_6CH_2OH \longrightarrow CH_3(CH_2)_4CH_2CH{=}CH_2 \underset{-H^+}{\overset{+H^+}{\rightleftharpoons}}$$

$$CH_3(CH_2)_4CH_2\overset{+}{C}HCH_3 \underset{+H^+}{\overset{-H^+}{\rightleftharpoons}} CH_3(CH_2)_4CH{=}CHCH_3 \rightleftharpoons \text{etc.}$$

Consequently acid-catalyzed dehydration is not a generally useful procedure for the conversion of primary alcohols to alkenes.

In contrast, secondary and tertiary alcohols are more easily dehydrated by acids.

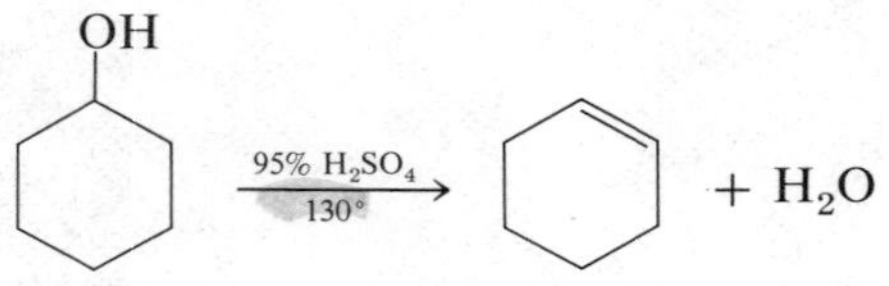

(83%)

cyclohexanol cyclohexene

$$CH_3CH_2\overset{OH}{\underset{CH_3}{C}}CH_3 \xrightarrow[90°]{46\%\ H_2SO_4} CH_3CH{=}C(CH_3)_2 + H_2O$$

(84%)

2-methyl-2-butanol 2-methyl-2-butene

$$CH_3CH_2CH_2\overset{OH}{C}HCH_3 \xrightarrow[95°]{62\%\ H_2SO_4} CH_3CH_2CH{=}CHCH_3 + H_2O$$

(65–80%)

2-pentanol *cis*- and *trans*-2-pentene

[2-Pentanol, 214 mL, is heated with a mixture of 200 mL of sulfuric acid and 200 mL of water and the alkene produced is distilled as formed. The distillate is washed, dried, and redistilled; yield, 65–80%. This product is mostly 2-pentenes. The small amount of 1-pentene also present can be removed by careful fractional distillation.]

In these cases dehydration probably occurs by the E1 mechanism, by way of intermediate carbocations.

(1) $CH_3CH_2CH_2\overset{\underset{\textstyle |}{OH}}{C}HCH_3 + H^+ \rightleftharpoons CH_3CH_2CH_2\overset{\underset{\textstyle |}{^+OH_2}}{C}HCH_3$

(2) $CH_3CH_2CH_2\overset{\underset{\textstyle |}{^+OH_2}}{C}HCH_3 \rightleftharpoons CH_3CH_2CH_2\overset{+}{C}HCH_3 + H_2O$

(3) $CH_3CH_2CH_2\overset{+}{C}HCH_3 \rightleftharpoons CH_3CH_2CH=CHCH_3 + H^+$

This method is especially suitable for relatively simple alcohols. In more complex cases, rearrangements may occur.

$$CH_3-\underset{\underset{\textstyle CH_3}{|}}{\overset{\overset{\textstyle CH_3}{|}}{C}}-\underset{\underset{\textstyle}{}}{\overset{\overset{\textstyle OH}{|}}{C}}H-CH_3 \xrightarrow[\Delta]{H^+} CH_3-\underset{\underset{\textstyle CH_3}{|}}{\overset{\overset{\textstyle CH_3}{|}}{C}}-\overset{+}{C}H-CH_3 \longrightarrow$$

$$CH_3-\underset{\underset{\textstyle CH_3}{|}}{\overset{+}{C}}-\overset{\overset{\textstyle CH_3}{|}}{C}H-CH_3 \xrightarrow{-H^+} CH_2=\underset{\underset{\textstyle CH_3}{|}}{C}-CHCH_3 + \underset{CH_3}{\overset{CH_3}{>}}C=C\underset{CH_3}{\overset{CH_3}{<}}$$

2,3-dimethyl-1-butene   2,3-dimethyl-2-butene
minor product   major product

A simple and effective procedure for dehydration of many alcohols, including primary alcohols, involves passing the vapors over alumina at 350–400°.

> Alumina, aluminum oxide, Al_2O_3, occurs naturally in crystalline form as ruby, sapphire, and corundum. Commercial alumina for laboratory use is a white powder which is available in many grades. It is highly insoluble in water and in organic solvents and has an extremely high melting point (over 2000°). It is used as an adsorbent in liquid chromatography, as a catalyst for some reactions, and as a catalyst support in other cases.

Alumina, like many aluminum salts, is a Lewis acid. The dehydration reaction probably occurs by some version of the E1 mechanism on the alumina surface.

$$ROH \qquad \overset{\overset{\textstyle R \quad H}{\diagdown\overset{+}{O}\diagup}}{\underset{\underset{\textstyle |}{Al}}{-Al-}} \qquad \overset{\overset{\textstyle OH}{|}}{\underset{\underset{\textstyle |}{}}{-Al^-}} + R^+ \qquad \overset{\overset{\textstyle ^+OH_2}{|}}{\underset{\underset{\textstyle |}{}}{-Al-}} + alkene\uparrow \longrightarrow \underset{\underset{\textstyle |}{}}{-Al-} + H_2O\uparrow$$

Accordingly, isomerization of olefins and rearrangments are common. These reactions can be suppressed by first treating the alumina with a base such as ammonia.

EXERCISE: 1-Cyclobutylethanol reacts with 60% aqueous sulfuric acid to give 1-methylcyclopentene. Write a mechanism, which shows each step, for this rearrangement reaction. If the rearrangement step is not clear, make a molecular model of the 1-cyclobutylethyl cation and perform the following operation: break the C_1—C_2 of the cyclobutane ring, and attach carbon 2 to the former cationic carbon. What is the structure of the product?

C. *Industrial Preparation of Alkenes*

Ethylene is an important item of commerce. It is used in large quantities for the manufacture of polyethylene and as an intermediate in the preparation of a host of other chemicals. It is obtained primarily as a cracking product in petroleum refining (Section 6.2). Although any hydrocarbon may be cracked to yield mainly ethylene, in the United States the primary material used for this purpose is ethane.

$$CH_3CH_3 \xrightarrow[\text{1 atm}]{\text{700–900°C}} CH_2{=}CH_2 + H_2$$

When higher hydrocarbons are submitted to the cracking process, significant amounts of propene are produced.

$$n\text{-}C_6H_{14} \xrightarrow{\text{700–900°C}} \underset{(15\%)}{CH_4} + \underset{(40\%)}{CH_2{=}CH_2} + \underset{(20\%)}{CH_3CH{=}CH_2} + \underset{(25\%)}{\text{other}}$$

A large amount of the propene produced in this country goes into the manufacture of polypropylene. Other important industrial alkenes are the butenes and 1,3-butadiene.

The 1979 industrial production of various alkenes is summarized in Table 11.5.

TABLE 11.5
1979 Production of Alkenes

Compound	Production, tons
$CH_2{=}CH_2$	14,595,000
$CH_3CH{=}CH_2$	7,150,000
$CH_2{=}CHCl$	3,770,000
$C_6H_5CH{=}CH_2$	3,740,000
$CH_2{=}CHCH{=}CH_2$	1,775,000

11.6
Reactions of Alkenes

In Appendix III, "Average Bond Energies," we find that the value for $C{=}C$ is 146 kcal mole^{-1}. This is 63 kcal mole^{-1} higher than the normal $C{-}C$ bond strength of 83 kcal mole^{-1}. The difference is reminiscent of the 65 kcal mole^{-1} required for rotation about the double bond in ethylene (Section 11.1) and may be considered roughly as the bond strength of the second or π-bond in ethylene. That is, the "second" bond of a double bond is substantially weaker than the first.

As we saw in Section 11.1, this view of alkenes as having two different kinds of bonds is simply a convenient way of visualizing a molecule and is mathematically equivalent to the "bent-bond" model. In the latter picture, as one of the two **equivalent** bonds breaks, the other actually becomes stronger. The total energy required to break one of the two bonds is thus 65 kcal mole^{-1}.

The reaction of this "weak" π-bond with a normal single bond to produce a molecule containing two new single bonds is generally a thermodynamically favorable process. For example, in gas phase reactions at 25°C:

$$CH_2{=}CH_2 + H_2 \longrightarrow CH_3CH_3 \qquad \Delta H° = -32.7 \text{ kcal mole}^{-1}$$
$$CH_2{=}CH_2 + Cl_2 \longrightarrow CH_2ClCH_2Cl \qquad \Delta H° = -43.2 \text{ kcal mole}^{-1}$$
$$CH_2{=}CH_2 + HBr \longrightarrow CH_3CH_2Br \qquad \Delta H° = -19.0 \text{ kcal mole}^{-1}$$
$$CH_2{=}CH_2 + H_2O \longrightarrow CH_3CH_2OH \qquad \Delta H° = -10.9 \text{ kcal mole}^{-1}$$

Not only do such *additions across a double bond* have favorable thermodynamics; many also have accessible pathways or reaction mechanisms. Such additions form an important part of the chemistry of alkenes.

A. Catalytic Hydrogenation

Even though the reaction is highly exothermic ($\Delta H° = -32.7$ kcal mole^{-1}), ethylene does not react with hydrogen at an appreciable rate without an appropriate catalyst. The conceptually simple "four-center" mechanism

cyclic four-membered
transition state

is apparently not an accessible mechanism. Such four-center mechanisms are rare because cyclic four-membered transition states, such as that shown, have unusually high energy. The high activation energy corresponds to an impractically slow reaction rate. The relatively high energies of such four-center transition states can be explained by molecular orbital concepts, and further discussion is deferred to Section 34.2.

The hydrogenation reaction does take place readily on the surface of some metals, particularly platinum, palladium, and nickel. These metals are known to coordinate with double bonds and form hydrides. The detailed reaction mechanism is complex and involves various types of metal–carbon bonds. A schematic representation that is suitable for our purposes is approximated as follows.

(catalyst surface)

Platinum is usually used as the black oxide known as "Adams' catalyst." The oxide is prepared by fusion of chloroplatinic acid, $H_2PtCl_6 \cdot 6H_2O$, hygroscopic red-brown crystals, or ammonium chloroplatinate, $(NH_4)_2PtCl_6$, yellow crystals, with sodium nitrate. It reacts readily with hydrogen gas even at low pressures to form a finely divided platinum metal catalyst. This is usually accomplished by stirring with a suitable inert solvent such as ethanol or acetic acid. The alkene is then added, and when the solution is stirred with the suspension of platinum under an atmosphere of hydrogen, hydrogen gas is absorbed rapidly. The hydrogen is usually contained in a gas buret so that the amount absorbed can be measured. The resulting mixture is filtered and the product is isolated from the filtrate. Only small amounts of platinum catalyst are required, but the filter paper residues are normally saved for recovery of the platinum, a rare and expensive material.

Palladium is usually used as a commercial preparation in which the finely divided metal is supported on a suitable inert surface, frequently charcoal (Pd/C) or barium sulfate (Pd/BaSO$_4$). Alkenes are normally hydrogenated in ethanol solution by stirring with Pd/C at room temperature under an atmosphere of hydrogen.

Nickel is usually used in a finely divided state called "Raney nickel." The catalyst is prepared by allowing nickel-aluminum alloy to react with aqueous sodium hydroxide. The aluminum dissolves and leaves the nickel as a finely divided suspension. Typical hydrogenations are conducted at moderately high pressures of hydrogen ($\sim$1000 psi).

Other hydrogenation catalysts used for specific purposes are rhodium, ruthenium, and copper–chromium oxide, but platinum, palladium, and nickel in their various forms are the most common. They are subject to "poisoning" by some compounds, notably sulfur containing compounds such as thiols and sulfides. These compounds bind firmly to the catalyst surface and destroy its catalytic activity.

Since the two hydrogens are added to the double bond from the surface of the metal, they are normally both added *to the same face of the double bond.* This type of addition is referred to as **syn,** just as the addition of two "pieces" of a reagent to opposite faces of a double bond is called **anti** addition. In the case of catalytic hydrogenation the alkene molecule is adsorbed to the catalyst surface with one face of the double bond coordinated to the surface; the two hydrogens are both added to this face.

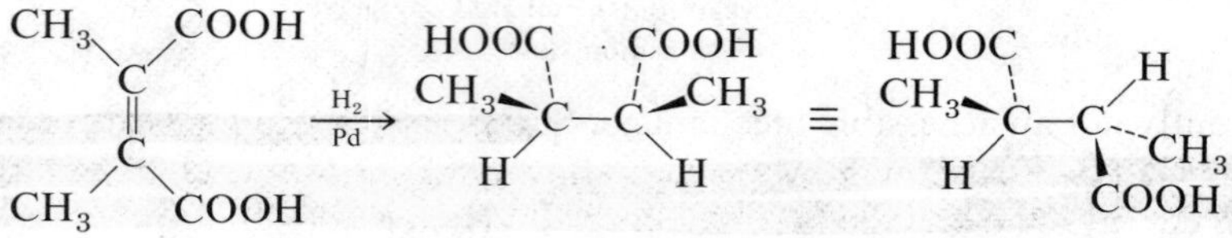

Although *syn* addition is the general rule, *anti* addition is sometimes observed. For example

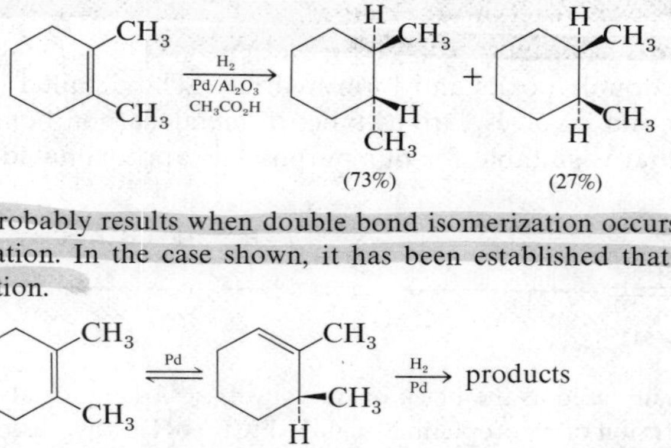

(73%) (27%)

Anti addition probably results when double bond isomerization occurs more rapidly than hydrogenation. In the case shown, it has been established that isomerization precedes reduction.

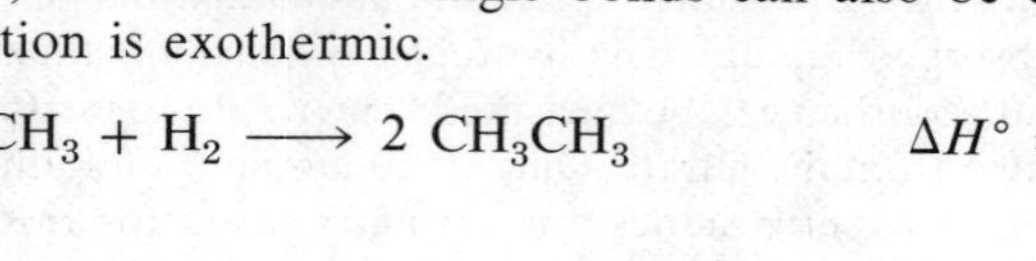

Palladium is particularly prone to catalyze double bond isomerization. Platinum, rhodium, or iridium should be used if isomerization is a problem.

In principle, carbon–carbon single bonds can also be cleaved by hydrogen, since the reaction is exothermic.

$$CH_3CH_2CH_2CH_3 + H_2 \longrightarrow 2\ CH_3CH_3 \qquad \Delta H° = -10.1 \text{ kcal mole}^{-1}$$

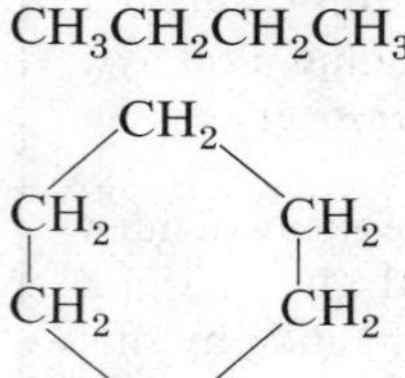

$$+ H_2 \longrightarrow CH_3(CH_2)_4CH_3 \qquad \Delta H° = -10.4 \text{ kcal mole}^{-1}$$

However, even though it is thermodynamically feasible, the reaction does not occur under ordinary conditions because alkanes are not effectively absorbed by the catalyst. Such carbon–carbon single bond cleavages *are* known with cyclopropanes and cyclobutanes. In these cases the ring strain of the small rings is released upon hydrogenation and the reactions are much more exothermic.

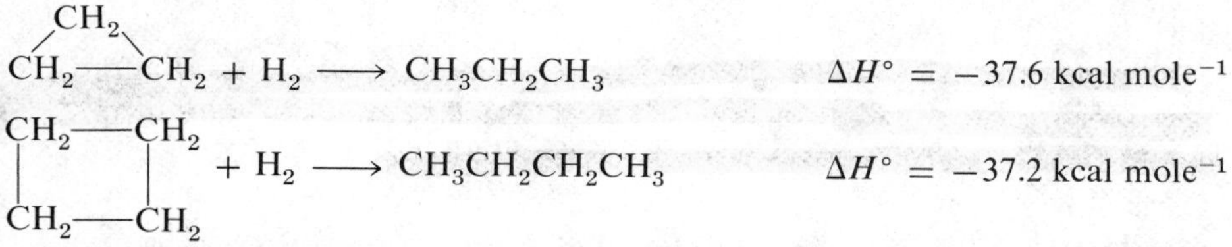

$$CH_2CH_2\text{—}CH_2 + H_2 \longrightarrow CH_3CH_2CH_3 \qquad \Delta H° = -37.6 \text{ kcal mole}^{-1}$$

$$\underset{CH_2\text{—}CH_2}{\overset{CH_2\text{—}CH_2}{|\qquad\quad|}} + H_2 \longrightarrow CH_3CH_2CH_2CH_3 \qquad \Delta H° = -37.2 \text{ kcal mole}^{-1}$$

In addition, the extra *p*-character in the small-ring C—C bonds (page 84) causes the molecules to behave somewhat like alkenes and be weakly absorbed by the catalyst. Even so, the conditions required for reaction are more drastic than those for hydrogenation of an alkene.

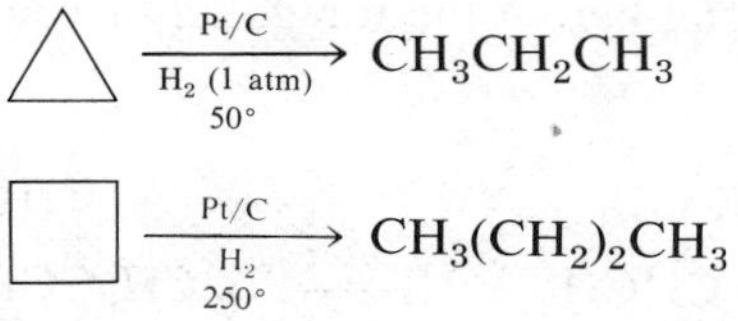

$$\triangle \xrightarrow[\substack{H_2 \text{ (1 atm)} \\ 50°}]{Pt/C} CH_3CH_2CH_3$$

$$\square \xrightarrow[\substack{H_2 \\ 250°}]{Pt/C} CH_3(CH_2)_2CH_3$$

EXERCISE: Hydrogenation of optically active 3,7-dimethyl-1-octene using platinum as catalyst gives optically active 2,6-dimethyloctane. If palladium is used as a catalyst, the 2,6-dimethyloctane is mostly racemic. Explain.

B. Electrophilic Additions

1. ADDITION OF HX. The region above and below a double bond is electron-rich because of the π-bond. Consequently double bonds have a tendency to act as Lewis bases and react with electrophilic reagents. An example is the reaction of 2-methylpropene with HCl. In the first step the double bond reacts with a proton to give a carbocation intermediate, which combines with chloride ion to give *t*-butyl chloride.

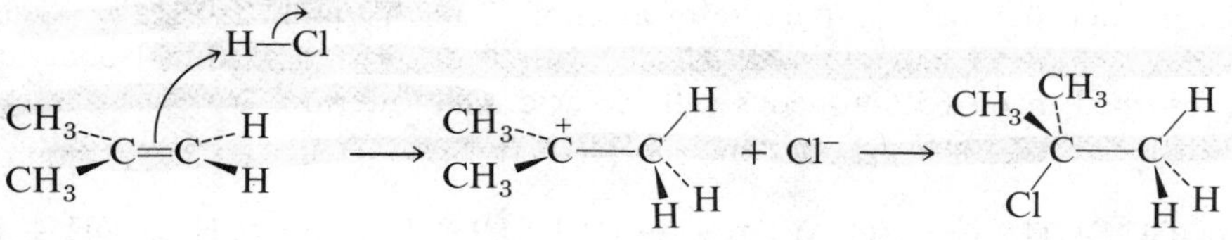

Some further examples of this reaction are

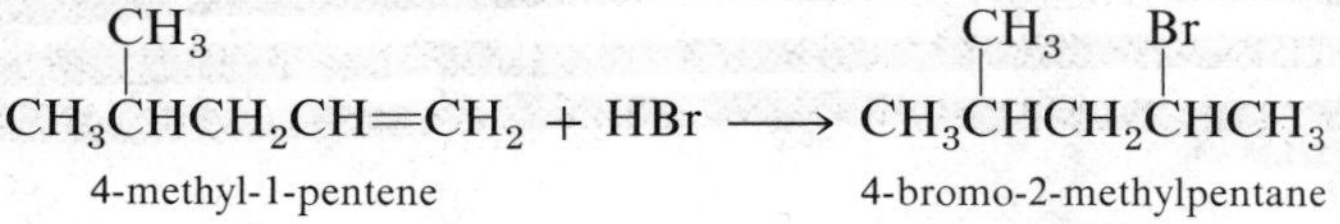

$$\underset{\text{4-methyl-1-pentene}}{\overset{\overset{\displaystyle CH_3}{|}}{CH_3CHCH_2CH=CH_2}} + HBr \longrightarrow \underset{\text{4-bromo-2-methylpentane}}{\overset{\overset{\displaystyle CH_3}{|}\quad\overset{\displaystyle Br}{|}}{CH_3CHCH_2CHCH_3}}$$

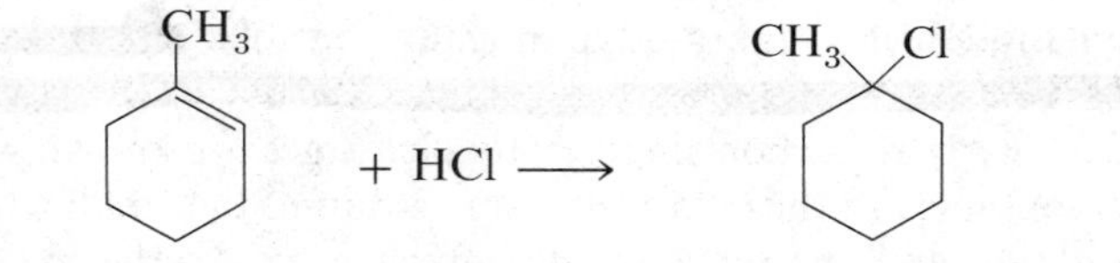

1-methylcyclohexene + HCl ⟶ 1-chloro-1-methylcyclohexane

With unsymmetrical alkenes the initial protonation occurs so as to afford the *more stable carbocation*. Since alkyl substituents stabilize carbocations, the proton adds to the less substituted carbon of the double bond.

This generalization is commonly referred to as **Markovnikov's rule.** It was formulated by Markovnikov long before the foregoing mechanistic interpretation was developed to explain it.

2-methyl-2-butene   2-iodo-2-methylbutane

If two intermediate carbocations of comparable stability can be formed, a mixture of products results.

$CH_3CH_2CH=CHCH_3 + HBr$

$[CH_3CH_2\overset{+}{C}HCH_2CH_3 \; Br^-] \longrightarrow CH_3CH_2CHBrCH_2CH_3$
3-bromopentane

$[CH_3CH_2CH_2\overset{+}{C}HCH_3 \; Br^-] \longrightarrow CH_3CH_2CH_2CHBrCH_3$
2-bromopentane

The addition of HX to a double bond is a significant reaction because of what it reveals about the general chemistry of alkenes, but it is not an important method for preparing the simpler alkyl halides. Better methods are generally available from alcohols.

2. ADDITION OF WATER. The hydration of alkenes is an important industrial method for the manufacture of alcohols. The hydration is usually accomplished by passing the alkene into a mixture of sulfuric acid and water. Isobutylene is absorbed in 60–65% aqueous sulfuric acid. The intermediate formed is undoubtedly the *t*-butyl cation, which reacts with water to give *t*-butyl alcohol.

$$(CH_3)_2C=CH_2 + H^+ \rightleftharpoons (CH_3)_3C^+ \xrightarrow{H_2O} (CH_3)_3C-\overset{+}{O}H_2 \rightleftharpoons (CH_3)_3COH + H^+$$

The reaction is the reverse of acid-catalyzed dehydration. Low temperatures and aqueous solution favor formation of the alcohol, whereas high temperatures and distillation of the alkene as it is formed shift the equilibrium toward the alkene. Under more vigorous conditions (Section 11.6.G) dimeric and polymeric products are produced.

Ethylene is also absorbed by sulfuric acid, but in this case 98% H_2SO_4 is required. The product is ethylsulfuric acid, which is hydrolyzed to ethyl alcohol in a separate step. This reaction may involve ethyl cation as an intermediate and is a rare example of the involvement of primary alkyl cations in solution.

$$H_2SO_4 + CH_2\!\!=\!\!CH_2 \longrightarrow CH_3CH_2{}^+ + HOSO_3{}^- \longrightarrow$$

$$CH_3CH_2OSO_2OH \xrightarrow{H_2O} CH_3CH_2OH + H_2SO_4$$

Although direct hydration is an important industrial process, it is seldom used as a laboratory procedure. Yields of alcohol are highly sensitive to reaction conditions, and more convenient laboratory reactions will be discussed later.

EXERCISE: Under strongly acidic conditions isomerization of double bonds frequently occurs. For example, if 1-butene is treated with concentrated sulfuric acid, 2-butanol is obtained on aqueous work-up. However, if the reaction is stopped short of completion and analyzed, the alkene is found to be a mixture of 1-butene, *cis*-2-butene, and *trans*-2-butene. Account for this observation mechanistically.

3. ADDITION OF HALOGENS. An important general reaction of double bonds is the addition of halogens.

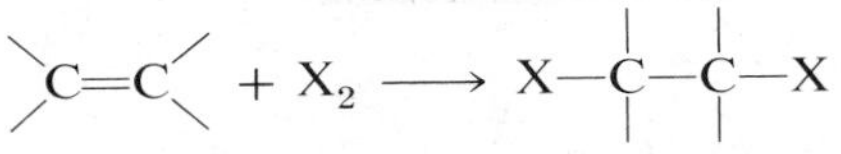

$$\diagup\!\!C\!\!=\!\!C\diagdown + X_2 \longrightarrow X\!-\!\overset{|}{C}\!-\!\overset{|}{C}\!-\!X \qquad X = Cl, Br$$

This reaction is rapid and serves as a simple diagnostic method for unsaturation. The reaction can be regarded as a nucleophilic displacement reaction on a halogen. The alkene is the nucleophile and halide ion is the leaving group.

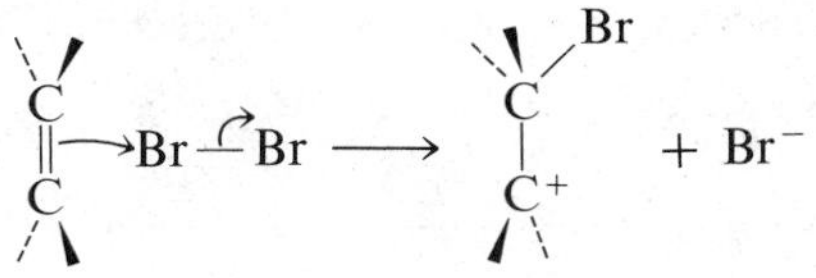

The resulting cation reacts with halide ion to give the observed product.

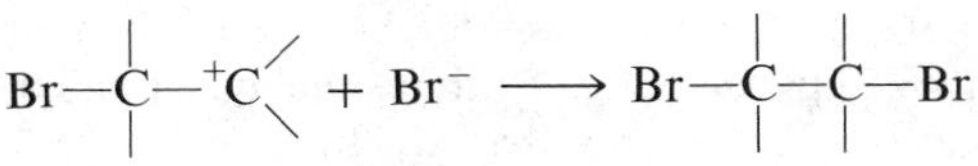

$$Br\!-\!\overset{|}{\underset{|}{C}}\!-\!{}^+C\diagdown + Br^- \longrightarrow Br\!-\!\overset{|}{\underset{|}{C}}\!-\!\overset{|}{\underset{|}{C}}\!-\!Br$$

The intermediate cation contains an electron-deficient cationic carbon and a halogen atom with nonbonding electron pairs. Consequently there is a tendency for overlap to produce a **cyclic halonium ion** as in Figure 11.21.

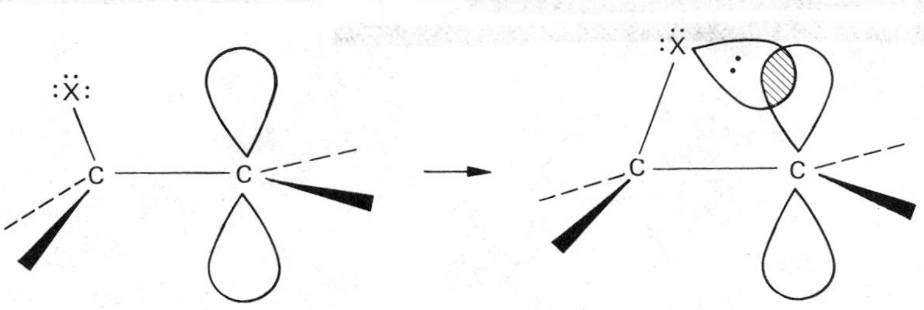

FIGURE 11.21. *Formation of cyclic halonium ion.*

The cyclic halonium ion may be written in Lewis form as

$$\overset{\ddot{X}^+}{\underset{C-C}{\diagdown}}$$

The advantage in terms of energy in forming such a structure is primarily the formation of an additional covalent bond. Furthermore, all of the atoms now have an octet electronic configuration. However, a price is paid for these gains. The bond angles in the three-membered ring structure are bent far from the desired tetrahedral geometry, and the positive charge is localized on the more electronegative halogen atom rather than on carbon.

In practice, the tendency of such a cation to exist in the cyclic form depends on the stability of the "open" carbocation. The intermediate formed from the addition of bromine to ethylene is best described as a symmetrical bromonium ion with relatively strong C—Br bonds. The alternative open form would be a highly unstable primary carbocation.

$$\overset{Br^+}{\underset{CH_2-CH_2}{\diagdown}} \quad \text{better than} \quad \overset{Br}{\underset{CH_2-\overset{+}{C}H_2}{\diagdown}}$$

The ion formed by addition of bromine to isobutylene is better described as a tertiary carbocation with a long and weak bond to bromine.

$$CH_3-\overset{\overset{\displaystyle \cdots Br}{|}}{\underset{\underset{\displaystyle CH_3}{|}}{\overset{+}{C}}}-CH_2$$

Cations such as these may be described in terms of three resonance structures: The actual ion is a composite or hybrid of the three structures A, B, and C.

$$\underset{A}{\overset{\ddot{X}}{\underset{|}{-\overset{+}{C}-C-}}} \longleftrightarrow \underset{B}{\overset{\overset{+}{X}}{\underset{|}{-C-C-}}} \longleftrightarrow \underset{C}{\overset{\ddot{X}}{\underset{|}{-C-C^+-}}}$$

If both A and C correspond to unstable carbocations, then structure B is a more important contributor to the actual structure of the ion. If either A or C corresponds to a relatively stable carbocation, then that structure contributes more and the ion has substantial carbocation character without as much halonium ion character.

The cyclic halonium ion intermediate has an important effect on the **stereochemistry** of halogen additions. When halide ion reacts with the cyclic ion, the reaction is a nucleophilic displacement reaction.

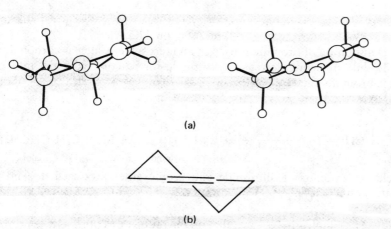

(a)

(b)

FIGURE 11.22. *Structure of cyclohexene:* (a) *stereo representation;* (b) *conventional symbolism* (*hydrogens attached to double bond are omitted for clarity*).

Since the nucleophile Br⁻ must approach carbon to the rear of the leaving group, the net result is *anti* **addition** of Br_2.

cyclopentene

$\xrightarrow[\substack{CCl_4 \\ -5°}]{Br_2}$

trans-1,2-dibromo-
cyclopentane

In additions to cyclohexene the initial product is the diaxial dibromide, which immediately undergoes ring flip to give the more stable diequatorial conformer.

$\xrightarrow{Br_2}$

trans-1,2-dibromocyclohexane

Note the conformation of cyclohexene. The double bond and the four atoms attached to it lie in one plane, as shown in Figure 11.22. The remaining two ring carbons lie above and below this plane in order to stagger the hydrogens.

The preference for **diaxial** addition in cyclohexenes is related mechanistically to the preference for diaxial elimination (Section 11.5.A).

When a solution of bromine is used in an inert solvent such as carbon tetrachloride, the only nucleophilic reagent available for reaction with the intermediate cation is bromide ion. In hydroxylic solvents the solvent itself is nucleophilic and can react in competition with the bromide ion.

$$C_6H_5CH{=}CH_2 \xrightarrow[\text{Br}_2\text{ in CH}_3\text{OH}]{\text{dilute solution of}} C_6H_5CHBr{-}CH_2Br + C_6H_5\overset{\overset{\displaystyle OCH_3}{|}}{C}HCH_2Br$$

minor   major

The relative amounts of dibromide and bromo ether produced depend on the concentration. Generally, for dilute solutions the product is almost exclusively the bromo ether.

Similarly, aqueous or aqueous alkaline solutions of chlorine or bromine produce the corresponding halo alcohols.

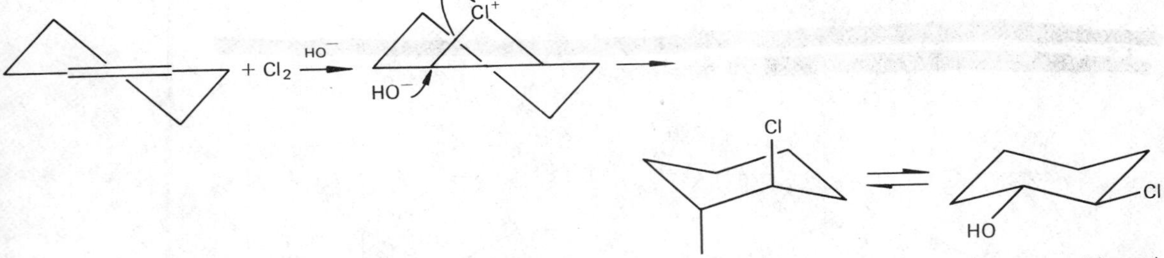

trans-2-chlorocyclohexanol

> A solution of chlorine in aqueous sodium hydroxide cooled with ice is added in portions to cyclohexene keeping the temperature at 15–20°. The mixture is saturated with salt and steam-distilled. The distillate is saturated with salt and extracted with ether. Distillation of the dried ether solution gives 70–73% yield of product.

$$\underset{CH_3}{\overset{CH_3}{>}}C{=}CH_3 + Br_2 \xrightarrow{H_2O} CH_3{-}\overset{\overset{\displaystyle OH}{|}}{\underset{\underset{\displaystyle CH_3}{|}}{C}}{-}CH_2Br$$

> Solutions of chlorine and bromine in water are in equilibrium with the corresponding hypohalous acids.
>
> $$X_2 + H_2O \rightleftharpoons H^+ + X^- + HOX$$
>
> A saturated solution of chlorine in water at 25° is 0.09 M and is one-third converted to chloride ion and hypochlorous acid. In saturated bromine water (0.2 M) 0.5% is converted to hypobromous acid. Thus, although the formation of halo alcohol may be formally regarded as an addition of HO—X across the double bond, the mechanism probably involves a reaction of an intermediate halonium ion with water or hydroxide ion as shown above.

Recall that the resulting halo alcohols provide a useful route to epoxides (Section 10.12).

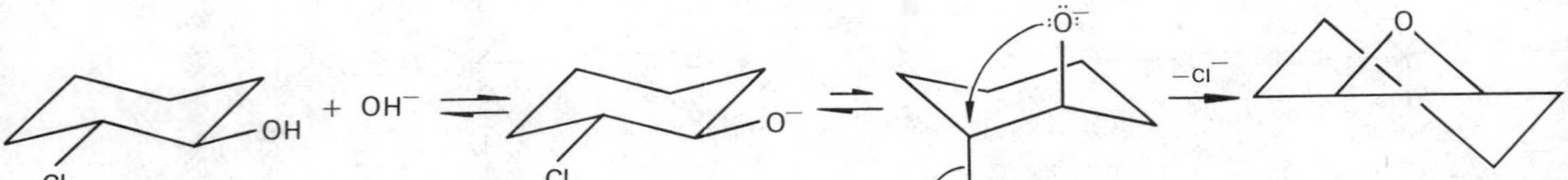

cyclohexene oxide

The 1,2-dihalides produced by the addition of halogens to alkenes are called **vicinal** dihalides (L., *vicinus,* near). They have many chemical properties in common with simple alkyl halides. For example, 1,2-dibromoethane readily enters into nucleophilic displacement reactions.

$$BrCH_2CH_2Br + 2\ CN^- \longrightarrow NCCH_2CH_2CN + 2\ Br^-$$

As with the simple monohalides, nucleophilic displacement is usually accompanied by some elimination, particularly when one or both of the halogens is attached to a secondary carbon. With strong bases elimination is the principal reaction.

$$\underset{\text{1,2-dichlorobutane}}{CH_3CH_2\overset{\overset{\displaystyle Cl}{|}}{C}HCH_2Cl} + CH_3O^- \longrightarrow \underset{\text{1-chloro-1-butene}}{CH_3CH_2CH{=}CHCl} + \underset{\text{2-chloro-1-butene}}{CH_3CH_2\overset{\overset{\displaystyle Cl}{|}}{C}{=}CH_2}$$

Such **dehydrohalogenations** are not generally useful ways to prepare haloalkenes because both isomers are usually produced. Dehydrohalogenation of both halogens is much more important as a method for preparing alkynes (Section 12.5).

EXERCISE: What product is produced when 2-methyl-2-butene reacts with (a) Br_2 in CCl_4, (b) Br_2 in CH_3CH_2OH, and (c) Br_2 in H_2O?

4. ADDITION OF MERCURIC ACETATE. Mercuric ion, Hg^{2+}, is an electrophilic reagent that adds to double bonds to form organomercury derivatives. Mercuric acetate in methanol or ethanol readily yields the corresponding alkoxyalkylmercuric acetate.

$$Hg(OAc)_2 + \ \overset{}{C}{=}\overset{}{C} \longrightarrow \underset{CH_3OH}{\overset{\overset{\displaystyle HgOAc}{|}}{C}{-}\overset{}{C}^+} + CH_3CO_2^- \longrightarrow \overset{\overset{\displaystyle HgOAc}{|}}{C}{-}\underset{\underset{\displaystyle OCH_3}{|}}{C}$$

Anti addition is the usual stereochemical result. Mercuric acetate or perchlorate in water gives the hydroxyalkylmercuric salt.

$$Hg^{2+} + \ \overset{}{C}{=}\overset{}{C} + H_2O \longrightarrow \overset{\overset{\displaystyle Hg^+}{|}}{C}{-}\underset{\underset{\displaystyle OH}{|}}{C} + H^+$$

These compounds are readily reduced with sodium borohydride, which replaces the C—Hg bond by C—H with liberation of free mercury. The intermediate organomercury compounds need not be isolated. The net result of mercuration in alcohol or water, followed by sodium borohydride reduction, is addition of alcohol or water to the alkene. The reduction is an excellent method for the synthesis of alcohols and ethers. **Addition follows the Markovnikov rule,** Hg^{2+} going to the less substituted carbon.

$$\underset{\text{1-hexene}}{n\text{-}C_4H_9CH{=}CH_2} \xrightarrow[\text{aq. THF}]{Hg(OAc)_2} \xrightarrow[\text{NaOH}]{NaBH_4} \underset{\underset{\text{2-hexanol}}{(96\%)}}{n\text{-}C_4H_9CHOHCH_3} + Hg$$

1-Hexene is added with stirring to an equivalent amount of mercuric acetate in 1:1 water–THF. After stirring for 10 min at 25°, aqueous NaOH is added, followed by a 0.5 M solution of $NaBH_4$ in 3 M NaOH. The organic layer is separated, dried and distilled to yield 2-hexanol.

The sodium borohydride used in this reaction is an important reagent in organic chemistry.

Sodium borohydride, $NaBH_4$, is the sodium salt of the borohydride ion, BH_4^-, a tetrahedral ion that can be regarded as derived from BH_3 and hydride ion, H^-. It is a white powder and dissolves in water to form stable solutions at basic pH. In acid the compound reacts rapidly to form hydrogen and sodium borate. Sodium borohydride is soluble in methanol and ethanol, but decomposes slowly in these solvents. It is appreciably soluble in diglyme (5.5 g per 100 g of solvent), but is almost insoluble in glyme or tetrahydrofuran. Sodium borohydride is a useful reducing agent for aldehydes and ketones (Section 13.8).

The reaction combination of mercuration and reduction is a useful laboratory alternative to acid-catalyzed hydration of olefins. Of course, it cannot compete with sulfuric acid in large-scale commercial productions.

EXERCISE: Write complete equations for the reactions of 2-methylpropene with (a) H_2/Pt, (b) HCl, (c) dil. H_2SO_4, (d) Br_2/CCl_4, (e) Cl_2/H_2O, (f) product of (e) with NaOH, and (g) $Hg(OAC)_2$ in water, followed by $NaBH_4$.

C. *Free Radical Additions*

The early literature of organic chemistry contained considerable disagreement on the mode of addition of HBr to terminal olefins. In some cases Markovnikov's rule appeared to hold; in other cases it did not. Often two chemists would add HBr to the same alkene and obtain contradictory results.

$$RCH{=}CH_2 + HBr \longrightarrow RCH_2CH_2Br \quad \text{or} \quad RCHCH_3 \overset{Br}{|}$$

In the 1930s this apparent dilemma was resolved when it was discovered that HBr (but *not* HCl or HI) can add to alkenes by two different mechanisms. Pure materials and pure solvents encourage addition by the electrophilic mechanism discussed in Section 11.6.B.1. This mechanism leads to normal Markovnikov addition.

$$RCH{=}CH_2 + HBr \xrightarrow{\text{(ionic mechanism)}} RCHCH_3 \overset{Br}{|}$$

Impure materials, oxygen, and some other additives were found to promote "abnormal" addition by a mechanism involving **free radical** intermediates.

$$RCH{=}CH_2 + HBr \xrightarrow[\text{mechanism}]{\text{(free radical}} RCH_2CH_2Br$$

The free radical mechanism starts with an initiation step that results in oxidation of HBr to bromine atoms.

$$\text{HBr} + \text{O}_2 \text{ (or other radical initiators)} \longrightarrow \text{Br·}$$

The bromine atom then adds to the alkene to give a free radical that continues the chain by abstracting hydrogen from a molecule of HBr. Both of the propagation steps are exothermic and have low activation energies.

$$\text{RCH}=\text{CH}_2 + \text{Br·} \longrightarrow \text{R}\overset{\cdot}{\text{C}}\text{HCH}_2\text{Br} \qquad\qquad \Delta H° = -9 \text{ kcal mole}^{-1}$$

$$\text{R}\overset{\cdot}{\text{C}}\text{HCH}_2\text{Br} + \text{HBr} \longrightarrow \text{RCH}_2\text{CH}_2\text{Br} + \text{Br·} \qquad \Delta H° = -7 \text{ kcal mole}^{-1}$$

Note that the bromine atom adds to the alkene in such a way as to give the more highly substituted (more stable) free radical. The overall outcome is thus **anti-Markovnikov** orientation. This abnormal addition or "peroxide effect" is a useful reaction with HBr, but is not significant with HCl or HI. The C—I bond is so weak that the addition of iodine atoms to double bonds is endothermic. It becomes exothermic only at elevated temperatures.

$$\text{RCH}=\text{CH}_2 + \text{I·} \longrightarrow \text{R}\overset{\cdot}{\text{C}}\text{HCH}_2\text{I} \qquad\qquad \Delta H° = +5 \text{ kcal mole}^{-1}$$

$$\text{R}\overset{\cdot}{\text{C}}\text{HCH}_2\text{I} + \text{HI} \longrightarrow \text{RCH}_2\text{CH}_2\text{I} + \text{I·} \qquad\quad \Delta H° = -24 \text{ kcal mole}^{-1}$$

The H—Cl bond is so strong that the second step in the sequence is endothermic and slow.

$$\text{RCH}=\text{CH}_2 + \text{Cl·} \longrightarrow \text{R}\overset{\cdot}{\text{C}}\text{HCH}_2\text{Cl} \qquad\qquad \Delta H° = -22 \text{ kcal mole}^{-1}$$

$$\text{R}\overset{\cdot}{\text{C}}\text{HCH}_2\text{Cl} + \text{HCl} \longrightarrow \text{RCH}_2\text{CH}_2\text{Cl} + \text{Cl·} \qquad \Delta H° = +8 \text{ kcal mole}^{-1}$$

Free radical chain reactions work best when both propagation steps are exothermic. An endothermic step corresponds to a slow and reversible reaction that breaks the chain.

Other compounds that have appropriate bond strengths can add to double bonds under free radical conditions. Examples include chlorine, bromine, hydrogen sulfide, thiols, and polyhaloalkanes.

$$\text{CH}_3\text{CH}_2\text{CH}=\text{CH}_2 + \text{H}_2\text{S} \xrightarrow[0°]{h\nu} \text{CH}_3\text{CH}_2\text{CH}_2\text{CH}_2\text{SH} + (\text{CH}_3\text{CH}_2\text{CH}_2\text{CH}_2)_2\text{S}$$
$$\qquad\qquad\qquad\qquad\qquad\qquad\qquad (68\%) \qquad\qquad\qquad (12\%)$$

$$(\text{CH}_3)_2\text{C}=\text{CH}_2 + \text{CH}_3\text{CH}_2\text{SH} \xrightarrow[\text{O}_2]{100°} (\text{CH}_3)_2\text{CHCH}_2\text{SCH}_2\text{CH}_3$$
$$\qquad\qquad\qquad\qquad\qquad\qquad\qquad\qquad (94\%)$$
$$\qquad\qquad\qquad\qquad\qquad\qquad\qquad \text{ethyl isobutyl sulfide}$$

Carbon tetrachloride and carbon tetrabromide react readily with olefins and free radical initiators to give 1:1 adducts. The propagation steps are

$$\text{RCH}=\text{CH}_2 + \cdot\text{CX}_3 \longrightarrow \text{R}\overset{\cdot}{\text{C}}\text{HCH}_2\text{CX}_3$$

$$\overset{\cdot}{\text{R}}\text{CHCH}_2\text{CX}_3 + \text{CX}_4 \longrightarrow \text{R}\overset{\overset{\text{X}}{|}}{\text{C}}\text{HCH}_2\text{CX}_3 + \cdot\text{CX}_3$$

$$CH_3(CH_2)_5CH{=}CH_2 + CCl_4 \xrightarrow[\Delta]{peroxide} CH_3(CH_2)_5CHClCH_2CCl_3$$

(75%)

1,1,1,3-tetrachlorononane

$$CH_3(CH_2)_5CH{=}CH_2 + CBr_4 \xrightarrow[75°]{h\nu} CH_3(CH_2)_5CHBrCH_2CBr_3$$

(88%)

1,1,1,3-tetrabromononane

In some cases, especially with ethylene itself, such reactions yield a mixture of **telomer** products in which the intermediate radicals have reacted with alkenes in the following way.

$$Cl_3CCH_2CH_2\cdot + CH_2{=}CH_2 \longrightarrow Cl_3CCH_2CH_2CH_2CH_2\cdot \xrightarrow{CH_2{=}CH_2}$$

$$Cl_3CCH_2CH_2CH_2CH_2CH_2CH_2\cdot \xrightarrow{etc.} Cl_3C(CH_2CH_2)_xCH_2CH_2\cdot \xrightarrow{CCl_4}$$

$$Cl_3C(CH_2CH_2)_xCH_2CH_2Cl + \cdot CCl_3$$

a telomer

$$\underset{\text{(100 atm)}}{CH_2{=}CH_2} + CCl_4 \xrightarrow[\underset{peroxide}{110°}]{H_2O} Cl(CH_2CH_2)_xCCl_3$$

9% $x = 1$

57% $x = 2$

24% $x = 3$

At this point it is instructive to review the meanings of the terms "Markovnikov addition" and "anti-Markovnikov addition." When first formulated, the Markovnikov rule simply stated that "in the addition of HX to an alkene, the H goes to the carbon already having the greater number of hydrogens."

Markovnikov addition

$$CH_3{-}CH{=}CH_2 + HX \longrightarrow CH_3{-}\overset{X}{\underset{|}{CH}}{-}\overset{H}{\underset{|}{CH_2}}$$

Anti-Markovnikov addition

$$CH_3{-}CH{=}CH_2 + HX \longrightarrow CH_3{-}\overset{H}{\underset{|}{CH}}{-}\overset{X}{\underset{|}{CH_2}}$$

We now recognize that electrophilic additions obey the Markovnikov rule simply because H^+ is the reagent that adds to the double bond first and *it always adds to the end that produces the more stable carbocation.*

$$CH_3CH{=}CH_2 + HX \longrightarrow CH_3\overset{+}{C}H{-}\overset{H}{\underset{|}{CH_2}} \quad (\text{not } CH_3\overset{H}{\underset{|}{C}}H{-}\overset{+}{C}H_2)$$

Free radical additions "disobey" the Markovnikov rule because in this case it is X· that adds first and *adds so as to produce the more stable free radical.*

$$CH_3CH{=}CH_2 + X\cdot \longrightarrow CH_3\overset{\cdot}{C}H{-}\overset{X}{\underset{|}{CH_2}} \quad (\text{not } CH_3\overset{X}{\underset{|}{C}}H{-}CH_2\cdot)$$

D. *Hydroboration*

Although the reaction of C=C double bonds with diborane was discovered less than three decades ago, it has become one of the most important reactions in the repertoire of the synthetic chemist.

> Diborane, B_2H_6, is a colorless, toxic gas that is spontaneously flammable in air. It is usually prepared by the reaction of sodium borohydride with boron trifluoride.
>
> $$3 \text{ NaBH}_4 + 4 \text{ BF}_3 \longrightarrow 2 \text{ B}_2\text{H}_6 + 3 \text{ NaBF}_4$$
>
> Borane itself, BH_3, is not known. In this compound boron has a sextet of electrons and is a Lewis acid. In ethers such as tetrahydrofuran or diglyme, common solvents for hydroboration reactions, diborane is readily soluble as an ether-monomer complex, $R_2O\text{-}BH_3$. Diborane has an unusual bridged structure because it is an electron-deficient compound. The 12 valence electrons are too few to provide enough normal two-electron bonds for an ethane-like structure with six B—H bonds. In the actual structure
>
>
> diborane
>
> four hydrogens and the two borons define a plane with four two-electron B—H bonds. The other two hydrogens lie above and below this plane and involve the unusual three-center two-electron bonds symbolized by ⅄ . Higher boron hydrides such as pentaborane, B_5H_9, hexaborane, B_6H_{10}, and decaborane, $B_{10}H_{14}$ are known. All are electron-deficient compounds with unusual structures involving bridged hydrogens and three-center bonds.

The B—H bond adds rapidly and quantitatively to many multiple bonds including C=C double bonds. With simple alkenes the product is a trialkylborane.

$$6 \text{ RCH=CH}_2 + \text{B}_2\text{H}_6 \longrightarrow 2 \text{ (RCH}_2\text{CH}_2)_3\text{B}$$

The addition appears to be dominated by steric considerations. The boron generally becomes attached to the less substituted and less sterically congested carbon. With highly substituted or hindered olefins, addition may stop at the mono- or dialkylborane stage. The reaction appears to involve initial coordination of BH_3 with the π-electrons of the double bond followed by formation of the C—H bond.

In cases where stereochemistry may be defined, exclusive *syn* addition is observed.

The alkylboranes are generally not isolated but are converted by subsequent reactions directly into desired products. The most important general reaction of alkylboranes is that with alkaline hydrogen peroxide.

$$\underset{\text{trans-2-methylcyclopentanol}}{\text{(cyclopentane with CH}_3\text{ and BR}_2)} \xrightarrow[\text{OH}^-]{\text{H}_2\text{O}_2} \text{(cyclopentane with CH}_3\text{ and OH)}$$

trans-2-methylcyclopentanol

Three separate processes are involved in the oxidation of alkylboranes to alcohols. In the first step, hydroperoxide anion adds to the electron-deficient boron atom.

$$\underset{\overset{|}{R}}{\overset{|}{R}}{-}B + {}^-\text{OOH} \longrightarrow \underset{\overset{|}{R}}{\overset{\overset{|}{R}}{R}}{-}\overset{-}{B}{-}O{-}OH$$

The resulting intermediate rearranges with loss of hydroxide ion. The driving force for the rearrangement is liberation of the stable anion OH^- and formation of the strong B—O bond.

$$\underset{\overset{|}{R}}{\overset{\overset{|}{R}}{R}}{-}\overset{-}{B}{-}O{-}OH \longrightarrow \underset{\overset{|}{R}}{\overset{|}{R}}{-}B{-}OR + OH^-$$

The B—O bond is much stronger than a B—C bond because of overlap of an oxygen *p*-orbital with its lone pair of electrons and the empty *p*-orbital on boron.

empty ──→ (orbital diagram) ──── full

--- B — O —

The oxygen lone pair becomes polarized toward boron in the sense indicated by resonance structures.

$$\text{>B}{-}\ddot{\text{O}}{-} \longleftrightarrow \text{>}\overset{-}{\text{B}}{=}\overset{..}{\text{O}}{}^+{-}$$

Hydrogen peroxide is available as the anhydrous liquid, b.p. 152°, or as aqueous solutions ranging from 3 to 90% in concentration. The compound is thermodynamically unstable with respect to water and oxygen, and high-strength solutions are explosively hazardous. The 3% solution is used medicinally as a topical antiseptic, but the 30% solution is commonly used in the organic laboratory. Even with the 30% reagent, experiments should be carried out behind safety shields and the material should be kept out of contact with skin and eyes.

The migration of an alkyl group, and its bonding electron pair, is analogous to the rearrangement of carbocations (Section 10.7.B).

$$\underset{\overset{|}{R}}{-}\overset{|}{C}{-}\overset{|}{\overset{+}{C}}{-} \longrightarrow \underset{\overset{|}{R}}{-}\overset{|}{\overset{+}{C}}{-}\overset{|}{C}{-}$$

The reaction of boranes with alkaline hydrogen peroxide is rapid and exothermic. The product R_2BOR reacts further by the same process to give a trialkyl borate ester.

$$R_2BOR + 2\,OOH^- \longrightarrow B(OR)_3 + 2\,OH^-$$

The borate ester is then hydrolyzed under the reaction conditions to the alcohol and sodium borate.

$$B(OR)_3 + OH^- + 3H_2O \longrightarrow 3\,ROH + B(OH)_4{}^-$$

> Alkyl borates can be prepared by heating a mixture of the alcohol and boric acid or boric anhydride, B_2O_3. The esters distill readily (trimethyl borate, b.p. 68°; triethyl borate, b.p. 120°). The esters are mild Lewis acids and are rapidly hydrolyzed by water.
>
> $$(RO)_3B + H_2O \rightleftharpoons (RO)_3\overset{-}{B}-\overset{+}{O}H_2 \rightleftharpoons (RO)_2\overset{\underset{\displaystyle OH}{|}}{B}-\overset{+}{O}\overset{\displaystyle H}{\underset{\displaystyle R}{\diagup}} \rightleftharpoons$$
>
> $$(RO)_2BOH + H_2O \rightleftharpoons \text{ etc.}$$
>
> Borate esters are not generally important in organic chemistry.

The net result of hydroboration and oxidation-hydrolysis is anti-Markovnikov hydration of a double bond. The reaction is a relatively simple and convenient laboratory procedure and has become an important synthetic reaction in organic chemistry.

1-methylcyclopentene   trans-2-methylcyclopentanol
                                                          (85%)

> Diborane prepared by reaction of sodium borohydride and boron trifluoride etherate in diglyme is swept by a stream of nitrogen into a solution of 1-methylcyclopentene in THF at 0°. Excess diborane is hydrolyzed by addition of ice. The reaction is completed by addition of aqueous sodium hydroxide followed slowly by 30% hydrogen peroxide. After stirring for an additional period the layers are separated, the aqueous phase is extracted with ether and the combined organic layers are dried and distilled to give 85% of trans-2-methylcyclopentanol.

EXERCISE: Write the structure, including stereochemistry where pertinent, for the alcohol produced by hydroboration-oxidation of each of the following alkenes.
(a) 4-methyl-1-pentene   (b) (E)-3-methyl-2-pentene
(c) 1,2-dideuteriocyclohexene

E. *Oxidation*

Alkenes are oxidized readily by potassium permanganate, $KMnO_4$, but the products depend on the reaction conditions.

> Potassium permanganate forms dark purple crystals that dissolve in water to give intense red solutions. In permanganate anion, $MnO_4{}^-$, manganese has an oxidation state of $+7$. As an oxidizing agent in basic solution, manganese is reduced to manganese dioxide, MnO_2, an insoluble brown compound that is frequently difficult to filter

because it tends to form colloidal suspensions. Treatment with SO_2 at this point forms the soluble $MnSO_4$. In acid solution reduction of permanganate to Mn^{2+} occurs. The two half-reactions are

$$3e^- + MnO_4^- + 2\,H_2O \longrightarrow MnO_2 + 4\,OH^-$$

$$5e^- + MnO_4^- + 8\,H^+ \longrightarrow Mn^{2+} + 4\,H_2O$$

In acid solution potassium permanganate is a strong reagent that attacks organic compounds almost indiscriminately. It will even oxidize HCl to Cl_2. Hence, in organic use potassium permanganate is almost always used in neutral or alkaline solutions in which MnO_2 is produced.

Cold dilute potassium permanganate reacts with double bonds to give vicinal diols, which are commonly called glycols.

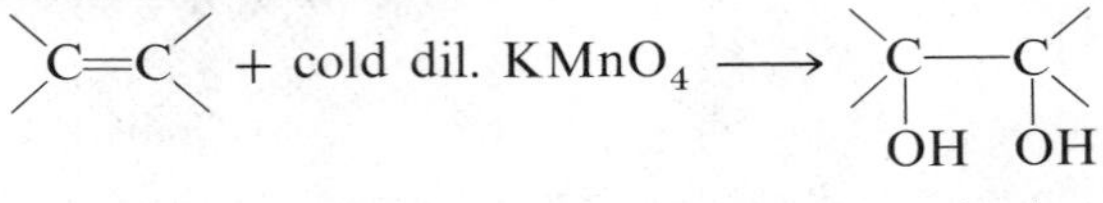

$$\text{C=C} + \text{cold dil. KMnO}_4 \longrightarrow \underset{\underset{\text{a glycol}}{\text{OH} \quad \text{OH}}}{\text{C--C}}$$

Reaction conditions need to be carefully controlled. Yields are variable and usually low. The reaction occurs with *syn* addition and is thought to involve an intermediate cyclic manganate ester that is rapidly hydrolyzed.

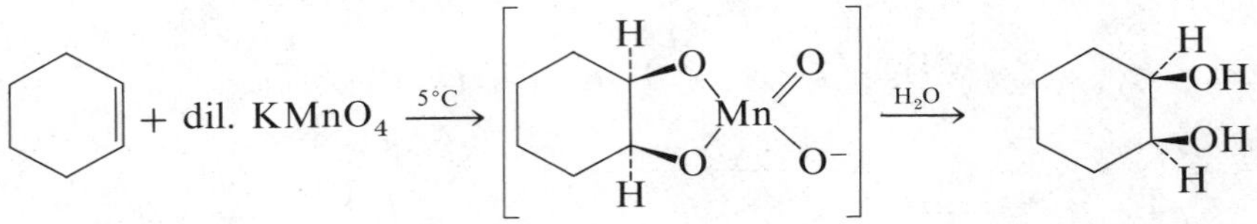

(30–33%)
cis-1,2-cyclohexanediol

The same overall reaction can be accomplished with osmium tetroxide, which forms isolable cyclic esters with alkenes.

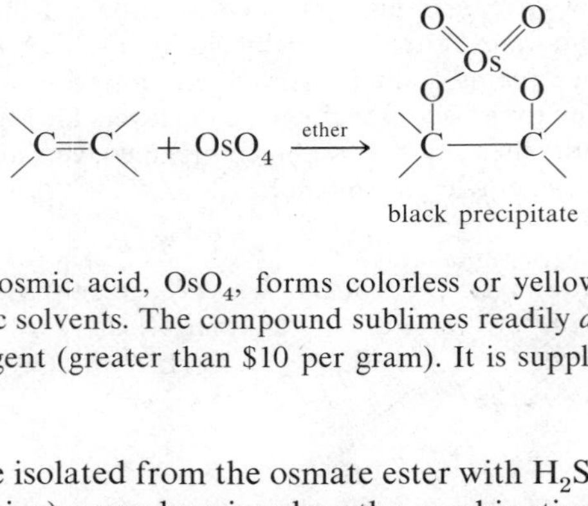

black precipitate

Osmium tetroxide, osmic acid, OsO_4, forms colorless or yellow crystals soluble in water and in organic solvents. The compound sublimes readily *and is highly toxic*. It is an expensive reagent (greater than $10 per gram). It is supplied commercially in small sealed tubes.

The *cis* diol can be isolated from the osmate ester with H_2S, but a more convenient (and less expensive) procedure involves the combination of hydrogen peroxide with a catalytic amount of osmium tetroxide. The osmate ester is formed but is converted by the peroxide to the *cis* diol. Osmium tetroxide is constantly regenerated, so that only a small amount need be used.

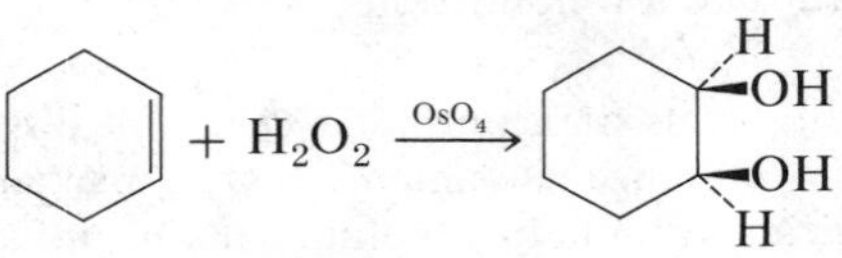

(58%)

When more concentrated solutions of potassium permanganate are used in the oxidation of alkenes, the initially formed glycol is oxidized further. The product is a mixture of ketones or carboxylic acids, depending on the extent of substitution of the double bond.

$$\underset{R'}{\overset{R}{>}}C=C\underset{H}{\overset{R''}{<}} \xrightarrow{\text{KMnO}_4} \underset{R'}{\overset{R}{>}}C=O + HO-\overset{O}{\overset{\|}{C}}R''$$

$$RCH=CH_2 \xrightarrow{\text{KMnO}_4} RCOOH + CO_2$$

This is not a common reaction in organic synthesis because the yields are usually low. Oxidative cleavage of the double bond can generally be accomplished in better yield by reaction with ozone.

Ozone, O_3, is an important constituent of the upper atmosphere where it is produced by action of solar ultraviolet radiation on atmospheric oxygen. Ozone, in turn, absorbs in the ultraviolet region of the spectrum and provides an important screen to limit the amount of this radiation that reaches the earth's surface. Ozone is thermodynamically unstable with respect to oxygen.

$$O_3 \longrightarrow \tfrac{3}{2}O_2 \qquad \Delta H° = -34 \text{ kcal mole}^{-1}$$

Ozone is produced in the laboratory with an "ozonator," a special apparatus in which an electrodeless discharge is induced in dry air passing through an alternating electric field. Ozone concentrations as high as 4% in air can be produced. The gas has a characteristic odor usually associated with electric arcs.

Reactions of alkenes with ozone are normally carried out by passing ozone-containing air through a solution of the alkene in an inert solvent at low temperatures (usually $-80°$). Reaction is rapid and completion of reaction is determined by testing the effluent gas with potassium iodide. Unreacted ozone reacts to give iodine. Suitable solvents for ozonizations include methylene chloride, alcohol, and ethyl acetate. The first formed addition product, the molozonide, rearranges rapidly, even at low temperatures, to the ozonide structure:

$$>C=C< \xrightarrow{O_3} \underset{\text{molozonide}}{\overset{O-O-O}{\underset{C---C}{<}}} \longrightarrow \underset{\text{ozonide}}{\overset{O}{\underset{O---O}{C}}}$$

In some cases polymeric structures are obtained. Some ozonides, especially the polymeric structures, decompose with explosive violence on heating; hence, the ozonides are generally not isolated but are decomposed directly to desired products.

Hydrolysis with water occurs readily to give carbonyl compounds and hydrogen peroxide.

$$\underset{O}{\overset{O-O}{C}} C< + H_2O \longrightarrow >C=O + O=C< + H_2O_2$$

Aldehydes are oxidized by hydrogen peroxide to carboxylic acids. Hence, *reduction* conditions are often used in decomposing the ozonides. Such conditions include zinc dust and acetic acid, catalytic hydrogenation, and dimethyl sulfide.

$$(CH_3)_2CHCH_2CH_2CH_2CH=CH_2 \xrightarrow{O_3} \xrightarrow[CH_3COOH]{Zn} (CH_3)_2CHCH_2CH_2CH_2CHO$$

6-methyl-1-heptene   5-methylhexanal (62%)

> 6-Methyl-1-heptene in methylene chloride at −78° is treated with ozone and is then added to a stirred mixture of zinc dust and 50% aqueous acetic acid. The mixture is refluxed for 1 hr and extracted with ether. Peroxides are removed from the ether with aqueous potassium iodide and the washed and dried solution is distilled to give 5-methylhexanal, b.p. 144°.

$$n\text{-}C_6H_{13}CH=CH_2 + O_3 \xrightarrow[-60°]{CH_3OH} \xrightarrow{(CH_3)_2S} n\text{-}C_6H_{13}CHO + (CH_3)_2SO + CH_2O$$

1-octene   heptanal (73%)   dimethyl sulfoxide (DMSO)

Treatment of the ozonide with sodium borohydride gives the corresponding alcohols.

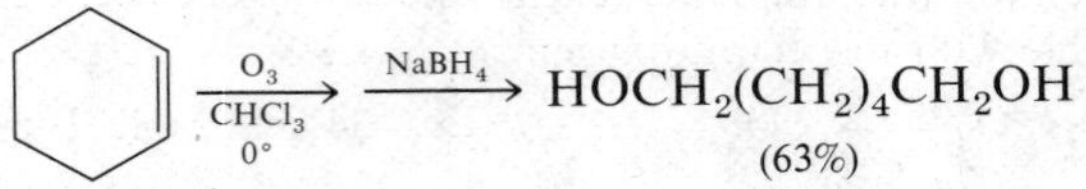

$$\text{(cyclohexene)} \xrightarrow[0°]{O_3 / CHCl_3} \xrightarrow{NaBH_4} HOCH_2(CH_2)_4CH_2OH$$

(63%)
1,6-hexanediol

Alkenes may also be oxidized by peroxycarboxylic acids. The product is an oxirane (Section 10.12.A).

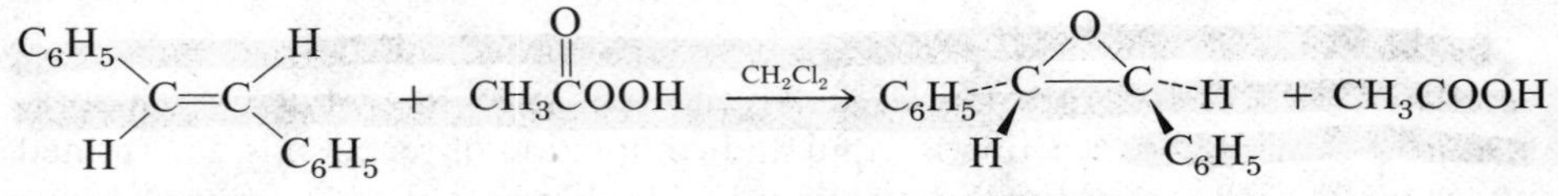

trans-stilbene   peroxyacetic acid   trans-stilbene oxide (78–83%)
trans-2,3-diphenyloxirane

Peroxycarboxylic acids have the general formula

$$R-\overset{\overset{\displaystyle O}{\|}}{C}-O-OH$$

Like hydrogen peroxide, H_2O_2, peroxycarboxylic acids are oxidizing agents and are often used for that purpose. Peroxycarboxylic acids are generally unstable and must be stored in the cold or, preferably, be prepared as needed. An important exception is 3-chloroperoxybenzoic acid, an exceptionally stable crystalline solid now available commercially. This reagent provides a simple and convenient one-step route to epoxides.

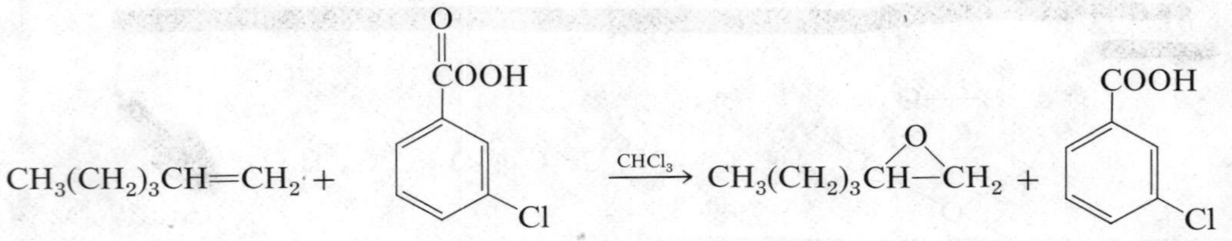

$CH_3(CH_2)_3CH=CH_2$ + 3-chloroperoxybenzoic acid $\xrightarrow{CHCl_3}$ $CH_3(CH_2)_3CH-CH_2$ (1-butyloxirane, 60%) + 3-chlorobenzoic acid

The mechanism of the reaction can be represented as a displacement reaction on an electrophilic oxygen by a nucleophilic alkene.

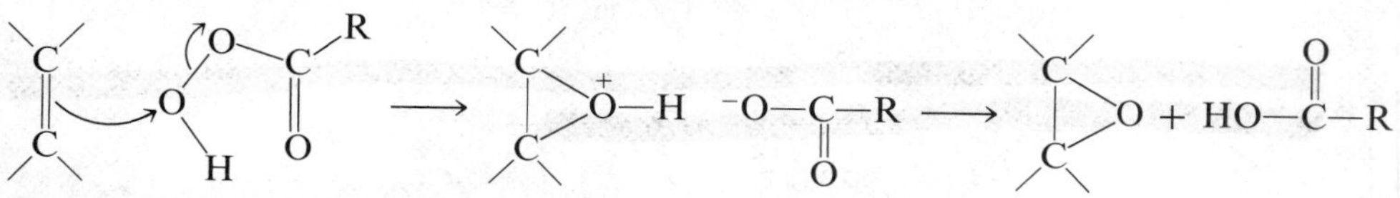

The reaction may actually take place in a single step.

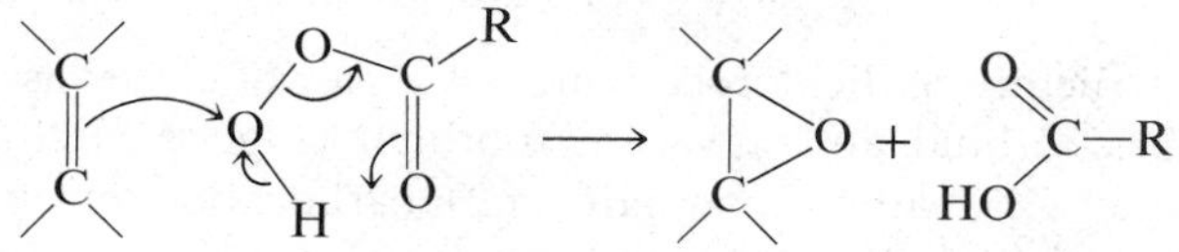

Recall that effective oxidation of the double bond may be achieved by the two-step combination of addition of aqueous halogen followed by base-catalyzed cyclization (page 310).

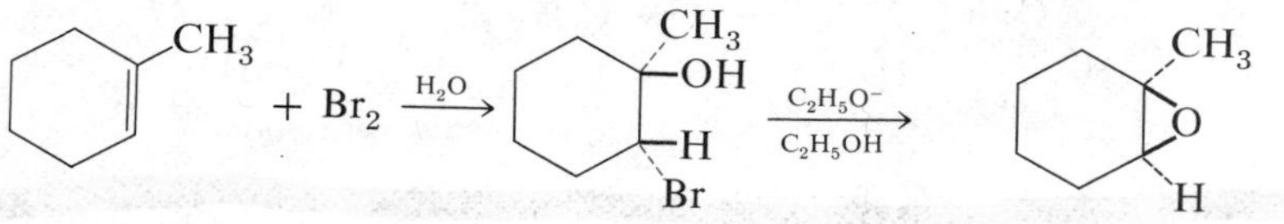

1-methylcyclohexene oxide

Ethylene oxide is a significant commercial item and is prepared industrially by the catalyzed air oxidation of ethylene.

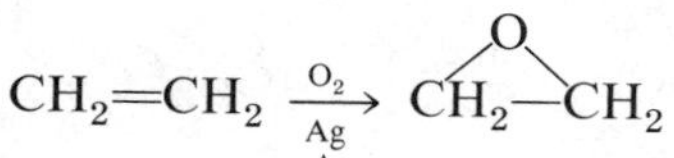

EXERCISE: How may cyclopentene be converted into each of the following compounds?

(a) cyclopentene oxide
(b) *cis*-1,2-cyclopentanediol
(c) $CH_2(CH_2CHO)_2$
(d) $CH_2(CH_2COOH)_2$
(e) $CH_2(CH_2CH_2OH)_2$

F. Addition of Carbenes and Carbenoids; Preparation of Cyclopropanes

Carbenes are reactive intermediates having the general formula $R_2C:$, in which carbon has only a sextet of electrons. Although carbenes are neutral species, the electron-deficient carbon is still "hungry" for electrons, and hence carbenes behave as **electrophiles**. One way in which carbenes may be generated is by the reaction of chloroform with a strong base. Because of the strong electron-attracting inductive effect of the three chlorines, chloroform is rather acidic ($pK \approx 25$). Thus, the trichloromethyl anion is formed in a small equilibrium concentration. If the reaction is carried out in water, the main fate of $Cl_3C:^-$ is protonation to regenerate chloroform. Another reaction, especially in less acidic solvents such as *t*-butyl alcohol, is loss of a chloride ion to give dichlorocarbene.

$$CHCl_3 + RO^- \rightleftharpoons ROH + \underset{\substack{\text{trichloromethyl}\\\text{anion}}}{Cl_3C:^-} \longrightarrow Cl^- + \underset{\text{dichlorocarbene}}{Cl_2C:}$$

If the reaction is carried out in the presence of an alkene, the carbene adds to the double bond to give a 1,1-dichlorocyclopropane.

$$(CH_3)_2C=CH_2 \xrightarrow[\substack{(CH_3)_3COH\\\Delta}]{\substack{CHCl_3\\(CH_3)_3COK}} \underset{\text{1,1-dichloro-2,2-dimethylcyclopropane}}{(CH_3)_2C\diagdown\overset{CCl_2}{\diagup}CH_2}$$

The electronic structure of dichlorocarbene has a pair of electrons in an approximately sp^2-hybrid orbital and a vacant p-orbital (Figure 11.23). That is, the carbene combines a carbanion lone pair and a carbocation vacancy on a single carbon.

Dichlorocarbene adds stereospecifically *syn* to many types of double bonds.

$$\underset{H}{\overset{C_2H_5}{\diagup}}C=C\underset{H}{\overset{CH_3}{\diagup}} \xrightarrow[\substack{(CH_3)_3COH\\\Delta}]{\substack{CHCl_3\\(CH_3)_3COK}} C_2H_5\cdots\overset{CCl_2}{C}\cdots CH_3$$

no *trans* isomer is formed

This reaction can also be applied to bromoform, $CHBr_3$, to yield the corresponding dibromocyclopropanes.

A related type of intermediate results when alkyllithium compounds are allowed to react with methylene chloride.

$$CH_3-\underset{\underset{CH_3}{|}}{C}=\underset{\underset{CH_3}{|}}{C}-CH_3 \xrightarrow[-35°]{\substack{C_4H_9Li\\CH_2Cl_2}} (CH_3)_2C\diagdown\overset{CHCl}{\diagup}C(CH_3)_2$$

(67%)
3-chloro-1,1,2,2-tetramethylcyclopropane

The reaction intermediate can be formulated as a chlorocarbene.

$$CH_2Cl_2 + C_4H_9Li \longrightarrow C_4H_{10} + LiCHCl_2$$

$$LiCHCl_2 \longrightarrow LiCl + \underset{\text{chlorocarbene}}{:CHCl}$$

$$\diagdown\!\!=\!\!\diagup + :CHCl \longrightarrow \underset{\substack{H\quad Cl}}{\bowtie}$$

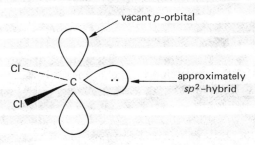

vacant *p*-orbital

Cl

C

..

approximately sp^2–hybrid

Cl

Figure 11.23. *Electronic structure of dichlorocarbene.*

Carbene itself, $H_2C:$, can be produced by photolysis or pyrolysis of diazomethane, CH_2N_2 (Section 18.7.A). However, a better way of adding $H_2C:$ to an alkene utilizes a mixture of diiodomethane and zinc dust (**Simmons–Smith reaction**). In practice, the zinc dust is usually activated by alloying it with a small amount of copper.

$$CH_3(CH_2)_4CH{=}CH_2 \xrightarrow[\substack{\text{ether} \\ \Delta}]{\substack{Zn(Cu) \\ CH_2I_2}} CH_3(CH_2)_4CH\overbrace{\qquad}^{CH_2}CH_2$$

1-heptene

(30%)
pentylcyclopropane

The reaction is applicable to many kinds of double bonds. Yields are generally only fair (30–70%), but the products are often difficult to prepare by alternative routes. The method appears to involve the formation of an organometallic species, iodomethylzinc (Section 15.9.). It is proposed that this species, which is sometimes termed a **carbenoid** because it behaves in some ways like a true carbene, adds to the double bond to give an adduct, which then eliminates zinc iodide.

$$CH_2I_2 + Zn(Cu) \longrightarrow ICH_2ZnI$$

$$\text{>C=C<} + ICH_2ZnI \xrightarrow[\Delta]{\text{ether}} \left[\begin{array}{c} I \;\; CH_2ZnI \\ -\overset{|}{\underset{|}{C}}-\overset{|}{\underset{|}{C}}- \end{array} \right] \longrightarrow -C\overbrace{\qquad}^{CH_2}C- + ZnI_2$$

EXERCISE: *cis*-2-Butene reacts with chlorocarbene to give two isomeric chlorodimethylcyclopropanes whereas *trans*-2-butene gives a single product. Explain.

G. *Polymerization*

Polymerization is the process wherein a small organic compound (a **monomer**) reacts with itself in such a way as to form a high molecular weight compound (a **polymer**). For example, ethylene can undergo polymerization to give polyethylene.

$$CH_2{=}CH_2 \longrightarrow XCH_2CH_2(CH_2CH_2)_nCH_2CH_2Y$$

ethylene

polyethylene

In the case of alkenes, polymerization amounts merely to the exchange of π-bonds for σ-bonds and is thermodynamically feasible. Polymerization may involve intermediate carbocations (**cationic polymerization**), free radicals (**radical polymerization**), or carbanions (**anionic polymerization**).

Cationic polymerization is not generally a practical method for preparing useful polymers. The process is used for the **dimerization** and **trimerization** of certain alkenes. As mentioned in Section 11.6.B.1, isobutylene is absorbed and hydrated by 60–65% aqueous sulfuric acid. Under more vigorous conditions (50% H_2SO_4 at 100°), the intermediate carbocation can react with alkene to form a new tertiary carbocation. Deprotonation of this new carbocation gives a mixture of alkenes known as "diisobutylenes."

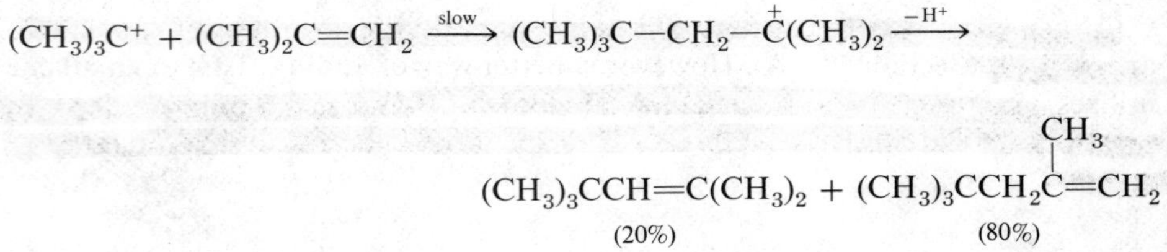

$$(CH_3)_3C^+ + (CH_3)_2C{=}CH_2 \xrightarrow{\text{slow}} (CH_3)_3C{-}CH_2{-}\overset{+}{C}(CH_3)_2 \xrightarrow{-H^+}$$

$$(CH_3)_3CCH{=}C(CH_3)_2 + (CH_3)_3CCH_2\overset{\overset{\displaystyle CH_3}{|}}{C}{=}CH_2$$

$$\text{(20\%)} \qquad\qquad\qquad \text{(80\%)}$$

"diisobutylenes"

Catalytic hydrogenation of this mixture gives 2,2,4-trimethylpentane, the so-called "isooctane" used as a standard for octane ratings of gasolines (Section 6.2).

Under still more vigorous conditions isobutylene reacts with sulfuric acid to produce a mixture of trimeric alkenes, "triisobutylenes."

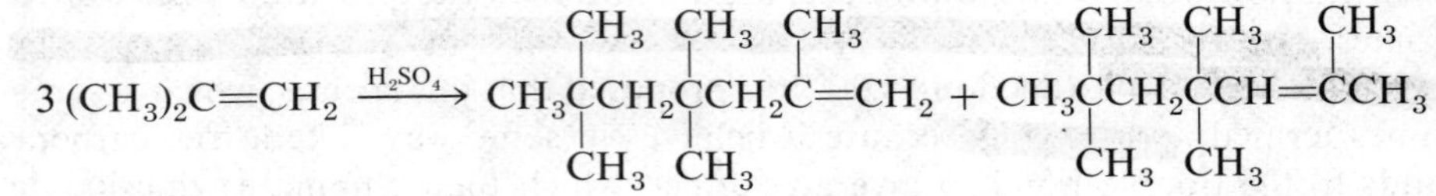

$$3\,(CH_3)_2C{=}CH_2 \xrightarrow{H_2SO_4} CH_3\overset{\overset{\displaystyle CH_3}{|}}{\underset{\underset{\displaystyle CH_3}{|}}{C}}CH_2\overset{\overset{\displaystyle CH_3}{|}}{\underset{\underset{\displaystyle CH_3}{|}}{C}}CH_2\overset{\overset{\displaystyle CH_3}{|}}{C}{=}CH_2 + CH_3\overset{\overset{\displaystyle CH_3}{|}}{\underset{\underset{\displaystyle CH_3}{|}}{C}}CH_2\overset{\overset{\displaystyle CH_3}{|}}{\underset{\underset{\displaystyle CH_3}{|}}{C}}CH{=}\overset{\overset{\displaystyle CH_3}{|}}{C}CH_3$$

"triisobutylenes"

Higher polymers and undesirable tars generally result from the reaction of other alkenes with strong hot acid.

In the absence of suitable alternative nucleophilic compounds to react with the carbocation intermediates, reaction with alkene is the only reaction mode possible. Reaction of isobutylene with a small amount of boron trifluoride occurs at low temperature to produce a high molecular weight polymer.

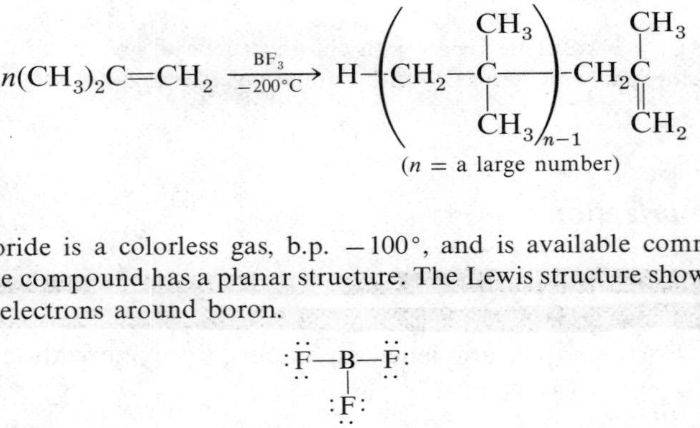

$$n(CH_3)_2C{=}CH_2 \xrightarrow[-200°C]{BF_3} H{-}\!\!\left(\!CH_2{-}\overset{\overset{\displaystyle CH_3}{|}}{\underset{\underset{\displaystyle CH_3}{|}}{C}}\!\right)_{\!\!n-1}\!\!\!\!CH_2\overset{\overset{\displaystyle CH_3}{|}}{\underset{\underset{\displaystyle CH_2}{\|}}{C}}$$

(n = a large number)

Boron trifluoride is a colorless gas, b.p. $-100°$, and is available commercially in cylinders. The compound has a planar structure. The Lewis structure shows that there are only six electrons around boron.

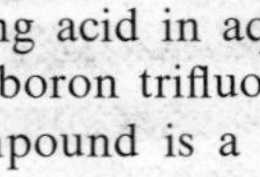

$$:\!\ddot{F}{-}B{-}\ddot{F}\!:$$
$$|$$
$$:\!\ddot{F}\!:$$

The tendency of boron to combine with an electron pair to form an octet is augmented by the electron-attracting character of the attached fluorines. Boron trifluoride is a strong Lewis acid. It reacts avidly with water to form a hydrate, $F_3\bar{B}{-}\overset{+}{O}H_2$, which is itself a strong acid but slowly hydrolyzes in water to form boric acid and HF. In fact, BF_3 has a strong affinity generally for oxygen, nitrogen, and fluorine. With HF it forms fluoboric acid, HBF_4, a strong acid in aqueous solution. With ethyl ether it forms the complex $(C_2H_5)_2O{:}BF_3$, boron trifluoride etherate, which can be formulated as $(C_2H_5)_2\overset{+}{O}{-}\bar{B}F_3$. This compound is a distillable liquid, b.p. $126°$, and is water-white when pure. We will encounter it as a useful acid catalyst.

Boron trifluoride does not react with alkenes in the rigorous absence of moisture. With traces of water, carbocations are produced. With isobutylene, for example, the intermediate salt $(CH_3)_3C^{+-}BF_3OH$ is produced in low concentration. The anion $F_3\bar{B}OH$ has low nucleophilicity, and the t-butyl cation is free to react with isobutylene to start the cationic polymerization.

In some cases the carbocation will abstract a tertiary hydrogen from an alkane. A reaction of this type is used to produce "isooctane" directly from isobutylene and isobutane.

$$(CH_3)_3CH + (CH_3)_2C{=}CH_2 \xrightarrow[-25°C]{HF} (CH_3)_3CCH_2CH(CH_3)_2$$

2,2,4-trimethylpentane
"isooctane"

A reasonable mechanism for this alkylation reaction is

$$(CH_3)_2C{=}CH_2 \xrightleftharpoons{HF} (CH_3)_3C^+ \xrightarrow{(CH_3)_2C{=}CH_2} (CH_3)_3CCH_2C^+(CH_3)_2$$

$$(CH_3)_3CCH_2C^+(CH_3)_2 + (CH_3)_3CH \rightleftharpoons (CH_3)_3CCH_2CH(CH_3)_2 + (CH_3)_3C^+$$

Under these conditions the dimeric carbocation does not react with more isobutylene, but instead abstracts hydrogen from isobutane to provide more *t*-butyl cation to continue the chain of reactions.

$$\left[\begin{array}{l}\text{Anhydrous hydrofluoric acid is a low-boiling (b.p. 19°), colorless liquid with a density} \\ \text{similar to that of water. It reacts with glass and is kept and handled in polyethylene} \\ \text{equipment. The pure liquid is a strong acid and is highly corrosive to tissues.}\end{array}\right]$$

Free radical polymerization may be initiated by the addition of many types of free radicals to an alkene double bond. The **telomerization** of ethylene in Section 11.6.C is an example. In that case the initiating group was $\cdot CX_3$. After the chain grows to four or five monomer units, the growing radical abstracts $X\cdot$ from CX_4 to give the telomer.

If no suitable radical addition reagent is available, the reaction of hydrocarbon radicals with alkenes can become the principal reaction to produce high molecular weight polymer chains. This reaction is an exceedingly important industrial process. Billions of pounds of polyethylene are made annually. The polymerization of ethylene requires high temperature and pressures.

$$n\ CH_2{=}CH_2 \xrightarrow[>100°;\ 15{,}000\ psi]{\text{trace } O_2 \text{ or peroxides}} Y{-}(CH_2CH_2)_n{-}Z$$

where n is a large number on the order of 1000

The end groups Y and Z depend on the initiators used and the termination reactions involved. The principal termination steps for ethylene polymerization are disproportionation and combination, as summarized in the following sequence of steps.

$$Y\cdot + CH_2{=}CH_2 \longrightarrow Y{-}CH_2{-}CH_2\cdot$$

$$YCH_2CH_2\cdot + CH_2{=}CH_2 \longrightarrow YCH_2CH_2CH_2CH_2\cdot \xrightarrow{etc.} Y(CH_2CH_2)_nCH_2CH_2\cdot$$

$$\left.\begin{array}{l}Y(CH_2CH_2)_nCH_2CH_2\cdot \\ \qquad + \\ Y(CH_2CH_2)_mCH_2CH_2\cdot\end{array}\right\}$$

$$\xrightarrow{\text{disproportionation}} Y(CH_2CH_2)_nCH_2CH_3 + Y(CH_2CH_2)_mCH{=}CH_2$$

$$\xrightarrow{\text{combination}} Y(CH_2CH_2)_nCH_2CH_2CH_2CH_2(CH_2CH_2)_mY$$

The product of this so-called "high-temperature polymerization" of ethylene does not have the simple linear structure shown. Ethyl and butyl groups are known to occur along the polymethylene chain, probably because of hydrogen abstraction reactions of the type

$$Y(CH_2CH_2)_n-\underset{\underset{\cdot CH_2}{|}}{\overset{\overset{CH_2}{\diagup}}{CH}}\overset{CH_2}{\underset{CH_2}{\diagdown}} \longrightarrow Y(CH_2CH_2)_n\overset{\overset{CH_2CH_2CH_2CH_3}{|}}{\underset{\cdot}{CH}}$$

chain grows
here

Linear polyethylene is made by an entirely different process described later.

Vinyl chloride, tetrafluoroethylene, and styrene are other important **monomers** used in free radical polymerizations. The Markovnikov addition of radicals to vinyl chloride applies with high specificity so that the product polymer has a complete head-to-tail structure.

$$Y\cdot \; + \; CH_2=CHCl \longrightarrow Y-CH_2\overset{\cdot}{CHCl} \longrightarrow \; \longrightarrow Y(CH_2CHCl)_nZ$$

polyvinyl chloride

> Vinyl chloride is manufactured on an enormous scale, mostly by dehydrochlorination of 1,2-dichloroethane (ethylene dichloride). In 1979 ethylene dichloride ranked seventeenth in chemicals produced in the United States. The only organic compounds produced in greater amount were ethylene, benzene, toluene, and propylene. The 1979 United States production of ethylene dichloride and vinyl chloride was 11.8 billion and 7.5 billion pounds, respectively. In 1974 the Occupational Safety and Health Administration concluded that vinyl chloride is a human carcinogen and set maximum limits to exposure.

Polyvinyl chloride is an extremely hard resin. In order to alter the physical properties of the polymer, low molecular weight liquids called **plasticizers** are added in the polymer formulation. Bis-2-ethylhexyl phthalate is one of the compounds added to polyvinyl chloride as a plasticizer. The resulting polymer has a tough leathery or rubber-like texture. It is used in plastic squeeze bottles, imitation leather upholstery, pipes, and so on.

Polytetrafluoroethylene or "Teflon" is a perfluoro polymer having great resistance to acids and organic solvents. It is used to coat "nonstick" frying pans and other cooking surfaces.

$$X-(CF_2CF_2)_n-Y$$

Teflon

> Two uses of the prefix **per-** are common in chemistry. One use designates a highly oxygenated compound that frequently, but not always, involves an O—O bond. Examples are hydrogen peroxide, peroxyacetic acid, permonosulfuric acid, H_2SO_5 ($HOSO_2OOH$, with O—O bond), and perchloric acid ($HOClO_3$, without an O—O bond). In its other use, per- refers to totally substituted, as in the examples perchloroethylene, $CCl_2=CCl_2$, and perfluoroalkane, C_nF_{2n+2}.

Polystyrene is an inexpensive plastic used to manufacture many familiar household items. It is a hard, colorless, somewhat brittle material.

$$\left(\begin{array}{c} C_6H_5 \\ | \\ CHCH_2 \end{array}\right)_n$$

> In the simple formulation of polystyrene, the end groups have been omitted. This simplification is common in the symbolism of polymer chemistry. The end groups constitute a minute portion of a high molecular weight polymer, although their character has a significant effect on the properties of the polymers.

In anionic polymerization, initiation is accomplished by addition of a nucleophile to a C=C double bond. Simple olefins are inert to most nucleophilic or basic reagents because most common anions are more stable than carbanions.

$$Y^- + CH_2{=}CH_2 \not\longrightarrow Y{-}CH_2CH_2^-$$

Only when the anion itself is an extremely powerful base will addition to the double bond occur. Amide ion (NH_2^-; pK of $NH_3 = 35$) is *not* generally a strong enough base for such a reaction.

t-Butyllithium (an organometallic reagent prepared from *t*-butyl chloride and lithium in ether at $-40°$) does react with ethylene.

$$(CH_3)_3C^- \; Li^+ + CH_2{=}CH_2 \longrightarrow (CH_3)_3CCH_2CH_2^- \; Li^+$$

Since primary carbanions are more stable than tertiary carbanions, the reaction as shown has favorable thermodynamics. This particular reaction is of limited use because *t*-butyllithium is such a highly reactive compound. For example, in ether it must be kept at rather low temperatures. At room temperature it rapidly decomposes ether, in part by an E2 reaction.

$$(CH_3)_3C^- \cdots H \cdots CH_2CH_2{-}OC_2H_5 \longrightarrow (CH_3)_3CH + CH_2{=}CH_2 + C_2H_5O^-$$

Other kinds of organometallic intermediates provide rapid polymerization. A particularly important example is a catalyst prepared from aluminum alkyls (R_3Al) and titanium tetrachloride. This catalyst polymerizes simple olefins by a mechanism that involves the carbanion character of the C—Al bond and the ability of the transition metal, titanium, to coordinate with the π-bonds of alkenes. The result is a rapid polymerization reaction that is used extensively with ethylene and propylene. Linear polyethylene, $(CH_2CH_2)_n$, prepared in this way, is more crystalline than high temperature polyethylene and has a higher density and melting point. The long chains of linear polyethylene can lie together in a regular manner in the solid without the defects in regularity imposed by the random branches of the high temperature polymer.

This example emphasizes how the properties of a polymer or macromolecular compound depend on molecular considerations of structure and interactions between chains. A polymer can have a regular structure characteristic of a crystal and at higher temperatures can melt to a viscous liquid. The liquid is viscous because the long chains form interpenetrating random coils and do not move freely past each other. An intermediate state is that of an amorphous glass, a solid in which the chains or coils are effectively frozen but not in the regular pattern typical of crystals. Polymers are characterized phenomenologically by two important types of temperatures that mark phase transitions, the crystalline melting point and the glass transition temperature (T_g).

PROBLEMS

1. (a) Give the structure and IUPAC name of each of the isomeric pentenes. Which ones are stereoisomers? Which ones are capable of optical activity?
  (b) Answer part (a) for the methylcyclopentenes.

2. Write the structure for each of the following names:
  (a) *trans*-3,4-dimethyl-2-pentene   (b) 4-methyl-3-penten-1-ol

(c) *cis*-3-ethyl-2-hexene (d) vinyl fluoride
(e) 1-bromocyclohexene (f) 3-methylcyclohexene
(g) (Z)-2-bromo-2-pentene (h) (E)-3-methyl-2-hexene
(i) *cis*-3-methylcyclooctene (j) *trans*-3-methylcyclooctene
(k) *trans*-3,4-dimethylcyclobutene (l) 2,3-dimethyl-2-butene

3. Each of the following names is incorrect. Give a correct name.
(a) 2-methylcyclopentene (b) 2-methyl-*cis*-3-pentene
(c) *trans*-1-butene (d) 1-bromoisobutylene
(e) 4-chlorocyclopentene (f) (E)-3-ethyl-3-pentene
(g) *trans*-pent-2-en-4-ol

4. Name each of the following structures. For stereoisomers, use both the *cis-trans* and E-Z nomenclature.

(a) 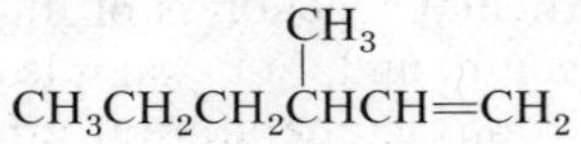 $CH_3CH_2CH_2\overset{\overset{\displaystyle CH_3}{|}}{CH}CH=CH_2$

(b) 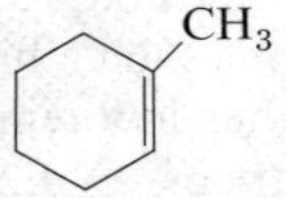 cyclohexene with CH_3

(c) $\underset{CH_3}{\overset{CH_3CH_2CH_2}{C}}=\underset{CH_2CH_3}{\overset{CH_3}{C}}$

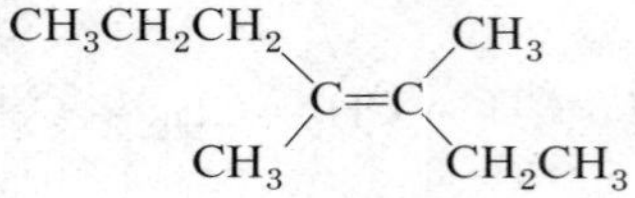

(d) 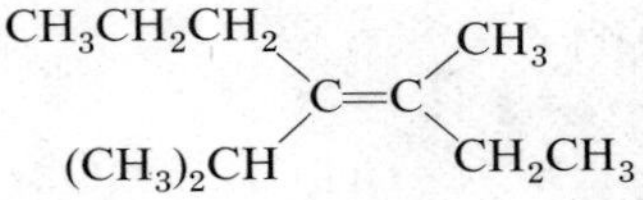 $\underset{(CH_3)_2CH}{\overset{CH_3CH_2CH_2}{C}}=\underset{CH_2CH_3}{\overset{CH_3}{C}}$

(e) 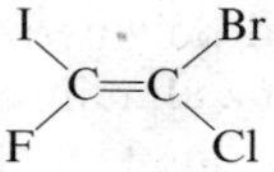 $\underset{F}{\overset{I}{C}}=\underset{Cl}{\overset{Br}{C}}$

(f) $ClCH_2CH_2CH=CH_2$

(g) 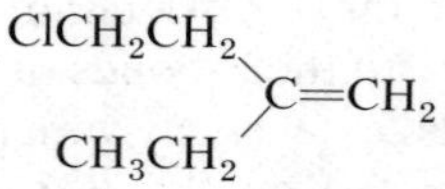 $\underset{CH_3CH_2}{\overset{ClCH_2CH_2}{C}}=CH_2$

(h) 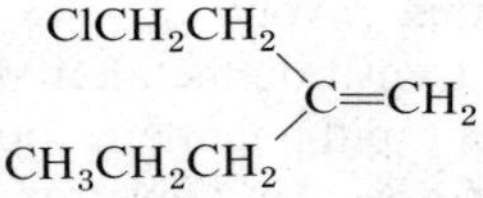 $\underset{CH_3CH_2CH_2}{\overset{ClCH_2CH_2}{C}}=CH_2$

(i) 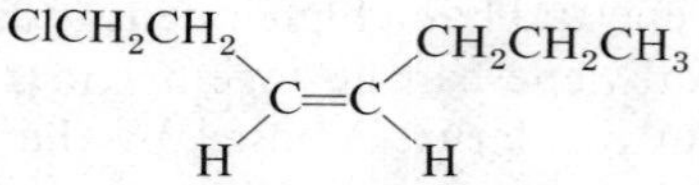 $\underset{H}{\overset{ClCH_2CH_2}{C}}=\underset{H}{\overset{CH_2CH_2CH_3}{C}}$

(j) $HOCH_2CH_2CH_2CH=CH_2$

(k) 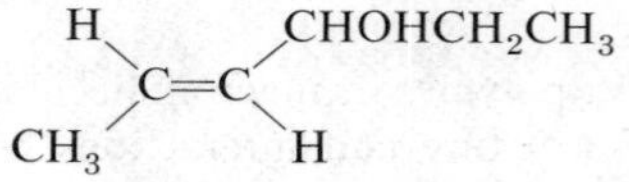 $\underset{CH_3}{\overset{H}{C}}=\underset{H}{\overset{CHOHCH_2CH_3}{C}}$

(l) 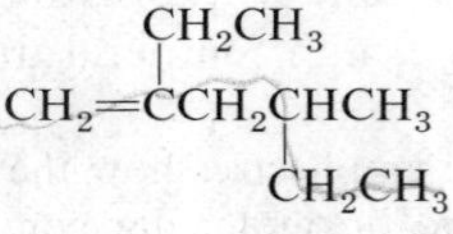 $CH_2=\overset{\overset{\displaystyle CH_2CH_3}{|}}{C}CH_2\overset{\overset{}{}}{C}HCH_3$ with CH_2CH_3

5. Give the structure and name of the principal organic product(s) produced from 3-ethyl-2-pentene under each of the following reaction conditions.
(a) $H_2/Pd-C$ (b) H_2O, Br_2
(c) $Cl_2/0°$ (d) cold dilute $KMnO_4$
(e) (i) B_2H_6; (ii) $NaOH-H_2O_2$ (f) (i) aq. $Hg(ClO_4)_2$; (ii) $NaBH_4$
(g) (i) O_3; (ii) Zn dust, aq. (h) HBr, inhibitor
     CH_3COOH
(i) HBr, peroxides (j) Br_2, dilute solution in CH_3OH
(k) peroxybenzoic acid (l) $CHBr_3$, *t*-BuOK, *t*-BuOH
(m) CH_2I_2, Zn(Cu)

6. Apply each of the reaction conditions in problem 5 to *cis*- and *trans*-3-hexene. For which reactions are the same products obtained from both stereoisomers? For which reactions do the products differ and how do they differ?

7. Apply each of the reaction conditions in problem 5 to cyclohexene. Specify stereochemistry where pertinent.

8. What is the principal organic product of each of the following reaction conditions? Specify stereochemistry where appropriate.

(a) $CH_3CH_2CH_2CHBrCH_3 \xrightarrow{(C_2H_5)_3CO^-}$

(b) 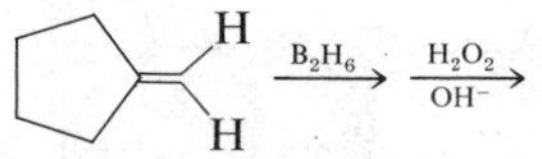 $\xrightarrow[\text{OH}^-]{B_2H_6 \quad H_2O_2}$

(c) 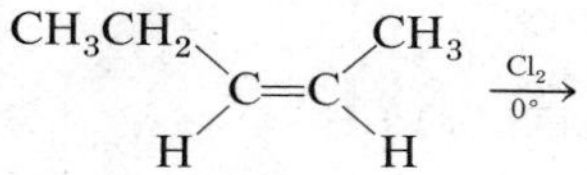 $\xrightarrow[0°]{Cl_2}$

(d) 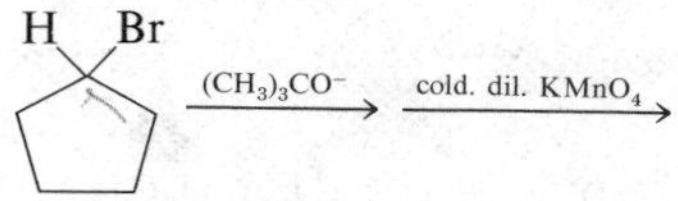 $\xrightarrow{(CH_3)_3CO^-} \xrightarrow{\text{cold. dil. KMnO}_4}$

(e) $CH_3CH_2CHClCD_2CH_3 \xrightarrow[\Delta]{\text{alc. KOH}}$

(f) 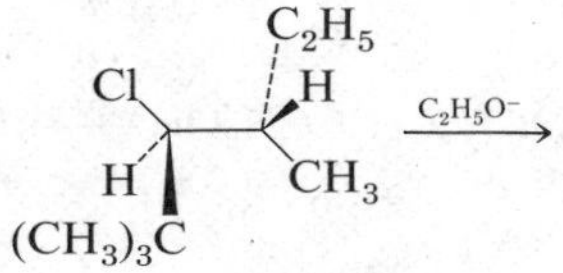 $\xrightarrow{C_2H_5O^-}$

(g) $\xrightarrow[H_2O]{Br_2} \xrightarrow[\Delta]{C_2H_5O^- \\ C_2H_5OH}$

(h) $\xrightarrow[-60°]{O_3 \quad (CH_3)_2S \\ CH_3OH}$

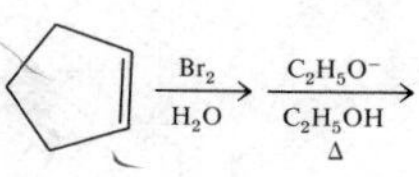

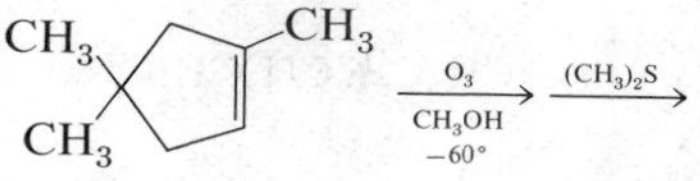

(i) 4-*t*-butylcyclohexene + Br_2/CCl_4

9. Show how one may accomplish each of the following transformations in a practical manner.

(a) $CH_3CHBrCH_3 \longrightarrow CH_3CH_2CH_2Br$

(b) $CH_3CHOHCH_3 \longrightarrow CH_3CH_2CH_2OH$

(c) 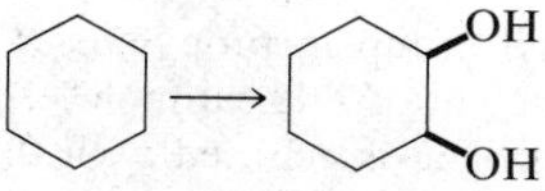
(*cis*)

(d)

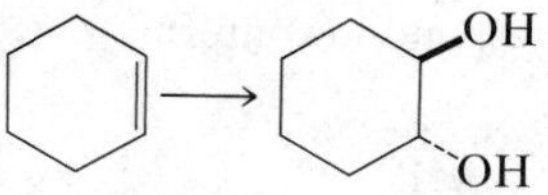

(e) $CH_3CH_2CHBrCH_2Br \longrightarrow CH_3CH_2CHBrCH_2CBr_3$

(f) 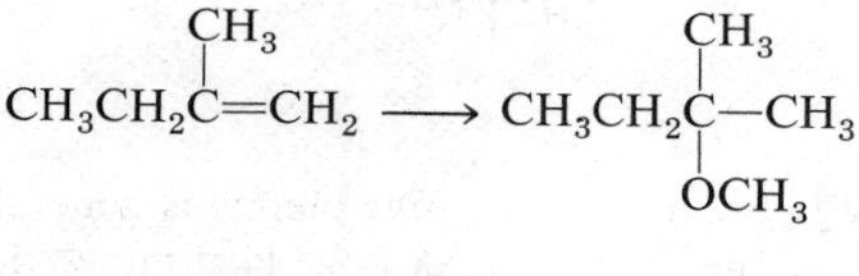 $CH_3CH_2\overset{\underset{\displaystyle CH_3}{|}}{C}=CH_2 \longrightarrow CH_3CH_2\overset{\underset{\displaystyle OCH_3}{|}}{\underset{}{C}}{-}CH_3$ (with CH₃ above)

(g) 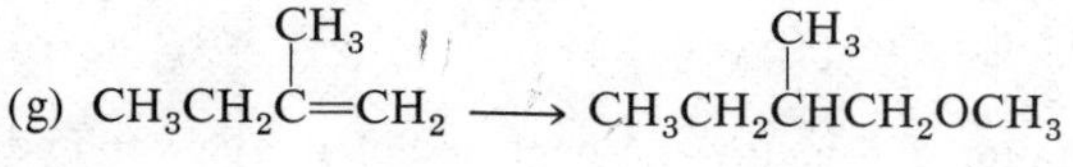 $CH_3CH_2\overset{\underset{\displaystyle CH_3}{|}}{C}=CH_2 \longrightarrow CH_3CH_2\overset{\underset{\displaystyle}{|}}{CH}CH_2OCH_3$ (with CH₃ above)

(h)

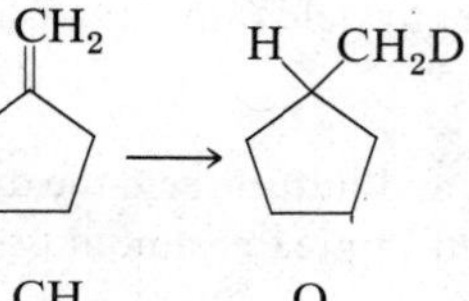

(i)

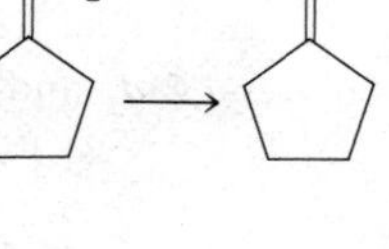

(j)

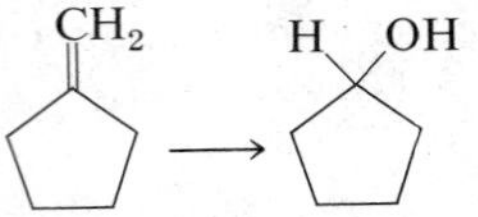

(k)

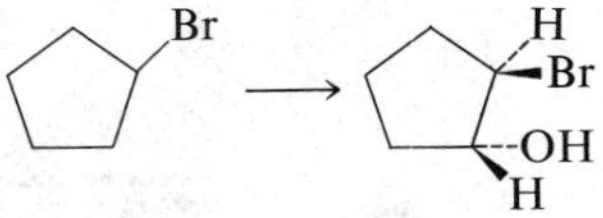

(l) $CH_3CH_2CH_2CH_2CH_2OH \longrightarrow CH_3CH_2CH_2CHBrCH_3$

(m) $CH_3CH_2\overset{\overset{\displaystyle CH_3}{|}}{\underset{\underset{\displaystyle D}{|}}{C}}CH_2OH \longrightarrow CH_3CH_2CH(CH_3)_2$

(n) $CH_3CH_2CH_2CH(CH_3)_2 \longrightarrow CH_3CH_2CH_2\overset{\overset{\displaystyle CH_3}{|}}{C}HCH_2OH$

(o)

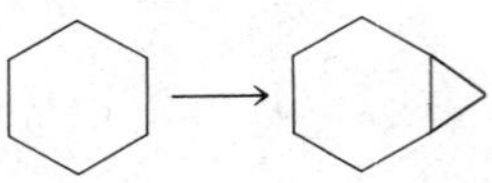

(p) 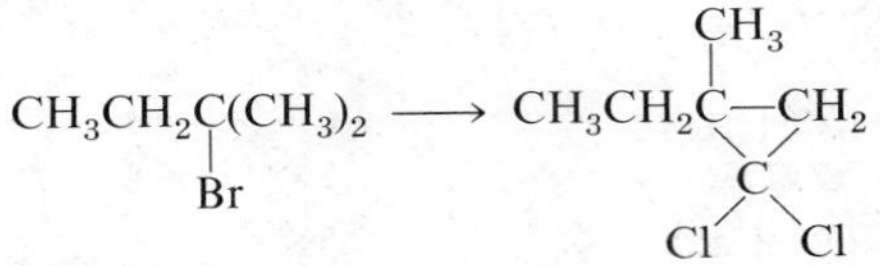 $CH_3CH_2\overset{\overset{\displaystyle}{|}}{\underset{\underset{\displaystyle Br}{|}}{C}}(CH_3)_2 \longrightarrow CH_3CH_2\overset{\overset{\displaystyle CH_3}{|}}{C}-CH_2$

10. The potential function for rotation of the methyl group in propylene is approximately that of a threefold barrier with a barrier height of 2.0 kcal mole^{-1}. The most stable conformation is A in which a methyl hydrogen is eclipsed with the double bond. The least stable conformation is B in which H-2 is eclipsed to a methyl hydrogen. Plot the energy of the system as a function of a 360° rotation of the methyl group. Identify the points along this plot that correspond to conformations A and B.

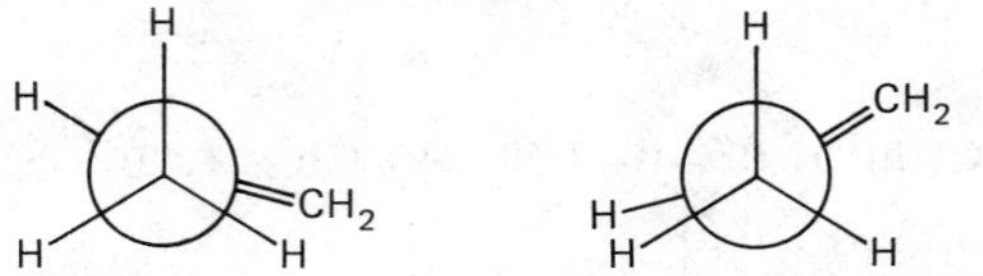

A   B

11. Although the difference in energy between *cis* and *trans* olefins is generally about 1 kcal mole^{-1}, for 4,4-dimethyl-2-pentene the *cis* isomer is 3.8 kcal mole^{-1} less stable than the *trans* isomer. Explain.

12. In the acid-catalyzed dehydration of 6-methyl-1,6-heptanediol, it is easy to find conditions that give smooth loss of one molecule of water to yield 6-methyl-5-hepten-1-ol. Explain.

13. In the formation of diisobutylenes from isobutylene and sulfuric acid the disubstituted olefin isomer, 2,4,4-trimethyl-1-pentene is produced in greater amount than the trisubstituted olefin, 2,4,4-trimethyl-2-pentene. Explain.

14. Compare the product of addition of bromine to ethene-*cis-1,2-d₂* and to ethene-*trans-1,2-d₂*.

$$
\begin{array}{cc}
\underset{H}{\overset{D}{\diagdown}}C=C\underset{H}{\overset{D}{\diagup}} & \underset{H}{\overset{D}{\diagdown}}C=C\underset{D}{\overset{H}{\diagup}} \\
\text{ethene-}cis\text{-}1,2\text{-}d_2 & \text{ethene-}trans\text{-}1,2\text{-}d_2
\end{array}
$$

On treatment with base each of the dibromides gives predominantly a single different dideuteriovinyl bromide. Show the structure in each case. (*Remember:* HBr is eliminated faster than DBr.)

15. The propagation steps for the radical addition of HY to propylene are

(a) $\quad CH_3CH{=}CH_2 + Y\cdot \longrightarrow CH_3\overset{\cdot}{C}HCH_2Y$

(b) $\quad CH_3\overset{\cdot}{C}HCH_2Y + HY \longrightarrow CH_3CH_2CH_2Y + Y\cdot$

$\quad\quad \overline{CH_3CH{=}CH_2 + HY \longrightarrow CH_3CH_2CH_2Y}$

$\Delta H°$ values are given in the table for steps (a) and (b) and for the net reaction with a number of reagents of the type HY in the gas phase. For which reagents is such a radical chain mechanism plausible?

Y:	F	Cl	Br	I	HS	HO	H_2N	CH_3	$(CH_3)_3C$
$\Delta H_a°$	-48	-22	-9	$+5$	-13	-32	-19	-26	-18
$\Delta H_b°$	$+41$	$+8$	-8	-24	-3	$+24$	$+8$	$+9$	-3
$\Delta H_{net}°$	-7	-14	-17	-18	-16	-8	-11	-17	-22

16. When isopropyl bromide is treated with sodium ethoxide in ethanol, propene and ethyl isopropyl ether are formed in a 3:1 ratio. If the hexadeuterioisopropyl bromide, $CD_3CHBrCD_3$, is used, $CD_3CH{=}CD_2$ and $(CD_3)_2CHOC_2H_5$ are formed in a ratio of 1:2. Explain.

17. The heat of hydrogenation, $\Delta H_{hydrog}°$, is defined as the enthalpy of the reaction of an alkene with hydrogen to give the alkane.

$$CH_2{=}CH_2 + H_2 = CH_3CH_3 \quad\quad \Delta H_{hydrog}° = -32.7 \text{ kcal mole}^{-1}$$

From the heats of formation given in Appendix I calculate heats of hydrogenation for a number of simple alkenes. Note that all monoalkyl ethylenes have about the same $\Delta H_{hydrog}°$ which is less (more positive) than that for ethylene. Explain. How would you expect $\Delta H_{hydrog}°$ to compare for isomeric *cis* and *trans* olefins?

18. Reaction of either 1-butene or 2-butene with HCl gives the same product, 2-chlorobutane, via the same carbocation, 2-butyl cation. Yet, the reaction of 1-butene is faster than that of 2-butene. Explain why, using simple energy diagrams. Using this explanation predict which is more reactive, *cis*-2-butene or *trans*-2-butene.

19. Consider a proposed free radical chain addition of HCN to $CH_3CH{=}CH_2$ to give *n*-propyl cyanide, $CH_3CH_2CH_2CN$. Use data in Appendix I and the following $DH°$ values:

$$\begin{array}{ll}
H{-}CN & 120 \text{ kcal mole}^{-1} \\
\underset{\underset{H}{|}}{CH_3CHCH_2CN} & 95 \text{ kcal mole}^{-1}
\end{array}$$

(a) Determine $\Delta H°$ for the net reaction, $CH_3CH{=}CH_2 + HCN = CH_3CH_2CH_2CN$

(b) Write the two chain-propagation steps for the proposed reaction and calculate $\Delta H°$ for each.

(c) Is the proposed reaction feasible? Explain.

20. Treatment of $C_7H_{15}Br$ with strong base gave an alkene mixture that was shown by careful gas chromatographic analysis and separation to consist of three alkenes, C_7H_{14}, C, D, and E. Catalytic hydrogenation of each alkene gave 2-methylhexane. Reaction of C with B_2H_6 followed by H_2O_2 and OH^- gave mostly an alcohol, F. Similar reaction of D or E gave approximately equal amounts of F and an isomeric alcohol G. What structural assignments can be made for C through G on the basis of these observations? What structural element is left undetermined by these data alone?

21. Propose a mechanism for each of the following reactions.

(a) 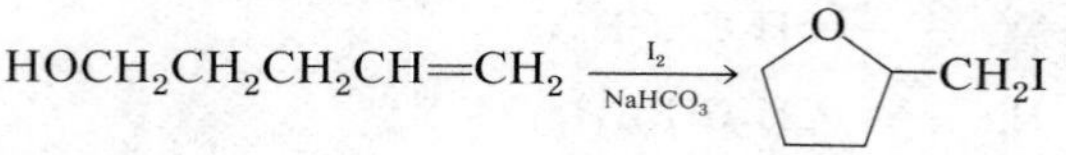 $HOCH_2CH_2CH_2CH=CH_2 \xrightarrow[NaHCO_3]{I_2}$ $-CH_2I$

(b) 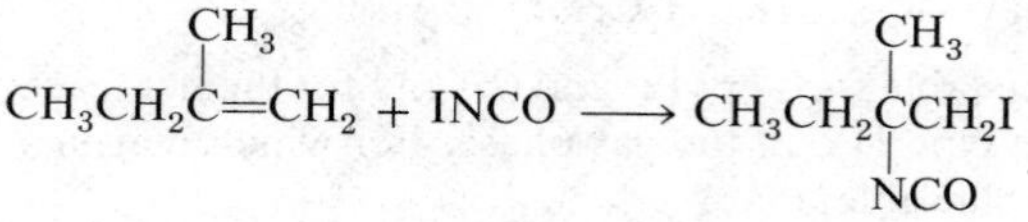
$$CH_3CH_2\overset{\underset{\displaystyle CH_3}{|}}{C}=CH_2 + INCO \longrightarrow CH_3CH_2\overset{\underset{\displaystyle NCO}{\overset{\displaystyle CH_3}{|}}}{C}CH_2I$$

(c) $CH_2-CH-CD_2Cl + CH_3O^- \xrightarrow[\Delta]{CH_3OH} CH_3OCH_2-CH-CD_2$

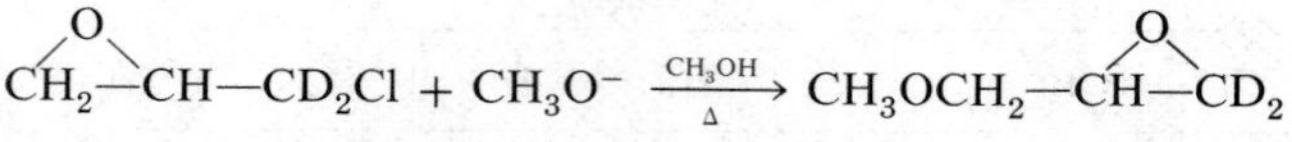

22. 3,3,5-Trimethyl-1-methylenecyclohexane reacts with *m*-chloroperoxybenzoic acid to give mainly the diastereomeric epoxide shown. However, when the same alkene is reacted first with Br_2 in water and then with aqueous base, the other diastereomer is the major product.

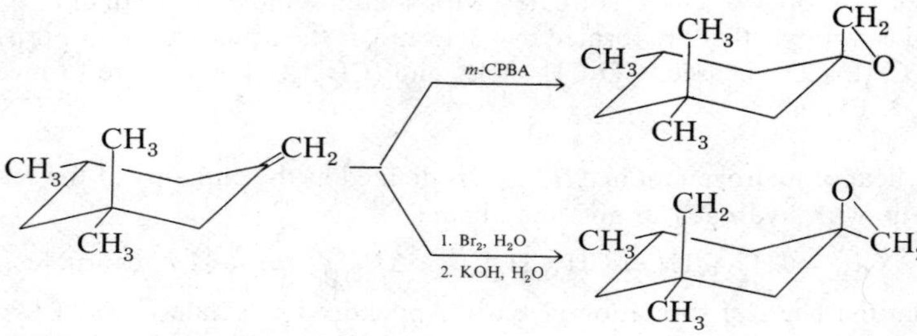

Explain.

23. Compound H, $C_{11}H_{24}O$, reacts with PBr_3 in ether at $0°$ to give I, $C_{11}H_{23}Br$. Treatment of I with potassium ethoxide in ethanol gives a mixture of J (major) and K (minor), both $C_{11}H_{22}$. Each of these compounds was treated first with ozone in $CHCl_3$ at $0°$, and then with $NaBH_4$. In each case a mixture of 3-methyl-1-butanol and 4-methyl-1-pentanol was produced. What are compounds H through K?

24. 3-Bromo-2,3-dimethylpentane was treated with KOH in refluxing ethanol. The resulting alkene mixture was separated by preparative gas chromatography into four fractions. The principal product was shown by nmr to be 2,3-dimethyl-2-pentene (no vinyl hydrogens). A second product was found by nmr to be 2-ethyl-3-methyl-1-butene (two vinyl hydrogens). The other two products have the following cmr chemical shifts.

compound L: 12.4, 17.8, 20.6, 28.5, 117.9, 141.0
compound M: 13.0, 13.1, 21.6, 37.4, 116.2, 141.5

What are the structures of compounds L and M?

25. Three compounds, all having the formula C_6H_{10}, are found to have the following cmr spectra. Propose structures for the three compounds.

compound N: 12.9, 124.9, 125.3
compound O: 17.6, 125.8, 132.3
compound P: 13.0, 18.0, 123.1, 127.4, 128.3, 130.2

26. 1-Chloro-3-methylcyclopentane was treated with potassium *t*-butoxide in *t*-butyl alcohol. Two isomeric alkenes were obtained. The cmr spectrum of the major product
had only four resonances, while the cmr spectrum of the minor product had six.
What are the structures of the major and minor products?

Chapter 12
Alkynes

12.1
Electronic Structure

Acetylene is known experimentally to have a linear structure. Its C≡C bond distance of 1.20 Å is the shortest C—C bond distance known. The C—H bond distance of 1.06 Å is shorter than that in ethylene (1.08 Å) or in ethane (1.10 Å) (Figure 12.1). These structural details are readily interpreted by an extension of the σ-π electronic structure of double bonds. In acetylene the σ-framework consists of C_{sp}-hybrid orbitals as indicated in Figure 12.2.

Recall (Section 11.1) that sp^2—s σ-bonds are shorter than are sp^3—s σ-bonds. The trend also holds for the sp—s bonds in acetylene. The effect of the amount of s-character in the C—H bond distance is shown graphically in Figure 12.3. Superimposed on the σ-electrons are two orthogonal π-electron systems as shown in Figure 12.4.

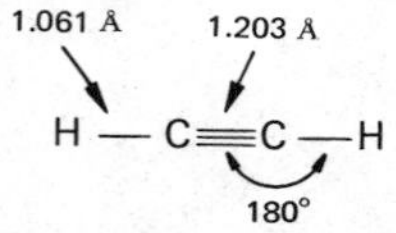

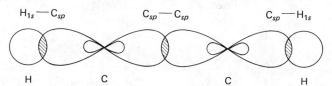

FIGURE 12.1. *Structure of acetylene.*

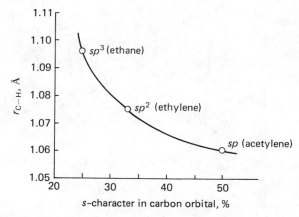

FIGURE 12.2. *σ-electronic framework of acetylene.*

FIGURE 12.3. *Relationship between C—H bond distance and the approximate amount of* s-*character in carbon orbital.*

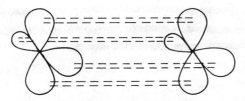

FIGURE 12.4. *π-systems of acetylene.*

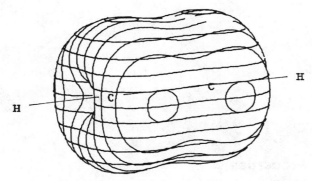

FIGURE 12.5. *π-electron density in the acetylene triple bond.*

The symbolic representations in Figure 12.4 are actually misleading because the electrons in two orthogonal *p*-orbitals form a cylindrically symmetrical torus or doughnut-like electron density distribution. A perspective view of the total π-electron density is seen in Figure 12.5.

12.2
Nomenclature

The simple alkynes are readily named in the common system as derivatives of acetylene.

$$CH_3C{\equiv}CH \qquad C_6H_5C{\equiv}CC_6H_5 \qquad F_3CC{\equiv}CH$$

methylacetylene   diphenylacetylene   trifluoromethylacetylene

In the IUPAC system the compounds are named as alkynes in which the final **-ane** of the parent alkane is replaced by the suffix **-yne.** The position of the triple bond is indicated by a number when necessary.

$$CH_3C{\equiv}CH \qquad (CH_3)_2CHC{\equiv}CH \qquad CH_3CH_2CH_2CHC{\equiv}CCH_3$$
$$CH_2CH_2CH_3$$

propyne   3-methyl-1-butyne   4-propyl-2-heptyne

If both -yne and -ol endings are used, the -ol is last and determines the numbering sequence.

$$HC{\equiv}CCH_2CH_2OH \qquad HOCH_2C{\equiv}CCH_2OH$$

3-butyn-1-ol   2-butyn-1,4-diol
(not but-4-ol-1-yne)

When both a double and triple bond are present, the hydrocarbon is named an **alkenyne** with numbers as low as possible given to the multiple bonds. In case of a choice, the double bond gets the lower number.

$$CH_3CH{=}CHC{\equiv}CH \qquad HC{\equiv}CCH_2CH{=}CH_2$$

3-penten-1-yne
(not 2-penten-4-yne)

1-penten-4-yne
(not 4-penten-1-yne)

In complex structures the alkynyl group is used as a modifying prefix.

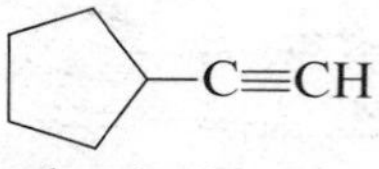

⬠—C≡CH

ethynylcyclopentane

EXERCISE: Name all ten isomeric pentynols (do not forget stereoisomerism).

12.3
Physical Properties

The physical properties of alkynes are similar to those of the corresponding alkenes. The lower members are gases with boiling points somewhat higher than the corresponding alkenes. Terminal alkynes have lower boiling points than isomeric internal alkynes (Table 12.1) and can be separated by careful fractional distillation.

A. Dipole Moments

The $CH_3{-}C{\equiv}$ bond in propyne is formed by overlap of a C_{sp^3}-hybrid orbital from the methyl carbon with a C_{sp}-hybrid from the acetylenic carbon. The bond is $C_{sp^3}{-}C_{sp}$. Since one orbital has more s-character than the other and is thereby more electronegative, the electron density in the resulting bond is not symmetri-

TABLE 12.1
Physical Properties of Alkynes

Compound	Boiling Point, °C	Melting Point, °C	d^{20}
ethyne (acetylene)	−84.0[a]	−81.5[b]	
propyne	−23.2	−102.7	
1-butyne	8.1	−122.5	
2-butyne	27	−32.3	
1-pentyne	39.3	−90.0	
2-pentyne	55.5	−101	
1-hexyne	71	−132	0.7152
2-hexyne	84	−88	0.7317
3-hexyne	81	−105	0.7231
phenylacetylene	143	−43	
diphenylacetylene	300	63.5	

[a] Sublimation temperature. [b] Under pressure.

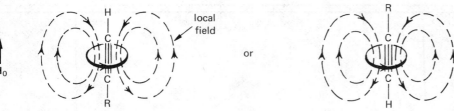

FIGURE 12.6. *Shielding of acetylenic protons by a triple bond in parallel orientation to the applied field.*

cal. The unsymmetrical electron distribution results in a dipole moment, larger than that observed for an alkene, but still relatively small.

$$CH_3CH_2C{\equiv}CH \qquad CH_3CH_2CH{=}CH_2 \qquad CH_3C{\equiv}CCH_3$$
$$\mu = 0.80 \text{ D} \qquad\qquad \mu = 0.30 \text{ D} \qquad\qquad \mu = 0 \text{ D}$$

Symmetrically disubstituted acetylenes, of course, have no net dipole moment.

B. *Nuclear Magnetic Resonance*

Protons attached directly to a triply bonded carbon resonate at $\delta = 2\text{–}3$ ppm. This is a much higher field resonance than that observed for vinyl protons. In fact, the resonance position for alkyne protons is only slightly downfield from the resonance position of alkane protons. The observed position is due to a deshielding effect of the "electronegative" triple bond superimposed on another effect due to magnetic anisotropy (see page 285) of the triple bond itself. Recall that a triple bond has a cylindrically symmetric sheath of π-electrons. As shown in Figure 12.6, the electrons in this torus can circulate in a magnetic field. This electronic motion induces a small local field (dashed lines in Figure 12.6). At the acetylenic proton the induced field opposes the applied field. Thus a higher applied field is required to bring this proton into resonance. The result of the induced field in this case is an effective shielding of the alkyne proton.

Actually, the diamagnetic shielding of acetylenic protons is a result of two factors. When the molecule is aligned perpendicular to $\mathbf{H}_0$, the acetylenic proton is deshielded, just as in the case of alkene hydrogens (Figure 12.7). This deshielding component is smaller than the shielding component diagrammed in Figure 12.6. When averaged over all possible orientations, the effect is a net diamagnetic shielding.

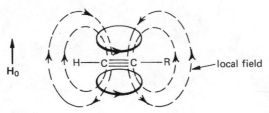

FIGURE 12.7. *Shielding of acetylenic protons for a triple bond perpendicular to the applied field.*

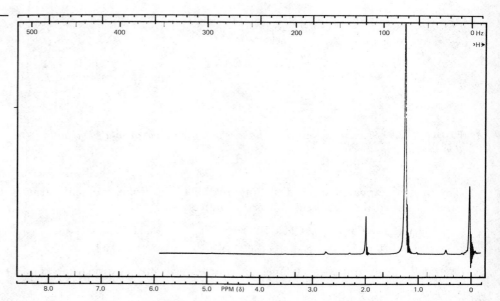

FIGURE 12.8. *Nmr spectrum of 3,3-dimethyl-1-butyne.*

This diamagnetic shielding diminishes the normal inductive effect. As a result of these two opposing effects, the resonance position of acetylenic protons is only slightly downfield from the resonance position of alkane protons. Similar effects operate on hydrogens bound to carbons adjacent to triple bonds, causing them to resonate about 1 ppm downfield from the corresponding alkane position.

The nmr spectra of 3,3-dimethyl-1-butyne and 1-hexyne are shown in Figures 12.8 and 12.9. Note that in the spectrum of 1-hexyne, the alkyne proton ($\delta = 1.7$ ppm) appears as a triplet with $J = 2.5$ Hz. This small splitting is the result of **long-range coupling** through the triple bond.

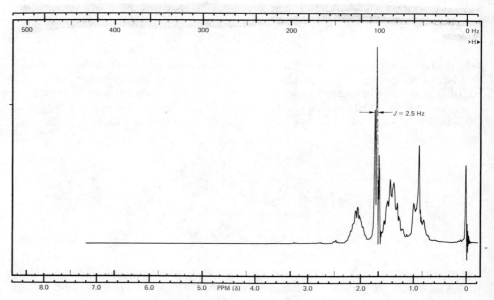

FIGURE 12.9. *Nmr spectrum of 1-hexyne.*

12.4
Acidity of Alkynes

The hydrogens in terminal alkynes are relatively acidic. Acetylene itself has a pK_a of about 25. It is a far weaker acid than water (pK_a 15.7) or the alcohols (pK_a 16–19), but it is much more acidic than ammonia (pK_a 35). Amide ion in liquid ammonia converts acetylene and other terminal alkynes into the corresponding carbanions.

$$RC{\equiv}CH + NH_2^- \rightleftharpoons RC{\equiv}C^- + NH_3$$

This reaction does not occur with alkenes or alkanes. Ethylene has a pK_a of about 44 and methane has a pK_a of about 50.

$$CH_2{=}CH_2 + NH_2^- \rightleftharpoons CH_2{=}CH^- + NH_3$$

$$CH_4 + NH_2^- \rightleftharpoons CH_3^- + NH_3$$

From the above pK_as we see that there is a vast difference in the stability of the carbanions $RC{\equiv}C^-$, $CH_2{=}CH^-$, and CH_3^-. This difference is explainable in terms of the character of the orbital occupied by the lone-pair electrons in the three anions. Methyl anion has a pyramidal structure with the lone-pair electrons in an orbital that is approximately sp^3 ($\frac{1}{4}s$ and $\frac{3}{4}p$). In vinyl anion the lone-pair electrons are in an sp^2-orbital ($\frac{1}{3}s$ and $\frac{2}{3}p$). In acetylide ion the lone pair is in an sp-orbital ($\frac{1}{2}s$ and $\frac{1}{2}p$).

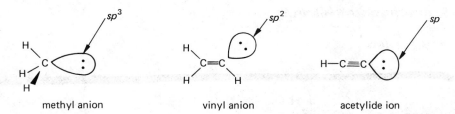

| methyl anion | vinyl anion | acetylide ion |

Electrons in s-orbitals are held, on the average, closer to the nucleus than they are in p-orbitals. This increased electrostatic attraction means that s-electrons have lower energy and greater stability than p-electrons. In general, the greater the amount of s-orbital in a hybrid orbital containing a pair of electrons, the less basic is that pair of electrons. Lower basicity corresponds to higher acidity of the conjugate acid.

| base strength ↑ | $CH_3{:}^-$ $CH_2{=}CH{:}^-$ $HC{\equiv}C{:}^-$ | CH_4 $CH_2{=}CH_2$ $HC{\equiv}CH$ | acid strength ↓ |

Alkynes are quantitatively deprotonated by alkyllithium compounds.

$$CH_3(CH_2)_3C{\equiv}CH + n\text{-}C_4H_9Li \longrightarrow CH_3(CH_2)_3C{\equiv}CLi + n\text{-}C_4H_{10}$$

The foregoing transformation is simply an acid-base reaction, with 1-hexyne being the acid and n-butyllithium being the base. Since the alkyne is a much stronger acid than the alkane (by over 20 pK units!), equilibrium lies essentially completely to the right. Alkyllithium compounds are available from the reaction of alkyl halides with lithium in a hydrocarbon solvent such as pentane or hexane.

$$n\text{-}C_4H_9Cl + 2 Li \xrightarrow{\text{hexane}} n\text{-}C_4H_9Li + LiCl\downarrow$$

A number of the simple alkyllithium compounds are commercially available. We shall study the reactions of alkyllithium compounds and other organometallic reagents in detail in Chapter 15. Some special uses of the alkynyllithium compounds will be considered in Section 12.5.C.

Terminal alkynes give insoluble salts with a number of heavy metal cations such as Ag^+ and Cu^+. The formation of the salts serves as a useful chemical diagnosis for the $RC{\equiv}CH$ function, but many of these salts are explosively sensitive when dry and should always be kept moist. The alkyne can be regenerated from the salt, and the overall process serves as a method for purifying terminal alkynes.

> Impure 1-hexyne is dissolved in 95% ethanol and aqueous silver nitrate is added. The white precipitate is filtered and washed with alcohol. On refluxing with sodium cyanide solution, the alkyne is regenerated and distilled. The cyanide converts silver cation to a stable complex.
>
> $$RC{\equiv}CAg + H_2O + 2CN^- \longrightarrow RC{\equiv}CH + OH^- + Ag(CN)_2{}^-$$

EXERCISE: Since 1-alkynes have substantially lower boiling points than internal alkynes, mixtures can be separated by careful fractional distillation. In practice, however, complete separation of such a mixture is difficult. Suggest a simple way in which the last traces of 1-pentyne might be removed from a sample of 2-pentyne.

12.5
Preparation of Alkynes

A. Acetylene

Acetylene itself is formed from the reaction of calcium carbide with water.

$$CaC_2 + H_2O \longrightarrow Ca(OH)_2 + HC{\equiv}CH$$

> Calcium carbide is a high-melting (m.p. 2300°) gray solid prepared by heating lime and coke in an electric furnace.
>
> $$CaO + 3C \longrightarrow CaC_2 + CO$$

This method was once an important industrial process for the manufacture of acetylene. However, the method has now been replaced by a process in which methane is pyrolyzed in a flow system with short contact time.

$$2 CH_4 \xrightarrow[\text{0.01-0.1 sec}]{\text{1500°C}} HC{\equiv}CH + 3 H_2$$

This reaction is endothermic at ordinary temperatures, but is thermodynamically favored at high temperatures.

At room temperature acetylene is thermodynamically unstable with respect to its elements, as shown by its large *positive* heat of formation ($\Delta H_f^\circ = +54.3$ kcal mole^{-1} at 25°).

$$C_2H_2 = 2\,C(s) + H_2 \qquad \Delta H^\circ = -54.3 \text{ kcal mole}^{-1}$$

This instability causes certain problems in the handling and storage of the material. When under pressure or in the presence of copper, it can convert to carbon and hydrogen with explosive violence. Although acetylene gas can be condensed readily (b.p. −84°), the liquid is similarly unstable. Since the gas is extremely soluble in acetone, commercial cylinders of acetylene contain pieces of pumice which are saturated with acetone. When the cylinder is filled, the acetylene mostly dissolves, giving a relatively stable solution. Acetylene is also appreciably soluble in water. A saturated aqueous solution at 25° and 1 atm pressure has a concentration of 0.05 M (0.13 g C_2H_2 per 100 mL).

B. *Elimination Reactions*

In principle, a triple bond can be introduced into a molecule by elimination of two molecules of HX from either a **geminal** (L., *geminus,* twin) or a **vicinal** (L., *vicinus,* near) dihalide.

$$-CBr_2-CH_2- \xrightarrow{-2\ HBr} -C{\equiv}C-$$
a *gem*-dibromide

$$-CHBr-CHBr- \xrightarrow{-2\ HBr} -C{\equiv}C-$$
a *vic*-dibromide

The dehydrohalogenation proceeds in stages, with the second molecule of HX being removed with greater difficulty than the first.

$$\left.\begin{array}{c} -CX_2CH_2- \\ \text{or} \\ -CHXCHX- \end{array}\right\} \xrightarrow{\text{faster}} -CX{=}CH- \xrightarrow{\text{slower}} -C{\equiv}C-$$

Typical reaction conditions for formation of alkynes involve the use of molten KOH, solid KOH moistened with alcohol, or concentrated alcoholic KOH solutions at temperatures of 100–200°. In practice, these conditions are so drastic that the method is only useful for the preparation of certain kinds of alkynes. Under these highly basic conditions the triple bond can **migrate** along a chain.

$$CH_3CH_2C{\equiv}CH \underset{\Delta}{\overset{\text{alc. KOH}}{\rightleftharpoons}} CH_3C{\equiv}CCH_3$$

Disubstituted alkynes are thermodynamically more stable than terminal alkynes (because of the preference for *s*-character in C—C bonds; Section 11.4). Consequently these conditions may be used only where such rearrangement is not possible.

$$C_6H_5CH{=}CHC_6H_5 \xrightarrow[\text{ether}]{Br_2} C_6H_5\overset{\overset{\displaystyle Br}{|}}{C}H-\overset{\overset{\displaystyle Br}{|}}{C}HC_6H_5 \xrightarrow[\Delta]{\text{alc. KOH}} C_6H_5C{\equiv}CC_6H_5$$

stilbene   (77–81%)   (66–69%)

Sodium amide is an effective strong base that is particularly appropriate for the preparation of 1-alkynes.

Sodium amide, $NaNH_2$, is a white solid prepared by the reaction of sodium with liquid ammonia. Sodium dissolves in liquid ammonia to give a blue solution of "solvated electrons."

$$Na \xrightarrow{NH_3} Na^+ + e^-(NH_3)$$

Such solutions are useful reducing agents. (One example is the reduction of alkynes in Section 12.6.A.) In the presence of small amounts of ferric ion, a reaction takes place with the liberation of hydrogen.

$$e^- + NH_3 \xrightarrow{Fe^{3+}} NH_2^- + \tfrac{1}{2} H_2$$

The net reaction is

$$2\,Na + 2\,NH_3 \xrightarrow{Fe^{3+}} 2\,NaNH_2 + H_2$$

In liquid ammonia, sodium amide is a strong base just as sodium hydroxide is in water.

Since NH_3 is much less acidic than water, sodium amide reacts quantitatively with water. Solutions of $NaNH_2$ in NH_3 readily absorb moisture from the atmosphere.

$$NH_2^- + H_2O \longrightarrow NH_3 + OH^-$$

In the organic laboratory, sodium amide is generally used as a solid suspension in some inert medium such as benzene or mineral oil or as a solution in liquid ammonia.

To dehydrohalogenate a dihalide, a suspension of sodium amide in mineral oil is heated to 150–165°. The dihalide is added slowly and a vigorous reaction ensues. Ammonia is evolved, and the sodium salt of the alkyne is formed. After cooling, the hydrocarbon is liberated by the addition of water.

$$\left.\begin{array}{l} RCX_2CH_3 \\ RCHXCH_2X \\ RCH_2CHX_2 \end{array}\right\} + 2\,NaNH_2 \longrightarrow RC\equiv CH \xrightarrow{NaNH_2} RC\equiv C^-Na^+ + NH_3$$

$$RC\equiv C^-Na^+ + H_2O \longrightarrow RC\equiv CH + Na^+OH^-$$

Since the reaction product is the salt of an alkyne, this method is useful for preparing terminal alkynes even when migration of the triple bond is possible.

$$n\text{-}C_{14}H_{29}CHBrCH_2Br \xrightarrow{NaNH_2} n\text{-}C_{14}H_{29}C\equiv CH$$
$$(65\%)$$

In fact, internal alkynes may be conveniently isomerized to terminal alkynes by the use of sodium amide at 150°.

$$n\text{-}C_5H_{11}C\equiv CCH_3 \xrightarrow[150°C]{NaNH_2} n\text{-}C_6H_{13}C\equiv C^-Na^+ + NH_3 \xrightarrow{H^+} n\text{-}C_6H_{13}C\equiv CH$$

EXERCISE: Suggest a way in which 2-pentene can be converted into (a) 2-pentyne and (b) 1-pentyne.

C. Displacement Reactions

Acetylide anions are highly nucleophilic and participate readily in S_N2 displacement reactions.

$$RC\equiv C:^- + R'X \longrightarrow RC\equiv CR' + X^-$$

This method is a useful general method for the preparation of certain types of alkynes. It is actually one of the few good methods we have encountered thus far for **lengthening a carbon chain.** The reaction may be carried out in liquid ammonia solution or in a polar aprotic solvent such as HMPT (hexamethylphosphoric triamide, Section 8.9.B). The acetylide anion is formed with sodium amide or with n-butyllithium.

Liquid ammonia is available commercially in cylinders. Although the compound boils at $-33°$, it has a relatively high heat of vaporization, due to extensive hydrogen bonding in the liquid. Because of this high heat of vaporization, boiling is a relatively slow process at room temperature. When using liquid ammonia, the material is kept in a normal reaction flask which is equipped with a type of trap or condenser containing dry ice ($-78°$). The liquid ammonia in the flask refluxes gently and condenses on the dry ice condenser.

The terminal alkyne is added to a solution of sodium amide in ammonia. After it has been converted into its salt, the alkyl halide is added. The mixture is stirred for a few hours, and water is then added. The hydrocarbon is separated from the aqueous ammonia layer and purified.

There is a fair amount of variety possible using this method. Acetylene itself may be alkylated either once to make a terminal alkyne or twice to make an internal alkyne.

$$HC\equiv CH + NaNH_2 \xrightarrow[-33°]{\text{liq. } NH_3} HC\equiv C^-Na^+ \xrightarrow{n\text{-}C_4H_9Br} CH_3(CH_2)_3C\equiv CH$$

(89%)

1-hexyne

$$HC\equiv CH \xrightarrow[\substack{\text{liq. } NH_3 \\ -33°}]{2\ NaNH_2} \xrightarrow{2\ n\text{-}C_3H_7Br} CH_3CH_2CH_2C\equiv CCH_2CH_2CH_3$$

(60–66%)

4-octyne

$$CH_3(CH_2)_3C\equiv CH + n\text{-}C_4H_9Li \xrightarrow[25°]{HMPT}$$

$$C_4H_{10}\uparrow + CH_3(CH_2)_3C\equiv C^-Li^+ \xrightarrow{CH_3(CH_2)_4Cl}$$

$$CH_3CH_2CH_2CH_2C\equiv CCH_2CH_2CH_2CH_2CH_3$$

(90%)

5-undecyne

Since acetylide ions are highly basic, they are also effective in E2 elimination reactions. For this reason the displacement reaction is only a good method for the synthesis of acetylenes when applied to primary halides that do not have branches close to the reaction center.

$$CH_3(CH_2)_3C\equiv C^-Li^+ + CH_3\overset{\overset{\displaystyle Br}{|}}{C}HCH_3 \xrightarrow[25°]{HMPT}$$

$$\begin{cases} CH_3CH_2CH_2CH_2C\equiv CCH(CH_3)_2 \\ \qquad\qquad (6\%) \\ \qquad\qquad + \\ \{CH_3CH_2CH_2CH_2C\equiv CH + CH_2=CHCH_3\} \\ \qquad\qquad (85\%) \end{cases}$$

$$CH_3(CH_2)_3C\equiv C^-Li^+ + CH_3\overset{\overset{\displaystyle CH_3}{|}}{C}HCH_2Br \xrightarrow[25°]{HMPT}$$

$$\begin{cases} CH_3CH_2CH_2CH_2C\equiv CCH_2CH(CH_3)_2 \\ \qquad\qquad (32\%) \\ \qquad\qquad + \\ \{CH_3CH_2CH_2CH_2C\equiv CH + (CH_3)_2C=CH_2\} \\ \qquad\qquad (68\%) \end{cases}$$

EXERCISE: Using *n*-butyl bromide as the only source of carbon, show how you might synthesize 3-octyne. Four steps are required (although the last two steps can be combined into a "one-pot" process).

12.6
Reactions of Alkynes

Many of the reactions of alkynes involve the triple bond in a manner analogous to comparable reactions of alkenes. However, just as a double bond is weaker than two single bonds, a triple bond is weaker still than three single bonds. This comparison is apparent in the average bond energies tabulated in Table 12.2. As a result the triple bond enters into some reactions not generally seen with alkenes.

TABLE 12.2
Average Bond Energies of
C—C Bonds

Bond	Average Bond Energy, kcal mole^{-1}
C—C	83
C=C	146
C≡C	200

A. Reduction

Hydrogenation of an alkyne to an alkane occurs readily with the same general catalysts used for the reduction of alkenes.

$$R-C\equiv C-R' + 2\ H_2 \xrightarrow{\text{Pt, Pd, Ni}} R-CH_2CH_2-R'$$

The first step in the reduction is a more exothermic reaction than is the second.

$$HC\equiv CH + H_2 \longrightarrow CH_2=CH_2 \qquad \Delta H° = -41.9\ \text{kcal mole}^{-1}$$

$$CH_2=CH_2 + H_2 \longrightarrow CH_3CH_3 \qquad \Delta H° = -32.7\ \text{kcal mole}^{-1}$$

The second reaction is so facile that, with most catalysts, it is not possible to stop the reduction at the alkene stage. However, with palladium or nickel, alkynes undergo hydrogenation extremely readily—faster than any other functional group. By taking advantage of this catalytic effect, one may accomplish the **partial hydrogenation** of an alkyne to an alkene. In practice, specially deactivated or **poisoned** catalysts are usually used. The most effective catalyst for this purpose is palladium metal which has been deposited in a finely divided state on solid BaSO$_4$ and then treated with quinoline (the actual poison). This catalyst is known as **Lindlar's catalyst.**

The function of the poison is to moderate the catalyst's activity to a point where triple bonds are still reduced at a reasonable rate but double bonds react only slowly. One can then readily stop the reduction after absorption of 1 mole of hydrogen and isolate the alkene in excellent yield.

Quinoline is a heterocyclic amine and is discussed in Chapter 32.

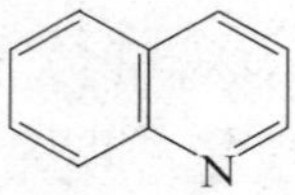

quinoline

It is isolated commercially from coal tar, but the commercial material contains trace amounts of sulfur compounds that are difficult to remove. Divalent sulfur compounds are such exceedingly powerful catalyst poisons that they completely inhibit the catalytic activity. For this reason, only pure synthetic quinoline may be used for this purpose.

An important aspect of these hydrogenations is the fact that hydrogen is delivered from the catalyst to the triple bond in such a manner as to generate a *cis* alkene.

$$C_2H_5C{\equiv}CC_2H_5 + H_2 \xrightarrow[\text{quinoline}]{Pd/BaSO_4}$$

An alternative synthetic method for the preparation of *cis* alkenes involves hydroboration (Section 12.6E).

Reduction of triple bonds can also be accomplished by treating the alkyne with sodium in liquid ammonia at $-33°$. This reduction produces exclusively the *trans* alkene.

$$C_4H_9C{\equiv}CC_4H_9 \xrightarrow[\text{liq. NH}_3]{Na} \xrightarrow{NH_4OH}$$

(80–90%)
trans-5-decene

The mechanism of this reaction involves the reduction of the triple bond by two electrons from sodium atoms. The first electron goes into an antibonding π-orbital to give a **radical anion.** This strongly basic species is protonated by ammonia to give a **vinyl radical,** which is reduced by another electron to give a **vinyl anion.** Final protonation of the vinyl anion by ammonia (acting as an acid) yields the *trans* alkene and amide ion.

Step 1. $R{-}C{\equiv}C{-}R + Na \longrightarrow Na^+ + \left[R{-}\ddot{C}{=}\dot{C}{-}R \right]^-$
radical anion

Step 2. $\left[R{-}\ddot{C}{=}\dot{C}{-}R \right]^- + NH_3 \longrightarrow NH_2^- +$
vinyl radical

Step 3. $+ Na \longrightarrow Na^+ +$
vinyl anion

Step 4. $+ NH_3 \longrightarrow NH_2^- +$

The stereochemistry of the final product is probably established in the reduction of the vinyl radical (step 3). The two vinyl radicals with the R groups *trans* or *cis* interconvert rapidly, but the *trans* form is preferred because of nonbonded interactions in the *cis* form. Since reduction of the two vinyl radicals probably proceeds at comparable rates and the *trans* form is present in much greater amount, the vinyl anion formed is mostly *trans*. The vinyl anion interconverts between *cis* and *trans* forms only relatively slowly and appears to protonate before it has a chance to isomerize.

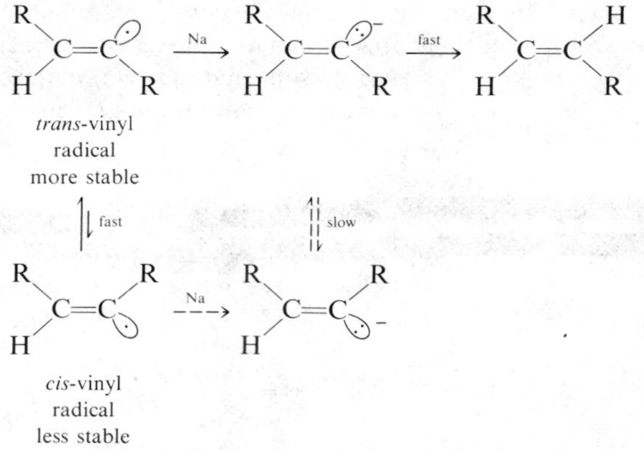

R—C=C (trans-vinyl radical more stable) → fast → R—C=C—R (H, R, H)

trans-vinyl
radical
more stable

fast ↓   slow

R—C=C (cis-vinyl radical less stable)

cis-vinyl
radical
less stable

Note that we have used the term **vinyl** in two different senses. It refers to the common name for the specific organic function, —CH=CH$_2$ (for example, vinyl chloride, CH$_2$=CHCl) but it is also used generically to refer to substitution at a carbon that is part of an alkene double bond (for example, CH$_3$CCl=CH$_2$, a vinyl or vinylic chloride).

Simple alkenes are not reduced by sodium in liquid ammonia, so it is easy to perform the selective reduction of an alkyne to an alkene by this method. It is important not to confuse a solution of Na in liquid NH$_3$ (which is actually a solution containing Na$^+$ ions and solvated electrons, e$^-$) with a solution of NaNH$_2$ in liquid NH$_3$ (which is a solution containing Na$^+$ ions and NH$_2^-$ ions). The former solution *reduces* alkynes. The latter solution does not reduce alkynes, but does deprotonate terminal alkynes.

By means of these several reactions it is possible to construct larger chains from smaller ones and to prepare either *cis* or *trans* alkenes with little contamination from the other. For example,

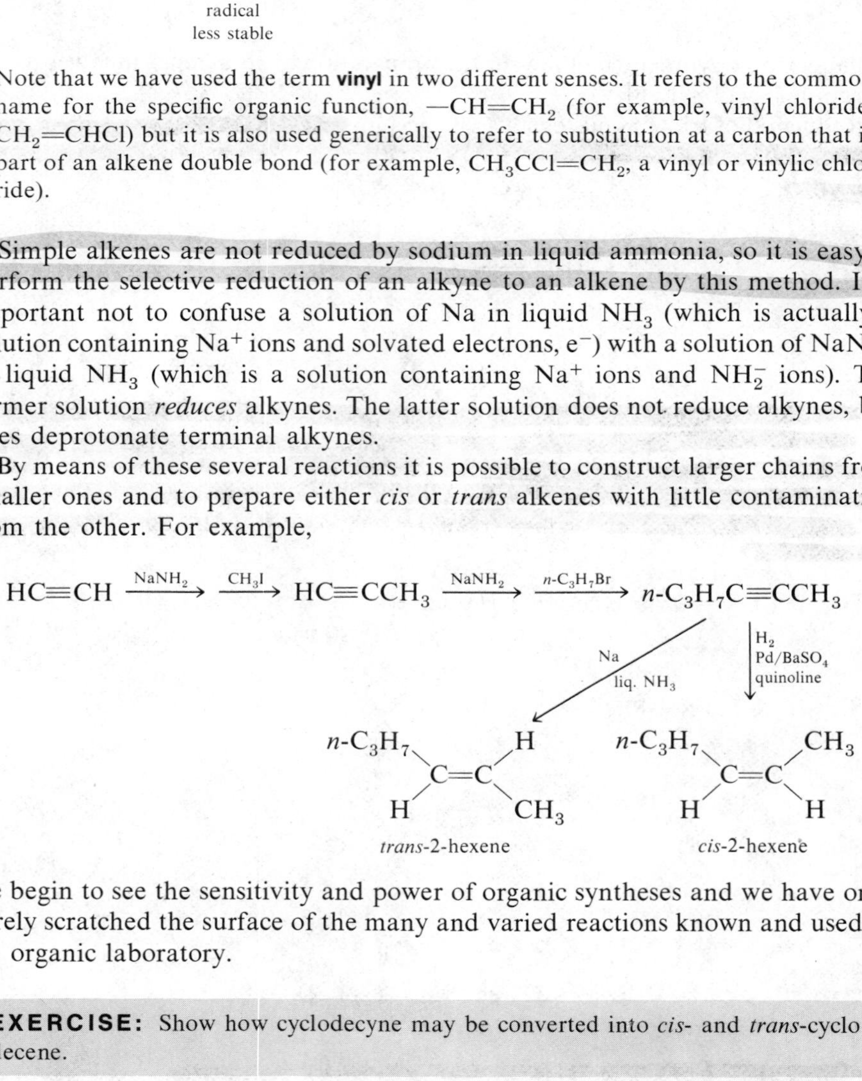

$$HC\equiv CH \xrightarrow{NaNH_2} \xrightarrow{CH_3I} HC\equiv CCH_3 \xrightarrow{NaNH_2} \xrightarrow{n\text{-}C_3H_7Br} n\text{-}C_3H_7C\equiv CCH_3$$

Na
liq. NH$_3$

H$_2$
Pd/BaSO$_4$
quinoline

n-C$_3$H$_7$, H
C=C
H, CH$_3$

trans-2-hexene

n-C$_3$H$_7$, CH$_3$
C=C
H, H

cis-2-hexene

We begin to see the sensitivity and power of organic syntheses and we have only barely scratched the surface of the many and varied reactions known and used in the organic laboratory.

EXERCISE: Show how cyclodecyne may be converted into *cis*- and *trans*-cyclodecene.

B. *Electrophilic Additions*

The triple bond reacts with HX and X$_2$ in much the same manner as does the double bond. The reaction goes in stages, and Markovnikov's rule is followed.

$$RC\equiv CH \xrightarrow{HX} RCX=CH_2 \xrightarrow{HX} RCX_2CH_3$$

$$RC\equiv CR' \xrightarrow{X_2} RCX=CXR' \xrightarrow{X_2} RCX_2CX_2R'$$

Some specific examples of such reactions are

$$CH_3(CH_2)_3C\equiv CH + HBr \xrightarrow[\substack{FeBr_3 \\ 15°}]{inhibitor} CH_3(CH_3)_3CBr=CH_2$$
$$(40\%)$$
$$\text{2-bromo-1-hexene}$$

$$CH_3C\equiv CH + Cl_2 \xrightarrow{60\text{-}70°} \underset{\substack{(20\%) \\ \text{(E)-1,2-dichloro-} \\ \text{1-propene}}}{\overset{CH_3\diagdown \qquad \diagup Cl}{\underset{Cl\diagup \qquad \diagdown H}{C=C}}} + \underset{\substack{(63\%) \\ \text{1,1,2,2-tetra-} \\ \text{chloropropane}}}{CH_3CCl_2CHCl_2}$$

Although addition across a triple bond is a more exothermic process than comparable addition across a double bond, alkynes are generally less reactive than
alkenes toward electrophilic reagents. This apparent anomaly is rationalized by
comparison of the intermediate carbocations produced from alkynes and alkenes.

$$RC\equiv CH + Y^+ \longrightarrow R\overset{+}{C}=CHY$$
$$\text{a vinyl cation}$$

$$RCH=CH_2 + Y^+ \longrightarrow R\overset{+}{C}HCH_2Y$$
$$\text{an alkyl cation}$$

The carbocation produced from the alkyne is a vinyl cation, $R\overset{+}{C}=CHY$, whose
electronic structure is shown in Figure 12.10. This type of carbocation is substantially less stable than an ordinary alkyl cation such as $R\overset{+}{C}HCH_2Y$, since the vacant
p-orbital belongs to an sp-hybridized carbon rather than to an sp^2-hybridized
carbon. Since a carbon that has sp-hybridization is more electronegative than one
having sp^2-hybridization, it is less tolerant of the positive charge. We shall return
to the subject of vinyl cations in Section 12.7.

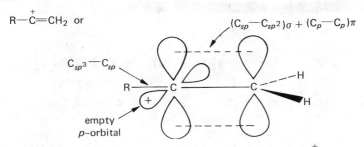

FIGURE 12.10. *Electronic structure of a vinyl cation,* $R\overset{+}{C}=CH_2$.

Once formed, the vinyl cation reacts with whatever nucleophiles are present.
For example, the overall reaction with HCl involves initial formation of the vinyl
cation, followed by its reaction with chloride ion.

$$RC\equiv CH + H^+ \longrightarrow R\overset{+}{C}=CH_2$$

$$R\overset{+}{C}=CH_2 + Cl^- \longrightarrow R-\overset{\overset{\textstyle Cl}{|}}{C}=CH_2$$

The initially formed vinyl halide also undergoes electrophilic addition. In this case the product is a *gem*-dihaloalkane.

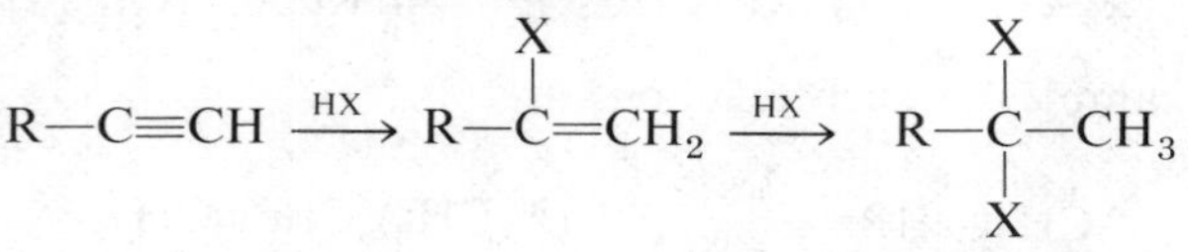

$$R-C{\equiv}CH \xrightarrow{HX} R-\overset{X}{\underset{}{C}}=CH_2 \xrightarrow{HX} R-\overset{X}{\underset{X}{C}}-CH_3$$

a *gem*-dihaloalkane

The addition can normally be stopped at the intermediate alkenyl halide stage.

The addition of halogen to a triple bond can also be stopped after the addition of 1 mole equivalent. The dihaloalkene produced generally has the *trans* structure.

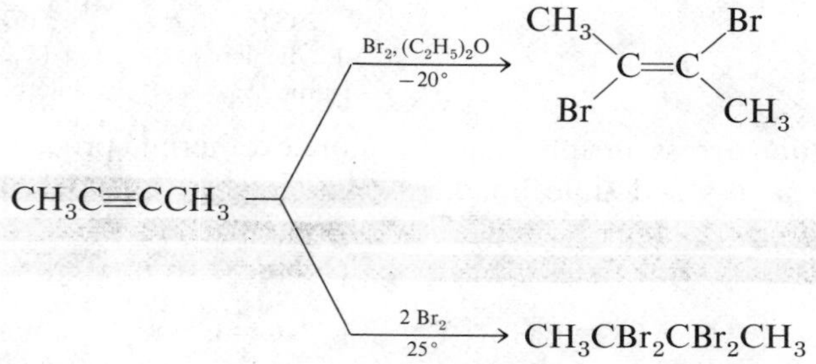

$$CH_3C{\equiv}CCH_3 \xrightarrow[]{}$$

$$\xrightarrow[-20°]{Br_2, (C_2H_5)_2O} \underset{Br}{\overset{CH_3}{C}}=\underset{CH_3}{\overset{Br}{C}}$$

$$\xrightarrow[25°]{2\ Br_2} CH_3CBr_2CBr_2CH_3$$

In reactions with aqueous sulfuric acid the intermediate carbocation reacts with water to produce an intermediate vinyl alcohol, $RC(OH){=}CH_2$. This reaction is poor with sulfuric acid alone, but is catalyzed by mercuric salts. Vinyl alcohols are unstable and rearrange immediately under the reaction conditions to give either an aldehyde or a ketone (Section 13.6).

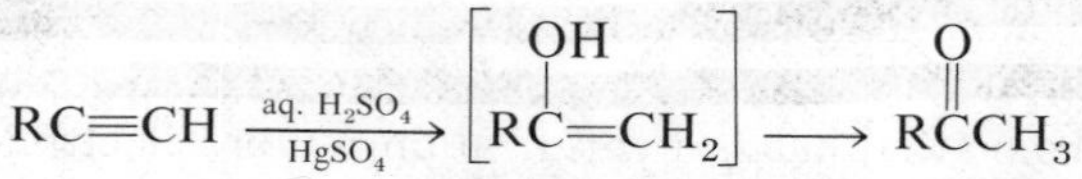

$$RC{\equiv}CH \xrightarrow[HgSO_4]{aq.\ H_2SO_4} \left[\overset{OH}{\underset{}{RC}}=CH_2 \right] \longrightarrow \overset{O}{\overset{\|}{RCCH_3}}$$

a vinyl alcohol   a ketone

Since the addition follows Markovnikov's rule, acetylene itself is the only alkyne that undergoes hydration to give an aldehyde.

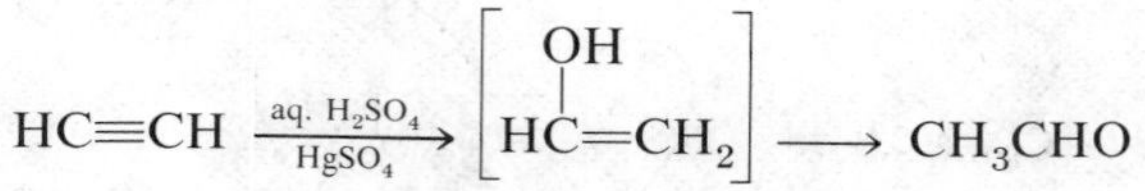

$$HC{\equiv}CH \xrightarrow[HgSO_4]{aq.\ H_2SO_4} \left[\overset{OH}{\underset{}{HC}}=CH_2 \right] \longrightarrow CH_3CHO$$

acetaldehyde

The reaction is a useful method for synthesizing a few special kinds of ketones. We shall discuss it further in that context in Section 13.5.C.

EXERCISE: Give the principal reaction product(s) for the reaction of 1-butyne, 2-butyne, and 2-pentyne with (a) HCl (1 eq.), (b) HCl (2 eq.), and (c) dil. H_2SO_4/Hg^{2+}.

C. Free Radical Additions

Radicals and atoms add to triple bonds just as they do to double bonds. Again, anti-Markovnikov orientation is observed.

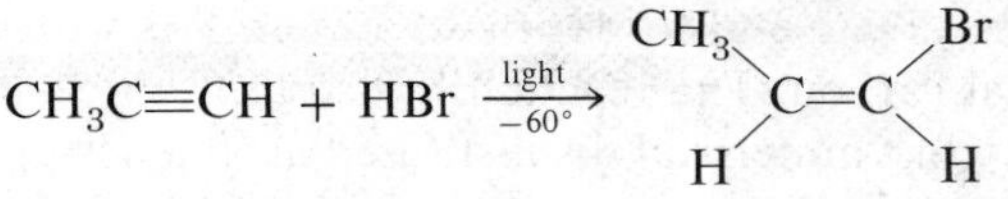

$$CH_3C{\equiv}CH + HBr \xrightarrow[-60°]{light}$$

CH$_3$ Br
   C=C
H   H

(88%)

(Z)-1-bromopropene

The reaction is a radical chain process involving the following propagation steps.

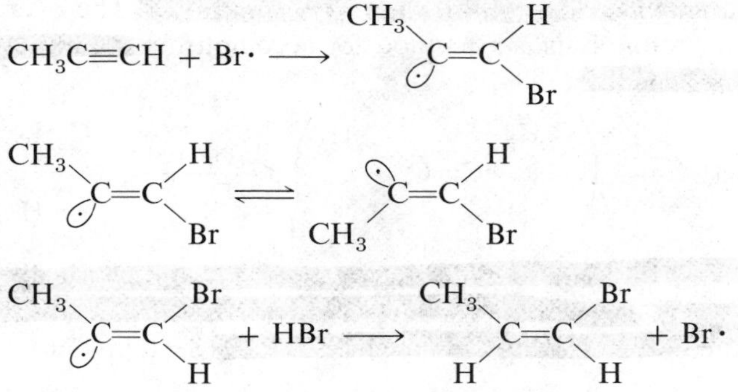

$$CH_3C{\equiv}CH + Br\cdot \longrightarrow$$

CH$_3$ H
   C=C
       Br

CH$_3$ H   H
   C=C   ⇌   C=C
       Br   CH$_3$ Br

CH$_3$ Br   CH$_3$ Br
   C=C   + HBr ⟶   C=C   + Br·
       H   H   H

As shown in this example, such reactions frequently give net *anti* addition, but there are many exceptions. In cases where *anti* addition is observed, it appears that the intermediate vinyl radical reacts with HBr from its more accessible side.

D. *Nucleophilic Additions*

Unlike simple alkenes, alkynes undergo nucleophilic addition reactions. For example, acetylene reacts with alkoxides in alcoholic solution to yield vinyl ethers. The reaction usually requires conditions of high temperature and pressure.

$$HC{\equiv}CH + RO^- \xrightarrow[\text{pressure}]{\underset{150°}{ROH}} ROCH{=}CH^- \xrightarrow{ROH} ROCH{=}CH_2 + RO^-$$

The reaction of a stable alkoxide ion to produce a less stable and more basic vinyl anion may seem surprising. However, the addition also involves the formation of a strong C—O bond at the expense of the relatively weak "third bond" of a triple bond. The net effect of stronger bonding is more than enough to compensate for the creation of a stronger base. The intermediate vinyl anion is immediately protonated by the alcohol solvent to regenerate the alkoxide ion.

E. *Hydroboration*

Hydroboration of alkynes is a useful laboratory process for the synthesis of several types of compounds. Diborane reacts with alkynes at 0° to produce the intermediate trivinylborane.

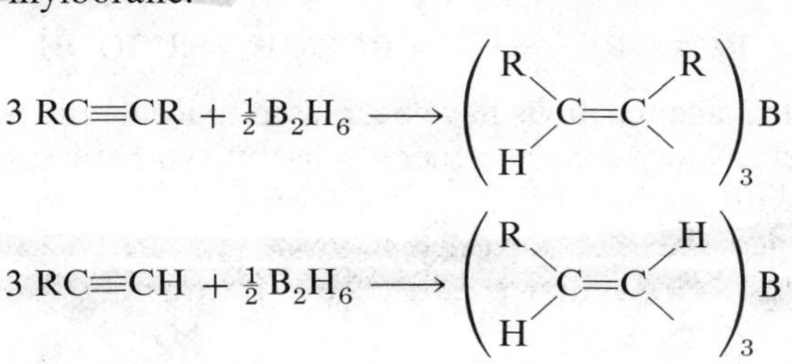

$$3\ RC{\equiv}CR + \tfrac{1}{2}B_2H_6 \longrightarrow \left(\begin{matrix} R & R \\ & C=C \\ H & \end{matrix} \right)_3 B$$

$$3\ RC{\equiv}CH + \tfrac{1}{2}B_2H_6 \longrightarrow \left(\begin{matrix} R & H \\ & C=C \\ H & \end{matrix} \right)_3 B$$

The reaction is generally useful for terminal alkynes. As with alkenes, the boron adds to the terminal carbon. The reaction is also useful with symmetrical disubstituted acetylenes. Unsymmetrical disubstituted acetylenes generally give a mixture of products. The net reaction is *syn* addition of H—BR$_2$ to the triple bond.

The resultant vinylboranes, like alkylboranes (Section 11.6.D), enter into several useful reactions. They undergo protonolysis to give the resulting alkene when treated with acetic acid. Protonolysis involves replacement of boron by the hydrogen of the carboxylic acid with *retention of configuration*. The overall process of hydroboration-protonolysis is a method for accomplishing *syn* hydrogenation of an alkyne to an alkene.

$$3\ C_2H_5C{\equiv}CC_2H_5 + \tfrac{1}{2} B_2H_6 \longrightarrow \left(\begin{array}{c} C_2H_5 \\ \diagdown \\ H \end{array} C{=}C \begin{array}{c} C_2H_5 \\ \diagup \\ \diagdown \end{array} \right)_3 B \xrightarrow{3\ CH_3CO_2H} 3 \begin{array}{c} C_2H_5 \\ \diagdown \\ H \end{array} C{=}C \begin{array}{c} C_2H_5 \\ \diagup \\ \diagdown H \end{array}$$

The C—B bond of a vinylborane can also be cleaved oxidatively with alkaline hydrogen peroxide. The initial product is a vinyl alcohol, which rearranges quantitatively to the corresponding aldehyde or ketone (Section 13.6).

$$n\text{-}C_4H_9C{\equiv}CH \xrightarrow{B_2H_6} \xrightarrow[\text{OH}^-]{H_2O_2} \left[\begin{array}{c} n\text{-}C_4H_9 \\ \diagdown \\ H \end{array} C{=}C \begin{array}{c} H \\ \diagup \\ \diagdown OH \end{array} \right] \longrightarrow n\text{-}C_4H_9CH_2\overset{\overset{\textstyle O}{\|}}{C}H$$

The overall effect of hydroboration-oxidation is that of hydration of the triple bond. Note that with terminal alkynes the *aldehyde* is formed (anti-Markovnikov hydration), whereas with direct H$_2$SO$_4$–HgSO$_4$ hydration the *ketone* is produced.

> **EXERCISE:** Write the principal organic reaction product of reaction of B$_2$H$_6$ followed by alkaline H$_2$O$_2$ with (a) 1-butyne, (b) 2-butyne, and (c) 2-pentyne.

F. *Oxidation*

Disubstituted acetylenes undergo oxidation by permanganate to yield 1,2-diketones and/or cleavage products. In a few cases, the reaction is a useful synthetic method for preparing 1,2-diketones.

$$CH_3(CH_2)_7C{\equiv}C(CH_2)_7COOH \xrightarrow[\substack{H_2O \\ pH\ 7.5}]{KMnO_4} CH_3(CH_2)_7\overset{\overset{\textstyle O}{\|}}{C}{-}\overset{\overset{\textstyle O}{\|}}{C}(CH_2)_7COOH$$
$$(92\text{--}96\%)$$

More vigorous conditions generally lead to cleavage in which mixtures of carboxylic acids result.

$$RC{\equiv}CR' \xrightarrow{\text{aq. } KMnO_4} RCOOH + R'COOH$$

Although such cleavage reactions have been used to locate the position of a triple bond in a molecule, they are not generally useful synthetic procedures because yields are frequently poor.

Terminal alkynes undergo an **oxidative coupling** reaction, often known as the Eglinton reaction. The reaction is carried out by treating the 1-alkyne with a cuprous salt in the presence of air or oxygen in a basic solvent such as pyridine. The coupled diyne is formed in high yield.

$$n\text{-}C_4H_9C\equiv CH \xrightarrow[\substack{\text{pyridine} \\ 60°}]{\text{CuCl}} n\text{-}C_4H_9C\equiv C-C\equiv CC_4H_9$$

(90%)
5,7-dodecadiyne

The oxidative coupling reaction has recently been used to prepare macrocyclic dimers, trimers, and so on, of diacetylenes.

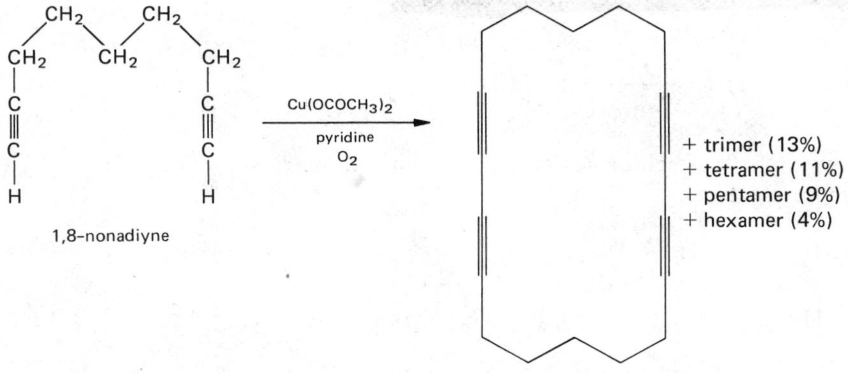

1,8–nonadiyne

$$\xrightarrow[\substack{\text{pyridine} \\ O_2}]{\text{Cu(OCOCH}_3)_2}$$

1,3,10,12–cyclooctadecatetrayne
(10%)

+ trimer (13%)
+ tetramer (11%)
+ pentamer (9%)
+ hexamer (4%)

For the preparation of unsymmetrical diynes, a process known as Cadiot–Chodkeivicz coupling may be employed. In this reaction a mixture of bromo- or chloroalkyne and an alkyne is treated with a cuprous salt. The haloalkyne may be prepared by reaction of the appropriate alkynyllithium with a halogen.

$$CH_2=CHC\equiv CH \xrightarrow[\substack{\text{ether} \\ -80°}]{n\text{-}C_4H_9Li} CH_2=CHC\equiv CLi \xrightarrow[-20°]{Br_2} CH_2=CHC\equiv C-Br$$

vinylacetylene
(but-3-en-1-yne)

1-bromobut-3-en-1-yne

$$CH_2=CHC\equiv C-Br + HC\equiv CCH_2CH_3 \xrightarrow[\substack{CH_3OH \\ CH_3NH_2 \\ 25°}]{\text{CuCl}} CH_2=CHC\equiv C-C\equiv CCH_2CH_3$$

(90%)
oct-1-en-3,5-diyne

A related reaction is the Nieuwland enyne synthesis. This dimerization reaction is usually carried out by treating the alkyne with a mixture of cuprous chloride, ammonium chloride, and HCl. The reaction is an important industrial process for the synthesis of vinylacetylene.

$$2\ HC\equiv CH \xrightarrow[\text{HCl}]{\text{CuCl, NH}_4\text{Cl}} HC\equiv C-CH=CH_2$$

vinylacetylene

In contrast to the Eglinton and Cadiot–Chodkeivicz reactions, no oxidation is involved in the Nieuwland synthesis. Rather, the C—H bond of one acetylene *adds* across the C≡C bond of another molecule.

EXERCISE: How may octa-1,7-dien-3,5-diyne be prepared, starting with acetylene?

12.7
Vinyl Halides

Alkenyl halides may be prepared by addition of 1 mole of hydrogen halide to an alkyne, often with the aid of a mild Lewis-acid catalyst.

$$CH_3C{\equiv}CH + HCl \xrightarrow{CuCl} CH_3CCl{=}CH_2$$

Alcoholic dehydrohalogenation of 1 mole of HX from a vicinal dihalide generally gives a mixture of possible haloalkenes.

$$CH_3CHClCH_2Cl \xrightarrow[\text{alcohol}]{KOH} CH_3CCl{=}CH_2 + \underset{H}{\overset{CH_3}{C}}{=}\underset{H}{\overset{Cl}{C}} + \underset{H}{\overset{CH_3}{C}}{=}\underset{Cl}{\overset{H}{C}}$$

However, the *vic*-dichlorides obtained from symmetrical olefins give good yields of single products.

$$\underset{H}{\overset{C_2H_5}{C}}{=}\underset{C_2H_5}{\overset{H}{C}} \xrightarrow{Cl_2} \underset{C_2H_5}{\overset{Cl}{\underset{H}{C}}}{-}\underset{Cl}{\overset{H\,C_2H_5}{C}} \xrightarrow[\substack{t\text{-BuOH} \\ 80°}]{t\text{-BuOK}} \underset{H}{\overset{C_2H_5}{C}}{=}\underset{Cl}{\overset{C_2H_5}{C}}$$

(E)-3-chloro-3-hexene

Haloalkenes in which the halogen is attached directly on the double bond have exceptionally low reactivity in S_N1 and S_N2 reactions. For example, 1-chloropropene is inert to potassium iodide in acetone under conditions where *n*-propyl chloride undergoes rapid S_N2 substitution.

$$CH_3CH_2CH_2Cl + I^- \longrightarrow CH_3CH_2CH_2I + Cl^- \qquad \text{fairly fast}$$

$$CH_3CH{=}CHCl + I^- \longrightarrow \text{no reaction}$$

Similarly, simple alkenyl halides do not give carbocations—S_N1 reactions for such halides are so slow that other kinds of reactions occur instead, such as addition to the multiple bond. The relative difficulty of ionizing a vinylic C—Cl bond is shown by the gas phase enthalpies.

	$\Delta H°$, *kcal mole*$^{-1}$
$CH_3Cl \longrightarrow CH_3^+ + Cl^-$	228
$CH_3CH_2Cl \longrightarrow CH_3CH_2^+ + Cl^-$	191
$CH_2{=}CHCl \longrightarrow CH_2{=}CH^+ + Cl^-$	223
$(CH_3)_2CHCl \longrightarrow (CH_3)_2CH^+ + Cl^-$	167

The difference between the energy required to form a vinyl cation and that needed for a simple primary carbocation is comparable to the difference between primary and secondary carbocations. Recall that secondary carbocations are common intermediates in many reactions but that simple primary carbocations are virtually unknown in solution. Primary vinyl cations are similarly unknown in S_N1 reactions; however, secondary vinyl cations of the type $RC{=}CH_2$ have been detected under special conditions. They are not important in most organic reactions of the simple vinyl halides.

This lack of reactivity is explained most simply as an increased difficulty in removing an atom with its pair of electrons from a bond to a vinyl orbital with its higher *s*-character than from a simple primary *sp*³-orbital. The increased strength of the vinyl-halogen bond compared to the ethyl-halogen bond is manifest also in the relative bond lengths and bond-dissociation energies as shown in the following examples.

	r_{C-X}, Å	$DH°$, kcal mole⁻¹
CH_3CH_2—Cl	1.78	81
CH_2=CH—Cl	1.72	88
CH_3CH_2—Br	1.94	68
CH_2=CH—Br	1.89	76

The *sp*²-carbon orbital involved in the vinyl halide bond is expected to produce a shorter and stronger bond than the ethyl *sp*³-orbital. However, an additional component leading to a still shorter and stronger bond is π-overlap between the π-orbital of the double bond and a lone-pair orbital of the halogen, as depicted in Figure 12.11.

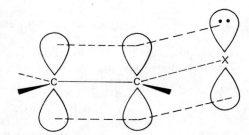

FIGURE 12.11. *π-orbital overlap in a vinyl halide.*

Such π-overlap can also be represented by resonance involving Lewis structures.

$$\left[CH_2{=}CH{-}\overset{..}{\underset{..}{X}}{:} \longleftrightarrow {}^{-}CH_2{-}CH{=}\overset{+}{\underset{..}{X}}{:} \right]$$

As a result of such overlap the C—X bond in a vinyl halide has *partial double bond character*. The amount of double bond character (that is, the contribution of resonance structure ⁻CH_2—CH=X⁺) is relatively small, but has a significant effect on reactivity.

EXERCISE: Using molecular models, construct a model of (3R,4S)-3,4-dichlorohexane. Convince yourself that if elimination of HCl occurs in an *anti* fashion, the product will be (E)-3-chloro-3-hexene, regardless of which chlorine is lost. Suggest a way in which 3-hexyne may be converted into (Z)-3-chloro-3-hexene (three steps are necessary).

PROBLEMS

1. Write out the structure corresponding to each of the following names.
   (a) methylisopropylacetylene
   (b) (R)-3-methyl-1-pentyne
   (c) vinylacetylene
   (d) 1-ethynylcyclohexanol
   (e) cyclodecyne
   (f) isobutylacetylene

(g) methyl ethynyl ether   (h) dodec-4-en-2-yne

(i) pent-2-yn-1-ol   (j) 3-methoxy-1-pentyne

2. Give an acceptable name for each of the following structures.

(a) $BrCH_2CH_2CH_2CH_2C{\equiv}CH$ (b) $(CH_3)_3CC{\equiv}CCH_2CH_3$

(c) $CH_2{=}CHI$ (d) 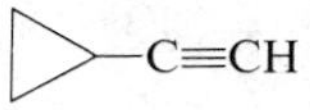 ▷—$C{\equiv}CH$

(e) $C_6H_5C{\equiv}CCH_2C(CH_3)_3$ (f) $BrC{\equiv}CCH_3$

(g) $HC{\equiv}CCH_2Br$ (h) 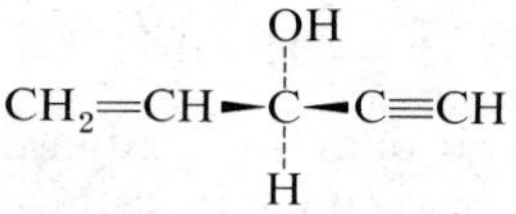 $CH_2{=}CH{-}\overset{\displaystyle OH}{\underset{\displaystyle H}{C}}{-}C{\equiv}CH$

(i) $HOCH_2C{\equiv}CCH_2OH$ (j) $(CH_3)_2CHC{\equiv}CCH_2CH_3$

(k) $CH_2{=}CHCH_2CH_2C{\equiv}CH$

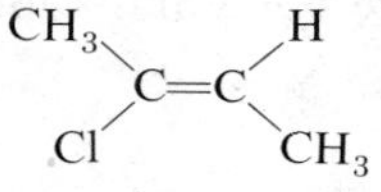

(m)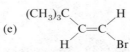

(l) $\underset{\displaystyle Cl}{\overset{\displaystyle CH_3}{\diagdown}}C{=}C\underset{\displaystyle CH_3}{\overset{\displaystyle H}{\diagup}}$

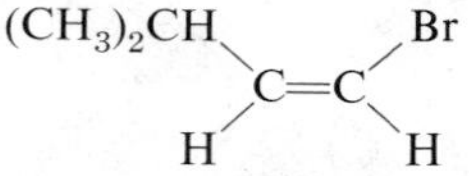

(n) $CH_3CHClC{\equiv}CCH_2CH_3$

3. Give the principal reaction product(s) for the reaction of 1-butyne with each of the following reagents. If no reaction is expected, so indicate with N.R.

(a) hot aq. $KMnO_4$ (b) H_2/Pt

(c) excess Br_2/CCl_4, $0°$ (d) aq. NaCl

(e) aq. $AgNO_3$ (f) 1. B_2H_6; 2. H_2O_2, NaOH

(g) aq. H_2SO_4; Hg^{2+} (h) 1. B_2H_6; 2. CH_3COOD

(i) H_2 (1 mole)/$Pd(BaSO_4)$–quinoline   (j) CuCl, NH_4Cl, HCl

(k) $CuCl_2$, pyridine, Δ

4. Answer problem 3 for 2-butyne.

5. Sketch the nmr spectrum expected for each of the following compounds. Indicate the number of peaks expected, the approximate δ-value for each group of peaks, and their area.

(a) $(CH_3)_3CC{\equiv}COCH_3$ (b) $CH_3C{\equiv}CH$

(c) $CH_3CHBrC{\equiv}CH$

(d) 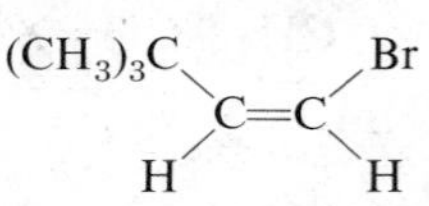 $\underset{\displaystyle H}{\overset{\displaystyle (CH_3)_3C}{\diagdown}}C{=}C\underset{\displaystyle H}{\overset{\displaystyle Br}{\diagup}}$

(e) $\underset{\displaystyle H}{\overset{\displaystyle (CH_3)_3C}{\diagdown}}C{=}C\underset{\displaystyle Br}{\overset{\displaystyle H}{\diagup}}$

6. Using propyne as the only source of carbon, devise practical syntheses for the following compounds.

(a) $(CH_3)_2CHBr$ (b) CH_3COCH_3

(c) CH_3CH_2CHO (d) $CH_3(CH_2)_4CH_3$

(e) 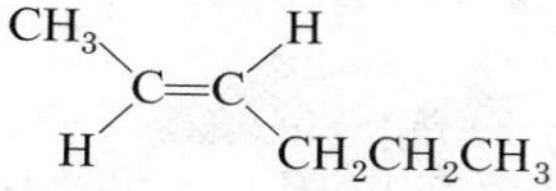 $\underset{\displaystyle H}{\overset{\displaystyle CH_3}{\diagdown}}C{=}C\underset{\displaystyle CH_2CH_2CH_3}{\overset{\displaystyle H}{\diagup}}$

(f) $CH_3CCl{=}CH_2$

(g) $CH_3CBr_2CBr_2D$ (h) $CH_3C{\equiv}C{-}C{\equiv}CCH_3$

7. Show how each of the following conversions may be accomplished in good yield. In each case, use only the indicated starting material as a source of carbon.

(a) $CH_3CH_2CH_2CH_3 \longrightarrow CH_3CH_2\overset{\overset{\displaystyle Cl}{|}}{C}HCH_3$ (containing *no* 1-chlorobutane)

(b) $CH_3CH_2CH_2CH=CH_2 \longrightarrow CH_3CH_2CH_2C\equiv CH$

(c) $CH_3CH_2CH_2Br \longrightarrow CH_3CH_2CH_2C\equiv CCH_3$

(d) $HC\equiv CH \longrightarrow CH_3CH_2CH_2CH_2OH$

(e) $HC\equiv CH \longrightarrow CH_3CH_2\overset{\overset{\displaystyle OCH_3}{|}}{C}HCH_2CH_2CH_3$

(f)
$$\underset{H}{\overset{C_6H_5}{\diagdown}}C=C\underset{C_6H_5}{\overset{H}{\diagup}} \longrightarrow \underset{H}{\overset{C_6H_5}{\diagdown}}C=C\underset{H}{\overset{C_6H_5}{\diagup}}$$

(g) $CH_3CH_2CH_2OH \longrightarrow CH_3COCH_3$

(h) $CH_3CH_2C\equiv CH \longrightarrow \underset{H}{\overset{CH_3CH_2}{\diagdown}}C=C\underset{H}{\overset{D}{\diagup}}$

(i) $HC\equiv CH \longrightarrow CH_3CH_2CH_2CH_3$

8. Show how each of the following conversions may be accomplished in good yield. In addition to the indicated starting material, other organic compounds may be used as necessary.

(a) $HC\equiv CH \longrightarrow CH_3CH_2\overset{\overset{\displaystyle OCH_3}{|}}{C}HCH_2CH_3$

(b) $CH_3\overset{\overset{\displaystyle OH}{|}}{C}HC\equiv CH \longrightarrow CH_3\overset{\overset{\displaystyle OH}{|}}{C}HC\equiv CC\equiv CCH_3$

(c) $HC\equiv CH \longrightarrow CH_3CH_2C\equiv CCH_2CH_2CH_3$

(d) $(CH_3)_2\overset{\overset{\displaystyle OH}{|}}{C}CH_2CH_3 \longrightarrow (CH_3)_2CHC\equiv CCH_3$

9. From isopentyl alcohol (3-methyl-1-butanol), acetylene, and any required straight-chain primary alcohols, derive a practical synthesis for 2-methylheptadecane, the sex attractant for the Tiger moth (page 97).

10. Muscalure, *cis*-9-tricosene, is the sex-attractant insect pheromone of the common housefly. Give a practical synthesis of this compound from acetylene and straight-chain alcohols.

11. The pK_as of ethane, ethylene, and acetylene are approximately 50, 44, and 25, respectively.

(a) The pK_a of NH_3 in liquid ammonia is approximately 35. What is the equilibrium constant for the following equilibrium in liquid ammonia for each of the above hydrocarbons.

$$RH + NH_2^- \rightleftharpoons R^- + NH_3$$

Assume that the above pK_as apply unchanged to a liquid ammonia solution. Note that this is a gross approximation but the above values are themselves approximations.

(b) If each hydrocarbon is treated with sodium amide in liquid ammonia and the resulting solution then treated with methyl iodide, different results are obtained. There is no reaction with ethane or ethylene, but acetylene gives a good yield of propyne. Explain this observation using the results in (a).

(c) The bond-dissociation energies, $DH°$, for the C—H bonds are ethane, 98; ethylene, 108; acetylene, 120 kcal mole^{-1}. The C—H bonds are progressively harder to break along this series, yet the compounds are increasingly acidic. Explain this apparent paradox.

12. Alkenyl chlorides react with a solution of sodium in liquid ammonia to replace the Cl by H with retention of configuration. This reaction may be regarded as a reduction with solvated electrons to produce a vinyl anion.

$$\underset{}{\overset{}{>}}C{=}C\overset{Cl}{\underset{}{<}} \xrightarrow{2\,e^-} \underset{}{\overset{}{>}}C{=}C^-\underset{}{<} + Cl^- \xrightarrow{NH_3} \underset{}{\overset{}{>}}C{=}C\overset{H}{\underset{}{<}} + NH_2^-$$

(a) What does the stereochemistry of the reaction reveal concerning the geometrical structure and configurational stability of the intermediate vinyl anion?

(b) This reaction is used in a sequence to invert the configuration of internal olefins. Show how *trans*-3-hexene may be converted to *cis*-3-hexene by use of this reaction as the final step.

13. The hydration of alkynes is catalyzed by Hg^{2+}. Write a mechanism for the hydration that accounts for this catalysis. (*Hint*: The mercuric ion adds to the triple bond to give a mercuricarbocation).

14. Compound A has the formula C_8H_{12} and is optically active. It reacts with hydrogen in the presence of platinum metal to give B, which has the formula C_8H_{18} and is optically inactive. Careful hydrogenation of A using H_2 and Lindlar's catalyst gives C, which has the formula C_8H_{14} and is optically active. Compound A reacts with sodium in ammonia to give D, which also has the formula C_8H_{14} and is optically inactive. What are compounds A through D?

15. Compound E has the formula C_7H_{12}. It reacts with dry HCl at $-20°$ to give F, $C_7H_{13}Cl$. Compound F reacts with potassium *t*-butoxide in *t*-butyl alcohol to give a small amount of E and mainly G, which as the formula C_7H_{12}. Ozonization of G gives cyclohexanone and formaldehyde. What are compounds E through G?

16. The reaction of (Z)-1,5-dibromo-1-pentene with ethanolic $NaOC_2H_5$ can give principally (Z)-1-bromo-5-ethoxy-1-pentene, 5-ethoxy-1-pentyne, or 2,5-diethoxy-1-pentene, depending on the reaction conditions. Explain. Why is 1,5-diethoxy-1-pentene not a principal product under any of these conditions?

17. Appendix II "Bond-Dissociation Energies," gives values of $DH°$ for $CH_2{=}CH—Cl$ and $CH_2{=}CH—Br$. Compare with the corresponding values for ethyl halides and explain any difference. Estimate $DH°$ for vinyl fluoride and vinyl iodide.

18. The proton-coupled cmr spectrum of 3-hexyne consists of a triplet at 13.2 ppm, a quartet at 15.6 ppm, and a singlet at 81.1 ppm. Which carbons are responsible for these resonances? Compare the values with those given for hexane (Table 9.4) and the 3-hexenes (Table 11.3). Suggest an explanation for the resonance value for the alkyne *sp*-hybridized carbon.

Chapter 13
Aldehydes and Ketones

13.1
Structure

Aldehydes and ketones are compounds containing the **carbonyl group** C=O. When two alkyl groups are attached to the carbonyl, the compound is a **ketone.** When two hydrogens, or one hydrogen and one alkyl group, are attached to the carbonyl, the compound is an **aldehyde.**

$$\overset{:\ddot{O}:}{R:C:R'} \qquad \overset{O}{\overset{\|}{R-C-R'}} \qquad RCOR'$$

Lewis structure   Kekulé structure   Condensed structure

The structure of formaldehyde, the simplest member of the class, is depicted below, along with its experimental bond lengths and bond angles. The carbon

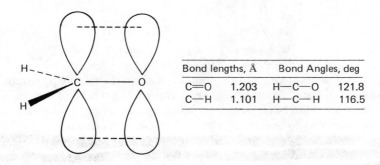

Bond lengths, Å		Bond Angles, deg	
C=O	1.203	H—C—O	121.8
C—H	1.101	H—C—H	116.5

atom is approximately sp^2-hybridized and forms σ-bonds to two hydrogens and one oxygen. The molecule is planar and the H—C—O and H—C—H bond angles are close to 120°, the idealized sp^2-angles. The remaining carbon p-orbital overlaps with the oxygen p_z-orbital, giving rise to a π-bond between these atoms. The oxygen atom also has two nonbonding electron pairs (the lone pairs) that occupy the remaining orbitals. A stereo representation of acetaldehyde is shown in Figure 13.1. Note the planarity of the carbonyl group. Also note that one C—H bond of the methyl group is eclipsed with the C=O bond and that the carbonyl C—H is staggered with respect to the other two C—H bonds.

Oxygen is more electronegative than carbon and attracts the bonding electrons

FIGURE 13.1. *Stereo representation of acetaldehyde.*

more strongly; that is, the higher nuclear charge on oxygen provides a greater attractive force than carbon. Accordingly, the C—O bond is polarized in the direction C^+—O^-. This effect is especially pronounced for the π-electrons and can be represented by the following resonance structures for formaldehyde.

$$
\left[
\begin{array}{c}
\overset{\displaystyle H}{\underset{\displaystyle H}{}} \overset{..}{:}C::\overset{..}{O}. \longleftrightarrow \overset{\displaystyle H}{\underset{\displaystyle H}{}} :\overset{+}{C}:\overset{..}{\underset{..}{O}}:^-
\end{array}
\right]
\quad \text{or} \quad
\left[
\begin{array}{c}
\overset{\displaystyle H}{\underset{\displaystyle H}{}}\!\!>\!C{=}O \longleftrightarrow \overset{\displaystyle H}{\underset{\displaystyle H}{}}\!\!>\!\overset{+}{C}{-}\bar{O}
\end{array}
\right]
$$

The actual structure is a composite of the normal octet structure, $CH_2{=}O$, and the polarized structure, $^+CH_2{-}O^-$, which corresponds to a carbocation oxide. The composite structure may be represented with dotted line symbolism which shows the partial charges in carbon and oxygen and the partial single bond character of the C—O bond.

$$
\overset{\displaystyle H}{\underset{\displaystyle H}{}}\!\!>\!\overset{\delta+}{C}\!\!\cdots\!\!\overset{\delta-}{O}
$$

One physical consequence of this bond polarity is that carbonyl compounds generally have rather high dipole moments. The experimental dipole moments of formaldehyde and acetone are 2.27 D and 2.85 D, respectively.

$$
\begin{array}{cc}
\overset{\displaystyle O}{\underset{\displaystyle H \quad H}{\overset{\|}{C}}}\uparrow & \overset{\displaystyle O}{\underset{\displaystyle CH_3 \quad CH_3}{\overset{\|}{C}}}\uparrow
\end{array}
$$

formaldehyde   acetone
dipole moment, 2.27 D   dipole moment, 2.85 D

The chemical consequences of this bond polarity will become apparent during our discussions of the reactions of carbonyl groups. We shall find that the positive carbon can react with bases and that much of the chemistry of the carbonyl function corresponds to that of a relatively stable carbocation.

The lone-pair electrons in the carbonyl oxygen have weakly basic properties. In acidic solution acetone acts as a Lewis base and is protonated to a small but significant extent.

$$
\overset{\displaystyle :O:}{\underset{}{CH_3{-}\overset{\|}{C}{-}CH_3}} + H^+ \rightleftharpoons \overset{\displaystyle :\overset{+}{O}H}{\underset{}{CH_3{-}\overset{\|}{C}{-}CH_3}}
$$

In fact, acetone is a much *weaker* Lewis base than is water. An acid strength corresponding to 82% sulfuric acid is required to give 50% protonation of acetone. This corresponds to an approximate pK_a for the conjugate acid of acetone of -7.2 (the approximate pK_a of H_3O^+ is -1.7). Even though the carbonyl group has only weakly basic properties, we shall find that this basicity plays an important role in the chemistry of aldehydes, ketones, and related compounds.

13.2
Nomenclature

A. Common Names

Traditionally, aldehyde names were derived from the name of the corresponding acid (Section 18.2) by dropping the suffix **-ic** (or **-oic**) and adding in its place

the suffix **-aldehyde.** These common names are still widely used for simpler aldehydes.

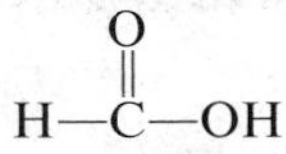

$$H-\overset{\overset{O}{\|}}{C}-OH \qquad H-\overset{\overset{O}{\|}}{C}-H$$

form**ic acid**   form**aldehyde**

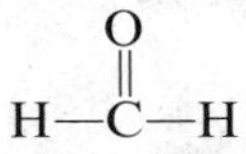

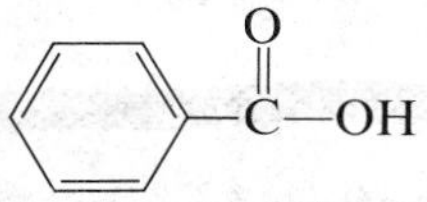

$$\text{benzoic acid} \qquad \text{benzaldehyde}$$

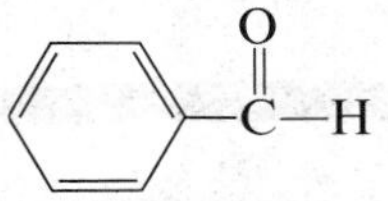

benz**oic acid**   benz**aldehyde**

Appendage groups are designated by the appropriate prefixes. The chain is la-
beled by using the Greek letters α, β, γ, and so on (see inside front cover), beginning with the carbon next to the carbonyl group.

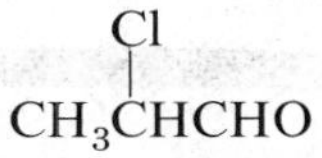

$$\overset{\overset{Cl}{|}}{CH_3CHCHO} \qquad \overset{\overset{Br}{|}}{CH_3CHCH_2CHO}$$

α-chloropropionaldehyde 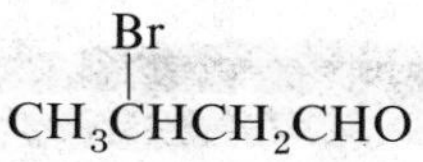 β-bromobutyraldehyde

The common names of ketones are derived by prefixing the word **ketone** by the
names of the two alkyl radical groups; the separate parts are separate words.

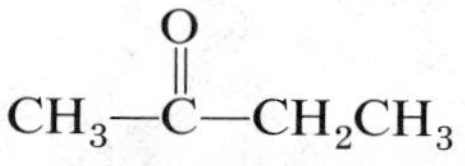

$$CH_3-\overset{\overset{O}{\|}}{C}-CH_2CH_3 \qquad CH_3CH_2\overset{\overset{CH_3}{|}}{CH}-\overset{\overset{O}{\|}}{C}-\overset{\overset{CH_3}{|}}{CH}CH_2CH_3$$

ethyl methyl ketone 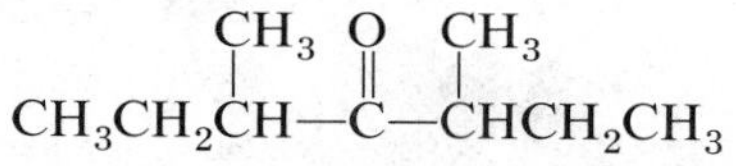   di-*sec*-butyl ketone

Dimethyl ketone has the additional trivial name **acetone,** which is universally
used.

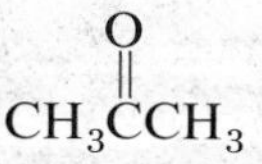

$$CH_3\overset{\overset{O}{\|}}{C}CH_3$$

acetone

As with aldehydes, appendages may be designated by a prefix using the Greek
letter notational system.

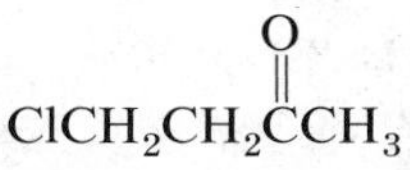

$$ClCH_2CH_2\overset{\overset{O}{\|}}{C}CH_3$$

β-chloroethyl methyl ketone

EXERCISE: Write the structures for each of the following ketones.
(a) isopropyl *n*-propyl ketone   (b) isobutyl neopentyl ketone
(c) γ-methoxybutyraldehyde   (d) β-bromobutyl methyl ketone
(e) *t*-butyl cyclohexyl ketone   (f) *sec*-butyl ethyl ketone

B. *IUPAC Names*

In the IUPAC system aldehyde names are derived from the name of the alkane
of the same carbon number. The final **-e** of the alkane is replaced by the suffix **-al.**
Since the carbonyl group is necessarily at the end of a chain, it is not necessary to

designate its position by a number, but as a suffix group it controls the numbering as the number 1 carbon. Note that the carbonyl carbon is the number 1 carbon in the IUPAC system, although it is given no designation in the common system.

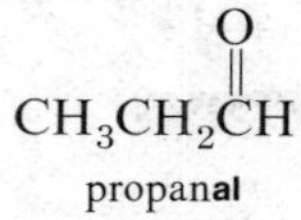

propan**al** 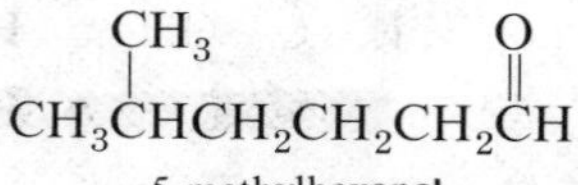   5-methylhexan**al**

More complicated aldehydes may be named using the suffix **-carbaldehyde.**

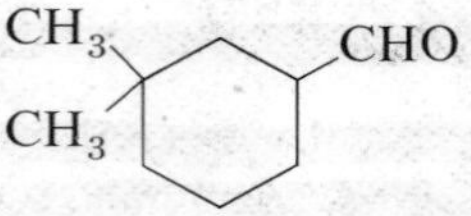

3,3-dimethylcyclohexane**carbaldehyde**

When it is necessary to name a compound as another functional group, the aldehyde grouping is designated **formyl.**

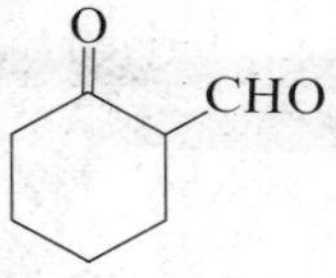

2-**formyl**cyclohexan**one**

The IUPAC names of ketones are derived from the name of corresponding alkane by replacing the final **-e** by **-one.** In acyclic ketones it is necessary to prefix the name by a number indicating which carbon along the longest chain is the carbonyl carbon. The longest chain containing the carbonyl group is numbered from the end that gives the carbonyl carbon the lower number. In cyclic ketones it is understood that the carbonyl carbon is number 1.

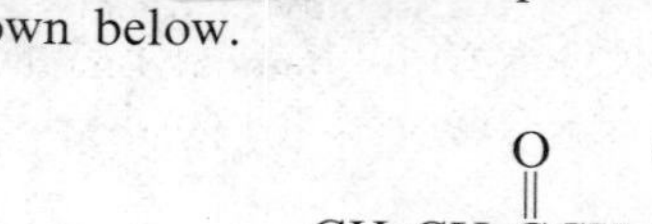

5-ethyl-3-heptan**one** 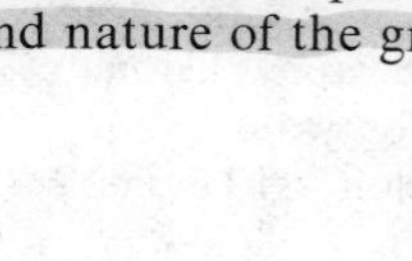   3-methylcyclohexan**one**

Occasionally it is necessary to name a molecule containing a carbonyl group as a derivative of a more important function. In such a case, the prefix **oxo-** is used, along with a number, to indicate the position and nature of the group. One such example is shown below.

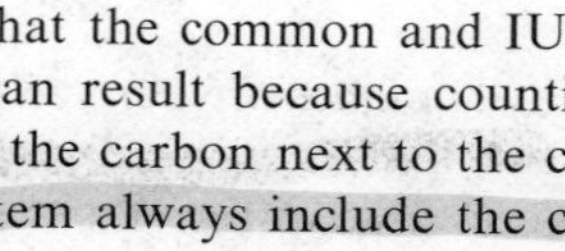

2-methyl-4-**oxo**hexanal

It is generally desirable that the common and IUPAC nomenclature systems not be mixed. Ambiguity can result because counting by Greek letters in the common system starts from the carbon next to the carbonyl group, whereas the numbers in the IUPAC system always include the carbonyl group.

ClCH$_2$CH$_2$CHO

correct:	β-chloropropionaldehyde or 3-chloropropanal
incorrect:	β-chloropropanal
allowed but not recommended:	3-chloropropionaldehyde

EXERCISE: What are the IUPAC names for each of the compounds in the exercise on page 359?

13.3
Physical Properties

Physical data for a number of aldehydes and ketones are collected in Tables 13.1 and 13.2. The boiling points at 1 atm for straight-chain aldehydes and methyl *n*-alkyl ketones are plotted in Figure 13.2 along with the corresponding data for straight-chain alkanes. As in other homologous series, there is a smooth increase in boiling point with increasing molecular weight. Aldehydes and ketones boil higher than alkanes of comparable molecular weights. This boiling-point elevation results from the interaction between dipoles.

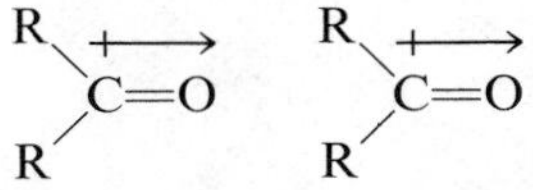

The discrepancy is largest with the simplest aldehyde, formaldehyde (mol. wt. 30, b.p. $-21°$), which boils 68° higher than ethane (mol. wt. 30, b.p. $-89°$). With higher members of the series, as the polar functional group becomes a smaller and

TABLE 13.1
Physical Properties of Some Aldehydes

Compound	Structure	Molecular Weight	Boiling Point, °C	Melting Point, °C
formaldehyde	HCHO	30	-21	-92
acetaldehyde	CH$_3$CHO	44	21	-121
propionaldehyde	CH$_3$CH$_2$CHO	58	49	-81
butyraldehyde	CH$_3$(CH$_2$)$_2$CHO	72	76	-99
valeraldehyde	CH$_3$(CH$_2$)$_3$CHO	86	103	-92
hexanal	CH$_3$(CH$_2$)$_4$CHO	100	128	-56
heptanal	CH$_3$(CH$_2$)$_5$CHO	114	153	-43
octanal	CH$_3$(CH$_2$)$_6$CHO	128	171	—
nonanal	CH$_3$(CH$_2$)$_7$CHO	142	192	—
decanal	CH$_3$(CH$_2$)$_8$CHO	156	209	-5
undecanal	CH$_3$(CH$_2$)$_9$CHO	170	—	-4
dodecanal	CH$_3$(CH$_2$)$_{10}$CHO	184	—	12

TABLE 13.2
Physical Properties of Some Ketones

Compound	Structure	Molecular Weight	Boiling Point, °C	Melting Point, °C	H_2O Solubility, wt. % (25°)
acetone	CH_3COCH_3	58	56	−95	∞
2-butanone	$CH_3CH_2COCH_3$	72	80	−86	25.6
2-pentanone	$CH_3(CH_2)_2COCH_3$	86	102	−78	5.5
3-pentanone	$CH_3CH_2COCH_2CH_3$	86	102	−40	4.8
2-hexanone	$CH_3(CH_2)_3COCH_3$	100	128	−57	1.6
3-hexanone	$CH_3(CH_2)_2COCH_2CH_3$	100	125	—	1.5
2-heptanone	$CH_3(CH_2)_4COCH_3$	114	151	−36	0.4
2-octanone	$CH_3(CH_2)_5COCH_3$	128	173	−16	—
2-nonanone	$CH_3(CH_2)_6COCH_3$	142	195	−7	—
2-decanone	$CH_3(CH_2)_7COCH_3$	156	210	14	—
2-undecanone	$CH_3(CH_2)_8COCH_3$	170	232	15	—
2-dodecanone	$CH_3(CH_2)_9COCH_3$	184	247	21	—

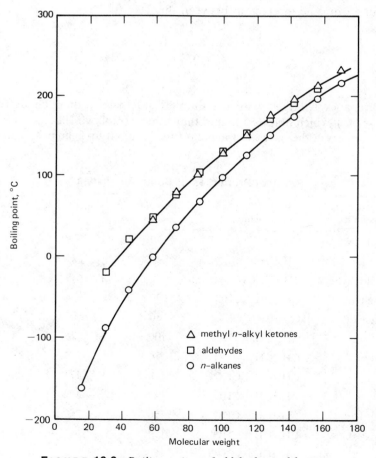

FIGURE 13.2. *Boiling points of aldehydes and ketones.*

smaller part of the molecule, the boiling point tends to come closer and closer to that of a corresponding alkane (see 2-dodecanone, mol. wt. 184, b.p. 247°; *n*-tridecane, mol. wt. 184, b.p. 235°).

13.4
Nuclear Magnetic Resonance

Nmr is an important technique for identifying aldehydes. The hydrogen attached to the carbonyl carbon gives rise to a characteristic band at very low field, usually around $\delta = 9.5$ ppm. In a magnetic field the circulating π-electrons produce an induced field that effectively deshields the aldehyde proton (Figure 13.3). That is, the induced field adds to the applied field in such a way that a smaller applied field is required to achieve resonance. The same phenomenon was discussed earlier with alkenes and accounts for the substantial downfield shift of vinyl protons (Section 11.3.B). In aldehydes the effect is greater and, in addition, the positive character of the carbonyl carbon provides a further downfield shift. The net result is a relatively large downfield resonance position for the aldehyde proton. Few other kinds of protons appear in this region; thus a peak at $\delta = 9.5$ is strongly indicative of the presence of a CHO function.

The same induced field that causes deshielding of a proton attached directly to a carbonyl group also produces significant deshielding of protons somewhat farther away at the α-carbon atoms. Typical chemical shifts for these protons are summarized in Table 13.3.

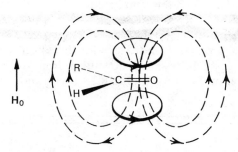

FIGURE 13.3. *The diamagnetic anisotropy of the carbonyl group.*

TABLE 13.3
Chemical Shifts for Aldehyde and Ketone Hydrogens

Hydrogen	Approximate Chemical Shift δ, ppm
$R-\overset{\displaystyle O}{\overset{\|}{C}}-H$	9.5
$R-\overset{\displaystyle O}{\overset{\|}{C}}-CH_3$	2.0
$R-\overset{\displaystyle O}{\overset{\|}{C}}-CH_2R$	2.2
$R-\overset{\displaystyle O}{\overset{\|}{C}}-CHR_2$	2.4

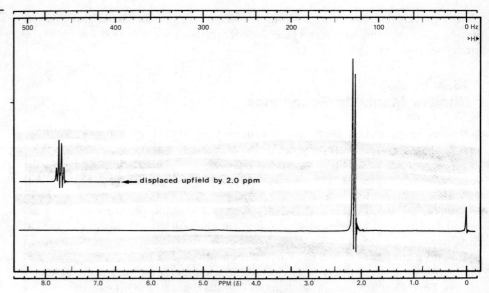

FIGURE 13.4. *Nmr spectrum of acetaldehyde,* CH_3CHO.

The spectra of acetaldehyde and 3-methyl-2-butanone shown in Figures 13.4 and 13.5 are characteristic. Note that the vicinal coupling constant in acetaldehyde is quite small, only 3 Hz.

The cmr chemical shifts for a few simple aldehydes and ketones are collected in Table 13.4. Note the extreme downfield shift of the carbonyl resonance—about 200 ppm from TMS. Carbonyl groups resonate further downfield than any other type of carbon. There are two main reasons for this effect. First, there is the fact that sp^2-hybridized carbons resonate downfield from sp^3-hybridized carbons. However, comparison of the shifts in Table 13.4 with those given for alkenes in

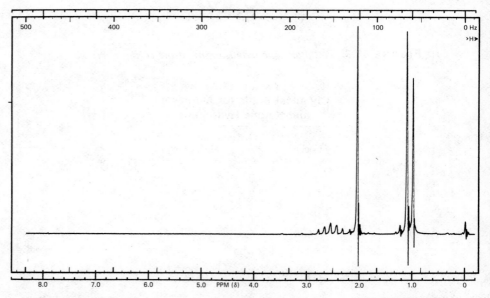

FIGURE 13.5. *Nmr spectrum of isopropyl methyl ketone,* $CH_3COCH(CH_3)_2$.

TABLE 13.4
Cmr Chemical Shifts of Aldehydes and Ketones

	C-1	C-2	C-3	C-4	C-5
ethanal	199.6	31.2			
propanal	201.8	36.7	5.2		
butanal	201.6	45.7	15.7	13.3	
propanone	30.2	205.1	30.2		
butanone	28.8	206.3	36.4	7.6	
2-pentanone	29.3	206.6	45.2	17.5	13.5
3-pentanone	7.3	35.3	209.3	35.3	7.3

Table 11.3 shows that aldehyde carbons resonate about 70 ppm downfield from alkene carbons of comparable substitution (compare propanal with 1-butene and butanal with 1-pentene). The reason for this additional shift is the polar nature of the carbonyl bond, which *deshields* the carbon nucleus even more. Because the carbonyl carbon has a long relaxation time, carbonyl resonances are normally much weaker than the resonances of other carbons, as shown in the cmr spectrum of 2-octanone (Figure 13.6).

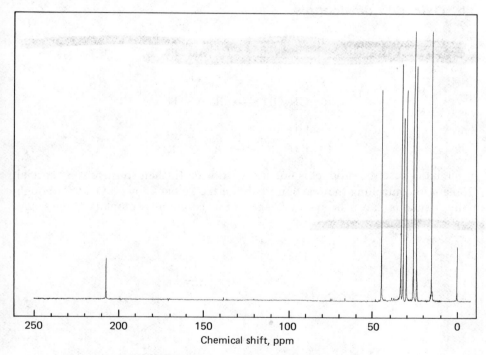

FIGURE 13.6. *Cmr spectrum of 2-octanone.*

Once a nucleus has been promoted from the α- to the β-spin state, there must be a way for it to give up the excess energy it acquired in the promotion so it can return to the α-spin state. Otherwise, absorption of rf power would stop as soon as all of the nuclei had been boosted to the higher energy state. In practice, the tiny amount of power that would be absorbed before "saturation" occurred would be very difficult to measure accurately. Fortunately, there are several "relaxation mechanisms" (ways in

which the "spin-flip" energy can be dissipated). The dominant relaxation mechanisms are different for different nuclei. For protons a major mechanism involves
motion of the molecule relative to other molecules in the sample. Although this
mechanism also contributes to relaxation of carbon nuclei, a more important one
involves interaction with hydrogens that are bonded *directly to the nucleus undergoing
relaxation*. As a consequence carbons that have no attached hydrogens have long
relaxation times. This is true for the carbonyl carbons of ketones (but not of aldehydes) and for other carbons that are fully substituted by alkyl groups ("quaternary
carbons"). We have already seen one other example of this effect. Note that the C-2
resonance in the cmr spectrum of 1,2,2-trichloropropane (Figure 9.27) is only about
one-sixth as intense as the resonances of the other two carbons.

13.5
Synthesis of Aldehydes and Ketones

The carbonyl group in aldehydes and ketones is one of the most important
functional groups. In this section we shall review several reactions that are good
methods for the synthesis of aldehydes and ketones.

A. *Oxidation of Alcohols*

As discussed in Section 10.7.E, aldehydes and ketones may be obtained by the
oxidation of primary and secondary alcohols, respectively.

$$R-CH_2OH \xrightarrow{[O]} R-\overset{\displaystyle O}{\overset{\|}{C}}-H$$

$$R-\overset{\displaystyle OH}{\overset{|}{C}H}-R' \xrightarrow{[O]} R-\overset{\displaystyle O}{\overset{\|}{C}}-R'$$

In the latter case the product is not easily oxidized further, so there is no special
problem in controlling the reaction to obtain the ketone in good yield. Although
many oxidants have been used, the most commonly employed ones are
chromium(VI) compounds.

$$+ \ Na_2Cr_2O_7 \ \xrightarrow[H_2SO_4]{H_2O}$$

menthol

(84%)
menthone

A mixture of 120 g of $Na_2Cr_2O_7$, 100 g of conc. H_2SO_4, and 600 mL of water is
prepared. To this solution is added 90 g of menthol (2-isopropyl-5-methylcyclohex-
anol). Heat is evolved, the temperature of the mixture rising to 55°. As soon as the
reaction is complete, the oily product layer is removed by ether extraction and distilled to obtain 75 g (84%) of menthone.

In the case of primary alcohols, this simple picture is clouded by the fact that
the product aldehyde may be further oxidized to a carboxylic acid. In most cases,

the primary alcohol undergoes oxidation *more rapidly* than the corresponding aldehyde. However, in aqueous solution the product aldehyde forms a hydrate, which is oxidized even more rapidly than the primary alcohol. Aldehyde hydrates are discussed thoroughly in Section 13.7.A.

$$R-CH_2OH \xrightarrow[fast]{[O]} R-\overset{\overset{\displaystyle O}{\|}}{C}-H \underset{H_2O}{\overset{[O]}{\underset{slow}{\rightleftarrows}}} \quad \begin{array}{c} R-\overset{\overset{\displaystyle O}{\|}}{C}-OH \\ \\ \underset{R-\overset{\overset{\displaystyle OH}{|}}{C}H-OH \xrightarrow{[O]}{fast} R-\overset{\overset{\displaystyle O}{\|}}{C}-OH}{} \end{array}$$

We mentioned previously (Section 10.7.E) that the preparation of aldehydes by oxidation of primary alcohols with aqueous chromic acid is limited to compounds of low molecular weight that can be distilled out as they are formed. In non-hydroxylic solvents, however, selective oxidation may be accomplished, and several oxidants that may be used in organic solvents have been developed for this this purpose. One such oxidant is the complex formed between chromium trioxide and 2 moles of the heterocyclic base pyridine. This material, bispyridinechromium(VI) oxide, is soluble in chloroform and dichloromethane.

$$n\text{-}C_7H_{15}CH_2OH + CrO_3(C_5H_5N)_2 \xrightarrow[\substack{25° \\ 1\ hr}]{CH_2Cl_2} n\text{-}C_7H_{15}CHO \\ (95\%)$$

A mixture of 6.0 g of chromium trioxide and 9.5 g of pyridine is prepared in 150 mL of dichloromethane. Octanol (1.32 g) is added to the deep red solution, and the resulting mixture is kept at 25° for 15 min. The reaction mixture is worked up to obtain 1.24 g (95%) of octanal.

B. *Oxidation of Alkenes*

Aldehydes and ketones may also be prepared by oxidative cleavage of C–C multiple bonds. A particularly useful reagent for this purpose is ozone (Section 11.6.E). Hydrolysis of the ozonide, usually under reductive conditions, results in the production of two carbonyl compounds.

6-methyl-1-heptene

(62%) 5-methylhexanal

With this example we introduce a new type of structural symbol that we shall use occasionally throughout the remainder of the book. These simplified structures are similar to the geometric shapes we use to symbolize cyclic compounds (page 79). The student should be careful to understand what these "line structures" mean, so as to avoid errors. The end of each line segment denotes a carbon with the proper number of attached hydrogens to fill out its tetravalence. Double and triple carbon–carbon bonds are denoted by two or three parallel lines. Other functional groups are written in *completely,* including the carbon and the hydrogens as pertinent. Sometimes one or more of the hydrogens are shown for additional clarity. The following examples should be examined carefully.

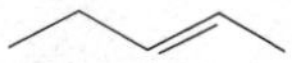

trans-2-pentene 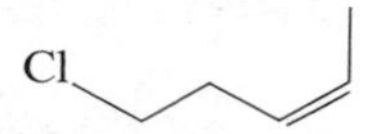   5-chloro-*cis*-2-pentene 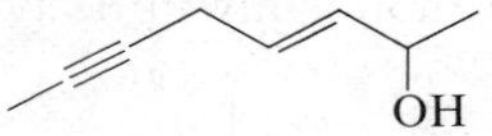   (E)-oct-3-en-6-yn-2-ol

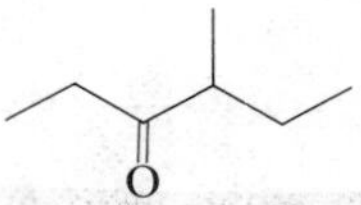

4-methyl-3-hexanone 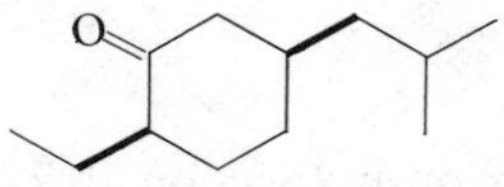   *cis*-2-ethyl-5-(2-methylpropyl)cyclohexanone

EXERCISE: Thumb back through the text and write a dozen structures at random in terms of the line diagrams introduced in this section. Check carefully to be sure that each line symbol corresponds correctly to the structure chosen.

C. *Hydration of Alkynes*

As discussed in Section 12.6, alkynes undergo hydration to yield an unstable vinyl alcohol that immediately rearranges to the corresponding ketone. The reaction is usually catalyzed by mercuric ion and sulfuric acid.

$$R-C\equiv C-R + H_2O \xrightarrow[H_2SO_4]{Hg^{2+}} \left[R-\overset{OH}{\underset{}{C}}=CH-R \right] \rightleftharpoons R-\overset{O}{\underset{}{C}}-CH_2R$$

The reaction is generally useful as a preparative method only when the alkyne is terminal, in which case a methyl alkyl ketone is always formed, or in cases where the molecule is symmetrical.

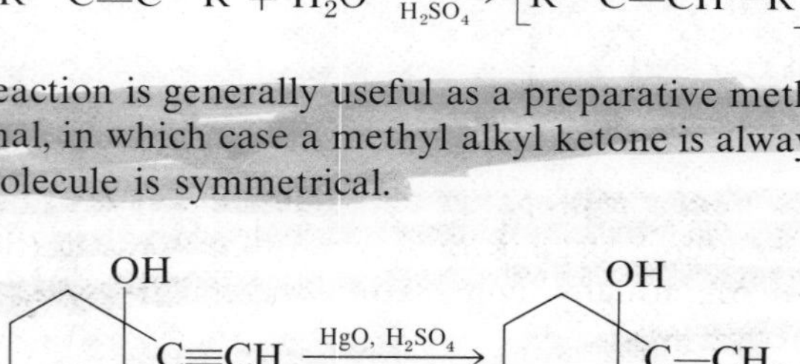

(84%)

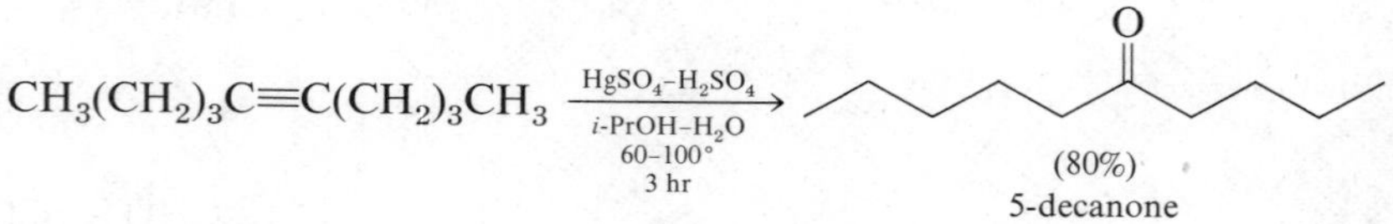

$$CH_3(CH_2)_3C\equiv C(CH_2)_3CH_3 \xrightarrow[\substack{i\text{-PrOH-H}_2O \\ 60-100° \\ 3 \text{ hr}}]{HgSO_4-H_2SO_4}$$

(80%)
5-decanone

Since the direct addition of water to a terminal alkyne always occurs in such a way that the hydroxy group becomes attached to the carbon bearing the alkyl group, the only alkyne that will yield an aldehyde upon hydration is acetylene itself. Indirect hydration of the triple bond, by the hydroboration route, yields the opposite result—terminal alkynes yield aldehydes (Section 12.6.E).

$$n\text{-}C_5H_{11}C\equiv CH \xrightarrow{B_2H_6} \xrightarrow[OH^-]{H_2O_2} n\text{-}C_6H_{13}CHO$$

EXERCISE: Give the principal product or products of hydration and hydroboration/H_2O_2 of each of the following alkynes. (a) 1-butyne, (b) 2-butyne, (c) 2-pentyne, and (d) 3-hexyne.

13.6 Enolization

The reactions of aldehydes and ketones can be divided into the following types.

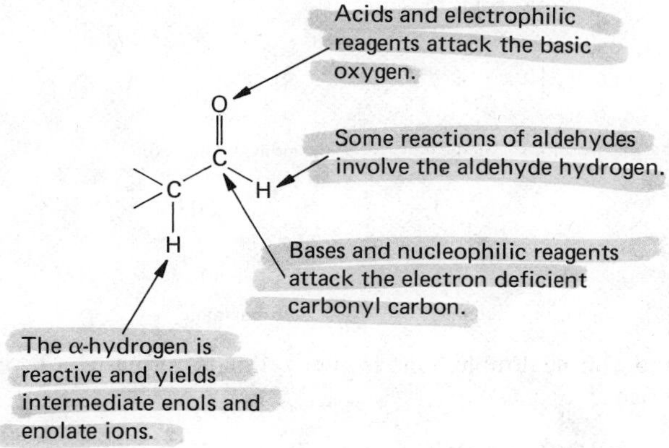

Acids and electrophilic reagents attack the basic oxygen.
Some reactions of aldehydes involve the aldehyde hydrogen.
Bases and nucleophilic reagents attack the electron deficient carbonyl carbon.
The α-hydrogen is reactive and yields intermediate enols and enolate ions.
O
C
C
H
H

A. *Keto–Enol Equilibria*

Aldehydes and ketones exist in solution as an equilibrium mixture of two isomeric forms, the keto form and the **enol** (from **-ene + -ol,** *unsaturated alcohol*) form. For simple aliphatic ketones there is very little of the enol form present at equilibrium, as shown by the following examples.

$$CH_3-\overset{O}{\overset{\|}{C}}-CH_3 \rightleftharpoons CH_3-\overset{OH}{\overset{|}{C}}=CH_2$$

(0.01%)

$$CH_3-\overset{O}{\overset{\|}{C}}-CH_2CH_3 \rightleftharpoons CH_3-\overset{OH}{\overset{|}{C}}=CH-CH_3$$

(0.12%)

O
OH

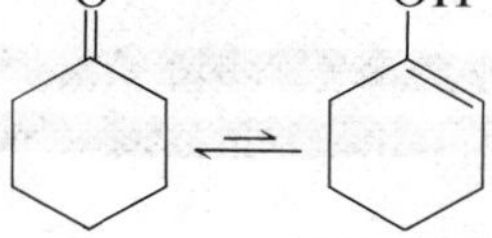

(1.2%)

This type of isomerisim, where the isomers differ only by the placement of a proton and the corresponding location of a double bond, is commonly referred to as **tautomerism.** The isomers are known as **tautomers.**

Strictly speaking, the tautomerism terminology is only used for this type of isomerism when there is a hetero atom such as nitrogen, oxygen, or sulfur present. In such cases, as exemplified in the following examples, the *rate of interconversion* of the isomers (tautomers) is relatively rapid.

$$CH_3-\overset{\overset{\displaystyle N-C_6H_5}{\|}}{C}-CH_3 \;\rightleftharpoons\; CH_2=\overset{\overset{\displaystyle NH-C_6H_5}{|}}{C}-CH_3$$

nitrosocyclohexane   cyclohexanone oxime

$$CH_3-\overset{\displaystyle O}{\underset{\underset{\displaystyle O^-}{+}}{N}} \;\rightleftharpoons\; CH_2=\overset{\displaystyle OH}{\underset{\underset{\displaystyle O^-}{+}}{N}}$$

nitromethane   aci-nitromethane

On the other hand, simple double bond isomerization is normally not considered as a case of tautomerism.

$$CH_3CH{=}CHCH_3 \;\rightleftharpoons\; CH_3CH_2CH{=}CH_2$$

Although this is purely a matter of semantics, the *rate* of the latter isomerization is also relatively slow.

Even though the percentage of enol tautomer at equilibrium is quite small, the enol is important in many reactions. As we shall soon see, many reactions of aldehydes and ketones occur by way of the unstable enol form.

Enolization is subject to both acid and base catalysis. In aqueous solutions the base is hydroxide ion. The base attacks a proton α to the carbonyl group to give an anion that is called an **enolate** ion. The enolate ion may be protonated on carbon, which regenerates the keto form, or on oxygen, which yields the enol form.

$$(1) \quad HO{:}^- + H{-}CH_2{-}\overset{\overset{\displaystyle O}{\|}}{C}CH_3 \;\rightleftharpoons\; \left[CH_2{=}\overset{\overset{\displaystyle O^-}{}}{C}CH_3 \;\longleftrightarrow\; {^-}\overset{\cdot\cdot}{C}H_2{-}\overset{\overset{\displaystyle O}{\|}}{C}CH_3 \right] + H_2O$$

$$(2) \quad \left[CH_2{=}\overset{\overset{\displaystyle O^-}{}}{C}CH_3 \;\longleftrightarrow\; {^-}\overset{\cdot\cdot}{C}H_2{-}\overset{\overset{\displaystyle O}{\|}}{C}CH_3 \right] + H_2O \;\rightleftharpoons\; CH_2{=}\overset{\overset{\displaystyle OH}{|}}{C}CH_3 + OH^-$$

Note that the first step in base-catalyzed enolization is formally analogous to E2 elimination. The "leaving group" may be considered to be the π-bond electron pair.

The first step of base-catalyzed enolization is an acid-base reaction, with acetone acting as a protic acid. The pK_a for acetone is approximately 20. Although acetone is an extremely weak acid compared to such familiar acids as HCl (pK_a -2.2), HF (pK_a $+3$), acetic acid (pK_a $+5$) or water (pK_a $+15.7$), we must remember that acidity is relative. If we compare acetone to ethane (pK_a estimated to be approximately $+50$), we see that the C—H bonds in acetone have considerably greater acidic properties than those in an alkane.

$$CH_3CH_3 \rightleftharpoons CH_3CH_2^- + H^+ \qquad K_a \approx 10^{-50}\ M$$

The reason for this enhanced acidity is apparent from a consideration of the conjugate bases produced by ionization of the two carbon acids. The anion produced from ethane has its negative charge localized on carbon. Since carbon is a fairly electropositive element, a carbanion is a high-energy species and the ionization that produces it is highly endothermic. On the other hand, the anion produced by ionization of acetone is not really a carbanion but a resonance hybrid of two structures.

$$CH_3\overset{\text{O}}{\overset{\|}{C}}CH_3 \rightleftharpoons H^+ + \left[\overset{\text{O}}{\underset{}{\overset{\|}{\overset{..}{C}H_2}}} - \overset{\text{O}}{\overset{\|}{C}}CH_3 \longleftrightarrow CH_2 = \overset{\text{O}^-}{\overset{|}{C}}CH_3 \right] \qquad K_a \approx 10^{-20}\ M$$

In one of the resonance structures the negative charge is borne by carbon, as in the ethyl anion. In the other the negative charge is on the more electronegative oxygen. Although both structures contribute to the resonance hybrid, that in which the negative charge is on oxygen is clearly dominating. It is important to point out once again that the two structures connected above by the double-headed arrow are *not isomers or tautomers* but *resonance structures*. The anion derived from acetone is neither one nor the other of the two indicated structures; it has the character of both. An alternative symbol that gives a better picture of the electronic distribution is

$$CH_3 - C \begin{smallmatrix} \diagdown \\ O^{\delta -} \\ \diagup \\ CH_2{}^{\delta -} \end{smallmatrix}$$

wherein we see that the negative charge is divided between the carbon and the oxygen. When this anion reacts with water, it can undergo protonation either on carbon, in which case the keto form results, or on oxygen, in which case the enol form is produced. The rate-limiting step for base-catalyzed enolization is usually the deprotonation step.

In neutral solution the principal base present is H_2O. Since H_2O is a much weaker base than OH^-, proton transfer from an aldehyde or ketone is not as rapid, and consequently enolization is slower. However, in acidic solution some of the weakly basic carbonyl groups are protonated. The protonated aldehyde or ketone loses a proton from carbon with much greater ease, even to such a weak base as H_2O. A carbonyl group is not very basic, and only a small amount of the protonated structure is present at equilibrium. The presence of the positive charge, however, greatly increases the rate of proton loss from carbon to solvent.

$$(1)\ \ H-CH_2-\overset{\text{O}}{\overset{\|}{C}}CH_3 + H_3O^+ \rightleftharpoons \left[H-CH_2-\overset{+\text{OH}}{\overset{\|}{C}}CH_3 \longleftrightarrow H-CH_2-\overset{\text{OH}}{\overset{|}{\underset{+}{C}}}CH_3 \right] + H_2O$$

$$(2)\ \ \left[H-CH_2-\overset{+\text{OH}}{\overset{\|}{C}}CH_3 \longleftrightarrow H-CH_2-\overset{\text{OH}}{\overset{|}{\underset{+}{C}}}CH_3 \right] + H_2O \rightleftharpoons CH_2=\overset{\text{OH}}{\overset{|}{C}}CH_3 + H_3O^+$$

In fact, the deprotonation of the protonated ketone (step 2) is analogous to the E1 elimination reaction, deprotonation of a carbocation.

$$H_2O \overset{\frown}{} H-CH_2\overset{+}{C}(CH_3)_2 \longrightarrow H_3O^+ + CH_2=C(CH_3)_2$$

In acid-catalyzed enolization the first step is a rapid equilibrium. Loss of a proton from carbon (step 2) is slower and is rate-determining.

Let us summarize. In aqueous solution aldehydes and ketones are in equilibrium with their corresponding enol forms. Interconversion of the enol and keto forms is catalyzed by either acid or base. At any given moment the vast majority of molecules are present as the more stable keto form. However, as we shall see, the small amount of enol tautomer present is involved as an important **intermediate** in many of the reactions of aldehydes and ketones.

One way in which the intermediate enols and enolates can be detected is by **deuterium exchange.** If one dissolves a ketone in D_2O containing DCl or Na^+OD^-, all of the α-hydrogens are exchanged for deuterium.

$$CH_3CH_2\overset{O}{\overset{\|}{C}}CH_2CH_3 + D_2O \xrightarrow[OD^-]{D^+ \text{ or}} CH_3CD_2\overset{O}{\overset{\|}{C}}CD_2CH_3 + H_2O$$

3-pentanone 3-pentanone-2,2,4,4-d_4

The amount of deuterium incorporation at equilibrium is related to the initial concentrations of the ketone and D_2O. In dilute solution the D_2O is present in large excess, and replacement of the α-hydrogens by deuterium is essentially complete. The *rate* of deuterium incorporation is proportional to the concentration of ketone and the catalyst, either D^+ or OD^-.

$$\text{rate}_{ex} = k[\text{ketone}][D^+] \quad \text{or} \quad k'[\text{ketone}][OD^-]$$

Such exchange reactions may be applied even when the aldehyde or ketone is not very soluble in water. Shaking such a compound with NaOD or DCl in D_2O for several hours results in virtually complete exchange.

2-methylcyclohexanone 2-methylcyclohexanone-2,6,6-d_3

Since the number of deuteriums is easily determined by mass spectrometry or by nmr, this reaction is a useful technique for counting the number of α-hydrogens.

EXERCISE: How many hydrogens are replaced by deuterium when each of the following compounds is treated with NaOD in D_2O?
(a) 2,2,4-trimethyl-3-pentanone   (b) 2-ethylbutanal
(c) 3-methylcyclopentanone   (d) *trans*-2-pentene

B. *Enolate Ions*

From the pK_a of acetone, +20, it is clear that in aqueous solution where the strongest base is OH^- the amount of enolate ion present is small.

$$CH_3\overset{\overset{O}{\parallel}}{C}CH_3 + OH^- \; \rightleftharpoons \; CH_2{=}\overset{\overset{O^-}{|}}{C}CH_3 + H_2O$$

pK_a 20   pK_a 15.7

$$K = \frac{[\text{enolate}][\cancel{H^+}]}{[\text{acetone}]} \cdot \frac{[H_2O]}{[OH^-][\cancel{H^+}]} = \frac{10^{-20}}{10^{-15.7}} \approx 10^{-4} \; M$$

However, if a ketone is treated with a much stronger base, it can be converted completely into the corresponding enolate ion.

pK_a 16.7   pK_a ~ 40

$$K = \frac{[\text{enolate}]}{[\text{cyclohexanone}]} \times \frac{[\text{diisopropylamine}]}{[\text{lithium diisopropylamide}]} = \frac{10^{-16.7}}{10^{-40}} \approx 10^{23}$$

Lithium diisopropylamide, $(i\text{-}C_3H_7)_2N^-Li^+$, is prepared by treating a solution of diisopropylamine in ether, THF, or 1,2-dimethoxyethane with n-butyllithium.

$$(i\text{-}C_3H_7)_2NH + n\text{-}BuLi \longrightarrow (i\text{-}C_3H_7)_2N^-Li^+ + C_4H_{10}$$

The base is widely used for converting ketones quantitatively into their corresponding lithium enolates.

$$CH_3\overset{\overset{O}{\parallel}}{-}C{-}CH_3 + (C_6H_5)_3C^-K^+ \; \rightleftharpoons \; CH_3{-}\overset{\overset{O^-K^+}{|}}{C}{=}CH_2 + (C_6H_5)_3CH$$

pK_a 20   pK_a 31.5

Potassium triphenylmethide, $(C_6H_5)_3C^-K^+$, is prepared by reaction of triphenyl-methane or chlorotriphenylmethane with potassium metal in an ether solvent such as glyme or THF.

$$(C_6H_5)_3CH + K \longrightarrow (C_6H_5)_3C^-K^+ + \tfrac{1}{2}H_2$$
$$(C_6H_5)_3CCl + 2\,K \longrightarrow (C_6H_5)_3C^-K^+ + KCl$$

The base gives intense blood-red solutions. Since enolate ions are colorless, a ketone may be converted into its enolate ion by titration with potassium triphenylmethide. So long as excess ketone remains, the solution is colorless. When sufficient base has been added to form the enolate ion completely, the red color of the triphenylmethide ion persists.

Solutions of enolate ions may be prepared and are quite stable if air and moisture are rigorously excluded. In many cases such enolate ions are valuable synthetic intermediates.

Enolate ions are ambident anions (Section 8.9). Just as they may undergo protonation on either carbon or oxygen, they may also react with other electrophilic species at either of these two centers. Two examples that illustrate this ambident character are the reactions with chlorotrimethylsilane and methyl iodide.

(95%)

> Chlorotrimethylsilane, $(CH_3)_3SiCl$, is a clear liquid, b.p. 57°, that has a number of uses in organic synthesis. The silicon–oxygen bond is strong ($DH° \approx 108$ kcal mole^{-1}). It is partly for this reason that chlorotrimethylsilane reacts with the ambient enolate ion exclusively on oxygen.

(98%)

> Potassium hydride, KH, is commercially available as a grey microcrystalline material dispersed in white mineral oil. Potassium hydride is insoluble in hydrocarbons, ethers, ammonia, and amines. For use, the mineral oil is washed away with an appropriate solvent and the hydride is used as a slurry. It is much more reactive than sodium hydride. It converts ketones to the potassium enolates in minutes at room temperature.

In one of the preceding reactions, reaction occurs exclusively on oxygen and in the other reaction occurs totally on carbon. Whether or not the oxygen or the carbon of an enolate ion is the site of reaction with an electrophile is determined by a number of factors. One important factor, which is illustrated by the examples just given, is the *reactivity of the electrophile*. In general, the more reactive the electrophile, the greater is the percentage of reaction at the oxygen atom. Other important factors are the nature of the solvent and the nature of the associated cation.

> The last two effects are both related to the **degree of association** of the enolate salt in solution. In an enolate ion the negative charge is borne mostly by the more electronegative oxygen atom. In most organic solvents the enolate salt exists in solution as an ion pair or as some higher aggregate of ions. In the ion pair the metal cation is in close juxtaposition to the *oxygen* of the enolate ion (it is actually partly *solvated* by this oxygen). The cation thus "protects" or "shields" the oxygen and tends to direct reactions of the ambient anion to the carbon. Polar aprotic solvents, such as dimethylformamide and dimethyl sulfoxide, solvate cations efficiently and tend to break up the ion pairs. The "bare" enolate ion has a greater tendency to undergo reaction on oxygen. With a given reagent the major amount of reaction may still be on carbon, but the percentage of reaction on oxygen is invariably higher in such solvents. The chemistry of such systems is rather complex and the details have not yet been fully worked out.
>
> Changing the metal cation from Li$^+$ to Na$^+$ to K$^+$ has a similar effect. Since the larger alkali metal ions require less solvation than does lithium, the corresponding enolate salts are more highly dissociated.

EXERCISE: The pK_a of acetone is about 20, and acetone contains about 0.01% of the enol form. Derive the pK_a of the enol form. How does this value compare with the pK_a values of alcohols?

C. Racemization

When (R)-3-phenyl-2-butanone is dissolved in aqueous ethanol that contains NaOH or HCl, the optical rotation of the solution gradually drops to zero. Reisolation from the reaction mixture yields a racemic mixture of the (R) and (S) enantiomers. The rate of racemization is proportional to the concentration of ketone and the concentration of NaOH or HCl. Clearly, racemization occurs by way of the intermediate enol form in which the former asymmetric carbon is planar and hence achiral.

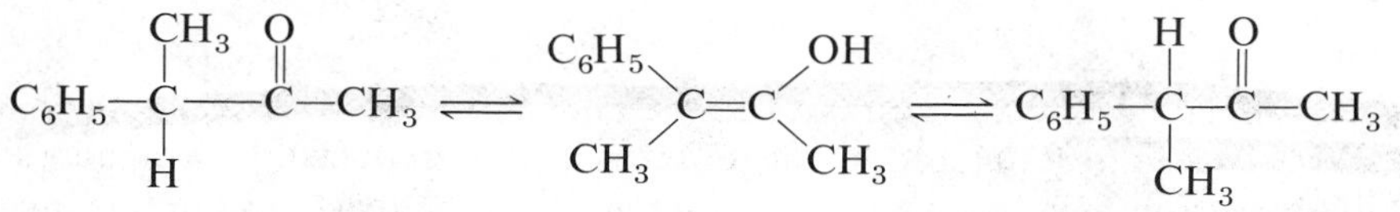

(R)-3-phenyl-2-butanone   (achiral)   (S)-3-phenyl-2-butanone

Since racemization involves the formation of the enol form, the rate of racemization is exactly equal to the rate of enolization.

$$\text{rate} = k[\text{ketone}][\text{H}^+] \quad \text{or} \quad k'[\text{ketone}][\text{OH}^-]$$

Furthermore, the rate of racemization is equal to the rate of deuterium incorporation because both reactions involve the same intermediate enol.

Note that racemization will occur *only* when the asymmetric carbon is α to the carbonyl group. If the aldehyde or ketone is chiral because of asymmetry at some other carbon, the enol form is also chiral, and enolization does not result in racemization.

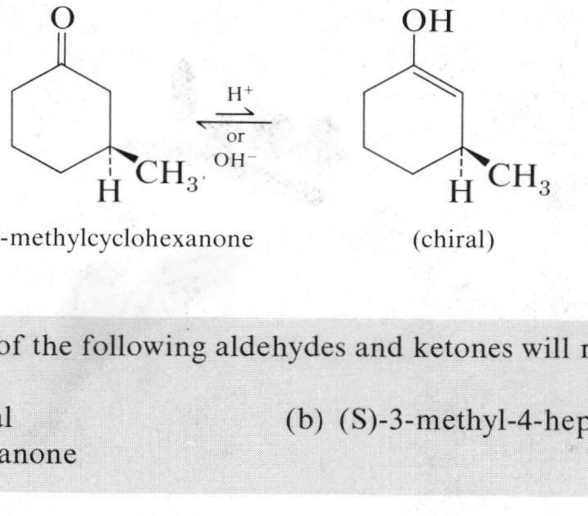

(S)-3-methylcyclohexanone   (chiral)

EXERCISE: Which of the following aldehydes and ketones will racemize in basic solution?
(a) (R)-2-methylbutanal   (b) (S)-3-methyl-4-heptanone
(c) (R)-4-methyl-2-hexanone

D. Halogenation

Aldehydes and ketones undergo acid- and base-catalyzed halogenation.

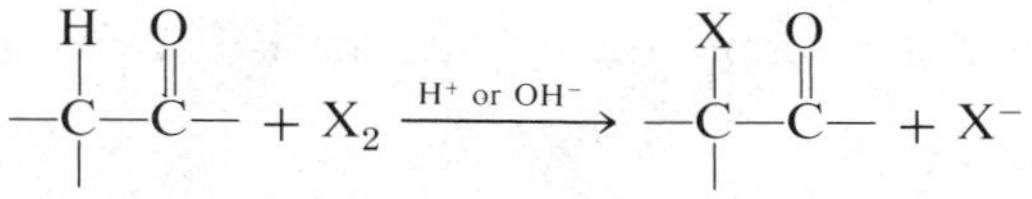

The reaction occurs with chlorine, bromine, and iodine.

$$\text{(cyclohexanone)} + Cl_2 \xrightarrow{H_2O} \text{(2-chlorocyclohexanone)} + HCl$$

(61–66%)

$$Br-\text{C}_6H_4-\overset{O}{\underset{\|}{C}}-CH_3 + Br_2 \xrightarrow[20°]{CH_3COOH} Br-\text{C}_6H_4-\overset{O}{\underset{\|}{C}}-CH_2Br + HBr$$

(69–77%)

Halogenation is often carried out without added catalyst. In this case the reaction is **autocatalytic** (catalyzed by one of the reaction products). In such an autocatalytic reaction there is no apparent reaction when the reactants are first mixed together. However, there is a slow uncatalyzed reaction that produces some HBr. As HBr is produced in the reaction, it catalyzes the formation of more HBr, which makes the reaction proceed even faster. In practice, autocatalytic reactions often have an **induction period** (a time when no apparent reaction is occurring), after which a rapid and vigorous reaction sets in and is soon over. In the base-catalyzed reaction, a full equivalent of base must obviously be used because an equivalent of HX is formed in the reaction. The actual species that reacts with the halogen is the enol form of the aldehyde or ketone; the purpose of the acid or base is simply to catalyze enolization. Acid-catalyzed halogenation is simply the normal electrophilic reaction of halogen with a double bond. The probable mechanism of the acid-catalyzed reaction is

$$(1) \quad -\overset{H}{\underset{|}{C}}-\overset{O}{\underset{\|}{C}}- + H^+ \underset{\text{fast}}{\rightleftharpoons} -\overset{H}{\underset{|}{C}}-\overset{\overset{+}{O}H}{\underset{\|}{C}}-$$

$$(2) \quad -\overset{H}{\underset{|}{C}}-\overset{\overset{+}{O}H}{\underset{\|}{C}}- \underset{\text{slow}}{\rightleftharpoons} \quad C=C\overset{OH}{\diagdown} \qquad enol$$

$$(3) \quad C=C\overset{OH}{\diagdown} + X_2 \underset{\text{fast}}{\rightleftharpoons} -\overset{X}{\underset{|}{C}}-\overset{\overset{+}{O}H}{\underset{\|}{C}}- + X^-$$

$$(4) \quad -\overset{X}{\underset{|}{C}}-\overset{\overset{+}{O}H}{\underset{\|}{C}}- \underset{\text{fast}}{\rightleftharpoons} -\overset{X}{\underset{|}{C}}-\overset{O}{\underset{\|}{C}}- + H^+$$

There is considerable evidence for this mechanism. The rate of the reaction depends only upon the concentration of the ketone and the acid; it is independent of halogen concentration. Chlorination and bromination occur at the same rate. Halogenation occurs at the same rate as does acid-catalyzed exchange of an α-proton for deuterium.

In the base-catalyzed reaction, the enolate ion is the probable intermediate.

$$(1) \quad -\overset{H}{\underset{|}{C}}-\overset{O}{\underset{\|}{C}}- + OH^- \rightleftharpoons \quad C=C\overset{O^-}{\diagdown} + H_2O$$

(2) $\left[\begin{array}{c} \diagup \\ C=C \diagdown \end{array} \overset{O^-}{} \longleftrightarrow \begin{array}{c} \diagup \\ C-C \diagdown \end{array} \overset{O}{} \right] + X_2 \rightleftharpoons \overset{X}{\underset{|}{-C}} - \overset{O}{\overset{\|}{C}} - + X^-$

The acid- and base-catalyzed halogenation reactions differ in several important aspects. In acid-catalyzed halogenation each successive halogenation step is normally slower than the previous one. Therefore it is usually possible to prepare a monohalo ketone in good yield by carrying out the halogenation under conditions of acid catalysis using one equivalent of halogen, as shown by the examples on page 376.

In the base-catalyzed reaction, each successive halogenation step is faster than the previous one, since the electron-attracting halogens increase the acidity of halogenated ketones; consequently base-catalyzed halogenation is not a generally useful method for preparation of a monohalo ketone.

Methyl ketones undergo base-catalyzed halogenation to give the trihalo ketone, which is normally not isolated.

$$R-\overset{O}{\overset{\|}{C}}-CH_3 + 3\,Br_2 + 3\,OH^- \longrightarrow R-\overset{O}{\overset{\|}{C}}-CBr_3 + 3\,Br^- + 3\,H_2O$$

Instead, the α,α,α-trihaloketone reacts further with hydroxide ion to give a carboxylate salt and the corresponding trihalomethane.

$$R-\overset{O}{\overset{\|}{C}}-CBr_3 + OH^- \longrightarrow RCO_2^- + CHBr_3 \qquad (13\text{-}1)$$

The overall reaction is known as the **haloform reaction.** The probable mechanism for the cleavage reaction is

(1) $R-\overset{O}{\overset{\|}{C}}-CBr_3 + OH^- \rightleftharpoons R-\overset{O^-}{\underset{\underset{OH}{|}}{C}}-CBr_3$

(2) $R-\overset{O^-}{\underset{\underset{OH}{|}}{C}}-CBr_3 \longrightarrow R-\overset{O}{\overset{\|}{C}}-OH + :CBr_3^-$

(3) $R-\overset{O}{\overset{\|}{C}}-OH + :CBr_3^- \underset{\text{fast}}{\rightleftharpoons} R-\overset{O}{\overset{\|}{C}}-O^- + HCBr_3$

The proposed mechanism is an example of a nucleophilic addition-elimination process. The addition of bases to a carbonyl group is treated in the next section. Furthermore, as we shall see in Chapters 18 and 19, the addition-elimination mechanism is important in the chemistry of carboxylic acid derivatives. In the present case the trihalomethyl anion is far more stable than methyl anion itself. (The pK_a of chloroform, $CHCl_3$, is about 25, and the other haloforms have comparable acidities.) Most of the time when hydroxide ion adds to the carbonyl group it comes right back off again, but sometimes the less stable CX_3^- ion comes off instead. This ion immediately becomes protonated so that the cleavage of the trihalomethyl ion, when it does occur, is irreversible. The net result is eventual cleavage of the halogenated ketone as shown in equation (13-1).

EXERCISE: The haloform reaction has been used as a method for converting a methyl ketone into a carboxylic acid. Show how 3,3-dimethylbutanoic acid, $(CH_3)_3CCH_2COOH$, could be prepared in this way. Note that the overall conversion amounts to an *oxidation*. What is it that gets reduced?

13.7
Carbonyl Addition Reactions

A. *Carbonyl Hydrates: Gem-diols*

The reaction of aldehydes and ketones with water to produce *gem*-diols is not a significant synthetic reaction, but it points up many of the important principles of reactions of carbonyl groups. Like the enolization reaction, the addition reaction with water involves consideration of both equilibrium and rate.

The equilibrium involved may be written as

$$\ce{C=O} + H_2O \overset{K}{\rightleftharpoons} \ce{C(OH)(OH)}$$

The magnitude of the equilibrium constant, K, is sensitive to the nature of the carbonyl group. In aqueous solution, the equilibrium constant is

$$K = \frac{\left[\ce{C(OH)(OH)}\right]}{\left[\ce{C=O}\right]}$$

(Note that the concentration of water, which is present in large excess, remains essentially constant. In such cases, it is usually not included in the definition of the equilibrium constant.) The equilibrium constant has a value of about 10^3 for formaldehyde, roughly 1 for other aldehydes such as propionaldehyde, and about 10^{-3} for ketones. Thus a solution of formaldehyde in water is almost all $CH_2(OH)_2$. Aqueous solutions of other aldehydes contain comparable amounts of $RCHO$ and $RCH(OH)_2$, and ketones are present almost wholly as the keto form, $RCOR'$.

$$CH_2{=}O + H_2O \overset{K}{\rightleftharpoons} HOCH_2OH \qquad K \approx 2 \times 10^3$$

$$\underset{\displaystyle \overset{\|}{O}}{CH_3CH_2CH} + H_2O \overset{K}{\rightleftharpoons} CH_3CH_2{-}\underset{\displaystyle \overset{|}{OH}}{\overset{OH}{CH}} \qquad K = 0.7$$

$$CH_3COCH_3 + H_2O \overset{K}{\rightleftharpoons} CH_3{-}\underset{\displaystyle \overset{|}{OH}}{\overset{OH}{C}}{-}CH_3 \qquad K \approx 0.002$$

This phenomenon can be explained with concepts that come from carbocation chemistry. Two important resonance structures can be written for a carbonyl group.

$$\left[\begin{array}{c} O \\ \parallel \\ -C- \end{array} \longleftrightarrow \begin{array}{c} O^- \\ \mid \\ -\overset{+}{C}- \end{array} \right]$$

The structure with the double bond is the more important because in it all atoms have complete octets. However, the other structure contributes to a significant extent. This dipolar structure has the character of a carbocation. Recall that the order of carbocation stability is secondary > primary > methyl.

The dipolar resonance structure of formaldehyde is analogous to a methyl cation.

$$\left[H-\overset{\overset{\displaystyle O}{\parallel}}{C}-H \longleftrightarrow H-\overset{\overset{\displaystyle O^-}{\mid}}{\underset{+}{C}}-H \right] \qquad H-\overset{\overset{\displaystyle H}{\mid}}{\underset{+}{C}}-H$$

formaldehyde   methyl cation

The dipolar resonance structure of a higher aldehyde is analogous to a primary carbocation, whereas that for a ketone is analogous to a secondary carbocation.

$$\left[CH_3CH_2-\overset{\overset{\displaystyle O}{\parallel}}{C}-H \longleftrightarrow CH_3CH_2-\overset{\overset{\displaystyle O^-}{\mid}}{\underset{+}{C}}-H \right] \qquad CH_3CH_2-\overset{\overset{\displaystyle H}{\mid}}{\underset{+}{C}}-H$$

propionaldehyde n-propyl cation

$$\left[CH_3-\overset{\overset{\displaystyle O}{\parallel}}{C}-CH_3 \longleftrightarrow CH_3-\overset{\overset{\displaystyle O^-}{\mid}}{\underset{+}{C}}-CH_3 \right] \qquad CH_3-\overset{\overset{\displaystyle H}{\mid}}{\underset{+}{C}}-CH_3$$

acetone   isopropyl cation

Just as isopropyl cation is more stable than n-propyl cation, acetone is more stable than propionaldehyde owing to the extra stabilization imparted by the dipolar resonance structure.

We can see this effect by an examination of heats of formation of isomeric aldehydes and ketones. Some of these comparisons are summarized in Table 13.5. Ketones are about 7 kcal mole^{-1} more stable than the isomeric aldehydes.

TABLE 13.5
Heats of Formation of Some Aldehydes and Ketones

Aldehydes	ΔH_f°, kcal mole^{-1} 25°, gas	Ketones	ΔH_f°, kcal mole^{-1} 25°, gas
HCHO	-26.0		
CH$_3$CHO	-39.7		
CH$_3$CH$_2$CHO	-45.5	CH$_3$COCH$_3$	-51.9
CH$_3$CH$_2$CH$_2$CHO	-48.9	CH$_3$CH$_2$COCH$_3$	-57.0
CH$_3$CH$_2$CH$_2$CH$_2$CHO	-54.5	CH$_3$CH$_2$CH$_2$COCH$_3$	-61.8
		CH$_3$CH$_2$COCH$_2$CH$_3$	-61.8

Consideration of the carbocation character of a carbonyl group has other corollaries as well. Consider the effect of a nearby polar substituent such as a chlorine, as in chloroacetaldehyde.

$$ClCH_2CHO$$

chloroacetaldehyde

The C—Cl dipole acts to destabilize the carbocation resonance structure. Hence, this structure contributes less to the overall resonance hybrid and the resonance hybrid is less stable as a result.

contributes less

No comparable effect operates on the corresponding hydrate. Thus, such a substituent shifts the equilibrium and the aldehyde is more hydrated. The equilibrium constant for trichloroacetaldehyde, "chloral," is about 3×10^4. This compound exists almost wholly as the hydrate.

[Chloral hydrate, $CCl_3CH(OH)_2$, is a crystalline solid, m.p. 57°, having a distinctive odor. Its narcotic effect has led to its illegal use as "knockout drops."]

The equilibrium between a carbonyl compound and its hydrate can also be described as the resultant of two rate constants.

$$CH_3\overset{O}{\overset{\|}{C}}CH_3 + H_2O \underset{k_{-1}}{\overset{k_1}{\rightleftharpoons}} CH_3\overset{OH}{\underset{OH}{\overset{|}{\underset{|}{C}}}}CH_3$$

The equilibrium constant is given by the ratio, $K = k_1/k_{-1}$. In the case of a ketone, such as acetone, the amount of hydrate present is so small that its rate of formation cannot be determined directly. The rate constant k_1 can be determined indirectly by an isotope exchange reaction. Water consists mostly of $H_2{}^{16}O$, but it also contains 0.20% of the heavy oxygen isotope as $H_2{}^{18}O$. Water enriched in the heavy isotope is available, and the rate of hydration can be followed as a rate of incorporation of ^{18}O into acetone. The ^{18}O content of the ketone can be determined by mass spectrometry (Chapter 17).

$$CH_3\overset{^{16}O}{\overset{\|}{C}}CH_3 + H_2{}^{18}O \underset{k_{-1}}{\overset{k_1}{\rightleftharpoons}} CH_3\overset{^{18}O}{\overset{\|}{C}}CH_3 + H_2{}^{16}O$$

This exchange reaction is slow in pure water, but is much faster in the presence of small amounts of acid or base. The exchange is both acid- and base-catalyzed. In the uncatalyzed reaction a molecule of water attacks the electron-deficient carbonyl carbon to produce an intermediate that undergoes rapid proton exchange to give the *gem*-diol.

The *gem*-diol decomposes to give back the ketone by an exact reversal of this sequence. If one of the oxygens in the diol is heavy oxygen, the dehydration process has an equal probability of losing the labeled oxygen in the leaving water or of retaining it in the ketone. However, water is a rather weakly basic reagent, and the carbonyl carbon is only slightly positive. Consequently the direct attack by water on the carbonyl carbon is a slow process.

Hydroxide ion is a much more basic reagent and its reaction with a carbonyl group is much faster than that of water.

$$
HO^- \overset{CH_3}{\underset{CH_3}{C}} = O \rightleftharpoons HO - \overset{CH_3}{\underset{CH_3}{C}} - O^- \overset{H_2O}{\rightleftharpoons} HO - \overset{CH_3}{\underset{CH_3}{C}} - OH + OH^-
$$

Note that the presence of hydroxide ion does not affect the *position* of equilibrium. It catalyzes the reverse reaction exactly as much as the forward reaction.

In the acid-catalyzed reaction the ketone oxygen is first protonated in a rapid equilibrium process exactly as in enolization.

$$
\overset{CH_3}{\underset{CH_3}{C}} = O + H^+ \rightleftharpoons \left[\overset{CH_3}{\underset{CH_3}{C}} = OH^+ \longleftrightarrow {}^+\overset{CH_3}{\underset{CH_3}{C}} - OH \right]
$$

In the protonated compound the carbonyl carbon has more positive charge than in the neutral ketone. One resonance structure is that of a hydroxycarbocation.

> A carbonyl group has dipolar character because oxygen is more electronegative than carbon and has greater attraction for electrons than carbon. The π-bond of the carbonyl group is relatively polarizable, and the electron density in this bond is displaced toward oxygen. In a protonated carbonyl group the oxonium ion oxygen is even more electronegative, and the electron density is displaced even more toward oxygen, leaving a more positive carbon.

The hydroxycarbocation reacts rapidly with water in a reaction that is analogous to the S_N1 solvolysis reaction involving carbocations (Section 8.12).

$$
H_2O + \left[{}^+\overset{CH_3}{\underset{CH_3}{C}} - OH \longleftrightarrow \overset{CH_3}{\underset{CH_3}{C}} = \overset{+}{O}H \right] \rightleftharpoons H_2\overset{+}{O} - \overset{CH_3}{\underset{CH_3}{C}} - OH \rightleftharpoons HO - \overset{CH_3}{\underset{CH_3}{C}} - OH + H^+
$$

$$
H_2O + {}^+\overset{CH_3}{\underset{CH_3}{C}} - CH_3 \rightleftharpoons H_2\overset{+}{O} - \overset{CH_3}{\underset{CH_3}{C}} - CH_3 \rightleftharpoons HO - \overset{CH_3}{\underset{CH_3}{C}} - CH_3 + H^+
$$

The reverse reaction of dehydration of the ketone hydrate is analogous to an El elimination of an alcohol.

EXERCISE: In this section and in Section 13.6, we have seen that several things are going on when an aldehyde or ketone is dissolved in aqueous solution. Some molecules are forming hydrates, while others are undergoing enolization. At any given time there are several neutral molecules and several anions or cations. Write equations showing the main processes that occur when acetone is dissolved in water at (a) pH 3 and (b) pH 11.

B. *Acetals and Ketals*

The equilibrium between carbonyl compounds and water is not a significant synthetic reaction because the *gem*-diols are generally unstable and readily dehydrate. However, the analogous reaction of alcohols has significant utility. The addition of 1 mole of an alcohol to the carbonyl group of an aldehyde or ketone yields a **hemiacetal** or a **hemiketal,** respectively.

$$CH_3CH_2CHO + CH_3OH \rightleftharpoons \underset{\text{a hemiacetal}}{CH_3CH_2\overset{\displaystyle OH}{\underset{\displaystyle |}{CH}}{-}OCH_3}$$

$$\underset{\text{a hemiketal}}{CH_3\overset{\displaystyle O}{\overset{\displaystyle \|}{C}}CH_3} + CH_3OH \rightleftharpoons CH_3\underset{\displaystyle OCH_3}{\overset{\displaystyle OH}{\underset{\displaystyle |}{\overset{\displaystyle |}{C}}}}CH_3$$

Addition of 2 moles of an alcohol, with the consequent formation of 1 mole of water, yields an **acetal** or a **ketal.**

$$CH_3CH_2CHO + 2\,CH_3OH \rightleftharpoons \underset{\text{an acetal}}{CH_3CH_2\overset{\displaystyle OCH_3}{\underset{\displaystyle OCH_3}{\overset{\displaystyle |}{\underset{\displaystyle |}{CH}}}}} + H_2O$$

$$\underset{\text{a ketal}}{CH_3\overset{\displaystyle O}{\overset{\displaystyle \|}{C}}CH_3} + 2\,CH_3OH \rightleftharpoons CH_3\overset{\displaystyle OCH_3}{\underset{\displaystyle OCH_3}{\overset{\displaystyle |}{\underset{\displaystyle |}{C}}}}CH_3 + H_2O$$

Formation of the hemiacetal or hemiketal is directly analogous to addition of water and is also subject to both acid and base catalysis. As with hydration, aldehydes give more of the addition product at equilibrium than do ketones.

Acid-catalyzed hemiacetal formation

(1) $CH_3\overset{\displaystyle O}{\overset{\displaystyle \|}{C}}H + H^+ \rightleftharpoons CH_3\overset{\displaystyle {}^+OH}{\overset{\displaystyle \|}{C}}H$

(2) $CH_3\overset{\displaystyle {}^+OH}{\overset{\displaystyle \|}{C}}H + CH_3OH \rightleftharpoons CH_3\underset{\displaystyle H}{\overset{\displaystyle OH}{\underset{\displaystyle |}{\overset{\displaystyle |}{CH}}}}\overset{\displaystyle +}{O}CH_3$

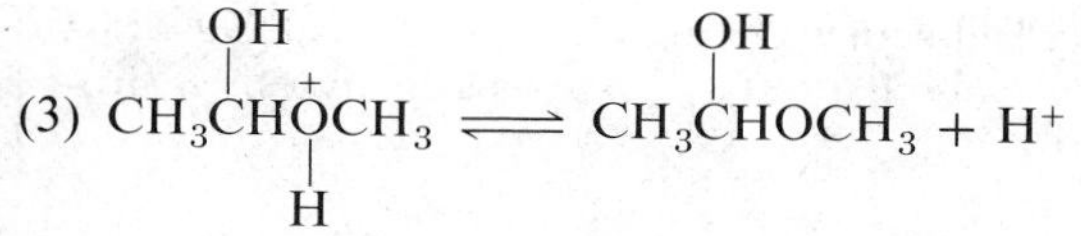

(3) $\underset{\text{H}}{\overset{\text{OH}}{CH_3\overset{+}{C}HOCH_3}} \rightleftharpoons \overset{\text{OH}}{CH_3CHOCH_3} + H^+$

Base-catalyzed hemiacetal formation

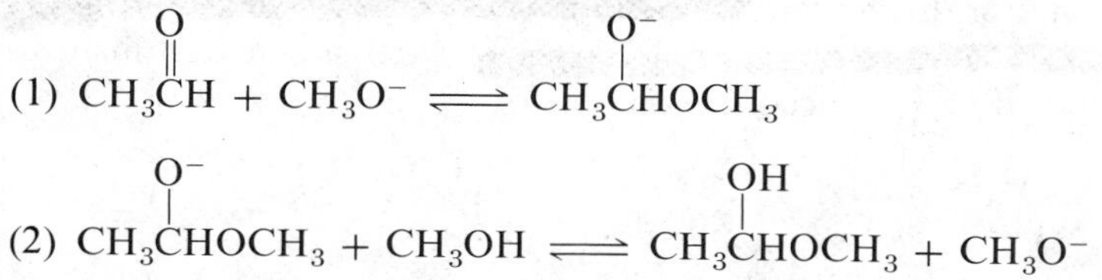

(1) $\overset{\text{O}}{\overset{\|}{CH_3CH}} + CH_3O^- \rightleftharpoons \overset{\text{O}^-}{CH_3CHOCH_3}$

(2) $\overset{\text{O}^-}{CH_3CHOCH_3} + CH_3OH \rightleftharpoons \overset{\text{OH}}{CH_3CHOCH_3} + CH_3O^-$

Like the hydrates, simple hemiacetals and hemiketals are generally not sufficiently stable for isolation.

Acetals and ketals are formed by way of the intermediate hemiacetal or hemiketal. Replacement of the —OH group by —OR is only acid-catalyzed.

Acid-catalyzed acetal formation

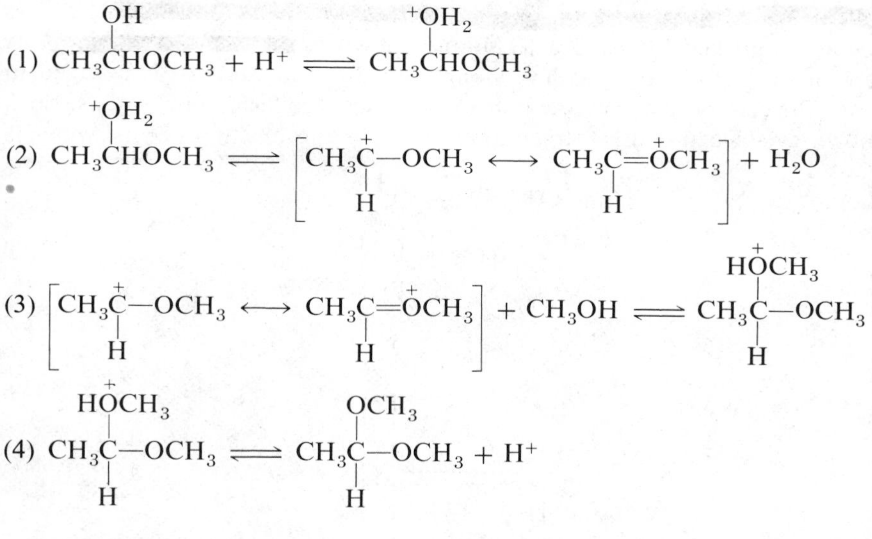

(1) $\overset{\text{OH}}{CH_3CHOCH_3} + H^+ \rightleftharpoons \overset{+\text{OH}_2}{CH_3CHOCH_3}$

(2) $\overset{+\text{OH}_2}{CH_3CHOCH_3} \rightleftharpoons \left[\underset{\text{H}}{CH_3\overset{+}{C}\text{—OCH}_3} \longleftrightarrow \underset{\text{H}}{CH_3C\overset{+}{=}OCH_3} \right] + H_2O$

(3) $\left[\underset{\text{H}}{CH_3\overset{+}{C}\text{—OCH}_3} \longleftrightarrow \underset{\text{H}}{CH_3C\overset{+}{=}OCH_3} \right] + CH_3OH \rightleftharpoons \underset{\text{H}}{CH_3\overset{\overset{+}{H}OCH_3}{C}\text{—OCH}_3}$

(4) $\underset{\text{H}}{CH_3\overset{\overset{+}{H}OCH_3}{C}\text{—OCH}_3} \rightleftharpoons \underset{\text{H}}{CH_3\overset{OCH_3}{C}\text{—OCH}_3} + H^+$

The net equilibrium that occurs when an aldehyde or ketone is treated with an alcohol and an acid catalyst is illustrated for acetone.

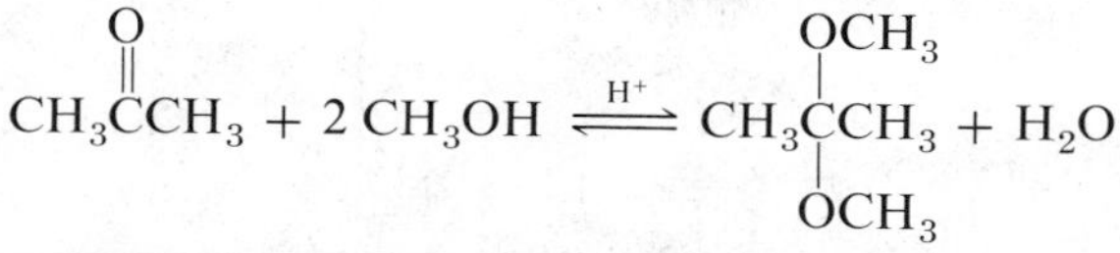

$$\overset{\text{O}}{\overset{\|}{CH_3CCH_3}} + 2\ CH_3OH \underset{}{\overset{H^+}{\rightleftharpoons}} \underset{OCH_3}{\overset{OCH_3}{CH_3CCH_3}} + H_2O$$

For simple aldehydes the overall equilibrium constant is favorable, and the acetal may be prepared simply by treating the aldehyde with two equivalents of alcohol and an acid catalyst.

$$CH_3CHO + 2\ CH_3CH_2OH \overset{H^+}{\rightleftharpoons} CH_3CH(OCH_2CH_3)_2 + H_2O$$

A mixture of 1305 mL of ethanol (21.7 moles), 500 g of acetaldehyde (11.4 moles) and 200 g of anhydrous $CaCl_2$ is placed in a 4 L bottle and kept at $25°$ for 1–2 days. At the end of this time the upper layer is washed with water and distilled to yield 790–815 g of 1,1-diethoxyethane, b.p. 101–103°. Note that $CaCl_2$ serves as a catalyst by hydrolyzing to give a small amount of HCl.

With larger aldehydes and with ketones, the equilibrium constant for acetalization or ketalization is generally unfavorable, more so for ketalization than for acetalization. For this reason the reaction is usually carried out using the alcohol as solvent to drive the equilibrium to the right. With aldehydes this usually allows the acetal to be produced in good yield.

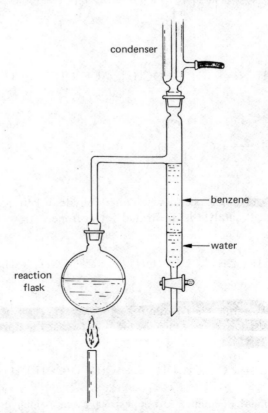

CHO
$+ CH_3OH \xrightarrow{HCl}$ CH(OCH$_3$)$_2$

NO$_2$

NO$_2$

3-nitrobenzaldehyde

(76–85%)
3-nitrobenzaldehyde
dimethyl acetal

With ketones the equilibrium lies even further to the left, and special techniques are used to remove water and drive the equilibrium to the right. This is usually accomplished by using a technique known as **azeotropic distillation**. A mixture of the ketone and alcohol, along with the acid catalyst, is refluxed in benzene. The condensate from the reflux condenser is collected in a water separator known as a Dean–Stark trap (Figure 13.7). As the reaction occurs, water is

condenser

benzene

water

reaction
flask

FIGURE 13.7. *The use of a water separator (Dean–Stark trap).*

formed and is distilled from the reaction flask as a binary azeotrope, which boils at 69° and is 91% benzene and 9% water. The distillate condenses in the condenser and runs down into the water separator where it forms two layers. At 20° the composition of the upper layer is 99.94% benzene and 0.06% water. The composition of the lower layer is 0.07% benzene and 99.93% water. As the water separator fills, the essentially dry benzene, being less dense, overflows and runs back into the reaction vessel. After a steady state is achieved, a normal reflux situation holds except that the water is trapped in the separator tube. This technique is a generally useful one for forcing equilibrium reactions to completion when one of the products is water. It may only be used when neither the ketone nor the alcohol is more volatile than benzene.

The acetal and ketal equilibria provide a fascinating illustration of the role of entropy in equilibria. As shown by the following examples, the formation of acetals and ketals is exothermic. The additional C—O bond formed is stronger than the second or π-bond of the carbonyl group.

$$\Delta H°, gas,$$
$$kcal\ mole^{-1}$$

$$
\left.
\begin{array}{ll}
CH_2O + 2\ CH_3OH \rightleftharpoons (CH_3O)_2CH_2 + H_2O & -19.0 \\
CH_3CHO + 2\ CH_3OH \rightleftharpoons (CH_3O)_2CHCH_3 + H_2O & -15.2 \\
CH_3COCH_3 + 2\ CH_3OH \rightleftharpoons (CH_3O)_2C(CH_3)_2 + H_2O & -11.7
\end{array}
\right\}\ . (13\text{-}2)
$$

These $\Delta H°$ values reflect the greater stability of ketones compared to aldehydes and show how formaldehyde is the least stable of all. However, note that even the ketone equilibrium is quite exothermic, yet the equilibrium constant is unfavorable. We must recall that the equilibrium constant is governed by $\Delta G°$, which depends on both enthalpy and entropy.

$$\Delta G° = \Delta H° - T\Delta S°$$

In these equilibria (13-2) three reactant molecules produce two product molecules. The resulting loss of the freedom of motion of one molecule corresponds to a negative entropy change. The formation of acetals from aldehydes is so exothermic that the equilibrium lies far to the right despite the unfavorable entropy change. In the formation of ketals, however, the entropy term dominates.

This unfavorable entropy effect is avoided by the use of a 1,2- or 1,3-diol to form a cyclic ketal as in (13-3) and (13-4).

(80%)

(13-3)

(83%)

(13-4)

Note in these cases that two reactant molecules produce two product molecules. The overall entropy change is approximately zero, and the exothermic enthalpy of the reaction results in favorable equilibria even for ketones.

Acetals and ketals are an important class of compounds in carbohydrate chemistry (Chapter 29). In other systems they are used principally to protect a carbonyl group during a synthetic scheme. Acetals and ketals are generally stable to basic conditions and are hydrolyzed back to carbonyl compounds in acidic solution. The following synthesis of 4-heptynal from 3-bromopropanal illustrates the use of an acetal as a **protecting group**.

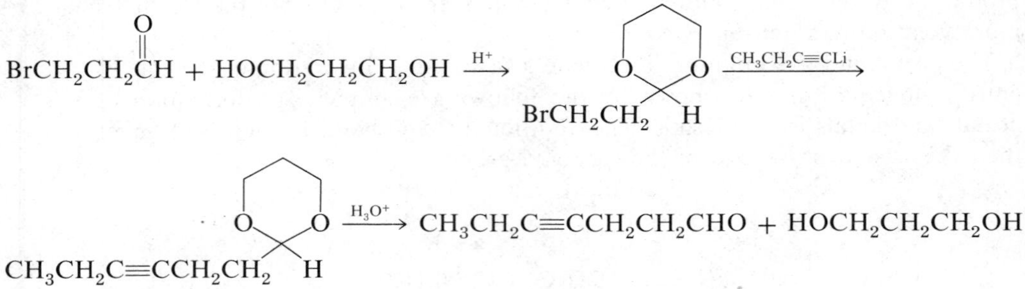

$$BrCH_2CH_2\overset{\displaystyle O}{\overset{\|}{C}}H + HOCH_2CH_2CH_2OH \xrightarrow{H^+}$$

$$\xrightarrow{CH_3CH_2C\equiv CLi}$$

$$\xrightarrow{H_3O^+} CH_3CH_2C\equiv CCH_2CH_2CHO + HOCH_2CH_2CH_2OH$$

In this case it is desired to replace Br by the 1-butynyl group. As we saw in Section 12.5, alkynyllithium compounds undergo ready alkylation by primary alkyl bromides. However, as we shall see in Section 15.6, organolithium reagents also react readily with the carbonyl group. Thus, direct replacement of bromine in the bromoaldehyde is not possible. Therefore the aldehyde is temporarily "protected" by conversion to the acetal, which is an ether and does not react with the organolithium reagent.

When acetals or ketals are pyrolyzed, one alcohol molecule is eliminated, and the product is an alkyl vinyl ether. Such compounds are related structurally to the enol form of the aldehyde or ketone, and they are commonly called **enol ethers**.

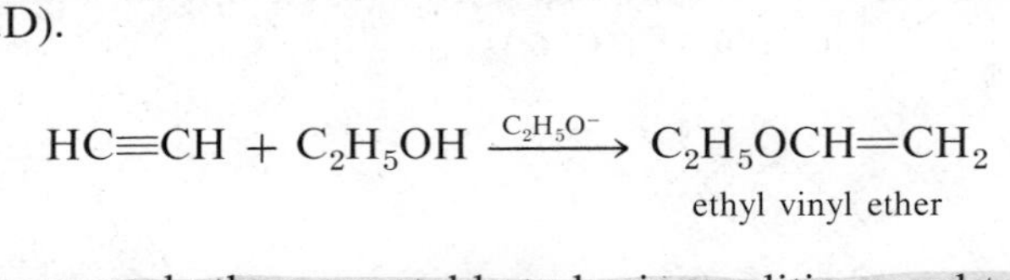

1-methoxycyclohexene

Enol ethers are also produced by the nucleophilic addition of alcohols to alkynes (Section 12.6.D).

$$HC\equiv CH + C_2H_5OH \xrightarrow{C_2H_5O^-} C_2H_5OCH=CH_2$$

ethyl vinyl ether

Like other ethers, enol ethers are stable to basic conditions and to basic reagents such as organolithium reagents. However, under acidic conditions they undergo rapid hydrolysis to give the aldehyde or ketone and alcohol.

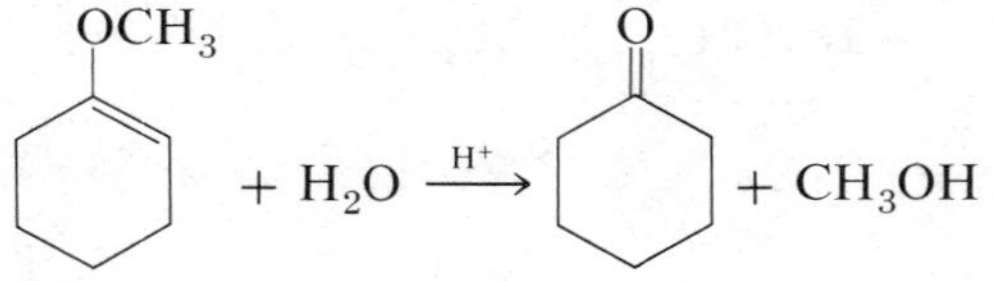

The mechanism of this ready hydrolysis involves the same type of intermediates as are involved in the formation and hydrolysis of acetals and ketals.

(1) $C_2H_5O-CH=CH_2 + H^+ \rightleftharpoons [C_2H_5O-\overset{+}{C}H-CH_3 \longleftrightarrow C_2H_5\overset{+}{O}=CH-CH_3]$

(2) $[C_2H_5O-\overset{+}{C}H-CH_3 \longleftrightarrow C_2H_5\overset{+}{O}=CH-CH_3] + H_2O \rightleftharpoons C_2H_5O-\overset{\overset{\displaystyle +OH_2}{|}}{C}H-CH_3$

(3) $C_2H_5O-\overset{\overset{\displaystyle +OH_2}{|}}{C}H-CH_3 \rightleftharpoons C_2H_5O-\overset{\overset{\displaystyle OH}{|}}{C}H-CH_3 + H^+$

(4) $C_2H_5O-\overset{\overset{\displaystyle OH}{|}}{C}H-CH_3 + H^+ \rightleftharpoons C_2H_5\overset{+}{\underset{\underset{\displaystyle H}{|}}{O}}-\overset{\overset{\displaystyle OH}{|}}{C}HCH_3$

(5) $C_2H_5\overset{+}{\underset{\underset{\displaystyle H}{|}}{O}}-\overset{\overset{\displaystyle OH}{|}}{C}HCH_3 \rightleftharpoons C_2H_5OH + [HO-\overset{+}{C}HCH_3 \longleftrightarrow H\overset{+}{O}=CHCH_3]$

(6) $[HO-\overset{+}{C}HCH_3 \longleftrightarrow H\overset{+}{O}=CHCH_3] \rightleftharpoons H\overset{\overset{\displaystyle O}{||}}{C}CH_3 + H^+$

Enol ethers are important intermediates in several important organic reactions (for example, Section 22.6.C).

Acetals and ketals are ethers and will form dangerous peroxides on exposure to air. Appropriate precaution should be taken when heating acetals and ketals that have had long exposure to oxygen.

There is a final important consequence to be discussed relative to the tendency of aldehydes to form acetals. On standing, aldehydes tend to form cyclic or polymeric acetals that are isomeric with several molecules of aldehyde. Formaldehyde itself is a gas that is available commercially as a 37% aqueous solution called formalin or as a solid polymer, paraformaldehyde. Formaldehyde for use in syntheses is normally obtained by heating the dry polymer.

[This linear polymer, $HO-(CH_2-O)_n-H$, forms the basis of some commercial plastics such as Delrin and Celcon. In these cases the terminal OH groups are "capped" with ester groups to prevent depolymerization or "unzipping" upon heating.]

Formaldehyde forms a cyclic trimer, trioxane, a solid having m.p. 64°, which can be sublimed unchanged. Paraldehyde, from acetaldehyde, is a liquid, b.p. 128°, that regenerates acetaldehyde on heating with a trace of acid.

trioxane   paraldehyde

Acetaldehyde also forms a cyclic tetramer, metaldehyde, a solid that sublimes readily. Other low molecular weight aldehydes form cyclic trimers related to paraldehyde. This kind of behavior is not shown by ketones.

EXERCISE: We shall see that aldehydes and ketones are *reduced* to alcohols by diborane. Outline a sequence of reactions that could be used to convert hept-6-en-2-one into 7-hydroxyheptan-2-one.

$$CH_2{=}CHCH_2CH_2CH_2\overset{\overset{\displaystyle O}{\|}}{C}CH_3 \longrightarrow HOCH_2CH_2CH_2CH_2CH_2\overset{\overset{\displaystyle O}{\|}}{C}CH_3$$

C. Reaction with Derivatives of Ammonia

Ammonia will react with aldehydes and ketones to form a compound containing the nitrogen analog of a carbonyl group. These compounds are called **imines.**

$$-\overset{\overset{\displaystyle O}{\|}}{C}- + NH_3 \rightleftharpoons \left[-\overset{\overset{\displaystyle OH}{|}}{\underset{\underset{\displaystyle NH_2}{|}}{C}}- \right] \rightleftharpoons H_2O + -\overset{\overset{\displaystyle NH}{\|}}{C}-$$

an imine

Imines derived from ammonia are an unimportant class of compounds. They hydrolyze rapidly even with water to generate carbonyl compounds.

The similar reaction with primary amines gives the more important substituted imine or **Schiff base.**

$$RCHO + R'NH_2 \rightleftharpoons RCH{=}NR' + H_2O$$

substituted imine
Schiff base

This type of reaction, in which two organic reagents are combined with the elimination of water, is generally referred to as a **condensation.** As in the case of the unsubstituted imines, most simple imines are fairly unstable compounds. They readily undergo hydrolysis back to the amine and carbonyl compound and are often prone to polymerization. However, when either the carbon or the nitrogen is substituted by a phenyl group, the resulting imine is generally rather stable.

(84–87%)

Imines prepared from aliphatic aldehydes and ketones and aliphatic amines are less stable than aromatic analogs and are somewhat more difficult to prepare. Since the equilibrium constant in this case is not as large as when there is a phenyl group attached to the C=N double bond, it is usually necessary to drive the reaction to completion by removal of water from the reaction mixture as it is formed, as in the formation of ketals. This is usually accomplished by azeotropic distillation, using benzene as solvent. An example of such a case is the condensation of cyclohexanone with *t*-butylamine.

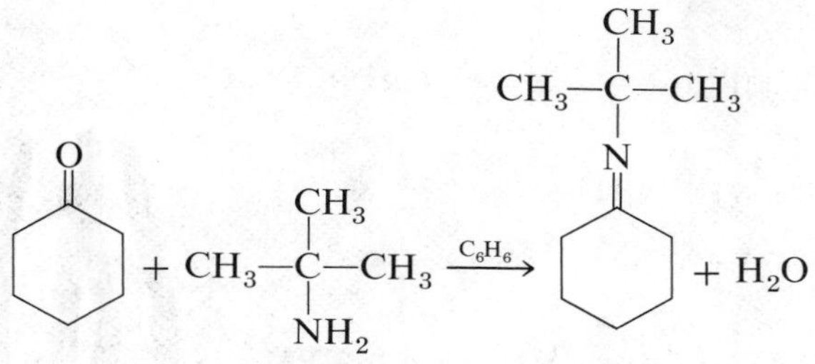

t-butylamine (85%)

Aldehydes and ketones also react with other ammonia derivatives to give analogous adducts. Common reagents are hydroxylamine (H_2NOH), hydrazine (H_2NNH_2), phenylhydrazine ($H_2NNHC_6H_5$), and semicarbazide ($H_2NNH\overset{O}{\overset{\|}{C}}NH_2$). Examples of such reactions follow. Unlike imines, the products of these reactions are generally stable.

$$CH_3-\overset{O}{\overset{\|}{C}}-CH_3 + H_2NOH \longrightarrow CH_3-\overset{NOH}{\overset{\|}{C}}-CH_3$$

hydroxylamine — an **oxime**, acetone oxime

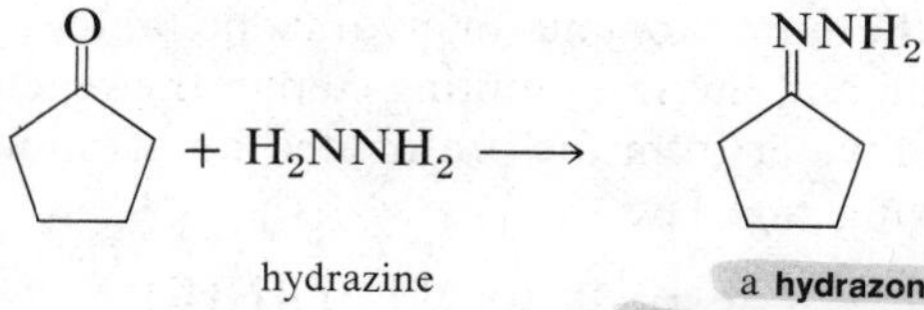

hydrazine — a **hydrazone**, cyclopentanone hydrazone

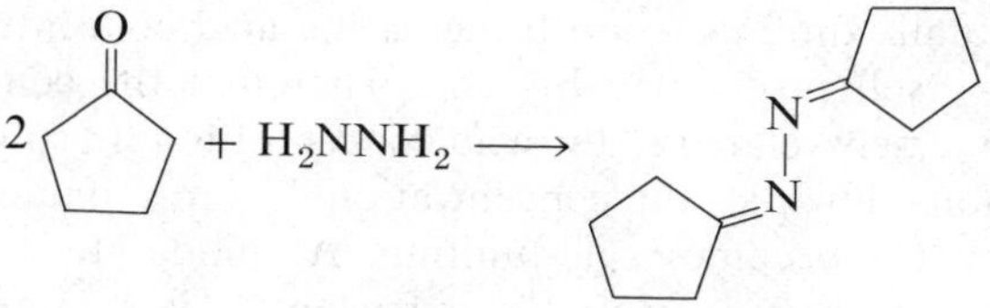

an **azine**
cyclopentanone azine

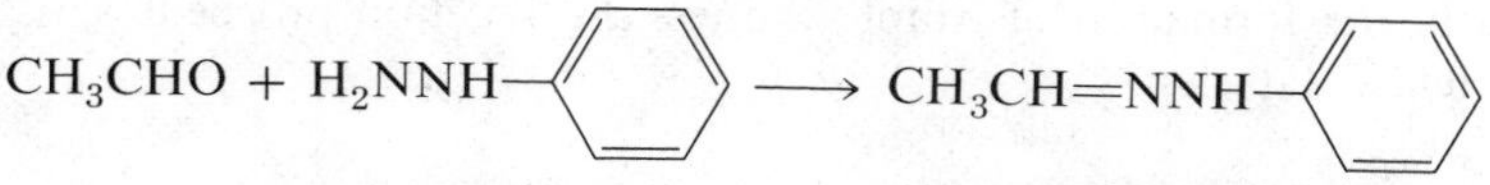

phenylhydrazine — a **phenylhydrazone**, acetaldehyde phenylhydrazone

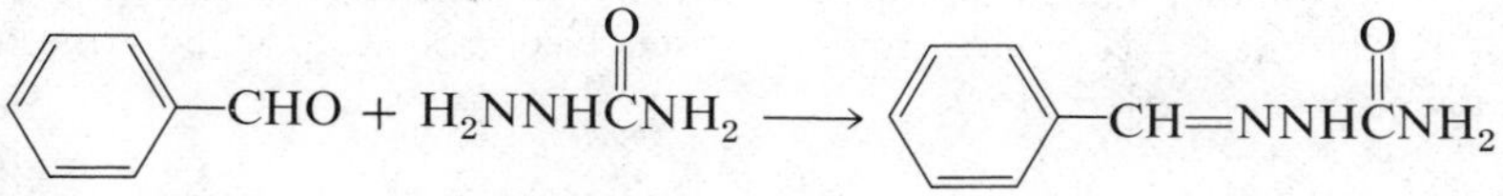

semicarbazide — a **semicarbazone**, benzaldehyde semicarbazone

The reactions of carbonyl compounds with substituted ammonia compounds are generally catalyzed by mild acid. The mechanism is directly analogous to the reactions discussed previously with water and alcohols.

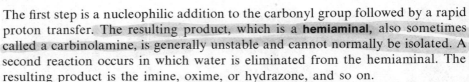

(1) $\ \diagup\!\!\!\diagdown C{=}O + R{-}\overset{..}{N}H_2 \;\rightleftharpoons\; \overset{\overset{-}{O}}{\underset{\ }{-C-\overset{+}{N}H_2R}}$

(2) $\overset{\overset{-}{O}}{\underset{\ }{-C-\overset{+}{N}H_2R}} \;\rightleftharpoons\; \overset{OH}{\underset{\ }{-C-NHR}}$

(3) $\overset{OH}{\underset{\ }{-C-NHR}} + H^+ \;\rightleftharpoons\; \overset{^+OH_2}{\underset{\ }{-C-NHR}}$

(4) $\overset{^+OH_2}{\underset{\ }{-C-NHR}} \;\rightleftharpoons\; \diagup\!\!\!\diagdown C{=}\overset{+}{N}HR + H_2O$

(5) $\diagup\!\!\!\diagdown C{=}\overset{+}{N}HR \;\rightleftharpoons\; \diagup\!\!\!\diagdown C{=}NR + H^+$

The first step is a nucleophilic addition to the carbonyl group followed by a rapid proton transfer. The resulting product, which is a **hemiaminal,** also sometimes called a carbinolamine, is generally unstable and cannot normally be isolated. A second reaction occurs in which water is eliminated from the hemiaminal. The resulting product is the imine, oxime, or hydrazone, and so on.

Step (3) is also rapid and the rate-limiting step in the reaction is generally step (4), the elimination of water from the protonated hemiaminal. The overall reaction obeys the following rate law.

$$\text{rate} = k[\text{ketone}][H^+][RNH_2]$$

Although the reaction is acid-catalyzed at moderate pH, at higher acid concentration the rate actually diminishes with increasing acid concentration because the nitrogen base is itself protonated by acid. Therefore the concentration of free nucleophile is inversely related to the acid concentration. In solutions having high acid concentrations (low pH) the concentration of unprotonated nitrogen base is so low that step (1) becomes rate-limiting. At moderate acid concentrations enough free nitrogen base is available that step (1) is a rapid equilibrium, yet enough acid is also available to catalyze step (4). For this reason the reaction is often run in the presence of a buffer such as sodium acetate. In some cases, particularly the formation of simple imines, the reaction proceeds satisfactorily without acid catalysis.

> **EXERCISE:** Write structures for the following derivatives of 2-pentanone: oxime, phenylhydrazone, hydrazone, semicarbazone, and Schiff base with aniline ($C_6H_5NH_2$). Note that there are two stereoisomers for each of these compounds (*cis* or *trans* about the C=N bond). How many stereoisomers exist for the azine of 2-pentanone?

D. *Addition of Acetylide Anions*

Just as acetylide ions act as nucleophilic reagents in S_N2 reactions (Section 11.5.C), so also do they react with carbonyl groups in aldehydes and ketones to form the corresponding alkynylcarbinols. In practice, the sodium salts of terminal alkynes are prepared from sodium amide in liquid ammonia and are treated with the carbonyl compound.

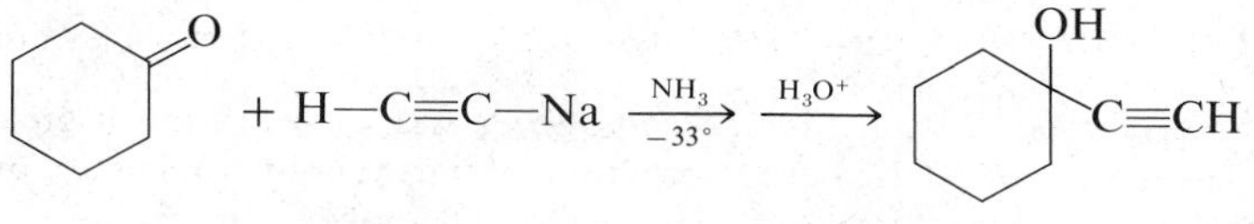

(65–75%)
1-ethynylcyclohexanol

A stream of dry acetylene is passed into a solution of 23 g of sodium in 1 L of liquid ammonia. After the sodium has been consumed, 98 g of cyclohexanone is added dropwise. The ammonia is allowed to evaporate, and the residue is treated with 400 mL of ice water and acidified with 50% H_2SO_4. The product is extracted with ether and distilled. The yield of 1-ethynylcyclohexanol is 81–93 g (65–75%).

EXERCISE: Show the principal reaction products from 1-ethynylcyclohexanol with (a) H_2SO_4/Hg^{2+}; (b) B_2H_6–alkaline-H_2O_2; (c) H_2/Lindlar catalyst; and (d) B_2H_6–CH_3COOH. Note how the reactions already learned can lead to compounds with more than one functional group.

E. Addition of HCN

Most acids will add to the carbonyl group to some extent, but usually the adducts are not stable. With HCl, for example, the equilibrium lies far to the left and α-chloro alcohols cannot be isolated.

$$CH_3\overset{O}{\overset{\|}{C}}CH_3 + HCl \rightleftharpoons CH_3\overset{OH}{\underset{Cl}{\overset{|}{\underset{|}{C}}}}CH_3$$

If such an α-chloro alcohol is produced by some other process, it immediately decomposes to give HCl and the corresponding aldehyde or ketone.

Hydrocyanic acid does add to many carbonyl compounds to give stable adducts. The product cyano alcohols are commonly called **cyanohydrins.** Since a strong C—C bond is formed in the adduct, the equilibrium often favors the product cyanohydrin.

$$CH_3\overset{O}{\overset{\|}{C}}CH_3 + HCN \rightleftharpoons CH_3\overset{OH}{\underset{CN}{\overset{|}{\underset{|}{C}}}}CH_3$$

acetone cyanohydrin

The relative stabilities of 1-chloro alcohols and cyanohydrins can be appreciated by comparing bond energies.

$$RCH{=}O + HY \rightleftharpoons RCH\underset{Y}{\overset{|}{-}}OH$$

$$\Delta H° = E(C{=}O) + E(H{-}Y) - E(O{-}H) - E(C{-}O) - E(C{-}Y)$$

The differences for $\Delta H°$ from one Y group to another are in the comparisons of $E(H—Y) - E(C—Y)$. To evaluate these bond-strength differences, compare the $DH°$ values of H—Y and CH_3—Y for Y = Cl and CN.

Y = Cl: $DH°(H—Cl) - DH°(CH_3—Cl) = 103 - 84 = 19$ kcal mole^{-1}
Y = CN: $DH°(H—CN) - DH°(CH_3—CN) = 120 - 112 = 8$ kcal mole^{-1}

The difference in bond strengths between H—Cl and C—Cl is much greater than between H—CN and C—CN; hence, formation of the cyanohydrin has a more favorable energy change than formation of a 1-chloroalkanol.

For aldehydes and most aliphatic ketones the equilibrium favors the adduct. For many aryl ketones and for some aliphatic ketones the equilibrium constant is small, and the reaction is not a useful one. The reaction is a typical nucleophilic addition, with the attacking nucleophile being CN$^-$. Addition is therefore catalyzed by base, which increases the cyanide concentration. The reaction can be carried out using liquid hydrogen cyanide (b.p. 26°) as the solvent.

$$ClCH_2CHO + HCN \longrightarrow ClCH_2\overset{\overset{\displaystyle OH}{|}}{C}HCN$$
$$(95\%)$$

Because of the high toxicity of HCN, procedures such as the foregoing are seldom used. A more common procedure is to generate the HCN *in situ* by the addition of HCl or H_2SO_4 to a mixture of the carbonyl compound and sodium or potassium cyanide.

$$CH_2O + KCN \xrightarrow[H_2O]{H_2SO_4} HOCH_2CN$$
$$(76-80\%)$$

A mixture of 130 g of potassium cyanide in 250 mL of H_2O and 170 mL of 37% formaldehyde solution (formalin) is prepared. To this solution is added a mixture of 57 mL of conc. H_2SO_4 and 173 mL of H_2O. The product, obtained by exhaustive extraction with ether, weighs 87–91 g (76–80%).

Under strongly basic conditions cyanohydrin formation may be reversed. The equilibrium is shifted by transformation of the weakly acidic HCN ($pK_a = 11$) into its conjugate base.

$$\rightleftharpoons \quad + \text{ HCN} \xrightarrow{\text{NaOH}} Na^+CN^- + H_2O$$

F. *The Aldol Condensation*

When acetaldehyde is treated with aqueous sodium hydroxide solution, 3-hydroxybutanal is formed in 50% yield.

$$2\ CH_3CHO \xrightarrow[5°,\ 4-5\ hr]{OH^-,\ H_2O} CH_3\overset{\overset{\displaystyle OH}{|}}{C}HCH_2CHO$$
$$(50\%)$$
$$\text{aldol}$$

The reaction is a general one for aldehydes that have a hydrogen α to the carbonyl group. The reaction is commonly known as the **aldol condensation,** from the trivial name of the simplest β-hydroxyaldehyde obtainable by this reaction. The term "aldol" is used both as a trivial name for 3-hydroxybutanal and as a generic name for β-hydroxyaldehydes and ketones in general. Under more vigorous conditions (base concentration, temperature), elimination of the β-hydroxy group occurs and an **α,β-unsaturated aldehyde** is produced.

$$2\ CH_3CH_2CH_2CHO \xrightarrow[\substack{80-100° \\ 3\ hr}]{1\ M\ NaOH,\ H_2O} \left[CH_3CH_2CH_2\overset{\displaystyle OH}{\underset{\displaystyle \underset{\displaystyle CH_3}{CH_2}}{CH}}CHCHO \right] \xrightarrow{-H_2O}$$

$$CH_3CH_2CH_2CH{=}\overset{\displaystyle CHO}{C}CH_2CH_3$$

(86%)
2-ethylhex-2-enal

The probable mechanism for the aldol condensation is illustrated for butyraldehyde.

(1) $CH_3CH_2CH_2CHO + OH^- \rightleftharpoons$

$$[CH_3CH_2CH{=}CH{-}O^- \longleftrightarrow CH_3CH_2\bar{C}H{-}CH{=}O] + H_2O$$

(2) $CH_3CH_2CH_2\overset{\displaystyle O}{\overset{\|}{C}}H + :\bar{C}H{-}CHO \rightleftharpoons CH_3CH_2CH_2\overset{\displaystyle O^-}{\underset{\displaystyle CH_2CH_3}{CH}}CHCHO$

hspace CH_2CH_3

(3) $CH_3CH_2CH_2\overset{\displaystyle O^-}{\underset{\displaystyle CH_2CH_3}{CH}}CHCHO + H_2O \rightleftharpoons CH_3CH_2CH_2\overset{\displaystyle OH}{\underset{\displaystyle CH_2CH_3}{CH}}CHCHO + OH^-$

The reaction is simply the nucleophilic addition of an enolate ion onto the carbonyl group of another, un-ionized, molecule. The slow, rate-limiting, step is usually this addition step. The dehydration of the β-hydroxyaldehyde or ketone formed by the above mechanism to produce an α,β-unsaturated carbonyl compound generally involves the enolate ion of the aldol product.

(4) $CH_3CH_2CH_2\overset{\displaystyle OH}{\underset{\displaystyle CH_2CH_3}{CH}}CHCHO + OH^- \rightleftharpoons CH_3CH_2CH_2\overset{\displaystyle OH}{\underset{\displaystyle CH_2CH_3}{CH}}CCHO + H_2O$

(5) $CH_3CH_2CH_2\overset{\displaystyle OH}{CH}{-}CCHO \rightleftharpoons CH_3CH_2CH_2CH{=}\underset{\displaystyle CH_2CH_3}{C}CHO + OH^-$

hspace CH_2CH_3

When a mixture of two aldehydes is treated with base, four aldol products are possible.

$$RCH_2CHO + R'CH_2CHO \longrightarrow \begin{cases} \overset{\displaystyle OH}{\underset{\displaystyle |}{}} \overset{\displaystyle R}{\underset{\displaystyle |}{}} \\ RCH_2CH-CHCHO \\ \\ \overset{\displaystyle OH}{\underset{\displaystyle |}{}} \overset{\displaystyle R'}{\underset{\displaystyle |}{}} \\ RCH_2CH-CHCHO \\ \\ \overset{\displaystyle OH}{\underset{\displaystyle |}{}} \overset{\displaystyle R}{\underset{\displaystyle |}{}} \\ R'CH_2CH-CHCHO \\ \\ \overset{\displaystyle OH}{\underset{\displaystyle |}{}} \overset{\displaystyle R'}{\underset{\displaystyle |}{}} \\ R'CH_2CH-CHCHO \end{cases}$$

In practice, a complex mixture usually results in such a situation. However, when one of the aldehydes cannot form an enolate ion or when one has a fairly unreactive carbonyl group, such a "mixed aldol condensation" is often feasible.

furfural + CH_3CH_2CHO $\xrightarrow{\text{NaOH, H}_2\text{O}}$ (72%)

A rather special case in which the mixed aldol condensation is useful employs formaldehyde as one component. The initial β-hydroxyaldehyde undergoes a further reaction with excess formaldehyde to give the diol and formate ion (a crossed Cannizzaro reaction; Section 13.8.D).

$$\underset{\text{isobutyraldehyde}}{CH_3\overset{\displaystyle CH_3}{\underset{\displaystyle |}{C}}HCHO} + CH_2O \xrightarrow{\text{NaOH, H}_2\text{O}} \underset{\substack{(90\%) \\ \text{2,2-dimethyl-1,3-propanediol}}}{CH_3-\overset{\displaystyle CH_3}{\underset{\displaystyle CH_2OH}{C}}-CH_2OH}$$

Although ketones also undergo the aldol condensation, the reaction in this case often requires rather special conditions. The overall reaction is an equilibrium process, and it appears that the equilibrium constant in most ketone aldol condensations is unfavorable. For example, acetone and its aldol product (4-hydroxy-4-methyl-2-pentanone, often known by the common name "diacetone alcohol") are in rapid equilibrium in the presence of base catalysts. The amount of the aldol product in the equilibrium in the presence of base catalysts. The amount of the aldol product in the equilibrium is only a few percent. However, if the product is removed from the basic catalyst as it is formed, the conversion can be accomplished in 80% yield.

$$2\ CH_3-\overset{\displaystyle O}{\overset{\displaystyle \|}{C}}-CH_3 \underset{}{\overset{\text{BaO}}{\rightleftharpoons}} CH_3-\overset{\displaystyle OH}{\underset{\displaystyle CH_3}{\overset{\displaystyle |}{C}}}-CH_2-\overset{\displaystyle O}{\overset{\displaystyle \|}{C}}-CH_3$$

"diacetone alcohol"

Intramolecular aldol condensation of diketones is an important method for the synthesis of cyclic compounds.

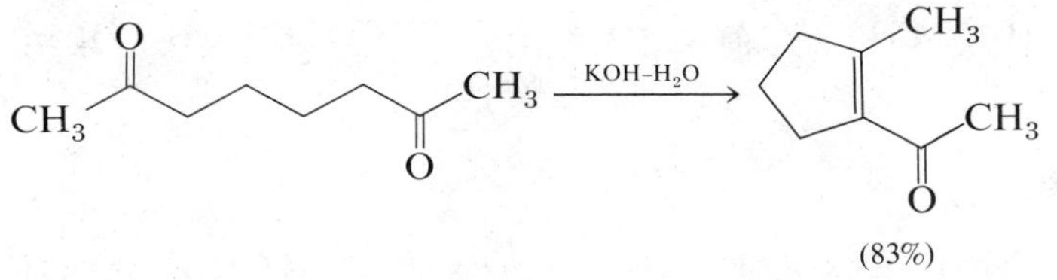

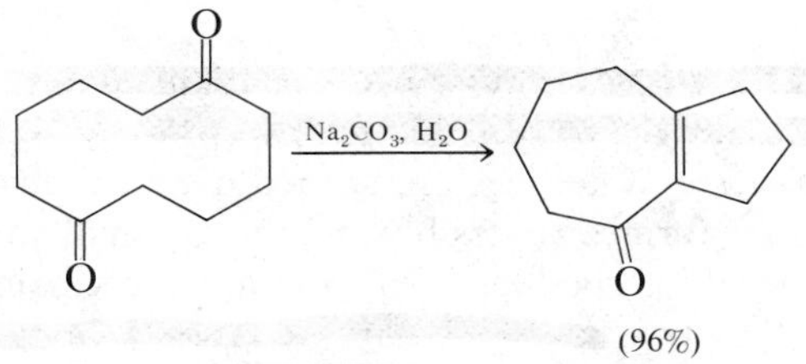

(83%)

(96%)

Since ketones undergo self-condensation much more slowly than aldehydes, mixed aldol condensations between a ketone and a nonenolizable aldehyde are usually clean. In order to assure the formation of a 1:1 adduct, an excess of ketone is often used, and the reaction is carried out under fairly mild conditions.

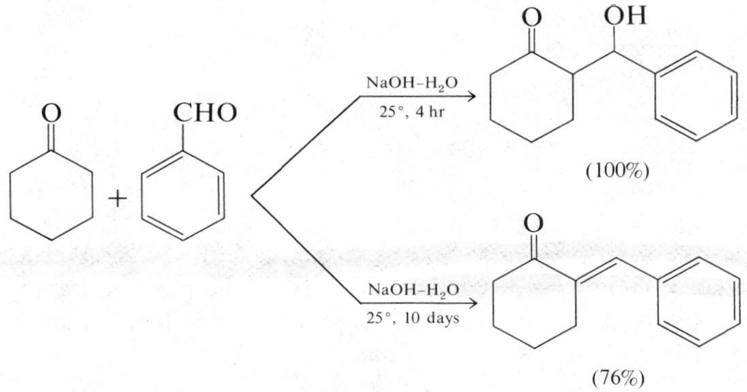

(100%)

(76%)

In cases where the aldehyde carbonyl group is fairly hindered, or when one side of the carbonyl group has only one hydrogen, 1:1 adducts form readily with little complication from more extensive reaction.

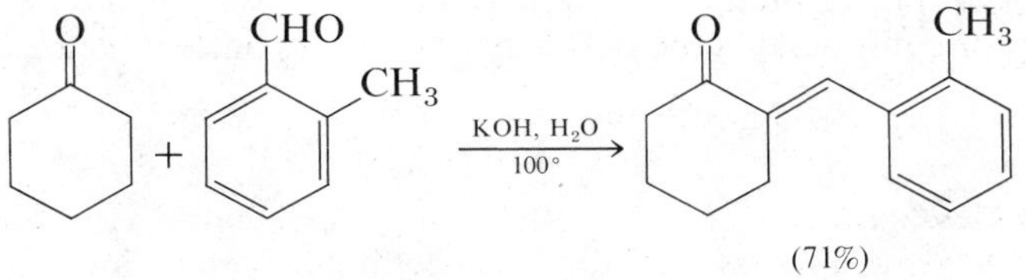

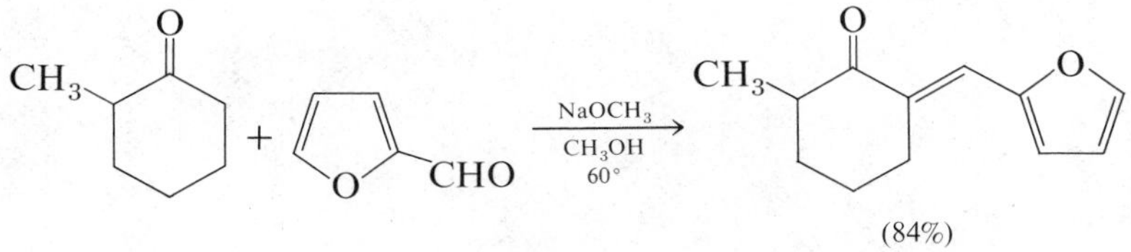

(71%)

(84%)

However, with excess (xs) aldehyde it is possible for reaction to occur at both sides of a ketone carbonyl group.

$$\text{xs } CH_3COCH_3 + C_6H_5CHO \xrightarrow[25°]{NaOH} CH_3COCH=CHC_6H_5$$

(65–78%)

benzalacetone

$$CH_3COCH_3 + \text{xs } C_6H_5CHO \xrightarrow[25°]{NaOH} C_6H_5CH=CH\overset{\overset{\displaystyle O}{\|}}{C}CH=CHC_6H_5$$

dibenzalacetone

The aldol condensations discussed up to this point are carried out using a base such as hydroxide ion or ethoxide ion in a protic solvent such as water or ethanol. Since aldehydes and ketones are a good deal less acidic than these solvents, the enolate ions are formed reversibly and in only small amounts (see Section 13.6.B). However, the aldol condensation may also be carried out in another manner. Recall that the strong base lithium diisopropylamide (LDA) converts a ketone completely into the corresponding enolate ion (see page 373). Since all of the ketone is consumed in this essentially irreversible reaction, aldol condensation does not occur. However, if an aldehyde is subsequently added to the cold enolate solution, a rapid and efficient mixed aldol condensation occurs. The initial product is the lithium salt of the β-hydroxy ketone. Aqueous work-up affords the aldol.

$$CH_3\overset{\overset{\displaystyle O}{\|}}{C}CH_3 \xrightarrow[THF, -78°]{LDA} CH_3\overset{\overset{\displaystyle O^-Li^+}{|}}{C}=CH_2 \xrightarrow{CH_3CH_2CHO}$$

$$CH_3\overset{\overset{\displaystyle O}{\|}}{C}CH_2\overset{\overset{\displaystyle O^-Li^+}{|}}{C}HCH_2CH_3 \xrightarrow[(work-up)]{H_2O} CH_3\overset{\overset{\displaystyle O}{\|}}{C}CH_2\overset{\overset{\displaystyle OH}{|}}{C}HCH_2CH_3$$

(85%)

4-hydroxy-2-hexanone

With unsymmetrical ketones, the strong base LDA always removes a proton from the *least sterically hindered position*; that is, from the position having the larger number of α-hydrogens. Thus, in such a case the mixed aldol condensation occurs specifically at one of the two α-positions.

$$CH_3\overset{\overset{\displaystyle O}{\|}}{C}CH_2CH_3 \xrightarrow[THF, -78°]{LDA} \xrightarrow{CH_3CHO} \xrightarrow{H_2O} CH_3\overset{\overset{\displaystyle OH}{|}}{C}HCH_2\overset{\overset{\displaystyle O}{\|}}{C}CH_2CH_3$$

(75%)

5-hydroxy-3-hexanone

This method of preforming the lithium enolate and subsequently adding the carbonyl receptor compound is the best way of carrying out a mixed aldol condensation.

EXERCISE: Show how each of the following compounds may be prepared from simple aldehydes and ketones utilizing the aldol condensation. For (e) two steps are required.

(a)

(b)

(c)

(d)

(e)

G. *Wittig Reaction*

Alkyl halides react with triphenylphosphine, $(C_6H_5)_3P$, by the S_N2 mechanism to give crystalline **phosphonium salts.**

$$(C_6H_5)_3P + CH_3CH_2Br \longrightarrow (C_6H_5)_3\overset{+}{P}CH_2CH_3\ Br^-$$

ethyltriphenylphosphonium bromide

Phosphine, PH_3, is the phosphorus analog of ammonia (Section 26.6). It is a poisonous gas and is usually spontaneously flammable because of the presence of impurities. Triphenylphosphine, $(C_6H_5)_3P$, is a commercially available crystalline solid, m.p. 80°. It is insoluble in water, but is soluble in most organic solvents. Although the pair of electrons on phosphorus is not appreciably basic, phosphines are generally reactive nucleophiles in S_N2 displacement reactions (Section 8.9).

Since phosphines are good nucleophiles and weak bases, competing elimination is not so important here as in other bimolecular substitutions. Consequently most primary and secondary alkyl halides give good yields of phosphonium salts.

(72%)
cyclohexyltriphenylphosphonium iodide

The alkyl proton α to the positive phosphorus may be removed by a strong base such as butyllithium or sodium hydride to give a neutral phosphorus compound called an **ylide** or **phosphorane.**

$$(C_6H_5)_3\overset{+}{P}CH_2CH_3\ Br^- + n\text{-}C_4H_9Li \xrightarrow{\text{ether}}$$

$$[C_6H_5)_3\overset{+}{P}\text{---}\overset{-}{C}HCH_3 \longleftrightarrow (C_6H_5)_3P{=}CHCH_3] + C_4H_{10} + LiBr$$

The formula $(C_6H_5)_3P{=}CHCH_3$ implies an expansion of the phosphorus octet and orbital overlap with phosphorus $3d$-atomic orbitals. Detailed quantum-mechanical studies show that the actual participation of such d-orbitals in bonding in phosphorus ylides is minor. Instead, the dipolar structure is stabilized by polarization of the electrons around phosphorus. Such polarization can be represented in terms of an induced dipole on phosphorus.

$$(C_6H_5)_3\overset{+}{\underset{\longleftrightarrow}{P}}\text{---}\overset{-}{C}HCH_3$$

Nevertheless, we will frequently use the simple pentacoordinate formula for convenience.

Ylides react rapidly with aldehydes and ketones, even at $-80°$, to give neutral products called **oxaphosphetanes.**

The mechanism of this addition involves an initial nucleophilic addition of the ylide carbon to the carbonyl group, giving a dipolar intermediate called a **betaine,** which then reacts further to give the oxaphosphetane. At $-80°$ the oxaphosphetane is stable in solution. Upon warming the solution to $0°$, it decomposes to give an alkene and triphenylphosphine oxide. The overall process, illustrated in equations (13-5) with acetone and the ylide derived from ethyl bromide, is called the **Wittig reaction.**

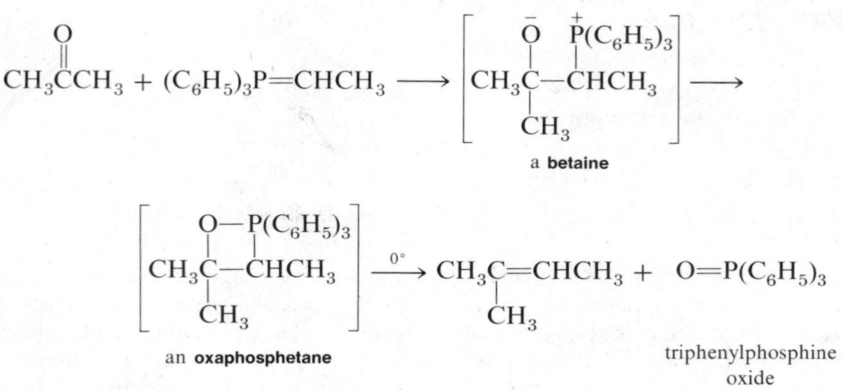

$$CH_3\overset{\text{O}}{\overset{\|}{C}}CH_3 + (C_6H_5)_3P=CHCH_3 \longrightarrow \left[CH_3\overset{\overset{\overset{-}{O}}{|}}{\underset{\underset{CH_3}{|}}{C}}-\overset{\overset{+}{P}(C_6H_5)_3}{\underset{}{C}}HCH_3 \right] \longrightarrow$$

a betaine

$$\left[CH_3\overset{\overset{O-P(C_6H_5)_3}{|}}{\underset{\underset{CH_3}{|}}{C}}-CHCH_3 \right] \xrightarrow{0°} CH_3\underset{\underset{CH_3}{|}}{C}=CHCH_3 + O=P(C_6H_5)_3 \quad (13\text{-}5)$$

an oxaphosphetane   triphenylphosphine
oxide

The Wittig reaction is an exceedingly useful method for the synthesis of alkenes. Although a mixture of *cis* and *trans* isomers often results, *only a single positional isomer is produced*. Consider, as an example, the synthesis of methylenecyclohexane. Dehydration of 1-methylcyclohexanol gives mainly 1-methylcyclohexene, since this isomer is more stable.

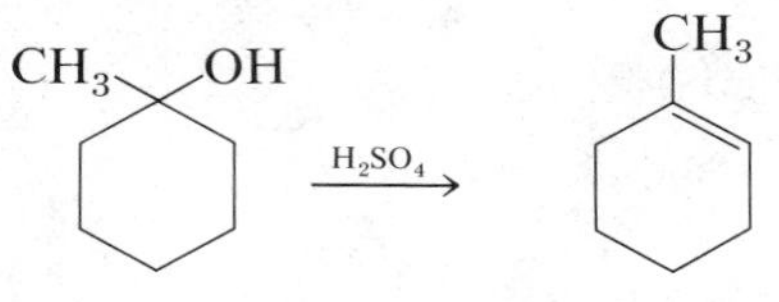

1-methylcyclohexene

The less stable isomer may be readily prepared from cyclohexanone by the Wittig reaction.

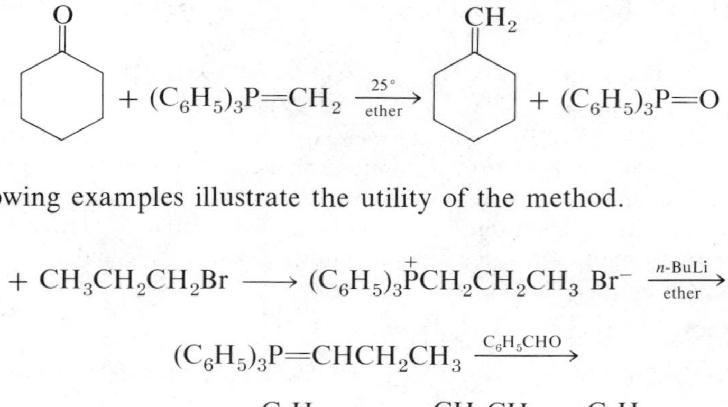

The following examples illustrate the utility of the method.

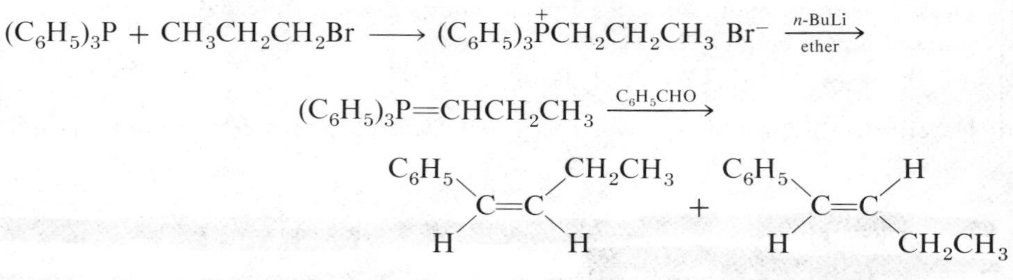

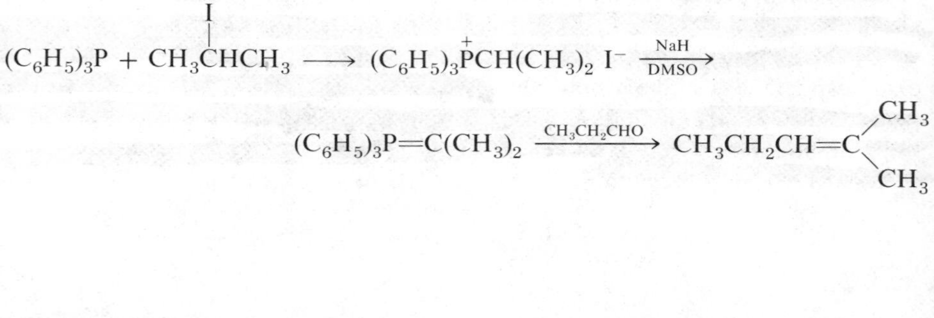

$$(C_6H_5)_3P + CH_3CH_2CH_2Br \longrightarrow (C_6H_5)_3\overset{+}{P}CH_2CH_2CH_3 \ Br^- \xrightarrow[\text{ether}]{n\text{-BuLi}}$$

$$(C_6H_5)_3P=CHCH_2CH_3 \xrightarrow{C_6H_5CHO}$$

$$(C_6H_5)_3P + CH_3\overset{\overset{I}{|}}{C}HCH_3 \longrightarrow (C_6H_5)_3\overset{+}{P}CH(CH_3)_2 \ I^- \xrightarrow[\text{DMSO}]{NaH}$$

$$(C_6H_5)_3P=C(CH_3)_2 \xrightarrow{CH_3CH_2CHO} CH_3CH_2CH=C\overset{\overset{CH_3}{\diagup}}{\underset{\underset{CH_3}{\diagdown}}{}}$$

13.8
Oxidation and Reduction

A. *Oxidation of Aldehydes and Ketones*

Aldehydes are oxidized to carboxylic acids with great ease. Oxidizing agents that have been used are Ag_2O, H_2O_2, $KMnO_4$, CrO_3, and peroxy acids.

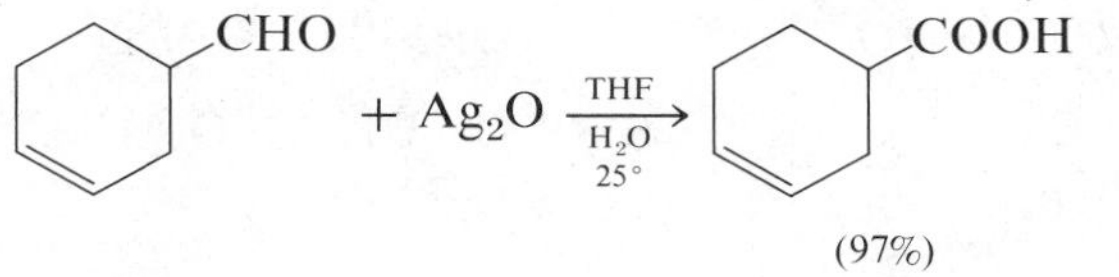

$$\text{(cyclohexenyl)CHO} + Ag_2O \xrightarrow[\substack{H_2O \\ 25°}]{THF} \text{(cyclohexenyl)COOH}$$

(97%)

$$\text{(o-hydroxyphenyl)CHO} + H_2O_2 \xrightarrow{acetone} \text{(o-hydroxyphenyl)COOH}$$

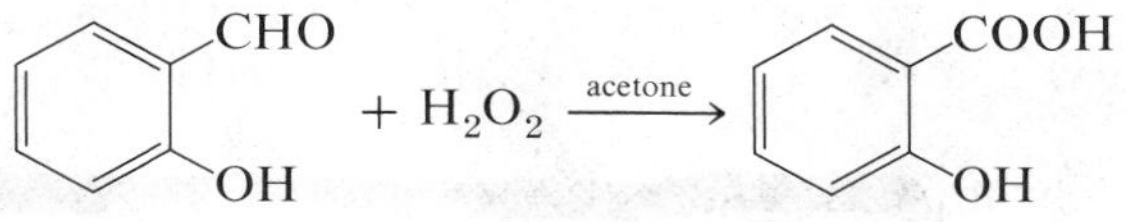

(100%)

$$n\text{-}C_6H_{13}CHO + CH_3\overset{O}{\overset{\|}{C}}OOH \longrightarrow n\text{-}C_6H_{13}COOH$$

peroxyacetic acid   (88%)

This oxidation is so facile that even atmospheric oxygen will bring it about. Most aldehyde samples that have been stored for some time before use are found to be contaminated with variable amounts of the corresponding carboxylic acid. In the case of oxidation by air **(autoxidation)** the initial oxidation product is a **peroxycarboxylic acid** (page 320). The peroxycarboxylic acid reacts with another molecule of aldehyde to give two carboxylic acid molecules.

$$RCHO + O_2 \longrightarrow R\overset{O}{\overset{\|}{C}}OOH$$

a peroxycarboxylic
acid

$$R\overset{O}{\overset{\|}{C}}OOH + RCHO \longrightarrow 2\ R\overset{O}{\overset{\|}{C}}OH$$

The initial oxidation (to the peroxycarboxylic acid stage) is a free radical chain process. A probable mechanism for this part of the reaction is as follows.

$$Y\cdot + R-\overset{O}{\overset{\|}{C}}-H \longrightarrow YH + R-\overset{O}{\overset{\|}{C}}\cdot \quad \Big\} \text{ initiation}$$

$$R-\overset{O}{\overset{\|}{C}}\cdot + O_2 \longrightarrow R-\overset{O}{\overset{\|}{C}}-O-O\cdot$$

$$R-\overset{O}{\overset{\|}{C}}-O-O\cdot + R-\overset{O}{\overset{\|}{C}}-H \longrightarrow R-\overset{O}{\overset{\|}{C}}-OOH + R-\overset{O}{\overset{\|}{C}}\cdot$$

propagation

The second stage, oxidation of the aldehyde by the initially formed peroxycarboxylic acid, is a special case of the **Baeyer–Villiger oxidation.** The probable course of this oxidation is shown in equations 13-6.

$$
\begin{aligned}
&\text{R—C—H} + \text{R—C—OOH} \rightleftharpoons \text{R—C—O—O—C—R} \\[2em]
&\text{R—C—O—O—C—R} \rightleftharpoons \text{R—C—OH} + {}^{-}\text{O—C—R} \\[2em]
&\text{R—C—OH} + {}^{-}\text{O—C—R} \longrightarrow 2\,\text{R—C—OH}
\end{aligned}
\tag{13-6}
$$

In contrast to aldehydes, ketones are oxidized only under rather special conditions. The Baeyer–Villiger oxidation, mentioned above, is one reaction in which a ketone undergoes oxidation. In this case the product is an ester.

$$
\text{CH}_3\text{CH}_2\overset{\text{O}}{\overset{\|}{\text{C}}}\text{CH}_2\text{CH}_3 \xrightarrow[\text{CH}_2\text{Cl}_2]{\text{CF}_3\text{CO}_3\text{H}} \text{CH}_3\text{CH}_2\overset{\text{O}}{\overset{\|}{\text{C}}}\text{—O—CH}_2\text{CH}_3
\tag{13-7}
$$
$$(78\%)$$

The mechanism of the reaction is believed to be similar to that outlined in equations (13-6) for the oxidation of an aldehyde except that the migrating group is an alkyl group rather than a hydrogen.

$$
\text{C}_2\text{H}_5\text{—}\underset{\text{C}_2\text{H}_5}{\overset{\text{OH}}{\text{C}}}\text{—O—O—C—CF}_3 \longrightarrow \text{C}_2\text{H}_5\overset{{}^{+}\text{OH}}{\text{—C—OC}_2\text{H}_5} + \text{CF}_3\text{CO}_2{}^{-} \longrightarrow
$$

$$
\text{C}_2\text{H}_5\overset{\text{O}}{\overset{\|}{\text{C}}}\text{OC}_2\text{H}_5 + \text{CF}_3\text{COOH}
\tag{13-8}
$$

The reaction is a preparatively useful one for the oxidation of certain ketones to esters. Cyclic ketones give cyclic esters (lactones; Section 28.5).

$$
+ \text{CH}_3\text{CO}_3\text{H} \xrightarrow[\substack{40° \\ 6.5\ \text{hr}}]{\text{CH}_3\text{COOEt}}
\tag{13-9}
$$
$$(90\%)$$

Symmetrical ketones, such as those illustrated in (13-7), (13-8), and (13-9), give only one product. Unsymmetrical ketones can give two oxidation products, and this is sometimes observed. When the two alkyl groups differ substantially, a clear selectivity can often be observed. The approximate order of *decreasing ease of migration* (the **migratory aptitude**) for various groups is hydrogen > phenyl > tertiary alkyl > secondary alkyl > primary alkyl > methyl. The following examples illustrate this selectivity.

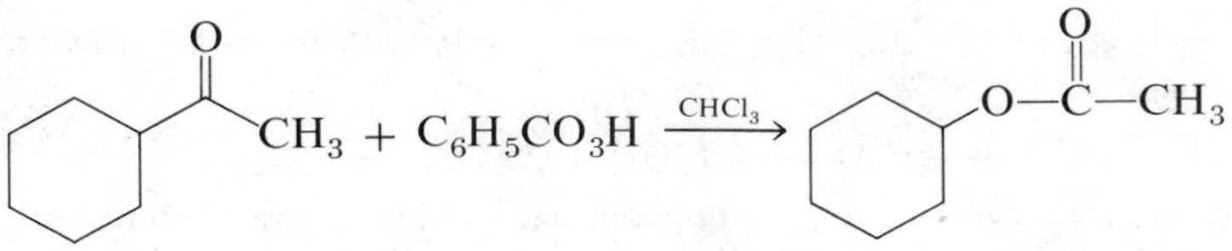

(67%)

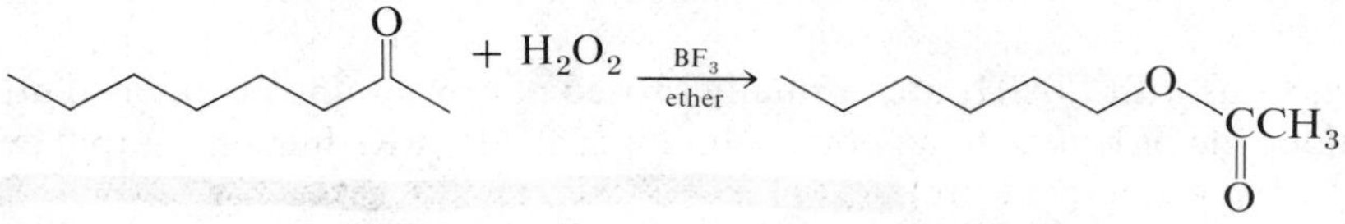

(62%)

Although ketones may be oxidized by other reagents, oxidative cleavage is seldom a useful preparative method. The conditions required for the oxidation of most ketones are sufficiently vigorous that complex mixtures result. The chief exception to this generalization is in symmetrical cyclic ketones, where the reaction can be useful. The oxidation of cyclohexanone by nitric acid is catalyzed by vanadium pentoxide. The product, adipic acid, is an important industrial chemical because it is one of the constituents of nylon 66 (Section 25.7.A).

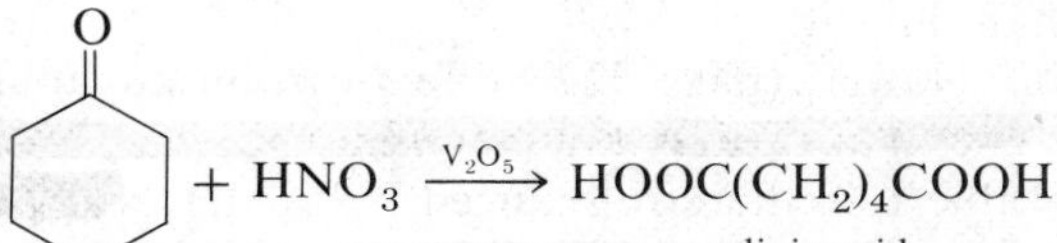

adipic acid

EXERCISE: Give the principal expected product of Baeyer–Villiger oxidation of (a) $C_6H_5COCH_3$, (b) cyclopentanone, (c) 3,3-dimethyl-2-butanone, and (d) methyl cyclohexyl ketone.

B. Metal Hydride Reduction

Aldehydes and ketones are easily reduced to the corresponding primary and secondary alcohols, respectively.

$$R-CHO \xrightarrow{[H]} R-CH_2OH$$

Many different reducing agents may be used. For laboratory applications the complex metal hydrides are particularly effective. Lithium aluminum hydride ($LiAlH_4$) is a powerful reducing agent that has been used for this purpose. purpose.

Lithium aluminum hydride, $LiAlH_4$, is a white salt-like compound prepared by reaction of lithium hydride with aluminum chloride.

$$4 \text{ LiH} + AlCl_3 \longrightarrow LiAlH_4 + 3 \text{ LiCl}$$

The compound is easily soluble in ethers such as ethyl ether, tetrahydrofuran and 1,2-dimethoxyethane (glyme). It has a clear relationship to sodium borohydride,

NaBH$_4$, and is a salt of Li$^+$ and AlH$_4^-$. It reacts avidly with traces of moisture to liberate hydrogen.

$$LiAlH_4 + 4 H_2O \longrightarrow LiOH + Al(OH)_3 + 4 H_2$$

All hydroxylic compounds (alcohols, carboxylic acids, and so on) react similarly. The dry crystalline powder must be used with care. It produces dust particles that are highly irritating to mucous membranes. It may inflame spontaneously while being crushed with a mortar and pestle and explodes violently when heated to about 120°.

Reactions with LiAlH$_4$ are normally carried out by adding an ether solution of the aldehyde or ketone to an ether solution of LiAlH$_4$. Reduction is rapid even at −78° (dry ice temperature). At the end of the reaction the alcohol is present as a mixture of lithium and aluminum salts and must be liberated by hydrolysis.

(90%)
cyclobutanol

The reagent also reduces many other oxygen- and nitrogen-containing functional groups, as illustrated in Table 13.5. The chief drawbacks of the reagent are its cost, which renders it useful only for fairly small-scale laboratory applications, and the hazards involved in handling it.

Sodium borohydride, NaBH$_4$ (page 312), offers certain advantages. This hydride is much less reactive than LiAlH$_4$ and is consequently more selective. Of the functional groups in Table 13.6 that are reduced by LiAlH$_4$, only aldehydes and ketones are reduced at a reasonable rate by NaBH$_4$. The reagent is moderately stable in aqueous and in alcoholic solution, expecially at basic pH's. The example below illustrates the selectivity that may be achieved with the reagent.

The carbonyl group is also reduced rapidly and quantitatively by diborane in ether or THF (page 315). The initial product is the ester of boric acid and an alcohol, a trialkyl borate. This material is rapidly hydrolyzed upon treatment with water.

$$(CH_3)_2CHCHO + B_2H_6 \xrightarrow{THF} [(CH_3)_2CHCH_2O]_3B \xrightarrow{H_2O} (CH_3)_2CHCH_2OH$$

(100%)   (100%)

EXERCISE: Suggest a way in which 1-octyne may be converted into 2-octanol.

C. Catalytic Hydrogenation

Aldehydes and ketones may also be reduced to alcohols by hydrogen gas in the presence of a metal catalyst (catalytic hydrogenation; Section 11.6.A). The chief

TABLE 13.6
Functional Groups Reduced by LiAlH$_4$

Functional Group	Product	Moles of LiAlH$_4$ Required
RCHO	RCH$_2$OH	0.25
R$_2$C=O	R$_2$CHOH	0.25
RCOOR′	RCH$_2$OH + R′OH	0.5
RCOOH	RCH$_2$OH	0.75
RCNH$_2$ (with =O above C)	RCH$_2$NH$_2$	1
RC≡N	RCH$_2$NH$_2$	0.5
RNO$_2$	RNH$_2$	1.5
RCl(Br, I)	RH	0.25

advantages of this method are that it is relatively simple to accomplish and usually affords a quantitative yield of product because no complicated work-up procedure is required. However, it suffers from the disadvantages that many of the catalysts used (Pd, Pt, Ru, Rh) are relatively expensive and that other functional groups (C=C, —C≡C—, —NO$_2$, —C≡N) also react.

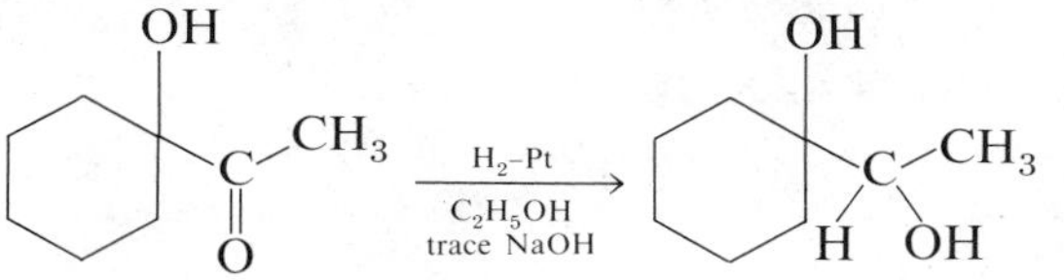

(100%)

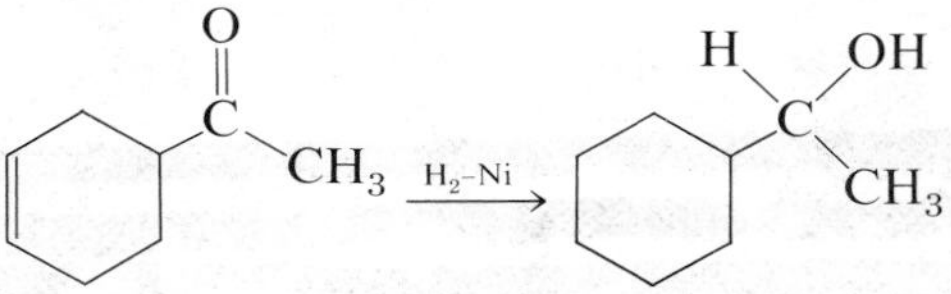

(96%)

D. *Cannizzaro Reaction*

When an aldehyde that has no hydrogens α to the carbonyl group is treated with concentrated aqueous base, a disproportionation reaction occurs. One-half of the aldehyde is reduced to a primary alcohol, and the other half is oxidized to the corresponding carboxylic acid.

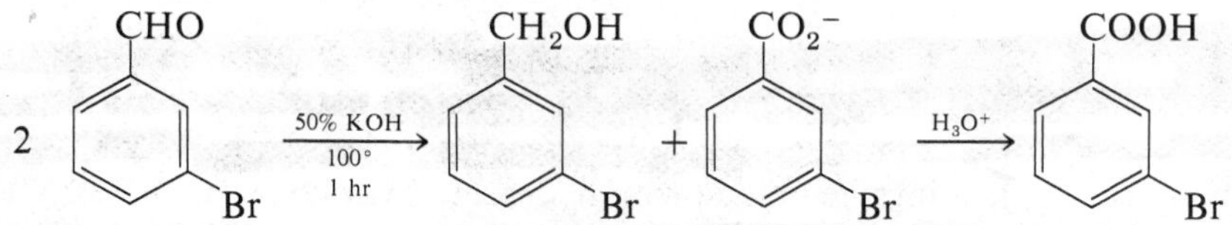

The reaction is known as the **Cannizzaro reaction** and is general for such aldehydes. It is not normally useful for preparative purposes because the maximum yield of either alcohol or acid is 50%. However, when an aldehyde is treated with aqueous base and formaldehyde, it is the formaldehyde, rather than the other aldehyde, that is oxidized. Such a reaction is known as a **crossed Cannizzaro reaction.** An interesting example of such a crossed Cannizzaro reaction combined with aldol condensation is the preparation of pentaerythritol from acetaldehyde and formaldehyde.

$$CH_3CHO + CH_2O \xrightarrow{OH^-} HOCH_2CH_2CHO \xrightarrow[OH^-]{CH_2O} HOCH_2CHCHO \xrightarrow[]{CH_2O}$$
$$\underset{CH_2OH}{|}$$

$$\underset{\overset{|}{CH_2OH}}{HOCH_2\overset{\overset{CH_2OH}{|}}{C}-CHO} \xrightarrow[OH^-]{CH_2O} \underset{\overset{|}{CH_2OH}}{HOCH_2\overset{\overset{CH_2OH}{|}}{C}CH_2OH} + HCO_2^-$$

(55–57%)
pentaerythritol

This reaction involves three successive aldol condensations of acetaldehyde with formaldehyde. The product has no α-hydrogens left and is reduced by the alkaline formaldehyde to give 2,2-bis-hydroxymethyl-1,3-propanediol (pentaerythritol).

The mechanism of the Cannizzaro reaction starts with addition of hydroxide ion to the carbonyl group. Most of the time it comes right back off again. Occasionally, however, H⁻ is expelled by transfer to another carbonyl group.

$$\underset{\overset{\parallel}{O}}{R-C-H} + {}^-OH \rightleftharpoons \underset{\overset{|}{OH}}{R-\overset{\overset{O^-}{|}}{C}-H}$$

$$\underset{\overset{|}{OH}}{R-\overset{\overset{O^-}{|}}{C}-H} \quad \underset{\overset{|}{H}}{\overset{\overset{O}{\parallel}}{C}-R} \longrightarrow \underset{\overset{|}{OH}}{R-\overset{\overset{O}{\parallel}}{C}} + \underset{\overset{|}{H}}{H-\overset{\overset{O^-}{|}}{C}-R} \longrightarrow RCO_2^- + HOCH_2R$$

The reaction is suitable only for aldehydes without an α-hydrogen because this H⁻-transfer process is slow and difficult. Any α-hydrogens present would react to give enolate ions and aldol condensation products far more rapidly than the Cannizzaro reaction.

F. Deoxygenation Reactions

Several methods are known whereby the oxygen of an aldehyde or ketone is replaced by two hydrogens.

$$\underset{\overset{\parallel}{O}}{R-C-R'} \longrightarrow R-CH_2-R'$$

One such method is the *Wolff–Kishner reduction*. The method involves first the formation of the hydrazone by reaction with hydrazine. At elevated temperatures with base the hydrazone loses nitrogen to give the hydrocarbon. The reaction is normally run by heating the ketone with hydrazine hydrate (Section 13.7.C) and sodium hydroxide in diethylene glycol, $HOCH_2CH_2OCH_2CH_2OH$, which has a b.p. of 245°. Alternatively, the reduction may be carried out in the polar aprotic solvent DMSO at 100°. The hydrazone forms, and water distills out of the mixture. On refluxing, nitrogen is evolved, and the product is isolated.

cyclononanone   (47%)
                                                                        cyclononane

The decomposition of the hydrazone involves an anionic intermediate. In the presence of base the hydrazone is in equilibrium with a double bond isomer. This isomer forms an anion with base, which loses nitrogen to produce the alkyl anion. This carbanion is exceedingly unstable and rapidly abstracts a proton from the solvent. Alkyl anions are rarely encountered in reactions because of the low acidity of hydrocarbons. In the present case they are formed only because the nitrogen also produced is an extremely stable molecule and its production provides the driving force for the reaction.

(1) $R_2C{=}N\ddot{N}H_2 + B^- \rightleftharpoons BH + [R_2C{=}N{-}\ddot{N}H^- \longleftrightarrow R_2\ddot{C}^-{-}N{=}NH]$

(2) $[R_2C{=}N{-}\ddot{N}H^- \longleftrightarrow R_2\ddot{C}^-{-}N{=}NH] + BH \rightleftharpoons R_2\overset{\overset{\displaystyle H}{|}}{C}{-}N{=}NH + B^-$

(3) $R_2CH{-}N{=}\ddot{N}H + B^- \rightleftharpoons R_2CH{-}N{=}\ddot{N}^- + BH$

(4) $R_2CH{-}N{=}\ddot{N}^- \xrightarrow{\text{slow}} R_2\ddot{C}H^- + N_2$

(5) $R_2\ddot{C}H^- + BH \xrightarrow{\text{fast}} R_2CH_2 + B^-$

An alternative procedure for the direct reduction of a carbonyl group to a methylene group involves refluxing the aldehyde or ketone with amalgamated zinc and hydrochloric acid (Clemmensen reduction).

Amalgamated zinc is zinc with a surface layer of mercury. It is prepared by treating zinc with an aqueous solution of a mercuric salt. Since zinc is higher on the electromotive force scale than mercury, it displaces mercury from its salts.

$$Zn + Hg^{2+} \longrightarrow Zn^{2+} + Hg$$

The mercury then alloys with the surface of zinc to produce an amalgam.

Reduction of the carbonyl compound occurs on the surface of the zinc, and, like many heterogeneous reactions, this reaction does not have a simple mechanism.
The Clemmensen reduction is suitable for compounds that can withstand treatment with hot acid. Many ketones are reduced in satisfactory yields.

$$C_6H_5COCH_3 \xrightarrow[\underset{\Delta}{\text{HCl}}]{\text{Zn(Hg)}} C_6H_5CH_2CH_3$$

acetophenone   (80%)
                                                            ethylbenzene

A different approach to such reductions involves the formation and reduction of dithioketals. Treatment of an aldehyde or ketone with 1,2-ethanedithiol, the sulfur analog of ethylene glycol (page 385), and a Lewis acid catalyst provides a dithioacetal or dithioketal.

$$\text{R}\overset{\displaystyle \overset{O}{\|}}{-}\text{C}-\text{R} + \text{HSCH}_2\text{CH}_2\text{SH} \xrightarrow{\text{BF}_3} \underset{\underset{\displaystyle \overset{|}{\text{R}} \quad \overset{|}{\text{R}}}{\text{C}}}{\text{S}\diagdown\diagup\text{S}} + \text{H}_2\text{O}$$

a dithioketal

$$\text{R}\overset{\displaystyle \overset{O}{\|}}{-}\text{C}-\text{H} + \text{HSCH}_2\text{CH}_2\text{SH} \xrightarrow{\text{ZnCl}_2} \underset{\underset{\displaystyle \overset{|}{\text{R}} \quad \overset{|}{\text{H}}}{\text{C}}}{\text{S}\diagdown\diagup\text{S}} + \text{H}_2\text{O}$$

a dithioacetal

When the dithioacetal or dithioketal is treated with Raney nickel, the carbon–sulfur bonds undergo **hydrogenolysis.** The overall result of the two-step sequence is replacement of the oxygen of the aldehyde or ketone by two hydrogens.

As it is normally prepared (page 304), Raney nickel is saturated with adsorbed hydrogen. Therefore, even though this reaction is a reduction, it is not necessary to use additional hydrogen. Reduction of a dithioacetal or dithioketal is a useful method for deoxygenation when it is desirable to carry out the operation under essentially neutral conditions.

EXERCISE: Which method is preferable for deoxygenation of each of the following aldehydes or ketones?

(a) $\text{BrCH}_2\text{CH}_2\text{CH}_2\text{CH}_2\text{CHO}$

(b) $(\text{CH}_3)_2\underset{\underset{\displaystyle \text{OH}}{|}}{\text{C}}\text{CH}_2\text{CHO}$

(c) 4-*n*-butylcyclohexanone

(d)

PROBLEMS

1. Provide common and IUPAC names for the following ketones and aldehydes.

(a) $\text{CH}_3\overset{\displaystyle \overset{O}{\|}}{\text{C}}\text{CH}_2\text{CH}_3$

(b) $\text{CH}_3\underset{\underset{\displaystyle \text{CH}_3}{|}}{\text{CH}}-\overset{\displaystyle \overset{O}{\|}}{\text{C}}\text{CH}_2\text{CH}_2\text{CH}_3$

(c) $\text{CH}_3\underset{\underset{\displaystyle \text{CH}_3}{|}}{\overset{\overset{\displaystyle \text{CH}_3}{|}}{\text{C}}}-\overset{\displaystyle \overset{O}{\|}}{\text{C}}\text{CH}_3$

(d) $\text{CH}_3\underset{\underset{\displaystyle \text{CH}_3}{|}}{\text{C}}\text{CH}_2\overset{\displaystyle \overset{O}{\|}}{\text{C}}\text{CH}_2\text{CH}_3$

(e)

(f) $(\text{CH}_3)_2\text{CHCHO}$

(g) $\text{CH}_3\text{CHBrCHO}$

(h) $\text{CH}_3\underset{\underset{\displaystyle \text{OCH}_3}{|}}{\text{CH}}\text{CH}_2\text{CHO}$

(i) $CH_2{=}CHCCH_2CH_3$ (with O above C)

(j) 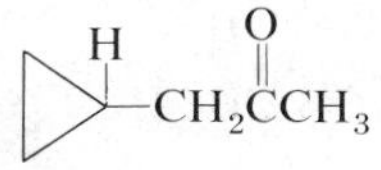 triangle with H, $-CH_2CCH_3$ (O above C)

2. Write IUPAC names for the following compounds.

(a) $CH_3CH_2CHCH_2CH_2CHO$ (with CH_3 above)

(b) $(CH_3)_2CHCH_2CCH_2CHCH_2CH_3$ (with O and CH_3 above)

(c) $CH_3CH_2CCH_2CH_2CHO$ (with O above)

(d) 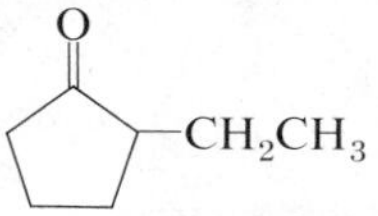 cyclopentane ring with $C{=}O$ and $-CH_2CH_3$

(e) $CH_3CHBrCH_2CCH_3$ (with O above)

(f) $(CH_3)_2C{=}CHCH_2CHO$

(g) 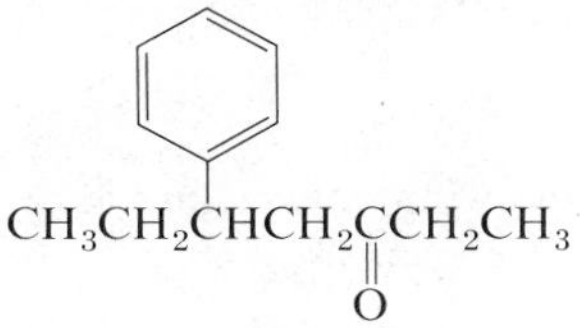 $CH_3CH_2CHCH_2CCH_2CH_3$ (with phenyl above, O below)

3. Write the structure of each of the following.
(a) methyl isobutyl ketone
(b) propionaldehyde diethyl acetal
(c) β-chlorobutyraldehyde
(d) 2,2-dimethylcyclopentanone
(e) cyclododecanone
(f) formaldehyde phenylhydrazone
(g) cyclohexanone oxime
(h) acetone semicarbazone

4. What is the product of the reaction of 4,4-dimethylcyclohexanone with each of the following reagents?
(a) Lithium diisopropylamide in THF, followed by CH_3CH_2Br
(b) Br_2 in acetic acid
(c) CH_3CO_3H
(d) $LiAlH_4$ in ether, followed by H_2O
(e) conc. HNO_3, V_2O_5
(f) $NaBH_4$ in C_2H_5OH
(g) $CH_3C{\equiv}C^-Na^+$ in liq. NH_3 at $-33°$, followed by H_2O
(h) KCN + aqueous sulfuric acid
(i) H_2NOH + sodium acetate and acetic acid
(j) ½ mole equivalent of H_2NNH_2 + sodium acetate and acetic acid
(k) 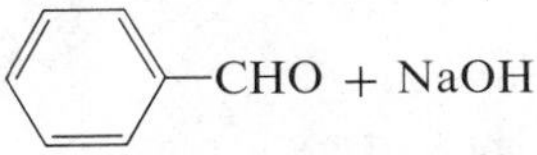 phenyl$-CHO$ + NaOH
(l) H_2NNH_2 + $Na^+{}^-OCH_2CH_2OCH_2CH_2OH$, $200°$
(m) zinc amalgam + hot conc. HCl
(n) NaOD in D_2O at $25°$
(o) $(C_6H_5)_3P{=}CHCH_2CH_3$
(p) $HSCH_2CH_2CH_2SH$, $ZnCl_2$

5. Show how each of the following compounds may be prepared from alkyl halides and alcohols containing four or fewer carbons.

(a) $CH_3CH_2CCH_2CH_2CH_3$ (with O above)

(b) $CH_3CHCH{=}CH_2$ (with CH_3 above)

(c) $CH_3CH_2CH_2CH{=}CCH_2OH$ (with CH_2CH_3 below)

(d) $CH_3CH_2CH{=}C$ (with CH_2CH_3 and CH_3)

(e) $CH_3CH_2CH_2CH_2\overset{\displaystyle OH}{\underset{\displaystyle |}{CH}}CH_2CH_2CH_2CH_2CH_3$

(f) $CH_3CH_2CH_2\overset{\displaystyle CH_2OH}{\underset{\displaystyle |}{\underset{\displaystyle CH_2OH}{C}}}CH_2OH$

6. Show how each of the following compounds may be prepared from pentanal. Any other reagents, organic or inorganic, may be used.

(a) $CH_3CH_2CH_2CH_2CH_3$

(b) $CH_3CH_2CH_2CH_2\overset{\displaystyle OH}{\underset{\displaystyle |}{CH}}C{\equiv}CH$

(c) $CH_3CH_2CH_2CH_2\overset{\displaystyle OH}{\underset{\displaystyle |}{CH}}CN$

(d) $CH_3CH_2CH_2CH_2COOH$

(e) $CH_3CH_2CH_2CH_2CH_2\overset{\displaystyle CH_3}{\underset{\displaystyle |}{CH}}CH_2CH_2CH_3$

(f) $CH_3CH_2CH_2CH_2CH_2Br$

(g) $CH_3CH_2CH_2CH_2CH_2OCH_2CH_2CH_2CH_2CH_3$ (h) $CH_3CH_2CH_2CH_2CH{=}CH_2$

7. Show how one may accomplish each of the following conversions in a practical manner using any necessary organic or inorganic reagents.

(a) $CH_3CH_2CH_2CHO \longrightarrow CH_3CH_2CH_2\overset{\displaystyle O}{\overset{\displaystyle \|}{C}}\overset{}{\underset{\displaystyle |}{\underset{\displaystyle Br}{CH}}}CH_2CH_3$

~~study~~ (b) $CH_3CH_2CH_2\overset{\displaystyle O}{\overset{\displaystyle \|}{C}}CH_3 \longrightarrow \overset{\displaystyle CH_3}{\underset{\displaystyle CH_3}{}}C{=}C\overset{\displaystyle CH_3}{\underset{\displaystyle CH_2CH_2CH_3}{}}$

(c) $CH_3CH_2CH_2CHO \longrightarrow CH_3CH_2\overset{\displaystyle CH_2OH}{\underset{\displaystyle |}{CH}}\underset{\displaystyle \underset{\displaystyle OH}{|}}{CH}CH_2CH_2CH_3$

(d) $CH_3CH_2CHO \longrightarrow CH_3\overset{\displaystyle CH_2OH}{\underset{\displaystyle |}{\underset{\displaystyle CH_2OH}{C}}}CH_2OH$

(e) $CH_3CH_2CH_2CH_2Br \longrightarrow CH_3CH_2CH_2CH_2CH_2CHO$

(f) $HC{\equiv}CH \longrightarrow CH_3CH_2CH_2\overset{\displaystyle O}{\overset{\displaystyle \|}{C}}CH_2CH_2CH_2CH_3$

8. In reactions of protonated acetone $CH_3\overset{\displaystyle {}^+OH}{\overset{\displaystyle \|}{C}}CH_3$, why does reaction with a base always occur at carbon rather than at the positive oxygen?

9. If the acid-catalyzed bromination of bromoacetone is analyzed *immediately after the bromine has all reacted,* the major product present is 1,1-dibromoacetone. If the reaction mixture is allowed to stir at room temperature for several hours and then analyzed, the sole product is found to be 1,3-dibromoacetone. Explain these observations with a reasonable mechanism for each step.

10. If 2-methylcyclohexanone is treated with 1 mole equivalent of potassium *t*-butoxide and methyl iodide in *t*-butyl alcohol, the reaction product has the composition shown. Explain.

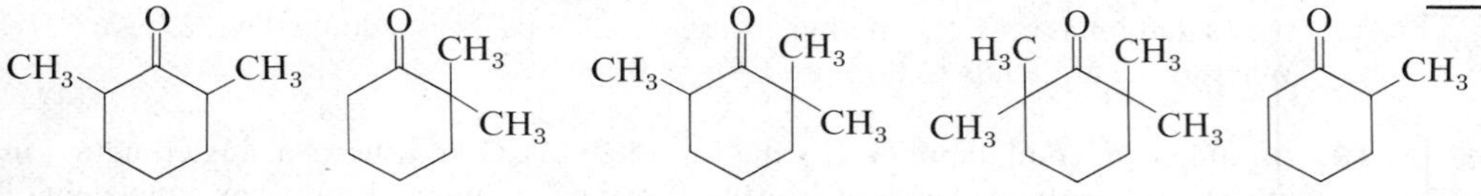

(9%)	(41%)	(21%)	(6%)	(23%)

11. Propose a mechanism for the following reaction.

$$\text{(acetal)} + Br_2 \xrightarrow{\text{trace } H^+} \text{(brominated product)}$$

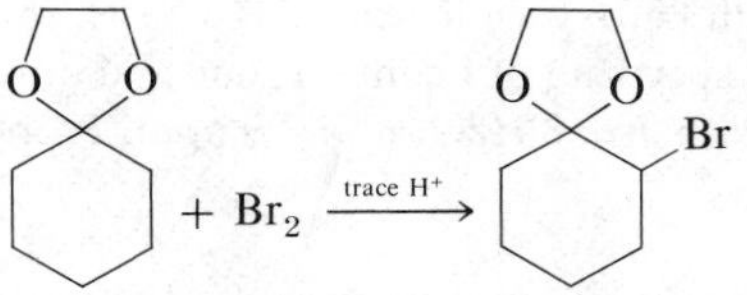

12. 1,1-Diethoxyethane hydrolyzes readily to acetaldehyde and ethanol in water containing some sulfuric acid. Write a complete reaction mechanism for this transformation including each significant intermediate and reaction step.

13. Is the equilibrium constant for the following equilibrium greater than, less than, or equal to unity? Explain briefly.

$$CF_3CHO + CH_3CH(OCH_3)_2 \rightleftharpoons CH_3CHO + CF_3CH(OCH_3)_2$$

14. When 4-hydroxybutanal is dissolved in methanol containing HCl, the following reaction occurs.

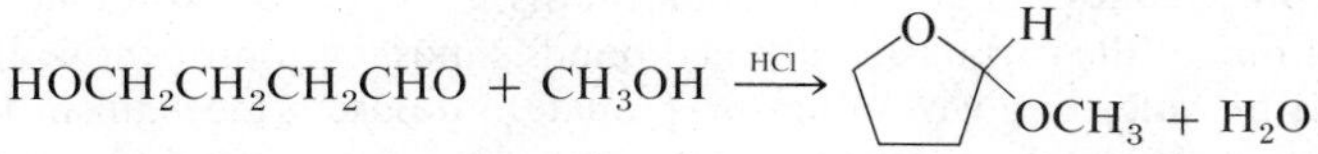

$$HOCH_2CH_2CH_2CHO + CH_3OH \xrightarrow{HCl} \text{(cyclic OCH}_3\text{ product)} + H_2O$$

(a) What type of compound is the product?
(b) Propose a mechanism for this reaction
Actually, 4-hydroxybutanal exists in solution mostly in the cyclic form.

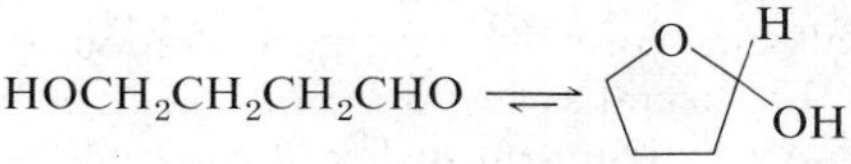

$$HOCH_2CH_2CH_2CHO \rightleftharpoons \text{(cyclic OH product)}$$

(c) What type of compound is this product?
(d) Propose a mechanism for the equilibrium.

15. Write a plausible reaction mechanism for the trimerization of acetaldehyde to paraldehyde with a trace of acid. How does this mechanism compare to the acid-catalyzed depolymerization of paraldehyde?

16. Undecanal is a sex attractant for the greater wax moth (*Galleria mellonella*). Show how to synthesize this compound efficiently from (a) 1-decanol and (b) 1-dodecanol.

17. Compound A, $C_7H_{16}O$, reacts with sodium dichromate in aqueous H_2SO_4 to give B, $C_7H_{14}O$. When B is treated with Na^+OD^- at 25° for several hours and then analyzed by mass spectroscopy, it is found to have a molecular weight of 116. Compound B is not oxidized by Ag_2O. What are A and B?

18. Compound C has the formula $C_{12}H_{20}$ and is optically active. It reacts with H_2 and Pt to give two isomers, D_1 and D_2, $C_{12}H_{22}$. Ozonolysis of C gives only E, $C_6H_{10}O$, which is optically active. Compound E reacts with hydroxylamine to give F, $C_6H_{11}NO$. When E is treated with DCl in D_2O for several hours and then analyzed by mass spectroscopy, it is found to have a molecular weight of 101. The nmr spectrum of E shows that it has only one methyl group, which appears as a doublet with $J = 6.5$ Hz. What are compounds C through F?

19. An unknown compound, G, has the formula C_6H_8O and shows a singlet methyl in the nmr. When treated with hydrogen gas and palladium it absorbs one equivalent of H_2 to give a product, H, with the formula $C_6H_{10}O$. The infrared spectrum of compound H shows that it has a carbonyl group. Compound H reacts with NaOD in D_2O to give a product shown by mass spectrometry to have the formula $C_6H_7D_3O$. Compound H reacts with peroxyacetic acid, CH_3CO_3H, to give I, which has the formula $C_6H_{10}O_2$. The nmr spectrum of I contains one and only one absorption due to a CH_3 group, a doublet with $J = 8$ Hz at $\delta = 1.9$ ppm. Propose structures for compounds G, H, and I.

20. *trans*-3-Isopropyl-6-methylcycloheptanone has the following proton-coupled cmr spectrum (letters in parentheses after each chemical shift indicate multiplicity of the resonance: s = singlet, d = doublet, t = triplet, q = quartet): 18.5(q), 18.9(q), 23.9(q), 32.0(t), 32.5(t), 34.9(d), 38.5(t), 41.9(d), 47.0(t), 51.6(t), and 213.4(s). From a reaction, samples of the *cis* and *trans* isomers of 3,6-dimethylcycloheptanone were obtained. These compounds have the following cmr spectra.

isomer J: 20.9(q), 29.2(d), 33.9(t), 50.8(t), 212.1(s)
isomer K: 23.7(q), 31.1(d), 37.9(t), 51.7(t), 212.1(s)

Which isomer has the *cis* structure, and which the *trans*?

21. Cyclopropanones undergo a reaction that is not common for other ketones—cleavage of one of the carbon-to-carbonyl bonds. Thus, cyclopropanone itself reacts with NaOH in water to give sodium propionate. Propose a mechanism for this unusual reaction. Predict the product that will result when 2-methylcyclopropanone is treated in the same manner. Treatment of 2-bromo-3-pentanone with NaOH in water yields, after acidification of the reaction mixture, 2-methylbutanoic acid. Propose a mechanism for this reaction (the **Favorskii reaction**).

22. Unsymmetrical alkynes normally undergo hydration to give a mixture of the two possible ketones. For example, hydration of 2-heptyne gives both 2-heptanone and 3-heptanone. However, treatment of hept-5-yn-2-one with aqueous sulfuric and a small amount of mercuric sulfate gives *only* heptan-2,5-dione. Suggest a mechanism that might account for this behavior.

23. Ethyl vinyl ether, $CH_2{=}CHOCH_2CH_3$, reacts with *n*-butanol and a trace of sulfuric acid in ether to give 1-*n*-butoxy-1-ethoxyethane. Treatment of this product with aqueous sulfuric acid affords a mixture of ethanol, *n*-butanol, and acetaldehyde. Propose a mechanism for each reaction. How might this chemistry be used to advantage in accomplishing the following conversion?

24. Each of the following compounds may be obtained by an intramolecular aldol condensation. Show the precursor in each case.

25. Bromination of 7-cholestanone gives 6-bromo-7-cholestanone with $J_{5H,6H} = 2.8$ Hz. This initial product slowly converts to the more stable 6-bromo stereoisomer that has $J_{5H,6H} = 11.8$ Hz. Assign stereostructures to the two isomers and explain their relative stabilities.

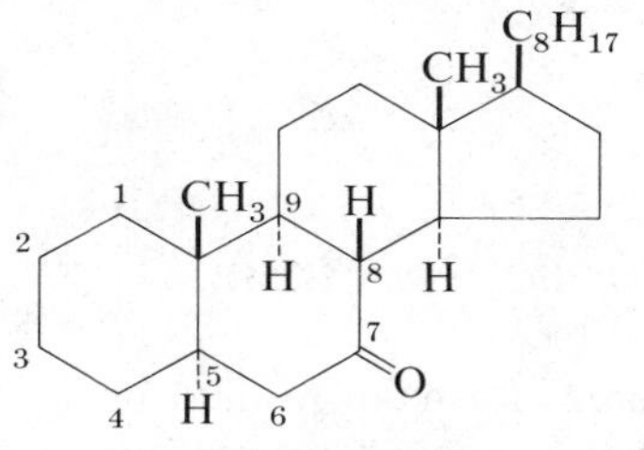

7-cholestanone

26. Ketones react with lithium diisopropylamide, LDA (page 373), in ether or THF to afford lithium enolates. 3-Methyl-2-butanone is treated with LDA in THF at $-70°$, followed by 2,2-dimethylpropanal. The sole product is 5-hydroxy-2,6,6-trimethyl-heptanone. Suggest an explanation for the selectivity observed in this aldol condensation.

27. When a mixture of an aldehyde or ketone and an α-halo ester is treated with a strong base, an α,β-epoxy ester is obtained.

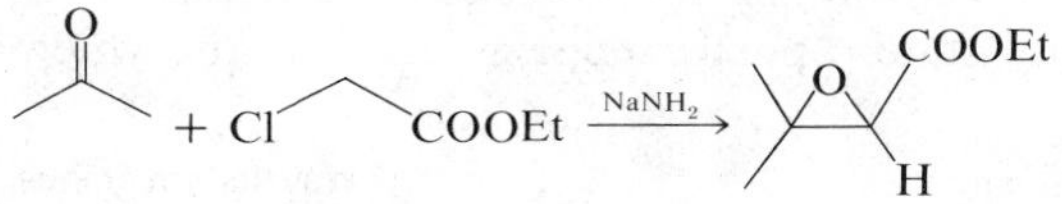

Propose a mechanism for this reaction **(Darzen's condensation).**

28. Treatment of an aldehyde or ketone with a mixture of ammonium chloride and sodium cyanide yields an α-amino nitrile **(Strecker synthesis).** For example,

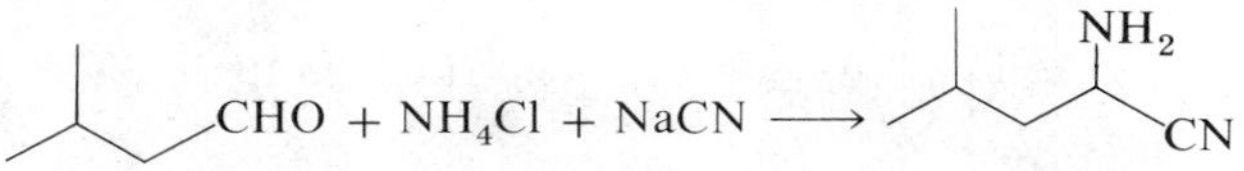

an α-amino nitrile

Propose a mechanism for the reaction. What product is expected when 2,6-heptanedione is subjected to the Strecker synthesis?

Chapter 14
Infrared Spectroscopy

14.1
The Electromagnetic Spectrum

There are many different forms of radiant energy—cosmic rays, x-rays, radio waves, visible light—that display wave properties. These apparently different types of radiation are known collectively as **electromagnetic radiation.** They are all considered as waves that travel at a constant velocity (the "speed of light," $c = 3.0 \times 10^{10}$ cm sec^{-1}) and differ in wavelength or frequency. The **electromagnetic spectrum** is diagrammed in Figure 14.1 along with the wavelength in centimeters of its various regions. The divisions between regions are arbitrary and are established in practice by the different instrumentation required to produce and record electromagnetic radiation in the different regions. As pointed out in Section 9.2, compounds may absorb radiant energy in various regions of the electromagnetic spectrum and thereby become excited from their ground state to a more energetic state. **Spectroscopy** is a technique whereby we measure the

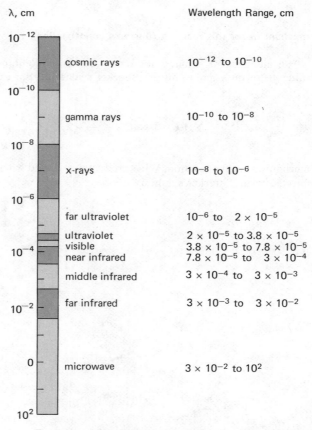

λ, cm		Wavelength Range, cm
10^{-12}	cosmic rays	10^{-12} to 10^{-10}
10^{-10}	gamma rays	10^{-10} to 10^{-8}
10^{-8}	x-rays	10^{-8} to 10^{-6}
10^{-6}	far ultraviolet	10^{-6} to 2×10^{-5}
	ultraviolet	2×10^{-5} to 3.8×10^{-5}
10^{-4}	visible	3.8×10^{-5} to 7.8×10^{-5}
	near infrared	7.8×10^{-5} to 3×10^{-4}
	middle infrared	3×10^{-4} to 3×10^{-3}
10^{-2}	far infrared	3×10^{-3} to 3×10^{-2}
0	microwave	3×10^{-2} to 10^2
10^2		

FIGURE 14.1. *The electromagnetic spectrum*

amount of radiation a substance absorbs at various wavelengths. From the **spectrum** of a compound we may often obtain useful information about the structure of the compound.

The relationship between the wavelength and frequency of radiation is given by

$$\nu = \frac{c}{\lambda}$$

where λ = wavelength in centimeters, ν = frequency in Hertz (Hz), and c = the velocity of light (2.998×10^{10} cm sec^{-1}). The relationship between energy and frequency is

$$\epsilon = h\nu$$

where ϵ = the energy of a photon and h = Planck's constant (6.6242×10^{-27} erg sec), or

$$E = Nh\nu$$

where E = the energy of an Avogadro number N of photons ($E = \epsilon \times 6.023 \times 10^{23}$). Thus, when a compound absorbs radiation of a given wavelength, each molecule absorbs an amount of energy ϵ, and each mole of the compound absorbs an amount of energy E. In organic chemistry energy is traditionally expressed in units of kilocalories per mole.

$$E \text{ (kcal mole}^{-1}\text{)} = \frac{2.857 \times 10^{-3}}{\lambda \text{ (cm)}}$$

Another system of units that is being adopted in many parts of the world is known as SI (Système International d'Unités). In this international system of units, the unit of energy is the joule, J; 1 cal $\equiv$ 4.184 J. The six fundamental units in SI are length = meters (m), mass = kilograms (kg), time = seconds (s), electrical current = amperes (A), temperature = degrees Kelvin (K), and luminosity = candelas (cd). These units are modified by 10^3 (kilo), 10^6 (mega), 10^{-3} (milli), 10^{-6} (micro), 10^{-9} (nano), 10^{-12} (pico). Many traditional units among chemists, such as calories (cal) or kilocalories (kcal) and centimeters (cm), are still in common use.

In this chapter, we are concerned with absorption of light in the **infrared** region of the electromagnetic spectrum. Wavelengths in this region are traditionally expressed in microns (μ), where $10^4 \mu$ = 1 cm. More commonly, another unit of measurement, the wavenumber, is used to describe infrared spectra. The wavenumber ($\tilde{\nu}$) is defined as the number of waves per centimeter and is expressed in units of cm^{-1} (reciprocal centimeters).

$$\tilde{\nu} = \frac{1}{\lambda}$$

By definition, the infrared region is split up into three parts (Figure 14.2): the **near infrared** ($\lambda = 0.78$–$3.0\,\mu$; $\tilde{\nu} = 12{,}820$–3333 cm^{-1}), the **middle infrared** ($\lambda = 3$–$30\,\mu$; $\tilde{\nu} = 3333$–333 cm^{-1}), and the **far infrared** ($\lambda = 30$–$300\,\mu$; $\tilde{\nu} = 333$–33 cm^{-1}). The near infrared region corresponds to energies in the range 37–10 kcal mole^{-1}. Since there are few absorptions of organic molecules in this range, it is seldom used for spectroscopic purposes. Radiation in the middle infrared

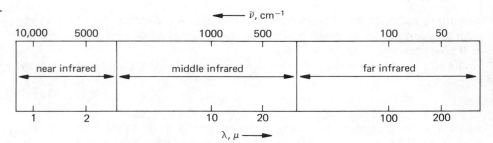

FIGURE 14.2. *Regions of the infrared spectrum. Notice that the scales are logarithmic.*

region has $E = 10$–1 kcal mole^{-1}, which corresponds to the differences commonly encountered between vibrational states. Spectroscopy in this region is extremely useful to the organic chemist. The far infrared region has $E = 1.0$–0.1 kcal mole^{-1}. This region has been little used for organic spectroscopy, again because few useful absorptions occur here.

14.2
Molecular Vibration

As discussed previously (Section 6.1), atoms in a molecule do not maintain fixed positions with respect to each other, but actually vibrate back and forth about an average value of the interatomic distance. This vibrational motion is quantized, as shown in the accompanying familiar diagram for a diatomic molecule (Figure 14.3). At room temperature most of the molecules in a given sample will be in the lowest vibrational state. However, absorption of light of the appropriate energy allows the molecule to become "excited" to the second vibrational level. In this level the amplitude of the molecular vibration is greater. In general, such absorption of an infrared light quantum can occur only if the dipole moment of the molecule is different in the two vibrational levels. The variation of the

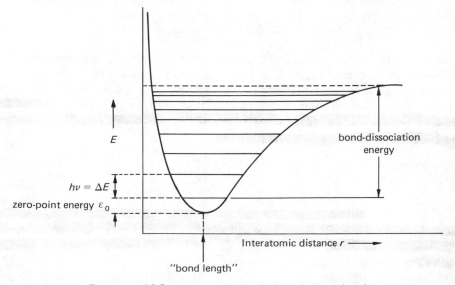

FIGURE 14.3. *Vibrational levels for vibrating bond.*

dipole moment with the change in interatomic distance during the vibration corresponds to an oscillating electric field that can interact with the oscillating electric field associated with electromagnetic radiation. The requirement that absorption of a vibrational quantum be accompanied by a change in dipole moment is known as a **selection rule.** Such a vibrational transition is said to be **infrared-active.** Vibrational transitions that do not result in a change of dipole moment of the molecule are not observed directly and are referred to as **infrared-inactive** transitions. Thus, carbon monoxide and iodine chloride absorb infrared light, but hydrogen, nitrogen, chlorine, and other symmetrical diatomics do not.

$$C\equiv O, \; I{-}Cl \qquad H_2, \; N_2, \; Cl_2$$
absorb in infrared   *do not absorb in infrared*

For more complex molecules, there are more possible vibrations. A nonlinear molecule containing n atoms has $3n - 6$ possible **fundamental vibrational modes.** Polyatomic molecules exhibit two distinct types of molecular vibration, **stretching** and **bending.** Vibrations of bonds involving hydrogens are especially significant because atoms of low mass tend to do a lot of moving compared to atoms of higher mass. The stretching and bending motions in a methylene group are diagrammed in Figure 14.4.

For polyatomic molecules of the size of typical organic compounds, the possible number of infrared absorption bands becomes very large, for example,

	Possible Infrared Absorption Bands
pentane	45
decane	90
triacontane	270

Many of these vibrations occur at the same frequency (that is, some vibrations are **degenerate**), and not all of the possible bands are generally seen as independent absorptions. However, additional bands, usually of low intensity, may occur as **overtones** (at approximately $\frac{1}{2}, \frac{1}{3}, \frac{1}{4}, \ldots$, and so on, the wavelength of the fundamental mode).

Stretching Vibrations

(a) symmetric

(b) asymmetric

Bending Vibrations

(a) scissoring (in-plane)

(b) rocking (in-plane)

(c) wagging (out-of-plane)

(d) twisting (out-of-plane)

FIGURE 14.4. *Some vibrational modes for the methylene group.*

Overtones may arise in two ways. If a molecule in the lowest or first vibrational state is excited to the third vibrational level, the energy required is almost twice that required for excitation to the second vibrational level. It is not exactly twice as much because the higher levels tend to lie closer together than the lower levels (see Figure 14.3).

Another type of overtone, commonly referred to as a **combination band,** occurs when a single photon has precisely the correct energy to excite two vibrations at once. For this to happen, the energy of the combination band must be the exact sum of the two independent absorptions.

As a result, the infrared spectrum of an organic compound is usually rather complex.

The spectrum of *n*-octane, shown in Figure 14.5, illustrates several features of an infrared spectrum. Note that the wavelength is plotted against the per cent transmittance of the sample. An absorption band is therefore represented by a "trough" in the curve; zero transmittance corresponds to 100% absorption of light of that wavelength.

The curve in Figure 14.5 is a spectrum of pure *n*-octane. The spectrum was measured on a Perkin–Elmer Model 735 Spectrometer using a cell 0.016 mm in length. Only four major absorption bands are apparent, at 2925, 1465, 1380, and 720 cm^{-1}. These four bands correspond to the C—H stretching vibrations, the CH_2 and CH_3 scissoring mode, the CH_3 rocking mode, and the CH_2 rocking mode, respectively. Figure 14.6 is a spectrum of the same sample measured in a cell 0.20 mm long. Since the amount of light absorbed is proportional to the number of molecules encountered by the beam as it passes through the sample, the longer cell allows absorption bands of low intensity to be observed. Many more bands can now be seen, especially in the region from 700 to 1300 cm^{-1}.

Because of its complexity, the spectrum cannot be analyzed completely. However, a peak-by-peak correspondence in the infrared spectra of two different samples is an excellent criterion of identity, as a comparison of the *n*-octane spectrum in Figure 14.6 with the *n*-heptane spectrum in Figure 14.7 readily shows. That is, the ir spectrum of *n*-heptane is similar to, but differs in significant respects from, that of *n*-octane.

Molecular vibrations are actually rather complex. Generally, all of the atoms in a molecule contribute to a vibration. Fortunately, however, some molecular vibrations can be treated by considering the motion of a few atoms relative to one

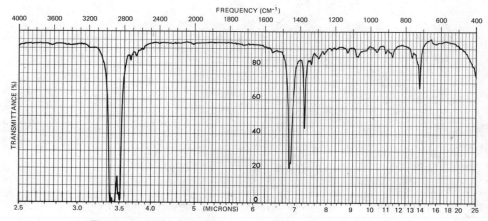

FIGURE 14.5. *Infrared spectrum of* n-*octane, 0.016 mm cell.*

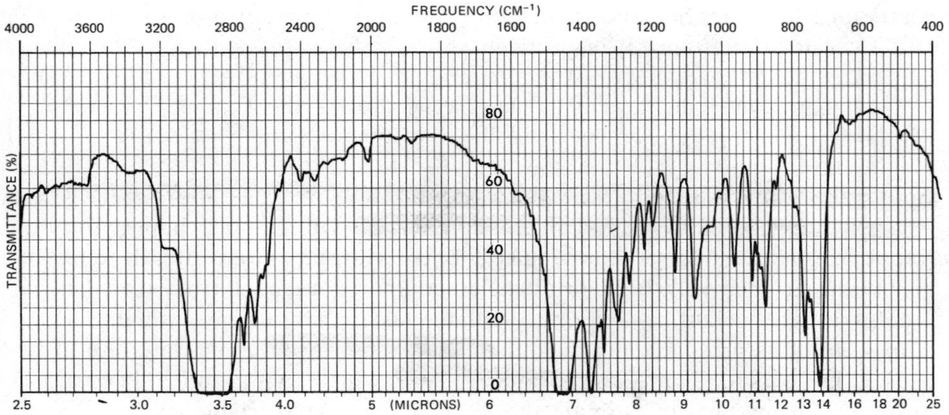

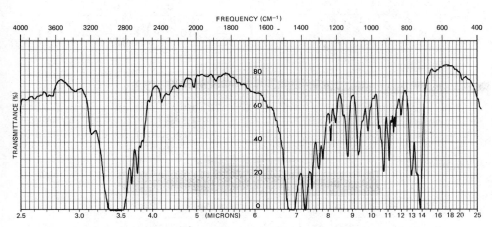

FIGURE 14.6. *Infrared spectrum of* n-*octane, 0.2 mm cell.*

FIGURE 14.7. *Infrared spectrum of* n-*heptane, 0.2 mm cell.*

another and ignoring the rest of the atoms in the molecule. For example, it is convenient to refer to the vibration of individual bonds. To a useful approximation (the **harmonic oscillator approximation**), the vibration frequency of a bond may be related to the masses of the vibrating atoms and the force constant, f, of the vibrating bond by equation (14-1). This equation corresponds to a simple Hooke's law model of two units coupled by a spring in which the force constant is the restoring force provided by the spring.

$$\tilde{\nu} = \frac{1}{2\pi c} \sqrt{\frac{f(m_1 + m_2)}{m_1 m_2}} \qquad (14\text{-}1)$$

where $\tilde{\nu}$ = vibrational frequency in cm^{-1} (wavenumber)

c = velocity of light in cm sec^{-1}
m_1 = mass of atom 1 in g
m_2 = mass of atom 2 in g
f = force constant in dyne cm^{-1} (g sec^{-2})

The larger the force constant, the higher the vibration frequency and the greater the energy spacings between vibrational quantum levels. The force constants for single, double, and triple bonds are approximately 5×10^5, 10×10^5, and 15×10^5 dynes cm^{-1}, respectively.

> Recall that 1 dyne is the force required to accelerate a 1 g mass 1 cm sec^{-2}. Therefore 1 dyne $= 1$ g cm sec^{-2}. The units of f, the force constant, are thus g sec^{-2}.

Force constants provide another measure of bond strength and generally are roughly proportional to bond-dissociation energies. On the other hand, vibration frequencies relate inversely to the masses of the vibrating atoms. Bonds to hydrogen occur at relatively high frequencies compared to bonds between heavier atoms—a light weight on a spring oscillates faster than a heavy weight.

In spite of its gross assumptions, the Hooke's law approximation is useful because it helps us to identify the *general region* in which a vibration will occur. For example, we may easily estimate the ^{12}C—^{1}H stretching frequency by

$$\tilde{\nu} = \frac{1}{2\pi \, 2.998 \times 10^{10} \text{ cm sec}^{-1}} \sqrt{\frac{5 \times 10^5 \text{ g sec}^{-2} \left(\frac{12}{6.023} + \frac{1}{6.023}\right) \times 10^{-23} \text{ g}}{\left(\frac{12}{6.023} \times 10^{-23} \text{ g}\right)\left(\frac{1}{6.023} \times 10^{-23} \text{ g}\right)}}$$

$$\tilde{\nu} = 3032 \text{ cm}^{-1}$$

The actual range for C—H absorptions is 2850–3000 cm^{-1}.

The approximate regions of the infrared spectrum where various bond vibrations are observed depend primarily on whether the bonds are single, double, triple or are bonds to hydrogen. These regions are summarized in Table 14.1.

TABLE 14.1
Approximate Values for Infrared Absorptions

Bond	General Absorption Region, cm^{-1}
C—C, C—N, C—O	800–1300
C=C, C=N, C=O	1500–1900
C≡C, C≡N	2000–2300
C—H, N—H, O—H	2850–3650

14.3
Characteristic Group Vibrations

As was pointed out in the previous section, the infrared spectrum of a polyatomic molecule is so complex that it is usually inconvenient to analyze it completely. However, extremely valuable information may be gleaned from the infrared spectrum of an organic compound using semiempirical methods. Consider the infrared spectra of 1-octene and 1-octadecene shown in Figures 14.8 and 14.9. Aside from the C—H stretching and bending vibrations at 2925, 1450, and 1370 cm^{-1}, we see several distinctive bands that do not appear in the spectra of typical alkanes (compare Figure 14.5). These new bands occur in the following general positions: 3080 cm^{-1}, 1640 cm^{-1}, 995 cm^{-1}, and 915 cm^{-1}. The Hooke's law approximation tells us that the band in the 3080 cm^{-1} region is the C—H stretch of the olefinic C—H bonds, and the 1640 cm^{-1} band is the C=C double bond stretching vibration. Other theoretical considerations suggest that the 995 cm^{-1} and 915 cm^{-1} bands are the olefinic C—H out-of-plane bending modes.

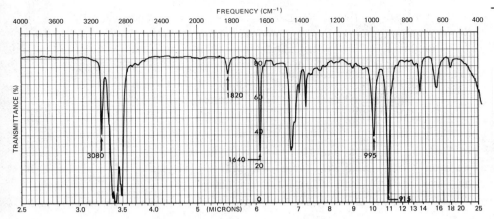

FIGURE 14.8. *Infrared spectrum of 1-octene.*

The weak band near 1820 cm^{-1} is an overtone of the fundamental band at 915 cm^{-1}.

The absorption bands mentioned above are characteristic of compounds containing a C=C double bond and may be used to determine unsaturation in an organic compound. Organic chemists use infrared spectroscopy in this semiempirical way. Most of the common functional groups give rise to characteristic absorption bands in defined regions of the infrared range. The chemist uses the presence or absence of a band in that region of the infrared spectrum as a diagnosis for the presence or absence of the corresponding functional group in this compound. One example will illustrate this point. The spectrum of 4-bromo-1-butene is shown in Figure 14.10. The spectrum is a fairly complex one, with a number of bands not characteristic of simple alkanes. The bands marked with arrows in Figure 14.10 are all due to vibrational transitions of the double bond. Note in particular the bands at 1842, 995, and 920 cm^{-1}. These bands are highly characteristic absorptions of a terminal vinyl group (R—CH=CH$_2$). The occurrence of these bands in the spectrum is taken as strong evidence for the presence of such a functional group in the molecule. (The bands at 3080 and 1640 cm^{-1}, although characteristic of alkenes, are not specific for compounds containing the group —CH=CH$_2$. As

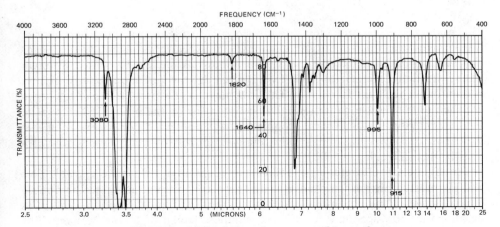

FIGURE 14.9. *Infrared spectrum of 1-octadecene.*

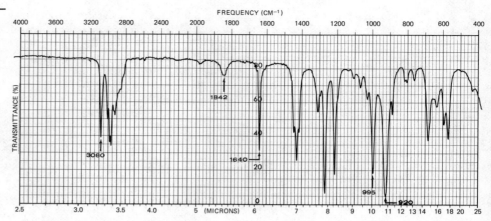

FIGURE 14.10. *Infrared spectrum of 4-bromo-1-butene,* $CH_2{=}CHCH_2CH_2Br$.

we shall see in a later section, other types of alkenes may also absorb in these regions.)

In the next few sections, we shall consider the characteristic group vibrations for various classes of compounds we have encountered.

14.4
Alkanes

As we saw previously, the major bands that appear in the infrared spectra of alkanes are those due to C—H stretching in the 2850–3000 cm^{-1} region, those due to CH_2 and CH_3 scissoring in the 1450–1470 cm^{-1} region, the band due to CH_3 rocking at about 1370–1380 cm^{-1}, and the CH_2 rocking bands at 720–725 cm^{-1}. These bands are of only limited diagnostic value because most alkanes contain all of these groupings. Some information may be culled from hydrocarbon spectra, however, if one looks at the fine details of the spectrum. For example, when a molecule has two methyl groups attached to the same carbon, the band at 1370–1380 cm^{-1} always appears as a doublet rather than as a single peak.

14.5
Alkenes

The alkene C—H stretching vibration occurs at higher wavenumber (shorter wavelength) than that due to an alkane C—H. Recall that alkene C—H bonds have greater *s*-character and are stronger than alkane C—H bonds. Stronger bonds are more difficult to stretch (higher force constant) and require greater energy or higher light frequency. Thus alkenes that have at least one hydrogen attached to the double bond normally absorb in the region 3050–3150 cm^{-1}. The relative intensity of this band, compared with the band for saturated C—H stretch, is roughly proportional to the relative numbers of the two types of hydrogens in the molecule.

The alkene C=C stretching mode occurs in the region 1645–1670 cm^{-1}. This band is most intense when there is only one alkyl group attached to the double bond. As more alkyl groups are added, the intensity of the absorption diminishes because the vibration now results in a smaller change of dipole moment. For trisubstituted, tetrasubstituted, and relatively symmetrical disubstituted alkenes,

TABLE 14.2
C—H Out-of-Plane Bending Absorptions of Alkenes

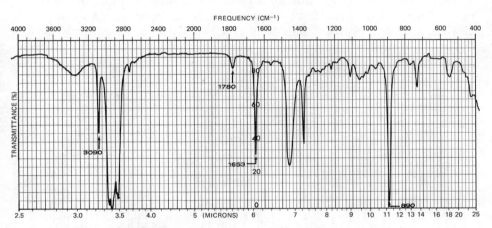

	Frequency cm^{-1}		Frequency cm^{-1}
R, H / C=C / H, H	**910** **990**	R, H / C=C / H, R	**970**
R, H / C=C / R, H	**890**	R, R / C=C / H, H	790–840
R, R / C=C / H, H	675–725	R, R / C=C / R, R	none

FIGURE 14.11. *Infrared spectrum of 2-methyl-1-heptene.*

the C—C stretching bond is often of such low intensity that it is not observable.

The C—H out-of-plane bending modes, which give rise to absorption bands in the region 700–1000 cm^{-1}, are useful for diagnosing the type of double bond present in some cases. The characteristic positions of these bands for various types of alkenes are summarized in Table 14.2. The most reliable bands for diagnostic purposes are those printed in boldface type.

The spectra that are reproduced in Figures 14.8–14.12 illustrate the characteristic vibrational bands of various types of alkenes. It is interesting to note that the C=C stretching band is absent from the spectrum of *trans*-4-octene because that vibration in this molecule results in no change in dipole moment. Also note that the C—H out-of-plane bending band gives rise to a characteristic overtone at about 1820 cm^{-1} for alkenes of the type R—CH=CH$_2$ and at 1780 cm^{-1} for alkenes of the type R$_2$C=CH$_2$, but not for other alkenes.

14.6
Alkynes

Terminal alkynes show a sharp C—H stretching band at 3300–3320 cm^{-1} and an intense C—H bending mode at 600–700 cm^{-1}. The C≡C stretch for terminal

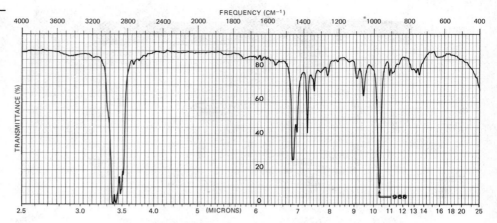

FIGURE 14.12. *Infrared spectrum of* trans-4-octene.

alkynes appears as a sharp absorption of moderate intensity at 2100–2140 cm⁻¹. For internal alkynes the C≡C stretch is a weak band occurring at 2200–2260 cm⁻¹; in hydrocarbons, this band is so weak that it is frequently not observed. Of course, if the molecule is symmetrical, the C≡C stretch is absent. These features are illustrated in the spectra shown in Figures 14.13 and 14.14.

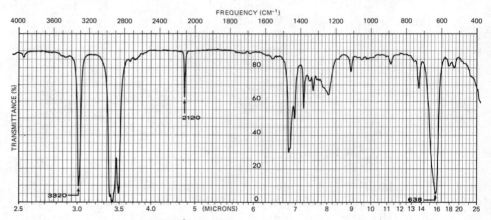

FIGURE 14.13. *Infrared spectrum of 1-octyne.*

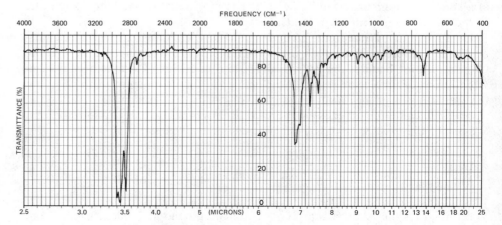

FIGURE 14.14. *Infrared spectrum of 2-octyne.*

14.7
Alkyl Halides

The characteristic absorption of alkyl halides is the band due to the C—X stretch. Typical positions for these bands are shown in Table 14.3. The C—Br and C—I stretching bands are used for diagnosis less than are the C—Cl and C—F bands because many of the commonly used spectrometers do not operate at wavelengths longer than about 700 cm⁻¹.

TABLE 14.3
Carbon–Halogen Stretching Bands

Bond	Frequency, cm⁻¹	Intensity
C—F	1000–1350	very strong
C—Cl	750–850	strong
C—Br	500–680	strong
C—I	200–500	strong

14.8
Alcohols and Ethers

The characteristic infrared bands of alcohols and ethers are the C—O stretch (1050–1200 cm⁻¹) and (for alcohols) the O—H stretch, which occurs in the 3200–3600 cm⁻¹ region. Although C—O stretching bands occur in a region of the spectrum where there are usually many other bands, they are relatively easy to identify because they are so intense. The infrared spectrum of *t*-butyl alcohol shown in Figure 14.15 is illustrative. Note the strong C—O stretch at 1200 cm⁻¹ and the extremely intense O—H stretch, which is centered at about 3360 cm⁻¹.

In Figure 14.16 are plotted the spectra of the O—H and C—H regions of *t*-butyl alcohol dissolved in carbon tetrachloride. (Carbon tetrachloride is a frequently used solvent for infrared studies because it is relatively inert chemically and is "transparent" to infrared light in most of the useful spectral regions.) Notice that in the first spectrum the 3440 cm⁻¹ O—H absorption is now accompanied by a sharp peak at 3620 cm⁻¹. As the solution is made more dilute, the 3620 cm⁻¹ band becomes more intense relative to the 3440 cm⁻¹ band. These two bands are both

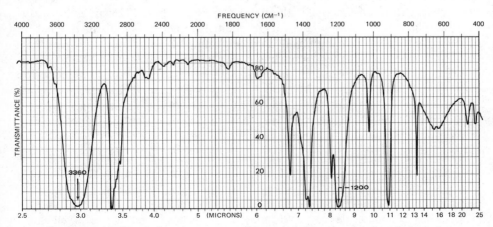

FIGURE 14.15. *Infrared spectrum of 2-methyl-2-propanol (t-butyl alcohol).*

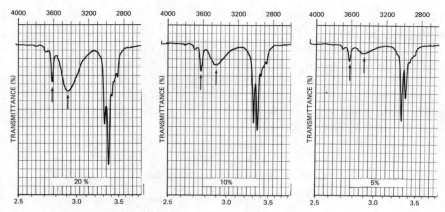

FIGURE 14.16. *Infrared spectra of various solutions of* t-*butyl alcohol in carbon tetrachloride.*

due to O—H stretch. The band at shorter wavelength (higher energy) is due to the stretching mode of the "free" hydroxy. The stretching mode of hydrogen-bonded or "associated" O—H bonds occurs at lower energy.

free hydroxy	[O—H]	$3620–3640 \text{ cm}^{-1}$
associated hydroxy	[O—H$\cdots$O]	$3250–3450 \text{ cm}^{-1}$

As the solution is made progressively more dilute, it is more likely that a molecule will exist in an unassociated state.

14.9
Aldehydes and Ketones

The characteristic infrared absorption for aldehydes and ketones is the band due to the C=O stretching vibration. Since the carbonyl group is highly polar, stretching of this bond results in a relatively large change in dipole moment. Consequently the carbonyl stretching band is an intense spectral feature. Because of its intensity, and also because it occurs in a region of the infrared spectrum commonly devoid of other absorptions, *the carbonyl stretch is perhaps the most reliable method for deducing the presence of such a functional group in a compound.* For simple saturated aldehydes the band occurs at about 1725 cm^{-1}. For saturated acyclic ketones the band occurs at about 1715 cm^{-1}. The distinctive nature of the C=O stretch is apparent in the spectrum of 2-heptanone shown in Figure 14.17. Since the carbonyl stretch is such an intense absorption, it often gives rise to a noticeable overtone in the $3400–3500 \text{ cm}^{-1}$ region. In 2-heptanone the carbonyl overtone occurs at 3440 cm^{-1} and may be seen in Figure 14.17. One must be cautious not to mistake this overtone for an OH absorption.

In cyclic ketones the exact stretching frequency depends on the size of the ring containing the carbonyl carbon. The magnitude of the effect is shown in Table 14.4.

The observed relationship between C=O stretching frequency and C—C=O bond angle has its origin in a "coupling" of the C=O stretch with that of the C—C bonds. For example, consider the hypothetical molecule depicted in Figure 14.18a in which the C—C=O bond angle is 90°. When the C=O bond is stretched, the carbon and oxygen move apart from one another. As shown in Figure 14.18b, this motion may

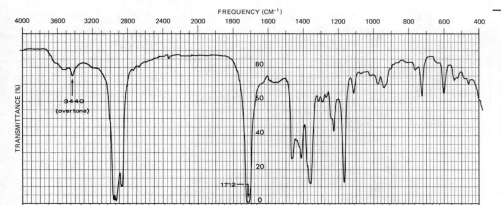

FIGURE 14.17. *Infrared spectrum of 2-heptanone.*

TABLE 14.4
Carbonyl Stretching Frequencies

Type of Ketone	Frequency, cm^{-1}
normal open chain	1715
cyclic	
three-membered	1850
four-membered	1780
five-membered	1745
six-membered	1715
seven-membered	1705

occur without seriously affecting the C—C bond length. However, the situation is very different in the hypothetical ketone shown in Figure 14.18c, where the C—C=O bond angle is 180°. When the C=O bond is stretched in this case, the C—C bond must be simultaneously *shortened* (Figure 14.18d). In a way, the C—C bond acts as a kind of "ballast" that resists stretching of the adjacent C=O bond. It is obvious from these simple extremes that the magnitude of coupling between the two bonds will relate to the C—C=O bond angle. Cyclohexanone and normal cyclic ketones have C—C=O bond angles of about 120°, the normal angle for sp^2-hybridization. Smaller ring size is associated with smaller internal bond angles and hence a larger C—C=O bond angle.

C
|↘90°
C=O

(a)

180°
C—C=O

(c)

C
/
C=O

(b)

C—C compressed   C=O stretched

C—C=O

(d)

FIGURE 14.18 *Illustrating the coupling of* C=O *and* C—C *stretching in ketones.*

The characteristic stretching frequencies of cyclic carbonyl groups apply to polycyclic systems as well. For example, the carbonyl group of camphor is part of a five-membered ring and the stretching frequency of 1740 cm^{-1} is characteristic of a cyclopentanone. The fact that the CO group in the bicyclic camphor is also part of a six-membered ring is not relevant—it is the character of ring strain that shows up in the spectrum. However, since ring-strain effects do depend somewhat on the detailed structure, variations of a few cm^{-1} from the values in Table 14.3 are to be expected.

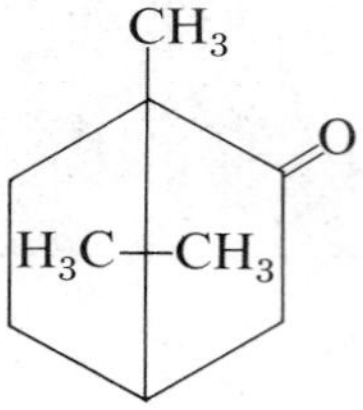

ν(CO) = 1740 cm^{-1}

camphor

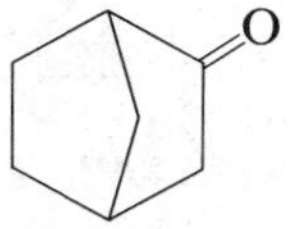

ν(CO) = 1751 cm^{-1}

norbornanone

The following bicyclic ketone is an apparent exception. Its C=O stretching band at 1731 cm^{-1} is far higher than expected for a cyclohexanone ring. A second look at the structure of this bicyclic ketone reveals, however, that its six-membered ring is that of a boat conformation rather than a chair.

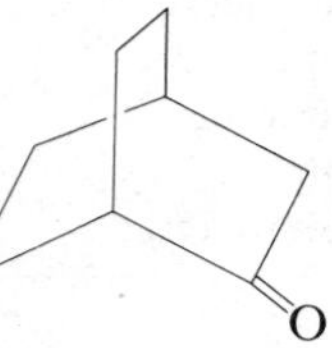

ν(CO) = 1731 cm^{-1}

2-bicyclo[2.2.2]octanone

In this section we have discussed the infrared spectra of aldehydes and ketones and have emphasized the utility of the carbonyl stretching band as a diagnostic tool for the carbonyl group. In later sections, we will discuss the spectra of carboxylic acids, anhydrides, acyl halides, esters, and amides, all of which contain a carbonyl group. In general, one may determine the precise nature of a carbonyl compound from the exact location of the C=O stretch and from a consideration of other bands in the spectrum.

EXERCISE: The carbonyl stretching frequency of ketene, $H_2C=C=O$, is 2150 cm^{-1}. Is this value consistent with the argument advanced to account for the magnitude of C=O stretching frequencies?

14.10
Summary: Principal Functional Group Absorptions

In this chapter, we have considered the infrared spectra of the classes of organic compounds taken up so far in this book. We have seen that the infrared spectra of organic compounds are so exceedingly complex that it is impractical to analyze a

427

Sec. 14.10

Summary:
Principal
Functional
Group
Absorptions

spectrum completely and assign each absorption to a given vibration. However, for each functional group there are characteristic absorptions that may be used empirically as a diagnosis for that particular functional group. Table 14.5 summarizes the infrared characteristics of alkanes, alkenes, alkynes, alkyl halides, alcohols, ethers, aldehydes, and ketones. The most useful bands are those printed in boldface type. As we consider other classes of compounds, we shall point out their characteristic infrared absorption bands. A further listing is given in Appendix VI.

TABLE 14.5
Principal Infrared Absorption

Class	Frequency, cm^{-1}	Intensity[a]	Assignment
1. Alkanes	2850–3000	s	C—H stretch
	1450–1470	s	
	1370–1380	s	CH$_2$ and CH$_3$ bend
	720–725	m	
2. Alkenes			
(a) RCH=CH$_2$	**3080–3140**	m	=C—H stretch
	1800–1860	m	overtone
	1645	m	C=C stretch
	990	s	
	910	s	C—H out-of-plane bend
(b) R$_2$C=CH$_2$	**3080–3140**	m	=C—H stretch
	1750–1800	m	overtone
	1650	m	C=C stretch
	890	s	C—H out-of-plane bend
(c) cis-RCH=CHR	3020	w	=C—H stretch
	1660	w	C=C stretch
	675–725	m	C—H out-of-plane bend
(d) trans-RCH=CHR	3020	w	=C—H stretch
	1675	vw	C=C stretch
	970	s	C—H out-of-plane bend
(e) R$_2$C=CHR	3020	w	=C—H stretch
	1670	w	C=C stretch
	790–840	s	C—H out-of-plane bend
(f) R$_2$C=CR$_2$	1670	vw	C=C stretch
3. Alkynes			
(a) RC≡CH	**3300**	s	≡C—H stretch
	2100–2140	m	C≡C stretch
	600–700	s	≡C—H bend
(b) RC≡CR	2190–2260	vw	C≡C stretch
4. Alkyl Halides			
(a) R—F	1000–1350	vs	C—F stretch
(b) R—Cl	750–850	s	C—Cl stretch
(c) R—Br	500–680	s	C—Br stretch
(d) R—I	200–500	s	C—I stretch
5. Alcohols			
(a) RCH$_2$OH	**3600**	var	free O—H stretch
	3400	s	bonded O—H stretch
	1050	s	C—O stretch
(b) R$_2$CHOH	**3600**	var	free O—H stretch
	3400	s	bonded O—H stretch
	1150	s	C—O stretch

TABLE 14.5 (continued)

Class	Frequency, cm^{-1}	Intensity[a]	Assignment
(c) R$_3$COH	**3600**	var	free O—H stretch
	3400	s	bonded O—H stretch
	1200	s	C—O stretch
6. Ethers	1070–1150	s	C—O stretch
7. Aldehydes	**1725**	s	C=O stretch
	2720, 2820	m	C—H stretch
8. Ketones			
(a) acyclic	**1715**	s	C=O stretch
(b) three-membered	**1850**	s	C=O stretch
(c) four-membered	**1780**	s	C=O stretch
(d) five-membered	**1740**	s	C=O stretch
(e) six-membered	**1715**	s	C=O stretch
(f) seven-membered	**1705**	s	C=O stretch

[a] vs = very strong, s = strong, m = medium, w = weak, vw = very weak, v = variable.

14.11
Instrumentation

An infrared spectrometer may be designed on either the single beam or the double beam principle. In a **single beam** spectrophotometer light from the radiation source (usually an oxide-coated ceramic rod that is heated electrically to about 1500°) is focused and passed through the sample, which is contained in a special cell. After passing through the sample, the emergent light beam is dispersed by a **monochromater** (either a prism or a diffraction grating) into its component wavelengths. The spectrum is scanned by slowly rotating the prism or grating. A **double beam** spectrophotometer operates on a similar principle except that the original light is split into two beams, one of which passes through the sample while the other passes through a reference cell. The instrument records the difference in intensity of these two beams. This type of instrument is especially useful when spectra are to be measured in solution. In such a case the reference cell contains pure solvent. Thus, if the solvent absorbs weakly in a given region of the spectrum, its absorption may be "canceled out." Since glass absorbs strongly in the useful infrared region, it cannot be used for the optical parts of a spectrophotometer. The prism and sample cell walls are usually fabricated from large NaCl or KBr crystals.

Modern research spectrophotometers are highly precise instruments that are both bulky and costly. Typical instruments cost from $8,000 to $30,000 and weigh 200–500 pounds. However, in 1969 two infrared spectrophotometers were packaged in space probes and sent to Mars! The purpose of this venture was to search for organic compounds such as methane in the atmosphere of the red planet. The spectrophotometers, which operated on the single beam principle, were designed and built at the University of California in Berkeley. They weighed 25 pounds each and occupied a volume only 1 ft^3 each. Using a sample length of about 6 miles (the effective depth of the Martian atmosphere), the spectrophotometers found water, carbon monoxide, and carbon dioxide–but no methane or other hydrocarbons. However, effective maximum levels, all in the parts per million range, were established for 39 other compounds. For example, NO_2, NH_3, SO_2, NO, O_3, CH_4, and CH_2O would have been detected had they been present in amounts as great as 4 ppm. An important spin-off of this space age research project has been the use of infrared as a method for analyzing for pollutants in the earth's atmosphere.

Infrared spectroscopy has recently found another interesting analytical use. A commercial instrument, called the Intoxalyzer, is used by police departments to analyze a person's breath for its alcohol content. The device is a single beam spectrophotometer that operates at a fixed wavelength (3.39 μ, 2950 cm^{-1}), corresponding to the C—H stretch of ethyl alcohol. Since exhaled breath rarely contains organic compounds other than alcohol, absorption at this wavelength is taken as a quantitative measure of drunkenness. A major disadvantage of this technique is that persons with certain illnesses exhale other organic compounds. For example, diabetics exhale acetone, which also absorbs at this wavelength. For this reason, crime labs that use the infrared method usually double-check their results with some other technique, such as gas chromatography.

PROBLEMS

1. Using the Hooke's law approximation, estimate $\tilde{\nu}$ for each of the following stretching vibrations. As force constants use 5×10^5, 10×10^5, and 15×10^5 dynes cm^{-1} for single, double, and triple bonds, respectively.

(a) O—H
(b) O—D
(c) C=C
(d) C≡C
(e) C—O
(f) C=O
(g) C≡N
(h) C—F

2. The acetylenic C—H stretch in 1-octyne occurs at 3350 cm^{-1}. Estimate the position of the C—D stretch in 1-deuterio-1-octyne and compare with the ^{13}C—H stretch in $CH_3(CH_2)_5C≡^{13}C—H$.

3. Identify the functional groups in each compound from the following infrared spectra. Note that the spectra in parts (f) and (g) were obtained using a different instrument than that used for the spectra in parts (a)–(e). This is a common situation. Although the spectra obtained with the two instruments have a different appearance, the characteristic absorptions are independent of the spectrometer used.

(a)

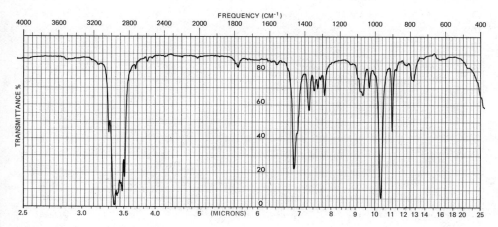

(b)

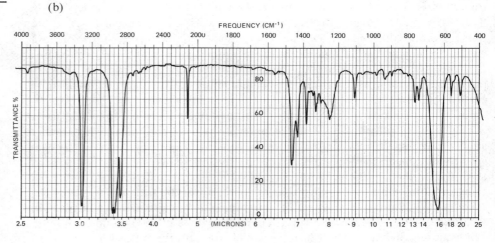

(c)

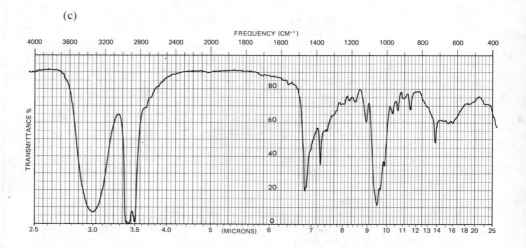

(d)

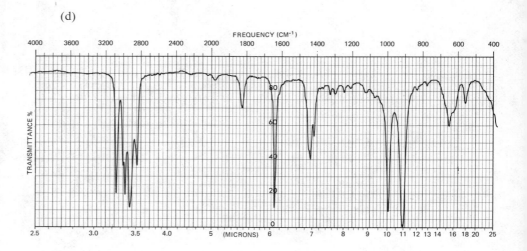

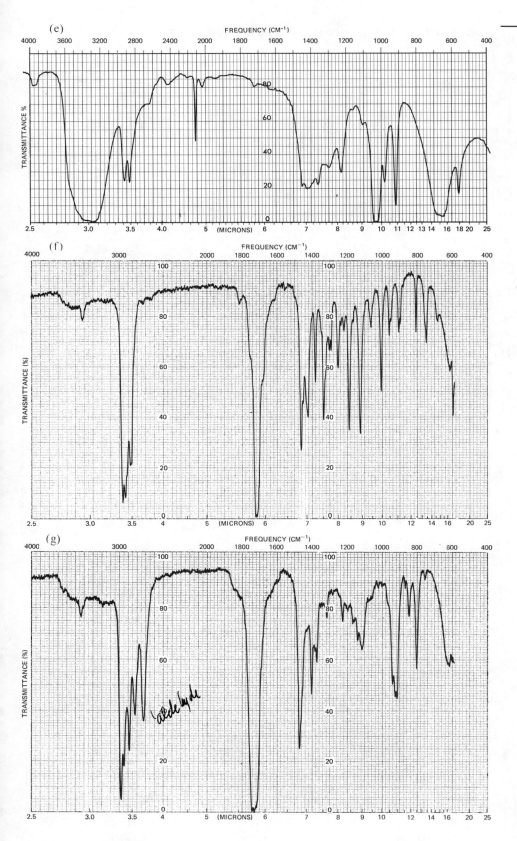

(e)

FREQUENCY (CM⁻¹)

(f)

FREQUENCY (CM⁻¹)

(g)

FREQUENCY (CM⁻¹)

4. Identify each of the following compounds from its ir and nmr spectra.

(a)

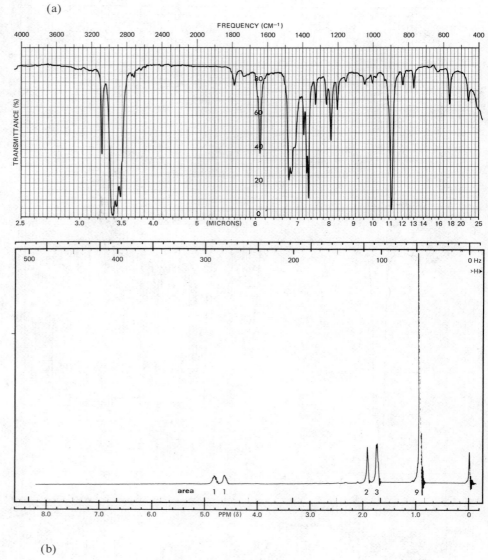

(b)

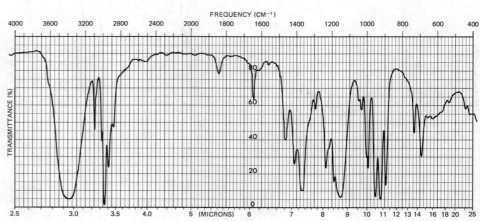

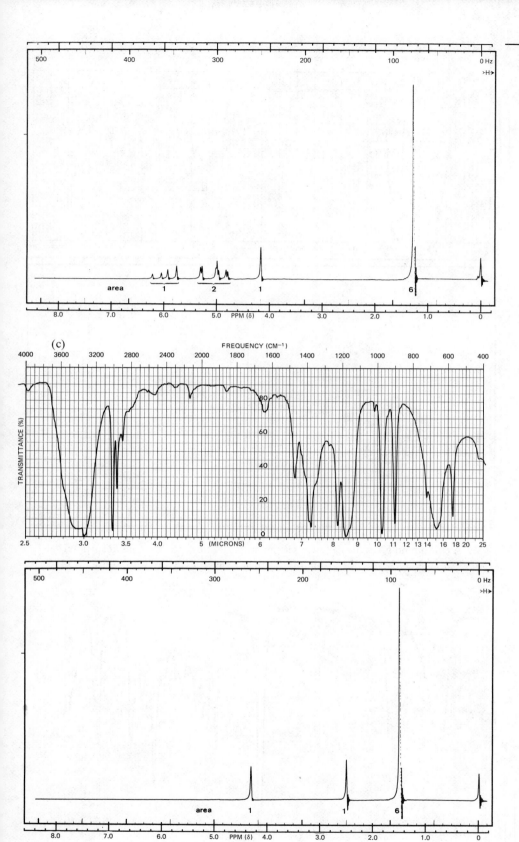

(d)

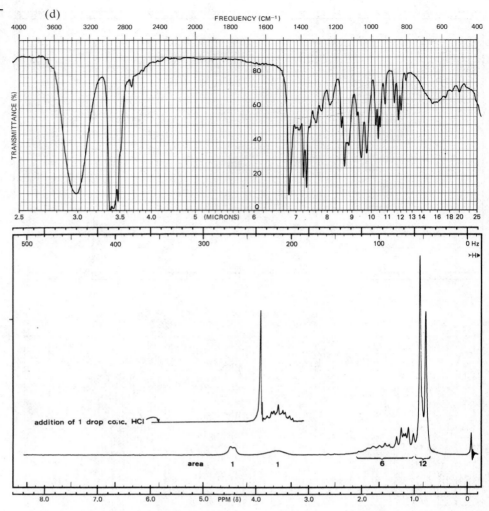

5. A compound gives the following ir spectrum. Its cmr spectrum has absorptions at 7.1, 8.5, 26.4, 30.7, 76.8, 212.7 ppm. Suggest a structure for the compound.

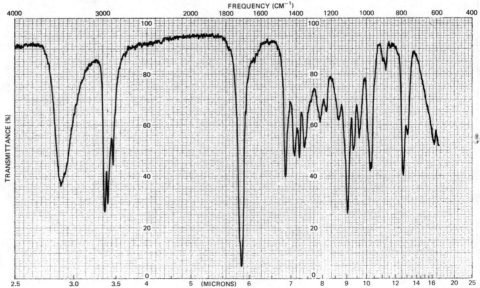

6. Identify the following isomeric compounds from their ir and cmr spectra.

(a) cmr: 7.9, 25.5, 33.5, 72.6 ppm

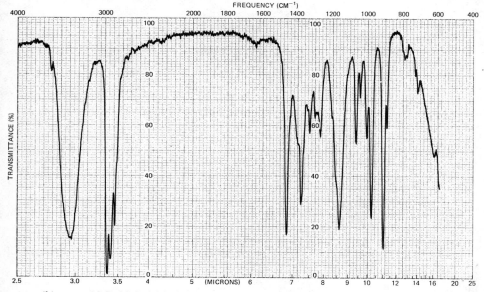

(b) cmr: 22.7, 23.5, 24.3, 25.1, 49.2, 65.5 ppm.

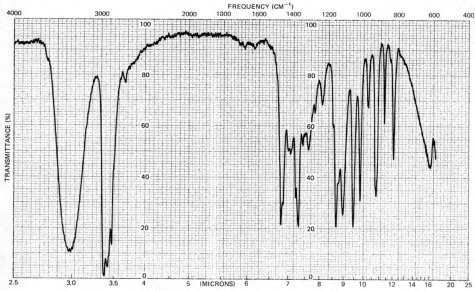

7. (a) For the heavier methyl halides one infrared frequency can be treated to an excellent approximation as a C—X stretching vibration. The position of this band is CH_3Cl, 732 cm^{-1}; CH_3Br, 611 cm^{-1}; CH_3I, 533 cm^{-1}. Find the corresponding C—X force constants and determine whether they are proportional to the corresponding $DH°$ (CH$_3$—X) values (see Appendix II).

(b) Astatine, element no. 85, is a halogen with no stable isotopes. The longest-lived isotope is ^{210}At with a half-life of 8.3 hr. At what value of $\tilde{\nu}$ would you expect to find the CH$_3$—At stretching band of methyl astatide?

(c) Methanethiol has a corresponding vibration (C—S stretch) at 705 cm^{-1}. Use your results from (a) to calculate the corresponding $DH°$ value. The experimental value for $DH°$ (CH$_3$—SH) is about 76 kcal mole^{-1}.

8. Identify the following compound from its ir and proton-decoupled cmr spectrum.

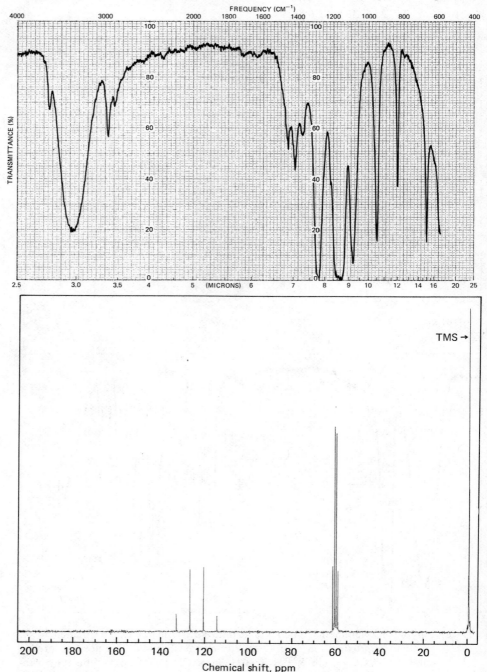

9. Dialkyl peroxides, ROOR, have an absorption in the region 820–1000 cm^{-1}, but this band is extremely weak and difficult to detect. Explain. Using the Hooke's law approximation, find the force constant for an O—O stretch of 900 cm^{-1}. How does it compare with the normal single bond f of 5×10^5? Explain.

10. For a harmonic oscillator the potential energy $E = f(r - r_0)^2$. In calculus form the radius of curvature is expressed as d^2E/dr^2. What is the relationship between the radius of curvature and the force constant f for a harmonic oscillator?

Chapter 15
Organometallic Compounds

15.1
Nomenclature

Organometallic compounds are substances in which an organic group is bonded directly to a metal, R—M. They are named by prefixing the name of the metal with the appropriate organic group name. The names are written as one word.

$$(CH_3)_3CLi \qquad (CH_3CH_2)_2Mg \qquad (CH_3)_3Al$$
$$\text{t-butyllithium} \qquad \text{diethylmagnesium} \qquad \text{trimethylaluminum}$$

$$(CH_3CH_2CH_2)_2Cd \qquad (CH_3CH_2)_2Zn \qquad (CH_3)_2Hg$$
$$\text{dipropylcadmium} \qquad \text{diethylzinc} \qquad \text{dimethylmercury}$$

$$CH_3Cu \qquad (CH_3)_4Si \qquad (CH_3CH_2)_4Pb$$
$$\text{methylcopper} \qquad \text{tetramethylsilicon} \qquad \text{tetraethyllead}$$

Compounds of boron, tin, and silicon are also named as derivatives of the simple hydrides: borane, BH_3; stannane, SnH_4; and silane, SiH_4. These compounds are indexed by *Chemical Abstracts* in this manner.

$$(CH_3CH_2)_3B \qquad (CH_3CH_2)_4Sn \qquad (CH_3)_3SiCH_2CH_3$$
$$\text{triethylborane} \qquad \text{tetraethylstannane} \qquad \text{ethyltrimethylsilane}$$

In some organometallic compounds, the valences of the metal are not all utilized in bonding to carbon but include bonds to inorganic atoms as well. Such compounds are named as organic derivatives of the corresponding inorganic salt.

$$CH_3CH_2MgBr \qquad CH_3HgCl \qquad CH_3CH_2AlCl_2$$
$$\text{ethylmagnesium} \qquad \text{methylmercuric chloride} \qquad \text{ethylaluminum}$$
$$\text{bromide} \qquad \qquad \text{dichloride}$$

In some cases, especially when the organic group contains other functional groups, it is convenient to name the metal as a substituent.

$$\overset{\qquad\qquad\qquad\qquad\quad OH}{NaC\equiv CCH_2CH_3 \qquad ClHgCH_2\overset{|}{C}HCH_3}$$
$$\text{1-sodio-1-butyne} \qquad \text{1-chloromecuri-2-propanol}$$

15.2
Structure

Since metals are **electropositive** elements, C—M bonds can have a high degree of **ionic character.** That is, dipolar resonance structures are often important contributors to the structure of such compounds.

$$\left[R—M \longleftrightarrow R^- \; M^+ \right] \equiv \overset{\delta-}{R}—\overset{\delta+}{M}$$

The degree of covalency of such C—M bonds depends markedly on the metal

TABLE 15.1
Electronegativity Values for Some Elements

Group	IA	IIA	IB	IIB	IIIA	IVA
	H					
	2.1					
	Li	Be			B	C
	1.0	1.5			2.0	2.5
	Na	Mg			Al	Si
	0.9	1.2			1.5	1.8
	K	Ca	Cu	Zn		Ge
	0.8	1.0	1.9	1.6		1.7
				Cd		Sn
				1.7		1.7
				Hg		Pb
				1.9		1.7

and is related to the **electronegativity** of the metal. Electronegativity is defined as the tendency of an element to attract electrons. A semiquantitative scale has been constructed in which each element is assigned an electronegativity value. On this scale a larger number signifies a greater affinity for electrons. When two elements of differing electronegativity are bonded, the bond will be polar with the "center of gravity" of electron density in the bond closer to the more electronegative element. The greater the difference in electronegativity, the more polar is the bond. Pauling electronegativities for carbon and several metals are listed in Table 15.1.

Because of the large differences in electronegativity between carbon and the alkali metals, alkyllithium, alkylsodium, and alkylpotassium compounds are highly ionic. Methyllithium exists as a tetramer in which the lithium cations and methyl anions occupy the corners of a distorted cube (Figure 15.1). Lithium–lithium distances are 2.67 Å and carbon–carbon distances are 3.67 Å. Methylsodium has a similar tetrameric structure. Methylpotassium displays almost classic salt-like behavior. Its crystal structure is similar to that of NaCl; each potassium ion is symmetrically surrounded by six methyl anions and vice versa.

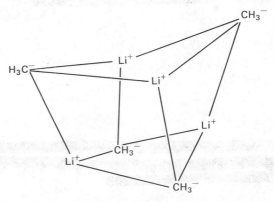

FIGURE 15.1. *Methyllithium tetramer. The lines are included to define geometry, not to signify bonds.*

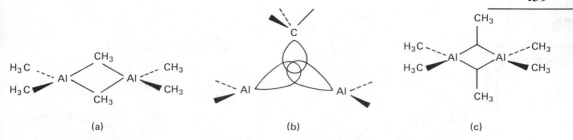

FIGURE 15.2. (a) *Trimethylaluminum dimer.* (b) *Three orbitals overlap to give a molecular orbital containing two electrons.* (c) *Structure showing the* ⊥ *symbol for a three-center two-electron bond.*

Other alkyllithiums form aggregates of tetramers, hexamers, and so on, that are usually soluble in organic solvents, even hydrocarbons. Some such aggregates have appreciable volatility. These aggregates can be thought of as ionic clusters with a hydrocarbon-like exterior.

With the less electropositive metals, such as Be, Mg, B, and Al, the carbon-metal bonds are polar but still partially covalent. The metals in these compounds generally do not have an inert gas configuration. This electron deficiency has several structural consequences. For example, **three-center two-electron** bonds are encountered commonly. We have previously seen one example of such bonding in the structure of diborane on page 315. Another example is trimethylaluminum, which exists as a dimer in which two aluminum atoms are bridged by two methyl groups (Figure 15.2a). The three-center bond may be thought of as arising by the simultaneous overlap of an orbital from each aluminum with an sp^3-orbital of carbon (Figure 15.2b). In this way both aluminum atoms realize the benefit of being surrounded by eight electrons, thus achieving an argon electronic configuration. The three-center bond is often symbolized by ⊥, as in Figure 15.2c.

Dialkylmagnesium compounds exist as polymeric structures in the solid phase or in hydrocarbon solvents (Figure 15.3a). In ether solvents monomeric dialkylmagnesium compounds are coordinated to two ether oxygens (Figure 15.3b). Alkylmagnesium halides, also known as **Grignard reagents,** are important organometallic compounds that have many uses in synthesis (Sections 15.4.A and 15.5.D). In dilute ether solution (about 0.1 M) Grignard reagents exist as monomers in which the magnesium is coordinated to two solvent molecules; their structures are similar to that of dialkylmagnesium (Figure 15.3b). However, in more concentrated solution (0.5–1 M) the principal species is a dimer in which two magnesium atoms are bridged by two bromines (Figure 15.4). In this structure

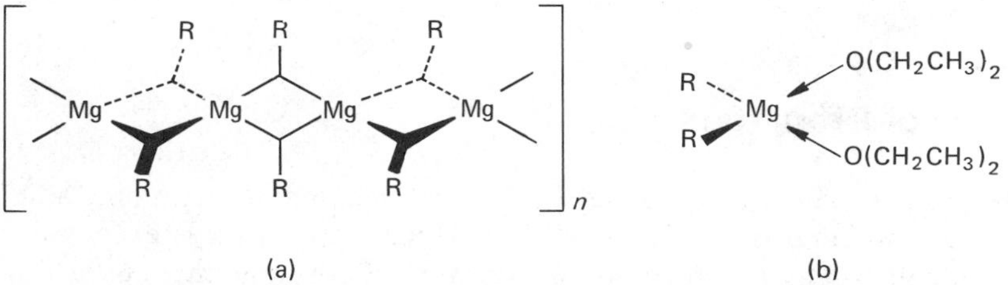

FIGURE 15.3. (a) *Dialkylmagnesium polymer.* (b) *Dialkylmagnesium coordinated to ether molecules. Note the use of arrows to show the dative or coordinative "donor" bonds.*

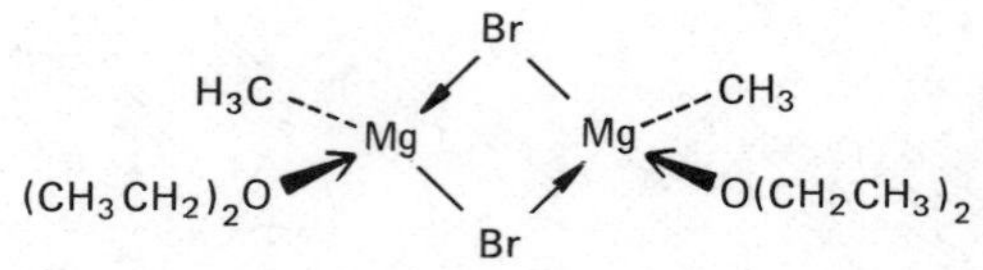

FIGURE 15.4. *Methylmagnesium bromide dimer in ether.*

each magnesium acquires its octet by additional coordination with one bromine from the other RMgBr and an ether oxygen. Note that the bridging bonds in this structure are *not* three-center two-electron bonds. Rather, the two bonds to each bromine are both two-center two-electron bonds. Such bonding is possible in this structure because the bromines, in contrast to the methyl groups in Figures 15.2 and 15.3, have unshared electron pairs to use in bonding.

In compounds of metals in groups IV and V (Si, Ge, Sn, Pb, Sb, Bi) there are sufficient electrons for the metals to engage in normal covalent bonding of the type with which we are familiar. Furthermore, since these metals tend to be much nearer to carbon in electronegativity than metals of groups I–III, the carbon–metal bonds are not very polar. Thus these organometallic compounds tend to resemble conventional organic compounds in their properties. For example, tetramethylsilicon and tetramethyllead are similar in structure to neopentane. The metal in each case has tetrahedral geometry (sp^3-hybridization) and makes covalent bonds to the four methyl groups.

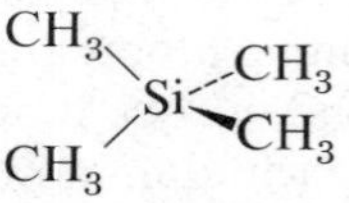

tetramethylsilicon
(tetramethylsilane)
b.p. 26.5°

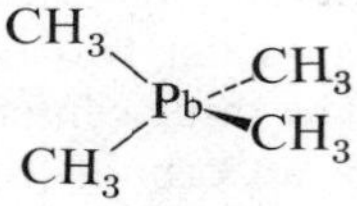

tetramethyllead
b.p. 110°

Trimethylantimony and trimethylbismuth have pyramidal structures in which the metal has an unshared electron pair, similar to that in ammonia (page 26).

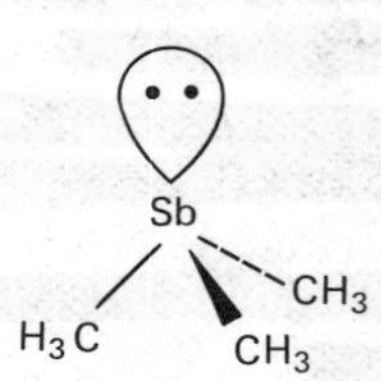

trimethylantimony
b.p. 80.6°

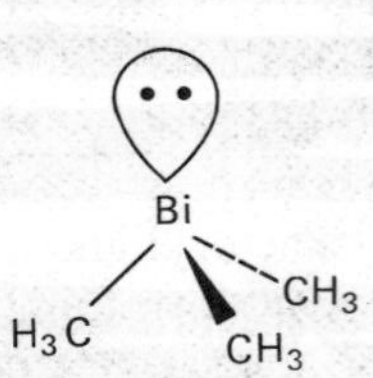

trimethylbismuth
b.p. 110°

15.3
Physical Properties

The melting points and boiling points of some simple organometallic compounds are summarized in Table 15.2. If sufficient caution is exercised, many organometallics may be prepared and handled in the same manner as other organic compounds. However, as we shall see, quite a few organometallics react vigorously with water or other protic compounds and with oxygen. Consequently,

TABLE 15.2
Physical Properties of Organometallic Compounds

Compound	Melting Point, °C	Boiling Point, °C
CH_3CH_2Li	95	subl. 95 (aggregated)
$(CH_3)_2Mg$	240	(probably polymeric)
$(CH_3)_3Al$	0	130
CH_3AlCl_2	73	97–100 (100 torr)
$(CH_3)_4Si$	—	26.5
$(CH_3CH_2)_4Si$	—	153
$(CH_3)_2Zn$	−42	46
$(CH_3)_2Cd$	−4.5	106
$(CH_3)_2Hg$	—	96
$(CH_3CH_2)_2Hg$	—	159
CH_3CH_2HgI	186	—
$(CH_3CH_2HgCl$	193	subl. 40
$(CH_3)_3Ga$	−19	56
$(CH_3)_3In$	89	subl. 89
$(CH_3)_3Te$	38.5	147
$(CH_3)_4Ge$	−88	43
$(CH_3)_4Sn$	−55	78
$(CH_3)_4Pb$	−27.5	110

care must be taken in performing such operations as distillation and recrystallization.

Many of the organometallic compounds in Table 15.2 decompose in water, but they are soluble in various inert aprotic organic solvents. Typical solvents are ethers and alkanes. Because of their solubility in convenient organic solvents and their extreme reactivity, a number of organometallic compounds used in organic syntheses are normally not purified, but are prepared and used in such solutions without isolation.

15.4
Preparation of Organometallic Compounds

A. Reaction of an Alkyl Halide with a Metal

Reaction of the metal with an alkyl halide is most generally used for the laboratory preparation of organolithium and organomagnesium compounds. The reaction is normally carried out by treating the metal with an ether or hydrocarbon solution of the alkyl halide.

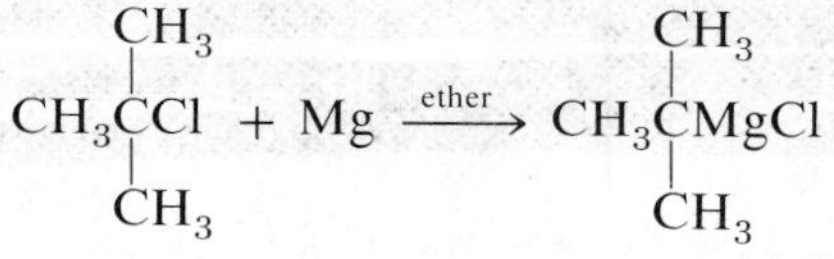

$$CH_3\underset{\underset{CH_3}{|}}{\overset{\overset{CH_3}{|}}{C}}Cl + Mg \xrightarrow{ether} CH_3\underset{\underset{CH_3}{|}}{\overset{\overset{CH_3}{|}}{C}}MgCl$$

A solution of 227 g of *t*-butyl chloride in 1300 mL of dry ether is stirred in contact with 61 g of magnesium turnings for 6–8 hr. A cloudy gray solution approximately 2 M in *t*-butylmagnesium chloride is obtained. This solution is used directly for further reactions.

This type of reaction is an example of a heterogeneous reaction—a reaction that occurs at the interface between two different phases. The alkyl halide in solution must react with magnesium on the surface of solid magnesium. In such reactions the surface area and its character are important. In the preparation of Grignard reagents the magnesium is usually in the form of metal shavings or turnings. The reaction mechanism consists of the following steps.

$$R\!-\!X + \overset{\}{\underset{\}{Mg}} \longrightarrow R\cdot\ + X\!-\!\overset{\}{\underset{\}{Mg}}$$

$$R\cdot\ + X\!-\!\overset{\}{\underset{\}{Mg}} \longrightarrow R\!-\!Mg\!-\!X$$

Reaction of RX at the magnesium surface produces an alkyl radical and a Mg—X species probably still associated with the metal surface. The resulting free radical, R·, then reacts with the ·MgX to produce the Grignard reagent, RMgX. The principal side reactions involve alternative reactions of organic radicals, mostly dimerization and disproportionation.

$$CH_3CH_2CH_2CH_2\cdot\ + CH_3CH_2CH_2CH_2\cdot\ \longrightarrow CH_3CH_2CH_2CH_2CH_2CH_2CH_2CH_3$$
$$CH_3CH_2CH_2CH_2\cdot\ + CH_3CH_2CH_2CH_2\cdot\ \longrightarrow CH_3CH_2CH_2CH_3 + CH_3CH_2CH\!=\!CH_2$$

However, for simple alkyl halides the yields of alkylmagnesium halide are high—frequently above 90%. The reaction works well with chlorides, bromides, and iodides. Reaction of alkyl chlorides is frequently somewhat sluggish and iodides are generally expensive. Hence, alkyl bromides are common laboratory reagents in Grignard syntheses.

A suitable solvent is essential for formation of the Grignard reagent because of the necessity for solvating the magnesium, as discussed in Section 15.2. The reaction is commonly carried out by adding an ether solution of the alkyl halide to magnesium turnings stirred in ether in a three-necked flask (Figure 15.5). The reaction is exothermic, particularly with bromides and iodides, and a reflux condenser is provided for returning the boiling ether. Anhydrous conditions must be maintained throughout the reaction since Grignard reagents react rapidly with traces of moisture.

Alkyllithium compounds are prepared in the same manner.

$$CH_3CH_2CH_2CH_2Br + 2Li \xrightarrow{\text{ether}} CH_3CH_2CH_2CH_2Li + LiBr$$

A solution of 68.5 g of *n*-butyl bromide in 300 mL of dry ether is added slowly to 8.6 g of lithium wire. The mixture is stirred at −10° for about 1 hr, during which time all of the lithium dissolves. The resulting ether solution of *n*-butyllithium is stored under nitrogen in a well-stoppered flask.

The reaction of alkyl halides with lithium is probably similar to that with magnesium. Organosodium and organopotassium compounds cannot be prepared in this manner. With these metals the chief product is an alkane (Wurtz reaction).

$$2\ CH_3CH_2CH_2Br + 2\ Na \longrightarrow CH_3CH_2CH_2CH_2CH_2CH_3 + 2\ NaBr$$

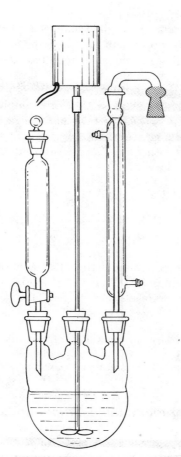

FIGURE 15.5. *Typical apparatus used for Grignard reactions. The three-necked flask carries a dropping funnel, stirring motor, and reflux condenser equipped with a drying tube.*

The alkane probably results when the initially formed alkylsodium or alkylpotassium displaces halide from a molecule of alkyl halide that has not yet reacted.

$$CH_3CH_2CH_2Br + 2\,Na \longrightarrow CH_3CH_2CH_2 :^- \; Na^+ + NaBr$$

$$CH_3CH_2CH_2 :^- + CH_3CH_2CH_2Br \longrightarrow CH_3CH_2CH_2CH_2CH_2CH_3 + Br^-$$

The second step presumably proceeds by the S_N2 mechanism. Alkylsodium compounds have been prepared by other methods, and they react with optically active alkyl halides with inversion of configuration.

$$CH_3CH_2Na + Cl\!-\!\underset{\underset{(CH_2)_5CH_3}{|}}{\overset{\overset{CH_3}{|}}{C}}\!-\!H \longrightarrow H\!-\!\underset{\underset{(CH_2)_5CH_3}{|}}{\overset{\overset{CH_3}{|}}{C}}\!-\!CH_2CH_3 + NaBr$$

(R)-2-chlorooctane $\qquad\qquad$ (R)-3-methylnonane

Note that in this example inversion has occurred at the reaction center. However, the reactant and product are both (R) because the leaving group has priority *a* and the incoming group has priority *b*.

The reactions of alkali metals with alkyl halides are heterogeneous surface reactions and may be more complex than the foregoing discussion would indicate. In some cases radical coupling reactions are required to explain the reaction products. For an example, see problem 15.

Although many other organometallic compounds may be prepared by direct reaction of the metal with an alkyl halide (for example, C_2H_5ZnI, CH_3HgCl), a more convenient and general laboratory method is metal exchange with an organolithium or organomagnesium compound (next section). However, the direct reaction of dihalides with zinc or magnesium can be used to prepare alkenes and cyclopropanes in good yield.

$$BrCH_2CH_2Br + Mg \longrightarrow MgBr_2 + CH_2{=}CH_2$$

$$BrCH_2CH_2CH_2Br + Zn \longrightarrow \triangle + ZnBr_2$$
$$(80\%)$$

Note that the probable intermediate in the latter reaction, $BrCH_2CH_2CH_2ZnBr$, is similar to the postulated intermediate in the Simmons–Smith procedure for preparation of cyclopropanes from alkenes (page 323).

EXERCISE: Write the structure and name of each organometallic compound derived from reactions of (a) methyl iodide, (b) cyclohexyl bromide, and (c) 2-chloro-2-methylbutane with magnesium and with lithium.

B. *Reaction of Organometallic Compounds with Salts*

One of the most useful methods for preparing most organometallic compounds in the laboratory is the exchange reaction of one organometallic with a salt to give a new organometallic and a new salt.

$$RM + M'X \rightleftharpoons RM' + MX \tag{15-1}$$

Although this is an equilibrium process, the equilibrium constant is dominated by the reduction potentials of the two metal cations. An abbreviated list of **standard reduction potentials** is contained in Table 15.3.

**TABLE 15.3
Standard Reduction Potentials**

Reaction	$E°$, volts
$Li^+ + e^- = Li$	-3.045
$Mg^{2+} + 2\,e^- = mg$	-2.370
$Al^{3+} + 3\,e^- = Al$	-1.660
$Si^{4+} + 4\,e^- = Si$	-0.840
$Zn^{2+} + 2\,e^- = Zn$	-0.763
$Cd^{2+} + 2\,e^- = Cd$	-0.402
$H^+ + e^- = \frac{1}{2}\,H_2$	0
$Sn^{4+} + 4\,e^- = Sn$	0.014
$Cu^+ + e^- = Cu$	0.522
$Hg^{2+} + 2\,e^- = Hg$	0.854

In general, reaction proceeds so that the more electropositive metal (more negative $E°$) exists as the more ionic inorganic salt. For example, Grignard reagents react readily with cadmium chloride to give organocadmium compounds and magnesium chloride.

$$2 \ CH_3CH_2MgCl + \underset{E° = -0.402}{CdCl_2} \xrightarrow{\text{ether}} (CH_3CH_2)_2Cd + \underset{E° = -2.370}{2 \ MgCl_2}$$

In a similar way, Grignard reagents may be used to prepare tetraalkylsilanes and dialkylmercury compounds.

$$4 \ CH_3MgCl + \underset{E° = -0.840}{SiCl_4} \longrightarrow (CH_3)_4Si + \underset{E° = -2.370}{4 \ MgCl_2}$$

$$\overset{\overset{\textstyle CH_3}{|}}{2 \ CH_3CH_2CHMgCl} + \underset{E° = 0.854}{HgCl_2} \longrightarrow \overset{\overset{\textstyle CH_3}{|}}{(CH_3CH_2CH)_2Hg} + \underset{E° = -2.370}{2 \ MgCl_2}$$

This reaction may be understood by considering the nature of the C—M bond in organometallic compounds. The standard reduction potential of a metal is a measure of its tendency to give up electrons and become an ion in aqueous solution. Metals with very negative reduction potentials give up electrons readily (that is, are good reducing agents). Metals with positive reduction potentials prefer to "hold on" to their electrons (that is, are poor reducing agents). The C—M bond has some covalent character, as do the M—M bonds in the free metal itself; in both cases the metal is formally neutral. In equilibria such as (15.1) the metal with the more positive reduction potential tends more to acquire electrons and prefers to exist as the organometallic compound. The metal with the more negative reduction potential prefers to exist as the ionic salt. Because lithium and magnesium are such highly electropositive elements and because their alkyl derivatives are so readily available, alkyllithium compounds and Grignard reagents are the usual starting materials in this synthetic route to other organometallics. Examples using alkyllithium compounds are

$$CH_3Li + \underset{E° = +0.522}{CuI} \longrightarrow CH_3Cu + \underset{E° = -3.045}{LiI}$$

$$4 \ n\text{-}C_4H_9Li + \underset{E° = +0.014}{SnCl_4} \longrightarrow (n\text{-}C_4H_9)_4Sn + \underset{E° = -3.045}{4 \ LiCl}$$

One final point should be specified. The standard reduction potentials given in Table 15.3 refer to the electrical potentials associated with conversion of a dilute aqueous solution of the metal salt to the free metal. Such potentials will be different for different metal salts in organic solvents, but the relative ranking of electropositive character generally remains unchanged.

EXERCISE: Predict the direction of equilibrium for each of the following cases.
(a) $2 \ (CH_3)_3Al + 3 \ ZnCl_2 \rightleftharpoons 3 \ (CH_3)_2Zn + 2 \ AlCl_3$
(b) $2 \ (CH_3)_2Hg + SiCl_4 \rightleftharpoons (CH_3)_4Si + 2 \ HgCl_2$
(c) $3 \ CuBr + (CH_3)_3Al \rightleftharpoons 3 \ CH_3Cu + AlBr_3$

C. *Metallation of C—H Bonds*

We have already encountered the metallation method for preparing an organo-metallic compound in Section 12.4. Recall that alkynes react with alkyllithium compounds to give the alkynyllithium and the appropriate alkane.

$$CH_3CH_2CH_2C{\equiv}CH + n\text{-}C_4H_9Li \longrightarrow CH_3CH_2CH_2C{\equiv}CLi + n\text{-}C_4H_{10}$$
$$\text{1-lithio-1-pentyne}$$

This reaction is simply an acid-base reaction, 1-pentyne being the acid and *n*-butyllithium the base. The equilibrium lies far to the right because 1-pentyne is a much stronger acid than *n*-butane. The reaction is a general method for replace-ment of a C—H bond by a carbon–metal bond, provided that the hydrocarbon to be metallated is considerably more acidic than the conjugate acid of the base to be used. Grignard reagents are also effective reagents for metallation of alkynes.

$$HC{\equiv}CH + CH_3CH_2MgBr \xrightarrow[\text{ether}]{\Delta} HC{\equiv}CMgBr + C_2H_6$$

$$\begin{array}{ccc} \text{ethylmagnesium} & & \text{ethynylmagnesium} \\ \text{bromide} & & \text{bromide} \end{array}$$

D. *Special Methods of Preparation*

We have seen in Chapter 11 that some organometallic compounds may be prepared by additions to alkenes. Hydroboration and oxymercuration are exam-ples.

$$6\ CH_2{=}CH_2 + B_2H_6 \longrightarrow 2\ (CH_3CH_2)_3B$$
$$\text{triethylborane}$$

$$CH_2CH{=}CH_2 + Hg(OAc)_2 + H_2O \longrightarrow \underset{\underset{\displaystyle OH}{|}}{CH_3CHCH_2HgOAc} + HOAc$$
$$\text{2-hydroxypropylmercuric}$$
$$\text{acetate}$$

Organoaluminum compounds may also be prepared by addition of compounds having an aluminum–hydrogen bond to alkenes or alkynes. Diisobutylaluminum hydride (DIBAL) is a useful reagent for this purpose.

$$\left(\underset{\underset{\displaystyle CH_3}{|}}{CH_3CHCH_2}\right)_2 AlH + CH_3CH_2C{\equiv}CH \longrightarrow \left(\underset{\underset{\displaystyle CH_3}{|}}{CH_3CHCH_2}\right)_2 Al \underset{H}{\overset{H}{\diagdown}} C{=}C \underset{CH_2CH_3}{\overset{H}{\diagup}}$$

$$\begin{array}{cc} \text{diisobutylaluminum} & \text{(E)-1-butenyldiisobutylaluminum} \\ \text{hydride} & \end{array}$$

Some organometallic compounds may be prepared by treating other organo-metallics with a free metal or by the disproportionation of two organometallic compounds.

$$(CH_3CH_2)_2Hg + 2\ Na \longrightarrow 2\ CH_3CH_2Na + Hg$$
$$(CH_3)_2Hg + 2\ CH_3CH_2Li \longrightarrow (CH_3CH_2)_2Hg + 2\ CH_3Li$$

However, these reactions are complicated and difficult to generalize. We shall not consider them further.

15.5
Reactions of Organometallic Compounds

A. *Hydrolysis*

Organometallic compounds in which the metal has an electronegativity value of about 1.7 or less (Table 15.1) react with water to give the hydrocarbon and a metal hydroxide. The more electropositive the metal is, the faster is the hydrolysis. Alkyllithium, alkylmagnesium, and alkylaluminum compounds react violently with water.

$$CH_3Li + H_2O \longrightarrow CH_4 + LiOH$$
$$CH_3CH_2MgBr + H_2O \longrightarrow CH_3CH_3 + HOMgBr$$
$$(CH_3)_3Al + 3 H_2O \longrightarrow 3 CH_4 + Al(OH)_3$$

Such compounds react similarly with other hydroxylic compounds, such as alcohols and carboxylic acids.

$$CH_3\!-\!\underset{\underset{CH_3}{|}}{\overset{\overset{CH_3}{|}}{C}}\!-\!Li + CH_3OH \longrightarrow CH_3\underset{\underset{}{}}{\overset{\overset{CH_3}{|}}{C}}HCH_3 + CH_3O^-Li^+$$

They also react with other compounds having relatively acidic hydrogens, such as thiols and amines.

Since the product of hydrolysis is an alkane, hydrolysis is not a very useful preparative reaction. However, it is important to recognize the limitation that this ready hydrolysis puts on the use of such organometallic compounds for other purposes. For example, it is not possible to prepare a Grignard reagent from an alkyl halide that also has an acidic hydrogen in the molecule, such as $ClCH_2CH_2CH_2OH$.

One important use for such hydrolysis reactions is **specific deuteriation.** When one carries out the hydrolysis with heavy water, deuterium oxide, the product is an alkane containing a deuterium at the position formerly occupied by the metal.

$$CH_3CH_2\underset{\underset{MgBr}{|}}{\overset{\overset{CH_3}{|}}{C}}CH_3 + D_2O \longrightarrow CH_3CH_2\underset{\underset{D}{|}}{\overset{\overset{CH_3}{|}}{C}}CH_3 + DOMgBr$$

2-methylbutane-*2-d*

Heavy water is now readily available, and this reaction is an excellent way of making hydrocarbons "labeled" with deuterium in a specific position. After reaction the magnesium salts are removed, and the ether solution of the labeled hydrocarbon is dried and distilled. We have already seen several examples of the use of labeled compounds in studies of reaction mechanisms.

Alkylzinc and alkylcadmium compounds also react with protic materials, but their reactions are not so violent. Compounds of silicon, tin, mercury, and lead are unaffected by water, but in acidic solution they also undergo hydrolysis.

$$(C_2H_5)_4Si + 4 HCl \xrightarrow{H_2O} 4 CH_3CH_3 + SiCl_4$$

EXERCISE: Suggest how each of the following deuterated compounds can be prepared from the corresponding hydrocarbon.
(a) $CH_3CH_2C\!\equiv\!CD$ (b) $(CH_3)_3CD$ (c) $(CH_3)_2CHCH_2D$

B. *Reaction with Halogens*

Most organometallic reagents react vigorously with chlorine and bromine (the reduction potential $Cl_2 + 2e^- = 2\,Cl^-$ is $+1.358$ volts).

$$RM + Cl_2 \longrightarrow RCl + M^+Cl^-$$

The reaction is not preparatively useful because the product is an alkyl halide and organometallic compounds are frequently derived from alkyl halides.

C. *Reaction with Oxygen*

Organic compounds of many metals react rapidly with oxygen. Some are so reactive that they spontaneously inflame in air, often with spectacular consequences.

> Alkylboranes burn with a brilliant green flame. In his graduate student days, one of the authors was briefly immersed in a sea of such green fire. Only the rapid reflexes of a lab partner with a fire extinguisher allowed the current textbook to come to fruition.

Because of this reactivity, it is common to carry out organometallic reactions under an inert atmosphere such as nitrogen or argon.

Controlled oxidation of Grignard reagents is one way of preparing alcohols. The first reaction that occurs when oxygen is bubbled into a Grignard solution is formation of the salt of a hydroperoxide.

$$RMgX + O_2 \longrightarrow R\!-\!O\!-\!O\!-\!MgX$$

With excess Grignard reagent the salt of the alcohol is formed.

$$ROOMgX + RMgX \longrightarrow 2\,ROMgX$$

On normal work-up with dilute acid, the alcohol is produced.

$$ROMgX + HCl \longrightarrow ROH + MgXCl$$

An example is the conversion of 1-bromo-4,4-dimethylpentane into the corresponding alcohol.

$$(CH_3)_3CCH_2CH_2CH_2MgBr \xrightarrow[\;]{O_2} \xrightarrow[H_2O]{NH_4Cl} (CH_3)_3CCH_2CH_2CH_2OH$$
(90%)
4,4-dimethyl-1-pentanol

Oxidation is a side reaction of all Grignard syntheses run in the presence of air. For this reason Grignard reagents are often handled under a nitrogen or argon atmosphere. In refluxing diethyl ether, a common solvent for Grignard syntheses, the ether vapor forms a suitable "blanket" of inert atmosphere.

The hydroperoxide is formed in good yield at low temperatures ($-70°$) and can be isolated by reaction with dilute acid, separating and drying the organic layer and evaporating.

$$RMgX + O_2 \xrightarrow{-78°} ROOMgX \xrightarrow{H_3O^+} ROOH$$

D. *Reaction with Carbonyl Compounds, Carbon Dioxide, and Epoxides*

One of the most useful techniques in organic chemistry for building up more complex molecules from simple ones involves the reaction of Grignard reagents and organolithium compounds with carbonyl compounds. Recall that the carbon–metal bonds in these highly reactive organometallic compounds are polarized in the sense $C^- M^+$. The negative carbon (carbanion) of Grignard reagents reacts readily and rapidly with the positive carbon of the carbonyl group of aldehydes and ketones. Some examples show the scope of this important reaction.

(64–69%)
cyclohexylmethanol

> The Grignard reagent is prepared from 26.7 g of magnesium turnings and 118.5 g of cyclohexyl chloride in 450 mL of dry ether. In a separate flask 50 g of dry paraformaldehyde (page 387) is heated to 180–200°, and the formaldehyde formed by depolymerization is carried by a stream of nitrogen gas into the Grignard solution. At the end of the reaction ice and dilute sulfuric acid are added, and the mixture is steam distilled. The distillate is extracted with ether and distilled at reduced pressure to yield 72.5–78.5 g of cyclohexylmethanol.

$$(CH_3)_2CHMgBr + CH_3CHO \xrightarrow{ether} \xrightarrow{H^+}_{H_2O} (CH_3)_2CHCHOHCH_3$$

(53–54%)
3-methyl-2-butanol

> A solution of 600 g of isopropyl bromide in ether is slowly added to a mixture of 146 g of dry magnesium turnings in ether. The Grignard solution is then cooled to $-5°$, and a solution of 200 g of acetaldehyde (prepared by heating paraldehyde (page 387) with a small amount of sulfuric acid) in ether is added. Ice and dilute sulfuric acid are added, and the mixture is extracted with ether. The dried extract is distilled to give 210–215 g of 3-methyl-2-butanol, b.p. 110–111.5°.

$$C_2H_5MgCl + (CH_3)_2CHCH_2COCH_3 \xrightarrow{ether} \xrightarrow{H^+}_{H_2O} (CH_3)_2CHCH_2\overset{\overset{\displaystyle OH}{|}}{\underset{\underset{\displaystyle CH_3}{|}}{C}}CH_2CH_3$$

3,5-dimethyl-3-hexanol

As shown by these examples, the reaction is useful for preparation of primary, secondary, and tertiary alcohols. Reaction of a Grignard reagent with formaldehyde gives a primary alcohol, other aldehydes yield secondary alcohols, and ketones lead to tertiary alcohols. Note that secondary and tertiary alcohols may generally be prepared by more than one combination of Grignard reagent and carbonyl component.

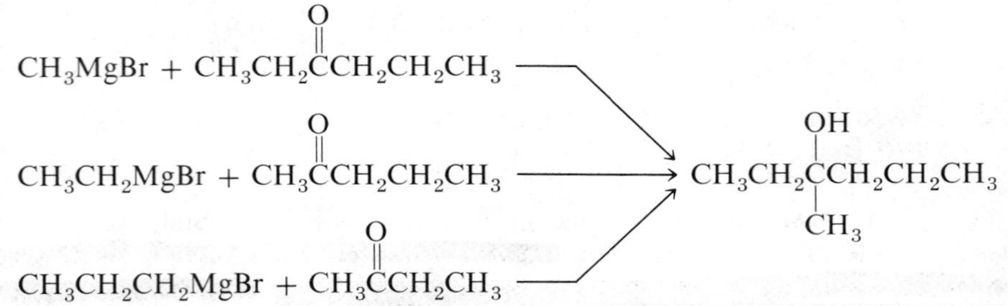

$$CH_3MgBr + CH_3CH_2\overset{\overset{\text{O}}{\|}}{C}CH_2CH_2CH_3 \longrightarrow$$

$$CH_3CH_2MgBr + CH_3\overset{\overset{\text{O}}{\|}}{C}CH_2CH_2CH_3 \longrightarrow CH_3CH_2\overset{\overset{\text{OH}}{|}}{\underset{\underset{\text{CH}_3}{|}}{C}}CH_2CH_2CH_3$$

$$CH_3CH_2CH_2MgBr + CH_3\overset{\overset{\text{O}}{\|}}{C}CH_2CH_3 \longrightarrow$$

The particular combination used is governed by such practical matters as cost and availability of reagents and ease of handling reactants.

For the preparation of many alcohols the Grignard reaction is a simple and straightforward process. However, side reactions are important and can dominate in sterically congested cases where the normal carbonyl addition reaction is slowed. One such side reaction is enolization.

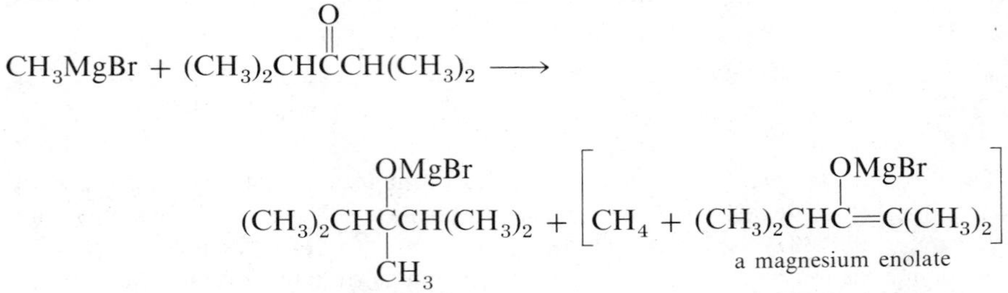

$$CH_3MgBr + (CH_3)_2CH\overset{\overset{\text{O}}{\|}}{C}CH(CH_3)_2 \longrightarrow$$

$$(CH_3)_2CH\overset{\overset{\text{OMgBr}}{|}}{\underset{\underset{\text{CH}_3}{|}}{C}}CH(CH_3)_2 + \left[CH_4 + (CH_3)_2CH\overset{\overset{\text{OMgBr}}{|}}{C}=C(CH_3)_2 \right]$$
$$\text{a magnesium enolate}$$

When water is added during normal work-up, the magnesium enolate is hydrolyzed to give back the starting ketone. In this side reaction the carbanionic carbon of the Grignard reagent is functioning as a base to produce the magnesium salt of the enolate ion.

> The reaction is probably more complex than this. The magnesium can function as a Lewis acid and coordinates with the carbonyl oxygen. The carbanionic carbon can react with the carbonyl carbon or an α-hydrogen. The reactions may occur via aggregates, and the magnesium and carbanion carbon may be different Grignard molecules. For simplicity, however, the reactions may be formulated as

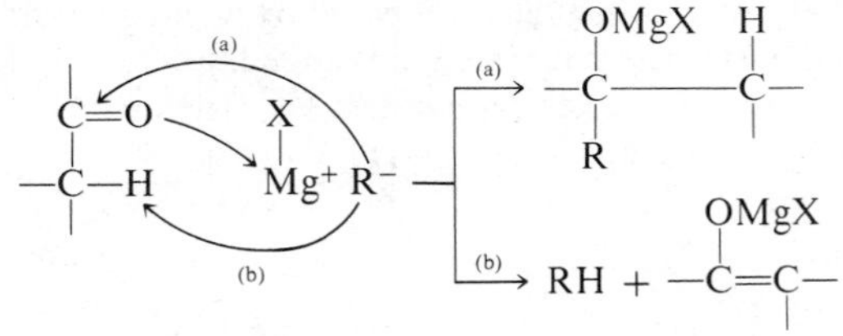

Another side reaction is important in hindered cases when the Grignard reagent has a β-hydrogen. In this reaction the carbonyl group is reduced and an alkene is formed.

$$(CH_3)_2CH\overset{\overset{\text{O}}{\|}}{C}CH(CH_3)_2 + (CH_3)_2CHMgBr \longrightarrow (CH_3)_2CH\overset{\overset{\text{OH}}{|}}{C}HCH(CH_3)_2 + CH_3CH=CH_2$$

> This reaction can be formulated as an alternative mode of reaction of a Grignard reagent coordinated to a carbonyl group.

$$\begin{array}{c} \text{C}=\text{O} \\ \quad\ \searrow \\ \text{H} \quad\ \stackrel{+}{\text{MgX}} \end{array} \xrightarrow{(c)} \begin{array}{c} \diagdown \\ \text{C}-\text{OMgX} \\ \diagup \quad | \\ \text{H} \end{array} + \begin{array}{c} \diagdown \quad\ \diagup \\ \text{C}=\text{C} \\ \diagup \quad\ \diagdown \end{array}$$

In such situations alkyllithium reagents are especially useful because they are more reactive than Grignard reagents and can be used at low temperature where the alternative reduction and enolization reactions are less important. A spectacular example of a hindered system prepared by an organolithium reaction is

$$(CH_3)_3CCOC(CH_3)_3 + (CH_3)_3CLi \xrightarrow[-70°]{\text{ether}} \xrightarrow{H_2O} \quad [(CH_3)_3C]_3COH$$

di-*t*-butyl ketone   *t*-butyllithium   (81%)

                                        3-*t*-butyl-2,2,4,4-tetramethyl-3-pentanol

The addition of lithium and sodium salts of alkynes to aldehydes and ketones (Section 13.7.D) is another example of this reaction.

$$CH_3C{\equiv}CH + NaNH_2 \longrightarrow CH_3C{\equiv}CNa \xrightarrow{CH_3\overset{O}{\overset{\|}{C}}CH_3} \xrightarrow{H_2O} CH_3C{\equiv}C\overset{OH}{\overset{|}{C}}(CH_3)_2$$

Grignard reagents also react readily with carbon dioxide. The initial product is the magnesium salt of a carboxylic acid (Chapter 17). This salt may be treated with dilute aqueous mineral acid to obtain the carboxylic acid. The reaction comprises one of the best methods for synthesizing carboxylic acids from alkyl halides.

$$CH_3CH_2\overset{Cl}{\overset{|}{C}}HCH_3 + Mg \xrightarrow{\text{ether}} CH_3CH_2\overset{MgCl}{\overset{|}{C}}HCH_3 \xrightarrow{CO_2}$$

$$CH_3CH_2\overset{CO_2^-\ ^+MgCl}{\overset{|}{C}}HCH_3 \xrightarrow[H_2SO_4]{H_2O} CH_3CH_2\overset{COOH}{\overset{|}{C}}HCH_3$$

Carbon dioxide gas is bubbled through a solution of *sec*-butylmagnesium chloride (prepared from 46 g of *sec*-butyl chloride and 13.4 g of magnesium in 400 mL of ether) at $-10°$. When CO_2 is no longer absorbed, the mixture is hydrolyzed with 25% aqueous H_2SO_4. Distillation of the crude product gives 40 g (80%) of 2-methylbutanoic acid.

A similar reaction occurs between carbon dioxide and an organolithium compound.

phenyllithium   benzoic acid

Grignard reagents undergo an interesting and useful reaction with ethyl orthoformate. The product is an acetal, which can be hydrolyzed to afford an aldehyde. The method is practical with Grignard reagents derived from either aliphatic or aromatic halides and provides a method for the conversion of a halide to an aldehyde.

(74%)

Ethyl orthoformate is an example of an orthoester; orthoesters bear the same relationship to esters as acetals do to aldehydes. Ethyl orthoformate is the most important member of this otherwise obscure class of compounds. It is a liquid, b.p. 146°, prepared by reaction of chloroform with ethanolic sodium ethoxide.

$$CHCl_3 + 3\ C_2H_5O^- \longrightarrow CH(OC_2H_5)_3 + 3\ Cl^-$$

The compound hydrolyzes readily in dilute acid to give first ethyl formate, then, upon further hydrolysis, formic acid.

$$CH(OC_2H_5)_3 + H_2O \xrightarrow{H^+} HCOOC_2H_5 + 2\ C_2H_5OH$$

$$HCOOC_2H_5 + H_2O \xrightarrow{H^+} HCOOH + C_2H_5OH$$

The mechanism of this reaction involves prior reaction of the ethyl orthoformate with the Lewis acid magnesium bromide, which is present in equilibrium with the Grignard reagent. Elimination of one of the ethoxy groups gives a relatively stable carbocation, which reacts with the Grignard reagent.

$$CH(OC_2H_5)_3 + MgBr_2 \rightleftharpoons Br_2\overset{-}{Mg}\overset{+}{O}C_2H_5 \rightleftharpoons$$
$$\underset{CH(OC_2H_5)_2}{|}$$

$$Br_2\overset{-}{Mg}OC_2H_5 + [C_2H_5O\overset{+}{-}\overset{+}{CH}-OC_2H_5 \longleftrightarrow C_2H_5\overset{+}{O}=CH-OC_2H_5] \xrightarrow{RMgBr}$$

$$RCH(OC_2H_5)_2$$

Finally, organolithium compounds react generally with epoxides. An example is the reaction of phenyllithium with 2-phenyloxirane to give 1,2-diphenylethanol. The reaction proceeds by the S_N2 mechanism, the phenyl "anion" being the attacking nucleophile and the oxide anion being the leaving group.

$$\underset{\text{2-phenyloxirane}}{C_6H_5\overset{O}{CH-}CH_2} + \underset{\text{phenyllithium}}{C_6H_5Li} \longrightarrow C_6H_5\overset{O^-\ Li^+}{CH}-CH_2C_6H_5 \xrightarrow{H_2O} \underset{\substack{(70-72\%)\\ \text{1,2-diphenylethanol}}}{C_6H_5\overset{OH}{CH}CH_2C_6H_5}$$

The leaving group is an RO^- and is a poor leaving group since it is the conjugate base of a *very weak* acid ROH (Section 8.10). However, a great deal of ring strain is relieved in the reaction (Section 10.12), and the activation energy is thereby lowered enough for reaction to occur readily.

A similar reaction occurs with Grignard reagents and ethylene oxide itself and is an important synthetic reaction since *the carbon chain is extended by two carbons in one step.*

$$R-Cl \xrightarrow[\text{ether}]{Mg} R-MgCl \xrightarrow[]{\overset{O}{CH_2-CH_2}} \xrightarrow{H^+} R-CH_2CH_2OH$$

$$CH_3CH_2CH_2CH_2Br \xrightarrow[\text{ether}]{Mg} \xrightarrow{\overset{O}{\overset{\triangle}{CH_2-CH_2}}} \xrightarrow{H^+} CH_3(CH_2)_3CH_2CH_2OH$$

(60–62%)
1-hexanol

Other epoxides are sometimes unsatisfactory for reaction with Grignard reagents because the magnesium halide generally present acts as a Lewis acid to promote a carbocation rearrangement to the more stable carbonyl isomer. The product then reacts with the Grignard reagent to give a different product.

$$(CH_3)_2\overset{O}{\overset{\triangle}{C-CH_2}} + CH_3CH_2MgBr \longrightarrow (CH_3)_2\overset{OMgBr}{\underset{|}{C}}CH_2CH_2CH_3$$

$$(CH_3)_2\overset{O}{\overset{\triangle}{C-CH_2}} + MgBr_2 \rightleftharpoons$$

$$\left[(CH_3)_2\overset{\overset{O\nearrow MgBr_2}{\triangle}}{C-\underset{H}{CH}} \longleftrightarrow (CH_3)_2\overset{+}{C}\underset{H}{\overset{\overline{O}\, MgBr_2}{-CH}} \right] \longrightarrow (CH_3)_2C\underset{H}{\overset{O\to MgBr_2}{-CH}}$$

$$(CH_3)_2CH\overset{O}{\overset{||}{C}}H + CH_3CH_2MgBr \longrightarrow (CH_3)_2CH\overset{OMgBr}{\underset{|}{C}}HCH_2CH_3$$

Such rearrangements do not generally occur with organolithium reagents because lithium cation is too weak as a Lewis acid.

EXERCISE: (a) Show how 1-bromobutane may be converted into each of the following compounds.
(i)   1-hexanol   (ii) 2-methyl-2-pentanol   (iii) 3-hexanol
(iv) 1-pentanol   (v) $CH_3CH_2CH_2CH_2COOH$
(b) Outline three different combinations of alkyl halide and ketone that can be used to prepare each of the following tertiary alcohols.
(i)   3-methyl-3-nonanol   (ii) 3,4,5-trimethyl-3-heptanol
(iii) 4-ethyl-2-methyl-4-octanol

E. Reaction with Other Organometallic Compounds

There are many reactions of one organometallic compound with another. Rather than going extensively into this fairly complex area, we shall simply mention one reaction, because the product is itself a reagent of considerable utility. Alkylcopper compounds react with alkyllithium compounds to give products known as **cuprates.** An example is the reaction of methyllithium with methylcopper to yield lithium dimethylcopper.

$$CH_3Li + CH_3Cu \longrightarrow (CH_3)_2CuLi$$

In practice, the cuprates can be prepared more conveniently by simply adding one equivalent of cuprous iodide to two equivalents of the alkyllithium compound.

Since Cu^+ has a much more positive reduction potential than lithium, methyl-copper is formed and then reacts with the second equivalent of methyllithium, giving the cuprate.

$$2\ CH_3CH_2CH_2Li + CuI \longrightarrow (CH_3CH_2CH_2)_2CuLi + LiI$$

The exact structures of these cuprates are not known, but in solution they appear to exist in a dimeric or tetrameric form.

15.6
Transition Metal Organometallic Compounds

Transition metal organometallic chemistry has been an area of incredible growth over the past three decades. Many such compounds are particularly important as catalysts in industrial processes, and research continues for new and better catalysts. The subject is at the interface between organic and inorganic chemistry. In this section we can provide only bare essentials of an introduction to this large, growing, and important field of chemistry.

The transition metals have a partially filled d-orbital shell. The maximum number of electrons in a filled valence shell for an element in the fourth period of the periodic table (the potassium–krypton row) is

$$4s^2 4p^6 3d^{10} = 18$$

Thus there is a tendency for these elements to surround themselves with 18 electrons and achieve an "inert" rare gas electronic configuration, just as elements in the second and third periods strive to achieve an eight-electron configuration. We shall see that a great many transition metal organometallic compounds have structures in which the metal is associated with either 16 or 18 electrons and that many reactions can be rationalized in terms of filling out the inert electron configuration. This "16- or 18-electron rule" provides a useful framework within which to organize the reactions of such compounds.

First we must take up the question of how to "count" the electrons in transition metal organometallic compounds. The abbreviated periodic table in Figure 15.6

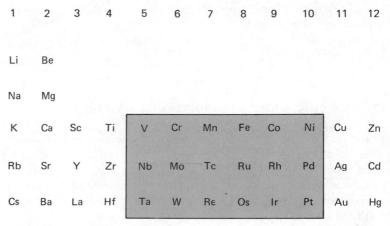

FIGURE 15.6. *Periodic table of the elements, showing the number of electrons in the valence shell for each element in the three transition series. The elements within the box most often obey the 16–18-electron rule.*

shows the elements of groups 1 and 2 and of the three transition series. The numbers at the top of each group correspond to the total number of electrons in the valence shell of an element in that group, regardless of whether the electrons reside in *s*-, *p*-, or *d*-orbitals in the atom itself. The transition elements within the box are those to which the 16- or 18-electron rule is most applicable, for reasons that we shall not take up.

In general, electron counting in transition metal compounds works exactly as it does for Lewis structures of second and third period elements except that dative or "donor" bonds are more common with metal systems. Recall that in such a bond two electrons are provided by a suitable donor atom or group, the **ligand.** Both electrons "belong" to the ligand for the purpose of determining formal charges, but do contribute to establishing a rare gas configuration. Common ligands of this type are $H_2\overset{..}{\underset{\times}{O}}$, $R_2\overset{..}{\underset{\times}{O}}$, $R_3\underset{\times}{N}$, $R_3\underset{\times}{P}$, $:\overset{..}{\underset{\times}{O}}{=}\underset{\times}{C}$, in which the donor electron pair is indicated by small $\times$s.

To illustrate the electron-counting procedure, consider the compound molybdenum hexacarbonyl. Six electrons come from the valence shell of the molybdenum atom (Figure 15.6). Each carbon monoxide ligand "donates" two more electrons. Thus the total electronic configuration about the molybdenum in $Mo(CO)_6$ is 18.

$$
\begin{array}{c}
O \\
C \\
OC\ \ \downarrow\ \ CO \\
Mo \\
OC\ \ \uparrow\ \ CO \\
C \\
O
\end{array}
\qquad
\begin{array}{rl}
6 & \text{electrons on Mo} \\
6 \times 2 = \underline{12} & \text{six CO ligands} \\
18 & \text{total electrons}
\end{array}
$$

molybdenum hexacarbonyl
(18 electrons)

None of the twelve carbonyl electrons is counted as "belonging" to Mo; hence, Mo has only the six valence electrons of its own and is formally netural. Note that these six electrons are not included in the symbolic representation. They would clutter up the diagram; rather, they are implied by the elemental symbol Mo.

A second example is methylmanganese pentacarbonyl. In this compound there are two types of bonds to the metal. The five carbon monoxide ligands each form a "donor" bond in the same manner as the ligands in $Mo(CO)_6$. However, the $Mn-CH_3$ bond may be viewed as a normal two-electron σ-bond, with one electron being contributed by each partner. Thus Mn has six valence electrons left that are not indicated in the structure—seven from a manganese atom (see Figure 15.6) minus one used in the bond to CH_3. The total electron count is arrived at as shown.

$$
\begin{array}{c}
O \\
C \\
OC\ \ \downarrow\ \ CO \\
Mn \\
OC\ \ \uparrow\ \ CH_3 \\
C \\
O
\end{array}
\qquad
\begin{array}{rl}
7 & \text{electrons on Mn} \\
-1 & \text{Mn electron used in the} \\
 & \quad Mn-CH_3\ \text{bond} \\
5 \times 2 = 10 & \text{five CO ligands} \\
\underline{2} & \text{the } Mn-CH_3 \text{ bonding pair} \\
18 & \text{total electrons}
\end{array}
$$

methylmanganese pentacarbonyl
(18 electrons)

In this compound the electrons "belonging" to Mn are the six Mn valence electrons, which are not used in bonding and are not shown on the structure, plus one-half of the electrons in the Mn—CH_3 σ-bond. Thus the metal is formally neutral.

Another kind of common ligand in transition metal organometallic compounds is the carbon–carbon multiple bond or π-bond of an alkene or alkyne. Accordingly, these groups usually are called π-donors to distinguish them from the σ-donors, the ligands discussed above with lone-pair electrons. The electron count works in the same way. For example, consider the following rhodium compound.

9	electrons on Rh
−1	Rh electron used in the Rh—H bond
$2 \times 2 = 4$	two phosphine ligands
2	CO ligand
2	π-bond of ethylene
2	the Rh—H bonding pair
18	total electrons

Again, the formal charge on the metal is zero.

The bonding of ethylene to transition metals ranges from rather weak to quite strong. A strong bond could perhaps be represented as

$$M \underset{CH_2}{\overset{CH_2}{\big<}}$$

It is easy to show that both representations give the same electron count.

In all of the examples discussed above the transition metal has achieved an 18-electron configuration and is said to be **coordinatively saturated.**

Transition metal organometallic compounds undergo a wide variety of reactions. Fortunately, almost all known reactions, including various steps in individual reaction mechanisms, can be grouped into five or six basic classes. We shall consider a few of these fundamental reaction types here.

1. Lewis Acid Association-Dissociation. An example is the reaction of tetrakis-(triphenylphosphine)nickel with HCl.

(18 electrons)   (18 electrons)

The forward reaction is association of the metal with the Lewis acid H^+. The reverse reaction is dissociation of the Lewis acid from the metal. In Lewis acid association-dissociation *the electron count of the metal is unchanged.* Another example is the reaction of manganese pentacarbonyl anion with methyl bromide. The reaction may be viewed as a nucleophilic displacement of bromide ion from the alkyl bromide that proceeds by the S_N2 mechanism. However, it also corresponds to association of the manganese with the Lewis acid "CH_3^+."

(18 electrons)   (18 electrons)

2. Lewis Base Association-Dissociation. This is the most common reaction in transition metal organometallic chemistry. An example is seen in the reaction of nickel tetracarbonyl with triphenylphosphine. The reaction mechanism consists of two steps. In the first, a CO ligand dissociates from the nickel, giving the intermediate nickel tricarbonyl. This species has an electron count of 16 and is **coordinatively unsaturated.** This electron-deficient intermediate reacts with triphenylphosphine by Lewis base **association.**

$$\text{Ni(CO)}_4 \longrightarrow \text{Ni(CO)}_3 + :\text{CO}$$
$$\text{(18 electrons)} \qquad \text{(16 electrons)}$$

$$\text{Ni(CO)}_3 + :\text{P(C}_6\text{H}_5)_3 \longrightarrow (\text{C}_6\text{H}_5)_3\text{PNi(CO)}_3$$
$$\text{(16 electrons)} \qquad \qquad \qquad \text{(18 electrons)}$$

In Lewis base association-dissociation *the electron count of the metal changes by ± 2.*

3. Oxidative Addition-Reductive Elimination. Many transition metal organometallic reactions involve the addition of a σ-bond to the metal. An example is the reaction of hydrogen with tris(triphenylphosphine)chloroiridium carbonyl. In this complex, the iridium has a 16-electron configuration. It exists in the coordinatively unsaturated form because the triphenylphosphine ligands are so large. The complex reacts with hydrogen to form two new iridium–hydrogen bonds, giving the 18-electron complex shown. The reaction is called **oxidative addition** because the formal oxidation state of the metal changes from +1 to +3.

The "oxidation state" of a metal in an organometallic compound may be defined in the following manner. Consider only the groups that are σ-bonded (H, R, Cl, CN, OH, etc.). The metal atom is considered to be more electropositive than any of these groups. Therefore it has a formal "oxidation number" of +1 for each such σ-bond. Donor ligands are ignored. The oxidation state of the metal is the sum of the oxidation numbers, minus one for each negative charge on the molecule and plus one for each positive charge on the complex. Some transition metal complexes contain more complex π-bonded ligands that cannot be treated in this simplified way, but they are beyond the scope of this introductory discussion.

The reverse of oxidative addition, dissociation of a hydrogen molecule with regeneration of the original iridium complex, is called **reductive elimination.** Note that the electron count of the metal changes by ± 2 in oxidative addition-reductive elimination.

4. Insertion-Deinsertion. Occasionally a donor ligand undergoes **insertion** into a σ-bond from the metal to another atom. In the process a donor bond and a σ-bond are traded for a new σ-bond. Thus the electron count of the metal decreases by two. In the reverse reaction the electron count increases by two. An example of insertion-deinsertion is the rearrangement of the 18-electron complex methylmanganese pentacarbonyl to the 16-electron complex acetylmanganese tetracarbonyl.

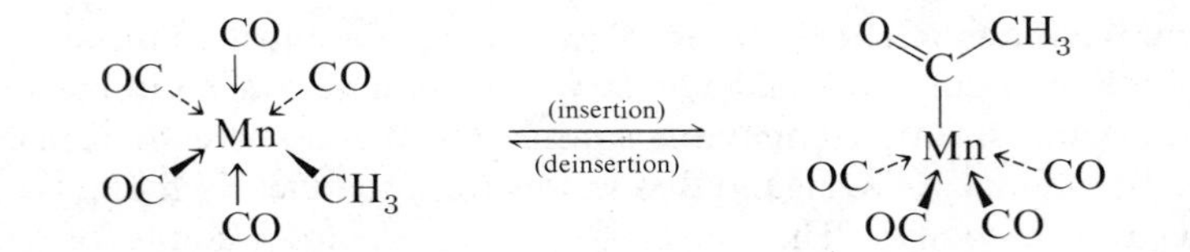

methylmanganese pentacarbonyl
(18 electrons)

acetylmanganese tetracarbonyl
(16 electrons)

Now let us briefly examine a few transition metal organometallic reactions and characterize them in terms of the foregoing fundamental transformations. Tris-(triphenylphosphine)carbonylrhodium hydride is an effective catalyst for the hydrogenation of alkenes.

$$CH_2=CH_2 + H_2 \xrightarrow{[(C_6H_5)_3P]_3RhHCO} CH_3CH_3$$

The reaction may be understood in terms of the following stepwise mechanism (for clarity of presentation, the triphenylphosphine ligands are symbolized by L).

(1)

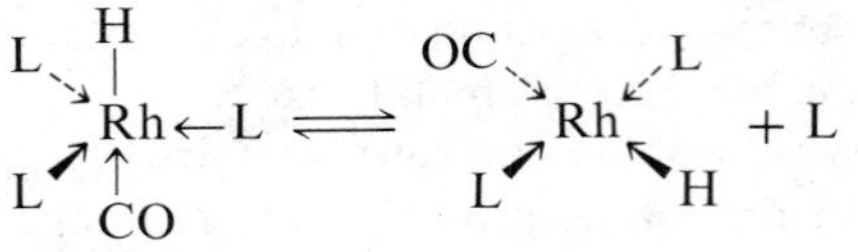

(18 electrons)   (16 electrons)

(2)

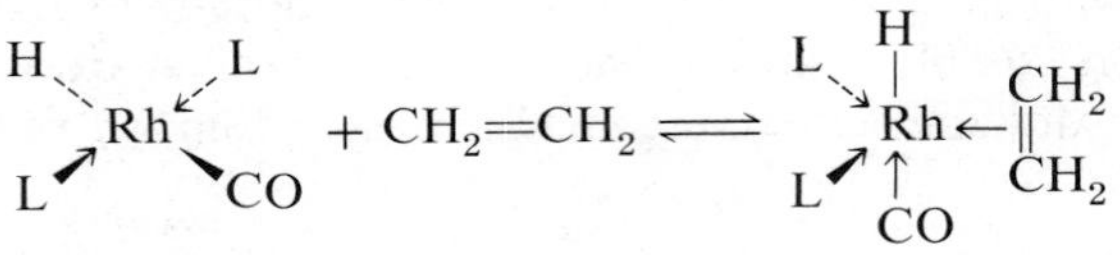

(16 electrons)   (18 electrons)

(3)

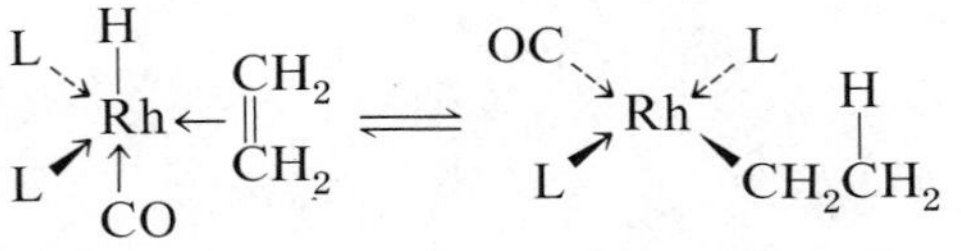

(18 electrons)   (16 electrons)

(4)

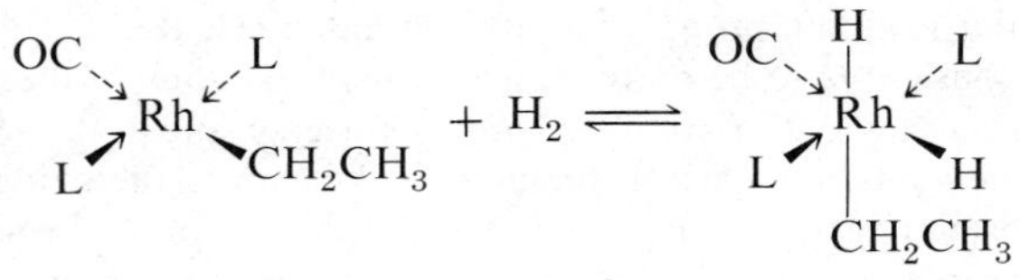

(16 electrons)   (18 electrons)

(5)

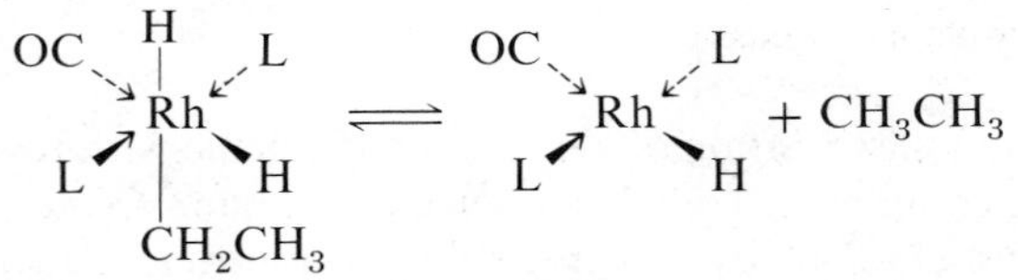

(18 electrons)   (16 electrons)

The mechanism consists of the following sequence of processes.

1. Lewis base dissociation.
2. Lewis base association.
3. Insertion.
4. Oxidative addition.
5. Reductive elimination.

The products of step 5 are the alkane molecule and the active catalyst, which recycles to step 1 and brings about the hydrogenation of another alkene molecule.

A related catalyst, tris(triphenylphosphine)rhodium chloride, also known as Wilkinson's catalyst, also brings about hydrogenation of alkenes by a mechanism similar to that above.

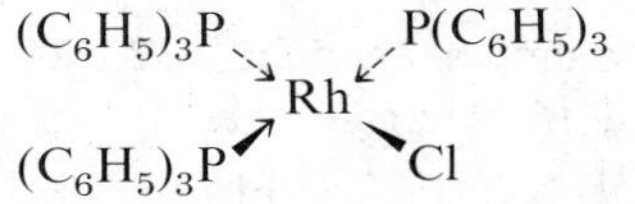

$$(C_6H_5)_3P \quad\quad P(C_6H_5)_3$$
$$Rh$$
$$(C_6H_5)_3P \quad\quad Cl$$

tris(triphenylphosphine)rhodium chloride
(Wilkinson's catalyst)

Wilkinson's catalyst also functions to bring about **decarbonylation** of aldehydes.

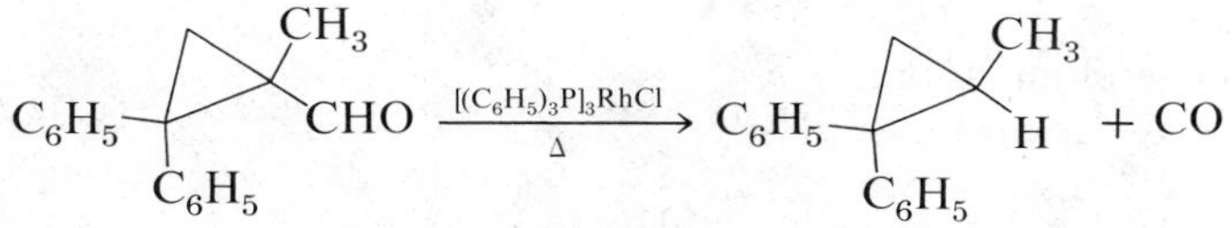

$$C_6H_5-\overset{CH_3}{\underset{C_6H_5}{C}}-CHO \xrightarrow[\Delta]{[(C_6H_5)_3P]_3RhCl} C_6H_5-\overset{CH_3}{\underset{C_6H_5}{C}}-H + CO$$

The reaction mechanism may be formulated as proceeding through the following steps.

(1) 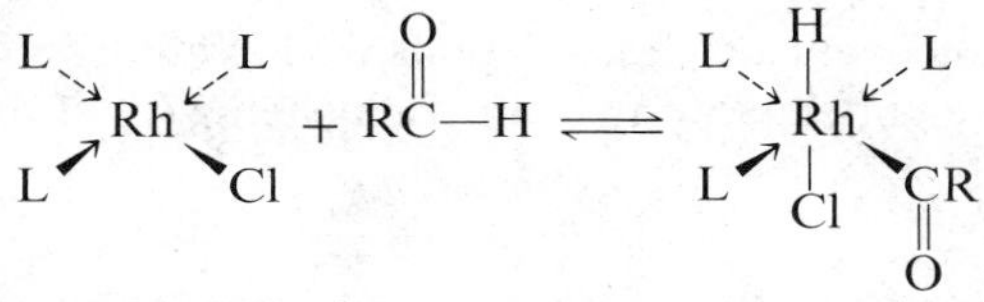
$$\begin{array}{c} L \quad L \\ Rh \\ L \quad Cl \end{array} + R\overset{O}{\overset{\|}{C}}-H \rightleftharpoons \begin{array}{c} L \quad H \quad L \\ Rh \\ L \quad Cl \quad CR \\ \quad\quad\quad \overset{\|}{O} \end{array}$$

(16 electrons)   (18 electrons)

(2) 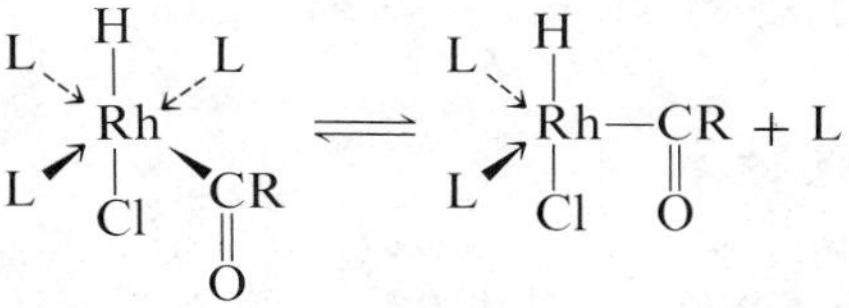
$$\begin{array}{c} L \quad H \quad L \\ Rh \\ L \quad Cl \quad CR \\ \quad\quad\quad \overset{\|}{O} \end{array} \rightleftharpoons \begin{array}{c} L \quad H \\ Rh-CR \\ L \quad Cl \quad O \end{array} + L$$

(18 electrons)   (16 electrons)

(3) 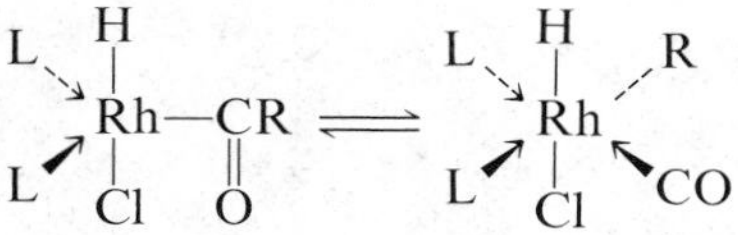
$$\begin{array}{c} L \quad H \\ Rh-CR \\ L \quad Cl \quad O \end{array} \rightleftharpoons \begin{array}{c} L \quad H \quad R \\ Rh \\ L \quad Cl \quad CO \end{array}$$

(16 electrons)   (18 electrons)

(4) 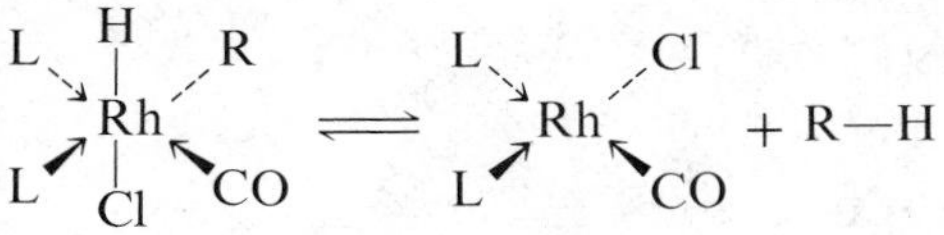
$$\begin{array}{c} L \quad H \quad R \\ Rh \\ L \quad Cl \quad CO \end{array} \rightleftharpoons \begin{array}{c} L \quad Cl \\ Rh \\ L \quad CO \end{array} + R-H$$

(18 electrons)   (16 electrons)

(5) 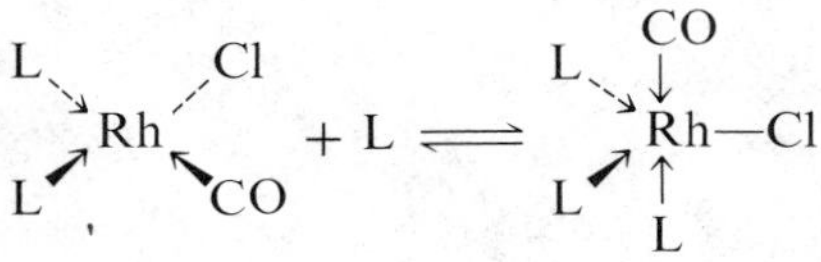
$$\begin{array}{c} L \quad Cl \\ Rh \\ L \quad CO \end{array} + L \rightleftharpoons \begin{array}{c} L \quad CO \\ Rh-Cl \\ L \quad L \end{array}$$

(16 electrons)   (18 electrons)

(6)

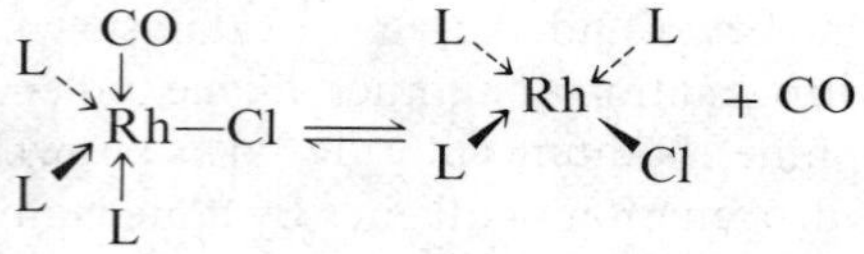

$$L_3Rh(CO)Cl \rightleftharpoons L_3RhCl + CO$$

(18 electrons)   (16 electrons)

Like the iridium complex shown on page 457, Wilkinson's catalyst exists in a coordinatively unsaturated 16-electron state. It can therefore undergo oxidative addition (which changes the electron count of the metal by +2) without prior dissociation of a donor ligand. Thus, the steps are

1. Oxidative addition.
2. Lewis base dissociation.
3. Deinsertion.
4. Reductive elimination.
5. Lewis base association.
6. Lewis base dissociation.

This mechanism is probably a little oversimplified, since there is evidence that free radical pairs may be involved in step 3. Nevertheless, the gross features are no doubt correct.

The simple structural ideas and mechanistic classifications we have introduced are but first steps toward a complete understanding of the rich chemistry of these fascinating compounds.

> **EXERCISE:** Review each structure given in this section and verify the "electron count" that has been assigned to the metal in each case.

PROBLEMS

1. Show how the following conversions may be accomplished.

(a) $(CH_3)_3CCH(CH_3)_2 \longrightarrow (CH_3)_3CCD(CH_3)_2$

(b) $CH_3CH_2CHClCH_3 \longrightarrow \left(CH_3CH_2\overset{\overset{\textstyle CH_3}{|}}{C}H\right)_4 Sn$

(c) $(CH_3)_4C \longrightarrow (CH_3)_3CCH_2MgCl$

(d) $CH_3CH_2Cl \longrightarrow (CH_3CH_2)_2Cd$

(e) $HC\equiv CH \longrightarrow (CH_3CH_2)CHC\equiv CD$

(f) $HC\equiv CH \longrightarrow CH_3CH_2C\equiv C\underset{\underset{\textstyle OH}{|}}{C}(CH_3)_2$

(g) $(CH_3)_3CCH_2Cl \longrightarrow (CH_3)_3CCH_2\overset{\overset{\textstyle OH}{|}}{C}HCH_3$

(h) 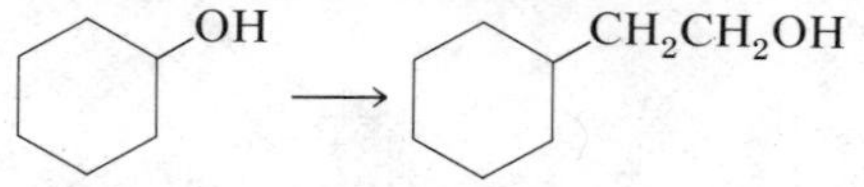 [cyclohexane with OH] $\longrightarrow$ [cyclohexane with CH_2CH_2OH]

(i) 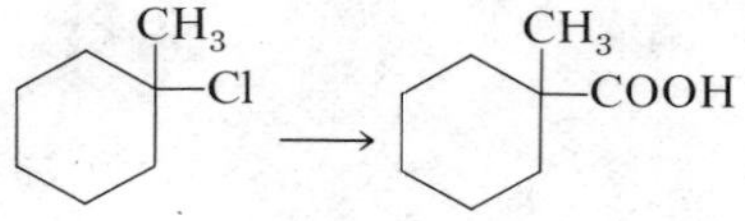 [cyclohexane with CH_3 and Cl] $\longrightarrow$ [cyclohexane with CH_3 and COOH]

2. Predict whether the equilibrium constant will be greater than or less than unity for each of the following reactions.
(a) $2 (CH_3)_3Al + 3 CdCl_2 \rightleftharpoons 3 (CH_3)_2Cd + 2 AlCl_3$
(b) $(CH_3)_2Hg + ZnCl_2 \rightleftharpoons (CH_3)_2Zn + HgCl_2$
(c) $2 (CH_3)_2Mg + SiCl_4 \rightleftharpoons (CH_3)_4Si + 2 MgCl_2$
(d) $CH_3Li + HCl \rightleftharpoons CH_4 + LiCl$
(e) $(CH_3)_2Zn + 2 LiCl \rightleftharpoons 2 CH_3Li + ZnCl_2$

3. (a) Dimethylberyllium has a polymeric structure analogous to that described for dimethylmagnesium in Figure 15.3a. However, di-*t*-butylberyllium is mono-meric. Suggest a reason for this difference. Predict the geometry of di-*t*-butyl-beryllium.
(b) Predict the geometry of dimethylmercury. The measured dipole moments of $(CH_3)_2Hg$ and $(C_2H_5)_2Hg$ are $\mu = 0.0$ D. Explain.

4. Trimethylborane reacts with methyllithium to give a product having the formula $C_4H_{12}BLi$. Propose a structure for this substance. What is the hybridization of boron? Describe the geometry of the species.

5. The reaction of alkyl halides with sodium metal was once used as a method for preparing certain alkanes (**Würtz** reaction). What do you think are the limitations of this reaction as a preparative method? Write out several alkanes that could be pre-pared this way in good yield.

6. Show how each of the following alcohols may be synthesized using Grignard rea-gents.

(a) $(CH_3CH_2CH_2)_3COH$

(b) $CH_3CH_2\overset{\displaystyle OH}{\underset{\displaystyle |}{C}}HCH_2CH_2CH_3$

(c) $CH_3CH_2CH_2\overset{\displaystyle OH}{\underset{\displaystyle |}{C}}HCH_2OH$

(d) $CH_3CH_2CH_2\overset{\displaystyle CH_3}{\underset{\displaystyle |}{\underset{\displaystyle CH_3}{C}}}CH_2CH_2OH$

(e) $(CH_3)_2\overset{\displaystyle OH}{\underset{\displaystyle |}{C}}CH_2CH_3$

7. Reaction of 2,2-diethyloxirane with ethylmagnesium bromide gives 4-ethyl-3-hexa-nol whereas reaction with ethyllithium gives 3-ethyl-3-hexanol. Explain.

8. Show how *sec*-butyl chloride can be converted into 5-methyl-1-heptanol by a route involving two Grignard steps.

9. Show how 2-chloro-2-methylbutane can be converted into each of the following compounds by a route involving at least one Grignard step.

(a) 2-methyl-2-butanol

(b) 2,2-dimethyl-1-butanol

(c) 3,3-dimethyl-1-pentanol

(d) 4,4-dimethyl-1-hexanol

(e) 4,4-dimethyl-3-hexanol

(f) $CH_3CH_2\overset{\displaystyle CH_3}{\underset{\displaystyle |}{\underset{\displaystyle CH_3}{C}}}COOH$

(g) $(CH_3CH_2\overset{\displaystyle CH_3}{\underset{\displaystyle |}{\underset{\displaystyle CH_3}{C}}})_2Hg$

(h) $CH_3CH_2\overset{\displaystyle D}{\underset{\displaystyle |}{C}}(CH_3)_2$

10. (a) Some transition metals form neutral complexes with carbon monoxide (metal carbonyls); for example, see nickel carbonyl on page 457. Chromium and iron also form stable carbonyls. Write the expected structures of these compounds.

(b) Other transition metals form stable metal carbonyl anions of the type $M(CO)_n^-$. Write the expected structures of the manganese and cobalt carbonyl anions and explain their stability.

11. The rate of exchange of triphenylphosphine for carbon monoxide in nickel carbonyl is inversely proportional to the pressure of carbon monoxide in contact with the reaction solution.

$$Ni(CO)_4 + (C_6H_5)_3P \longrightarrow (C_6H_5)_3PNi(CO)_3 + CO$$

Explain.

12. Hydroformylation of alkenes (the "oxo reaction") is an important industrial process. For example, ethylene reacts with hydrogen and carbon monoxide in the presence of hydridocobalt tetracarbonyl to give propionaldehyde.

$$CH_2{=}CH_2 + CO + H_2 \xrightarrow[100-120]{HCo(CO)_4} CH_3CH_2CHO$$

$$(72\%)$$

Propose a mechanism which is consistent with the "16- or 18-electron rule." Classify each step in your mechanism by reaction type (i.e., Lewis acid association, insertion, etc.).

13. Write a plausible mechanism for the hydrogenation of ethylene with Wilkinson's catalyst. Wilkinson's catalyst also brings about the decarbonylation of acyl halides.

$$CH_3CH_2CH_2CH_2\overset{\overset{\displaystyle O}{\|}}{C}Cl \xrightarrow{[(C_6H_5)_3P]_3RhCl} CH_3CH_2CH_2CH_2Cl + CO$$

Propose a mechanism for this reaction.

14. An alternative mechanism for coupling two alkyl halides by sodium metal (page 443) is dimerization of the free radicals produced during formation of the organosodium compound.

$$2\ RCl + 2Na \longrightarrow 2\ R\cdot + 2\ NaCl$$

$$2\ R\cdot \longrightarrow R{-}R$$

This mechanism has been tested by treating (R)-2-chlorooctane with sodium. The product is a mixture of (S,S)- and (R,S)-7,8-dimethyltetradecane.

(a) What does this result show about the free radical coupling mechanism above? Remember that free radicals are effectively planar and thus achiral.

(b) One of the dimethyltetradecanes is meso and one is optically active. Which is which?

(c) Are the two products necessarily formed in equal amounts? Explain.

(d) Does this experiment rigorously exclude either mechanism? Explain.

In another experiment aimed at probing the mechanism of the Würtz reaction, a 50:50 mixture of *n*-pentyl iodide and neopentyl iodide was treated with metallic sodium. The coupling products, *n*-decane, 2,2-dimethyloctane, and 2,2,5,5-tetramethylhexane, were formed in a ratio of 1.2:1.7:1.1.

(e) With which mechanism is this result most compatible?

15. The reaction of organometallic compounds with halogens may be regarded as an electrophilic substitution in distinction to the more common nucleophilic substitution. The stereochemistry of one such reaction, involving a carbon-tin bond, has been studied with the following sequence.

When trineopentyltin chloride is treated with sodium metal a salt is formed.

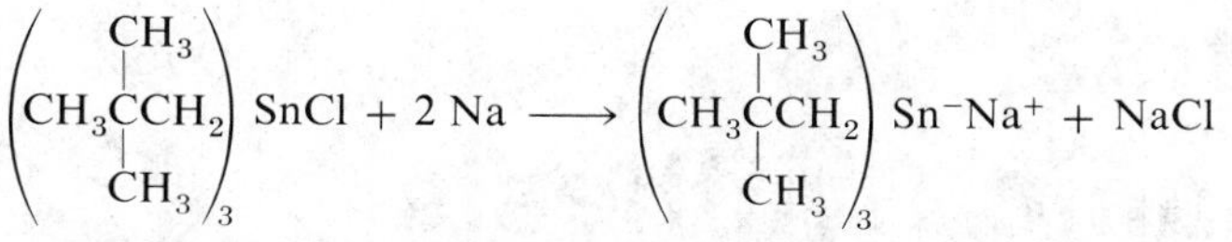

$$\left(\underset{\underset{CH_3}{|}}{\overset{\overset{CH_3}{|}}{CH_3CCH_2}}\right)_3 SnCl + 2\,Na \longrightarrow \left(\underset{\underset{CH_3}{|}}{\overset{\overset{CH_3}{|}}{CH_3CCH_2}}\right)_3 Sn^-Na^+ + NaCl$$

This salt is treated with (R)-2-bromobutane to give *sec*-butyltrineopentyltin (15-2).

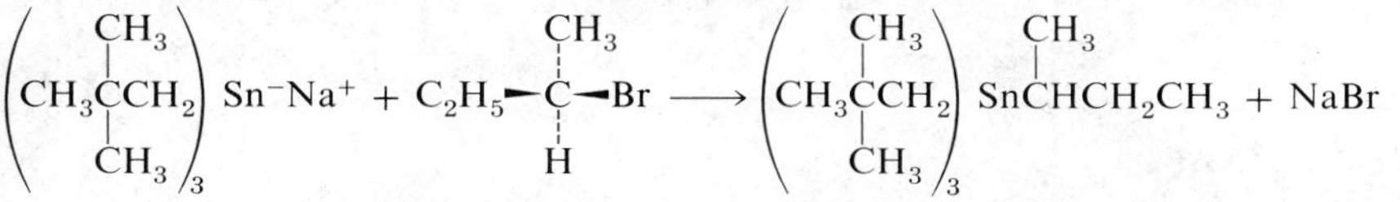

$$\left(\underset{\underset{CH_3}{|}}{\overset{\overset{CH_3}{|}}{CH_3CCH_2}}\right)_3 Sn^-Na^+ + C_2H_5\!-\!\!\underset{\underset{H}{|}}{\overset{\overset{CH_3}{|}}{C}}\!\!-\!Br \longrightarrow \left(\underset{\underset{CH_3}{|}}{\overset{\overset{CH_3}{|}}{CH_3CCH_2}}\right)_3 Sn\overset{\overset{CH_3}{|}}{C}HCH_2CH_3 + NaBr \quad (15\text{-}2)$$

Treatment of the tetraalkyltin compound with bromine gives (R)-2-bromobutane and trineopentyltin bromide (15-3).

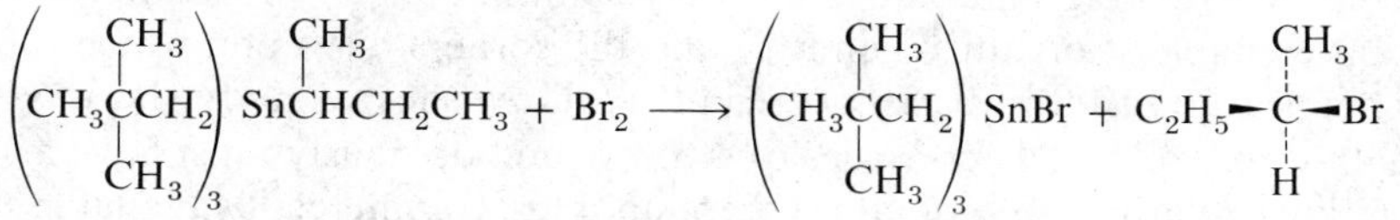

$$\left(\underset{\underset{CH_3}{|}}{\overset{\overset{CH_3}{|}}{CH_3CCH_2}}\right)_3 Sn\overset{\overset{CH_3}{|}}{C}HCH_2CH_3 + Br_2 \longrightarrow \left(\underset{\underset{CH_3}{|}}{\overset{\overset{CH_3}{|}}{CH_3CCH_2}}\right)_3 SnBr + C_2H_5\!-\!\!\underset{\underset{H}{|}}{\overset{\overset{CH_3}{|}}{C}}\!\!-\!Br \quad (15\text{-}3)$$

(a) What is the probable mechanism of reaction (15-2)?

(b) Comment on the overall stereochemistry of the two steps.

(c) Why is it the *sec*-butyl group, rather than one of the neopentyl groups, that reacts in (15-3)?

Chapter 16
Organic Synthesis

16.1
Introduction

Organic synthesis is the preparation of a desired organic compound from a commercially available material, usually by some multistep procedure. It is an important element of organic chemistry and the cornerstone upon which the organic chemical industry is based. A scientist who wishes to study the physical, chemical, or physiological properties of a compound obviously must have a sample of it. Since relatively few organic compounds are commercially available from chemical suppliers, the scientist often must synthesize the desired material. In this chapter we shall show how to approach such a problem using the chemistry we have learned so far.

16.2
Considerations in Synthesis Design

The goal in any synthesis is to obtain a pure sample of the desired product by the most efficient and convenient procedure possible. For this reason one usually strives to use reactions that can reliably be expected to give only a single product and avoids reactions that will give a mixture of products. It is also important to plan a synthesis that entails the fewest possible steps. This is necessary both in terms of the amount of time consumed in an overly long route and in the ultimate yield that may be realized. A ten-step synthesis averaging 80% yield per step will give an overall yield of only 10.7%.

In planning a synthesis, three interrelated factors are involved.

1. Construction of the Proper Carbon Skeleton. In a sense, reactions that result in formation of a new C—C bond are the most important reactions in organic chemistry because these reactions allow us to build up more complicated structures. Carbon–carbon bond-forming reactions that we have encountered are summarized as follows.

(a) Reaction of primary alkyl halides with cyanide (Section 8.6).

$$RCH_2X + CN^- \longrightarrow RCH_2CN$$

(b) Addition of HCN to aldehydes and ketones (Section 13.7.E).

$$R_2C{=}O + HCN \longrightarrow R_2\overset{\displaystyle OH}{\underset{\displaystyle |}{C}}CN$$

(c) Reaction of primary alkyl halides with acetylide ions (Sections 8.3 and 12.5.C).

$$RCH_2X + {}^-C{\equiv}CR' \longrightarrow RCH_2C{\equiv}CR'$$

(d) Addition of acetylide ions to aldehydes and ketones (Section 13.7.D).

$$R_2C=O + {}^-C\equiv CR' \longrightarrow \xrightarrow{H^+} R_2\overset{\overset{\displaystyle OH}{|}}{C}C\equiv CR'$$

(e) Reactions of Grignard reagents and alkyllithium compounds with carbon dioxide, aldehydes, and ketones (Section 15.5.E).

$$RMgX + CO_2 \longrightarrow \xrightarrow{H^+} RCOOH$$

$$RMgX + R'CHO \longrightarrow \xrightarrow{H^+} R-\overset{\overset{\displaystyle OH}{|}}{C}H-R'$$

$$RMgX + R_2'C=O \longrightarrow \xrightarrow{H^+} R-\overset{\overset{\displaystyle OH}{|}}{\underset{\underset{\displaystyle R'}{|}}{C}}-R'$$

(f) Alkylation of enolate ions with primary alkyl halides (Section 13.6.B).

$$R_2CH-\overset{\overset{\displaystyle O}{\|}}{C}R' \xrightarrow{base} R_2C=\overset{\overset{\displaystyle O^-}{|}}{C}R' \xrightarrow{R''CH_2X} R_2\overset{\overset{\displaystyle O}{\|}}{\underset{\underset{\displaystyle CH_2R''}{|}}{C}}-\overset{\displaystyle }{C}R'$$

(g) The Wittig reaction (Section 13.7.G).

$$R_2C=O + R'CH=P(C_6H_5)_3 \longrightarrow R_2C=CHR'$$

(h) The aldol condensation (Section 13.7.F).

$$2\,RCH_2CHO \longrightarrow RCH_2\overset{\overset{\displaystyle HO}{|}}{C}H\overset{\overset{\displaystyle R}{|}}{C}HCHO \longrightarrow RCH_2CH=\overset{\overset{\displaystyle R}{|}}{C}CHO$$

$$R\overset{\overset{\displaystyle O}{\|}}{C}CH_3 \xrightarrow{LDA} R\overset{\overset{\displaystyle O^-}{|}}{C}=CH_2 \xrightarrow{R'CHO} R\overset{\overset{\displaystyle O}{\|}}{C}CH_2\overset{\overset{\displaystyle OH}{|}}{C}HR'$$

A more complete summary of C—C bond-forming reactions is given in Appendix II of the Study Guide.

2. Placement of Desired Functional Groups in Their Proper Place. This aspect of a synthesis involves the introduction, removal, or interconversion of functional groups. We have encountered a great many such reactions. Rather than summarize them all, we shall only give a few illustrative examples.

(a) Introduction of a functional group.

$$R-\overset{\overset{\displaystyle R'}{|}}{C}H-R'' + Br_2 \xrightarrow{h\nu} R-\overset{\overset{\displaystyle R'}{|}}{\underset{\underset{\displaystyle Br}{|}}{C}}-R''$$

(b) Removal of a functional group.

$$R-\overset{\overset{\displaystyle O}{\|}}{C}-R' + H_2NNH_2 \xrightarrow[\underset{\Delta}{(HOCH_2CH_2)_2O}]{NaOH} RCH_2R'$$

$$-\overset{\overset{\displaystyle H}{|}}{\underset{|}{C}}-\overset{\overset{\displaystyle OH}{|}}{\underset{|}{C}}- \xrightarrow[(-H_2O)]{H^+} {}_{>}C=C_{<} \xrightarrow{\underset{Pd}{H_2}} -\overset{\overset{\displaystyle H}{|}}{\underset{|}{C}}-\overset{\overset{\displaystyle H}{|}}{\underset{|}{C}}-$$

(c) Interconversion of functional groups.

$$\underset{\overset{|}{-}\text{CH}\overset{|}{-}}{\overset{\text{OH}}{}} \quad \underset{\text{reduction}}{\overset{\text{oxidation}}{\rightleftarrows}} \quad \underset{-\text{C}-}{\overset{\text{O}}{\parallel}}$$

$$-\text{CH}_2\text{Br} \quad \underset{\text{PBr}_3}{\overset{\text{OH}^-}{\rightleftarrows}} \quad -\text{CH}_2\text{OH}$$

Insofar as is possible, one should choose reactions for building up the carbon skeleton so that the least amount of subsequent functional group manipulation is necessary. A complete summary of functional group interconversions is included in Appendix II of the Study Guide.

3. Control of Stereochemistry Where Relevant. When more than one stereo-isomer of the desired product is possible, it is necessary to design a synthesis that will yield only that isomer. In such cases it is important to use reactions that are **stereoselective**—that is, reactions yielding largely one stereoisomer when two or more might result. The stereoselective reactions we have encountered are summarized as follows.

(a) S_N2 displacement reactions of secondary halides (Section 8.5).

$$\underset{\overset{\displaystyle H}{\underset{\displaystyle R'}{}}}{\overset{\displaystyle R}{}}\!\!\!\diagdown\!\!\text{C}\!-\!\text{X} + \text{Y}^- \longrightarrow \text{Y}\!-\!\text{C}\!\!\!\diagup\overset{\displaystyle R}{\underset{\displaystyle R'}{}}\!\!\!H + \text{X}^-$$

(b) Catalytic hydrogenation of alkynes (Section 12.6.A).

$$\text{R}\!-\!\text{C}\!\equiv\!\text{C}\!-\!\text{R}' \xrightarrow[\text{cat.}]{\text{H}_2} \underset{\overset{\displaystyle R}{}}{\overset{\displaystyle H}{}}\!\!\!\diagdown\!\text{C}\!=\!\text{C}\!\!\diagup\overset{\displaystyle H}{\underset{\displaystyle R'}{}}$$

(c) Metal ammonia reduction of alkynes (Section 12.6.A).

$$\text{R}\!-\!\text{C}\!\equiv\!\text{C}\!-\!\text{R}' \xrightarrow[\text{NH}_3]{\text{Na}} \underset{\overset{\displaystyle R}{}}{\overset{\displaystyle H}{}}\!\!\!\diagdown\!\text{C}\!=\!\text{C}\!\!\diagup\overset{\displaystyle R'}{\underset{\displaystyle H}{}}$$

(d) Oxidation of alkenes with osmium tetroxide (Section 11.6.E).

$$+ \text{H}_2\text{O}_2 \xrightarrow{\text{OsO}_4}$$

(e) Addition of halogens to alkenes (Section 11.6.B).

$$+ \text{Br}_2 \longrightarrow$$

(f) Bimolecular elimination of alkyl halides (Section 11.5.A)

$$\underset{\overset{\displaystyle R}{\underset{\displaystyle R'}{}}}{\overset{\displaystyle H}{}}\!\!\!\diagdown\!\!\text{C}\!-\!\text{C}\!\!\!\diagup\overset{\displaystyle H}{\diagup}\!\!\!\underset{\displaystyle Br}{\overset{\displaystyle R''}{}} \xrightarrow{t\text{-BuO}^-} \underset{\overset{\displaystyle R'}{}}{\overset{\displaystyle R}{}}\!\!\!\diagdown\!\text{C}\!=\!\text{C}\!\!\diagup\overset{\displaystyle H}{\underset{\displaystyle R''}{}}$$

(g) Hydroboration (Section 11.6D).

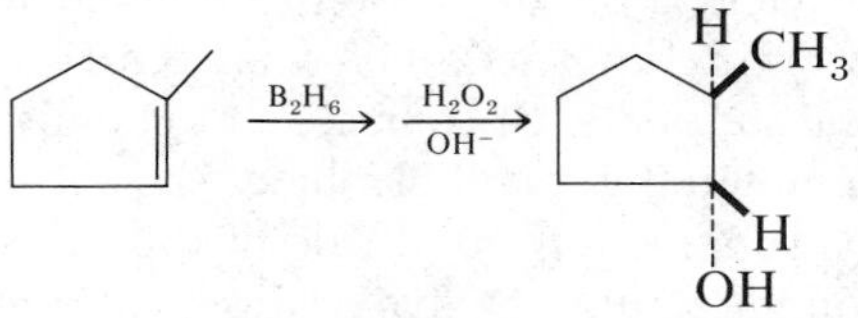

(h) Epoxidation of alkenes (Section 11.6.E).

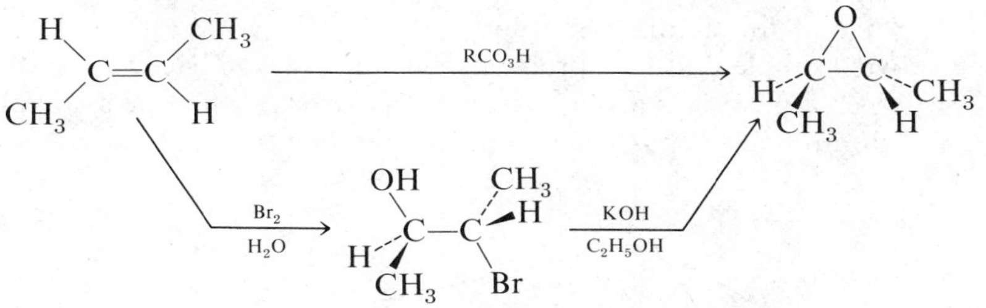

(i) Ring opening of epoxides (Section 10.12).

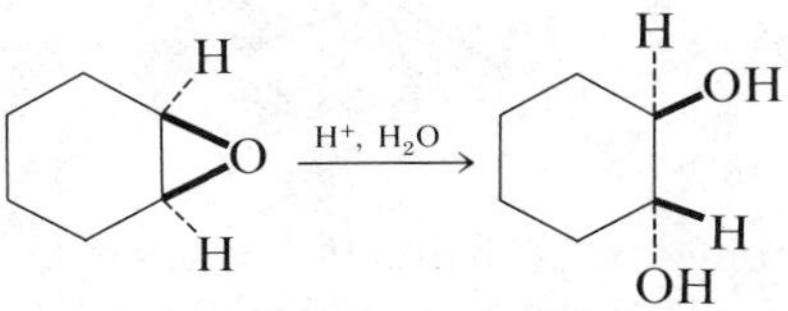

(k) Cyclopropanation (Section 11.6.F).

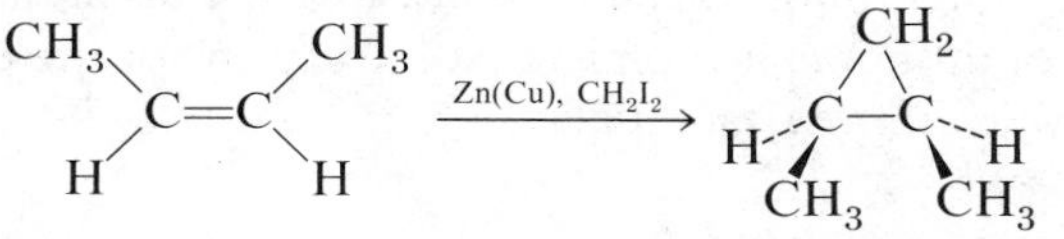

Sometimes it will not be possible to control a synthesis so that only the desired stereoisomer is produced because a method that will accomplish that goal is lacking. In such a case the next best solution is to prepare the mixture of isomers and separate the desired isomer. At this point in our study of organic synthesis we shall only touch on this subject of stereoselectivity, but we will return to the topic in Section 34.8.

EXERCISE: Consider the methods available for the general conversion of RY to RZ in which Y and Z are the eight groups given below. Set up a 8 × 8 matrix with Y down the side and Z across the top. Mark with a minus sign those conversions for which no *general* reaction sequence is presently known to you. Mark with a 1 those interconversions of functional groups that can be generally accomplished by a simple reaction process we have studied. Finally, mark with a + those interconversions that can be accomplished in a sequence of two or more reaction steps. Y and Z: H, Br, OH, CH_2OH, CHO, COOH, CN, $COCH_3$.

16.3
Planning a Synthesis

In planning a synthesis, one works from the product *backward*. Remember that the goal is to connect the desired product with some commercially available starting material by a series of reactions each of which will, insofar as possible, give a single product in high yield. For this reason, the practicing chemist usually acquires a fairly good working knowledge of what types of starting materials are available from chemical suppliers. Of course, if the chemist is not sure whether or not a possible starting material is available, he or she checks for its availability in the catalogs of various suppliers. A good rule of thumb is that most monofunctional aliphatic compounds containing five or fewer carbons may be purchased. A great many others are also available, but this type of information is acquired only by experience. For the purpose of learning how to design a synthesis, we shall follow the "five or fewer carbons" rule.

For relatively simple synthetic problems, one may reason backward in this way and soon arrive at possible starting materials that are known to be available. The result of such an analysis has been called a synthetic tree.

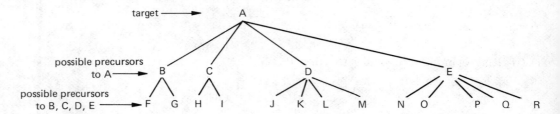

For more complex problems the synthetic tree soon becomes unwieldy. In fact, current research is directed toward the application of computers to synthetic design. In most cases the practicing chemist solves such problems by a combination of logical analysis and intuition. The synthetic tree is built until the chemist recognizes, by insight or intuition, a complete path from one of the possible intermediates to an available starting material.

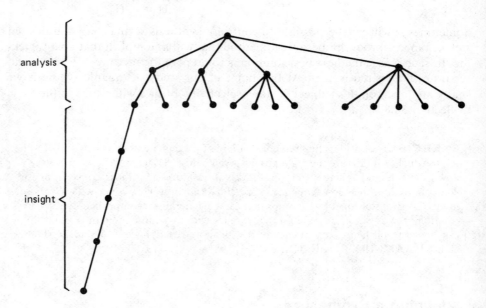

The importance of insight and intuition, relative to analytical reasoning, should not be underestimated for science in general and synthetic design in particular.

Nevertheless, insight and intuition cannot function in the absence of a body of facts—in the present case, a thorough knowledge of organic reactions.

The best way to illustrate synthesis design is to demonstrate with a few specific simple examples.

Example 16-1. Plan a synthesis of 4-methyl-1-pentanol.

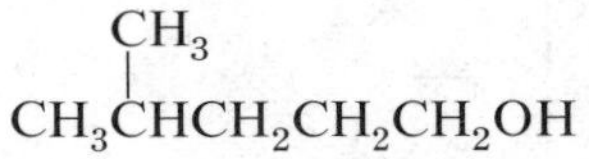

$$\underset{\underset{\displaystyle CH_3}{|}}{CH_3CHCH_2CH_2CH_2OH}$$

Since the product contains six carbons, we must build up the skeleton from a simpler starting material. In principle, there are a number of ways in which this may be done.

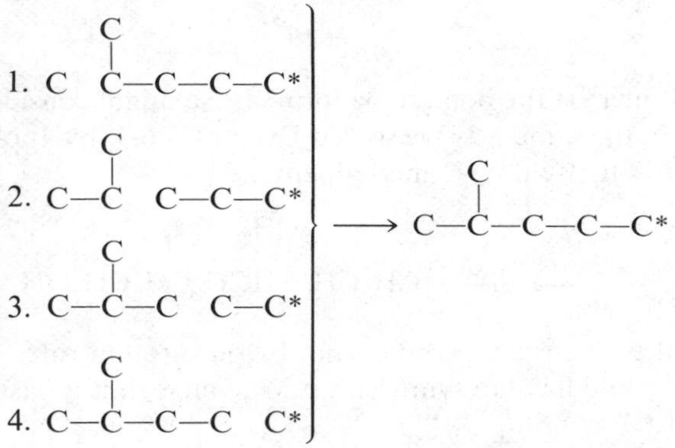

1. $\overset{\displaystyle C}{\underset{}{}}$ C C—C—C—C*

2. C—C C—C—C*

3. C—C—C C—C*

4. C—C—C—C C*

$\longrightarrow$ C—C—C—C—C*

The carbon marked with an asterisk indicates the desired point of functionality, in this case, a hydroxy group. Of the various combinations, No. 4 is most attractive, since the carbon–carbon bond to be created is adjacent to the functional group in the final product. In planning a synthesis, *it is generally most productive to look first at combinations in which the functional group is close to the carbon–carbon bonds that will be formed in the synthesis.* We immediately recognize that 4-methyl-1-pentanol is readily available by reaction of the Grignard reagent derived from 1-chloro-3-methylbutane with formaldehyde.

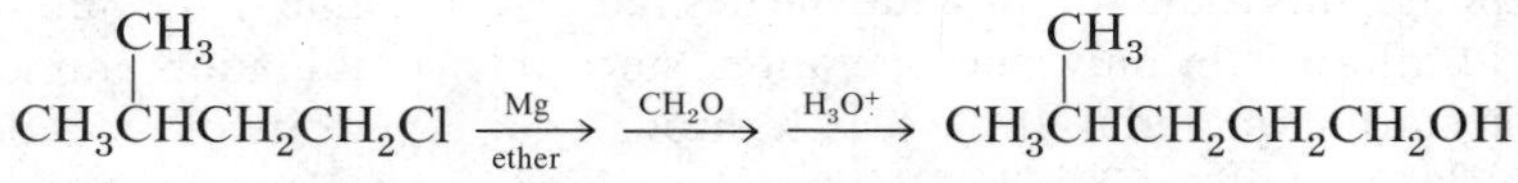

$$\underset{\underset{\displaystyle CH_3}{|}}{CH_3CHCH_2CH_2Cl} \xrightarrow[\text{ether}]{Mg} \xrightarrow{CH_2O} \xrightarrow{H_3O^+} \underset{\underset{\displaystyle CH_3}{|}}{CH_3CHCH_2CH_2CH_2OH}$$

Example 16-2. Plan a synthesis of 4-methyl-1-hexanol.

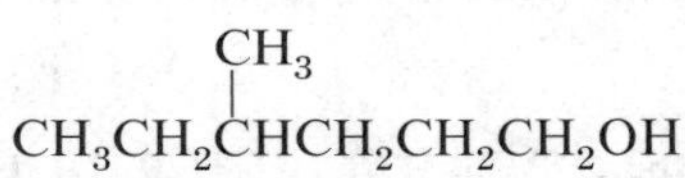

$$\underset{\underset{\displaystyle CH_3}{|}}{CH_3CH_2CHCH_2CH_2CH_2OH}$$

In this example, the desired product contains seven carbons. Analysis of the carbon skelton gives the following combinations.

$$
\begin{array}{l}
\qquad\qquad\quad\ \overset{\displaystyle C}{\underset{|}{\ }} \\[-4pt]
1.\ \ C\ \ C-C-C-C-C^* \\[10pt]
\qquad\qquad \overset{\displaystyle C}{\underset{|}{\ }} \\[-4pt]
2.\ \ C-C\ \ C-C-C-C^* \\[10pt]
\qquad\qquad \overset{\displaystyle C}{\underset{|}{\ }} \\[-4pt]
3.\ \ C-C-C-C-C-C^* \\[10pt]
\qquad\qquad \overset{\displaystyle C}{\underset{|}{\ }} \\[-4pt]
4.\ \ C-C-C\ \ C-C-C^* \\[10pt]
\qquad\qquad \overset{\displaystyle C}{\underset{|}{\ }} \\[-4pt]
5.\ \ C-C-C-C\ \ C-C^* \\[10pt]
\qquad\qquad \overset{\displaystyle C}{\underset{|}{\ }} \\[-4pt]
6.\ \ C-C-C-C-C\ \ C^*
\end{array}
\qquad\longrightarrow\qquad
\begin{array}{l}
\qquad\quad \overset{\displaystyle C}{\underset{|}{\ }} \\[-4pt]
C-C-C-C-C-C^*
\end{array}
$$

In No. 6 the functional group is nearest the bond to be formed. We might consider solving this synthetic problem in the same way we solved Example 16-1, by application of the Grignard synthesis to 1-chloro-3-methylpentane.

$$
\underset{\displaystyle CH_3CH_2\overset{\textstyle CH_3}{\overset{|}{C}}HCH_2CH_2Cl}{}\ \xrightarrow[\text{ether}]{Mg}\ \xrightarrow{CH_2O}\ \xrightarrow{H_3O^+}\ CH_3CH_2\overset{\textstyle CH_3}{\overset{|}{C}}HCH_2CH_2CH_2OH
$$

However, this starting material contains six carbons and, by our ground rules, is not readily available. Thus we would have to synthesize it. Of course, this is easily accomplished as follows.

$$
CH_3CH_2\overset{\textstyle CH_3}{\overset{|}{C}}HCH_2Cl\ \xrightarrow[\text{ether}]{Mg}\ \xrightarrow{CH_2O}\ \xrightarrow{H_3O^+}
$$

$$
CH_3CH_2\overset{\textstyle CH_3}{\overset{|}{C}}HCH_2CH_2OH\ \xrightarrow[\text{pyridine}]{SOCl_2}\ CH_3CH_2\overset{\textstyle CH_3}{\overset{|}{C}}HCH_2CH_2Cl
$$

Thus our overall synthesis of 4-methyl-1-hexanol would require three separate operations beginning with 1-chloro-2-methylbutane.

Although this route will provide the desired product, note that combination No. 5 could afford it in only one operation, since this combination employs a five-carbon starting material. If we search through our repertoire of reactions for one in which the unit —CH₂CH₂OH may be joined to another molecule, we recall that Grignard reagents react with ethylene oxide to produce exactly this result. Thus we can prepare the Grignard reagent of 1-chloro-2-methylbutane and allow it to react with ethylene oxide to obtain 4-methyl-1-hexanol.

$$
CH_3CH_2\overset{\textstyle CH_3}{\overset{|}{C}}HCH_2Cl\ \xrightarrow[\text{ether}]{Mg}\ \xrightarrow{\overset{O}{CH_2-CH_2}}\ \xrightarrow{H_3O^+}\ CH_3CH_2\overset{\textstyle CH_3}{\overset{|}{C}}HCH_2CH_2CH_2OH
$$

This example illustrates a second rule which should be observed when planning a synthesis. *It is generally more productive to add a large fragment in a single reaction than to add several smaller fragments sequentially.* One should therefore examine all of the various carbon–carbon bonds that might be formed, even though they are not immediately adjacent to the desired functional group. In some cases a reaction may exist that allows the formation of such a remote bond.

Example 16-3. Plan a synthesis of 3-methyl-1-pentene.

$$\overset{\displaystyle CH_3}{\underset{\displaystyle |}{CH_3CH_2CHCH{=}CH_2}}$$

First, write out the combinations which would produce the skeleton.

$$\left.\begin{array}{l} \text{1. } C \quad \overset{\displaystyle C}{\underset{\displaystyle |}{C}} \overset{*}{-}\overset{*}{C}-C-C \\[2ex] \text{2. } C{-}C \quad \overset{\displaystyle C}{\underset{\displaystyle |}{C}} \overset{*}{-}\overset{*}{C}-C \\[2ex] \text{3. } C{-}C{-}\overset{\displaystyle C}{\underset{\displaystyle |}{C}} \quad \overset{*}{C}{-}\overset{*}{C} \\[2ex] \text{4. } C{-}C{-}\overset{\displaystyle C}{\underset{\displaystyle |}{C}}{-}\overset{*}{C} \quad \overset{*}{C} \end{array}\right\} \longrightarrow C{-}C{-}\overset{\displaystyle C}{\underset{\displaystyle |}{C}}{-}C{-}C$$

The functional group in our desired product is a carbon–carbon double bond. We should now review the methods whereby this functional group can be introduced into a molecule. Briefly, they are

(a) Dehydration of an alcohol (Sections 10.7 and 11.5).

$$\overset{\displaystyle OH}{\underset{\displaystyle |}{RCHCH_3}} \xrightarrow{H^+} RCH{=}CH_2 + H_2O$$

(b) Elimination of an alkyl halide (Section 11.5).

$$RCH_2CH_2X + R'O^- \longrightarrow RCH{=}CH_2 + X^- + R'OH$$

(c) Hydrogenation of an alkyne (Section 12.6).

$$RC{\equiv}CH \xrightarrow[\substack{\text{Lindlar's}\\\text{catalyst}}]{H_2} RCH{=}CH_2$$

(d) Wittig reaction (Section 13.7).

$$RCHO + (C_6H_5)_3P{=}CH_2 \longrightarrow RCH{=}CH_2 + (C_6H_5)_3P{=}O$$

For the problem under consideration, alcohol dehydration is not suitable, since carbocation rearrangements could result in a mixture of products (Section 12.5.B). Base-catalyzed elimination of an alkyl halide is a possibility. The required halide would be 1-chloro-3-methylpentane, which could be prepared as discussed in Example 16-2. This would provide a three-step synthesis of the desired alkene.

$$\overset{\displaystyle CH_3}{\underset{\displaystyle |}{CH_3CH_2CHCH_2Cl}} \longrightarrow \overset{\displaystyle CH_3}{\underset{\displaystyle |}{CH_3CH_2CHCH_2CH_2OH}} \dashrightarrow$$

$$\overset{\displaystyle CH_3}{\underset{\displaystyle |}{CH_3CH_2CHCH_2CH_2Cl}} \longrightarrow \overset{\displaystyle CH_3}{\underset{\displaystyle |}{CH_3CH_2CHCH{=}CH_2}}$$

At first glance, partial hydrogenation of 3-methyl-1-pentyne is attractive, since the necessary alkyne appears to be available by alkylation of acetylene with 2-bromo-butane.

$$CH_3CH_2\overset{\overset{\displaystyle Br}{|}}{C}HCH_3 + NaC\equiv CH \longrightarrow CH_3CH_2\overset{\overset{\displaystyle CH_3}{|}}{C}HC\equiv CH \longrightarrow CH_3CH_2\overset{\overset{\displaystyle CH_3}{|}}{C}HCH=CH_2$$

However, recall that acetylide ions cause *elimination* of secondary alkyl halides (Section 12.5.C).

$$CH_3CH_2\overset{\overset{\displaystyle Br}{|}}{C}HCH_3 + NaC\equiv CH \longrightarrow CH_3CH_2CH=CH_2 + HC\equiv CH + NaBr$$
(plus other isomers)

Therefore, this two-step synthesis of 3-methyl-1-pentane is not applicable. This example illustrates an important aspect of synthesis design. After writing out a possible route to a desired product, carefully review the chemistry that would be involved to see if there are any subtle structural features that render the route inoperable for the specific carbon skeleton under consideration.

The final method for creation of a carbon–carbon double bond is the Wittig reaction. In this case the reaction of 2-methylbutanol with methylenetriphenyl-phosphorane would yield the desired alkene in one step.

$$CH_3CH_2\overset{\overset{\displaystyle CH_3}{|}}{C}HCHO \longrightarrow CH_3CH_2\overset{\overset{\displaystyle CH_3}{|}}{C}HCH=CH_2$$

This is clearly a more efficient synthesis of 3-methyl-1-pentene than either of the two routes considered heretofore, since only one step is involved. Even so, the practicing chemist might not choose this method, for practical reasons. The required reagent is prepared by reaction of methyltriphenylphosphonium bromide with a strong base such as *n*-butyllithium.

$$(C_6H_5)_3\overset{+}{P}-CH_3Br^- \xrightarrow{n-C_4H_9Li} (C_6H_5)_3P=CH_2$$

Both of these reactants are rather expensive. Furthermore, the molecular weight of the phosphonium salt is 357. Thus, to carry out the reaction on a one-mole scale (which would provide a maximum of 82 g of alkene), one must employ 357 g of phosphonium salt. Consequently, for preparation of a large quantity of 3-methyl-1-pentene, the chemist would probably use the longer Grignard route rather than the one-step Wittig reaction. For preparation of only a small amount of material, where cost is of less importance, one might well use the Wittig procedure.

Example 16-4. Plan a synthesis of *cis*-2-methyl-5-decene.

$$CH_3\overset{\overset{\displaystyle CH_3}{|}}{C}HCH_2CH_2 \overset{H}{\diagdown}\underset{\diagup}{C}=\underset{\diagdown}{C}\overset{H}{\diagup} CH_2CH_2CH_2CH_3$$

Here stereochemistry is important because only the *cis* isomer is desired. In this case that consideration dominates the planning, since we have at this point only one method available for the stereospecific production of a *cis* alkene.

$$\underset{\overset{|}{CH_3}}{CH_3CHCH_2CH_2C\equiv CCH_2CH_2CH_2CH_3} \xrightarrow[\text{Lindlar's}]{H_2} \underset{\overset{|}{CH_3}}{CH_3CHCH_2CH_2} \underset{CH_2CH_2CH_2CH_3}{\overset{\overset{H}{\diagup} \underset{}{C=C} \overset{H}{\diagdown}}{}}$$

Other methods for introducing the functional group would give a mixture of positional and/or stereoisomers. The required alkyne contains 11 carbons and must be built up from smaller fragments. If we consider making bonds closest to the functional group, we have

$$\underset{\overset{|}{C}}{C-C-C-C} + \overset{*}{C}-\overset{*}{C} + C-C-C-C \longrightarrow \underset{\overset{|}{C}}{C-C-C-C-\overset{*}{C}-\overset{*}{C}-C-C-C-C}$$

The reaction that allows us to accomplish this desired conversion is the S_N2 displacement of primary alkyl halides with the acetylene anion.

$$HC\equiv CH \xrightarrow[NH_3]{NaNH_2} \xrightarrow{\overset{CH_3}{\overset{|}{CH_3CHCH_2CH_2Br}}} \underset{\overset{|}{CH_3}}{CH_3CHCH_2CH_2C\equiv CH} \xrightarrow[NH_3]{NaNH_2}$$

$$\xrightarrow{BrCH_2CH_2CH_2CH_3} \underset{\overset{|}{CH_3}}{CH_3CHCH_2CH_2C\equiv CCH_2CH_2CH_2CH_3}$$

Example 16-5. Plan a synthesis of 2-ethylhexanoic acid.

$$\underset{\overset{|}{CH_3CH_2}}{CH_3CH_2CH_2CH_2CHCOOH}$$

To simplify the problem, consider only the combinations involving five carbons or smaller fragments.

$$
\begin{array}{l}
1.\ \ C-C-C \quad \underset{\overset{|}{C-C}}{C-\overset{*}{C}-C} \\[4mm]
\qquad\qquad \underset{\overset{|}{C-C}}{} \\[1mm]
2.\ \ C-C-C-C \quad C-\overset{*}{C}
\end{array}
\Bigg\} \longrightarrow \underset{\underset{\overset{|}{C-C}}{}}{C-C-C-C-\overset{*}{C}-C}
$$

Combination No. 1 is not practical, since we know no simple methods for making a bond so far from a functional group. At this point, also, we do not have any reactions in our repertoire that allow us to attach an alkyl group to C-2 of a carboxylic acid. However, we can form such a bond adjacent to the aldehyde functional group by using the aldol condensation (Section 13.7.F).

$$2\ CH_3CH_2CH_2CHO \xrightarrow[\Delta]{KOH,\ H_2O} \underset{\overset{|}{CH_3CH_2}}{CH_3CH_2CH_2CH=CCHO}$$

With this realization, we recognize a very efficient synthesis of 2-ethylhexanoic acid.

$$\underset{\overset{|}{CH_3CH_2}}{CH_3CH_2CH_2CH=CCHO} \xrightarrow{Ag_2O} \underset{\overset{|}{CH_3CH_2}}{CH_3CH_2CH_2CH=CHCOOH} \xrightarrow[Pd/C]{H_2} \underset{\overset{|}{CH_3CH_2}}{CH_3CH_2CH_2CH_2CHCOOH}$$

This example illustrates yet another aspect of synthesis design. Look for "hidden functionality." In this case it is important to notice that although the carbon–

carbon double bond does not appear in the ultimate product, the most effective synthesis proceeds through an intermediate containing this functional group. This example also demonstrates the importance of the order of reactions. A sequence involving first hydrogenation and then oxidation could result in reduction of the aldehyde group to an alcohol function that is more difficult to oxidize to a carboxylic acid.

Example 16-6. Plan a synthesis of 4-methylnonane.

$$CH_3CH_2CH_2CH_2CH_2\overset{\overset{\displaystyle CH_3}{|}}{C}HCH_2CH_2CH_3$$

Here we have a compound that has no functional group. Since it has ten carbons, we would like to assemble it from two five-carbon building blocks. Thus the questions are how to accomplish the following combination and how to get rid of any functional groups that might be present in our intermediates.

$$C-C-C-C-C \quad \overset{\overset{\displaystyle C}{|}}{C}-C-C-C \longrightarrow C-C-C-C-C-\overset{\overset{\displaystyle C}{|}}{C}-C-C-C$$

First, let us review the methods at our disposal for removal of a functional group. We have learned three such methods—hydrogenation of a multiple bond, removal of the oxygen from an aldehyde or ketone, and removal of halogen by formation and hydrolysis of an organometallic reagent.

(a) Hydrogenation of alkenes or alkynes (Sections 11.6 and 12.6).

$$\underset{\diagup}{\overset{\diagdown}{C}}=\underset{\diagdown}{\overset{\diagup}{C}} + H_2 \xrightarrow{\text{cat.}} -\overset{\overset{\displaystyle H}{|}}{C}-\overset{\overset{\displaystyle H}{|}}{C}-$$

$$-C\equiv C- + 2\,H_2 \xrightarrow{\text{cat.}} -\overset{\overset{\displaystyle H}{|}}{\underset{\underset{\displaystyle H}{|}}{C}}-\overset{\overset{\displaystyle H}{|}}{\underset{\underset{\displaystyle H}{|}}{C}}-$$

(b) Deoxygenation of aldehydes and ketones (Section 13.0).

$$-\overset{\overset{\displaystyle O}{\|}}{C}- \xrightarrow[\substack{\text{Wolff–Kishner} \\ \text{reduction}}]{\text{Clemmensen or}} -\overset{\overset{\displaystyle H}{|}}{\underset{\underset{\displaystyle H}{|}}{C}}-$$

(c) Hydrolysis of organometallic compounds (Section 15.5).

$$-\overset{|}{\underset{|}{C}}-Cl \xrightarrow[\text{ether}]{\text{Mg}} -\overset{|}{\underset{|}{C}}-MgCl \xrightarrow{H_2O} -\overset{|}{\underset{|}{C}}-H$$

There are a number of ways we might use these reactions in a synthesis of 4-methylnonane. For example, we could make the Grignard reagent from 1-chloropentane, add it to 2-pentanone, and dehydrate the resulting tertiary alcohol. A mixture of $C_{10}H_{20}$ isomers is expected to result, but this is of no consequence, since all of the isomers will give 4-methylnonane upon hydrogenation.

$$CH_3CH_2CH_2CH_2CH_2Cl \xrightarrow[\text{ether}]{Mg} \xrightarrow{CH_3\overset{O}{\overset{||}{C}}CH_2CH_2CH_3} CH_3CH_2CH_2CH_2CH_2\overset{\overset{\displaystyle CH_3}{|}}{\underset{\underset{\displaystyle OH}{|}}{C}}CH_2CH_2CH_3 \xrightarrow[\Delta]{H_2SO_4}$$

$$\left\{ \begin{array}{c} CH_3CH_2CH_2CH_2CH{=}\overset{\overset{\displaystyle CH_3}{|}}{C}CH_2CH_2CH_3 \\ (\textit{cis and trans}) \\ + \\ CH_3CH_2CH_2CH_2CH_2\overset{\overset{\displaystyle CH_3}{|}}{C}{=}CHCH_2CH_3 \\ (\textit{cis and trans}) \\ + \\ CH_3CH_2CH_2CH_2CH_2\overset{\overset{\displaystyle CH_2}{||}}{C}CH_2CH_2CH_3 \end{array} \right\} \xrightarrow[Pd/C]{H_2} CH_3CH_2CH_2CH_2CH_2\overset{\overset{\displaystyle CH_3}{|}}{C}HCH_2CH_2CH_3$$

16.4
Protecting Groups

It often happens in the design of a synthesis that one wishes to carry out a transformation on one functional group with a reagent that would also react with some other functional group present in the same molecule. This may often be done by temporarily **protecting** one of the functional groups by changing it into another functional group that is unreactive to the reagent in question. For example, suppose it is desirable to carry out the following conversion.

$$CH_3\overset{O}{\overset{||}{C}}CH_2\overset{\overset{\displaystyle CH_3}{|}}{\underset{\underset{\displaystyle CH_3}{|}}{C}}CH_2Br \longrightarrow CH_3\overset{O}{\overset{||}{C}}CH_2\overset{\overset{\displaystyle CH_3}{|}}{\underset{\underset{\displaystyle CH_3}{|}}{C}}CH_2COOH$$

The obvious method of employing the corresponding Grignard reagent cannot be used here because the bromide has a carbonyl group in the molecule and Grignard reagents react with ketones. The problem can be circumvented by first transforming the ketone into a ketal.

$$CH_3\overset{O}{\overset{||}{C}}CH_2\overset{\overset{\displaystyle CH_3}{|}}{\underset{\underset{\displaystyle CH_3}{|}}{C}}CH_2Br + HOCH_2CH_2OH \xrightarrow[(-H_2O)]{H^+} CH_3{-}\overset{\overset{\displaystyle CH_2{-}CH_2}{\overset{|\quad\;\;|}{\overset{\displaystyle O\quad\;\; O}{\diagdown\;\;\diagup}}}}{C}{-}CH_2\overset{\overset{\displaystyle CH_3}{|}}{\underset{\underset{\displaystyle CH_3}{|}}{C}}CH_2Br$$

Since a ketal is an ether and ethers do not react with Grignard reagents, the Grignard synthesis may now be carried out. After the Grignard step, the ketal is hydrolyzed back to the ketone by treatment with acid and excess water.

$$CH_3{-}\overset{\overset{\displaystyle CH_2{-}CH_2}{\overset{|\quad\;\;|}{\overset{\displaystyle O\quad\;\; O}{\diagdown\;\;\diagup}}}}{C}{-}CH_2\overset{\overset{\displaystyle CH_3}{|}}{\underset{\underset{\displaystyle CH_3}{|}}{C}}CH_2Br \xrightarrow[\text{ether}]{Mg} \xrightarrow{CO_2} \xrightarrow{H_3O^+} CH_3\overset{O}{\overset{||}{C}}CH_2\overset{\overset{\displaystyle CH_3}{|}}{\underset{\underset{\displaystyle CH_3}{|}}{C}}CH_2COOH$$

In this example the ketal group is a protecting group. It is a good protecting group for aldehydes and ketones because it is easily introduced and removed and, since it is an ether, is stable to many reagents.

A method often used for the protection of primary alcohols is to convert them into *t*-butyl ethers. For example, suppose it is desired to convert 3-bromopropanol into 3-deuterio-1-propanol. Again, the Grignard reagent might be used, except for the fact that the hydroxy function would interfere. However, recall that primary and secondary alcohols may be transformed into *t*-butyl ethers by treatment with isobutylene and acid (Section 10.11.A).

$$
\underset{}{BrCH_2CH_2CH_2OH} + \underset{}{CH_3}\overset{CH_3}{\underset{}{C}}{=}CH_2 \xrightarrow{H_2SO_4} BrCH_2CH_2CH_2O\overset{CH_3}{\underset{CH_3}{\overset{|}{\underset{|}{C}}}}CH_3
$$

The hydroxy group has now been protected by conversion into an ether function, which is stable to Grignard reagents. The Grignard reagent is prepared and hydrolyzed with D_2O, and the product is hydrolyzed with aqueous sulfuric acid to obtain the deuterated propanol.

$$
BrCH_2CH_2CH_2O\overset{CH_3}{\underset{CH_3}{\overset{|}{\underset{|}{C}}}}CH_3 \xrightarrow[ether]{Mg} BrMgCH_2CH_2CH_2O\overset{CH_3}{\underset{CH_3}{\overset{|}{\underset{|}{C}}}}CH_3 \xrightarrow{D_2O}
$$

$$
DCH_2CH_2CH_2O\overset{CH_3}{\underset{CH_3}{\overset{|}{\underset{|}{C}}}}CH_3 \xrightarrow[H_2O]{H_2SO_4} \underset{\text{3-deuterio-1-propanol}}{DCH_2CH_2CH_2OH} + CH_3\overset{CH_3}{\underset{}{C}}{=}CH_2
$$

Another protecting group commonly used for alcohols is the trimethylsilyl ether, which is formed by treating the alcohol with trimethylchlorosilane and an organic base such as triethylamine.

$$
-\overset{|}{\underset{|}{C}}-OH + (CH_3)_3SiCl \xrightarrow{(C_2H_5)_3N} -\overset{|}{\underset{|}{C}}-O-Si(CH_3)_3
$$

The silyl ether grouping is stable to most neutral and basic conditions. Upon treatment with mild aqueous acid, the alcohol is regenerated.

$$
-\overset{|}{\underset{|}{C}}-O-Si(CH_3)_3 \xrightarrow[25°]{H_3O^+} -\overset{|}{\underset{|}{C}}-OH + (CH_3)_3SiOH
$$

16.5
Industrial Syntheses

Organic compounds are synthesized for two fundamentally different kinds of reasons. On the one hand, we may need a specific compound in order to study its properties or to use it for further research purposes. For such purposes relatively small amounts generally suffice, and cost is not an important criterion—within limits. On the other hand, a compound may have commercial significance, and for such purposes economic factors take on a vital importance. The cost of a medicinal used in small quantity where no other product will work is clearly of a different magnitude than that of a polymeric building material that must compete with wood and steel.

Most of the reactions and syntheses we have studied are useful for understanding the chemistry of different kinds of functional groups and for the laboratory preparation of various compounds. Few of these reactions, however, are suitable for the industrial preparation of compounds, some of which are produced in the amount of millions of pounds a year. An important distinction is the following: the reactions and laboratory preparations that we study have generality. These methods, with minor modifications, are suitable for the preparation of whole classes of compounds. On the other hand, many industrial preparations are specific. They apply for making one and only one specific compound. Many such reactions are gas-phase catalytic processes with the precise catalysts and reaction conditions carefully worked out. An important advantage of such processes is that they are continuous rather than batch processes. Ideally, in a continuous process the reactants are fed into one end of a chemical plant and the product comes out at the other end without stopping, ready for marketing or for the next step. In practice, this ideal is rarely achieved. Catalysts lose their efficiency with time and need to be replaced. By-products build up and need to be cleaned out.

Much research in the chemical industry is devoted to discovering new products with useful properties. But much other research is devoted to existing products, improving processes to obtain these products cheaper and more efficiently, and, in many cases, purer. Research in *process development* is a fascinating area of its own that requires special talents of creativity, patience, and chemical knowledge. Close attention in process development is also given to the environment and to energy conservation. A suitable industrial process must involve a minimum of waste products that require disposal.

P R O B L E M S

1. Plan a synthesis for each of the following compounds from starting materials containing five or fewer carbons.
  (a) 1-hexanol
  (b) 2-hexanol
  (c) 3-hexanol
  (d) 1-heptyne
  (e) *trans*-3-heptene
  (f) 5-methyl-5-nonanol
  (g) 2,6-dimethyl-2-heptanol
  (h) 1-cyclopentyl-2-methylpropene

2. Plan a synthesis for each of the following compounds from starting materials containing five or fewer carbons. For these compounds, the most efficient synthesis will involve removal of a functional group from an intermediate.
  (a) 5-(3-methylbutyl)-2,8-dimethylnonane
  (b) 1-cyclopentylpentane
  (c) 2,9-dimethyldecane
  (d) 2,3,6-trimethylheptane
  (e) 4-methyl-1-heptene

3. Plan a synthesis for each of the following compounds from monofunctional starting materials containing five or fewer carbons. In each case, the final product may be racemic; it is not necessary to plan the synthesis using optically active reactants. Relative stereochemistry *is* important.
  (a) *trans*-2-hexene
  (b) *trans*-2-butylcyclopentanol
  (c) *cis*-4-octene
  (d) (3R,4R)-hexane-3,4-diol
  (e) (2R,3S)-3-methyl-2-heptanol
  (f)
  (g) *trans*-1,2-diethylcyclopropane

4. Show how each of the following *optically active* compounds may be prepared from optically active starting materials containing five or fewer carbons.
(a) (S)-4-methyl-1-hexanol   (b) (R)-2-methylpentanenitrile
(c) (R)-4,4-dimethyl-2-pentanol

5. Show how one may carry out each of the following conversions in good yield.

(a)

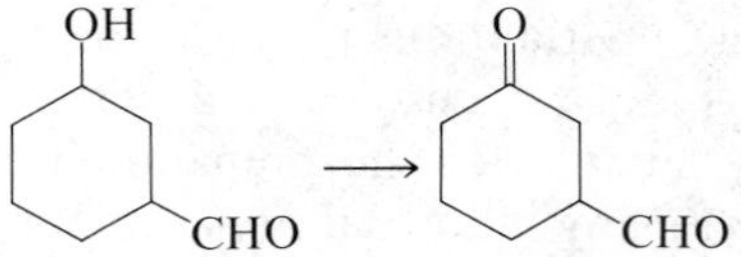

(b) $BrCH_2CH_2CH_2CH_2CHO \longrightarrow CH_2{=}CHCH_2CH_2CHO$

(c) 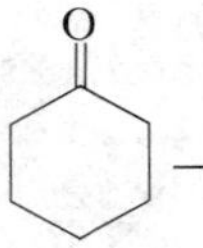 $\longrightarrow HOCH_2CH_2CH_2CH_2CH_2\overset{\overset{\displaystyle OH}{|}}{C}HCH_3$

6. Plan a synthesis for each of the following compounds from starting materials containing five or fewer carbons. Difunctional starting materials may be used if necessary.

(a) *cis*-2,2-dimethyl-3-octene
(b) $CH_3CH_2CH_2CH_2\overset{\overset{\displaystyle O}{||}}{C}CH_2CH_2CHO$
(c) 5-hydroxy-2-hexanone
(d) 4-ethyl-2-methyl-2,4-octadiene
(e) 2-methyloct-2-en-5-yn-4-ol
(f) 3-ethylhept-6-en-4-yn-3-ol
(g) 6-methylheptan-1,4-diol

Chapter 17
Mass Spectrometry

17.1
Introduction

When a beam of electrons of energy greater than the ionization potential [about 8–13 electron volts (eV) (185–300 kcal mole^{-1}) for most compounds] is passed through a sample of an organic compound in the vapor state, ionization of some molecules occurs. In one form of ionization, one of the valence electrons of the molecule is "knocked out," leaving behind a **radical cation.**

$$e^- + CH_4 \longrightarrow 2\,e^- + [CH_4]^{+}$$

In CH_4 eight valence electrons bond the four hydrogens to carbon. The symbol $[CH_4]^{+}$ represents a structure in which *seven* valence electrons bond the four hydrogens to carbon. The $+$ sign shows that the species has a net positive charge. The · signifies that the species has an odd number of electrons. If a mixture of compounds is bombarded with electrons, a mixture of radical cations differing in mass will obviously be produced.

$$CH_4,\ C_2H_6 \longrightarrow [CH_4]^{+},\ [C_2H_6]^{+}$$
$$m/e:\qquad\qquad\qquad 16\qquad 30$$

where $m/e \equiv$ mass-to-charge ratio.

In practice, even when a pure substance is bombarded with electrons, a mixture of cations is produced. This process, termed fragmentation, will be discussed later. For example, methane may give cations with masses of 16, 15, 14, 13, and 12.

$$CH_4 \longrightarrow [CH_4]^{+},\ CH_3{}^+,\ [CH_2]^{+},\ CH^+,\ [C]^{+}$$
$$m/e:\qquad 16\qquad\quad 15\qquad\quad 14\qquad 13\qquad 12$$

Lewis structures for these cations are

$$
\left[\begin{array}{c} H \\ H\!:\!\overset{..}{C}\!\cdot\!H \\ H \end{array}\right]^{+}
\qquad
\begin{array}{c} H \\ H\!:\!\overset{..}{C}{}^+ \\ H \end{array}
\qquad
\left[\begin{array}{c} H \\ H\!:\!\overset{..}{C}\cdot \end{array}\right]^{+}
\qquad
H\!:\!\overset{..}{C}{}^+
\qquad
\left[\,:\!\overset{..}{C}\,\right]^{+}
$$

Note that the cations having an even number of hydrogens are **radical cations,** whereas the cations having an odd number of hydrogens are normal **carbocations.**

A **mass spectrometer** is an instrument that is designed to ionize gaseous molecules, separate the ions produced on the basis of their mass-to-charge ratio, and record the relative number of different ions produced. A **mass spectrum** is a plot of the data obtained from the mass spectrometer. It is customary for the mass-to-charge ratio (m/e) to be plotted as the abscissa and the relative number of ions of relative intensity (height of each peak) to appear as the ordinate. Mass spectrometry differs from spectroscopy in that no absorption of light is involved. Nevertheless, it has been called a "spectroscopy" because the mass "spectrum" resembles other kinds of spectra.

17.2
Instrumentation

The simplest and most common mass spectrometer currently in use is based on **single focusing magnetic deflection.** A schematic sketch of a semicircular instrument based on this principle is shown in Figure 17.1. The sample vapor is introduced at point *a*, usually at very low pressure (10^{-6} to 10^{-7} torr). A low pressure is necessary to minimize the number of collisions between ions and un-ionized molecules. Such collisions lead to reactions that produce new ions not related to the starting compound. Such ions can lead to errors in interpretation of the data. As the sample vapor enters the ionizing chamber (see enlarged insert), it passes through the electron beam *b* where ionization occurs. The resulting ions pass between two charged plates *c*, where there is a difference in potential of several thousand volts. In this region, the ions are accelerated and pass through a slit *d* into the magnetic field.

The radius of the path followed by an ion of mass *m* in a magnetic field is proportional to its charge *e* and the accelerating potential (V). The potential energy of the accelerated ion (eV) equals its kinetic energy.

$$eV = \tfrac{1}{2}mv^2 \tag{17-1}$$

In a magnetic field of strength **H** the ion will experience a centripedal force **H**ev, which is balanced by a centrifugal force mv^2/r, where v is the velocity of the ion and r is the radius of the circular path followed by the ion.

$$\mathbf{H}ev = \frac{mv^2}{r} \tag{17-2}$$

$$r = \frac{mv}{e\mathbf{H}} \tag{17-3}$$

As a collection of ions of different mass enters the magnetic field region (f), each ion will follow a circular path described by equation (17-3). Ions of larger m/e will follow a path of greater radius, and ions of lesser m/e will follow a path of smaller radius. In the example diagrammed in Figure 17.2 the ions of $m/e = y$ are passing through the slit (g) and impinging upon the ion collector (i).

Elimination of the velocity term from equations (17-2) and (17-3) gives

$$\frac{m}{e} = \frac{\mathbf{H}^2 r^2}{2V} \tag{17-4}$$

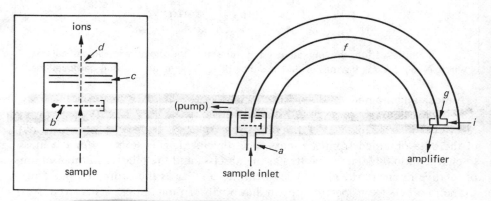

FIGURE 17.1. *Single-focusing magnetic deflection mass spectrometer.*

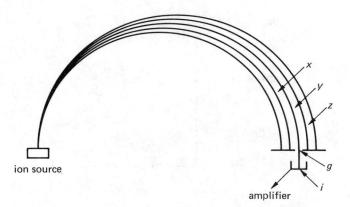

FIGURE 17.2. *Ions with* m/e = y *are focused on the collector* i *through slit* g.

This relationship shows that for an ion of given mass-to-charge ratio (m/e), the radius of deflection r can be increased either by increasing V, the accelerating voltage, or by decreasing **H**, the magnetic field strength. For example, in the case diagrammed in Figure 17.2 a slight increase in V will cause the radius of deflection of all of the ions to increase somewhat. In Figure 17.3 ions of $m/e = y$ no longer pass through the slit and into the collector, but ions of $m/e = x$ do.

Note that the same effect might have been obtained by decreasing **H** slightly or by moving the collector slit slightly to the left. In actual practice this last technique is not technically feasible. Scanning of the spectrum is achieved by either **magnetic** or **electrical scanning.** In the first technique the accelerating voltage V is kept constant while the magnetic field strength **H** is increased slowly. In the latter case **H** is kept constant while V is decreased slowly. When either of these techniques is used, ions of progressively higher m/e attain the necessary radius of deflection to pass through the collector slit (g) and into the ion collector (i).

As the ions enter the collector, they impinge upon a photomultiplier tube where a minute current is produced. The magnitude of this current is proportional to the intensity of the ion beam. The current produced is amplified and fed to a recorder. Some mass spectrometers use a recording oscillograph to record the spectrum. The current from the amplifier is fed to a mirror galvanometer. As the mirror is deflected, it reflects a narrow beam of light onto a sheet of photographic paper

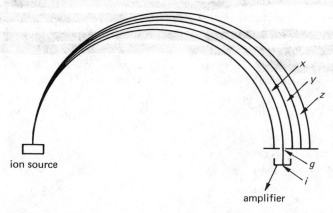

FIGURE 17.3. *At higher* V *or lower* **H,** *ions with* m/e = x *are now focused on the collector* i.

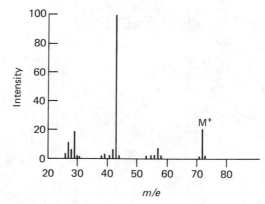

FIGURE 17.4. *Mass spectrum of 2-butanone.*

that is moving through the recorder at a constant rate. If ultraviolet light is used, the photographic chart does not even have to be developed before examination.

In actual practice the mirror galvanometer usually has five mirrors arranged in such a manner that five scans of differing amplitude are obtained simultaneously. The relative sensitivities of the five simultaneous scans are usually $1:3:10:30:100$. It is more convenient to plot the mass spectrum in the form of a bar graph. The most intense peak (the "base peak") is assigned the arbitrary value of 100, and all other peaks are given their proportionate value. Such a bar graph may be constructed manually, by measuring and plotting all of the peaks in a spectrum, or it may be done automatically by computer. A mass spectrum recorded in this manner is shown in Figure 17.4.

17.3
The Molecular Ion: Molecular Formula

The molecular weight of a compound is one datum that can usually be obtained by visual inspection of a mass spectrum. Although the radical cations produced by the initial electron impact usually undergo extensive fragmentation to give cations of smaller m/e (next section), the particle of highest m/e generally (but not always) corresponds to the ionized molecule, and m/e for this particle (called the **molecular ion** and abbreviated M^+) gives the molecular weight of the compound.

If the spectrum is measured with a "high-resolution" spectrometer, it is possible to determine a unique molecular formula for any peak in a mass spectrum, including the molecular ion. This is possible because atomic masses are not integers. For example, consider the molecules CO, N_2, and C_2H_4, all of which have a **nominal mass** of 28. The actual masses of the four atomic particles are H = 1.007825, C = 12.000000 (by definition), N = 14.003050, O = 15.994914. Therefore, the actual masses of CO, N_2 and C_2H_4 are

^{12}C	12.0000	$^{14}N_2$	28.0061	$^{12}C_2$	24.0000
^{16}O	15.9949			$^1H_4$	4.0314
	27.9949				28.0314

Since a high-resolution spectrometer can readily measure mass with an accuracy of better than 1 part in 10,000, the above three masses are readily distinguishable, as shown in Figure 17.5.

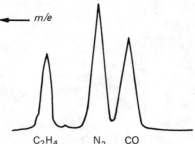

Figure 17.5. *High resolution mass spectrum of a mixture of ethylene, nitrogen, and carbon monoxide.*

Because the mass spectrometer measures the exact m/e for each ion and because most of the elements commonly found in organic compounds have more than one naturally occurring isotope, a given peak will usually be accompanied by several isotope peaks. Table 17.1 shows the common isotopes of some of the elements.

Consider the molecular ion derived from methane. Most of the methane molecules are $^{12}C^1H_4$ and have the nominal mass 16. However, a few molecules are either $^{13}C^1H_4$ or $^{12}C^2H^1H_3$ and have the nominal mass 17. An even smaller number of molecules will have *both* a ^{13}C and an ^2H or will have two ^2H isotopes and will have the nominal mass 18. An exact expression for the ratio $(M + 1)/(M)$ can be derived from probability mathematics but is rather complex. The theoretical intensities of the various isotope peaks may be looked up in special tables compiled for this purpose. However, the contributions of ^2H and ^{17}O to $(M + 1)/(M)$ are relatively small and the ratio is given to a satisfactory approximation for most compounds having few N and S atoms by equation 17-5.

$$\frac{M + 1}{M} = \frac{0.01107}{0.98893}c + 0.00015h + 0.00367n + 0.00037o + 0.0080s \quad (17\text{-}5)$$

where M = intensity of the molecular ion (ions containing no heavy isotopes)
    $M + 1$ = intensity of the molecular ion + 1 peak (ions containing one ^{13}C, ^2H, ^{15}N, ^{17}O, or ^{33}S)
c, h, n, o, s = number of carbons, hydrogens, nitrogens, oxygens, sulfurs.

TABLE 17.1
Natural Abundance of Common Isotopes

Element		Abundance, %		
hydrogen	99.985 ^1H	0.015 ^2H		
carbon	98.893 ^{12}C	1.107 ^{13}C		
nitrogen	99.634 ^{14}N	0.366 ^{15}N		
oxygen	99.759 ^{16}O	0.037 ^{17}O	0.204 ^{18}O	
sulfur	95.0 ^{32}S	0.76 ^{33}S	4.22 ^{34}S	0.014 ^{36}S
fluorine	100 ^{19}F			
chlorine	75.53 ^{35}Cl	24.47 ^{37}Cl		
bromine	50.54 ^{79}Br	49.46 ^{81}Br		
iodine	100 ^{127}I			

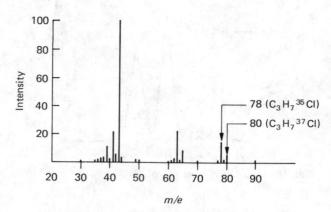

FIGURE 17.6. *Mass spectrum of 2-chloropropane.*

Using this relationship, we may readily estimate the intensity of the M + 1 peak in mass spectrum of methane.

$$\frac{M + 1}{M} = 0.01119(1) + 0.00015(4) = 0.01179$$

Thus, the peak at m/e 17 in the mass spectrum of methane should be approximately 1.18% as intense as the peak at m/e 16.

A similar relationship may be derived for calculation of the intensity of the M + 2 peak. However, in order to obtain an exact figure, a lengthy computation is required. For most compounds the M + 2 peak is small and is not especially useful. However, for compounds containing chlorine or bromine, the M + 2 isotopic peak is substantial. The characteristic doublets observed in the mass spectra of compounds containing chlorine and bromine are an excellent way of diagnosing for the presence of these elements, as shown in Figures 17.6 and 17.7.

One use to which isotope peaks may be put is in approximating the molecular formula of the parent ion in the mass spectrum of an unknown compound. However, one must exercise caution when applying the foregoing computations. First, the M + 1 peak is generally much less intense than the parent ion. Unless the parent ion is a fairly strong one, its isotope peak may be too weak to measure accurately. Second, intermolecular proton transfer reaction can give M + 1 peaks that are *not due to isotopes*. Third, the presence of a small amount of impurity with a strong peak at M + 1 of the sample will interfere with accurate measurement.

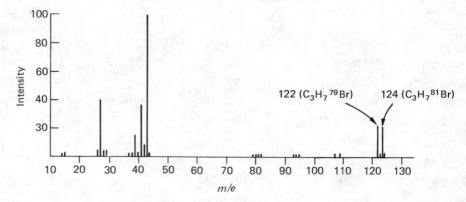

FIGURE 17.7. *Mass spectrum of 1-bromopropane.*

EXERCISE: Estimate the intensity of the M + 1 peak for each of the following compounds.

(a) $C_8H_{16}O_4$ (b) $C_{11}H_{16}N_2$ (c) $C_{13}H_{20}$

(d) $C_{60}H_{122}$ (e) CH_3I (f) C_2F_6

17.4
Fragmentation

A. *Simple Bond Cleavage*

When an electron collides with a molecule in the ionizing chamber of the mass spectrometer, ionization will occur if the impinging electron transfers to the molecule an amount of energy equal to or greater than its ionization potential. The ionization potentials for several organic molecules are given in Table 17.2. When the colliding electron has *excess energy,* a part of this excess energy will normally be carried away by the radical cation produced in the collision. If the molecular ion gains enough surplus energy, bond cleavage (fragmentation) may occur, with the resultant formation of a new cation and a free radical. Typically, the electron beams employed in the ionization process have an energy of 50–70 eV (1150–1610 kcal mole^{-1}). Since this is far in excess of the typical bond energies encountered in organic compounds (50–130 kcal mole^{-1}), fragmentation is normally extensive.

Consider the case of the simplest hydrocarbon, methane. The mass spectrum of methane is shown in Figure 17.8 in bar graph form as well as tabular form. Note that the base peak (most intense peak) corresponds to the molecular ion (m/e 16). Note also the monoisotopic peak at m/e 17 (M + 1), which has an intensity 1.11% that of the molecular ion, within 0.07% of the intensity predicted by theory. Examination of the mass spectrum reveals that cations are also produced and measured that have m/e values of 15, 14, 13, 12, 2, and 1. The following modes of fragmentation may be postulated to explain these various cationic fragments. Initial ionization supplies the molecular ion, with m/e 16.

$$CH_4 + e^- \longrightarrow [CH_4]^{\ddagger} + 2e^-$$
$$m/e\ 16$$

Some of these ions move into the accelerating region and are passed into the magnetic field. However, since they possess a large amount of excess energy, many undergo fragmentation prior to entering the magnetic field, giving a methyl cation (m/e 15) and a hydrogen atom.

TABLE 17.2
Ionization Potentials

Compound	Ionization Potential, electron volts (eV)
benzene	9.25
aniline	7.70
acetylene	11.40
ethylene	10.52
methane	12.98
methanol	10.85
methyl chloride	11.35

m/e	Intensity
1	3.4
2	0.2
12	2.8
13	8.0
14	16.0
15	86.0
16	100.0
17	1.11

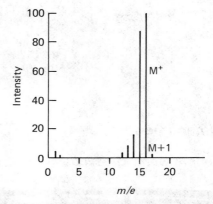

FIGURE 17.8. *Mass spectrum of methane.*

$$[CH_4]^{+\cdot} \longrightarrow CH_3^+ + H\cdot$$
$$m/e\ 15$$

Occasionally this cleavage occurs in such a way as to produce a methyl radical and a bare proton (m/e 1).

$$[CH_4]^{+\cdot} \longrightarrow CH_3\cdot + H^+$$
$$m/e\ 1$$

The fragment CH_3^+ can be accelerated, deflected, and collected as a cation of m/e 15, or it too may undergo fragmentation, giving a hydrogen atom and a new radical cation of m/e 14.

$$CH_3^+ \longrightarrow [CH_2]^{+\cdot} + H\cdot$$
$$m/e\ 14$$

Similar events give rise to particles of m/e 13 and 12.

$$[CH_2]^{+\cdot} \longrightarrow CH^+ + H\cdot$$
$$m/e\ 13$$

$$CH^+ \longrightarrow [C]^{+\cdot} + H\cdot$$
$$m/e\ 12$$

Occasionally an ion may eject an ionized hydrogen molecule, giving rise to the weak peak at m/e 2.

$$[CH_4]^{+\cdot} \longrightarrow CH_2 + [H_2]^{+\cdot}$$
$$m/e\ 2$$

More complicated alkanes give very complicated spectra, with peaks at virtually every value of m/e. However, most of these fragment peaks are of extremely low intensity. The more intense fragment peaks have m/e values of M − 15, M − 29, M − 43, M − 57, and so on, corresponding to scission of the hydrocarbon chain at various places along its length. The spectrum of *n*-dodecane, plotted in Figure 17.9, is illustrative.

There is a reasonably intense molecular ion (4% of the base peak) at m/e 170. The peak at m/e 155, corresponding to loss of CH_3 (M − 15) is so weak as not to be noticeable. However, the peaks at m/e 141 (M − 29), 127 (M − 43), and so on,

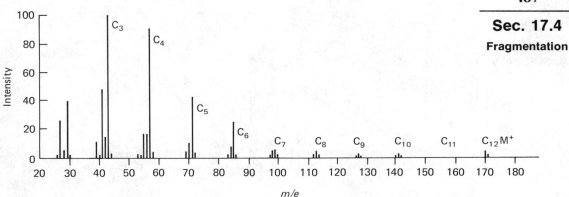

FIGURE 17.9. *Mass spectrum of* n-*dodecane.*

are apparent. Note that intensity decreases regularly as mass increases, after the particle of m/e 43 (corresponding to $C_3H_7^+$). The modes of fragmentation responsible for the spectrum of *n*-dodecane are indicated in Figure 17.10.

When there is a branch point in the chain, an unusually large amount of fragmentation will occur there because a more stable carbocation results. Thus, in 2-methylpentane, loss of C_3H_7 or CH_3 is much greater than loss of C_2H_5, since the former modes give secondary carbocations, whereas the latter gives a primary carbocation.

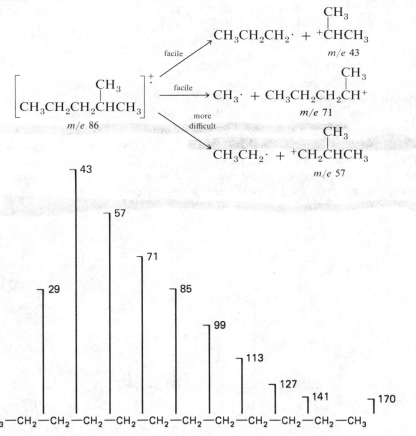

FIGURE 17.10. *Fragmentation of* n-*dodecane.*

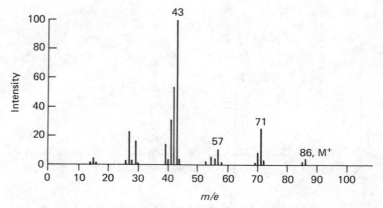

FIGURE 17.11. *Mass spectrum of 2-methylpentane.*

The spectrum of 2-methylpentane, plotted in Figure 17.11, illustrates this behavior.

On the other hand, the isomeric hydrocarbon 3-methylpentane can cleave in three ways so as to give a secondary carbocation. Two of these cleavages amount to loss of C_2H_5. Correspondingly, the M − 29 peak in its spectrum, shown in Figure 17.12, is the most intense peak.

$$
\left[\begin{array}{c} CH_3 \\ | \\ CH_3CH_2CHCH_2CH_3 \end{array} \right]^{+\cdot} \longrightarrow CH_3CH_2\cdot + \overset{CH_3}{\underset{}{^+CHCH_2CH_3}}
$$

m/e 86   *m/e 57*

Note that 3-methylpentane cannot undergo a simple cleavage to give a particle with m/e 43. The peak in its spectrum with this value must arise by a process involving some sort of skeletal rearrangement.

The mode of fragmentation in the preceding discussion is common in mass spectrometry. A radical cation undergoes bond cleavage in such a manner as to give the *most stable cationic fragment.* What we know about the relative stabilities of various cations from other areas of organic chemistry may often be used to predict how fragmentation will occur in a mass spectrometer. The case of the methylpentanes is a good example of this principle. In Chapter 8 we discussed the S_N1 reactions of alkyl halides to give carbocationic intermediates and found a

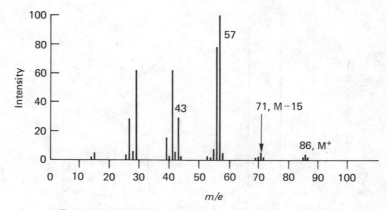

FIGURE 17.12. *Mass spectrum of 3-methylpentane.*

reactivity order tertiary > secondary > primary. From this order, and other data, we concluded that tertiary carbocations are more stable than secondary ones, which are, in turn, more stable than primary carbocations. Although these results are in solution and mass spectrometry occurs in the vapor phase, we can use our qualitative knowledge of carbocation stabilities to "interpret" the fragmentation pattern of hydrocarbons.

[Some of the enthalpy data for ionization of alkyl chlorides given in Table 8.12 on page 175 were actually obtained by mass spectrometric methods.]

In alkanes with a quaternary carbon fragmentation to give tertiary carbocations is so facile that such hydrocarbons frequently give no detectable parent peak.

EXERCISE: What are the principal fragments expected for 3,3-dimethylheptane?

B. Two-Bond Cleavage, Elimination of a Neutral Molecule

Some compounds give extremely weak molecular ion peaks. This tends to happen when some form of fragmentation is particularly easy. Such behavior is typical of alcohols, which often give no detectable molecular ion whatsoever. The spectrum of 2-methyl-2-butanol in Figure 17.13 illustrates this phenomenon.

The molecular ion, which would appear at m/e 88, is not observed. Instead, sizable peaks are observed at m/e values of 73 (M − 15) and 59 (M − 29), corresponding to cleavage of the radical ion so as to give stable oxonium ions.

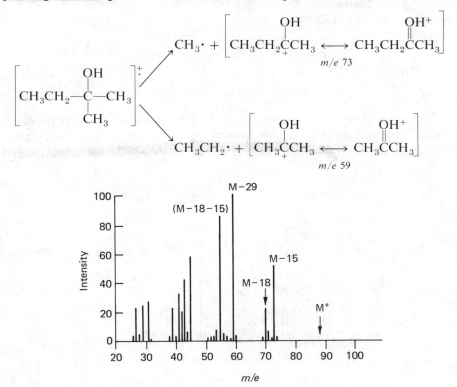

FIGURE 17.13. *Mass spectrum of 2-methyl-2-butanol.*

In addition, there is a substantial peak at m/e 70, corresponding to *loss of water* from the molecular ion. This type of fragmentation, in which a radical cation expels a neutral molecule, giving a new radical cation, is common with alcohols and ethers.

$$\left[\begin{array}{c} OH \\ | \\ CH_3CH_2-C-CH_3 \\ | \\ CH_3 \end{array} \right]^{+\cdot} \xrightarrow{-H_2O} \left[CH_3CH=C \begin{array}{c} CH_3 \\ \diagdown \\ \diagup \\ CH_3 \end{array} \right]^{+\cdot} \quad or \quad \left[\begin{array}{c} CH_3 \\ | \\ CH_3CH_2C=CH_2 \end{array} \right]^{+\cdot}$$

m/e 70 $\qquad\qquad$ m/e 70

Of course, these new molecular ions can undergo fragmentation of the type first discussed. The peak at m/e 55 probably arises from such a stepwise path.

$$\left[\begin{array}{c} OH \\ | \\ CH_3CH_2C-CH_3 \\ | \\ CH_3 \end{array} \right]^{+\cdot} \xrightarrow{-H_2O} \left[\begin{array}{c} CH_3 \\ | \\ CH_3CH_2C=CH_2 \end{array} \right]^{+\cdot} \xrightarrow{-CH_3\cdot} \left[\begin{array}{c} CH_3 \\ | \\ {}^+CH_2-C=CH_2 \end{array} \longleftrightarrow \begin{array}{c} CH_3 \\ | \\ CH_2=C-CH_2{}^+ \end{array} \right]$$

m/e 70 $\qquad\qquad\qquad\qquad$ m/e 55

[The m/e 55 fragment is a substituted allyl cation, a conjugated carbocation to be discussed in Chapter 21.]

Another example of such behavior is 3-methyl-1-butanol, whose mass spectrum is plotted in Figure 17.14. This is a particularly dramatic example. Again the molecular ion is nonexistent. The exact mechanism of the fragmentation in this case is unclear. There is evidence that one of the hydrogens that is lost with the hydroxy group comes from the methyl group. The nature of the C_5H_{10} radical cation produced is not known.

$$\left[\begin{array}{c} CH_3 \\ | \\ CH_3CHCH_2CH_2OH \end{array} \right]^{+\cdot} \longrightarrow \begin{array}{c} CH_3 \\ | \\ \dot{C}H_2CHCH_2CH_2\overset{+}{O}\diagup{}^H{}_{\diagdown H} \end{array} \longrightarrow \left[C_4H_{10} \right]^{+\cdot} + H_2O$$

m/e 70

There is one other type of fragmentation, also involving expulsion of a neutral molecule, that we shall introduce at this point. The spectrum of butyraldehyde is plotted in Figure 17.15. The most striking thing about the spectrum is the fact that

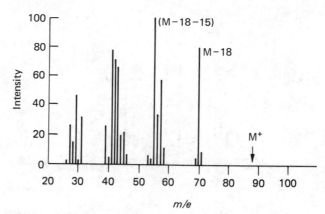

FIGURE 17.14. *Mass spectrum of 3-methyl-1-butanol.*

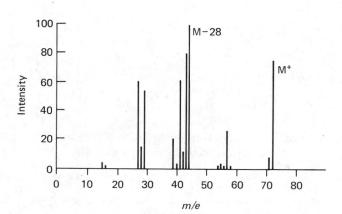

FIGURE 17.15. *Mass spectrum of butyraldehyde.*

the base peak (m/e 44) is an even number. Thus it must correspond to expulsion of a molecule, rather than a radical, from the molecular ion. Extensive studies suggest that this fragment arises in the following way.

$$
\left[\begin{array}{c} H \\ CH_2 \\ CH_2 \qquad O \\ \qquad CH \\ CH_2 \end{array}\right]^{\ddot{+}} \longrightarrow \begin{array}{c} CH_2 \\ \| \\ CH_2 \end{array} + \left[\begin{array}{c} H \\ O \\ CH \\ CH_2 \end{array}\right]^{\ddot{+}}
$$

$$m/e\ 72 \qquad\qquad m/e\ 44$$

There is some evidence which suggests that this fragmentation may involve two distinct steps, transfer of a hydrogen atom to the carbonyl oxygen from the γ-carbon followed by scission of the α, β-bond.

$$
\left[\begin{array}{c} H \\ CH_2 \quad O \\ CH_2 \quad CH \\ CH_2 \end{array}\right]^{\ddot{+}} \longrightarrow \begin{array}{c} \cdot \\ CH_2 \quad O^+ \\ CH_2 \quad CH \\ CH_2 \end{array} \longrightarrow \begin{array}{c} CH_2 \\ \| \\ CH_2 \end{array} + \left[\begin{array}{c} HO \\ CH \\ CH_2 \end{array}\right]^{\ddot{+}}
$$

This rearrangement reaction is called a **McLafferty rearrangement.** It can provide useful information concerning the structure of isomeric aldehydes and ketones. For example, 2-methylbutanal and 3-methylbutanal both undergo the rearrangement. In the former case one observes an intense peak at m/e 58, but in the latter the rearrangement peak occurs at m/e 44.

$$
\left[\begin{array}{c} H \\ CH_2 \\ CH_2 \qquad O \\ \qquad CH \\ CH \\ CH_3 \end{array}\right]^{\ddot{+}} \longrightarrow \begin{array}{c} CH_2 \\ \| \\ CH_2 \end{array} + \left[\begin{array}{c} H \\ O \\ CH \\ CH \\ CH_3 \end{array}\right]^{\ddot{+}}
$$

$$m/e\ 86 \qquad\qquad m/e\ 58$$

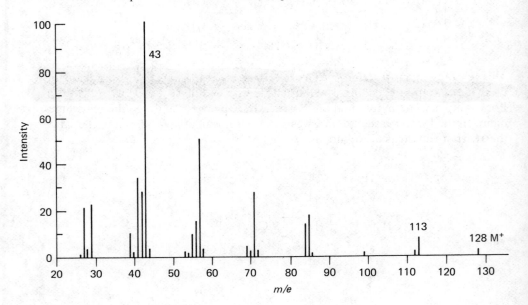

An additional fragmentation common to ketones is cleavage of a bond to the carbonyl group to give a cation of the oxonium ion type.

$$\left[\begin{array}{c} O \\ \parallel \\ R-C-R' \end{array} \right]^{+} \quad \xrightarrow{-R'} \quad \left[R-C\equiv\ddot{O}^{+} \longleftrightarrow R-\overset{+}{C}=\ddot{O} \right]$$

> **EXERCISE:** Write equations showing the four principal fragmentation products expected in the mass spectrum of 2-methyl-4-heptanone. There are two different McLafferty rearrangement ions and two different α-cleavages leading to oxonium ions.

PROBLEMS

1. (a) Estimate the relative intensity of the peaks at m/e 112, 114, and 116 in the mass spectrum of 1,2-dichloropropane.

    (b) A compound shows a molecular ion at m/e 138 with a ratio of $(M + 1)/M$ of 0.111. Show how this piece of information can be used to distinguish among the three formulas $C_{10}H_{18}$, $C_8H_{10}O_2$, and $C_8H_{14}N_2$.

2. The principal fragmentation mode of amines (RCH_2NH_2) is cleavage of the C_1-C_2 bond. For example, the most abundant fragment in the mass spectrum of tripropylamine, $(CH_3CH_2CH_2)_3N$, is at m/e 114. Suggest a reason that accounts for this favorable fragmentation pathway.

3. An unknown compound contains only carbon and hydrogen. Its mass spectrum is shown. Propose a structure for the compound.

4. The following mass spectra are of 2,2-dimethylpentane, 2,3-dimethylpentane, and 2,4-dimethylpentane. Assign structures on the basis of the mass spectra.

(a)

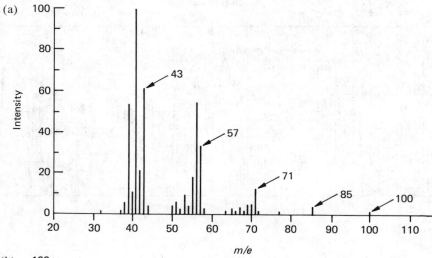

(b)

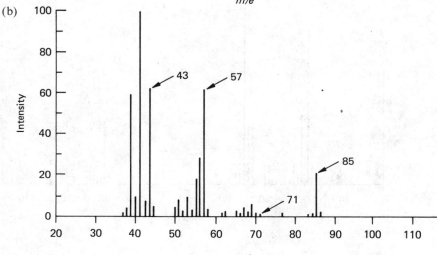

(c)

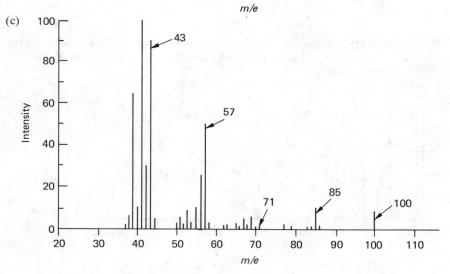

5. Identify the following compound from its ir and mass spectra.

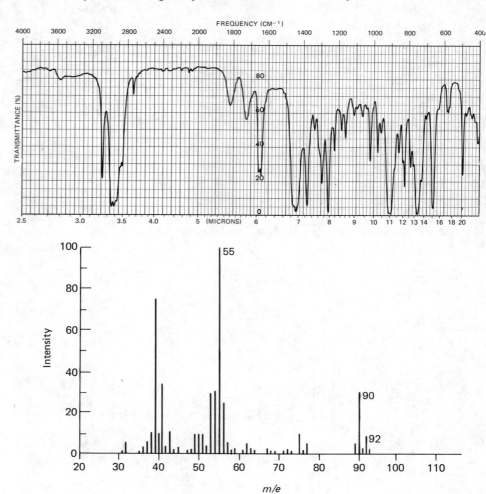

6. Two hydrocarbons were isolated from blue-green algae. The mass spectra of the two hydrocarbons, shown below, provided a clue to their structures. Suggest structures for the two hydrocarbons.

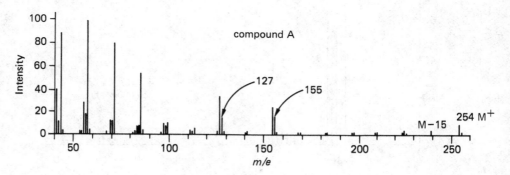

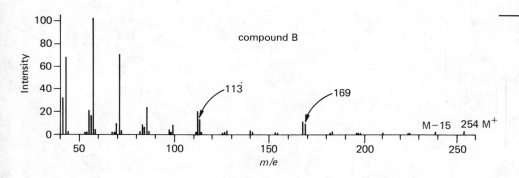

compound B

7. A compound has infrared absorption at 1710 cm^{-1}. Its mass spectrum is shown below. Suggest a structure of the compound.

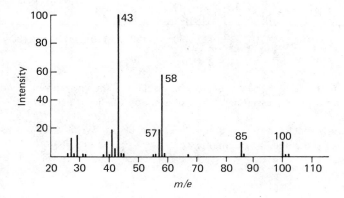

8. Propose a structure for the following compound from its mass spectrum. The ir spectrum shows a strong absorption at 1710 cm^{-1}.

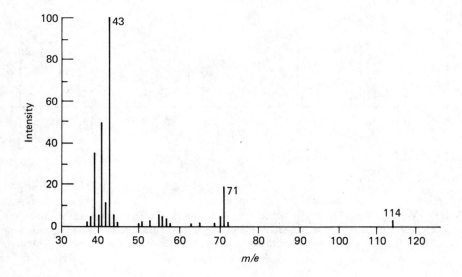

9. An unknown compound shows strong ir absorption at 3400 cm^{-1}. Its mass spectrum is shown below.

(a) Propose a possible structure for the compound.

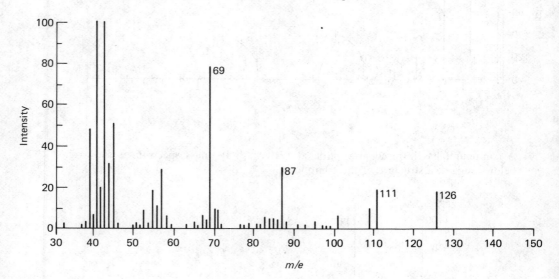

(b) The ir and nmr spectra of the compound are shown in problem 4d, Chapter 14. Confirm your assignment by examination of these spectra.

10. The mass spectrum of 2-octanone is shown below. Write mechanisms showing the origin of the principal fragments.

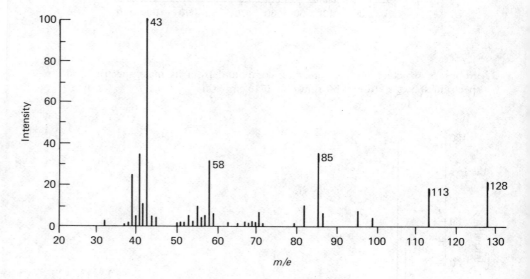

11. The ionization potential of 2-methylbutane is 10.35 eV or 238.7 kcal mole^{-1} (1 eV $\equiv$ 23.06 kcal mole^{-1}); hence, ΔH_f° for 2-methylbutane cation is obtained from ΔH_f° of 2-methylbutane as $-36.9 + 238.7 = 201.8$ kcal mole^{-1}. From the following ΔH_f° values given for possible fragmentation products of the radical cation, calculate ΔH° for each of the fragmentation reactions shown.

	ΔH_f°, kcal mole^{-1}		ΔH_f°, kcal mole^{-1}
CH_3^+	260	$CH_3\cdot$	34
$C_2H_5^+$	219	$C_2H_5\cdot$	26
$(CH_3)_2CHCH_2^+$	205	$(CH_3)_2CHCH_2\cdot$	13.5
$(CH_3)_2CH^+$	190	$(CH_3)_2CH\cdot$	17.5
$CH_3CH_2\overset{+}{C}HCH_3$	192	$CH_3CH_2\overset{\cdot}{C}HCH_3$	12

$$[(CH_3)_2CHCH_2CH_3]^{\dot{+}} \quad
\begin{aligned}
&\longrightarrow (CH_3)_2CHCH_2\cdot \; + CH_3^+ \\
&\longrightarrow (CH_3)_2CHCH_2^+ + CH_3\cdot \\
&\longrightarrow (CH_3)_2CH\cdot \; + C_2H_5^+ \\
&\longrightarrow (CH_3)_2CH^+ + C_2H_5\cdot \\
&\longrightarrow CH_3\cdot \; + CH_3\overset{+}{C}HCH_2CH_3 \\
&\longrightarrow CH_3^+ + CH_3CHCH_2CH_3
\end{aligned}$$

On the basis of these results, what will be the relative intensity order of the fragment cations?

Chapter 18
Carboxylic Acids

18.1
Structure

Carboxylic acids are distinguished by the functional grouping COOH. Four ways of writing this grouping, referred to as the **carboxy group,** are shown.

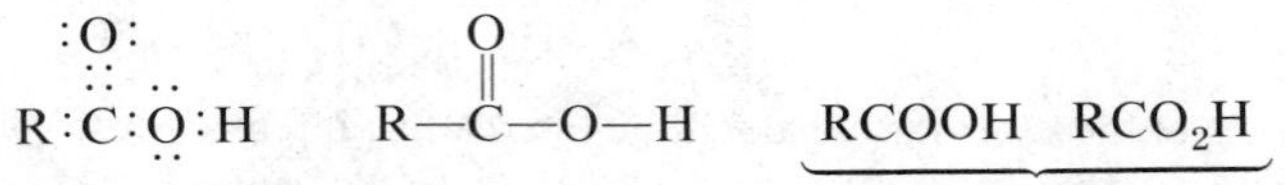

$$R\!:\!\overset{:O:}{\underset{..}{C}}\!:\!\overset{..}{\underset{..}{O}}\!:\!H \qquad R\!-\!\overset{\overset{O}{\|}}{C}\!-\!O\!-\!H \qquad RCOOH \quad RCO_2H$$

Lewis structure	Kekulé structure	condensed structures

Either an organic group or a hydrogen may be attached to the carboxy group.

The carbon atom in a carboxy group uses three hybrid orbitals to bond to the oxygen of the OH group, to the carboxy oxygen, and to hydrogen or an organic radical. These three orbitals are approximately sp^2-hybrids that lie in one plane. The remaining p-orbital on the carbon forms a π-bond to a p-orbital on the carboxy oxygen. There are two distinct C–O bond distances, corresponding to C=O and C—O. The bond angles and bond lengths of formic acid, as determined by microwave spectroscopy, are shown in Figure 18.1. Note that the bond angles around the carboxy carbon are only approximately those expected for sp^2-hybridization. The array HCOO is planar, and the hydroxy hydrogen lies outside of this plane.

In the solid and liquid phases, as well as in the vapor phase at moderately high pressure, carboxylic acids exist largely in the dimeric form depicted.

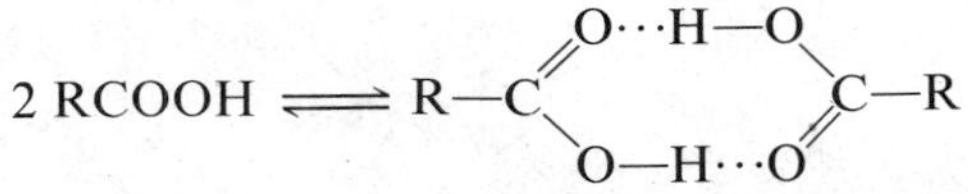

$$2\,RCOOH \;\rightleftharpoons\; R\!-\!C\!\!\underset{O-H\cdots O}{\overset{O\cdots H-O}{<\qquad>}}\!\!C\!-\!R$$

For formic acid in the vapor phase, $\Delta H°$ for the dimerization has been determined to be -14 kcal mole^{-1}. The factor that stabilizes the dimeric form is undoubtedly the reciprocal hydrogen bonding shown in the diagram.

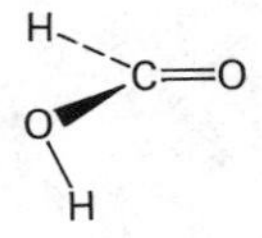

Bond Lengths, Å		Bond Angles, deg	
C=O	1.202	H—C=O	124.1
C—O	1.343	O—C=O	124.9
C—H	1.097	H—C—O	111.0
O—H	0.972	H—O—C	106.3

FIGURE 18.1. *Structure of formic acid.*

18.2
Nomenclature

Two systems of nomenclature are currently in use for carboxylic acids, and the student should be acquainted with both. Since many of the simpler acids are naturally occurring and were discovered early in the history of organic chemistry, they have well-entrenched "common" names. At the 1892 IUPAC Congress it was agreed to derive the name of a carboxylic acid systematically from that of the normal alkane having the same number of carbon atoms by dropping the ending **-e** and adding the suffix **-oic acid.** The common and IUPAC names for the first ten straight-chain acids, as well as other selected examples, are given in Table 18.1. The name used by *Chemical Abstracts* in indexing is printed in bold type.

Past caproic acid, the even-numbered carboxylic acids are the most important because it is only the even-numbered acids that occur in nature. Carboxylic acids are **biosynthesized** (built up by living organisms) by the combination of acetic acid units. Since acetic acid is a two-carbon building block, most of the naturally occurring acids have an even number of carbon atoms in the chain.

For naming a substituted carboxylic acid in the IUPAC system, the longest carbon chain *containing the carboxy group* is numbered from 1 to *n*, beginning with the carboxy carbon. The name of this parent straight-chain carboxylic acid is then prefixed by the names of the various substituents.

$$\underset{7\quad6\quad5\quad4\quad3\quad2\quad1}{CH_3CHCH_2CH_2CH_2CHCOOH}$$

2-hydroxy-6-methylheptanoic acid

$$\underset{6\quad5\quad4\quad3\quad2\quad1}{CH_3(CH_2)_{11}CHCH_2CH_2CH_2CHCOOH}$$

6-methyl-2-propyloctadecanoic acid

TABLE 18.1
Nomenclature of Carboxylic Acids

Compound	Common Name	IUPAC Name
HCOOH	**formic acid**	methanoic acid
CH_3COOH	**acetic acid**	ethanoic acid
CH_3CH_2COOH	propionic acid	**propanoic acid**
$CH_3(CH_2)_2COOH$	butyric acid	**butanoic acid**
$CH_3(CH_2)_3COOH$	valeric acid	**pentanoic acid**
$CH_3(CH_2)_4COOH$	caproic acid	**hexanoic acid**
$CH_3(CH_2)_5COOH$	enanthic acid	**heptanoic acid**
$CH_3(CH_2)_6COOH$	caprylic acid	**octanoic acid**
$CH_3(CH_2)_7COOH$	pelargonic acid	**nonanoic acid**
$CH_3(CH_2)_8COOH$	capric acid	**decanoic acid**
$CH_3(CH_2)_{10}COOH$	lauric acid	**dodecanoic acid**
$CH_3(CH_2)_{12}COOH$	myristic acid	**tetradecanoic acid**
$CH_3(CH_2)_{14}COOH$	palmitic acid	**hexadecanoic acid**
$CH_3(CH_2)_{16}COOH$	stearic acid	**octadecanoic acid**

When using common names, the chain is labeled α, β, γ, δ, and so on, beginning with the carbon adjacent to the carboxy carbon (see table of Greek letters inside the front cover).

As in the case of aldehydes and ketones, it is desirable not to mix IUPAC and common nomenclature (page 360).

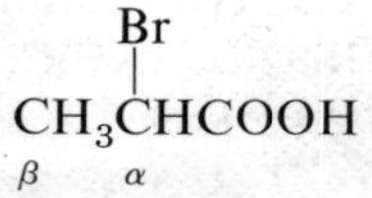

α-bromopropionic acid
2-bromopropanoic acid
(not 2-bromopropionic acid)

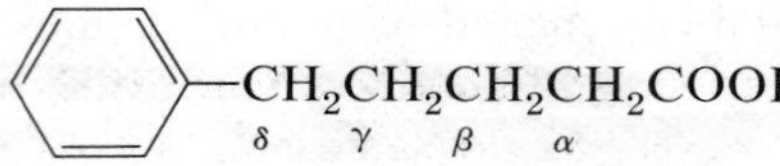

δ-phenylvaleric acid
5-phenylpentanoic acid
(not δ-phenylpentanoic acid)

It is not always possible or convenient to name a carboxylic acid in the foregoing way. This is the case with cyclic acids.

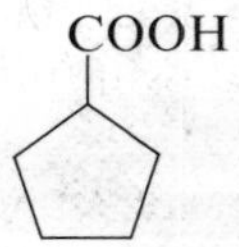

cyclopentanecarboxylic acid 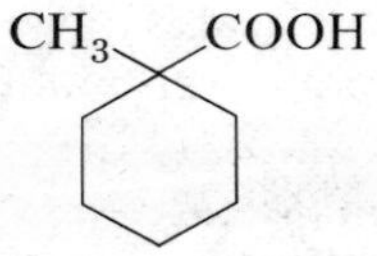   1-methylcyclohexanecarboxylic acid

In rare cases it may be necessary to name a compound containing a carboxy group as a derivative of some other function. Then the COOH group is designated "carboxy." One such example is

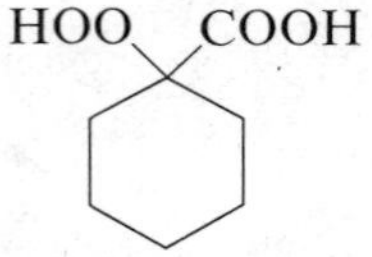

1-carboxycyclohexyl hydroperoxide

EXERCISE: What is the IUPAC name for each of the following carboxylic acids?

(a) $ClCH_2CH_2CH_2CH_2COOH$

(b) $(CH_3)_3CCH_2CH_2CH_2COOH$

(c) $CH_3CH_2\overset{\overset{\displaystyle CH_3}{|}}{C}HCH_2\overset{\overset{\displaystyle I}{|}}{C}HCOOH$

(d) $CH_2{=}CHCH{=}CHCOOH$

18.3
Physical Properties

Table 18.2 lists the melting point, boiling point, and water solubility of the first 18 straight-chain carboxylic acids. The boiling points of the straight-chain acids are plotted against molecular weight in Figure 19.5 on page 531. The boiling points of carboxylic acids are higher than expected for their molecular weights because of hydrogen bonding. The lower molecular weight acids are liquids at room temperature. The first four acids are fully miscible with water in all proportions. As the chain length is increased, the water solubility steadily decreases.

TABLE 18.2
Physical Properties of Carboxylic Acids

Acid	Melting Point, °C	Boiling Point, °C (760 mm)	Solubility in H_2O g/100 ml, 20°C
formic	8.4	101	∞
acetic	16.6	118	∞
propanoic	−21	141	∞
butanoic	−5	164	∞
pentanoic	−34	186	4.97
hexanoic	−3	205	0.968
heptanoic	−8	223	0.244
octanoic	17	239	0.068
nonanoic	15	255	0.026
decanoic	32	270	0.015
undecanoic	29	280	0.0093
dodecanoic	44	299	0.0055
tridecanoic	42	312	0.0033
tetradecanoic	54	251 (100 mm)	0.0020
pentadecanoic	53	257 (100 mm)	0.0015
hexadecanoic	63	267 (100 mm)	0.00072
heptadecanoic	63	—	0.00042
octadecanoic	72	—	0.00029

18.4
Acidity

A. *Ionization*

We saw in Section 4.5 that compounds containing the functional group

$-C\overset{\displaystyle O}{\underset{\displaystyle OH}{\diagup}}$ are weakly acidic; in fact, it is this property from which the class

derives its name. When acetic acid is dissolved in water, the equilibrium shown in equation (18-1) exists.

$$CH_3-\overset{\displaystyle O}{\overset{\|}{C}}-OH + H_2O \rightleftharpoons CH_3-\overset{\displaystyle O}{\overset{\|}{C}}-O^- + H_3O^+ \qquad (18\text{-}1)$$

The equilibrium constant for this reaction, denoted as K_a or the "acid dissociation tion constant" has the magnitude

$$K_a = \frac{[CH_3CO_2^-][H^+]}{[CH_3COOH]} = 1.8 \times 10^{-5} \ M \qquad (18\text{-}2)$$

Remember that the concentration of H_2O, which remains essentially invariant for dilute solutions (at 55.5 M), is not carried in the denominator of the expression for K_a. More correctly, equation (18-2) is an expression for the equilibrium

$$CH_3COOH \rightleftharpoons CH_3CO_2^- + H^+$$

The exact equilibrium expression for equation (18-1) is

$$K = \frac{[CH_3CO_2^-][H_3O^+]}{[CH_3COOH][H_2O]} = 3.25 \times 10^{-7}$$

It follows that $K_a = [H_2O] \times K = 1.8 \times 10^{-5}\ M$.

The corresponding $pK_a = -\log (1.8 \times 10^{-5}) = 5 - \log (1.8) = 4.74$.

Dissociation constants of this magnitude put the carboxylic acids in the class of relatively weak acids. For example, a 0.1 M aqueous solution of acetic acid is only 1.3% dissociated into ions. Strong acids, such as HCl and H_2SO_4, are completely dissociated in dilute aqueous solution. Nevertheless, the carboxylic acids are distinctly acidic—their aqueous solutions have the characteristic sour taste of hydronium ion. Although the carboxylic acids are weak acids compared with mineral acids, they are much stronger than alcohols. Recall that ethanol has a dissociation constant of about 10^{-16}; ethanol is only 10^{-11} as strong an acid as acetic acid.

The question immediately arises, "Why is acetic acid more acidic than ethanol?" The answer lies mostly in the relative stability of the negative charge of the anion. In ethoxide ion the negative charge is concentrated on a single oxygen atom; ethoxide ion is basic because this concentrated charge provides strong attraction for a proton. In acetate ion, however, the charge on the carboxy group is divided between two oxygens. Acetate ion is not adequately represented by a single Lewis structure. A second and equivalent structure may be written that differs only in the position of electrons.

$$\left[CH_3-C\overset{\displaystyle \ddot{O}:}{\underset{\displaystyle \ddot{O}:^-}{}} \longleftrightarrow CH_3-C\overset{\displaystyle \ddot{O}:^-}{\underset{\displaystyle \ddot{O}:}{}} \right]$$

Acetate ion is described as a resonance hybrid of these two principal structures (Section 2.4). The hybrid structure may also be written as

$$CH_3-C\overset{\displaystyle O^{\frac{1}{2}-}}{\underset{\displaystyle O^{\frac{1}{2}-}}{}}$$

This symbol emphasizes that only half of a negative charge resides on each oxygen. The attraction for a proton is therefore reduced.

Another way of describing this phenomenon is wholly in terms of energy. The energies required to remove a proton from ethanol and acetic acid in the dilute gas phase are

$$CH_3CH_2OH \rightleftharpoons CH_3CH_2O^- + H^+ \qquad \Delta H^\circ = 379\ \text{kcal mole}^{-1}$$
$$CH_3COOH \rightleftharpoons CH_3CO_2^- + H^+ \qquad \Delta H^\circ = 346\ \text{kcal mole}^{-1}$$

It takes a large amount of energy to separate charges because of their Coulombic attraction. The energy required for dissociation of acetic acid is substantially less than that for ethanol because in acetate ion the negative charge is attracted by two oxygen nuclei and the reduced charge density has less internal Coulombic repulsion. This energy effect in acetate ion is sometimes referred to as "resonance" stabilization or as a resonance energy (Chapter 20).

An alternative description can be given in terms of the overlap of atomic orbitals. Each oxygen contributes a p-orbital that can overlap in π-bonding to a p-orbital of the carboxy carbon, as illustrated in Figure 18.2. The resulting three-

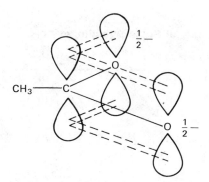

FIGURE 18.2. *π-orbital interactions in acetate ion.*

center π-molecular orbital system has four electrons with excess electron density put on the end oxygens. Such multicenter π-molecular orbital systems are characteristic of **conjugated systems** and are discussed in more detail in Chapter 20.

In either representation, the two C—O bonds in a carboxylate ion are equivalent. The C—O bond length is 1.26 Å, which is between the bond length of 1.20 Å for C=O and 1.34 Å for C—O in formic acid (see Figure 18.1).

> **EXERCISE:** Formic acid has a dissociation constant, $K_a = 1.77 \times 10^{-4}$ M. What is its pK_a? Calculate the approximate concentration of formate ions in a solution nominally 0.1 M in formic acid.

B. *Inductive Effects*

In Section 10.4 we found that electronegative groups whose bonds to carbon are highly polar have important effects on the acidity of alcohols. We saw that this effect could be interpreted in terms of the electrostatic interaction of a bond dipole with the negative charge on oxygen in the alkoxide ion. Substituent groups also affect the acidity of carboxylic acids. Because carboxylic acids are more acidic than alcohols, it is easier to determine their dissociation constants. Consequently, a wealth of such quantitative acidity data is available. Some of these results are summarized in Table 18.3.

Atoms that have high electronegativity tend to withdraw electron density from carbon and have a marked acid-strengthening effect. Chloroacetic acid is 1.9 pK_a units more acidic than acetic acid. The C—Cl bond dipole is oriented in such a way that the positive end is closer to the negative charge on the carboxy group than is the negative end. Electrostatic attraction exceeds the repulsion and the negative charge of the anion is therefore stabilized by the chlorine.

$$Cl \overset{\nwarrow}{\underset{}{}} CH_2—CO_2^-$$

We used the same explanation to interpret the effect of a chlorine substituent on the acidity of ethanol (page 239).

$$Cl \overset{\nwarrow}{\underset{}{}} CH_2—CH_2O^-$$

The acid-strengthening effect, in this case of 1.6 pK_a units, is quite similar to that

TABLE 18.3
Acidity of Some Substituted Acetic Acids

Acid	K_a, M	pK_a
CH_3COOH	1.8×10^{-5}	4.74
FCH_2COOH	2.6×10^{-3}	2.59
F_3CCOOH	0.59	0.23
$ClCH_2COOH$	1.4×10^{-3}	2.86
$Cl_2CHCOOH$	5.5×10^{-2}	1.26
Cl_3CCOOH	0.23	0.64
$BrCH_2COOH$	1.3×10^{-3}	2.90
ICH_2COOH	6.7×10^{-4}	3.18
$HOCH_2COOH$	1.5×10^{-4}	3.83
CH_3OCH_2COOH	2.9×10^{-4}	3.54
$CH_2{=}CHCH_2COOH$	4.5×10^{-5}	4.35
$HC{\equiv}CCH_2COOH$	4.8×10^{-4}	3.32
CH_3CH_2COOH	1.3×10^{-5}	4.87
$NCCH_2COOH$	3.4×10^{-3}	2.46
$C_6H_5CH_2COOH$	4.9×10^{-5}	4.31

in chloroacetic acid. In both anions the negative charge is two atoms away from
the C—Cl bond.

Carbon-carbon double and triple bonds have a significant electron-attracting
effect that is reflected in the enhanced acidity of vinylacetic and ethynylacetic
acids. An sp^2-hybridized carbon orbital with its greater s-character is effectively
more electronegative than an sp^3-orbital. Recall that alkenes and alkynes have
small but significant dipole moments (Sections 11.3.A and 12.3.A).

The higher alkanoic acids are somewhat less acidic than acetic acid. Alkyl
groups show a small but significant electron-donating inductive effect in appropriate systems in solution.

*The inductive effect of remote substituents falls off dramatically with increased
distance from the charged center.* This effect is expected because electrostatic interactions between charges are inversely proportional to the distance between them.
This effect is exemplified by the chlorobutanoic acids whose acidity constants are
shown in Table 18.4. Beyond a few methylene groups, the effect becomes negligible.

EXERCISE: Using graph paper, plot the pK_as of chloroacetic acid and trichloroacetic acid versus the pK_as of the corresponding fluoroacetic acids. Note that the pK_a
of acetic acid falls on the line. What is the predicted pK_a of difluoroacetic acid?

C. Salt Formation

There is one more aspect of acidity of carboxylic acids that we should consider.
In Section 18.3 we saw that carboxylic acids of more than five carbons are essentially insoluble in water. Beyond this point, the polar portion of the molecule

TABLE 18.4
Acidity of Butanoic Acids

Acid	K_a, M	pK_a
$\overset{\text{Cl}}{\underset{|}{\text{CH}_3\text{CH}_2\text{CHCOOH}}}$	139×10^{-5}	2.86
$\overset{\text{Cl}}{\underset{|}{\text{CH}_3\text{CHCH}_2\text{COOH}}}$	8.9×10^{-5}	4.05
$\overset{\text{Cl}}{\underset{|}{\text{CH}_2\text{CH}_2\text{CH}_2\text{COOH}}}$	3.0×10^{-5}	4.52
$\text{CH}_3\text{CH}_2\text{CH}_2\text{COOH}$	1.5×10^{-5}	4.82

(COOH) becomes less important than the nonpolar hydrocarbon tail ($-$R). Now consider the reaction of a carboxylic acid such as dodecanoic acid with a strong base like hydroxide ion.

$$CH_3(CH_2)_{10}COOH + OH^- \overset{K}{\rightleftharpoons} H_2O + CH_3(CH_2)_{10}CO_2^- \qquad (18\text{-}3)$$

The equilibrium constant for reaction (18-3) may be derived as follows.

$$K_a = \frac{[CH_3(CH_2)_{10}CO_2^-][H^+]}{[CH_3(CH_2)_{10}COOH]} = 1.3 \times 10^{-5} \, M \qquad (18\text{-}4)$$

$$K_w = [H^+][OH^-] = 10^{-14} \, M^2 \qquad (18\text{-}5)$$

Rearranging (18-5), we have

$$[H^+] = \frac{10^{-14}}{[OH^-]} \, M \qquad (18\text{-}6)$$

Substituting (18-6) into (18-4) and expanding, we have

$$K = \frac{[CH_3(CH_2)_{10}CO_2^-]}{[CH_3(CH_2)_{10}COOH][OH^-]} = 1.3 \times 10^9 \, M^{-1} \qquad (18\text{-}7)$$

Equation (18-7) is merely the equilibrium expression for reaction (18-3). The large value of K shows that the reaction proceeds to completion. Thus dodecanoic acid is converted by aqueous sodium hydroxide completely into the salt, sodium dodecanoate. Note that the anions of carboxylic acids are named by dropping **-ic** from the name of the parent acid and adding the suffix **-ate**. Although dodecanoic acid is a neutral molecule, sodium dodecanoate is a salt. Dissolution of this salt gives an anion and a cation, which can be solvated by water. It is not surprising that the solubility of sodium dodecanoate (1.2 g per 100 mL) is much greater than that of dodecanoic acid itself (0.0055 g per 100 mL).

D. *Soaps*

The sodium and potassium salts of long-chain carboxylic acids ("fatty acids") are obtained by the reaction of natural fats with sodium or potassium hydroxide. These salts, referred to as soaps, have the interesting and useful ability to solubilize nonpolar organic substances. This phenomenon can easily be understood if one considers the structure of such a salt.

$$CH_3CH_2CH_2CH_2CH_2CH_2CH_2CH_2CH_2CH_2CH_2CH_2CH_2CH_2CH_2CO_2^- \, K^+$$

The molecule has a polar ionic region and a large nonpolar hydrocarbon region. In aqueous solution a number of carboxylate ions tend to cluster together so that the hydrocarbon tails are close to each other, thus reducing their energy by the attractive van der Waals forces enjoyed by normal hydrocarbons. The surface of the sphere-like cluster is then occupied by the highly polar —CO_2^- groups. These polar groups face the medium, where they may be solvated by H_2O or paired with a cation. The resulting spherical structure, called a **micelle,** is depicted in cross section in Figure 18.3. The wavy lines in the figure represent the long hydrocarbon chains of the salt molecules.

Organic material not normally soluble in water (such as butter or motor oil) may "dissolve" in the hydrocarbon interior of a micelle. The overall process of soap solubilization is diagrammed schematically in Figure 18.4.

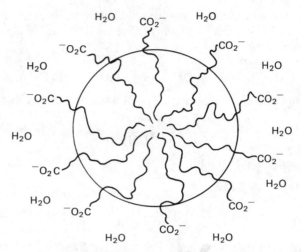

FIGURE 18.3. *Cross section of a micelle.*

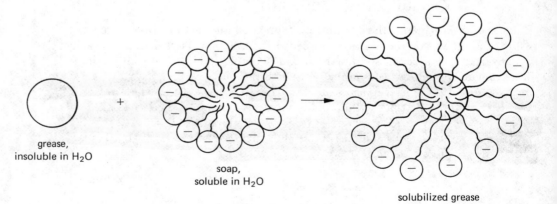

grease, insoluble in H_2O

soap, soluble in H_2O

solubilized grease

FIGURE 18.4. *Schematic diagram of soap solubilization.*

There are natural bacteria that can metabolize soaps. This degradation is most rapid when there are no branches in the hydrocarbon chain of the soap molecule. Since the naturally occurring fatty acids are all unbranched compounds, soaps derived from natural fats are said to be **biodegradable.** Before 1933 all cleaning materials were soaps. In that year the first synthetic detergents were marketed. Detergents have the useful property of not forming the hard "scum" that often results from the use of a soap with hard water. This scum is actually the insoluble magnesium and calcium salts of the fatty acid. The first detergents were compounds called **alkylbenzene-sulfonates.** Like soaps, they had a large nonpolar hydrocarbon tail and a polar end.

$$R \overset{}{\underset{}{\bigcirc}}\!\!-SO_3^- \; K^+$$

R = branched alkyl chain

However, being branched compounds, these early detergents were not rapidly biodegradable. Since the materials could not be completely metabolized by the bacteria that operate in sewage treatment plants, they were passed into natural waterways with the treated sewage. They often reappeared as foam or suds on the surface of lakes and rivers. After an intensive research project, the detergent industry in 1965 introduced **linear alkanesulfonate detergents.**

$$CH_3(CH_2)_n CH_2 SO_3^- \; K^+$$

Since the new detergents are straight-chain compounds, they can be metabolized rapidly by the natural bacteria.

18.5
Spectroscopy

A. *Nuclear Magnetic Resonance*

The resonance positions for various types of hydrogens in carboxylic acids are summarized in Table 18.5. Hydrogens attached to the number 2 carbon of a carboxylic acid resonate at roughly the same place as do the analogous hydrogens in aldehydes and ketones. The very low-field resonance of the carboxy proton is associated with the dimeric hydrogen-bonded structure discussed in Section 18.1. The spectrum of 2-methylpropanoic acid is shown in Figure 18.5.

TABLE 18.5
Chemical Shifts of Carboxylic Acid Hydrogens

Type of Hydrogen	Chemical Shift, δ, ppm
CH_3COOH	2.0
RCH_2COOH	2.36
$R_2CHCOOH$	2.52
$RCOOH$	about 10–13

B. *Infrared*

One of the characteristic absorptions of acids is the $C{=}O$ stretch, which occurs in the region 1710–1760 cm^{-1}. The exact position and appearance of this absorption depends on the physical state in which the measurement is made. In pure liquids or in the solid state the $C{=}O$ stretch occurs as a broad band at about 1710 cm^{-1}. In dilute solution (CCl_4, $CHCl_3$) the absorption band is narrower and

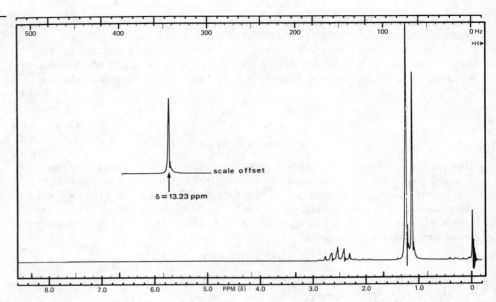

FIGURE 18.5. *Nmr spectrum of 2-methylpropanoic acid,* $(CH_3)_2CHCOOH$.

occurs at about 1760 cm^{-1}. The OH stretch occurs as a series of weak bands from 2500 to 3000 cm^{-1}. The spectrum of hexanoic acid in Figure 18.6 illustrates these bands.

EXERCISE: Review the characteristic spectral features of the following classes of compounds: alcohols, ethers, alkenes, alkynes, aldehydes, ketones, and carboxylic acids. Make a list showing the principal nmr chemical shifts and the characteristic ir absorptions for each of the following simple compounds: 1-propanol, di-*n*-propyl ether, 1-pentene, 1-pentyne, butanal, 3-pentanone, and propanoic acid. Is ir or nmr the more reliable method for ascertaining which functional group is present in a molecule? Explain why nmr is more reliable for determining *where* the functional group is in the molecule.

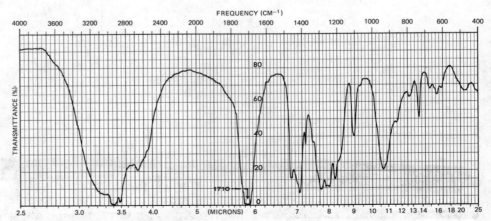

FIGURE 18.6. *Infrared spectrum of hexanoic acid.*

18.6
Synthesis

Carboxylic acids are most often prepared in the laboratory by

1. Hydrolysis of nitriles.
2. Carbonation of organometallic compounds.
3. Oxidation of primary alcohols or aldehydes.
4. Decarboxylation of substituted malonic acids.

In this section we will discuss the first three of these methods; decarboxylation of malonic acids will be discussed in Chapter 27.

A. *Hydrolysis of Nitriles*

Nitriles are compounds containing the functional group —C≡N (Section 20.6). They undergo hydrolysis to give 1 mole of carboxylic acid and 1 mole of ammonia.

$$R—C≡N + 2 H_2O \longrightarrow RCOOH + NH_3$$

In neutral solution, the reaction is immeasurably slow. However, the reaction is strongly catalyzed by either acid or base. Typical examples are the acidic hydrolysis of phenylacetonitrile and the basic hydrolysis of *n*-decyl cyanide.

phenylacetonitrile   phenylacetic acid

A mixture of 1150 mL water, 840 mL conc. H_2SO_4, and 700 g phenylacetonitrile is heated at reflux for 3 hr. Phenylacetic acid (630 g, 78% yield) is obtained when the reaction mixture is poured into cold water.

$$CH_3(CH_2)_9CN + 2 H_2O \xrightarrow[\substack{C_2H_5OH \\ reflux \\ 77\ hr}]{KOH} \xrightarrow{HCl} CH_3(CH_2)_9COOH + NH_3$$

n-decyl cyanide   undecanoic acid

A mixture of 27 g of *n*-decyl cyanide, 200 g of 20% ethanolic KOH, and 50 mL of water is refluxed for 77 hr, during which time ammonia is evolved. The solvent is evaporated, and the residue is treated with conc. HCl. After washing with water, 24 g of undecanoic acid (80%), m.p. 29°, is obtained.

As we saw in Chapter 8, nitriles are conveniently prepared from primary alkyl halides by treatment with cyanide ion. Carboxylic acids may therefore be prepared from alkyl halides by conversion to the nitrile, which is then hydrolyzed.

$$CH_3CH_2CH_2CH_2Br + NaCN \xrightarrow{CH_3SOCH_3} CH_3CH_2CH_2CH_2CN \xrightarrow[\substack{H_2O \\ ethylene \\ glycol}]{KOH}$$

n-butyl bromide   (92%)
                                               n-butyl cyanide

$$\xrightarrow{H^+} CH_3CH_2CH_2CH_2COOH$$

(90%)
pentanoic acid

The mechanism of the hydrolysis reaction will be discussed in Section 19.6.

B. *Carbonation of Organometallic Reagents*

Alkyl halides may also be converted into carboxylic acids by formation of an organometallic reagent, which is then allowed to react with carbon dioxide. The product is the salt of a carboxylic acid. Treatment of this salt with aqueous mineral acid liberates the free acid. Grignard reagents are commonly used (Section 15.5).

$$(CH_3)_3CCl \xrightarrow[\text{ether}]{Mg} (CH_3)_3CMgCl \xrightarrow{CO_2} (CH_3)_3CCOOMgCl \xrightarrow{H_3O^+} (CH_3)_3CCOOH$$

(63%)

2,2-dimethylpropanoic acid
pivalic acid

The Grignard method of converting an alkyl halide to the corresponding carboxylic acid must be used when the alkyl halide is tertiary, as in the foregoing example, or is otherwise unreactive in S_N2 reactions. With primary and unbranched secondary alkyl halides the nitrile displacement method discussed in the previous section may also be used.

C. *Oxidation of Primary Alcohols or Aldehydes*

The third generally useful method for preparing carboxylic acids involves oxidation of aldehydes (obtained in the aldol condensation; Section 13.7.F) or primary alcohols (obtained, for example, by hydroboration of terminal alkenes; Section 11.6.D). A useful oxidizing agent for this purpose is potassium permanganate.

$$CH_3(CH_2)_3\underset{\overset{|}{CH_2CH_3}}{C}HCH_2OH + KMnO_4 \xrightarrow[25°]{H_2O,\ NaOH} \xrightarrow{H_2SO_4} CH_3(CH_2)_3\underset{\overset{|}{CH_2CH_3}}{C}HCOOH$$

2-ethyl-1-hexanol 2-ethylhexanoic acid

$$CH_3(CH_2)_3\underset{\overset{|}{CH_2CH_3}}{C}HCHO + KMnO_4 \xrightarrow{\text{same as above}}$$

(78%)

2-ethylhexanal

The initial product in oxidation of a primary alcohol is the corresponding aldehyde. However, with aqueous permanganate the aldehyde undergoes subsequent oxidation *more rapidly* than the primary alcohol, so it is normally not observed in the reaction mixture.

Another reagent that may be used for the oxidation of either primary alcohols or aldehydes to carboxylic acids is nitric acid. Although this procedure works well on simple compounds, it is rather vigorous and cannot be used with compounds containing acid-sensitive functional groups elsewhere in the molecule.

$$HOCH_2CH_2CH_2CH_2CH_2CHO + HNO_3 \xrightarrow[0°]{H_2O} HOOCCH_2CH_2CH_2CH_2COOH$$

(75%)

6-hydroxyhexanal hexanedioic acid

$$ClCH_2CH_2CHO + \text{fuming } HNO_3 \xrightarrow{35°} ClCH_2CH_2COOH$$

(65%)

3-chloropropanal 3-chloropropanoic acid

An extremely mild and selective reagent for the oxidation of aldehydes to carboxylic acids is silver oxide suspended in aqueous base. Although the method usually affords the desired acid in excellent yield, it is expensive to carry out on a large scale owing to the cost of silver oxide, unless one reclaims and recycles the silver metal.

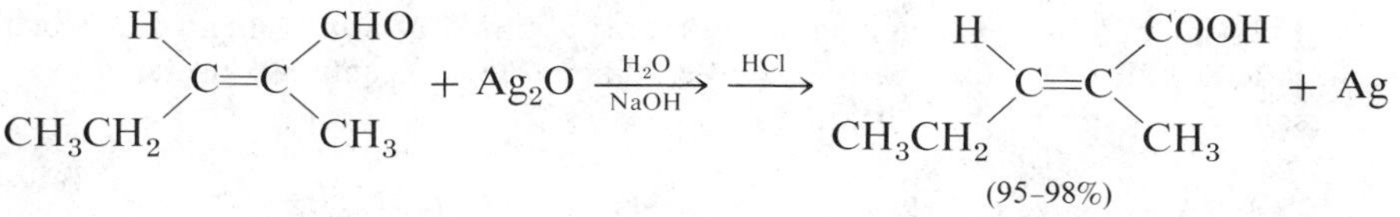

$$\underset{CH_3CH_2}{\overset{H}{>}} C = C \underset{CH_3}{\overset{CHO}{<}} + Ag_2O \xrightarrow[NaOH]{H_2O} \xrightarrow{HCl} \underset{CH_3CH_2}{\overset{H}{>}} C = C \underset{CH_3}{\overset{COOH}{<}} + Ag$$

(95–98%)

Note the advantageous use of Ag_2O in this case because the double bond is sensitive to stronger oxidizing agents.

> Silver oxide is a brown solid that has only slight solubility in water. It is usually prepared as needed by mixing a solution of silver nitrate with sodium hydroxide. The precipitate may be filtered, washed with water, and used as an aqueous suspension.

EXERCISE: How may each of the following conversions be accomplished most effectively?
(a) $(CH_3CH_2CH_2)_3CBr \longrightarrow (CH_3CH_2CH_2)_3CCOOH$
(b) $HOCH_2CH_2CH_2CH_2CH_2CH_2Br \longrightarrow HOCH_2CH_2CH_2CH_2CH_2CH_2COOH$
(c) $CH_3CH_2CH_2CH_2CH=CH_2 \longrightarrow CH_3CH_2CH_2CH_2CH_2COOH$

18.7
Reactions

The chemistry of carboxylic acids may be divided mechanistically into four categories.

1. Reactions involving the acidic O—H bond.
2. Reactions occurring in the hydrocarbon side chain.
3. Reactions occurring *at* the carboxy carbon atom.
4. One-carbon degradations.

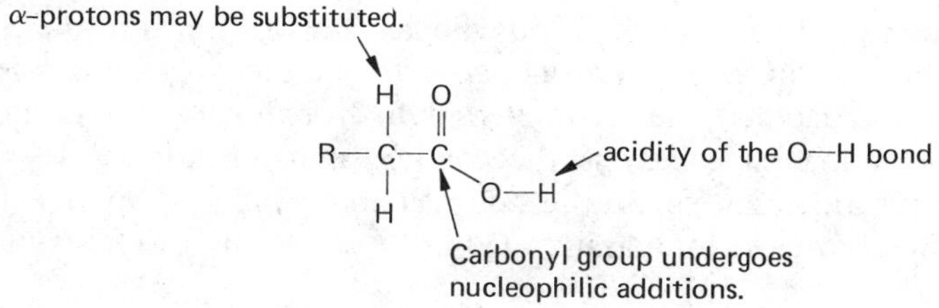

α–protons may be substituted.

$$R - \underset{\underset{H}{|}}{\overset{\overset{H}{|}}{C}} - \overset{\overset{O}{\|}}{C} \underset{O-H}{}$$

acidity of the O—H bond

Carbonyl group undergoes
nucleophilic additions.

A. Reactions Involving the OH Bond

We have already seen one important reaction of carboxylic acids involving the OH bond—the reaction with bases to give salts.

$$CH_3CH_2CH_2COOH + Na^+OH^- \longrightarrow CH_3CH_2CH_2CO_2^- Na^+ + H_2O$$

Another important reaction involving this bond is the reaction of carboxylic acids with diazomethane. The products of this reaction are the **methyl ester** and nitrogen.

$$RCOOH + CH_2N_2 \longrightarrow RCOOCH_3 + N_2$$

> Diazomethane is a yellow gas boiling at about 0°. It is highly toxic and, under certain conditions, explosive. Diazomethane is another example of a compound for which multiple Kekulé or Lewis structures can be written. The molecule is considered to be a resonance hybrid of the following forms.
>
> $$[:\overset{-}{C}H_2 - \overset{+}{N} \equiv N: \quad \longleftrightarrow \quad CH_2 = \overset{+}{N} = \overset{..}{\overset{..}{N}}:]$$

The reaction of diazomethane with carboxylic acids probably involves the following steps.

(1) $R - \overset{O}{\overset{||}{C}} - O - H + :\overset{-}{C}H_2 - \overset{+}{N} \equiv N: \longrightarrow R - \overset{O}{\overset{||}{C}} - \overset{..}{\overset{..}{O}}:^- + CH_3 - \overset{+}{N} \equiv N:$

(2) $R - \overset{O}{\overset{||}{C}} - \overset{..}{O}:^- + \overset{+}{C}H_3 - N \equiv N: \longrightarrow R - \overset{O}{\overset{||}{C}} - OCH_3 + :N \equiv N:$

The first step is a simple acid-base reaction; the moderately acidic carboxylic acid transfers a proton to the basic carbon atom of diazomethane. The pair of ions thus formed immediately reacts, probably by the S_N2 mechanism; carboxylate ion is the entering nucleophile and nitrogen is the leaving group.

Because of the toxicity and danger of explosion, diazomethane reactions are almost never carried out on a large scale. However, because of the convenience of the procedure (yields are usually quantitative and the only by-product is a gas), it is frequently used for the small-scale conversion of an acid into its methyl ester, especially when the acid is a relatively precious one.

(100%)

The sodium or silver salts of carboxylic acids also enter into the S_N2 reaction with alkyl halides, in a manner analogous to the second step of the preceding reaction with diazomethane. However, dehydrohalogenation is an important competing reaction, especially with secondary and tertiary halides (Chapters 8 and 11). In certain cases, where the side reaction is blocked, the reaction is sufficiently clean as to be of preparative value, as shown by the following example.

(93%)

EXERCISE: Write equations for the reactions of 3-methylpentanoic acid with (a) diazomethane in ether and (b) sodium hydroxide, followed by methyl iodide in aqueous dioxane.

B. *Reactions Involving the Hydrocarbon Side Chain*

Carboxylic acids undergo the normal reactions of alkanes, as modified by the presence of the carboxy group, in the hydrocarbon chain of the molecule. For example, butanoic acid undergoes combustion and free radical chlorination.

$$CH_3CH_2CH_2COOH \begin{array}{c} \xrightarrow{O_2} CO_2 + H_2O \\ \\ \xrightarrow[\substack{heat\ or \\ light}]{Cl_2} \text{mixture of mono- and} \\ \text{polychlorobutanoic acids} \end{array}$$

Since these reactions are not selective for any particular position along the chain, they generally have no preparative utility.

The indiscriminate nature of such free radical reactions is demonstrated by the light-initiated chlorination of butanoic acid in CCl_4 at 25°.

$$CH_3CH_2CH_2COOH \xrightarrow[h\nu]{Cl_2} \begin{cases} ClCH_2CH_2CH_2COOH \\ \qquad\quad (31\%) \\ \qquad\quad Cl \\ \qquad\quad | \\ CH_3CHCH_2COOH \\ \qquad\quad (64\%) \\ \qquad\qquad\quad Cl \\ \qquad\qquad\quad | \\ CH_3CH_2CHCOOH \\ \qquad\quad (5\%) \end{cases}$$

One reaction of the aliphatic chain that does have utility is the reaction of carboxylic acids with phosphorus tribromide and bromine. This reaction is sometimes known as the **Hell–Volhard–Zelinsky** reaction, after its discoverers.

$$3\ CH_3CH_2CH_2COOH + PBr_3 + 3\ Br_2 \longrightarrow 3\ CH_3CH_2\overset{\overset{\displaystyle Br}{|}}{C}\overset{\overset{\displaystyle O}{\|}}{H}CBr + H_3PO_3 + 3\ HBr$$

Note that the reaction is *positionally selective*-only the hydrogen on C-2 is replaced. This is not a free radical halogenation reaction. The overall result, α-bromination, is accomplished by a sequence of steps. The key step involves the reaction of bromine with the **enol form** of the corresponding **acyl bromide**. Phosphorus tribromide facilitates the reaction by reacting with the carboxylic acid to yield the acyl bromide (Section 18.7.C.2), which undergoes **enolization** much more readily than the acid itself.

$$3\ RCH_2COOH + PBr_3 \longrightarrow 3\ RCH_2\overset{\overset{\displaystyle O}{\|}}{C}Br + H_3PO_3$$

$$RCH_2\overset{\overset{\displaystyle O}{\|}}{C}Br \underset{}{\overset{H^+}{\rightleftharpoons}} RCH=\overset{\overset{\displaystyle OH}{|}}{C}Br$$
$$\text{keto form} \qquad\qquad \text{enol form}$$

$$RCH=\overset{\overset{\displaystyle OH}{|}}{C}Br + Br_2 \longrightarrow R\overset{\overset{\displaystyle}{|}}{\underset{\underset{\displaystyle Br}{|}}{C}}H\overset{\overset{\displaystyle O}{\|}}{C}Br + HBr$$

The reaction is completely analogous to the acid-catalyzed bromination of ketones (Section 13.6.D). It is only necessary to use a catalytic amount of PBr$_3$ because the product acyl bromide enters into the following equilibrium with the starting acid.

$$\underset{\underset{\displaystyle Br}{|}}{RCHCBr} + RCH_2COH \rightleftharpoons \underset{\underset{\displaystyle Br}{|}}{RCHCOH} + RCH_2CBr$$

Under these conditions the product that is isolated is the α-bromo carboxylic acid itself rather than the α-bromo acyl bromide.

$$CH_3CH_2CH_2COOH + Br_2 \xrightarrow{PBr_3} \underset{\underset{\displaystyle Br}{|}}{CH_3CH_2CHCOOH} + HBr$$

(82%)

Since phosphorus reacts rapidly with bromine to give PBr$_3$, the reaction is often carried out by simply heating the carboxylic acid with a mixture of phosphorus and bromine. In the Hell–Volhard–Zelinsky reaction one normally begins with a carboxylic acid and ends up with the α-bromo carboxylic acid, as illustrated in the last example. However, it is important to remember that the crucial reaction, introduction of the bromine into the molecule, is actually a reaction of the intermediate acyl bromide.

EXERCISE: Write a detailed mechanism, showing each step, for the reaction of phosphorus tribromide, bromine, and acetic acid to give bromoacetic acid.

C. Reactions Occurring at the Carbonyl Carbon

As in aldehydes and ketones, the carbonyl group in carboxylic acids is polarized. That is, the bonding electrons have higher density in the neighborhood of the oxygen than at the carbon.

$$\left[\underset{\displaystyle R-\overset{\displaystyle \overset{O}{\|}}{C}-OH}{} \longleftrightarrow R-\overset{\displaystyle \overset{O^-}{|}}{\underset{+}{C}}-OH \right]$$

It is reasonable, then, to expect that *nucleophilic additions* to the carboxy group would occur. As with aldehydes and ketones, both *base-catalyzed* and *acid-catalyzed* nucleophilic additions are observed.

1. BASE-CATALYZED NUCLEOPHILIC ADDITIONS. With carboxylic acids base-catalyzed nucleophilic additions are rare, and with good reason. Consider the reaction of acetic acid with sodium methoxide. Since methoxide ion is a strong base (the pK_a of methanol is about 16) and acetic acid is a moderately strong acid (pK_a about 5), the simple acid-base equilibrium below lies strongly to the right ($K \approx 10^{11}$) and is established very rapidly.

$$CH_3-\overset{\displaystyle \overset{O}{\|}}{C}-OH + CH_3O^- \underset{K}{\rightleftharpoons} CH_3-\overset{\displaystyle \overset{O}{\|}}{C}-O^- + CH_3OH$$

In other words, the acetic acid is converted immediately and quantitatively into acetate ion and the methoxide into methanol. Even in the presence of excess methoxide ion no further reaction occurs, since the acetate carbonyl is less electrophilic. That is, nucleophilic addition to the carbonyl would require that two anions combine to give a dianion. Since like charges repel one another, this reaction is unlikely.

$$CH_3-\overset{\overset{\displaystyle O}{\|}}{C}-O^- + CH_3O^- \;\not\!\longrightarrow\; CH_3-\overset{\overset{\displaystyle O^-}{|}}{\underset{\underset{\displaystyle O^-}{|}}{C}}-OCH_3$$

Even so, several base-catalyzed nucleophilic additions of carboxylic acids are known. As we shall see, each involves rather special conditions.

The most common reaction of this type is the reaction of carboxylic acids with ammonia or amines to give **amides.** When ammonia is bubbled through butanoic acid at 185°, butanamide is obtained in 85% yield. The reaction involves two stages. At room temperature, or even below, butanoic acid reacts with the weak base ammonia to give the salt ammonium butanoate.

$$CH_3CH_2CH_2COOH + NH_3 \xrightarrow{25°} CH_3CH_2CH_2CO_2^- \; NH_4^+$$

This salt is perfectly stable at normal temperatures. However, pyrolysis of the salt results in the elimination of water and formation of the amide.

$$\underset{\text{ammonium butanoate}}{CH_3CH_2CH_2CO_2^- \; NH_4^+} \xrightarrow{185°} \underset{\text{butanamide}}{CH_3CH_2CH_2\overset{\overset{\displaystyle O}{\|}}{C}NH_2} + H_2O$$

The reaction occurs only because ammonium butanoate, being the salt of a weak acid and a weak base, is in equilibrium with a significant amount of ammonia and butanoic acid. The actual dehydration step is probably the result of nucleophilic addition of ammonia to the carbonyl group of butanoic acid itself.

$$R-\overset{\overset{\displaystyle O}{\|}}{C}-O^- \; NH_4^+ \rightleftharpoons R-\overset{\overset{\displaystyle O}{\|}}{C}-OH + NH_3$$

$$R-\overset{\overset{\displaystyle O}{\|}}{C}-OH + :NH_3 \rightleftharpoons R-\overset{\overset{\displaystyle O^-}{|}}{\underset{\underset{\displaystyle NH_3^+}{|}}{C}}-OH$$

$$R-\overset{\overset{\displaystyle O^-}{|}}{\underset{\underset{\displaystyle NH_3^+}{|}}{C}}-OH \rightleftharpoons R-\overset{\overset{\displaystyle O^-}{|}}{\underset{\underset{\displaystyle NH_2}{|}}{C}}-OH_2^+$$

$$R-\overset{\overset{\displaystyle O}{|}}{\underset{\underset{\displaystyle NH_2}{|}}{C}}-OH_2^+ \rightleftharpoons R-\overset{\overset{\displaystyle O}{\|}}{C}-NH_2 + H_2O$$

Another nucleophilic addition to the carboxylate group that is of some interest is in the reduction of carboxylic acids by lithium aluminum hydride.

$$\underset{\text{CH}_3\text{O}}{\overset{\text{CH}_3\text{O}}{\diagdown}}\text{—COOH} + \text{LiAlH}_4 \xrightarrow[\text{ether}]{} \xrightarrow[\text{H}_2\text{O}]{\text{H}_2\text{SO}_4} \underset{\text{CH}_3\text{O}}{\overset{\text{CH}_3\text{O}}{\diagdown}}\text{—CH}_2\text{OH}$$

(93%)

The first step in this reaction is an acid-base reaction, giving the lithium salt of the acid, hydrogen gas, and aluminum hydride.

$$\text{RCOOH} + \text{LiAlH}_4 \longrightarrow \text{RCOOLi} + \text{H}_2 + \text{AlH}_3$$

The lithium carboxylate is then reduced further, eventually to the salt of the corresponding primary alcohol. Tetrahydroaluminate ion, AlH_4^-, is so reactive, it reduces even a carboxylate ion.

$$\text{RCO}_2^- + \text{AlH}_4^- \longrightarrow \text{R—}\underset{\text{H}}{\overset{\text{O}^-}{\underset{|}{\overset{|}{\text{C}}}}}\text{—O}^- + \text{AlH}_3$$

The reaction is undoubtedly assisted by the Lewis-acid character of aluminum salts, which reduce the effective negative charge on oxygen. The remaining steps in the reduction are still more complex, but undoubtedly also involve lithium and aluminum salts. For example, further reaction of the bis-alkoxide dianion could involve expulsion of O^{2-} as an aluminum oxide with formation of an intermediate aldehyde. The aldehyde is then rapidly reduced to the alcohol with lithium aluminum hydride.

2. ACID-CATALYZED NUCLEOPHILIC ADDITIONS. Although base-catalyzed nucleophilic additions to the carboxy group are relatively rare, acid-catalyzed additions are quite common. Carboxylic acids react readily with alcohols in the presence of catalytic amounts of mineral acids to yield compounds called **esters** (Chapter 19). The process is called **esterification.**

$$\text{CH}_3\text{COOH} + \text{C}_2\text{H}_5\text{OH} \overset{\text{H}^+}{\rightleftharpoons} \text{CH}_3\text{COOC}_2\text{H}_5 + \text{H}_2\text{O} \qquad (18\text{-}8)$$

Unlike most of the reactions we have encountered, this one has an equilibrium constant of relatively low magnitude. The experimental equilibrium constant for the reaction of acetic acid with ethanol is

$$K_{eq} = \frac{[\text{CH}_3\text{COOC}_2\text{H}_5][\text{H}_2\text{O}]}{[\text{CH}_3\text{COOH}][\text{C}_2\text{H}_5\text{OH}]} = 3.38$$

As in any equilibrium process, the reaction may be driven in one direction by controlling the concentration of either the reactants or products (Le Châtelier's principle). For reaction (18-8) the equilibrium constant tells us that an equimolar mixture of acetic acid and ethanol will eventually reach equilibrium to give a mixture containing 0.35 mole each of acetic acid and ethanol and 0.65 mole each

TABLE 18.7
Equilibrium Compositions

	CH_3COOH	$+$ C_2H_5OH	$\rightleftharpoons$ $CH_3COOC_2H_5$	$+$ H_2O
at start	1	1	0	0
at equilibrium	0.35	0.35	0.65	0.65
at start	1	10	0	0
at equilibrium	0.03	9.03	0.97	0.97
at start	1	100	0	0
at equilibrium	0.007	99.007	0.993	0.993

of ethyl acetate and water. Of course, the same equilibrium mixture will be obtained if one starts with equimolar quantities of ethyl actate and water.

If we increase the concentration of either reactant relative to the other, the reaction will be driven to the right and the equilibrium mixture will contain proportionately more ethyl acetate and water. Table 18.7 shows the equilibrium compositions that will be achieved starting with various mixtures of acetic acid and ethanol.

Similar results will obviously be obtained by increasing the acetic acid concentration rather than the ethanol concentration. In a practical situation, when one wants to prepare an ester, it is desirable to obtain the maximum yield of pure product. It is often done as suggested in the preceding paragraph—by using a large excess of one of the reactants. For economic reasons, the reactant chosen is usually the less expensive of the two.

The mechanism of the acid-catalyzed esterification reaction has been studied thoroughly. All of the experimental facts are consistent with a mechanism involving the following steps (illustrated for acetic acid and methanol).

$$
(1)\quad CH_3\!-\!\overset{\overset{\displaystyle O}{\|}}{C}\!-\!OH + H^+ \rightleftharpoons CH_3\!-\!\overset{\overset{\displaystyle OH^+}{\|}}{C}\!-\!OH
$$

$$
(2)\quad CH_3\!-\!\overset{\overset{\displaystyle OH^+}{\|}}{C}\!-\!OH + CH_3\overset{..}{\underset{..}{O}}H \rightleftharpoons CH_3\!-\!\overset{\overset{\displaystyle OH}{|}}{\underset{\underset{\displaystyle +}{\displaystyle H\overset{}{O}CH_3}}{C}}\!-\!OH
$$

$$
(3)\quad CH_3\!-\!\overset{\overset{\displaystyle OH}{|}}{\underset{\underset{\displaystyle +}{\displaystyle H\overset{}{O}CH_3}}{C}}\!-\!OH \rightleftharpoons CH_3\!-\!\overset{\overset{\displaystyle OH}{|}}{\underset{\displaystyle OCH_3}{C}}\!-\!OH + H^+
$$

$$
(4)\quad CH_3\!-\!\overset{\overset{\displaystyle OH}{|}}{\underset{\displaystyle OCH_3}{C}}\!-\!OH + H^+ \rightleftharpoons CH_3\!-\!\overset{\overset{\displaystyle OH}{|}}{\underset{\displaystyle OCH_3}{C}}\!-\!OH_2^+
$$

$$
(5)\quad CH_3\!-\!\overset{\overset{\displaystyle :\overset{..}{O}H}{|}}{\underset{\displaystyle OCH_3}{C}}\!-\!OH_2^+ \rightleftharpoons CH_3\!-\!\overset{\overset{\displaystyle \overset{..}{O}H^+}{\|}}{C}\!-\!OCH_3 + H_2O
$$

$$(6)\ CH_3\overset{\overset{\displaystyle OH^+}{\|}}{C}\!-\!OCH_3 \rightleftharpoons CH_3\overset{\overset{\displaystyle O}{\|}}{C}\!-\!OCH_3 + H^+$$

Steps (1), (3), (4), and (6) are rapid proton-transfer steps—simple acid-base reactions. Although we show "bare" protons in each case, they are actually solvated by some Lewis base, which may be methanol, water, or any of the other oxygenated species present. In steps (2) and (5), C—O bonds are formed or broken. These steps have higher activation energies than the proton-transfer steps.

The foregoing mechanism is an extremely important one in organic chemistry. As mentioned previously, it is based on a large amount of experimental data. Two of the more important experiments involved the use of ^{18}O-labeled materials. The first of these interesting experiments demonstrated that the oxygen–carbonyl bond is broken during the esterification process. Benzoic acid was treated in the presence of HCl with methanol enriched in ^{18}O. The water produced in the reaction was isolated and shown to be normal $H_2{}^{16}O$.

$$C_6H_5\overset{\overset{\displaystyle O}{\|}}{C}\!-\!OH + CH_3{}^{18}OH \overset{H^+}{\rightleftharpoons} C_6H_5\overset{\overset{\displaystyle O}{\|}}{C}\!-\!{}^{18}OCH_3 + H_2{}^{16}O$$

This experiment rules out mechanisms such as the following, in which the oxygen in the water produced comes from the alcohol.

$$C_6H_5\overset{\overset{\displaystyle O}{\|}}{C}\!-\!\overset{\cdot\cdot}{\underset{\cdot\cdot}{O}}H + CH_3\!-\!\overset{+}{\underset{\cdot\cdot}{O}}H_2 \overset{/\!/}{\longrightarrow} C_6H_5\overset{\overset{\displaystyle O}{\|}}{C}\!-\!\underset{\underset{\displaystyle H}{|}}{\overset{+}{O}}CH_3 + H_2O$$

The second important labeling experiment showed that a symmetrical **intermediate** intervenes in the process. Ethyl benzoate enriched in ^{18}O in the carbonyl oxygen was hydrolyzed with HCl and normal water. The reaction was stopped short of completion, and the recovered ethyl benzoate was analyzed. It was found that exchange of ^{18}O in the ester by ^{16}O had occurred. Although hydrolysis occurs approximately five times faster than exchange, this experiment demonstrates that an intermediate is formed that can go on to give acid or reverse to give exchanged ester. Mechanisms such as the following, which is analogous to the S_N2 displacement in saturated systems, are definitely ruled out.

$$R\!-\!\underset{\underset{\displaystyle H}{|}}{\overset{\overset{\displaystyle O}{\|}}{C}}\!-\!\overset{+}{O}R + H_2O \longrightarrow \left[H_2\overset{\delta+}{O}\cdots\underset{\underset{\displaystyle R}{|}}{\overset{\overset{\displaystyle O}{\|}}{C}}\cdots\overset{\delta+}{O}R \right]^{\ddagger} \longrightarrow H_2\overset{+}{O}\!-\!\overset{\overset{\displaystyle O}{\|}}{C}\!-\!R + ROH$$

Note that the accepted mechanism involves simply an acid-catalyzed addition of an alcohol to the carbonyl group and is completely analogous to the similar reactions with aldehydes and ketones to form intermediate hemiacetals (Section 13.7.A and B).

Acid-catalyzed esterification is an important method for the preparation of carboxylic acid esters (Section 19.5). It is a general reaction for acids. Occasionally, however, the reaction is very slow or the equilibrium constant is very unfavorable. This situation often occurs when the acid is extensively branched near the carboxy group. In such cases esters may be prepared in another way. When an acid is dissolved in concentrated sulfuric acid, the initially formed oxonium ion dissociates to form an **acylium ion.**

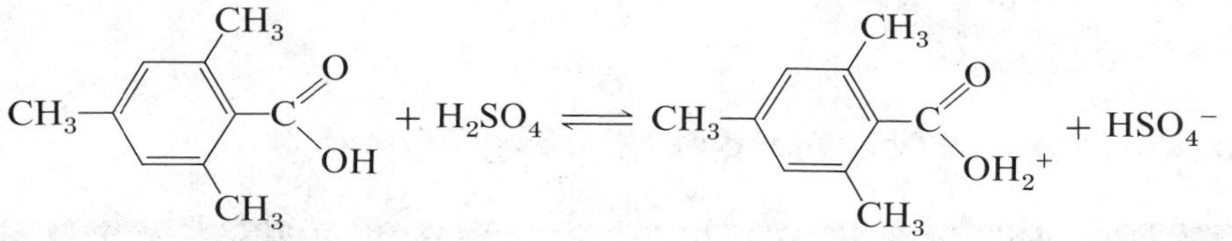

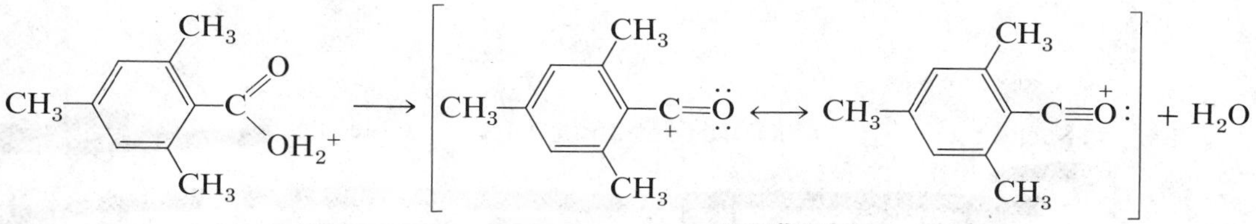

an **acylium ion**

The water produced reacts with more sulfuric acid to give hydronium ion and bisulfate ion.

$$H_2O + H_2SO_4 \longrightarrow H_3O^+ + HSO_4^-$$

Since the solution contains only weakly nucleophilic species (H_2SO_4 and HSO_4^-), the acylium ion remains in solution. If this solution is rapidly diluted with an alcohol, immediate reaction occurs to give an ester.

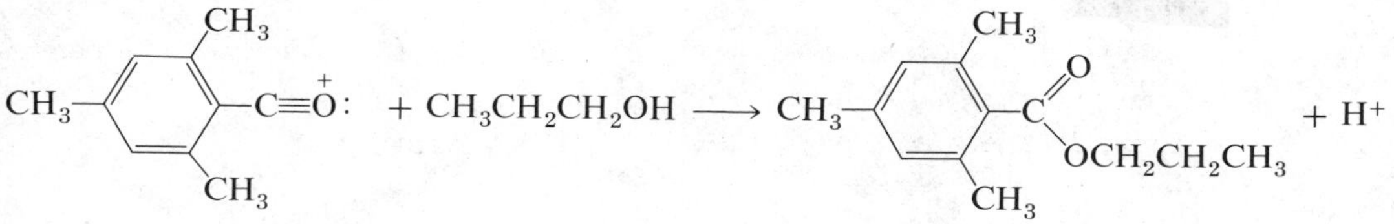

Carboxylic acids react with thionyl chloride, phosphorus pentachloride, and phosphorus tribromide in the same way that alcohols do (Section 10.7.D). The products are **acyl halides** (Chapter 19).

$$CH_3-\overset{O}{\overset{\|}{C}}-OH + Cl-\overset{O}{\overset{\|}{S}}-Cl \longrightarrow CH_3-\overset{O}{\overset{\|}{C}}-Cl + HCl + SO_2$$

acetyl chloride

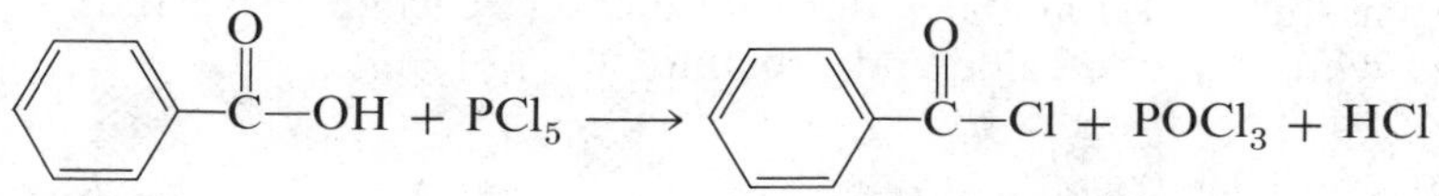

benzoyl chloride

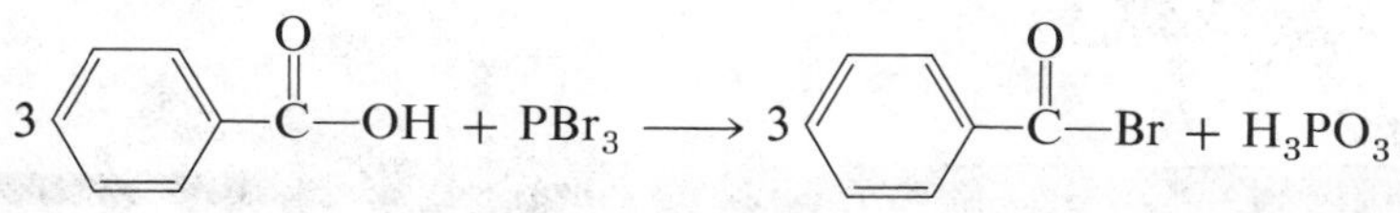

benzoyl bromide

EXERCISE: Write the equations for the reactions of hexanoic acid with (a) ethanol and a catalytic amount of concentrated sulfuric acid, (b) concentrated sulfuric acid, followed by addition of the sulfuric acid solution to ice-cold ethanol, (c) thionyl chloride, (d) phosphorus pentachloride, and (e) phosphorus tribromide.

D. One-Carbon Degradation of Carboxylic Acids

Carboxylic acids undergo several reactions in which the carboxy group is replaced by halogen.

$$RCOOH \longrightarrow RX$$

Such reactions, in which carbons are lost from a molecule, are called "degradations."

In the **Hunsdiecker reaction** the silver salt of a carboxylic acid, prepared by treating the acid with silver oxide, is treated with a halogen. Bromine is the usual reagent, but iodine may also be used. Carbon dioxide is evolved and the corresponding alkyl halide is obtained, usually in fair to good yield.

$$CH_3\overset{O}{\overset{\|}{O}}CCH_2CH_2CH_2CH_2CO^- \ Ag^+ \ \xrightarrow[CCl_4]{Br_2} \ CH_3\overset{O}{\overset{\|}{O}}CCH_2CH_2CH_2CH_2Br + AgBr + CO_2$$

(65–68%)

The reaction appears to proceed by a free radical path and may be formulated as follows.

$$(1) \quad R-\overset{O}{\overset{\|}{C}}-O^- \ Ag^+ + Br_2 \longrightarrow R-\overset{O}{\overset{\|}{C}}-OBr + AgBr$$

$$(2) \quad R-\overset{O}{\overset{\|}{C}}-OBr \longrightarrow R-\overset{O}{\overset{\|}{C}}-O\cdot + Br\cdot \qquad \text{initiation step}$$

$$(3) \quad R-\overset{O}{\overset{\|}{C}}-O\cdot \longrightarrow R\cdot + CO_2$$

$$(4) \quad R\cdot + R-\overset{O}{\overset{\|}{C}}-OBr \longrightarrow RBr + R-\overset{O}{\overset{\|}{C}}-O\cdot$$

propagation steps

In a useful modification of the Hunsdiecker reaction the carboxylic acid is treated with mercuric oxide and bromine.

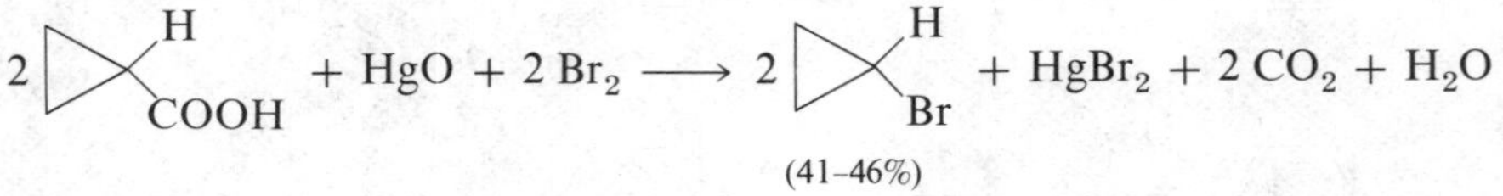

$$2 \ \triangleright\!\!\!<\overset{H}{\underset{COOH}{}} + HgO + 2 \ Br_2 \longrightarrow 2 \ \triangleright\!\!\!<\overset{H}{\underset{Br}{}} + HgBr_2 + 2 \ CO_2 + H_2O$$

(41–46%)

In the **Kochi reaction** the carboxylic acid is treated with lead tetraacetate and lithium chloride; the product is an alkyl chloride.

$$\text{[cyclobutyl]}-COOH + Pb(O\overset{O}{\overset{\|}{C}}CH_3)_4 + LiCl \longrightarrow \text{[cyclobutyl]}-Cl + CO_2 + LiPb(O\overset{O}{\overset{\|}{C}}CH_3)_3 + CH_3COOH$$

(100%)

The Hundsdiecker and Kochi reactions complement each other, the former giving best results with primary alkyl carboxylic acids and the latter being preferred for secondary and tertiary alkyl carboxylic acids.

EXERCISE: Write equations showing how (a) hexanoic acid can be converted into 1-bromopentane and (b) 1-bromopentane can be converted into hexanoic acid.

18.8
Occurrence of Carboxylic Acids

Carboxylic acids are widespread in nature, both as such and in the form of esters. Partly because they are easily isolated as salts, they were among the earliest known organic compounds.

Formic acid was first discovered in 1670 by the distillation of ants. Its name comes from the Latin word for ant, *formica*. Formic acid is partially responsible for the irritation resulting from the sting of the red ant and the stinging nettle.

Acetic acid is a product of fermentation. The characteristic taste of sour wine is due to the oxidation of ethanol to acetic acid. Vinegar is a dilute solution of acetic acid. Although the acid has been known in the form of vinegar since antiquity, it was first isolated in pure form by Stahl in 1700. Pure acetic acid is known as glacial acetic acid. This term arises from the relatively high melting point of the compound (17 °C, 63 °F). In earlier times, when buildings were not heated as they are now, pure acetic acid was commonly observed to be a solid at "room temperature." Acetic acid is also one of the products of pyrolysis of wood (destructive distillation).

Butyric acid is responsible for the sharp odor of rancid butter. It was first isolated from this source. Caproic acid also has a penetrating unpleasant odor described as "goat-like." Indeed, its name, as well as those of caprylic acid and capric acid, is derived from the Latin word for goat, *caper*. These acids and their esters are widespread in nature.

Juvenile hormone and juvabione are examples of carboxylic acids that occur in nature in the form of their methyl esters.

juvenile hormone   juvabione

These compounds are associated with the pupal development of various insects. Such compounds offer some promise as insect-control agents. Since they are highly species-specific and leave no residues, they have obvious advantages over other commonly used pesticides.

PROBLEMS

1. Give the IUPAC name for each of the following compounds.

(a) $CH_3CH_2\overset{\overset{\displaystyle CH_3}{|}}{C}HCH_2COOH$

(b) $(CH_3)_3CCOOH$

(c) $BrCH_2CH_2CH_2COOH$

(d) ICH_2COOH

(e) $CH_3CH_2\overset{\overset{\displaystyle OH}{|}}{C}HCOOH$

(f) $CH_3(CH_2)_8COOH$

(g) 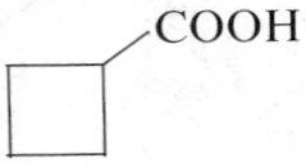 a cyclobutane ring with a COOH substituent

(h) $CH_3\overset{\overset{\displaystyle OCH_3}{|}}{C}HCH_2COOH$

(i) $(CH_3)_2CH\overset{\overset{\displaystyle }{|}}{C}HCH_2COOH$
$\qquad\quad\overset{\displaystyle |}{CH_3}$

2. Write out the correct structure for each of the following names.
   (a) β-chlorobutyric acid
   (b) hexanoic acid
   (c) γ-methoxyvaleric acid
   (d) cyclopentanecarboxylic acid
   (e) 3-phenylpropanoic acid
   (f) *cis*-2-pentenoic acid
   (g) (E)-4-hydroxypent-2-enoic acid
   (h) α-chloro-β-bromopropionic acid

3. Give the products in each of the following reactions of cyclohexanecarboxylic acid.
   (a) LiAlH₄ in ether, then dilute hydrochloric acid
   (b) conc. H₂SO₄, followed by ice-cold ethanol
   (c) P + Br₂, heat, then water
   (d) diazomethane in ether
   (e) isopropyl alcohol (excess), trace of H₂SO₄
   (f) ammonia, 200°
   (g) methylamine (CH₃NH₂), 200°
   (h) SOCl₂, heat
   (i) PBr₃, heat
   (j) Pt/H₂, room temperature
   (k) dilute aqueous sodium hydroxide at room temperature
   (l) silver hydroxide, followed by bromine in carbon tetrachloride
   (m) Pb(OAc)₄ + LiCl

4. Two general methods for converting alkyl halides to carboxylic acids are displacement by cyanide ion, followed by hydrolysis, and conversion to the Grignard reagent, followed by carbonation with carbon dioxide. Which method is superior for each of the following transformations? In which cases would a protecting group facilitate the conversion? Explain why.
   (a) $(CH_3)_3CCl \longrightarrow (CH_3)_3CCOOH$
   (b) $BrCH_2CH_2Br \longrightarrow HOOCCH_2CH_2COOH$
   (c) $CH_3COCH_2CH_2CH_2Br \longrightarrow CH_3COCH_2CH_2CH_2COOH$
   (d) $(CH_3)_3CCH_2Br \longrightarrow (CH_3)_3CCH_2COOH$
   (e) $CH_3CH_2CH_2CH_2Br \longrightarrow CH_3CH_2CH_2CH_2COOH$
   (f) $HOCH_2CH_2CH_2CH_2Br \longrightarrow HOCH_2CH_2CH_2CH_2COOH$

5. Show how neopentane may be converted into each of the following compounds.

(a) $(CH_3)_3CCH_2COOH$

(b) $(CH_3)_3CCHBrCOOH$

6. Show how butanal may be converted into each of the following compounds.

(a) $CH_3CH_2CH_2COOH$

(b) $CH_3CH_2CH_2CH=\overset{\overset{\displaystyle COOH}{|}}{C}CH_2CH_3$

(c) $CH_3CH_2CH_2\overset{\overset{\displaystyle OH}{|}}{C}HCOOH$

(d) $CH_3CH_2CH_2CH_2COOH$

7. Show how 3-methylbutanoic acid may be converted into each of the following compounds.

(a) $(CH_3)_2CHCH_2CH_2OH$

(b) $(CH_3)_2CHCH_2CH_2COOH$

(c) $(CH_3)_2CHCH_2CH_2\overset{\overset{\displaystyle Br}{|}}{C}HCH_2Br$

(d) $(CH_3)_2C=CHCOOCH_3$

(e) $(CH_3)_2CHCH_2\overset{\overset{\displaystyle O}{||}}{C}CH_3$

(f) $(CH_3)_2CHCH_2\overset{\overset{\displaystyle O}{||}}{C}NH_2$

(g) $(CH_3)_2CHCH_2Br$

8. Show how each of the following transformations may be accomplished in a practical manner.

(a) 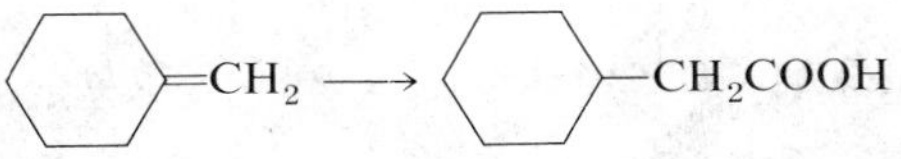 ⬡$=CH_2 \longrightarrow$ ⬡$-CH_2COOH$

(b) $(CH_3)_3CCH=CH_2 \longrightarrow (CH_3)_3CCOOH$

(c) $CH_3COCH_2CH_2CH_2\overset{\overset{\displaystyle CH_3}{|}}{\underset{\underset{\displaystyle CH_3}{|}}{C}}Br \longrightarrow CH_3COCH_2CH_2CH_2\overset{\overset{\displaystyle CH_3}{|}}{\underset{\underset{\displaystyle CH_3}{|}}{C}}COOH$

(d) $CH_3CH_2COOH \longrightarrow CH_3CH_2CH_2COOH$

(e) $CH_3CH_2CH_2COOH \longrightarrow CH_3CH_2COOH$

(f) $CH_3CH_2CH_2COOH \longrightarrow CH_3CH_2CH_2N_3$

9. The dissociation constant of acetic acid is 1.8×10^{-5} M. Calculate the percent dissociation when the following amounts of acetic acid are made up to 1 L with water at 25°.

(a) 0.1 mole   (b) 0.01 mole   (c) 0.001 mole

10. The following dissociation constants are given. Calculate the corresponding pK_a values.

(a) $(CH_3)_2CHCH_2CH_2COOH$; $K_a = 1.4 \times 10^{-5}$ M

(b) 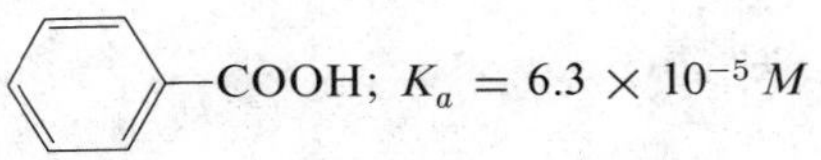 ⬡$-COOH$; $K_a = 6.3 \times 10^{-5}$ M

(c) $Cl_2CHCOOH$; $K_a = 5.5 \times 10^{-2}$ M

(d) Cl_3CCOOH; $K_a = 0.23\ M$

(e) $CH_3CONHCH_2COOH$; $K_a = 2.1 \times 10^{-4}\ M$

11. In each of the following pairs, which is the stronger *base?* Explain briefly.
(a) $CH_3CH_2O^-$; $CH_3CO_2^-$
(b) $ClCH_2CH_2CO_2^-$; $CH_3CH_2CH_2CO_2^-$
(c) $ClCH_2CH_2CO_2^-$; $CH_3CHClCO_2^-$
(d) $FCH_2CO_2^-$; $F_2CHCO_2^-$
(e) $HC{\equiv}CCH_2CO_2^-$; $CH_3CH_2CH_2CO_2^-$
(f) Cl^-; $CH_3CO_2^-$

12. The two carboxy groups in 3-chlorohexanedioic acid are not equivalent and have different dissociation constants. Which carboxy group is the more acidic?

13. From the progression of acidity constants for chlorobutanoic acids in Table 18.4 and the pK_a of 3-cyanobutanoic acid, 4.44, estimate the pK_a of 2-cyanobutanoic acid.

14. When propanoic acid is refluxed with some sulfuric acid in water enriched with $H_2^{18}O$, ^{18}O gradually appears in the carboxylic acid group. Write the mechanism for this reaction, showing each intermediate in the reaction pathway.

15. When 5-hydroxyhexanoic acid is treated with a trace of sulfuric acid in benzene solution, the following reaction occurs.
(a) Propose a mechanism for the reaction.
(b) The equilibrium constant for this process is much larger than that normally observed for an esterification reaction. Explain.

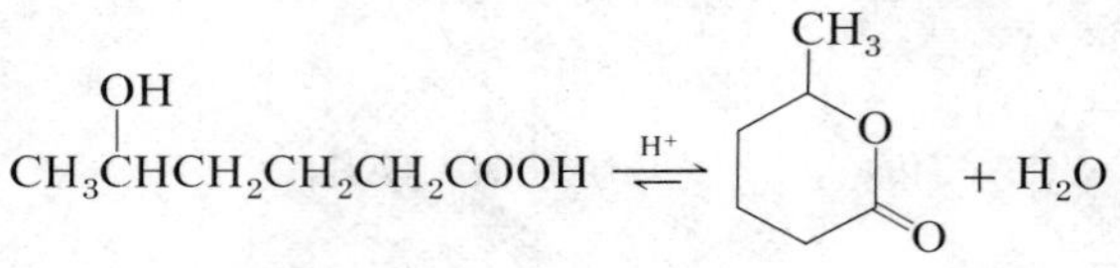

$$\underset{\substack{|\\ OH}}{CH_3CHCH_2CH_2CH_2COOH} \overset{H^+}{\rightleftharpoons} \text{(structure)} + H_2O$$

16. On refluxing with D_2O containing a strong acid, propanoic acid is slowly converted to CH_3CD_2COOD. Write a plausible mechanism for this reaction.

17. Values of heats of formation, ΔH_f°, for the ideal gas state at 25° are given in the table that follows for several compounds. Calculate ΔH° for the following equilibrium in the gas phase.

$$CH_3COOH + C_2H_5OH \rightleftharpoons CH_3COOC_2H_5 + H_2O$$

In the liquid phase, ΔH° for this equilibrium is -0.9 kcal mole^{-1}. Why is there such a difference between the two values?

Compound	ΔH_f° at 25°, kcal mole^{-1}
CH_3COOH	-103.3
C_2H_5OH	-56.2
$CH_3COOC_2H_5$	-106.3
H_2O	-57.8

18. The following reaction is exothermic in the gas phase.

$$CH_3CO_2^- + ClCH_2COOH \rightleftharpoons CH_3COOH + ClCH_2CO_2^- \qquad \Delta H^\circ = -13\ \text{kcal mole}^{-1}$$

(a) Explain briefly why this reaction is exothermic.

(b) Perform a simple calculation to determine whether the electrostatic interaction of a C—Cl dipole with a carboxylate anion has the proper magnitude to account for this energy difference. For this purpose treat $ClCH_2CO_2^-$ as having the following structure in which the CCCl plane is perpendicular to the OCO plane.

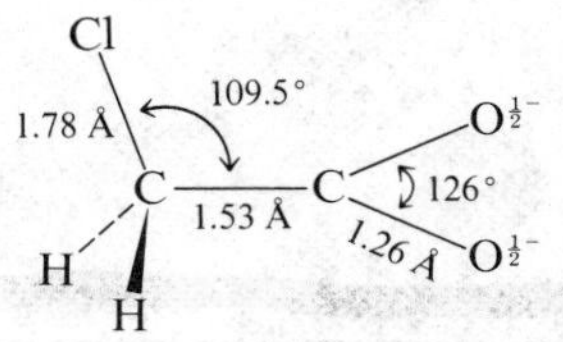

in which θ is the angle between the dipole and the charge. For a charge q of one electron, $\mu = 1$ D, $r = 1$ Å, and $\theta = 0°$, the energy E is 69 kcal mole^{-1}. Consider the effect of the chlorine to be that of a point dipole of 1.9 D. The electrostatic energy for a point dipole and a charge is given by

$$E = \frac{q\mu \cos \theta}{r^2}$$

Chapter 19
Derivatives of Carboxylic Acids

19.1
Structure

Functional group derivatives of carboxylic acids are those compounds that are transformed into carboxylic acids by simple hydrolysis. The most common such derivatives are **esters,** in which the hydroxy group is replaced by an alkoxy group.

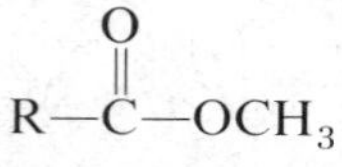

$$R-\overset{\overset{\displaystyle O}{\|}}{C}-OCH_3$$

an ester

Amides are compounds in which the hydroxy group is replaced by an amino group. The nitrogen of the amino group may bear zero, one, or two alkyl groups.

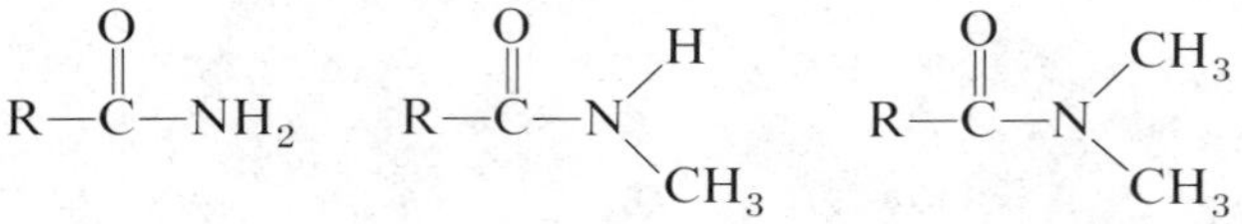

$$R-\overset{\overset{\displaystyle O}{\|}}{C}-NH_2 \qquad R-\overset{\overset{\displaystyle O}{\|}}{C}-\overset{\overset{\displaystyle H}{|}}{N}-CH_3 \qquad R-\overset{\overset{\displaystyle O}{\|}}{C}-\overset{\overset{\displaystyle CH_3}{|}}{N}-CH_3$$

amides

Acyl halides are derivatives in which the carboxy OH is replaced by a halogen atom; **acyl chlorides** and **acyl bromides** are the most commonly encountered acyl halides.

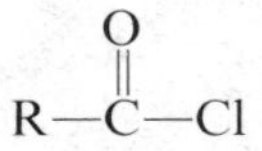

$$R-\overset{\overset{\displaystyle O}{\|}}{C}-Cl \qquad R-\overset{\overset{\displaystyle O}{\|}}{C}-Br$$

an acyl chloride 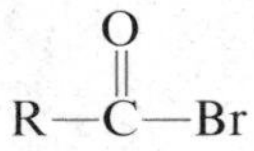   an acyl bromide

Acid anhydrides are molecules in which one molecule of water has been removed from two molecules of a carboxylic acid. The only acyclic anhydride of general importance is acetic anhydride.

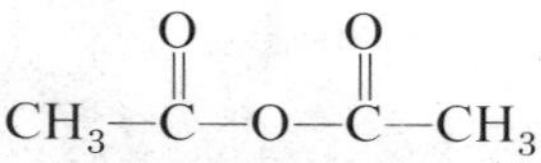

$$CH_3-\overset{\overset{\displaystyle O}{\|}}{C}-O-\overset{\overset{\displaystyle O}{\|}}{C}-CH_3$$

acetic anhydride

In a strict sense, **nitriles** are functional derivatives of carboxylic acids because they may be hydrolyzed to carboxylic acids (Section 18.6.A).

$$R-CN$$

a nitrile

The simplest ester, methyl formate, may be considered a simple derivative of formic acid in which the OH group is replaced by an OCH_3 group. Correspondingly, the molecular geometry of methyl formate is similar to that of formic acid. Experimental bond lengths and bond angles, determined by microwave spectroscopy, are given in Figure 19.1.

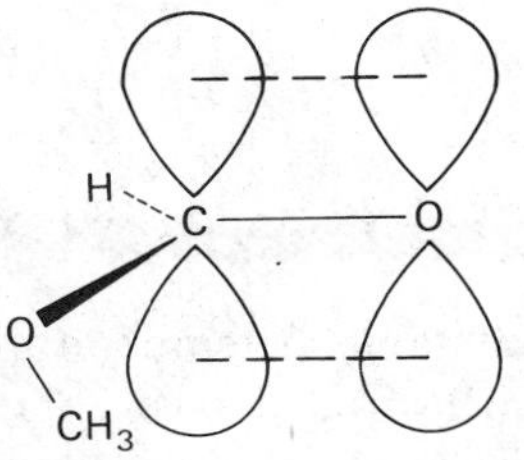

Bond Lengths, Å		Bond Angles, deg	
C=O	1.200	H—C=O	124.95
C(=O)—O	1.334	O—C=O	125.87
C(H$_3$)—O	1.437	H—C—O	109.18
C(=O)—H	1.101	CH$_3$—O—C	114.78

FIGURE 19.1. *Structure of methyl formate.*

Note that the C_{sp^2}—O σ-bond is considerably shorter than the C_{sp^3}—O σ-bond. Two factors are apparently important in accounting for this bond shortening. In Section 11.1 we saw that because of the difference in "length" of various hybrid orbitals, C_{sp^3}—C_{sp^3} σ-bonds are longer than C_{sp^2}—C_{sp^2} σ-bonds. This factor is probably also important in methyl formate. Another factor involves the dipolar resonance form (19-1) as a contributor to the structure of methyl formate.

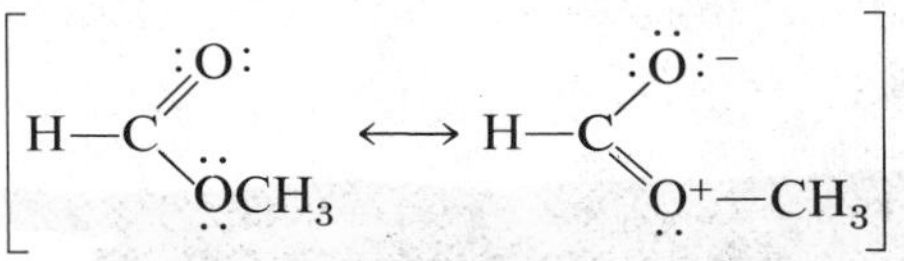

$$\left[\begin{array}{c} H-C \overset{:O:}{\underset{\ddot{O}CH_3}{}} \end{array} \longleftrightarrow \begin{array}{c} H-C \overset{:\ddot{O}:^-}{\underset{O^+-CH_3}{}} \end{array} \right] \qquad (19\text{-}1)$$

To the extent that this form contributes to the actual structure of the molecule, the C_{sp^2}—O σ-bond will be shorter because it has some **double bond character.** This latter factor is especially important in amides.

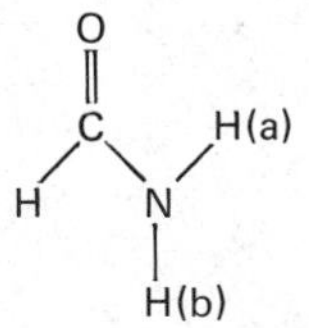

Bond Lengths, Å		Bond Angles, deg	
C=O	1.193	H—C=O	122.97
C—N	1.376	H—C—N	113.23
C—H	1.102	N—C=O	123.80
N—H(a)	1.014	C—N—H(a)	117.15
N—H(b)	1.002	C—N—H(b)	120.62
		H—N—H	118.88

FIGURE 19.2. *Structure of formamide.*

Microwave measurements on formamide indicate the structure shown in Figure 19.2. The entire molecule is planar. Note that the two hydrogens attached to nitrogen are distinguishable. The barrier to rotation about the C—N bond has been measured experimentally and is found to be 18 kcal mole^{-1}. A high degree of double bond character in this bond, as indicated in the dipolar resonance form (19-2), has been invoked to explain this relatively high rotational barrier.

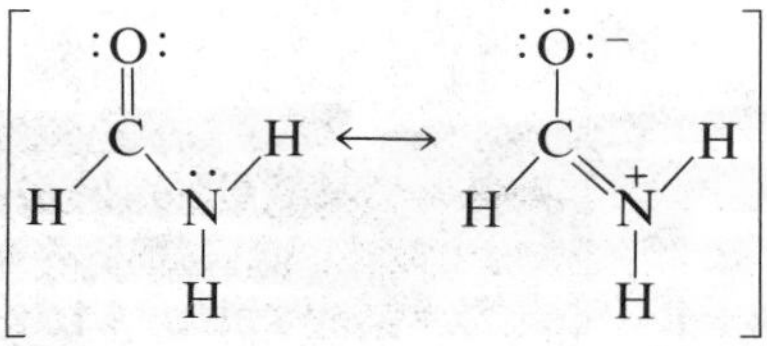

$$\left[\begin{array}{c} :O: \\ \| \\ C \\ H \diagup \ddot{N} \diagdown H \\ | \\ H \end{array} \longleftrightarrow \begin{array}{c} :\ddot{O}:^- \\ | \\ C \\ H \diagup \overset{+}{N} \diagdown H \\ | \\ H \end{array} \right] \qquad (19\text{-}2)$$

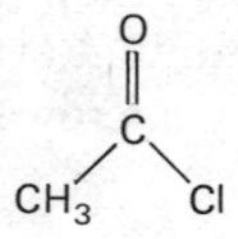

Bond Lengths, Å		Bond Angles, deg	
C=O	1.192	C—C=O	127.08
C—C	1.499	C—C—Cl	112.66
C—Cl	1.789	O=C—Cl	120.26
C—H	1.083		

FIGURE 19.3. *Structure of acetyl chloride.*

Because of the high barrier to rotation about the C—N bond, amides have a relatively rigid structure.

Since the simplest acyl chloride, formyl chloride, is not stable at temperatures above −60°, its structural parameters have not been measured. However, the bond lengths and bond angles have been determined for acetyl chloride (Figure 19.3). The C—Cl bond is not appreciably shorter than the analogous bond in methyl chloride (1.784 Å), suggesting that dipolar resonance structures are not particularly important in the case of acyl halides.

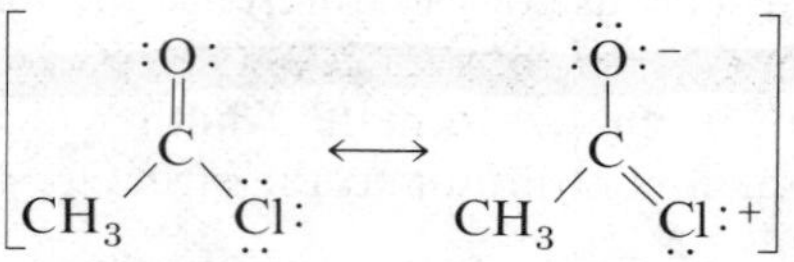

As judged by the $-\overset{O}{\underset{}{C}}-Y$ bond distances, the importance of such dipolar resonance structures decreases in the order NR_2, OR, Cl.

Acetonitrile (methyl cyanide) has a structure analogous to that of propyne. The nitrile carbon is *sp*-hybridized. It forms a $C_{sp^3}-C_{sp}$ σ-bond to the methyl group and a $C_{sp}-N_{sp}$ σ-bond to nitrogen. The two remaining *p*-orbitals on carbon overlap with two *p*-orbitals on nitrogen, giving rise to a typical axially symmetric triple bond. The lone pair is in an *sp*-orbital on nitrogen (Figure 19.4).

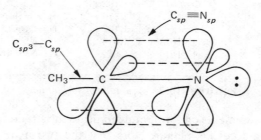

FIGURE 19.4. *Orbital structure of acetonitrile.*

19.2
Nomenclature

Esters are named in the following way. The first word of the name is the stem name of the alkyl group attached to oxygen. The second word of the name is the name of the parent acid with the suffix **-ic** replaced by **-ate**. This nomenclature applies for both common and IUPAC names of acids.

$$H-\overset{\overset{\displaystyle O}{\|}}{C}-OCH_3 \qquad CH_3-\overset{\overset{\displaystyle O}{\|}}{C}-OCH_2CH_3 \qquad \overset{\overset{\displaystyle CH_3}{|}}{CH_3CH}-\overset{\overset{\displaystyle O}{\|}}{C}-OCH(CH_3)_2$$

methyl form**ate**   ethyl acet**ate**   isopropyl isobutyr**ate**
isopropyl 2-methylpropano**ate**

Amides are named by dropping the suffix **-ic** or **-oic** from the name of the parent acid and adding the suffix **-amide**.

$$H-\overset{\overset{\displaystyle O}{\|}}{C}-NH_2 \qquad CH_3\overset{\overset{\displaystyle O}{\|}}{C}-NH_2 \qquad CH_3CH_2CH_2\overset{\overset{\displaystyle O}{\|}}{C}-NH_2$$

form**amide**   acet**amide**   butan**amide**

A substituted nitrogen is indicated by prefixing the name of the simple amide by N-, followed by the name of the substituent group.

$$CH_3CH_2\overset{\overset{\displaystyle O}{\|}}{C}-N\overset{\displaystyle H}{\underset{\displaystyle CH_3}{}} \qquad \overset{\displaystyle CH_3}{CH_3CHCH_2}\overset{\overset{\displaystyle O}{\|}}{C}-N\overset{\displaystyle CH_3}{\underset{\displaystyle CH_3}{}}$$

N-methylpropanamide   N,N,3-trimethylbutanamide

Acyl halides are named in a similar manner. In this case the suffix **-ic** is replaced by the suffix **-yl,** and the halide name is added as a second word. (Note that for acyl halides, the "o" of the ending -oic is retained.)

$$CH_3\overset{\overset{\displaystyle O}{\|}}{C}-Cl \qquad \overset{\displaystyle CH_3}{CH_3CH_2CHCH_2}\overset{\overset{\displaystyle O}{\|}}{C}-Br \qquad CH_3CH_2\overset{\overset{\displaystyle O}{\|}}{C}-F$$

acet**yl chloride**   3-methylpentano**yl bromide**   propano**yl fluoride**

Anhydrides are named by adding **anhydride** to the name of the corresponding carboxylic acid.

$$CH_3-\overset{\overset{\displaystyle O}{\|}}{C}-O-\overset{\overset{\displaystyle O}{\|}}{C}-CH_3 \qquad \overset{\displaystyle CH_3}{CH_3CH_2CHCH_2}-\overset{\overset{\displaystyle O}{\|}}{C}-O-\overset{\overset{\displaystyle O}{\|}}{C}-CH_2\overset{\displaystyle CH_3}{CHCH_2CH_3}$$

acetic anhydride   3-methylpentanoic anhydride

For mixed anhydrides, the parent name of each acid is given, followed by the word **anhydride.**

$$H-\overset{\overset{\displaystyle O}{\|}}{C}-O-\overset{\overset{\displaystyle O}{\|}}{C}-CH_3$$

acetic formic anhydride

Nitriles are named in the IUPAC system as alkanenitriles. Simple nitriles are usually referred to as derivatives of the corresponding carboxylic acid, by replacing the suffic **-ic** or **-oic** by the suffix **-onitrile.**

$$CH_3CN \qquad CH_3CH_2CN \qquad BrCH_2CH_2CH_2CH_2CH_2CN$$

aceto**nitrile**   propane**nitrile**   6-bromohexane**nitrile**
(ethane**nitrile**)

For functional derivatives of carboxylic acids that are named as alkanecarboxylic acids (page 499) the suffix **-carboxylic acid** is replaced by **-carboxamide, -carbonyl halide,** or **-carbonitrile.**

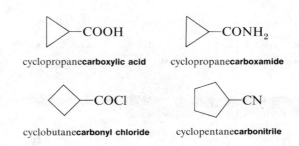

cyclopropane**carboxylic acid**   cyclopropane**carboxamide**

cyclobutane**carbonyl chloride**   cyclopentane**carbonitrile**

Occasionally it is necessary to name an acid derivative function as a derivative of some other functional stem. The group names for the various radicals as prefixes are given in Table 19.1.

TABLE 19.1
Functional Group Names

Radical	Group Name as Prefix
—COOCH$_3$	methoxycarbonyl
—COOCH$_2$CH$_3$	ethoxycarbonyl
—CONH$_2$	carbamoyl
—COCl	chloroformyl
—COBr	bromoformyl
—CN	cyano

Examples of such names are

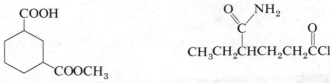

3-methoxycarbonylcyclohexanecarboxylic acid   4-carbamoylhexanoyl chloride

It is important to remember that *Chemical Abstracts* does not always follow the IUPAC nomenclature conventions. For example, the current indices of *Chemical Abstracts* use the IUPAC prefix methoxycarbonyl- for —COOCH$_3$, but use chlorocarbonyl- instead of chloroformyl- for —COCl. A user should always consult the Index Guide before searching a recent *Chemical Abstracts* index for a compound by name.

EXERCISE: Write IUPAC names for the methyl ester, amide, acyl chloride, anhydride, and nitrile corresponding to each of the following carboxylic acids.
(a) CH$_3$CH$_2$CH$_2$CH$_2$CH$_2$COOH   (b) (CH$_3$)$_2$CHCH$_2$CH$_2$COOH
(c) CH$_2$=CHCH$_2$CH$_2$COOH   (d) BrCH$_2$CH$_2$COOH
Write IUPAC names for the following ester and amide.
(e) (CH$_3$)$_2$CHCH$_2$CH$_2$COOCH$_2$CH(CH$_3$)$_2$ (f)· (CH$_3$)$_3$CCH$_2$CH$_2$CON(CH$_2$CH$_3$)$_2$

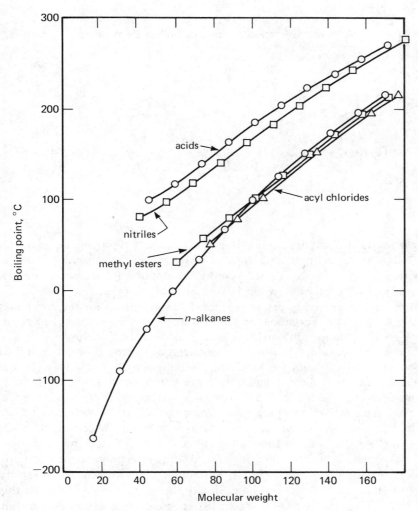

FIGURE 19.5. *Boiling points of various compounds.*

19.3
Physical Properties

In Figure 19.5 the boiling points of straight-chain acids, nitriles, methyl esters, and acyl chlorides are plotted against molecular weight. For comparison the boiling point curve for *n*-alkanes is also given. It can readily be seen that esters and acyl halides have approximately the boiling points expected for hydrocarbons of the same molecular weight. This correspondence indicates that the main attractive forces in the condensed phase for these compounds are the relatively weak van der Waals forces.

On the other hand, carboxylic acids and nitriles boil much higher than hydrocarbons of equivalent weight. In these compounds polar intermolecular forces must also be important in the liquid state.

$$R\!-\!C\!\equiv\!N\!:$$
$$\longmapsto$$

$$:\!N\!\equiv\!\overset{+}{C}\!-\!R$$
$$\longleftarrow$$

TABLE 19.2
Physical Properties for Acetamide Derivatives

	Molecular Weight	Melting Point, °C	Boiling Point, °C
$CH_3-\overset{O}{\overset{\|}{C}}-NH_2$	59	82	221
$CH_3-\overset{O}{\overset{\|}{C}}-NHCH_3$	73	28	204
$CH_3-\overset{O}{\overset{\|}{C}}-N(CH_3)_2$	87	-20	165

For carboxylic acids hydrogen bonding in the condensed phase is an added factor. In addition to supplying enough energy to overcome the normal van der Waals attractive forces, additional energy must be supplied to overcome the "extra" polar attractive forces. The result is a higher boiling point.

All methyl and ethyl esters, acyl chlorides, and nitriles for the straight-chain acids lower than tetradecanoic are liquids at room temperature. Simple anhydrides above nonanoic anhydride are solid at room temperature.

Amides, in particular, show the strong effects of hydrogen bonding in the condensed phase. The melting points and boiling points of acetamide, N-methylacetamide, and N,N-dimethylacetamide are tabulated in Table 19.2. Note that acetamide boils 215° higher than a comparable alkane. Dimethylacetamide still boils 95° higher than a comparable alkane, but it has a boiling point almost exactly the same as that for an acid or nitrile of comparable weight. A similar downward trend is seen in the melting points of the three compounds.

The explanation of these interesting trends lies in the phenomenon of hydrogen bonding. Acetamide, with two hydrogens attached to nitrogen, is extensively hydrogen-bonded in both the solid and liquid phases. In methylacetamide there is only one N—H, and therefore the hydrogen bonding is less extensive. Finally, dimethylacetamide cannot engage in hydrogen bonding at all, since it has no hydrogens attached to nitrogen.

Esters, amides, nitriles, acyl halides, and anhydrides are generally soluble in common organic solvents (ether, chloroform, benzene, and so on). Acetonitrile, dimethylformamide, and dimethylacetamide are miscible with water in all proportions. Because of their polar, aprotic nature, these compounds are excellent solvents. Ethyl acetate, which is only sparingly soluble in water, is also a common solvent. Because of its excellent solvent properties, ethyl acetate is a common constituent of many brands of paint remover and is also used as a fingernail polish remover. It may easily be recognized by its characteristic "fruity" odor.

19.4
Spectroscopy

A. *Nuclear Magnetic Resonance*

The chemical shifts of protons in the vicinity of the carbonyl group have similar resonance positions regardless of the exact nature of the compound. Typical val-

TABLE 19.3
Chemical Shifts in Compounds of the Type R—Y

Y	Chemical Shift in δ		
	CH_3Y	RCH_2Y	CH_3CH_2Y
—CHO	2.20	2.40	1.08
—COOH	2.10	2.36	1.16
—COOCH$_3$	2.03	2.13	1.12
—COCl	2.67		
—CONH$_2$	2.08	2.23	1.13
—CN	2.00	2.28	1.14

ues are summarized in Table 19.3. A typical example is the spectrum of methyl propanoate, shown in Figure 19.6.

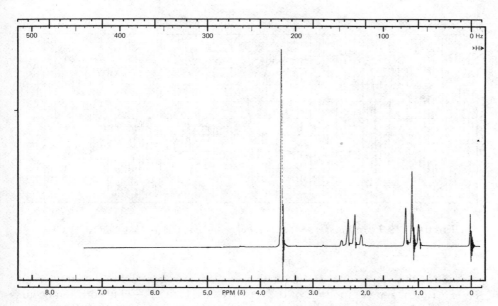

FIGURE 19.6. *Nmr spectrum of methyl propanoate,* $CH_3CH_2COOCH_3$.

B. *Infrared*

In Chapter 14 we saw that the characteristic absorption of aldehydes and ketones is the C=O stretch that occurs in the 1710–1825 cm^{-1} region. Other compounds containing the carbonyl group also absorb in this general region. The exact position of the absorption depends on the nature of the functional group. Typical values are listed in Table 19.4.

In addition to the carbonyl stretch, these derivatives have other useful infrared absorptions. Esters have a characteristic carbonyl C—O single bond stretch in the 1050–1250 cm^{-1} region. Both the C=O stretch at 1738 cm^{-1} and the C—O single bond stretch at 1170 cm^{-1} are clearly seen in the spectrum of ethyl octanoate in Figure 19.7.

TABLE 19.4
Carbonyl Stretching Bands of
Carboxylic Acid Derivatives in Solution

Functional Group	C=O Stretch, cm^{-1}
$-\overset{\overset{O}{\|}}{C}-OR$	1735
$-\overset{\overset{O}{\|}}{C}-Cl$	1800
$-\overset{\overset{O}{\|}}{C}-O-\overset{\overset{O}{\|}}{C}-$	1820 and 1760 (two peaks)
$-\overset{\overset{O}{\|}}{C}-NR_2$	1650–1690

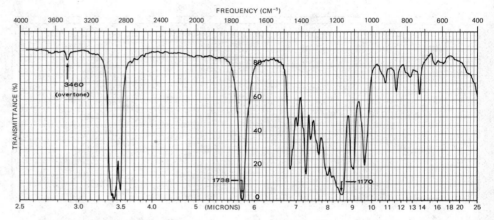

FIGURE 19.7. *Infrared spectrum of ethyl octanoate,* $CH_3(CH_2)_6COO_2CH_2CH_3$.

Amides that have one or two hydrogens on nitrogen show a characteristic N—H stretch. For compounds of the type $RCONH_2$ the N—H absorption occurs as two peaks at 3400 and 3500 cm^{-1}. For RCONHR compounds the N—H stretch comes at 3440 cm^{-1}.

The typical absorption of nitriles is the C≡N stretch at 2110–2160 cm^{-1}. Note that this absorption occurs in the general region where the C≡C triple bond absorbs (Section 14.6).

Recall that the exact stretching frequency for cyclic ketones is related to ring size. This same effect is seen with cyclic esters, which are called **lactones** (Section 28.5), and with cyclic amides, which are called **lactams** (Section 30.5).

$$CH_3\overset{\overset{O}{\|}}{C}OCH_3$$
1736 cm^{-1} 1735 cm^{-1} 1770 cm^{-1} 1800 cm^{-1}

$$CH_3\overset{\overset{\displaystyle O}{\|}}{C}NH_2$$

1680 cm^{-1} 1670 cm^{-1} 1700 cm^{-1} 1745 cm^{-1}

EXERCISE: Sketch the expected nmr spectra of (a) ethyl propanoate and (b) isopropyl acetate.

19.5
Basicity of the Carbonyl Oxygen

As in the cases of aldehydes, ketones (Section 13.6), and carboxylic acids (Section 18.7), the carbonyl oxygen of carboxylic acid derivatives has basic properties. The conjugate acid, an oxonium salt, plays an important role as an intermediate in acid-catalyzed reactions of all of these types of compounds. The actual basicity of the lone-pair electrons of the carbonyl oxygen depends markedly on the nature of the radical attached to the carbonyl group. This basicity is generally expressed quantitatively in terms of the acidity or pK_a of the conjugate acid.

$$\overset{\overset{\displaystyle +OH}{\|}}{RCY} \;\rightleftharpoons\; \overset{\overset{\displaystyle O}{\|}}{RCY} + H^+$$

$$K_a = \frac{[RCOY][H^+]}{\left[\overset{\overset{\displaystyle +OH}{\|}}{RCY}\right]}$$

$$pK_a = -\log K_a$$

Some pK_a values are summarized in Table 19.5. Most protonated carbonyl compounds are strong acids, stronger than H_3O^+ and comparable in acidity to sulfuric acid (p$K_a = -5.2$). That is, the carbonyl compounds themselves are weak bases in water, in a class with bisulfate ion. Some of the individual structural effects warrant comment.

Alcohols are generally a little weaker as bases than water itself, and ethers are weaker bases still. These variations are probably the result of solvation differences. The fewer the number of protons on an oxonium oxygen, the less the amount of solvation stabilization by hydrogen bonds to water. If the acid structure is less stable for whatever reason, the conjugate base has lower basicity.

The carbonyl oxygen of aldehydes and ketones is less basic than an alcohol or ether oxygen by several powers of ten. The lone-pair electrons of the carbonyl oxygen may be considered to be approximately sp^2 in character. These electrons have greater s-character than the lone pairs of alcohol oxygens. Hence, the lone-pair electrons of carbonyl oxygens are more tightly held. As a result, the carbonyl group as a conjugate base is more stable, and the corresponding protonated carbonyl is more acidic. This system is exactly analogous to the corresponding hydrocarbon cases. Recall that ethylene is more acidic than ethane (Section 12.4). In a protonated carbonyl group the O—H bond is described approximately as O$_{sp^2}$—H.

TABLE 19.5
Acidities of Protonated Compounds

Compound	Conjugate Acid	pK_a of Conjugate Acid
CH_3CONH_2	$CH_3\overset{\displaystyle +OH}{\underset{\displaystyle \|}{C}}NH_2$	0.0
H_2O	H_3O^+	-1.7
CH_3OH	$CH_3\overset{+}{O}H_2$	-2.2
$(CH_3CH_2)_2O$	$(CH_3CH_2)_2\overset{+}{O}H$	-3.6
CH_3COOH	$CH_3\overset{\displaystyle +OH}{\underset{\displaystyle \|}{C}}OH$	-6
$CH_3COOC_2H_5$	$CH_3\overset{\displaystyle +OH}{\underset{\displaystyle \|}{C}}OC_2H_5$	-6.5
CH_3COCH_3	$CH_3\overset{\displaystyle +OH}{\underset{\displaystyle \|}{C}}CH_3$	-7.2
CH_3CHO	$CH_3\overset{\displaystyle +OH}{\underset{\displaystyle \|}{C}}H$	~-8
CH_3COCl	$CH_3\overset{\displaystyle +OH}{\underset{\displaystyle \|}{C}}Cl$	~-9
CH_3CN	$CH_3C\equiv\overset{+}{N}H$	-10.1

In a protonated alcohol or ether the O—H bond involves an oxygen orbital that has greater *p*-character.

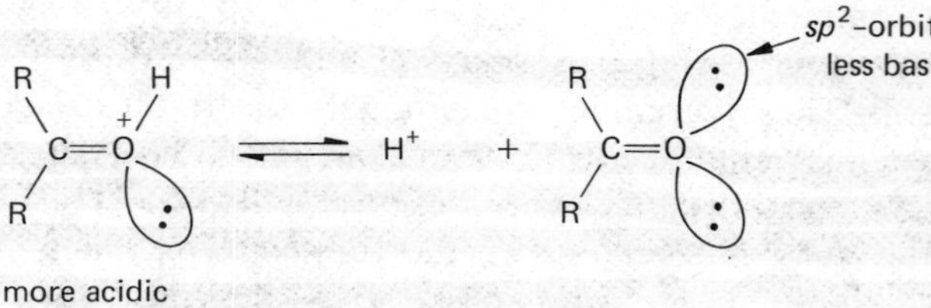

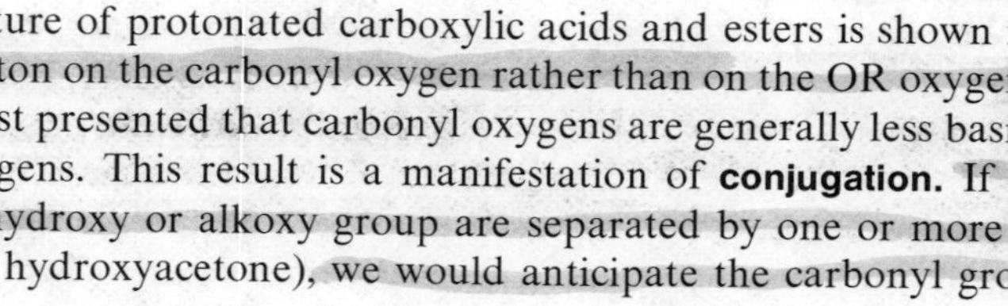

The structure of protonated carboxylic acids and esters is shown in Table 19.5 with the proton on the carbonyl oxygen rather than on the OR oxygen, despite the argument just presented that carbonyl oxygens are generally less basic than singly bonded oxygens. This result is a manifestation of **conjugation.** If the carbonyl group and hydroxy or alkoxy group are separated by one or more carbons (for example, in hydroxyacetone), we would anticipate the carbonyl group to be the less basic. That is, in an acidic medium the hydroxy or alkoxy group is protonated to a greater degree than is the carbonyl group.

$$\underset{\text{acetonyloxonium ion (more)}}{CH_3\overset{\overset{\displaystyle O}{\|}}{C}CH_2\overset{+}{O}H_2}$$

$$\underset{\text{hydroxyacetone}}{CH_3\overset{\overset{\displaystyle O}{\|}}{C}CH_2OH} + H^+ \rightleftharpoons \underset{\text{hydroxyacetonium ion (less)}}{CH_3\overset{\overset{\displaystyle +}{\overset{\displaystyle O}{\|}H}}{C}CH_2OH}$$

> Note that both protonated isomers are more acidic than their monofunctional counterparts: acetonyloxonium ion is more acidic than propyloxonium ion, $CH_3CH_2CH_2OH_2^+$, and hydroxyacetonium ion is more acidic than acetonium ion. In both cases the substituent has an electron-attracting inductive effect that results in an increase in acidity, just as in the case of substituted acetic acids (Section 18.4.B).

When the OH or OR group is attached directly to the carbonyl group, electron density on the singly bonded oxygen can "leak over" to the electron-attracting carbonyl group, as symbolized by the resonance structures.

$$\left[R-\overset{\overset{\displaystyle O}{\|}}{C}-OR' \longleftrightarrow R-\overset{\overset{\displaystyle O^-}{|}}{C}=\overset{+}{O}R' \right]$$

The actual electronic structure of the carboxylic acid or ester group may be represented as

$$R-\overset{\overset{\displaystyle O^{\delta-}}{\vdots}}{C}\overset{\delta+}{\cdots}OR'$$

The partial positive charge or oxonium character of the OR group makes this oxygen less basic than an ether oxygen. The partial negative charge on the carbonyl oxygen makes it more basic than a ketone oxygen. This argument does not mean that the alternative protonated compound cannot exist. It does say that this protonated compound, an acyloxonium ion, is much more acidic than a simple oxonium ion.

$$R-\overset{\overset{\displaystyle O}{\|}}{C}-\overset{+}{O}\overset{\displaystyle H}{\underset{\displaystyle H}{}} \qquad \text{more acidic than} \qquad \underset{\text{oxonium ion}}{RCH_2-\overset{+}{O}\overset{\displaystyle H}{\underset{\displaystyle H}{}}}$$

On the other hand, the carbonyl-protonated carboxylic grouping is stabilized by conjugation; the positive charge is distributed between the two oxygens, much as the negative charge is distributed in acetate ion (Section 18.4.A).

$$\left[R-\overset{\overset{\displaystyle +OH}{\|}}{C}-OH \longleftrightarrow R-\overset{\overset{\displaystyle OH}{|}}{C}=\overset{+}{O}H \longleftrightarrow R-\overset{\overset{\displaystyle OH}{|}}{\underset{+}{C}}-OH \right] \text{ less acidic than } R-\overset{\overset{\displaystyle +OH}{\|}}{C}-R'$$

These same considerations apply to an even greater extent in the case of amides. Ammonia is much more basic than water: $pK_a(NH_4^+) = 9.5$; $pK_a(H_3O^+) = -1.7$. The nitrogen in an amide is far less basic than that in ammonia because of the important contribution of the dipolar resonance structure.

$$\left[\begin{array}{ccc} \underset{\displaystyle R-\overset{\displaystyle O}{\overset{\displaystyle \|}{C}}-NH_2} {} & \longleftrightarrow & \underset{\displaystyle R-\overset{\displaystyle O^-}{\overset{\displaystyle |}{C}}=\overset{+}{N}H_2} {} \end{array}\right] \equiv \quad R-\overset{O^{\delta-}}{\overset{\vdots}{C}}\cdots\overset{\delta+}{N}H_2$$

That is, the nitrogen in an amide already has some of the character of an ammonium ion. If this nitrogen becomes protonated, the resonance stabilization of the amide is lost.

$$R-\overset{O}{\overset{\|}{C}}-\overset{+}{N}H_3$$

(no lone pair for conjugation with carbonyl group)

The situation is represented schematically in Figure 19.8. Note that the O-protonated amide is greatly stabilized by resonance.

$$\left[\begin{array}{ccccc} R-\overset{+OH}{\overset{\|}{C}}-NH_2 & \longleftrightarrow & R-\overset{OH}{\overset{|}{C}}-\overset{+}{N}H_2 & \longleftrightarrow & R-\overset{OH}{\overset{|}{C}}=\overset{+}{N}H_2 \end{array}\right]$$

In fact, as shown in Table 19.5, the O-protonated amide is almost 100 times less acidic than H_3O^+.

Finally, note that the triply bonded nitrogen in nitriles is far less basic than ammonia. That is, the protonated nitrile, $R-C\equiv\overset{+}{N}H$, is some 20 powers of ten more acidic than NH_4^+. Again, this difference points up the powerful effect that s-character has on the relative stability of lone-pair electrons. The lone-pair electrons in a nitrile are approximately sp in character; those in ammonia are almost sp^3 in character. Recall the effect of a corresponding change in hydrocarbons: acetylene is far more acidic than methane (Section 12.4).

$HC\equiv C-H$	CH_3-H	
$pK_a \approx 25$	$pK_a \approx 50$	$\Delta pK_a \approx 25$
$RC\equiv\overset{+}{N}-H$	$\overset{+}{N}H_3-H$	
$pK_a = -10$	$pK_a = 9.5$	$\Delta pK_a \approx 20$

EXERCISE: In this section we showed that the basicity of the lone-pair electrons on carbonyl groups of various carboxylic acid derivatives is related to the importance of the dipolar resonance structure. Consider how the following bond lengths agree with this interpretation.

CH_3-NH_2	1.47 Å	$CH_3\overset{O}{\overset{\|}{C}}-NH_2$	1.36 Å
CH_3-OCH_3	1.42 Å	$CH_3\overset{O}{\overset{\|}{C}}-OCH_3$	1.36 Å
CH_3-F	1.38 Å	$CH_3\overset{O}{\overset{\|}{C}}-F$	1.37 Å

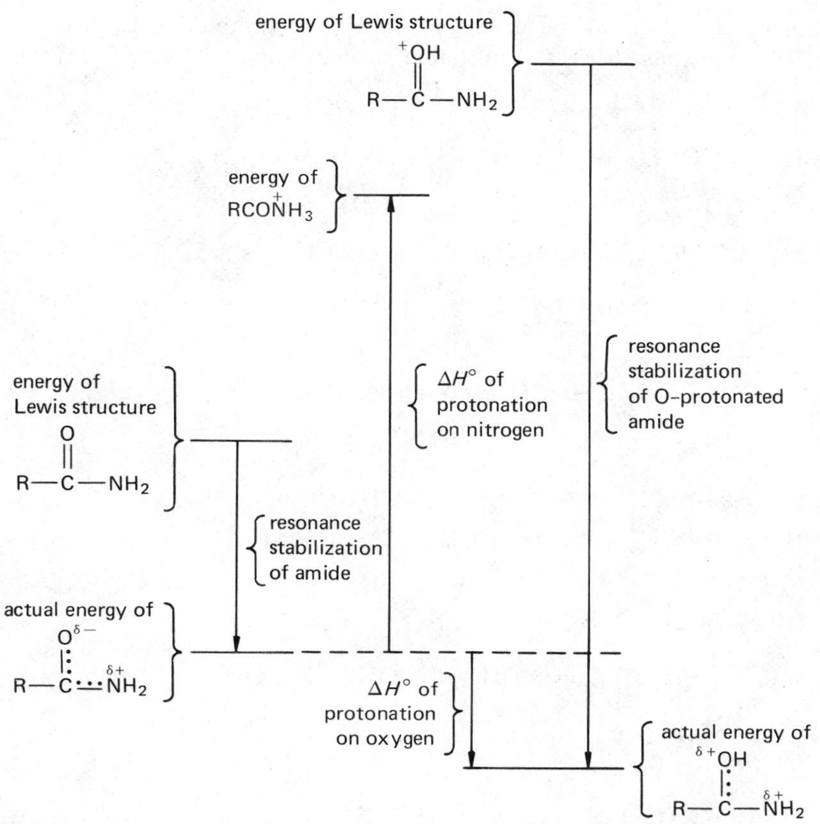

539

Sec. 19.6

Hydrolysis,
Nucleophilic
Addition-
Elimination

energy of Lewis structure

$$\overset{+OH}{\underset{\parallel}{R—C—NH_2}}$$

energy of
$RCO\overset{+}{N}H_3$

energy of
Lewis structure

$$\overset{O}{\underset{\parallel}{R—C—NH_2}}$$

resonance
stabilization
of O-protonated
amide

$\Delta H°$ of
protonation
on nitrogen

resonance
stabilization
of amide

actual energy of

$$\overset{O^{\delta-}}{\underset{}{R—C\cdots\overset{\delta+}{N}H_2}}$$

$\Delta H°$ of
protonation
on oxygen

actual energy of

$$\overset{\delta+OH}{\underset{}{R—C—\overset{\delta+}{N}H_2}}$$

FIGURE 19.8. *Illustrating why N-protonated amides are more acidic than O-protonated amides, even though H_3N: is more basic than H_2O:. The schematic given is for the system*

$$RCONH_2 + H_3O^+ \rightleftharpoons H_2O + \left\{ \overset{+OH}{\underset{\parallel}{RCNH_2}} \text{ or } \overset{O}{\underset{\parallel}{RCNH_3^+}} \right\}$$

19.6
Hydrolysis, Nucleophilic Addition-Elimination

The most characteristic reaction of the functional derivatives of carboxylic acids is **hydrolysis,** the reaction with water to give the carboxylic acid itself.

$$\overset{O}{\underset{\parallel}{R—C—X}} + H_2O \longrightarrow \overset{O}{\underset{\parallel}{R—C—OH}} + HX$$

X = halogen, OR, NR$_2$, OOCR

Esters react with water to yield the corresponding carboxylic acid and alcohol.

$$\overset{O}{\underset{\parallel}{CH_3C—OCH_3}} + H_2O \rightleftharpoons CH_3COOH + CH_3OH$$

The reaction is generally slow, but is strongly catalyzed by acid. The acid-catalyzed hydrolysis of esters of primary and secondary alcohols is simply the reverse of the acid-catalyzed esterification reaction (Section 18.7). The reaction is an equilibrium process, but can be driven practically to completion by using a large excess of water (page 516). The probable mechanism involves the following steps.

$$(1) \quad CH_3-\overset{\overset{\displaystyle O}{\|}}{C}-OCH_3 + H^+ \rightleftharpoons CH_3-\overset{\overset{\displaystyle +OH}{\|}}{C}-OCH_3$$

$$(2) \quad CH_3-\overset{\overset{\displaystyle +OH}{\|}}{C}-OCH_3 + H_2O \rightleftharpoons CH_3-\underset{\underset{\displaystyle +OH_2}{|}}{\overset{\overset{\displaystyle OH}{|}}{C}}-OCH_3$$

$$(3) \quad CH_3-\underset{\underset{\displaystyle +OH_2}{|}}{\overset{\overset{\displaystyle OH}{|}}{C}}-OCH_3 \rightleftharpoons CH_3-\underset{\underset{\displaystyle OH}{|}}{\overset{\overset{\displaystyle OH}{|}}{C}}-OCH_3 + H^+$$

$$(4) \quad CH_3-\underset{\underset{\displaystyle OH}{|}}{\overset{\overset{\displaystyle OH}{|}}{C}}-OCH_3 + H^+ \rightleftharpoons CH_3-\underset{\underset{\displaystyle OH}{|}}{\overset{\overset{\displaystyle OH\ H}{|}}{C}}-\overset{+}{O}CH_3$$

$$(5) \quad CH_3-\underset{\underset{\displaystyle OH}{|}}{\overset{\overset{\displaystyle OH\ H}{|}}{C}}-\underset{+}{O}CH_3 \rightleftharpoons CH_3-\overset{\overset{\displaystyle +OH}{\|}}{C}-OH + CH_3OH$$

$$(6) \quad CH_3-\overset{\overset{\displaystyle +OH}{\|}}{C}-OH \rightleftharpoons CH_3-\overset{\overset{\displaystyle O}{\|}}{C}-OH + H^+$$

Let us examine the role of the acid catalyst in the preceding reaction. In neutral water the preponderant nucleophile present is water. Even though the carbonyl double bond is polarized, water is not a sufficiently strong nucleophile to add to it at a reasonable rate. Furthermore, addition of water to methyl acetate would produce an intermediate bearing both a positive and a negative charge. Since charge separation requires electrostatic energy, this type of addition is exceptionally slow.

$$CH_3-\overset{\overset{\displaystyle O}{\|}}{C}-OCH_3 + H_2O \rightleftharpoons CH_3-\underset{\underset{\displaystyle +OH_2}{|}}{\overset{\overset{\displaystyle O^-}{|}}{C}}-OCH_3$$

In the presence of mineral acids, the ester may be protonated (Section 19.5).

$$CH_3-\overset{\overset{\displaystyle O}{\|}}{C}-OCH_3 + H^+ \rightleftharpoons \left[CH_3-\overset{\overset{\displaystyle +OH}{\|}}{C}-OCH_3 \longleftrightarrow CH_3-\overset{\overset{\displaystyle OH}{|}}{C}=\overset{+}{O}CH_3 \right]$$
$$\text{resonance-stabilized cation}$$

Since there is a very large excess of water molecules, and since the ester carbonyl is actually less basic than water (Table 19.5), only a small percentage of the ester is protonated at moderate acid concentration. However, the carbonyl carbon in the protonated species is much more electrophilic and reacts much faster with the weak nucleophile water than does the unprotonated ester. Furthermore, addition now involves no charge separation.

541

Sec. 19.6

**Hydrolysis,
Nucleophilic
Addition-
Elimination**

$$CH_3-\overset{\overset{\displaystyle OH^+}{\|}}{C}-OCH_3 + H_2\ddot{O} \rightleftharpoons CH_3-\overset{\overset{\displaystyle OH}{|}}{\underset{\underset{\displaystyle +OH_2}{|}}{C}}-OCH_3$$

In some cases acid-catalyzed ester hydrolysis involves cleavage of the *alkyl-oxygen* bond rather than the *acyl-oxygen* bond. Such is the case with *t*-butyl acetate (19-3).

$$CH_3\overset{\overset{\displaystyle O}{\|}}{C}-O-\overset{\overset{\displaystyle CH_3}{|}}{\underset{\underset{\displaystyle CH_3}{|}}{C}}CH_3 + H_2O \overset{H^+}{\rightleftharpoons} CH_3COOH + CH_3\overset{\overset{\displaystyle CH_3}{|}}{\underset{\underset{\displaystyle CH_3}{|}}{C}}-OH \quad (19\text{-}3)$$

Although the products are the same in both cases, the different mechanisms may be demonstrated by the labeling experiments (19-4) and (19-5).

$$CH_3\overset{\overset{\displaystyle O}{\|}}{C}-{}^{18}OCH_3 + H_2O \longrightarrow CH_3\overset{\overset{\displaystyle O}{\|}}{C}-OH + CH_3{}^{18}OH \quad (19\text{-}4)$$

$$CH_3\overset{\overset{\displaystyle O}{\|}}{C}-{}^{18}O-\overset{\overset{\displaystyle CH_3}{|}}{\underset{\underset{\displaystyle CH_3}{|}}{C}}CH_3 + H_2O \longrightarrow CH_3\overset{\overset{\displaystyle O}{\|}}{C}-{}^{18}OH + CH_3\overset{\overset{\displaystyle CH_3}{|}}{\underset{\underset{\displaystyle CH_3}{|}}{C}}-OH \quad (19\text{-}5)$$

The probable mechanism for this reaction involves the following steps.

$$(1) \quad CH_3\overset{\overset{\displaystyle O}{\|}}{C}-O-\overset{\overset{\displaystyle CH_3}{|}}{\underset{\underset{\displaystyle CH_3}{|}}{C}}CH_3 + H^+ \rightleftharpoons CH_3\overset{\overset{\displaystyle +OH}{\|}}{C}-O-\overset{\overset{\displaystyle CH_3}{|}}{\underset{\underset{\displaystyle CH_3}{|}}{C}}CH_3$$

$$(2) \quad CH_3\overset{\overset{\displaystyle +OH}{\|}}{C}-O-\overset{\overset{\displaystyle CH_3}{|}}{\underset{\underset{\displaystyle CH_3}{|}}{C}}CH_3 \rightleftharpoons CH_3\overset{\overset{\displaystyle OH}{|}}{C}=O + CH_3\overset{\overset{\displaystyle CH_3}{|}}{\underset{\underset{\displaystyle CH_3}{|}}{C}}{}^+$$

$$(3) \quad CH_3\overset{\overset{\displaystyle CH_3}{|}}{\underset{\underset{\displaystyle CH_3}{|}}{C}}{}^+ + H_2O \rightleftharpoons CH_3\overset{\overset{\displaystyle CH_3}{|}}{\underset{\underset{\displaystyle CH_3}{|}}{C}}-\overset{+}{O}H_2$$

$$(4) \quad CH_3\overset{\overset{\displaystyle CH_3}{|}}{\underset{\underset{\displaystyle CH_3}{|}}{C}}-\overset{+}{O}H_2 \rightleftharpoons CH_3\overset{\overset{\displaystyle CH_3}{|}}{\underset{\underset{\displaystyle CH_3}{|}}{C}}-OH + H^+$$

It is reasonable that *t*-butyl acetate should react by this mechanism whereas methyl acetate does not. The reaction is simply an acid-catalyzed S_N1 process in which the nucleophile water replaces the nucleophile acetic acid. The reactive intermediate is the *t*-butyl cation. Esters of other alcohols that give rise to relatively stable carbocations also undergo hydrolysis by this mechanism.

Ester hydrolysis is also strongly catalyzed by hydroxide ion. Since the carboxylic acid product neutralizes one equivalent of hydroxide, it is actually necessary to employ stoichiometric amount of base. That is, hydroxide ion is actually a reagent instead of just a catalyst. The products are the salt of the carboxylic acid and the corresponding alcohol.

$$CH_3(CH_2)_8CH{=}\overset{\overset{\displaystyle CH_3}{|}}{C}COOCH_3 + KOH \xrightarrow[\underset{\Delta}{H_2O}]{C_2H_5OH}$$

$$CH_3(CH_2)_8CH{=}\overset{\overset{\displaystyle CH_3}{|}}{C}CO_2{}^- \ K^+ \xrightarrow{H_2SO_4} CH_3(CH_2)_8CH{=}\overset{\overset{\displaystyle CH_3}{|}}{C}COOH$$
$$(68{-}83\%)$$

> To a solution of 20 g of methyl (E)-2-methyl-2-dodecenoate in 100 mL of 95% aqueous ethanol is added 8.8 g of potassium hydroxide. The solution is refluxed for 1.5 hr, concentrated to 40 mL, and acidified by the addition of 5 N sulfuric acid. The product is isolated by extraction with petroleum ether. After removal of the petroleum ether, the crude product is distilled to yield 18 g of pure acid, m.p. 29.5–32.5°.

The probable mechanism for base-catalyzed hydrolysis is illustrated below with methyl acetate.

$$(1)\ CH_3\overset{\overset{\displaystyle O}{\|}}{C}{-}OCH_3 + OH^- \rightleftharpoons CH_3\underset{\underset{\displaystyle OH}{|}}{\overset{\overset{\displaystyle O^-}{|}}{C}}{-}OCH_3$$

$$(2)\ CH_3\underset{\underset{\displaystyle OH}{|}}{\overset{\overset{\displaystyle O^-}{|}}{C}}{-}OCH_3 \rightleftharpoons CH_3\overset{\overset{\displaystyle O}{\|}}{C}{-}OH + CH_3O^-$$

$$(3)\ CH_3\overset{\overset{\displaystyle O}{\|}}{C}{-}OH + CH_3O^- \longrightarrow CH_3\overset{\overset{\displaystyle O}{\|}}{C}{-}O^- + CH_3OH$$

In basic aqueous solution two nucleophiles are present, H_2O and OH^-. As we saw above, H_2O is a poor nucleophile and therefore reacts slowly with the carbonyl carbon. On the other hand, OH^- is a much stronger nucleophile and adds more rapidly to the carbonyl carbon.

After addition has taken place, elimination of a nucleophile from the tetrahedral intermediate can occur. Elimination of hydroxide ion merely reverses the initial addition step. However, elimination of methoxide ion gives a new species—acetic acid. Since acetic acid is a weak acid and methoxide ion is a strong base, a rapid acid-base reaction then occurs, yielding acetate ion and methanol. Because of the great difference in acidity between acetic acid ($pK_a \approx 5$) and methanol ($pK_a \approx 16$), this last step is essentially irreversible (K for the last reaction $\approx 10^{11}$). Thus basic hydrolysis of esters differs from acid-catalyzed hydrolysis in that the equilibrium constant for the overall reaction is very large and it is sufficient to use only one equivalent of water in order for the reaction to proceed to completion.

It is interesting to compare the C—O bond-forming step in the acid-catalyzed and base-catalyzed mechanisms. In the former case the "weak" nucleophile H_2O

543

Sec. 19.6

**Hydrolysis,
Nucleophilic
Addition-
Elimination**

adds to the "strongly" electrophilic bond $C=\overset{+}{O}H$

$$CH_3\overset{\overset{+}{O}H}{\underset{}{C}}-OCH_3 + \overset{..}{\overset{..}{O}}H_2 \rightleftharpoons CH_3\overset{OH}{\underset{\underset{+}{O}H_2}{C}}-OCH_3$$

In the latter case the "strong" nucleophile OH^- adds to the "weakly" electrophilic bond $C=O$

$$CH_3\overset{O}{\underset{}{C}}-OCH_3 + :\overset{..}{O}H^- \rightleftharpoons CH_3\overset{O^-}{\underset{OH}{C}}-OCH_3$$

Either case is better than the case where the "weak" nucleophile H_2O adds to the "weakly" electrophilic bond $C=O$.

The most rapid hydrolysis would involve *both* acid and base catalysis.

$$CH_3\overset{\overset{+}{O}H}{\underset{}{C}}-OCH_3 + OH^- \longrightarrow CH_3\overset{OH}{\underset{OH}{C}}-OCH_3$$

In aqueous solution this mechanism is not observed for a simple reason. In acidic solution, where the concentration of $C=\overset{+}{O}H$ species is appreciable, the concentration of OH^- is very small. In basic solution, where the concentration of OH^- is appreciable, the concentration of $C=\overset{+}{O}H$ is low.

Amides undergo hydrolysis to 1 mole of carboxylic acid and 1 mole of ammonia or amine. The reaction is catalyzed by acid or base. Amides undergo hydrolysis much more slowly than esters, and therefore more vigorous conditions are normally required.

(88–90%)

> A mixture of 600 g of 2-phenylbutanamide, 1 L of water, and 400 mL of conc. sulfuric acid is refluxed for 2 hr. After cooling the mixture and diluting with 1 L of water, the oily organic layer is separated and distilled to yield 530–554 g of 2-phenylbutanoic acid, m.p. 42°.

Both the acid- and base-catalyzed reactions are essentially irreversible. In the former case ammonium ion is produced; in the latter case a carboxylate ion is formed.

$$R\overset{O}{\underset{}{C}}-NH_2 + H_2O \overset{H^+}{\rightleftharpoons} R\overset{O}{\underset{}{C}}-OH + NH_3 \overset{H^+}{\longrightarrow} NH_4^+$$

$$R\overset{O}{\underset{}{C}}-NH_2 + H_2O \overset{OH^-}{\rightleftharpoons} R\overset{O}{\underset{}{C}}-OH + NH_3 \overset{OH^-}{\longrightarrow} RCO_2^-$$

Acyl halides and anhydrides undergo hydrolysis with great ease. Acetyl chloride reacts with water to give acetic acid and hydrogen chloride, whereas acetic anhydride gives two equivalents of acetic acid.

$$\underset{\underset{O}{\parallel}}{CH_3C}-Cl + H_2O \longrightarrow \underset{\underset{O}{\parallel}}{CH_3C}-OH + HCl$$

$$\underset{\underset{O}{\parallel}}{CH_3C}-O-\underset{\underset{O}{\parallel}}{CCH_3} + H_2O \longrightarrow 2\,\underset{\underset{O}{\parallel}}{CH_3C}-OH$$

As with esters and amides, hydrolysis of acyl halides and anhydrides is subject to acid or base catalysis. However, both acyl halides and anhydrides react much more rapidly than esters, and uncatalyzed hydrolysis occurs readily, provided the reaction mixture is homogeneous. Since most acyl halides and anhydrides are only sparingly soluble in water, hydrolysis often appears to be slow. However, if a solvent is used that dissolves both water and the organic reactant, hydrolysis is rapid.

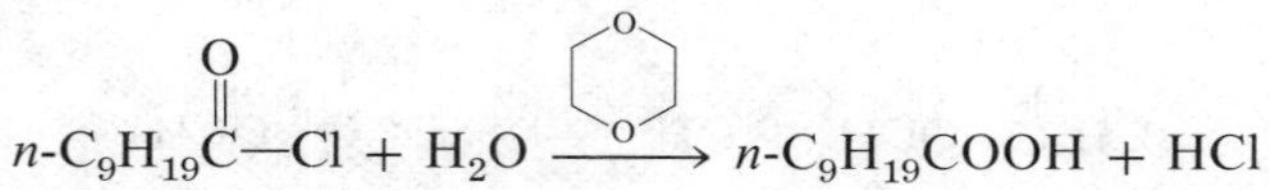

$$n\text{-}C_9H_{19}\underset{\underset{O}{\parallel}}{C}-Cl + H_2O \xrightarrow{} n\text{-}C_9H_{19}COOH + HCl$$

Unlike the other types of compounds we have discussed in this chapter, nitriles do not contain the carbonyl group. They are considered to be functional derivatives of carboxylic acids because exhaustive hydrolysis of the nitrile (cyano) group yields the carboxy group.

$$RCN + H_2O \longrightarrow RCOOH + NH_3$$

The actual conditions required for this hydrolysis, which proceeds by way of the intermediate amide, are quite severe.

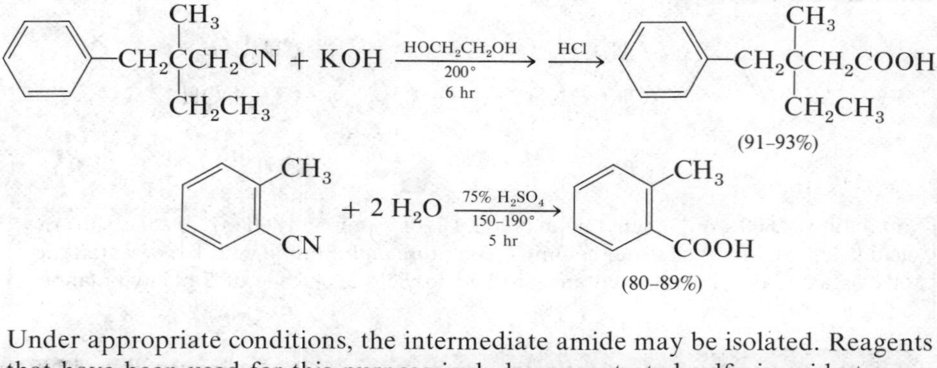

(91–93%)

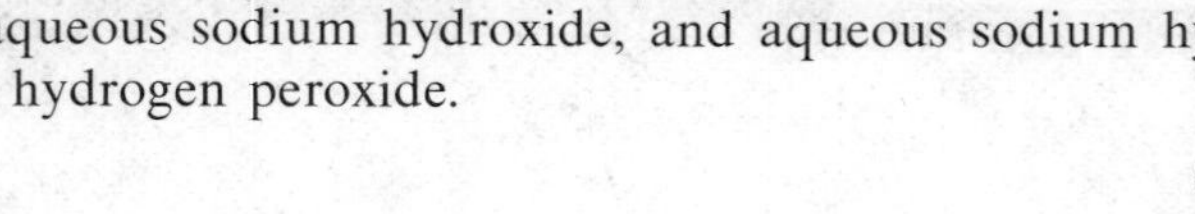

(80–89%)

Under appropriate conditions, the intermediate amide may be isolated. Reagents that have been used for this purpose include concentrated sulfuric acid at room temperature, aqueous sodium hydroxide, and aqueous sodium hydroxide containing 6–12% hydrogen peroxide.

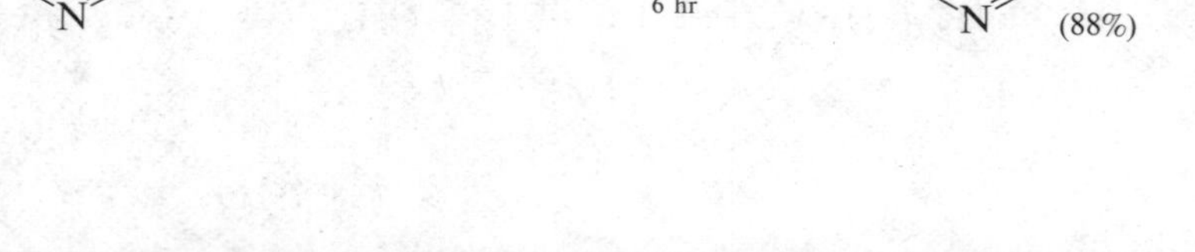

(88%)

545

Sec. 19.6

Hydrolysis,
Nucleophilic
Addition-
Elimination

The mechanism for hydrolysis of the nitrile grouping is probably similar to that observed in the hydrolysis of esters, amides, and acyl halides. The polarized $C \equiv N$ triple bond is in many respects similar to a carbonyl group. Hydration of the neutral bond is extremely slow, but the poor nucleophile water adds easily to the conjugate acid.

$$R-C\equiv\overset{+}{N}-H + H_2O \rightleftharpoons R-\overset{\overset{+}{O}H_2}{\underset{|}{C}}=NH \rightleftharpoons R-\overset{\overset{OH}{|}}{C}=NH + H^+$$

The resulting neutral species is simply an unstable tautomer of a primary amide and is converted to the amide by a protonation-deprotonation sequence.

$$R-\overset{\overset{OH}{|}}{C}=NH + H^+ \rightleftharpoons \left[R-\overset{\overset{OH}{|}}{C}=\overset{+}{N}H_2 \longleftrightarrow R-\overset{\overset{+OH}{\|}}{C}-NH_2 \right] \rightleftharpoons R-\overset{\overset{O}{\|}}{C}-NH_2 + H^+$$

In a similar manner, nucleophilic hydroxide ion will add to the neutral nitrile grouping to yield, after protonation, the unstable amide tautomer.

$$R-C\equiv N: + OH^- \rightleftharpoons R-\overset{\overset{OH}{|}}{C}=\overset{..}{\underset{..}{N}}{}^- \underset{OH^-}{\overset{H_2O}{\rightleftharpoons}} R-\overset{\overset{OH}{|}}{C}=NH$$

In hydrolysis of an ester, acyl halide, or anhydride the overall result is replacement of the nucleophile —OH for one of the nucleophiles —OR, —X, or —OCOR. Thus hydrolyses of these compounds, as well as some of the reactions of carboxylic acids themselves (Section 18.7.C), are but further examples of **nucleophilic substitution** (Chapter 8). However, as we have seen in this section, the mechanism for these nucleophilic substitution reactions is different from the substitution mechanisms we have previously encountered (S_N1, S_N2). At this point it is instructive to examine the three distinct ways in which the bond-breaking and bond-making operations of a nucleophilic substitution process may be timed.

1. Bond breaking may occur first, followed in a subsequent step by bond making.

$$R-Y \rightleftharpoons R^+ + Y^-$$
$$R^+ + :N^- \longrightarrow R-N$$

This sequence of steps is involved in the S_N1 process for substitution in alkyl halides (Section 8.12). In the case of carboxylic acid derivatives, the intermediate carbocation is called an **acylium ion.**

$$R-\overset{\overset{O}{\|}}{C}-Y \rightleftharpoons :Y^- + \left[R-\overset{+}{C}=\overset{..}{O}: \longleftrightarrow R-C\equiv\overset{+}{O}: \right]$$

acylium ion

$$R-\overset{+}{C}=\overset{..}{O}: + :N^- \longrightarrow R-\overset{\overset{O}{\|}}{C}-N$$

Only relatively few reactions of carboxylic acids and their derivatives occur by this mechanism. One example of a reaction proceeding by this path was seen on page 519.

2. Bond breaking and bond making may occur more or less synchronously.

$$R-Y + :N^- \longrightarrow \left[\overset{\delta-}{N}\cdots R \cdots \overset{\delta-}{Y}\right]^{\ddagger} \longrightarrow N-R + :Y^-$$

This is the familiar S_N2 mechanism (Sections 8.4–8.10). The synchronous mechanism is rare in the chemistry of carboxylic acid derivatives. It has been suggested by a few workers that some reactions of acyl halides may occur by this path, but actual evidence for the mechanism is sparse.

3. Bond making may occur first, followed in a subsequent step by bond breaking. This mechanism is not possible in the case of simple alkyl halides, since it would require the intervention of a pentacoordinate carbon. However, the mechanism is the most common one in the chemistry of carboxylic acids and their derivatives.

$$R-\overset{\overset{O}{\|}}{C}-Y + :N^- \rightleftharpoons R-\overset{\overset{O^-}{|}}{\underset{N}{C}}-Y$$

$$R-\overset{\overset{O^-}{|}}{\underset{N}{C}}-Y \rightleftharpoons R-\overset{\overset{O}{\|}}{C}-N + :Y^-$$

In this case *addition* to the carbonyl group occurs first, giving an intermediate in which the former carbonyl carbon now has sp^3-hybridization. This intermediate then decomposes by ejection of a nucleophile, restoring the carbonyl group. This is an extremely important mechanism. Almost all nucleophilic substitution reactions of carboxylic acids and their derivatives occur by this pathway, the so-called **nucleophilic addition-elimination** mechanism.

EXERCISE: Write equations showing all steps in the acid- and base-catalyzed hydrolysis mechanisms for acetamide.

19.7
Other Nucleophilic Substitution Reactions

The nucleophilic substitution mechanism discussed in the previous section is general. Carboxylic acids and their derivatives react with nucleophiles other than water in the same manner. In this section we shall consider some of these other reactions.

A. Reaction with Alcohols

Acyl halides react with alcohols to yield esters and the corresponding hydrohalic acid.

$$\underset{\text{acetyl chloride}}{CH_3\overset{\overset{O}{\|}}{C}-Cl} + \underset{\text{isobutyl alcohol}}{CH_3\overset{\overset{CH_3}{|}}{C}HCH_2OH} \longrightarrow \underset{\text{isobutyl acetate}}{CH_3\overset{\overset{O}{\|}}{C}-OCH_2\overset{\overset{CH_3}{|}}{C}HCH_3} + HCl$$

Such reactions are usually carried out in the presence of some base that serves to neutralize the HX formed in the reaction.

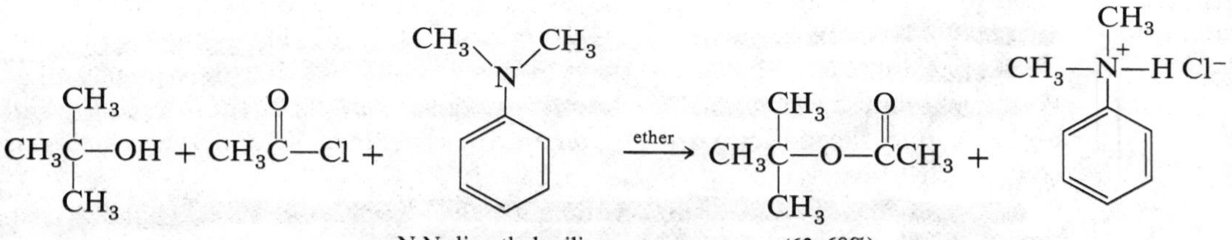

$$\underset{\underset{\displaystyle CH_3}{|}}{\overset{\overset{\displaystyle CH_3}{|}}{CH_3C}}-OH + CH_3\overset{\overset{\displaystyle O}{\|}}{C}-Cl + \underset{\text{N,N-dimethylaniline}}{\left[\text{N,N-dimethylaniline}\right]} \xrightarrow{\text{ether}} \underset{\underset{\displaystyle CH_3}{|}}{\overset{\overset{\displaystyle CH_3}{|}}{CH_3C}}-O-\overset{\overset{\displaystyle O}{\|}}{C}CH_3 + \underset{(63\text{–}68\%)}{}$$

To a refluxing solution of 114 g of *t*-butyl alcohol and 202 g of N,N-dimethylaniline in 200 mL of dry ether is added dropwise 124 g of acetyl chloride. After addition of all the acyl chloride, the mixture is cooled in an ice bath, and the solid N,N-dimethylaniline hydrochloride is removed by filtration. The ether layer is extracted with aqueous sulfuric acid to remove excess amine and worked up to yield 110–119 g of *t*-butyl acetate, b.p. 95–98°.

Other bases commonly used for this purpose are triethylamine and pyridine.

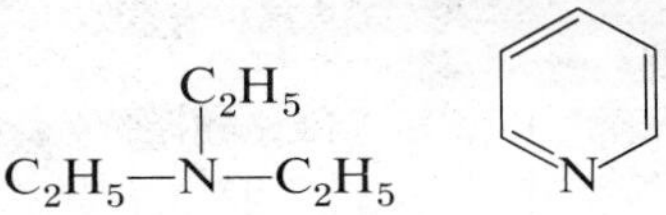

$$\underset{\text{triethylamine}}{C_2H_5-\overset{\overset{\displaystyle C_2H_5}{|}}{N}-C_2H_5} \qquad \underset{\text{pyridine}}{}$$

Since acyl halides are readily available from the corresponding carboxylic acids (Section 18.7.C), the following sequence is often used for the preparation of esters.

$$RCOOH \xrightarrow[\text{PCl}_3]{\overset{SOCl_2}{\text{or}}} R\overset{\overset{\displaystyle O}{\|}}{C}-Cl \xrightarrow{R'OH} R\overset{\overset{\displaystyle O}{\|}}{C}-OR'$$

Anhydrides also react readily with alcohols. The product is 1 mole of ester and 1 mole of the carboxylic acid corresponding to the anhydride used.

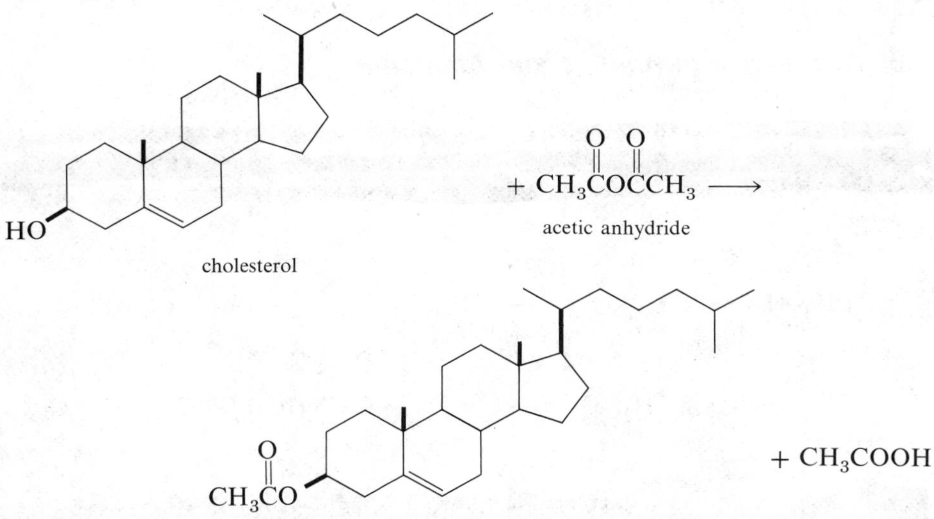

cholesterol

$$+ CH_3\overset{\overset{\displaystyle O}{\|}}{C}O\overset{\overset{\displaystyle O}{\|}}{C}CH_3 \longrightarrow$$

acetic anhydride

cholesteryl acetate

$$+ CH_3COOH$$

$$\left[\begin{array}{l} \text{A mixture of 5 g of cholesterol and 7.5 mL of acetic anhydride is boiled for 1 hr. The} \\ \text{mixture is cooled and filtered to yield 5 g of cholesteryl acetate, m.p. 114–115°.} \end{array} \right]$$

The mechanism of the reaction is the same as that for the reaction of an alcohol with an acyl halide; the leaving group in this case is the carboxylate anion rather than a halide ion. This reaction is an important method for the preparation of acetates, since acetic anhydride is an inexpensive reagent and the reaction is convenient to carry out.

Esters undergo reaction with alcohols to give a new ester and a new alcohol. The reaction is catalyzed by either acid or base and is called **transesterification.**

$$CH_3\overset{\overset{\displaystyle O}{\|}}{C}-OCH_2CH_3 + CH_3OH \underset{}{\overset{H^+ \text{ or } OCH_3^-}{\rightleftharpoons}} CH_3\overset{\overset{\displaystyle O}{\|}}{C}-OCH_3 + C_2H_5OH$$

The mechanism for the transesterification process involves steps identical to those given for acid-catalyzed and base-catalyzed ester hydrolysis, with one significant exception. In base-catalyzed transesterification, step (3) of the mechanism on page 542 cannot occur because the free carboxylic acid is never formed. Thus, base-catalyzed transesterification is subject to the same equilibrium conditions that apply to the acid-catalyzed reaction.

Amides react with alcohols under acidic conditions to yield the corresponding ester and an ammonium salt.

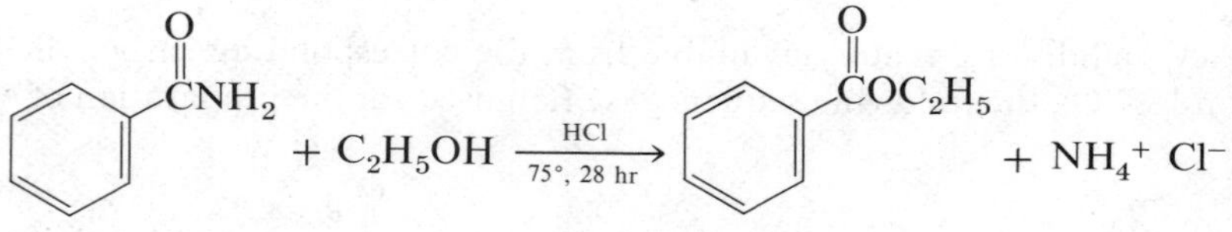

(52%)

The base-catalyzed equivalent of this process requires such drastic conditions that it is rarely encountered.

B. Reaction with Amines and Ammonia

Acyl halides react with ammonia or amines that have at least one hydrogen bound to nitrogen to give amides. Since one equivalent of hydrogen chloride is formed in the reaction, two equivalents of ammonia or the amine must be used.

$$CH_3CH_2\overset{\overset{\displaystyle O}{\|}}{C}-Cl + 2 NH_3 \longrightarrow CH_3CH_2\overset{\overset{\displaystyle O}{\|}}{C}-NH_2 + NH_4^+ Cl^-$$

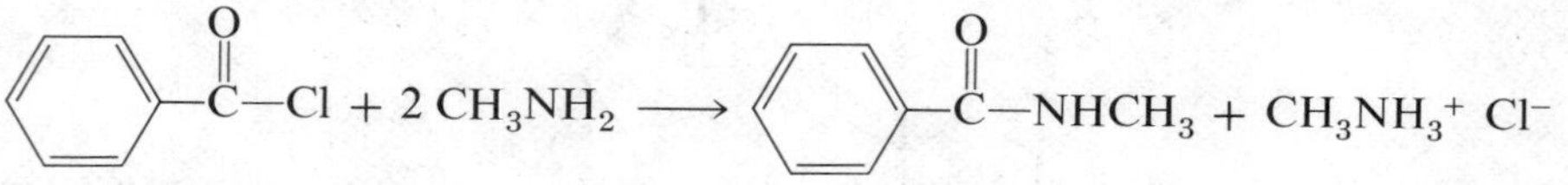

Alternatively, sodium hydroxide may be used to neutralize the HCl in a procedure known as the **Schotten–Baumann** method.

549

Sec. 19.7

Other
Nucleophilic
Substitution
Reactions

$$
\text{(C}_6\text{H}_5)\text{—C(=O)—Cl} + \text{piperidine-NH} + \text{NaOH} \xrightarrow{\text{H}_2\text{O}} \text{(C}_6\text{H}_5)\text{—C(=O)—N(piperidine)} + \text{Na}^+\text{Cl}^-
$$

The acyl chloride used is generally insoluble in water and reacts slowly with the sodium hydroxide. The organic amine, however, dissolves in the acid chloride and reacts rapidly. The amine hydrochloride also produced dissolves in the aqueous phase and reacts rapidly with hydroxide ion to regenerate the amine.

Amines that contain three alkyl groups bound to nitrogen cannot react with acyl halides to give amides, since they have no replaceable hydrogen. However, such amines do react with acyl halides, giving highly reactive **acylammonium salts.**

$$
\text{CH}_3\overset{O}{\overset{\|}{C}}\text{—Cl} + \text{CH}_3\text{CH}_2\overset{\text{CH}_2\text{CH}_3}{\overset{|}{\text{N}}}\text{CH}_2\text{CH}_3 \longrightarrow \text{CH}_3\overset{O}{\overset{\|}{C}}\text{—}\overset{\text{C}_2\text{H}_5}{\overset{|}{\overset{+}{\text{N}}}}\text{—C}_2\text{H}_5 \ \text{Cl}^-
$$
$$
\underset{\text{C}_2\text{H}_5}{}
$$

Acylammonium salts are difficult to isolate because they are so extremely reactive. They react instantaneously with water, alcohols, or other nucleophilic species.

$$
\text{CH}_3\overset{O}{\overset{\|}{C}}\text{—}\overset{\text{C}_2\text{H}_5}{\overset{|}{\overset{+}{\text{N}}}}\text{—C}_2\text{H}_5 \ \text{Cl}^- + \text{H}_2\text{O} \longrightarrow \text{CH}_3\text{COOH} + \text{H—}\overset{\text{C}_2\text{H}_5}{\overset{|}{\overset{+}{\text{N}}}}\text{—C}_2\text{H}_5 \ \text{Cl}^-
$$

$$
\text{CH}_3\overset{O}{\overset{\|}{C}}\text{—}\overset{\text{C}_2\text{H}_5}{\overset{|}{\overset{+}{\text{N}}}}\text{—C}_2\text{H}_5 \ \text{Cl}^- + \text{CH}_3\text{OH} \longrightarrow \text{CH}_3\overset{O}{\overset{\|}{C}}\text{—OCH}_3 + \text{H—}\overset{\text{C}_2\text{H}_5}{\overset{|}{\overset{+}{\text{N}}}}\text{—C}_2\text{H}_5 \ \text{Cl}^-
$$

It is likely that acylammonium ions are intermediates in esterification reactions in which triethylamine or pyridine is used to neutralize HCl (page 547).

The reaction of ammonia and amines with anhydrides follows a similar course; the products are 1 mole of amide and 1 mole of carboxylic acid. Since the liberated acid reacts to form a salt with the ammonia or the amine, it is necessary to employ an excess of that reactant.

$$
2 \ \text{CH}_3\text{CH}_2\text{NH}_2 + \text{CH}_3\overset{O}{\overset{\|}{C}}\text{—O—}\overset{O}{\overset{\|}{C}}\text{CH}_3 \longrightarrow
$$
$$
\text{CH}_3\text{CH}_2\text{NH—}\overset{O}{\overset{\|}{C}}\text{CH}_3 + \text{CH}_3\text{CH}_2\text{NH}_3^+ \ \text{CH}_3\text{CO}_2^-
$$

As in the analogous reaction of amines with acyl halides, one may carry out the reaction in the presence of one equivalent of tertiary amine.

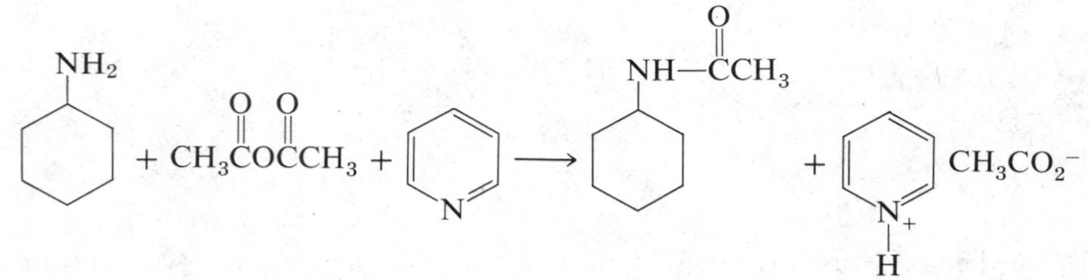

Esters also react with ammonia and amines to yield the corresponding amide and the alcohol of the ester. The reaction is synthetically useful in cases where the corresponding acyl halide or anhydride is unstable or not easily available. An interesting example of such a case is

$$CH_3CH(OH)COOC_2H_5 + NH_3 \xrightarrow[24\ hr]{25^\circ} CH_3CH(OH){-}\overset{O}{\overset{\|}{C}}NH_2 + C_2H_5OH$$

(70–74%)

In this case the acyl halide method for preparing the amide may not be used. Since the molecule contains an OH group, which will react rapidly with an acyl halide, an attempt to prepare the acyl halide would lead to a polymer.

$$CH_3CH(OH)COOH + SOCl_2 \longrightarrow \left[CH_3CH(OH)COCl\right] \longrightarrow \left({-}OCH(CH_3){-}\overset{O}{\overset{\|}{C}}{-}O{-}CH(CH_3){-}\overset{O}{\overset{\|}{C}}{-}\right)_n$$

Amides and nitriles also react with ammonia and amines under suitable conditions. However, these reactions are rarely encountered, and we shall not discuss them here.

C. *Reaction of Acyl Halides and Anhydrides with Carboxylic Acids and Carboxylate Salts. Synthesis of Anhydrides*

A mixture of an acid anhydride and a carboxylic acid undergoes equilibration when heated.

$$R\overset{O}{\overset{\|}{C}}O\overset{O}{\overset{\|}{C}}R + 2\,R'COOH \overset{\Delta}{\rightleftharpoons} R'\overset{O}{\overset{\|}{C}}O\overset{O}{\overset{\|}{C}}R' + 2\,RCOOH$$

The reaction is preparatively useful when the anhydride is acetic anhydride. In this case, acetic acid can be distilled off as it is formed because it is the most volatile component in the equilibrium.

b.p. 142° | benzoic acid | (72–74%) benzoic anhydride | b.p. 117°

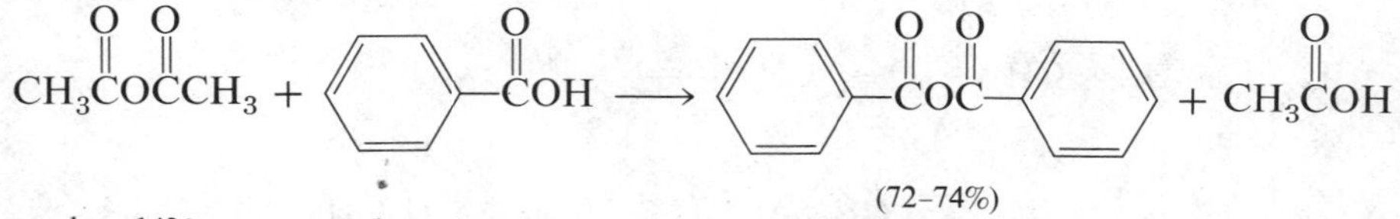

The only carboxylic acid derivatives that undergo a useful reaction with carboxylate salts are acyl halides. The product is an anhydride.

551

Sec. 19.7

Other
Nucleophilic
Substitution
Reactions

$$C_6H_{13}\overset{O}{\underset{\|}{C}}Cl + C_6H_{13}\overset{O}{\underset{\|}{C}}O^-\ Na^+ \xrightarrow{H_2O} \left(C_6H_{13}\overset{O}{\underset{\|}{C}}\right)_2O + Na^+Cl^-$$

| heptanoyl | sodium | heptanoic |
| chloride | heptanoate | anhydride |

The procedure may also be used to prepare unsymmetrical anhydrides.

$$CH_3\overset{O}{\underset{\|}{C}}Cl + H\overset{O}{\underset{\|}{C}}O^-\ Na^+ \xrightarrow[\substack{0° \\ 24\ hr}]{THF} CH_3\overset{O}{\underset{\|}{C}}O\overset{O}{\underset{\|}{C}}H + Na^+Cl^-$$

(60%)
acetic formic
anhydride

D. Reaction with Organometallic Compounds

Acyl halides react with various organometallic reagents to give ketones. When using a Grignard reagent, best results are obtained if the reaction is carried out at low temperature. Anhydrous ferric chloride is often added as a catalyst.

(71%)

$$CH_3\overset{O}{\underset{\|}{C}}Cl + CH_3CH_2CH_2CH_2MgCl \xrightarrow[\substack{FeCl_3 \\ -70°}]{ether} CH_3\overset{O}{\underset{\|}{C}}CH_2CH_2CH_2CH_3$$

(72%)

If excess Grignard reagent is used, the product ketone reacts further, giving a tertiary alcohol after hydrolysis.

$$R{-}\overset{O}{\underset{\|}{C}}Cl \xrightarrow{R'MgX} R{-}\overset{O}{\underset{\|}{C}}{-}R' \xrightarrow{R'MgX} R{-}\overset{O^-\ ^+MgX}{\underset{R'}{\overset{|}{\underset{|}{C}}}}{-}R' \xrightarrow{H_2O} R{-}\overset{OH}{\underset{R'}{\overset{|}{\underset{|}{C}}}}{-}R'$$

In fact, the reaction is a useful method for preparing ketones only because acyl halides are so very reactive.

Some of the less reactive organometallic compounds still react rapidly with acyl halides, but react with ketones only sluggishly or not at all. Such is the case with lithium organocuprates, which are obtained by treating an organolithium compound with cuprous iodide (Section 15.5.E).

$$2\ CH_3Li + CuI \xrightarrow{ether} LiCu(CH_3)_2 + LiI$$

The cuprate reacts rapidly with acyl halides and aldehydes, slowly with ketones, and not at all with esters, amides, and nitriles.

$$(CH_3)_3\overset{\overset{\displaystyle O}{\|}}{C}CCl \ + \ (CH_3)_2CuLi \ \xrightarrow[-78°]{ether} \ (CH_3)_3\overset{\overset{\displaystyle O}{\|}}{C}CCH_3$$

2,2-dimethylpropanoyl
chloride

(60%)
3,3-dimethyl-2-butanone

$$CH_3\overset{\overset{\displaystyle CH_3}{|}}{CH}CH_2\overset{\overset{\displaystyle O}{\|}}{C}Cl + \left(CH_3\overset{\overset{\displaystyle CH_3}{|}}{C}=CH\right)_2CuLi \ \xrightarrow[-5°]{ether} \ CH_3\overset{\overset{\displaystyle CH_3}{|}}{CH}CH_2\overset{\overset{\displaystyle O}{\|}}{C}CH=C\overset{\overset{\displaystyle CH_3}{}}{\underset{CH_3}{}}$$

3-methylbutanoyl
chloride

(70%)
2,6-dimethylhept-2-en-4-one

Since esters are generally less reactive than ketones, the preparation of ketones
by nucleophilic substitution on an ester is generally unsatisfactory. The method is
only useful when the product ketone is sterically hindered.

$$CH_3\overset{\overset{\displaystyle CH_3}{|}}{\underset{\underset{\displaystyle CH_3}{|}}{C}}COOC_2H_5 + CH_3\overset{\overset{\displaystyle CH_3}{|}}{\underset{\underset{\displaystyle CH_3}{|}}{C}}Li \ \xrightarrow{ether} \ (CH_3)_3\overset{\overset{\displaystyle O}{\|}}{C}CC(CH_3)_3$$

(80%)

In less hindered cases the only isolable product is the tertiary alcohol.
The reaction of esters with two equivalents of a Grignard reagent or an
alkyllithium is a useful method for the synthesis of tertiary alcohols in which two
of the alkyl groups are equivalent.

$$2\ C_6H_5MgBr + CH_3\overset{\overset{\displaystyle O}{\|}}{C}OC_2H_5 \ \longrightarrow \ \xrightarrow{H_2O} \ C_6H_5\overset{\overset{\displaystyle OH}{|}}{\underset{\underset{\displaystyle CH_3}{|}}{C}}C_6H_5$$

1,1-diphenylethanol

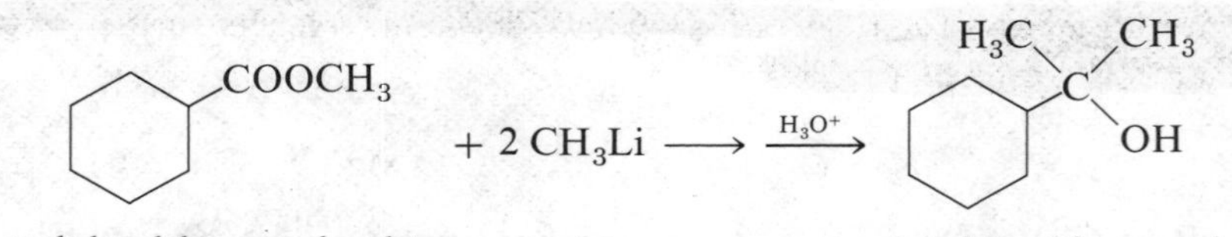

methyl cyclohexanecarboxylate   2-cyclohexyl-2-propanol

If an ester of formic acid is used, the product is a secondary alcohol.

$$2\ CH_3CH_2CH_2CH_2MgBr + H\overset{\overset{\displaystyle O}{\|}}{C}OC_2H_5 \ \xrightarrow{ether}$$

$$\xrightarrow{H_3O^+} \ CH_3CH_2CH_2CH_2\overset{\overset{\displaystyle OH}{|}}{C}HCH_2CH_2CH_2CH_3$$

(85%)
5-nonanol

Carbonate esters yield tertiary alcohols in which all three of the carbinol alkyl
groups come from the organometallic reagent.

553

Sec. 19.7

Other
Nucleophilic
Substitution
Reactions

$$3\ C_2H_5MgBr + CH_3O\overset{\overset{\displaystyle O}{\|}}{C}OCH_3 \longrightarrow \xrightarrow{H_3O^+} C_2H_5-\overset{\overset{\displaystyle OH}{|}}{\underset{\underset{\displaystyle C_2H_5}{|}}{C}}-C_2H_5$$

$$\text{(85\%)}$$

dimethyl carbonate   triethylcarbinol

Dialkyl carbonates are generally prepared by reaction of alcohols with phosgene, $COCl_2$.

$$COCl_2 + 2\ CH_3CH_2OH \longrightarrow CH_3CH_2O\overset{\overset{\displaystyle O}{\|}}{C}OCH_2CH_3 + 2\ HCl$$

[
Phosgene is a highly toxic colorless gas, b.p. 7.6°, having a distinctive odor. It was used in World War I as a war gas. This compound is the diacid chloride of carbonic acid and reacts accordingly. It is hydrolyzed by water to give carbon dioxide and HCl.

$$COCl_2 + 2\ H_2O \longrightarrow 2\ HCl + [CO(OH)_2] \longrightarrow CO_2 + H_2O$$

Dimethyl carbonate is a colorless liquid, b.p. 90°, and is available commercially.
]

Nitriles undergo a synthetically useful reaction with Grignard reagents or alkyllithium reagents. The initial product of addition of the organometallic to the triple bond is the salt of an imine (Section 13.7.C). This imine salt, even though it still has a C=N double bond, does not easily react further with the organometallic reagent because the resulting product would be a species with *two* negative charges on the same atom. When dilute acid is added at the end of the reaction, the imine salt is protonated to yield the imine, which undergoes rapid hydrolysis to the corresponding ketone.

$$R-C{\equiv}N + CH_3Li \longrightarrow R-\overset{\overset{\displaystyle N^-\ Li^+}{\|}}{C}-CH_3$$

$$\nearrow\!\!\!\!/\!\!\!\!/\quad R-\overset{\overset{\displaystyle Li^+\ N^{2-}\ Li^+}{|}}{\underset{\underset{\displaystyle CH_3}{|}}{C}}-CH_3 \quad \text{(not formed)}$$

$$\searrow^{H_3O^+}\quad R-\overset{\overset{\displaystyle NH}{\|}}{C}-CH_3 \xrightarrow{H_3O^+} R-\overset{\overset{\displaystyle O}{\|}}{C}-CH_3$$

The method serves as a convenient method for the preparation of ketones from nitriles. With acetonitrile itself higher yields result when ether is replaced by benzene as the solvent. Some specific examples are

$$CH_3OCH_2CN + C_6H_5MgBr \longrightarrow \xrightarrow{H_3O^+} CH_3OCH_2\overset{\overset{\displaystyle O}{\|}}{C}C_6H_5$$

$$\text{(78\%)}$$

+ $C_6H_5MgBr \longrightarrow \xrightarrow{H_3O^+}$

$$\text{(73\%)}$$

EXERCISE: Make a 5 × 5 matrix with acetic anhydride, acetyl chloride, ethyl acetate, acetamide, and acetonitrile on one side and ethanol, ammonia, water, methylmagnesium bromide, and sodium acetate on the other side. Each of the 25 intersections of the matrix represents a possible reaction, that is, acetyl chloride + ethanol, methyl acetate + ethanol, and so on. Which of the combinations represent reactions we have discussed in this section? For each of the important reactions we have discussed, be sure you know the details—acid catalysis, base catalysis, uncatalyzed, reaction mechanism, and so on.

19.8 Reduction

Acyl halides may be reduced to aldehydes or to primary alcohols.

$$\underset{\text{R—CCl}}{\overset{O}{\parallel}} \xrightarrow{\text{[H]}} \underset{\text{R—CH}}{\overset{O}{\parallel}} \xrightarrow{\text{[H]}} RCH_2OH$$

The selective reduction of an acyl halide is one of the most useful ways of preparing aldehydes. Such selective reduction is possible because acyl halides are generally more reactive than the product aldehydes. One procedure for accomplishing the selective reduction is catalytic hydrogenation; the method is called a **Rosenmund reduction.** The acyl halide is hydrogenated in the presence of a catalyst such as palladium deposited on barium sulfate. As in the reduction of alkynes to alkenes, a "regulator" or "catalyst poison" (page 344) is frequently added in order to moderate the effectiveness of the catalyst and thereby inhibit subsequent reduction of the product aldehyde.

$$\underset{\text{CH}_3\text{OCCH}_2\text{CH}_2\text{CCl}}{\overset{O\quad\quad O}{\parallel\quad\quad\parallel}} + H_2 \xrightarrow[\substack{\text{quinoline}\\\text{sulfur}\\\text{xylene, } 110°}]{\text{Pd/BaSO}_4} \underset{\text{CH}_3\text{OCCH}_2\text{CH}_2\text{CH}}{\overset{O\quad\quad O}{\parallel\quad\quad\parallel}}$$
$$(65\%)$$

Another reagent that has found use for the selective reduction of an acyl halide to an aldehyde is lithium tri-*t*-butoxyaluminohydride, LiAl(*t*-C₄H₉O)₃H.

In general, acyl halides are reduced more rapidly than aldehydes. However, lithium aluminum hydride is so extremely reactive that a selective reduction is difficult to accomplish. The tri-*t*-butoxy derivative, prepared by treating the hydride with three equivalents of *t*-butyl alcohol in ether, is less reactive and therefore more selective.

$$\text{LiAlH}_4 + 3\,t\text{-C}_4\text{H}_9\text{OH} \longrightarrow \text{Li}(t\text{-C}_4\text{H}_9\text{O})_3\text{AlH} + 3\,\text{H}_2$$

If excess reducing agent is used, the product aldehyde is reduced further to a primary alcohol.

$$(1)\ \underset{\text{R—C—Cl}}{\overset{O}{\parallel}} + \text{Li}(t\text{-C}_4\text{H}_9)_3\text{AlH} \xrightarrow{\text{faster}} \underset{\underset{\text{H}}{\mid}}{\overset{\text{O}^-}{\mid}} R\text{—C—Cl} + \text{Li}^+ + (t\text{-C}_4\text{H}_9)_3\text{Al}$$

(2) $\underset{\underset{H}{\overset{O^-}{\overset{|}{R-C-Cl}}}}{} \rightleftharpoons \underset{\overset{O}{\overset{\parallel}{R-C-H}}}{} + Cl^-$

(3) $\underset{\overset{O}{\overset{\parallel}{R-C-H}}}{} + Li(t\text{-}C_4H_9)_3AlH \xrightarrow{\text{slower}} RCH_2O^- + Li^+ + (t\text{-}C_4H_9)_3Al$

With the more reactive lithium aluminum hydride, it is difficult to achieve selectivity.

Since esters are generally less reactive than aldehydes, they cannot be selectively reduced to the aldehyde stage. However, the reduction of an ester to a primary alcohol is an important preparative method. The most generally used reducing agents are lithium aluminum hydride and lithium borohydride.

$$\text{C}_6\text{H}_5\text{-}\overset{\overset{O}{\parallel}}{\text{C}}\text{OC}_2\text{H}_5 + LiAlH_4 \longrightarrow \xrightarrow{H_2O} \text{C}_6\text{H}_5\text{-}CH_2OH + C_2H_5OH$$
$$(90\%)$$

$$CH_3(CH_2)_{14}COOCH_2CH_2CH_2CH_3 + LiBH_4 \xrightarrow[\Delta]{THF} \xrightarrow{H_2O} CH_3(CH_2)_{15}OH + CH_3CH_2CH_2CH_2OH$$
$$(95\%)$$

> Lithium borohydride, $LiBH_4$, is a white solid, m.p. 284°, which is extremely hygroscopic. It is prepared by the reaction of sodium borohydride (page 344) with lithium chloride in ethanol.
>
> $$NaBH_4 + LiCl \longrightarrow LiBH_4 + NaCl$$
>
> It is a more reactive reducing agent than sodium borohydride, but is less reactive than lithium aluminum hydride. It is much more soluble in ether (4 g per 100 mL) than is sodium borohydride.

Esters are also reduced by sodium in ethanol (the **Bouveault–Blanc reaction**). Before the discovery of lithium aluminum hydride, this was the most common laboratory method for reducing esters, and it is still an important method for large-scale preparations where reagent cost is a concern.

$$n\text{-}C_{11}H_{23}COOC_2H_5 + Na \xrightarrow[\Delta]{C_2H_5OH} n\text{-}C_{11}H_{23}CH_2OH + C_2H_5OH$$
$$\text{ethyl dodecanoate} \qquad\qquad (65\text{--}75\%)$$
$$\text{1-dodecanol}$$

The reaction mechanism is not known in detail, but undoubtedly involves electron transfer from sodium to the carbonyl group as a first step. The reaction is *not* a reduction by hydrogen liberated from the reaction of sodium with ethanol.

Reduction of amides having at least one hydrogen on nitrogen with lithium aluminum hydride in ether or THF provides the corresponding primary or secondary amines.

$$\text{C}_6\text{H}_5\text{-}OCH_2\overset{\overset{O}{\parallel}}{\text{C}}NH_2 + LiAlH_4 \xrightarrow[\Delta]{\text{ether}} \xrightarrow{H_2O} \text{C}_6\text{H}_5\text{-}OCH_2CH_2NH_2$$
$$(80\%)$$

$$CH_3(CH_2)_{10}\overset{\overset{\textstyle O}{\|}}{C}NHCH_3 + LiAlH_4 \xrightarrow[\Delta]{ether} \xrightarrow{H_2O} CH_3(CH_2)_{10}CH_2NHCH_3$$
$$(81\text{–}95\%)$$

A solution of 38 g of lithium aluminum hydride in 1800 mL of dry ether is placed in a 5 L three-necked flask equipped with a condenser and a mechanical stirrer. The solution is gently refluxed while 160 g of N-methyldodecanamide is slowly added over a period of 3 hr. The mixture is refluxed an additional 2 hr, then stirred overnight. The reaction mixture is worked up by the addition of 82 mL of water. After filtration to remove the solid aluminum and lithium salts, the ether is evaporated, and the residue is distilled to yield 121–142 g of N-methyldodecylamine.

Although the exact mechanism of this reaction has not been elucidated, the first step is probably reaction of the strongly basic LiAlH$_4$ with the weakly acidic NH bond, giving the lithium salt of the amide.

$$R\overset{\overset{\textstyle O}{\|}}{C}NH_2 + LiAlH_4 \longrightarrow R\overset{\overset{\textstyle O}{\|}}{C}NH^- \, Li^+ + AlH_3 + H_2$$

Aluminum hydride may then add to the carbonyl group.

$$R\overset{\overset{\textstyle O}{\|}}{C}NH^- \, Li^+ + AlH_3 \longrightarrow R\overset{\overset{\textstyle O-AlH_2}{|}}{\underset{\underset{\textstyle H}{|}}{C}}NH^- \, Li^+$$

This tetrahedral intermediate may decompose by elimination of H$_2$AlO$^-$, which is actually not a bad leaving group (recall that aluminum is an amphoteric metal; H$_3$AlO$_3$ is a protonic acid).

$$R\overset{\overset{\textstyle OAlH_2}{|}}{\underset{\underset{\textstyle H}{|}}{C}}NH^- \, Li^+ \longrightarrow R\underset{\underset{\textstyle H}{|}}{C}=NH + H_2AlO^- \, Li^+$$

The resulting imine is now reduced by another hydride (from AlH$_4^-$ or H$_2$AlO$^-$).

$$R\overset{\overset{\textstyle NH}{\|}}{C}H + AlH_4^- \text{ (or } H_2AlO^-) \longrightarrow R\overset{\overset{\textstyle NH-Al\diagdown}{|}}{\underset{\underset{\textstyle H}{|}}{C}}H$$

Upon aqueous work-up, the N—Al bond is hydrolyzed to liberate the amine.

$$RCH_2-NH-Al\diagdown + H_2O \longrightarrow RCH_2NH_2 + HO-Al\diagdown$$

N,N-Dialkylamides may also be reduced to amines by lithium aluminum hydride.

$$CH_3\overset{\overset{\textstyle O}{\|}}{C}N(CH_2CH_3)_2 + LiAlH_4 \xrightarrow[\text{reflux}]{ether} \xrightarrow{H_2O} (CH_3CH_2)_3N$$
$$\text{N,N-diethylacetamide} \qquad\qquad\qquad\qquad\qquad (50\%)$$
$$\text{triethylamine}$$

With disubstituted amides the reduction may generally be controlled so that the aldehyde may be obtained. This occurs when the initial tetrahedral intermediate is sufficiently stable so that it survives until all of the hydride has been consumed.

If one wishes to prepare an aldehyde in this manner, it is necessary to keep the amide in excess by slowly adding the reducing agent to it. Several modified hydrides have been used for this purpose. One reagent that is particularly useful with simple dimethylamides is lithium triethoxyaluminohydride, prepared *in situ* by the reaction of three equivalents of ethanol with one equivalent of lithium aluminum hydride.

$$LiAlH_4 + 3\ C_2H_5OH \xrightarrow{\text{ether}} LiAlH(OC_2H_5)_3 + 3\ H_2$$

$$(CH_3)_3C\overset{\overset{\displaystyle O}{\|}}{C}N(CH_3)_2 + LiAlH(OC_2H_5)_3 \xrightarrow{\text{ether}} \xrightarrow{H_2O} (CH_3)_3C\overset{\overset{\displaystyle O}{\|}}{C}H$$

(88%)
2,2-dimethylpropanal
pivalaldehyde

The C≡N triple bond in nitriles is successfully reduced by a number of reagents. The most generally useful methods are lithium aluminum hydride reduction and catalytic hydrogenation.

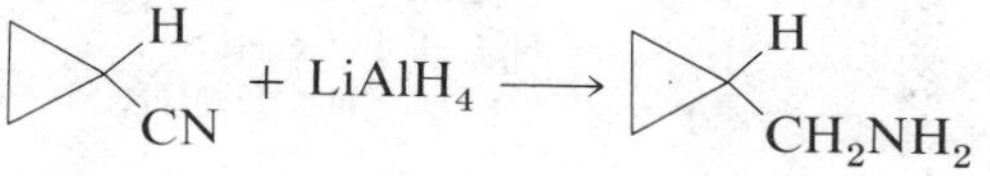

(79%)

$$CH_3(CH_2)_{11}CN + LiAlH_4 \longrightarrow CH_3(CH_2)_{12}NH_2$$

(90%)

$$C_6H_5CH_2CN + 2\ H_2 \xrightarrow[\substack{2000\ psi \\ NH_3}]{\substack{Ni \\ 120°}} C_6H_5CH_2CH_2NH_2$$

(83–87%)

If the reduction is carried out under carefully controlled conditions, the initially formed imine salt may often be obtained. Hydrolysis of this salt gives the corresponding aldehyde.

$$RC≡N \xrightarrow{\text{``H''}} R\overset{\overset{\displaystyle N^-}{\|}}{C}{-}H \xrightarrow{H_3O^+} R\overset{\overset{\displaystyle O}{\|}}{C}{-}H$$

A particularly useful reagent for this purpose is diisobutylaluminum hydride (DIBAL).

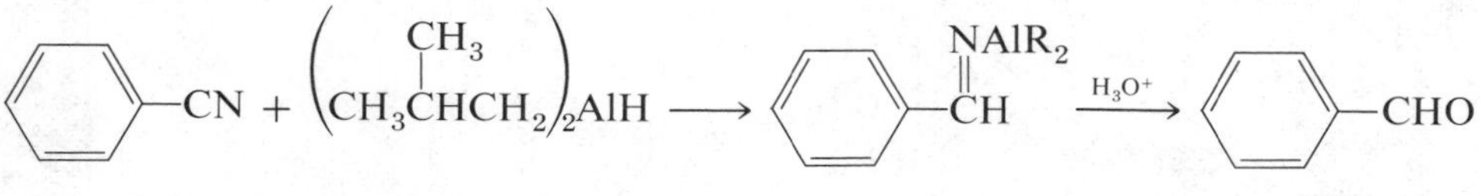

benzonitrile   DIBAL   (90%)
benzaldehyde

Diisobutylaluminum hydride, DIBAL, is a clear colorless liquid that boils at 140° at 4 torr. Like many other alkylaluminum and alkylboron compounds, it is flammable in air. It is supplied commercially in small tanks or lecture bottles. It is normally used as a benzene solution under an inert atmosphere such as nitrogen or argon.

Nitriles may also be converted into aldehydes by a method known as the **Stephen reduction.** In this reaction the reducing agent is stannous chloride and hydrogen chloride in an inert solvent such as ether or ethyl acetate.

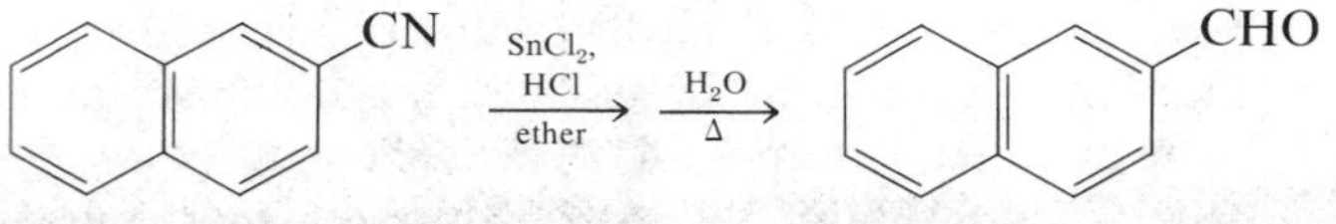

naphthalene-2-carbonitrile   naphthalene-2-carboxaldehyde

The reaction involves reduction to an intermediate immonium salt which is readily hydrolyzed.

$$RCN \longrightarrow (RCH{=}NH_2{}^+)_2 \; SnCl_6{}^{2-} \xrightarrow{\;H_2O\;} RCHO$$

Although the Stephen reduction is successful with some aliphatic aldehydes, it is most useful when the group attached to the nitrile is an aromatic ring.

EXERCISE: Show how one may accomplish each of the following transformations.
(a) $CH_3CH_2CH_2COCl$ $\longrightarrow$ $CH_3CH_2CH_2CHO$
(b) $CH_3CH_2CH_2CN$ $\longrightarrow$ $CH_3CH_2CH_2CHO$
(c) $CH_3CH_2CH_2COOCH_3$ $\longrightarrow$ $CH_3CH_2CH_2CH_2OH$
(d) $CH_3CH_2CH_2CONH_2$ $\longrightarrow$ $CH_3CH_2CH_2CH_2NH_2$
(e) $CH_3CH_2CH_2CON(CH_3)_2$ $\longrightarrow$ $CH_3CH_2CH_2CHO$
(f) $CH_3CH_2CH_2CN$ $\longrightarrow$ $CH_3CH_2CH_2CH_2NH_2$

19.9
Acidity of the α-Protons

Like the α-protons of aldehydes and ketones (Section 13.6), protons adjacent to the carbonyl group in carboxylic acid derivatives are weakly acidic. Table 19.6 lists the pK_a values for some representative compounds. Recall that the main reason for the acidity of aldehydes and ketones relative to the alkanes is the fact that the resulting anion is resonance-stabilized. In fact, the resonance contributor with the negative charge on oxygen (the enolate ion) is the more important struc-

TABLE 19.6

Compound	pK_a of **H**
$CH_3\overset{O}{\overset{\|}{C}}Cl$	~16
$CH_3\overset{O}{\overset{\|}{C}}H$	17
$CH_3\overset{O}{\overset{\|}{C}}CH_3$	20
$CH_3\overset{O}{\overset{\|}{C}}OCH_3$	25
CH_3CN	25
$CH_3\overset{O}{\overset{\|}{C}}N(CH_3)_2$	~30
CH_3CH_3	~50

ture because of the greater electronegativity of oxygen. This stabilization of the anion greatly reduces the $\Delta H°$ for the ionization process and is responsible for the fact that acetone is approximately 30 powers of ten more acidic than ethane.

$$CH_3CH_3 \rightleftharpoons H^+ + \underset{\text{charge localized on carbon}}{CH_3CH_2^-}$$

$$CH_3\overset{O}{\overset{\|}{C}}CH_3 \rightleftharpoons H^+ + \left[CH_3\overset{O}{\overset{\|}{C}}CH_2^- \longleftrightarrow CH_3\overset{O^-}{\overset{|}{C}}{=}CH_2\right]$$

charge delocalized—mainly on oxygen

The anions obtained upon deprotonation of esters, amides, acyl halides, anhydrides, and nitriles are also stabilized by delocalization of the negative charge onto the carbonyl oxygen. Consequently these compounds also act as weak acids; representative pK_as are summarized in Table 19.6.

The α-proton acidity manifests itself in the chemistry of carboxylic acid derivatives in several ways. For example, if ethyl acetate is dissolved in deuterioethanol that contains a catalytic amount of sodium ethoxide, exchange of the α-protons by deuterium occurs.

$$CH_3\overset{O}{\overset{\|}{C}}OC_2H_5 + C_2H_5OD \xrightleftharpoons{C_2H_5O^-} CD_3\overset{O}{\overset{\|}{C}}OC_2H_5 + C_2H_5OH$$

Ethyl acetate is a much weaker acid than ethanol (pK_a 15.9), so the above equilibrium is established relatively slowly. For example, a solution of ethyl acetate in deuterioethanol containing 0.1 *M* sodium ethoxide is only 50% exchanged after two weeks at 25°. For this reason, if one wishes to exchange the α-protons of an ester, it is necessary to reflux the solution for several hours.

If an ester is treated with a sufficiently strong base, it can be completely converted to the corresponding anion. For example, *t*-butyl acetate reacts with lithium isopropylcyclohexylamide in THF at −78° to give *t*-butyl lithioacetate, which may be isolated as a white crystalline solid.

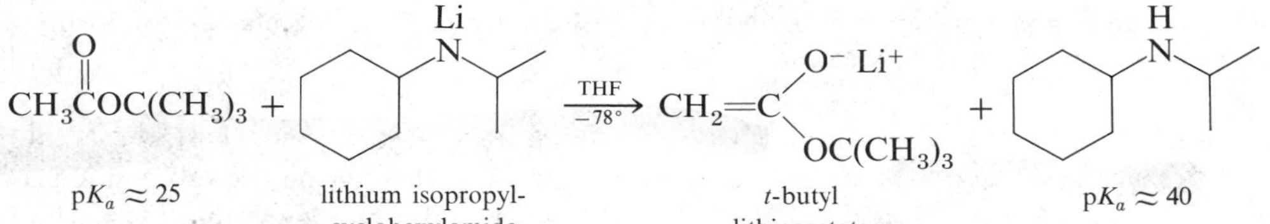

p$K_a \approx 25$ — lithium isopropylcyclohexylamide — *t*-butyl lithioacetate — p$K_a \approx 40$

Such ester anions are strong bases and are also good nucleophiles. They undergo reactions similar to those of the corresponding enolate ions derived from ketones. Examples are S_N2 displacements with primary alkyl halides and additions to aldehyde and ketone carbonyl groups.

$$\underset{\text{ethyl hexanoate}}{CH_3CH_2CH_2CH_2CH_2COOC_2H_5} \xrightarrow[\text{THF},\ -78°]{(i\text{-}C_3H_7)_2NLi} \xrightarrow{CH_3I} \underset{\substack{(83\%)\\ \text{ethyl 2-methylhexanoate}}}{CH_3CH_2CH_2CH_2\overset{CH_3}{\overset{|}{C}}HCOOC_2H_5}$$

$$CH_3COOC(CH_3)_3 \xrightarrow[\substack{THF \\ -78°}]{(i\text{-}C_3H_7)_2NLi} \xrightarrow{H_2O}$$

HO   CH$_2$COOC(CH$_3$)$_3$

(93%)

t-butyl (1-hydroxycyclopentyl)acetate

A related reaction results when an aldehyde or ketone is treated with an α-halo ester and zinc in an inert solvent.

$$+ \ BrCH_2COOC_2H_5 + Zn \xrightarrow{toluene} \xrightarrow{H_3O^+}$$

OH

CH$_2$COOC$_2$H$_5$

(70%)

ethyl (1-hydroxycyclohexyl)acetate

This reaction, which is known as the **Reformatsky reaction,** involves the formation of an intermediate organozinc compound that may be regarded as the anion of the ester, closely associated with a zinc cation.

$$BrCH_2COOR + Zn \longrightarrow CH_2{=}\overset{\overset{\displaystyle O^- ZnBr^+}{|}}{C}{-}OR$$

Ester and nitrile anions also undergo another reaction that is an important synthetic procedure, the **Claisen condensation.** In this reaction the ester anion condenses with an un-ionized ester molecule to give a β-keto ester.

$$2 \ CH_3\overset{\overset{\displaystyle O}{\|}}{C}OC_2H_5 \xrightarrow[\text{C}_2\text{H}_5\text{OH}]{\text{NaOC}_2\text{H}_5} \xrightarrow{H_3O^+} CH_3\overset{\overset{\displaystyle O}{\|}}{C}CH_2\overset{\overset{\displaystyle O}{\|}}{C}OC_2H_5$$
(75%)

> A mixture of 1.2 moles of ethyl acetate and 0.2 mole of alcohol-free sodium ethoxide is heated at 78° for 8 hr. The mixture is then cooled to 10°, and 36 g of 33% aqueous acetic acid is slowly added. The aqueous layer is washed with ether, and the combined organic layers are dried and distilled to give ethyl acetoacetate, b.p. 78–80° (16 torr), in 75–76% yield.

The product in this example, ethyl 3-oxobutanoate, is known by the trivial name **ethyl acetoacetate,** or simply **acetoacetic ester.** For this reason the self-condensation of esters is sometimes called the **acetoacetic ester condensation,** even when other esters are involved.

The reaction is mechanistically similar to the aldol condensation (Section 13.7.F) in that the conjugate base of the ester is a reactive intermediate.

$$(1) \ CH_3\overset{\overset{\displaystyle O}{\|}}{C}OC_2H_5 + C_2H_5O^- \ \rightleftharpoons \ ^-CH_2\overset{\overset{\displaystyle O}{\|}}{C}OC_2H_5 + C_2H_5OH$$

$$(2) \ ^-CH_2\overset{\overset{\displaystyle O}{\|}}{C}OC_2H_5 + CH_3\overset{\overset{\displaystyle O}{\|}}{C}OC_2H_5 \ \rightleftharpoons \ CH_3\underset{\underset{\displaystyle OC_2H_5}{|}}{\overset{\overset{\displaystyle O^-}{|}}{C}}CH_2\overset{\overset{\displaystyle O}{\|}}{C}OC_2H_5$$

(3) $\underset{\underset{\displaystyle OC_2H_5}{|}}{CH_3\overset{\displaystyle O^-}{\overset{|}{C}}CH_2\overset{\displaystyle O}{\overset{||}{C}}OC_2H_5} \;\rightleftharpoons\; CH_3\overset{\displaystyle O}{\overset{||}{C}}CH_2\overset{\displaystyle O}{\overset{||}{C}}OC_2H_5 + C_2H_5O^-$

(4) $CH_3\overset{\displaystyle O}{\overset{||}{C}}CH_2\overset{\displaystyle O}{\overset{||}{C}}OC_2H_5 + C_2H_5O^- \;\rightleftharpoons\; CH_3\overset{\displaystyle O^-}{\overset{|}{C}}{=}CH\overset{\displaystyle O}{\overset{||}{C}}OC_2H_5 + C_2H_5OH$

As we shall see in Section 27.7 C, 1,3-dicarbonyl compounds are fairly strong carbon acids; the pK_a of acetoacetic ester itself is about 11. Thus the equilibrium for the last step in this mechanism lies far to the right. Since the pK_a of ethanol is about 16, K for step (4) is about 10^5. This final, essentially irreversible, step provides the driving force for the Claisen condensation. A dramatic illustration is provided by attempted Claisen condensation with ethyl 2-methylpropanoate.

$$2\,(CH_3)_2CH\overset{\displaystyle O}{\overset{||}{C}}OC_2H_5 \;\xrightarrow[C_2H_5OH]{NaOC_2H_5}\;\xrightarrow{H_3O^+}\; (CH_3)_2CH\overset{\displaystyle O}{\overset{||}{C}}{-}\underset{\underset{\displaystyle CH_3}{|}}{\overset{\overset{\displaystyle CH_3}{|}}{C}}{-}\overset{\displaystyle O}{\overset{||}{C}}OC_2H_5$$

$$(0\%)$$

In this case there are no protons in the normally acidic position between the two carbonyl groups; hence, the final step in the mechanism cannot occur. The overall equilibrium constant for steps (1) through (3) in the mechanism is apparently too small for condensation to be observed in the absence of step (4). The most acidic proton available in the hypothetical product is the proton at C-4, which is a normal proton α to a ketone carbonyl; its pK_a is therefore about 20. If a much stronger base is used to catalyze the reaction, this proton can be removed and reaction can now be observed.

$$(CH_3)_2CHCOOC_2H_5 + (C_6H_5)_3C^-Na^+ \;\longrightarrow\; \xrightarrow[(work\text{-}up)]{H_3O^+}\; (CH_3)_2CH\overset{\displaystyle O}{\overset{||}{C}}{-}\underset{\underset{\displaystyle CH_3}{|}}{\overset{\overset{\displaystyle CH_3}{|}}{C}}{-}COOC_2H_5$$

$$(45\text{-}60\%)$$

The base in the foregoing example, sodium triphenylmethide, is much more basic than ethoxide ion. It is the conjugate base of the weak carbon acid triphenylmethane (p$K_a = 31.5$), and it will be discussed in detail in Section 22.5. Since it is such a strong base, K for the last step in the Claisen condensation is very large (10^{11}), even though a normal ketone is being deprotonated.

$$(CH_3)_2CH\overset{\displaystyle O}{\overset{||}{C}}{-}\underset{\underset{\displaystyle CH_3}{|}}{\overset{\overset{\displaystyle CH_3}{|}}{C}}{-}COOC_2H_5 + (C_6H_5)_3C^- \;\overset{K}{\rightleftharpoons}$$

$$\text{p}K_a \approx 20$$

$$(CH_3)_2C{=}\underset{\underset{\displaystyle CH_3}{|}}{\overset{\overset{\displaystyle O^-}{|}}{C}}{-}\underset{\underset{\displaystyle CH_3}{|}}{\overset{\overset{\displaystyle CH_3}{|}}{C}}{-}COOC_2H_5 + (C_6H_5)_3CH$$

$$\text{p}K_a = 31.5$$

Mixed Claisen condensations between two esters are successful when one of the esters has no α-hydrogens, as shown by the following examples.

$$CH_3\overset{O}{\underset{||}{C}}OC_2H_5 + H\overset{O}{\underset{||}{C}}OC_2H_5 \xrightarrow[C_2H_5OH]{NaOC_2H_5} \xrightarrow{H_3O^+} H\overset{O}{\underset{||}{C}}CH_2\overset{O}{\underset{||}{C}}OC_2H_5$$

(79%)

ethyl formylacetate

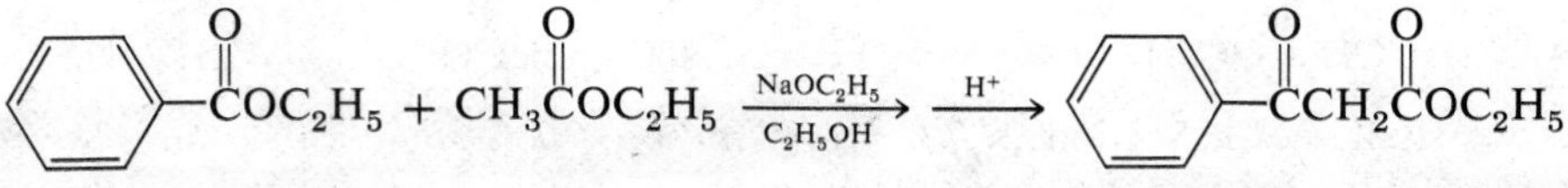

$$\underset{}{\bigcirc}\overset{O}{\underset{||}{C}}OC_2H_5 + CH_3\overset{O}{\underset{||}{C}}OC_2H_5 \xrightarrow[C_2H_5OH]{NaOC_2H_5} \xrightarrow{H^+} \underset{}{\bigcirc}\overset{O}{\underset{||}{C}}CH_2\overset{O}{\underset{||}{C}}OC_2H_5$$

(55–70%)

Thus the overall result of a mixed Claisen condensation is given by the general equation

$$RCOOC_2H_5 + H-\overset{R'}{\underset{R''}{C}}-COOC_2H_5 \longrightarrow R\overset{O}{\underset{||}{C}}-\overset{R'}{\underset{R''}{C}}-COOC_2H_5$$

Best results are obtained when R has no α-hydrogens and either R′ or R″ is hydrogen.

The Claisen condensation is used extensively in biological reactions to build up and degrade chains. Nature, however, does not use alcoholic sodium ethoxide as a reaction solvent (despite attempts of some persons to provide a suitable alcoholic medium). Instead of a normal ester, the biochemical processes make use of a thioester together with enzyme catalysts specific for each step. The key ingredient is coenzyme A or CoA in the form of the thioacetate ester or acetyl CoA.

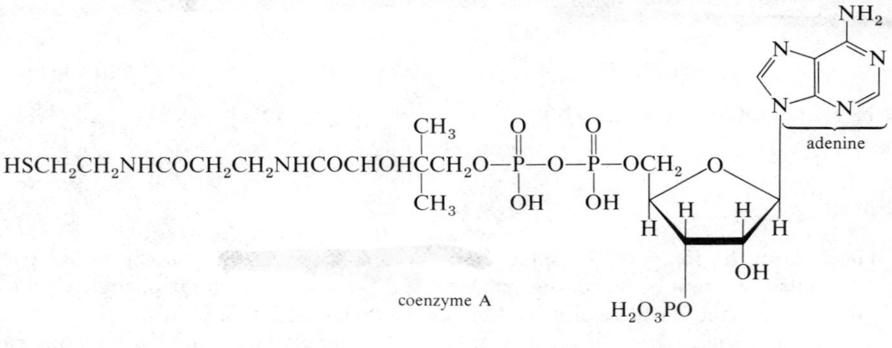

$$HSCH_2CH_2NHCOCH_2CH_2NHCOCHOHC\overset{CH_3}{\underset{CH_3}{|}}CH_2O-\overset{O}{\underset{OH}{\overset{||}{P}}}-O-\overset{O}{\underset{OH}{\overset{||}{P}}}-OCH_2$$

adenine

coenzyme A

H_2O_3PO

$$2\ CH_3CO-CoA \rightleftharpoons CH_3COCH_2CO-CoA + CoA$$
$$\Updownarrow$$
$$CH_3CHOHCH_2CO-CoA$$
$$\Updownarrow$$
$$CH_3CH=CHCO-CoA$$
$$\Updownarrow$$
$$CH_3CH_2CH_2CO-CoA \rightleftharpoons \text{higher fatty acids}$$

The initial condensation is of the Claisen type and is followed by reduction and dehydration reactions to give a butyrate ester than can react with acetyl CoA to build up higher fatty acids. Note that these acids are built up and degraded two carbons at

563

Sec. 19.10

Reactions
of Amides That
Occur on
Nitrogen

a time. A consequence of this mechanism is that fatty acids almost always have an even number of carbons in the chain.

EXERCISE: Show how ethyl acetate may be converted into each of the following compounds.

(a) ethyl hexanoate

(b) ethyl 3-hydroxyhexanoate

(c) $\overset{\overset{\textstyle O}{\|}}{H}CCH_2COOCH_2CH_3$

(d) $CD_3COOCH_2CH_3$

19.10
Reactions of Amides That Occur on Nitrogen

The N—H bonds of amides are also acidic. In fact, since the negative charge in the resulting **amidate** anion is shared between oxygen and nitrogen, rather than oxygen and carbon, they are substantially more acidic than ketones; acetamide has $pK_a \sim 15$.

$$R-\overset{\overset{\textstyle O}{\|}}{C}-NH_2 \rightleftharpoons H^+ + \left[R-\overset{\overset{\textstyle O}{\|}}{C}-NH^- \longleftrightarrow R-\overset{\overset{\textstyle O^-}{|}}{C}=NH \right]$$

an **amidate** anion

The lability of the amide N—H bond is reflected in many of the reactions of amides. One of these is dehydration. A primary amide may be converted into the corresponding nitrile by treatment with an efficient dehydrating agent such as P_2O_5, $POCl_3$, $SOCl_2$, or acetic anhydride.

$$CH_3CH_2CH_2\overset{\overset{\textstyle CH_3}{|}}{C}H \overset{\overset{\textstyle CH_3}{|}}{C}H-\overset{\overset{\textstyle O}{\|}}{C}-NH_2 \xrightarrow[\substack{benzene \\ 80°}]{SOCl_2} CH_3CH_2CH_2\overset{\overset{\textstyle CH_3}{|}}{C}H \overset{\overset{\textstyle CH_3}{|}}{C}H-CN + SO_2 + HCl$$

2,3-dimethylhexanamide   (90%)
2,3-dimethylhexanenitrile

If a primary amide is treated with bromine in the presence of aqueous base, an interesting reaction occurs.

$$\text{—}CH_2\overset{\overset{\textstyle O}{\|}}{C}NH_2 + Br_2 \xrightarrow[H_2O]{NaOH} \text{—}CH_2NH_2$$

phenylacetamide   benzylamine

This reaction, known as the **Hofmann degradation of amides,** is discussed in Section 24.6.H.

EXERCISE: Consider the four compounds ethyl bromide, propanoic acid, propanamide, and propanenitrile. Make a 4 × 4 matrix with the four compounds on each side of the matrix. There are twelve intersections of the matrix representing the conversion of one compound to another (ignore the four that connect a compound with itself). We have now learned reactions that correspond to eight of these twelve possibilities. Review these eight methods.

19.11
Pyrolytic Eliminations

Several types of organic compounds undergo elimination to give alkenes when heated to relatively high temperatures. Such **pyrolytic eliminations** are often preparatively useful. In this section we shall discuss the pyrolytic elimination of esters and xanthates.

$$R-\overset{\overset{O}{\|}}{C}-OCH_2CHR_2' \xrightarrow{\Delta} RCOOH + CH_2=CR_2'$$

an ester

$$RS-\overset{\overset{S}{\|}}{C}-OCH_2CHR_2' \xrightarrow{\Delta} RSH + COS + CH_2=CR_2'$$

a xanthate

When a carboxylic ester is heated to 300–500°, elimination occurs to give the carboxylic acid and an alkene. The elimination may be accomplished by heating the ester in the liquid phase or by passing the gaseous ester through an electrically heated vapor-phase reactor.

$$CH_3\overset{\overset{O}{\|}}{C}OCH_2CH_2CH_2CH_3 \xrightarrow{500°} CH_2=CHCH_2CH_3 + CH_3COOH$$
$$(100\%)$$

The elimination is believed to occur by a concerted mechanism involving a six-center transition state.

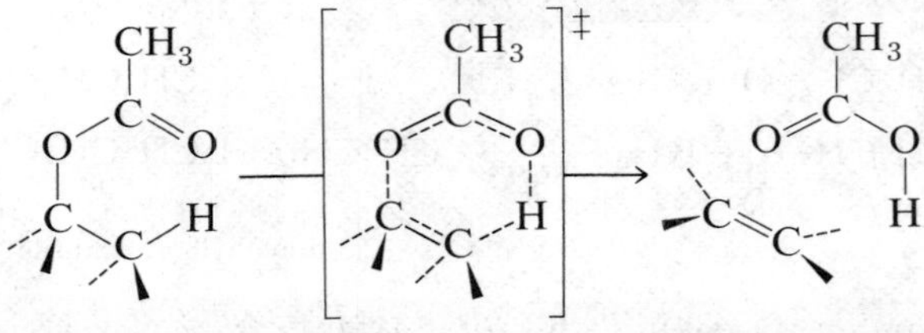

When the alkyl group of the ester has two or more β-hydrogens that may be lost, mixtures are obtained. For simple esters the elimination is nearly statistical.

$$CH_3-\overset{\overset{\overset{O}{\|}}{O}\overset{\|}{C}CH_3}{CH}-CH_2CH_3 \xrightarrow{500°} CH_2=CHCH_2CH_3 + CH_3CH=CHCH_3$$

observed ratio: 57%   43%
(statistical: 60%   40%)

Xanthate salts are prepared by treating an alcohol with carbon disulfide and sodium hydroxide.

$$ROH + CS_2 + NaOH \longrightarrow RO\overset{\overset{S}{\|}}{C}S^- \; Na^+$$

When the xanthate salt is treated with methyl iodide, S_N2 displacement occurs and a **xanthate ester** is formed.

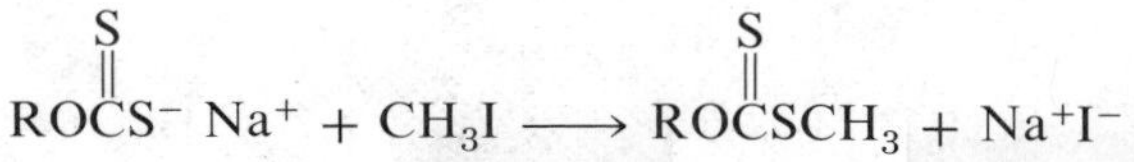

$$\underset{\text{S}}{ROC S^-}\,Na^+ + CH_3I \longrightarrow \underset{\text{S}}{ROC SCH_3} + Na^+I^-$$

Xanthates undergo pyrolytic elimination at somewhat lower temperatures than esters. The method, called **Chugaev reaction,** has been widely used as method of dehydrating alcohols.

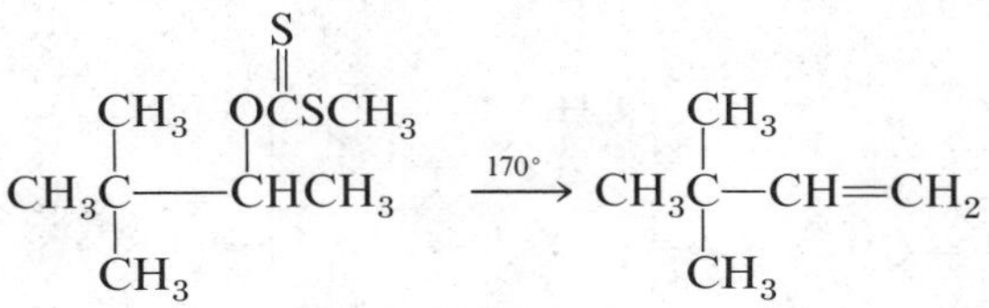

methyl 3,3-dimethyl-
2-butylxanthate

(71%)
3,3-dimethyl-1-butene

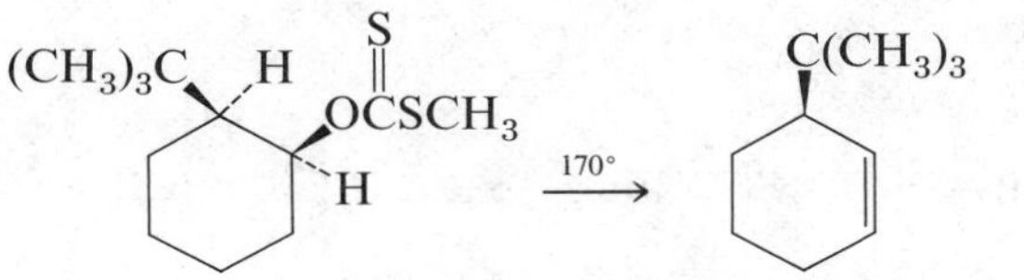

(61%)
3-*t*-butylcyclohexene

Recall that the E2 elimination of alkyl halides to give alkenes is an *anti* elimination (Section 12.5.A). That is, the hydrogen and halogen depart from opposite sides of the molecule.

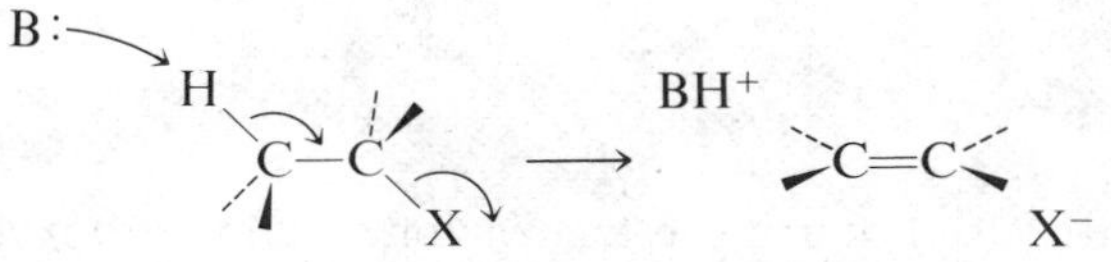

In contrast, ester and xanthate pyrolyses are *syn* eliminations; the leaving groups depart from the same side of the molecule. This stereochemistry was first demonstrated with an elegant labeling experiment by Curtin and Kellom in 1953. It was shown that 1,2-diphenylethyl acetate undergoes pyrolysis to give only *trans*-1,2-diphenylethylene (*trans*-stilbene).

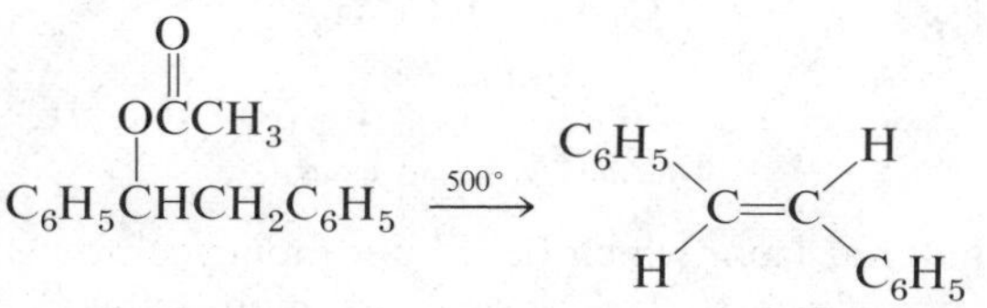

1,2-diphenylethyl acetate

trans-stilbene

The two deuterium-labeled analogs shown in (19-6) and (19-7) were then prepared and pyrolyzed. The (R,R) compound gives *trans*-stilbene, which contains one deuterium atom per molecule, and the (R,S) compound gives a product that has no deuterium.

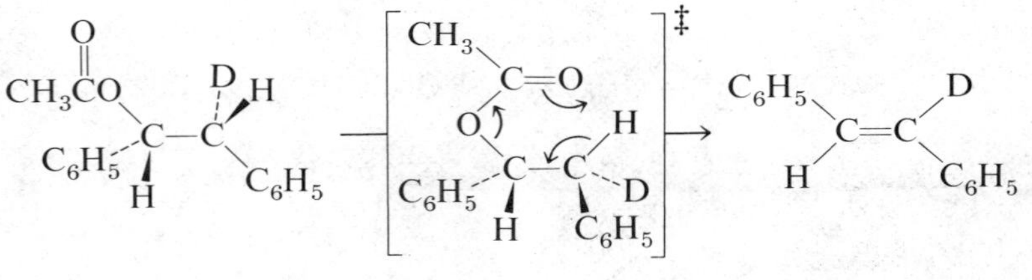

$$\text{CH}_3\overset{\text{O}}{\overset{\|}{\text{CO}}}\underset{\text{C}_6\text{H}_5}{\overset{\text{D}}{\underset{\text{H}}{\text{C}}}}\underset{\text{C}_6\text{H}_5}{\overset{\text{H}}{\text{C}}} \longrightarrow \left[\begin{array}{c} \text{CH}_3 \\ \text{C}=\text{O} \end{array} \right]^{\ddagger} \longrightarrow \underset{\text{H}}{\overset{\text{C}_6\text{H}_5}{\text{C}}}=\underset{\text{C}_6\text{H}_5}{\overset{\text{D}}{\text{C}}} \quad (19\text{-}6)$$

(R,R)

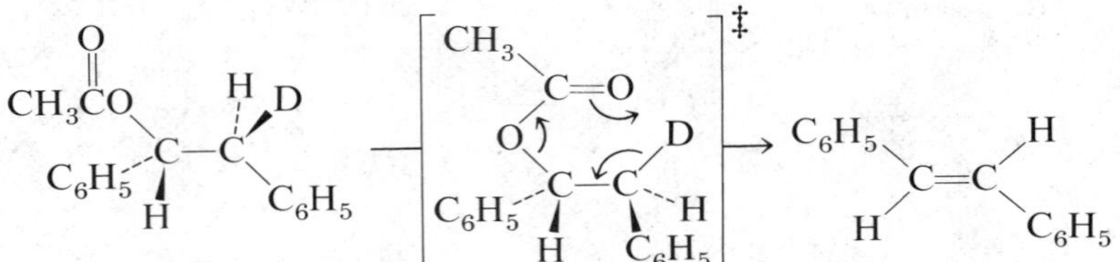

$$\text{CH}_3\overset{\text{O}}{\overset{\|}{\text{CO}}}\underset{\text{C}_6\text{H}_5}{\overset{\text{H}}{\underset{\text{H}}{\text{C}}}}\underset{\text{C}_6\text{H}_5}{\overset{\text{D}}{\text{C}}} \longrightarrow \left[\begin{array}{c} \text{CH}_3 \\ \text{C}=\text{O} \end{array} \right]^{\ddagger} \longrightarrow \underset{\text{H}}{\overset{\text{C}_6\text{H}_5}{\text{C}}}=\underset{\text{C}_6\text{H}_5}{\overset{\text{H}}{\text{C}}} \quad (19\text{-}7)$$

(R,S)

EXERCISE: Write the structure of the alkene produced from each of the following reaction sequences.

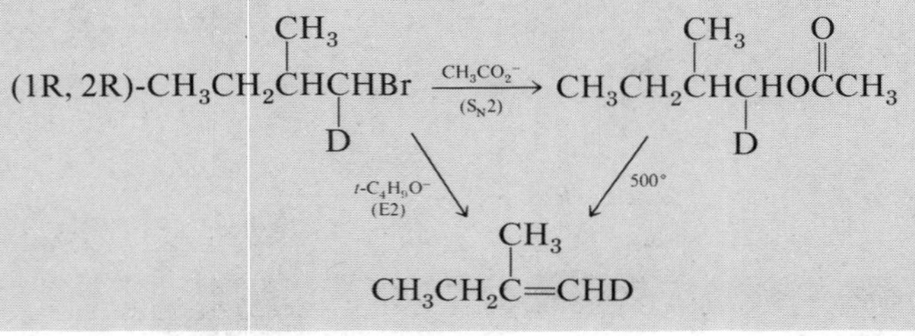

$$\text{(1R, 2R)-CH}_3\text{CH}_2\underset{\text{D}}{\overset{\text{CH}_3}{\text{CHCHBr}}} \xrightarrow[\text{(S}_\text{N}2)]{\text{CH}_3\text{CO}_2^-} \text{CH}_3\text{CH}_2\underset{\text{D}}{\overset{\text{CH}_3}{\text{CHCHO}}}\overset{\text{O}}{\overset{\|}{\text{CCH}_3}}$$

$$\underset{\text{(E2)}}{\overset{t\text{-C}_4\text{H}_9\text{O}^-}{\searrow}} \qquad \overset{500°}{\swarrow}$$

$$\text{CH}_3\text{CH}_2\underset{}{\overset{\text{CH}_3}{\text{C}}}=\text{CHD}$$

19.12
Waxes and Fats

A. *Waxes*

Waxes are naturally occurring esters of long-chain carboxylic acids (C_{16} or greater) with long-chain alcohols (C_{16} or greater). They are low-melting solids that have a characteristic "waxy" feel. We present three examples. **Spermaceti** is a wax that separates from the oil of the sperm whale upon cooling. It is mainly cetyl palmitate (cetyl alcohol $\equiv$ 1-hexadecanol) and melts at 42–47°.

$$\text{C}_{15}\text{H}_{31}\overset{\text{O}}{\overset{\|}{\text{C}}}\text{OC}_{16}\text{H}_{33}$$

spermaceti
cetyl palmitate
1-hexadecyl hexadecanoate

Beeswax is the material from which bees build honeycomb cells. It melts at 60–82° and is a mixture of esters. Hydrolysis yields mainly the C_{26} and C_{28} straight-chain carboxylic acids and the C_{30} and C_{32} straight-chain primary alcohols.

$$\text{C}_{25\text{-}27}\text{H}_{51\text{-}55}\overset{\text{O}}{\overset{\|}{\text{C}}}\text{OC}_{30\text{-}32}\text{H}_{61\text{-}65}$$

beeswax

Carnauba wax occurs as the coating on Brazilian palm leaves. It has a high melting point (80–87°) and is impervious to water. It is widely used as an ingredient in automobile and floor polish. It is a mixture of esters of the C_{24} and C_{28} carboxylic acids and the C_{32} and C_{34} 1-alkanols. Other components are present in smaller amounts.

B. *Fats*

Fats are naturally occurring esters of long-chain carboxylic acids and the triol glycerol (1,2,3-propanetriol). They are also called **glycerides.**

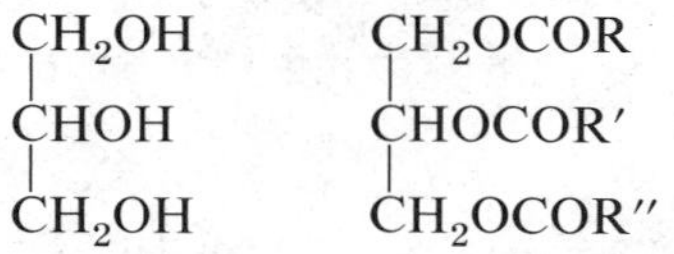

$$
\begin{array}{ll}
CH_2OH & CH_2OCOR \\
| & | \\
CHOH & CHOCOR' \\
| & | \\
CH_2OH & CH_2OCOR''
\end{array}
$$

glycerol   a fat or glyceride
1,2,3-propanetriol

Hydrolysis of fats yields glycerol and the component carboxylic acids. The straight-chain carboxylic acids that may be obtained from fats are frequently called **fat acids** or **fatty acids.** Fatty acids may be saturated or unsaturated. The most common saturated fatty acids are lauric acid, myristic acid, palmitic acid, and stearic acid.

$$CH_3(CH_2)_{10}COOH \quad CH_3(CH_2)_{12}COOH \quad CH_3(CH_2)_{14}COOH \quad CH_3(CH_2)_{16}COOH$$

| lauric acid | myristic acid | palmitic acid | stearic acid |
| dodecanoic acid | tetradecanoic acid | hexadecanoic acid | octadecanoic acid |

The most important unsaturated fatty acids have 18 carbon atoms, with one or more double bonds. Examples are oleic acid, linoleic acid, and linolenic acid.

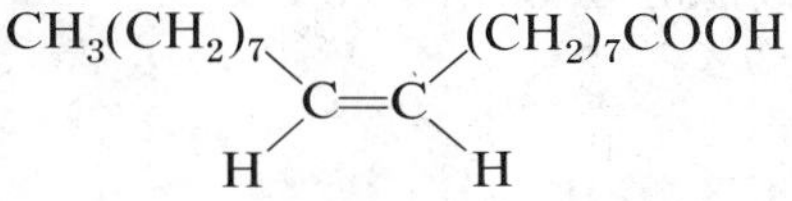

oleic acid
(Z)-9-octadecenoic acid

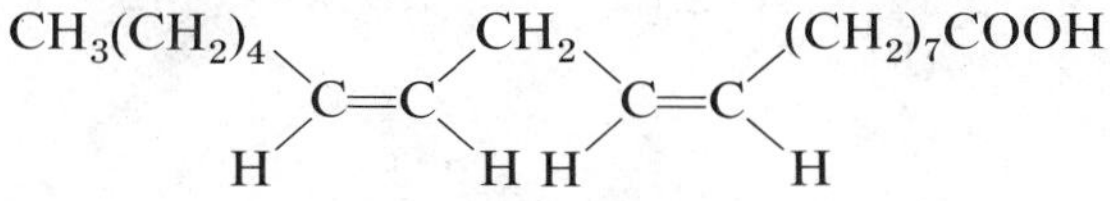

linoleic acid
(Z,Z)-9,12-octadecadienoic acid

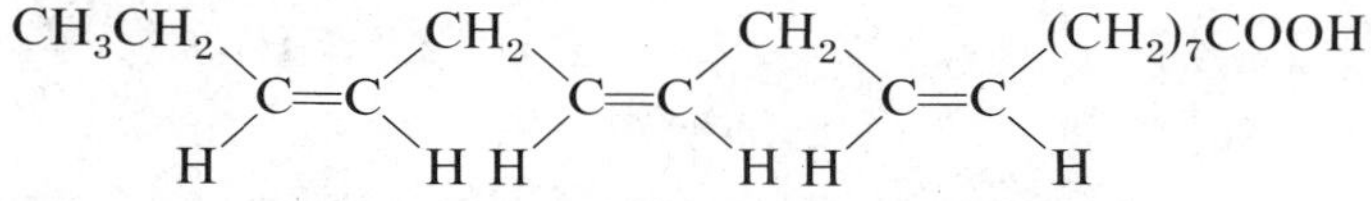

linolenic acid
(Z,Z,Z)-9,12,15-octadecatrienoic acid

Almost invariably, the double bonds have the *cis* or (Z) configuration. A significant exception is eleostearic acid, which has one *cis* and two *trans* double bonds.

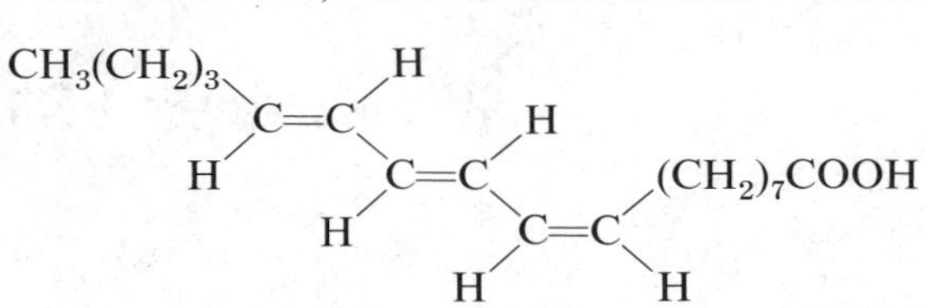

eleostearic acid
(Z,E,E)-9,11,13-octadecatrienoic acid

Natural fats are generally complex mixtures of triesters of glycerol **(triglycerides).** In general, the secondary hydroxy group is esterified with C_{18} acids, and the primary hydroxy groups are esterified with either C_{18} or other fatty acids. For example, hydrolysis of palm oil yields 1–3% myristic acid, 34–43% palmitic acid, 3–6% stearic acid, 38–40% oleic acid, and 5–11% linoleic acid. Some natural fats yield large amounts of a single fatty acid on hydrolysis; tung oil yields 2–6% stearic acid, 4–16% oleic acid, 1–10% linoleic acid, and 74–91% eleostearic acid.

Fats undergo the typical reactions of esters. An important commercial reaction of fats is alkaline hydrolysis. The product fatty acid salts are used as soaps (Section 18.4.D).

$$\begin{array}{l} CH_2OCOR \\ | \\ CH_2OCOR \\ | \\ CH_2OCOR \end{array} + 3\ NaOH \xrightarrow{H_2O} \begin{array}{l} CH_2OH \\ | \\ CHOH \\ | \\ CH_2OH \end{array} + 3\ RCO_2^-\ Na^+$$

The alkaline hydrolysis of an ester is often referred to as **saponification,** from this process.

The melting point of a fat depends on the amount of unsaturation in the fatty acids. Fats with a preponderance of unsaturated fatty acids have melting points below room temperature and are called **oils.** Fats with little unsaturation are solid at normal temperatures. For the manufacture of soaps and for certain food uses, solid fats are preferable to oils. The melting point of a natural fat may be increased by hydrogenation. Industrially, the process is called **hardening.** Vegetable oils, such as cottonseed and peanut, are often hardened to the consistency of lard by partial hydrogenation.

Phosphoglycerides are fats in which glycerol is esterified to two fatty acids and to phosphoric acid. Such monophosphate esters are called **phosphatidic acids.** Most phosphatidic acids contain one saturated and one unsaturated fatty acid. Because different acids are esterified at the C-1 and C-3 hydroxy groups of glycerol, phosphatidic acids are chiral; the absolute configuration at the C-2 ester link is (R).

a phosphatidic acid

Free phosphatidic acids are rare in nature. Usually the phosphoric acid moiety is esterified to a second alcohol component. Important examples are **phosphatidyl ethanolamine** and **phosphatidyl choline.**

phosphatidyl ethanolamine

phosphatidyl choline

Phosphoglycerides are important biomolecules and occur widely in plants and animals. They are often referred to collectively as "phospholipids."

> **Lipid** is a term that has been used to describe the group of natural substances which are soluble in hydrocarbons and insoluble in water. It includes fats, waxes, phosphoglycerides, natural hydrocarbons, and so on. Most biochemists reserve the term lipid for natural compounds that yield fatty acids upon hydrolysis.

Phosphoglycerides have long nonpolar "tails" and a small, highly polar "head."

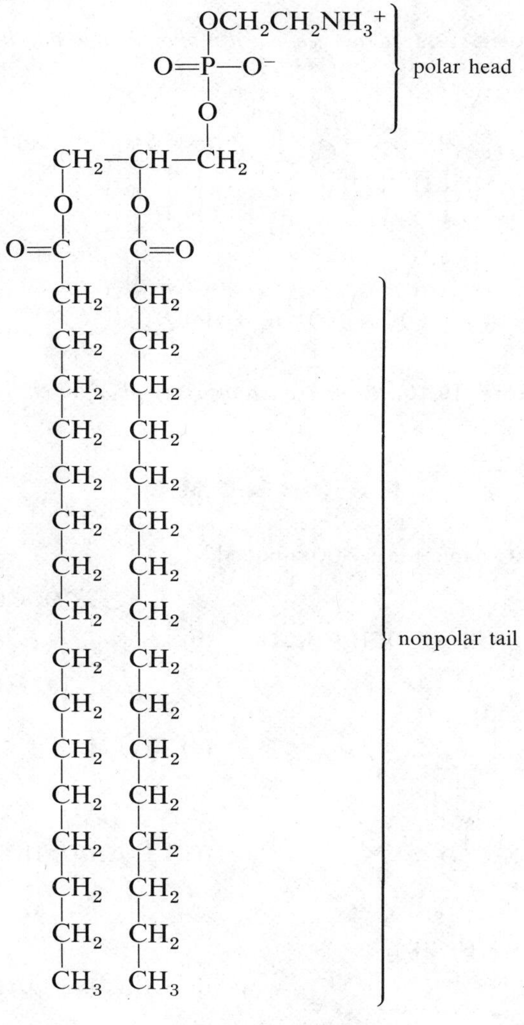

In aqueous solution, they disperse to form micelles in the same way soaps do (Section 18.4.D). The nonpolar tails cluster together in the middle of the micelle, leaving the polar heads exposed to the aqueous environment (Figure 19.9). Phospholipids also form **bilayers,** particularly at the interface between two aqueous surfaces. In such bilayers the hydrocarbon tails cluster toward one another, leaving a layer of polar heads on either side of the bilayer exposed to the aqueous phases (Figure 19.10). Such bilayers appear to form the fundamental framework of natural membranes.

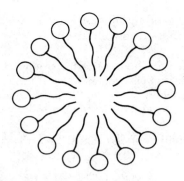

FIGURE 19.9. *Cross section of a phospholipid micelle.*

(H_2O)

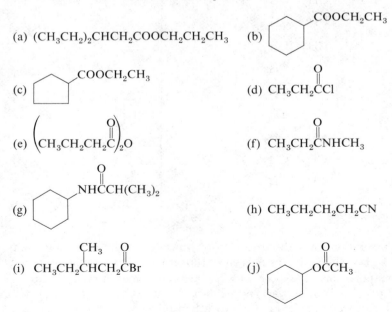

(H_2O)

FIGURE 19.10. *Cross section of a phospholipid bilayer.*

PROBLEMS

1. Write the IUPAC name for each compound.

(a) $(CH_3CH_2)_2CHCH_2COOCH_2CH_2CH_3$

(b)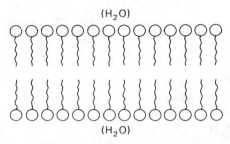

(c) cyclopentane COOCH₂CH₃

(d) $CH_3CH_2\overset{O}{\overset{\|}{C}}Cl$

(e) $\left(CH_3CH_2CH_2\overset{O}{\overset{\|}{C}}\right)_2O$

(f) $CH_3CH_2\overset{O}{\overset{\|}{C}}NHCH_3$

(g) cyclohexyl $NH\overset{O}{\overset{\|}{C}}CH(CH_3)_2$

(h) $CH_3CH_2CH_2CH_2CN$

(i) $CH_3CH_2\overset{CH_3}{\underset{|}{C}}HCH_2\overset{O}{\overset{\|}{C}}Br$

(j) cyclohexyl $O\overset{O}{\overset{\|}{C}}CH_3$

2. Write a structure that corresponds to each name.
(a) N,3-diethylhexanamide
(b) N,N-dimethylformamide
(c) ethyl butanoate
(d) methyl 3-chloropropanoate

(e) 5-methylhexanenitrile   (f) acetic formic anhydride
. (g) cyclohexanecarboxamide   (h) 3-methylcyclopentanenitrile
(i) cyclobutyl formate   (j) 3-formylhexanoic acid
(k) ethyl acetoacetate   (l) N-bromoacetamide
(m) butanoyl bromide   (n) propanoic anhydride

3. In a dilute solution of acetic acid in 0.1 M aqueous HCl, what percentage of the acetic acid is present as $CH_3CO_2^-$ and what percentage is present as $CH_3C(OH)_2^+$?

4. The pK_a of $CH_3\overset{\overset{O}{\|}}{C}OH_2^+$ may be estimated to be approximately -12. Calculate the ratio of $CH_3\overset{\overset{OH^+}{\|}}{C}OH$ to $CH_3\overset{\overset{O}{\|}}{C}OH_2^+$ in an acidic solution of acetic acid.

5. A neutral compound, $C_7H_{13}O_2Br$, does not give an oxime or phenylhydrazone derivative. The infrared spectrum shows bands at 2850–2950 cm^{-1} but none above 3000 cm^{-1}. Another strong band is at 1740 cm^{-1}. The nmr shows the following pattern: $\delta = 1.0$ ppm (3H) triplet; $\delta = 1.3$ ppm (6H) doublet; $\delta = 2.1$ ppm (2H) mult; $\delta = 4.2$ ppm (1H) triplet; $\delta = 4.6$ ppm (1H) mult. Deduce the structure and assign their bands.

6. Identify the compound having the following ir, nmr, and mass spectra.

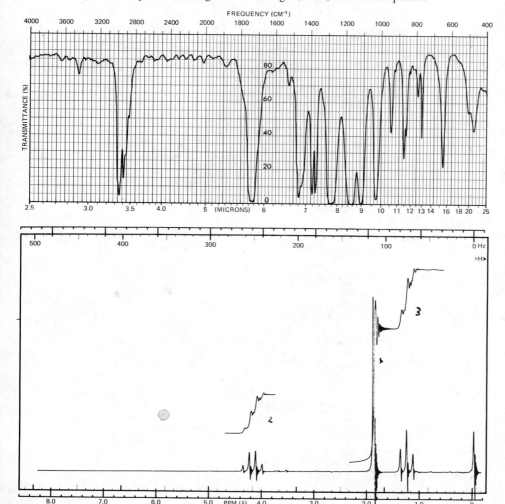

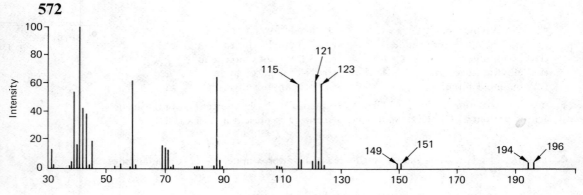

7. What are the organic products of each of the following reactions or sequences of reactions?

(a) $CH_3CH_2CH_2OH + CH_3\overset{\overset{\displaystyle O}{\|}}{C}Cl \longrightarrow \overset{500°}{\longrightarrow}$

(b) 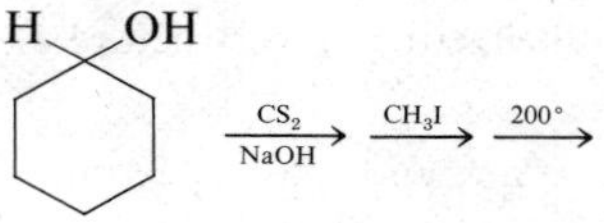 [cyclohexanol structure] $\xrightarrow[NaOH]{CS_2} \xrightarrow{CH_3I} \xrightarrow{200°}$

(c) 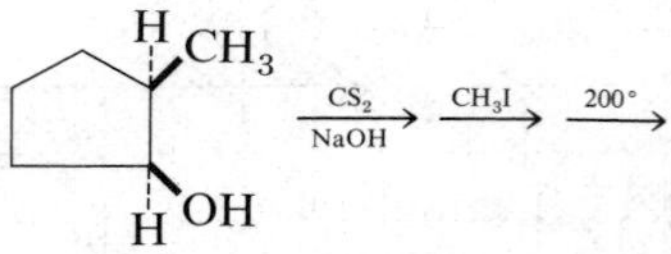 [cyclopentane with CH₃ and OH structure] $\xrightarrow[NaOH]{CS_2} \xrightarrow{CH_3I} \xrightarrow{200°}$

(d) $CH_3CH_2CH_2Cl \xrightarrow[\substack{ethanol \\ 75°}]{NaCN} \xrightarrow[\substack{H_2O \\ 100°}]{NaOH}$

(e) 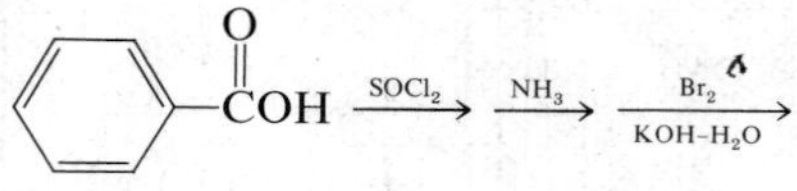 [benzene ring]$\overset{\overset{\displaystyle O}{\|}}{C}OH \xrightarrow{SOCl_2} \xrightarrow{NH_3} \xrightarrow[KOH-H_2O]{Br_2 \, \Delta}$

(f) 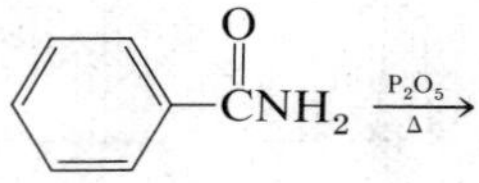 [benzene ring]$\overset{\overset{\displaystyle O}{\|}}{C}NH_2 \xrightarrow[\Delta]{P_2O_5}$

(g) $(CH_3)_3C\overset{\overset{\displaystyle O}{\|}}{C}NH_2 \xrightarrow[\substack{D_2O \\ 25°, \, 5 \, min}]{0.1 \, M \, NaOD}$

(h) 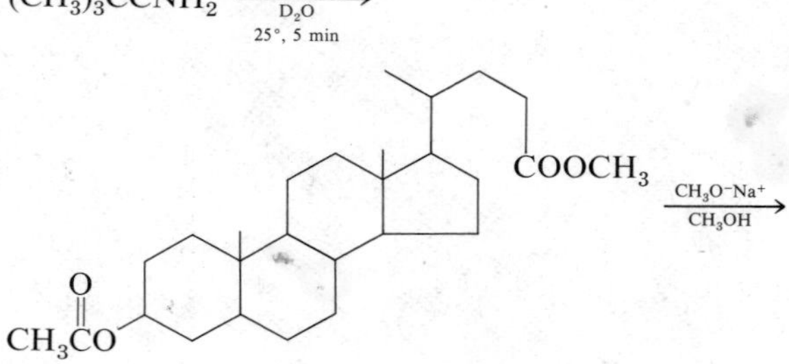 [steroid structure with COOCH₃ and CH₃CO–O] $\xrightarrow[CH_3OH]{CH_3O^-Na^+}$

(i) 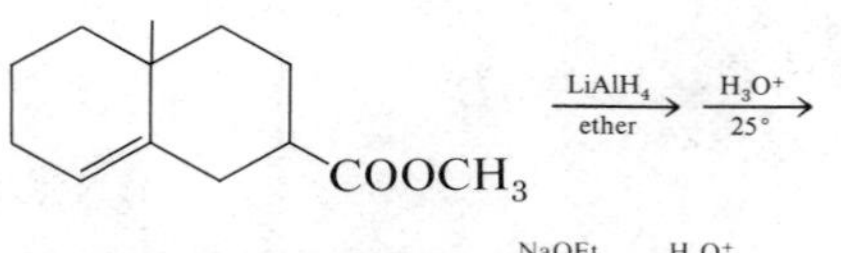 [bicyclic structure with COOCH₃] $\xrightarrow[ether]{LiAlH_4} \xrightarrow[25°]{H_3O^+}$

(j) $CH_3CH_2COOCH_2CH_3 \xrightarrow[EtOH]{NaOEt} \xrightarrow{H_3O^+}$

(k) $BrCH_2COOCH_3 + CH_3CH_2CHO + Zn \xrightarrow{benzene}$

(l) $CH_3CH_2COOCH_2CH_3 \xrightarrow[THF, \, -78°]{(i\text{-}C_3H_7)_2NLi} \xrightarrow{CH_3CH_2CHO} \xrightarrow{H_2O}$

8. Show how butanoic acid may be converted into each of the following compounds. More than one step may be required in some cases.

(a) $CH_3CH_2CH_2\overset{\overset{\displaystyle O}{\|}}{C}Cl$

(b) $\left(CH_3CH_2CH_2\overset{\overset{\displaystyle O}{\|}}{C}\right)_2O$

(c) $CH_3CH_2CH_2CH_2O\overset{\overset{\displaystyle O}{\|}}{C}CH_2CH_2CH_3$

(d) $CH_3CH_2CH_2\overset{\overset{\displaystyle O}{\|}}{C}N(CH_3)_2$

(e) $CH_3CH_2CH_2C{\equiv}N$

(f) $CH_3CH_2CH_2CH_2NH_2$

(g) $CH_3CH_2CH_2Cl$

(h) $CH_3CH_2CH_2CH_2N(CH_2CH_3)_2$

(i) $CH_3CH_2CH_2\overset{\overset{\displaystyle O}{\|}}{C}CH_2\overset{\overset{\displaystyle CH_3}{|}}{C}HCH_3$

(j) $CH_3CH_2CH_2\overset{\overset{\displaystyle CH_3}{|}}{\underset{\underset{\displaystyle CH_3}{|}}{C}}OH$

(k) $CH_3CH_2CH_2CHO$

(l) $CH_3CH_2CH_2CH_2OH$

(m) $CH_3CH_2CH_2\overset{\overset{\displaystyle OH}{|}}{C}H\underset{\underset{\displaystyle CH_3CH_2}{|}}{C}HCH_2OH$

(n) $CH_3CH_2CH_2\overset{\overset{\displaystyle OH}{|}}{C}H\underset{\underset{\displaystyle CH_2CH_3}{|}}{C}HCOOC_2H_5$

(o) $CH_3CH_2CH_2CO\underset{\underset{\displaystyle CH_2CH_3}{|}}{C}HCOOC_2H_5$

9. The following reaction sequence is observed.

$$\underset{(S)\text{-}(+)}{C_6H_5CH_2\overset{\overset{\displaystyle CH_3}{|}}{C}HCOOH} \xrightarrow[\substack{(2)\ NH_3 \\ (3)\ Br_2,\ OH^-}]{(1)\ SOCl_2} \underset{(-)}{C_6H_5CH_2\overset{\overset{\displaystyle CH_3}{|}}{C}HNH_2}$$

Using considerations of reaction mechanism, derive whether the configuration of the final product is (R) or (S).

10. Predict the product of each of the following reactions. Rationalize your predictions.

(a) $H_2N\!-\!\overset{\overset{\displaystyle O}{\|}}{C}\!-\!Cl + CH_3O^- \longrightarrow$

(b) $CH_3O\!-\!\overset{\overset{\displaystyle O}{\|}}{C}\!-\!Cl + H_2N^- \longrightarrow$

11. How may each of the following compounds be prepared from monofunctional compounds containing five or fewer carbons.

(a) $(CH_3)_2CHCH_2CH_2CH_2COOCH_3$

(b) $CH_3CH_2CH_2CH(OH)CH_2COOCH_2CH_3$

(c) $(CH_3)_3C\overset{\overset{\displaystyle O}{\|}}{C}CH_2COOCH_2CH_3$

(d) 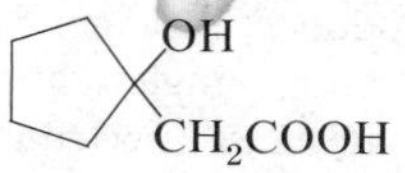 $\overset{\displaystyle OH}{\underset{\displaystyle CH_2COOH}{}}$

12. N-Methylacetamide reacts with triethyloxonium tetrafluoroborate to give a salt, $C_5H_{12}NO^+\ BF_4^-$. When this salt is treated with sodium bicarbonate, a compound $C_5H_{11}NO$ is produced.

$$CH_3\overset{\overset{\displaystyle O}{\|}}{C}NHCH_3 + (C_2H_5)_3O^+\ BF_4^- \longrightarrow C_5H_{12}NO^+\ BF_4^- \xrightarrow{NaHCO_3} C_5H_{11}NO$$

(a) What are the structures of the two compounds?

(b) Rationalize the formation of the salt in mechanistic terms.

(c) Predict the product that will be obtained if the salt is treated with aqueous acid.

13. Write a mechanism that explains the following reaction (the Ritter reaction).

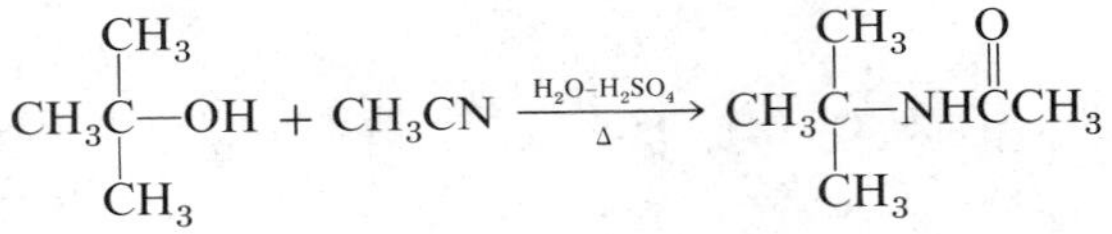

$$CH_3\overset{\displaystyle CH_3}{\underset{\displaystyle CH_3}{C}}{-}OH + CH_3CN \xrightarrow[\Delta]{H_2O-H_2SO_4} CH_3\overset{\displaystyle CH_3}{\underset{\displaystyle CH_3}{C}}{-}NH\overset{\displaystyle O}{C}CH_3$$

What product is expected when 2-methyl-2,4-pentanediol is treated with acetonitrile and aqueous sulfuric acid?

14. Explain why the pyrolysis of *cis*-2-methylcyclohexyl acetate gives only 3-methylcyclohexene, whereas *trans*-2-methylcyclohexyl acetate gives a mixture of 1-methylcyclohexene and 3-methylcyclohexene.

15. The elimination of a carboxylic acid from an ester is generally an endothermic process, for example,

$$CH_3CH_2CH_2CH_2OCOCH_3 \longrightarrow$$
$$CH_3CH_2CH{=}CH_2 + CH_3COOH \quad \Delta H° = +12.6 \text{ kcal mole}^{-1}$$

Yet the pyrolytic elimination is a useful preparative reaction. How do you account for this? Why is the pyrolytic elimination carried out at high temperature (300–500°)?

16. Orthoesters are compounds that have three alkoxy groups attached to the same carbon, for example, ethyl orthoacetate, $CH_3C(OC_2H_5)_3$. When ethyl orthoacetate is treated with dilute aqueous acid, ethyl acetate is obtained. Explain with a mechanism.

$$CH_3C(OC_2H_5)_3 + H_2O \xrightarrow{H^+} CH_3COOC_2H_5 + 2\ C_2H_5OH$$

17. The most intense peaks in the mass spectra of methyl pentanoate and methyl 2-methylbutanoate are m/e 74 and m/e 88, respectively. Explain.

18. Treatment of diethyl adipate, $C_2H_5OOC(CH_2)_4COOC_2H_5$, with sodium ethoxide in ethanol, followed by neutralization with aqueous acid, yields a compound having the formula $C_8H_{12}O_3$. Propose a structure for this product, and write a mechanism that accounts for its formation. This reaction is a general reaction of certain diesters (the Dieckmann condensation) and will be discussed in Section 27.6.

19. A solution of methyl cyclohexyl ketone in chloroform is treated with peroxybenzoic acid for 16 hr at 25°. The reaction mixture is worked up to obtain A, which has infrared absorption at 1740 cm^{-1} and 1250 cm^{-1}. The nmr spectrum of A shows a sharp three-proton singlet at $\delta = 2.0$ ppm and a one-proton multiplet at $\delta = 4.8$ ppm. The mass spectrum of A shows a molecular ion at m/e 142 and the most intense peak at m/e 43. What is A and how is it formed?

20. (R)-2-Butanol, $[\alpha]_D = -13.5°$, reacts with methanesulfonyl chloride to give a methanesulfonate. Treatment of the methanesulfonate with aqueous sodium hydroxide affords 2-butanol having $[\alpha]_D = +13.5°$.

(a) From this result, what conclusions may you draw regarding the mechanism of the hydrolysis?

(b) How may (S)-2-octanol be converted into (R)-2-methoxyoctane?

Chapter 20
Conjugation

20.1
Allylic Systems

A. *Allylic Cations*

When 2-buten-1-ol is treated with hydrogen bromide at 0°, a mixture of about 3 parts of 1-bromo-2-butene to 1 part of 3-bromo-1-butene is produced. A comparable mixture is produced when 3-buten-2-ol is treated with HBr under the same conditions.

$$CH_3CH = CHCH_2OH$$
$$CH_3CHOHCH = CH_2$$
$$\xrightarrow[0°]{HBr} CH_3CH = CHCH_2Br + CH_3CHBrCH = CH_2$$
$$3:1$$

Similarly, when 1-chloro-3-methyl-2-butene is hydrolyzed in water containing silver oxide at room temperature, a mixture of alcohols consisting of 15% of 3-methyl-2-buten-1-ol and 85% of 2-methyl-3-buten-2-ol is produced. Essentially the same mixture is obtained by the reaction of 3-chloro-3-methyl-1-butene with silver oxide in water.

$$(CH_3)_2C = CHCH_2Cl$$
$$\underset{Cl}{(CH_3)_2CCH = CH_2}$$
$$\xrightarrow[\substack{H_2O \\ 25°}]{Ag_2O} (CH_3)_2C = CHCH_2OH + \underset{(85\%)}{(CH_3)_2\overset{OH}{C}CH = CH_2} + AgCl$$
$$(15\%)$$

> Silver cation catalyzes the formation of carbocations from alkyl halides. The Ag$^+$ tends to coordinate with the leaving halide group; that is, it provides a potent "pull" that contributes to the driving force of the reaction.
>
> $$RX + Ag^+ \longrightarrow [R \cdots X \cdots Ag]^+ \longrightarrow R^+ + AgX\downarrow$$
> $$\text{(transition state)}$$
>
> Furthermore, silver chloride, bromide, and iodide are highly insoluble salts and remove the halide ion from further equilibration reactions. The net reaction of an alkyl halide with silver oxide is generally written as
>
> $$2 RX + Ag_2O + H_2O \longrightarrow 2 ROH + 2 AgX\downarrow$$
>
> However, since these reactions often involve carbocation intermediates, other reactions such as rearrangements and eliminations frequently occur. Because of these reaction possibilities and the high cost of the reagent, the reaction of silver oxide with alkyl halides has little preparative significance and is used mainly to study the properties of carbocation intermediates.

These observations are explained by the formation of an intermediate cation in which the positive charge is delocalized over two carbons.

$$CH_3CH=CHCH_2\overset{+}{O}H_2$$

$$\overset{+}{O}H_2$$
$$CH_3\overset{|}{C}HCH=CH_2$$

$$\xrightarrow{-H_2O} [CH_3CH=CH\overset{+}{C}H_2 \longleftrightarrow CH_3\overset{+}{C}HCH=CH_2] \xrightarrow{Br^-}$$

$$CH_3CH=CHCH_2Br + CH_3\overset{Br}{\underset{|}{C}}HCH=CH_2$$

$$(CH_3)_2C=CHCH_2Cl$$

$$Cl$$
$$(CH_3)_2\overset{|}{C}CH=CH_2$$

$$\xrightarrow{-Cl^-} [(CH_3)_2C=CH\overset{+}{C}H_2 \longleftrightarrow (CH_3)_2\overset{+}{C}CH=CH_2] \xrightarrow{H_2O}$$

$$\overset{+}{O}H_2$$
$$(CH_3)_2C=CHCH_2\overset{+}{O}H_2 + (CH_3)_2\overset{|}{C}CH=CH_2$$

The intermediate carbocations in the foregoing reactions are described as resonance hybrids of two important structures. The simplest such cation is the 2-propen-1-yl or **allyl cation**.

$$[CH_2=CH-\overset{+}{C}H_2 \longleftrightarrow \overset{+}{C}H_2-CH=CH_2] \equiv \overset{\frac{1}{2}+}{CH_2}\cdots CH \cdots \overset{\frac{1}{2}+}{CH_2}$$
allyl cation

Various symbols may be used to describe the hybrid electronic structure of the allyl cation. The two resonance structures in brackets show that the positive charge in the cation is divided equally between the two indicated positions. The alternative structure with dotted bonds shows that each C—C bond has an order of $1\frac{1}{2}$ and that each end carbon has $\frac{1}{2}$ of a positive charge.

The grouping $CH_2=CHCH_2-$ is called the **allyl group**, just as CH_3CH_2- is called the ethyl group. The group name is used in naming many compounds containing the allyl group.

$$CH_2=CHCH_2Cl \qquad CH_2=CHCH_2OH \qquad CH_2=CHCH_2O\overset{O}{\overset{\|}{C}}CH_3$$
allyl chloride $\qquad\qquad$ allyl alcohol $\qquad\qquad$ allyl acetate

Substituents may be indicated by the use of Greek letters.

$$\underset{\gamma}{CH_2}=\underset{\beta}{CH}\underset{\alpha}{\overset{\overset{CH_3}{|}}{CH}}Cl \qquad \underset{CH_3}{\overset{CH_3}{\diagdown}}\underset{\gamma}{C}=\underset{\beta}{CH}\underset{\alpha}{CH_2}OH$$

α-methylallyl chloride $\qquad\qquad$ γ,γ-dimethylallyl alcohol
3-chloro-1-butene $\qquad\qquad\qquad$ 3-methyl-2-buten-1-ol

One important structural feature of allyl cation is that all of the atoms lie in one plane in such a way that the empty *p*-orbital of the carbocation can overlap with the π-orbital of the double bond. The electron density in the double bond is shared in the manner indicated by the symbols

$$[CH_2=CH-\overset{+}{C}H_2 \longleftrightarrow \overset{+}{C}H_2-CH=CH_2]$$

It is important to recall that resonance structures are used to symbolize alternative configurations of electron density. The geometry of the nuclei remains *precisely* the same in all resonance structures. The symbol $CH_2=CH-\overset{+}{C}H_2$ would normally indicate a C=C having a short distance and a C—C having a longer distance. However, the alternative structure $\overset{+}{C}H_2$—CH=CH$_2$ contributes equally to the actual structure symbolized by dotted lines as $(CH_2\cdots CH\cdots CH_2)^+$. Allyl cation has two equivalent C—C bonds of equal length. The parent ion has been detected in the gas phase and is inferred as an intermediate in some solution reactions. The structure has not been determined experimentally, but sophisticated quantum-mechanical calculations show the C—C bond length to be between those for single and double bonds. Similarly, the two terminal carbon atoms share the positive charge equally.

A stereo representation of the planar structure of allyl cation and the corresponding orbital description are shown in Figure 20.1. The two electrons in the π-orbital are in a molecular orbital extending over all three carbon atoms (Section 20.1.E).

Since the positive charge is spread over a larger volume, we expect allyl cation to be more stable than a simple primary alkyl cation. This expectation is confirmed by a comparison of gas phase enthalpies of ionization of alkyl chlorides.

$$RCl \longrightarrow R^+ + Cl^-$$

R	$\Delta H°$, kcal mole^{-1}
CH_3CH_2Cl	191
CH_2=$CHCH_2Cl$	173
$(CH_3)_2CHCl$	167

In fact, *allyl cation is roughly comparable in relative stability to a secondary alkyl cation.* Similarly, methylallyl cation has the positive charge spread between a secondary and primary position, and the net stabilization is comparable to that of a tertiary alkyl cation.

$$[CH_3\overset{+}{C}H—CH=CH_2 \longleftrightarrow CH_3CH=CH—\overset{+}{C}H_2] \equiv CH_3\overset{\delta+}{CH}\cdots CH\cdots\overset{\delta+}{CH_2}$$

When an allylic cation reacts with a nucleophilic reagent, it can react at either positive center and generally produces a mixture of products. As a result, reac-

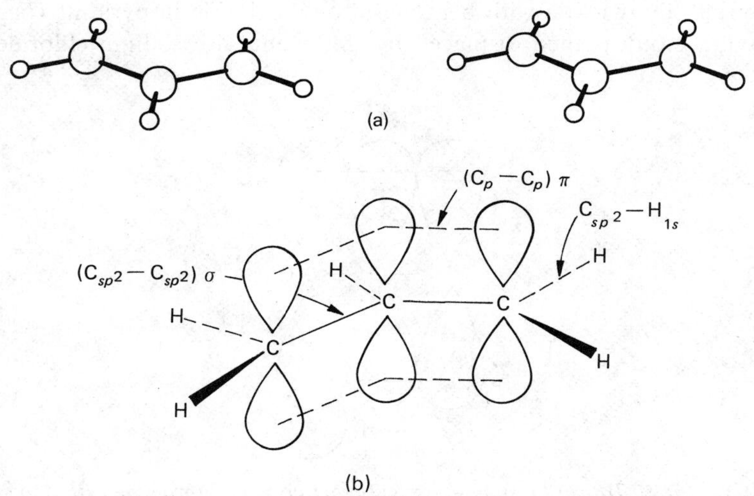

(a)

(b)

FIGURE 20.1. *Allyl cation:* (a) *stereo structure;* (b) *orbital description.*

tions that proceed by way of allyl cations often appear to give "rearranged" products. Such reactions are called **allylic rearrangement**. For example, in the first reaction we encountered in this chapter, 2-buten-1-ol reacts with HBr to give 1-bromo-2-butene, a "normal" product. However, the reaction also gives 3-bromo-1-butene, a product of allylic rearrangement. Allylic rearrangement can be observed even with the parent system by labeling one carbon with radioactive ^{14}C.

$$CH_2{=}CH^{14}CH_2OH \xrightarrow{\text{SOCl}_2} CH_2{=}CH^{14}CH_2Cl + {}^{14}CH_2{=}CHCH_2Cl$$

(ratio depends on reaction conditions)

> **EXERCISE:** Write the structures for all of the isomeric products expected from (R)-2-chloro-(E)-3-hexene with silver oxide in aqueous dioxane at 25°.

B. S_N2 Reactions

In addition to forming carbocations relatively easily, allylic halides and alcohols also undergo substitution by the S_N2 mechanism more readily than analogous saturated systems. For example, allyl bromide undergoes bimolecular substitution about 40 times more rapidly than does ethyl bromide. This enhanced reactivity is apparently due to the fact that the double bond stabilizes the transition state for substitution and therefore lowers the activation energy for the process (Figure 20.2). Recall (Figure 8.5) that the S_N2 transition state can be viewed as a p-orbital on the central carbon that simultaneously overlaps with orbitals from the entering and leaving nucleophiles. In an allylic system this p-orbital also participates in π-overlap with the adjacent double bond. Thus, even though allylic systems are prone to react by the S_N1 mechanism, because they form carbocations easily, it is possible by a careful choice of reaction conditions to cause them to react *without allylic rearrangement* by way of the S_N2 mechanism.

$$CH_3CH{=}CHCH_2OH \xrightarrow[\text{pyridine}]{C_6H_5SO_2Cl} CH_3CH{=}CHCH_2OSO_2C_6H_5 \xrightarrow{Cl^-} CH_3CH{=}CHCH_2Cl$$

but-2-en-1-ol   1-chloro-2-butene

In the foregoing example, the allylic hydroxy group is first converted into a good leaving group by reaction with benzenesulfonyl chloride in pyridine (page 255). This leaving group is then displaced by treatment with sodium chloride. Since

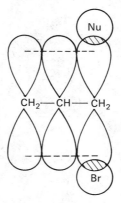

FIGURE 20.2. *Transition state of an S_N2 reaction of allyl bromide with a nucleophile,* Nu.

substitution occurs by the S_N2 mechanism, only 1-chloro-2-butene is obtained. If substitution is accomplished by treatment of the alcohol with HCl, a mixture of 1-chloro-2-butene and 3-chloro-1-butene is obtained, analogously to the reaction with HBr shown on page 575.

The reactivity of allylic compounds in displacement reactions is sufficiently high that they even react with Grignard reagents.

$$CH_2=CHCH_2Br + \overset{\bigtriangleup}{} -MgBr \longrightarrow \overset{\bigtriangleup}{} -CH_2CH=CH_2 + MgBr_2$$

(70%)

> Cyclopentylmagnesium bromide is prepared from 745 g of bromocyclopentane and 125 g of magnesium in 3 L of anhydrous ether. The mixture is refluxed, and 605 g of allyl bromide is added slowly. The mixture is stirred for 2 hr, and cold 6 N HCl is added. The ether layer is separated, washed, dried, and distilled to yield 70% of allylcyclopentane, 3-cyclopentyl-1-propene, b.p. 121–125°.

The reaction of Grignard reagents with allyl bromide is a good method for preparing 1-alkenes. The corresponding reaction with normal saturated alkyl halides does not occur.

EXERCISE: Show how 4,4-dimethyl-1-pentene and 4-methyl-1-hexene may be synthesized starting with alkyl halides.

C. Allylic Anions

The allyl Grignard reagent may be prepared by treating an allyl halide with magnesium.

$$CH_2=CHCH_2X + Mg \xrightarrow{\text{ether}} CH_2=CHCH_2MgX$$

However, since allyl halides react with Grignard reagents to give 1-alkenes, care must be taken in preparing the allyl Grignard reagent. If the reaction is attempted at too high concentration, a large amount of the coupling product, 1,5-hexadiene or "biallyl," is formed.

$$CH_2=CHCH_2MgX + CH_2=CHCH_2X \longrightarrow CH_2=CHCH_2CH_2CH=CH_2 + MgX_2$$

1,5-hexadiene
biallyl

The allyl Grignard reagent can be prepared in good yield by minimizing the further S_N2 reaction of the Grignard reagent with allyl halide. This result may be accomplished by adding a dilute solution of the allyl halide in ether slowly to a large excess of vigorously stirred magnesium. This technique is an example of the **dilution principle.** The rate of reaction of an alkyl halide with magnesium depends on the concentration of the alkyl halide and the surface area of the magnesium.

$$\text{rate}_1 = k_1[CH_2=CHCH_2X][Mg \text{ surface}]$$

The rate of the displacement step depends on the concentrations of allyl halide and Grignard reagent.

$$\text{rate}_2 = k_2[CH_2=CHCH_2X][CH_2=CHCH_2MgX]$$

When the solution is diluted, the concentrations of the allyl halide and the Grignard reagent decrease, but the surface area of the magnesium remains unchanged. Dilution retards both reactions, but it slows the second reaction more than the first.

Two isomeric allylic halides give Grignard reagents with indistinguishable properties.

$$CH_3CHBrCH=CH_2 \xrightarrow{Mg} C_4H_7MgBr \xleftarrow{Mg} CH_3CH=CHCH_2Br$$

$$\downarrow \text{aq } H^+$$

$$CH_3CH_2CH=CH_2 + CH_3CH=CHCH_3$$
$$(57\%) \qquad\qquad (27\% \text{ cis})$$
$$(16\% \text{ trans})$$

The allylic Grignard reagent undergoes rapid isomerization.

$$CH_3CH=CHCH_2MgBr \underset{}{\overset{\text{fast}}{\rightleftharpoons}} CH_3\underset{\underset{MgBr}{|}}{CH}CH=CH_2$$

Although Grignard reagents apparently have substantial C—Mg covalent bonding, we have seen that in many reactions the reagents behave as carbanion salts, $[R-MgX \longleftrightarrow \bar{R} \ \overset{+}{M}gX]$. Similarly, allylic Grignard reagents have a high degree of ionic character involving the magnesium cation salt of an allylic anion;

$$[CH_2=CH-CH_2-MgX \longleftrightarrow CH_2=CH-\bar{C}H_2 \ \overset{+}{M}gX \longleftrightarrow {}^-CH_2-CH=CH_2 \ \overset{+}{M}gX]$$

The negative charge is spread between two **conjugated** carbons. This spreading of ionic character facilitates rearrangement of the magnesium. In the reaction with an acid, protonation may occur at either of the carbons where there is negative charge. To summarize, the two isomeric allylic halides 1-bromo-2-butene and 3-bromo-1-butene can give *two* isomeric allylic Grignard reagents. The two isomers are in *rapid equilibrium* by allylic rearrangement of the magnesium. The net result on hydrolysis is protonation at either negative center of the corresponding allyl anion to give a mixture of product alkenes, 1-butene and 2-butene in the present example.

The spreading of charge in an allylic anion is a stabilizing mechanism; such anions are more stable than simple unconjugated anions. The allylic hydrogen of propene, for example, is substantially more acidic than any hydrogen in ethane or propane.

$$CH_2=CHCH_3 + B:^- \longrightarrow BH + [CH_2=CH-\overset{..}{\bar{C}}H_2 \longleftrightarrow \overset{..}{\bar{C}}H_2-CH=CH_2]$$
more acidic $\qquad\qquad\qquad\qquad\qquad\qquad$ delocalized allyl anion

$$CH_3CH_3 + B:^- \longrightarrow BH + CH_3\overset{..}{\bar{C}}H_2$$
less $\qquad\qquad\qquad$ localized
acidic $\qquad\qquad\quad$ ethyl anion

We have encountered similar kinds of conjugated or resonance-stabilized anions before: carboxylate ion (Section 18.4.A), enolate ion (Section 13.6.A), amidate ion (Section 19.10), even nitrite ion (Section 8.9.A).

$$\left[O=C-\bar{O} \longleftrightarrow \bar{O}-C=O \atop {|\atop R} \qquad {|\atop R} \right] \qquad \left[R_2\bar{C}-C=O \longleftrightarrow R_2C=C-O^- \atop {|\atop R} \qquad\qquad {|\atop R} \right]$$

carboxylate ion $\qquad\qquad\qquad\qquad\qquad$ enolate ion

$$\left[\begin{array}{ccc} \overset{\displaystyle O}{\underset{\displaystyle \|}{}} & & \overset{\displaystyle O^-}{\underset{\displaystyle |}{}} \\ R{-}C{-}NH^- & \longleftrightarrow & R{-}C{=}NH \end{array} \right] \qquad [O{=}N{-}O^- \longleftrightarrow {}^-O{-}N{=}O]$$

amidate ion   nitrite ion

All such ions are described in the same way. The resonance structures provide a way of representing a rather complex electronic distribution by means of the formal symbolism of Lewis structures. The actual structure is a composite of the resonance structures.

The mathematics of linear combinations is especially useful for describing this situation. In this approach, the wave function of a molecule Ψ is represented as a linear combination of simpler wave functions ψ.

$$\Psi = a\psi_a + b\psi_b + c\psi_c + \cdots$$

For many molecules a single structure provides a satisfactory representation. For example, methane is well represented by a single Lewis structure ψ_a, and

$$\Psi = \psi_a$$

Other possible Lewis structures, such as

$$\psi_b = \overset{\displaystyle H}{\underset{\displaystyle H}{H{-}\overset{|}{\underset{|}{C}}{:}^-H^+}}$$

are so unlikely that their contribution to the linear combination is negligible; that is, b for methane is a very small number.

In allyl cation and anion, carboxylate ion, and nitrite ion, the two resonance structures are equivalent and must have equal coefficients in the linear combination.

$$\Psi(CH_2{\cdots}CH{\cdots}CH_2)^+ = a\psi_a(CH_2{=}CH{-}\overset{+}{C}H_2) + b\psi_b(\overset{+}{C}H_2{-}CH{=}CH_2)$$

where $a = b$.

In enolate ions the two structures are not equivalent. The structure with negative charge on the electronegative oxygen is more stable than that with the charge on carbon. Hence these two structures enter into the linear combination with unequal coefficients.

$$\Psi(CH_2{\cdots}CH{\cdots}O)^- = a\psi_a(CH_2{=}CH{-}O^-) + b\psi_b(\overset{-}{C}H_2{-}CH{=}O)$$

where $a > b$. We say that ψ_a contributes more than ψ_b; hence, the amount of negative charge on oxygen in Ψ, the resonance hybrid, is greater than that on carbon.

Resonance structures whose coefficients are very small contribute so little to the actual structure of the resonance hybrid that their contribution is generally neglected. It is this important property that allows us to represent even complex organic molecules by what is really a rather simple symbolism.

We have already encountered a number of neutral functions that have electronic structures closely related to that of allyl anion. In all such systems a lone pair of electrons is conjugated with a multiple bond.

$$\left[\begin{array}{ccc} \overset{\displaystyle :\overset{..}{O}:}{\underset{\displaystyle \|}{}} & & \overset{\displaystyle :\overset{..}{O}:^-}{\underset{\displaystyle |}{}} \\ R{-}C{-}\overset{..}{\underset{..}{O}}H & \longleftrightarrow & R{-}C{=}\overset{+}{\underset{..}{O}}H \end{array} \right] \qquad \left[\begin{array}{ccc} \overset{\displaystyle :\overset{..}{O}:}{\underset{\displaystyle \|}{}} & & \overset{\displaystyle :\overset{..}{O}:^-}{\underset{\displaystyle |}{}} \\ R{-}C{-}\overset{..}{N}H_2 & \longleftrightarrow & R{-}C{=}\overset{+}{N}H_2 \end{array} \right]$$

carboxylic acid   amide

These cases involve nonequivalent resonance structures. The dipolar structures involve charge separation and are generally less stable than the normal Lewis structures. Hence the dipolar structures generally contribute less to the overall resonance hybrids. But they do contribute, and we have seen (Section 19.1) how consideration of such structures is essential to the understanding of the chemistry of such functional groups; the normal Lewis structures alone provide an inadequate description of the actual electronic structures of such groups.

> **EXERCISE:** What products are expected from treatment of 3-chloro-1-pentene with magnesium in ether, followed by carbon dioxide, then dilute aqueous sulfuric acid?

D. *Allylic Radicals*

Allylic radicals are also stabilized by resonance.

$$[CH_2{=}CH{-}\dot{C}H_2 \longleftrightarrow \dot{C}H_2{-}CH{=}CH_2]$$

The odd-electron character is spread between two carbons, and this radical is more stable than a simple alkyl radical. This increased stability is reflected in the relatively low bond-dissociation energy of bonds conjugated to a double bond.

$$DH°, kcal\ mole^{-1}$$

$$CH_3CH_2{-}H \qquad 98$$

$$CH_2{=}CHCH_2{-}H \qquad 87$$

Advantage can be taken of this low bond-dissociation energy of allylic C—H bonds in free radical halogenation—but only under special circumstances because of the alternative reaction path of addition to the double bond. One method for accomplishing **allylic bromination** is with the reagent N-bromosuccinimide. This material is available commercially and is prepared by bromination of the cyclic imide of succinic acid (Section 27.6).

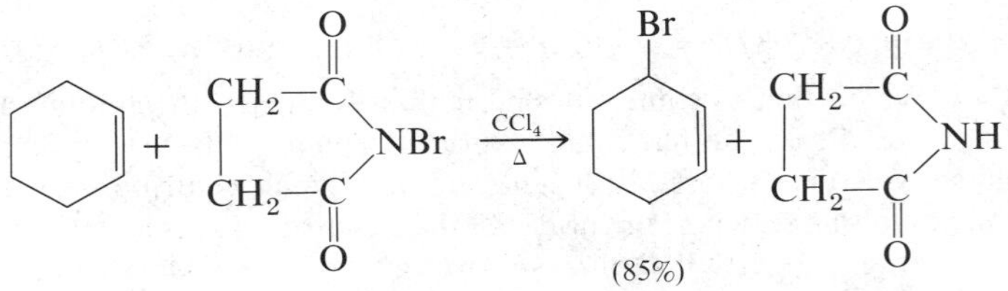

N-bromosuccinimide   succinimide

The reaction is not simple. It is important to use a medium in which N-bromosuccinimide is insoluble. Carbon tetrachloride is commonly used for this purpose. Reaction occurs in part on the surface of the N-bromosuccinimide, although the active reagent appears to be bromine formed in dilute solution from the reaction of traces of acid and moisture with the bromoimide.

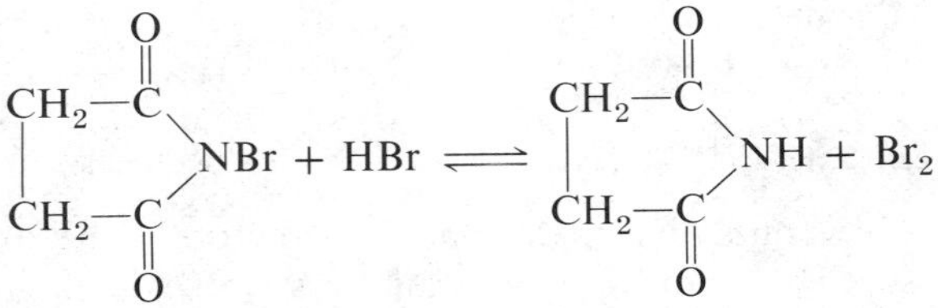

The bromine is then involved in free radical chain bromination of the allylic hydrogen. Under these conditions of high dilution no significant addition of bromine to the double bond occurs.

One of the reasons for using a nonpolar solvent such as CCl_4 in this reaction is that the normal addition of Br_2 to a double bond is an ionic reaction.

$$\underset{C}{\overset{C}{\|}} + Br_2 \longrightarrow \underset{C}{\overset{C}{|}}\overset{+}{Br} + Br^-$$

In the absence of a suitable solvent to solvate these ions, one or more excess bromine molecules are required for this role.

$$\underset{C}{\overset{C}{\|}} + 2\,Br_2 \longrightarrow \underset{C}{\overset{C}{|}}\overset{+}{Br} + Br_3^-$$

Thus the reaction kinetics has a relatively high order in bromine and the ionic addition has a low reaction rate when bromine is kept in low concentration.

A free radical initiator or light is often used to promote the reaction. Because the reaction intermediate is a resonance-stabilized radical, two products can be obtained in unsymmetrical cases.

$$[R\overset{\cdot}{C}H-CH=CHR' \longleftrightarrow RCH=CH-\overset{\cdot}{C}HR'] \xrightarrow{Br_2}$$

$$\overset{\displaystyle Br}{\underset{\displaystyle |}{RCH}}-CH=CHR' + RCH=CH-\overset{\displaystyle Br}{\underset{\displaystyle |}{C}HR'}$$

Allyl chloride is prepared commercially in large quantity by the direct free radical chlorination of propylene at high temperature. At higher temperatures the normal *addition* of chlorine atom to the double bond becomes unfavorable for entropic reasons, and hydrogen abstraction is the principal reaction. In the addition reaction two species become one and freedom of motion is lost, whereas the entropy change in hydrogen abstraction is small. Entropy considerations are especially important at high temperatures.

$$CH_3CH=CH_2 + Cl_2 \xrightarrow{500°} ClCH_2CH=CH_2 + HCl$$

$$Cl_2 \rightleftharpoons 2\,Cl\cdot$$

$$CH_2=CHCH_3 + Cl\cdot \longrightarrow HCl + CH_2=CH\overset{\cdot}{C}H_2$$

$$CH_2=CH\overset{\cdot}{C}H_2 + Cl_2 \longrightarrow Cl\cdot + CH_2=CHCH_2Cl$$

Allyl alcohol is prepared from the chloride by hydrolysis.

EXERCISE: Write the equations for the reactions of the following alkenes with N-bromosuccinimide in refluxing carbon tetrachloride. (a) 2-methylpropene, (b) cyclopentene, (c) 2-pentene.

E. *Molecular Orbital Description of Allylic Systems*

The resonance description of allylic conjugation involves alternative descriptions of bonding by pairs of electrons in normal Lewis structures.

The asterisk represents a positive or negative charge or an odd electron. Most of the electrons bond the skeleton of the compound and are not involved in the resonance stabilization or conjugation. In the molecular orbital picture these electrons form the σ-bonding framework of the molecule. Conjugation is a phenomenon associated more generally with the π-electron system. The π-system consists of p-orbitals overlapping to form π-bonds above and below the plane of the atoms that define the allylic system (Figure 20.3).

The involvement of three p-orbitals in this manner clearly gives greater bonding, and it is not difficult to understand the stabilization that such orbital overlap bestows on allyl cation. However, according to the Pauli principle, only two electrons of opposite spin can be associated with any single orbital. What are we to do with the third and fourth π-electrons of allyl radical and anion?

Three p_z-orbitals overlapping, as in allyl, generate *three* different molecular orbitals, each having its own energy. One is most bonding and is occupied by two electrons of opposite spin in allyl cation. The third electron of allyl radical must be put into the second molecular orbital. The fourth electron of allyl anion can also be put into this second molecular orbital. The third molecular orbital has high energy and is not involved in bonding in any of these compounds. These relationships are shown in Figure 20.4.

The π-molecular orbitals can be regarded as having molecular orbital quantum numbers or their equivalent in nodes. When p-orbitals overlap as in allyl, they do so in such a way as to generate one molecular orbital having no nodes, a second having one node, and a third having two nodes.

Since these molecular orbitals are made up of p-orbitals overlapping in a π-fashion, all of the molecular orbitals have one other node, the nodal plane of the component p-orbitals. This plane is illustrated in Figure 20.3. The nodes referred to above are nodes in addition to this nodal plane.

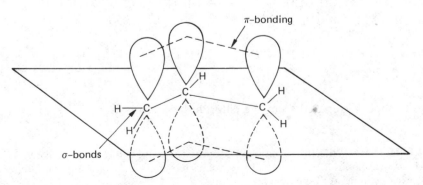

FIGURE 20.3. *σ- and π-bonds in allyl systems.*

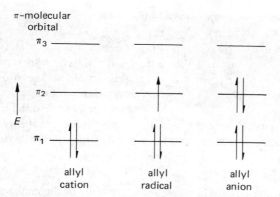

π-molecular
orbital

E

allyl
cation

allyl
radical

allyl
anion

FIGURE 20.4. π-molecular orbital energies in allyl.

These molecular orbitals for allyl are shown in Figure 20.5. Recall that when functions of the same sign overlap, electron density is put in the overlap region between the nuclei, and bonding results. Electron density does not exist at a node. The overlap of two wave functions of opposite sign creates a node in the overlap

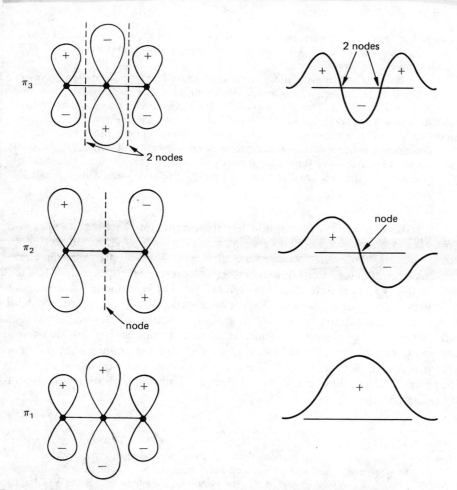

FIGURE 20.5. π-molecular orbitals of allyl.

FIGURE 20.6. Standing waves of a linear system.

region and signifies a region devoid of electron density. The absence of such electron density to counter nuclear repulsion produces **antibonding**. Hence, the first allyl π-molecular orbital, π_1, has no nodes and is completely bonding. In π_2 there is a node going through the middle carbon. The two remaining p-orbital wave functions are so far apart that overlap is small and this molecular orbital is approximately **nonbonding**. The highest molecular orbital, π_3, has two nodes and is antibonding. In general, the greater the number of nodes, the higher the energy of an orbital and the lower the stability. The greatest stability (lowest energy) results when electrons are associated as far as possible with the most bonding molecular orbitals.

These molecular orbitals can be described analytically by the mathematical functions

$$\pi_1 = \frac{1}{2}p_1 + \frac{\sqrt{2}}{2}p_2 + \frac{1}{2}p_3$$

$$\pi_2 = \frac{\sqrt{2}}{2}p_1 - \frac{\sqrt{2}}{2}p_3$$

$$\pi_3 = \frac{1}{2}p_1 - \frac{\sqrt{2}}{2}p_2 - \frac{1}{2}p_3$$

in which p_1, p_2, and p_3 are the mathematical functions for the three p-atomic orbitals. Note that the node at the middle carbon of π_2 means simply that the coefficient of p_2 in this molecular orbital is zero.

It is instructive to compare these molecular orbitals with the standing waves of other linear systems, such as a vibrating violin string or the sound waves in a pipe organ, as in Figure 20.6. The lowest-energy wave has no nodes and is either positive throughout or negative throughout. The next lowest wave has one node, and the third has two nodes. Note the close resemblence to the π-molecular orbitals of allyl. The correspondence reaffirms the common properties of all waves. Moreover, the smooth continuous nature of wave functions allows us to understand why the end coefficients in π_1 and π_3 of allyl are smaller than the middle coefficient and why the middle coefficient is zero in π_2.

This molecular orbital description has important uses in understanding why more than one electron pair can be associated with a single group of overlapping atomic orbitals. The approach is subject to more or less accurate numerical quantum-mechanical calculation. But the molecular orbital description does not lend itself to the type of simple structural symbolism that is so facile with resonance structures. Accordingly, the use of resonance structures is important and widespread in the qualitative understanding and prediction of charge distributions and reactivities of conjugated systems. Resonance structures also provide an accurate accounting of all of the electrons. It is important to recognize that resonance structures and molecular orbitals are complementary views of the same reality. They offer different dissections of total electron-density distributions that help us to understand the whole.

EXERCISE: Pentadienyl anion, $CH_2=CH-CH=CH-CH_2^-$, has its negative charge divided among three carbons. Write the corresponding resonance structures. There are now five π-molecular orbitals (MO) with 0 to 4 nodes. Sketch the five π-MOs ranked in order of expected energy. How many π-electrons are involved and which of the π-MOs are occupied?

20.2
Dienes

A. Structure and Stability

Double bonds separated by one or more carbon atoms react more or less independently. The heats of hydrogenation of such double bonds are essentially those of independent units. For example, $\Delta H°$ for the reaction of 1,5-hexadiene with hydrogen is exothermic by 60 kcal mole^{-1}, exactly twice that for the reaction of 1-hexene with hydrogen. (Recall that the heat of hydrogenation of an alkene is $\Delta H°$ for the reaction   alkene + H$_2$ $\longrightarrow$ alkane.)

$$CH_2\!=\!CHCH_2CH_2CH\!=\!CH_2 + 2\,H_2 \longrightarrow$$
$$CH_3CH_2CH_2CH_2CH_2CH_3 \quad \Delta H°_{hydrog} = -60\ kcal\ mole^{-1}$$
$$CH_3CH_2CH_2CH_2CH\!=\!CH_2 + H_2 \longrightarrow$$
$$CH_3CH_2CH_2CH_2CH_2CH_3 \quad \Delta H°_{hydrog} = -30\ kcal\ mole^{-1}$$

Heats of hydrogenation for other alkenes and dienes are included in Table 20.1.

TABLE 20.1
Heats of Hydrogenation

	$\Delta H°_{hydrog}$, kcal mole^{-1}
$CH_3CH_2CH\!=\!CH_2$	-30.2
$CH_3CH_2CH_2CH\!=\!CH_2$	-29.8
$CH_3CH_2CH_2CH_2CH\!=\!CH_2$	-30.0
$CH_2\!=\!CH\!-\!CH\!=\!CH_2$	-56.5
$CH_2\!=\!CHCH_2CH\!=\!CH_2$	-60.4
$CH_2\!=\!CHCH_2CH_2CH\!=\!CH_2$	-60.0

Note that 1,3-butadiene is a significant exception to the preceding generalization. Its hydrogenation is about 4 kcal mole^{-1} *less exothermic* than for the other two dienes. This compound is an example of a **conjugated diene,** a diene in which the two double bonds are separated by a single bond. Dienes in which two or more single bonds separate the double bonds are called **unconjugated dienes.** The double bonds in unconjugated dienes are called **isolated double bonds.**

Conjugated dienes are significantly more stable than would be expected for a compound with completely independent double bonds. This relatively small but significant difference is attributed to two factors, which are shown in Figure 20.7.

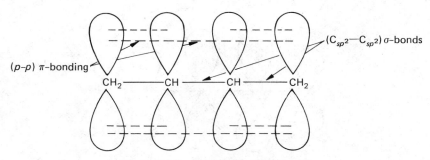

FIGURE 20.7. *Structure of 1,3-butadiene.*

First, the two double-bond distances are essentially normal, but the C_2—C_3 single bond is shorter than the 1.54 Å distance normally associated with C—C single bonds. This decreased bond length results in part from the increased s-character

$$CH_2 \xlongequal{1.34 Å} CH \xrightarrow{1.48 Å} CH {=\!=} CH_2$$

of the carbon orbitals comprising this bond; the single bond between the double bonds may be described approximately as C_{sp^2}—C_{sp^2}. This shorter bond is somewhat stronger than C—C bonds having less s-character. Second, the p_z-orbitals on carbons 2 and 3 can also overlap to give some double-bond character to the C_2—C_3 single bond. This factor also contributes some additional stability to the conjugated double-bond system. However, this overlap is much less than those between C-1 and C-2 and between C-3 and C-4 carbons because of the greater distance between the C-2 and C-3 p-orbitals.

The π-system of 1,3-butadiene consists of four overlapping p-orbitals which

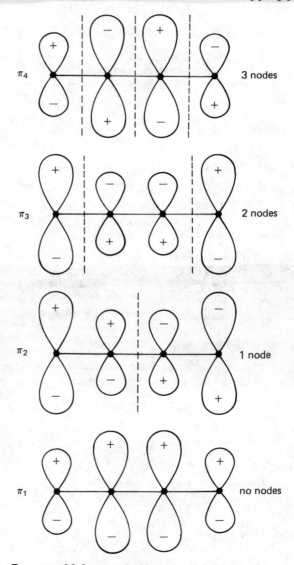

FIGURE 20.8. π-molecular orbitals of 1,3-butadiene.

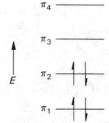

FIGURE 20.9. *Relative energies of π-molecular orbitals of 1,3-butadiene.*

generate four molecular orbitals. These four orbitals are shown schematically in Figure 20.8. The lowest-energy orbital, π_1, has no nodes and is bonding between carbons 1 and 2, between carbons 2 and 3, and between carbons 3 and 4. The second MO, π_2, has one node. Therefore, this orbital is bonding between carbons 1 and 2 and between carbons 3 and 4 but is antibonding between the center carbons. The four electrons associated with the π-system are in molecular orbitals π_1 and π_2 (Figure 20.9). Therefore, the π-bonding between carbons 2 and 3 produced by π_1 is partially offset by the antibonding nature of π_2 in this region of the molecule. The antibonding character of π_2 does not quite cancel the bonding character of π_1, since π_1 is of lower energy than π_2. However, there is little net π-bonding between the center carbons, and the net electronic structure is given to a reasonable approximation by the normal Lewis structure

$$H_2C::CH:CH::CH_2$$

B. *Addition Reactions*

The conjugated character of 1,3-dienes is shown in two-step addition reactions. Such additions are almost invariably initiated at the end of a chain of conjugation to produce a resonance-stabilized allylic intermediate rather than a nonconjugated intermediate.

$$CH_2=CH-CH=CH_2 + HBr$$

$$\overset{+}{C}H_2-\underset{\underset{H}{|}}{C}H-CH=CH_2$$

$$\left[CH_2-\underset{\underset{H}{|}}{\overset{+}{C}}H-CH=CH_2 \longleftrightarrow CH_2-CH=CH-\overset{+}{C}H_2 \right] + Br^-$$

This intermediate reacts in a second step to give a mixture of products characteristic of the intermediate allylic system.

$$[CH_3\overset{+}{C}H-CH=CH_2 \longleftrightarrow CH_3CH=CH-\overset{+}{C}H_2] + Br^- \longrightarrow$$

$$\underset{\underset{Br}{|}}{CH_3CHCH=CH_2} + CH_3CH=CHCH_2Br$$

(80%)   (20%)
α-methylallyl   crotyl bromide
bromide

A further example is seen in the addition of bromine to 1,3-butadiene.

$$CH_2{=}CH{-}CH{=}CH_2 + Br_2 \xrightarrow[-15°]{\text{hexane}} \begin{bmatrix} CH_2{=}CH{-}\overset{+}{C}HCH_2Br \\ \updownarrow \\ \overset{+}{C}H_2{-}CH{=}CHCH_2Br \end{bmatrix} Br^- \longrightarrow$$

$$BrCH_2CH{=}CHCH_2Br + CH_2{=}CHCHBrCH_2Br$$
$$\text{54\%} \qquad\qquad\qquad \text{46\%}$$

[In this case, the allylic carbocation is sufficiently stabilized that the effect of a cyclic
bromonium ion is minimized (see Section 11.6.B.3).]

When the mixture of dibromides in the last example is warmed to 60°, the composition of the mixture changes to one consisting of 90% of (E)-1,4-dibromo-2-butene. This compound is easy to isolate in a pure state because it is a solid, m.p. 54°, and crystallizes readily.

$$\left.\begin{aligned} BrCH_2CH{=}CHCH_2Br \quad &(46\%) \\ + \\ CH_2{=}CHCHBrCH_2Br \quad &(54\%) \end{aligned}\right\} \xrightarrow{60°} \underset{(90\%)}{\overset{BrCH_2}{\underset{H}{}}\!\!\diagup\!\! \overset{}{C}{=}C\!\!\diagup\!\!\overset{H}{\underset{CH_2Br}{}}}$$

Thus, (E)-1,4-dibromo-2-butene is the most stable product, but it is formed at a rate comparable to the rate of formation of the other isomer.

This example illustrates an important concept in organic chemistry, **kinetic versus thermodynamic control.** In the addition reaction, the product composition is determined by the relative rates of reaction of the nucleophilic reagent at the two positions of positive charge. These relative reactivities need not, and generally do not, reflect the relative thermodynamic stabilities of the products. In the pres-

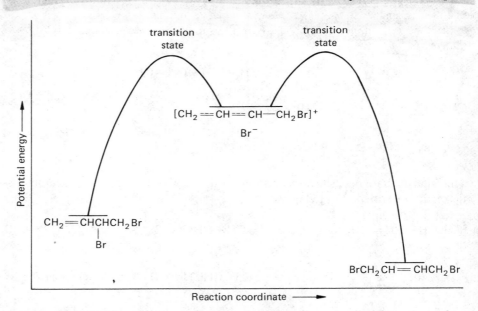

FIGURE 20.10. *Kinetic and thermodynamic effects in the formation of dibromobutenes.*

ent case, the reaction of the intermediate carbocation with bromide ion occurs approximately equally at both cationic centers; the reaction shows little selectivity. However, 1,4-dibromo-2-butene has a disubstituted double bond and is somewhat more stable than 3,4-dibromo-1-butene, which has a monosubstituted double bond (Section 11.4). Under conditions where the dibromides can react further to reform the carbocation, the more stable isomer predominates. Such a process provides a mechanism for establishing equilibrium.

The situation is illustrated in Figure 20.10. The two alternative transition states derived from the intermediate carbocation have comparable energies and give the alternative products at approximately equal rates. Actually, the rate of formation of 3,4-dibromo-1-butene, which involves reaction at the more positive secondary carbocation, is a little faster than the formation of the 1,4-dibromo isomer. However, the 1,4-isomer is the more stable; it reforms the carbocation less readily than the 3,4-isomer. Hence, at equilibrium, some of the 3,4-isomer is converted to 1,4-isomer, and the latter predominates.

The contrast between kinetic and thermodynamic control is important and will be encountered from time to time in our further study of organic chemistry. Another example is found in the reaction products of butadiene with hydrogen bromide. As shown above, 3-bromo-1-butene (α-methylallyl bromide) is the dominant product of the addition reaction. However, the equilibrium mixture consists of only 15% of α-methylallyl bromide and 85% of 1-bromo-2-butene (crotyl bromide). Once again, equilibrium favors the more highly substituted double bond. On prolonged reaction or treatment with strong Lewis acids such as ferric bromide, the equilibrium mixture is produced (Figure 20.11).

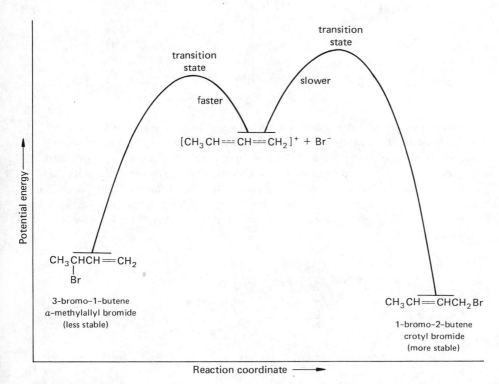

FIGURE 20.11. *Kinetic and thermodynamic effects in the formation of bromobutenes.*

EXERCISE: The reaction of 1,3-butadiene with aqueous bromine is expected to afford three isomeric bromobutenols (see page 310). Give the structures of these three products and predict the principal product of the reaction of each isomer with sodium hydroxide (page 321).

C. 1,2-Dienes: Allenes

1,2-Propadiene, $CH_2=C=CH_2$, has the trivial name of allene. Both double bonds in this hydrocarbon are especially short; the bond distance of 1.31 Å is between those for the double bond in ethylene, 1.34 Å, and the triple bond in acetylene, 1.20 Å. The electronic structure can be represented in terms of two double-bond systems *at right angles* as in Figure 20.12. Note that the central carbon is sp-hybridized. The additional s-character in these $C=C$ double bonds accounts for the rather short length.

One especially interesting feature of allenes, which results from the nonplanar character of the molecule, is that suitably substituted allenes are chiral and can be obtained as optically active enantiomers. For example, penta-2,3-diene has no plane of symmetry. Its mirror images are not superimposable, and this hydrocarbon is capable of existence in $(+)$ and $(-)$ enantiomers.

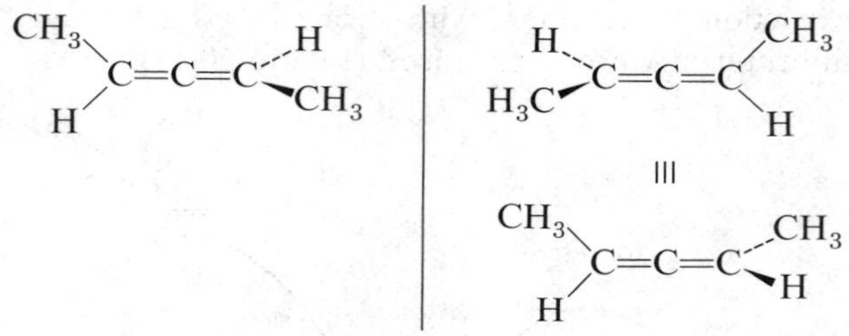

mirror images of penta-2,3-diene

A stereo representation of such an allene is given in Figure 20.13.

Molecules with **cumulated** double bonds, as in allene, do not constitute an important class of compounds. They are generally difficult to prepare and can frequently be isomerized to more stable dienes. Allene, for example, with $\Delta H_f^\circ = 45.9$ kcal mole^{-1} is 1.6 kcal mole^{-1} *less* stable than propyne with $\Delta H_f^\circ = 44.3$ kcal mole^{-1}; and 1,2-butadiene (methylallene, $CH_3CH=C=CH_2$, $\Delta H_f^\circ = 38.3$ kcal mole^{-1}), though slightly more stable than 1-butyne ($\Delta H_f^\circ = 39.5$ kcal mole^{-1}),

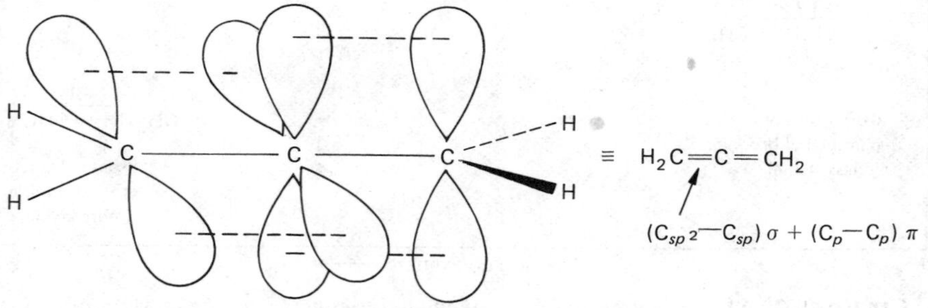

FIGURE 20.12. *Orbital structure of allene.*

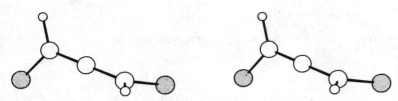

FIGURE 20.13. *Stereo representation of a 1,3-disubstituted allene.*

is almost 4 kcal mole^{-1} less stable than 2-butyne ($\Delta H_f^\circ = 34.7$ kcal mole^{-1}) and almost 13 kcal mole^{-1} less stable than 1,3-butadiene ($\Delta H_f^\circ = 26.1$ kcal mole^{-1}).

EXERCISE: Use molecular models to show that 1,3-dichloropropa-1,2-diene is chiral and that 1,4-dichlorobuta-1,2,3-triene is not.

D. *Preparation of Dienes*

Many dienes can be prepared in much the same way as monoenes except that two functional groups are involved. Some examples are

$$(CH_3)_2\overset{OH}{\underset{}{C}}\text{---}\overset{OH}{\underset{}{C}}(CH_3)_2 \xrightarrow[420\text{-}470^\circ]{Al_2O_3} CH_2{=}\overset{}{\underset{CH_3}{C}}\text{---}\overset{}{\underset{CH_3}{C}}{=}CH_2$$

(79–85%)

2,3-dimethyl-1,3-butadiene

$$CH_3COOCH_2CH_2CH_2CH_2CH_2OOCCH_3 \xrightarrow{575^\circ} 2\ CH_3COOH + CH_2{=}CHCH_2CH{=}CH_2$$

(63–71%)

1,4-pentadiene

Many other synthetic methods are known, but most require difunctional compounds (to be studied later) as starting materials.

Allylic halides are useful for preparing both conjugated and unconjugated dienes. Displacement by a vinylmagnesium halide on an allyl halide is another route to 1,4-dienes (see Section 20.1B).

$$CH_2{=}CHCH_2Cl + CH_2{=}CHMgBr \xrightarrow{ether} CH_2{=}CHCH_2CH{=}CH_2$$

The Wittig reaction (Section 13.7.G) with allylphosphoranes gives 1,3-dienes.

$$CH_2{=}CHCH_2Cl + P(C_6H_5)_3 \longrightarrow CH_2{=}CHCH_2\overset{+}{P}(C_6H_5)_3\ Cl^-$$

allyltriphenylphosphonium chloride

$$CH_2{=}CHCH_2\overset{+}{P}(C_6H_5)_3 \xrightarrow{LiOC_2H_5} [CH_2{=}CH\overset{-}{C}H\overset{+}{P}(C_6H_5)_3] \xrightarrow{C_6H_5CHO}$$

$$CH_2{=}CHCH{=}CHC_6H_5 + (C_6H_5)_3PO$$

1-phenyl-1,3-butadiene

Another use of allylic intermediates is exemplified in one preparation of 2-methyl-1,3-butadiene (isoprene).

$$CH_3COCH_3 + HC\equiv CH \xrightarrow{\text{KOH}} (CH_3)_2\overset{\displaystyle OH}{\underset{\displaystyle |}{C}}C\equiv CH \xrightarrow{\text{Pd, H}_2}$$

2-methyl-3-butyn-2-ol

$$(CH_3)_2\underset{\displaystyle \underset{\displaystyle OH}{|}}{C}CH=CH_2 \xrightarrow[\Delta]{\text{Al}_2\text{O}_3} CH_2=\overset{\displaystyle CH_3}{\overset{\displaystyle |}{C}}-CH=CH_2$$

2-methyl-3-buten-2-ol   isoprene

E. *Diene Polymers*

Butadiene is probably the most important of the dienes. It is prepared commercially in large quantity by the catalytic dehydrogenation of butane or of butane–butene mixtures.

$$CH_3CH_2CH_2CH_3 \xrightarrow[\text{catalyst}]{600°} CH_2=CH-CH=CH_2 + 2 H_2$$

Butadiene is a gas at room temperature, b.p. $-4.5°$, m.p. $-113°$. Large amounts are used as an intermediate in organic syntheses. For example, its reaction with chlorine is used to prepare large quantities of 1,4-dichloro-2-butene, an important intermediate in one route to nylon (Section 27.6.B). However, most butadiene is used directly, often in conjunction with one or more other monomers, to produce polymers.

A copolymer of 4–5 moles of butadiene to 1 mole of styrene, $C_6H_5CH=CH_2$, is an **elastomer** known as the synthetic rubber Buna S or GRS. In this free radical polymerization, the butadiene adds by *cis*- and *trans*-1,4- and 1,2-additions. The repeating units in the polymer are

$$\underset{\displaystyle \underset{\displaystyle C_6H_5}{|}}{-CH}-CH_2- \qquad -CH_2CH=CHCH_2- \qquad \underset{\displaystyle \underset{\displaystyle CH=CH_2}{|}}{-CH_2-CH}-$$

cis and *trans*

Butadiene is also polymerized with Ziegler–Natta catalysts (alkylaluminum and titanium chloride) or by alkali metal catalysts based on alkylsodium formulations or lithium dispersions. Some of these methods are highly specific and give either *cis*- or *trans*-1,4- or 1,2-addition.

Neoprene is a synthetic elastomer obtained by the free radical polymerization of chloroprene, 2-chloro-1,3-butadiene, which in turn is prepared by a route starting with acetylene (page 351).

$$2 HC\equiv CH \xrightarrow{\text{CuCl}} CH_2=CHC\equiv CH \xrightarrow{\text{HCl}} CH_2=\overset{\displaystyle Cl}{\overset{\displaystyle |}{C}}-CH=CH_2$$

vinylacetylene   chloroprene

Neoprene has unique properties, such as resistance to oils, oxygen, and heat.

2-Methyl-1,3-butadiene, isoprene, also polymerizes under the influence of acids or Ziegler–Natta catalysts to a polyisoprene with rubber-like properties.

$$n \; CH_2=\overset{\displaystyle CH_3}{\overset{\displaystyle |}{C}}-CH=CH_2 \longrightarrow \left(CH_2\overset{\displaystyle CH_3}{\overset{\displaystyle |}{C}}=CHCH_2\right)_n$$

isoprene   polyisoprene
2-methyl-1,3-butadiene

The double bonds in this synthetic polyisoprene are both *cis* and *trans*.
Natural rubber consists mostly of polyisoprene with *cis*-1,4 head-to-tail units.

The natural latex is not a useful elastomer or rubber but requires a **vulcanization** process. One such process was discovered by Charles Goodyear in 1839 and involves heating the raw rubber with sulfur. The process appears to involve addition of sulfur units to the double bonds with the production of **crosslinks** between the polymer chains. Because of these crosslinks, the polymer resists distortion and tends to return to its original shape.

> Elasticity has important structural requirements. If a polymer has regular repeating units, regions of the polymer may pack together by van der Waals forces in a manner similar to crystals. Such polymers are more or less crystalline and tend to be hard solids. Polymers that have flexible and irregular chains tend to be less rigid, but such a polymer is not an elastomer unless it returns to its original shape when stress is removed. Hence, elastomers tend to have flexible chains with varying amounts of crosslinking.

Some plants produce a polyisoprene with *trans*-1,4-isoprene units.

This material, known as gutta-percha, is a harder and less elastomeric natural polymer than rubber.

20.3
Unsaturated Carbonyl Compounds

A. *Unsaturated Aldehydes and Ketones*

Compounds having both a carbonyl group and a double bond are known as unsaturated aldehydes or ketones. As with dienes, the two centers of unsaturation can be conjugated or unconjugated and the conjugated isomers are generally more stable. For example,

$$CH_2=CHCH_2CHO \longrightarrow CH_3CH=CHCHO \qquad \Delta H° = -6 \text{ kcal mole}^{-1}$$

unconjugated isomer   conjugated isomer

$$CH_2=CHCH_2CH=CH_2 \longrightarrow CH_3CH=CHCH=CH_2 \qquad \Delta H° = -7 \text{ kcal mole}^{-1}$$

unconjugated isomer   conjugated isomer

The stabilizing effect of conjugation in unsaturated carbonyl compounds is approximately of the same magnitude as that in the corresponding dienes. It is explained in the same manner in terms of the stabilizing effect of the central C_{sp^2}—C_{sp^2} bond and by the overlap of p-orbitals to give π-bonding.

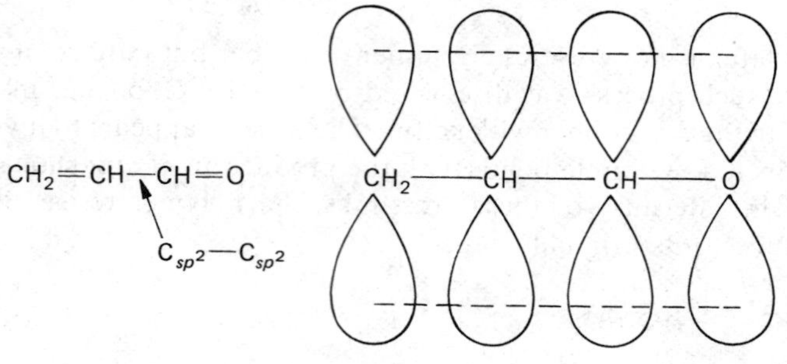

$$CH_2\!\!=\!\!CH\!-\!\!CH\!\!=\!\!O$$

$$C_{sp^2}\!-\!C_{sp^2}$$

Because of the greater stability of the conjugated unsaturated aldehydes and ketones, an isolated double bond will tend to **move into conjugation** if a suitable pathway is available. This migration of double bonds is especially facile for double bonds that are β,γ to the carbonyl group by acid- and base-catalyzed reactions that involve the intermediate enol.

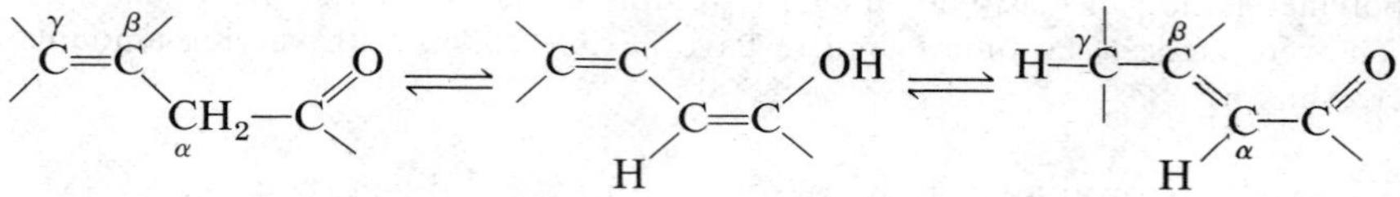

β,γ-unsaturated isomer (less stable) $\quad$ enol form $\quad$ α,β-unsaturated isomer (more stable)

For example, the methylene hydrogens in 3-butenal (vinylacetaldehyde) are appreciably acidic because they are α to the carbonyl group and give rise to a resonance-stabilized enolate ion. However, in this enolate ion a further resonance structure can be written that shows that negative charge is also spread to the γ-carbon.

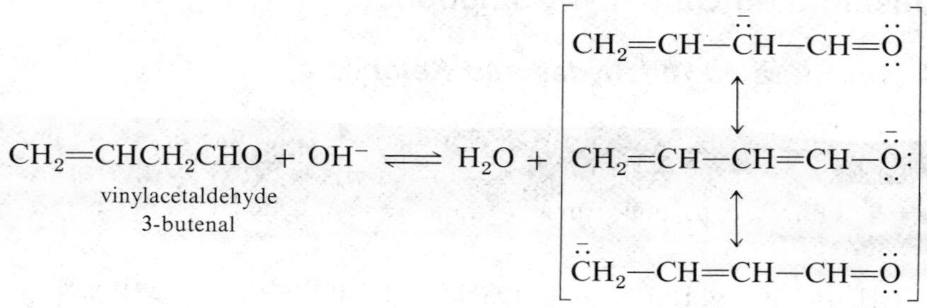

$$CH_2\!\!=\!\!CHCH_2CHO + OH^- \rightleftharpoons H_2O + \left[\begin{array}{c} CH_2\!\!=\!\!CH\!-\!\ddot{C}H\!-\!CH\!\!=\!\!\ddot{O} \\ \updownarrow \\ CH_2\!\!=\!\!CH\!-\!CH\!\!=\!\!CH\!-\!\ddot{O}: \\ \updownarrow \\ \ddot{C}H_2\!-\!CH\!\!=\!\!CH\!-\!CH\!\!=\!\!\ddot{O} \end{array} \right]$$

vinylacetaldehyde
3-butenal

The delocalized anion can be protonated by water on oxygen to give the enol form, on the α-carbon to regenerate vinylacetaldehyde, or at the γ-carbon to generate crotonaldehyde.

$$\left[\begin{array}{c} CH_2=CH-CH=CH-\ddot{\overset{..}{O}}:^- \\ \updownarrow \\ CH_2=CH-\overset{..}{\overset{-}{C}}H-CH=O \\ \updownarrow \\ \ddot{\overset{-}{C}}H_2-CH=CH-CH=O \end{array}\right] + H_2O \rightleftharpoons$$

$$\begin{array}{c} CH_2=CH-CH=CH-OH \\ \text{enol form} \\ \rightleftharpoons CH_2=CH-CH_2-CH=O \\ \text{vinylacetaldehyde} \\ CH_3-CH=CH-CH=O \\ \text{crotonaldehyde} \end{array}$$

All of these compounds are interconverted by base, but at equilibrium the most stable isomer, the conjugated crotonaldehyde, predominates to the extent of about 99.99% (Figure 20.14).

One way in which these interconversions can be demonstrated is by deuterium exchange. The enolate ion derived from crotonaldehyde or vinylacetaldehyde can react with D_2O at either of the carbons that bear the negative charge.

$$\left[\begin{array}{c} ^-CH_2-CH=CHCHO \\ \updownarrow \\ CH_2=CH-\overset{-}{C}HCHO \end{array}\right] \xrightarrow{D_2O} \begin{array}{c} CH_2D-CH=CHCHO \\ CH_2=CH-CHDCHO \end{array}$$

Repeated reaction with base to re-form the enolate ion and reaction with D_2O eventually produce the tetradeuterio compound $CD_3CH=CDCHO$.

The acid-catalyzed interconversions involve the intermediate enol in exact analogy to simple aldehydes and ketones (Section 13.6). Rapid and reversible protonation occurs at the carbonyl oxygen to form a hydroxycarbocation, which can lose a proton from carbon to form an enol. This enol is also a diene and can

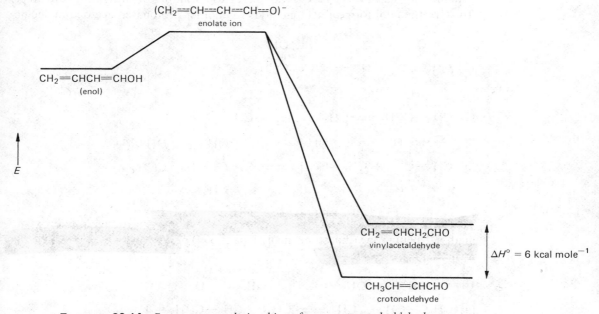

FIGURE 20.14. *Some energy relationships of an unsaturated aldehyde.*

reprotonate at either the α- or the γ-carbon to produce the β,γ- or α,β-unsaturated carbonyl, respectively.

$$CH_2{=}CH{-}CH_2{-}\overset{O}{\overset{\|}{C}}H \underset{}{\overset{H^+}{\rightleftharpoons}} \left[\begin{array}{c} CH_2{=}CH{-}CH_2{-}\overset{\overset{+}{O}H}{\overset{\|}{C}}H \\ \updownarrow \\ CH_2{=}CH{-}CH_2{-}\overset{OH}{\underset{+}{C}H} \end{array} \right] \overset{-H^+}{\rightleftharpoons} CH_2{=}CH{-}CH{=}\overset{OH}{C}H \overset{H^+}{\rightleftharpoons}$$

a dienol

$$\left[CH_3\overset{+}{C}H{-}CH{=}\overset{OH}{C}H \longleftrightarrow CH_3CH{=}CH{-}\overset{OH}{\underset{+}{C}H} \longleftrightarrow CH_3CH{=}CH{-}\overset{\overset{+}{O}H}{\overset{\|}{C}}H \right] \overset{-H^+}{\underset{H^+}{\rightleftharpoons}}$$

$$CH_3CH{=}CHCHO$$

α,β-Unsaturated aldehydes and ketones are often obtained directly in the condensation of aldehydes and ketones under basic conditions (the aldol condensation, Section 13.7.G). They may also be obtained under acidic conditions. For example, the acid-catalyzed condensation of acetone produces 4-methylpent-3-en-2-one, commonly called "mesityl oxide."

$$2\ CH_3COCH_3 \xrightarrow{H_2SO_4} CH_3\overset{CH_3}{\overset{|}{C}}{=}CHCOCH_3$$

mesityl oxide
4-methylpent-3-en-2-one

The mechanism of this condensation involves a number of straightforward steps. To start, the enol form of acetone adds to another protonated acetone molecule.

(1) $CH_3\overset{O}{\overset{\|}{C}}CH_3 + H^+ \rightleftharpoons CH_3\overset{+OH}{\overset{\|}{C}}CH_3$

(2) $CH_3{-}\overset{+OH}{\overset{\|}{C}}CH_3 \rightleftharpoons H^+ + CH_2{=}\overset{OH}{C}CH_3$

(3) $CH_3{-}\overset{+OH}{\overset{\|}{C}} \overset{}{\underset{CH_3}{}} + CH_2{=}\overset{:OH}{C}CH_3 \rightleftharpoons CH_3\overset{OH}{\underset{CH_3}{\overset{|}{C}}}{-}CH_2{-}\overset{+OH}{\overset{\|}{C}}CH_3$

The resulting oxonium ion can lose a proton from oxygen to give 4-hydroxy-4-methyl-2-pentanone (diacetone alcohol, page 394) or from carbon to give the enol form of diacetone alcohol.

(4) $CH_3\overset{OH}{\underset{CH_3}{\overset{|}{C}}}{-}CH_2{-}\overset{+OH}{\overset{\|}{C}}CH_3 \rightleftharpoons CH_3\overset{OH}{\underset{CH_3}{\overset{|}{C}}}{-}CH_2{-}\overset{O}{\overset{\|}{C}}CH_3 + H^+$

(4′) $\underset{\underset{CH_3}{|}}{\overset{\overset{OH}{|}}{CH_3C}}-CH_2-\overset{\overset{+OH}{|}}{CCH_3} \rightleftharpoons \underset{\underset{CH_3}{|}}{\overset{\overset{OH}{|}}{CH_3C}}-CH=\overset{\overset{OH}{|}}{CCH_3} + H^+$

The latter species is an enol form of a ketone and is unstable relative to the ketone; it is present in only low concentration. Protonation on the tertiary hydroxy gives an oxonium ion that readily eliminates water to form a new cation. The cation is resonance-stabilized with the positive charge spread over oxygen and two carbons. The oxonium ion structure is the more important structure because all atoms have octet configurations.

(5) $\underset{\underset{CH_3}{|}}{\overset{\overset{OH}{|}}{CH_3C}}-CH=\overset{\overset{OH}{|}}{CCH_3} + H^+ \rightleftharpoons \underset{\underset{CH_3}{|}}{\overset{\overset{+OH_2}{|}}{CH_3C}}-CH=\overset{\overset{OH}{|}}{CCH_3}$

(6) $\underset{\underset{CH_3}{|}}{\overset{\overset{+OH_2}{|}}{CH_3C}}-CH=\overset{\overset{OH}{|}}{CCH_3} \rightleftharpoons$

$H_2O + \left[\underset{\underset{CH_3}{|}}{\overset{\overset{+}{}}{CH_3C}}-CH=\overset{\overset{OH}{|}}{CCH_3} \longleftrightarrow \underset{\underset{CH_3}{|}}{CH_3C}=CH-\overset{\overset{OH}{|}}{\underset{+}{C}CH_3} \longleftrightarrow \underset{\underset{CH_3}{|}}{CH_3C}=CH-\overset{\overset{+OH}{||}}{C}CH_3\right]$

Loss of the proton from oxygen gives the product mesityl oxide.

(7) $\underset{\underset{CH_3}{|}}{CH_3C}=CH-\overset{\overset{+OH}{||}}{C}CH_3 \rightleftharpoons \underset{\underset{CH_3}{|}}{CH_3C}=CH-\overset{\overset{O}{||}}{C}CH_3 + H^+$

It should be emphasized that in this reaction sequence, simple alkyl cations are not involved. *Every organic cationic intermediate either is an oxonium ion or has an oxonium ion resonance form.*

Recall from Section 13.7.F that mixed aldol condensations are successful ways of preparing α,β-unsaturated ketones in cases where one carbonyl compound cannot form an enol or enolate (because it has no α-protons) and/or has an especially reactive carbonyl group. These conditions are met by aromatic aldehydes, since the aldehyde function is generally more reactive than a ketone carbonyl and aromatic aldehydes have no α-protons. Condensations of aromatic ketones with aromatic aldehydes generally give good yields of enones, which are usually nicely crystalline substances. Such enones are sometimes referred to as "chalcones."

acetophenone   benzaldehyde   chalcone
                                                    (benzalacetophenone)

> Benzaldehyde (460 g) is added all at once to an ice-cold solution of 520 g of acetophenone and 218 g of NaOH in a mixture of 1960 mL of water and 1225 mL of ethanol. The mixture is stirred mechanically for several hours, during which time the product separates as light yellow crystals. Filtration affords 770 g of chalcone, m.p. 55–57°.

The simplest conjugated unsaturated carbonyl compound is propenal, CH_2=CHCHO, commonly known as acrolein. This compound is a liquid, b.p. 53°, having a powerful, pungent odor. It may be prepared by a special reaction in which the readily available triol, glycerol, is heated with sulfuric acid or potassium acid sulfate.

$$HOCH_2CHOHCH_2OH \xrightarrow[200-230°]{KHSO_4} CH_2=CHCHO$$

glycerol   acrolein
1,2,3-propanetriol   propenal

Most of us are familiar with the odor of acrolein because a similar dehydration occurs thermally when fats burn or decompose on a hot surface. Recall that fats are esters of glycerol (Section 19.12).

α,β-Unsaturated aldehydes and ketones are also available by oxidation of the corresponding unsaturated alcohols, which in turn are frequently available by Grignard syntheses. The oxidation requires mild conditions in order not to oxidize the double bond. One reagent that is specific for allylic alcohols is manganese dioxide in a specially active form obtained by treatment of manganese sulfate with base and potassium permanganate.

$$CH_2=CHCH_2OH \xrightarrow[\substack{petroleum\ ether \\ reflux}]{MnO_2} CH_2=CHCHO$$

allyl alcohol   (99%)
acrolein

retinol   (80%)
(vitamin A₁)   retinal

α,β-Unsaturated aldehydes and ketones undergo many of the reactions expected separately for the double bond and carbonyl functions. The C=O group forms normal derivatives such as oximes, phenylhydrazones, and so on (Section 13.7.C).

$$CH_3CH=CH\overset{\text{O}}{\overset{\|}{C}}CH_3 + H_2NOH \longrightarrow CH_3CH=CH\overset{\text{NOH}}{\overset{\|}{C}}CH_3$$

an oxime

The aldehyde group is oxidized under mild conditions to a carboxylic acid (Section 13.8.A).

$$CH_2=CHCHO + Ag_2O \longrightarrow CH_2=CHCOOH + 2\ Ag$$

Bromine can be added to the double bond (Section 11.6.B.3)

$$C_6H_5CH=CHCOCH_3 + Br_2 \xrightarrow[10-20°]{CCl_4} C_6H_5CHBrCHBrCOCH_3$$

(52–57%)

However, some reactions are unique to the conjugated system. Additions may occur across the ends of the conjugated system or to either one of the double bonds, just as in the case of conjugated dienes (Section 20.2.B). Additions that occur to a single double bond are called **1,2-additions** or **normal additions.** Additions that occur across the ends of the conjugated system are called **1,4-additions** or **conjugate additions.**

1,2-additions

$$C=C-C=O + X-Y \longrightarrow \overset{\overset{X}{|}\overset{Y}{|}}{C-C-C=O} \quad or \quad \overset{\overset{X}{|}\overset{Y}{|}}{C=C-C-O}$$

1,4-additions

$$C=C-C=O + X-Y \longrightarrow \overset{\overset{X}{|}\overset{Y}{|}}{C-C=C-O}$$

> Do not be confused by the terms *1,2-addition* and *1,4-addition*. The numbers do not refer to the carbon numbers in any given compound. The terms mean that the addition is to the 1 and 2 positions or the 1 and 4 positions of a conjugated system. For example, the addition of Br_2 to the double bond in pent-3-en-2-one is an example of a 1,2-addition.
>
> $$\underset{54321}{CH_3CH=CH-\overset{\overset{\textstyle O}{\|}}{C}CH_3} + Br_2 \longrightarrow CH_3\overset{\overset{\textstyle Br}{|}}{CH}-\overset{\overset{\textstyle Br}{|}}{CH}-\overset{\overset{\textstyle O}{\|}}{C}CH_3$$

Cyanide ion, which normally adds to the C=O bond in aldehydes and ketones (Section 13.7.F), frequently adds instead to a C=C bond that is conjugated with a carbonyl group.

$$C_6H_5CH=CH-\overset{\overset{\textstyle O}{\|}}{C}C_6H_5 + KCN \xrightarrow{CH_3CO_2H} C_6H_5\overset{\overset{\textstyle CN}{|}}{C}HCH_2\overset{\overset{\textstyle O}{\|}}{C}C_6H_5$$

(93–96%)

The reaction appears to be 1,2-addition to the double bond. However, the mechanism actually involves 1,4-addition of HCN to the conjugated system. The initial product of the 1,4-addition is an enol, which tautomerizes to the observed keto form.

$$(1)\ C_6H_5CH=CH-\overset{\overset{\textstyle O}{\|}}{C}C_6H_5 + H^+ \rightleftharpoons C_6H_5CH=CH-\overset{\overset{\textstyle {}^+OH}{\|}}{C}C_6H_5$$

(2) $C_6H_5CH=CH-\overset{\overset{+OH}{\|}}{C}C_6H_5 + CN^- \rightleftharpoons C_6H_5\overset{\overset{CN}{|}}{C}H-CH=\overset{\overset{OH}{|}}{C}C_6H_5$

(3) $C_6H_5\overset{\overset{CN}{|}}{C}H-CH=\overset{\overset{OH}{|}}{C}C_6H_5 + H^+ \rightleftharpoons C_6H_5\overset{\overset{CN}{|}}{C}H-CH_2-\overset{\overset{+OH}{\|}}{C}C_6H_5$

(4) $C_6H_5\overset{\overset{CN}{|}}{C}H-CH_2-\overset{\overset{+OH}{\|}}{C}C_6H_5 \rightleftharpoons C_6H_5\overset{\overset{CN}{|}}{C}HCH_2\overset{\overset{O}{\|}}{C}C_6H_5 + H^+$

A particularly effective method for accomplishing the 1,4-addition of HCN to
α,β-unsaturated ketones employs triethylaluminum as a catalyst. The procedure
gives high yields of the conjugate adduct even when the enone is highly substituted at the β-position.

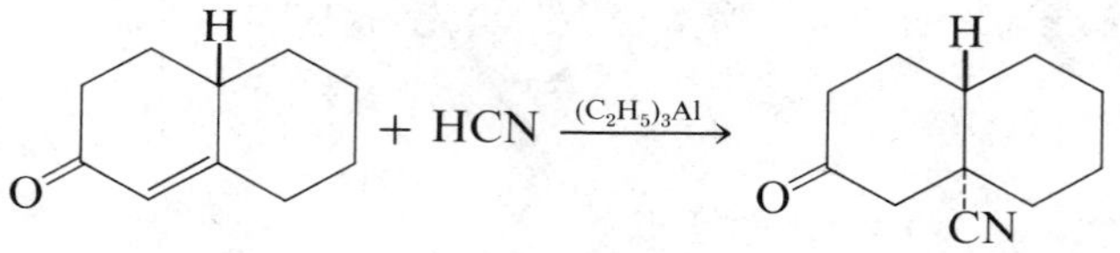

(85%)

Organometallic compounds may add either 1,2 or 1,4. Grignard reagents show
variable behavior depending on the structure of the conjugated system. The most
important factor in determining whether the addition is 1,2 or 1,4 seems to be
steric hindrance. Most α,β-unsaturated aldehydes undergo normal 1,2-addition to
the carbonyl group.

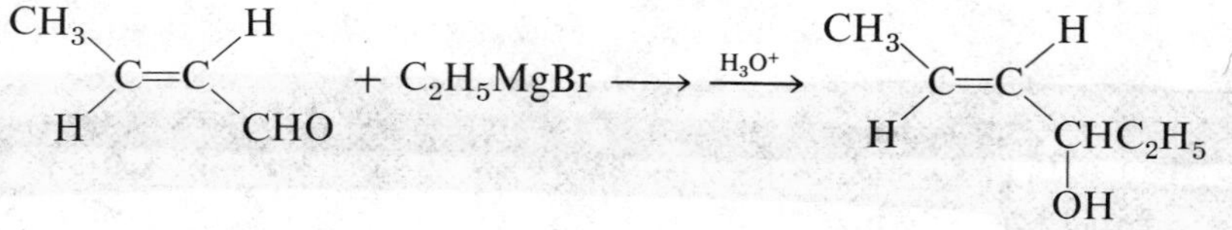

(70%)
trans-4-hexen-3-ol

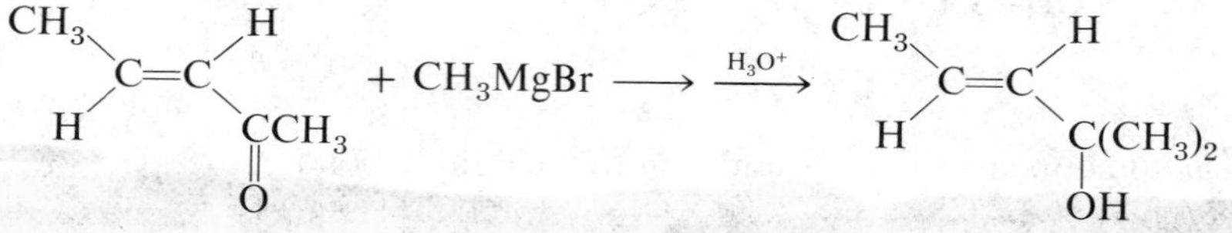

(80%)
2-methyl-*trans*-3-penten-2-ol

However, many other α,β-unsaturated ketones give substantial amounts also of
the 1,4-adduct.

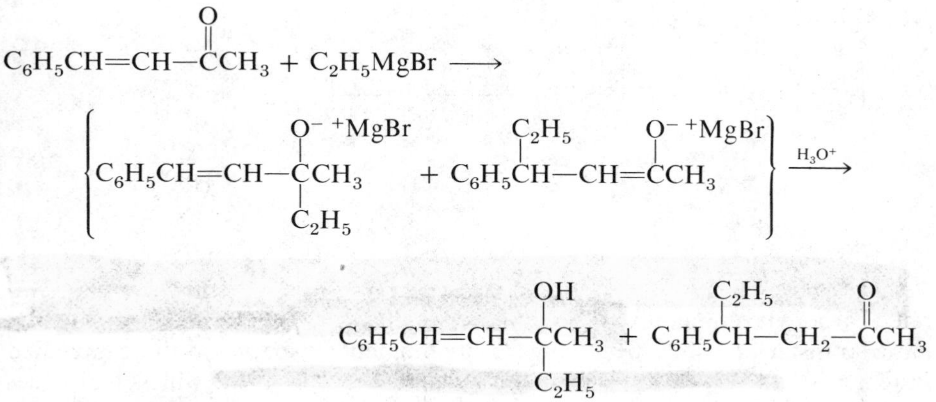

$$C_6H_5CH=CH-\overset{\overset{\displaystyle O}{\|}}{C}CH_3 + C_2H_5MgBr \longrightarrow$$

$$\left\{ \begin{array}{l} C_6H_5CH=CH-\overset{\overset{\displaystyle O^- \ ^+MgBr}{|}}{\underset{\underset{\displaystyle C_2H_5}{|}}{C}}CH_3 \quad + \quad C_6H_5\overset{\overset{\displaystyle C_2H_5}{|}}{C}H-CH=\overset{\overset{\displaystyle O^- \ ^+MgBr}{|}}{C}CH_3 \end{array} \right\} \xrightarrow{H_3O^+}$$

$$C_6H_5CH=CH-\overset{\overset{\displaystyle OH}{|}}{\underset{\underset{\displaystyle C_2H_5}{|}}{C}}CH_3 \quad + \quad C_6H_5\overset{\overset{\displaystyle C_2H_5}{|}}{C}H-CH_2-\overset{\overset{\displaystyle O}{\|}}{C}CH_3$$

(37%)	(57%)
1,2-adduct	1,4-adduct

In some cases the 1,4-adduct is almost the sole product.

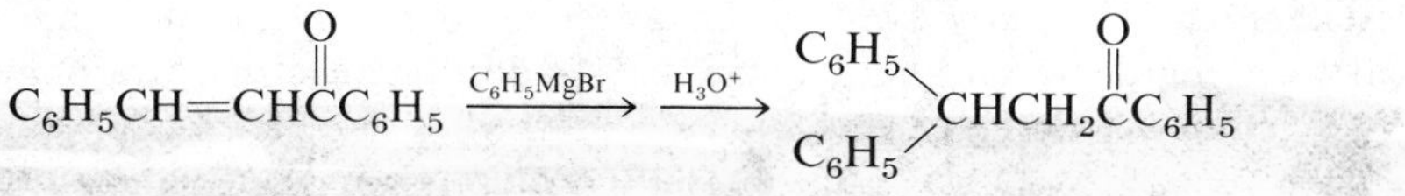

$$C_6H_5CH=CH\overset{\overset{\displaystyle O}{\|}}{C}C_6H_5 \xrightarrow{C_6H_5MgBr} \xrightarrow{H_3O^+} \overset{\displaystyle C_6H_5}{\underset{\displaystyle C_6H_5}{>}}CHCH_2\overset{\overset{\displaystyle O}{\|}}{C}C_6H_5$$

(96%)
1,3,3-triphenyl-1-propanone

Organolithium compounds show a much greater tendency to engage in 1,2-addition.

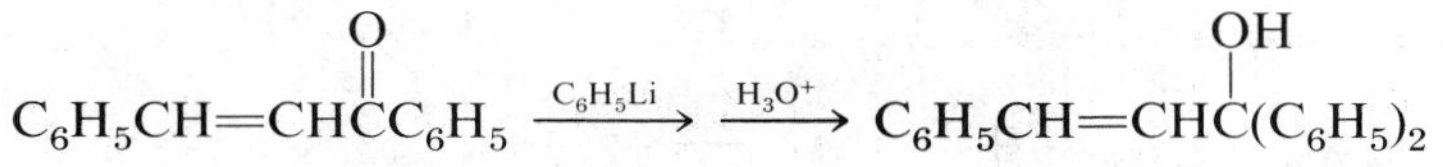

$$C_6H_5CH=CH\overset{\overset{\displaystyle O}{\|}}{C}C_6H_5 \xrightarrow{C_6H_5Li} \xrightarrow{H_3O^+} C_6H_5CH=CH\overset{\overset{\displaystyle OH}{|}}{C}(C_6H_5)_2$$

(75%)

When one wants to maximize 1,2-addition, it is customary to utilize the organolithium reagent.

On the other hand, 1,4-addition may be achieved by using lithium dialkylcuprates, which are readily prepared from the corresponding alkyllithium reagent and cuprous iodide (Section 15.5.E).

$$2 \ CH_3Li + CuI \longrightarrow (CH_3)_2CuLi$$

These reagents add exclusively in a 1,4-fashion.

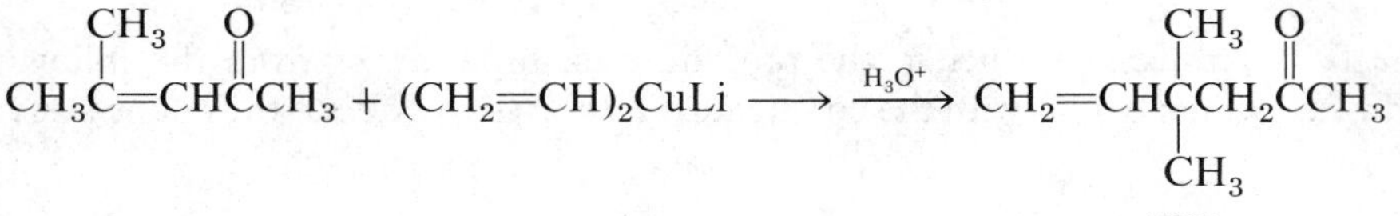

$$CH_3\overset{\overset{\displaystyle CH_3}{|}}{C}=CH\overset{\overset{\displaystyle O}{\|}}{C}CH_3 + (CH_2=CH)_2CuLi \longrightarrow \xrightarrow{H_3O^+} CH_2=CH\overset{\overset{\displaystyle CH_3}{|}}{\underset{\underset{\displaystyle CH_3}{|}}{C}}CH_2\overset{\overset{\displaystyle O}{\|}}{C}CH_3$$

(72%)

The same result may often be achieved by forming the dialkylcuprate *in situ* from the corresponding Grignard reagent. In practice, it is only necessary to use a catalytic amount of cuprous bromide or iodide.

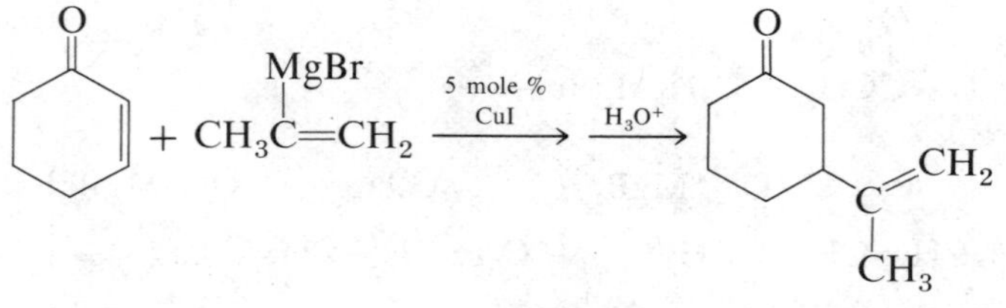

(68%)

The mechanism for the 1,4-additions of organometallic reagents to α,β-unsaturated carbonyl compounds is not completely understood, but it appears to involve initial transfer of an electron from an organometallic species to the conjugated system with the formation of an intermediate radical anion, which undergoes further reactions. Compounds such as the cuprates are effective because Cu^+ is readily oxidized to Cu^{2+}. Organolithium and organomagnesium compounds do not react by this mechanism, since neither metal is easily oxidized to a higher valence state. The 1,4-addition reactions that are observed with Grignard reagents arise from traces of transition metal impurities, such as Cu^+ and Fe^{2+}, which are present in commercially available magnesium. If highly purified magnesium is used to prepare the Grignard reagent, 1,4-addition reactions are not observed.

Reduction of α,β-unsaturated carbonyl compounds can also involve either the C=C or the C=O double bond. Lithium aluminum hydride reduction of most such compounds gives the highest amount of simple carbonyl reduction.

$$CH_3CH{=}CHCHO + LiAlH_4 \xrightarrow{\text{ether}} \xrightarrow{H_2O} CH_3CH{=}CHCH_2OH$$

(82%)

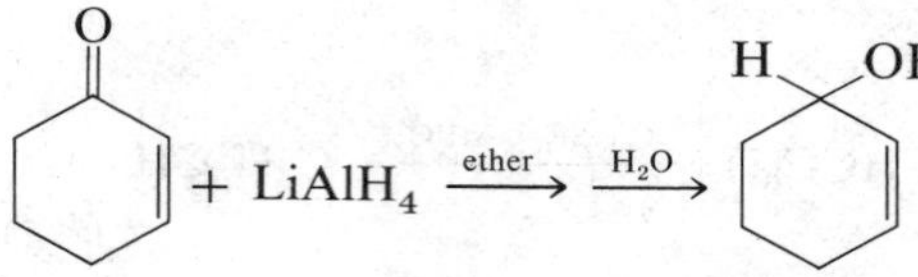

(97%)

In contrast, sodium borohydride in ethanol produces substantial 1,4-addition.

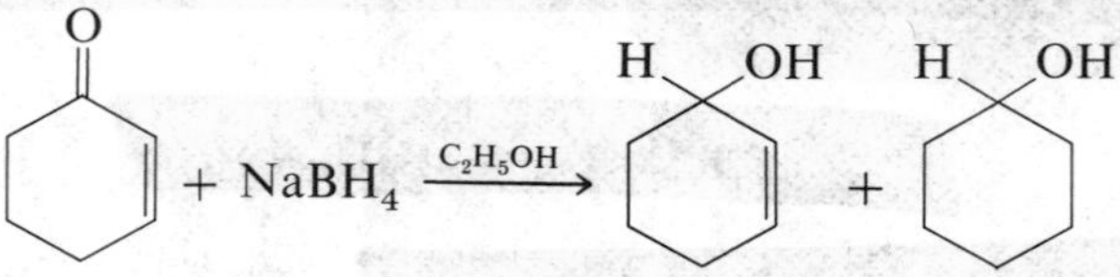

(59%)   (41%)

The fully reduced product in the preceding example arises from the following pathway, which begins with the conjugate addition of hydride to the enone system.

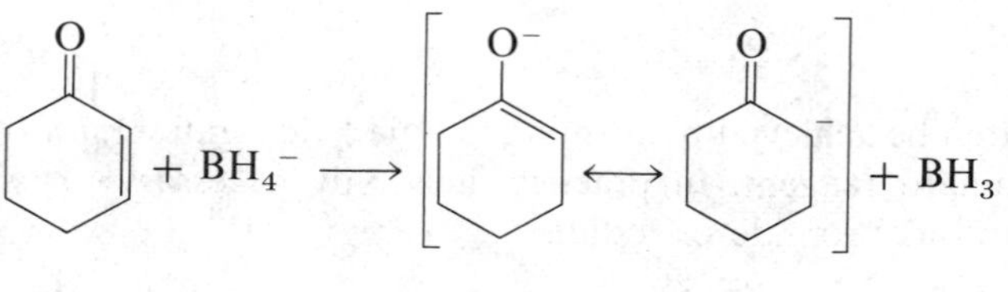

(2) 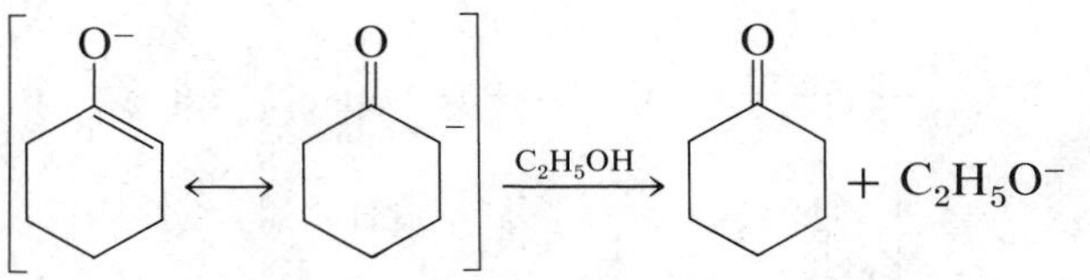 $\left[\text{structure} \leftrightarrow \text{structure} \right] \xrightarrow{\text{C}_2\text{H}_5\text{OH}} \text{structure} + \text{C}_2\text{H}_5\text{O}^-$

(3) structure $+ \text{BH}_4^- \longrightarrow$ structure $+ \text{BH}_3$

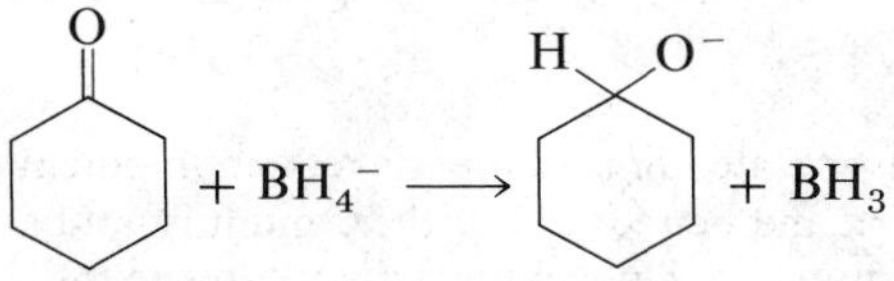

(4) 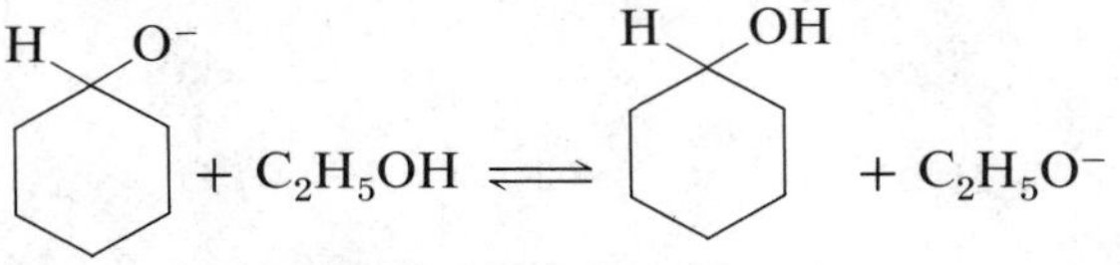 structure $+ \text{C}_2\text{H}_5\text{OH} \rightleftharpoons$ structure $+ \text{C}_2\text{H}_5\text{O}^-$

The double bond of such a conjugated system may generally be reduced cleanly by either of two procedures, catalytic hydrogenation or lithium–ammonia reduction.

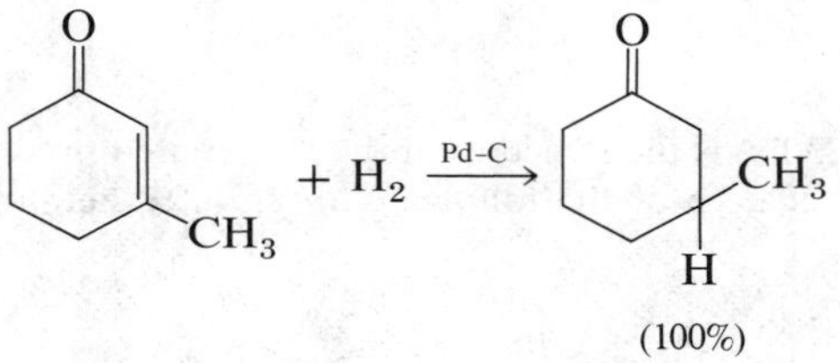

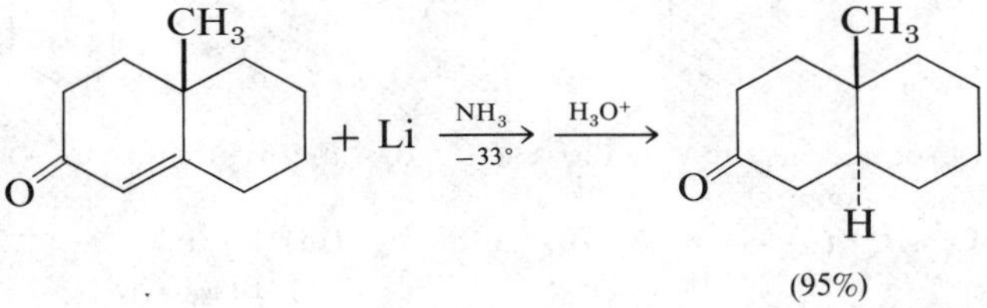

structure $+ \text{H}_2 \xrightarrow{\text{Pd–C}}$ structure

(100%)

structure $+ \text{Li} \xrightarrow[-33°]{\text{NH}_3} \xrightarrow{\text{H}_3\text{O}^+}$ structure

(95%)

The mechanism of the latter reduction is similar to that seen earlier in the reduction of alkynes to alkenes (Section 12.6.A). The first step involves addition of an electron to the conjugated system, giving a resonance-stabilized radical anion. This radical anion protonates on carbon giving a radical, which is reduced by another electron with the formation of an enolate ion.

(1) 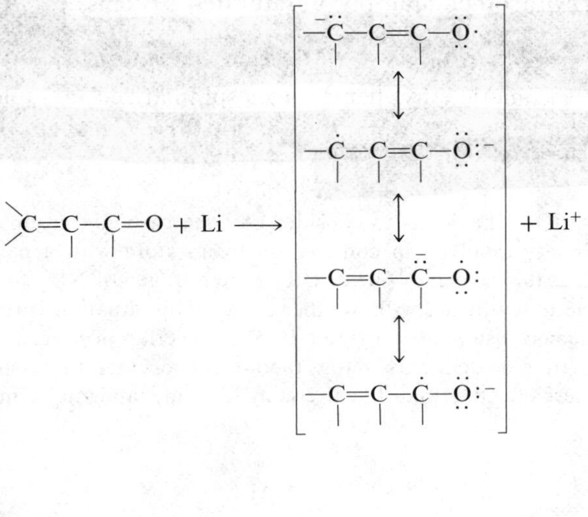 $\text{C}=\text{C}-\text{C}=\text{O} + \text{Li} \longrightarrow$ $\left[\text{resonance structures} \right] + \text{Li}^+$

(2) $\left[\overset{|}{\underset{|}{C}}\cdots\overset{|}{\underset{|}{C}}\cdots\overset{|}{\underset{|}{C}}\cdots\overset{..}{\underset{..}{O}} \right]^{-} + NH_3 \longrightarrow -\overset{H}{\underset{|}{C}}-\overset{|}{\underset{|}{C}}=\overset{|}{\underset{|}{C}}-\overset{..}{\underset{.}{O}}: + NH_2^{-}$

(3) $-\overset{H}{\underset{|}{C}}-\overset{|}{\underset{|}{C}}=\overset{|}{\underset{|}{C}}-\overset{..}{\underset{..}{O}}: + Li \longrightarrow \left[-\overset{H}{\underset{|}{C}}-\overset{|}{\underset{|}{C}}=\overset{|}{\underset{|}{C}}-\overset{..}{\underset{..}{O}}:^{-} \longleftrightarrow -\overset{H}{\underset{|}{C}}-\overset{\overset{\bar{..}}{}}{\underset{|}{C}}-\overset{|}{\underset{|}{C}}=O \right] + Li^{+}$

Under the conditions of the reaction, the enolate ion is stable. Its reduction potential is too high for it to accept another electron and be reduced further, and it is not basic enough to be protonated by a weak acid ammonia. Upon aqueous workup, the enolate ion is protonated to give the ketone.

$$\left[\overset{>}{\underset{}{C}}H\cdots\overset{|}{\underset{|}{C}}\cdots\overset{|}{\underset{|}{C}}\cdots O \right]^{-} + H_3O^{+} \longrightarrow \overset{>}{\underset{}{C}}H-\overset{|}{\underset{|}{C}}H-\overset{|}{\underset{|}{C}}=O + H_2O$$

The partial reduction of a conjugated enone in this reaction is a particularly impressive example of the special properties of conjugated systems, since isolated C=C double bonds are *not* reduced by lithium in ammonia and isolated C=O double bonds *are* reduced by the reagent.

EXERCISE: (a) What is the structure of the deuteriated product formed by base-catalyzed exchange of each of the following unsaturated ketones with excess D_2O?

(b) Write the equations showing the reaction of 4-methylpent-3-en-2-one with each of the following reagents.
(i) n-C_4H_9Li (ii) n-C_4H_9MgBr, CuBr   (iii) H_2/Pd
(iv) Li–NH_3 (v) HCN, $(C_2H_5)_3Al$ (vi) Br_2–CCl_4

B. *Unsaturated Carboxylic Acids and Derivatives*

Both conjugated and unconjugated unsaturated carboxylic acids and acid derivatives are known. As with other multiply unsaturated systems, conjugation provides added stabilization, but the magnitude of this stabilization is rather small, substantially smaller than for dienes or unsaturated aldehydes and ketones. The heats of formation of isomeric ethyl pentenoates summarized in Table 20.2 demonstrate this point. In other words, the carboxylic function is less effective in conjugation than is a simple carbonyl group.

One way of rationalizing this behavior is to consider that the carbonyl group in a carboxylic function is already involved in conjugation to an atom with a pair of electrons to donate—such as the oxygen of OH or OR or the nitrogen of NH_2. Such a carbonyl group is less able to conjugate with another group. This situation is representative of a **cross-conjugated** system as illustrated by the π-overlap of p-orbitals in Figure 20.15. Much qualitative evidence, as well as theoretical considerations, shows that cross conjugation is less effective than linear conjugation in stabilizing a molecule.

TABLE 20.2
Heats of Formation of Ethyl Pentenoates

Isomer	ΔH_f°, kcal mole^{-1}
$CH_2{=}CHCH_2CH_2COOC_2H_5$	-92.1 ± 0.6
$\begin{array}{cc} CH_3 & CH_2COOC_2H_5 \\ \diagdown & \diagup \\ C{=}C \\ \diagup & \diagdown \\ H & H \end{array}$	-92.6 ± 0.9
$\begin{array}{cc} CH_3 & H \\ \diagdown & \diagup \\ C{=}C \\ \diagup & \diagdown \\ H & CH_2COOC_2H_5 \end{array}$	-93.2 ± 0.7
$\begin{array}{cc} CH_3CH_2 & COOC_2H_5 \\ \diagdown & \diagup \\ C{=}C \\ \diagup & \diagdown \\ H & H \end{array}$	-94.3 ± 0.7
$\begin{array}{cc} CH_3CH_2 & H \\ \diagdown & \diagup \\ C{=}C \\ \diagup & \diagdown \\ H & COOC_2H_5 \end{array}$	-94.2 ± 0.9

As a result of the reduced effectiveness of conjugation in unsaturated acid functions, in some compounds the unconjugated isomer may be the more stable. For example, 4-methylpent-3-enoic acid, with a trisubstituted double bond, is more stable than 4-methylpent-2-enoic acid, which has a conjugated π-system but has only a disubstituted double bond.

$$\underset{\text{4-methylpent-3-enoic acid}}{\overset{\displaystyle CH_3}{CH_3\overset{|}{C}{=}CHCH_2COOH}} \rightleftharpoons \underset{\text{4-methylpent-2-enoic acid}}{\overset{\displaystyle CH_3}{CH_3\overset{|}{C}HCH{=}CHCOOH}}$$

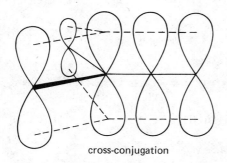

cross-conjugation

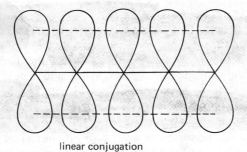

linear conjugation

FIGURE 20.15. *Comparison of linear conjugation and cross-conjugation.*

Unsaturated carboxylic acids and their derivatives may be prepared by many of the same routes appropriate for the saturated analogs. An example is the sequence $RX \rightarrow RCOOH$, in which R contains a double bond.

$$CH_2{=}CHCH_2Cl \xrightarrow{\text{CuCN}} CH_2{=}CHCH_2CN \xrightarrow[\Delta]{\text{aq. HCl}} CH_2{=}CHCH_2COOH$$

allyl chloride

(52–62%)

vinylacetic acid
but-3-enoic acid

Another example involves the oxidation of a methyl ketone with hypohalite (Section 13.6.D).

$$(CH_3)_2C{=}CH\overset{\overset{\displaystyle O}{\|}}{C}CH_3 \xrightarrow[\text{aq. dioxane}]{\text{KOCl}} \xrightarrow{\text{H}^+} CHCl_3 + (CH_3)_2C{=}CHCOOH$$

(49–53%)

3-methylbut-2-enoic acid

α,β-Unsaturated acids and derivatives are also available by elimination of HX from α-halo acids and esters.

$$CH_3(CH_2)_9\overset{\overset{\displaystyle CH_3}{|}}{C}HCOOH \xrightarrow[\text{Br}_2]{\text{PBr}_3} CH_3(CH_2)_9\overset{\overset{\displaystyle CH_3}{|}}{C}BrCOBr \xrightarrow{\text{CH}_3\text{OH}} CH_3(CH_2)_9\overset{\overset{\displaystyle CH_3}{|}}{\underset{\underset{\displaystyle Br}{|}}{C}}COOCH_3$$

2-methyldodecanoic acid

$\downarrow$ (CH₃)₃COK
(CH₃)₃COH

$\downarrow$ quinoline
160–170°

$$CH_3(CH_2)_9\overset{\overset{\displaystyle CH_2}{\|}}{C}COOC(CH_3)_3 \qquad CH_3(CH_2)_8CH{=}\overset{\overset{\displaystyle CH_3}{|}}{C}COOCH_3$$

(70–85%)

$\downarrow$ alc KOH

$\downarrow$ H⁺

$\downarrow$ alc KOH

$\downarrow$ H⁺

$$CH_3(CH_2)_9\overset{\overset{\displaystyle CH_2}{\|}}{C}COOH \qquad CH_3(CH_2)_8CH{=}\overset{\overset{\displaystyle CH_3}{|}}{C}COOH$$

2-decylpropenoic acid

(68–83%)

2-methyldodec-2-enoic acid

Note in the foregoing examples that the direction of elimination can sometimes be controlled by the choice of basic reagent used. Although both of these eliminations occur by the E2 mechanism, recall that the bulky reagent potassium *t*-butoxide tends to abstract primary hydrogens (Section 11.5.A).

Certain condensation reactions of aldehydes and ketones provide especially valuable routes to α,β-unsaturated acids and esters. One such method is a variation of the Wittig reaction (Section 13.7.G). Instead of a phosphonium salt, this variant, called the **Horner–Emmons reaction,** utilizes a phosphonic acid derivative. The preparation and chemistry of such phosphonic esters will be discussed in Section 26.7. An example of the Horner–Emmons reaction is the preparation of ethyl cyclohexylideneacetate.

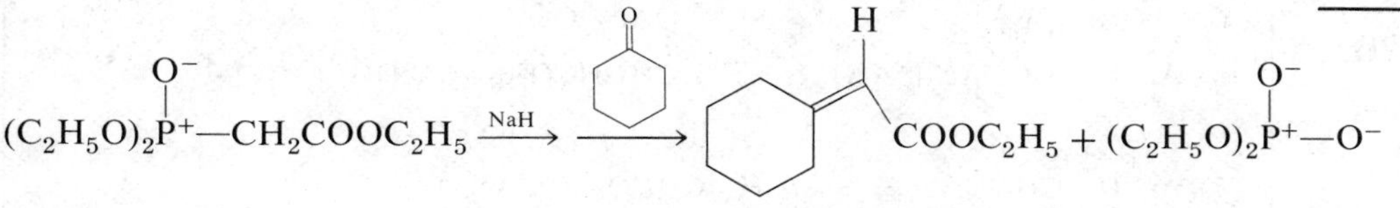

$$(C_2H_5O)_2\overset{O^-}{\underset{}{P^+}}\!\!-\!CH_2COOC_2H_5 \xrightarrow{\text{NaH}}$$

ethyl cyclohexylideneacetate structure $+ (C_2H_5O)_2\overset{O^-}{\underset{}{P^+}}\!\!-\!O^-$

(77%)

triethyl phosphonoacetate   ethyl cyclohexylideneacetate   diethyl phosphate ion

The mechanism of this condensation is analogous to that of the Wittig reaction. The protons α to the ester carbonyl are also made more acidic by the adjacent positive phosphorus and are more acidic than normal ester protons. A proton is rapidly removed by the strong base NaH. The resulting anion adds to the carbonyl group of the ketone, giving an intermediate that eliminates diethyl phosphate and the α,β-unsaturated ester.

Other condensation reactions are unique for aldehydes that do not have α-hydrogens. An example is the **Perkin reaction** in which an aromatic aldehyde is heated with an acid anhydride and the corresponding carboxylate salt. Acetic anhydride and sodium acetate are most commonly used.

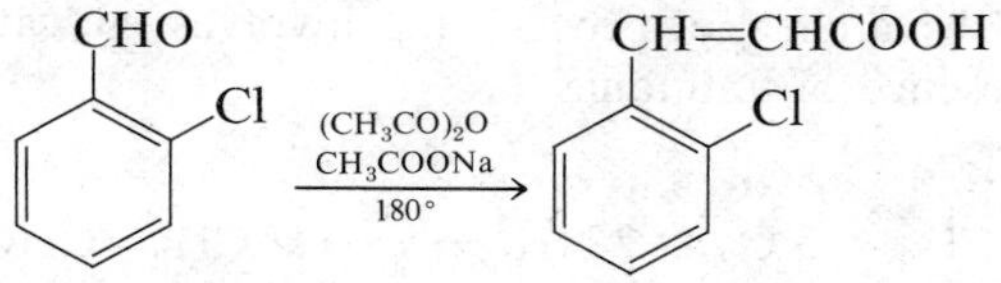

$$\text{o-chlorobenzaldehyde} \xrightarrow[\substack{CH_3COONa \\ 180°}]{(CH_3CO)_2O} \text{o-chlorocinnamic acid}$$

(72%)

o-chlorocinnamic acid

The reaction is a typical base-catalyzed condensation in which the enolate ion of an acid anhydride is an intermediate.

The Perkin reaction appears to proceed by way of the following interesting mechanism.

$$(1)\quad CH_3\overset{O}{\overset{\|}{C}}\overset{O}{\overset{\|}{C}}CH_3 + AcO^- \rightleftharpoons {}^-\!:CH_2\overset{O}{\overset{\|}{C}}\overset{O}{\overset{\|}{C}}CH_3 + AcOH$$

$$(2)\quad Ar\overset{O}{\overset{\|}{C}}H + {}^-\!:CH_2\overset{O}{\overset{\|}{C}}OAc \rightleftharpoons Ar\overset{O^-}{\overset{|}{C}}H\!-\!CH_2\overset{O}{\overset{\|}{C}}OAc$$

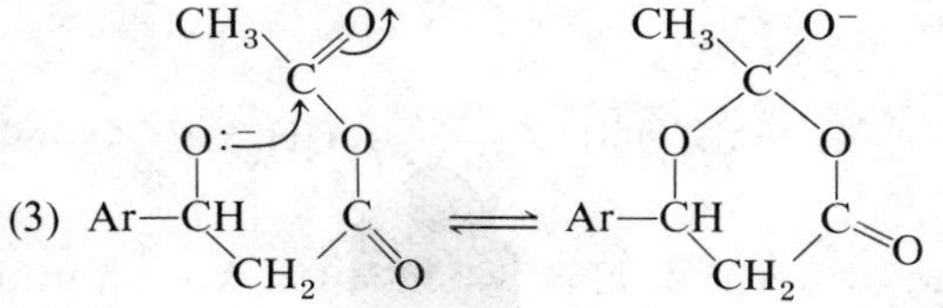

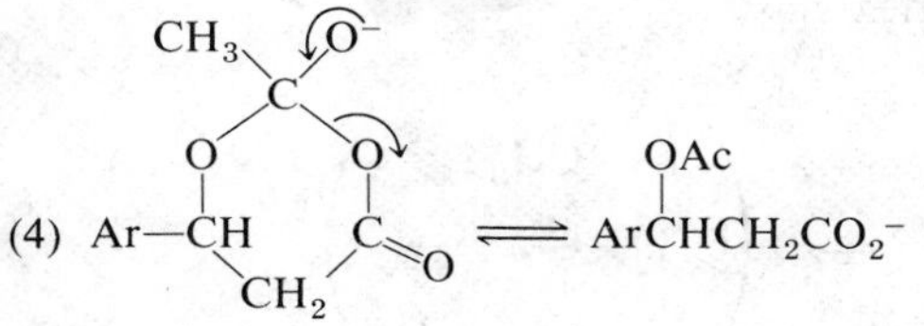

$$(3)\quad Ar\!-\!CH \cdots \rightleftharpoons Ar\!-\!CH \cdots$$

$$(4)\quad Ar\!-\!CH \cdots \rightleftharpoons Ar\overset{OAc}{\overset{|}{C}}HCH_2CO_2^-$$

$$(5)\quad Ar\overset{OAc}{\overset{|}{C}}HCH_2CO_2^- + Ac_2O \rightleftharpoons Ar\overset{OAc}{\overset{|}{C}}HCH_2\overset{O}{\overset{\|}{C}}OAc + AcO^-$$

$$(6) \quad \underset{\substack{| \\ OAc}}{ArCH}\underset{\substack{\\ }}{CH_2}\overset{\substack{O \\ ||}}{C}OAc + AcO^- \rightleftharpoons Ar\underset{\substack{| \\ OAc}}{CH}\underset{\substack{| \\ }}{\bar{C}H}\overset{\substack{O \\ ||}}{C}OAc + AcOH$$

$$(7) \quad Ar\underset{\substack{| \\ OAc}}{CH}\overset{\frown}{\underset{\substack{\\ }}{CH}}\overset{\substack{O \\ ||}}{C}OAc \rightleftharpoons ArCH{=}CH\overset{\substack{O \\ ||}}{C}OAc + AcO^-$$

$$(8) \quad ArCH{=}CH\overset{\substack{O \\ ||}}{C}OAc + AcOH \rightleftharpoons ArCH{=}CHCOOH + Ac_2O$$

The simplest unsaturated carboxylic acid is propenoic acid, $CH_2{=}CHCOOH$, commonly known as acrylic acid, a liquid having b.p. 141.6°. The corresponding nitrile, $CH_2{=}CHCN$, acrylonitrile, is an important industrial material that is made in large quantity for use in synthetic fibers and polymers; its 1979 production in the United States was 1,010,000 tons. Acrylonitrile is also a liquid, b.p. 78.5°. It was once prepared industrially by addition of HCN to acetylene.

$$HC{\equiv}CH + HCN \xrightarrow[NH_4Cl]{CuCl} CH_2{=}CHCN$$

It is now prepared by a cheaper process that involves the catalytic oxidation of propene in the presence of ammonia.

$$2\,CH_3CH{=}CH_2 + 3\,O_2 + 2\,NH_3 \xrightarrow[catalyst]{450°} 2\,CH_2{=}CHCN + 6\,H_2O$$

Free radical polymerization of acrylonitrile in aqueous solution gives a polymer that can be spun to give the textile Orlon or Acrilan.

$$CH_2{=}CHCN \longrightarrow \left(CH_2{-}\underset{\substack{| \\ CN}}{CH}\right)_n$$

Orlon, Acrilan

Copolymerization with butadiene and styrene gives an inexpensive plastic terpolymer known as ABS.

The methyl ester of α-methylacrylic acid is also an important monomer. It is prepared from acetone by the sequence

$$CH_3COCH_3 + HCN \longrightarrow (CH_3)_2\underset{\substack{| \\ OH}}{C}CN \xrightarrow{H_2SO_4} CH_2{=}\underset{\substack{| \\ CH_3}}{C}CONH_2 \xrightarrow[H_2SO_4]{CH_3OH} CH_2{=}\underset{\substack{| \\ CH_3}}{C}COOCH_3$$

Poly(methyl methacrylate) prepared by free radical polymerization is a stiff transparent plastic known as Lucite or Plexiglas.

$$CH_2{=}\underset{\substack{| \\ CH_3}}{C}{-}COOCH_3 \longrightarrow \left(CH_2{-}\underset{\substack{| \\ COOCH_3}}{\overset{\substack{CH_3 \\ |}}{C}}\right)_n$$

poly(methyl methacrylate)
Lucite, Plexiglas

Note that with both acrylonitrile and methyl methacrylate free radical polymerization involves almost exclusively head-to-tail combination of the monomers.

Unsaturated acids and esters are widespread in nature. Ricinoleic acid is a derivative of stearic acid, $CH_3(CH_2)_{16}COOH$, and is obtained from castor oil. Other important unsaturated octadecanoic acids widespread as glyceryl esters in fats are oleic acid, linoleic acid, and linolenic acid (Section 19.12).

$$CH_3(CH_2)_5CHOHCH_2CH{=}CH(CH_2)_7COOH$$
ricinoleic acid

$$CH_3(CH_2)_7CH{=}CH(CH_2)_7COOH$$
oleic acid

$$CH_3(CH_2)_4CH{=}CHCH_2CH{=}CH(CH_2)_7COOH$$
linoleic acid

$$CH_3CH_2CH{=}CHCH_2CH{=}CHCH_2CH{=}CH(CH_2)_7COOH$$
linolenic acid

Linseed oil is an example of a **drying oil** and contains a high percentage of linolenic acid. On exposure to air, the highly unsaturated chain reacts with oxygen and crosslinks to give a tough transparent polymer. Oil-based paint is a combination of drying oil with suspended pigment. Varnish also contains such drying oils and also involves the formation of a tough waterproof film by oxygen-promoted free radical crosslinking.

C. Ketenes

The compound $CH_2{=}C{=}O$ is known as ketene and is the carbonyl analog of allene. It is a toxic gas, b.p. $-48°$, and is prepared by the pyrolysis of acetone at high temperatures.

$$CH_3COCH_3 \xrightarrow{700°} CH_4 + CH_2{=}C{=}O$$

The reaction appears to be a free radical chain decomposition.

Substituted ketenes are prepared by treatment of acyl halides with triethylamine or by treatment of α-halo acyl halides with zinc.

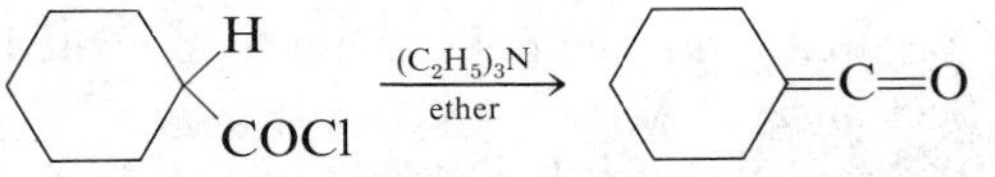

(32%)
pentamethyleneketene

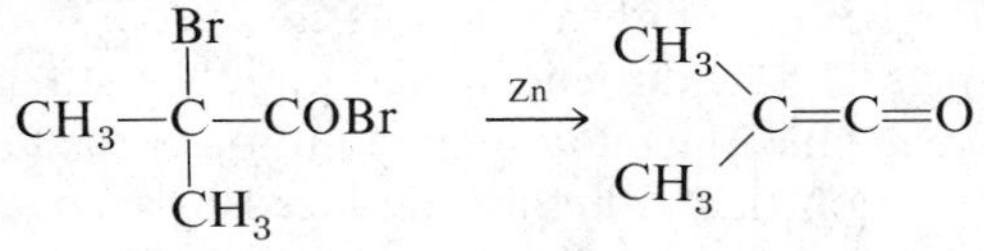

α-bromoisobutyryl bromide

(46–54%)
dimethylketene

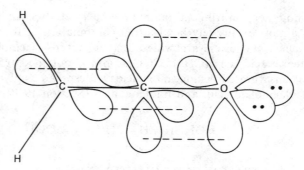

FIGURE 20.16. *Orbital structure of ketene.*

Ketenes react as "super anhydrides." With water they give carboxylic acids, and with alcohols they give esters. One commercial synthesis of acetic anhydride involves the combination of acetic acid with ketene.

$$CH_3COOH + CH_2{=}C{=}O \longrightarrow (CH_3CO)_2O$$

The United States production of acetic anhydride in 1979, mostly by this reaction, was 755,000 tons.

The ketene group is extremely reactive and is a relatively unimportant functional group. Ketenes, like allenes, have two π-systems at right angles. The two double bonds are *not* conjugated (Figure 20.16).

20.4
Higher Conjugated Systems

Many compounds having more than two conjugated double bonds are known. In such systems each double bond alternates with a single bond to allow extensive π-overlap of p-orbitals. One example is *trans*-1,3,5-hexatriene, a liquid, b.p. 79°.

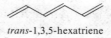

trans-1,3,5-hexatriene

Another is retinol (vitamin A_1), an alcohol with five conjugated double bonds that we encountered in Section 20.3.A.

Despite the stabilization that such highly conjugated compounds derive from their extensive π-electronic systems, they are generally *more* reactive, not less reactive, than their nonconjugated isomers. The reason for this apparent paradox is simply that the intermediate radicals or ions are even more stabilized by conjugation. For example, 1,3,5-hexatriene reacts rapidly with acids, bromine, free radicals, and other reagents. The addition of a proton to the terminal carbon atom gives a pentadienyl cation that is highly stabilized by resonance. That is, in this carbocation the positive charge is distributed among three carbons.

$$CH_2=CH-CH=CH-CH=CH_2 + H^+ \longrightarrow$$

$$
\left[
\begin{array}{c}
CH_2=CH-CH=CH-\overset{+}{C}H-CH_3 \\
\updownarrow \\
CH_2=CH-\overset{+}{C}H-CH=CH-CH_3 \\
\updownarrow \\
\overset{+}{C}H_2-CH=CH-CH=CH-CH_3
\end{array}
\right]
\equiv
\overset{\delta+}{C}H_2\cdots CH\cdots \overset{\delta+}{C}H\cdots CH\cdots \overset{\delta+}{C}H-CH_3
$$

The resonance stabilization of such carbocation, carbanion, and free radical intermediates and of the transition states leading to them is much greater than the stabilization afforded by π-overlap in the starting polyenes. As a result, such polyenes are highly reactive. Exposure to air or light is often sufficient to initiate free radical chain polymerization.

> **EXERCISE:** Sketch the six π-molecular orbitals of 1,3,5-hexatriene. Which of these MOs are occupied? Use these MOs to consider the bonding character between adjacent carbons and compare with a normal Lewis structure for this conjugated hydrocarbon. How many position isomers are expected from the rate-controlled addition of one mole of bromine to 1,3,5-hexatriene? Which isomer is expected to predominate at equilibrium?

20.5
The Diels–Alder Reaction

Conjugated dienes undergo a **cycloaddition reaction** with certain multiple bonds to form cyclohexenes and related compounds. This reaction, which is an exceedingly important synthetic method, is called the **Diels–Alder reaction.** The simplest Diels–Alder reaction is the reaction of 1,3-butadiene and ethylene to yield cyclohexene.

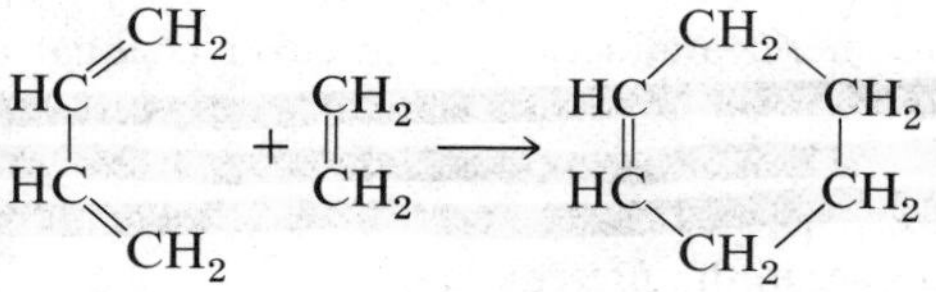

Although this simple example is known, it is *very slow* and only occurs under conditions of heat and pressure. However, the Diels–Alder reaction is facilitated by the presence of electron-donating groups on the diene component and by the presence of electron-attracting groups on the monoene component, often referred to as the "dienophile."

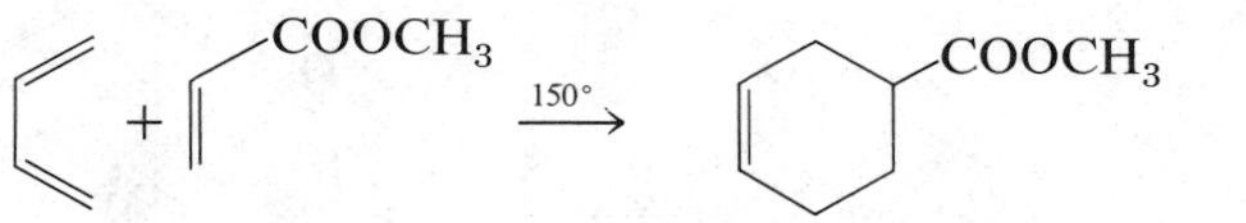

diene dienophile   methyl 4-cyclohexenecarboxylate

The reaction is a particularly versatile method for the preparation of cyclohexane derivatives of varied sorts. Another example is

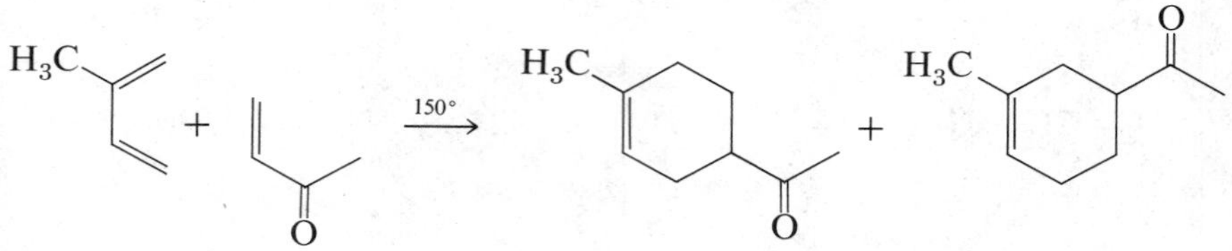

| isoprene | methyl vinyl ketone | | (70%) methyl 4-methyl-3-cyclohexenyl ketone | | (30%) methyl 3-methyl-3-cyclohexenyl ketone |

As shown by the foregoing example, many Diels–Alder reactions can give two isomers, depending on the orientation of the diene and the dienophile. The two different orientations are sometimes called head-to-head and head-to-tail.

Head-to-head orientation

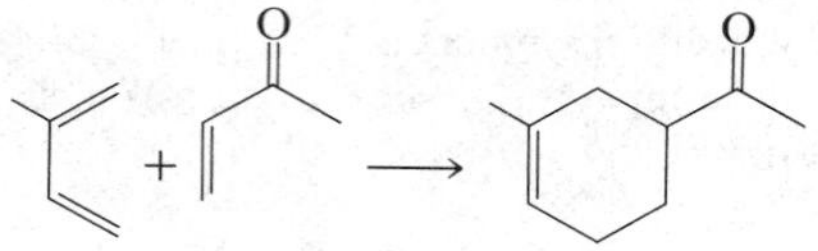

(30%)

Head-to-tail orientation

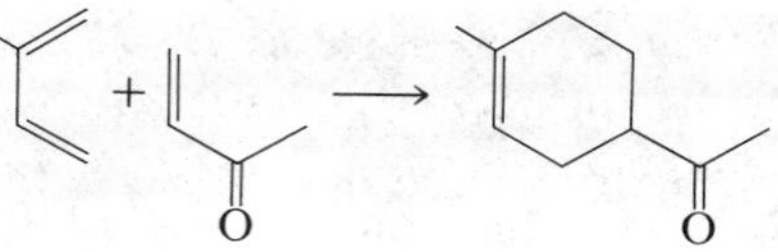

(70%)

In general, the two isomeric products are formed in unequal amounts, as in the present example. A useful way to predict which isomer will predominate is to treat the reaction as though it proceeds in two steps by way of free radical intermediates. Then examine the four hypothetical diradical intermediates and decide which is the most stable. For example, in the present case

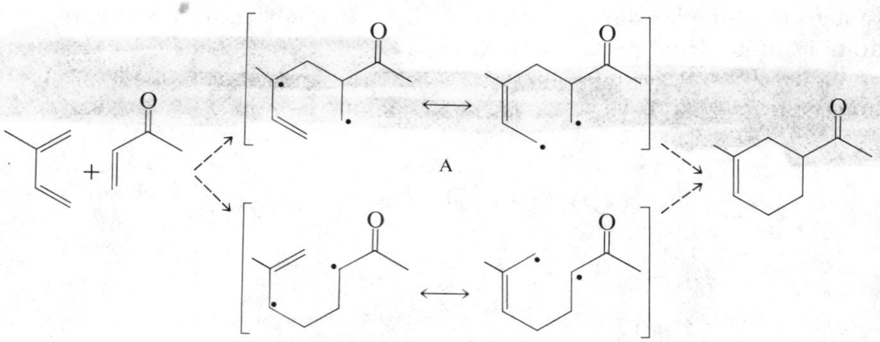

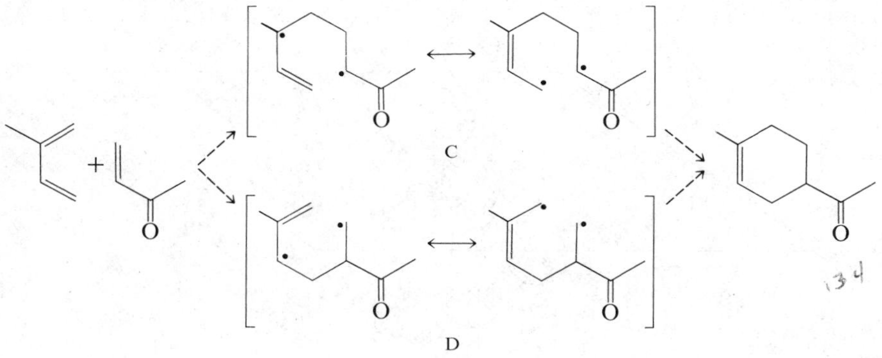

Of the four hypothetical intermediates, C is the most stable; one radical is primary–tertiary allylic, and the other is stabilized by the adjacent carbonyl group. In A, there is also a primary–tertiary allylic radical, but the other odd electron is primary and is not stabilized by an adjacent carbonyl group.

It is important to remember that this method is simply a technique for predicting which product will predominate; it does not mean that the reaction actually proceeds by way of discrete diradical intermediates.

The reaction has wide scope because multiple bonds other than C=C may be used

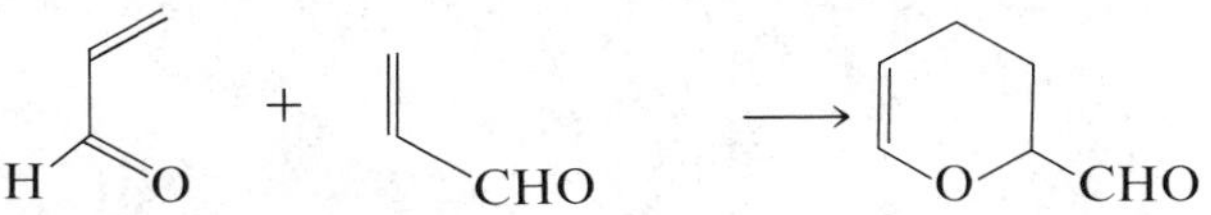

acrolein

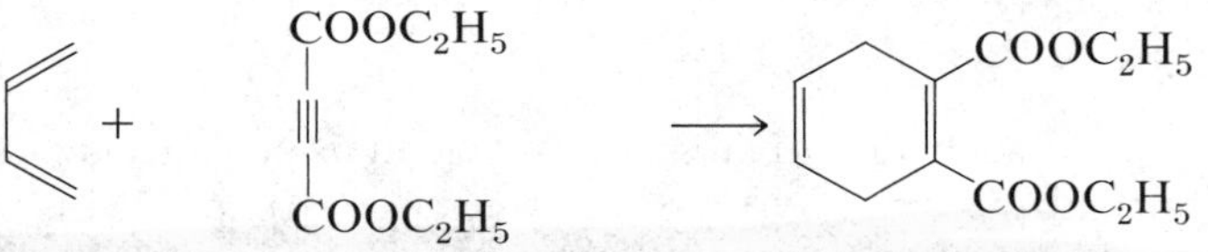

diethyl acetylenedicarboxylate   diethyl cyclohexa-1,4-diene-1,2-dicarboxylate

The Diels–Alder reaction is also known as a thermal cycloaddition reaction. The mechanism of the reaction involves σ-overlap of the π-orbitals of the two unsaturated systems, as illustrated in Figure 20.17. The Diels–Alder reaction involves specifically four π-electrons on one system and two on another and is therefore referred to as a 4 + 2 cycloaddition. A remarkable fact is that although such 4 + 2 reactions are common and general, analogous 2 + 2 and 4 + 4 thermal cycloadditions do not normally occur. The requirement of *six* electrons in the cyclic transition state in Figure 20.17 is now well understood and involves properties of molecular orbitals that we will not review here (this topic is discussed in Section 34.2). We can mention, however, that the matter is closely related to the fact that the benzene ring, which has special properties of stability to be discussed in Chapter 22, also has six electrons in a cyclic π-system.

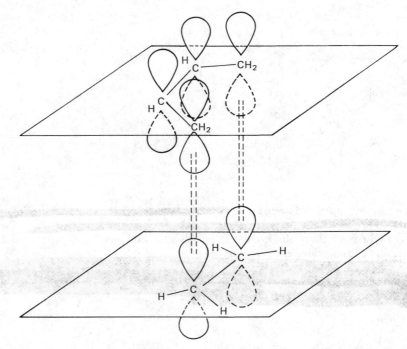

FIGURE 20.17. *Transition state of a Diels–Alder reaction.*

$$\begin{matrix} CH_2 \\ \| \\ CH_2 \end{matrix} + \begin{matrix} CH_2 \\ \| \\ CH_2 \end{matrix} \;\xrightarrow[\;\Delta\;]{/\!/}\; \square$$

2 + 2

4 + 4

The net result of such a non-Diels–Alder cycloaddition can sometimes be accomplished, but only by other types of reactions (see Section 34.2).

The transition state of the Diels–Alder reaction depicted in Figure 20.17 has stereochemical consequences. The following examples illustrate that the reaction is a *syn* addition with respect to both the diene and the dienophile.

When both the diene and dienophile are suitably substituted, a further stereochemical feature arises because the reactants may approach each other in two distinct orientations. The substituent on the dienophile may be directed away from the diene (*exo* approach) or toward the diene (*endo* approach), resulting in two stereoisomers.

Exo *approach*

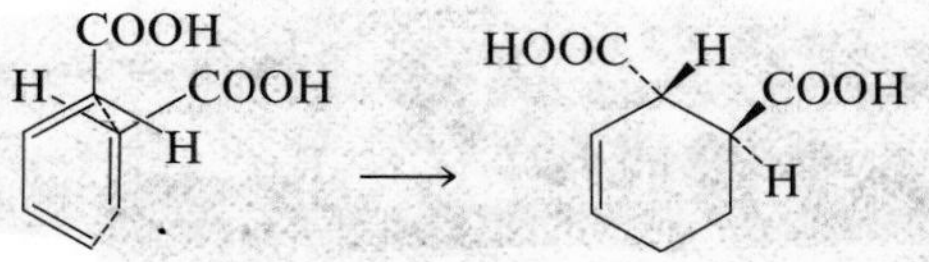

Endo *approach*

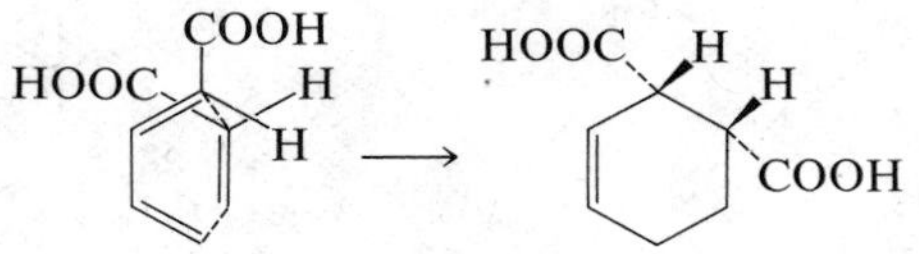

The Diels–Alder reaction shows a general preference for *endo* approach, but the *endo/exo* ratio depends on the reaction conditions, such as temperature and solvent polarity.

The direct product of a Diels–Alder reaction is a cyclohexene or 1,4-cyclohexadiene derivative, and these compounds can generally be converted to many other cyclohexane derivatives.

The use of cyclic dienes in the Diels–Alder reaction leads to bicyclic products. An especially important diene of this type is cyclopentadiene.

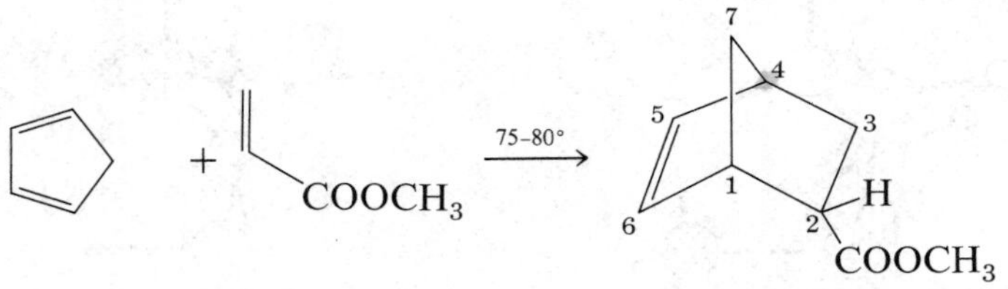

cyclopentadiene   methyl acrylate

(79–91%)
methyl bicyclo[2.2.1]-
hept-5-ene-*endo*-2-carboxylate

Bicyclic compounds constitute an important group of organic structures. The number of cycles in such a compound is determined by the minimum number of ring bonds that must be broken to obtain an acyclic compound. Bicyclic structures have in common

1. Two bridgehead atoms.
2. Three arms connecting the two bridgehead atoms.

Bicyclic compounds are named as derivatives of the alkane corresponding to the total number of carbons in both ring skeletons. The term **bicyclo** is appended as a prefix together with the numbers of carbons in each of the three connecting arms inserted in brackets.

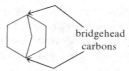

bicyclo[2.2.1]heptane   bicyclo[4.4.0]decane   bicyclo[4.1.0]heptane

bridgehead
carbons

The numbering system used to assign substituents starts at a bridgehead position, proceeds along the *longest* arm to the other bridgehead position, and continues along the next longest arm. The bridgehead position chosen to start the numbering is that which gives the lower substituent number.

7,7-dimethylbicyclo[2.2.1]heptan-2-ol

trans-1,6-dichlorobicyclo[4.3.0]nonane

7,7-dicyanobicyclo[4.1.0]hepta-2,4-diene

In many bicyclic structures the stereochemistry at the bridgehead positions is established by steric constraints. There is only a single bicyclo[2.2.1]heptane, for

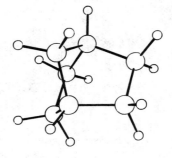

(a)

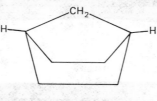

(b)

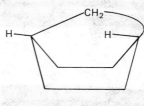

(c)

FIGURE 20.18. *Bicyclo[2.2.1]heptane (norbornane):* (a) *stereo view;* (b) *conventional perspective drawing;* (c) *hypothetical* trans *isomer*—too strained to exist.

example, that in which the methylene bridge is joined *cis* at the 1,4-positions of a boat cyclohexane. The corresponding compound with a *trans* attachment is too strained to exist (Figure 20.18). In other systems both *cis* and *trans* ring fusions can occur and are specified appropriately in the nomenclature. Some bicyclic systems have a further aspect of stereochemistry that must be noted. For example, the Diels–Alder reaction between cyclopentadiene and methyl acrylate, discussed on page 617, could have given two stereoisomeric products, one in which the methoxycarbonyl group is *cis* to the two-carbon bridge (called the **endo** isomer) or one in which this group is *cis* to the one-carbon bridge (called the **exo** isomer). Diels–Alder reactions of cyclopentadiene generally produce the *endo* isomer.

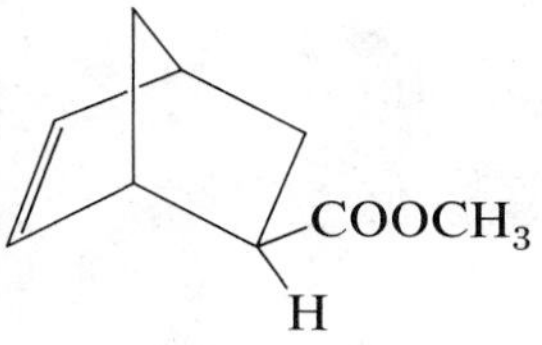

methyl bicyclo[2.2.1]-
hept-5-ene-*exo*-2-carboxylate

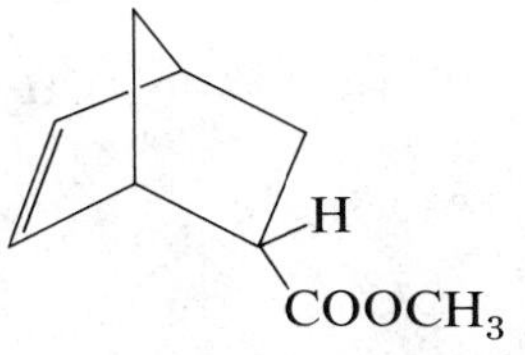

methyl bicyclo[2.2.1]-
hept-5-ene-*endo*-2-carboxylate

Cyclopentadiene is a low-boiling hydrocarbon, b.p. 46°, available commercially as a dimer that can be readily cracked thermally. The dimer boils at 170°. When the free monomer has been prepared by slow distillation of the dimer, it must be used immediately as it re-dimerizes on standing. The dimerization reaction is a Diels–Alder reaction in which one molecule acts as the diene and another takes the role of the dienophile. The *endo* dimer is produced.

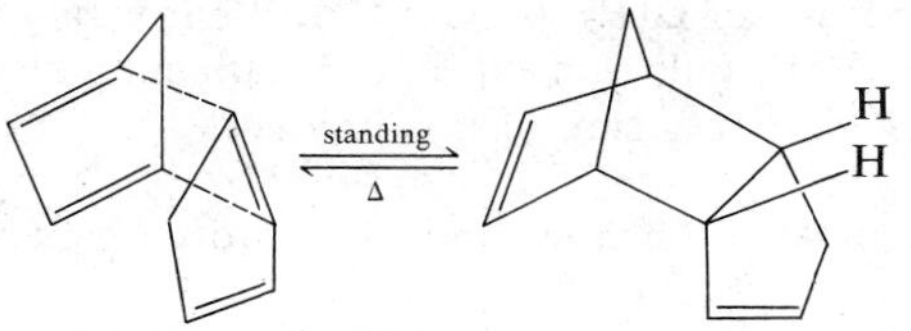

endo addition

dicyclopentadiene

EXERCISE: Write the equations for the Diels–Alder reactions of cyclopentadiene with (a) vinyl acetate, (b) acrylic acid, and (c) dimethyl acetylenedicarboxylate, $CH_3OOCC\equiv CCOOCH_3$. What is the name of each product?

PROBLEMS

1. Draw all of the important resonance structures for each of the following allylic type carbocations.

(a) $CH_2=CH-CH=CH-CH_2^+$ ≡

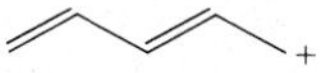

(b) $CH_2=CH-\overset{\overset{\displaystyle CH_2^+}{|}}{C}=CH_2$ ≡

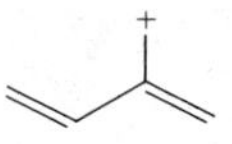

(c)

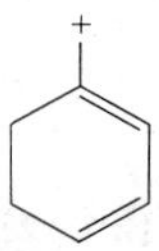

2. For each of the following pairs of allylic resonance structures, determine which is the more stable and contributes more to the resonance hybrid.

(a) $[^+CH_2—CH=CHCH_3 \longleftrightarrow CH_2=CH—\overset{+}{C}HCH_3]$

(b) $[CF_3\overset{-}{C}H—CH=CHCH_3 \longleftrightarrow CF_3CH=CH—\overset{-}{C}HCH_3]$

(c) $[CF_3\overset{+}{C}H—CH=CHCH_3 \longleftrightarrow CF_3CH=CH—\overset{+}{C}HCH_3]$

(d) $[CH_2=CH—O^- \longleftrightarrow {}^-CH_2—CH=O]$

(e) $\left[CH_3\overset{\overset{O}{\|}}{C}—NH_2 \longleftrightarrow CH_3\overset{\overset{O^-}{|}}{C}=\overset{+}{N}H_2 \right]$

(f) $[CH_2=CH—OCH_3 \longleftrightarrow {}^-CH_2—CH=\overset{+}{O}CH_3]$

(g)

3. (a) A common procedure for measuring ^{14}C is to obtain it in the form of carbon dioxide, which is passed into aqueous barium hydroxide and precipitated as barium carbonate. The white product is dried and pressed into a pellet, which is counted with a Geiger counter. Given $Ba^{14}CO_3$ as starting material, present a practical synthesis of $CH_2=CH^{14}CH_2OH$. How would you show that allylic rearrangement occurs when this labeled allyl alcohol reacts with thionyl chloride (page 578)?

(b) Allylic rearrangement can also be demonstrated with deuterium labeling. Present a practical synthesis of $CH_2=CHCD_2OH$. What would the nmr spectrum look like? What product would you expect on treatment with thionyl chloride? What is the expected nmr spectrum of this product?

4. Illustrate the use of allylic halides and Grignard reagents by preparation of the following alkenes.

(a) $CH_3CH_2CH_2CH=CH_2$ (b) $C_6H_5CH_2CH=CH_2$

(c) $CH_2=CHCH_2CH=CH_2$ (d) $(CH_3)_3CCH_2CH=CH_2$

5. What is the principal product of reaction of each of the following alcohols with activated MnO_2?

(a) $CH_3CH=CHCH_2OH$ (b) $CH_3CH_2CH_2CH_2OH$

(c) $HOCH_2CH_2CH=CHCH_2OH$ (d)

(e)

(f) $CH_3C\equiv CCHOHCH_3$

6. The nmr spectrum at $-80°$ of the Grignard reagent prepared from 4-bromo-2-methyl-2-butene shows the following signals: $\delta = 0.6$ ppm, doublet (2H); $\delta = 1.6$ ppm, doublet (6H); $\delta = 5.6$ ppm, triplet (1H). Which structure for the Grignard reagent best fits this nmr spectrum?

$$(CH_3)_2C=CHCH_2MgBr \quad \text{or} \quad CH_2=CHC(CH_3)_2MgBr$$

On warming the solution to room temperature, the doublet at $\delta = 1.6$ ppm first broadens and then becomes a sharp singlet. How do you interpret this behavior?

7. The reaction of 1-octene with N-bromosuccinimide in carbon tetrachloride with a small amount of benzoyl peroxide, $(C_6H_5COO)_2$, gives a mixture of 17% of 3-bromo-1-octene, 44% of *trans*-1-bromo-2-octene, and 39% of *cis*-1-bromo-2-octene. Account for these products with a reaction mechanism showing all significant intermediates.

8. Allylic chlorination can be accomplished by the use of *t*-butyl hypochlorite, a reagent prepared by passing chlorine into an alkaline solution of *t*-butyl alcohol.
(a) Write a plausible mechanism for this reaction.
(b) An example of the use of $(CH_3)_3COCl$ in allylic chlorination is

$$\underset{\text{H}}{\overset{(CH_3)_3C}{>}}C{=}C\underset{\text{CH}_3}{\overset{\text{H}}{<}} \xrightarrow[-78°]{\underset{hv}{(CH_3)_3COCl}} \underset{\text{H}}{\overset{(CH_3)_3C}{>}}C{=}C\underset{\text{CH}_2\text{Cl}}{\overset{\text{H}}{<}} + (CH_3)_3CCHClCH{=}CH_2$$

$$(93\%) \qquad\qquad (7\%)$$

Write a reasonable reaction mechanism.
(c) In the example in (b), note that none of the *cis* isomer is obtained. If we start with the *cis* olefin, the reaction takes the following course.

$$\underset{(CH_3)_3C}{\overset{\text{H}}{>}}C{=}C\underset{\text{CH}_3}{\overset{\text{H}}{<}} \xrightarrow[-78°]{\underset{hv}{(CH_3)_3COCl}} \underset{(CH_3)_3C}{\overset{\text{H}}{>}}C{=}C\underset{\text{CH}_2\text{Cl}}{\overset{\text{H}}{<}} + (CH_3)_3CCHClCH{=}CH_2$$

$$(76\%) \qquad\qquad (24\%)$$

What does this experiment reveal concerning the configurational stability about the C—C bond in an allyl radical, at least at $-78°$?
(d) Why was it necessary to do the experiment with both the *cis* and *trans* olefins? Could the conclusion in (c) have been derived from the results of part (b) alone?
(e) The rotation barrier in allyl radical is estimated to be about 10 kcal mole^{-1}. Explain why such a barrier exists.

9. When 1,3-butadiene is allowed to react with hydrogen chloride in acetic acid at room temperature, there is produced a mixture of 22% 1-chloro-2-butene and 78% 3-chloro-1-butene. On treatment with ferric chloride or on prolonged treatment with hydrogen chloride, this mixture is converted to 75% 1-chloro-2-butene and 25% 3-chloro-1-butene. Explain.

10. Reaction of butadiene with carbon tetrachloride at 110° in the presence of peroxides gives a 23% yield of 1,5,5,5-tetrachloro-2-pentene. Write a plausible mechanism for its formation.

11. Reaction of 2,3-dimethyl-1,3-butadiene with Cl_2 in carbon tetrachloride in the dark at $-20°$ gives 45% of the expected product, 1,4-dichloro-2,3-dimethyl-2-butene, in addition to 54% of A and 1% of B.
Compound A shows mass spectral parent peaks at m/e 118 and 116, with an intense fragment peak at m/e 81. The nmr spectrum shows singlets at $\delta = 1.90$ ppm (3H) and $\delta = 4.20$ ppm (2H) and four peaks at $\delta = 6.06, 6.19, 6.22, 6.30$ ppm (4H).
Compound B shows mass spectral parent peaks at m/e 118 and 116. The nmr spectrum shows singlets at $\delta = 1.78$ ppm (3H), 1.85 ppm (3H), and 6.20 ppm (1H), and two peaks at $\delta = 5.08, 5.00$ ppm (2H).
Deduce the structures of A and B and write a plausible mechanism for their formation.

12. On heating 2-buten-1-ol with dilute sulfuric acid, a mixture of three structurally different isomeric ethers is produced of the type $(C_4H_7)_2O$. Give the structures of these ethers [do not count *cis-trans* or (R,S) isomers] and write a plausible reaction mechanism for their formation.

13. In the reaction of 1,3-cyclopentadiene with hydrogen chloride at 0°, no significant amount of 4-chlorocyclopentene is produced. Explain.

14. (a) When 1-pentyne is treated with 4 N alcoholic potassium hydroxide at 175°, it is converted slowly into an equilibrium mixture of 1.3% 1-pentyne, 95.2% 2-pentyne, and 3.5% 1,2-pentadiene. Calculate $\Delta G°$ differences between these isomers for the equilibrium composition. Write a reasonable reaction mechanism showing all intermediates in the equilibrium reaction.

   (b) Sodium amide in liquid ammonia is a stronger basic system than alcoholic KOH, yet even prolonged treatment of 1-pentyne by $NaNH_2$ in liquid ammonia leads to recovery of the 1-pentyne essentially unchanged. Explain.

15. Reaction of 3-methyl-1,2-butadiene with Cl_2 under free radical conditions gives a good yield of 3-chloro-2-methyl-1,3-butadiene. Write a reasonable reaction mechanism.

16. One convenient preparation of acrolein involves the treatment of glycerol, $CH_2OHCHOHCH_2OH$, with sulfuric acid. Write a plausible reaction mechanism.

17. 5-Methylcyclopent-2-en-1-one reacts with refluxing aqueous sodium hydroxide to give 2-methylcyclopent-2-en-1-one. Explain with a mechanism.

18. Consider the following equilibria between α,β- and β,γ-isomers. For R = CH_3 the equilibrium constant is about unity. For R = $(CH_3)_3C$, however, the equilibrium is displaced substantially in the α,β-direction. Explain this result. The use of molecular models may be helpful.

$$
\underset{\underset{CH_3}{|}}{CH_3CH_2CHCH=CCOOC_2H_5} \rightleftharpoons \underset{\underset{R}{|}\quad\underset{CH_3}{|}}{CH_3CH_2C=CHCHCOOC_2H_5}
$$

R = CH_3:   (45%)   (55%)
R = $C(CH_3)_3$:   (86%)   (14%)

19. The gas phase enthalpy of ionization ($\Delta H°$ for RCl → R^+ + Cl^-) for *trans*-1-chloro-2-butene is about 161 kcal mole^{-1} and is significantly lower than that for 3-chloro-2-methyl-1-propene, 169 kcal mole^{-1}. Explain, using resonance structures where desirable.

20. Show how each of the following conversions may be accomplished.

(a)

(b)

(c)

(d)

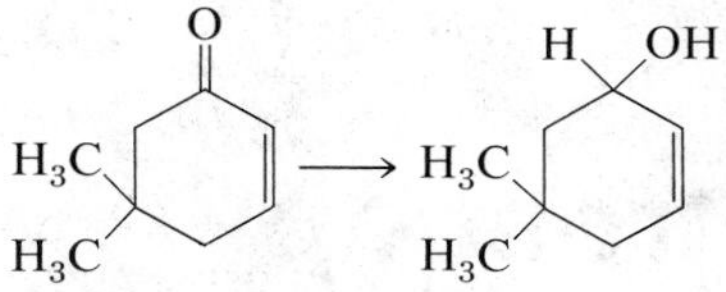

(e)

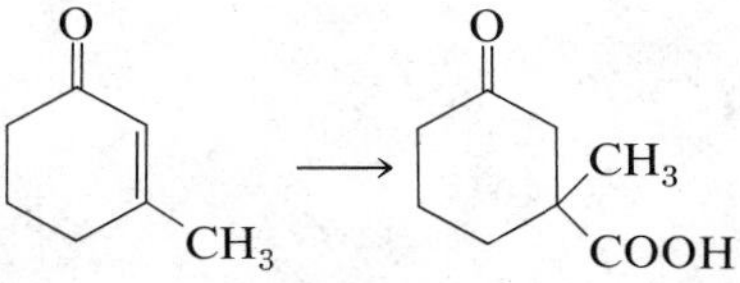

21. Propose a plausible mechanism for the reaction of ketene with water to give acetic acid.

22. What diene and dienophile produce the following Diels–Alder adducts?

(a)

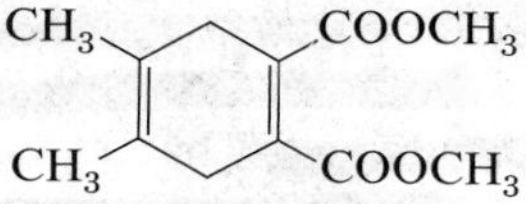

(b)

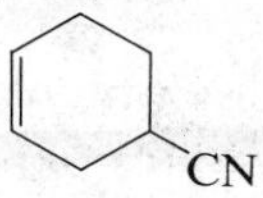

(c)

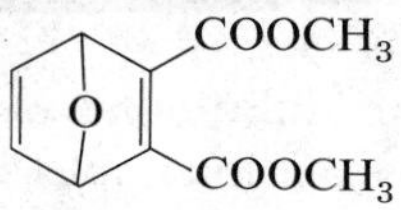

(d)

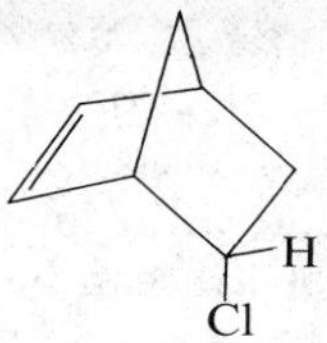

(e)

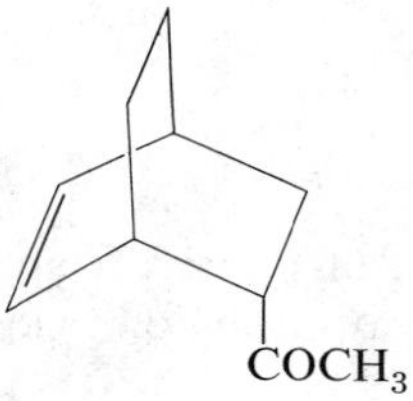

(f)

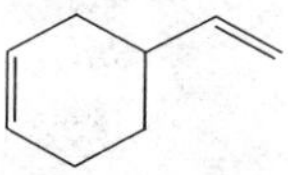

(g)

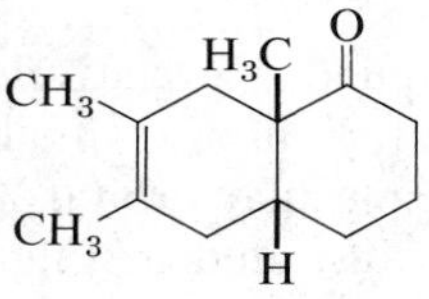

(h)

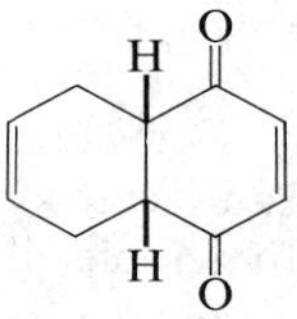

23. For each of the following Diels–Alder reactions, predict the orientation and stereochemistry.

(a)

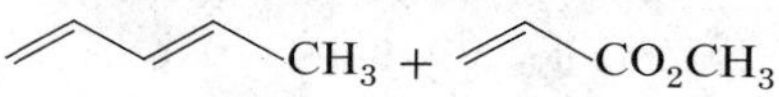

(b)

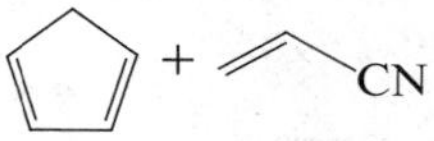

(c)

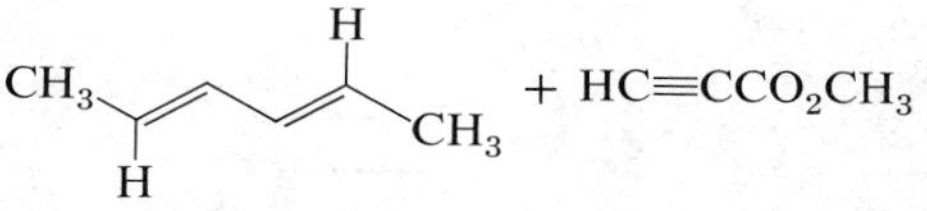

Chapter 21
Ultraviolet Spectroscopy

21.1
Electronic Transitions

A molecule can absorb a quantum of microwave radiation (about 1 cal mole^{-1}) and change from one rotational state to another. Vibrational energy changes are associated with light quanta in the infrared region of the spectrum (about 3–10 kcal mole^{-1}). A change in the electronic energy of a molecule requires light in the visible (40–70 kcal mole^{-1}) or ultraviolet (70–300 kcal mole^{-1}) regions. The energies required for such **electronic transitions** are of the magnitude of bond strengths because the electrons involved are valence electrons. That is, the energy of light quanta in this region of the electromagnetic spectrum is sufficient to **excite** an electron from a bonding to an antibonding state.

The resulting **excited electronic states,** in contrast to the **ground electronic state,** are often difficult to describe by resonance symbolism. In simple diatomic molecules an excited state can sometimes be described rather simply. For example, one excited state of LiF can be described by the process

$$Li^+F^- \xrightarrow{\ h\nu\ } Li\cdot F\cdot$$

$$\underset{\text{state}}{\underset{\text{ground}}{\quad}} \qquad \underset{\text{state}}{\underset{\text{excited}}{\quad}}$$

Absorption of a photon is accompanied by a shift in electron density from fluorine to lithium, and the resulting excited state resembles two atoms held in close proximity.

The excited states of polyatomic molecules are not usually described so simply. Fortunately, molecular orbital concepts can often be applied in a relatively simple and straightforward way. For example, the electronic transition of methane involves the excitation of an electron from a bonding molecular orbital, σ, to the corresponding antibonding molecular orbital, σ*, as illustrated in Figure 21.1.

Recall that the bonding molecular orbital between two atoms is formed by the positive overlap of two hybrid orbitals and is symbolized as in Figure 21.2. The corresponding antibonding molecular orbital, σ*, is produced by the negative overlap of the hybrid orbitals. Note that this negative overlap produces an additional node between the nuclei and reduces the electron density that is so essential for covalent bonding. In the excited state, the electron in σ* partially cancels the

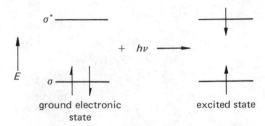

FIGURE 21.1. *Ground and excited states.*

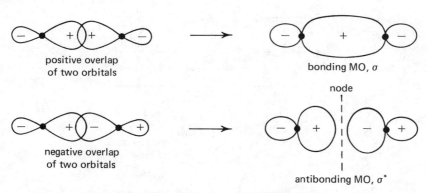

positive overlap
of two orbitals

bonding MO, σ

node

negative overlap
of two orbitals

antibonding MO, σ^*

FIGURE 21.2. *Bonding and antibonding molecular orbitals.*

bonding provided by the remaining electron in σ; hence, the energy required for excitation is of the order of magnitude of bond strengths.

The bonding molecular orbitals of methane, and of alkanes generally, are relatively stable and low in energy. Excitation of an electron requires light of high energy with a wavelength about 150 nm (1500 Å) or less. Light in this region is strongly absorbed by the oxygen in air, and spectroscopic measurements of such compounds require special instruments in which air is completely excluded. This region of the light spectrum is called the vacuum ultraviolet and is unimportant in routine organic laboratory studies.

Wavelengths of light above about 200 nm are not absorbed by air, and it is this region that is most important for organic chemists. The range of about 200–400 nm is called the ultraviolet; the visible region of the spectrum ranges from wavelengths of about 400 nm (violet light) to about 750 nm (red light). The energy of such light is insufficient to affect most σ-bonds, but it is in the range of π-electron energies, especially for conjugated systems. That is, ultraviolet-visible spectroscopy is an important spectroscopic tool for the study of conjugated multiple bonds. The π-molecular orbitals of such conjugated systems extend over several atoms. The highest occupied or least bonding of such molecular orbitals already have at least one node. Electronic excitation generally involves the transition of an electron to a molecular orbital having an additional node, and, as a general rule, the more nodes an electron has in a wave function, the less energy it takes to add another node.

21.2
$\pi \to \pi^*$ Transitions

Absorption of light that produces excitation of an electron from a bonding π to an antibonding π^*-molecular orbital is referred to as a $\pi \to \pi^*$ transition. For example, 1,3-butadiene has an intense absorption band at 217 nm (usually written as λ_{max} 217 nm; that is, lambda max = 217 nanometers) that results from the excitation of an electron from π_2 to π_3 (Figure 21.3). Recall that π_2 has one node and π_3 has two (Figure 20.8). 1,3,5-Hexatriene absorbs at longer wavelength, $\lambda_{max} = 258$ nm. It takes less energy to excite an electron from π_3, the highest occupied π-molecular orbital of hexatriene, which has two nodes, to π_4, which has three nodes. *The longer the chain of conjugation, the longer the wavelength of the absorption band.* For example, the lowest energy $\pi \to \pi^*$ transition of 1,3,5,7-

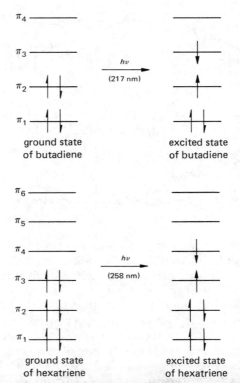

FIGURE 21.3. *Electronic states of butadiene,* $CH_2{=}CHCH{=}CH_2$, *and 1,3,5-hexatriene,* $CH_2{=}CHCH{=}CHCH{=}CH_2$.

octatetraene occurs at the still longer wavelength of 304 nm, whereas ethylene itself absorbs in the vacuum ultraviolet at 175 nm.

> Compounds generally have many excited electronic states, but organic chemists are mostly concerned with the lowest or more stable states, since these are the states that are accessible with the energies of ultraviolet and visible light. Many of these states can be described in terms of electron transitions that involve other than just the highest occupied and lowest vacant molecular orbitals. For example, other electronic states of butadiene arise from the electronic transition $\pi_1 \rightarrow \pi_3$ or $\pi_1 \rightarrow \pi_4$, but such transitions occur in the vacuum ultraviolet. However, benzene has three absorption bands rather close together at 264, 207, and 179 nm that involve excited states that do not differ much in energy from each other—the corresponding energies are 108, 138, and 160 kcal mole^{-1}, respectively (see Section 22.3.C).

The highly conjugated hydrocarbon *trans-β*-carotene, with eleven double bonds in conjugation, has two intense long-wavelength absorptions in alkane solution at 483 nm and 453 nm. These absorptions are in the visible region of the spectrum and correspond to blue to blue-green light. Since light of this color is absorbed by the compound, *β*-carotene appears yellow to orange in solution. For further aspects of color, see Section 34.3.A.

cis-β-Carotene has two absorption peaks at essentially the same wavelengths but with weaker intensities. This result is quite general; π-systems that are prevented from achieving coplanarity show significant changes from coplanar analogs, particularly in absorption intensities. In *cis-β*-carotene the two groups on the same side of the double bond sterically interfere with each other.

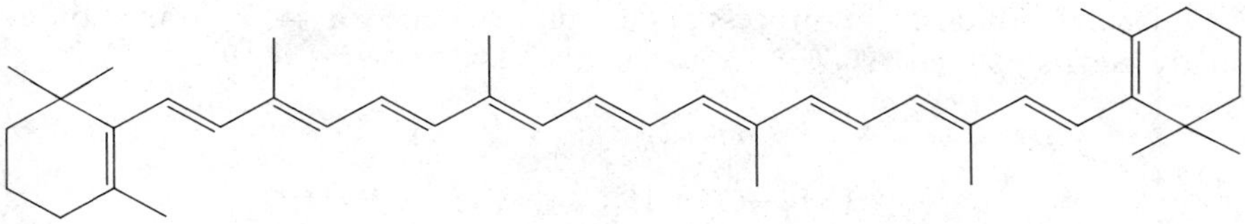

trans-β-carotene

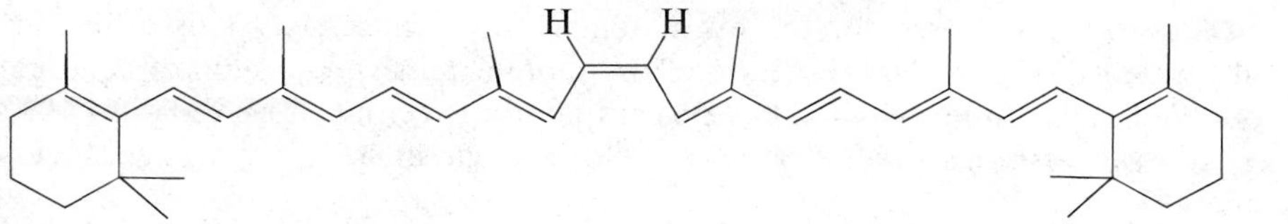

cis-β-carotene

α,β-Unsaturated aldehydes and ketones also have high-intensity absorptions resulting from the transition of an electron from π_2 to π_3. These $\pi \rightarrow \pi^*$ transitions occur at almost exactly the same wavelength as those for the corresponding dienes.

$$CH_2{=}CHCH{=}CH_2 \qquad CH_2{=}CHCHO$$
$$\lambda_{max} \text{ 217 nm} \qquad\qquad \lambda_{max} \text{ 218 nm}$$

As in the case of polyenes, the wavelength of the light absorbed by unsaturated carbonyl compounds increases as the chain of conjugation increases. The effect is illustrated in Table 21.1.

TABLE 22.1
Spectra of Some Polyene Aldehydes

Aldehyde	λ_{max}, nm
$CH_3CH{=}CHCHO$	220
$CH_3CH{=}CHCH{=}CHCHO$	270
$CH_3(CH{=}CH)_3CHO$	312
$CH_3(CH{=}CH)_4CHO$	343
$CH_3(CH{=}CH)_5CHO$	370
$CH_3(CH{=}CH)_6CHO$	393
$CH_3(CH{=}CH)_7CHO$	415

21.3
$n \rightarrow \pi^*$ Transitions

Carbonyl groups have another characteristic absorption that is associated with the lone-pair electrons on oxygen. Since these electrons are bound to only a single atom, they are not held as tightly as σ-electrons, and they can also be excited to

π^*-molecular orbitals. The process results in a so-called $n \rightarrow \pi^*$ transition and usually occurs at relatively long wavelength.

	$CH_3-\overset{\overset{\displaystyle O}{\|}}{C}-CH_3$	$CH_3\overset{\overset{\displaystyle O}{\|}}{C}CH{=}CH_2$
$n \rightarrow \pi^*$:	λ_{max} 270 nm	λ_{max} 324 nm
$\pi \rightarrow \pi^*$:	187 nm	219 nm

The π-system of methyl vinyl ketone is more extended than that of acetone, and less energy is required for the excitation in the former case. This difference is illustrated in Figure 21.4. Because of the more extensive π-system of conjugated double bonds of methyl vinyl ketone compared to acetone, both the $n \rightarrow \pi^*$ and $\pi \rightarrow \pi^*$ transitions of methyl vinyl ketone occur at longer wavelength (lower energy).

One important distinguishing characteristic of $n \rightarrow \pi^*$ transitions results from the critical feature that the lone-pair electrons tend to be concentrated in a different region of space from the π-electrons (Figure 21.5). Although $n \rightarrow \pi^*$ transitions often occur at lower energy (longer wavelength) than $\pi \rightarrow \pi^*$ transitions, they are *less probable*. A given quantum of $n \rightarrow \pi^*$ light must encounter many more molecules before it is absorbed than is the case for $\pi \rightarrow \pi^*$ light quanta. This difference shows up experimentally in an absorption spectrum as an intensity difference; $\pi \rightarrow \pi^*$ absorptions are generally much more intense ("strong absorption") than $n \rightarrow \pi^*$ absorptions ("weak absorption"). The difference in intensity is two to three orders of magnitude.

The intensity is expressed as an **extinction coefficient** ϵ. The amount of light absorbed depends on the extinction coefficient and the number of molecules in the light path. The number of molecules depends on the concentration of the solution and the path length of the absorption cell. The amount of light that

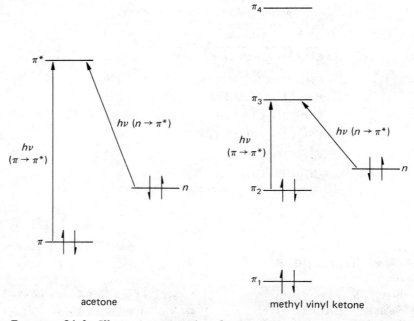

FIGURE 21.4. *Illustrating* $\pi \rightarrow \pi^*$ *and* n $\rightarrow \pi^*$ *transitions in two ketones.*

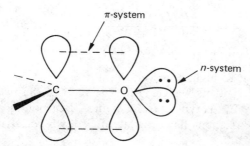

FIGURE 21.5. *Lone pairs* (n-*system*) *and* π-*system of a carbonyl group.*

passes through a solution (transmittance) is given by Beer's law

$$\log \frac{I_0}{I} = \epsilon cd$$

where I_0 is the intensity of the light before it encounters the cell, I is the intensity of the light emerging from the cell, c is the concentration in moles per liter, and d is the path length in centimeters.

As an example, the spectrum of two concentrations of mesityl oxide, $(CH_3)_2C{=}CHCOCH_3$, in the same 1 cm cell (a common path length) is shown in Figure 21.6. A highly dilute solution is used for the $\pi \rightarrow \pi^*$ absorption at 235 nm. The extinction coefficient for this transition is calculated as

$$\epsilon = \frac{\log I_0/I}{cd} = \frac{1.18}{(9.37 \times 10^{-5})(1)} = 12{,}600 \text{ L mole}^{-1} \text{ cm}^{-1}$$

In this dilute solution the absorption due to the $n \rightarrow \pi^*$ transition is so weak it is barely discernible. A more concentrated solution gives greater absorption and, from the second curve in Figure 21.6, we may calculate ϵ for this transition at 326 nm to be

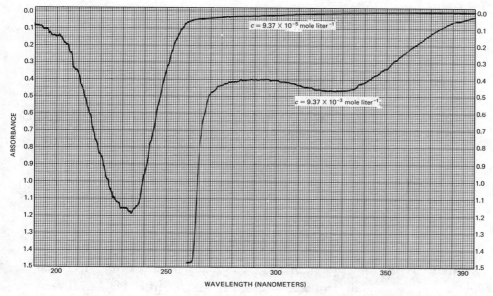

FIGURE 21.6. *Ultraviolet absorption spectra of mesityl oxide,* $(CH_3)_2C{=}CHCOCH_3$.

$$\epsilon = \frac{\log I_0/I}{cd} = \frac{0.47}{(9.37 \times 10^{-3})(1)} = 50$$

(Note that the units of ϵ are usually omitted.)

This concentration is so high, however, that the $\pi \to \pi^*$ transition absorbs light essentially completely at its wavelength. The ratio of the two extinction coefficients, $12{,}600/50 = 252$, is typical for unsaturated carbonyl compounds. In general, the $\pi \to \pi^*$ transitions have ϵ of about 10^4, whereas ϵ for $n \to \pi^*$ transitions are about 10–100.

EXERCISE: A solution of 0.00731 g of crotonic acid, $CH_3CH{=}CHCOOH$, in 10 mL of methanol was prepared. The ultraviolet spectrum was measured in a 1 cm cell. Although the $\pi \to \pi^*$ transition was off scale, the $n \to \pi^*$ transition at 250 nm showed an absorbance of 0.77. A 1 mL aliquot of this solution was diluted to 100 mL and the spectrum was recorded again. The $\pi \to \pi^*$ transition was seen clearly at 200 nm with an absorbance of 0.86. Calculate the extinction coefficients for the two transitions.

21.4
Alkyl Substituents

We saw in Section 21.2 that 1,3-butadiene has a $\pi \to \pi^*$ transition at 217 nm. Alkyl-substituted butadienes have the same π-system, but their absorption spectra vary significantly. Examples are

	λ_{max}, nm
$CH_2{=}CHCH{=}CH_2$	217
$CH_2{=}\overset{\overset{\textstyle CH_3}{\textstyle \vert}}{C}CH{=}CH_2$	220
$CH_3CH{=}CHCH{=}CH_2$	223.5
$CH_2{=}\overset{\overset{\textstyle CH_3}{\textstyle \vert}}{C}{-}\overset{\overset{\textstyle CH_3}{\textstyle \vert}}{C}{=}CH_2$	226
$CH_3CH{=}CHCH{=}CHCH_3$	227

Each methyl group increases the wavelength of the absorption peak by about 5 nm. A similar effect shows up with unsaturated carbonyl compounds.

	λ_{max}, nm ($\pi \to \pi^*$)
$CH_2{=}CHCOCH_3$	219
$CH_3CH{=}CHCOCH_3$	224
$(CH_3)_2C{=}CHCOCH_3$	235

This effect arises from the overlap of σ-bonds in the alkyl substituent with the π-system. The resulting **hyperconjugation** is symbolized in Figure 21.7.

Hyperconjugation is much less effective than conjugation. Adding a methyl

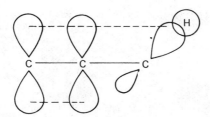

FIGURE 21.7. *Illustration of hyperconjugation between a C—H bond and a double bond.*

group to butadiene has little more than 10% of the effect of adding another vinyl group. Nevertheless, even this small effect has been useful for structure proofs, and extensive empirical correlations have been compiled for determining the actual effect of an alkyl substituent in various positions of dienes, trienes, and unsaturated carbonyl compounds. Some correlations of this type **(Woodward–Fieser rules)** are summarized in Table 21.2 for use with conjugated dienes, trienes and α,β-unsaturated ketones.

TABLE 21.2

**Empirical Parameters for $\pi \rightarrow \pi^*$ Transitions of Conjugated Systems in Ethanol
(Woodward–Fieser Rules)**

Parent System	λ_{max}, nm	Substituent Corrections	
Polyenes			
		double bond[a]	+30
	214	alkyl group	+5
		exocyclic C=C double bond[b]	+5
	253	OR groups	0
(in six-membered ring)		Cl, Br	+5
α,β-Unsaturated ketones			
	215	double bond[a]	+30
		alkyl group	+10
		exocyclic C=C double bond[b]	+5

[a] The double bond must be attached at the end of the π-system to produce a linear conjugated system.

[b] An exocyclic double bond is attached to a ring at one end, for example,

Some examples of the use of this table follow.

parent (circled)	214
1 exocyclic double bond (a)	5
3 alkyl substituents (b)	15
calcd λ_{max}	234 nm
expl λ_{max}	234 nm

[a]CH_3

parent (circled)	253
4 alkyl substituents (a)	20
calcd λ_{max}	273 nm
expl λ_{max}	262 nm

[a]$CH(CH_3)_2$

parent (circled)	215
1 exocyclic double bond (a)	5
3 alkyl groups (b)	30
calcd λ_{max}	250 nm
expl λ_{max}	251 nm

Note that the parent is a ketone and therefore includes alkyl group (c). Note also that the correction for an exocyclic double bond does not apply to the carbonyl group.

parent (circled)	214
double bond (a)	30
3 alkyl groups (b)	15
calcd λ_{max}	259 nm
expl λ_{max}	263 nm

The rules reproduced here have been simplified to serve as an introduction to the general concept of the use of such empirical generalizations in ultraviolet spectroscopy. Note that the only substituents that are relevant are those attached directly to the conjugated π-system. Alkyl groups and double bonds elsewhere in the molecule have little or no effect on the transitions of the π-system. Extended correlations of this type are available for other parents and for other types of substituent and solvent corrections. The general approach constitutes one of the most useful applications of ultraviolet spectroscopy in structure proofs.

EXERCISE: Use the Woodward–Fierer rules to predict the ultraviolet absorptions for (a) 3-methyl-1,3-pentadiene, (b) 1,2-dimethylcyclohexa-1,3-diene, (c) hex-4-en-3-one, and (d) 3-methylcyclohex-2-en-1-one.

21.5
Other Functional Groups

Alcohols and ethers do not have conjugated π-systems and are transparent in the normal ultraviolet and visible regions. Ethanol and ether are common solvents for recording ultraviolet spectra. Sulfides, however, have relatively intense absorption at about 210 nm with a weaker band at about 230 nm. These absorptions are probably associated with transition of a lone-pair electron in sulfur to a sulfur $3d$-orbital.

The carbonyl group in carboxylic acid derivatives is significantly different from that in ketones. Alkanoic acids have a low-intensity band about 200–210 nm, anhydrides absorb at somewhat longer wavelength, and the acid chlorides are still longer at about 235 nm.

Simple acetylenes absorb in the vacuum ultraviolet. Conjugated triple bonds show the type of absorption in the accessible ultraviolet expected for extended π-systems. The C≡N group of nitriles also absorbs at short wavelength, below 160 nm.

The simple alkyl fluorides and chlorides have no absorption maxima in the normal ultraviolet region. Alkyl bromides and iodides, however, do have λ_{max} in the region about 250–260 nm. These absorptions are attributed to transition of a lone-pair electron to an antibonding σ^*-orbital. C—Br and C—I bonds are sufficiently weak that the corresponding σ^*-orbitals have low enough energy to give transition energies in this ultraviolet region.

21.6
Photochemical Reactions

An excited state has more electronic energy than the ground state, and such states are generally rather short-lived. This excess energy is generally dissipated within less than 10^{-7} sec. One important way in which this energy is removed is by conversion of the electronic energy to vibrational and rotational energy. That is, the energy of moving electrons is converted in part to that of moving nuclei. Such energy, in turn, may simply be distributed as translational energy to other colliding molecules, in which case the net result has been the conversion of light to heat.

Alternatively, the vibrational energy may suffice to cause rearrangements or to break bonds. We saw one example of bond-breaking in the light-initiated chlorination of alkanes (Section 6.3.A).

$$Cl_2 \xrightarrow{h\nu} 2\ Cl\cdot$$

This reaction could have been expressed as

$$Cl_2 \xrightarrow{h\nu} Cl_2^* \longrightarrow 2\ Cl\cdot$$

in which Cl_2^* refers to an electronically excited state of Cl_2. The light promotes an electron to a Cl—Cl σ^*-orbital. An electron in an antibonding orbital produces a weaker bond than when such an orbital is vacant and generally gives rise to a lower bond-dissociation energy.

Many examples of different types of photochemical reactions are known for organic compounds, but we will only discuss one at this point (for further examples see Section 34.4). In the electronically excited state of an alkene the double bond is generally weaker than in the ground state, and *cis-trans* isomerization is more facile.

trans-stilbene   *cis*-stilbene

This type of photochemical reaction is involved in the chemistry of vision. Vitamin A_1 (retinol) is an alcohol that is oxidized enzymatically to vitamin A aldehyde (retinal), a *cis* form of which combines with a protein, opsin, to produce the light-sensitive compound rhodopsin. This compound is contained in the rods of the retina and absorbs at 500 nm. Absorption of light quanta of this wavelength results in conversion to the *trans* isomer. This isomerization is accompanied by a conformational change that excites the nerve cell and produces a separation into opsin and *trans*-retinal. The *trans* aldehyde is converted to the *cis* form by an enzyme, retinal isomerase, and the cycle starts anew. A wavelength of 500 nm corresponds to the blue-green region of the light spectrum and suggests why the rods are so sensitive to light of this color. Only a few light quanta are required to give a visual response to the dark-adapted eye. Bright light causes temporary impairment of vision because it depletes the rhodopsin and time is required for the protein to be reconstructed via the retinal isomerase cycle.

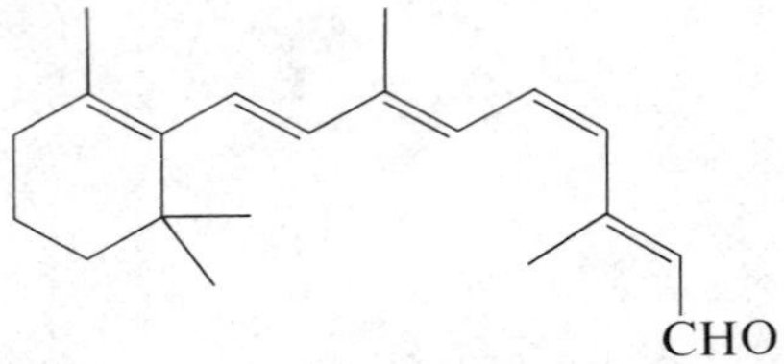

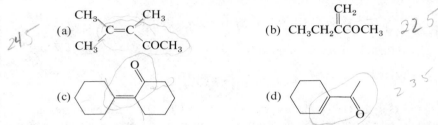

cis-retinal 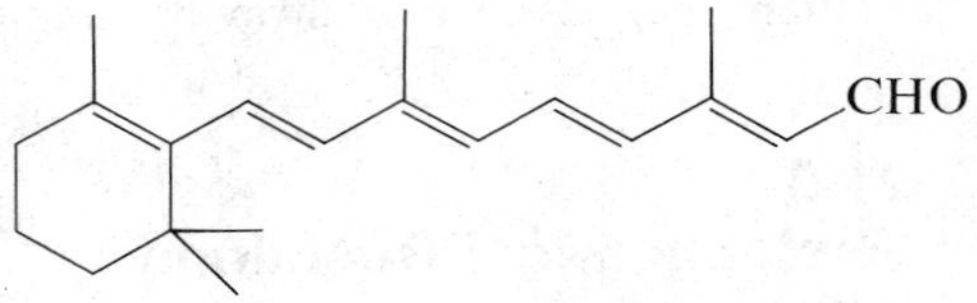   trans-retinal

PROBLEMS

1. The labels fell off four bottles of ketones known to have the structures below. Measurement of the ultraviolet spectra of the contents of the four bottles gave λ_{max} at 221, 233, 249, and 258 nm. Assign structures to the appropriate λ_{max}.

(a)
$$\begin{array}{c} CH_3 \\ CH_3 \end{array} C = C \begin{array}{c} CH_3 \\ COCH_3 \end{array}$$

(b) $CH_3CH_2\overset{\overset{\displaystyle CH_2}{\|}}{C}COCH_3$

(c)

(d)

2. Alkyl bromides and iodides are normally stored in the dark or in dark bottles. On exposure to light they slowly turn brown or violet, respectively. Give an explanation for this phenomenon based on a reasonable photochemical mechanism.

3. Which of the following compounds would be suitable as solvents for recording normal ultraviolet spectra of substrates? Explain briefly.

methanol   perfluoropropane   1-chlorobutane
ethyl ether   ethyl iodide   methylene bromide
methyl butyl sulfide   benzene   cyclohexane
acetonitrile

4. Steroids are important biomolecules that are discussed in more detail in Section 34.6.B. They have the carbon skeleton

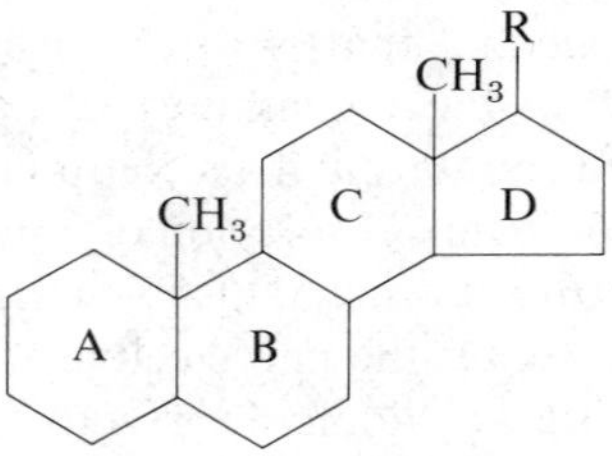

with the rings labeled as indicated. Many such steroid compounds are known. In the following examples only the A and B rings are indicated and significant. Using the empirical parameters in Table 21.2, calculate the expected λ_{max} and compare with the experimental values given.

(a) A B

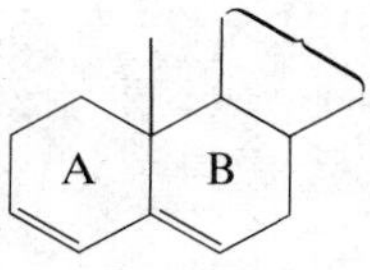

234 nm

(b)

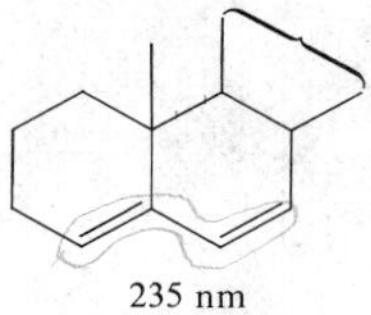

235 nm

(c)

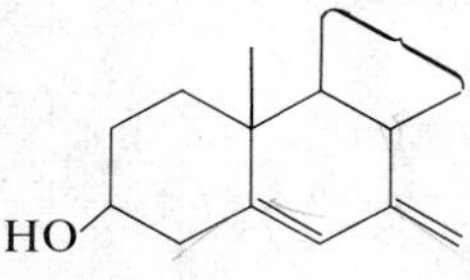

HO

236 nm

(d)

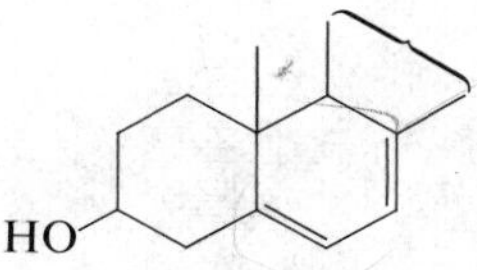

HO

282 nm

(e)

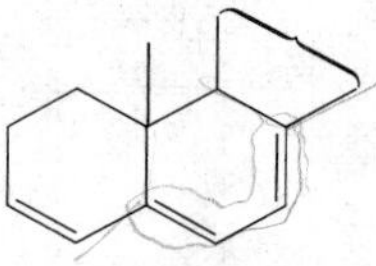

315 nm

(f)

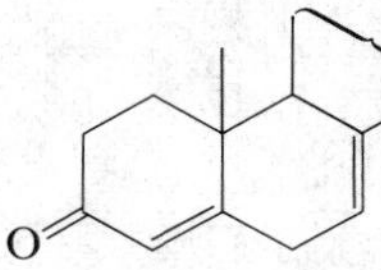

O

235 nm

(g)

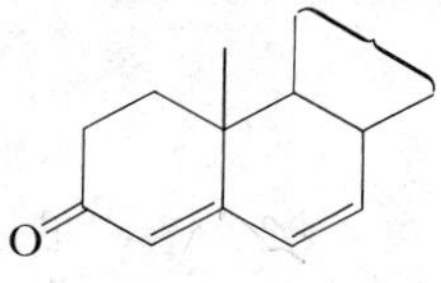

O

284 nm

(h)

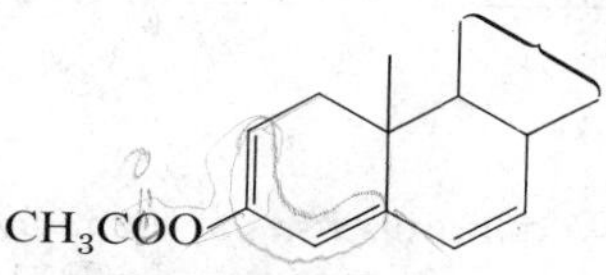

CH_3COO

306 nm

(i)

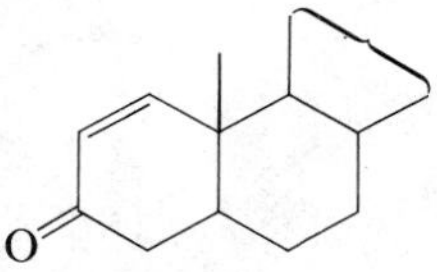

O

230 nm

(j)

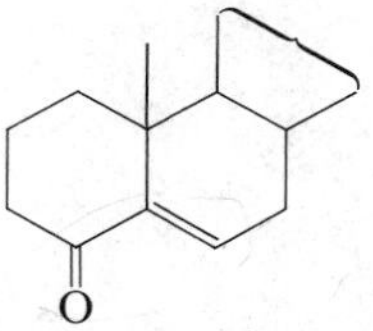

O

241 nm

5. A number of simple conjugated polyenes, $H(CH{=}CH)_n H$, are now known up to $n = 10$; λ_{max} in nanometers corresponding to values of n are as follows: 2, 217; 3, 268; 4, 304; 5, 334; 6, 364; 7, 390; 8, 410; 10, 447. A crude model of such a conjugated π-system is that of an electron in a box having the dimensions of the π-system. A quantum-mechanical treatment of such a model suggests that $1/\lambda_{max}$ should be approximately a linear function of $1/n$. Test this prediction with the data given and try to interpolate to find λ_{max} for the missing polyene with $n = 9$.

6. The natural product α-cyperone is a member of a large class of compounds called terpenes discussed in Section 34.6.A. α-Cyperone was originally assigned the structure shown. Based on the ultraviolet absorption at λ_{max} 251 nm found for α-cyperone, determine whether this structure is reasonable, and if not, propose an alternative with the same carbon skeleton.

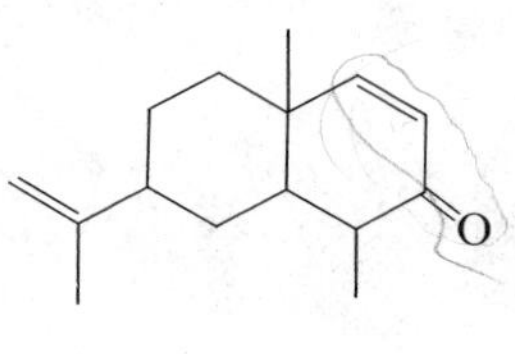

7. The ultraviolet spectrum of 3,6,6-trimethylcyclohex-2-en-1-one, is shown below. The concentration is 1.486×10^{-5} g mL^{-1} in ethanol. Calculate ϵ and compare λ_{max} with the value predicted by Woodward's rules.

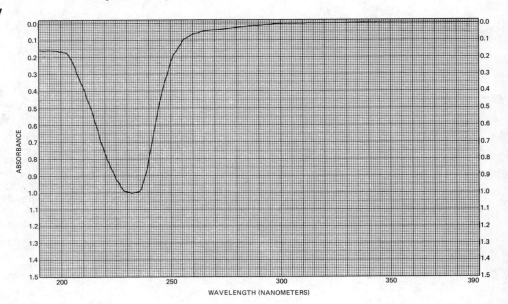

WAVELENGTH (NANOMETERS)

8. Woodward's rules apply to cyclopentenones if the parent α,β-unsaturated ketone is assigned a base value of 212 nm.

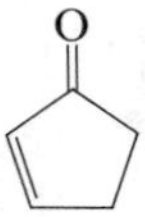

λ_{max} 212 nm

Calculate the λ_{max} expected for 3-methylcyclopent-2-enone.

Chapter 22
Benzene and the
Aromatic Ring

22.1
Benzene

A. The Benzene Enigma

The hydrocarbon now known as benzene was first isolated by Michael Faraday in 1825 from an oily condensate that deposited from illuminating gas. Faraday determined that it has equal numbers of carbons and hydrogens and named the new compound "carbureted hydrogen." In 1834 Mitscherlich found that the same hydrocarbon may be produced by pyrolysis with lime of benzoic acid, which had been isolated from gum benzoin. By vapor density measurements, Mitscherlich established the molecular formula to be C_6H_6. He named the compound benzin, but other influential chemists protested that this name implied a relationship to alkaloids such as quinine. Finally, the German name benzol, based on the German öl, oil, was adopted. In France and England, the name **benzene** was adopted, to avoid confusion with the typical alcohol ending.

> During the early history of benzene, Laurent proposed the name pheno (Gk., *phainein*, to shine) in keeping with the discovery of the material in illuminating gas. Although the name never gained acceptance, it persists in **phenyl,** the name of the C_6H_5- group.

Other preparations of benzene followed these early discoveries, and it was soon recognized that benzene is the parent hydrocarbon of a whole family of organic compounds. The physical properties of benzene (b.p. 80.1°, m.p. 5.5°) are consistent with its molecular formula of C_6H_6. For example, cyclohexane, C_6H_{12}, has b.p. 80.7° and m.p. 6.5°. A six-carbon saturated alkane would have the formula C_6H_{14}. Therefore, benzene must have four double bonds and/or rings. Yet, it does not exhibit the high reactivity of typical polyenes. In fact, it is remarkably inert to many reagents. For example, it does not react with aqueous potassium permanganate or with bromine water. It does not even react with concentrated sulfuric acid in the cold. It is stable to air and tolerates free-radical initiators. It may be used as a solvent for Grignard reagents and alkyllithium compounds. All of these properties are totally inconsistent with such C_6H_6 structures as

$$CH_2=C=\overset{\overset{\displaystyle H}{|}}{C}-\overset{\overset{\displaystyle H}{|}}{C}=C=CH_2 \quad \text{or} \quad CH_3C\equiv C-C\equiv CCH_3$$

The fact that benzene has a formula that suggests a polyene structure but does not behave at all like other polyenes was a dilemma for nineteenth century chemists. Furthermore, new compounds were continually being discovered that were structurally related to benzene. It was clear that there is something fundamentally different about benzene and its derivatives. As a group, the benzene-like compounds were called **aromatic** compounds because many of them have characteristic aromas. The Kekulé theory of valence, first proposed in 1859, allowed acceptable structures to be written for aliphatic compounds such as ethane and ethylene, but at first it did not appear to be applicable to aromatic compounds.

In 1865 Kekulé suggested a regular hexagon structure for benzene with a hydrogen attached at each corner of a hexagonal array of carbons.

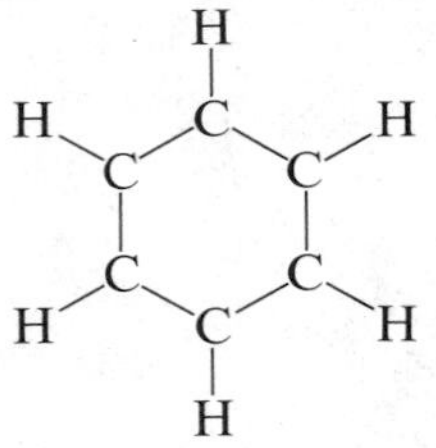

However, this structure violates the tetravalence of carbon inherent in his theory. He later modified his structure to treat benzene as an equilibrating mixture of cyclohexatrienes. However, this structure does not account for the nonolefinic character of benzene.

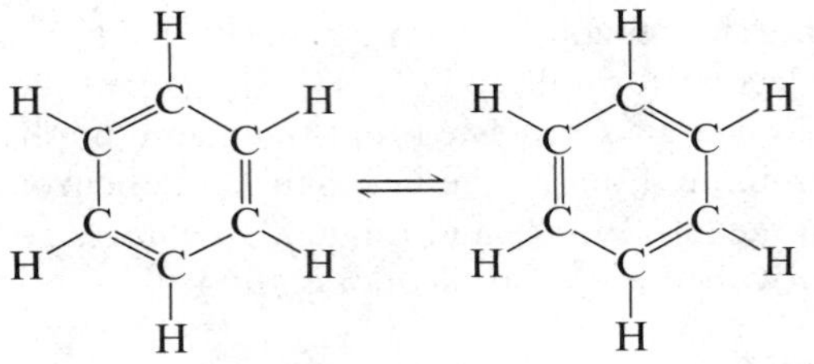

cyclohexatriene

Other attempts by nineteenth century chemists to explain the benzene problem only emphasize the frustrations inherent in the limited theory of the day. One such example was Armstrong's centroid formula in which the fourth valence of each carbon is directed toward the center of the ring.

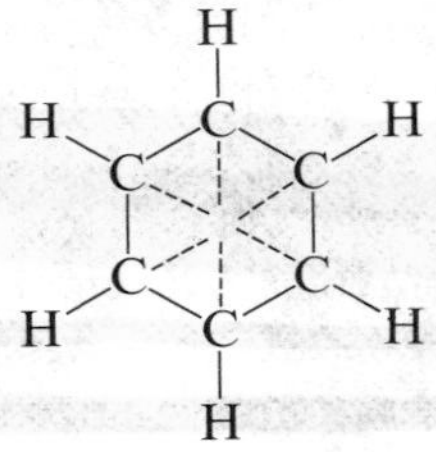

Ladenburg, in 1879, proposed an interesting structure that would solve the problem of why benzene displays no polyene properties. In the Ladenburg proposal, benzene was treated as a tetracyclic compound with no double bonds.

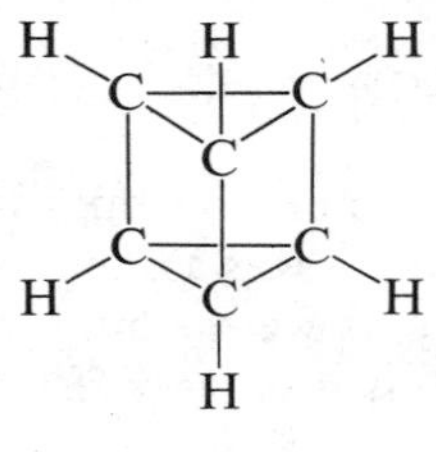

Ladenburg's representation of benzene, while not the structure of benzene, is a perfectly valid structure for an organic compound. It has come to be known as "Ladenburg benzene" or "prismane." After considerable effort, prismane was finally synthesized in 1973 by organic chemists at Columbia University. Upon heating to 90°, it isomerizes to benzene.

Only with the advent of modern wave mechanics did the structure of benzene take its place within a unified electronic theory. The x-ray crystal structure of benzene shows that the compound does indeed have a regular hexagonal structure as Kekulé had originally suggested. The C—C bond distance of 1.40 Å is intermediate between those for a single bond (1.54 Å) and a double bond (1.33 Å). In a regular hexagon the bond angles are all 120°, and this suggests the involvement of sp^2-hybrid orbitals. We can now recognize the "fourth valence," which was so difficult for nineteenth century chemists to explain, as being π-bonds from p-orbitals *extending equally around the ring*, as in Figure 22.1.

In resonance language, we may depict benzene by two equivalent resonance structures.

Note the important difference in meaning between this formulation and that of equilibrating cyclohexatrienes. Cyclohexatriene would have alternating single and double bonds, and the chemical equilibrium between the two alternative structures requires the movement of nuclei.

In the resonance structures the C—C distances remain the same. The resulting resonance hybrid may be written with dotted lines to indicate the partial-double-bond character of the benzene bonds.

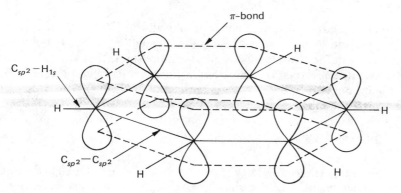

FIGURE 22.1. *Orbital structure of benzene.*

B. *Resonance Energy of Benzene*

From an examination of the heat of hydrogenation of benzene it is possible to estimate how much more stable benzene is compared to a hypothetical "cyclohexatriene." This imaginary quantity is called the **resonance energy** of benzene. The heat of hydrogenation of the double bond in cyclohexene is -28.4 kcal mole^{-1}. That for one double bond in 1,3-cyclohexadiene is -26.5 kcal mole^{-1}.

$$+ \ H_2 \longrightarrow \qquad \Delta H^\circ = -28.4 \text{ kcal mole}^{-1}$$

$$+ \ H_2 \longrightarrow \qquad \Delta H^\circ = -26.5 \text{ kcal mole}^{-1}$$

By a simple extrapolation, we might expect the heat of hydrogenation of a 1,3,5-cyclohexatriene with alternating single and double bonds to be about -24.5 kcal mole^{-1}.

$$+ \ H_2 \longrightarrow \qquad \Delta H^\circ \approx -24.5 \text{ kcal mole}^{-1}(?)$$

Benzene can in fact be hydrogenated but only with difficulty. It hydrogenates slowly under conditions where simple alkenes react rapidly. When hydrogenation does occur, it generally goes all the way and cyclohexane results. The heat of hydrogenation for the complete reduction of benzene to cyclohexane is -49.3 kcal mole^{-1}.

$$+ \ 3 \ H_2 \longrightarrow \qquad \Delta H^\circ = -49.3 \text{ kcal mole}^{-1}$$

Since the heat of hydrogenation of 1,3-cyclohexadiene to cyclohexane is -54.9 kcal mole^{-1}, the heat of hydrogenation of benzene to 1,3-cyclohexadiene is $-49.3 - (-54.9) = +5.6$ kcal mole^{-1}; the process is actually endothermic!

$$+ \ H_2 \longrightarrow \qquad \Delta H^\circ = +5.6 \text{ kcal mole}^{-1}$$

These energy relationships are shown graphically in Figure 22.2.

By comparison with the actual heat of hydrogenation of one bond in benzene, we find that benzene is about 30 kcal mole^{-1} more stable than it would be if it had the cyclohexatriene structure. This stabilization energy defines the resonance energy of benzene; that is, the resonance energy is the difference in energy between the real benzene and that of a single principal Lewis resonance structure. Other derivations of this quantity give somewhat different values; one commonly used number is 36 kcal mole^{-1}.

Actually, the true resonance energy of benzene should not be referred to a cyclohexatriene with alternating bonds of different lengths. Rather, it should be referred to a hypothetical model having the geometry of benzene but with π-overlap allowed only between alternating bonds. Such a structure requires the deformation of cyclohexatriene—stretching the double bonds and compressing the single bonds

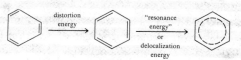

Various estimates have been made of this distortion energy, but one estimated value of about 30 kcal mole^{-1} appears to be reasonable. This would make the actual resonance energy of benzene about 60 kcal mole^{-1}. To distinguish between these different energy quantities, this number of about 60 kcal mole^{-1} is referred to as a **delocalization energy** because it is the energy liberated when electrons are allowed to *delocalize* or *relax* from a hypothetical compound with the benzene geometry having the electrons constrained to alternating single and double bonds to the electronic structure of benzene itself. The value of about 30 kcal mole^{-1}, derived from heats of hydrogenation, is referred to as the **empirical resonance energy.**

Furthermore, this use of the term *resonance* should not be confused with the resonance that occurs in, for example, nmr (nuclear magnetic resonance) in which resonance refers to a matching of the frequency of an irradiating electromagnetic beam with the energy difference between two nuclear spin states in a magnetic field. However, both kinds of resonance are in fact related to the resonance phenomena of vibrations that allowed Joshua's horn to bring down the walls of Jericho.

The benzene ring can also conjugate with other π-electron groups and provide additional stabilization. Comparison of some heats of hydrogenation shows that the benzene ring is less effective in this regard than a double bond.

$$CH_3CH_2CH{=}CH_2 + H_2 \longrightarrow CH_3CH_2CH_2CH_3 \qquad \Delta H^\circ = -30.2 \text{ kcal mole}^{-1}$$
$$CH_2{=}CH{-}CH{=}CH_2 + H_2 \longrightarrow CH_2{=}CHCH_2CH_3 \qquad \Delta H^\circ = -26.3 \text{ kcal mole}^{-1}$$

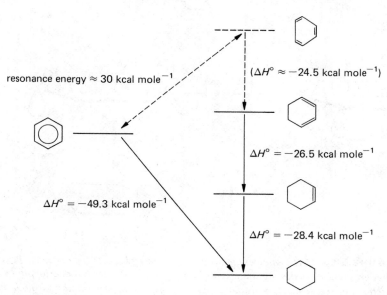

FIGURE 22.2. *Estimation of the resonance energy of benzene.*

C. *Molecular Orbital Theory of Benzene*

The π-system of benzene is made up of six overlapping p-orbitals on carbon, which will therefore give rise to six π-molecular orbitals. The lowest, most stable molecular orbital, ψ_1, has no nodes and consists of all six p-orbitals overlapping around the ring. The next two molecular orbitals, ψ_2 and ψ_3, are not as bonding as ψ_1 and have higher energy. Each has one node; the node in ψ_2 is at right angles to that in ψ_3. These two molecular orbitals have identical energies and are therefore said to be **degenerate**. The three molecular orbitals designated as ψ_1, ψ_2, and ψ_3 are the occupied molecular orbitals in benzene; one pair of electrons can be put in each to accommodate all six π-electrons of benzene. The three remaining π-molecular orbitals have less bonding and higher energy; they are not occupied by electrons. The relative energies of all six π-molecular orbitals and the molecular orbitals themselves are represented in Figure 22.3.

A characteristic feature of this molecular orbital pattern is that the lowest-lying molecular orbital is a single molecular orbital; thereafter the molecular orbitals occur in pairs of equal energy until only one highest-lying level is left. These

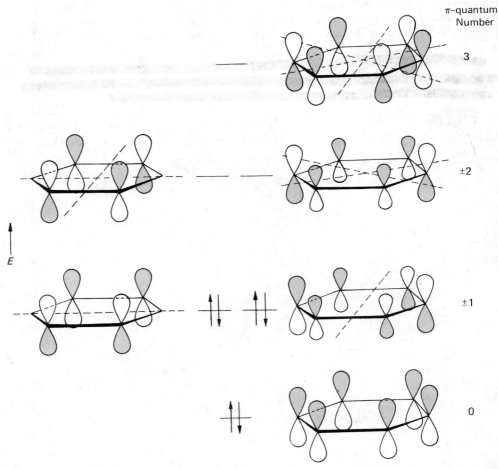

FIGURE 22.3. *The π-molecular orbitals and energy levels for benzene. Positive lobes are shaded. Nodes are indicated by dashed lines. Recall that overlapping wave functions of the same sign are bonding; wave functions separated by a node have opposite signs and are antibonding.*

molecular orbital levels could be identified by quantum numbers, 0, ± 1, ± 2, and so on. Each quantum number represents a **shell** of orbitals, much as we have $2s$- and $2p$-shells in atomic structure. In benzene the ± 1 shell is filled, and we can attribute the stability of benzene to this filled-shell structure in much the same way as the noble gases (helium, neon, argon, and so on) have stability associated with filled atomic orbital shells.

It is important to distinguish between the π-electronic system of benzene as symbolized commonly by a set of six p-orbitals overlapping as in Figure 22.1 and the molecular orbitals as symbolized in Figure 22.3. The necessity for the distinction comes about because the π-electronic system contains six electrons. According to the Pauli principle (Section 2.5), no two electrons can have the same quantum numbers. The six electrons can be divided into two groups of three based on electronic spin (a quantum number, $\pm\frac{1}{2}$), but each electron in the set of three must then belong to a different orbital. In the case of benzene such orbitals are the π-molecular orbitals characterized by quantum numbers of 0, $+1$, and -1. The pattern of six electrons in a ring gives benzene a special character known as **aromatic stability.** Examples of other aromatic systems, which are not based on the benzene ring, are discussed later in this chapter. Note that this use of the term *aromatic* has nothing whatsoever to do with smell. Although the term was first used to describe a class of compounds that had strong odors, it is now recognized that the special property setting benzene and related compounds apart from aliphatic compounds is the filled molecular orbital shell. Thus, the term "aromatic" has come to be associated in organic chemistry with the general class of compounds possessing cyclic π-electron systems that have this special stabilizing electronic character.

Many such "aromatic" compounds are known. They include derivatives of benzene in which one or more groups are attached to the ring. Examples are toluene and benzoic acid.

toluene   benzoic acid

There are polycyclic benzenoid compounds in which two or more benzene rings are fused together; examples are naphthalene and coronene.

naphthalene   coronene

There are also aromatic heterocyclic compounds, compounds in which one or more atoms other than carbon participate in the cyclic conjugated ring. Examples are pyridine and pyrrole.

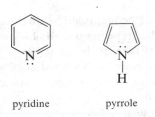

pyridine   pyrrole

All of these cases involve cyclic systems of six π-electrons.

D. *Symbols for the Benzene Ring*

Finally, the symbolism used for the benzene ring deserves further comment. We have discussed the electronic structure of benzene in terms of resonance structures and molecular orbital theory and with reference to hypothetical formulations of cyclohexatriene. We have used symbolic representations of molecular orbitals and various symbols based on hexagons. These symbols are all in common use in various contexts and may be summarized as follows.

The molecular orbital representation in Figure 22.3 is especially useful for understanding the high stability of the benzene ring, but it is too complex and cumbersome a symbolism for normal use. The hexagon with an inscribed circle is a simple and commonly used representation of the aromatic π-system and is especially useful for the representation of aromatic structures.

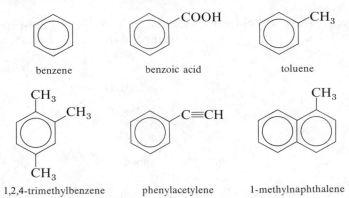

benzene   benzoic acid   toluene

1,2,4-trimethylbenzene   phenylacetylene   1-methylnaphthalene

However, this symbol has an important disadvantage in not allowing an accurate accounting of electrons; that is, it does not correspond to a Lewis structure. In all of our other structural representations, a bond symbolized by a straight line corresponds to two electrons. No such simple correspondence applies to the inscribed circle; for example, the circle in benzene corresponds to six π-electrons, whereas the two circles in naphthalene correspond to a total of ten π-electrons.

The alternating-double-bonds symbol does allow a simple and accurate accounting of electrons and does correspond to a Lewis structure.

This symbol is used frequently to represent the benzene ring, and the student must be wary not to read this symbol as that of cyclohexatriene—that is, as a

cyclic polyene. Generally, this symbol is used as a shorthand for a resonance hybrid of Kekulé structures.

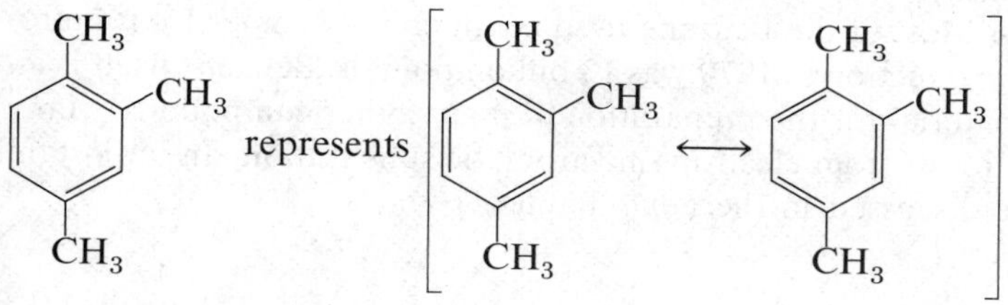

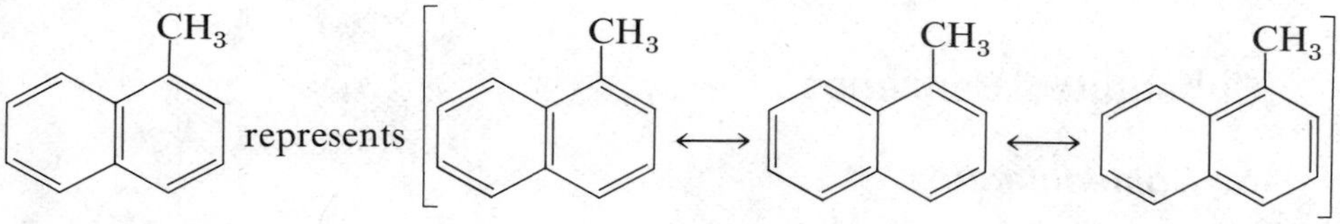

This is the symbol for benzene rings used generally throughout this textbook. We shall see later in this chapter and the next how this symbol lends itself readily to following the mechanisms of reactions at the benzene ring.

EXERCISE: Write the two Kekulé structures for benzene as Lewis structures showing all valence electrons.

E. *Formation of Benzene*

The high stability of the benzene ring is further demonstrated by reactions that produce this ring system. Cyclohexane rings can be **dehydrogenated** with suitable reagents or catalysts.

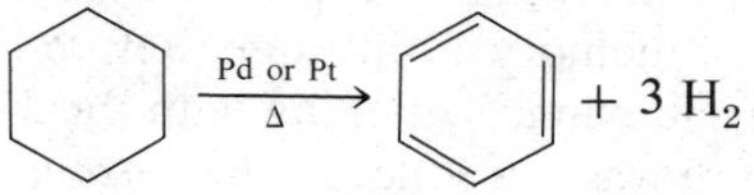

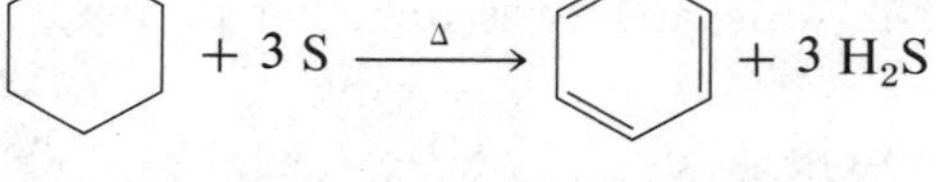

Dehydrogenation with cyclization can be accomplished from aliphatic hydrocarbons.

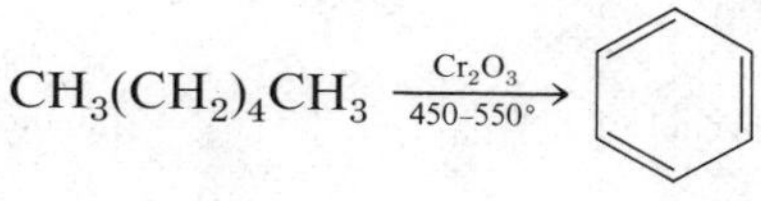

(20%)

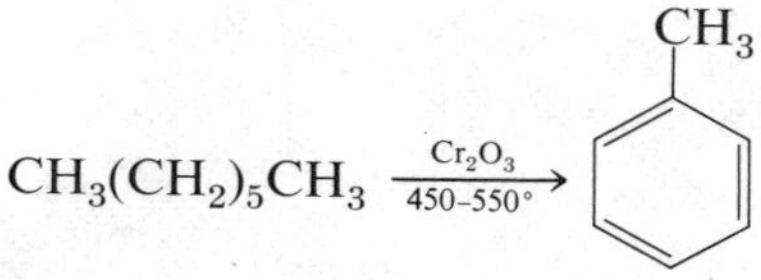

(36%)

Such reactions form the basis of the **hydroforming process** of petroleum refining. Gasoline fractions are heated with platinum catalysts **(platforming)** to produce mixtures of aromatic hydrocarbons by cyclization and dehydrogenation of aliphatic hydrocarbons. Most of the benzene used commercially comes from petroleum. United States production in 1979 was 13 billion pounds. Benzene itself is an important starting material for the preparation of many other compounds. Many of these compounds result from electrophilic aromatic substitution, an important reaction that will be discussed in the next chapter.

22.2
Substituted Benzenes

A. *Nomenclature*

Benzene derivatives are named in a systematic manner by combining the substituent prefix with the word benzene. The names are written as one word with no spaces. Since benzene has sixfold symmetry, there is only one monosubstituted benzene for each substituent, and no position number is necessary.

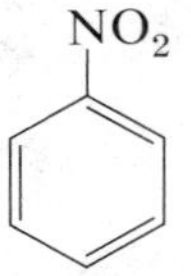

nitrobenzene 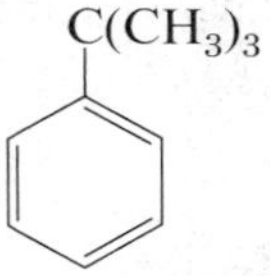   *t*-butylbenzene 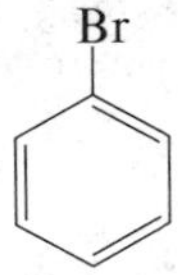   bromobenzene

A number of monosubstituted benzene derivatives have special names that are in such common use that they have IUPAC sanction. We shall refer to the following twelve compounds by their IUPAC-approved common names; thus the student should commit them to memory at this time. Before 1978 *Chemical Abstracts* also used these names for indexing. Beginning with the 1978 indices, however, a new system of nomenclature was introduced. The *Chemical Abstracts* names currently in use are shown in brackets. Although a scientist must be aware of these indexing names, they are not in widespread use for other purposes.

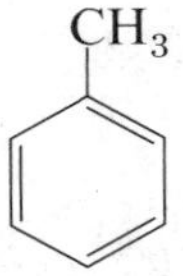

toluene
[methylbenzene]

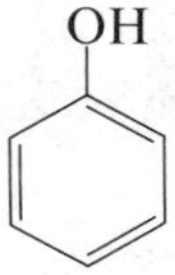

phenol
[phenol]

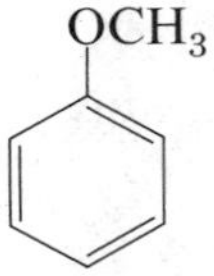

anisole
[methoxybenzene]

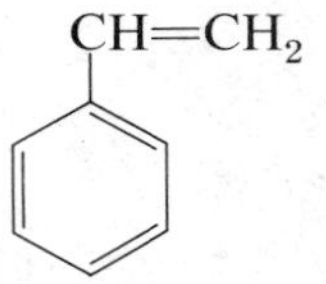

styrene
[ethenylbenzene]

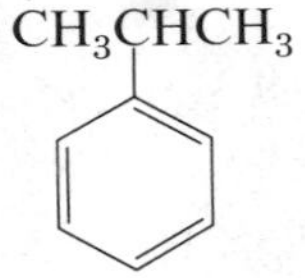

cumene
[1-methylethylbenzene]

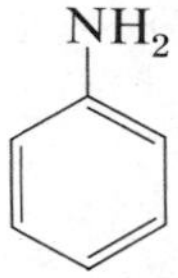

aniline
[benzenamine]

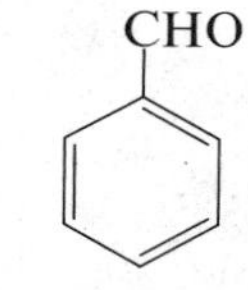

benzaldehyde
[benzaldehyde]

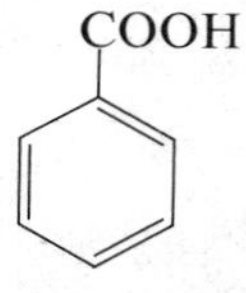

benzoic acid
[benzoic acid]

acetophenone
[1-phenylethanone]

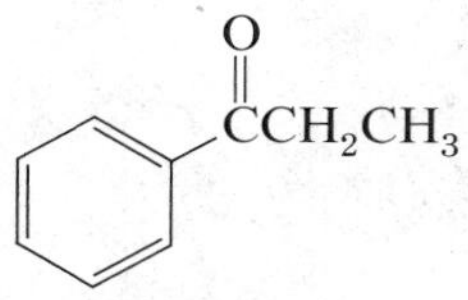

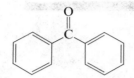

propiophenone
[1-phenylpropanone]

benzophenone
[diphenylmethanone]

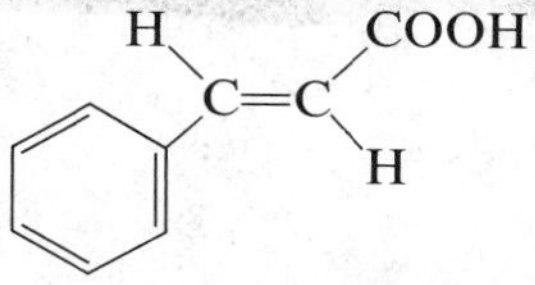

cinnamic acid
[(E)-3-phenylpropenoic acid]

When there are two or more substituents, some specification of position is required. The numbering system is straightforward.

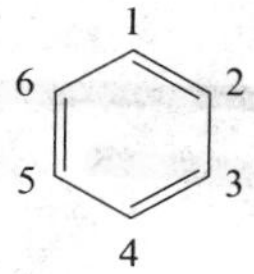

For disubstituted benzene derivatives, the three possible isomers are named using the Greek prefixes **ortho-, meta-,** and **para-** (often shortened to **o-, m-,** and **p-**).

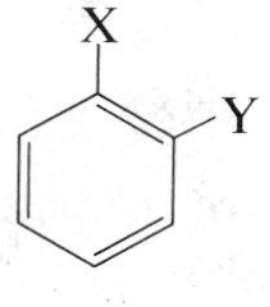

ortho- or *o-* 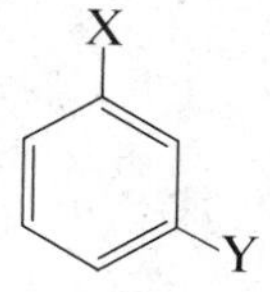   *meta-* or *m-* 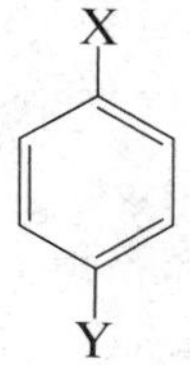   *para-* or *p-*

The following examples illustrate the use of these prefixes.

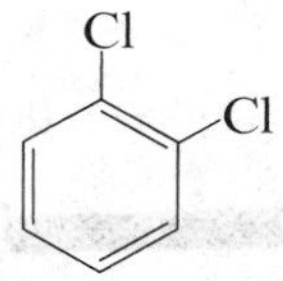

ortho-dichlorobenzene
o-dichlorobenzene

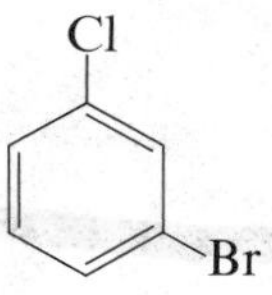

meta-bromochlorobenzene
m-bromochlorobenzene

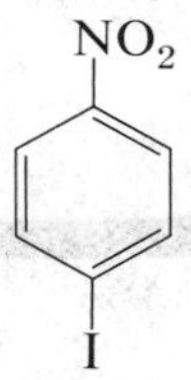

para-iodonitrobenzene
p-iodonitrobenzene

Note that the substituent prefixes are ordered alphabetically. When one of the substituents corresponds to a monosubstituted benzene that has a special name, the disubstituted compound is named as a derivative of that parent.

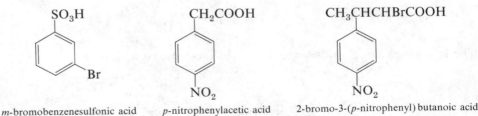

p-nitrotoluene *m*-chlorophenol *o*-bromoanisole

However, if the substituent introduced is the same as one already present, the compound is named as a derivative of benzene.

p-divinylbenzene   1,2,3-trimethylbenzene
(not *p*-vinylstyrene) (not 3-methyl-*o*-xylene)

When two different substituents in a disubstituted compound each correspond to a special parent, there is a difficulty in deciding which one to name as a derivative of the other. In such cases, the substituent that is normally treated as a suffix takes precedence. Some examples are

p-methylanisole *o*-aminobenzaldehyde *p*-aminophenol *m*-isopropylacetophenone

In many other cases, the systematic IUPAC name of the parent is in common use, and derivatives are named accordingly.

m-bromobenzenesulfonic acid *p*-nitrophenylacetic acid 2-bromo-3-(*p*-nitrophenyl) butanoic acid

For polysubstituted benzenes, the numbering system should be used.

1,3,5-tribromobenzene 2,4-dinitroanisole 2,4,6-trichlorophenol

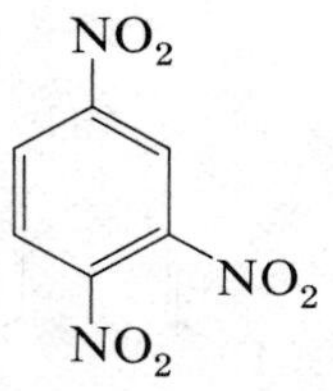

1,2,4-trinitrobenzene
(not 1,3,4-trinitrobenzene;
the lower numbers are used)

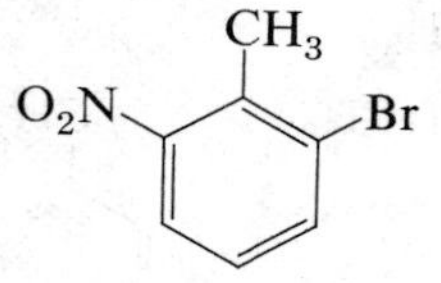

2-bromo-6-nitrotoluene
(prefixes are alphabetic)

Some di- and polysubstituted benzenes have common or trivial names that are widely used and should be learned by the student. Some of these special names follow; others will be brought up in subsequent chapters dealing with the chemistry of such compounds.

p-xylene

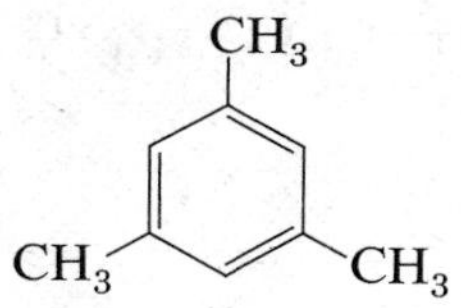

mesitylene
1,3,5-trimethylbenzene

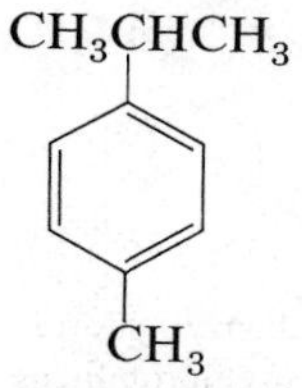

p-cymene
1-isopropyl-4-methylbenzene

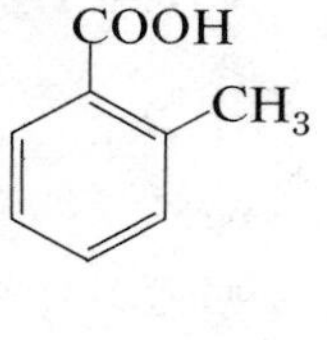

o-toluic acid
o-methylbenzoic
acid

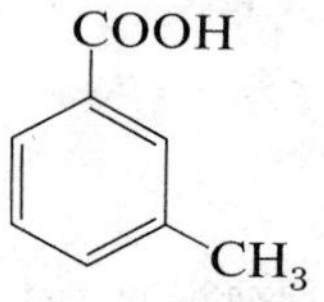

m-toluic acid
m-methylbenzoic
acid

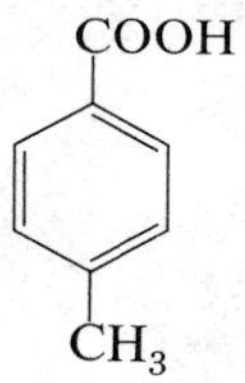

p-toluic acid
p-methylbenzoic
acid

Aromatic hydrocarbons have the generic name of **arene.** Accordingly, for many purposes the abbreviation Ar, for aryl, is used just as R is used for alkyl; thus, the symbol ArR refers to arylalkanes. We have already learned that for benzene itself the term **phenyl** is used. Examples of names employing this prefix are

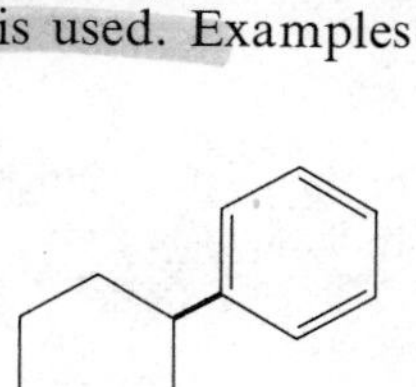

trans-2-phenylcyclohexane-
carboxylic acid

β-phenylpropionaldehyde
3-phenylpropanal

Similarly, derivatives of toluene, the xylenes, and mesitylene, where the additional substituent is attached to the ring, may be named by using the prefixes **tolyl-, xylyl-,** or **mesityl-.**

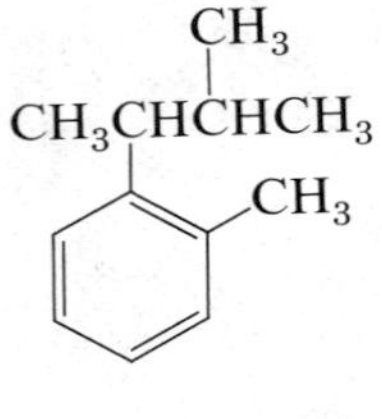

CH₃
CH₃CHCHCH₃
CH₃
2-methyl-3-*o*-tolylbutane

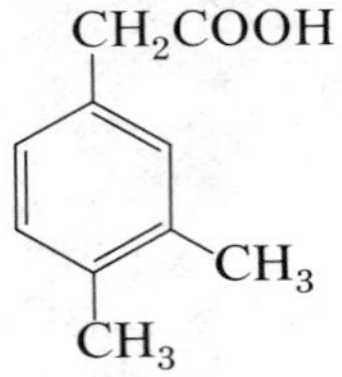

CH₂COOH
CH₃
CH₃
3,4-xylylacetic acid

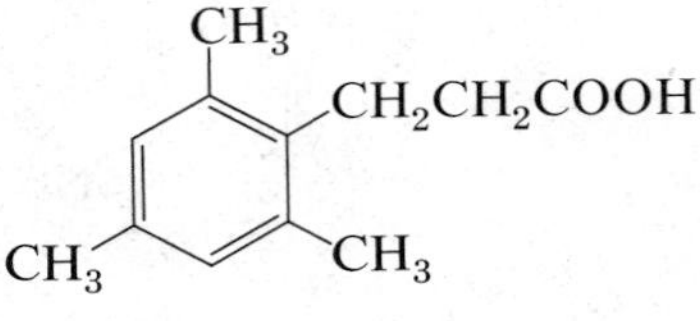

CH₃
CH₂CH₂COOH
CH₃
CH₃
3-mesitylpropanoic acid

Certain other group names are used for derivatives of these hydrocarbons when the substituent is attached to a side chain. Note in some of these cases how Greek letters are used to define the side-chain position relative to the benzene ring. The prefix **benzyl** for the phenylmethyl group is especially important.

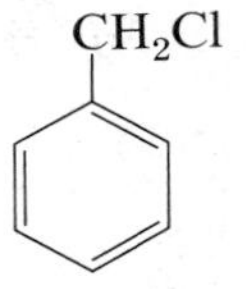

CH₂Cl

benzyl chloride
α-chlorotoluene

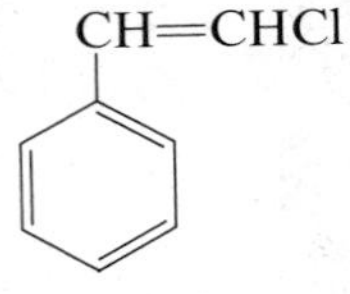

CH=CHCl

β-styryl chloride
β-chlorostyrene

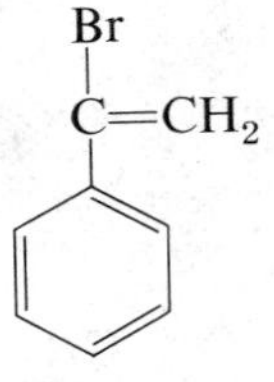

Br
C=CH₂

α-styryl bromide
α-bromostyrene

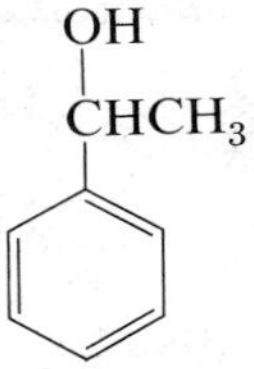

OH
CHCH₃

α-phenethyl alcohol

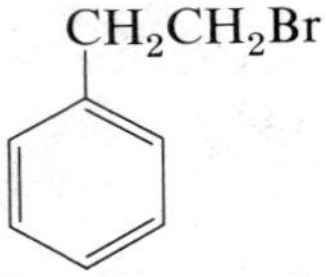

CH₂CH₂Br

β-phenethyl bromide

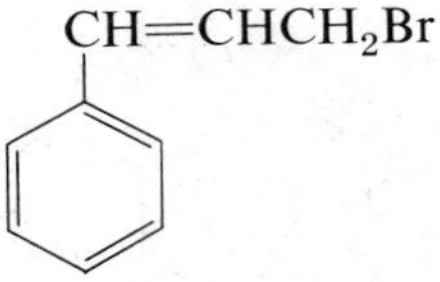

CH=CHCH₂Br

cinnamyl bromide

O
H   OCCH₃
C

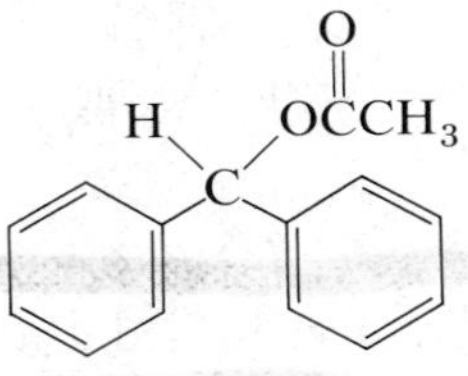

benzhydryl acetate
diphenylmethyl acetate

EXERCISE: Write the structures and names of the twelve methyl and polymethyl benzenes and the six (monochlorophenyl)propanoic acids.

B. *Körner's Absolute Method*

One of the classical methods for distinguishing *ortho-*, *meta-*, and *para*-disubstituted benzenes is based on a simple logical corollary of a hexagonally symmetrical benzene and is known as Körner's absolute method. The method depends on the following simple principles applied to disubstituted benzenes where both substituents are the same.

1. In a *para*-disubstituted benzene all four hydrogens are equivalent. Further substitution can only lead to a single trisubstituted benzene.

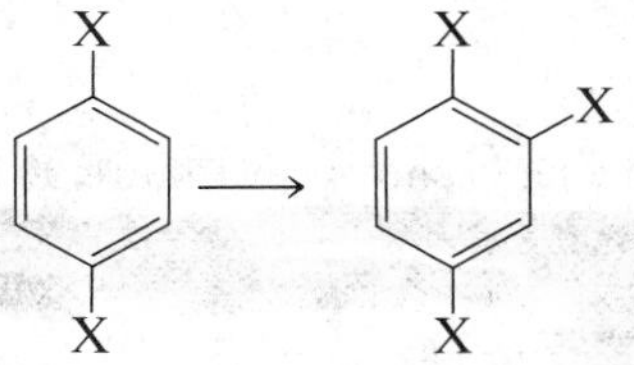

para   1,2,4-trisubstituted

2. In an *ortho*-disubstituted benzene there are two types of hydrogen, and further substitution can, in principle, lead to two isomeric trisubstituted benzenes.

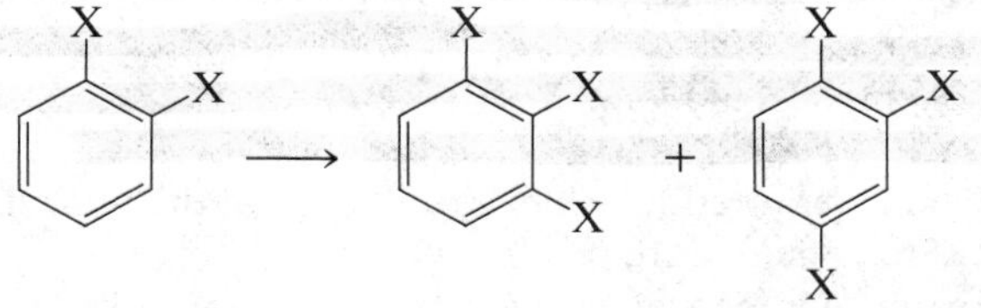

ortho   1,2,3-trisubstituted   1,2,4-trisubstituted

3. In a *meta*-disubstituted benzene there are three nonequivalent hydrogens, and further substitution can lead to three different trisubstituted benzenes.

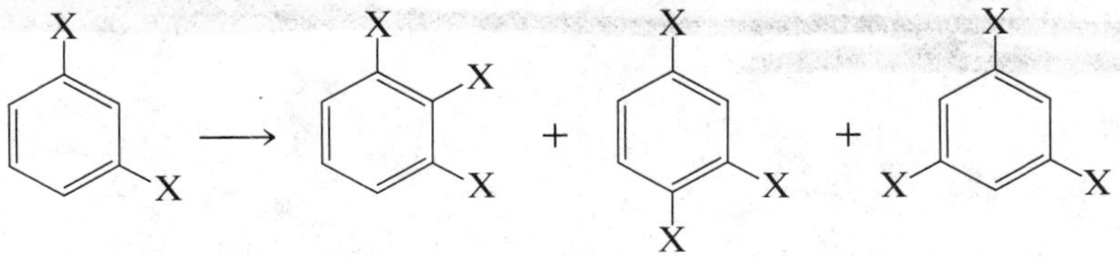

meta   1,2,3-trisubstituted   1,2,4-trisubstituted   1,3,5-trisubstituted

Furthermore, only one trisubstituted isomer, the 1,2,4-, is given by all three disubstituted compounds. In practice, the application of this method requires careful work because the possible products are generally not formed in comparable amounts. The existing groups on a benzene ring provide orientation specificity, to be discussed in detail later, that directs an incoming group to given positions. In practice, some isomers may be formed in such small quantity as to defy detection. This was especially true in earlier years when actual isolation was required. Nevertheless, by careful work the structures of many disubstituted benzenes were established early in the history of modern organic chemistry. By interconversion of functional groups, the structures of many other substituted benzenes could also be assigned.

Körner's absolute method is now only of historical interest as a valuable example of chemical logic. It is now generally feasible to convert an unknown benzene derivative into one of the many known derivatives or to use spectroscopic methods, primarily nmr and ir.

EXERCISE: Show how Körner's absolute method could be used to distinguish among the three possible trisubstituted benzenes (for example, 1,2,3-, 1,2,4-, and 1,3,5-trimethylbenzene).

22.3 Spectra

A. *Nmr Spectra*

The nmr spectrum of benzene is shown in Figure 22.4. Since the six hydrogens are equivalent, the spectrum consists of a single line. The unusual feature of the spectrum is the position of the singlet, $\delta = 7.27$ ppm. Recall that δ for olefinic protons is generally about 5 ppm (Section 11.3.B).

The downfield shift of the benzene hydrogens results from the cyclic nature of the π-electrons of the aromatic ring. This cyclic electronic system can be likened to a circular wire, which in a magnetic field produces a current around the ring. This current is exactly analogous to the current induced in the π-electrons of a double bond (Figure 22.5), except that in benzene this electron current extends around the ring rather than being localized in one double bond. The resulting **ring current** has an induced magnetic field that *adds to* the externally applied field at the protons, just as in the related case of olefinic hydrogens (Figure 22.5). Since a smaller applied field is required to achieve resonance at the nucleus of the proton, the net result is a downfield shift.

The effect is greater for a benzene ring than for a simple alkene, in part because the benzene ring has six π-electrons in the cycle. The proton resonance of substituted benzene rings occurs generally in the region of $\delta = 7$–8, a region in which few other kinds of proton resonances occur. Nmr peaks in this region are diagnostic for aromatic protons. Substituents have a normal type of effect: electronegative substituents generally cause a downfield shift, and electron-donating groups usually produce an upfield shift.

| δ for aromatic H, ppm: | 6.95 | 7.27 | 7.34 | 7.98 |

In some monosubstituted benzenes all five benzenoid hydrogens have approximately the same chemical shift, and the aromatic protons appear as a singlet. This is usually the case when the substituent is an alkyl group or some other group having approximately the same electronegativity as carbon. An example is toluene, the nmr spectrum of which is shown in Figure 22.6.

Also note in Figure 22.6 that the methyl group attached to the benzene ring resonates at $\delta = 2.32$ ppm, about 1.4 ppm downfield from the resonance position of a methyl group in an alkane. The main cause of this downfield shift is the diamagnetic anisotropy or ring current of the aromatic ring. The effect is not as great with the methyl group as it is for hydrogens directly attached to the ring because the methyl hydrogens are farther from the circulating electrons than the benzenoid hydrogens.

When the substituent in a monosubstituted benzene is sufficiently electronegative or electropositive relative to carbon, the *ortho, meta,* and *para* hydrogens have significantly different chemical shifts and the nmr spectrum becomes more complex. Such a spectrum is shown by nitrobenzene (Figure 22.7).

The spectra of disubstituted benzenes can sometimes be rather complex. In

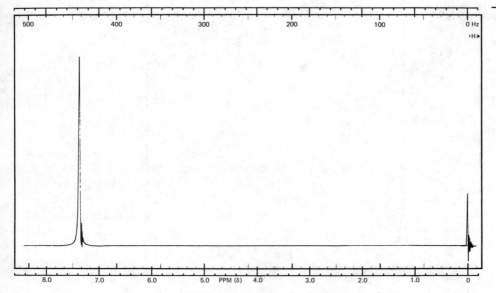

FIGURE 22.4. *Nmr spectrum of benzene.*

p-dichlorobenzene the four benzenoid hydrogens are equivalent, and the nmr spectrum is a sharp singlet (Figure 22.8). On the other hand, there are 24 lines in the nmr spectrum of *o*-dichlorobenzene (Figure 22.9). Some of these signals are of low intensity, and others are so close together that they appear merged if instrument resolution is inadequate. The analysis of such complex splitting patterns is beyond the scope of this text, but the student should know that such spectra can be analyzed and interpreted by experts to give structural information.

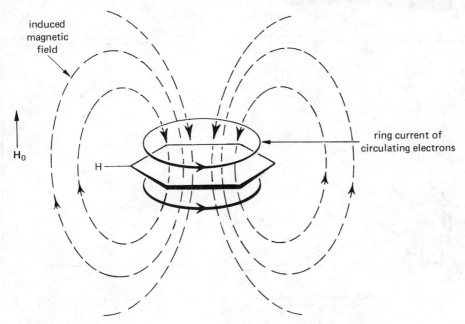

FIGURE 22.5. *Effect of ring current in benzene π-system increases effective magnetic field at the proton.*

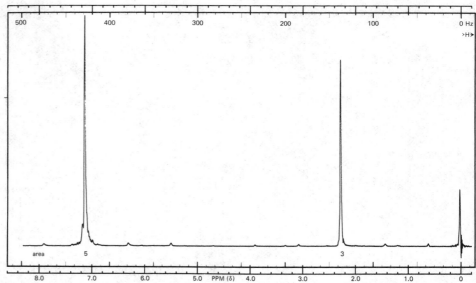

FIGURE 22.6. *Nmr spectrum of toluene,* $C_6H_5CH_3$.

When the two substituents are of different electronegativity, the nmr spectra are sometimes sufficiently simple as to be interpretable by a "first-order" approximation. An example is the spectrum of *p*-nitrotoluene, shown in Figure 22.10. To a first approximation, the benzenoid region in *p*-nitrotoluene may be regarded as a pair of doublets arising from coupling between the hydrogens on C-2 and C-3 with $J = 8$ Hz. Each doublet has an intensity of 2, relative to 3 for the methyl group because the hydrogens at C-2 and C-6 are equivalent and the hydrogens at C-3 and C-5 are equivalent.

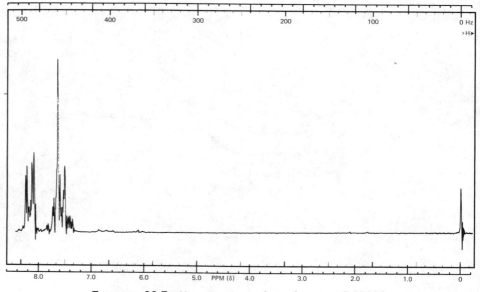

FIGURE 22.7. *Nmr spectrum of nitrobenzene,* $C_6H_5NO_2$.

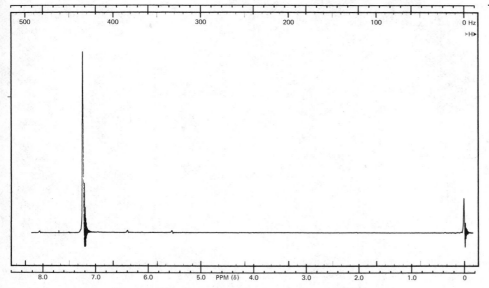

FIGURE 22.8. *Nmr spectrum of* p-*dichlorobenzene,* $p\text{-}Cl_2C_6H_4$.

$J = 8$ Hz

The cmr chemical shifts for benzene and several simple alkyl benzenes are collected in Table 22.1. The data point out an important difference between nmr and cmr spectra. *Diamagnetic anisotropy effects are of only minor importance in*

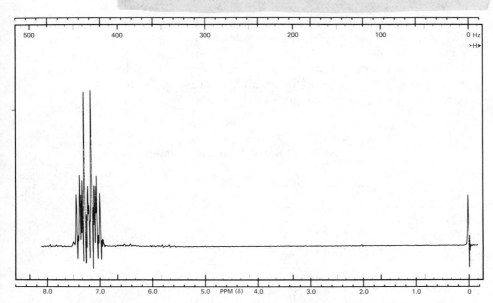

FIGURE 22.9. *Nmr spectrum of* o-*dichlorobenzene,* $o\text{-}Cl_2C_6H_4$.

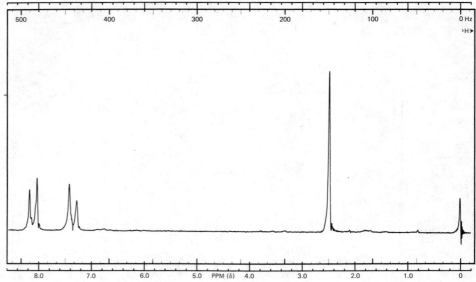

FIGURE 22.10. *Nmr spectrum of* p-*nitrotoluene,* p-$NO_2C_6H_4CH_3$.

determining carbon chemical shifts. For comparison, the chemical shifts of two
alkenes analogous to two of the aromatic compounds listed in Table 22.1 are
shown below.

Thus, the carbon chemical shifts of aromatic compounds for both the sp^2-carbons
of the ring itself and the sp^3-carbons bonded to the ring are similar to those of a
comparable alkene.

However, cmr spectroscopy can be useful in determining the substitution pattern on a benzene ring. Note, for example, that o-, m-, and p-xylene have cmr
spectra consisting of four, five, and three signals, respectively.

TABLE 22.1
Cmr Spectra of Some Aromatic Compounds

Compound	Aromatic Resonances					Side-Chain Resonances		
	C-1	C-2	C-3	C-4	C-5	C-1′	C-2′	C-3′
benzene	128.7							
toluene	137.8	129.3	128.5	125.6		21.3		
o-xylene	136.4		129.9	126.1		19.6		
m-xylene	137.5	130.1		126.4	128.3	21.3		
p-xylene	134.5	129.1				20.9		
mesitylene	137.6	127.4				21.2		
ethylbenzene	144.1	128.1	128.5	125.9		29.3	16.8	
n-propylbenzene	142.5	128.7	128.4	125.9		38.5	25.2	14.0

EXERCISE: Sketch the expected nmr and cmr spectra of (a) 1-chloro-3-nitroben-
zene and (b) 1-chloro-4-nitrobenzene.

B. *Infrared Spectra*

Aromatic C—H bonds have a stretching frequency at about 3030 cm^{-1}. This is
the same region as alkene C—H stretching bonds. Both bonds are close to
C_{sp^2}—H_{1s} in character. Aromatic C—H bonds also give rise generally to a series of
low-intensity combination and overtone bands at 2000–1660 cm^{-1} (5–6 μ). A further set of one or more intense bands in the 700–900 cm^{-1} region results from
out-of-plane bending. These last absorptions are especially valuable because
neighboring C—H's couple together to give bands whose absorption frequency is
characteristic of the number of vicinal or adjacent C—H's in the ring. For example, four adjacent C—H's, as in an *ortho*-disubstituted benzene, give absorption at
about 750 cm^{-1}, whereas three adjacent C—H's, as in a *meta*-disubstituted benzene, give absorption at somewhat higher frequency. Table 22.2 summarizes the
one or two ir absorption bands generally found in this region for different substitution patterns on the benzene ring.

C. *Ultraviolet Spectra*

The π-system of benzene has two highest occupied molecular orbitals, ψ_2 and
ψ_3, and two lowest vacant molecular orbitals, ψ_4 and ψ_5 (Figure 22.3). We might
expect to see four kinds of $\pi \rightarrow \pi^*$ transitions: $\psi_2 \rightarrow \psi_4$, $\psi_2 \rightarrow \psi_5$, $\psi_3 \rightarrow \psi_4$,
and $\psi_3 \rightarrow \psi_5$. These four transitions all correspond to the same energy, and for
this type of situation there is a breakdown in our simple picture of an electronic
excitation as involving the transition from one molecular orbital to another. Several low-lying excited states of benzene exist that we would have to describe as
various composites of the four simple transitions described above. An adequate
treatment of the ultraviolet spectrum of benzene requires a rather complex quantum-mechanical discussion, which we will not develop. The longest-wavelength
absorption of benzene gives a series of sharp bands centered at 255 nm with

TABLE 22.2
Infrared Absorption of
Benzene Derivatives

Substitution	Out-of-Plane Bending Frequencies, cm^{-1}
mono-	770–730, 710–690
ortho-	770–735
meta-	810–750, 710–690
para-	840–810
1,2,3-	780–760, 745–705
1,3,5-	865–810, 730–675
1,2,4-	825–805, 885–870
1,2,3,4-	810–800
1,2,4,5-	870–855
1,2,3,5	850–840
penta-	870

$\epsilon = 230$, a relatively low intensity for a $\pi \rightarrow \pi^*$ transition. This low value results from the high symmetry of benzene, which gives this absorption a relatively low probability. Such an absorption is called **symmetry-forbidden.**

A vinyl group attached to a benzene ring constitutes a conjugated system. Styrene has two principal absorption bands: λ_{max} 244 nm with $\epsilon = 12,000$ and λ_{max} 282 nm with $\epsilon = 450$. The more intense band is a polyene type of $\pi \rightarrow \pi^*$ transition, whereas the less intense band corresponds to a substituted benzene.

1,2-Diphenylethylene (stilbene) allows another comparison of *cis* and *trans* isomers (page 626).

trans-stilbene
$\lambda_{max} = 295$ nm
$\epsilon = 27,000$

cis-stilbene
$\lambda_{max} = 280$ nm
$\epsilon = 13,500$

trans-Stilbene has no significant steric interactions. The compound has an extended coplanar π-system. In *cis*-stilbene, however, the two phenyl groups are on the same side of the double bond and sterically interfere with each other. The rings cannot both be coplanar with the double bond, and π-conjugation is not as effective as it is in the *trans* isomer. The result is a small change in λ_{max} but a large decrease in the extinction coefficient.

Ultraviolet spectroscopy is a useful method for determining the overall π-electronic system of a molecule, particularly with polycyclic aromatic hydrocarbons (Chapter 31), but this method is not generally useful for determining the substitution pattern in simple substituted benzenes.

22.4
Dipole Moments in Benzene Derivatives

Methyl chloride has a dipole moment of 1.94 D in the gas phase. The experimental measurement of the dipole moment gives only its magnitude and not its direction. Nevertheless, there is no doubt that the dipole moment in methyl chloride is oriented from carbon to chlorine.

$$CH_3 - Cl$$
$$\overset{\longrightarrow}{\vdash}$$
$$1.94 \text{ D}$$

This orientation agrees with quantum-mechanical calculations and with spectroscopic interpretations of related compounds. Chlorobenzene has a dipole moment of 1.75 D in the gas phase. The direction of the dipole is undoubtedly also from carbon to chlorine

$$\overset{\longrightarrow}{\vdash}$$
$$1.75 \text{ D}$$

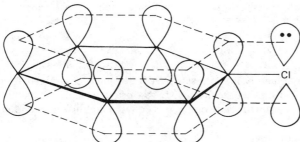

FIGURE 22.11. *Conjugation of a chlorine lone pair with the benzene π-system. Actually, a chlorine 3p-orbital is involved, and the additional node is omitted for simplicity.*

The magnitude of the dipole moment of chlorobenzene is smaller than that of methyl chloride for two reasons. The C—Cl bond in methyl chloride may be represented approximately as C_{sp^3}—Cl_p. The bond in chlorobenzene is approximately C_{sp^2}—Cl_p. The higher s-character of the benzene orbital makes it more electronegative than an sp^3-orbital; hence, the electronegativity difference with the more electronegative chlorine orbital is reduced. The second contribution to the reduced dipole moment in chlorobenzene results from conjugation of one of the chlorine lone pairs with the benzene π-system, illustrated in Figure 22.11. The lone pair is actually part of the π-system and may be represented in terms of resonance structures by

The effect of conjugation is small; the ionic structures contribute only a slight amount to the overall electronic structure. This slight conjugation, however, is equivalent to a dipole moment for the π-system, μ_π, oriented in the opposite direction from that associated with the C—Cl σ-bond, μ_σ. The net dipole moment is the vector sum and is less than that of μ_σ alone.

Dipole moments of some other benzene derivatives in the gas phase are summarized in Table 22.3. The dipole moments of multiply substituted benzenes are generally close to the vector sum of the constituent dipoles. *p*-Dichlorobenzene has a net dipole moment of zero because the two component C—Cl dipoles oppose and cancel each other.

$$\text{net } \mu = 1.75 + (-1.75) = 0$$

Because of the geometry of the hexagonal benzene ring, *ortho* and *meta* vector sums are given simply as

$$\mu = (\mu_1{}^2 + \mu_2{}^2 \pm \mu_1\mu_2)^{1/2} \qquad \begin{array}{l} + \ ortho \\ - \ meta \end{array}$$

This equation generally is quite satisfactory for *meta* groups but frequently inadequate for *ortho* groups. *Ortho* groups are so close to each other that electronic effects are mutually perturbed. For example, this equation applied to *o*- and *m*-dichlorobenzene gives

$$\mu_o = [1.75^2 + 1.75^2 + (1.75)(1.75)]^{1/2} = 3.03 \text{ D}$$
$$\mu_m = [1.75^2 + 1.75^2 - (1.75)(1.75)]^{1/2} = 1.75 \text{ D}$$

The *meta* result is close to the experimental value of 1.68 D, but the calculated *ortho* value is substantially higher than the experimental value of 2.52 D.

Toluene has a small but distinct dipole moment of 0.37 D. We note from the data in Table 22.3 that the dipole moment of *p*-chlorotoluene is approximately that of the sum of the dipole moments of toluene and chlorobenzene; hence, both component dipoles are operating in the same direction.

$$CH_3 - \underset{\longrightarrow}{} - Cl$$

$$\overset{0.37 \text{ D}}{\longrightarrow} \qquad \overset{1.75 \text{ D}}{\longrightarrow}$$

$$0.37 + 1.75 = 2.12 \text{ D} \qquad (\text{experimental } \mu = 2.21 \text{ D})$$

The dipole moment of toluene results in part from the character of the C_{methyl}—C_{ring} bond. This bond can be described approximately as C_{sp^3}—C_{sp^2}. The sp^2-orbital is more electronegative than the sp^3-orbital and produces an electronic displacement corresponding to the direction of the dipole moment indicated for toluene.

$$CH_3 - \underset{\longrightarrow}{\bigcirc}$$

TABLE 22.3
Dipole Moments of Substituted Benzenes

Compound	μ, D (gas phase)	Compound	μ, D (gas phase)
C_6H_6	0	p-$C_6H_4Cl_2$	0
C_6H_5F	1.63	o-$CH_3C_6H_4Cl$	1.57
C_6H_5Cl	1.75	p-$CH_3C_6H_4Cl$	2.21
C_6H_5Br	1.72	o-$CH_3C_6H_4F$	1.35
C_6H_5I	1.71	m-$CH_3C_6H_4F$	1.85
$C_6H_5CH_3$	0.37	p-$CH_3C_6H_4F$	2.01
$C_6H_5NO_2$	4.28	m-$ClC_6H_4NO_2$	3.72
o-$C_6H_4Cl_2$	2.52	p-$ClC_6H_4NO_2$	2.81
m-$C_6H_4Cl_2$	1.68		

The same approach applied to nitrobenzene derivatives shows that the direction of the dipole in nitrobenzene is away from the benzene ring.

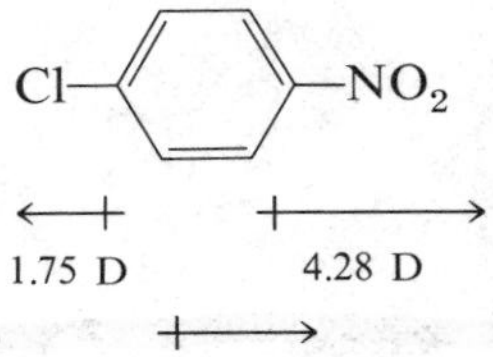

$$Cl-\langle\text{ring}\rangle-NO_2$$

$\xleftarrow{\quad}+\qquad\quad+\xrightarrow{\quad\quad}$

1.75 D   4.28 D

$+\xrightarrow{\quad\quad}$

$\mu_{net} = 4.28 - 1.75 = 2.53$ D   (experimental $\mu = 2.81$ D)

The direction thus derived for the dipole moment in nitrobenzene is what we would have expected from the electronic structure of the nitro group

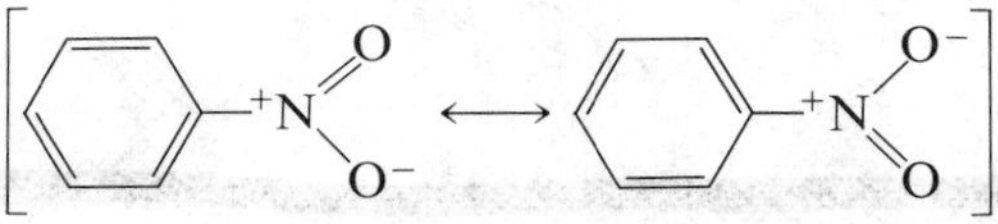

$$\left[\langle\text{ring}\rangle-\overset{+}{N}\begin{matrix}O\\O^-\end{matrix} \longleftrightarrow \langle\text{ring}\rangle-\overset{+}{N}\begin{matrix}O^-\\O\end{matrix} \right]$$

The relatively high magnitude of μ for nitrobenzene also follows from the formal charges required in the Lewis structures for the nitro group.

EXERCISE: Calculate the dipole moment expected for *p*-fluorotoluene by vector addition and compare with the experimental result in Table 22.3.

22.5
Side-Chain Reactions

A. *Free Radical Halogenation*

Benzene itself does not undergo the type of free radical chlorination typical of alkanes. The bond-dissociation energy of the phenyl–hydrogen bond is rather high ($DH° = 110$ kcal mole^{-1}), undoubtedly because the bond involved is C_{sp^2}—H_{1s} and has extra *s*-character. Consequently, the hydrogen transfer reaction is endothermic.

$$C_6H_5—H + Cl\cdot \rightleftharpoons C_6H_5\cdot + HCl \qquad \Delta H° = +7 \text{ kcal mole}^{-1}$$

Instead, chlorine atoms tend to add to the ring with the ultimate formation of a hexachlorocyclohexane.

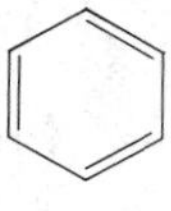

$$\langle\text{ring}\rangle + Cl_2 \xrightarrow{h\nu} C_6H_6Cl_6$$

benzene hexachloride
1,2,3,4,5,6-hexachlorocyclohexane

Eight geometric isomers are possible. The so-called γ-isomer (gammexane, lindane) has insecticidal properties and constitutes 18% of the mixture.

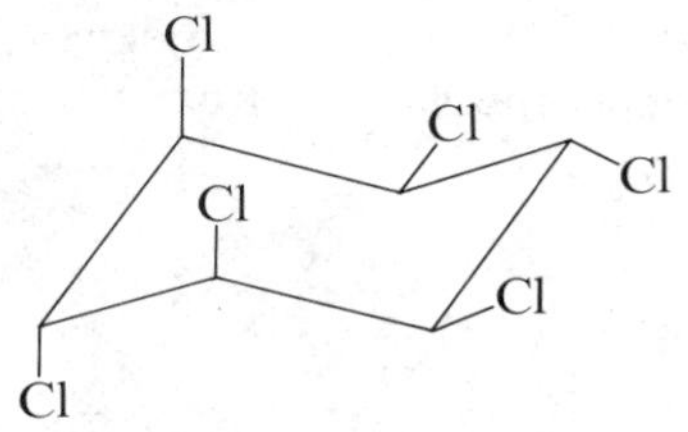

γ-benzene hexachloride

Reaction of the benzene hexachlorides with hot alcoholic potassium hydroxide gives 1,2,4-trichlorobenzene.

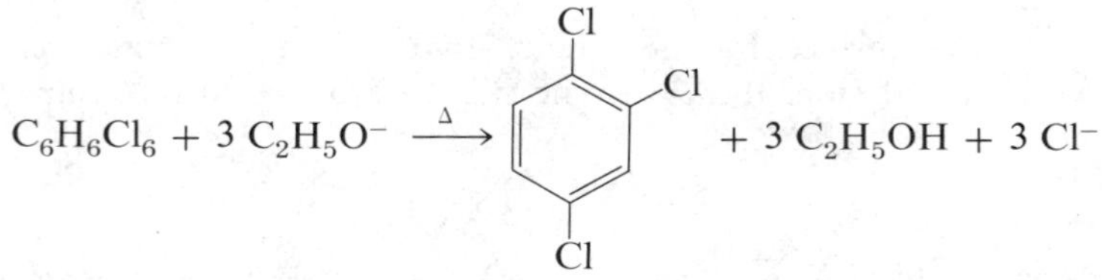

$$C_6H_6Cl_6 + 3\ C_2H_5O^- \xrightarrow{\Delta} \qquad + 3\ C_2H_5OH + 3\ Cl^-$$

In contrast, toluene undergoes smooth free radical chlorination on the methyl group to give benzyl chloride. Benzyl chloride undergoes further halogenation to give benzal chloride and benzotrichloride.

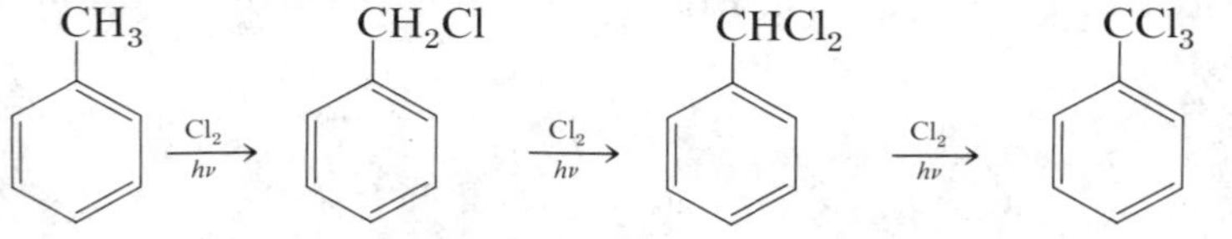

benzyl chloride   benzal chloride   benzotrichloride

The extent of chlorination may be controlled by monitoring the amount of chlorine used.

The reaction of toluene with chlorine atoms occurs exclusively at the methyl group because of the low bond–dissociation energy of the benzyl–hydrogen bond ($DH° = 87$ kcal mole^{-1}).

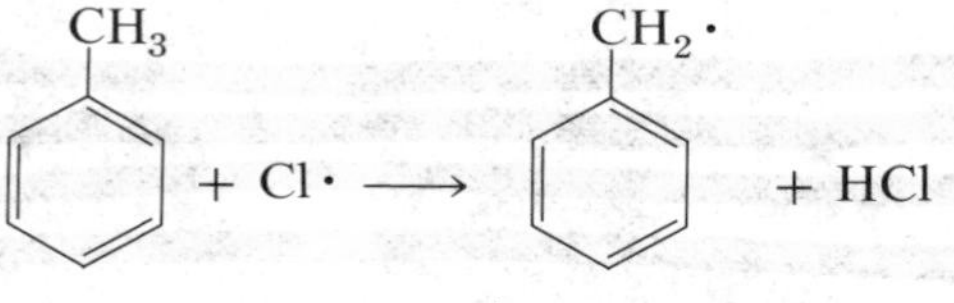

$$+ Cl\cdot \longrightarrow \qquad + HCl \qquad \Delta H° = -16 \text{ kcal mole}^{-1}$$

benzyl radical

The benzyl radical is especially stable for the same reason the allyl radical is stabilized (Section 20.1.D). Delocalization of the odd electron into the ring spreads out and diffuses the free radical character of the molecule. This conjugation can be represented by orbital overlap between the carbon $2p$-orbital containing the odd electron and the ring π-system, as in Figure 22.12.

Alternatively, the conjugated system can be represented by resonance structures.

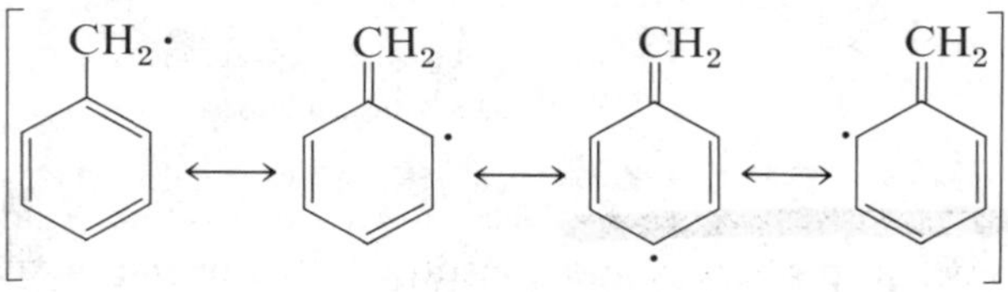

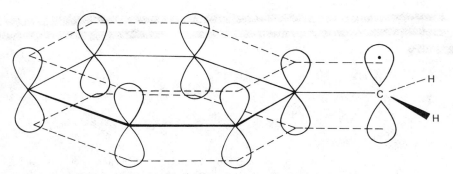

FIGURE 22.12. *Delocalization of the benzyl radical.*

In the next step of the chain halogenation reaction, benzyl radical reacts with chlorine to regenerate a chlorine atom, which then continues the chain.

Note that benzyl radical reacts exclusively at the exocyclic position. The *ortho* and *para* positions do have odd-electron character, but reaction at these positions produces a chloride that does not have the aromatic stability of a benzene ring. The resulting effect on the thermodynamics of reaction is substantial.

$$\Delta H° = -14 \text{ kcal mole}^{-1}$$

$$\Delta H° \approx +24 \text{ kcal mole}^{-1}$$

The free radical chain bromination of toluene is exactly analogous and is a suitable route to benzyl bromide. With xylene the two methyl groups undergo successive halogenation.

(48–53%)

α,α'-dibromo-*o*-xylene

(74–80%)

α,α,α',α'-tetrabromo-*o*-xylene

With the higher alkylbenzenes chlorination is limited in its synthetic utility because reaction along the alkyl chain occurs in addition to reaction at the benzylic position.

CH$_2$CH$_3$ + Cl$_2$ $\xrightarrow{h\nu}$ CHClCH$_3$ +   CH$_2$CH$_2$Cl

(56%)   (44%)

α-chloroethylbenzene   β-chloroethylbenzene

1-chloro-1-phenylethane   1-chloro-2-phenylethane

However, bromination occurs exclusively at the benzylic position.

CH$_2$CH$_3$ + Br$_2$ $\xrightarrow{h\nu}$ CHBrCH$_3$

This difference in behavior again reflects the greater reactivity of chlorine atoms compared to bromine; recall that bromine generally is a more selective reagent than chlorine (Section 6.3.B).

EXERCISE: Show how cumene could be converted into 2-phenylpropene.

B. *Benzylic Displacement and Carbocation Reactions*

The reactions of phenylalkyl systems are more or less comparable to those of analogous alkyl systems—halides undergo displacements, eliminations, formation of Grignard reagents, and so on. However, when the halogen is α to a benzene ring, the compounds are especially reactive. Benzyl halides are generally at least 100 times as reactive as ethyl halides in S$_N$2 displacement reactions. This high reactivity is attributed to conjugation of the ring π-electrons in the transition state (Figure 22.13). Accordingly, such displacement reactions are straightforward and facile. Some examples are

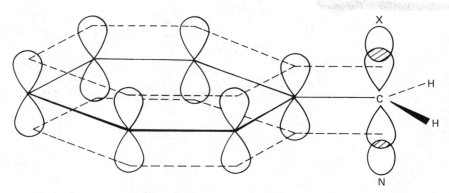

FIGURE 22.13. *Transition state for S$_N$2 reaction with a benzyl halide.*

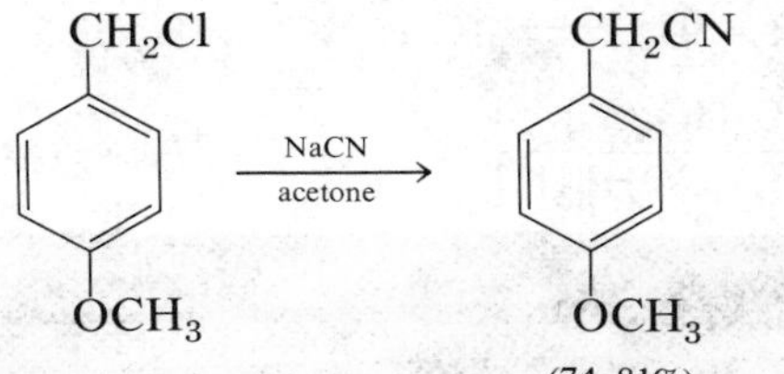

(74–81%)
p-anisylacetonitrile

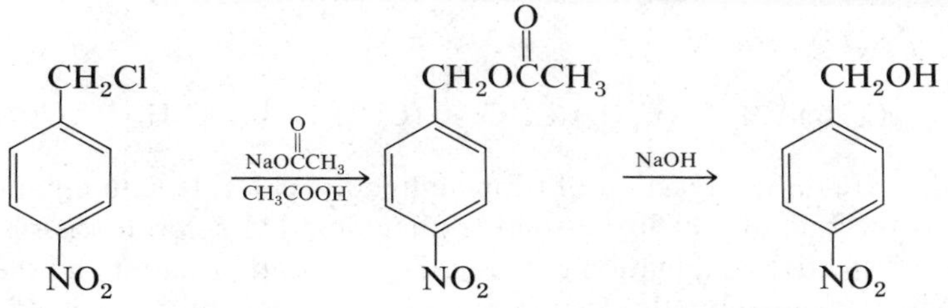

p-nitrobenzyl chloride

(64–71%)
p-nitrobenzyl alcohol

Benzylic compounds also react rapidly by the S_N1 mechanism because of the relative stability of the benzyl cation.

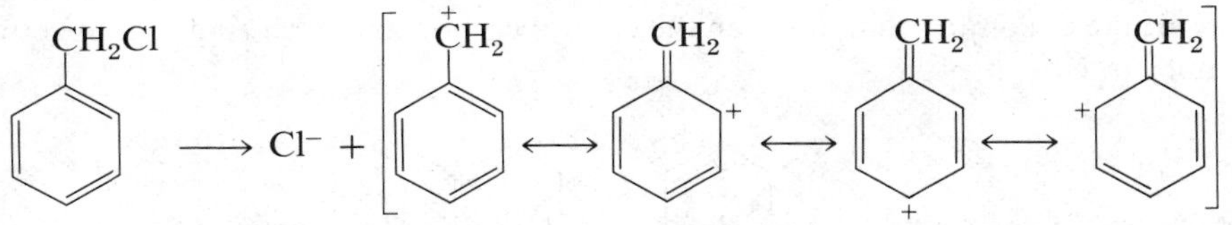

In fact, the gas-phase enthalpy of ionization of benzyl chloride is more comparable to that of secondary alkyl chlorides than to that of primary alkyl chlorides.

	$\Delta H°$, $kcal\ mole^{-1}$
$C_6H_5CH_2Cl \rightleftharpoons C_6H_5CH_2^+ + Cl^-$	154
$C_2H_5Cl \rightleftharpoons C_2H_5^+ + Cl^-$	191
$(CH_3)_2CHCl \rightleftharpoons (CH_3)_2CH^+ + Cl^-$	167
$(CH_3)_3CCl \rightleftharpoons (CH_3)_3C^+ + Cl^-$	151

When the carbocation center is conjugated with two or three benzene rings, the positive charge is distributed to a still greater extent. For example, triphenylmethyl cation has ten resonance structures in which the charge is spread to six *ortho* and three *para* positions. Consequently, triphenylmethyl chloride ionizes readily and shows exceptional reactivity. A liquid sulfur dioxide solution is colored yellow and conducts electricity because of the triphenylmethyl cations and chloride ions present.

$$(C_6H_5)_3CCl \xrightleftharpoons{SO_2} (C_6H_5)_3C^+ + Cl^-$$
(yellow color)

$$K = \frac{[(C_6H_5)_3C^+][Cl^-]}{[(C_6H_5)_3CCl]} = 4 \times 10^{-5}\ M$$

Similarly, triphenylmethanol is converted into substantial amounts of triphenylmethyl cation in strong aqueous sulfuric acid.

$$(C_6H_5)_3COH + H^+ \rightleftharpoons (C_6H_5)_3C^+ + H_2O$$

$$K = \frac{[(C_6H_5)_3C^+]}{[(C_6H_5)_3COH][H^+]} = 2 \times 10^{-7} M^{-1}$$

To give some idea of relative magnitudes of reactivity, benzyl chloride undergoes S_N1-type reactions much more slowly than t-butyl chloride, diphenylmethyl chloride is 10^1 to 10^3 times faster than t-butyl chloride, and triphenylmethyl chloride is 10^6 to 10^7 times more reactive than t-butyl chloride.

Order of S_N1 reactivity

$$(C_6H_5)_3CCl > (C_6H_5)_2CHCl > (CH_3)_3CCl > C_6H_5CH_2Cl$$

In fact, the rate of S_N1 reaction of triphenylmethyl chloride with ethanol is comparable to the rate at which the solid triphenylmethyl chloride dissolves.

The most effective conjugation between the carbocation center and the benzene π-electrons in triphenylmethyl cation requires that the whole molecule be coplanar. In this type of structure, however, the *ortho* hydrogens of the phenyl groups are only about 0.5 Å apart; the resulting steric repulsion forces the rings to tilt apart. The actual structure of triphenylmethyl cation is that of a three-bladed propeller, as shown by the stereo plot of tris-(*p*-aminophenyl)methyl cation in Figure 22.14. This twisting of the phenyl groups does somewhat diminish the magnitude of conjugation between the central carbon and each ring, but the effect is not large.

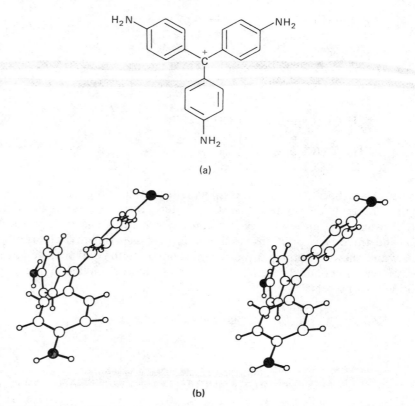

(a)

(b)

FIGURE 22.14. *Tris*(p-aminophenyl)methyl cation: (a) conventional structure; (b) stereo structure.

The high reactivity of benzylic compounds in displacement and carbocation reactions is also seen in reactions of benzyl alcohols.

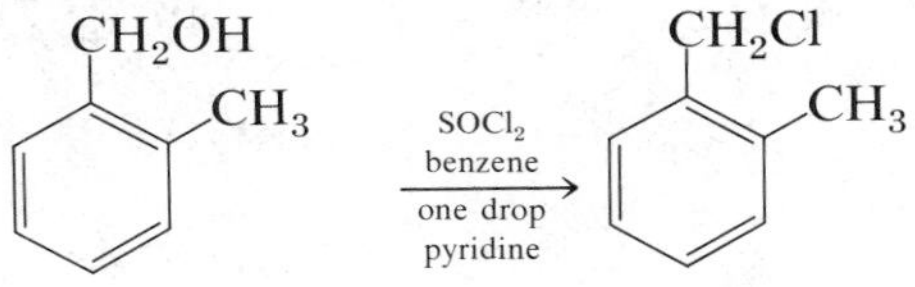

o-methylbenzyl alcohol $\xrightarrow[\substack{\text{benzene} \\ \text{one drop} \\ \text{pyridine}}]{\text{SOCl}_2}$ α-chloro-o-xylene (75–89%)

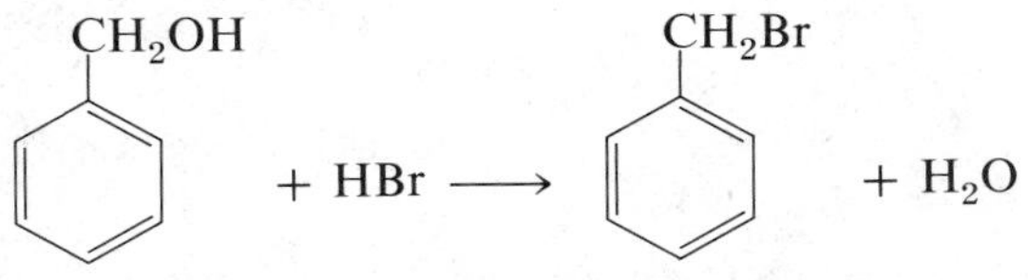

benzyl alcohol + HBr ⟶ benzyl bromide + H₂O

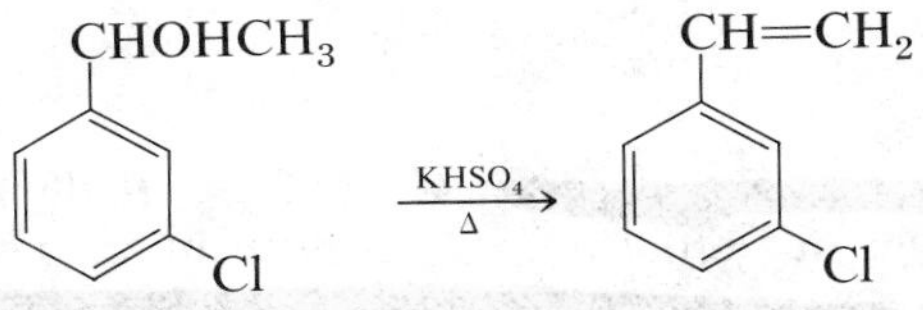

1-(m-chlorophenyl)ethanol $\xrightarrow[\Delta]{\text{KHSO}_4}$ m-chlorostyrene (80–82%)

EXERCISE: Write the resonance structures for m- and p-methylbenzyl cations and predict which of the isomeric cations is the more stable.

C. Oxidation

Like allylic alcohols (Section 20.3.A), benzylic alcohols may be oxidized to corresponding aldehydes or ketones using manganese dioxide, a reagent that does not bring about the oxidation of normal alcohols.

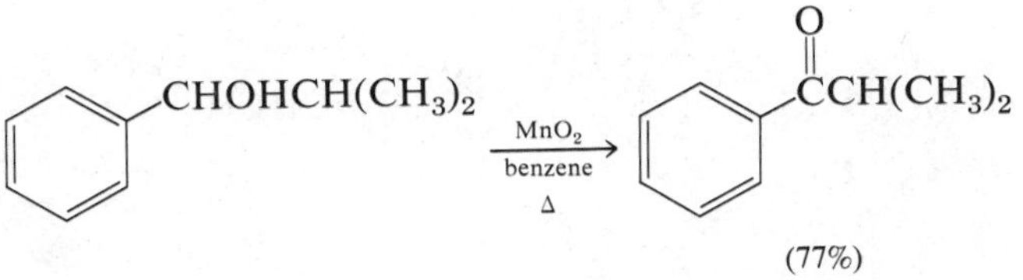

$\xrightarrow[\substack{\text{benzene} \\ \Delta}]{\text{MnO}_2}$ isobutyrophenone (77%)

The benzene ring is rather stable to oxidizing agents, and under appropriate conditions side-chain alkyl groups are oxidized instead. Sodium dichromate in aqueous sulfuric acid or acetic acid is a common laboratory procedure, but aqueous nitric acid and potassium permanganate have also been used.

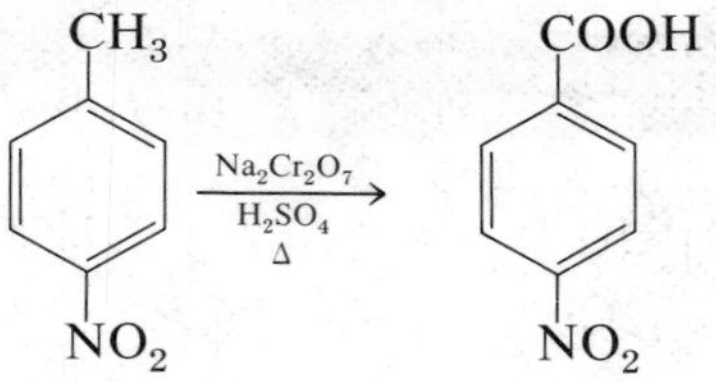

$$\underset{NO_2}{\underset{CH_3}{\bigcirc}} \xrightarrow[\substack{H_2SO_4 \\ \Delta}]{Na_2Cr_2O_7} \underset{NO_2}{\underset{COOH}{\bigcirc}}$$

(82–86%)
p-nitrobenzoic acid

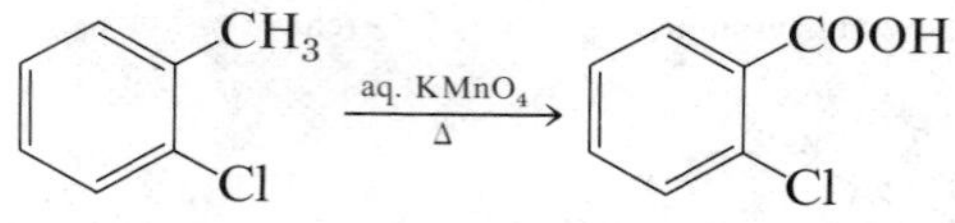

$$\underset{Cl}{\underset{CH_3}{\bigcirc}} \xrightarrow[\Delta]{aq.\ KMnO_4} \underset{Cl}{\underset{COOH}{\bigcirc}}$$

(76–78%)
o-chlorobenzoic acid

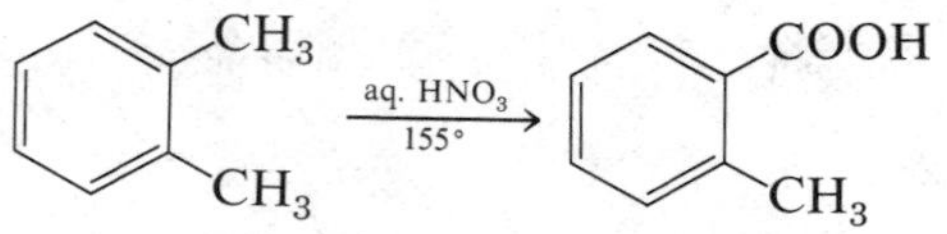

$$\underset{CH_3}{\underset{CH_3}{\bigcirc}} \xrightarrow[155°]{aq.\ HNO_3} \underset{CH_3}{\underset{COOH}{\bigcirc}}$$

(53–55%)
o-toluic acid

The detailed reaction mechanisms by which these oxidations occur are rather complex. They involve numerous intermediates including chromate and permanganate esters, but they also appear to involve an intermediate benzyl cation.

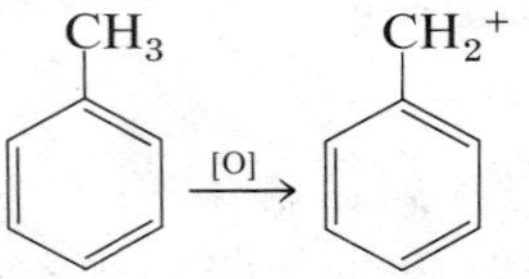

$$\underset{}{\underset{CH_3}{\bigcirc}} \xrightarrow{[O]} \underset{}{\underset{CH_2{}^+}{\bigcirc}}$$

As we have seen, this carbocation is relatively stable because of conjugation of the positive charge with the benzene ring. Reaction with water yields benzyl alcohol, which can oxidize further. Larger side chains can also be oxidized completely so long as there is one benzylic hydrogen for the initial oxidation. Cleavage reactions of larger side chains probably involve the formation of an intermediate alkene.

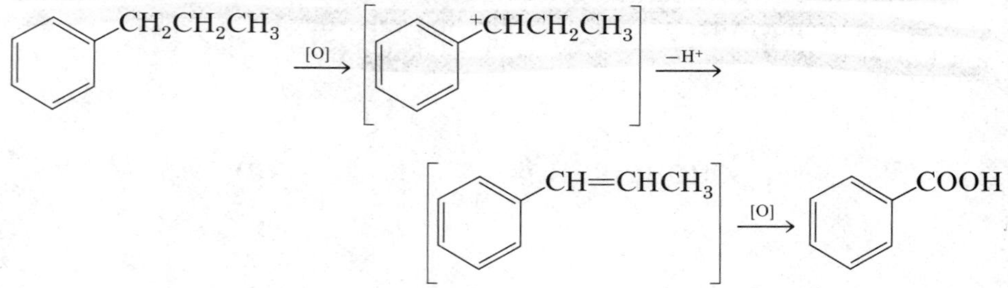

$$\underset{}{\underset{CH_2CH_2CH_3}{\bigcirc}} \xrightarrow{[O]} \left[\underset{}{\underset{{}^+CHCH_2CH_3}{\bigcirc}}\right] \xrightarrow{-H^+}$$

$$\left[\underset{}{\underset{CH=CHCH_3}{\bigcirc}}\right] \xrightarrow{[O]} \underset{}{\underset{COOH}{\bigcirc}}$$

benzoic acid

The more extensive oxidation required in these reactions often results in lower yields so that they are not so useful for laboratory preparations as they are for structural identification. When there is no benzylic hydrogen, the side chain resists oxidation. For example, vigorous conditions are required for the oxidation of *t*-butylbenzene, and the product is trimethylacetic acid, the product of oxidation of the benzene ring.

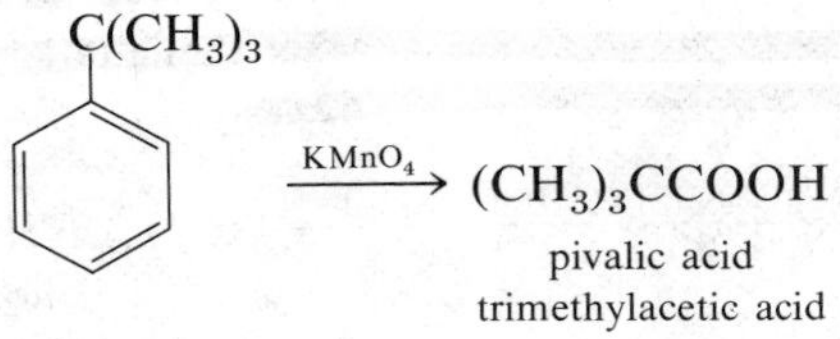

$$\overset{C(CH_3)_3}{} \xrightarrow{KMnO_4} (CH_3)_3CCOOH$$

pivalic acid
trimethylacetic acid

Oxidation of side-chain methyl groups is an important industrial route to aromatic carboxylic acids. The most important oxidzing agent for such reactions is air.

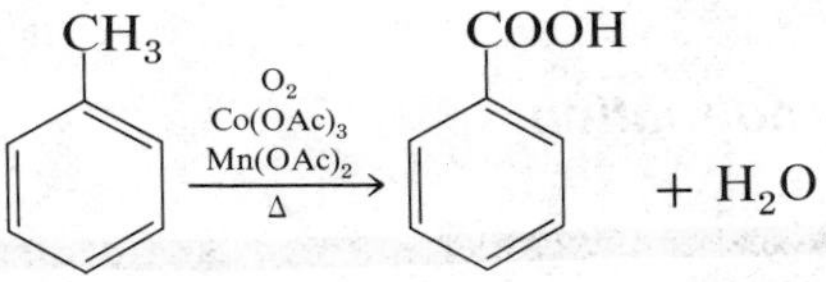

$$\overset{CH_3}{} \xrightarrow[\Delta]{\substack{O_2 \\ Co(OAc)_3 \\ Mn(OAc)_2}} \overset{COOH}{} + H_2O$$

An important industrial reaction of this general type is the oxidation of *p*-xylene to the dicarboxylic acid (Section 27.6).

EXERCISE: Suggest two different two-step ways in which toluene can be converted into benzyl alcohol.

D. *Acidity of Alkylbenzenes*

Benzyl anion is stabilized by delocalization of the negative charge into the benzene ring.

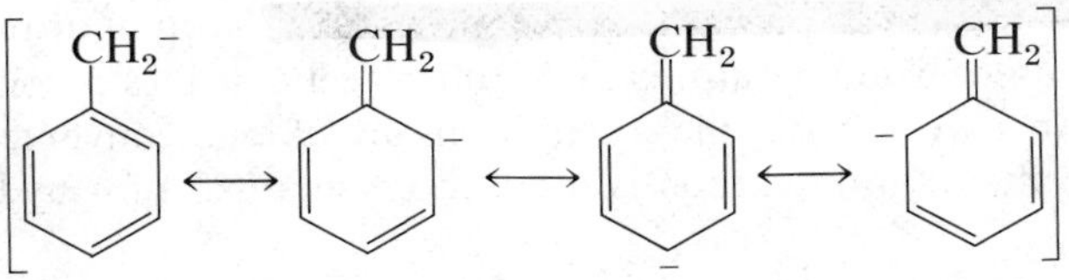

As a result, toluene is more acidic than the alkanes. Its pK_a is about 41 compared to a value of about 50 for ethane. Toluene is still a very weak acid and is not significantly converted to the anion even with $NaNH_2$ in liquid ammonia. As

TABLE 22.4
Acidity of Some Hydrocarbons

Hydrocarbon	Conjugate Base	pK_a
ethane	$CH_3CH_2^-$	~50
benzene	$C_6H_5^-$	43
toluene	$C_6H_5CH_2^-$	41
diphenylmethane	$(C_6H_5)_2CH^-$	34
triphenylmethane	$(C_6H_5)_3C^-$	31.5

additional benzene rings are added, however, the acidity increases markedly. Some relevant pK_a data are summarized in Table 22.4.

Di- and triphenylmethane are sufficiently acidic to be converted significantly to the corresponding carbanions with sodium amide. Hence, synthesis may be accomplished with these anions. One example is

$$(C_6H_5)_2CH_2 \xrightarrow[\text{liq. NH}_3]{\text{NaNH}_2} \xrightarrow[(C_2H_5)_2O]{\text{CH}_3(\text{CH}_2)_2\text{CH}_2\text{Br}} (C_6H_5)_2CH(CH_2)_3CH_3$$

(92%)
1,1-diphenylpentane

22.6
Reduction

A. *Catalytic Hydrogenation*

Benzene rings are substantially more resistant to catalytic hydrogenation than alkenes or alkynes. In molecules that contain both a double bond and a benzene ring, the double bond may be preferentially hydrogenated without difficulty.

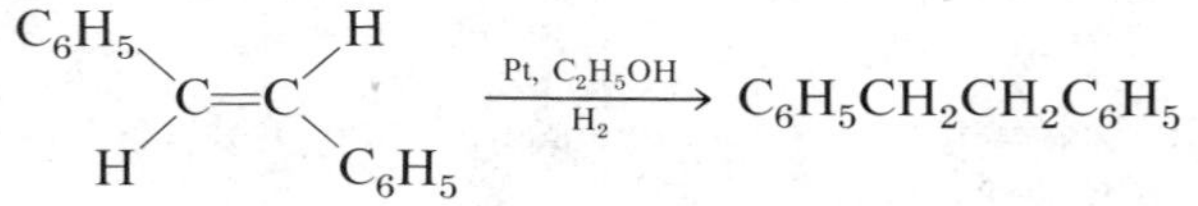

trans-stilbene

bibenzyl
1,2-diphenylethane

Hydrogenation of the benzene ring occurs under more vigorous conditions and yields the corresponding cyclohexane. It is generally impractical to stop the reaction at an intermediate stage, since cyclohexadienes and cyclohexenes hydrogenate more readily than benzenes. Dialkylbenzenes tend to give predominantly the *cis*-dialkylcyclohexane, although the exact stereochemistry of the reduction depends on the reaction conditions and catalysts used. Platinum or palladium catalysts may be used at temperatures near 100°; acetic acid is a common solvent. Nevertheless, reactions under these conditions are often inconveniently slow, and ruthenium or rhodium on carbon is often more successful for hydrogenation of aromatic rings.

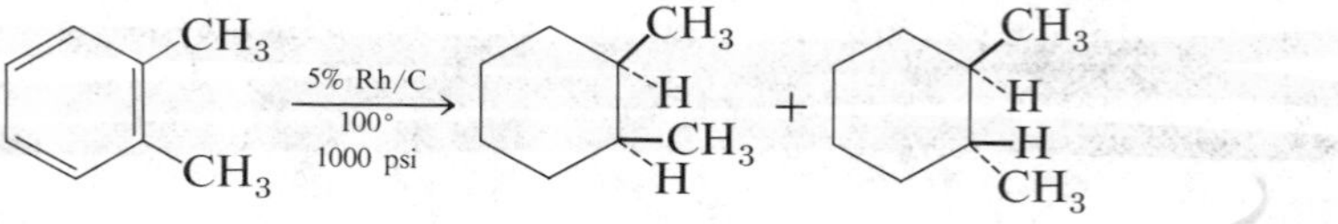

about 9:1

With aromatic aldehydes and ketones the functional group undergoes reduction faster than the ring.

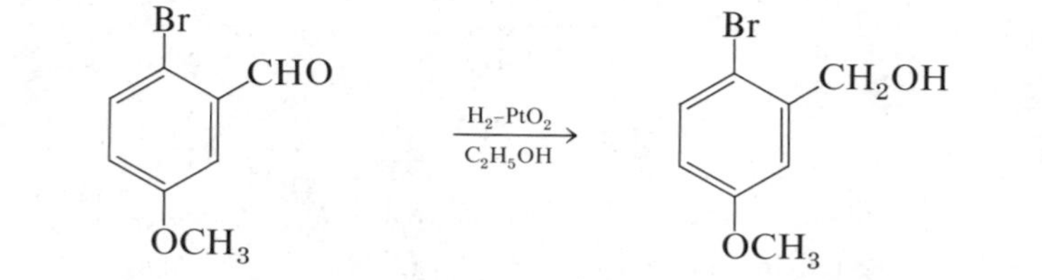

2-bromo-5-methoxybenzaldehyde

(83%)
2-bromo-5-methoxybenzyl alcohol

B. *Hydrogenolysis of Benzylic Groups*

Benzylic halides can be converted to hydrocarbons in the same ways that suffice for converting alkyl halides to alkanes—formation and hydrolysis of a Grignard reagent or by reduction with lithium aluminum hydride. Benzylic alcohols may be converted into the corresponding halide or sulfonate ester and thence into the hydrocarbon.

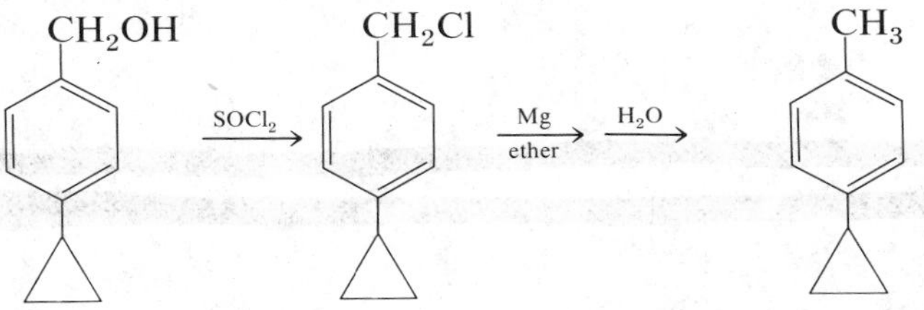

p-cyclopropyltoluene

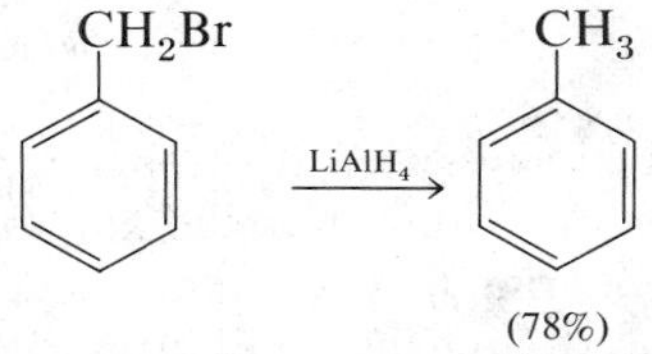

benzyl bromide   toluene

These reactions differ from comparable reactions of normal alkyl halides only in that they occur a little more rapidly since the reactants are benzylic (Section 22.5.B).

Benzylic alcohols may be reduced directly to the corresponding hydrocarbon by treatment with hydrogen in the presence of palladium and a small amount of perchloric acid.

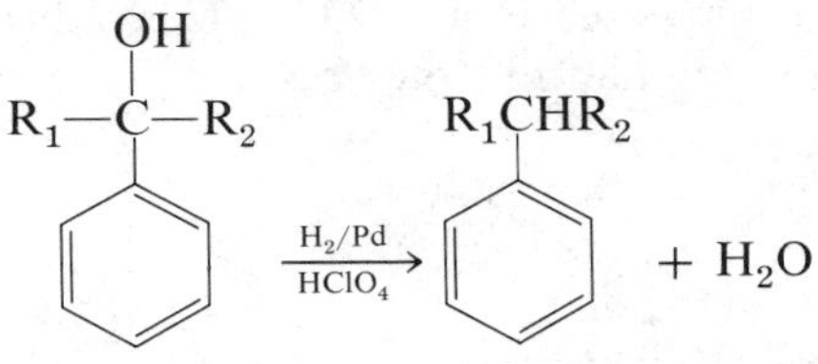

Since the carbonyl group in aromatic aldehydes and ketones is usually hydrogenated more rapidly than the benzene ring, the carbonyl group in such compounds can be *completely removed* by the use of two equivalents of hydrogen.

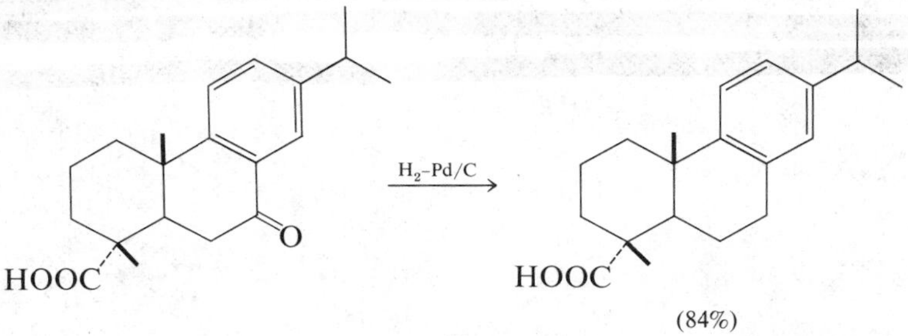

(84%)

This type of process, in which hydrogen breaks a single bond, is known as **hydrogenolysis.** Another example is

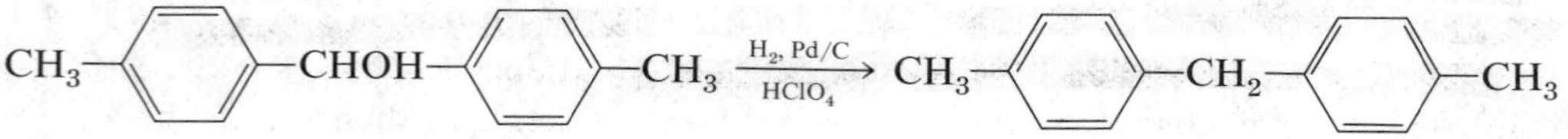

$$CH_3-\!\!\!\bigcirc\!\!\!-CHOH-\!\!\!\bigcirc\!\!\!-CH_3 \xrightarrow[HClO_4]{H_2,\ Pd/C} CH_3-\!\!\!\bigcirc\!\!\!-CH_2-\!\!\!\bigcirc\!\!\!-CH_3$$

di-(*p*-tolyl)methanol   di-(*p*-tolyl)methane

C. *Birch Reduction*

Aromatic rings can be reduced by alkali metals in a mixture of liquid ammonia and alcohol. The product of this reduction is an unconjugated cyclohexadiene.

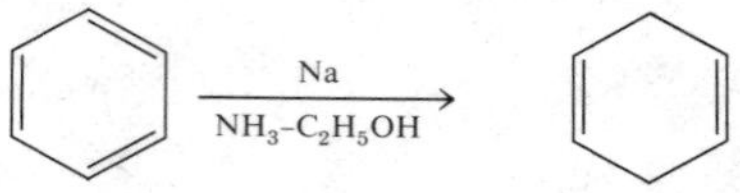

$$\bigcirc \xrightarrow[NH_3-C_2H_5OH]{Na} \bigcirc$$

1,4-cyclohexadiene

Recall that a similar reduction is used to prepare *trans* alkenes from alkynes (Section 12.6.A). A solution of sodium in liquid ammonia contains solvated electrons, which add to a benzene ring to give a radical anion. Benzene and alkylbenzenes are not readily reduced, and the equilibrium lies far to the left.

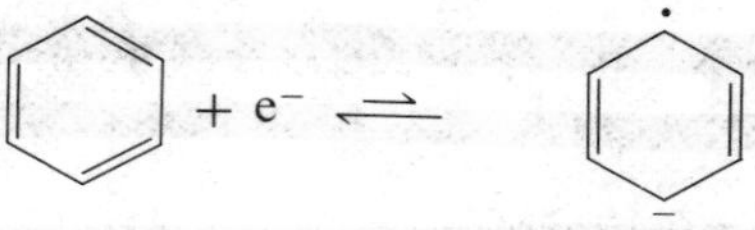

$$\bigcirc + e^- \rightleftharpoons \bigcirc$$

benzene radical anion

Note that the radical anion has seven electrons in the benzene π-system. The extra electron has added to the lowest empty molecular orbital in benzene, an antibonding π-molecular orbital (see Figure 22.3). The radical anion is still highly conjugated, and many resonance structures can be written for it, some of which are

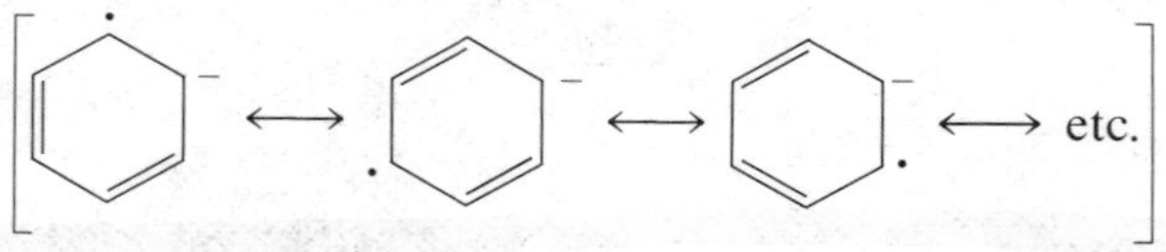

$$\left[\bigcirc \longleftrightarrow \bigcirc \longleftrightarrow \bigcirc \longleftrightarrow \text{etc.} \right]$$

Because the odd electron is in an antibonding orbital, benzene radical anion is less stable than benzene. The ion reacts readily with proton donors. Ammonia itself is too weakly acidic to react, but ethanol is a sufficiently strong acid to protonate the radical anion. The resulting cyclohexadienyl radical immediately reacts with another solvated electron to form the corresponding cyclohexadienyl anion.

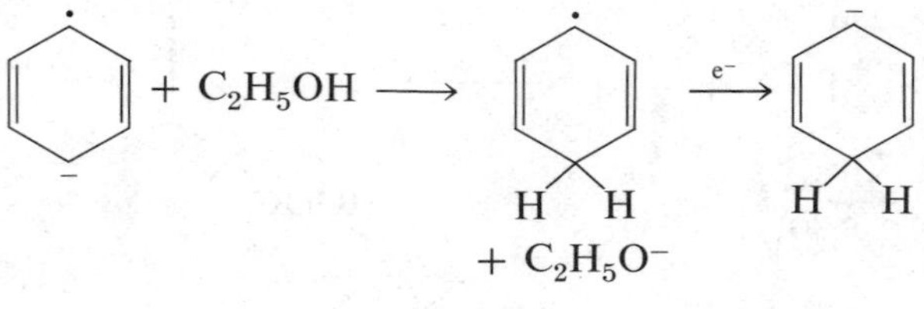

$$\bigcirc + C_2H_5OH \longrightarrow \underset{H\ \ H}{\bigcirc} \xrightarrow{e^-} \underset{H\ \ H}{\bigcirc}$$

$$+\ C_2H_5O^-$$

This anion is a strong base and reacts immediately with ethanol to give 1,4-cyclohexadiene and ethoxide ion.

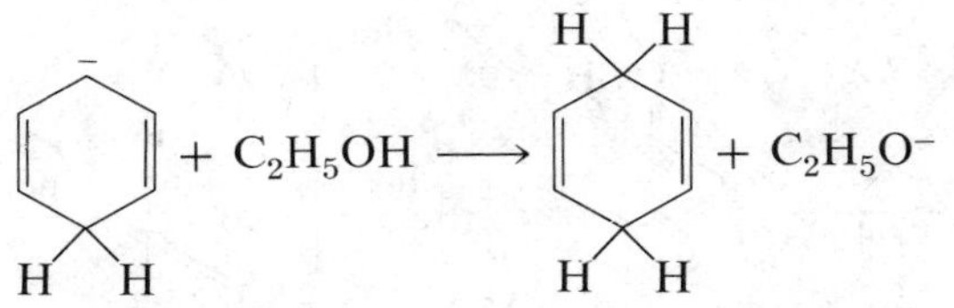

Cyclohexadienyl anion is a conjugated carbanion of the allylic type, and the negative charge is correspondingly distributed over several carbons, as indicated by the resonance structures

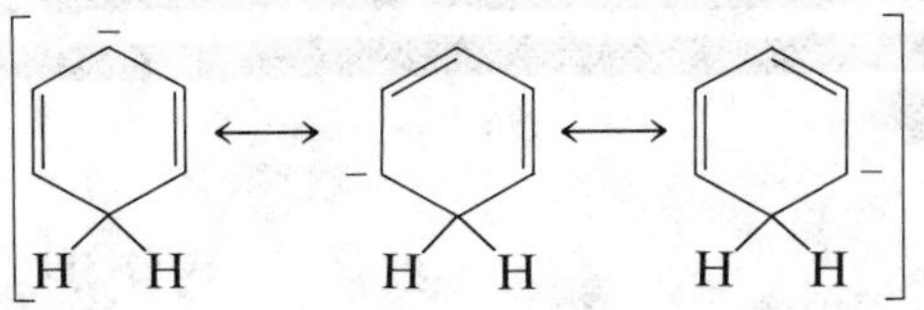

Protonation at the central carbon is much faster than at the end carbons of the conjugated chain. This result is quite general, even though the product of protonation at the terminal carbon of the conjugated chain produces a conjugated diene. The reason is not readily apparent, although various more or less sophisticated explanations have been given for this unusual effect.

Since the product contains isolated double bonds, no further reduction takes place, and the cyclohexadiene may be isolated in good yield. On prolonged contact with base, the carbanion is re-formed, allowing eventual isomerization of the unconjugated diene to the conjugated isomer, which is rapidly reduced to the monoene.

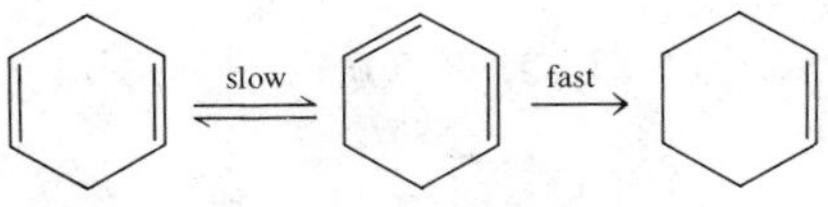

Thus, by a proper choice of solvent and temperature one may reduce the benzene ring to either the 1,4-cyclohexadiene or the cyclohexene.

With substituted benzenes a single product is often formed in good yield; for example,

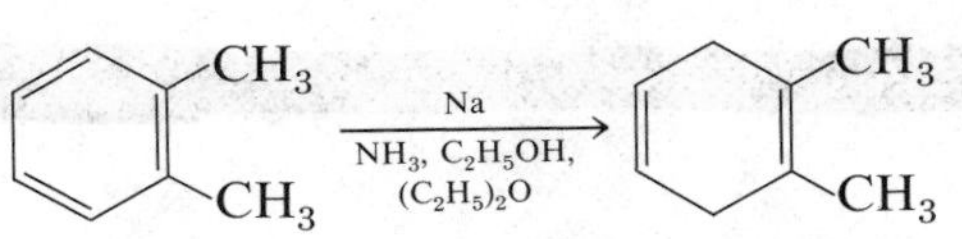

(77–92%)
1,2-dimethyl-1,4-cyclohexadiene

[Sodium is added in pieces to a mixture of liquid ammonia, ether, ethanol, and *o*-xylene cooled in a dry ice bath. The ammonia is allowed to evaporate, water is added, and the washed and dried organic layer is distilled.]

Birch reduction is particularly useful with anisole and alkylanisoles. Addition of hydrogen always occurs in such a way that an enol ether is produced. Hydrolysis occurs readily (Section 13.7.B) to give a β,γ-unsaturated ketone. Under the acidic conditions of the hydrolysis, the double bond moves into conjugation with the carbonyl group (Section 20.3.A).

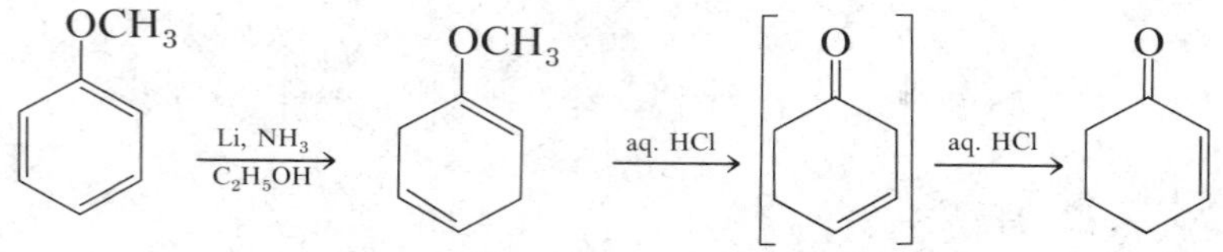

anisole   (84%) 1-methoxycyclo-hexa-1,4-diene   3-cyclo-hexenone   2-cyclo-hexenone

In contrast to the Birch reductions of toluene and anisole, which provide 2-substituted 1,4-cyclohexadienes, reduction of benzoic acid gives the completely unconjugated product.

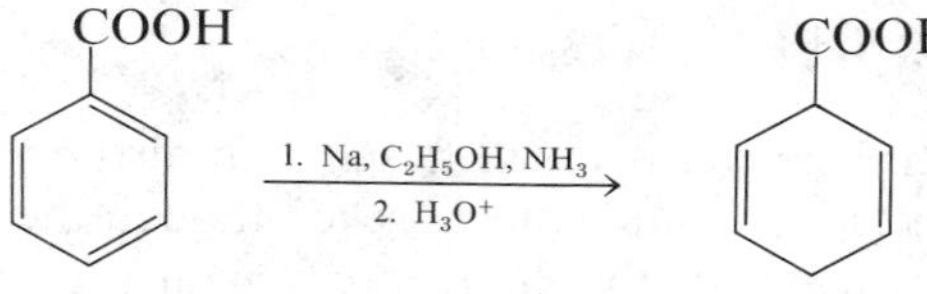

(89–95%)
2,5-cyclohexadiene-1-carboxylic acid

The initial reduction product, before work-up, is a type of enolate ion (Sections 13.9 and 19.9). If an alkyl halide is added to the ammonia solution of this enolate ion, good yields of alkylation product may be obtained.

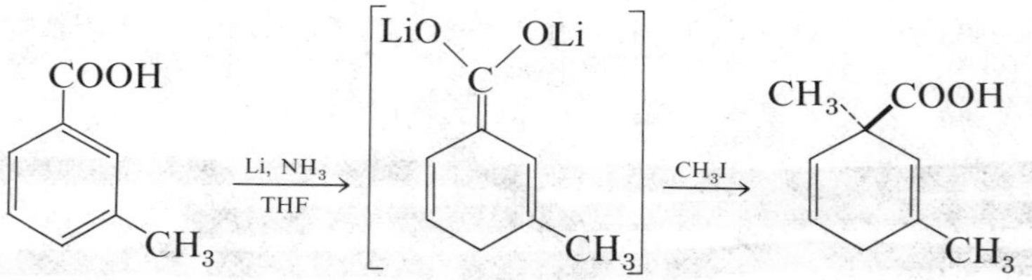

(95%)

Similar results may be obtained in the Birch reduction–alkylation of aromatic ketones.

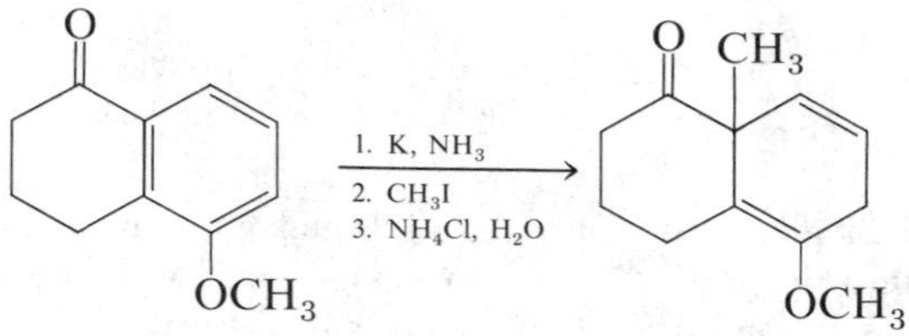

(93%)

Double bonds conjugated to the aromatic ring can also be reduced by alkali metals in liquid ammonia.

(80%)

Although the benzene ring can also be reduced by lithium in ammonia, it is not as reactive as the conjugated double bond. Thus, selective reduction is possible, as in the foregoing example. Of course, the Birch reduction cannot generally be applied to systems that contain other easily reducible functions such as halogens, nitro groups, or carbonyl functions.

EXERCISE: What reaction products, if any, would you expect in the reactions of *cis*-2-butene, 2-phenylpropene, and 4-phenyl-1-butene with H_2–Pd/C and with Na/NH$_3$–C$_2$H$_5$OH and from treatment of α- and β-phenylethanols with H_2–Pd/C in C$_2$H$_5$OH containing a small amount of perchloric acid?

22.7
Aromaticity

A. *Cyclooctatetraene: The Hückel 4n + 2 Rule*

Cyclooctatetraene is a well-known hydrocarbon; it is a liquid, b.p. 152°, that shows all of the chemistry typical of conjugated polyenes. It polymerizes on exposure to light and air and reacts readily with acids, halogens, and other reagents. In other words, it shows none of the "aromatic" stabilization associated with benzene. If cyclooctatetraene had the structure of a planar regular octagon analogous to the hexagon of benzene, we could write two resonance structures of the benzene Kekulé type.

We would therefore anticipate a significant amount of resonance energy for such a structure. Why, then, is cyclooctatetraene not an "aromatic" compound? The π-molecular orbital energy-level pattern is shown in Figure 22.15. Six of the eight π-electrons are put into the three lowest molecular orbital levels, but the one pair left is not enough to fill the next shell. Thus, planar, octagonal cyclooctatetraene has an incomplete orbital shell and would therefore not be expected to have the special stability characteristic of benzene.

Note that the last two electrons in Figure 22.15 are placed with the same spin, one in each of the degenerate orbitals. This arrangement is a consequence of Hund's rule, just as in atomic structure. Two electrons of the same spin are prevented from close approach by the Pauli principle—two electrons with the same quantum numbers cannot occupy the same region of space. Two electrons of opposite spin stay apart only because of electrostatic repulsion; hence, such a system has higher net energy and is less stable than one in which the electrons have the same spin.

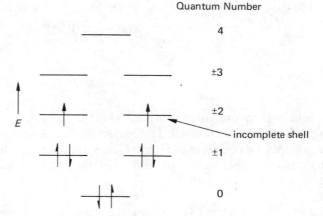

FIGURE 22.15. *The π-molecular orbital energy level pattern for a planar octagonal cyclooctatetraene.*

In fact, the structure of cyclooctatetraene is in keeping with this analysis. The molecule is tub-shaped and has bond lengths characteristic of alternating single and double bonds (Figure 22.16).

As a result of the tub shape, the π-orbitals of adjacent double bonds are twisted with respect to one another and overlap is greatly reduced. That is, looking down any single bond, the π-orbitals are almost at right angles to each other. In short, because of the instability associated with an incomplete orbital shell in the planar

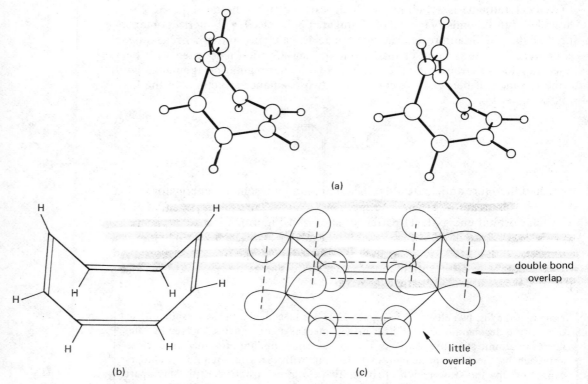

FIGURE 22.16. *Cyclooctatetraene:* (a) *stereo representation;* (b) *Kekulé structure;* (c) *π-orbital structure.*

octagonal geometry, cyclooctatetraene prefers a nonplanar structure in which the alternating double bonds are effectively not conjugated with each other!

Incomplete atomic orbital shells are associated with a relative ease of gaining or losing electrons to form an ion having a filled-shell electronic configuration; for example,

$$Li \cdot \longrightarrow Li^+ + e^-$$

$$:\ddot{F}\cdot\ + e^- \longrightarrow\ :\ddot{F}:^-$$

The same behavior is seen with incomplete molecular orbital shells. Cyclooctatetraene reacts readily with alkali metals in ether solvents to form alkali metal salts of cyclooctatetraene dianion.

cyclooctatetraene
dianion

The dianion has the planar structure of a regular octagon with C—C bond distances of 1.4 Å, quite similar to the C—C bond distances in benzene! Cyclooctatetraene dianion has ten π-electrons, just enough to fill the molecular orbital shell with π-quantum numbers of ± 2. This reaction of cyclooctatetraene provides a remarkable demonstration of the usefulness of simple molecular orbital concepts. Cyclooctatetraene is not an aromatic system; in fact, a planar octagonal cyclooctatetraene could even be described as **antiaromatic.** However, the dianion, with two more π-electrons, is definitely an aromatic system.

The foregoing discussion illustrates a general principle. For all cyclic π-electronic systems, successive molecular orbitals above the lowest level can be characterized by quantum numbers of $\pm n$, where n is any integer.

> The absolute value of the quantum number n indicates the number of nodal planes that bisect the ring. Alternatively, and equivalently, we can consider the quantum number to represent the angular momentum of an electron circling round the ring. The lowest level then corresponds to an electron having zero angular momentum. Thereafter, the momentum can be represented clockwise or counterclockwise about the ring; hence, above zero the quantum numbers come as $\pm$ integer pairs.

To summarize, a filled orbital shell corresponds to a relatively stable electronic configuration. Examples in atomic orbitals are the filled $1s$-shell of helium and the filled $2p$-shell of neon. Similarly, filled π-molecular orbital shells give the stability associated with "aromatic" systems and bestow that stabilization commonly known as aromatic character or aromaticity. It takes two electrons of opposite spin to fill the lowest π-molecular orbital level for which $n = 0$. Thereafter, four electrons are required to give a filled π-molecular orbital shell. That is, filled shells are associated with a total of $4n + 2$ electrons, or two ($n = 0$), six ($n = 1$), ten ($n = 2$), fourteen ($n = 3$), and so on, electrons. This rule is known as the **Hückel 4n + 2 rule** after the late Erich Hückel, the German theoretical chemist who first developed the rule in the mid-1930s.

Many examples of compounds are now known to which the Hückel rule can be applied. The results are truly remarkable for such a simple rule; a vast amount of experimental chemistry can be summarized by the generalization that *those mon-*

ocyclic π systems with $4n + 2$ electrons show relative stability compared to acyclic analogs. Furthermore, those monocyclic systems with other than $4n + 2$ electrons appear to be destabilized relative to acyclic analogs and can be said to have "antiaromatic" character. In succeeding sections we will summarize some of the experimental evidence for several values of n.

> **EXERCISE:** Construct a molecular model of cyclooctatetraene. Place the model in the "tub" conformation described on page 676 and note the lack of overlap between the adjacent double-bond π-orbitals.

B. *Two-Electron Systems*

One two-electron cyclic π-system is obviously ethylene, a well-known and relatively stable compound. However, another cyclic π-system with two electrons is cyclopropenyl cation, a rather stable carbocation.

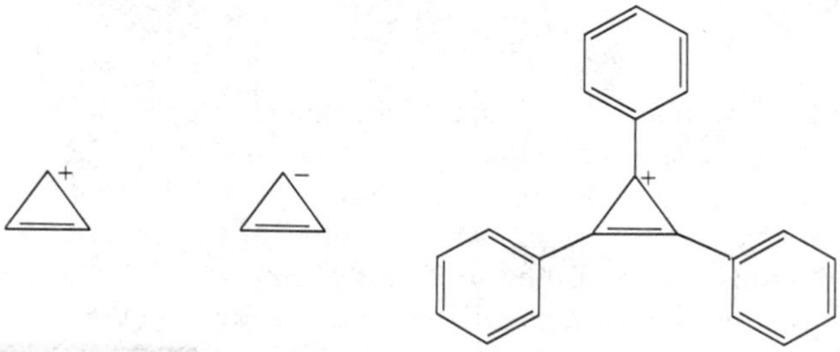

cyclopropenyl
cation

cyclopropenyl
anion

triphenylcyclopropenyl
cation

Triphenylcyclopropenyl cation is such a stable carbocation that many of its salts can be isolated and stored in bottles. On the other hand, the cyclopropenyl anion is unknown. The acidity of the methylene group in the known hydrocarbon cyclopropene has been deduced from several experiments to be *less* than that of alkanes.

The generalization that cyclic two-electron systems show relative stability makes its appearance in some subtle ways. For example, compounds containing the cyclopropenone ring system have unusually high dipole moments; this result is explained on the basis that the dipolar resonance structure contributes more to the resonance hybrid of cyclopropenone because it embodies the "aromatic" cyclopropenyl cation.

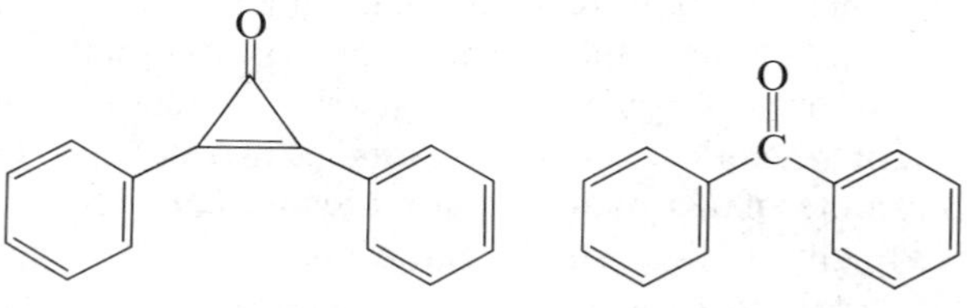

$\mu = 5.08$ D
diphenylcyclopropenone

$\mu = 2.97$ D
benzophenone

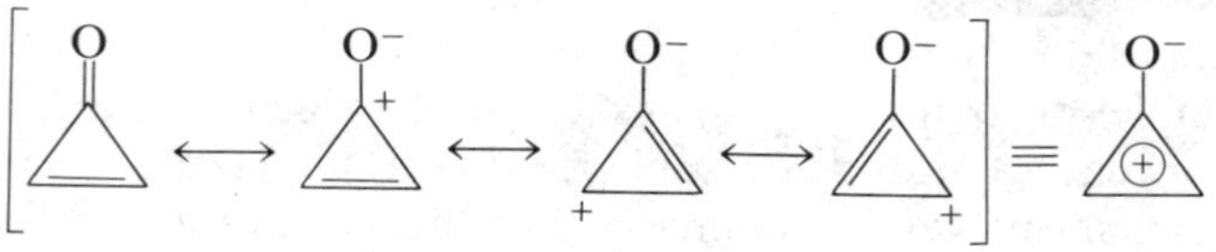

As with the structure of cyclooctatetraene dianion on page 677, this example illustrates the use of an inscribed circle to represent an aromatic cycle of $4n + 2$ electrons.

C. *Six-Electron Systems*

Cyclobutadiene has a cyclic π-system with four electrons and does not fit the $4n + 2$ rule; accordingly, we would expect it to have antiaromatic character. Cyclobutadiene is a known but very reactive hydrocarbon. It can be captured only at very low temperatures. Under most conditions it has but a fleeting existence and yields only dimeric products. This same reactivity is characteristic of various substituted derivatives, for example

CH$_3$—CH$_3$—Cl → (Li-Hg) → tetramethylcyclobutadiene →

Cyclic π-systems with six electrons fit the $4n + 2$ rule, and the most important such cycle is, of course, benzene. Some other six-electron cycles are ions. Cyclopentadienyl anion is a rather stable carbanion whose conjugate acid, cyclopentadiene, is an unusually acidic hydrocarbon with a pK_a of 16. Recall that such a value is far lower than that of triphenylmethane ($pK_a = 31.5$). In fact, cyclopentadiene is comparable in acidity to water and the alcohols.

$$\text{cyclopentadiene} \xrightleftharpoons{pK_a = 16} \text{cyclopentadienyl anion} + H^+$$

By contrast, cycloheptatriene is a nonacidic hydrocarbon; it appears to be less acidic than the open-chain heptatriene.

cycloheptatriene $CH_3CH{=}CHCH{=}CHCH{=}CH_2$
                          heptatriene

The cycloheptatrienyl anion has seven equivalent resonance structures of the type shown and would be expected to have a well-distributed negative charge. But it also has an incomplete molecular orbital shell with its eight π-electrons, and this property conveys antiaromatic character (Figure 22.17).

cycloheptatrienyl anion

Figure 22.17 also shows that the situation is reversed for the corresponding cations. Cyclopentadienyl cation is highly reactive and difficult to prepare. It has only four π-electrons. On the other hand, cycloheptatrienyl cation has six π-electrons and is a remarkably stable carbocation. It is readily prepared by oxidation of cycloheptatriene and many of its salts are stable crystalline compounds.

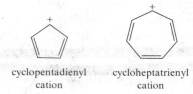

cyclopentadienyl
cation

cycloheptatrienyl
cation

EXERCISE: Of the two ketones shown below, one is highly reactive and one is unusually stable. Which is which?

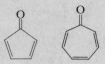

D. *Ten-Electron Systems*

The Hückel $4n + 2$ rule says that monocyclic π-electron systems with ten electrons will have filled π-molecular orbital shells with the highest occupied molecular orbital having quantum numbers of ± 2. We have encountered one such system in cyclooctatetraene dianion. A related system is cyclononatetraenyl anion. This anion is a known system and gives evidence of having a planar nonagon structure, despite the high angle strain in such a ring system.

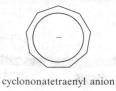

cyclononatetraenyl anion

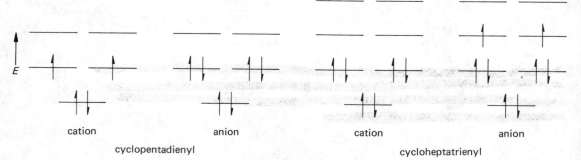

cation   anion   cation   anion

cyclopentadienyl   cycloheptatrienyl

FIGURE 22.17. *Molecular orbital energy levels for cyclopentadienyl and cycloheptatrienyl ions showing filled molecular orbital shells for six π-electrons.*

A neutral ten-π-electron hydrocarbon homologous to benzene would be cyclo-decapentaene. The planar all-*cis* structure has highly strained bond angles. The alternative structure with two *trans* double bonds cannot achieve planarity because of interaction between the two interior hydrogens. As a result, cyclo-decapentaene does not have the expected aromatic stability, but is instead a highly reactive hydrocarbon.

all-*cis* or all-(Z)-cyclodecapentaene   (Z,Z,E,Z,E)-cyclodecapentaene

An unusual hydrocarbon has been prepared in which the two interior hydrogens of cyclodecapentaene have been replaced by a bridging methylene group. This hydrocarbon cannot have a completely coplanar π-system, but enough cyclic overlap occurs to give the compound significant aromatic character.

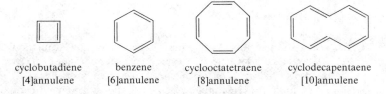

bicyclo[4.4.1]undeca-1,3,5,7,9-pentaene

E. *Larger Cyclic π-Systems*

Cyclobutadiene, benzene, and cyclooctatetraene are the first three members of a family of monocyclic $(CH)_n$ compounds known as **annulenes.** Cyclobutadiene is [4]annulene, benzene is [6]annulene, and cyclooctatetraene is [8]annulene. Other fully conjugated cyclic polyenes are named in an analogous fashion.

| cyclobutadiene | benzene | cyclooctatetraene | cyclodecapentaene |
| [4]annulene | [6]annulene | [8]annulene | [10]annulene |

A number of larger annulenes are known that further confirm the generality of the $4n + 2$ rule. Cyclododecahexaene, [12]annulene, is a polyolefinic compound that reacts with alkali metals to give a dianion that, with fourteen electrons, follows the $4n + 2$ rule. The interesting hydrocarbon depicted in Figure 22.18 has a fourteen-electron π-system and is a stable molecule that has the properties of an aromatic system. The periphery indicated on the right of Figure 22.18 is a [14]annulene that follows the $4n + 2$ rule.

Cyclohexadecaoctaene, [16]annulene, with sixteen π-electrons, does not fit the

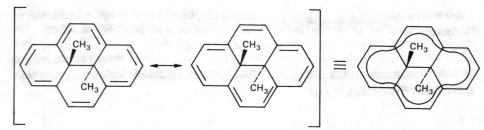

FIGURE 22.18. trans-*10b,10c-Dimethyl-10b,10c-dihydropyrene.*

$4n + 2$ rule. It has polyolefinic behavior, but reacts with alkali metals to form the aromatic cyclic dianion with eighteen π-electrons.

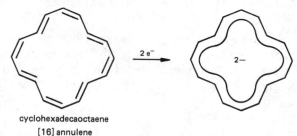

cyclohexadecaoctaene
[16] annulene

The corresponding neutral eighteen-π-electron hydrocarbon, cyclooctadecanonaene, [18]annulene, has been synthesized as a relatively stable brown-red compound. X-ray structure analysis shows that the bonds have equal length; however, this result has been questioned by some theoretical calculations, and the aromaticity of this hydrocarbon is still controversial.

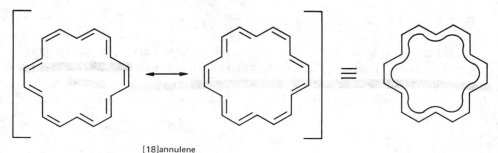

[18]annulene

EXERCISE: Which of the following systems is expected to show aromaticity in the Hückel sense? (a) cycloeicosadecaene, (b) cyclodoeicosaundecaene, (c) cyclooctatetraene radical cation (which is seen in the mass spectrum of cyclooctatetraene), (d) cyclobutadiene dianion, (e) [26]annulene, (f) cycloundecapentaenyl cation.

F. Metallocenes

Dicyclopentadienyliron, $(C_5H_5)_2Fe$, was first reported in 1951. This remarkable compound is an orange solid, is stable even at elevated temperatures (up to 500°), and is easily prepared by the reaction of cyclopentadienyl anion with ferrous chloride.

$$2\,C_5H_5^- + Fe^{2+} \longrightarrow (C_5H_5)_2Fe$$

A solution of $FeCl_2 \cdot 4H_2O$ in dimethyl sulfoxide is added dropwise with stirring and under nitrogen to a mixture of powdered KOH and cyclopentadiene in dimethoxyethane (DME). When reaction is complete, the mixture is added to ice and hydrochloric acid, filtered, washed and dried. The product may be purified by sublimation. Yield 89–98%; m.p. 173–4°.

The structure of this compound and many derivatives has been shown to be that of two planar pentagonal rings in parallel planes sharing a common axis and with an iron atom centered between them (Figure 22.19). In the crystal the rings are staggered as shown, but there is actually essentially free rotation of the rings about the fivefold axis. From Figure 15.6 (page 455) it can be seen that Fe^{2+} has six valence electrons. Each cyclopentadienyl ring contributes six π-electrons for a total around iron of eighteen; that is, this unusual organometallic compound satisfies the transition metal eighteen-electron rule.

Alternatively, we could regard the compound in terms of a neutral iron atom, with eight valence electrons outside the argon shell, bound to two cyclopentadienyl radicals with five π-electrons each. We still satisfy the eighteen-electron rule.

The compound has been given the name **ferrocene** and is now known to be just one representative of a large class of related compounds in which one or more cyclopentadienyl rings is π-*bonded* to a central transition metal. Some of these compounds are of the *sandwich* type similar to ferrocene. Examples are ruthenocene, $(C_5H_5)_2Ru$, and cobalticinium cation, $(C_5H_5)_2Co^+$. Others are *half-sandwich* compounds that include other ligands, such as

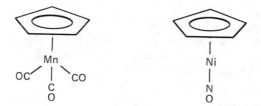

An important source of ring–metal bonding in these compounds is the overlap of π-molecular orbitals of the cyclopentadienyl rings with suitable *d*-orbitals of the central metal. One such interaction is shown in Figure 22.20. In this figure the vertical dashed line corresponds to a node perpendicular to the page; that is, the *p*-orbitals of the two cyclopentadienyl rings correspond to a π-MO with a quantum number of 1 just as in the benzene MO shown in Figure 22.3. Because the node runs through one of the ring carbons, the *p*-orbital on that carbon cannot participate. The *d*-orbital shown on iron shares the same nodal plane. Accordingly, the signs of the various orbital lobes correspond and give relatively high resultant overlap and bonding.

FIGURE 22.19. *Structure of ferrocene.*

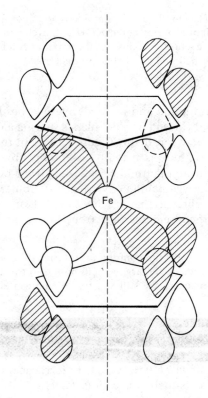

FIGURE 22.20. *One of the interactions between ring π-MOs and metal d-orbitals that contribute to bonding in ferrocene.*

Related organometallic compounds are known with other π-ligands. Examples are dibenzenechromium, cyclobutadieneiron tricarbonyl, and bis(cyclooctatetraene)uranium (uranocene).

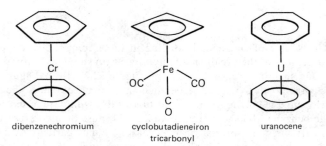

| dibenzenechromium | cyclobutadieneiron tricarbonyl | uranocene |

Note that some of the π-systems used correspond to Hückel 4n ring systems. For these cases it is convenient to consider that the ring has sufficient additional electrons to convert it to the next higher 4n + 2 system and that the central metal is appropriately positively charged. Note also that a number of compounds of these types are known that do not conform to the eighteen-electron rule.

Many of these compounds serve as important reagents and catalysts. Ferrocene itself has a rich and diverse chemistry. The whole field is one of active modern research and is still growing rapidly.

EXERCISE: Review all of the compounds cited in this section to determine which ones, if any, are exceptions to the eighteen-electron rule.

PROBLEMS

1. Write structures corresponding to each of the following names.
   (a) *m*-fluoroanisole
   (b) 2,4,6-tribromobenzoic acid
   (c) 2,4-dinitrotoluene
   (d) α-bromomesitylene
   (e) *m*-divinylbenzene
   (f) *p*-cyanophenylacetylene
   (g) *o*-diisopropylbenzene
   (h) 2-bromo-6-chloroaniline

2. Give an acceptable name for each of the following structures.

(a) 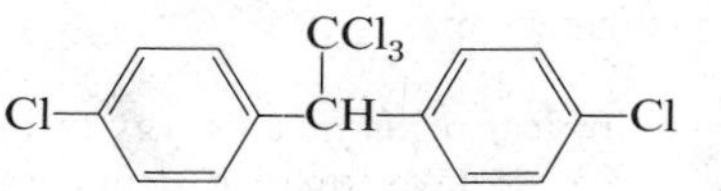 (DDT)

(b)

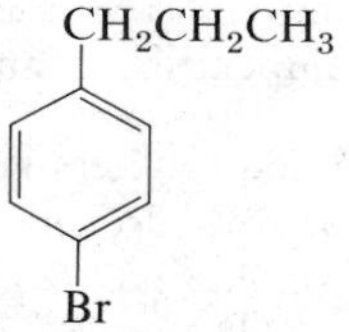

(c)

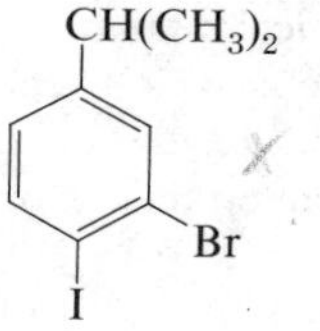

(d)

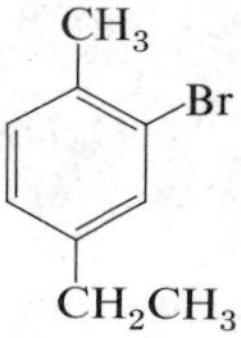

(e)

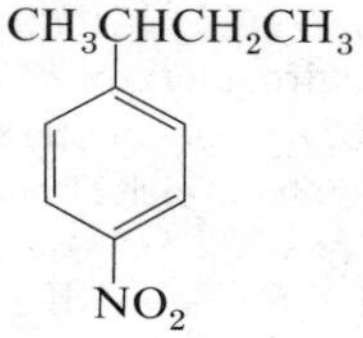

(f)

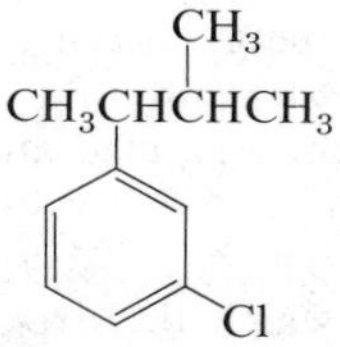

(g)

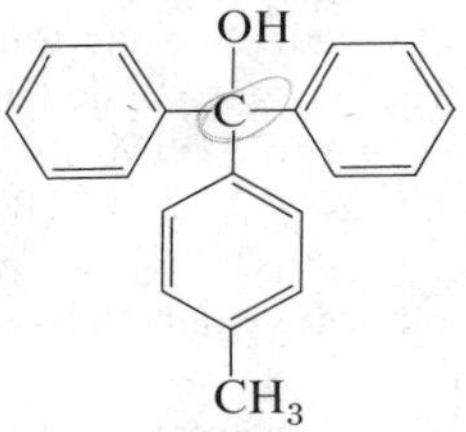

(h)

(i)

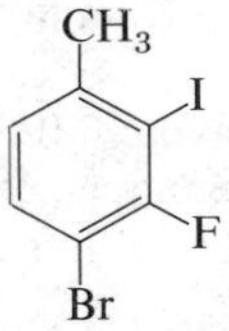

(j)

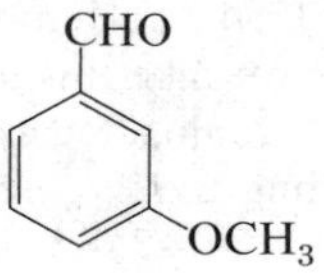

(k)

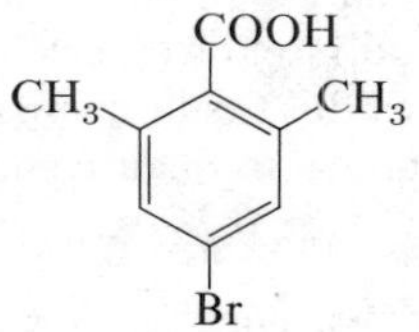

(l)

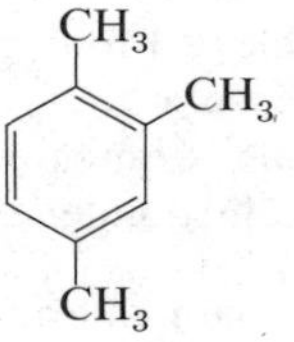

3. In Kekulé's day, one puzzling aspect of his dynamic theory for benzene was provided by 1,2-dimethylbenzene. According to his theory, there should be two distinct such compounds, one with a double bond between the two methyl-substituted carbons and one with a single bond in this position.

Only a single 1,2-dimethylbenzene is known, however.
(a) Does Landenburg's formula solve this problem?
(b) Explain with modern resonance theory.

4. When passed through a hot tube, acetylene gives fair amounts of benzene. What is $\Delta H°$ for the reaction

$$3 \ HC\equiv CH \longrightarrow$$

The entropy change for this reaction is $\Delta S° = -79.7$ eu. How do you explain the negative sign of this entropy change? Calculate $\Delta G°$ for the reaction at 25°. Where does the equilibrium lie at room temperature? This reaction does not occur spontaneously at room temperature. Can you give a reason?

5. (a) A common method for estimating the empirical resonance energy of benzene is to take the heat of hydrogenation of one Kekulé resonance structure as three times that of cyclohexene. What value of the empirical resonance energy does this procedure yield? Note how the exact value of the empirical resonance energy depends so markedly on the model used for a hypothetical system.
   (b) The heat of hydrogenation of cyclooctene to cyclooctane is -23.3 kcal mole^{-1}. That for 1,3,5,7-cyclooctatetraene is -100.9 kcal mole^{-1}. Use the procedure in (a) to calculate an empirical resonance energy for cyclooctatetraene. How do you interpret this result?
   (c) Another method for estimating empirical resonance energies makes use of Appendix III, "Average Bond Energies." In this table $E(C-H) = 99$, $E(C-C) = 83$, $E(C=C) = 146$ kcal mole^{-1}. Calculate the total bond energy of a hypothetical cyclohexatriene. This energy is the so-called heat of atomization, the heat required to dissociate a molecule into all of its constituent separated atoms. For benzene, this heat is actually $\Delta H°_{atom.} = +1318$ kcal mole^{-1}. What value for the empirical resonance energy results?
      Apply this same method to cyclooctatetraene, for which the experimental $\Delta H°_{atom.} = +1713$ kcal mole^{-1}.
   (d) The heat of combustion of cyclooctatetraene (reactant and products in the gas phase) is $\Delta H°_{comb.} = -1054.7$ kcal mole^{-1}. Derive the value of $\Delta H°_{comb.}$ per C—H group in C_8H_8. If we take this number to be a kind of group equivalent heat for a CH group in a conjugated system having little resonance energy, one can derive a corresponding $\Delta H°_{comb.}$ for a nonconjugated cyclohexatriene. The actual $\Delta H°_{comb.}$ for benzene is -757.5 kcal mole^{-1}. What is the derived empirical resonance energy of benzene?
   (e) Thermochemical measurements on [18]annulene give a derived heat of atomization, $\Delta H°_{atom.} = +3890$ kcal mole^{-1}. From the method in part (c), calculate the corresponding empirical resonance energy. How does this result compare with expectations from the $4n + 2$ rule? Compare the empirical resonance energy per C—H unit of [18]annulene and [6]annulene.

6. Consider the possible free radical chain chlorination of benzene.

$$C_6H_6 + Cl\cdot \longrightarrow C_6H_5\cdot + HCl$$

$$C_6H_5\cdot + Cl_2 \longrightarrow C_6H_5Cl + Cl\cdot$$

From the data in Appendix I calculate $\Delta H°$ for each reaction. Use these results to explain why this method is *not* a satisfactory way of preparing chlorobenzene.

7. A compound, $C_{10}H_{14}$, has the following nmr and ir spectra. Determine the structure of the compound.

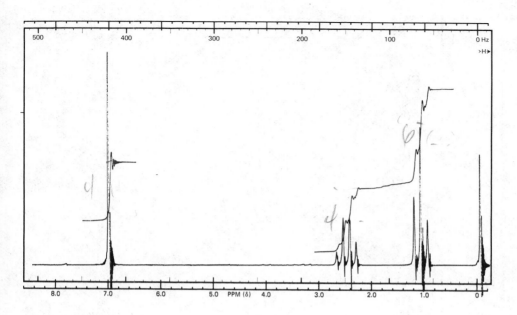

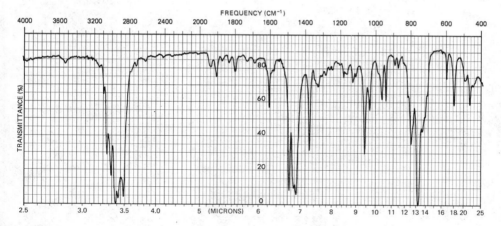

8. One general reaction of halobenzenes is the formation of Grignard reagents (Section 15.4.A). Thus, bromobenzene gives a Grignard reagent, which is hydrolyzed by water to give benzene.

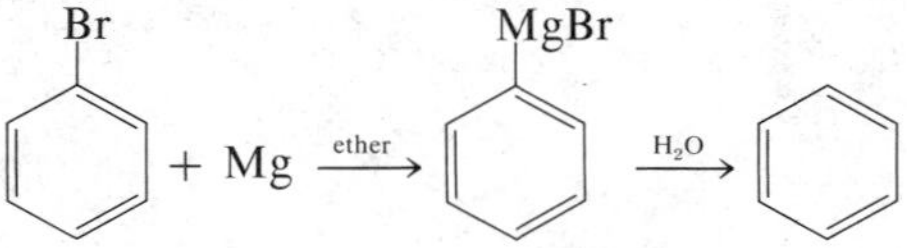

All six isomers of bromodimethylbenzene, $BrC_6H_3(CH_3)_2$, are known. Show how the above reaction, with Körner's absolute method, may be used to establish the structures of the three isomeric dimethylbenzenes.

9. Give the principal product of the following reactions or reaction sequences.

(a) 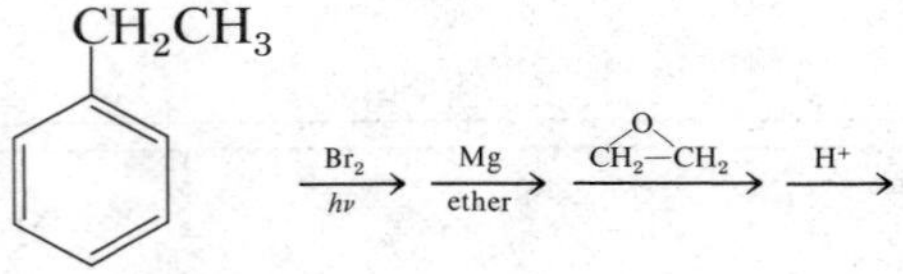
$$\underset{hv}{\overset{Br_2}{\longrightarrow}} \underset{ether}{\overset{Mg}{\longrightarrow}} \overset{\overset{O}{\overset{}{\diagup\diagdown}}}{\underset{}{CH_2-CH_2}} \overset{H^+}{\longrightarrow}$$

CH₂CH₃ (on benzene ring)

(b)
$$\underset{C_2H_5OH}{\overset{Na, NH_3}{\longrightarrow}}$$

CH₃ ... CH₃ (para on benzene ring)

(c)
$$\underset{hv}{\overset{Br_2 \ (1 \ mole)}{\longrightarrow}} \underset{\Delta}{\overset{CH_3OH}{\longrightarrow}}$$

CH(CH₃)₂ (on benzene ring)

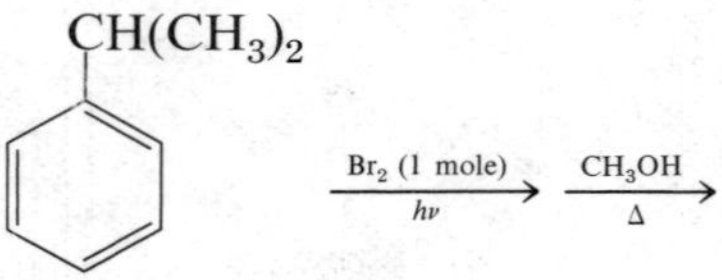

(d) 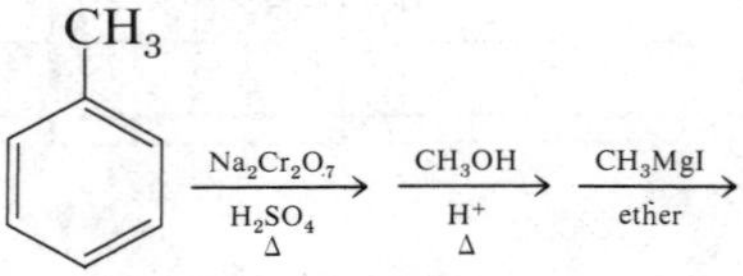
$$\underset{\underset{\Delta}{H_2SO_4}}{\overset{Na_2Cr_2O_7}{\longrightarrow}} \underset{\underset{\Delta}{H^+}}{\overset{CH_3OH}{\longrightarrow}} \underset{ether}{\overset{CH_3MgI}{\longrightarrow}}$$

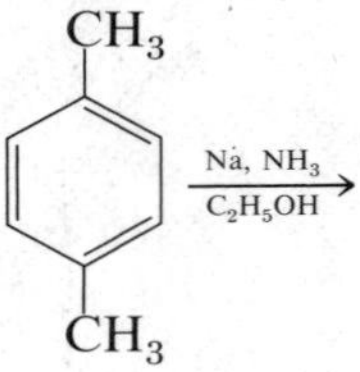

CH₃ (on benzene ring)

(e) 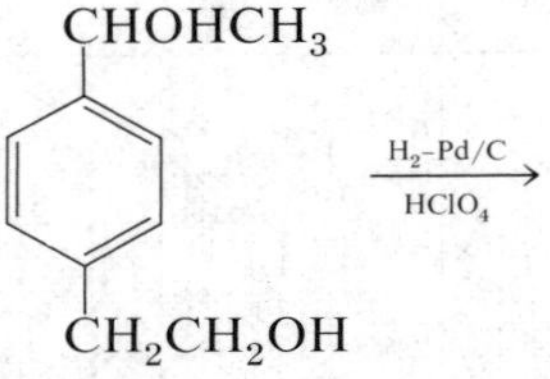
$$\underset{HClO_4}{\overset{H_2-Pd/C}{\longrightarrow}}$$

CHOHCH₃ ... CH₂CH₂OH (para on benzene ring)

(f) 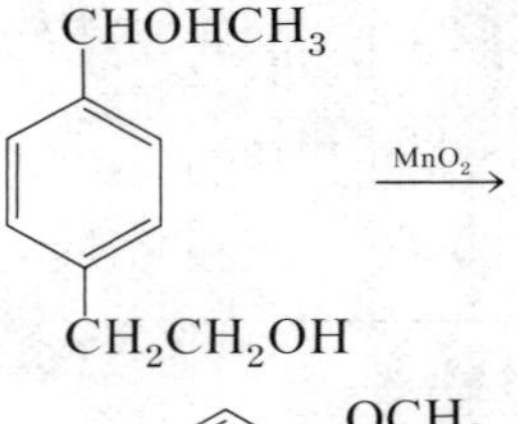
$$\overset{MnO_2}{\longrightarrow}$$

CHOHCH₃ ... CH₂CH₂OH (para on benzene ring)

(g) 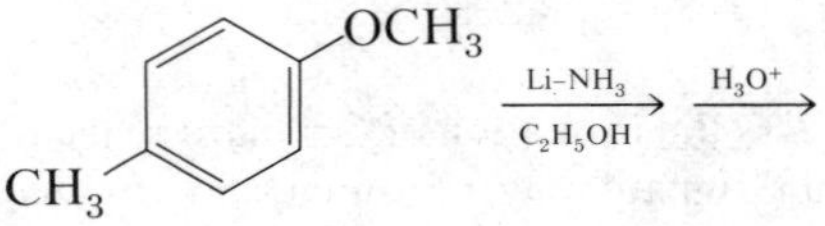
$$\underset{C_2H_5OH}{\overset{Li-NH_3}{\longrightarrow}} \overset{H_3O^+}{\longrightarrow}$$

OCH₃ ... CH₃ (para on benzene ring)

10. Show how one may synthesize each of the following compounds, starting with benzene, toluene, xylene, or ethylbenzene.

(a) 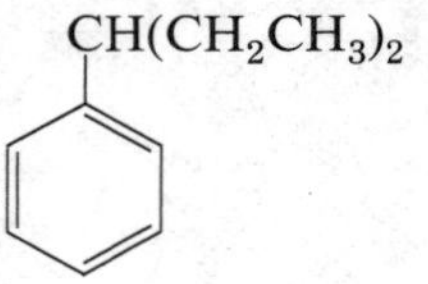 CH(CH₂CH₃)₂ (on benzene ring)

(b) 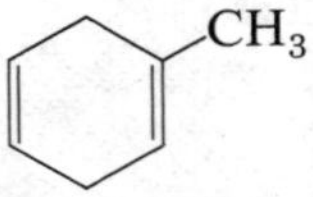 CH₃ (on cyclohexene ring)

(c) 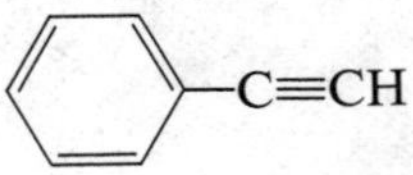 benzene ring—C≡CH

(d) [benzene ring]—CH_2CH_2—[benzene ring]

(e) [benzene ring with CH_2D at top and CH_3 at bottom]

(f) [benzene ring with $CH_2C≡CH$]

11. (a) Write out the steps of the free radical chain bromination of toluene to give benzyl bromide.

   (b) From $\Delta H_f°$ and $DH°$ values listed in Appendices I and II, calculate $\Delta H°$ for the reactions in part (a).

   (c) Compare these values with those for ethane and the tertiary position of isobutane. How feasible are these brominations?

12. Write structures for the eight possible benzene hexachlorides. Which one is capable of optical isomerism? Which one is slowest to react in E2 elimination reactions?

13. o-Phthalyl alcohol, 1,2-bis(hydroxymethyl)benzene, on treatment with acid, gives the corresponding cyclic ether.

[benzene ring with CH_2OH and CH_2OH substituents] $\xrightarrow{H^+}$ [bicyclic structure with O]

Give a reasonable mechanism for this reaction.

14. Which of the following hydrocarbons is expected to be the most acidic? Why?

[cyclopentadiene with $=CH_3$] [cyclopentadiene with CH_3 and $=CH_2$] [cyclopentadiene with CH_3 and $=CH_2$]

15. Each of the following compounds is known for a given value of x. Determine x for each case.

   (a) $(C_5H_5)Co(CO)_x$

   (b) $(C_6H_6)Mn(CO)_x{}^+$

   (c) $(C_5H_5)Mn(C_6H_5C≡CC_6H_5)(CO)_x$

   (d) $(C_6H_6)Mo(CO)_x$

   (e) $(C_5H_5)V(CO)_x$

   (f) $(C_7H_7)Mo(CO)_x{}^+$

   (g) $(C_5H_5)FeI(CO)_x$

16. Ferrocene undergoes many electrophilic substitution reactions and behaves in this regard as an aromatic system more reactive than benzene. The substitution intermediate can be described as a butadiene-type ligand associated with a *ferric* central metal.

   (a) Write the structure of the substitution intermediate in detail and determine whether it is a sixteen- or eighteen-electron organometallic.

   (b) If a substituted ferrocene is used, the second substituent can either enter the same ring as the first or the other ring. If it enters the same ring as the first substituent, how many isomers are possible? If it enters the other ring, how many isomers are possible with and without free rotation of the rings?

Chapter 23
Electrophilic Aromatic Substitution

In the previous chapter we encountered aromatic compounds for the first time and learned something of the special properties of the aromatic ring. In this chapter we will look at the most important reaction of the aromatic ring—substitution of one electrophile (usually a proton) by another electron-deficient species. The reaction applies to a number of different electrophilic reagents and provides an important route to many substituted aromatic compounds. Moreover, when applied to benzene systems already containing one or more substituents, the reaction shows specificity and reactivity effects which are readily rationalized by theory. Thus, this chapter provides an especially integrated combination of theory and synthesis.

23.1
Halogenation

Alkenes react rapidly with bromine even at low temperatures to give the product of *addition* of bromine.

$$CH_2{=}CH_2 + Br_2 \longrightarrow BrCH_2CH_2Br \qquad \Delta H^\circ = -29.2 \text{ kcal mole}^{-1}$$

The reaction is highly *exothermic* because two C—Br bonds are substantially more stable than a Br—Br bond and the second bond of a double bond. The corresponding addition reaction of benzene is slightly *endothermic*.

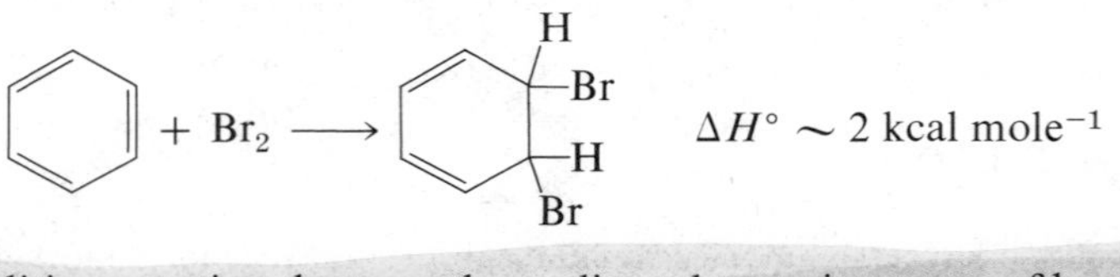

$$\Delta H^\circ \sim 2 \text{ kcal mole}^{-1}$$

Such an addition reaction destroys the cyclic π electronic system of benzene. Note that the difference in ΔH° for the two cases is approximately the resonance energy of benzene.

Benzene does react with bromine, but the reaction requires the use of appropriate Lewis acids such as ferric bromide. The product of the reaction is the result of *substitution* rather than addition.

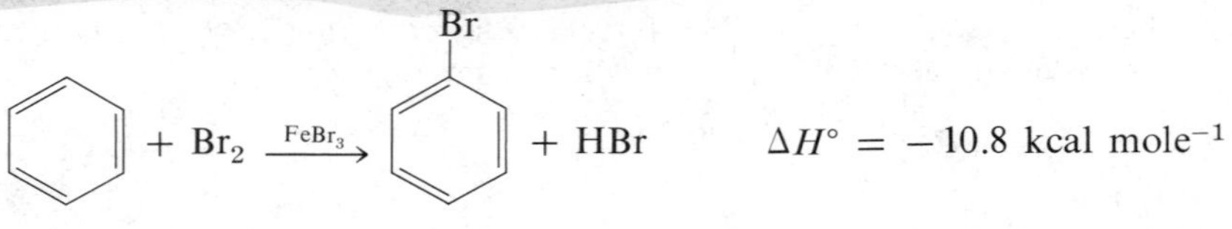

$$\Delta H^\circ = -10.8 \text{ kcal mole}^{-1}$$

Sixty grams of bromine is added slowly to a mixture of 33 g of benzene and 2 g of iron filings. The mixture is warmed for an additional half hour; the red vapors of bromine should no longer be visible. Water is added, and the washed and dried organic layer is distilled to give 40 g of bromobenzene, b.p. 156°.

In this procedure, the iron reacts rapidly with bromine to give ferric bromide. Anhydrous ferric halides are Lewis acids and react avidly with bases such as water.

$$FeX_3 + H_2O \rightleftharpoons H_2O \cdot FeX_3 \longrightarrow \text{higher hydrates}$$

The anhydrous salts are difficult to keep pure and are frequently made from the elements as needed, as in the procedure given above for the bromination of benzene.

It is the Lewis-acid character of ferric salts that allows them to function as catalysts in this reaction. Aluminum halides are used frequently for the same purpose. Recall that the first step in the bromination of an alkene is a displacement by the alkene as a nucleophile on bromine with bromide ion as a leaving group (Section 11.6.B.3).

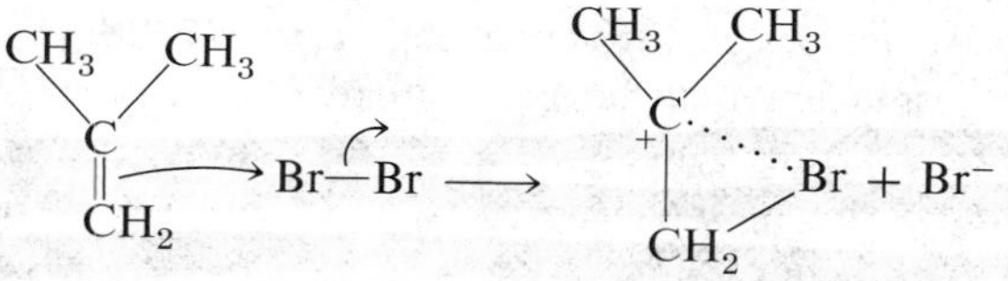

In nonpolar solvents, the leaving bromide ion requires additional solvation by bromine (page 583).

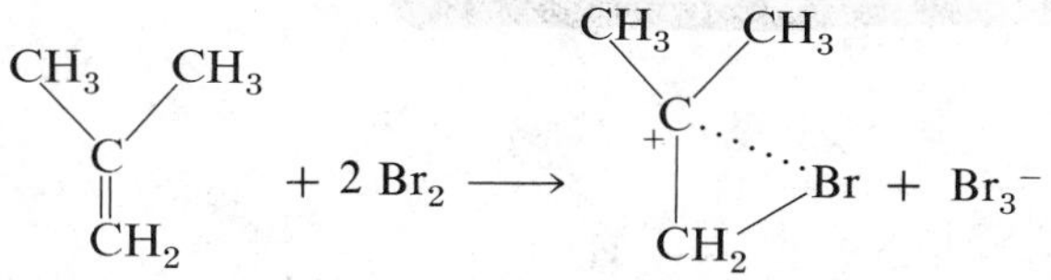

In this case, a bromine molecule serves as a mild Lewis acid to help pull bromide ion from bromine. This type of "pull" is provided more powerfully by a stronger Lewis acid such as ferric bromide.

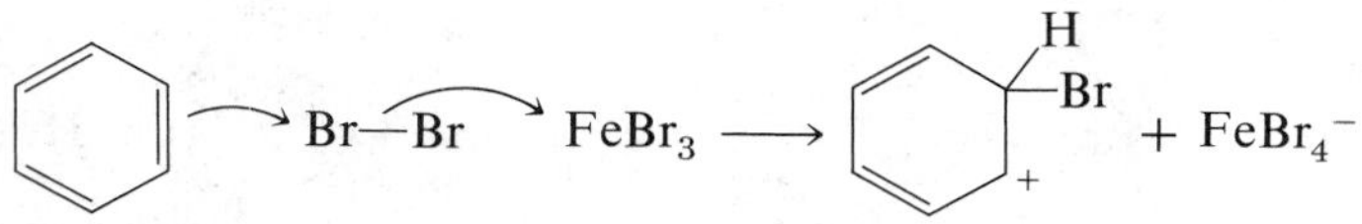

Benzene is a much weaker nucleophilic reagent than a simple alkene and requires a more electrophilic reagent for reaction.

The intermediate in the bromination of benzene is a conjugated carbocation. Its structure may be expressed by three Lewis structures

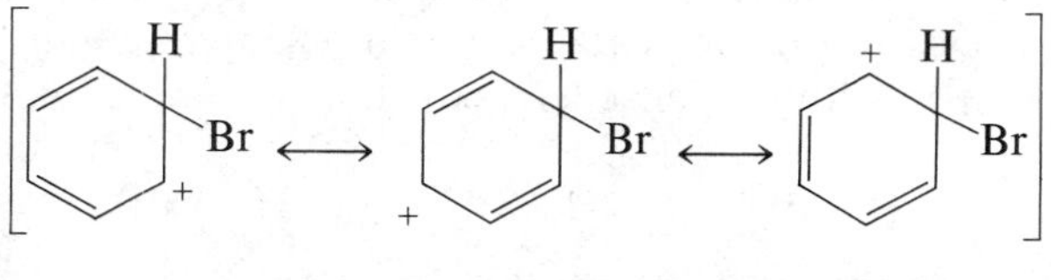

The resulting structure is that of an approximately tetrahedral carbon attached to a planar pentadienyl cation, as shown by the stereo representation in Figure 23.1.

This resonance-stabilized pentadienyl cation is often symbolized by using a

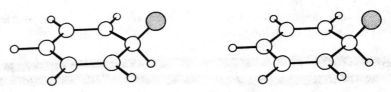

FIGURE 23.1. *Stereo representation of the intermediate in the bromination of benzene.*

dotted line to indicate that the positive charge is delocalized over the three positions indicated in the foregoing resonance structures.

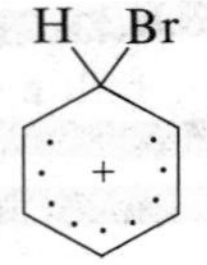

Again, however, this symbol conveys no accounting of electrons. The student is urged to use Lewis structures exclusively at this stage in order to understand more fully the electron displacements that occur in reactions.

Carbocations can generally react with a nucleophilic reagent, rearrange, or lose a proton. Reaction of the pentadienyl cation intermediate with a nucleophile would give a product without the benzene-ring resonance. Consequently, this type of reaction is rarely observed in electrophilic aromatic reactions. Rearrangements are significant only in some special cases to be discussed later. The only important reaction of our bromination intermediate is loss of a proton to restore the cyclic π-system and yield the substitution product.

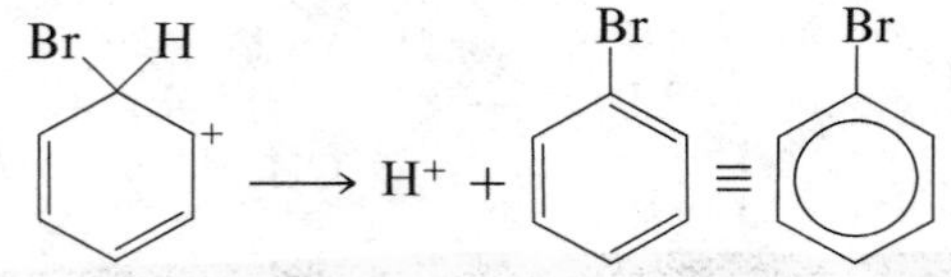

The overall reaction sequence is

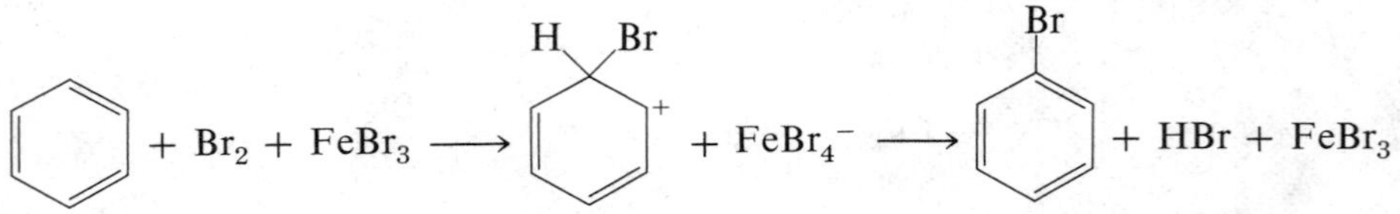

The first step is rate-determining as indicated by the energy profile shown in Figure 23.2. The experimental evidence for this reaction mechanism comes from many studies of chemical kinetics, isotope effects, and structural effects.

The reaction mechanism for bromination of benzene is general for other electrophilic aromatic substitutions as well. Reaction occurs with an electron-deficient (electrophilic) species to give a pentadienyl cation intermediate, which loses a proton to give the substituted benzene product.

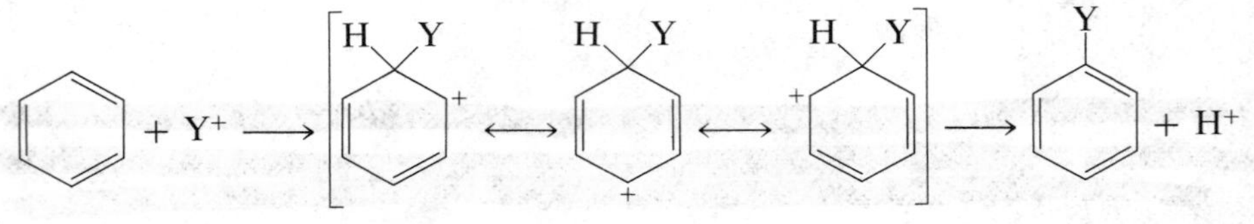

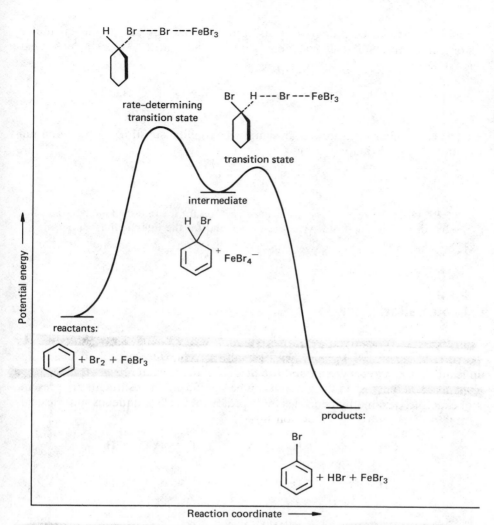

FIGURE 23.2. *Reaction profile for bromination of benzene.*

Chlorination is directly analogous to bromination.

The last example allows an important comparison of these electrophilic halogenations and the free radical halogenations of methylbenzenes that we encountered in Section 22.5.A. Recall that under free radical conditions substitution occurs in the side chain, principally at the benzylic position.

Iodobenzene can be prepared by using iodine and an oxidizing agent under acidic conditions. Suitable oxidizing agents include nitric acid or arsenic acid (H_3AsO_4).

$$C_6H_6 + I_2 + HNO_3 \xrightarrow{50°} NO + NO_2 + C_6H_5I$$

$$(86–87\%)$$

The reaction involves a normal aromatic electrophilic substitution by iodonium ion obtained by oxidation

$$I_2 \xrightarrow{-2\ e^-} 2\ I^+$$

EXERCISE: Write all of the resonance structures for the intermediate free radical produced from reaction of *p*-xylene with Br· and for the intermediate carbocation produced from the reaction of *p*-xylene with Br+.

23.2
Protonation

Benzene is an extremely weak base, much weaker than an alkene. Benzene is only slightly protonated in concentrated sulfuric acid, whereas isobutylene is significantly protonated even in sulfuric acid containing water. Some protonation does occur, however, and the amount can be significant if substituents are present. For example, hexamethylbenzene is 50% protonated in 90% aqueous sulfuric acid.

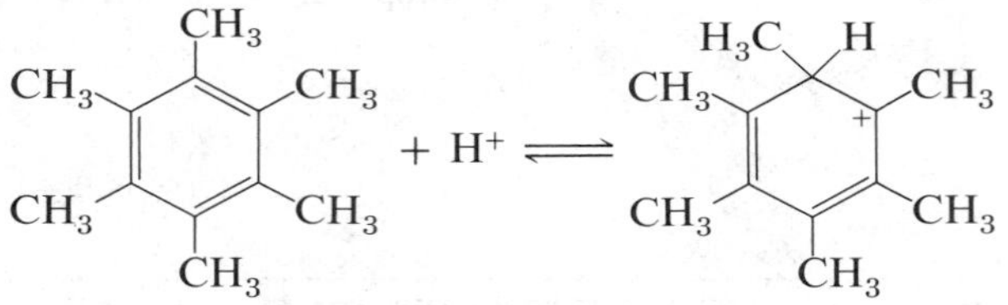

Protonation of benzene can be detected by hydrogen isotope exchange reactions in acid. If benzene is stirred for several days at room temperature with 80% aqueous sulfuric acid containing deuterium or tritium, the isotope distributes between the benzene and the aqueous acid.

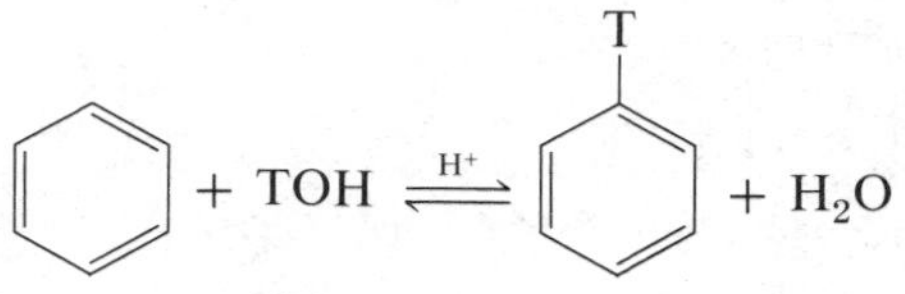

benzene-*t*

Tritium is normally used as a radioactive **tracer isotope.** It is typically used in a ratio of less than one part per million of ordinary hydrogen. Therefore, in an exchange process such as this, it is unlikely that a given molecule will have more than one tritium bound to it. The radioactivity of tritium can be measured by a sensitive instrument called a **liquid scintillation counter;** hence, tritium incorporation can be precisely measured by using only a small amount of the isotope. On the other hand, deuterium is used as a **macroscopic isotope.** Incorporation must be monitored by much less sensitive analytical techniques, such as nmr or mass spectrometry. The exchange reaction will give mixtures of deuteriated benzenes containing varying numbers of deuterium atoms attached to the ring. The amount of deuterium incorporation will depend on the relative amounts of ^{1}H and ^{2}H isotopes in the hydrogen "pool." If a large excess of D_2SO_4 and D_2O is used, benzene-d_6, C_6D_6, can be obtained.

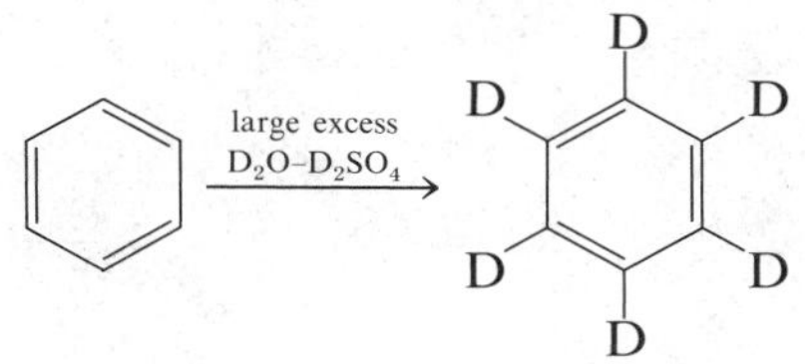

benzene-d_6

The exchange reaction is a simple type of electrophilic aromatic substitution reaction in which the electrophilic reagent is D^+.

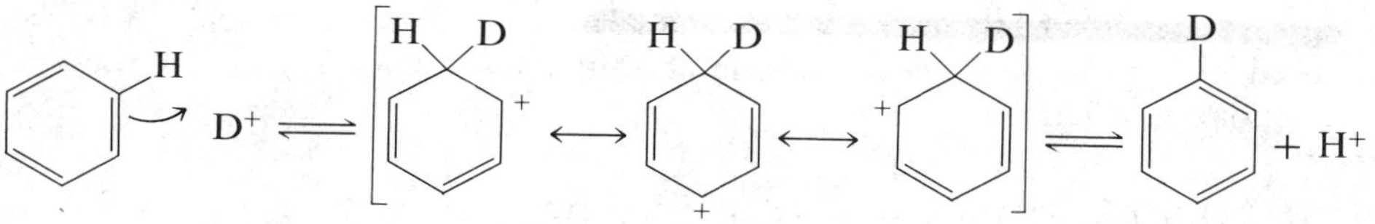

The intermediate pentadienyl cation undergoes only one significant reaction—loss of a proton (or a deuteron). This reaction, which regenerates the aromatic π-system, is much faster than its reaction with water. With ordinary carbocations, reaction with a nucleophilic species is an important reaction, but with such carbocations, elimination of a proton does not have the extraordinary driving force of the formation of an aromatic ring.

EXERCISE: What product would you expect from treatment of benzene with excess D_2O–D_2SO_4 containing some tritiated water?

23.3
Nitration

The reaction of alkenes with nitric acid is not a generally useful reaction. Addition of nitric acid to the double bond is accompanied by more or less oxidation. However, benzene is quite stable to most oxidizing agents, and its reaction with nitric acid is an important organic reaction. Actually, the nitrating reagent generally used is a mixture of concentrated nitric acid and sulfuric acid.

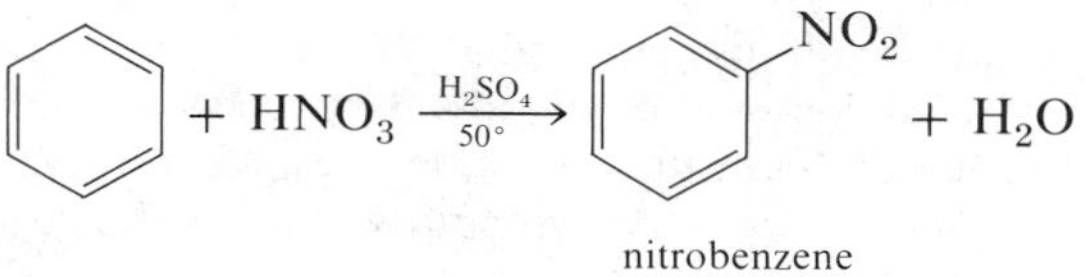

nitrobenzene

To a flask containing 65 g of benzene is added a mixture of 110 mL of conc. H_2SO_4 and 85 mL of conc. HNO_3. The acid mixture is added in portions so that the temperature does not exceed 50°. After all of the acid has been added, the reaction mixture is cooled and the oily nitrobenzene layer is separated, washed, and distilled. The yield of pure product is 85–88 g (83–86%).

The nitro group is an important functional group in aromatic chemistry because it may be converted into many other functional groups. The nitration reaction thus provides a route to many substituted aromatic compounds. The chemistry of the nitro group will be detailed in Section 25.1. Many properties of the nitro group can be interpreted on the basis of a resonance hybrid of two Lewis structures.

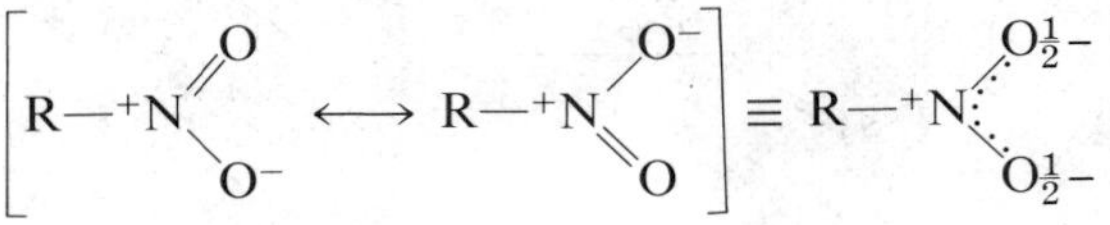

In these structures, the O—N—O system is seen to have an allylic anion type of π-system.

In a mixture of nitric and sulfuric acids, an equilibrium is established in which many species are present. One of these species is the nitronium ion, NO_2^+, which has been detected by spectroscopic methods. In the mixture of acids, it is produced by a process in which sulfuric acid functions as an acid and nitric acid functions as a base.

$$(1) \quad H_2SO_4 + HONO_2 \rightleftharpoons H_2\overset{+}{O}NO_2 + HSO_4^-$$

$$(2) \quad H_2\overset{+}{O}NO_2 + H_2SO_4 \rightleftharpoons H_3O^+ + {}^+NO_2 + HSO_4^-$$

$$\overline{2\,H_2SO_4 + HONO_2 \rightleftharpoons H_3O^+ + NO_2^+ + 2\,HSO_4^-}$$

The structure of nitronium ion is known from spectroscopic measurements. It is related to the isoelectronic compound carbon dioxide. The molecule is linear and is a powerful electrophilic reagent.

$$O=\overset{+}{N}=O$$

nitronium ion

It reacts directly with benzene to give a pentadienyl cation intermediate.

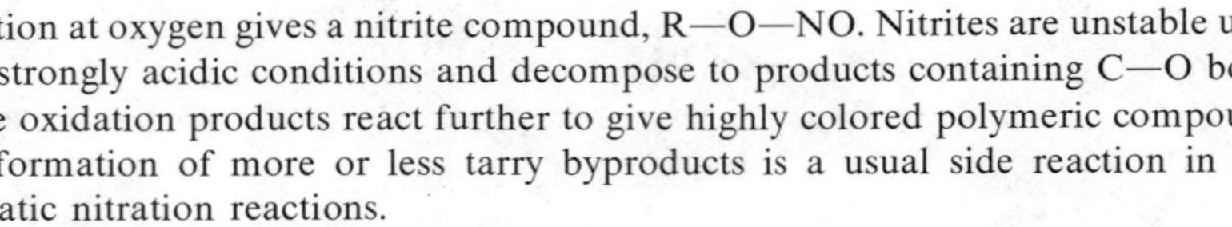

Note that reaction occurs on nitrogen rather than oxygen.

Reaction at oxygen gives a nitrite compound, R—O—NO. Nitrites are unstable under such strongly acidic conditions and decompose to products containing C—O bonds. These oxidation products react further to give highly colored polymeric compounds. The formation of more or less tarry byproducts is a usual side reaction in most aromatic nitration reactions.

Aromatic nitro compounds are important intermediates for the synthesis of other aromatic deviatives. The most important reaction of the nitro group is reduction. We shall return to this subject in Section 24.6.C.

23.4
Friedel–Crafts Reactions

A. *Acylations*

The electrophile in electrophilic aromatic substitution can also be a carbocation. Such reactions are called **Friedel–Crafts reactions.** The most useful version of the reaction is Friedel–Crafts **acylation** in which the entering electrophile is an acyl group, RCO—, derived from a carboxylic acid derivative, usually an acyl halide or anhydride. The carbonyl group in such acid derivatives is sufficiently basic that formation of a complex occurs with strong Lewis acids such as aluminum chloride.

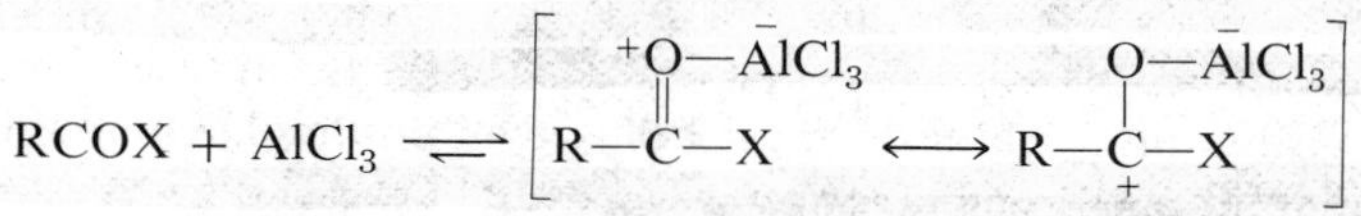

$$RCOX + AlCl_3 \rightleftharpoons \left[\begin{array}{c} {}^+O-\bar{A}lCl_3 \\ \| \\ R-C-X \end{array} \longleftrightarrow \begin{array}{c} O-\bar{A}lCl_3 \\ | \\ R-\overset{+}{C}-X \end{array} \right]$$

Aluminum chloride, $AlCl_3$, can be prepared by the direct reaction of aluminum with chlorine or hydrogen chloride. The anhydrous compound is available as a white powder that fumes in air and has the strong odor of HCl from reaction with atmospheric moisture. It can be sublimed and is soluble in many organic solvents such as benzene, nitrobenzene, and carbon tetrachloride. In benzene solution aluminum chloride exists as a dimer, Al_2Cl_6. Anhydrous aluminum chloride reacts vigorously with water with evolution of HCl. It is a strong Lewis acid that forms complexes with most oxygen-containing compounds. In laboratory use it is kept in tightly sealed bottles, and the fine powder is handled in air as little as possible, preferably in a hood.

The carbocation character of a carbonyl carbon is tremendously enhanced by coordination to aluminum chloride, and in many cases the complex itself is sufficiently electrophilic to react with aromatic rings. In other cases, the complex exists in equilibrium with a small amount of the corresponding **acylium ion,** which is an even more powerful electrophile.

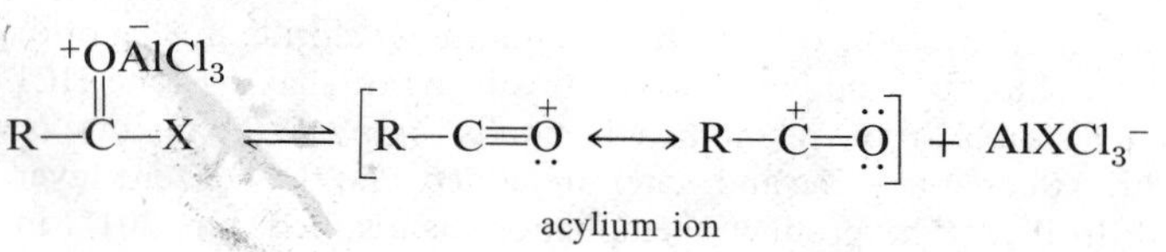

$$\begin{array}{c} {}^+O\bar{A}lCl_3 \\ \| \\ R-C-X \end{array} \rightleftharpoons \left[R-C\equiv\overset{+}{O} \longleftrightarrow R-\overset{+}{C}=\overset{..}{\overset{..}{O}} \right] + AlXCl_3^- \qquad (23\text{-}1)$$

acylium ion

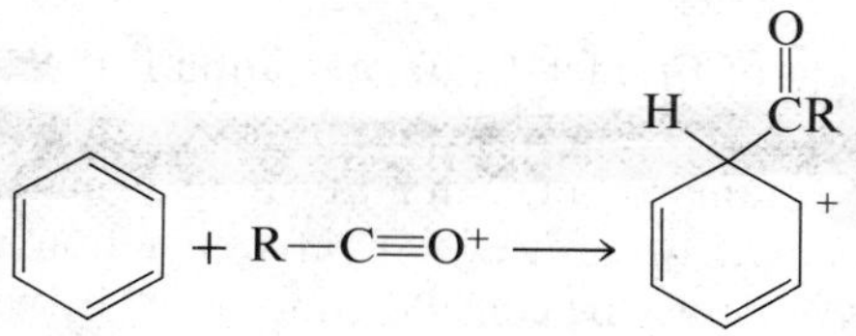

$$\qquad (23\text{-}2)$$

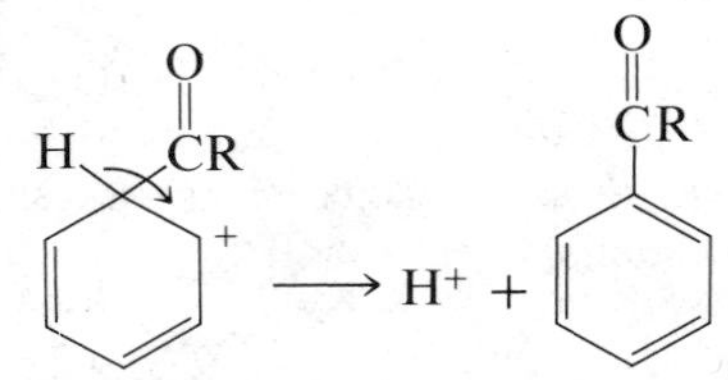

$$\text{(23-3)}$$

As shown in equations (23-1)–(23-3), the mechanism for reaction of the acylium ion with benzene is completely analogous to that of other electrophilic reagents. The final product is an aromatic ketone whose carbonyl group is sufficiently basic to be complexed completely by aluminum chloride.

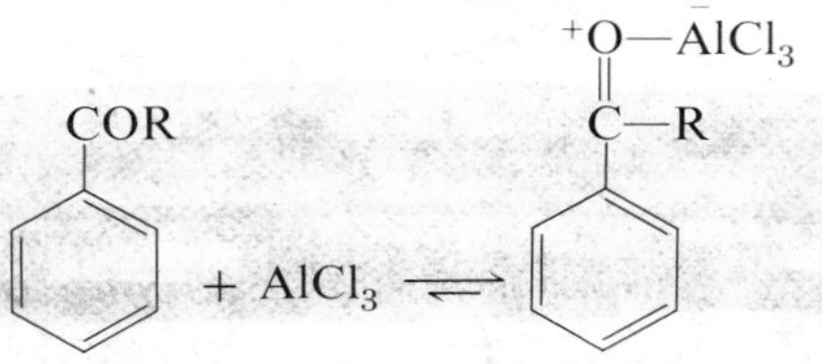

It is this complex that is the actual reaction product. The work-up procedure involves treatment with water or dilute hydrochloric acid to decompose the complex and dissolve the aluminum salts. The liberated ketone remains in the organic layer and is isolated by crystallization or distillation. Because it complexes with the product, aluminum chloride must be used in equimolar amounts. Furthermore, the complexed ketone is resistant to further reaction so that high yields of pure product are readily available by this reaction. Friedel–Crafts acylation is an important and useful reaction in aromatic chemistry. An example is the acylation of benzene with acetyl chloride.

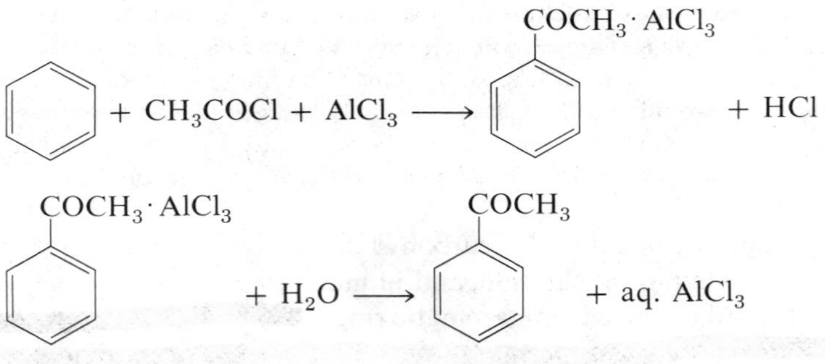

$$+ CH_3COCl + AlCl_3 \longrightarrow \quad COCH_3 \cdot AlCl_3 \quad + HCl$$

$$COCH_3 \cdot AlCl_3 \quad + H_2O \longrightarrow \quad COCH_3 \quad + \text{aq. } AlCl_3$$

acetophenone

> To a cooled mixture of 40 g of anhydrous aluminum chloride in 88 g of dry benzene, 29 g of acetyl chloride is added slowly with stirring or shaking. The HCl evolved is absorbed in a suitable trap. When the addition is complete, the mixture is warmed to 50° for 1 hr. After cooling, ice and water are added, and the benzene layer is washed, dried, and distilled. The product acetophenone is distilled, b.p. 201°, in a yield of 27 g.

An interesting variant of Friedel–Crafts acylation is the **Gatterman–Koch** aldehyde synthesis, the reaction of an aromatic hydrocarbon with carbon monoxide and hydrogen chloride in the presence of a Lewis acid such as aluminum chloride. The reaction is equivalent to a Friedel–Crafts acylation with formyl chloride, HCOCl, and may conveniently be considered to occur by way of the electrophilic intermediate $HCO^+AlCl_4^-$.

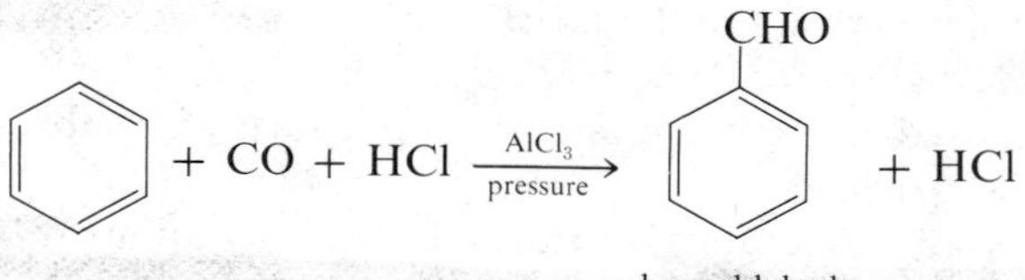

$$\text{benzene} + CO + HCl \xrightarrow[\text{pressure}]{AlCl_3} \text{(CHO benzene)} + HCl$$

benzaldehyde

The reaction is primarily of industrial importance.

B. *Alkylations*

Benzene undergoes Friedel–Crafts **alkylation** when treated with an alkyl halide and a Lewis-acid catalyst such as $FeBr_3$ or $AlCl_3$. An example is the reaction of benzene with *t*-butyl chloride to give *t*-butylbenzene.

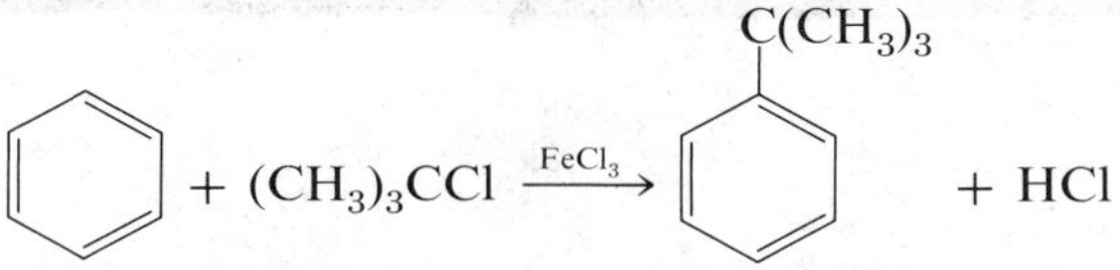

$$\text{benzene} + (CH_3)_3CCl \xrightarrow{FeCl_3} \text{C}(CH_3)_3\text{benzene} + HCl$$

> To a mixture of 25 g of anhydrous ferric chloride (or 50 g of anhydrous aluminum chloride) in 200 mL of benzene cooled to 10° is added slowly and with stirring 50 g of *t*-butyl chloride. The mixture is allowed to warm to room temperature, and when no further HCl is evolved, ice, water, and dilute hydrochloric acid are added. The organic layer is washed, dried, and distilled to give *t*-butylbenzene, b.p. 168.5°.

In this Friedel–Crafts alkylation the attacking electrophile is the *t*-butyl cation, which is produced in the reaction of *t*-butyl chloride with $FeCl_3$. In the absence of other nucleophiles, this electrophilic species reacts with the aromatic ring.

$$(CH_3)_3CCl + FeCl_3 \rightleftharpoons (CH_3)_3C^+FeCl_4^-$$

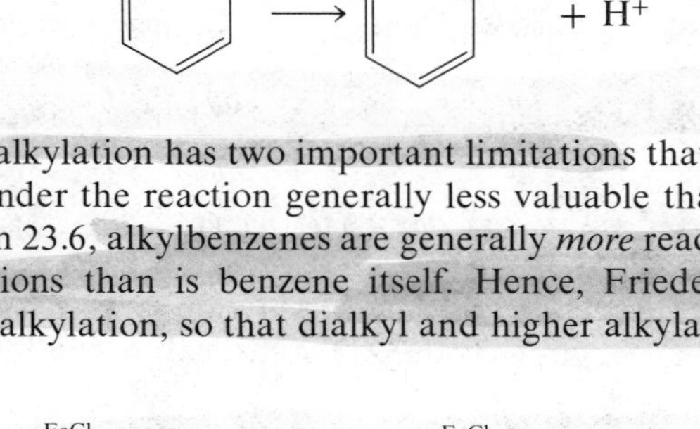

$$\text{benzene} + (CH_3)_3C^+ \longrightarrow \text{[(CH_3)_3C, H arenium ion]}$$

$$\text{[(CH_3)_3C, H arenium ion]} \longrightarrow \text{C(CH_3)_3 benzene} + H^+$$

Friedel–Crafts alkylation has two important limitations that severely restrict its usefulness and render the reaction generally less valuable than acylation. As we shall see in Section 23.6, alkylbenzenes are generally *more* reactive in electrophilic substitution reactions than is benzene itself. Hence, Friedel–Crafts alkylation tends to give overalkylation, so that dialkyl and higher alkylated by-products are formed.

$$C_6H_6 \xrightarrow[(CH_3)_3CCl]{FeCl_3} C_6H_5C(CH_3)_3 \xrightarrow[\substack{(CH_3)_3CCl \\ \text{(faster)}}]{FeCl_3} C_6H_4[C(CH_3)_3]_2$$

The only practical way of controlling such additional reactions is to keep benzene in large excess. This approach is practical with benzene itself, an inexpensive compound often used as a solvent, but it is impractical with substituted benzenes, which are more expensive.

Another important limitation of Friedel–Crafts alkylations relates to an alternative reaction of many carbocations, particularly in the absence of reactive nucleophiles, namely, rearrangement to isomeric carbocations. Isopropyl chloride or bromide reacts normally with aluminum chloride and benzene to give isopropylbenzene.

However, 1-chloropropane also gives isopropylbenzene under these conditions. Rearrangement to the secondary carbocation is essentially complete.

Primary alkyl halides are less reactive than secondary or tertiary halides, and higher temperatures are normally required. Under some conditions, the rearrangement of primary systems is only partial. Under these conditions, a displacement reaction by benzene on the alkyl halide coordinated with the Lewis acid competes with carbocation rearrangement. It should be emphasized, however, that at least some rearrangement always occurs with suitable primary systems and greatly limits the utility of this reaction.

Friedel–Crafts alkylations may also be accomplished with alcohols and a catalyst such as aluminum chloride or boron trifluoride. The reaction has the same limitations as the alkyl halide reactions in requiring a large excess of benzene and in giving rearrangement products in suitable cases. In addition, one reaction product is water, which coordinates with Lewis acids. Thus, with alcohols a stoichiometric amount of Lewis acid is required.

Note that equilibration of the isomeric pentyl cations is rapid compared to alkylation of benzene.

$$CH_3\overset{+}{C}HCH_2CH_2CH_3 \rightleftharpoons CH_3CH_2\overset{+}{C}HCH_2CH_3 \rightleftharpoons CH_3CH_2CH_2\overset{+}{C}HCH_3$$

Since both cations are secondary, they are of comparable stability and the observed product ratio (2:1) reflects the 2:1 statistical bias in favor of the 2-pentyl cation.

Alkylation reactions can also be accomplished with alkenes. Typical catalysts used, HF–BF_3 and HCl–$AlCl_3$, generate carbocations in the usual way.

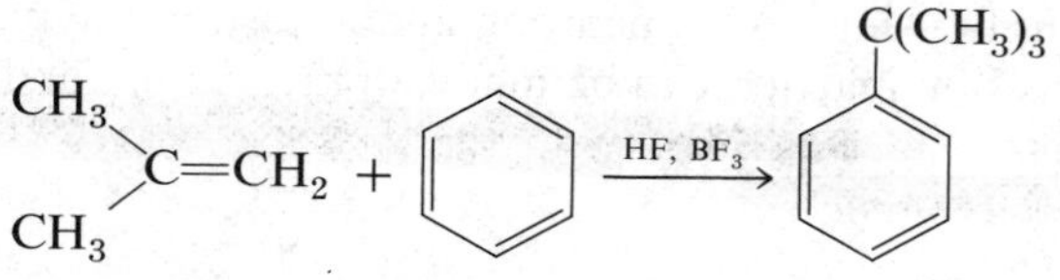

$$(CH_3)_2C{=}CH_2 + HF + BF_3 \rightleftharpoons (CH_3)_3C^+\, BF_4^-$$

This reaction is used industrially to prepare alkylbenzenes, but it is not an important laboratory reaction.

A reaction that is closely related to Friedel–Crafts alkylation is **chloromethylation,** the reaction of aromatic rings with formaldehyde, hydrogen chloride, and a Lewis acid such as zinc chloride.

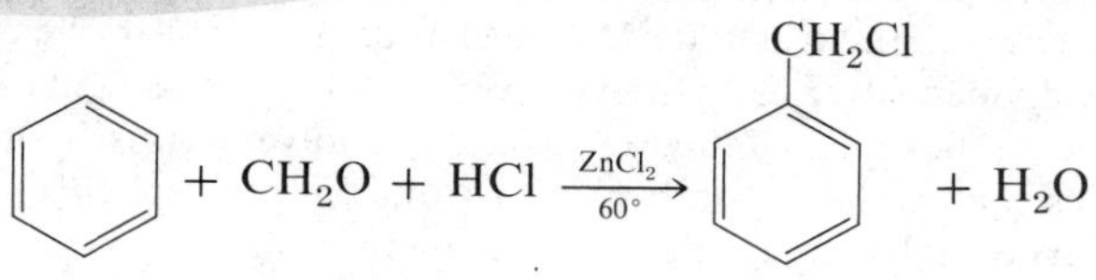

(79%)

The reaction is an electrophilic aromatic substitution, probably by the oxonium ion formed by coordination of the formaldehyde with the Lewis acid.

$$[H_2C{=}\overset{+}{O}{-}\overset{-}{Z}nCl_2 \longleftrightarrow H_2\overset{+}{C}{-}O{-}\overset{-}{Z}nCl_2]$$

The resulting reagent will react with aromatic rings that are at least as reactive as benzene. The product of electrophilic aromatic substitution by the coordinated aldehyde is the corresponding alcohol, but this alcohol is benzylic, and in the presence of $ZnCl_2$–HCl it is converted rapidly to the corresponding chloride.

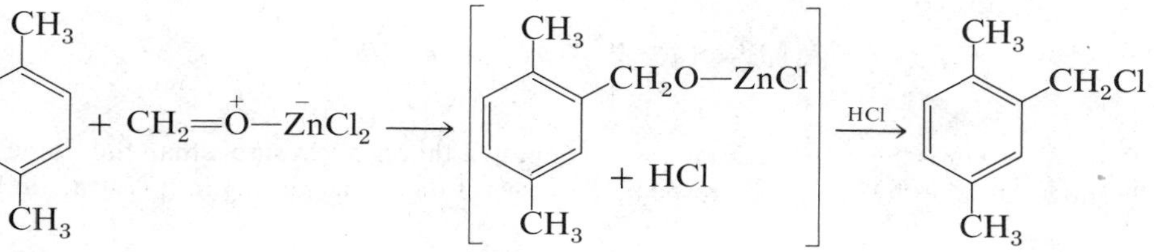

[The chloromethylation reaction must be conducted in an efficient hood and with extreme care. Under these reaction conditions bis-chloromethyl ether, $ClCH_2OCH_2Cl$, is produced. This compound is a potent carcinogen and should be avoided whenever possible.]

An important side reaction in chloromethylation reactions is reaction of the product, which is a reactive alkyl halide, with the starting aromatic compound. For example, chloromethylation of benzene gives some diphenylmethane, which arises from reaction of the initial product, benzyl chloride, with benzene.

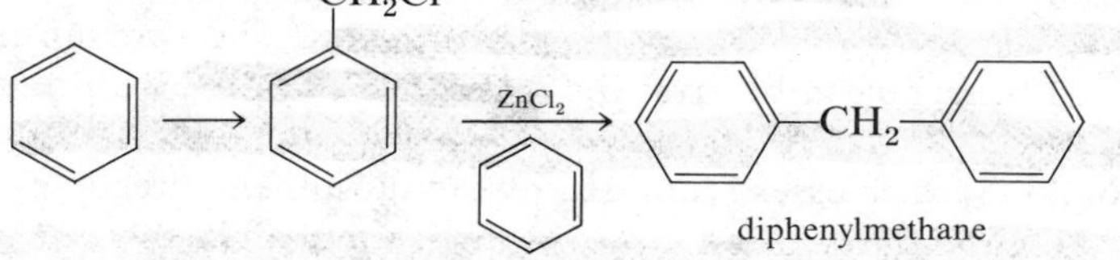

diphenylmethane

Of course, by simply using excess benzene and adjusting the reaction conditions appropriately, this side reaction can be made a practical method for the synthesis of diphenylmethane. A related reaction is the Friedel–Crafts alkylation of benzene with carbon tetrachloride.

$$\text{(benzene)} + CCl_4 \xrightarrow[\Delta]{AlCl_3} (C_6H_5)_3CCl$$

(84–86%)
triphenylmethyl chloride

Triphenylmethyl chloride, or "trityl" chloride, is a colorless crystalline solid, m.p. 111–112°, which forms the starting point of a fascinating chapter of organic chemistry. In experiments reported in 1900 Moses Gomberg treated triphenylmethyl chloride with finely divided silver in an inert atmosphere to obtain a white solid hydrocarbon formulated as hexaphenylethane. In organic solvents this hydrocarbon gives yellow solutions that rapidly absorb oxygen from the atmosphere. These solutions contain a relatively stable free radical, triphenylmethyl.

$$(C_6H_5)_3C\!-\!C(C_6H_5)_3 \rightleftharpoons 2\,(C_6H_5)_3C\cdot \xrightarrow{O_2} (C_6H_5)_3COOC(C_6H_5)_3$$

bis-triphenylmethyl
peroxide

The stability of triphenylmethyl radical stems from π-conjugation in the radical and steric hindrance in the dimer. Equilibrium studies showed the central bond in the dimer to have a strength of only 11 kcal mole^{-1}. A most remarkable epilogue to this story was provided by recent structural studies based primarily on nmr evidence that show that the hydrocarbon considered to be hexaphenylethane for the better part of a century does not have this structure at all, but is instead the product of dimerization at one *para* position.

$$2\,(C_6H_5)_3C\cdot \rightleftharpoons (C_6H_5)_3C\!-\!\underset{H}{\overset{}{\bigcirc}}\!=\!C\!\overset{C_6H_5}{\underset{C_6H_5}{}}$$

In other words, hexaphenylethane is so congested that it is less stable than the isomer shown despite the loss of the resonance energy of one benzene ring in this structure!

EXERCISE: Write the equations illustrating the reaction of benzene with (a) butanoyl chloride + aluminum chloride, (b) 2-methylpropene + HF + BF$_3$, (c) 2-butanol + H$_2$SO$_4$, (d) formaldehyde + HCl + ZnCl$_2$, (e) 2-chloro-2-methylbutane + AlCl$_3$.

23.5
Orientation in Electrophilic Aromatic Substitution

Benzene can give only a single monosubstituted product in electrophilic aromatic substitution. However, substitution on a compound that already has a group attached to the ring can give three products. The two substituents in a disubstituted benzene can be arranged *ortho, meta,* or *para* with respect to each other. These three isomers are generally *not* formed in equal amounts. The product distribution in such cases is affected by the substituent already present on the ring. With some groups further substitution gives mainly the *ortho-* and *para-*

703

Sec. 23.5

**Orientation in
Electrophilic
Aromatic
Substitution**

disubstituted products. Examples are seen in the products formed from bromination of bromobenzene or anisole or in the nitration of toluene.

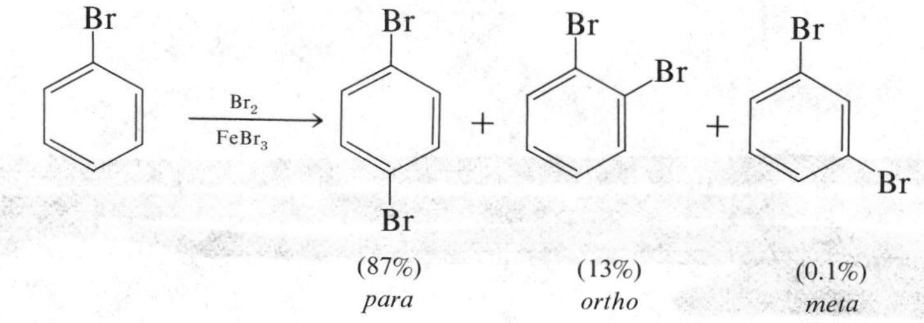

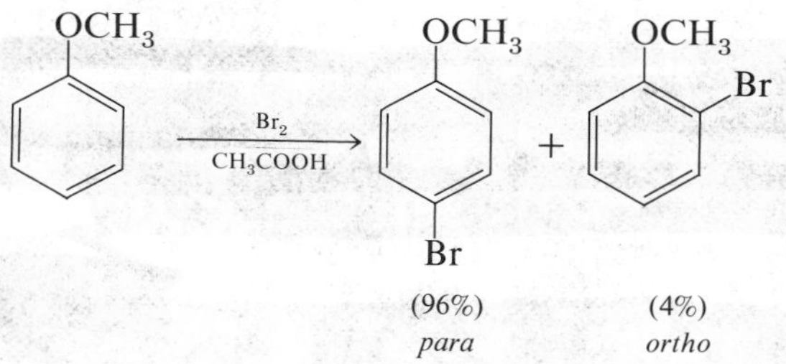

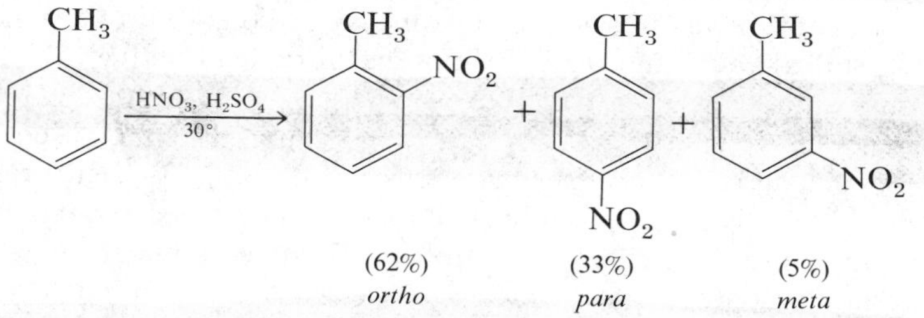

(87%)
para

(13%)
ortho

(0.1%)
meta

(96%)
para

(4%)
ortho

(62%)
ortho

(33%)
para

(5%)
meta

On the other hand, further substitution on some substituted benzenes gives essentially all *meta*-disubstituted product, with little or none of the *ortho* and *para* products. Examples are bromination and nitration of nitrobenzene and nitration of methyl benzoate.

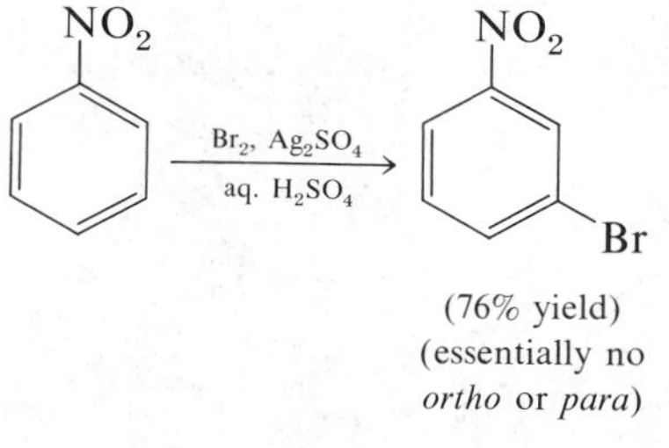

(76% yield)
(essentially no
ortho or *para*)

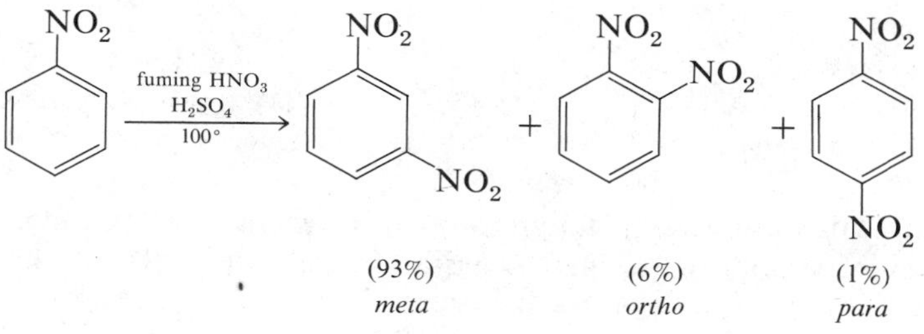

(93%)
meta

(6%)
ortho

(1%)
para

(81–85% yield)

As the foregoing examples suggest, substituent groups can be divided into two categories, those that are **ortho,para directors** and those that are **meta directors.** Bromo, methyl, and methoxy groups are *ortho,para* directors and nitro and ester groups are *meta* directors. Note that *ortho* and *para* are produced together, although the *ortho/para* ratio may vary with different groups and under different reaction conditions with the same group.

Substituent groups may also be characterized with respect to their effect on the *rate* of further substitution reactions. Some substituents cause the aromatic ring to be *less reactive* than benzene itself; these groups are said to be **deactivating.** An example is the nitro group, which is a powerful deactivating group. Nitrobenzene is much less reactive than benzene, as shown by the conditions required for nitration of benzene (page 695) and nitrobenzene (page 703). When two deactivating groups are attached to the ring, even more drastic conditions are necessary for further substitution. For example, *m*-dinitrobenzene may be converted into 1,3,5-trinitrobenzene in 45% yield by heating 60 g of the dinitrobenzene with 1 kg of fuming sulfuric acid and 0.5 kg of fuming nitric acid at 100° for 5 days.

Carbonyl groups are also deactivating groups. Thus, the products of Friedel–Crafts acylation are *less reactive* than the starting material. Consequently, it is quite easy to achieve monosubstitution in such reactions. In fact, Friedel–Crafts acylation is generally not applicable at all to aromatic rings that contain a strongly deactivating group. For example, nitrobenzene does not react under Friedel–Crafts acylation conditions and is even used as a solvent for such reactions!

Other groups are **activating;** electrophilic substitution on rings containing these groups is more rapid than with benzene. As we mentioned in our discussion of Friedel–Crafts alkylation, alkyl groups are activating groups, and hence alkylbenzenes are more reactive than benzene itself. This effect has an important consequence, since it means that the product of Friedel–Crafts alkylation is *more reactive* than the starting material. Thus, it is difficult to avoid the formation of overalkylation by-products.

The activating effect of alkyl groups can also be seen in the conditions required for nitration of mesitylene, which may be compared with the conditions used for the nitration of benzene itself (page 695).

mesitylene

(74–76%)
nitromesitylene

Mesitylene with its three activating groups is so reactive that it undergoes halogenation even without the normal Lewis-acid catalyst (see Section 23.1).

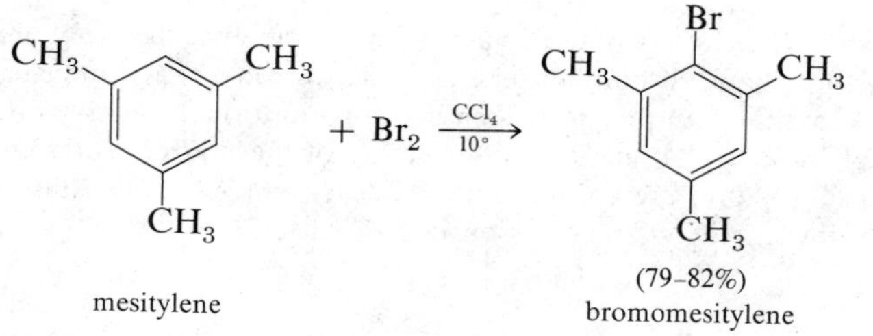

705

Sec. 23.5

**Orientation in
Electrophilic
Aromatic
Substitution**

mesitylene

(79–82%)
bromomesitylene

In such a case, the reaction may be considered to be a displacement reaction on halogen with the ring acting as a nucleophile.

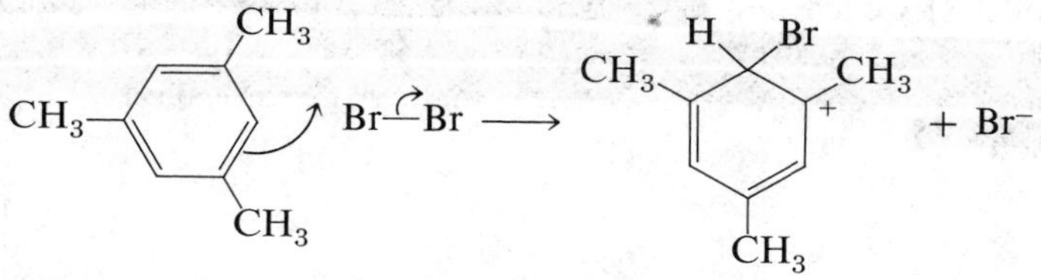

Mesitylene is especially reactive because the intermediate produced is highly stabilized; all three of the usual resonance structures correspond to tertiary carbocations:

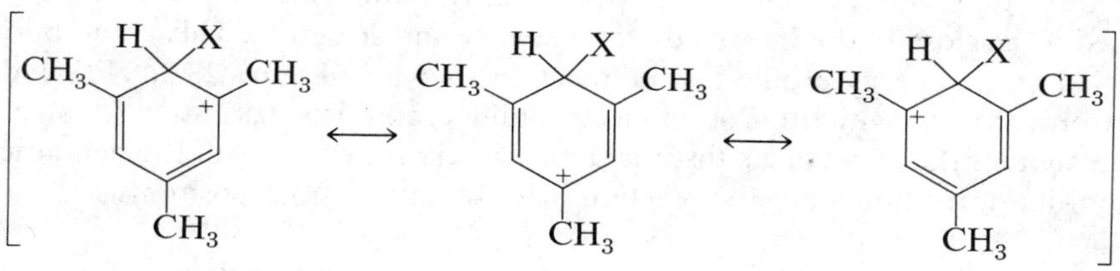

We will elaborate on the foregoing rationale for the activating effect of alkyl groups in Section 23.6 and 23.7. For the present, however, it is useful to note that substituent groups may be grouped into three different classes with regard to whether they are activating or deactivating and whether they are *ortho,para*-directing or *meta*-directing.

1. ***Ortho,para-directing and activating.*** Functional groups in this category include R (alkyl), NH_2, NR_2, and NHCOR (amino, alkylamino, and amide). OH, OR, and OCOR (hydroxy, alkoxy, and ester).
2. ***Ortho,para-directing and deactivating.*** The most important functional groups in this category are the halogens, F, Cl, Br, and I.
3. ***Meta-directing and deactivating.*** This group includes NO_2 (nitro), SO_3H (sulfonic acid), and all carbonyl compounds: COOH, COOR, CHO, and COR (carboxylic acids, esters, aldehydes, and ketones).

Note that all activating groups are *ortho,para* directors and all *meta* directors are deactivating. These generalizations derive from many experimental observations and form a set of empirical and useful rules. However, these rules are also subject to a consistent and satisfying interpretation by the modern theory of organic chemistry. This theory has its basis in the electron-donating and electron-attracting character of different functional groups, as discussed in the next section.

EXERCISE: Make a six-by-four matrix. On one side put the aromatic compounds toluene, acetophenone, bromobenzene, anisole, nitrobenzene, and acetanilide. On the other side put the four reactions bromination, nitration, Friedel–Crafts alkylation, and Friedel–Crafts acylation. Each intersection of the matrix represents a reaction. What products (if any) are formed in each reaction? Which reactions occur more rapidly than the analogous reaction on benzene itself?

23.6
Theory of Orientation in Electrophilic Aromatic Substitution

In Section 23.1 we learned that the mechanism of electrophilic aromatic substitution involves combination of a positive or electrophilic species with a pair of π-electrons of the benzene ring to form an intermediate having a pentadienyl cation structure.

The transition state leading to the pentadienyl cation intermediate may be assumed to be close to the intermediate in energy and structure. This assumption is implied in the energy profile for the reaction shown in Figure 23.3. The modern electronic theory of orientation in electrophilic aromatic substitution involves an assessment of the effect of a substituent on the relative energies of the pentadienyl cation-like transition state for reaction at different possible positions.

[The consideration that the transition states resemble the intermediate in structure is
another application of the Hammond postulate (page 115).]

For example, reaction at the *ortho* position of toluene gives rise to a transition state that resembles the intermediate

Two of the structures are those of secondary carbocations, but the third corresponds to a more stable tertiary carbocation. As a result, this intermediate and *hence also the transition state that leads to it* are more stable—have lower energy—than the corresponding intermediate and transition state for benzene in which all three resonance structures are those of secondary carbocations. The *ortho* position of toluene is therefore expected to be more reactive than a single position of benzene.

[This argument must be put on a per-hydrogen basis. Without specific orientation
preferences, statistics alone would give a reactivity ratio for benzene/*ortho*/*meta*/*para*
of 6:2:2:1.]

707

Sec. 23.6

Theory of
Orientation in
Electrophilic
Aromatic
Substitution

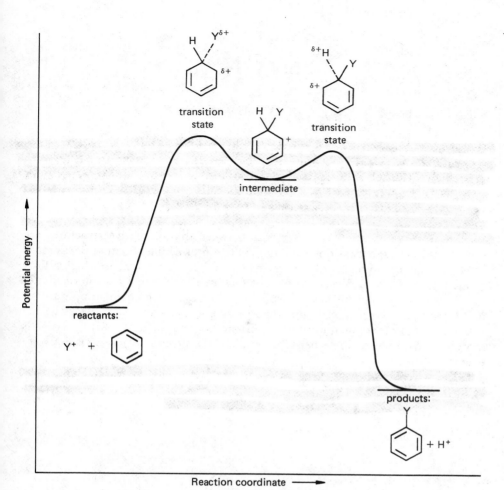

FIGURE 23.3. *Energy profile for electrophilic substitution on benzene.*

Reaction at the *meta* position gives rise to the following resonance structures.

All three structures are those of secondary carbocations. Each structure is stabilized slightly by the C_{methyl}–C_{ring} dipole.

Correspondingly, the *meta* position of toluene is expected to be somewhat more reactive than a benzene position but not nearly so reactive as in *ortho* position.

Finally, we apply this approach to the *para* position to generate the resonance structures

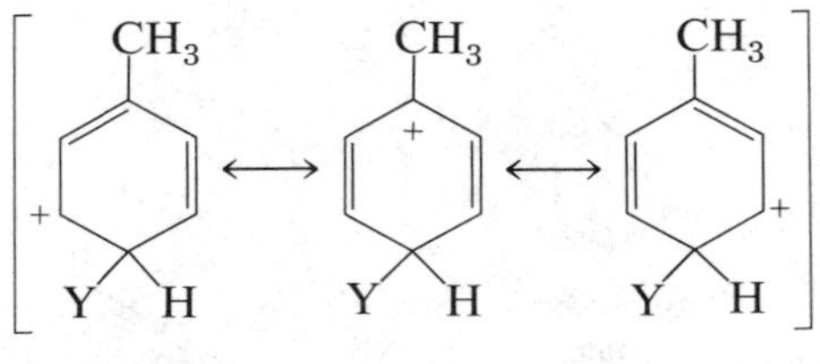

Here again we find two secondary carbocation structures and one tertiary carbocation. The overall energy of the transition state is comparable to that for *ortho* substitution. Indeed, this approach does not distinguish between preference for *ortho* relative to *para* substitution, but does indicate why substituents divide into the two broad groups of *ortho,para* and *meta* directors.

The resulting energy profile for reaction at toluene is compared with that for benzene in Figure 23.4. The differences between the alternative structures are somewhat less in the transition state than in the intermediate because the amount of positive charge to be distributed is greater in the intermediate in which a fully formed C—Y bond is developed. We have chosen to examine the intermediates but only for the convenience of symbolism. The same arguments apply to the developing positive charge on the benzene ring in the transition states. The net result is that of predominant *ortho,para* orientation; although the *meta* position is more reactive than a single benzene position, the *ortho* and *para* positions are even more so.

We next apply this approach to the corresponding reaction at the *ortho, para,* and *meta* positions of nitrobenzene and derive three sets of resonance structures for the intermediates (and transition states) involved.

ortho

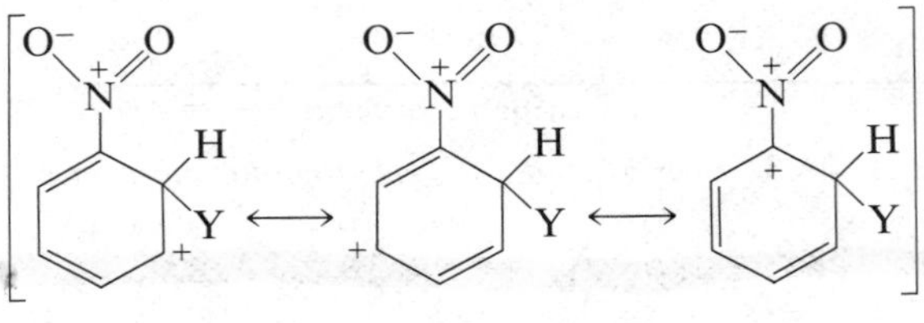

para

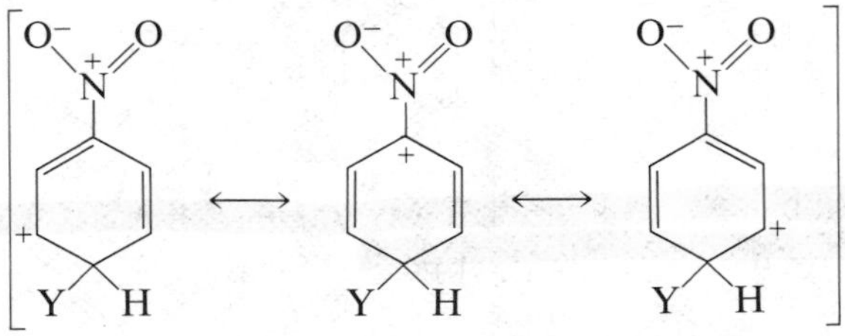

meta

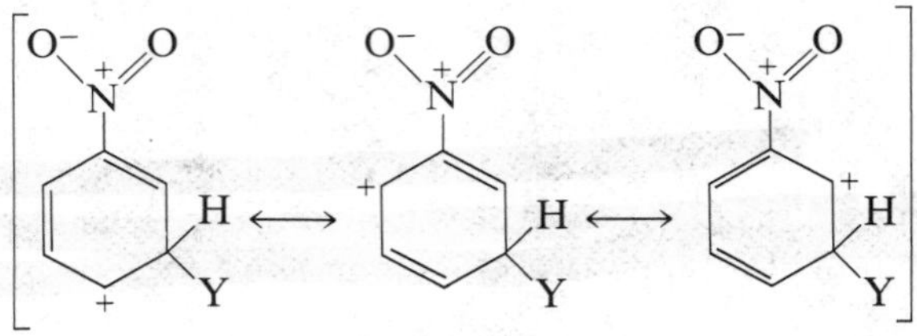

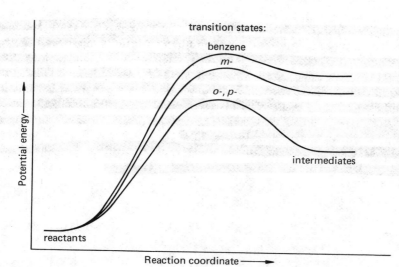

FIGURE 23.4. *Energy profile for reaction at* ortho, para, *and* meta *positions of toluene compared to benzene.*

All of these structures involve the electrostatic repulsion of the carbocation charge with the strong dipole of the nitro group.

$$(+) \quad \overset{+\longrightarrow}{C-NO_2}$$

That is, every one of these structures is substantially less stable than the corresponding structure for reaction at benzene; hence, all positions in nitrobenzene are expected to be deactivated relative to benzene. For reaction at the *ortho* and *para* positions, however, one structure is that of a carbocation right next to the positive nitrogen of the nitro group. This structure in each case is of such high

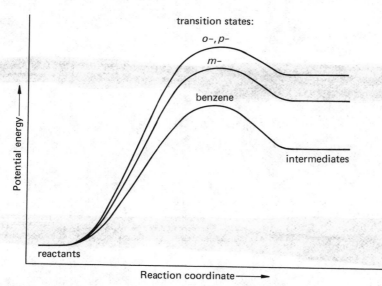

FIGURE 23.5. *The intermediate derived from* meta *reaction of nitrobenzene is formed less readily than that from attack at benzene itself but more readily than that from reaction at the* ortho *or* para *position.*

energy compared to the other structures, in which the positive charges are separated by one or more atoms, that it contributes very little to the overall resonance hybrid. The *meta* reaction involves only structures in which the positive charges are separated; thus, although the transition state for *meta* substitution is of higher energy than for reaction at benzene, it is of lower energy than those for reaction at the *ortho* or *para* positions. We can phrase this result another way: the *meta* reaction is deactivated less than *ortho* or *para* reaction. The corresponding reaction profiles are summarized in Figure 23.5.

These principles apply generally to other types of substituents. For anisole, we may write the same sets of three resonance structures.

ortho

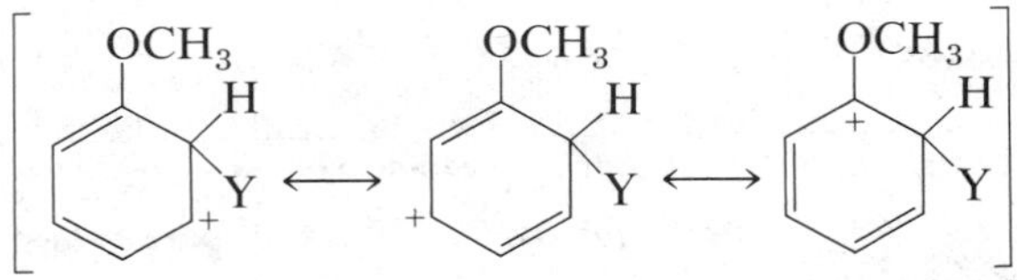

para

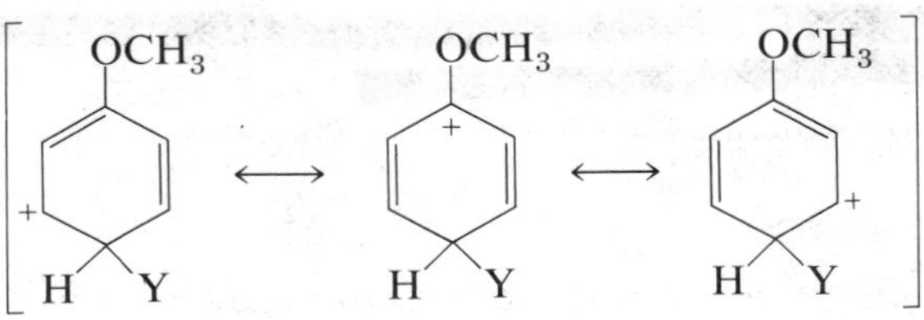

meta

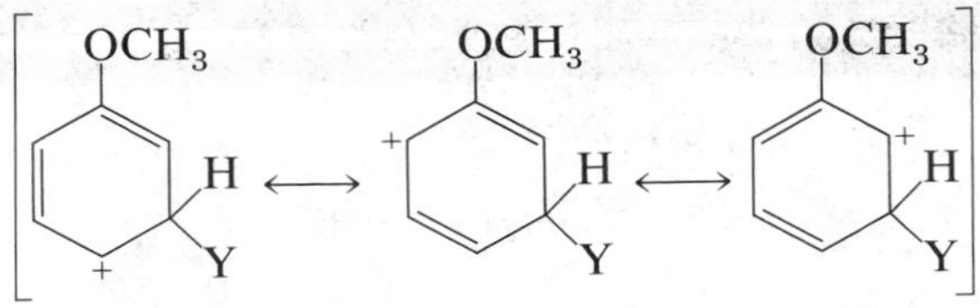

All of these structures are expected to be somewhat destabilized by the electrostatic interaction with the C—O dipole, but reaction at the *ortho* and *para* positions also corresponds to oxonium ions.

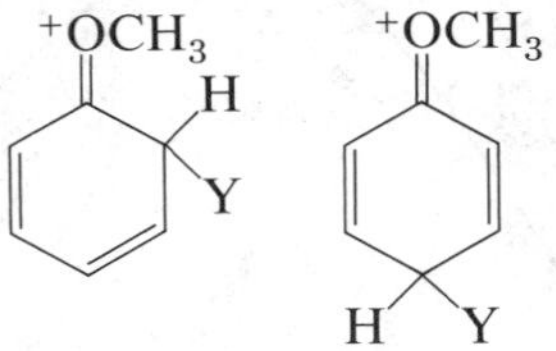

These additional structures greatly stabilize the intermediates and the transition states leading to them. Similar structures are involved in acid-catalyzed reactions of carbonyl compounds.

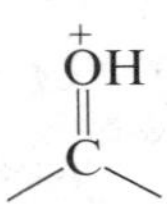

711

Sec. 23.6
**Theory of
Orientation in
Electrophilic
Aromatic
Substitution**

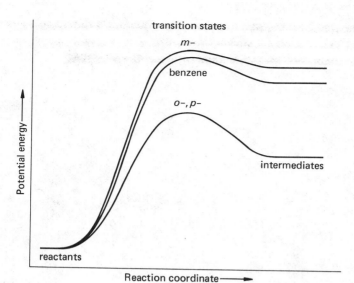

FIGURE 23.6. *Reaction profile for reaction at the* ortho,para, *and* meta *positions of anisole compared to benzene. The same figure would apply to reaction of phenol and aniline.*

The oxonium ion structures so dominate the system that the *ortho* and *para* positions of anisole are highly activated compared to benzene. We shall see in Chapter 30 that electrophilic substitution reactions at these positions in phenols and phenyl ethers and esters are accomplished under rather mild conditions. On the other hand, reaction at the *meta* positions is expected to be somewhat less facile than in benzene. The resulting reaction profile is shown in Figure 23.6.

Reactions at the *ortho* and *para* positions of aromatic amines involve related immonium ion structures, for example,

Consequently these positions are also highly activated relative to benzene.

Let us now apply the procedure to a halobenzene. Reaction at the *meta* position gives the three structures.

meta

All three structures are strongly destabilized by electrostatic interaction of the positive charge with the C—X dipole.

$$(+) \quad \overset{+\longrightarrow}{C-X}$$

Accordingly, the *meta* position in the halobenzene is strongly deactivated relative to benzene. Reactions at the *ortho* and *para* positions involve similar carbocation structures destabilized by interaction with the C—X dipole.

ortho

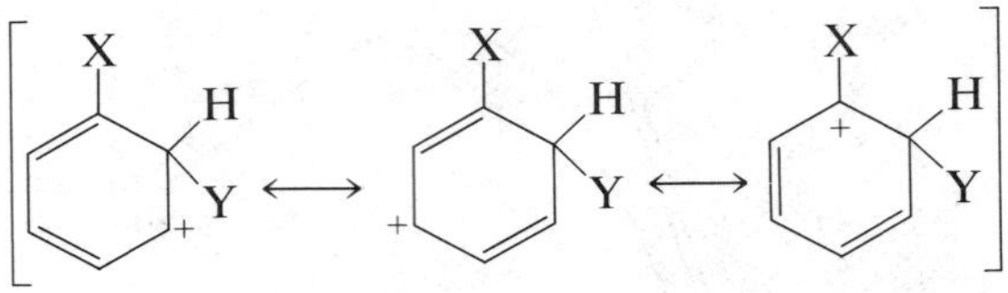

para

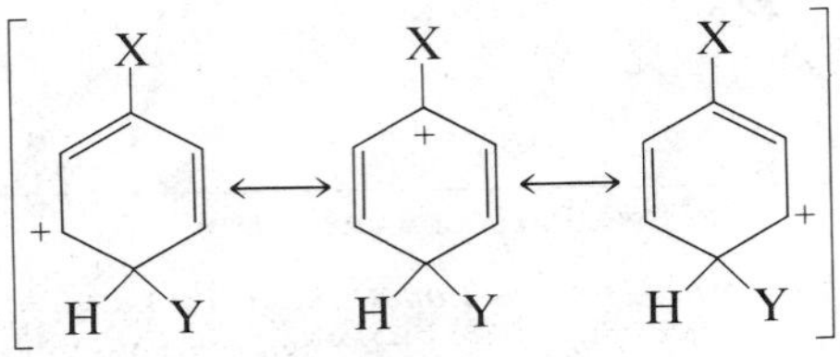

In both cases, however, one structure is that of an α-halocarbocation in which interaction with a halogen lone pair is possible to give the halonium ion structures.

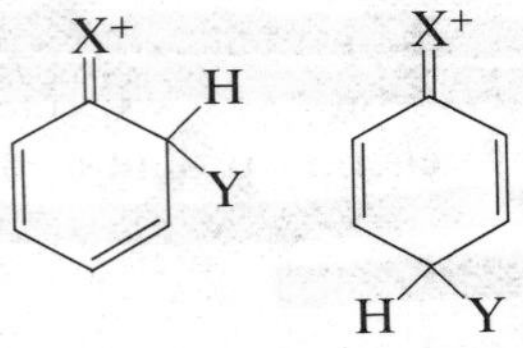

Such halonium ion structures are not nearly as stable as related oxonium and immonium ions. In practice, the additional contribution of such structures does not compensate for the deactivating effect of the C—X dipoles on the other structures, but it does make reaction at the *ortho* and *para* positions far more facile than at the *meta* position. The corresponding reaction profile is shown in Figure 23.7.

EXERCISE: Use the theory developed in this section to predict the orientation specificity of vinyl and formyl groups, —CH=CH$_2$ and —CHO.

23.7
Quantitative Reactivities: Partial Rate Factors

Nitration reactions have been studied extensively for many aromatic compounds, and relative reactivities at different positions have been determined. Furthermore, by studying the reaction of a mixture of benzene and some other compound, it is often possible to determine the quantitative reactivities of various positions relative to a benzene position. These statistically corrected relative reactivities are known as partial rate factors.

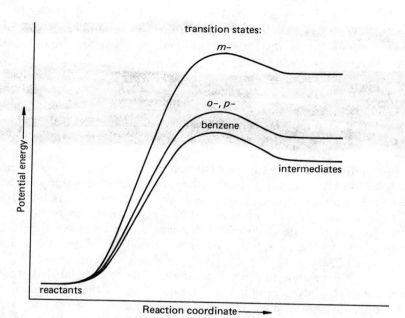

FIGURE 23.7. *Energy profile for reaction of a halobenzene compared to benzene.*

For example, the reaction of equimolar amounts of toluene and benzene with a small amount of nitric acid in acetic anhydride at 30° gives one part of nitrobenzene to 27 parts of nitrotoluenes. The nitrotoluenes formed are 58.1% *ortho*, 3.7% *meta,* and 38.2% *para*. The partial rate factors, f_i, are calculated as follows.

$$f_o = (\text{fraction } o)\left(\tfrac{\text{no. benzene positions}}{\text{no. } o \text{ positions}}\right)\left(\tfrac{\text{toluene reactivity}}{\text{benzene reactivity}}\right)$$
$$= (0.581)(\tfrac{6}{2})(27) = 47$$
$$f_m = (0.037)(\tfrac{6}{2})(27) = 3$$
$$f_p = (0.382)(\tfrac{6}{1})(27) = 62$$

Note that the *meta* position is more reactive than a benzene position, as predicted by the theory developed in Section 23.6.

Partial rate factors for nitration of several substituted benzenes are summarized in Table 23.1. Some effects are clearly apparent in these results. For example, a *t*-butyl group has much the same effect as a methyl group in the *meta* and *para* positions, but at the *ortho* position *t*-butylbenzene is much less reactive than tolu-

TABLE 23.1
Partial Rate Factors for Nitration

ene. The difference is clearly to be attributed to steric hindrance caused by the bulky *t*-butyl group. At the distant *meta* and *para* positions the size of the alkyl group has little effect.

The halobenzenes follow the theory outlined in Section 23.6. All positions are less reactive than benzene, but the *meta* positions are more strongly deactivated than *ortho* and *para*. The chloromethyl group is of especial interest since the stabilizing effect of an alkyl group is superimposed on the deactivating effect of the C—Cl dipole. The result is a net *ortho,para* orientation with a little net deactivation.

But the quantitative data are sparse. The amounts formed of some isomers are so minute as to defy detection even by modern gas chromatography analytical methods. One approach to obtaining quantitative reactivity results for all positions in a given molecule, even when they differ greatly in reactivity, has been to study the simplest possible electrophilic aromatic substitution reaction, the replacement of one hydrogen isotope by another. Examples of this reaction were given in Section 23.2.

In principle, it is possible to prepare a variety of specifically labeled aromatic compounds and to study quantitatively the rate of loss of the hydrogen isotope under a consistent set of acidic conditions. A comparison of the rates of replacement of deuterium by hydrogen (protodedeuteration) of specifically deuteriated anisoles with the corresponding rate for deuteriobenzene in aqueous perchloric acid gives the results displayed in Table 23.2. These results demonstrate the high

TABLE 23.2
**Relative Rates of Protodedeuteration in Aqueous
Perchloric Acid**

| 1 | 6 × 10⁴ | 0.3 | 2 × 10⁴ |

reactivity of the *ortho* and *para* positions compared to benzene and the lower reactivity in the *meta* position. These same relative rates are expected to correspond approximately to nitration as well and imply that nitration of anisole gives only a few parts per million of *m*-nitro product. This minute quantity is extremely difficult to detect directly in the product mixture.

EXERCISE: Using the data in Table 23.1, calculate the percent composition of the three chloronitrobenzenes formed by nitration of chlorobenzene.

23.8
Effects of Multiple Substituents

The relative rates of replacement of tritium by hydrogen (protodetritiation) in trifluoroacetic acid for toluene and the dimethylbenzenes compared to benzene

TABLE 23.3
Relative Rates of Protodetritiation in Trifluoroacetic Acid

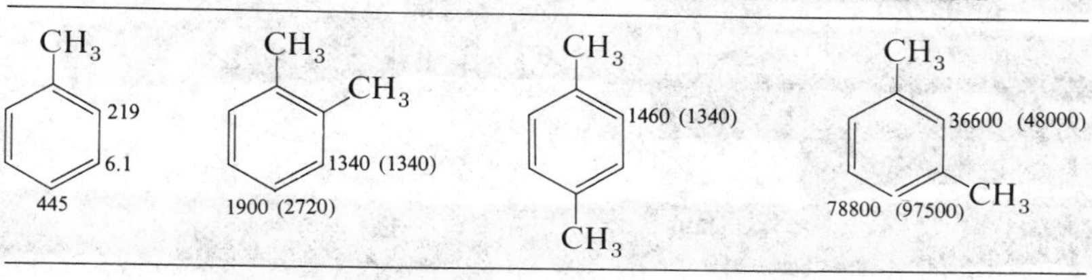

are summarized in Table 23.3. The energy effects of two methyl groups are approximately additive compared to the effect of one methyl group in toluene. For example, the 3-position in o-dimethylbenzene is *ortho* to one methyl and *meta* to the other. The predicted reactivity is therefore (219)(6.1) = 1340, which agrees exactly with the experimental reactivity.

> The product of the two partial rate factors is taken because it is the activation energy quantities that are additive
>
> $$\Delta E^{\ddagger} \text{ (3-position in } o\text{-dimethylbenzene)} = \Delta E^{\ddagger}(o\text{-}) + \Delta E^{\ddagger}(m\text{-})$$
>
> The energies are related to the logarithms of the rate constants.
>
> $$RT \ln k(\text{3-position}) = RT \ln (o\text{-}) + RT \ln (m\text{-})$$
> $$\log f \text{ (3-position)} = \log f_o + \log f_m = \log f_o f_m$$
> $$f \text{ (3-position)} = f_o f_m$$

The relative reactivities of the toluene positions were used as partial rate factors to derive the predicted reactivities of the dimethylbenzenes given in parentheses in Table 23.3. The approximate agreement can be generalized to electrophilic substitution reactions of polysubstituted benzenes. That is, the net orientation effects of two or more substituents can be predicted approximately by examining the effects of each substituent separately. If all substituents orient preferentially to the same positions, such positions are strongly preferred. For example, nitration of the following disubstituted benzenes gives the percentage of nitration at each position as indicated.

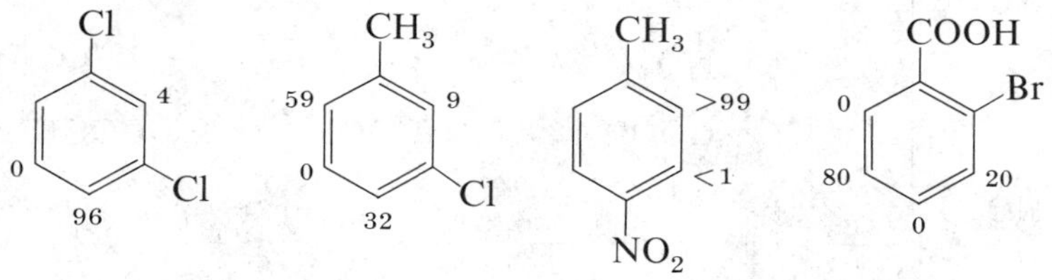

In m-chlorotoluene, the 5-position is *meta* both to chlorine and to methyl, and no significant reaction occurs at this position. The other positions are all *ortho* or *para* to both groups, and a disagreeable mixture results. In p-nitrotoluene, however, the highly favored 2-position is *ortho* to the *ortho,para*-directing methyl and *meta* to the *meta*-directing nitro.

If the groups already present have conflicting orientation preferences, it is helpful to divide substituents into three classes:

1. Strongly activating *ortho,para* directors, such as OR and NR$_2$.
2. Alkyl groups and halogens.
3. All *meta* directors.

If two substituents belong to different classes, the orientation effect of the superior class dominates. The following nitration results are examples.

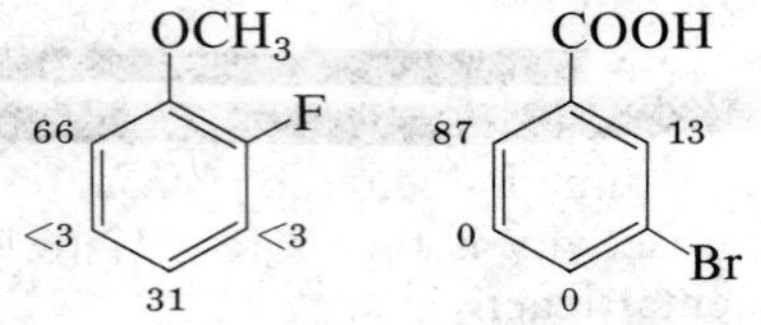

Note that the effects of all *ortho,para* directors dominate over *meta* directors.

Finally, if both substituents are in the same class, all bets are off and horrible mixtures can be anticipated. The following nitration results are examples.

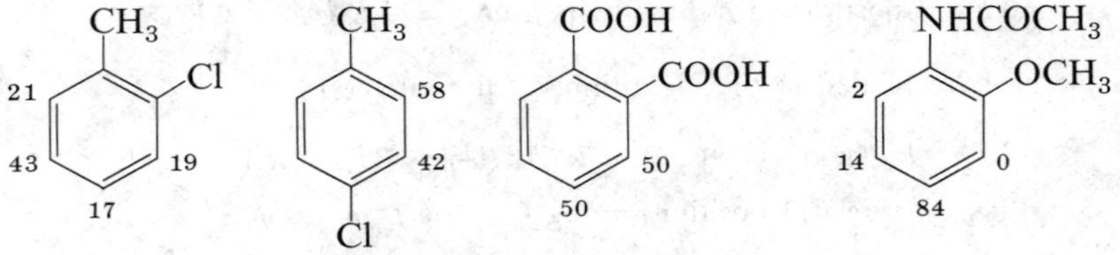

In our subsequent studies of the reactions of functional groups on benzene rings, we shall see that many syntheses can be accomplished by aromatic substitution reactions combined with functional group transformations. In such sequences, the order in which reactions are accomplished is of great importance because of the orientation preferences of different groups. One example will demonstrate this point.

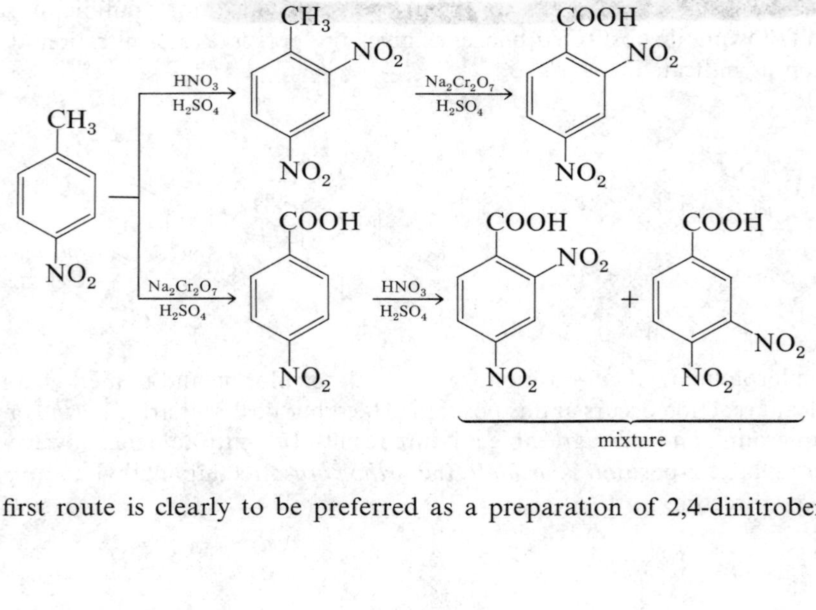

The first route is clearly to be preferred as a preparation of 2,4-dinitrobenzoic acid.

717

Sec. 23.9
Synthetic
Utility of
Electrophilic
Aromatic
Substitution

23.9
Synthetic Utility of Electrophilic Aromatic Substitution

In this chapter we have seen how some important functional groups can be introduced directly into benzene and many of its derivatives by electrophilic substitution reactions. Important examples of such synthetically useful reactions are halogenation, nitration, and Friedel–Crafts acylation. In Chapter 26 we will learn of an additional useful reaction, sulfonation. Each of these functional groups can serve as a substrate for additional electrophilic substitution reactions, or the group can be converted to other functional groups. Halogens can be converted via lithium or Grignard reagents to a variety of functions. These and other reactions of aromatic halides will be discussed in Chapter 30. Nitro compounds can be reduced to amines, which in turn can be transformed to many different groups as detailed in Chapter 25. Aromatic ketones can participate in the usual reactions of carbonyl groups. One important such reaction is reduction of the carbonyl group to a methylene group by either the Wolff–Kishner or the Clemmensen method (Section 13.8.E).

(82%)
hexylbenzene

(77%)
octadecylbenzene

Since Friedel–Crafts *acylation* is generally a clean, high-yield reaction, the combination of acylation and reduction is generally to be preferred to Friedel–Crafts *alkylation*.

In our discussion of electrophilic substitution reactions we have considered the effect on orientation and reactivity of other substituents already on the benzene ring. We now need to consider how electrophilic substitution may be used in a practical sense to prepare polysubstituted benzenes. We want especially to consider how the process may be used in some cases to provide practical syntheses of pure compounds.

Remember that the goal of any chemical synthesis is generally to prepare **one pure compound** for some purpose. Therefore, whenever possible, one must use reactions that do not give mixtures of isomers. When there is no known method that provides only one isomer, a synthesis may still be acceptable if the desired isomer is produced in substantial amounts (hopefully as the *major* product) and if it may be separated in some way from the unwanted isomers.

Some electrophilic substitution reactions fit the first criterion; that is, one of the possible isomers is produced almost exclusively. Substitution on *meta*-orienting compounds usually falls into this category. Thus, the following substitution reactions are good preparative reactions.

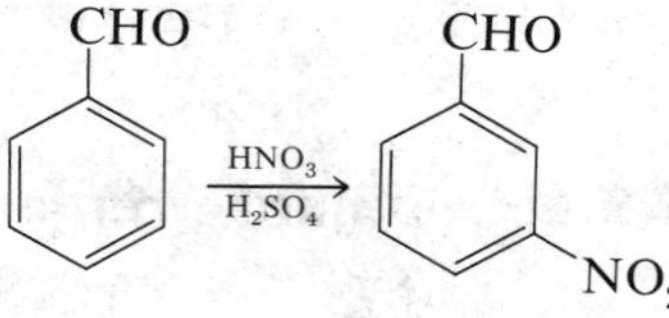

(75–84%)

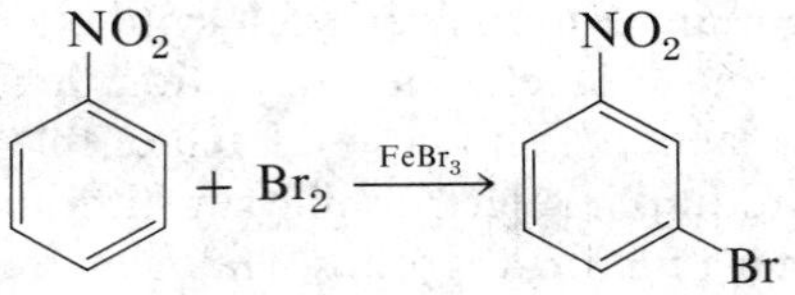

(60–75%)

Recall that the Friedel–Crafts acylation reaction often *does not work when the ring already contains a meta-directing group*. Thus, *m*-nitroacetophenone may be prepared by nitration of acetophenone but *not* by Friedel–Crafts acylation of nitrobenzene.

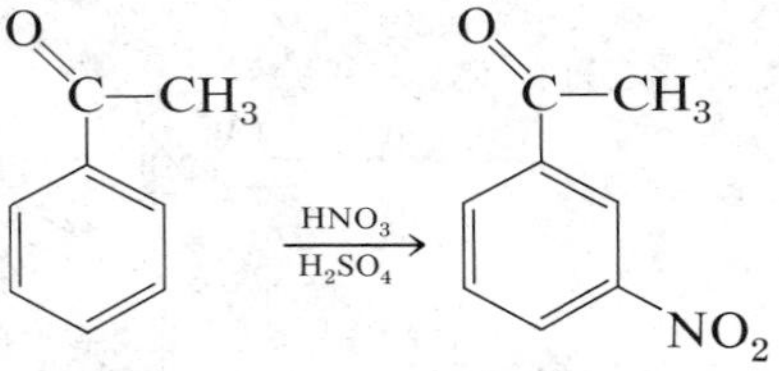

acetophenone   (55%)

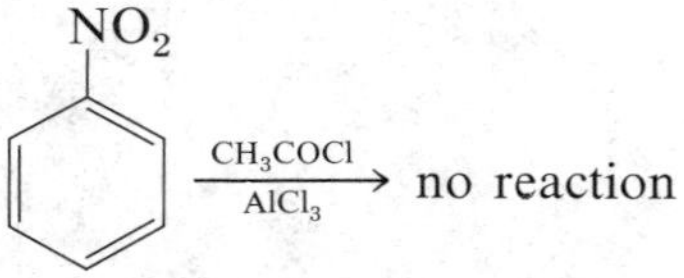

In many of these reactions, a few percent of the *ortho* and *para* isomers are produced. However, if the major isomer is crystalline, as is usually the case, it may easily be purified by recrystallization.

When the substituent already in the ring is an *ortho,para* director, mixtures invariably result, as we have seen in previous sections. In such cases direct electrophilic substitution is less satisfactory as a synthetic method. However, some benzene derivatives may still be obtained in this manner, particularly the *para* isomers. Because of its symmetrical nature, the *para* isomer usually has a significantly higher melting point than the *ortho* or *meta* isomer. Some representative data are summarized in Table 23.4. Recall that a higher melting point represents a more stable crystal lattice and lower solubility. Consequently the higher melting *para* isomer may often be crystallized from the mixture of *ortho* and *para* products of direct substitution. It is generally *not possible* to isolate the *ortho* isomer in a pure state by this technique.

719

Sec. 23.9

Synthetic
Utility of
Electrophilic
Aromatic
Substitution

TABLE 23.4
Melting Points of Disubstituted Benzenes

Substituents	Melting Point °C		
	ortho	*meta*	*para*
Br, Br	7	−7	87
Cl, Cl	−17	−25	53
Br, Cl	−12	−22	68
CH$_3$, Br	−26	−40	29
CH$_3$, NO$_2$	−10	16	55
Br, NO$_2$	43	56	127
Cl, NO$_2$	35	46	84
Br, COOH	150	155	255
Cl, COOH	142	158	243
OH, Br	6	33	66

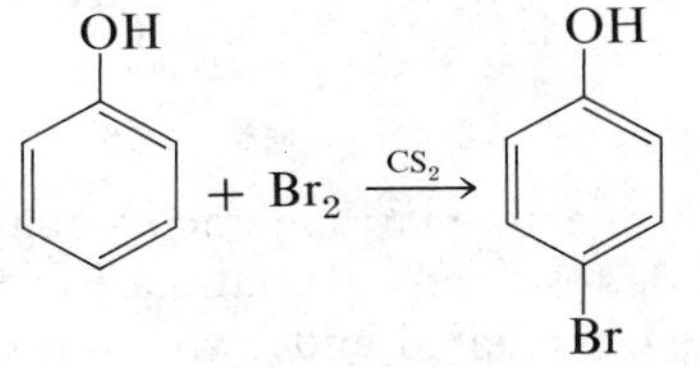

(80–84%)
m.p. 66°

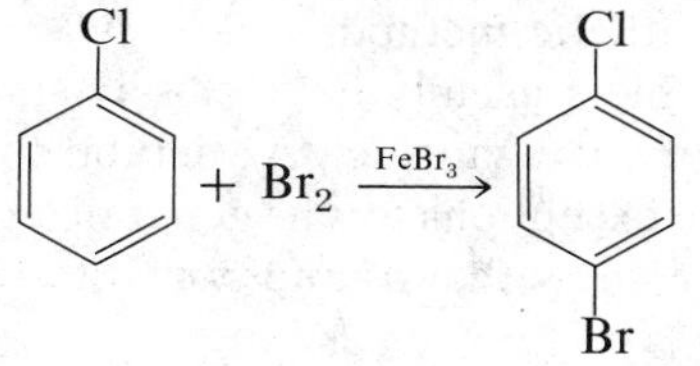

(60%)
m.p. 68°

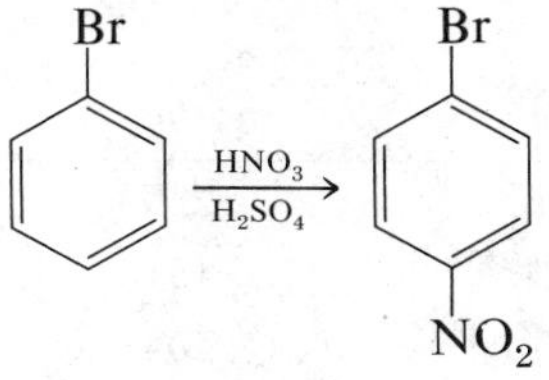

(72%)
m.p. 127°

Another useful generalization is that the acylating agent obtained by coordination of aluminum chloride with an acyl halide behaves as a rather bulky reagent. Consequently Friedel–Crafts acylation reactions tend to give almost completely *para* products, which are usually easy to separate from the small amounts of other isomers.

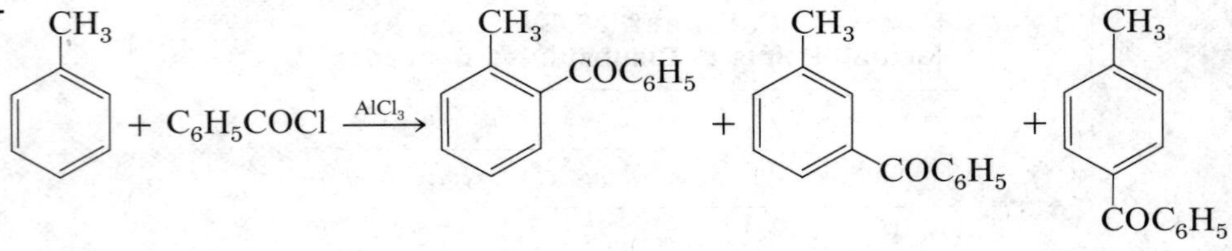

(9%)
o-methyl
benzophenone

(1%)
m-methyl-
benzophenone

(90%)
p-methyl-
benzophenone

On the other hand, Friedel–Crafts alkylations tend to be rather nonspecific. Often the orientations appear to be quite unusual. For example, under mild conditions with aluminum chloride in acetonitrile, isopropylation of toluene gives predominantly the expected *ortho* and *para* products, but there is also a substantial amount of *meta* product.

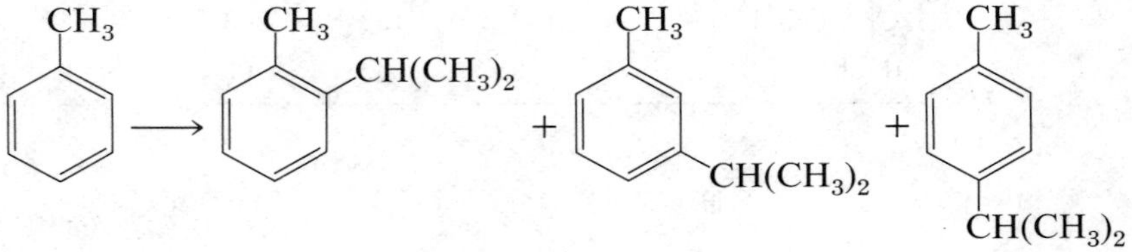

(63%)

(12%)

(25%)

This unusual behavior is due to rearrangements that alkylbenzenes undergo under the conditions of Friedel–Crafts alkylation. For example, if the foregoing mixture of isopropyltoluenes is treated under vigorous conditions with $AlCl_3$ and HCl, the product is exclusively the *meta* isomer. Although we shall not go into these rearrangements in detail, the student should be aware that Friedel–Crafts alkylations are often complicated and difficult to predict. Thus the reaction has less general utility as a synthetic method.

Since *ortho* and *para* isomers usually have closely similar boiling points, fractional distillation is usually *not* a satisfactory method for separation of such isomer mixtures, but there are exceptions to this generalization. Some representative data collected in Table 23.5 show that *o*- and *p*-nitrotoluenes differ sufficiently in

TABLE 23.5
Boiling Points of Disubstituted Benzenes

Substituents	Boiling Point, °C		
	ortho	*meta*	*para*
Br, Br	225	218	219
Cl, Cl	181	173	174
Br, Cl	204	196	196
CH_3, Br	182	184	184
CH_3, Cl	159	162	162
Br, NO_2	258	265	256
Cl, NO_2	246	236	242
CH_3, NO_2	220	233	238
NO_2, NO_2	319	291	299
OCH_3, NO_2	277	258	274

boiling point to be separable by fractional distillation. On the other hand, the melting points of the bromotoluenes are too low for effective crystallization, and their boiling points are too close for simple fractionation; hence, the bromination of toluene is *not* a satisfactory route to any of the bromotoluenes.

In summary, direct electrophilic substitution is a useful synthetic method as such if only one isomer is produced or if the mixture can be conveniently separated by physical means. To predict whether such a reaction will be useful, the chemist must consider both the mechanism of the reaction—that is, what the isomer distribution is expected to be—and the probable physical properties of the expected products. We shall see in future chapters that the utility of electrophilic substitution may be extended by modification and interrelation of functional groups and by a technique in which one or more positions on the ring are temporarily deactivated or blocked.

EXERCISE: In this chapter we have learned how Br and $COCH_3$ groups can be introduced into a number of aromatic compounds by bromination and Friedel–Crafts acylation, respectively. Review the transformations that you have already learned of $ArBr$ to $ArCH_2CH=CH_2$, $ArCOOH$, $ArCR_2OH$ and of $ArCOCH_3$ to $ArCOOH$, $ArCH_2CH_3$, $ArC(CH_3)_2OH$. How is each of these transformations affected by the presence in the aryl group of each of the following functions: CHO, COOH, Br, NO_2?

PROBLEMS

1. Benzene can be iodinated with iodine and an oxidizing agent such as nitric acid or hydrogen peroxide. The actual electrophilic reagent in this reaction is probably $IOH_2{}^+$, which may be regarded as I^+ bound to a water molecule. Write a balanced equation for the generation of this intermediate from I_2 and H_2O_2. Include this as part of an overall mechanism for the reaction of I_2 and H_2O_2 with benzene to give iodobenzene, C_6H_5I.

2. (a) The chloromethylation reaction of benzene with formaldehyde

$$C_6H_6 + CH_2O + HCl \xrightarrow{\text{ZnCl}_2} C_6H_5CH_2Cl + H_2O$$

benzyl chloride

could involve as the principal electrophilic reagent either $CH_2=\overset{+}{O}-\overset{-}{Z}nCl_2$ or $\overset{+}{C}H_2Cl$. Write complete reaction mechanisms using both intermediates. Note that under these reaction conditions, benzyl alcohol, $C_6H_5CH_2OH$, reacts rapidly with $ZnCl_2$ and HCl (Lucas reagent) to give benzyl chloride.

(b) In such chloromethylation reactions a carcinogenic agent, bis-chloromethyl ether, $ClCH_2OCH_2Cl$, is produced as a by-product. Write a plausible mechanism for formation of this compound from formaldehyde, HCl, and $ZnCl_2$, showing each intermediate involved.

3. Benzene can be mercurated to give phenylmercuric acetate, $C_6H_5HgOOCCH_3$, with mercuric acetate in acetic acid containing some perchloric acid as an acid catalyst. The electrophilic reagent involved is probably $\overset{+}{H}gOOCCH_3$. Write a complete reaction mechanism.

4. Biphenyl, C_6H_5—C_6H_5, may be considered as a benzene with a phenyl substituent. Show why this hydrocarbon is expected to direct to the *ortho,para* positions, using resonance structures.

5. Use resonance structures to show why the COOH group in benzoic acid is a *meta* director.

6. Indicate the principal mononitration product or products expected from each of the following compounds.

(a) 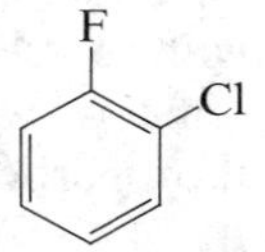 [structure: benzene with F and Cl] (b) 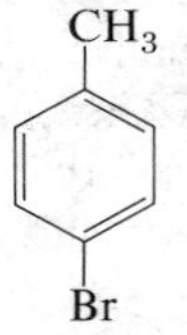 [structure: benzene with CH_3 and Br] (c) 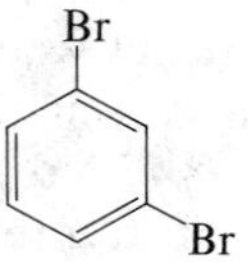 [structure: benzene with Br and Br] (d)  [structure: benzene with NH_2 and CH_3]

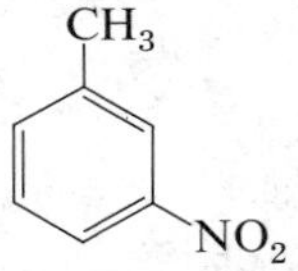

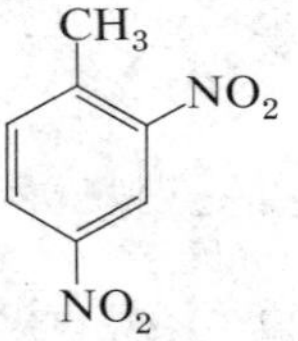

(e) [structure: benzene with CH_3 and NO_2] (f) [structure: benzene with CH_3, NO_2, NO_2] (g) 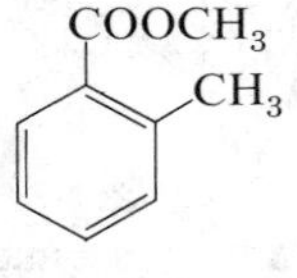 [structure: benzene with $COOCH_3$ and CH_3]

(h) 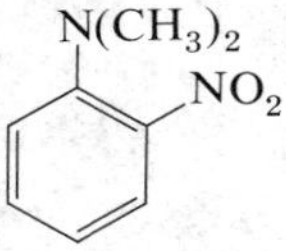 [structure: benzene with $N(CH_3)_2$ and NO_2] (i) 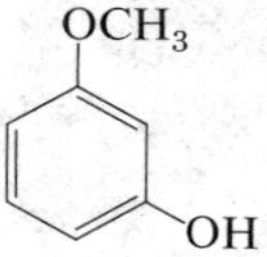 [structure: benzene with OCH_3 and OH] (j) 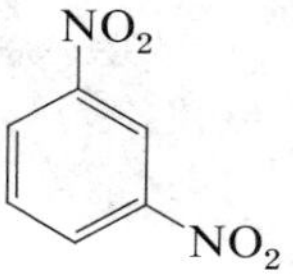 [structure: benzene with NO_2 and NO_2]

7. (a) Toluene is 605 times as reactive as benzene toward bromination in aqueous acetic acid. The bromotoluenes produced are 32.9% *ortho*, 0.3% *meta*, and 66.8% *para*. Calculate the partial rate factors.

(b) The partial rate factors for chlorination of toluene are *ortho*, 620; *meta*, 5.0; *para*, 820. Calculate the isomer distribution in chlorination of *m*-xylene (*m*-dimethylbenzene). The experimental result is 77% 4-, 23% 2-, and ~0% 5-.

(c) The partial rate factors for chlorination of chlorobenzene are *ortho*, 0.1; *meta*, 0.002; *para*, 0.41. Calculate the isomer distribution in chlorination of *p*-chlorotoluene (the experimental result is 77% 2,4-dichlorotoluene and 23% 3,4-dichlorotoluene).

8. Which of the following compounds can probably be prepared in a pure state from benzene by using two successive electrophilic substitution reactions? For each compound, write out the reaction sequence and describe how the intermediates and products would be purified.

(a) 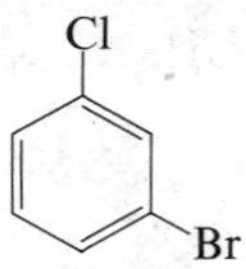 [structure: benzene with Cl and Br] (b) 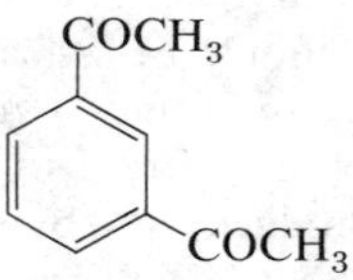 [structure: benzene with $COCH_3$ and $COCH_3$] (c) 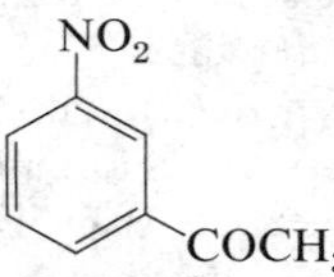 [structure: benzene with NO_2 and $COCH_3$]

(d) 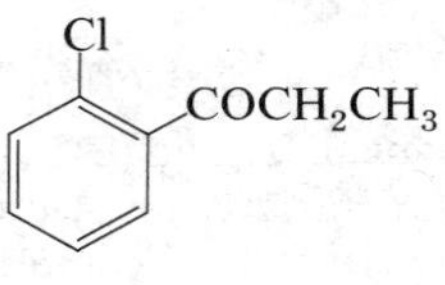 [structure: benzene with Cl and $COCH_2CH_3$] (e)  [structure: benzene with Cl and NO_2] (f) 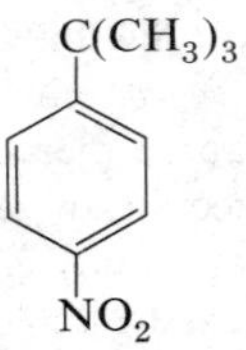 [structure: benzene with $C(CH_3)_3$ and NO_2]

9. Which of the following compounds can probably be prepared in a pure state by electrophilic substitution on a disubstituted benzene? Outline the method in each case.

(a)

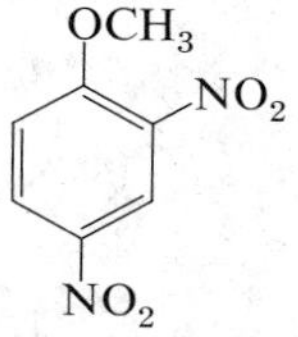

(b)

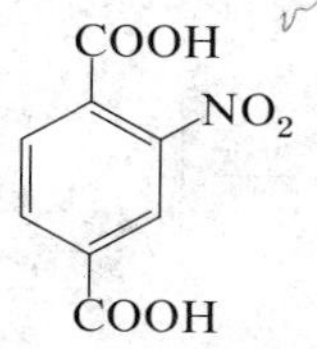

(c)

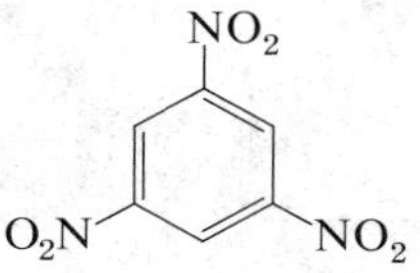

(d)

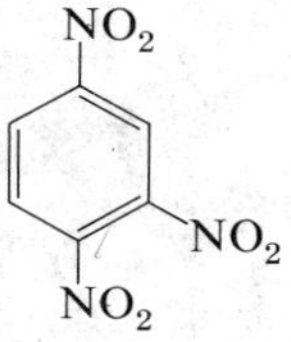

(e)

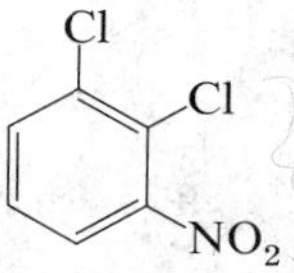

(f)

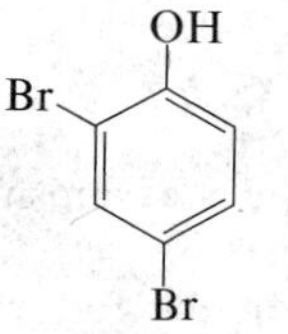

(g)

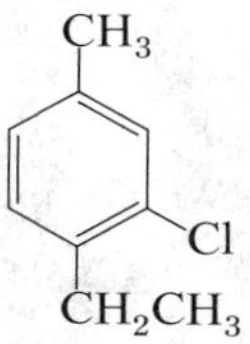

(h)

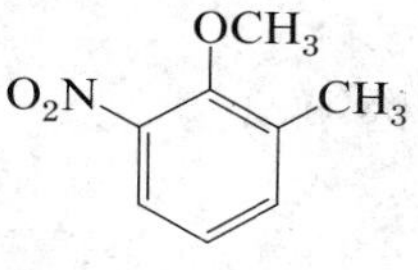

(i)

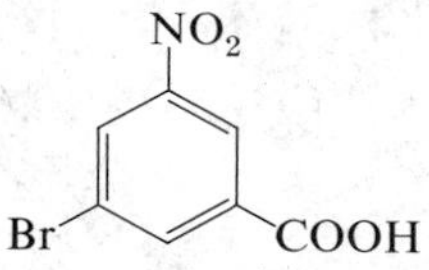

(j)

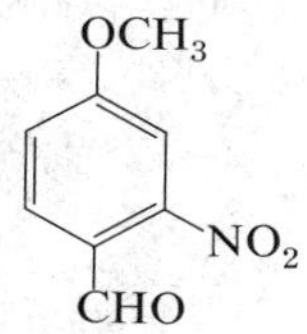

10. Toluene is *ortho,para*-directing, whereas trifluoromethylbenzene, $C_6H_5CF_3$, is *meta*-directing. Explain.

11. (a) The reaction of benzene with isobutyl alcohol and BF_3 gives primarily *t*-butylbenzene. Explain.

(b) The reaction of 2-butanol and BF_3 with benzene at $0°$ gives 2-phenylbutane in good yield. When 2-butanol-*2-d*, $CH_3CDOHCH_2CH_3$, is used, a mixture of deuterated compounds is obtained that includes major amounts of

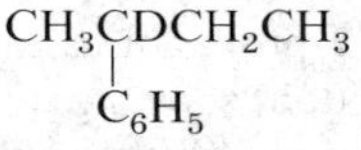

$$\underset{\underset{C_6H_5}{|}}{CH_3CDCH_2CH_3} \quad \text{and} \quad \underset{\underset{C_6H_5}{|}}{CH_3CHCHDCH_3}$$

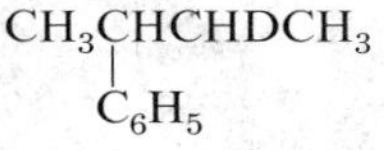

Explain.

(c) By contrast, the reaction of $CD_3CHOHCH_3$ with benzene and BF_3 gives the following compound in good yield.

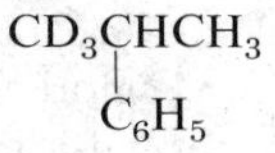

$$\underset{\underset{C_6H_5}{|}}{CD_3CHCH_3}$$

with no deuterium scrambling. How do you account for this difference?

(d) 2-Propanol-*1-d_3*, $CD_3CHOHCH_3$, has an asymmetric carbon and significant optical activity. According to the mechanism of the alkylation reaction, what do you expect for the steric course of the reaction of this optically active alcohol with benzene and BF_3?

12. When a solution of *p*-di-(3-pentyl)benzene in benzene is treated with aluminum chloride at $25°$, a rapid transfer of a pentyl group occurs to give monopentylbenzene. The reaction product is approximately one part of 2-phenylpentane and two parts of 3-phenylpentane. Account for these results with a reasonable reaction mechanism.

13. Show how all three nitrobenzoic acids may be prepared from toluene.

14. On heating with aqueous sulfuric acid, styrene reacts to form a dimer in good yield.

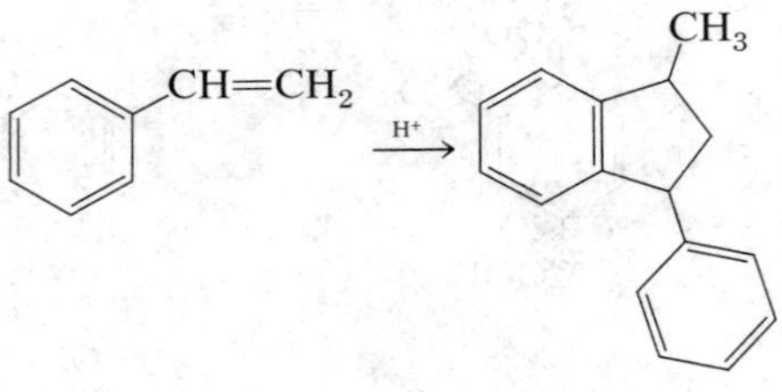

(77–81%)

Write a reasonable mechanism, showing all intermediates involved.

15. Show how each of the following compounds may be prepared from benzene or toluene in a practical manner.

(a) [structure with CH_3 and Br]

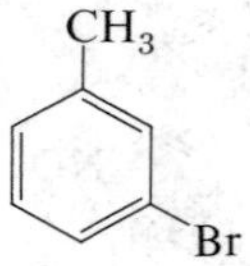

(b) [structure with $CH(CH_2CH_3)_2$]

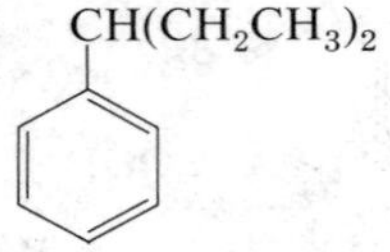

(c) [structure with CH_2CH_3 and NO_2]

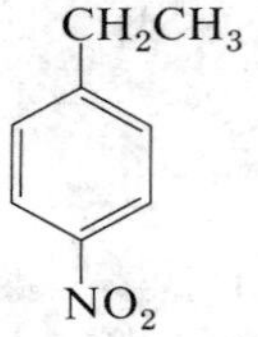

(d) [structure with CH_2CH_3 and Br]

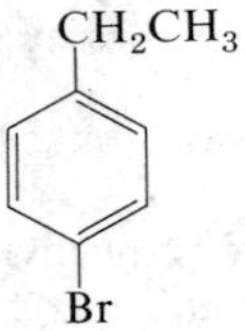

(e) [structure with $CH_2\overset{OH}{\underset{}{C}}CH_2CH_3$ and CH_3]

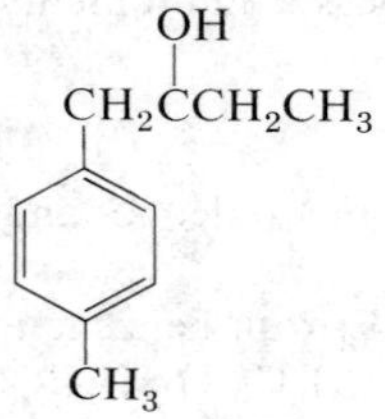

(f) [structure with $COOCH_3$ and Br]

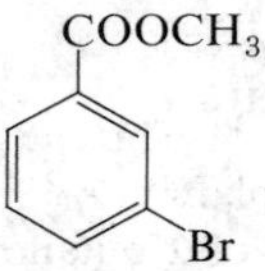

16. The solvolysis reaction of 2-chloro-2-phenylpropane in aqueous acetone is an S_N1 carbocation process that yields 2-phenyl-2-propanol as the principal product.
   (a) Write out the mechanism of this reaction, showing any intermediates involved.
   (b) The rate of reaction depends markedly on substituents in the phenyl group. The order of reactivity given by p-substituents is

$$CH_3O > CH_3 > H > NO_2$$

Explain, using resonance structures.

17. The dissociation of triarylmethyl chlorides into ions in liquid sulfur dioxide solution has been studied quantitatively, and a number of dissociation constants have been measured for the equilibrium

$$Ar_3CCl \overset{K}{\rightleftharpoons} Ar_3C^+ + Cl^-$$

Rank the following compounds in order of increasing K and explain. (m-$ClC_6H_4)_3CCl$, (p-$O_2NC_6H_4)_3CCl$, (m-$CH_3C_6H_4)_3CCl$, (p-$CH_3C_6H_4)_3CCl$, (p-$CH_3OC_6H_4)_3CCl$, $(C_6H_5)_3CCl$

18. (a) Solvolysis of 2-methyl-2-phenylpropyl tosylate in acetic acid gives primarily a mixture of 2-methyl-1-phenyl-2-propyl acetate and 2-methyl-1-phenylpropene. Write a reasonable reaction mechanism.
   (b) Solvolyses of 3-phenyl-2-butyl tosylates in acetic acid also give mixtures of alkenes and esters. The ester product of solvolysis of the optically active (2S,3R) diastereomer is racemic (equal amounts of (2S,3R)-3-phenyl-2-butyl acetate and (2R,3S)-3-phenyl-2-butyl acetate, whereas that from the (2R,3R) tosylate is the optically active ester (2R,3R)-3-phenyl-2-butyl acetate. Provide a reasonable explanation.

Chapter 24
Amines

24.1
Structure

Amines are compounds in which one or more alkyl or aryl groups are attached to nitrogen. They may be considered the organic relatives of ammonia in the same way that alcohols and ethers are related to water.

$$H_2O \qquad ROH \qquad R_2O$$
water $\qquad$ alcohols $\qquad$ ethers

$$NH_3 \qquad \underline{RNH_2 \quad R_2NH \quad R_3N}$$
ammonia $\qquad\qquad$ amines

Amines are classified as **primary, secondary,** or **tertiary,** according to the number of alkyl or aryl groups joined to the nitrogen. Note that these descriptive adjectives are used here to denote the *degree of substitution* on nitrogen, not the nature of the substituent groups. In secondary and tertiary amines the alkyl or aryl groups may be the same or different.

Some Primary Amines

$$CH_3NH_2 \qquad (CH_3)_2CHNH_2 \qquad (CH_3)_3CNH_2$$

Some Secondary Amines

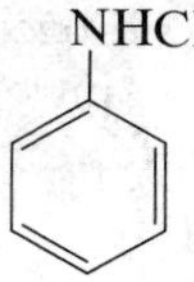

$$NHCH_3$$

$$(CH_3CH_2)_2NH \qquad CH_3CH_2NH$$
$$\overset{\displaystyle CH_3}{|}$$

Some Tertiary Amines

$$N(CH_3)_2$$

$$CH_3\diagdown \underset{N}{} \diagup CH_3$$

$$(CH_3)_3N$$

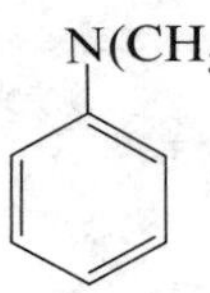

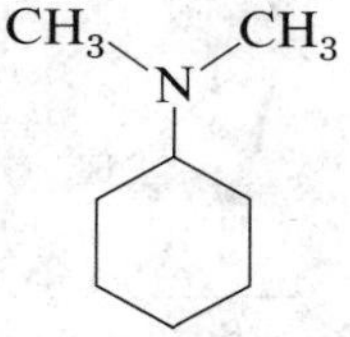

Quaternary ammonium compounds are related to simple inorganic ammonium salts. Again the four groups joined to nitrogen in the ammonium ion may be the same or different.

Some Quaternary Ammonium Compounds

$(CH_3CH_2)_2\overset{+}{N}(CH_3)_2 \; Br^-$ $\overset{+}{N}(CH_3)_3 \; Br^-$

$(CH_3)_4\overset{+}{N} \; Cl^-$

$CH_3CH_2CH_2\overset{+}{N}(CH_3)_3 \; OH^-$

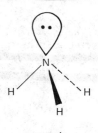

Recall that ammonia has a pyramidal shape. The N—H bond length is 1.008 Å, and the HNH bond angle is 107.3°. The hybridization of nitrogen is approxi-

ammonia

mately *sp³*. It forms three approximately *sp³-s* σ-bonds to hydrogen and has a **nonbonding electron pair** that occupies the other approximately *sp³*-orbital. Amines have similar structures, as shown in Figure 24.1.

Consider the following progression of bond lengths (in Å).

CH_3—CH_3	1.531	H—CH_3	1.085
CH_3—NH_2	1.474	H—NH_2	1.012
CH_3—OH	1.427	H—OH	0.957
CH_3—F	1.385	H—F	0.917

As we proceed along the first row of the periodic table, the increasing nuclear charge causes the electron orbitals to shrink and result in shorter bonds.

The nonbonding electron pair is extremely important in the chemistry of amines, since it is responsible for the typical basic and nucleophilic properties of these compounds. Amines having an aryl group attached to nitrogen are characterized by somewhat larger HNH and HNC angles; that is, the nitrogen is more nearly planar than in alkylamines. We will discuss the reason for this difference in Section 24.4.

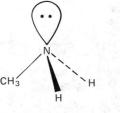

Bond Length, Å		Bond Angle, deg	
NH	1.011	HNH	105.9
CN	1.474	HNC	112.9

(a)

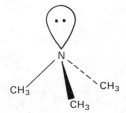

Bond Length, Å		Bond Angle, deg	
CN	1.47	CNC	108

(b)

FIGURE 24.1. *Simple amine structure:* (a) *methylamine;* (b) *trimethylamine.*

Because of the pyramidal geometry, an amine with three different groups joined to nitrogen is chiral (alternatively, amines may be regarded as approximately tetrahedral with the nonbonding pair being the fourth "group"). Recall

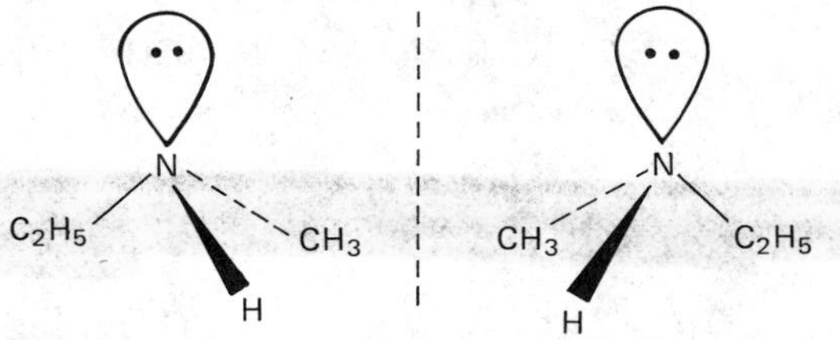

that enantiomeric carbon compounds may be separated and that the individual enantiomers are quite stable because it is necessary to break and re-form bonds to interconvert them. In contrast, the two enantiomers of a chiral amine are readily interconvertible by a process known as **nitrogen inversion.** For simple amines the activation energy required for inversion is rather small, on the order of 6 kcal mole^{-1}. In the planar transition state for inversion the nitrogen has sp^2-hybridization with the lone pair in the p_z-orbital.

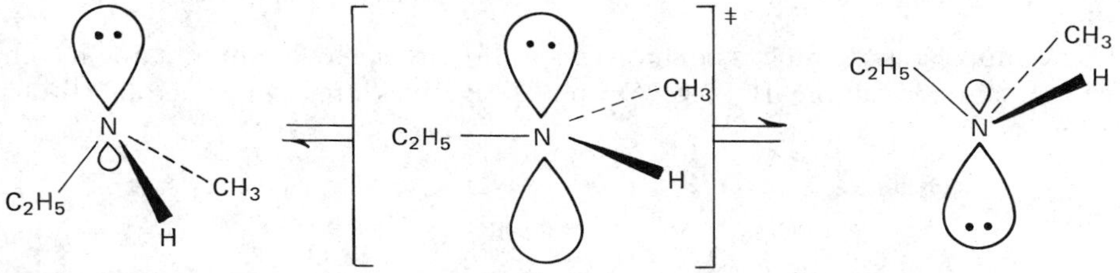

For quaternary ammonium compounds such inversion is not possible, and chiral ions may be separated into enantiomers that are relatively stable.

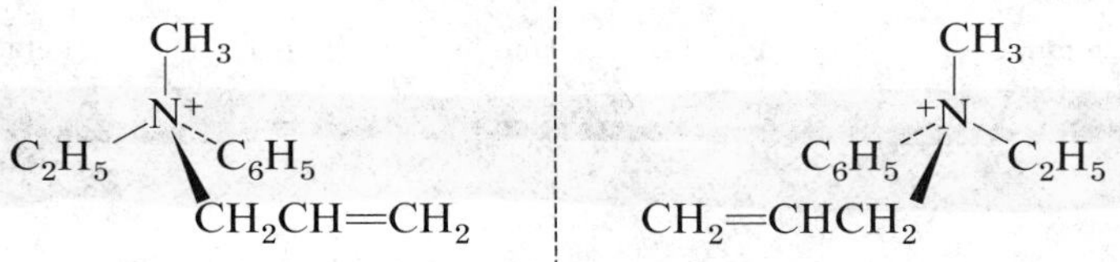

24.2
Nomenclature of Amines

Like most other classes of organic compounds, amines have been named in several ways. Simple amines are usually referred to by common names, which are derived by using the suffix **-amine,** preceded by the name or names of the alkyl groups. The names are written as one word.

$$CH_3NH_2 \qquad (CH_3CH_2)_2NH \qquad (CH_3CH_2CH_2)_3N$$

methylamine · · · · · · · diethylamine · · · · · · tri-*n*-propylamine

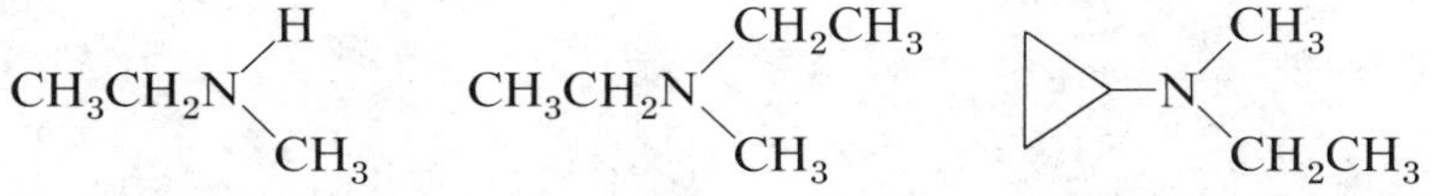

ethylmethylamine · · · · · diethylmethylamine · · · · cyclopropylethyl-
· methylamine

Under the IUPAC rules, amines are named as derivatives of a parent hydrocarbon by using the prefix **amino-** to designate the group —NH_2.

$$CH_3NH_2 \qquad CH_3CH_2\overset{\displaystyle CH_3}{\underset{\displaystyle}{C}}HNH_2 \qquad (CH_3)_2CHCH_2\overset{\displaystyle NH_2}{\underset{\displaystyle}{C}}HCH_2CH_3$$

aminomethane   2-aminobutane   4-amino-2-methylhexane

In this system secondary and tertiary amines are named by using a compound prefix that includes the names of all but the largest alkyl group.

$$CH_3CH_2N(CH_3)_2 \qquad (CH_3)_2CHCH_2\overset{\displaystyle CH_3}{\underset{\displaystyle}{C}}HN(CH_2CH_3)_2$$

dimethylaminoethane   4-(diethylamino)-2-methylpentane

$$CH_3CH_2\overset{\displaystyle CH_3}{\underset{\displaystyle}{C}}HN\overset{\displaystyle CH_3}{\underset{\displaystyle CH_2CH_3}{}}\qquad CH_3CH_2\overset{\displaystyle CH_3}{\underset{\displaystyle}{C}}HCH_2CH_2N\overset{\displaystyle H}{\underset{\displaystyle CH_3}{}}$$

2-(ethylmethylamino)butane   1-(methylamino)-3-methylpentane

The simplest arylamine is **aniline.** This well-entrenched common name has the official sanction of the IUPAC. Simple derivatives are named as substituted anilines.

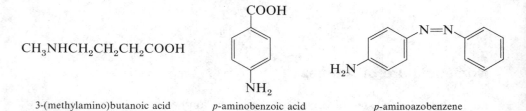

aniline   *m*-bromoaniline   *p*-nitroaniline   N,N-dimethylaniline

When it is necessary to name a compound containing the amino group as a derivative of some other function, the prefix **amino-** is employed.

$$CH_3NHCH_2CH_2CH_2COOH$$

3-(methylamino)butanoic acid   *p*-aminobenzoic acid   *p*-aminoazobenzene

A number of aromatic amines have trivial names that have received IUPAC sanction. Some of the more important examples are

p-toluidine   *m*-anisidine   *o*-phenetidine
p-aminotoluene   *m*-methoxyaniline   *o*-ethoxyaniline

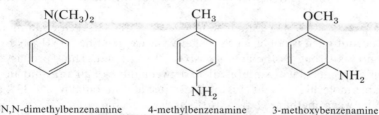

sulfanilic acid
p-aminobenzenesulfonic acid

anthranilic acid
o-aminobenzoic acid

Chemical Abstracts has recently adopted a new system for naming amines that is more rational than either the common or IUPAC system and will probably gain universal acceptance for the nomenclature of these compounds. In this system amines are named in the same manner as are alcohols (Section 10.2). The name of the alkane is modified by replacing the final **-e** by the suffix **-amine.**

$CH_3CH_2CH_2NH_2$ $(CH_3)_2CHCH_2CH_2CH_2CH_2NH_2$ $(CH_3)_2CHNH_2$
propan**amine**   5-methylhexan**amine**   2-propan**amine**

For secondary and tertiary amines the parent alkane is taken to be the alkyl group with the longest chain. If two alkyl groups are "tied" by this criterion, the parent alkane is the one with the greater number of substituents. The remaining alkyl groups are named as substituents by using the prefix **N-** to indicate that they are attached to nitrogen.

$(CH_3CH_2)_2NH$ $CH_3CH_2CH_2NHCH_2CH(CH_3)_2$ $(CH_3CH_2CH_2)_2NCH_3$
N-ethylethanamine   2-methyl-N-propylpropanamine   N-methyl-N-propylpropanamine

The *Chemical Abstracts* name for aniline is benzenamine; derivatives are named accordingly.

N,N-dimethylbenzenamine   4-methylbenzenamine   3-methoxybenzenamine

In this book we shall use common names for simple amines such as methylamine, triethylamine, and di-*n*-propylamine. Because it is so widely used in the chemical literature, we shall retain the name aniline for the simplest aromatic amine and name derivatives as substituted anilines. For more complex amines, we shall use the *Chemical Abstracts* system.

EXERCISE: What are the *Chemical Abstracts* names for the amines depicted on pages 727–28?

24.3
Physical Properties of Amines

A. *Colligative Properties*

The melting points, boiling points, and densities of some simple amines are collected in Table 24.1. As with other classes of compounds, certain trends are

TABLE 24.1
Physical Properties of Amines

	Molecular Weight	Melting Point, °C	Boiling Point, °C	Density
Primary Amines				
CH_3NH_2	31	−94	−6.3	0.6628
$CH_3CH_2NH_2$	45	−81	16.6	0.6829
$CH_3CH_2CH_2NH_2$	59	−83	47.8	0.7173
$CH_3CH_2CH_2CH_2NH_2$	73	−49	77.8	0.7414
Secondary Amines				
$(CH_3)_2NH$	45	−93	7.4	0.6804
$(CH_3CH_2)_2NH$	73	−48	56.3	0.7056
$(CH_3CH_2CH_2)_2NH$	101	−40	110	0.7400
$(CH_3CH_2CH_2CH_2)_2NH$	129	−60	159	0.7670
Tertiary Amines				
$(CH_3)_3N$	59	−117	2.9	0.6356
$(CH_3CH_2)_3N$	101	−114	89.3	0.7256
$(CH_3CH_2CH_2)_3N$	143	−94	155	0.7558
$(CH_3CH_2CH_2CH_2)_3N$	185		213	0.7771

evident in the properties. All three properties increase with molecular weight as a consequence of the greater intermolecular attraction with the larger members in the series.

Like alcohols, the lower amines show the effect of hydrogen bonding (Section 10.3). Since nitrogen is not as electronegative as oxygen, the N—H···N hydrogen bond is not as strong as the analogous O—H···O bond. Thus, primary amines have boiling points that are intermediate between those of alkanes and alcohols of comparable molecular weight (Figure 24.2), just as ammonia, b.p. −33°, is intermediate between methane, b.p. −161°, and water, b.p. 100°.

Hydrogen bonding is more important with primary than with secondary amines and is not possible at all with tertiary amines. Thus, a primary amine always boils higher than a secondary or tertiary amine of the same molecular weight (Figure 24.2).

B. *Spectroscopic Properties*

1. INFRARED SPECTRA. The characteristic infrared absorptions of amines are associated with the N—H and C—N bonds. Typical bands are summarized in Table 24.2. For diagnostic purposes, the C—N absorptions are not very useful because these bands occur in a spectral region that normally also contains many bands for other types of compounds. Particularly useful absorptions are the weak N—H stretching bands of primary amines, the N—H bending mode of primary amines, and the N—H wagging mode for primary and secondary amines. The N—H stretch of secondary amines is so weak that it is often not observed. Infrared spectroscopy is useful in diagnosing the presence of a tertiary amino group only in an indirect sense; if an amine does not show ir absorption bands charac-

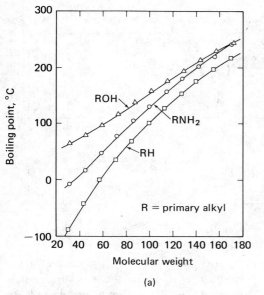

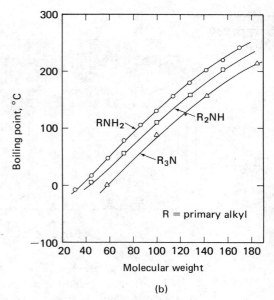

FIGURE 24.2. *Boiling points:* (a) *alkanes, alcohols, and primary amines;* (b) *primary, secondary, and tertiary amines.*

teristic for primary or secondary amines, it may be concluded that it is tertiary. An example is the spectrum of *n*-hexylamine, which is shown in Figure 24.3.

2. NUCLEAR MAGNETIC RESONANCE SPECTRA. Since nitrogen is more electronegative than carbon, the protons near the amino group are deshielded. The downfield shifts are not as pronounced as in the case of alcohols and ethers (Sections 10.5 and 10.9). As with alcohols and ethers, the exact chemical shift is dependent upon whether the protons are part of a CH_3, a CH_2, or a CH group.

$$CH_3NR_2 \qquad R'CH_2NR_2 \qquad R'_2CHNR_2$$

δ, ppm: 2.2   2.4   2.8

Protons β to nitrogen are affected to a much smaller extent; they are normally seen in the range $\delta = 1.1–1.7$ ppm.

Protons bound directly to the nitrogen in primary and secondary amines may resonate anywhere in the region from $\delta = 0.6$ ppm to $\delta = 3.0$ ppm. The exact

TABLE 24.2
Infrared Spectra of Amines

Frequency, cm^{-1}	Intensity	Assignment	Compound Type
3500, 3400 (doublet)	weak	N—H stretching	primary
3310–3350	very weak	N—H stretching	secondary
1580–1650	medium to strong	N—H bending	primary
1020–1250	weak to medium	C—N stretching	primary, secondary
666–909	medium to strong	N—H wagging	primary, secondary

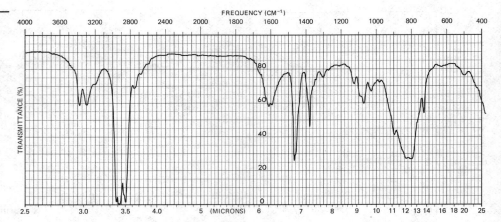

FIGURE 24.3. *Infrared spectrum of* n-*hexylamine.*

resonance position is dependent on the purity of the sample, the nature of the solvent, the concentration, and the temperature at which the measurement is made. As with alcohols, coupling of the type H—C—N—H is generally not observed because of proton exchange. The spectrum of di-*n*-propylamine is shown in Figure 24.4.

3. MASS SPECTRA. Mass spectrometry is often a useful technique for establishing the presence of nitrogen in an organic compound. Hydrocarbons and oxygen-containing compounds always have *even molecular weights,* but compounds containing an odd number of nitrogen atoms have odd molecular weights. Thus, the molecular ion of a monoamine is always odd, and such an observation can be used to conclude that a molecule contains an odd number of nitrogens. Unfortunately, the parent molecular ions of simple amines tend to have low intensity because there is a favorable fragmentation pathway. For this reason, the absence of an odd molecular ion does not establish that a molecule is not a monoamine.

The chief fragmentation mode is cleavage of the C_α—C_β bond to give an alkyl

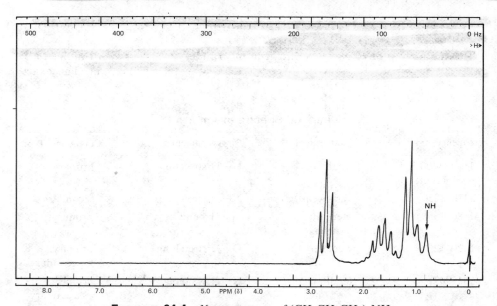

FIGURE 24.4. *Nmr spectrum of* $(CH_3CH_2CH_2)_2NH$.

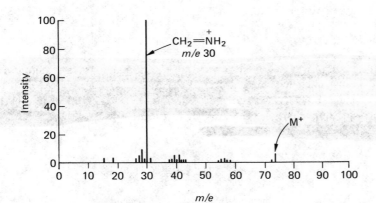

FIGURE 24.5. *Mass spectrum of isobutylamine,* $(CH_3)_2CHCH_2NH_2$.

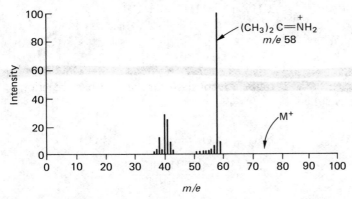

FIGURE 24.6. *Mass spectrum of* t-*butylamine,* $(CH_3)_3CNH_2$.

radical and an **immonium ion.** Primary amines that are not branched at the α-carbon give a strong fragment with $m/e = 30$.

$$[CH_3CH_2CH_2NH_2]^+ \longrightarrow CH_3CH_2\cdot + CH_2{=}NH_2^+$$
$$m/e = 30$$

When the amine is branched at the α-carbon, an analogous cleavage occurs, leading to a homologous immonium ion; loss of the larger group is preferred.

$$\begin{bmatrix} CH_3 \\ | \\ CH_3CH_2CH_2CHNH_2 \end{bmatrix}^+ \longrightarrow CH_3CH_2CH_2\cdot + CH_3CH{=}NH_2^+$$
$$m/e = 44$$

These cleavage patterns are illustrated by the spectra of isobutylamine and *t*-butylamine shown in Figures 24.5 and 24.6.

EXERCISE: There are eight amines having the formula $C_4H_{11}N$. Write their structures and make a list of the characteristic ir, nmr, and mass spectral features expected for each.

24.4
Basicity

Because of the presence of a nonbonding electron pair on nitrogen, amines are Lewis bases just like alcohols and ethers (Sections 10.7.B and 19.5). Nitrogen is not as electronegative as oxygen, and amines have a greater tendency to react with a proton than alcohols. Looking at it another way, alkyloxonium ions are more acidic than alkylammonium ions.

$$CH_3OH + H^+ \rightleftharpoons CH_3\overset{+}{O}H_2$$

less	more
basic	acidic

$$CH_3NH_2 + H^+ \rightleftharpoons CH_3\overset{+}{N}H_3$$

more	less
basic	acidic

Since amines are much more basic than water, aqueous solutions of amines have basic properties. The equilibrium constant for the acid-base reaction of an amine with water is called K_b.

$$RNH_2 + H_2O \overset{K_b}{\rightleftharpoons} RNH_3^+ + OH^-$$

$$K_b = \frac{[RNH_3^+][OH^-]}{[RNH_2]}$$

As usual in such expressions, the concentration of water is not included in the equilibrium expression because it is present in large excess and is essentially constant (Section 18.4). It is often convenient to refer to the dissociation constant of the corresponding ammonium ion when comparing base strengths of amines. This equilibrium constant, like other dissociation constants, is called K_a (Section 4.5).

$$RNH_3^+ + H_2O \overset{K_a}{\rightleftharpoons} RNH_2 + H_3O^+$$

$$K_a = \frac{[RNH_2][H_3O^+]}{[RNH_3^+]}$$

K_b for an amine and K_a for the corresponding ammonium ion are related by expression (24-1)

or

$$K_a K_b = 10^{-14} \ M^2$$
$$pK_a + pK_b = 14$$

(24-1)

[The student should derive this relationship by using the foregoing equations and the relationship for the dissociation of water, $K_w = [H_3O^+][OH^-] = 10^{-14} \ M^2$.]

The pK_bs for some typical amines and the pK_as for the corresponding ammonium ions are collected in Table 24.3, together with the pK_b for ammonia and the pK_a

TABLE 24.3
Basicity of Some Alkylamines

Amine	pK_b, 25°	Conjugate Acid	pK_a, 25°
NH_3	4.76	NH_4^+	9.24
CH_3NH_2	3.38	$CH_3NH_3^+$	10.62
$CH_3CH_2NH_2$	3.36	$CH_3CH_2NH_3^+$	10.64
$(CH_3)_3CNH_2$	3.32	$(CH_3)_3CNH_3^+$	10.68
$(CH_3)_2NH$	3.27	$(CH_3)_2NH_2^+$	10.73
$(CH_3CH_2)_2NH$	3.06	$(CH_3CH_2)_2NH_2^+$	10.94
$(CH_3)_3N$	4.21	$(CH_3)_3NH^+$	9.79
$(CH_3CH_2)_3N$	3.25	$(CH_3CH_2)_3NH^+$	10.75

for ammonium ion for reference. Notice that the simple alkylammonium ions all have pK_as in the range 10–11 and are therefore slightly *less acidic* than NH_4^+ itself. In other words, amines are only slightly more basic than NH_3.

It is important to distinguish between K_a for the dissociation of NH_4^+ and K_a for NH_3 itself and not to confuse them. Ammonia itself is an extremely weak acid; the pK_a for NH_3 is about 35 and the pK_b for NH_2^-, correspondingly, is −21. Analogous anions derived by deprotonation of amines are known and are useful reagents for some organic reactions. Because amines are such feeble acids, powerful bases are needed for deprotonation; alkyllithium compounds are commonly used.

$$(CH_3CH_2CH_2)_2NH + \textit{n-}C_4H_9Li \longrightarrow (CH_3CH_2CH_2)_2N^-Li^+ + C_4H_{10}$$

dipropylamine	butyllithium	lithium	butane
$pK_a \sim 40$		dipropylamide	$pK_a \sim 50$

In aqueous solution, arylamines are *substantially less basic* than alkylamines. Correspondingly, the acidity of anilinium ion is substantially greater than that of alkylammonium ions.

$$K_a = 2.5 \times 10^{-5} M \qquad pK_a = 4.60$$

$$(CH_3)_2CHNH_3^+ \rightleftharpoons (CH_3)_2CHNH_2 + H_3O^+$$
$$K_a = 2.5 \times 10^{-12} M \qquad pK_a = 11.60$$

Aliphatic amines have basicity comparable to dilute solutions of sodium hydroxide; the basicity of aniline is comparable to sodium acetate.

The reduced basicity of aniline compared to aliphatic amines may be attributed in part to the electron-attracting inductive effect of a phenyl group; for example, phenylacetic acid ($pK_a = 4.31$) is more acidic than acetic acid ($pK_a = 4.76$). However, this effect is small compared to the effect of delocalization of the nitrogen lone pair into the benzene ring.

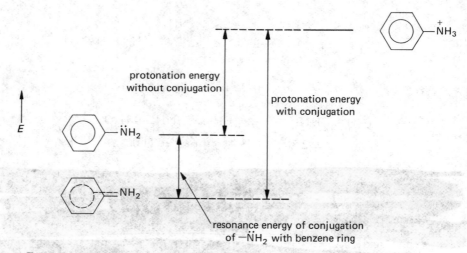

This delocalization renders the lone-pair less accessible for bonding. Alternatively and equivalently, this delocalization effect can be expressed as a resonance stabilization of the amine that is not present in the ammonium ion. This energy effect is illustrated in Figure 24.7. The resonance energy of conjugation results in displacement of the protonation equilibrium toward the amine.

Ammonia itself and amines generally have a pyramidal structure (Section 24.1); the H—N—H bond angle in ammonia is 107.1°. The most effective conjugation of the nitrogen lone pair with the benzene ring would be obtained for a lone pair in a *p*-orbital parallel to the *p*-orbitals of the aromatic π-system. However, lone pairs are generally more stable in orbitals having some *s*-character. In the case of aniline, an energy compromise is reached in which the lone-pair orbital has more *p*-character than in ammonia but in which the orbital retains some *s*-character. As a result, the NH_2 group in aniline is still pyramidal but with a larger H—N—H bond angle (113.9°) than in ammonia. The H—N—H plane intersects the plane of the benzene ring at an angle of 39.4°. The orbital structure of aniline is represented in Figure 24.8.

Substituents on the aniline ring affect basicity in ways that are generally interpretable with the principles of substituent effects discussed previously. Table 24.4 summarizes the pK_a values of a number of substituted anilinium ions.

Ortho substituents sometimes give unexpected results because of steric effects; for example, *o*-methylaniline is less basic than aniline, whereas in the *meta* and *para* positions a methyl substituent exerts its typical electron-donating effect to give enhanced basicity. Bromo, chloro, iodo, and CF_3 groups show normal electron-attracting inductive effects that decrease the basicity of aniline. The nitro group has an especially potent effect in the *para* position that is attributed to direct conjugation with the amino group.

FIGURE 24.7. *Conjugation with the phenyl ring decreases the basicity of the amino group in aniline.*

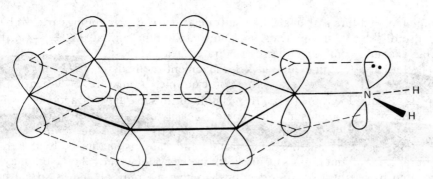

FIGURE 24.8. *The partially pyramidal amino group in aniline can still conjugate with the phenyl π-system.*

Because of their basic properties, amines form salts with acids. Since these salts are ionic compounds, they are usually water-soluble, even in cases where the corresponding amine is insoluble in water.

$$CH_3(CH_2)_9NH_2 + HCl \longrightarrow CH_3(CH_2)_9\overset{+}{N}H_3 \; Cl^-$$

$$\begin{array}{cc} \textit{n-decylamine} & \textit{n-decylammonium chloride} \\ \text{(insoluble in } H_2O) & \text{(soluble in } H_2O) \end{array}$$

Even though aromatic amines are only one-millionth as basic as alkylamines (see Tables 24.3 and 24.4), they are still protonated even in dilute acidic solutions. For example, aniline is essentially completely protonated in 0.1 M HCl solution (pH = 1). Hence, although aniline is only slightly soluble in water, it dissolves

TABLE 24.4
pK_as of Anilinium Ions

Substituent	pK_a, 25°		
	ortho	*meta*	*para*
H	4.60	4.60	4.60
benzoyl			2.17
bromo	2.53	3.58	3.86
chloro	2.65	3.52	3.98
cyano	0.95	2.75	1.74
fluoro	3.20	3.57	4.65
iodo	2.60	3.60	3.78
methoxy	4.52	4.23	5.34
methyl	4.44	4.72	5.10
nitro	−0.26	2.47	1.00
trifluoromethyl		3.20	2.75

completely in dilute hydrohalic and sulfuric acids. The nitroanilines are less basic but also dissolve in strong acids. 2,4-Dinitroanilinium ion has $pK_a = -4.4$; this amine is soluble only in rather concentrated acids.

The basicity of amines provides a convenient method for separating amines from neutral organic compounds. For example, a mixture of *n*-decylamine (b.p. 221°) and dodecane (b.p. 216°) is difficult to separate by fractional distillation. The two compounds may be separated easily by *extracting* the mixture with sufficient 10% aqueous hydrochloric acid to convert all of the amine into the ammonium salt. The alkane, being insoluble in water, is unaffected by this treatment, and the ammonium salt dissolves in the water layer. The layers may be separated by use of a separatory funnel to give the pure alkane. A strong base such as sodium hydroxide is then added to the aqueous solution to neutralize the ammonium salt and liberate the free amine. The water-insoluble amine now forms a second layer that may be separated.

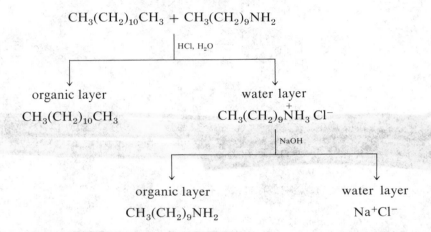

Amines also form salts with carboxylic acids. Again the salts are ionic and are often water-soluble.

$$CH_3\overset{O}{\overset{\|}{C}}OH + CH_3NH_2 \longrightarrow CH_3\overset{O}{\overset{\|}{C}}O^- \ H_3\overset{+}{N}CH_3$$

acetic acid   methylamine   methylammonium acetate

This salt-forming reaction is often used as a method for **resolving** racemic mixtures of organic acids.

> The student should review the basic principles of stereochemistry in Chapter 7. **Resolution** is the term used to describe the separation of two enantiomers from each other.

Consider a racemic mixture of α-hydroxypropionic acids (lactic acids). Recall that the two enantiomers have identical physical properties and cannot be separated by crystallization or distillation techniques. The mixture will react with methylamine to give a racemic mixture of methylammonium lactates, which also cannot be separated by physical methods.

$$
\begin{array}{c}
\underset{\text{(R)-lactic acid}}{\overset{\displaystyle \text{COOH}}{\underset{\displaystyle \text{CH}_3}{\text{H}-\text{C}-\text{OH}}}} \\
+ \\
\underset{\text{(S)-lactic acid}}{\overset{\displaystyle \text{COOH}}{\underset{\displaystyle \text{CH}_3}{\text{HO}-\text{C}-\text{H}}}}
\end{array}
\;+\; \text{CH}_3\text{NH}_2 \longrightarrow
\begin{array}{c}
\underset{\text{methylammonium (R)-lactate}}{\overset{\displaystyle \text{CO}_2^- \;\; \text{CH}_3\overset{+}{\text{N}}\text{H}_3}{\underset{\displaystyle \text{CH}_3}{\text{H}-\text{C}-\text{OH}}}} \\
+ \\
\underset{\text{methylammonium (S)-lactate}}{\overset{\displaystyle \text{CO}_2^- \;\; \text{CH}_3\overset{+}{\text{N}}\text{H}_3}{\underset{\displaystyle \text{CH}_3}{\text{HO}-\text{C}-\text{H}}}}
\end{array}
$$

However, consider the situation when one enantiomer of a chiral amine is used to form the salt.

$$
\begin{array}{c}
\underset{\text{(R)-lactic acid}}{\overset{\displaystyle \text{COOH}}{\underset{\displaystyle \text{CH}_3}{\text{H}-\text{C}-\text{OH}}}} \\
+ \\
\underset{\text{(S)-lactic acid}}{\overset{\displaystyle \text{COOH}}{\underset{\displaystyle \text{CH}_3}{\text{HO}-\text{C}-\text{H}}}}
\end{array}
\;+\; \underset{\text{(S)-1-phenylethylamine}}{\overset{\displaystyle \text{NH}_2}{\underset{\displaystyle \text{H}}{\text{C}_6\text{H}_5-\text{C}-\text{CH}_3}}}
\longrightarrow
\begin{array}{c}
\underset{\text{(S)-1-phenylethylammonium (R)-lactate}}{\overset{\displaystyle \text{NH}_3^+ \qquad \text{CO}_2^-}{\underset{\displaystyle \text{H} \qquad\quad \text{CH}_3}{\text{C}_6\text{H}_5-\text{C}-\text{CH}_3 \; \text{H}-\text{C}-\text{OH}}}} \\
\\
\underset{\text{(S)-1-phenylethylammonium (S)-lactate}}{\overset{\displaystyle \text{NH}_3^+ \qquad \text{CO}_2^-}{\underset{\displaystyle \text{H} \qquad\quad \text{CH}_3}{\text{C}_6\text{H}_5-\text{C}-\text{CH}_3 \; \text{HO}-\text{C}-\text{H}}}}
\end{array}
$$

The two salts are now *diastereomeric* rather than enantiomeric, and they have different physical properties. For example, the (S,R) salt may be more soluble in some solvents than the (S,S) salt. Because of this difference in solubility, the two salts can be separated by fractional crystallization. Each of the diastereomeric salts can then be treated with a strong acid such as hydrochloric or sulfuric acid to liberate the free carboxylic acid. Acidification of the (S,R) salt gives enantiomerically pure (R)-lactic acid, whereas similar treatment of the (S,S) salt gives pure (S)-lactic acid.

$$
\begin{array}{c}
\text{(S)-acid} \\
+ \\
\text{(R)-acid}
\end{array}
\;+\; \text{(S)-amine} \longrightarrow
\begin{array}{c}
\text{(S,S)-salt} \\
+ \\
\text{(S,R)-salt}
\end{array}
\xrightarrow{\text{separate}}
\begin{array}{cc}
\text{(S,S)-salt} & + & \text{(S,R)-salt} \\
\downarrow{\scriptstyle \text{HCl}} & & \downarrow{\scriptstyle \text{HCl}} \\
\text{(S)-acid} & & \text{(R)-acid} \\
+ & & + \\
\text{(S)-ammonium} & & \text{(S)-ammonium} \\
\text{chloride} & & \text{chloride}
\end{array}
$$

Of course, in order to use this technique for resolution, suitable optically active amines must be available. Fortunately, a number of such compounds are readily available and relatively inexpensive. A particularly useful source of such resolving agents is the class of naturally occurring amines called alkaloids, which occur in nature in only one enantiomeric form. Examples are strychnine and brucine (Section 34.6). Another frequently used resolving agent is 1-phenyl-2-propanamine (amphetamine). Although not a natural product, synthetic amphetamine is readily available in both enantiomeric forms.

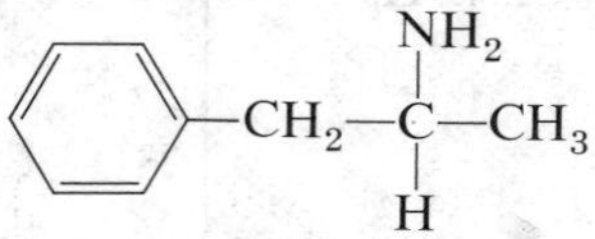

amphetamine

EXERCISE: Using the data in Tables 24.3 and 24.4, calculate equilibrium constants for the following equilibria.

(a) 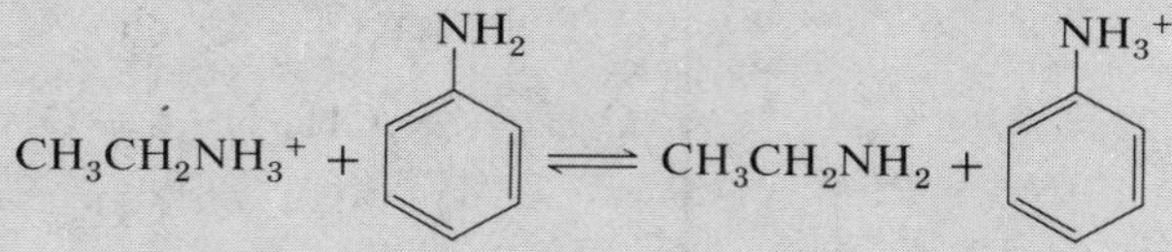 $CH_3CH_2NH_3^+$ + [aniline] $\rightleftharpoons$ $CH_3CH_2NH_2$ + [anilinium]

(b) 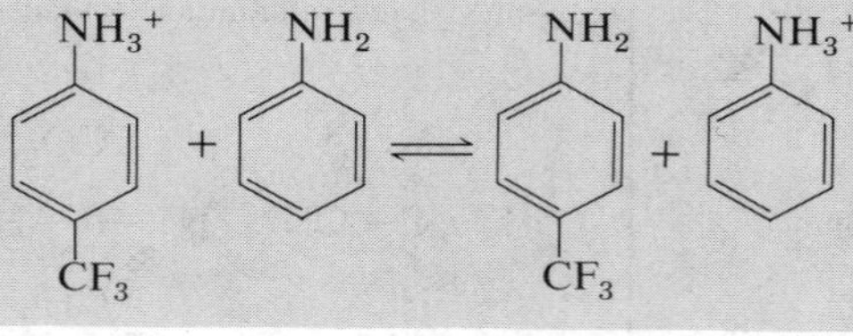 [p-CF₃-anilinium] + [aniline] $\rightleftharpoons$ [p-CF₃-aniline] + [anilinium]

24.5
Quaternary Ammonium Compounds

A. Tertiary Amines as Nucleophiles

Recall that there is a correlation between Lewis basicity and the nucleophilicity of a species (Section 8.9). Amines are more basic than alcohols or ethers, and they are also more nucleophilic. For example, a mixture of diethyl ether and methyl iodide does not react under ordinary conditions, but triethylamine and methyl iodide react violently at room temperature. If the reaction is carried out in a solvent to moderate its vigor, the product, which is a tetraalkylammonium iodide, may be obtained in good yield.

$$(C_2H_5)_2O + CH_3I \xrightarrow{25°} NR$$

$$(C_2H_5)_3N + CH_3I \xrightarrow[\text{ether}]{25°} (C_2H_5)_3\overset{+}{N}CH_3 \ I^-$$

methyltriethylammonium iodide

Such compounds, which have four alkyl groups replacing the four hydrogens of the ammonium ion, are called **quaternary ammonium compounds.** Since they are ionic, they are generally water-soluble and have fairly high melting points. They often decompose at the melting point.

$$(CH_3)_4N^+ \ Cl^-$$

tetramethylammonium chloride
m.p. 420°

$$(CH_3CH_2CH_2)_4N^+ \ Br^-$$

tetrapropylammonium bromide
m.p. 252°

Quaternary ammonium compounds are important as intermediates in some reactions that we shall encounter and also occur in nature. Acetylcholine, which is important in the neural transport system of mammals, is an example.

$$\underset{\text{acetylcholine bromide}}{CH_3\overset{\displaystyle O}{\overset{\|}{C}}OCH_2CH_2\overset{\displaystyle \overset{+}{C}H_3}{\underset{\displaystyle CH_3}{\overset{|}{N}}}CH_3 \ \ Br^-}$$

Quaternary ammonium hydroxides are as basic as alkali hydroxides. They decompose on heating (Hofmann degradation; Section 24.7.E) and find use as base catalysts in organic systems.

B. *Phase-Transfer Catalysis*

In Section 11.6.F we learned that chloroform reacts with strong bases to form dichlorocarbene, which can then add to double bonds to give dichlorocyclopropane derivatives. If a solution of cyclohexene in chloroform is stirred with 50% aqueous sodium hydroxide, only small yields of the cyclopropane are formed. The hydroxide ion stays in the aqueous phase, and the only reaction that occurs is at the interface between the organic and aqueous phases. However, if we add a small amount of benzyltriethylammonium chloride to the mixture, a reaction occurs and 7,7-dichlorobicyclo[4.1.0]heptane can be isolated in 72% yield.

(72%)

To explain what has happened, we need to recognize that although the quaternary ammonium compound is a salt soluble in water, it also has a large organic group that provides solubility in organic solvents *as an ion pair*. Although the chloride was used because of its availability and greater convenience in handling, it is swamped by the large excess of hydroxide ion in the aqueous solution. Benzyltriethylammonium hydroxide ion pairs extract in part into the chloroform layer. In this medium the hydroxide ion is especially reactive because of the reduced hydrogen bonding. Dichlorocarbene is produced in the chloroform solution, in which there also is a high concentration of cyclohexene.

$$C_6H_5CH_2NEt_3^+(aq.) + OH^-(aq.) \rightleftharpoons C_6H_5CH_2NEt_3^+ \ OH^-(CHCl_3)$$

$$C_6H_5CH_2NEt_3^+ \ OH^- + CHCl_3 \longrightarrow H_2O + C_6H_5CH_2NEt_3^+ \ CCl_3^-$$

$$C_6H_5CH_2NEt_3^+ \ CCl_3^- \longrightarrow :CCl_2 + C_6H_5CH_2NEt_3^+ \ Cl^-$$

The benzyltriethylammonium chloride ion pairs extract into the aqueous phase, where the ammonium ion can again pick up a hydroxide ion and begin the cycle anew. The key to the procedure is the solubility of the quaternary ammonium salt in both water and organic solvents.

$$C_6H_5CH_2N(CH_2CH_3)_3{}^+ \ OH^-$$

soluble in both water and
organic solvents

The general technique is called **phase-transfer catalysis** and can be applied to a number of different types of reaction. The general procedure is to use concentrated solutions with an aqueous and an organic phase. The quaternary ammonium salt used need only have organic groups that are sufficiently large to provide solubility in organic solvents. Among the ones commonly used are tetrabutylammonium, methyltrioctylammonium, and hexadecyltrimethylammonium salts. Some additional examples of applications of phase-transfer catalysis are

$$CH_3(CH_2)_7CH{=}CH_2 \text{ (benzene soln.)} \xrightarrow[\substack{\text{aq. KMnO}_4 \\ 40-50°}]{(CH_3(CH_2)_6CH_2)_3\overset{+}{N}CH_3 \ Cl^-} CH_3(CH_2)_7COOH$$
$$(91\%)$$

$$C_6H_5CH_2COCH_3 + CH_3(CH_2)_3Br \xrightarrow[\substack{50\% \text{ aq. NaOH}}]{C_6H_5CH_2\overset{+}{N}Et_3 \ OH^-} CH_3COCHCH_2CH_2CH_2CH_3$$
$$\overset{|}{C_6H_5}$$
$$(90\%)$$

$$CH_3(CH_2)_9Br \xrightarrow[\substack{\text{aq. NaSCN} \\ 100°}]{(n\text{-}C_6H_{13})_3\overset{+}{N}CH_3 \ Cl^-} CH_3(CH_2)_9SCN$$
$$(100\%)$$

Note that the examples include alkylation, oxidation, and displacement reactions, that the anions are not restricted to hydroxide ion, and that various temperatures may be used.

EXERCISE: For each of the above examples what species are in the aqueous phase? In the organic phase? Which species are passing from one phase to another?

24.6
Synthesis of Amines

A. *Direct Alkylation of Ammonia or Other Amines*

In Section 8.7 it was mentioned that ammonia reacts with primary alkyl halides by the S_N2 mechanism to give an alkylammonium halide. In principle, this type of displacement reaction might be used as a way of synthesizing primary amines.

$$CH_3CH_2Br + NH_3 \longrightarrow CH_3CH_2NH_3{}^+ \ Br^- \xrightarrow{\text{NaOH}} CH_3CH_2NH_2$$

In practice, this method is not very useful because of the side reactions that occur. The product alkylammonium ion is fairly acidic and may transfer a proton to a molecule of ammonia that has not yet reacted to give the primary amine and the ammonium ion. Since the primary amine is also nucleophilic, it may undergo further reaction giving a secondary amine. By similar equilibria and further alkylation, the tertiary amine and even the quaternary ammonium compound may be formed. The actual result is a complex mixture even when equivalent molar amounts of ammonia and alkyl halide are used.

$$RBr + NH_3 \longrightarrow RNH_3^+ \; Br^-$$
$$RNH_3^+ + NH_3 \rightleftharpoons RNH_2 + NH_4^+$$
$$RNH_2 + RBr \longrightarrow R_2NH_2^+ \; Br^-$$
$$R_2NH_2^+ + NH_3 \rightleftharpoons R_2NH + NH_4^+$$
$$R_2NH + RBr \longrightarrow R_3NH^+ \; Br^-$$
$$R_3NH^+ + NH_3 \rightleftharpoons R_3N + NH_4^+$$
$$R_3N + RBr \longrightarrow R_4N^+ \; Br^-$$

The "overalkylation" may be suppressed by using a large excess of ammonia or the amine being alkylated. This ploy is only practical in cases where the amine is relatively inexpensive and sufficiently volatile that the unreacted excess may be easily removed. An example is the preparation of *n*-butylamine by the reaction of *n*-butyl bromide with ammonia.

$$CH_3CH_2CH_2CH_2Br + NH_3 \longrightarrow \overset{NaOH}{\longrightarrow} CH_3CH_2CH_2CH_2NH_2$$
$$(47\%)$$

A solution of 300 g of NH_3 (20 moles) in 8 L of 90% aqueous ethanol is prepared. *n*-Butyl bromide is added slowly until 1507 g (11 moles) has been added. The reaction mixture is stirred at 25° for 48 hr and then made basic with aqueous NaOH. Fractional distillation of the organic layer gives 388 g of *n*-butylamine (47%) along with some di-*n*-butylamine and tri-*n*-butylamine, which have higher boiling points.

Secondary and tertiary amines may also be prepared this way, but the yields are again often low due to overalkylation. Also, if the amine is not readily available or is expensive, it is undesirable to use it in excess. In many cases where a pure primary, secondary, or tertiary amine is desired, direct alkylation is not a practical synthetic method. Several indirect methods have been devised to accomplish this purpose, and we shall study some of them in later parts of this section.

We saw in the previous section that aromatic amines are much less basic than alkylamines. They are also less nucleophilic, and their reactions with alkyl halides require somewhat more vigorous conditions. Since these amines are less reactive nucleophiles, it is easier to achieve monoalkylation, as illustrated with the following synthesis of N-benzylaniline.

(85–87%)
N-benzylaniline

The sodium bicarbonate serves to neutralize the HCl that is produced in the reaction.

B. *Indirect Alkylation: The Gabriel Synthesis*

Pure primary amines can be prepared in good yield by a method called the **Gabriel synthesis.** This method involves the alkylation of a "protected" form of ammonia. The compound phthalimide (Section 27.6) is prepared from ammonia and the dicarboxylic acid phthalic acid. Imides have acidic properties because the negative charge of the conjugate base is delocalized over both oxygens and the nitrogen. The pK_a of phthalimide is 8.3. In aqueous basic solution the compound is converted almost completely into the anion.

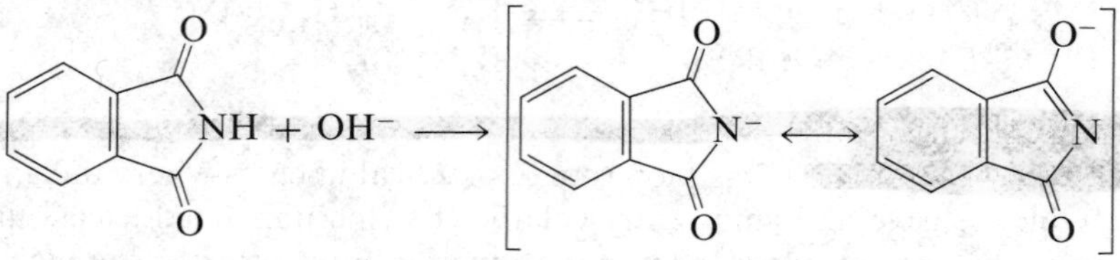

phthalimide

The phthalimide anion has nucleophilic properties and can enter into displacement reactions with alkyl halides. Reaction could in principle take place on either oxygen or nitrogen, but since nitrogen is more nucleophilic, it occurs mostly on nitrogen. Further alkylation cannot occur because there are no acidic protons. The product is an N-alkylphthalimide, and hydrolysis gives the amine and phthalic acid.

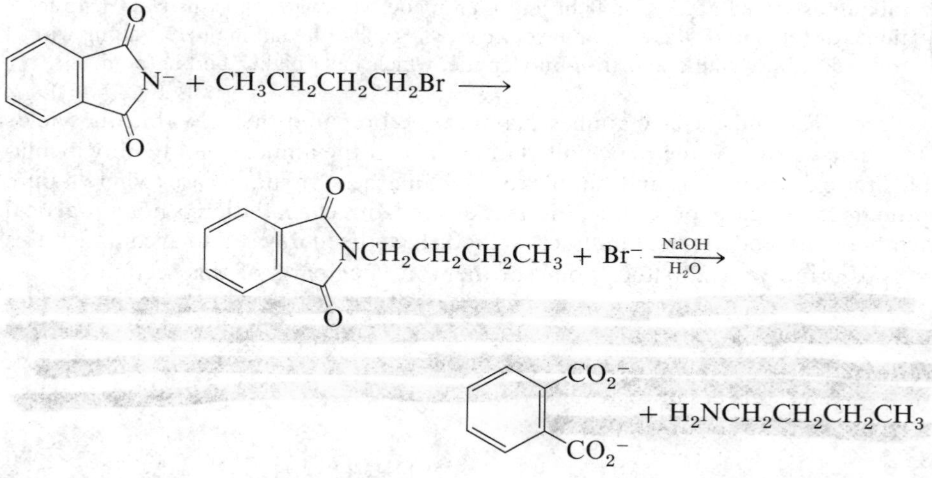

The best solvent for the alkylation appears to be dimethylformamide, $HCON(CH_3)_2$. The Gabriel synthesis is frequently used in the preparation of α-amino carboxylic acids, and we shall encounter it again in that context in Chapter 29.

In many cases, the alkaline hydrolysis of the alkylphthalimide is very slow. An alternative procedure for recovering the amine is called the Ing–Manske modification and involves the use of hydrazine. The product is the amine and a cyclic diamide of hydrazine.

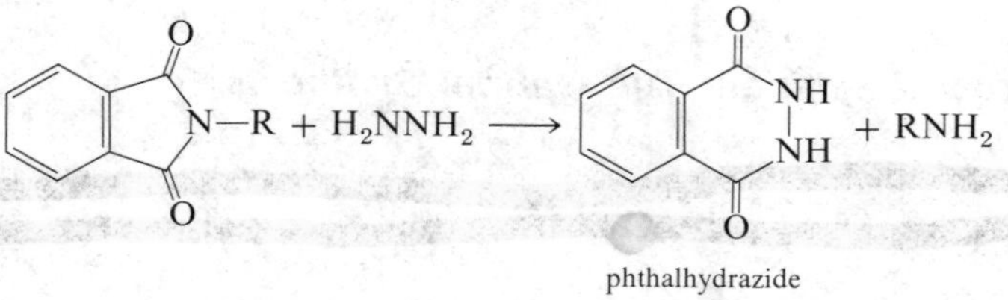

phthalhydrazide

EXERCISE: Explain why the Gabriel synthesis cannot be used to prepare each of the following primary amines. (a) neopentylamine, (b) *t*-butylamine, (c) di-*n*-propyl-amine.

C. Reduction of Nitro Compounds

Nitro compounds undergo ready reduction to yield primary amines. Because aromatic nitro compounds of a wide variety are available from nitration of aromatic compounds (Chapter 23), this method constitutes the most general synthesis of aromatic amines. Reduction may be accomplished by catalytic hydrogenation or by the use of chemical reducing agents in acidic solution.

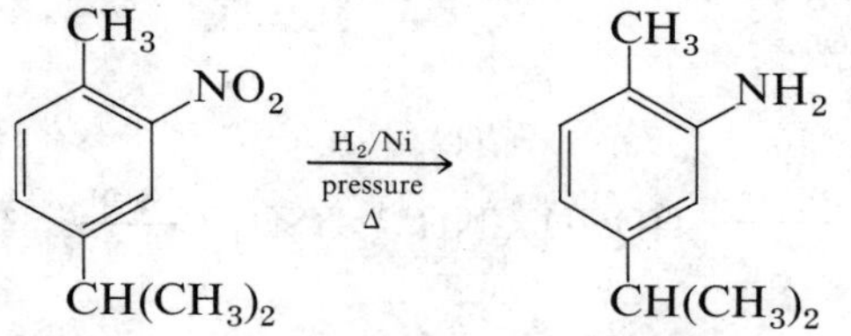

(87–90%)
2-methyl-5-isopropylaniline

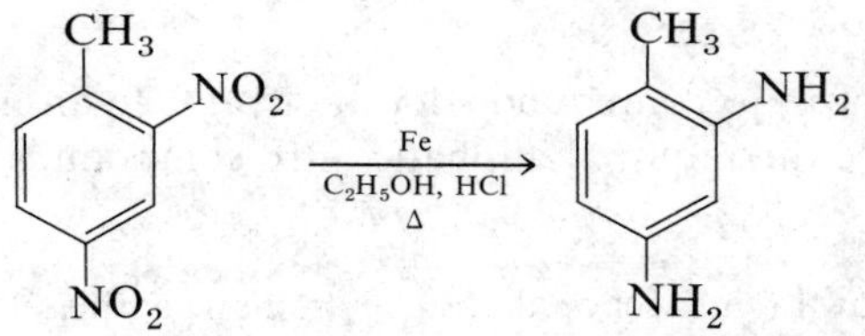

(74%)
2,4-diaminotoluene

Many chemical reducing agents have been used for the conversion of aromatic nitro groups to amines. Among the most common are metals and acid, usually iron or zinc and dilute hydrochloric acid. Stannous chloride, $SnCl_2$, and hydrochloric acid are an especially useful combination when other reducible groups, such as carbonyl groups, are present.

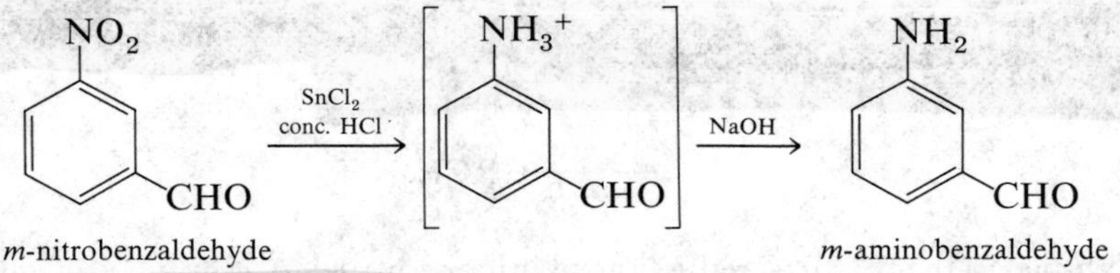

m-nitrobenzaldehyde   *m*-aminobenzaldehyde

Sodium or ammonium sulfide or hydrosulfide is useful for reducing one of the nitro groups in dinitro compounds.

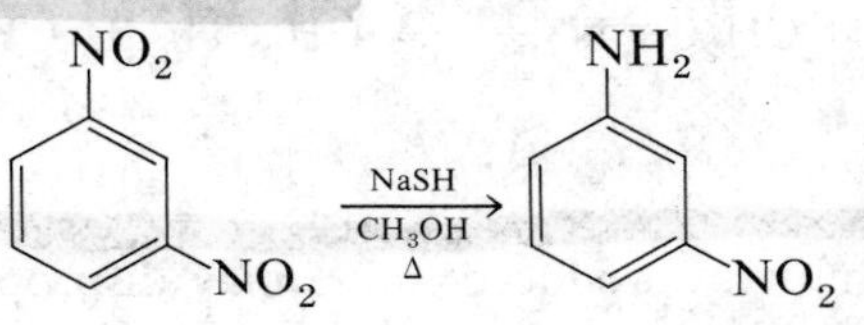

(79–85%)
m-dinitrobenzene   *m*-nitroaniline

The method can be applied to unsymmetrical dinitro compounds as well, and selective reductions are sometimes possible.

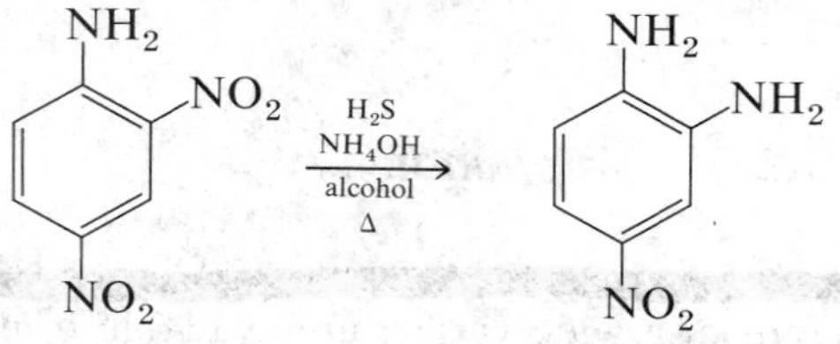

(52–58%)

2,4-dinitroaniline 1,2-diamino-4-nitrobenzene

Aliphatic nitro compounds may also be reduced to primary amines. An especially efficient reducing agent is a mixture of iron powder and ferrous sulfate in aqueous acidic solution.

$$HOCH_2C(CH_3)_2NO_2 \xrightarrow[H_2O,\ H_2SO_4]{Fe,\ FeSO_4} HOCH_2C(CH_3)_2NH_3^+\ HSO_4^- \xrightarrow{NaOH} HOCH_2C(CH_3)_2NH_2$$

(90%)

2-amino-2-methyl-1-propanol

However, reduction of nitro compounds does not provide an important route to aliphatic amines, since the required aliphatic nitro compounds are not common.

EXERCISE: We saw in this section that some aromatic amines are readily synthesized from aromatic compounds by the two-step sequence of (1) nitration and (2) reduction of the —NO_2 group to —NH_2. Show how this method may be applied to the synthesis of 2,4-dimethylaniline. Explain why the method is not applicable for the synthesis of *m*-methoxyaniline or *o*-aminobenzaldehyde.

D. *Reduction of Nitriles*

Nitriles are reduced by hydrogen and a catalyst or by lithium aluminum hydride in an ether solvent to give primary amines.

$$RC{\equiv}N \xrightarrow[\text{or } LiAlH_4]{H_2\text{-cat.}} RCH_2NH_2$$

An example is the $LiAlH_4$ reduction of butanenitrile to *n*-butylamine in high yield.

$$CH_3CH_2CH_2C{\equiv}N \xrightarrow{LiAlH_4} CH_3CH_2CH_2CH_2NH_2$$

(86%)

In the catalytic hydrogenation procedure, secondary amines are often produced as by-products. The initially produced imine may disproportionate by reaction with some of the primary amine already produced in the reduction to give a new imine. Hydrogenation of this imine gives the secondary amine.

(1) $RCN + H_2 \longrightarrow RCH{=}NH$
(2) $RCH{=}NH + H_2 \longrightarrow RCH_2NH_2$
(3) $RCH{=}NH + RCH_2NH_2 \rightleftharpoons RCH{=}NCH_2R + NH_3$
(4) $RCH{=}NCH_2R + H_2 \longrightarrow RCH_2NHCH_2R$

This side reaction may be suppressed by carrying out the hydrogenation in the presence of excess NH_3, which forces equilibrium (3) to the left.

$$RCN \xrightarrow[\text{EtOH,NH}_3]{\text{H}_2\text{-Pd/C}} RCH_2NH_2$$

Secondary amine formation may also be minimized by carrying out the reaction in acetic anhydride as solvent. The primary amine produced is rapidly converted into the amide.

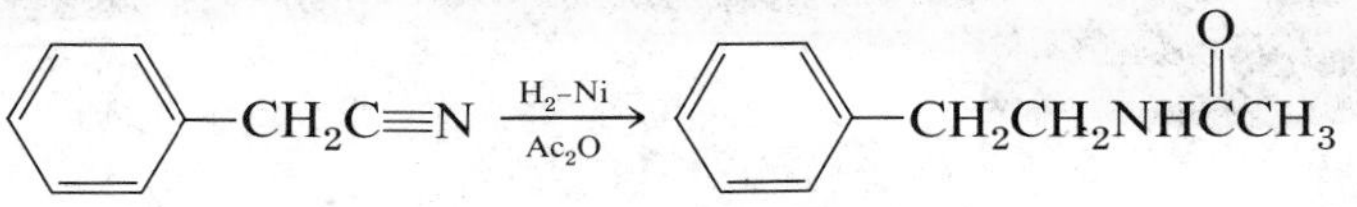

(97%)

The amine may then be obtained by hydrolysis of the amide. Since nitriles are easily available by several methods, many primary amines may be synthesized by this procedure.

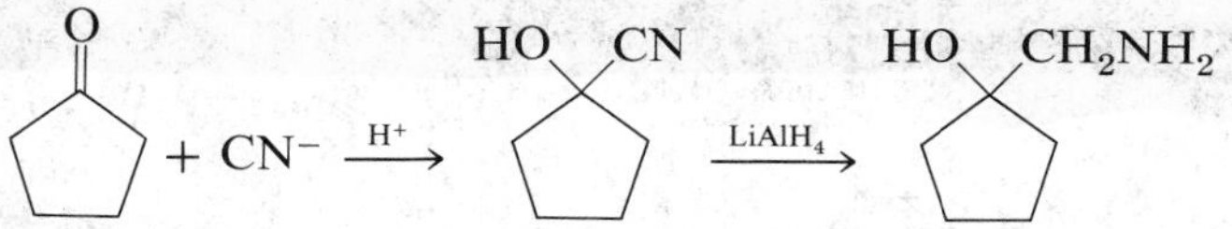

Notice that the cyano group, CN^-, is thus synthetically equivalent to $-CH_2NH_2$.

EXERCISE: Write equations showing how the following conversions may be accomplished.

(a) $(CH_3)_2CHCH_2CH_2Br \longrightarrow (CH_3)_2CHCH_2CH_2CH_2NH_2$

(b) $(CH_3)_2CHCHO \longrightarrow (CH_3)_2CHCHCH_2NH_2$ with OH on the indicated carbon

E. Reduction of Oximes

Aldoximes and ketoximes, which may be prepared from aldehydes or ketones by reaction with hydroxylamine (Section 13.7.C), are reduced by lithium aluminum hydride or hydrogen to primary amines. Since oximes are easily generated in high yield, this is a useful synthetic method.

$$CH_3CH_2CH_2\overset{O}{\overset{\|}{C}}CH_3 + H_2NOH \longrightarrow CH_3CH_2CH_2\overset{NOH}{\overset{\|}{C}}CH_3 \xrightarrow{\underset{\text{C}_2\text{H}_5\text{OH}}{\text{H}_2\text{-Ni}}} CH_3CH_2CH_2\overset{NH_2}{\overset{|}{C}}HCH_3$$

(85%)

(80%)

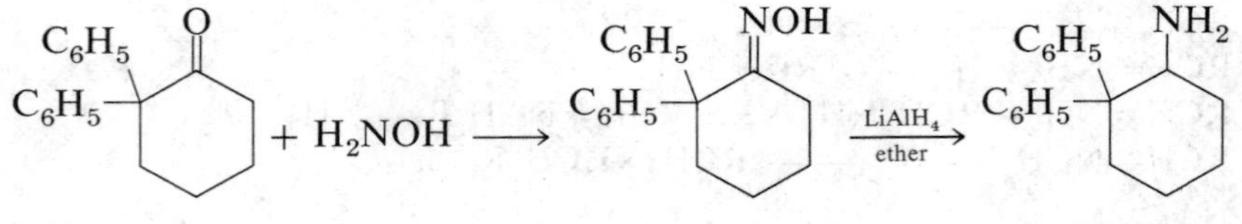

F. *Reduction of Imines: Reductive Amination*

Ammonia and primary amines condense with aldehydes and ketones to give imines (Section 13.7.C). In the case of ammonia, the imines are unstable and cannot be isolated. However, if a mixture of a carbonyl compound and ammonia is treated with hydrogen and a suitable hydrogenation catalyst, the C=N bond of the unstable imine is reduced and an amine results. The process is often called "reductive amination."

(89%)

benzaldehyde benzylamine

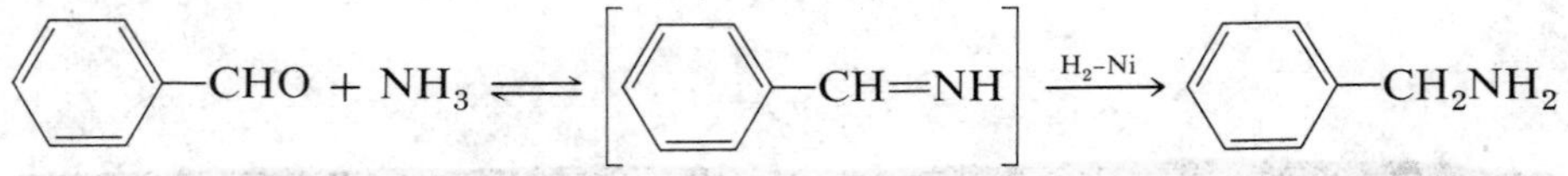

A significant side reaction complicates the reductive amination method. As the primary amine begins to build up, it may condense with the starting aldehyde to give a different imine. Reduction of this imine gives a secondary amine.

dibenzylamine

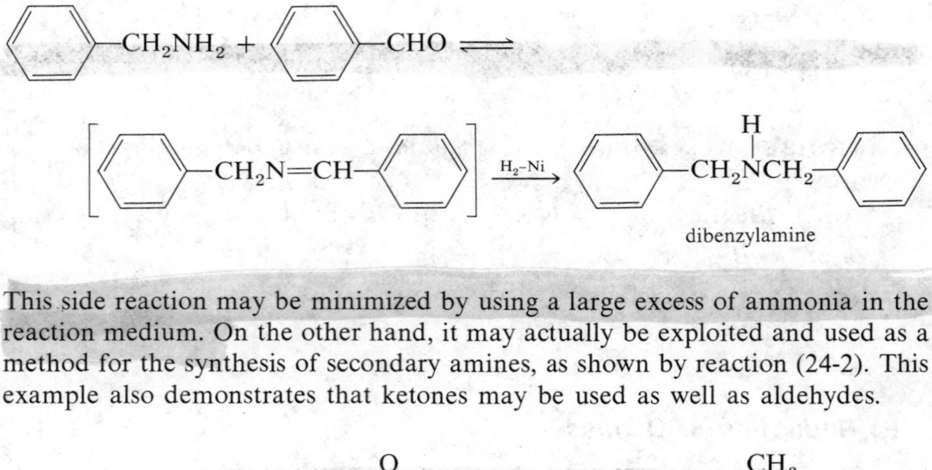

This side reaction may be minimized by using a large excess of ammonia in the reaction medium. On the other hand, it may actually be exploited and used as a method for the synthesis of secondary amines, as shown by reaction (24-2). This example also demonstrates that ketones may be used as well as aldehydes.

$$HOCH_2CH_2NH_2 + CH_3\overset{O}{\overset{\|}{C}}CH_3 \xrightarrow[C_2H_5OH]{H_2\text{-}Pt} HOCH_2CH_2NH\overset{CH_3}{\overset{|}{C}}HCH_3 \quad (24\text{-}2)$$

(95%)

One version of reductive amination, which is frequently employed for the synthesis of tertiary amines where at least one of the alkyl groups is methyl, is the **Eschweiler-Clarke** reaction. Instead of hydrogen, the reducing agent is formic acid, which is oxidized to carbon dioxide.

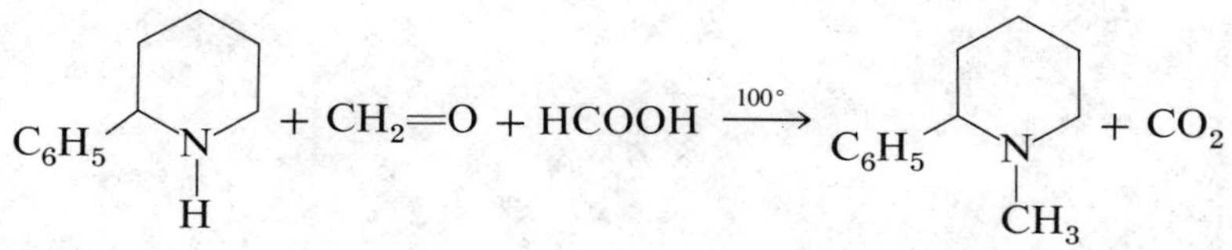

$$C_6H_5-\underset{\underset{H}{|}}{N}\text{(piperidine)} + CH_2=O + HCOOH \xrightarrow{100°} C_6H_5-\underset{\underset{CH_3}{|}}{N}\text{(piperidine)} + CO_2 \quad (24\text{-}3)$$

(94%)

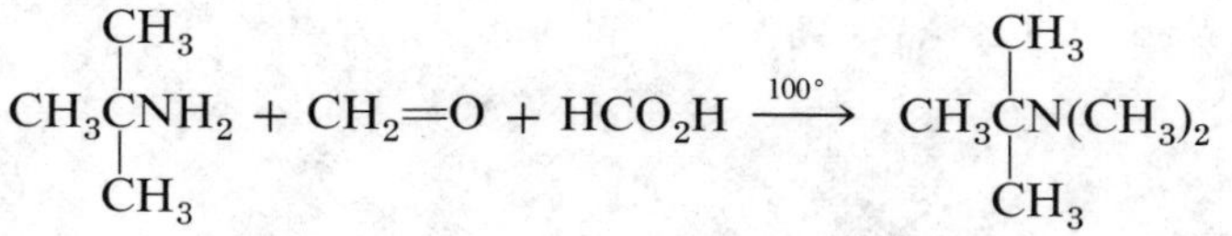

$$CH_3\underset{\underset{CH_3}{|}}{\overset{\overset{CH_3}{|}}{C}}NH_2 + CH_2=O + HCO_2H \xrightarrow{100°} CH_3\underset{\underset{CH_3}{|}}{\overset{\overset{CH_3}{|}}{C}}N(CH_3)_2 \quad (24\text{-}4)$$

(95%)
t-butyldimethylamine

As shown by equations (24-3) and (24-4), the reaction proceeds in excellent yield. The intermediate that is reduced is an **immonium ion,** and the reduction may be visualized as follows.

$$R_2NH + CH_2=O \underset{}{\overset{H^+}{\rightleftharpoons}} R_2N-CH_2OH$$

$$R_2N-CH_2OH \underset{}{\overset{H^+}{\rightleftharpoons}} R_2\overset{+}{N}=CH_2 + H_2O$$

$$R_2\overset{+}{N}=CH_2 + H-\overset{\overset{O}{\|}}{C}-O-H \longrightarrow R_2N-CH_3 + CO_2 + H^+$$

An earlier version of this reaction, called the **Leuckart** reaction, gives lower yields, but is more general. It can be used to prepare primary, secondary, or tertiary amines. In this method, the ketone is heated with a formate salt or a formamide at 180–200°.

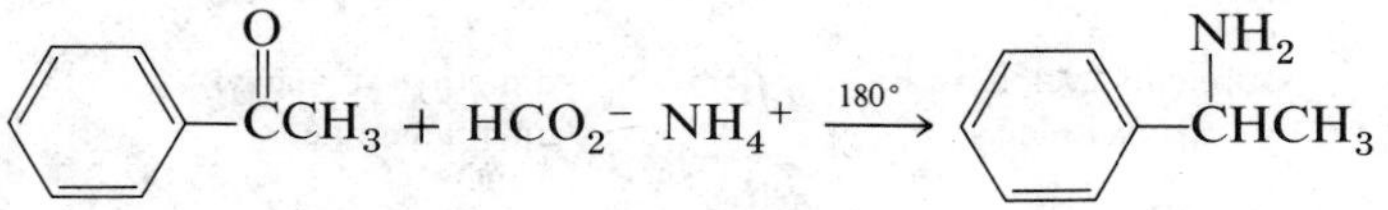

$$\text{C}_6\text{H}_5-\overset{\overset{O}{\|}}{C}CH_3 + HCO_2^-\ NH_4^+ \xrightarrow{180°} \text{C}_6\text{H}_5-\underset{\underset{NH_2}{|}}{C}HCH_3$$

(60–66%)

$$\text{(cyclooctanone)} + H\overset{\overset{O}{\|}}{C}N(CH_3)_2 \xrightarrow{200°} \text{(cyclooctane-N(CH}_3)_2\text{)}$$

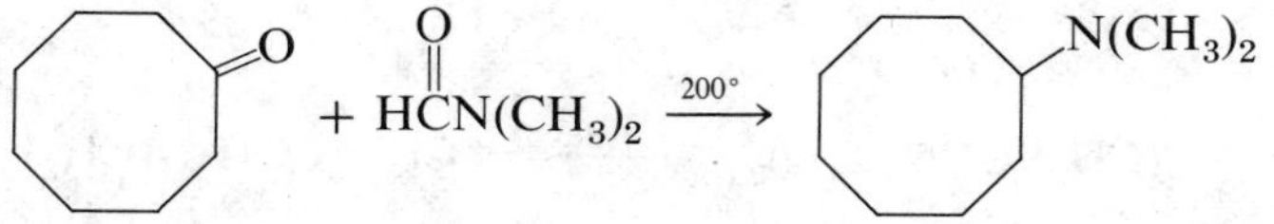

(75%)
(dimethylamino)cyclooctane

EXERCISE: What are the products of each of the following reactions?

(a) $\underset{\underset{CH_3}{}}{\overset{\overset{CHO}{}}{\text{C}_6\text{H}_4}}$ + NH$_3$ (excess) $\xrightarrow{H_2/Ni}$

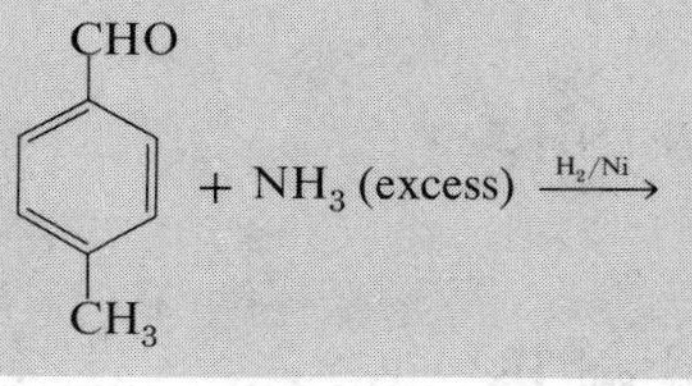

(b) $CH_3CH_2CH_2NH_2$ + [cyclopentanone] $\xrightarrow{H_2/Pt}$

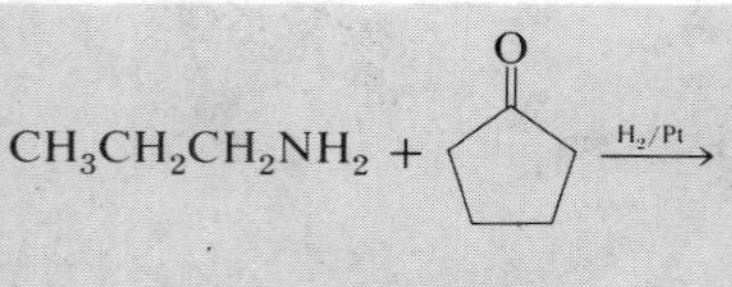

(c) [benzene ring]$CH_2\overset{\overset{O}{\|}}{C}CH_3$ + $H\overset{\overset{O}{\|}}{C}N(CH_3)_2$ $\xrightarrow{200°}$

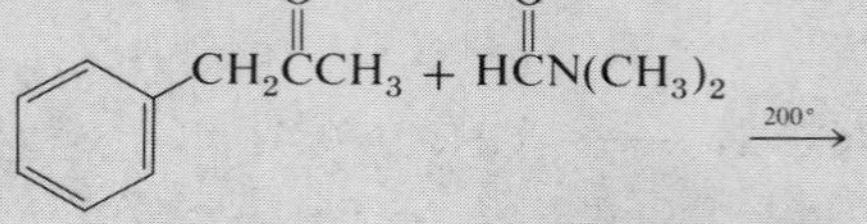

(d) [pyrrolidine ring with N-H]CH_3 + $CH_2{=}O$ + $HCOOH$ $\xrightarrow{100°}$

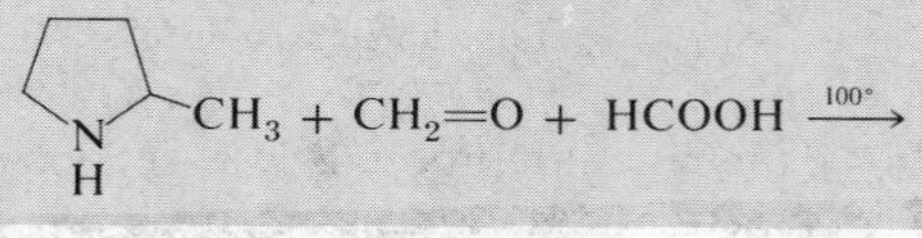

G. *Reduction of Amides*

Amides are reduced by lithium aluminum hydride in refluxing ether to give amines (Section 19.8). The reduction is unusual in that a C=O group is reduced to CH_2. Yields are generally good.

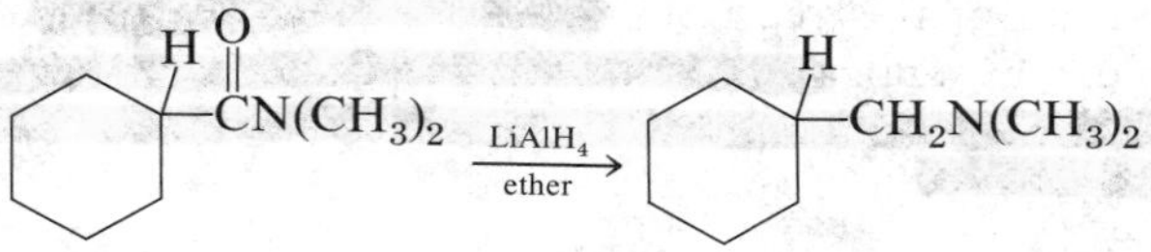

[structure: cyclohexane with H and $\overset{\overset{O}{\|}}{C}N(CH_3)_2$] $\xrightarrow[\text{ether}]{LiAlH_4}$ [structure: cyclohexane with H and $CH_2N(CH_3)_2$]

N,N-dimethylcyclohexane-
carboxamide

(88%)
N,N-dimethylcyclohexyl-
methanamine

Diborane, B_2H_6, may also be used as the reducing agent.

$(CH_3)_3C\overset{\overset{O}{\|}}{C}N(CH_3)_2$ $\xrightarrow[\text{THF}]{B_2H_6}$ $(CH_3)_3CCH_2N(CH_3)_2$

N,N,2,2-tetramethylpropanamide

(79%)
N,N,2,2-tetramethylpropanamine

The method also serves as a method to prepare primary or secondary amines, depending on the structure of the amide used.

$R\overset{\overset{O}{\|}}{C}NH_2$ $\xrightarrow{LiAlH_4}$ RCH_2NH_2

$R\overset{\overset{O}{\|}}{C}NHR'$ $\xrightarrow{LiAlH_4}$ RCH_2NHR'

EXERCISE: The two-step sequence of (1) acylation of an amine and (2) reduction of the resulting amide to an amine constitutes a method for the indirect addition of an alkyl group to nitrogen. Show how this sequence of reactions can be used to accomplish the following transformations.

(a) $CH_2CH_2CH_2CH_2NH_2 \longrightarrow CH_3CH_2CH_2CH_2NHCH_2CH(CH_3)_2$

(b) $(CH_3)_2CHNH_2 \longrightarrow (CH_3)_2CHNCH_2CH_2CH_2CH_3$
$\qquad\qquad\qquad\qquad\qquad\quad |$
$\qquad\qquad\qquad\qquad\quad CH_2CH_3$

H. Preparation of Amines from Carboxylic Acids: The Hofmann, Curtius, and Schmidt Rearrangements

These three reactions all accomplish the same overall process—conversion of a carboxylic acid to a primary amine with *loss of the carboxy carbon* of the acid.

$$\underset{\text{R}}{\overset{\overset{\displaystyle O}{\|}}{R-C-OH}} \longrightarrow R-NH_2 + CO_2$$

Examples of the three reactions are shown.

Hofmann

$$CH_3(CH_2)_7CH_2CONH_2 + Cl_2 + OH^- \longrightarrow CH_3(CH_2)_7CH_2NH_2$$
$$\text{decanamide} \qquad\qquad\qquad\qquad\qquad \underset{\text{nonanamine}}{(66\%)}$$

Curtius

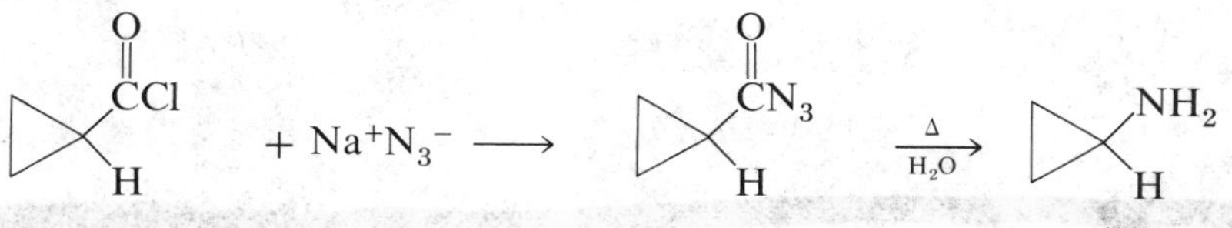

cyclopropanecarbonyl   cyclopropanecarbonyl   cyclopropanamine
chloride   azide   (60%)

Schmidt

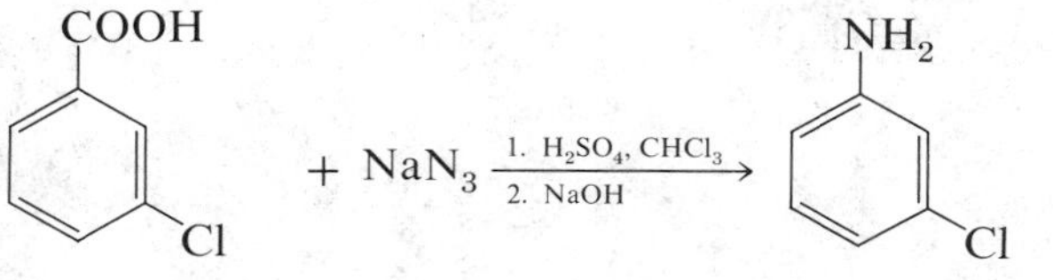

m-chlorobenzoic acid m-chloroaniline
(75%)

Although the Hofmann rearrangement begins with an amide, the Curtius reaction with an acyl azide, and the Schmidt reaction with an acid, the three reactions are related mechanistically. In each case, the crucial intermediate is probably the same, an **acyl nitrene.**

The probable mechanism for the Hofmann rearrangement is outlined as follows. The first two steps are simply the base-catalyzed halogenation of the amide, which is mechanistically related to the base-catalyzed halogenation of ketones (Section 13.6.D).

$$\text{(1)} \quad \overset{O}{\overset{\|}{R\overset{}{C}NH_2}} + OH^- \rightleftharpoons \overset{O}{\overset{\|}{R\overset{}{C}NH^-}} + H_2O$$

$$\text{(2)} \quad \overset{O}{\overset{\|}{R\overset{}{C}NH^-}} + Cl_2 \rightleftharpoons \overset{O}{\overset{\|}{R\overset{}{C}NHCl}} + Cl^-$$

The N-chloroamide is more acidic than the starting amide and also reacts with base to give the corresponding anion. This intermediate loses chloride ion to give a highly reactive intermediate called a **nitrene.** Nitrenes are neutral molecules in which the nitrogen has only six electrons; they are structurally similar to carbenes (Section 11.6).

$$\text{(3)} \quad \overset{O}{\overset{\|}{R\overset{}{C}\overset{..}{N}HCl}} + OH^- \rightleftharpoons \overset{O}{\overset{\|}{R\overset{}{C}\overset{..}{\overset{..}{N}}Cl}} + H_2O$$

$$\text{(4)} \quad \overset{O}{\overset{\|}{R\overset{}{C}\overset{..}{N}Cl}} \rightleftharpoons \underset{\substack{\text{an acyl} \\ \text{nitrene}}}{\overset{O}{\overset{\|}{R\overset{}{C}}}{-}\overset{..}{\overset{..}{N}}} + Cl^-$$

Acyl nitrenes undergo a rapid rearrangement to give a compound called an **isocyanate,** which is similar to an allene (Section 20.2.C) or a ketene (Section 20.3.C).

$$\text{(5)} \quad \underset{\text{an isocyanate}}{R{-}\overset{O}{\overset{\|}{C}}{-}\overset{..}{\overset{..}{N}}:} \longrightarrow R{-}\overset{..}{\overset{..}{N}}{=}C{=}O$$

Like ketenes, isocyanates react rapidly with water; the products in this case are **carbamic acids,** which are thermally unstable. Decarboxylation of the carbamic acid occurs to give the amine and carbon dioxide.

$$\text{(6)} \quad \underset{\text{an alkyl isocyanate}}{R{-}N{=}C{=}O} + H_2O \longrightarrow \underset{\text{a carbamic acid}}{\overset{O}{\overset{\|}{RNH\overset{}{C}OH}}}$$

$$\text{(7)} \quad \overset{O}{\overset{\|}{RNH\overset{}{C}OH}} \longrightarrow RNH_2 + CO_2$$

If the Hofmann rearrangement is carried out in alcohol solution rather than in water, steps (6) and (7) are not possible. Instead, the isocyanate adds the alcohol to give the ester of the carbamic acid, which is stable and may be isolated.

$$\text{(6')} \quad R{-}N{=}C{=}O + CH_3OH \longrightarrow \underset{\text{a methyl carbamate}}{\overset{O}{\overset{\|}{RNH\overset{}{C}OCH_3}}}$$

In the Curtius reaction, the acyl azide loses nitrogen upon heating to give the acyl nitrene directly. The remaining steps (5)–(7) are the same. Acyl azides are potentially explosive, and the decomposition is therefore somewhat hazardous to carry out.

$$\underset{\underset{\displaystyle O}{\|}}{R-C}-\overset{\displaystyle\cdot\cdot}{\underset{\displaystyle\cdot\cdot}{N}}-\overset{+}{N}\equiv N\colon \xrightarrow{\Delta} \underset{\underset{\displaystyle O}{\|}}{R-C}-\overset{\cdot\cdot}{N}\colon + N_2$$

The Schmidt reaction also proceeds by way of the acyl azide, which is formed by reaction of the carboxylic acid with hydrazoic acid, HN_3, under the acidic conditions of the reaction.

$$\underset{\underset{\displaystyle O}{\|}}{R-C}-OH + HN_3 \xrightarrow{H_2SO_4} \underset{\underset{\displaystyle O}{\|}}{R-C}-N_3 + H_2O$$

EXERCISE: Write equations showing the preparation of 3-methylpentanamine and o-bromoaniline by the Hofmann, Curtius, and Schmidt rearrangements.

24.7
Reactions of Amines

Certain important reactions of amines that have already been presented will not be discussed further here. These are the reactions with protons (Section 24.4) and with alkyl halides (Sections 24.5 and 24.6.A).

A. Formation of Amides

Recall that ammonia and primary and secondary amines react with acyl halides and acid anhydrides to give amides (Section 19.4.B). If an acyl halide is used, the hydrohalic acid produced will neutralize an additional equivalent of amine.

$$\underset{\underset{\displaystyle O}{\|}}{CH_2{=}CHCCl} + 2\ CH_3NH_2 \longrightarrow \underset{\underset{\displaystyle O}{\|}}{CH_2{=}CHCNHCH_3} + CH_3NH_3{}^+\ Cl^-$$

Aromatic amines can also be converted to amides with acyl chlorides or anhydrides. We shall find that this is frequently a useful procedure because the amide group is less strongly activating than the amino group in electrophilic aromatic substitution reactions. Thus the high reactivity of the amine can be moderated by conversion into an amide. For this *moderating* purpose the acetyl group is used most often and is usually introduced with acetic anhydride. Acetanilides can also be prepared by direct heating of aniline or a substituted aniline with acetic acid, but this process is slower.

acetanilide

Acetanilide is a colorless crystalline solid, m.p. 114°, that behaves as a neutral compound under normal conditions. The basic character of aniline is reduced by the acetyl group. The conjugate acids of amides have pK_as of about 0 to $+1$; that is, they are extensively protonated in 10% sulfuric acid.

$$\underset{RNHCCH_3}{\overset{\overset{+OH}{\parallel}}{}} \underset{K_a}{\rightleftharpoons} H_3O^+ + RNHCOCH_3 \qquad K_a \sim 0.1\text{--}1\ M$$

Acetanilide is a somewhat weaker base, just as aniline is a weaker base than aliphatic amines; the pK_a of the conjugate acid of acetanilide is about -1 to -2. It is protonated by strong sulfuric acid solutions. Acetanilide is also a weak acid with a pK_a estimated to be about 15. It is not appreciably soluble in dilute aqueous alkali hydroxides and requires more basic conditions to form the conjugate anion.

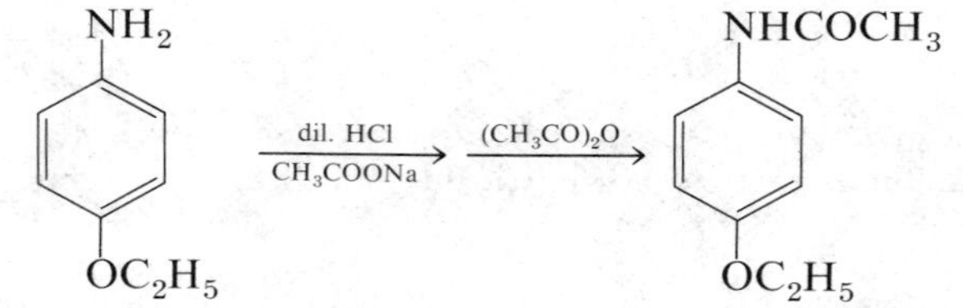

p-phenetidine
p-ethoxyaniline

phenacetin
p-ethoxyacetanilide

Phenacetin has been used as an analgesic and in mixture with aspirin and caffeine as formerly popular over-the-counter analgesic pills (the "P" in APC). It has since been removed from such compositions because of suspected side effects.

The mixture of HCl and sodium acetate creates a buffered medium that keeps the amine in solution as the ammonium salt in equilibrium with a small amount of free amine. The free amine reacts rapidly with acetic anhydride to form the amide in high yield and in a pure state. Acetic anhydride hydrolyzes slowly under these conditions. The amide can be hydrolyzed back to the amine by heating with alcoholic HCl.

B. *Reaction with Nitrous Acid*

Amines undergo interesting reactions with nitrous acid, HNO_2. The reaction products depend on whether the amine is primary, secondary, or tertiary, and whether it is aromatic or aliphatic.

Secondary amines, either aromatic or aliphatic, give N-nitroso compounds, which are usually yellow.

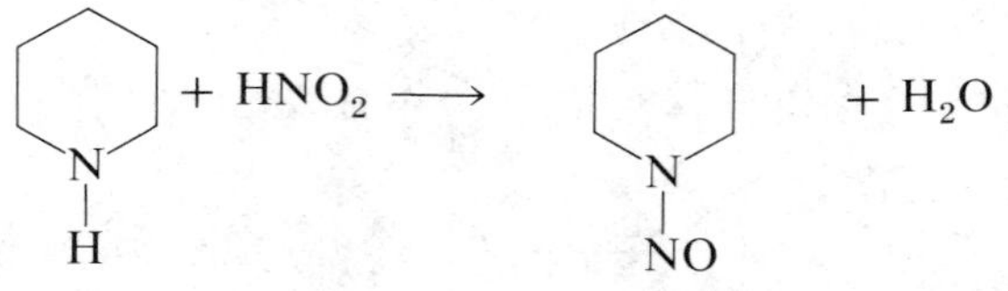

piperidine

N-nitrosopiperidine
(yellow)

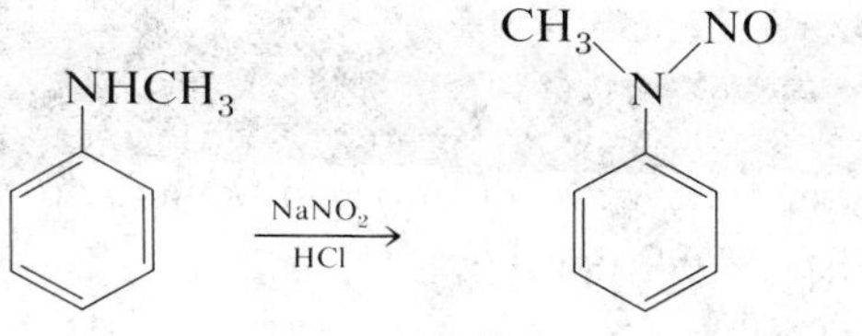

N-nitroso-N-methylaniline

The reaction mechanism may be thought of in terms of the following steps.

$1°$ $\longrightarrow$ $Ar-N\equiv N^{\oplus}$

$2°$ $\longrightarrow$ $N-N=O$

$3°$, $Ar \longrightarrow$ electrophilic substitution

(1) $HO-N=O + H^+ \rightleftharpoons H_2\overset{+}{O}-N=O$

(2) $H_2\overset{+}{O}-N=O \rightleftharpoons H_2O + {}^+\overset{..}{N}=\overset{..}{O}$

(3) $R_2\overset{..}{N}H + NO^+ \rightleftharpoons R_2\overset{+}{N}\overset{\displaystyle N=O}{\underset{\displaystyle H}{\diagup}}$

(4) $R_2\overset{+}{N}\overset{\displaystyle N=O}{\underset{\displaystyle H}{\diagup}} \rightleftharpoons R_2N-N=O + H^+$

Since tertiary amines have no proton on nitrogen, step (4) is blocked. Thus no overall reaction occurs by this pathway. However, if the tertiary amine is aromatic, an alternative reaction path is available. The mild electrophile NO$^+$ (the **nitrosonium ion**) attacks the highly reactive aromatic ring, and electrophilic aromatic substitution occurs (Section 24.7.C). Substitution occurs only at the *para* position.

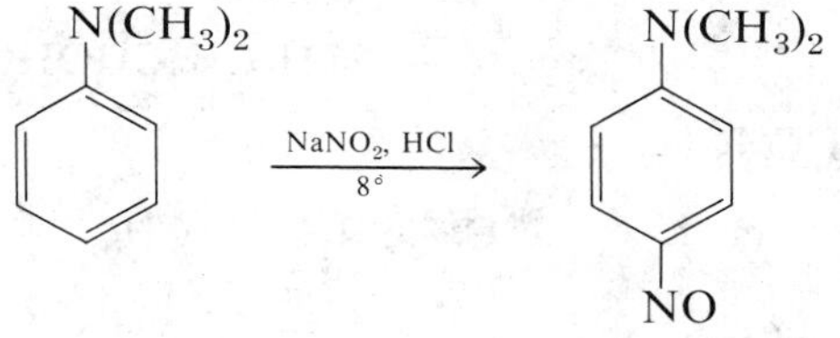

(80–89%)
p-nitrosodimethylaniline

With primary amines, the reaction proceeds further, and a diazonium compound is produced.

(5) $R-\underset{\displaystyle ..}{\overset{\displaystyle H}{N}}-N=O + H^+ \rightleftharpoons \left[R-\underset{\displaystyle ..}{\overset{\displaystyle H}{N}}-N=\overset{+}{O}H \longleftarrow R-\underset{\displaystyle +}{\overset{\displaystyle H}{N}}=N-OH \right]$

(6) $R-\underset{\displaystyle +}{\overset{\displaystyle H}{N}}=N-OH \rightleftharpoons R-N=N-OH + H^+$

(7) $R-\overset{..}{N}=\overset{..}{N}-OH + H^+ \rightleftharpoons R-\overset{..}{N}=\overset{..}{N}-OH_2{}^+$

(8) $R-\overset{..}{N}=\overset{..}{N}-OH_2{}^+ \rightleftharpoons H_2O + R-\overset{+}{N}\equiv\overset{..}{N}$

alkanediazonium cation

Alkanediazonium compounds are exceedingly unstable and decompose, even at low temperatures, to give nitrogen and carbocations, which then react by the normal pathways of substitution, elimination, and rearrangement. An example is the reaction of *n*-butylamine with nitrous acid, generated *in situ* from sodium nitrite and aqueous hydrochloric acid.

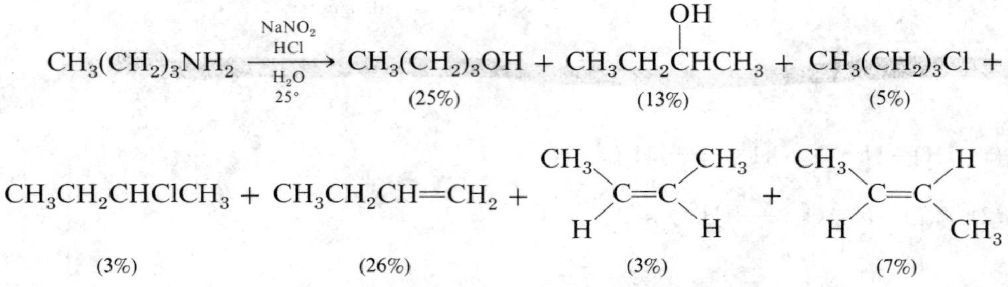

$$CH_3(CH_2)_3NH_2 \xrightarrow[\substack{H_2O \\ 25°}]{\substack{NaNO_2 \\ HCl}} CH_3(CH_2)_3OH + \underset{(13\%)}{CH_3CH_2\overset{\displaystyle OH}{\overset{|}{C}}HCH_3} + \underset{(5\%)}{CH_3(CH_2)_3Cl} +$$
$$\underset{(25\%)}{}$$

$$\underset{(3\%)}{CH_3CH_2CHClCH_3} + \underset{(26\%)}{CH_3CH_2CH=CH_2} + \underset{(3\%)}{\overset{CH_3}{\underset{H}{}}C=C\overset{CH_3}{\underset{H}{}}} + \underset{(7\%)}{\overset{CH_3}{\underset{H}{}}C=C\overset{H}{\underset{CH_3}{}}}$$

The primary alkyl products are derived from displacement reactions by water or chloride ion.

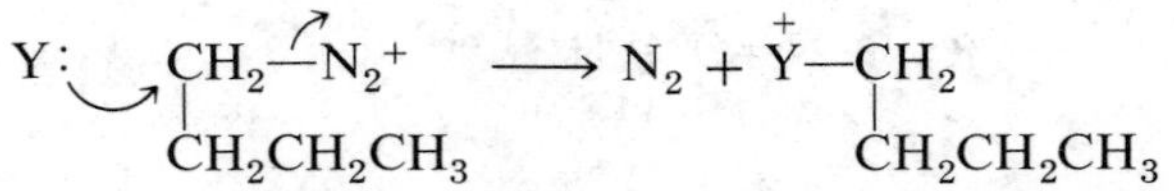

$$Y: \overset{\curvearrowright}{\underset{CH_2CH_2CH_3}{CH_2-N_2^+}} \longrightarrow N_2 + \overset{+}{\underset{CH_2CH_2CH_3}{Y-CH_2}}$$

Elimination of a proton and rearrangements are important alternative pathways.

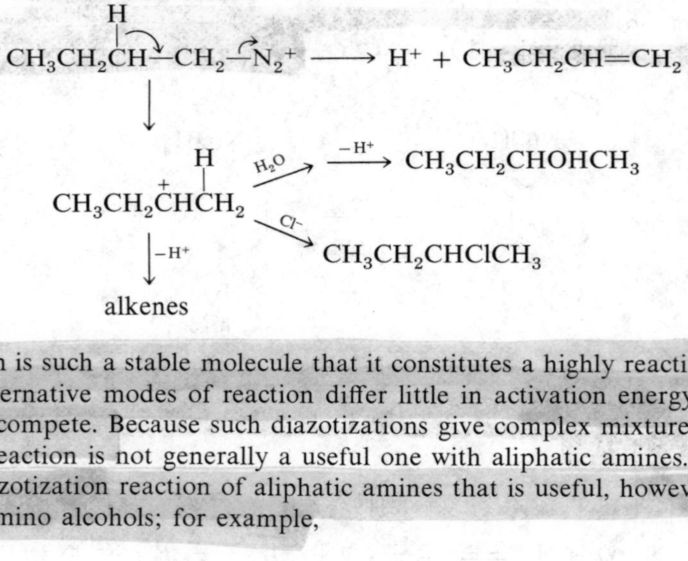

$$CH_3CH_2\overset{\displaystyle H}{\overset{|}{C}}H\overset{\curvearrowright}{-}CH_2\overset{\curvearrowright}{-}N_2^+ \longrightarrow H^+ + CH_3CH_2CH=CH_2$$

$$\downarrow$$

$$\underset{CH_3CH_2\overset{+}{C}HCH_2}{\overset{\displaystyle H}{\overset{|}{}}} \overset{H_2O}{\underset{Cl^-}{\nearrow}} \xrightarrow{-H^+} CH_3CH_2CHOHCH_3$$

$$\searrow CH_3CH_2CHClCH_3$$

$$\downarrow -H^+$$

alkenes

Nitrogen is such a stable molecule that it constitutes a highly reactive leaving group. Alternative modes of reaction differ little in activation energy so many pathways compete. Because such diazotizations give complex mixtures of products, the reaction is not generally a useful one with aliphatic amines.

One diazotization reaction of aliphatic amines that is useful, however, is that with 1,2-amino alcohols; for example,

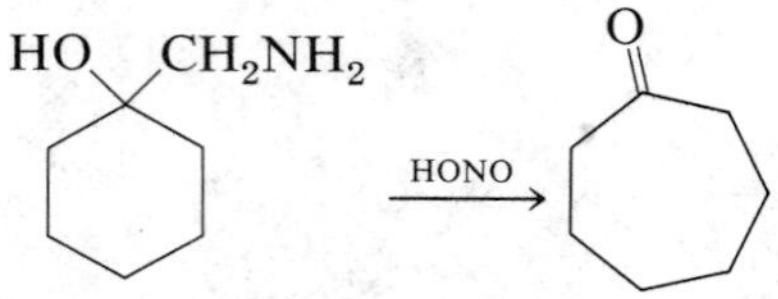

This reaction is a useful procedure for ring expansion (page 251) and succeeds because rearrangement produces a highly stable carbocation—the protonated ketone.

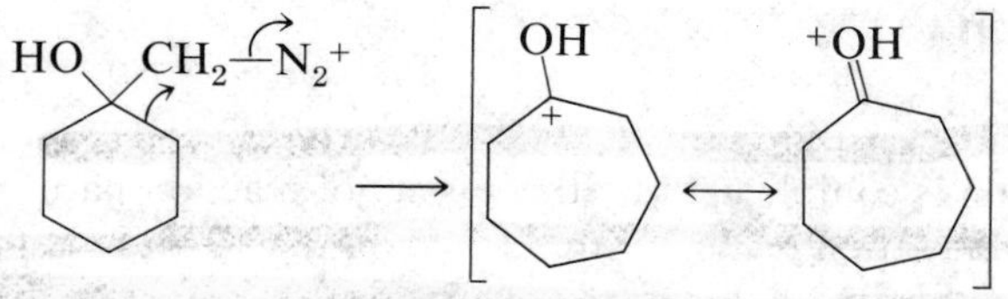

The diazonium ions produced by the reaction of primary aromatic amines are more stable than alkanediazonium ions. Aqueous solutions of these **arenediazonium ions** are stable at ice-bath temperatures.

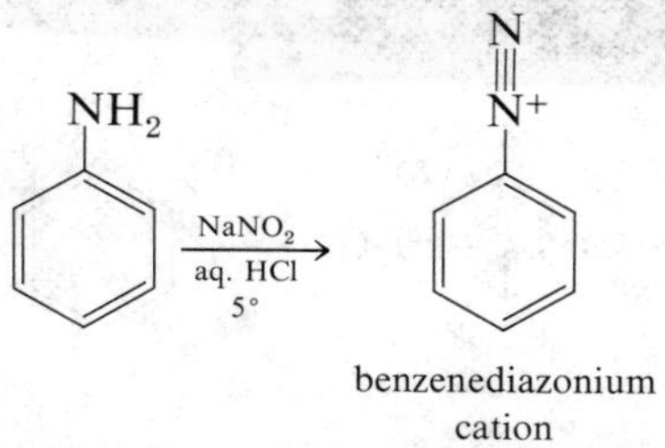

benzenediazonium
cation

The overall process of converting a primary aromatic amine into an arenediazonium salt is called **diazotization.** As we shall see in Chapter 25, diazotization provides access to a wide variety of substituted aromatic compounds.

The mechanism of diazotization formulated on page 755 is probably oversimplified. In solution nitrous acid is in equilibrium with several other species, such as dinitrogen trioxide, the anhydride of nitrous acid.

$$2 \text{ HONO} \rightleftharpoons \text{H}_2\text{O} + \text{N}_2\text{O}_3$$

With halide ions, the equilibria contain nitrosyl halides, which are the mixed anhydrides of nitrous acid and hydrohalic acids.

$$\text{HONO} + \text{HX} \rightleftharpoons \text{ONX} + \text{H}_2\text{O}$$

The actual nitrosating agent in many cases is probably N_2O_3, although in solutions containing halide ion the corresponding nitrosyl halide may also play a role. Nitrous acid itself is an intermediate oxidation state of nitrogen and disproportionates to nitric oxide and nitric acid.

$$3 \text{ HONO} \longrightarrow 2 \text{ NO} + \text{H}_3\text{O}^+ + \text{NO}_3^-$$

The rate of this reaction is temperature-dependent and is also strongly dependent on the concentration of nitrous acid—the rate is proportional to $[\text{HONO}]^4$. Consequently, nitrous acid solutions are usually kept cold and dilute and are used immediately.

EXERCISE: Write equations showing the reaction product(s) expected from treatment of each of the following amines with an aqueous solution of NaNO_2 and HCl at 0°.
(a) *n*-propylamine   (b) di-*n*-propylamine   (c) tri-*n*-propylamine
(d) aniline   (e) N-ethylaniline   (f) N,N-diethylaniline

C. *Oxidation*

Because of their basic nature, amines are oxidized with ease. With primary amines, oxidation is complicated by the variety of reaction paths that are available. Few useful oxidation reactions are known for this class. Secondary amines are easily oxidized to hydroxylamines. Again yields are generally poor due to over-oxidation.

$$R_2NH + H_2O_2 \longrightarrow R_2NOH + H_2O$$

Tertiary amines are oxidized cleanly to tertiary amine oxides. Useful oxidants are H_2O_2 or organic peroxyacids, RCO_3H.

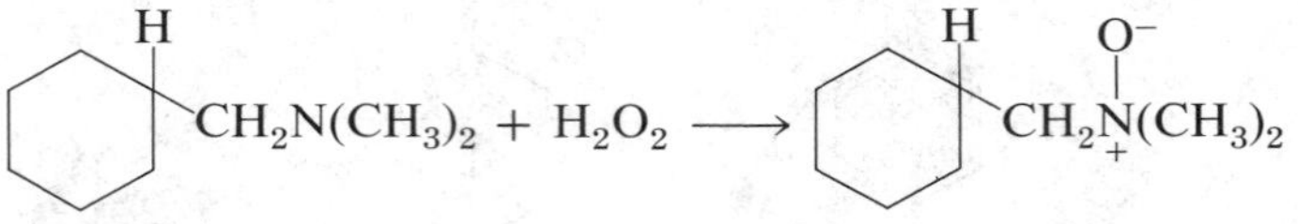

(90%)

A mixture of 49 g of N,N-dimethylcyclohexylmethanamine, 45 mL of methanol, and 120 g of 30% aqueous H_2O_2 is kept at room temperature for 36 hr. The excess H_2O_2 is destroyed by the addition of a small amount of colloidal platinum. The solution is then filtered and evaporated to obtain the crude amine oxide in greater than 90% yield.

Amine oxides fall into the class of organic compounds for which no completely uncharged Lewis structure may be written. The Lewis electron-dot representation of trimethylamine oxide shows that both the oxygen and the nitrogen have octet configurations and that they bear (−) and (+) formal charges, respectively.

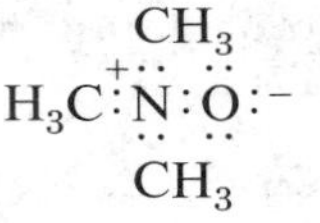

trimethylamine oxide

Aromatic amines are readily oxidized by a variety of oxidizing agents as well as by air. As a result, the oxidation of other functional groups cannot usually be carried out as satisfactorily if amino groups are also present.

The nature of amine oxidations is demonstrated by oxidation of *p*-bis(dimethyl-amino)benzene, which gives a relatively stable radical cation called Wurster's blue.

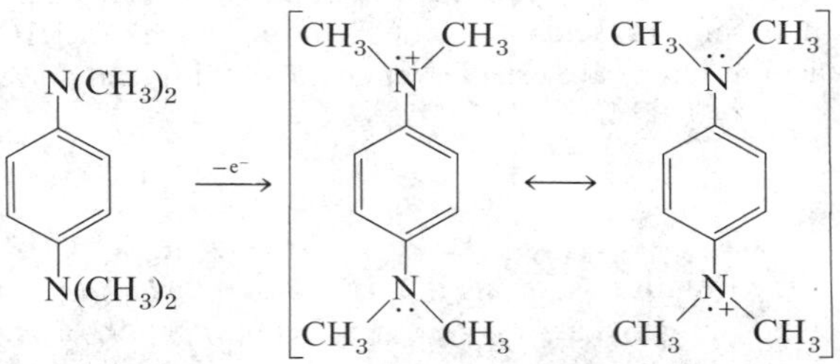

Wurster's blue

Radical cations were encountered previously as gas-phase species in mass spectrometry (Chapter 17). Wurster's blue is an example of a radical cation that is stable in solution. Other examples are now known to be important intermediates in various oxidation reactions. For example, the radical cation formed from aniline reacts further with aniline to produce highly colored polymeric compounds. On treatment with acidic potassium dichromate, aniline gives a black insoluble dye, aniline black, that is difficult to characterize. A proposed structure for the compound is

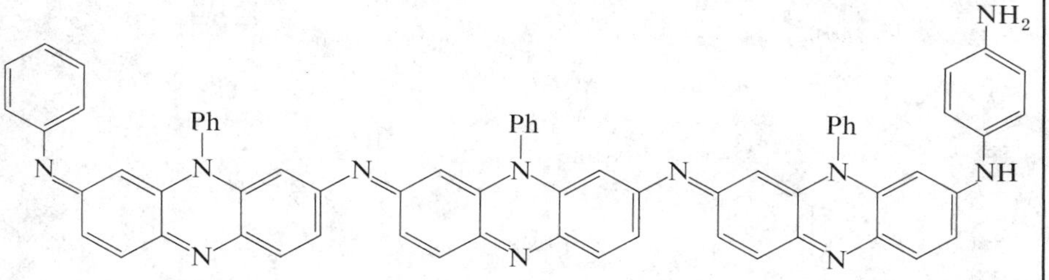

Further oxidation gives some *p*-benzoquinone, a compound we will consider in detail in Chapter 30.

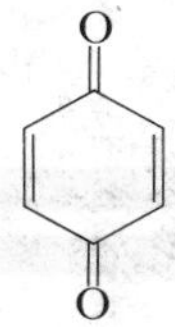

p-benzoquinone

The facile oxidation of amines often gives rise to undesired byproducts in the course of other preparations, but there are some oxidation reactions of the amino group itself that have some preparative significance. One example is the direct oxidation to a nitro group by the action of trifluoroperoxyacetic acid.

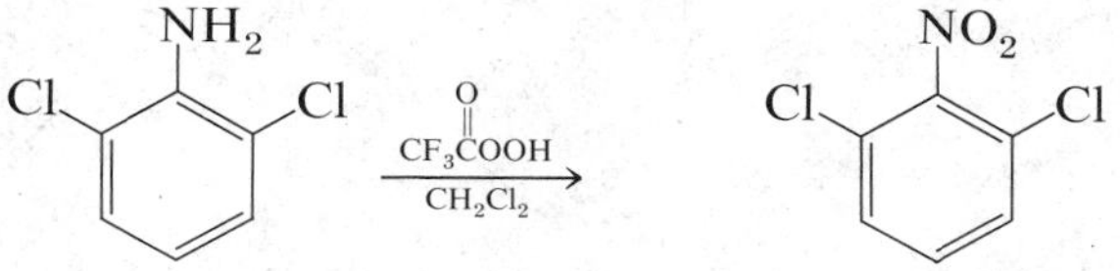

2,6-dichloroaniline   2,6-dichloronitrobenzene

Trifluoroperoxyacetic acid is prepared as needed by stirring trifluoroacetic anhydride in CH_2Cl_2 with 90% hydrogen peroxide. The amine to be oxidized is added to the resulting solution and heated to reflux. Trifluoroperoxyacetic acid is another of the reagents that find limited but important use in organic chemistry. It is the best reagent, for example, for the direct oxidation of an arylamine to the nitroarene.

This oxidation reactions works best with arylamines that have electron-attracting groups such as halogen, nitro, cyano, and so on.

D. *Electrophilic Aromatic Substitution*

Aromatic amines are highly activated toward substitution in the ring by electrophilic reagents. Reaction with such amines generally occurs under rather mild conditions. For example, halogenation is so facile that all unsubstituted *ortho* and *para* positions become substituted.

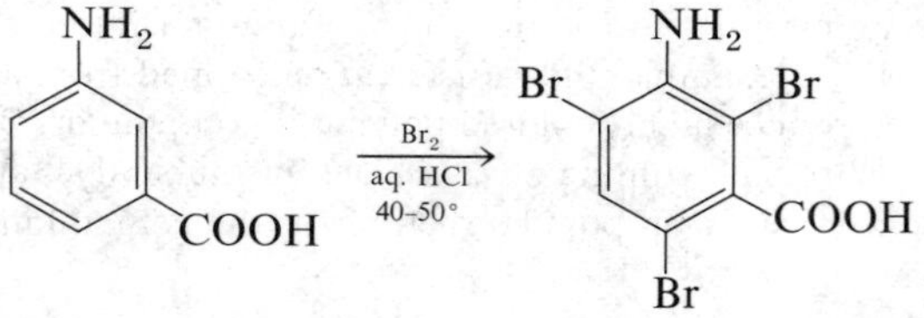

m-aminobenzoic acid → 3-amino-2,4,6-tribromo-benzoic acid

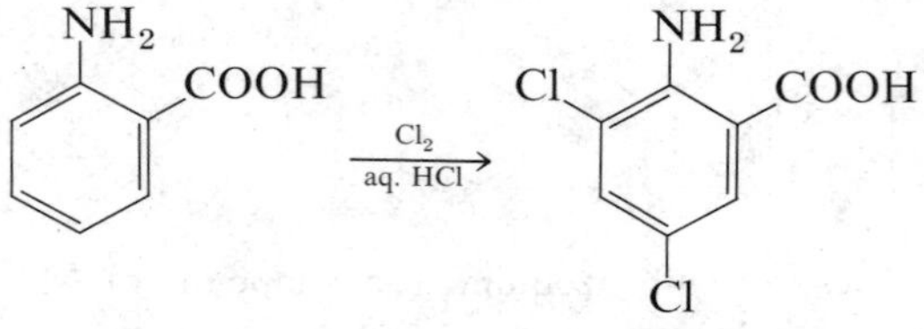

Br₂ / aq. HCl / 40–50°

anthranilic acid → 2-amino-3,5-dichloro-benzoic acid (69–78%)

Cl₂ / aq. HCl

Nitration of aromatic primary amines is not generally a useful reaction because nitric acid is an oxidizing agent and amines are sensitive to oxidation. A mixture of aniline and nitric acid can burst into flame. Nitration of tertiary aromatic amines can be accomplished conveniently and in good yield; a satisfactory method is nitration in acetic acid.

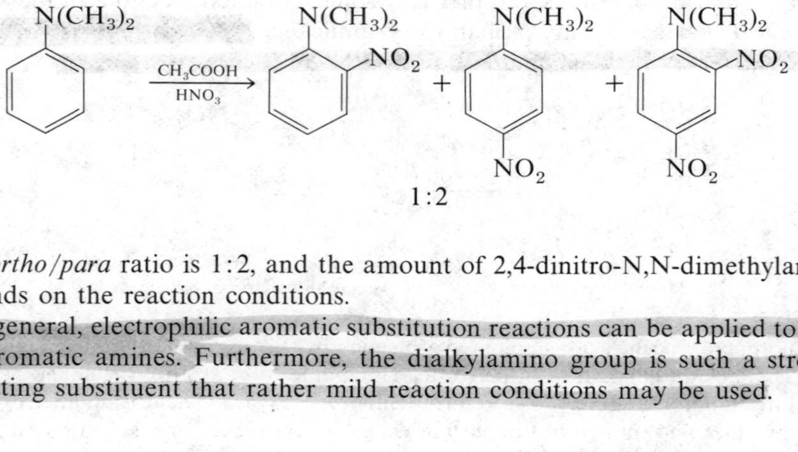

1:2

The *ortho/para* ratio is 1:2, and the amount of 2,4-dinitro-N,N-dimethylaniline depends on the reaction conditions.

In general, electrophilic aromatic substitution reactions can be applied to tertiary aromatic amines. Furthermore, the dialkylamino group is such a strongly activating substituent that rather mild reaction conditions may be used.

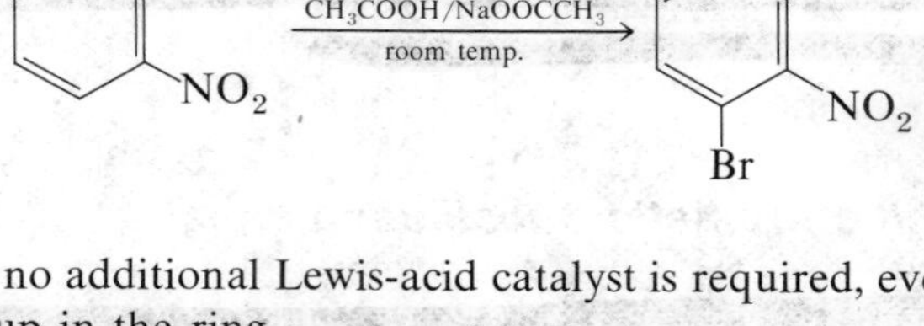

Br₂ / CH₃COOH/NaOOCCH₃ / room temp.

In this example, no additional Lewis-acid catalyst is required, even with a deactivating nitro group in the ring.

Friedel–Crafts acylations can also be accomplished under mild conditions, for example,

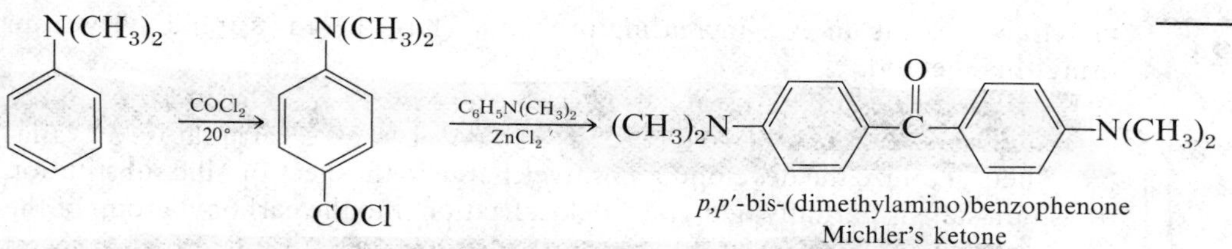

p,p'-bis-(dimethylamino)benzophenone
Michler's ketone
(a dye intermediate)

Aromatic tertiary amines can undergo a useful variant of Friedel–Crafts acylation, the **Vilsmeier reaction,** the treatment of a reactive aromatic ring with dimethylformamide and phosphorus oxychloride to introduce an aldehyde group. The reaction does not apply to simple benzene hydrocarbons because it requires activating substituents.

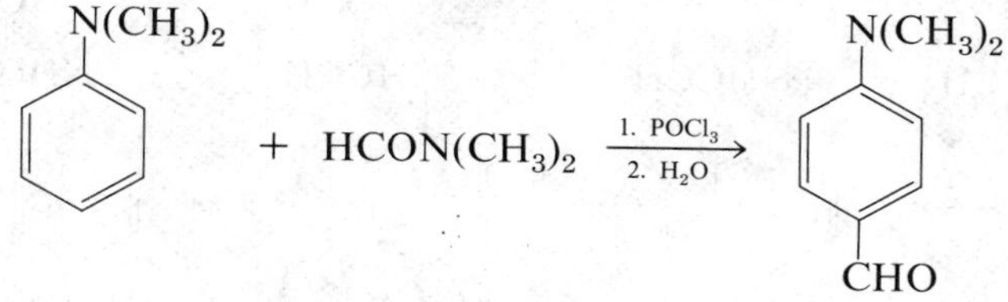

dimethylaniline dimethylformamide p-dimethylaminobenz-
(DMF) aldehyde

The electrophilic reagent in the Vilsmeier reaction is a chloroimmonium ion, which is formed in the following manner.

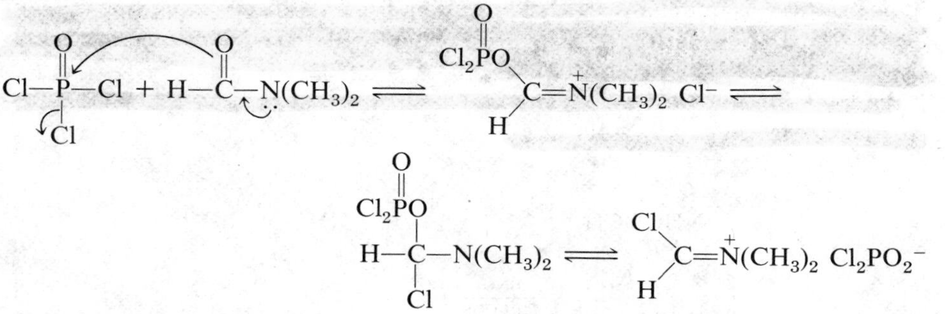

This electrophilic species reacts only with aromatic rings that contain highly activating groups such as OH and NR_2. The initial product, an α-chloroamine, hydrolyzes rapidly during work-up to afford the aldehyde.

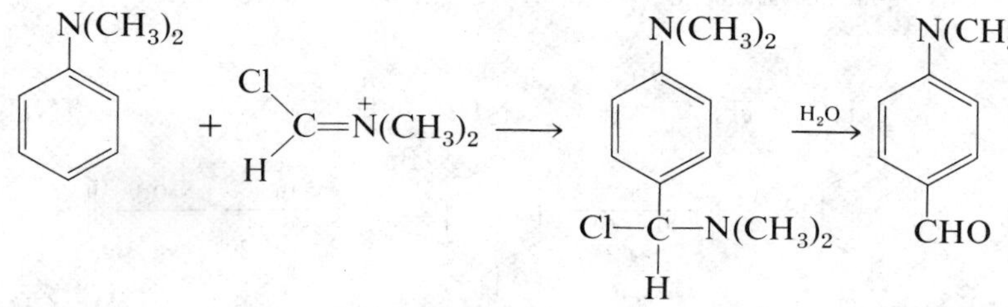

The Vilsmeier reaction has been an important industrial process, particularly for formylation of reactive heterocyclic compounds. However, large quantities of by-product phosphorus compounds are produced, and the disposal of these waste materials presents an ever-increasing problem. Thus, there is a trend away from using this method.

The reactivity of aromatic amines to electrophilic reagents is reduced when the amino group is converted into an amide. The nitrogen lone pair that is so available to help stabilize the developing positive charge in the electrophilic substitution transition state is partially tied up by delocalization into the carbonyl group of the amide. Amide groups are still *ortho,para* directors, but they are not nearly as activating as amino groups and electrophilic reactions on amide derivatives are readily controlled. Another way of understanding the moderating effect of the acyl group on nitrogen is by consideration of the resonance structures for the intermediate that results from electrophilic attack *para* to the amide group.

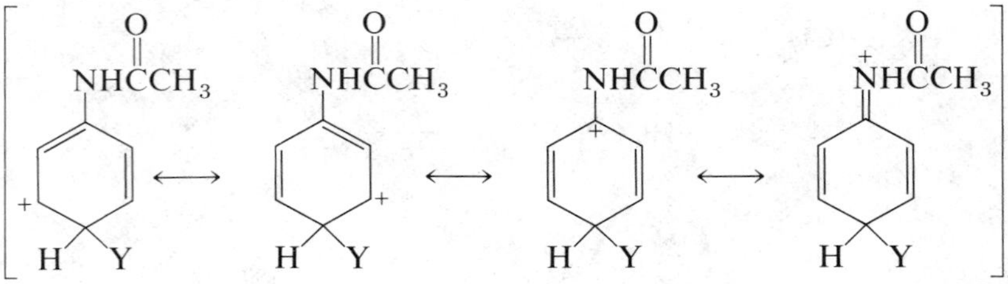

The fact that the developing positive charge can be placed on the nitrogen still causes attack at this position (or at the *ortho* position) to be more favorable than attack at the *meta* position. However, this resonance structure is not quite as important as it is in the intermediate resulting from attack at the *para* position of the corresponding amine because the positive nitrogen is now adjacent to the positive carbon of the carbonyl group.

Acetanilides are widely used as substrates for electrophilic aromatic substitution reactions. A few examples are

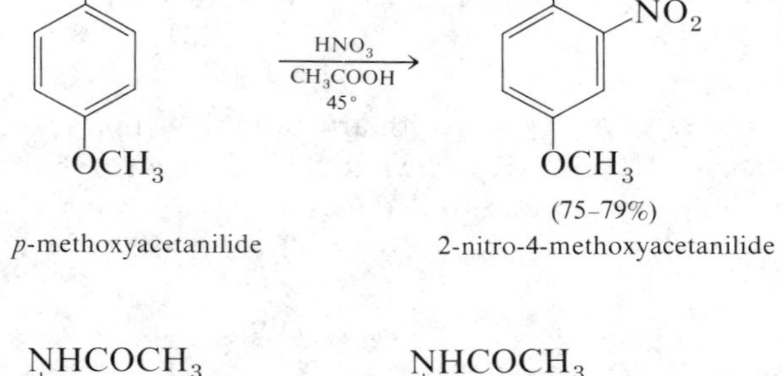

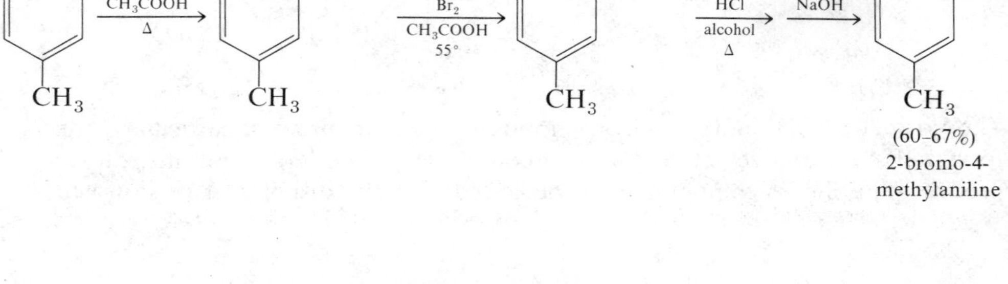

p-methoxyacetanilide → 2-nitro-4-methoxyacetanilide (75–79%)

(60–67%)
2-bromo-4-methylaniline

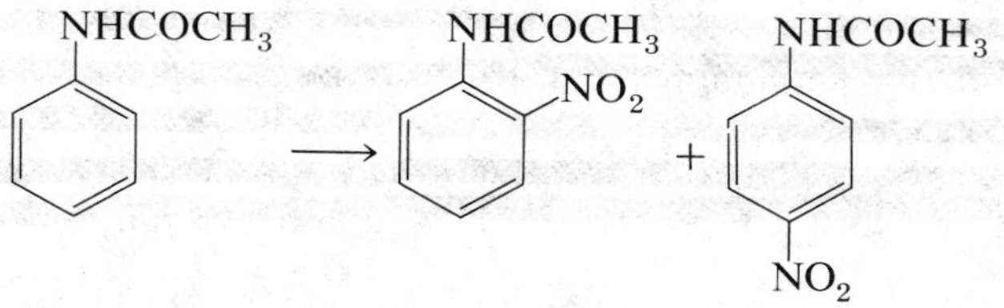

NHCOCH$_3$ NHCOCH$_3$ NHCOCH$_3$

| 90% aq. HNO$_3$, $-20°$ | (23%) | (77%) |
| HNO$_3$, Ac$_2$O, 20° | (68%) | (30%) |

The last examples show that the *ortho/para* ratio can sometimes be altered by choosing the proper reaction conditions.

E. *Elimination of the Amino Group: The Cope and Hofmann Elimination Reactions*

Simple amines undergo neither base-catalyzed nor acid-catalyzed elimination reactions. In the former case, the leaving group would be NH$_2^-$, which is the conjugate base of a very weak acid, ammonia (pK_a = 35). In the latter case, the leaving group NH$_3$ is still not a very good one, since its conjugate acid, the ammonium ion, NH$_4^+$, has a relatively low acidity (pK_a = 9.24).

$$B:^- + H-\overset{|}{\underset{|}{C}}-\overset{|}{\underset{|}{C}}-NH_2 \;\not\longrightarrow\; BH + \;\overset{}{\underset{}{>}}C=C\overset{}{\underset{}{<}} + \;^-:NH_2$$

$$B: + H-\overset{|}{\underset{|}{C}}-\overset{|}{\underset{|}{C}}-\overset{+}{N}H_3 \;\not\longrightarrow\; BH^+ + \;\overset{}{\underset{}{>}}C=C\overset{}{\underset{}{<}} + \;:NH_3$$

However, quaternary ammonium hydroxides do undergo elimination upon being heated.

$$CH_3CH_2\overset{+}{N}(CH_2CH_3)_3\ OH^- \overset{\Delta}{\longrightarrow} CH_2{=}CH_2 + N(CH_2CH_3)_3 + H_2O$$

tetraethylammonium hydroxide   triethylamine

Elimination proceeds by the E2 mechanism with hydroxide ion as the attacking base.

$$HO^- + H-CH_2-CH_2-\overset{+}{N}R_3 \longrightarrow \left[\overset{\delta-}{HO}\cdots H \cdots CH_2 \cdots CH_2 \cdots \overset{\delta+}{N}R_3 \right]^{\ddagger} \longrightarrow$$

$$HOH + CH_2{=}CH_2 + NR_3$$

The elimination reaction itself is the final step in a process known as **Hofmann exhaustive methylation** or **Hofmann degradation.** In this process a primary, secondary, or tertiary amine is first treated with enough methyl iodide to convert it into the quaternary ammonium iodide. Then the iodide is converted into the hydroxide with silver oxide and water. The elimination reaction to give the alkene is finally effected by heating the dry quaternary ammonium hydroxide at 100° or higher. In the process, the C—N bond is broken and an amine and an alkene are produced.

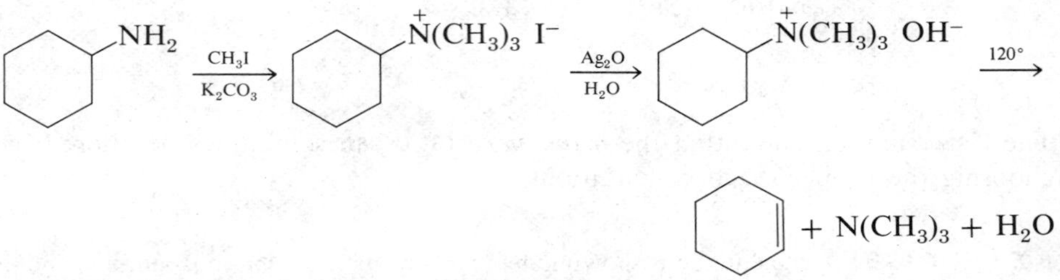

If the amine is cyclic, then the product is an amino alkene.

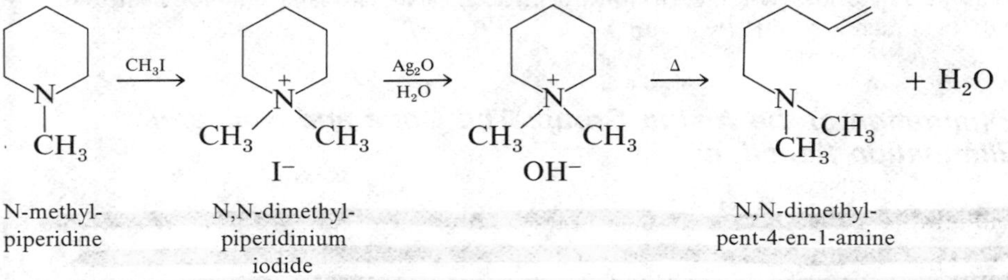

N-methyl-
piperidine

N,N-dimethyl-
piperidinium
iodide

N,N-dimethyl-
pent-4-en-1-amine

The process may be repeated with the product of (24-6) to yield a diene, liberating the nitrogen as trimethylamine (24-7).

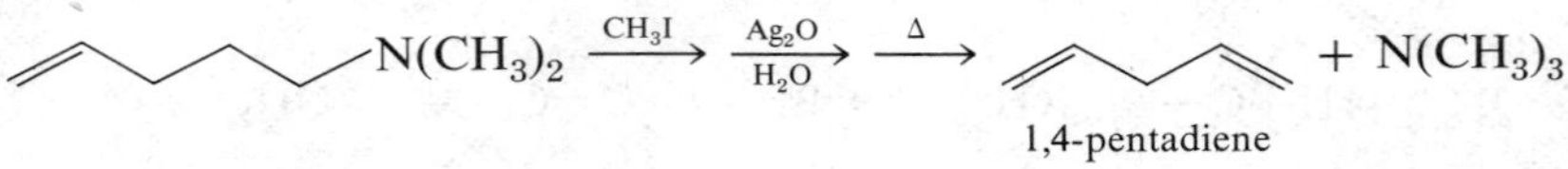

1,4-pentadiene

For bicyclic amines in which the nitrogen is a part of both rings, three cycles are required to remove the nitrogen completely.

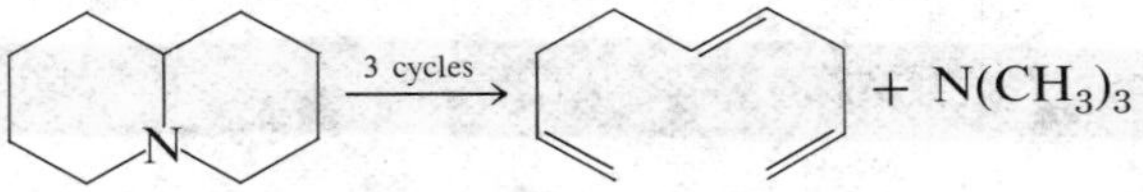

A side reaction that may occur is S_N2 displacement. This is seldom a significant reaction except in cases where there are no β-hydrogens.

$$(CH_3)_3\overset{+}{N}{-}CH_3 \quad OH^- \xrightarrow[\Delta]{H_2O} (CH_3)_3N + CH_3OH$$

When the quaternary ammonium hydroxide has two or more different β-hydrogens, more than one alkene may be formed in the elimination. Unlike normal E2 eliminations with alkyl halides, the Hofmann elimination gives predominantly the less highly substituted alkene, as comparison of (24-8) and (24-9) shows.

$$CH_3CH_2\overset{\overset{\displaystyle Br}{|}}{C}HCH_3 \xrightarrow[\Delta]{\underset{EtOH}{NaOEt}} CH_3CH{=}CHCH_3 + CH_3CH_2CH{=}CH_2 \quad (24\text{-}8)$$

(81%)   (19%)

$$\underset{(5\%)}{\underset{|}{\overset{\overset{+}{N}(CH_3)_3}{CH_3CH_2\overset{|}{C}HCH_3}}}\ OH^- \longrightarrow \underset{(5\%)}{CH_3CH=CHCH_3} + \underset{(95\%)}{CH_3CH_2CH=CH_2} \quad (24\text{-}9)$$

This type of behavior is also seen when the ammonium compound contains two different alkyl groups that may be lost as the alkene.

$$CH_3CH_2CH_2CH_2\overset{\overset{CH_3}{|}}{\underset{\underset{CH_3}{|}}{N^+}}\overset{CH_3}{\overset{|}{CH_2CHCH_3}}\ OH^- \overset{\Delta}{\longrightarrow}
\begin{cases}
CH_3CH_2CH=CH_2 + (CH_3)_2NCH_2\overset{\overset{CH_3}{|}}{C}HCH_3 \quad (64\%) \\
\quad\quad\quad\quad\quad\quad\quad\quad + \\
\overset{\overset{CH_3}{|}}{CH_3}C=CH_2 + (CH_3)_2NCH_2CH_2CH_2CH_3 \quad (36\%)
\end{cases}$$

For ammonium hydroxides having only simple alkyl groups, such as the foregoing examples, the mode of elimination may be generalized as follows (**Hofmann rule**): *"In the decomposition of quaternary ammonium hydroxides, the hydrogen is lost most easily from CH_3, next from RCH_2—, and least easily from R_2CH—."* The direction of elimination in the Hofmann reaction is probably governed mostly by steric factors. The generality of the rule is shown by the following additional examples.

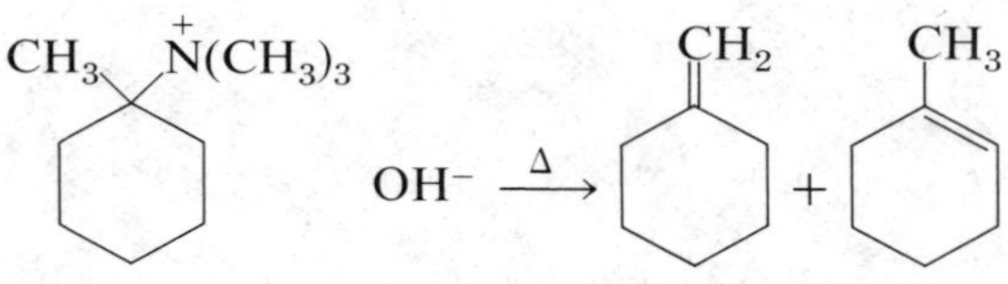

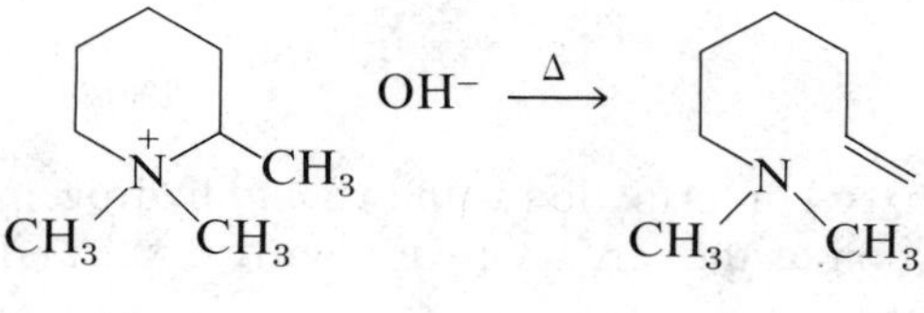

One way in which steric effects contribute to this preference for the less substituted olefin is by their effect on the populations of different conformations. For example, in the 2-butyl case (24-9), the most stable conformation may be represented by a Newman projection of the C_2—C_3 bond as

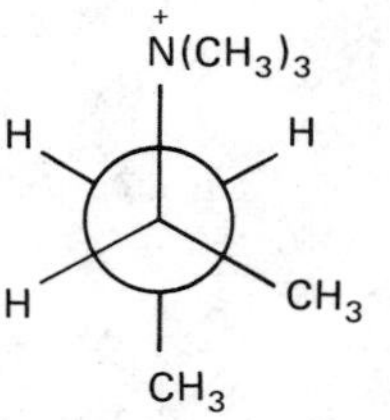

but this conformation has no *anti* hydrogen at C_3. *Anti* elimination can only occur in the two conformations

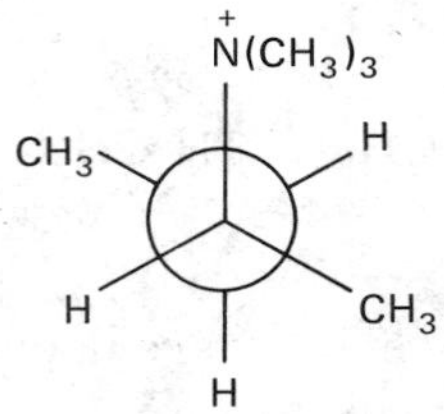

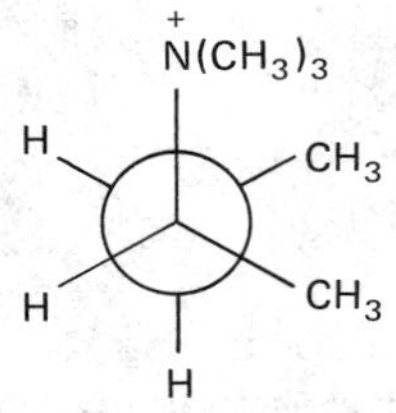

Both of these conformations have a methyl group *gauche* to the bulky trimethyl-ammonium group—this group is comparable in size to *t*-butyl. Hence, the populations of these conformations are small. On the other hand, all conformations with respect to the C_1—C_2 bond have an *anti* hydrogen.

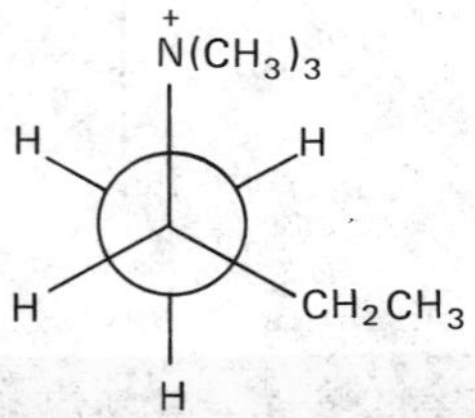

Although removal of a proton from C-3 would be faster because a more stable disubstituted ethylene results (Section 1.5.A), the population of conformations with *anti* hydrogen at C-3 is so small that the inherently slower reaction at C-1 dominates. An interesting example is the elimination reaction of the cyclohexane compound.

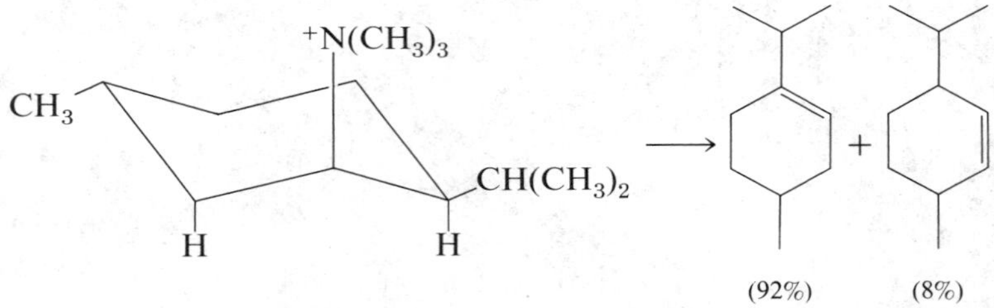

(92%)   (8%)

This compound has a fixed conformation with two *anti* hydrogens, one secondary and one tertiary. Reaction at the tertiary hydrogen is faster and gives the more highly substituted olefin.

When electron-withdrawing groups are attached to one of the β-carbons, the Hofmann rule is not followed.

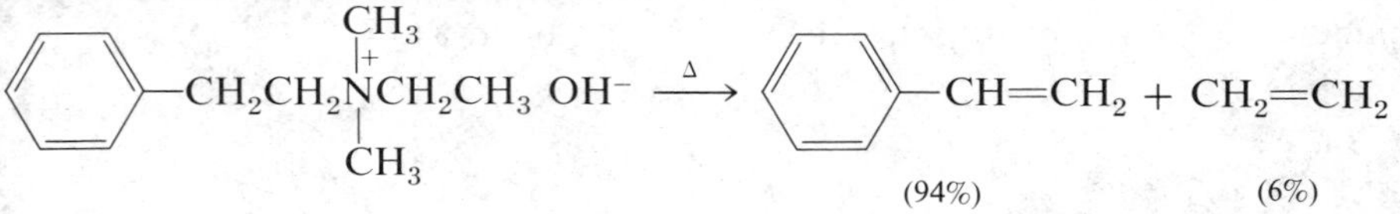

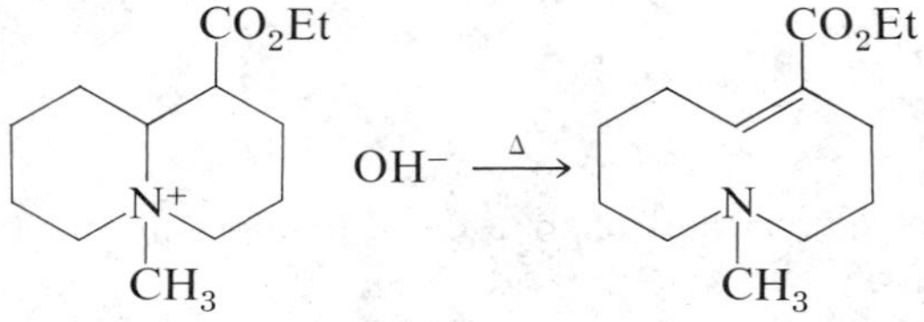

(94%)   (6%)

Elimination of the amino group may also be brought about by the thermal elimination of amine oxides (Section 24.7.C). These compounds undergo elimination when heated to 150–200°, provided that there is at least one hydrogen β to the nitrogen. The reaction is called the **Cope elimination** and is a useful alternative to the Hofmann degradation as a method for removing nitrogen from a compound. It is also useful as a preparative method for certain alkenes.

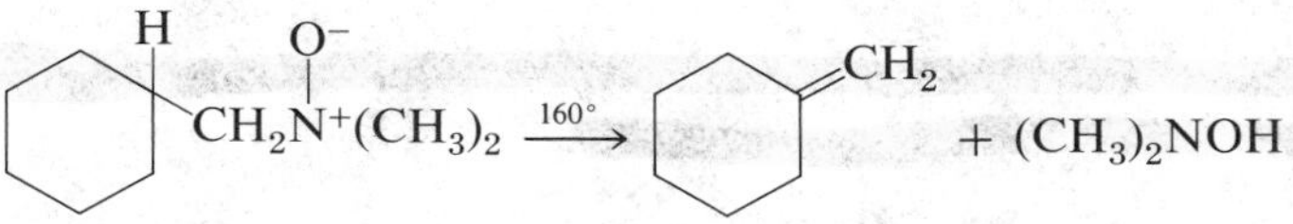

$$\text{(cyclohexyl)}\overset{\text{H}}{\underset{}{\text{C}}}\text{H}_2\overset{\text{O}^-}{\underset{}{\text{N}^+}}(\text{CH}_3)_2 \xrightarrow{160°} \text{(cyclohexane)}=\text{CH}_2 + (\text{CH}_3)_2\text{NOH}$$

Crude N,N-dimethylcyclohexylmethanamine oxide (about 50 g) is placed in a flask that has been evacuated to a pressure of about 10 torr. The liquefied amine oxide is heated at 160° for 2 hr. Water is added, and the alkene layer is separated and distilled to obtain 30 g of methylenecyclohexane (98%).

Amine oxide pyrolysis is similar mechanistically to ester pyrolysis (Section 19.11). The mechanism is a sort of internal E2 process in which the oxide oxygen acts as the attacking base, abstracting the β-proton in a concerted reaction.

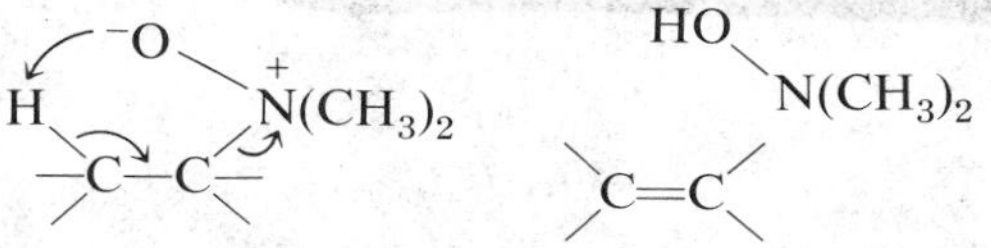

$$\underset{\text{H}}{\overset{\text{O}}{\underset{=\text{C}-\text{C}=}{\overset{+}{\text{N}(\text{CH}_3)_2}}}} \qquad \underset{\text{C}=\text{C}}{\overset{\text{HO}\quad\text{N}(\text{CH}_3)_2}{}}$$

This mechanism is supported by experiments that clearly show the elimination to be *syn* (see Section 19.11).

EXERCISE: Write equations showing the product or products expected from the following reactions.
(a) N,N-dimethyl-2-pentanamine + methyl iodide; silver hydroxide; heat
(b) N,N-diethyl-1-octanamine + methyl iodide; silver hydroxide; heat
(c) triethylamine + hydrogen peroxide; heat
(d) (1R,2S)-1-deuterio-N,N,2-trimethyl-1-butanamine + hydrogen peroxide; heat

24.8
Enamines

Enamines are compounds in which an amino group is attached directly to a C=C double bond. They are the nitrogen analogs of enols.

$$\underset{\text{an enol}}{\overset{\text{OH}}{\underset{}{\text{C}=\text{C}}}} \qquad \underset{\text{an enamine}}{\overset{\text{NH}_2}{\underset{}{\text{C}=\text{C}}}}$$

Like enols, enamines are generally unstable and undergo rapid conversion into the imine tautomer.

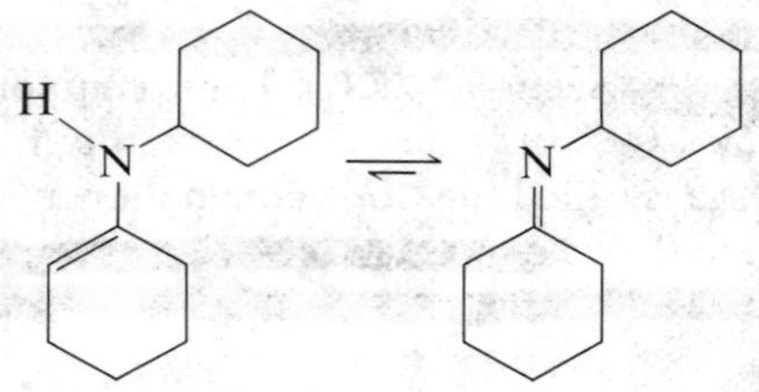

When the nitrogen of an enamine is tertiary, such tautomerism cannot occur, and the enamine may be isolated and handled.

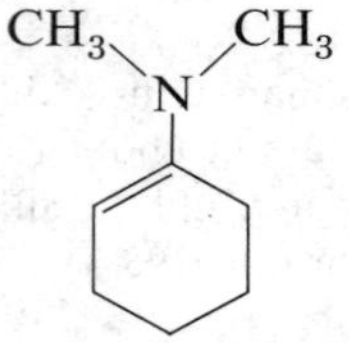

N,N-dimethyl-1-cyclohexenamine

Tertiary enamines are prepared by reaction of a secondary amine with an aldehyde or ketone. Water must be removed as it is formed in order to shift the equilibrium to the enamine product (page 384). Cyclic secondary amines are commonly used.

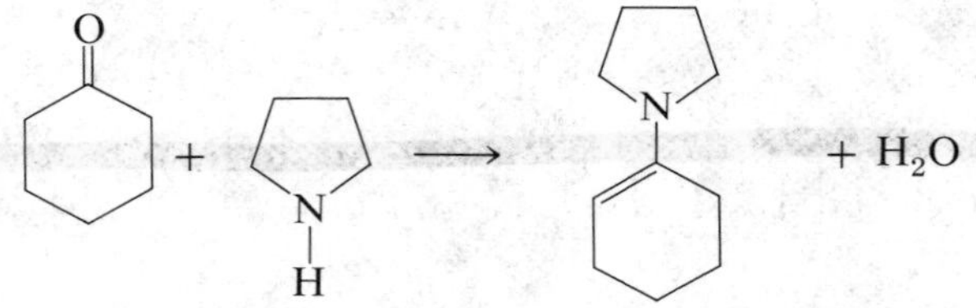

pyrrolidine   N-(1-cyclo-
hexenyl)pyrrolidine

The mechanism for enamine formation is similar to the mechanism for formation of an imine; the reaction is subject to both acid and base catalysis (Section 13.7.C).

(1) [ring with O] $+ H^+ \rightleftharpoons$ [ring with OH^+]

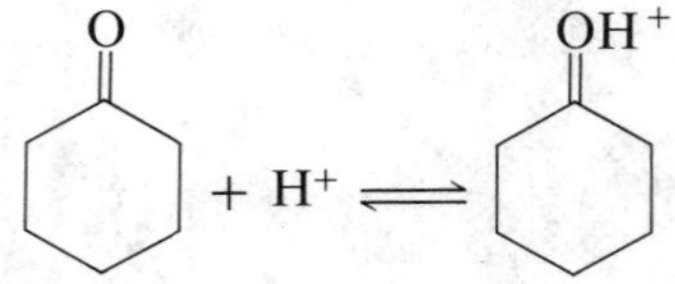

(2) [ring with OH^+] $+ R_2NH \rightleftharpoons$ [ring with HO, $\overset{+}{N}HR_2$]

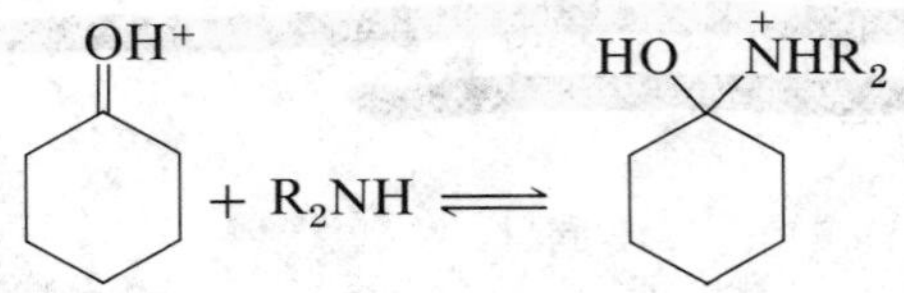

(3) [ring with HO, $\overset{+}{N}HR_2$] $\rightleftharpoons$ [ring with HO, NR_2] $+ H^+$

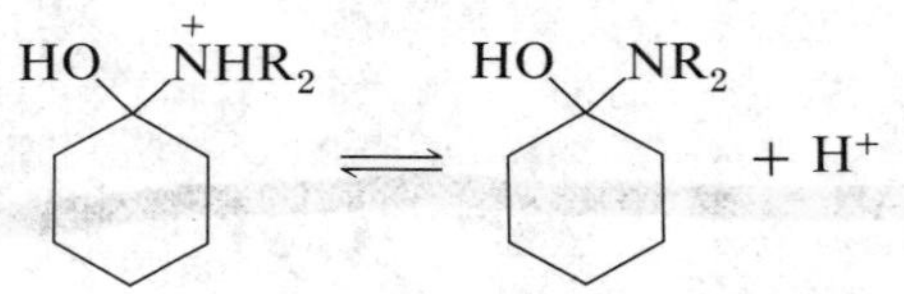

(4)

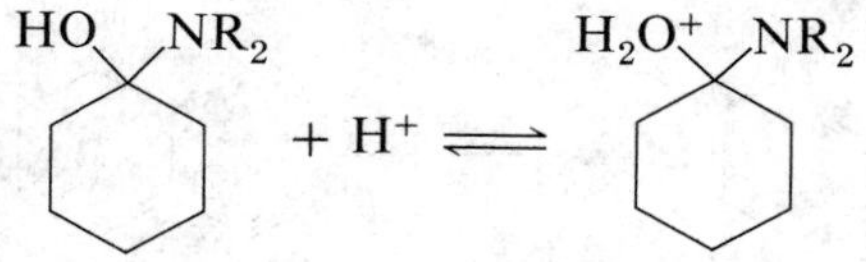

(5)

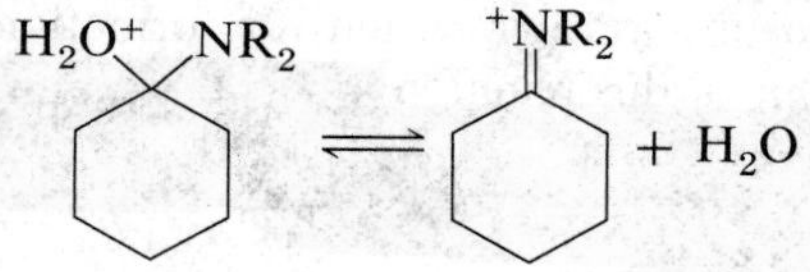

(6)

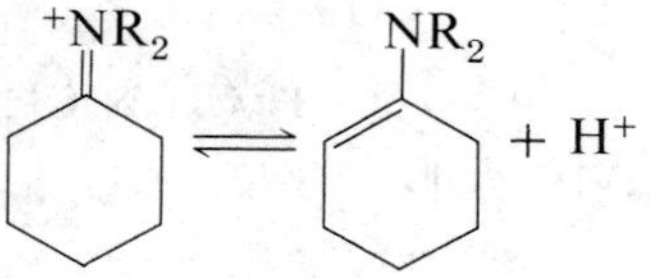

The products are sensitive to aqueous acid and revert to the carbonyl compound and the amine in dilute acid.

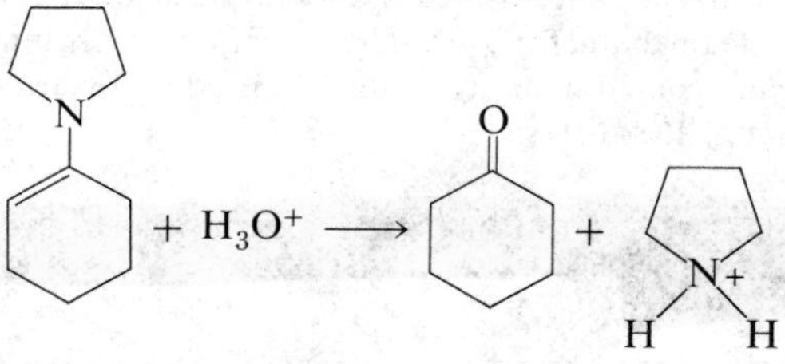

Enamines are useful intermediates in some reactions because the β-carbon of the double bond has nucleophilic character, as shown in the following resonance structures.

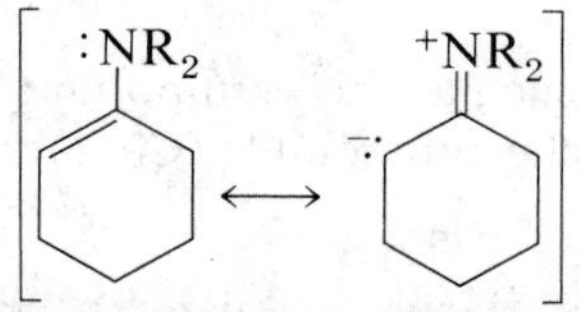

Reaction occurs rapidly with reactive alkyl halides to give alkylated immonium compounds, which undergo facile hydrolysis to give the alkylated ketone.

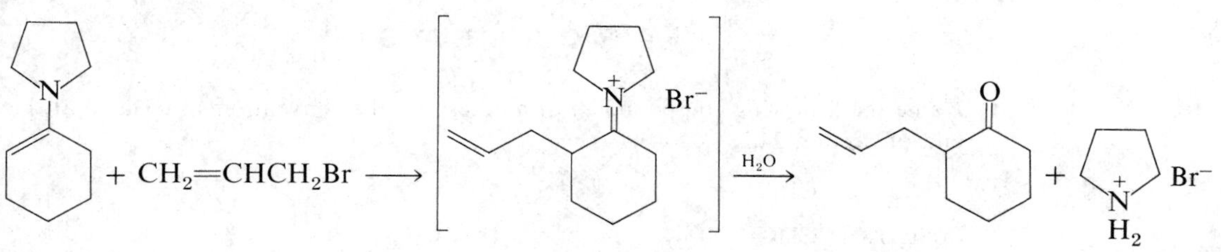

$$+ \ CH_2\!\!=\!\!CHCH_2Br \ \longrightarrow$$

2-allylcyclohexanone

Enamines may also be prepared by the dehydrogenation of tertiary amines. Dehydrogenation is accomplished by oxidation to an immonium ion with mercuric acetate, followed by deprotonation of the immonium ion with sodium bicarbonate.

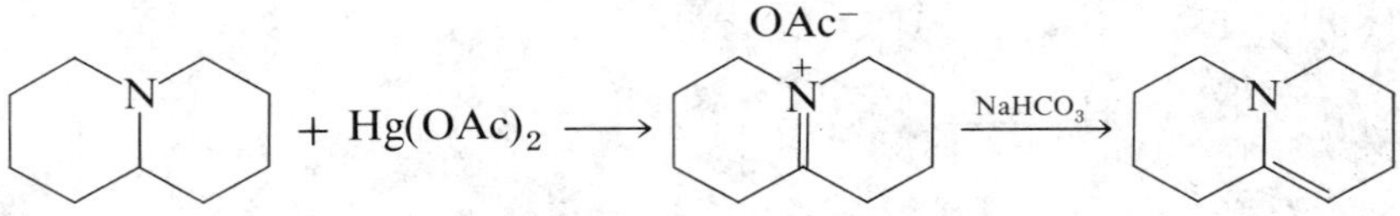

$$\text{structure} + Hg(OAc)_2 \longrightarrow \text{structure (OAc}^-) \xrightarrow{NaHCO_3} \text{structure}$$

Immonium ions are intermediates in an important reaction for formation of carbon–carbon bonds, the **Mannich reaction.** In this reaction, a secondary amine, a ketone, or other carbonyl compound that can undergo easy enolization, and an aldehyde (often formaldehyde) combine to give a β-amino ketone.

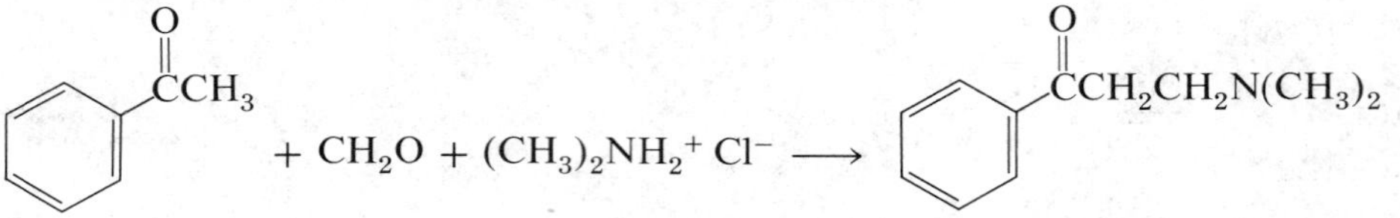

$$\underset{\text{acetophenone}}{\text{(ring)}\overset{O}{\overset{\|}{C}}CH_3} + CH_2O + (CH_3)_2NH_2{}^+ Cl^- \longrightarrow \underset{\beta\text{-(N,N-dimethylamino)propiophenone}}{\text{(ring)}\overset{O}{\overset{\|}{C}}CH_2CH_2N(CH_3)_2}$$

> A mixture of 60 g of acetophenone, 52.7 g of dimethylamine hydrochloride, 19.8 g of paraformaldehyde, 1 mL of concentrated hydrochloric acid, and 80 mL of 95% ethanol is heated on a steam bath for 2 hr. The warm yellow solution is diluted with 400 mL of acetone and cooled in an ice bath. The product separates as large crystals, 72–77 g (68–72%), m.p. 138–141°.

The Mannich reaction involves an intermediate immonium ion, which reacts with the enol form of the ketone.

$$CH_2{=}O + R_2\overset{+}{N}H_2 \longrightarrow CH_2{=}\overset{+}{N}R_2 + H_2O$$

$$\underset{}{\overset{:\ddot{O}H}{\underset{|}{R'C}}{=}CH_2} + CH_2{=}\overset{+}{N}R_2 \longrightarrow \underset{}{\overset{:\overset{+}{O}H}{\underset{\|}{R'C}}{-}CH_2{-}CH_2{-}\ddot{N}R_2}$$

We shall see that chemistry such as this is important in connection with methods for preparation of heterocyclic compounds (Chapter 32).

EXERCISE: Show how the Mannich reaction may be used to prepare 4-methyl-5-(N,N-dimethylamino)-3-pentanone.

PROBLEMS

1. Name the following compounds. For amines, use the convention enunciated at the end of Section 24.2.

(a) $CH_3CH_2\overset{\overset{\displaystyle CH_3}{|}}{C}HCH_2NH_2$

(b) $(CH_3)_3N$

(c) $CH_2{=}CHCH_2N\overset{\displaystyle CH_3}{\underset{\displaystyle CH_2CH_3}{\big\langle}}$

(d) 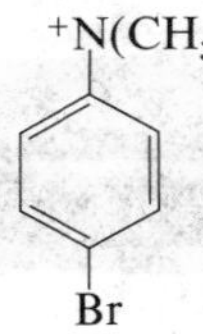 $\text{(}^+N(CH_3)_3\ Cl^-\text{ ring with }Br\text{)}$

(e) 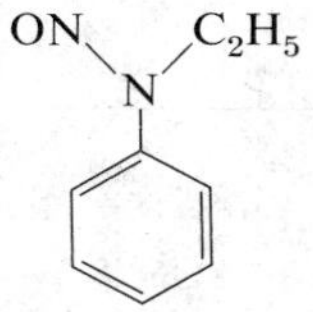 ON‑N(C₂H₅)‑C₆H₅

(f) 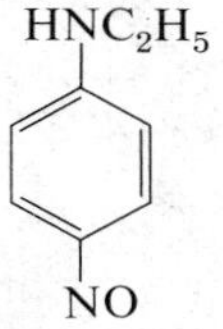 HNC₂H₅ ... NO

(g) $CH_3CH_2\overset{+}{N}(CH_3)_3\ I^-$

(h) $CH_3CH_2CH_2N^+(CH_3)_2$ with O^-

(i) $(CH_3CH_2)_2CHN(CH_3)_2$

(j) 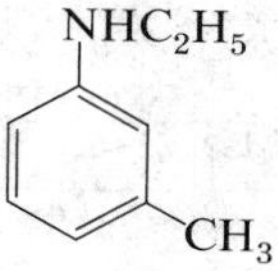 NHC₂H₅ ... CH₃

(k) 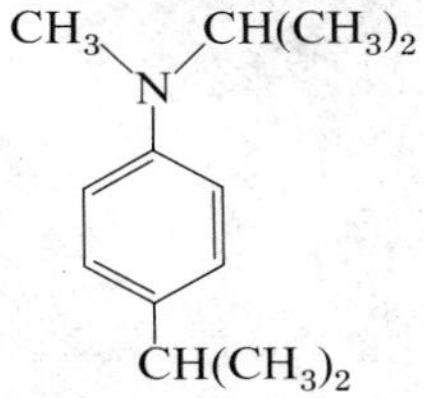 CH₃‑N‑CH(CH₃)₂ on ring with CH(CH₃)₂

(l) 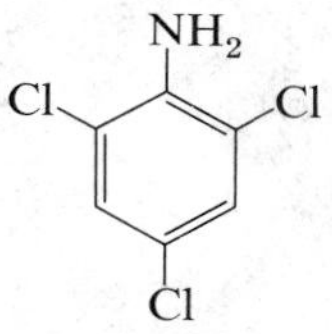 NH₂ with Cl, Cl, Cl substituents

2. Quaternary ammonium salts that have four different groups on nitrogen are chiral and may be resolved. The optically active allylethylmethylphenylammonium halides racemize slowly in solution. The rate of racemization is temperature-dependent and is faster for the iodide than for the bromide. Propose a mechanism for the racemization.

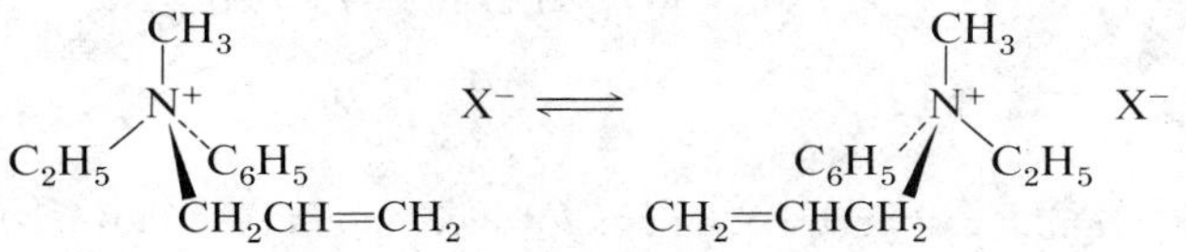

$$\begin{array}{ccc} CH_3 & & CH_3 \\ | & & | \\ N^+ & X^- \rightleftharpoons & N^+ \quad X^- \\ C_2H_5 / | \backslash C_6H_5 & & C_6H_5 / | \backslash C_2H_5 \\ CH_2CH{=}CH_2 & CH_2{=}CHCH_2 & \end{array}$$

3. The nmr, ir, and mass spectra of an unknown compound are shown below. Propose a structure for the compound.

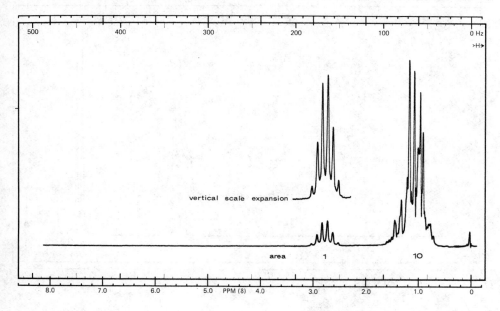

vertical scale expansion

area   1   10

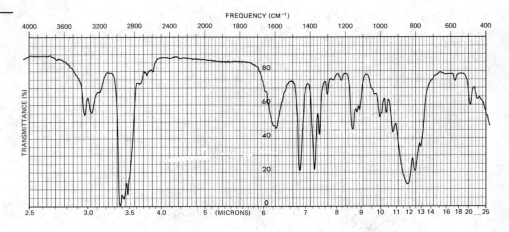

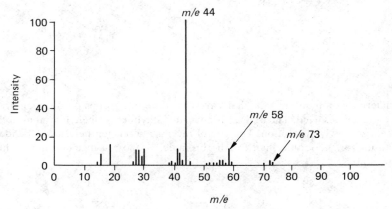

4. The nmr, ir, and mass spectra of an unknown compound are shown below. Propose a structure for the compound.

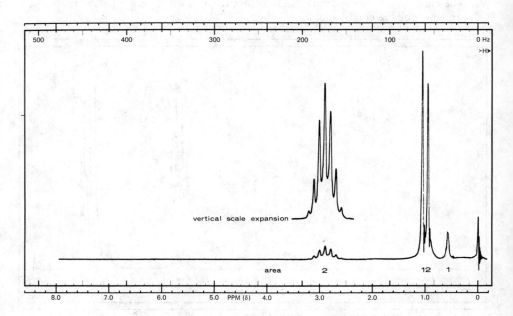

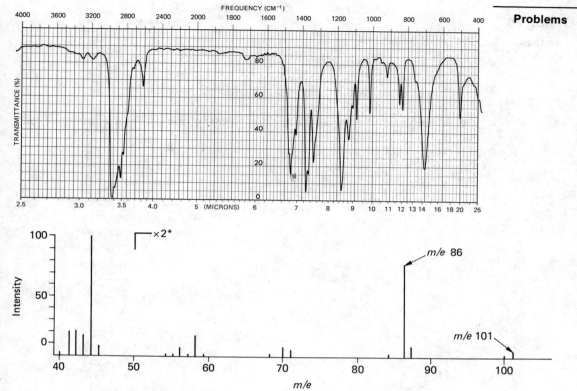

*The vertical scale is expanded by a factor of 2 above m/e 50.

5. Although the inversion barrier for trimethylamine is only 6 kcal mole^{-1}, that for the heterocyclic tertiary amine N-methylaziridine is about 19 kcal mole^{-1}. Propose an explanation.

$$\begin{array}{c} CH_2 \\ | \quad\quad N-CH_3 \\ CH_2 \end{array}$$

N-methylaziridine

6. Consider a solution of methylamine in water.
(a) At what pH are the CH_3NH_2 and $CH_3NH_3^+$ concentrations exactly equal?
(b) Calculate the $[CH_3NH_2]/[CH_3NH_3^+]$ ratio at pH 6, 8, 10, and 12.

7. Consider the reaction of methylamine with acetic acid.

$$CH_3NH_2 + CH_3COOH \overset{K}{\rightleftharpoons} CH_3NH_3^+ + CH_3CO_2^-$$

(a) Using the data in Tables 18.2 and 24.3, calculate K.
(b) At what pH does $[CH_3CO_2^-] = [CH_3NH_3^+]$?

8. (a) Propose a method for separating a mixture of cyclohexanecarboxylic acid, tributylamine, and decane.
(b) Alcohols react with phthalic anhydride to give monophthalate esters

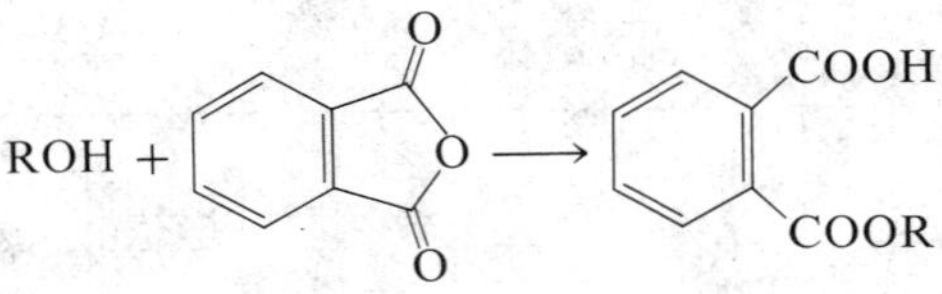

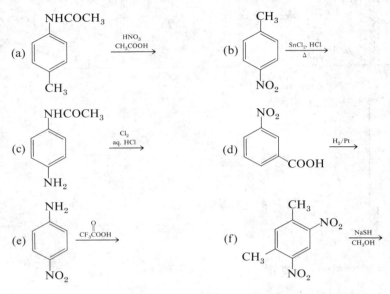

$$ROH +$$

Suggest a method for the resolution of racemic 2-octanol.

9. What is (are) the principal organic product(s) of each of the following reactions?

(a) [structure: NHCOCH₃ and CH₃ on benzene ring] $\xrightarrow[\text{CH}_3\text{COOH}]{\text{HNO}_3}$

(b) [structure: CH₃ and NO₂ on benzene ring] $\xrightarrow[\Delta]{\text{SnCl}_2, \text{HCl}}$

(c) [structure: NHCOCH₃ and NH₂ on benzene ring] $\xrightarrow[\text{aq. HCl}]{\text{Cl}_2}$

(d) [structure: NO₂ and COOH on benzene ring] $\xrightarrow{\text{H}_2/\text{Pt}}$

(e) [structure: NH₂ and NO₂ on benzene ring] $\xrightarrow{\text{CF}_3\text{COOH}}$

(f) [structure: CH₃, NO₂, CH₃, NO₂ on benzene ring] $\xrightarrow[\text{CH}_3\text{OH}]{\text{NaSH}}$

10. Suggest a sequence of reactions involving the Mannich condensation and the Hofmann elimination that can be used to convert acetone into methyl vinyl ketone.

11. Compare the behavior toward aqueous nitrous acid of each of the following compounds.

(a) 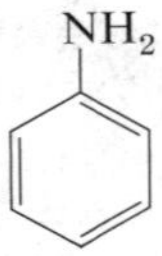 NH₂ [benzene ring]

(b) 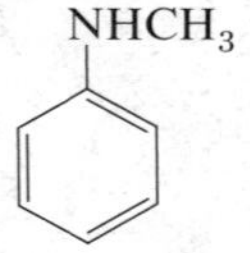 NHCH₃ [benzene ring]

(c) 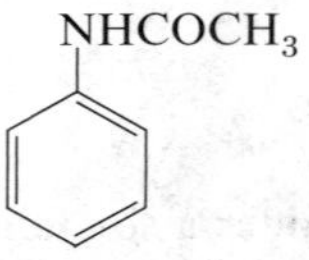 NHCOCH₃ [benzene ring]

(d) 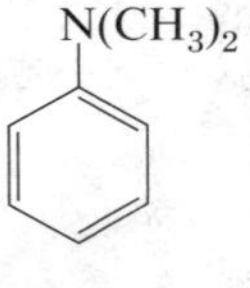 N(CH₃)₂ [benzene ring]

(e) 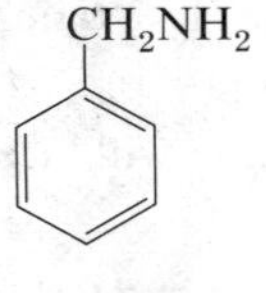 CH₂NH₂ [benzene ring]

(f) 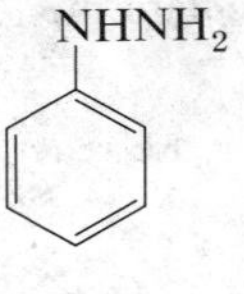 NHNH₂ [benzene ring]

(g) 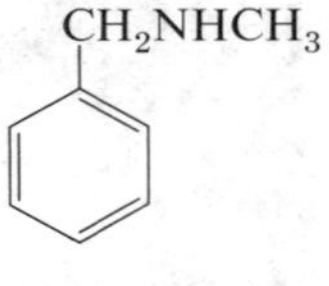 CH₂NHCH₃ [benzene ring]

(h) 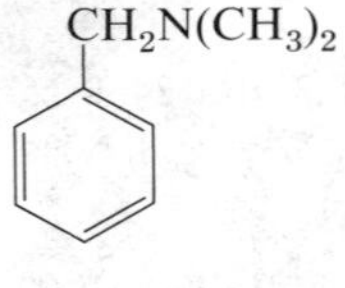 CH₂N(CH₃)₂ [benzene ring]

(i) 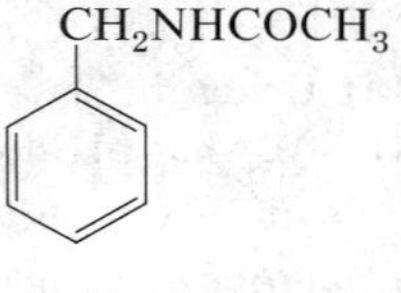 CH₂NHCOCH₃ [benzene ring]

12. The dipole moment of p-(N,N-dimethylamino)benzonitrile, 6.60 D, is substantially greater than the sum of the dipole moments of N,N-dimethylaniline, 1.57 D, and benzonitrile, 3.93 D. Explain.

13. Write out the mechanism for bromination of N,N-dimethylaniline in the *para* position with Br_2 and show why this compound is so much more reactive than benzene.

14. Although *o*-methylaniline ($pK_a = 4.44$) is a somewhat weaker base than aniline ($pK_a = 4.60$), *o*-methyl-N,N-dimethylaniline ($pK_a = 6.11$) is a much stronger base than N,N-dimethylaniline ($pK_a = 5.15$). Give a rational explanation.

15. In each of the following pairs of compounds, which is the more *basic* in aqueous solution? Give a brief explanation.

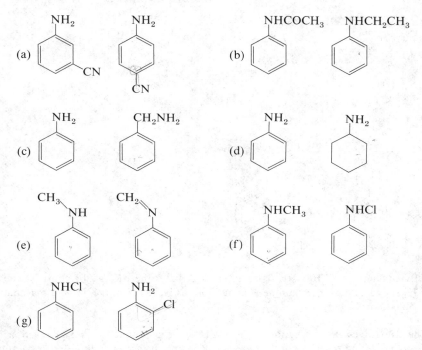

16. Propose a synthesis of each of the following compounds from alcohols containing five or fewer carbon atoms.

(a) $CH_3(CH_2)_4NH_2$

(b) $CH_3(CH_2)_4N(CH_3)_2$

(c) $CH_3CH_2CH_2\overset{\underset{|}{CH_3}}{N}CH_2CH_2CH_3$

(d) $(CH_3)_2CHCH_2CH_2NHCH_2CH_3$

17. Propose a method for the stereospecific conversion of
(a) (R)-2-octanol into (S)-2-octylamine
X (b) (R)-2-octanol into (R)-2-octylamine

18. Propose a synthesis for each of the following compounds.

(a) $H_2N\overset{\underset{|}{CH_3}}{C}H\overset{\underset{|}{}}{C}HCH_2CH_3$ (mixture of diastereomers)
$\overset{\underset{|}{OH}}{}$

(b) 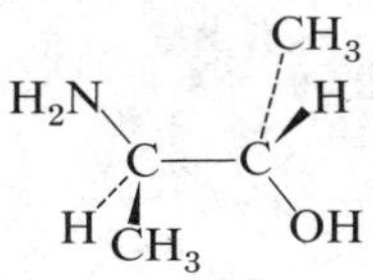 [structure of C–C with H_2N, CH_3, H, CH_3, OH groups]   (indicated diastereomer only)

(c)

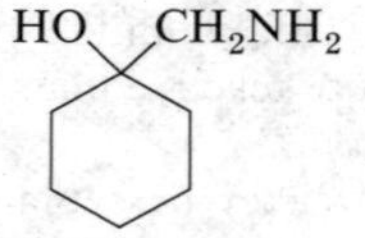

(d) $CH_3CH_2\overset{\overset{\displaystyle O}{\|}}{C}CHCH_2N(CH_2CH_3)_2$
$\qquad\quad\underset{CH_3}{|}$

(e) 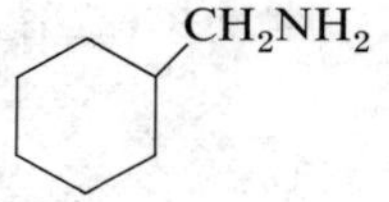 ![]   (*Hint:* See Section 20.5.)

(f) 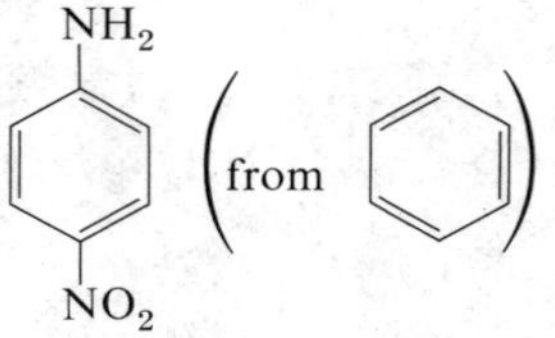 ![] (from ![])

(g) 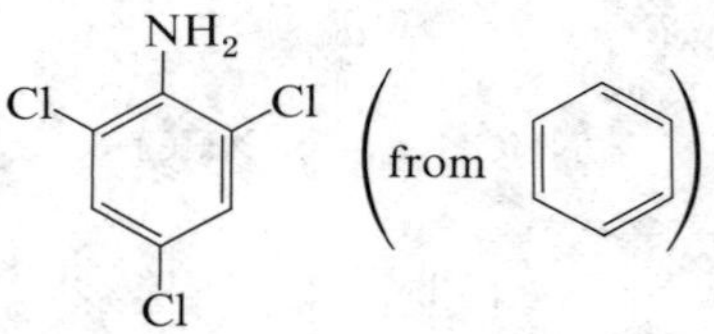 ![] (from ![])

19. Diphenylamine, $(C_6H_5)_2NH$, is a rather weak base; the pK_a of the conjugate acid, 0.79, shows that diphenylamine is about 10^{-4} as basic as aniline. Give a reasonable explanation.

20. The trimethylanilinium cation, $C_6H_5N(CH_3)_3^+$ is prepared by an S_N2 reaction of dimethylaniline with a methyl halide or sulfonate. The compound undergoes a number of electrophilic substitution reactions such as nitration. Write out the resonance structures involved for reaction at the *meta* and *para* positions and determine whether the trimethylammonium group is activating or deactivating, and *ortho,para-* or *meta-*directing.

21. Show how to accomplish each of the following conversions.
(a) $CH_2{=}CHCO_2C_2H_5 \longrightarrow H_2NCH_2CH_2CH_2CH_2N(CH_3)_2$

(b) 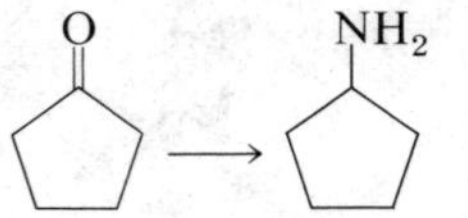 ![] $\longrightarrow$![]

(c) 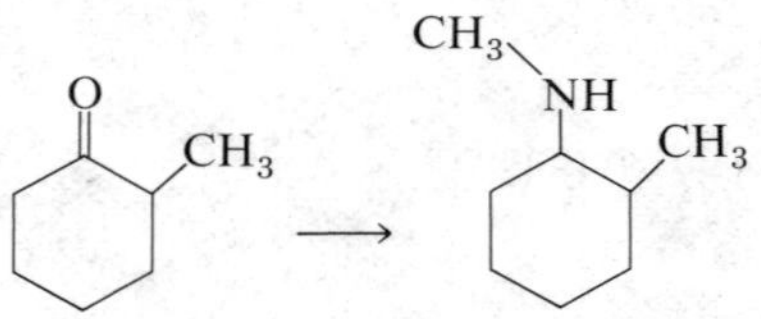 ![] $\longrightarrow$![]

(d) 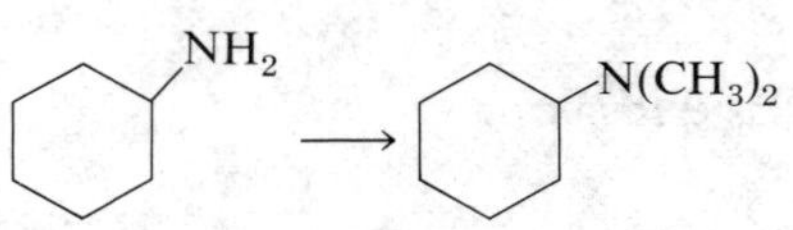 ![] $\longrightarrow$![]

(e)

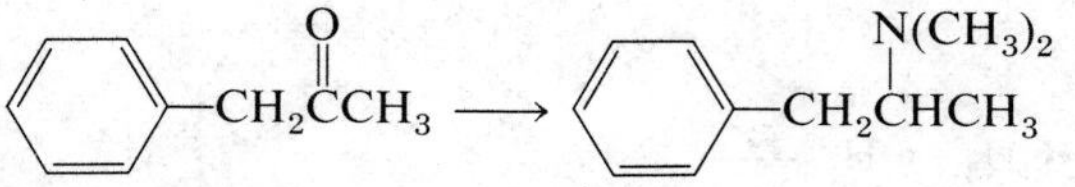

22. What is the expected product when piperidine is subjected to each of the following sets of reactions?

(a)

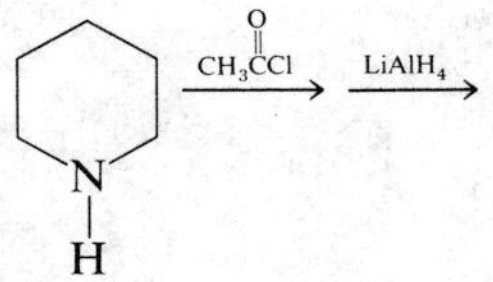

piperidine

(b)

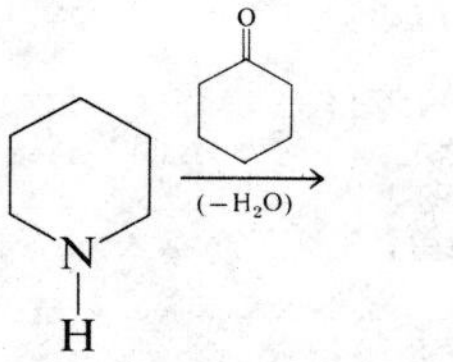

(c)

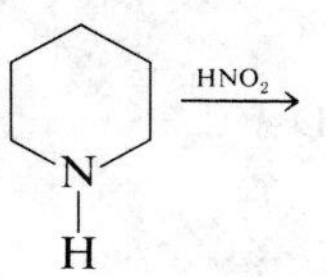

(d)

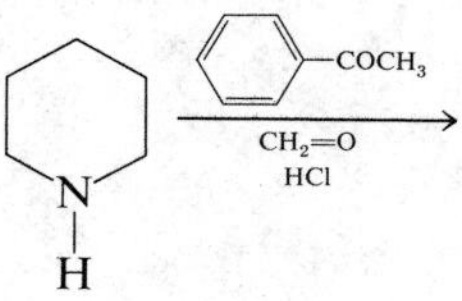

(e)

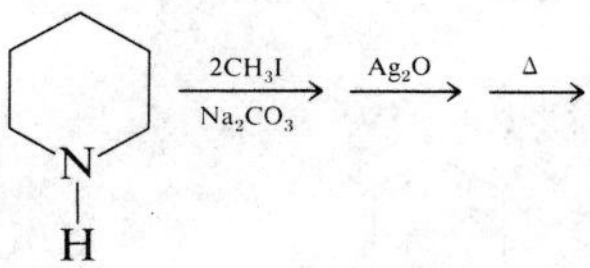

(f)

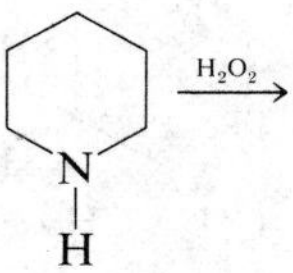

23. *cis*-2-Butene is subjected to the following sequence of reactions.

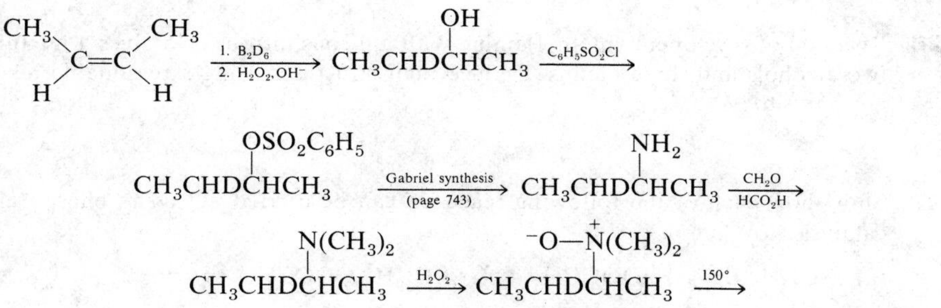

$$CH_3CHDCHCH_3 \xrightarrow{C_6H_5SO_2Cl}$$

$$\underset{CH_3CHDCHCH_3}{OSO_2C_6H_5} \xrightarrow[\text{(page 743)}]{\text{Gabriel synthesis}} \underset{CH_3CHDCHCH_3}{NH_2} \xrightarrow[HCO_2H]{CH_2O}$$

$$\underset{CH_3CHDCHCH_3}{N(CH_3)_2} \xrightarrow{H_2O_2} \underset{CH_3CHDCHCH_3}{^-O-\overset{+}{N}(CH_3)_2} \xrightarrow{150°}$$

1-butene + *cis*-2-butene + *trans*-2-butene

Two of the butene isomers produced in the pyrolysis contain one atom of deuterium per molecule and the other isomer contains only hydrogen. Which isomer contains no deuterium? Explain.

24. Predict the major product in each of the following elimination reactions.

(a) 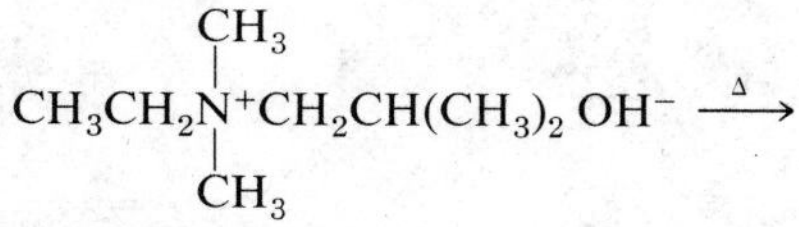 $CH_3CH_2\overset{\underset{\displaystyle CH_3}{|}}{\underset{\underset{\displaystyle CH_3}{|}}{N^+}}CH_2CH(CH_3)_2 \ OH^- \xrightarrow{\Delta}$

(b) 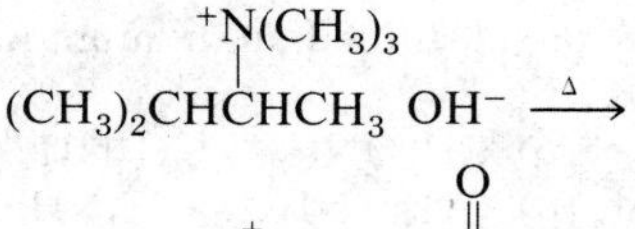 $(CH_3)_2CH\overset{\overset{\displaystyle +N(CH_3)_3}{|}}{C}HCH_3 \ OH^- \xrightarrow{\Delta}$

(c) $(CH_3CH_2)_3\overset{+}{N}CH_2CH_2\overset{\overset{\displaystyle O}{\|}}{C}CH_3 \ OH^- \xrightarrow{\Delta}$

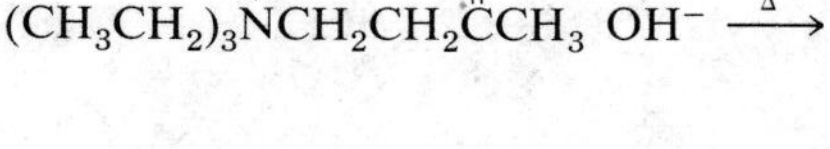

(d) 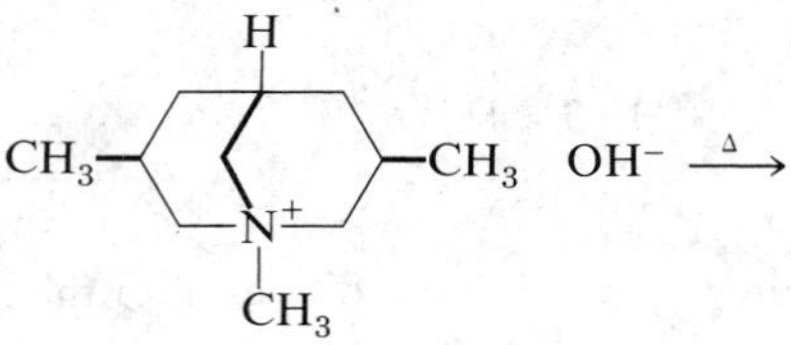 CH_3- $-CH_3$ $OH^- \xrightarrow{\Delta}$

(e)

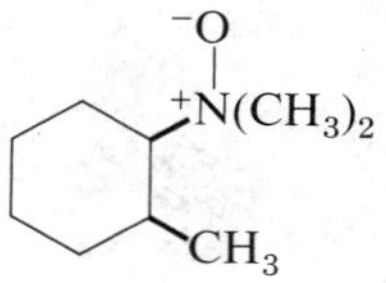

25. Show how to accomplish each of the following conversions.

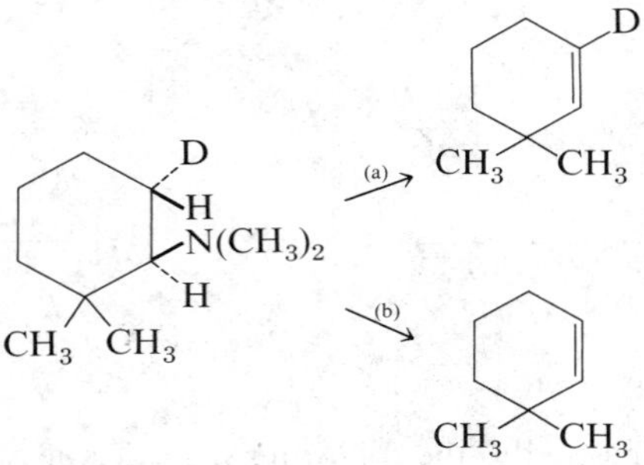

26. Reaction of (cyclopentylmethyl)amine with aqueous nitrous acid gives a mixture of two alcohols and three olefins. Deduce their structures using a reasonable reaction mechanism.

27. Show how each of the following reactions can be carried out using phase-transfer catalysis.

(a) $n\text{-}C_{10}H_{21}Br + NaOOCCH_3 \longrightarrow n\text{-}C_{10}H_{21}OOCCH_3$

(b) $3,4\text{-}(CH_3)_2C_6H_3COCH_3 + NaOD + D_2O \longrightarrow 3,4\text{-}(CH_3)_2C_6H_3COCD_3$

(c) $(CH_3)_2CH(CH_2)_4OH + (CH_3)_2SO_4 \longrightarrow (CH_3)_2CH(CH_2)_4OCH_3$

(d) $(CH_3)_2CHCHO + C_6H_5CH_2Cl \longrightarrow C_6H_5CH_2C(CH_3)_2CHO$

(e) $C_6H_5CH{=}CH_2 + CHCl_3 \xrightarrow{Cl_2} C_6H_5$

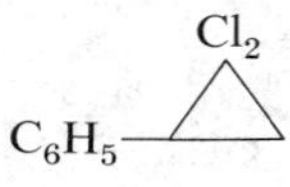

28. Which member of each of the following pairs of substituted ammonium ions is the more acidic? Explain briefly.

(a) $ClCH_2CH_2NH_3^+$; $CH_3CH_2CH_2NH_3^+$ (b) $CH_3ONH_3^+$; $CH_3NH_3^+$

(c) $CH_3CONH_3^+$; $CH_3CH_2NH_3^+$ (d) $CH_2{=}NH_2^+$; $CH_3NH_3^+$

(e) $CH_3OOCCH_2NH_3^+$; $CH_3CH_2CH_2NH_3^+$ (f) $CH_2{=}CHNH_3^+$; $CH_3CH_2NH_3^+$

29. Granatine, $C_9H_{17}N$, is an alkaloid that occurs in pomegranate. Two stages of the Hofmann exhaustive methylation remove the nitrogen and yield a mixture of cyclo-octadienes identified by catalytic hydrogenation to cyclooctane. The ultraviolet spectrum of the mixture shows the absence of the conjugated diene, 1,3-cyclooctadiene. Deduce the structure of granatine.

30. N-Chloroacetanilide is converted to a mixture of 32% o-chloroacetanilide and 68% p-chloroacetanilide in the presence of HCl. The use of ^{36}Cl-enriched HCl finds the isotopic Cl incorporated into the product. The reaction of acetanilide with chlorine under the same reaction conditions gives the same product composition. Write a reaction mechanism for the rearrangement of N-chloroacetanilide (Orton rearrangement) to account for these facts.

Chapter 25
Other Nitrogen Functions

The amino group is the most important functional group of nitrogen. Neverthe-less, a number of other nitrogen-containing functional groups are known, and some have been mentioned from time to time in this text. Several of these other nitrogen functional groups are listed in Table 25.1 and will be discussed in this chapter. We shall see that they are of varying importance in organic chemistry. Some of these functional groups have special importance when they are attached to aromatic rings. An example is the diazonium group, a particularly significant aromatic functional group that is useful for the preparation of a wide variety of other compounds.

25.1
Nitro Compounds

Nitroalkanes are relatively rare, although a few of the simpler ones are com-mercially available. Examples are nitromethane, which is used as a high-power fuel in racing engines, nitroethane, and 2-nitropropane.

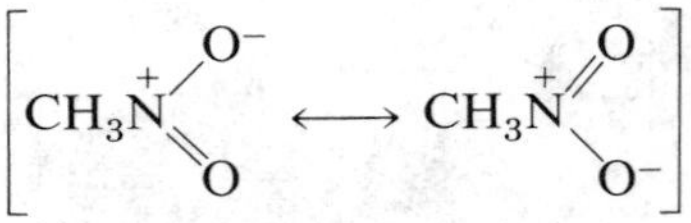

$$\left[CH_3\overset{+}{N}\overset{O^-}{\underset{O}{\diagup}} \longleftrightarrow CH_3\overset{+}{N}\overset{O}{\underset{O^-}{\diagup}} \right] \equiv CH_3NO_2 \qquad CH_3CH_2NO_2 \qquad (CH_3)_2CHNO_2$$

nitromethane   nitroethane   2-nitropropane

Aromatic nitro compounds are much more common because they are easily pre-pared by the electrophilic nitration of aromatic compounds (Chapter 23).

TABLE 25.1
Functional Groups Containing Nitrogen

Structure	Name	Example .	
R—NO$_2$	nitro	C$_6$H$_5$NO$_2$	nitrobenzene
R—NCO	isocyanate	C$_6$H$_5$NCO	phenyl isocyanate
R—NHCOOR′	urethane, carbamate	C$_6$H$_5$NHCOOCH$_3$	methyl N-phenyl-carbamate
R—NHCONH—R′	urea	H$_2$NCONH$_2$	urea
R—N$_3$	azide	CH$_3$CH$_2$N$_3$	ethyl azide
R—N=N—R′	azo	C$_6$H$_5$N=NC$_6$H$_5$	azobenzene
R—$\overset{O^-}{\underset{\vert}{N^+}}$=N—R	azoxy	C$_6$H$_5$$\overset{O^-}{\underset{\vert}{N^+}}$=NC$_6H_5$	azoxybenzene
R—NHNH$_2$	hydrazine, diazine	C$_6$H$_5$NHNH$_2$	phenylhydrazine
R$_2$C=N$_2$	diazo	CH$_2$=N$_2$	diazomethane
R—N$_2$$^+$	diazonium	C$_6$H$_5$N$_2$$^+$ Cl$^-$	benzenediazonium chloride

A. *Nitroalkanes*

Nitroalkanes are prepared industrially by the free radical nitration of alkanes (see problem 11, page 120).

$$CH_4 + HNO_3 \xrightarrow{400°} CH_3NO_2 + H_2O$$

Some nitro compounds may be prepared in the laboratory by the displacement of alkyl halides with nitrite ion. Since nitrite is an ambident anion, some alkyl nitrite is usually produced as a by-product (Section 8.9).

$$\underset{}{CH_3(CH_2)_5\overset{I}{C}HCH_3} + NaNO_2 \longrightarrow \underset{(58\%)}{CH_3(CH_2)_5\overset{NO_2}{C}HCH_3} + \underset{(30\%)}{CH_3(CH_2)_5\overset{ONO}{C}HCH_3}$$

Yields of nitroalkane are higher when silver nitrite is used, but this added economy is tempered by the cost of the silver salt.

$$CH_3(CH_2)_6CH_2I + AgNO_2 \longrightarrow \underset{(83\%)}{CH_3(CH_2)_6CH_2NO_2} + \underset{(11\%)}{CH_3(CH_2)_6CH_2ONO}$$

The most striking chemical property of nitroalkanes is their acidity. The pK_a of nitromethane is 10.2, that of nitroethane is 8.5, and that of 2-nitropropane is 7.8. 2-Nitropropane is so acidic that it is 50% ionized at pH 7.8! Like carboxylic acids and ketones, nitro compounds owe their acidity to the fact that the conjugate base is resonance-stabilized.

$$CH_3\overset{+}{N}\underset{O^-}{\overset{O}{}} \longrightarrow H^+ + \left[\ddot{:}CH_2-\overset{+}{N}\underset{O^-}{\overset{O}{}} \longleftrightarrow \ddot{:}CH_2-N\underset{O}{\overset{O^-}{}} \longleftrightarrow CH_2{=}\overset{+}{N}\underset{O^-}{\overset{O^-}{}} \right]$$

methylnitronate ion

The anions derived from nitroalkanes are nucleophilic and enter into typical nucleophilic reactions. One particularly useful reaction is analogous to the aldol condensation.

$$CH_3(CH_2)_7CHO + CH_3NO_2 \xrightarrow[\text{EtOH}]{\text{NaOH}} \underset{(80\%)}{CH_3(CH_2)_7\overset{OH}{C}HCH_2NO_2}$$

1-nitro-2-decanol

Since nitro compounds are so acidic, only weakly basic catalysts are required. In the case of aromatic aldehydes, dehydration of the initial β-hydroxynitro compound usually results.

$$\text{C}_6\text{H}_5{-}CHO + CH_3NO_2 \xrightarrow[25°]{n\text{-}C_5H_{11}NH_2} \text{C}_6\text{H}_5{-}CH{=}CHNO_2$$

(75%)
β-nitrostyrene

Another general reaction of nitro compounds is reduction to the corresponding amine (Section 24.6.C). Reduction of β-hydroxynitroalkanes produced by the aldol condensation provides a convenient method for preparing 1,2-amino alcohols (page 746).

B. *Nitroarenes*

The preparation of nitroarenes by electrophilic nitration reactions has been discussed previously. Some nitroarenes are available by oxidation of the corresponding amine (Section 24.7.C). We shall see later in this chapter that another preparation from amines is via the diazonium group.

Nitrobenzene and related nitro compounds are generally high-boiling liquids. Nitrobenzene is a pale yellow oil, b.p. 210–211°, having a characteristic odor of almonds. It was used at one time in shoe polish, but is readily absorbed through the skin and is poisonous.

2,4,6-Trinitrotoluene, TNT, is an important explosive. It is relatively insensitive to shock and is used with a detonator. It melts at 81°, and it can be poured as the melt into containers such as bombs and hand grenades. 1,3,5-Trinitrobenzene is less sensitive than TNT to shock and has more explosive power, but is more difficult to prepare. Direct introduction of the third nitro group into toluene is assisted by the methyl group. Small amounts of 1,3,5-trinitrobenzene are prepared by oxidation of TNT to trinitrobenzoic acid.

2,4,6-trinitrotoluene   2,4,6-trinitrobenzoic acid
                                         (57–69%)

> Sodium dichromate (540 g) is added in portions to a solution of 360 g of technical grade trinitrotoluene in 1960 mL of concentrated sulfuric acid at such a rate as to keep the temperature in the range 45–55°. After addition is complete, the mixture is stirred 2 hr at 50° and then poured onto 4 kg of crushed ice. The yield of purified acid is 230–280 g.

2,4,6-Trinitrobenzoic acid is a strong acid ($pK = 0.7$) whose anion decomposes on heating to give carbon dioxide and a phenyl anion that is stabilized by the electron-attracting inductive effect of the three nitro groups.

1,3,5-trinitrobenzene

C. *Reactions of Nitroarenes*

The nitro group is relatively stable to many reagents. It is generally inert to acids and most electrophilic reagents; hence, it may be present in a ring when reactions with such reagents are used. The nitro group is also stable to most oxidizing agents, but it reacts with Grignard reagents and other strongly basic compounds such as lithium aluminum hydride. The most important reaction of the nitro group in aromatic compounds is reduction, but the reduction product depends on the reaction conditions used. Catalytic hydrogenation and reduction in acidic media yield the corresponding amine (Section 24.6.C).

$$ArNO_2 \longrightarrow ArNH_2$$

Reduction of the nitro group actually proceeds in a series of two-electron steps. In acid the intermediate compounds cannot be isolated, but are reduced rapidly in turn.

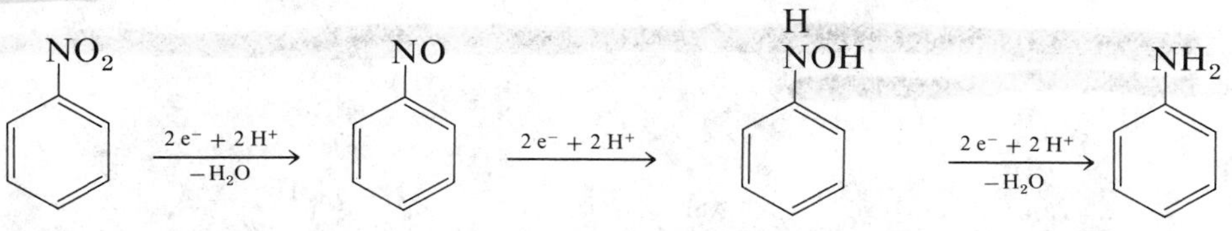

| nitrobenzene | nitrosobenzene | phenylhydroxylamine | aniline |

In neutral media a higher reduction potential is required, and reduction is readily stopped at the hydroxylamine stage.

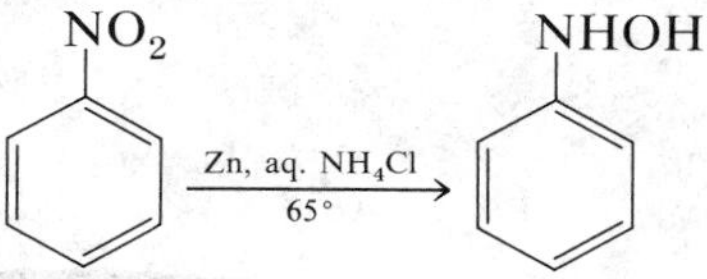

(62–68%)

Aromatic hydroxylamines are relatively unimportant compounds. Phenylhydroxylamine is a water-soluble, crystalline solid, m.p. 82°, that deteriorates in storage. It may be oxidized to nitrosobenzene.

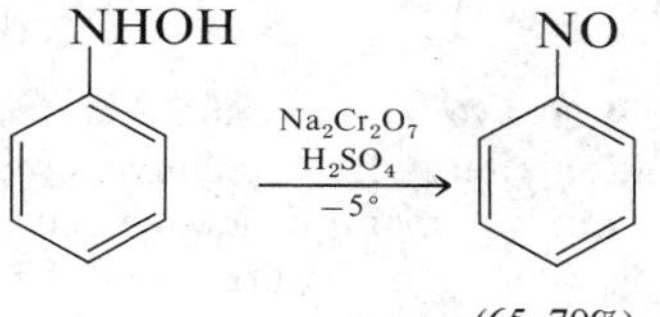

(65–70%)

Both hydroxylamino and nitroso groups are readily reduced to amines by chemical reduction in acid or by catalytic hydrogenation.

Reduction in basic media gives binuclear compounds.

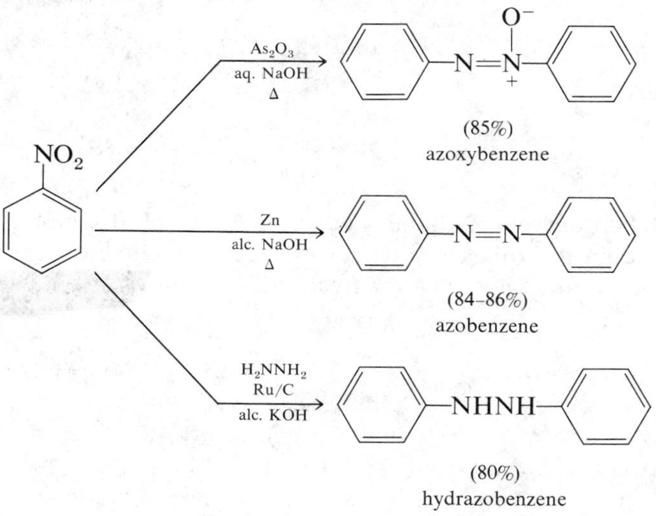

(85%)
azoxybenzene

(84–86%)
azobenzene

(80%)
hydrazobenzene

All of these compounds are reduced to aniline under acidic conditions. They may also be interconverted by the following reactions.

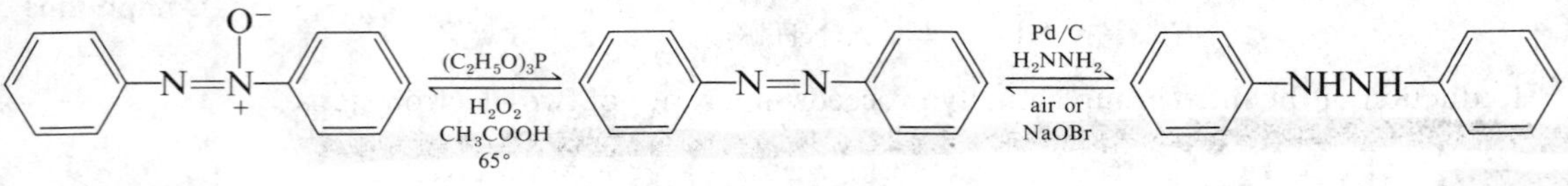

These binuclear compounds may best be considered to arise by condensation reactions during reduction.

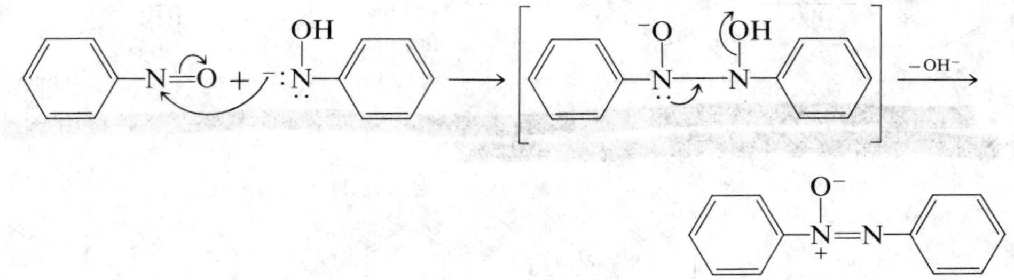

In fact, azoxybenzene can be prepared by the base-catalyzed condensation of phenylhydroxylamine with nitrosobenzene. The azoxy function is the least important functional group among these compounds. Azobenzene is a bright orange-red solid; although this parent compound has only limited significance, the azo linkage is an important component of azo dyes (Section 34.3).

Many azo compounds show *cis-trans* isomerism. The *trans* isomer is generally the more stable, and the activation energy for the conversion is sufficiently low that the *cis* isomer is generally not seen. For example, azobenzene can be converted in part to the *cis* isomer by photolysis, but the activation energy required to convert back to *trans*-azobenzene is only 23–25 kcal mole^{-1} in various solvents. This reaction has a half-life on the order of hours at room temperature.

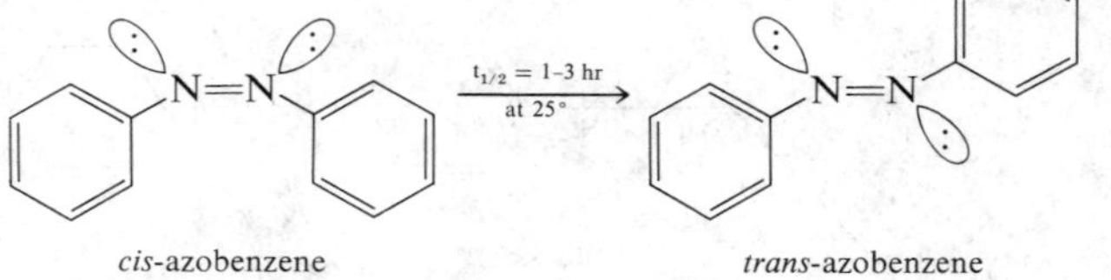

cis-azobenzene   *trans*-azobenzene

Hydrazobenzene or 1,2-diphenylhydrazine is a colorless solid that air-oxidizes on standing to azobenzene. It is significant principally because of a rearrangement that it undergoes in strongly acidic solution, the **benzidine rearrangement.**

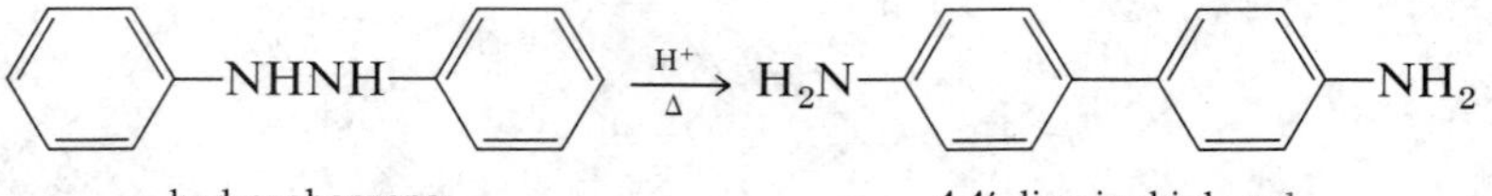

hydrazobenzene   4,4'-diaminobiphenyl
(benzidine)

This remarkable reaction involves the mono- or diprotonated salt in which bonding occurs between the *para* positions as the N—N bond is broken.

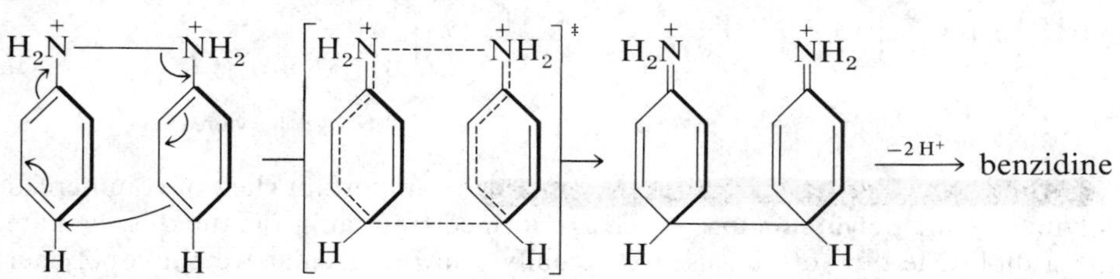

transition state

$$\xrightarrow{-2\,H^+}\ benzidine$$

Benzidine has had important uses as an intermediate in dye manufacture, but the compound has recently been demonstrated to be carcinogenic. Its place was taken by dichlorobenzidine, which now also appears to be a carcinogen.

Azoxybenzene, azobenzene, and hydrazobenzene are all conveniently reduced to aniline with sodium hydrosulfite.

$$C_6H_5-N{=}N-C_6H_5 + 2\,Na_2S_2O_4 + 2\,H_2O \longrightarrow 2\,C_6H_5NH_2 + 4\,NaHSO_3$$

> Sodium hydrosulfite, or sodium dithionite, is a useful reagent in neutral or alkaline solution. At acid pHs it decomposes with the liberation of sulfur. It is especially useful in the reductive cleavage of the azo groups in azo dyes. These dyes are generally water-soluble, and it suffices simply to add sodium hydrosulfite to an aqueous solution until the color of the dye has been discharged. The products are the corresponding amines.

EXERCISE: Review the preparation of *o*-nitrotoluene by the nitration of toluene (Section 23.5). Apply each of the reactions discussed in this section to *o*-nitrotoluene.

25.2
Isocyanates, Carbamates, and Ureas

We have encountered alkyl isocyanates previously as intermediates in the Hofmann rearrangement of amides (Section 24.6.H). They may also be prepared by S_N2 displacement of alkyl halides with cyanate ion. This ion is ambident and reacts preferentially at the nitrogen end.

$$CH_3CH_2CH_2CH_2Br + Na^+\,NCO^- \longrightarrow CH_3CH_2CH_2CH_2NCO$$

n-butyl isocyanate

Isocyanates react with water to give N-alkyl carbamic acids, which are unstable and spontaneously lose carbon dioxide to give the corresponding amine.

$$R-NCO + H_2O \longrightarrow [R-NHCOOH] \longrightarrow RNH_2 + CO_2$$

Isocyanates give carbamate esters with alcohols and ureas with amines.

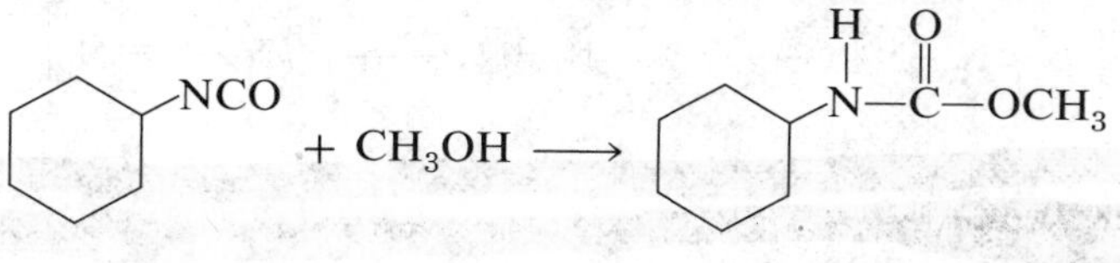

methyl N-cyclohexylcarbamate

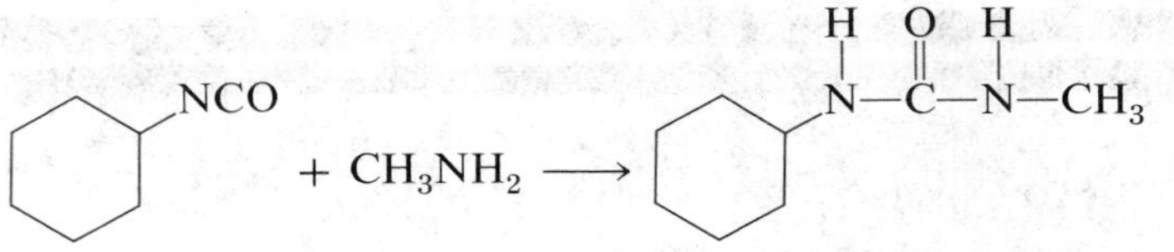

N-methyl-N'-cyclohexylurea

Carbamate esters are also called **urethanes.** An important class of commercial polymers is the **polyurethanes,** which are formed from an aromatic diisocyanate and a diol. One type of diol used is actually a low molecular weight copolymer made from ethylene glycol and adipic acid. When this polymer, which has free hydroxy end groups, is mixed with the diisocyanate, a larger polymer is produced.

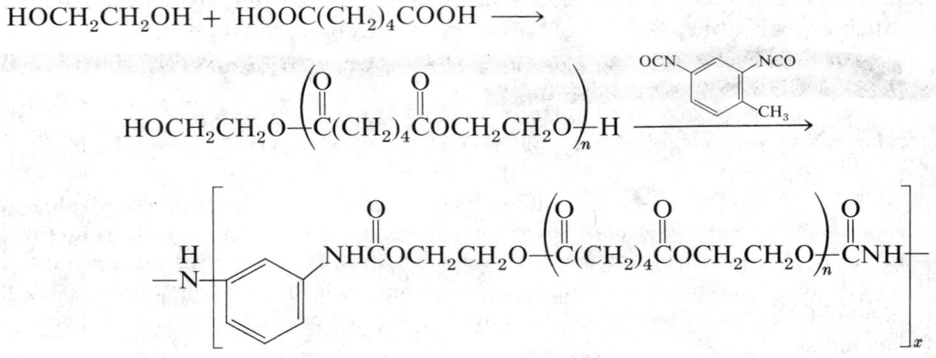

$$HOCH_2CH_2OH + HOOC(CH_2)_4COOH \longrightarrow$$

a polyurethane

In the manufacturing process, a little water is mixed in with the diol. Some of the diisocyanate reacts with water to give an aromatic diamine and carbon dioxide. The carbon dioxide forms bubbles that are trapped in the bulk of the polymer as it solidifies. The result is a spongy product called **polyurethane foam.**

25.3
Azides

Organic azides are compounds with the general formula RN_3. They are related to the inorganic acid, hydrazoic acid, HN_3. Azide ion, N_3^-, is a resonance hybrid of the following important dipolar structures.

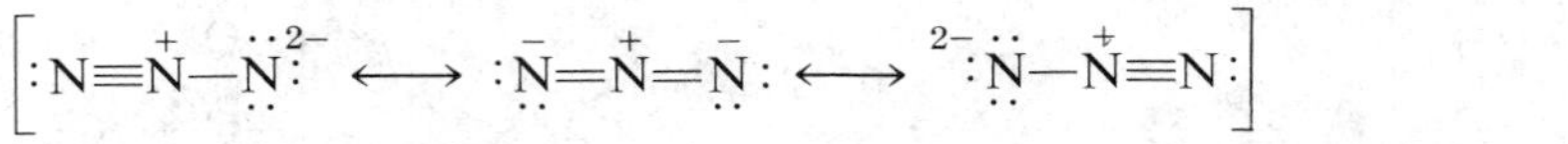

$$\left[:N{\equiv}\overset{+}{N}{-}\overset{..}{\underset{..}{N}}:^{2-} \longleftrightarrow :\overset{-}{\underset{..}{N}}{=}\overset{+}{N}{=}\overset{-}{\underset{..}{N}}: \longleftrightarrow {}^{2-}:\overset{..}{\underset{..}{N}}{-}\overset{+}{N}{\equiv}N: \right]$$

This anion is relatively nonbasic for anionic nitrogen (the pK_a of HN_3 is 11) and is a good nucleophile. Alkyl azides are best prepared by nucleophilic displacement on alkyl halides.

$$CH_3CH_2CH_2CH_2Br + N_3^- \xrightarrow[H_2O]{CH_3OH} CH_3CH_2CH_2CH_2N_3 + Br^-$$

A mixture of 34.5 g of NaN_3, 68.5 g of *n*-butyl bromide, 70 mL of water, and 25 mL of methanol is refluxed for 24 hr. The *n*-butyl azide separates as an oily layer. It is dried and distilled behind a safety barricade to obtain 40 g of pure *n*-butyl azide (90%).

Azide ion is also sufficiently nucleophilic to open the epoxide ring. β-Hydroxy-alkyl azides may be prepared in this way.

$$CH_3CH\!-\!CH_2 + N_3^- \xrightarrow[25°, \, 24 \, hr]{H_2O} CH_3\overset{OH}{\underset{|}{C}}HCH_2N_3$$

(70%)

Acyl azides may be prepared from acyl halides and azide ion (Section 24.6.H).

$$R\overset{O}{\overset{\|}{C}}Cl + N_3^- \longrightarrow R\overset{O}{\overset{\|}{C}}N_3 + Cl^-$$

Alkyl azides are reduced by lithium aluminum hydride or by catalytic hydrogenation to give the corresponding amines. The two-step process of (1) displacement of halide ion by azide ion, and (2) reduction of the resulting azide provides a convenient synthesis of pure primary amines. Also, since azide ion is relatively nonbasic yet still highly nucleophilic, its substitution/elimination ratio is high, even with secondary and β-branched alkyl halides (see Section 8.13).

$$CH_3CH_2CH_2\overset{Br}{\underset{|}{C}}HCH_3 + Na^+N_3^- \xrightarrow[H_2O]{C_2H_5OH} CH_3CH_2CH_2\overset{N_3}{\underset{|}{C}}HCH_3 \xrightarrow{LiAlH_4} CH_3CH_2CH_2\overset{NH_2}{\underset{|}{C}}HCH_3$$

2-pentyl azide

Both alkyl and acyl azides are thermally unstable and lose nitrogen on heating. In some cases, particularly when the nitrogen content of the molecule is higher than about 25%, the decomposition can occur with explosive violence. Decomposition of alkyl azides gives a complex mixture of products. Acyl azides decompose to the acyl nitrene, which rearranges to an **isocyanate** (Schmidt and Curtius rearrangements; Section 24.6.H).

$$R\!-\!\overset{O}{\overset{\|}{C}}\!-\!N_3 \longrightarrow N_2 + R\!-\!\overset{O}{\overset{\|}{C}}\!-\!\overset{..}{\underset{..}{N}} \longrightarrow R\!-\!N\!=\!C\!=\!O$$

an isocyanate

EXERCISE: Compare the Lewis structure of an isocyanate with related structures of an azide. What is the expected product of S_N2 displacement of benzyl chloride with sodium cyanate in methanol solution?

25.4
Diazo Compounds

Diazo compounds have the general formula $R_2C\!=\!N_2$. The electronic structure of diazomethane, the simplest diazo compound, shows that the carbon has nucleophilic properties.

$$\left[\ddot{\ ^-}CH_2-\overset{+}{N}\equiv N: \longleftrightarrow CH_2=\overset{+}{N}=N:^- \right]$$

diazomethane

We have encountered diazomethane previously as a reagent for converting carboxylic acids into methyl esters (Section 18.7.A).

$$\underset{\substack{\parallel \\ O}}{R\overset{O}{C}OH} + CH_2N_2 \longrightarrow \underset{\substack{\parallel \\ O}}{R\overset{O}{C}OCH_3} + N_2$$

Diazomethane is prepared by treating an N-methyl-N-nitrosoamide with concentrated potassium hydroxide solution.

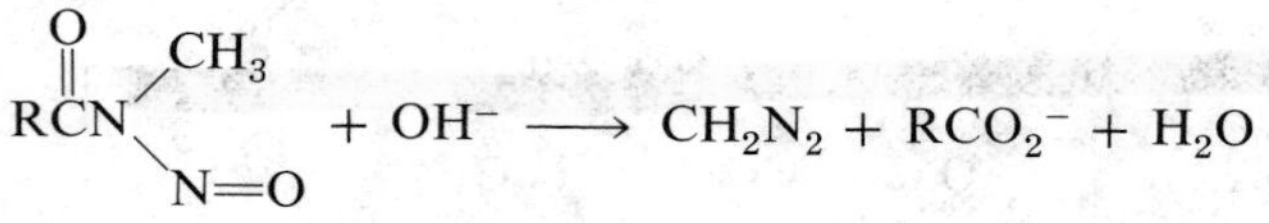

$$\underset{\substack{\parallel \\ O}}{R\overset{O}{C}N}\overset{CH_3}{\underset{N=O}{\diagdown}} + OH^- \longrightarrow CH_2N_2 + RCO_2^- + H_2O$$

The preparation is carried out in a two-phase mixture consisting of ether and aqueous KOH. The diazomethane dissolves in the ether as it is formed and it is generally used as an ether solution.

Other diazo compounds are also known. α-Diazo ketones and α-diazo esters are relatively stable since the carbonyl group can delocalize the carbanionic electron pair.

$$\left[\underset{\substack{\parallel \\ O}}{R-\overset{O}{C}-\ddot{C}H-\overset{+}{N}\equiv N:} \longleftrightarrow \underset{\substack{| \\ O^-}}{R-\overset{O^-}{C}=CH-\overset{+}{N}\equiv N:} \longleftrightarrow \underset{\substack{\parallel \\ O}}{R-\overset{O}{C}-CH=\overset{+}{N}=N:^-} \right]$$

This type of diazo compound is conveniently prepared by the reaction of diazomethane with an acyl halide.

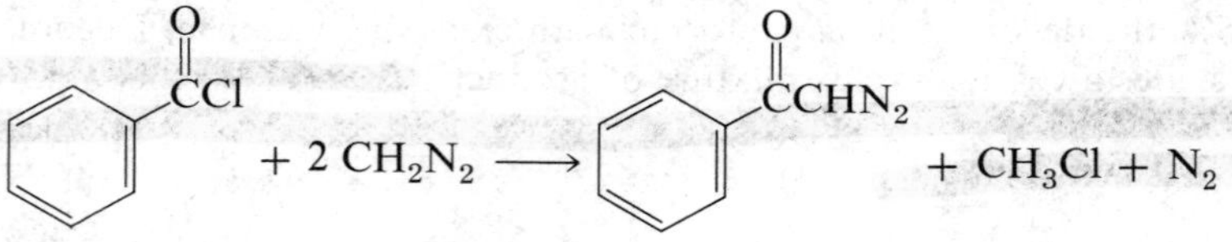

Excess diazomethane must be used to react with the HCl that is produced in the reaction. If only one equivalent of diazomethane is used, a chloromethyl ketone results.

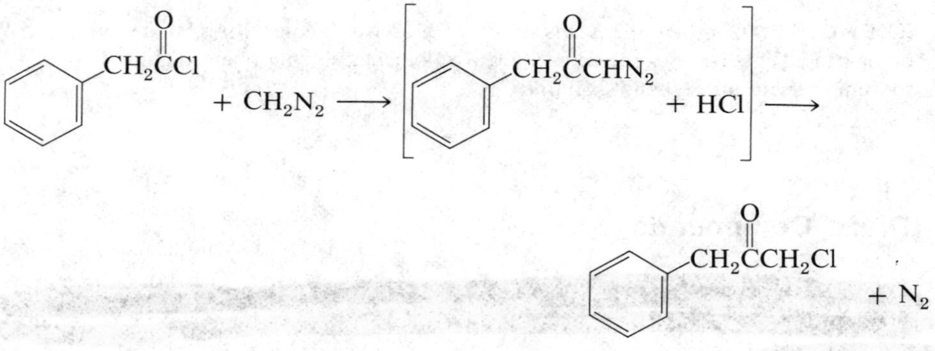

(83–85%)

The reaction of diazomethane with acyl halides is another reaction that shows the nucleophilic nature of the carbon in this compound. The mechanism of the reaction may be visualized as follows.

$$(1) \quad R-\overset{\overset{\displaystyle O}{\|}}{C}-Cl + \:\!\ddot{C}H_2-\overset{+}{N}\equiv N: \; \rightleftharpoons \; R-\overset{\overset{\displaystyle O^-}{|}}{\underset{\underset{\displaystyle Cl}{|}}{C}}-CH_2-\overset{+}{N}\equiv N:$$

$$(2) \quad R-\overset{\overset{\displaystyle O^-}{|}}{\underset{\underset{\displaystyle Cl}{|}}{C}}-CH_2-\overset{+}{N}\equiv N: \; \rightleftharpoons \; R-\overset{\overset{\displaystyle O}{\|}}{C}-CH_2-\overset{+}{N}\equiv N: + Cl^-$$

$$(3) \quad R-\overset{\overset{\displaystyle O}{\|}}{C}-CH_2-\overset{+}{N}\equiv N: \; \rightleftharpoons \; R-\overset{\overset{\displaystyle O}{\|}}{C}-CHN_2 + H^+$$

If there is no excess diazomethane to react with the proton liberated in step (3), the chloride ion displaces nitrogen to give the chloro ketone.

$$(4) \quad Cl^- + R-\overset{\overset{\displaystyle O}{\|}}{C}-CH_2-\overset{+}{N}\equiv N: \; \longrightarrow \; R-\overset{\overset{\displaystyle O}{\|}}{C}-CH_2Cl + :N\equiv N:$$

Like azides, diazo compounds lose nitrogen either thermally or when irradiated with ultraviolet light. The decomposition is catalyzed by copper powder. The initial product is a carbene (Section 11.6.F), which then reacts further. In the case of α-diazo ketones, the resulting acylcarbene rearranges to give a ketene (see the Curtius rearrangement; Section 24.6.H).

$$C_6H_5-\overset{\overset{\displaystyle N_2}{\|}}{C}-\overset{\overset{\displaystyle O}{\|}}{C}-C_6H_5 \; \xrightarrow[-N_2]{110°} \; \left[C_6H_5-\overset{\cdot\cdot}{C}\text{-}\overset{\overset{\displaystyle O}{\|}}{C}-C_6H_5 \right] \; \longrightarrow \; \underset{C_6H_5}{\overset{C_6H_5}{\diagdown}}C=C=O$$

(65%)

diphenylketene

The carbenes derived from some diazo compounds may be trapped by reaction with an alkene; the products are cyclopropane derivatives.

$$\bigcirc\!\!\!= + N_2CHCOOC_2H_5 \; \xrightarrow[Cu]{\Delta} \; \text{(bicyclic structure)}\overset{H}{\underset{COOC_2H_5}{\diagup}} + N_2$$

ethyl 7-bicyclo[4.1.0]-
heptanecarboxylate

EXERCISE: Show how propanoic acid can be converted into the following compounds by a route using diazomethane. (a) methyl propanoate; (b) 1-diazo-2-butanone; (c) 1-chloro-2-butanone; (d) ethylketene (a reactive ketene that rapidly dimerizes).

25.5
Diazonium Salts

In Section 24.7.B we found that aromatic amines react with nitrous acid in aqueous solution to give solutions of arenediazonium salts, which are moderately stable if kept cold.

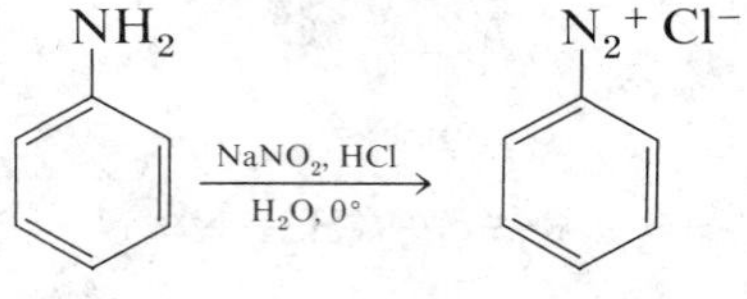

aniline

benzenediazonium chloride

These unstable compounds comprise an important class of synthetic intermediates. In a sense, they are the "Grignard reagents" of aromatic chemistry, since they can be used in the synthesis of such a wide variety of other aromatic compounds. In this section we shall discuss the chemistry of this useful group of compounds.

A. *Acid-Base Equilibria of Arenediazonium Ions*

In acid solution arenediazonium salts have the diazonium ion structure with a linear C—N—N bond system. The diazonium ion has a π-system that can conjugate with the aromatic π-system (Figure 25.1). This conjugation is responsible in part for the relative stability of these compounds. Recall that aliphatic diazonium ions are not at all stable and generally react immediately upon formation (Section 24.7.B).

Arenediazonium ions behave as dibasic acids. The two steps in the equilibria are represented as

$$\text{Ar}-\overset{+}{\text{N}}\equiv\text{N} + 2\,\text{H}_2\text{O} \overset{K_1}{\rightleftharpoons} \text{Ar}-\text{N}=\text{NOH} + \text{H}_3\text{O}^+$$

arenediazohydroxide

$$\text{Ar}-\text{N}=\text{NOH} + \text{H}_2\text{O} \overset{K_2}{\rightleftharpoons} \text{Ar}-\text{N}=\text{N}-\text{O}^- + \text{H}_3\text{O}^+$$

arenediazotate ion

The diazonium ions represent an unusual class of dibasic acids in that $K_2 \gg K_1$; that is, the arenediazohydroxide is present only in small amount. For the phenyl group, equal concentrations of benzenediazonium ion and benzenediazotate are present at a pH of 11.9. Even in neutral solutions with pH $= 7$, the diazonium ions are generally the most predominant species present.

Benzenediazotate ion exists in *syn* and *anti* forms, like other compounds containing C=N and N=N bonds. The *anti* form is the more stable, but the less stable *syn* isomer is that formed first by reaction of the diazonium cation with hydroxide ion.

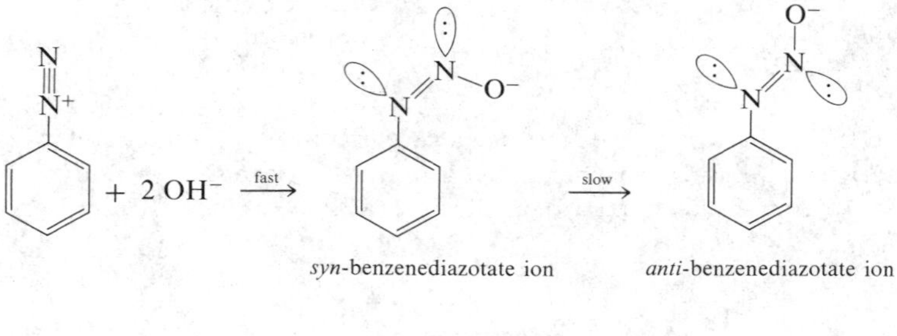

syn-benzenediazotate ion

anti-benzenediazotate ion

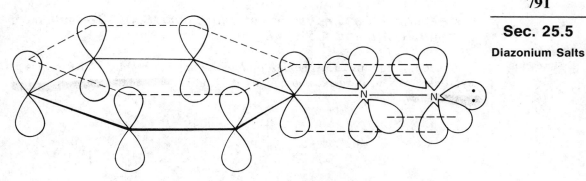

FIGURE 25.1. *Orbital structure of benzenediazonium cation.*

Alkanediazonium ions react by S_N2 displacement, S_N1 formation of carbocations, E1 and E2 eliminations, and carbocation rearrangements (Section 24.7.B). None of these pathways is readily available to arenediazonium ions. The most likely reaction, formation of an aryl cation, is limited by the high energy of these species. In phenyl cation, the empty orbital has approximately sp^2-hybridization and cannot conjugate with the π-electronic system (Figure 25.2). For example, the enthalpy of formation in the gas phase of phenyl cation from chlorobenzene is about the same as that for the ionization of vinyl chloride and is almost as high as the enthalpy of formation of methyl cation from methyl chloride.

$$C_6H_5Cl \longrightarrow C_6H_5^+ + Cl^- \qquad \Delta H^\circ = 223 \text{ kcal mole}^{-1}$$
$$CH_2{=}CHCl \longrightarrow C_2H_3^+ + Cl^- \qquad \Delta H^\circ = 223 \text{ kcal mole}^{-1}$$
$$CH_3Cl \longrightarrow CH_3^+ + Cl^- \qquad \Delta H^\circ = 227 \text{ kcal mole}^{-1}$$

Nevertheless, many reactions of diazonium ions are the reactions expected if aryl cations were intermediates. In many preparative methods, however, the aryl cation is not free; it is combined as a complex with a metal, often copper. Other reactions involve free radical intermediates. All of the following types of compounds can be prepared by appropriate reactions of the arenediazonium ions, ArN_2^+.

ArOH	ArCN	Ar—Ar′
ArI	ArF	ArNHNH₂
ArSH	ArNO₂	ArN₃
ArCl	ArH	ArN=NAr′
ArBr	Ar—Ar	

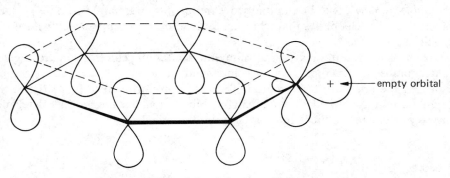

FIGURE 25.2. *Orbital diagram of phenyl cation.*

B. *Thermal Decomposition of Diazonium Salts; Formation of ArOH, ArI, and ArSH*

Aqueous solutions of arenediazonium ions are not stable. Nitrogen gas is evolved slowly in the cold and rapidly on heating. The net reaction is that of hydrolysis.

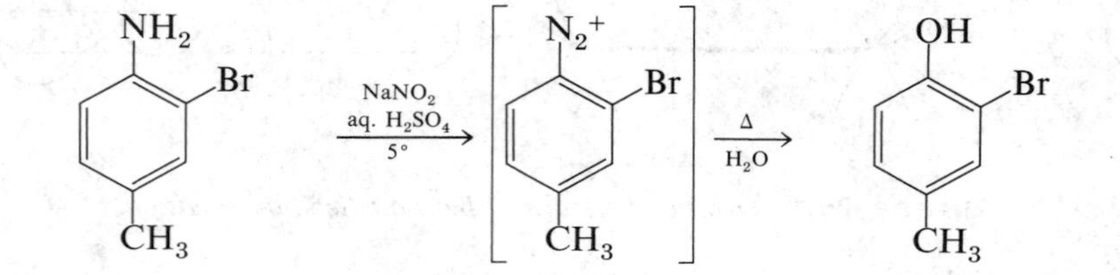

2-bromo-4-methylaniline   (80–92%)
2-bromo-4-methylphenol

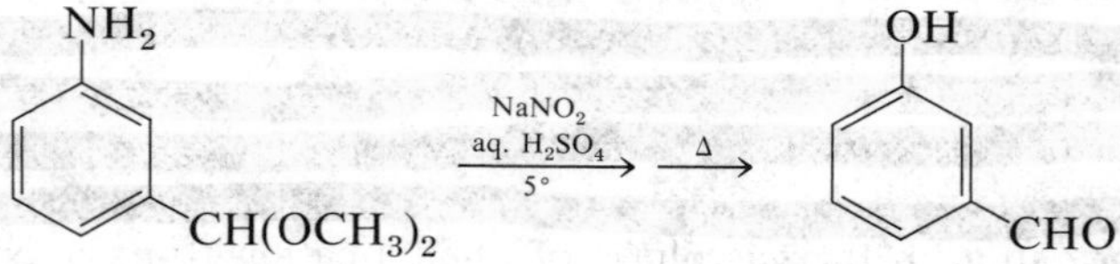

m-aminobenzaldehyde   *m*-hydroxybenzaldehyde
dimethyl acetal

[In this example, note that the hot aqueous acid also causes hydrolysis of the acetal function to the aldehyde.]

Hydrolysis of the diazonium ion appears to be an S_N1 type of process involving the aryl cation.

$$ArN_2^+ \longrightarrow N_2 + Ar^+$$

The aryl cation forms despite its high energy because of the great stability of nitrogen; that is, the formation of N_2 is a powerful driving force for the decomposition of diazonium ions. The aryl cation intermediate is highly reactive and reacts rapidly with water to form the corresponding phenol.

$$Ar^+ + H_2O \longrightarrow Ar-OH_2^+ \rightleftharpoons ArOH + H^+$$

However, the aryl cation also reacts with other nucleophiles that may be present—such as halide ion. If HCl is used in the diazotization, some chloroarene is also produced.

$$Ar^+ + Cl^- \longrightarrow ArCl$$

For this reason sulfuric acid is normally used as the acid in diazotizations in which the diazonium salt is to be thermally decomposed. Sulfate and bisulfate ions are much poorer nucleophiles than chloride ion and do not compete well with water for the aryl cation.

When aqueous solutions of diazonium ions containing chloride ion are allowed to decompose, the rate of reaction is independent of the chloride ion concentration, but the amount of chloroarene formed is proportional to [Cl⁻]. Thus, the rate-determining step does not depend on chloride ion, but the product-determining steps do. This result is interpretable by the scheme

$$Ar-N_2^+ \xrightarrow{\text{slow}} N_2 + Ar^+ \begin{array}{c} \xrightarrow{H_2O} ArOH + H_3O^+ \\ \xrightarrow{Cl^-} ArCl \end{array}$$

The competition of chloride ion with water for the intermediate aryl cation is usually inadequate for this method to be a successful preparation of chloroarenes; the Sandmeyer reaction (Section 25.5.C) is generally better.

Highly nucleophilic anions can compete successfully with water for the intermediate aryl cation and lead to satisfactory syntheses. A useful example is iodide ion.

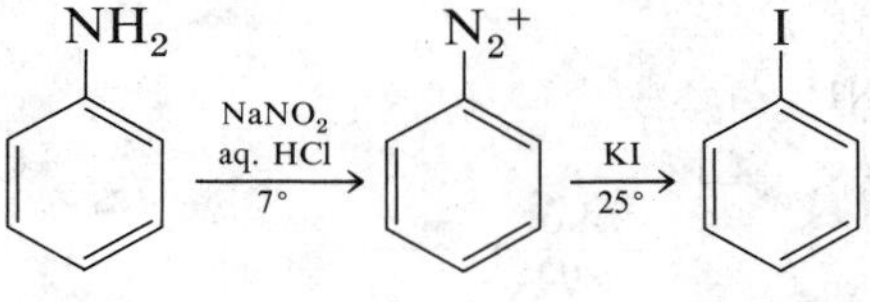

(74–76%)

In this case HCl is used for the diazotization, and the resulting solution contains Cl^-; nevertheless, the more nucleophilic I^- dominates the reaction. The diazotization could have been accomplished with hydriodic acid, but this acid is far more expensive than HCl.

Reaction of aqueous diazonium salts with HS^- or with metal polysulfides has been used for preparation of thiophenols, but violent reactions and explosions have been reported and the method is not recommended. An alternative route involves potassium ethyl xanthate, $KSCSOC_2H_5$, which is available commercially from the reaction of potassium ethoxide with carbon disulfide.

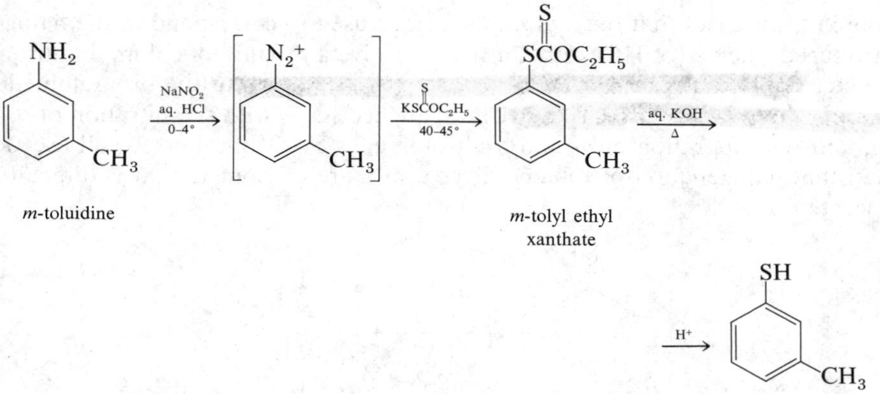

m-toluidine

m-tolyl ethyl
xanthate

(63–75%)
m-methylthiophenol
m-thiocresol

This reaction involves a variation of the hydrolysis reaction above. Reaction of the diazonium ion with the ethyl xanthate ion gives first the diazoxanthate.

$$Ar—N=N—\overset{\overset{\displaystyle S}{\|}}{S}COC_2H_5$$

which then decomposes with liberation of nitrogen by an ion-pair or radical mechanism. Even this method is hazardous because the intermediate diazoxanthate can detonate and should be allowed to decompose as formed. The use of traces of nickel—even a nichrome stirrer—has been recommended to facilitate the controlled decomposition.

EXERCISE: Write equations showing how benzene may be converted into phenol and into iodobenzene.

C. *The Sandmeyer Reaction: Preparation of ArCl, ArBr, and ArCN*

Decomposition of diazonium salts is catalyzed by cuprous salts. In laboratory practice, the cold diazonium solution is added dropwise to a hot suspension of cuprous bromide, chloride, or cyanide to give the corresponding aromatic product in a method known as the **Sandmeyer reaction.** Some examples of Sandmeyer reactions are

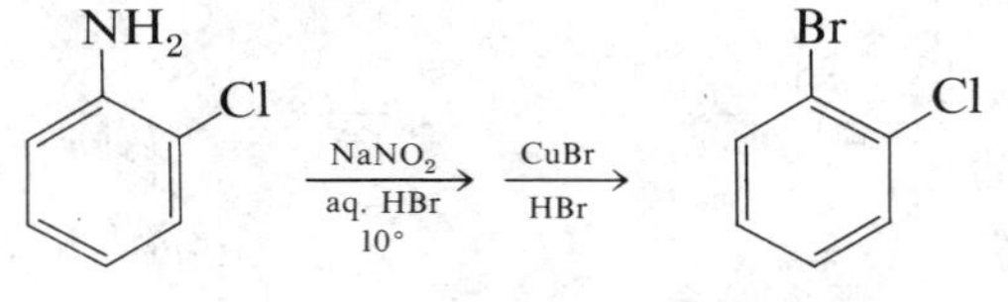

o-chloroaniline

(89–95%)
o-bromochlorobenzene

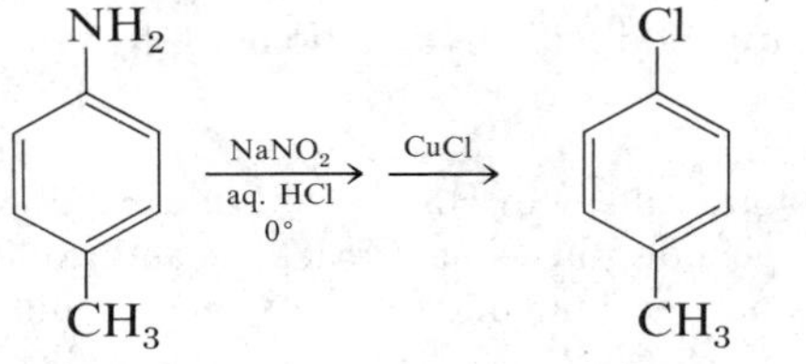

p-toluidine

(70–79%)
p-chlorotoluene

Note in these cases that the hydrohalic acid is used to correspond to the halogen introduced; the use of HCl with CuBr would give a mixture of chloro and bromo product. Incidentally, this route to *o*- or *p*-chlorotoluene (or the bromotoluenes) gives the individual pure isomers. The sequence starts with the nitration of toluene and the separation of *o*- and *p*-nitrotoluenes by distillation. Recall (Section 23.9) that halogenation of toluene gives a mixture of isomers that is difficult to separate.

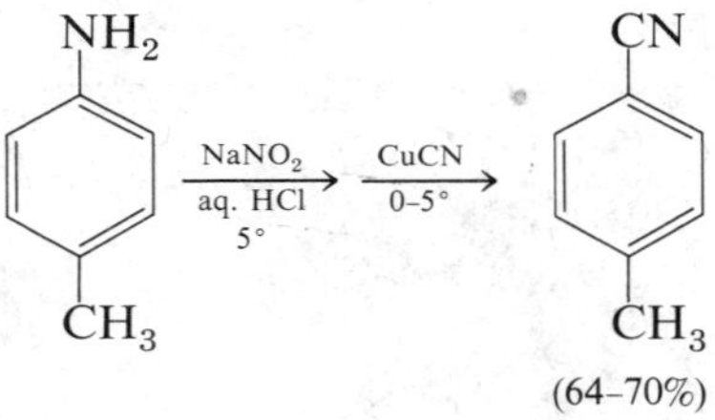

(64–70%)
p-tolunitrile

Note that the decomposition with CuCN occurs even in the cold.

> Cupric chloride is normally obtained as a blue-green hydrate, $CuCl_2 \cdot 2H_2O$. This color is characteristic of many cupric salts. Cupric chloride and bromide are readily soluble in water. Cuprous bromide and chloride are white insoluble powders prepared by reducing an aqueous solution of cupric sulfate and sodium bromide or chloride with sodium bisulfite. The cuprous halide precipitates as a white powder, which is filtered and used directly. On standing in air the white cuprous salts darken by oxidation. Cuprous cyanide is prepared by treating an aqueous suspension of cuprous chloride with sodium cyanide. Cuprous cyanide is also insoluble in water, but dissolves in excess sodium cyanide with formation of a complex, $Cu(CN)_2^-$. This solution is used directly in the Sandmeyer reaction.

The aromatic nitriles prepared by the Sandmeyer reaction can, of course, be hydrolyzed to carboxylic acids, reduced to benzylamines, treated with Grignard

reagents to produce ketones, and so on. Consequently, the diazonium salts provide an entry to a host of aromatic compounds.

EXERCISE: Write equations showing the conversion of benzene into the following compounds by multistep routes involving a Sandmeyer reaction as one step.
(a) chlorobenzene, (b) bromobenzene, (c) benzoic acid, (d) benzylamine, (e) benzamide.

D. *Preparation of Fluoro- and Nitroarenes*

Some diazonium salts are fairly stable and can be isolated and handled. One
such salt is the fluoborate. This salt is prepared by diazotization with sodium
nitrate and fluoboric acid. The diazonium fluoborate usually precipitates and is
filtered. The isolated salt is useful in two significant reactions.

In one reaction a suspension of the diazonium fluoborate in aqueous sodium
nitrite is treated with copper powder. Nitrogen is evolved, and the corresponding
nitro compound is produced.

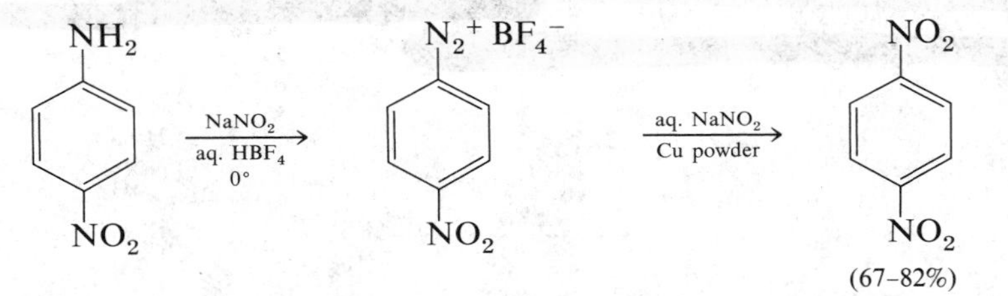

p-nitroaniline p-nitrobenzenediazonium fluoborate   (67–82%) p-dinitrobenzene

This example is similar to the Sandmeyer reaction, which was discussed in the
previous section. Note that in this case copper powder, rather than a cuprous salt,
is used to bring about decomposition of the diazonium ion. This variant is sometimes called the **Gatterman reaction.** The Gatterman modification can also be
used to prepare aryl halides, but it is not as useful for this purpose as is the
Sandmeyer reaction. The diazonium ion method for effecting the conversion
$ArNH_2 \longrightarrow ArNO_2$ is an alternative to oxidation with trifluoroperoxyacetic
acid (Section 24.7.C). Both routes are successful with amines containing electron-
attracting groups, but only the diazonium reaction can be applied to amines with
electron-donating groups.

The isolated diazonium fluoborate can be decomposed thermally either as the
dry salt or in an inert solvent such as THF to provide a satisfactory preparation of
aryl fluorides **(Schiemann reaction).**

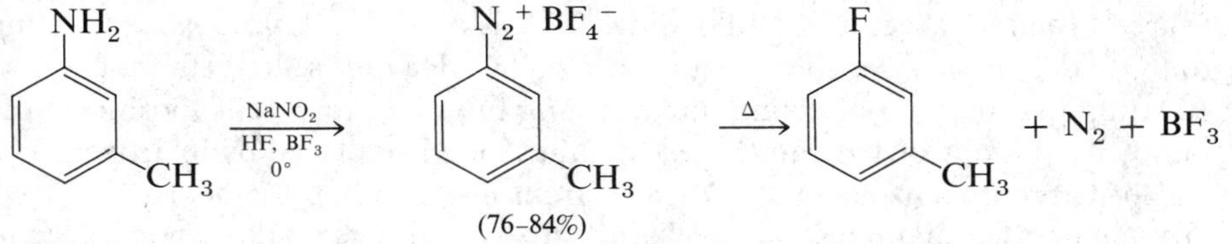

m-toluidine   (76–84%) m-methylbenzenediazonium fluoborate m-fluorotoluene

An improved procedure makes use of hexafluorophosphoric acid, HPF_6. The corresponding diazonium salts are less soluble than the fluoborates and are obtained in generally higher yield. The dry salt is thermally decomposed to form the aryl fluoride.

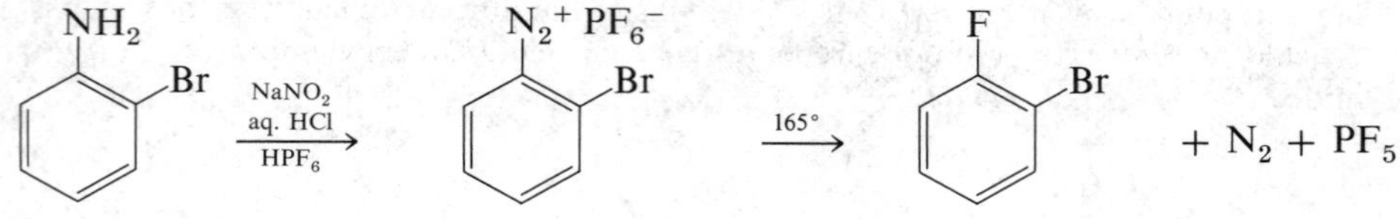

o-bromoaniline   *o*-bromobenzenediazonium   *o*-bromofluorobenzene
hexafluorophosphate

E. *Replacement of the Diazonium Group by Hydrogen*

One of the most useful reactions of diazonium salts is the replacement of the diazonium group by hydrogen. This reaction allows use of the amino group to direct the orientation of an electrophilic aromatic substitution reaction, after which the amino group is removed. The most generally useful reagent for the reaction is hypophosphorous acid, H_3PO_2.

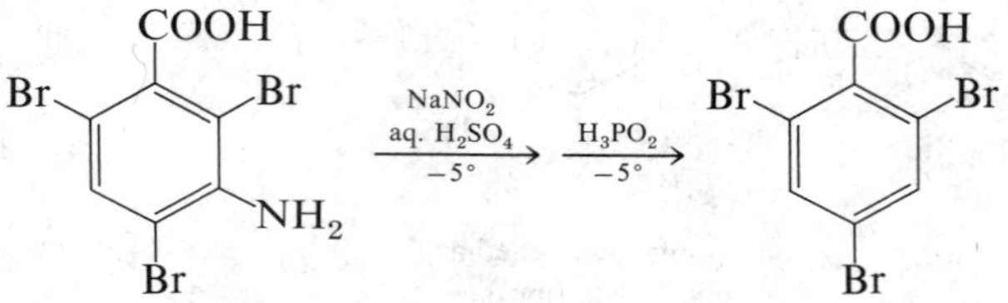

3-amino-2,4,6-tribromo-   2,4,6-tribromo-
benzoic acid   benzoic acid

Hypophosphorous acid is a low-melting (m.p. 26.5°) crystalline solid having a structure with two P—H bonds,

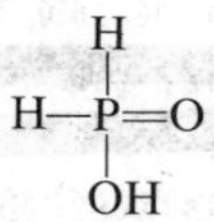

$$H-\overset{\overset{\displaystyle H}{|}}{P}=O$$
$$|$$
$$OH$$

Salts are prepared by treating white phosphorus with alkali or alkaline earth hydroxides. The free acid can be liberated from the water-soluble calcium salt with sulfuric acid. Aqueous solutions are available commercially and can be used directly. The monobasic acid has $pK_a = 1.2$ and is a powerful reducing agent.

In the foregoing example, the amino group is used to direct the facile introduction of three bromines (Section 24.7.D). Since NH_2 is a powerful *ortho,para*-directing group, reaction occurs at the positions indicated. Having served its function of activating the ring and directing the incoming bromines to specific positions, the amino group is diazotized and the diazonium group is replaced by hydrogen. The yield of tribromobenzoic acid is 70–80% from *m*-aminobenzoic acid.

In some cases diamines can be "tetrazotized," and reaction can be accomplished at both diazonium groups. An example is provided by the following preparation of 3,3′-dimethylbiphenyl (biphenyls will be discussed in Section 32.2).

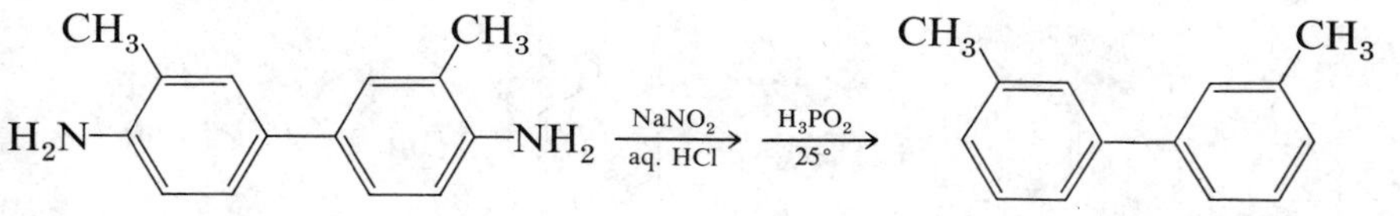

4,4′-diamino-3,3′-dimethylbiphenyl   3,3′-dimethylbiphenyl

EXERCISE: Write equations showing how benzene may be converted into 1,3,5-trichlorobenzene.

F. Arylation Reactions

Arylation provides a convenient preparation of unsymmetrical biaryls in which the diazonium group is replaced by an aromatic ring. Diazotization is carried out in the usual way except that a minimum of water is used. The solution is made basic, and the resulting concentrated aqueous solution is stirred in the cold with a liquid aromatic compound.

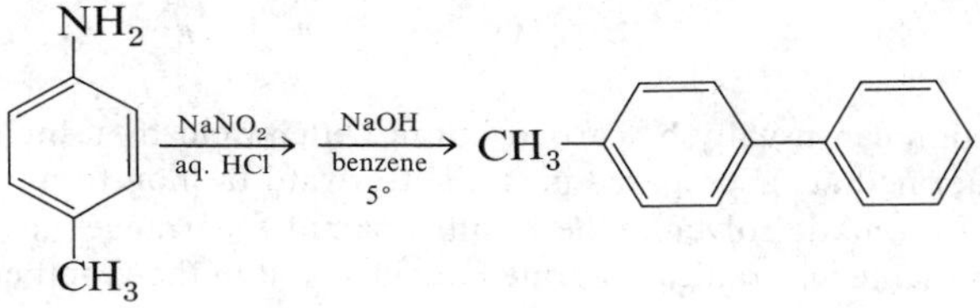

(21%)
4-methylbiphenyl

The reaction is called the **Gomberg–Bachmann** reaction. Yields are generally low, but the starting materials are usually readily available, and there are few other methods for the synthesis of such biaryls (Section 32.2).

The mechanism of the Gomberg-Bachmann reaction involves free radicals. Recall that in basic solution the diazonium salt is in equilibrium with the covalent diazohydroxide (Section 25.5.A). This material is extracted into the organic phase, where homolytic fission of the C—N and N—O bonds occur. The resulting products of this homolysis are an aryl radical, a nitrogen molecule, and a hydroxy radical. The driving force for simultaneous rupture of two bonds is the formation of the highly stable nitrogen molecule.

$$\text{Ar}-\text{N}=\text{N}-\text{OH} \longrightarrow \text{Ar}\cdot + \text{N}_2 + \cdot\text{OH}$$

The aryl radicals produced are, of course, highly reactive intermediates. They react rapidly with the aromatic ring of the solvent.

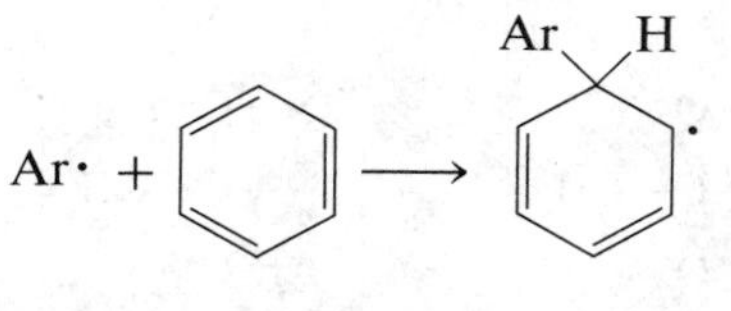

The resulting radical reacts with some other radical in solution to afford the final product. The overall result is substitution of a hydrogen atom of the solvent by the aryl radical corresponding to the beginning arylamine.

$$Ar \overset{H}{\underset{\cdot}{\bigcirc}} + \cdot OH \longrightarrow \overset{Ar}{\bigcirc} + H_2O$$

Although the mechanism involves intermediate free radicals, it does not appear to be a radical chain reaction. The reaction is somewhat different than most free radical reactions in that the concentration of radicals becomes rather high. The typical low yields result from the many alternative reactions that are available to the intermediate radicals.

Because the aromatic substitution involves radicals, the normal orientation rules do not hold; almost all substituents tend to give *ortho* and *para* orientation, and mixtures of products are common. In using this method to prepare an unsymmetrical biaryl, it is best to start with the substituents on the diazonium ring and to keep the ring to be added as simple as possible.

$$\overset{N_2^+}{\bigcirc} \xrightarrow[\text{toluene}]{\text{NaOH}} \bigcirc\!\!-\!\!\bigcirc\!\!-CH_3$$

(66% *o*-, 19% *m*-, 14% *p*-)

The same reaction can usually be carried out in better yield by using an N-nitrosoamide. This intermediate is prepared in straightforward fashion from the amine and is heated in an aromatic solvent. The N-nitrosoamide rearranges to a diazo ester, which forms the same aryl radical intermediate involved in the Gomberg–Bachmann reaction.

$$\overset{NH_2}{\underset{NO_2}{\bigcirc}} \xrightarrow[\text{AcOH}]{(CH_3CO)_2O} \left[\overset{NHCOCH_3}{\underset{NO_2}{\bigcirc}} \right] \xrightarrow[\text{CH}_3\text{COOH}]{N_2O_3} \overset{\overset{NO}{|}}{\underset{NO_2}{\overset{NCOCH_3}{\bigcirc}}} \xrightarrow[\text{benzene}]{NaOOCCH_3}$$

N-nitroso-*m*-nitro-acetanilide

$$\underset{NO_2}{\bigcirc\!\!-\!\!\bigcirc}$$

(56–60%)
3-nitrobiphenyl

The rearrangement of the intermediate nitrosoamide can be regarded as an intramolecular transesterification.

$$\left[\begin{matrix} N\!=\!O \\ | \\ N\!-\!\overset{O}{\underset{\|}{C}}\!-\!CH_3 \\ | \\ Ar \end{matrix} \longleftrightarrow \begin{matrix} N\!-\!O^- \\ | \\ {}^+N\!-\!\overset{O}{\underset{\|}{C}}\!-\!CH_3 \\ | \\ Ar \end{matrix} \right] \longrightarrow \left[\begin{matrix} N\!-\!O \\ | \\ {}^+N\!-\!\overset{O}{\underset{\|}{C}}\!-\!CH_3 \\ | \\ Ar \quad O^- \end{matrix} \right] \longrightarrow$$

$$Ar\!-\!N\!=\!N\!-\!O\!-\!\overset{O}{\underset{\|}{C}}CH_3 \longrightarrow Ar\cdot + N_2 + \cdot\overset{O}{\underset{\|}{O}}CCH_3$$

EXERCISE: 4-Ethylbiphenyl may be prepared by the Gomberg–Bachmann reaction beginning with 4-ethylaniline and benzene or beginning with aniline and ethylbenzene. Write the equations for these two syntheses, and explain why one combination is superior to the other.

G. Diazonium Ions as Electrophiles: Azo Compounds

The diazonium cation bears a resemblance to some other species that are known as intermediates in electrophilic aromatic substitution reactions. Compare

$$Ar-\overset{+}{N}\equiv N: \qquad R-\overset{+}{C}\equiv \overset{..}{O}: \qquad \overset{+}{O}\equiv N:$$

Arenediazonium ions can react as electrophilic reagents in aromatic substitutions, but they are such mild reagents that only the most activated rings can be used. In practice, such reactions are limited primarily to aromatic amines and phenols. This lack of reactivity was already apparent in the chemistry of the diazonium salts already discussed. They do not react with the mild base, water,

$$Ar-\overset{+}{N}\equiv N + H_2O \rightleftharpoons Ar-N=N-\overset{+}{O}H_2$$

and water is a stronger base than most aromatic rings. Note that the central nitrogen is analogous to an ammonium salt and that the reaction with bases that does occur is exclusively at the terminal nitrogen.

$$Ar-\overset{+}{N}\equiv N + {}^-OH \rightleftharpoons Ar-N=N-OH$$

Reaction with primary aromatic amines occurs in a related manner, with the amino nitrogen as a base, to give a derivative of a **triazene**.

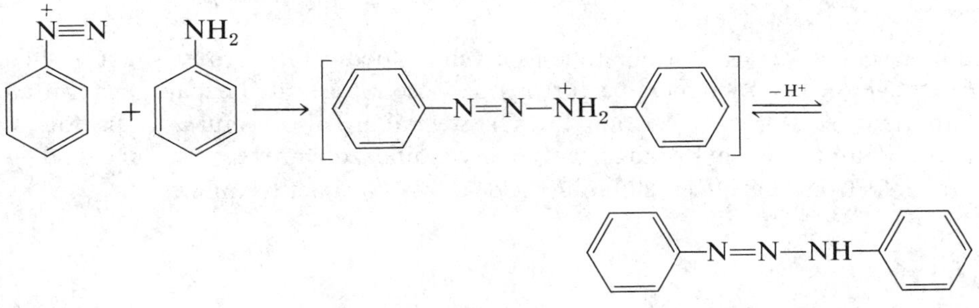

(82–85%)
diazoaminobenzene
1,3-diphenyltriazene

Diazoaminobenzene reacts with excess aniline and acid to give the aromatic substitution product *p*-aminoazobenzene.

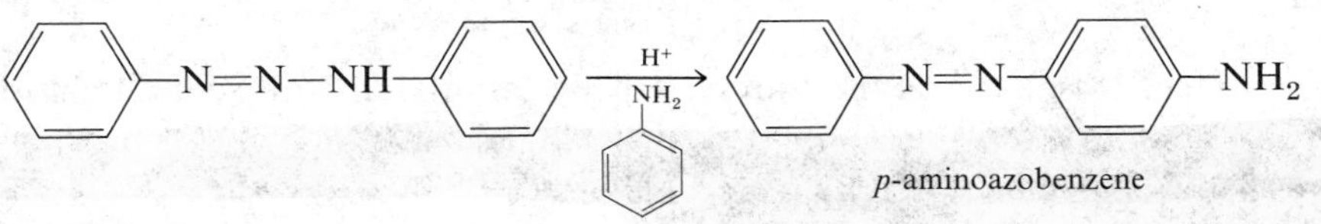

p-aminoazobenzene

This reaction is usually assumed to involve a reversal of diazoaminobenzene formation followed by the slower reaction of benzenediazonium cation with the *para* position of aniline.

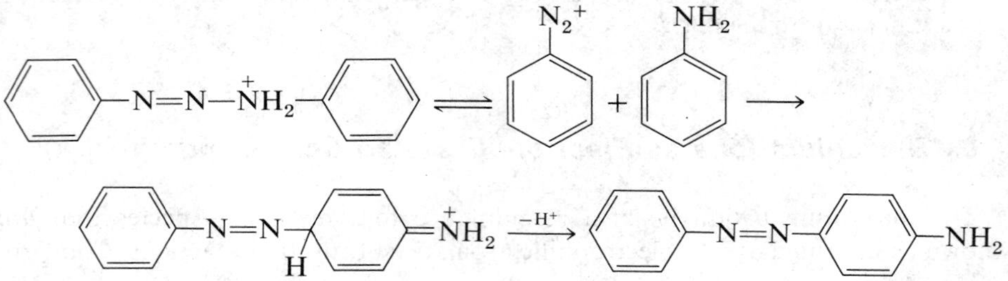

This mechanism incorporates the frequently encountered distinction between kinetically and thermodynamically controlled reactions and explains many features of the reaction; however, the reaction also produces a variety of minor by-products that suggest a more complex reaction mechanism.

With N,N-dimethylaniline the triazene is so unstable that it is not observed, and the net observed reaction is that of aromatic substitution.

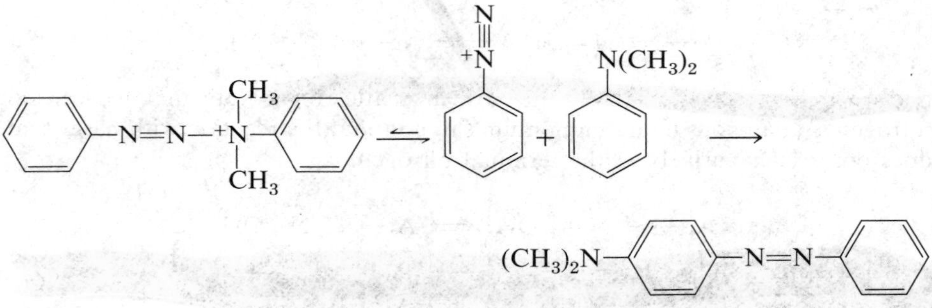

p-dimethylaminoazobenzene
butter yellow

Butter yellow was used at one time as a yellow food coloring, but as a suspected carcinogen it is no longer used for this purpose. Substituted azoarenes form an important class of dyes (Section 33.2.C). Several are also useful as indicators in the laboratory. Methyl orange, *p*-dimethylaminoazobenzene-*p'*-sulfonic acid, is prepared from diazotized sulfanilic acid and N,N-dimethylaniline.

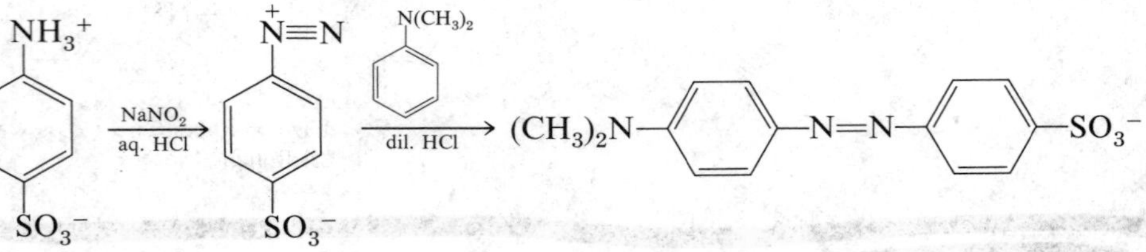

sulfanilic acid

methyl orange

The product is isolated as the sodium salt by salting out with sodium chloride. Note that these so-called "coupling reactions" of diazonium salts occur almost exclusively at the *para* position. Reaction occurs generally at the *ortho* position only when the *para* position is blocked.

The azo group has nitrogen lone pairs and is expected to show basic properties. However, each of these lone pairs is in an approximately sp^2-orbital and is less basic than an amino lone pair in which the orbital has less s-character. Furthermore, the adjacent nitrogen further reduces the basicity. As a result the azo group in azobenzene itself is a rather weak base; the pK_a of the protonated compound is -2.5.

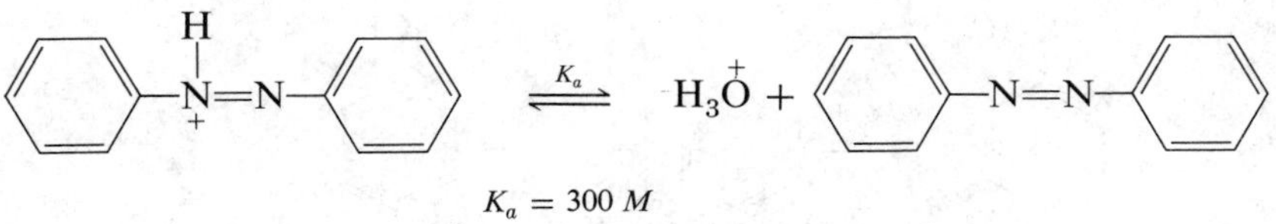

$$K_a = 300 \ M$$

Azobenzene itself is protonated only in rather strong acid. Methyl orange has a pK_a of 3.5; this value refers to the protonated azo group, not to the dimethylamino or sulfonic acid groups. At pHs much above 3.5, methyl orange is in the yellow azo form. At pHs lower than 3.5 it is present instead in the red protonated form.

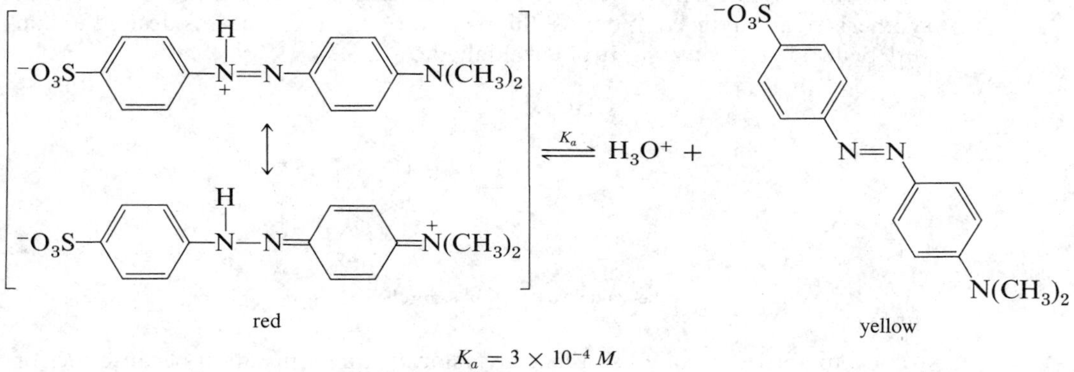

red   yellow

$$K_a = 3 \times 10^{-4} \ M$$

Note that the azo-protonated form of methyl orange is stabilized by the p-N(CH$_3$)$_2$ group. This stabilization renders the protonated form less acidic than protonated azobenzene.

H. Synthetic Utility of Arenediazonium Salts

We saw in Chapter 23 that the nitro group may be introduced with ease into a wide variety of aromatic compounds. We have also seen that the nitro group may be reduced to the amino group by several reliable methods (Section 24.6.C) and that the resulting aromatic amines may be converted into arenediazonium salts (Section 24.7.B). In this section we have learned how the diazonium group may be replaced by —OH, —I, —SH, —Cl, —Br, —F, —NO$_2$, —F, —H, and —Ar. Thus, *any of these functions can be introduced into an aromatic ring at any position that can be nitrated.*

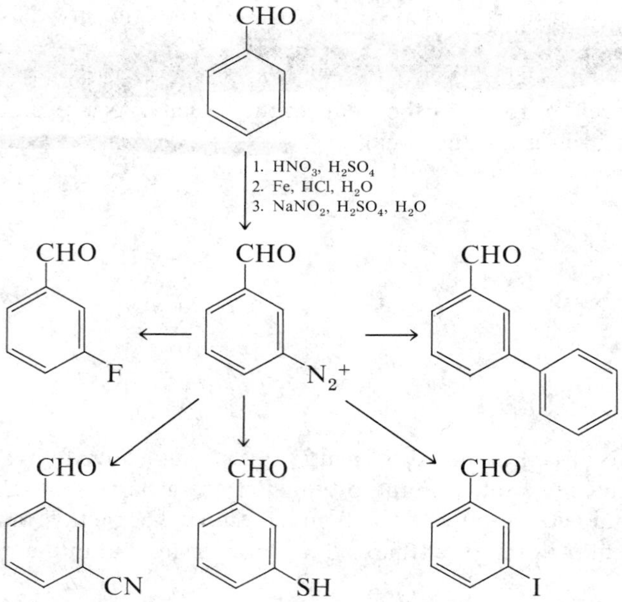

CHO
[benzaldehyde structure]

1. HNO₃, H₂SO₄
2. Fe, HCl, H₂O
3. NaNO₂, H₂SO₄, H₂O

[reaction scheme with central diazonium CHO/N₂⁺ compound radiating to products: m-F-benzaldehyde, biphenyl-carbaldehyde, m-CN-benzaldehyde, m-SH-benzaldehyde, m-I-benzaldehyde]

But diazonium chemistry can be used in another advantageous manner for the synthesis of aromatic compounds. Suppose, for example, that it is desired to convert benzaldehyde into *p*-nitrobenzaldehyde.

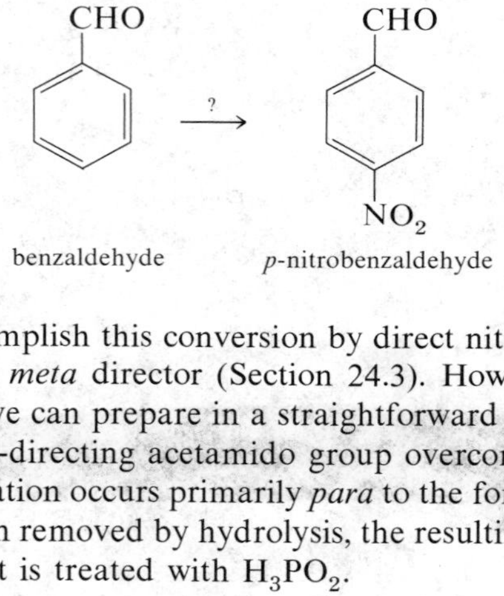

[benzaldehyde → p-nitrobenzaldehyde with "?" over arrow]

benzaldehyde   *p*-nitrobenzaldehyde

We are unable to accomplish this conversion by direct nitration because the formyl group is a strong *meta* director (Section 24.3). However, in *m*-acetamido-benzaldehyde, which we can prepare in a straightforward synthesis from benzaldehyde, the *ortho,para*-directing acetamido group overcomes the *meta*-directing formyl group, and nitration occurs primarily *para* to the formyl group. The amide protecting group is then removed by hydrolysis, the resulting amine is diazotized, and the diazonium salt is treated with H_3PO_2.

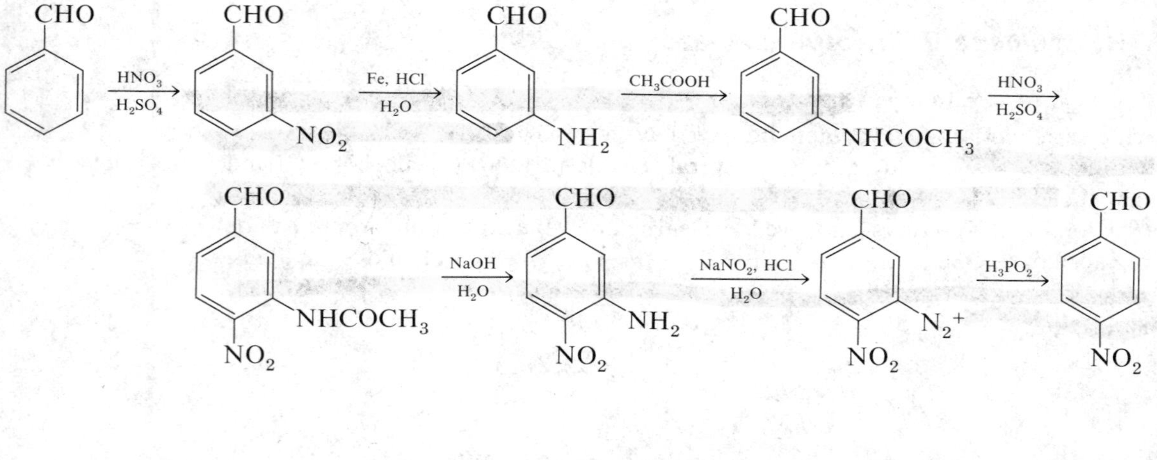

[reaction sequence:]
CHO $\xrightarrow[H_2SO_4]{HNO_3}$ CHO(m-NO₂) $\xrightarrow[H_2O]{Fe, HCl}$ CHO(m-NH₂) $\xrightarrow{CH_3COOH}$ CHO(m-NHCOCH₃) $\xrightarrow[H_2SO_4]{HNO_3}$

CHO(NHCOCH₃, NO₂) $\xrightarrow[H_2O]{NaOH}$ CHO(NH₂, NO₂) $\xrightarrow[H_2O]{NaNO_2, HCl}$ CHO(N₂⁺, NO₂) $\xrightarrow{H_3PO_2}$ CHO(p-NO₂)

With these examples, we see the central position that the versatile diazonium function holds in aromatic synthesis.

PROBLEMS

1. What is the principal product obtained from *p*-toluenediazonium cation with each of the following reagents?

(a) I^-
(b) CuCN
(c) OH^- (cold)
(d) H_2O (hot)
(e) CuBr
(f) $NaNO_2$, Cu powder
(g) aq. NaOH, benzene, 5°
(h) (1) $NaBF_4$; (2) heat
(i) H_3PO_2
(j) CuCl
(k) N,N-diethylaniline
(l) (1) HPF_6; (2) heat

2. Each of the following compounds is a significant dye intermediate. Give a practical laboratory preparation for each starting with benzene or toluene.

(a)
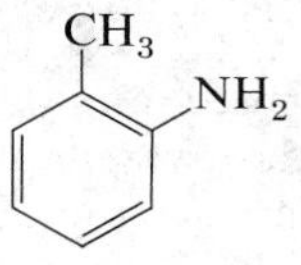
(b)
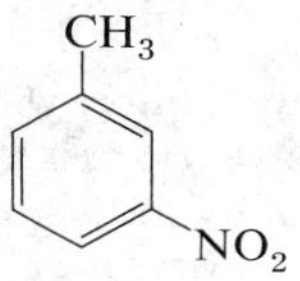
(c)
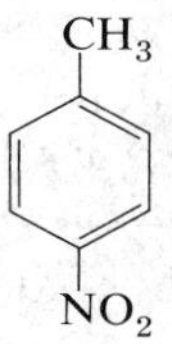

(d)
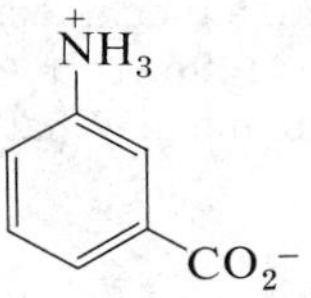
(e)
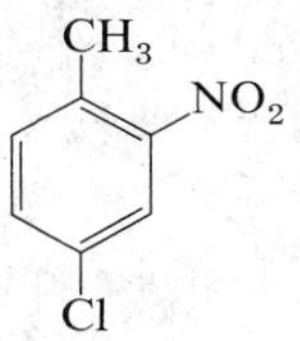
(f)
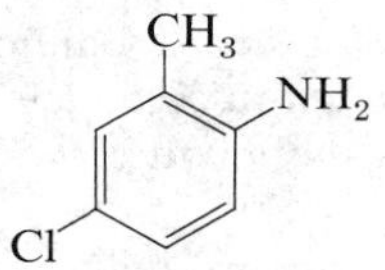

(g)
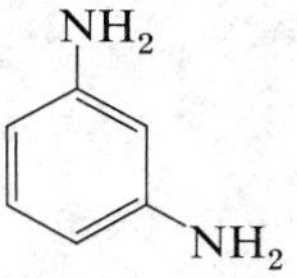
(h)
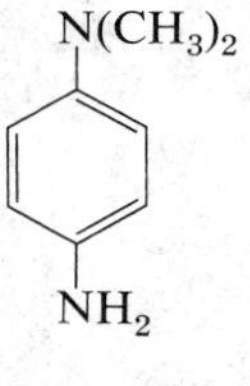
(i)
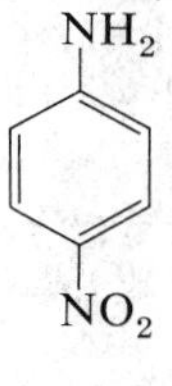

3. Show how each of the following conversions can be accomplished in a practical manner.

(a)
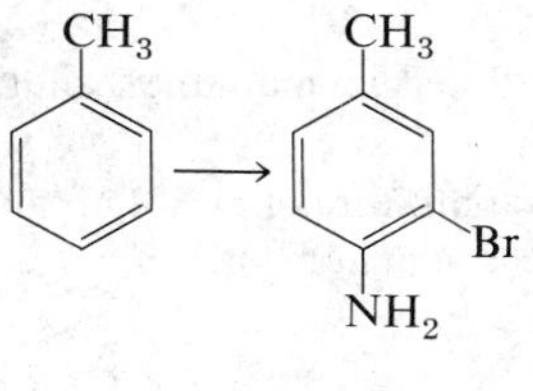
(b)
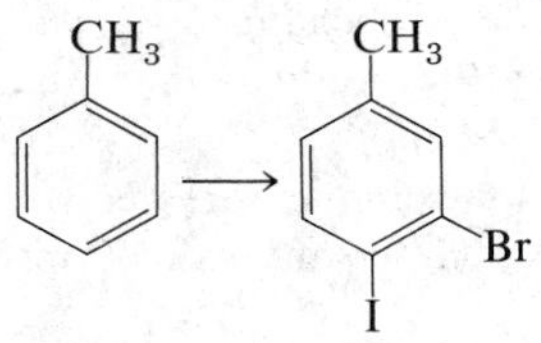

(c)
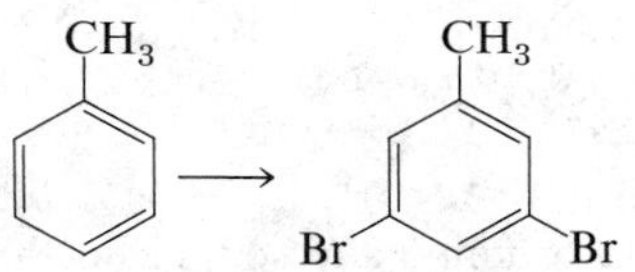
(d)
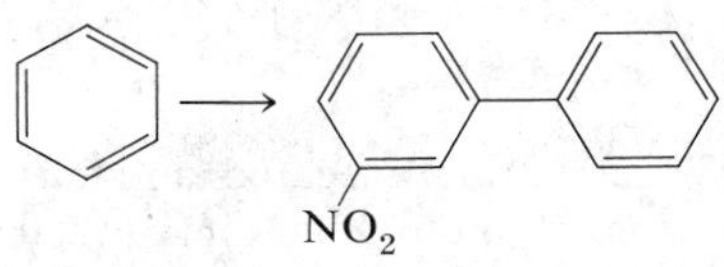

(e)

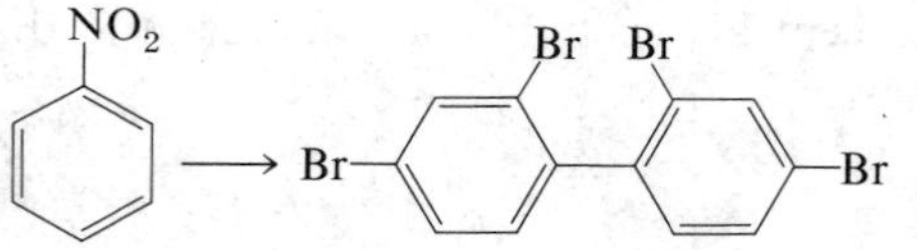

(f)

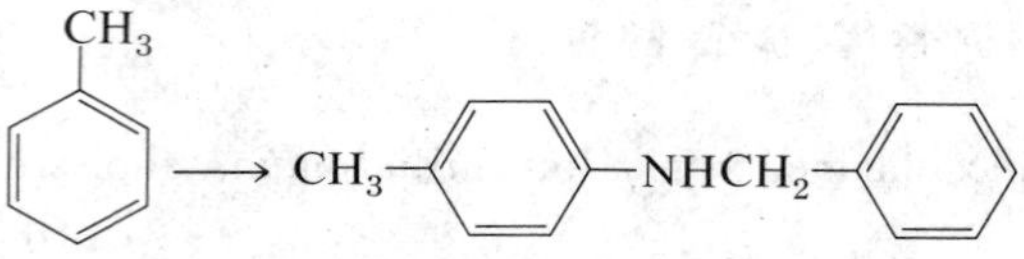

(g)

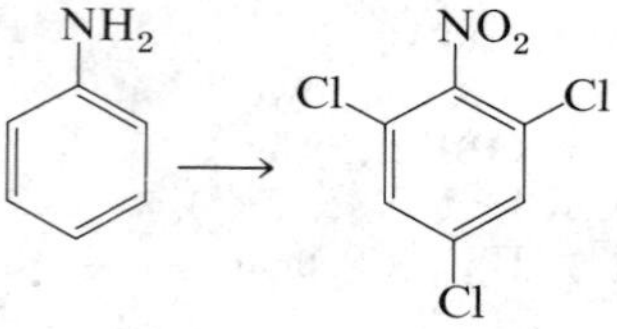

(h)

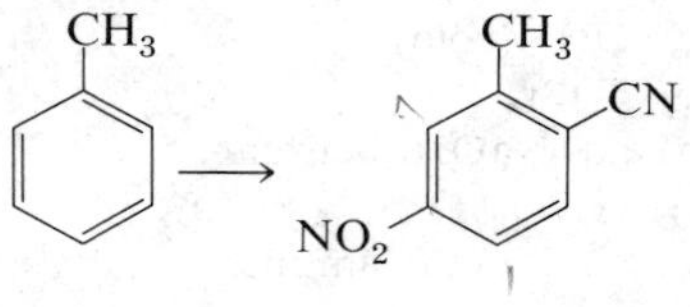

4. A small amount of methyl orange is added to a solution containing equimolar amounts of acetic acid and sodium acetate. Is this solution yellow or red?

5. A diazonium salt prepared from p-nitroaniline, when decomposed in nitrobenzene, gives a 69% yield of 4,4'-dinitrobiphenyl. That is, reaction of the aryl radical formed from the diazonium salt occurs primarily at the *para* position of the nitrobenzene. Give a reasonable explanation of this orientation behavior.

6. A student tried to prepare p-bromocumene by diazotizing p-aminocumene with a mixture of aqueous sodium nitrite and hydrochloric acid at $0°$ followed by reaction with hot cuprous bromide, but a mixture of products was obtained. What was the nature of the mixture?

7. Give the principal reduction product from m-nitrotoluene under each of the following conditions.
(a) Zn, alc. NaOH   (b) Pt/H$_2$
(c) Zn, aq. NH$_4$Cl   (d) SnCl$_2$, HCl
(e) H$_2$NNH$_2$, Ru/C, alc. KOH   (f) As$_2$O$_3$, aq. NaOH

8. When α-diazoketones are irradiated or heated in aqueous solution, the product obtained is a carboxylic acid.

$$\underset{\substack{\|\\ O}}{R\overset{O}{C}CHN_2} \xrightarrow[H_2O]{h\nu \text{ or } \Delta} RCH_2COOH$$

(a) Propose a mechanism for the transformation.
(b) Predict the product when diazoacetone is irradiated in methanol solution.

9. When cyclohexanone is treated with diazomethane, a mixture of cycloheptanone and methylenecyclohexane oxide is produced. Propose a mechanism.

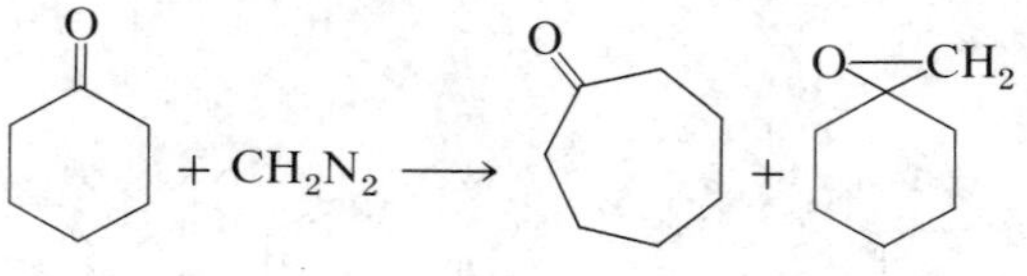

$$+ CH_2N_2 \longrightarrow \quad + $$

10. When cyclohexanecarboxamide is treated with bromine and sodium methoxide in methanol, the product obtained is methyl N-cyclohexylcarbamate.

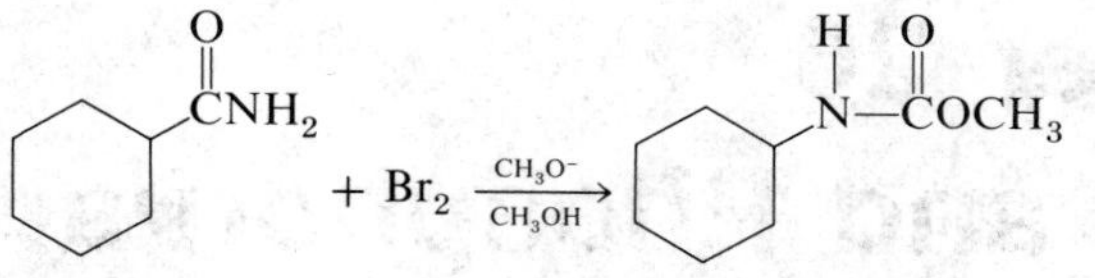

Rationalize with a plausible mechanism.

11. When ethyl N-cyclohexylcarbamate is refluxed with 1 M potassium hydroxide in methanol for 100 hr, the only product obtained is the methyl ester in 95% yield.

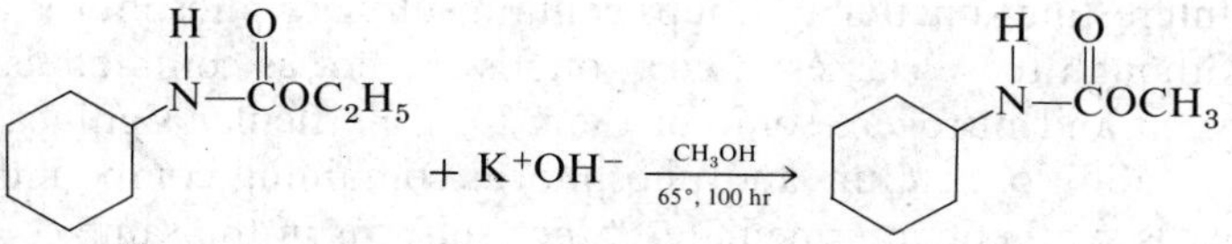

Explain why no cyclohexylamine is produced.

12. Thermal decomposition of N-nitrosoacetanilide in carbon tetrachloride yields chlorobenzene. Write a reasonable mechanism for this reaction.

Chapter 26
Sulfur and Phosphorus Compounds

Several interesting functional groups contain sulfur or phosphorus as a central element. Although these classes of compounds are not as common as those containing oxygen and nitrogen, some of them have particular significance in biochemistry. In addition, sulfur- and phosphorus-containing compounds have important uses as synthetic intermediates. Since sulfur is in the same column of the periodic table as oxygen, we expect to find similarities in the chemistry of analogous sulfur and oxygen functions. Similarly, we expect comparable chemistry of analogous nitrogen and phosphorus compounds. To some extent this correspondence is observed. However, the third-period elements are both less electronegative and more polarizable than their second-period relatives and these differences result in significant quantitative differences in chemistry. Furthermore, sulfur and phosphorus have higher oxidation states available and can form some compounds that have no counterparts in oxygen and nitrogen chemistry. We have already encountered some of these functional groups (for example, sulfonate esters, phosphorus ylides). In this chapter we shall consider all of the important sulfur and phosphorus functions in a systematic fashion.

26.1
Thiols and Sulfides

Thiols, RSH, and sulfides, R_2S, bear an obvious relationship to alcohols and ethers. In the IUPAC system the alkane name is combined with the suffix **-thiol** in the same way that alcohols are named as alkanols. One difference is that the final **-e** of the alkane name is retained in naming thiols.

$$\underset{\text{ethanethiol}}{CH_3CH_2SH} \qquad \underset{\text{2-butanethiol}}{CH_3CH_2\overset{\overset{\displaystyle CH_3}{|}}{C}HSH}$$

A common name is that of alkyl mercaptan, analogous to alkyl alcohol, but the mercaptan nomenclature is falling into disuse in favor of IUPAC systematic names.

Sulfides are commonly named in a manner analogous to the common nomenclature of ethers. The two alkyl group names are followed by the word **sulfide.**

$$\underset{\text{dimethyl sulfide}}{CH_3SCH_3} \qquad \underset{\text{ethyl propyl sulfide}}{CH_3CH_2SCH_2CH_2CH_3}$$

In the IUPAC system sulfides are named as alkylthioalkanes. The prefix **alkylthio-** is analogous to **alkoxy-** and refers to a group RS—. As with ethers, the larger of the two alkyl groups is taken as the stem.

$$CH_3SCH_2\overset{\overset{\displaystyle CH_3}{|}}{C}HCH_2CH_3$$

2-methyl-1-(methylthio)butane

$$CH_3\overset{\overset{\displaystyle CH_3}{|}}{C}HSCH_2CH_2CH_2CH_3$$

1-isopropylthiobutane

The IUPAC system for naming sulfides is only used in practice for complex structures that are not conveniently named as dialkyl sulfides.

The principal structural differences between methanethiol and methanol are that the C—S bond is about 0.4 Å longer than the C—O bond and the C—S—H bond angle is relatively more acute than the C—O—H angle. In thiols, as in H_2S itself, sulfur uses orbitals that are rich in p-character for bonding. The barrier to rotation about the C—S bond is identical to that about the C—O bond in methanol, 1.1 kcal mole^{-1}.

Bond Distances, Å		Bond Angles, deg	
C—H	1.10	H—C—H	110.2
C—S	1.82	H—C—S	108
S—H	1.33	C—S—H	100.3

A similar structure is found for dimethyl sulfide. Again, the C—S—C angle is relatively small, corresponding to C—S bonds in which sulfur uses a high percentage of its $3p$-orbitals.

Bond Distances, Å		Bond Angles, deg	
C—H	1.09	H—C—H	109.5
C—S	1.80	H—C—S	106.7
		C—S—C	98.9

Thiols have boiling points that are almost normal for their molecular weight; they generally boil somewhat higher than the corresponding chlorides. For example, ethanethiol has b.p. 37° compared to ethyl chloride, b.p. 13°. Thiols are stronger acids than alcohols, just as H_2S is a stronger acid than water. The pK_a of ethanethiol, 10.60, indicates that the compound is completely converted to its anion by hydroxide ion.

$$C_2H_5SH + OH^- \rightleftharpoons C_2H_5S^- + H_2O$$
$$\text{p}K_a\ 10.6 \qquad\qquad\qquad \text{p}K_a\ 15.7$$

Although the thiols are more acidic than alcohols, sulfur is less electronegative than oxygen. Hence, thiols have lower dipole moments (CH_3SH, $\mu = 1.26$ D; CH_3OH, $\mu = 1.71$ D) and hydrogen bonding between thiol molecules is much weaker than for alcohols. However, hydrogen-bonding from the acidic SH protons to water oxygen is significant, and the thiols have some water solubility.

The most impressive property of thiols is their odor. Their intensely disagreeable odors discourage use as laboratory reagents. Thiols contribute to the characteristic odors of skunk and onion. Two methanethiol esters are known to give the urine of many persons a distinctive odor after eating asparagus. The nose is more sensitive than any laboratory instrument in detecting ethanethiol; one part in 50 billion parts of air can be detected. Small amounts are included in heating gas, which is otherwise almost odorless, as an effective warning device against leaks. The lower molecular weight sulfides have similarly repugnant odors.

Thioaldehydes and thioketones are also known. The lower molecular weight compounds are red liquids with intensely obnoxious odors. Although they may be prepared in a monomeric form, they rapidly polymerize to give cyclic trimers.

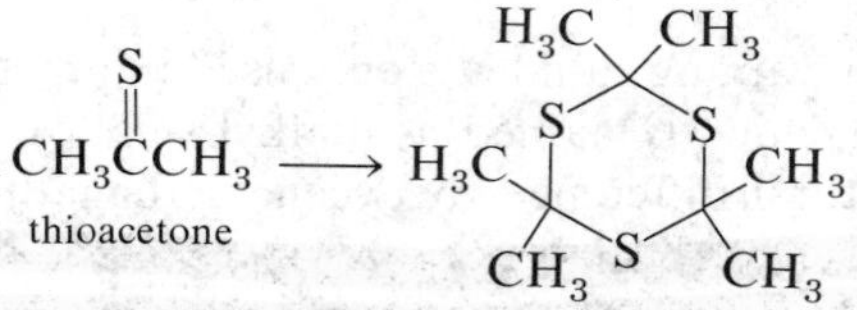

$$\underset{\text{thioacetone}}{CH_3\overset{\displaystyle S}{\overset{\|}{C}}CH_3} \longrightarrow$$

Thioacids, RCOSH, and thioesters, RCOSR′, are more common.

$$\underset{\text{thioacetic acid}}{CH_3\overset{\displaystyle O}{\overset{\|}{C}}SH} \qquad \underset{\textit{t}\text{-butyl thiopropanoate}}{CH_3CH_2\overset{\displaystyle O}{\overset{\|}{C}}SC(CH_3)_3}$$

Note that the stable tautomeric form of thioacids is the one with C=O and S—H groups and not the alternative structure with C=S and O—H groups. A particularly important example of a thioester is acetyl coenzyme A, "acetyl CoA," an important intermediate used by nature in the biosynthesis of numerous organic compounds (for examples, see Sections 19.9 and 34.7).

EXERCISE: Give the structure corresponding to each of the following names. (a) ethyl isopropyl sulfide; (b) butyl mercaptan; (c) 3-methylthiooctane; (d) diphenyl sulfide; (e) 3-pentanethiol; (f) thioacetaldehyde; (g) ethyl thioacetate.

26.2
Preparation of Thiols and Sulfides

Thiols can be prepared from alkyl halides by displacement with hydrosulfide ion, HS⁻, in ethanol solution.

$$CH_3(CH_2)_{16}CH_2I + Na^+SH^- \xrightarrow{C_2H_5OH} \underset{\text{1-octadecanethiol}}{CH_3(CH_2)_{16}CH_2SH} + Na^+Br^-$$

In preparing thiols by this method it is necessary to employ a large excess of hydrosulfide because of the equilibrium

$$HS^- + RSH \rightleftharpoons H_2S + RS^-$$

The thiol anion produced by this equilibrium is itself a good nucleophile and can react with the alkyl halide to give the corresponding sulfide.

$$RS^- + RBr \longrightarrow RSR + Br^-$$

The use of a large excess of hydrosulfide makes its reaction with the alkyl halide more probable and maximizes the yield of thiol.

For this reason HS⁻ in such displacement reactions has been almost exclusively replaced by thiourea.

$$\underset{\text{thiourea}}{H_2N-\overset{\displaystyle S}{\overset{\|}{C}}-NH_2}$$

Thiourea, a commercially available solid, m.p. 178°, is soluble in water and alcohols. The sulfur is nucleophilic in S_N2 displacement reactions on alkyl halides. The product salt is readily hydrolyzed to the alkanethiol.

$$CH_3(CH_2)_{11}Br + H_2N\overset{\overset{\displaystyle S}{\|}}{C}-NH_2 \xrightarrow[\Delta]{95\% \text{ ethanol}} CH_3(CH_2)_{11}S-\overset{\overset{\displaystyle \overset{+}{N}H_2}{\|}}{C}NH_2 \ Br^-$$
n-dodecyl bromide

$$CH_3(CH_2)_{11}S-\overset{\overset{\displaystyle \overset{+}{N}H_2}{\|}}{C}NH_2 \ Br^- \xrightarrow[\Delta]{\text{aq. NaOH}} CH_3(CH_2)_{11}SH + H_2N\overset{\overset{\displaystyle O}{\|}}{C}NH_2$$
(79–83%)
1-dodecanethiol

This route avoids formation of sulfides. The other product of the hydrolysis is urea, H_2NCONH_2 (Section 25.2).

Thiols can also be prepared by reaction of Grignard reagents with sulfur.

$$(CH_3)_3CMgBr + S_8 \longrightarrow (CH_3)_3CSMgBr \xrightarrow{HCl} (CH_3)_3CSH + MgBrCl$$
t-butyl mercaptan
2-methyl-2-propanethiol

Both symmetrical and unsymmetrical sulfides can be prepared by S_N2 displacement of alkylthio anions on alkyl halides or sulfonates.

$$CH_3CH_2SH + OH^- \xrightarrow{H_2O} CH_3CH_2S^-$$

$$CH_3CH_2S^- + (CH_3)_2CHCH_2Br \longrightarrow (CH_3)_2CHCH_2SCH_2CH_3$$
(95%)
ethyl isobutyl sulfide

This general method for preparing dialkyl sulfides is directly analogous to the Williamson ether synthesis (Section 10.10.A).

EXERCISE: Show how the following conversions may be accomplished.

(a) $(CH_3)_3CCH_2Br \longrightarrow (CH_3)_3CCH_2SH$

(b) $CH_3CH_2CH_2OH \longrightarrow (CH_3CH_2CH_2)_2S$

26.3
Reactions of Thiols and Sulfides

One of the characteristic similarities between the thiol chemistry and alcohol chemistry is that both classes of compounds are weak acids. However, just as HCl is a stronger acid in aqueous solution than HF, RSH compounds are substantially more acidic than their ROH counterparts. The pK_a of ethanethiol, 10.6, tells us that the compound is half-ionized at pH 10.6, which corresponds roughly to the pH of a 5% sodium carbonate solution. Thus, unlike alcohols, thiols may be converted essentially quantitatively into the corresponding anions in aqueous solution.

$$CH_3CH_2SH \qquad CH_3CH_2OH$$
ethanethiol ethanol
$pK_a = 10.6$ $pK_a = 15.7$

Thiols are readily oxidized to **disulfides.** The disulfide bond is weak and is easily reduced to give the thiol.

$$2 \; RSH \underset{\text{reduction}}{\overset{\text{oxidation}}{\rightleftharpoons}} RS\text{—}SR$$
$$\text{a disulfide}$$

Mild oxidizing agents suffice for the oxidation. Iodine is often used for this purpose.

$$2 \; CH_3CH_2SH + I_2 \longrightarrow CH_3CH_2SSCH_2CH_3 + 2 \; HI$$
$$\text{diethyl disulfide}$$

A commonly used reducing agent for regeneration of the thiol is lithium in liquid ammonia.

$$CH_3CH_2SSCH_2CH_3 \xrightarrow[\text{NH}_3]{\text{Li}} \xrightarrow[\text{(work-up)}]{H_3O^+} 2 \; CH_3CH_2SH$$

The facile thiol–disulfide redox system is especially important in biological systems. The disulfide link occurs in proteins and hormones, and the redox reaction itself plays an important role in molecular biology. It has been suggested that this reaction may be involved in the mechanism of memory in the brain. An interesting natural product is thioctic acid, which contains a cyclic disulfide link. Thioctic acid is an essential substance for the growth of various organisms and has been used for the treatment of liver disease and as an antidote for the toxin of poisonous *Amanita* mushrooms.

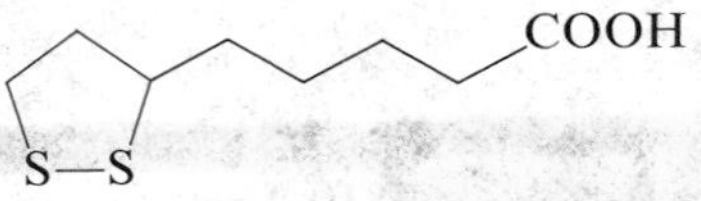

thioctic acid

The sulfur–sulfur bond of disulfides is susceptible to cleavage by nucleophiles.

$$N\!:^- + \; RS\text{—}SR \longrightarrow N\text{—}SR + RS\!:^-$$

A synthetically useful version of this reaction is found in the thiolation of ketone and ester enolates.

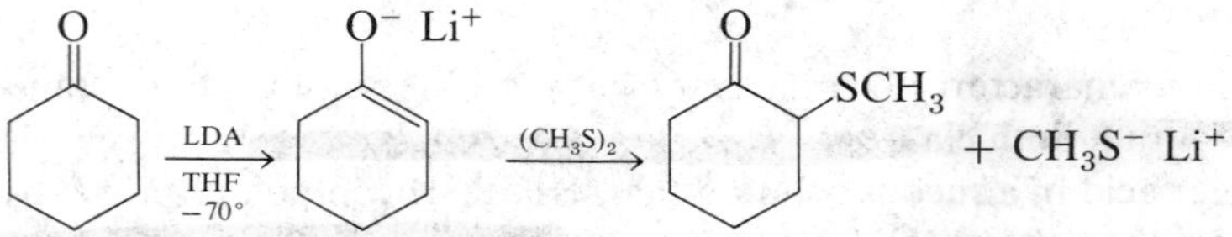

Strong oxidants such as potassium permanganate or hot nitric acid oxidize thiols or disulfides to the corresponding sulfonic acids. If a mixture of chlorine and nitric acid is used as the oxidant, the corresponding sulfonyl chloride is obtained. Sulfonyl chlorides have the same relationship to sulfonic acids as acyl halides have to carboxylic acids. Thus they are hydrolyzed to sulfonic acids by water.

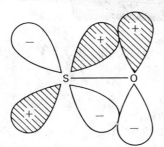

di-*o*-nitrophenyl disulfide   *o*-nitrobenzenesulfonyl chloride   *o*-nitrobenzenesulfonic acid

Sulfides are also easily oxidized. The initial oxidation product is a **sulfoxide.**
Further oxidation of the sulfoxide yields a **sulfone.**

$$CH_3SCH_3 + H_2O_2 \longrightarrow CH_3-\overset{\displaystyle O^-}{\underset{}{S^+}}-CH_3 \xrightarrow{RCO_3H} CH_3-\overset{\displaystyle O^-}{\underset{\displaystyle O^-}{S^{2+}}}-CH_3$$

or   *or*

$$CH_3-\overset{\displaystyle O}{S}-CH_3 \qquad\qquad CH_3-\overset{\displaystyle O}{\underset{\displaystyle O}{S}}-CH_3$$

dimethyl sulfoxide   dimethyl sulfone
(DMSO)

> Sulfur compounds such as these are frequently represented for convenience as having
> S=O double bonds and an expanded octet around sulfur. However, the sulfur–
> oxygen bond is not a double bond in the same sense as carbon–carbon or carbon–
> oxygen double bonds in which the "second bond" is viewed as arising from π-overlap
> of atomic *p*-orbitals. Because sulfur has *d*-orbitals of rather low energy, it has long
> been speculated that sulfur–oxygen "double" bonds result from overlap of an oxygen
> *p*-orbital with a sulfur *d*-orbital.
>
> However, sophisticated quantum-mechanical calculations do not confirm this hy-
> pothesis. The exact nature of such sulfur "double bonds" remains unclear. Because of
> their highly polar nature, classical coulombic attraction of the partially positive sulfur
> for the partially negative oxygen provides much of the bonding. It also appears that
> the high polarizability of the sulfur valence electrons is involved.

We have previously encountered dimethyl sulfoxide as an important solvent in
the class of dipolar aprotic solvents (Section 8.9). It is a relatively inexpensive
colorless liquid that is miscible with water in all proportions. It is prepared industrially by the NO_2-catalyzed air-oxidation of dimethyl sulfide, a by-product produced in tonnage quantities in the sulfite pulping process for paper manufacture.

$$CH_3SCH_3 + \tfrac{1}{2} O_2 \xrightarrow{NO_2} \overset{\overset{\displaystyle O}{\|}}{CH_3SCH_3}$$

dimethyl sulfide   dimethyl sulfoxide (DMSO)

Dimethyl sulfoxide owes its utility to the fact that it readily dissolves many inorganic salts as well as most organic compounds. Thus it is an excellent solvent for reactions such as displacements on alkyl halides. However, the peculiar solvent properties of DMSO result in a particular hazard that is associated with its use. The compound diffuses through the skin almost instantaneously and carries any solutes along with it. For this reason, one should always wear impermeable gloves when working with DMSO solutions.

Note that sulfoxides are similar in electronic structure to amine oxides (Section 24.7.C). Like amine oxides, sulfoxides undergo thermal elimination to give alkenes (Section 24.7.E). The process is a *syn* elimination (Section 19.11).

$$CH_3(CH_2)_4CH_2\overset{\overset{\displaystyle SCH_3}{|}}{C}HCN \xrightarrow{H_2O_2} CH_3(CH_2)_4CH_2\overset{\overset{\displaystyle O}{\overset{\displaystyle \diagdown S}{\diagdown}}CH_3}{|}{C}HCN \xrightarrow{150°} CH_3(CH_2)_4CH=CHCN + [CH_3SOH]$$
(96%)

The sulfur-containing by-product in this reaction is a **sulfenic acid.** Sulfenic acids are unstable and undergo disproportionation to the corresponding **sulfinic acid** and disulfide, which are the actual by-products of a sulfoxide elimination.

$$3\ CH_3SOH \longrightarrow CH_3SSCH_3 + CH_3SO_2H + H_2O$$

methanesulfenic acid   dimethyl disulfide   methanesulfinic acid

The sulfur in dialkyl sulfides is nucleophilic. Consequently, sulfides react readily with alkyl halides by the normal S_N2 mechanism to produce **trialkylsulfonium salts,** which are usually hygroscopic solids.

$$(CH_3)_2\overset{..}{S}: + CH_3 \overset{..}{-}I \longrightarrow (CH_3)_2\overset{..}{S}{}^+\!\!-CH_3\ I^-$$

trimethylsulfonium iodide

Like other S_N2 displacements, the reaction works best with primary halides.

$$(CH_3)_2CHCH_2Br + (CH_3)_2S \longrightarrow (CH_3)_2CHCH_2\overset{+}{S}(CH_3)_2\ Br^-$$

isobutyldimethylsulfonium bromide

When trialkylsulfonium salts are heated, the reaction reverses. Halide ion acts as the nucleophile, and the dialkyl sulfide is the leaving group. The driving force for reaction is vaporization of the volatile products.

$$Br:^- + CH_3\!-\!S^+(CH_3)_2 \xrightarrow{\Delta} CH_3Br\uparrow + (CH_3)_2S\uparrow$$

Nature makes extensive use of this S_N2 reaction. The compound *S*-adenosylmethionine is a methylating agent in biochemical S_N2 reactions, which are catalyzed by appropriate enzymes. It can be regarded as the body's equivalent of methyl iodide.

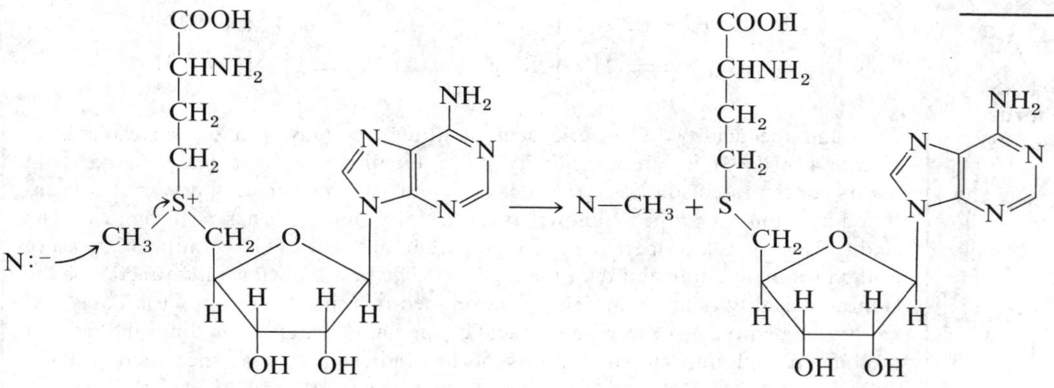

$$\text{S-adenosylmethionine} \longrightarrow \text{N—CH}_3 + \text{S}$$

S-adenosylmethionine

Like 1,2-and 1,3-diols, the analogous 1,2- and 1,3-dithiols react with aldehydes and ketones under conditions of acid catalysis to give cyclic thioacetals and thioketals.

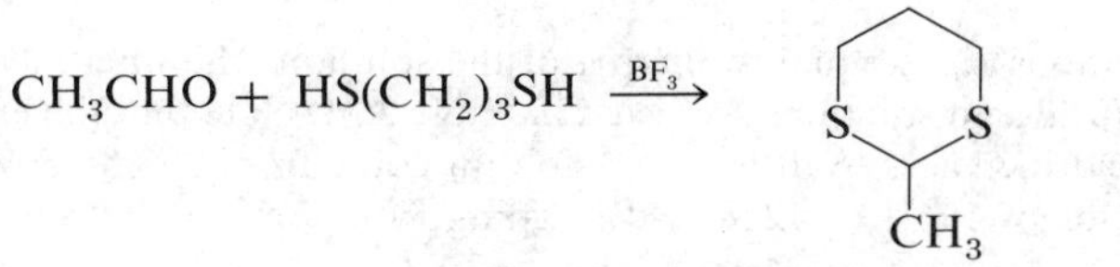

$$\text{CH}_3\text{CHO} + \text{HS(CH}_2)_3\text{SH} \xrightarrow{\text{BF}_3} \underset{\underset{\text{CH}_3}{}}{\overset{}{\text{S}\quad\text{S}}} \qquad (26\text{-}1)$$

1,3-propanedithiol   2-methyl-1,3-dithiane

The thioacetals that are prepared in this way using 1,3-propanedithiol are useful as synthetic reagents (Section 26.8). The carbon–sulfur bond can be reductively cleaved by certain reagents, the most common of which is Raney nickel. The products of the reaction are the hydrocarbons formed by hydrogenolysis of each C—S bond.

$$\text{R—S—R}' + \text{H}_2 \xrightarrow{\text{Raney Ni}} \text{RH} + \text{R}'\text{H}$$

Desulfurization of thioacetals and thioketals provides a method for net deoxygenation of aldehydes and ketones (Section 13.8.F).

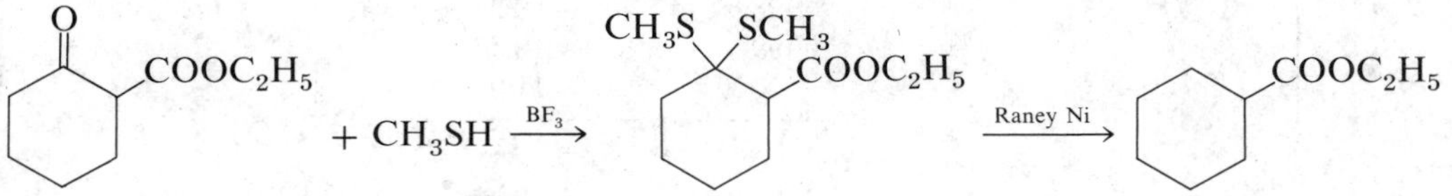

$$\text{(cyclohexanone-COOC}_2\text{H}_5\text{)} + \text{CH}_3\text{SH} \xrightarrow{\text{BF}_3} \text{(CH}_3\text{S, SCH}_3\text{-COOC}_2\text{H}_5\text{)} \xrightarrow{\text{Raney Ni}} \text{(cyclohexane-COOC}_2\text{H}_5\text{)}$$

(75%)

EXERCISE: (a) Write equations illustrating the following reaction sequences beginning with 1-butanethiol.
(i) 1. NaOH, CH$_3$Br; 2. H$_2$O$_2$; 3. 150°   (ii) 1. I$_2$, KI; 2. Li, NH$_3$
(iii) 1. NaOH, CH$_3$CH$_2$I; 2. CH$_3$I
(b) Write a reasonable reaction mechanism for equation (26-1).

26.4
Sulfate Esters

Sulfuric acid, H$_2$SO$_4$, is a strong dibasic inorganic acid with p$K_1 \approx -5$ and p$K_2 = 1.99$.

$$H_2SO_4 \rightleftharpoons H^+ + HSO_4^- \qquad K_a \approx 1.4 \times 10^5 \; M$$

$$HSO_4^- \rightleftharpoons H^+ + SO_4^{2-} \qquad K_a = 1.0 \times 10^{-2} \; M$$

Although the acidity of sulfuric acid in dilute aqueous solution corresponds to $K_1 \approx 1.4 \times 10^5 \; M$, its effective acidity or "protonating power" increases markedly in highly concentrated solutions. In order to quantify the protonating power of concentrated solutions, a property known as an "acidity function" has been defined. The acidity function is a property of a given medium that provides a quantitative measure of the proton donating ability of the medium. The best-known acidity function is the Hammett acidity function, which was developed by using a series of weak bases that are protonated only in exceedingly "acidic" media. For a given medium the ratio of protonated and unprotonated forms of the indicator base is measured, usually spectrophotometrically. The Hammett acidity function, H_0, is defined as the negative logarithm of the ratio of protonated and unprotonated species.

$$H_0 = -\log \frac{[BH^+]}{[B]}$$

In dilute solutions H_0 is equal to the pH of the solution. Hammett acidity functions for some sulfuric acid solutions are listed in Table 26.1. Note that the effective acidity of sulfuric acid increases by $10^{1.59}$ or 39-fold in going from 30% to 50% sulfuric acid and by a factor of $10^{3.11}$ or 1288-fold in going from 70% to 90% sulfuric acid.

TABLE 26.1
H_0 for Sulfuric Acid–Water Mixtures

% H_2SO_4	H_0
5	−0.02
10	−0.43
30	−1.82
50	−3.41
70	−5.92
90	−9.03
95	−9.73
98	−10.27
99	−10.57
100	−11.94

Both mono- and diesters of sulfuric acid are known. Like sulfuric acid itself, the sulfuric acid esters are often considered as resonance hybrids involving an expanded sulfur octet.

$$\left[RO-\overset{O^-}{\underset{O^-}{\overset{|}{\underset{|}{S^{2+}}}}}-OH \longleftrightarrow RO-\overset{O^-}{\underset{O}{\overset{|}{\underset{\|}{S^+}}}}-OH \longleftrightarrow RO-\overset{O}{\underset{O^-}{\overset{\|}{\underset{|}{S^+}}}}-OH \longleftrightarrow RO-\overset{O}{\underset{O}{\overset{\|}{\underset{\|}{S}}}}-OH \right]$$

For convenience, we shall only use the Kekulé structure represented as having two S=O bonds.

Diesters of sulfuric acid are named by combining the alkyl group name(s) with the word "sulfate" just as though they were salts of sulfuric acid.

$$\begin{array}{cc} \overset{\displaystyle O}{\underset{\displaystyle O}{\|}} & \overset{\displaystyle O}{\underset{\displaystyle O}{\|}} \\ CH_3OSOCH_3 & CH_3OSOCH(CH_3)_2 \end{array}$$

dimethyl sulfate   methyl isopropyl sulfate

Monoesters are named as alkylsulfuric acids.

$$\begin{array}{cc} \overset{\displaystyle O}{\underset{\displaystyle O}{\|}} & \overset{\displaystyle O}{\underset{\displaystyle O}{\|}} \\ CH_3OSOH & CH_3CH_2OSOH \end{array}$$

methylsulfuric acid   ethylsulfuric acid

Dialkyl sulfates are highly polar compounds and generally have rather high boiling points. Their water solubility is surprisingly low (Table 26.2). Alkylsulfuric acids are approximately as acidic as sulfuric acid itself.

$$CH_3OSOH \rightleftharpoons H^+ + CH_3OSO^-$$

They readily form inorganic salts, which are named as metal alkyl sulfates.

$$CH_3OSO_3^- \; Na^+ \qquad (CH_3CH_2OSO_3^-)_2 \; Ba^{2+}$$

sodium methyl sulfate   barium ethyl sulfate

The monoesters are rarely encountered as reagents in organic chemistry. Ethylsulfuric acid is an intermediate in the industrial hydration of ethylene to give ethanol.

$$CH_2{=}CH_2 + H_2SO_4 \xrightarrow{0°} CH_3CH_2OSO_3H \xrightarrow{H_2O} CH_3CH_2OH + H_2SO_4$$

Dimethyl sulfate and diethyl sulfate are encountered rather more frequently as organic reagents. Both diesters are readily available, inexpensive materials. They are prepared commercially from the corresponding alcohol and sulfuric acid.

$$2 \; ROH + H_2SO_4 \longrightarrow ROSO_2OR + H_2O$$

Since alkylsulfuric acids are such strong acids, the alkyl sulfate ion is a good leaving group, roughly comparable to iodide ion. Hence, dimethyl sulfate and diethyl sulfate readily enter into S_N2 displacement processes (Section 8.10).

$$(CH_3)_2CHCH_2O^- + CH_3OSO_2OCH_3 \longrightarrow (CH_3)_2CHCH_2OCH_3 + CH_3OSO_3^-$$

They are used in organic chemistry mainly for this purpose—as alkylating agents.

TABLE 26.2
Physical Properties of Dialkyl Sulfates

	Melting Point, °C	Boiling Point, °C	Solubility in H_2O, g/100 ml
$CH_3OSO_2OCH_3$	−27	188	2.8
$CH_3CH_2OSO_2OCH_2CH_3$	−25	210	very low

26.5
Sulfonic Acids

Recall that **sulfonic acids** contain the functional group —SO_3H joined to carbon. They are named as alkane**sulfonic acids** or arene**sulfonic acids.**

$$CH_3\overset{\displaystyle O}{\underset{\displaystyle O}{S}}OH \qquad (CH_3)_2CH\overset{\displaystyle O}{\underset{\displaystyle O}{S}}OH \qquad CF_3\overset{\displaystyle O}{\underset{\displaystyle O}{S}}OH$$

methanesulfonic acid   2-propanesulfonic acid   trifluoromethanesulfonic acid

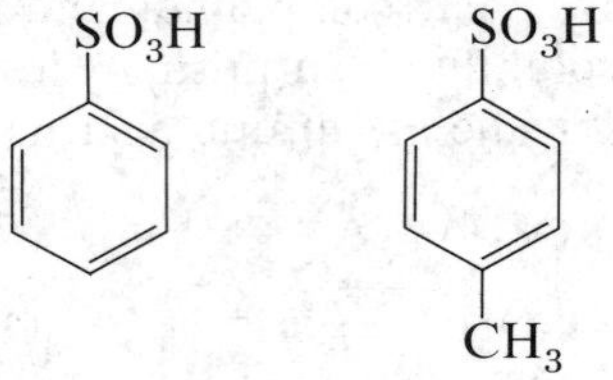

benzenesulfonic
acid   *p*-toluenesulfonic
acid

Sulfonic acids are strong acids, as strong as typical mineral acids.

$$CH_3CH_2\overset{\displaystyle O}{\underset{\displaystyle O}{S}}OH \rightleftharpoons H^+ + CH_3CH_2\overset{\displaystyle O}{\underset{\displaystyle O}{S}}O^-$$

Because of the inductive effect of the fluorines, trifluoromethanesulfonic acid is much more acidic and, indeed, is one of the strongest acids known.

A. *Alkanesulfonic Acids*

Alkanesulfonic acids can be prepared by nucleophilic displacement of alkyl halides with bisulfite ion, an ambident anion. Because of the greater nucleophilicity of sulfur, alkylation occurs primarily on sulfur rather than on oxygen. The initial product is the salt of the sulfonic acid, which is converted into the sulfonic acid by treatment with strong acid.

$$(CH_3)_2CHCH_2CH_2Br + HO-\overset{\displaystyle O}{\underset{\displaystyle \cdot\cdot}{S}}-O^-\,Na^+ \xrightarrow{H_2O} (CH_3)_2CHCH_2CH_2\overset{\displaystyle O}{\underset{\displaystyle O}{S}}O^-\,Na^+$$

(96%)
sodium 3-methylbutanesulfonate

Sodium salts of α-hydroxysulfonic acids are obtained by the addition of sodium bisulfite to aldehydes and some ketones.

$$CH_3\overset{\overset{\displaystyle O}{\|}}{C}H + NaHSO_3 \longrightarrow CH_3\overset{\overset{\displaystyle OH}{|}}{C}H\overset{\overset{\displaystyle O}{\|}}{\underset{\underset{\displaystyle O}{\|}}{S}}O^- Na^+$$

(89%)
sodium 1-hydroxyethanesulfonate

$$CH_3\overset{\overset{\displaystyle O}{\|}}{C}CH_3 + NaHSO_3 \longrightarrow (CH_3)_2\overset{\overset{\displaystyle OH}{|}}{C}SO_3^- Na^+$$

(59%)
sodium 2-hydroxy-2-propanesulfonate

Sulfonic acid esters are best prepared via the sulfonyl chlorides, which are obtained from the sodium sulfonate by treatment with phosphorus pentachloride (PCl$_5$) or thionyl chloride (SOCl$_2$).

$$CH_3CH_2\overset{\overset{\displaystyle O}{\|}}{\underset{\underset{\displaystyle O}{\|}}{S}}O^- Na^+ + PCl_5 \longrightarrow CH_3CH_2\overset{\overset{\displaystyle O}{\|}}{\underset{\underset{\displaystyle O}{\|}}{S}}Cl + POCl_3 + NaCl$$

ethanesulfonyl chloride

$$CH_3CH_2\overset{\overset{\displaystyle O}{\|}}{\underset{\underset{\displaystyle O}{\|}}{S}}Cl + CH_3O^- Na^+ \longrightarrow CH_3CH_2\overset{\overset{\displaystyle O}{\|}}{\underset{\underset{\displaystyle O}{\|}}{S}}OCH_3 + NaCl$$

methyl ethanesulfonate

As with the alkyl sulfates, the alkanesulfonates are potent alkylating agents because the sulfonate ion is a reactive leaving group. One class of alkanesulfonates in common use is the esters of methanesulfonic acid, which are prepared from methanesulfonyl chloride ("mesyl chloride"), an inexpensive commercial material. Methanesulfonates, frequently called "mesylates," are used in substitution and elimination processes in the same way as alkyl halides.

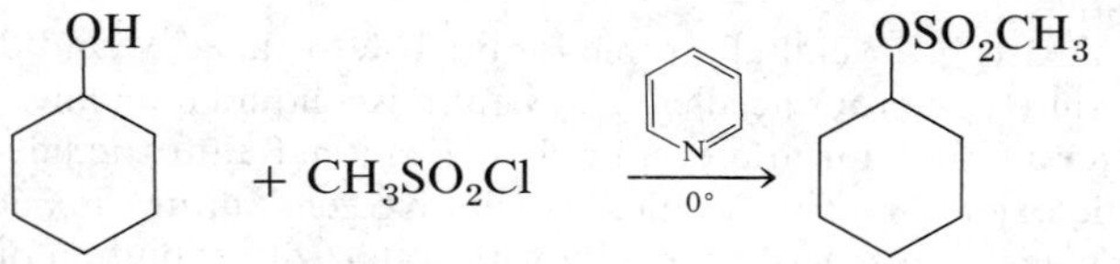

methanesulfonyl chloride   cyclohexyl
"mesyl chloride"   methanesulfonate

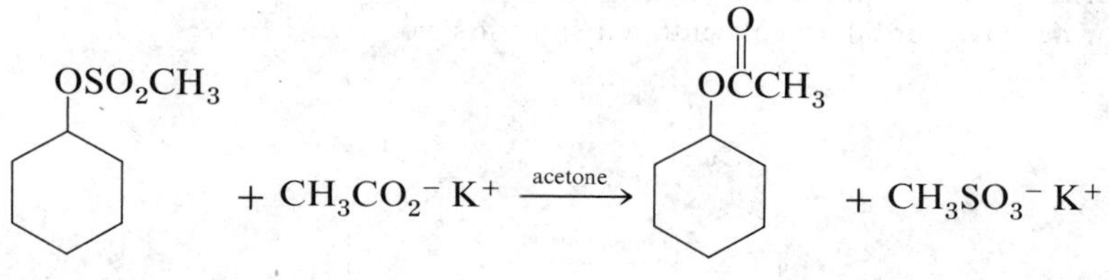

EXERCISE: Write equations showing how propyl propanesulfonate can be prepared from organic compounds that do not contain sulfur.

B. *Arenesulfonic Acids*

Aromatic sulfonic acids are more common than the aliphatic acids because of their availability through electrophilic sulfonation reactions.

1. ELECTROPHILIC AROMATIC SULFONATION. In the reactions of alkenes with sulfuric acid, the acid acts primarily as a protonating reagent to produce a carbocation that reacts with any nucleophile present (Sections 11.6.B and 11.6.F). We have seen that benzene itself undergoes protonation in strong sulfuric acid (Section 23.2). However, unless such a reaction is followed by means of a hydrogen isotope, it remains an *invisible* reaction.

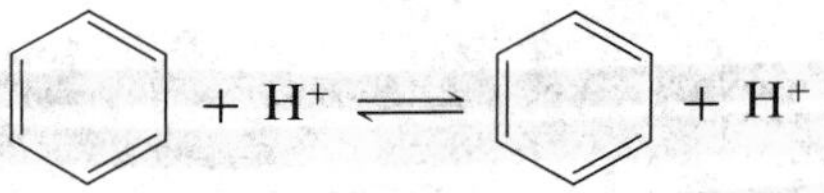

With strong sulfuric acid, double bonds can react with the sulfur trioxide present. Such reactions of alkenes frequently result in oxidation of the organic material with concomitant reduction of sulfur trioxide to sulfur dioxide. Such reactions are not usually useful reactions of alkenes, although some exceptional cases do exist. The reaction of benzene with sulfur trioxide is a useful and important reaction. Sulfur trioxide is an electrophilic reagent, and it reacts with benzene to give benzenesulfonic acid, the product of a sulfonation reaction. The reaction is usually carried out with a solution of sulfur trioxide in sulfuric acid, known as fuming sulfuric acid.

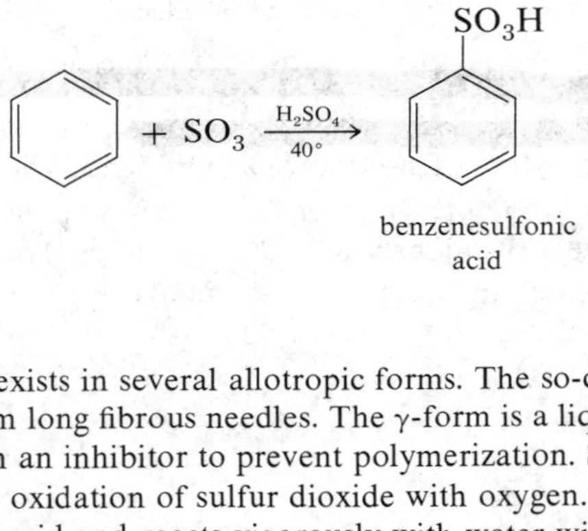

benzenesulfonic
acid

Sulfur trioxide, SO_3, exists in several allotropic forms. The so-called α- and β-forms are polymers that form long fibrous needles. The γ-form is a liquid monomer available commercially with an inhibitor to prevent polymerization. Sulfur trioxide is prepared by the catalytic oxidation of sulfur dioxide with oxygen. Sulfur trioxide is the anhydride of sulfuric acid and reacts vigorously with water with evolution of much heat. The reaction with heavy water, D_2O, is used to prepare D_2SO_4. Sulfuric acid is prepared commercially by dissolving sulfur trioxide in sulfuric acid to produce "fuming sulfuric acid." Commercial fuming sulfuric acid contains 7–8% of SO_3. Dilution with water gives ordinary concentrated sulfuric acid.

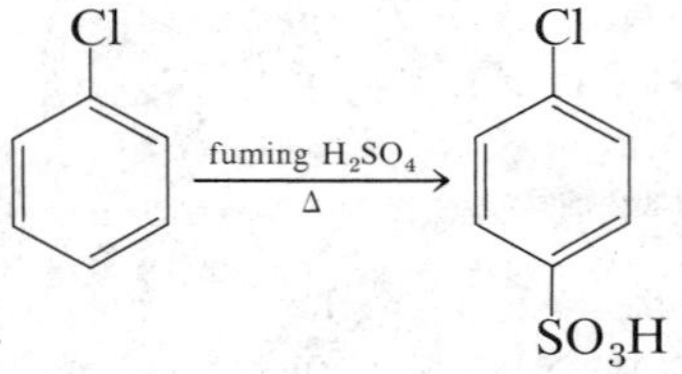

| (43%) | (4%) | (53%) |

Note that milder reaction conditions suffice to bring about sulfonation of toluene, which is more reactive than either chlorobenzene or benzene owing to the electron-donating effect of the methyl group. The isomer distribution often depends on the exact experimental conditions; for example, the product composition from toluene at 100° is 13% *ortho*, 8% *meta*, and 79% *para*. The sulfonation reaction is reversible, and the product depends on whether the reaction conditions favor kinetic or thermodynamic control. In the sulfonation of toluene at low temperature, the reaction product is the product of kinetic control; that is, the product composition reflects relative energies of transition states. At higher temperatures the reverse reaction has a significant rate, and the reaction takes on the aspects of an equilibrium.

$$\text{Ar—H} + SO_3 \rightleftharpoons \text{Ar—SO}_3\text{H}$$

important at high
temperatures

The sulfonic group is a rather bulky group, and steric hindrance interaction with *ortho* substituents is significant. At equilibrium the relatively unhindered *p*-toluenesulfonic acid dominates over *o*-toluenesulfonic acid. Such steric-hindrance effects are much less evident at the transition state for sulfonation; hence, *o*- and *p*-toluenesulfonic acids are formed at comparable rates.

According to the principle of microscopic reversibility (page 107), the back reaction must be the exact reverse of the forward reaction. The forward reaction is a reaction with sulfur trioxide to form a dipolar neutral intermediate that loses a proton to form the arenesulfonate ion.

Consequently, the reverse reaction involves reaction by a proton at a ring carbon of the sulfonate ion. This reverse reaction is faster for a more hindered *ortho*-substituted sulfonic acid because steric congestion effects are relieved in the dipolar intermediate.

The reversal of sulfonation can be carried out by heating the sulfonic acid in dilute aqueous sulfuric acid. In this way the sulfonic acid group can serve as a protecting group to direct aromatic substitution into other positions. An example of this strategy is provided by the following preparation of pure *o*-bromophenol.

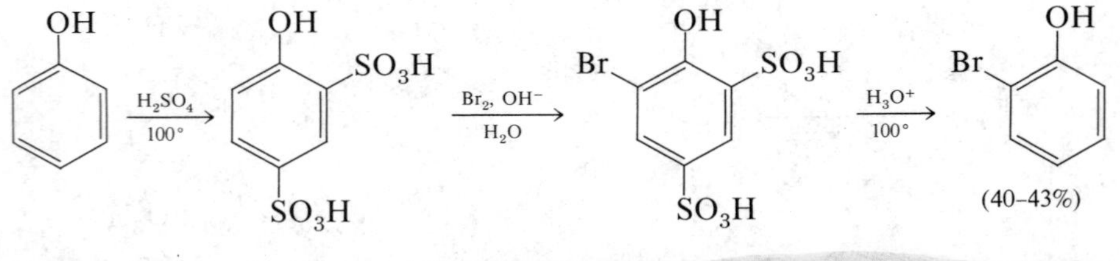

The sulfonic acid function is deactivating and strongly *meta*-directing in electrophilic aromatic substitution reactions. This is easy to understand in light of the Lewis structure of benzenesulfonic acid, which has a doubly positive sulfur adjacent to the ring. Benzene can be disulfonated by the use of hot 20% fuming sulfuric acid (sulfuric acid containing 20% sulfur trioxide).

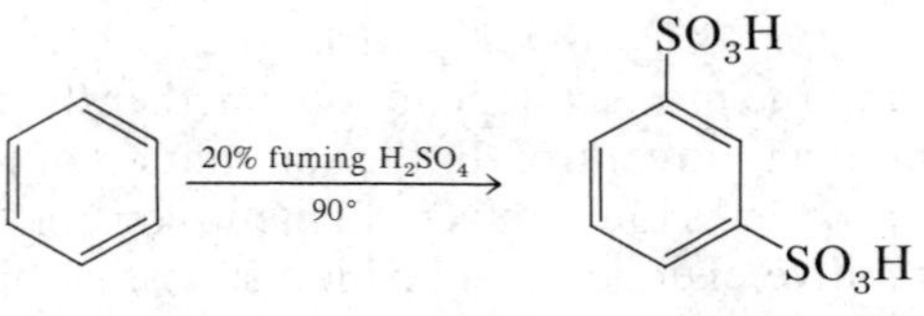

(90%)
m-benzenedisulfonic acid

As was shown by the foregoing example of sulfonation of phenol, aromatic amines and phenols undergo sulfonation readily. With aniline sulfonation occurs first on the nitrogen; the initial product must be heated to give the ring-sulfonated product.

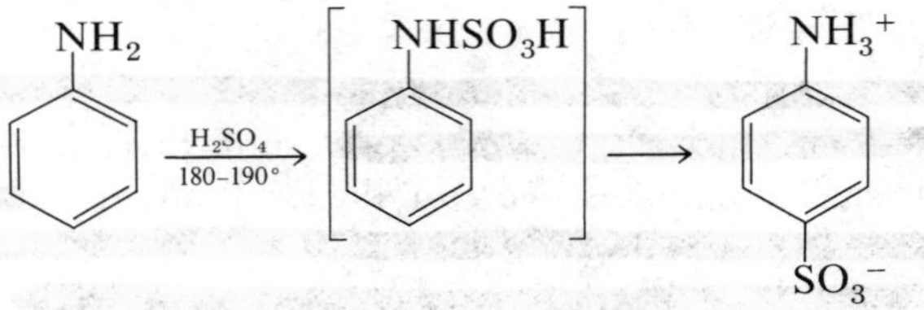

sulfanilic acid

Sulfanilic acid contains an acidic and a basic group in the same molecule and exists in the zwitterionic or inner-salt form, as shown.

Arenesulfonyl chlorides may be prepared by sulfonation of aromatic compounds using chlorosulfonic acid, $ClSO_3H$. An example is chlorosulfonation of acetanilide.

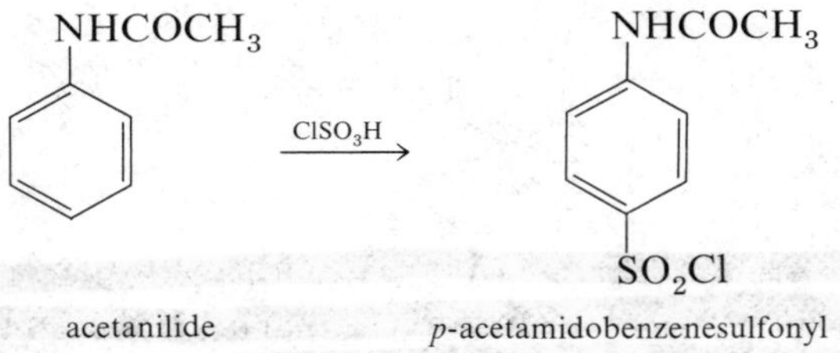

acetanilide   *p*-acetamidobenzenesulfonyl chloride

Arenesulfonic acids are strong acids, about as strong as hydrochloric acid. They are completely dissociated in aqueous solution; indeed, the resulting solubility presents problems in isolations. Consequently, the products of sulfonation reactions are usually isolated as salts. The sodium salts, like all sodium salts, are water-soluble, but they are generally not as soluble as sodium sulfate or sodium chloride. The solubility products of sodium arenesulfonates are sufficiently low that these salts can be "salted out" by addition of more soluble sodium salts to an aqueous solution.

$$ArSO_3^- + Na^+ \rightleftharpoons ArSO_3^- Na^+\downarrow$$

The sodium salts of benzenesulfonic acid containing a long alkyl side chain behave as detergents. The sulfonate end is hydrophilic and dissolves in water. The alkane end is hydrophobic and fat-soluble, and the combination serves to emulsify fatty materials (Section 18.4.D). At one time the alkane side chain was made by the carbocation polymerization of propylene to give a tetrameric olefin that was used to alkylate benzene; this alkylate was then sulfonated to give the product, which was widely used in many common household detergents.

$$CH_3CH{=}CH_2 \longrightarrow CH_3\overset{\overset{\displaystyle CH_3}{|}}{C}H(CH_2\overset{\overset{\displaystyle CH_3}{|}}{C}H)_2CH{=}\overset{\overset{\displaystyle CH_3}{|}}{C}H \xrightarrow[\text{benzene}]{AlCl_3}$$

$$CH_3\overset{\overset{\displaystyle CH}{|}}{C}H(CH_2\overset{\overset{\displaystyle CH_3}{|}}{C}H)_2CH_2\overset{\overset{\displaystyle CH_3}{|}}{C}H{-}\bigcirc \longrightarrow CH_3(\overset{\overset{\displaystyle CH_3}{|}}{C}HCH_2)_3\overset{\overset{\displaystyle CH_3}{|}}{C}H{-}\bigcirc{-}SO_3^- Na^+$$

The widespread use of large quantities of this material caused problems in the purification of sewage effluent because the branched chains were only slowly biodegradable. The detergent industry has now completely replaced this product with one prepared from a straight-chain C_{12}–C_{15} alkane. The hydrocarbon is chlorinated and used for Friedel–Crafts alkylation of benzene followed by sulfonation to give a product that has the linear character necessary for rapid biodegradability by bacteria.

$$\begin{array}{c} CH_3(CH_2)_x \\ \underset{CH_3(CH_2)_y}{\overset{|}{C}H}{-}\bigcirc{-}SO_3^- Na^+ \end{array}$$

modern household detergent

p-Toluenesulfonic acid is readily available as the crystalline monohydrate, $C_7H_7SO_3^-H_3O^+$, m.p. 105°. It is prepared by salting out the sulfonation product from toluene with concentrated hydrochloric acid. Alternatively, the barium salt is treated with the stoichiometric amount of sulfuric acid, and the insoluble barium sulfate is filtered. The filtrate is a strong acid, and concentrated solutions will dehydrate cellulose (filter paper!) just like sulfuric acid. Indeed, *p*-toluenesulfonic acid is used as an acid catalyst in place of sulfuric acid in cases where the oxidizing character of sulfuric acid is deleterious.

EXERCISE: *p*-Chlorobenzenesulfonic acid is a relatively inexpensive, commercially available material. Suggest a way in which it may be used to prepare 2,6-dinitrochlorobenzene.

2. REACTIONS OF ARENESULFONIC ACIDS. The sulfonate group in aromatic sulfonic acid can be replaced by **nucleophilic aromatic substitution** reactions. The conditions are drastic: fusion with alkali hydroxide or other salts at temperatures of 200–350°. For example,

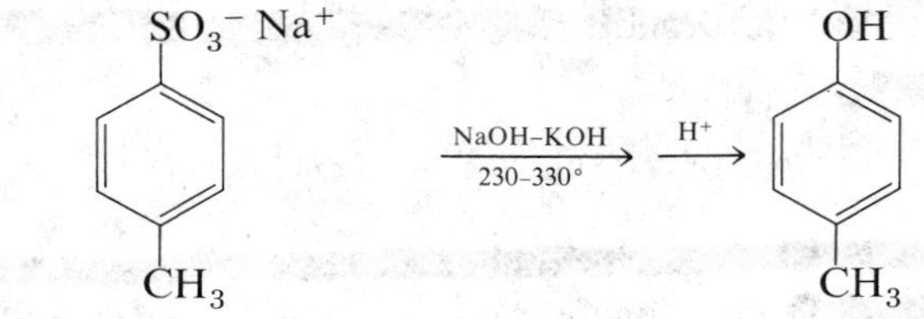

sodium *p*-toluenesulfonate

p-cresol
(63–72%)

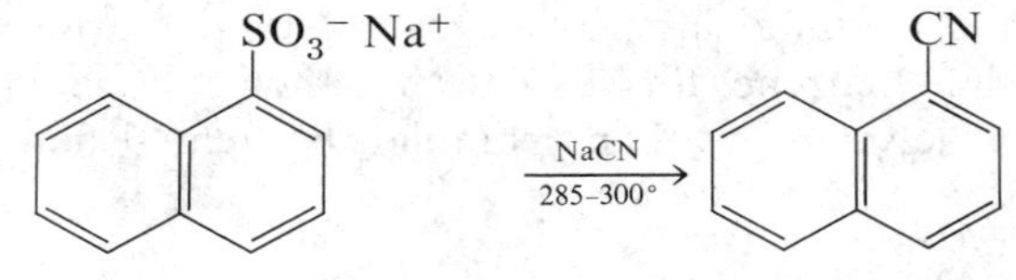

sodium 1-naphthalenesulfonate

1-naphthonitrile
(60–70%)

The method is used primarily for preparing phenols and nitriles. The reaction conditions required are tolerated by few other functional groups; hence, the scope of the reaction is limited. Other nucleophilic aromatic substitution reactions will be considered in Section 30.2.

An unusual feature of this reaction is the fact that it takes place in a fused-salt medium. Such media are highly polar liquids composed almost wholly of ions. In general, neutral organic compounds are not soluble in such ionic media; hence, we have encountered them only rarely in organic reactions. Fused-salt media are useful in organic chemistry only when the organic compound is itself a salt.

As with the aliphatic sulfonic acids, one of the most important reactions of aromatic sulfonic acids is conversion to the corresponding sulfonyl chloride. This reaction is most conveniently carried out on the sodium salt by treatment with PCl_5 or $POCl_3$. The product can be distilled or crystallized from benzene: benzenesulfonyl chloride, b.p. 251.5°; *p*-toluenesulfonyl chloride, m.p. 68°. Alternatively, the acid chloride may be prepared by direct sulfonation with chlorosulfonic acid as discussed on page 820.

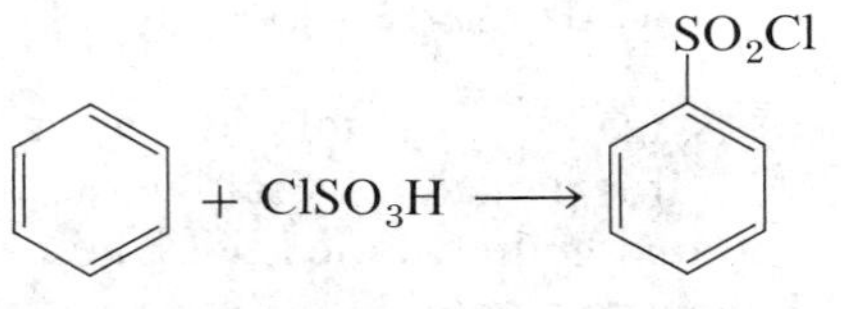

p-Toluenesulfonyl chloride (tosyl chloride) is used to prepare tosyl esters from alcohols. The procedure involves combining the reagents with excess pyridine at room temperature. Pyridinium chloride separates from solution; the mixture is then added to dilute hydrochloric acid, and the product tosylate is filtered or extracted into ether.

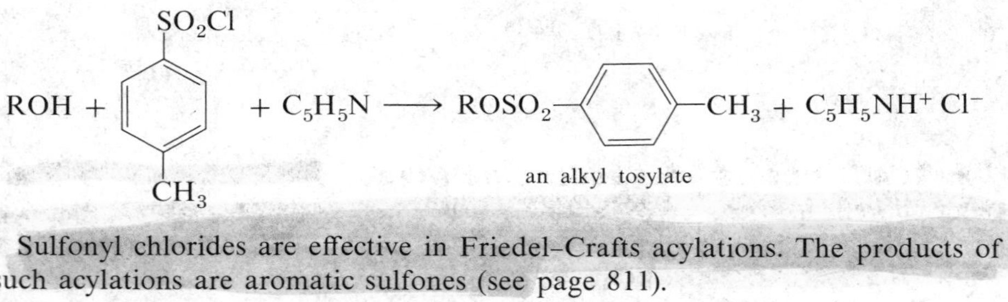

an alkyl tosylate

Sulfonyl chlorides are effective in Friedel–Crafts acylations. The products of such acylations are aromatic sulfones (see page 811).

$$C_6H_5SO_2Cl \xrightarrow[C_6H_6]{AlCl_3} C_6H_5SO_2C_6H_5$$

phenyl sulfone

Reaction of sulfonyl chlorides with ammonia or amines gives the corresponding sulfonamides. Many such compounds have found important medicinal use as antibacterial agents. Examples are sulfanilamide (*p*-aminobenzenesulfonamide), sulfadiazine [*p*-amino-N-(2-pyrimidyl)benzenesulfonamide], and sulfathiazole [*p*-amino-N-(2-thiazolyl)benzenesulfonamide].

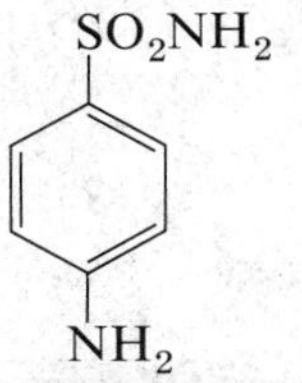

sulfanilamide 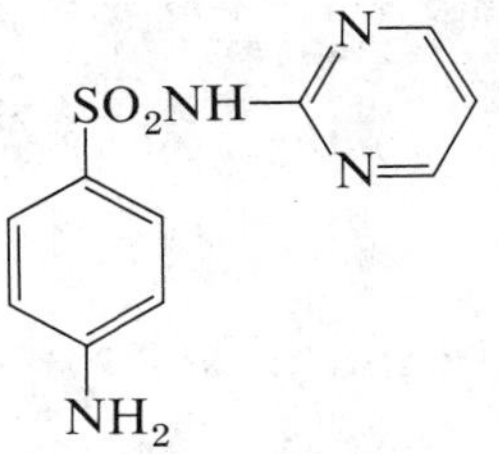   sulfadiazine 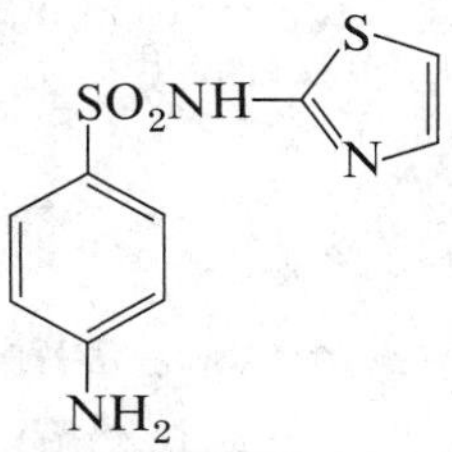   sulfathiazole

They are no longer used in human medicine because of side reactions and the development of antibiotics, but they are still used in veterinary medicine.

Reduction of a sulfonyl chloride with zinc and water gives a sulfinic acid (page 812).

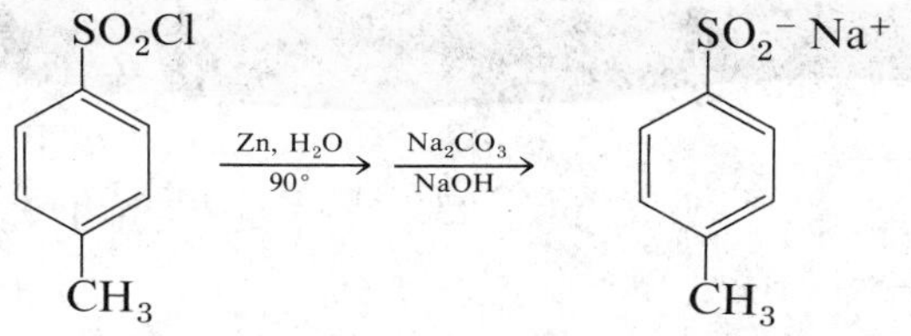

(64%)

sodium *p*-toluenesulfinate

Sulfinic acids are a generally unimportant class of compounds, but the salts show one interesting feature. As nucleophilic reagents reacting with alkyl halides, reaction occurs on sulfur rather than oxygen to produce a sulfone; that is, sulfinates are another example of ambident anions (Section 8.9; see also page 817).

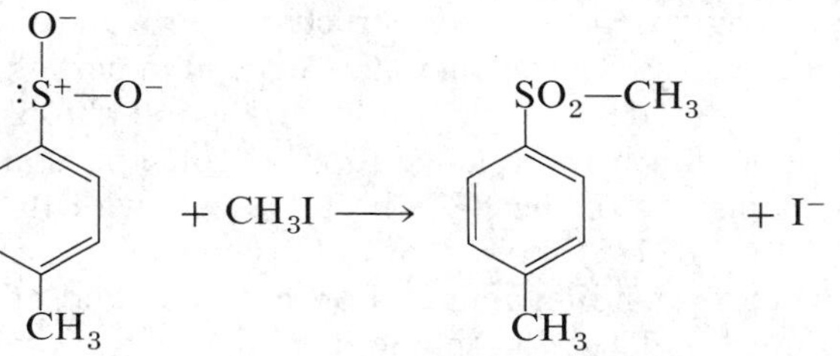

methyl *p*-tolyl sulfone

Under more strongly reducing conditions sulfonyl chlorides give the corresponding thiols.

EXERCISE: (a) Sulfonation of bromobenzene gives *p*-bromobenzenesulfonic acid, which is isolated as the sodium salt. This salt is frequently called sodium brosylate, and the brosylate group is sometimes used instead of the analogous tosylate group. How can sodium brosylate be converted into brosyl chloride? Give the reaction product resulting from treatment of *p*-bromobenzenesulfonyl chloride with each of the following sets of reagents.

(i)   *n*-C$_4$H$_9$OH, pyridine   (ii) toluene, AlCl$_3$

(iii) 1. Zn, water; 2. Na$_2$CO$_3$; 3. C$_6$H$_5$CH$_2$Cl   (iv) Zn, H$_2$SO$_4$

  (b) Sulfonium salts with three different groups, RR′R″S$^+$, are chiral and have inversion barriers sufficiently high (25–30 kcal mole^{-1}) that optically active enantiomers can be obtained at room temperature. Review each sulfur-containing functional group discussed so far in this chapter and give an example of each one that is also expected to be chiral.

26.6
Phosphines and Phosphonium Salts

Phosphines are the phosphorus analogs of amines. They are named by appending the suffix **-phosphine** to the stem names of the alkyl groups attached to phosphorus.

$$CH_3CH_2PH_2 \qquad (CH_3CH_2CH_2CH_2)_2PH \qquad (C_6H_5)_3P$$

ethylphosphine   di-*n*-butylphosphine   triphenylphosphine

Tertiary phosphines are the most important. They are conveniently prepared by reaction of Grignard reagents with phosphorus trichloride.

tricyclohexylphosphine

Phosphines are more highly pyramidal than amines, but there is considerable variation in the bond angles about phosphorus. Trimethylphosphine has a CPC angle of 99°, somewhat expanded from the HPH angle of 93° in phosphine itself. Triphenylphosphine has a CPC angle of 103°.

Since phosphines are analogs of amines, we expect the phosphorus lone pair to show characteristic basic properties. Phosphines do act as Lewis bases, but the base strength is strongly dependent on structure, particularly on the degree of substitution on phosphorus. The changes may depend in part on bond-angle variations between the pyramidal phosphine and tetrahedral phosphonium salt. Representative pK_a values for the protonated forms of some phosphines are collected in Table 26.3. Comparison of the pK_as in Table 26.3 with the pK_as of amines given in Table 24.3 (page 735) shows that tertiary phosphines are about 100-fold *less basic* than the corresponding amines. For secondary and primary phosphines the difference is much greater. Phosphine itself, PH$_3$, shows no basic properties whatsoever; the pK_a of PH$_4^+$ has been estimated to be −14! Like the analogous amines, phenylphosphines are less basic than the alkyl compounds.

Even though phosphines are considerably less basic than amines, the phosphorus is highly polarizable, and phosphines are highly nucleophilic and readily participate in S$_N$2 reactions.

$$(C_6H_5)_3P + C_6H_5CH_2Cl \longrightarrow (C_6H_5)_3\overset{+}{P}CH_2C_6H_5 \ Cl^-$$

benzyltriphenylphosphonium chloride

TABLE 26.3
Acidity of
Phosphonium Ions

R_3PH^+	pK_a
$(C_2H_5)_3PH^+$	8.69
$(n\text{-}C_4H_9)_3PH^+$	8.43
$(C_4H_9)_2PH_2{}^+$	4.51
$(CH_3)_2PH_2{}^+$	3.91
$i\text{-}C_4H_9PH_3{}^+$	-0.02
$n\text{-}C_8H_{17}PH_3{}^+$	0.43
$(C_6H_5)_3PH^+$	2.73
$(C_6H_5)_2PH_2{}^+$	0.03

Phosphonium salts are most important for their use as reagents in the Wittig synthesis of alkenes (Section 13.7.G).

Tertiary phosphines also act as good donor ligands toward metals and are commonly used in the preparation of organometallic complexes such as tris-triphenyl-phosphinerhodium(I) chloride (Section 15.5).

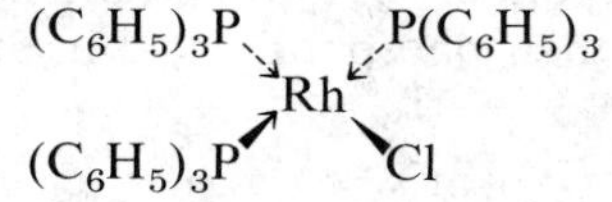

EXERCISE: Review the discussion of the Wittig reaction in Section 13.7.G. Write the equations illustrating the use of benzyltriphenylphosphonium chloride to prepare several alkenes. How could the triphenylphosphine that is used in this sequence be prepared?

26.7
Phosphate and Phosphonate Esters

Phosphorus forms several oxyacids. The most common one is orthophosphoric acid, more commonly called simply phosphoric acid, H_3PO_4. When orthophosphoric acid is heated above 210°, it loses water with the formation of pyrophosphoric acid, which may be regarded as an anhydride of phosphoric acid.

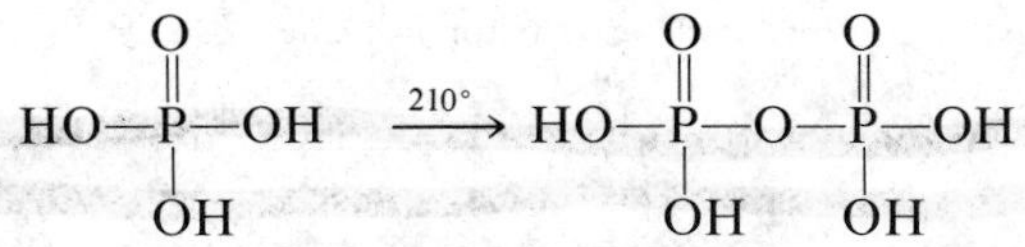

orthophosphoric acid   pyrophosphoric acid

"Polyphosphoric acid" (PPA) is a mixture of phosphoric anhydrides that is prepared by heating H_3PO_4 with phosphorus pentoxide, P_2O_5. It consists of about 55% tripoly-phosphoric acid, the remainder being H_3PO_4 and higher polyphosphoric acids.

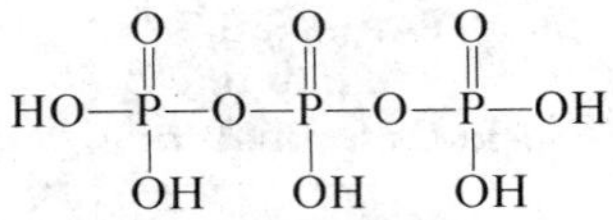

$$HO\overset{\overset{\displaystyle O}{\|}}{\underset{\underset{\displaystyle OH}{|}}{P}}\!-\!O\!-\!\overset{\overset{\displaystyle O}{\|}}{\underset{\underset{\displaystyle OH}{|}}{P}}\!-\!O\!-\!\overset{\overset{\displaystyle O}{\|}}{\underset{\underset{\displaystyle OH}{|}}{P}}\!-\!OH$$

tripolyphosphoric acid

Polyphosphoric acid is commonly used as an acid catalyst in some organic reactions.

Orthophosphoric acid is a tribasic acid having $pK_1 = 2.15$, $pK_2 = 7.20$, and $pK_3 = 12.38$.

$$H_3PO_4 \rightleftharpoons H^+ + H_2PO_4^- \qquad K_a = 7.1 \times 10^{-3}\ M$$
$$H_2PO_4^- \rightleftharpoons H^+ + HPO_4^{2-} \qquad K_a = 6.3 \times 10^{-8}\ M$$
$$HPO_4^{2-} \rightleftharpoons H^+ + PO_4^{3-} \qquad K_a = 4.2 \times 10^{-13}\ M$$

It may form mono-, di-, and triesters.

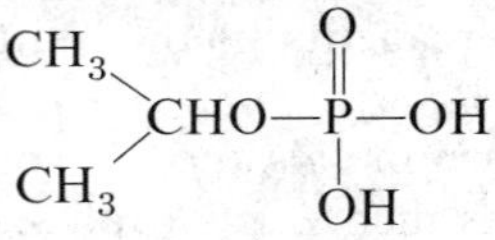

isopropyl phosphate 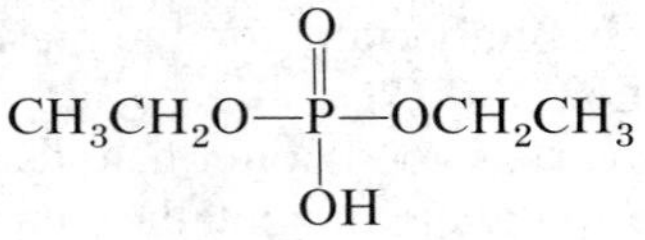   diethyl phosphate 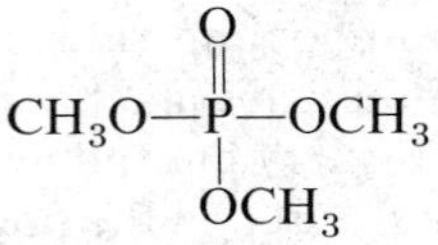   trimethyl phosphate

The mono- and diesters still contain OH groups and have acidic properties. They are actually stronger acids than phosphoric acid itself.

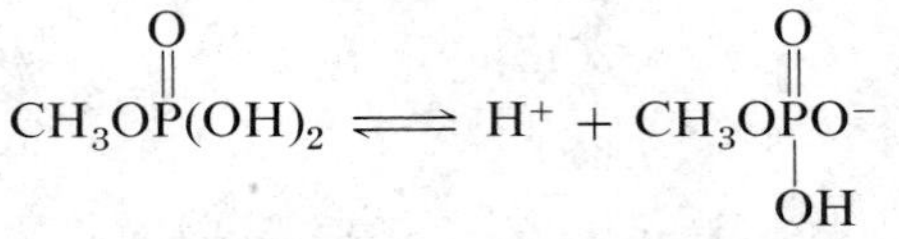

$$CH_3OP(OH)_2 \rightleftharpoons H^+ + CH_3O\overset{\overset{\displaystyle O}{\|}}{\underset{\underset{\displaystyle OH}{|}}{P}}O^- \qquad pK_a = 1.54$$

methyl phosphate

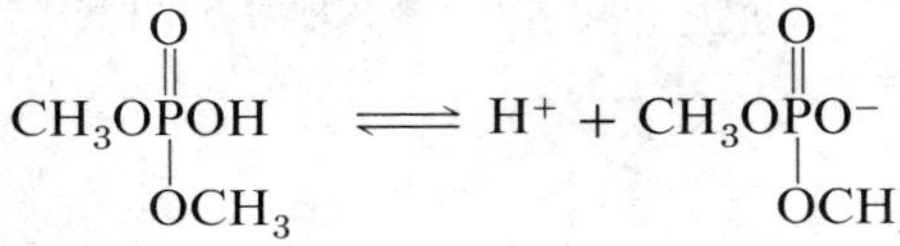

$$CH_3O\overset{\overset{\displaystyle O}{\|}}{\underset{\underset{\displaystyle OCH_3}{|}}{P}}OH \rightleftharpoons H^+ + CH_3O\overset{\overset{\displaystyle O}{\|}}{\underset{\underset{\displaystyle OCH_3}{|}}{P}}O^- \qquad pK_a = 1.29$$

dimethyl phosphate

Analogous esters are possible for pyrophosphoric acid, but the most common are the monoesters.

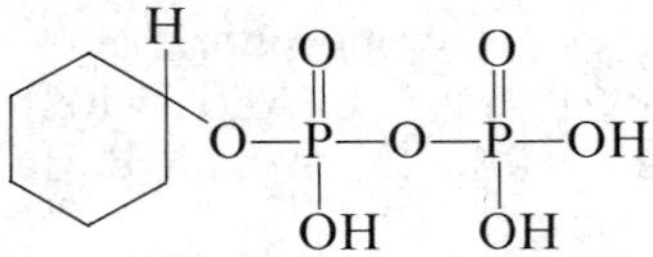

cyclohexyl pyrophosphate

Phosphate triesters are commonly prepared from the alcohol and phosphorus oxychloride, which is the acyl halide corresponding to phosphoric acid.

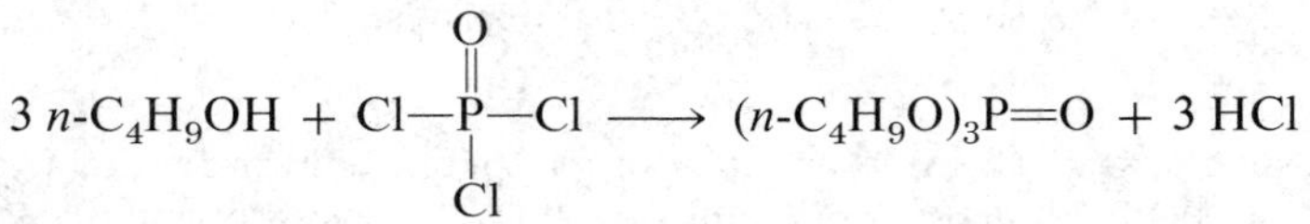

$$3\ n\text{-}C_4H_9OH + Cl\overset{\overset{\displaystyle O}{\|}}{\underset{\underset{\displaystyle Cl}{|}}{P}}Cl \longrightarrow (n\text{-}C_4H_9O)_3P{=}O + 3\ HCl$$

tributyl phosphate

Compounds prepared by the replacement of two of the chlorines of $POCl_3$ (phosphorochloridates) may be used in a similar way to prepare mixed phosphates.

$$(t\text{-}C_4H_9O)_2\overset{\overset{\textstyle O}{\|}}{P}\!\!-\!\!Cl \;+\; \underset{\substack{\text{cyclohexyl} \\ \text{(ring with H and OH)}}}{\bigcirc}\!\!-\!\!OH \;\longrightarrow\; \bigcirc\!\!-\!\!O\!\!-\!\!\overset{\overset{\textstyle O}{\|}}{P}(OC_4H_9\text{-}t)_2 \;+\; HCl$$

di-*t*-butyl
phosphorochloridate

cyclohexyl
di-*t*-butyl phosphate

The only reaction of phosphate esters that we shall consider here is hydrolysis. Hydrolysis may be either acid- or base-catalyzed and may involve either C—O or P—O bond rupture. Under basic conditions hydrolysis occurs mainly by an addition-elimination mechanism, similar to that involved in the hydrolysis of carboxylic acid esters.

$$CH_3O\!\!-\!\!\overset{\overset{\textstyle O}{\|}}{\underset{\underset{\textstyle OCH_3}{|}}{P}}\!\!-\!\!OCH_3 \;+\; OH^- \;\rightleftharpoons\; CH_3O\!\!-\!\!\overset{\overset{\textstyle HO}{\diagdown}\overset{\textstyle O^-}{\diagup}}{\underset{\underset{\textstyle OCH_3}{|}}{P}}\!\!-\!\!OCH_3 \;\rightleftharpoons$$

$$CH_3O\!\!-\!\!\overset{\overset{\textstyle OH}{|}}{\underset{\underset{\textstyle O}{\|}}{P}}\!\!-\!\!OCH_3 \;+\; CH_3O^- \;\longrightarrow\; CH_3O\!\!-\!\!\overset{\overset{\textstyle O}{\|}}{\underset{\underset{\textstyle O^-}{|}}{P}}\!\!-\!\!OCH_3 \;+\; CH_3OH$$

The first alkyl group of a trialkyl phosphate is hydrolyzed most easily with the second and third groups being hydrolyzed rather more sluggishly.

Under acidic conditions C—O bond cleavage is the predominant mode of hydrolysis, although P—O rupture may also be observed. Cleavage of the C—O bonds may occur by either the S_N2 or the S_N1 mechanism, the former being preferred with primary alkyl phosphates and the latter with tertiary systems.

$$S_N2:\quad CH_3O\!\!-\!\!\overset{\overset{\textstyle O}{\|}}{\underset{\underset{\textstyle OCH_3}{|}}{P}}\!\!-\!\!OCH_3 \;+\; H^+ \;\rightleftharpoons\; CH_3O\!\!-\!\!\overset{\overset{\textstyle OH}{|}}{\underset{\underset{\textstyle O\text{---}CH_3}{|}}{P^+}}\!\!-\!\!OCH_3 \;\rightleftharpoons$$

$$H_2\overset{..}{O}:\!\!-$$

$$CH_3O\!\!-\!\!\overset{\overset{\textstyle OH}{|}}{\underset{\underset{\textstyle O}{\|}}{P}}\!\!-\!\!OCH_3 \;+\; CH_3OH \;+\; H^+$$

$$S_N1:\quad t\text{-}C_4H_9O\!\!-\!\!\overset{\overset{\textstyle O}{\|}}{\underset{\underset{\textstyle OH}{|}}{P}}\!\!-\!\!OH \;+\; H^+ \;\rightleftharpoons\; t\text{-}C_4H_9O\!\!-\!\!\overset{\overset{\textstyle OH}{|}}{\underset{\underset{\textstyle OH}{|}}{P^+}}\!\!-\!\!OH \;\rightleftharpoons$$

$$O\!\!=\!\!\overset{\overset{\textstyle OH}{|}}{\underset{\underset{\textstyle OH}{|}}{P}}\!\!-\!\!OH \;+\; t\text{-}C_4H_9^+ \;\xrightarrow{\;H_2O\;}\; t\text{-}C_4H_9OH \;+\; H^+$$

As a leaving group in such substitution reactions, phosphate is about as good as bromide ion. Alkyl pyrophosphates are somewhat better leaving groups, being about 100-fold more effective as leaving groups than iodide ion. The pyrophosphate group is an important leaving group in nucleophilic substitution reactions that occur in nature. A number of compounds are built up by plants (**biosynthesized**) from acetic acid units. By a series of enzyme-catalyzed steps, acetic acid is transformed into the compound isopentenyl pyrophosphate, which undergoes enzyme-catalyzed isomerization to γ,γ-dimethylallyl pyrophosphate.

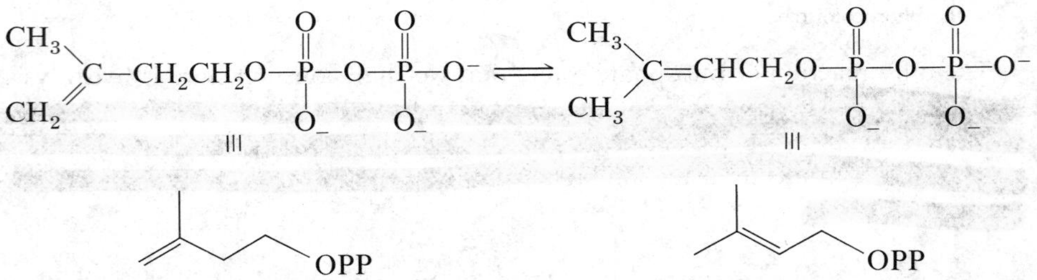

isopentenyl pyrophosphate γ,γ-dimethylallyl pyrophosphate

Since the pyrophosphate ion is such a good leaving group, γ,γ-dimethylallyl pyrophosphate readily ionizes to give the allylic cation.

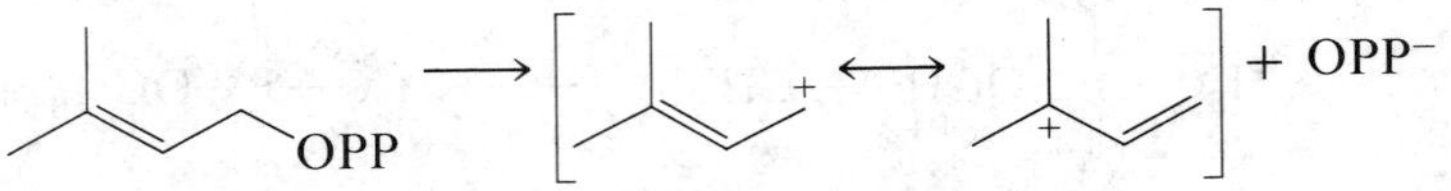

The dimethylallyl cation so produced reacts with isopentenyl pyrophosphate to give a new carbocation, which eliminates to give geranyl pyrophosphate.

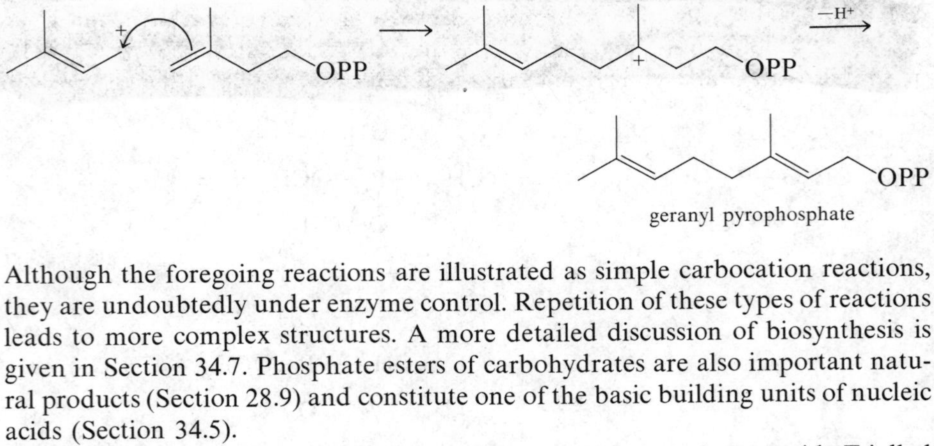

geranyl pyrophosphate

Although the foregoing reactions are illustrated as simple carbocation reactions, they are undoubtedly under enzyme control. Repetition of these types of reactions leads to more complex structures. A more detailed discussion of biosynthesis is given in Section 34.7. Phosphate esters of carbohydrates are also important natural products (Section 28.9) and constitute one of the basic building units of nucleic acids (Section 34.5).

Phosphorous acid, H_3PO_3, is a less important oxyphosphorus acid. Trialkyl phosphites are generally prepared by treatment of phosphorus trichloride with the alcohol and pyridine.

$$PCl_3 + 3\ C_2H_5OH + 3\ C_5H_5N \longrightarrow (C_2H_5O)_3P + 3\ C_5H_5NH^+\ Cl^-$$
triethyl phosphite

Phosphonic acids contain the functional group $-PO_3H_2$ attached to carbon. They are named as alkylphosphonic acids.

$$CH_3\overset{\displaystyle O}{\underset{\displaystyle OH}{\overset{\|}{P}}}{-}OH \qquad CH_3CH_2CH_2\overset{\displaystyle O}{\underset{\displaystyle OH}{\overset{\|}{P}}}{-}OH$$

methylphosphonic acid   1-propylphosphonic acid

The most common derivatives are the diesters.

$$CH_3CH_2\overset{\displaystyle O}{\underset{\displaystyle OCH_3}{\overset{\|}{P}}}{-}OCH_3 \qquad CH_3CH_2\overset{\displaystyle O}{\underset{\displaystyle OCH_2CH_3}{\overset{\|}{P}}}{-}OCH_2CH_3$$

dimethyl ethylphosphonate   diethyl ethylphosphonate

Dialkyl phosphonates are best prepared from trialkyl phosphites by a reaction known as the **Arbuzov–Michaelis reaction.** For example, when trimethyl phosphite is heated at 200° with a catalytic amount of methyl iodide, dimethyl methylphosphonate is produced in virtually quantitative yield.

$$(CH_3O)_3P\colon \xrightarrow[200°]{CH_3I} \quad CH_3\overset{\displaystyle O}{\overset{\|}{P}}(OCH_3)_2$$

trimethyl phosphite   dimethyl methylphosphonate

The reaction mechanism involves two successive S_N2 processes. In the first step the trialkyl phosphite displaces iodide from methyl iodide, giving an alkyltrialkoxyphosphonium salt.

(1) $(CH_3O)_3P\colon + CH_3{-}I \longrightarrow (CH_3O)_3P^+{-}CH_3 \; I^-$

methyltrimethoxyphosphonium iodide

The liberated iodide ion attacks one of the methoxy groups in a second S_N2 process, displacing the neutral dialkyl phosphonate.

(2) $I^- + CH_3{-}O{-}\overset{\displaystyle OCH_3}{\underset{\displaystyle OCH_3}{\overset{\displaystyle |}{\underset{\displaystyle |}{P^+}}}}{-}CH_3 \longrightarrow CH_3I + CH_3{-}\overset{\displaystyle O}{\underset{\displaystyle OCH_3}{\overset{\|}{P}}}{-}OCH_3$

dimethyl methylphosphonate

Since the alkyl halide is regenerated in the second step, only a catalytic amount is required in order to initiate reaction.

The Arbuzov–Michaelis reaction has been applied to the synthesis of numerous dialkyl phosphonates. If a full equivalent of alkyl halide is used, dialkyl phosphonates having different groups attached to oxygen and phosphorus may be prepared.

$$(CH_3CH_2O)_3P\colon + BrCH_2COOC_2H_5 \xrightarrow{200°} C_2H_5O\overset{\displaystyle O}{\overset{\|}{C}}CH_2\overset{\displaystyle O}{\overset{\|}{P}}(OC_2H_5)_2 + CH_3CH_2Br$$

triethyl phosphite   ethyl bromoacetate   triethyl phosphonoacetate

Monoalkyl phosphonates are readily obtained from the diesters by alkaline hydrolysis. Hydrolysis of the second group is more difficult.

$$CH_3CH_2CH_2CH_2\overset{O}{\overset{\|}{P}}(OCH_3)_2 + NaOH \xrightarrow[\Delta]{C_2H_5OH}$$

dimethyl butylphosphonate

$$CH_3CH_2CH_2CH_2\overset{O}{\overset{\|}{\underset{OCH_3}{P}}}O^-Na^+ \xrightarrow{dil.\ H_2SO_4} CH_3CH_2CH_2CH_2\overset{O}{\overset{\|}{\underset{OCH_3}{P}}}OH$$

sodium methyl butylphosphonate   methyl butylphosphonate

EXERCISE: Write equations illustrating the preparation of the following phosphate and phosphonate esters using $POCl_3$ or PCl_3 as the source of phosphorus.

(a) $CH_3O\overset{O}{\overset{\|}{P}}(OC_2H_5)_2$

(b) $(i\text{-}C_4H_9O_3)_3PO$

(c) $CH_3CH_2CH_2\overset{O}{\overset{\|}{P}}(OC_2H_5)_2$

(d) $CH_3CH_2\overset{O}{\overset{\|}{\underset{OH}{P}}}OCH_2CH_3$

26.8
Sulfur- and Phosphorus-Stabilized Carbanions

Sulfur- and phosphorus-containing functional groups stabilize adjacent carbanions to varying degrees, depending on the exact nature of the function. This property gives rise to carbanionic reagents that have important uses in organic synthesis.

The most well-known and useful compounds in this class are the phosphorus ylides, or Wittig reagents (Section 13.7.G). These compounds are formed by reaction of alkyl phosphonium salts with a strong base such as n-butyllithium.

$$(C_6H_5)_3\overset{+}{P}-CH_3 + n\text{-}C_4H_9Li \longrightarrow (C_6H_5)_3\overset{+}{P}-CH_2{}^- + n\text{-}C_4H_{10}$$

The bonding in such ylides is still a subject of controversy. At one time, it was thought that the ylide P–C bond could be described as a double bond resulting from overlap of the carbon p-orbital with a phosphorus d-orbital. However, quantum-mechanical calculations suggest that such p–d double bonds are relatively unimportant. As in the case of the S–O bond (page 811), a large part of the bonding in ylides is electrostatic—attraction of the positive phosphorus for the negative carbon. However, this cannot be the complete answer, since alkylammonium salts are *not* converted into ylides under conditions that suffice to form phosphorus ylides. Electrostatic bonding in the nitrogen ylide should be as important as in a phosphorus ylide.

$$R_3\overset{+}{N}-CH_2{}^-$$

an ammonium ylide

The higher polarizability of the phosphorus valence electrons is probably in-

831

Sec. 26.8

Sulfur- and
Phosphorus-
Stabilized
Carbanions

volved in stabilizing the dipolar ylide structure. Such polarization can be viewed as an induced dipole on phosphorus.

$$R_3\overset{+}{P}-CH_2^-$$

Whatever the exact explanation for the bonding in phosphorus ylides, it is quantitatively significant. Although pK_as for simple phosphonium salts have not been measured, salts that also contain a keto or ester function attached to the acidic position are substantially more acidic than 1,3-diketones and 1,3-keto esters (Section 27.7.B.2).

$$(C_6H_5)_3\overset{+}{P}-CH_2\overset{O}{\overset{\|}{C}}C_6H_5 \qquad (C_6H_5)_3\overset{+}{P}-CH_2CO_2C_2H_5$$
$$\text{p}K_a = 6.0 \qquad\qquad\qquad \text{p}K_a = 9.1$$

From these pKs, it may be concluded that the $(C_6H_5)_3P-$ substituent stabilizes a carbanion slightly more than does a carbonyl group!

A simple phosphonate may be deprotonated by a strong base such as *n*-butyl-lithium, and the resulting carbanion adds to aldehydes and ketones. Aqueous work-up of this initial adduct affords the β-hydroxyphosphonate.

$$CH_3\overset{O}{\overset{\|}{P}}(OC_2H_5)_2 + n\text{-}C_4H_9Li \longrightarrow {}^-CH_2\overset{O}{\overset{\|}{P}}(OC_2H_5)_2 \xrightarrow{CH_3CHO}$$

diethyl methanephosphonate

$$CH_3\overset{O^-}{\overset{|}{C}}HCH_2\overset{O}{\overset{\|}{P}}(OC_2H_5)_2 \xrightarrow{H_2O} CH_3\overset{OH}{\overset{|}{C}}HCH_2\overset{O}{\overset{\|}{P}}(OC_2H_5)_2$$

diethyl 2-hydroxypropanephosphonate

Phosphonates that have a carbonyl group attached to the α-carbon are more acidic, having pK_as of about 15.

$$C_2H_5OOCCH_2\overset{O}{\overset{\|}{P}}(OC_2H_5)_2$$

triethyl phosphonoacetate
estimated p$K_a \approx 15$

Comparison of this value with the pK_as of malonic ester and acetoacetic ester (Table 27.6, page 875) shows that the dialkylphosphono group, $(RO)_2PO-$, is not quite as effective as a carbonyl group at stabilizing an adjacent carbanion. Anions of such activated phosphonates add to aldehydes and ketones, and the resulting products eliminate dialkylphosphonate ion to give α,β-unsaturated esters or ketones **(Horner–Emmons** reaction: Section 20.3.B).

$$(C_2H_5O)_2\overset{O}{\overset{\|}{P}}CH_2CN \xrightarrow{NaH} \xrightarrow{CH_3O-\!\!\!\langle\ \rangle\!\!\!-CHO} CH_3O-\!\!\!\langle\ \rangle\!\!\!-CH\!\!=\!\!CHCN$$

(88%)

Sulfur also stabilizes adjacent carbanions. An example is thioanisole, which may be deprotonated by *n*-butyllithium. The resulting lithium compound may be alkylated by primary alkyl halides and also reacts with aldehydes and ketones.

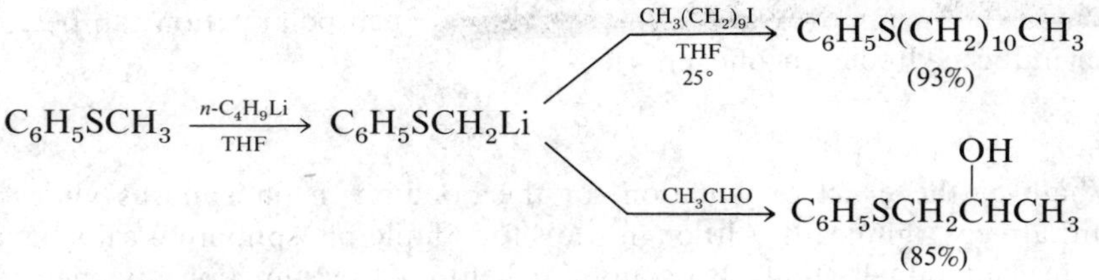

$$\text{C}_6\text{H}_5\text{SCH}_3 \xrightarrow[\text{THF}]{n\text{-C}_4\text{H}_9\text{Li}} \text{C}_6\text{H}_5\text{SCH}_2\text{Li}$$

$$\xrightarrow[\substack{\text{THF}\\25°}]{\text{CH}_3(\text{CH}_2)_9\text{I}} \text{C}_6\text{H}_5\text{S}(\text{CH}_2)_{10}\text{CH}_3$$
(93%)

$$\xrightarrow{\text{CH}_3\text{CHO}} \text{C}_6\text{H}_5\text{SCH}_2\overset{\overset{\displaystyle\text{OH}}{|}}{\text{CH}}\text{CH}_3$$
(85%)

The acetal protons in dithioacetals are activated by two sulfur atoms and are correspondingly more acidic than simple sulfides. An extensively studied member of this class is 1,3-dithiane, the product of reaction of 1,3-propanedithiol and formaldehyde. The pK_a of 1,3-dithiane has been determined to be 31.5.

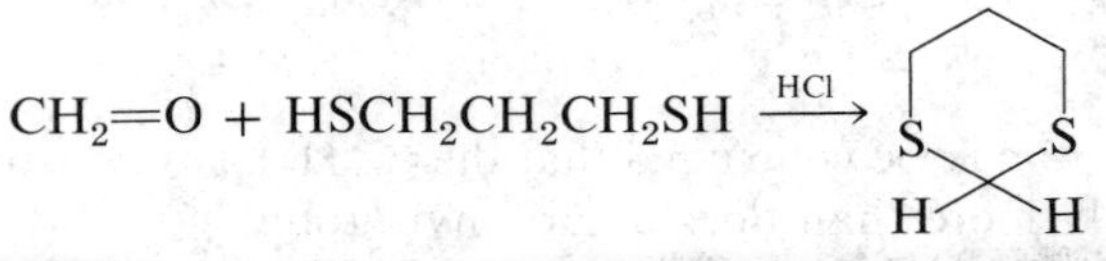

$$\text{CH}_2{=}\text{O} + \text{HSCH}_2\text{CH}_2\text{CH}_2\text{SH} \xrightarrow{\text{HCl}}$$

1,3-dithiane

1,3-Dithiane is readily converted into the 1,3-dithianyl anion by treatment with *n*-butyllithium.

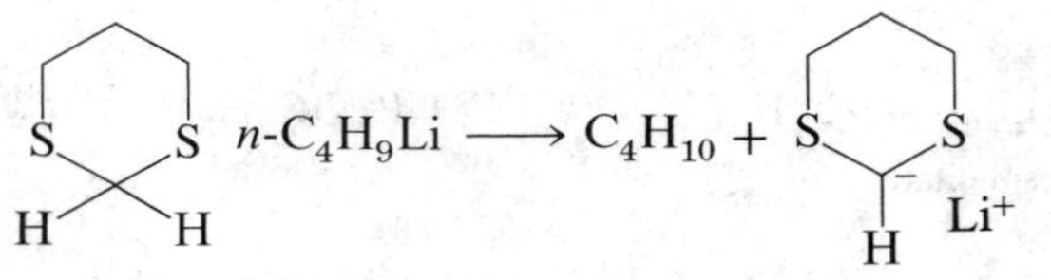

$$n\text{-C}_4\text{H}_9\text{Li} \longrightarrow \text{C}_4\text{H}_{10} + $$

Analogous anions may be produced from other 1,3-dithianes.

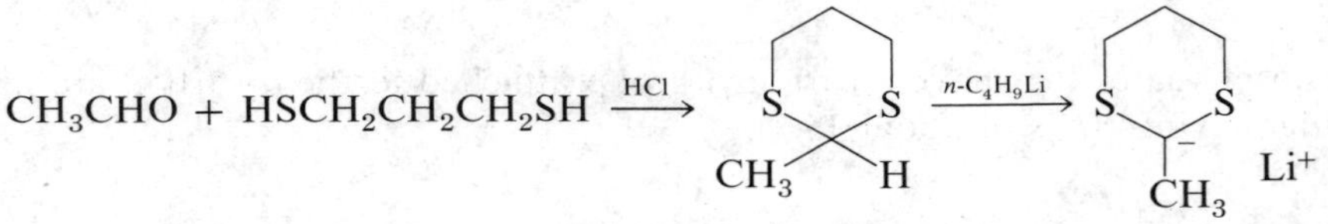

$$\text{CH}_3\text{CHO} + \text{HSCH}_2\text{CH}_2\text{CH}_2\text{SH} \xrightarrow{\text{HCl}} \xrightarrow{n\text{-C}_4\text{H}_9\text{Li}}$$

The 1,3-dithianyl anions are nucleophilic and undergo addition to aldehyde and ketone carbonyl groups.

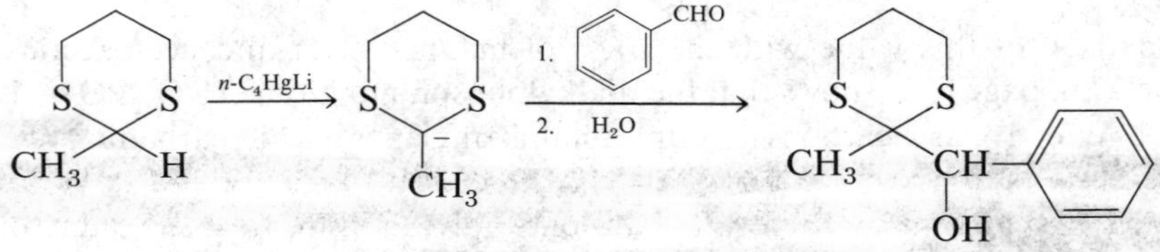

$$\xrightarrow{n\text{-C}_4\text{HgLi}} \xrightarrow[\text{2. }\text{H}_2\text{O}]{\text{1.}}$$
(91%)

The resulting thioacetals are stable to normal hydrolytic conditions, but hydrolyze easily when treated with mercuric chloride in aqueous acetonitrile.

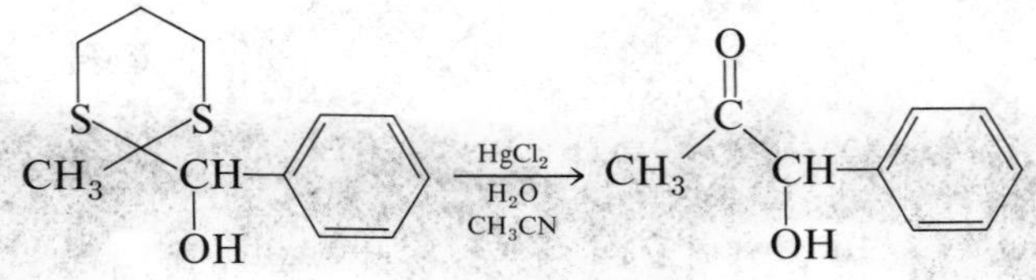

$$\xrightarrow[\substack{\text{H}_2\text{O}\\\text{CH}_3\text{CN}}]{\text{HgCl}_2}$$
(90%)

833

Sec. 26.8

**Sulfur- and
Phosphorus-
Stabilized
Carbanions**

The overall process constitutes one method for the synthesis of α-hydroxy ketones (Section 27.4.A).

The dithiane method may also be used as a method of preparing certain β-hydroxy ketones. If the dithianyl anion is used as a nucleophile to open an epoxide ring (Section 15.5.D), hydrolysis of the resulting hydroxy dithiane affords the β-hydroxy ketone.

(81%)

(83%)
2-hydroxy-4-nonanone

Like phosphonium salts, the analogous sulfonium salts are acidic. Because the sulfur bears a positive charge, which aids in stabilizing the conjugate base, sulfonium salts are much more acidic than simple sulfides or even dithioacetals. One way in which the enhanced acidity of sulfonium salts is shown is by base-catalyzed exchange of the acidic protons in deuterated media such as D_2O.

$$(CH_3)_3S^+ \ I^- \xrightarrow{OD^-, \ D_2O} (CD_3)_3S^+ \ I^-$$

Although exact pK_as of such sulfonium ions have not been measured, they are probably on the order of 25. If a strong base such as n-butyllithium is used, the sulfonium salt can be converted completely into ylides, which are analogous to the phosphonium ylides.

$$(CH_3)_3S^+ \ I^- + n\text{-}C_4H_9Li \xrightarrow[-70°]{THF} (CH_3)_2\overset{+}{S}\text{—}\overset{-}{C}H_2 + C_4H_{10} + LiI$$

Sulfonium ylides are unstable at temperatures higher than 0°. However, at lower temperatures they add to aldehydes and ketones. The initial product is a zwitterion that behaves differently from the zwitterion produced by addition of a phosphonium ylide to a carbonyl compound (Section 13.7.G). In this case, the alkoxide ion acts as the nucleophile in an *intramolecular S_N2 process* and dimethyl sulfide is the leaving group. The product is an epoxide.

(93%)

Sulfoxides and sulfones (page 811) are also relatively acidic; the pK_as of dimethyl sulfoxide and dimethyl sulfone are 35 and 31, respectively, in DMSO solution.

$$CH_3\overset{\displaystyle O}{\underset{\displaystyle O}{S}}CH_3 \qquad CH_3\overset{\displaystyle O}{S}CH_3$$

dimethyl sulfone	dimethyl sulfoxide
$pK_a = 31$ (in DMSO)	$pK_a = 35$ (in DMSO)

The anions derived from sulfones react with electrophiles in the same manner as the other sulfur-stabilized anions we have considered. An example is the reaction of the dimethyl sulfone anion with carboxylic acid esters to give β-keto sulfones.

$$CH_3SO_2CH_3 \xrightarrow[\text{DMSO}]{\text{NaH}} CH_3SO_2CH_2^- \xrightarrow{CH_3(CH_2)_{14}COOCH_3} CH_3(CH_2)_{14}\overset{\displaystyle O}{C}CH_2\overset{\displaystyle O}{\underset{\displaystyle O}{S}}CH_3$$

(83%)

This reaction is analogous to the mixed Claisen condensation (Section 19.9). Note that dimethyl sulfoxide is used as a solvent even though it is also acidic. The difference of 4 pK units between dimethyl sulfone and DMSO suffices for synthetic usefulness.

EXERCISE: Write equations illustrating the synthesis of each of the following compounds by a route involving the use of an organosulfur or organophosphorus reagent.

(a) 1-nonene

(b) $CH_3CH_2CH_2CH_2CH\overset{\displaystyle O}{-\!\!-}CH_2$

(c) 4-hydroxy-3-hexanone

(d) 4-hydroxy-2-hexanone

(e) $CH_3CH_2CH_2CH{=}CHCN$

(f) $CH_3SCH_2CH_2C_6H_5$

PROBLEMS

1. Give the structure corresponding to each of the following names.

    (a) ethyl neopentyl sulfide

    (b) isobutyl mercaptan

    (c) 2-methylthiocyclopentanone

    (d) dibutyl disulfide

    (e) cyclohexanethiol

    (f) isobutylsulfuric acid

    (g) diethyl sulfate

    (h) methyl p-nitrobenzenesulfonate

    (i) tributyl phosphate

    (j) diethyl ethylphosphonate

    (k) cyclohexyl methanesulfonate

    (l) triphenyl phosphite

    (m) allyl pyrophosphate

2. Give the IUPAC name corresponding to each of the following structures.

    (a) $(CH_3)_2CHCH_2\overset{\displaystyle SH}{\underset{\displaystyle |}{C}H}CH_3$

    (b) $(CH_3)_2CHCHCH_2CH_3$
                            $\underset{\displaystyle SCH_3}{\overset{\displaystyle |}{}}$

(c) 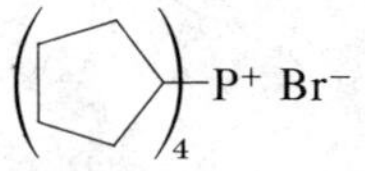 (cyclopentyl)$_4$—P$^+$ Br$^-$

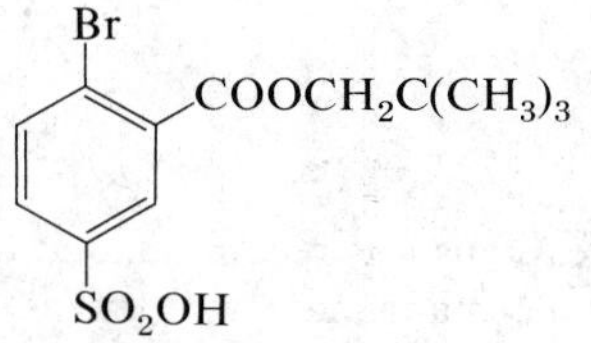

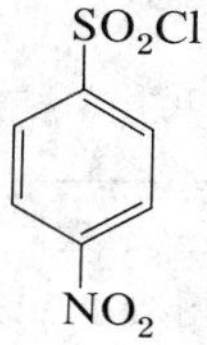

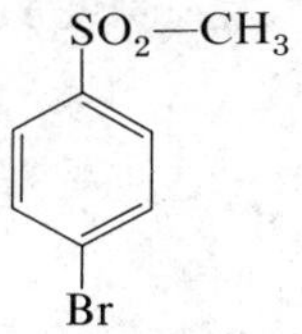

(d) 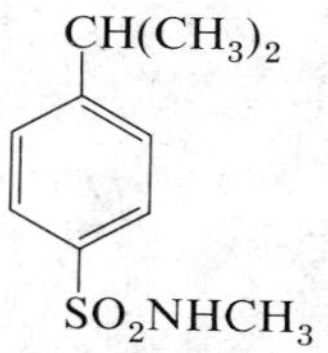 4-isopropyl-C$_6$H$_4$—SO$_2$NHCH$_3$

(e) 2-Br, 5-SO$_2$OH-C$_6$H$_3$—COOCH$_2$C(CH$_3$)$_3$

(f) 4-NO$_2$-C$_6$H$_4$—SO$_2$Cl

(g) 4-Br-C$_6$H$_4$—SO$_2$—CH$_3$

(h) (cyclopropyl)—S—S—(cyclopropyl)

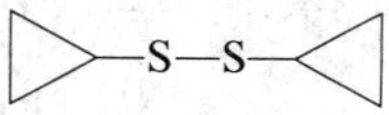

(i) CH$_3$CH$_2$$\overset{\text{O}}{\overset{\|}{\text{P}}}(OEt)_2$

3. Complete the following reactions.

(a) (CH$_3$CH$_2$CH$_2$CH$_2$O)$_3$P + CH$_3$CH$_2$CH$_2$CH$_2$Br $\xrightarrow[\text{(trace)}]{200°}$

(b) t-BuO$^-$K$^+$ + CH$_3$OSO$_2$OCH$_3$ $\xrightarrow[\Delta]{t\text{-BuOH}}$

(c) CH$_3$CH$_2$CH$_2$$\overset{\text{O}}{\overset{\|}{\text{P}}}$(OCH$_3$)$_2$ + NaOH $\xrightarrow[\Delta]{C_2H_5OH}$ $\xrightarrow[0°]{H_2SO_4}$

(d) (C$_2$H$_5$O)$_2$$\overset{\text{O}}{\overset{\|}{\text{P}}}$Cl + (CH$_3$)$_2$CHCH$_2CH_2$OH $\longrightarrow$

(e) POCl$_3$ + CH$_3$OH (excess) $\longrightarrow$

(f) CH$_3$CH$_2$S$^+$(CH$_3$)$_2$ I$^-$ $\xrightarrow{\Delta}$

(g) (CH$_3$)$_3$CCH$_2$S$^+$(CH$_2$CH$_3$)$_2$ Br$^-$ $\xrightarrow{\Delta}$

(h) 1-[$^+$S(CH$_2$CH$_3$)$_2$]-1-CH$_3$-cyclohexane   Cl$^-$ $\xrightarrow{\Delta}$

(i) 2-SO$_3$H-naphthalene $\xrightarrow[\text{fuse}]{\text{NaOH–KOH}}$

(j) C$_6$H$_5$SO$_2$$^-$ + C$_6$H$_5$CH$_2$Cl $\longrightarrow$

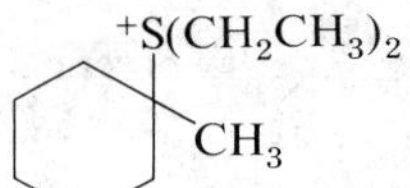

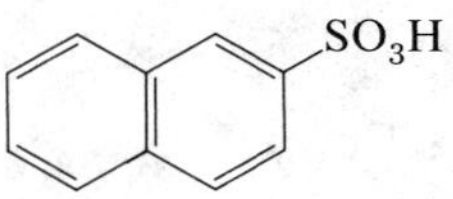

(k) HS(CH$_2$)$_4$SH $\xrightarrow{I_2, KI}$

(l) HS(CH$_2$)$_3$SH + C$_6$H$_5$—CHO $\xrightarrow{HCl}$

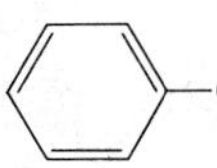

(m) 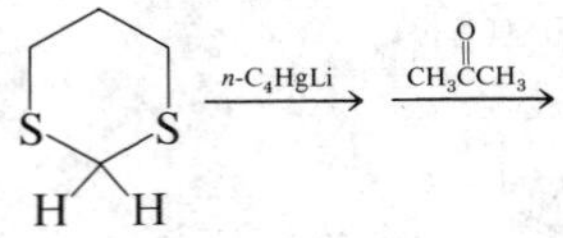 $\xrightarrow{n\text{-}C_4HgLi}$ $\xrightarrow{CH_3\overset{O}{\overset{\|}{C}}CH_3}$

(n) 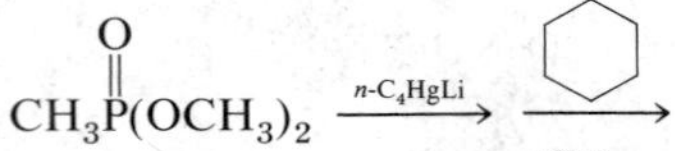 $CH_3\overset{O}{\overset{\|}{P}}(OCH_3)_2$ $\xrightarrow{n\text{-}C_4HgLi}$

$\xrightarrow{CHO}$

(o) $(CH_3)_3S^+ Br^-$ $\xrightarrow{n\text{-}C_4HgLi}$ $\longrightarrow$

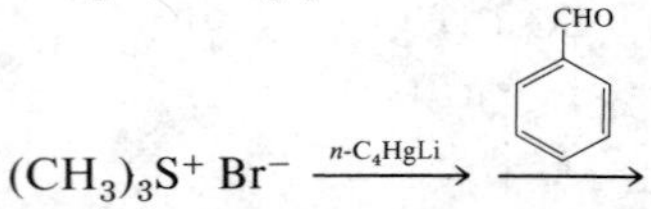

4. To minimize the odor released by traces of mercaptans it is recommended that reaction vessels be rinsed with aqueous potassium permanganate as soon as possible. Nitric acid may also be used for this purpose. The thiols are oxidized to alkanesulfonic acids, and the sulfides are oxidized to sulfones. Write balanced equations for

   (a) $CH_3SH + KMnO_4 \xrightarrow{H_2O} CH_3SO_3K + MnO_2 + KOH$

   (b) $(CH_3)_2S + KMnO_4 \xrightarrow{H_2O} CH_3SO_2CH_3 + MnO_2 + KOH$

   (c) $(CH_3)_2S + HNO_3 \longrightarrow CH_3SO_2CH_3 + H_2O + NO_2$

5. The following reaction sequences are impractical. Determine what is wrong in each case.

   (a) $(CH_3)_3CO^-K^+ \xrightarrow{PCl_3} [(CH_3)_3C-O]_3P \xrightarrow[200°]{(CH_3)_3CBr} (CH_3)_3C\overset{O}{\overset{\|}{P}}(OC(CH_3)_3)_2$

   (b)

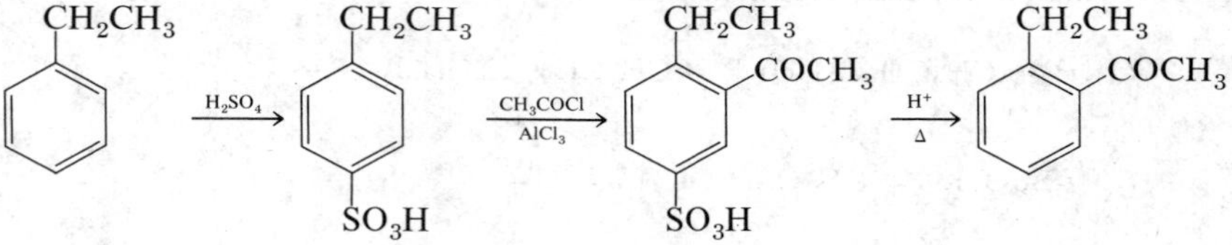

   (c)

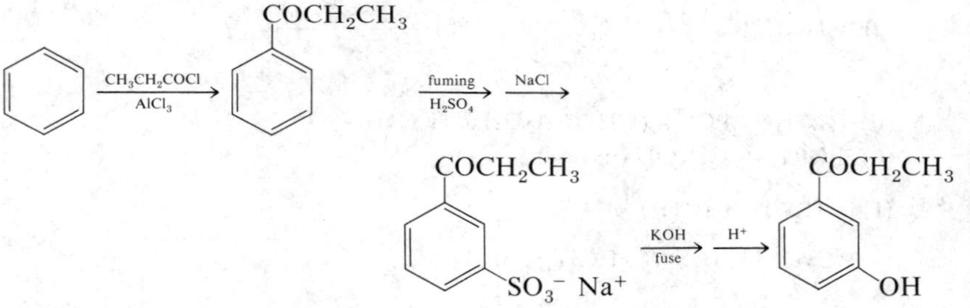

6. Show how to accomplish each of the following conversions in a practical manner.
   (a) $(CH_3)_3CBr \longrightarrow (CH_3)_3CSCH_3$
   (b) $CH_3CH_2CH_2CH_2OH \longrightarrow (CH_3CH_2CH_2CH_2)_2S_2$

   (c)

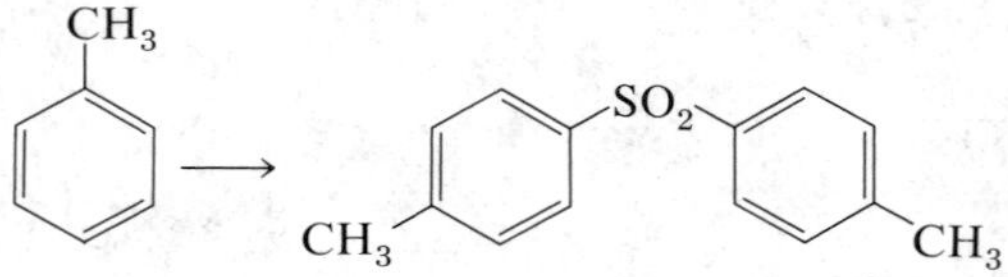

   (d)

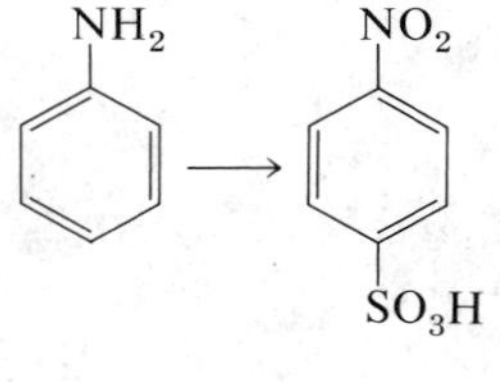

(e)

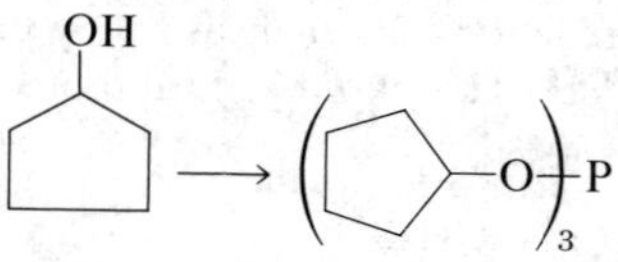

OH

$\longrightarrow$

$\left(\bigcirc\!\!-\!\!O \right)_{\!3}\!\!-\!\!P$

(f)

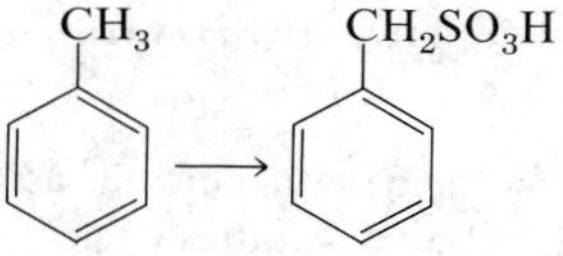

CH₃

$\longrightarrow$

CH₂SO₃H

(g)

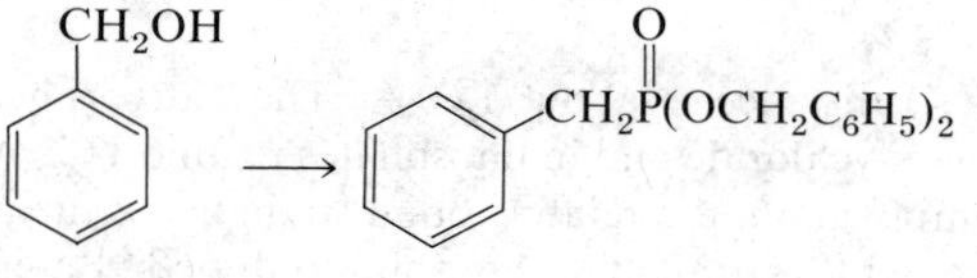

CH₂OH

$\longrightarrow$

O
‖
–CH₂P(OCH₂C₆H₅)₂

(h)

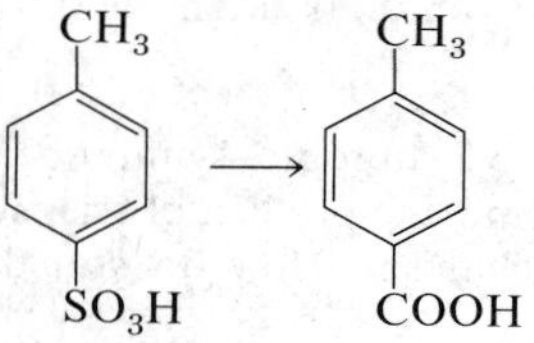

CH₃

SO₃H

$\longrightarrow$

CH₃

COOH

(i)

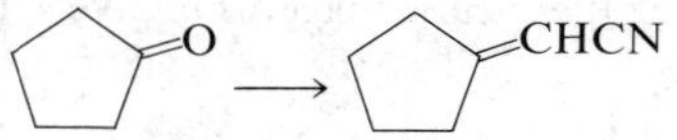

$\bigcirc\!\!=\!\!O$ $\longrightarrow$ $\bigcirc\!\!=\!\!CHCN$

(j)

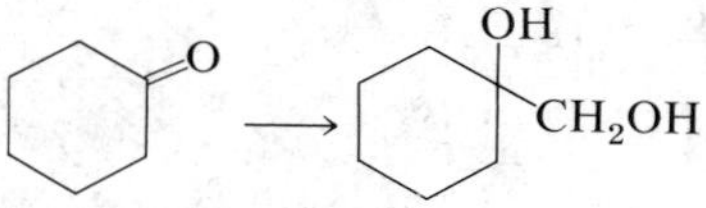

$\bigcirc\!\!=\!\!O$ $\longrightarrow$

OH
CH₂OH

(k)

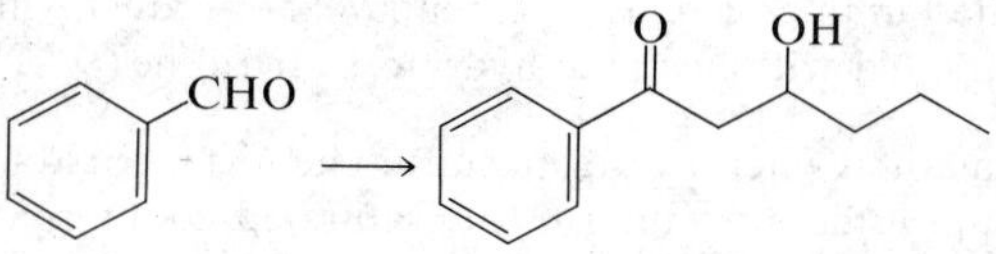

CHO

$\longrightarrow$

O   OH
‖ |

(l)

S
O   O

$\longrightarrow$

S –CH₂CH₂CH₃
O   O

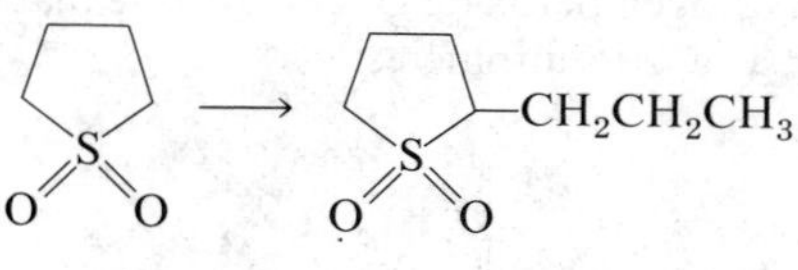

7. Mustard gas or bis(β-chloroethyl) sulfide, $(ClCH_2CH_2)_2S$, is an oily liquid that was used extensively as a poison gas in World War I. It is a deadly vesicant that causes blindness and numerous other effects. The active agent is actually the cyclic sulfonium salt

$$ClCH_2CH_2\!\!-\!\!\overset{+}{S}\!\!\underset{CH_2}{\overset{CH_2}{\diagup}}$$

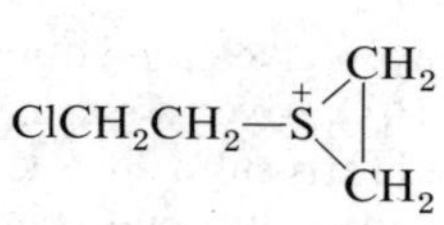

which reacts with nucleophilic materials in the body. The formation of the cyclic sulfonium salt can be regarded as an internal or intramolecular S_N2 displacement reaction. Write out this reaction mechanism. What mechanism does this process suggest for the subsequent reaction of the cyclic sulfonium salt with nucleophilic reagents?

8. Thiols are used as inhibitors in free radical reactions. In such use they end up as disulfides. The bond-dissociation energy of CH_3S—H is about 75 kcal mole^{-1}. Calculate $\Delta H°$ for the reaction

$$CH_4 + CH_3S \cdot \rightleftharpoons CH_3 \cdot + CH_3SH$$

In which direction does the equilibrium lie? Explain how CH_3SH works as an inhibitor.

9. Compare the relative acidities of benzenesulfonic acid with benzoic acid and of benzenesulfonamide with benzamide. Predict the relative acidities of acetophenone and phenyl methyl sulfone.

10. The normal single bond P—O distance is about 1.60 Å. The dative phosphorus–oxygen bond as in phosphorus oxychloride and in phosphates is almost 0.2 Å shorter. What does this comparison imply about the relative bond strengths of the two types of P–O bonds? The rearrangement of trimethyl phosphite to dimethyl methylphosphonate is exothermic by 47 kcal mole^{-1}. What do you think might be the principal driving force for this reaction?

11. One of the products of the reaction of sodium 3,5-dibromo-4-aminobenzenesulfonate with aqueous bromine is 2,4,6-tribromoaniline, the product of an **ipso-** substitution reaction. Write a reasonable reaction mechanism. How do you think *ipso* substitution is defined?

12. β-Keto sulfoxides react with aqueous acids or other acidic reagents to give α-hydroxy sulfides **(Pummerer rearrangement)**.

$$C_6H_5\overset{O}{\overset{\|}{C}}CH_2\overset{O}{\overset{\|}{S}}CH_3 \xrightarrow[\substack{DMSO \\ 25°}]{HCl, H_2O} C_6H_5\overset{O}{\overset{\|}{C}}CHSCH_3$$
$$\underset{OH}{|}$$
(80%)

Propose a mechanism for the Pummerer rearrangement. What product will result if such a β-keto sulfoxide is treated with acetic anhydride in pyridine?

13. Suggest a general synthetic method whereby organosulfur chemistry could be used to convert saturated ketones and esters into their α,β-unsaturated counterparts.

14. Oxidation of diphenyl disulfide with hydrogen peroxide gives a dioxide that could be formulated as either a disulfoxide or as a thiosulfonic ester.

$$C_6H_5\overset{O}{\overset{\|}{S}}-\overset{O}{\overset{\|}{S}}C_6H_5 \qquad C_6H_5S-\overset{O}{\underset{\underset{O}{\|}}{\overset{\|}{S}}}C_6H_5$$

a disulfoxide　　　　a thiosulfonic ester

Based on thermodynamic analogies available in Appendix I, which formulation do you think is correct?

15. Oxidation of trithioformaldehyde gives a mixture of two bis-sulfoxides, A and B. Further oxidation of A with H_2O_2 gives a single tris-sulfoxide, C, whereas further oxidation of B gives a mixture of tris-sulfoxides, C and D. What are the structures of the bis-sulfoxides, A and B, and the tris-sulfoxides, C and D?

trithioformaldehyde

Chapter 27
Difunctional Compounds

27.1
Introduction

To a first approximation the chemical properties of difunctional compounds are a summation of those of the individual functions. For example, cyclohex-2-en-1-one is a difunctional compound that undergoes typical alkene reactions (catalytic hydrogenation) and normal ketone reactions (reduction by lithium aluminum hydride.

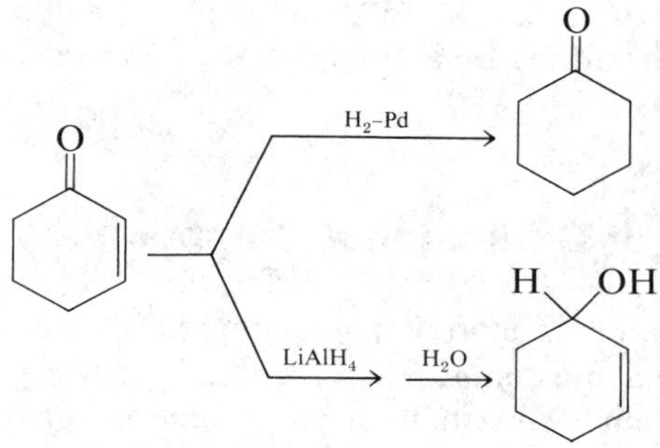

However, in many cases the two functional groups interact in such a way as to give the compound chemical properties that are not observed with the simple monofunctional compounds. In cyclohex-2-en-1-one the two functional groups form a conjugated system (Chapter 20), so it undergoes some special reactions, such as 1,4-addition of lithium dimethylcuprate (Section 20.3.A).

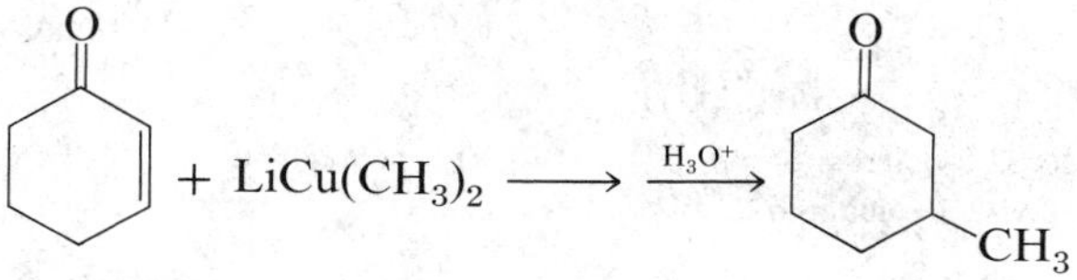

This is a special reaction of the difunctional compound because neither simple alkenes nor simple ketones react with the reagent.

In other cases the chemical properties of a difunctional compound are similar to those of a corresponding monofunctional compound in a qualitative sense but not in a quantitative sense. An example is the reaction of allyl bromide with azide ion. The reaction is a normal S_N2 replacement of a primary halide, but since the organic group is allylic, the reaction is over 50 times faster than it is with propyl bromide (Section 20.1.A).

$$CH_2{=}CHCH_2Br + N_3^- \xrightarrow{\text{faster}} CH_2{=}CHCH_2N_3 + Br^-$$

$$CH_3CH_2CH_2Br + N_3^- \xrightarrow{\text{slower}} CH_3CH_2CH_2N_3 + Br^-$$

In still other cases two functional groups in a molecule may enter into a chemical reaction *with each other.* An example is the intramolecular S_N2 reaction leading to cyclic ethers (problems 14 and 15 on page 184).

$$HOCH_2CH_2CH_2CH_2Br + OH^- \rightleftharpoons {}^-OCH_2CH_2CH_2CH_2Br \longrightarrow \underset{O}{\bigcirc} + Br^-$$

As may be seen in the foregoing examples, we have already encountered the reactions of a number of difunctional compounds, mainly in the study of conjugated systems (Chapter 20). In this chapter, we shall take up a few specific types of difunctional compounds, pointing out some of the unique chemistry which results from the cooperation or interaction of the two functional groups. Specific difunctional compounds we shall consider at this time are those containing the functional groups —OH and C=O: diols, diketones, dicarboxylic acids, hydroxy aldehydes, hydroxy ketones, hydroxy acids, and keto acids. The chemistry of these difunctional compounds forms a necessary foundation for our study of the chemistry of carbohydrates (Chapter 28). In Chapter 29 we shall consider another large and important class of difunctional compounds, amino acids.

27.2
Nomenclature of Difunctional Compounds

Recall that most simple monofunctional compounds are named in such a way that the ending of the name denotes the functional group: acetic **acid,** 3-penta**ol,** cyclohexan**one,** 1-but**ene.** Alkyl halides are exceptions to this generalization, in that they are considered as derivatives of the parent alkane, for example, 2-chloroheptane. When a compound contains two like functional groups, it is generally named in the same way except that the typical group suffix is combined with **di-** to indicate the presence of two groups. Numbers are used to locate the positions of the groups on the carbon skeleton.

(2E,5Z)-7-methyl-2,5-octa**diene** or
trans,cis-7-methyl-2,5-octa**diene**

$$CH_3C{\equiv}C-C{\equiv}CH$$

1,3-penta**diyne**

$$CH_3\overset{O}{\overset{\|}{C}}CH_2\overset{O}{\overset{\|}{C}}CH_3$$

2,4-pentane**dione**

cis-1,3-cyclohexane**diol**

$$\underset{\underset{Br}{|}}{CH_3}CHCH_2\underset{\underset{CH_3}{|}}{CH}CH_2CH_2Br$$

1,5-**dibromo**-3-methylhexane

trans-1,2-cyclopentane**dicarboxylic acid**

Diols are sometimes called **glycols.** This is a trivial nomenclature widely used in the chemical industry, particularly for some of the simpler diols, which are important commercial items. For example, ethylene glycol (1,2-ethanediol) is the most widely used antifreeze additive for automobile radiators.

The aliphatic dicarboxylic acids having up to ten carbons in their chains have

TABLE 27.1
Names of Some Dicarboxylic Acids

n	Formula	Common	IUPAC
C_2	$\overset{O\ O}{\underset{}{HOC\ COH}}$	oxalic acid	ethanedioic acid
C_3	$HOCCH_2COH$	malonic acid	propanedioic acid
C_4	$HOC(CH_2)_2COH$	succinic acid	butanedioic acid
C_5	$HOC(CH_2)_3COH$	glutaric acid	pentanedioic acid
C_6	$HOC(CH_2)_4COH$	adipic acid	hexanedioic acid
C_7	$HOC(CH_2)_5COH$	pimelic acid	heptanedioic acid
C_8	$HOC(CH_2)_6COH$	suberic acid	octanedioic acid
C_9	$HOC(CH_2)_7COH$	azelaic acid	nonanedioic acid
C_{10}	$HOC(CH_2)_8COH$	sebacic acid	decanedioic acid

common names that are used extensively in the chemical literature (Table 27.1). Although these names are no longer used for indexing purposes, the student should be aware of them since they were uniformly used before about 1975. The benzenedicarboxylic acids are known as phthalic, isophthalic, and terephthalic acids. The last-named acid is a highly important industrial material; it forms one of the building blocks of the synthetic fiber known as polyester, Dacron, or Terylene.

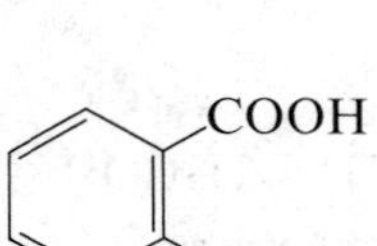

phthalic acid
1,2-benzenedicarboxylic acid

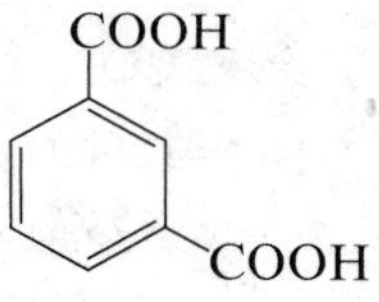

isophthalic acid
1,3-benzenedicarboxylic acid

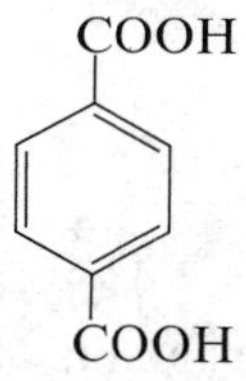

terephthalic acid
1,4-benzenedicarboxylic acid

Two unsaturated aliphatic diacids that have widely used common names are maleic and fumaric acids.

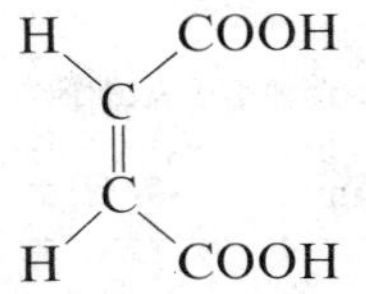

maleic acid
cis-butenedioic acid

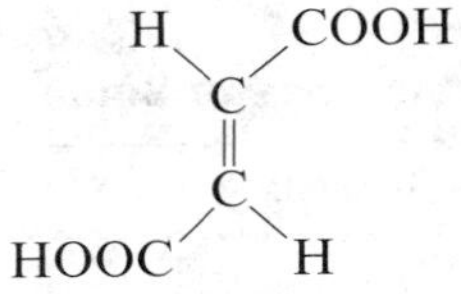

fumaric acid
trans-butenedioic acid

Aldehydes and functional derivatives corresponding to common diacids are frequently named as derivatives of the acids, in the same manner as is used to name simple aldehydes and functional derivatives.

$$HCCH_2CH_2CH$$

$$HCCH_2CCH_2CH$$

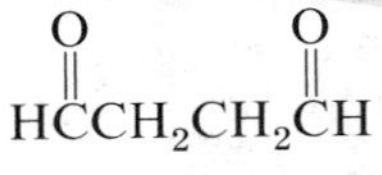

succinaldehyde

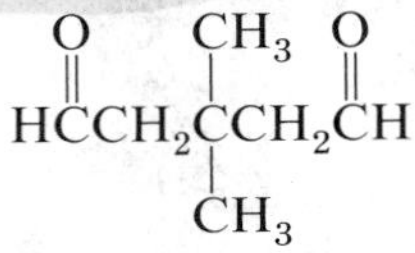

β,β-dimethylglutaraldehyde

$$ClCCCl$$

oxalyl chloride

$$CH_3OCCH_2CH_2CH_2CH_2COCH_3$$

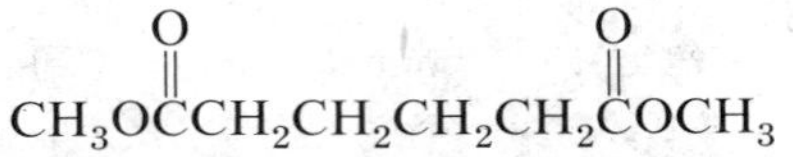

dimethyl adipate

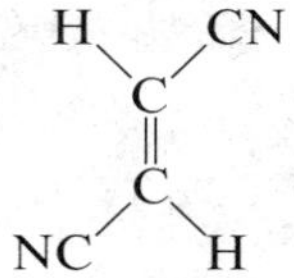

fumaronitrile

When a compound contains two different functional groups, one of the groups (the principal function) is usually expressed in the ending of the name and the other as a prefix.

$$HOCH_2CH_2COOH$$

3-hydroxypropan**oic acid**
β-hydroxypropionic acid

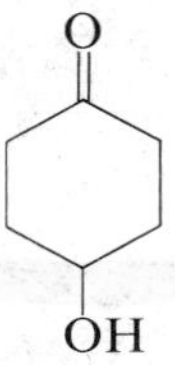

4-hydroxycyclohexan**one**

Alkenes and alkynes are exceptions in that the double or triple bond cannot be expressed as a prefix. For compounds containing a multiple bond and another functional group, two suffixes are used.

$$CH_2{=}CHCH_2CH_2OH$$
3-buten-1-ol

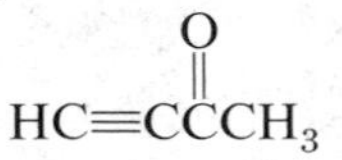

$$HC{\equiv}CCCH_3$$
3-butyn-2-one

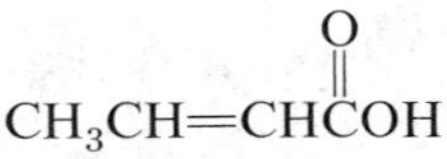

$$CH_3CH{=}CHCOH$$
2-butenoic acid

In the naming of a difunctional compound, a choice must be made as to which group is the principal function. The generally accepted order is carboxylic acid, sulfonic acid, ester, acyl halide, amide, nitrile, aldehyde, ketone, alcohol, thiol, amine, alkyne, alkene. Since alkenes and alkynes cannot be designated by prefixes, they are always indicated by a second suffix, which is placed *before* the final suffix of any function higher in the order. Table 27.2 contains a listing of the

TABLE 27.2
Functional Groups as Prefixes and Suffixes

Group	Prefix	Suffix
—COOH	carboxy-	-oic acid -carboxylic acid
—SO$_3$H	sulfo-	-sulfonic acid
—COOR	alkoxycarbonyl-	-carboxylate
—COCl	chloroformyl-	-oyl chloride -carbonyl chloride
—CONH$_2$	carbamoyl-	-amide -carboxamide
—CN	cyano-	-nitrile -carbonitrile
—CHO	formyl- oxo-	-al -carboxaldehyde -carbaldehyde
$\overset{\text{O}}{\underset{}{\overset{\|}{-\text{C}-}}}$	oxo- (IUPAC) keto- (common)	-one
—OH	hydroxy-	-ol
—SH	mercapto-	-thiol
—NH$_2$	amino-	-amine
—C≡C—	—	-yne
—C=C—	—	-ene
—Cl	chloro-	—

common functions with the appropriate prefix and suffix used to designate each one. Some examples of difunctional compound names follow.

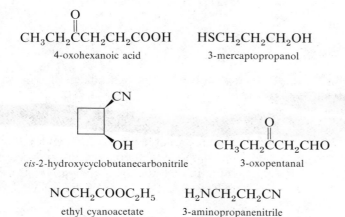

$$\underset{\text{4-oxohexanoic acid}}{\text{CH}_3\text{CH}_2\overset{\overset{\text{O}}{\|}}{\text{C}}\text{CH}_2\text{CH}_2\text{COOH}}$$ 4-oxohexanoic acid

HSCH$_2$CH$_2$CH$_2$OH
3-mercaptopropanol

cis-2-hydroxycyclobutanecarbonitrile

$$\underset{\text{3-oxopentanal}}{\text{CH}_3\text{CH}_2\overset{\overset{\text{O}}{\|}}{\text{C}}\text{CH}_2\text{CHO}}$$ 3-oxopentanal

NCCH$_2$COOC$_2$H$_5$
ethyl cyanoacetate

H$_2$NCH$_2$CH$_2$CN
3-aminopropanenitrile

EXERCISE: Write the structures of a six-carbon compound containing each pairwise combination of the following functional groups: —OH, C=C, C=O (both aldehyde and ketone), COOH. Assign a name to each of your fifteen structures.

**27.3
Diols**

A. *Preparation of Diols*

1,2-Diols are usually prepared from the corresponding alkene by the net addition of two hydroxy groups to the double bond **(hydroxylation).** Direct hydroxylation may be accomplished by oxidation of the alkene with $KMnO_4$ or OsO_4 (Section 11.6.E). Overall hydroxylation may be achieved by conversion of the alkene to an epoxide (Section 11.6.E), which is then hydrolyzed to the 1,2-diol (Section 10.12).

Both hydroxylation processes are stereospecific. Direct hydroxylation with $KMnO_4$ or OsO_4 results in *syn* addition of the two hydroxy groups to the double bond. *Syn* hydroxylation of a cyclic alkene having a *cis* double bond affords a *cis* diol; for example, *cis*-cyclooctene yields *cis*-1,2-cyclooctenediol.

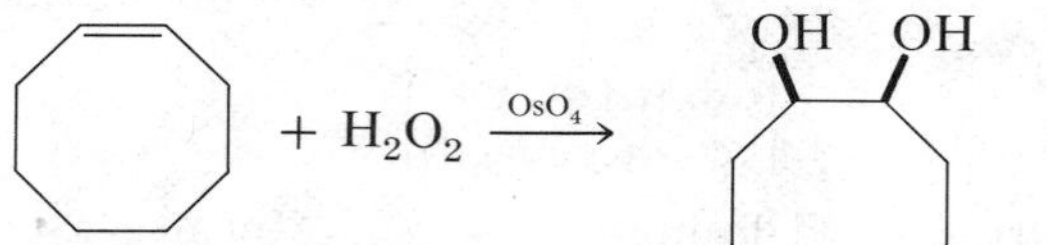

cis-cyclooctene   *cis*-1,2-cyclooctanediol

If the double bond in the ring is *trans,* then *syn* hydroxylation affords the *trans* diol.

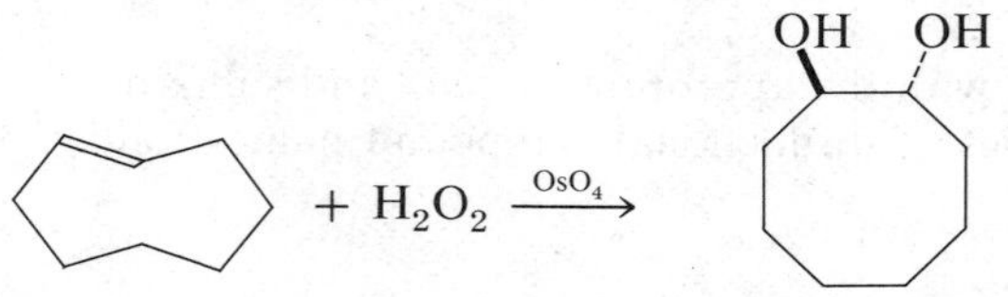

trans-cyclooctene   *trans*-1,2-cyclooctanediol

Recall that stable *trans* isomers only exist for cyclic alkenes having eight or more atoms in the ring (Section 11.4).

The two-step procedure for hydroxylation of an alkene is also stereospecific, but this process produces net *anti* addition of the two hydroxy groups to the double bond. For example, cyclohexene (*cis* double bond) reacts with peroxyacetic acid to give cyclohexene oxide. This compound undergoes acid-catalyzed ring opening by the S_N2 mechanism, resulting in *inversion of configuration* at one of the two C—O bonds. Thus, the overall result is formation of the *trans* 1,2-diol.

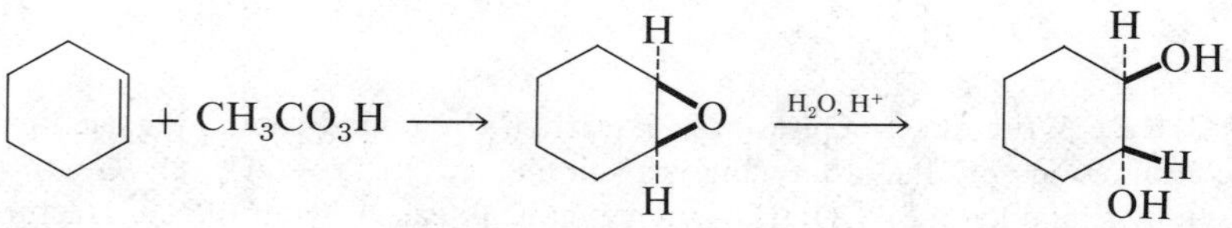

cyclohexene oxide   *trans*-1,2-cyclohexanediol

With acyclic alkenes, different diols result from addition to the *cis* or *trans* isomer of the alkene. For a symmetrical alkene, such as 2-butene, *syn* hydroxylation of the *cis* isomer gives a *meso* diol.

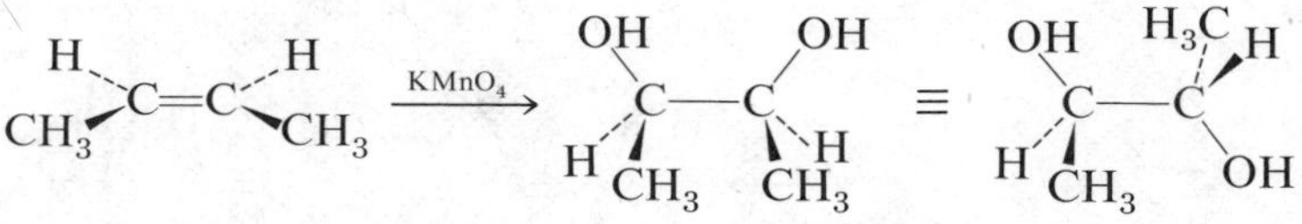

meso-2,3-butanediol

Syn hydroxylation of the *trans* isomer gives a 50:50 mixture of two enantiomeric diols. These two products arise from addition of the reagent to the two faces of the planar alkene molecule. Since the reagent is achiral, the transition states leading to the two products are enantiomeric and equal in energy. The product is therefore a **racemic mixture** (Section 7.4). To distinguish this mixture of enantiomers from the *meso* diol, it is frequently designated as (±) or *dl* (meaning an equimolar or racemic mixture of the dextrorotatory and levorotatory enantiomers).

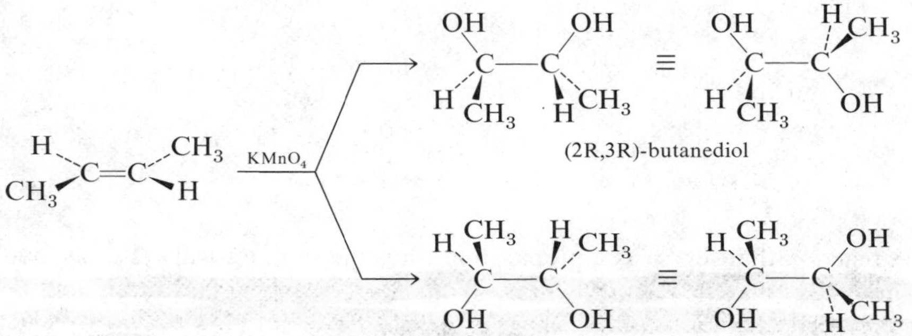

(2R,3R)-butanediol

(2S,3S)-butanediol
[mixture = (±)-2,3-butanediol]

For *anti* hydroxylation, the situation is just exactly reversed—the *cis* alkene gives the (±)-diol and the *trans* alkene affords the *meso* diol. When the acyclic alkene is not symmetrical, both *syn* and *anti* additions of each isomer produce racemic mixtures. The following equations illustrate the *syn* hydroxylation of the two isomers of 2-pentene.

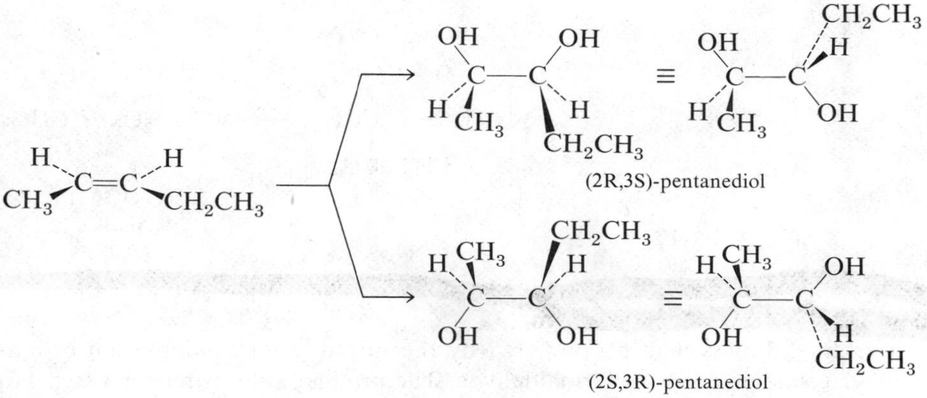

(2R,3S)-pentanediol

(2S,3R)-pentanediol

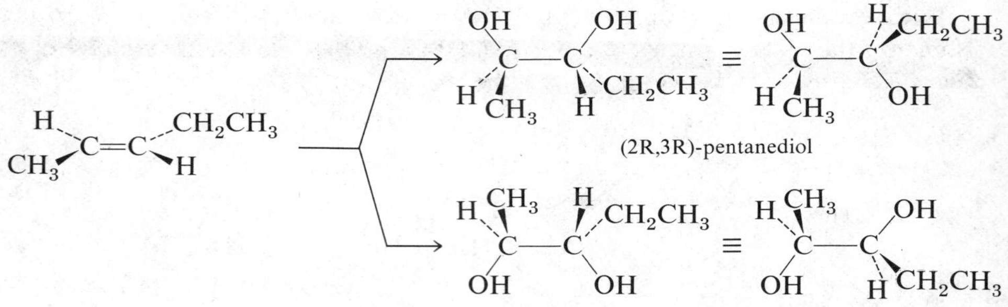

(2R,3R)-pentanediol

(2S,3S)-pentanediol

> The (2R,3S) and (2S,3R) isomers of 2,3-pentanediol are called **erythro** isomers, and the (2R,3R) and (2S,3S) isomers are called **threo** isomers. These names derive from carbohydrate chemistry (Chapter 28) and are frequently used for other simple difunctional compounds. When a compound contains two asymmetric carbons that have two identical attached groups and a third that differs, *the isomer that would be meso if the third groups were identical is the erythro isomer.* The other isomer is the *threo* isomer.

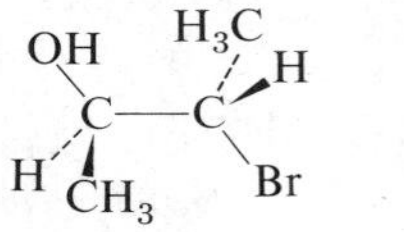

erythro-3-bromo-2-butanol 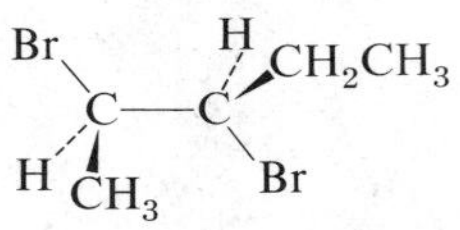   *threo*-2,3-dibromopentane

A reaction that serves as a preparation for some symmetrical 1,2-diols involves the **reductive dimerization** of ketones. The reducing agent is generally an electropositive metal, such as sodium or magnesium. The reaction occurs by electron transfer from the metal to the ketone to produce a **ketyl**, or **radical anion.** Dimerization of two radical anions affords the dianion of a 1,2-diol, which is hydrolyzed in a separate step to the diol itself.

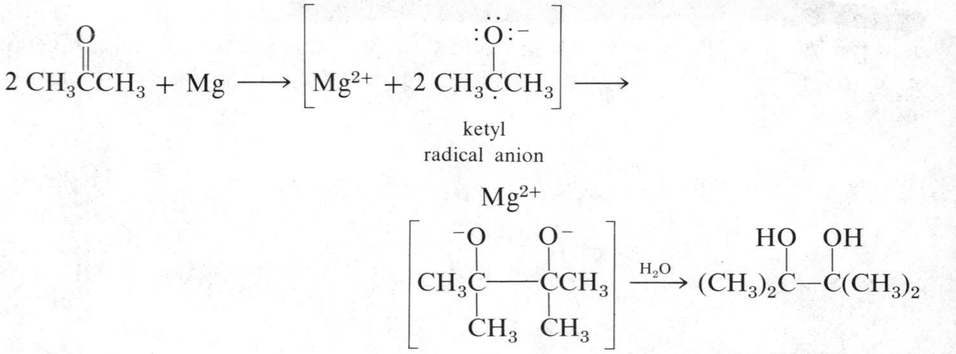

ketyl
radical anion

(43–50%)
pinacol

Because the diol produced from acetone has the trivial name "pinacol," this reaction is called the **pinacol reaction.**

Certain 1,3-diols may be prepared by the mixed aldol condensation of α,α-dialkylacetaldehydes with formaldehyde. The product aldol is reduced to a 1,3-diol by the Cannizzaro reaction (Sections 13.7.F and 13.8.D).

$$(CH_3)_2CHCHO + CH_2O \xrightarrow{KOH} HOCH_2\overset{\overset{\displaystyle CH_3}{|}}{\underset{\underset{\displaystyle CH_3}{|}}{C}}CH_2OH$$

Other types of diols are generally prepared by reduction of the appropriate dicarbonyl compounds, as indicated by the following examples.

$$C_2H_5O\overset{\overset{\displaystyle O}{\|}}{C}(CH_2)_8\overset{\overset{\displaystyle O}{\|}}{C}OC_2H_5 \xrightarrow{LiAlH_4} HO(CH_2)_{10}OH$$
(75%)
1,10-decanediol

$$CH_3\overset{\overset{\displaystyle O}{\|}}{C}(CH_2)_3\overset{\overset{\displaystyle O}{\|}}{C}CH_3 + NaBH_4 \xrightarrow{C_2H_5OH} CH_3\overset{\overset{\displaystyle OH}{|}}{C}H(CH_2)_3\overset{\overset{\displaystyle OH}{|}}{C}HCH_3$$
(80%)
2,6-heptanediol

EXERCISE: Write equations for the overall *anti* hydroxylation of *cis*- and *trans*-2-butene and of *cis*- and *trans*-2-pentene using the sequence of reagents (1) peroxyacetic acid and (2) aqueous acid. Clearly illustrate the stereochemistry at each step of the process.

B. Reactions of Diols

1. DEHYDRATION: THE PINACOL REARRANGEMENT. Dehydration of 1,2-diols under acid catalysis is frequently accompanied by skeletal rearrangement. For example, pinacol (2,3-dimethylbutane-2,3-diol, page 846) reacts with sulfuric acid to give *t*-butyl methyl ketone, which has the trivial name "pinacolone."

$$CH_3\overset{\overset{\displaystyle OH}{|}}{\underset{\underset{\displaystyle CH_3}{|}}{C}}\text{---}\overset{\overset{\displaystyle OH}{|}}{\underset{\underset{\displaystyle CH_3}{|}}{C}}CH_3 \xrightarrow{H_2SO_4} CH_3\overset{\overset{\displaystyle CH_3}{|}}{\underset{\underset{\displaystyle CH_3}{|}}{C}}\text{---}\overset{\overset{\displaystyle O}{\|}}{C}CH_3$$
pinacol   pinacolone

The mechanism of this **pinacol rearrangement** involves 1,2-migration of a methyl group and its bonding electron pair from one carbinyl position to an adjacent electron-deficient center (Section 10.7.B). The driving force for the rearrangement is formation of a stable oxonium ion, the conjugate acid of a ketone.

$$(CH_3)_2\overset{\overset{\displaystyle OH}{|}}{C}\text{---}\overset{\overset{\displaystyle OH}{|}}{C}(CH_3)_2 + H^+ \rightleftharpoons (CH_3)_2\overset{\overset{\displaystyle OH}{|}}{C}\text{---}\overset{\overset{\displaystyle \overset{+}{O}H_2}{|}}{C}(CH_3)_2 \rightleftharpoons$$

$$CH_3\overset{\overset{\displaystyle :OH}{|}}{\underset{\underset{\displaystyle CH_3}{|}}{C}}\text{---}\overset{\overset{\displaystyle CH_3}{}}{\underset{\underset{\displaystyle CH_3}{}}{\overset{+}{C}}} \rightleftharpoons CH_3\overset{\overset{\displaystyle \overset{+}{O}H}{\|}}{C}\text{---}C(CH_3)_3 \rightleftharpoons CH_3\overset{\overset{\displaystyle O}{\|}}{C}C(CH_3)_3 + H^+$$

tertiary carbocation   oxonium ion   pinacolone

By combining the pinacol reaction with this acid-catalyzed rearrangement process, interesting and unusual compounds may be prepared.

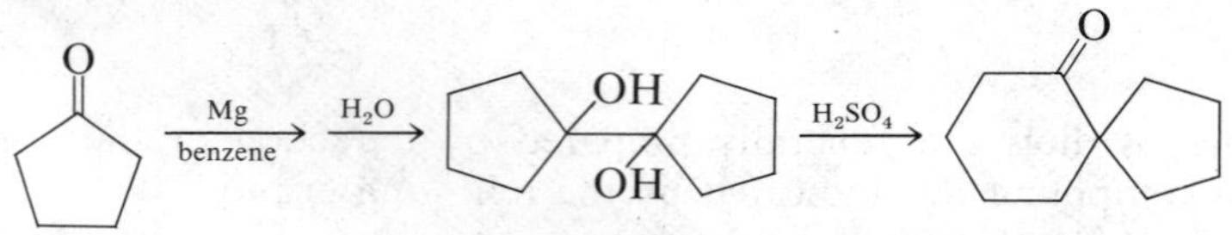

spiro[4.5]decan-6-one

Bicyclic compounds having one carbon common to both rings are **spiro** compounds.
The nomenclature is based on the following scheme.

$$(CH_2)_n C \quad (CH_2)_m \equiv \text{spiro}[n.m]\text{alkane}$$

Numbering starts next to the common carbon and proceeds around the smaller ring
first.

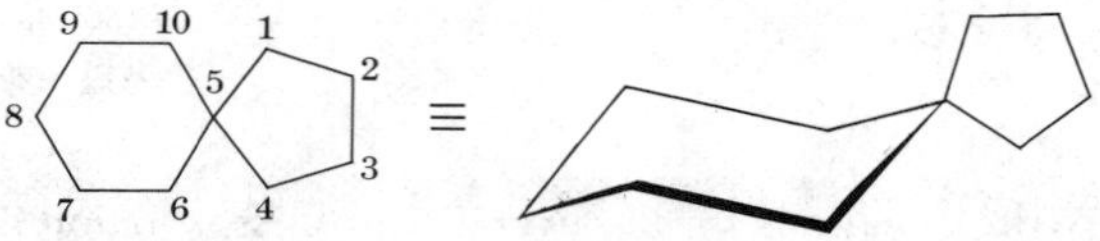

spiro[4.5]decane

Dehydration of 1,4- and 1,5-diols often leads to the formation of cyclic ethers,
particularly when one of the hydroxy groups is tertiary.

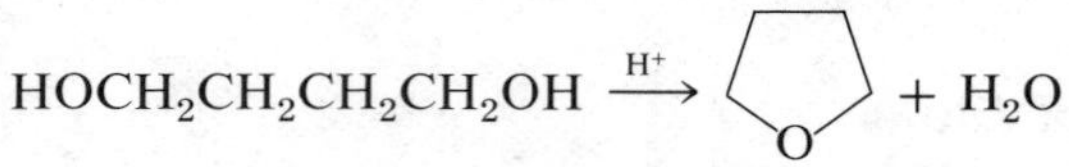

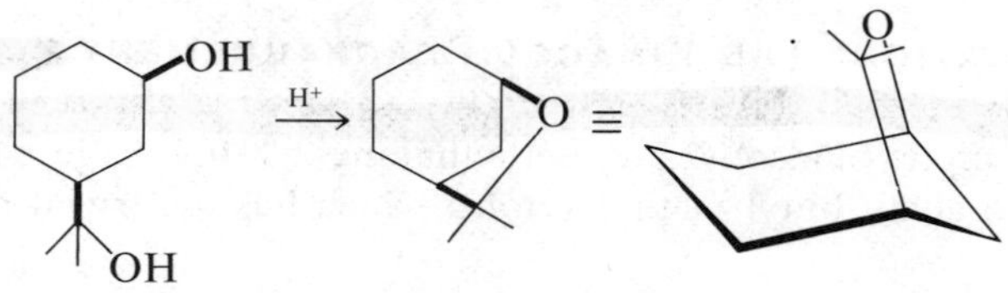

$$HOCH_2CH_2CH_2CH_2OH \xrightarrow{H^+} \bigcirc + H_2O$$

In the first case the reaction undoubtedly occurs by intramolecular nucleophilic
displacement on the initially formed oxonium ion.

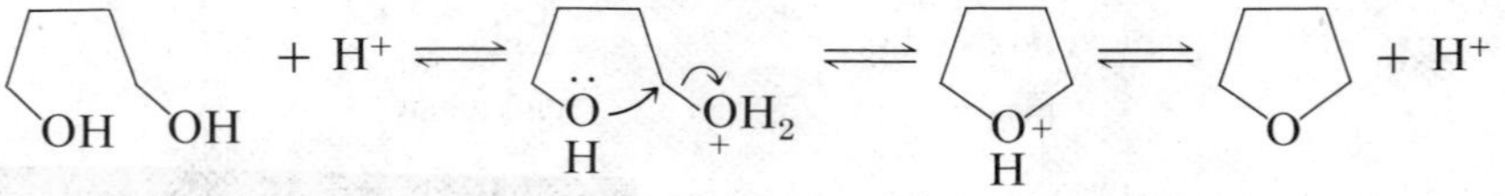

The second reaction probably involves the formation of a tertiary carbocation
which is trapped by the secondary hydroxy group.

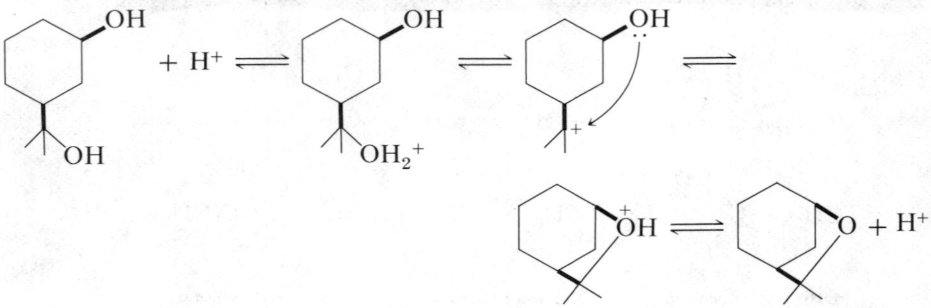

In this case the cyclization is possible because the two groups are *cis*. The *trans* analog cannot give a cyclic product.

EXERCISE: What are the principal products of the pinacol rearrangements of the following diol when $Y=CH_3$ and when $Y=NO_2$.

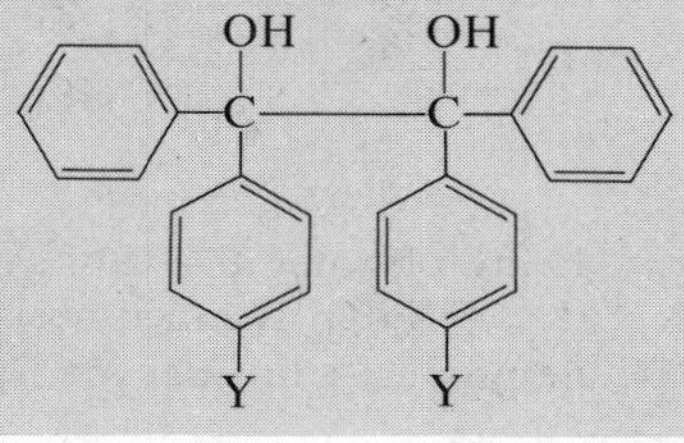

---

2. **OXIDATION.** 1,2-Diols undergo easy cleavage of the C—C bond joining the two hydroxy carbons when treated with periodic acid or lead tetraacetate. Combined with the hydroxylation process, this oxidation constitutes a method for the cleavage of alkenes that is complementary to ozonolysis (Section 11.6.E).

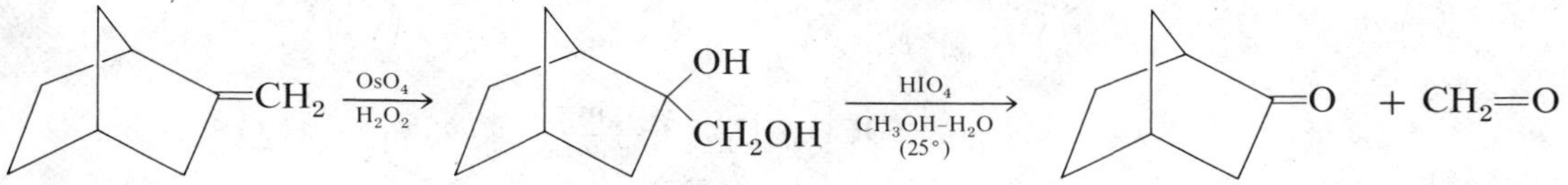

(85%)
bicyclo[2.2.1]heptan-2-one
norbornanone

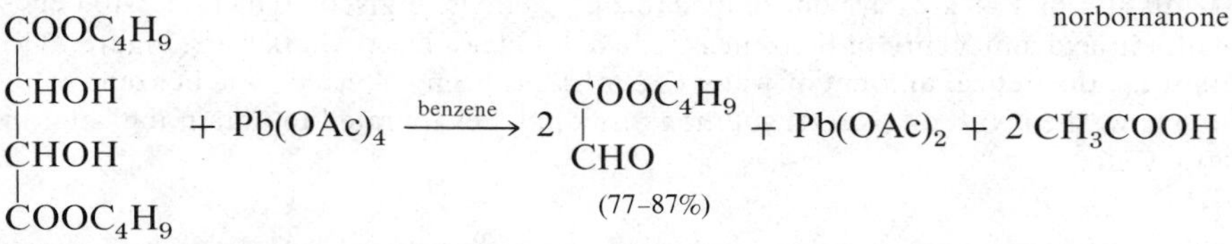

(77–87%)

The mechanism of periodic oxidation has been studied in detail and involves the formation of a cyclic diester of periodic acid. Decomposition of this cyclic diester yields the two carbonyl fragments and iodic acid.

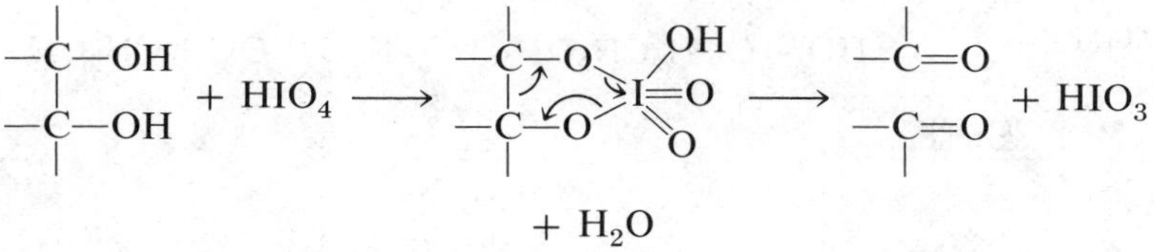

The lead tetraacetate oxidation is believed to occur by fragmentation of an alkoxy lead compound.

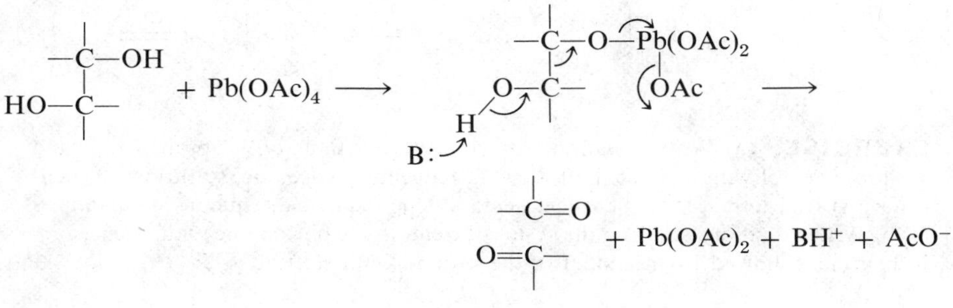

Various procedures have been developed in which alkene hydroxylation and the diol cleavage reactions are combined into one operation. One such reaction (the Lemieux–Johnson reaction) involves treating an alkene with sodium periodate and a catalytic amount of osmium tetroxide.

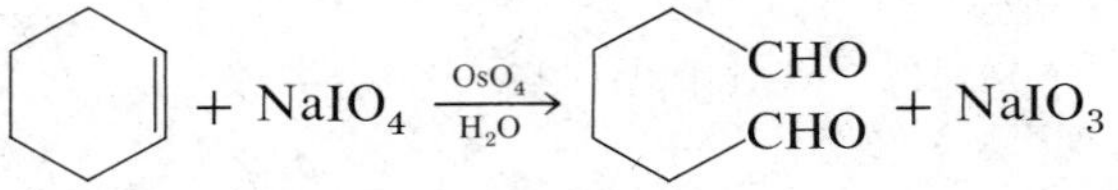

$$\text{[cyclohexene]} + NaIO_4 \xrightarrow[H_2O]{OsO_4} \text{[cyclohexane with CHO, CHO]} + NaIO_3$$

> A mixture of 15 mL of ether, 15 mL of water, 0.41 g of cyclohexene, and 0.065 g of OsO_4 is stirred at 25° while 2.32 g of $NaIO_4$ is added over a period of 40 min. After an additional 80 min at 25°, the product adipaldehyde is isolated in 77% yield.

3. KETAL AND ACETAL FORMATION. Certain 1,2- and 1,3-diols react with aldehydes and ketones in the presence of acid catalysts to form cyclic acetals or ketals (Section 13.7.B).

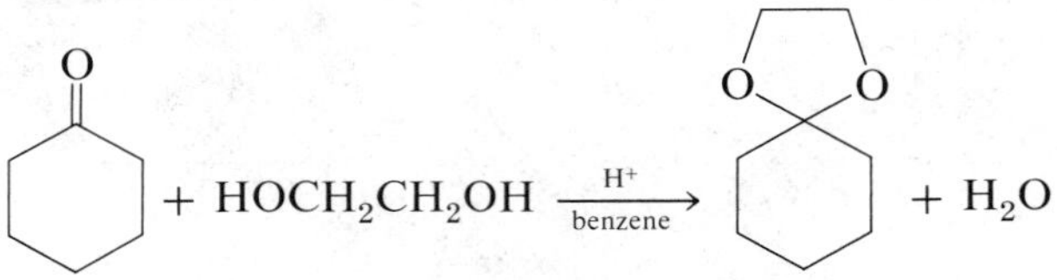

$$\text{[cyclohexanone]} + HOCH_2CH_2OH \xrightarrow[benzene]{H^+} \text{[cyclic ketal]} + H_2O$$

> A mixture of 11.8 g of cyclohexanone, 8.2 g of ethylene glycol, 0.05 g of p-toluenesulfonic acid and 50 mL of benzene is refluxed under a Dean–Stark trap (Figure 13.7) until the theoretical amount of water (2.2 mL) has been collected. The benzene solution is washed with dil. NaOH solution, dried, and evaporated to obtain the ketal in 80% yield.

The probable mechanism for this reaction is identical to that given in Section 13.7.B for ketal formation except that the addition of the second alcohol group is an *intramolecular* process.

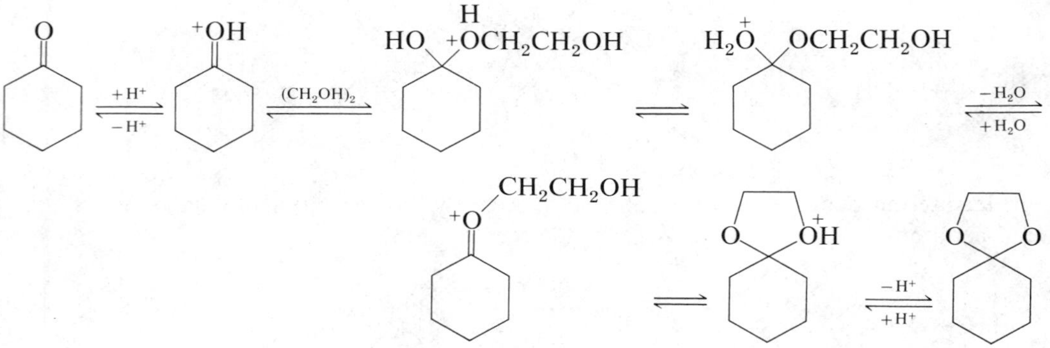

EXERCISE: (a) Write equations showing the reactions of 1,2-propanediol and 1,3-propanediol with each of the following reagents. (i) acetone, sulfuric acid, benzene; (ii) lead tetraacetate, benzene, heat; (iii) periodic acid, aqueous methanol.
(b) Write equations showing the result of treating cyclohexanone with magnesium in benzene followed by treatment of the product with sulfuric acid.

27.4
Hydroxy Aldehydes and Ketones

A. *Synthesis of Hydroxy Aldehydes and Ketones*

α-Hydroxy ketones result from the treatment of esters with sodium in an inert solvent such as ether or benzene. Such compounds are called **acyloins,** and the reaction is called the **acyloin condensation.** The initial product of the reaction is the disodium salt of an enediol, which is hydrolyzed to give the acyloin.

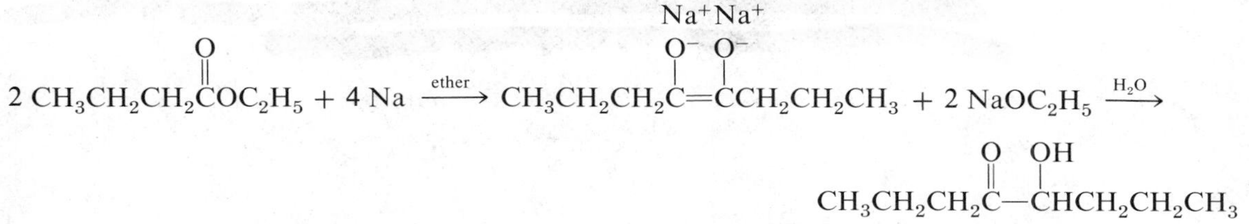

$$2\ CH_3CH_2CH_2\overset{O}{\overset{\|}{C}}OC_2H_5 + 4\ Na \xrightarrow{\text{ether}} CH_3CH_2CH_2\overset{\overset{\displaystyle Na^+Na^+}{\overset{\displaystyle O^-\ O^-}{|\ \ \ |}}}{C}{=}CCH_2CH_2CH_3 + 2\ NaOC_2H_5 \xrightarrow{H_2O}$$

$$CH_3CH_2CH_2\overset{O}{\overset{\|}{C}}{-}\overset{OH}{\overset{|}{C}}HCH_2CH_2CH_3$$

The acyloin condensation is a useful method for the synthesis of ring compounds, particularly for medium-sized rings (8–13 members). In such cases, the reaction must be carried out under conditions of high dilution to suppress intermolecular reactions.

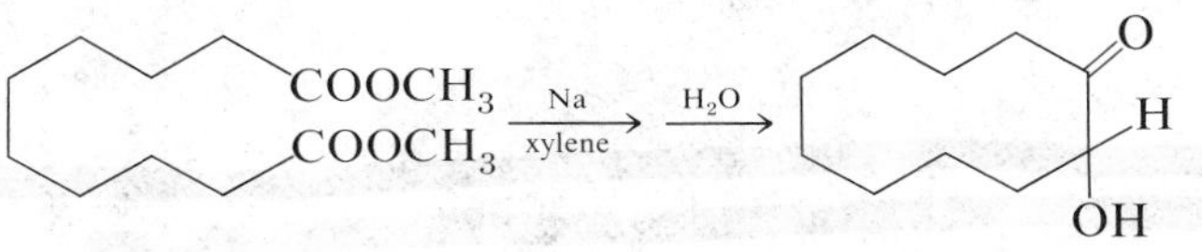

(66%)
2-oxocyclodecanol

The acyloin condensation is related mechanistically to the pinacol reaction (page 846) in that electron transfer from sodium to the ester carbonyl produces an intermediate ketyl. The chief side reaction is the Claisen condensation (Section 19.9), which stems from the alkoxide ion produced as a by-product in the reaction.

Aromatic aldehydes are converted into acyloins by sodium cyanide in aqueous ethanol. The reaction is called the **benzoin condensation,** and cyanide ion is a specific catalyst.

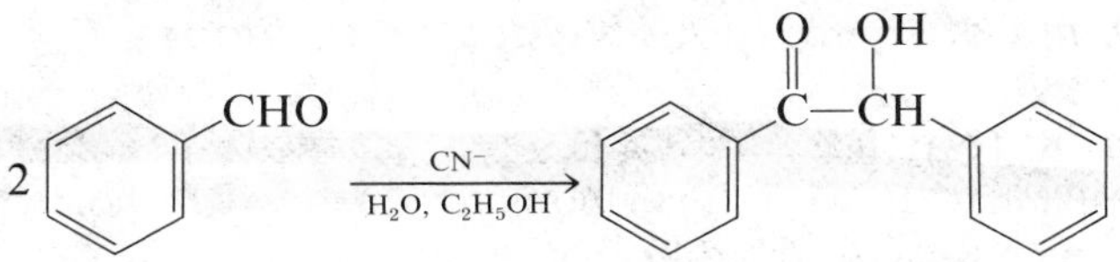

benzaldehyde

(92%)
benzoin

The catalyst functions by first adding to the carbonyl group to form the cyanohydrin (see Section 13.7.E). The former aldehyde hydrogen is now α to a cyano group and is sufficiently acidic to be removed by a base. The resulting carbanion then adds to another molecule of aldehyde to give an intermediate cyano diol. Elimination of cyanide ion yields the acyloin and regenerates the catalyst.

$$RCHO + CN^- \rightleftharpoons \underset{\text{RCHCN}}{\overset{O^-}{\mid}} \overset{H_2O}{\underset{-OH}{\rightleftharpoons}} \underset{\text{RCHCN}}{\overset{OH}{\mid}} \overset{OH^-}{\underset{H_2O}{\rightleftharpoons}} \underset{\text{RC—CN}}{\overset{OH}{\mid}} \overset{RCHO}{\rightleftharpoons}$$

$$\underset{\underset{CN}{\mid}}{\overset{HO \quad O^-}{\underset{\mid}{RC—CHR}}} \overset{H_2O}{\underset{-OH}{\rightleftharpoons}} \underset{\underset{CN}{\mid}}{\overset{HO \quad OH}{\underset{\mid}{RC—CHR}}} \overset{-OH}{\underset{H_2O}{\rightleftharpoons}} \underset{\underset{CN}{\mid}}{\overset{-O \quad OH}{\underset{\mid}{RC—CHR}}} \rightleftharpoons \overset{O \quad OH}{\underset{\mid}{RC—CHR}} + CN^-$$

Although the acyloin and benzoin condensations produce the same type of product, it is important to remember that they involve *entirely different mechanisms*.

The most general synthesis of β-hydroxy aldehydes and ketones is the aldol condensation (Section 13.7.F). Recall that simple aldehydes condense to form β-hydroxy aldehydes when treated with cold aqueous base.

$$2\ CH_3CHO \xrightarrow[\substack{H_2O \\ 0°}]{NaOH} \underset{\overset{\mid}{\text{OH}}}{CH_3CHCH_2CHO}$$

Under more forcing conditions, such as are necessary to accomplish the initial condensation with aldehydes of more than six carbons, the β-hydroxy aldehyde undergoes dehydration to give the α,β-unsaturated aldehyde.

heptanal

$$\xrightarrow[\substack{EtOH \\ reflux}]{NaOC_2H_5}$$

(70%)
2-pentyl-2-nonenal

Mixed aldol condensations can be performed by first converting a ketone into the lithium enolate, which is then treated with an aldehyde. Hydrolysis of the initially formed alkoxide with water affords the β-hydroxy ketone, usually in good yield.

$$\underset{\overset{\parallel}{O}}{CH_3CCH_3} \xrightarrow[\text{THF, }-70°]{(i\text{-}C_3H_7)_2NLi} \quad \xrightarrow{H_2O}$$

(88%)
4-hydroxy-4-phenyl-2-butanone

B. *Reactions of Hydroxy Aldehydes and Ketones*

1. DEHYDRATION. β-Hydroxy aldehydes and ketones undergo acid-catalyzed dehydration more easily than normal alcohols. The following examples illustrate the magnitude of the differences.

Relative Rate

$$\underset{\overset{\mid}{\text{OH}}}{CH_3CHCH_2CH_2CH_3} \xrightarrow{H_2SO_4} \text{mixture of pentenes} \qquad 1$$

$$\underset{\overset{\mid}{\text{OH}}}{CH_3CHCH_2}\underset{\overset{\parallel}{\text{O}}}{CCH_3} \xrightarrow{H_2SO_4} CH_3CH{=}CHCCH_3 \qquad {>}10^5$$

Recall that the dehydration of a secondary or tertiary alcohol involves the formation of an intermediate carbocation; the rate of formation of this intermediate determines the rate of dehydration (Section 10.7.B).

(1)

$$-\overset{\text{H}}{\underset{|}{\text{C}}}-\overset{\text{OH}}{\underset{|}{\text{C}}}-\ +\ \text{H}^+\ \underset{\text{fast}}{\rightleftharpoons}\ -\overset{\text{H}}{\underset{|}{\text{C}}}-\overset{\overset{+}{\text{O}}\text{H}_2}{\underset{|}{\text{C}}}-$$

(2)

$$-\overset{\text{H}}{\underset{|}{\text{C}}}-\overset{\overset{+}{\text{O}}\text{H}_2}{\underset{|}{\text{C}}}-\ \underset{\text{slow}}{\rightleftharpoons}\ -\overset{\text{H}}{\underset{|}{\text{C}}}-\text{C}^{+}\big<\ +\ \text{H}_2\text{O}$$

(3)

$$-\overset{\text{H}}{\underset{|}{\text{C}}}-\text{C}^{+}\big<\ \underset{\text{fast}}{\rightleftharpoons}\ \big>\text{C}=\text{C}\big<\ +\ \text{H}^+$$

β-Hydroxy ketones undergo dehydration by a different mechanism, involving the *enol* form of the ketone. The rate-determining step is formation of the enol. Elimination of water from the protonated enol gives a stable oxonium ion, which is simply the protonated form of the α,β-unsaturated ketone.

(1)

$$-\overset{\text{OH}}{\underset{|}{\text{C}}}-\text{CH}-\overset{\text{O}}{\overset{\|}{\text{C}}}-\ \underset{\text{slow}}{\rightleftharpoons}\ -\overset{\text{OH}}{\underset{|}{\text{C}}}-\overset{\text{OH}}{\underset{|}{\text{C}}}=\text{C}-$$

(2)

$$-\overset{\text{OH}}{\underset{|}{\text{C}}}-\overset{\text{OH}}{\underset{|}{\text{C}}}=\text{C}-\ +\ \text{H}^+\ \underset{\text{fast}}{\rightleftharpoons}\ -\overset{\overset{+}{\text{O}}\text{H}_2}{\underset{|}{\text{C}}}-\overset{\text{OH}}{\underset{|}{\text{C}}}=\text{C}-$$

(3)

$$-\overset{\overset{+}{\text{O}}\text{H}_2}{\underset{|}{\text{C}}}-\overset{\text{OH}}{\underset{|}{\text{C}}}=\text{C}-\ \underset{\text{fast}}{\rightleftharpoons}\ \left[\ \big>\overset{+}{\text{C}}-\overset{\text{OH}}{\underset{|}{\text{C}}}=\text{C}-\ \longleftrightarrow\ \big>\text{C}=\text{C}-\overset{^+\text{OH}}{\overset{\|}{\text{C}}}-\ \right]$$

(4)

$$\big>\text{C}=\text{C}-\overset{\overset{+}{\text{O}}\text{H}}{\overset{\|}{\text{C}}}-\ \underset{\text{fast}}{\rightleftharpoons}\ \big>\text{C}=\text{C}-\overset{\text{O}}{\overset{\|}{\text{C}}}-\ +\ \text{H}^+$$

Normal alcohols do not undergo dehydration under basic conditions, as shown by the fact that *t*-butyl alcohol solutions of potassium *t*-butoxide are quite stable (Section 10.7.A). However, β-hydroxy aldehydes and ketones undergo dehydration fairly easily under basic conditions. In this case the dehydration actually proceeds via the enolate ion (Section 13.6.B).

$$-\overset{\text{OH}}{\underset{|}{\text{C}}}-\text{CH}-\overset{\text{O}}{\overset{\|}{\text{C}}}-\ +\ \text{OH}^-\ \rightleftharpoons\ -\overset{\text{OH}}{\underset{|}{\text{C}}}-\overset{\text{O}^-}{\underset{|}{\text{C}}}=\text{C}-$$

$$-\overset{\text{OH}}{\underset{|}{\text{C}}}-\overset{\text{O}^-}{\underset{|}{\text{C}}}=\text{C}-\ \rightleftharpoons\ \big>\text{C}=\text{C}-\overset{\text{O}}{\overset{\|}{\text{C}}}-\ +\ \text{OH}^-$$

In contrast to the easy dehydration of β-hydroxy carbonyl compounds, α-hydroxy ketones undergo acid-catalyzed dehydration with even *more* difficulty than normal alcohols. In this case, the intermediate carbocation would be destabilized by the inductive effect of the adjacent carbonyl group.

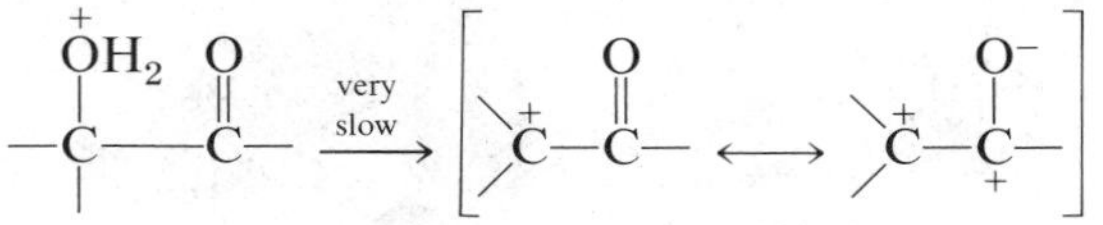

An example is the preparation of 3-methylbut-3-en-2-one by heating a mixture of the α-hydroxy ketone and *p*-toluenesulfonic acid in an oil bath at 150°. These conditions are far more vigorous than required for dehydration of normal tertiary-alcohols.

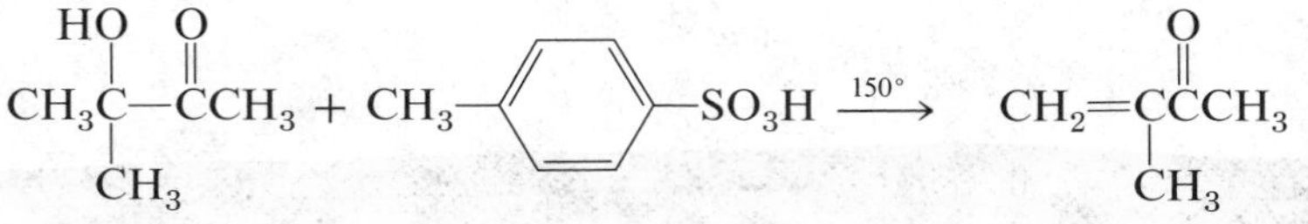

(74%)
3-methylbut-3-en-2-one

2. CYCLIC HEMIACETALS AND HEMIKETALS. Hydroxy aldehydes and ketones usually exist, to some extent, in a cyclic hemiacetal or hemiketal form (Section 13.7.B), particularly when the ring is five- or six-membered.

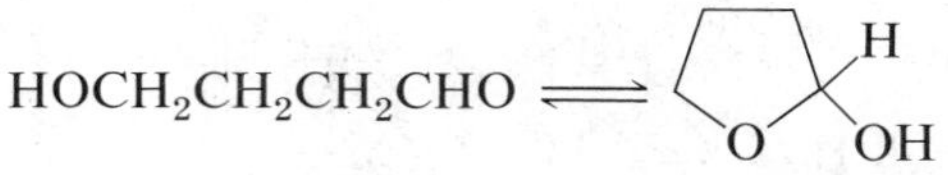

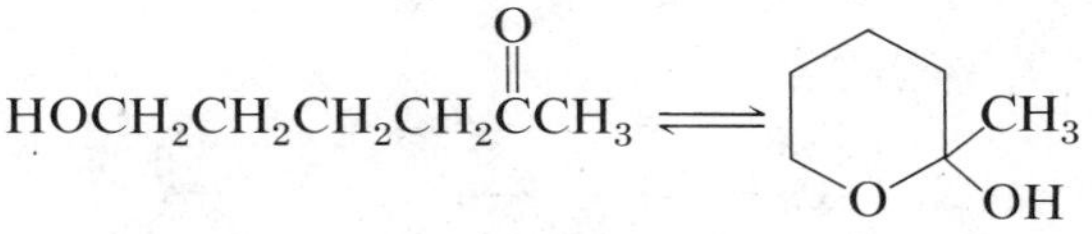

$$HOCH_2CH_2CH_2CHO \rightleftharpoons$$

$$HOCH_2CH_2CH_2CH_2\overset{O}{\overset{\|}{C}}CH_3 \rightleftharpoons$$

The data in Table 27.3 show that the cyclic form predominates with 4- and 5-hydroxy aldehydes. The formation of a cyclic hemiacetal from such a hydroxy carbonyl compound is subject to acid or base catalysis, as in the formation of acetals by intermolecular reaction (Section 13.7.B). However, when five- or six-membered rings are involved, the cyclization is so facile that it occurs even under neutral conditions. Thus, any reaction that would nominally give a 4- or 5-hydroxy carbonyl compound will yield an equilibrium of the open-chain and ring-closed isomers.

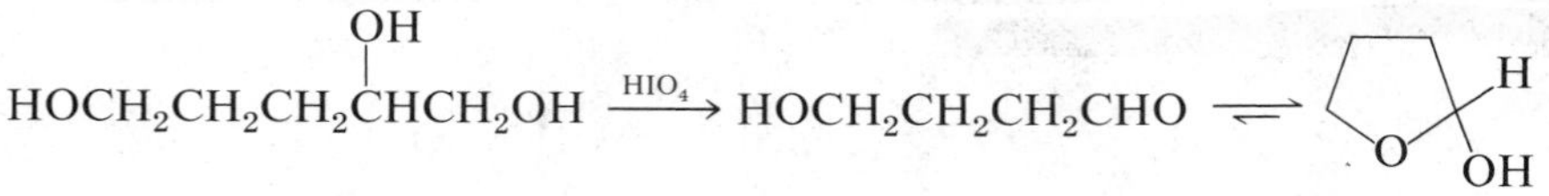

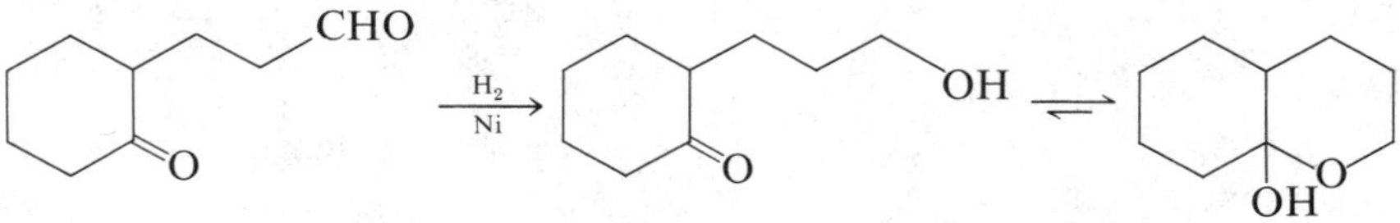

$$HOCH_2CH_2CH_2\overset{OH}{\overset{|}{C}H}CH_2OH \xrightarrow{HIO_4} HOCH_2CH_2CH_2CHO \rightleftharpoons$$

TABLE 27.3

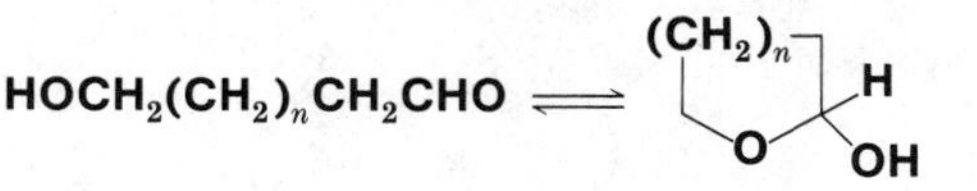

$$HOCH_2(CH_2)_nCH_2CHO \rightleftharpoons$$

n	Ring Size	Percent Free Aldehyde
1	5	11
2	6	6
3	7	85

Like noncyclic hemiacetals, these compounds react with alcohols under acid catalysis to give acetals or ketals.

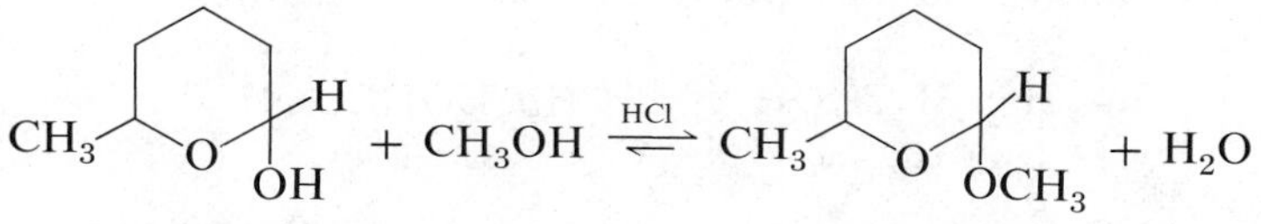

$$CH_3 + CH_3OH \underset{HCl}{\rightleftharpoons} CH_3 + H_2O$$

Since there is usually a small amount of the open-chain hydroxy carbonyl compound in equilibrium with the cyclic hemiacetal form, solutions of such compounds can show reactions of either form, as the following examples show.

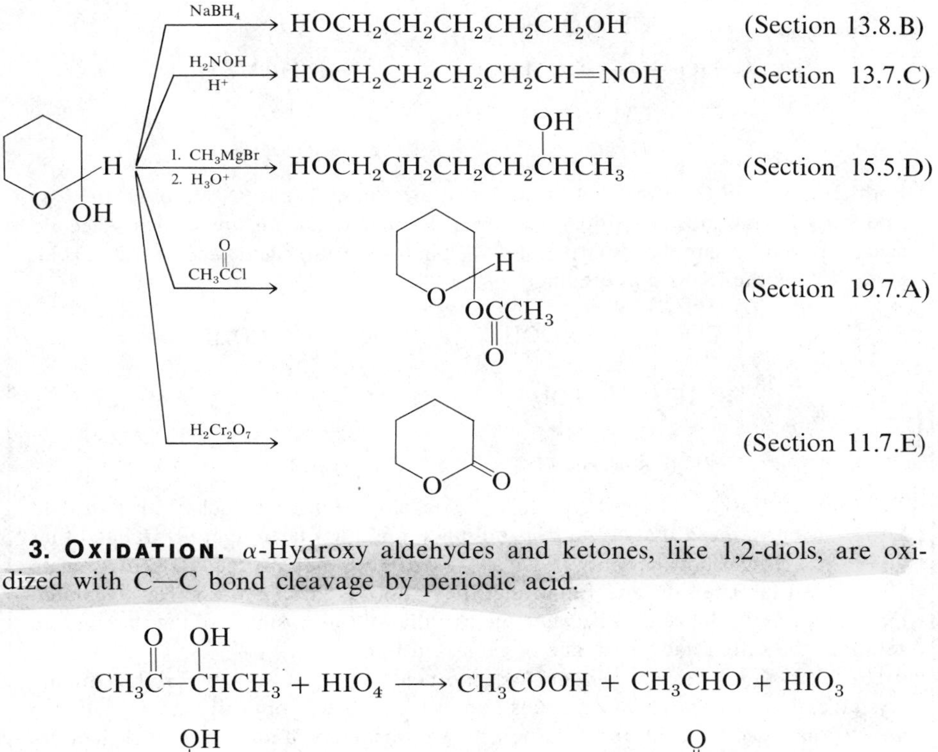

$\xrightarrow{NaBH_4}$ $HOCH_2CH_2CH_2CH_2CH_2OH$ (Section 13.8.B)

$\xrightarrow[H^+]{H_2NOH}$ $HOCH_2CH_2CH_2CH_2CH=NOH$ (Section 13.7.C)

$\xrightarrow[2.\ H_3O^+]{1.\ CH_3MgBr}$ $HOCH_2CH_2CH_2CH_2\overset{\overset{OH}{|}}{C}HCH_3$ (Section 15.5.D)

$\xrightarrow{CH_3\overset{\overset{O}{\|}}{C}Cl}$ (Section 19.7.A)

$\xrightarrow{H_2Cr_2O_7}$ (Section 11.7.E)

3. OXIDATION. α-Hydroxy aldehydes and ketones, like 1,2-diols, are oxidized with C—C bond cleavage by periodic acid.

$$CH_3\overset{\overset{O}{\|}}{C}-\overset{\overset{OH}{|}}{C}HCH_3 + HIO_4 \longrightarrow CH_3COOH + CH_3CHO + HIO_3$$

$$CH_3CH_2\overset{\overset{OH}{|}}{C}HCHO + HIO_4 \longrightarrow CH_3CH_2CHO + H\overset{\overset{O}{\|}}{C}OH + HIO_3$$

The reaction constitutes a useful method for structure determination in the carbohydrate field (Chapter 28).

EXERCISE: Write reasonable mechanisms for the five reactions of the cyclic hemi-acetal of 5-hydroxypentanal shown on page 855. Clearly indicate each step.

27.5
Hydroxy Acids

Many hydroxy acids are important in nature and, as would be expected, have trivial names that are in common use. Glycolic acid is a constituent of cane-sugar juice. Lactic acid is produced by the fermentation of lactose; it was first isolated from sour milk. Other important hydroxy acids are dicarboxylic acids. Malic acid occurs in fruit juices. Tartaric acid has been known since antiquity as the monopotassium salt (cream of tartar) deposited in the lees of wine. The hydroxy acids with asymmetric carbons are optically active in nature.

$$HOCH_2COOH$$
glycolic acid

$$CH_3CHOHCOOH$$
lactic acid

$$\overset{\displaystyle OH}{\underset{\displaystyle |}{HOOCCH_2CHCOOH}}$$
malic acid

$$\overset{\displaystyle OH \quad OH}{\underset{\displaystyle | \quad\ |}{HOOCCH-CHCOOH}}$$
tartaric acid

$$\overset{\displaystyle CH_2COOH}{\underset{\displaystyle CH_2COOH}{HO-C-COOH}}$$
citric acid

mandelic acid

Both the (+) and (−) forms of tartaric acid are found in nature, although the (+) acid is by far the more common. Two optically inactive forms are known. **Racemic acid,** m.p. 206°, is simply a mixture of (+)- and (−)-tartaric acid. *meso*-Tartaric acid, m.p. 140°, is the (R,S) diastereomer.

(+)-(R,R)-tartaric acid

meso-tartaric acid

Tartaric acid played an important role in the development of stereochemistry. In 1848 Louis Pasteur noticed that crystals of sodium ammonium tartrate are chiral and that all of the crystals show chirality in the same sense. He then proceeded to investigate 19 different tartrate salts and found that they all gave chiral crystals. He postulated that there was a relationship between the chirality of the crystals and the fact that, in solution, the salts rotate the plane of polarized light.

However, there was a problem. **Racemic acid,** obtained as a by-product in the crystallization of tartaric acid, was also known, and it was optically inactive. It was known that racemic acid and tartaric acid were isomers, and Mitscherlich had reported that crystals of sodium ammonium tartrate and sodium ammonium racemate were identical in all respects except that the tartrate gave a dextrorotatory solution whereas the racemate gave an optically inactive solution.

Pasteur repeated Mitscherlich's work on sodium ammonium racemate and was disappointed to discover that Mitscherlich had been correct and that crystals of the racemate salt are indeed chiral. Upon closer examination, however, he noticed that the crystals are not all chiral in the same sense. In his words, "the hemihedral faces which in the tartrate are all turned one way are in the racemate inclined sometimes to the right and sometimes to the left." In short, the racemate salt gives a mixture of nonsuperimposable mirror image crystals. Using a pair of tweezers, Pasteur carefully separated the left-handed from the right-handed crystals, dissolved each in water, and measured their optical rotations. To his great excitement, he discovered that one solution was dextrorotatory and the other was levorotatory. When he converted the separated salts back to the free acids, he found that one was identical with natural (+)-tartaric acid and that the other was a new tartaric acid isomer, identical in all respects save the sign of its optical rotation. Pasteur had accomplished the first resolution—separation of a racemate into its component enantiomers.

Pasteur's work paved the way for an understanding of stereoisomerism. He made the important suggestion that since the crystals of the enantiomeric salts show handedness, the molecules themselves might also show handedness—and this before the idea of chemical bonds had even been conceived.

Citric acid is a hydroxytricarboxylic acid that is widespread in nature and is especially prevalent, as its trivial name implies, in the juice of citrus fruits. These hydroxy acids are related to carbohydrates (Chapter 28) and to amino acids (Chapter 29).

A. Synthesis of Hydroxy Acids

α-Hydroxy acids are most commonly prepared by hydrolysis of α-halo acids. Recall that α-halo acids are readily available by Hell–Vollhard–Zelinsky bromination of carboxylic acids (Section 18.7.B). Thus the two-step sequence provides a way to introduce the hydroxy group at the α-position of a carboxylic acid.

$$CH_3CH_2COOH \xrightarrow[P]{Br_2} CH_3CHBrCOOH \xrightarrow{OH^-} CH_3CHOHCO_2^-$$

β-Hydroxy acids cannot be prepared by hydrolysis of the corresponding β-halo acid, since these compounds undergo elimination in base to give the unsaturated acids.

$$CH_3CHBrCH_2COOH \xrightarrow{OH^-} CH_3CH{=}CHCO_2^-$$

α-Hydroxy acids are also generally available by hydrolysis of cyanohydrins, which result from the reaction of HCN with aldehydes or ketones (Section 13.7.E). Since the addition of HCN to a carbonyl group is reversed by the strongly basic conditions necessary to hydrolyze a nitrile to an acid, the hydrolysis is done under acidic conditions.

$$CH_3CH_2CHO \xrightarrow{HCN} CH_3CH_2\underset{\underset{OH}{|}}{\overset{\overset{CH_3}{|}}{C}}CN \xrightarrow[25°]{conc.\ HCl} CH_3CH_2\underset{\underset{OH}{|}}{\overset{\overset{CH_3}{|}}{C}}COOH$$

(55%)
2-methyl-2-hydroxybutanoic acid

(70%)
α-hydroxyphenylacetic acid
mandelic acid

β-Hydroxy acids and their derivatives are available by methods analogous to the aldol condensation (Section 19.9).

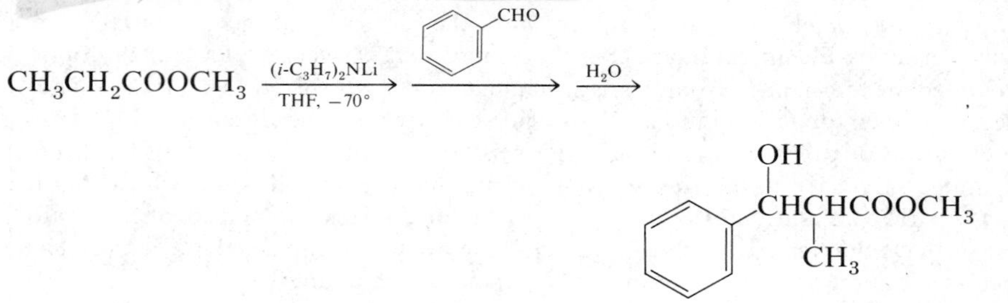

$$CH_3CH_2COOCH_3 \xrightarrow[\text{THF, } -70°]{(i\text{-}C_3H_7)_2NLi} \xrightarrow{C_6H_5CHO} \xrightarrow{H_2O}$$

OH
|
C₆H₅CHCHCOOCH₃
|
CH₃

(95%)
methyl 3-hydroxy-2-methyl-3-phenylpropanoate

A variety of hydroxy acids are available by hydrolysis of lactones, which may be obtained by the Baeyer–Villiger oxidation of cyclic ketones (Section 13.8.A).

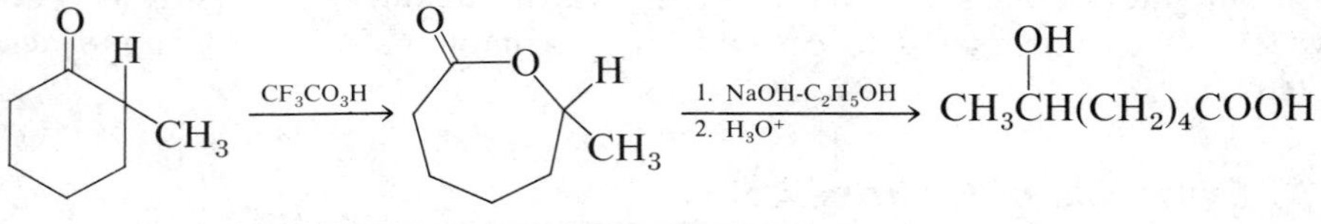

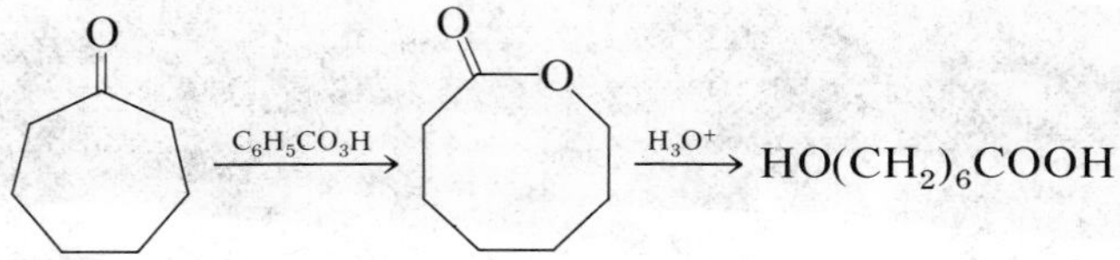

$$\xrightarrow[\text{2. H}_3O^+]{\text{1. NaOH-C}_2H_5OH} \quad CH_3\overset{\overset{\displaystyle OH}{|}}{C}H(CH_2)_4COOH$$

$$\xrightarrow{H_3O^+} HO(CH_2)_6COOH$$

EXERCISE: Show how 2-hydroxypentanoic acid may be prepared from (a) pentanoic acid and (b) butanal. Outline a synthesis of 3-hydroxypentanoic acid.

B. Reactions of Hydroxy Acids

1. FORMATION OF LACTONES. Recall that carboxylic acids react with alcohols under acid catalysis to yield esters (Section 18.7.C).

$$\overset{\displaystyle O}{\overset{\|}{R}COH} + R'OH \underset{}{\overset{H^+}{\rightleftharpoons}} \overset{\displaystyle O}{\overset{\|}{R}COR'} + H_2O$$

A hydroxy acid contains both of these functional groups, and thus it can undergo intramolecular esterification to yield a cyclic ester, which is called a **lactone**.

$$HOCH_2CH_2CH_2COOH \overset{H^+}{\rightleftharpoons} \quad + H_2O$$

γ-hydroxybutyric acid   γ-butyrolactone

Lactonization, like normal esterification, is an equilibrium process. Only when the lactone has a five- or six-membered ring is there a substantial amount of lactone present under equilibrium conditions, as shown by the data in Table 27.4. The

TABLE 27.4
Hydrolytic Equilibria of Lactones

Lactone Formula	Equilibrium Composition	
	Hydroxy Acid, %	Lactone, %
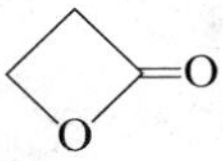	100	0
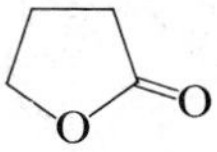	27	73
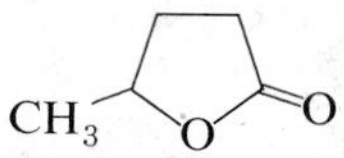	5	95
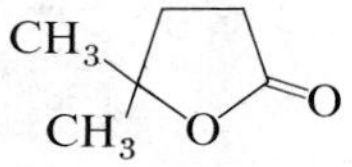	2	98
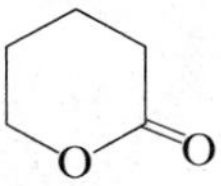	91	9
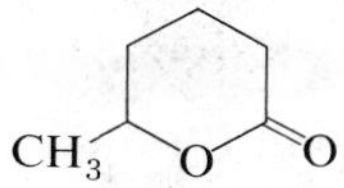	79	21
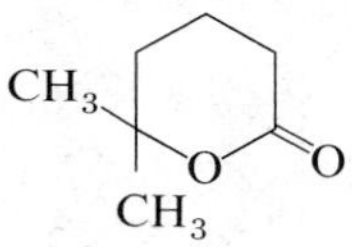	75	25
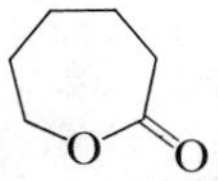	~100	~0

data in Table 27.4 also reveal that alkyl substitution on the ring increases the amount of lactone present at equilibrium.

Although the larger lactones do not exist to any appreciable extent in *equilibrium* with the free hydroxy acids, such lactones may be prepared under the proper conditions. It is necessary to treat the hydroxy acid with acid under conditions where the water formed in the reaction is removed so as to shift the unfavorable equilibrium toward the lactone. It is also necessary to operate in very dilute solution so as to minimize the intermolecular esterification reaction, which leads to a polymer.

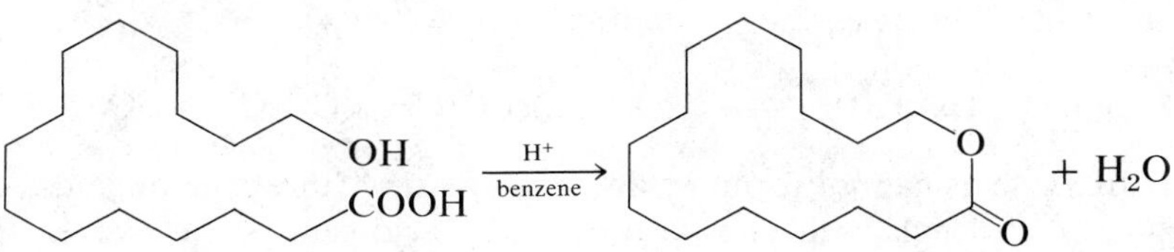

15-hydroxypentadecanoic acid
(0.007 *M*)

(100%)
15-hydroxypentadecanoic
acid lactone

γ-Lactones and δ-lactones form from the hydroxy acids so readily that it is often not necessary even to add acid to catalyze the intramolecular esterification; mere traces of acid in the solvent or on the glassware suffice to bring about lactonization. Thus, in any reaction that would yield a 4- or 5-hydroxy acid, the corresponding lactone is often the isolated product.

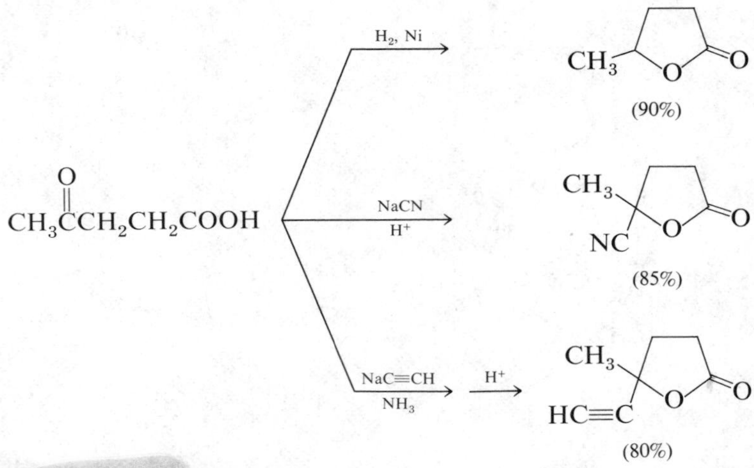

Lactones may also result from reactions of other substituted carboxylic acids, as shown by the following examples.

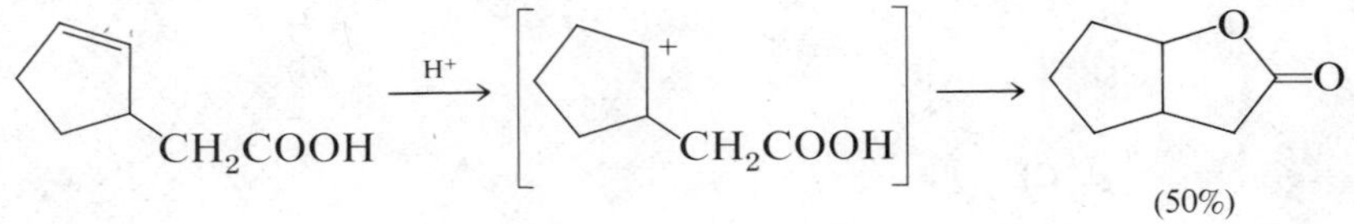

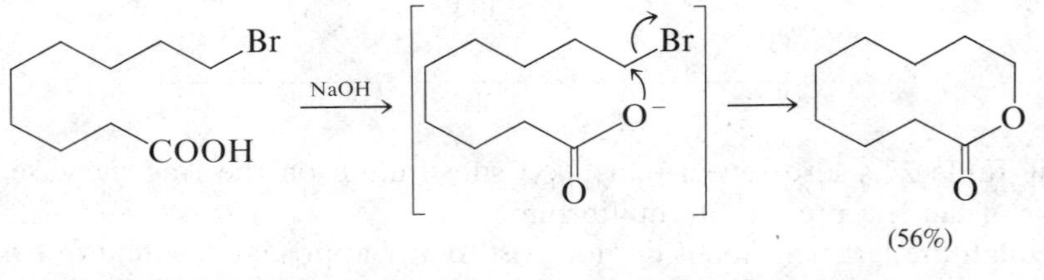

The second of these reactions must be done under high dilution to suppress intermolecular displacement reactions.

2. POLYMERIZATION; LACTIDES. As discussed in the previous section, 4- and 5-hydroxy acids react rapidly in an intramolecular process to afford lactones. Other hydroxy acids, which cannot form five- or six-membered rings, undergo polymerization unless the reaction is carried out under high dilution conditions.

$$n \text{ HO(CH}_2)_8\overset{\text{O}}{\overset{\|}{\text{C}}}\text{OH} \longrightarrow \text{HO(CH}_2)_8\overset{\text{O}}{\overset{\|}{\text{C}}}\text{O(CH}_2)_8\overset{\text{O}}{\overset{\|}{\text{C}}}\text{O(CH}_2)_8\overset{\text{O}}{\overset{\|}{\text{C}}}\text{O. . .}$$

α-Hydroxy acids cannot form a stable lactone ring (three-membered), so they undergo intermolecular self-esterification under acid catalysis. However, the initial dimeric product is now a form of 5-hydroxy acid, so lactonization occurs. The product, which is a dilactone containing two molecules of the original α-hydroxy acid, is called a lactide.

$$2 \overset{\overset{\displaystyle OH}{|}}{CH_3CHCOOH} \xrightarrow[-H_2O]{H^+} \overset{\overset{\displaystyle HO}{|} \overset{\displaystyle O}{\parallel}}{CH_3CHCOCHCOOH} \underset{\overset{|}{CH_3}}{} \xrightarrow{-H_2O}$$

$$\text{lactic acid} \qquad\qquad\qquad\qquad\qquad \text{lactide}$$

3. DEHYDRATION. Like β-hydroxy aldehydes and ketones, β-hydroxy acids and their derivatives undergo dehydration easily under acidic conditions. The mechanism is similar to that for dehydration of the other β-hydroxy carbonyl compounds discussed previously (Sections 20.3.A and 27.4.B). Since conjugation of a double bond with an acid or ester carbonyl group is less stabilizing than with an aldehyde or ketone carbonyl (Section 20.3), mixtures of the α,β-unsaturated and β,γ-unsaturated acids often result from dehydration of a β-hydroxy acid.

$$\overset{\overset{\displaystyle OH}{|}}{CH_3CH_2CCH_2COOC_2H_5} \xrightarrow[\Delta]{KHSO_4} \underset{\overset{|}{CH_3}}{}$$

$$\overset{\overset{\displaystyle CH_3}{|}}{CH_3CH_2C}{=}CHCOOC_2H_5 \; + \; \overset{\overset{\displaystyle CH_3}{|}}{CH_3CH}{=}CCH_2COOC_2H_5$$

$$\text{(57\%)} \qquad\qquad\qquad\qquad\qquad \text{(43\%)}$$

EXERCISE: What are the principal reactions expected when each of the following hydroxy acids is treated with acid?
(a) 2-hydroxybutanoic acid
(b) 3-hydroxybutanoic acid
(c) 4-hydroxybutanoic acid

27.6
Dicarboxylic Acids

The simple aliphatic dicarboxylic acids are fairly widespread in nature and crystallize readily from aqueous solutions. Consequently, they are easy to isolate and were among the earliest known organic compounds. Oxalic acid occurs in many plants, such as rhubarb, usually as the potassium salt. The insoluble calcium salt is found in plant cells and in some calculi, which are stony deposits found in the human body. The acid is poisonous. Succinic acid occurs in fossils, fungi, lichens, and amber. It was first isolated in 1546 from the distillate of amber. Glutaric acid occurs in sugar beets and is also found in the aqueous extract of crude wool. Adipic acid may also be isolated from sugar beets, but it is normally synthesized from cyclohexane and its derivatives, as discussed in the next section.

A. Synthesis of Dicarboxylic Acids

Several dicarboxylic acids may be prepared by methods involving the hydrolysis of nitriles. For example, malonic acid is prepared from chloroacetic acid via cyanoacetic acid. The displacement reaction and the alkaline hydrolysis are carried out in one operation, and the product is isolated in about 80% yield.

$$ClCH_2COOH + NaOH + NaCN \xrightarrow{H_2O} [NCCH_2CO_2^- \ Na^+] \xrightarrow{NaOH}{H_2O}$$

$$Na^+ \ ^-O_2CCH_2CO_2^- \ Na^+ \xrightarrow{HCl} HOOCCH_2COOH$$
$$(77\text{–}82\%)$$

A similar example is the synthesis of glutaric acid by the acid-catalyzed hydrolysis of 1,3-dicyanopropane.

$$BrCH_2CH_2CH_2Br \xrightarrow{NaCN}{H_2O} NCCH_2CH_2CH_2CN \xrightarrow{HCl}{H_2O} HOOCCH_2CH_2CH_2COOH$$
$$(82\%) \qquad\qquad\qquad (84\%)$$
$$\text{glutaric acid}$$

Succinic acid derivatives are often available by conjugate addition of cyanide to α,β-unsaturated esters (Section 20.3.A.). Hydrolysis of the β-cyano acid yields the corresponding succinic acid.

$$CH_3CH{=}CHCOOC_2H_5 + NaCN \xrightarrow{H_2O} CH_3\overset{\overset{\displaystyle CN}{|}}{C}HCH_2COOH \xrightarrow{Ba(OH)_2}{H_2O}$$

$$\xrightarrow{H_3O^+} CH_3\overset{\overset{\displaystyle COOH}{|}}{C}HCH_2COOH$$
$$(66\text{–}70\%)$$
$$\alpha\text{-methylsuccinic acid}$$
$$2\text{-methylbutanedioic acid}$$

Certain diacids are conveniently prepared by the oxidation of cyclic alkenes or ketones. This is particularly true for adipic acid derivatives because cyclohexane derivatives are generally readily available. Examples are

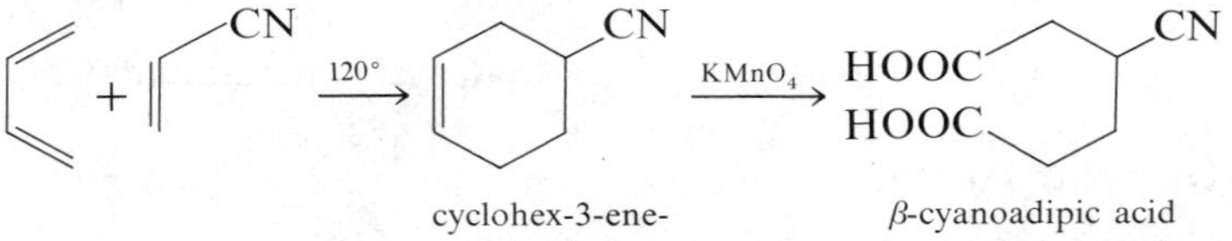

cyclohex-3-ene-
carbonitrile

β-cyanoadipic acid
3-cyanohexanedioic
acid

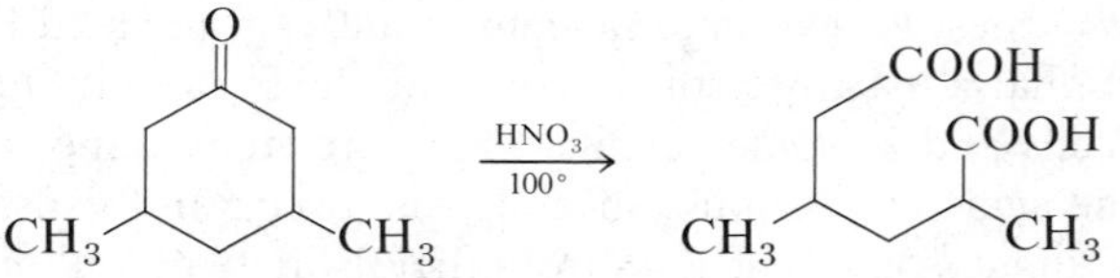

3,5-dimethylcyclohexanone

α,β'-dimethyladipic acid
2,4-dimethylhexanedioic acid

(60%)

Adipic acid is manufactured on a large scale by several methods, one of which is the oxidation of cyclohexane or cyclohexene. The United States production of adipic acid in 1979 was almost 2 billion pounds, and essentially all of it was used for making nylon and derived polymers (Section 27.7.A).

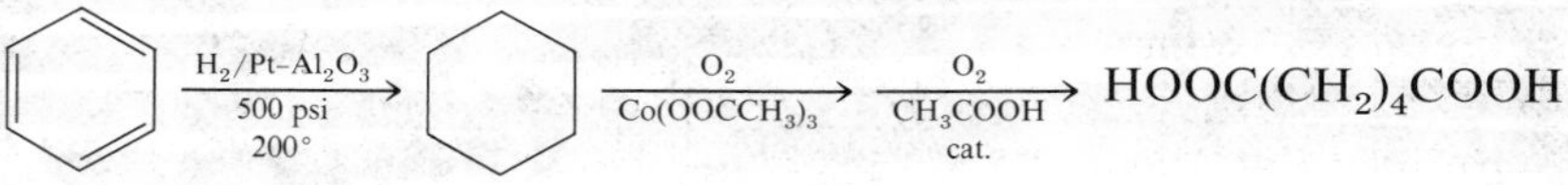

One acid that is usually considered to be an inorganic acid of carbon is carbonic acid. However, important organic derivatives are known. The diacyl chloride, phosgene, $COCl_2$, is prepared commercially by allowing CO and Cl_2 to react in the presence of a catalyst. Phosgene reacts with alcohols to give dialkyl carbonates.

$$COCl_2 + 2\ ROH \longrightarrow RO\overset{\overset{\displaystyle O}{\|}}{C}OR + 2\ HCl$$
dialkyl
carbonate

The diamide of carbonic acid, urea, H_2NCONH_2, is a metabolic product that has important commercial and historical significance in organic chemistry (Chapter 1).

On a tonnage basis, the most important dicarboxylic acid by far is terephthalic acid, which is prepared on an industrial scale by the air oxidation of *p*-xylene (Section 22.5.C). The oxidation of the two side chains occurs in stages, first to *p*-toluic acid, which in turn is oxidized to terephthalic acid.

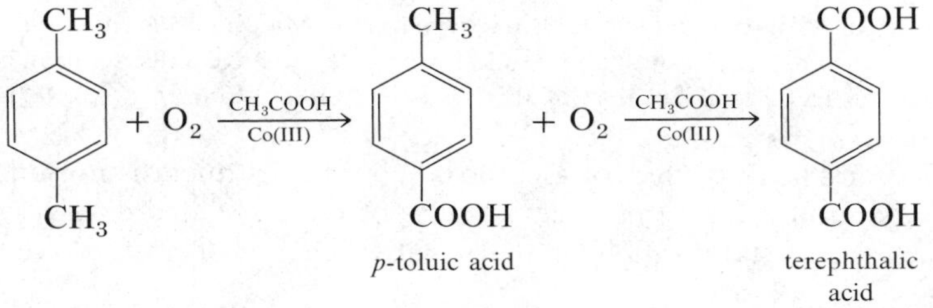

p-toluic acid   terephthalic acid

The first step proceeds much more easily than the second. Indeed, the oxidation of *p*-toluic acid requires such high temperatures that oxidation of the acetic acid solvent becomes a significant cost concern and corrosion of reaction vessels is a problem. United States production of terephthalic acid and dimethyl terephthalate in 1979 was more than 7 billion pounds.

Phthalic acid is another important industrial product. Various high boiling esters, particularly the bis-2-ethylhexyl ester, are widely used as plasticizers. Phthalic acid is prepared commercially by the oxidation of naphthalene (Section 33.3.C) or *o*-xylene. It loses water readily on heating to produce phthalic anhydride, a compound with a characteristic odor that forms long colorless needles on sublimation.

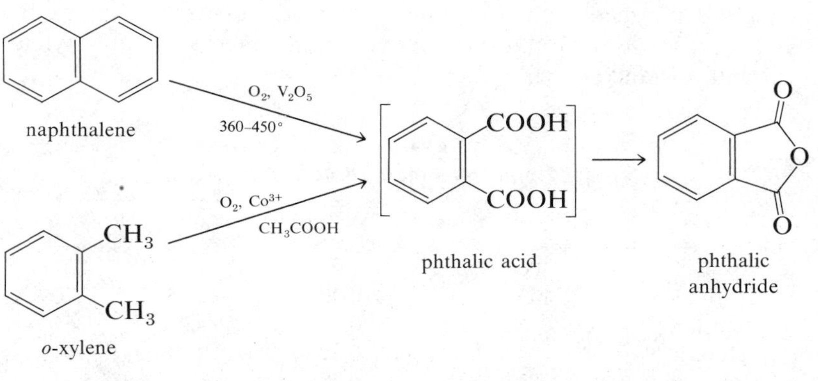

naphthalene   phthalic acid   phthalic anhydride

o-xylene

Phthalic anhydride is used in the manufacture of glyptal resins, highly cross-linked, infusible polyesters prepared by heating the anhydride with glycerol. Potassium hydrogen phthalate is a well-characterized compound available in pure anhydrous form. It is used as a primary standard in titrations with bases.

B. *Reactions of Dicarboxylic Acids and Their Derivatives*

1. ACIDITY OF DICARBOXYLIC ACIDS. The dicarboxylic acids are dihydric acids and are characterized by two dissociation constants, K_1 and K_2.

$$HOOC(CH_2)_n COOH \xrightarrow{K_1} HOOC(CH_2)_n CO_2^- + H^+$$

$$HOOC(CH_2)_n CO_2^- \xrightarrow{K_2} {}^-O_2C(CH_2)_n CO_2^- + H^+$$

The dissociation constants for several diacids are summarized in Table 27.5. If we treat the COOH group as a substituent in acetic acid, YCH_2COOH, the higher acidity of malonic acid compared to acetic acid ($pK_a = 4.76$) indicates that the COOH group acts as an electron-attracting inductive group.

$$\left[\begin{array}{l} \text{Be careful of statistical effects in this comparison. Malonic acid has two COOH} \\ \text{groups which can lose a proton and would be expected to have a dissociation constant} \\ \text{twice that of acetic acid because of this statistical effect alone.} \end{array} \right]$$

The acid-strengthening effect of a carboxylic acid substituent is not unexpected. All carbonyl groups have this effect because of the associated dipole which provides electrostatic stabilization of the negative charge of a carboxylate anion.

electrostatic attraction

$\overleftarrow{\qquad +}$

$HOOC\sim\!\sim\!\sim CO_2^-$

On the other hand, K_2 for a dicarboxylic acid is generally less than the dissociation constant of acetic acid. The presence of a carboxylate ion substituent reduces the acidity of an acid. This effect is clearly associated with the electrostatic repulsion of two negative charges in the dicarboxylate ion.

electrostatic repulsion

$^-O_2C\sim\!\sim\!\sim CO_2^-$

As expected for such a phenomenon, both the acid-strengthening effect of a carboxylic acid substituent and the acid-weakening effect of a carboxylate anion diminish with distance down a chain.

TABLE 27.5
Acidity of Alkanedioic Acids

Acid	$K_1 \times 10^{-5} M$	$K_2 \times 10^{-5} M$	pK_1	pK_2
oxalic	5400	5.4	1.27	4.27
malonic	140	0.20	2.85	5.70
succinic	6.2	0.23	4.21	5.64
glutaric	4.6	0.39	4.34	5.41
adipic	3.7	0.39	4.43	5.41

> The second dissociation constant of oxalic acid seems anomolous by this comparison. Oxalate monoanion is *more* acidic than a neutral alkanoic acid despite the high electrostatic repulsion inherent in the oxalate dianion. This exception to the above-mentioned generalization is probably a solvation phenomenon and is associated with the high charge density on the oxalate dianion.

2. FORMATION OF POLYESTERS AND POLYAMIDES. Two of the most characteristic reactions of carboxylic acids are the formation of esters and amides (Section 18.7). Dicarboxylic acids may be converted into diesters and diamides in a straightforward fashion. When a diol or diamine is used, a polymer results. For example, Dacron is prepared by the reaction of dimethyl terephthalate and ethylene glycol. In one industrial process, the two reactants are heated together and methanol is distilled from the reactor.

$$x \text{ CH}_3\text{OOC}\text{---}\langle\bigcirc\rangle\text{---}\text{COOCH}_3 + x \text{ HOCH}_2\text{CH}_2\text{OH} \xrightarrow{\Delta}$$

$$\left(\text{OOC}\text{---}\langle\bigcirc\rangle\text{---}\text{COOCH}_2\text{CH}_2\right)_x + 2x \text{ CH}_3\text{OH}\uparrow$$

Polymers such as this are called **condensation polymers,** since they are formed by a reaction in which the monomers condense with the elimination of a small by-product molecule, in this case methanol. They are distinguished from **addition polymers** such as polyethylene and polystyrene (Section 11.6.G) in which polymerization results from the stepwise addition of monomer molecules to the growing polymer chain in such a way that no by-product molecules are produced.

Polyamides are an important class of commercial polymers. The best known is nylon 6,6, which is a **copolymer** (a polymer made up of two different monomers) formed from 1,6-hexanediamine and adipic acid. The polymer is manufactured by heating an equimolar mixture of the two monomers at 270° at a pressure of about 10 atm.

$$\text{H}_2\text{N(CH}_2)_6\text{NH}_2 + \text{HOOC(CH}_2)_4\text{COOH} \xrightarrow[\text{10 atm}]{270°}$$

$$\left(\text{NH(CH}_2)_6\text{NH}\overset{\overset{\displaystyle O}{\|}}{\text{C}}\text{(CH}_2)_4\overset{\overset{\displaystyle O}{\|}}{\text{C}}\right)_n + \text{H}_2\text{O}$$

nylon 6,6

The two monomers employed in the synthesis of nylon 6,6 have been produced commercially from benzene. The synthesis of adipic acid was outlined in the previous section. The 1,6-hexanediamine can be produced from adipic acid by converting it into the diamide, which is dehydrated to 1,4-dicyanobutane. Reduction of the dinitrile gives the diamine.

$$\text{HOOC(CH}_2)_4\text{COOH} \xrightarrow[\Delta]{\text{NH}_3} \text{H}_2\text{N}\overset{\overset{\displaystyle O}{\|}}{\text{C}}\text{(CH}_2)_4\overset{\overset{\displaystyle O}{\|}}{\text{C}}\text{NH}_2 \xrightarrow{-\text{H}_2\text{O}}$$

$$\text{NC(CH}_2)_4\text{CN} \xrightarrow[\substack{130° \\ 2000 \text{ psi}}]{\text{H}_2-\text{Ni}} \text{H}_2\text{N(CH}_2)_6\text{NH}_2$$

The diamine is also prepared from butadiene by the following route.

$$CH_2=CHCH=CH_2 \xrightarrow{Cl_2} \left[\underset{\underset{Cl}{|}}{CH_2=CHCHCH_2Cl} + ClCH_2CH=CHCH_2Cl \right] \xrightarrow[\text{CuCN}]{CN^-}$$

$$NCCH_2CH=CHCH_2CN \xrightarrow[\text{NH}_3]{H_2\text{-cat.}} H_2N(CH_2)_6NH_2$$

Another form of nylon is nylon 6, which is produced by polymerization of the amino acid 6-aminohexanoic acid. The actual monomer used is the cyclic amide caprolactam.

$$H_2N(CH_2)_5COOH \qquad \text{or}$$

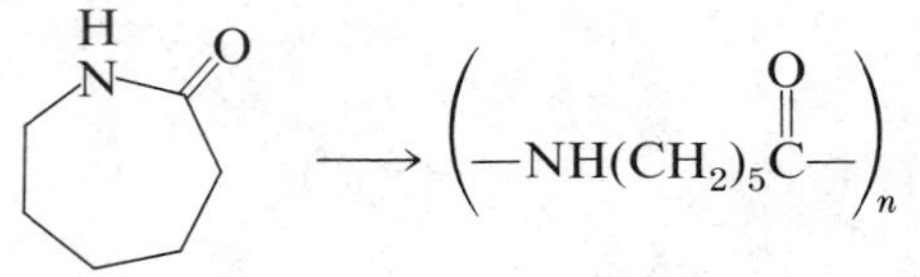

6-aminohexanoic
acid   caprolactam   nylon 6

3. BEHAVIOR ON HEATING. Anhydrous oxalic acid can be sublimed by careful heating, but at higher temperatures it decomposes to carbon dioxide and formic acid. Formic acid also decomposes under these conditions to carbon monoxide and water.

Malonic acid decarboxylates smoothly on heating to give acetic acid.

$$HOOCCH_2COOH \xrightarrow{150°} CO_2 + CH_3COOH$$

This reaction is general for all substituted malonic acids and for β-keto acids as well. The mechanism may involve a cyclic six-center transition state similar to that discussed previously for the Diels–Alder reaction (Section 20.5). The initial product is an enol, which rapidly tautomerizes to acetic acid.

Decarboxylation

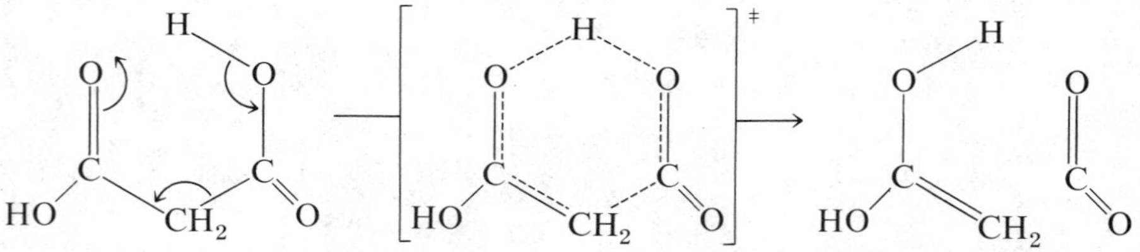

Diels–Alder Reaction

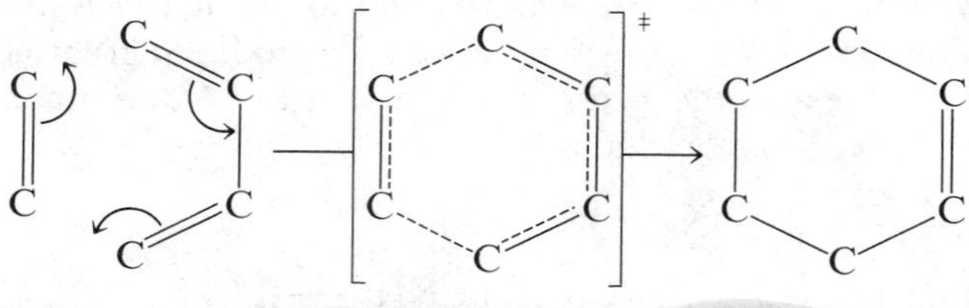

Decarboxylation of substituted malonic acids is a frequently used process for the synthesis of carboxylic acids (Section 27.7.C.3). Typical decarboxylation conditions involve heating the diacid at 120–180° for several hours.

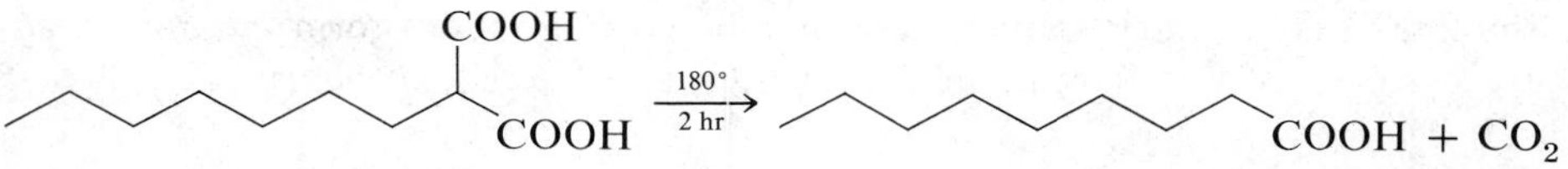

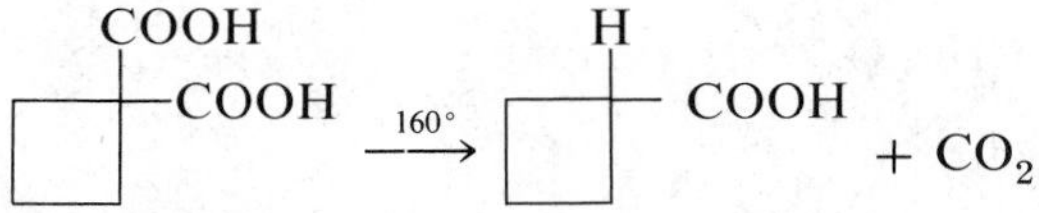

$$\xrightarrow[\text{2 hr}]{180^\circ}$$

$$\xrightarrow{160^\circ}$$

Succinic and glutaric acids lose water on heating to give cyclic anhydrides.

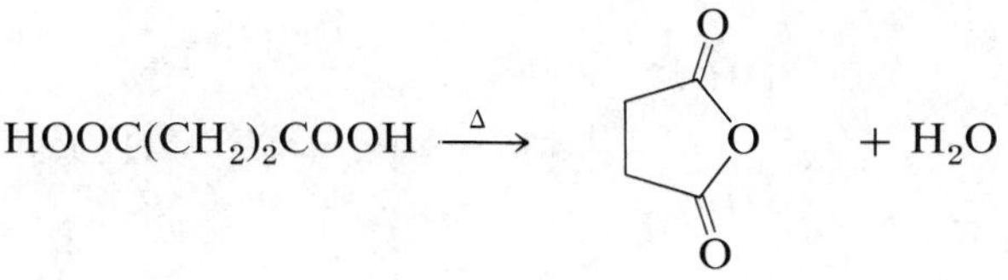

$$\text{HOOC(CH}_2)_2\text{COOH} \xrightarrow{\Delta}$$

$$+ \text{ H}_2\text{O}$$

succinic anhydride

However, the preparation of these anhydrides is best accomplished by heating with acetyl chloride or acetic anhydride. These reagents react with the water produced by the dehydration.

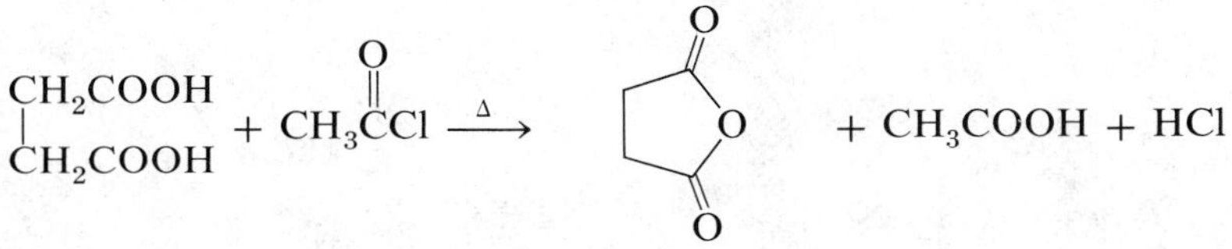

$$\begin{array}{c}\text{CH}_2\text{COOH}\\|\\\text{CH}_2\text{COOH}\end{array} + \text{CH}_3\overset{\text{O}}{\overset{\|}{\text{C}}}\text{Cl} \xrightarrow{\Delta} \qquad + \text{ CH}_3\text{COOH} + \text{HCl}$$

succinic acid   (95%)
                      succinic anhydride

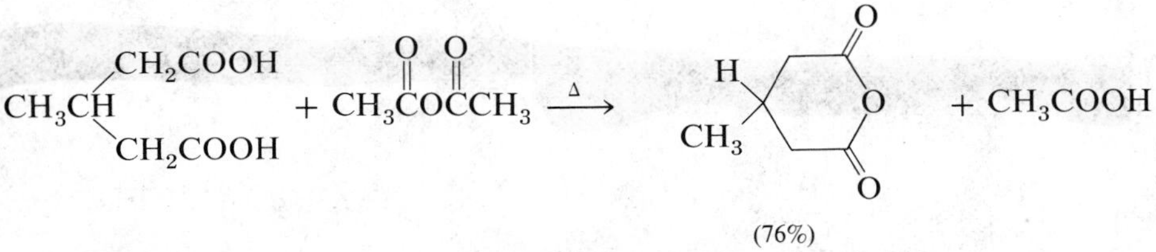

$$\text{CH}_3\text{CH}\begin{array}{c}\text{CH}_2\text{COOH}\\\\\text{CH}_2\text{COOH}\end{array} + \text{CH}_3\overset{\text{O}}{\overset{\|}{\text{C}}}\text{O}\overset{\text{O}}{\overset{\|}{\text{C}}}\text{CH}_3 \xrightarrow{\Delta} \qquad + \text{ CH}_3\text{COOH}$$

β-methylglutaric acid   (76%)
                              β-methylglutaric anhydride

The easy dehydration of phthalic acid to phthalic anhydride was mentioned in Section 27.6.A. Another example that shows the generality of cyclic anhydride formation is

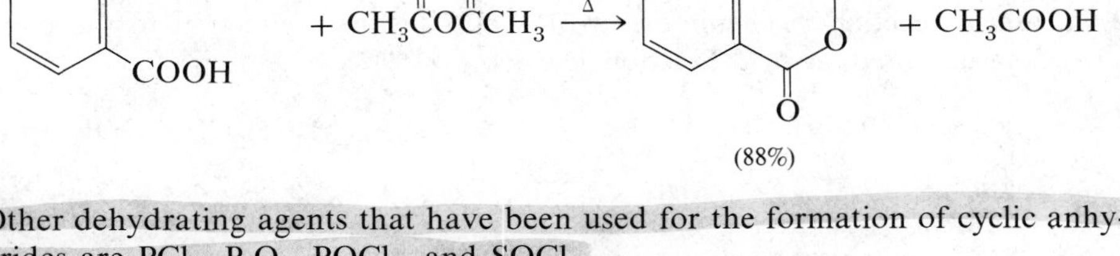

$$+ \text{ CH}_3\overset{\text{O}}{\overset{\|}{\text{C}}}\text{O}\overset{\text{O}}{\overset{\|}{\text{C}}}\text{CH}_3 \xrightarrow{\Delta} \qquad + \text{ CH}_3\text{COOH}$$

(88%)

Other dehydrating agents that have been used for the formation of cyclic anhydrides are PCl_5, P_2O_5, POCl_3, and SOCl_2.

Succinic and glutaric acid and their derivatives also form cyclic **imides** with ammonia and primary amines. Five-membered ring imides form the most readily; pyrolysis of the diammonium salt often gives excellent yields.

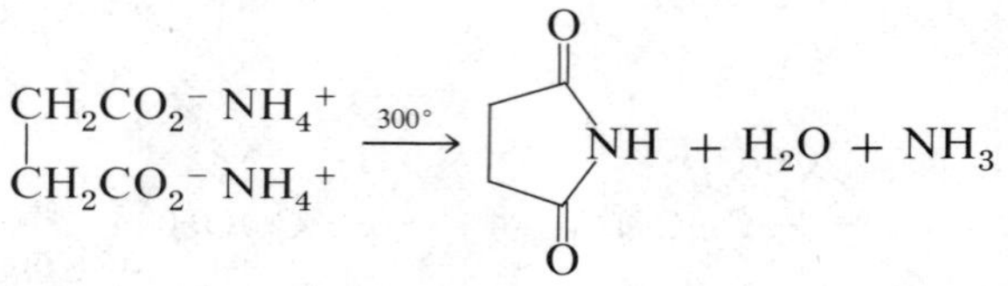

ammonium succinate   succinimide
        (83%)

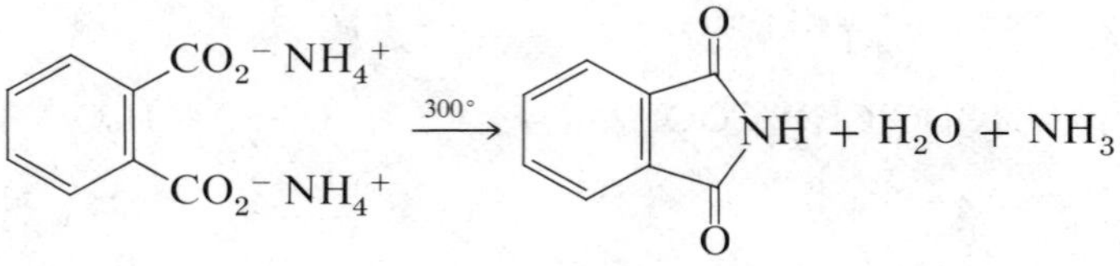

ammonium phthalate   phthalimide
        (97%)

Six-membered ring imides form less readily; a convenient method of preparation involves pyrolysis of the monoamide of the corresponding dicarboxylic acid, as illustrated by the following example.

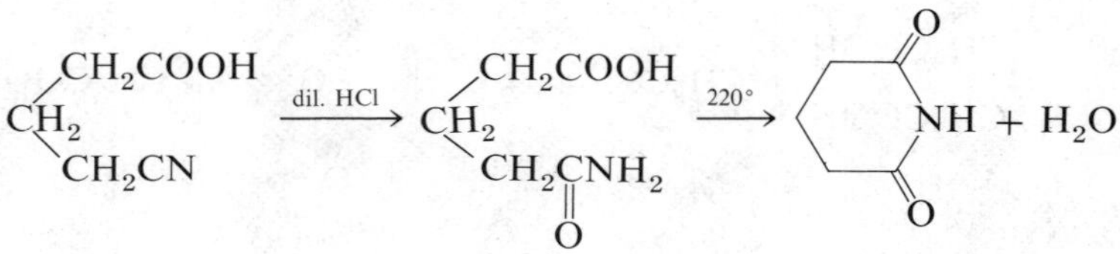

(65% overall)
glutarimide

The cyclic anhydrides derived from dicarboxylic acids are effective reactants in Friedel–Crafts acylations (Section 23.4).

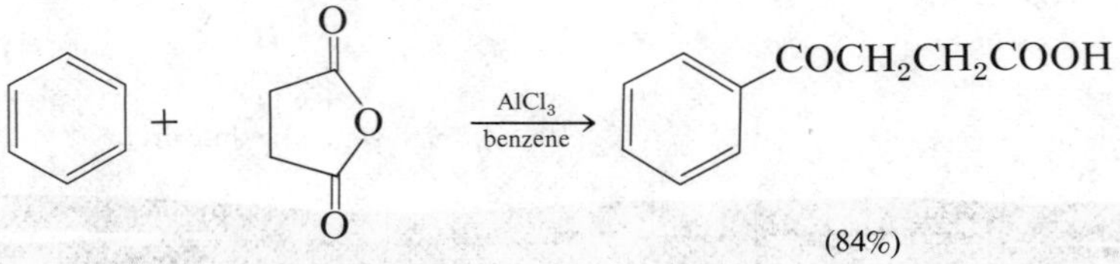

succinic anhydride   4-phenyl-4-oxobutanoic acid
           (84%)

The products of these acylations are themselves difunctional compounds and undergo the characteristic reactions of ketones and carboxylic acids generally. One useful reaction is Clemmensen or Wolff–Kishner reduction to the corresponding arylalkanoic acid (Section 13.8.E).

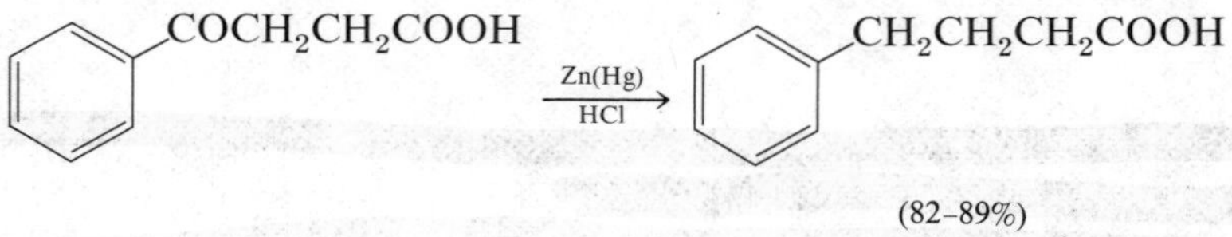

(82–89%)
4-phenylbutanoic acid

An important property of such phenyl-substituted acids is the ease with which they undergo intramolecular Friedel–Crafts acylation reactions to form five- and six-membered cyclic ketones.

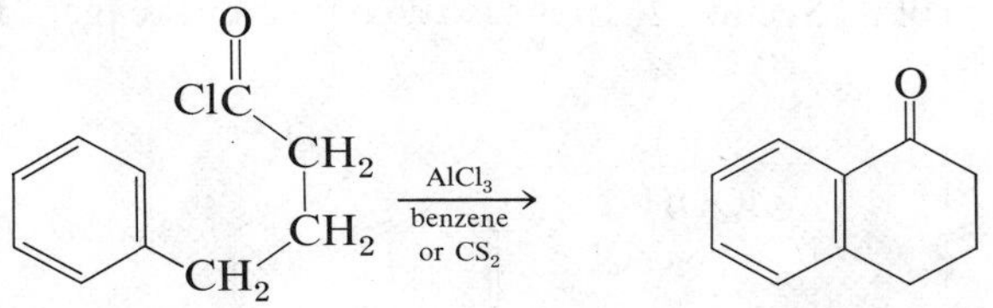

(70–90%)
1-oxo-1,2,3,4-tetrahydronaphthalene
α-tetralone

Such cyclizations are generally excellent preparative methods when the product is a five- or six-membered ring ketone. Commonly used reagents are $AlCl_3$ with the acid chloride and sulfuric acid, polyphosphoric acid, or liquid hydrogen fluoride with the free acid.

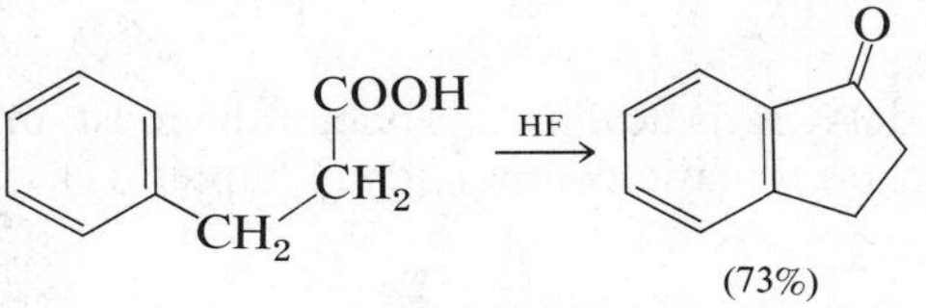

(73%)
1-indanone

> The carboxylic acid is weighed into a polyethylene beaker, and—in an efficient hood—liquid HF is added from an inverted tank previously cooled to 5° (use polyethylene or rubber gloves). The mixture is stirred, and the HF is allowed to evaporate over the course of several hours. The residue is mixed with aqueous Na_2CO_3 and extracted with benzene. The product is obtained by distillation or crystallization. Yields are typically 70–90%. This is one of the best and simplest procedures for Friedel–Crafts cyclizations.

> Anhydrous HF is a low-boiling liquid, b.p. 19°, available in cylinders. It is highly corrosive to glass and tissue and must be handled with due caution. The liquid is an excellent solvent for oxygen-containing organic compounds (hydrogen bonding). It does not attack polyethylene or Teflon, and these polymers make suitable reaction vessels. Because of the etching of glass windows, it is generally best to use one specific hood in a laboratory for HF reactions. The vapors should not be inhaled, and the material causes severe burns on contact with skin.

Hot sulfuric acid is useful for some special cases. Reaction of phthalic anhydride with benzene gives a keto acid that cyclizes in sulfuric acid to give anthraquinone in high yield.

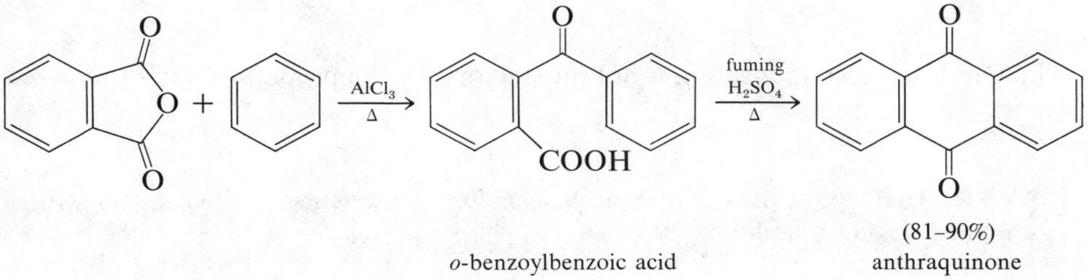

o-benzoylbenzoic acid   (81–90%)
anthraquinone

This reaction is an apparent exception to the generalization that Friedel–Crafts acylations do not occur on aromatic ketones. However, in this case both reactants are in the
same molecule, and ring formation provides an additional driving force.

Polyphosphoric acid (PPA, Section 26.2) is also a convenient reagent for carrying
out such cyclizations.

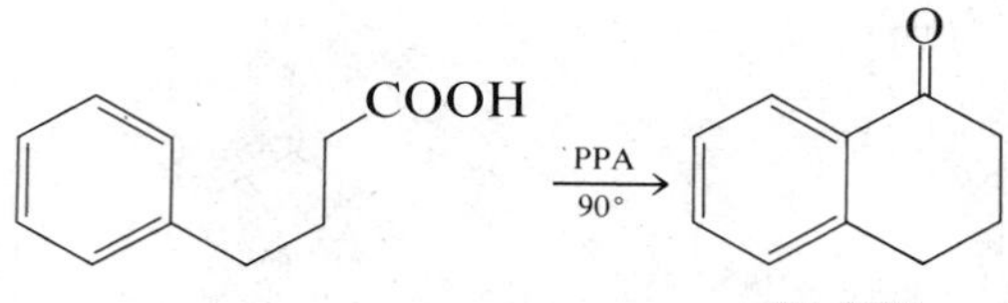

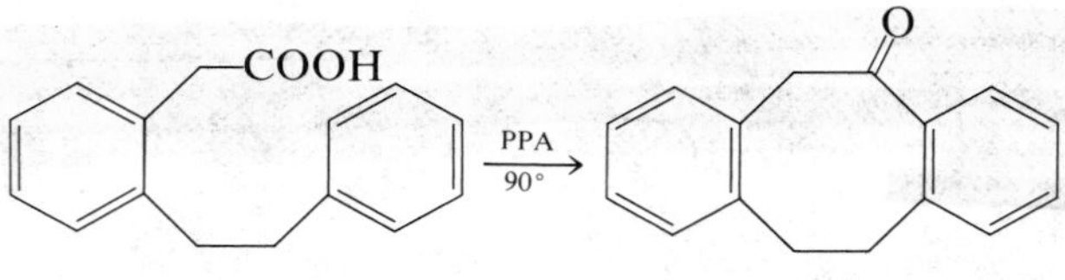

(75–86%)

(93%)

Intramolecular Friedel–Crafts acylation is one of the most important methods for
synthesis of polycyclic aromatic compounds (Chapter 31).

EXERCISE: (a) There are four isomeric diacids having the formula $C_5H_8O_4$.
Write their structures and predict the product of heating each isomer at 200° for 2 hr.
(b) Write the equations illustrating the following reaction sequence starting with
anisole: 1. succinic anhydride, $AlCl_3$; 2. Zn(Hg), HCl; 3. HF.

4. DIECKMANN CONDENSATION. Adipic and pimelic acid esters undergo
a cyclic Claisen condensation (Section 19.9) known as a **Dieckmann condensation;**
the products of such reactions are five- and six-membered cyclic β-keto esters.

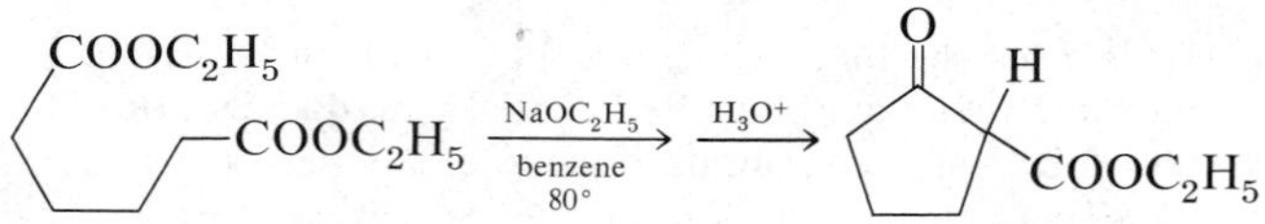

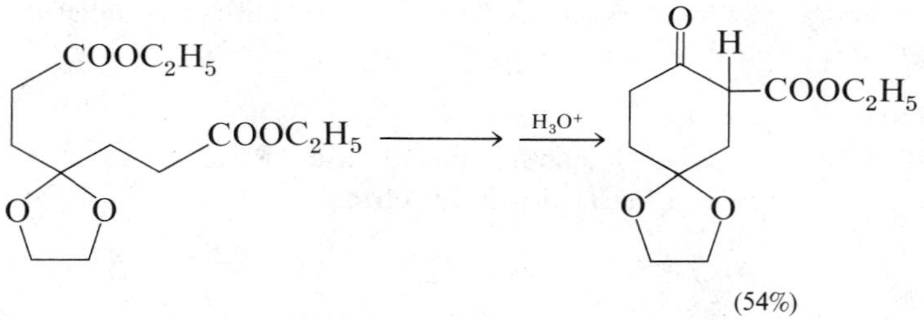

(80%)

(54%)

The Dieckmann condensation is not satisfactory for the preparation of other sized
rings.

EXERCISE: Write the stepwise mechanism for the reaction of diethyl adipate with
sodium ethoxide in refluxing benzene.

871

Sec. 27.7

Diketones,
Keto Aldehydes,
Keto Acids,
and Keto Esters

27.7
Diketones, Keto Aldehydes, Keto Acids, and Keto Esters

This diverse group of difunctional compounds is best considered together because their chemistry is dominated by the interaction of two carbonyl groups in each case. As we shall see, the 1,2-compounds, though fairly rare, do have some interesting aspects in their chemistry. The most important group of dicarbonyl compounds are the 1,3-isomers because of their importance in synthesis. Other dicarbonyl compounds show chemical behavior that is simply that of the monofunctional counterparts except that the presence of two functional groups in the same molecule allows intramolecular reactions, leading to the formation of ring compounds.

A. Synthesis

α-Diketones may be obtained by the mild oxidation of α-hydroxy ketones, which are available by the acyloin condensation (Section 27.4.A). Since the product α-diketones are also susceptible to oxidation (with cleavage of the carbonyl–carbonyl bond), especially mild oxidants must be used. Cupric acetate is especially effective.

(88%)

α-Diketones and α-keto aldehydes are also available by the direct oxidation of simple ketones with selenium dioxide.

(60%)

$$\underset{}{C_6H_5\overset{O}{\overset{\|}{C}}CH_3} + SeO_2 \xrightarrow[\substack{dioxane\\100°}]{H_2O} C_6H_5\overset{O}{\overset{\|}{C}}CHO$$
(70%)

Selenium dioxide, SeO_2, is a white, crystalline material that melts at 340°. It is prepared by oxidizing selenium metal with nitric acid. Although it is rather high melting, it has a substantial vapor pressure at moderate temperatures (12.5 torr at 70°). The yellowish green vapor has a pungent odor. In the body it is reduced to selenium metal, which may produce liver damage. Prolonged occupational exposure to selenium or SeO_2 leads to a garlic odor of breath and sweat.

1,3-Dicarbonyl compounds are almost uniformly prepared by some version of the Claisen condensation. In Section 19.9 we saw that esters react with base to give β-keto esters.

$$2\ CH_3COOC_2H_5 \xrightarrow{C_2H_5O^-Na^+} \xrightarrow{H_3O^+} CH_3\overset{O}{\overset{\|}{C}}CH_2COOC_2H_5$$

β-Diketones and β-keto aldehydes may be prepared by a **mixed Claisen condensation** using a ketone and an ester.

$$CH_3\overset{O}{\overset{\|}{C}}CH_3 + CH_3\overset{O}{\overset{\|}{C}}OEt \xrightarrow[\text{ether}]{\text{NaH}} \xrightarrow{H_3O^+} CH_3\overset{O}{\overset{\|}{C}}CH_2\overset{O}{\overset{\|}{C}}CH_3$$
$$(85\%)$$

$$C_6H_5\overset{O}{\overset{\|}{C}}OEt + C_6H_5\overset{O}{\overset{\|}{C}}CH_3 \xrightarrow[\text{benzene}]{\text{NaNH}_2} \xrightarrow{H_3O^+} C_6H_5\overset{O}{\overset{\|}{C}}CH_2\overset{O}{\overset{\|}{C}}C_6H_5$$
$$(73\%)$$

[These equations illustrate the use of a "shorthand" convention we shall employ for
the remaining examples in this chapter. In order to conserve space and give the
equations a less cluttered appearance, the abbreviations Et for C_2H_5 and Me for CH_3
will be used. Thus EtOH and MeOH represent ethanol and methanol, respectively.]

When ethyl formate is used in a mixed Claisen condensation, the product is a
β-keto aldehyde, which exists almost entirely in the enolic form (Section 27.7.B).

$$(75\%)$$

The mixed Claisen condensation of ketones and esters works well because ketones are considerably more acidic than are esters (Section 19.9). Thus, in the
basic medium, the ketone is deprotonated to a larger extent than the ester.

$$CH_3\overset{O}{\overset{\|}{C}}CH_3 + EtO^- \rightleftharpoons CH_3\overset{O^-}{\overset{|}{C}}{=}CH_2 + EtOH \qquad K \approx 10^{-4}$$

$$CH_3\overset{O}{\overset{\|}{C}}OEt + EtO^- \rightleftharpoons CH_2{=}\overset{O^-}{\overset{|}{C}}OEt + EtOH \qquad K \approx 10^{-9}$$

Of course, once the ketone enolate is formed, it may react with another unionized
ketone molecule (aldol condensation) or with the ester. However, the aldol condensation of ketones is usually thermodynamically unfavorable (page 394), and
this reaction is only a minor side reaction.

$$CH_3\overset{O^-}{\overset{|}{C}}{=}CH_2 + CH_3\overset{O}{\overset{\|}{C}}CH_3 \rightleftharpoons CH_3\overset{O}{\overset{\|}{C}}CH_2\overset{O^-}{\overset{|}{C}}(CH_3)_2$$

On the other hand, the Claisen condensation is driven by the all-important final
deprotonation of the acidic product. Thus the β-diketone is formed in high yield.

Cyclic β-didetones are formed by intramolecular Claisen condensation of 1,4-
and 1,5-keto esters. The reaction is a useful method for the formation of five- and
six-membered rings. This reaction is clearly analogous to the Dieckmann condensation (Section 27.6.B).

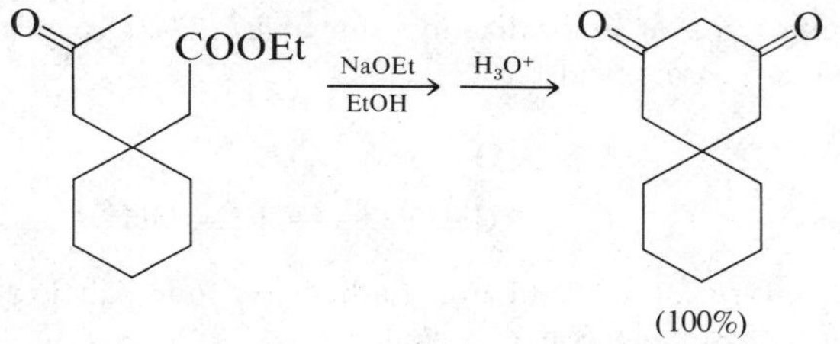

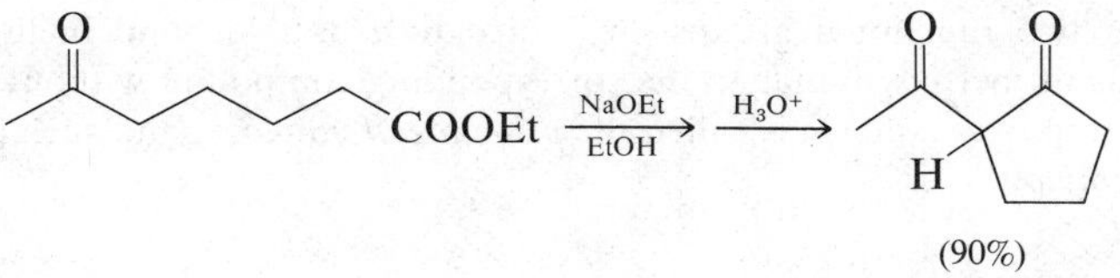

873

Sec. 27.7

Diketones,
Keto Aldehydes,
Keto Acids,
and Keto Esters

(100%)

(90%)

> **EXERCISE:** Outline syntheses of the following dicarbonyl compounds. (a) 3,4-hexanedione; (b) 2,4-pentanedione; (c) 3-oxobutanal.

B. *Properties*

1. KETO-ENOL EQUILIBRIA. Recall that simple ketones exist very largely in the keto form with but a trace of the enol (vinyl alcohol) form present at equilibrium (Section 13.6.A).

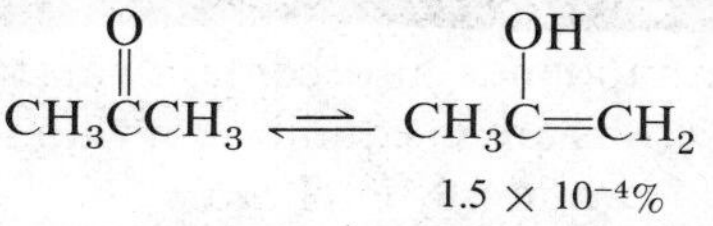

$$1.5 \times 10^{-4}\%$$

In contrast, 1,2- and 1,3-dicarbonyl compounds often contain a large amount of enol form in equilibrium with the dicarbonyl form. For example, 2,4-pentanedione is a mixture of 84% dione and 16% enolic form in aqueous solution. In hexane solution the compound exists almost entirely in the enolic form.

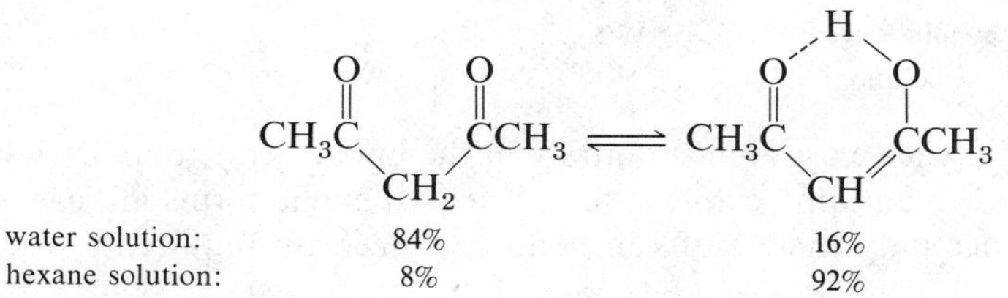

(27-1)

water solution:	84%	16%
hexane solution:	8%	92%

One important reason for this phenomenon is the ability of the enol to form an intramolecular hydrogen bond as shown in (27-1). Such intramolecular hydrogen bonds are especially favorable when six-membered rings are formed. The enolic form also benefits from resonance stabilization in a way not available to the dicarbonyl compound itself.

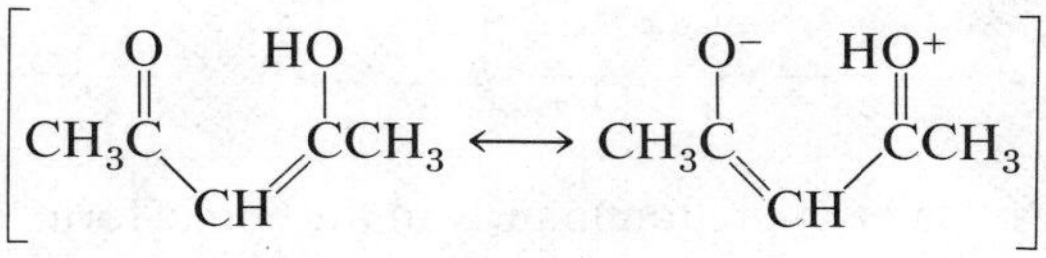

(27-2)

Note that the type of delocalization shown in (27-2) is precisely the kind that is involved in carboxylic acids.

$$
\underset{\text{O}}{\overset{\text{O}}{\parallel}}\text{RC—OH} \quad \longleftrightarrow \quad \underset{\text{O}^-}{\overset{}{}}\text{RC}\overset{+}{=}\text{OH}
$$

In the enolic form of a 1,3-diketone, a hydroxy lone pair is delocalized *through the double bond* into the carbonyl oxygen.

Whenever two functional groups are joined to a double bond in this way, the molecule has properties similar to the corresponding compound without the double bond. This empirical concept is called the *principle of vinylogy,* and such compounds are called *vinylogs.*

$$
\underset{\text{O}}{\overset{\text{O}}{\parallel}}\text{CH}_3\overset{\text{OH}}{\overset{|}{\text{C}}}\text{CCH}=\text{CCH}_3 \quad \text{is a vinylog of} \quad \text{CH}_3\overset{\text{O}}{\overset{\parallel}{\text{C}}}\text{—OH}
$$

$$
\text{CH}_3\overset{\text{O}}{\overset{\parallel}{\text{C}}}\text{CH}=\overset{\text{NH}_2}{\overset{|}{\text{C}}}\text{CCH}_3 \quad \text{is a vinylog of} \quad \text{CH}_3\overset{\text{O}}{\overset{\parallel}{\text{C}}}\text{—NH}_2
$$

Note that the percentage of enol form at equilibrium is higher in nonpolar aprotic solvents because in such solvents the intramolecular hydrogen bond is most beneficial. In protic solvents the dicarbonyl compound itself as well as the enol can hydrogen-bond to solvent molecules, and the ability of the enol to form an intramolecular hydrogen bond provides no extra stabilization.

Other 1,3-dicarbonyl compounds also contain substantial amounts of enolic forms in solution. β-Keto esters are in equilibrium with significant amounts of the form in which the ketone carbonyl is enolized.

water solution:	90%		10%
hexane solution:	51%		49%

β-Keto aldehydes exist almost entirely in the enolic form; both carbonyl groups are enolized to an appreciable extent. The two enolic forms are easily interconvertible, since only small shifts in bond distances are required.

0%   76%   24%
(CCl₄ solution)

Cyclic 1,3-diketones also exist predominantly in the enolic form, even though they cannot participate in intramolecular hydrogen bonding for reasons of geometry.

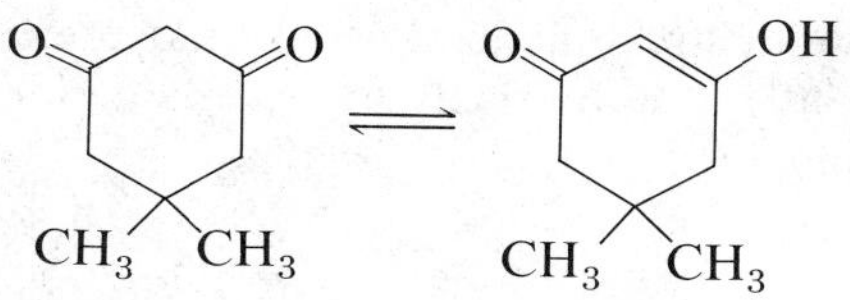

875

Sec. 27.7

Diketones,
Keto Aldehydes,
Keto Acids,
and Keto Esters

5%   95%

(water solution)

1,2-Diketones also show enhanced amounts of enol form. The main driving force for enolization in this case is relief of the electrostatic repulsion that occurs when the two electrophilic carbonyl groups are adjacent to each other.

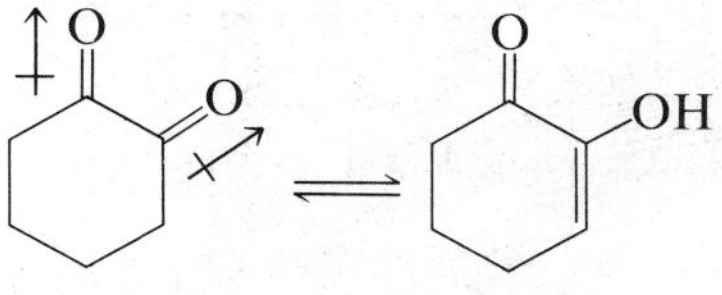

minor   major

2. 1,3-DICARBONYL COMPOUNDS AS CARBON ACIDS. 1,3-Dicarbonyl compounds, which have a hydrogen bound to the carbon between the two carbonyl groups, are much stronger acids than normal aldehydes, ketones, or esters because the charge in the resulting enolate ion can be delocalized into both carbonyl groups.

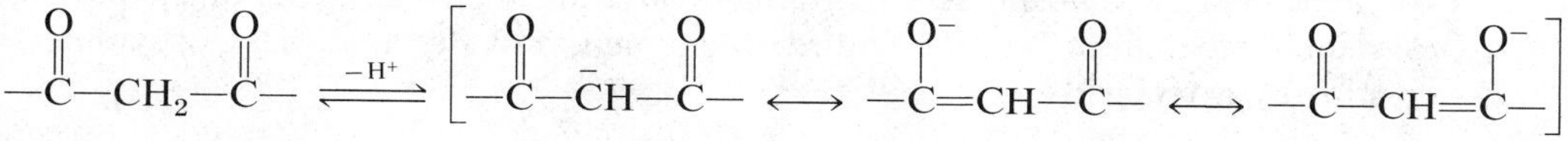

Some typical pK_as for such systems are contained in Table 27.6. The acidities of

TABLE 27.6
Acidity of β-Dicarbonyl Compounds

Compound	pK_a
$NCCH_2COCH_3$	9
$CH_3CCH_2CCH_3$	9
$CH_3CCH_2COCH_3$	11
$CH_3CCHCCH_3$ $\quad\quad\;\; CH_3$	11
$NCCH_2CN$	11
$CH_3OCCH_2COCH_3$	13

1,3-dicarbonyl compounds are sufficiently high that they are converted to their conjugate bases essentially quantitatively by hydroxide ion in water or by alkoxide ion in alcoholic solvent.

$$CH_3\overset{\overset{\displaystyle O}{\|}}{C}CH_2\overset{\overset{\displaystyle O}{\|}}{C}OCH_3 + CH_3O^-Na^+ \overset{K}{\rightleftharpoons} CH_3\overset{\overset{\displaystyle Na^+}{\underset{\displaystyle O^-}{|}}}{C}=CH\overset{\overset{\displaystyle O}{\|}}{C}OCH_3 + CH_3OH$$

$$\text{p}K_a \ 11 \qquad\qquad\qquad\qquad\qquad\qquad\qquad\qquad \text{p}K_a \ 16$$

$$K = \frac{\left[CH_3\overset{\overset{\displaystyle O^-}{|}}{C}=CHCOOCH_3\right]\left[CH_3OH\right]}{\left[CH_3\overset{\overset{\displaystyle O}{\|}}{C}CH_2COOCH_3\right]\left[CH_3O^-\right]} \approx 10^5$$

As we shall see, these easily accessible carbanions are valuable synthetic intermediates.

C. Reactions

1. The Benzilic Acid Rearrangement. α-Diketones undergo an interesting rearrangement reaction upon being treated with strong base. The reaction is called the **benzilic acid rearrangement** after the trivial name of the parent system. Note that the benzilic acid rearrangement provides a synthesis of symmetrical α-hydroxy acids.

$$C_6H_5-\overset{\overset{\displaystyle O}{\|}}{C}-\overset{\overset{\displaystyle O}{\|}}{C}-C_6H_5 \xrightarrow[\substack{H_2O-C_2H_5OH \\ 100°}]{KOH} \xrightarrow{H_3O^+} (C_6H_5)_2\overset{\overset{\displaystyle OH}{|}}{C}COOH$$

$$\text{benzil} \qquad\qquad\qquad\qquad\qquad\qquad \underset{\substack{(95\%) \\ \text{benzilic acid}}}{}$$

The mechanism of this rearrangement involves the addition of hydroxide ion to one of the carbonyl groups. This initially formed adduct is probably ionized further to a dianion, which undergoes a molecular rearrangement to give the dianion of the α-hydroxy acid. The involvement of a dianion explains the strongly basic conditions required.

$$R-\overset{\overset{\displaystyle O}{\|}}{C}-\overset{\overset{\displaystyle O}{\|}}{C}-R + OH^- \rightleftharpoons HO-\overset{\overset{\displaystyle O^-}{|}}{\underset{\underset{\displaystyle R}{|}}{C}}-\overset{\overset{\displaystyle O}{\|}}{C}-R$$

$$HO-\overset{\overset{\displaystyle O^-}{|}}{\underset{\underset{\displaystyle R}{|}}{C}}-\overset{\overset{\displaystyle O}{\|}}{C}-R + OH^- \rightleftharpoons {}^-O-\overset{\overset{\displaystyle O^-}{|}}{\underset{\underset{\displaystyle R}{|}}{C}}-\overset{\overset{\displaystyle O}{\|}}{C}-R + H_2O$$

$$^-O-\overset{\overset{\displaystyle O^-}{|}}{\underset{\underset{\displaystyle R}{|}}{C}}-\overset{\overset{\displaystyle O}{\|}}{C}-R \longrightarrow {}^-O-\overset{\overset{\displaystyle O}{\|}}{C}-\overset{\overset{\displaystyle O^-}{|}}{\underset{\underset{\displaystyle R}{|}}{C}}-R$$

$$-O-\overset{\overset{\displaystyle O}{\|}}{C}-\overset{\overset{\displaystyle O^-}{|}}{\underset{\underset{\displaystyle R}{|}}{C}}-R + H_2O \; \rightleftharpoons \; -O\overset{\overset{\displaystyle O}{\|}}{C}-\overset{\overset{\displaystyle OH}{|}}{C}R_2 + OH^-$$

When the α-diketone is cyclic, the rearrangement serves as a method of **ring contraction.**

(80%)

2. DECARBOXYLATION OF β-KETO ACIDS.

β-Keto acids undergo thermal decarboxylation in the same manner as do 1,3-diacids (Section 27.6.B). In this case milder conditions suffice to bring about decarboxylation; 2-ethyl-3-oxohexanoic acid has a half-life of only 15 min at 50°.

$$CH_3CH_2CH_2\overset{\overset{\displaystyle O}{\|}}{C}\underset{\underset{\displaystyle CH_2CH_3}{|}}{CH}COOH \xrightarrow[\substack{H_2O \\ 2 \text{ hr}}]{50°} CH_3CH_2CH_2\overset{\overset{\displaystyle O}{\|}}{C}CH_2CH_2CH_3 + CO_2$$

The mechanism may involve a concerted, six-center transition state as depicted on page 866 for the decarboxylation of malonic acid. The initial product in the case of a β-keto acid is the enol form of the ketone. This mechanism is consistent with the resistance of bridgehead bicyclic β-keto acids to decarboxylation; the product would be a highly strained bridgehead olefin.

EXERCISE: Write the equations illustrating the application of the following sequence of operations to ethyl propanoate: (1) sodium ethoxide in refluxing ethanol, (2) work-up with cold 10% aqueous HCl, (3) refluxing 6 N HCl. Note that this sequence of reactions results in the formation of a symmetrical ketone. What starting material is required to synthesize 5-nonanone by this method?

3. ALKYLATION OF 1,3-DICARBONYL COMPOUNDS: THE MALONIC ESTER AND ACETOACETIC ESTER SYNTHESES.

The anions of 1,3-dicarbonyl compounds are nucleophiles and may take part in S_N2 displacement reactions with alkyl halides. Diethyl malonate and ethyl acetoacetate are inexpensive commercial compounds that are often alkylated in this manner. Hydrolysis of the alkylated product followed by decarboxylation of the resulting β-diacid or β-keto acid provides an important general synthesis of acids and methyl ketones. The overall processes are called the **malonic ester synthesis** or the **acetoacetic ester synthesis.** Some examples are given in (27-3) and (27-4).

$$CH_2(COOEt)_2 \xrightarrow[\text{EtOH}]{\text{NaOEt}} \xrightarrow{n\text{-}C_4H_9Br} \underset{\substack{(80–90\%) \\ \text{diethyl } n\text{-butylmalonate}}}{n\text{-}C_4H_9CH(COOEt)_2} \xrightarrow[115°]{\text{conc. HCl}} \underset{\text{hexanoic acid}}{n\text{-}C_4H_9CH_2COOH} \quad (27\text{-}3)$$

$$CH_2(COOEt)_2 \xrightarrow[\text{EtOH}]{\text{NaOEt}} \xrightarrow{(CH_3)_2CHI} (CH_3)_2CHCH(COOEt)_2 \xrightarrow[115°]{\text{conc. HCl}}$$
$$\underset{\text{diethyl isopropylmalonate}}{(70\text{-}75\%)}$$

$$(CH_3)_2CHCH_2COOH \quad (27\text{-}4)$$

3-methylbutanoic acid

The initially formed alkylmalonic ester may be alkylated again with the same alkyl halide or with a different one to widen the scope of the procedure.

$$CH_2(COOEt)_2 \xrightarrow[\text{EtOH}]{\text{NaOEt}} \xrightarrow{n\text{-}C_4H_9Br} n\text{-}C_4H_9CH(COOEt)_2 \xrightarrow[\text{EtOH}]{\text{NaOEt}} \xrightarrow{CH_3Br}$$

$$\underset{}{CH_3CH_2CH_2CH_2\overset{\overset{\displaystyle CH_3}{|}}{C}(COOEt)_2} \xrightarrow[115°]{\text{conc. HCl}} CH_3CH_2CH_2CH_2\overset{\overset{\displaystyle CH_3}{|}}{C}HCOOH$$
2-methylhexanoic acid

The overall synthetic result of alkylation and decarboxylation of malonic ester is an alkyl or dialkylacetic acid.

$$CH_2(COOEt)_2 \xrightarrow{\text{NaOEt}} \xrightarrow{R'X} R'CH(COOEt)_2 \xrightarrow{\text{NaOEt}} \xrightarrow{R''X}$$

$$R'R''C(COOEt)_2 \xrightarrow[\Delta]{H^+} \underset{R''}{\overset{R'}{>}}CHCOOH$$

A principal limitation in the synthetic sequence is that the alkylation process is an S_N2 reaction; E2 elimination is an expected side reaction whose importance depends on the structure of RX. If a suitable dihalide is used in the reaction, 2 moles of malonic ester can be added to both ends of a chain; alternatively, intramolecular alkylation in the second step leads to a cyclic diester.

$$CH_2(COOEt)_2 \xrightarrow[\text{EtOH}]{\text{2 NaOEt}} \xrightarrow{Br(CH_2)_3Cl} \boxed{\underset{}{\overset{\text{COOEt}}{-}\overset{}{\underset{}{-}\text{COOEt}}}} \xrightarrow[\text{EtOH}]{\text{KOH}} \xrightarrow[\Delta]{H_3O^+} \boxed{\overset{\text{H}}{\underset{}{-}}\text{COOH}}$$

(42-44%)
cyclobutanecarboxylic acid

If ethyl acetoacetate is used as the starting material, the combination of alkylation, hydrolysis, and decarboxylation provides a synthesis of various methyl alkyl ketones.

$$\underset{}{CH_3\overset{\overset{\displaystyle O}{\|}}{C}CH_2COOEt} \xrightarrow[\text{EtOH}]{\text{NaOEt}} \xrightarrow{n\text{-}BuBr} CH_3\overset{\overset{\displaystyle O}{\|}}{C}\underset{\underset{\displaystyle C_4H_9}{|}}{C}HCOOEt \xrightarrow[H_2O]{\text{NaOH}} \xrightarrow[25°]{H_2SO_4}$$
(69-72%)

$$CH_3\overset{\overset{\displaystyle O}{\|}}{C}CH_2CH_2CH_2CH_2CH_3$$
(61% overall)
2-heptanone

879

Sec. 27.7

**Diketones,
Keto Aldehydes,
Keto Acids,
and Keto Esters**

$$CH_3\overset{O}{\underset{\parallel}{C}}CH_2COOEt \xrightarrow[\text{EtOH}]{\text{NaOEt}} \xrightarrow{\text{ClCH}_2\text{COOEt}} CH_3\overset{O}{\underset{\parallel}{C}}\underset{\underset{CH_2COOEt}{|}}{CHCOOEt} \xrightarrow[100°]{\text{conc. HCl}}$$

(56–62%)

$$CH_3\overset{O}{\underset{\parallel}{C}}CH_2CH_2COOH$$

(50% overall)
4-oxopentanoic acid

Again the starting β-keto ester may be alkylated successively with two different alkyl halides. After hydrolysis and decarboxylation, the product is a ketone that is branched at the α-carbon.

$$CH_3\overset{O}{\underset{\parallel}{C}}CH_2COOEt \xrightarrow[\text{EtOH}]{\text{NaOEt}} \xrightarrow{\text{CH}_3\text{I}} \xrightarrow[\text{EtOH}]{\text{NaOEt}} \xrightarrow{\text{CH}_2=\text{CHCH}_2\text{Br}}$$

$$CH_3\overset{O}{\underset{\parallel}{C}}-\underset{\underset{CH_2CH=CH_2}{|}}{\overset{\overset{CH_3}{|}}{C}}COOEt \xrightarrow[\text{H}_2\text{O}]{\text{NaOH}} \xrightarrow{\text{H}_2\text{SO}_4} CH_3\overset{O}{\underset{\parallel}{C}}\underset{\underset{CH_3}{|}}{CHCH_2CH=CH_2}$$

3-methylhex-5-en-2-one

The net result is summarized in (27-5) and is again subject to the usual limitations of S_N2 reactions.

$$CH_3COCH_2COOEt \xrightarrow{R'X} \xrightarrow{R''X} \xrightarrow[\Delta]{H^+} CH_3\overset{O}{\underset{\parallel}{C}}\underset{\underset{R''}{\diagdown}}{\overset{\overset{R'}{\diagup}}{CH}} \qquad (27\text{-}5)$$

Other β-diketones and β-keto esters may be alkylated in the same manner. The following examples show the tremendous utility of these reactions in building up complicated organic structures.

1 equiv NaOH
CH₃I, H₂O

(70%)
2-methyl-1,3-cyclohexanedione

2 equiv NaOMe
CH₃I, MeOH

(80%)
2,2-dimethyl-1,3-cyclohexanedione

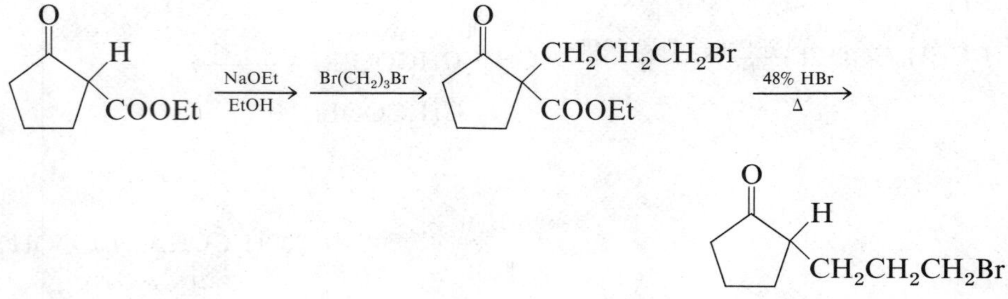

(60%)
2-(3-bromopropyl)cyclopentanone

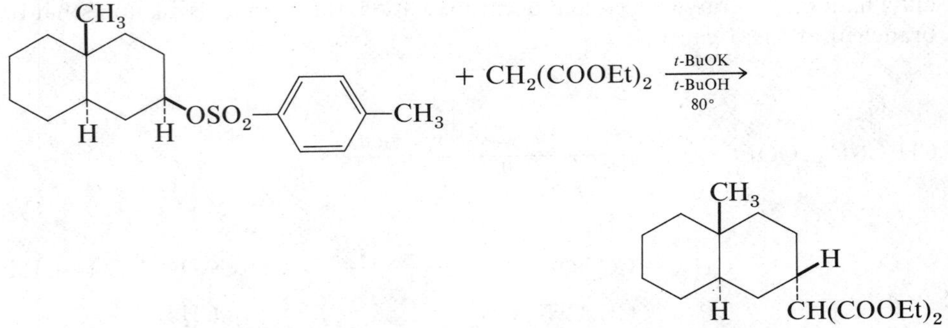

(65%)

EXERCISE: Write equations showing all steps in the syntheses of the following compounds by the malonic ester or acetoacetic ester syntheses.

(a) $CH_2{=}CHCH_2CH_2\overset{\overset{\displaystyle O}{\|}}{C}CH_3$

(b) $(CH_3)_2CHCH_2CH_2CH_2COOH$

---

4. THE KNOEVENAGEL CONDENSATION. We have seen that malonic ester is more acidic than normal ketones and aldehydes; that is, in equilibrium with base the carbanion from malonic ester is present in large excess over that from the ketone or aldehyde. The consequence is a condensation of malonic ester with the carbonyl group of the ketone or aldehyde. This result should be contrasted with the condensation of a ketone with a monoester (Section 27.7.A).

$$RCOR' + {}^-CH(COOEt)_2 \rightleftharpoons R\overset{\overset{\displaystyle O^-}{|}}{\underset{\underset{\displaystyle R'}{|}}{C}}{-}CH(COOEt)_2$$

Several further equilibria occur.

$$RR'\overset{\overset{\displaystyle O^-}{|}}{C}CH(COOEt)_2 + EtOH \rightleftharpoons RR'\overset{\overset{\displaystyle OH}{|}}{C}CH(COOEt)_2 + EtO^-$$

$$RR'\overset{\overset{\displaystyle OH}{|}}{C}CH(COOEt)_2 + EtO^- \rightleftharpoons RR'\overset{\overset{\displaystyle OH}{|}}{C}\overset{\text{—}}{C}(COOEt)_2 + EtOH$$

The next step is an elimination reaction that pulls the whole equilibrium.

881

Sec. 27.7

**Diketones,
Keto Aldehydes,
Keto Acids,
and Keto Esters**

$$\underset{|}{\overset{OH}{RR'CC(COOEt)_2}} \longrightarrow OH^- + RR'C=C(COOEt)_2$$

The net result is a condensation *catalyzed* by base.

$$RR'CO + CH_2(COOEt)_2 \xrightarrow{\text{base}} H_2O + RR'C=C(COOEt)_2$$

In the Claisen condensations discussed previously, the base was a reagent required in stoichiometric quantity. In the present case, catalytic amounts suffice and, in fact, the reaction works well with such weak base catalysts as amines. The overall reaction works best for aldehydes, although ketones have been used. Some examples are

$$(CH_3)_2CHCH_2CHO + CH_2(COOEt)_2 \xrightarrow[\text{benzene}]{\text{piperidine}} (CH_3)_2CHCH_2CH=C(COOEt)_2$$
$$\Delta$$
$$(78\%)$$

$$(80\%)$$

A variation of this reaction makes use of malonic acid with an amine catalyst. In this case some of the carboxylate anion is present, but since amines are weak bases (Section 24.4), the fraction of carboxylate anion is small. The carbanion derived from the α-hydrogen is also present because the carbonyl groups of the carboxylic acid stabilize the carbanion just as they stabilize the ester carbonyls. The resulting condensation products decarboxylate on heating to give α,β-unsaturated acids. The basic catalyst frequently used in this reaction is the tertiary amine pyridine.

$$\underset{\text{heptanal}}{CH_3(CH_2)_5CHO} + CH_2(COOH)_2 \xrightarrow[\Delta]{\text{pyridine}} [CH_3(CH_2)_5CH=C(COOH)_2] \xrightarrow{-CO_2}$$

$$CH_3(CH_2)_5CH=CHCOOH$$
$$(75-85\%)$$
$$\text{non-2-enoic acid}$$

This version of the Knoevenagel condensation works well with all aldehydes. It can be employed with ketones, but yields are generally low.

$$(CH_3CH_2)_2CO + CH_2(COOH)_2 \xrightarrow[\Delta]{\overset{\text{pyridine}}{\text{piperidine}}} \xrightarrow{-CO_2} (CH_3CH_2)_2C=CHCOOH$$
$$(35\%)$$

EXERCISE: The Knoevenagel condensation provides a synthesis of unsaturated malonic esters or of α,β-unsaturated acids. Outline syntheses of the following compounds.
(a) $CH_3CH_2CH=CHCOOH$
(b) $CH_3CH_2CH=C(COOC_2H_5)_2$

5. THE MICHAEL ADDITION. In Section 20.3.A., we saw that α,β-unsaturated carbonyl compounds may react with such nucleophiles as cyanide ion and Grignard reagents either by 1,2- or 1,4-addition. The 1,4-addition of a carbanion to an α,β-unsaturated carbonyl system is called a **Michael addition.** It is a common and useful reaction. For example, when a mixture of 2-cyclohexen-1-one and diethyl malonate is treated with a catalytic amount of sodium ethoxide in ethanol, the following addition reaction occurs.

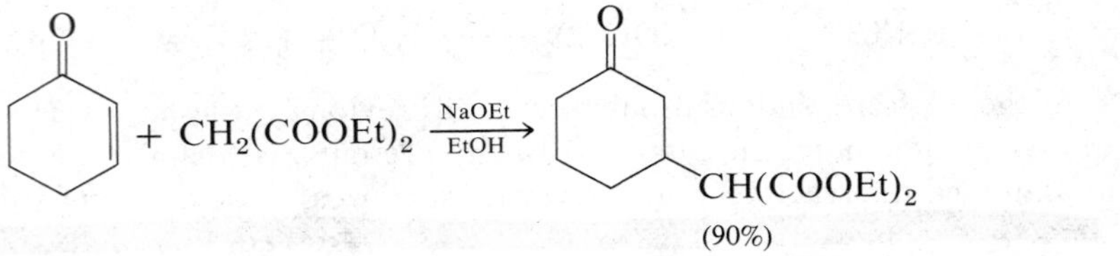

$$+ \ CH_2(COOEt)_2 \xrightarrow[\text{EtOH}]{\text{NaOEt}} \quad CH(COOEt)_2$$
(90%)

The mechanism of the Michael addition is illustrated with diethyl malonate and acrolein; the product is obtained in 50% yield.

(1) $CH_2(COOEt)_2 + EtO^- \rightleftharpoons \ ^-:CH(COOEt)_2 + EtOH$

(2) $HC\overset{O}{-}CH=CH_2 + \ ^-:CH(COOEt)_2 \rightleftharpoons HC\overset{O^-}{=}CH-CH_2-CH(COOEt)_2$

(3) $CH\overset{O^-}{=}CHCH_2CH(COOEt)_2 + EtOH \rightleftharpoons HC\overset{O}{C}H_2CH_2CH(COOEt)_2 + EtO^-$

A Michael addition such as this is similar to the alkylation of a carbanion by an alkyl halide—with one important exception. In the alkylation with an alkyl halide, a stoichimetric amount of base is consumed; in the Michael addition, the base functions as a catalyst. Thus only a small amount of base need be used in Michael additions, and the process is reversible. The driving force for the reaction is the formation of a new C—C single bond at the expense of the π-bond of the unsaturated carbonyl compound; this driving force is essentially the same as that of all additions to a double bond.

Michael additions are observed between carbon acids containing an acidic proton and a variety of α,β-unsaturated carbonyl systems. A few representative examples are

$$CH_2(COOEt)_2 + CH_2=CHCCH_3 \xrightarrow[\substack{\text{EtOH} \\ 0°}]{\text{NaOEt}} CH_3CCH_2CH_2CH(COOEt)_2$$
(71%)

$$CH_3CCH_2CCH_3 + CH_2=CHCN \xrightarrow[\substack{t\text{-BuOH} \\ 25°}]{\text{Et}_3N} CH_3CCHCCH_3$$
$$CH_2CH_2CN$$
(77%)

$$+ \ CH_2=CHCOOEt \xrightarrow[\substack{\text{EtOH} \\ 0°}]{\text{NaOEt}} \quad CH_2CH_2COOEt$$

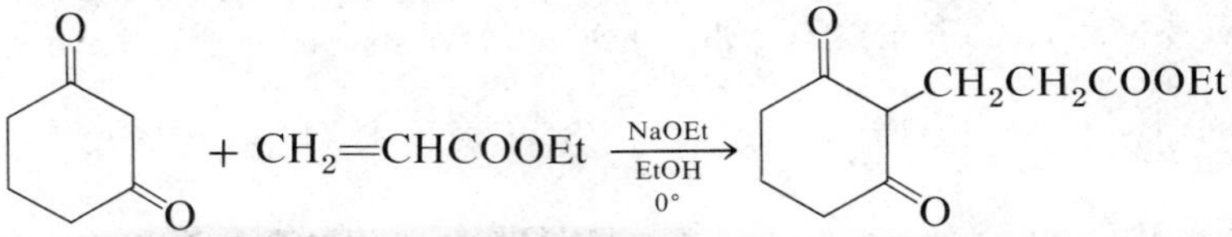

If an excess of the α,β-unsaturated carbonyl component is used, it is possible to achieve **dialkylation.**

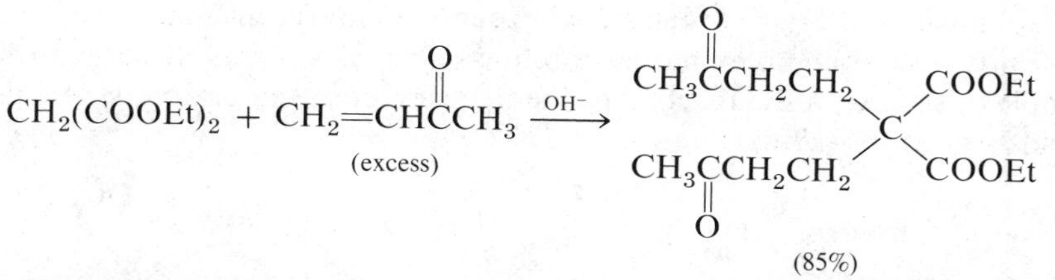

$$CH_2(COOEt)_2 + CH_2{=}CHCCH_3 \xrightarrow{OH^-}$$

(excess)

(85%)

The Michael addition constitutes a useful method for the synthesis of 1,5-dicarbonyl systems. When diethyl malonate or acetoacetic ester is used as the adding group, the product may be hydrolyzed and decarboxylated to obtain the alkylated acid or ketone.

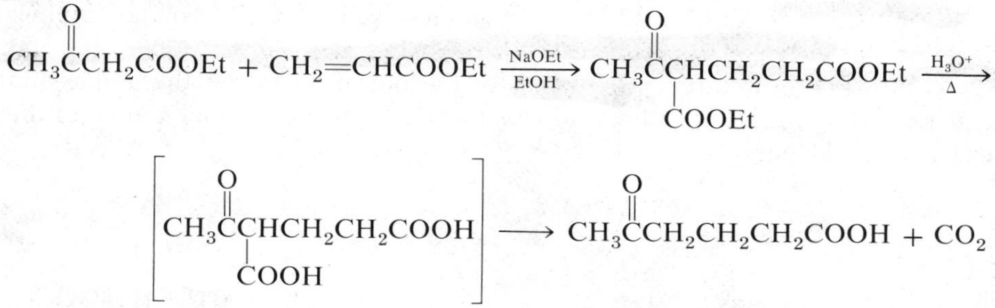

$$CH_3CCH_2COOEt + CH_2{=}CHCOOEt \xrightarrow[EtOH]{NaOEt} CH_3CCHCH_2CH_2COOEt \xrightarrow[\Delta]{H_3O^+}$$

$$\qquad\qquad\qquad\qquad\qquad\qquad\qquad\qquad\qquad\qquad\qquad\qquad COOEt$$

$$\left[CH_3CCHCH_2CH_2COOH \atop \qquad\quad COOH \right] \longrightarrow CH_3CCH_2CH_2CH_2COOH + CO_2$$

Although the Michael addition is most successful when the carbon acid is relatively acidic, such as a 1,3-dicarbonyl compound, the reaction also occurs with simple ketones.

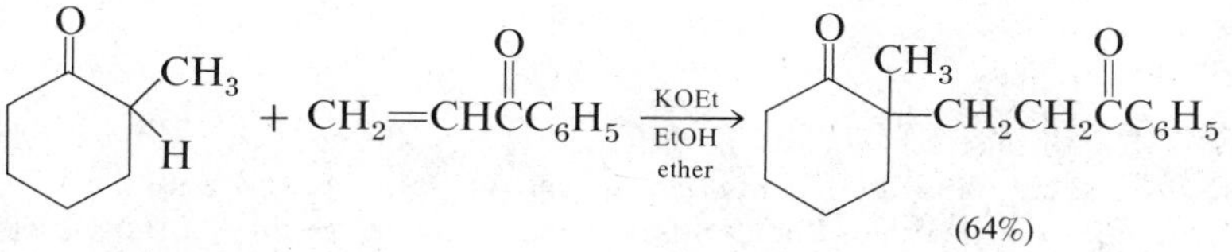

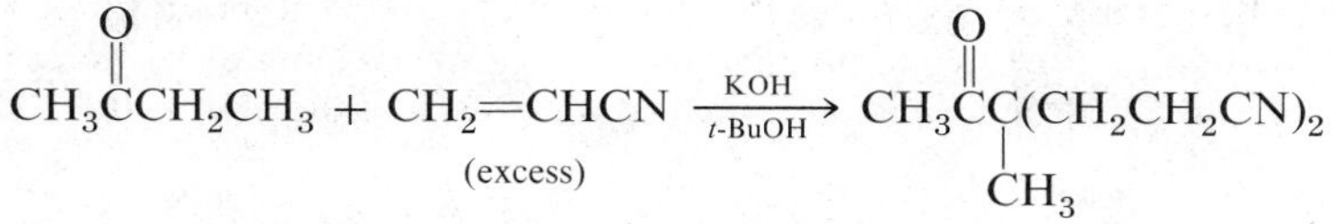

(64%)

$$CH_3CCH_2CH_3 + CH_2{=}CHCN \xrightarrow[t\text{-BuOH}]{KOH} CH_3CC(CH_2CH_2CN)_2$$

(excess)

$$\qquad\qquad\qquad\qquad\qquad\qquad\qquad\qquad\qquad\qquad CH_3$$

A useful variant of the Michael addition occurs with methyl vinyl ketone and its derivatives. The initially formed 1,5-diketone undergoes a subsequent intramolecular aldol condensation to yield a cyclohexenone ring. The process is essentially a combination of the Michael reaction and aldol condensation and is called **Robinson annelation.**

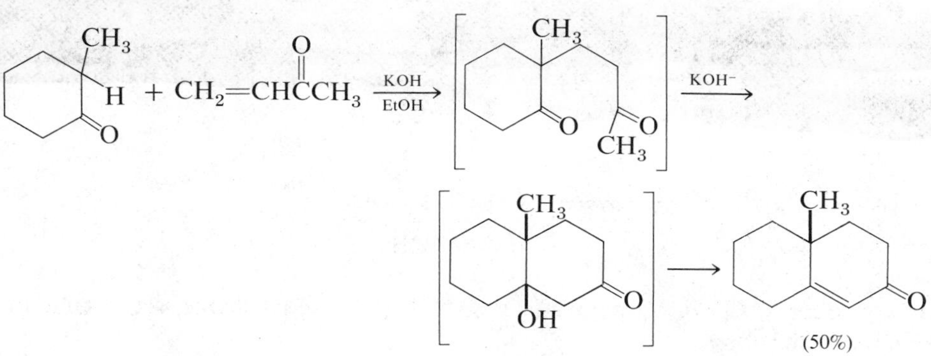

(50%)

The Robinson annelation sequence has been very useful in building up the carbon framework of complex natural products such as steroids (Section 33.5). An example of the use of the reaction as the first step in the laboratory synthesis of a steroid is given in (27-6).

$$+ CH_3CH_2\overset{O}{\underset{\|}{C}}CH{=}CH_2 \xrightarrow[\text{EtOH}]{\text{OH}^-} CH_3 \qquad (27\text{-}6)$$

Enamines, which are formed by the reaction of ketones with secondary amines (Section 24.8), readily react with α,β-unsaturated nitriles, aldehydes, ketones, and esters by Michael addition. Hydrolysis of the product removes the amine and generates the product of indirect Michael addition of the original ketone to the electrophilic alkene.

> **EXERCISE:** Make a 3 × 3 matrix with methyl acrylate, acrolein, and methyl vinyl ketone on one side and diethyl malonate, acetoacetic ester, and 2-ethoxycarbonyl-cyclopentanone on the other. Each of the nine intersections of the matrix represents a Michael addition reaction. Assuming that the reactants are used in a ratio of 1:1 in each case, write the structures of the nine reaction products. Now write the products expected from the hydrolysis and decarboxylation of each of the nine initial adducts.

6. REVERSE CLAISEN CONDENSATION. As we saw in Section 19.9, the driving force for the Claisen condensation is provided by the conversion of the product β-keto ester into its conjugate base; when the product has no protons between the two carbonyl groups, the overall equilibrium constant is unfavorable. Such a fully alkylated system is said to be a nonenolizable β-keto ester. Nonenolizable β-keto esters may be prepared by the alkylation of a β-keto ester, as discussed in Section 19.9. At the end of the alkylation, the base has all been consumed and the product is stable. If a catalytic amount of sodium ethoxide is now added, a reverse Claisen condensation occurs.

$$CH_3\overset{O}{\underset{\|}{C}}CH_2COOEt \xrightarrow[\text{EtOH}]{2\,\text{NaOEt}} \xrightarrow{2\,CH_3I} CH_3\overset{O}{\underset{\|}{C}}{-}\underset{\underset{CH_3}{|}}{\overset{\overset{CH_3}{|}}{C}}COOEt \xrightarrow[\text{EtOH}]{\text{NaOEt}}$$

$$(CH_3)_2CHCOOEt + CH_3COOEt$$

885

Sec. 27.7

**Diketones,
Keto Aldehydes,
Keto Acids,
and Keto Esters**

Only a catalytic amount of sodium ethoxide is required for the reverse Claisen because no base is consumed in the overall reaction.

If sodium hydroxide is used, a full equivalent of base is consumed because one of the products is acetic acid.

The reverse Claisen condensation is a common side-reaction that is observed when nonenolizable β-keto esters are hydrolyzed under basic conditions; it competes with hydrolysis especially well when the β-keto ester is cyclic. For this reason it is usually best to hydrolyze such compounds under acidic conditions.

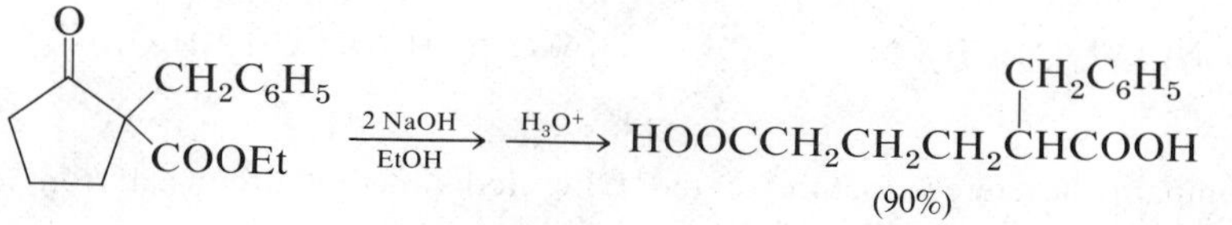

$$\xrightarrow[\text{EtOH}]{\text{2 NaOH}} \xrightarrow{\text{H}_3\text{O}^+} \text{HOOCCH}_2\text{CH}_2\text{CH}_2\text{CHCOOH} \quad (27\text{-}7)$$
$$(90\%)$$

Not only is the reaction an annoying side reaction, but also it can be used as synthesis of dicarboxylic acids, as shown in equation (27-7). The same principles apply to nonenolizable β-diketones. An example of such a reaction, which yields a keto ester, is given in (27-8).

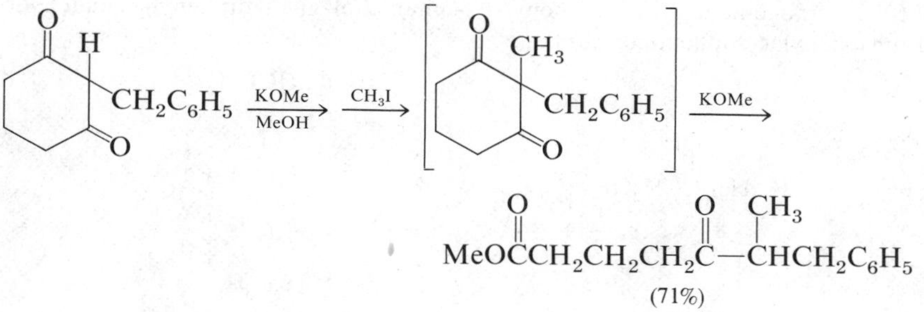

$$\text{MeOCCH}_2\text{CH}_2\text{CH}_2\text{C}-\text{CHCH}_2\text{C}_6\text{H}_5 \quad (27\text{-}8)$$
$$(71\%)$$

EXERCISE: Show how 1,3-cyclopentanedione can be transformed into 5-methyl-4-oxohexanoic acid by a two-step synthesis involving a reverse Claisen condensation as one of the steps.

PROBLEMS

1. Name the following compounds.

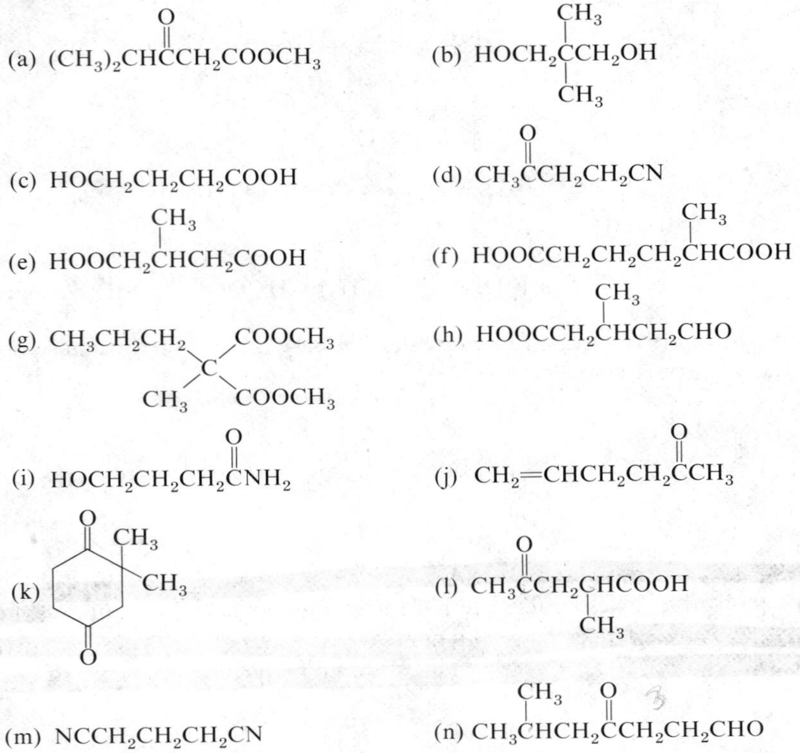

(a) $(CH_3)_2CHCCH_2COOCH_3$ (with $\overset{O}{\overset{\|}{C}}$)

(b) $HOCH_2\overset{\overset{\displaystyle CH_3}{|}}{\underset{\underset{\displaystyle CH_3}{|}}{C}}CH_2OH$

(c) $HOCH_2CH_2CH_2COOH$

(d) $CH_3\overset{O}{\overset{\|}{C}}CH_2CH_2CN$

(e) $HOOCH_2\overset{\overset{\displaystyle CH_3}{|}}{C}HCH_2COOH$

(f) $HOOCCH_2CH_2CH_2\overset{\overset{\displaystyle CH_3}{|}}{C}HCOOH$

(g) $CH_3CH_2CH_2\overset{\displaystyle COOCH_3}{\underset{\displaystyle CH_3}{\overset{|}{\underset{|}{C}}}}COOCH_3$

(h) $HOOCCH_2\overset{\overset{\displaystyle CH_3}{|}}{C}HCH_2CHO$

(i) $HOCH_2CH_2CH_2\overset{O}{\overset{\|}{C}}NH_2$

(j) $CH_2{=}CHCH_2CH_2\overset{O}{\overset{\|}{C}}CH_3$

(k) (cyclohexanone ring with $\overset{O}{\|}$ at top and bottom, CH_3, CH_3 substituents)

(l) $CH_3\overset{O}{\overset{\|}{C}}CH_2\overset{\overset{\displaystyle}{}}{\underset{\underset{\displaystyle CH_3}{|}}{C}}HCOOH$

(m) $NCCH_2CH_2CH_2CN$

(n) $CH_3\overset{\overset{\displaystyle CH_3}{|}}{C}HCH_2\overset{O}{\overset{\|}{C}}CH_2CH_2CHO$

2. Compare the stereostructure of the 1,2-cyclodecanediol produced from *cis-* and *trans*-cyclodecene by each of the following reactions.
(a) 1. HCO_3H; 2. aqueous NaOH
(b) OsO_4, H_2O_2
(c) 1. aqueous Br_2; 2. aqueous NaOH

3. Pinacol rearrangement of 1,1-diphenyl-1,2-ethanediol gives diphenylacetaldehyde and not phenylacetophenone. Explain.

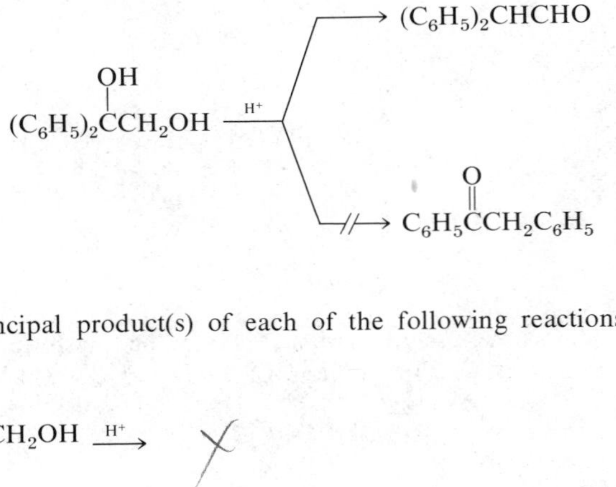

$(C_6H_5)_2\overset{\overset{\displaystyle OH}{|}}{C}CH_2OH \xrightarrow{\text{H}^+}$

$\longrightarrow (C_6H_5)_2CHCHO$

$\xrightarrow{//}\ C_6H_5\overset{O}{\overset{\|}{C}}CH_2C_6H_5$

4. Give the principal product(s) of each of the following reactions or reaction sequences.

(a) 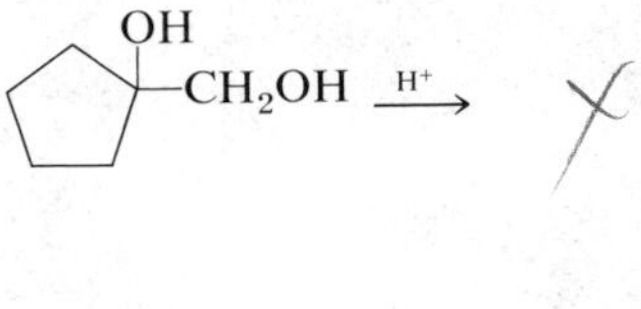 (cyclopentane ring with OH and $-CH_2OH$) $\xrightarrow{\text{H}^+}$ X

(b) $\underset{\overset{\displaystyle |}{CH_3}}{\overset{\overset{\displaystyle CH_3}{\displaystyle |}}{HOCH_2CCHOHCH_2OH}}$ + HIO_4 $\longrightarrow$

(c) $(CH_3)_2CHCHO + HOCH_2CH_2OH \xrightarrow{H^+}$

(d) $HO(CH_2)_6COOH \xrightarrow{\Delta}$

(e) $CH_3CH_2CHOHCOOH \xrightarrow{\Delta}$

(f) $HOCH_2CH_2CH_2COOH \xrightarrow{\Delta}$

(g) $+ EtOOC(CH_2)_3COOEt \xrightarrow[EtOH]{EtONa} \xrightarrow[\Delta]{dil\ H^+} C_{11}H_{10}O_2$

(h) $\underset{\displaystyle CH_3}{\overset{\displaystyle D}{\underset{\displaystyle |}{\overset{\displaystyle |}{H-\!\!-\!\!OH}}}}$ $\xrightarrow[pyridine]{C_6H_5SO_2Cl}$ $\xrightarrow{CH_2(COOEt)_2}{EtONa/EtOH}$ $\xrightarrow[\Delta]{dil\ H^+}$ $C_4H_7DO_2$
optically
active

(i) $2\ CH_2(COOEt)_2 + C_6H_5CHO \xrightarrow[EtOH]{EtONa} \xrightarrow[\Delta]{dil\ H^+} C_{11}H_{12}O_4$

(j) $C_6H_5COCH_2COOEt \xrightarrow[EtOH]{EtONa} \xrightarrow{CH_3CH_2I} \xrightarrow[EtOH]{EtONa} \xrightarrow{CH_3I} \xrightarrow[\Delta]{H^+} C_{11}H_{14}O$

(k) $EtOOCCOOEt + CH_3COOEt \xrightarrow[EtOH]{EtONa}$

5. *cis*-1,2-Cyclopentanediol reacts with benzaldehyde, C_6H_5CHO, in the presence of HCl and anhydrous magnesium sulfate to give two compounds with the formula $C_{12}H_{14}O_2$. What are the structures of the two compounds? What is the function of the magnesium sulfate? Explain why *trans*-1,2-cyclopentanediol does not undergo a similar reaction.

6. (a) Write alternative Lewis structures for periodic acid, $HOIO_3$, and iodic acid, $HOIO_2$, making use of $^-O\!-\!I^+$ or $O\!=\!I$ bonds. Be careful to count electrons and assign formal charges properly; note that iodic acid has a lone pair of electrons on iodine. Follow the changes in electron pairs symbolized in the cyclic mechanism on page 849.

(b) *cis*-1,2-Cyclopentanediol is oxidized to glutaraldehyde (1,5-pentanedial) by periodic acid much more rapidly than the *trans* isomer. Explain.

(c) Suggest a method whereby periodic acid could be used to distinguish between 1,2,3-pentanetriol and 1,2,4-pentanetriol.

7. (a) 4,5-Dihydroxypentanal exists in solution largely as a cyclic hemiacetal. From the data in Table 27.3 predict which form will predominate at equilibrium.

(b) When 4,5-dihydroxypentanal is treated with silver oxide and the resulting dihydroxyhexanoic acid is treated with acid, a lactone results. What is the structure of the lactone (see Table 27.4)?

8. Propose mechanisms for the following reactions.

(a) 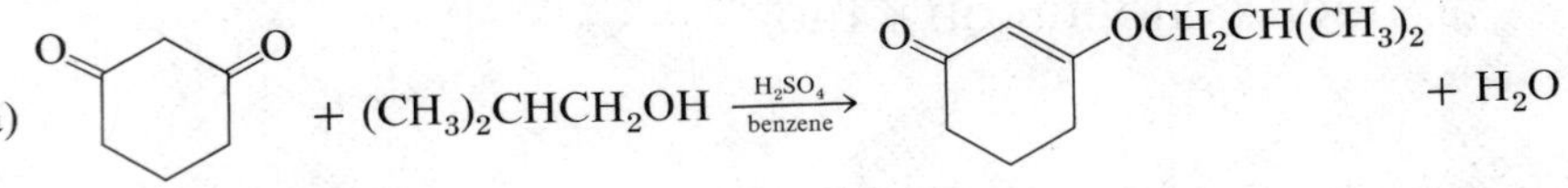 + $(CH_3)_2CHCH_2OH$ $\xrightarrow[\text{benzene}]{H_2SO_4}$ + H_2O

(b) 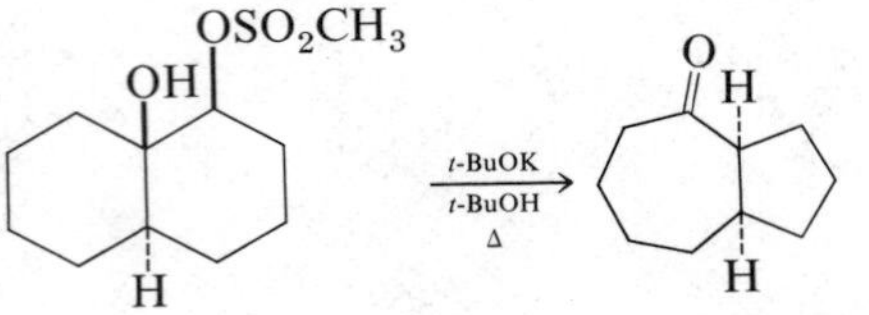 $\xrightarrow[\substack{t\text{-BuOH} \\ \Delta}]{t\text{-BuOK}}$

(c) 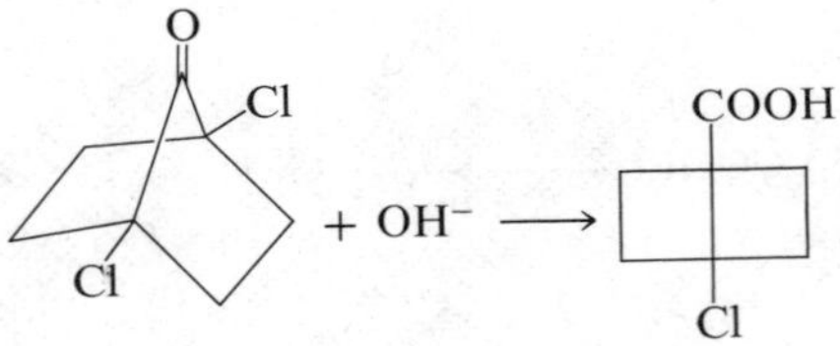 + OH^- $\longrightarrow$

(d) 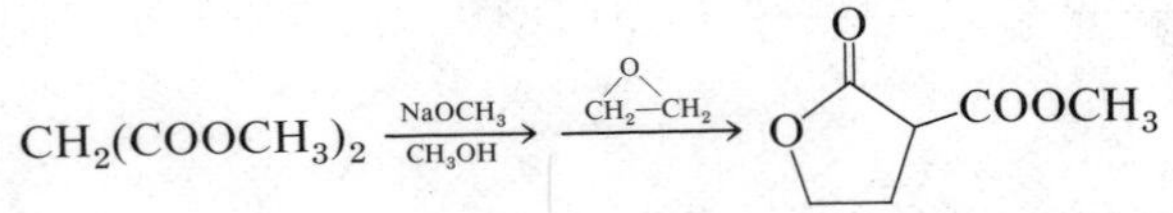 $CH_2(COOCH_3)_2$ $\xrightarrow[\text{CH}_3\text{OH}]{\text{NaOCH}_3}$

(e) 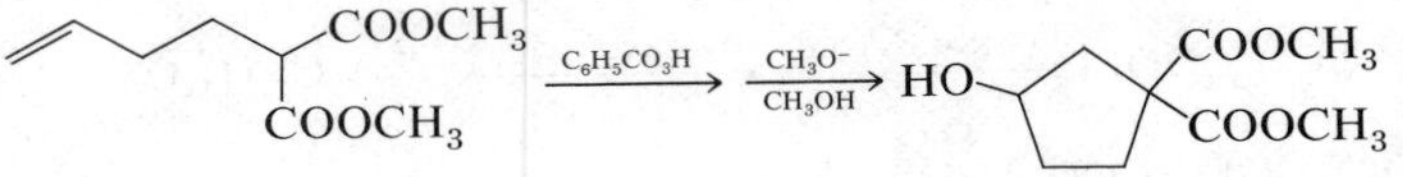 $\xrightarrow[\text{CH}_3\text{OH}]{C_6H_5CO_3H}$ $\xrightarrow[\text{CH}_3\text{OH}]{\text{CH}_3\text{O}^-}$

✓(f) $p\text{-}CH_3OC_6H_4C(C_6H_5)_2CH_2OH + SOCl_2 \longrightarrow (C_6H_5)_2C{=}CHC_6H_4OCH_3\text{-}p$

9. When ethyl acetoacetate is treated with 1,3-dibromopropane and 2 moles of sodium ethoxide in ethanol, a product (A) is produced that has the formula $C_9H_{14}O_3$. Compound A has an infrared spectrum that shows only one carbonyl absorption and no OH bond. Suggest a structure for A and rationalize its formation.

10. Give the expected product for each reaction sequence.

(a) 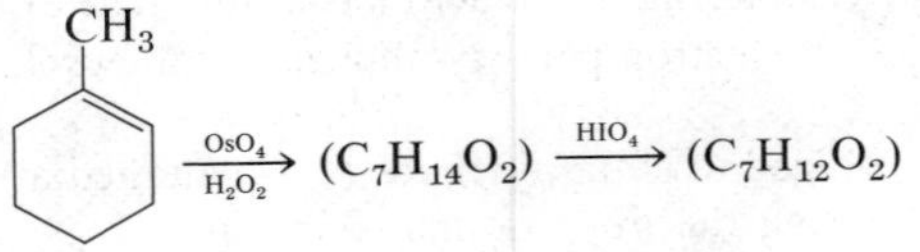 $\xrightarrow[\text{H}_2\text{O}_2]{\text{OsO}_4}$ $(C_7H_{14}O_2)$ $\xrightarrow{\text{HIO}_4}$ $(C_7H_{12}O_2)$

(b) $CH_3CH_2\overset{\displaystyle O}{\overset{\|}{C}}CH_2CH_3$ $\xrightarrow[\text{benzene}]{\text{Mg}}$ $\xrightarrow{\text{H}_2\text{O}}$ $(C_{10}H_{22}O_2)$ $\xrightarrow[\Delta]{\text{H}_3\text{O}^+}$ $(C_{10}H_{20}O)$

(c) 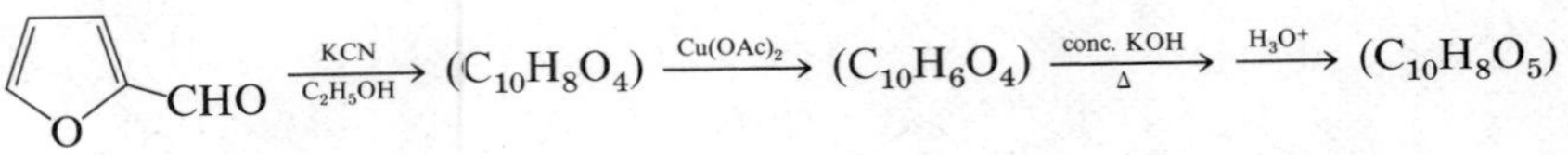 $\xrightarrow[\text{C}_2\text{H}_5\text{OH}]{\text{KCN}}$ $(C_{10}H_8O_4)$ $\xrightarrow{\text{Cu(OAc)}_2}$ $(C_{10}H_6O_4)$ $\xrightarrow[\Delta]{\text{conc. KOH}}$ $\xrightarrow{\text{H}_3\text{O}^+}$ $(C_{10}H_8O_5)$

(d) 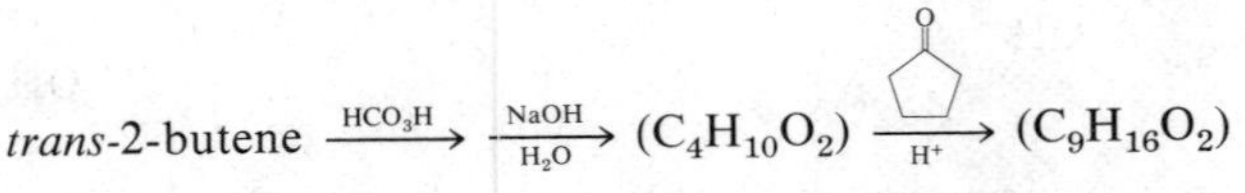 *trans*-2-butene $\xrightarrow{\text{HCO}_3\text{H}}$ $\xrightarrow[\text{H}_2\text{O}]{\text{NaOH}}$ $(C_4H_{10}O_2)$ $\xrightarrow[\text{H}^+]{}$ $(C_9H_{16}O_2)$

11. When 2-methyl-2-methoxycarbonylcyclopentanone is treated with sodium methoxide in refluxing methanol and the solution is then neutralized, 5-methyl-2-

methoxycarbonylcyclopentanone is produced. Write a mechanism for the reaction and explain.

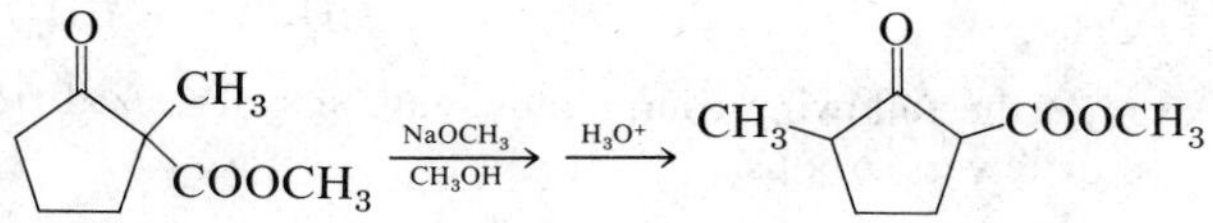

12. Show how one may accomplish each of the following conversions in a practical manner.

(a) $CH_3COCH_3 \longrightarrow (CH_3)_3CCOOH$

(b) $CH_3CH_2COOH \longrightarrow CH_3CH_2CHOHCHCOOH$ (with CH_3 on the CHCOOH carbon)

(c) $CH_3-\!\!\bigcirc\!\!-CHO \longrightarrow CH_3-\!\!\bigcirc\!\!-CHOH-CHOH-\!\!\bigcirc\!\!-CH_3$

(d)

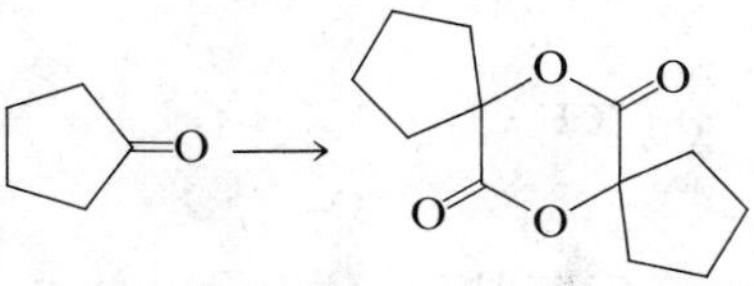

(e) $CH_3CH_2COOH \longrightarrow CH_3CH_2CHOHCHOHCH_2CH_3$

(f)

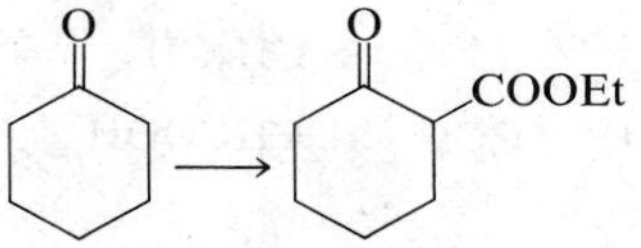

(g) $C_6H_5CH_2COOEt \longrightarrow C_6H_5CH(COOEt)_2$

(h) $C_6H_5CH(COOEt)_2 \longrightarrow HOOCCHCH_2COOH$ (with C_6H_5 on the CH)

13. Show how each of the following compounds may be prepared starting with diethyl malonate.

(a) $(CH_3)_2CHCH_2CH_2COOH$

(b) $CH_2\!\!=\!\!CHCH_2CHCOOH$ (with CH_3)

(c) $CH\!\!=\!\!CCH_2CH_2COOH$ (with CH_3)

(d)

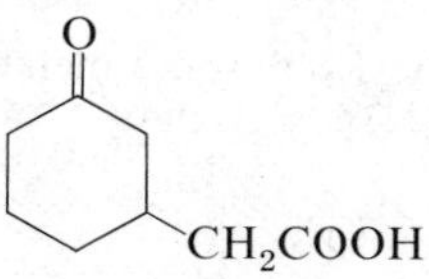

(e) $HOOCCH_2CH_2CH_2COOH$

(f)

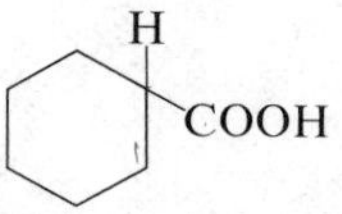

14. Show how each of the following compounds may be prepared starting with ethyl acetoacetate.

(a) $CH_3CCH_2CH_2CH_2CH(CH_3)_2$ (with O double bond)

(b) $CH_3CCH_2CH_2CH_2COOH$ (with O double bond)

(c) $CH_3\overset{O}{\overset{\|}{C}}CHCH_2CH_2CH_3$
$\quad\quad\underset{CH_2CH_3}{|}$

(d) $CH_3\overset{O}{\overset{\|}{C}}CHCH\underset{CH_3}{|}CH_2CH\underset{CH_3}{|}\overset{O}{\overset{\|}{C}}CH_3$

15. Show how each of the following compounds may be synthesized from compounds containing five or fewer carbons.

(a) $HOOCCH_2\underset{\underset{CH_2CH_3}{|}}{CH}COOH$

(b) 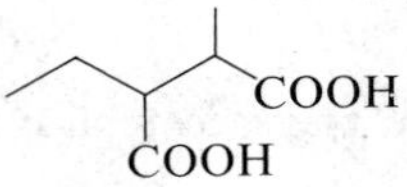 a structure with ethyl, methyl branch, COOH and COOH groups

(c) $(CH_3)_2N\overset{O}{\overset{\|}{C}}CH_2CH_2CH_2COOH$

(d) 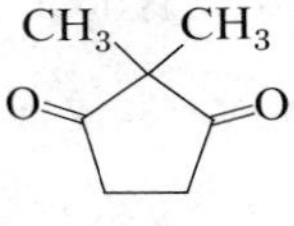 3,3-dimethylcyclopentane-1,2-dione with two CH_3 groups, two O

(e) $HOOCCH_2CH_2\overset{O}{\overset{\|}{C}}CH(CH_3)_2$

(f) 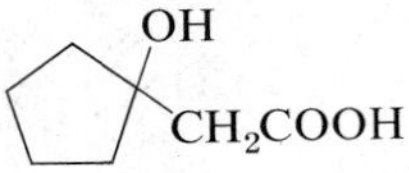 cyclopentane with OH and CH_2COOH

(g) $(CH_3)_2CH\overset{O}{\overset{\|}{C}}-\overset{O}{\overset{\|}{C}}CH(CH_3)_2$

(h) $(CH_3CH_2)_3C\overset{O}{\overset{\|}{C}}CH_2CH_3$

(i) $CH_3\overset{O}{\overset{\|}{C}}CH_2CH_2CH_2CH_2COOH$

(j) $CH_3CH_2CH_2\overset{O}{\overset{\|}{C}}\underset{\underset{CH_3}{|}}{C}HCH_2CH_3$

(k) $CH_3CH_2CH_2CH_2\overset{O}{\overset{\|}{C}}COOH$

(l) $HOOCCH_2CH_2\underset{\underset{CH_2CH_3}{|}}{CH}COOH$

(m) 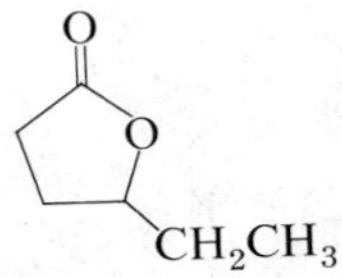 lactone ring with CH_2CH_3

(n) 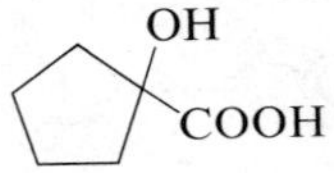 cyclopentane with OH and COOH

(o) 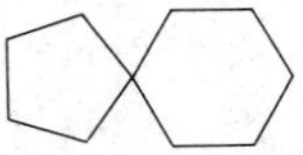 spiro bicyclic structure

16. Suggest a procedure for the dialkylation of diethyl malonate with benzyl chloride using phase-transfer catalysis. What is the final product of hydrolysis of this product and heating in acid?

17. Show how the Robinson annelation may be used to prepare each compound.

(a) 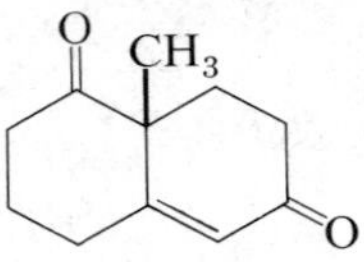 bicyclic structure with O, CH_3

(b) 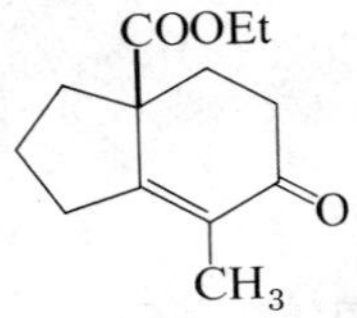 bicyclic structure with COOEt, CH_3, O

(c) 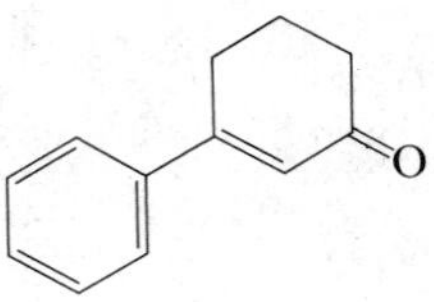 structure with phenyl ring and cyclohexenone

(d) 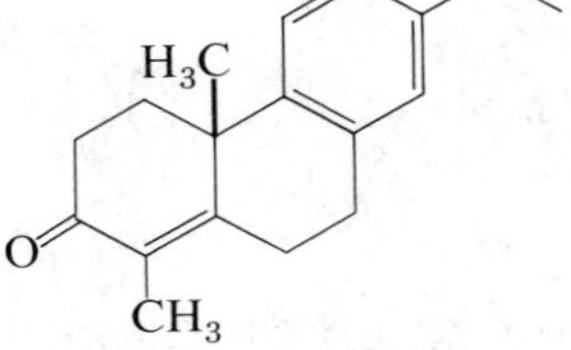 tricyclic structure with H_3C, isopropyl, O, CH_3

18. 1-Phenylethane-*1*-*d*, $C_6H_5CHDCH_3$ (F), has been prepared in optically active form. This compound is particularly interesting because its asymmetry is due entirely to the isotopic difference between H and D; nevertheless, the magnitude of its rotation has the relatively high value of $[\alpha]_D \pm 0.6°$. The absolute configuration of F is related to the known configuration of mandelic acid by the following sequence of reactions.

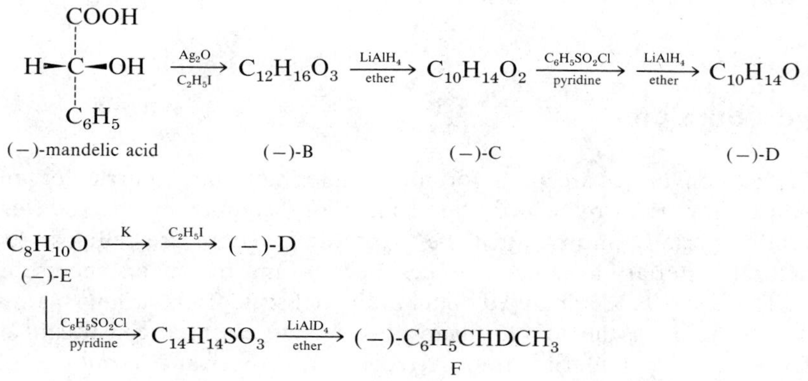

$$
\begin{array}{c}
\text{COOH} \\
| \\
\text{H}\!-\!\overset{|}{\text{C}}\!-\!\text{OH} \\
| \\
\text{C}_6\text{H}_5
\end{array}
\xrightarrow[\text{C}_2\text{H}_5\text{I}]{\text{Ag}_2\text{O}}
\text{C}_{12}\text{H}_{16}\text{O}_3
\xrightarrow[\text{ether}]{\text{LiAlH}_4}
\text{C}_{10}\text{H}_{14}\text{O}_2
\xrightarrow[\text{pyridine}]{\text{C}_6\text{H}_5\text{SO}_2\text{Cl}}
\xrightarrow[\text{ether}]{\text{LiAlH}_4}
\text{C}_{10}\text{H}_{14}\text{O}
$$

(−)-mandelic acid   (−)-B   (−)-C   (−)-D

$$
\text{C}_8\text{H}_{10}\text{O} \xrightarrow{\text{K}} \xrightarrow{\text{C}_2\text{H}_5\text{I}} (-)\text{-D}
$$
(−)-E

$$
\xrightarrow[\text{pyridine}]{\text{C}_6\text{H}_5\text{SO}_2\text{Cl}} \text{C}_{14}\text{H}_{14}\text{SO}_3 \xrightarrow[\text{ether}]{\text{LiAlD}_4} (-)\text{-C}_6\text{H}_5\text{CHDCH}_3
$$
                                                        F

Deduce the absolute configuration of (−)-F and the structure and configuration of each intermediate, B through E, in the sequence. Assign the proper R,S notation to each structure B through F.

19. The following reaction is similar to the acyloin condensation. Propose a mechanism for the reaction.

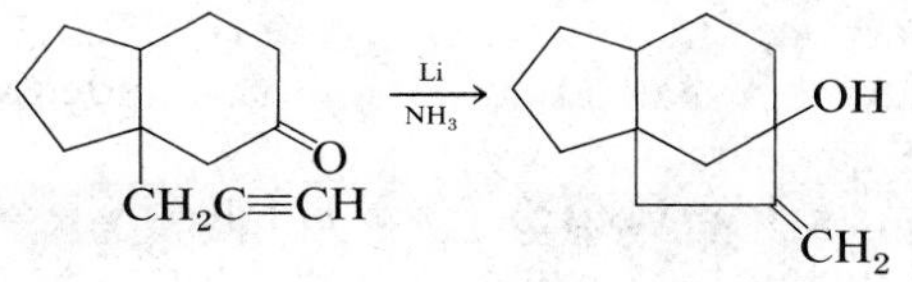

Chapter 28
Carbohydrates

28.1
Introduction

The carbohydrates are an important group of naturally occurring organic compounds. They are extremely widespread in plants, comprising up to 80% of the dry weight. Especially important in the vegetable kingdom are cellulose (the chief structural material of plants), starches, pectins, and the sugars sucrose and glucose. These sugars are obtained on an industrial scale from various plant sources. In higher animals, the simple sugar glucose is an essential constituent of blood and occurs in a polymeric form as glycogen in the liver and in muscle. Carbohydrates also occur in a bound form in adenosine triphosphate, which is a key material in biological energy storage and transport systems, and in the nucleic acids, which control the production of enzymes and the transfer of genetic information.

The term **carbohydrate** is used loosely to characterize the whole group of natural products that are related to the simple sugars. The name first arose because the simple sugars, such as glucose ($C_6H_{12}O_6$), have molecular formulas that appear to be "hydrates of carbon," that is, $C_6H_{12}O_6 = (C \cdot H_2O)_6$. Although subsequent structural investigations revealed that this simple-minded view was erroneous, the term carbohydrate has persisted.

Sugars, also called **saccharides,** are the simplest type of carbohydrate. An example is glucose, which is the cyclic hemiacetal form of one of the diastereomers of 2,3,4,5,6-pentahydroxyhexanal. As we shall see in a later section, although glucose exists almost entirely in the cyclic form, in solution it appears to be in equilibrium with a minute amount of the noncyclic pentahydroxyaldehyde form. The structure of naturally occurring glucose is shown as an example of a simple sugar. The open-chain form is drawn as a Fischer projection formula (Section 7.5), which unambiguously specifies the stereochemistry at each of the four asymmetric carbons.

glucose

As is generally true for natural products, *the carbohydrates occur in optically active form,* and only one enantiomer is found in nature. Glucose is an example of a **monosaccharide,** a term that means glucose is not hydrolyzable into smaller units.

Maltose is an example of a **disaccharide**; upon hydrolysis under mildly acidic conditions, maltose yields two equivalents of the monosaccharide glucose.

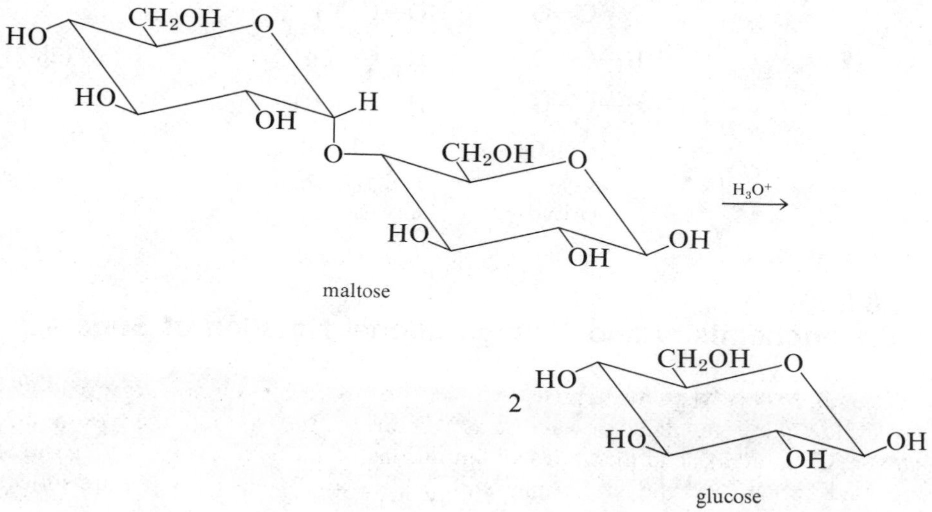

maltose

$\xrightarrow{H_3O^+}$

glucose

A **trisaccharide** yields three monosaccharides on hydrolysis, a **tetrasaccharide** four, and so forth. **Oligosaccharide** is a general term applied to sugar polymers containing up to eight units. **Polysaccharide** refers to polymers in which the number of subunits is greater than eight; the natural polysaccharides generally consist of 100–3000 subunits.

The monosaccharides are also characterized in terms of the number of carbons in the chain and the nature of the carbonyl group, aldehyde or ketone. Glucose, which is a six-carbon aldehyde, is a **hexose,** which specifies the number of carbons, and an **aldose,** which shows that it is an aldehyde. It is completely characterized by the general term **aldohexose.** Other aldoses are glyceraldehyde, an **aldotriose,** erythrose, an **aldotetrose,** and arabinose, an **aldopentose.** The structures of these examples are shown in Fischer projections as their open-chain forms in (28-1).

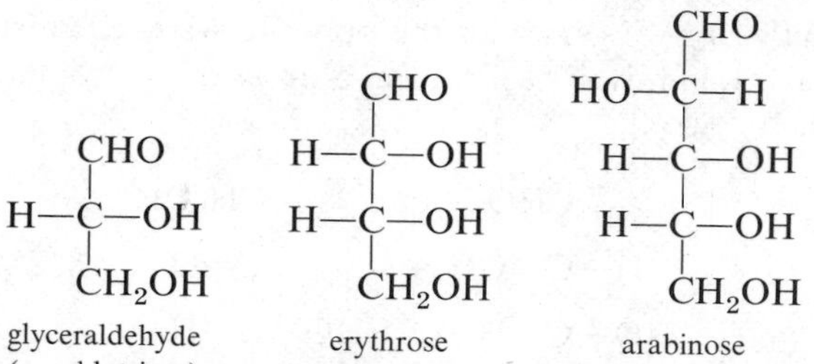

glyceraldehyde (an aldotriose) erythrose (an aldotetrose) arabinose (an aldopentose) (28-1)

Most of the naturally occurring sugars are derived from the aldoses, and the most widespread are the aldohexoses and the aldopentoses.

A few important saccharides are **ketoses,** meaning that they contain a ketone, rather than an aldehyde, carbonyl group. Fructose is an example of a **ketohexose,** a six-carbon pentahydroxy ketone. An example of a **ketopentose** is ribulose. Both compounds are shown (28-2) in open-chain form as Fischer projections.

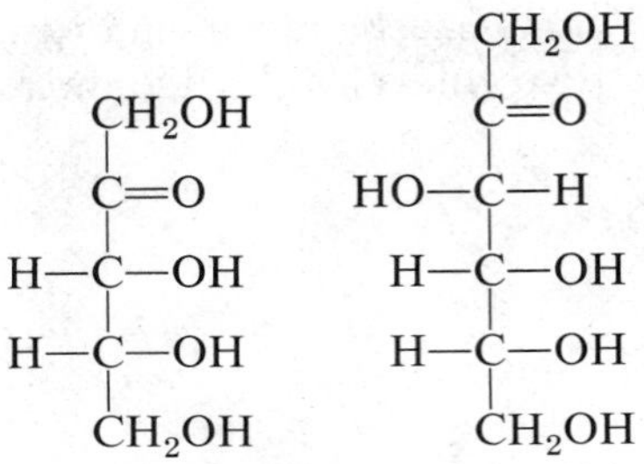

$$
\begin{array}{ccc}
& & CH_2OH \\
& & | \\
CH_2OH & & C{=}O \\
| & & | \\
C{=}O & HO{-}C{-}H \\
| & | \\
H{-}C{-}OH & H{-}C{-}OH \\
| & | \\
H{-}C{-}OH & H{-}C{-}OH \\
| & | \\
CH_2OH & CH_2OH \\
\text{ribulose} & \text{fructose} \\
\text{(a ketopentose)} & \text{(a ketohexose)}
\end{array}
\tag{28-2}
$$

28.2
Stereochemistry and Configurational Notation of Sugars

The simplest polyhydroxy aldehyde is the compound 2,3-dihydroxypropanal or glyceraldehyde. The molecule has one asymmetric center, so there are two enantiomers (Chapter 7). The absolute configurations of the glyceraldehyde enantiomers are known, and the enantiomer with $[\alpha]_D = +8.7°$ has the structure shown in (28-3) on the left (Fischer projection).

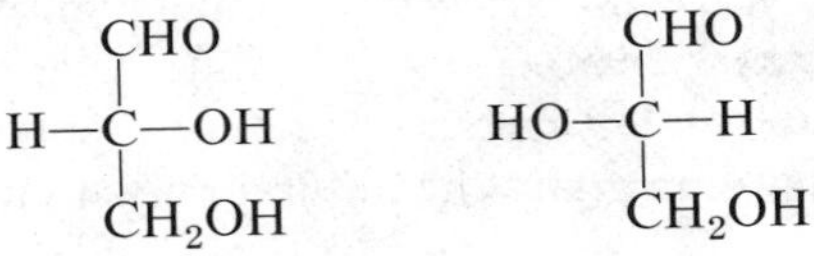

$$
\begin{array}{cc}
CHO & CHO \\
| & | \\
H{-}C{-}OH & HO{-}C{-}H \\
| & | \\
CH_2OH & CH_2OH \\
\text{D-(+)-glyceraldehyde} & \text{L-(−)-glyceraldehyde} \\
\text{R-(+)-glyceraldehyde} & \text{S-(−)-glyceraldehyde}
\end{array}
\tag{28-3}
$$

This enantiomer may be distinguished from the other by calling it (+)-glyceraldehyde, meaning "the dextrorotatory enantiomer of glyceraldehyde." A complete description, which specifies the absolute stereochemistry of the enantiomer, is R-(+)-glyceraldehyde (Section 7.3). In the carbohydrate field, it is customary to use another and older system of configurational notation, wherein this enantiomer is called D-(+)-glyceraldehyde. Under this convention, all D-*sugars have the same stereochemistry as* D-(+)-*glyceraldehyde at the asymmetric carbon most distant from the carbonyl group.* Sugars with the opposite stereochemistry at this center are members of the L-family. Thus, natural glucose is D-(+)-glucose and fructose is D-(−)-fructose.

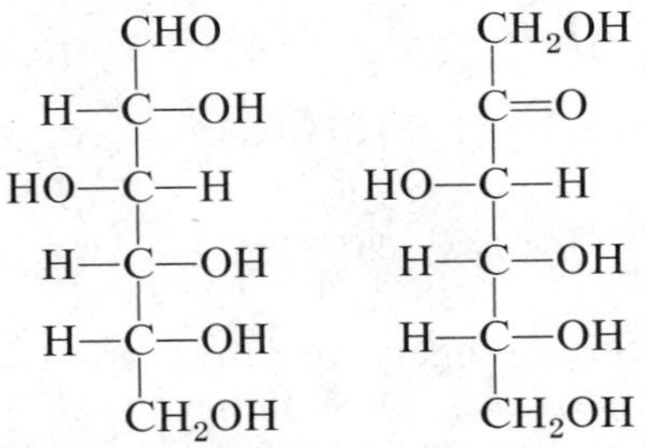

$$
\begin{array}{cc}
CHO & CH_2OH \\
| & | \\
H{-}C{-}OH & C{=}O \\
| & | \\
HO{-}C{-}H & HO{-}C{-}H \\
| & | \\
H{-}C{-}OH & H{-}C{-}OH \\
| & | \\
H{-}C{-}OH & H{-}C{-}OH \\
| & | \\
CH_2OH & CH_2OH \\
\text{D-(+)-glucose} & \text{D-(−)-fructose}
\end{array}
$$

Recall that for a compound with n asymmetric centers, there are 2^n possible optical isomers. Thus, there are 2 aldotrioses, 4 aldotetroses, 8 aldopentoses, and

895

Sec. 28.2
Stereochemistry
and
Configurational
Notation
of Sugars

16 aldohexoses. Half of these compounds belong to the D-family (are related to D-glyceraldehyde) and half belong to the L-family. To avoid cumbersome names, each isomer has been given a trivial name; that is, D-(+)-glucose is (2R,3S,4R,5R)-2,3,4,5,6-pentahydroxyhexanol. Fischer projections depicting the complete D-family of the aldoses in their open-chain forms are shown in Table 28.1. Note that abbreviated Fischer projections are used—the asymmetric carbons are omitted. The student should thoroughly review Section 7.5 on the use of these projection formulas. In writing Fischer projections of carbohydrates, the convention is observed that the main chain is always written from top to bottom with CHO at the top and CH$_2$OH at the bottom.

Although the naturally occurring sugars generally belong to the D-family shown in Table 28.1, an equal number of compounds have the L-configuration. Each D-sugar has an enantiomeric L-counterpart.

Recall that a molecule with two or more asymmetric atoms may be achiral if it has a plane of symmetry. Such compounds are called *meso* compounds (page 136). It often happens in carbohydrate chemistry that a chiral compound undergoes a chemical reaction to yield a *meso* product. For example, consider the reduction of the aldotetrose D-(−)-erythrose by sodium borohydride. The product is *meso*-1,2,3,4-butanetetrol (erythritol). The same compound would be produced by the reduction of L-(+)-erythrose.

On the other hand, the aldotetroses D-(−)-threose and L-(+)-threose each yield an optically active butanetetrol on reduction.

The formation of a *meso* compound can be a powerful piece of information for use in determining the relative stereochemistry of a compound. For example, the fact that erythrose undergoes reduction to give a *meso* tetrol proves that its two asymmetric centers are either (R,R) or (S,S). Conversely, since threose gives a tetrol that is optically active, it must have either the (R,S) or (S,R) configuration.

TABLE 28.1
The D-Family Aldoses

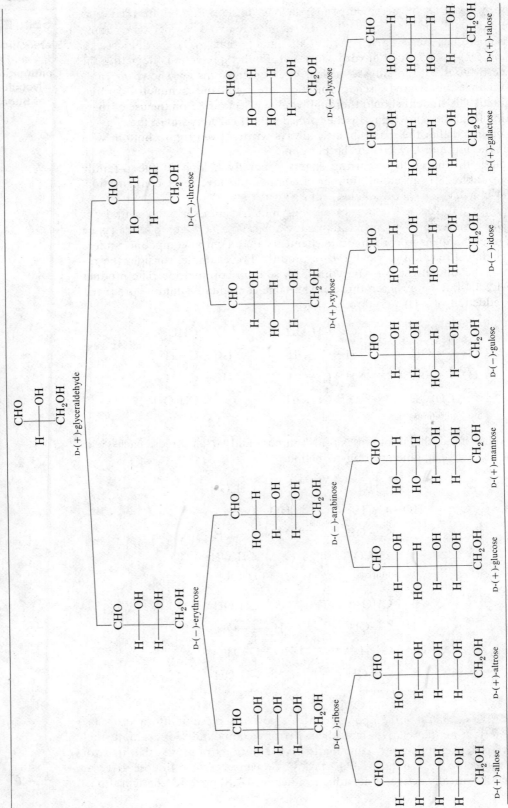

897

Sec. 28.3

**Cyclic
Hemiacetals:
Anomerism:
Glycosides**

EXERCISE: Write Fischer projections for the products of $NaBH_4$ reduction of D-(+)-galactose, L-(−)-xylose, and D-(+)-mannose. Are any of these products achiral?

28.3
Cyclic Hemiacetals: Anomerism; Glycosides

In the last chapter (Section 27.4.B.2), we learned that 4- and 5-hydroxy aldehydes and ketones exist mainly in the form of a cyclic hemiacetal or hemiketal.

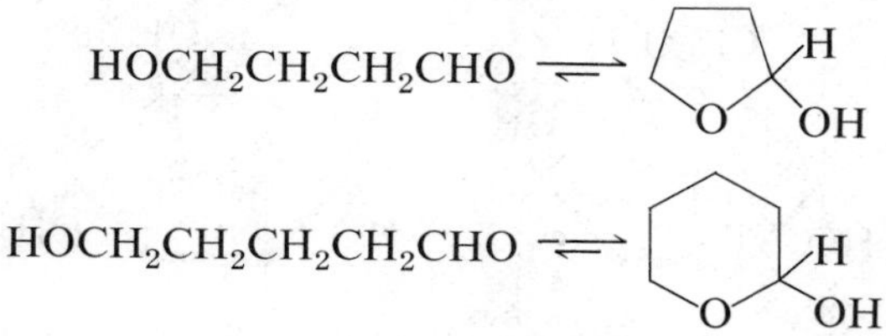

$$HOCH_2CH_2CH_2CHO \rightleftharpoons$$

$$HOCH_2CH_2CH_2CH_2CHO \rightleftharpoons$$

It is not surprising then that the sugars also exist in such a cyclic form. Although
either the five- or six-membered hemiacetal structure is possible, almost all of the
simple sugars exist in the six-membered ring form (see Table 27.3, page 855). Note
that when the hemiacetal is formed, *the former aldehyde carbon becomes asymmetric.* Thus, there are two cyclic forms of glucose.

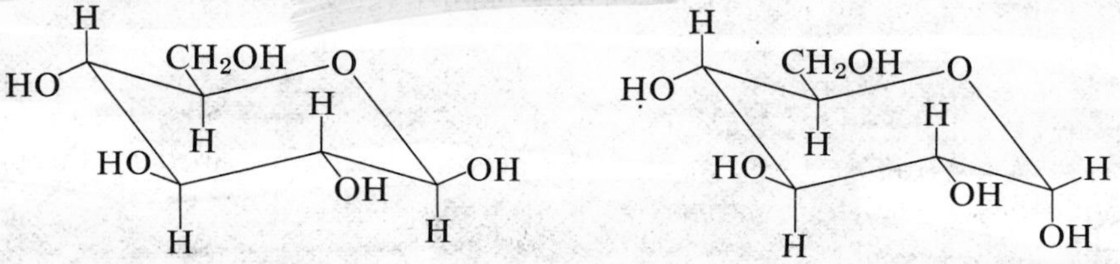

β-D-glucose   α-D-glucose

The two cyclic isomers of glucose differ only in the stereochemistry at C-1, the
acetal carbon (former aldehyde carbon). Such isomers are called **anomers,** and
the acetal carbon (or the ketal carbon in the case of a cyclic ketose) is called the
anomeric carbon. The two anomers are commonly differentiated by the Greek
letters α and β; for example, α-D-glucose, β-D-glucose. For the aldohexoses, the β
anomer is the one that has the OH at C-1 and the CH_2OH at C-5, *cis* with respect
to each other on the ring.

The Fischer projection formulas shown in Table 28.1 are a convenient way in
which to represent the open-chain form of sugars. Modified Fischer projections
have frequently been used to depict the cyclic hemiacetal form. For example, the
two D-glucose anomers may be represented as

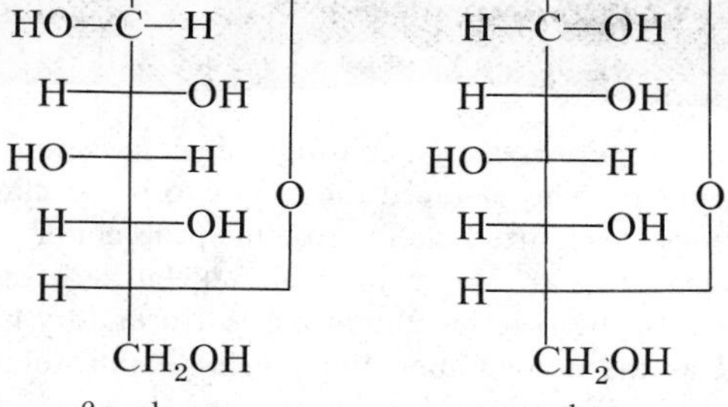

β-D-glucose   α-D-glucose

Note the convention used to represent stereochemistry at anomeric carbon. The OH at C-1 in the β-anomer is written to the left and that in the α-anomer is written to the right.

Because these modified projections lead to awkward drawings of bond lengths, which offend the sensibilities of many chemists, Haworth introduced an alternate projection formula, which is used extensively by sugar chemists. In a Haworth projection, the sugar ring is written as a planar hexagon with the oxygen in the upper right vertex. Substituents are indicated by straight lines through each vertex, either above or below the plane. The OH at the anomeric carbon is up in the β-anomer and down in the α-anomer. Hydrogens attached to the ring are omitted.

β-D-glucose α-D-glucose

There is a simple way to convert a Fischer projection to a Haworth projection, or vice versa. The OH groups that project to the *left* in a Fischer projection project *up* in a Haworth projection. In this book, we shall use Fischer projections to depict open-chain sugars and the more accurate chair representations to depict cyclic forms.

The six-membered ring form of a sugar is called a **pyranose** from the name of the simplest heterocyclic compound containing such a ring, pyran. Thus, β-D-glucose is a pyranose form, and it may be completely described by the name β-D-glucopyranose. Although the free sugars normally do not exist as the five-membered ring form, numerous derivatives are known that have such a structure. They are called **furanoses** from the name of the parent heterocyclic compound, furan.

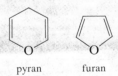

pyran   furan

Pure β-D-glucose has an optical rotation $[\alpha]_D = +18.7°$; the α-anomer has $[\alpha]_D = +112°$. Both anomers have been isolated in pure crystalline states. If either pure anomer is dissolved in water, the optical rotation of the solution gradually changes until it reaches an equilibrium value of $+52.7°$. This phenomenon, which was first observed in 1846, results from the interconversion of the two anomers in solution and is called **mutarotation.** At equilibrium, the solution contains 63.6% of the more stable β-anomer and 36.4% of the α-anomer.

The phenomenon of anomerism caused considerable confusion for the early workers in carbohydrate chemistry, who believed the sugars to be acyclic compounds. After their cyclic structure was recognized, there arose the problem of how to name the two anomers for each sugar. In 1909, a system of nomenclature based purely on optical rotation was adopted. In the D series, the more dextrorotatory member of a pair of anomers is defined as the α-D-anomer and the less dextrorotatory anomer is the β-D-anomer. When reliable methods for determining the stereochemistry at C-1 be-

899

Sec. 28.3
Cyclic
Hemiacetals:
Anomerism:
Glycosides

came available, it turned out that all α-anomers have the same absolute configuration at C-1.

Interconversion of the two anomers is subject to both acid and base catalysis and occurs by the normal mechanism for acetal formation and hydrolysis (Sections 13.7.B and 27.4.B). The open-chain form is probably an intermediate in the process. The mechanism of acid-catalyzed interchange of the α- and β-anomers of glucose is outlined as follows (all ring hydrogens except the one on the anomeric carbon are omitted for clarity).

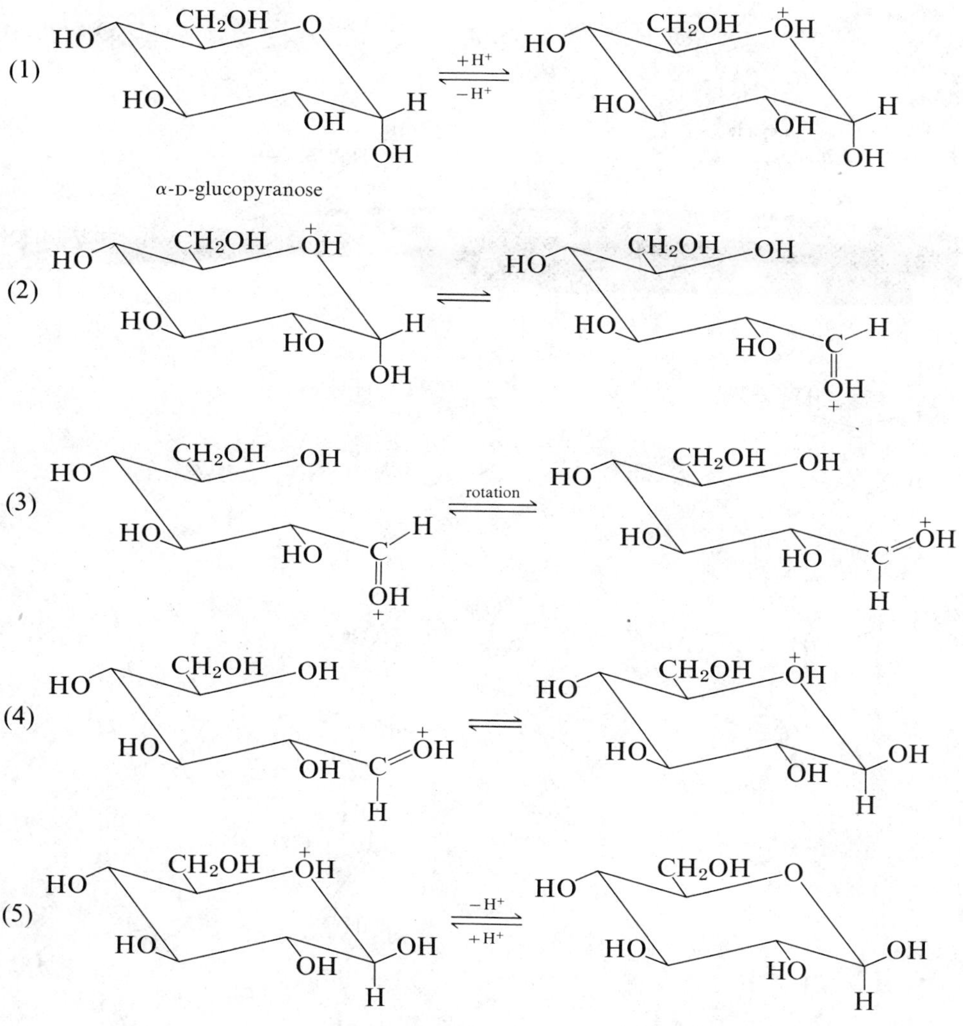

α-D-glucopyranose

β-D-glucopyranose

When an aldose is dissolved in an alcohol and the solution is treated with a mineral acid catalyst, a cyclic acetal is produced (Section 27.4.B). In carbohydrate chemistry such cyclic acetals are called **glycosides**. A glycoside derived from glucose is a **glucoside,** one derived from mannose is a **mannoside,** and so on. Like the hemiacetal, these cyclic acetals may exist in both α- and β-anomeric forms as shown in (28-4) for the methyl mannosides.

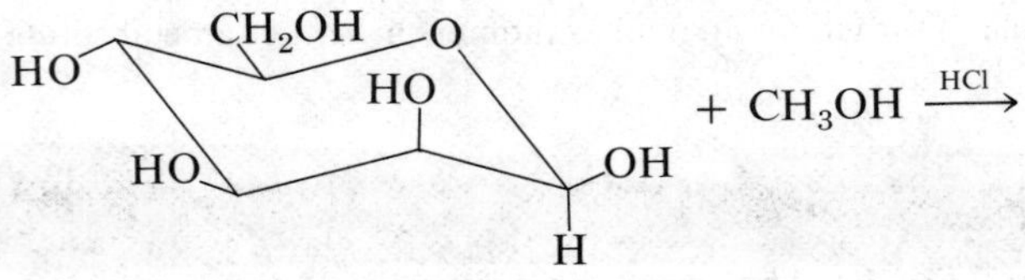

β-D-mannose

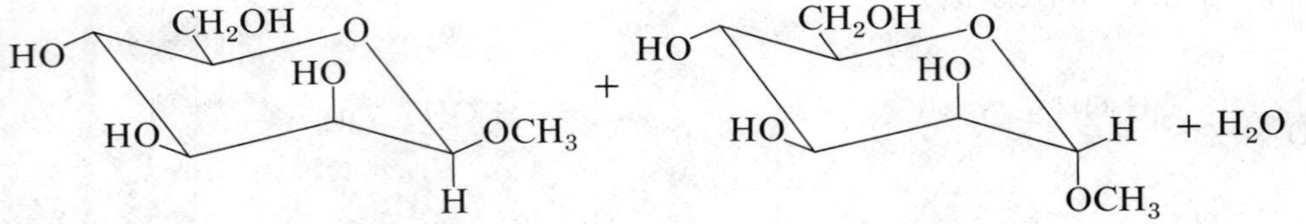

methyl β-D-mannoside
methyl β-D-mannopyranoside

methyl α-D-mannoside
methyl α-D-mannopyranoside

Glycosides form only under acid catalysis; the mechanism for the formation of the methyl galactosides is outlined as follows.

(1)

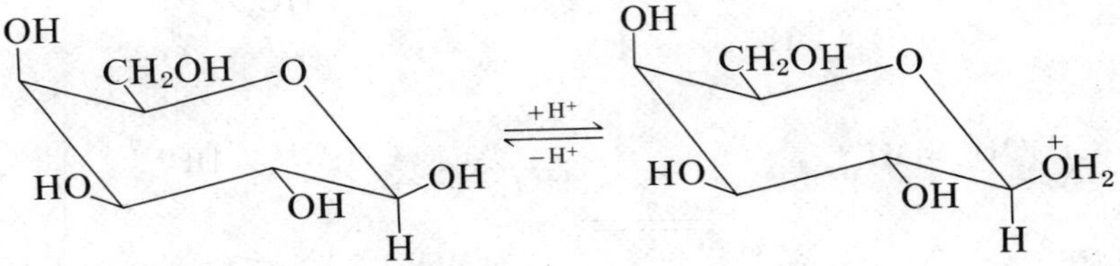

β-D-galactose

(2)

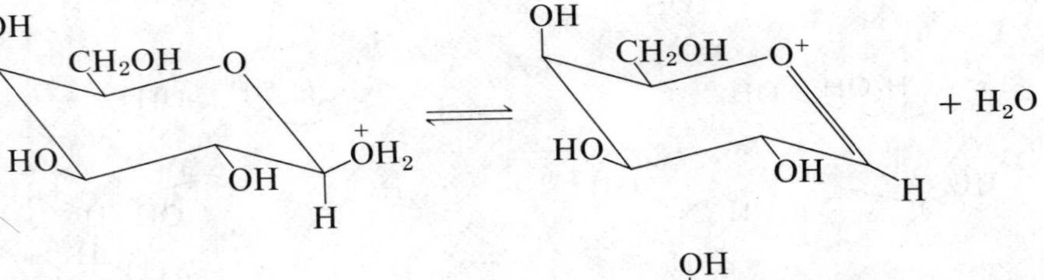

(3)

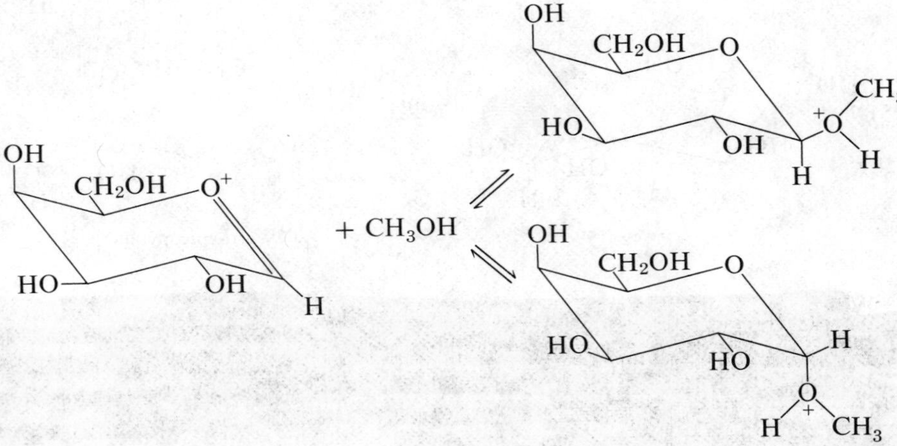

901

Sec. 28.3

**Cyclic
Hemiacetals:
Anomerism:
Glycosides**

(4)

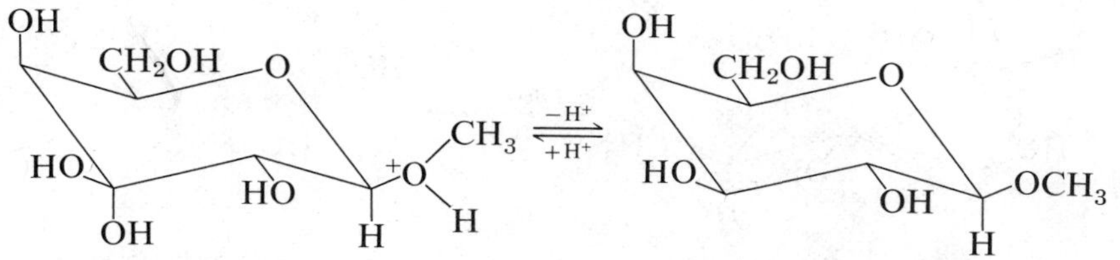

methyl β-D-galactoside
methyl β-D-galactopyranoside

The formation of glycosides is a reversible process under acidic conditions. If a glycoside is treated with an acid catalyst in aqueous solution where water is present in excess, the equilibrium shifts and hydrolysis occurs. Of course, under acidic conditions a mixture of the anomeric sugars results.

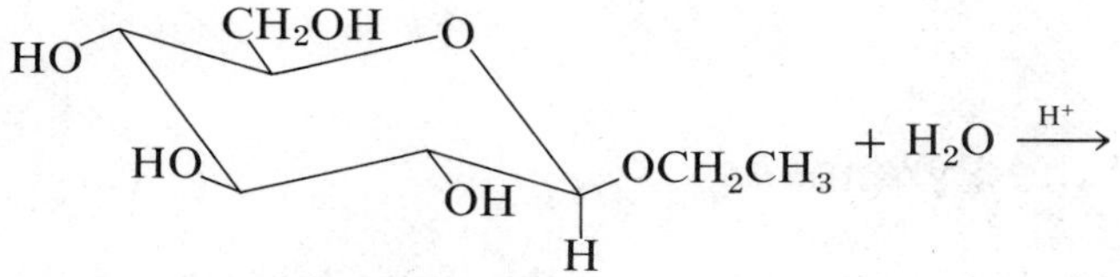

ethyl β-D-glucoside

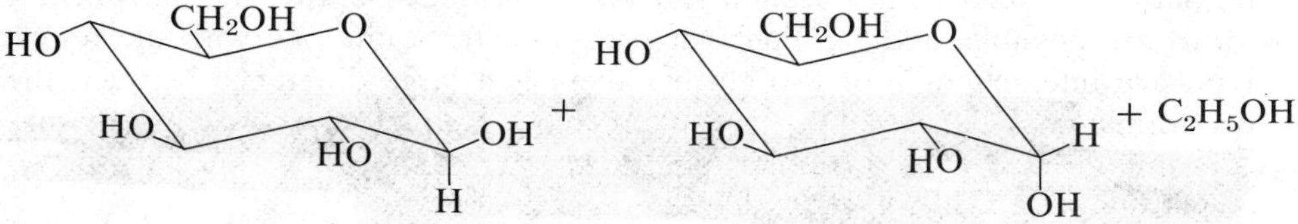

β-D-glucose   α-D-glucose

The hydrolysis of glycosides may also be brought about by certain **enzymes.** Enzymes are complex natural products, mainly protein in nature (Chapter 29), which function as catalysts in biological reactions. They are extremely potent catalysts, often speeding up reactions by factors as large as 10^{10}. They also show remarkable structural specificity, as shown by the present example. Methyl α-D-glucopyranoside is hydrolyzed in the presence of an enzyme, isolated from yeast, called α-glucosidase. This particular enzyme only catalyzes the hydrolysis of α-glucoside linkages; methyl β-D-glucopyranoside is unaffected by it.

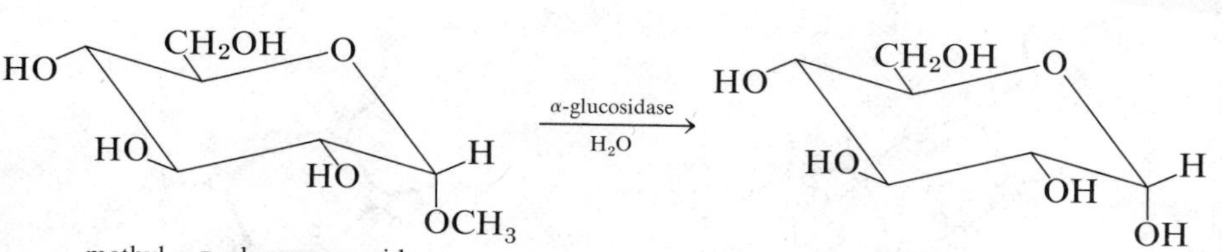

methyl α-D-glucopyranoside   α-D-glucopyranose

Another enzyme, β-glucosidase, from almonds, has opposite properties; it only catalyzes the hydrolysis of β-glucosides.

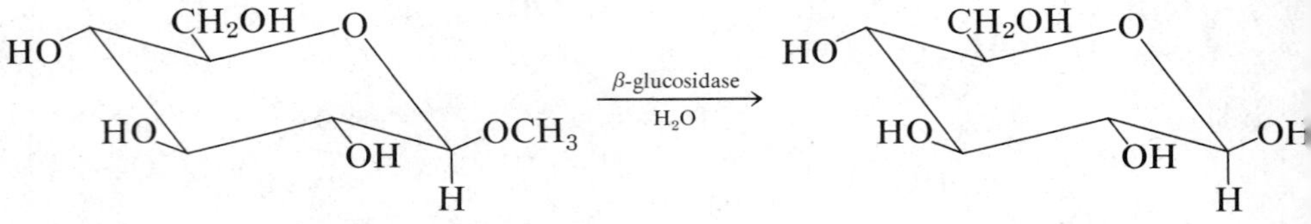

methyl β-D-glucopyranoside   β-D-glucopyranose

Similar enzymes are known that specifically catalyze the hydrolysis of α- and β-galactosides (α- and β-galactosidase) and other glycosidic bonds. These enzymes are very useful in determining the stereochemistry of the glycoside links in oligosaccharides and polysaccharides (Sections 28.7 and 28.8).

EXERCISE: Write Haworth and chair perspective formulas for α-D-altrose. Using the chair perspective formula, write all of the steps for mutarotation to β-D-altrose.

28.4
Conformations of the Pyranoses

As has been tacitly implied in the structures used thus far in this chapter, the pyranose forms of sugars exist in a chair conformation similar to the stable conformation of cyclohexane (Section 5.7). As in cyclohexane, two alternative chair forms are possible, and the one that predominates is the one with the fewest repulsive interactions. For β-D-glucose there is a large difference between the two forms. In one form all five substituents are in equatorial positions, whereas they are all axial in the other conformation; the difference between these two conformations has been estimated to be 6 kcal mole^{-1}.

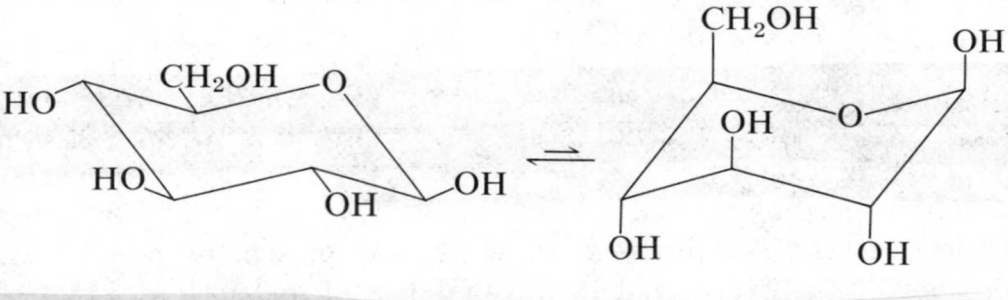

Of the eight D-aldohexoses, glucose is the only one that can have all five substituents equatorial. It is no accident that glucose is the most abundant natural monosaccharide. A stereo structure of β-D-glucose is given in Figure 28.1.

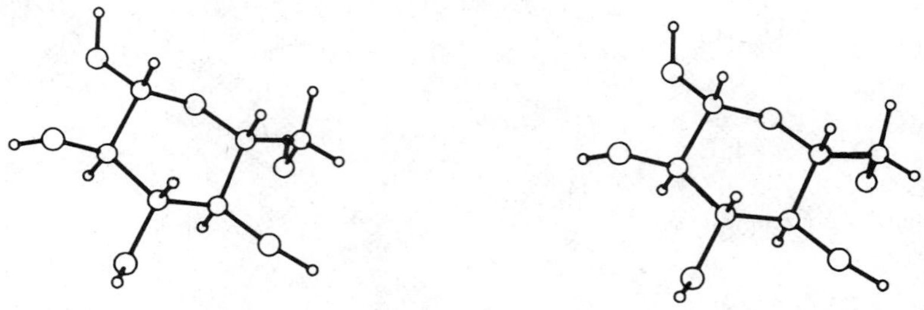

FIGURE 28.1. *Stereo structure of β-D-glucose.* [*Reproduced with permission from* Molecular Structure and Dimensions, *International Union of Crystallography,* 1972.]

If one remembers that β-D-glucose has all substituents equatorial, it is easy to write conformational structures for the other aldohexoses by simply referring to Table 28.1. For example, D-allose differs from D-glucose only in the configuration at C-3. Thus, β-D-allose is

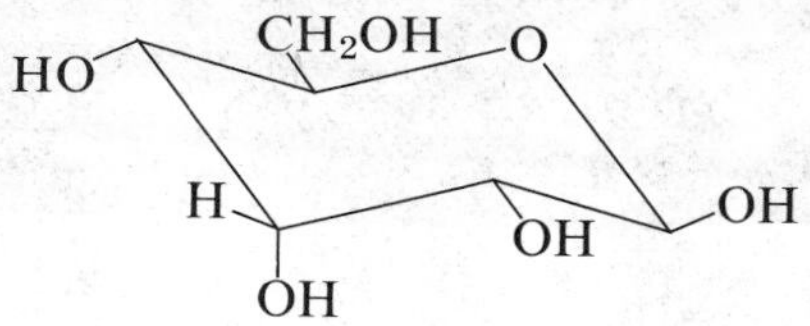

β-D-allose

For most of the aldohexoses, the more stable conformation is the one with the CH_2OH group in an equatorial position. However, in a few cases the two conformations are nearly equal in energy and substantial amounts of both may be present at equilibrium. For α-D-idose, the stable conformation is that in which the CH_2OH group is axial.

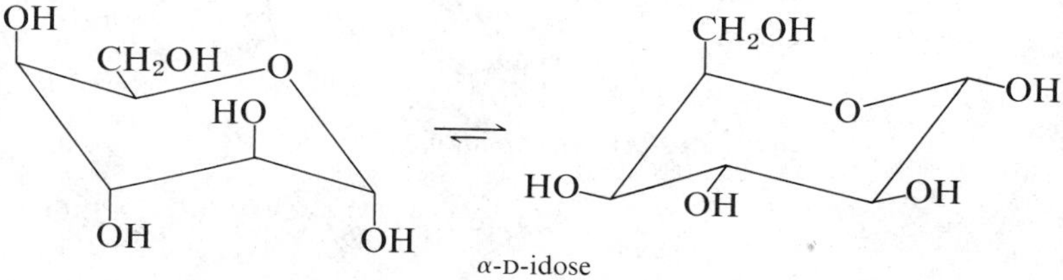

α-D-idose

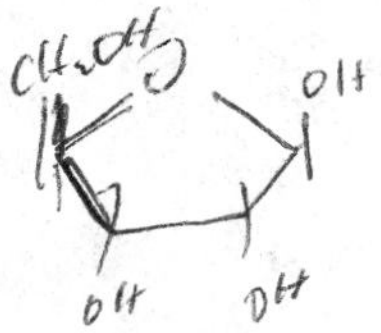

EXERCISE: Write chair perspective formulas for β-D-gulose and α-D-talose.

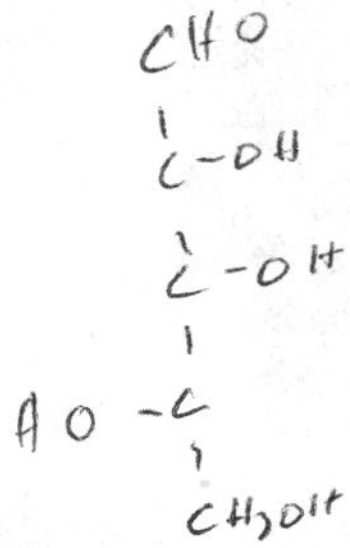

28.5
Reactions of Monosaccharides

A. *Ether Formation*

In Section 28.3 we discussed the formation of glycosides in which the OH group at the anomeric carbon is replaced by an alkoxy group under mildly acidic conditions. The remaining hydroxy groups are unaffected by this process because such a process would involve a primary or secondary carbocation, rather than the far more stable oxonium ion that is involved in glycoside formation.

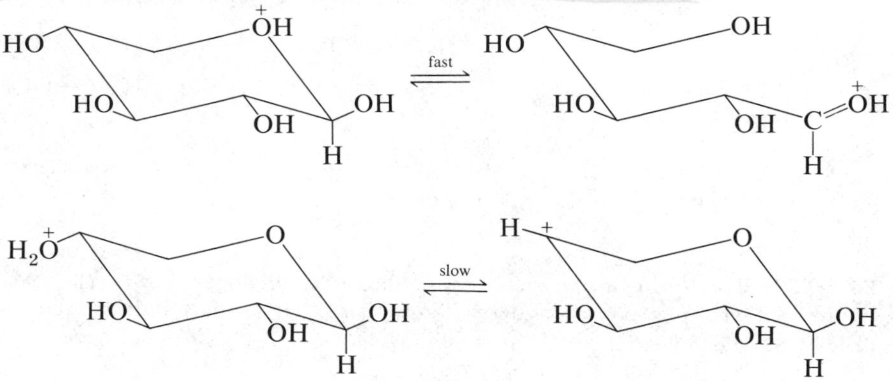

The other hydroxy groups can be converted into ethers by an application of the Williamson ether synthesis (Section 10.10.A). The most common ethers are the methyl ethers, which are prepared by treating the sugar with 30% aqueous sodium hydroxide and dimethyl sulfate, or with silver oxide and methyl iodide. Since the free aldehyde form of an aldose is not stable to strongly basic conditions, it is customary to **protect** the anomeric carbon by converting the sugar into the methyl glycoside. The glycoside linkage can then be cleaved by mild acid hydrolysis because the normal ether linkages are stable under these conditions.

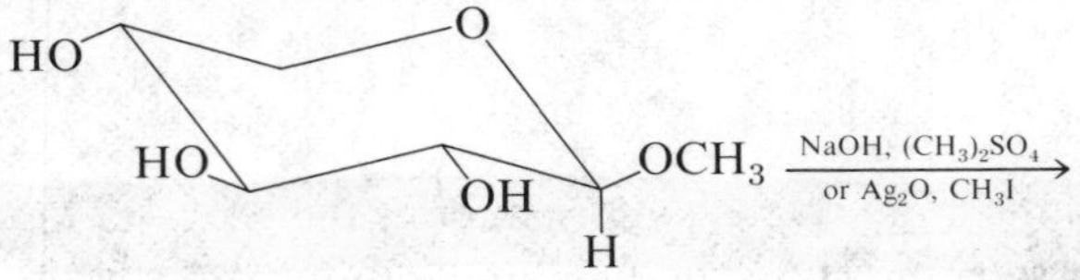

methyl β-D-xyloside

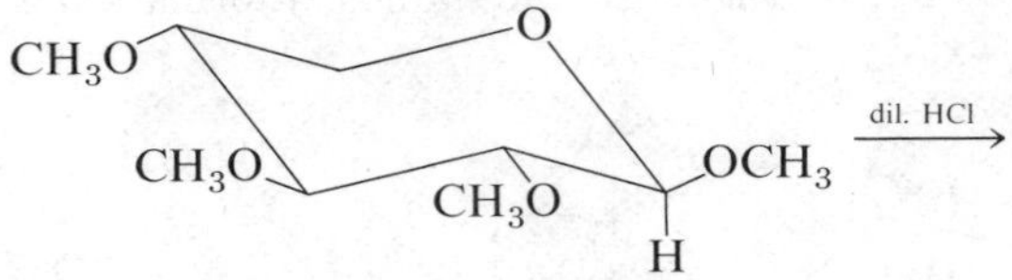

$\xrightarrow{\text{NaOH, (CH}_3)_2\text{SO}_4 \text{ or Ag}_2\text{O, CH}_3\text{I}}$

methyl 2,3,4-tri-O-methyl-β-D-xyloside

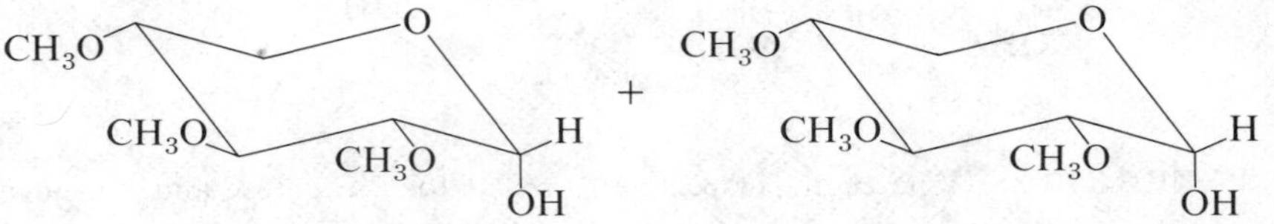

$\xrightarrow{\text{dil. HCl}}$

2,3,4-tri-O-methyl-β-D-xylose   +   2,3,4-tri-O-methyl-α-D-xylose

Methylation can be a useful method for determining the size of the acetal ring in a glycoside. For example, oxidation of the above mixture of anomeric tri-O-methylxylose yields a 2,3,4-trimethoxyglutaric acid, thus establishing that the original methyl xyloside had the pyranose structure.

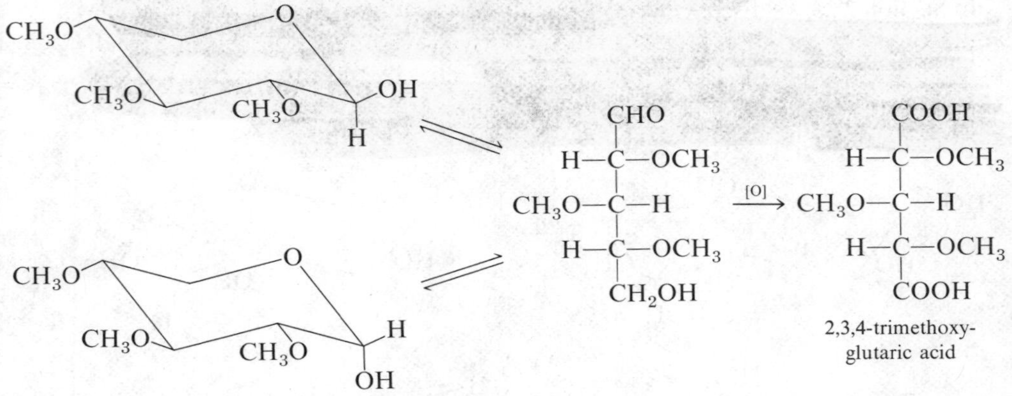

$$
\begin{array}{c}
\text{CHO} \\
\text{H---C---OCH}_3 \\
\text{CH}_3\text{O---C---H} \\
\text{H---C---OCH}_3 \\
\text{CH}_2\text{OH}
\end{array}
\xrightarrow{[O]}
\begin{array}{c}
\text{COOH} \\
\text{H---C---OCH}_3 \\
\text{CH}_3\text{O---C---H} \\
\text{H---C---OCH}_3 \\
\text{COOH}
\end{array}
$$

2,3,4-trimethoxy-glutaric acid

EXERCISE: Using chair perspective formulas, write equations showing the reaction of methyl β-D-glucopyranoside with Ag$_2$O and CH$_3$I. What products are produced when this material is treated with aqueous HCl?

B. *Formation of Cyclic Acetals and Ketals*

Recall that 1,2- and 1,3-diols condense with aldehydes and ketones to form cyclic acetals or ketals (Section 27.3.B). If the diol is itself cyclic, the acetal or ketal forms only when the two OH groups are *cis,* for geometric reasons.

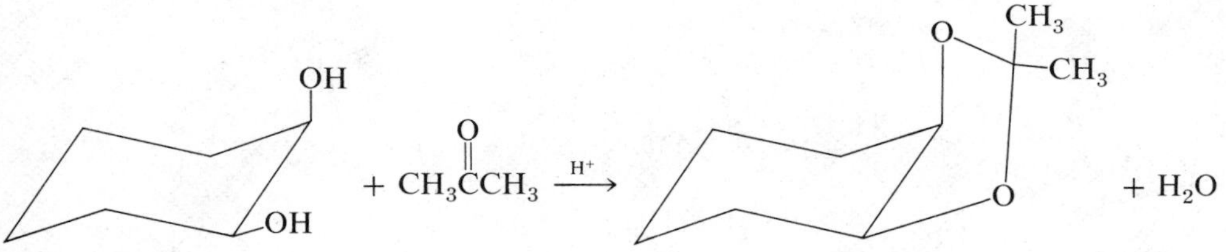

Since sugars are polyhydroxy compounds, they also undergo this reaction. The reaction is often complicated by the fact that the ring size in the product is not the same as it is in the free sugar. This usually occurs when the more stable pyranose form does not have *cis* vicinal hydroxy groups, but the furanose form does. Thus galactose reacts with acetone to give the diketal shown because in the α-form, which is present under the acidic conditions of the reaction, there are two pairs of *cis* vicinal OH groups.

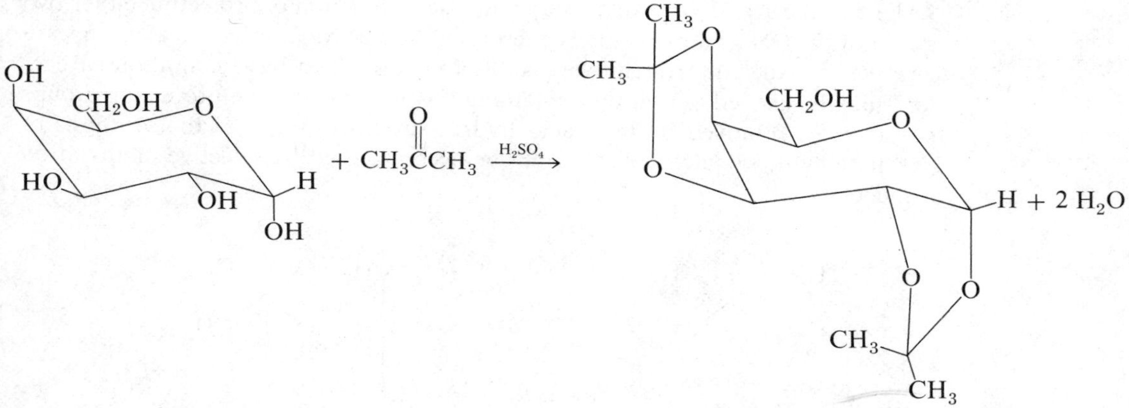

α-D-galactose

1,2:3,4-di-O-isopropylidine-
α-D-galactopyranoside

Glucose, on the other hand, reacts by way of the furanose form.

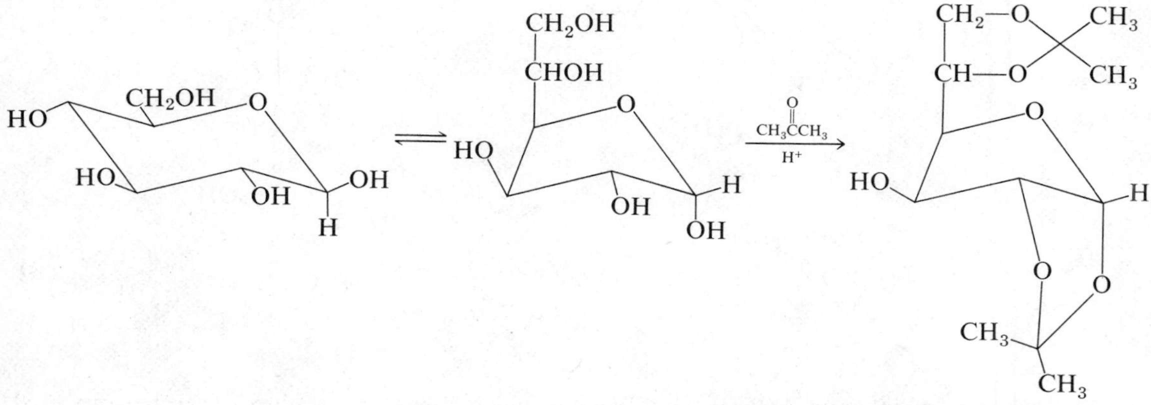

1,2:5,6-di-O-isopropylidine-
α-D-glucofuranose

Similar condensations occur with aldehydes. Benzaldehyde shows a tendency to form six-membered ring acetals. Thus, benzaldehyde reacts with methyl α-D-galactoside to give the 4,6-benzylidine derivative.

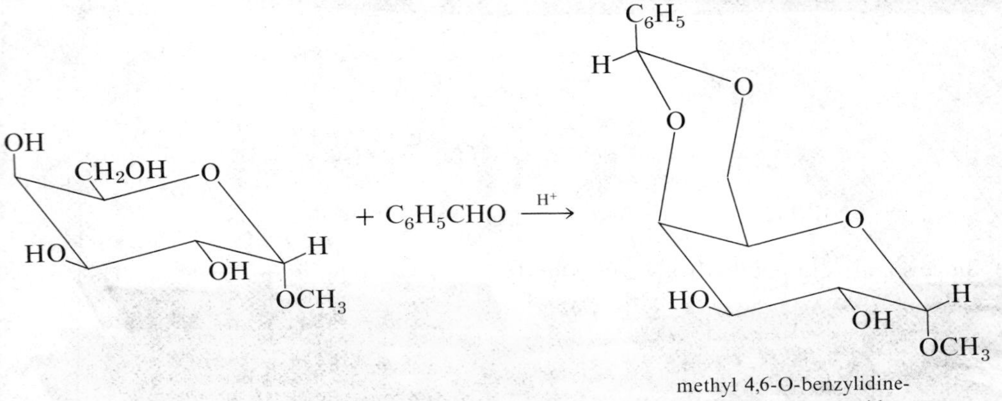

methyl 4,6-O-benzylidine-
α-D-galactopyranoside

These cyclic acetals and ketals serve the useful function of protecting either two or four of the OH groups normally present in the free sugar. The acetal groups are sensitive to acid, but are relatively stable to neutral and basic conditions. Reactions may be carried out on the remaining OH groups, and the protecting groups may then be removed by mild acid hydrolysis. An example is the synthesis of 3-O-methylglucose, a feat that cannot be accomplished by selective methylation of glucose itself.

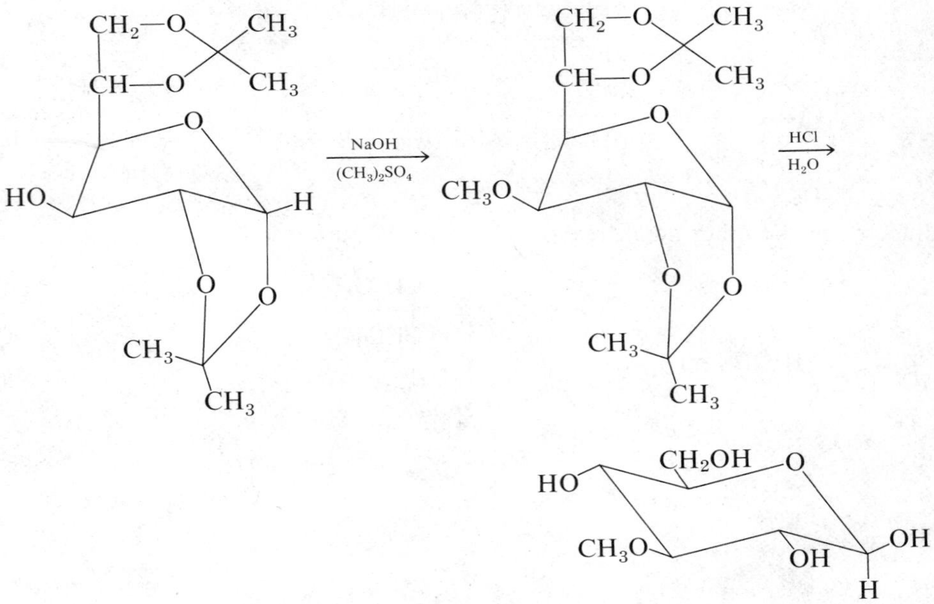

Another example is the following, which shows how D-glucose can be converted into D-allose by inversion of stereochemistry at C-3.

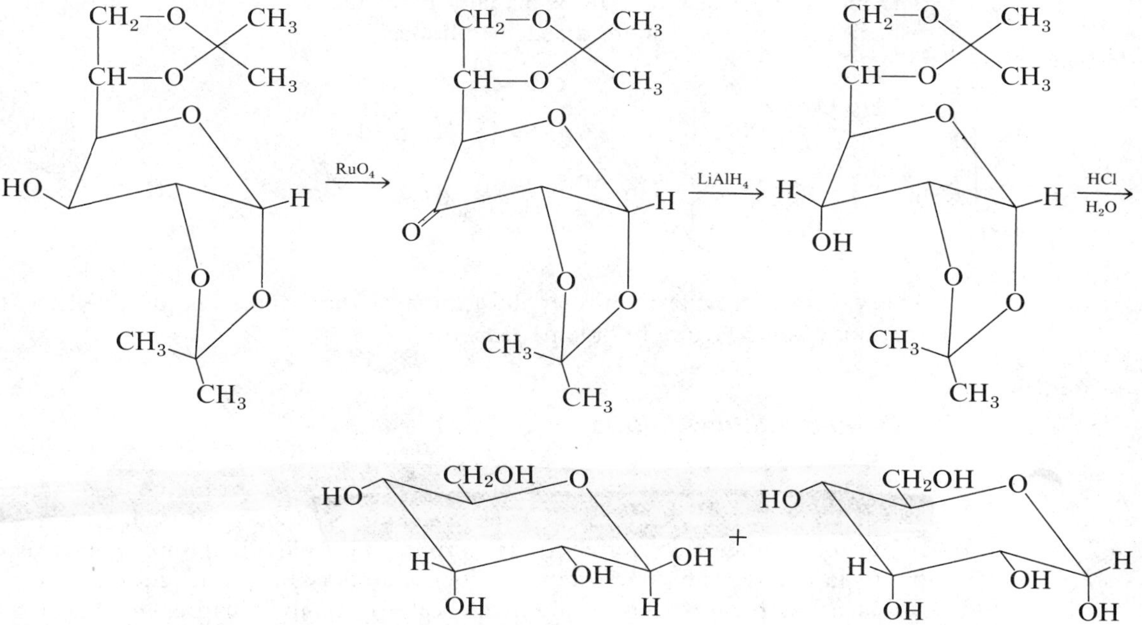

CH$_2$—O   CH$_3$
CH—O   CH$_3$
O
HO   H
O   O
CH$_3$
CH$_3$

$\xrightarrow{\text{RuO}_4}$

CH$_2$—O   CH$_3$
CH—O   CH$_3$
O
O   H
O
CH$_3$
CH$_3$

$\xrightarrow{\text{LiAlH}_4}$

CH$_2$—O   CH$_3$
CH—O   CH$_3$
O
H   H
OH
O
CH$_3$
CH$_3$

$\xrightarrow[\text{H}_2\text{O}]{\text{HCl}}$

HO   CH$_2$OH   O
H   OH   OH
OH   H
β-D-allose

+

HO   CH$_2$OH   O
H   H
OH   OH
OH
α-D-allose

EXERCISE: The reaction product of glucose and acetone, 1,2:5,6-di-O-isopropylidine-α-D-glucofuranose, is a valuable synthetic intermediate, as shown by the examples in this section. Suggest a way in which this intermediate might be used in the synthesis of 3-deoxyglucose (glucose lacking the hydroxy group at C-3).

C. *Esterification*

The hydroxy groups in sugars may be esterified by normal methods (Section 19.7.A). The most common procedure to form acetates uses acetic anhydride and a mild basic catalyst such as sodium acetate or pyridine. At low temperature acetylation in pyridine occurs more rapidly than interconversion of the anomers; at 0° either α-D- or β-D-glucose gives the corresponding pentaacetate. At higher temperatures the anomers interconvert rapidly, and the β-pentaacetate is produced preferentially, since the equatorial OH of the β-anomer reacts more rapidly than does the axial OH of the α-anomer.

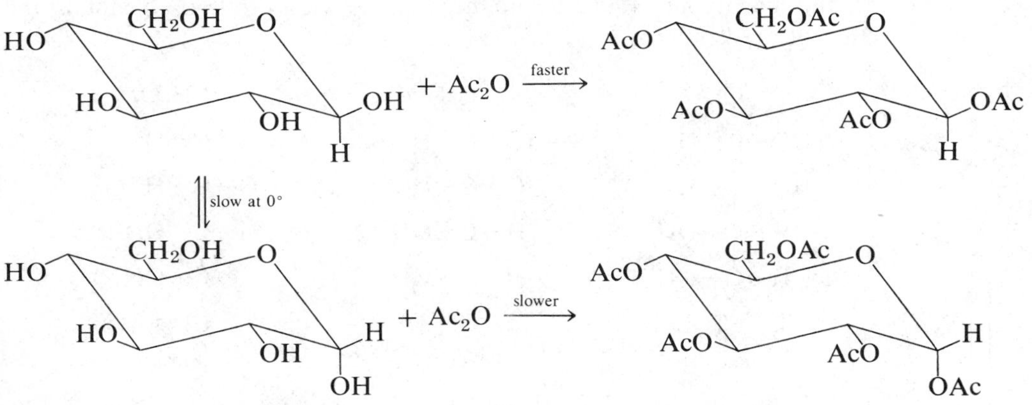

HO   CH$_2$OH   O
HO   OH   OH
H

+ Ac$_2$O $\xrightarrow{\text{faster}}$

AcO   CH$_2$OAc   O
AcO   AcO   OAc
H

$\updownarrow$ slow at 0°

HO   CH$_2$OH   O
HO   OH   H
OH

+ Ac$_2$O $\xrightarrow{\text{slower}}$

AcO   CH$_2$OAc   O
AcO   AcO   H
OAc

The more stable pentaacetate is actually the α-form, but equilibrium is established only under still more drastic conditions.

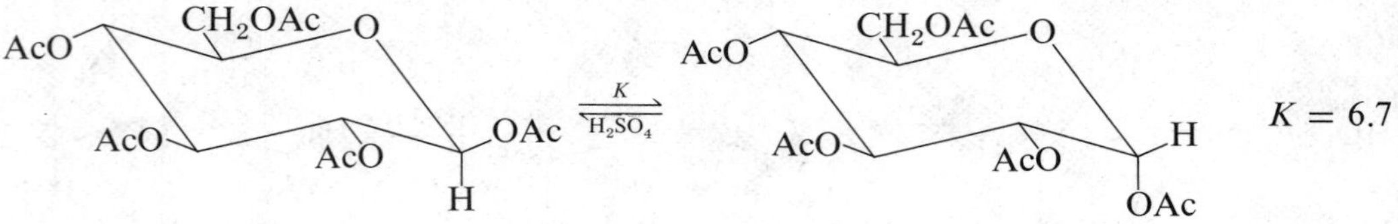

$$K = 6.7$$

This example provides a further illustration of the importance of kinetic and thermodynamic factors in organic reactions.

D. *Reduction: Alditols*

Monosaccharides may be reduced by various methods to the corresponding polyalcohols, which as a class are called **alditols**. Reduction of D-glucose gives D-glucitol (D-glucitol is referred to as D-sorbitol in the older literature), which also occurs in nature. The same compound is produced by the reduction of L-gulose. D-Glucitol is prepared on an industrial scale by catalytic hydrogenation of D-glucose over a nickel catalyst. The reduction probably occurs on the small amount of open-chain form that is present in equilibrium with the cyclic form. As the open-chain form is removed in this way, the equilibrium continually shifts until all of the sugar is reduced.

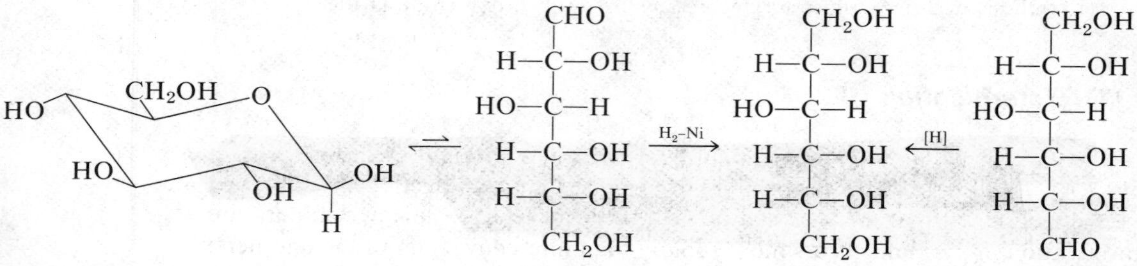

D-Mannitol, produced by the sodium borohydride reduction of D-mannose, is widespread in nature, occurring in such varied sources as olives, marine algae, onions, and mushrooms. It is also produced, along with a little D-glucitol, in the reduction of the ketohexose D-fructose.

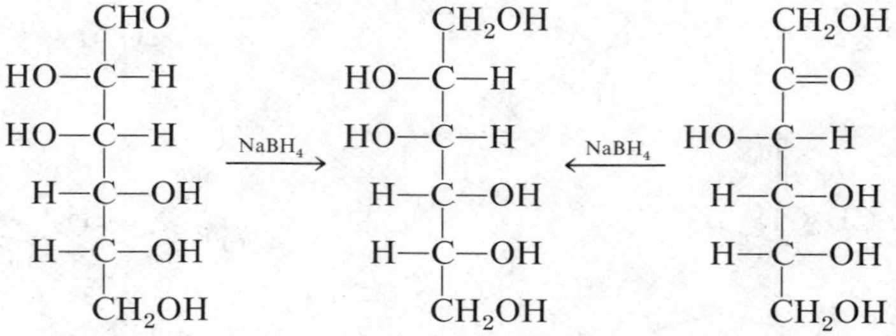

E. *Oxidation: Aldonic and Saccharic Acids*

Sugars are oxidizable in several ways. In the aldoses, the most susceptible group is the aldehyde group. For preparative purposes, the most convenient method employs bromine in a buffered solution at pH 5–6. Yields of the polyhydroxy carboxylic acids **(aldonic acids)** are usually in the range 50–70%. With glucose, yields as high as 95% have been achieved.

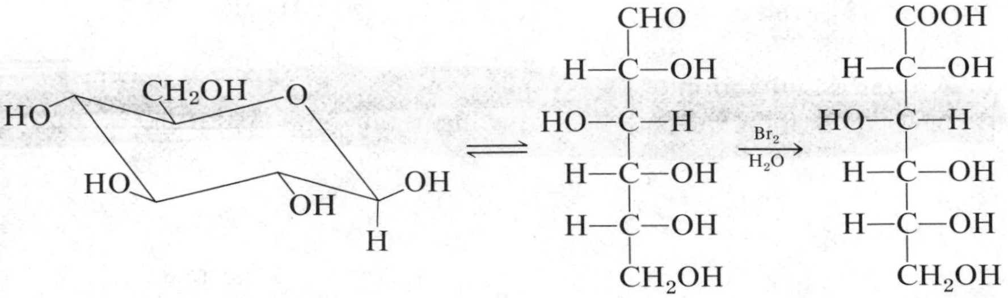

D-glucose   D-gluconic acid

Since they are 4-hydroxyalkanoic acids, the aldonic acids lactonize readily (Section 27.4.B). Although either a five- or six-membered lactone might, in principle, be formed, the more stable lactones are those containing a five-membered ring (see Table 27.4).

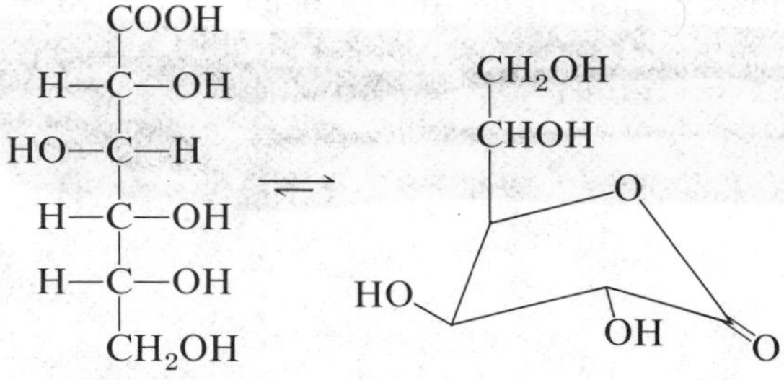

D-gluconic acid   D-gluconic acid
γ-lactone

The easy oxidation of aldoses provides a basis for an analytical method that has been widely used in sugar chemistry. Two examples are Fehling's test, employing cupric ion as the oxidant, and Tollens' test, in which silver ion is the oxidant. In the Fehling reaction (28-5) the presence of a potential aldehyde group is shown by the formation of cupric oxide as a brick-red precipitate. In Tollens' test (28-6) the silver ion is reduced to metallic silver, which deposits in the form of a mirror on the inside of the test tube.

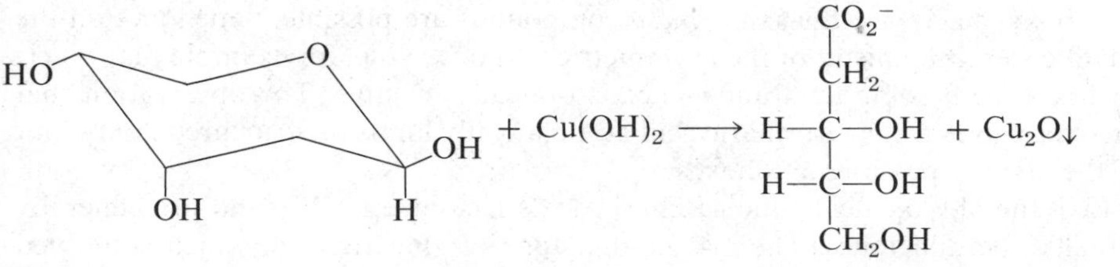

$$\beta\text{-D-2-deoxyribose} \quad + \quad Cu(OH)_2 \longrightarrow \quad + \quad Cu_2O\downarrow \qquad (28\text{-}5)$$

β-D-2-deoxyribose

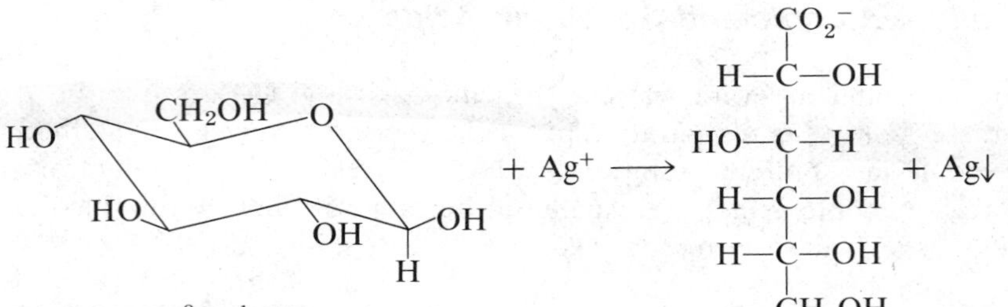

$$\beta\text{-D-glucose} + Ag^+ \longrightarrow \begin{array}{c} CO_2^- \\ H-C-OH \\ HO-C-H \\ H-C-OH \\ H-C-OH \\ CH_2OH \end{array} + Ag\downarrow \qquad (28\text{-}6)$$

If the sugar is in the form of a glycoside, then the anomeric carbon is protected under basic conditions, and the sugar is stable to these mild oxidizing conditions.

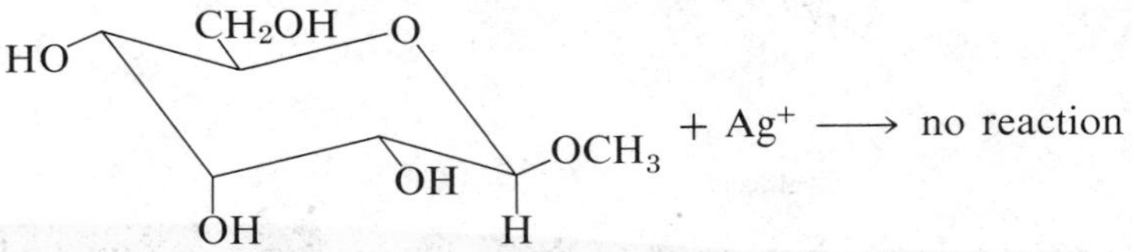

methyl β-D-allopyranoside $+ Ag^+ \longrightarrow$ no reaction

Such compounds are called **nonreducing sugars**; sugars that do reduce basic solutions of Cu^{2+} or Ag^+ are called **reducing sugars.**

Under more vigorous oxidizing conditions, one or more hydroxy groups may be oxidized. The primary OH groups are attacked most readily and are generally oxidized all the way to the carboxylic acid stage. The product is a polyhydroxy dicarboxylic acid called a **saccharic acid.** A convenient oxidizing agent for the preparation of saccharic acids is aqueous nitric acid.

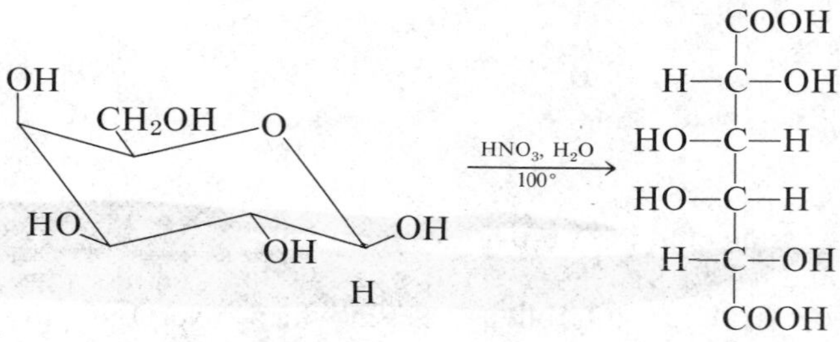

β-D-galactose $\xrightarrow[100°]{HNO_3, H_2O}$ mucic acid (galactaric acid)

The saccharic acids have proven to be useful in unraveling the puzzle of the relative configuration of the aldoses. Since the two ends of the chain in such a dicarboxylic acid are the same, *meso* compounds are possible, depending on the relative stereochemistry of the asymmetric carbons. Note, for example that mucic acid is a *meso* compound and hence is optically inactive. The observation that galactose gives a *meso* saccharic acid automatically limits its structure to only four of the sixteen possible aldohexoses.

Like the aldonic acids, the saccharic acids lactonize readily and are generally found to be dilactones. The 1,4:3,6-dilactone of glucaric acid, which is derived from D-glucose, is shown in (28-7).

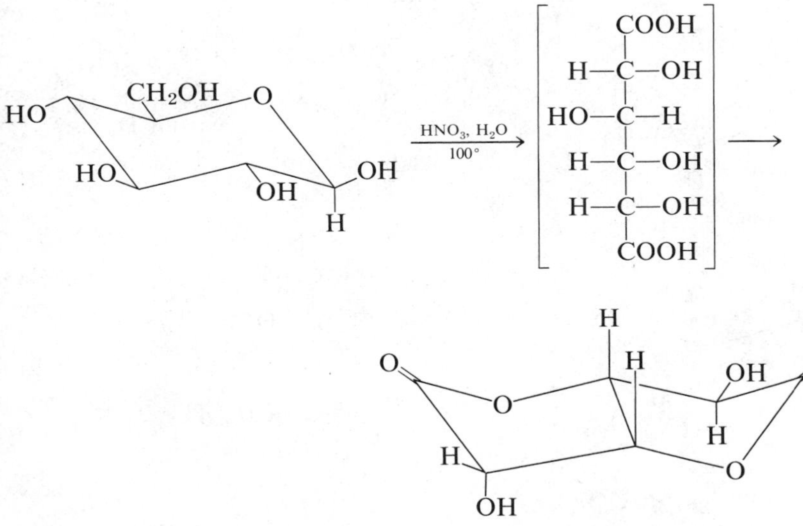

$$\xrightarrow[\text{100°}]{\text{HNO}_3,\ \text{H}_2\text{O}}$$

$$
\begin{bmatrix}
\text{COOH} \\
\text{H}-\text{C}-\text{OH} \\
\text{HO}-\text{C}-\text{H} \\
\text{H}-\text{C}-\text{OH} \\
\text{H}-\text{C}-\text{OH} \\
\text{COOH}
\end{bmatrix}
\longrightarrow
$$

(28-7)

EXERCISE: Write Fischer projection formulas for the aldonic and saccharic acids derived from D-galactose, D-mannose, and D-xylose. Are any of these acids achiral? What are the principal lactones expected from the three aldonic acids?

F. Oxidation by Periodic Acid

Like other vicinal diols, sugars are cleaved by periodic acid (Section 27.3.B). For example, methyl 2-deoxyribopyranoside reacts with one equivalent of periodic acid to give the dialdehyde shown in (28-8).

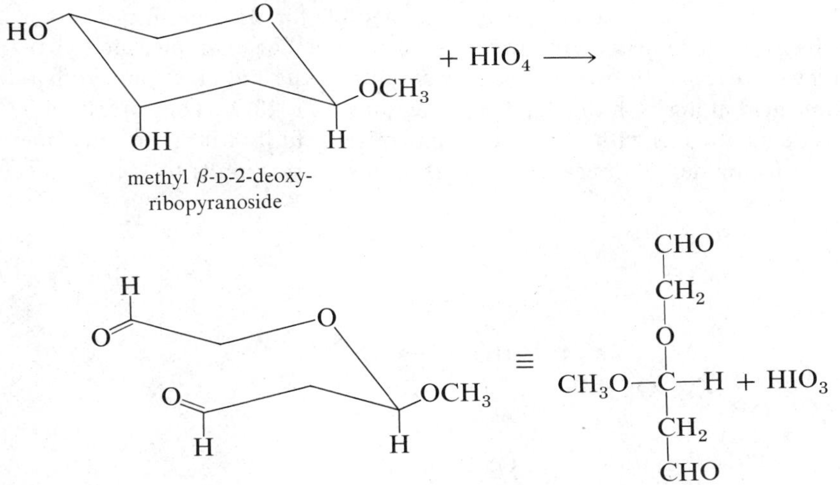

$+ \text{HIO}_4 \longrightarrow$

methyl β-D-2-deoxy-
ribopyranoside

$\equiv$

$$
\begin{array}{c}
\text{CHO} \\
\text{CH}_2 \\
\text{O} \\
\text{CH}_3\text{O}-\text{C}-\text{H} \\
\text{CH}_2 \\
\text{CHO}
\end{array}
+ \text{HIO}_3 \quad (28\text{-}8)
$$

When there are more than two adjacent hydroxy groups, the initially formed α-hydroxy aldehyde undergoes further oxidation (Section 27.4.B).

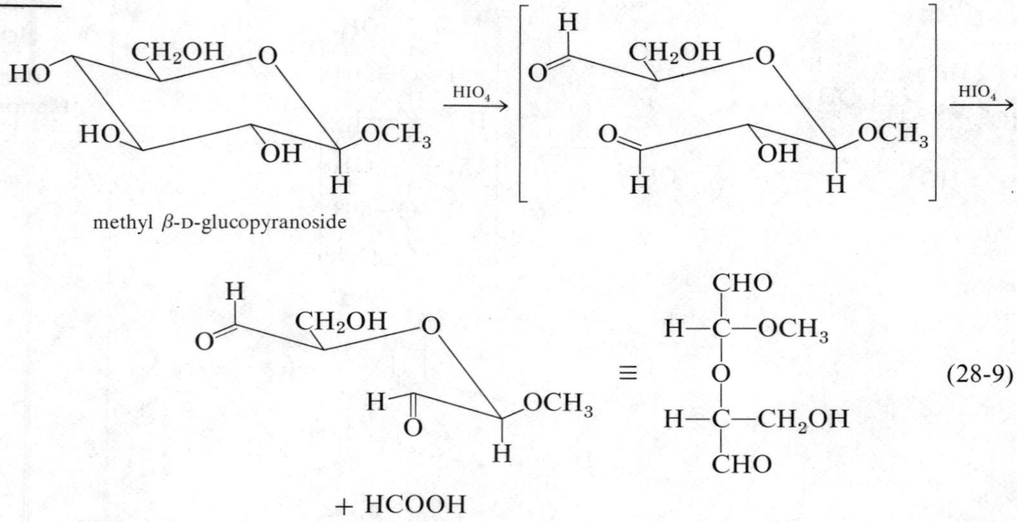

methyl β-D-glucopyranoside

$$\equiv$$

(28-9)

+ HCOOH

As shown in reaction (28-9), each time an α-hydroxy aldehyde is cleaved, one equivalent of formic acid is produced. With the free sugars it is the open-chain form that is oxidized, and complete oxidation occurs. Glucose yields five equivalents of formic acid and one equivalent of formaldehyde, which arises from C-6.

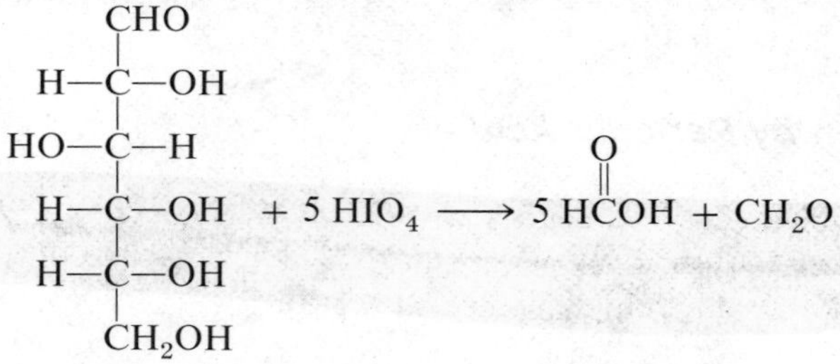

(28-10)

Periodate oxidation has been applied as a method for determining whether glycosides have the furanose or the pyranose structure. For example, methyl D-glucopyranoside reacts with two equivalents of the reagent and gives one equivalent of formic acid along with the dialdehyde as shown in (28-9). The corresponding furanoside also reacts with two equivalents of reagent, but it yields only one equivalent of formaldehyde because the carbon lost corresponds to C-6.

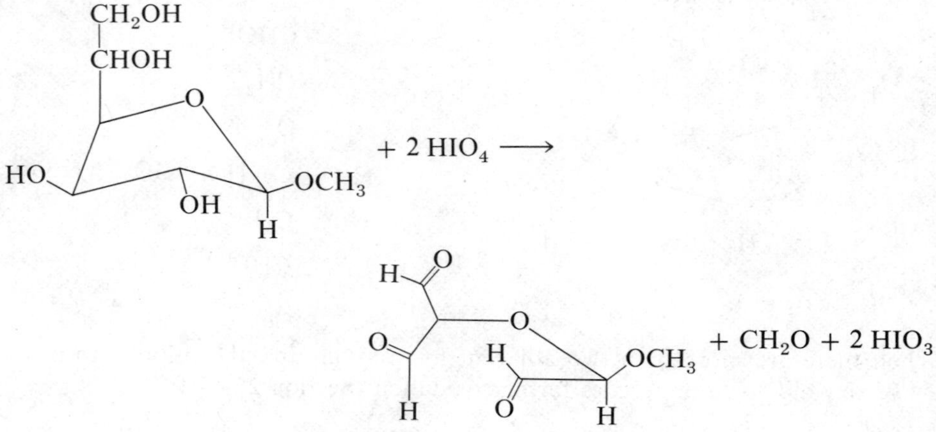

Periodic acid has also been used to determine the configuration at the anomeric carbon in the pyranosides. Note that oxidation of methyl β-D-glucopyranoside yields a dialdehyde that contains two asymmetric carbons corresponding to C-1 and C-5 in the glucoside itself; C-3 is lost as formic acid, and C-2 and C-4 become achiral. The other D-aldohexoses differ from D-glucose only in the relative configuration at these three centers. Thus, the methyl β-D-glucosides of all of the aldohexoses give this same dialdehyde on oxidation. One asymmetric carbon in the product is determined by the D-configuration, whereas the other is determined by the β-glycoside linkage.

methyl β-D-galactopyranoside

$$+ 2\ HIO_4 \longrightarrow$$

$$
\begin{array}{c}
CHO \\
H-C-OCH_3 \\
O \\
H-C-CH_2OH \\
CHO
\end{array}
\quad +\ HCOOH\ +\ 2\ HIO_3
$$

EXERCISE: How many equivalents of periodic acid are required for complete oxidation of each of the following sugars? For each case, indicate how many equivalents of formaldehyde and formic acid are produced and write the structure of any dialdehyde that remains.
(a) methyl 4,6-O-benzylidine-α-D-galactopyranoside (page 906)
(b) methyl β-D-xyloside (page 904)
(c) L-($-$)-threitol (page 895)

G. *Phenylhydrazones and Osazones*

Because of their polyhydroxy nature, sugars are rather difficult to isolate and purify. They are extremely water-soluble and tend to form viscous syrups that crystallize poorly. Naturally occurring examples of such syrups are honey and molasses. These properties caused severe problems in working with sugars in the nineteenth century, before the advent of today's powerful spectroscopic methods of analysis. At that time the only way to ascertain the identity or nonidentity of two compounds was to compare melting points. In 1884 Emil Fischer introduced the use of phenylhydrazine as a reagent in sugar chemistry and opened up a new vista in the subject. Fischer found that a monosaccharide, such as glucose, will react by way of its open-chain form with phenylhydrazine in acetic acid to give a normal phenylhydrazone (Section 13.7.C). However, the initially formed phenylhydrazone reacts further with two more equivalents of phenylhydrazine to yield a derivative called an **osazone.**

$$
\begin{array}{c}
\text{CHO} \\
\text{H}-\text{C}-\text{OH} \\
\text{HO}-\text{C}-\text{H} \\
\text{H}-\text{C}-\text{OH} \\
\text{H}-\text{C}-\text{OH} \\
\text{CH}_2\text{OH}
\end{array}
\xrightarrow[\text{HOAc}]{\text{C}_6\text{H}_5\text{NHNH}_2}
\begin{array}{c}
\text{CH}=\text{NNHC}_6\text{H}_5 \\
\text{H}-\text{C}-\text{OH} \\
\text{HO}-\text{C}-\text{H} \\
\text{H}-\text{C}-\text{OH} \\
\text{H}-\text{C}-\text{OH} \\
\text{CH}_2\text{OH}
\end{array}
\xrightarrow[\text{HOAc}]{2\,\text{C}_6\text{H}_5\text{NHNH}_2}
$$

D-glucose   D-glucose
                   phenylhydrazone

$$
\begin{array}{c}
\text{CH}=\text{NNHC}_6\text{H}_5 \\
\text{C}=\text{NNHC}_6\text{H}_5 \\
\text{HO}-\text{C}-\text{H} \\
\text{H}-\text{C}-\text{OH} \\
\text{H}-\text{C}-\text{OH} \\
\text{CH}_2\text{OH}
\end{array}
\quad + \text{C}_6\text{H}_5\text{NH}_2 + \text{NH}_3 + \text{H}_2\text{O}
$$

D-glucose
phenylosazone

The osazones were found to be bright yellow crystalline materials with characteristic crystal forms. However, the osazones soon proved to be more valuable
than they appeared at first sight. Notice that the chirality at C-2 in D-glucose is
lost upon formation of glucose phenylosazone. Thus, *D-mannose gives the same
phenylosazone as does D-glucose,* thus proving that the two aldohexoses have the
same absolute stereochemistry at C-3, C-4, and C-5. Furthermore, the ketohexose,
D-fructose, also gives glucose phenylosazone, thereby establishing that it also has

$$
\begin{array}{c}
\text{CHO} \\
\text{H}-\text{C}-\text{OH} \\
\text{HO}-\text{C}-\text{H} \\
\text{H}-\text{C}-\text{OH} \\
\text{H}-\text{C}-\text{OH} \\
\text{CH}_2\text{OH}
\end{array}
\longrightarrow
\begin{array}{c}
\text{CH}=\text{NNHC}_6\text{H}_5 \\
\text{C}=\text{NNHC}_6\text{H}_5 \\
\text{HO}-\text{C}-\text{H} \\
\text{H}-\text{C}-\text{OH} \\
\text{H}-\text{C}-\text{OH} \\
\text{CH}_2\text{OH}
\end{array}
\longleftarrow
\begin{array}{c}
\text{CHO} \\
\text{HO}-\text{C}-\text{H} \\
\text{HO}-\text{C}-\text{H} \\
\text{H}-\text{C}-\text{OH} \\
\text{H}-\text{C}-\text{OH} \\
\text{CH}_2\text{OH}
\end{array}
$$

D-glucose   D-glucose   D-mannose
                   phenylosazone

$$
\begin{array}{c}
\text{CH}_2\text{OH} \\
\text{C}=\text{O} \\
\text{HO}-\text{C}-\text{H} \\
\text{H}-\text{C}-\text{OH} \\
\text{H}-\text{C}-\text{OH} \\
\text{CH}_2\text{OH}
\end{array}
$$

D-fructose

this configuration at C-3, C-4, and C-5 (and, incidentally, that its carbonyl group is at C-2). Notice that osazones are bis-phenylhydrazones and that the reaction with phenylhydrazine stops at this stage; that is, further reaction at the C-3 hydroxy group does not normally occur.

EXERCISE: D-(+)-Sorbose is a ketohexose that gives the same osazone as do the aldohexoses D-(−)-gulose and D-(−)-idose. What is the structure of D-(+)-sorbose?

H. *Chain Extension: The Kiliani–Fischer Synthesis*

When D-glyceraldehyde is treated with HCN, a mixture of two cyanohydrins is produced. Both cyanohydrins have the (R) configuration at C-3, corresponding to the same configuration at C-2 in D-glyceraldehyde. They differ only in the configuration at C-2, the new asymmetric carbon. Hydrolysis of the two cyanohydrins yields the same aldonic acids as are produced by the mild oxidation of the aldotetroses D-erythrose and D-threose. Since glyceraldehyde is a chiral molecule, the transition states leading to the two cyanohydrins are diastereomeric rather than enantiomeric, and the two products are not produced in equal amounts (see Section 7.6).

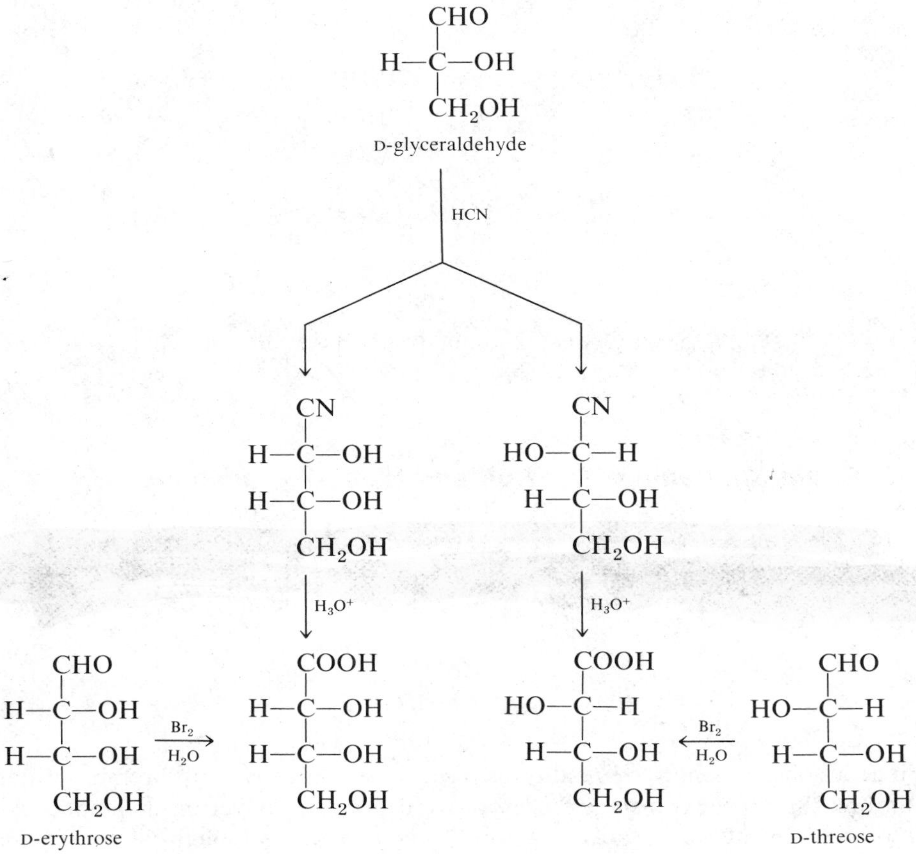

The cyanohydrin chain-lengthening procedure has been applied extensively in sugar chemistry and has come to be known as the **Kiliani–Fischer synthesis.** Fischer discovered that the aldonic acids produced by hydrolysis of the cyanohydrins lactonize on heating to aldonolactones. He also discovered that these lactones may be reduced with sodium amalgam at pH 3.0–3.5 to give a new aldose. A more modern method involves reduction of the lactone with aqueous sodium borohydride at pH 3–4. The complete Kiliani–Fischer synthesis provides a method for converting an aldopentose into an aldohexose or an aldohexose into an aldoheptose. The synthesis always provides two diastereomers, usually in unequal amounts, which differ only in their configuration at the new C-2 (old C-1). An example is D-arabinose, which yields a mixture of D-glucose and D-mannose.

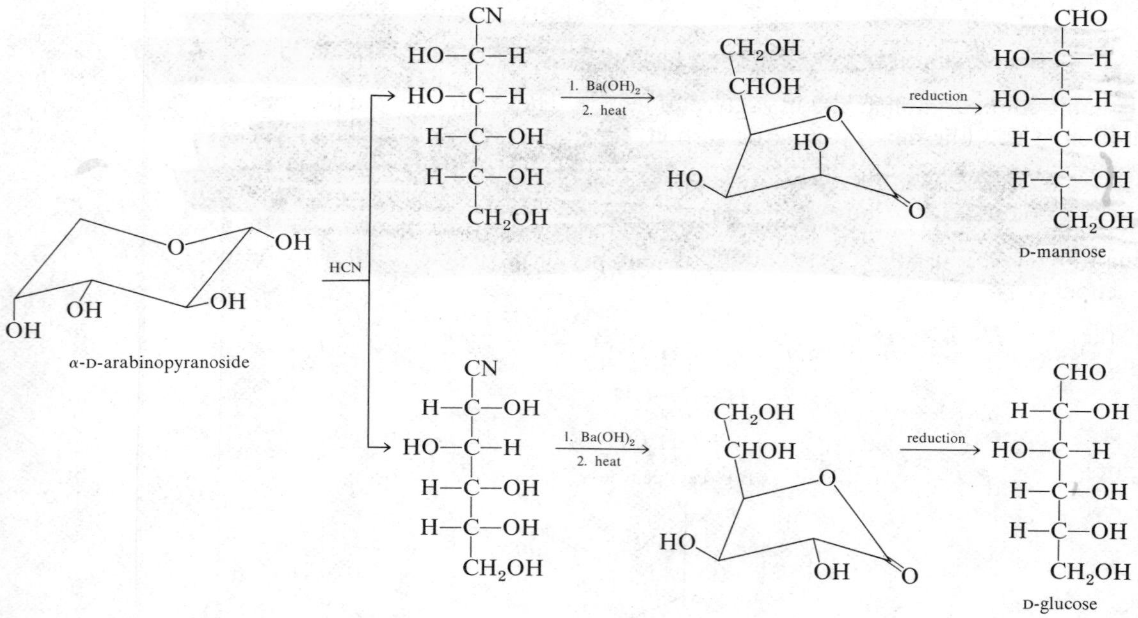

EXERCISE: What two aldohexoses result from application of the Kiliani–Fischer synthesis to (a) D-(−)-ribose and (b) D-(+)-xylose?

I. *Chain Shortening: The Ruff and Wohl Degradations*

In 1896 Ruff discovered that the calcium salts of aldonic acids are oxidized by hydrogen peroxide, the reaction being catalyzed by ferric salts. The oxidation occurs with cleavage of the C_1—C_2 bond and the product is the lower aldose.

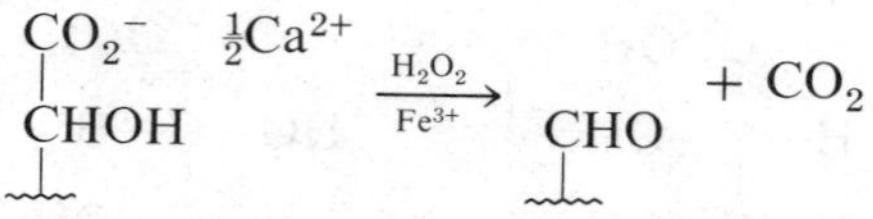

Since aldohexoses may be readily oxidized to aldonic acids by bromine water (Section 28.5.E), the two-stage process provides a way of converting an aldohexose into an aldopentose; it is called the **Ruff degradation.** Although yields are not high, the Ruff degradation has been a useful technique for the synthesis of certain aldopentoses. Since asymmetry is lost at C-2, two aldohexoses that differ only in chirality at this center yield the same aldopentose.

```
   CHO                      CO2- ½Ca2+
 H—C—OH                   H—C—OH
HO—C—H     Br2   Ca(OH)2  HO—C—H      H2O2
 H—C—OH    ——→   ——→       H—C—OH      Fe3+
 H—C—OH    H2O             H—C—OH              ↘
   CH2OH                     CH2OH                   CHO
D-glucose                calcium D-gluconate      HO—C—H
                                                   H—C—OH
   CHO                      CO2- ½Ca2+           H—C—OH
HO—C—H                   HO—C—H                   CH2OH
HO—C—H     Br2   Ca(OH)2  HO—C—H      H2O2       (40–50%)
 H—C—OH    ——→   ——→       H—C—OH      Fe3+       D-arabinose
 H—C—OH    H2O             H—C—OH              ↗
   CH2OH                     CH2OH
D-mannose                calcium D-mannonate
```

Unfortunately, the process is not very useful for the conversion of aldopentoses into aldotetroses because of low yields.

Another process, called the **Wohl degradation,** accomplishes the same overall conversion, shortening the aldose chain by the removal of C-1. The Wohl degradation is essentially the reverse of the Kiliani–Fischer synthesis. The aldose is first converted into its oxime by treatment with hydroxylamine (Section 13.7.C). When the resulting polyhydroxy oxime is heated with acetic anhydride and sodium acetate, all of the hydroxy groups are acetylated and the oxime group is *dehydrated* to a cyano group. The product is the acetate ester of a cyanohydrin. The ester groups are removed by treatment with base. Under the basic conditions of hydrolysis, the cyanohydrin is decomposed to the corresponding aldehyde. Again, the process does not give especially high yields, but it is applicable to pentoses as well as to hexoses.

```
   CHO                CH=NOH              C≡N
 H—C—OH             H—C—OH             H—C—OAc
 H—C—OH    H2NOH    H—C—OH    Ac2O     H—C—OAc    NaOCH3
 H—C—OH    ——→      H—C—OH    ——→      H—C—OAc    ——→
   CH2OH    ——→       CH2OH    NaOAc      CH2OAc     CHCl3
D-ribose

                    ⌈   CN    ⌉              CHO
                    │ H—C—OH  │  NaOCH3    H—C—OH
                    │ H—C—OH  │  ——→       H—C—OH    + NaCN + CH3OH
                    │ H—C—OH  │              CH2OH
                    ⌊   CH2OH ⌋          D-erythrose
```

EXERCISE: What tetrose results from application of the Wohl degradation to D-(−)-lyxose? Which other pentose also affords this tetrose?

28.6
Relative Stereochemistry of the Monosaccharides: The Fischer Proof

In the late nineteenth century organic chemists were faced with a puzzling mystery regarding the structures of the monosaccharides. A number of compounds had been isolated that were known to have the same formula and that had the same connectivity. That is, the available evidence showed that glucose, galactose, and mannose were all 2,3,4,5,6-pentahydroxyhexanals. The Le Bel–Van't Hoff theory of stereoisomerism provided an explanation for this phenomenon. The challenging question was, which relative arrangement of the four asymmetric carbons corresponds to glucose, which to mannose, and so on?

The challenge was taken up by Emil Fischer, who succeeded in establishing the correct stereostructures for D-glucose, D-mannose, D-fructose, and D-arabinose in 1891. The structure proof consists of an elegant series of logical deductions and has come to be known as "the Fischer proof." We will present a modernized version of the proof here because it typifies the method that has been used to establish the structures of all the sugars.

At the outset Fischer realized that he could establish the *relative configuration* of the various asymmetric centers in a sugar, but that he had no way to determine the *absolute configuration* of any of the compounds. In order to understand this distinction, consider the four aldotetroses D- and L-threose and D- and L-erythrose. All four compounds are oxidized by nitric acid to saccharic acids. The enantiomeric D- and L-threoses give enantiomeric D- and L-2,3-dihydroxysuccinic acids, which are called D-tartaric acid and L-tartaric acid. However, both D- and L-erythrose are oxidized by nitric acid to a saccharic acid that is an optically inactive diacid, *meso*-tartaric acid. *Meso*-tartaric acid owes its optical inactivity to an internal symmetry plane, and hence the relative configuration of its two asymmetric carbons is fixed. It follows that one of the erythroses has the (R,R) configuration and the other has the (S,S) configuration. However, there was no way to tell which is which.

$$
\begin{array}{ccc}
\text{CHO} & & \text{COOH} \\
| & & | \\
\text{HO}-\text{C}-\text{H} & \xrightarrow{\text{HNO}_3} & \text{HO}-\text{C}-\text{H} \\
| & & | \\
\text{H}-\text{C}-\text{OH} & & \text{H}-\text{C}-\text{OH} \\
| & & | \\
\text{CH}_2\text{OH} & & \text{COOH} \\
\text{D-threose} & & \text{D-tartaric acid}
\end{array}
$$

$$
\begin{array}{ccc}
\text{CHO} & & \text{COOH} \\
| & & | \\
\text{H}-\text{C}-\text{OH} & \xrightarrow{\text{HNO}_3} & \text{H}-\text{C}-\text{OH} \\
| & & | \\
\text{HO}-\text{C}-\text{H} & & \text{HO}-\text{C}-\text{H} \\
| & & | \\
\text{CH}_2\text{OH} & & \text{COOH} \\
\text{L-threose} & & \text{L-tartaric acid}
\end{array}
$$

919

Sec. 28.6

**Relative
Stereochemistry
of the
Monosaccharides:
The Fischer
Proof**

```
        CHO
     H—C—OH        HNO₃
     H—C—OH
        CH₂OH
     D-erythrose                        COOH
                                     H—C—OH
                                     - - - - - - - - - - - -   symmetry plane
        CHO                          H—C—OH
    HO—C—H          HNO₃               COOH
    HO—C—H                          meso-tartaric acid
        CH₂OH
     L-erythrose
```

The question of the absolute configuration was not settled until 1954, when Bijvoet determined the absolute configuration of a salt of D-tartaric acid by an x-ray crystallographic technique known as anomalous dispersion.

1. Fischer started by arbitrarily choosing what is now called the (R) configuration for the stereochemistry at C-5 in D-glucose. He had a 50% chance of being correct in this assignment, but it has no bearing on the rest of the proof because the other centers were to be determined relative to C-5. In 1954 Bijvoet's work showed that Fischer had actually made the correct choice in an absolute sense. The structures of the eight aldohexoses having the (R) configuration at C-5 are shown in Table 28.2 and are designated **1** through **8**.

2. It was known that glucose and mannose give the same osazone (Section 28.5.G). Therefore the two compounds have the same configuration at C-3, C-4, and C-5; they differ only at C-2. The two compounds must be **1** and **2, 3** and **4, 5** and **6,** or **7** and **8.**

TABLE 28.2
The (5R)-Aldohexoses

```
     CHO              CHO              CHO              CHO
  H—C—OH          HO—C—H           H—C—OH           HO—C—H
  H—C—OH           H—C—OH          HO—C—H           HO—C—H
  H—C—OH           H—C—OH           H—C—OH           H—C—OH
  H—C—OH           H—C—OH           H—C—OH           H—C—OH
     CH₂OH            CH₂OH            CH₂OH            CH₂OH
      1                2                3                4


     CHO              CHO              CHO              CHO
  H—C—OH          HO—C—H           H—C—OH           HO—C—H
  H—C—OH           H—C—OH          HO—C—H           HO—C—H
 HO—C—H          HO—C—H          HO—C—H          HO—C—H
  H—C—OH           H—C—OH           H—C—OH           H—C—OH
     CH₂OH            CH₂OH            CH₂OH            CH₂OH
      5                6                7                8
```

3. *Both* D-glucose and D-mannose are oxidized by nitric acid to *optically active* saccharic acids (Section 28.5.E). The aldohexoses with structures **1** and **7** would give *meso*-saccharic acids. Therefore D-glucose and D-mannose must be either **3** and **4** or **5** and **6**. (Remember that the two compounds differ only at C-2, so eliminating **1** also eliminates **2**, and eliminating **7** also eliminates **8**.)

4. Kiliani–Fischer chain extension of the aldopentose D-arabinose yields both D-glucose and D-mannose (Section 28.5.H). Therefore, D-arabinose has the same configuration at its C-2, C-3, and C-4 as D-glucose and D-mannose at C-3, C-4, and C-5, respectively. D-Arabinose must be either

$$
\begin{array}{c}
\text{CHO} \\
\text{HO—C—H} \\
\text{H—C—OH} \\
\text{H—C—OH} \\
\text{CH}_2\text{OH} \\
\textbf{9}
\end{array}
\xrightarrow{\text{Kiliani–Fischer}} \textbf{3 and 4}
$$

or

$$
\begin{array}{c}
\text{CHO} \\
\text{H—C—OH} \\
\text{HO—C—H} \\
\text{H—C—OH} \\
\text{CH}_2\text{OH} \\
\textbf{10}
\end{array}
\xrightarrow{\text{Kiliani–Fischer}} \textbf{5 and 6}
$$

However, oxidation of D-arabinose gives an *optically active* diacid. The saccharic acid derived from aldopentose **10** would be *meso,* so D-arabinose must be **9**, and D-glucose and D-mannose must be **3** and **4**.

$$
\begin{array}{c}
\text{CHO} \\
\text{HO—C—H} \\
\text{H—C—OH} \\
\text{H—C—OH} \\
\text{CH}_2\text{OH} \\
\textbf{9, D-arabinose}
\end{array}
\xrightarrow{\text{HNO}_3}
\begin{array}{c}
\text{COOH} \\
\text{HO—C—H} \\
\text{H—C—OH} \\
\text{H—C—OH} \\
\text{COOH} \\
\text{optically active}
\end{array}
$$

or

$$
\begin{array}{c}
\text{CHO} \\
\text{H—C—OH} \\
\text{HO—C—H} \\
\text{H—C—OH} \\
\text{CH}_2\text{OH} \\
\textbf{10}
\end{array}
\xrightarrow{\text{HNO}_3}
\begin{array}{c}
\text{COOH} \\
\text{H—C—OH} \\
\text{---HO—C—H---} \quad \text{symmetry plane} \\
\text{H—C—OH} \\
\text{COOH} \\
\textit{meso}
\end{array}
$$

5. Fischer developed a method that allowed him to *interchange* the two ends of an aldose chain. The method is fairly involved, and we need not go into the chemical details here. However, consider the results when the method is applied to the aldohexoses that have structures **3** and **4**.

```
   CHO                CH₂OH              CHO
 H—C—OH            H—C—OH           HO—C—H
HO—C—H            HO—C—H           HO—C—H
 H—C—OH     --->    H—C—OH      ≡    H—C—OH
 H—C—OH             H—C—OH          HO—C—H
   CH₂OH              CHO               CH₂OH

     3            rotate structure 180°
```

```
   CHO                CH₂OH              CHO
HO—C—H            HO—C—H           HO—C—H
HO—C—H            HO—C—H           HO—C—H
 H—C—OH     --->    H—C—OH      ≡    H—C—OH
 H—C—OH             H—C—OH           H—C—OH
   CH₂OH              CHO               CH₂OH

     4            rotate structure 180°
```

When C-1 and C-6 are interchanged, compound **3** gives a *different aldohexose*. However, when the same operation is performed on compound **4,** the final product is the same as the starting material. Fischer applied his method to D-glucose and discovered that a new aldohexose was produced, which he named L-gulose. The proof was complete; D-glucose must have structure **3** and D-mannose must have structure **4!**

> **EXERCISE:** Referring to Table 28.1, choose any aldohexose except glucose or mannose. Work through the reactions involved in the Fischer proof and write the structures of the compounds that would be involved if glucose had the structure you have chosen. Which experiments show that the chosen aldohexose is not glucose?

28.7
Oligosaccharides

Oligosaccharides (Gr., *oligos,* a few) are polysaccharides that yield from two to eight monosaccharide units upon hydrolysis. The most common are the disaccharides, which are dimers composed of two monosaccharides. The two monosaccharides may be the same or different. Disaccharides are joined by a glycoside linkage from the OH group of one monosaccharide to the anomeric carbon of the other.

A simple example of a disaccharide is maltose, which is produced by the enzymatic hydrolysis of starch (Section 28.8). Maltose contains two D-glucose units, both in the pyranose form. The C-4 hydroxy group of one glucose is bound by an α-glycoside bond to the anomeric carbon of the other unit. The disaccharide is obtained in crystalline form in which the hydroxy group on the other anomeric carbon is β, but it mutarotates in solution to a mixture of the α- and β-forms.

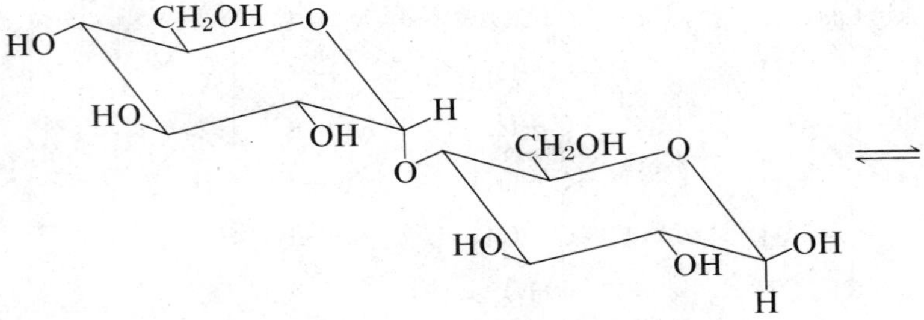

4-O-(α-D-glucopyranosyl)-β-D-glucopyranose
β-maltose

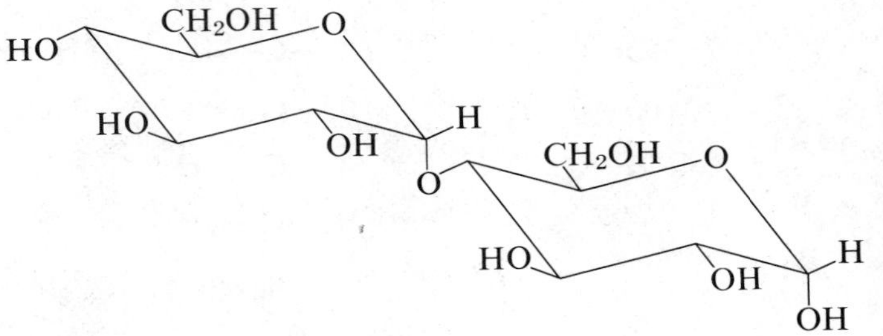

4-O-(α-D-glucopyranosyl)-α-D-glucopyranose
α-maltose

In maltose, one of the glucose units has its aldehyde carbon firmly bound in the glycosidic linkage to the other unit. However, the carbonyl group of the second ring is only in the hemiacetal form, and it may therefore undergo normal carbonyl reactions, just as the monosaccharides do. Thus maltose is oxidized by Tollens' reagent and by Fehling's solution and is a reducing sugar. Disaccharides undergo most of the same reactions as do the monosaccharides. For example, maltose is oxidized by bromine water to maltobionic acid.

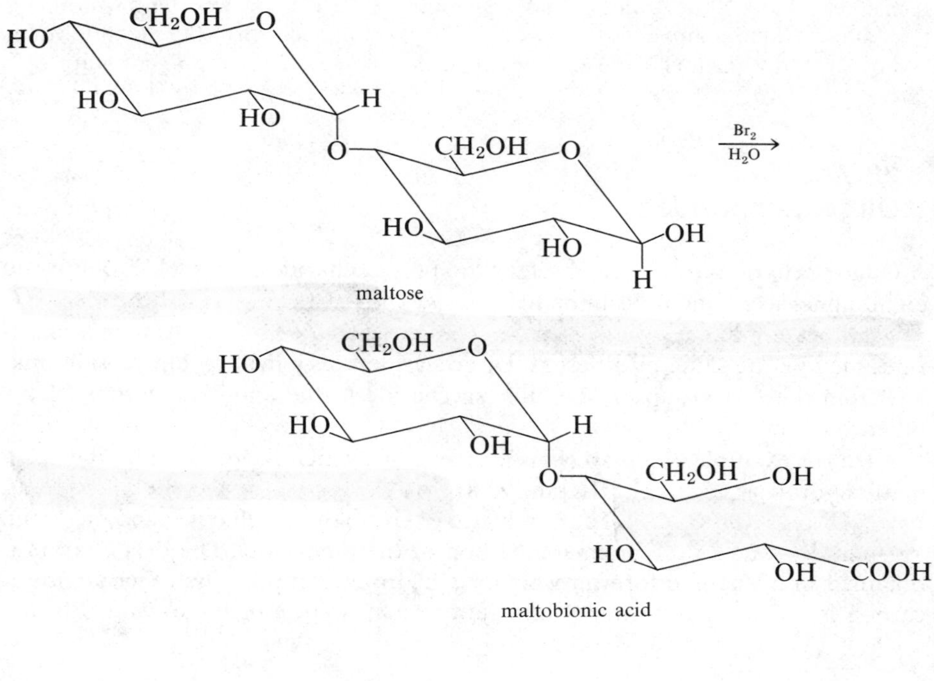

maltose

maltobionic acid

The structure of maltose is shown by the following observations. Methylation of maltobionic acid gives octa-O-methylmaltobionic acid, which is hydrolyzed under acidic conditions. The products are 2,3,4,6-tetra-O-methyl-D-glucose and 2,3,5,6-tetra-O-methylgluconic acid. Thus both rings in maltose exist in the pyranose form.

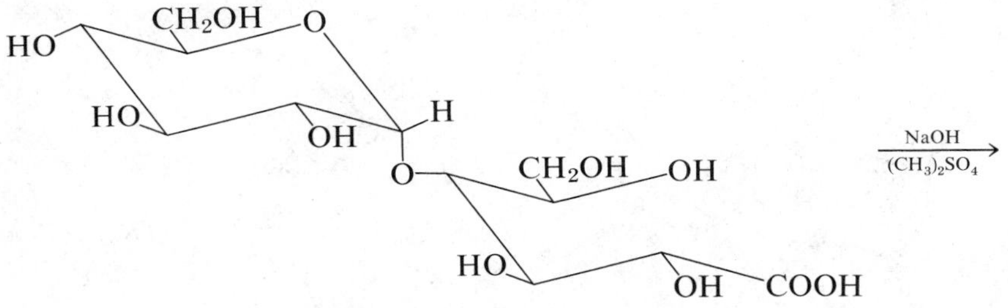

maltobionic acid

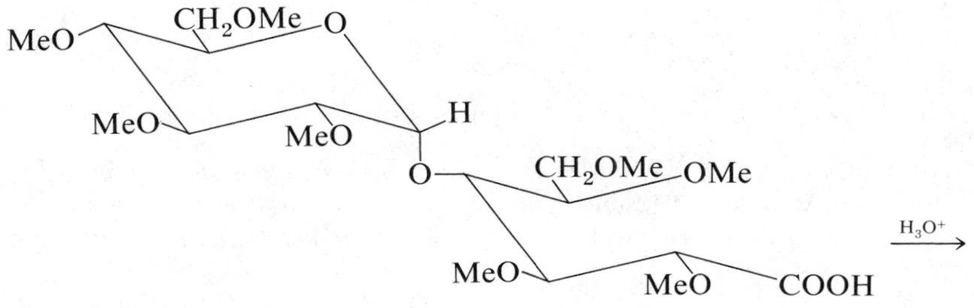

octa-O-methylmaltobionic acid

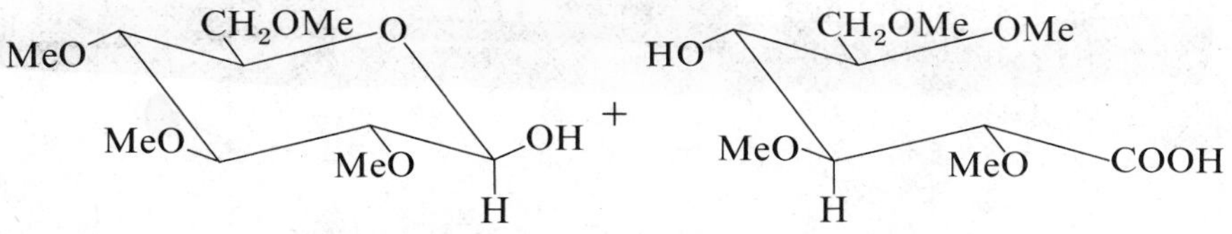

2,3,4,6-tetra-O-methyl-β-D-glucose   2,3,5,6-tetra-O-methyl-D-gluconic acid

The only remaining question is the stereochemistry of the glycoside link in maltose. This is established by the fact that the yeast enzyme α-glycosidase catalyzes the hydrolysis of maltose, whereas β-glycosidase does not affect the disaccharide (Section 28.3). Thus the glycoside linkage in maltose is α.

Cellobiose is a disaccharide that is obtained by the partial hydrolysis of cellulose (Section 28.8). Its chemical properties are very similar to those of maltose. Like maltose, cellobiose is a reducing sugar; it is oxidized by bromine to cellobionic acid, which is methylated to octa-O-methylcellobionic acid. Hydrolysis of the latter gives the same two products as hydrolysis of octa-O-methylmaltobionic acid. The only difference is that the hydrolysis of cellobiose is catalyzed by β-glucosidase and not by α-glucosidase. Therefore cellobiose is isomeric with maltose and contains a β-glycosidic linkage. The structure of β-cellobiose is shown in Figure 28.2.

Lactose is an example of a disaccharide in which the two monosaccharide units are different. It constitutes about 5% by weight of mammalian milk. It is produced commercially from whey, which is obtained as a by-product in the manufacture of cheese. Evaporation of the whey at temperatures below 95° causes the less soluble α-anomer to precipitate. Hydrolysis of lactose affords one equivalent of glucose and one equivalent of galactose. Application of the methylation technique to

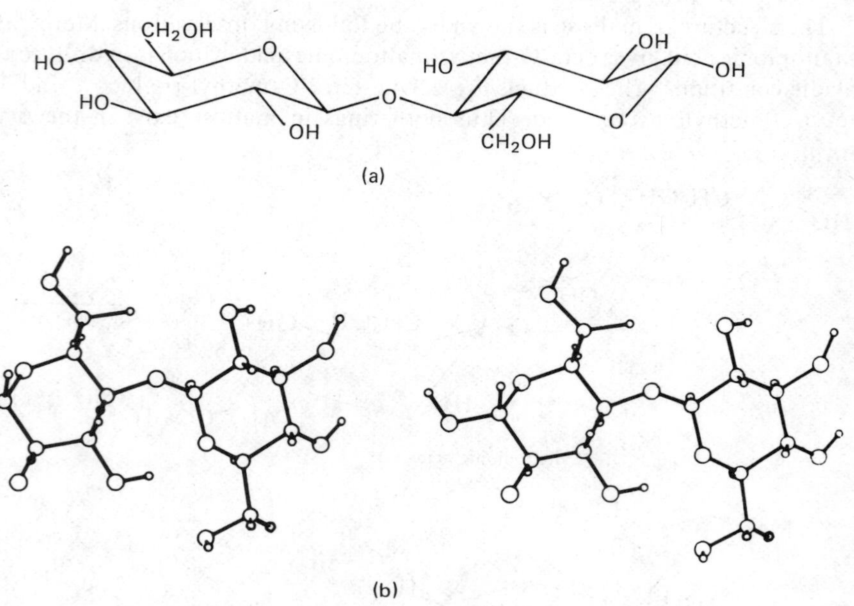

(a)

(b)

FIGURE 28.2. *β-Cellobiose, 4-O-(β-D-glucopyranosyl)-β-D-glucopyranose:* (a) *conventional structure;* (b) *stereo structure.* [*Part* (b) *reproduced with permission from* Molecular Structure and Dimensions. *International Union of Crystallography, 1972.*]

lactobionic acid shows that the galactose unit is bound in the glycoside form and that both rings are pyranoses. The hydrolysis of lactose is catalyzed by an enzyme called β-galactosidase, which is specific for the hydrolysis of β-galactoside links.

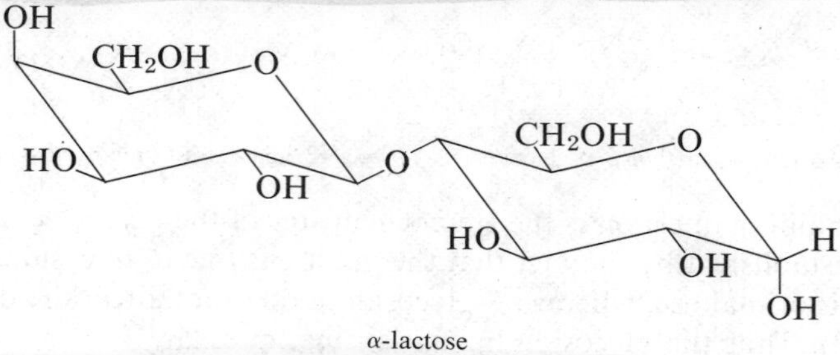

α-lactose
4-O-(β-D-galactopyranosyl)-α-D-glucopyranose

Sucrose is one of the most widespread sugars in nature. It is produced commercially from sugar cane and sugar beets. It is a disaccharide composed of one D-glucose and D-fructose unit; these are joined by an acetal linkage *between the two anomeric carbons.* The glucose unit is in the pyranose form and the fructose is in the furanose form. The structure of sucrose is given in Figure 28.3. Since both anomeric carbons are bound in the acetal form, sucrose is a *nonreducing sugar.*

Acidic hydrolysis of sucrose yields an equimolar mixture of D-glucose and D-fructose. Sucrose itself is dextrorotatory, having an optical rotation $[\alpha]_D = +66°$. D-Glucose is also dextrorotatory (the equilibrium mixture of α- and β-anomers has $[\alpha]_D = +52.5°$), but D-fructose is strongly levorotatory (the equilibrium mixture of fructose isomers has $[\alpha]_D = -92.4°$). In the early days of carbohydrate chemistry D-glucose was known as "dextrose" and D-fructose was called "levulose," terms that were derived from the signs of rotation of the two monosaccharides.

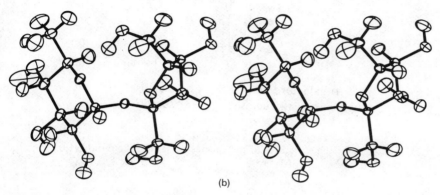

(a)

(b)

FIGURE 28.3. *Sucrose, α-D-glucopyranosyl-βD-fructofuranoside or β-D-fructo-furanosyl-α-D-glucopyranoside:* (a) *conventional structure;* (b) *stereo structure. The stereo structure also illustrates the way in which modern x-ray structure determinations are presented. The noncircular shapes of atoms, called "thermal ellipsoids," represent thermal motions.* [*Part* (b) *reproduced with permission from G. M. Brown and H. A. Levy,* Acta Cryst., **B29:**790 *(1973).*]

In the process of hydrolysis the dextrorotatory sucrose solution becomes levorotatory because an equimolar mixture of D-glucose and D-fructose has $[\alpha]_D = -20°$. This commonly encountered mixture is called "invert sugar" from the inversion in the sign of rotation that occurs during its formation. A number of organisms, including honeybees, have enzymes that catalyze the hydrolysis of sucrose. These enzymes are usually called **invertases** and are specific for the β-D-fructofuranoside linkage. Honey is largely a mixture of D-glucose, D-fructose, and sucrose. It has been shown that the equilibrium mixture resulting from hydrolysis of sucrose contains 32% β-D-glucopyranose, 18% α-D-glucopyranose, 34% β-D-fructopyranose, and 16% β-D-fructofuranose. Note that although glucose exists only in pyranose forms, fructose exists as a mixture of pyranose and furanose forms.

Raffinose is an example of a trisaccharide. It is a very minor constituent in sugar beets (0.01–0.02%) and is obtained as a by-product in the isolation of sucrose from this source. Raffinose is nonreducing and on hydrolysis yields one equivalent each of D-galactose, D-glucose, and D-fructose. If the hydrolysis is catalyzed by the enzyme α-galactosidase, the products are galactose and sucrose.

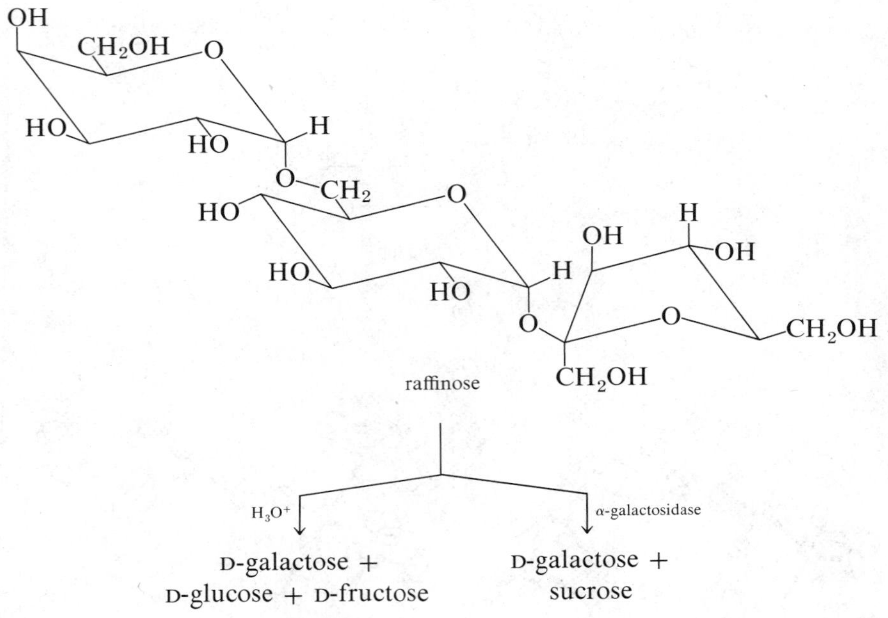

raffinose

D-galactose +
D-glucose + D-fructose

H_3O^+

α-galactosidase

D-galactose +
sucrose

EXERCISE: Write the structures of the species present in the equilibrium mixture that results from hydrolysis of sucrose.

28.8
Polysaccharides

Polysaccharides differ from the oligosaccharides only in the number of monosaccharide units that make up the molecule. The majority of the natural polysaccharides contain from 80 to 100 units, but some materials have much larger molecular weights; cellulose, for example, has an average of about 3000 glucose units per molecule. Polysaccharides may have a linear structure in which the individual monosaccharides are joined one to the other by glycosidic bonds, or they may be branched. A branched polysaccharide has a linear backbone, but additional OH groups on some of the monosaccharide units are involved in glycosidic bonding to another chain of sugars. A few cyclic polysaccharides are also known. The three types are illustrated schematically in Figure 28.4.

Cellulose is probably the single most abundant organic compound on the earth. It is the chief structural component of plant cells. For example, it comprises 10–20% of the dry weight of leaves, about 50% of the weight of tree wood and bark, and about 90% of the weight of cotton fibers, from which pure cellulose is most easily obtained. Filter paper is a source of almost pure cellulose in the laboratory.

Structurally, cellulose is a polymer of D-glucose in which the individual units are linked by β-glucoside links from the anomeric carbon of one unit to the C-4 hydroxy of the next unit. It may be hydrolyzed by 40% aqueous hydrochloric acid to give D-glucose in 95% yield. Partial hydrolysis, which may be brought about by enzymatic methods, yields the disaccharide cellobiose (Section 28.7). It is a linear polysaccharide, the isolated form containing an average of 3000 units per chain, corresponding to an average molecular weight of about 500,000. Some degradation occurs during the isolation; the actual "native cellulose" as it exists in plants

linear polysaccharide

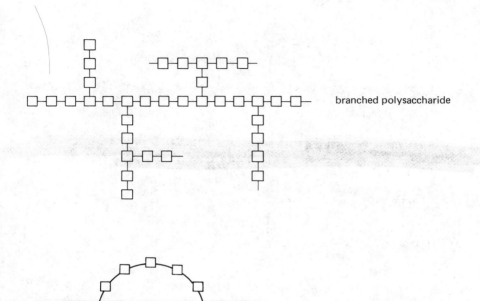

branched polysaccharide

cyclic polysaccharide

FIGURE 28.4. *Types of polysaccharides.*

may contain as many as 10,000–15,000 glucose units per chain, corresponding to a molecular weight of 1.6–2.4 million. The strength of wood derives principally from hydrogen bonds of one chain with hydroxy groups of neighboring chains.

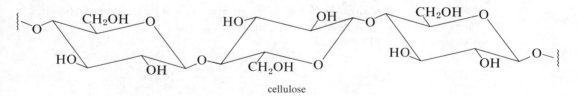

cellulose

Although the higher animals do not have enzymes that can catalyze the degradation of cellulose to glucose, such enzymes **(cellulases)** are common in microorganisms. Cellulases produced by the microflora that reside in the digestive tracts of herbivorous animals permit these animals to utilize cellulose as a food source.

Various chemically modified forms of cellulose have long been used in commercial applications. Cellulose may be nitrated by a mixture of HNO_3 and H_2SO_4. The product is a partially degraded cellulose in which some of the free OH groups have been converted into nitrate esters. The average number of nitrate ester groups per glucose unit is variable and depends on the composition of the nitrating mixture and the reaction time. Highly nitrated cellulose, in which 2.5–2.7 OH groups per glucose unit are nitrated, has explosive properties and has been used in the manufacture of blasting powder. Nitrated cellulose possessing a lower nitrogen content (2.1–2.5 ONO_2 groups per glucose unit) is used in the preparation of plastics (celluloid) and lacquers.

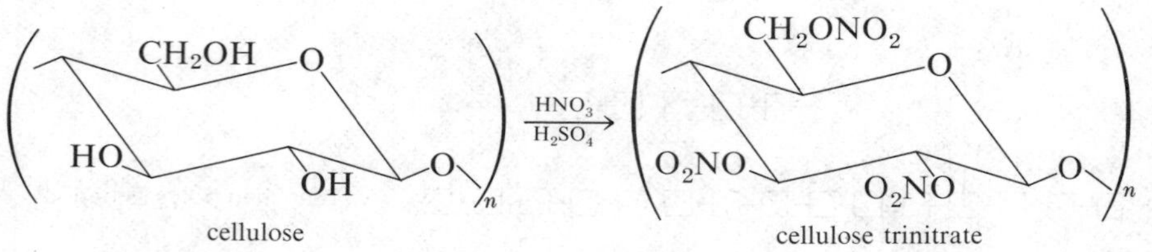

cellulose → cellulose trinitrate

$\xrightarrow[H_2SO_4]{HNO_3}$

Cellulose acetate, prepared by acetylation of cellulose with acetic anhydride and sulfuric acid, is used in the manufacture of rayon.

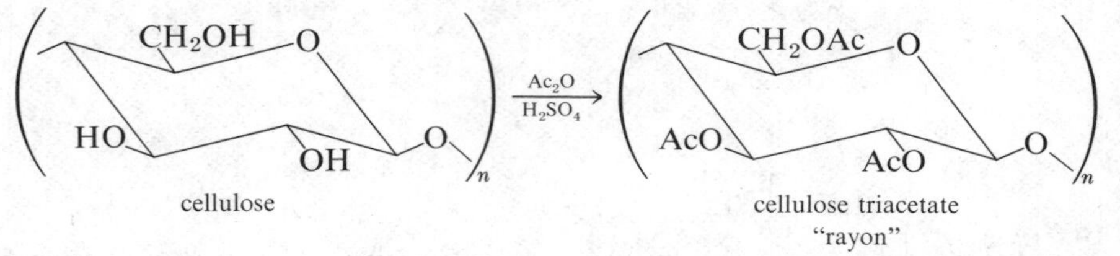

cellulose → cellulose triacetate "rayon"

$\xrightarrow[H_2SO_4]{Ac_2O}$

Starch is the second most abundant polysaccharide and occurs in both the vegetable and animal kingdoms. It is the chief source of carbohydrate for humans and is therefore of considerable economic importance. The polysaccharide is deposited in the plant in the form of small insoluble particles called starch granules. Like cellulose, the term starch is a general one; there is a considerable variety in the nature of the starch molecules produced by a given plant. Natural starch may be separated into two gross fractions called **amylose** and **amylopectin.**

Like cellulose, starch yields only D-glucose on hydrolysis. Although amylose appears to be essentially unbranched, amylopectin has a highly branched structure. Both types of starch have very high molecular weights, corresponding to many thousands of glucose units per molecule. The main chain consists of D-glucose units bound through the C-4 OH group as in cellulose but with the glucoside bond having the α-configuration. In the branched form, the branches appear to be to the C-6 OH group.

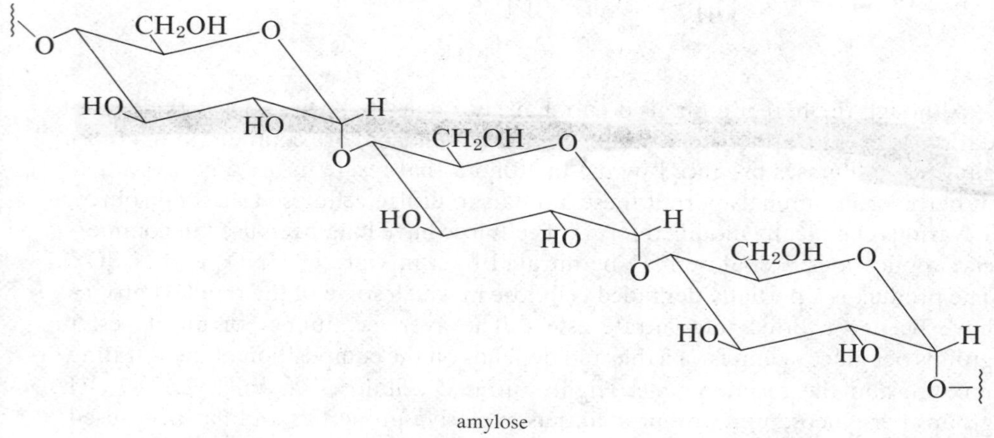

amylose

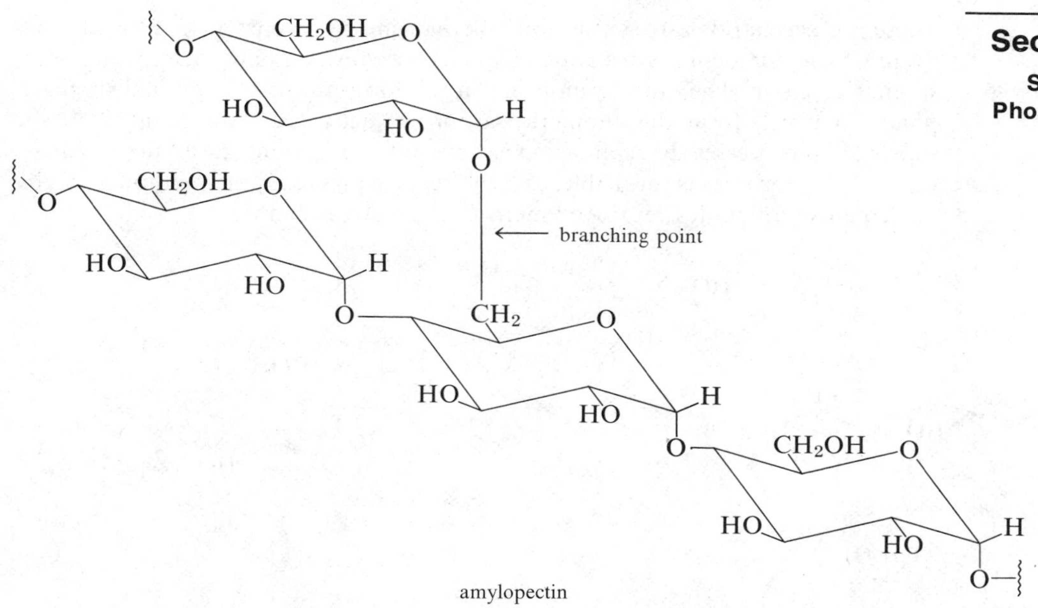

amylopectin

Partial hydrolysis of starch yields the disaccharide maltose (Section 28.7).

Glycogen is a polysaccharide that is structurally similar to starch. It is the form in which animals store glucose for further use. It is found in most tissues, but the best source is liver or muscle. Glycogen has a structure similar to that of amylopectin but is more highly branched.

28.9
Sugar Phosphates

The sugar phosphates are a class of carbohydrates that is particularly important in living systems. The chemistry of phosphate esters was discussed in Section 26.7. Sugar phosphates are intermediates in many metabolic processes, such as the degradation of glycogen to lactic acid in muscle (glycolysis), the fermentation of sugars to alcohol, and the biosynthesis of carbohydrates in plants by the process of photosynthesis. They are also constituents of ribonucleic and deoxyribonucleic acids (RNA and DNA), which are of utmost importance in the transfer of genetic information.

Typical sugar phosphates, which are known to be involved in the biosynthesis and biodegradation of the polysaccharides glycogen and starch, are α-D-glucopyranosyl phosphate and D-glucose 6-phosphate.

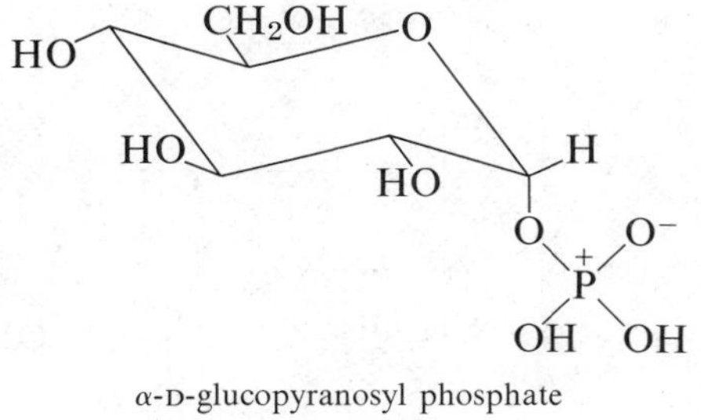

α-D-glucopyranosyl phosphate

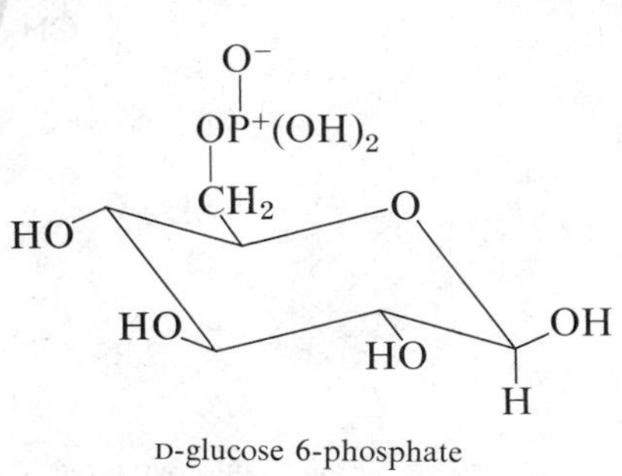

D-glucose 6-phosphate

These polysaccharides are synthesized in organisms by an enzyme-catalyzed process in which glucose units are added in a stepwise fashion onto the growing polysaccharide chain. The glucose units are in the form of α-D-glucopyranosyl phosphate, in which form the anomeric carbon is "activated" toward nucleophilic substitution processes (phosphate ion is a much better leaving group than hydroxide ion). The process is reversible, and the reverse process is the method whereby the organism degrades, or depolymerizes, the polysaccharide.

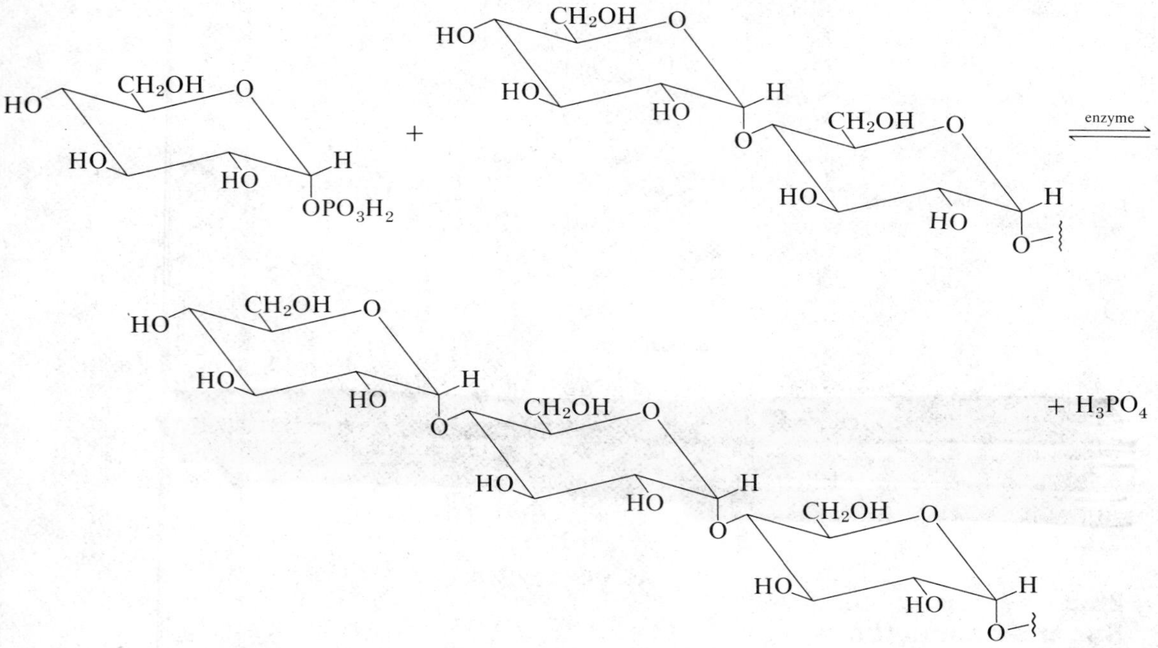

Similar enzymes catalyze the formation and cleavage of the $1 \rightarrow 6$ glycosidic bond by way of D-glucose 6-phosphate.

α-D-Glucopyranosyl phosphate is also involved in the formation and degradation of sucrose.

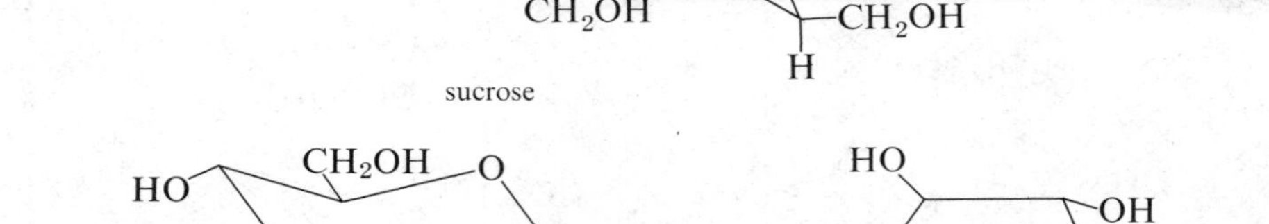

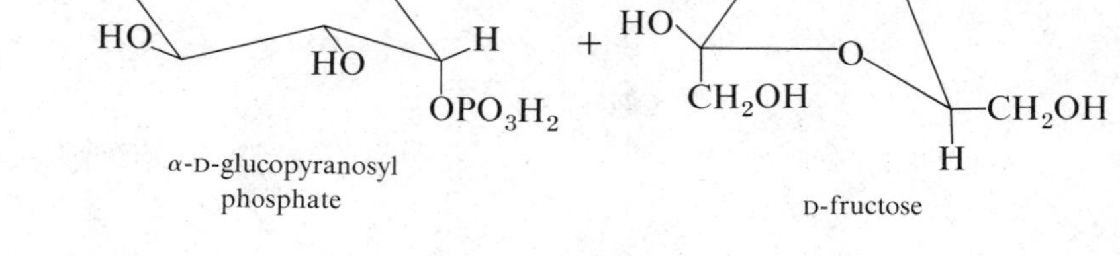

sucrose

α-D-glucopyranosyl phosphate

D-fructose

28.10
Natural Glycosides

Sugars are often found to occur in organisms in the form of glycosides. Hydrolysis of a glycoside yields the sugar (the **glycosyl residue**) and the alkyl or aryl group to which it is bound (the **aglycon**). There are many types of glycosides, and we shall only give a few examples here.

Amygdalin was one of the first glycosides to be discovered. It occurs in bitter almonds and is a glycoside formed from the disaccharide gentiobiose and the cyanohydrin of benzaldehyde. Almonds contain an enzyme that catalyzes the conversion of amygdalin to HCN, benzaldehyde, and two molecules of D-glucose.

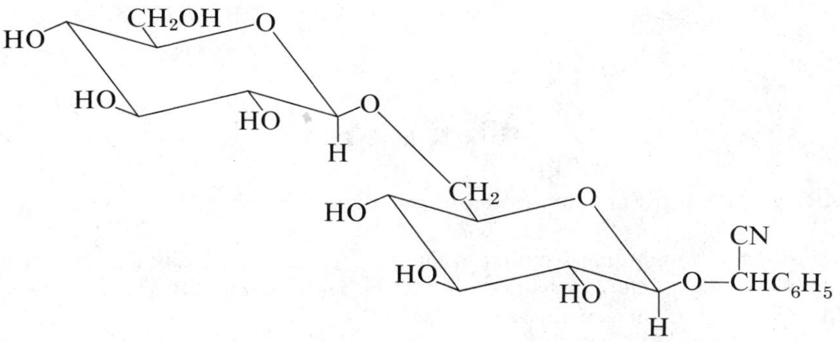

amygdalin

Amygdalin is the chief constituent of laetrile, which is reputed to be effective in the treatment of cancer. Considerable controversy has surrounded this substance, which has been banned by the United States Food and Drug Administration on the grounds that it is ineffective. In spite of this prohibition, a lively black-market in the drug has persisted, and a number of State legislatures legalized its use in 1978. In 1979 the FDA announced that it was reopening investigations of the possible efficacy of the substance.

Another natural glycoside is peonin, which is responsible for the color of the dark red peony (Section 34.3).

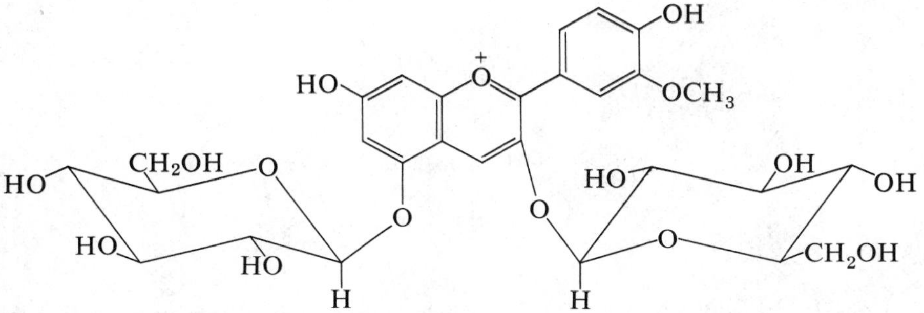

peonin

A number of naturally occurring antibiotics contain sugars bound as glycosides. The glycosyl groups often have unusual structures. An example is erythromycin A, a widely used antibiotic. The aglycon is a fourteen-membered lactone containing four hydroxy groups, two of which are bound to the rare monosaccharides cladinose and desosamine.

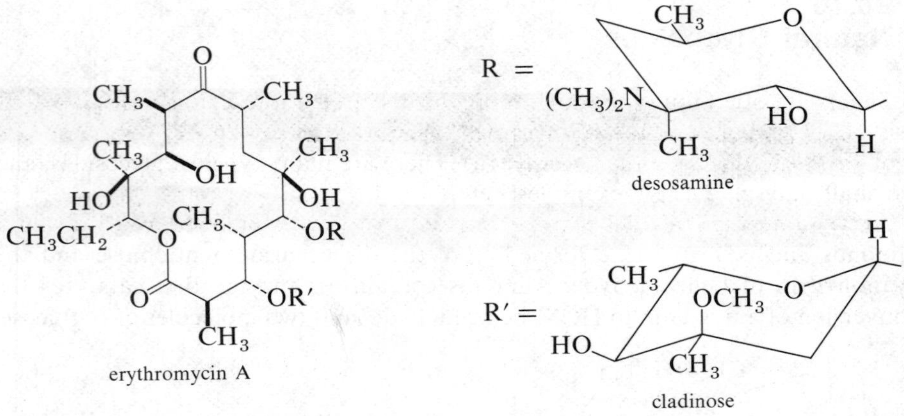

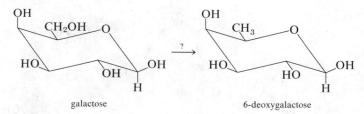

erythromycin A

R = desosamine

R' = cladinose

PROBLEMS

1. Assign (R) and (S) notations to all of the aldoses in Table 28.1.

2. Construct a "family tree," similar to that in Table 28.1, that contains the structures for all of the D-ketoses having six or fewer carbons. Identify which aldoses and ketoses will give the same osazones.

3. Using Table 28.1, identify all of the aldoses that give *meso* saccharic acids on oxidation by nitric acid.

4. 5-Hydroxyheptanal exists in two cyclic hemiacetal forms. Write three-dimensional structures for the two compounds. Which is more stable? Write a mechanism for interconversion of the two forms under conditions of acid catalysis and base catalysis.

5. (a) Draw three-dimensional projection structures for the two conformations of β-D-xylopyranose. Predict which conformation predominates in solution.
   (b) Answer part (a) for α-D-arabinopyranose.

6. Suggest a method for the synthesis of 6-deoxygalactose from galactose. (*Hint:* See page 905.)

galactose   ⟶?   6-deoxygalactose

7. Consider the addition of HCN to 2-methylcyclopentanone. Write structures for the two diastereomeric cyanohydrins that can result. Construct molecular models for the two compounds. Assuming that addition of cyanide ion to the protonated ketone is the rate-limiting step, which diastereomeric cyanohydrin should form faster?

8. (a) Write equations that show the application of the Kiliani–Fischer synthesis to each of the D-aldotetroses. Which aldopentoses are obtained from D-threose and which from D-erythrose?
   (b) Answer part (a) for the aldopentoses.

9. Write equations that show the application of the Ruff degradation to each of the D-aldohexoses. Which aldopentoses are obtained from each aldohexose?

10. Under the proper conditions D-glucose reacts with benzaldehyde to give 2,4-O-benzylidine-D-glucose.

2,4-O-benzylidine-D-glucose

This compound is reduced to 2,4-O-benzylidine-D-glucitol, which reacts with periodic acid to give the benzylidine derivative of an aldopentose. Hydrolysis of the latter compound gives the aldopentose. What are the structure and name of the aldopentose?

11. Complete acid- or base-catalyzed hydrolysis of one class of nucleic acids yields a D-aldopentose, A, phosphoric acid, and several purine and pyrimidine bases. Nitric acid oxidation of A yields a *meso* diacid, B. Treatment of A with hydroxylamine forms the oxime, C, which upon treatment with acetic anhydride is converted into an acetylated cyanohydrin, D. Hydrolysis of compound D gives an aldotetrose, E, which is oxidized by nitric acid to a *meso* diacid, F. What are the structures of compounds A through F?

12. A disaccharide, G, $C_{11}H_{20}O_{10}$, is hydrolyzable by α-glucosidase, yielding D-glucose and a D-pentose. The disaccharide does not reduce Fehling's solution. Methylation of G with dimethyl sulfate in NaOH yields a heptamethyl ether, H, which upon acid hydrolysis yields 2,3,4,6-tetra-O-methyl-D-glucose and a pentose tri-O-methyl ether, I. Oxidation of I by bromine water yields 2,3,4-tri-O-methyl-D-ribonic acid. Assign structures to compounds G through I compatible with these data.

13. A naturally occurring compound, J, has the formula $C_7H_{14}O_6$. It is nonreducing and does not mutarotate. Compound J is hydrolyzed by aqueous HCl to compound K, $C_6H_{12}O_6$, a reducing sugar. Oxidation of K with dilute HNO_3 gives an optically inactive diacid, L ($C_6H_{10}O_8$). Ruff degradation of K gives a new reducing sugar, M ($C_5H_{10}O_5$), which is oxidized by dilute HNO_3 to an optically active diacid, N ($C_5H_8O_7$). Compound J is treated successively with NaOH and dimethyl sulfate, aqueous HCl, and hot nitric acid. From the product mixture, one may isolate α,β-dimethoxysuccinic acid and α-methoxymalonic acid.

α,β-dimethoxysuccinic acid α-methoxymalonic acid

(a) Give structures for compounds J through N.
(b) What structural ambiguity exists, if any?

14. An aldopentose, O, is oxidized to a diacid, P, which is optically active. Compound O is also degraded to an aldotetrose, Q, which undergoes oxidation to an optically inactive diacid, R. Assuming that O has the D-configuration (4R), what are the structures of O through R?

15. Aldohexose S is reduced by sodium borohydride (NaBH$_4$) to an optically inactive alditol, T. Ruff degradation of S gives an aldopentose, U, which is oxidized by nitric acid to an optically active saccharic acid, V. What are compounds S through V, assuming them to be D-sugars?

16. Oxidation of aldohexose W by nitric acid gives an optically active saccharic acid, X. Ruff degradation of W gives an aldopentose, Y, which yields an optically inactive diacid, Z, on nitric acid oxidation. When compound W is subjected to a series of reactions that exchange C-1 and C-6, the same aldohexose is obtained. Assuming them to be D-sugars, what are compounds W through Z?

17. The optical rotations for the α- and β-anomers of D-mannose are $[\alpha]_D = +29.3°$ and $[\alpha]_D = -17.0°$, respectively. In water solution each form mutarotates to an equilibrium value of $[\alpha]_D = +14.2°$. Calculate the percentage of each anomer present at equilibrium.

18. The disaccharide melibiose is hydrolyzed by dilute acid to a mixture of D-glucose and D-galactose. Melibiose is a reducing sugar and is oxidized by bromine water to melibionic acid, which is methylated by sodium hydroxide and dimethyl sulfate to octa-O-methylmelibionic acid. Hydrolysis of the latter gives a tetra-O-methylgluconic acid, AA, and a tetra-O-methylgalactose, BB. Compound AA is oxidized by nitric acid to tetra-O-methylglucaric acid. Compound BB is also obtained by the acidic hydrolysis of methyl 2,3,4,6-tetra-O-methylgalactopyranoside. Melibiose is hydrolyzed by an α-galactosidase from almonds. What is the structure of melibiose?

19. The trisaccharide gentianose is hydrolyzed by acid to two equivalents of D-glucose and one of D-fructose. Partial acid hydrolysis yields D-fructose and gentibiose (page 931). The enzymes of almond emulsion cleave gentianose into D-glucose and sucrose. What is the structure of gentianose?

20. Write Haworth projections for the following saccharides.
(a) α-D-galactopyranose
(b) methyl β-D-mannoside
(c) α-maltose
(d) β-cellobiose

21. 1,2:5,6-Di-O-isopropylidine-α-D-glucofuranose reacts with aqueous acetic acid to give 1,2-O-isopropylidine-α-D-glucofuranose. That is, the 5,6-ketal is hydrolyzed selectively. This selective hydrolysis of the 5,6-ketal in compounds of this series is normal. Making use of this general reaction, along with other reactions discussed in this chapter, outline methods for the conversion of D-glucose into the following products.
(a) D-(+)-xylose
(b) *meso*-2,4-dihydroxyglutaric acid
(c) L-(+)-ribose

22. It is not possible to convert glyceraldehyde into its ketal with acetone by the direct reaction of the two compounds in the presence of acid. However, D-mannitol (page 908) reacts smoothly with acetone in the presence of zinc chloride to form a bis-ketal, 1,2:5,6-di-O-isopropylidine-D-mannitol. How might this observation be used to achieve the indirect synthesis of 2,3-O-isopropylidine-D-glyceraldehyde?

Chapter 29
Amino Acids, Peptides, and Proteins

29.1
Introduction

Amino acids constitute a particularly important class of difunctional compounds because they are the building blocks from which proteins are constructed. Since the two functional groups in an amino acid are, respectively, basic and acidic, the compounds are **amphoteric** and actually exist as **zwitterions** or **inner salts.** For example, glycine, the simplest amino acid, exists mostly in the form shown, rather than as aminoacetic acid.

$$\overset{+}{H_3}NCH_2CO_2^- \;\rightleftharpoons\; H_2NCH_2COOH$$

glycine	glycine
zwitterion form	amino acid form

Amino acids owe their important place in nature to the fact that they may form amide bonds between two molecules. Such a linkage is also called a **peptide bond,** and the resulting compounds are called **peptides.** For example, the peptide formed from two molecules of glycine is glycylglycine, a **dipeptide.** Like glycine, it is amphoteric and exists as a zwitterion.

$$\overset{+}{H_3}NCH_2\overset{\displaystyle O}{\overset{\|}{C}}-NHCH_2CO_2^-$$

peptide bond

glycylglycine
a dipeptide

Higher peptides are also possible; a **tripeptide** contains three amino acid building blocks, a **tetrapeptide** four, and so on.

$$\overset{+}{H_3}NCH_2\overset{\displaystyle O}{\overset{\|}{C}}NHCH_2\overset{\displaystyle O}{\overset{\|}{C}}NHCH_2CO_2^-$$

glycylglycylglycine
a tripeptide

As more units are added to the chain, a polymer of any length may be achieved. Such polymers are called, as a class, **polypeptides. Proteins** are special types of polypeptides composed primarily of about 20 different specific amino acids. They are large molecules, with molecular weights from 6000 to more than 1,000,000 (from about 50 to more than 8000 amino acids per molecule).

29.2
Structure, Nomenclature, and Physical Properties of Amino Acids

Most of the important natural amino acids are α-amino acids; that is, the amino group occurs at the position adjacent to the carboxy function.

$$\overset{+}{N}H_3$$
$$CH_3\overset{|}{C}HCO_2^-$$

alanine
an α-amino acid

$$H_3\overset{+}{N}CH_2CH_2CO_2^-$$

β-alanine
a β-amino acid

The important natural amino acids are listed in Table 29.1, along with the three-letter code that is conventionally used as an abbreviation for the name of each. The structures are all written in the amino acid form, rather than as zwitterions, since alternative zwitterionic structures are possible for some.

The inner-salt nature of the amino acids results in physical properties that are somewhat different from the properties normally found in organic compounds. Zwitterions are highly polar substances for which intermolecular electrostatic attractions lead to rather strong crystal lattice structures. Consequently, melting

TABLE 29.1
Common Amino Acids

$R-\overset{NH_2}{\underset{|}{C}}HCOOH$	Name	Abbreviation
$H-\overset{NH_2}{\underset{|}{C}}HCOOH$	glycine	Gly
$CH_3-\overset{NH_2}{\underset{|}{C}}HCOOH$	alanine	Ala
$CH_3\overset{CH_3}{\underset{|}{C}}H-\overset{NH_2}{\underset{|}{C}}HCOOH$	valine	Val
$CH_3\overset{CH_3}{\underset{|}{C}}HCH_2-\overset{NH_2}{\underset{|}{C}}HCOOH$	leucine	Leu
$CH_3CH_2\overset{CH_3}{\underset{|}{C}}H-\overset{NH_2}{\underset{|}{C}}HCOOH$	isoleucine	Ile
$CH_3SCH_2CH_2-\overset{NH_2}{\underset{|}{C}}HCOOH$	methionine	Met
proline structure	proline	Pro
phenylalanine structure $-CH_2-\overset{NH_2}{\underset{|}{C}}HCOOH$	phenylalanine	Phe

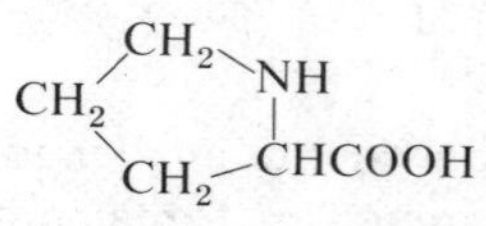

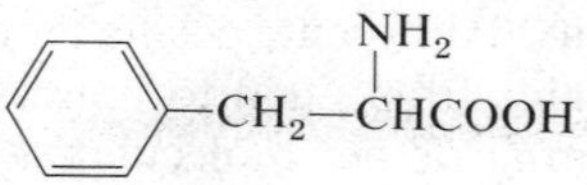

937

Sec. 29.2

Structure,
Nomenclature,
and Physical
Properties of
Amino Acids

TABLE 29.1 (continued)

$R-\underset{\underset{NH_2}{|}}{CH}COOH$	Name	Abbreviation

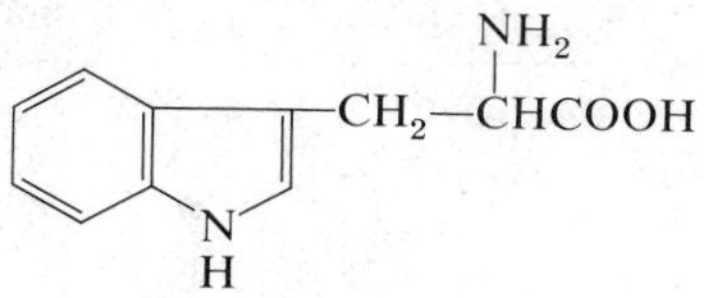

Structure	Name	Abbreviation
indole-$CH_2-\underset{\underset{NH_2}{|}}{CH}COOH$	tryptophan	Trp
$HOCH_2-\underset{\underset{NH_2}{|}}{CH}COOH$	serine	Ser
$CH_3\underset{\underset{OH}{|}}{CH}-\underset{\underset{NH_2}{|}}{CH}COOH$	threonine	Thr
$HSCH_2-\underset{\underset{NH_2}{|}}{CH}COOH$	cysteine	Cys
$HO-C_6H_4-CH_2-\underset{\underset{NH_2}{|}}{CH}COOH$	tyrosine	Tyr
$H_2N\overset{\overset{O}{\|}}{C}CH_2-\underset{\underset{NH_2}{|}}{CH}COOH$	asparagine	Asn
$H_2N\overset{\overset{O}{\|}}{C}CH_2CH_2-\underset{\underset{NH_2}{|}}{CH}COOH$	glutamine	Gln
$HO\overset{\overset{O}{\|}}{C}CH_2-\underset{\underset{NH_2}{|}}{CH}COOH$	aspartic acid	Asp
$HO\overset{\overset{O}{\|}}{C}CH_2CH_2-\underset{\underset{NH_2}{|}}{CH}COOH$	glutamic acid	Glu
$H_2NCH_2CH_2CH_2CH_2-\underset{\underset{NH_2}{|}}{CH}COOH$	lysine	Lys
$H_2N\overset{\overset{NH}{\|}}{C}NHCH_2CH_2CH_2-\underset{\underset{NH_2}{|}}{CH}COOH$	arginine	Arg
imidazole-$CH_2-\underset{\underset{NH_2}{|}}{CH}COOH$	histidine	His

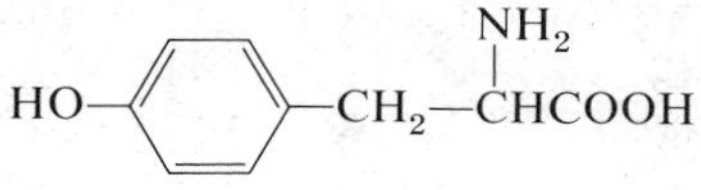

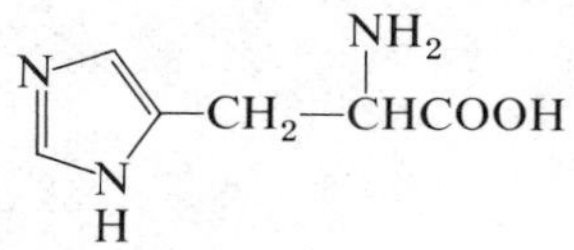

points are generally high. Most amino acids decompose instead of melting, and it is customary to record decomposition points (Table 29.2). In general, decomposition points are dependent on the rate of heating of the sample and are not reliable physical properties. Most of the amino acids are only sparingly soluble in water, again as a consequence of the strong intermolecular forces acting in the crystal

TABLE 29.2
Physical Properties of Amino Acids

Amino Acid	Decomposition Point, °C	Water Solubility, g/100 ml H_2O at 25	$[\alpha]_D^{25}$	pK_1	pK_2	pK_3
glycine	233	25		2.35	9.78	
alanine	297	16.7	+8.5	2.35	9.87	
valine	315	8.9	+13.9	2.29	9.72	
leucine	293	2.4	−10.8	2.33	9.74	
isoleucine	284	4.1	+11.3	2.32	9.76	
methionine	280	3.4	−8.2	2.17	9.27	
proline	220	162	−85.0	1.95	10.64	
phenylalanine	283	3.0	−35.1	2.58	9.24	
tryptothan	289	1.1	−31.5	2.43	9.44	
serine	228	5.0	−6.8	2.19	9.44	
threonine	225	very	−28.3	2.09	9.10	
cysteine			+6.5	1.86	8.35	10.34
tyrosine	342	0.04	−10.6	2.20	9.11	10.07
asparagine	234	3.5	−5.4	2.02	8.80	
glutamine	185	3.7	+6.1	2.17	9.13	
aspartic acid	270	0.54	+25.0	1.99	3.90	10.00
glutamic acid	247	0.86	+31.4	2.13	4.32	9.95
lysine	225	very	+14.6	2.16	9.20	10.80
arginine	244	15	+12.5	1.82	8.99	13.20
histidine	287	4.2	−39.7	1.81	6.05	9.15

lattice. Exceptions are glycine, alanine, proline, lysine, and arginine, which are all quite soluble in water.

With the exception of glycine, all of the common amino acids are chiral molecules. The naturally occurring compounds all have the same absolute configuration at the asymmetric carbon. As with carbohydrates, it is traditional to use the D- and L-nomenclature with amino acids. Natural amino acids belong to the L-series (Figure 29.1). The stereo structure of L-proline is shown in Figure 29.2. Optical rotations for the natural L-amino acids are given in Table 29.2.

EXERCISE: (a) Commit to memory the names, structures, and abbreviations of the 20 common amino acids in Table 29.1.
(b) Assign (R) and (S) stereochemical descriptors to L-alanine, L-serine, and L-cysteine (see Section 7.3).

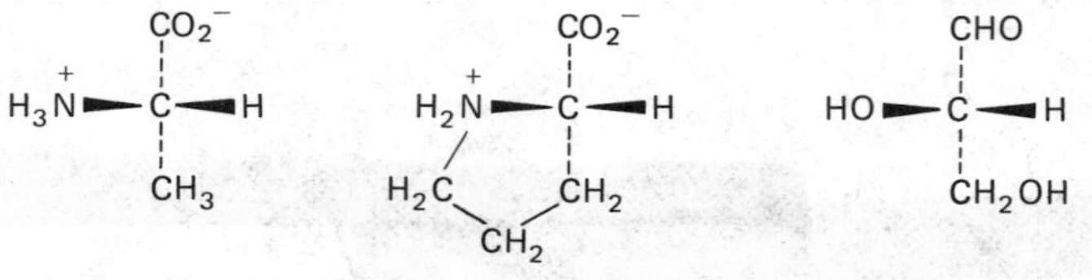

L-alanine   L-proline   L-glyceraldehyde

FIGURE 29.1. *The relationship of L-alanine and L-proline to L-glyceraldehyde.*

FIGURE 29.2. *Stereo structure of L-proline.* [*Reproduced with permission from* Molecular Structure and Dimensions. *International Union of Crystallography, 1972.*]

29.3
Acid-Base Properties of Amino Acids

Amino acids show both acidic and basic properties **(amphoterism).** In acidic solution, the amino acid is completely protonated and exists as the conjugate acid.

$$H_3\overset{+}{N}CH_2CO_2^- + H^+ \longrightarrow H_3\overset{+}{N}CH_2COOH$$

The titration curve for glycine hydrochloride is shown in Figure 29.3. The salt behaves as a typical dihydric acid.

$$H_3\overset{+}{N}CH_2COOH \underset{}{\overset{K_1}{\rightleftharpoons}} H^+ + H_3\overset{+}{N}CH_2CO_2^-$$

$$H_3\overset{+}{N}CH_2CO_2^- \underset{}{\overset{K_2}{\rightleftharpoons}} H^+ + H_2NCH_2CO_2^-$$

$$K_1 = \frac{[H^+][H_3\overset{+}{N}CH_2CO_2^-]}{[H_3\overset{+}{N}CH_2COOH]}$$

$$K_2 = \frac{[H^+][H_2NCH_2CO_2^-]}{[H_3\overset{+}{N}CH_2CO_2^-]}$$

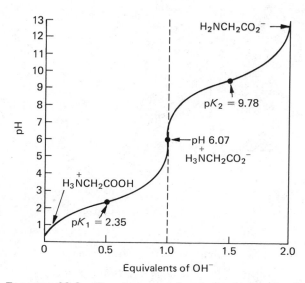

FIGURE 29.3. *Titration curve for glycine hydrochloride.*

When the hydrochloride has been half-neutralized, $[H_3\overset{+}{N}CH_2COOH] = [H_3\overset{+}{N}CH_2CO_2^-]$. The pH of the solution at this point is equal to pK_1. This first dissociation constant refers to ionization of the COOH, which is the more acidic of the two acidic groups in the dibasic acid. Note that glycine is substantially more acidic than acetic acid, which has $pK_a = 4.76$, because of the large inductive effect of the $-\overset{+}{N}H_3$ group (see Sections 10.4 and 18.4).

After one equivalent of base has been added, the chief species in solution is the zwitterionic form of the amino acid itself. The pH of the solution at this point is simply the pH of a solution of the amino acid in pure water. This point represents the pH at which the solubility of the amino acid is at a minimum, and is called the **isoelectric point.**

Addition of a further half-equivalent of base corresponds to half-neutralization of the acid $H_3\overset{+}{N}CH_2CO_2^-$. At this point $[H_3\overset{+}{N}CH_2CO_2^-] = [H_2NCH_2CO_2^-]$, and the pH of the solution is equal to pK_2, the dissociation constant for the protonated amino group. Note that pK_2 for glycine, 9.78, is slightly lower than that for the conjugate acid of methylamine, which has $pK_a = 10.4$.

$$H_3\overset{+}{N}CH_2CO_2^- \rightleftharpoons H^+ + H_2NCH_2CO_2^- \qquad pK_a = 9.8$$

$$H_3\overset{+}{N}CH_3 \rightleftharpoons H^+ + H_2NCH_3 \qquad pK_a = 10.4$$

Thus, the ammonium group of glycine is slightly more acidic than the methyl-ammonium ion.

It may seem surprising that a carboxylate anion, with its negative charge, should make an ammonium cation *more* acidic. The explanation probably has to do with solvation effects. A dipolar or zwitterionic compound with its charges sufficiently separated can have both ionic centers solvated in a normal fashion—as if the charges were on separate molecules. When the two charges are close together, however, solvation becomes less efficient. As shown in Figure 29.4, solvent dipoles (a) and (b) provide normal solvation. Solvent dipoles (c) and (d), however, provide a stabilizing Coulombic attraction to one charge but repulsion with the other. As a result, a dipolar system of this type is less stabilized by solvation and is less favored in an equilibrium.

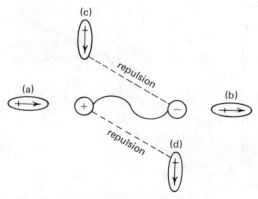

Solvation effects in a zwitterion

FIGURE 29.4. *Solvation effects in a zwitterion.*

As shown in Table 29.2, most of the amino acids show similar values of pK_1 and pK_2. Aspartic and glutamic acids each have an additional carboxy group, with pK_as of 3.90 and 4.32, respectively.

$pK_2 = 3.90$ →

$\overset{+}{N}H_3$ ← $pK_3 = 10.00$

$HOOCCH_2\overset{|}{C}HCOOH$ ← $pK_1 = 1.99$

aspartic acid

$pK_2 = 4.32$ →

$\overset{+}{N}H_3$ ← $pK_3 = 9.95$

$HOOCCH_2CH_2\overset{|}{C}HCOOH$ ← $pK_1 = 2.13$

glutamic acid

Lysine has two amino groups with pK_as of 9.20 and 10.8. The more basic group is probably the one more remote from the carboxy group. Consequently, the zwitterionic form of lysine is probably

$$H_3\overset{+}{N}CH_2CH_2CH_2CH_2\overset{\overset{\displaystyle NH_2}{|}}{C}HCO_2^-$$

lysine

Arginine contains the strongly basic guanadino group, corresponding to pK_a 13.2. It exists in the following zwitterionic form

$$H_2N\overset{\overset{\displaystyle \overset{+}{N}H_2}{\|}}{C}NHCH_2CH_2CH_2\overset{\overset{\displaystyle NH_2}{|}}{C}HCO_2^-$$

arginine

Guanidines are compounds of the general formula

$$RNH\overset{\overset{\displaystyle NH}{\|}}{C}NH_2$$

Protonation of the guanidino group on the imino nitrogen results in a cation that is highly resonance-stabilized. Guanidines are among the strongest organic bases.

$$RNH\overset{\overset{\displaystyle NH}{\|}}{C}NH_2 + H^+ \rightleftharpoons \left[RNH-\overset{\overset{\displaystyle \overset{+}{N}H_2}{\|}}{C}-NH_2 \longleftrightarrow R\overset{+}{N}H=\overset{\overset{\displaystyle NH_2}{|}}{C}-NH_2 \longleftrightarrow RNH-\overset{\overset{\displaystyle NH_2}{|}}{C}=\overset{+}{N}H_2 \right]$$

The high pK_3 for arginine shows that the guanidino group is half-protonated even at pH 13.2.

Tyrosine and histidine also contain other titratable groups, corresponding to the pK_as of 10.07 and 6.05, respectively. These ionization constants refer to the phenolic hydroxy in tyrosine and the imidazole ring in histidine. We shall discuss these two types of weak acids in later sections. The third titratable group in cysteine is the SH, which has a pK_a of 10.34, a normal value for a mercaptan.

EXERCISE: What is the principal organic species present in an aqueous solution of glycine at (a) pH 2, (b) pH 4, (c) pH 8, and (d) pH 11?

29.4
Synthesis of Amino Acids

A. Commercial Availability

All of the common amino acids are available from chemical suppliers in optically active form. Table 29.3 lists the prices per 100 g of the amino acids quoted by various suppliers in 1979. These prices reflect several factors.

All of the racemic amino acids are synthetic and are prepared commercially by methods to be outlined later in this section. The prices of the synthetic amino acids reflect both the ease of synthesis and the demand for the various compounds. Note that the most expensive racemic amino acid is the cyclic compound proline, which cannot be easily prepared by the standard methods that serve for the other amino acids.

Some of the available L-amino acids are isolated from natural sources; this is generally true when the price is lower than that for the racemate. The relatively low price of glutamic acid is a consequence of the fact that monosodium glutamate (MSG) is widely used as flavor enhancer in food preparation. The L-amino acid is prepared by a fermentation process in tonnage quantities, and its low price reflects this volume. Some of the commercially available L-amino acids and all of the D-enantiomers are prepared by resolution of the synthetic racemates. Their high costs result from the additional expenses incurred in the resolution process (see Section 29.4.F).

TABLE 29.3
Prices of Amino Acids

Amino Acid	Price per 100 g, $		
	L-enantiomer	D-enantiomer	Racemate
glycine	—	—	1.60
alanine	15.90	114.00	4.50
valine	12.50	108.00	4.50
leucine	13.00	206.00	29.10
isoleucine	70.40	—	22.00
methionine	14.00	66.00	4.00
proline	25.30	2360.00	273.00
phenylalanine	23.10	261.40	13.40
tryptophan	23.50	260.00	29.20
serine	64.00	115.00	16.50
threonine	30.80	25.00	24.68
cysteine	9.40[a]	2030.00[a]	170.00[a]
tyrosine	3.95	560.00	36.00
asparagine	8.70[b]	20.00[b]	8.70[b]
glutamine	10.50	733.00	—
aspartic acid	3.00	52.00	2.90
glutamic acid	2.50	132.50	7.70[c]
lysine	4.30[a]	772.00[a]	12.00[a]
arginine	5.25[a]	900.00[a]	94.00[a]
histidine	11.90	12.50[a]	26.00

[a] Hydrochloride salt.
[b] Hydrate.
[c] Sodium salt.

B. Amination of α-Halo Acids

α-Halo acids are available by the Hell–Volhard–Zelinsky halogenation of carboxylic acids (Section 18.7.B). Recall that the direct alkylation of ammonia or an amine is not generally a satisfactory method for preparing amines owing to the overalkylation problem (Section 24.5). The reaction is somewhat better for preparing α-amino acids because the amino group in the product amino acids is less basic (by about 0.8 pK_a unit) than the amine itself. Thus the second alkylation reaction is now slower than the first. A number of α-amino acids may be prepared in this way.

$$\underset{\overset{|}{CH_3}\overset{}{C}HCOOH}{\overset{Br}{}} + NH_4{}^+\ OH^- \xrightarrow[\substack{25° \\ 4\ days}]{H_2O} \underset{\overset{|}{CH_3}\overset{}{C}HCO_2{}^-}{\overset{\overset{+}{N}H_3}{}} + NH_4{}^+\ Br^-$$

> α-Bromopropionic acid (153 g) is added to 5.8 L of concentrated aqueous ammonia and the resulting solution is kept at room temperature for 4 days. The solution is evaporated to dryness and extracted with warm absolute ethanol to remove ammonium bromide. The amino acid is obtained as a white crystalline mass. Yield, 50 g (56%).

EXERCISE: Write equations for the syntheses of phenylalanine, valine, and leucine starting with the corresponding carboxylic acids. What special problems arise in the application of this method to the synthesis of serine or tyrosine?

C. Alkylation of N-Substituted Aminomalonic Esters

An especially useful general method for the synthesis of α-amino acids involves a variation of the malonic ester synthesis (Section 27.7.C). Diethyl malonate may be monobrominated to yield a bromide that enters into the S_N2 reaction with the potassium salt of phthalimide to give N-phthalimidomalonic ester.

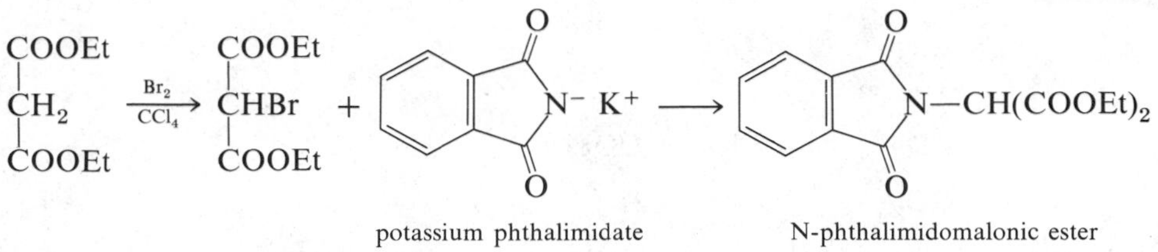

potassium phthalimidate   N-phthalimidomalonic ester

The ester may be alkylated by a variety of alkyl halides or α,β-unsaturated carbonyl compounds. Vigorous acid hydrolysis causes hydrolysis of both ester groups and the phthalimido group and decarboxylation of the resulting malonic acid. The product is a racemic α-amino acid.

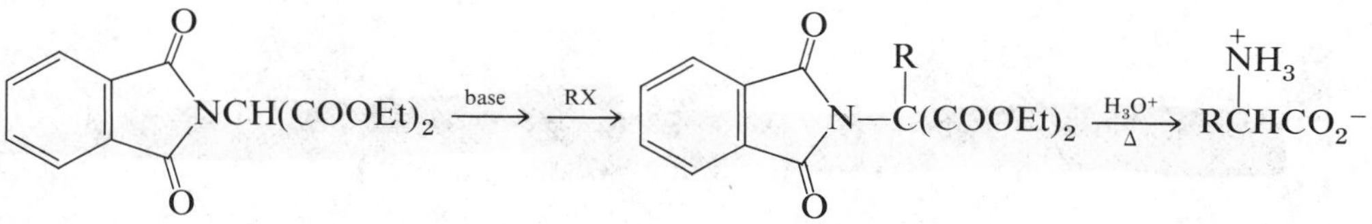

A few specific examples are given.

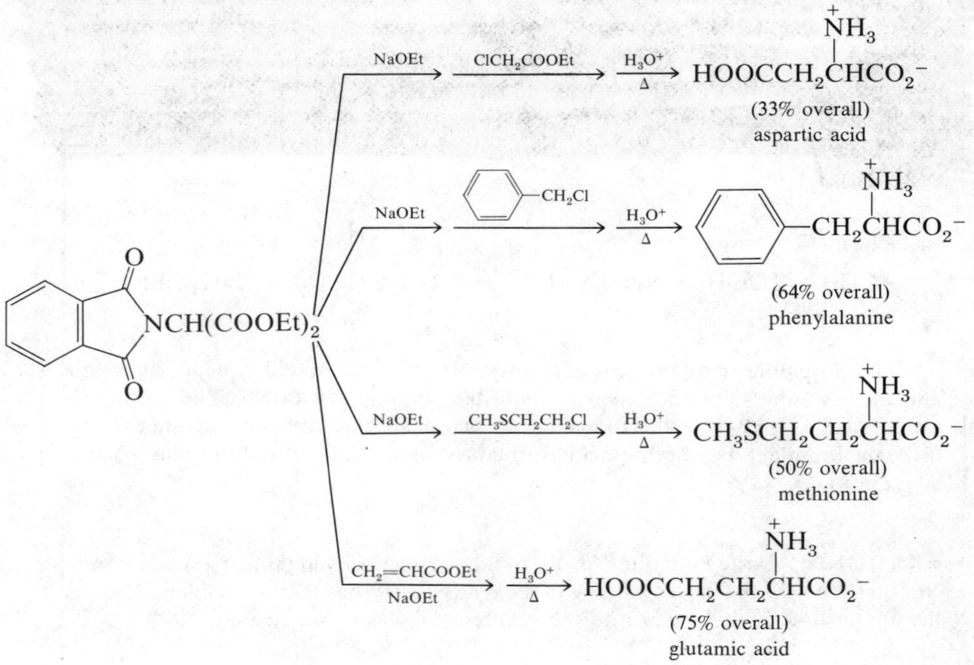

Other procedures similar to this are also useful. The best method utilizes the N-acetamido rather than the N-phthalimido derivative. The starting material is prepared from malonic ester in the following way: Treatment of the diester with nitrous acid gives a nitroso derivative, which tautomerizes to the oxime, and hydrogenation of the oxime in acetic anhydride solution gives acetamidomalonic ester.

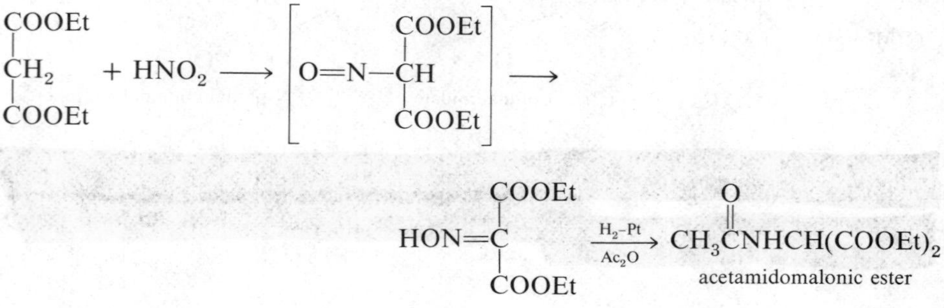

The acetamidomalonic ester is alkylated, and the resulting product is hydrolyzed and decarboxylated to obtain the amino acid.

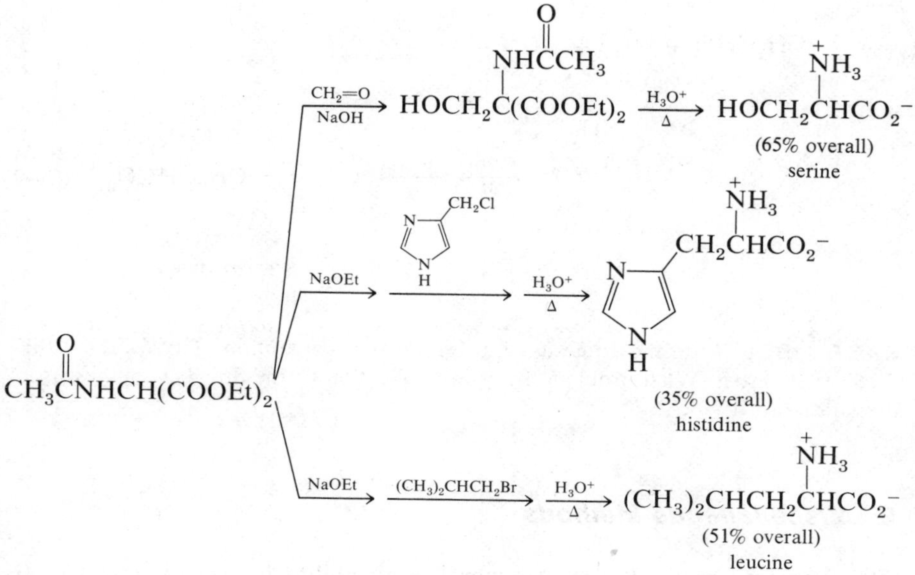

$$\xrightarrow[\text{NaOH}]{\text{CH}_2=\text{O}} \underset{\overset{|}{\text{NHCCH}_3}}{\overset{\overset{\displaystyle O}{\parallel}}{\text{HOCH}_2\text{C(COOEt)}_2}} \xrightarrow[\Delta]{\text{H}_3\text{O}^+} \underset{\overset{|}{\overset{+}{\text{NH}_3}}}{\text{HOCH}_2\text{CHCO}_2^-}$$

(65% overall)
serine

$$\xrightarrow{\text{NaOEt}} \text{[imidazole-CH}_2\text{Cl]} \xrightarrow[\Delta]{\text{H}_3\text{O}^+} \underset{\overset{|}{\overset{+}{\text{NH}_3}}}{\text{CH}_2\text{CHCO}_2^-}$$

(35% overall)
histidine

$$\text{CH}_3\overset{\overset{\displaystyle O}{\parallel}}{\text{C}}\text{NHCH(COOEt)}_2$$

$$\xrightarrow{\text{NaOEt}} \xrightarrow{(\text{CH}_3)_2\text{CHCH}_2\text{Br}} \xrightarrow[\Delta]{\text{H}_3\text{O}^+} \underset{\overset{|}{\overset{+}{\text{NH}_3}}}{(\text{CH}_3)_2\text{CHCH}_2\text{CHCO}_2^-}$$

(51% overall)
leucine

EXERCISE: Write the equations illustrating the syntheses of tyrosine, alanine, and valine using both phthalimidomalonic ester and acetamidomalonic ester.

D. Strecker Synthesis

Another method of some generality for the preparation of α-amino acids is the hydrolysis of α-amino nitriles, which are available by the treatment of aldehydes with ammonia and HCN (Strecker synthesis).

$$\text{RCHO} + \text{NH}_3 + \text{HCN} \longrightarrow \underset{\overset{|}{\text{NH}_2}}{\text{RCHCN}} \xrightarrow{\text{H}_3\text{O}^+} \underset{\overset{|}{\overset{+}{\text{NH}_3}}}{\text{RCHCO}_2^-}$$

The mechanism of formation of the α-amino nitrile probably involves the addition of HCN to the imine, which is formed by condensation of the aldehyde with ammonia.

$$\text{RCHO} + \text{NH}_3 \rightleftharpoons \text{H}_2\text{O} + \text{RCH}=\text{NH} \xrightarrow{\text{HCN}} \underset{\overset{|}{\text{NH}_2}}{\text{RCHCN}}$$

Examples of amino acids that have been prepared by the Strecker synthesis are given in (29-1) and (29-2).

$$\text{CH}_2=\text{O} + \text{NH}_4^+ \text{ CN}^- \xrightarrow{\text{H}_2\text{SO}_4} \text{H}_2\text{NCH}_2\text{CN} \xrightarrow[\Delta]{\overset{\text{BaO}}{\text{H}_2\text{O}}} \text{H}_3\overset{+}{\text{N}}\text{CH}_2\text{CO}_2^- \quad (29\text{-}1)$$

(42%)
glycine

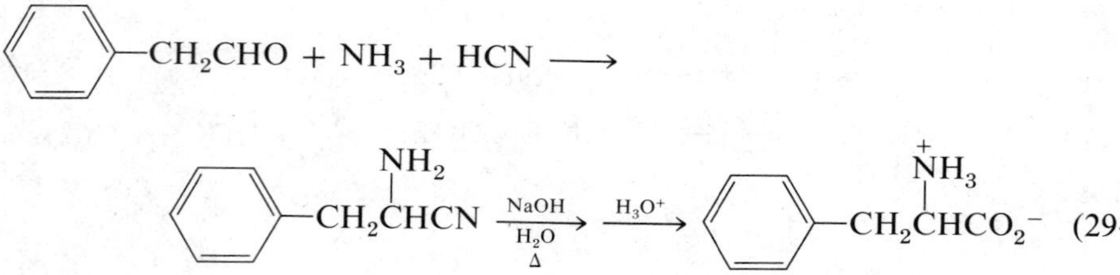

$$\text{—CH}_2\text{CHO} + \text{NH}_3 + \text{HCN} \longrightarrow$$

(29-2)

(74%)
phenylalanine

EXERCISE: Write the equations illustrating the preparation of tyrosine by the Strecker synthesis. What problem arises in application of the Strecker synthesis for preparation of lysine?

E. *Miscellaneous Methods*

The foregoing methods are of general applicability for the synthesis of the simpler amino acids, either natural or unnatural. Some of the more complicated structures must be prepared in other ways. For example, the heterocyclic amino acid proline has been synthesized by the following route.

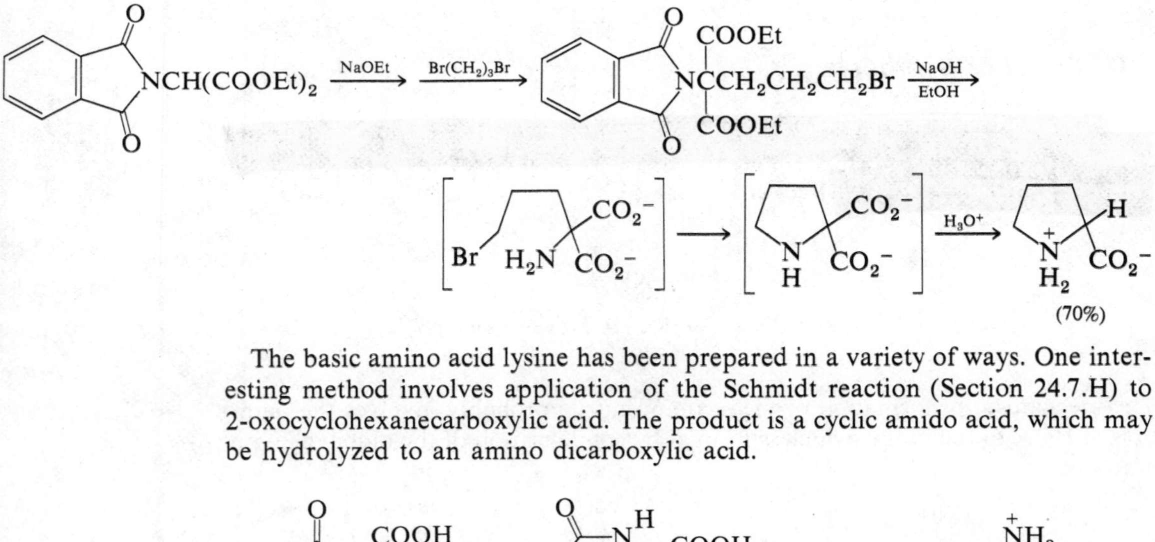

(70%)

The basic amino acid lysine has been prepared in a variety of ways. One interesting method involves application of the Schmidt reaction (Section 24.7.H) to 2-oxocyclohexanecarboxylic acid. The product is a cyclic amido acid, which may be hydrolyzed to an amino dicarboxylic acid.

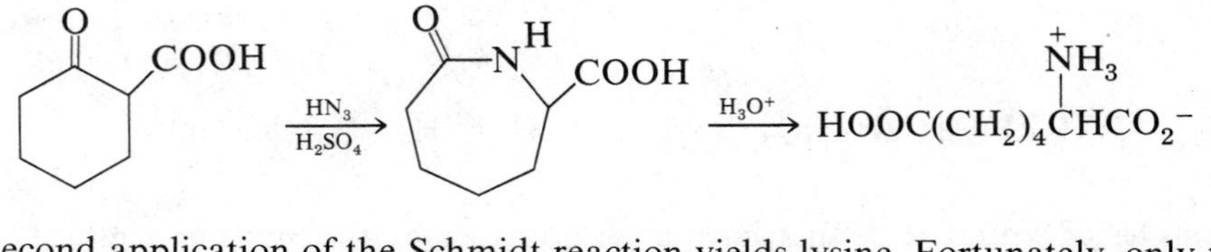

A second application of the Schmidt reaction yields lysine. Fortunately, only the carboxy group that is not α to the amino group reacts; in fact, α-amino acids fail to react at all in the Schmidt reaction.

(74%)

The Schmidt reaction, introduced in Section 24.7.H as a reaction of carboxylic acids, also may be applied to ketones. It is a general method for the conversion of ketones to amides.

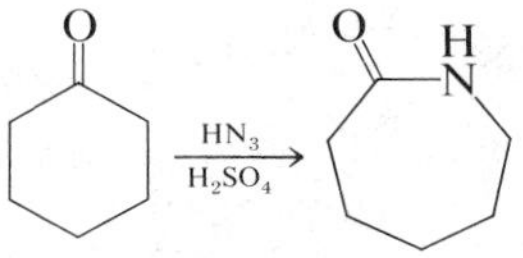

A probable mechanism for the conversion is

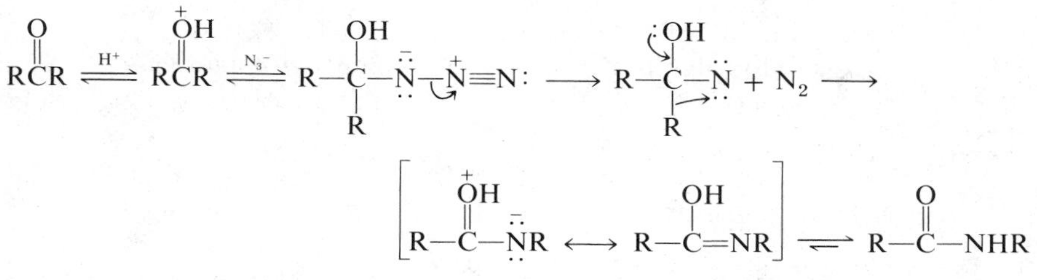

The Schmidt reaction may also be used for the synthesis of simple amino acids if it is applied to an alkylated malonic acid.

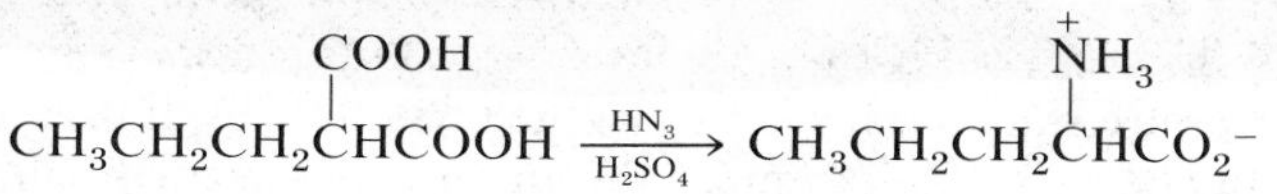

> **EXERCISE:** Show how phenylalanine, leucine, and glutamic acid could be prepared starting with malonic ester and using the Schmidt reaction in one step.

F. Resolution

Amino acids that are synthesized by the methods outlined in the preceding sections are obtained as racemates. It is usually desirable to have one of the two enantiomers, usually the L-enantiomer. For this reason a good deal of attention has been paid to the problem of **resolving** racemic amino acids.

One method that may be used for the resolution of amino acids involves converting them into diastereomeric salts (Section 24.4). The amino group is usually converted into an amide so that the material is not amphoteric. For example, alanine reacts with benzoyl chloride in aqueous base to give N-benzoylalanine, which is a typical acid.

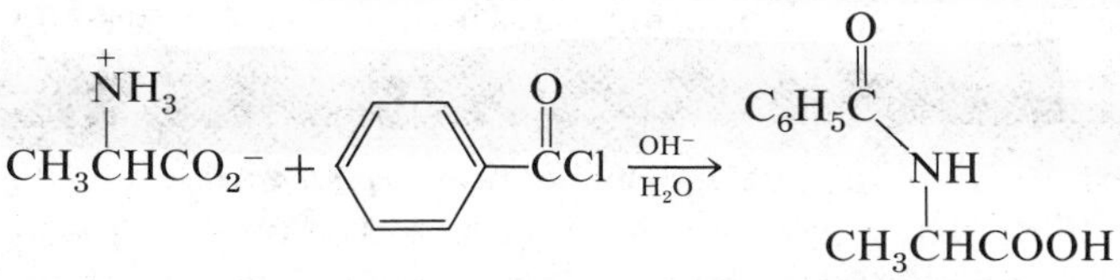

benzoyl chloride   N-benzoylalanine

The racemic N-benzoylalanine is resolved in the normal way (Section 24.4) with brucine or strychnine. If brucine is used, it is the brucine salt of D-alanine that is less soluble. If strychnine is used, the strychnine salt of L-alanine crystallizes. Acidification of the salts yields the D- and L-enantiomers of N-benzoylalanine.

Basic hydrolysis then affords the pure enantiomeric amino acids. The process is outlined schematically as follows.

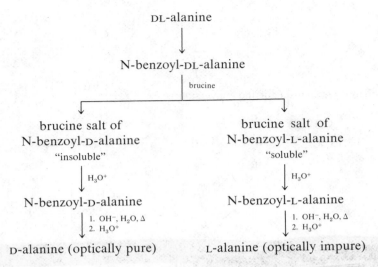

In cases such as that outlined, the enantiomer that forms the less soluble salt is usually obtained in an optically pure state. The other enantiomer is usually obtained in an impure state because some of the less soluble salt invariably remains in solution. In the case diagrammed, the impure N-benzoyl-L-alanine may be treated with strychnine to give the insoluble strychnine salt. In this way both enantiomers may be obtained in an optically pure state.

For all its simplicity, the **method of diastereomeric salts** suffers from several severe drawbacks. The less soluble diastereomeric salt is usually contaminated with the other salt, and several tedious recrystallizations may be required in order to purify it. These repetitive crystallizations are wasteful of both time and material, which may often be quite valuable. There is no way to predict which chiral base will give well-defined crystals with a given amino acid or which enantiomer of the amino acid will form the less soluble salt.

Various biological procedures are much more useful for the routine large-scale resolution of amino acids. The success of biological resolution stems from the fact that organisms are generally capable of utilizing only one enantiomer of a racemic substance. Thus, if a racemic amino acid is fed to an animal or microorganism, one enantiomer is consumed. The unreacted enantiomer may then be isolated from the culture medium in the case of microorganisms or from the urine of the animal. Since L-enantiomers are utilized by almost all organisms, this method is useful for preparing optically pure D-enantiomers.

In practice, the procedure of using the whole animal for resolution is of only limited value. A more useful adaptation of the basic principle employs the use of crude enzyme preparations that catalyze some reaction on only one enantiomer. An example is the resolution of DL-leucine by *hog renal acylase,* an enzyme isolated from hog kidneys. The enzyme functions as a catalyst for the hydrolysis of amide linkages and is specific for amides of L-amino acids. For resolution, the racemic amino acid is first converted into the N-acetyl derivative, which is then incubated with a small amount of the crude enzyme preparation. The enzyme catalyzes hydrolysis of N-acetyl-L-leucine to the amino acid, leaving N-acetyl-D-leucine unchanged. The two enantiomers are easily separable, since one is acidic and the other is amphoteric.

DL-leucine $\xrightarrow{Ac_2O}$ N-acetyl-DL-leucine $\xrightarrow[\text{acylase}]{\text{hog renal}}$ $\left\{ \begin{array}{c} \text{L-leucine} \\ + \\ \text{N-acetyl-D-leucine} \end{array} \right.$

A suspension of 17.3 g of N-acetyl-DL-leucine in 1 L of water is adjusted to pH 7.0 with NH_4OH solution, and 0.012 g of hog renal acylase powder is added. The mixture is agitated at 38° for 24 hr. The mixture is acidified with 10 mL of acetic acid, filtered, and evaporated under vacuum to a volume of about 50 mL. Upon addition of ethanol, L-leucine crystallizes. The semipure amino acid is recrystallized from ethanol–water to give 5 g (80%) of optically pure L-leucine.

The filtrates from the foregoing process are acidified to pH 2 with HCl and chilled, whereupon N-acetyl-D-leucine crystallizes. One recrystallization from water gives 7 g (80%) of optically pure product. It may be hydrolyzed by refluxing with 2 N HCl to obtain pure D-leucine.

Another biological process employs enzymes that specifically *destroy* one of the enantiomers. An example of such an enzyme is L-*amino acid oxidase,* which is obtained from rattlesnake venom. The enzyme catalyzes the oxidation of the amino acid to an α-keto acid and ammonia and is specific for L-enantiomers.

$$O_2 + \overset{\overset{+}{N}H_3}{\underset{|}{R\,CHCO_2^-}} \xrightarrow[\text{oxidase}]{\text{amino acid}} \overset{O}{\underset{\|}{R\,CCOOH}} + NH_3$$

Analogous enzymes, such as hog kidney D-amino acid oxidase, perform the same function on the other enantiomer. The chief advantages of this method are its simplicity (it is not necessary to convert the amino acid into a derivative prior to resolution) and the extremely high stereospecificity. The chief drawbacks are the relative unavailability of the necessary enzymes and the fact that one of the two enantiomers is destroyed.

29.5
Reactions of Amino Acids

A. *Esterification*

The carboxy group of an amino acid may be esterified in the normal way. Methyl, ethyl, and benzyl esters are employed extensively as intermediates in the synthesis of peptides (Section 29.6). The methyl and ethyl esters are normally prepared by treating a suspension of the amino acid in the appropriate alcohol with anhydrous hydrogen chloride. The amino acid ester is isolated as the crystalline hydrochloride salt.

$$\text{C}_6\text{H}_5\text{—CH}_2\overset{\overset{+}{N}H_3}{\underset{|}{\text{CHCO}_2^-}} \xrightarrow[\text{CH}_3\text{OH}]{\text{HCl}} \text{C}_6\text{H}_5\text{—CH}_2\overset{\overset{+}{N}H_3\ \ Cl^-}{\underset{|}{\text{CHCOOCH}_3}}$$

(90%)
phenylalanine methyl
ester hydrochloride

Benzyl esters are often prepared using benzenesulfonic acid as the catalyst. The water produced in the reaction is removed by azeotropic distillation, thus avoiding the use of a large excess of benzyl alcohol.

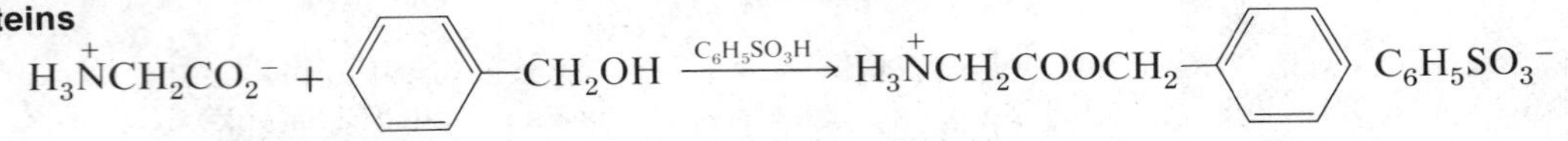

$$H_3\overset{+}{N}CH_2CO_2^- + \text{C}_6\text{H}_5\text{-CH}_2\text{OH} \xrightarrow{\text{C}_6\text{H}_5\text{SO}_3\text{H}} H_3\overset{+}{N}CH_2COOCH_2\text{-C}_6\text{H}_5 \quad \text{C}_6\text{H}_5\text{SO}_3^-$$

(90%)
glycine benzyl ester
benzenesulfonate

As we shall see later, the benzyl esters are especially useful derivatives because they may be converted back to acids by nonhydrolytic methods. For example, glycine benzyl ester reacts with hydrogen in the presence of palladium to give glycine and toluene.

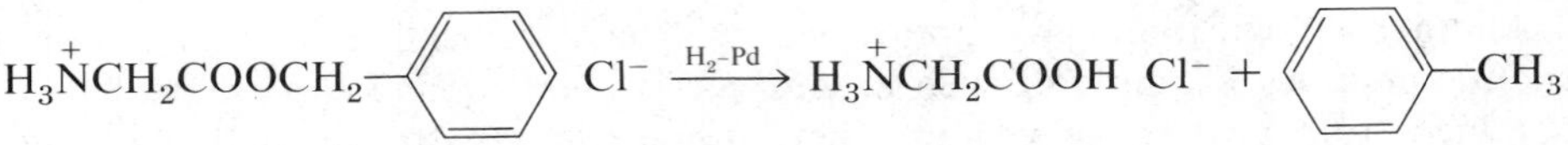

$$H_3\overset{+}{N}CH_2COOCH_2\text{-C}_6\text{H}_5 \quad Cl^- \xrightarrow{\text{H}_2\text{-Pd}} H_3\overset{+}{N}CH_2COOH \quad Cl^- + \text{C}_6\text{H}_5\text{-CH}_3$$

This is another example of hydrogenolysis, cleavage of a σ-bond by hydrogen (Section 22.6.B).

B. *Amide Formation*

Acylation of the amino group in amino acids is best carried out under basic conditions, so that a substantial concentration of the free amino form is present. An example is N-benzoylation under Schotten–Baumann conditions (Section 19.7.B).

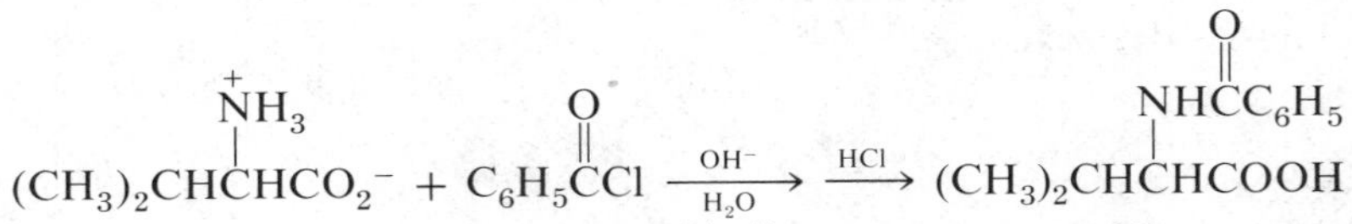

$$(CH_3)_2CHCHCO_2^- + C_6H_5\overset{O}{\overset{\|}{C}}Cl \xrightarrow[\text{H}_2\text{O}]{\text{OH}^-} \xrightarrow{\text{HCl}} (CH_3)_2CHCHCOOH$$
$$\underset{\overset{|}{\overset{+}{N}H_3}}{} \qquad \qquad \underset{2\text{ hr, }4°}{} \qquad \underset{\overset{|}{NHCC_6H_5}}{}$$

(80%)
N-benzoylvaline

Amides may also be prepared by reaction with acetic anhydride.

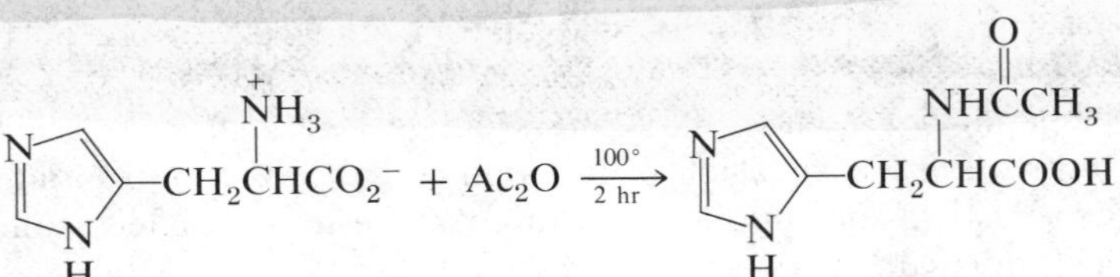

$$\text{imidazole-}CH_2\overset{+}{\underset{\overset{|}{N}H_3}{C}}HCO_2^- + Ac_2O \xrightarrow[2\text{ hr}]{100°} \text{imidazole-}CH_2\overset{NHCCH_3}{\underset{\overset{\|}{O}}{C}}HCOOH$$

(80%)
N-acetylhistidine

Sulfonamides result from reaction with arenesulfonyl halides.

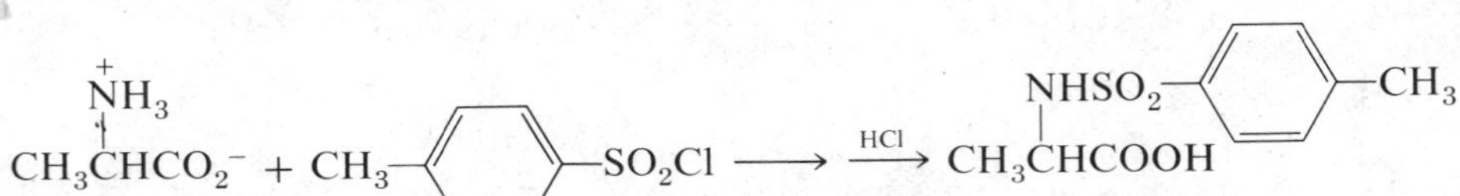

$$CH_3\overset{+}{\underset{\overset{|}{N}H_3}{C}}HCO_2^- + CH_3\text{-C}_6\text{H}_4\text{-SO}_2Cl \xrightarrow{\text{HCl}} CH_3\overset{NHSO_2\text{-C}_6\text{H}_4\text{-CH}_3}{C}HCOOH$$

(70%)

EXERCISE: Write equations illustrating the conversion of phenylalanine into N-acetylphenylalanine methyl ester.

C. *Ninhydrin Reaction*

When an aqueous solution of an α-amino acid is treated with triketohydrindene hydrate (ninhydrin), a violet color is produced.

ninhydrin

violet

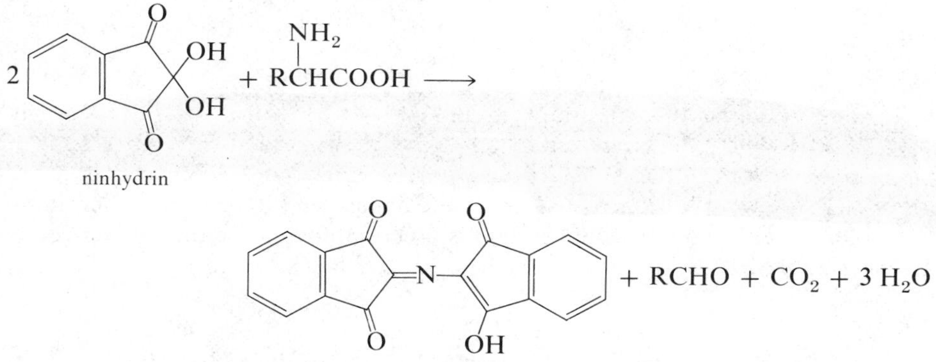

The reaction mechanism is complicated, and we shall not go into it here. Note that the amino acid contributes only a nitrogen to the violet-colored product. The manner in which this nitrogen becomes separated from the amino acid is known and is an example of a reaction of α-amino acids that is biologically important.

In the first stages of the ninhydrin reaction, the amino acid is oxidized to give an α-imino acid. In aqueous solution this imino acid is hydrolyzed to an α-keto acid and ammonia. The ammonia reacts further to give the violet pigment, and the α-keto acid decarboxylates to give an aldehyde.

$$\mathrm{R\overset{NH_2}{\overset{|}{C}}HCOOH} \xrightarrow{[O]} \mathrm{R\overset{NH}{\overset{\|}{C}}COOH} \xrightarrow{H_2O} \mathrm{NH_3} + \mathrm{R\overset{O}{\overset{\|}{C}}COOH} \longrightarrow \mathrm{RCHO} + \mathrm{CO_2} \qquad (29\text{-}3)$$

The process of **oxidative deamination,** outlined in (29-3), is an important pathway for the biodegradation of α-amino acids.

The ninhydrin reaction is important because it may be used as an analytical method for α-amino acids. The violet solutions show a significant absorption at 570 nm, and the intensity of the absorption is proportional to the amount of α-amino acid present. The reaction is not given by proline, in which the α-amino group is secondary. Another reaction occurs with proline to give a product that can be assayed at another wavelength.

29.6 Peptides

A. *Structure and Nomenclature*

Peptides, also called polypeptides, are amino acid polymers containing from 2 to about 50 individual units. The individual amino acids are connected by amide linkages from the amino group of one unit to the carboxy group of another. Unless a polypeptide is cyclic, it will contain a free $-NH_3^+$ group (the N-terminal end) and a free $-CO_2^-$ group (the C-terminal end).

a polypeptide

By convention, peptide structures are always written with the N-terminal unit on the left and the C-terminal unit on the right. They are named by prefixing the name of the C-terminal unit with the group names of the other amino acids, beginning with the N-terminal unit. Since the names tend to become rather unintelligible, a shorthand notational system is used employing the three-letter codes given in Table 29.1.

glycylalanine
Gly-Ala

glycylphenylalanylglycine
Gly-Phe-Gly

glycylserylphenylalanylglycine
Gly-Ser-Phe-Gly

The stereo structure of Gly-Phe-Gly is shown in Figure 29.5.

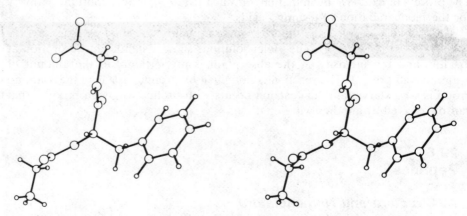

FIGURE 29.5. *Stereo structure of Gly-Phe-Gly.* [*Reproduced with permission from* Molecular Structure and Dimensions. *International Union of Crystallography, 1972.*]

Peptides are formed by partial hydrolysis of proteins, which are amino acid polymers of much higher molecular weight (more than 50 amino acid units). Upon hydrolysis of a protein, some amide linkages are broken and a complex mixture of peptides results. Complete hydrolysis gives a mixture of amino acids. Some peptides are also important natural products. An example is the nonapeptide bradykinin, which occurs in blood plasma and is involved in the regulation of blood pressure.

Arg-Pro-Pro-Gly-Phe-Ser-Pro-Phe-Arg

bradykinin

The central feature of the polypeptide chain is the succession of amide linkages. Recall from our previous study (Chapter 19) that the C—N bond in an amide has a high degree of "double bond character" resulting from delocalization of the nitrogen lone pair into the carbonyl group. This delocalization reduces the basicity of the nitrogen and causes restricted rotation about the C—N bond.

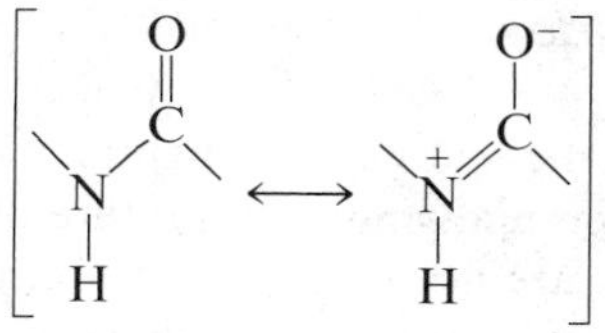

The restricted rotation has an important effect on the three-dimensional structure of proteins, as we shall see later.

The only other type of covalent bond between amino acids in proteins and peptides is the disulfide linkage between two cysteine units.

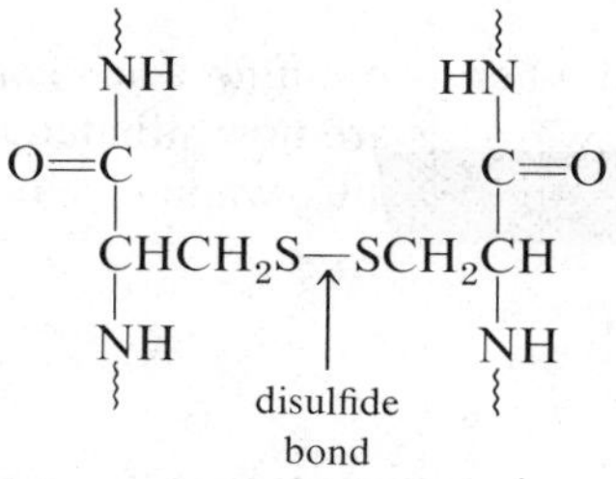

Recall that disulfides, R—S—S—R, are formed by the mild oxidation of thiols (page 810). The disulfide linkage is easily reduced to regenerate the thiols.

$$2 \text{ RSH} \underset{\text{reduction}}{\overset{\text{oxidation}}{\rightleftarrows}} \text{RSSR}$$

When such a disulfide bond occurs between two cysteine residues in the same chain, a "loop" results, as in the posterior pituitary hormone oxytocin. If the cysteine units are in different chains, the disulfide link may bind the two chains together, as in the A and B chains of insulin (Figure 29.6).

Like the simpler amino acids, peptides are amphoteric compounds, since they usually still contain a free α-amino and a free α-carboxy group; they exist as zwitterions. The pK_as for the two functions in a few simple peptides are listed in Table 29.4. Also included are the isoelectric points, pH$_1$, the pH at which the peptide is least soluble in aqueous solution.

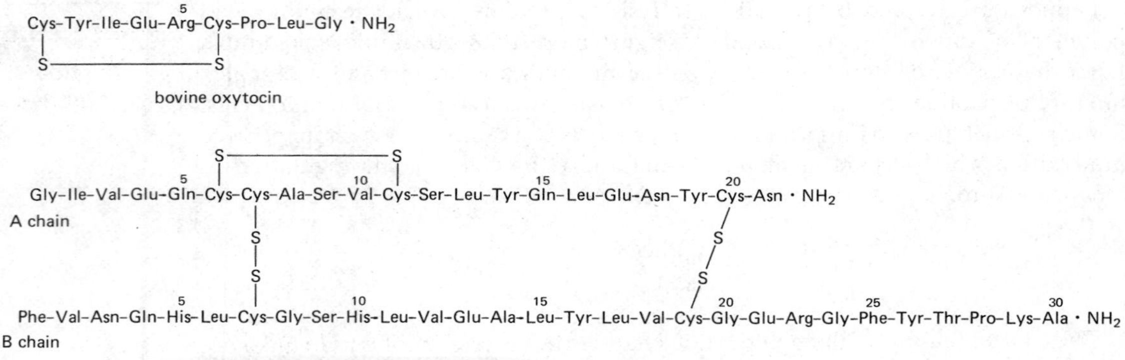

$$\underset{5}{\text{Cys-Tyr-Ile-Glu-Arg-Cys-Pro-Leu-Gly}} \cdot NH_2$$

```
Cys-Tyr-Ile-Glu-Arg-Cys-Pro-Leu-Gly · NH₂
 |                   |
 S──────────────────S
```

bovine oxytocin

```
            S────────────S
            |            |
   5        |   10       |     15            20
Gly-Ile-Val-Glu-Gln-Cys-Cys-Ala-Ser-Val-Cys-Ser-Leu-Tyr-Gln-Leu-Glu-Asn-Tyr-Cys-Asn · NH₂
A chain                  |                              |
                         S                              S
                         |                             /
                         S                            S
                         |                           /
   5                     |  10                      /  20          25              30
Phe-Val-Asn-Gln-His-Leu-Cys-Gly-Ser-His-Leu-Val-Glu-Ala-Leu-Tyr-Leu-Val-Cys-Gly-Glu-Arg-Gly-Phe-Tyr-Thr-Pro-Lys-Ala · NH₂
B chain
```

bovine insulin

FIGURE 29.6. *Amino acid sequence and disulfide bridges of bovine oxytocin and bovine insulin. The N-terminal units are at the left and the C-terminal units are at the right. All three C-terminal units occur as amides,* —$CONH_2$.

B. Synthesis of Peptides

The simplest method for the synthesis of peptides is the polymerization of an amino acid. The resulting **homopolymer** is a mixture of peptides of variable chain length. Such homopolymers are not found in nature, but the synthetic ones have been useful in understanding some of the physical and spectral properties of proteins.

$$H_3\overset{+}{N}CH_2CO_2^- \longrightarrow H_3\overset{+}{N}CH_2\overset{O}{\overset{\|}{C}}\left(NHCH_2\overset{O}{\overset{\|}{C}}\right)_n NHCH_2CO_2^-$$

polyglycine

The first product formed when two amino acids condense is a dipeptide. The terminal amino and carboxy groups are now situated so that they can interact to form a six-membered ring diamide. Thus, when glycine is heated, the cyclic dimer 2,5-diketopiperazine is produced.

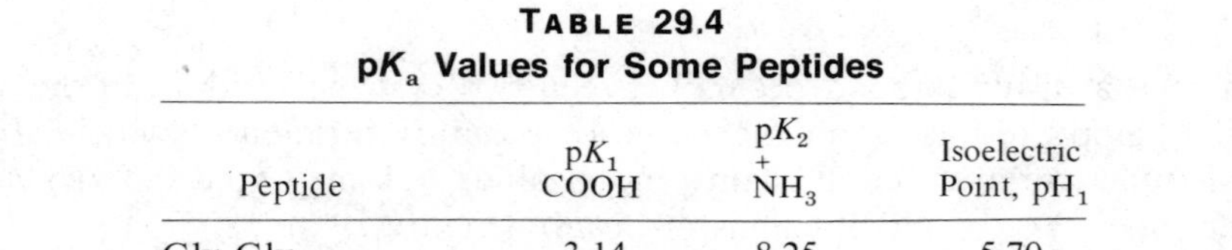

$$2\ H_3\overset{+}{N}CH_2CO_2^- \xrightarrow{\text{slow}} H_3\overset{+}{N}CH_2\overset{O}{\overset{\|}{C}}NHCH_2CO_2^- \xrightarrow{\text{fast}}$$

2,5-diketopiperazine
"glycine anhydride"

TABLE 29.4
pK_a Values for Some Peptides

Peptide	pK_1 COOH	pK_2 $\overset{+}{N}H_3$	Isoelectric Point, pH_1
Gly-Gly	3.14	8.25	5.70
Gly-Ala	3.15	8.23	5.69
Ala-Gly	3.17	8.18	5.68
Gly-Gly-Gly	3.23	8.09	5.66
Ala-Ala-Ala-Ala	3.42	7.94	5.68

Piperazine is a heterocyclic diamine, which is numbered as shown. It is the nitrogen analog of 1,4-dioxane (page 244).

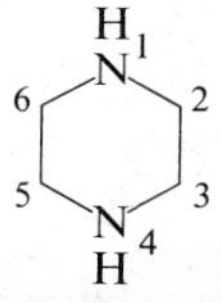

piperazine 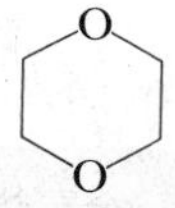   1,4-dioxane

Polymerization probably involves initial diketopiperazine formation, followed by opening of the ring by another amino acid or by the growing polymer chain.

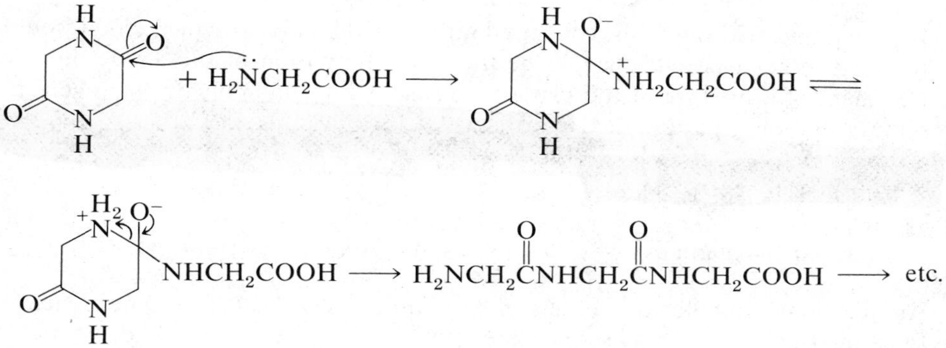

$$+ \text{H}_2\text{NCH}_2\text{COOH} \longrightarrow \overset{+}{\text{N}}\text{H}_2\text{CH}_2\text{COOH} \rightleftharpoons$$

$$\text{NHCH}_2\text{COOH} \longrightarrow \text{H}_2\text{NCH}_2\overset{O}{\overset{\|}{\text{C}}}\text{NHCH}_2\overset{O}{\overset{\|}{\text{C}}}\text{NHCH}_2\text{COOH} \longrightarrow \text{etc.}$$

Diketopiperazines are best prepared from the corresponding amino acid esters. These compounds are stable indefinitely as the hydrochloride salts, but the free amines spontaneously dimerize to give the 2,5-diketopiperazine.

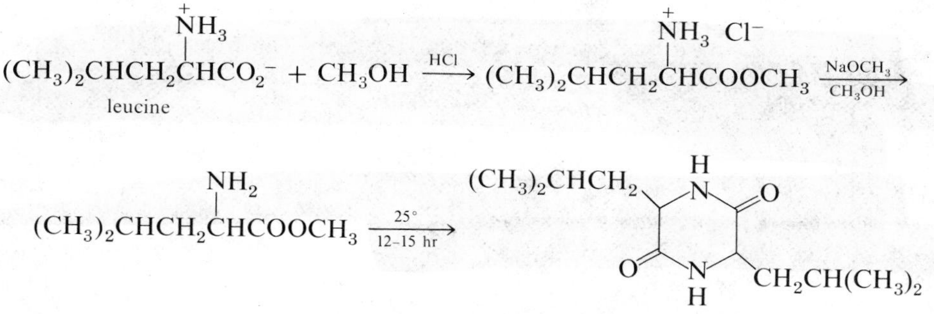

$$(\text{CH}_3)_2\text{CHCH}_2\overset{\overset{+}{\text{N}}\text{H}_3}{\underset{|}{\text{CH}}}\text{CO}_2^- + \text{CH}_3\text{OH} \xrightarrow{\text{HCl}} (\text{CH}_3)_2\text{CHCH}_2\overset{\overset{+}{\text{N}}\text{H}_3 \ \text{Cl}^-}{\underset{|}{\text{CH}}}\text{COOCH}_3 \xrightarrow[\text{CH}_3\text{OH}]{\text{NaOCH}_3}$$

leucine

$$(\text{CH}_3)_2\text{CHCH}_2\overset{\text{NH}_2}{\underset{|}{\text{CH}}}\text{COOCH}_3 \xrightarrow[\text{12–15 hr}]{25°}$$

(60% overall)
leucine anhydride

Hydrolysis of one of the amide bonds in a 2,5-diketopiperazine is one method for preparing simple dipeptides.

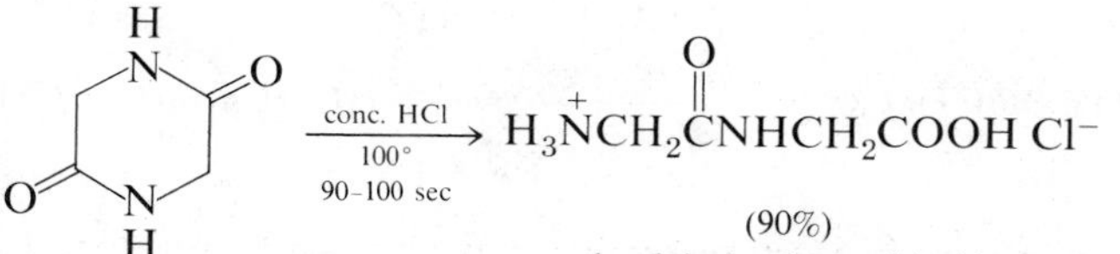

$$\xrightarrow[\substack{100° \\ 90–100 \text{ sec}}]{\text{conc. HCl}} \text{H}_3\overset{+}{\text{N}}\text{CH}_2\overset{O}{\overset{\|}{\text{C}}}\text{NHCH}_2\text{COOH } \text{Cl}^-$$

(90%)
glycylglycine hydrochloride

The rational synthesis of peptides is a challenging task that has only been solved in the past few decades. In order to illustrate the difficulty, consider the synthesis of the simple dipeptide glycylalanine from glycine and alanine. The problem is to form an amide linkage between the carboxy group of glycine and the amino group of alanine.

$$\underset{}{H_3\overset{+}{N}CH_2CO_2^-} + \underset{}{H_3\overset{+}{N}\overset{\overset{\displaystyle CH_3}{|}}{C}HCO_2^-} \dashrightarrow H_3\overset{+}{N}CH_2\overset{\overset{\displaystyle O}{\|}}{C}-NH\overset{\overset{\displaystyle CH_3}{|}}{C}HCO_2^-$$

The normal method for converting a carboxylic acid into an amide is to activate the carboxy group by converting it to an acyl halide and then to add the amine.

$$RCOOH \xrightarrow{SOCl_2} R\overset{\overset{\displaystyle O}{\|}}{C}Cl \xrightarrow{R'NH_2} R\overset{\overset{\displaystyle O}{\|}}{C}NHR'$$

But an amino acid cannot be converted into an acyl halide; polymerization would result. Another possibility would be the direct formation of the amide link by treatment of a mixture of the two amino acids with some dehydrating agent to remove the water produced. However, such a direct approach will give a mixture of four different dipeptides. Furthermore, each of these dipeptides can react further to give higher peptides.

$$\text{glycine} + \text{alanine} \xrightarrow{-H_2O} \text{Gly-Gly} + \text{Ala-Ala} + \text{Gly-Ala} + \text{Ala-Gly}$$

An additional complication enters in with amino acids that have other reactive functional groups, such as serine, threonine, lysine, and aspartic acid.

The general method that has been developed to avoid these difficulties involves the use of **protecting groups.** Protecting groups have been developed for both the amino and carboxy groups, as well as for the other groups that occur in the side chains of the various amino acids. A suitable protecting group must fulfill several criteria.

1. The protecting group must be easy to introduce into the molecule.
2. It must protect the functional group under conditions of amide formation.
3. It must be removable under conditions that leave the newly created amide link intact.

Carboxy groups are normally protected by conversion into the methyl, ethyl, or benzyl ester. Since esters are hydrolyzed more easily than amides, the protecting group can be removed by alkaline hydrolysis.

$$\sim\!\overset{\overset{\displaystyle O}{\|}}{C}NH\overset{\overset{}{|}}{\underset{\underset{\displaystyle R}{|}}{C}}HCOOCH_3 \xrightarrow[H_2O]{OH^-} \xrightarrow{H_3O^+} \sim\!\overset{\overset{\displaystyle O}{\|}}{C}NH\overset{\overset{}{|}}{\underset{\underset{\displaystyle R}{|}}{C}}HCOOH + CH_3OH$$

Benzyl esters may be cleaved by hydrogenolysis (Section 22.6.B).

$$\sim\!\overset{\overset{\displaystyle O}{\|}}{C}NH\overset{\overset{}{|}}{\underset{\underset{\displaystyle R}{|}}{C}}HCOOCH_2\!-\!\!\bigcirc \xrightarrow{H_2\text{-}Pd/C} \sim\!\overset{\overset{\displaystyle O}{\|}}{C}NH\overset{\overset{}{|}}{\underset{\underset{\displaystyle R}{|}}{C}}HCOOH + CH_3\!-\!\!\bigcirc$$

Of the many amino protecting groups that have been developed, we shall discuss only two, the benzyloxycarbonyl (**"carbobenzoxy," Cbz**) and the **t-butoxycarbonyl (Boc)** groups. The benzyloxycarbonyl group is introduced by treating the amino acid with benzyl chloroformate in alkaline solution.

$$\text{H}_3\overset{+}{\text{N}}\text{CH}_2\text{CO}_2^- + \text{C}_6\text{H}_5-\text{CH}_2\text{O}\overset{\text{O}}{\overset{\|}{\text{C}}}\text{Cl} \xrightarrow[\substack{\text{H}_2\text{O}\\5°\\30\text{ min}}]{\text{NaOH}} \xrightarrow{\text{H}_3\text{O}^+} \text{C}_6\text{H}_5-\text{CH}_2\text{O}\overset{\text{O}}{\overset{\|}{\text{C}}}\text{NHCH}_2\text{COOH}$$

glycine   benzyl chloroformate

(70–80%)
benzyloxycarbonylglycine
Cbz-Gly

Benzyl chloroformate is the half benzyl ester, half acyl chloride of carbonic acid. It is prepared by treating benzyl alcohol with phosgene.

$$\text{C}_6\text{H}_5-\text{CH}_2\text{OH} + \text{COCl}_2 \longrightarrow \text{C}_6\text{H}_5-\text{CH}_2\text{O}\overset{\text{O}}{\overset{\|}{\text{C}}}\text{Cl}$$

(95–99%)

The new C—N linkage in a benzyloxycarbonyl amino acid is part of a carbamate grouping (Sections 19.10 and 24.8.D). Like amides, carbamates hydrolyze with great difficulty. However, the benzyl–oxygen bond can be cleaved by catalytic hydrogenolysis, yielding the unstable carbamic acid, which undergoes decarboxylation (Section 24.6.H).

$$\text{C}_6\text{H}_5-\text{CH}_2\text{O}\overset{\text{O}}{\overset{\|}{\text{C}}}\text{NHCHCOOH} \xrightarrow{\text{H}_2-\text{Pd}}$$
$$\text{R}$$

$$\text{C}_6\text{H}_5-\text{CH}_3 + \left[\text{HO}\overset{\text{O}}{\overset{\|}{\text{C}}}\text{NHCHCOOH}\right] \longrightarrow \text{CO}_2 + \text{H}_3\overset{+}{\text{N}}\text{CHCO}_2^-$$
$$\text{R} \qquad\qquad \text{R}$$

The *t*-butoxycarbonyl group is introduced by treating the amino acid with di-*t*-butyl dicarbonate.

$$\left(\begin{array}{c}\text{CH}_3\ \text{O}\\ \text{CH}_3\text{CO}-\text{C}\\ \text{CH}_3\end{array}\right)_2\!\!\!\text{O} + \text{H}_3\overset{+}{\text{N}}\overset{\text{CH}_3}{\text{CHCO}_2^-} \xrightarrow{\text{Et}_3\text{N}} \text{CH}_3\overset{\text{CH}_3}{\underset{\text{CH}_3}{\text{C}}}-\text{O}\overset{\text{O}}{\overset{\|}{\text{C}}}\text{NH}\overset{\text{CH}_3}{\text{CHCOOH}}$$

di-*t*-butyl dicarbonate

t-butoxycarbonylalanine
Boc-Ala

Di-*t*-butyl dicarbonate is commercially available. *t*-Butyl chloroformate is too unstable to use in the reaction. Another reagent that has been used is *t*-butyl azidoformate. However, because of its toxicity and explosive nature, it is no longer commonly used.

The *t*-butoxycarbonyl group is removed by treating the protected amino acid or peptide with hydrogen chloride in acetic acid or dioxane as solvent.

$$\text{\textit{t}-BuO}\overset{\text{O}}{\overset{\|}{\text{C}}}\text{NH}\overset{\text{R}}{\text{CHCOOH}} \xrightarrow{\text{HCl}} \text{H}_3\overset{+}{\text{N}}\overset{\text{R}}{\text{CHCOOH}}\ \text{Cl}^-$$

The initial reaction is cleavage of the alkyl–oxygen bond (Section 19.6). The resulting carbamic acid then decarboxylates, giving the amine.

$$\underset{\overset{+}{\text{OH}}}{\underset{\|}{(CH_3)_3COCNHR}} \longrightarrow (CH_3)_3C^+ + HOCNHR$$

$$\underset{\overset{O}{\|}}{HOCNHR} \longrightarrow CO_2 + H_2NR$$

The most generally useful coupling reagent is **dicyclohexylcarbodiimide (DCC),** a commercially available reagent that is prepared from cyclohexylamine and carbon disulfide by the route indicated.

$$2 \bigcirc{-}NH_2 + CS_2 \longrightarrow \bigcirc{-}\underset{\overset{S}{\|}}{NHCNH}{-}\bigcirc \xrightarrow{\text{HgO}}$$

$$\bigcirc{-}N{=}C{=}N{-}\bigcirc$$

(86%)
dicyclohexylcarbodiimide
DCC

DCC is an effective catalyst for condensation of carboxylic acids with alcohols and amines. DCC activates the free carboxy group of the N-protected amino acid. An equimolar mixture of a carboxylic acid, an amine, and DCC results in the corresponding amide and the highly insoluble N,N′-dicyclohexylurea.

$$RCOOH + R'NH_2 + C_6H_{11}N{=}C{=}NC_6H_{11} \longrightarrow \underset{\overset{O}{\|}}{RCNHR'} + \underset{\overset{O}{\|}}{C_6H_{11}NHCNHC_6H_{11}}$$

N,N′-dicyclohexylurea

The probable mechanism for the DCC coupling reaction is outlined as follows. Addition of the carboxylic acid to the diimide gives the ester of isourea, an O-acylisourea.

$$\underset{\overset{O}{\|}}{RCOH} + R'N{=}C{=}NR' \longrightarrow \underset{\overset{O}{\|}}{R{-}C}{-}O{-}\underset{\overset{NHR'}{|}}{C}{=}NR'$$

an O-acylisourea

The intermediate O-acylisourea is an *activated carboxylic acid derivative* similar in reactivity to an anhydride or an acyl halide. Nucleophilic substitution by the amine yields the amide and the dialkylurea.

$$\underset{\overset{O}{\|}}{RC}{-}O{-}\underset{\overset{NHR'}{|}}{C}{=}NR' + R''NH_2 \rightleftharpoons \left[\underset{\overset{NHR''}{|}}{\overset{\overset{OH}{|}}{RC}{-}O{-}\underset{\overset{NHR'}{|}}{C}{=}NR'} \right] \rightleftharpoons$$

$$\left[\underset{\overset{NHR''}{|}}{\overset{\overset{O^-}{|}}{RC}{-}O{-}\underset{\overset{NHR'}{|}}{C}{=}NHR'} \right] \longrightarrow \underset{\overset{O}{\|}}{RCNHR''} + \underset{\overset{O}{\|}}{R'NHCNHR'}$$

An example of the synthesis of a dipeptide utilizing this method is the synthesis of threonylalanine (Thr-Ala) from benzyloxycarbonylthreonine and alanine benzyl ester.

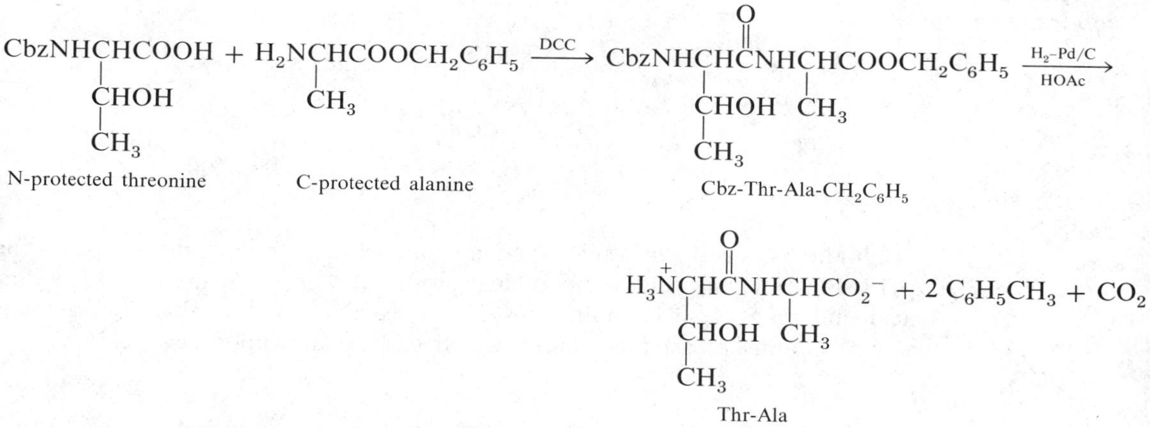

$$CbzNHCHCOOH + H_2NCHCOOCH_2C_6H_5 \xrightarrow{DCC} CbzNHCHCNHCHCOOCH_2C_6H_5 \xrightarrow[HOAc]{H_2-Pd/C}$$
$$\underset{\underset{CH_3}{|}}{\overset{|}{CHOH}} \qquad \underset{CH_3}{|}$$

N-protected threonine   C-protected alanine

Cbz-Thr-Ala-CH$_2$C$_6$H$_5$

$$H_3\overset{+}{N}CHCNHCHCO_2^- + 2\ C_6H_5CH_3 + CO_2$$

Thr-Ala

Another widely used method for coupling the two protected amino acids involves activation of the carboxy group by conversion into an acyl azide. The N-protected amino acid ester is treated with hydrazine to give the acyl hydrazide.

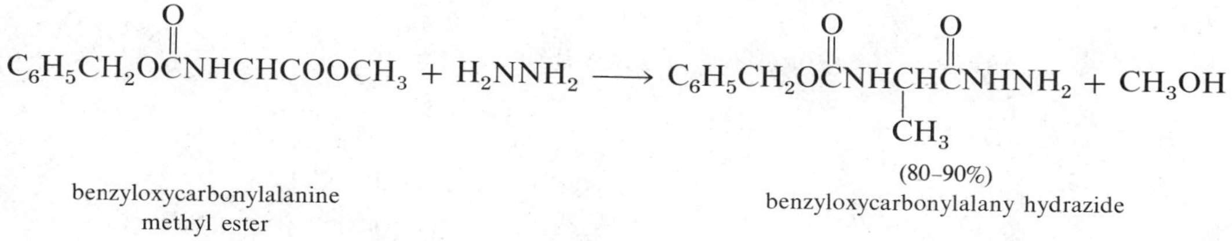

$$C_6H_5CH_2OCNHCHCOOCH_3 + H_2NNH_2 \longrightarrow C_6H_5CH_2OCNHCHCNHNH_2 + CH_3OH$$

benzyloxycarbonylalanine
methyl ester

(80–90%)
benzyloxycarbonylalany hydrazide

The hydrazide is treated with nitrous acid to give the acyl azide.

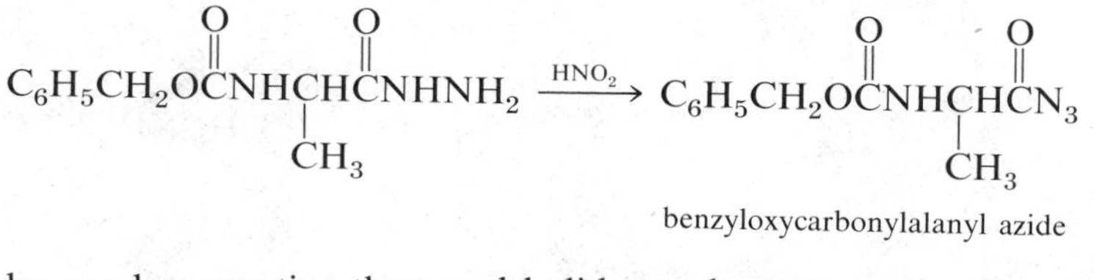

$$C_6H_5CH_2OCNHCHCNHNH_2 \xrightarrow{HNO_2} C_6H_5CH_2OCNHCHCN_3$$

benzyloxycarbonylalanyl azide

Acyl azides are less reactive than acyl halides and more reactive than esters. A typical nucleophilic substitution occurs when the acyl azide is treated with an amino acid in alkaline solution.

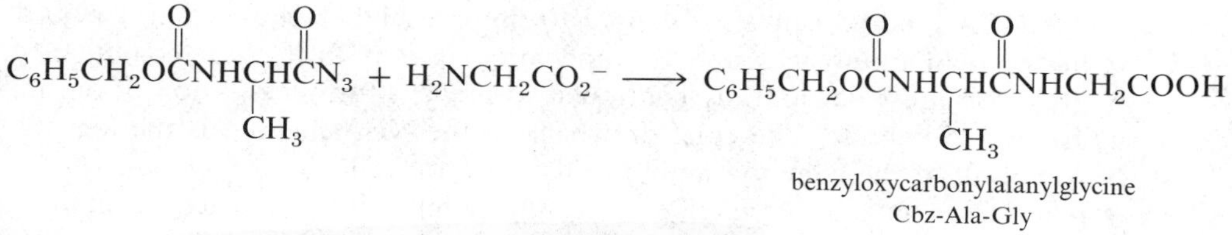

$$C_6H_5CH_2OCNHCHCN_3 + H_2NCH_2CO_2^- \longrightarrow C_6H_5CH_2OCNHCHCNHCH_2COOH$$

benzyloxycarbonylalanylglycine
Cbz-Ala-Gly

When using the acyl azide coupling, it is not necessary to employ a carboxy protecting group.

Another elegant method for peptide synthesis involves the use of amino acid N-carboxy anhydrides **(NCAs** or **Leuchs' anhydrides)**. The additional carbonyl group serves the dual function of protecting the nitrogen and activating the carbonyl.

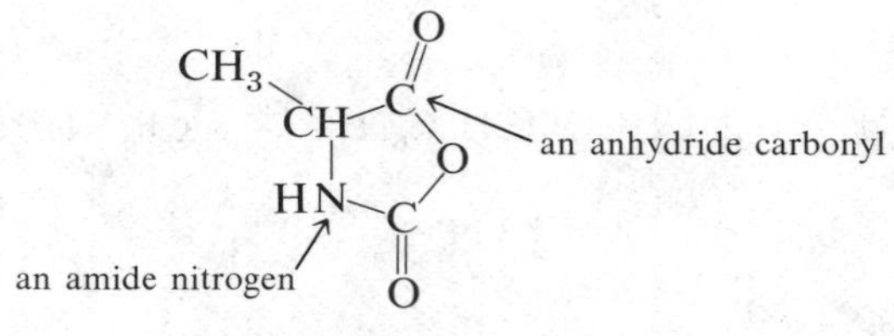

alanine NCA

When the NCA is slowly added to an aqueous solution of an amino acid at 0° and pH 10, an excellent yield of dipeptide is produced. The amino group of the amino acid undergoes a nucleophilic substitution reaction on the anhydride carbonyl group, yielding a carbamic acid, which decarboxylates upon acidification.

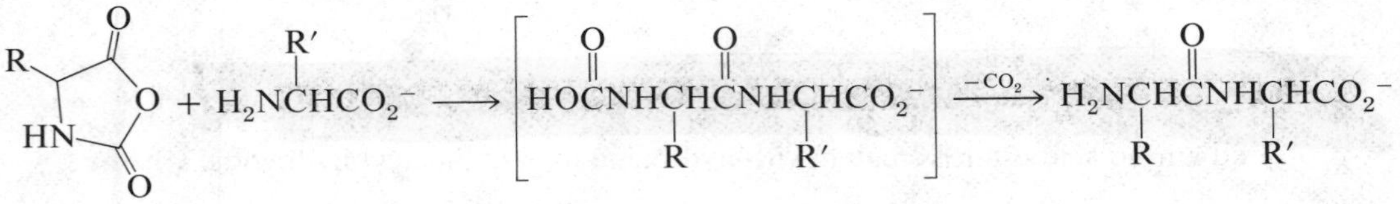

Examples of dipeptides prepared in this manner are

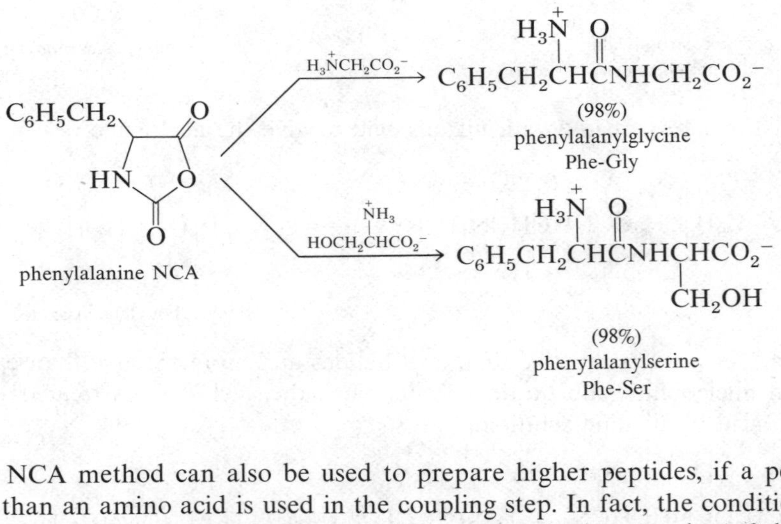

The NCA method can also be used to prepare higher peptides, if a peptide rather than an amino acid is used in the coupling step. In fact, the conditions of the reaction must be carefully controlled to prevent further reaction of the initially formed peptide. The chief drawback of the NCA method is the fact that NCAs of all of the common amino acids are not available.

Thus far we have discussed peptide synthesis only with amino acids containing no other reactive groups. When there is another functional group present in the molecule, it too must be protected until after the peptide has been formed. Typical protecting groups are benzyloxycarbonyl for the second amino group in lysine and benzyl for the sulfur in cysteine.

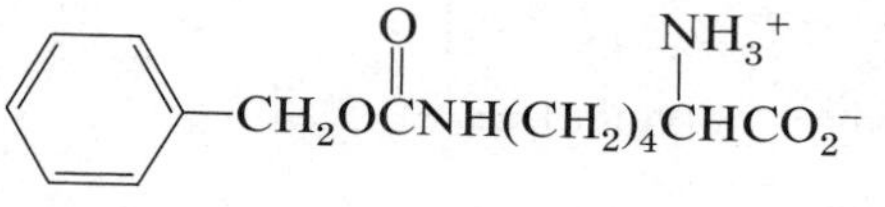

$$C_6H_5-CH_2OCNH(CH_2)_4\overset{\overset{+NH_3}{|}}{C}HCO_2^-$$

ε-benzyloxycarbonyllysine

$$C_6H_5-CH_2SCH_2\overset{\overset{+NH_3}{|}}{C}HCO_2^-$$

S-benzylcysteine

Both protecting groups are removable by cleavage with anhydrous acids such as hydrogen bromide in acetic acid. The second carboxy group in aspartic acid or glutamic acid is usually protected as a methyl or benzyl ester.

A recent development that has revolutionized peptide synthesis is the **solid-phase technique** introduced by R. B. Merrifield of Rockefeller University. In the Merrifield method the peptide or protein is synthesized throughout a swollen cross-linked polymer network that is insoluble and can be recovered by filtration. The polymer used is polystyrene (page 326) in which some of the benzene rings are substituted by —CH₂Cl groups. The polystyrene used is cross-linked with about 1% of divinylbenzenes. The particle sizes range from 20 to 70 μ in diameter.

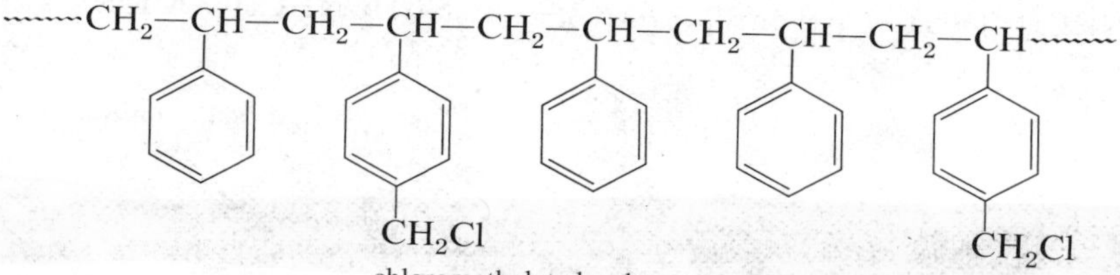

chloromethylated polystyrene

Typically, about one out of every 10–100 phenyl groups is chloromethylated.

The C-terminal amino acid of the desired peptide is bound to the polymer by shaking a solution of the N-protected amino acid salt in an organic solvent such as DMF with the insoluble polymer. The product is an amino acid ester in which the alkoxy group of the ester is the polymer itself.

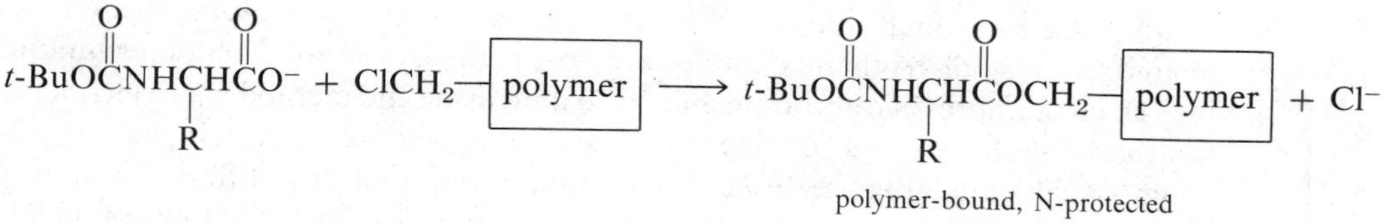

polymer-bound, N-protected
amino acid

Excess reagents are removed by filtration, the insoluble polymer-bound amino ester is washed and the Boc group is removed by treatment with acid.

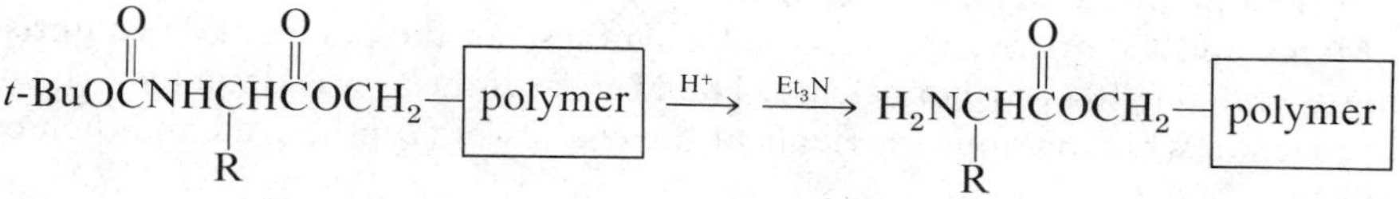

polymer-bound amino acid

A solution of an N-protected amino acid is then added with a coupling agent, and the heterogeneous mixture is shaken until coupling is complete.

$$t\text{-BuOCNHCHCOOH} + \text{H}_2\text{NCHCOCH}_2\!-\!\boxed{\text{polymer}} \xrightarrow{\text{DCC}}$$

N-protected amino acid   polymer-bound amino acid

$$t\text{-BuOCNHCHCNHCHCOCH}_2\!-\!\boxed{\text{polymer}}$$

N-protected, polymer-bound dipeptide

The polymer, now bound to an N-protected dipeptide, is again filtered and washed, and a strong anhydrous acid, usually trifluoroacetic acid, is added to remove the protecting group.

$$t\text{-BuOCNHCHCNHCHCOCH}_2\!-\!\boxed{\text{polymer}} \xrightarrow{\text{H}^+}\xrightarrow{\text{Et}_3\text{N}} \text{H}_2\text{NCHCNHCHCOCH}_2\!-\!\boxed{\text{polymer}}$$

polymer-bound dipeptide

The process may now be repeated to add the third amino acid, and so on. At the end of the synthesis, the peptide is removed from the resin by treatment with anhydrous hydrogen fluoride. At the same time, all side-chain protecting groups are also removed. This final cleavage step does not affect the amide linkages of the peptide chain. The synthetic peptide is then purified by a suitable chromatographic method.

The great advantages of the solid-phase technique are the ease of operation and the high overall yield. Since the growing peptide chain is bound to the highly insoluble polystyrene resin, no mechanical losses are entailed in the intermediate isolation and purification stages. Furthermore, since the method involves the repetitive use of a small number of similar operations, the synthesis is easily automated. Almost all synthetic peptides are now made by the solid-phase technique.

The most impressive accomplishment to date using the method was Merrifield's synthesis of the enzyme bovine pancreatic ribonuclease, a protein having 124 amino acid units (Figure 29.7). The synthesis required 369 different chemical reactions, including coupling and various deprotection steps. A total of 11,931 operations of the automated peptide synthesis machine were entailed, including injection of reagents, filtrations, washings, and so on. The overall yield of synthetic protein, before cleavage from the resin, was 17%, corresponding to a yield of >99% for each step.

At the same time that the Merrifield synthesis was announced, another group, at Merck and Company, also reported a synthesis of the polypeptide comprising amino acids 21–104 of ribonuclease. The Merck synthesis was done by a classical approach in which various segments of the chain were synthesized and then coupled together.

EXERCISE: Write out all of the steps in a rational synthesis of the pentapeptide Ala-Val-Phe-Ala-Ala. As N-protecting groups use the benzyloxycarbonyl group, and for coupling use DCC.

Lys-Glu-Thr-Ala-Ala-Ala-Lys-Phe-Glu-Arg-Glu-His-Met-Asp (5, 10)
Ser 15

Met-Glu-Asn-Cys-Tyr-Asn-Ser-Ser-Ser-Ala-Ala-Ser-Thr-Ser (25, 20)

30 Met
Lys
Ser
Arg
Asn
35 Leu
Thr
Lys
Asp
Arg
Cys 40
Lys
Pro
Val-Asn

Glu-Arg-Cys-Asp-Thr-Ile-Ser-Met-Thr-Ser-Tyr-Ser (85, 80, 75)
Glu-Tyr
Thr-Asn-Cys 70
Glu Cys 65
Gly-Asn Lys Ala
Val
Asn
Lys
Glu 60
Ser
Cys

Thr
Gly
Ser
Ser 90
Lys
Tyr
Pro
Asn
Cys 95
Ala
Tyr-Lys-Thr-Thr-Glu-Ala-Asn-Lys-His-Ile-Ile-Val-Val (100, 105)
Ala
Val 110 Cys
Ala

124 Val-Ser-Ala-Asp-Phe-His-Val-Pro-Val-Tyr-Pro-Asn-Gly-Glu (120, 115)

Thr-Phe-Val-His-Glu-Ser-Leu-Ala-Asp-Val-Glu (45, 50, 55)

FIGURE 29.7. *Amino acid sequence of bovine pancreatic ribonuclease A. The bonds between Cys units, 26—84, 40—95, 58—110, and 65—72 are disulfide linkages. Note how these linkages lock the chain into a folded conformation.*

C. *Structure Determination*

1. AMINO ACID ANALYSIS. The first step in determining the structure of a polypeptide or protein is the cleavage of any disulfide bridges that may be present. This reaction is commonly done by oxidizing the substance with peroxyformic acid, which converts the two cysteine units into cysteic acid units. If the compound contains no disulfide bridges, this step is not necessary.

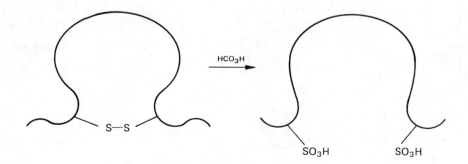

The next analytical step is to determine the total amino acid composition. The material is subjected to total hydrolysis by some suitable method, typically heating with 6 N HCl at 112° for 24–72 hr. The hydrolyzate is then purified and analyzed by a chromatographic technique. The analytical method currently in use employs a commercial instrument called an **amino acid analyzer.** The mixture of amino acids is chromatographed on an ion exchange column with an aqueous buffer solution as eluent. The effluent from the column is automatically mixed with ninhydrin solution, and the presence of an amino acid is indicated by the typical violet color produced in the reaction (Section 29.5.C). The effluent is monitored at appropriate wavelengths with a visible spectrometer, and the absorbance is plotted by a recorder as a function of time. By comparing the chromatogram of an unknown mixture with that of a mixture of known composition, the analyst may arrive at a quantitative analysis of his mixture. The chromatogram determined for a standard mixture of amino acids is shown at the right in Figure 29.8. The left curve is a chromatogram of hydrolyzed bradykinin (page 953).

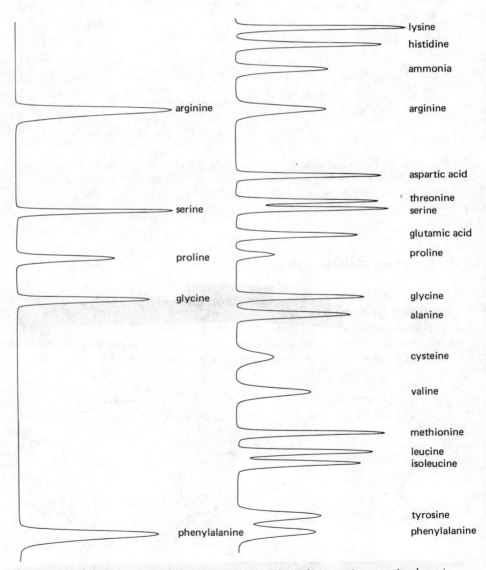

FIGURE 29.8. *Amino acid analyzer traces. The right curve is an equimolar mixture. The left curve is the analysis of a sample of hydrolyzed bradykinin.*

2. IDENTIFICATION OF THE N-TERMINAL AMINO ACID. There are two methods available for identifying the amino acid unit that occupies the N-terminal position in the polypeptide chain. The first is called the **Sanger method.** The —NH$_2$ group in amino acids and peptides reacts with 2,4-dinitrofluorobenzene to form yellow 2,4-dinitrophenyl (DNP) derivatives. The reaction, illustrated for glycine, is an example of aromatic nucleophilic substitution (see Section 30.3.A).

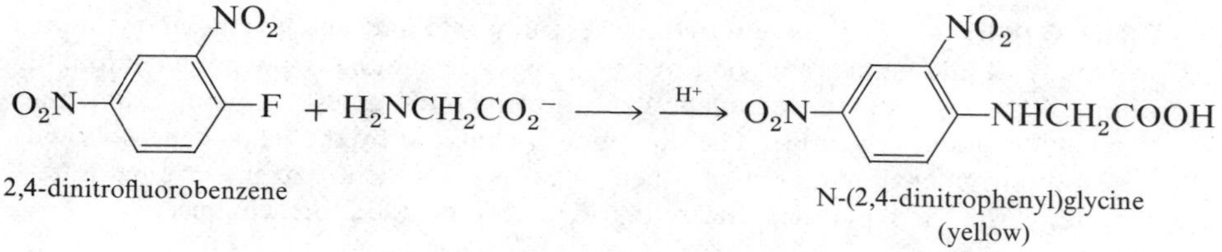

$$O_2N\text{—}\underset{\text{2,4-dinitrofluorobenzene}}{\overset{\displaystyle NO_2}{\bigcirc}}\text{—}F + H_2NCH_2CO_2^- \xrightarrow{\;H^+\;} O_2N\text{—}\underset{\substack{\text{N-(2,4-dinitrophenyl)glycine}\\ \text{(yellow)}}}{\overset{\displaystyle NO_2}{\bigcirc}}\text{—}NHCH_2COOH$$

If the Sanger reaction is carried out on a peptide, the only α-amino group that undergoes the reaction is the free group on the N-terminal end. Total hydrolysis of the DNP-labeled peptide then gives a mixture of amino acids, only one of which is labeled with the DNP function on the α-amino group. By knowing which amino acid bears the label, the investigator knows which amino acid is at the N-terminal end of the peptide.

The other technique for N-terminal analysis, which is actually more useful, is called the **Edman degradation.** In the Edman degradation, the peptide is allowed to react with phenylisothiocyanate, C$_6$H$_5$N=C=S. The terminal NH$_2$ group reacts to form the phenylthiocarbamoyl derivative of the peptide. The labeled peptide is then treated with anhydrous HCl in an organic solvent. Although these conditions do not hydrolyze the amide linkages, the labeled amino acid undergoes a cyclization reaction, giving a phenylthiohydantoin. In the process the end group also becomes separated from the remainder of the peptide chain.

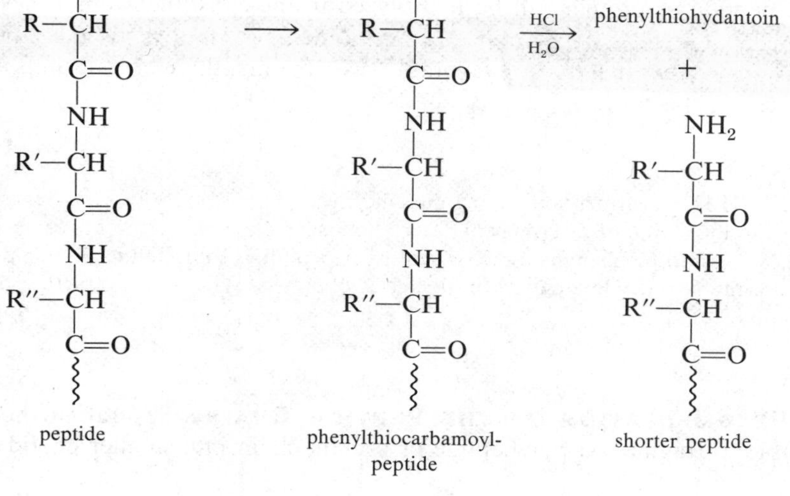

peptide → phenylthiocarbamoyl-peptide → phenylthiohydantoin + shorter peptide

The substituted phenylthiohydantoin produced can be identified chromatographically by comparing it with known materials. Furthermore, the degraded peptide can be isolated and subjected to another cycle of the Edman degradation to identify the new N-terminal unit. The process has been automated and has been used to identify the first 60 amino acids in whale myoglobin, a protein that contains 153 amino acids in its chain.

EXERCISE: A tetrapeptide shows serine, valine, alanine, and glycine upon amino acid analysis. Edman degradation gives a tripeptide and the N-phenylhydantoin of valine. A second Edman degradation on the tripeptide gives a dipeptide and the N-phenylhydantoin of serine. The dipeptide is analyzed by the Sanger method and N-(2,4-dinitrophenyl)alanine is obtained. What is the structure of the original tetrapeptide? Write the equations illustrating the entire degradation sequence.

3. IDENTIFICATION OF THE C-TERMINAL AMINO ACID. The C-terminal amino acid may be identified by hydrolyzing with the enzyme **carboxypeptidase,** which specifically catalyzes the hydrolysis of the C-terminal amide link in a peptide or protein chain.

Thus, when the material is incubated with carboxypeptidase, the first free amino acid to appear in solution is the one that occupies the C-terminal position. Of course, once that amino acid has been removed from the chain, the enzyme continues to function and goes to work on the next residue, and so on. Eventually, the entire peptide or protein will be hydrolyzed to the constituent amino acids. By measuring the rate of appearance of amino acids in the hydrolyzate, one may identify the C-terminal unit. In favorable cases the first three or four units may be identified in this way.

EXERCISE: A tripeptide having the empirical composition Phe, Gly, Ser is subjected to the action of carboxypeptidase. The first free amino acid to appear in solution is found to be phenylalanine. When the tripeptide is subjected to Edman degradation, the N-phenylhydantoin of glycine is obtained. What is the structure of the tripeptide?

4. FRAGMENTATION OF THE PEPTIDE CHAIN. Several methods are available to fragment a polypeptide or protein chain into smaller peptides. The

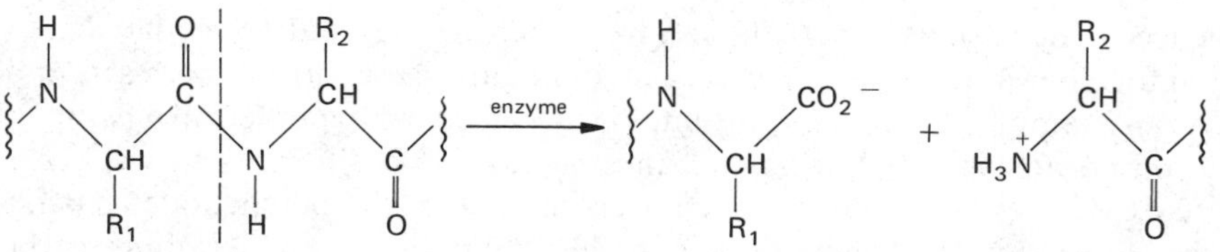

Enzyme	R_1
trypsin	Lys, Arg
chymotrypsin	Phe, Trp, Tyr
pepsin	Phe, Trp, Tyr, Leu, Asp, Glu

FIGURE 29.9. *Specificity of proteases.*

most useful method is enzymatic hydrolysis. There are several enzymes available, called **proteases,** that catalyze hydrolysis of the peptide chain, usually at specific positions. For example, the enzyme **trypsin,** which occurs in the intestines of mammals, causes cleavage of peptide bonds only when the carbonyl group is part of a lysine or arginine unit. In a similar way, **chymotrypsin,** another intestinal enzyme, catalyzes hydrolysis of phenylalanine, tryptophan, and tyrosine positions. **Pepsin,** a gastric protease, is much less specific, causing rupture of the chain wherever there is phenylalanine, tryptophan, tyrosine, leucine, aspartic acid, or glutamic acid (Figure 29.9). Abnormal cleavage is sometimes observed.

Another useful method for selective cleavage of polypeptide chains employs **cyanogen bromide,** BrCN. This reagent cleaves the chain only at the carbonyl group of methionine units; the methionine is converted into a C-terminal homoserine lactone unit.

homoserine lactone
unit

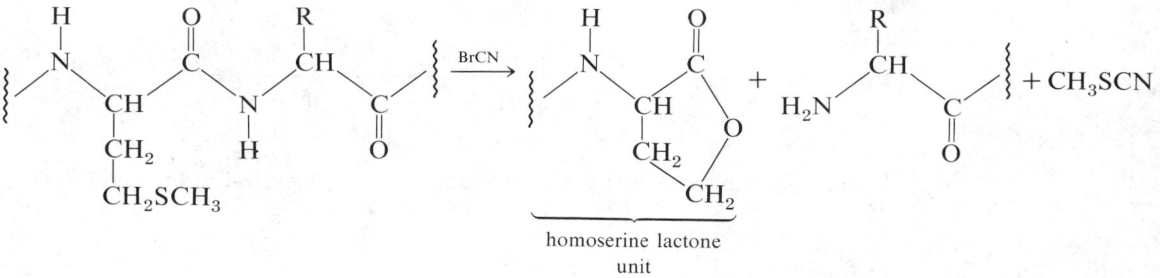

Partial degradation of the polypeptide chain, using one of the aforementioned methods, is a crucial step in determining the proper amino acid sequence of the molecule. Usually the purified polypeptide or protein is first incubated with trypsin, the most selective protease. The resulting mixture of peptide fragments is chromatographed, and the pure fragments are isolated. The peptides produced will usually contain from 2 to about 20 amino acid units. If the polypeptide chain is very long and if it contains relatively few lysine and arginine units, much larger fragments may be produced. The purified fragments are then analyzed for total amino acid content and subjected to repetitive Edman degradation to determine their structures.

The process is then repeated using a different cleavage method, usually cyanogen bromide. This second set of peptide fragments is then analyzed and sequenced. The various peptide blocks from the two degradation methods are then fitted together to produce a structure that unequivocally satisfies both sets of data.

As an example of the reasoning employed, consider a hypothetical eicosapeptide (20 amino acid units) having the amino acid composition ($Gly_2Ala_4Leu_4Phe_3$ $TrpLys_2Met_2SerArg$). End-group analysis shows that the polypeptide has alanine at the N-terminus (Sanger method) and phenylalanine at the C-terminus (carboxypeptidase). The material is hydrolyzed with trypsin to give four fragments: a tripeptide, two pentapeptides, and a heptapeptide. The four peptide fragments are each sequenced by repetitive Edman degradation and found to have the following structures.

I   Trp-Phe-Arg

II   Ala-Leu-Gly-Met-Lys

III   Leu-Gly-Leu-Leu-Phe

IV   Ala-Ala-Ser-Met-Ala-Phe-Lys

At this point the investigator knows that fragment III must correspond to the last five amino acids in the chain because trypsin does not cleave a chain at a phenylalanine carbonyl. Furthermore, fragment II or IV must correspond to the N-terminal end, but it is not possible with this information alone to write a unique complete sequence.

The intact polypeptide is then cleaved with cyanogen bromide, and the fragments are isolated, purified, and sequenced as before. Three fragments are produced, having the structures

V   Ala-Leu-Gly-Met

VI   Ala-Phe-Lys-Leu-Gly-Leu-Leu-Phe

VII   Lys-Trp-Phe-Arg-Ala-Ala-Ser-Met

The four fragments in the first degradation and the three fragments in the second are then ordered in an overlapping way to arrive at an unambiguous structure.

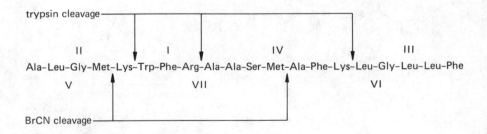

In practice, identification of the complete sequence of a complicated polypeptide or protein is rarely as simple as this example, and the actual process is usually tedious and time-consuming. It often happens that almost the entire sequence is elucidated, but the exact positions of a few amino acids remain doubtful. The general process is still being improved, and routine sequencing will undoubtedly become less tedious and more efficient during the next few years.

The first major structure determination was that of insulin (page 954) by Sanger in 1953. The next significant accomplishment in this area was the sequencing of adrenocorticotrophin (40 amino acid units), the hormone produced in the anterior pituitary gland that stimulates the adrenal cortex.

$$\overset{5}{\text{Ser-Tyr-Ser-Ser-Met-Glu-His-Phe-Arg-Trp-Gly-Lys-Pro-Val-}}$$

$$\overset{15}{\text{Gly-Lys-Lys-Arg-Arg-Pro-Val-Lys-Val-Tyr-Pro-Asn-Gly-}}$$

$$\overset{30}{\text{Ala-Glu-Asp-Glu-Ser-Ala-Glu-Ala-Phe-Pro-Leu-Glu-Phe}}$$

human adrenocorticotrophin

Using techniques such as these, such large proteins as bovine chymotrypsinogen (245 amino acid units) and glyceraldehyde 3-phosphate dehydrogenase (333 amino acid units) have been sequenced.

EXERCISE: What is the structure of a pentapeptide that gives Gly-Ala, Leu-Phe, Leu-Leu, and Ala-Leu upon partial hydrolysis and the N-phenylhydantoin of glycine upon Edman degradation? What products will be obtained from incubation of this pentapeptide with pepsin?

29.7
Proteins

A. *Molecular Shape*

Proteins serve two important biological functions. On the one hand, they serve as structural material. The structural proteins tend to be **fibrous** in nature. That is, the long polypeptide chains are lined up more or less parallel to each other and are bonded one to another by hydrogen bonds. Depending on the actual three-dimensional structure of the individual protein molecule and its interaction with other similar molecules, a variety of structural forms may result. Examples are the protective tissues such as hair, skin, nails, and claws (α- and β-keratins), connective tissues such as tendon (collagen), and the contractile material of muscle (myosin). Fibrous proteins are usually insoluble in water.

The other important function of proteins is as biological catalysts and regulators. They are responsible for increasing and regulating the speed of biochemical reactions and the transport of various materials throughout an organism. The catalytic proteins **(enzymes)** and transport proteins tend to be **globular** in nature. The polypeptide chain is folded around itself in such a way as to give the entire molecule a rounded shape. Each globular protein has its own characteristic geometry, which is a result of interactions between different sites on the chain. The intrachain interactions may be of four types: disulfide bridging, hydrogen bonding, dipolar interactions, or van der Waals attraction.

The globular proteins are water-soluble. Sometimes each molecule of a globular protein consists of a single long polypeptide chain twisted about and folded back upon itself. In other cases the molecule is composed of several subunits. Each subunit is a single polypeptide chain that has adopted its own unique three-dimensional geometry. Several of the subunits are then bonded together by secondary forces (hydrogen bonding and van der Waals attraction) to give the total globular unit.

Globular proteins often carry a nonprotein molecule (the **prosthetic** group) as a part of their structure. The prosthetic group may be covalently bonded to the polypeptide chain, or it may be held in place by other forces.

B. *Factors That Influence the Molecular Shape*

As we saw in the previous section, proteins are amino acid polymers containing more than about 50 individual units per chain. A polymer may be **linear** or **branched** (Section 11.6.G).

$$\sim\sim\sim CH_2-CH_2-CH_2-CH_2-CH_2-CH_2-CH_2-CH_2-CH_2-CH_2-CH_2\sim\sim\sim$$

linear polyethylene

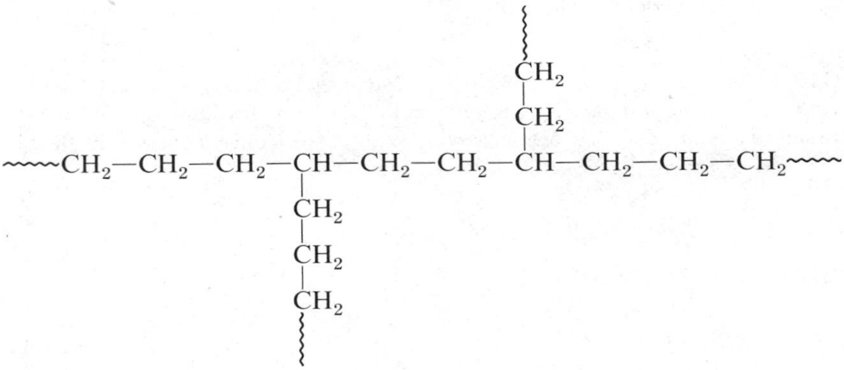

$$\sim\sim\sim CH_2-CH_2-CH_2-CH-CH_2-CH_2-CH-CH_2-CH_2-CH_2\sim\sim\sim$$

branched polyethylene

Proteins are linear polymers. The backbone of the polymer chain is the repeating unit

$$-NHCHCO-$$
$$\overset{\displaystyle R}{|}$$

In addition to the rigidity imparted to the polymer chain by the restricted rotation about the amide bond (Sections 19.1 and 29.6), the three-dimensional structure of the macromolecule is determined by two factors.

1. SECONDARY INTER- OR INTRACHAIN BONDING. Disulfide bridges between cysteine units in separate chains lead to a type of cross-linking. An example is seen in insulin (Figure 29.6), in which the A and B chains are bonded together by two disulfide links. When the two cysteine units are in the same chain, as in oxytocin, the A chain of insulin (Figure 29.6), or ribonuclease (Figure 29.7), disulfide bridging results in loops in the chain.

Hydrogen bonding is another type of secondary bonding that may occur between two different chains or between different regions of the same chain. Although hydrogen bonds are inherently weak (about 5 kcal mole^{-1} per hydrogen bond), a polypeptide chain contains many C=O and N—H groups that may engage in such bonding. The total amount of bonding that results from many small interactions is substantial and plays an important role in the actual shape or conformation of the molecule. Reciprocal hydrogen bonding may occur between the C=O and N—H groups of different chains and thus bind them together. Intrachain hydrogen bonding causes the chain to fold back on itself in some specific fashion.

hydrogen bonds
polypeptide chain

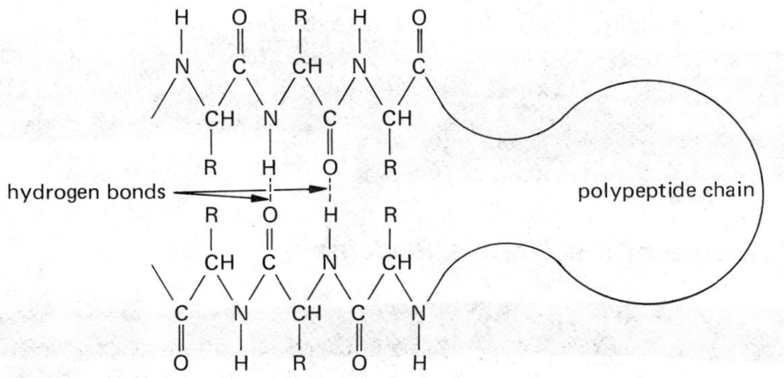

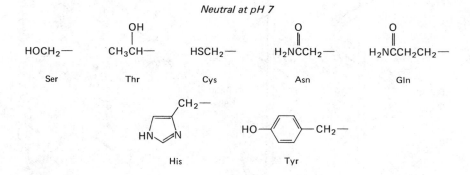

2. Electronic and Steric Properties of the Side-Chain Groups. Some of the side-chain groups that project from the polypeptide backbone are nonpolar (Figure 29.10). In globular proteins, these nonpolar groups are found to be about equally distributed between the interior and the surface of the molecule.

Gly
Ala
Val
Leu
Ile
Phe
Trp
Met
Pro

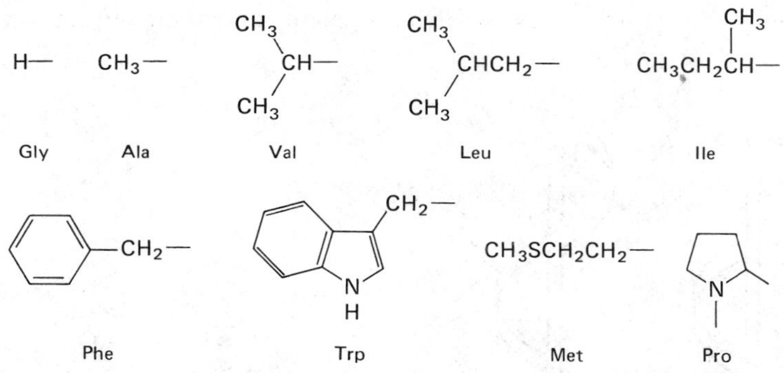

Figure 29.10. *Nonpolar side chains.*

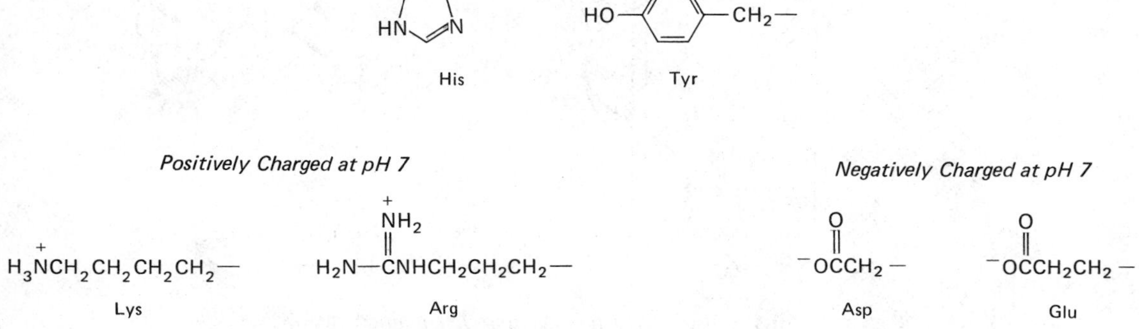

Neutral at pH 7
Ser
Thr
Cys
Asn
Gln
His
Tyr
Positively Charged at pH 7
Lys
Arg
Negatively Charged at pH 7
Asp
Glu

Figure 29.11. *Polar side chains.*

Other side chains are polar and can hydrogen-bond to water molecules. Since the globular proteins exist mainly in aqueous solutions, the polar side chains are found mainly on the outer surface of the molecule. The polar or **hydrophilic** side chains are listed in Figure 29.11. Some are neutral, and others bear either a negative or a positive charge at neutral pH.

C. *Structure of the Fibrous Proteins*

The most important type of conformation found in fibrous proteins is the α-helix. In this structure the polypeptide chain coils about itself in a spiral manner. The spiral or helix is held together by intrachain hydrogen bonding. The α-helix is "right-handed" and has a pitch of 5.4 Å or 3.6 amino acid units (Figure 29.12). Although a right-handed α-helix can form from either D- or L-amino acids (but not from DL), the right-handed version is more stable with the natural L-amino acids. A dramatic demonstration of the α-helix is shown by the stereo representation of polyalanine in Figure 29.13.

Not all polypeptide chains can form a stable α-helix. The stability of the coil is governed by the nature of the side-chain groups and their sequence along the

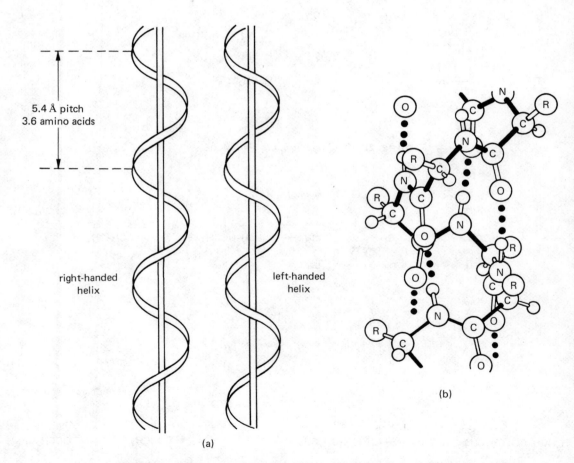

5.4 Å pitch
3.6 amino acids

right-handed
helix

left-handed
helix

(b)

(a)

FIGURE 29.12. (a) *Right-handed and left-handed helix. Note that ordinary screws are right-handed helices.* (b) *Diagram of a peptide α-helix.*

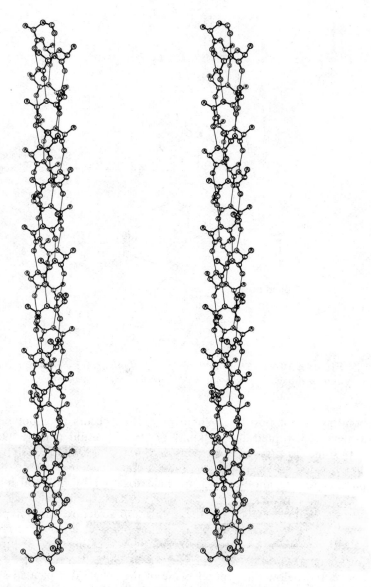

FIGURE 29.13. *Stereo representation of polyalanine. [Courtesy of C.K. Johnson, Oak Ridge National Laboratory.]*

chain. Polyalanine, where the side chains are small and uncharged, forms a stable α-helix. However, polylysine does not. At pH 7 the terminal amino groups in the lysine side chains are all protonated. Electrostatic repulsion between the neighboring ammonium groups disrupts the regular coil and forces polylysine to adopt a **random coil** conformation. At pH 12 the lysine amino groups are uncharged, and the material spontaneously adopts the α-helical structure. In a similar way, polyglutamic acid exists as a random coil at pH 7, where the terminal carboxy groups are ionized, and as an α-helix at pH 2, where they are uncharged.

Proline is particularly interesting. Since the α-amino group in proline is part of a five-membered ring, rotation about the C—N bond is impossible. Furthermore, the amide nitrogen in polyproline has no hydrogens, and intrachain hydrogen

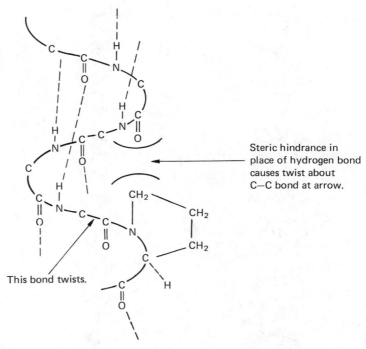

Steric hindrance in
place of hydrogen bond
causes twist about
C—C bond at arrow.

This bond twists.

FIGURE 29.14. *Showing the origin of a kink in an α-helix at proline. The proline unit in a peptide chain has no N—H for hydrogen bonding.*

bonding is not possible. Wherever proline occurs in a polypeptide chain, the α-helix is disrupted and a "kink" or "bend" results (Figure 29.14).

In some cases, such as the keratins of hair and wool, several α-helices coil about one another to produce a **superhelix.** In other cases, the helices are lined up parallel to one another and are held together by intercoil hydrogen bonding.

Another type of conformation found in the fibrous proteins is the β- or **pleated sheet** structure of β-keratin (silk). In the β-structure, the polypeptide chains are extended in a "linear" or zigzag arrangement. Neighboring chains are bonded together by reciprocal interchain hydrogen bonding. The result is a structure resembling a pleated sheet (Figure 29.15). Side-chain groups extend alternately above and below the general plane of the sheet. The pleated-sheet structure results in the side-chain groups being fairly close together. For this reason side chains that are bulky or have like charges disrupt the arrangement. In the β-keratin of silk fibroin 86% of the amino acid residues are glycine, alanine, and serine, all of which have small side chains.

D. Structure of the Globular Proteins

Globular proteins are designed to be soluble in the aqueous body fluids. They also must have a unique structure that creates an **active site** where the catalytic or transport function of the protein is carried out. The specific coiling that produces the proper geometry of the protein results from a delicate interplay of all the forces we have discussed up until now. Some folding is stabilized by interchain disulfide bridges. The molecule tends to orient itself so that the nonpolar side chains lie inside the bulk of the structure where they attract each other by van der

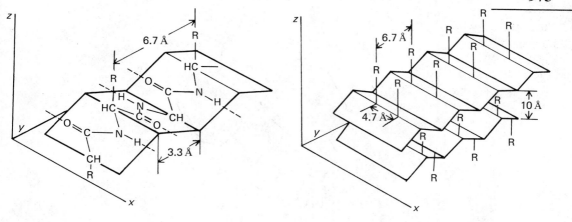

extended polypeptide chain   array of chains to give pleated sheets

FIGURE 29.15. *Schematic diagrams of pleated-sheet structure of polypeptides. The peptide bonds lie in the plane of the pleated sheet; the side chains lie above and below the sheet alternately. The polypeptide chains are held together by interchain hydrogen bonds, shown as dotted lines.*

Waals forces. The polar side chains tend to be on the surface of the molecule where they can hydrogen-bond to the solvent molecules and confer the necessary water solubility. Further coiling and compacting of the structure result from interchain hydrogen bonds between the amide linkages inside the bulk of the molecule. Some segments of the polypeptide chain may have the typical α-helical structure, and others may be random coil. In other cases the chain may fold back on itself in the β- or pleated-sheet fashion. A schematic representation of a globular protein is shown in Figure 29.16.

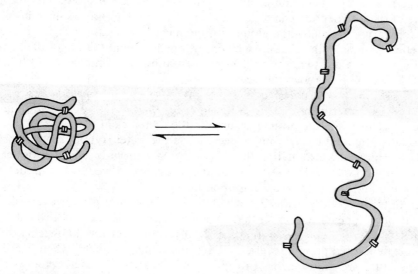

FIGURE 29.16. *Schematic diagram of a globular protein with intrachain bonds (hydrogen bonds, van der Waals forces, and so on), showing reversible denaturation to random coil chain. [Adapted with permission from S. J. Baum,* Introduction to Organic and Biological Chemistry, *2nd ed. Macmillan Publishing Co., Inc., 1978.]*

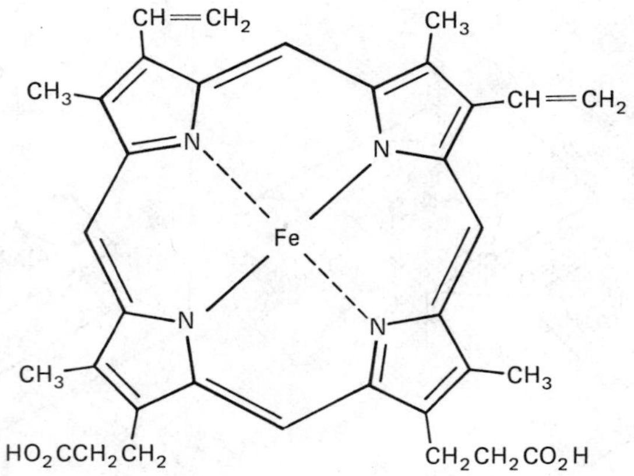

FIGURE 29.17. *Hemin, the prosthetic group of hemoglobin and myoglobin.*

If the protein contains a prosthetic group, that group will be imbedded at some point within the overall three-dimensional structure of the protein, either covalently bonded to the polypeptide chain or simply held by secondary forces. An example of a prosthetic group is heme, which is found in hemoglobin and myoglobin (Figure 29.17). In these proteins, both of which are oxygen carriers—myoglobin in muscle and hemoglobin in the bloodstream—the function of the prosthetic group is to bind an oxygen molecule. In both cases the polypeptide chain folds in such a way as to leave a hydrophobic "pocket" into which the heme just fits. The heme pocket is equipped with a histidine situated in such a way that its imidazole nitrogen can act as a fifth ligand for the ferrous ion in the center of the heme molecule. The prosthetic group is further held in its pocket by hydrogen bonding between the two propionic acid side chains and other appropriate side chains within the pocket.

The stereo representation of myoglobin in Figure 29.18 shows only the backbone of the polypeptide chain and the heme; substituent groups have been deleted for clarity. Note how the globular protein coils up on itself. There are several α-helical regions in the chain. An extensive one is seen at the top of the molecule and is viewed almost end-on in this representation. The imidazole "fifth ligand" (not shown) is just above the heme.

Under proper conditions the delicate three-dimensional structure of globular proteins may be disrupted. This process is called **denaturation.** Denaturation commonly occurs when the protein is subjected to extremes in temperature or pH. It is usually attended by a dramatic decrease in the water solubility of the protein. An example is the coagulation that results when skim milk is heated or acidified (denaturation of lactalbumin). A similar process is involved in the hardening of the white and the yolk of an egg upon heating.

Until fairly recently it was believed that denaturation was an irreversible process. It now appears, however, that in some cases the process is reversible. The reverse process is called **renaturation.** Many cases are now known in which a soluble denatured protein reverts to its natural folded geometry when the pH and temperature are adjusted back to the point where the native protein is stable. Thus the three-dimensional structure of a protein seems to be a natural consequence of the specific amino acid sequence in its polypeptide chain; that is, the unique conformation of each protein is simply the most stable structure that molecule can have under biological conditions.

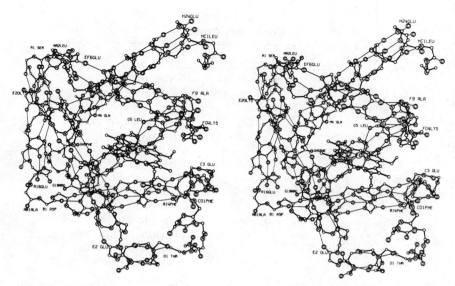

FIGURE 29.18. *Stereo representation of myoglobin (side-chain substituents deleted).* [*Courtesy of C. K. Johnson, Oak Ridge National Laboratory.*]

E. *Biological Function of Proteins—An Overview*

Although a complete discussion of the biological function of proteins is beyond the scope of this book, we shall give a brief tabular summary of the topic here. As discussed previously, one important function is structural. Some examples of structural proteins are listed in Table 29.5.

The regulatory proteins serve an immense variety of purposes. A few examples are given in Table 29.6.

The regulatory proteins serve an immense variety of purposes. We have already encountered several enzymes, compounds that act as catalysts for biological reactions. Others function to carry important reactants from one site in the organism to another. Still other members of this group serve to store materials that will later be metabolized for production of energy. Other important regulatory functions are handled by the proteins of the immune system, which protect the organism from the effects of foreign substances, and by the hormones, which pass messages from one point in the organism to another. A few examples of specific regulatory proteins are given in Table 29.6.

TABLE 29.5
Structural Proteins

Example	Function
α-keratin	structural component of skin, hair, feathers, nails
collagen	connective tissues, tendon, bone, cartilage
fibroin	silk of spider webs, cocoons
sclerotin	exoskeletons of insects
myosin	stationary component of muscle
actin	contractile component of muscle

**TABLE 29.6
Regulatory Proteins**

Example	Type	Function
carboxypeptidase	enzyme	catalyzes hydrolysis of polypeptide chains
trypsin	enzyme	catalyzes hydrolysis of polypeptide chains
hemoglobin	transport	carries oxygen in the bloodstream
myoglobin	transport	carries oxygen in muscles
cytochrome c	transport	carries electrons
ovalbumin	storage	food storage in egg white
casein	storage	milk protein
antibodies	protective	form insoluble complexes with foreign substances in the bloodstream
insulin	hormone	regulates the metabolism of glucose

PROBLEMS

1. For each of the following compounds write the structure of the principal ionic species present in aqueous solution at pH 2, 7, and 12.
   (a) isoleucine
   (b) aspartic acid
   (c) lysine
   (d) glycylglycine (Gly-Gly)
   (e) lyslyglycine (Lsy-Gly)
   (f) alanylaspartylvaline (Ala-Asp-Val)

2. Show how the isoelectric point of an amino acid can be computed from pK_1 and pK_2.

3. The pK_as for β-alanine and 4-aminobutanoic acid are shown below. Compare these values with the pK_as for the α-amino acids in Table 29.2 and explain the differences.

$$\overset{+}{H_3N}CH_2CH_2CO_2^- \qquad \overset{+}{H_3N}CH_2CH_2CH_2CO_2^-$$
$$\beta\text{-alanine} \qquad\qquad \text{4-aminobutanoic acid}$$
$$pK_1 = 3.55 \qquad\qquad pK_1 = 4.03$$
$$pK_2 = 10.24 \qquad\qquad pK_2 = 10.56$$

What are the isoelectric points for these two amino acids?

4. The dipeptide Gly-Asp has three known pK_a values: 2.81, 4.45, and 8.60. Associate each pK_a with the appropriate functional group in the structure of this peptide. Give a practical synthesis of this peptide starting with the amino acids.

5. Propose syntheses for the following amino acids.

(a) $CH_3CH_2CH_2CH_2\overset{\overset{+}{N}H_3}{\underset{|}{C}H}CO_2^-$

(b) 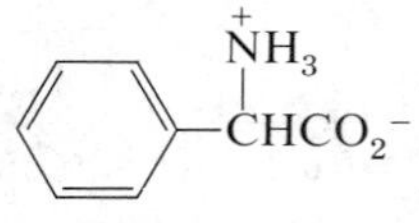 ⬡—$\overset{\overset{+}{N}H_3}{\underset{|}{C}H}CO_2^-$

(c) $(CH_3)_3C\overset{\overset{+}{N}H_3}{\underset{|}{C}H}CO_2^-$

(d) $CH_3CH_2\overset{\overset{+}{N}H_3}{\underset{\underset{CH_3}{|}}{C}}CO_2^-$

(e) 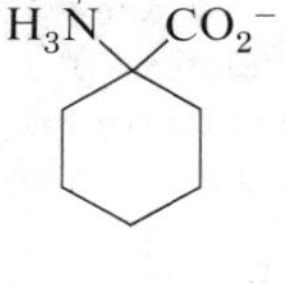 $H_3\overset{+}{N}$ CO_2^- (on cyclohexane)

(f) 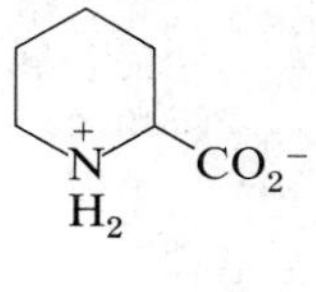 (piperidine ring with) $\overset{+}{\underset{H_2}{N}}$—$CO_2^-$

6. The following isotopically labeled amino acids are desired for biochemical research. Show how each may be prepared. The only acceptable sources of ^{14}C are $Ba^{14}CO_3$ and $Na^{14}CN$. The ^{14}C-labeled atom is marked with an asterisk in each case. Deuterated compounds may be prepared using D_2O, $LiAlD_4$, or D_2.

$$\overset{+}{N}H_3$$
(a) $CH_3CH{-}\overset{*}{C}O_2{}^-$

$$\overset{+}{N}H_3$$
(b) $CH_3{-}\overset{*}{C}HCO_2{}^-$

$$\overset{+}{N}H_3$$
(c) $CD_3CHCO_2{}^-$

$$\overset{+}{N}H_3$$
(d) (phenyl ring)$-CH_2\overset{}{C}DCO_2{}^-$

$$\overset{+}{N}H_3$$
(e) $^*CH_3SCH_2CH_2CHCO_2{}^-$

$$\overset{+}{N}H_3$$
(f) $(CD_3)_2CHCHCO_2{}^-$

$$\overset{+}{N}H_3$$
(g) $HOO\overset{*}{C}CH_2CHCO_2{}^-$

$$NH_2$$
(h) $H_3\overset{+}{N}CD_2CH_2CH_2CH_2CHCO_2{}^-$

7. Glycine undergoes acid-catalyzed esterification more slowly than does propanoic acid. Explain.

8. Explain why the benzoyl group cannot be used as a N-protecting group for peptide synthesis.

9. In 1914 Maillard reported a study of the polymerization of glycine. The amino acid was heated in glycerol solution. The main product of the reaction was found to be 2,5-diketopiperazine. A polypeptide fraction was produced in low yield. The predominant peptides in this fraction were found to be the even peptides tetraglycine and hexaglycine. Explain.

10. In 1903 Emil Fischer introduced a rational method for the stepwise construction of peptides. The process, known as the α-haloacyl halide method, is outlined below in a synthesis of glycylglycine.

$$\underset{O}{\overset{O}{\|}} \qquad \underset{O}{\overset{O}{\|}}$$

$BrCH_2CBr + H_2NCH_2CO_2{}^- \longrightarrow BrCH_2CNHCH_2CO_2{}^- \xrightarrow[\Delta]{NH_3} H_3\overset{+}{N}CH_2CNHCH_2CO_2{}^-$

(a) Show how the α-haloacyl halide method can be used to synthesize glycyl-L-alanine and glycylglycyl-L-alanine.
(b) Which α-haloacyl halides would be used to add alanyl or valyl units?
(c) If the method is applied to L-alanine using the acyl halides in part (b), what will the products be?
(d) What would be the chief problem in applying the α-haloacyl halide method to a fairly complex polypeptide such as Ala-Val-Phe-Ala-Ala?

11. Write out all the steps in a synthesis of the hexapeptide Gly-Ala-Pro-Ala-Ala-Val. As N-protecting groups use either the benzyloxycarbonyl or t-butoxycarbonyl groups. For coupling use DCC.

12. Propose a synthesis for the hexapeptide in problem 11 using N-carboxy anhydrides. Assume that the necessary anhydrides are commercially available.

13. Propose a synthesis of the pentapeptide Ala-Lys-Glu-Gly-Gly. Note that the terminal amino group of lysine and the terminal carboxy group of glutamic acid must be protected.

14. Propose a synthesis of the decapeptide Ala-Val-Phe-Ala-Ala-Ala-Val-Phe-Ala-Ala.

15. Pyroglutamic acid, pyroGlu, is a cyclic lactam obtained by heating glutamic acid.

pyroGlu

This derivative of proline occurs in an important tripeptide, thyrotropin-releasing hormone, TRH, which occurs in brain tissue. It also occurs in the anterior lobe of the pituitary gland where it stimulates the secretion of several other hormones. TRH has been shown to be pyroglutamylhystidylprolineamide. Write out the structure of TRH. A sensitive assay method has been developed that makes use of synthetic hormone. Propose a synthesis of TRH from pyroglutamic acid and other required reagents.

16. The cyanogen bromide method for cleavage of peptide chains involves reaction of the nucleophilic sulfur of a methionine unit with the carbon of BrCN. Write out the complete reaction mechanism.

17. What polypeptide fragments will be produced when ribonuclease A (Figure 29.7) is partially hydrolyzed with (a) trypsin, (b) chymotrypsin, or (c) cyanogen bromide. Assume that the disulfide bonds are cleaved prior to hydrolysis.

18. Gastrins are heptadecapeptide (17 amino acid units) hormones that stimulate the secretion of gastric acid in the stomach of mammals. Feline gastrin has the empirical amino acid composition ($Ala_2Asp_1Gly_2Glu_5Leu_1Met_1Phe_1Pro_1Trp_2Tyr_1$). The peptide was digested with chymotrypsin and four peptide fragments were isolated. The four fragments were sequenced and found to be

I   Glu-Gly-Pro-Trp
II Gly-Trp
III Met-Asp-Phe
IV Leu-Glu-Glu-Glu-Glu-Ala-Ala-Tyr

End-group analysis revealed that the N-terminal unit is Glu and the C-terminal unit is Phe. What two structures for feline gastrin are compatible with the foregoing evidence?

19. Porcine pancreatic secretory trypsin inhibitor I is a protein containing 56 amino acid units. Acidic hydrolysis, followed by amino acid analysis, gave the following empirical composition: $Asp_4Thr_6Ser_6Glu_7Pro_5Gly_4Ala_1Val_4Cys_6Ile_3Leu_2Tyr_2Lys_4Arg_2$. (*Note:* Complete hydrolysis does not distinguish Gln from Glu or Asn from Asp.) After cleavage of disulfide bridges, the protein was digested with trypsin. Nine fragments were isolated and purified by chromatography. The nine fragments were each sequenced by repetitive Edman degradation. Eight of the fragments were found to be

T-1   Lys
T-2   Arg
T-3   Ser-Gly-Pro-Cys
T-4   Thr-Ser-Pro-Gln-Arg
T-5   Gln-Thr-Pro-Val-Leu-Ile-Gln-Lys
T-6   Ser-Asn-Glu-Cys-Val-Leu-Cys-Ser-Glu-Asn-Lys
T-7   Ile-Tyr-Asn-Pro-Val-Cys-Gly-Thr-Asp-Gly-Ile-Thr-Tyr
T-8   Glu-Ala-Thr-Cys-Thr-Ser-Glu-Val-Ser-Gly-Cys-Pro-Lys

The ninth fragment contained 24 amino acid units and had the empirical composition $Asp_4Thr_2Ser_2Glu_2Pro_1Gly_2Val_2Cys_3Ile_2Leu_1Tyr_2Lys_1$. Seven cycles of Edman degradation showed that the N-terminal end of fragment T-9 had the composition

T-9   Ile-Tyr-Asn-Pro-Val-Cys-Gly···

Edman degradation of the intact protein showed the N-terminal unit to be Thr. The C-terminal residue was shown to be Cys.

The protein was then digested with chymotrypsin and three peptide fragments were isolated. The three chymotryptic fragments were each subjected to total hydrolysis and analyzed for amino acid composition. They were also subjected to three cycles of Edman degradation to identify the N-terminal sequence and incubated with carboxypeptidase to identify the C-terminal unit. The partial structures of the three fragments were found to be

Ch-1   Thr-Ser-Pro($Thr_2Ser_2Glu_3Pro_1Gly_1Ala_1Val_1Cys_2Ile_1Lys_1Arg_1$)Tyr
Ch-2   Asn-Pro-Val($Asp_1Thr_2Gly_2Cys_1Ile_1$)Tyr
Ch-3   Ser-Asn-Glu($Asp_1Thr_1Ser_2Glu_3Pro_2Gly_1Val_2Cys_2Ile_1Leu_2Lys_3Arg_1$)Cys

The intact protein was then treated with methyl isothiocyanate. This reagent modifies the lysine side chains so that they are not cleaved by trypsin. The modified protein was digested with trypsin and three fragments were isolated. The three fragments were isolated, hydrolyzed, and analyzed and shown to have the following empirical compositions:

*T-1 $Thy_1Ser_1Pro_1Arg_1Glu_1$
*T-2 $Thr_1Ser_1Glu_2Pro_2Gly_1Val_1Cys_1Ile_1Leu_1Lys_1$
*T-3 $Asp_4Thr_4Ser_4Glu_4Pro_2Gly_3Ala_1Val_3Cys_5Ile_2Leu_1Tyr_2Lys_3Arg_1$

From the data, what is the complete amino acid sequence of the protein?

Chapter 30
Aromatic Halides, Phenols, Phenyl Ethers, and Quinones

30.1
Introduction

When halogen is in the side chain of an alkyl-substituted benzene, the chemistry is essentially the same as for any alkyl halide. As we saw in Section 22.5, enhanced reactivity is seen in the displacement and carbocation reactions of *benzylic* halides, since the phenyl group stabilizes either an S_N2 transition state or a cationic center to which it is directly attached. Side-chain halides are prepared by methods analogous to those used for the preparation of normal alkyl halides, mainly from corresponding alcohols (Section 10.7.D). Recall that free radical halogenation is especially useful for preparation of benzylic halides (Section 22.5.A).

When halogen is attached directly to the benzene ring, there is some special chemistry, in regard to both preparations and reactions, that is different from the normal chemistry of alkyl halides. We shall take up these unique reactions pertaining to aryl halides in Sections 30.2 and 30.3. Since several of the important reactions of aryl halides give rise to phenols, it is convenient to study that important family of organic compounds next, and we shall do so in Sections 30.4–30.6. Finally, we shall take up the quinones in Section 30.7, since the chemistry of this class of compounds is intimately wound up with the chemistry of phenols.

30.2
Preparation of Halobenzenes

One important preparation of ring halides is by electrophilic aromatic substitution (Section 23.1). The active reagent in these reactions is an actual or incipient halonium ion. For suitably activated rings the halogen alone may be used as the reagent.

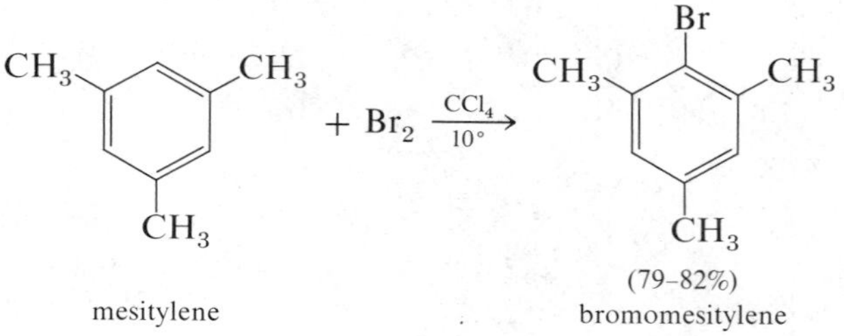

mesitylene   bromomesitylene
(79–82%)

For less reactive rings a Lewis acid such as a ferric salt is used to catalyze the reaction.

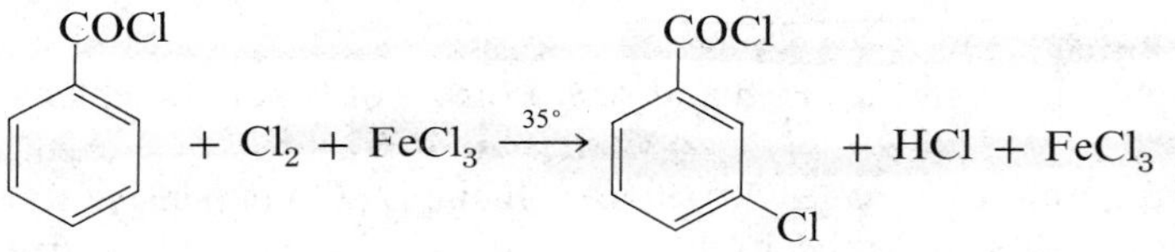

$$\text{(COCl-benzene)} + Cl_2 + FeCl_3 \xrightarrow{35°} \text{(COCl, Cl-benzene)} + HCl + FeCl_3$$

Direct electrophilic aromatic substitution is most useful as a method for synthesis of halobenzenes when the product is essentially a single compound. This is usually the case when the benzene ring contains a *meta*-directing group or when the structure of the starting material is such that only a single product can be produced. Thus bromination of *p*-nitrotoluene gives only one product, as does bromination of *p*-xylene.

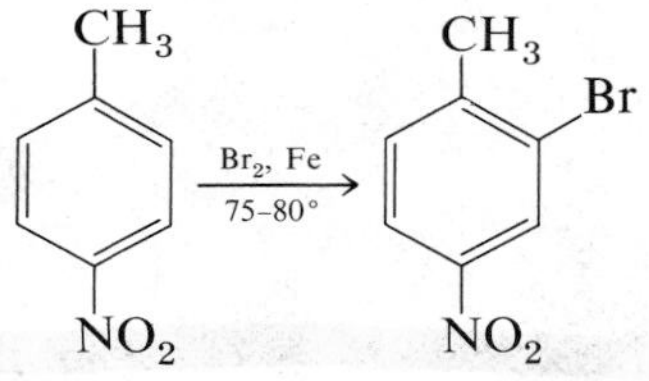

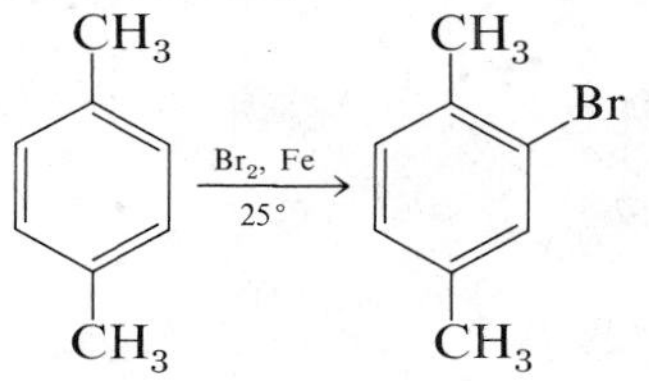

$$\text{(CH}_3\text{, NO}_2\text{-benzene)} \xrightarrow[75-80°]{Br_2, \ Fe} \text{(CH}_3\text{, Br, NO}_2\text{-benzene)}$$

$$\text{(CH}_3\text{, CH}_3\text{-benzene)} \xrightarrow[25°]{Br_2, \ Fe} \text{(CH}_3\text{, Br, CH}_3\text{-benzene)}$$

When the ring contains a single *ortho,para*-directing group, a mixture of products generally results. If the group is very large, as in the case of *t*-butyl, the amount of *ortho* product is small, and the *para*-disubstituted product may be obtained in good yield.

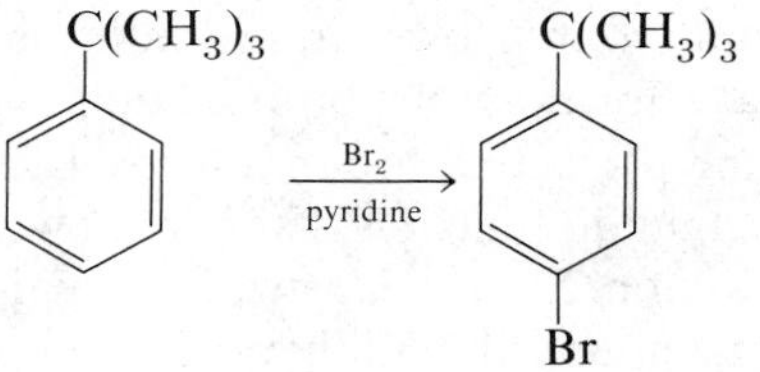

$$\text{(C(CH}_3)_3\text{-benzene)} \xrightarrow[\text{pyridine}]{Br_2} \text{(C(CH}_3)_3\text{, Br-benzene)}$$
(94%)

However, in most such cases direct halogenation gives a mixture of isomeric products that is difficult to separate. For example, bromination of toluene gives a mixture consisting of 65% *p*-bromotoluene and 35% *o*-bromotoluene.

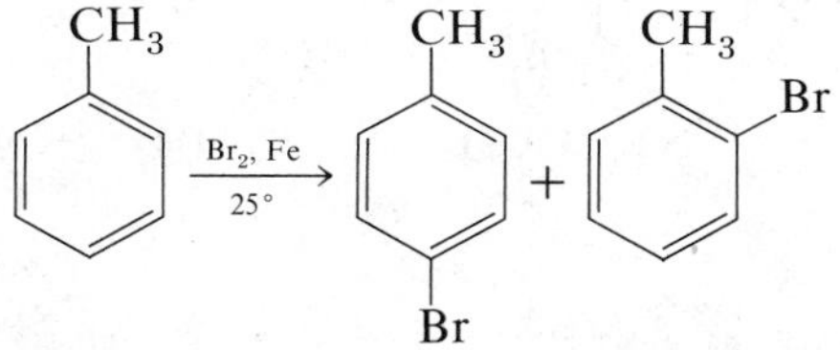

$$\text{(CH}_3\text{-benzene)} \xrightarrow[25°]{Br_2, \ Fe} \text{(CH}_3\text{, Br-benzene)} + \text{(CH}_3\text{, Br-benzene)}$$
(65%)   (35%)

984

Chap. 30

**Aromatic
Halides,
Phenols,
Phenyl Ethers,
and Quinones**

It is in situations like this that arenediazonium salts are useful (Section 25.5). For example, although the *ortho* and *para* isomers of bromotoluene cannot easily be separated by distillation, the nitrotoluenes are readily separated by this method (see Table 23.5, page 720). Since toluene is an inexpensive starting material, direct nitration followed by fractional distillation represents an economical method for the preparation of both the *ortho* and *para* isomers.

(55%)   (45%)
b.p. 238°   b.p. 220°

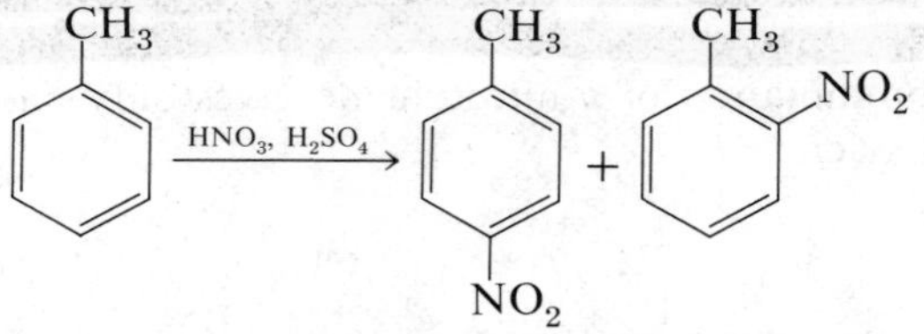

Both isomers may be converted into the corresponding bromotoluenes by reduction of nitro to amino (Section 24.6.C), followed by Sandmeyer reaction (Section 25.5C). For example,

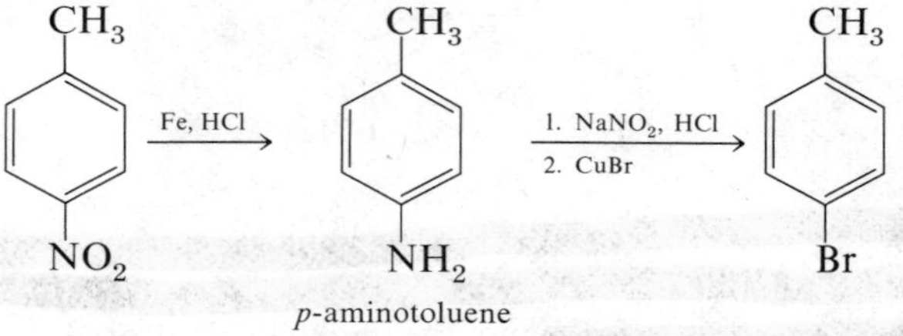

p-aminotoluene

The foregoing example illustrates the indirect *replacement* of a nitro group by halogen via the diazonium group. However, arenediazonium chemistry may also be utilized in another way. For example, bromination of *p*-acetamidotoluene followed by hydrolysis of the protecting group affords 4-amino-3-bromotoluene in good purity, since the acetamido function is a more powerful *ortho,para*-directing group than is methyl (Section 23.8). Diazotization of this product, followed by reduction of the arenediazonium salt with hypophosphorous acid, affords *m*-bromotoluene.

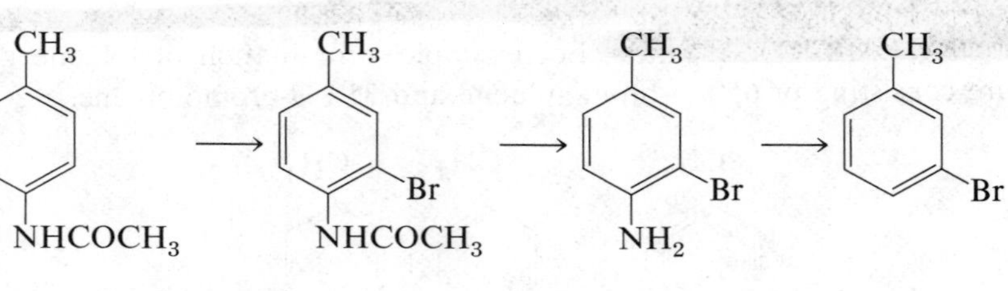

EXERCISE: Write equations illustrating the preparation of the three chlorobromobenzenes starting with benzene.

30.3
Reactions of Halobenzenes

A. *Substitution Reactions*

The halogen of an aryl halide can be replaced by other nucleophiles. However, such substitution reactions do *not* occur by the S_N2 mechanism. Like vinyl halides (Section 12.7), aryl halides cannot achieve the geometry necessary for a backside displacement; the ring shields the rear of the C—X bond.

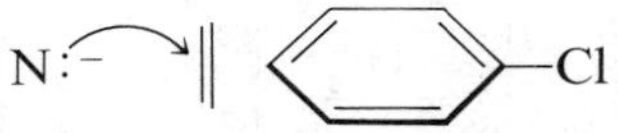

Instead, nucleophilic substitution occurs by two other mechanisms, **addition-elimination** and **elimination-addition**.

1. THE ADDITION-ELIMINATION MECHANISM. Aryl halides that have electron-attracting groups in the positions *ortho* and *para* to the halogen undergo substitution under rather mild conditions. The most effective groups are nitro and carbonyl.

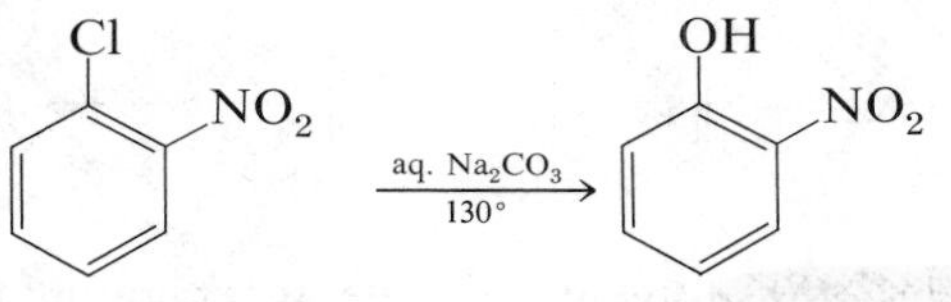

o-nitrochlorobenzene   *o*-nitrophenol

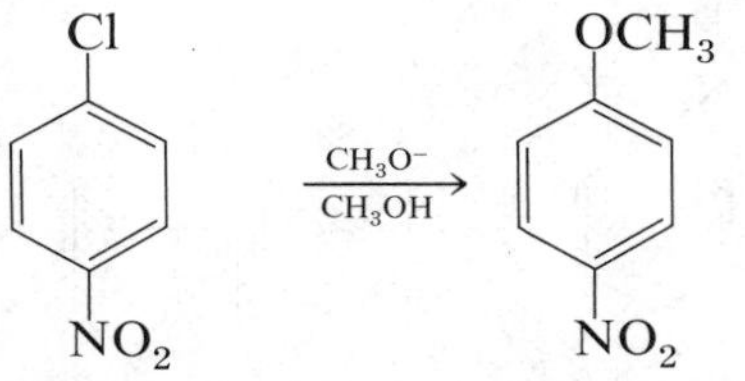

p-nitrochlorobenzene   *p*-nitroanisole

With two nitro groups in conjugating positions, this type of displacement reaction is quite facile:

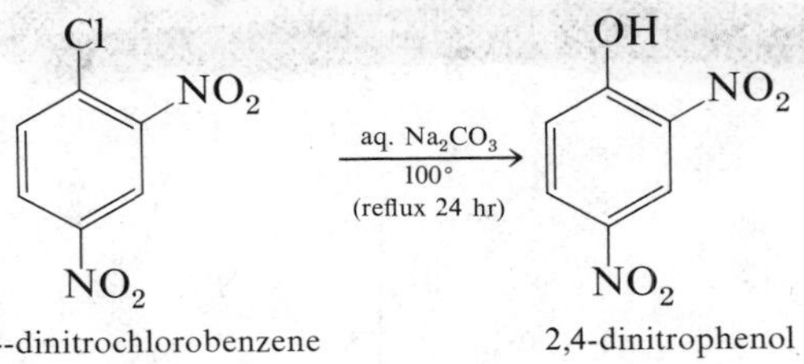

2,4-dinitrochlorobenzene   2,4-dinitrophenol

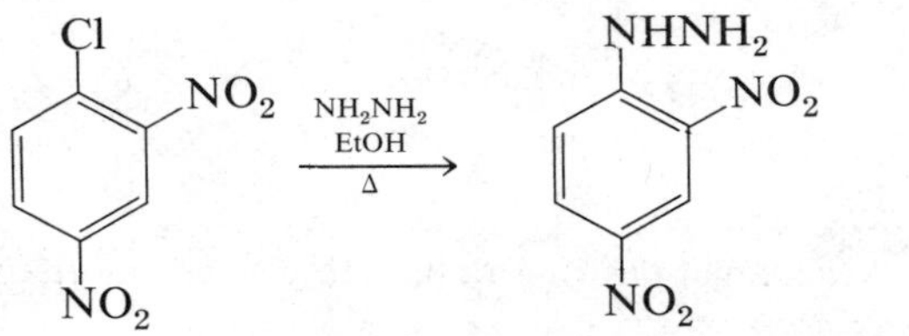

(30-1)

2,4-dinitrophenylhydrazine

986

Chap. 30

**Aromatic
Halides,
Phenols,
Phenyl Ethers,
and Quinones**

> 2,4-Dinitrophenylhydrazine, prepared as shown in (30-1), is a common reagent used for preparing the corresponding 2,4-dinitrophenylhydrazone derivatives of aldehydes and ketones (Section 13.7.C). These derivatives are usually crystalline compounds with well-defined melting points and are useful for characterizing aldehydes and ketones; the 2,4-dinitrophenylhydrazones are commonly abbreviated as DNPs.

The mechanism of these substitution reactions involves two steps, an addition followed by an elimination. It is analogous to the nucleophilic addition-elimination mechanism so prevalent in the chemistry of carboxylic acids and derivatives of carboxylic acids (Chapter 19).

$$R\overset{\displaystyle O}{\underset{}{\overset{\|}{C}}}X + N{:}^- \rightleftharpoons \left[R\overset{\displaystyle O^-}{\underset{N}{\overset{|}{\underset{|}{C}}}}X\right] \rightleftharpoons R\overset{\displaystyle O}{\underset{}{\overset{\|}{C}}}N + X{:}^-$$

In the first step the attacking nucleophile adds to the benzene ring to give a resonance-stabilized pentadienyl anion.

The pentadienyl anion can eject the nucleophile, regenerating the reactants, or it can eject halide ion, giving the substitution product.

However, even a conjugated pentadienyl anion is not sufficiently stable for this mechanism to operate with such simple aryl halides as chlorobenzene or *o*-bromotoluene. Electron-attracting groups provide further resonance stabilization of the anion, thus lowering its energy enough for it to be formed as a reaction intermediate. As the foregoing resonance structures show, the nitro or carbonyl groups are most effective when they are *ortho* or *para* to the leaving group.

EXERCISE: Write equations illustrating the conversion of chlorobenzene to *p*-methoxyaniline.

2. THE ELIMINATION-ADDITION MECHANISM; BENZYNE. After the foregoing discussion it may seem surprising that one commercial preparation of phenol involves heating chlorobenzene itself with aqueous sodium hydroxide.

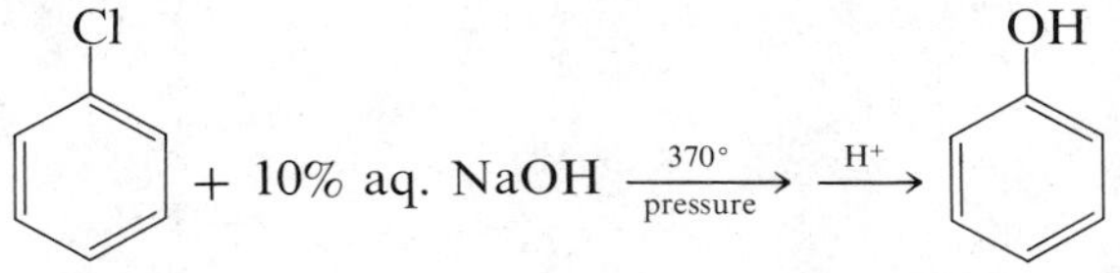

$$\text{(chlorobenzene)} + 10\% \text{ aq. NaOH} \xrightarrow[\text{pressure}]{370°} \xrightarrow{H^+} \text{(phenol)}$$

However, this reaction is not a simple displacement of chloride by hydroxide. *o*-Chlorotoluene, for example, gives not only *o*-methylphenol in this reaction but also *m*-methylphenol.

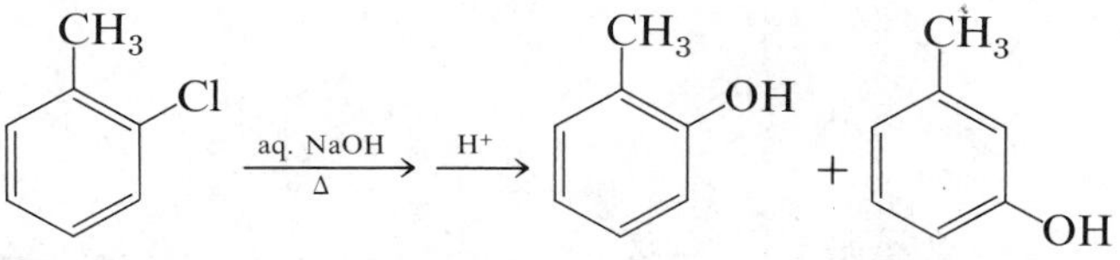

An analogous reaction occurs under milder conditions with amide ion in liquid ammonia.

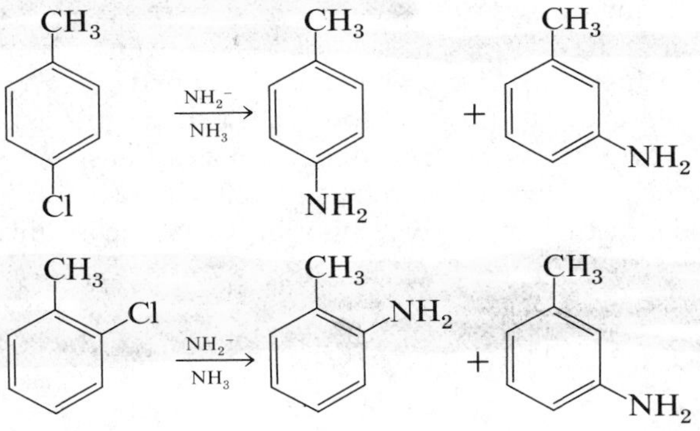

The remarkable feature of these reactions is that the entering group substitutes not only at the position of the displaced halide but also at the ring position adjacent to the original halide. Even iodobenzene shows this behavior, as has been demonstrated using ^{14}C-labeled materials.

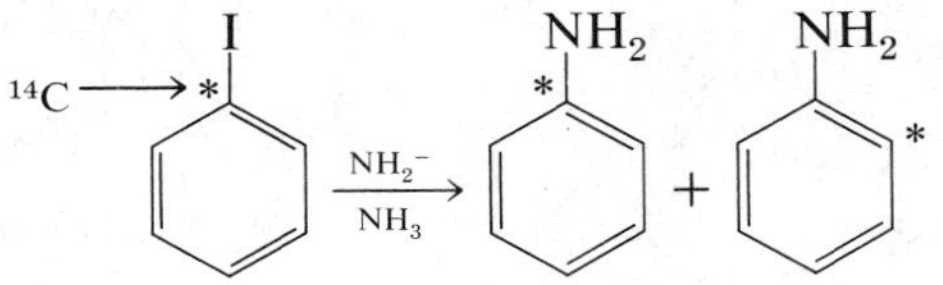

These results are rationalized by the involvement as a reactive intermediate of the product of an elimination reaction—dehydrobenzene or "benzyne."

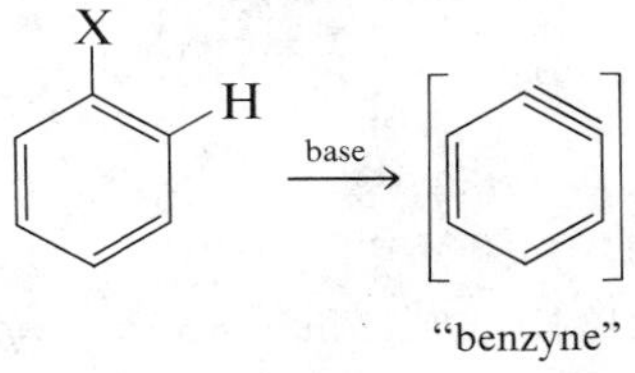

"benzyne"

The detailed mechanism involves a series of steps. Benzene itself is a weak acid, but its pK_a of 43 corresponds to a much higher acidity than the alkanes (pK_as $\sim$ 50). The close proximity of an electronegative halogen renders an adjacent hydrogen sufficiently acidic that it is removed by a strong base such as NH_2^-.

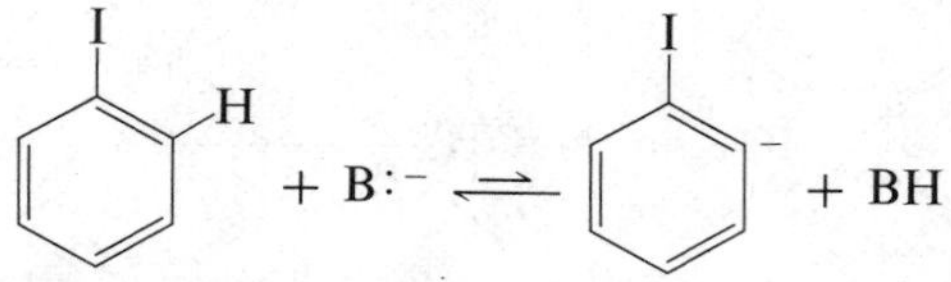

$$\text{[structure]} + \text{B}:^- \rightleftharpoons \text{[structure]}^- + \text{BH}$$

The intermediate iodophenyl anion can itself pick up a proton to regenerate the original iodobenzene, *or it can lose iodide ion.*

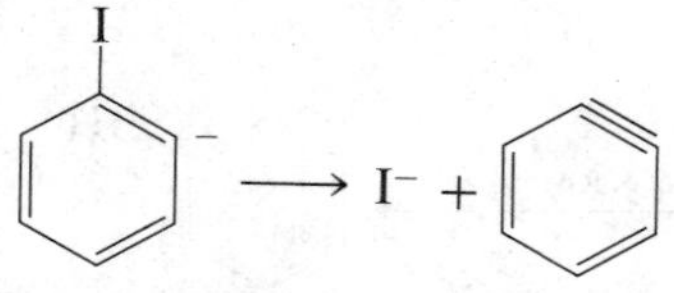

$$\text{[structure]}^- \longrightarrow \text{I}^- + \text{[structure]}$$

The driving force for this reaction is the formation of a stable halide ion. The "benzyne" generated is a very reactive intermediate. The "triple bond" in benzyne is highly strained. Recall that the two carbons in acetylene are *sp*-hybridized and that the H—C—C angles are 180°. That is, the two carbons of an alkyne *as well as the two atoms directly attached to the triple bond* comprise a linear array. For geometric reasons this is impossible when the triple bond is in a small ring. In fact, the smallest stable cycloalkyne is cyclooctyne, and even it is less stable than acyclic alkynes. Nevertheless, benzyne does form as a transient reaction intermediate. It has even been detected spectroscopically by using special techniques.

The electronic structure of benzyne may be visualized readily as a distorted acetylene. A triple bond has two π-bonds, as shown in Figure 30.1. One π-bond is constructed from *p*-orbitals perpendicular to the plane of the paper (dotted line in Figure 30.1); the other π-bond derives from overlap of *p*-orbitals in the plane of

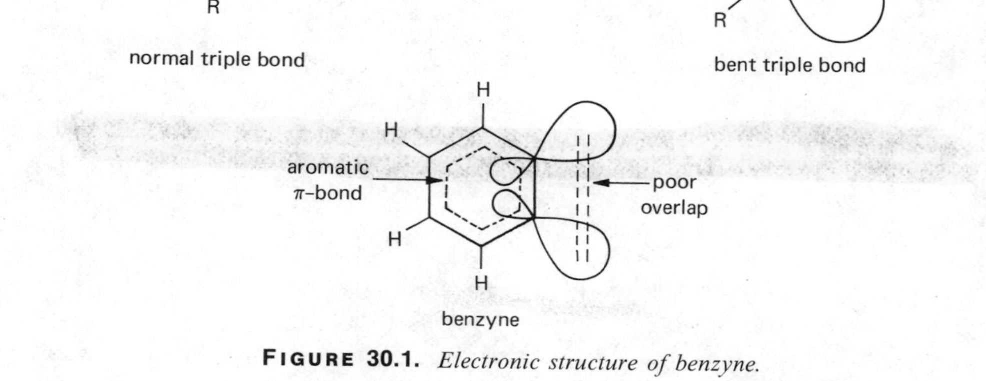

FIGURE 30.1. *Electronic structure of benzyne.*

the page, as illustrated. When the triple bond is distorted from linearity, one π-bond is essentially unchanged, but the other π-bond now involves hybrid orbitals directed away from each other with consequent reduced overlap. Reduced overlap means a weaker and more reactive bond. Benzyne is related in this sense to the distorted acetylenes. The two orbitals shown in Figure 30.1 provide inefficient overlap and a weak, reactive bond. The resulting strained triple bond reacts readily with any available nucleophilic reagent *at either end of the triple bond*.

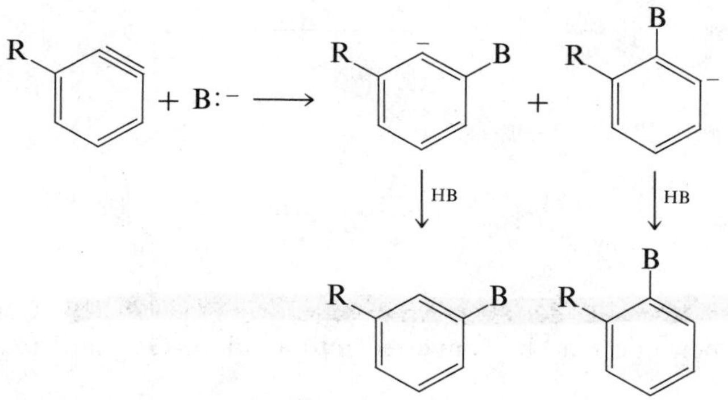

Benzyne may also be generated by decomposition of the diazonium salt produced by diazotization of anthranilic acid. The diazonium compound is an inner salt and is insoluble. The dry salt will detonate and must be kept moist and handled with care. The controlled decomposition in ethylene chloride provides the unusual strained hydrocarbon, biphenylene.

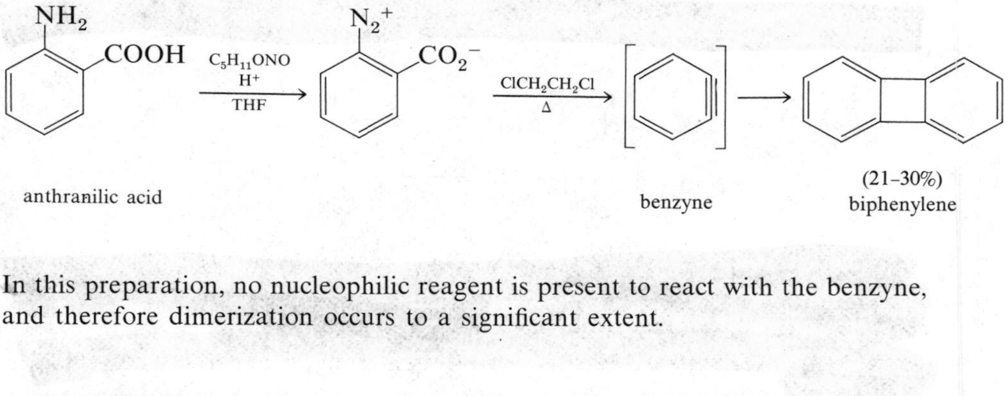

anthranilic acid   benzyne   biphenylene
                                                                          (21–30%)

In this preparation, no nucleophilic reagent is present to react with the benzyne, and therefore dimerization occurs to a significant extent.

EXERCISE: What products are produced by treatment of *o*-, *m*-, and *p*-bromotoluene with KNH_2 in liquid NH_3?

B. *Metallation*

Aryl bromides form Grignard reagents in a normal fashion, and these derivatives undergo all the usual reactions of Grignard reagents; they react with aldehydes and ketones, CO_2, D_2O, and so on.

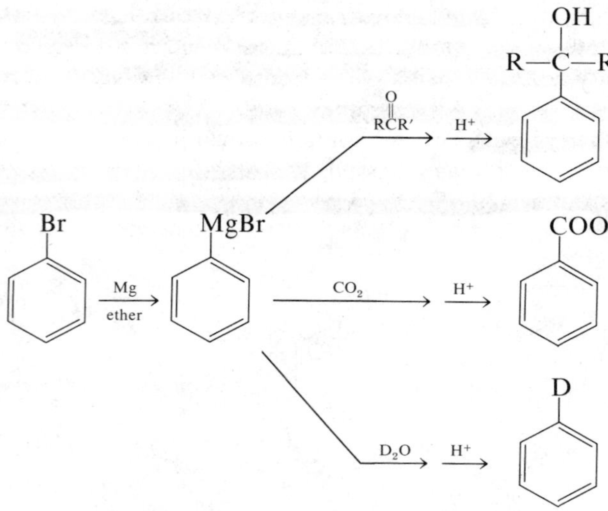

However, aryl chlorides do not react with magnesium in ether. Consequently a bromochlorobenzene may be converted into a chloro-Grignard reagent.

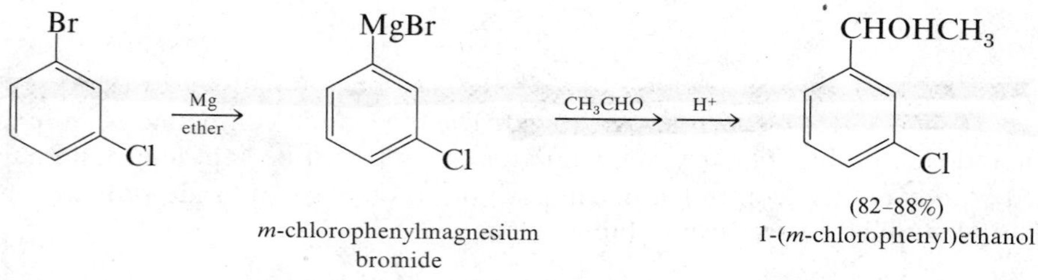

m-chlorophenylmagnesium bromide

(82–88%)
1-(*m*-chlorophenyl)ethanol

Grignard reagents *can* be produced from aryl chlorides by using tetrahydrofuran (THF) as the solvent.

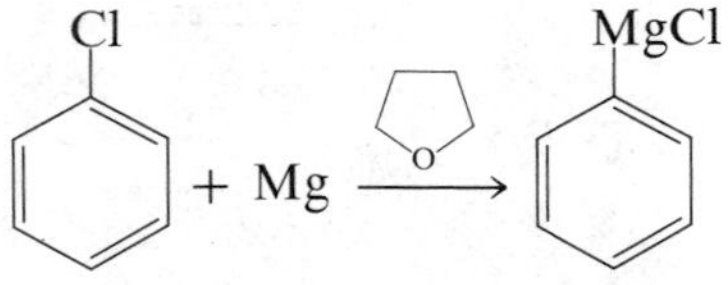

Of course, the formation of the Grignard reagent is successful only if no functional group is present that will react with such reagents: examples of such reactive groups are NO_2, NO, COR, SO_3R, CN, OH, and NH_2.

Aryllithium reagents may be prepared from lithium metal and the aryl chloride or bromide.

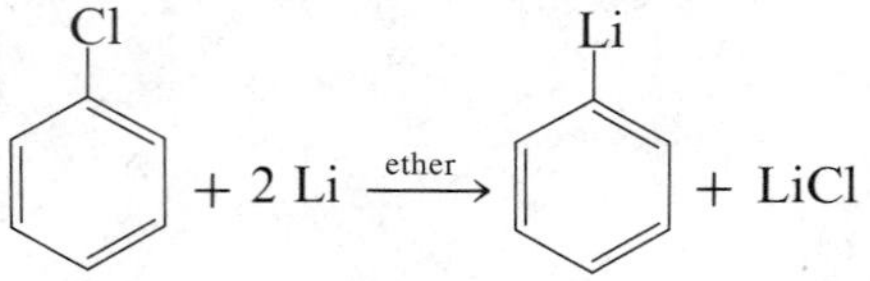

The lithium reagents generally undergo the same reactions as the Grignard reagents. Furthermore, aryllithiums can be prepared by **transmetallation** of an aryl bromide or iodide with an alkyllithium.

(85%)

With aryl bromides or iodides the reaction is rapid, even at low temperature. The reaction may be regarded as a displacement reaction on halogen to form the lithium salt of a more stable anion.

$$\text{ArX} + \text{R}^- \text{Li}^+ \longrightarrow \text{Ar}^- \text{Li}^+ + \text{RX}$$

This reaction is most successful with aromatic bromides and iodides; the chlorides do not react as cleanly or rapidly.

When a mixture of an alkyl halide and an aryl halide is treated with sodium in ether, coupling occurs to give the arylalkane. The reaction, known as the **Wurtz–Fittig** reaction, is sometimes an excellent preparative method.

$$\text{C}_6\text{H}_5\text{Br} + \text{CH}_3(\text{CH}_2)_3\text{CH}_2\text{Br} + 2\,\text{Na} \xrightarrow[20°]{\text{ether}} \text{C}_6\text{H}_5(\text{CH}_2)_4\text{CH}_3 + 2\,\text{NaBr}$$

(65–70%)

In this reaction the sodium reacts first with the aryl bromide to form the arylsodium, which then reacts with the alkyl bromide by the S_N2 mechanism.

$$\text{ArBr} + 2\,\text{Na} \longrightarrow \text{ArNa} + \text{NaBr}$$
$$\text{ArNa} + \text{RX} \longrightarrow \text{ArR} + \text{NaX}$$

Under the proper conditions, the yields are quite good; this reaction is related to, but is considerably better than, the Wurtz reaction of alkyl halides with sodium (Section 15.4.A). Like all organometallic reactions, the Wurtz–Fittig reaction is only applicable to compounds not having highly reactive functional groups (hydroxy, carbonyl, nitro, and so on).

A reaction that is superficially related to the Wurtz reaction is the **Ullmann** reaction, in which two molecules of an aryl iodide are coupled by heating with copper powder. The product is a **biaryl**, a compound in which two benzene rings are joined together.

2,2'-dinitrobiphenyl

The chemistry of biaryls will be detailed in a subsequent chapter (Chapter 32). The Ullmann reaction works well with chlorides, bromides, and iodides and is facilitated by electron-attracting groups such as NO_2 and CN functions. The reaction involves the formation of an arylcopper intermediate that undergoes a free-radical-like coupling, probably while still coordinated to copper.

A method of coupling aryl halides that is complementary to the Ullmann reaction involves formation of the diarylcuprate (Section 15.5.E), which is oxidized by oxygen at low temperature.

992

Chap. 30

**Aromatic
Halides,
Phenols,
Phenyl Ethers,
and Quinones**

lithium diphenylcuprate

(75%)
biphenyl

The cuprate method avoids the high temperatures of the Ullmann reaction, but it is only applicable with arenes not having functional groups that react with the aryllithium intermediate.

EXERCISE: Show how *p*-bromotoluene may be converted into *p*-toluic acid (page 649), 4,4′-dimethylbiphenyl, and 4-*n*-propyltoluene.

30.4
Nomenclature of Phenols and Phenyl Ethers

Compounds having a hydroxy group directly attached to a benzene ring are called **phenols.** The term phenol is also used for the parent compound, hydroxybenzene. Hydroxybenzene may be regarded as an enol (Section 13.6), as implied by the name phenol, from **ph**enyl + **enol.** However, unlike simple ketones, which are far more stable than their corresponding enols, the tautomeric equilibrium for phenol lies far on the side of the enol form. The reason for this difference is the resonance energy of the aromatic ring, which provides an important stabilization of the enol form.

$$CH_3-\overset{O}{\overset{\|}{C}}-CH_3 \rightleftharpoons CH_3-\overset{OH}{\overset{|}{C}}=CH_2 \qquad \Delta H° \approx +14 \text{ kcal mol}^{-1}$$

cyclohexa-2,4-dien-1-one   phenol

$$\Delta H° \approx -16 \text{ kcal mol}^{-1}$$

Since the functional group occurs as a suffix in phen**ol,** many compounds containing an aromatic hydroxy group are named as derivatives of the parent compound phenol, as illustrated by the following IUPAC names. Although *Chemical Abstracts* utilizes the IUPAC-approved name phenol for the parent compound, substituted phenols are indexed as derivatives of **benzenol.** The *Chemical Abstracts* names are given in brackets for the following compounds. We shall not use the name benzenol in this text.

m-bromophenol
[3-bromobenzenol]

p-*t*-butylphenol
[4-(1,1-dimethylethyl)benzenol]

2,4-dinitrophenol
[2,4-dinitrobenzenol]

Suffix groups such as sulfonic acid and carboxylic acid take priority, and when these groups are present, the hydroxy group is used as a modifying prefix.

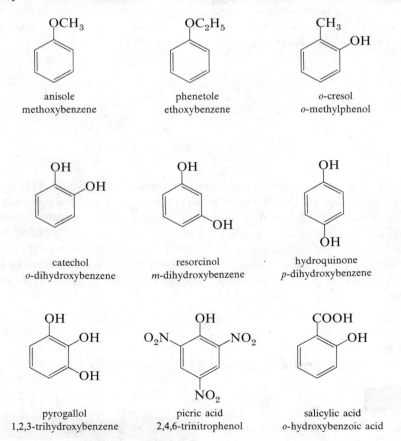

p-hydroxybenzoic acid   m-hydroxybenzaldehyde   2,4-dihydroxybenzene-
sulfonic acid

Phenyl ethers are named in the IUPAC system as alkoxyarenes, although the "ether" nomenclature is used for some compounds.

phenyl ether   t-butoxybenzene   p-benzyloxytoluene

Phenols and their ethers are widespread in nature, and, as is usual for such compounds, trivial names abound. Many of these names are in such common use that they should be learned.

anisole   phenetole   o-cresol
methoxybenzene   ethoxybenzene   o-methylphenol

catechol   resorcinol   hydroquinone
o-dihydroxybenzene m-dihydroxybenzene p-dihydroxybenzene

pyrogallol   picric acid   salicylic acid
1,2,3-trihydroxybenzene 2,4,6-trinitrophenol o-hydroxybenzoic acid

994

Chap. 30

Aromatic
Halides,
Phenols,
Phenyl Ethers,
and Quinones

Note that compounds with more than one hydroxy group are named with the hydroxy prefix. Terms such as "phen-diol" or "benzene-triol" are *not* used.

Other trivial names are common, and the student can expect to find them in the current literature.

phloroglucinol   gallic acid   guaiacol

a urushiol

Urushiols are the active constituents of the allergenic oils of poison ivy, sumac, and oak. They are C-3 alkylated catechols in which the side chain can be saturated or can contain up to three double bonds. In poison ivy and poison sumac the side chain contains fifteen carbons; in poison oak it contains seventeen carbons.

Lignins are complex natural products that occur together with cellulose in the "woody" part of plants such as shrubs and trees. Because lignins are high molecular weight polymers, their exact structures are not known. They are composed of three basic building blocks: coniferyl alcohol, sinapyl alcohol, and *p*-coumaryl alcohol.

coniferyl alcohol   sinapyl alcohol   *p*-coumaryl alcohol

Different plants apparently have lignins of different composition. Guaiacyl lignins are derived mainly from coniferyl alcohol. Guaiacyl-syringyl lignins are also derived principally from coniferyl alcohol, but contain significant amounts of sinapyl alcohol. *p*-Coumaryl alcohol is a minor constituent (1–5%) of all lignins. The actual polymeric structures of the lignins contain a variety of types of linkages between the phenolic monomers, as shown by the part structure in Figure 30.2. There also seem to be glycosidic bonds between the lignin polymer and the cellulose units in the wood.

30.5
Preparation and Properties of Phenols and Ethers

Phenols are generally crystalline compounds with distinctive odors. Phenol itself melts at 40.9°, but is often found to be semiliquid because of the presence of water, which lowers the melting point; it is completely liquefied by the addition

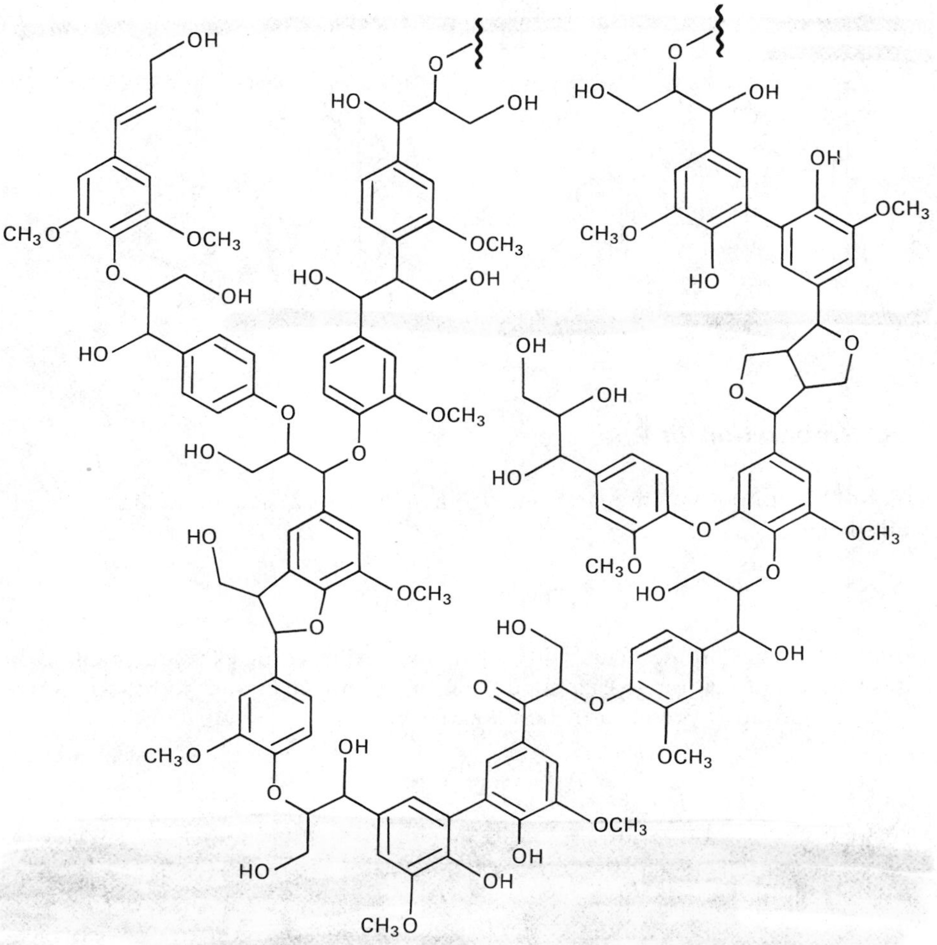

995

Sec. 30.5

**Preparation
and Properties
of Phenols
and Ethers**

FIGURE 30.2. *Typical structure of a softwood lignin. The existence of the types of linkages illustrated has been established, but they are arranged in this partial structure in an arbitrary manner for illustrative purposes.*

of 8% of water. It is soluble to the extent of 6.7 g per 100 mL of cold water and is totally miscible with hot water. The lower alkylphenols are sparingly soluble in water; for example, *o*-cresol dissolves to the extent of 2.5 g per 100 mL of water at 25°. Phenol and the cresols are widely used in commercial disinfectants. Phenol turns pink on exposure to air because of oxidation. The sensitivity of phenols to air oxidation is enhanced by the presence of more than one hydroxy group and by alkali. An alkaline solution of pyrogallol rapidly removes oxygen from a stream of air, and it is often used for this purpose. The oxidation of phenols to quinones will be discussed in Section 30.7.B. Phenols are sufficiently acidic that they are caustic toward flesh and are poisonous.

The lower phenyl ethers are liquids; for example, anisole boils at 154°. Unlike phenols, the ethers are essentially insoluble in water. They lack the hydroxy group of phenol, which can hydrogen-bond to water oxygens. The ether oxygens have relatively low basicity and form only weak hydrogen bonds to water hydrogens. The low basicity of the oxygens of phenyl ethers compared to aliphatic ethers stems from conjugation of a lone pair with the aromatic ring. The same phenome-

996

Chap. 30

**Aromatic
Halides,
Phenols,
Phenyl Ethers,
and Quinones**

non is responsible for the reduced basicity of aromatic amines compared to aliphatic amines.

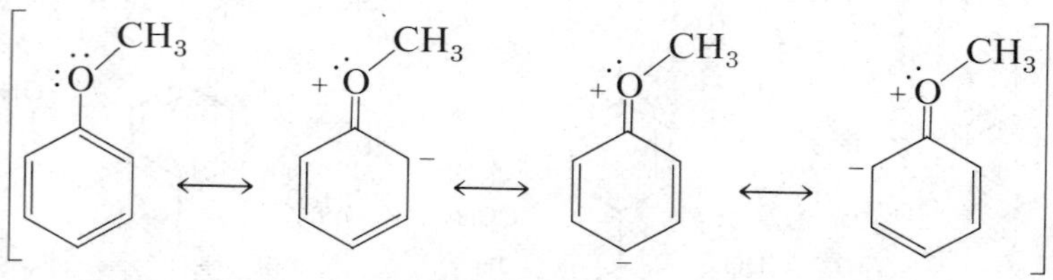

Aryl ethers are also more stable to oxidation than phenols.

A. *Preparation of Phenols*

All of the important preparations of phenols involve reactions that have already been discussed.

$$ArSO_3^- \xrightarrow[\Delta]{NaOH} ArOH$$

Fusion of arenesulfonic acids with alkali hydroxide is an excellent method in cases where sulfonation gives a good yield of sulfonic acid and no base-sensitive functional group is present (Section 26.2.B).

$$ArCl \xrightarrow[\Delta]{NaOH} ArOH$$

The hydrolysis of haloarenes with alkali at high temperature is a commercial preparation, but is not suitable for general laboratory use because of the involvement of benzyne intermediates and the formation of mixtures of isomeric phenols (Section 30.3.A). An exception occurs in those systems that contain strongly electron-attracting groups, such as nitro groups, conjugated with the halogen. In such cases nucleophilic aromatic substitution proceeds smoothly to give the hydroxy compound (Section 30.3.A).

$$ArN_2^+ \xrightarrow[\Delta]{H_2O} ArOH + H_3O^+ + N_2$$

The hydrolysis of arenediazonium salts is a good route to many phenols and has been discussed in Section 25.3.

An important industrial preparation of phenol involves the oxidation of cumene, an inexpensive hydrocarbon that may be prepared by alkylation of benzene (Section 23.4).

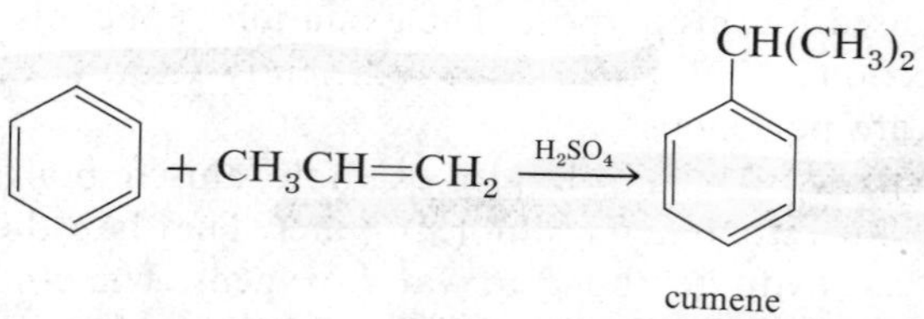

cumene

Cumene is air-oxidized (Section 10.6.B) to obtain cumene hydroperoxide.

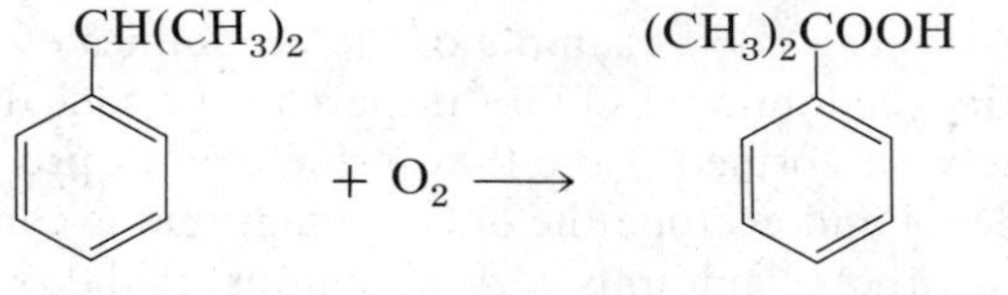

cumene hydroperoxide

The oxidation is a typical free radical chain process. It is especially facile because the intermediate cumyl radical is tertiary and benzylic.

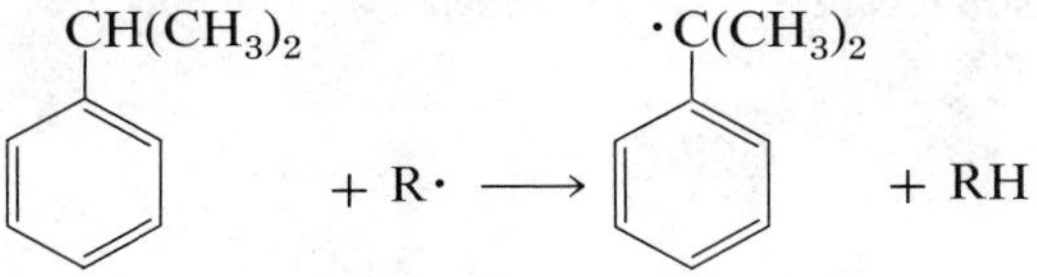

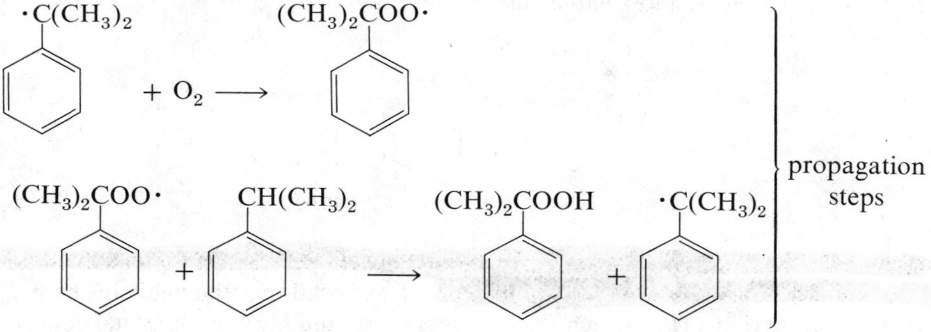

The cumene hydroperoxide is treated with sulfuric acid to obtain phenol and acetone.

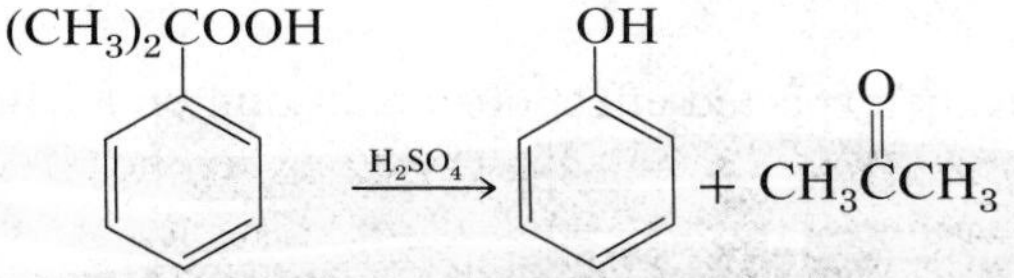

This interesting fragmentation probably occurs by the following mechanism.

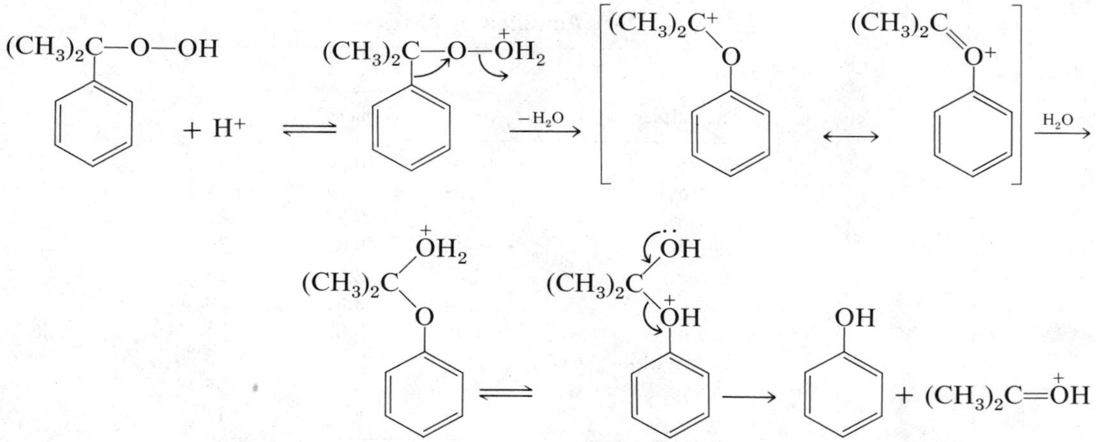

998

Chap. 30

Aromatic
Halides,
Phenols,
Phenyl Ethers,
and Quinones

The cumene process is a good example of the economics of industrial organic synthesis. It owes its economic feasibility in part to the fact that *two* important commercial products are formed. Note that in the overall process from benzene and propene to phenol and acetone the only reagent that is consumed is oxygen. In fact, the overall process amounts to a circuitous oxidation of both starting hydrocarbons. The sulfuric acid that is used both in the alkylation and in decomposition of the hydroperoxide is recycled.

EXERCISE: Write equations illustrating four ways by which benzene may be converted into phenol.

B. *Acidity of Phenols*

Phenol has $pK_a = 10.00$. The pK_as of some substituted phenols are summarized in Tables 30.1; most values are in the range from 8 to 10.

$$K_a = 1.0 \times 10^{-10} \, M \qquad pK_a = 10.00$$

Phenols are generally several orders of magnitude less acidic than carboxylic acids, but are far more acidic than alcohols. If we recall that the value of the pK_a correspond to that pH at which the conjugate acid and base are in equal concentrations, pK_as of 8–10 imply that phenols will dissolve in dilute alkali hydroxide solutions (pH 12–14); however, water-insoluble phenols will not dissolve in aqueous sodium bicarbonate (pH $\sim$ 6–7). Carboxylic acids dissolve in aqueous bicarbonate.

Table 30.1 shows the expected effect of substitution on acidity. Electron-donating groups on the ring are acid-weakening. Halogens increase acidity, and strongly electron-attracting groups such as cyano and nitro have pronounced acid-strengthening effects. With these substituents, the negative charge in the anion can be delocalized onto the oxygen or nitrogen of the substituent.

TABLE 30.1
Acidities of Phenols

Substituent	pK_a (25°)		
	ortho	*meta*	*para*
H	10.00	10.00	10.00
methyl	10.29	10.09	10.26
fluoro	8.81	9.28	9.81
chloro	8.48	9.02	9.38
bromo	8.42	8.87	9.26
iodo	8.46	8.88	9.20
methoxy	9.98	9.65	10.21
methylthio		9.53	9.53
cyano			7.95
nitro	7.22	8.39	7.15

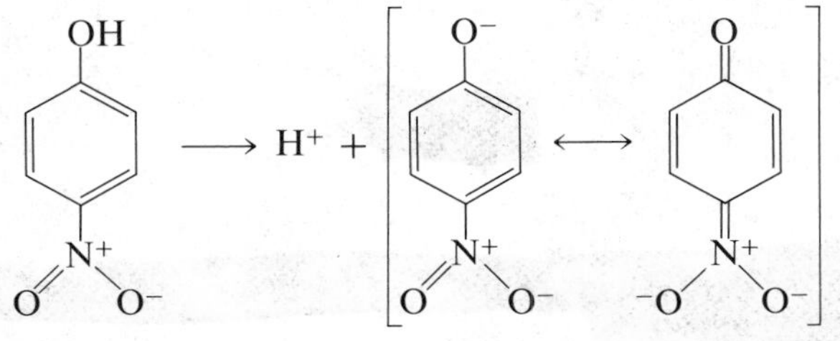

999

Sec. 30.5

**Preparation
and Properties
of Phenols
and Ethers**

Note that a nitro group is more acid-strengthening when it is *ortho* or *para* to the hydroxy group. Dinitrophenols are comparable to carboxylic acids in acidity; for example, the pK_a of 2,4-dinitrophenol is 4.09. Picric acid, 2,4,6-trinitrophenol, has $pK_a = 0.25$ and is a rather strong acid, comparable to trifluoroacetic acid.

EXERCISE: (a) Using graph paper, plot the pK_as of phenols from Table 30.1 versus the pK_as of corresponding anilinium ions from Table 24.4. Explain the result.
(b) Calculate the pH of a 0.1 M solution of phenol in water. What is the pH of a solution containing 0.1 M phenol and 0.1 M sodium phenolate?

C. *Preparation of Ethers*

Alkyl phenyl ethers can be prepared by the Williamson synthesis—the S_N2 reaction of phenoxide ions with alkyl halides. As is usually the case with S_N2 reactions, this preparation works best for primary halides and is least successful with tertiary halides. The reaction can be carried out in water, acetone, dimethyl-formamide, or even alcohol. Because of the great difference in acidities of alcohols and phenols, equilibrium (30-2) lies far to the left.

$$ArO^- + ROH \rightleftharpoons ArOH + RO^- \tag{30-2}$$

Thus, reaction of the alkyl halide with alkoxide ion is not an important side reaction. Some examples of preparations of aryl ethers are

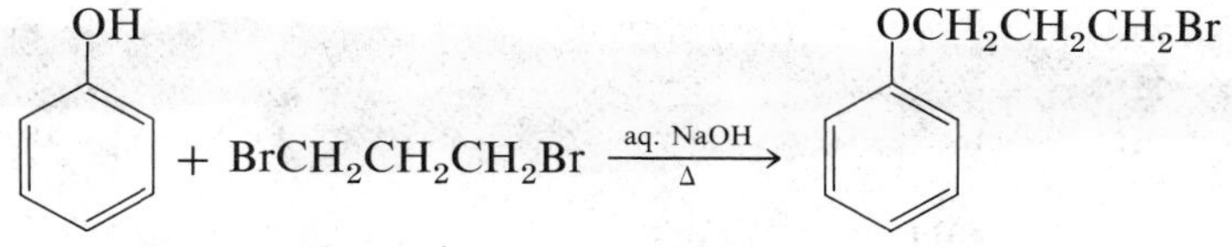

(84–85%)
1-bromo-3-phenoxypropane

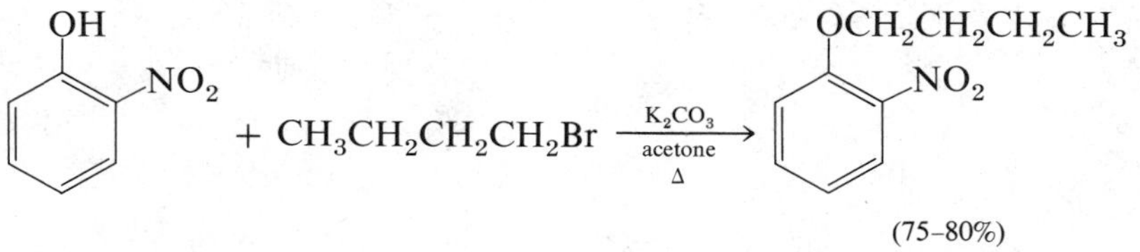

(75–80%)
1-butoxy-2-nitrobenzene

For the preparation of aryl methyl ethers, dimethyl sulfate is especially convenient.

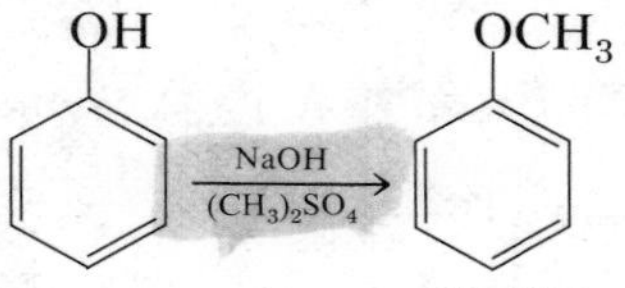

1000

Chap. 30

**Aromatic
Halides,
Phenols,
Phenyl Ethers,
and Quinones**

(72–75%)

Ethers can also be prepared by nucleophilic aromatic substitution in suitable cases (see Section 30.3.A).

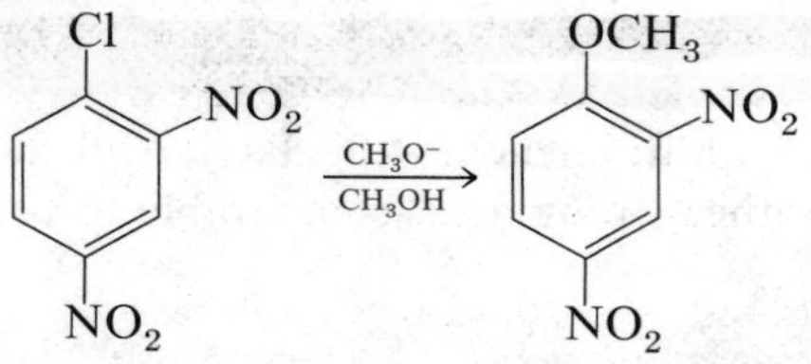

2,4-dinitroanisole

EXERCISE: Write the equation illustrating the synthesis of allyl phenyl ether, $C_6H_5OCH_2CH{=}CH_2$.

30.6
Reactions of Phenols and Ethers

A. *Esterification*

Phenols can be converted to esters but *not* by direct reaction with carboxylic acids. Although the esterification equilibrium is exothermic for alcohols, it is slightly endothermic for phenols. For example, in the gas phase

$$C_6H_5OH + CH_3COOH \rightleftharpoons C_6H_5OOCCH_3 + H_2O \qquad \Delta H° = +1.5 \text{ kcal mole}^{-1}$$

$$C_2H_5OH + CH_3COOH \rightleftharpoons C_2H_5OOCCH_3 + H_2O \qquad \Delta H° = -4.6 \text{ kcal mole}^{-1}$$

This example again demonstrates the significant differences between alcohols and phenols. Aryl esters can be prepared by allowing the phenol to react with an acid chloride or anhydride under basic or acid catalysis. Some typical examples are

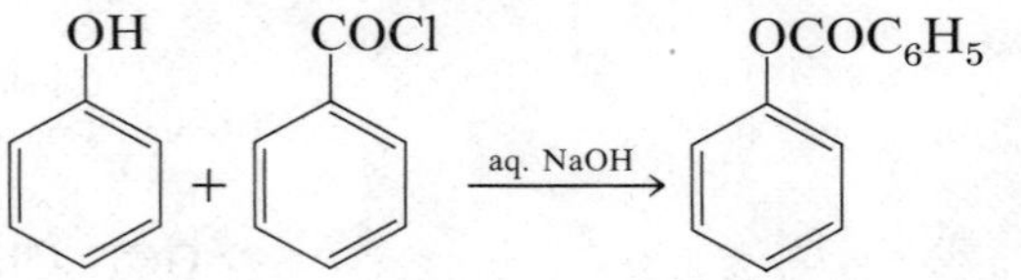

phenyl benzoate

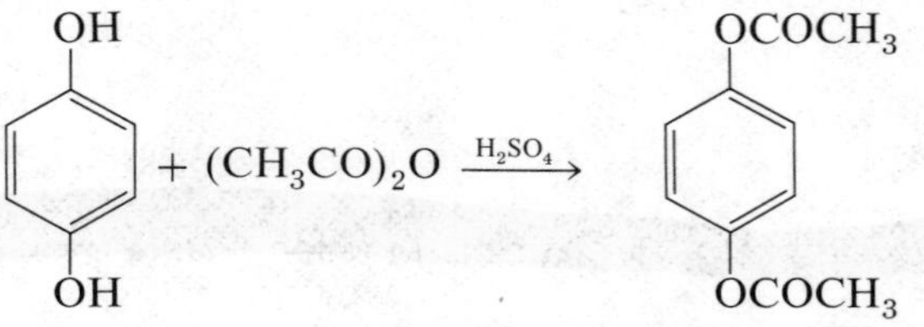

hydroquinone diacetate

One of the best-known aromatic acetates is acetylsalicylic acid, or aspirin, which is prepared by the acetylation of salicylic acid.

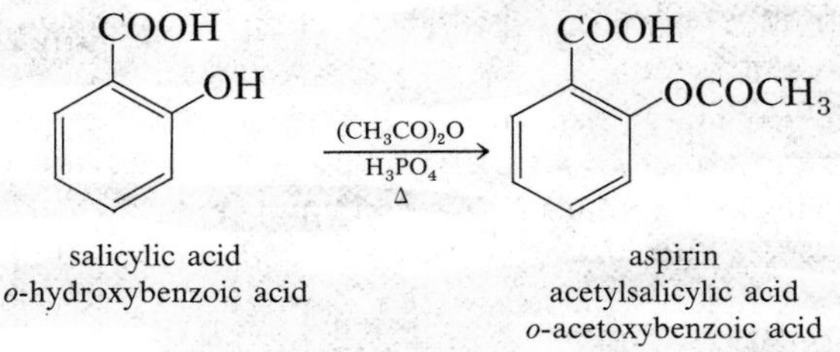

salicylic acid
o-hydroxybenzoic acid

aspirin
acetylsalicylic acid
o-acetoxybenzoic acid

Aspirin is widely used, primarily for its analgesic effect but also as an antipyretic and antirheumatic. It is not so innocuous a drug as one might imagine from its widespread use and ready availability. Repeated use may cause gastrointestinal bleeding, and large doses can provoke a host of reactions including vomiting, diarrhea, vertigo, and hallucinations. The average dose is 0.3–1 g; single doses of 10–30 g can be fatal.

B. *Reactions of Phenolate Ions*

Phenolate ions may be considered as enolate ions, and many of the reactions of phenolate ions point up this relationship. It is convenient to distinguish such reactions from those of the conjugate acids, the phenols. Many of the reactions of phenols resemble those of the corresponding ethers and may be considered together.

1. HALOGENATION. The reaction of an aqueous solution of phenol with bromine gives a precipitate of 2,4,4,6-tetrabromocyclohexa-2,5-dienone. This precipitate is normally washed with aqueous sodium bisulfite to generate 2,4,6-tribromophenol. The reactive form of phenol in this process is the phenolate ion. As successive bromines are introduced into the ring, the products are progressively more acidic, and a greater fraction of the phenol is present in the phenolate form. Thus each bromine is attached more rapidly than the previous one, until the product is no longer a phenol. The overall process is similar to the bromination of a ketone under basic conditions (Section 13.6.D).

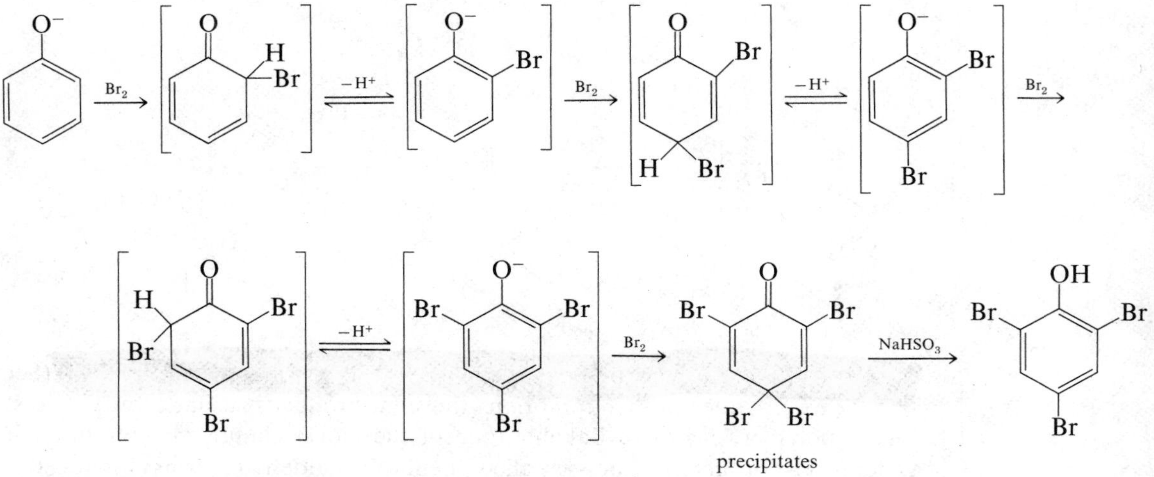

precipitates

1002

Chap. 30

Aromatic
Halides,
Phenols,
Phenyl Ethers,
and Quinones

Corresponding reactions occur with chlorine and iodine and with other phenols. The net reaction is halogenation of all available *ortho* and *para* positions. Halogenation of phenol is possible under acid conditions, and the incorporation of successive halogens can be controlled (Section 30.6.D). Here also we see an analogy to the acid-catalyzed, as well as the base-catalyzed, halogenation of carbonyl compounds (Section 13.6.D).

2. CONDENSATION WITH ALDEHYDES. A characteristic reaction of enolate ions from aldehydes and ketones is the condensation with other carbonyl groups as in the aldol condensation (Section 13.7.F). A similar reaction occurs with phenolate ions. Phenol reacts with formaldehyde in the presence of dilute alkali to give a mixture of *o*- and *p*-hydroxybenzyl alcohols.

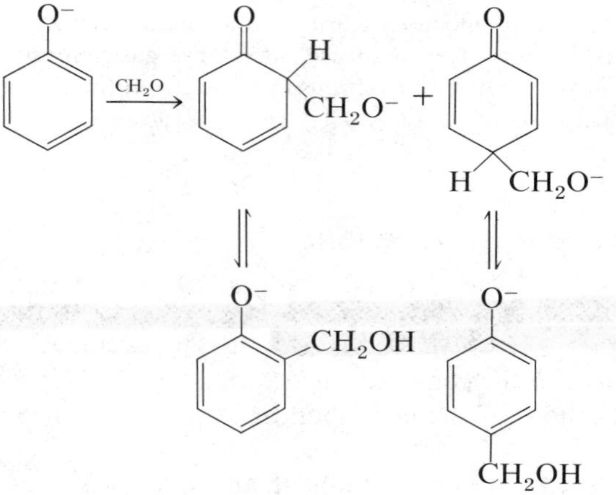

The reaction is difficult to control because of further condensations that lead to a polymeric product.

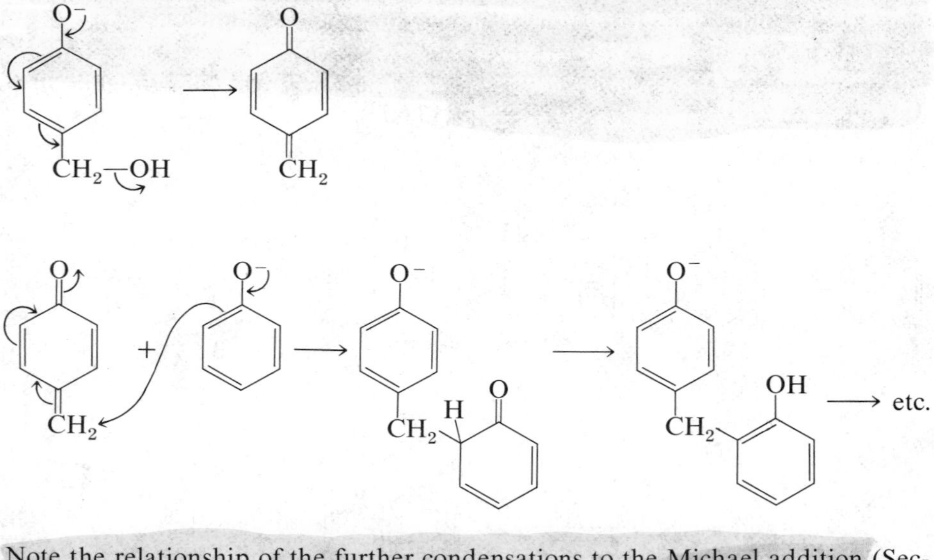

Note the relationship of the further condensations to the Michael addition (Section 27.7.C). Under proper conditions the final product is a dark, brittle, cross-linked polymer known as Bakelite, one of the oldest commercial plastics. The general class of such polymers is called phenol-formaldehyde resins (Figure 30.3).

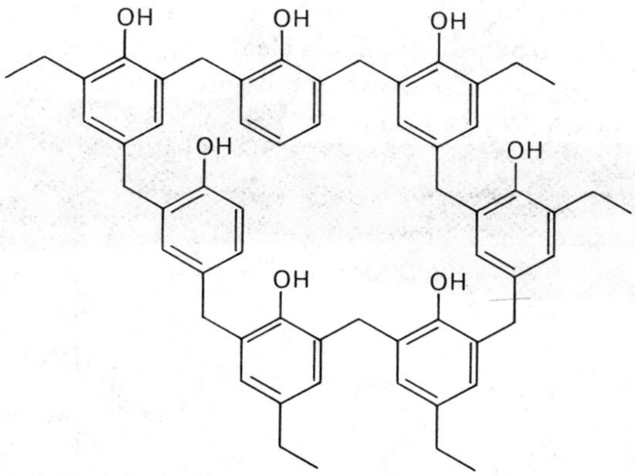

FIGURE 30.3. *Partial structure of a phenol-formaldehyde resin.*

3. KOLBE SYNTHESIS. The reaction of carbanions with carbon dioxide to give carboxylate salts (Section 15.5.D) has its counterpart in the reaction of phenolate ions with CO_2.

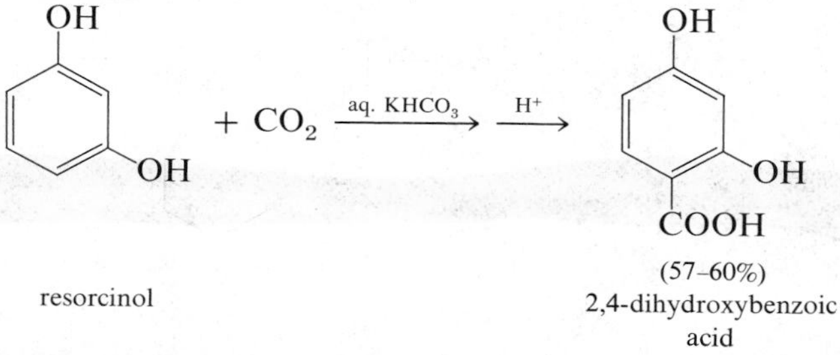

resorcinol

(57–60%)
2,4-dihydroxybenzoic
acid

The reaction of phenols with carbon dioxide under basic conditions is called the **Kolbe synthesis** and is important as a method for preparing analogs of aspirin. The product depends on the manner in which the reaction is carried out. Carbonation of sodium phenolate at relatively low temperature gives sodium salicylate. However, the reaction is reversible, and best yields are obtained only if carried out under pressure.

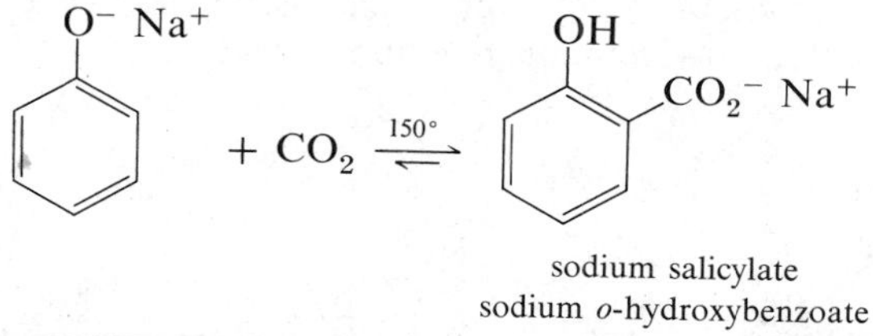

sodium salicylate
sodium *o*-hydroxybenzoate

Under more severe conditions isomerization to the more stable *para* isomer occurs. Thus potassium salicylate smoothly isomerizes to the *para* isomer at 240°.

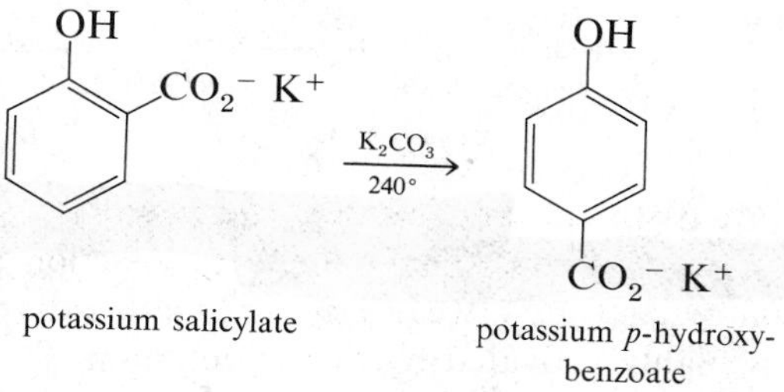

potassium salicylate

potassium *p*-hydroxy-
benzoate

1004

Chap. 30

**Aromatic
Halides,
Phenols,
Phenyl Ethers,
and Quinones**

Although the Kolbe synthesis may be written simply as the enolate condensation of phenolate ion with carbon dioxide, it is clear that coordination phenomena are involved. However, these mechanistic details are not yet fully understood.

4. REIMER–TIEMANN REACTION. The **Reimer–Tiemann reaction** is the reaction of a phenol with chloroform in basic solution to give a hydroxybenzaldehyde. Reaction occurs primarily in an *ortho* position unless both are blocked. The reaction mechanism involves the prior formation of dichlorocarbene by the reaction of chloroform with alkali (pages 321–322).

$$CHCl_3 + OH^- \rightleftharpoons CCl_3 \longrightarrow \quad :CCl_2$$
$$\text{dichlorocarbene}$$

The dichlorocarbene then reacts with the phenolate ion to give a dichloromethyl compound, which rapidly hydrolyzes.

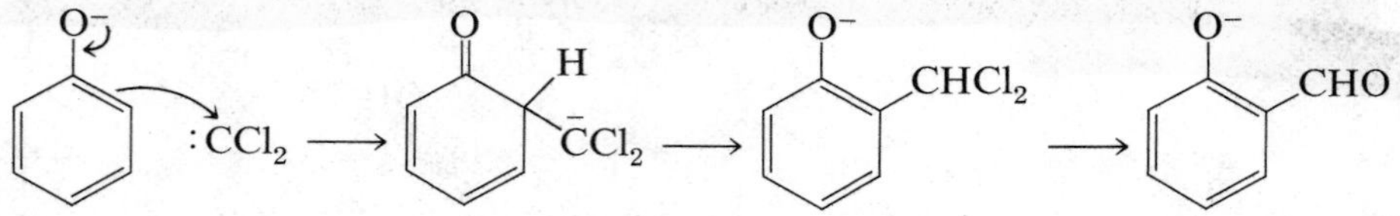

The final hydrolysis reaction is facilitated by the phenoxide ion in the following way.

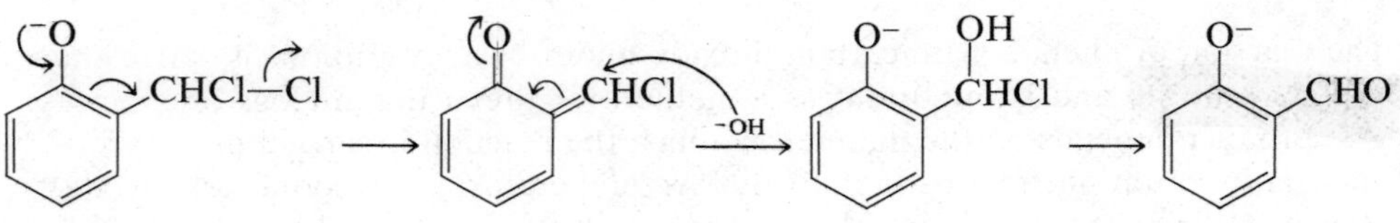

The essential correctness of the overall mechanism is revealed by an interesting by-product of the Reimer–Tiemann reaction on *p*-cresol.

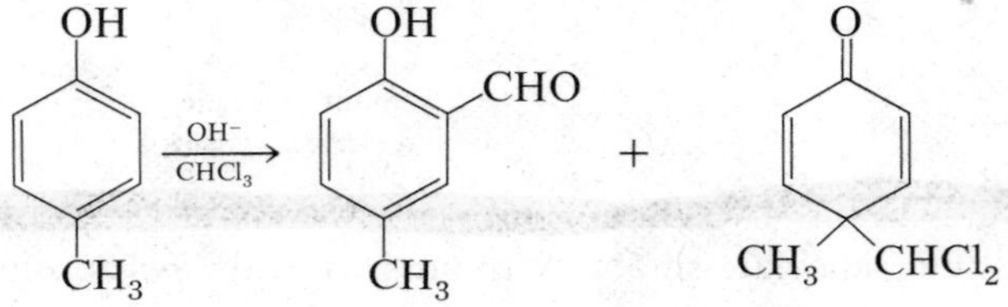

2-hydroxy-5-methyl-
benzaldehyde

4-methyl-4-dichloromethyl-
cyclohex-2,5-dienone

(5%)

5. DIAZONIUM COUPLING. Phenols react in basic solution with diazonium salts to give the corresponding arylazophenols. The reaction is an electrophilic aromatic substitution reaction by a weak electrophile, the diazonium ion, on an aromatic ring that is highly activated by the oxide anion.

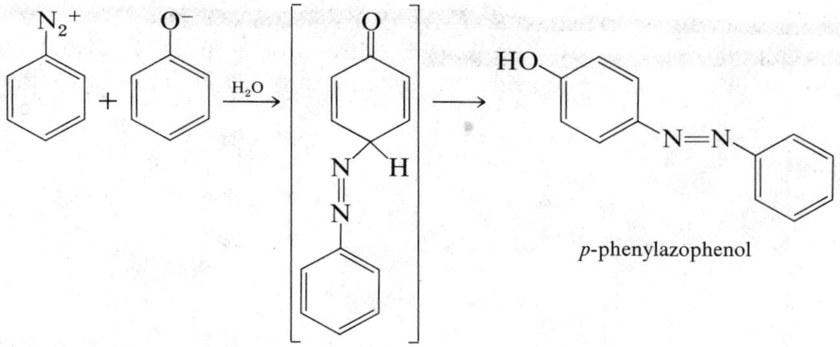

p-phenylazophenol

The product is almost exclusively the *para* isomer; the *ortho* isomer is formed to the extent of only 1%. Arylazophenols constitute an important class of azo dyes. Further examples will be discussed in Section 36.1.C.

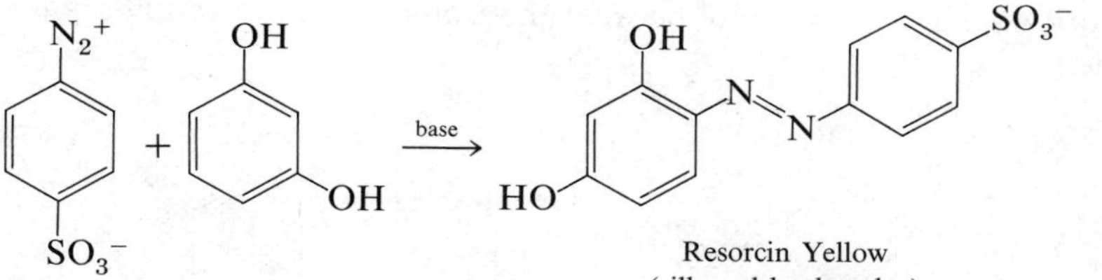

Resorcin Yellow
(silk and leather dye)

EXERCISE: Write equations showing the reactions (if any) of *o*-cresol with the following reagents.
(a) acetic acid, H_2SO_4
(b) acetic anhydride, H_2SO_4
(c) 1. Br_2–H_2O; 2. $NaHSO_3$
(d) NaOH, CO_2, 150°
(e) KOH, $CHCl_3$
(f) $C_6H_5N_2^+$ Cl^-

C. *Electrophilic Substitutions on Phenols and Phenyl Ethers*

In acidic solutions electrophilic substitutions occur on the un-ionized phenol. Such substitutions are still rather facile because of the activating nature of the hydroxy group, which is a strong *ortho,para* director. Reaction at one of these positions gives an intermediate cation that is essentially a protonated ketone.

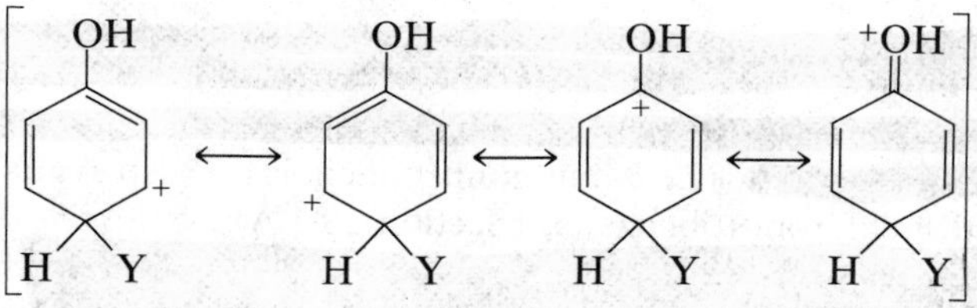

The last structure shows the role of an oxygen lone pair in stabilizing the intermediate and the transition state leading to it. Exactly the same phenomenon applies to the aromatic ethers; that is, alkoxy groups are also powerful *ortho,para* directors. Consequently, for many electrophilic aromatic substitution reactions, phenols and ethers can be considered together. The principal difference between the two groups of compounds lies in the greater water solubility of the phenols. Many electrophilic reactions of phenols can be carried out in aqueous solutions.

1006

Chap. 30

**Aromatic
Halides,
Phenols,
Phenyl Ethers,
and Quinones**

The phenol ring is sufficiently reactive that reaction occurs readily even with such feeble electrophiles as nitrous acid. Nitrosation is usually carried out in aqueous solution or in acetic acid; the principal product is the *p*-nitrosophenol.

15:1

p-nitrosophenol   *o*-nitrosophenol

Phenol is nitrated by dilute aqueous nitric acid, even at room temperature. Nitration of phenol also yields large amounts of tarry by-products produced by oxidation of the ring. Nevertheless, nitration is a satisfactory method for preparing both *o*- and *p*-nitrophenol because the isomers can be readily separated and purified.

(30–40%)   (15%)
o-nitrophenol   *p*-nitrophenol

o-Nitrophenol has lower solubility and higher volatility because of the **chelation** or intramolecular hydrogen bonding between the hydroxy group and the nitro group. Chelation (Gk., *chele,* claw) refers to formation of a ring by coordination with a pair of electrons.

Because the acceptor hydrogen of *o*-nitrophenol is involved in chelation, it is not available for hydrogen bonding to solvent water molecules. The resulting lower solubility and higher volatility are such that *o*-nitrophenol can be steam-distilled from the reaction mixture. The *o*- and *p*-nitrophenols are also available by hydrolysis of *o*- and *p*-chloronitrobenzenes (Section 30.3.A).

2,4-Dinitrophenol can be prepared by dinitration of phenol, using somewhat stronger nitric acid than is used for mononitration. However, a more convenient preparation of this phenol involves dinitration of chlorobenzene and hydrolysis of the resulting 2,4-dinitrochlorobenzene (Section 30.3.A).

Picric acid, 2,4,6-trinitrophenol, is prepared by treating phenol with concentrated sulfuric acid at 100°, followed by nitric acid, first in an ice bath, then at higher temperature. The first reaction that occurs in this sequence is disulfonation of the ring to give 4-hydroxybenzene-1,3-disulfonic acid. This substance is nitrated at the remaining *ortho* position by cold nitric acid. At higher temperature, sulfonic acid groups are replaced by nitro groups.

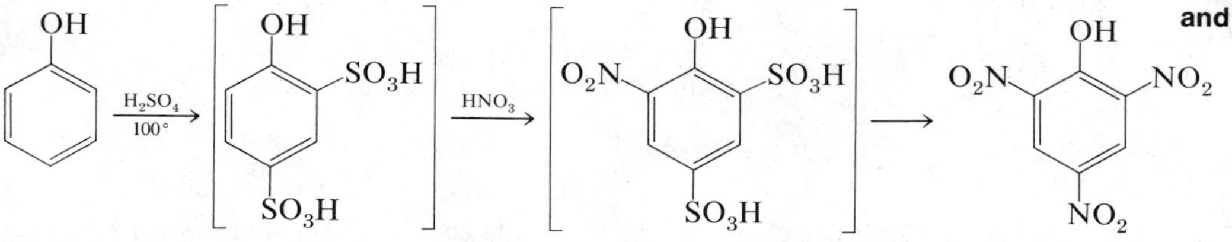

4-hydroxybenzene-
1,3-disulfonic acid

(90%)
picric acid

The reaction in which a sulfonic acid group is replaced by a nitro group is not uncommon in electrophilic aromatic substitutions but occurs more often as a side reaction rather than the main reaction. The mechanism is exactly the same as for substitution of a proton, except that a different cation is lost.

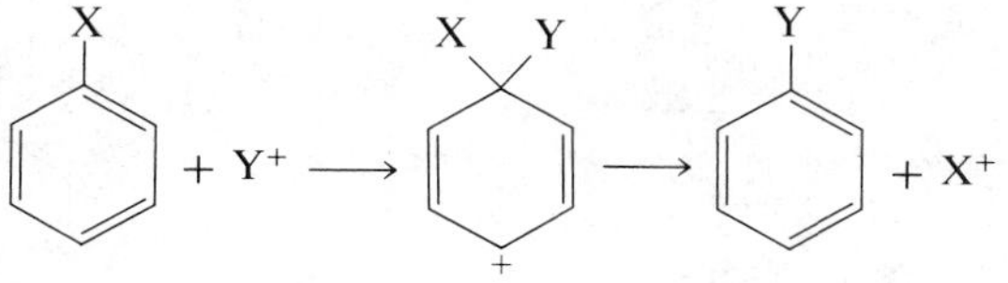

The reaction is best when strong *ortho,para*-directing groups such as —OR and —NR$_2$ are present and with functions that form relatively stable electrophilic molecules. The sulfonic acid group is prone to such replacement because it is lost as a neutral molecule, SO$_3$.

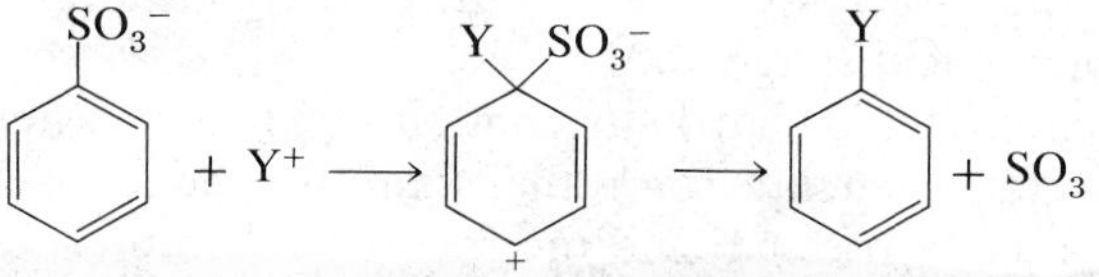

Picric acid forms yellow crystals, m.p. 123°. It explodes at temperatures above 300° and was once used as a synthetic dye. It is now important principally because of the molecular complexes it forms with many compounds, especially with polycyclic aromatic hydrocarbons and their derivatives. These complexes are of the "charge-transfer" type to be discussed in Section 30.7. Such picric acid complexes are called picrates. They can be crystallized and are useful for purification purposes. On treatment with base, the picric acid component is converted to the picrate ion, which does not form complexes; thus the other component of the complex is readily recovered.

Anisole is readily nitrated to give a mixture of *o*- and *p*-nitroanisole. The composition of the mixture depends on the reaction conditions; nitric and sulfuric acids give more *para* substitution, whereas nitration in acetic anhydride gives more *ortho*.

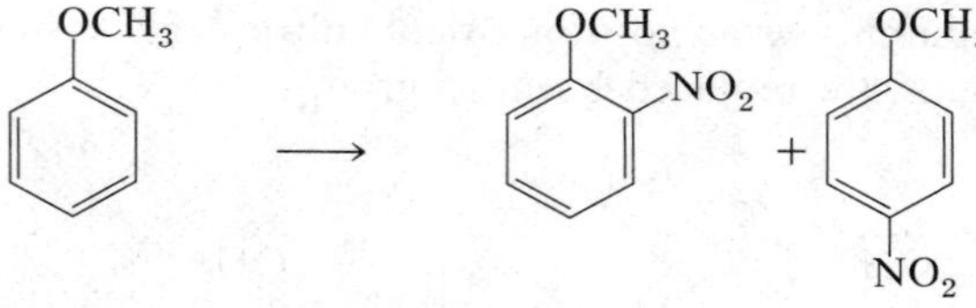

1008

Chap. 30

**Aromatic
Halides,
Phenols,
Phenyl Ethers,
and Quinones**

HNO$_3$–H$_2$SO$_4$, 45°	(31%)	(67%)
HNO$_3$–Ac$_2$O, 10°	(71%)	(28%)

These compounds are also available from the corresponding chloronitrobenzenes by substitution with methoxide ion (Section 30.3.A).

Monosulfonation of phenol gives an equimolar mixture of *ortho* and *para* substitution products if the reaction is carried out at room temperature, but predominantly *p*-hydroxybenzenesulfonic acid if steam-bath temperature is used. The behavior is typical for sulfonation reactions and is due to the easy reversibility of sulfonation (Section 26.5.B).

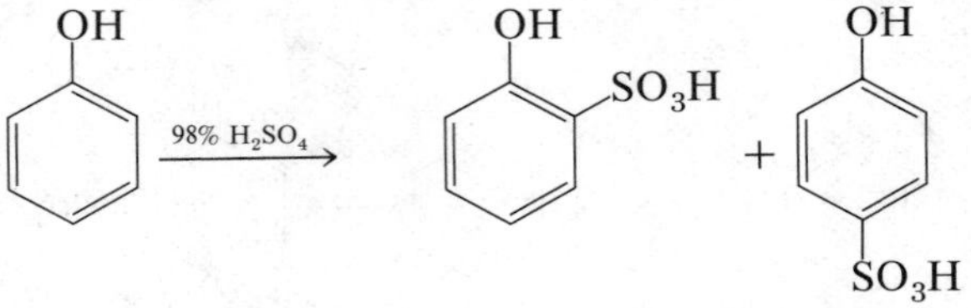

20°	(49%)	(51%)
100°	(10%)	(90%)

With concentrated sulfuric acid the disulfonic acid is formed. This product can be isolated as the sodium salt or can be used directly for further reactions as in the preparation of picric acid (page 1007).

We saw in Section 30.6.B that halogenation of phenol in neutral solution involves reaction of the phenolate ion rather than the phenol itself. In acidic solution phenolate ion is suppressed, and the free phenol is involved in electrophilic halogenation. By a proper choice of reaction conditions one, two, or three halogens can be introduced into the available *ortho* and *para* positions.

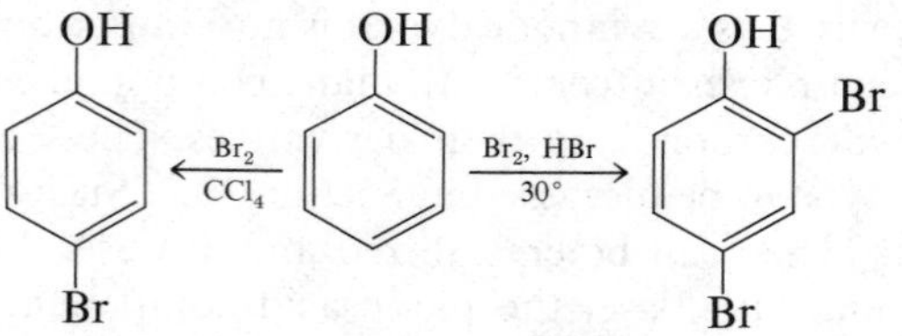

Anisole behaves in an analogous manner.

Anisole functions as an excellent substrate for Friedel–Crafts acylation. Mild reaction conditions suffice because the alkoxy group is highly activating.

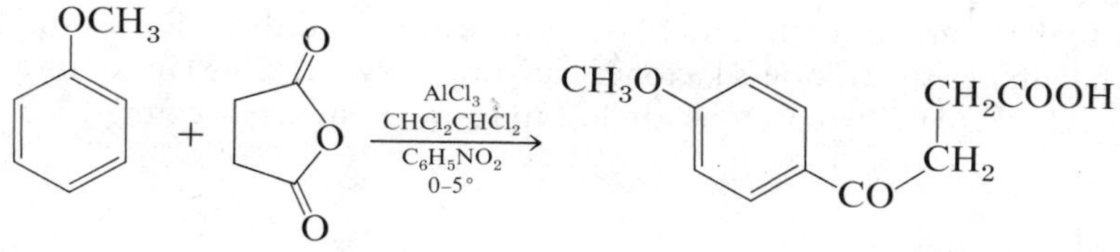

succinic anhydride	(85%) β-(p-methoxybenzoyl)-propionic acid

Primary alkoxybenzenes also work well in Friedel–Crafts acylation.

Direct Friedel–Crafts acylation of phenol is generally unsatisfactory. However, there are some exceptions. For example, treatment of phenol with acetic acid and boron trifluoride affords *p*-hydroxyacetophenone in excellent yield. Almost none of the *ortho* isomer is produced.

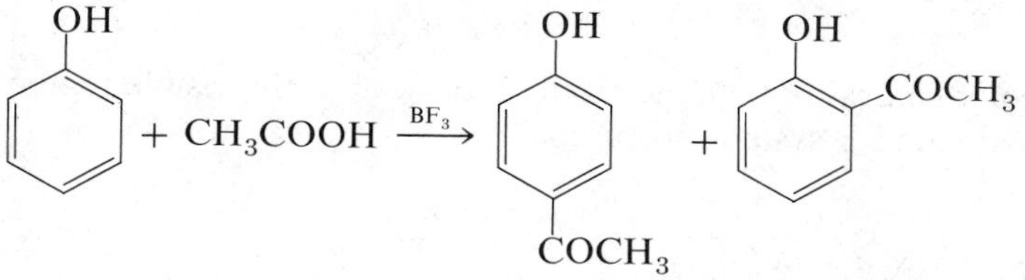

(95%)   (trace)

Many other phenol acylations are known, but most proceed in low yield, partly because of competing esterification of the hydroxy group.

An alternative acylation method that avoids the problem of esterification is the **Houben–Hoesch synthesis.** In this reaction the phenol is treated with a nitrile instead of an acyl halide or anhydride. The usual catalyst is zinc chloride.

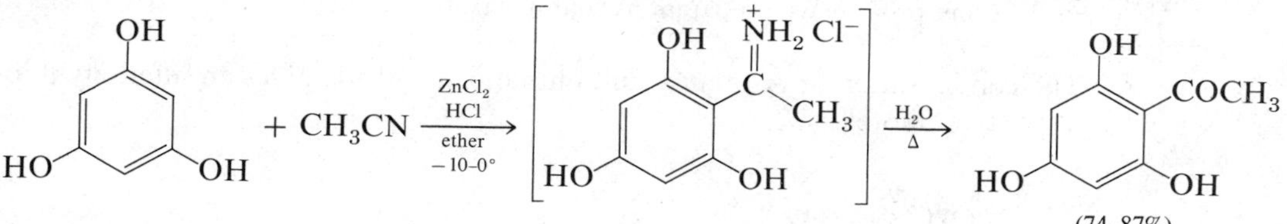

phloroglucinol   (74–87%)
2,4,6-trihydroxyacetophenone

The reaction is generally not as useful for monohydric phenols as it is for di- and polyhydric phenols; it may also be applied to many of the phenyl ethers.

Phenols undergo a special Friedel–Crafts acylation with phthalic anhydride and sulfuric acid or zinc chloride. In this case two molecules of phenol condense with one molecule of phthalic anhydride to give triarylmethane derivatives known as **phthaleins.**

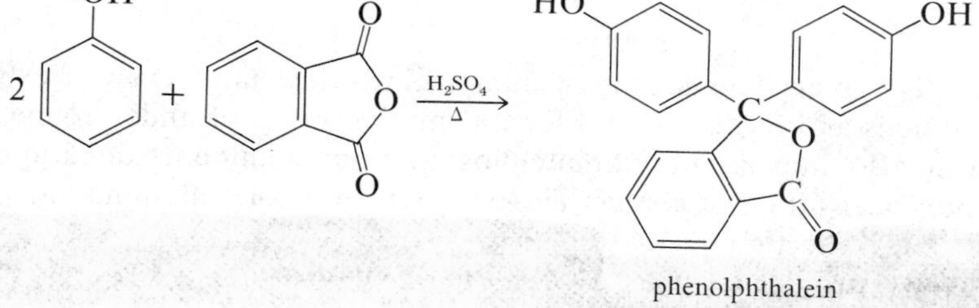

(30-3)

phenolphthalein
(colorless lactone form)

1010

Chap. 30

Aromatic
Halides,
Phenols,
Phenyl Ethers,
and Quinones

The phthaleins are an important class of indicators and dyes. For example, phenolphthalein has the colorless lactone structure shown in (30-3) in solutions below pH 8.5. Above pH 9 two protons are lost to form an intensely colored red dianion (30-4).

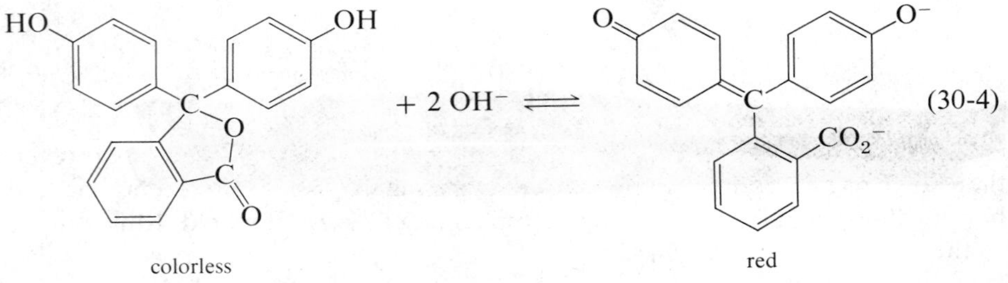

$$+\ 2\ OH^- \rightleftharpoons \qquad (30\text{-}4)$$

colorless

red

The red color comes from an electronic transition in the visible region associated with the extended π-system of the ion.

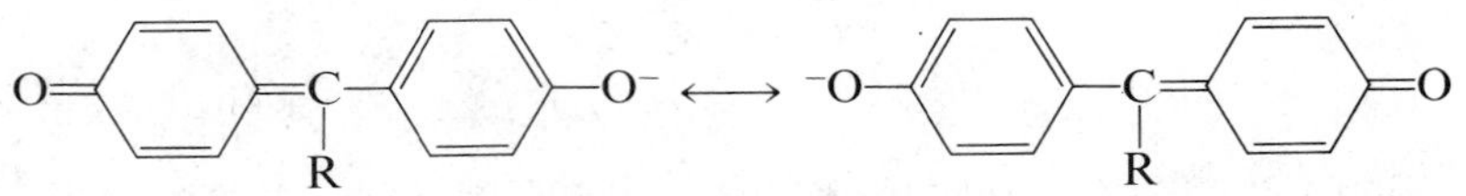

Highly conjugated anions or cations of this type are invariably highly colored.

[Phenolphthalein is used medicinally as a laxative and is the principal active ingredient in some proprietary preparations sold as laxative agents.]

The condensation of resorcinol and phthalic anhydride gives an intensely fluorescent dye, fluorescein.

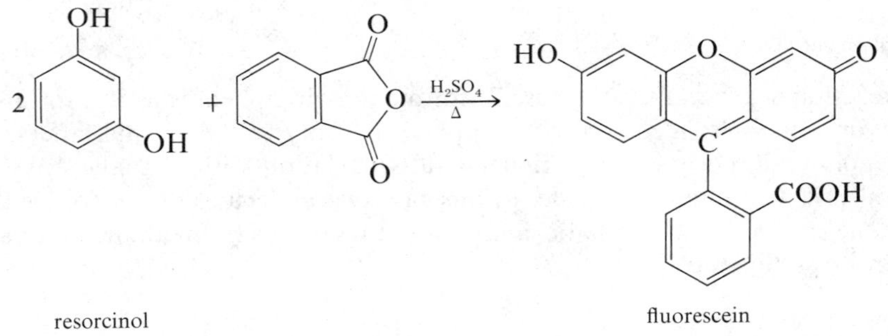

resorcinol

fluorescein

The yellowish green fluorescence of fluorescein is detectable even in extremely dilute solutions and has been used for tracing the course of underground rivers. Fluorescein also finds use in ophthalmology; a minute amount added to the eye assists the visual fitting of contact lenses under ultraviolet illumination.

Phenyl esters (Section 30.5.A) undergo a Lewis-acid-catalyzed rearrangement that amounts to intramolecular Friedel–Crafts acylation. The reaction is known as the **Fries rearrangement** and is carried out by heating the ester with aluminum chloride, often with no solvent.

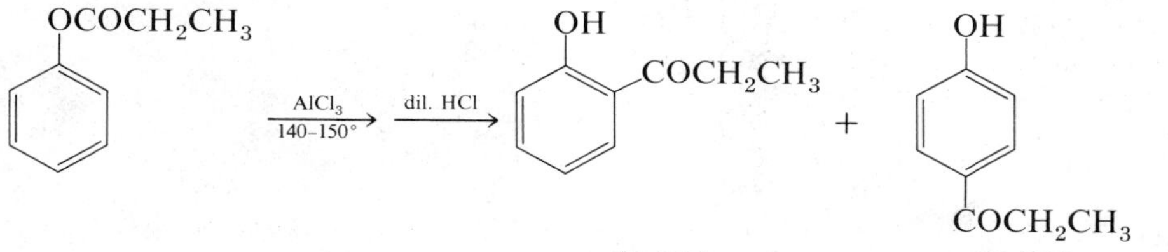

phenyl propionate → o-hydroxypropiophenone (32–35%) + p-hydroxypropiophenone (45–50%)

The two products can be conveniently separated by fractional distillation. In many cases a single product is formed in good yield.

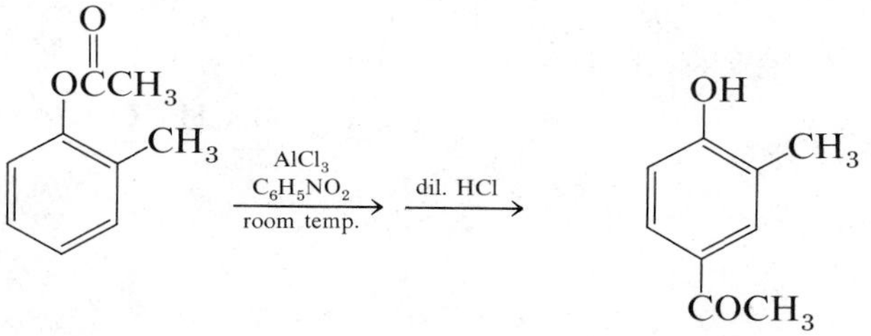

o-tolyl acetate → 3-methyl-4-hydroxyacetophenone (80–85%)

EXERCISE: Write equations showing the reactions of phenol with the following reagents.
(a) $NaNO_2$, HCl, 0°
(b) HNO_3, H_2O, 25°
(c) conc. H_2SO_4, 100°
(d) Br_2, CCl_4, 25°
(e) acetic acid, BF_3
(f) phthalic anhydride, H_2SO_4, Δ
(g) 1. acetic anhydride, H_2SO_4;
    2. $AlCl_3$, 150°

D. Reactions of Ethers

Alkyl aryl ethers are cleaved by acids just as are dialkyl ethers. Hydrobromic or hydriodic acid is commonly used.

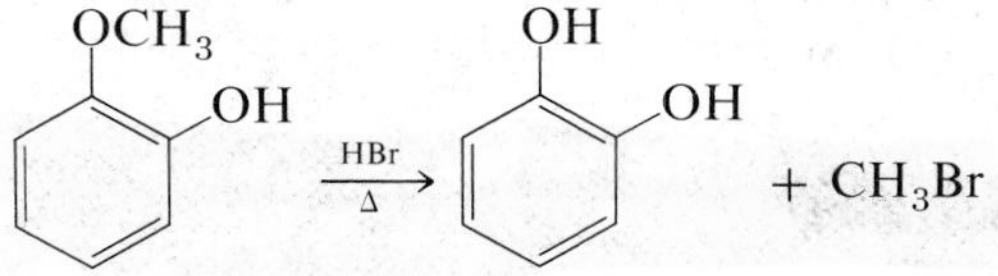

$+ CH_3Br$

The reaction mechanism is the same as for aliphatic ethers; the protonated ether undergoes S_N1 or S_N2 cleavage. Because the phenyl group is not susceptible to either S_N1 or S_N2 reaction, cleavage of the aliphatic C—O bond always occurs.

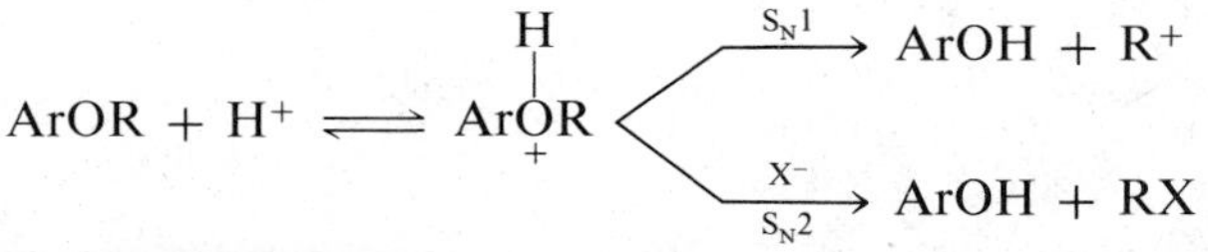

When R is a tertiary alkyl group, the ether cleavage is especially facile; cleavage occurs by the S_N1 mechanism.

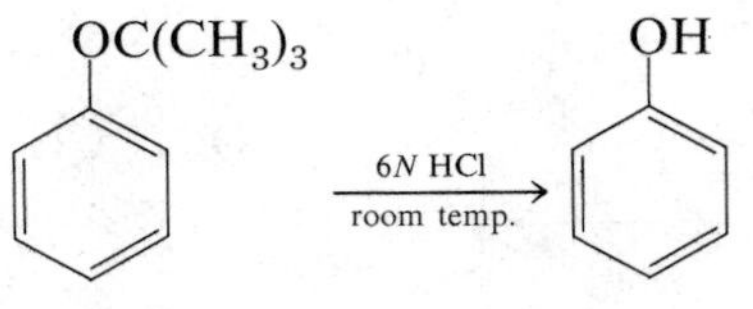

1012

Chap. 30

Aromatic
Halides,
Phenols,
Phenyl Ethers,
and Quinones

OC(CH$_3$)$_3$ OH

$$\xrightarrow[\text{room temp.}]{6N\ \text{HCl}}$$

t-butyl
phenyl ether

Allyl aryl ethers undergo a fascinating and useful rearrangement reaction upon being heated to about 200°. The reaction is called the **Claisen rearrangement** and results in apparent migration of the allyl group to the *ortho* position on the benzene ring.

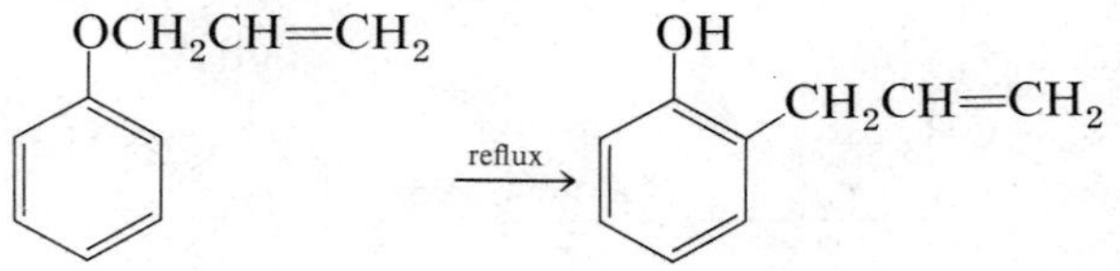

OCH$_2$CH=CH$_2$ OH
         CH$_2$CH=CH$_2$

$$\xrightarrow{\text{reflux}}$$

allyl phenyl ether   (73%)
         o-allylphenol

If both *ortho* positions are occupied by substituents, rearrangement occurs to the *para* position.

The reaction mechanism is known to involve a concerted formation of a C—C bond between the *ortho* carbon and the terminal position of the allyl group as the C—O bond is broken.

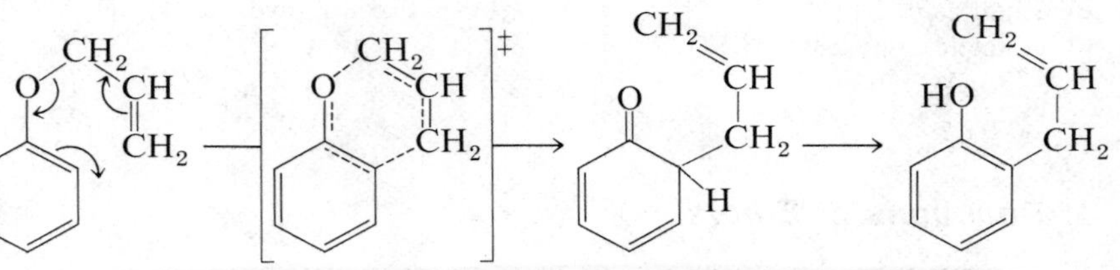

The transition state has the six-electron aromatic conjugation characteristic of benzene. A related type of transition state was noted earlier in the Diels–Alder reaction (Section 20.5). An orbital description of the Claisen rearrangement transition state is shown in Figure 30.4.

Note that the γ-carbon of the allyl group becomes attached to the benzene ring. The allylic rearrangement is observable with an unsymmetrical allyl group.

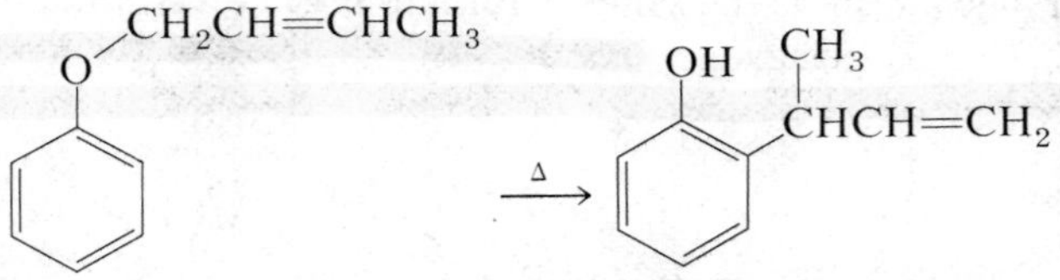

O—CH$_2$CH=CHCH$_3$ OH   CH$_3$
         CHCH=CH$_2$

$$\xrightarrow{\Delta}$$

         o-(α-methylallyl)phenol

If no *ortho* hydrogen is available, enolization to the phenol cannot occur, and a second rearrangement occurs to the *para* position.

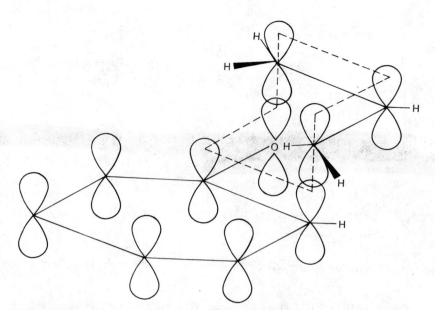

FIGURE 30.4. *Transition state for Claisen rearrangement. Dashed line shows the orbital interactions important in the reaction zone only.*

Note that two successive allylic rearrangements restore the original orientation of the allylic group. In the second rearrangement the transition state involves a six-membered ring of carbon atoms.

Both types of rearrangements have their counterparts in aliphatic systems. The rearrangement of allyl ethers of enols is another example of a Claisen rearrangement.

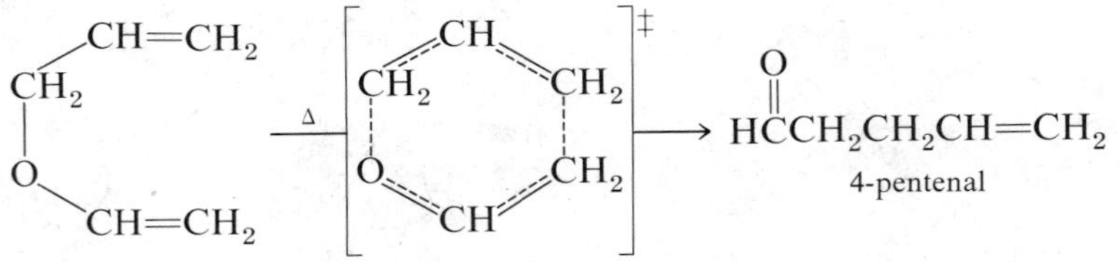

1014

Chap. 30

Aromatic
Halides,
Phenols,
Phenyl Ethers,
and Quinones

$$CH_2=CH-CH_2-O-CH_2-CH=CH_2 \xrightarrow{\Delta} [transition state] \longrightarrow HCCH_2CH_2CH=CH_2$$

4-pentenal

The all-carbon rearrangement is known as the **Cope rearrangement** and is a general thermal reaction of 1,5-dienes.

$$CH_2=CHCH-CHCH=CH_2 \xrightarrow{\Delta} CH_3CH=CHCH_2CH_2CH=CHCH_3$$

3,4-dimethyl-1,5-hexadiene   2,6-octadiene

The reaction can be observed for 1,5-hexadiene itself with deuterium labeling.

$$CD_2=CHCH_2CH_2CH=CD_2 \rightleftharpoons CH_2=CHCD_2CD_2CH=CH_2$$

The mechanism of these **sigmatropic rearrangements** is discussed in more detail in Section 34.2.

EXERCISE: Write the equations showing the following reaction sequences starting with *p*-cresol.
(a) 1. NaOH, $CH_3CH=CHCH_2Br$; 2. 200°
(b) 1. NaOH, $CH_3OSO_3CH_3$; 2. HNO_3, Ac_2O, 10°; 3. HBr, 100°

30.7
Quinones

A. *Nomenclature*

Quinones are cyclohexadiendiones, but they are named as derivatives of aromatic systems: benzoquinones are derived from benzene, toluquinones from toluene, naphthoquinones from naphthalene, and so on. "Quinone" is used both as a generic term and as a common name for *p*-benzoquinone.

Many quinones and especially hydroxyquinones occur in nature. Some examples are

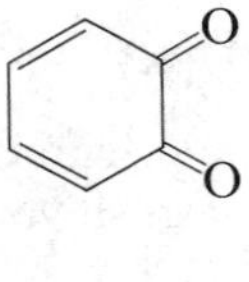

o-benzoquinone 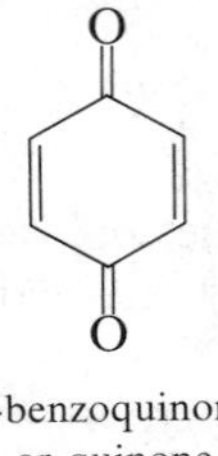 *p*-benzoquinone toluquinone   fumigatin
       or quinone   2-methyl-1,4- 3-hydroxy-2-methoxy-5-
              benzoquinone methyl-1,4-benzoquinone
                    (antibiotic substance)

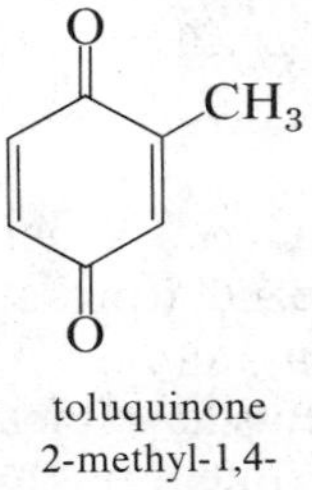

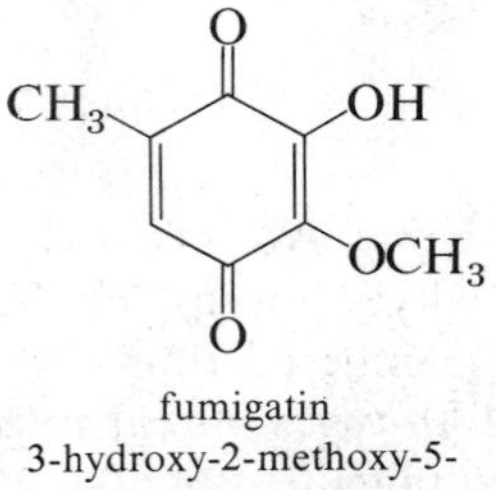

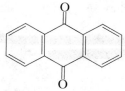

1,4-naphthoquinone   1,2-naphthoquinone   2,6-naphthoquinone

9,10-anthraquinone   9,10-phenanthraquinone   phthiocol
2-hydroxy-3-methyl-
1,4-naphthoquinone
(antibiotic substance)

Hydroxynaphthoquinones and hydroxyanthraquinones are especially common, either free or as glycosides. Many natural pigments have quinone structures.

alizarin   juglone   rhein
(madder root)   (walnut shells)   (rhubarb)

Quinone structures are frequently associated with color and the structural units

and

are referred to as "quinoid" structures.

B. *Preparation*

The only important method for the preparation of quinones is oxidation of aromatic hydroxy and amino compounds. Substituted phenols or aniline derivatives can be used with some oxidizing agents. For example, p-benzoquinone can be prepared by oxidation of benzene or aniline with a variety of oxidizing agents, but the usual laboratory preparation involves the oxidation of hydroquinone.

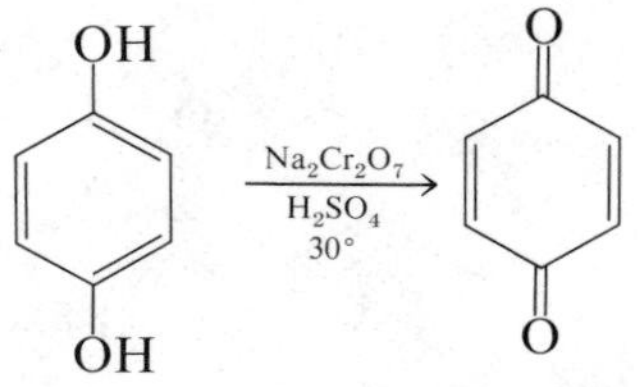

1016

Chap. 30

**Aromatic
Halides,
Phenols,
Phenyl Ethers,
and Quinones**

hydroquinone   (86–92%)
quinone

p-Benzoquinone forms yellow crystals, m.p. 115.7°, that are slightly soluble in water and can be sublimed or steam-distilled.

Aminophenols are easily oxidized to quinones, and this route constitutes one of the best methods for the preparation of some substituted quinones.

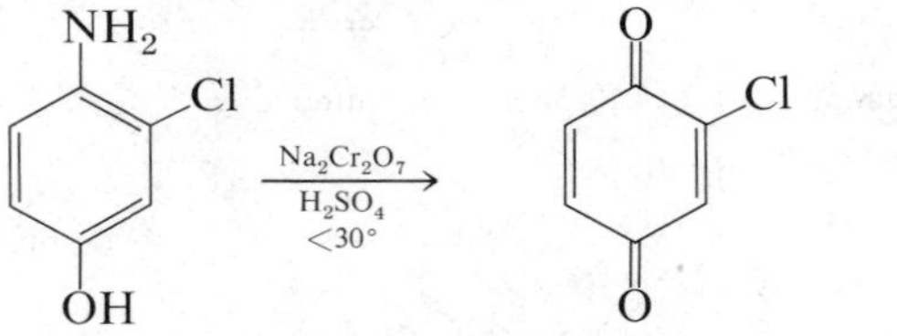

2-chloro-1,4-benzoquinone

Other oxidizing agents that are used for the preparation of quinones are ferric ion, nitrogen tetroxide (N_2O_4), and sodium chlorate–vanadium pentoxide. Many other oxidizing agents have also been used, and the best one for any given compound must be determined by experiment. For example, the preparation indicated in (30-5) makes advantageous use of nitric acid.

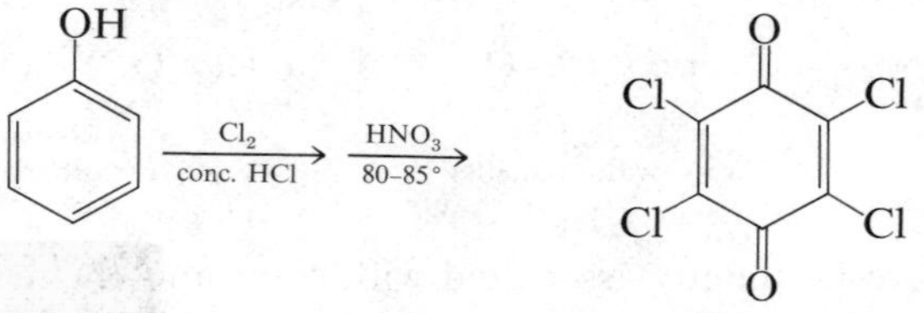

(30-5)

(60%)
2,3,5,6-tetrachloro-1,4-benzoquinone
chloranil

Chloranil is commercially available and has several uses in organic chemistry that we will encounter later (Sections 30.7.D and 31.3.B).

The oxidation of *o*-dihydroxybenzenes to *o*-quinones can be carried out with silver oxide in ether.

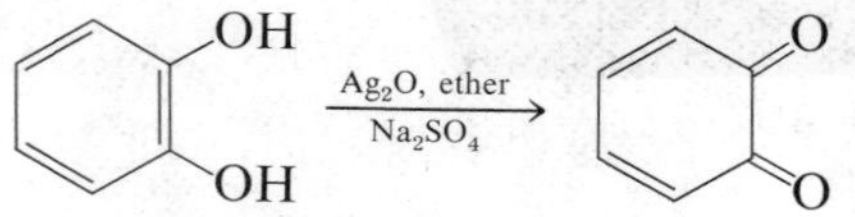

(30-6)

catechol   *o*-benzoquinone

o-Benzoquinone forms red crystals that are water-sensitive; anhydrous sodium sulfate is used in the preparation (30-6) to remove the water formed in the oxidation.

In many cases phenyl ethers and esters undergo oxidation to the corresponding quinone with loss of the alkyl or acyl group. An example is the oxidation of 2,6-di-*t*-butyl-4-methoxyphenyl propionate by ceric ammonium nitrate, CAN.

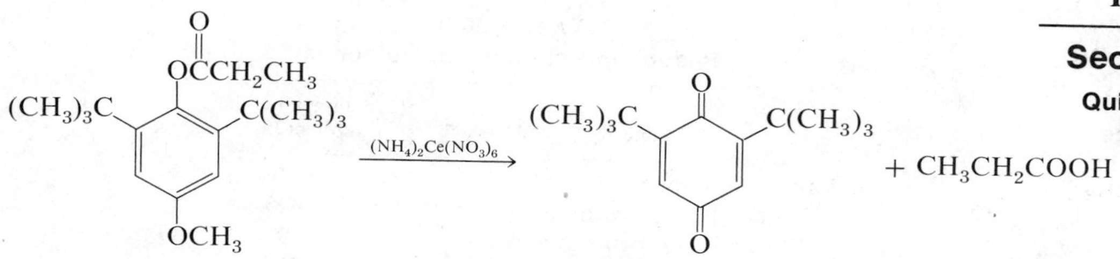

$$(CH_3)_3C \qquad C(CH_3)_3 \qquad \xrightarrow{(NH_4)_2Ce(NO_3)_6} \qquad (CH_3)_3C \qquad C(CH_3)_3 \qquad + \ CH_3CH_2COOH$$

C. Reduction-Oxidation Equilibria

Just as 1,2- and 1,4-dihydroxybenzenes are readily oxidized to quinones, the quinones are readily reduced back to dihydroxy compounds. This reduction can be carried out chemically.

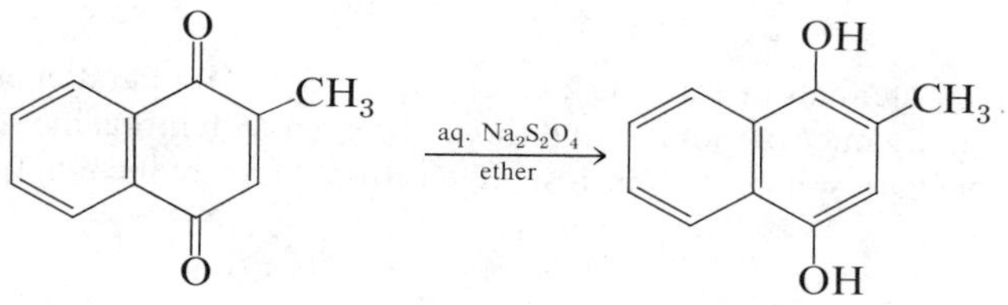

2-methyl-1,4-naphthoquinone

(95%)
2-methyl-1,4-dihydroxy-
naphthalene

However, the most important aspect of this redox system is that it is electrochemically reversible.

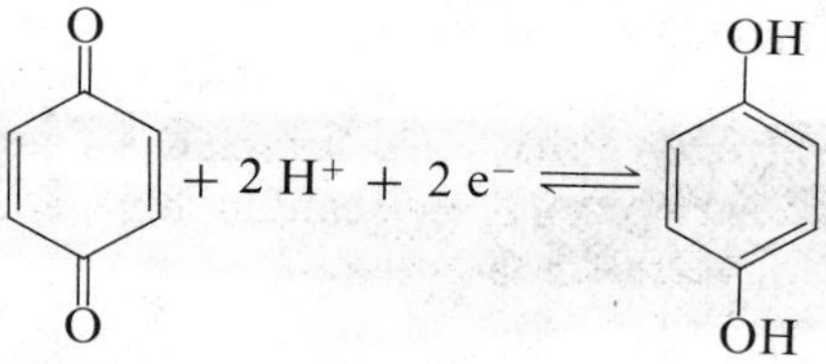

$$+ \ 2\,H^+ + 2\,e^- \ \rightleftharpoons$$

The electrical potential of this cell is given by the Nernst equation (30-7)

$$E = E^\circ + \frac{2.303\ RT}{n\mathcal{F}} \log \frac{[\text{quinone}][H^+]^2}{[\text{hydroquinone}]} \qquad (30\text{-}7)$$

in which $\mathcal{F}$ is the Faraday. At 25° equation (30-7) may be written as (30-8), in which the electrical potential is given in volts.

$$E^{25^\circ} = E^\circ - 0.059\ \text{pH} + 0.0296 \log \frac{[\text{quinone}]}{[\text{hydroquinone}]} \qquad (30\text{-}8)$$

The standard potential E° is that given at unit hydrogen ion concentration and equal concentrations of quinone and hydroquinone. Some values of E° are listed in Table 30.2. The more positive the value of the potential, the more readily the quinone is reduced. Note that electron-donating groups such as methyl and hydroxy stabilize the quinone form relative to the hydroquinone and result in lowering the reduction potential; electron-attracting groups such as halogen have the opposite effect.

1018

Chap. 30

**Aromatic
Halides,
Phenols,
Phenyl Ethers,
and Quinones**

TABLE 30.2
Reduction Potentials of Quinones

Quinone	Reduction Potential $E°$, volts (25°)
1,4-benzoquinone	0.699
2-methyl-1,4-benzoquinone	0.645
2-hydroxy-1,4-benzoquinone	0.59
2-bromo-1,4-benzoquinone	0.715
2-chloro-1,4-benzoquinone	0.713
1,2-benzoquinone	0.78
1,4-naphthoquinone	0.47
1,2-naphthoquinone	0.56
9,10-anthraquinone	0.13
9,10-phenanthraquinone	0.44

The reduction potentials in Table 30.2 allow one to see a clear parallel between redox phenomena in organic compounds and those observed with inorganic species. Recall that oxidation corresponds to the loss of electrons, and reduction to the gain of electrons.

$$Fe^{2+} \underset{\text{reduction}}{\overset{\text{oxidation}}{\rightleftarrows}} Fe^{3+} + e^-$$

The more electron-rich a species is, the easier is its oxidation and the more difficult is its reduction. In Table 30.2 we see that the electron-attracting substituents chloro and bromo do indeed cause the quinone to be reduced more easily (more positive reduction potential). Similarly, the electron-donating substituents hydroxy and methyl cause the quinone to be reduced less easily.

The reduction of quinone occurs in two one-electron steps. The product of the first step is a radical anion that can be detected in dilute solution by the technique of electron spin resonance spectroscopy.

Electron spin resonance, esr, is closely related to nuclear magnetic resonance. Electrons, like protons, have spin, and in molecules with an odd number of electrons (radicals) the resulting net electronic spin is aligned with or against an applied magnetic field. With commercial magnets the energy difference between the two states is in the microwave region of electromagnetic radiation. The resulting esr spectra have been extremely useful for detecting small concentrations of radicals, and the details of the spectra provide important information about the electronic structures of radicals. These details, however, are beyond the scope of an introductory textbook.

The same radical anions are produced by one-electron oxidations of hydroquinone dianions.

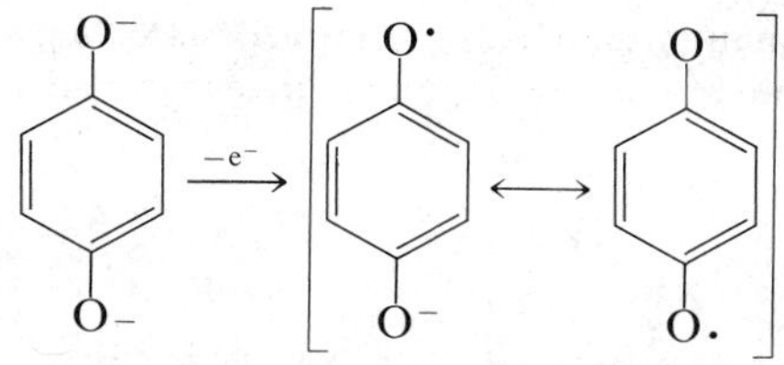

Phenoxide ions are also subject to one-electron oxidation to give the correspond-ing neutral radicals.

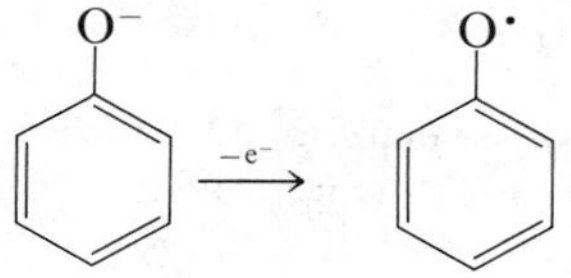

phenoxy radical

Such radicals are involved in many of the reactions of phenols including reactions of naturally occurring phenols.

Vitamin E, α-tocopherol, is a phenol that is widespread in plant materials. It appears to have several functions in animals, but one important function seems to be as a radical scavanger. The corresponding phenoxy radical is less reactive and less damag-ing to body constituents. Free radicals have been implicated in the aging process.

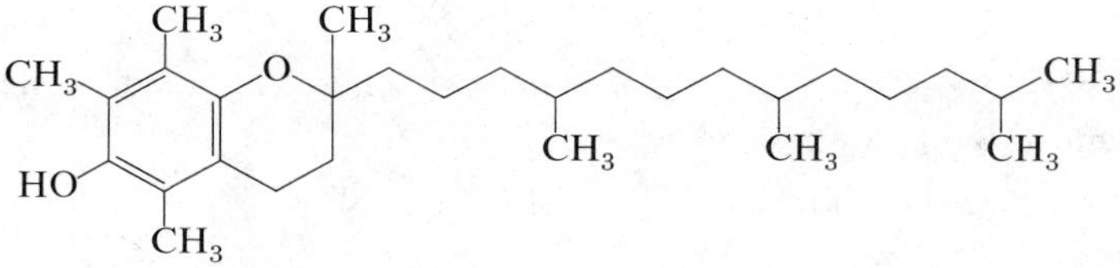

α-tocophenol
vitamin E

Quinone–hydroquinone redox systems have a number of important uses. Hy-droquinone itself, for example, is an important photographic developer.

Silver bromide crystals that have become photoactivated by exposure to light are reduced by hydroquinone. The photoactivated silver bromide is reduced to black silver metal and the hydroquinone is oxidized to p-benzoquinone. The residual silver bromide is then removed by "hypo," sodium hyposulfite, which forms a soluble complex with silver cation. The result is a black image where the silver bromide emulsion was exposed to light. Some developer formulas include p-methylamino-phenol, usually as the sulfate (Elon, Metol), which also oxidizes to p-benzoquinone.

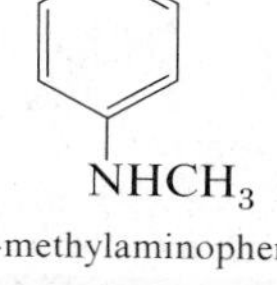

p-methylaminophenol

The oxidation-reduction of hydroquinone and quinone derivatives play an im-portant role in physiological redox processes.

1020

Chap. 30
**Aromatic
Halides,
Phenols,
Phenyl Ethers,
and Quinones**

Vitamin K is actually many vitamins; for example, K_1, K_2, K_3, and so on. They are all related to 1,4-naphthoquinone or compounds that are oxidized to it. For example, vitamin $K_{2(30)}$ is

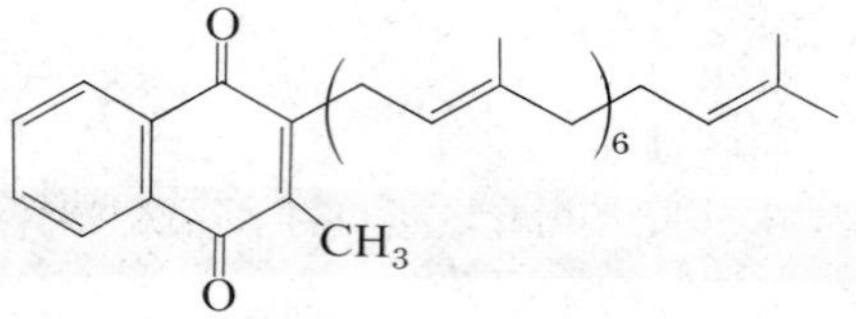

Others vary in the length of the side chain. The K vitamins are present in blood as coagulation factors.

A related series of compounds is coenzyme Q, which occurs in many kinds of cells with $n = 6$, 8, or 10 ($n = 10$ in mammalian cells); indeed, when first discovered, it was called **ubiquinone** because it was so ubiquitous in cells. Coenzyme Q is involved in electron-transport systems, and the long isoprenoid chain is undoubtedly designed to promote fat solubility.

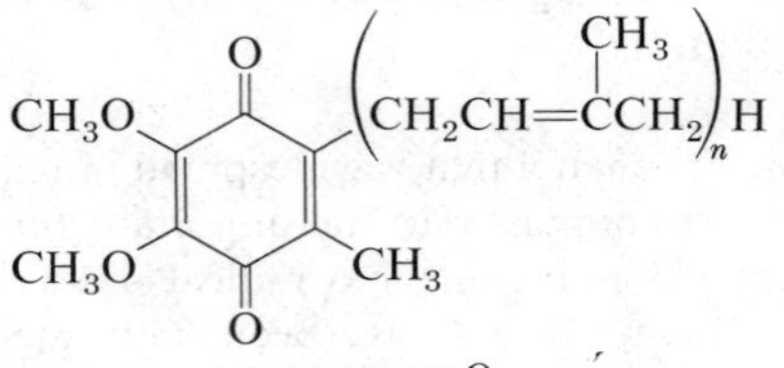

coenzyme Q

EXERCISE: Using the reduction potentials in Table 30.2, predict direction of equilibrium for the following reaction.

$$\left.\begin{array}{c} \text{1,4-benzoquinone} \\ + \\ \text{2-chlorohydroquinone} \end{array}\right\} \rightleftharpoons \left\{\begin{array}{c} \text{hydroquinone} \\ + \\ \text{2-chloro-1,4-benzoquinone} \end{array}\right.$$

D. *Charge-Transfer Complexes*

An equimolar mixture of *p*-benzoquinone and hydroquinone forms a dark green crystalline molecular complex, "quinhydrone," having a definite melting point of 171°. This material dissolves in hot water, and the solution is largely dissociated into its components.

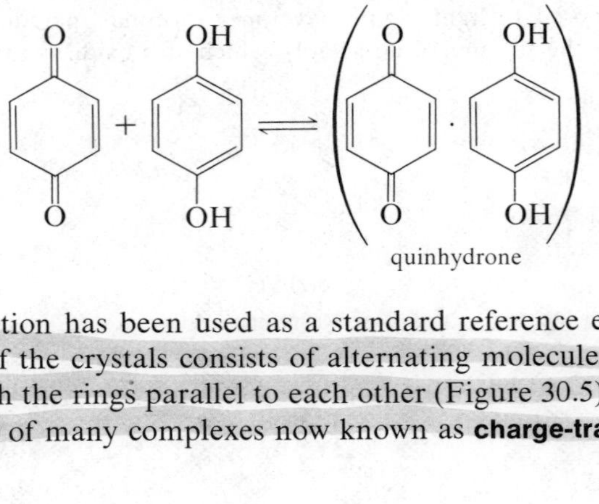

quinhydrone

The buffered solution has been used as a standard reference electrode.

The structure of the crystals consists of alternating molecules of quinone and hydroquinone with the rings parallel to each other (Figure 30.5). This complex is only one example of many complexes now known as **charge-transfer** complexes.

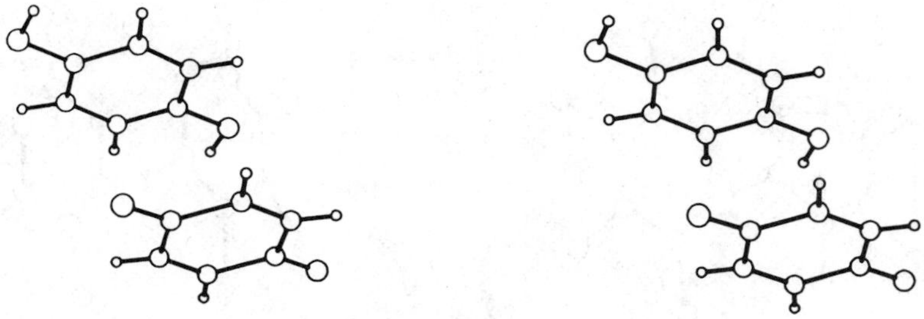

FIGURE 30.5. *Stereo diagram of quinhydrone.* [*Adapted with permission from* Molecular Structure and Dimensions. *International Union of Crystallography, 1972.*]

Such complexes are characterized by one component that is electron-rich (the donor) and another component that is strongly electron-attracting (the acceptor); hence, such complexes are also known as donor-acceptor complexes. In resonance language the complexes are characterized as a hybrid of two resonance structures.

$$[D:A \longleftrightarrow \overset{+}{D} \cdot \overset{-}{A} \cdot]$$

The second structure, the "charge-transfer structure," makes only a small contribution to the total electronic structure; that is, this second structure provides the bonding that holds the two components together, but the bond strength involved is only a few kcal mole^{-1}.

In molecular orbital language the donors have a filled high-energy orbital in which electrons are held rather loosely; that is, this "highest occupied" molecular orbital has a low ionization potential. The acceptors have a relatively low-lying vacant molecular orbital; in fact, common acceptors frequently form radical anions readily on one-electron reduction in which the electron enters this "lowest vacant" orbital. In the complex there is some overlap of the highest occupied orbital of the donor with the lowest vacant orbital of the acceptor that results in transfer of some electron density from donor to acceptor. The amount of charge transferred is small and corresponds typically to a small fraction (~ 0.05) of an electron. The molecular orbital approach is entirely equivalent to the resonance interpretation.

Typical donors that form charge-transfer complexes are benzene rings with electron-donating groups such as —OH, —OCH$_3$, —N(CH$_3$)$_2$, —CH$_3$, and so on. Common acceptors are compounds with several nitro groups, such as 1,3,5-trinitrobenzene and picric acid or quinones. Especially potent are quinones with additional electron-attracting groups; chloranil (tetrachloro-*p*-benzoquinone) is an important example. The structure of the complex formed from hexamethylbenzene and chloranil is shown in Figure 30.6; this complex has a bond strength of about 5 kcal mole^{-1}. Compounds with several CN groups are also used as acceptors. Some examples are tetracyanoethylene and 2,3-dicyano-1,4-benzoquinone.

Charge-transfer complexes are often intensely colored. The color is associated with an electronic transition in which a substantial fraction of an electron is transferred from donor to acceptor. Charge-transfer interactions are now recognized as being important in other solid-state structures in which the interactions are weaker. Furthermore, many reactions and reaction mechanisms are now recognized to involve charge-transfer phenomena; however, a detailed treatment of such phenomena must be deferred to advanced organic chemistry texts.

1022

Chap. 30

**Aromatic
Halides,
Phenols,
Phenyl Ethers,
and Quinones**

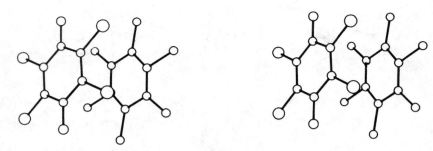

FIGURE 30.6. *Stereo diagram of hexamethylbenzene–chloranil complex.* [*Adapted
with permission from* Molecular Structure and Dimensions. *International Union
of Crystallography, 1972.*]

E. *Reactions of Quinones*

Quinones are α,β-unsaturated carbonyl compounds and show double bond reactions typical of such structures. One significant reaction is addition of hydrogen
chloride.

chlorohydroquinone

This reaction is simply an acid-catalyzed conjugate addition (see Section 20.3.A).

Amines add readily. 1,4-Benzoquinone reacts with aniline to give 2,5-dianilino-
1,4-benzoquinone and hydroquinone.

This reaction provides an interesting contrast to the addition of HCl, where the
product is the chlorohydroquinone. In this case the group entering the hydroquinone ring is *electron-donating*. Thus the initial product of conjugate addition of
aniline to 1,4-benzoquinone, 2-anilinohydroquinone, is rapidly oxidized by an
equivalent of 1,4-benzoquinone, which is reduced to hydroquinone.

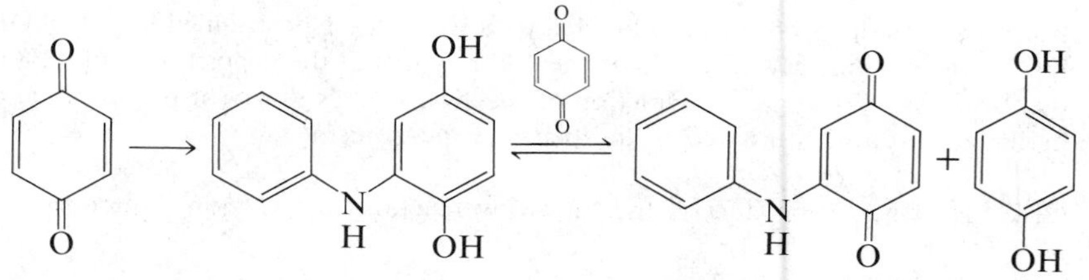

This equilibrium lies far to the right because of the reduction potentials of the two quinones. The 2-anilino-1,4-benzoquinone then undergoes a second conjugate addition resulting in the formation of 2,5-dianilinohydroquinone, which is similarly oxidized to produce the isolated product.

Quinones also function as potent dienophiles in the Diels–Alder reaction (Section 20.5). An example is the reaction of 1,4-benzoquinone with butadiene, which occurs in acetic acid solution at room temperature. The product may be isolated or it may be treated with HCl, whereupon tautomerization to the more stable hydroquinone form occurs.

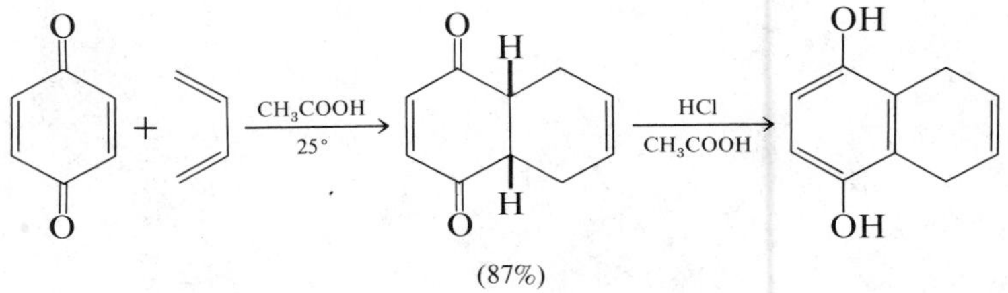

(87%)

EXERCISE: Write equations illustrating the reaction of 2,3-dimethyl-1,4-benzo-quinone with (a) HCl, (b) CH_3NH_2, and (c) 2-methyl-1,3-butadiene (25°).

PROBLEMS

1. Write structures for each of the following names.
   (a) *m*-cresol
   (b) benzyne
   (c) 3-chloro-1,2-benzoquinone
   (d) *o*-methoxyphenol
   (e) picric acid
   (f) benzyl phenyl ether
   (g) 3-(*o*-hydroxyphenyl)pentanoic acid
   (h) *p*-isobutylphenol
   (i) 2-methoxy-1,4-naphthoquinone
   (j) 2,5-dichloro-1,4-benzoquinone

2. The Sanger method for identifying the N-terminal amino acid of a peptide involves reaction with 2,4-dinitrofluorobenzene (Section 29.6.C). Write a plausible mechanism for this reaction.

3. When 2,4,6-trinitroanisole is treated with methoxide ion in methanol, a red anion having the composition $(C_8H_8O_8N_3)^-$ is produced. Such anions are called Meisenheimer complexes after the chemist who first suggested the correct structure. What structure do you think he suggested? One of Meisenheimer's experiments compared the product of reaction of 2,4,6-trinitroanisole with ethoxide ion with the product of 2,4,6-trinitrophenyl ethyl ether with methoxide ion. What do you think he found?

4. The reaction of chlorobenzene with hot aqueous sodium hydroxide actually goes in part by way of a benzyne intermediate and in part by nucleophilic aromatic substitu-

1024

Chap. 30

**Aromatic
Halides,
Phenols,
Phenyl Ethers,
and Quinones**

tion. Reaction of chlorobenzene labeled with ^{14}C at the 1-position with $4\,M$ NaOH at 340° gives phenol in which 58% of the ^{14}C remains at the 1-position and 42% is at the 2-position. Calculate the fraction of reaction going by way of nucleophilic aromatic substitution compared to the benzyne mechanism.

5. Give the principal product of the following reactions or reaction sequences:

(a)

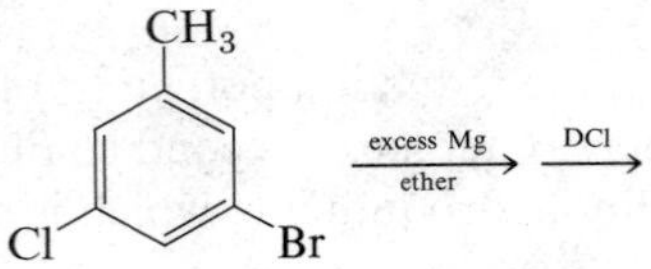

$$\xrightarrow[\text{ether}]{\text{excess Mg}} \xrightarrow{\text{DCl}}$$

(b)

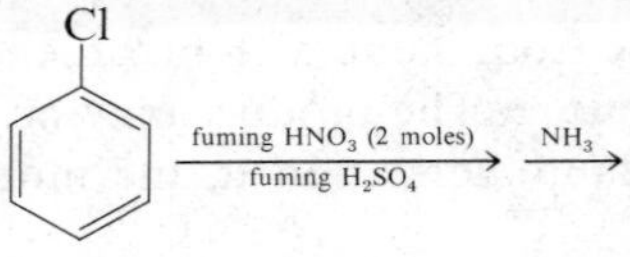

$$\xrightarrow[\text{fuming H}_2\text{SO}_4]{\text{fuming HNO}_3\ (2\ \text{moles})} \xrightarrow{\text{NH}_3}$$

(c)

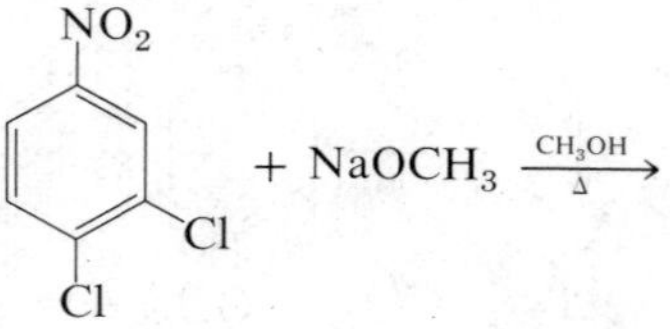

$+ \text{NaOCH}_3 \xrightarrow[\Delta]{\text{CH}_3\text{OH}}$

(d)

$$\xrightarrow[h\nu]{\text{Br}_2} \xrightarrow[\text{ether}]{\text{Mg}} \xrightarrow{\overset{O}{\text{CH}_2-\text{CH}_2}} \xrightarrow{\text{H}^+}$$

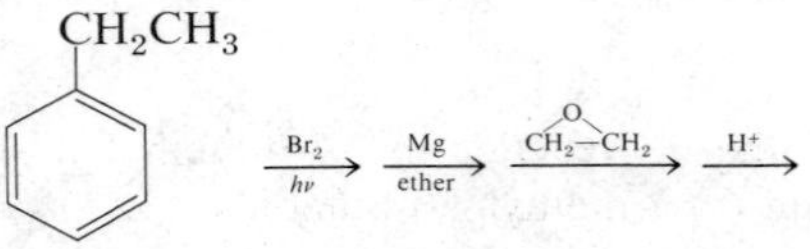

6. Write the principal reaction product or products, if any, of o-cresol with the following reagents.

(a) $(CH_3O)_2SO_2$, NaOH
(b) $Na_2S_2O_4$
(c) $SnCl_2$, HCl
(d) $Na_2Cr_2O_7$, H_2SO_4
(e) $C_6H_5N_2^+$, aq. NaOH
(f) CH_3COOH, H_2SO_4, Δ
(g) aq. NH_3
(h) 98% H_2SO_4, 25°
(i) $KMnO_4$, Δ
(j) cold dilute HNO_3
(k) Br_2, H_2O
(l) $(CH_3CO)_2O$
(m) $CHCl_3$, aq. NaOH
(n) HONO
(o) $LiAlH_4$
(p) HNO_3 (2 moles) in CH_3COOH
(q) HBr, Δ
(r) CO_2, K_2CO_3, 240°
(s) Br_2 (1 mole), CCl_4

7. Write the principal reaction product or products, if any, when 2-methylanisole is subjected to the following conditions.

(a) $Na_2S_2O_4$
(b) conc. H_2SO_4
(c) HBr, Δ
(d) Br_2, CH_3COOH
(e) HNO_3 (2 moles) in CH_3COOH
(f) $LiAlH_4$
(g) $(CH_3CO)_2O$
(h) CH_3COCl, $ZnCl_2$
(i) phthalic anhydride, $C_6H_5NO_2$, $AlCl_3$, 0°
(j) Na, liquid NH_3

8. Each of the following phenol or quinone derivatives has the common or trivial name shown and is a compound of some significance. Provide the IUPAC name and show how each may be synthesized from the indicated starting material.

(a)

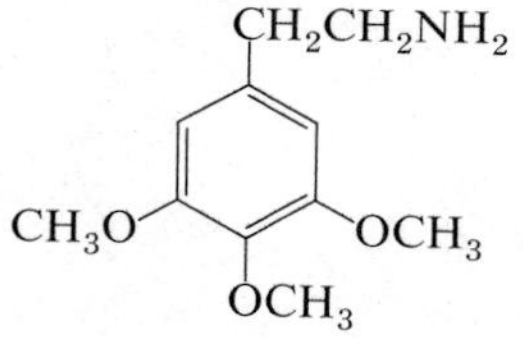

CH₂CH₂NH₂ / CH₃O / OCH₃ / OCH₃

Mescaline is the active ingredient in peyote (mescal buttons) and is used as a psychotomimetic (that is, mimics psychosis) drug. Prepare from gallic acid.

CrO, or SOCl₂, CN⁻ — LiAlH₄, NH₃OCOCH₃

(b)

OH CH₃ / O₂N / CHCH₂CH₃ / NO₂

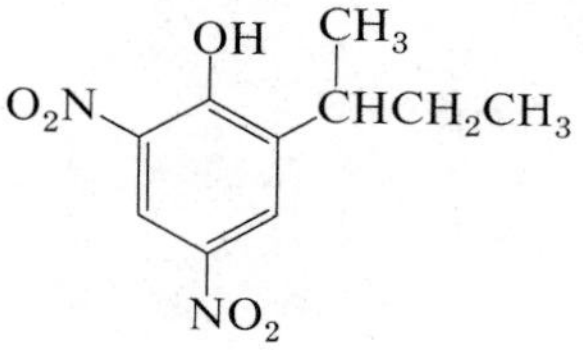

The ester of this phenol with 3-methyl-but-2-enoic acid is **binapacryl,** which is used as a fungicide and mitocide. Synthesize the phenolic portion shown from phenol.

(c) 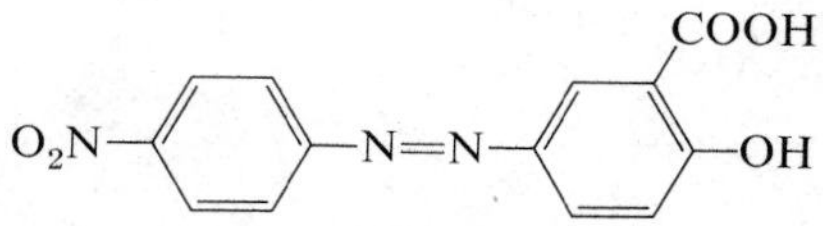 O₂N—〈 〉—N=N—〈 〉 COOH / OH

Alizarine yellow R is used as an indicator in alkaline solutions. The color changes from yellow to red over the pH range from 10 to 12. Synthesize from aniline.

(d)

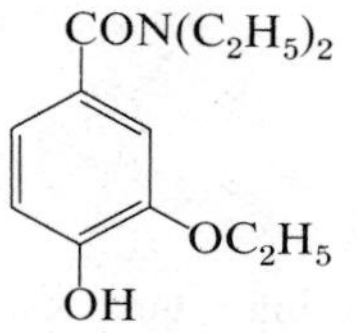

CON(C₂H₅)₂ / OC₂H₅ / OH

Anacardiol is a drug that acts as a stimulant on the central nervous system. Prepare from catechol.

(e)

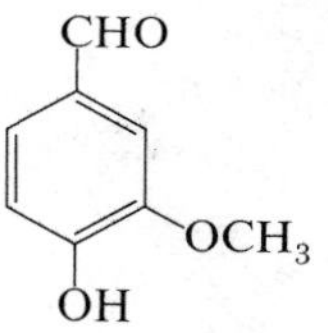

CHO / OCH₃ / OH

Vanillin occurs naturally in vanilla and other plant materials and is used as a flavoring agent. Prepare from guaiacol.

(f)

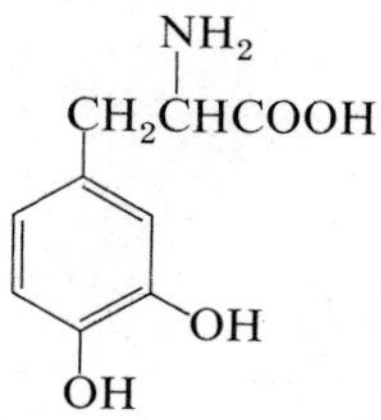

NH₂ / CH₂CHCOOH / OH / OH

Dopa having levorotation **(L-dopa)** is found in some beans. It is used in a treatment of Parkinson's disease. Prepare the racemic compound from catechol.

(g)

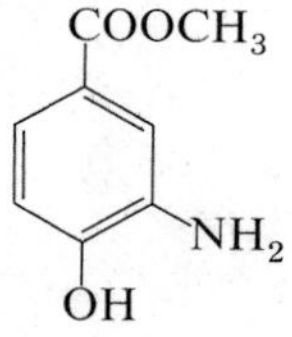

COOCH₃ / NH₂ / OH

Orthocaine is used as a surface anesthetic. Prepare from phenol.

9. Show how each of the following conversions may be accomplished.

(a)

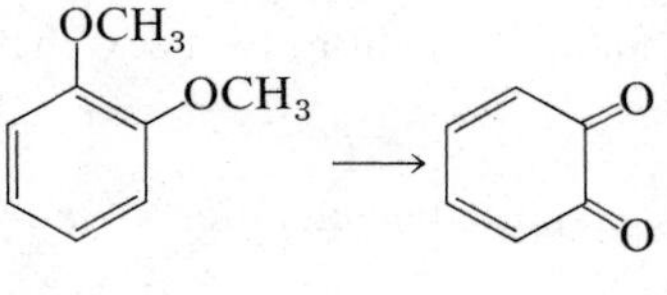

OCH₃ / OCH₃ ⟶ O / O

1026

Chap. 30

**Aromatic
Halides,
Phenols,
Phenyl Ethers,
and Quinones**

(b)

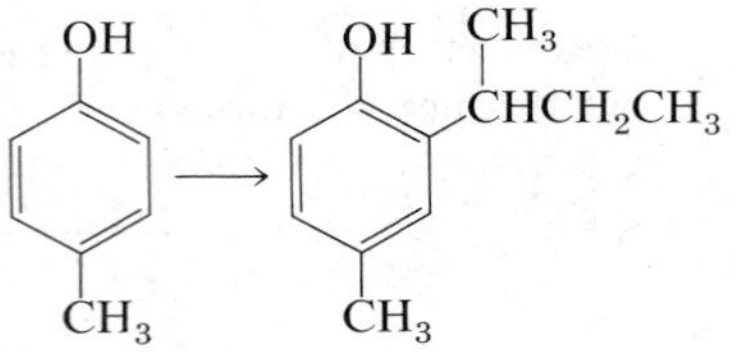

(c)

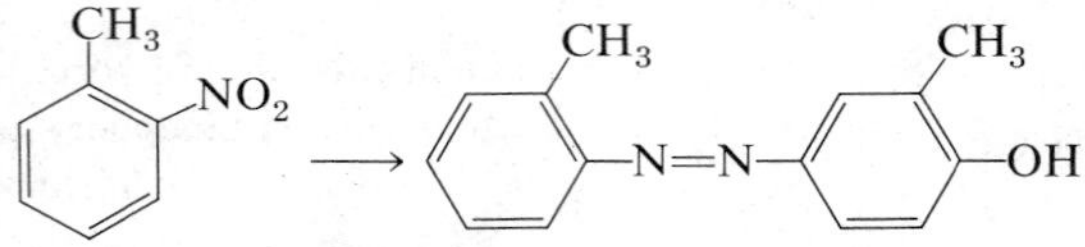

(d)

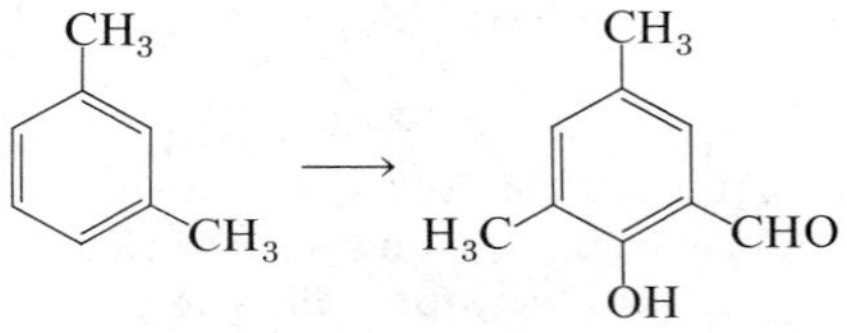

(e)

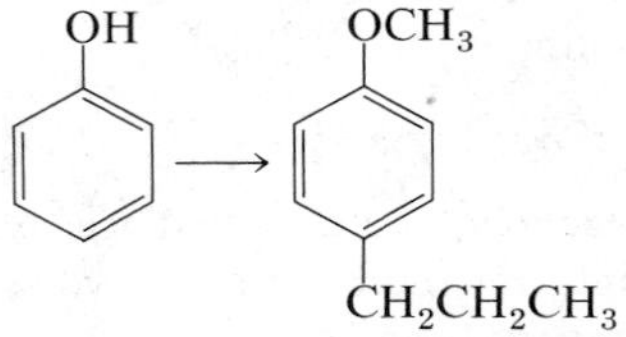

10. What is the principal organic product of each of the following sequences?

(a) $C_6H_5CH=CHCH_2Cl \xrightarrow{C_6H_5O^-} \xrightarrow{200°}$

(b)

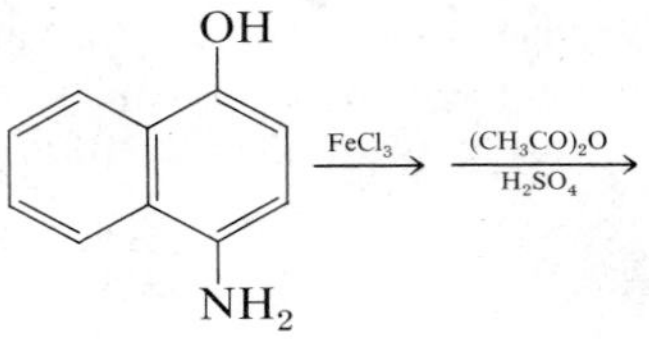

$\xrightarrow{FeCl_3} \xrightarrow[H_2SO_4]{(CH_3CO)_2O}$

(c) $OHC-\underset{\underset{CH_3}{|}}{CH}-\underset{\underset{CH_3}{|}}{CH}-CHO \xrightarrow{CH_2=P(C_6H_5)_3 \ (2 \ moles)} \xrightarrow{\Delta}$

(d) $(CH_3)_2C=CHCH_2OCH=CH_2 \xrightarrow{\Delta} \xrightarrow{CH_2=P(C_6H_5)_3} \xrightarrow{\Delta}$

(e) $C_6H_5NH_2 \xrightarrow[HCl]{NaNO_2}$

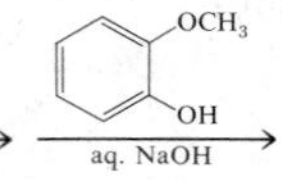

$\xrightarrow{aq. \ NaOH}$

(f)

$\xrightarrow{(CH_3CO)_2O} \xrightarrow[\Delta]{AlCl_3}$

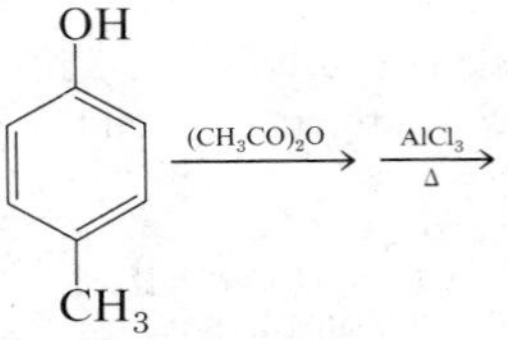

11. 2,6-Dichlorophenol is present in some ticks and is thought to be a sex pheromone. Devise a practical synthesis from phenol.

12. In Table 23.2 (page 714) the relative rates of protodedeuteration of o-, m-, and *p*-anisole-*d* are summarized. Show how each of these deuterated anisoles can be prepared uncontaminated by the other isomers.

13. Another component of urushiol, the active constituent of the irritating oil of poison ivy, is 3-pentadecyl-1,2-dihydroxybenzene. Synthesize this compound from catechol (be careful in handling the product!).

14. When *n*-butyl benzenesulfonate is heated with an ethanolic solution of potassium benzyloxide, the product is a mixture of *n*-butyl ethyl ether and *n*-butyl benzyl ether. However, if the isomeric salt, potassium *p*-methylphenolate, is used, the product is almost exclusively *n*-butyl *p*-methylphenyl ether. Explain.

15. Acetanilide is oxidized in the body by oxygen and a hydroxylase enzyme to *p*-hydroxyacetanilide. Show how this compound can be synthesized from phenol.

16. 2,2-Bis-(*p*-hydroxyphenyl)propane or "Bisphenol A," used commercially in the manufacture of epoxy resins and as a fungicide, is prepared by the reaction of phenol with acetone in acid.

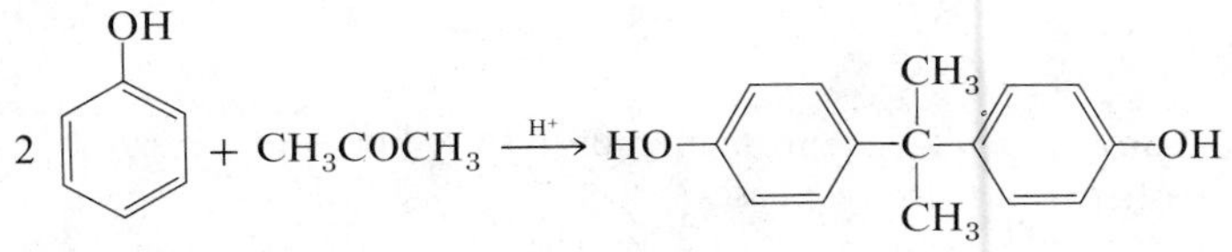

Bisphenol A

Write out the mechanism of this reaction, showing all intermediates involved.

17. The sulfonation of *p*-cymene (1-methyl-4-isopropylbenzene) gives the 2-sulfonic acid. Is this the expected orientation? Explain. Use this fact to synthesize carvacrol, 2-methyl-5-isopropylphenol, from *p*-cymene. Carvacrol is found in the essential oils from thyme, marjoram, and summer savory. It has a pleasant thymol-like odor.

18. 4-Hexylresorcinol is used medicinally as an antiseptic. Suggest a practical synthesis from resorcinol.

19. Rank each group in order of increasing acidity.
 (a) phenol, 3-acetylphenol, 4-acetylphenol
 (b) *p*-dimethylaminomethylphenol, *p*-dimethylaminophenol, trimethyl-(*p*-hydroxy-phenyl)ammonium ion
 (c) 2-hydroxy-1,4-benzoquinone, 2,5-dimethoxyphenol, 4-hydroxy-1,2-benzoquinone

20. A student attempted the following synthesis of *o*-methoxybenzyl alcohol from *o*-cresol, but got essentially no yield. What went wrong?

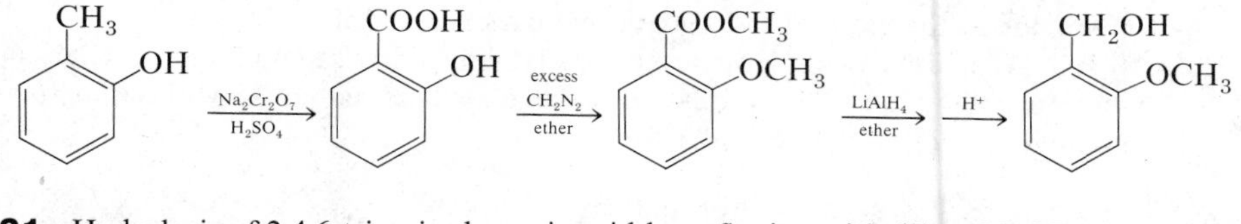

21. Hydrolysis of 2,4,6-triaminobenzoic acid by refluxing with dilute NaOH gives phloroglucinol.

1028

Chap. 30

**Aromatic
Halides,
Phenols,
Phenyl Ethers,
and Quinones**

Write a reasonable mechanism for this reaction. (*Hint:* The reaction involves the nonaromatic keto forms.)

22. *o*-Phenylazophenol is readily separable from the *para* isomer by steam distillation. Give a reasonable explanation for the greater volatility of the *ortho* isomer. Can this explanation also be used to explain the greater volatility of *o*-hydroxypropiophenone compared to the *para* isomer?

23. Write out the mechanism of both steps involved in the Hoesch reaction of phloroglucinol with acetonitrile, shown on page 1009. The corresponding reaction of phenol gives mostly the imino ether hydrochloride,

$$C_6H_5OCCH_3$$
$$\underset{\overset{||}{\underset{NH_2^+}{}} Cl^-}{}$$

Write the mechanism for the formation of this product. How do you account for the difference between these two cases?

24. (a) Write out a reasonable mechanism for the sulfuric acid-catalyzed condensation of phenol with phthalic acid. Be sure to show all intermediates.
(b) Phenol does not form diphenyl ether with sulfuric acid, yet the condensation of resorcinol with phthalic anhydride to give fluorescein includes the formation of an ether link from two phenolic hydroxy groups. Give a reasonable explanation.

25. Equimolar mixtures of *p*-benzoquinone with hydroquinone and of 2-chloro-1,4-benzoquinone with chlorohydroquinone in the same buffer solution are contained in separate beakers. The beakers are connected by a salt bridge, and the potential difference between them is measured. What is this potential difference? Which beaker constitutes the negative end (cathode) of this battery?

26. Allyl chloride labeled with ^{14}C is allowed to react with the anion from 2-methyl-6-allylphenol to form the corresponding ether. When this ether is heated, the Claisen rearrangement product 2-methyl-4,6-diallylphenol is formed. More than half, but not all, of the ^{14}C is found in the allyl group in the 4-position. Explain.

27. Vinyl ethers are readily available by an alcohol exchange reaction using commercially available ethyl vinyl ether.

$$ROH + C_2H_5OCH{=}CH_2 \xrightarrow{Hg^{2+}} ROCH{=}CH_2 + C_2H_5OH$$

(a) Write a mechanism that accounts for the mercuric ion catalysis. (*Hint:* The Hg^{2+} first adds to the enol ether to give a mercuricarbocation.)
(b) If ROH is an allyl alcohol, the resulting allyl vinyl ether can undergo the Claisen rearrangement. Show how this sequence may be used in a synthesis of (1-methyl-2-cyclohexenyl)acetaldehyde.

28. Tri-*o*-cresyl phosphate, (*o*-CH$_3$C$_6$H$_4$O)$_3$PO, is used as a gasoline additive. Suggest a preparation.

29. Diazotization of 2,4-dinitroaniline in aqueous solution is accompanied by some conversion to phenols in which a nitro group is replaced by a hydroxy group.

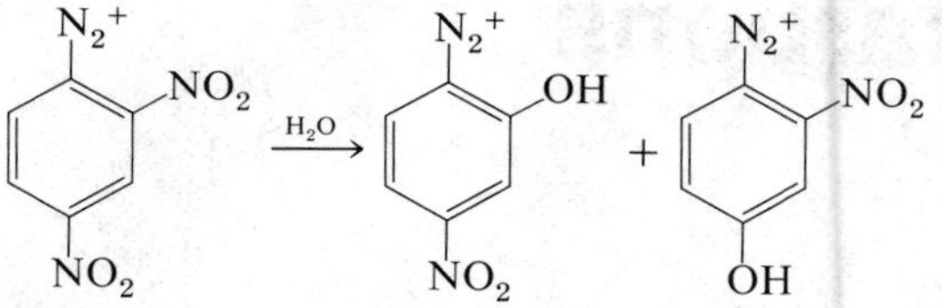

Give a reasonable mechanism for this reaction.

30. In the chlorination of aniline in aqueous solution some chlorophenol is produced. Write a reasonable mechanism.

Chapter 31
Polycyclic Aromatic Hydrocarbons

31.1
Nomenclature

Polycyclic aromatic hydrocarbons may be dissected into two broad classes: the biaryls and the condensed benzenoid hydrocarbons. The latter class is by far the larger and more important group.

The biaryls are benzenoid compounds in which two rings are linked together by a single bond. The parent system of this class is biphenyl. In numbering the ring positions, the rings are considered to be joined at the 1-position, and the two rings are distinguished by the use of primes.

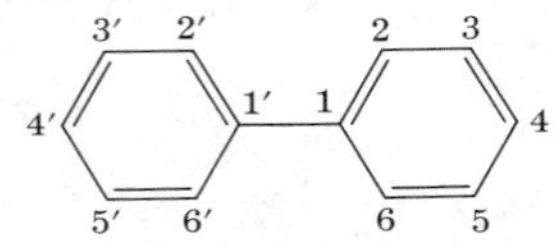

biphenyl

Simple derivatives can be named by use of *ortho, meta, para* nomenclature.

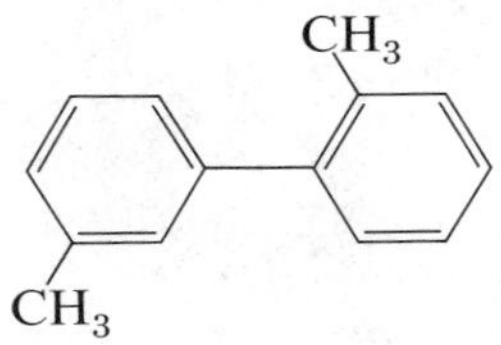

o,m'-dimethylbiphenyl
2,3'-dimethylbiphenyl

More complex compounds are named using numbers. Again, substituents in one ring are designated by the use of primes.

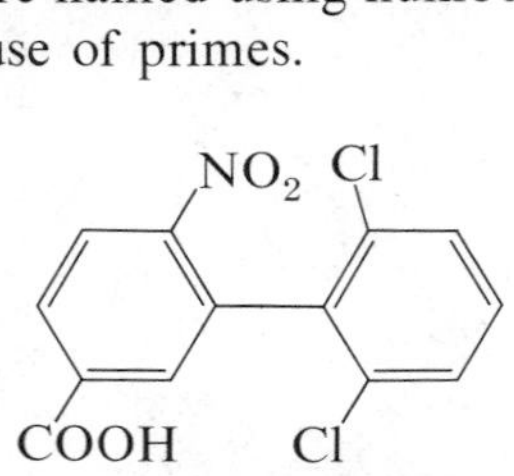

2',6'-dichloro-6-nitrobiphenyl-3-carboxylic acid

The condensed benzenoid compounds are characterized by two or more benzene rings **fused** or superimposed together at *ortho* positions in such a way that each pair of rings shares two carbons. The simplest members of this group are naphthalene, with two rings, and anthracene and phenanthrene, with three rings. In the IUPAC systematic names all carbons that may bear a substituent are numbered. Carbons that are part of a ring junction are denoted by a lowercase a or b following the number of the immediately preceding carbon. The numbering systems for naphthalene, anthracene, and phenanthrene are

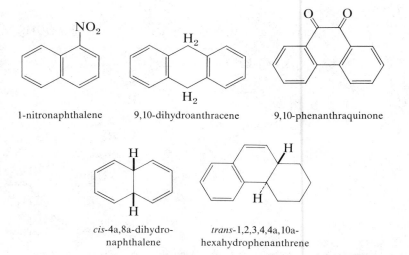

naphthalene   anthracene   phenanthrene

Derivatives are named using these numbering systems.

1-nitronaphthalene   9,10-dihydroanthracene   9,10-phenanthraquinone

cis-4a,8a-dihydro-   *trans*-1,2,3,4,4a,10a-
naphthalene   hexahydrophenanthrene

EXERCISE: Name the following compounds.

31.2
Biphenyl

A. *Synthesis*

Biphenyl is prepared commercially by the pyrolysis of benzene.

$$2 \, C_6H_6 \xrightarrow{\Delta} C_6H_5{-}C_6H_5 + H_2$$

It is a colorless crystalline solid with a melting point of 70°. Substituted biphenyls are prepared by electrophilic aromatic substitution reactions on the parent hydrocarbon (Section 31.2.C) or from benzene derivatives using reactions we have already studied. One of the most useful methods is the benzidine rearrangement (page 784).

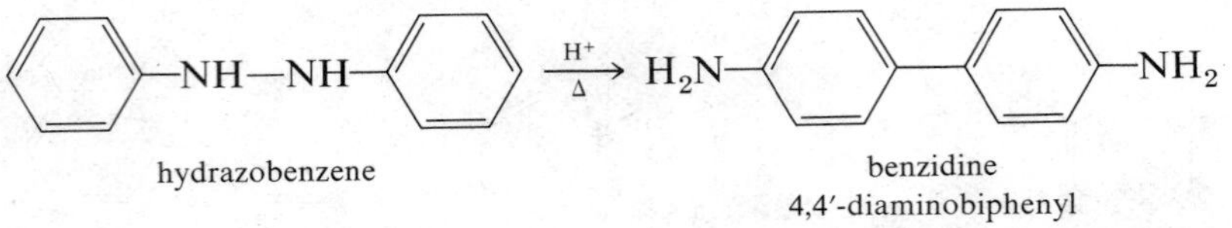

hydrazobenzene

benzidine
4,4′-diaminobiphenyl

The amino groups in benzidine can be converted to many other functional groups by way of the bis-diazonium salt. Thus, benzidine and substituted benzidines are useful intermediates for the preparation of a variety of symmetrical biphenyls.

(80%)
2,2′-dimethyl-5,5′-dimethoxybiphenyl

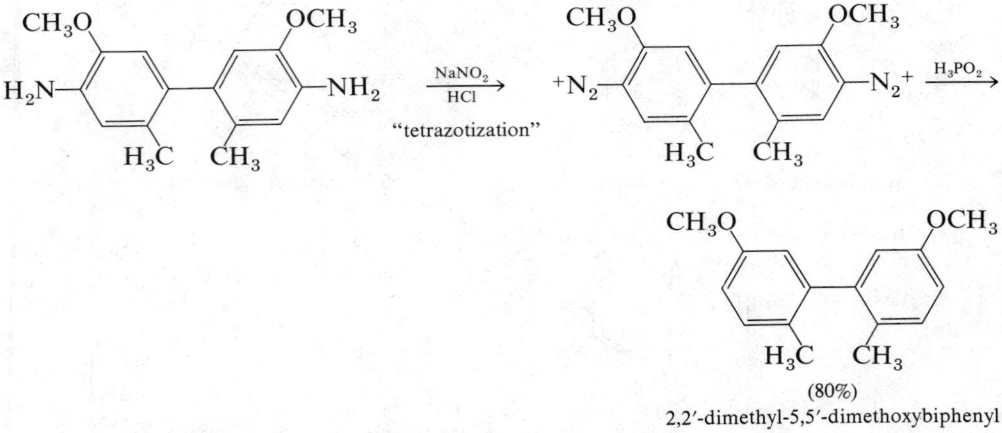

The Ullmann reaction (Section 30.3.B) is also useful for the preparation of symmetrically substituted biphenyls, as is the oxidation of certain lithium diarylcuprates (Section 30.3.B). The Gomberg–Bachmann reaction (Section 25.6) is suitable for the preparation of some unsymmetrical biphenyls.

EXERCISE: Propose syntheses of (a) *m,m′*-dimethylbiphenyl and (b) *m*-methylbiphenyl beginning with *o*-nitrotoluene.

B. *Structure*

In the crystal both benzene rings of biphenyl lie in the same plane. However, in solution and in the vapor phase the two rings are twisted with respect to each other by an angle of about 45° (Figure 31.1). This twisting is the result of steric interactions between the 2,2′ and 6,6′ pairs of hydrogens (Figure 31.2). The magnitude of these repulsions is relatively small, only a few kcal mole^{-1}, and in the crystal is less than the stablization obtained by stacking biphenyls together in coplanar arrays. Of course, these crystal-packing forces do not exist in the vapor phase, and the twisting of the rings causes greater separation of the hydrogens.

These repulsion effects are enhanced by *ortho* substituents larger than hydrogen. When the groups are sufficiently large, rotation of the phenyl rings with respect to each other is hindered or prevented. For example, 6,6′-dinitrobiphenyl-

(a)

~45°

(b)

FIGURE 31.1. *Structure of biphenyl in the solution or vapor phase:* (a) *stereo representation;* (b) *perspective diagram.*

2,2'-dicarboxylic acid can be resolved into its enantiomers and each enantiomer is stable indefinitely (Figure 31.3). The nitro and carboxylic acid groups are so bulky that they cannot pass by each other, and rotation about the bond joining the two rings is prevented.

If the bulky nitro groups are replaced by the smaller fluorine atoms, the resulting compound, 6,6'-difluorobiphenyl-2,2'-dicarboxylic acid, can still be obtained in optically active form. However, the compound racemizes readily; that is, the enantiomers are readily interconverted. The racemization process involves squeezing the fluorines past the adjacent COOH groups via a planar transition state.

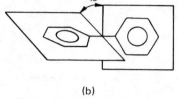

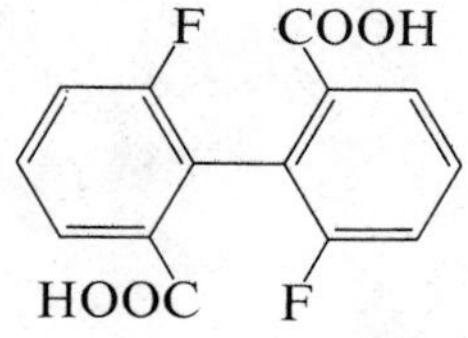

F   COOH

HOOC   F

6,6'-difluorobiphenyl-2,2'-dicarboxylic acid

This transition state is congested and requires bending bonds. The process takes energy and is measurably slow. On the other hand, all attempts to resolve bi-

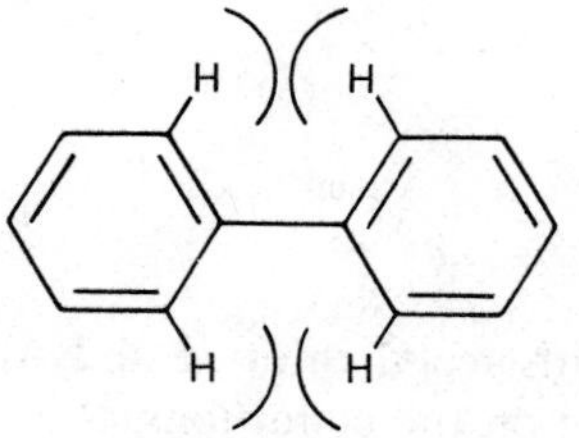

H) (H

H) (H

FIGURE 31.2. *Steric interactions between* ortho-*hydrogens in biphenyl.*

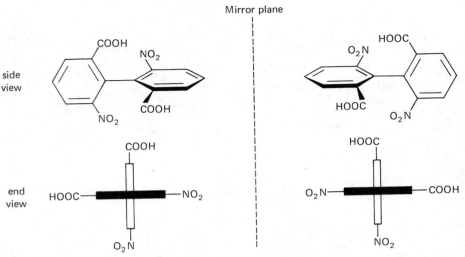

FIGURE 31.3. *Enantiomers of 6,6'-dinitrobiphenyl-2,2'-dicarboxylic acid.*

phenyl-2,2'-dicarboxylic acid (diphenic acid) have failed. The process of slipping a small hydrogen past the carboxylic acid group is so facile that racemization of enantiomers occurs rapidly.

diphenic acid

> **EXERCISE:** Construct a molecular model of a biphenyl having different substituents in the 2,2'- and 6,6'-positions. Demonstrate that the molecule is chiral so long as the two rings are not coplanar.

C. *Reactions*

Biphenyl undergoes electrophilic aromatic substitution more readily than benzene; a phenyl substituent is activating and is an *ortho,para* director. Nitration in acetic anhydride solution gives primarily 2-nitrobiphenyl, but most other substitution reactions give primarily *para* orientation. Bromination, for example, gives almost wholly 4-bromobiphenyl, and excess reagent leads readily to 4,4'-dibromobiphenyl. Typical partial rate factors are

$$C_6H_5-\begin{array}{c}130\\ \\ 3900\end{array} \qquad \text{for Br}_2 \text{ in 50\% aqueous acetic acid}$$

Friedel–Crafts acylation with acetyl chloride and AlCl$_3$ yields 4-acetyl- and 4,4'-diacetylbiphenyl depending on the conditions. Reaction with succinic anhydride is another example.

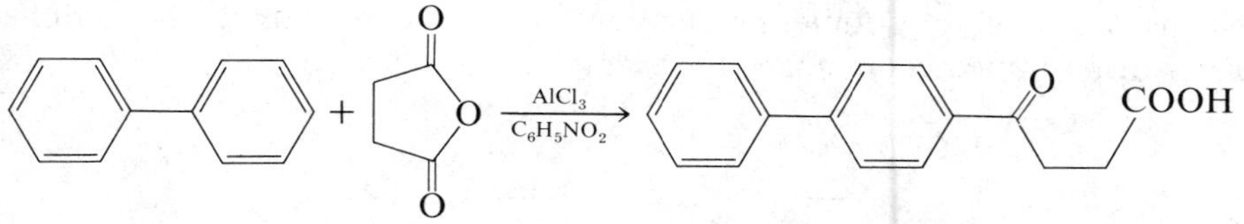

(70–90%)
4-oxo-4-(4-biphenylyl)butanoic
acid

In general, 4-substituted and 4,4'-disubstituted biphenyls can often be prepared by way of electrophilic substitution reactions of biphenyl. Other derivatives are constructed from benzene compounds by way of the syntheses described in Section 31.2.A.

EXERCISE: Write the resonance structures for the intermediate cation resulting from attack of Br^+ at the *para* position and at the *meta* position of biphenyl. Explain why a phenyl substituent is *ortho,para*-directing.

D. *Related Compounds*

The terphenyls have three benzene rings linked together. All three possible isomers, *ortho, meta,* and *para,* are known. Note how the greater symmetry of the *para* isomer confers a much higher melting point.

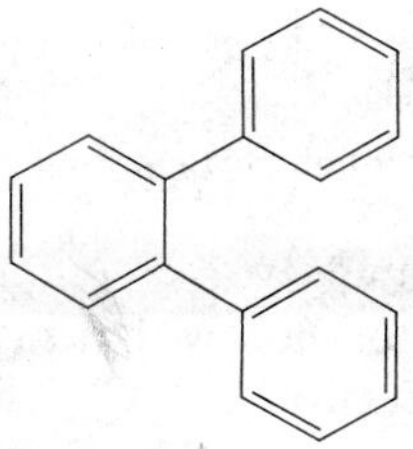

o-terphenyl m.p. 57°	*m*-terphenyl m.p. 87°	*p*-terphenyl m.p. 171°

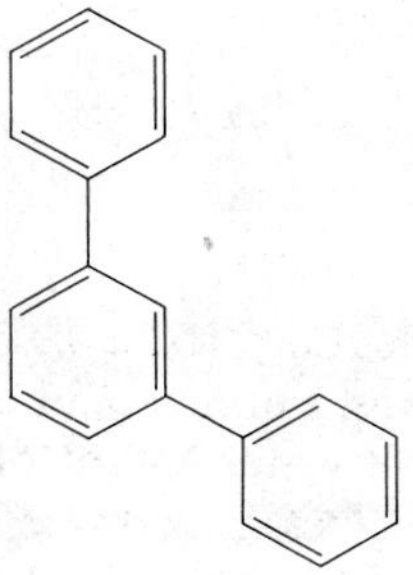

Many of the higher polyphenyls are known, especially for the *para* isomers. *p*-Quaterphenyl has four phenyl groups linked and melts at 320°. These compounds are generally such insoluble materials that they are difficult to work with.

Fluorene is a biphenyl in which two *ortho* positions are linked by a methylene group. It is obtained commercially from coal tar.

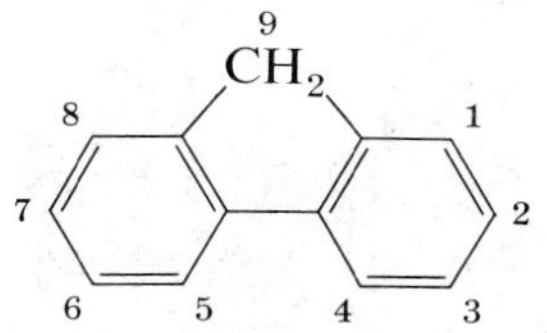

fluorene

The 2- and 7-positions correspond to the *para* positions of biphenyl and are, accordingly, the most reactive positions in electrophilic aromatic substitution reactions. Most such substitutions on fluorene give predominantly the 2-substituted or 2,7-disubstituted compounds, for example,

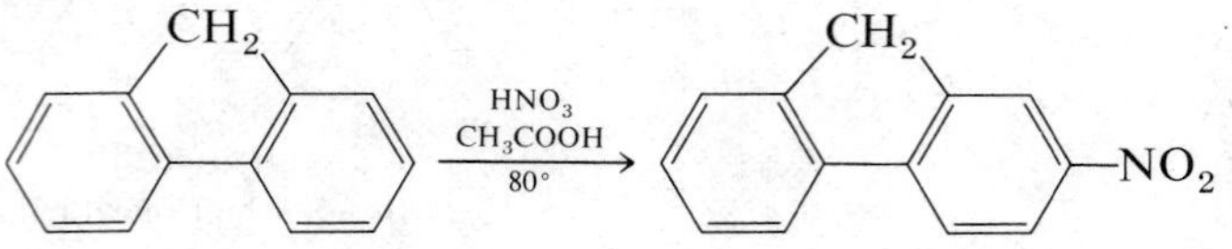

(79%)
2-nitrofluorene

The methylene group is an important center for other reactions. Oxidation gives the corresponding yellow ketone, fluorenone.

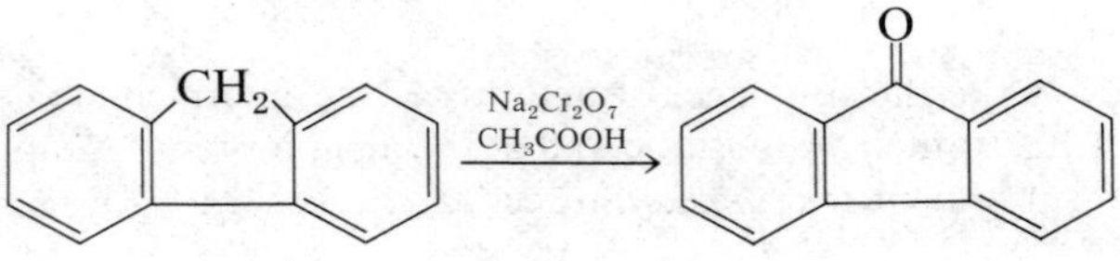

fluorenone

One of the especially interesting aspects of the chemistry of fluorene is its relatively high acidity. The pK_a value of 23 puts the methylene group of this hydrocarbon in the same range as ketones and esters. Alkali metal salts can be prepared by melting with potassium hydroxide or by treatment with butyllithium.

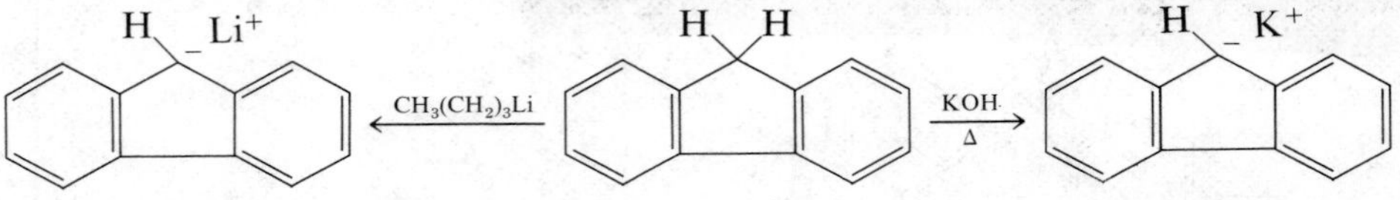

9-fluorenyllithium

9-fluorenylpotassium

The reason for this remarkably high acidity is related to the central five-membered ring structure. Cyclopentadiene is a highly acidic hydrocarbon with a pK_a of 16—an acidity comparable to that of water and alcohols (Section 22.7.C). The resulting "aromatic" character of this electronic system stabilizes the anion relative to the hydrocarbon. The anion is less basic and the hydrocarbon is more acidic than would be expected in the absence of such stabilization.

If one or both double bonds in cyclopentadiene are replaced by benzene rings, the corresponding anion has reduced stability relative to its conjugate acid because the delocalization of negative charge disrupts the benzene conjugation.

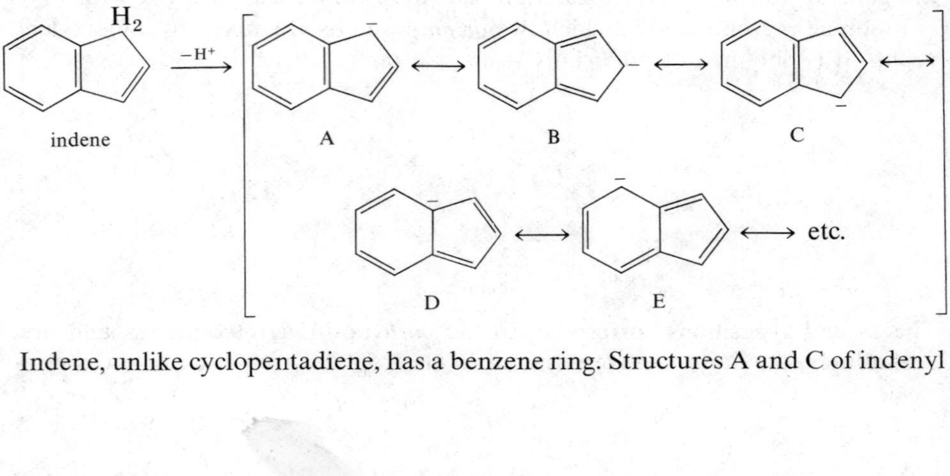

indene   A   B   C

D   E   etc.

Indene, unlike cyclopentadiene, has a benzene ring. Structures A and C of indenyl

TABLE 31.1
Acidities of Some Hydrocarbons

Formula	Name	pK_a
	cyclopentadiene	16
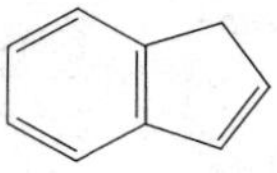	indene	20
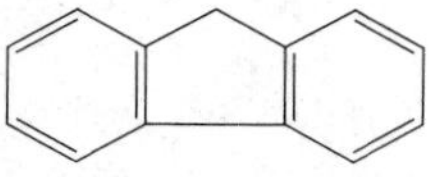	fluorene	23
$(C_6H_5)_3CH$		31.5
$(C_6H_5)_2CH_2$		34
$C_6H_5CH_3$		41

anion also have benzene rings, but the other structures, B, D, E, and so on, have no benzene rings and are expected to be much less stable. The same principles apply to fluorene and fluorenyl anion. The corresponding pK_a values are summarized in Table 31.1 and compared with several other hydrocarbons for reference.

> **EXERCISE:** Suggest a way in which *p*-terphenyl might be synthesized starting with biphenyl and benzene and any necessary inorganic reagents. (*Hint:* See Section 31.2.A.)

31.3
Naphthalene

A. *Structure and Occurrence*

Naphthalene is a colorless crystalline hydrocarbon, m.p. 80°. It sublimes readily and is isolated in quantity from coal tar.

> Coal tar is obtained from the conversion of bituminous coal to coke. The coal is heated in the absence of oxygen, giving gas and a distillate boiling over a wide range. The low-boiling range contains benzene, toluene, and xylenes. A fraction boiling at 195–230°, called naphthalene oil, yields crude naphthalene on cooling. The higher-boiling coal tar is a black odiferous complex mixture containing many polycyclic hydrocarbons and heterocyclic compounds.

Naphthalene is the parent hydrocarbon of the series of fused benzene polycyclic structures. X-ray analysis shows it to have the structure shown in Figure 31.4. The bonds are not all of the same length, but are close to the benzene value of 1.397 Å. Naphthalene can be considered to be resonance hybrid of three structures.

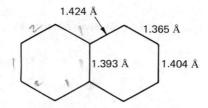

FIGURE 31.4. *Structure of naphthalene.*

Accordingly, it has an empirical resonance energy of about 60 kcal mole^{-1}, a value somewhat greater than that of benzene (Section 22.1.B).

Substituted naphthalenes are named using the numbering system given in Section 31.1. Monosubstituted naphthalenes are often named using α- and β-nomenclature for the 1- and 2-positions, respectively, for example,

α-methylnaphthalene
1-methylnaphthalene

β-naphthol
2-naphthol

Reduced naphthalenes are widespread in nature, particularly in terpenes and steroids (Section 34.6). The fully reduced form, decahydronaphthalene, has the trivial name decalin. Two diastereomeric forms are possible, in which the hydro-

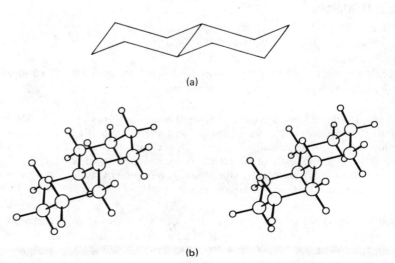

(a)

(b)

FIGURE 31.5. trans-*Decalin:* (a) *conventional representation;* (b) *stereo structure.*

gens at the ring juncture carbons are either *cis* or *trans*. The decalins may also be named using the systematic nomenclature for bicyclic compounds (page 617).

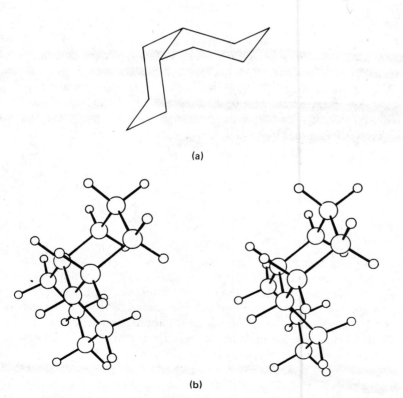

cis-bicyclo[4.4.0]decane
(*cis*-decalin)

trans-bicyclo[4.4.0]decane
(*trans*-decalin)

In both *cis*- and *trans*-decalin the two cyclohexane rings are each in the chair conformation. In *trans*-decalin the two chair cyclohexanes are fused together in such a way that each ring comprises two equatorial substituents on the other one (Figure 31.5). In *cis*-decalin the cyclohexane chairs are joined together as equatorial and axial substituents (Figure 31.6).

EXERCISE: From the resonance structures on page 1038, determine the fractional double bond character of all of the different C–C bonds in naphthalene. A pure C_{sp^2}—C_{sp^2} single bond is expected to have a length of about 1.50 Å. Using this value and the bond lengths of ethylene and benzene, draw a smooth curve for bond length as a function of double bond character. Calculate the bond lengths expected for the different bonds in naphthalene using this curve and compare with the experimental values in Figure 31.4.

(a)

(b)

FIGURE 31.6. cis-*Decalin:* (a) *conventional representation;* (b) *stereo structure.*

B. *Synthesis*

The naphthalene ring system can be prepared from suitable benzene derivatives (Section 27.6.B). The following sequence is an example.

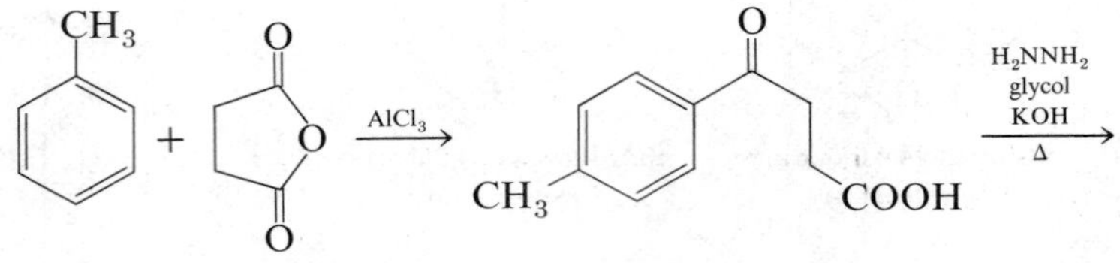

toluene	succinic anhydride		4-*p*-tolyl-4-oxo-butanoic acid

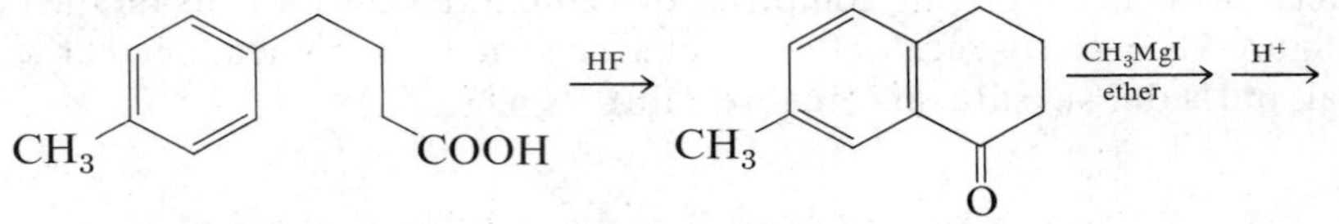

4-*p*-tolylbutanoic acid   1-oxo-7-methyl-1,2,3,4-tetrahydronaphthalene

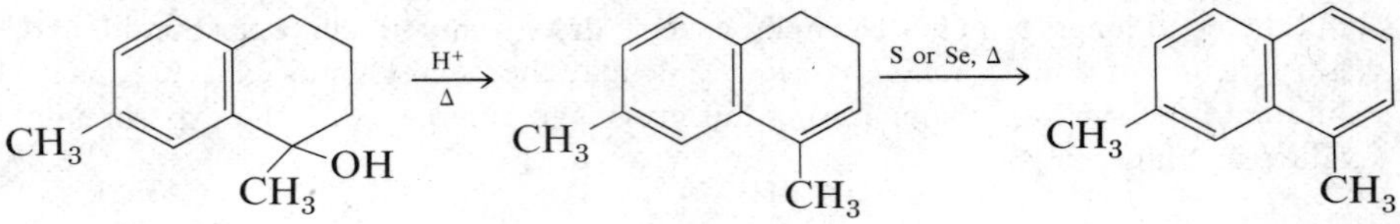

1,7-dimethyl-1,2,3,4-tetrahydro-1-naphthol	4,6-dimethyl-1,2-dihydronaphthalene	1,7-dimethylnaphthalene

The last step in the foregoing synthesis of 1,7-dimethylnaphthalene illustrates the **aromatization** of a hydroaromatic compound. The driving force is formation of the stable aromatic ring. When sulfur or selenium is used for aromatization, it is concomitantly reduced to H_2S or H_2Se, respectively. Palladium metal can also be used to catalyze the aromatization. In this case, which amounts to the reverse of catalytic hydrogenation, hydrogen is eliminated.

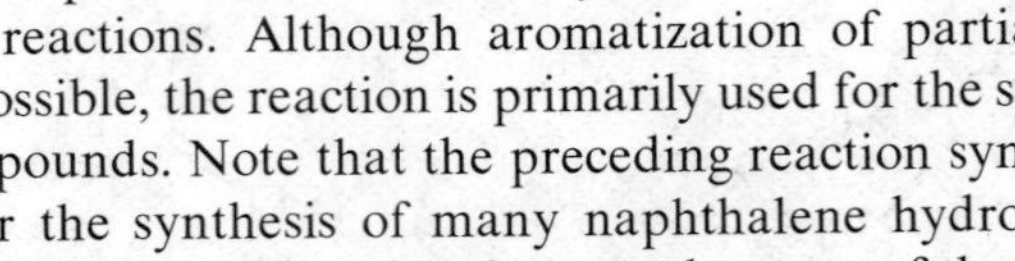

naphthalene

Tetrachloro-*p*-benzoquinone, chloranil, may also be used as a reagent for such dehydrogenation reactions. Although aromatization of partially hydrogenated benzenes is also possible, the reaction is primarily used for the synthesis of polycyclic aromatic compounds. Note that the preceding reaction synthesis is subject to wide variation for the synthesis of many naphthalene hydrocarbons. It is less useful for the introduction of functional groups because of the sensitivity of most groups to several of the reactions involved.

Another important way of building up the second ring makes use of the Diels–Alder reaction (Section 20.5). *p*-Benzoquinone is an excellent dienophile and reacts with a wide range of conjugated dienes.

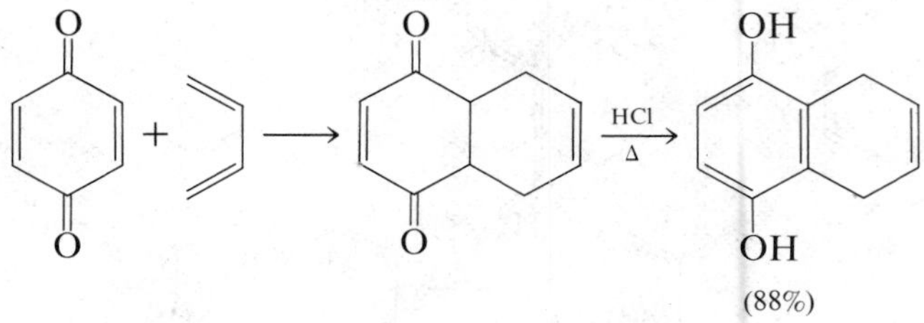

(88%)

The product is a dihydroxydihydronaphthalene that can be converted in good yield to 1,4-naphthoquinone.

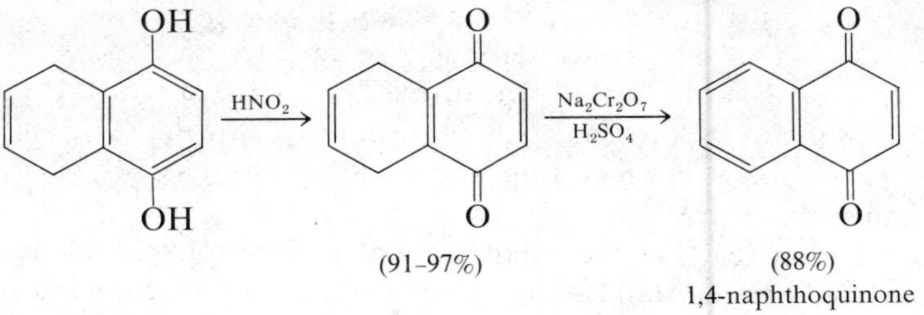

(91–97%) (88%)

1,4-naphthoquinone

EXERCISE: Outline syntheses of (a) 1-propyl-7-methoxynaphthalene and (b) 2,3-dimethyl-1,4-naphthoquinone.

C. Reactions of Naphthalene

1. ELECTROPHILIC SUBSTITUTION. Naphthalene undergoes a number of the usual electrophilic aromatic substitution reactions such as nitration, halogenation, sulfonation, and Friedel–Crafts acylation. The 1-position is the more reactive, for example,

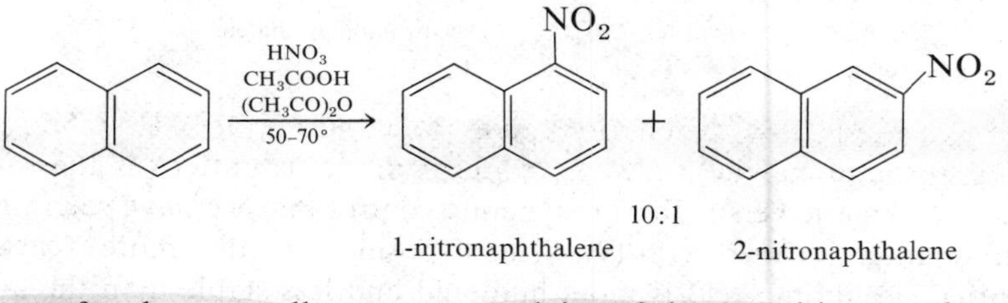

10:1

1-nitronaphthalene 2-nitronaphthalene

The reason for the generally greater reactivity of the 1-position can be seen by examination of the resonance structures for the two transition states or the intermediates resulting from them.

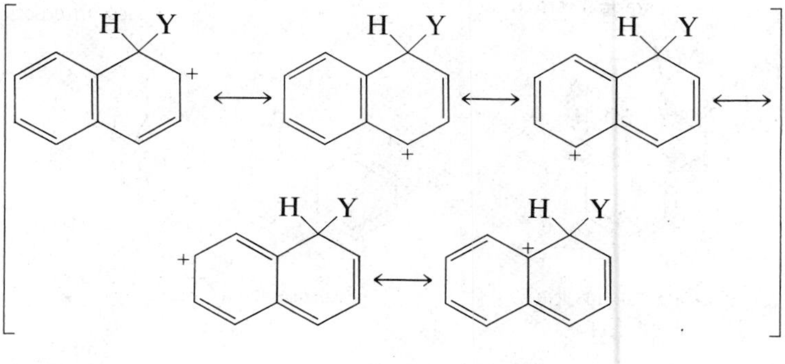

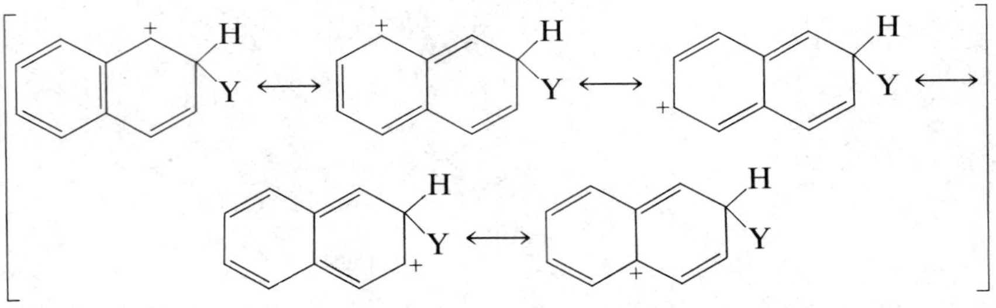

In both cases the positive charge can be distributed to five different positions, but these carbocation structures are not equivalent in energy. In the α case the first two structures still have an intact benzene ring and are consequently much more stable than the remaining three structures. The first two structures contribute much more to the overall resonance hybrid. In the β case, however, only the first structure has an intact benzene ring; the resulting resonance hybrid has higher energy than in the α case.

In the nitration reaction the small amount of 2-nitronaphthalene formed is readily removed by recrystallization; hence, the nitration reaction is a satisfactory route to 1-nitronaphthalene. More vigorous nitration conditions give mixtures of 1,5- and 1,8-dinitronaphthalenes. Since the nitro group is a deactivating group, the second nitro group enters the other ring.

Bromination is also an excellent reaction and gives substantially pure 1-bromonaphthalene.

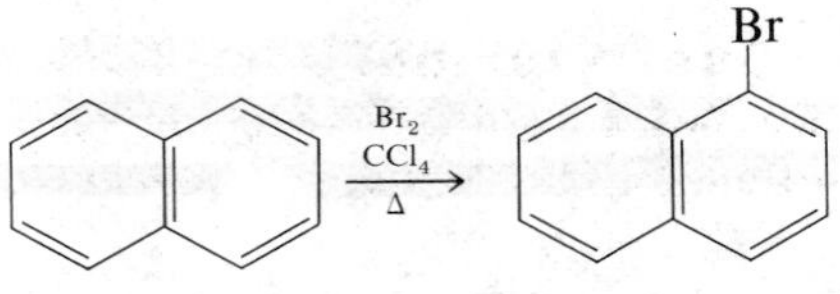

(72–75%)
1-bromonaphthalene

Sulfonation under mild conditions gives the 1-sulfonic acid. However, at higher temperature naphthalene-2-sulfonic acid results. This pattern is the same phenomenon of kinetic versus thermodynamic control that we have seen previously for sulfonations (pages 818–19). The 1-position is the more reactive, but 1-naphthalenesulfonic acid is more hindered and less stable than the 2-acid because the bulky sulfonic acid group is within the van der Waals radius of the 8-hydrogen.

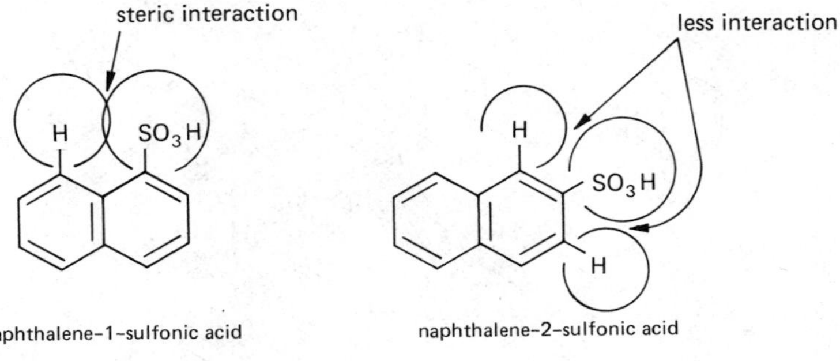

naphthalene–1–sulfonic acid

naphthalene–2–sulfonic acid

Under conditions where the sulfonation reaction is reversible, the 2-acid is the dominant product.

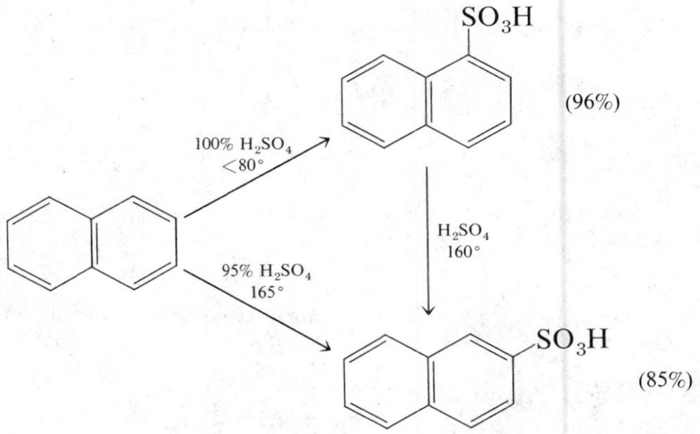

More vigorous sulfonation conditions give di- and trisulfonic acids whose structures are highly dependent on the precise reaction conditions. Several such reaction conditions have been worked out in detail to lead primarily to individual isomers for the preparation of several naphthalenedisulfonic acids that are useful intermediates for the preparation of dyes.

Friedel-Crafts acylation reactions also frequently give mixtures. In general, use of $AlCl_3$ with CS_2 as solvent gives predominantly the α-product, but separation from the β-product also produced can be difficult or impractical.

3:1
difficult to separate

(81%)
1-benzoylnaphthalene

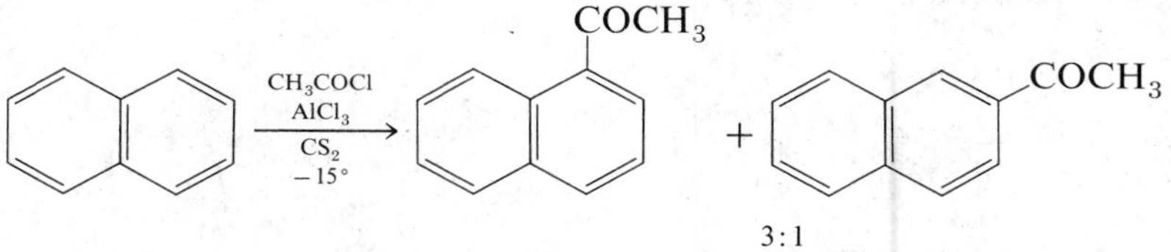

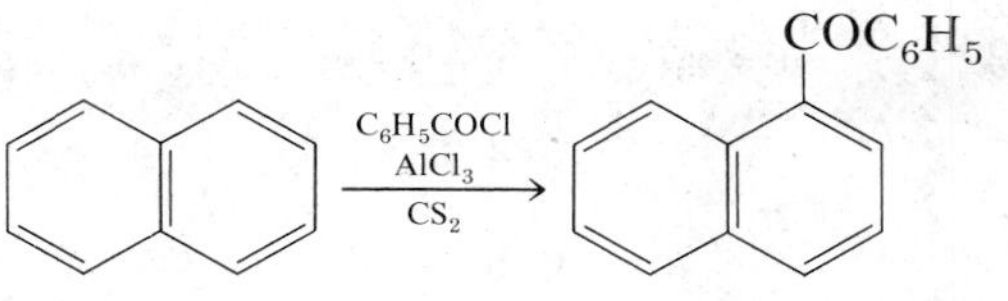

Use of nitrobenzene as solvent generally leads to the β-isomer. These generalizations are only approximate. The reaction products depend on the reaction conditions and the concentrations of reagents. These reactions are not simple, and the nature of the rate-determining step can differ for α- and β-reactions. Some useful specific examples are

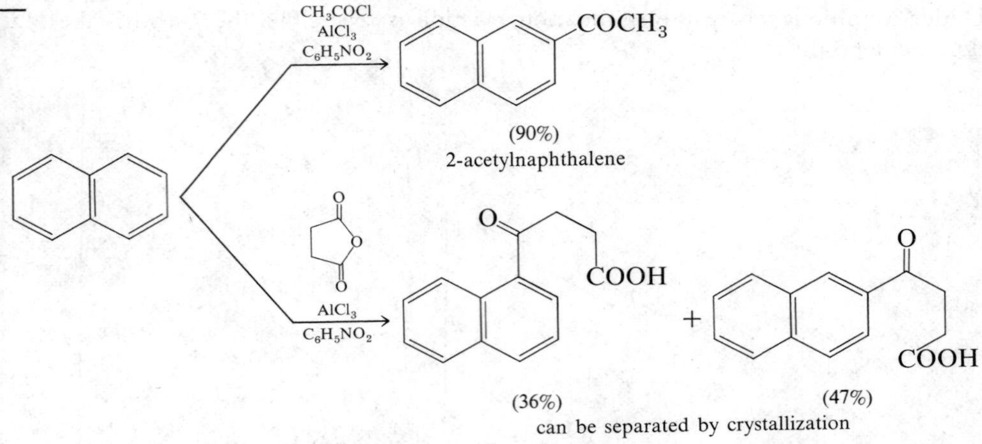

(90%)
2-acetylnaphthalene

(36%)   (47%)
can be separated by crystallization

2. OXIDATION. Under many oxidation conditions naphthalene is oxidized to
1,4-naphthoquinone, but the yields are frequently rather poor.

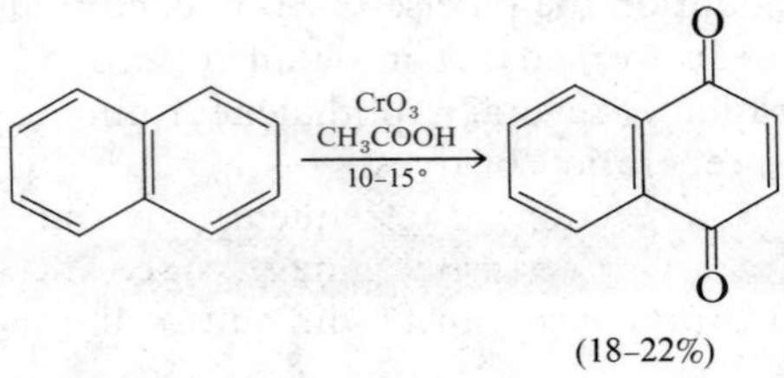

(18–22%)

More vigorous oxidation results in loss of one ring and constitutes one commercial
preparation of phthalic anhydride.

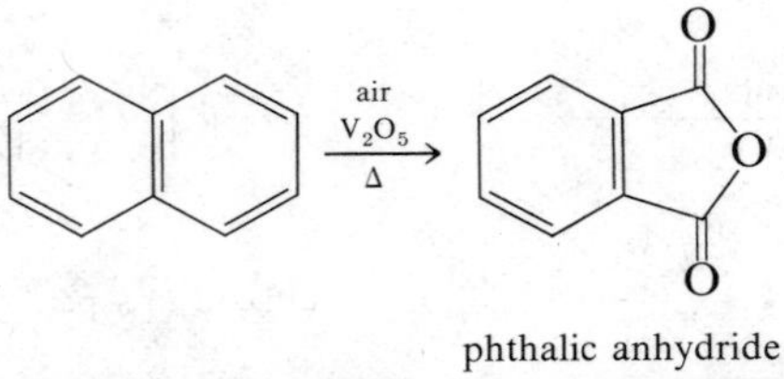

phthalic anhydride

When a substituted naphthalene is oxidized, either ring may be destroyed depending
on the substituent present. For example,

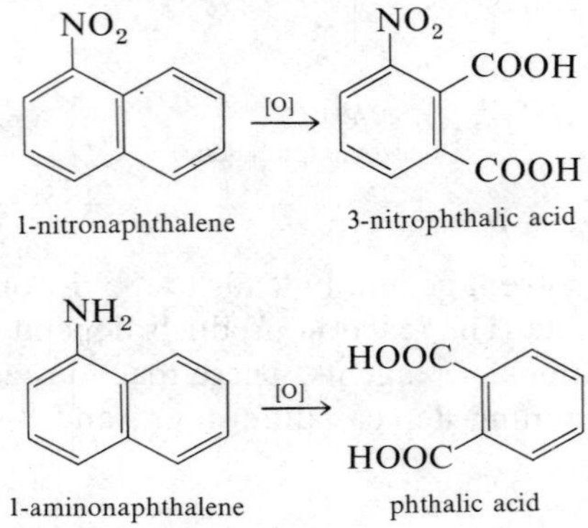

1-nitronaphthalene   3-nitrophthalic acid

1-aminonaphthalene   phthalic acid

1045

Here again we see a connection between electron density and the ease of oxidation (see page 1018). In both compounds it is the more electron-rich ring that is oxidized.

3. REDUCTION. Birch reduction (Section 22.6.C) of naphthalene yields 1,4-dihydronaphthalene. Note that in this product an isolated double bond is produced that does not reduce further.

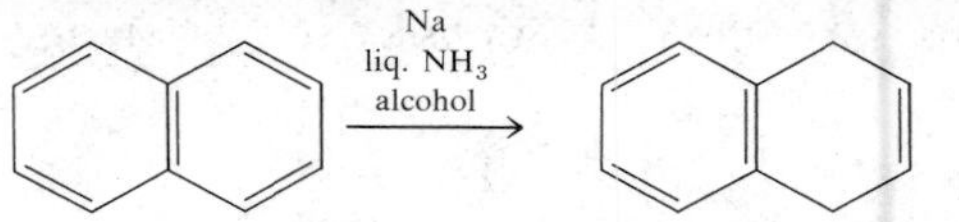

$$\text{naphthalene} \xrightarrow[\text{alcohol}]{\substack{\text{Na} \\ \text{liq. NH}_3}} \text{1,4-dihydronaphthalene}$$

1,4-dihydronaphthalene

Catalytic hydrogenation gives either 1,2,3,4-tetrahydronaphthalene (tetralin) or decahydronaphthalene (decalin) depending on catalyst or conditions.

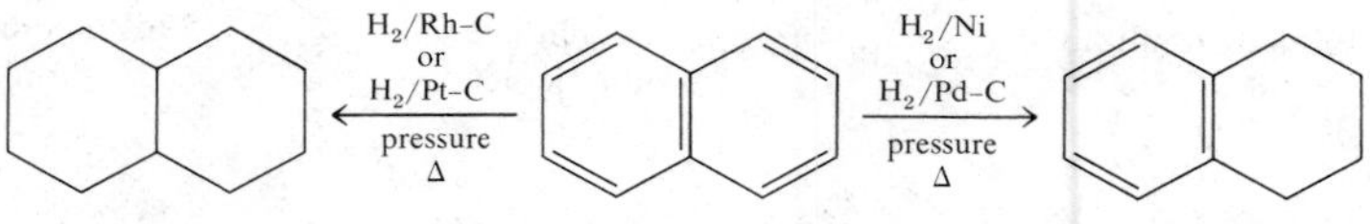

$$\text{decalin} \xleftarrow[\substack{\text{pressure} \\ \Delta}]{\substack{\text{H}_2/\text{Rh–C} \\ \text{or} \\ \text{H}_2/\text{Pt–C}}} \text{naphthalene} \xrightarrow[\substack{\text{pressure} \\ \Delta}]{\substack{\text{H}_2/\text{Ni} \\ \text{or} \\ \text{H}_2/\text{Pd–C}}} \text{tetralin}$$

decalin   tetralin

cis-Decalin (Figure 31.6) is the predominant product of complete hydrogenation. Tetralin and the decalins are high-boiling liquids that find some use as solvents.

EXERCISE: Write the equations illustrating the conversion of naphthalene into each of the following.
(a) 1-bromonaphthalene   (b) 1-nitronaphthalene
(c) 2-naphthalenesulfonic acid   (d) 1,4-dihydronaphthalene
(e) 1-naphthalenesulfonic acid   (f) phthalic anhydride
(g) cis-decalin

D. Substituted Naphthalenes

Functional groups on a naphthalene ring behave more or less as their benzenoid analogs. For example, nitro groups can be reduced to amines, and bromides can be converted to Grignard or lithium reagents.

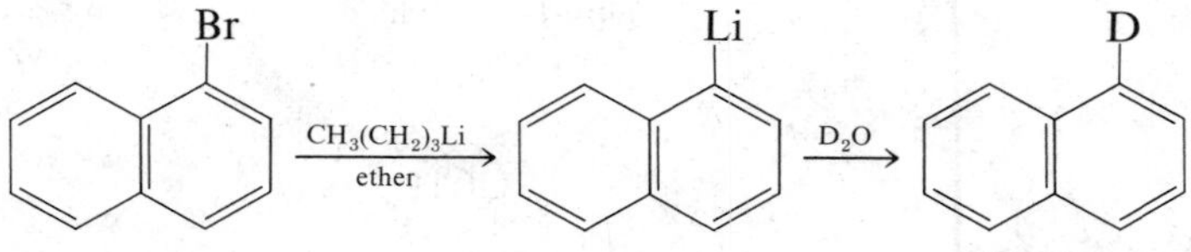

$$\text{Br} \xrightarrow[\text{ether}]{\text{CH}_3(\text{CH}_2)_3\text{Li}} \text{Li} \xrightarrow{\text{D}_2\text{O}} \text{D}$$

naphthalene-1-d

An especially useful reaction is the fusion of the sulfonic acids with sodium or potassium hydroxide.

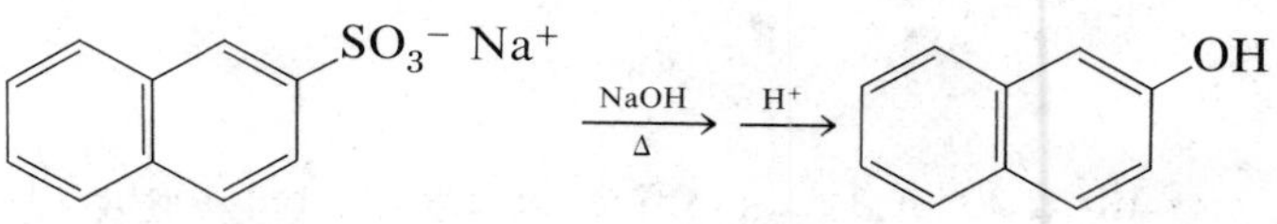

$$\text{SO}_3^- \text{ Na}^+ \xrightarrow[\Delta]{\text{NaOH}} \xrightarrow{\text{H}^+} \text{OH}$$

2-naphthol

Since both naphthalenesulfonic acids are available by sulfonation under different conditions (pages 1042–43), this reaction provides a route to either α-naphthol or β-naphthol.

In the further electrophilic substitution reactions of monosubstituted naphthalenes some simple generalizations can be made.

1. *Meta*-directing substituents in either the 1- and 2-position generally direct to the 5- and 8-positions, the α-positions of the other ring. Examples:

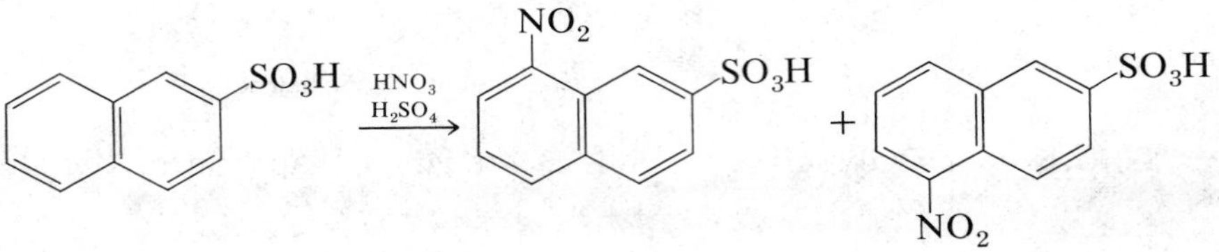

2-naphthalenesulfonic
acid   8-nitronaphthalene-
2-sulfonic acid   5-nitronaphthalene-
2-sulfonic acid

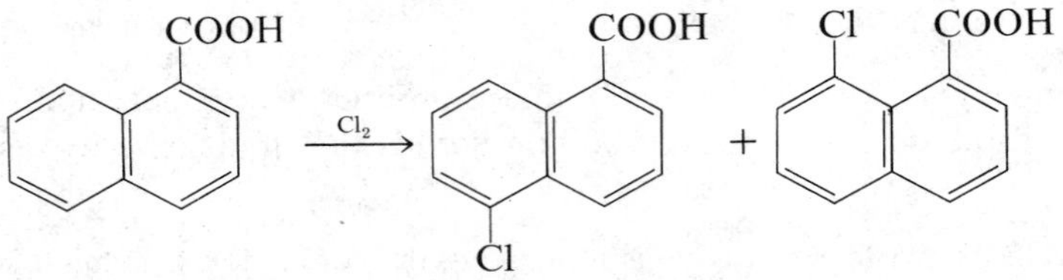

1-naphthoic acid   5-chloro-1-naphthoic
acid   8-chloro-1-naphthoic
acid

2. *Ortho,para*-directing groups in the 1-position direct principally to the 4-position, but also occasionally to the 2-position as well. Examples:

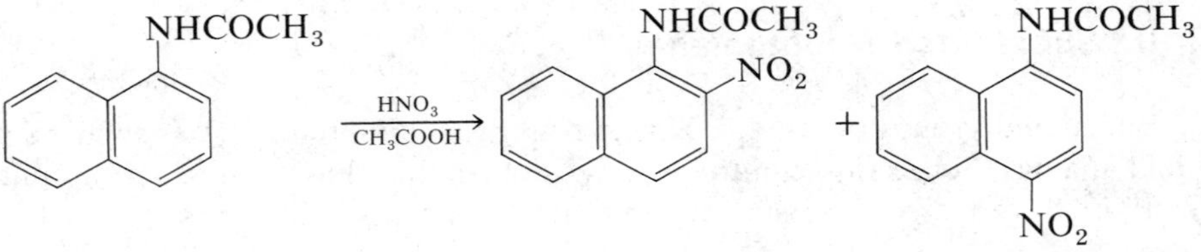

2-nitro-1-acetylamino-
naphthalene   4-nitro-1-acetylamino-
naphthalene

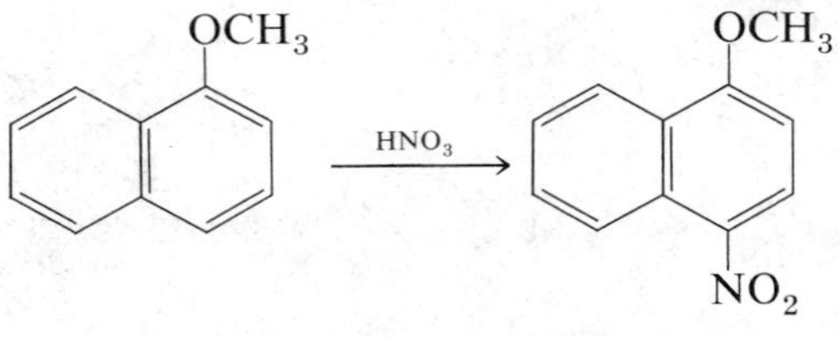

4-nitro-1-methoxy-
naphthalene

3. *Ortho,para* directors in the 2-position generally direct to the 1-position. Examples:

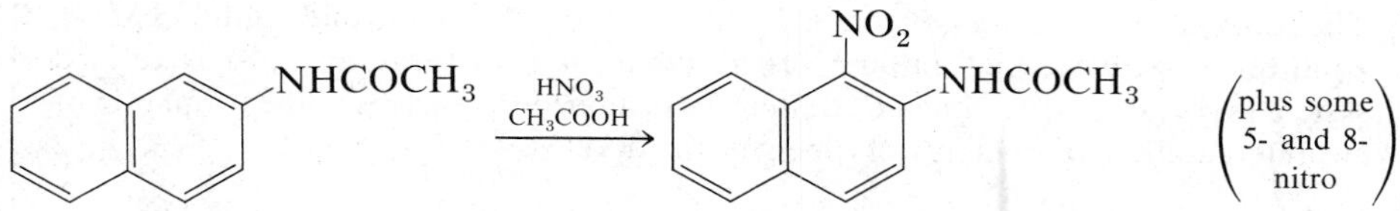

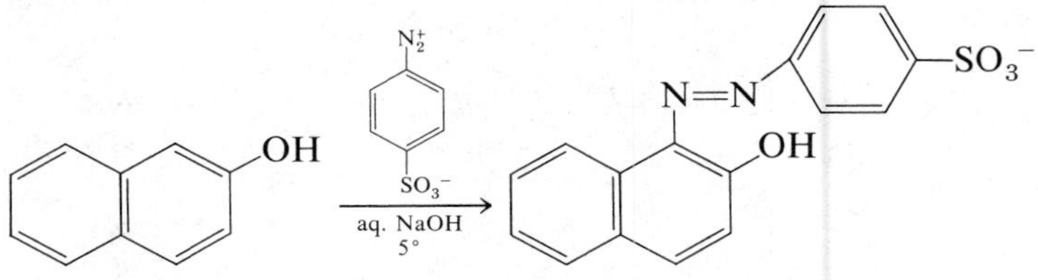

Orange II

Exceptions to these generalizations are not uncommon, especially in Friedel–Crafts acylations and sulfonation.

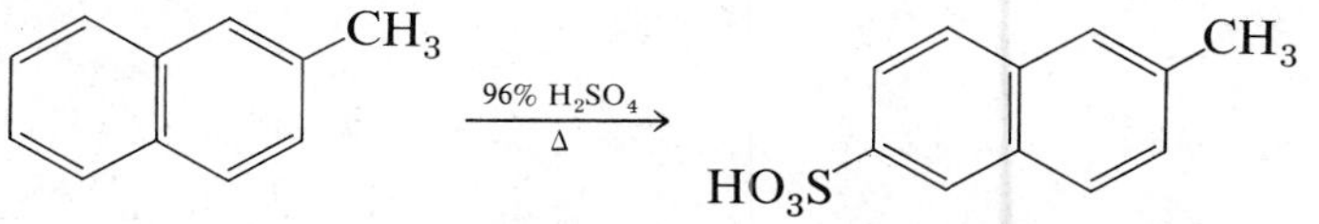

(80%)
6-methylnaphthalene-2-sulfonic acid

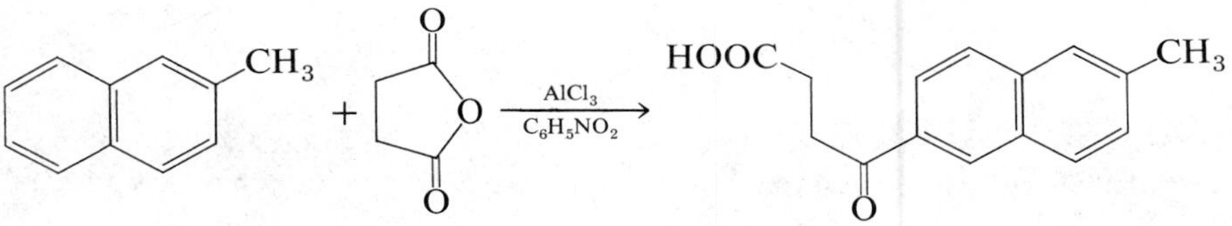

(60–79%)

One of the important reactions in naphthalene chemistry, the **Bucherer** reaction, involves the interconversion of naphthols and aminonaphthalenes and does not apply generally in benzene chemistry.

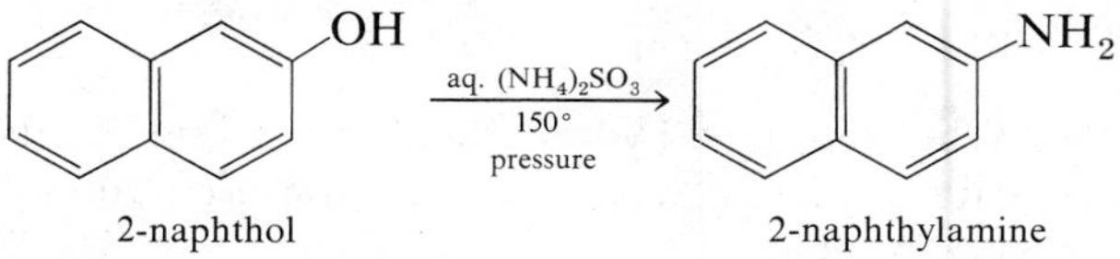

2-naphthol   2-naphthylamine

2-Naphthol is readily available from 2-naphthalenesulfonic acid; hence, the Bucherer reaction provides a simple route to 2-aminonaphthalene, which, in turn, can be converted to many other functions via the diazonium ion.

[2-Naphthylamine is a powdery solid that at one time was widely used as an important intermediate in dye chemistry. This amine is carcinogenic and is no longer used.]

The reaction is reversible and also provides a hydrolytic route from amine to naphthol.

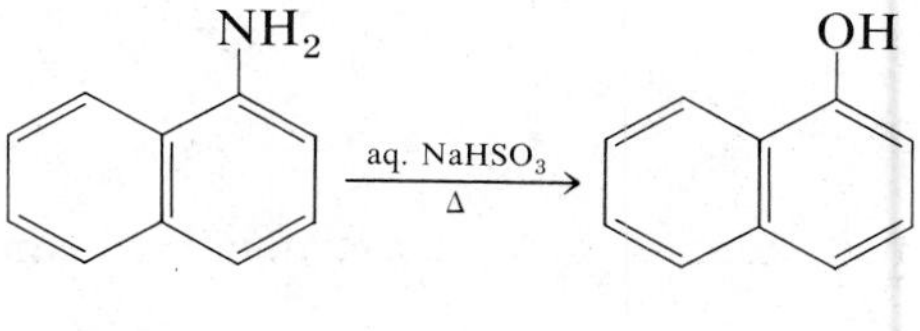

The sulfite or bisulfite ion is essential in this reaction. The amine and naphthol are in equilibrium with a small amount of the imine or keto form, an α,β-unsaturated system that undergoes conjugate addition by bisulfite ion much as in the formation of bisulfite addition compounds of aldehydes and ketones (pages 816–17).

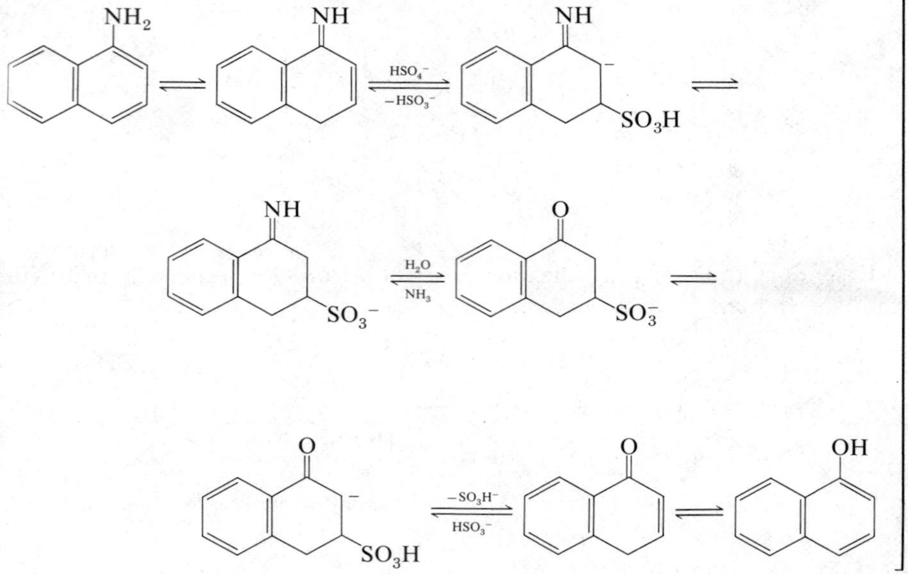

EXERCISE: Write the most stable resonance structure for nitration of 2-methoxy-naphthalene at each of the seven free positions. Explain why nitration occurs primarily at C-1.

31.4
Anthracene and Phenanthrene

A. Structure and Stability

The isomeric tricyclic benzenoid hydrocarbons differ significantly in thermodynamic stability; the linear system, anthracene, is almost 6 kcal mole^{-1} less stable than the angular system, phenanthrene.

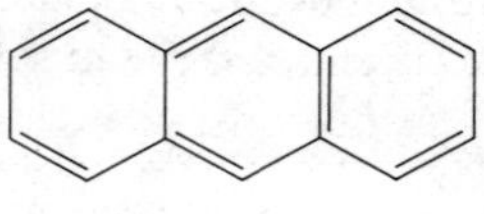

anthracene

$$\Delta H_f^\circ = +55.2 \text{ kcal mole}^{-1}$$

phenanthrene

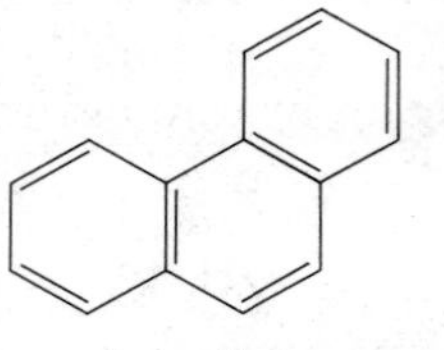

$$\Delta H_f^\circ = +49.5 \text{ kcal mole}^{-1}$$

The empirical resonance energies show a corresponding change; one set of values is 84 kcal mole^{-1} for anthracene and 91 kcal mole^{-1} for phenanthrene. The empirical resonance energy of benzene calculated in the same way is 36 kcal mole^{-1}. The resonance energies of anthracene and phenanthrene are not much more than that of two benzene rings; that is, the third ring contributes relatively little additional resonance stabilization. We shall see that this characteristic is reflected in the reactivities of these hydrocarbons.

B. *Preparation of Anthracenes and Phenanthrenes*

Anthracene and phenanthrene are both available from coal tar in grades that are suitable for most reactions. Commercial material requires extensive further treatment to obtain the pure hydrocarbons. When pure, anthracene (m.p. 216°) is colorless and exhibits a beautiful blue fluorescence. This fluorescence is diminished or altered by impurities in the commercial material. Phenanthrene also is a colorless crystalline solid (m.p. 101°), but it does not fluoresce.

Both ring systems can be built up from simpler compounds. Anthracene and many derivatives are available from phthalic anhydride and benzene compounds via benzoylbenzoic acid (page 869).

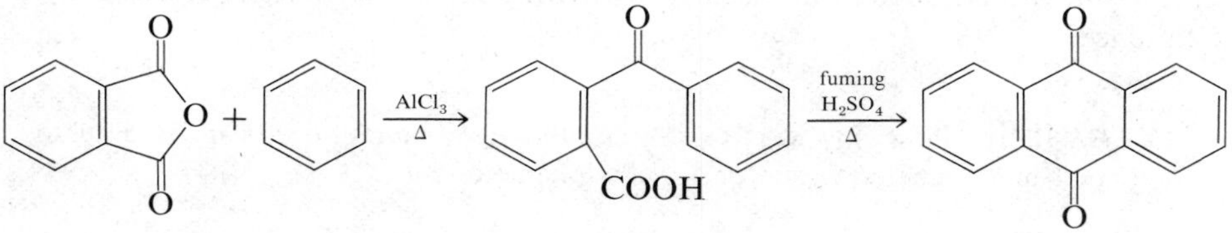

benzoylbenzoic acid   anthraquinone

Anthraquinones can be reduced directly to anthracene by several reducing agents, but the use of sodium borohydride and boron fluoride etherate is convenient.

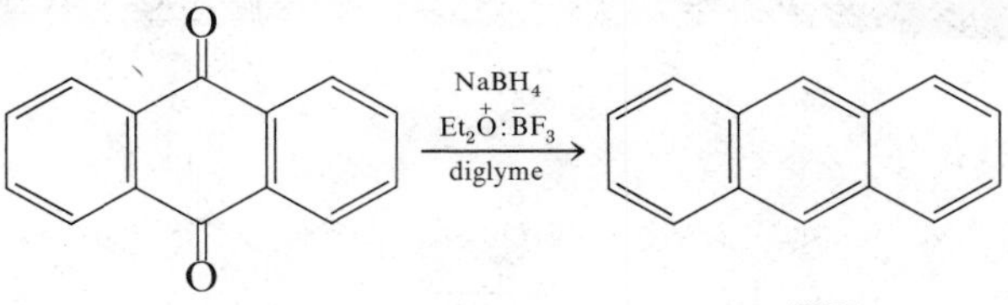

(73%)

The phenanthrene ring system can be built up from naphthalene.

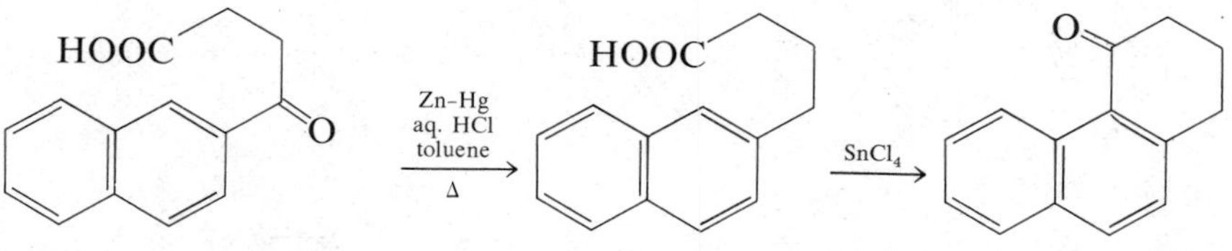

(91%)   (88%)

β-(2-naphthoyl)propionic acid   4-(2-naphthoyl)-   4-oxo-1,2,3,4-
4-oxo-4-(2-naphthyl)butanoic acid   butanoic acid   tetrahydro-
                                                                 phenanthrene

Note that the cyclization goes exclusively to the 1-position of naphthalene (Section 31.3.C). A similar sequence starting from the 1-substituted naphthalene also gives the phenanthrene ring system.

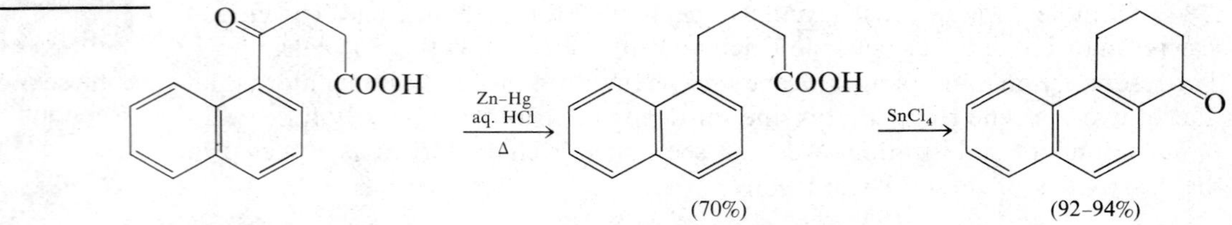

β-(1-naphthoyl)propionic acid
4-oxo-4-(1-naphthyl)butanoic acid

4-(1-naphthyl)butanoic acid (70%)

1-oxo-1,2,3,4-tetrahydro-
phenanthrene (92–94%)

The cyclic ketones can be converted to phenanthrene by successive reduction, dehydration, and dehydrogenation.

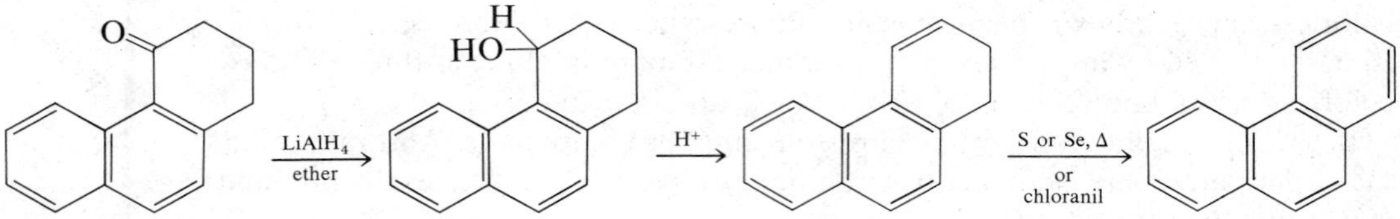

Many substituted phenanthrenes may be synthesized by variations of this general sequence.

EXERCISE: Show how the two β-(naphthoyl) propionic acids can be used to prepare 1-methylphenanthrene or 4-methylphenanthrene.

C. Reactions

Anthracene and phenanthrene undergo ready oxidation to the corresponding quinones.

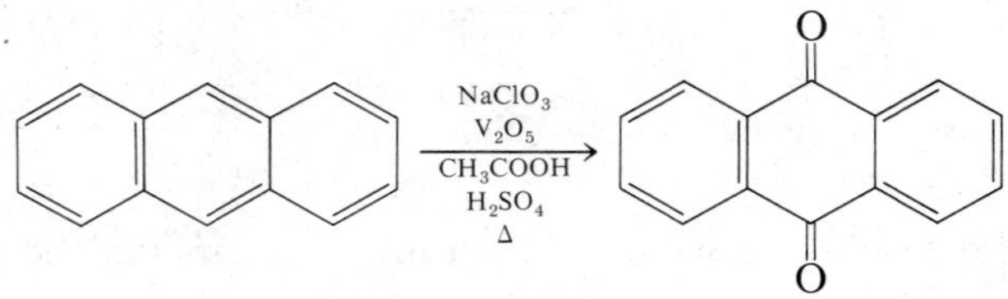

(88–91%)
9,10-anthraquinone

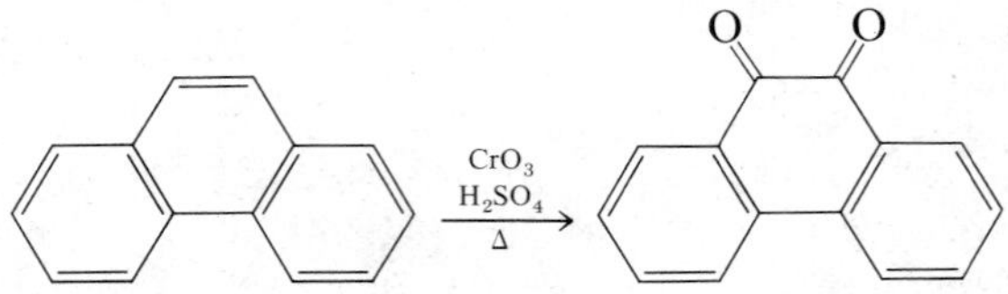

(44–48%)
9,10-phenthraquinone

Anthraquinone can be partially reduced to give anthrone.

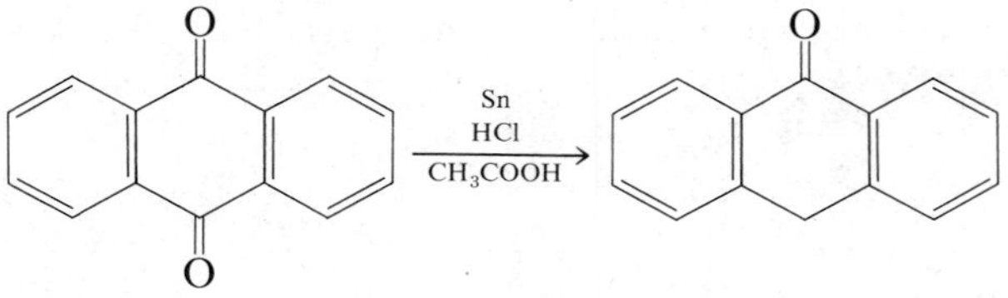

(82%)
anthrone

Anthrone is the keto form of 9-anthranol; both isomers can be isolated, but anthrone is the stable form.

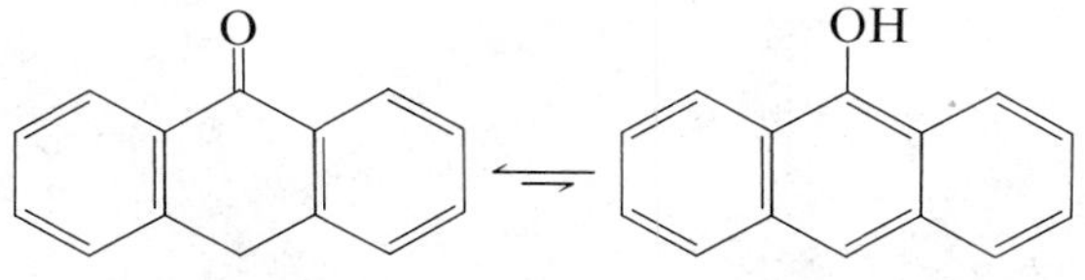

anthrone   9-anthranol

Both anthracene and phenanthrene can be reduced readily to dihydro compounds.

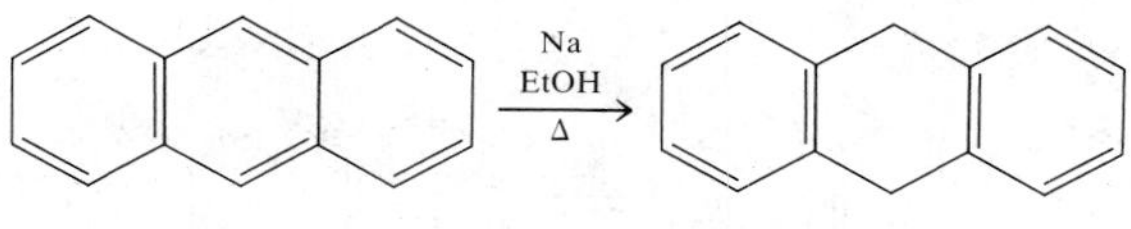

(75–79%)
9,10-dihydroanthracene

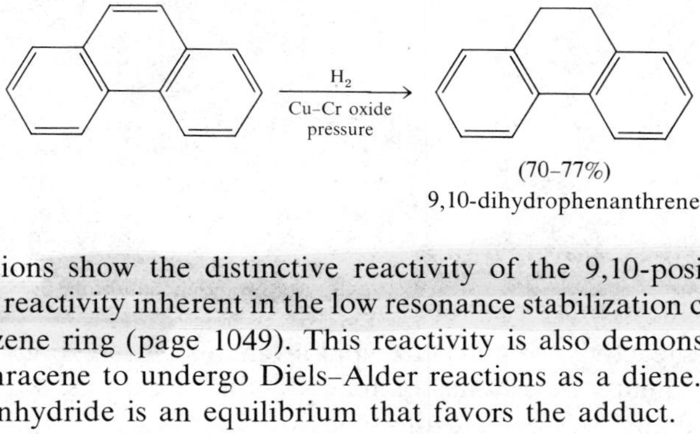

(70–77%)
9,10-dihydrophenanthrene

These reactions show the distinctive reactivity of the 9,10-positions of both compounds, a reactivity inherent in the low resonance stabilization contributed by the third benzene ring (page 1049). This reactivity is also demonstrated by the ability of anthracene to undergo Diels–Alder reactions as a diene. The reaction with maleic anhydride is an equilibrium that favors the adduct.

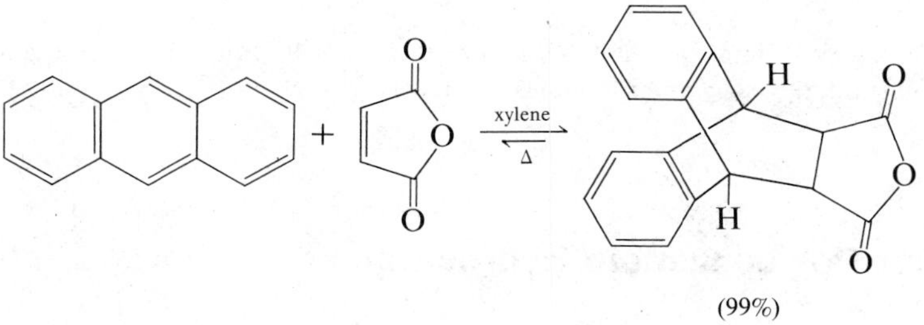

maleic
anhydride

(99%)

A novel reaction of this type is with benzyne (pages 987–89) to give the unusual hydrocarbon triptycene.

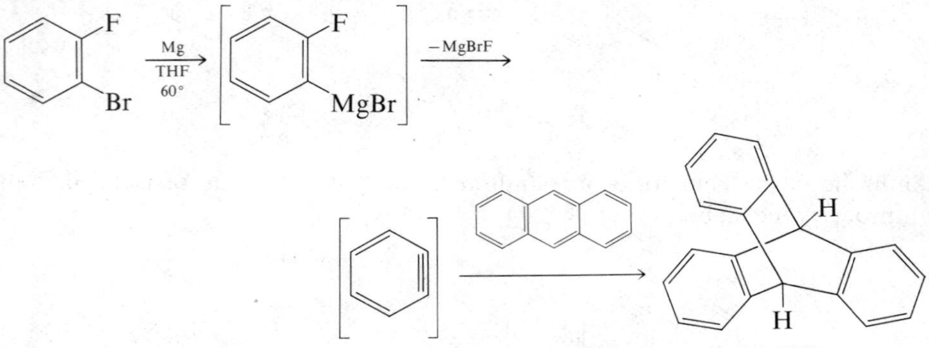

(28%)
triptycene

Electrophilic aromatic substitution reactions with anthracene and phenanthrene occur most readily in the 9-position and frequently give disubstituted products.

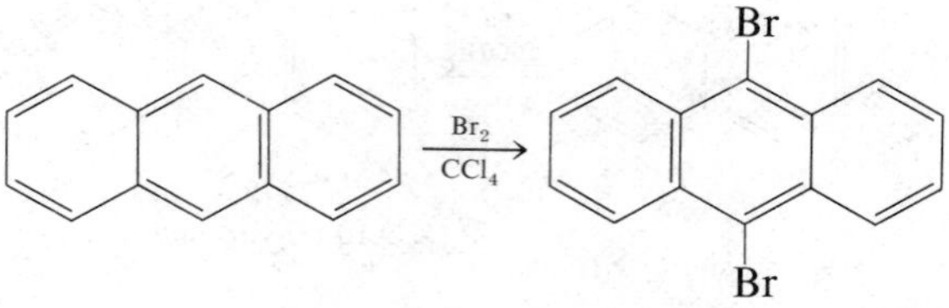

(83–88%)
9,10-dibromoanthracene

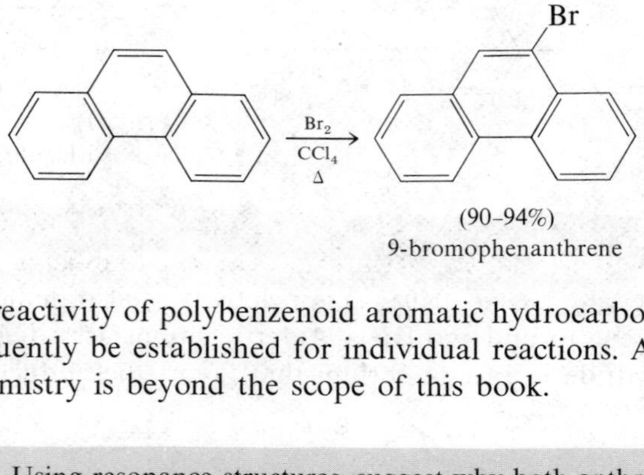

(90–94%)
9-bromophenanthrene

Because of the reactivity of polybenzenoid aromatic hydrocarbons, special conditions must frequently be established for individual reactions. A detailed discussion of this chemistry is beyond the scope of this book.

EXERCISE: Using resonance structures, suggest why both anthracene and phenanthrene undergo electrophilic attack primarily at C-9.

31.5
Higher Polybenzenoid Hydrocarbons

A large number of polybenzenoid hydrocarbons are known, and some are relatively important. Some multi-ring systems, with their established common names and their numbering systems, are

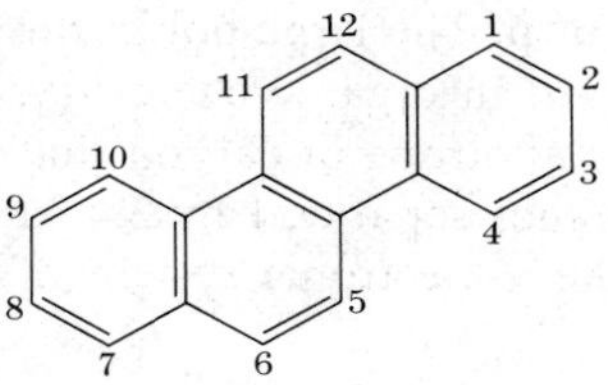

chrysene

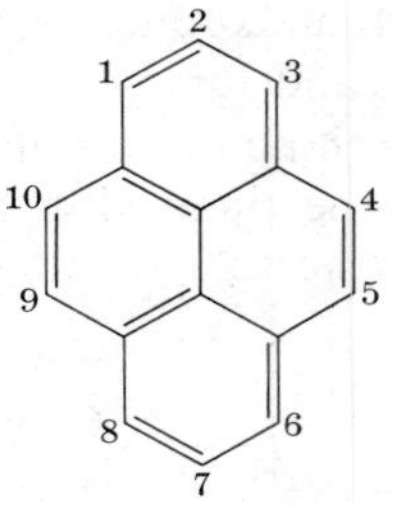

pyrene

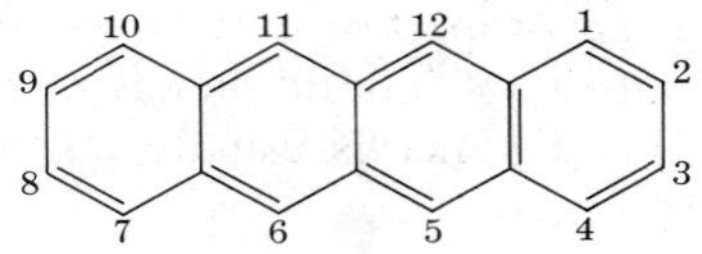

tetracene

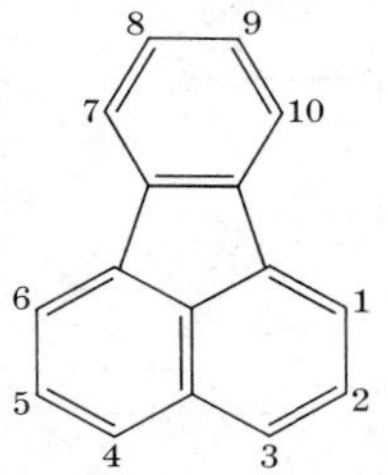

fluoranthene

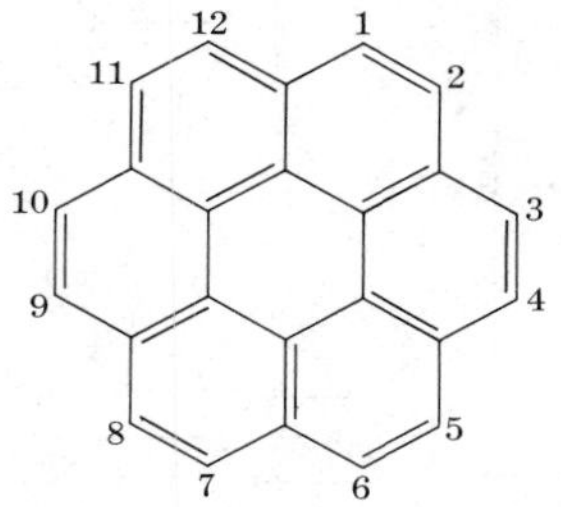

coronene

Some of these hydrocarbons are available from coal tar; others are prepared from simpler systems by building up rings in the manner shown in the preceding section for anthracene and phenanthrene.

Tetracene is an orange compound that shows much of the chemistry of anthracene. Oxidation and reduction occur readily at the 5- and 12-positions, and the hydrocarbon reacts readily as a Diels–Alder diene. Higher linear **acenes** are known: pentacene, with five fused benzene rings in a row, is blue; hexacene is green; and heptacene is a deep greenish black. The higher linear acenes are reactive, air-sensitive, and difficult to obtain pure.

Chrysene is similar to phenanthrene in its reactions: it can be oxidized to the 5,6-quinone.

Pyrene is among the most important of these hydrocarbons. Pyrene undergoes the usual electrophilic aromatic substitution reactions such as halogenation, nitration, Friedel–Crafts acylation, and so on. These reactions occur exclusively at the 1-position.

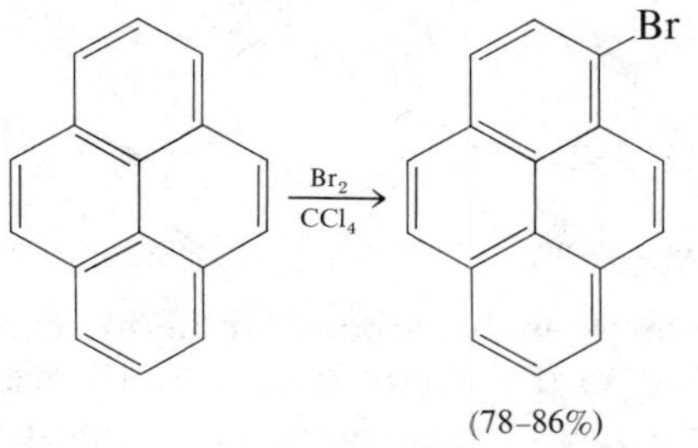

(78–86%)
1-bromopyrene

Two numbering systems have been used for pyrene, and care must be taken in reading the literature, particularly the older literature, to establish which system has been used. The system shown here is the accepted IUPAC numbering, but even today references will be found with the older nomenclature.

Polycyclic systems much larger than coronene are known. The large polycyclic hydrocarbons have low solubility, and few are significant in organic chemistry. Their properties start to approach those of graphite, an allotrope of carbon that consists of infinite planes of benzene rings with the planes separated by 3.4 Å. This distance is usually taken as the total width of the π-electronic system of benzene.

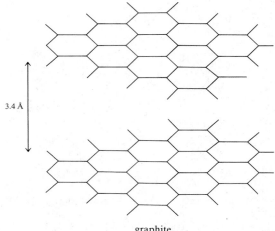

3.4 Å

graphite

A number of polycyclic aromatic hydrocarbons are named as **benz-** derivatives of simpler systems. The position of fusion of the benz-ring is represented by a lowercase italic letter that designates the side around the periphery of the parent system used for the fusion. For example, the sides of anthracene are lettered starting with side *a* between positions 1 and 2.

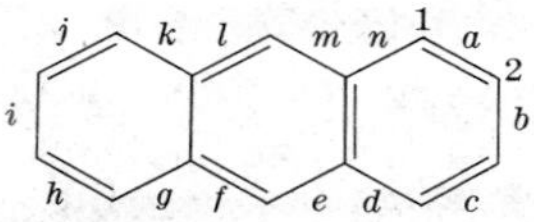

In this way the following hydrocarbons are derived.

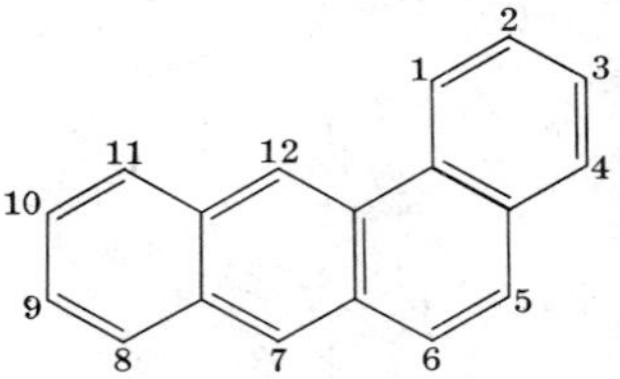

benz[*a*]anthracene 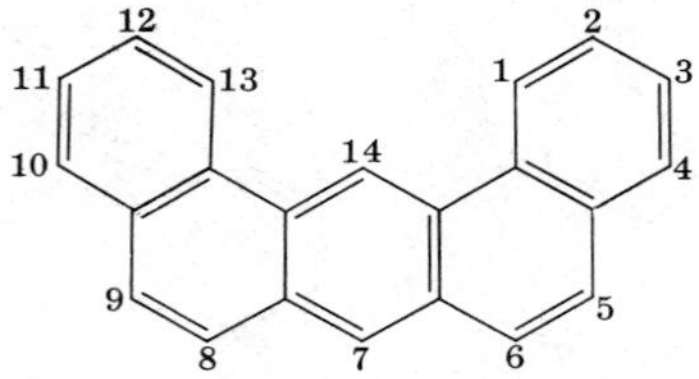   dibenz[*a,j*]anthracene

Some polycyclic aromatic hydrocarbons are highly carcinogenic compounds. Minute amounts painted on the skin of mice will produce skin tumors (epithelioma) in the course of a few months. Some of the most potent of the carcinogenic hydrocarbons are dibenz[*a,h*]anthracene (but not dibenz[*a,c*]anthracene), benzo[*a*]pyrene (but not benzo[*e*]pyrene), dibenzo[*a,i*]pyrene, and benzo[*b*]fluoranthene. These compounds occur in coal tar and in soot. A high incidence of scrotal cancer in chimney sweeps was noticed in England as early as 1775. All of these carcinogenic hydrocarbons have been detected in minute quantity in tobacco smoke.

The way in which these polycyclic aromatic hydrocarbons produce malignant tumors has been actively investigated for several decades, and the overall mechanism is now fairly well understood. The actual carcinogens turn out to be metabolic products of the polycyclic hydrocarbons. After the hydrocarbon enters the organism, it is enzymatically oxidized. This oxidation is a normal cellular function and is intended to render the hydrocarbon more water-soluble so that it may be eliminated from the organism. An example is benzo[a]pyrene, which gives rise to the highly carcinogenic diol epoxide shown.

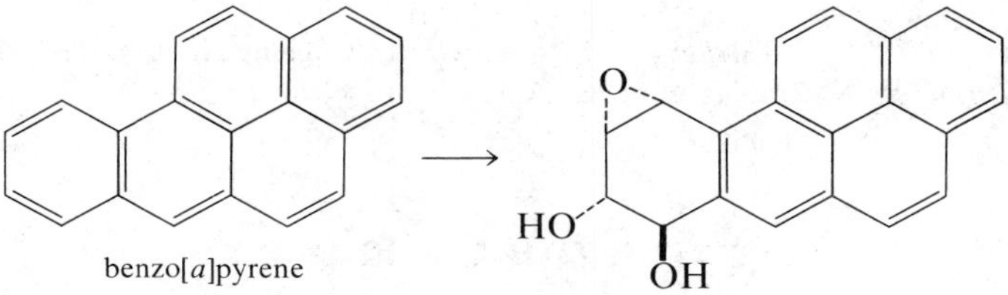

benzo[a]pyrene

The diol epoxides are carcinogenic because they alkylate cellular DNA, causing mutations and an eventual loss of the cell's ability to undergo controlled replication.

Benzo[c]phenanthrene presents a further aspect of interesting chemistry.

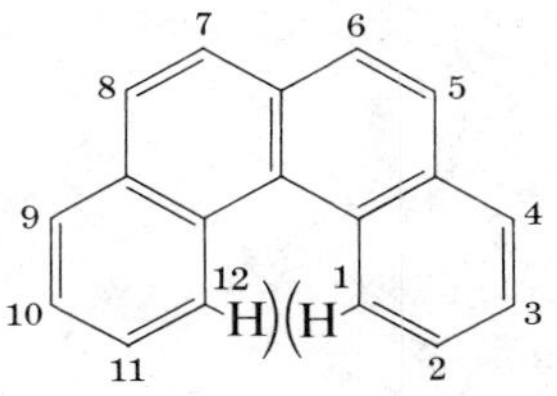

benzo[c]phenanthrene

The hydrogen atoms at the 1- and 12-positions interact significantly, and the molecule is forced to twist somewhat from coplanarity. With two additional benzene rings we obtain the spirally *fused* hydrocarbon hexahelicene.

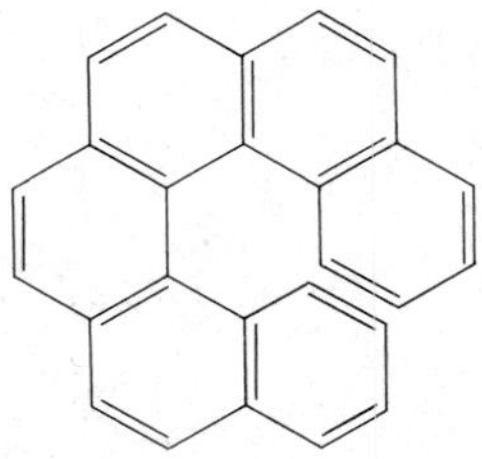

hexahelicene

If this molecule were planar, two sets of CH groups would have to exist in the same space. In practice the hydrocarbon adopts a spiral structure that is also chiral. The enantiomers of this hydrocarbon have been obtained and have enormous optical rotations, $[\alpha]_D$ 3700°. The spiral structure has been demonstrated experimentally by the x-ray structure determination of 2-methylhexahelicene as shown in the stereo plot in Figure 31.7.

EXERCISE: Write the structures of the hydrocarbons named on page 1054.

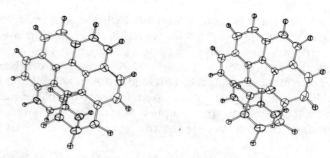

FIGURE 31.7. *Stereo structure of 2-methylhexahelicene.* [*Reproduced with permission from K. N. Trueblood et al.,* Acta Cryst., **B29**:223 (1973).]

PROBLEMS

1. Show a practical synthesis of each of the following compounds starting with a suitable benzene derivative.

(a)

(b)

(c)

(d)

(e) Br — —Br

(f)

(g)

(h)

2. 6,6'-Dinitrobiphenyl-2,2'-dicarboxylic acid can be prepared by the Ullmann reaction on 2-iodo-3-nitrobenzoic acid. Give a reasonable preparation of this compound from available materials.

3. Substitution reactions of 2-methylnaphthalene with bulky electrophilic reagents tend to occur at the 6-position. Explain why this position is preferred to the sterically equivalent 7-position.

4. Write out the mechanism for the conversion of 2-naphthol to 2-aminonaphthalene showing every intermediate involved in the Bucherer reaction.

5. The heat of formation of naphthalene, ΔH_f°, is 36.1 kcal mole^{-1}; ΔH_f° for *trans*-decalin is -43.5 kcal mole^{-1}.
   (a) Calculate the heat of hydrogenation of naphthalene to *trans*-decalin.
   (b) Using the heat of hydrogenation of cyclohexene as a comparison standard, estimate the heat of hydrogenation of naphthalene in the absence of any conjugation stabilization.
   (c) Compare (a) and (b) to derive the corresponding empirical resonance energy of naphthalene.

6. Cadinene, $C_{15}H_{24}$, is a sesquiterpene (Section 34.6.A) occurring in the essential oils of junipers and cedars. Dehydrogenation gives the naphthalene hydrocarbon cadalene.

cadinene   cadalene

   (a) What is the IUPAC name for cadalene?
   (b) Give a rational synthesis of cadalene from toluene and any necessary aliphatic compounds.

7. Provide a practical synthesis of each of the following compounds from naphthalene:
   (a) 2-bromonaphthalene   (b) 1-methylnaphthalene
   (c) 1-isopropylnaphthalene   (d) 1-naphthyl propyl ketone
   (e) 2-phenylnaphthalene   (f) 1,2-naphthoquinone
   (g) 1-naphthoic acid   (h) 2-naphthoic acid
   (i) ethyl α-naphthoylacetate

8. Allyl β-naphthyl ether undergoes the Claisen rearrangement to give exclusively 1-allyl-2-naphthol. Give a reasonable explanation for the decided preference of this reaction over the alternative reaction to 3-allyl-2-naphthol.

9. The difference in empirical resonance energies of anthracene and phenanthrene can be accounted for on the basis of resonance structures. Whereas we can write two benzenoid resonance structures for benzene (page 639) and three for naphthalene (page 1038), there are four structures for anthracene and five for phenanthrene.
   (a) Write out both sets of resonance structures for anthracene and phenanthrene.
   (b) For each of the five different C—C bonds in anthracene compare the number of resonance structures in which each is single or double and determine the fraction of double bond character (bond order). Compare with the bond lengths predicted using the curve you constructed for the exercise on page 1039 with the experimental values determined by x-ray crystal structure techniques as

bond distances in Å

10. Give the expected dominant product or products in mononitration of each of the following compounds.

(a)

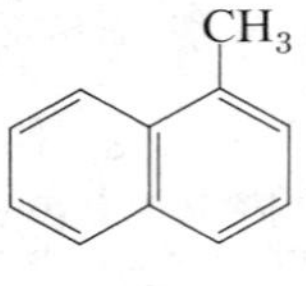

(b)

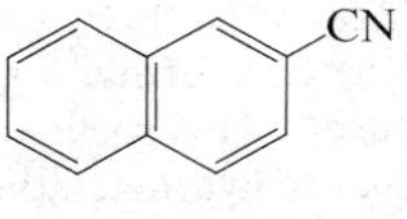

(c)

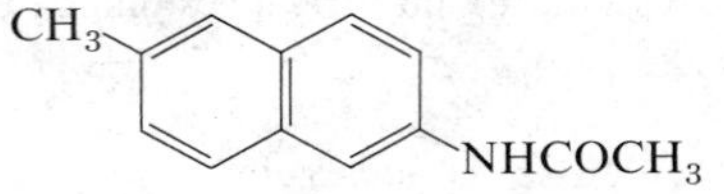

(d)

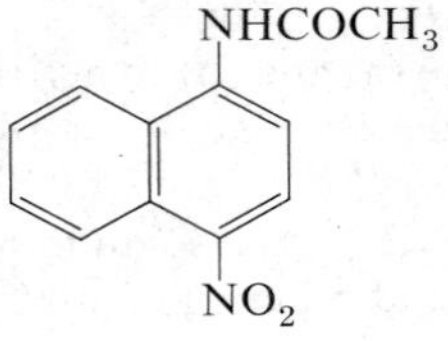

(e)

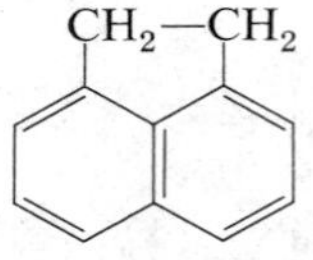

(f)

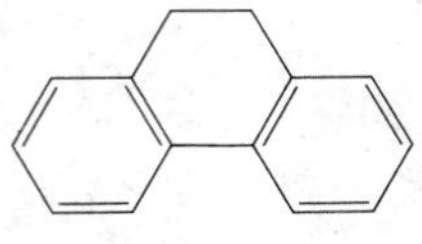

(g) 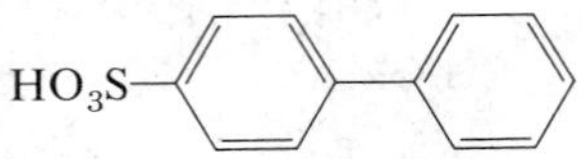 HO_3S—

(h)

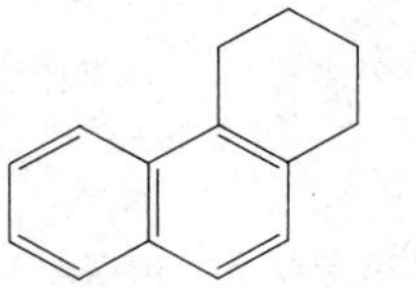

(i)

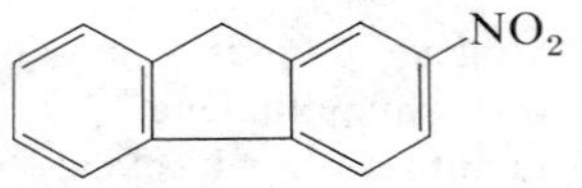

(j)

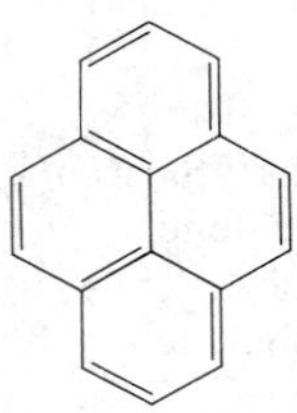

(k)

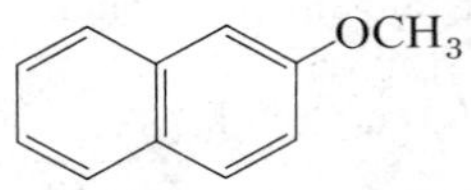

(l)

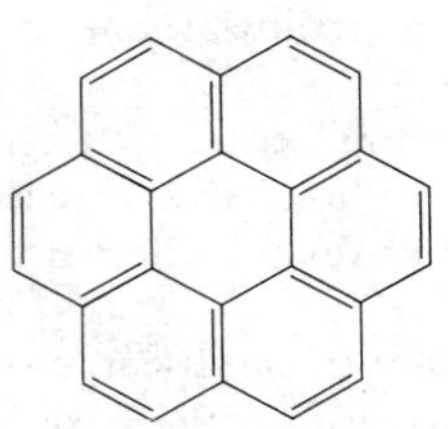

11. Acetylation of phenanthrene with acetyl chloride and $AlCl_3$ in nitrobenzene gives primarily 3-acetylphenanthrene. 2-Acetylphenanthrene is best prepared by Friedel–Crafts acetylation of 9,10-dihydrophenanthrene (note that this hydrocarbon is a biphenyl compound and the 2-position corresponds to the *para* position of biphenyl) followed by dehydrogenation with Pd/C. Show how to prepare each of the following phenanthrene derivatives.
(a) 2- and 3-phenanthrenecarboxylic acid
(b) 2- and 3-aminophenanthrene
(c) 2- and 3-bromophenanthrene
(d) phenanthrene-*2-d* and phenanthrene-*3-d*

12. (a) Starting from naphthalene or either of the monomethylnaphthalenes show how to prepare all five possible methylphenanthrenes. (*Note:* Some of these are harder than others.)
(b) α-Methylsuccinic anhydride reacts with naphthalene and $AlCl_3$ in nitrobenzene to give about equal amounts of 4-oxo-4-(1-naphthyl)-2-methylbutanoic acid and 4-oxo-4-(2-naphthyl)-2-methylbutanoic acid. These acids can be separated and

used as starting materials for problem (a). Which of the methylphenanthrenes can be prepared in this way?

13. Show how anthraquinone can be prepared from 1,4-naphthoquinone.

14. The following methyl derivatives have been shown to be carcinogenic. Supply an adequate name for each compound.

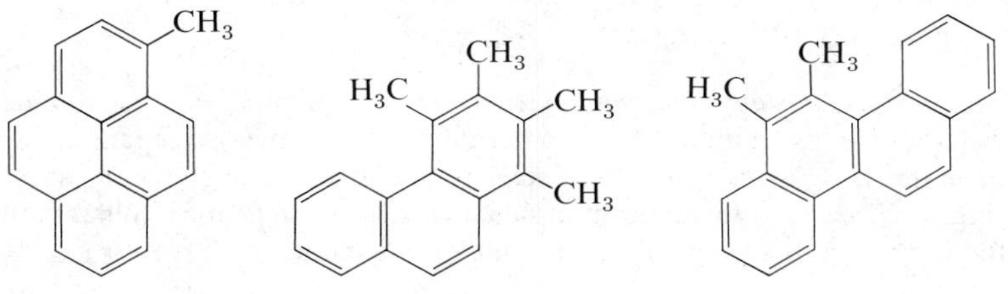

15. The acidity of fluorene is sufficiently high that it will undergo condensation reactions as do esters in alcoholic sodium ethoxide. Show how such condensation reactions can be utilized for the preparation of the following compounds.
(a) fluorene-9-carboxylic acid
(b) 9-methylfluorene-9-carboxylic acid
(c) 9-benzoylfluorene
(d) fluorene-9-carboxaldehyde

16. Suggest a procedure using phase-transfer catalysis for the alkylation of fluorene with *n*-butyl bromide.

17. (a) Write a reasonable mechanism for the following reaction showing all intermediates involved.

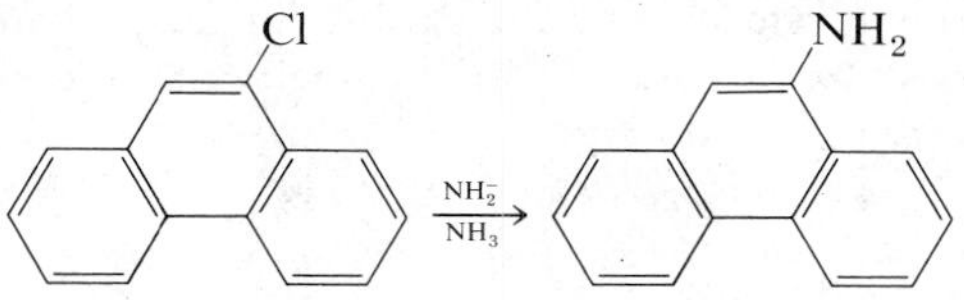

(b) On the basis of this mechanism, what would be the course of reaction for 2-methyl-9-chlorophenanthrene?

18. Give a reasonable mechanism for the following reaction, showing all intermediates involved.

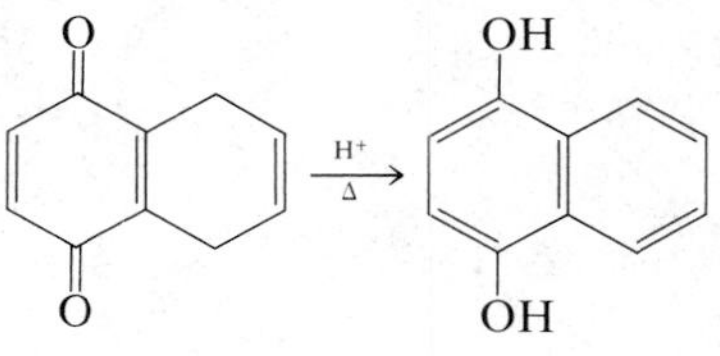

19. 2,6-Naphthoquinone is reduced more readily than 1,2-naphthoquinone. Explain.

20. Steganone is a naturally occurring biphenyl, which was shown by x-ray analysis to have the following structure.

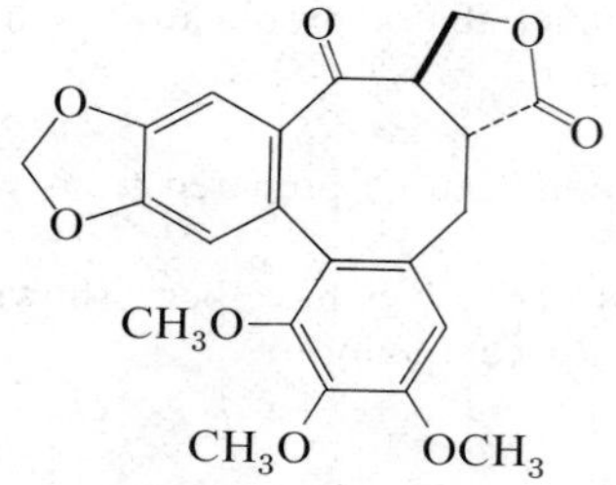

steganone

A synthesis of steganone was carried out, and a product was obtained that was different from steganone. This isomeric product, named isosteganone, was shown to have the same gross structure as steganone and was also shown to have the lactone ring fused *trans* to the eight-membered ring, as in steganone. What is the nature of the difference between steganone and isosteganone?

21. Triptycene (page 1052) is a triarylmethane formally similar to triphenylmethane. Triphenylmethane is a relatively acidic hydrocarbon with a pK_a of 31.5, whereas the pK_a of triptycene is at least 10 units higher. Explain.

22. How many stereoisomers exist for 1,3-bis-(2-bromo-6-methylphenyl)benzene? Write their structures. Which are chiral?

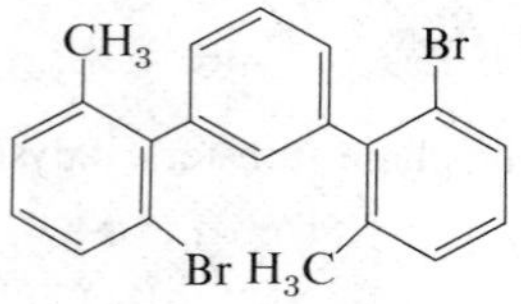

23. The perinaphthenyl cation is a relatively stable carbocation in which the positive charge can be distributed to six equivalent positions. Write resonance structures showing this equivalency. Use the stability of the perinaphthenyl cation to explain why the 1-position of pyrene is exceptionally reactive in electrophilic substitution reactions.

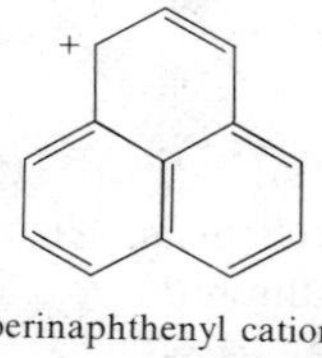

perinaphthenyl cation

Chapter 32
Heterocyclic Compounds

32.1
Introduction

Heterocycles are cyclic compounds in which one or more ring atoms are not carbon (that is, *hetero* atoms). Although heterocyclic compounds are known that incorporate many different elements into cyclic structures (for example, N, O, S, B, Al, Si, P, Sn, As, Cu), we shall consider only some of the more common systems in which the hetero atom is N, O, or S.

Heterocycles are conveniently grouped into two classes, nonaromatic and aromatic. The nonaromatic compounds have physical and chemical properties that are typical of the particular hetero atom. Thus, tetrahydrofuran and 1,4-dioxane are typical ethers, while 1,3,5-trioxane behaves as an acetal.

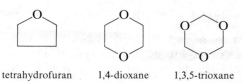

    tetrahydrofuran   1,4-dioxane   1,3,5-trioxane

Pyrrolidine and piperidine are typical secondary amines and the bicyclic compound quinuclidine is a tertiary amine.

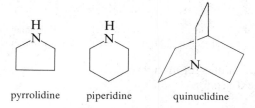

    pyrrolidine   piperidine   quinuclidine

Since the chemistry of these compounds parallels the chemistry of their acyclic relatives, we shall treat them here only briefly.

The aromatic heterocycles include such compounds as pyridine, where nitrogen replaces one of the CH groups in benzene, and pyrrole, in which the aromatic sextet is supplied by the four electrons of the two double bonds and the lone pair on nitrogen.

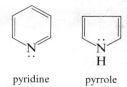

    pyridine   pyrrole

Other aromatic heterocycles contain more than one hetero atom, and still others contain fused aromatic rings. Examples that we will treat in more detail later include

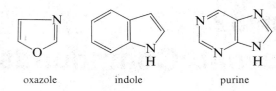

oxazole   indole   purine

The nomenclature of these heterocyclic series is a vast sea of special names for individual ring systems and trivial names for individual compounds. In the course of developing the chemistry of some important groups of compounds we will treat the associated nomenclature. There is only one naming scheme common to all of these compounds, and it, unfortunately, is used only in cases where alternative nomenclature based on special names is awkward. This scheme is based on the corresponding hydrocarbon. The compound formed by replacing a carbon by a hetero atom is named by an appropriate prefix: **aza** for nitrogen, **oxa** for oxygen, and **thia** for sulfur. Some examples are

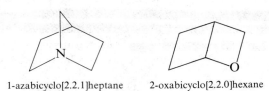

1-azabicyclo[2.2.1]heptane   2-oxabicyclo[2.2.0]hexane

Saturated monocyclic rings are named according to ring size as 3-, **-irane**; 4-, **-etane**; 5-, **-olane**; and 6-, **-ane**. Even this system does not apply to nitrogen-containing rings and finds only limited use in common practice. Some examples of this nomenclature are

1,3-dithiane   oxolane   1,3-dioxolane
(used commonly)   (rarely used)   (used commonly)

The commonly used names for monocyclic rings with a single hetero atom will be discussed in the next section.

32.2
Nonaromatic Heterocycles

A. *Nomenclature*

Names in common use of some fully saturated heterocycles containing only one hetero atom are shown below.

oxirane   thiirane   aziridine

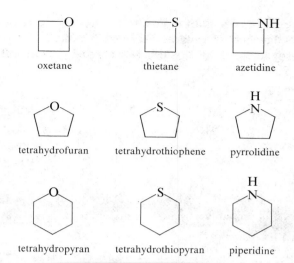

oxetane   thietane   azetidine

tetrahydrofuran   tetrahydrothiophene   pyrrolidine

tetrahydropyran   tetrahydrothiopyran   piperidine

In naming substituted derivatives, the ring is numbered beginning with the hetero atom.

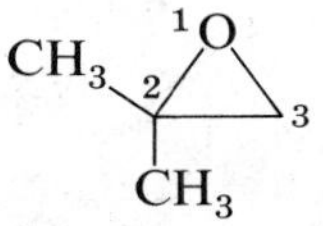

2,2-dimethyloxirane 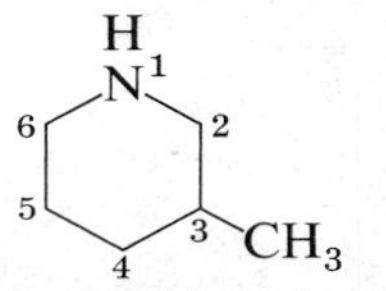   3-methylpiperidine 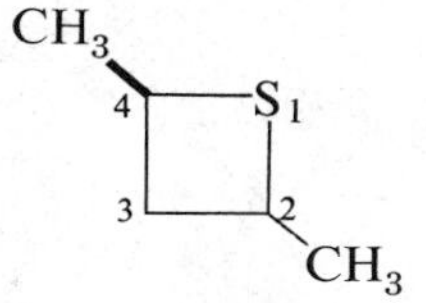   *trans*-2,4-dimethylthietane

B. *Three-Membered Rings*

The common three-membered heterocycles are ethylene oxide (oxirane), ethyleneimine (aziridine), and ethylene sulfide (thiirane).

ethylene oxide   ethyleneimine   ethylene sulfide
oxirane   aziridine   thiirane

Ethylene oxides have been discussed previously (Sections 10.12A and 11.6.E). Recall that the two most general syntheses are the oxidation of alkenes with peroxyacids and the base-catalyzed cyclization of halohydrins (page 265).

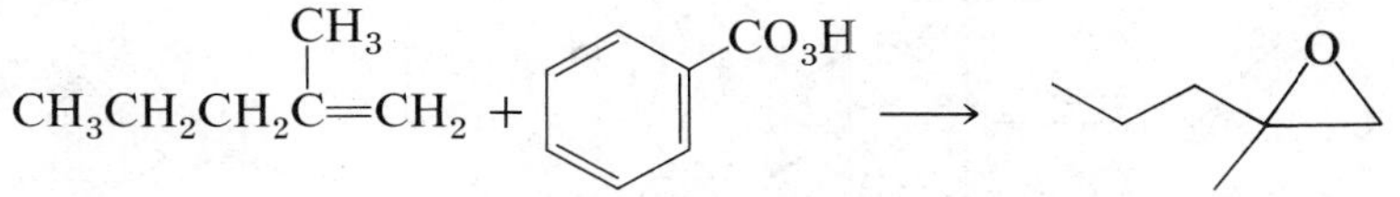

2-methyl-2-propyloxirane

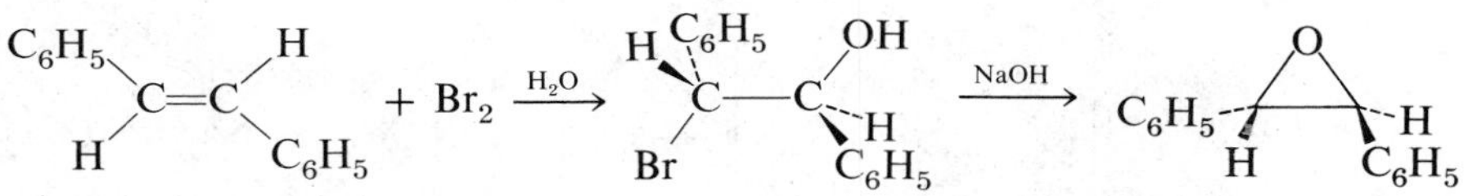

C_6H_5 $C=C$ C_6H_5 + Br_2 $\xrightarrow{H_2O}$ ... $\xrightarrow{NaOH}$

trans-2,3-diphenyloxirane

Aziridines are most commonly prepared by related cyclization reactions. A classical method (the **Wenker synthesis**) consists of converting a β-amino alcohol into a β-amino hydrogen sulfate, which is cyclized by treatment with a strong base.

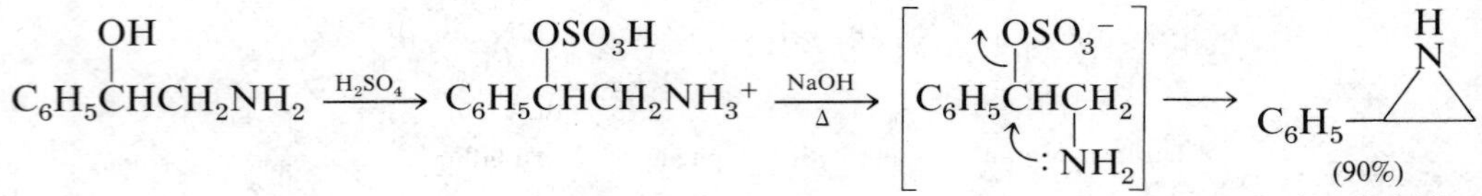

2-phenylaziridine
(90%)

They may also be prepared by cyclization of β-haloalkylamines and their derivatives. An example is the conversion of an alkene into an aziridine via the iodo isocyanate and iodo carbamate.

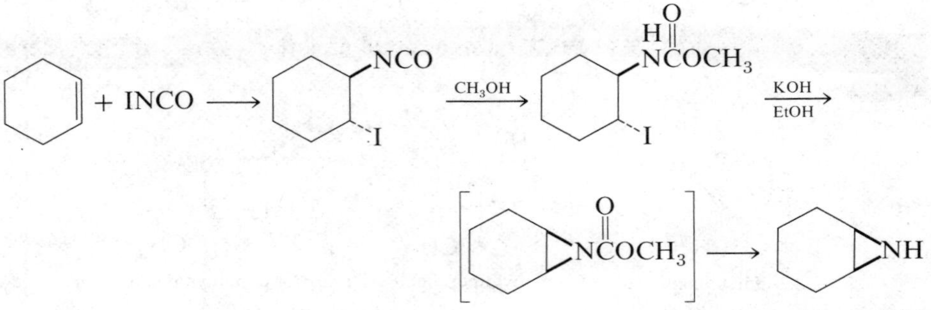

cyclohexeneimine

Under basic conditions, the NH of the carbamate grouping is partially deprotonated ($pK_a \sim 15$). The resulting anion accomplishes an internal S_N2 displacement, leading to the cyclized carbamate, which hydrolyzes to the aziridine.

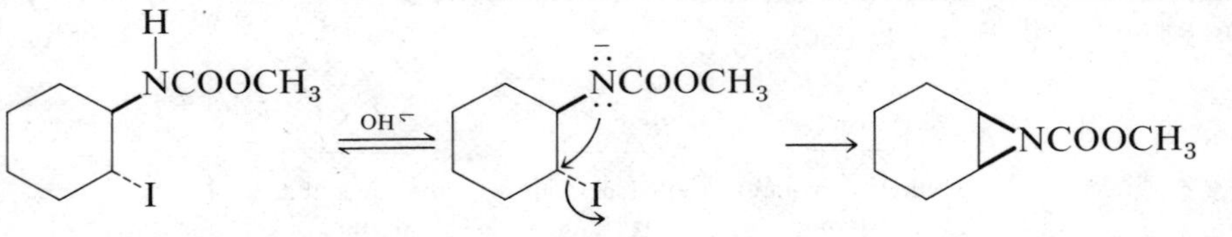

Thiiranes are most conveniently prepared from the corresponding oxirane. An especially useful method involves treating the epoxide with sodium thiocyanate. The reaction is formulated as shown in (32-1).

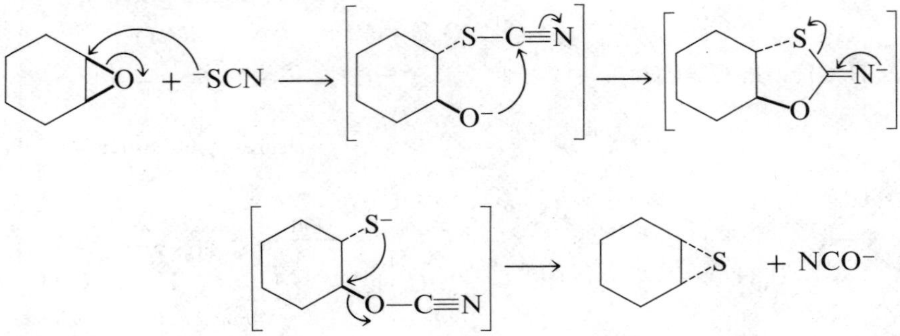

cyclohexene sulfide

The most striking chemical property of the three-membered heterocycles is their extraordinary reactivity. This enhanced reactivity has its origin in the relief of ring strain that occurs when the ring is cleaved. Recall that ethylene oxide is much more reactive than normal ethers and undergoes ring opening by dilute acid or by base (Section 10.12.A).

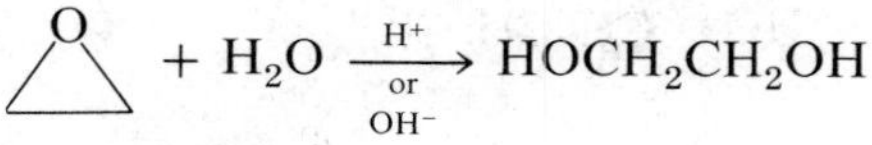

$$\triangle\!\!\!\!O + H_2O \xrightarrow[\substack{or \\ OH^-}]{H^+} HOCH_2CH_2OH$$

Similar reactivity is observed with aziridines and with thiiranes.

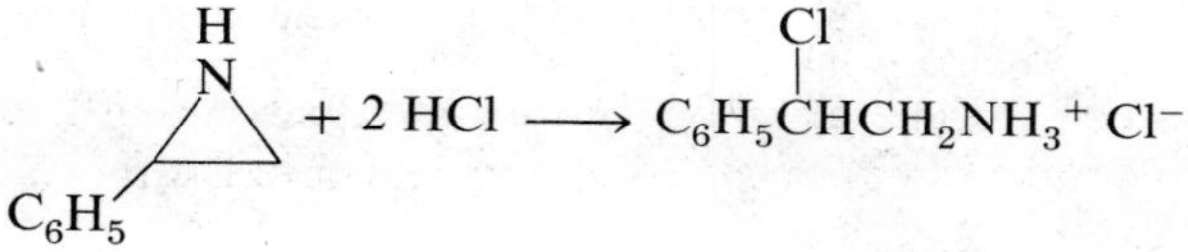

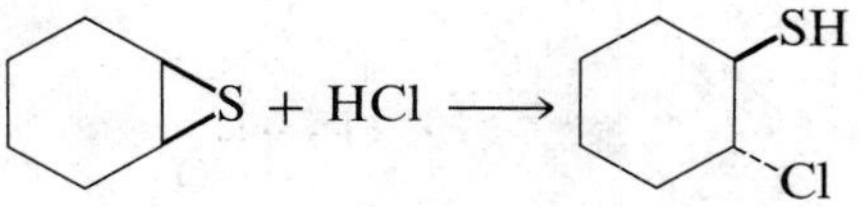

$$\text{+ 2 HCl} \longrightarrow C_6H_5\overset{\overset{\textstyle Cl}{|}}{C}HCH_2NH_3^+\;Cl^-$$

$$+ \;HCl \longrightarrow$$

EXERCISE: From the data in Appendix I calculate and compare $\Delta H°$ for the two reactions

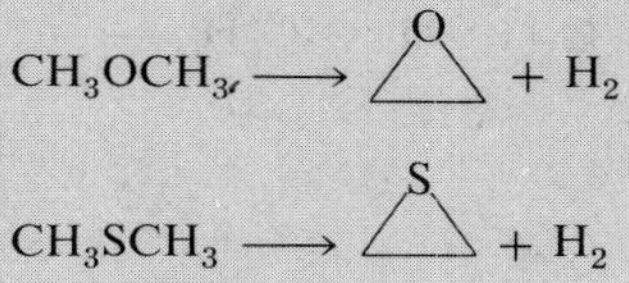

$$CH_3OCH_3 \longrightarrow \triangle\!\!\!\!O + H_2$$

$$CH_3SCH_3 \longrightarrow \triangle\!\!\!\!S + H_2$$

How can you rationalize the result?

C. Four-Membered Rings

The four-membered-ring heterocycles are rarer, mainly because of the greater difficulty of preparing four-membered rings (Section 8.14).

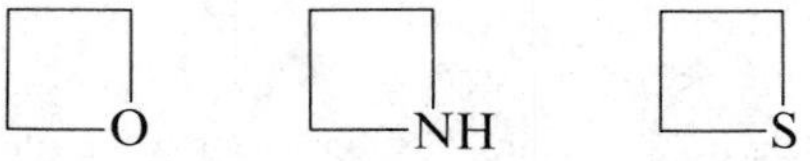

oxetane   azetidine   thietane

In some favorable cases the rings may be formed by direct ring closure, but yields in such reactions are often low.

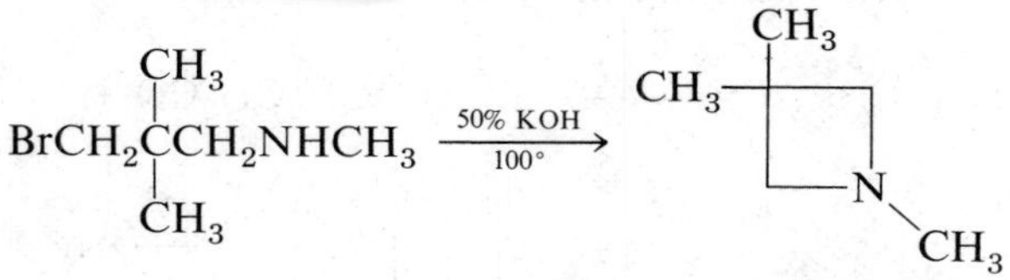

$$\underset{\underset{\textstyle CH_3}{|}}{\overset{\overset{\textstyle CH_3}{|}}{BrCH_2CCH_2NHCH_3}} \xrightarrow[100°]{50\% \text{ KOH}}$$

(80%)
1,3,3-trimethylazetidine

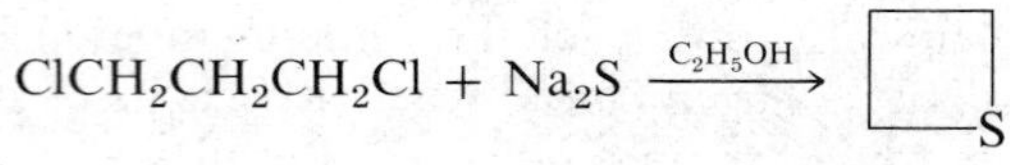

$$ClCH_2CH_2CH_2Cl + Na_2S \xrightarrow{C_2H_5OH}$$

(20–30%)
thietane

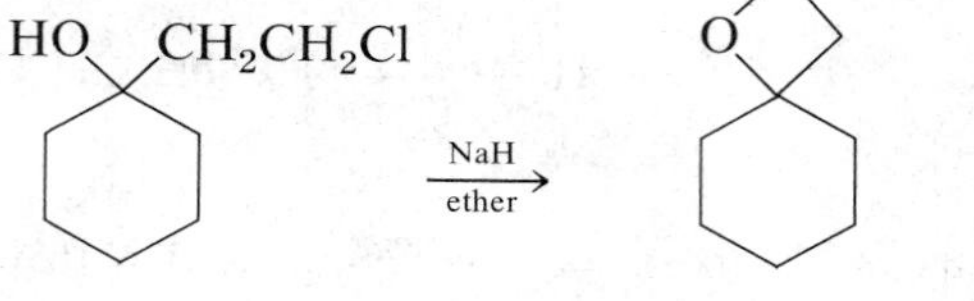

$$\xrightarrow[\text{ether}]{\text{NaH}}$$

(55%)
1-oxaspiro[5.3]nonane

Another approach to generation of the four-membered-ring heterocycles is by the union of two double bonds, a cycloaddition reaction. Examples are the formation of β-lactones and β-lactams.

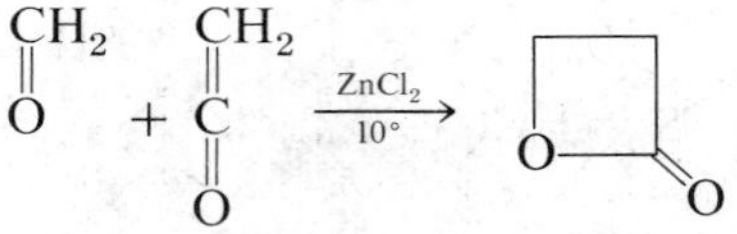

$$\underset{O}{\overset{CH_2}{\|}} + \underset{\overset{\|}{O}}{\overset{CH_2}{\underset{C}{\|}}} \xrightarrow[10°]{ZnCl_2}$$

(88%)
β-propiolactone

$$(C_6H_5)_2C=C=O + C_6H_5CH=NC_6H_5 \longrightarrow$$

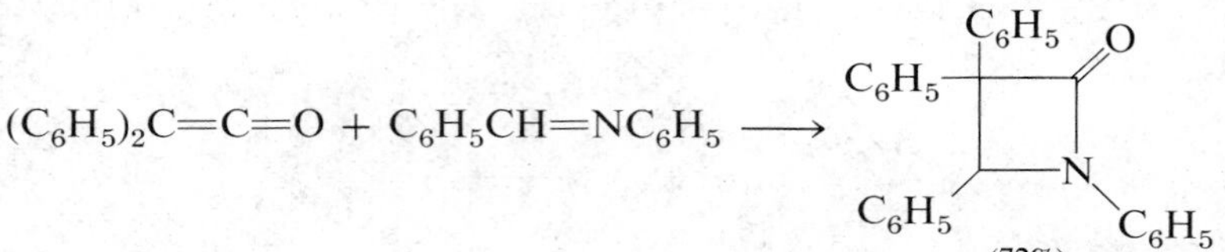

(72%)
1,3,3,4-tetraphenyl-
azetidin-2-one

The cycloaddition reactions of these examples are *not* analogous to Diels–Alder reactions. Diels–Alder reactions are examples of 4 + 2 cycloaddition reactions and involve a cyclic six-electron transition state related to benzenoid aromatic systems (Section 34.2). An analogous transition state for the present 2 + 2 cycloadditions has only four electrons and is not aromatic.

six-electron aromatic
transition state for
4 + 2 Diels–Alder
reaction

four-electron transition
state for a concerted
2 + 2 cycloaddition—
nonaromatic

Instead, stepwise mechanisms are involved.

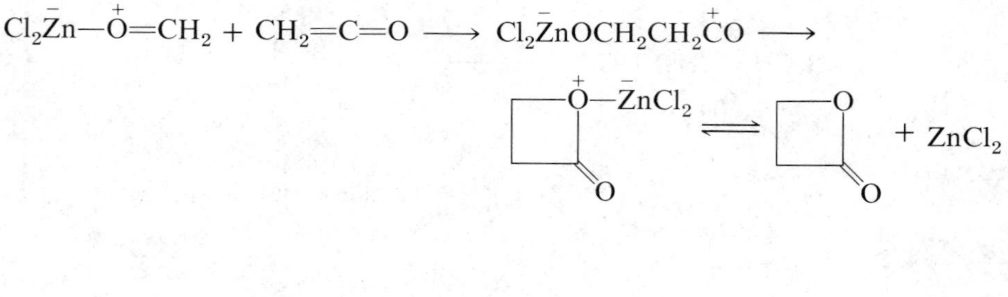

$$Cl_2\bar{Z}n-\overset{+}{O}=CH_2 + CH_2=C=O \longrightarrow Cl_2\bar{Z}nOCH_2CH_2\overset{+}{C}O \longrightarrow$$

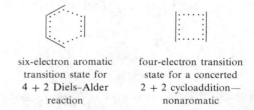

$$+ ZnCl_2$$

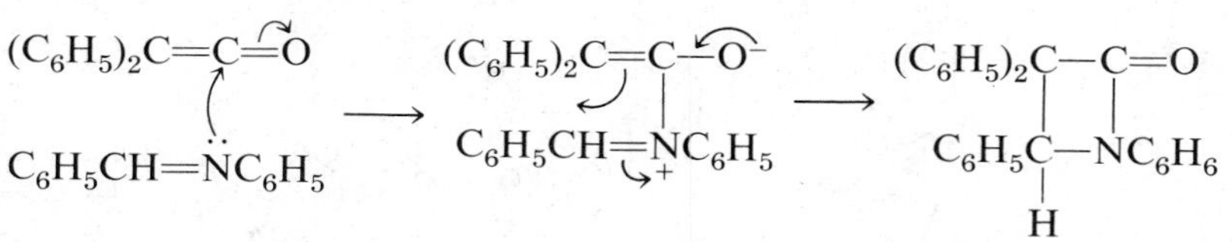

$$(C_6H_5)_2C=C-\overset{\frown}{O} \longrightarrow (C_6H_5)_2C=C-O^- \longrightarrow (C_6H_5)_2C-C=O$$

$$C_6H_5CH=\overset{..}{N}C_6H_5 \qquad C_6H_5CH=\overset{+}{N}C_6H_5 \qquad C_6H_5\overset{|}{C}-NC_6H_6$$

$$\overset{|}{H}$$

The stepwise processes also rationalize the observed modes of addition.

Like the three-membered ring analogs, oxetanes, azetidines, and thietanes are susceptible to acid-catalyzed ring-opening reactions.

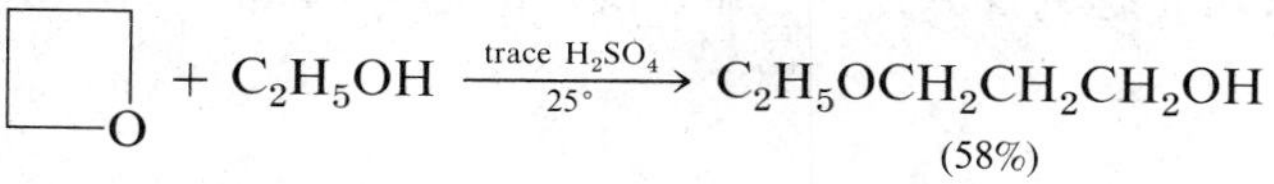

$$\square_O + C_2H_5OH \xrightarrow[25°]{\text{trace } H_2SO_4} C_2H_5OCH_2CH_2CH_2OH \qquad (32\text{-}2)$$
$$(58\%)$$

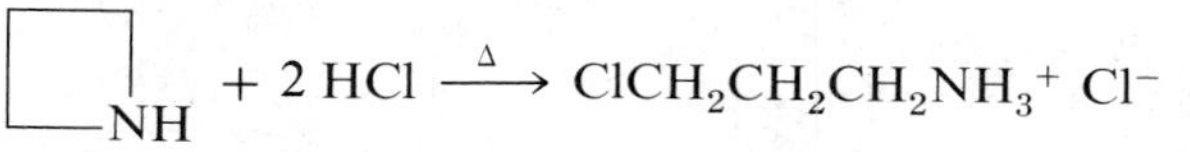

$$\square_{NH} + 2 \text{ HCl} \xrightarrow{\Delta} ClCH_2CH_2CH_2NH_3^+ \text{ Cl}^- \qquad (32\text{-}3)$$

They are also more reactive than their open-chain relatives in nucleophile reactions but are much less reactive than the analogous three-membered ring compounds. Note the strenuous conditions required in the example (32-4).

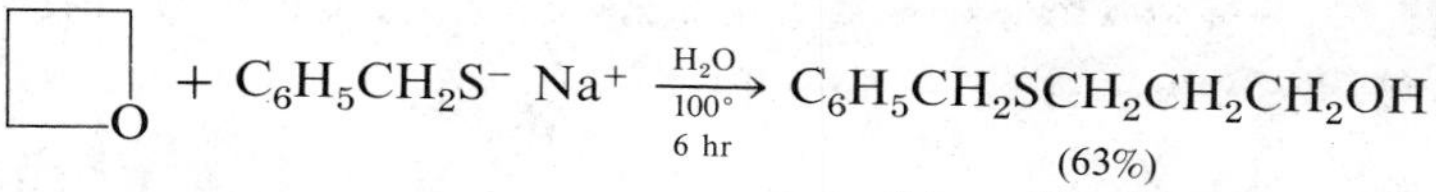

$$\square_O + C_6H_5CH_2S^- \text{ Na}^+ \xrightarrow[\substack{6 \text{ hr}}]{H_2O \\ 100°} C_6H_5CH_2SCH_2CH_2CH_2OH \qquad (32\text{-}4)$$
$$(63\%)$$
3-benzylthio-1-propanol

EXERCISE: Write equations showing each step in reactions (32-2), (32-3), and (32-4).

D. *Five- and Six-Membered Rings*

One source of the saturated five-membered-ring heterocycles is reduction of available aromatic compounds derived from furan and pyrrole.

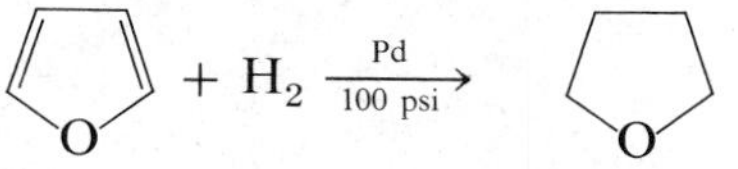

$$\text{furan} + H_2 \xrightarrow[100 \text{ psi}]{Pd} \text{tetrahydrofuran}$$
$$(100\%)$$

furan   tetrahydrofuran

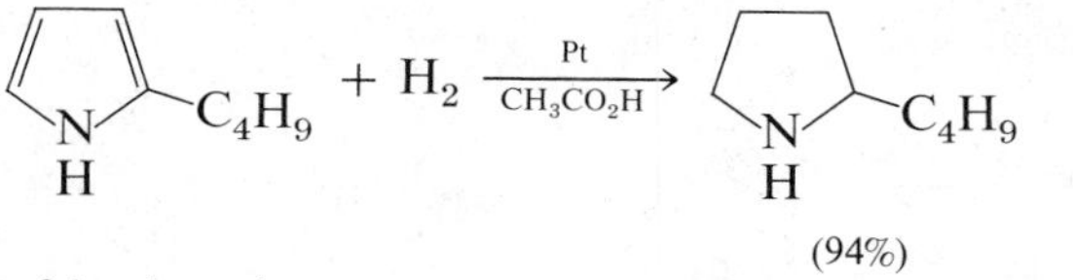

$$\text{2-butylpyrrole} + H_2 \xrightarrow[CH_3CO_2H]{Pt} \text{2-butylpyrrolidine}$$
$$(94\%)$$

2-butylpyrrole   2-butylpyrrolidine

Many piperidine derivatives may be prepared by hydrogenation of the corresponding pyridine.

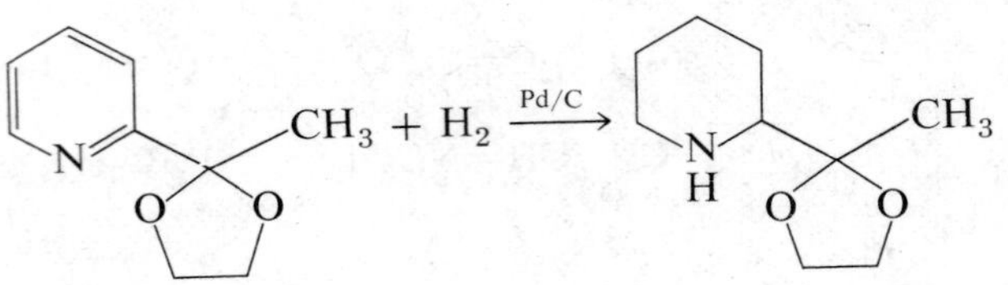
CH3 + H2
Pd/C
CH3

(77%)

The pyridine ring is more easily reduced than indole or pyrrole. For example, the selective reduction (32-5) may be carried out in high yield.

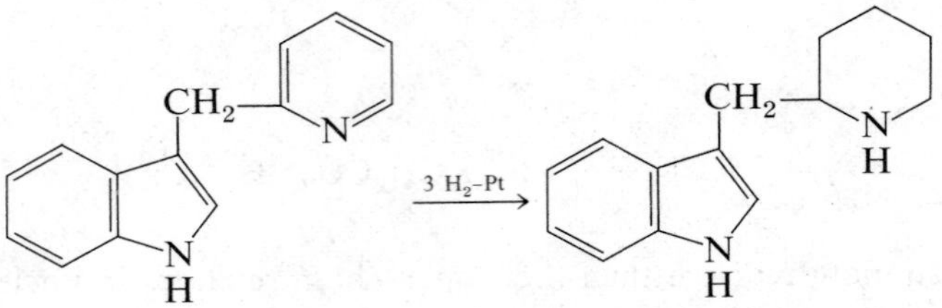
CH2
3 H2–Pt
CH2

(32-5)

Aside from reduction of aromatic heterocycles, the main synthetic route to the five- and six-membered-ring saturated compounds is by ring closure of suitable difunctional compounds. Some examples are

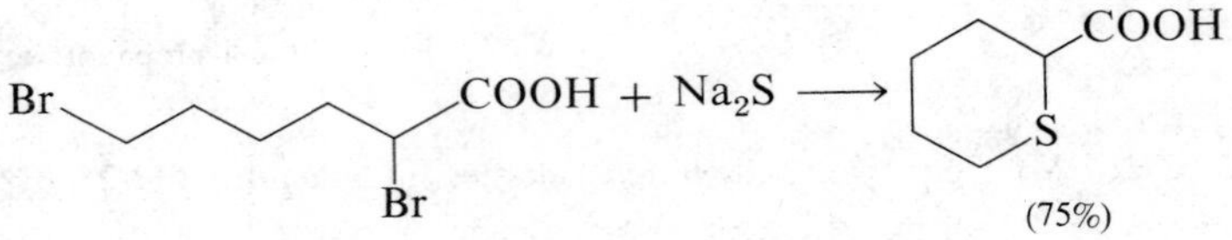
Br
COOH + Na2S
Br
COOH
S

(75%)

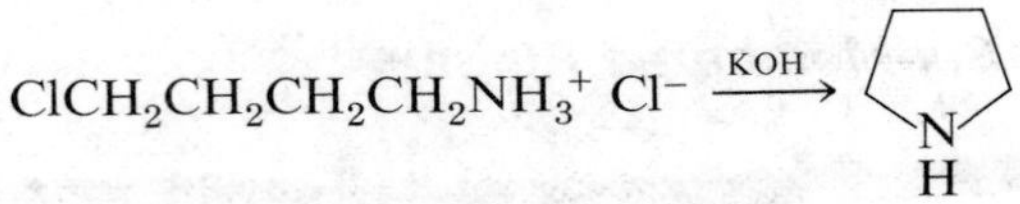
ClCH2CH2CH2CH2NH3+ Cl−
KOH

(50%)

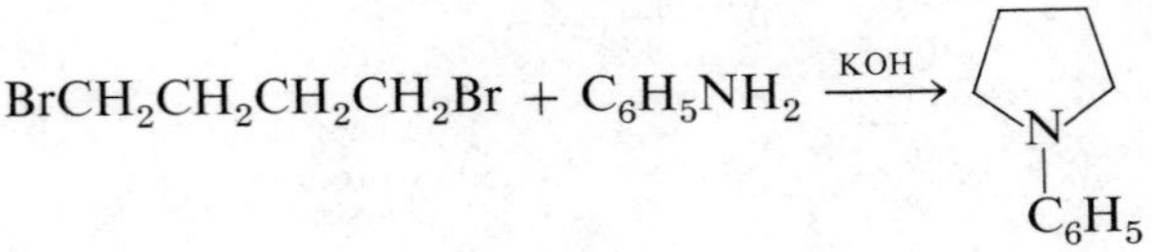
BrCH2CH2CH2CH2Br + C6H5NH2
KOH
C6H5

(100%)

Cyclic esters (lactones) and amides (lactams) are also examples of heterocyclic compounds.

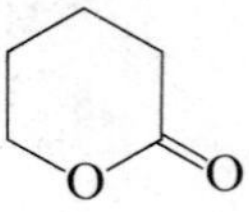

δ-valerolactone

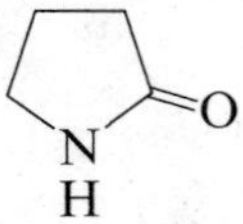

γ-butyrolactam
(2-pyrrolidone)

32.3
Furan, Pyrrole, and Thiophene

A. *Structure and Properties*

The structures of these three heterocycles would suggest that they have highly reactive diene character.

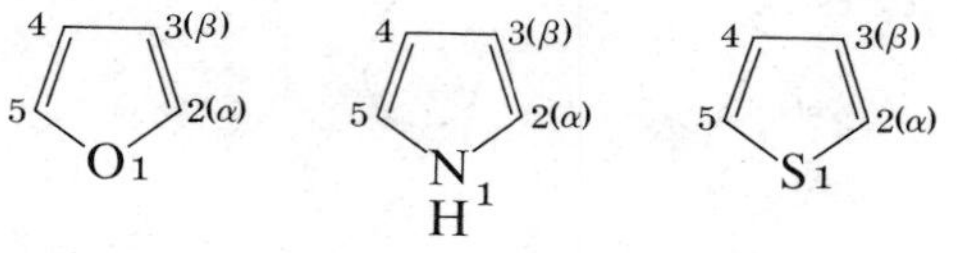

furan   pyrrole   thiophene

However, like benzene, they have many chemical properties that are not typical of dienes. They undergo substitution rather than addition reactions, and they show the effect of a ring current in their nmr spectra. In short, these heterocycles have characteristics associated with aromaticity.

From an orbital point of view, pyrrole has a planar pentagonal structure in which the four carbons and the nitrogen have sp^2-hybridization. Each ring atom forms two sp^2—sp^2 σ-bonds to its neighboring ring atoms, and each forms one sp^2—s σ-bond to a hydrogen. The remaining p_z-orbitals on each ring atom overlap to form a π-molecular system in which the three lowest molecular orbitals are bonding. The six π-electrons (one for each carbon and two for nitrogen) fill the three bonding orbitals and give the molecule its aromatic character. Pyrrole (Figure 32.1) is isoelectronic with cyclopentadienyl anion, an unusually stable carbanion that also has a cyclic π-electronic system with six electrons (Section 22.7.C).

Furan and thiophene have similar structures. In these cases the second lone pair on the heteroatom may be considered to occupy an sp^2-orbital that is perpendicular to the π-system of the ring (Figure 32.2).

The aromatic character of these heterocycles may also be expressed by using resonance structures, which show that a pair of electrons from the hetero atom is delocalized around the ring.

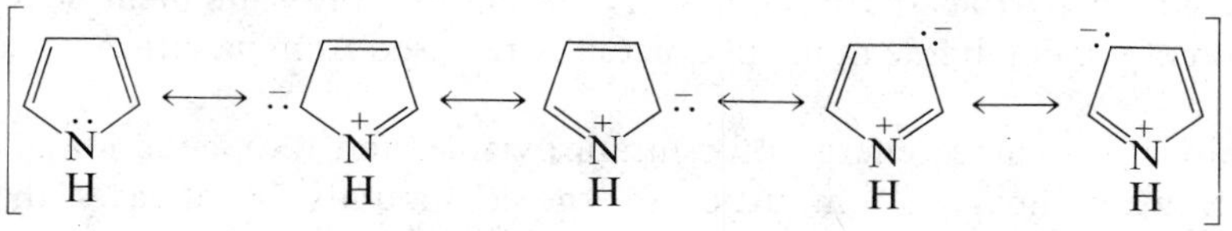

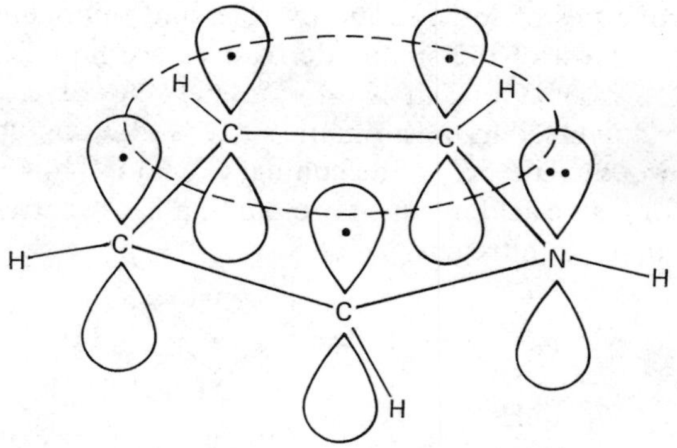

FIGURE 32.1. *Orbital structure of pyrrole.*

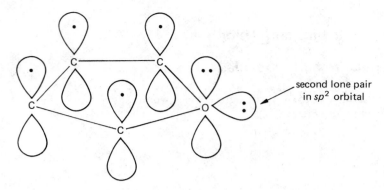

FIGURE 32.2. *Orbital structure of furan.*

This delocalization of the lone-pair electrons away from the hetero atom can be inferred from the dipole moments of these aromatic heterocycles and their nonaromatic counterparts.

1.73 D	1.90 D	1.58 D

0.70 D	0.51 D	1.81 D

In the saturated compounds, the hetero atom is at the negative end of the dipole. In the aromatic heterocycles the dipole moment associated with the π-system opposes the σ-moment. As a result the net dipole moment of furan and thiophene is reduced. In pyrrole the π-moment is larger than the σ-moment so that the direction of the net dipole moment is actually reversed from its saturated counterpart!

Empirical resonance energies for furan, pyrrole, and thiophene may be computed from the heats of combustion for the compounds. In all cases, there is a substantial stabilization energy, although of considerably smaller magnitude than for benzene.

The physical properties of some furan, pyrrole, and thiophene derivatives are listed in Table 32.1. Most of the simple derivatives are liquids.

Although pyrrole is an amine, it is an extremely nonbasic one because the nitrogen lone pair is involved in the aromatic sextet and is therefore less available for bonding to a proton. The pK_a of its conjugate acid is -4.4. In fact, this pK_a corresponds to a conjugate acid in which protonation has occurred predominantly on carbon rather than on nitrogen.

$$pK_a = -4.4$$

TABLE 32.1
Physical Properties of Some Furans,
Pyrroles, and Thiophenes

Compound	Melting Point, °C	Boiling Point, °C
furan	−86	31
2-chlorofuran	—	78
3-chlorofuran	—	79
2-methylfuran	—	63
3-methylfuran	—	66
2-nitrofuran	29	135
pyrrole	—	131
1-methylpyrrole	—	115
2-methylpyrrole	—	148
3-methylpyrrole	—	143
1-phenylpyrrole	62	234
2-phenylpyrrole	129	271
thiophene	−38	84
2-chlorothiophene	−72	128
2-methylthiophene	−63	113
3-methylthiophene	−69	115
2-nitrothiophene	46	225
2,4-dimethylthiophene	—	141

Pyrrole compounds occur widely in living systems. One of the more important pyrrole compounds is the porphyrin **hemin,** the prosthetic group of hemoglobin and myoglobin (see Figure 29.16). A number of simple alkylpyrroles have played an important role in the elucidation of the porphyrin structures. Drastic reduction of hemin gives a complex mixture from which the four pyrroles—hemopyrrole, cryptopyrrole, phyllopyrrole, and opsopyrrole—have been isolated.

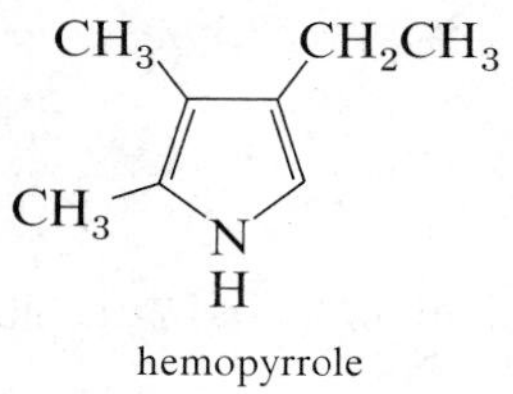

hemopyrrole

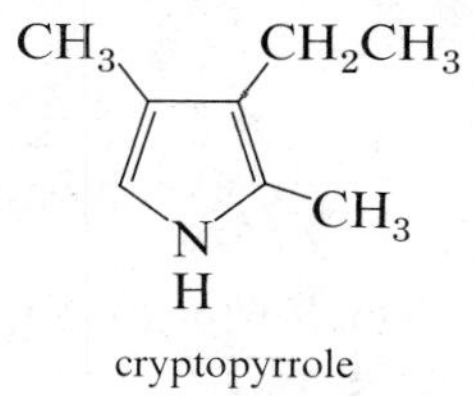

cryptopyrrole

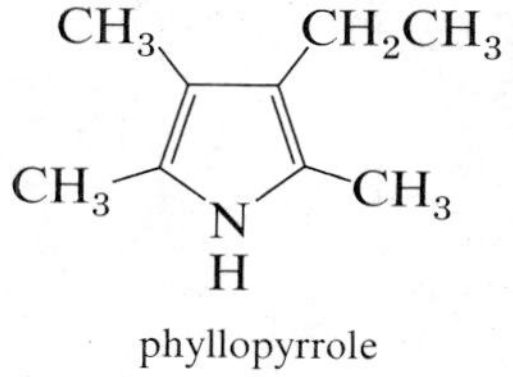

phyllopyrrole

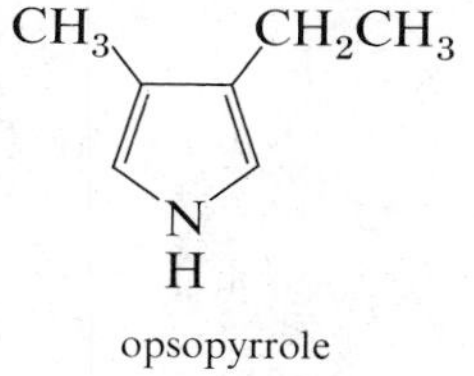

opsopyrrole

The function of hemoglobin in an organism is to transport oxygen; 1 g of hemoglobin absorbs 1.35 mL of oxygen at STP, corresponding to exactly one molecule of O_2 per atom of iron. The oxygen binds to the hemoglobin molecule in the vicinity of the iron, and the binding constant is proportional to the partial pressure of oxygen. In the lungs where the partial pressure of oxygen is high, hemoglobin binds oxygen. In the tissues served by the bloodstream the oxyhemoglobin dissociates back into O_2 and hemoglobin, which returns to the lungs for another load. Carbon monoxide is a poison because it forms a tight complex with the iron of hemoglobin and prevents the iron from binding to oxygen.

The porphyrins are derivatives of porphine, a tetrapyrrole heterocycle, and occur as metal complexes in the active sites of a number of enzymes. The porphine nucleus contains a conjugating system of eighteen π-centers, indicated as a heavy line in the structure shown. This system obeys the $4n + 2$ rule and is therefore an aromatic cycle.

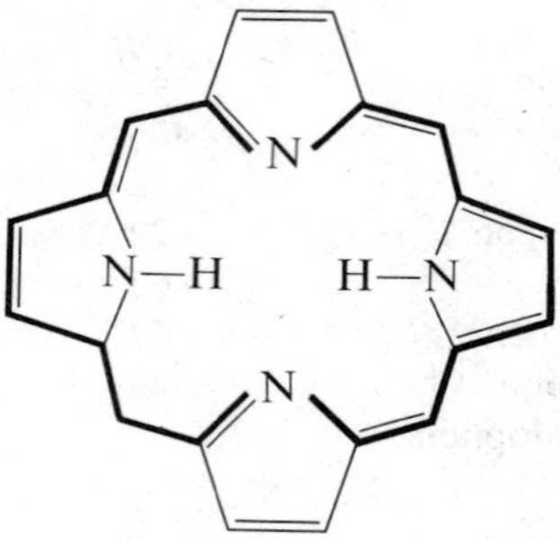

porphine

EXERCISE: One approach to calculating empirical resonance energies is to compare the heats of hydrogenation of furan, pyrrole, and thiophene to tetrahydrofuran, pyrrolidine, and tetrahydrothiophene, respectively, with the heat of hydrogenation of cyclopentadiene to cyclopentane. Apply this method with the data in Appendix I. Rationalize the results using resonance structures.

B. *Synthesis*

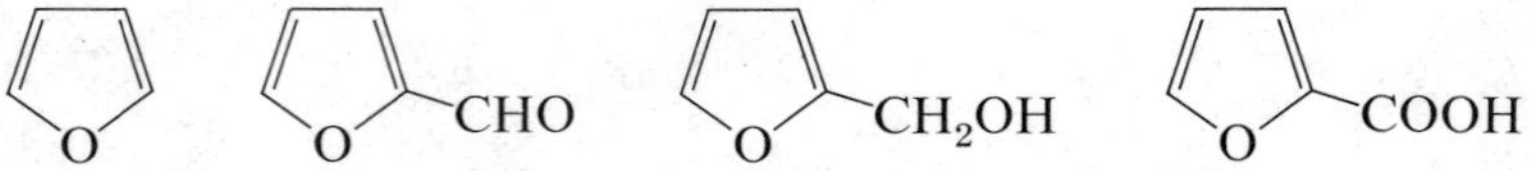

Furan, 2-furaldehyde (furfural), 2-furylmethanol, and 2-furoic acid are all inexpensive commercial items.

furan   2-furaldehyde   2-furylmethanol   2-furoic acid
furfural

The ultimate source of these heterocycles is furfural, which is obtained industrially by the acid hydrolysis of the polysaccharides of oat hulls, corn cobs, or straw. These polysaccharides are built up from pentose units. Dehydration of the pentose may be formulated as follows.

$$
\begin{array}{c}
\text{CHO} \\
\text{H}-\text{C}-\text{OH} \\
\text{HO}-\text{C}-\text{H} \\
\text{H}-\text{C}-\text{OH} \\
\text{CH}_2\text{OH}
\end{array}
\underset{-\text{H}^+}{\overset{+\text{H}^+}{\rightleftharpoons}}
\begin{array}{c}
\overset{+}{\text{CHOH}} \\
\text{H}-\text{C}-\text{OH} \\
\text{HO}-\text{C}-\text{H} \\
\text{H}-\text{C}-\text{OH} \\
\text{CH}_2\text{OH}
\end{array}
\underset{+\text{H}^+}{\overset{-\text{H}^+}{\rightleftharpoons}}
\begin{array}{c}
\text{CHOH} \\
\text{C}-\text{OH} \\
\text{HO}-\text{C}-\text{H} \\
\text{H}-\text{C}-\text{OH} \\
\text{CH}_2\text{OH}
\end{array}
\underset{-\text{H}^+}{\overset{+\text{H}^+}{\rightleftharpoons}}
$$

$$
\begin{array}{c}
\text{CHOH} \\
\text{C}-\text{OH} \\
\text{H}_2\overset{+}{\text{O}}-\text{C}-\text{H} \\
\text{H}-\text{C}-\text{OH} \\
\text{CH}_2\text{OH}
\end{array}
\underset{+\text{H}_2\text{O}}{\overset{-\text{H}_2\text{O}}{\rightleftharpoons}}
\left[
\begin{array}{c}
\text{CHOH} \\
\text{C}-\text{OH} \\
\overset{+}{\text{C}}-\text{H} \\
\text{H}-\text{C}-\text{OH} \\
\text{CH}_2\text{OH}
\end{array}
\longleftrightarrow
\begin{array}{c}
\overset{+}{\text{CHOH}} \\
\text{C}-\text{OH} \\
\text{C}-\text{H} \\
\text{H}-\text{C}-\text{OH} \\
\text{CH}_2\text{OH}
\end{array}
\right]
\underset{+\text{H}^+}{\overset{-\text{H}^+}{\rightleftharpoons}}
$$

$$
\begin{array}{c}
\text{CHO} \\
\text{C}-\text{OH} \\
\text{C}-\text{H} \\
\text{H}-\text{C}-\text{OH} \\
\text{CH}_2\text{OH}
\end{array}
\underset{-\text{H}^+}{\overset{+\text{H}^+}{\rightleftharpoons}}
\begin{array}{c}
\text{CHO} \\
\text{C}-\text{OH} \\
\text{C}-\text{H} \\
\text{HC}-\overset{+}{\text{OH}}_2 \\
\text{CH}_2\text{OH}
\end{array}
\underset{+\text{H}_2\text{O}}{\overset{-\text{H}_2\text{O}}{\rightleftharpoons}}
\left[
\begin{array}{c}
\text{CHO} \\
\overset{+}{\text{C}}-\text{OH} \\
\text{C}-\text{H} \\
\text{C}-\text{H} \\
\text{CH}_2\text{OH}
\end{array}
\longleftrightarrow
\begin{array}{c}
\text{CHO} \\
\text{C}-\text{OH} \\
\text{C}-\text{H} \\
\text{H}-\overset{+}{\text{C}} \\
\text{CH}_2\text{OH}
\end{array}
\right]
\rightleftharpoons
$$

$$
\begin{array}{c}
\text{CHO OH} \\
\text{C} \\
\text{C}-\text{H} \\
\text{C}-\text{H} \\
\text{CH}_2 \\
\end{array}
\overset{+}{\text{OH}}
\underset{+\text{H}^+}{\overset{-\text{H}^+}{\rightleftharpoons}}
\begin{array}{c}
\text{CHO} \\
\text{HC} \quad \text{C}-\text{OH} \\
\text{HC} \quad \text{CH}_2 \\
\text{O}
\end{array}
\underset{-\text{H}^+}{\overset{+\text{H}^+}{\rightleftharpoons}}
\begin{array}{c}
\text{CHO} \\
\text{OH}_2{}^+ \\
\text{O}
\end{array}
\underset{+\text{H}_2\text{O}}{\overset{-\text{H}_2\text{O}}{\rightleftharpoons}}
\begin{array}{c}
\text{CHO} \\
\overset{+}{\text{O}}
\end{array}
\overset{-\text{H}^+}{\longrightarrow}
\underset{\text{O}}{\overset{}{\bigcirc}}-\text{CHO}
$$

(100%)

Pyrrole is prepared commercially by the fractional distillation of coal tar or by passing a mixture of furan, ammonia, and steam over a catalyst at 400°.

Thiophene is prepared industrially by passing a mixture of butane, butenes, or butadiene and sulfur through a reactor heated at 600° for a contact time of about 1 sec.

$$
n\text{-C}_4\text{H}_{10} + \text{S} \xrightarrow{600°} \underset{\text{S}}{\bigcirc\!\!\!\!\bigcirc} + \text{H}_2\text{S}
$$

Substituted furans, pyrroles, and thiophenes may be prepared by electrophilic substitution on one of the available materials discussed or by a variety of cyclization reactions. The most general is the **Paal–Knorr** synthesis, in which a 1,4-dicarbonyl compound is heated with a dehydrating agent, ammonia, or an inorganic sulfide to produce the furan, pyrrole, or thiophene, respectively.

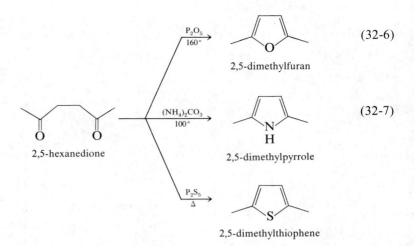

$$(32\text{-}6)$$

2,5-dimethylfuran

$$(32\text{-}7)$$

2,5-hexanedione

2,5-dimethylpyrrole

2,5-dimethylthiophene

Another general method for the synthesis of substituted pyrroles is the **Knorr pyrrole synthesis,** the condensation of an α-aminoketone with a β-keto ester. The method is illustrated in a synthesis of diethyl 3,5-dimethylpyrrole-2,4-dicarboxylate.

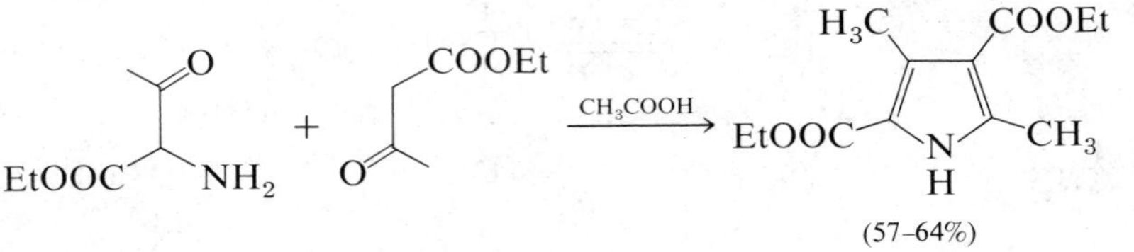

(57–64%)

The probable mechanism of the Knorr synthesis is

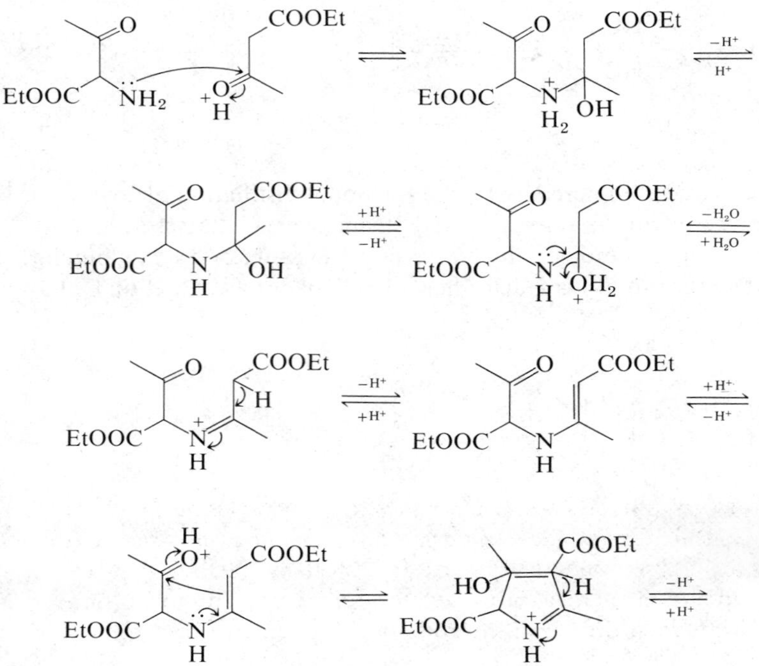

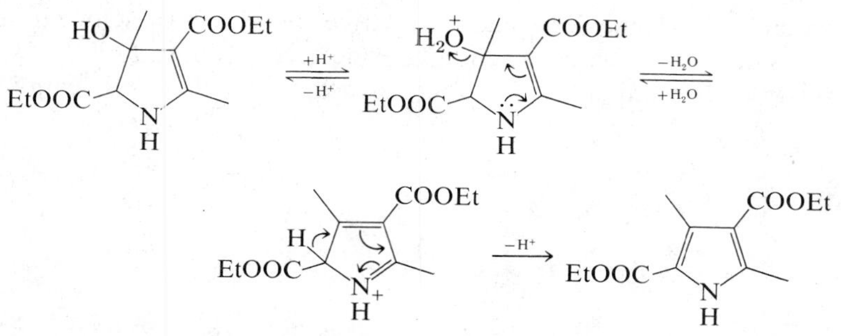

Notice how each individual step involves either an oxonium ion or an ammonium ion. No unstabilized carbocation is involved.

EXERCISE: (a) The mechanism detailed on page 1073 is an extremely important example in heterocyclic chemistry and deserves careful study. For each of the neutral intermediates shown in the reaction pathway examine alternative sites of protonation and show why each of these protonated compounds is less likely to lead to a carbocation intermediate than the intermediate shown.
(b) Write reasonable mechanisms for reactions (32-6) and (32-7).

C. Reactions

The most typical reaction of furan, pyrrole, and thiophene is electrophilic substitution. All three heterocycles are much more reactive than benzene, the reactivity order being

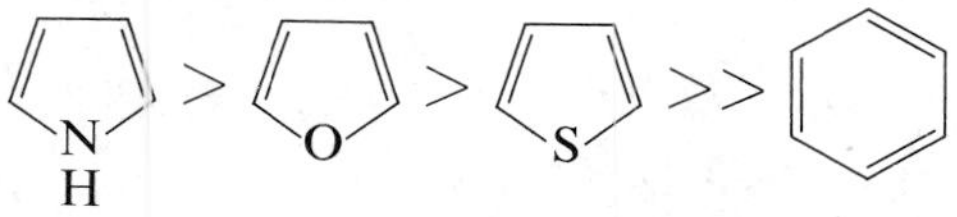

To give some idea of the magnitude of this reactivity order, partial rate factors (reactivities relative to benzene) for tritium exchange with trifluoroacetic acid (page 715) for thiophene are

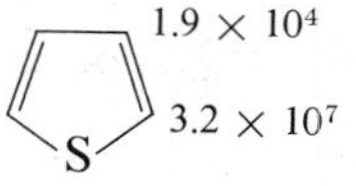

Because of this high reactivity, even mild electrophiles suffice to cause reaction. Substitution occurs predominantly at the α-position (C-2).

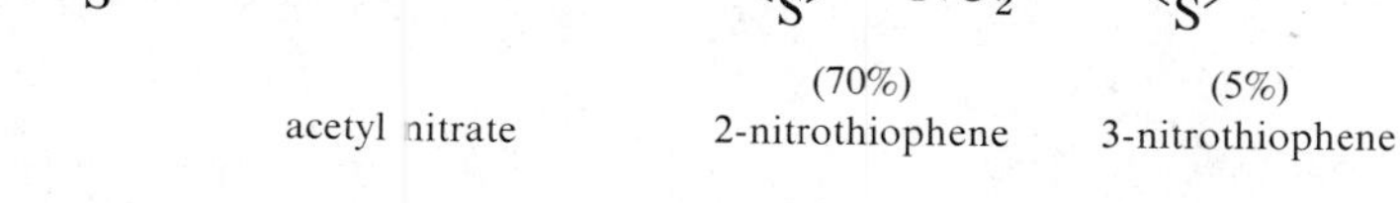

acetyl nitrate   2-nitrothiophene   3-nitrothiophene

This orientation is understandable in terms of the mechanism of electrophilic aromatic substitution. The α/β ratio is determined by the relative energies of the transition states leading to the two isomers. As in the case of substituted benzenes

(Section 23.1), we may estimate the relative energies of these two transition states by considering the actual reaction intermediates produced by attack at the α- or β-position. The important resonance structures for these two cations are

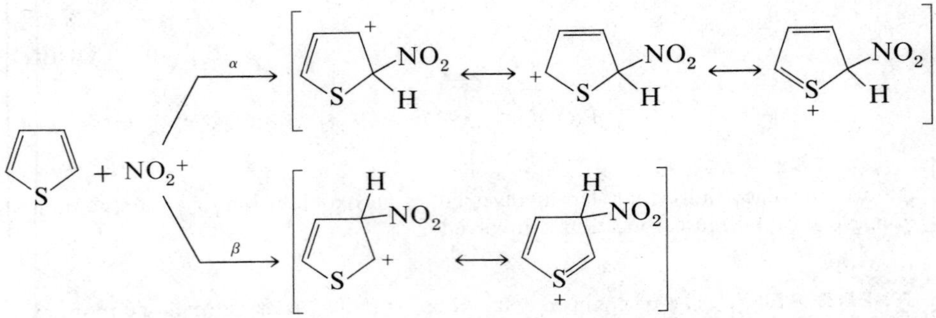

The most important of these structures are the two with the positive charge on sulfur because, in these two sulfonium cation structures, all atoms have octets of electrons. Nevertheless, as the sets of resonance structures show, the charge on the cation resulting from attack at the α-position is more extensively delocalized than that for the cation resulting from attack at the β-position. The following examples further demonstrate the generality of α-attack.

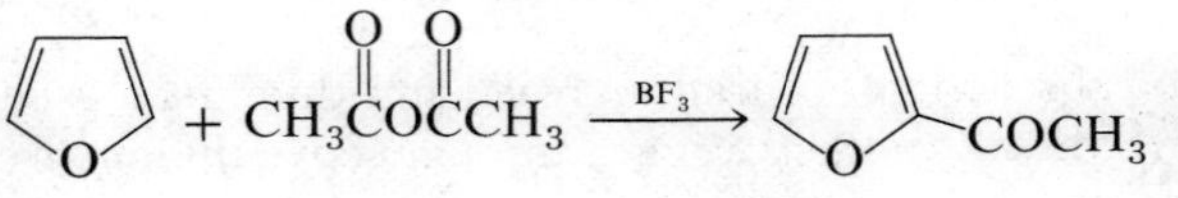

(75–92%)
2-acetylfuran

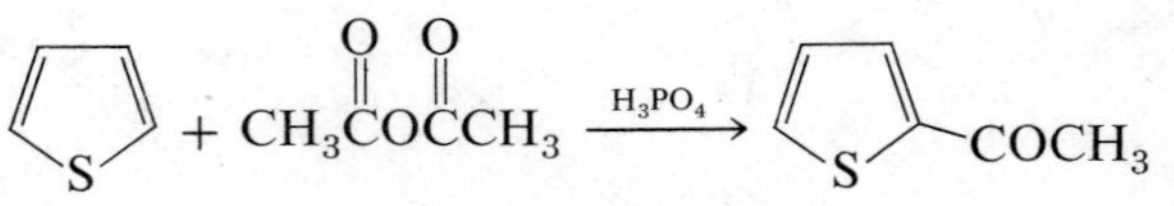

(94%)
2-acetylthiophene

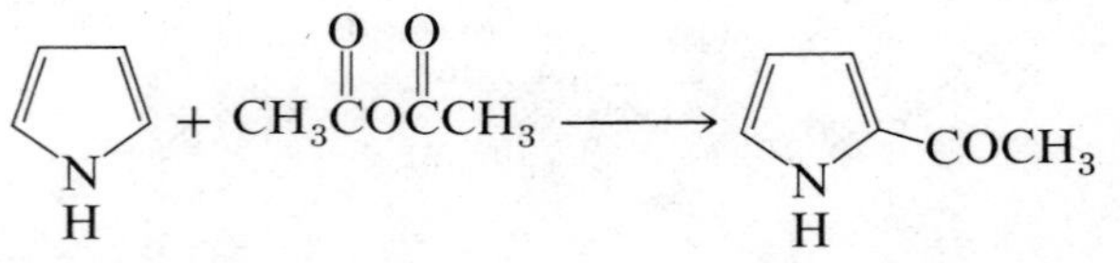

(60%)
2-acetylpyrrole

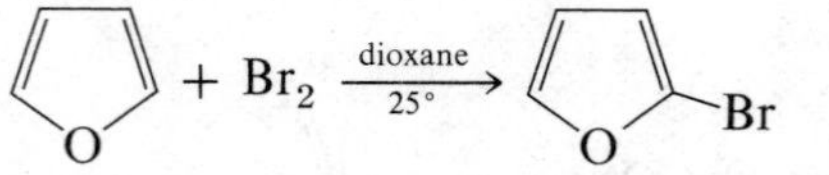

(75%)
2-bromofuran

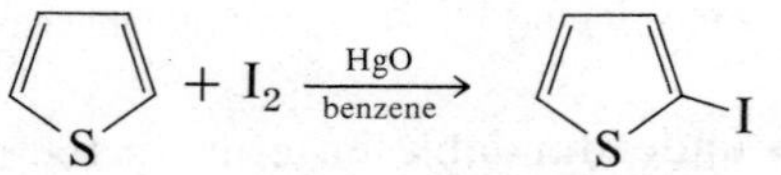

(75%)
2-iodothiophene

1077

Sec. 32.4

Condensed
Furans,
Pyrroles,
and Thiophenes

In the last example, note that 2-iodothiophene is the sole product of iodination, even though the reaction is carried out in benzene as solvent; that is, thiophene is so much more reactive than benzene that no significant amount of iodobenzene is formed.

Hydrolytic ring opening is a typical reaction of furans. In essence, the reaction is the reverse of the Paal-Knorr synthesis. Careful hydrolysis of furans can lead to the corresponding 1,4-dicarbonyl compounds in good yield.

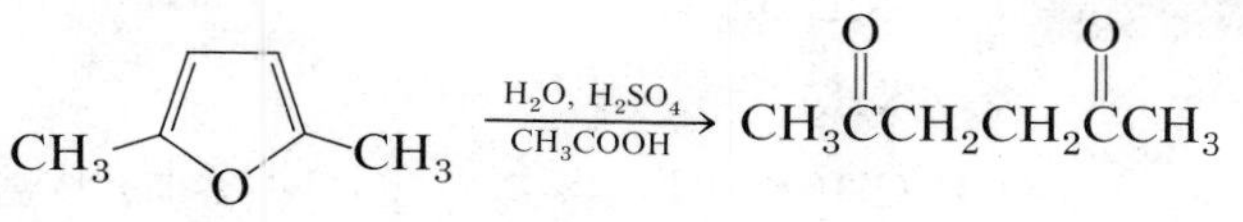

(90%)

The hydrolysis is initiated by electrophilic attack on the ring.

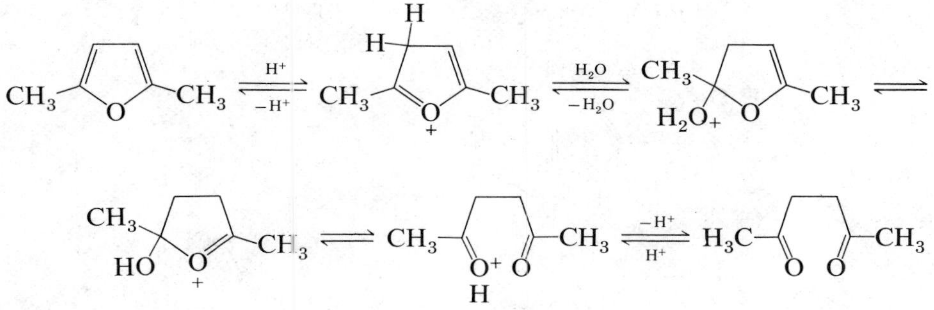

Pyrroles are polymerized by even dilute acids, probably by a mechanism such as the following.

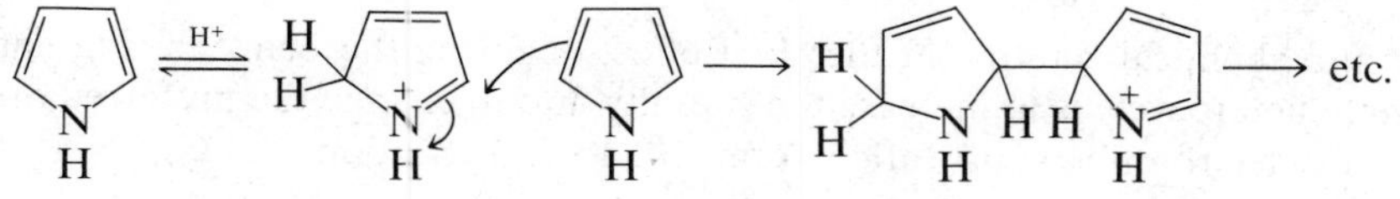

Thiophenes are more stable and do not undergo hydrolysis.

> **EXERCISE:** Give reasonable syntheses, starting with the unsubstituted heterocycle, of (a) 1-(2-pyrryl)-1-propanone, (b) furan-2-d, and (c) 2-(chloromethyl)thiophene.

32.4
Condensed Furans, Pyrroles, and Thiophenes

A. Structure and Nomenclature

Benzofuran, indole, and benzothiophene are related to the monocyclic heterocycles in the same way that naphthalene is a benzo derivative of benzene. As with the simple heterocycles, the rings are numbered beginning with the heteroatom; carbazole is an exception.

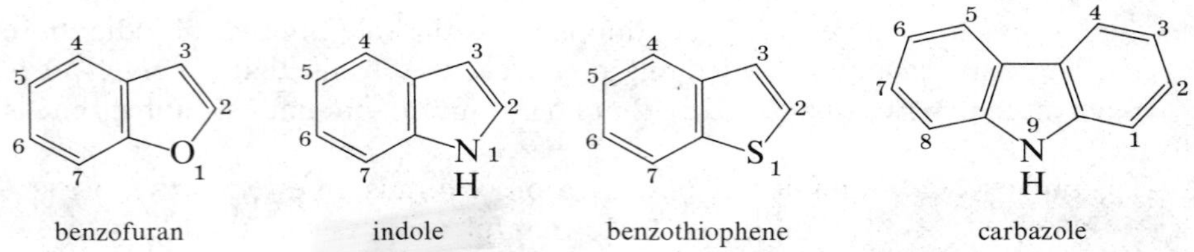

benzofuran   indole   benzothiophene   carbazole

Of the four systems, indoles are by far the most important. Many natural products have indole structures (see Section 34.5).

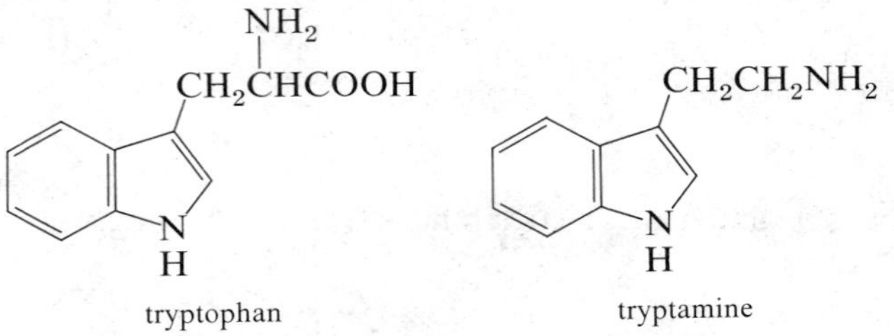

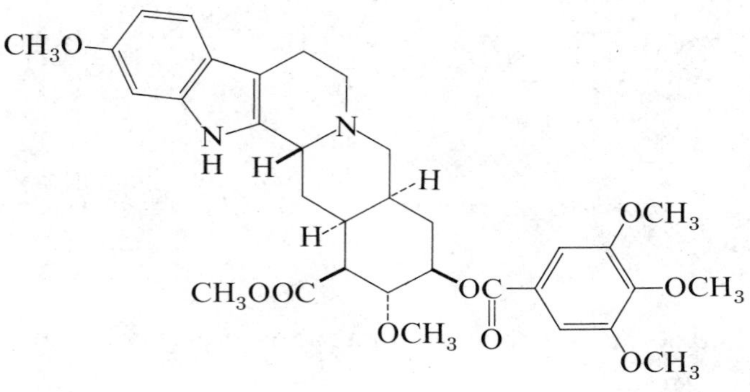

tryptophan   tryptamine

reserpine

From a chemical standpoint the chief effect of fusing the benzene ring onto the simple heterocycle is to increase the stability and to change the preferred orienta-. tion in electrophilic substitution from C-2 to C-3 (Section 32.4.C).

B. Synthesis

The most general synthesis of indoles is the **Fischer indole synthesis,** in which the phenylhydrazone of an aldehyde or ketone is treated with a catalyst such as BF_3, $ZnCl_2$, or polyphosphoric acid (PPA).

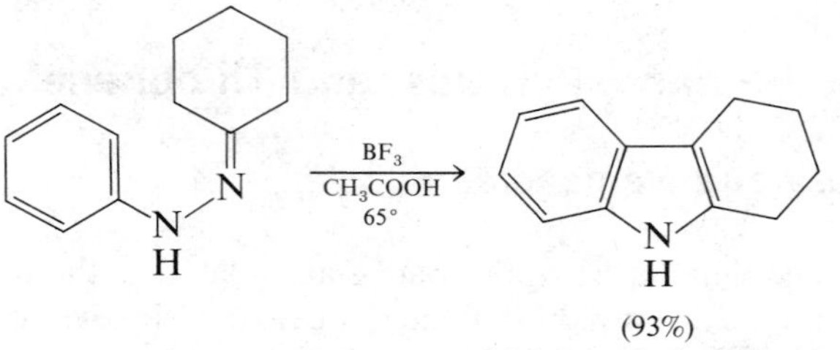

(93%)
1,2,3,4-tetrahydrocarbazole

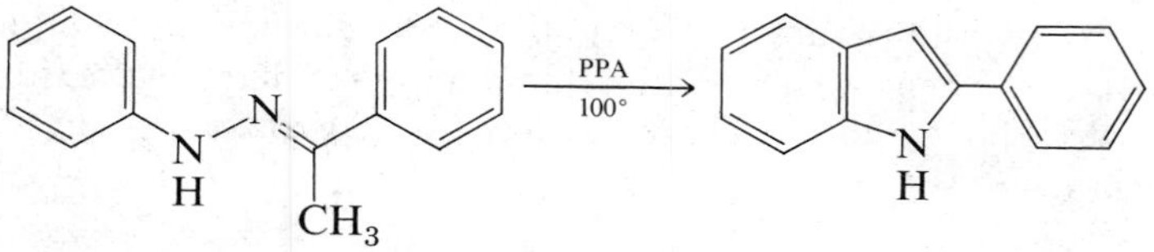

1079

Sec. 32.4

**Condensed
Furans,
Pyrroles,
and Thiophenes**

(73%)
2-phenylindole

The mechanism of the Fischer synthesis has been the subject of much study. The available evidence is in accord with a pathway involving a benzidine-like rearrangement (Section 25.1).

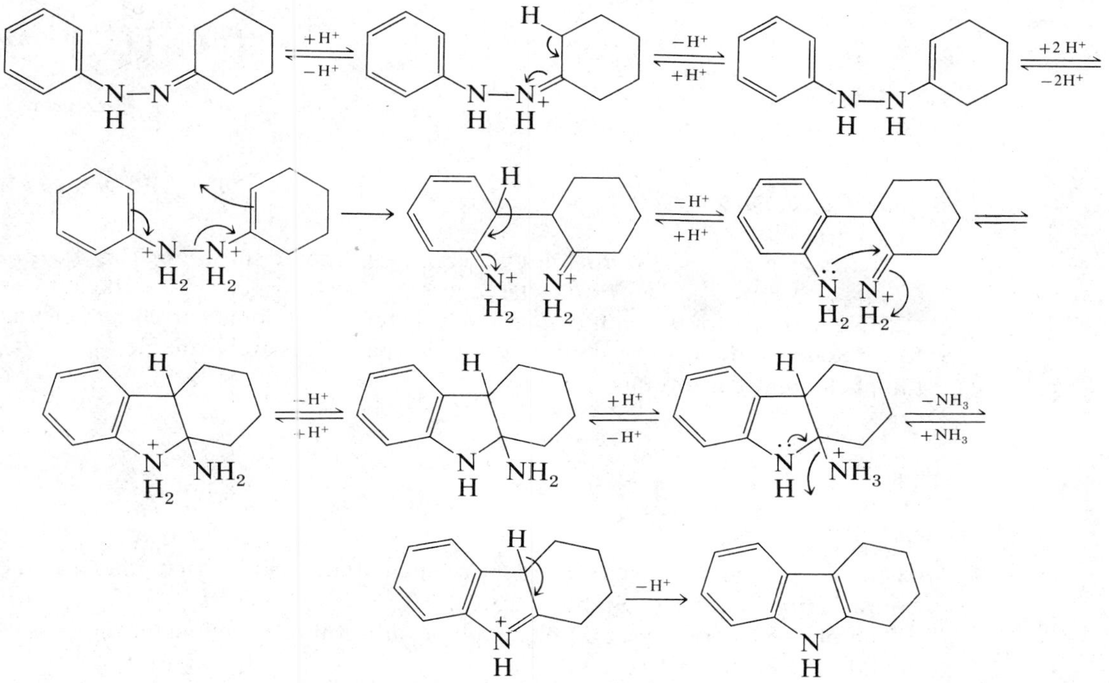

The reaction fails with the phenylhydrazone of acetaldehyde and thus cannot be used to prepare indole itself. However, the phenylhydrazone of pyruvic acid does react to yield indole-2-carboxylic acid, which can be decarboxylated to give indole.

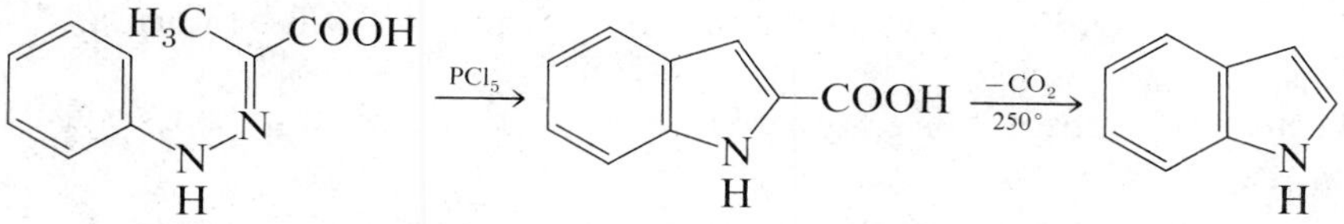

Benzofuran is prepared from coumarin, which in turn is prepared from salicylaldehyde by the Perkin synthesis (Section 20.3.B).

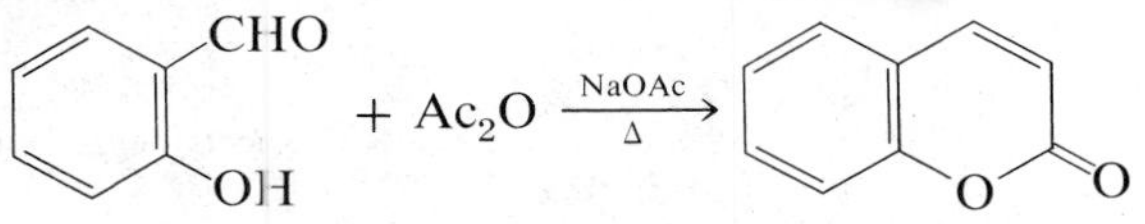

coumarin

Coumarin reacts with bromine to give a dibromide that undergoes an interesting ring contraction when treated with ethanolic potassium hydroxide to give benzofuran-2-carboxylic acid. The acid may be decarboxylated by distillation over calcium oxide.

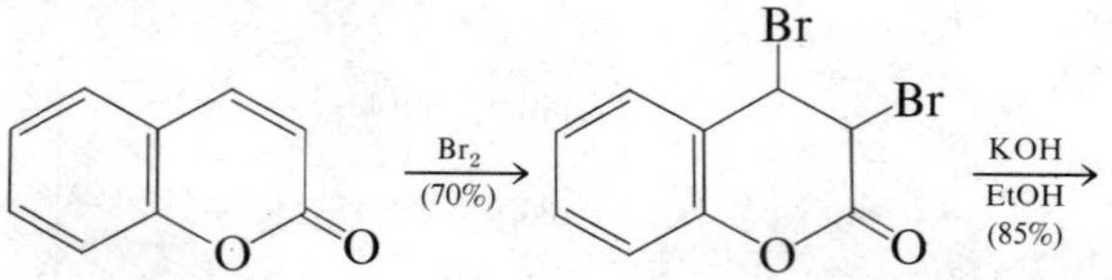

coumarin   coumarilic acid   benzofuran

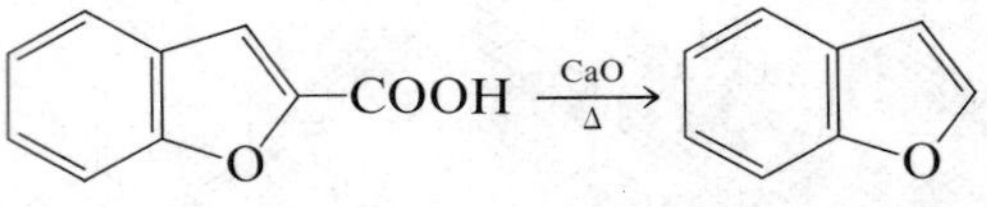

C. Reactions

All three condensed heterocycles undergo electrophilic substitution in the heterocyclic ring rather than in the benzene ring. However, each is markedly less reactive than the corresponding monocyclic heterocycle. Some partial rate factors for protodetritiation with trifluoroacetic acid (page 715) are available for benzothiophene and benzofuran.

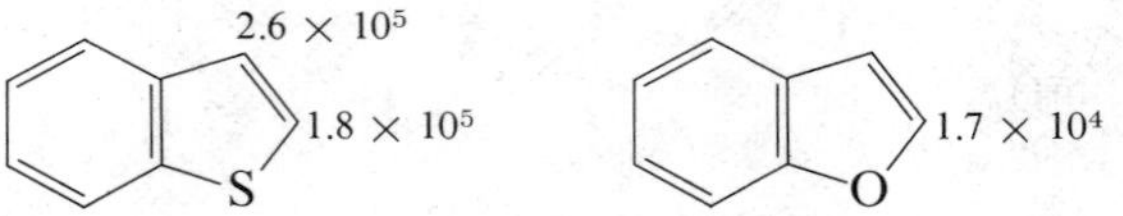

These values are at least two orders of magnitude smaller than that for the α-position of thiophene (Section 32.3.C).

The preferred orientation in electrophilic substitution reactions in these compounds can be summarized as follows.

1. In benzofuran the most reactive position is C-2.
2. In benzothiophene C-2 and C-3 have comparable reactivities, with C-3 being somewhat the more reactive.
3. In indole the most reactive position is C-3.

The way in which these generalizations apply in practice will be illustrated with some specific examples.

Electrophilic substitution in benzofuran occurs predominantly at C-2, just as in furan itself.

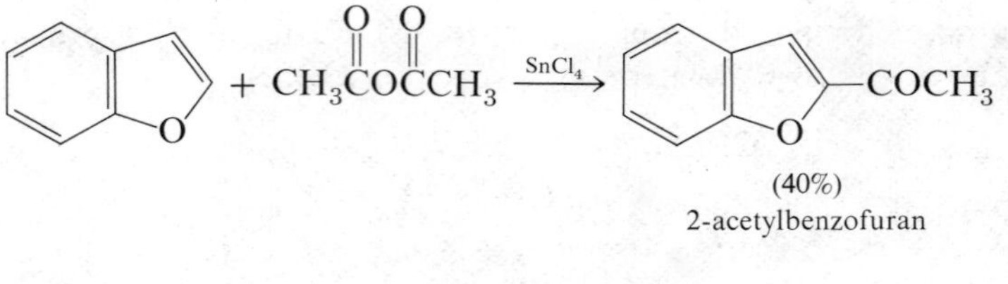

(40%)
2-acetylbenzofuran

If the 2-position is occupied, reaction occurs at C-3.

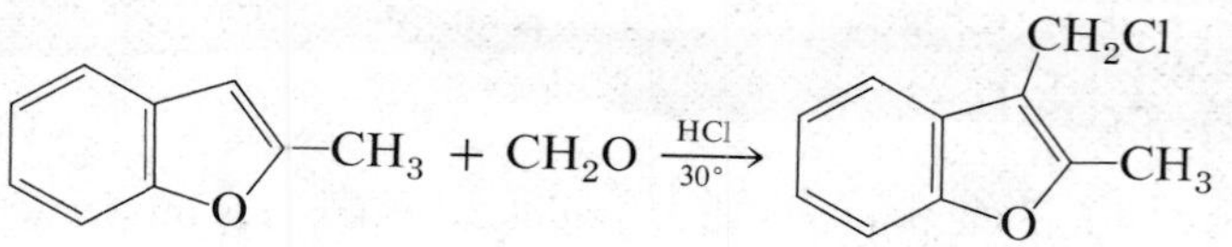

1081

Sec. 32.4

**Condensed
Furans,
Pyrroles,
and Thiophenes**

(70%)
2-methyl-3-chloromethyl-
benzofuran

The preferred reaction at C-3 in indole and benzothiophene is illustrated by

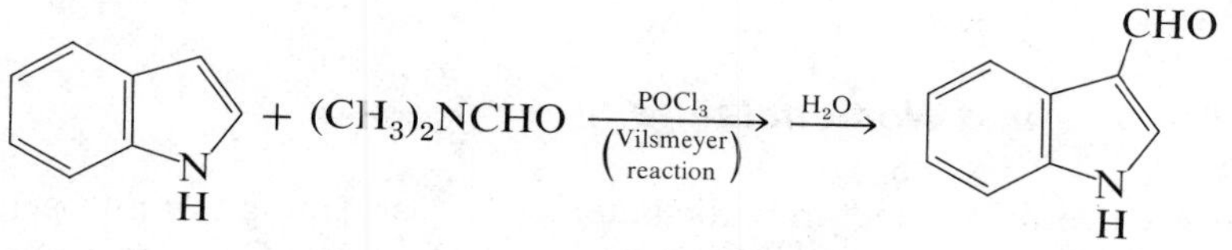

(97%)
indole-3-carboxaldehyde

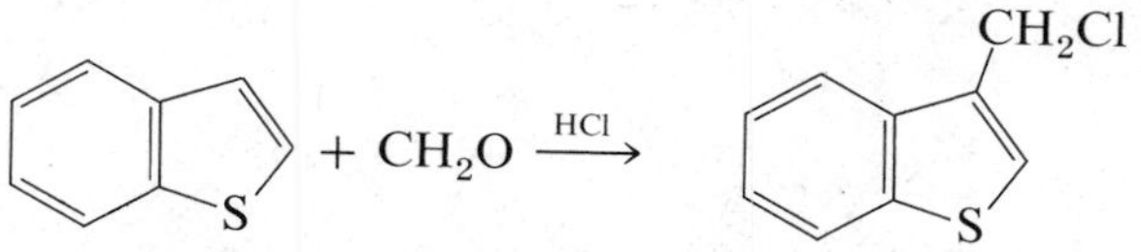

(56%)
3-chloromethylbenzothiophene

With benzothiophene, other isomers are usually produced as well, but are not always detected or isolated.

These orientation specificities can be rationalized by considering the intermediate ions produced by attack at C-2 and C-3. Reaction at C-2 gives a carbocation in which the charge is distributed to the benzene ring and to the hetero atom; however, the structure with the charge on the hetero atom no longer has a benzene ring. In contrast, reaction at C-3 does not permit effective distribution of charge around the benzene ring, but the electron pair on the hetero atom is utilized efficiently without disruption of the benzene resonance.

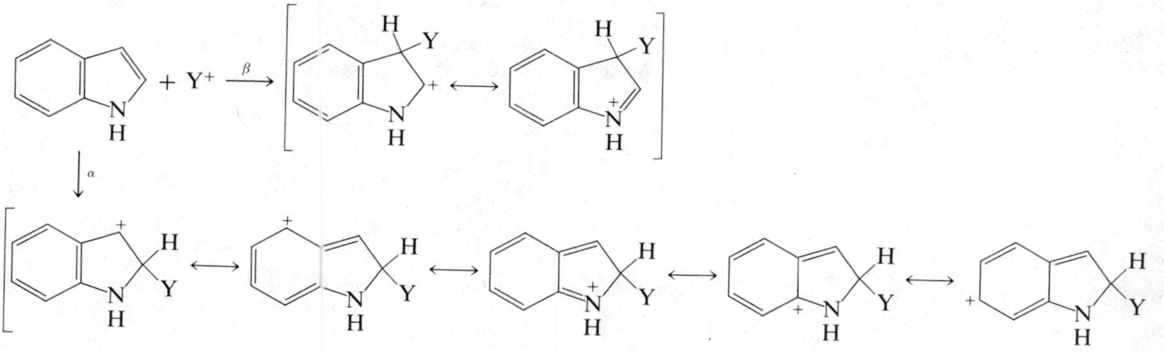

The relative reactivities depend on the balance of these contracting effects. The experimental results suggest that for indole the direct involvement of the basic nitrogen lone pair is much more important than conjugation with the benzene ring, whereas with benzofuran the oxygen lone pair is less basic and the involvement of the benzene ring now dominates. In the case of benzothiophene the two effects are roughly comparable in magnitude.

EXERCISE: From the principles and examples developed in this section, work out whether the 2- or 3-position of carbazole is the more reactive in electrophilic substitution.

32.5
Azoles

A. *Structure and Nomenclature*

Azoles are five-membered aromatic heterocycles containing two nitrogens, one nitrogen and one oxygen, or one nitrogen and one sulfur. They are named and numbered as shown. They may be considered as aza analogs of furan, pyrrole, and thiophene in the same way that pyridine is an aza analog of benzene (see Section 32.6).

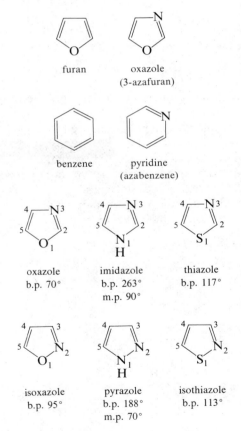

furan

oxazole
(3-azafuran)

benzene

pyridine
(azabenzene)

oxazole
b.p. 70°

imidazole
b.p. 263°
m.p. 90°

thiazole
b.p. 117°

isoxazole
b.p. 95°

pyrazole
b.p. 188°
m.p. 70°

isothiazole
b.p. 113°

From a molecular orbital standpoint the azoles are similar to the simpler aromatic heterocycles. For example, in imidazole each carbon and nitrogen may be considered to be sp^2-hybridized. One nitrogen makes two sp^2—sp^2 σ-bonds to carbon and one sp^2—s σ-bond to hydrogen. The other nitrogen has its lone pair in the third sp^2-orbital. The π-molecular orbital system is made up from the p_z-

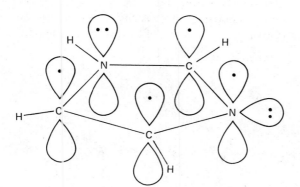

FIGURE 32.3. *Orbital structure of imidazole.*

orbitals from each ring atom (Figure 32.3). Six π-electrons (one from each carbon and from one nitrogen, two from the other nitrogen) complete the aromatic shell.

An examination of the physical properties of the simple azoles reveals that imidazole and pyrazole have anomalously high boiling points. They are also the only simple azoles that are solids at room temperature. These properties clearly result from intermolecular hydrogen bonding. With imidazole the hydrogen bonding is of a linear polymer, whereas pyrazole seems to exist largely as the dimer.

Like pyridine (pK_a 5.2; Section 32.6.A), thiazole (pK_a 2.4), pyrazole (pK_a 2.5), and oxazole (pK_a 0.8) are weak bases. As in pyridine, the nitrogen lone pair is in an sp^2-orbital. Recall that greater s-character of lone-pair electrons is associated with heightened stability and lower basicity (Section 12.4). A similar trend is seen with nitrogen acids (Table 32.2).

The higher s-character of the pyridine lone pair compared to aliphatic amines is

TABLE 32.2
Hybridization and Acidity

Orbital	Carbon Acid	pK_a	Nitrogen Acid	pK_a
sp	$HC{\equiv}CH$	25	$CH_3C{\equiv}\overset{+}{N}H$	-10
sp^2	$CH_2{=}CH_2$	44	pyridine $N^+{-}H$	5
sp^3	$CH_3{-}CH_3$	50	$(CH_3)_3\overset{+}{N}H$	10

sufficient to account for a decrease in basicity of several powers of ten. In pyrazole, thiazole, and isoxasole the basicity of the nitrogen lone pair is further reduced by the presence of the other hetero atom.

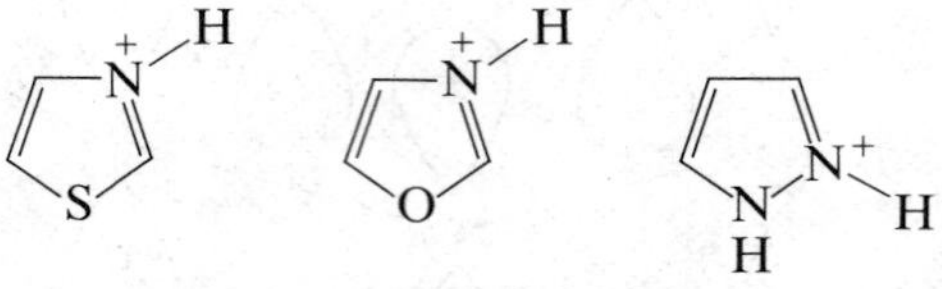

pK_a 2.4   pK_a 0.8   pK_a 2.5

In marked contrast to these results, imidazole seems to be abnormally basic for a compound with sp^2-hybridized nitrogen (pK_a 7.0). The enhanced basicity of imidazole is presumably due to the symmetry of the conjugate acid and the consequent resonance stabilization.

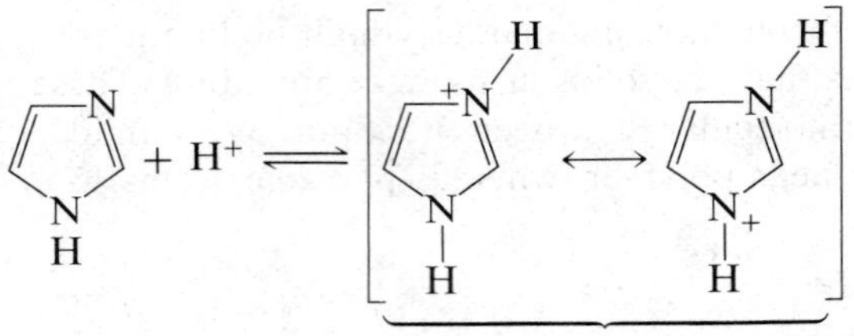

pK_a 7.0

Its pK_a of 7.0 means that imidazole is half-protonated in neutral water. As a result the basicity of imidazole plays an important role in biological processes. The imidazole ring in the amino acid histidine is often involved as a proton acceptor in the active site of enzymes.

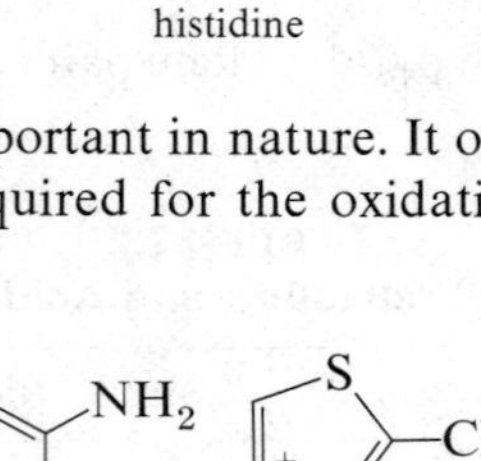

histidine

The thiazole ring is also important in nature. It occurs, for example, in vitamin B$_1$, thiamine, a coenzyme required for the oxidative decarboxylation of α-keto acids.

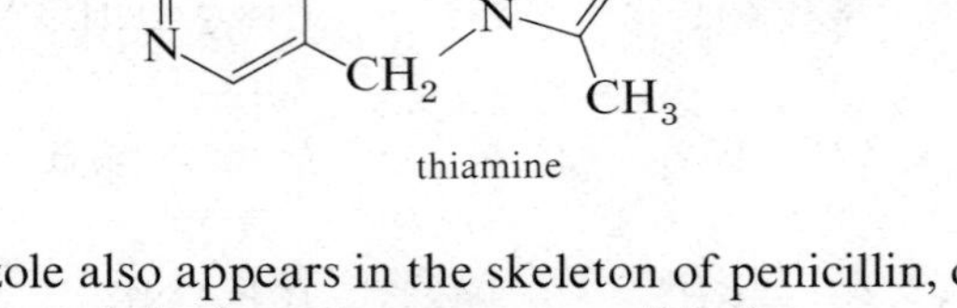

thiamine

A tetrahydrothiazole also appears in the skeleton of penicillin, one of the first and still most important of the broad-spectrum antibiotics.

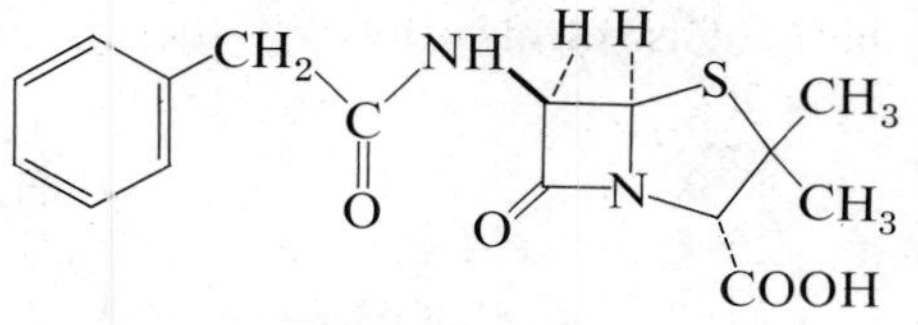

benzylpenicillin

B. *Synthesis*

Pyrazoles and isoxazoles may be synthesized by the reaction of hydrazine or hydroxylamine with a 1,3-dicarbonyl compound or its equivalent.

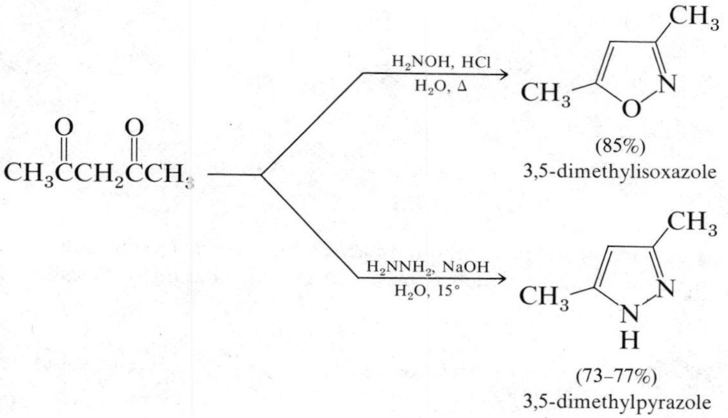

$$CH_3CCH_2CCH_3$$

$\xrightarrow[\text{H}_2\text{O, }\Delta]{\text{H}_2\text{NOH, HCl}}$

(85%)
3,5-dimethylisoxazole

$\xrightarrow[\text{H}_2\text{O, }15°]{\text{H}_2\text{NNH}_2\text{, NaOH}}$

(73–77%)
3,5-dimethylpyrazole

The reaction proceeds through an oxime or hydrazone, which undergoes cyclization.

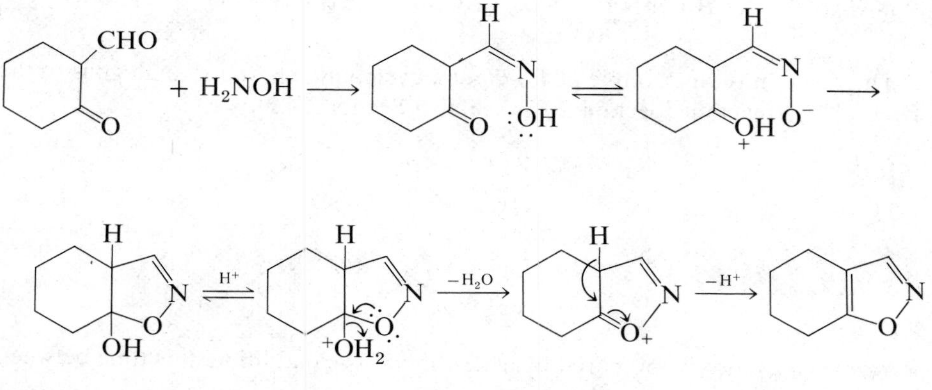

If a substituted hydrazine is used, a 1-substituted pyrazole results.

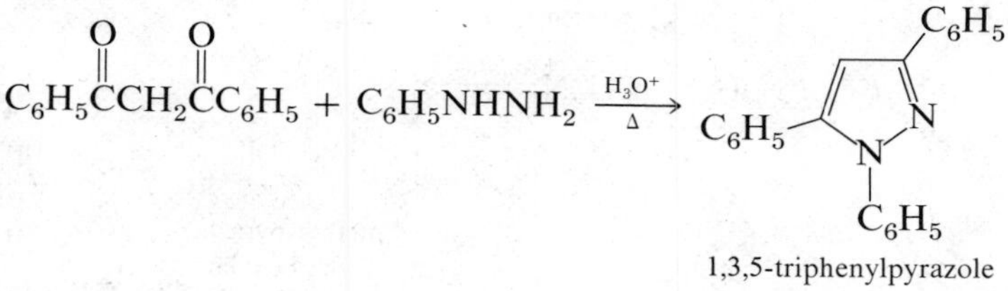

$$C_6H_5CCH_2CC_6H_5 + C_6H_5NHNH_2 \xrightarrow[\Delta]{H_3O^+}$$
(32-8)

1,3,5-triphenylpyrazole

An alternative synthesis of isoxazoles involves the cycloaddition of a nitrile oxide to an acetylene.

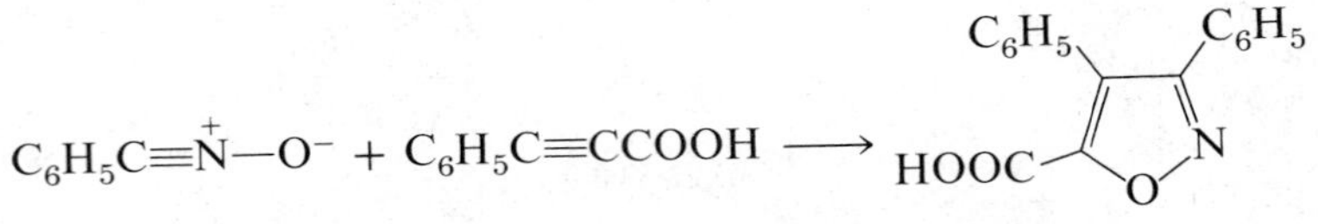

$$C_6H_5C\overset{+}{\equiv}N\text{—}O^- + C_6H_5C\equiv CCOOH \longrightarrow$$

benzonitrile oxide

3,4-diphenylisoxazole-5-carboxylic acid

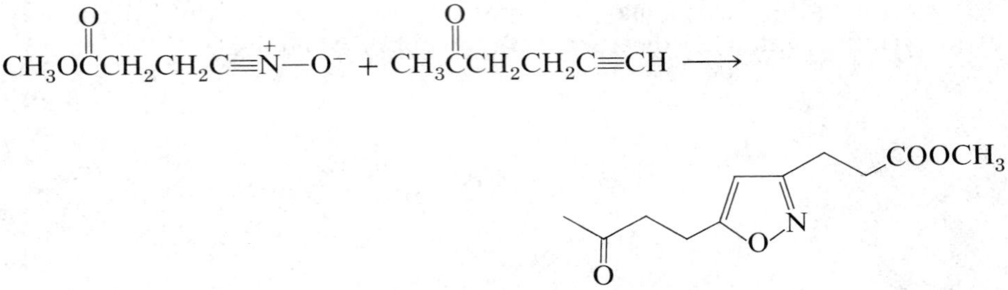

$$CH_3O\overset{O}{\overset{\|}{C}}CH_2CH_2C\overset{+}{\equiv}N\text{—}O^- + CH_3\overset{O}{\overset{\|}{C}}CH_2CH_2C\equiv CH \longrightarrow$$

(50%)

> Nitrile oxides are unstable compounds generated *in situ* by the dehydration of a hydroxamic acid chloride, which is prepared by chlorination of an aldoxime. 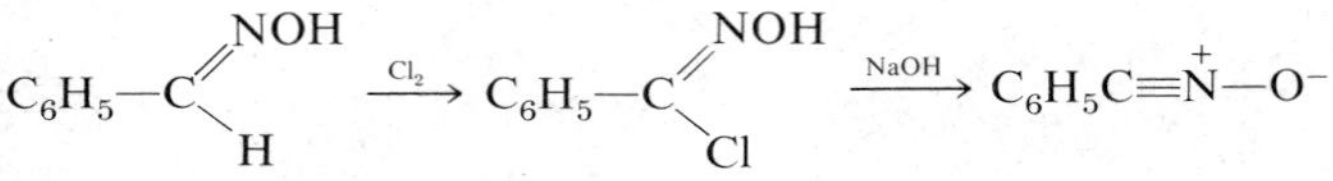
>
> $$C_6H_5\text{—}\overset{NOH}{\underset{H}{C}} \xrightarrow{Cl_2} C_6H_5\text{—}\overset{NOH}{\underset{Cl}{C}} \xrightarrow{NaOH} C_6H_5C\overset{+}{\equiv}N\text{—}O^-$$
>
> An alternative preparation involves dehydration of a nitroalkane.
>
> $$CH_3CH_2\overset{O}{\underset{+}{\overset{\|}{N}}}\diagdown_{O^-} + C_6H_5NCO \longrightarrow CH_3C\overset{+}{\equiv}N\text{—}O^-$$

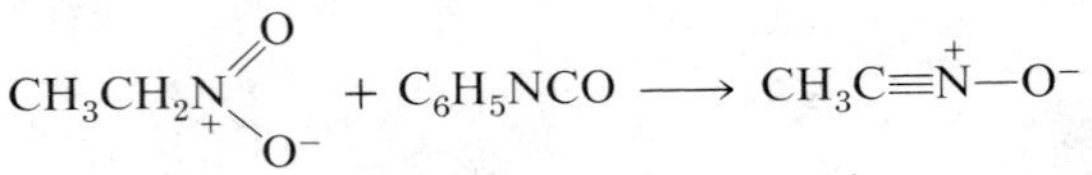

The reaction is an example of **1,3-dipolar cycloaddition** and is analogous to the Diels–Alder reaction (Section 20.5; see also Section 34.2).

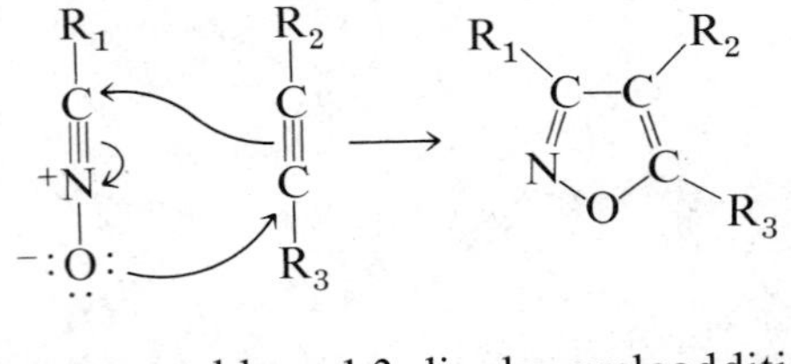

Pyrazoles may also be prepared by a 1,3-dipolar cycloaddition, this time between diazomethane and an acetylene.

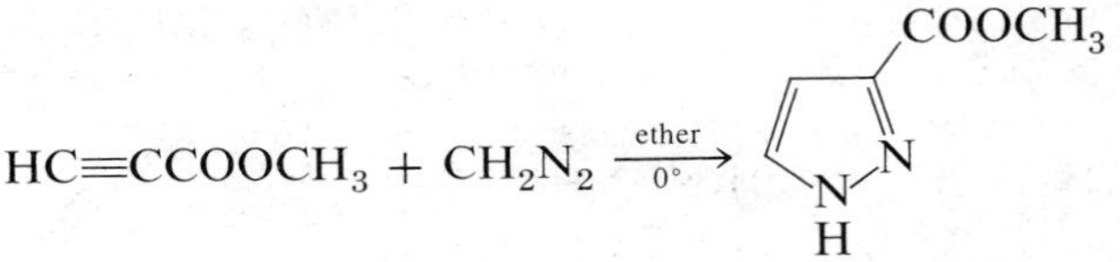

$$HC\equiv CCOOCH_3 + CH_2N_2 \xrightarrow[0°]{ether}$$

(80%)

methyl pyrazole-3-carboxylate

The reaction is formulated in a completely analogous manner.

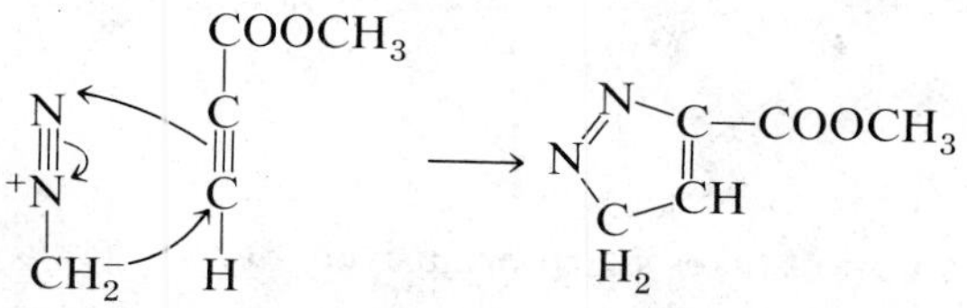

The initially formed product tautomerizes to the more stable aromatic system.

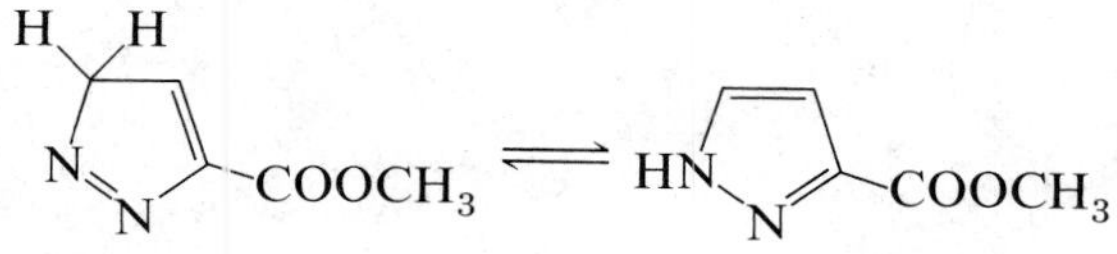

3-Substituted pyrazoles bearing a proton on nitrogen are in rapid equilibrium with the 5-isomers.

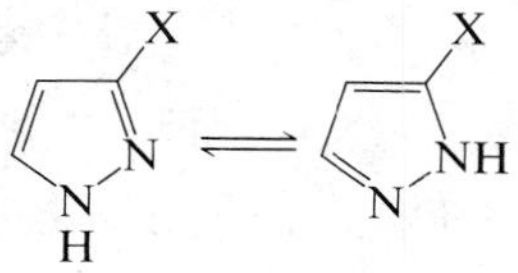

The most general synthesis of the 1,3-azoles is the dehydration of 1,4-dicarbonyl compounds, a form of Paal–Knorr cyclization.

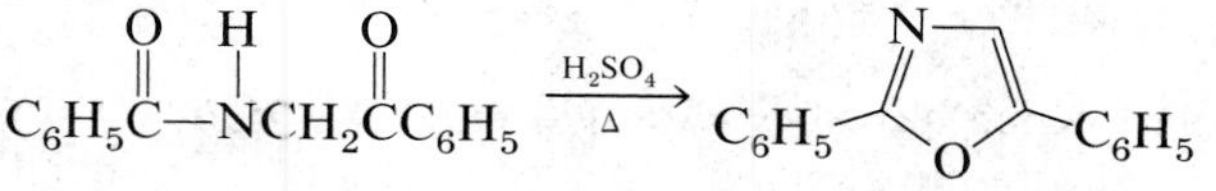

$$C_6H_5\overset{O}{\overset{\|}{C}}-\overset{H}{\underset{}{N}}CH_2\overset{O}{\overset{\|}{C}}C_6H_5 \xrightarrow[\Delta]{H_2SO_4} C_6H_5-\underset{O}{\overset{N}{\diagup}}-C_6H_5 \qquad (32\text{-}9)$$

2,5-diphenyloxazole

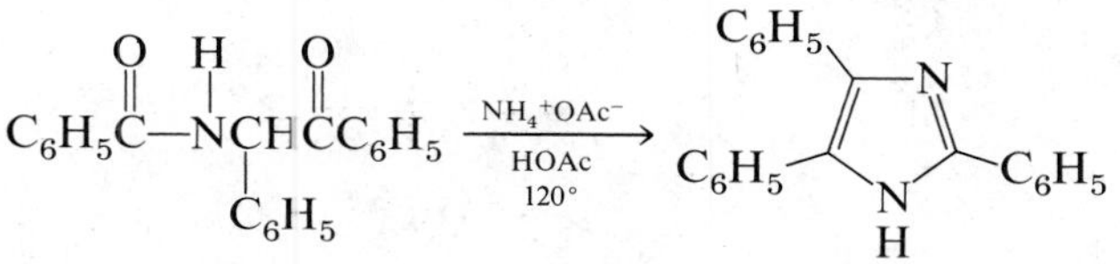

$$C_6H_5\overset{O}{\overset{\|}{C}}-\underset{\underset{C_6H_5}{|}}{\overset{H}{N}}\overset{O}{\overset{\|}{CH}}\overset{O}{\overset{\|}{C}}C_6H_5 \xrightarrow[\substack{HOAc \\ 120°}]{NH_4^+OAc^-} \underset{\overset{N}{H}}{\overset{C_6H_5}{\underset{C_6H_5}{}}}-C_6H_5 \qquad (32\text{-}10)$$

(93%)
2,4,5-triphenylimidazole

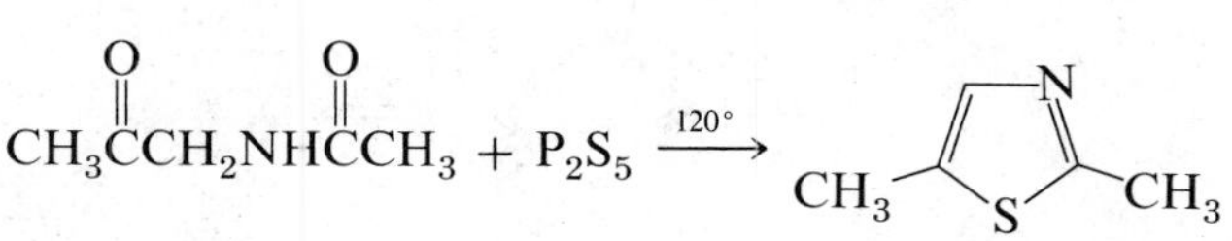

$$CH_3\overset{O}{\overset{\|}{C}}CH_2NH\overset{O}{\overset{\|}{C}}CH_3 + P_2S_5 \xrightarrow{120°} CH_3-\underset{S}{\overset{N}{\diagup}}CH_3$$

2,5-dimethylthiazole

EXERCISE: Write reasonable reaction mechanisms for reactions (32-8), (32-9), and (32-10).

C. Reactions

The azoles are markedly less reactive than furan, pyrrole, and thiophene. The reduced reactivity is due to the electronegative azole nitrogen. For the 1,2-azoles

the reactivity order is

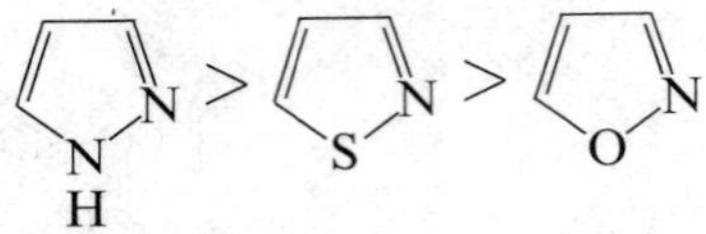

Electrophilic substitution takes place exclusively at C-4.

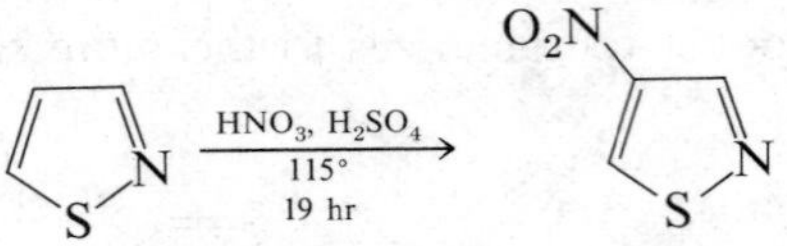

(97%)
4-nitroisothiazole

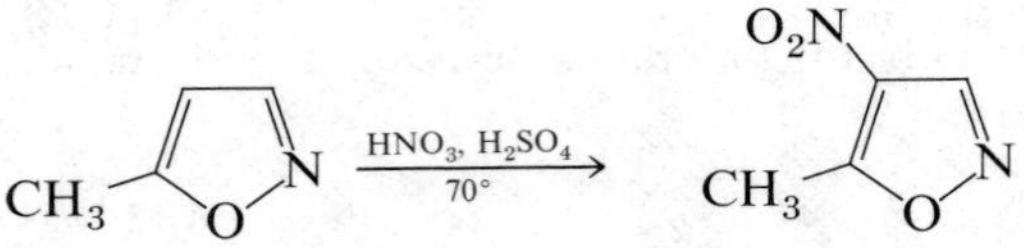

5-methylisoxazole   4-nitro-5-methylisoxazole

For the 1,3-azoles the reactivity order seems to be

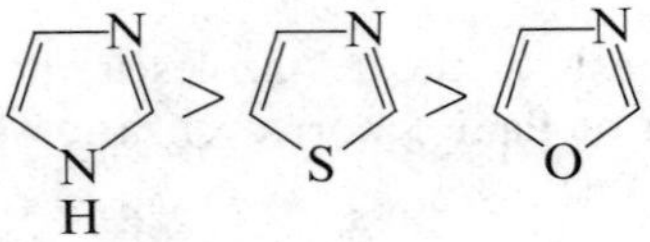

For imidazoles, which have been studied most extensively, substitution occurs preferentially at C-4 (equivalent to C-5 by tautomerism) rather than at C-2.

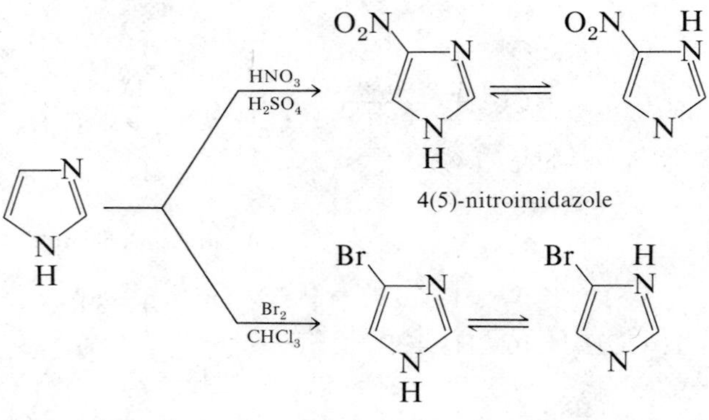

4(5)-nitroimidazole

4(5)-bromoimidazole

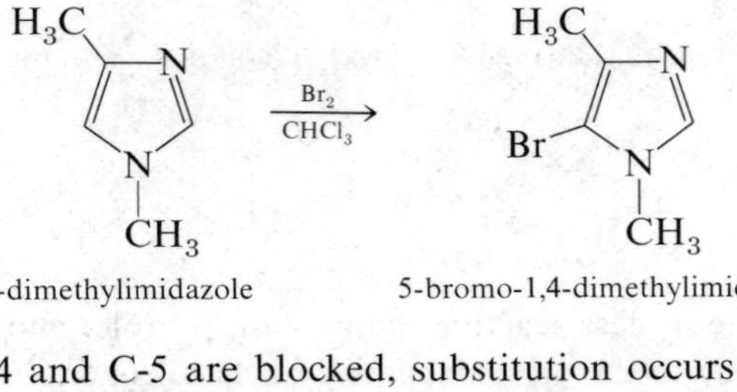

1,4-dimethylimidazole   5-bromo-1,4-dimethylimidazole

When both C-4 and C-5 are blocked, substitution occurs at C-2.

H₃C

4-methyl-5-nitroimidazole

$\xrightarrow[\text{CHCl}_3]{\text{Br}_2}$

2-bromo-4-methyl-5-nitroimidazole

EXERCISE: We learned in Section 32.3.C that the positions in thiophene α to the sulfur are much more reactive than the β- positions toward electrophilic substitution. Explain why the 5-position of thiazole (α to the sulfur) is *less* reactive than the 4-position (β to the sulfur).

32.6
Pyridine

A. *Structure and Physical Properties*

Pyridine is an analog of benzene in which one of the CH units is replaced by nitrogen (Figure 32.4). The nitrogen lone pair is located in an sp^2-hybrid orbital that is perpendicular to the π-system of the ring. The effect on the basicity of the nitrogen (pK_a 5.2) has been discussed in Section 32.5.A. Various values have been deduced for the empirical resonance energy of pyridine, but it would appear to be roughly comparable to benzene. The resonance stabilization is shown by the two equivalent Kekulé structures and the three zwitterionic forms with negative charge on nitrogen.

The surplus negative charge on nitrogen is manifest in the dipole moment of pyridine, which is substantially greater than that of piperidine, the nonaromatic

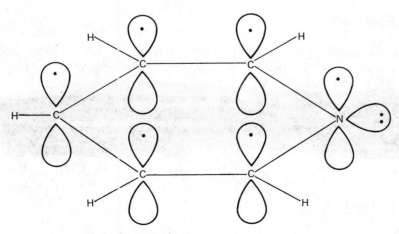

FIGURE 32.4. *Orbital structure of pyridine.*

analog. That is, the π-moment is in the same direction as the σ-moment and the net moment is additive.

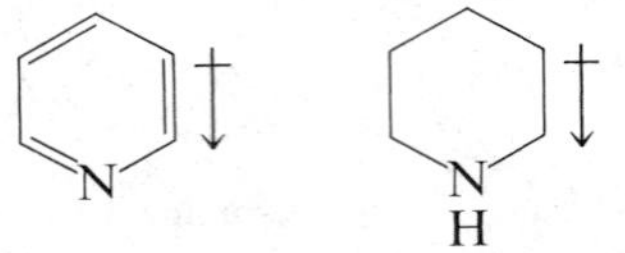

2.26 D 1.17 D

As the charged resonance structures and the dipole moment show, the ring in pyridine is relatively electron-deficient. This deficiency is reflected in many of the reactions of pyridine (Section 32.6.C).

Alkylpyridines have the trivial names of picolines (C_6H_7N), lutidines (C_7H_9N), or collidines ($C_8H_{11}N$).

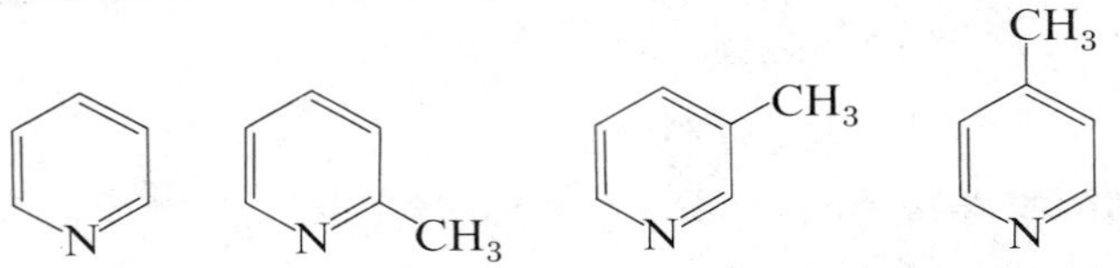

pyridine b.p. 115° | α-picoline b.p. 128° | β-picoline b.p. 144° | γ-picoline b.p. 144°

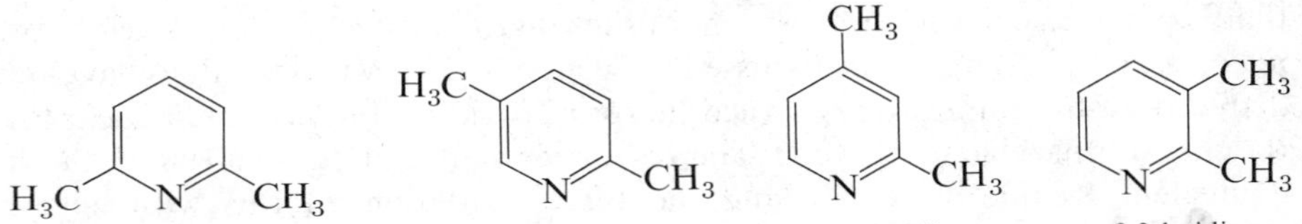

2,6-lutidine b.p. 144° | 2,5-lutidine b.p. 157° | 2,4-lutidine b.p. 157° | 2,3-lutidine b.p. 163°

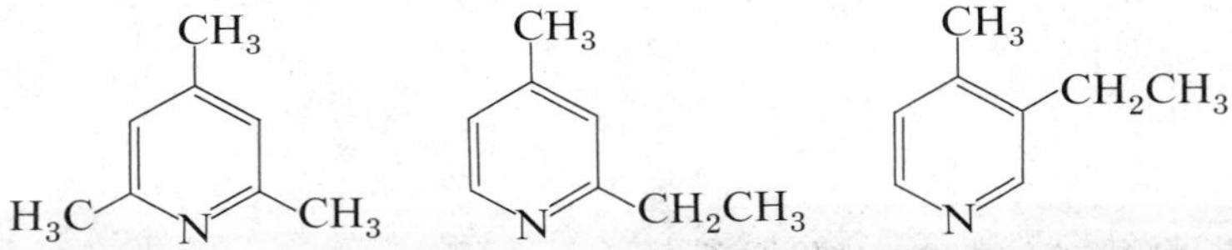

sym-collidine b.p. 170° | α-collidine b.p. 179° | β-collidine b.p. 196°

B. *Synthesis*

Pyridine itself and most of the simpler alkylpyridines are available from coal tar distillates. Several syntheses are available for deriving substituted pyridines from other compounds.

The most general technique for constructing the ring is the **Hantzsch pyridine synthesis.** Although numerous variations are known, the simplest consists of the condensation of a β-keto ester with an aldehyde and ammonia. The product is a 1,4-dihydropyridine, which is subsequently aromatized by oxidation.

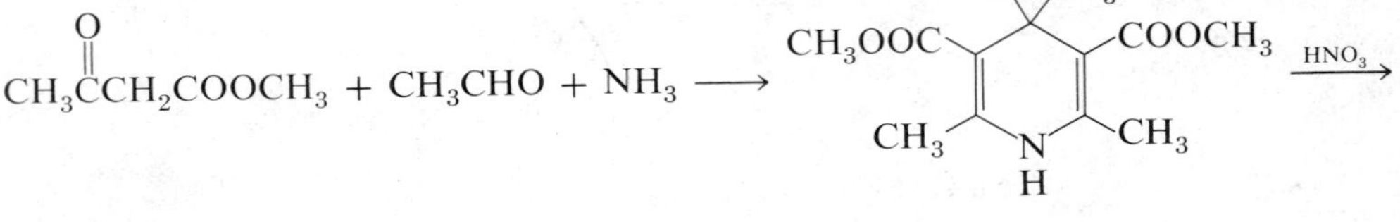
$CH_3CCH_2COOCH_3 + CH_3CHO + NH_3 \longrightarrow$
H CH3
CH3OOC
COOCH3
HNO3
CH3
N
H
CH3

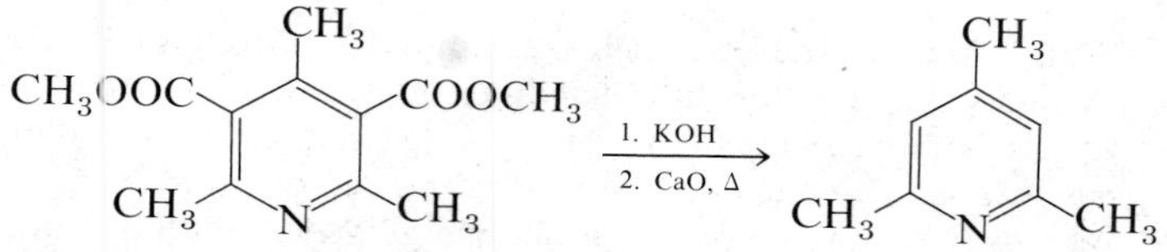

$$\begin{array}{c}\text{CH}_3\text{OOC, CH}_3, \text{COOCH}_3\text{ pyridine with CH}_3, \text{N}, \text{CH}_3\end{array} \xrightarrow[\text{2. CaO, }\Delta]{\text{1. KOH}} \text{CH}_3\text{-pyridine-CH}_3$$

A reasonable mechanism for the Hantzsch reaction is outlined as follows. The first step is probably a Knoevenagel condensation of the aldehyde (Section 27.7.C) with the β-keto ester.

(1) $\text{CH}_3\text{CHO} + \text{CH}_3\overset{\text{O}}{\overset{\|}{\text{C}}}\text{CH}_2\text{COOCH}_3 \longrightarrow$

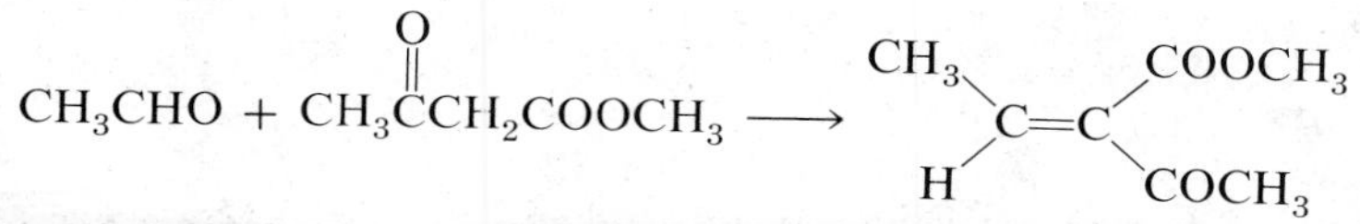

A part of the β-keto ester also condenses with ammonia to form an enamine.

(2) $\text{CH}_3\overset{\text{O}}{\overset{\|}{\text{C}}}\text{CH}_2\text{COOCH}_3 + \text{NH}_3 \longrightarrow \text{CH}_3\overset{\text{NH}_2}{\overset{|}{\text{C}}}=\text{CHCOOCH}_3 + \text{H}_2\text{O}$

The unsaturated keto ester produced in step (1) then undergoes a condensation with the enamine produced in step (2).

(3)

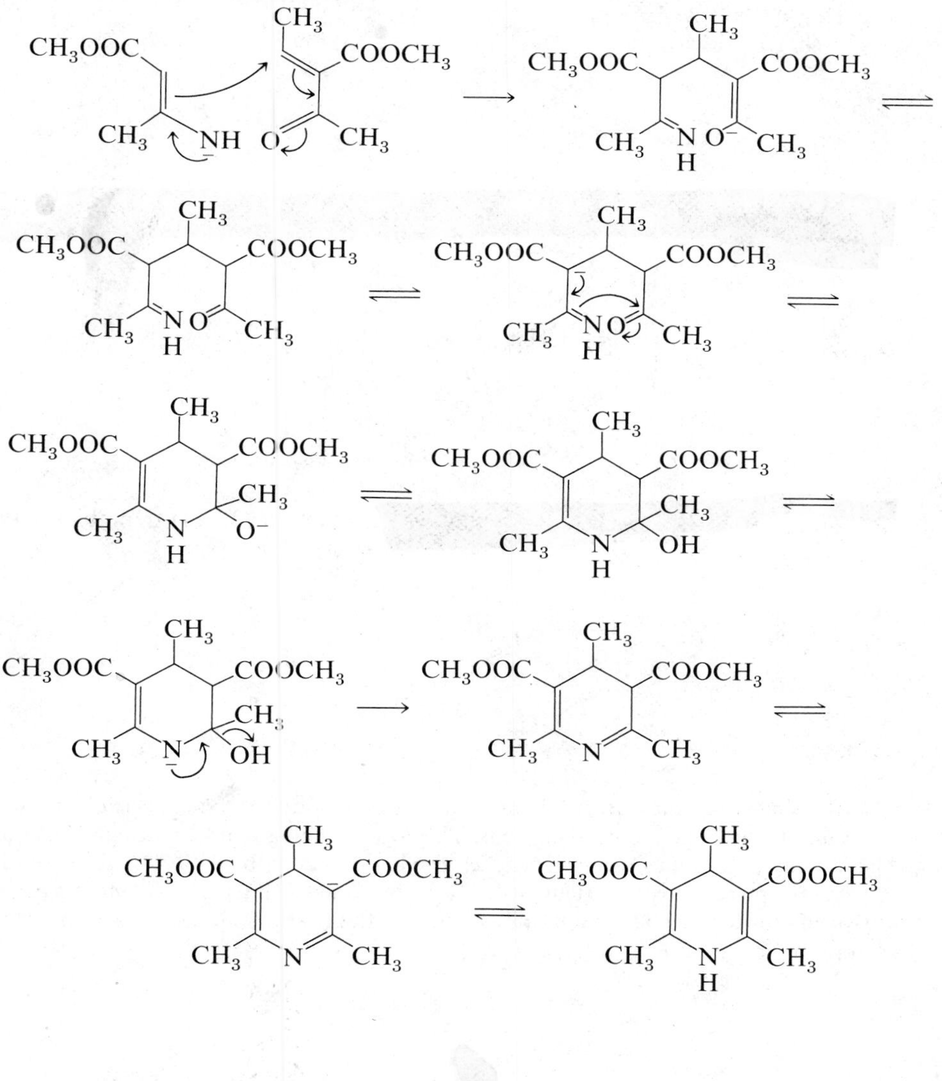

EXERCISE: The Hantzsch pyridine synthesis is usually run with ammonia in alcohol solution, mildly basic conditions. The carbanion intermediates in the reaction are enolate ions with the negative charge stabilized on oxygen and nitrogen. The reaction sequence above involves a number of protonations and deprotonations. What is the base used for deprotonations? Where do the protons come from in the protonations? Trace the course of each protonation and deprotonation involved, and write resonance structures to show the stabilization of negative charge in each carbanion intermediate.

C. *Reactions*

The nitrogen lone pair has basic and nucleophilic properties, although both are diminished by the hybridization effect. Pyridines form salts with acids and are widely used as catalysts and "acid scavengers" in reactions where strong acids are produced.

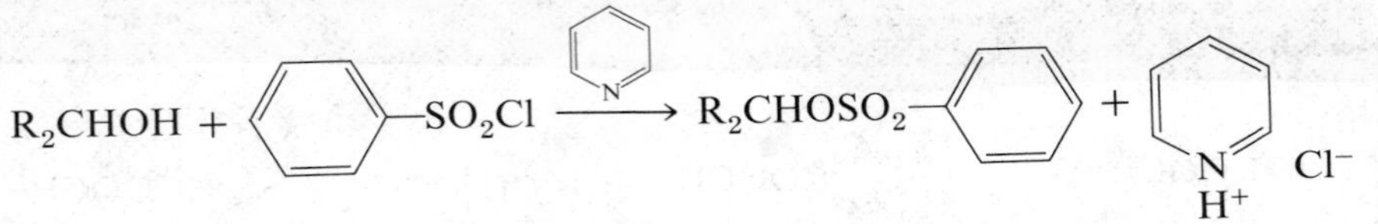

pyridinium chloride

The nitrogen may be alkylated by primary alkyl halides, leading to N-alkylpyridinium salts.

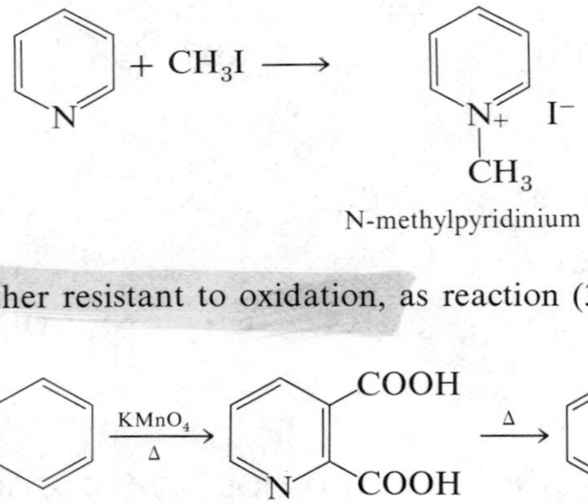

N-methylpyridinium iodide

Pyridines are rather resistant to oxidation, as reaction (32-11) demonstrates.

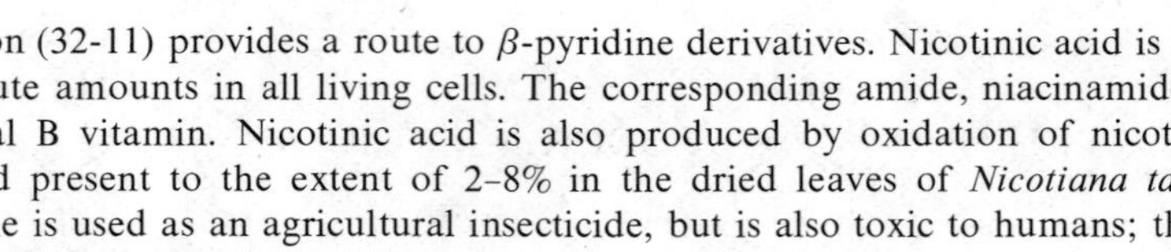

(32-11)

quinoline — (65–70%) quinolinic acid — nicotinic acid (niacin)

Reaction (32-11) provides a route to β-pyridine derivatives. Nicotinic acid is present in minute amounts in all living cells. The corresponding amide, niacinamide, is an essential B vitamin. Nicotinic acid is also produced by oxidation of nicotine, an alkaloid present to the extent of 2–8% in the dried leaves of *Nicotiana tabacum*. Nicotine is used as an agricultural insecticide, but is also toxic to humans; the fatal dose by ingestion is about 40 mg.

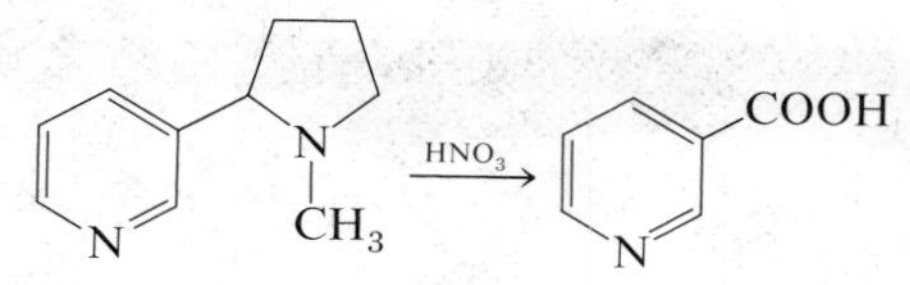

nicotine

Because of its resistance to oxidation, pyridine can even be used as a solvent for chromium trioxide oxidations (Sarrett procedure; see also Section 13.5.A). However, under the proper conditions the nitrogen can be oxidized to the N-oxide, as are other tertiary amines (Section 24.7.C).

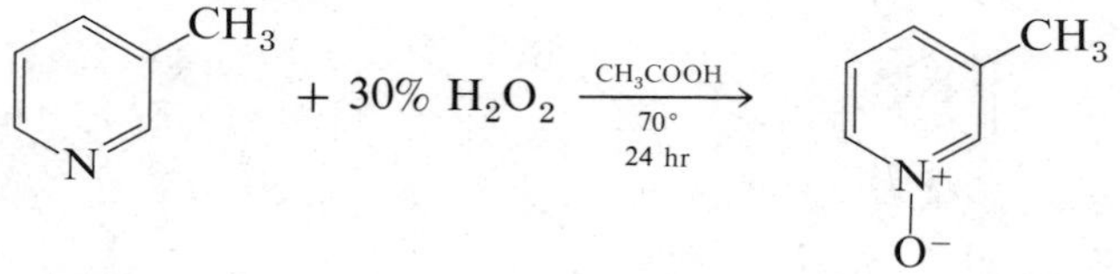

(75%)
3-methylpyridine N-oxide

As we shall see later, pyridine N-oxides are important synthetic intermediates. Pyridine is resistant to electrophilic aromatic substitution conditions, not only because of the electron-deficient ring but also because under the acidic conditions of such reactions the nitrogen is protonated or complexed with a Lewis acid. In general, pyridine is *less* reactive in such reactions than trimethylaniliniun ion.

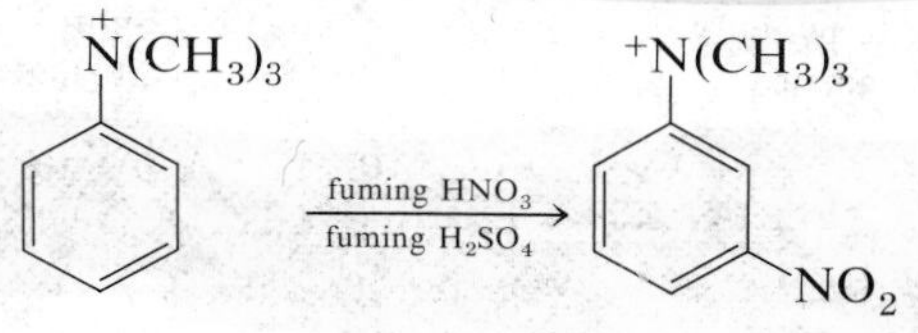

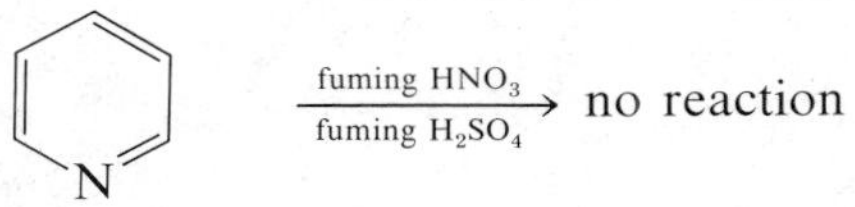

Substitution is achieved only under the most drastic conditions, for example

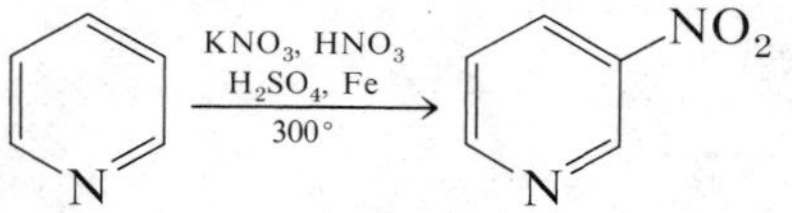

(22%)
3-nitropyridine

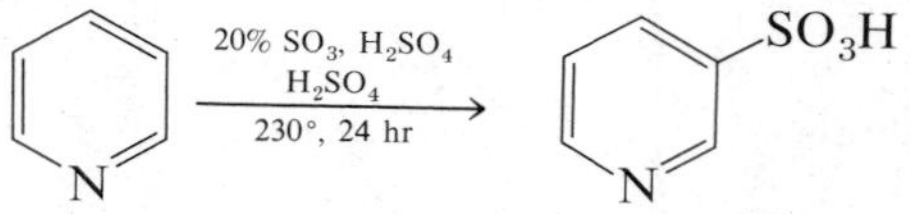

(71%)
pyridine-3-sulfonic acid

Alkyl and amino groups activate the ring toward electrophilic substitution. In the alkylpyridines the ring nitrogen directing influence predominates (C-3 or C-5 attack) regardless of the position of alkylation.

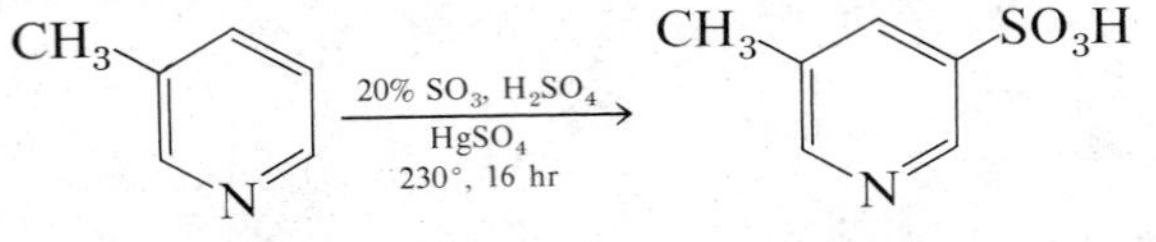

β-picoline → 5-methylpyridine-3-sulfonic acid

Conditions: 20% SO₃, H₂SO₄, HgSO₄, 230°, 16 hr

Amino groups, either free or acylated, govern the position of further substitution (*ortho* or *para* to the amino).

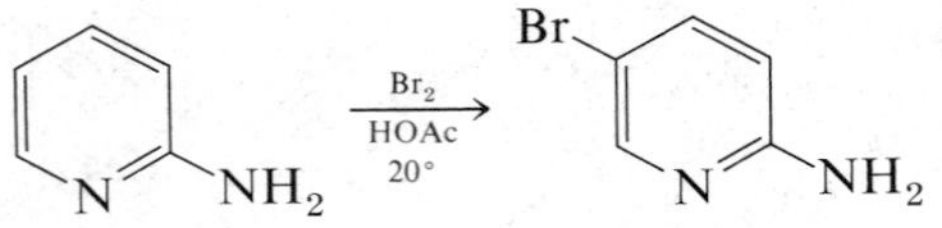

2-aminopyridine → 5-bromo-2-amino-pyridine (90%)

Conditions: Br₂, HOAc, 20°

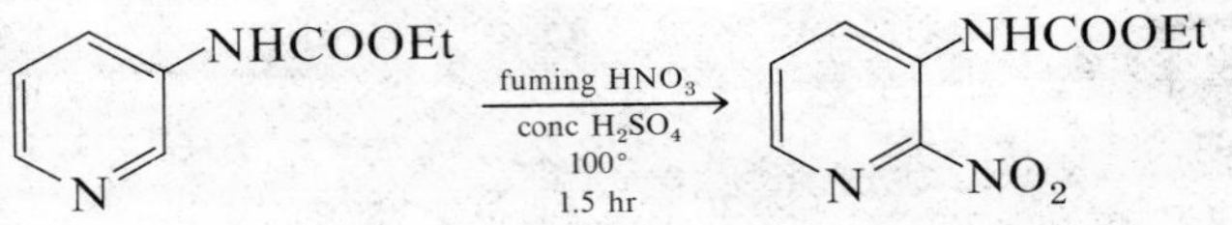

ethyl N-(3-pyridyl)-carbamate → ethyl N-(2-nitro-3-pyridyl)-carbamate (61%)

Conditions: fuming HNO₃, conc H₂SO₄, 100°, 1.5 hr

The predominant 3-substitution in pyridine is explainable in terms of the resonance structures of the intermediate ions, and the corresponding transition states, produced by electrophilic attack at the three positions.

Attack at C-2

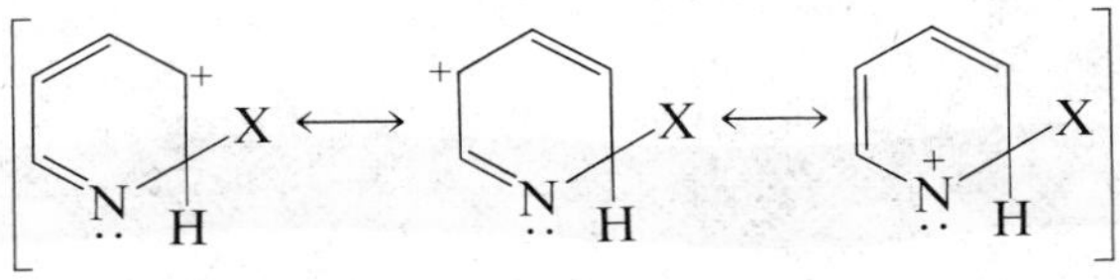

Attack at C-4

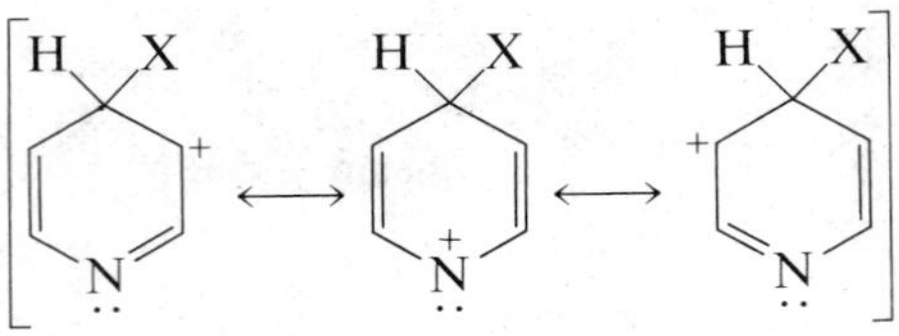

Attack at C-3

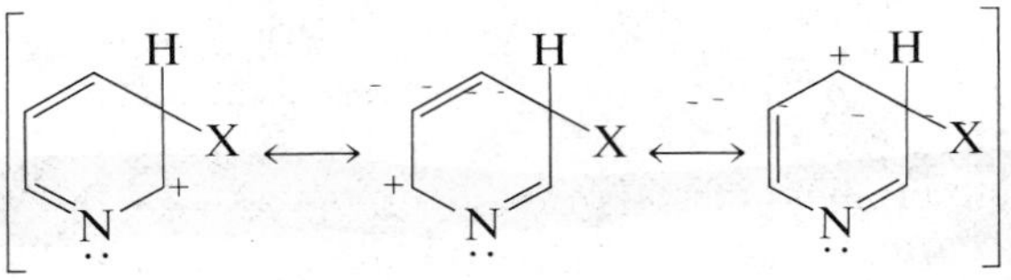

Compared with the ion produced with benzene, all three ions are destabilized by the inductive effect of the nitrogen, especially if it is protonated or coordinated with a Lewis acid. However, the situation is much worse when attack is at C-2 or C-4 than at C-3. In the two former cases one of the structures of the intermediate ion has the positive charge on an electron-deficient nitrogen. Thus the situation in pyridine is similar to that in nitrobenzene. Electrophilic attack is retarded at all positions, but especially at C-2 and C-4.

Pyridine N-oxides undergo electrophilic substitution somewhat more readily. Reaction generally occurs at C-4.

$$\underset{\underset{O^-}{N+}}{\bigcirc} \xrightarrow[\substack{H_2SO_4 \\ 90°, 14 \text{ hr}}]{\text{fuming NHO}_3} \underset{\underset{O^-}{N+}}{\overset{NO_2}{\bigcirc}}$$

(90%)

The N-oxide can often be used as an "activated" form of the pyridine. Treatment of the substituted N-oxide with PCl_3 results in deoxygenation.

$$\underset{\underset{O^-}{N+}}{\overset{NO_2}{\bigcirc}} + PCl_3 \xrightarrow{\Delta} \underset{N}{\overset{NO_2}{\bigcirc}} + POCl_3$$

The electron-deficient nature of the pyridine ring is also manifest in the ease with which pyridines undergo **nucleophilic substitution.** A particularly useful and unusual example is the synthesis of aminopyridines by the reaction of a pyridine with an alkali metal amide **(Chichibabin reaction).**

$$\underset{N}{\bigcirc} + NaNH_2 \xrightarrow[\text{reflux}]{C_6H_5N(CH_3)_2} \underset{N}{\bigcirc}\underset{NH^- Na^+}{} \xrightarrow{H_2O} \underset{N}{\bigcirc}\underset{NH_2}{}$$

(70–80%)
2-aminopyridine

Attack occurs at C-2 or C-6 unless both positions are occupied, in which case substitution can occur at C-4. The reaction is initiated by attack by the nucleophile at C-2 or C-4. Attack occurs at these positions because the negative charge can be delocalized onto the ring nitrogen.

$$\underset{N}{\bigcirc} + NH_2^- \longrightarrow \left[\underset{\underset{NH_2}{N}}{\overset{H}{\bigcirc}} \longleftrightarrow \underset{\underset{NH_2}{N}}{\overset{H}{\bigcirc}} \longleftrightarrow \underset{\underset{NH_2}{N}}{\overset{H}{\bigcirc}} \right]$$

The second step is elimination of hydride ion, which reacts with the amino-pyridine to give H_2. The driving force for the elimination of hydride ion is, of course, the formation of the aromatic cycle.

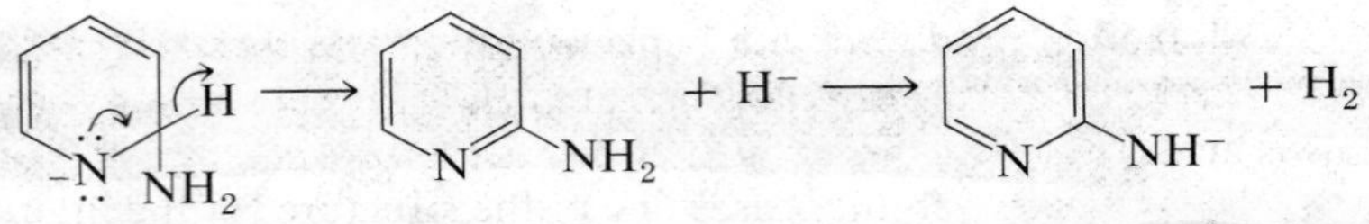

Chichibabin-like reactions are also observed with organolithium compounds.

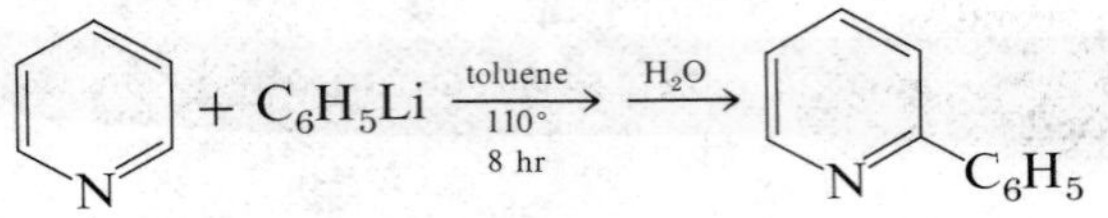

(50%)
2-phenylpyridine

Diazotization of the 2- and 4-aminopyridines yields the 2- and 4-hydroxy-pyridines, which exist completely in the keto form (for example, α-pyridone, γ-pyridone).

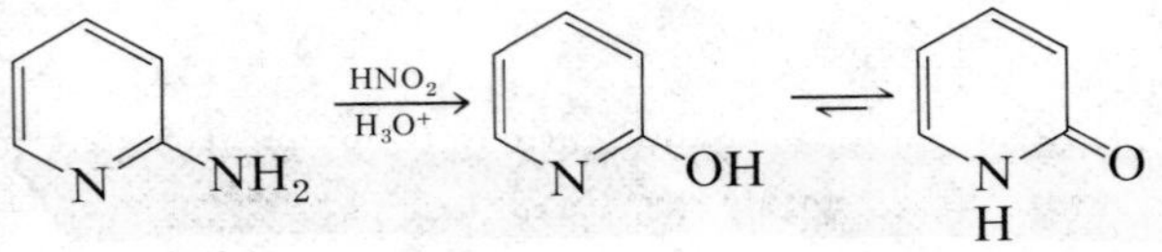

α-pyridone

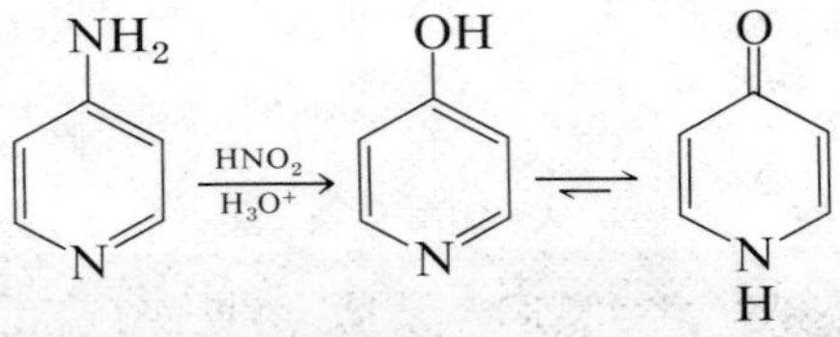

γ-pyridone

N-Alkylated α-pyridones may be produced by the ferricyanide oxidation of N-alkylpyridinium salts.

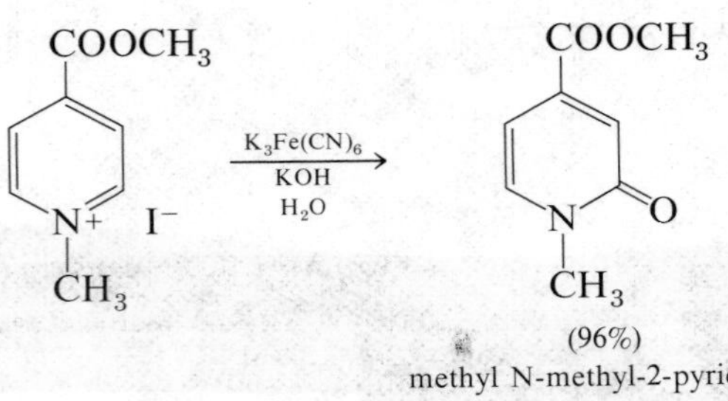

(96%)
methyl N-methyl-2-pyridone-
4-carboxylate

The oxidation occurs by initial addition of hydroxide to give a dihydropyridine, which is then oxidized to the pyridone.

Even though the simple pyridones exist in the tautomeric form with hydrogen attached to nitrogen (amide form), they still have extensive aromatic character, as shown by the important dipolar resonance structure.

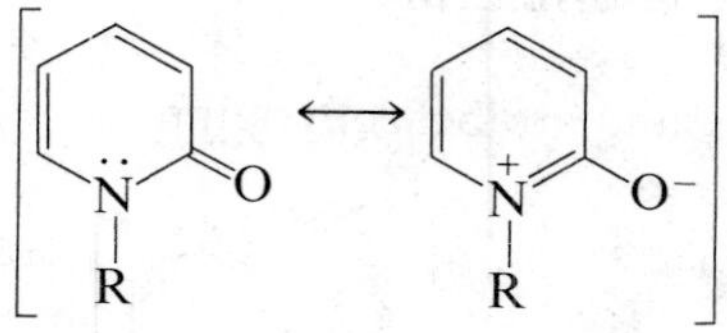

A final feature of the pyridine ring is of interest. The methyl groups in α- and γ-picoline are comparable in acidity to methyl ketones and readily undergo base-catalyzed reactions.

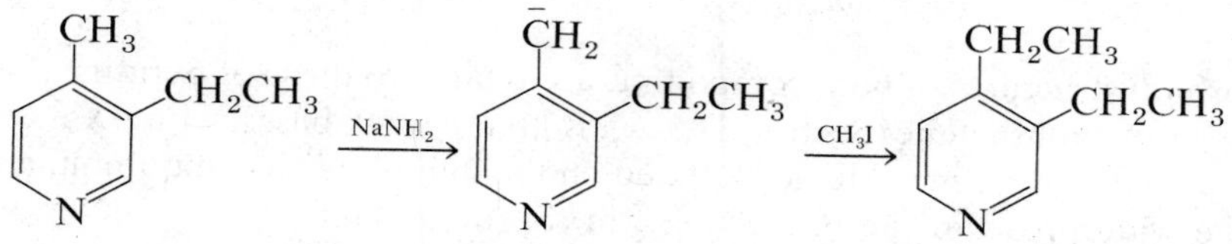

(80%)
3,4-diethylpyridine

The enhanced acidity at these positions is again attributed to delocalization of negative charge in the intermediate anion into the ring and especially onto the nitrogen.

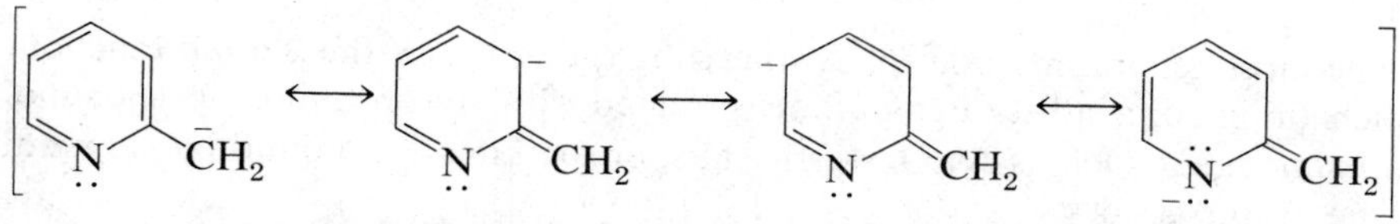

This side-chain acidity is enhanced in the N-alkylpyridinium compounds.

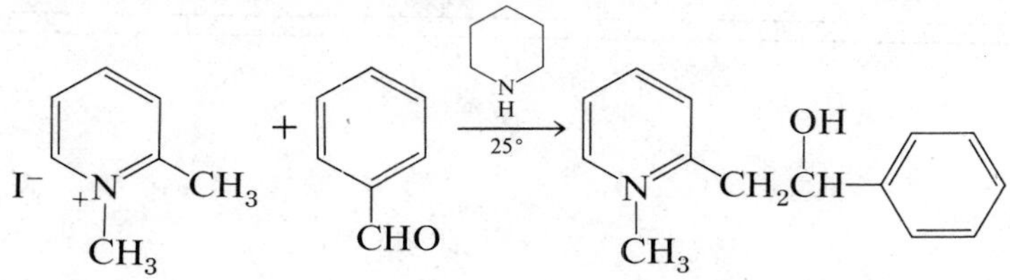

(60%)

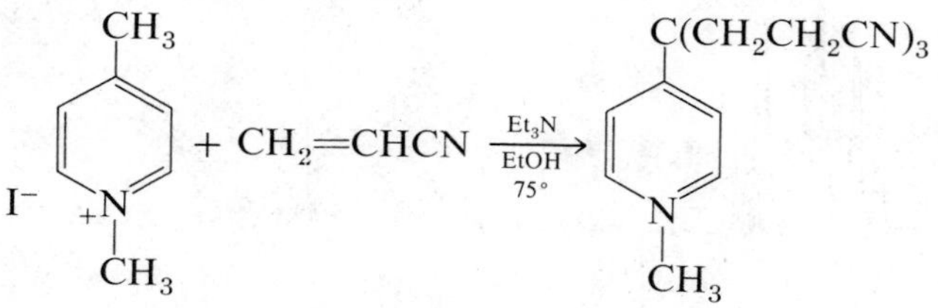

(92%)

EXERCISE: Explain why 3-methylpyridine is less acidic than 4-methylpyridine. Draw resonance structures showing that both positive and negative charges are delocalized to the 2- and 4-positions of pyridine N-oxide.

32.7
Quinoline and Isoquinoline

A. *Structure and Nomenclature*

Quinoline and isoquinoline are benzopyridines.

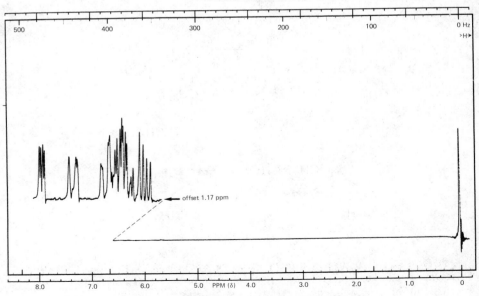

quinoline
b.p. 238°
pK_a 4.8

isoquinoline
b.p. 243°
pK_a 5.4

The orbital structures of both compounds are related to those of pyridine (Section 32.6.C) and naphthalene (Section 31.3.A). Both are weak bases, with pK_as comparable to that of pyridine. Alkaloids based on the quinoline and isoquinoline skeleton are widespread in the plant kingdom (Section 34.6.C).

The nmr spectrum of quinoline is shown in Figure 32.5. Note the downfield position of the resonances, a consequence of the electron-attracting effect of the pyridine nitrogen.

B. *Synthesis*

The most general method for synthesizing quinolines is the **Skraup reaction,** in which aniline or a substituted aniline is treated with glycerol, sulfuric acid, and an oxidizing agent such as As_2O_5, ferric salts, or the nitro compound corresponding to the amine used.

offset 1.17 ppm

FIGURE 32.5. *Nmr spectrum of quinoline.*

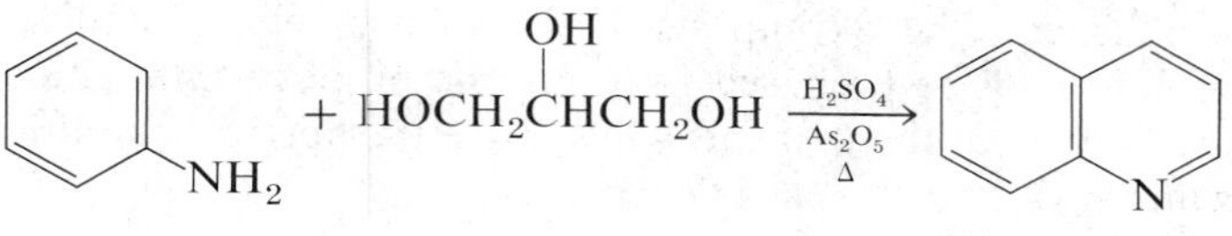

$$(84\text{--}91\%)$$

The mechanism of the Skraup reaction probably involves initial dehydration of the glycerol to give acrolein, which undergoes a 1,4-addition by the aniline. The resulting β-anilinopropionaldehyde is then cyclized to a dihydroquinoline, which is finally oxidized to give the product.

(1)

(2)

(3)

(4)

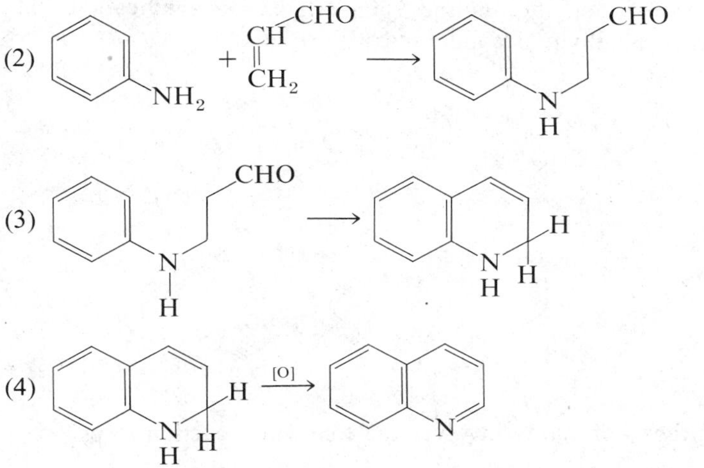

Identical results are obtained if an α,β-unsaturated ketone or aldehyde is substituted for the glycerol.

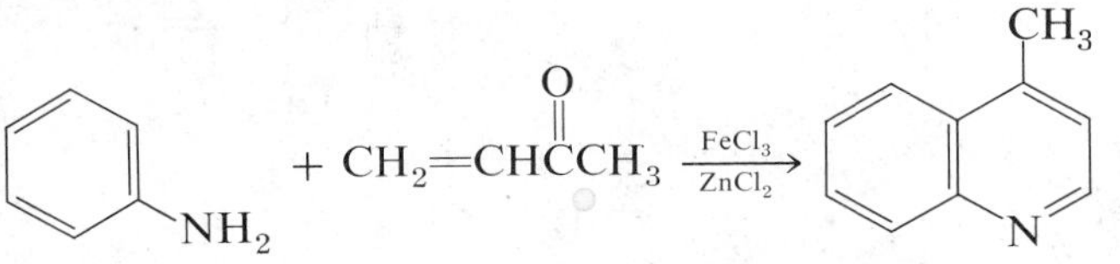

(73%)
lepidine

If a saturated aldehyde is used, an initial aldol condensation occurs to give an α,β-unsaturated aldehyde that engages in the normal condensation. **(Döbner–Miller reaction).**

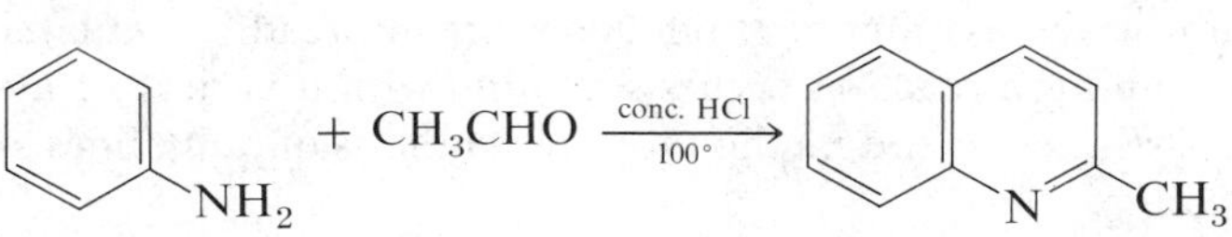

(32%)
quinaldine

In some of these cases an oxidizing agent is not included; in these cases unsaturated reaction intermediates probably serve as oxidizing agents, but this point has not been established. The Skraup synthesis is extremely versatile; almost any desired quinoline may be prepared by using the proper combination of aniline and aldehyde, so long as the reagents will survive the hot acid conditions. A more complex example is

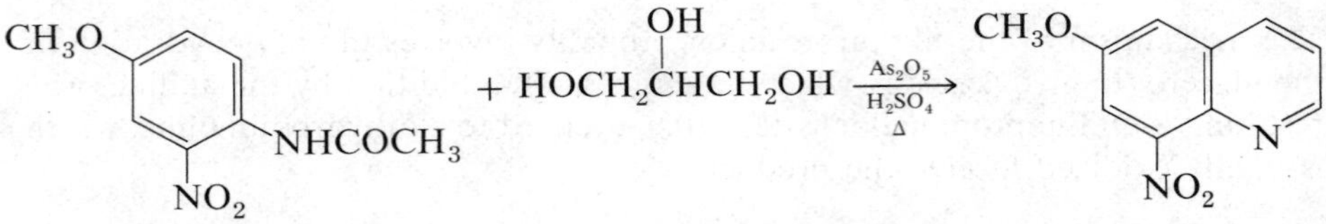

(67%)
6-methoxy-8-nitroquinoline

A second general preparation of quinolines is the **Friedländer synthesis.** In this method an *o*-aminobenzaldehyde is condensed with a ketone.

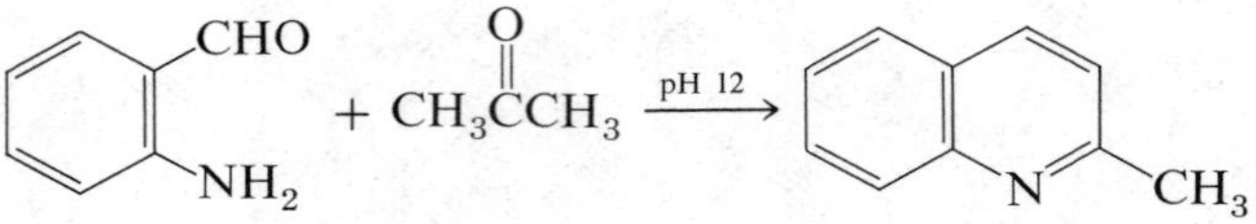

(85%)

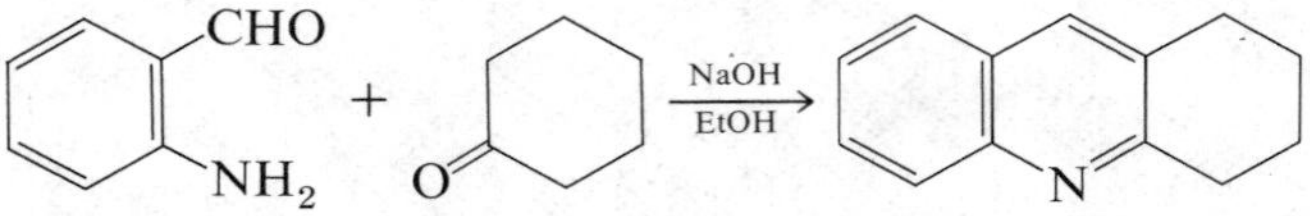

The Friedländer synthesis probably involves the following reaction steps.

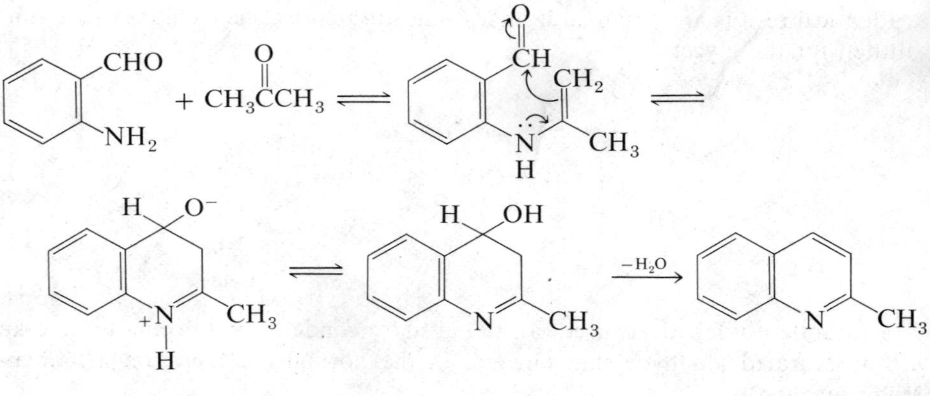

Although substituted *o*-aminobenzaldehydes are not readily available, the parent compound is, and the reaction occurs smoothly with a variety of aldehydes and ketones. It constitutes a good method for the synthesis of quinolines substituted in the pyridine ring.

Isoquinolines are most easily prepared by a reaction known as the **Bischler–Napieralski synthesis.** An acyl derivative of a *β*-phenylethylamine is treated with a dehydrating agent to give a dihydroisoquinoline, which is dehydrogenated to the isoquinoline.

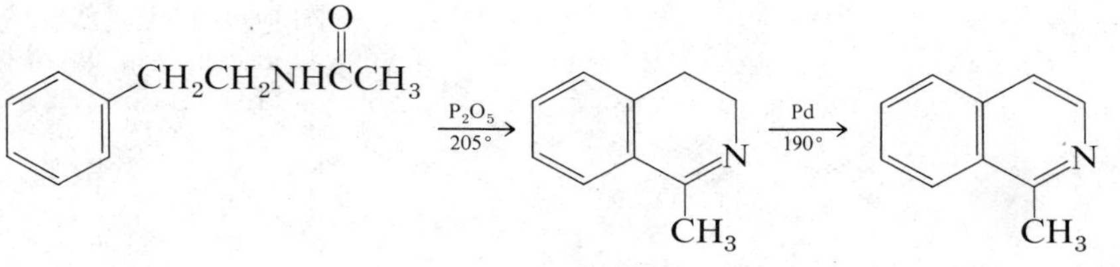

(83%)   (83%)
1-methylisoquinoline

EXERCISE: What quinoline is obtained from a Skraup synthesis using *p*-toluidine and phenyl vinyl ketone? Write a complete reaction mechanism showing all intermediates up to the dihydroquinoline.

C. Reactions

Quinoline and isoquinoline are considerably more reactive than pyridine in electrophilic substitution reactions. For reactions carried out in strongly acidic solution, reaction occurs on the protonated form, and substitution occurs in the benzene ring at C-5 and C-8.

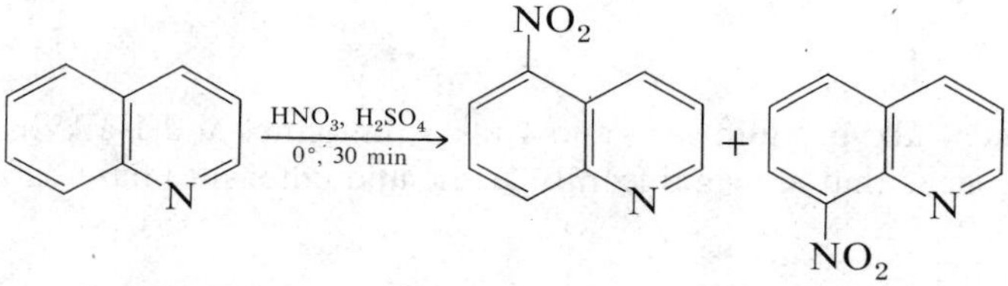

(52%)   (48%)
5-nitroquinoline   8-nitroquinoline

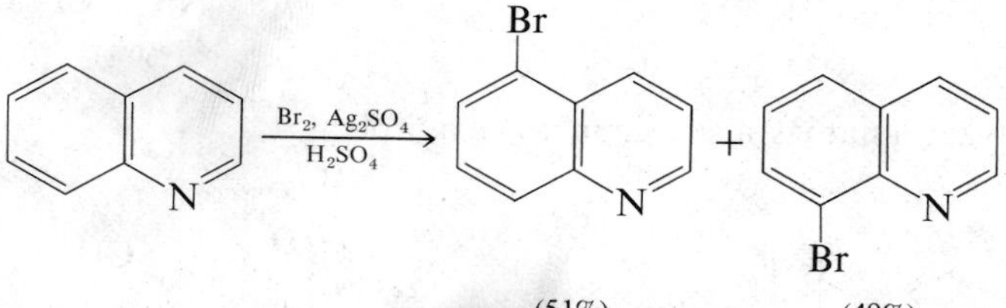

(51%)   (49%)
5-bromoquinoline   8-bromoquinoline

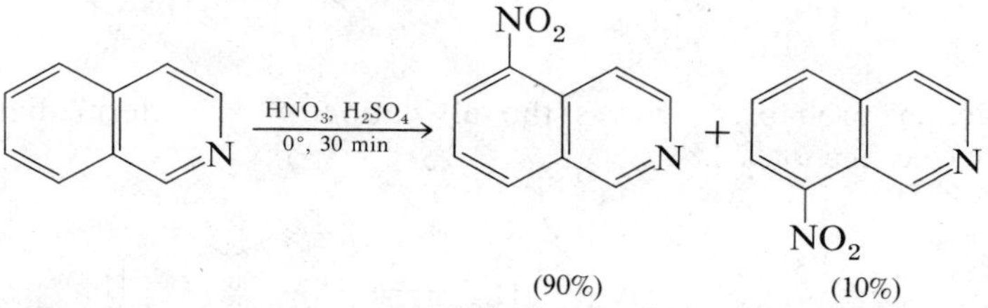

(90%)   (10%)
5-nitroisoquinoline   8-nitroisoquinoline

As with pyridine N-oxide, quinoline N-oxide undergoes nitration considerably more easily; reaction occurs at C-4.

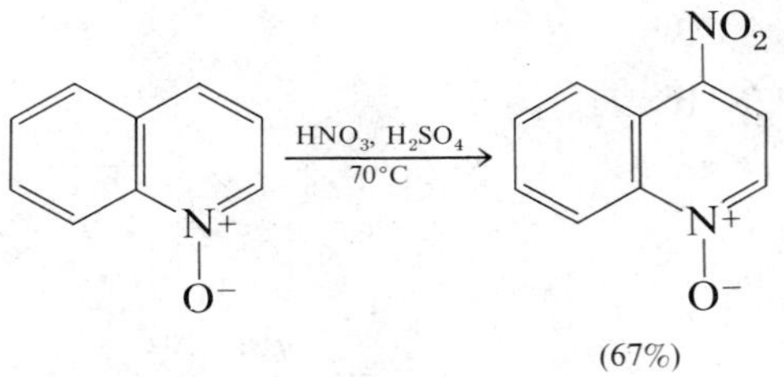

(67%)

Both quinoline and isoquinoline readily undergo nucleophilic substitution reactions of the Chichibabin type.

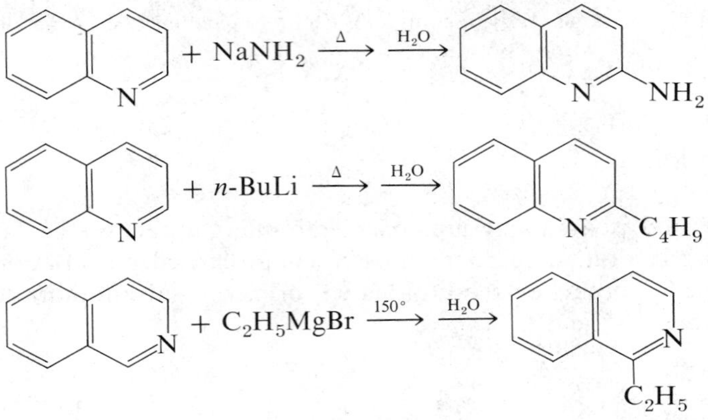

Like 2- and 4-alkylpyridines, 2- and 4-alkylquinolines and 1-alkylisoquinolines have α-hydrogens that are significantly acidic and enter into base-catalyzed reactions.

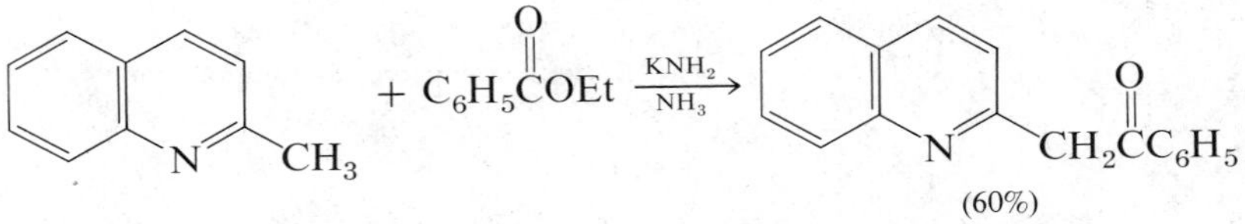

(60%)

Acid-catalyzed analogs also occur, for example,

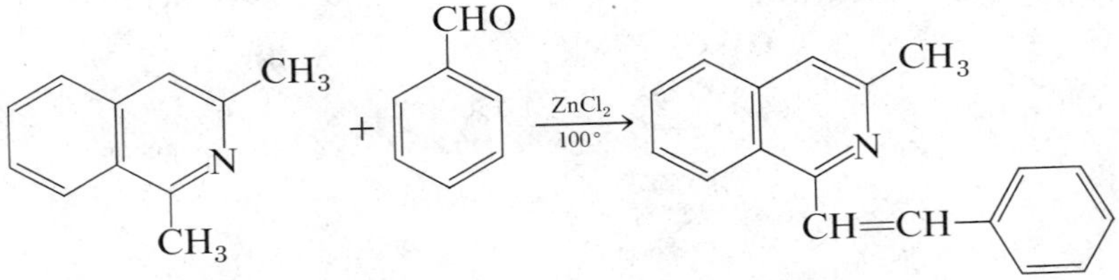

This reaction probably involves the alkylation of an intermediate enamine tautomer (see Section 24.8).

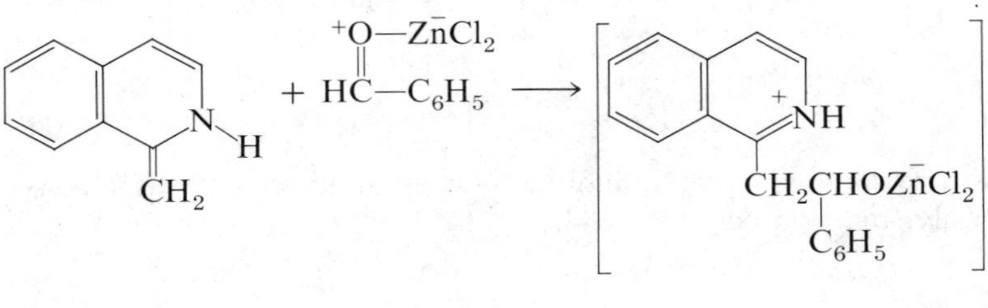

EXERCISE: Give the expected product from quinoline in each of the following reactions: (a) C_6H_5Li; (b) CH_3I, CH_3CN; (c) 2 Br_2, Ag^+, H_2SO_4; (d) 30% H_2O_2, CH_3COOH, 70°; and (e) $KMnO_4$, heat.

32.8 Diazines

A. *Structure and Occurrence*

In this section, we shall take a brief look at another class of heterocycles, the diazines. The three types of diazabenzenes are

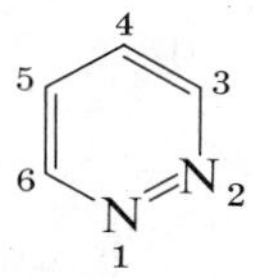

pyridazine
b.p. 208°
pK_a 2.3

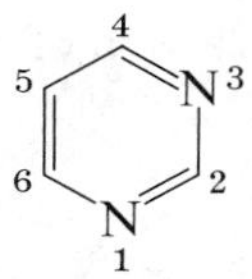

pyrimidine
b.p. 134°
pK_a 1.3

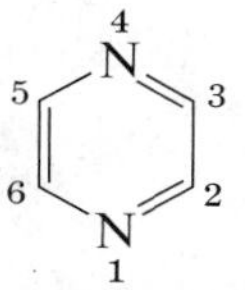

pyrazine
b.p. 118°
pK_a 0.7

In addition to these three diazines, the bicyclic tetraaza compound, purine, as an important heterocyclic system.

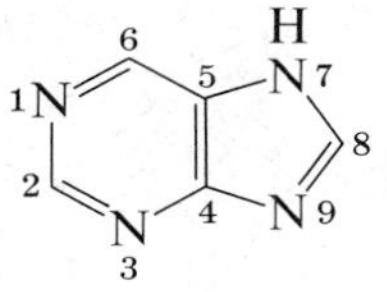

purine
m.p. 217°
pK_a 2.3

These ring systems, particularly that of pyrimidine, occur commonly in natural products. The pyrimidines cytosine, thymine, and uracil are especially important because they are components of nucleic acids, as are the purine derivatives adenine and guanine (Section 33.5).

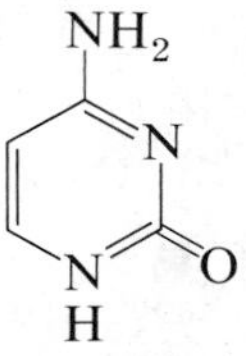

cytosine

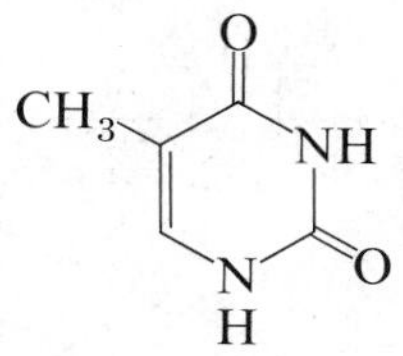

thymine

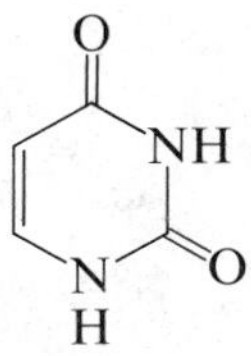

uracil

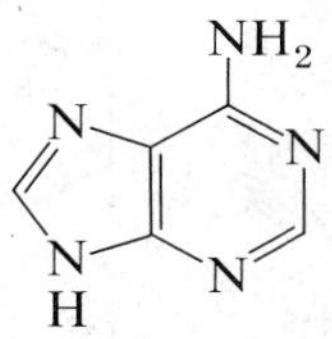

adenine

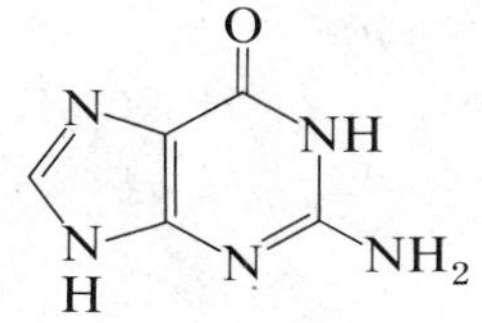

guanine

The purine nucleus also occurs in such compounds as caffeine (coffee and tea) and theobromine (cacao beans).

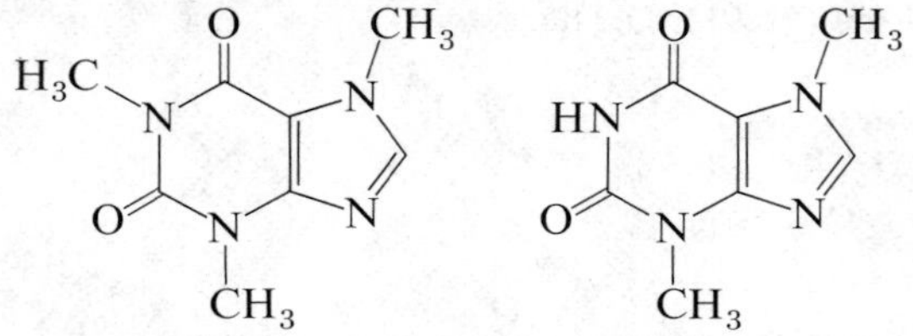

caffeine   theobromine

B. *Synthesis*

Pyridazines are prepared by the reaction of hydrazine with 1,4-dicarbonyl compounds.

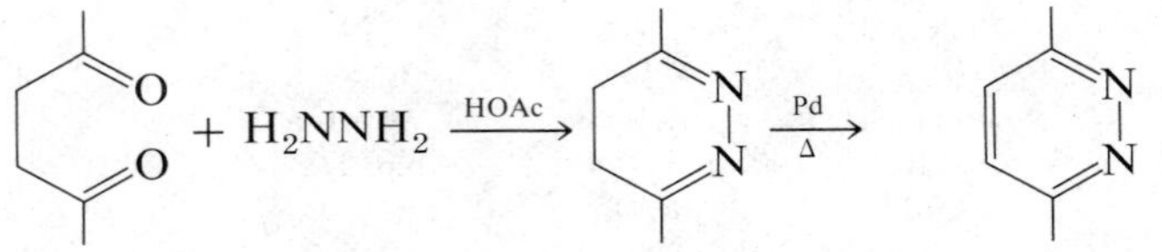

2,5-dimethylpyridazine

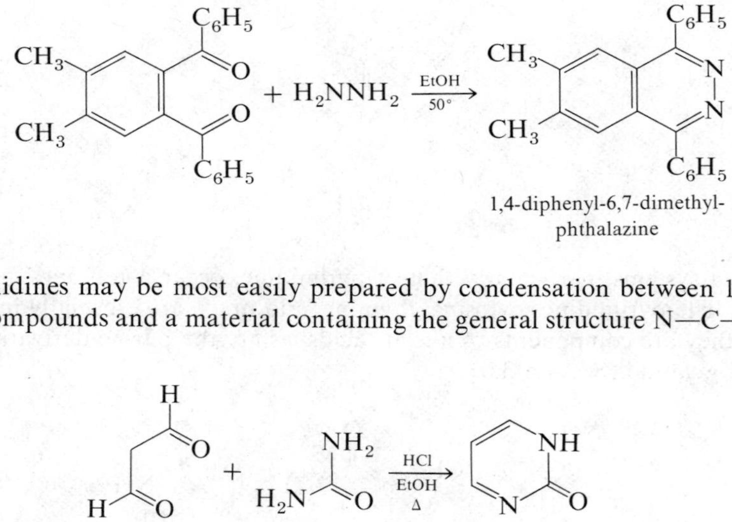

1,4-diphenyl-6,7-dimethyl-
phthalazine

Pyrimidines may be most easily prepared by condensation between 1,3-dicarbonyl compounds and a material containing the general structure N—C—N, such as urea.

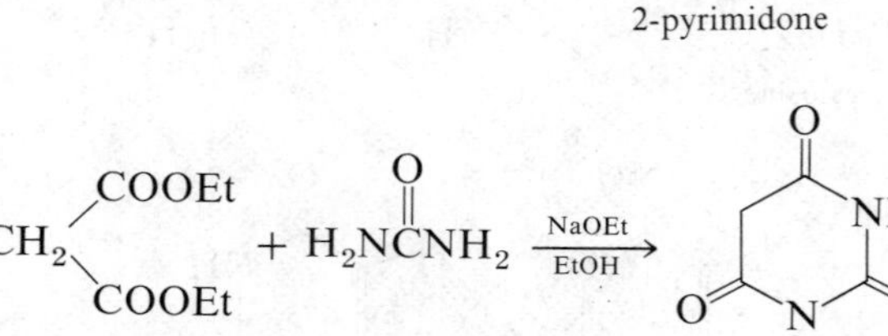

(73%)
2-pyrimidone

(72–78%)
barbituric acid

Note that the C-2 oxygenated pyrimidine, like the C-2 oxygenated pyridine, exists in the keto form.

Pyrazines result from the dimerization of α-aminocarbonyl compounds. The initial dihydropyrazines may be oxidized to obtain the pyrazine.

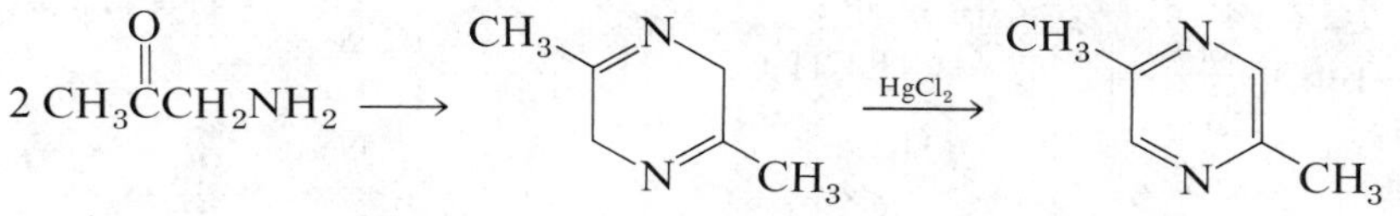

2,5-dimethylpyrazine

Pyrazines are also obtained from the condensation of 1,2-diamines with 1,2-dicarbonyl compounds. When 1,2-diaminobenzene (*o*-phenylenediamine) is used, the product is a benzopyrazine (quinoxaline). The reaction has been used as a diagnostic test for such 1,2-dicarbonyl compounds.

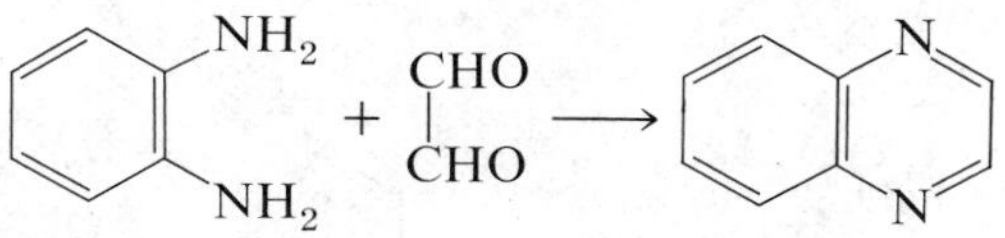

(85–90%)
quinoxaline

C. Reactions

Because of the second nitrogen in the ring system, the diazines are even less reactive than pyridine toward electrophilic substitution. When activating groups are present on the ring, such substitutions may occur.

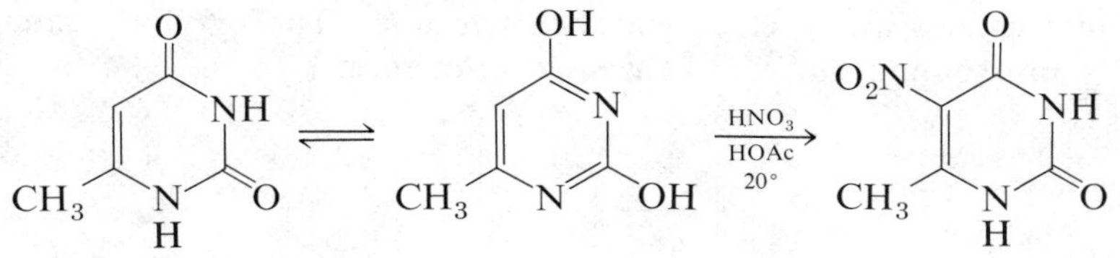

As with quinoline and isoquinoline, attack on the benzodiazines occurs in the benzene ring.

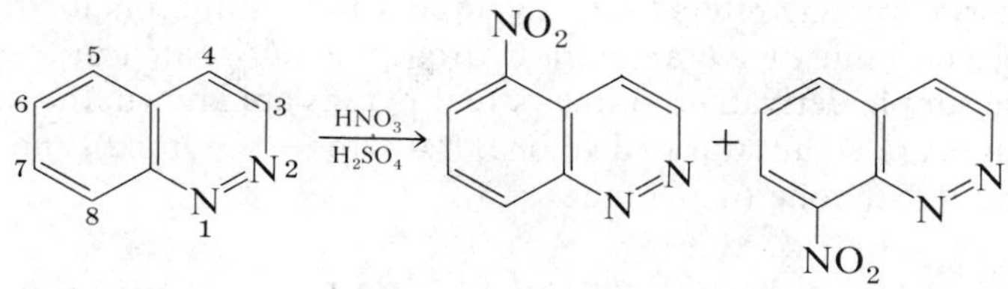

cinnoline   5-nitrocinnoline   8-nitrocinnoline

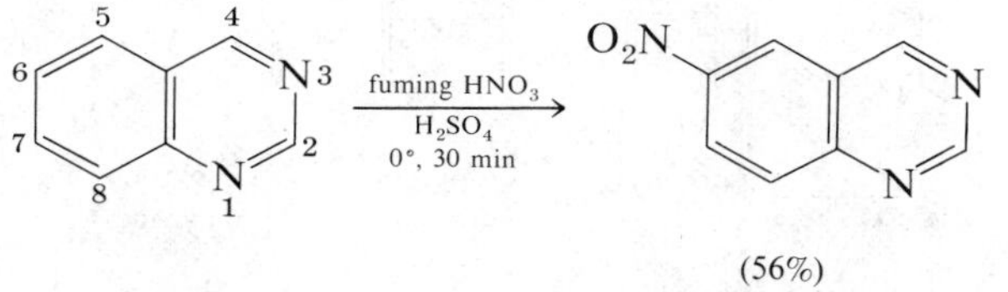

(56%)
6-nitroquinazoline

quinazoline

Many other reactions of the diazines and their benzo derivatives are similar to those observed with pyridine, quinoline, and isoquinoline. The following reactions illustrate some of these similarities.

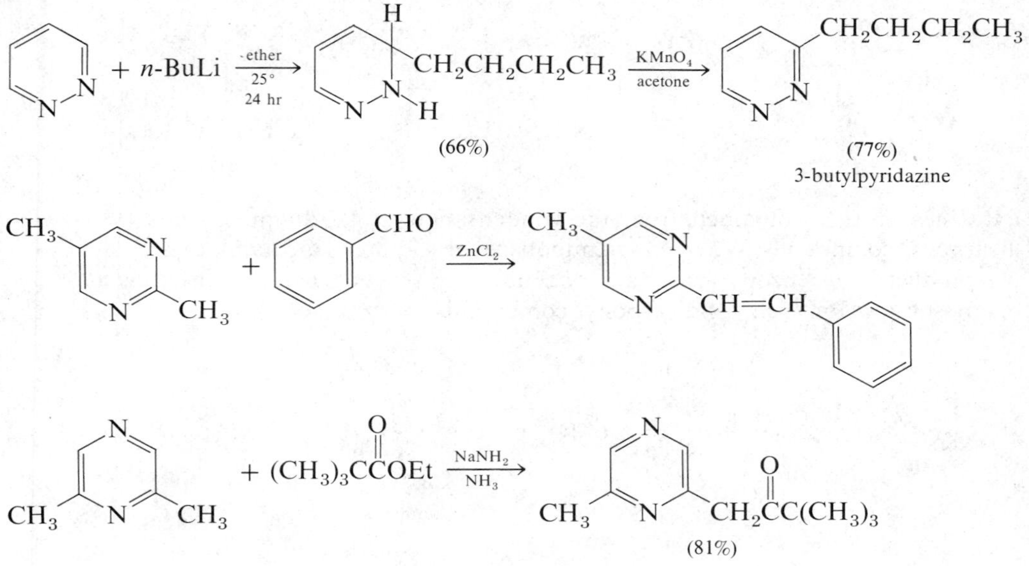

(66%)

(77%)
3-butylpyridazine

(81%)

32.9
Pyrones and Pyrylium Salts

Pyrylium cations are isoelectronic with pyridines. The pyrylium cation ring system is an oxonium salt with benzenoid resonance.

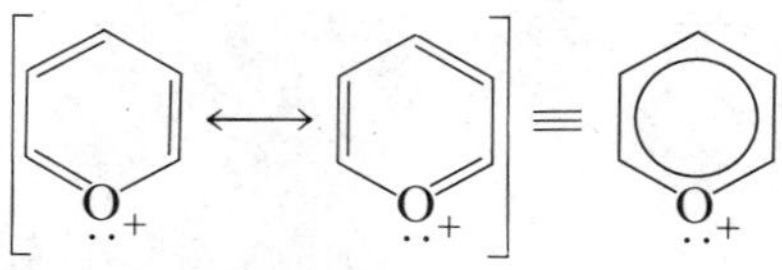

pyrylium cation

The parent neutral ring system is that of pyran-2-H or pyran-4-H. In this nomenclature the term "pyran" refers to the hypothetical neutral aromatic ring system; the real molecules must have an extra hydrogen as designated in the name. Although some simple derivatives of the parent pyrans are known, the most important derivatives are an unsaturated ketone, 1,4-pyrone (γ-pyrone), and an unsaturated lactone, 1,2-pyrone (α-pyrone).

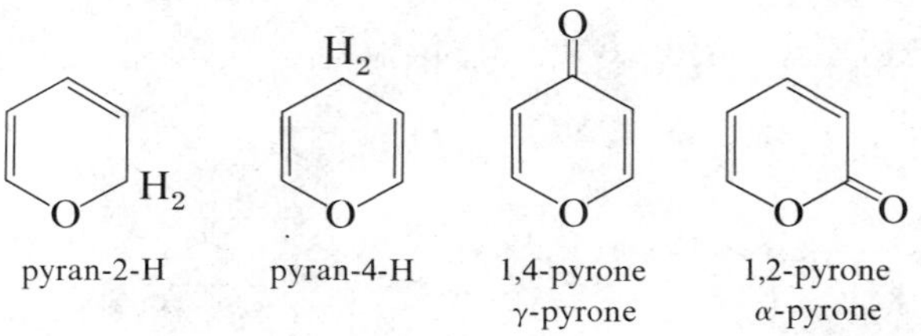

pyran-2-H   pyran-4-H   1,4-pyrone   1,2-pyrone
                          γ-pyrone   α-pyrone

Pyrone and pyrylium salt structures are widespread in nature (Section 34.3.B).

A. *Pyrones*

Treatment of malic acid with fuming sulfuric acid produces 1,2-pyrone-5-carboxylic acid, which on heating to 650° decarboxylates to α-pyrone.

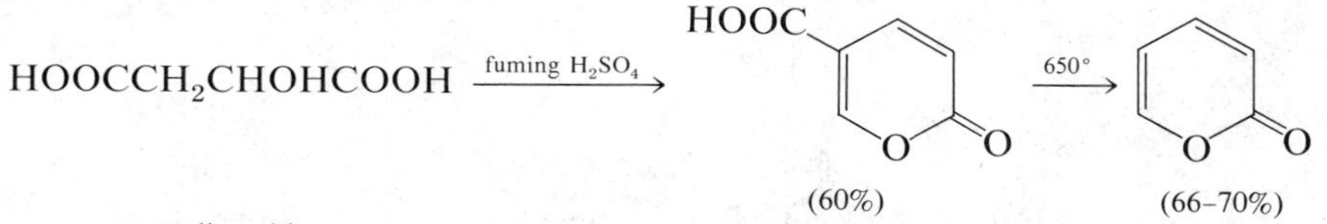
HOOCCH$_2$CHOHCOOH
fuming H$_2$SO$_4$
HOOC
O
O
650°
O
O

malic acid — (60%) coumalic acid — (66–70%) α-pyrone

γ-Pyrones are available by intramolecular cyclization of 1,3,5-triketones.

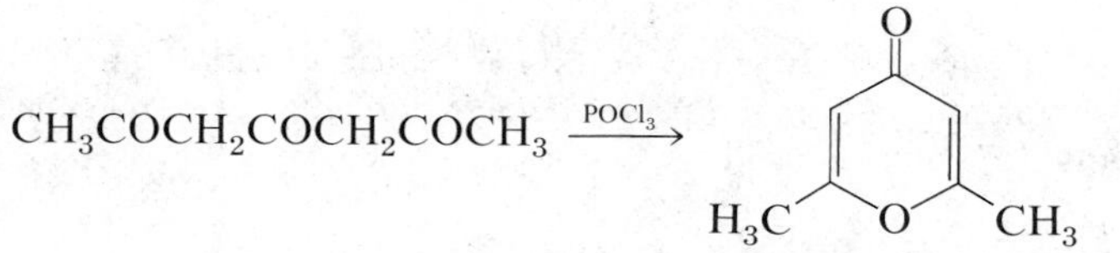
CH$_3$COCH$_2$COCH$_2$COCH$_3$
POCl$_3$
O
H$_3$C
O
CH$_3$

2,6-dimethyl-1,4-pyrone

The required triketones are prepared as needed by condensation of a β-diketone with an ester.

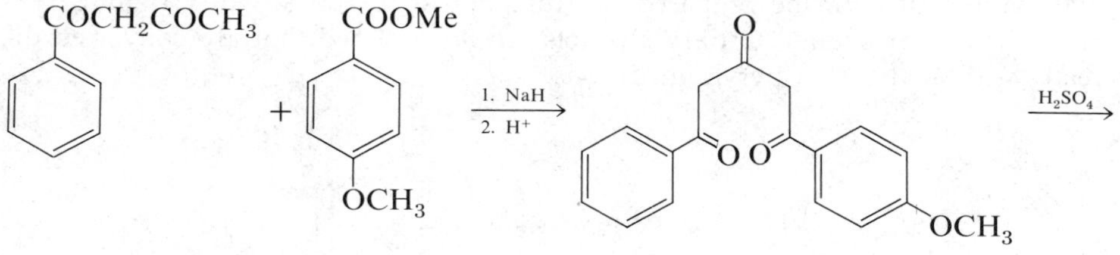
COCH$_2$COCH$_3$
COOMe
+
OCH$_3$
1. NaH
2. H$^+$
O
O O
OCH$_3$
H$_2$SO$_4$
O
O
OCH$_3$

(77–86%)

(88–98%)

α-Pyrones are dienes and partake readily in Diels–Alder reactions with appropriate dienophiles. The intermediate product readily decarboxylates.

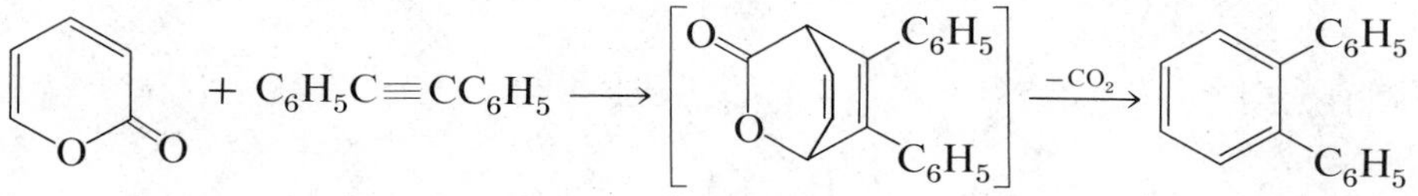
O
O
+ C$_6$H$_5$C≡CC$_6$H$_5$
O
O
C$_6$H$_5$
C$_6$H$_5$
−CO$_2$
C$_6$H$_5$
C$_6$H$_5$

The pyrones react with ammonia and primary amines under mild conditions to give the corresponding pyridones.

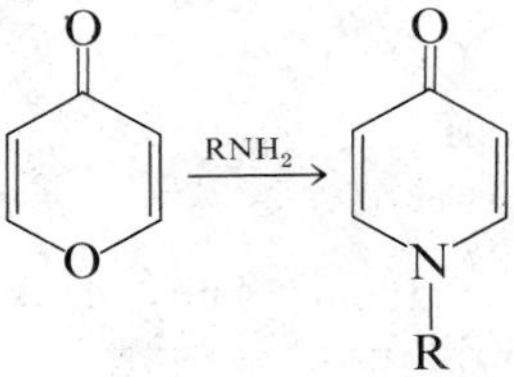
O
O
RNH$_2$
O
N
R

Pyrones are relatively basic and salts can frequently be isolated. The pK_a of 2,6-dimethyl-1,4-pyrone is 0.4; that is, this pyrone is about as basic as trifluoroacetate ion. The conjugate acid of a pyrone can be considered as a hydroxypyrylium salt.

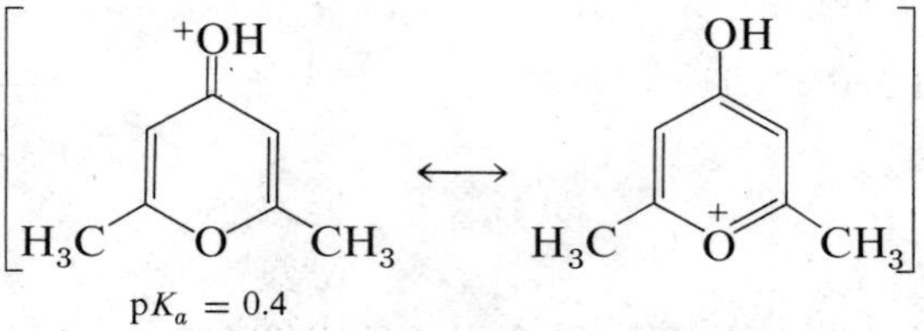

pK_a = 0.4

EXERCISE: The first step in the reaction of malic acid with fuming sulfuric acid is decarbonylation to give 3-oxopropanoic acid. Write out a reasonable mechanism for the formation of coumalic acid.

B. Pyrylium Salts

Pyrylium salts can be prepared by treatment of pyrones with Grignard reagents. The intermediate tertiary alcohols are not isolated, but are converted directly with acid to the pyrylium cations.

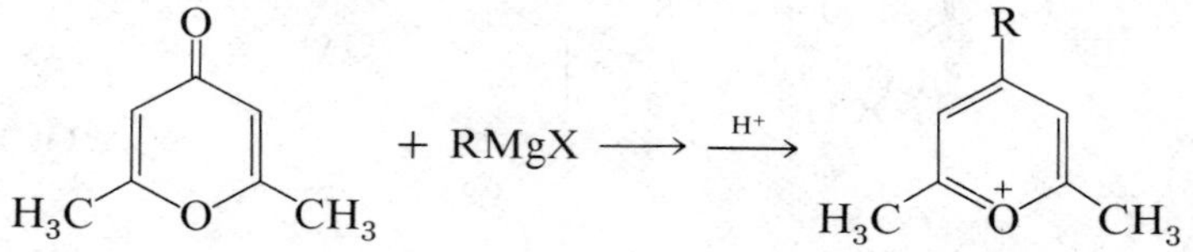

Alternatively, they are prepared by acid-catalyzed condensations of α,β-unsaturated ketones. Some examples are

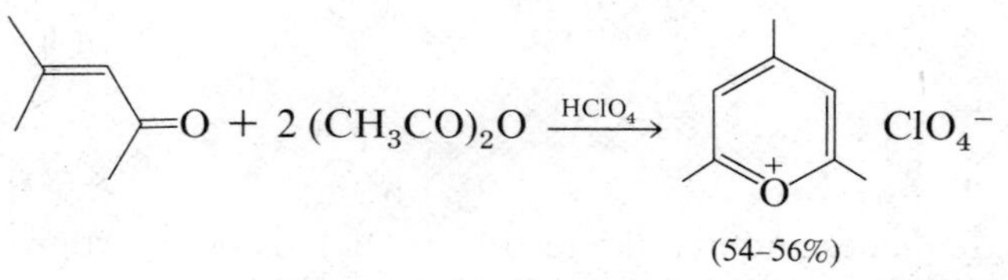

$$(32\text{-}12)$$

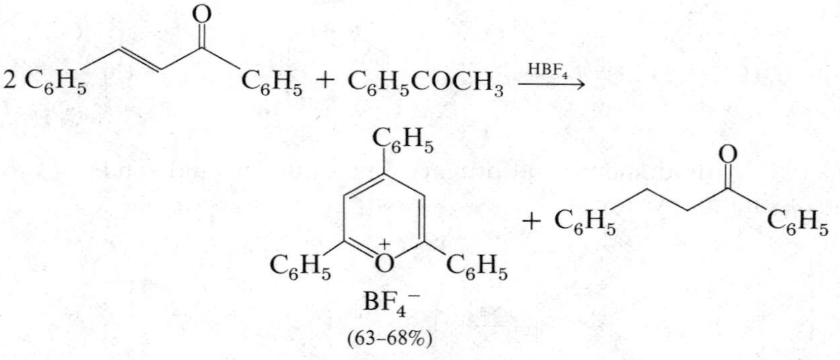

(54–56%)

$$(32\text{-}13)$$

(63–68%)

Pyrylium salts can be useful reagents. They react with nucleophilic reagents in the α-position. One such example is

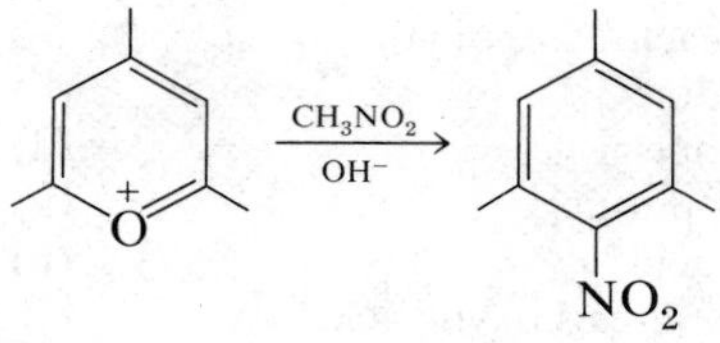

$$\text{(32-14)}$$

EXERCISE: Write reasonable reaction mechanisms for reactions (32-12), (32-13), and (32-14).

PROBLEMS

1. Name each of the following compounds.

(a)

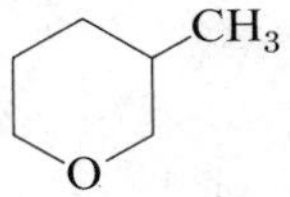

(b)

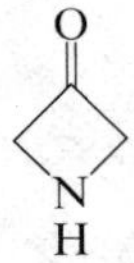

(c)

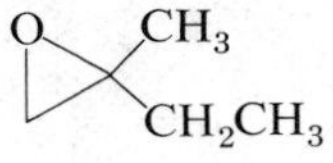

(d)

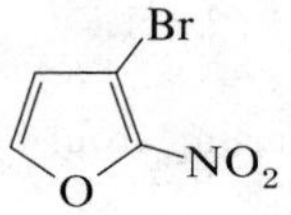

(e)

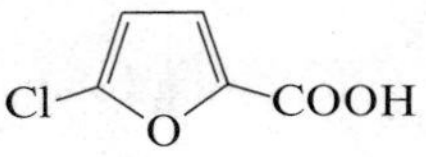

(f)

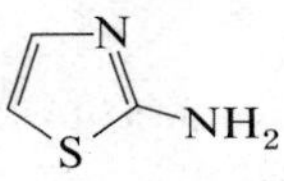

(g)

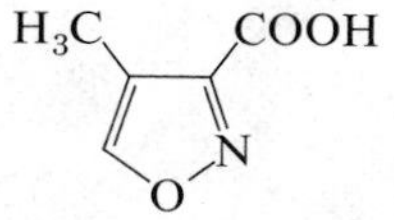

(h)

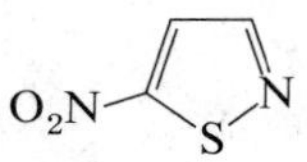

(i)

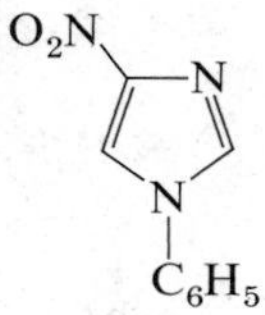

(j)

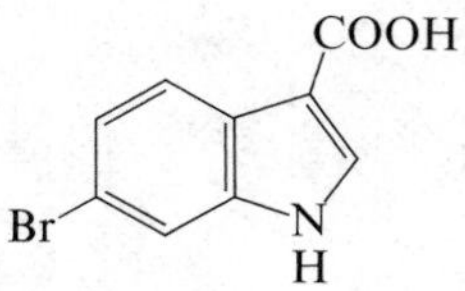

(k)

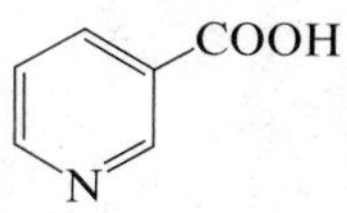

(l)

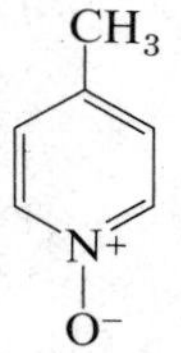

(m)

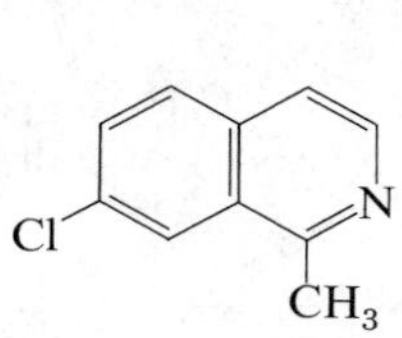

(n)

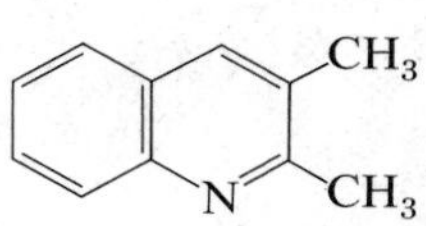

(o)

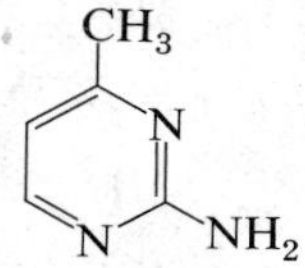

(p)

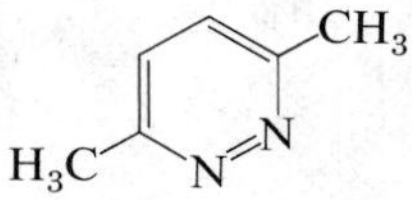

(q)

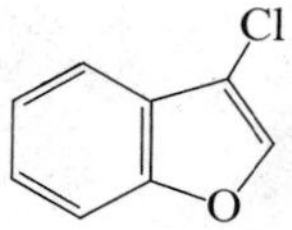

(r)

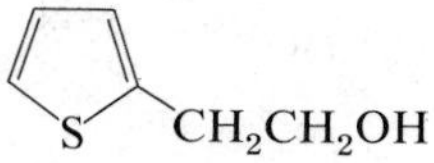

(s)

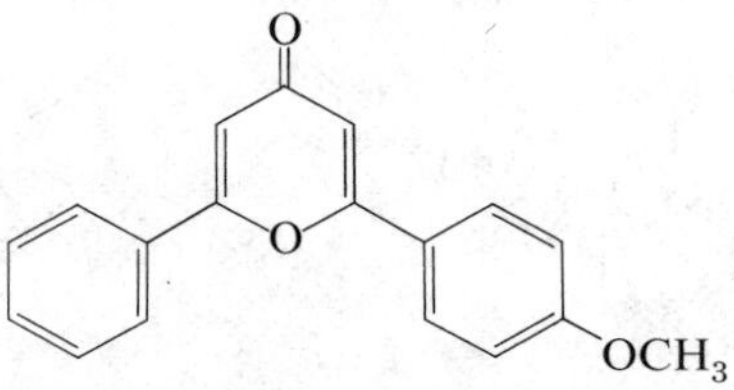

(t)

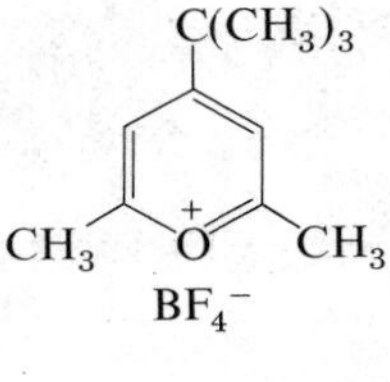

2. Write a structure for each compound.
- (a) 1,2-diphenylaziridine
- (b) 2,5-dihydrofuran
- (c) 1-methyl-2-pyridone
- (d) 8-bromoisoquinoline
- (e) 7-methyl-6-aminopurine
- (f) 2-aminopurine
- (g) (3-indolyl)acetic acid
- (h) 4-nitroquinoline-1-oxide
- (i) 4-chlorothiophene-2-carboxylic acid
- (j) 2-methyl-5-phenylpyrazine
- (k) 5-nitroquinoline-2-carboxylic acid
- (l) 2-nitrothiazole
- (m) 3-cyanoisoxazole
- (n) 4,6-dimethyl-1,2-pyrone

3. Outline a synthesis for each of the following compounds.

(a) C_6H_5 —   (b) 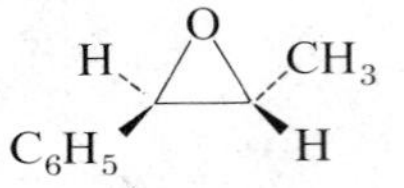   (c)

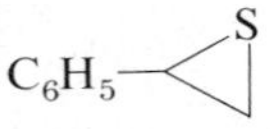

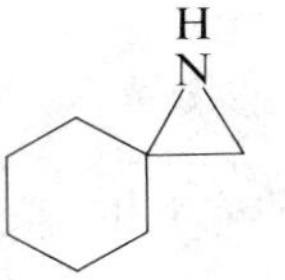

(d) 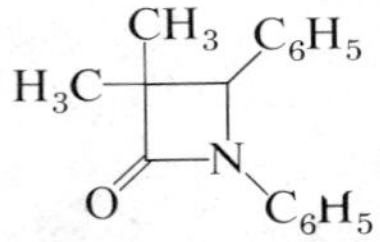   (e) 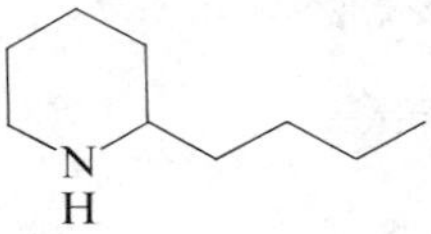   (f)

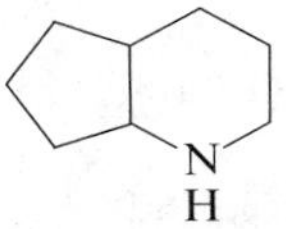

4. Outline a synthesis for each of the following compounds, starting from nonheterocyclic precursors.

(a)   (b)

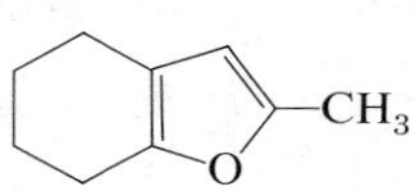

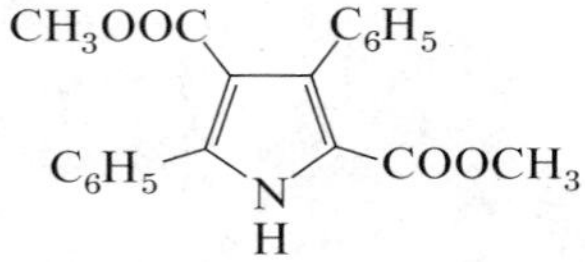

(c) 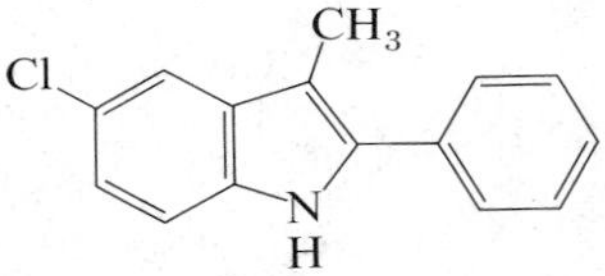   (d)

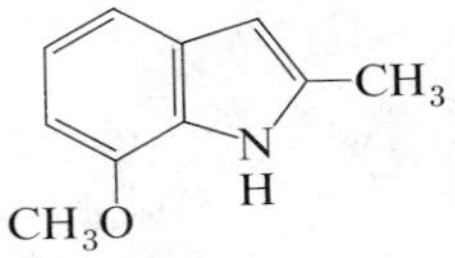

(e) 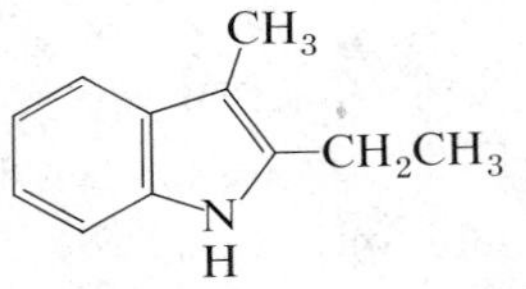   (f) 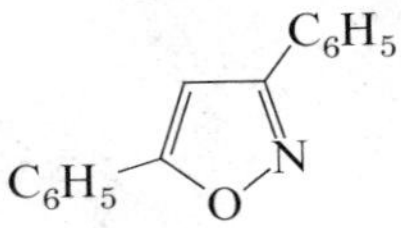   (g)

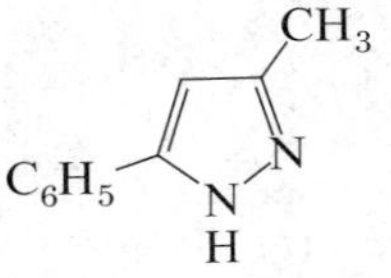

(h) 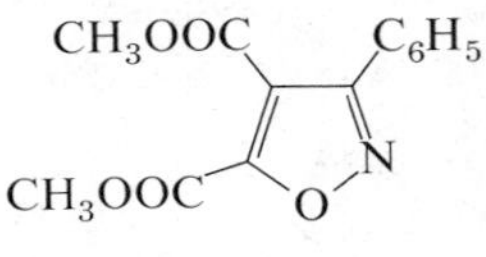   (i) 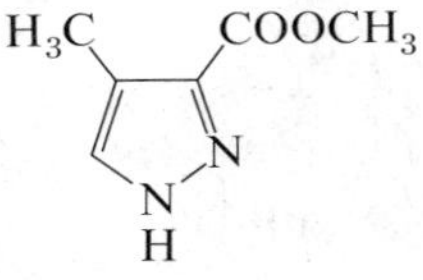   (j)

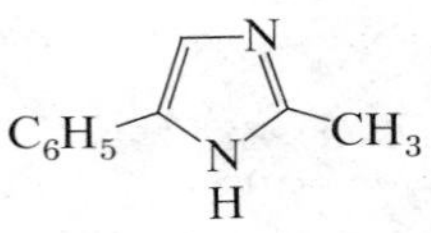

(k) 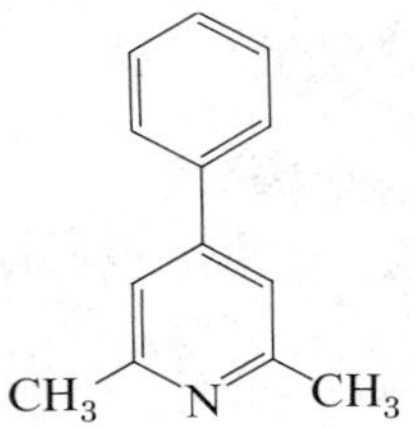   (l) 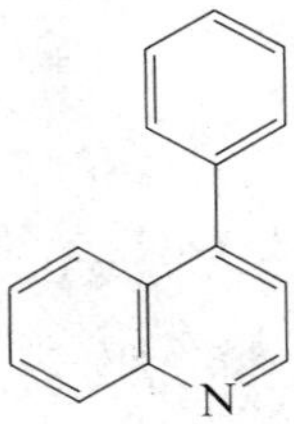   (m)

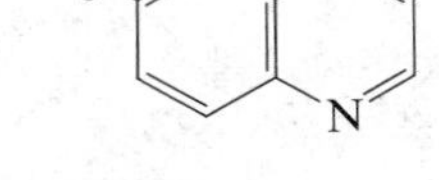

(n)

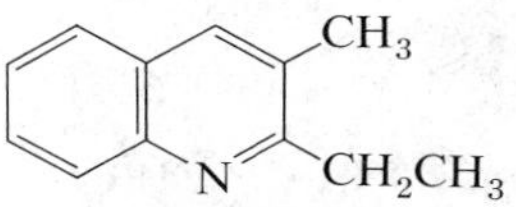

(o)

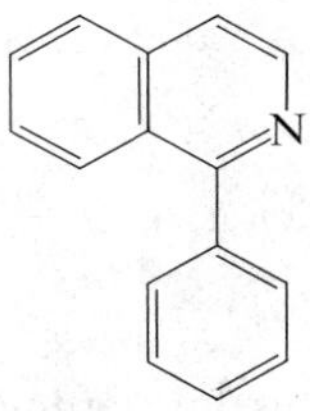

(p)

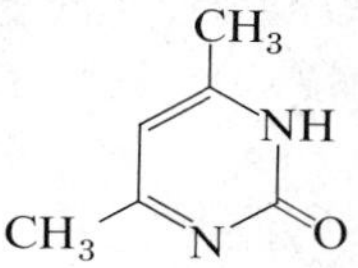

(q)

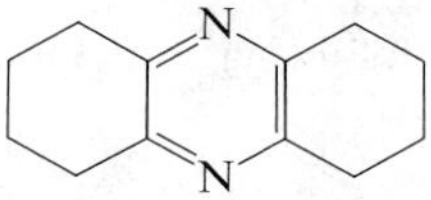

(r)

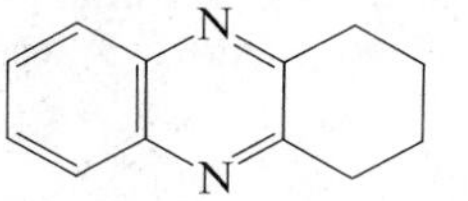

(s)

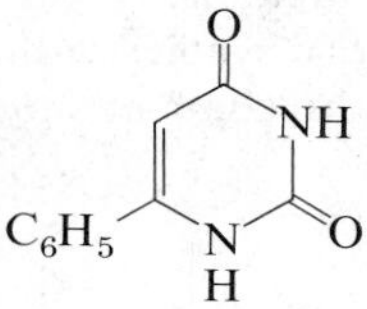

(t)

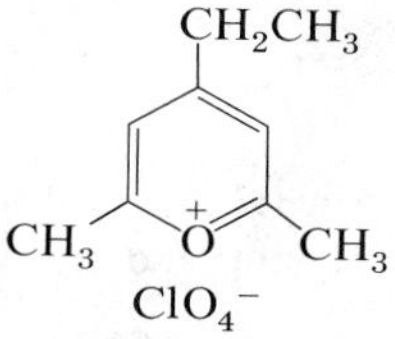

(u)

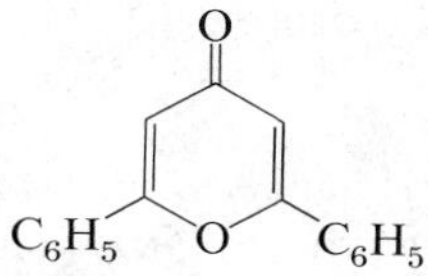

5. Outline a synthesis for each of the following compounds from the corresponding unsubstituted or alkyl-substituted heterocyclic system.

(a)

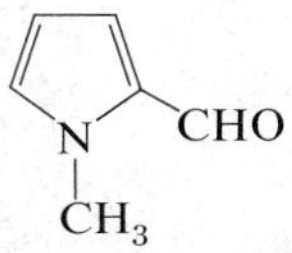

(b)

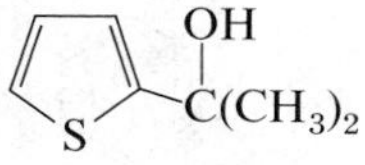

(c)

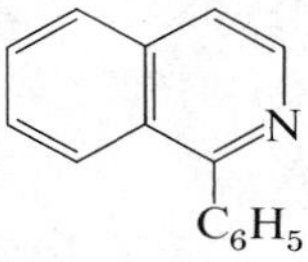

(d)

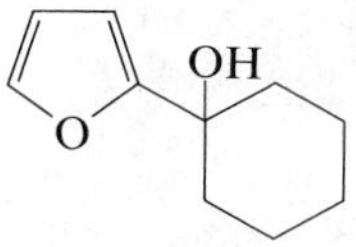

(e)

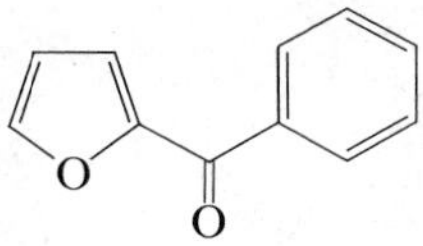

(f)

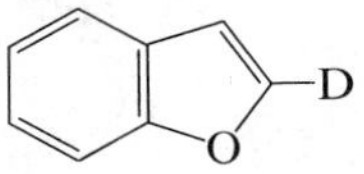

(g)

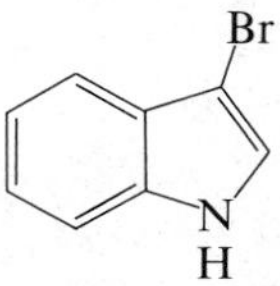

(h)

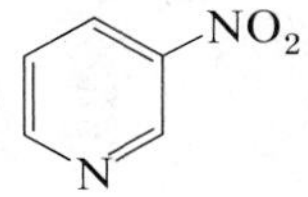

(i)

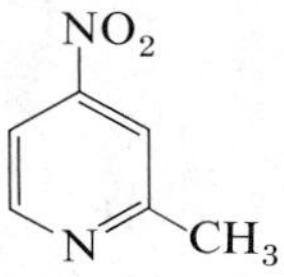

(j)

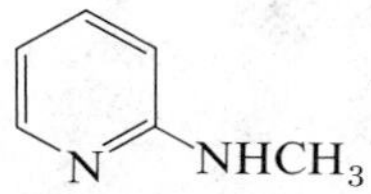

(k)

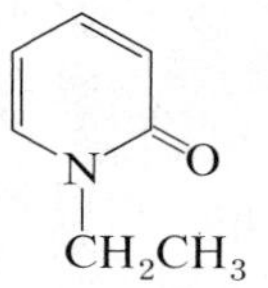

(l)

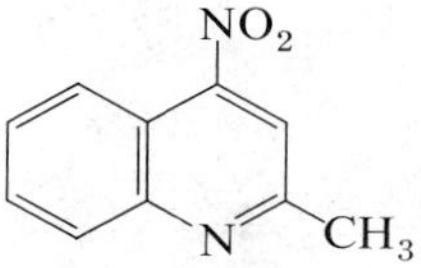

(m)

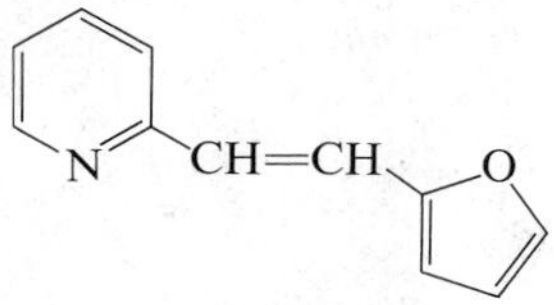

(n)

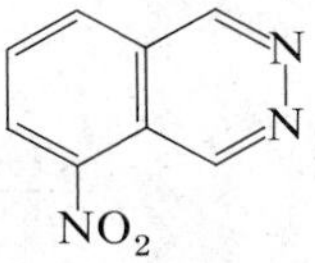

6. Write a reasonable mechanism that explains the following reaction.

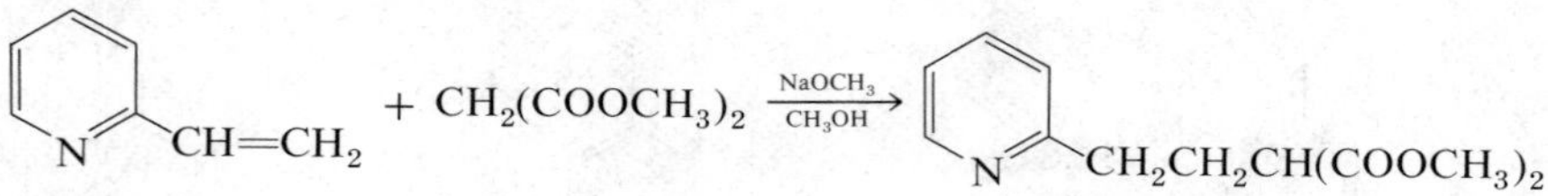

7. *o*-Aminobenzaldehyde is a useful starting material in the Friedländer synthesis of quinolines.
   (a) Synthesize this compound from available materials.
   (b) Use *o*-aminobenzaldehyde in the synthesis of the following compounds.

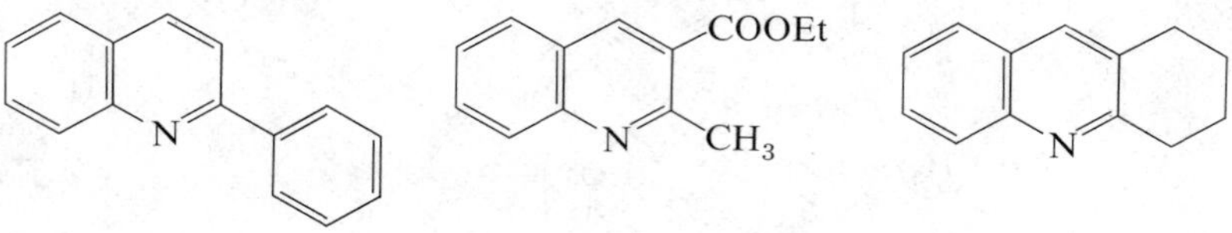

   (c) The mechanism of the Friedländer synthesis given on page 1100 was abbreviated. Write out the complete mechanism, showing all of the intermediates involved.

8. The pyridine ring is so inert that Friedel–Crafts reactions fail completely. Suggest a method to synthesize phenyl 3-pyridyl ketone.

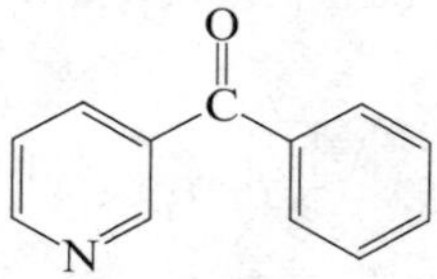

9. Predict the major product from each of the following reactions.

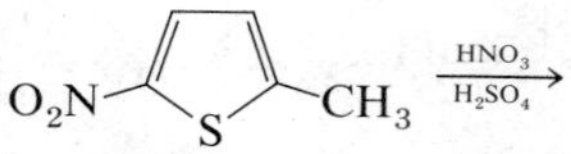

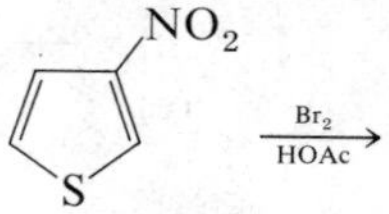

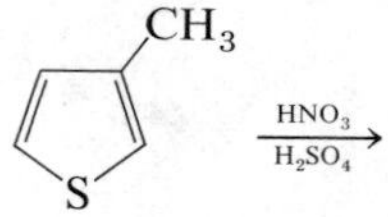

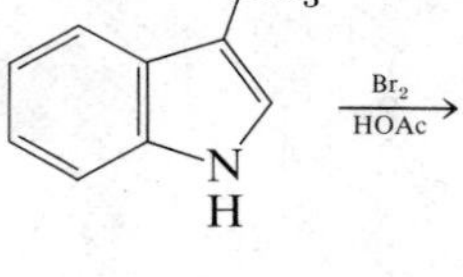

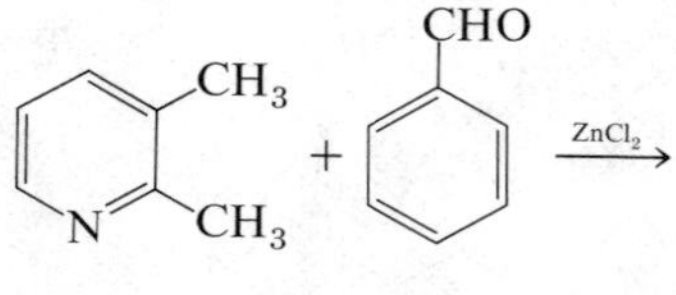

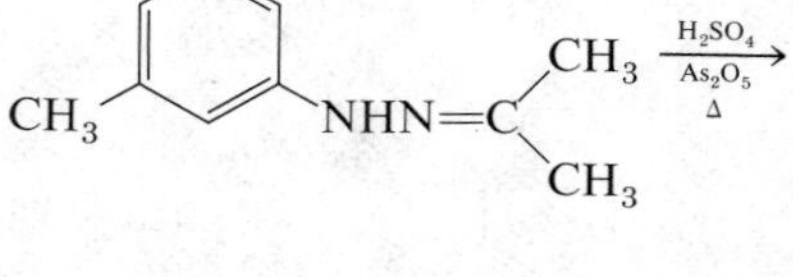

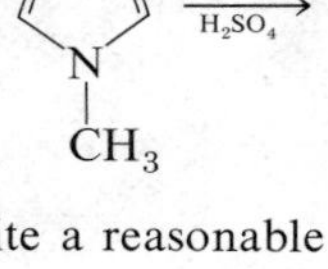

10. Write a reasonable mechanism for the following reaction.

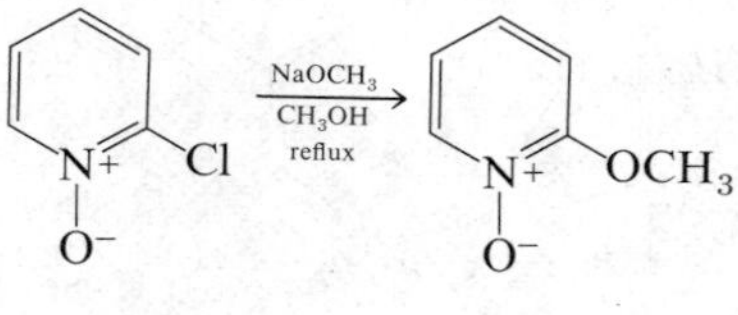

11. Pyridine N-oxide reacts with benzyl bromide to give N-benzyloxypyridinium bromide. Treatment of this salt with strong base gives benzaldehyde (92%) and pyridine. Rationalize with a reasonable mechanism.

12. Write a mechanism, showing all steps, that explains the following reaction.

$$CH_3CH_2\overset{\overset{\displaystyle O}{\|}}{C}CH{=}CHCl + H_2NOH \xrightarrow[\Delta]{EtOH}$$

(60%)   (40%)

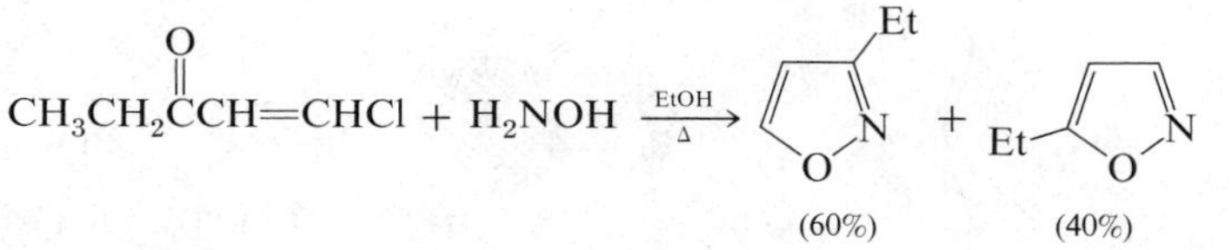

13. Pyrrole reacts with ethylmagnesium bromide, followed by methyl iodide, to give a mixture of 2- and 3-methylpyrrole. Rationalize this result, using resonance structures where desirable.

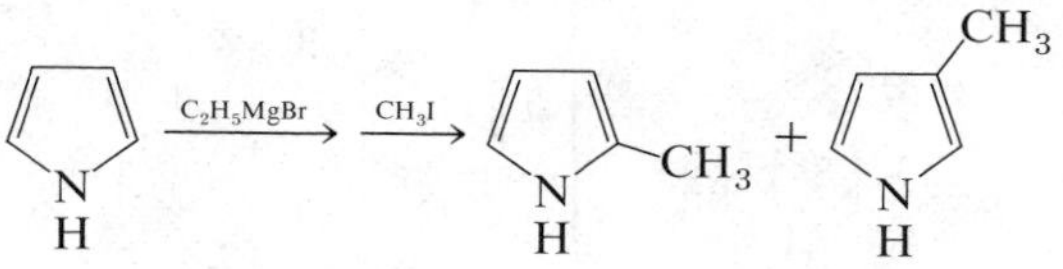

14. Write a mechanism for the following reaction in which a furan is produced.

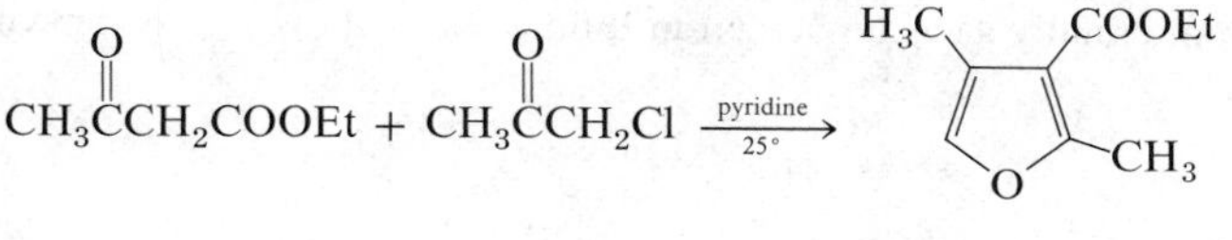

$$CH_3\overset{\overset{\displaystyle O}{\|}}{C}CH_2COOEt + CH_3\overset{\overset{\displaystyle O}{\|}}{C}CH_2Cl \xrightarrow[25°]{pyridine}$$

15. Explain why the methyl protons in 1-methylisoquinoline are more acidic than the methyl protons in 3-methylisoquinoline.

16. Heats of formation ΔH_f° are pyridine, $+34.6$, and piperidine, -11.8 kcal mole^{-1}. Before these data can be used to estimate the empirical resonance energy of pyridine, we need a value for the heat of hydrogenation of a C=N double bond. Data for several compounds have recently become available that suggest a value of -21 kcal mole^{-1} for ΔH° for the reaction

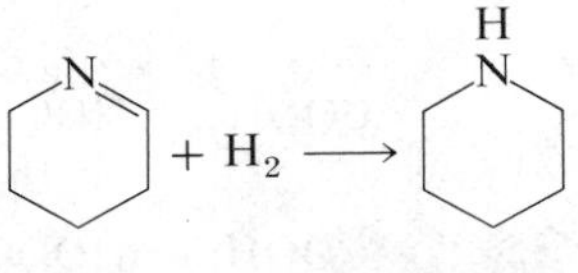

(a) Use this information together with corresponding results for the heat of hydrogenation of cyclohexene (Appendix I) and derive an empirical resonance energy for pyridine.

(b) An alternative method for calculating the empirical resonance energy is to compare the experimental heat of atomization with that obtained by use of a table of average bond energies, such as that in Appendix III. Compare the value you calculate by this method with the commonly quoted value for the resonance energy of pyridine of 23 kcal mole^{-1}. To get some insight into the source of the discrepancy, compare the calculated and observed heats of atomization of piperidine. How accurate are the results expected from the use of average bond energies?

17. Dipole moments of furan, thiophene, and pyrrole were discussed in Section 32.3.A, and the assignments of directions of the dipoles were presented. Given the following dipole moment data, deduce the directions assigned on page 1070.

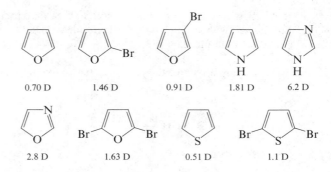

0.70 D   1.46 D   0.91 D   1.81 D   6.2 D

2.8 D   1.63 D   0.51 D   1.1 D

18. Given the following dipole moments, deduce the direction of the dipole moment of pyridine. Compared to the dipole moment of piperidine, is this direction reasonable?

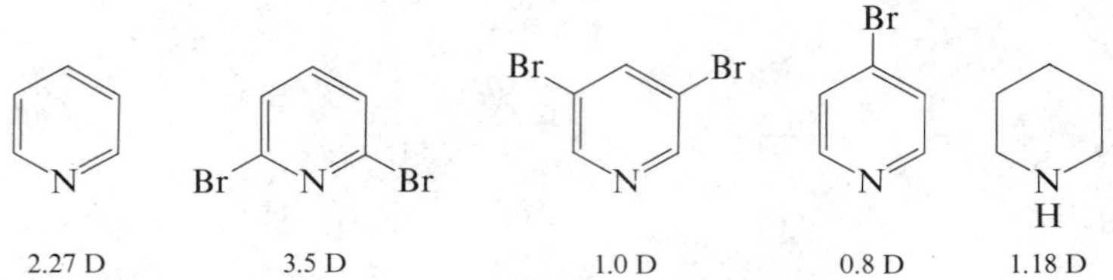

2.27 D   3.5 D   1.0 D   0.8 D   1.18 D

19. Umbelliferone is a coumarin derivative present as a glucoside in many plants. It is used commercially as a sun-screen in lotions. Show how it may be synthesized from resorcinol.

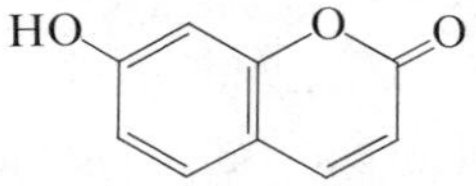

20. Papaverine, $C_{20}H_{21}O_4N$, is an alkaloid present in opium and is used as a muscle relaxant. It is nonaddicting, but is classified as a narcotic. Reaction with excess hydriodic acid gives 4 moles of CH_3I and shows the presence of $4\ CH_3O$ groups (Zeisel determination). Oxidation with $KMnO_4$ gives first a ketone, $C_{20}H_{19}O_5N$; continued oxidation gives a mixture from which the compounds shown below were isolated and identified. Deduce the structure of papaverine, and interpret the reactions described.

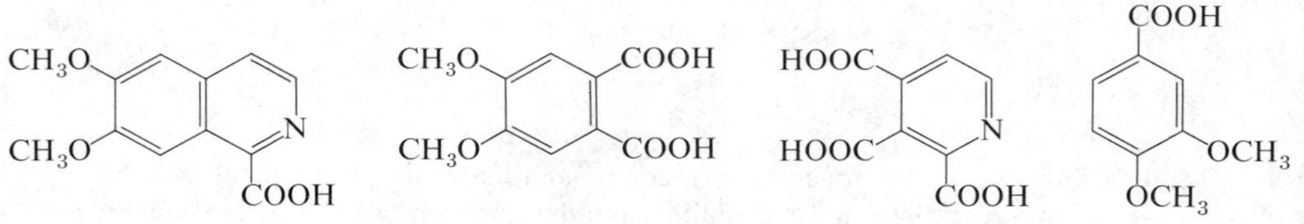

21. Write a mechanism for the base-catalyzed chlorination of an oxime to give a hydroxamic acid chloride (page 1086).

22. Azulene is an isomer of naphthalene that is characterized by its brilliant blue color. 4,6,8-Trimethylazulene is formed in 43–49% yield when 2,4,6-trimethylpyrillium perchlorate is treated with cyclopentadienylsodium in THF. Write a reasonable reaction mechanism for this reaction.

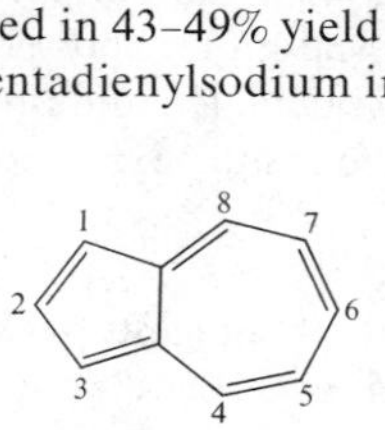

azulene

Chapter 33
The Chemical Literature

33.1
Research Journals

The total knowledge of chemistry is contained in hundreds of thousands of books and journals that are known collectively as **the literature.** New knowledge is communicated to the world for the first time as a **paper** or **communication** in a **research journal.** There are perhaps 10,000 journals that publish original articles on chemical topics, but only about 50 are of general interest to most chemists. Some journals, such as the *Journal of the American Chemical Society,* publish articles in all branches of chemistry. Others, such as the *Journal of Organic Chemistry,* only publish articles dealing with a specific area. A partial listing of typical journals that would be of interest to an organic chemist, with the normal abbreviation printed in italic type, is given. The language(s) used in each journal is also indicated.

1. *Angewandte Chemie* (German)
2. *Angewandte Chemie International Edition in English* (English)
3. *Justus Liebig's Annalen der Chemie* (German)
4. *Bulletin of the Chemical Society of Japan* (English)
5. *Canadian Journal of Chemistry* (English, French)
6. *Chemische Berichte* (German)
7. *Journal of the Chemical Society, Chemical Communications* (English)
8. *Collection of Czechoslovak Chemical Communications* (English)
9. *Comptes rendus hebdomadaries, Series C* (French)
10. *Helvetica Chimica Acta* (German, French, English)
11. *Journal of the American Chemical Society* (English)
12. *Journal of the Chemical Society, Dalton Transactions* (English)
13. *Journal of the Chemical Society, Perkin Transactions* (English)
14. *Journal of Heterocyclic Chemistry* (English)
15. *Journal of the Indian Chemical Society* (English)
16. *Journal of Medicinal Chemistry* (English)
17. *Journal of Organometallic Chemistry* (English, German, French)
18. *Journal of Organic Chemistry* (English)
19. *Synthesis* (English)
20. *Synthetic Communications* (English)
21. *Tetrahedron* (English, German, French)
22. *Tetrahedron Letters* (English, German, French)

An original article in a research journal may be in the form of a **full paper,** a **note,** or a **communication.** A full paper is a complete report on a research project, with full experimental details and interpretation. It is always accompanied by a short abstract, written by the authors. A note is a final report on a project of smaller scope. It includes experimental details, but has no abstract. A communication is a preliminary report on a finding of unusual significance. Communications are extremely concise, usually less than 1000 words, and have little or no experimental detail. In most cases a communication will be followed later by a full paper after the project has been completed. Some journals, such as *J. Am. Chem. Soc.* and *J. Org. Chem.*, publish both papers and communications, and others, such as *Tetrahedron Lett.* and *J. Chem. Soc., Chem.Commun.*, publish only communications. Research articles are documented with references to the literature, to other research articles, and to books. The traditional form for such a **literature citation** is Author(s), *journal abbreviation,* **volume number,** page number (year). For example,

H. O. House and B. M. Trost, *J. Org. Chem.,* **30,** 2052 (1965).

However, in 1979, the American Chemical Society journals recommended a new form for literature citations. In the newly recommended form the authors last names are given first, followed by first names and initials, followed by the *journal abbreviation,* **year,** *volume number,* and page number. For example,

House, H. O.; Trost, B. M., *J. Org. Chem.,* **1965,** *30,* 2052.

At the present time, both formats are in use.

If a practicing chemist is to keep abreast of the developments in his or her field, it is essential that he or she peruse a number of research journals regularly as they appear. All of the journals listed above appear periodically, usually weekly, semi-monthly, or monthly. Most chemists regularly scan the tables of contents of a dozen or so journals that publish articles in areas of interest to them.

33.2
Books and Review Articles

The original research journals comprise the **primary literature** of chemistry; they are the ultimate source that must be consulted for authoritative information on any subject. A second category of chemical literature is classed as **secondary literature.** The secondary literature consists of reference books and review articles in which the primary literature is collated and interpreted.

A. *Handbooks*

There are a number of excellent handbooks that compile data about individual organic compounds. The most extensive and most useful is the *Handbuch der Organischen Chemie,* commonly known as *Beilstein,* after its first editor. *Beilstein* is a multivolume handbook that lists all known organic compounds, together with their physical properties, methods of preparation, chemical properties, and any other available information. The main disadvantage of *Beilstein* is that it is not up to date. All of the literature through 1929 is completely covered, and certain classes of compounds are covered through 1959. We shall consider the use of *Beilstein* in Section 33.4.

The Handbook of Chemistry and Physics, published by CRC Press, Inc., Boca Raton, Florida, is revised regularly. It contains a useful collection of data and a copy may be found on the desk of almost all practicing chemists. The most important table for organic chemists is "Physical Constants of Organic Compounds," which occupies a major portion of the book. This table contains the name, formula, color, and several important physical properties for several thousand common organic compounds. Compounds are listed alphabetically by the IUPAC names. A similar volume is Lange's *Handbook of Chemistry,* McGraw-Hill Book Company, New York.

The Dictionary of Organic Compounds, edited by Heilbron, Cook, Bunbury, and Hey, is a five-volume handbook published by Oxford University Press, New York. It contains names, formulas, physical properties, and references for about 40,000 organic compounds. Compounds are listed alphabetically and there is no index.

The Merck Index of Chemicals and Drugs is published periodically by Merck and Company, Rahway, New Jersey. It concentrates on compounds of medicinal importance, but covers most simple organic compounds, whether or not they have significant physiological properties. In addition to names and formulas, the *Merck Index* lists physical properties, methods of synthesis, physiological properties, and medicinal uses and also gives the generic and trade names for all compounds that are used as drugs.

B. *Review Articles*

A review article is a survey of a single limited topic. For example, a chemist may assemble all of the information available on a topic by reading the original research articles and condense the information into a review article. frequently with his or her own interpretation. There are several periodicals that specialize in publishing review articles. A few important to organic chemists are

1. *Chemical Reviews* (English)
2. *Chemical Society Reviews* (English)
3. *Angewandte Chemie* (German)
4. *Angewandte Chemie International Edition in English* (English)
5. *Fortschritte der Chemischen Forschung* (German)
6. *Reviews of Pure and Applied Chemistry* (English)
7. *Synthesis* (English)
8. *Organometallic Chemistry Reviews* (English)
9. *Accounts of Chemical Research* (English)

In addition to review journals such as these, there are a number of open-ended serial publications that are published at somewhat irregular intervals in hardbound form. These books are similar in content and format to the normal review journals. A few examples are

1. *Advances in Carbohydrate Chemistry*
2. *Advances in Free Radical Chemistry*
3. *Advances in Photochemistry*
4. *Progress in Physical Organic Chemistry*
5. *Advances in Physical Organic Chemistry*
6. *Organic Reactions*
7. *Progress in Organic Chemistry*
8. *Progress in Stereochemistry*
9. *Topics in Stereochemistry*

Organic Reactions is a particularly important reference source for organic chemists. It is published approximately yearly and contains review articles on general reactions, for example, "The Wittig Reaction" and "The Clemmensen Reaction." The articles are accompanied by extensive tables of applications of the reaction.

C. *Monographs*

There are a large number of excellent books available that provide in-depth surveys of specific areas. The number of such monographs is far too great even to attempt to list here, and the student is referred to the card catalog in his or her own library. Several examples, merely to indicate the types of topics covered, are

1. H. C. Brown, *Hydroboration*. Benjamin, New York, 1962.
2. E. L. Eliel, *Stereochemistry of Carbon Compounds*. McGraw-Hill, New York, 1962.
3. H. O. House, *Modern Synthetic Reactions*. Benjamin, New York, 1971.
4. A. Streitwieser, Jr., *Molecular Orbital Theory for Organic Chemists*. Wiley, New York, 1961.
5. N. J. Turro, *Molecular Photochemistry*. Benjamin, New York, 1965.

D. *Books Covering Methods and Reagents*

There are several useful books that are devoted to synthetic methods or to reagents used in organic reactions. *Organic Syntheses* is published by John Wiley & Sons, New York. It is a collection of procedures for the preparation of specific compounds. The work has appeared annually since 1921. The procedures for each ten-year period are collected in cumulative volumes, of which five now exist. The procedures in *Organic Syntheses* are submitted by any chemist who wishes to do so and are then tested in the laboratory of a member of the editorial board. Although the methods given pertain to specific compounds, an attempt is made to include procedures that have general applicability. For this reason *Organic Syntheses* is a useful source of model procedures when the chemist wishes to carry out a new preparation The cumulative volumes are each thoroughly indexed, and there is a collective index for the five cumulative volumes.

Theilheimer, *Synthetic Methods of Organic Chemistry*, S. Karger Verlag, Basel, is an annual compilation of synthetic methods. It is organized by way of a system based upon types of bond formations or bond cleavages. There is an index with each volume and a cumulative index after each fifth volume.

Reagents for Organic Synthesis by L. F. Fieser and M. Fieser (Wiley, New York) is an exceedingly useful compendium of reagents and catalysts used in organic chemistry. In addition to the main volume, seven supplements are now available. The work gives information on how each reagent is prepared, commercial suppliers, and references to its uses.

33.3
Abstract Journals

Abstract journals are periodicals that publish short abstracts of articles that have appeared in the original research journals. There are currently two such publications devoted to the original chemical literature, *Chemical Abstracts* and *Referativnyl Zhurnal* (Russian). A German abstract journal, *Chemisches Zentralblatt,* ceased publication in 1970, but it is frequently useful for retrieving information published before that date.

Chemical Abstracts is published weekly by the American Chemical Society and includes abstracts in English of nearly every paper that contains chemical information, regardless of the original language. Abstracts appear from 3 to 12 months after the appearance of the original paper. The abstracts are grouped into 80 sections, of which sections 21–34 pertain to organic chemistry. Each individual abstract is preceded by the authors' names, the authors' address, the journal citation, and the language of the original article.

Although many chemists use *Chemical Abstracts* routinely to keep abreast of a broad area of chemistry, it is most useful because of its indexes. From its beginning in 1907 until 1961 there were annual indexes. Since 1962 there have been semiannual indexes, covering the periods January–June and July–December. From 1907 until 1956 there were published additional ten-year indexes. Since 1957 the cumulative indexes have appeared at five-year intervals. The most recent complete index is the *Ninth Collective Index,* covering the period 1972–1976. Each annual and collective index has a subject index and an author index. Formula indexes for the periods 1920–1946, 1947–1956, 1957–1961, 1962–1966, 1967–1971, and 1972–1976 are also available. The most useful of the indexes is the Subject Index. Each compound referred to in any paper abstracted during the index period is listed alphabetically. Following each entry is an abstract number.

To search *Chemical Abstracts* for information concerning a given compound, one looks up the name of the compound in each of the collective indexes and then in the semiannual indexes that have appeared since the last collective index. The abstract numbers are then used to locate the abstracts, and these are scanned. If it appears from an abstract that the original paper contains information of use, the original paper is consulted.

For example, suppose we wish to know what has been published regarding the carcinogenic (tumor-producing) properties of the hydrocarbon benz[*a*]anthracene during the period 1967–1971.

benz[*a*]anthracene

Consulting the *Eighth Collective Index,* which covers the period, we find the listing

Benz[*a*]anthracene [*56 – 55 – 3*]

[*a*]

Following this listing, there are a number of indexed topics, in alphabetical order. A portion is shown.

bond length and heat of dissocn. of, **69:** 35282v
bond length in, **72:** 6380g
bond lengths and energy levels of, **75:** 5052b
bond order and localization energy of, in excited
    state, polarizability in relation to, **68:** 82565d
bromination of, mechanism of, **66:** 37049m
bud induction by, in tobacco callus culture, **74:**
    2887w
butoxyphenyl siloxanes contg., degradation of, **72:**
    133523f
carbonate dehydratase inhibition by, **67:** 115102k
carcinogenesis by, two-stage process in relation to,
    73: 33378q
carcinogenic activity of
    67: 98520a; **72:** 10910x
    activation energy of elec. conduction in relation
        to, **70:** 42062x
    mol. orbitals in relation to, **72:** 64737q
carcinogenic activity of hydrocarbons and, **72:**
    53064s
carcinogenic and menadione reductase-inducing
    activities of, **68:** 47948j
as carcinogen in human environment, **69:** 25669n
cell invasiveness in presence of, **68:** 103288h
cell protection against dimethylbenzanthracene by,
    aryl hydroxylase formation in, **71:** 122020t
cheese contamination with, from wax coating, **72:**
    109888h
chem. shifts of, mol. orbital calcn. in relation to, **73:**
    55252w

The number after each topic indicates the *Chemical Abstracts* volume number and abstract number where the information will be found. For example, the listing **67**:98520a means that abstract 98520 in volume **67** contains information on the carcinogenic activity of benz[*a*]anthracene. Going to volume **67** of the abstracts (1967), we find the following abstract.

98520a The carcinogenic activities in mice of compounds related to benz[a]anthracene. E. Boyland and P. Sims (Roy. Cancer Hosp., London). *Int. J. Cancer* 2(5), 500–4(1967)(Eng). The carcinogenic activities of 18 aromatic hydrocarbons and their metabolic intermediates were compared after 3–10 s.c. injections of 1 mg. into C57 black mice. The monohydroxymethyl derivs. of 7,12-dimethylbenz[*a*]anthracene and some related compds. were active carcinogens, but were much less so than the parent hydrocarbon. Epoxides formed at the 5,6-bond (K-region) of chrysene, benz[*a*]anthracene, 7-methylbenz[*a*]anthracene, and dibenz[*a,h*]anthracene produced tumors when given at high dose levels, but were not as active as the parent hydrocarbons. The epoxide derived from phenanthrene was inactive. All of the compds. were prepd. by known methods with the exception of 7,12-dimethylbenz[*a*]anthracene, dibenz[*a,h*]anthracene, and chrysene which were obtained com. and 7-(diacetoxymethyl)benz[*a*]anthracene which was prepd. by heating benz[*a*]anthracene-7-carboxaldehyde under reflux with Ac_2O for 6 hrs. It sepd. from EtOH in needles, m. 196°. CTJN

If we desire more complete information, we may consult the original article, which was published in the *International Journal of Cancer Research,* volume **2,** on page 500, in 1967.

In order to use *Chemical Abstracts* efficiently, one must have a good command of organic nomenclature. All compounds are listed as derivatives of a parent compound, for example,

$CH_3CH(Cl)CH_2CH_2COOH$ ≡ valeric acid, 4-chloro

≡ 2-cyclohexen-1-one, 3-chloro, 2-phenyl

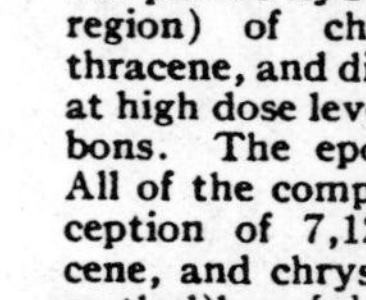

Note that *Chemical Abstracts* does not always use IUPAC nomenclature. At the beginning of each collective index, there is an extensive section dealing with the system of nomenclature used in indexing. In cases where it is not clear which name is used for indexing a particular compound, the formula index is useful. However, the formula index is much more tedious to use because one must often sift through an extensive list of isomers. It is also less reliable than the subject index, since omissions are more frequent. However, because of the frequency with which the compilers of *Chemical Abstracts* have changed nomenclature systems in recent years, the formula index has become an indispensable aid in searching for more complex structures.

Beginning with the issuance of the volume indexes covering the January–June 1972 period, the Subject Index has been issued in three parts: the Chemical Substance Index, the General Subject Index, and the Index Guide. All references to distinct, definable chemical substances are collected in the Chemical Substance Index, and all entries pertinent to any other topics (concepts, processes, organism names, diseases, reactions, generalized classes of compounds, and so on) are found in the General Subject Index. The Index Guide serves to guide the user quickly and efficiently to the proper headings in these two indexes. The Index Guide can be used to find a Chemical Substance Index name for trivial, commercial, and other nonsystematically named substances. It represents a compilation of indexing cross-references, preferred index headings, synonyms, and general index notes on thousands of chemical terms and names and should be consulted before using either the Chemical Substance or General Subject Index. The Index Guide is supplemented annually to cover additions and changes that may occur within a volume indexing period.

33.4
Beilstein

Beilstein's *Handbuch der Organischen Chemie* is shelved in the reference section of most chemical libraries. There have been four editions of the work and the first three are obsolete. The fourth edition (*vierte Auflage*) consists of a main series (*das Hauptwerk*) and three supplementary series (*erstes, zweites,* and *drittes Ergänzungswerk*). The periods covered by the various series are

Main series	antiquity–1909
First supplement	1910–1919
Second supplement	1920–1929
Third supplement	1930–1949 (incomplete)

A fourth supplement, covering the period 1950–1959, has begun to appear but it is also incomplete. The main series consists of 27 volumes (*Bands*), each bound as a separate book. Each supplementary series also consists of 27 volumes, and entries in the supplements are cross-referenced to the main series. Volumes in the supplementary series are sometimes bound as more than one book, and in some cases two or more volumes are bound together.

Compounds are grouped into three major divisions, in the following manner.

Division	Volumes
Acyclishe reihe,	
Acyclic compounds	1–4
Isocyclishe reihe,	
Carbocyclic compounds	5–16
Heterocyclishe reihe,	
Heterocyclic compounds	17–27

There is a fourth minor division—carbohydrates, rubber-like compounds, and carotenoids—contained in volumes 30 and 31, which only appears with the main series.

Volumes 28 and 29 are a subject index (*Generalsachregister*) and a formula index (*Generalformelregister*), respectively. The most recent indexes are part of the second supplement and they cover the main series and the first and second supplements, that is, through 1929. Earlier versions of the indexes are obsolete. One cannot rely completely on the indexes, because only representative compounds are indexed. However, they are useful to obtain rapidly the approximate location of a compound in the handbook, particularly for heterocyclic compounds. The index listing gives the volume and page numbers where the compound will be found. Bold type indicates the volume number and normal type indicates the page number; supplementary series page numbers are preceded by the appropriate Roman numeral. For example, volume **28** of the second supplementary series contains the listing

Indol **20**, 304, I 121, II 196; **21** II 567.

Thus, we find indole listed on p. 304 in volume **20** of the main series, on p. 121 of volume **20** of the first supplementary series, and on p. 196 of volume **20** of the second supplementary series. The final entry refers to a correction, which appeared on p. 567, at the end of volume **21** of the second supplementary series. We find the same listing in volume **29** of the second supplementary series, which is the formula index, under C_8H_7N, the formula of indole.

Although the *Beilstein* indexes are useful, one should become familiar with the basic organizational system of the handbook if it is to be used to best advantage. In each of the first two major divisions—acyclic compounds and carbocyclic compounds—compounds are listed according to the following order of basic classes.

1. Hydrocarbons (*Kohlenwasserstoffe*), RH.
2. Hydroxy compounds (*Oxyverbindungen*), ROH.
3. Carbonyl compounds (*Oxoverbindungen*), $R_2C{=}O$.
4. Carboxylic acids (*Carbonsäuren*), RCOOH.
5. Sulfinic acids (*Sulfinsäuren*), RSO_2H.
6. Sulfonic acids (*Sulfonsäuren*), RSO_3H.
7. Selenium acids (*Seleninsäuren* and *Selenosäuren*), $RSeO_2H$ and $RSeO_3H$
8. Amines (*Amine*), RNH_2, R_2NH, R_3N.
9. Hydroxylamines (*Hydroxylamine*), RNHOH.
10. Hydrazines (*Hydrazine*), $RNHNH_2$.
11. Azo compounds (*Azo-Verbindungen*), RN=NH.

Following these basic classes, there are a further 27 rare classes, which we shall not list.

The handbook begins with acyclic hydrocarbons; the very first entry is methane, CH_4. After all of the derivatives of methane have been listed, one finds
ethane, followed by its derivatives, and so on, through all the hydrocarbons having the empirical formula C_nH_{2n+2}. When all alkanes and their substitution derivatives have been listed, hydrocarbons with the formula C_nH_{2n} follow, beginning
with ethylene (C_2H_4), and going on up in carbon number. Next are listed hydrocarbons with the formula C_nH_{2n-2}. In this section we find alkynes and dienes; the
first entry is acetylene, C_2H_2. The following group of compounds has the general
formula C_nH_{2n-4}, then C_nH_{2n-6}, and so on. Thus, within a class of compounds,
such as hydrocarbons, compounds are listed in order of *increasing unsaturation*.
The general formula for the compounds listed on a given pair of pages is printed
at the top of the left-hand page.

After all hydrocarbons and their derivatives have been listed, the hydroxy compounds are listed. In the acyclic division, the first hydroxy compound is methanol,
CH_3OH, which has the empirical formula $C_nH_{2n+2}O$. Following the alcohols of
this formula, one finds alcohols with the formula $C_nH_{2n}O$, and so on. When all
mono alcohols have been listed, the diols are listed, beginning with the
$C_nH_{2n+2}O_2$ compounds. Next come the triols, tetraols, and so on. When the alcohols have been exhausted, the aldehydes and ketones are listed, and so on down
the list of classes of compounds.

Polyfunctional compounds are indexed *under the class that occurs last in the
listing*. For example, hydroxycarboxylic acids are indexed under carboxylic acids,
amino sulfonic acids under amines, and so on. When three or more of the basic
functional groups are present, the same rule applies; a hydroxy amino acid will be
found under the amines.

Following each compound in the handbook, one will find its derivatives. The
derivatives are of three types and are listed in the following order.

1. FUNCTIONAL DERIVATIVES. These compounds are derivatives of the
basic functional group and are hydrolyzable (in principle) to the parent compound. For example, dimethyl ether and methyl nitrate are both considered as
functional derivatives of methyl alcohol and are indexed after it.

$$CH_3OCH_3 \xrightarrow{H_2O} 2\ CH_3OH$$

$$CH_3ONO_2 \xrightarrow{H_2O} CH_3OH + HNO_3$$

2. SUBSTITUTION DERIVATIVES. These are compounds in which a C—H
has been replaced by C—X, C—NO, C—NO$_2$, or C—N$_3$. They are listed in the
order

1. Halides
   (a) Fluorides, such as CH_3F.
   (b) Chlorides, such as CH_3Cl.
   (c) Bromides, such as CH_3Br.
   (d) Iodides, such as CH_3I.
2. Nitroso derivatives, such as CH_3NO.
3. Nitro derivatives, such as CH_3NO_2.
4. Azido derivatives, such as CH_3N_3.

When there is more than one of the same group attached to the basic compound,
the polysubstituted compounds follow the monosubstituted compound. For example, the fluorinated methanes appear in the order CH_3F, CH_2F_2, CHF_3, CF_4.
When two different substitution groups are present, the compound is listed under

the group that *occurs last* in the foregoing list. Thus, fluorochloromethane, CH_2FCl, appears immediately after methyl chloride, CH_3Cl; in effect, CH_2FCl is considered as a substitution derivative of CH_3Cl. Likewise, chloronitromethane, $ClCH_2NO_2$, follows nitromethane in the listing. One must be careful not to confuse substitution derivatives with functional derivatives. For example, methyl hypochlorite, CH_3OCl, is listed with the functional derivatives of methyl alcohol because, in principle, it is hydrolyzable to methyl alcohol.

$$CH_3OCl \xrightarrow{H_2O} CH_3OH + HOCl$$

3. SULFUR AND SELENIUM COMPOUNDS. These compounds are listed as replacement derivatives under the corresponding oxygen compound. For example, methyl mercaptan, CH_3SH, and dimethyl selenide, $(CH_3)_2Se$, are listed last under the derivatives of methyl alcohol. Similarly, dithioacetic acid, CH_3CS_SH, is found among the final listings that follow acetic acid.

A similar organization is followed with the carbocyclic compounds. For heterocyclic compounds, the same scheme is used, but there is an additional division into *hetero numbers*. Most practicing chemists do not use the hetero numbers but, rather, rely on the subject or formula index to locate the parent heterocycle in the handbook. One must remember that many familiar compounds not normally thought of as heterocyclic compounds indeed are. For example, succinic anhydride will be found in the third division, as a dicarbonyl derivative of the heterocycle tetrahydrofuran.

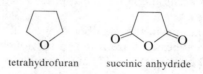

tetrahydrofuran   succinic anhydride

PROBLEMS

1. Look up the following compounds in Beilstein's *Handbuch der Organischen Chemie*. Record the melting point and/or boiling point, the system number, and the page number in the main series and each supplementary series when the compound is found.

(a) O_2N-⬡$-SeBr$ (with NO_2)

(b) naphthalene-$COOH$

(c) $HO-$⬡$-COOH$

(d) $ClCH_2CH_2CH_2OH$

(e) ⬠$-(CH_2)_{12}COOCH_3$

(f) $CH_2{=}C$ with CN_3 (C=O) and CH_2CN_3 (C=O)

(g)

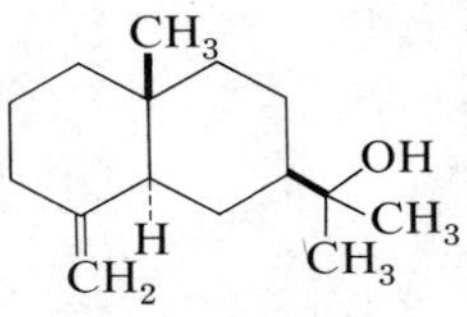

β-eudesmol

(h)

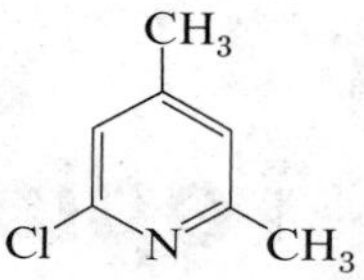

2. Do a complete literature search for each of the following compounds using Beilstein to cover the literature up to the time of the most recent Beilstein reference and the *Chemical Abstracts* formula indices to cover the literature since that time. Indicate where you looked (with time periods covered) and what references you found. If the original research journals are available in your library, scan the pertinent articles and record any physical constants (e.g., melting point, color, crystal form, spectra). List the references you find in one of the two formats discussed on p. 1116, being careful to use the correct journal abbreviations as used by the American Chemical Society journals [see *Chemical Abstracts Service Source Index* (1907–1974 Cumulative)].

(a)

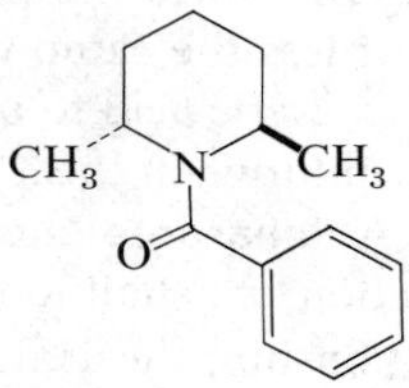

(b) $CH_2{=}CH{-}\underset{\underset{CH_3}{|}}{\overset{\overset{CH_3}{|}}{C}}{-}COCH_3$

(c) $CH_3CH_2CH_2CH_2CH_2\overset{\overset{N_3}{|}}{C}HCH_3$

(d)

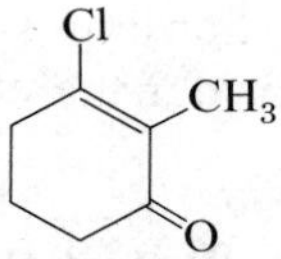

Chapter 34
Special Topics

34.1
The Hammett Equation: An Example of a Linear Free Energy Relationship

One of the important areas of research in modern organic chemistry is the elucidation of reaction mechanisms. Because we cannot see the intermediate structures along the path from reactants to products, we must deduce the mechanism of a reaction from indirect evidence. Physical organic chemists make use of many tools in gathering evidence that can be used to draw inferences about the mechanisms of reactions. One of the most useful ideas is the concept that a given structural feature will affect related reactions in more or less the same way. For example, if replacement of a hydrogen by chlorine causes acetic acid to become a stronger acid, then it is reasonable that introduction of a chlorine at the α-position of propanoic acid will also result in enhanced acidity. A **linear free energy relationship** is simply a quantitation of this idea. In this section, we shall look briefly at the first and most important linear free energy relationship, the Hammett $\sigma\rho$ equation. Since the relationship is based on the acidities of aromatic carboxylic acids, we must first consider the effect of substituents on the acidity of benzoic acid.

A. *Acidity of Substituted Benzoic Acids*

Phenylacetic acid ($K_a = 4.9 \times 10^{-5}$ M, pK_a = 4.31) is somewhat more acidic than acetic acid ($K_a = 1.8 \times 10^{-5}$ M, pK_a = 4.74). The sp^2-hybrid carbons of the benzene ring are effectively more electronegative than sp^3-hybrid carbons. This electron-attracting effect of a phenyl group was noted earlier in the dipole moment of toluene (Section 22.4). The phenyl group behaves as a normal type of substituent in that the acid-strengthening effect is attenuated rapidly down an alkyl chain, as shown by the acidity data displayed in Table 34.1.

Extrapolating in the other direction, putting the benzene ring still closer to the COOH group, we would expect benzoic acid to be much stronger acid than acetic acid. Benzoic acid ($K_a = 6.3 \times 10^{-5}$ M, pK_a = 4.20) is somewhat stronger than acetic acid, but the difference is much less than an extrapolation from Table 34.1

TABLE 34.1
Acidity of Phenylalkanoic Acids

	K_a, M	pK_a
CH_3COOH	1.8×10^{-5}	4.74
C_6H_5COOH	6.3×10^{-5}	4.20
$C_6H_5CH_2COOH$	4.9×10^{-5}	4.31
$C_6H_5CH_2CH_2COOH$	2.2×10^{-5}	4.66
$C_6H_5CH_2CH_2CH_2COOH$	1.8×10^{-5}	4.76

would suggest. The reason for the difference lies in the conjugation of the benzene ring with the carboxy group. This conjugation is less effective in the negatively charged carboxylate ion and renders the carboxy group effectively less acidic (compare α,β-unsaturated acids, Section 20.3.B).

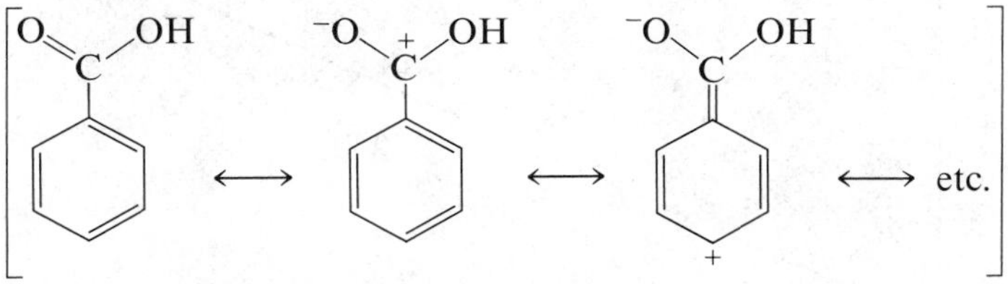

Quantitative acidity measurements have been obtained for a variety of substituted benzoic acids because of the theoretical significance of aromatic chemistry and especially of the effect of structure on reactivity in these geometrically rigid and well-defined compounds. A number of the available pK_a measurements for *ortho-*, *meta-*, and *para-*substituted benzoic acids are summarized in Table 34.2 and compared with the corresponding substituted acetic acids.

Interesting comparisons may be made by plotting the pK_as of substituted benzoic acids against those of the corresponding acetic acids. Figure 34.1 shows such a plot for *meta-*substituted benzoic acids. A fair linear correlation is obtained, with a slope of about 0.2. That is, substituent effects in the acetic acid series are more pronounced than at the *meta* position of benzoic acid. This is reasonable, since the substituent is closer to the carboxylate in the former case.

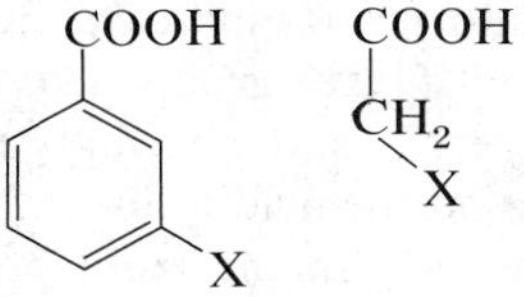

TABLE 34.2
Acidities of Substituted Benzoic and Acetic Acids

Substituent Y	Y—CH$_2$COOH	Y—C$_6$H$_4$COOH		
		ortho	*meta*	*para*
		pK_a at 25°		
H	4.74	4.20	4.20	4.20
CH$_3$	4.87	3.91	4.27	4.38
C$_2$H$_5$	4.82	3.79	4.27	4.35
F	2.59	3.27	3.86	4.14
Cl	2.85	2.92	3.83	3.97
Br	2.90	2.85	3.81	3.97
I	3.18	2.86	3.85	4.02
CN	2.47	3.14	3.64	3.55
CF$_3$	3.06		3.77	3.66
HO	3.83	2.98	4.08	4.57
CH$_3$O	3.57	4.09	4.09	4.47
C$_6$H$_5$	4.31	3.46	4.14	4.21
NO$_2$		2.21	3.49	3.42

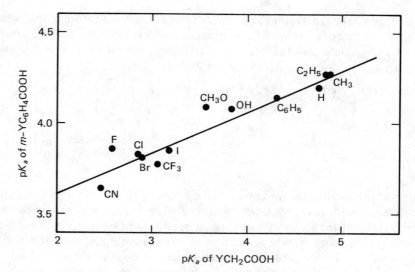

FIGURE 34.1. *Comparison of acidities of* meta-*substituted benzoic acids to those of corresponding substituted acetic acids.*

We have seen previously (Sections 10.4 and 18.4.B) that inductive effects diminish dramatically with distance.

A similar plot of the pK_as of *para*-substituted benzoic acids (Figure 34.2) shows much more scatter. Indeed, the points seem to form no consistent pattern. Electronegative atoms such as the halogens do increase the acidity of both acids, but the oxygen atoms in hydroxy and alkoxy groups cause substantial *decreases* in the acidity of benzoic acid. Furthermore, the strongly electron-attracting groups CN and CF_3 have a greater effect on the acidity of benzoic acid when they are in the *para* position than when they are in the *meta* position, even though the *para* position is one carbon farther removed from the carboxy group.

Reason is restored when we recognize that the *para* position is directly conju-

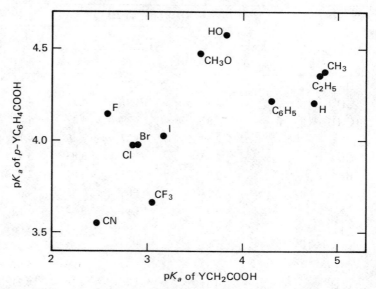

FIGURE 34.2. *Comparison of acidities of* para-*substituted benzoic acids to those of corresponding substituted acetic acids.*

1129

Sec. 34.1

**The Hammett
Equation:
An Example
of a Linear
Free Energy
Relationship**

gated with the reaction center, whereas the *meta* position is not. The pronounced acid-weakening effect of RO and HO groups can be interpreted on the basis of conjugative electron donation, which stabilizes the acid.

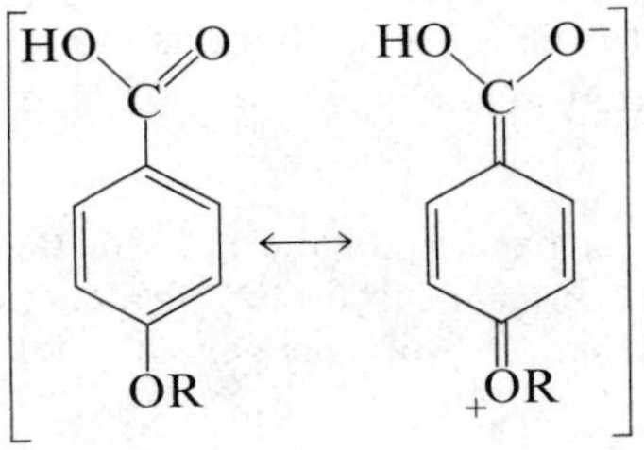

Comparable delocalization in the conjugate base is less important because the carboxylate group would then bear a double negative charge.

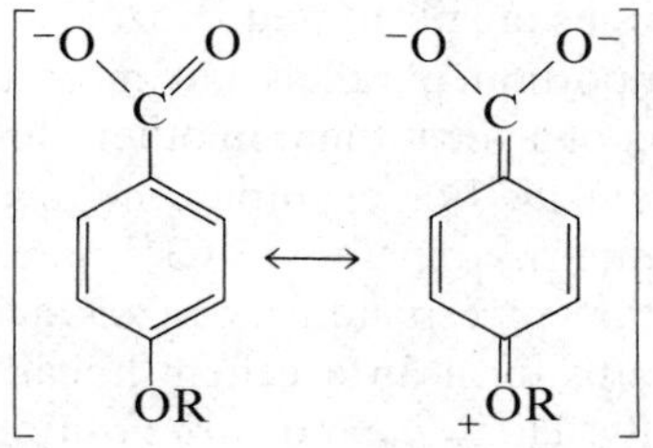

Since the RO group stabilizes the acid more than it does the anion, *acidity is decreased*.

A similar analysis of the enhanced acid-strengthening effect of the electron-attracting groups is possible. For example, the benzene π-electrons are rather polarizable. That is, they can be displaced toward an electron-attracting group or away from an electron-repelling group. In *p*-cyanobenzoic acid, this polarization provides a mechanism for electrostatic stabilization of the anion form. Thus, ionization is facilitated.

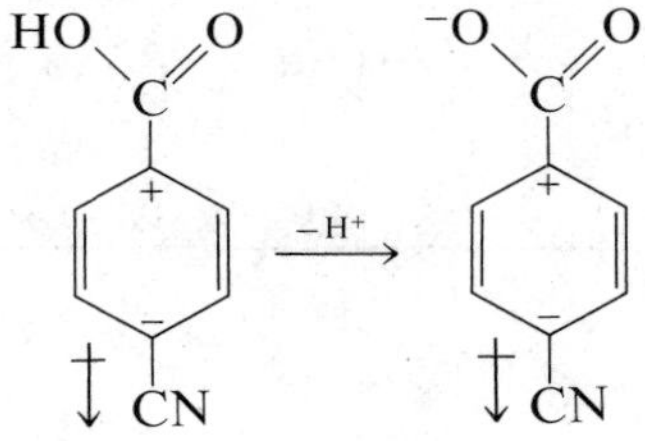

In the *meta* isomer such polarization can also operate, but it is not as effective in stabilizing the carboxylate as it is in the *para*-substituted benzoic acid.

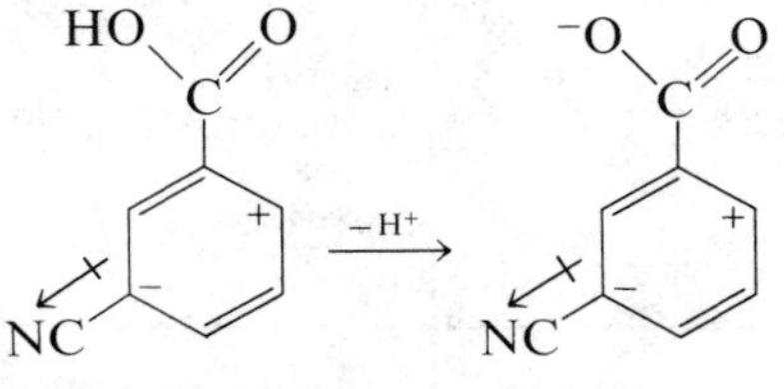

Ortho substituents present an additional complication. The steric hindrance provided by the close proximity of substituent and carboxy group prevents the carboxy group from achieving complete coplanarity with the benzene ring. That is, the carboxy group in these compounds is no longer as conjugated with the benzene ring, and the differential stabilization of the carboxylic acid relative to the carboxylate ion is lost. Consequently, all *ortho* substituents, including alkyl groups, cause an increase in the acidity of benzoic acid. Because steric effects are involved as well as electronic effects, there is no useful correlation with the acidities of other substituted acids.

EXERCISE: Using graph paper, plot the pK_as of the *meta*-substituted benzoic acids against the pK_as of the *meta*-substituted phenols (Table 30.1). Note that the points for the comparable *para* substituents generally fall off the line.

B. *The Hammett σρ Equation*

Comparison of the acidities of substituted benzoic acids with substituted acetic acids has shown some important parallels but more important differences. We next inquire as to the effect of a substituent in other phenyl compounds compared with the effect in benzoic acid. For example, the acidities of some substituted phenylacetic acids are summarized in Table 34.3. Figure 34.3 shows a plot of these acidities compared with the corresponding benzoic acids.

The *meta* and *para* groups form an excellent linear correlation, but the *ortho* groups deviate substantially. The *meta* and *para* positions are sufficiently removed from the center of reaction that only electronic effects are important; the *ortho* position involves steric effects as well. The important conclusion to be drawn from Figure 34.3 is that, in the *meta* and *para* positions, the electronic effects of a substituent in one system are proportional to the effects in another. The slope of the line in Figure 34.3, 0.46, shows that a given substituent has only half the effect in modifying the acidity of phenylacetic acid that it has on benzoic acid.

This type of interrelationship has been demonstrated for a wide variety of reactions and equilibria and has been formulated as equation (34-1) by Professor L. P. Hammett. Equation (34-1) is now known as the Hammett equation.

$$\log K_i / K_H = \rho \sigma_i \qquad (34\text{-}1)$$

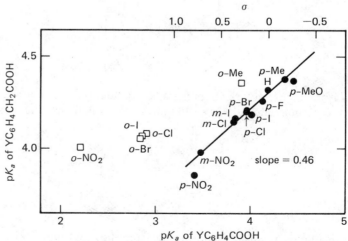

FIGURE 34.3. *Comparison of acidities of substituted phenylacetic acids to those of corresponding substituted benzoic acids.*

1131

Sec. 34.1

**The Hammett
Equation:
An Example
of a Linear
Free Energy
Relationship**

TABLE 34.3
Acidities of Substituted Phenylacetic Acids

| Substituent, Y | pK_a of Y—$C_6H_4CH_2COOH$ | | |
	ortho	meta	para
H	4.31	4.31	4.31
CH_3	4.35		4.37
F			4.25
Cl	4.07	4.14	4.19
Br	4.05		4.19
I	4.04	4.16	4.18
NO_2	4.00	3.97	3.85
CH_3O			4.36

In this equation, K_i is the equilibrium or rate constant given by substituent i, compared with that for the unsubstituted compound (substituent = H). The Greek letter sigma, σ_i, is a number characteristic of the substituent, whereas rho, ρ, is a number characteristic of the reaction. For the acidities of benzoic acids in water at 25°, ρ is *defined* as unity; thus, σ_i-values are given directly as the pK_a difference, pK_a(benzoic acid) − pK_a(substituted benzoic acid). Table 34.4 summarizes σ-values for a number of *meta* and *para* substituents. A negative σ-value signifies an electron-donating group; a positive σ-value signifies an electron-attracting group. The larger the magnitude of σ, the greater is the effect of the substituent.

Figure 34.3 corresponds to a Hammett plot with $\rho = 0.46$. The magnitude of ρ indicates the sensitivity of a given equilibrium or reaction to a given substituent. A positive ρ-value signifies that the equilibrium or reaction is aided by electron-

TABLE 34.4
Hammett Substituent Constants, σ

Group	σ_m	σ_p
CH_3	−0.069	−0.170
$C(CH_3)_3$	−0.10	−0.197
C_6H_5	0.06	−0.01
CF_3	0.43	0.54
CN	0.56	0.660
$COCH_3$	0.376	0.502
NH_2	−0.16	−0.66
NO_2	0.710	0.778
OCH_3	0.115	−0.268
OH	0.121	−0.37
F	0.337	0.062
Cl	0.373	0.227
Br	0.391	0.232
I	0.352	0.18

attracting substituents. For ionization of phenylacetic acids, the ρ of $+0.46$ indicates that a given electron-attracting substituent facilitates ionization but has only 0.46 of the effect that the same substituent has in facilitating ionization of benzoic acid.

As another example, the rates of hydrolysis of substituted methyl benzoates by hydroxide ion in aqueous acetone solution form a straight line when plotted against the Hammett σ-constants with a slope (ρ) of 2.23. This result means that substituent groups that facilitate ionization of benzoic acid also facilitate hydrolysis of methyl benzoate. Thus we conclude that the transition state for ester hydrolysis has substantial negative charge (positive ρ indicates stabilization by electron-attracting groups). This result is consistent with our view (Chapter 19) that ester hydrolysis proceeds through an anionic tetrahedral intermediate.

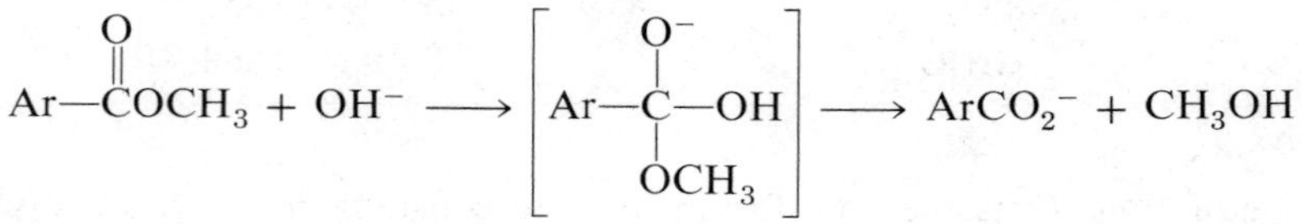

$$\text{Ar}-\overset{\overset{\displaystyle O}{\|}}{\text{C}}\text{OCH}_3 + \text{OH}^- \longrightarrow \left[\text{Ar}-\overset{\overset{\displaystyle O^-}{|}}{\underset{\underset{\displaystyle OCH_3}{|}}{\text{C}}}-\text{OH}\right] \longrightarrow \text{ArCO}_2^- + \text{CH}_3\text{OH}$$

Conversely, the reaction of substituted dimethylanilines with methyl iodide in aqueous acetone to give the trimethylanilinium iodide has $\rho = -3.30$. In this case electron-donating substituents help to stabilize the developing positive charge close to the ring and lead to a negative ρ-value.

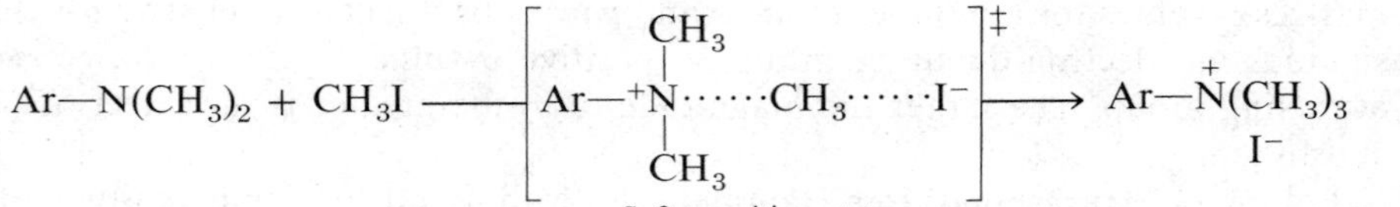

$$\text{Ar}-\text{N}(\text{CH}_3)_2 + \text{CH}_3\text{I} \longrightarrow \left[\text{Ar}-\overset{\overset{\displaystyle CH_3}{|}}{\underset{\underset{\displaystyle CH_3}{|}}{\overset{+}{\text{N}}}}\cdots\text{CH}_3\cdots\text{I}^-\right]^{\ddagger} \longrightarrow \text{Ar}-\overset{+}{\text{N}}(\text{CH}_3)_3 \quad \text{I}^-$$

S_N2 transition state

EXERCISE: (a) Plot the pK_as of substituted phenols from Table 30.1 versus the corresponding σ values and determine ρ. Make a comparable plot for the pK_as of anilinium ions using the data of Table 24.4. Why does a given substituent have a greater effect on ionization of a phenol or an anilinium ion that it does on benzoic acid? Explain using resonance structures where desirable.

(b) The pK_as of m- and p-iodoxybenzoic acids in water at $25°$ are 3.50 and 3.44, respectively. Calculate σ_m and σ_p for the iodoxy group, $-\text{IO}_2$. Give a rationalization of the approximate magnitude on the basis of Lewis structures of the $-\text{IO}_2$ group.

(c) Give the sign of ρ for the reaction

$$\text{ArCH}_2\text{NMe}_2 + \text{CH}_3\text{I} \longrightarrow \text{ArCH}_2\text{N}(\text{CH}_3)_3^+ \text{ I}^-$$

(d) Compare the expected magnitude of ρ for part (c) with that for the following reaction.

$$\text{ArCH}_2\text{CH}_2\text{NMe}_2 + \text{CH}_3\text{I} \longrightarrow \text{ArCH}_2\text{CH}_2\text{N}(\text{CH}_3)_3^+ \text{ I}^-$$

(e) What is the sign of ρ for the reaction

$$\text{ArSO}_2\text{OCH}_3 + \text{C}_2\text{H}_5\text{O}^- \longrightarrow \text{ArSO}_3^- + \text{CH}_3\text{OC}_2\text{H}_5$$

(f) The first-order rate constants for solvolysis is isopropyl esters of substituted benzenesulfonic acids in 50% aqueous ethanol at $25°$ are

Substituent	$10^5 k$, sec^{-1}
H	2.50
p-CH$_3$	1.47
p-Br	7.30
m-NO$_2$	34.2
p-NO$_2$	44.4

Calculate ρ for this reaction and predict the solvolysis rate of isopropyl m-trifluoromethylbenzenesulfonate.

(g) For the diazonium ion–diazotate ion equilibrium expressed as

$$\text{ArN}_2^+ \underset{}{\overset{K}{\rightleftharpoons}} \text{ArN}_2\text{O}^- + 2\text{ H}^+$$

we learned (problem 25-14) that $\frac{1}{2}\log[\text{ArN}_2\text{O}^-]/[\text{ArN}_2^+] = \text{pH} - \frac{1}{2}\text{p}K$. Some values of $\frac{1}{2}\text{p}K$ for substituted benzenediazo compounds at 25° are as follows: p-NO$_2$, 9.44; m-Cl, 10.70; H, 11.90; m-CH$_3$, 12.12. Plot these values against the appropriate σ values and calculate ρ for this equilibrium. Does ρ have the sign you would have expected?

34.2
Pericyclic Transition States

In our study of organic chemistry we have encountered a number of reactions that proceed through cyclic transition states. Examples are the Diels–Alder reaction (Section 20.5) and the Claisen and Cope rearrangements (Section 30.6.D). These transition states were related to benzene and the facility of these reactions was compared with the aromatic character of benzene. The comparison is accurate and, indeed, these and many other reactions, including a number we have already studied, can be treated as cyclic electronic systems *to which the Hückel* $4n + 2$ *rule can be applied*. In this section we shall demonstrate this generality and show its application to several important types of reaction.

A. *Electrocyclic Reactions*

cis-1,3,5-Hexatriene undergoes a facile transformation on heating to give 1,3-cyclohexadiene. This type of isomerization is known as an **electrocyclic** reaction. The reaction can be perceived as proceeding through a cyclic six-membered transition state.

cis-1,3,5-hexatriene   1,3-cyclohexadiene

Moreover, if we follow the electrons involved by the conventional symbolism of a curved arrow for each pair of electrons involved, we find that three arrows are required. That is, six electrons participate in the transformation and hint at the involvement of $4n + 2$ in some manner.

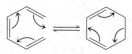

There is a complication, however, with regard to stereochemistry. In the open-chain hexatriene best π-overlap of the double bonds is achieved when all six carbons and eight hydrogens lie in the same plane—including both terminal CH_2 groups. In the cyclohexadiene, however, the two methylene groups form part of a ring and are no longer coplanar. That is, in the transformation from hexatriene to cyclohexadiene, the terminal methylene groups must rotate out of coplanarity.

In principle, these rotations can take two possible modes. They can both rotate in the same sense when viewed from the same direction **(conrotatory motion)** or in the opposite sense **(disrotatory motion)**

$$H-C-H \quad H-C-H \qquad H-C-H \quad H-C-H$$

conrotatory motion   disrotatory motion

In the simple case of hexatriene itself, these alternative modes of rotation cannot be distinguished, but substituted compounds would lead to different isomers. Consider the (E,Z,E)-1,6-dimethyl compound as an example. If both end groups rotate in opposite directions in disrotatory fashion, the product is the *cis*-dimethyl-cyclohexadiene. If they rotate together in conrotatory fashion, the product is the *trans*-dimethylcyclohexadiene.

(E,Z,E)-octa-2,4,6-triene $\xrightarrow{\text{disrotatory}}$ *cis*-5,6-dimethylcyclohexa-1,3-diene

$\xrightarrow{\text{conrotatory}}$ *trans*-5,6-dimethylcyclohexa-1,3-diene

The reaction involves the conversion of the two terminal p-orbitals from π-bonding to σ-bonding. At this point it is important to examine the signs of the orbital wave functions to determine whether the orbital overlaps involved are bonding (positive) or antibonding (negative) (Figure 34.4). In the starting hexatriene the p-orbitals have signs assigned to the wave functions to provide the most positive overlaps and lead to a set of π-molecular orbitals that describe the electronic structure. Note that for clarity only the overlap (dotted line) at the top is shown. In disrotatory motion both terminal CH_2 groups rotate so that the positive lobes interact to give positive overlap throughout, exactly as in the related benzene system included for comparison in Figure 34.4. Even though the orbitals in

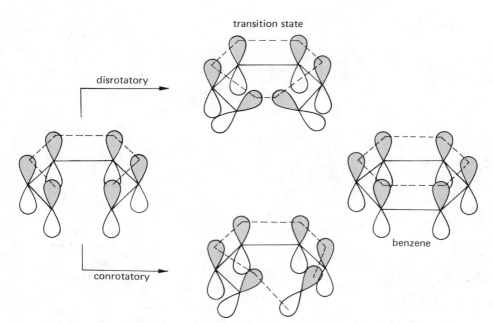

FIGURE 34.4. *Orbital interactions involved in disrotatory and conrotatory ring closure of 1,3,5-hexatriene, compared to benzene. Shaded lobes represent positive wave functions; unshaded lobes represent negative wave functions.*

the transition state for electrocyclic reaction are not aligned exactly as in benzene, the all-important overlap characteristics are the same in both; that is, the orbital overlaps are all positive around the ring.

This result is to be contrasted with the pattern for conrotatory motion. In this case the positive lobe of one terminal *p*-orbital starts to overlap with the negative lobe of the other terminal *p*-orbital. The disrotatory transition state clearly more closely resembles orbital interactions in benzene and, indeed, the electrocyclic reaction of hexatrienes is completely disrotatory. The product of the dimethyl case shown is *exclusively* the *cis*-dimethylcyclohexadiene, even though this product is thermodynamically less stable than the alternative *trans* structure.

The substantial difference in activation energies for the two cases is directly related to the stabilization energies of aromatic systems having cyclic π-systems with $4n + 2$ electrons. Recall that a cyclic system of *p*-orbitals gives rise to a pattern of molecular orbitals in a distinctive manner. There is a lowest-lying molecular orbital with zero nodes and thereafter the molecular orbitals occur as degenerate pairs having one, two, and so on, nodes (Section 22.7.). In the benzene-like transition state for disrotatory ring closure in Figure 34.4, this symmetry is disturbed so that the higher molecular orbitals no longer occur as degenerate pairs, but the molecular orbitals within each pair still have energies that are close together. As such they still represent orbital "shells" that provide aromatic-like stability when filled. The molecular orbital energies of the disrotatory transition state of Figure 34.4 are illustrated in Figure 34.5a.

The negative overlap required for the conrotatory ring closure gives rise to an entirely different pattern of molecular orbital energies. The negative overlap constitutes a node; hence, there cannot be any molecular orbital with zero nodes. Instead, we find a pair of molecular orbitals of similar energy with one node each, a higher pair with two nodes, and so on. This pattern is illustrated in Figure 34.5b.

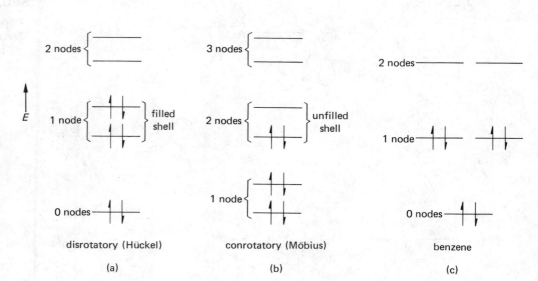

FIGURE 34.5. *Energy level diagrams for alternative transition states for ring closure of hexatriene compared to that for benzene.*

Six electrons leave the second shell unfilled, a condition that represents relative instability.

It is convenient to have names for these two possible patterns of molecular orbital levels. The pattern for disrotatory closure is a **Hückel molecular orbital system** and gives filled molecular orbital shells with $4n + 2$ electrons. The conrotatory pattern in Figure 34.5b is frequently referred to as a **Möbius molecular orbital system,** and has the important characteristic of giving filled molecular orbital shells with $4n$ electrons.

This name derives from the topology of a **Möbius strip.** A Möbius strip is formed by taking a circular band, cutting in one place, giving one twist, and rejoining at the cut. The resulting strip has no inside nor outside! Both are joined in one continuous manner (Figure 34.6). In a Hückel molecular orbital system, the *p*-orbitals are set up with a positive "top" and negative "bottom." In the Möbius system, the negative overlap joins the "top" and the "bottom" in a manner that resembles the joining of the inside and outside of a Möbius strip.

strip having
inside and outside

Möbius strip
inside is the outside

FIGURE 34.6. *Illustrating the surfaces of a Möbius strip.*

Note that this point also emphasizes the difference between setting up atomic orbitals, such as *p*-orbitals, to overlap in a given fashion (basis functions) and the set of molecular orbitals that results from such overlaps. If we start with *n* interacting atomic orbitals, we must end up with *n* molecular orbitals. The energies of the molecular orbitals depend on how the starting atomic orbitals overlap. We may summarize this discussion as follows: A set of *p*-orbitals overlapping in a cyclic manner with zero (or an even number) of negative overlaps gives rise to a Hückel pattern of molecular orbital energy levels to which quantum numbers can be assigned as 0, ± 1, ± 2, and so on. Cyclic interaction of a set of p-orbitals with one (or an odd number) of negative overlaps gives rise to a set of molecular orbitals having the Möbius pattern of energy levels to which quantum numbers can be assigned as ± 1, ± 2, and so on.

Let us now apply these principles to the corresponding electrocyclic ring closure of 1,3,5,7-octatetraene to 1,3,5-cyclooctatriene. In contrast to the (E,Z,E)-dimethylhexatriene case discussed previously, the (E,Z,Z,E)-dimethyloctatetraene compound shown (34-2) gives, as the first product of thermal electrocyclic reaction, *exclusively* the *trans*-dimethylcyclooctatriene, the product of *conrotatory* motion!

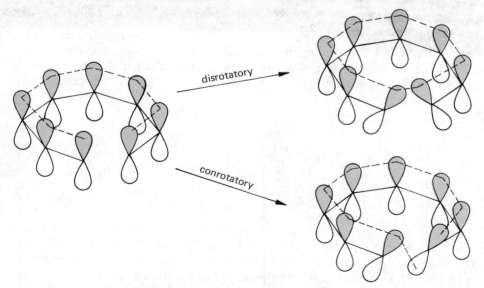

(34-2)

cis-7,8-dimethylcycloocta-1,3,5-triene

(E,Z,Z,E)-deca-2,4,6,8-tetraene

trans-7,8-dimethylcyclooctа-1,3,5-triene
exclusive product

To see why this system changes so dramatically from the hexatriene case, we again look at the orbital overlaps involved (Figure 34.7). In disrotatory motion the overlaps involved are again all positive and give rise to a Hückel pattern of molecular orbital energy levels. But the eight electrons now involved no longer fit the $4n + 2$ rule. The result is the instability associated with an unfilled orbital shell. On the other hand, the conrotatory transition state gives rise to a Möbius pattern of molecular orbital levels. The eight electrons fill the first two shells and have the

FIGURE 34.7. *Orbital overlaps involved in cyclization of octatetraene.*

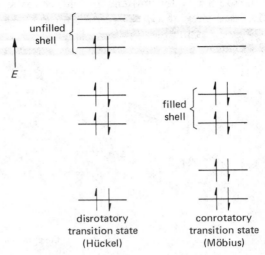

FIGURE 34.8. *Energy level pattern of molecular orbitals for cyclization of octatetraene.*

stability associated with filled orbital shells (Figure 34.8); that is, conrotatory ring closure of octatetraene involves a transition state that has Möbius aromatic character and is energetically more favorable than the Hückel antiaromatic character of the transition state for disrotatory ring closure.

This result may be generalized. *Those thermal electrocyclic reactions that involve 4n + 2 electrons react with disrotatory motion* so that the orbitals involved can overlap in the Hückel sense. *Those thermal electrocyclic reactions that involve 4n electrons react with conrotatory motion* so that the orbitals involved can overlap in the Möbius sense. These generalizations hold whether the reaction involved is that of ring closure or ring opening (principle of microscopic reversibility).

The thermal ring opening of cyclobutenes provide a further example. On heating, *cis*-3,4-dimethylcyclobutene is smoothly converted to (E,Z)-hexa-2,4-diene.

cis-3,4-dimethyl-
cyclobutene

(E,Z)-hexa-2,4-diene

The reaction involves a four-electron cycle. The filled-shell molecular orbital system of the transition state thus requires Möbius overlap and conrotatory motion (Figure 34.9). Disrotatory ring opening would give a Hückel cyclic system, which, with four electrons, would be antiaromatic (Figure 34.9).

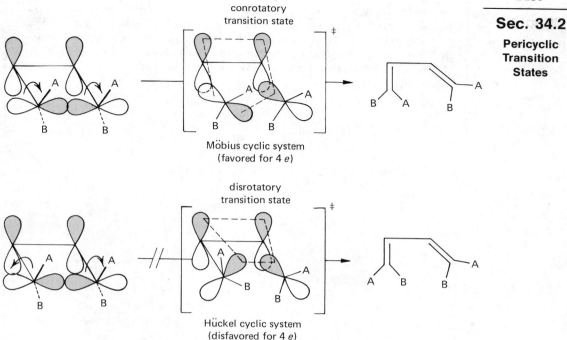

FIGURE 34.9. *Orbital interactions for electrocyclic ring openings of cyclobutene.*

If this example is approached from the microscopically reverse direction, ring closure of butadiene, we obtain the equivalent result (Figure 34.10). This transition state looks superficially different from that derived from ring opening (Figure 34.9); that is, the + and − labeling of different lobes is different, but both pictures have one negative overlap and correspond to a Möbius cyclic system. Moreover, they give rise to precisely the same set of four molecular orbitals. In the approach shown here, it is only necessary to determine whether a given transition state is a Möbius or Hückel cyclic system. Other approaches based on detailed consideration of the symmetries of individual molecular orbitals may also be applied, but the present $4n$ versus $4n + 2$ approach is exactly equivalent and simpler to use in practice.

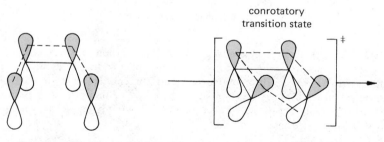

FIGURE 34.10. *Orbital interactions for ring closure of butadiene.*

The foregoing considerations lead to the following generalizations for thermal electrocyclic reactions.

$4n$ electrons (4, 8, 12, etc.)	conrotatory motion
$4n + 2$ electrons (2, 6, 10, 14, etc.)	disrotatory motion

These rules apply to thermal reactions only. For photochemical reactions, the rules are usually exactly opposite, because electronic excited states have some important symmetry differences from ground states that we will not explore here. Furthermore, photochemical electrocyclic reactions are often not concerted reactions (Section 34.4). Thermal and photochemical reactions can sometimes be combined to give interesting results, for example,

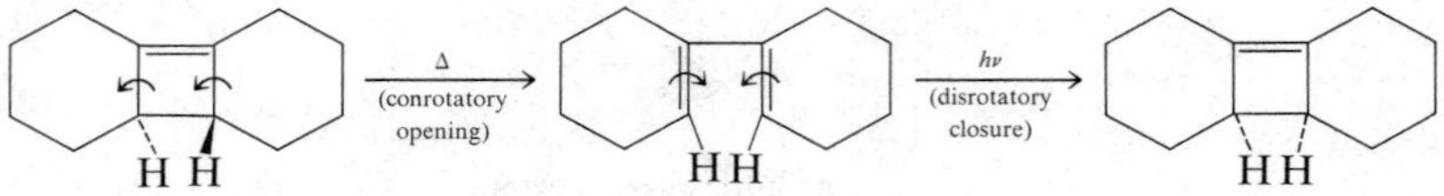

Some additional examples (34-3) and (34-4) follow.

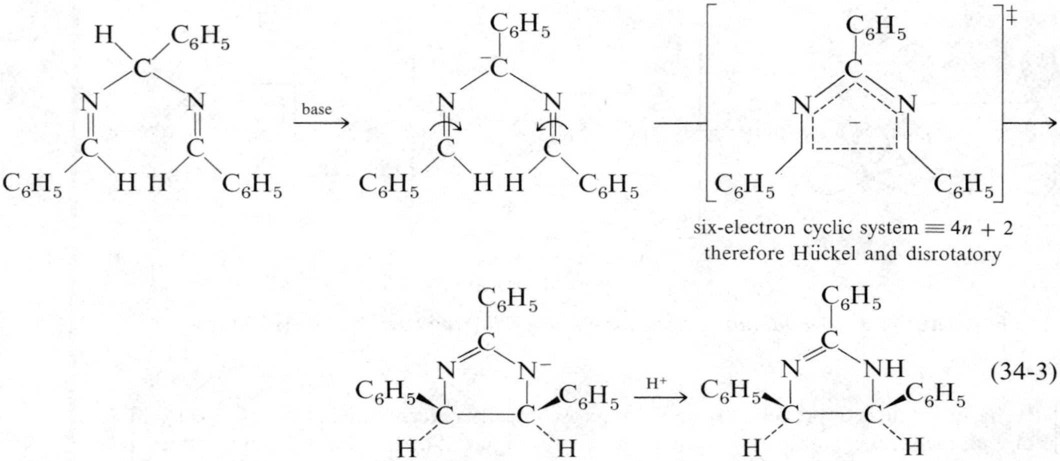

six-electron cyclic system ≡ $4n + 2$
therefore Hückel and disrotatory

cis product
amarine

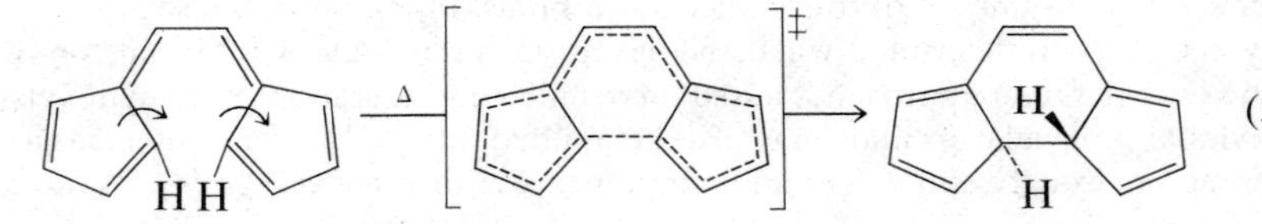

(34-3)

twelve-electron cyclic system ≡ $4n$
therefore Möbius and conrotatory

(34-4)

EXERCISE: (a) The theoretically aromatic hydrocarbon [10]annulene has been the goal of numerous synthetic efforts. One attempted synthesis should have produced the (Z,E,Z,Z,E) stereoisomer. However, an isomer of 9,10-dihydronaphthalene was obtained instead. What was the stereochemistry of this product, *cis* or *trans*?

(b) Of the two stereoisomers of each of the following compounds, which is expected to be the more thermally stable?

    (i) bicyclo[4.2.0]octa-2,4-diene
    (ii) bicyclo[4.2.0]oct-7-ene

(c) In each of the following thermal electrocyclic reactions, predict whether the product is *cis* or *trans*.

    (i) (E,E)-hepta-2,5-dien-4-yl cation to 4,5-dimethylcyclopent-2-en-1-yl cation
    (ii) *cis*-bicyclo[5.2.0]nona-2,8-diene to bicyclo[4.3.0]nona-2,4-diene
    (iii) (E,E,E)-5-phenylnona-2,5,7-trien-4-yl anion to 1-phenyl-4,5-dimethyl-cyclohepta-2,6-dien-1-yl anion

B. Cycloaddition Reactions

The Diels–Alder reaction is an example of a cycloaddition reaction.

Diels–Alder reactions involve the formation of two σ-bonds and one π-bond from the two π-bonds of a diene and the π-bond of a monoene. Accordingly, the common Diels–Alder reaction can be referred to as a $(_4\pi + {_2}\pi)$ cycloaddition reaction. If we label the orbitals involved, we can see that the normal Diels–Alder reaction involves a Hückel cyclic electronic system with six electrons; that is, the transition state is aromatic (Figure 34.11).

On the other hand, $(_2\pi + {_2}\pi)$ cycloaddition reactions are unknown. Ethylene, for example, does not dimerize, even under high pressure, despite the large favorable enthalpy for forming cyclobutane.

$$2\ CH_2{=}CH_2 \longrightarrow \begin{array}{c} CH_2{-}CH_2 \\ |\qquad\ | \\ CH_2{-}CH_2 \end{array} \qquad \Delta H° = -18.2 \text{ kcal mole}^{-1}$$

Cycloaddition in the $(_2\pi + {_2}\pi)$ manner involves a transition state of the antiaromatic Hückel type (four electrons $\neq 4n + 2$). Instead, ethylene undergoes linear polymerization to form polyethylene. Examples of thermal cycloadditions to form cyclobutanes are known, but in virtually every case the mechanism appears to involve a diradical intermediate rather than a concerted reaction via a cyclic transition state.

$$2\ CH_2{=}CHCN \xrightarrow[\Delta]{\text{pressure}} [NC\dot{C}HCH_2CH_2\dot{C}HCN] \longrightarrow \underset{CN}{\overset{CN}{\square}}$$

However, many photochemical $(_2\pi + {_2}\pi)$ cycloadditions are known. In photochemical reactions, a $4n$ cyclic transition state is acceptable, although many of these reactions may also involve diradical intermediates. One example of a photochemical cycloaddition occurs when cinnamic acid is exposed to sunlight.

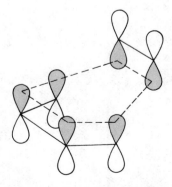

FIGURE 34.11. *Transition state for $(_4\pi + {_2}\pi)$ cycloaddition reaction; all overlaps are positive.*

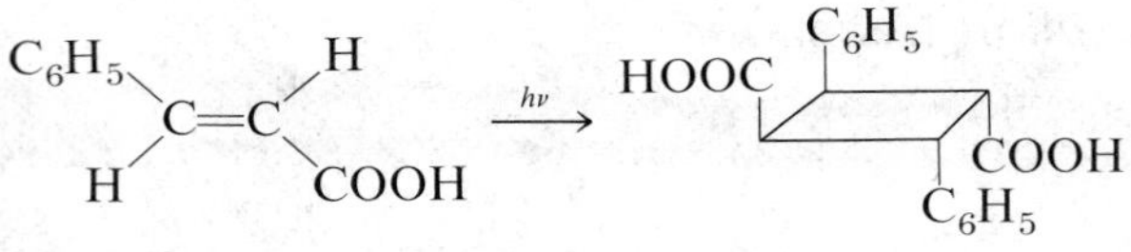

trans-cinnamic acid   α-truxillic acid

Many variations of $(_4\pi + {}_2\pi)$ cycloadditions are known. Two further examples are

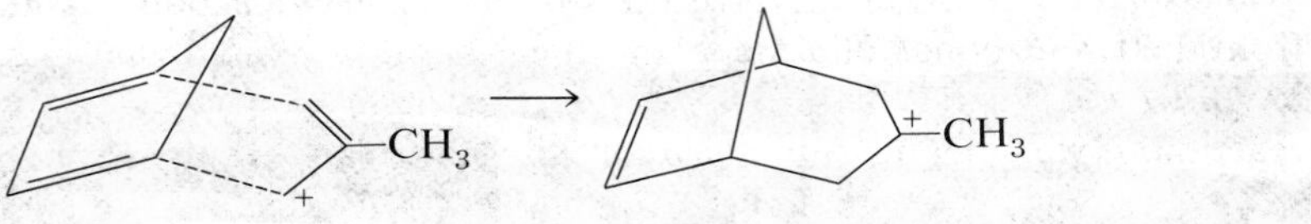

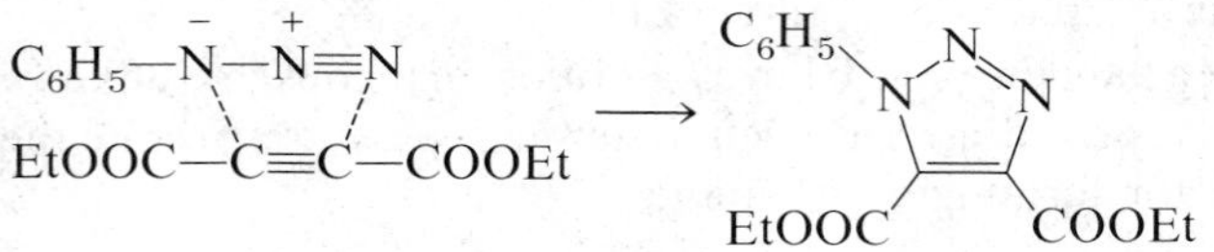

Other examples of dipolar cycloaddition reactions were discussed in an earlier chapter (Section 32.5.B); these examples are all of the $(_4\pi + {}_2\pi)$ type.

Still other examples involve higher $4n + 2$ cyclic electronic transition states.

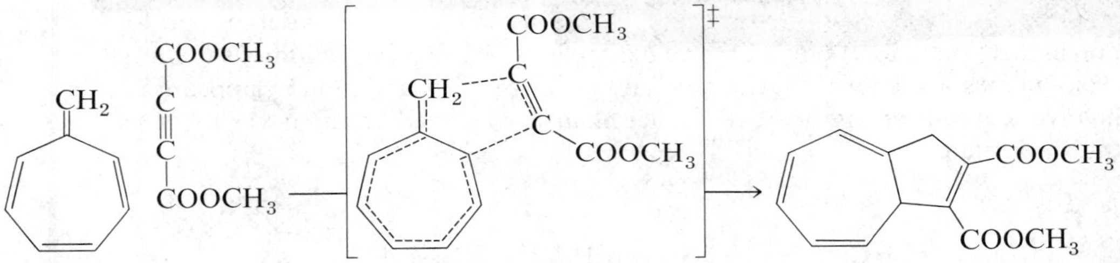

$(_8\pi + {}_2\pi)$ cycloaddition
$\equiv 4n + 2$ cyclic transition state

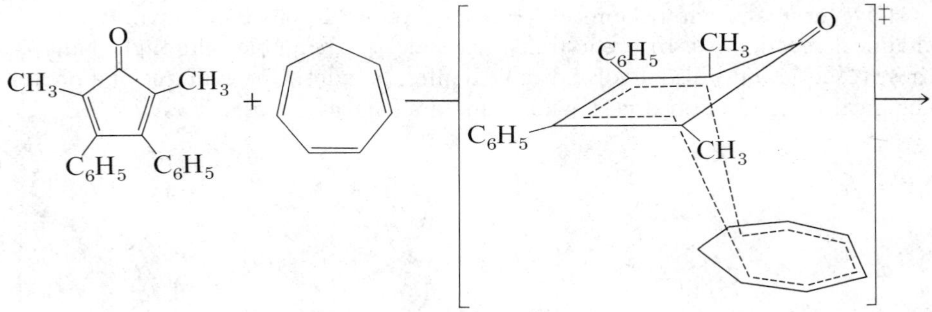

$(_8\pi + {}_6\pi)$ cycloaddition
$\equiv 4n + 2$ cyclic transition state

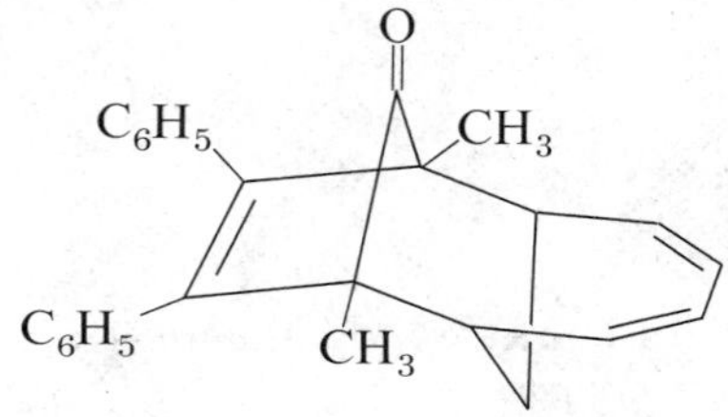

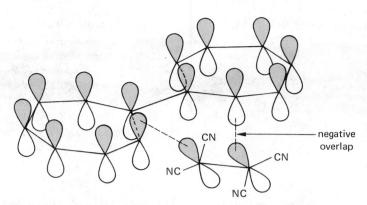

FIGURE 34.12. *Orbital interactions for* trans *addition of tetracyanoethylene to heptafulvalene.*

A remarkable exception would appear to be reaction (34-5), which involves a $(_{14}\pi + {_2}\pi)$ cycloaddition and does not fit the $4n + 2$ rule.

$$\text{heptafulvalene} \quad \text{tetracyanoethylene} \quad \quad \quad (34\text{-}5)$$

However, note that the product is the result of *anti* addition. The corresponding transition state (Figure 34.12) involves a negative overlap that would correspond to a Möbius cyclic electronic system, a favorable transition state for a 16-electron cyclic system!

In principle, ethylene could dimerize if the positive lobes of one molecule could overlap with the opposite lobes of another (Figure 34.13). The resulting transition state has one negative overlap and creates a Möbius molecular orbital system. However, the steric constraints of such a four-membered cyclic system mean that

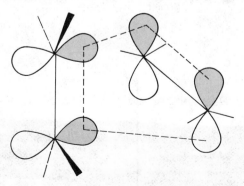

FIGURE 34.13. *Orbital interactions for the hypothetical* trans *or* $(_2\pi_a + {_2}\pi_s)$ *cycloaddition of one ethylene with another.*

it possesses energy that is too high for it to compete with alternative modes of reaction.

Addition to the same side of a π-system is called **suprafacial** and is symbolized with a subscript s; addition to opposite sides of a π-system is called **antarafacial** and is symbolized with a subscript a. Hence, the normal Diels–Alder reaction is an example of a $(_4\pi_s + _2\pi_s)$ cycloaddition. Reaction (34-5) of heptafulvalene with tetracyanoethylene discussed previously is an example of a $(_{14}\pi_a + _2\pi_s)$ cyclo-addition. In general, $(_p\pi_s + _q\pi_s)$ cycloadditions are thermally "allowed" when $p + q = 4n + 2$, whereas $(_p\pi_a + _q\pi_s)$ thermal cycloadditions, a somewhat rarer breed, are thermally allowed when $p + q = 4n$.

EXERCISE: (a) Consider the electrocyclization of 1,3,5-hexatriene as an intramolecular $(_4\pi_s + _2\pi_s)$ cycloaddition (Diels–Alder reaction) in which one of the double bonds "adds" across the two ends of the remaining diene system. What is the stereochemistry predicted by this method? Apply this type of reasoning to the cyclization of 1,3-butadiene to cyclobutene, viewing it as a $(_2\pi_a + _2\pi_s)$ cycloaddition.

(b) For each of the following cycloaddition reactions determine whether the reaction is thermally allowed for the stereochemistry shown.

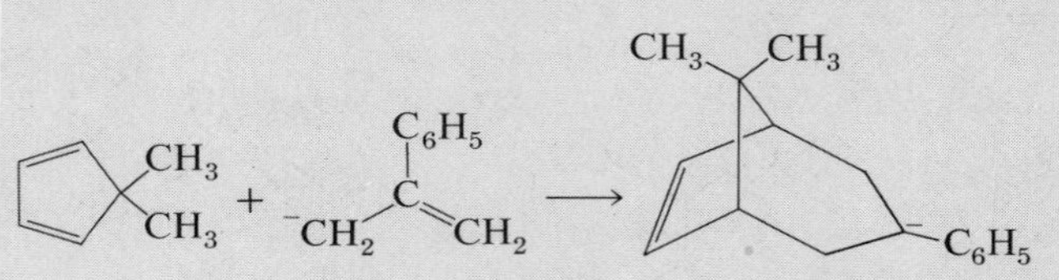

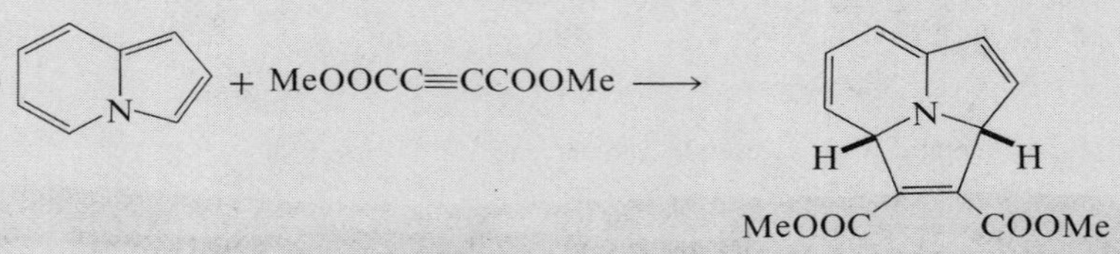

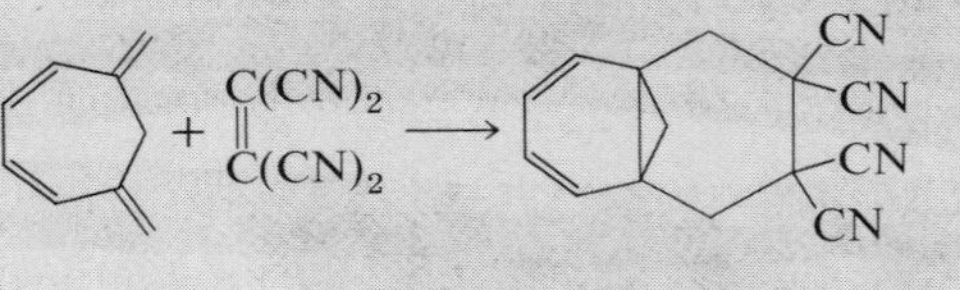

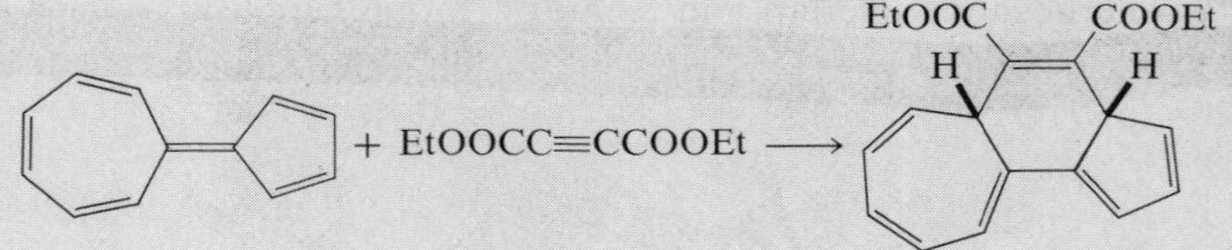

C. Sigmatropic Rearrangements

A **sigmatropic rearrangement** of order $[p,q]$ is defined as a concerted rearrangement in which a bond between two conjugated systems is broken at the same time that a new one is formed at the pth atom of one of the systems and the qth atom of the other. For example, the Claisen rearrangement (Section 30.6.D) is an example of a [3,3] sigmatropic rearrangement.

The Cope rearrangement is also a [3,3] sigmatropic rearrangement.

We have alluded earlier to the benzene-like or aromatic character of such transition states. Indeed, we expect such aromatic character whenever $4n + 2$ electrons are involved. One well-known example that fits this pattern is the familiar carbocation rearrangement, a reaction that could be termed a two-electron [2,1] sigmatropic rearrangement.

The orbital picture of the transition state follows in straightforward fashion (Figure 34.14). Only positive orbital overlaps are involved; hence, the system is of the Hückel type and has aromatic stabilization for two electrons. Note that the migrating group is still pyramidal and that the cyclic system is derived from hybrid orbitals rather than pure *p*-orbitals.

Such rearrangements are common for carbocations but do not occur for carbanions. The 1,2-rearrangement of carbanions would be described as a four-electron [2,1] sigmatropic rearrangement and would be antiaromatic.

Another sigmatropic rearrangement that involves a Hückel six-electron transition state is a 1,5-hydrogen shift, a [5,1] sigmatropic rearrangement. An example

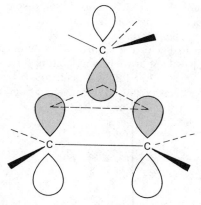

FIGURE 34.14. *Orbital interactions in a [2.1]sigmatropic rearrangement.*

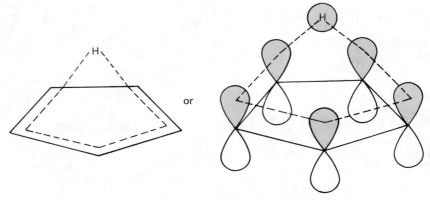

FIGURE 34.15. *Orbital interactions for a [5,1]sigmatropic rearrangement.*

of this reaction is afforded by 5-methyl-1,3-cyclopentadiene, which rearranges readily, first to 1-methyl-1,3-cyclopentadiene and then to an equilibrium mixture that contains 2-methyl-1,3-cyclopentadiene.

The [5,1] sigmatropic character of the transition state is apparent when written as in Figure 34.15. All overlaps involved are positive; hence, the system is Hückel and aromatic for six electrons.

The benzidine rearrangement (Section 25.1.C) is an example of a [5,5] sigmatropic rearrangement and is aromatic for a Hückel ten-electron cycle.

In this section we have only touched on what has become a vast and growing field, a subject of much current research in organic chemistry. The unifying concepts that have brought such order to a hitherto mysterious collection of experimental results are only a few years old. The vitality and growth of modern organic chemistry are emphasized by the realization that the interpretation of pericyclic transition states given in these pages was unknown in the organic chemistry of only two decades ago.

EXERCISE: (a) Although *trans*-1,2-divinylcyclopropane is a relatively stable compound, the *cis* isomer rearranges rapidly at 0°. Suggest a reason for the high reactivity of this isomer. What product does it give?

(b) Allyl acetate is treated with the strong base lithium diisopropylamide. The resulting enolate solution is kept at room temperature for 2 hr and then quenched with dilute acid. The product is found to be 4-pentenoic acid. Propose a mechanism.

(c) Determine whether each of the following sigmatropic reactions is thermally allowed for the stereochemistry shown.

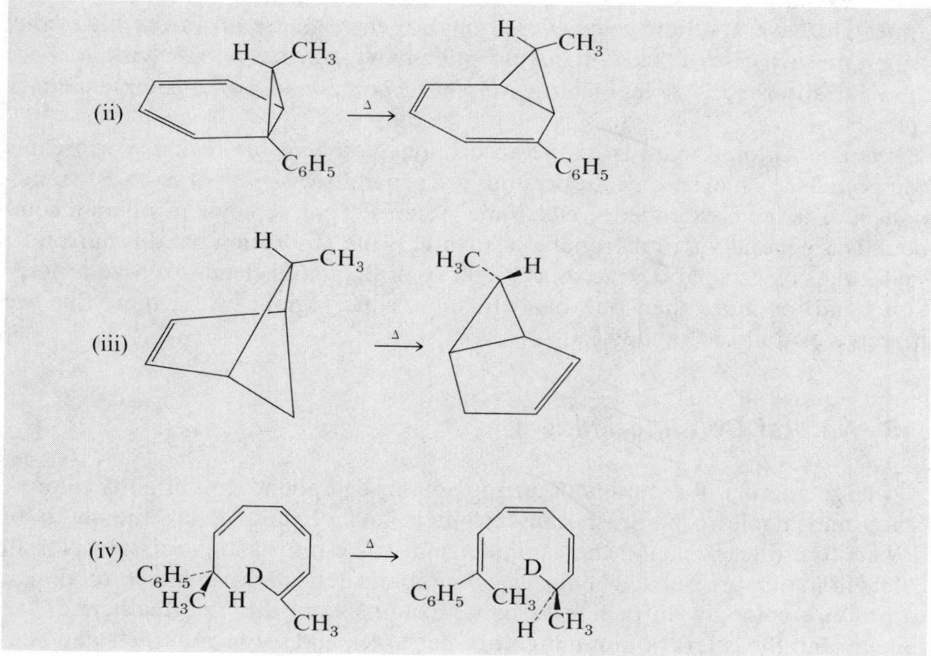

(ii)

(iii)

(iv)

34.3
Organic Coloring Matters

A. *Color*

As discussed in Chapter 21, all compounds can be excited electronically by electromagnetic radiation. For most organic compounds such transitions are in the ultraviolet region of the spectrum, and such compounds are white and colorless. However, when an electronic transition is in the visible range (about 400–750 nm), the compound will appear to us as colored. The color perceived for different wavelengths of light are summarized in Table 34.5.

Light of a given wavelength is perceived as the indicated color. However, if that wavelength is *absorbed,* we perceive the complementary color. Some compounds

TABLE 34.5
Color and Wavelength

Wavelength of light, nm	Color	Complementary color
400–430	violet	green-yellow
430–480	blue	yellow
480–490	green-blue	orange
490–510	blue-green	red
510–530	green	purple
530–570	yellow-green	violet
570–580	yellow	blue
580–600	orange	green-blue
600–680	red	blue-green
680–750	purple	green

appear to have a yellow color even though their λ_{max} are all in the ultraviolet region. In such cases a "tail" of an absorption band stretches into the visible. Since the light absorbed is violet or blue, we see the compound as the complementary color of yellow.

Intensely colored materials have absorptions in the visible region. For organic compounds, such electronic absorptions are generally $\pi \rightarrow \pi^*$ or $n \rightarrow \pi^*$ transitions and involve extended π-electronic systems. That is, color in organic compounds is generally a property of π-structure. If the absorption band is narrow or sharp, the color will appear to us as bright or brilliant and clean. A broad absorption band, or more than one band in the visible region, gives colors that we perceive as dull or "muddy."

B. *Natural Coloring Matters*

A large number of naturally occurring organic compounds are brightly colored. Some find a role in nature because of their color. Examples are the colors of flowers to attract bees and the camouflage of some insects and animals. Not all colors in nature are due to chemicals. Some colors depend on architectural design to produce color by diffraction of light. Examples are the blue feathers of the bluejay and the colors of hummingbirds, peacocks, and some butterflies and beetles. More often, however, specific organic compounds are involved. A few of the important classes of compounds with representative examples will be summarized here.

Anthocyanins provide much of the color of the plant world. They are responsible for the red color of buds and young shoots and for the purple of autumn leaves as the green chlorophyll decomposes with the approach of winter. Their colors depend in part on the pH of their environment. For example, the blue cornflower and red rose have the same anthocyanin, cyanin. The blue color is that of the potassium salt. Anthocyanins are actually present as glycosides. Hydrolysis gives the corresponding anthocyanidins; that is, anthocyanidins are the aglycones of anthocyanins.

Only three anthocyanidins are important: cyanidin (crimson to blue-red flowers, cherries, cranberries), pelargonidin (pelargonium, geranium), and delphinidin (delphinium, pansy, grape). These compounds have the following structures.

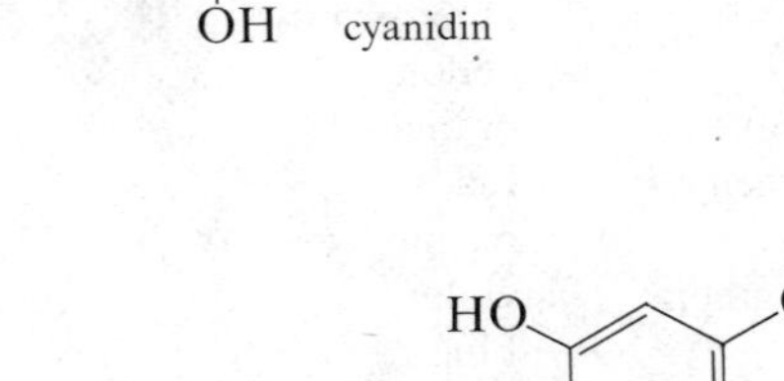

cyanidin

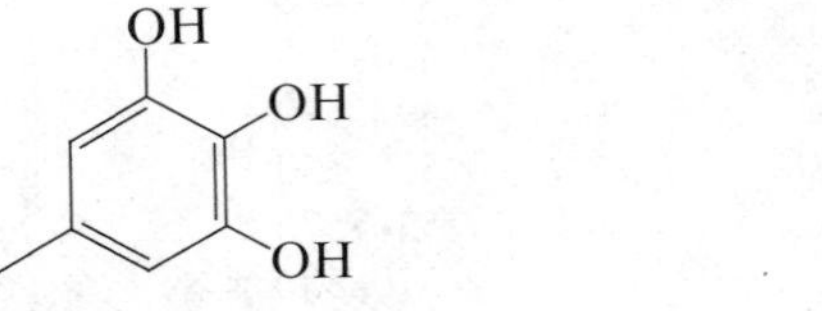

pelargonidin

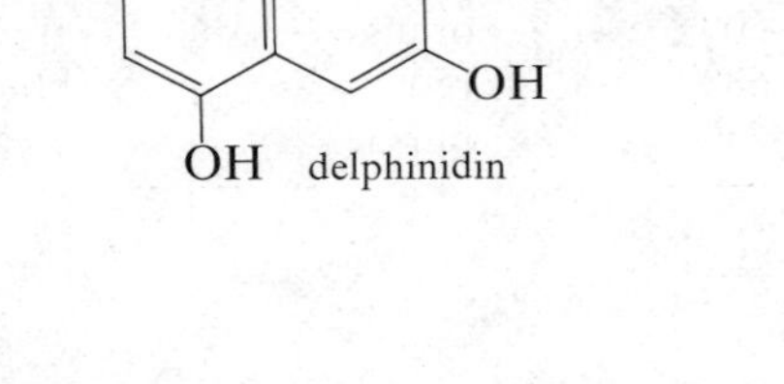

delphinidin

These pyrylium salts (Section 32.9) are generally considered as derivatives of a parent structure, flavone, a nucleus that is widespread in nature.

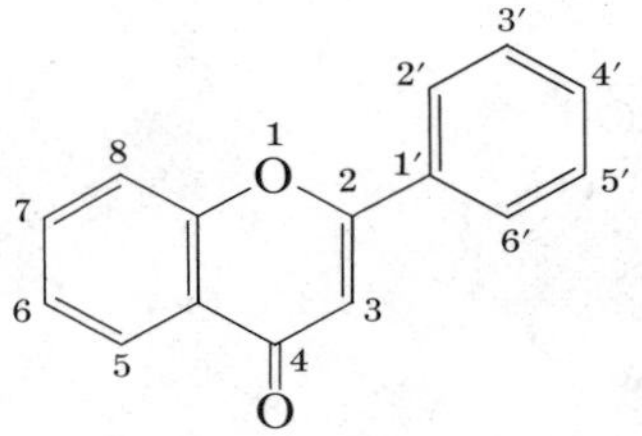

flavone

In the corresponding glycosides, the sugar units are attached at the 3- and 5-positions.

Carotenoids are also widespread in nature from bacteria and fungi to vegetable and animal life. Examples are β-carotene, which is a precursor for vitamin A, and lycopene, which occurs in tomatoes and ripe fruit.

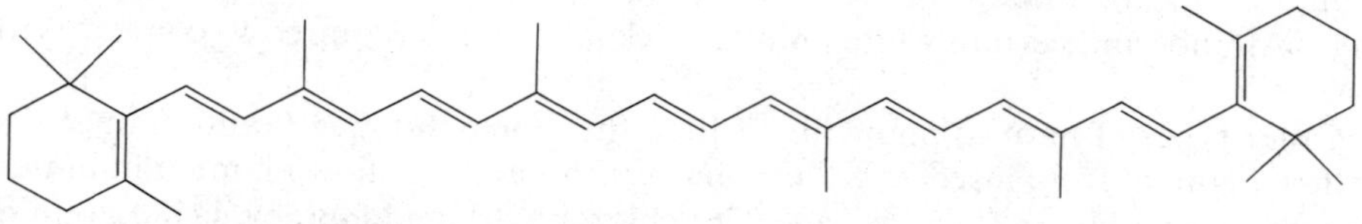

β-carotene

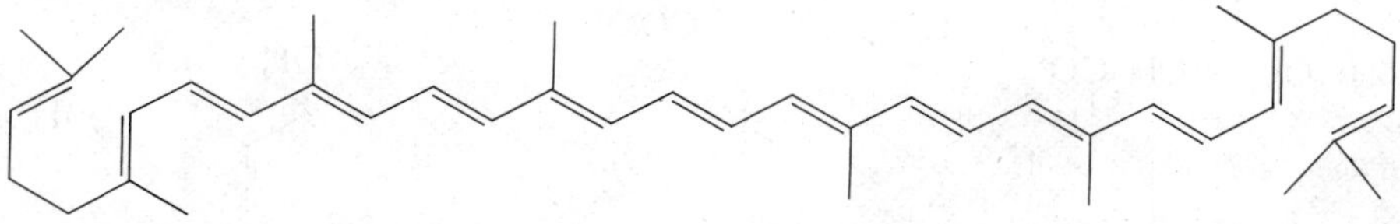

lycopene

The color of these hydrocarbons clearly comes from $\pi \rightarrow \pi^*$ transitions of the long conjugated system. Note that these compounds are terpenes (Section 34.6.A). Carotenoids also occur in marine biology, for example, in the skins of fish, sea stars, anemones, corals, and crustaceans, frequently in combination with proteins. Denaturation of the protein in boiling water frees the carotenoid and unmasks its color as in the red color of boiled lobster.

Naphthoquinones and **anthraquinones** also occur in both animal and vegetable worlds. Some examples were given in Section 30.7.A. Echinochrome is a polyhydroxynaphthoquinone that occurs as a red pigment in the sea urchin and sand dollar.

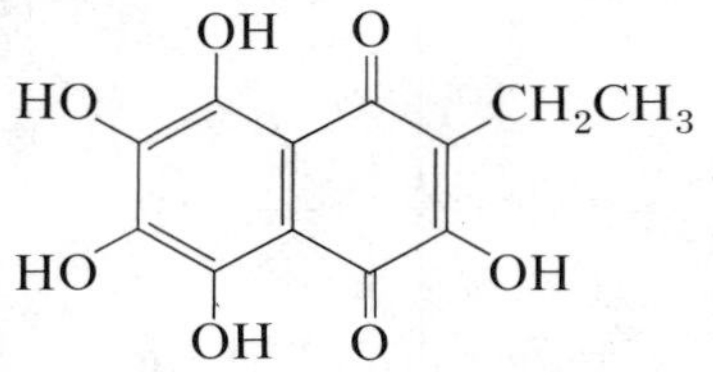

echinochrome

Cochineal is a dried female insect, *Coccus cacti* L., used for a red coloring in food products, cosmetics, and pigments. The principal constituent is carminic acid, a polyhydroxyanthraquinone attached to glucose.

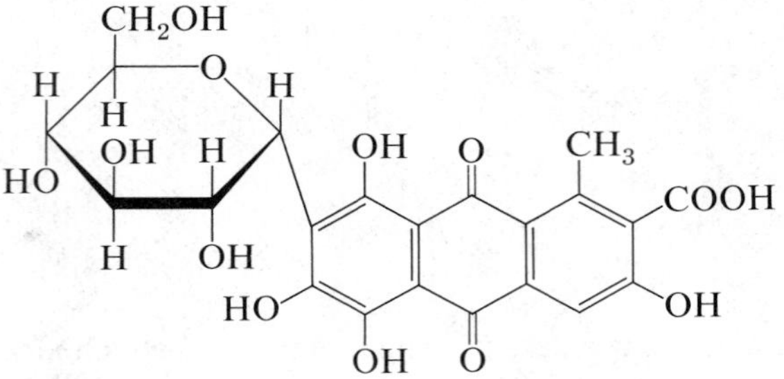

Melanins are complex quinoidal compounds derived from the oxidation and polymerization of tyrosine. Melanins occur in such varied places as feathers, hair, eyes, and the ink of cephalopods. They occur in the skin of all humans, except albinos, and are responsible for the varied skin coloration among the races of man. Albinos and certain white animals lack an enzyme required to convert tyrosine.

Other types of natural pigments include the **ommochromes** (xanthommatin, a yellow pigment from insect eyes), **pterins** (xanthopterin, yellow pigment in butterfly wings), **porphyrins** (hemin, page 976; chlorophyll), and **indigoids** (indigotin or indigo; occurs as glucoside in many plants; used as a blue dye; see next section).

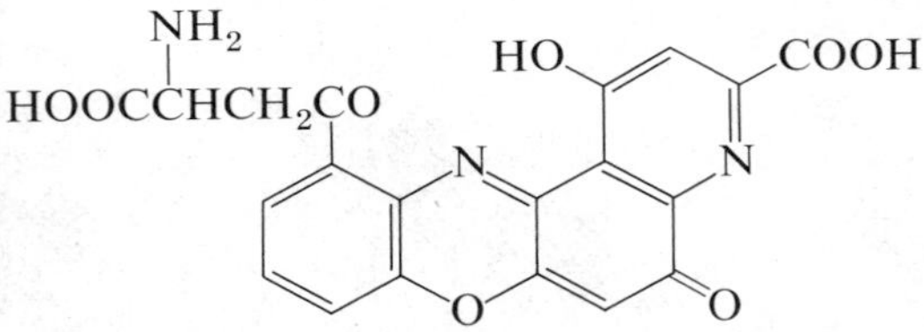

xanthommatin

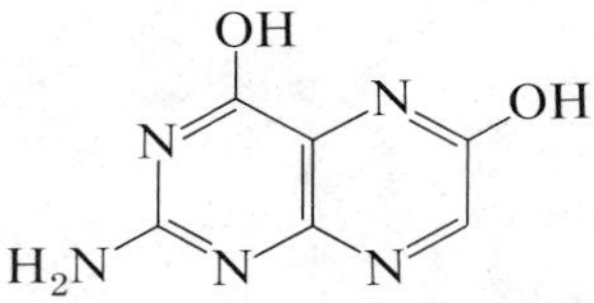

xanthopterin

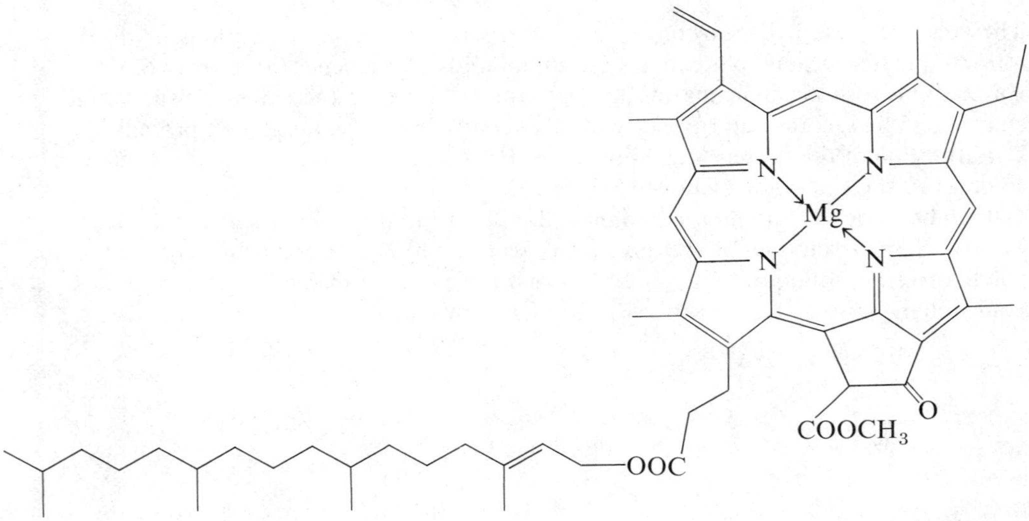

chlorophyll a

C. Dyes and Dyeing

Dyes are coloring matters that will bind in some manner to a substrate, usually a fiber or cloth, and are fast to light and to washing. Dyes have been known to man for thousands of years. Early dyes were entirely of natural origin, but common dyes in use today are almost all synthetic. Different methods were and still are used for combining the dye with the fiber. Some of the principal categories follow.

Vat dyes are exemplified by indigo, a highly insoluble blue compound known to the ancient world. It is also the *woad* of ancient Britain. A warm suspension of indigo with other materials was allowed to ferment for several days. This process produced a reduced and soluble "leuco" compound that is colorless. The material to be dyed was immersed in this solution and then exposed to air to reoxidize the leuco base. Indigo is now produced synthetically and is reduced to the leuco form with sodium hydrosulfite. It can be oxidized by exposure to air or more quickly by use of an oxidizing agent such as sodium perborate. The insoluble blue pigment so produced is "locked" within the fiber.

indigotin
indigo

leuco form

Mordant dyes are used in conjunction with a mordant (L. *mordere,* to bite), usually a metal salt that forms an insoluble complex or "lake" with the dye. The dye is applied to fiber or cloth that has been pretreated with a metal salt. An example known to the ancient world was the extract of the madder root, which was mordanted with aluminum salts to produce a color known as turkey red. Other metal salts give different colors. The actual dye that coordinates with the metal is alizarin. Alizarin was first synthesized in 1869, and shortly thereafter synthetically manufactured material drove the natural product from the market with important economic repercussions.

alizarin mordanted with Al^{3+}

Direct dyes can be applied to the fiber directly from an aqueous solution. This process is especially applicable to wool and silk. These fibers are proteins that incorporate both acidic and basic groups that can combine with **basic** and **acid dyes,** respectively. An example is **mauve,** the dye that started the modern synthetic dyestuff industry but is no longer used.

mauve

William Henry Perkin was a student at the Royal College of Chemistry when, in 1856 in his home laboratory, at the age of 18, he treated aniline sulfate with sodium dichromate and obtained a black precipitate from which he extracted a purple compound. This material showed promise as a dye and he resigned his position to manufacture it. The product was successful and cloth dyed with mauve was even worn by Queen Victoria. Not long afterward, additional synthetic dyes were synthesized by German chemists and the synthetic dyestuff industry gradually became a German industry. By the time of World War I, almost all of the world production of synthetic dyes was German.

Perkin's success in the discovery of mauve was based on the fact that his "aniline" was impure. It was prepared by nitrating and reducing "benzene" that contained substantial amounts of toluene!

Mauve is an example of a basic dye that can ion-pair with acidic centers of the fiber. An example of a direct acid dye is benzopurpurin 4B, whose sulfonate ion groups can pair with cationic centers in the fiber.

benzopurpurin 4B

Disperse dyes are used as aqueous dispersions of finely divided dyes or colloidal suspensions that form solid solutions of the dye within the fiber. They are especially useful for polyester synthetic fibers. These fibers have no acid or basic groups for use with direct dyes and are sensitive to hydrolysis in the strongly alkaline conditions of vat dyeing. Disperse dyes tend to have important limitations. They frequently lack fastness to washing, tend to sublime out on ironing, and are subject to fading with NO_2 or ozone in the atmosphere, a condition known as *gas fading*.

Dyes are also classified on the basis of chemical structure. These structures frequently contain a functional group that is principally involved in the $n \to \pi^*$ and $\pi \to \pi^*$ transitions that give rise to the color. Examples of such groups, called **chromophores,** are the azo group, $-N{=}N-$, the carbonyl groups in quinones, and extended chains of conjugation. Some of the principal chemical classes of dyes follow.

Azo dyes form the largest chemical class of dyestuffs. These dyes number in the thousands. They consist of a diazotized amine coupled to an amine or a phenol and have one or more azo linkages. An example of a diazo dye, a dye with two azo groups, is direct blue 2B, prepared by coupling tetrazotized benzidine with H-acid (8-amino-1-naphthol-3,6-disulfonic acid) in alkaline solution. If H-acid is coupled with a diazonium ion in dilute acid solution, reaction occurs next to the amino group. Both positions can be coupled to different diazonium salts.

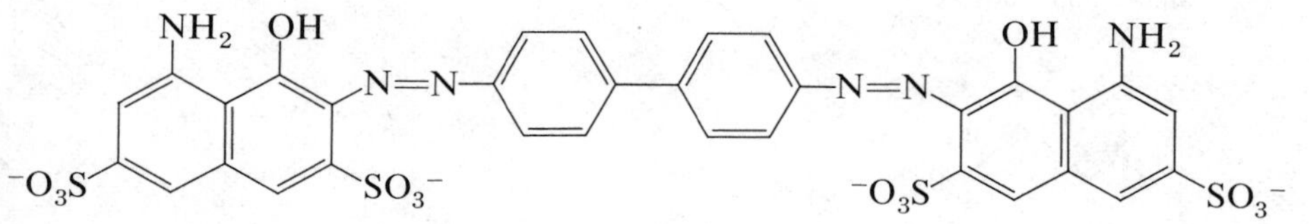

direct blue 2B

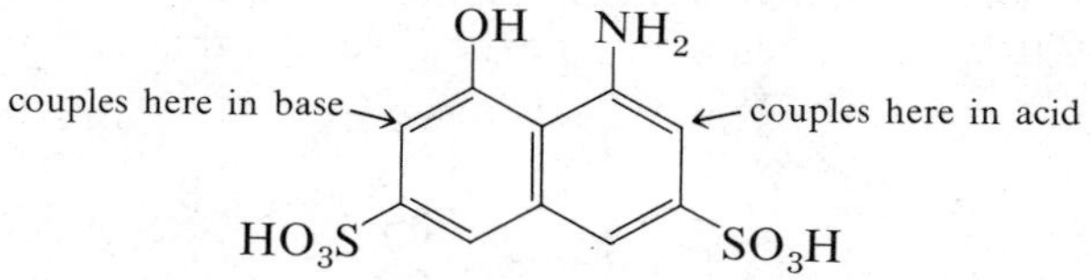

H-acid

Triphenylmethane dyes are derivatives of triphenylmethyl cation. They are basic dyes for wool or silk or for suitably mordanted cotton. Malachite green is a typical example that is prepared by condensing benzaldehyde with dimethylaniline and oxidizing the intermediate leuco base.

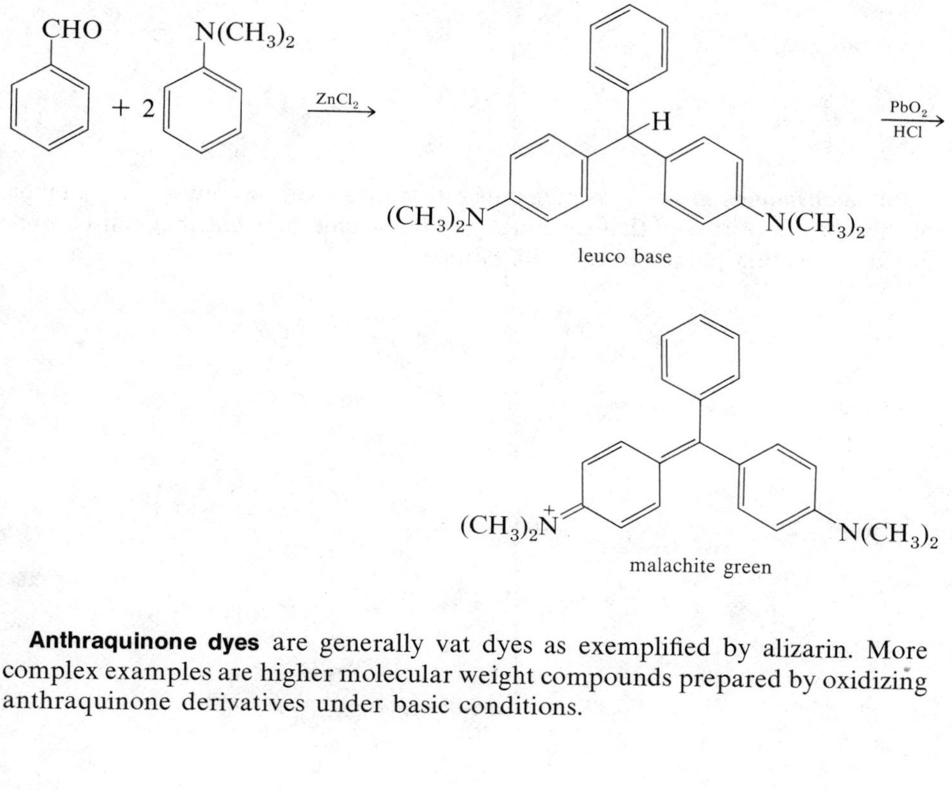

leuco base

malachite green

Anthraquinone dyes are generally vat dyes as exemplified by alizarin. More complex examples are higher molecular weight compounds prepared by oxidizing anthraquinone derivatives under basic conditions.

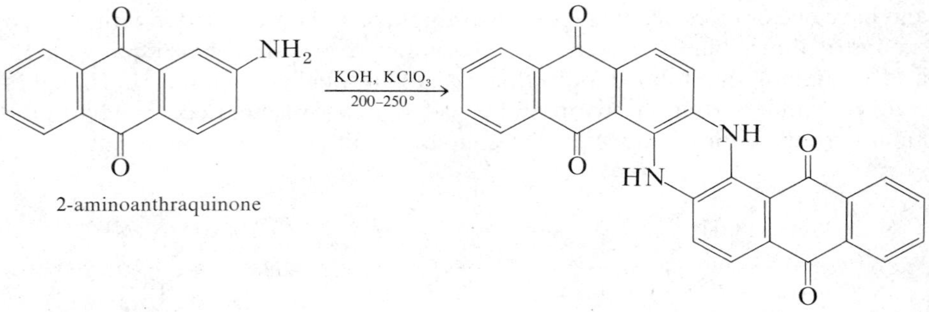

2-aminoanthraquinone

KOH, KClO₃
200–250°

indanthrone
indanthrene blue R

Indigoid dyes are also vat dyes, as represented by indigo itself. A dibromo-indigo has historical interest as the Tyrian purple of the ancient world. This dye was laboriously isolated from a family of mollusks (*Murex*). Its use was restricted to the wealthy. It is now a relatively inexpensive dye that still finds some use.

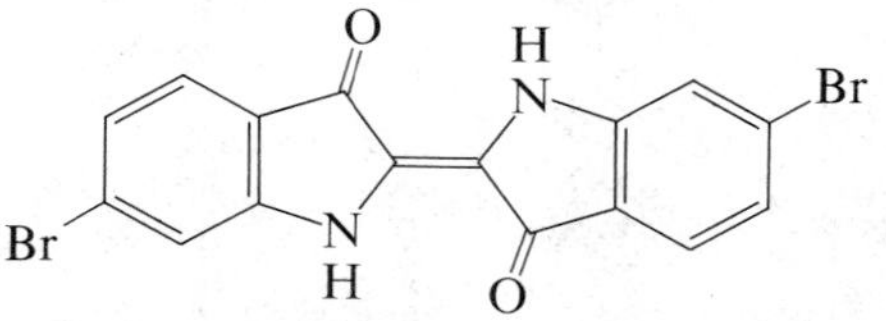

Tyrian purple
6,6′-dibromoindigo

Azine dyes are derivatives of phenoxazine, phenothiazine, or phenazine. Mauve and aniline black (page 759) are derivatives of phenazine. Methylene blue is a thiazine derivative used as a bacteriological stain.

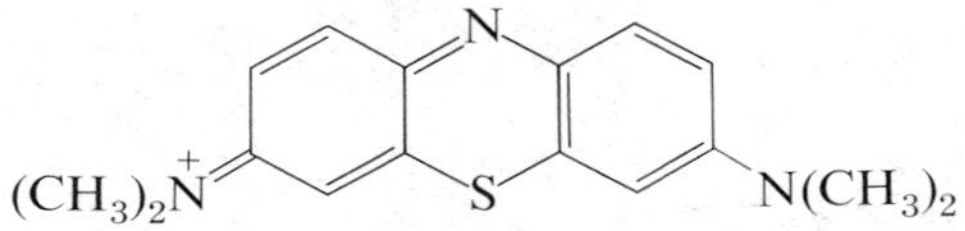

methylene blue

Phthalocyanines are used as pigments rather than dyes. An important member of this class is copper phthalocyanine, a brilliant blue pigment that can be prepared by heating phthalonitrile with copper.

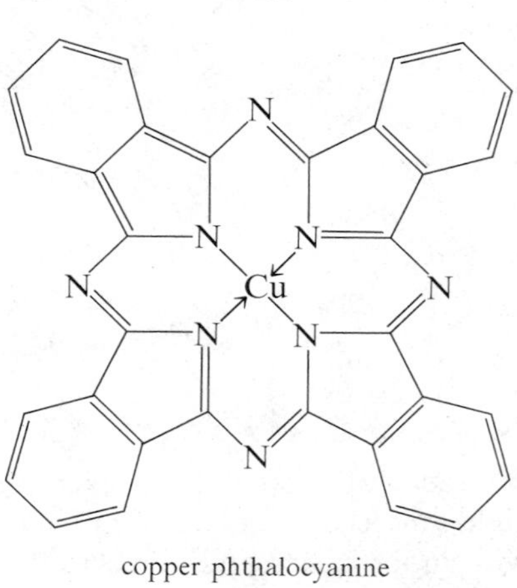

copper phthalocyanine

34.4
Photochemistry

A. *Electronically Excited States*

Most organic molecules have an even number of electrons with all electrons paired. Within each pair, the opposing electron spins cancel, and the molecule has no net electronic spin. Such an electronic structure is called a **singlet** state. When a ground-state singlet absorbs a photon of sufficient energy, it is converted to an excited singlet state. The process is a **vertical transition;** that is, electronic excitation is so fast that the excited state has the same geometry of bond distances and bond angles. The most stable geometry of the excited state often differs from that of the ground state with the result that the excited electronic state is often formed in an excited vibrational state as well. These relationships are illustrated in Figure 34.16. A vertical transition of the type illustrated is also known as a **Franck–Condon transition.**

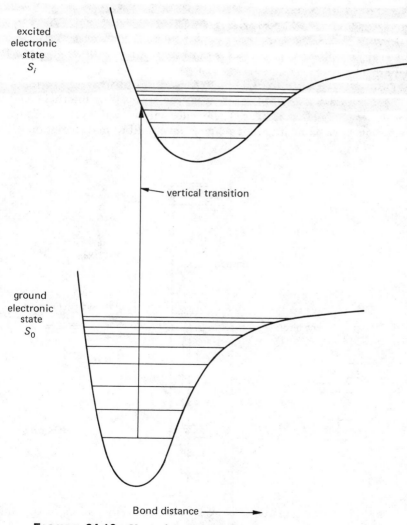

FIGURE 34.16. *Vertical or Franck–Condon electronic transition.*

Figure 34.16 is simplified in showing energy levels as a function of a single bond coordinate. For real molecules, there are many bonds and the resulting pattern of energy levels is multidimensional and complex. Furthermore, there are many excited singlet states, which may be represented collectively as S_i. The lowest excited singlet state is then represented as S_1.

The first formed excited state generally gives up its extra vibrational energy to other bonds in the molecule and falls to the lowest vibrational level of this state. The time required for this process is very short, about 10^{-13} sec, the time required for a single vibration. In the next step, this electronic excited state gives up more energy to other vibrational modes in the molecule or in collisions with solvent and becomes an excited vibrational level of the lowest excited state, S_1. This excited vibrational state again gives up vibrational energy until it reaches the lowest vibrational level of S_1. The change from S_i to S_1 is called **internal conversion** and takes about 10^{-11} sec, or about 10^2 vibrations.

The lifetime of the S_1 state in its lowest vibrational level is longer, about 10^{-8} to 10^{-7} sec. During this time one of four possibilities can occur. These alternatives are discussed below with the help of a Jablonski diagram (Figure 34.17).

1. S_1 can emit a photon and undergo an electronic transition to the ground state, a process called **fluorescence.** Because the excited state has lost energy before fluorescence occurs, the fluorescence photon has less energy than the originally exciting photon. Fluorescent light has longer wavelength than the light required for the original excitation.

2. S_1 can give up energy to other vibrations or to solvent and become an excited vibrational state of the ground state, S_0. This is an internal conversion and is a nonradiative process. The excited vibrational level again gives energy to its environment until it achieves an equilibrium distribution with the

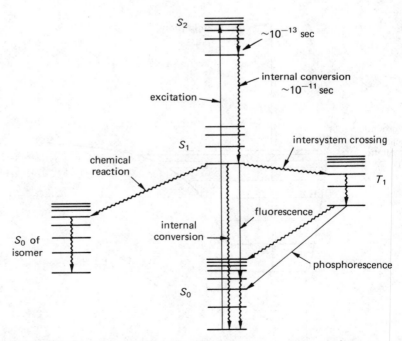

FIGURE 34.17. *Jablonski diagram. Nonradiative processes are shown as wavy lines.*

lowest vibrational level. The net result of all of these changes is the conversion of the original light quantum into heat.

3. S_1 can undergo internal conversion to an excited vibrational level of the ground state of a different compound, an isomer whose lowest vibrational level of S_0 corresponds to a different geometry than our starting material. Alternatively, S_1 can react on collision with another molecule. In either case, we have achieved a photochemical reaction.

4. S_1 can undergo **intersystem crossing** to the triplet state, T_1. A triplet state is one in which one electron spin has been changed so that the molecule has two electrons that cannot pair. The lowest triplet state is usually of higher energy than the ground state (the rare exceptions are compounds with "ground-state triplets"). Nevertheless, T_1 is of lower energy than S_1. Electrons with the same spin tend to stay apart because of the Pauli principle. As a result, the electrostatic energy of electronic repulsion is less than in comparable singlet states.

In many compounds, the switching of an electronic spin is an improbable process, and triplet states are not important in the photochemistry of such compounds. In certain other compounds, particularly $\pi \rightarrow \pi^*$ states of polycyclic aromatic hydrocarbons and $n \rightarrow \pi^*$ states of many ketones, the process of intersystem crossing is more probable. The process can take as little as 10^{-9} sec. Since the lifetime of S_1 is generally in the range of 10^{-7} to 10^{-8} sec, in such cases almost all of the excited states intersystem cross to T_1.

Triplet states are fairly long-lived, with lifetimes of greater than 10^{-5} sec, in some cases up to a second or so. One reason for such long lifetimes is that conversion to S_0 again requires switching an electronic spin. The rate at which intersystem crossing takes place depends in large part on the energy difference between the two states. The energy difference between S_1 and T_1 is generally much less than between T_1 and S_0; hence, the latter intersystem crossing is much less probable.

Triplet states themselves have four possible ways for shedding their excess energy.

1. T_1 can thermally decay to S_0. The net result in this case is again the conversion of a light quantum to heat.

2. T_1 can emit a photon. This process is called phosphorescence. As in the case of fluorescence, the wavelength of the phosphorescent light is longer than that of the initially exciting light. Phosphorescence is a low-probability event because of the change in electron spin required. It is generally observed only at low temperature for which thermal events have been slowed.

3. T_1 can convert to T_1 of an isomeric molecule or it can intersystem cross to S_0 of an isomer. Either process results in a photochemical isomerization. Alternatively, T_1 can react on collision with another molecule to provide a photochemical reaction.

4. T_1 can transfer its electronic spin to another molecule and become converted to S_0. This process is called *triplet energy transfer* and is symbolized as

$$T_1 + S_0' \longrightarrow S_0 + T_1'$$

This reaction is actually an equilibrium governed by the usual free energy requirements, but it is usually important only when the reaction shown is exothermic.

B. *Photochemical Reactions*

Some photochemical reactions involve simple isomerizations. An example is the *cis-trans* interconversion of alkenes. In this case the excited state involves a twisted double bond (an example is Figure 11.7), and conversion to the ground state has an equal probability of going *cis* and *trans*. If a wavelength of light can be chosen at which the *trans* isomer absorbs and the *cis* does not, a *trans* isomer can be converted completely to the *cis*. In other cases a photochemical **stationary state** is achieved.

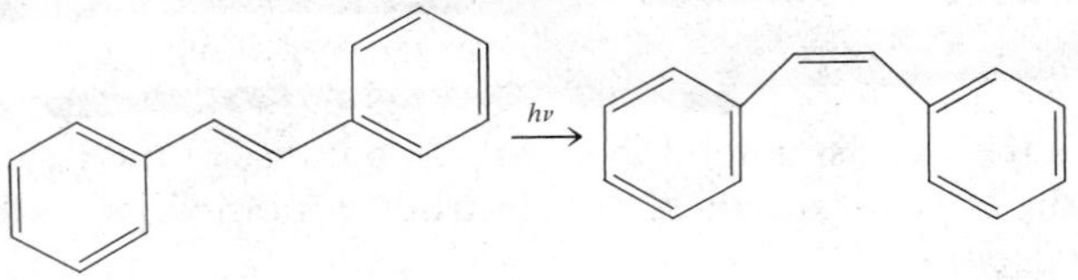

trans-stilbene   *cis*-stilbene

The breaking of bonds is a common photochemical reaction. We saw an early example in the dissociation of chlorine to initiate free radical chain reactions (Section 6.3.A). This reaction is also common for ketones and is called a **Norrish type I** reaction.

$$CH_3COCH_3 \xrightarrow{h\nu} CH_3CO\cdot + CH_3\cdot$$
$$CH_3CO\cdot \longrightarrow CH_3\cdot + CO$$
$$2\, CH_3\cdot \longrightarrow CH_3CH_3$$

For cyclic ketones, the photochemical product is a **diradical** that can undergo further reactions.

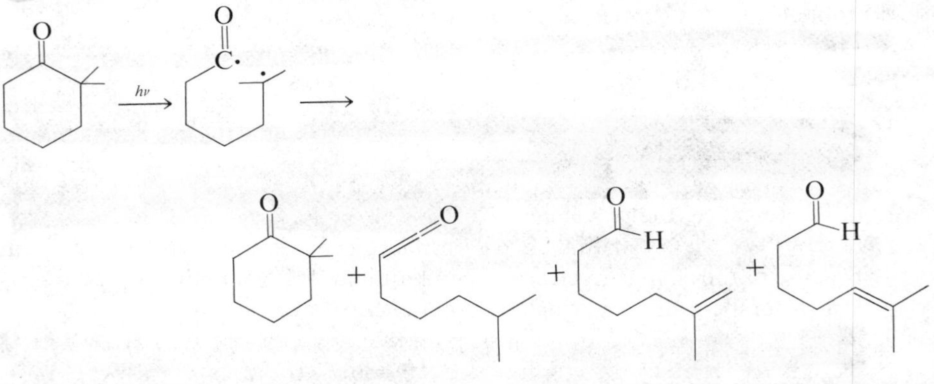

Another reaction that ketones undergo involves intramolecular hydrogen atom transfer via a six-membered-ring transition state to form another diradical. This reaction is called a **Norrish type II** process.

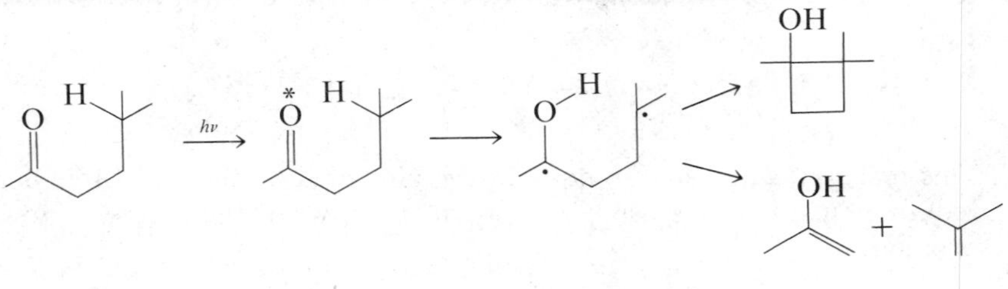

⌈ Note the use of an asterisk to indicate an excited *state*. We have previously used this
| symbol to refer to an antibonding orbital. The asterisk symbolism is used commonly
⌊ in both contexts. ⌋

α,β-Unsaturated carbonyl compounds can undergo photochemical dimerization with the formation of a four-membered ring.

The formation of four-membered rings is also common with dienes.

1,3-cycloheptadiene   bicyclo[3.2.0]hept-6-ene

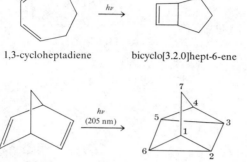

norbornadiene   quadricyclene
bicyclo[2.2.1]-   tetracyclo[2.2.1.0^{2,6}.0^{3,5}]-
hepta-2,5-diene   heptane

Quadricyclene is an unusual hydrocarbon given its trivial name from its tetracyclic nature. The compound is considered to be tetracyclic because four C—C
bonds must be broken to obtain an acyclic system. Note how the systematic name
is derived from that of the bicyclic parent, bicyclo[2.2.1]heptane. The two additional bridges are both "zero-carbon" bridges, and join C-2 to C-6 and C-3 to C-5,
respectively. These additional bridges are included within the bracket that specifies the nature of the cyclic skeleton as the locants 0^{2,6}. 0^{3,5}. This last reaction is
a slow process and the compound is formed in low yield. Most of the light quanta
absorbed end up as heat, and the **quantum yield** is low. The quantum yield is
defined as the number of product molecules divided by the number of light
quanta absorbed.

This reaction becomes much more efficient by triplet energy transfer.

(95%)

Norbornadiene is transparent to light of 313 nm wavelength. The ketone, however, absorbs this light in an $n \rightarrow \pi^*$ transition. Intersystem crossing occurs readily to T_1 of the ketone. On colliding with a molecule of norbornadiene, the ketone

T_1 gives up its triplet character and is converted back to the ground state. Norbornadiene is converted to its T_1, which undergoes the changes in geometry required to intersystem cross to the ground state of quadricyclene. In this example, the acetophenone functions as a **triplet sensitizer.**

Sensitizers work in the same fashion in photography. The direct photoexcitation of a grain of silver bromide by light (page 1019) is an inefficient process and long exposure times would be required. In modern photography, a light photon first excites a dye molecule and its excited state transfers its excitation energy to a silver bromide grain that is then developed as usual. The dye functions as the sensitizer.

Some photochemical reactions involve rather deep-seated rearrangements.

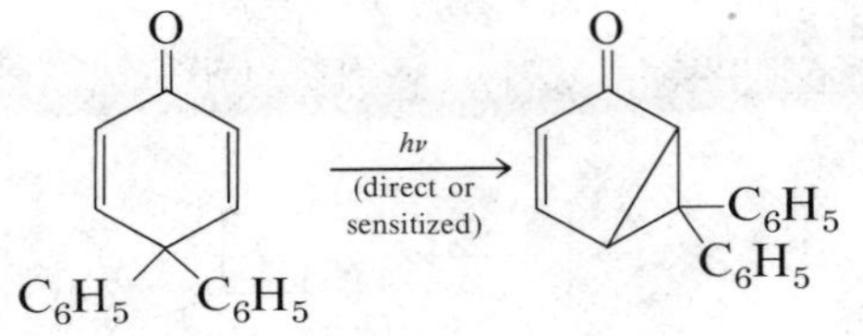

This reaction can be rationalized by the following bond-switching process in the excited state.

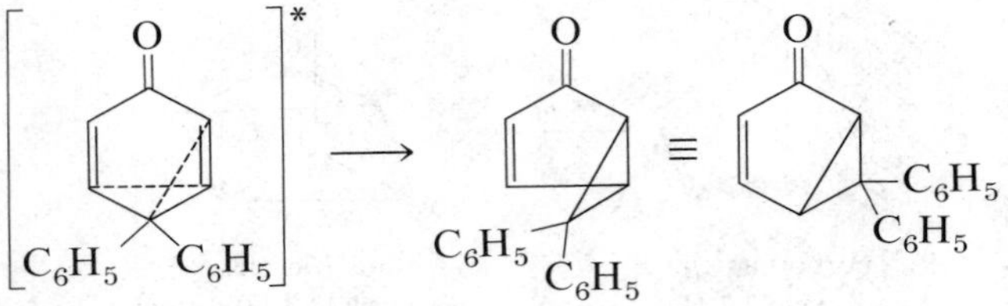

Mercury vapor is often used as a photosensitizer.

$$Hg \xrightarrow[\text{(253.7 nm)}]{h\nu} Hg^*$$

$$Hg^* + A \longrightarrow Hg + A^*$$

Small amounts of mercury vapor suffice to make the following reactions preparatively useful.

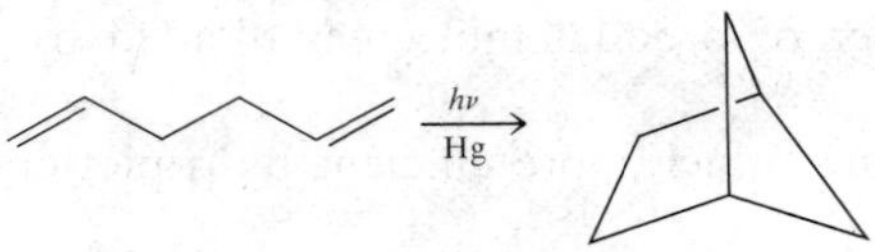

bicyclo[2.1.1]hexane

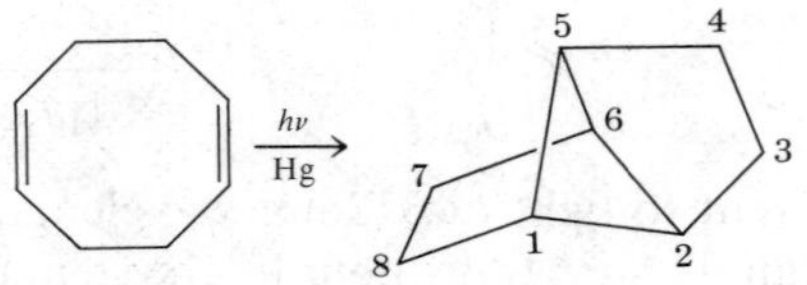

tricyclo[3.3.0.0^{2,6}]octane

34.5
Nucleic Acids

Nucleic acids are important biomolecules that play a crucial role in the storage of genetic information and in protein biosynthesis. There are two types of nucleic acid, ribonucleic acid (RNA) and deoxyribonucleic acid (DNA). Both are biopolymers in which the repeating monomer units are called **nucleotides.** A nucleotide, in turn, is a complex molecule made up of one unit each of phosphate, a pentose, and a heterocyclic base. For each class of nucleic acid there are four main nucleotide monomers. In RNA the pentose is ribose and the heterocyclic base is a pyrimidine, uracil or cytosine, or a purine, adenine or guanine.

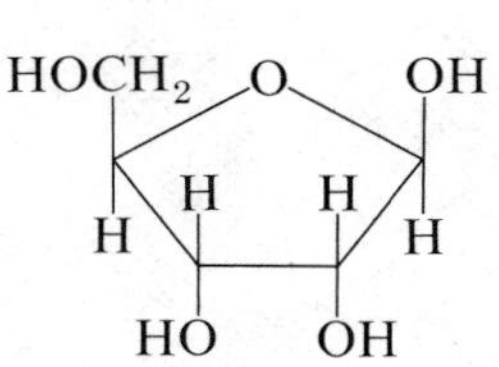

ribose 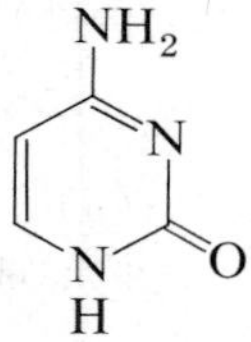   cytosine 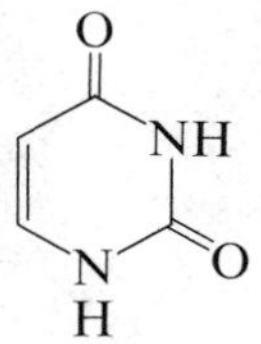   uracil

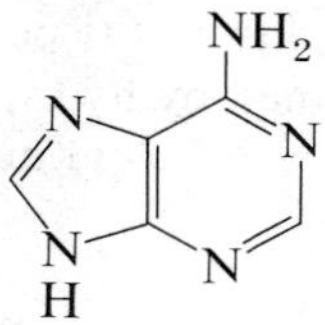

adenine 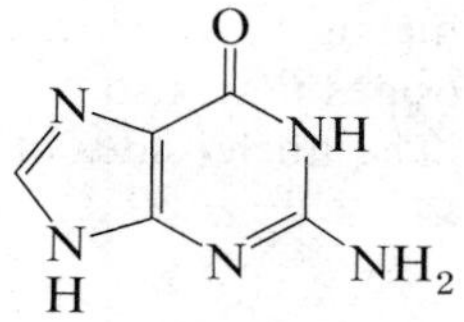   guanine

The nucleotides themselves are called uridylic acid, cytidylic acid, and so on.

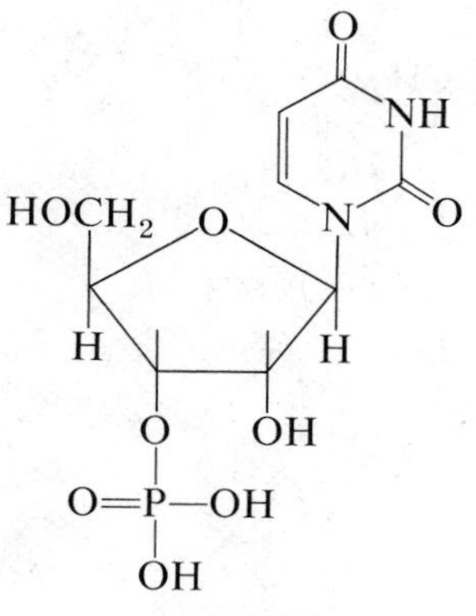

uridylic acid 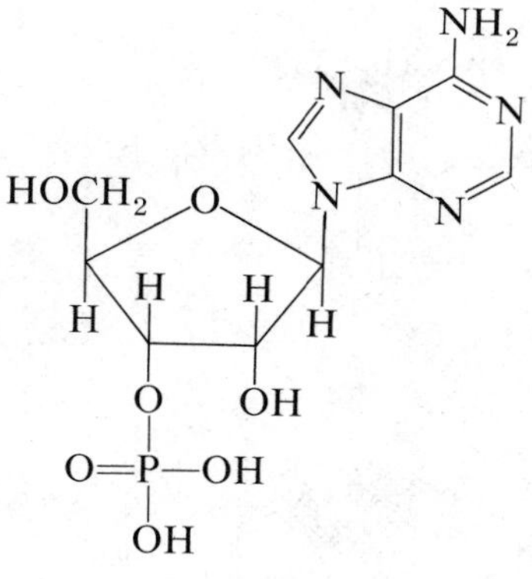   adenylic acid

In DNA the pentose is 2-deoxyribose. The heterocyclic bases are the same, except that thymine replaces uracil.

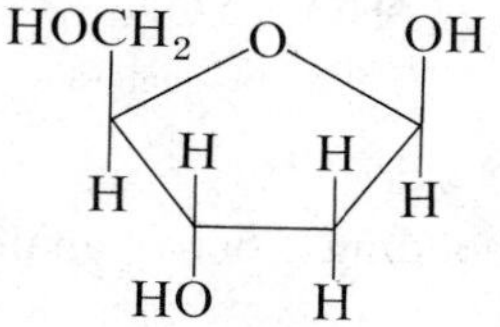

2-deoxyribose 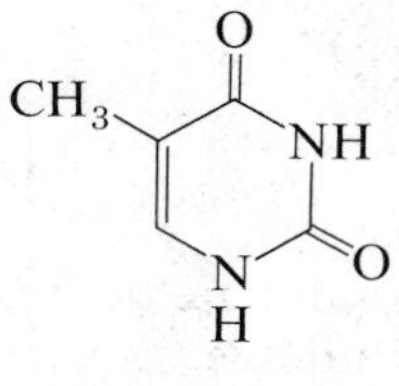   thymine

The nucleotides are called 2-deoxythymidylic acid, 2-deoxycytidylic acid, and so on.

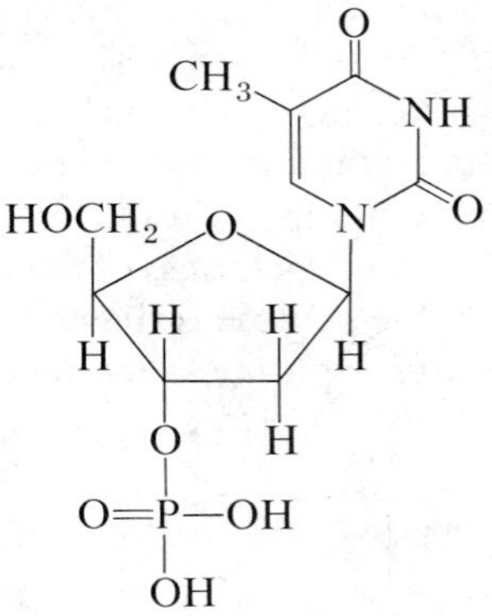

2-deoxythymidylic acid

Base-catalyzed hydrolysis of a nucleotide removes the phosphate group and yields a **nucleoside,** which is a glycoside formed from the pentose and the heterocyclic base. The nucleosides may also be obtained by hydrolysis of the appropriate nucleic acid itself. The nucleosides of RNA are cytidine, uridine, adenosine, and guanosine.

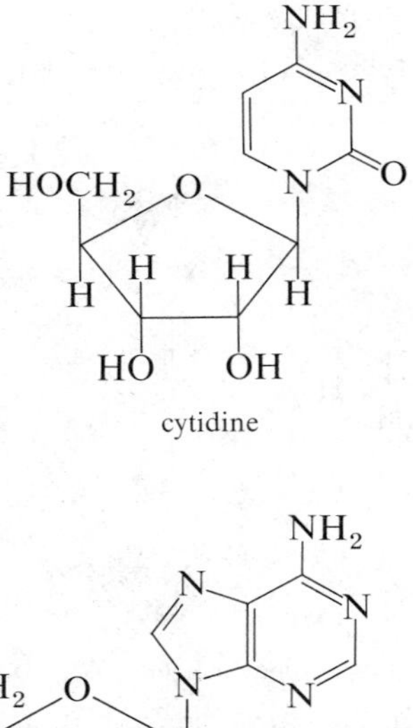

cytidine

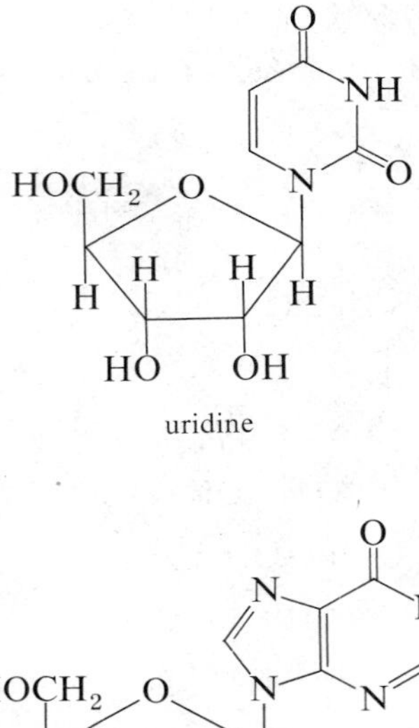

uridine

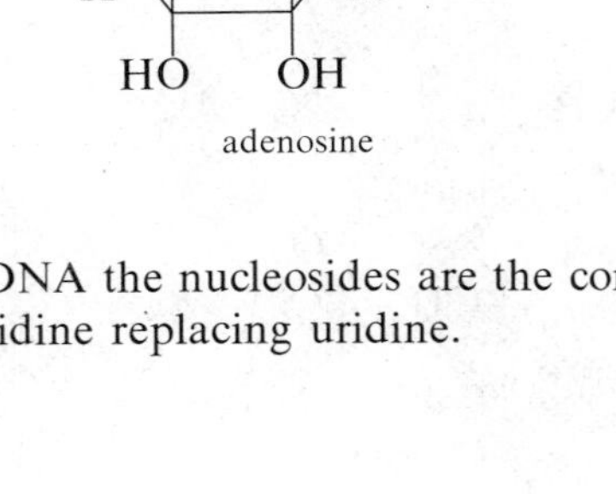

adenosine

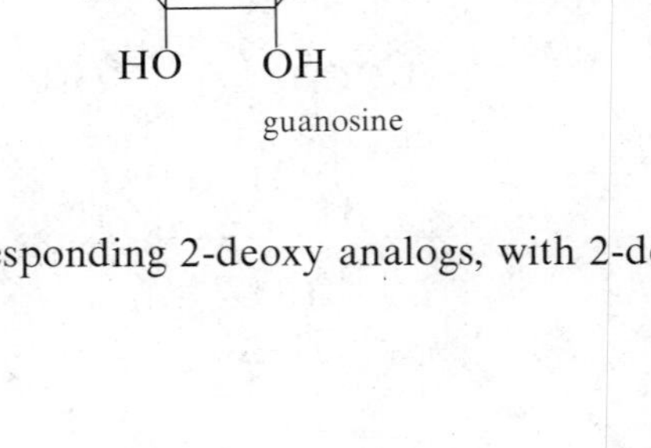

guanosine

For DNA the nucleosides are the corresponding 2-deoxy analogs, with 2-deoxythymidine replacing uridine.

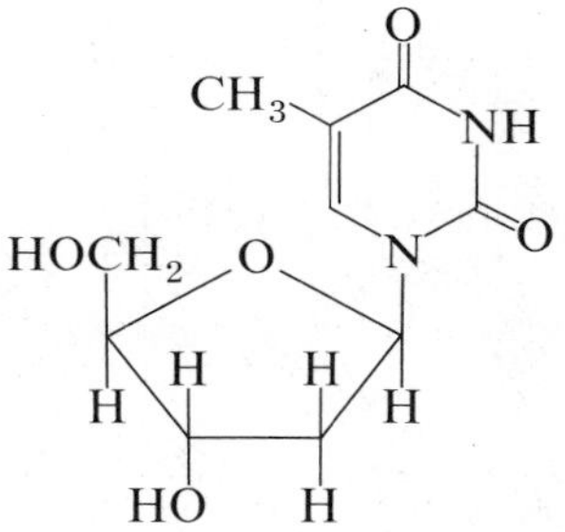

2-deoxythymidine

The nucleic acids may be extremely large molecules, with molecular weights up to 4 billion. The nucleic acid backbone is a copolymer of phosphoric acid and either ribose or 2-deoxyribose molecules, with one of the four heterocycles, adenine, guanine, cytosine, and uracil (or thymine), linked to C-1 of each of the pentose units (see Figure 34.18).

DNA occurs in the nuclei of all cells and is the molecule in which genetic information is stored. In the precise sequence of purine and pyrimidine bases along its phosphodiester backbone, it carries the information necessary for the exact duplication of the cell and, in fact, for the construction of the entire organism. DNA is actually a double-stranded helix of two individual molecules about 20 Å apart. The two chains are held together by reciprocal hydrogen bonding between pairs of bases in opposite positions in the two chains. The molecular geometry is such that adenine forms a strong reciprocal bond to thymine (AT) and guanine to cytosine (GC).

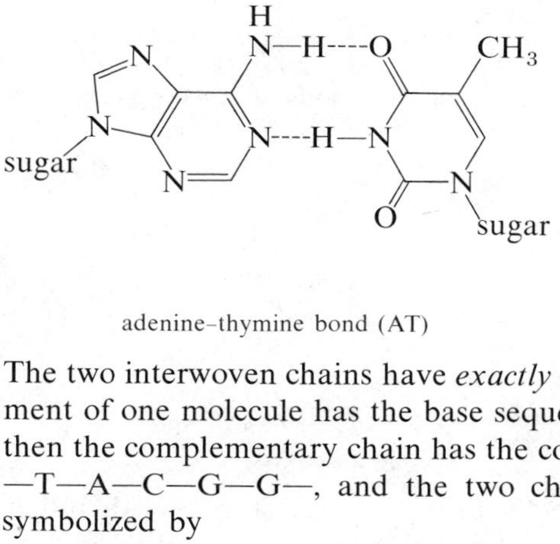

adenine–thymine bond (AT) 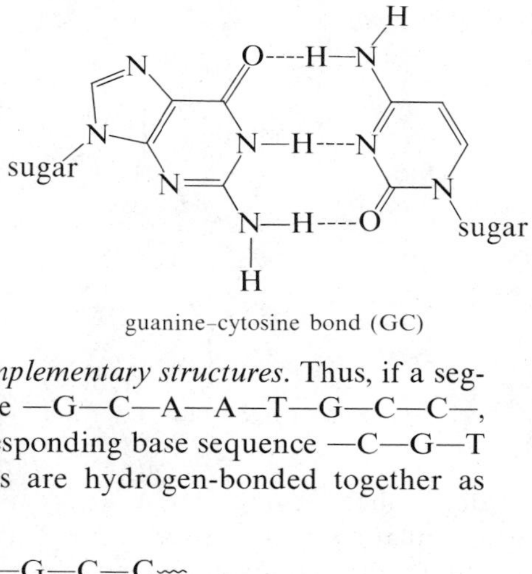   guanine–cytosine bond (GC)

The two interwoven chains have *exactly complementary structures*. Thus, if a segment of one molecule has the base sequence —G—C—A—A—T—G—C—C—, then the complementary chain has the corresponding base sequence —C—G—T—T—A—C—G—G—, and the two chains are hydrogen-bonded together as symbolized by

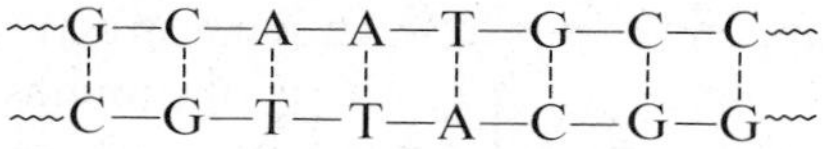

Genetic information is passed from cell to cell in a process called **DNA replication.** As one of the essential steps in cell mitosis, the nuclear DNA helices separate. Each of the separate chains then functions as a template upon which another chain exactly complementary to itself is constructed. The end result is that one DNA double helix is transformed into two. In this way each daughter cell acquires an identical set of nuclear DNA molecules.

RNA is similar to DNA except that the pentose units are ribose instead of

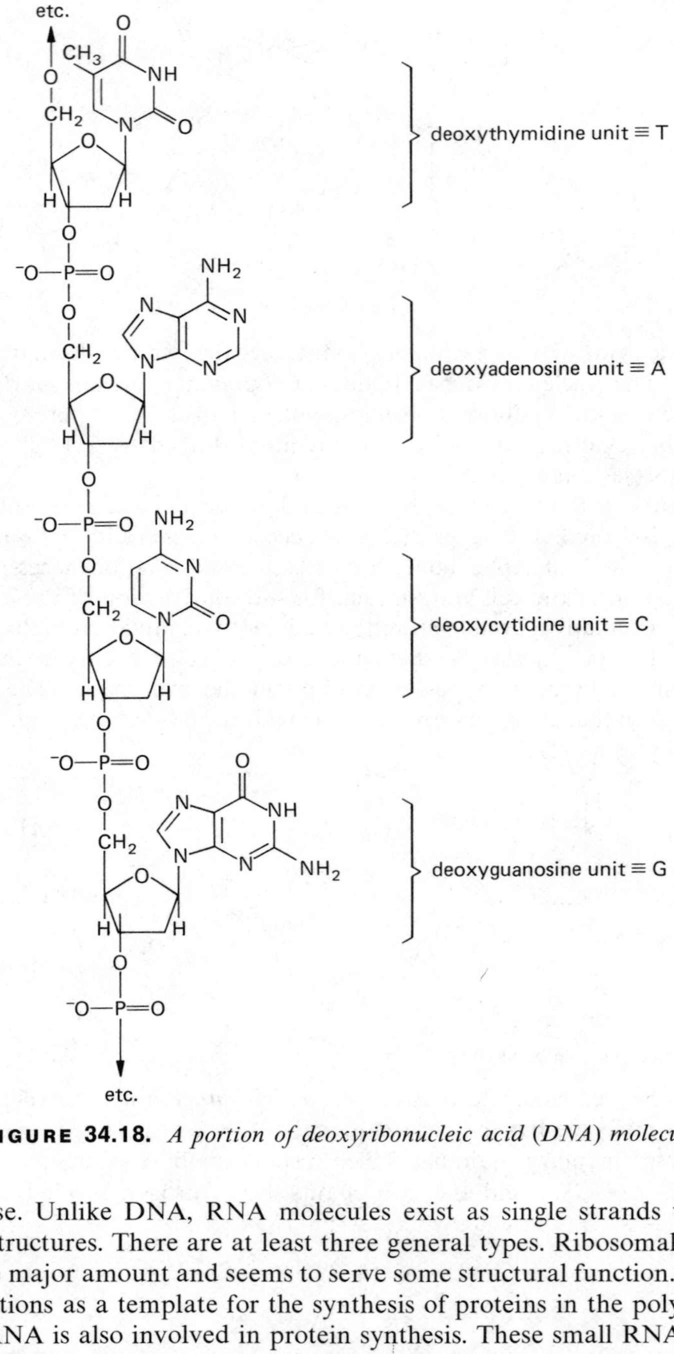

FIGURE 34.18. *A portion of deoxyribonucleic acid (DNA) molecule.*

deoxyribose. Unlike DNA, RNA molecules exist as single strands with rather irregular structures. There are at least three general types. Ribosomal RNA constitutes the major amount and seems to serve some structural function. Messenger RNA functions as a template for the synthesis of proteins in the polyribosomes. Transfer RNA is also involved in protein synthesis. These small RNA molecules bind to individual amino acids and guide them into place on the growing protein chain.

Many fascinating details of DNA and RNA chemistry have been worked out over the last twenty years. Much of the story of the genetic code and how it functions has been unraveled, and much more remains to be learned. Although we cannot even scratch the surface of this intriguing story here, suffice it to say that the unique hydrogen bonding that is possible between the complementary purines and pyrimidines is crucial to life as we know it.

1165

Sec. 34.6

Natural
Products:
Terpenes,
Steroids,
and Alkaloids

34.6
Natural Products: Terpenes, Steroids, and Alkaloids

Metabolism is the collection of chemical processes by which an organism creates and maintains its substance and obtains energy in order to grow and function. Almost all of these chemical processes involve organic compounds and reactions and naturally fall under the purview of the organic chemist. The metabolic processes of various organisms are varied and complex. In fact, the study of these processes is the subject of an entire discipline of science—biochemistry. However, many of the end products of metabolism are readily isolable organic compounds and have historical importance in organic chemistry. These compounds are grouped together under the broad heading of **natural products.**

There are many different classes of naturally occurring compounds, and some have already been encountered in this book. Some, such as fats (Section 19.12), carbohydrates (Chapter 28), proteins (Chapter 29), and nucleic acids (Section 34.5), have obvious roles in the functioning of organisms. These natural products, together with a relatively small number of related substances, occur in almost all organisms; they are called **primary metabolites.** The processes whereby they are produced are called **primary metabolic processes.** That is, most living organisms, regardless of species, produce the common sugars and sugar derivatives, the common fatty acids, and the simple carboxylic acids.

A second class of natural products is called **secondary metabolites.** They are not necessarily of secondary importance to the organism, but their distribution in nature tends to be much more species-dependent. They are the product of **secondary metabolic** processes of the organism. Examples of such secondary metabolites are the terpenes, the steroids, and the alkaloids. Because of the central role these natural products have played in the development of organic chemistry, we shall briefly examine their structures at this point.

A. *Terpenes*

The **terpenes** are a class of organic compounds that are the most abundant components of the **essential oils** of many plants and flowers. Essential oils are obtained by distilling the plants with water; the oil that separates from the distillate usually has highly characteristic odors identified with the plant origin. In the days of alchemists this procedure was common. The resulting mixture of organic compounds was thought to be the *essence* of the plant, hence the term essential oil. As we shall see in Section 34.7, terpenes are synthesized by organisms ("biosynthesized") from acetic acid by way of the important biological intermediate isopentenyl pyrophosphate. Terpene structures may generally be dissected into several "isoprene units."

isopentenyl pyrophosphate   an isoprene unit

Compounds derived from a single isoprene unit are rare in nature. However, compounds composed of two isoprene units are common. These materials are called **monoterpenes.** The simplest acyclic example is geraniol, a constituent of the oil of geranium. A related monoterpene is citronellal, which is responsible for the characteristic aroma of lemon oil.

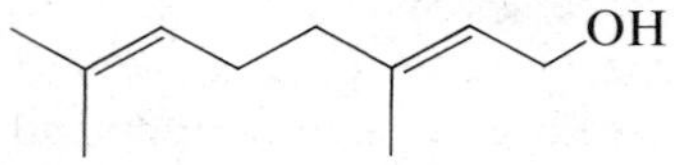

geraniol 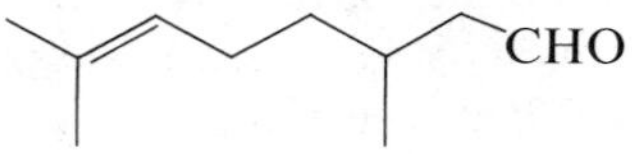   citronellal

Menthol (peppermint oil), β-pinene (turpentine), and camphor (obtained from the wood and leaves of the camphor tree) are examples of cyclic monoterpenes.

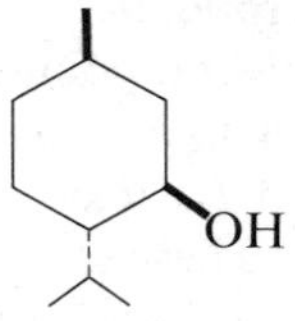

menthol   β-pinene 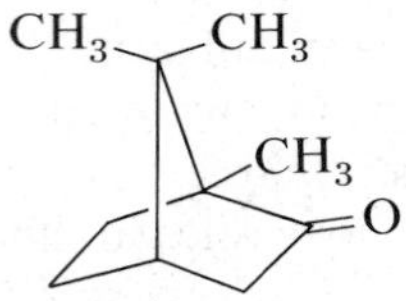   camphor

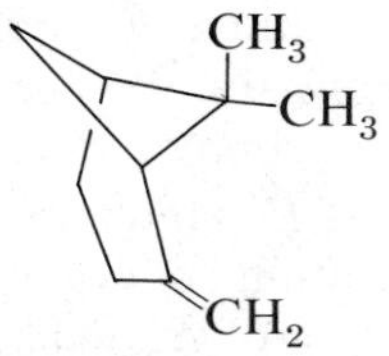

Sesquiterpenes are C_{15} compounds that are composed of three isoprene units. Farnesol, which occurs in the essential oils of rose, acacia, and cyclamen, may be regarded as the parent acyclic alcohol. It has the characteristic odor of lily of the valley and is used in perfumery. A simple monocyclic sesquiterpene is bisabolene, which is found in the oils of bergamot and myrrh.

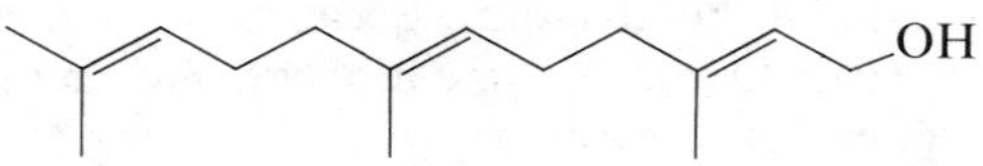

farnesol 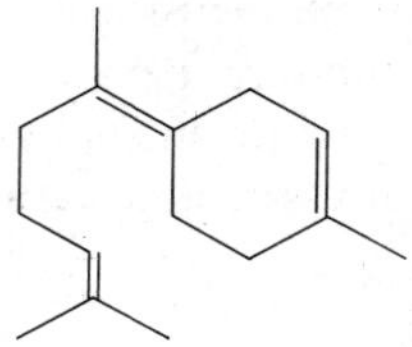   bisabolene

Most terpenes have cyclic or polycyclic structures. A tremendous variety of fascinating structures are known. Examples are nootkatone (aroma of grapefruit), β-santalol (sandalwood oil), guaiol (guaiacum wood), and copaene (copaiba balsam oil).

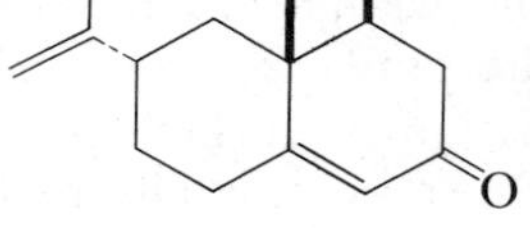

nootkatone   β-santalol

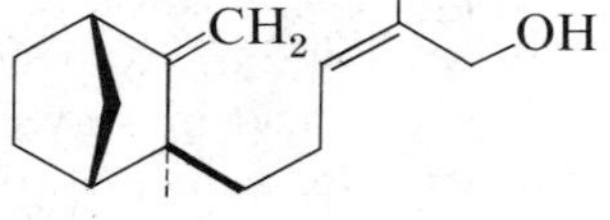

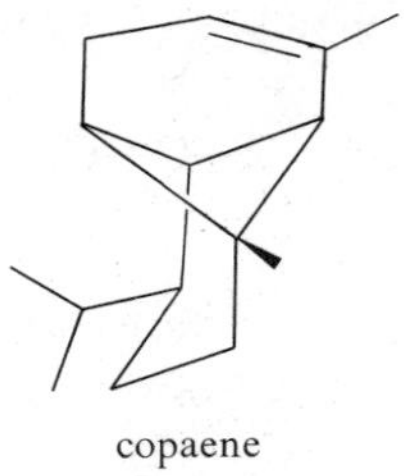

copaene 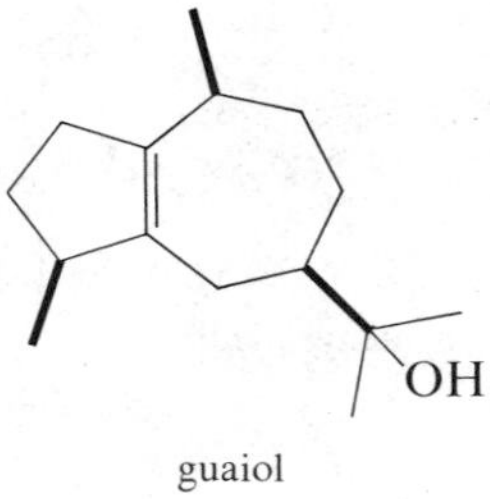   guaiol

Diterpenes, C_{20} compounds composed of four isoprene units, include the important visual factor retinal (vitamin A aldehyde) and abietic acid, a component of rosin, the nonvolatile exudate of coniferous trees.

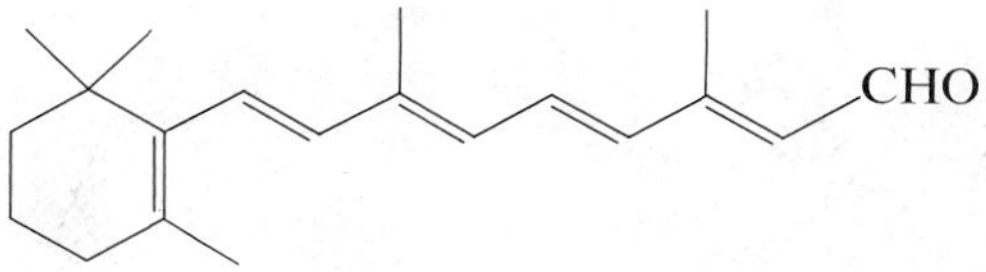

retinal

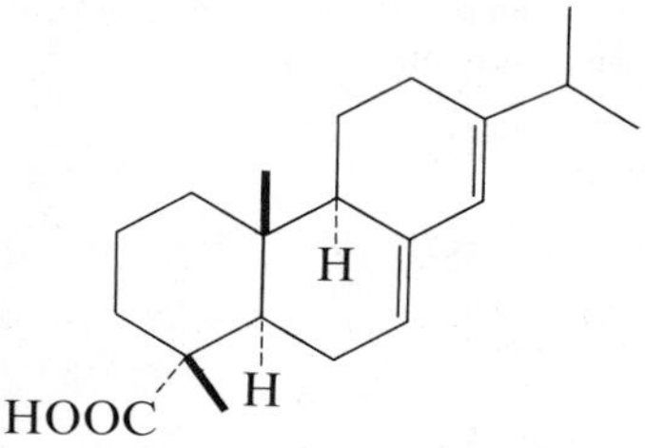

abietic acid

Terpenes having 25 carbons (sesterterpenes) are rare. However, the C_{30} compounds, or **triterpenes,** are common. An interesting example is squalene, a high-boiling viscous oil that is found in large quantities in shark liver oil. It may be isolated in smaller amounts from olive oil, wheat germ oil, rice bran oil, and yeast, and it is an intermediate in the biosynthesis of steroids (Section 34.6.B).

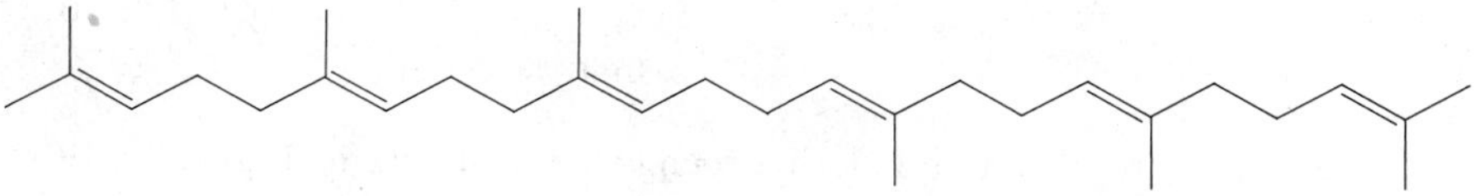

squalene

Other triterpenes are polycyclic, such as β-amyrin, a major constituent of the resin of the Manila elemi tree, and cyclolaudenol, a component of the neutral fraction of opium.

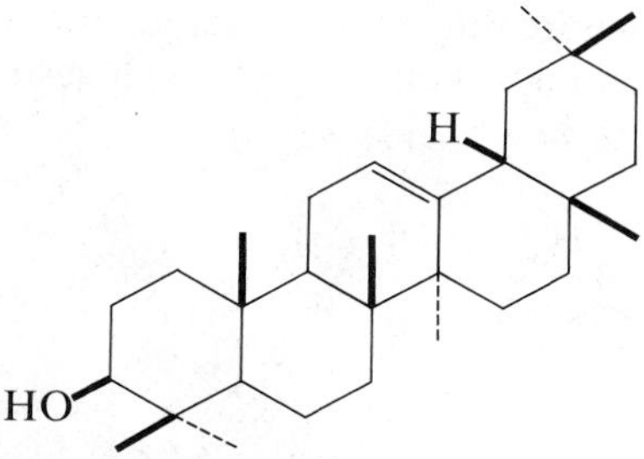

β-amyrin

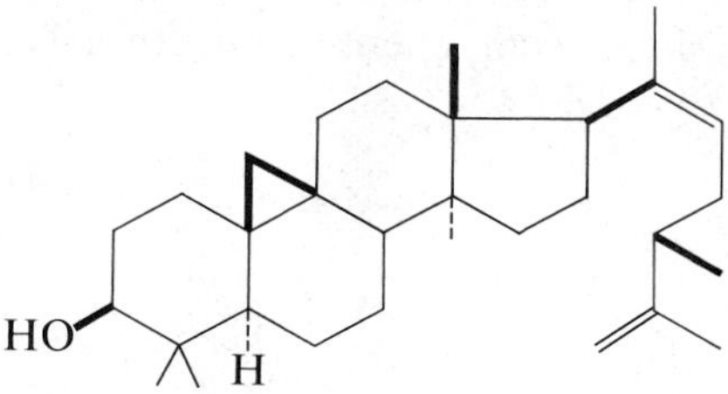

cyclolaudenol

Natural rubber and gutta percha (Section 20.2.E) are polyterpenes, being made up of a large number of isoprene units.

The elucidation of the structures of the terpenes has provided a fascinating and important chapter in organic chemistry that really started only about a half century ago. Early terpene research led to the recognition of skeletal rearrangements that were among the first examples of carbocation rearrangements. A particularly important example is the camphene hydrochloride–isobornyl chloride rearrangement, which we can recognize as a simple 1,2-aklyl rearrangement.

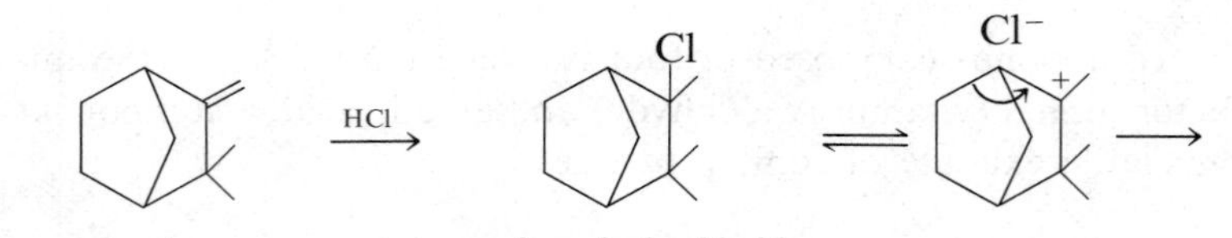

camphene
(many essential oils)

camphene hydrochloride

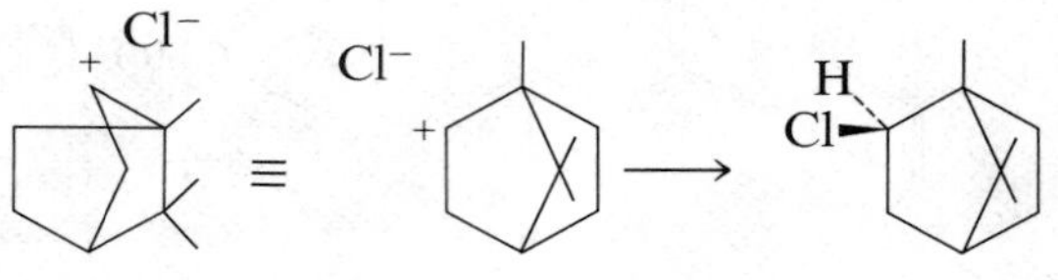

isobornyl chloride

Similar rearrangements are widespread in terpene chemistry. A further example is the rearrangement of longifolene to longifolene hydrochloride.

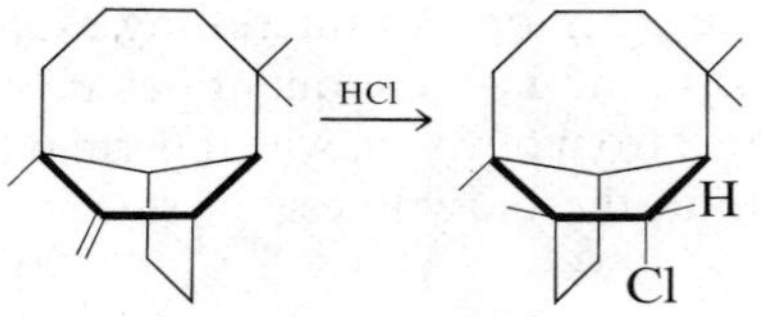

longifolene

longifolene
hydrochloride

The stereo structure of longifolene hydrochloride is shown in Figure 34.19.

Synthetic routes to the simpler terpenes are now available, but many of the more complex polycyclic terpenes provide synthetic challenges that intrigue present-day synthetic research chemists.

B. *Steroids*

Steroids are tetracyclic natural products that are related to the terpenes in that they are biosynthesized by a similar route. An important example is cholesterol, the major component of human gall stones (Gk., *chole,* bile).

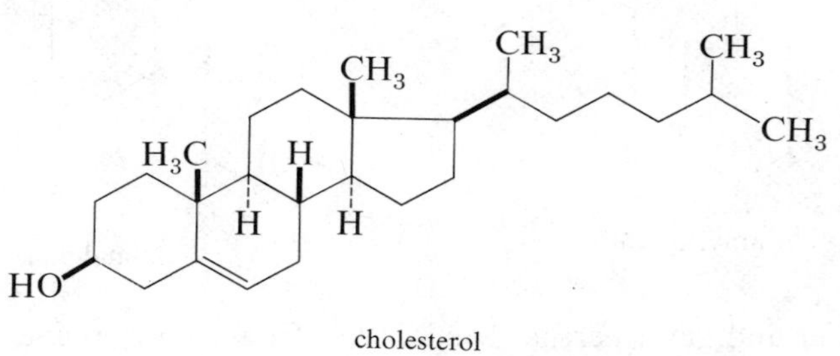

cholesterol

Actually, cholesterol is present in some amount in all normal animal tissues, but it is concentrated in the brain and in the spinal cord. The total amount present in a 180 lb person is 240 g, about $\frac{1}{2}$ lb! It is present partly as the free alcohol and partly esterified with fatty acids.

The structure of cholesterol illustrates the basic steroid skeleton, which is that of a hydrogenated 1,2-cyclopentenophenanthrene having two methyl substituents at C-10 and C-13 and an additional side chain at C-17. The stereochemistry at the

1169

Sec. 34.6

**Natural
Products:
Terpenes,
Steroids,
and Alkaloids**

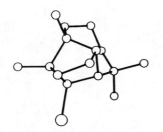

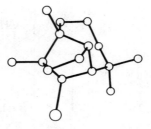

FIGURE 34.19. *Stereo structure of longifolene hydrochloride. Note that hydrogens are not shown.* [*Reproduced with permission from* Molecular Structures and Dimensions. *International Union of Crystallography, 1972.*]

various asymmetric carbons is almost invariably that shown, and in subsequent examples we shall not indicate stereochemistry unless it differs from the usual.

1,2-cyclopentenophenanthrene

general steroid ring structure

Other steroids are also common constituents of animal tissues and play important roles in normal biological process. Cholic acid, deoxycholic acid, and chenodeoxycholic acid occur in the bile duct.

cholic acid

deoxycholic acid

chenodeoxycholic acid

The bile acids exist as amides of the amino acid glycine, H_2NCH_2COOH, or the aminosulfonic acid taurine, $H_2NCH_2CH_2SO_3H$. The sodium salts have a large hydrocarbon region and a highly polar region and function in the intestinal tract as emulsifying agents to promote the absorption of fats. They are a type of biological "soap" (Section 18.4.D).

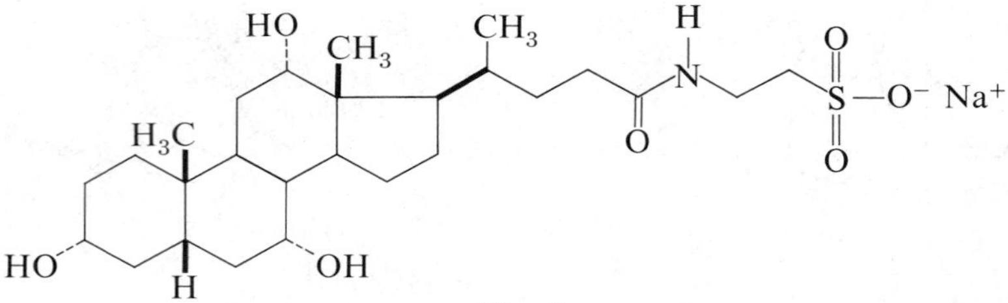

a bile salt

Estrone, progesterone, testosterone, and androsterone are steroid sex hormones.

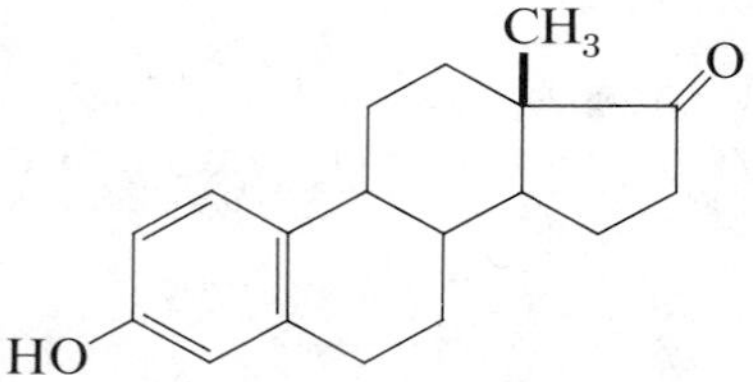

estrone

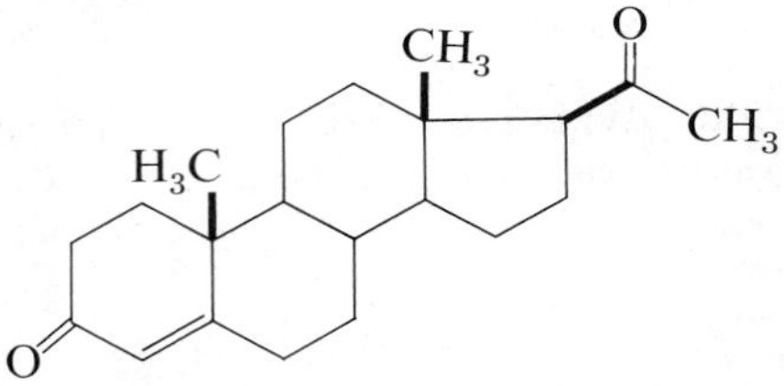

progesterone

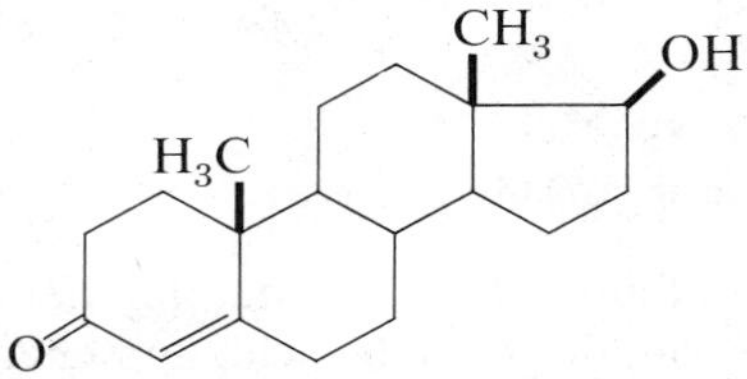

testosterone

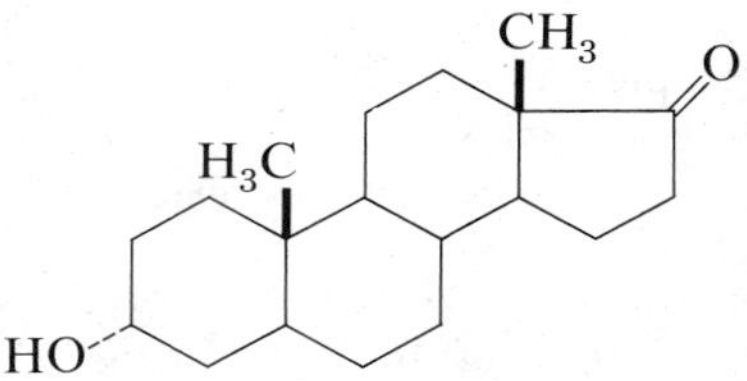

androsterone

Estrone is an example of an **estrogen,** or female sex hormone. Estrogens are secreted by the ovary and are responsible for the typical female sexual characteristics. Progesterone is another type of female sex hormone. It is also produced in the ovary and is the progestational hormone of the placenta and corpus luteum. Testosterone and androsterone are **androgens,** or male sex hormones. They are produced in the testes and are responsible for the typical male sexual characteristics.

One of the most dramatic achievements of synthetic organic chemistry, and one that has already had profound impact on the history and mores of human societies, has been the development of "the pill." Actually, there are a number of different oral contraceptives in use. They are mainly synthetic steroids that interfere in some way with the normal estrus or progestational cycle in the female. One example is norethindrone, also known by the trade name Norlutin.

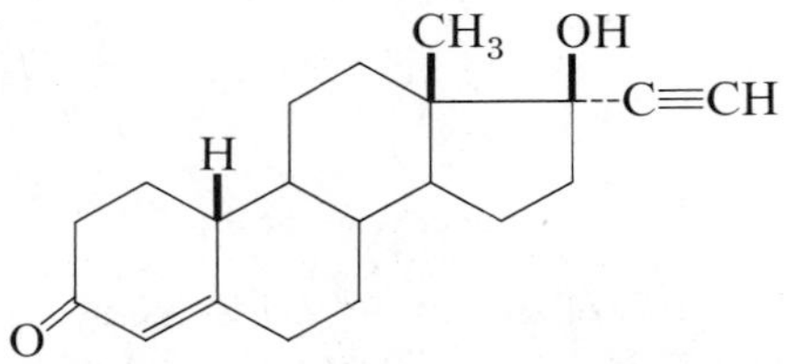

norethindrone
Norlutin

Steroids are widespread in the plant kingdom as well as in animals. One exam-

1171

Sec. 34.6

**Natural
Products:
Terpenes,
Steroids,
and Alkaloids**

ple is digitalis, a preparation made from the dried seeds and leaves of the purple foxglove. Historically, digitalis was used as a poison and as a medicine in heart therapy. The active agents in digitalis are **cardiac glycosides,** complex molecules built up from a steroid and several carbohydrates. Hydrolysis of digitoxin, one of the cardiac glycosides from digitalis, yields the steroid digitoxigenin.

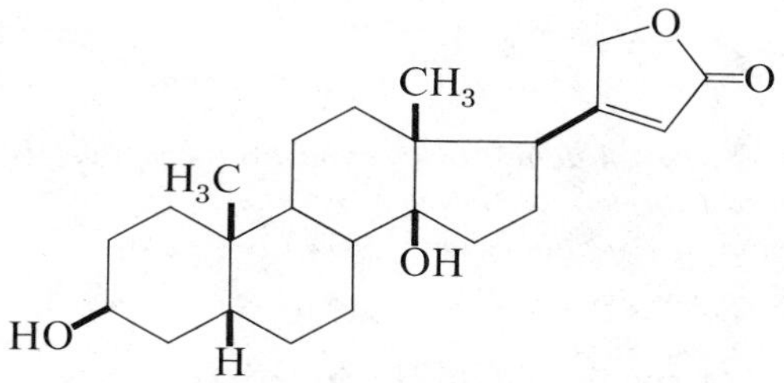

digitoxigenin

C. Alkaloids

Alkaloids constitute a class of basic, nitrogen-containing plant products that have complex structures and possess significant pharmacological properties. The name alkaloid, or "alkali-like," was first proposed by the pharmacist W. Meissner in the early nineteenth century before anything was known about the chemical structures of the compounds.

The first alkaloid isolated in a pure state was morphine, by Sertürner in 1805. The compound occurs in poppies and is responsible for the physiological effect of opium.

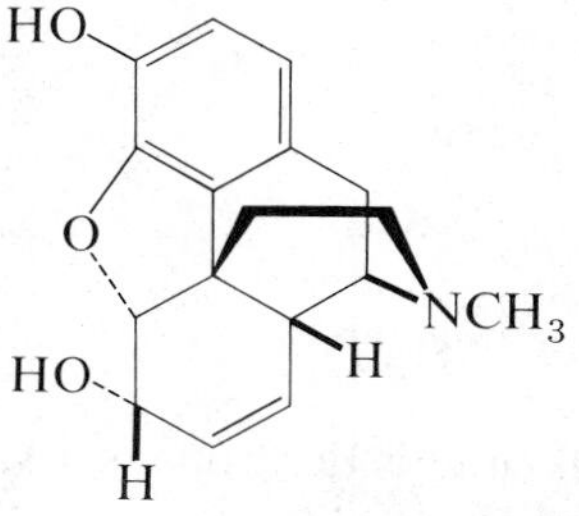

morphine

Other members of the morphine family are the O-methyl derivative codeine and the diacetyl derivative heroin.

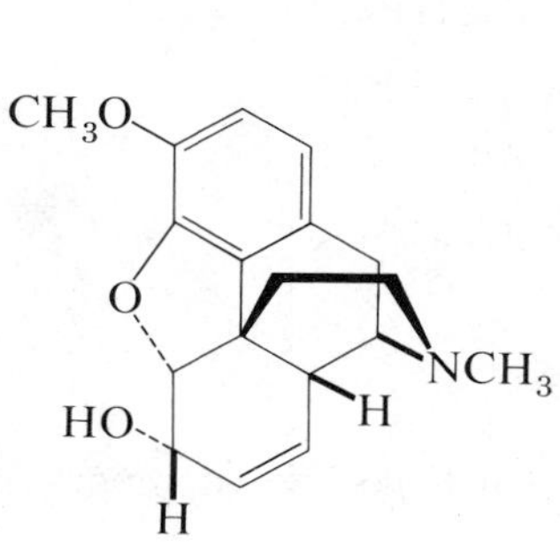

codeine

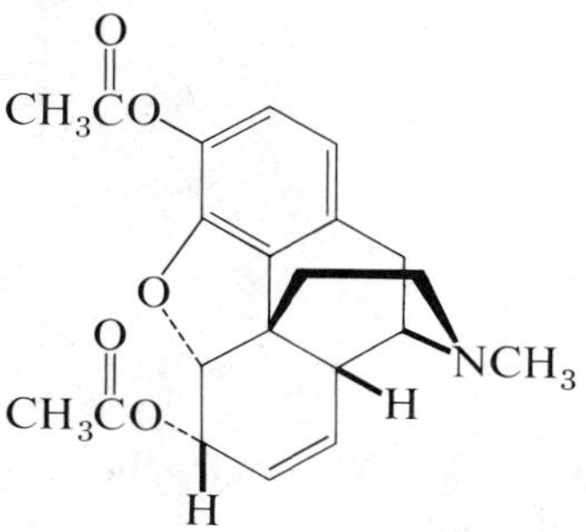

heroin

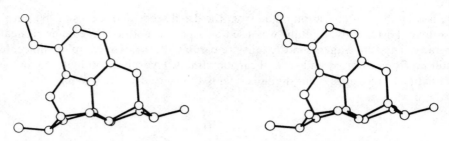

FIGURE 34.20. *Stereo structure of codeine hydrobromide. Hydrogens are not shown.* [*Reproduced with permission from* Molecular Structure and Dimensions. *International Union of Crystallography, 1972.*]

The stereo structure of codeine hydrobromide is shown in Figure 34.20.

Another common family of rather simple alkaloids is related to phenylethylamine. An example is mescaline, which occurs in several species of cactus. It is the active principle of mescal buttons, which were once used by some American Indians in religious rites. It has more recently gained notoriety as an illegal hallucinogen. Recent studies have shown that virtually all "mescaline" in street sales is actually LSD (lysergic acid diethylamide), which is an even more potent hallucinogen.

mescaline

lysergic acid diethylamide
"LSD"

Note that both of these hallucinogens contain a β-phenylethylamine grouping, as does amphetamine (page 740).

Another representative alkaloid is the tropane alkaloid cocaine, which has important anesthetic properties.

cocaine

Quinine is an alkaloid from cinchona bark, which has had an important use as an antimalarial agent.

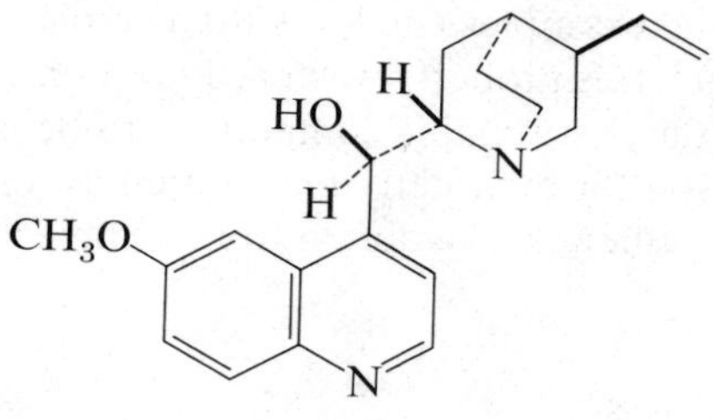

1173

Sec. 34.6

Natural
Products:
Terpenes,
Steroids,
and Alkaloids

quinine

Nicotine is the chief alkaloid of the tobacco plant.

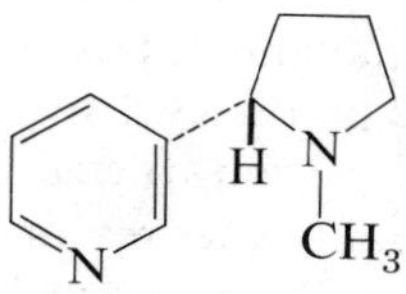

nicotine

Strychnine and brucine are intricate heptacyclic alkaloids that have been used as rodent poisons. They also find use as resolving agents in organic chemistry (page 738), since they are inexpensive and optically active and form well-defined salts with a variety of organic acids.

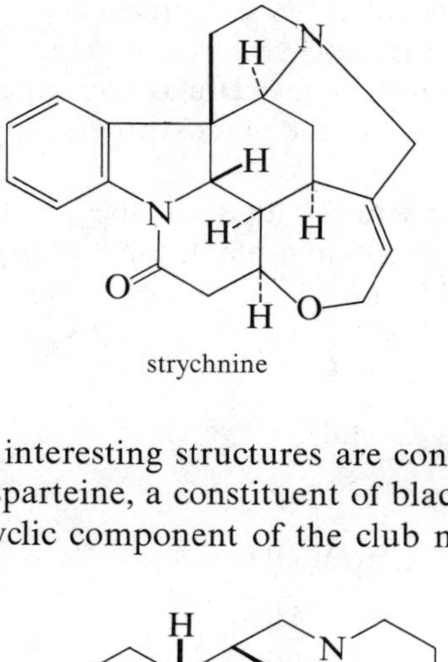

strychnine

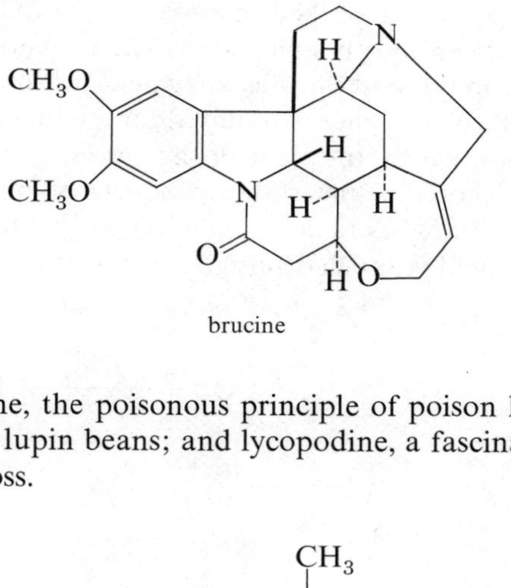

brucine

Other interesting structures are coniine, the poisonous principle of poison hemlock; sparteine, a constituent of black lupin beans; and lycopodine, a fascinating tetracyclic component of the club moss.

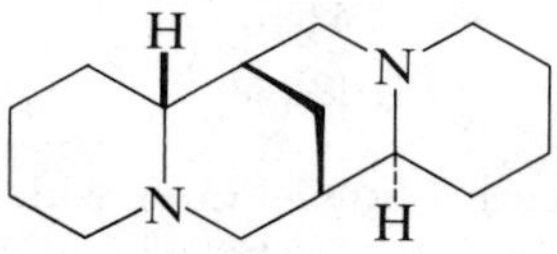

sparteine

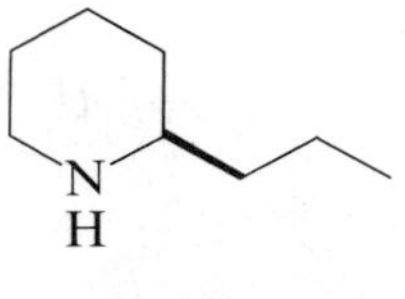

coniine

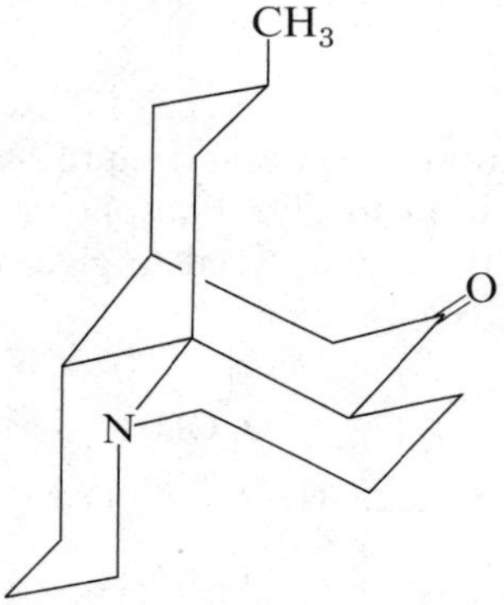

lycopodine

As is obvious from the examples cited, alkaloids typically have potent physiological properties. In fact, this property was partly responsible for the fact that the alkaloids were among the first organic compounds to be isolated in a pure state. Of course, another reason for their early recognition as discrete chemical entities is the fact that they are rather easy to obtain from complex plant material because of their basic nature.

34.7
Biosynthesis

In the previous section, we surveyed the classes of secondary plant metabolites known as terpenes, steroids, and alkaloids. It is not within the scope of this book to go into full details on metabolic processes. However, the elucidation of the pathways by which natural products are constructed by organisms **(biosynthesized)** is a major area of current research. For that reason we shall survey briefly one area of biosynthesis, that of the terpenes and steroids.

The basic raw material of the higher plants is carbon dioxide. This basic building unit is reduced in a process called **photosynthesis,** to give simple sugars and sugar derivatives.

$$n\,CO_2 + n\,H_2O \longrightarrow (CH_2O)_n + n\,O_2 \qquad (34\text{-}6)$$

The process described in equation (34-6) is wonderfully complex, and its details have been mostly elucidated. The conversion is catalyzed by the green pigment chlorophyll and various enzymes, and numerous other cellular constituents are involved. The energy required for the functioning of the photosynthetic apparatus is supplied by the light of the sun.

One of the first stable products of photosynthesis that can be identified is 3-phosphoglyceric acid (3-PGA). If $^{14}CO_2$ is "fed" to a plant, the 3-phosphoglyceric acid is labeled initially in the carboxy group.

$$^{14}CO_2 \xrightarrow{\text{photosynthesis}}$$

$$\begin{array}{c} ^{14}\text{COOH} \\ | \\ \text{H}-\text{C}-\text{OH} \\ | \\ \text{CH}_2\text{OPOH} \\ | \\ \text{OH} \end{array}$$

3-phosphoglyceric acid

This substance is isomerized to 2-phosphoglyceric acid (2-PGA), which undergoes dehydration to give phosphoenolpyruvic acid (PEP). The phosphoenolpyruvic acid is then hydrolyzed to give pyruvic acid and phosphate ion.

$$\begin{array}{c} \text{COOH} \\ | \\ \text{H}-\text{C}-\text{OH} \\ | \\ \text{CH}_2\text{OPO}_3\text{H}_2 \end{array} \rightleftharpoons \begin{array}{c} \text{COOH} \\ | \\ \text{H}-\text{C}-\text{OPO}_3\text{H}_2 \\ | \\ \text{CH}_2\text{OH} \end{array} \longrightarrow \begin{array}{c} \text{HOOC} \quad \text{OP(OH)}_2 \\ \diagdown \quad || \\ \text{C} \quad \text{O} \\ || \\ \text{C} \\ \diagup \quad \diagdown \\ \text{H} \quad \text{H} \end{array} \longrightarrow \begin{array}{c} \text{O} \\ || \\ \text{CH}_3\text{CCOOH} \end{array}$$

3-PGA   2-PGA   PEP   pyruvic acid

⌈ All of these reactions, like almost all biological reactions, are mediated by enzymes.
Enzymes function as catalysts and merely lower the activation energy for a reaction,
thereby allowing it to occur more rapidly. An enzyme cannot make a thermodynamically unfavorable process favorable. All biological processes are inherently exothermic. However, just as in any multistep reaction, a given step may be endothermic. In
⌊ writing reactions in this section, we shall omit the enzymes from the equations. ⌋

Pyruvic acid, PEP, 2-PGA, and 3-PGA are all involved in a number of primary
metabolic processes that we shall not consider. We shall look only at one reaction
of pyruvic acid, a reaction that leads us further down the path from CO_2 to the
terpenes and steroids. In a complex series of steps, which are generally understood, the α-keto acid is oxidatively decarboxylated and coupled with the —SH
group of coenzyme A to give a substance known as **acetyl coenzyme A,** or
$CH_3COSCoA$.

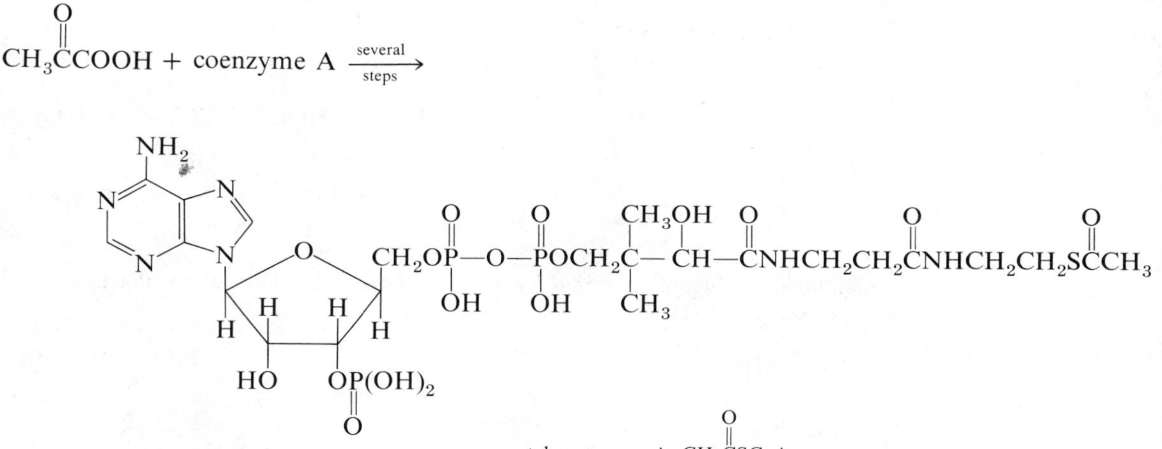

$$\underset{O}{\overset{O}{CH_3\overset{\parallel}{C}COOH}} + \text{coenzyme A} \xrightarrow[\text{steps}]{\text{several}}$$

acetyl coenzyme A, $CH_3\overset{O}{\overset{\parallel}{C}}SCoA$

Acetyl coenzyme A is a **thiolester.** In essence, it is the biological equivalent of
ethyl acetate. The —SR group serves two functions; it activates the acetyl group
for nucleophilic attack, and it renders the —CH_3 protons more acidic for
enolization.

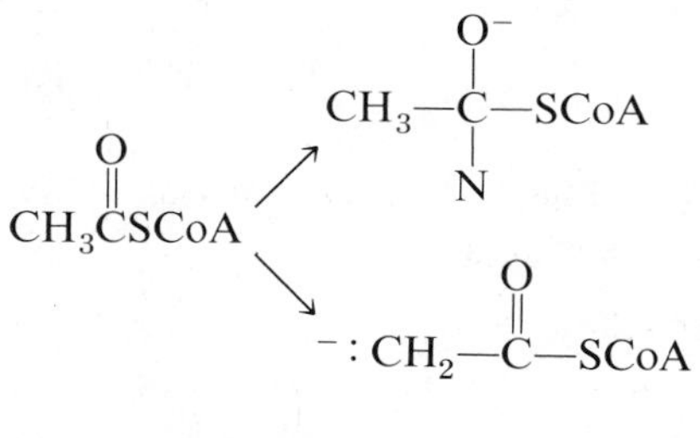

We have already seen in Section 19.9 how these properties are utilized in the
formation of fatty acids. The first step of this process is a biological version of the
Claisen condensation, affording acetoacetyl coenzyme A.

$$2\ CH_3\overset{O}{\overset{\parallel}{C}}SCoA \rightleftharpoons CH_3\overset{O}{\overset{\parallel}{C}}CH_2\overset{O}{\overset{\parallel}{C}}SCoA + CoASH$$

acetoacetyl coenzyme A

This substance reacts with another molecule of acetyl CoA in a form of aldol condensation to give β-hydroxy-β-methylglutaryl CoA.

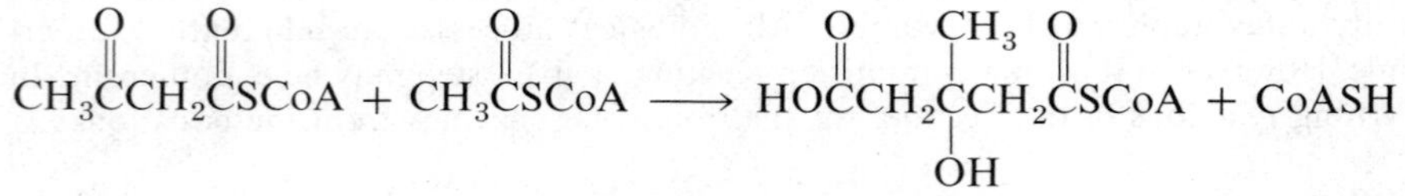

$$CH_3\overset{O}{\overset{\|}{C}}CH_2\overset{O}{\overset{\|}{C}}SCoA + CH_3\overset{O}{\overset{\|}{C}}SCoA \longrightarrow HO\overset{O}{\overset{\|}{C}}CH_2\overset{CH_3}{\underset{OH}{\overset{|}{C}}}CH_2\overset{O}{\overset{\|}{C}}SCoA + CoASH$$

β-hydroxy-β-methylglutaryl CoA

The thiolester grouping is reduced, first to a hydroxy aldehydo acid, then to a dihydroxy acid, **mevalonic acid.**

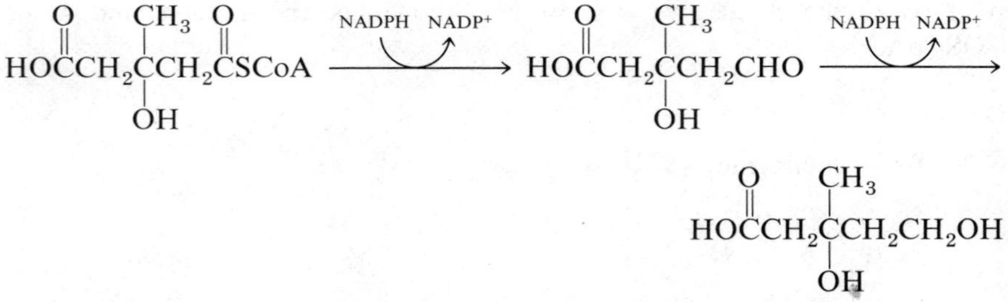

$$HO\overset{O}{\overset{\|}{C}}CH_2\overset{CH_3}{\underset{OH}{\overset{|}{C}}}CH_2\overset{O}{\overset{\|}{C}}SCoA \xrightarrow{\text{NADPH} \quad \text{NADP}^+} HO\overset{O}{\overset{\|}{C}}CH_2\overset{CH_3}{\underset{OH}{\overset{|}{C}}}CH_2CHO \xrightarrow{\text{NADPH} \quad \text{NADP}^+}$$

$$HO\overset{O}{\overset{\|}{C}}CH_2\overset{CH_3}{\underset{OH}{\overset{|}{C}}}CH_2CH_2OH$$

(R)-3,5-dihydroxy-3-methylpentanoic acid
mevalonic acid

The biological reducing agent in these two reductions is **nicotinamide adenine dinucleotide phosphate, NADPH.**

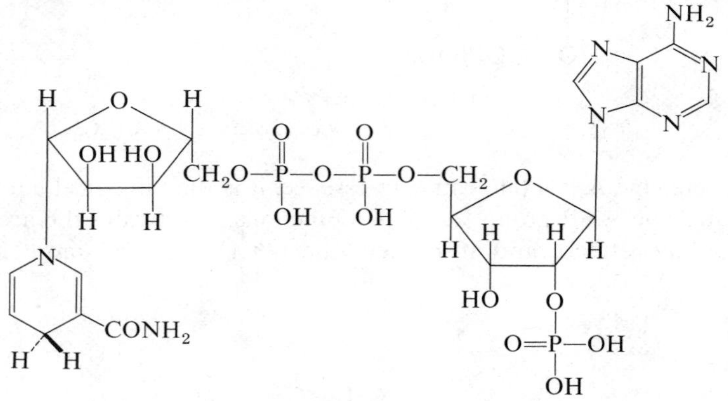

NADPH

It is classified as a coenzyme and functions with the aid of an enzyme. It functions as a reducing agent in the following manner.

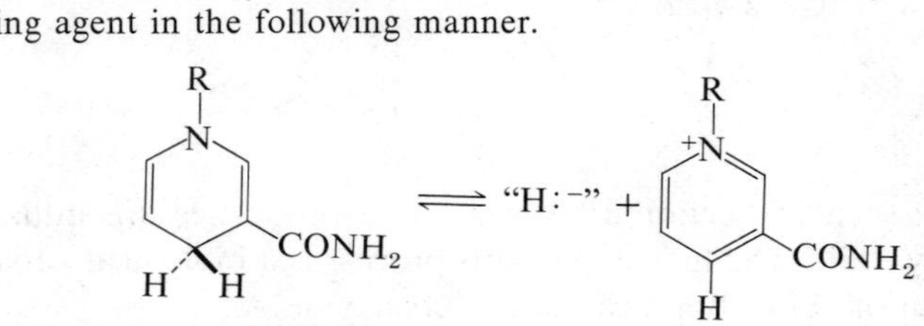

NADPH $\rightleftharpoons$ "H:⁻" + NADP⁺

Biological reactions involving oxidation or reduction are conventionally written as indicated, the curved arrow symbolizing that NADPH is consumed and NADP⁺ is produced in the reaction.

Mevalonic acid is converted, in two steps, into 5-pyrophosphomevalonic acid.

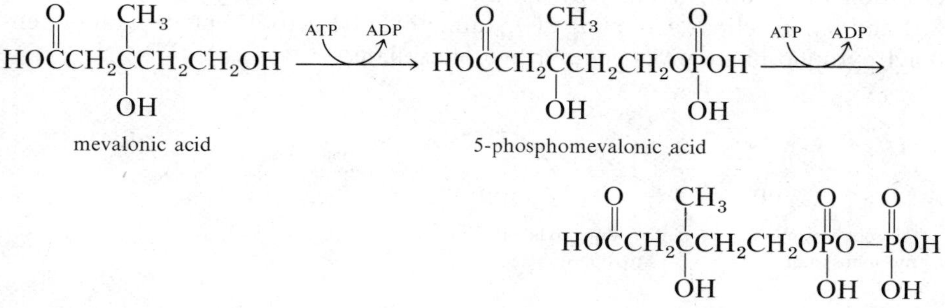

mevalonic acid

5-phosphomevalonic acid

5-pyrophosphomevalonic acid

Phosphorylation is accomplished with the aid of another coenzyme, adenosine triphosphate (ATP). The by-product of each phosphorylation step is adenosine diphosphate (ADP).

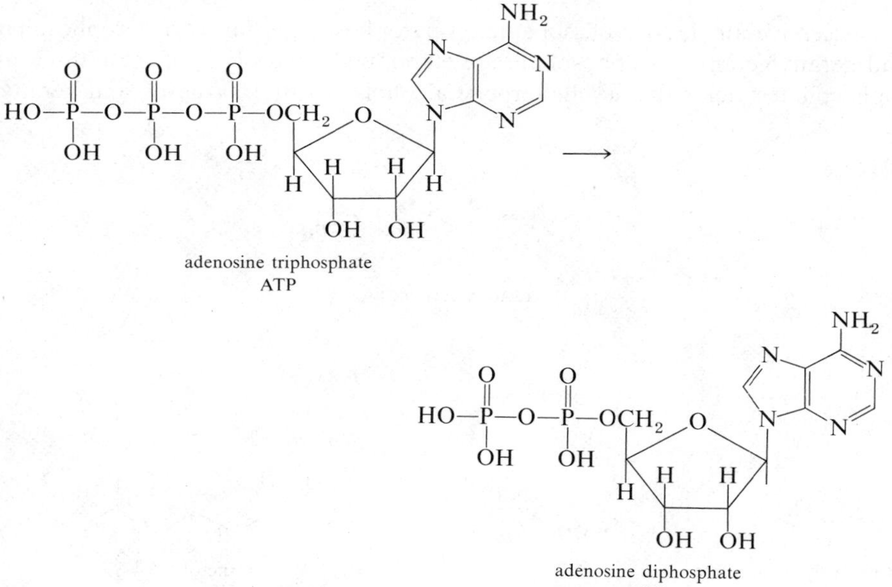

adenosine triphosphate
ATP

adenosine diphosphate
ADP

5-Pyrophosphomevalonic acid is phosphorylated once again to yield 3-phospho-5-pyrophosphomevalonic acid, which undergoes "decarboxylative elimination" to yield isopentenyl pyrophosphate.

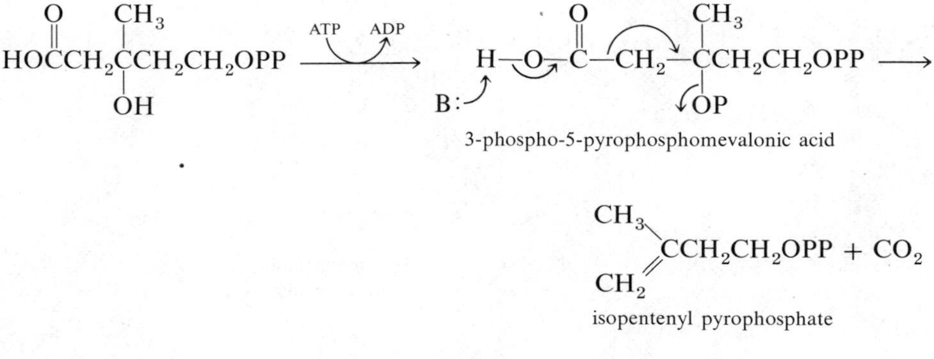

3-phospho-5-pyrophosphomevalonic acid

isopentenyl pyrophosphate

Isopentenyl pyrophosphate is the biological "isoprene unit." It undergoes isomerization to γ,γ-dimethylallyl pyrophosphate, which is highly activated toward S_N1 and S_N2 reactions. Coupling of γ,γ-dimethylallyl pyrophosphate with isopentenyl pyrophosphate gives geranyl pyrophosphate.

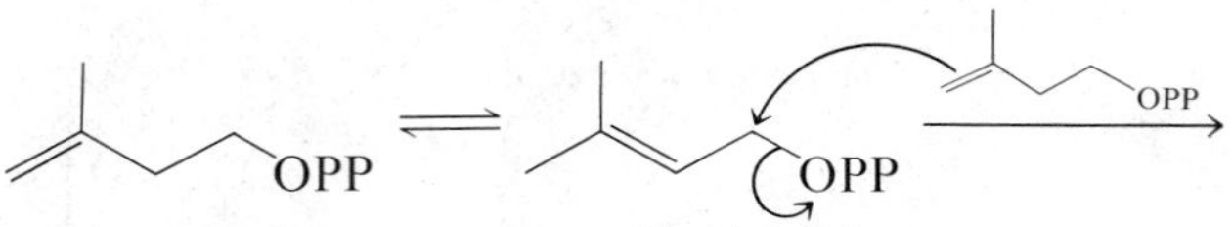

isopentenyl
pyrophosphate

γ,γ-dimethylallyl
pyrophosphate

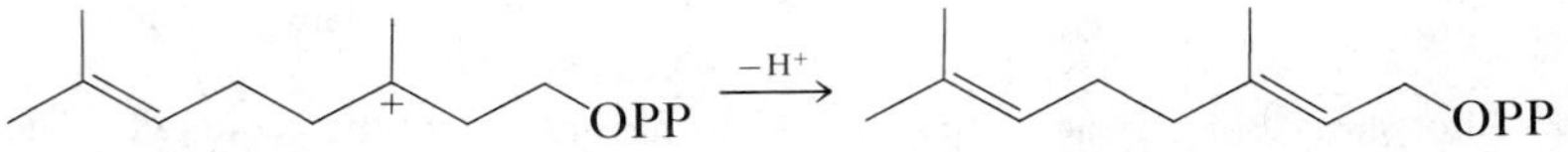

geranyl pyrophosphate

By successive alkylations of isopentenyl pyrophosphate, farnesyl pyrophosphate and geranylgeranyl pyrophosphate are elaborated. Hydrolysis of these three pyrophosphates yields the acyclic terpene alcohols geraniol, farnesol, and geranylgeraniol.

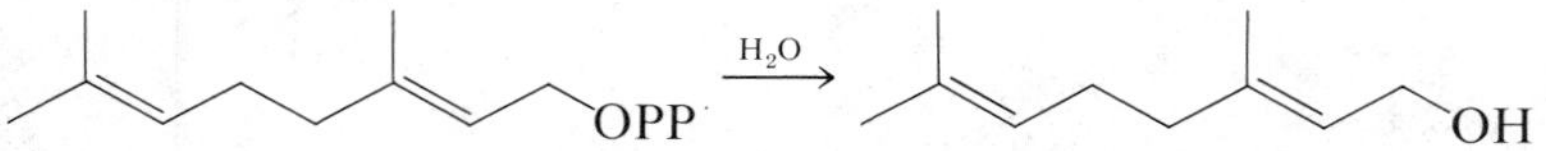

geranyl pyrophosphate

geraniol
(a monoterpene)

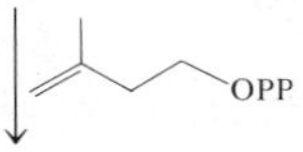

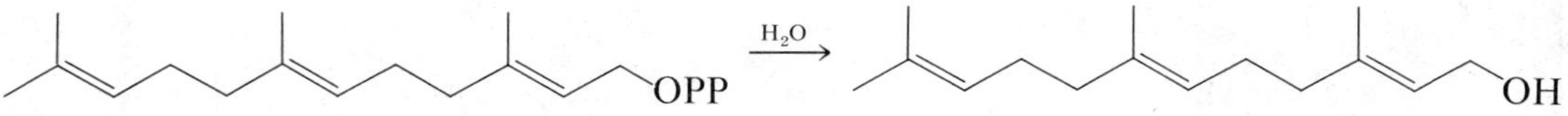

farnesyl pyrophosphate

farnesol
(a sesquiterpene)

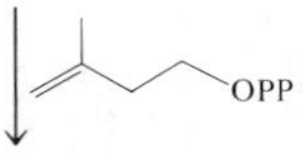

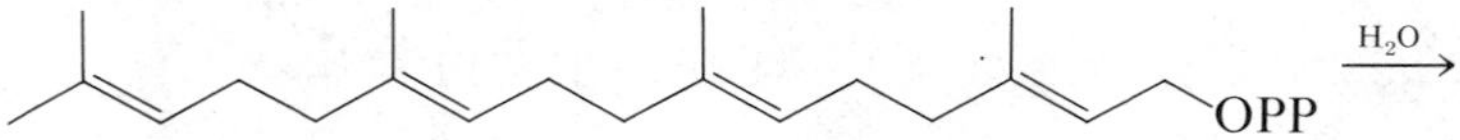

geranylgeranyl pyrophosphate

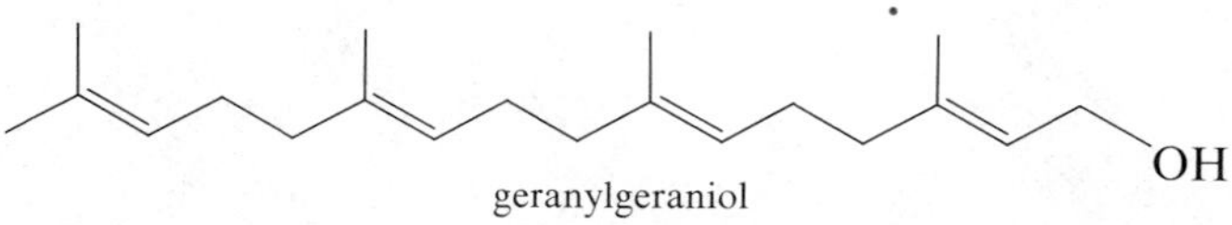

geranylgeraniol
(a diterpene)

The three acyclic pyrophosphates undergo a marvelous assortment of subsequent reactions to yield the cyclic monoterpenes, sesquiterpenes, and diterpenes that are so widespread in the plant kingdom. For example, double bond isomerization, followed by two successive enzyme-catalyzed cyclizations, converts geranyl pyrophosphate into α-pinene, the familiar fragrant principle of turpentine.

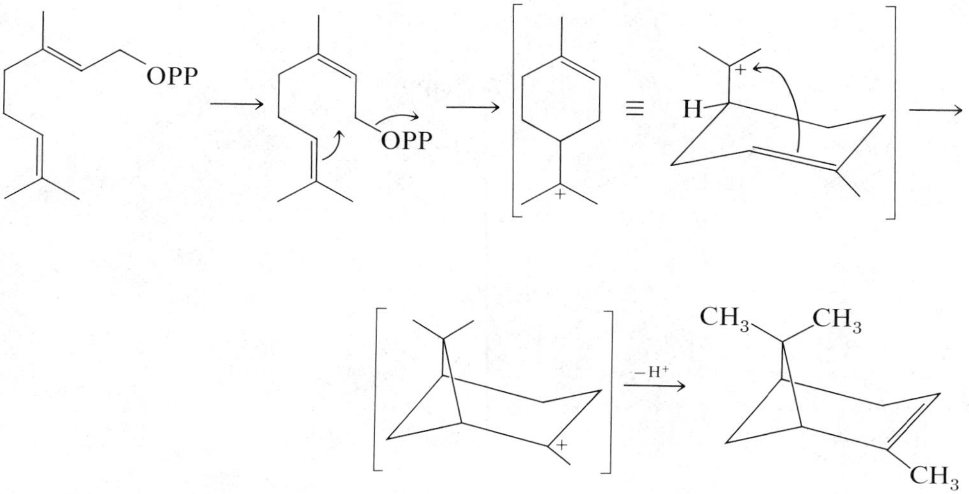

α-pinene

The following sequence of steps may be envisioned from farnesyl pyrophosphate to valencene, which is responsible for the aroma and taste of Valencia oranges.

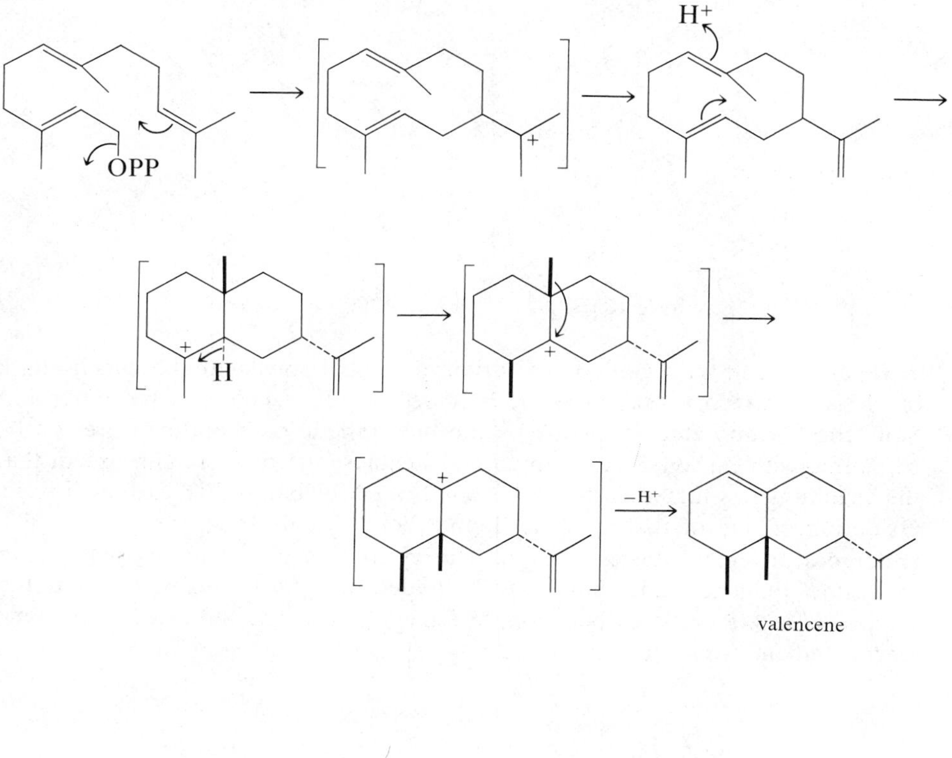

valencene

Similar cyclizations yield the diterpenes. The foregoing metabolic paths are common to many plants. Differences occur mainly in the ways in which geranyl pyrophosphate and farnesyl pyrophosphate are converted into the various terpenes.

Farnesyl pyrophosphate undergoes another biological reaction—reductive coupling. The initial product of this coupling process is a cyclopropylmethyl pyrophosphate, presqualene alcohol pyrophosphate, which is then converted into squalene.

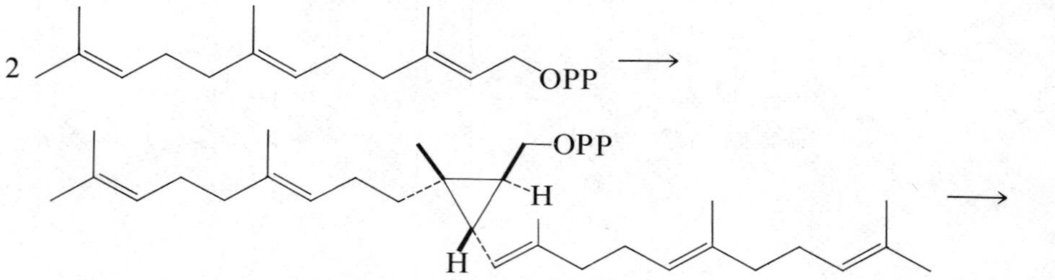

presqualene alcohol pyrophosphate

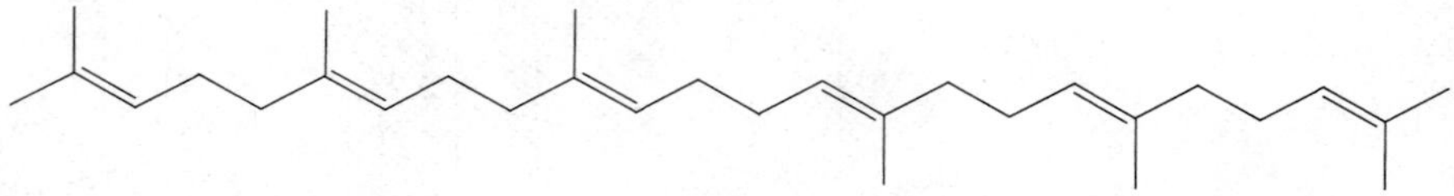

squalene

The exact mechanism whereby presqualene alcohol pyrophosphate is converted into squalene is still unknown, but it may reasonably be formulated as follows.

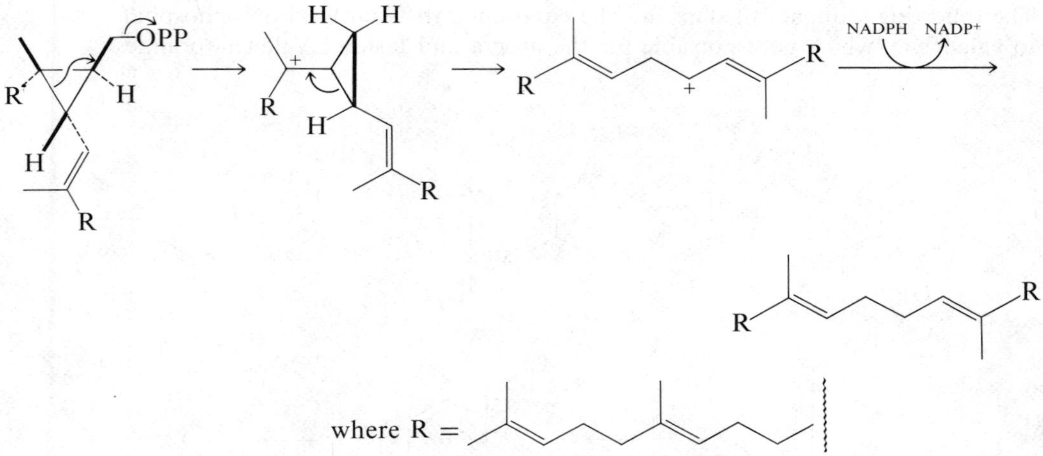

where R =

Squalene was once regarded as a curious terpenoid polyene that occurs in shark oil. It is now recognized to be the precursor to the whole family of steroids. Squalene is elaborated by plants by the biosynthetic path outlined previously, beginning with carbon dioxide. In animals its biosynthesis is similar, except that the beginning raw materials are more complicated foodstuffs than carbon dioxide. Oxidation of one of the squalene double bonds yields squalene oxide, which undergoes an enzymatic cyclization. The cyclization is a complex reaction in which the squalene oxide chain must be folded in a precise manner. The initial product is a tetracyclic carbocation, which undergoes the indicated "backbone rearrangement" to afford lanosterol.

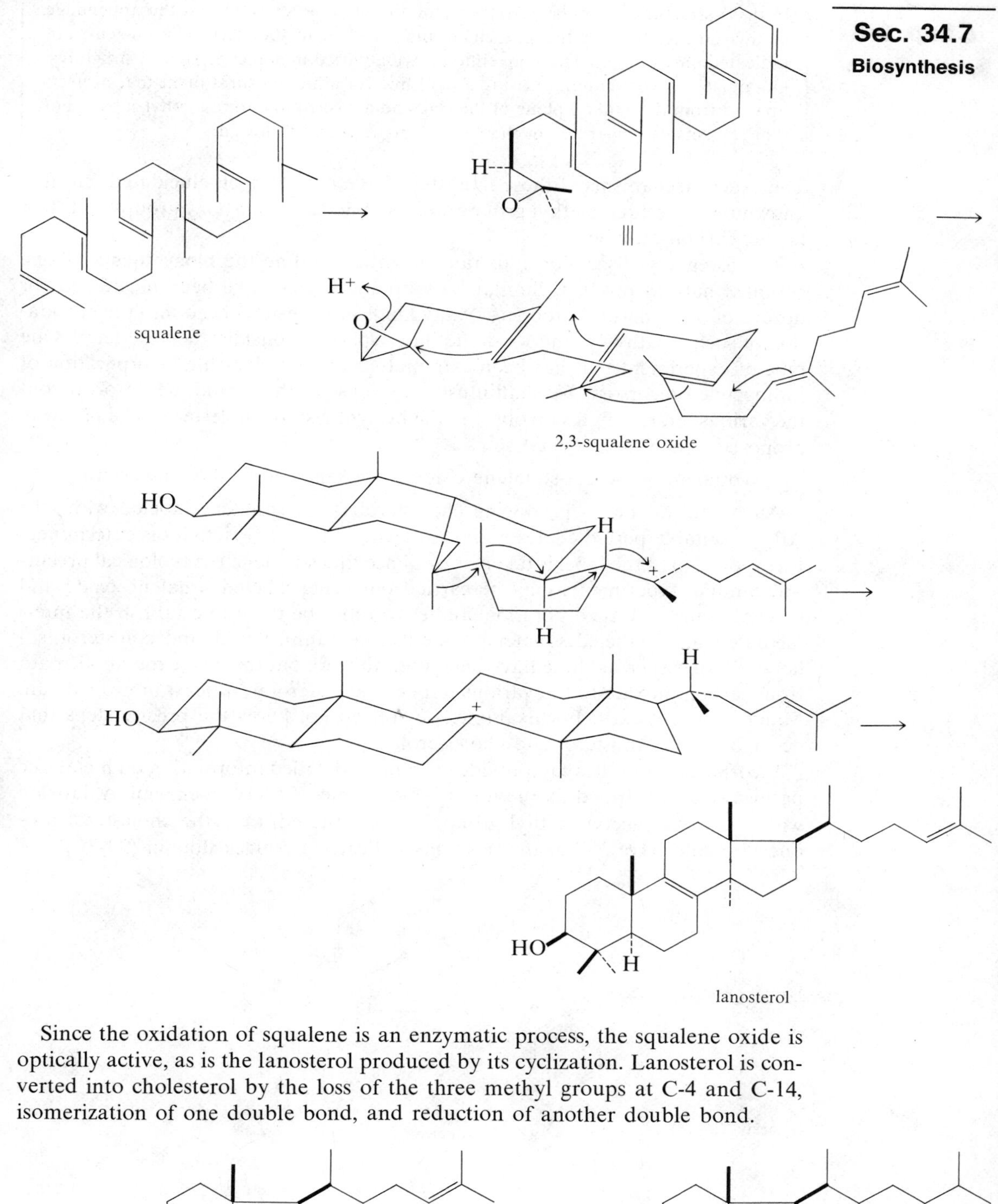

squalene

2,3-squalene oxide

lanosterol

Since the oxidation of squalene is an enzymatic process, the squalene oxide is optically active, as is the lanosterol produced by its cyclization. Lanosterol is converted into cholesterol by the loss of the three methyl groups at C-4 and C-14, isomerization of one double bond, and reduction of another double bond.

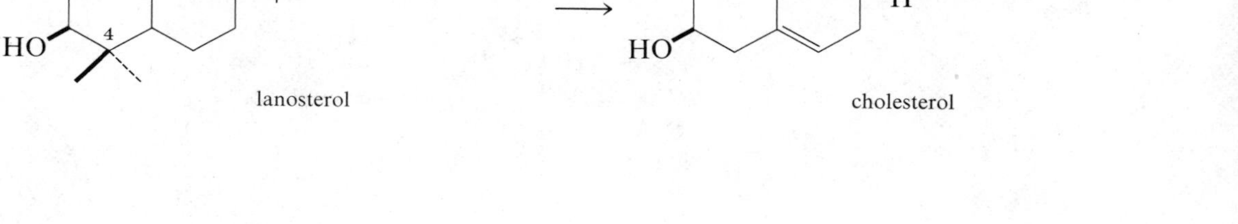

lanosterol   cholesterol

In these structural formulas for lanosterol and cholesterol, note that the appendage methyl groups, both in the side chain and attached to the tetracyclic nucleus, are indicated only as lines. This convention is widely used in depicting steroid structures. Of course, the convention of using a bold line for a methyl substituent that projects upward from the general plane of the ring and a dashed line for a methyl substituent that projects downward from the general plane is still followed.

The exact mechanisms of these final steps have not yet been elucidated, but it is known that the three methyl groups are first oxidized to carboxy groups and then lost as carbon dioxide.

The foregoing discussion illustrates in broad outline the **biosynthesis** of one complex natural product. Similar biosynthetic routes have been elucidated for dozens of other natural products. Many kinds of experiments go into the elucidation of such a pathway, and we do not have space to consider them in detail. One type of experiment that has been extremely useful involves the incorporation of radioactive precursors. We shall illustrate the use of this technique in sorting out the various intermediates involved in the biosynthesis of cholesterol. The pathway proposed above for the final steps is

squalene $\longrightarrow$ 2,3-squalene oxide $\longrightarrow$ lanosterol $\longrightarrow$ cholesterol

An organism that is producing cholesterol is fed squalene labeled with ^{14}C. After a suitable period cholesterol is isolated, and its radioactivity is determined. Incorporation of the label is taken as evidence that squalene is a biological precursor. Similar experiments may be carried out using labeled squalene oxide and labeled lanosterol. Incorporation studies can also be done to establish the intermediate stages; labeled squalene oxide can be administered, and lanosterol isolated. Studies such as these have been used to work out the entire metabolic path from carbon dioxide to the various terpenes and steroids, at least in gross detail. Many gaps still exist. For example, we still do not know the precise steps that occur between lanosterol and cholesterol.

Incorporation studies can provide even more detailed information on metabolic pathways, as illustrated by the following example. If acetyl coenzyme A, labeled with ^{14}C in the acetyl methyl group, is administered, then the cholesterol produced should have ^{14}C in the positions indicated by black dots in (34-7) if the

$$^{14}CH_3CSCoA \longrightarrow$$

acetyl CoA

squalene

(34-7)

cholesterol

biosynthetic path outlined in this section is correct. This experiment has been done, and the cholesterol produced was subjected to a lengthy multistep degradation. The degradation was carried out in such a way that each of the 27 carbons could be examined separately for radioactivity. The label was found precisely where it is predicted to be on the basis of the biosynthetic hypothesis. In particular, the experiment ruled out an earlier hypothesis regarding the possible biosynthesis of cholesterol, which may be summarized as

Note the different labeling pattern at C-7, C-12, and C-13 that would result from this model. In fact, the degradation revealed that C-7 and C-13 are labeled, whereas C-12 is not, thus refuting this earlier hypothesis.

34.8
Stereoselective Synthesis

In the preceding section we considered the question of how organisms synthesize complex organic molecules. We saw that the chemical reactions involved are the same reactions we use in the laboratory—aldol condensations, Claisen condensations, oxidations and reductions, and so on. Only the reagents and catalysts are different. Organisms make use of highly complex reagents to achieve rapid and specific reactions. An example is the biological reducing agent nicotinamide adenine dinucleotide phosphate (NADPH)—the organic chemist uses $NaBH_4$ or $LiAlH_4$. Although $NaBH_4$ and $LiAlH_4$ are much simpler molecules than NADPH, they are not nearly as specific. Under the influence of various enzyme catalysts, NADPH can perform selective reductions that chemists cannot yet hope to achieve with their laboratory reducing agents.

In fact, one of the major goals of organic chemists who specialize in synthetic methods is to duplicate or surpass the finess with which nature can construct complicated molecules. In many areas, we are far from that goal, although significant advances have been made. For example, in 1973 a team of organic chemists headed by the late Professor R. B. Woodward of Harvard University and by Professor Albert Eschenmoser at the Swiss Federal Institute of Technology completed a laboratory synthesis of vitamin B_{12} (Figure 34.21).

The synthesis was a notable achievement. In terms of effort expended, it is undoubtedly the most impressive organic synthesis ever done. More than 90 separate synthetic reactions were required, and dozens of highly trained chemists labored for years to bring it to completion. However, compared with the ease with which nature elaborates this complex molecule, the organic chemist must still take second place.

However, chemists *have* been able to synthesize many interesting molecules that are not produced by nature (perhaps because nature has no need for them).

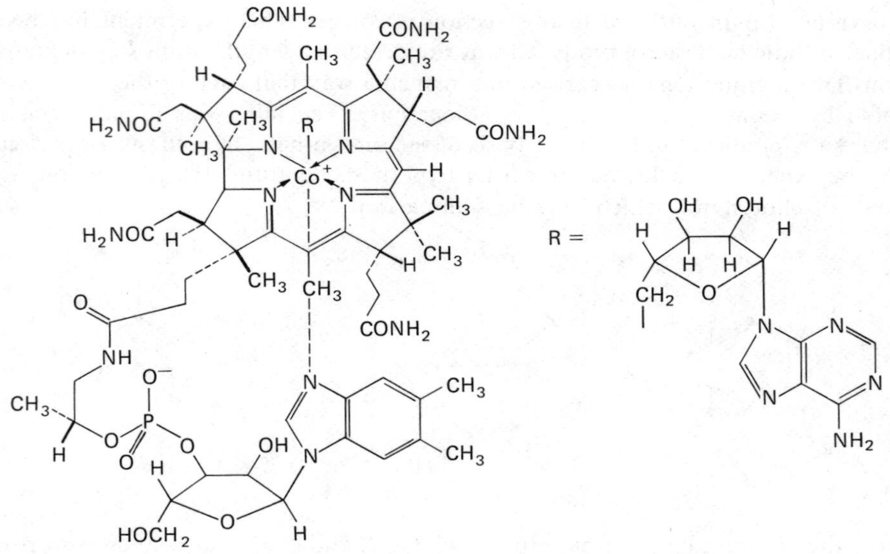

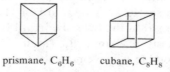

FIGURE 34.21. *Vitamin B_{12}:* (a) *structure;* (b) *stereo structure.* [*Courtesy of C. K. Johnson, Oak Ridge National Laboratory*].

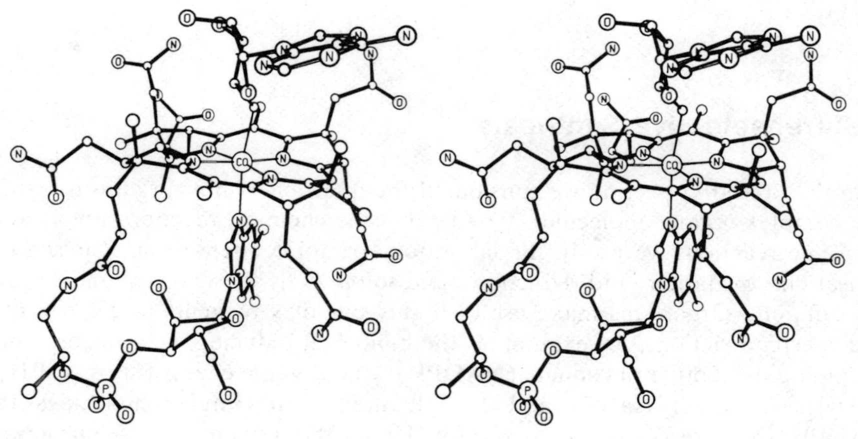

Examples are the highly strained hydrocarbons prismane and cubane.

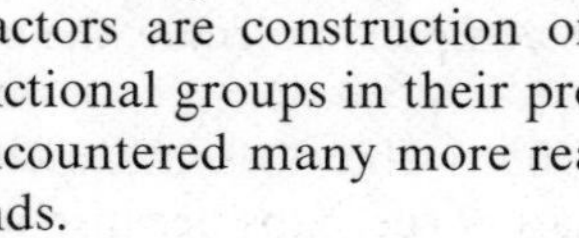

prismane, C_6H_6 cubane, C_8H_8

As we pointed out in Chapter 16 when we first surveyed the problem of designing the synthesis of an organic compound, several interrelated factors must be considered. Two of these factors are construction of the carbon skeleton and placement of the desired functional groups in their proper positions. In the intervening chapters, we have encountered many more reactions that can be used to accomplish both of these ends.

The third major consideration in synthetic design that we mentioned in Chapter 16 is stereochemical control. At the time, we had studied relatively few reactions that could be used to establish stereochemistry. Our repertoire is now more suited for a discussion of that problem, and we shall examine it briefly. In general, stereochemistry may be controlled in two ways: thermodynamically or kinetically.

A. Thermodynamic Control

If the stereoisomer we wish to synthesize is the more stable stereoisomer, we may carry out the synthesis in such a way that the product, or one of its precursors, may equilibrate with its stereoisomers. For example, suppose we wish to synthesize methyl trans-4-t-butylcyclohexanecarboxylate. Since both substituents are equatorial in the trans stereoisomer, it is more stable than the cis isomer.

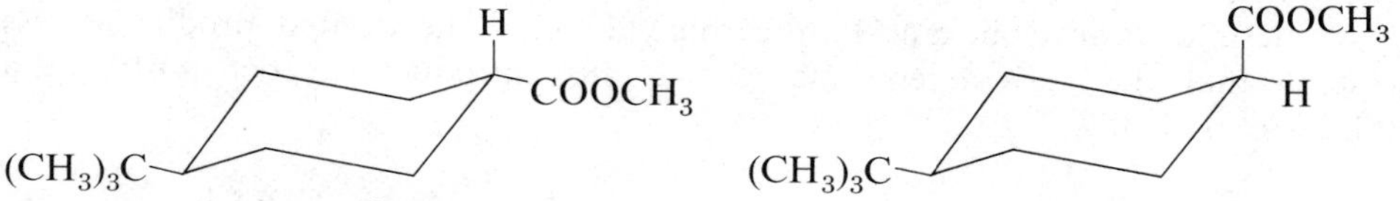

trans stereoisomer
(more stable)

cis stereoisomer
(less stable)

We may take advantage of this information in our synthesis. A reasonable synthesis of the compound, beginning with phenol, is outlined (34-8).

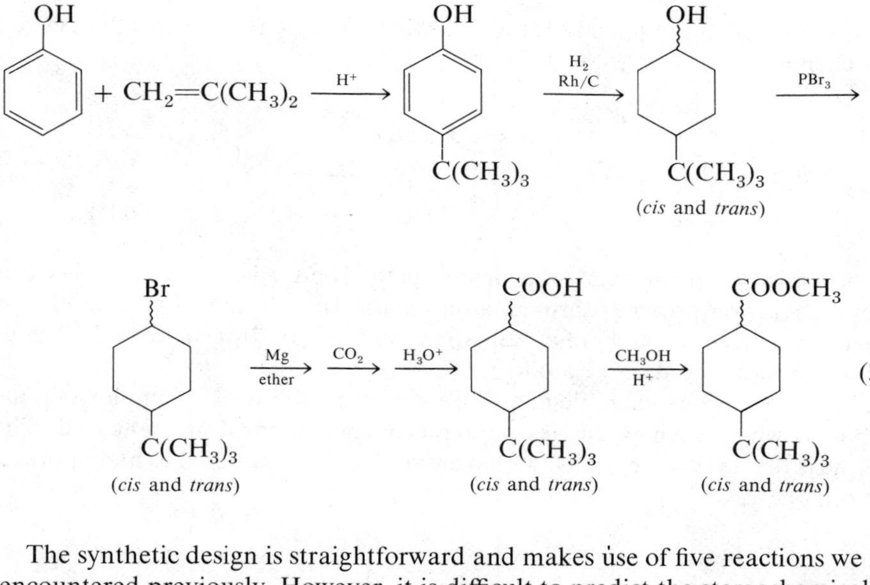

(34-8)

The synthetic design is straightforward and makes use of five reactions we have encountered previously. However, it is difficult to predict the stereochemical outcome, and a mixture of cis and trans isomers will probably result. Since the more stable trans isomer is the desired product, obtaining a mixture of stereoisomers presents no problem. One may simply equilibrate the product by treating the final mixture with sodium methoxide in anhydrous methanol. The more stable trans isomer predominates at equilibrium.

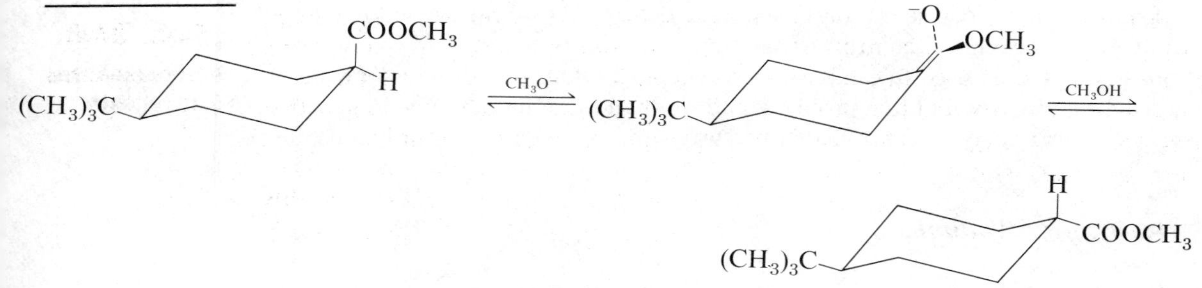

B. *Kinetic Control*

In cases where the desired stereoisomer is *not* the more stable one, we must carry out the synthesis in such a way that the less stable stereoisomer is produced more rapidly than the more stable stereoisomer. As an example, consider the synthesis of *cis*-cyclohexane-1,2-dicarboxylic acid. The desired product is less stable than the *trans* stereoisomer, so we must introduce stereochemistry by a kinetic method.

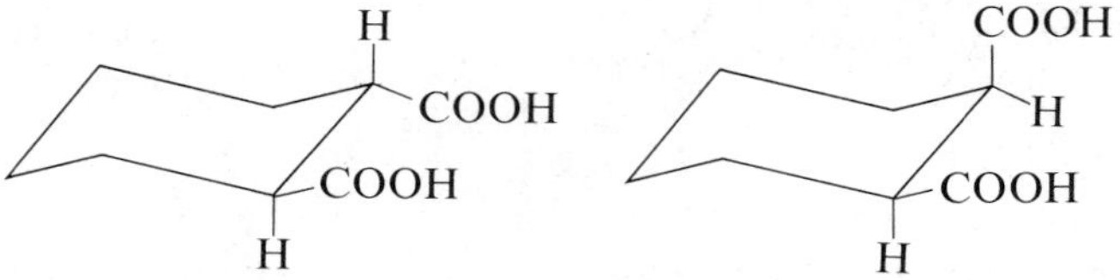

trans stereoisomer
(more stable)

cis stereoisomer
(less stable)

We may synthesize the diacid by the following route, which assures the desired stereochemistry.

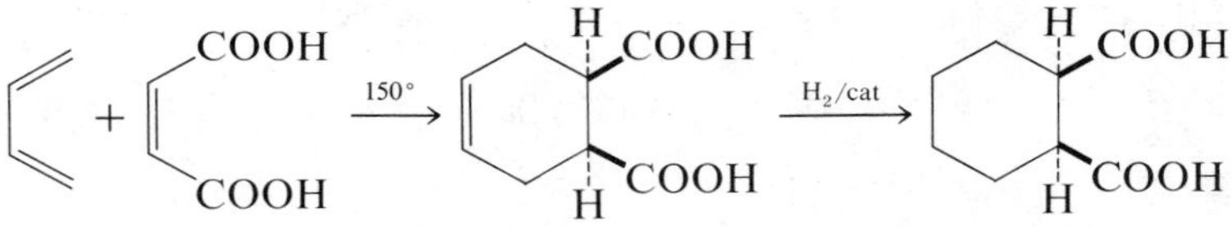

Stereochemistry is established in the first step, the Diels–Alder reaction. Since this pericyclic reaction proceeds through an aromatic $4n + 2$ transition state, the *cis* isomer is produced much more rapidly than the *trans*. Hydrogenation of the double bond affords the desired product.

Let us consider one example of the laboratory synthesis of a complex organic compound where such principles of stereochemical control are practiced. The example is that of estrone, which we have previously encountered as an important sex hormone.

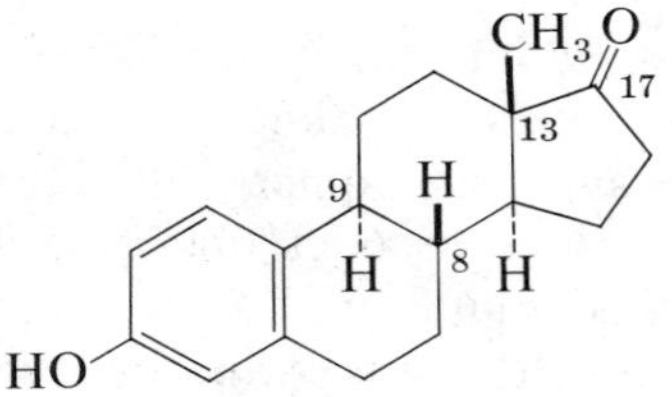

estrone

The molecule has four asymmetric centers, and there are therefore sixteen stereoisomers (Section 7.6). Only the isomer shown has significant physiological activity. There have been numerous syntheses of estrone, but by far the most efficient is one discovered by the Russian chemist I. V. Torgov and perfected by various other workers. In fact, the Torgov synthesis of estrone is so efficient that it is used as a commercial preparation for the steroid.

The Torgov synthesis begins with 6-methoxy-1-tetralone, a substance readily available from naphthalene. The ketone is treated with vinylmagnesium bromide to give an allylic alcohol.

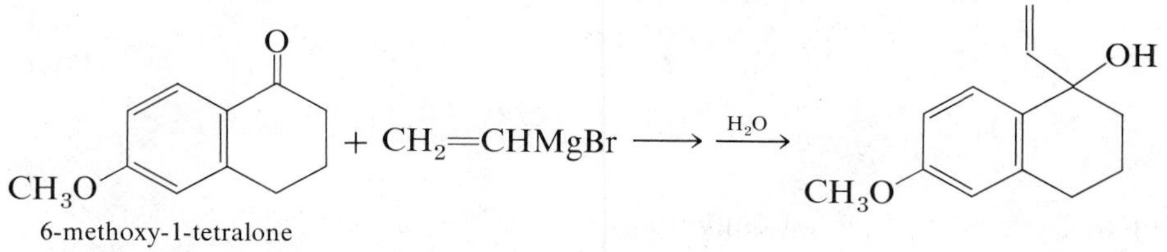

6-methoxy-1-tetralone

When this alcohol is heated in xylene with 2-methyl-1,3-cyclopentanedione and acetic acid, a crystalline tetracyclic ketone is produced in 70% yield.

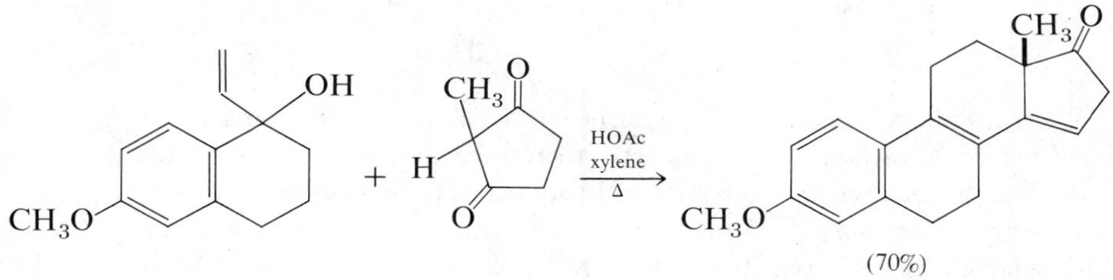

(70%)

The mechanism of this reaction involves an acid-catalyzed ionization of the allylic alcohol (Section 20.1.A), which reacts with the enol form of the 1,3-diketone.

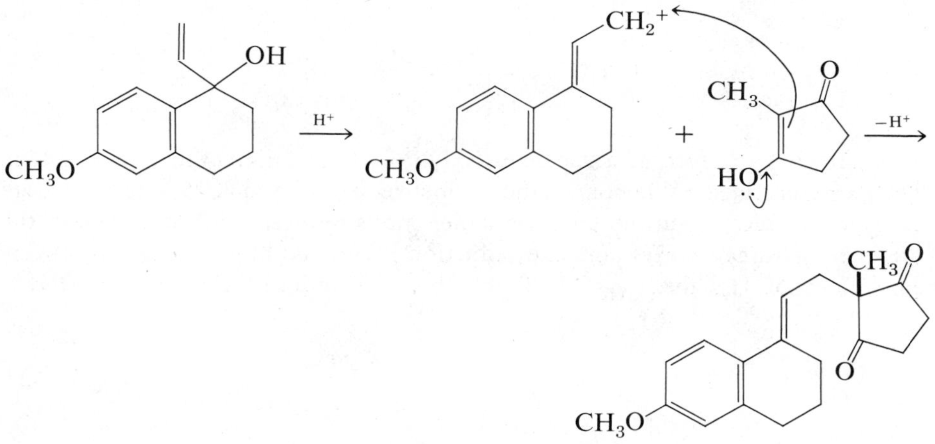

The double bond then undergoes acid-catalyzed isomerization, and a type of Friedel–Crafts alkylation reaction follows.

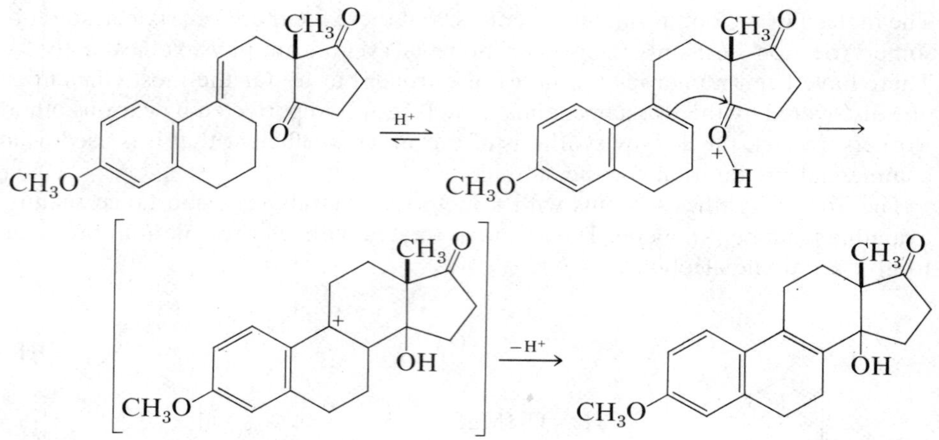

Finally, the allylic alcohol dehydrates to give the diene.

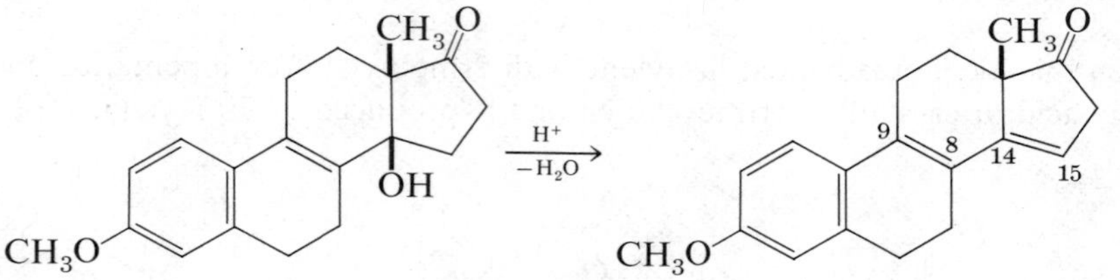

With the carbon skeleton of estrone elaborated and the functional groups in their proper position, it remains only to reduce the two double bonds in such a way that the proper stereoisomer is produced. At this point we should mention that estrone is *not* the most stable of the sixteen stereoisomers. Rather, 14-isoestrone is the most stable isomer. Therefore, chirality must be imparted to C-14 by some kinetic method.

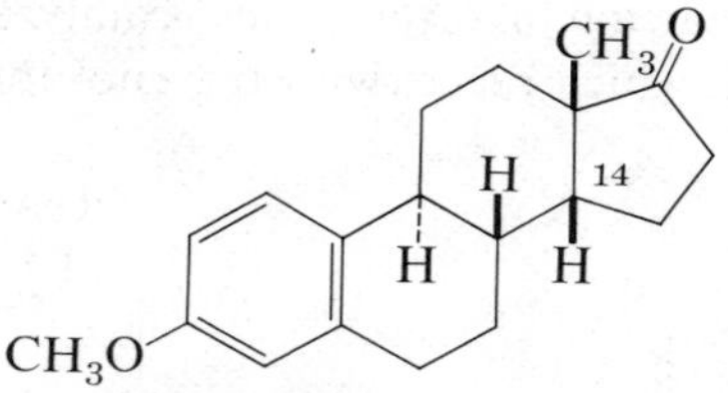

14-isoestrone

This is readily achieved by a selective catalytic hydrogenation of the diene system. The more reactive double bond is the one between C-14 and C-15, since it is more easily accessible. When the tetracyclic dienone is reduced with hydrogen in the presence of Raney nickel, and the reduction is stopped after the uptake of one equivalent of H_2, the C_{14}–C_{15} double bond is reduced cleanly and stereoselectivally.

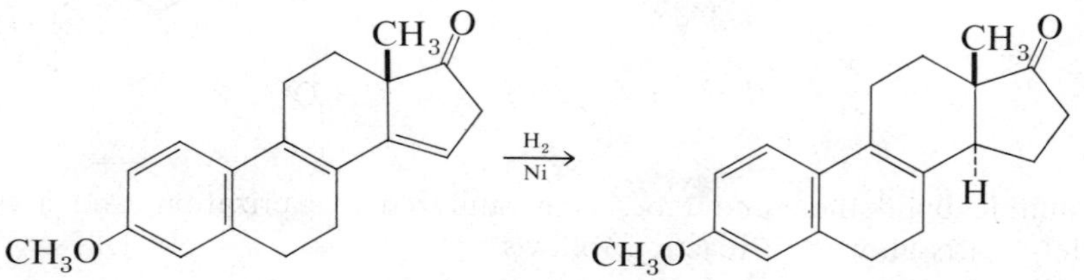

In this case, the methyl group, which projects over the top of the nearly flat molecule, shields that side of the molecule from the surface of the catalyst. Therefore, the less stable stereoisomer is produced more rapidly, and stereoselectivity is achieved.

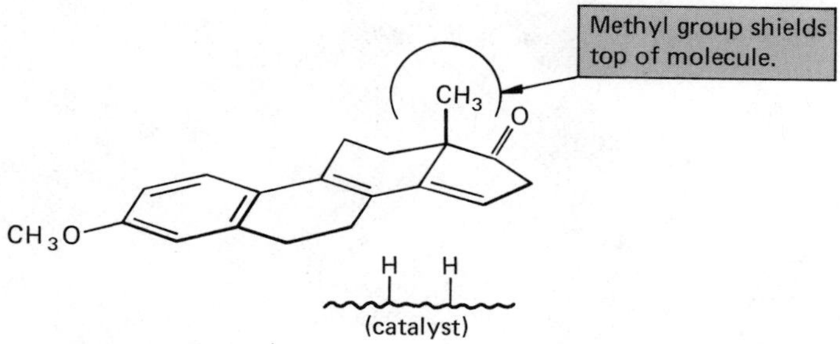

The remaining double bond between C-8 and C-9 may be reduced under conditions of thermodynamic control because the desired stereochemistry at these centers is the thermodynamically favored stereochemistry. This is achieved by treating the compound with potassium in liquid ammonia. Both the ketone and the double bond are reduced in this reaction, but the important feature is the stereochemistry. The alkali metal–ammonia reducing medium has the important property of producing the most stable reduction product (for example, see Section 12.6.A). Thus the stereochemistry at C-8 and C-9 is that desired.

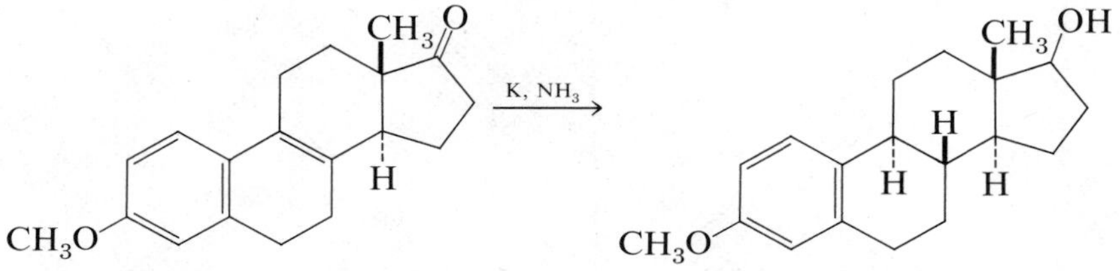

The secondary alcohol is oxidized back to a ketone, and the aryl methyl ether is cleaved (Section 30.6.D) to complete the synthesis.

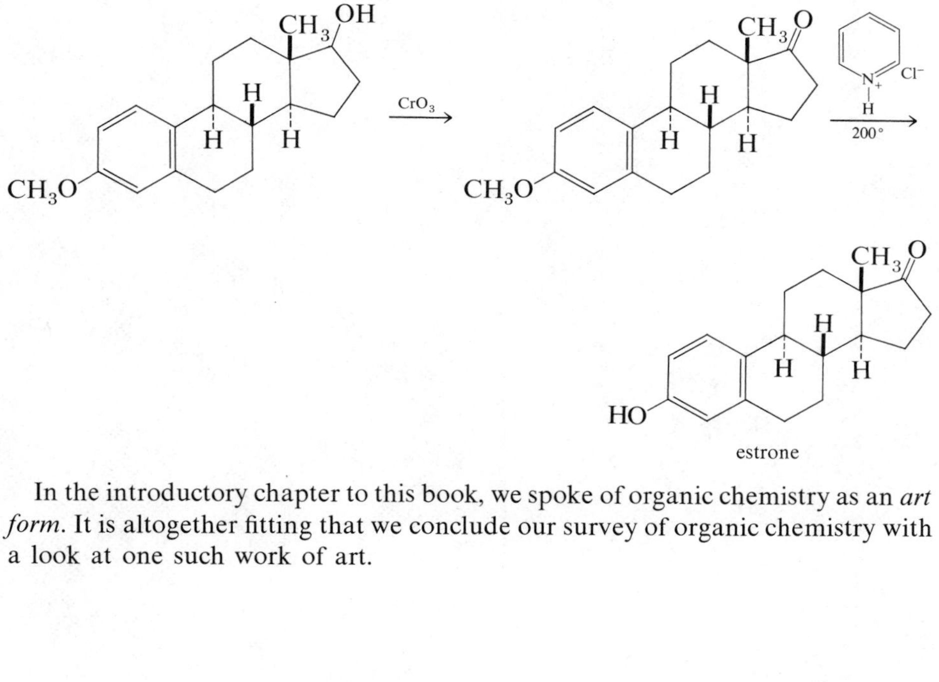

estrone

In the introductory chapter to this book, we spoke of organic chemistry as an *art form*. It is altogether fitting that we conclude our survey of organic chemistry with a look at one such work of art.

Appendix I
Heats of Formation
ΔH_f° (gas, 25°C), kcal mole^{-1}

Alkanes

methane	—17.9	2,2-dimethylpropane	—40.3
ethane	—20.2	hexane	—39.9
propane	—24.8	2-methylpentane	—41.8
butane	—30.4	3-methylpentane	—41.1
2-methylpropane	—32.4	2,2-dimethylbutane	—44.5
pentane	—35.1	2,3-dimethylbutane	—42.6
2-methylbutane	—36.9		

Cycloalkanes

cyclopropane	12.7	methylcyclopentane	—25.3
cyclobutane	6.8	methylcyclohexane	—37.0
cyclopentane	—18.4	ethylcyclohexane	—41.0
cyclohexane	—29.5	1,1-dimethylcyclohexane	—43.2
cycloheptane	—28.2	cis-1,2-dimethylcyclohexane	—41.1
cyclooctane	—29.7	trans-1,2-dimethylcyclohexane	—43.0
cyclononane	—31.7	cis-1,3-dimethylcyclohexane	—44.1
cyclodecane	—36.9	trans-1,3-dimethylcyclohexane	—42.2
		cis-1,4-dimethylcyclohexane	—42.2
		trans-1,4-dimethylcyclohexane	—44.1

Alkenes

ethylene	12.5	2-methyl-1-butene	— 8.6
propene	4.9	2-methyl-2-butene	—10.1
1-butene	— 0.2	cyclobutene	37.5
cis-2-butene	— 1.9	cyclopentene	8.2
trans-2-butene	— 3.0	cyclohexene	— 1.1
2-methylpropene	— 4.3	1-methylcyclohexene	—10.3
1-pentene	— 5.3	cycloheptene	— 2.2
cis-2-pentene	— 7.0	cyclooctene	— 6.5
trans-2-pentene	— 7.9		

Alkynes and Polyenes

acetylene	54.3	cis-1,3-pentadiene	19.1
propyne	44.4	trans-1,3-pentadiene	18.1
1-butyne	39.5	1,4-pentadiene	25.3
2-butyne	34.7	2-methyl-1,3-butadiene	18.1
allene	45.6	cyclopentadiene	31.9
1,2-butadiene	38.8	1,3-cyclohexadiene	25.4
1,3-butadiene	26.1	1,3,5,7-cyclooctatetraene	71.1
1,2-pentadiene	33.6		

Aromatic Hydrocarbons

benzene	19.8	styrene	35.3
toluene	12.0	naphthalene	36.1
o-xylene	4.6	1,2,3,4-tetrahydronaphthalene	7.3
m-xylene	4.1	anthracene	55.2
p-xylene	4.3	9,10-dihydroanthracene	38.2
ethylbenzene	7.1	phenanthrene	49.5

Alcohols

methanol	—48.1	*t*-butyl alcohol	—74.7
ethanol	—56.2	cyclopentanol	—58.0
allyl alcohol	—29.6	cyclohexanol	—68.4
1-propanol	—61.2	benzyl alcohol	—24.0
2-propanol	—65.1	ethylene glycol	—93.9

Ethers

dimethyl ether	—44.0	1,1-dimethoxyethane	— 93.3
ethylene oxide	—12.6	2,2-dimethoxypropane	—101.9
tetrahydrofuran	—44.0	anisole	— 17.3
diethyl ether	—60.3		

Aldehydes and Ketones

formaldehyde	—26.0	butanal	—49.0
acetaldehyde	—39.7	cyclopentanone	—46.0
propionaldehyde	—45.5	cyclohexanone	—54.0
acetone	—51.9	benzaldehyde	— 8.8
2-butenal	—24.0		

Other Oxygen Compounds

formic acid	— 90.6	benzoic acid	— 70.1
acetic acid	—103.3	acetic anhydride	—137.1
vinyl acetate	— 75.5	furan	— 8.3
ethyl acetate	— 106.3	phenol	— 23.0

Nitrogen Compounds

methylamine	— 5.5	pyridine	34.6
dimethylamine	— 4.7	piperidine	—11.8
trimethylamine	— 5.7	aniline	20.8
ethylamine	—11.4	benzonitrile	51.5
acrylonitrile	44.1	dimethylformamide	—45.8
acetonitrile	21.0	acetanilide	—30.8
propionitrile	12.1	methyl nitrite	—15.8
pyrrole	25.9	nitromethane	—17.9
pyrrolidine	— 0.8	glycine	—93.7

Halogen Compounds

methyl chloride	—20.6	bromobenzene	25.2
methylene chloride	—23.0	chlorobenzene	12.2
chloroform	—24.6	acetyl chloride	—58.4
carbon tetrachloride	—25.2	methyl fluoride	— 56.8
vinyl chloride	8.6	methyl bromide	— 9.1
ethyl chloride	—26.1	methyl iodide	3.4
n-propyl chloride	—31.0	ethyl bromide	—15.2
isopropyl chloride	—33.6	benzyl chloride	4.5

Sulfur Compounds

methanethiol	—5.4	thiirane	19.7
ethanethiol	—11.0	dimethyl sulfoxide	—36.1
dimethyl sulfide	—8.9	dimethyl sulfone	—89.1
dimethyl disulfide	—5.6	thiophene	27.6
thiophenol	26.9	tetrahydrothiophene	—8.1

Inorganic Compounds

CO_2	—94.05	NH_3	—10.9
H_2O	—57.80	CO	—26.42
H_2S	—4.8	H_2NNH_2	22.7
SO_2	—71.0	O_3	34.0
HCl	—22.1	NO_2	7.9
Br_2	7.4	HF	—65.0
HBr	— 8.7	HNO_3	—32.1
I_2	14.9	HNO_2	—18.4
HI	6.3	H_2O_2	—32.53
H_2O_2	—32.5	NO	21.6
		HCN	31.2

Atoms and Radicals

H	52.1	$CH_3\cdot$	34
Li	38.4	$C_2H_5\cdot$	26
C	170.9	$(CH_3)_2CH\cdot$	18
N	113.0	$(CH_3)_3C\cdot$	7
O	59.6	$CH_2{=}CH\cdot$	68
F	18.9	$CH_2{=}CHCH_2\cdot$	40
Cl	28.9	$C_6H_5CH_2\cdot$	47
Br	26.7	$C_6H_5\cdot$	80
I	25.5	$CH_3CO\cdot$	—5
S	65.7	$CH_3CO_2\cdot$	—50
Na	25.8	$CH_3O\cdot$	4
		$C_2H_5O\cdot$	—5
		$HCO\cdot$	9
		$HOOC\cdot$	53
		$CH_3SO_2\cdot$	—61
		$CH_3COCH_2\cdot$	—6

Appendix II
Bond-Dissociation Energies

$DH°$, kcal mole^{-1} for A—B Bonds

A	(52.1) H	(18.9) F	(29.0) Cl	(26.7) Br	(25.5) I	(9.3) OH	(40.1) NH$_2$	(34) CH$_3$	(26) C$_2$H$_5$	(18) i-C$_3$H$_7$	(7) t-C$_4$H$_9$	(78) C$_6$H$_5$	(99) CN
(34) CH$_3$	104	109	84	70	56	91	80	88	85	84	81	100	112
(26) C$_2$H$_5$	98	107	81	68	53	91	77	85	82	80	77	97	113
(21) n-C$_3$H$_7$	98	107	81	68	53	91	78	85	82	80	77	97	112
(18) i-C$_3$H$_7$	95	106	80	68	53	92	78	84	80	78	73	95	111
(7) t-C$_4$H$_9$	91	108	79	65	50	91	76	81	77	73	68	90	
(78) C$_6$H$_5$	110	125	95	80	64	110	97	100	97	95	90	113	125
(47) C$_6$H$_5$CH$_2$	87		72	57	47	80		74	71	70		88	
(40) allyl	87		69	55	42	79		74	71	70	66		
(−5) CH$_3$CO	87	120	82	67	50	108		81	77	75	71	93	
(−5) C$_2$H$_5$O	103					43		81	81			99	
(68) CH$_2$=CH	108		88	76				97	94	93	86	111	123
(52.1) H	104.2	135.8	103.2	87.5	71.3	119.2	103.2	104	98	95	91	110	120

Numbers in parentheses are the heats of formation, $\Delta H_f°$, for the corresponding atom or radical.

Appendix III
Average Bond Energies

Average Bond Energies, kcal mole^{-1}

H	C	N	O	F	Si	S	Cl	Br	I	
104	99	93	111	135	76	83	103	87	71	H
	83[a]	73[b]	86[c]	116[d]	72	65	81	68	52	C
		39	53[e]	65			46			N
			47	45	108		52	48	56	O
				37	135					F
					53		91	74	56	Si
						60	61	52		S
							58			Cl
								46		Br
									36	I

[a] C=C 146, C≡C 200.
[b] C=N 147, C≡N 213.
[c] C=O 176 (aldehydes), 179 (ketones).
[d] In CF_4.
[e] In nitrites and nitrates.

Appendix IV
Table of Acid Dissociation Constants

Acidities of Inorganic Acids at 25°

Name	Formula	pK_a
ammonium ion	$NH_4{}^+$	9.24
boric acid	H_3BO_3	9.24
carbon dioxide	CO_2	6.35[a]
cyanic acid	HOCN	3.46
hydrazinium ion	$H_2N\overset{+}{N}H_3$	7.94
hydrazoic acid	HN_3	4.68
hydriodic acid	HI	-5.2
hydrobromic acid	HBr	-4.7
hydrochloric acid	HCl	-2.2
hydrocyanic acid	HCN	9.22
hydrofluoric acid	HF	3.18
hydrogen peroxide	H_2O_2	11.65
hydrogen selenide	H_2Se	3.89 (11.0)[b]
hydrogen sulfide	H_2S	6.97 (12.9)[b]
hydroxylammonium ion	$H_3\overset{+}{N}OH$	5.95
hypobromous acid	HOBr	8.6
hypochlorous acid	HOCl	7.53
hypophosphorus acid	H_3PO_2	1.2
nitric acid	$HONO_2$	-1.3
nitrous acid	HONO	3.23
periodic acid	H_3IO_5	1.55 (8.27)[b]
phosphoric acid	$(HO)_3PO$	2.15 (7.20, 12.38)[b]
sulfuric acid	$(HO)_2SO_2$	~ -5.2 (1.99)[b]
sulfurous acid	$(HO)_2SO$	1.8 (7.2)[b]
thiocyanic acid	HCNS	-1.9

[a] For the equilibrium $CO_2(aq) = H^+(aq) + HCO_3{}^-(aq)$.
[b] Second and third acidity constants in parentheses.

Acidities of Organic Acids at 25°

Acid	pK_a	Acid	pK_a
$CH_3\overset{\underset{\displaystyle \|}{O}}{\overset{+}{N}}OH$	-11.9	$(CH_3)_2C{=}\overset{H}{\underset{+}{N}}OH$	-1.9
$C_6H_5\overset{\underset{\displaystyle \|}{O}}{\overset{+}{N}}OH$	-11.3	CH_3SO_3H	~ -1.2
$C_6H_5\overset{+}{C}{\equiv}NH$	-10.5	$CH_3\overset{\overset{\displaystyle OH}{\|}}{C}{=}\overset{+}{N}H_2$	~ 0
$CH_3\overset{+}{C}{\equiv}NH$	-10.1	$CH_3(CH_2)_3\overset{+}{P}H_3$	0
$CH_3\overset{\overset{\displaystyle H}{\|}}{C}{=}\overset{+}{O}H$	~ -8	$(CH_3)_2\overset{+}{S}OH$	0
$C_6H_5\overset{\overset{\displaystyle OH}{\|}}{C}{=}\overset{+}{O}H$	-7.3	CF_3COOH	0.2
$(CH_3)_2C{=}\overset{+}{O}H$	-7.2	picric acid (2,4,6-trinitrophenol)	0.25
$C_6H_5\overset{\overset{\displaystyle H}{\|}}{C}{=}\overset{+}{O}H$	-7.1	N-hydroxypyridinium	0.79
$CH_3\overset{+}{S}H_2$	-6.8	$(C_6H_5)_2\overset{+}{N}H_2$	0.8
$C_6H_5\overset{+}{O}H_2$	-6.7	$O_2N{-}C_6H_4{-}\overset{+}{N}H_3$	1.00
$C_6H_5\overset{\overset{\displaystyle H}{+}}{O}CH_3$	~ -6.5	$C_6H_5\overset{+}{N}H{=}\overset{\overset{\displaystyle OH}{\|}}{C}C_6H_5$	2.17
$CH_3\overset{\overset{\displaystyle OC_2H_5}{\|}}{C}{=}\overset{+}{O}H$	-6.5	$CH_3\overset{+}{P}H_3$	~ 2.5
$C_6H_5N{=}\overset{\overset{\displaystyle OH}{\|}}{\underset{+}{N}}C_6H_5$	-6.45	$O_2N{-}C_6H_4{-}COOH$	3.42
$C_6H_5\overset{\overset{\displaystyle +OH}{\|}}{C}CH_3$	-6.2	$CH_2(NO_2)_2$	3.57
$CH_3\overset{\overset{\displaystyle +OH}{\|}}{C}OH$	-6.1	$(CH_3)_2\overset{+}{P}H_2$	3.91
$(CH_3)_2\overset{+}{S}H$	-5.4	2,4-dinitrophenol	4.09
$(CH_3)_3C\overset{+}{O}H_2$	-3.8	$C_6H_5\overset{+}{N}H_3$	4.60
$(CH_3CH_2)_2\overset{+}{O}H$	-3.6	$(CH_3)_3\overset{+}{N}OH$	4.7
$(CH_3)_2CH\overset{+}{O}H_2$	-3.2	CH_3COOH	4.74
$C_6H_5N{=}\overset{\overset{\displaystyle H}{\|}}{\underset{+}{N}}C_6H_5$	-2.9		
$C_2H_5\overset{+}{O}H_2$	-2.4		
$CH_3\overset{+}{O}H_2$	-2.2		
$C_6H_5\overset{\overset{\displaystyle OH}{\|}}{C}{=}\overset{+}{N}H_2$	-2.0		

Acidities of Organic Acids at 25° (continued)

Acid	pK_a	Acid	pK_a
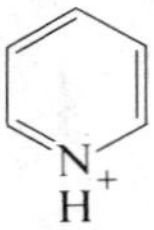 (pyridinium)	5.29	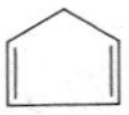 (cyclopentadiene)	16.0
$(CH_3CO)_3CH$	5.85	$C_6H_5COCH_3$	16
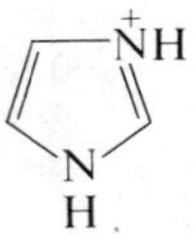 (imidazolium)	7.0	$(CH_3)_3COH$	18
		CH_3COCH_3	20
O_2N—⟨ ⟩—OH	7.15	(indene)	20
C_6H_5SH	7.8	(fluorene)	23
$(CH_3)_3\overset{+}{P}H$	8.65	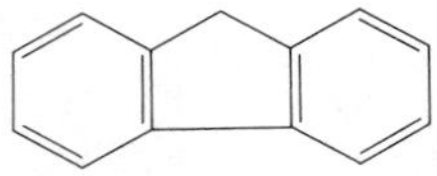 $CH_3SO_2CH_3$	23
$(CH_3CO)_2CH_2$	9	$CH_3COOC_2H_5$	24.5
$(CH_3)_3\overset{+}{N}H$	9.79	$HC{\equiv}CH$	~25
C_6H_5OH	10.00	CH_3CN	~25
CH_3NO_2	10.21	$(C_6H_5)_3CH$	31.5
CH_3CH_2SH	10.60	$(C_6H_5)_2CH_2$	34
$CH_3\overset{+}{N}H_3$	10.62	$C_2H_5NH_2$	~35
$(CH_3)_2\overset{+}{N}H_2$	10.73	$C_6H_5CH_3$	41
$CH_3COCH_2COOC_2H_5$	11	⟨ ⟩—H	43
$CH_2(CN)_2$	11.2	$CH_2{=}CH_2$	44
CF_3CH_2OH	12.4	△	46
$CH_2(COOC_2H_5)_2$	13.3	CH_4	~49
$(CH_3SO_2)_2CH_2$	14	C_2H_6	~50
CH_3OH	15.5	⬡ (cyclohexane)	~52
C_2H_5OH	15.9	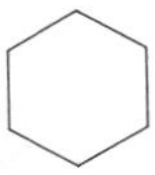	

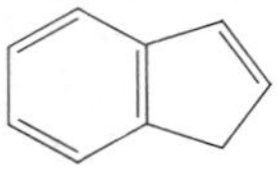

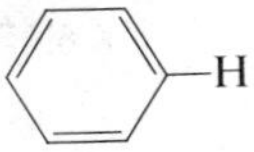

Appendix V
Proton Chemical Shifts

Proton Chemical Shifts, δ, ppm, for C—H

Y	CH_3Y	CH_3—C—Y	CH_3—C—C—Y	R—CH_2—Y	RCH_2—C—Y	R_2CH—Y
H	0.23	0.9	0.9	0.9	1.3	1.3
CH=CH$_2$	1.71	1.0		2.0		1.7
C≡CH	1.80	1.2	1.0	2.1	1.5	2.6
C$_6$H$_5$	2.35	1.3	1.0	2.6	1.7	2.9
F	4.27	1.2		4.4		
Cl	3.06	1.5	1.1	3.5	1.8	4.1
Br	2.69	1.7	1.1	3.4	1.9	4.2
I	2.16	1.9	1.0	3.2	1.9	4.2
OH	3.39	1.2	0.9	3.5	1.5	3.9
OR	3.24	1.2	1.1	3.3	1.6	3.6
OAc	3.67	1.3	1.1	4.0	1.6	4.9
CHO	2.18	1.1	1.0	2.4	1.7	2.4
COCH$_3$	2.09	1.1	0.9	2.4	1.6	2.5
COOH	2.08	1.2	1.0	2.3	1.7	2.6
NH$_2$	2.47	1.1	0.9	2.7	1.4	3.1
NHCOCH$_3$	2.71	1.1	1.0	3.2	1.6	4.0
SH	2.00	1.3	1.0	2.5	1.6	3.2
CN	1.98	1.4	1.1	2.3	1.7	2.7
NO$_2$	4.29	1.6	1.0	4.3	2.0	4.4

Proton Chemical Shifts for Y—H

Group Type	δ, ppm
ROH	0.5–5.5
ArOH	4–8
RCOOH	10–13
R_2C=NOH	7.4–10.2
RSH	0.9–2.5
ArSH	3–4
RSO$_3$H	11–12
RNH$_2$, R$_2$NH	0.4–3.5
ArNH$_2$, ArRNH	2.9–4.8
RCONH$_2$	5.0–6.5
RCONHR	6.0–8.2
RCONHAr	7.8–9.4

Appendix VI
Infrared Bands

TABLE A
Characteristic Stretching Frequencies

Bond	$\tilde{\nu}$, cm^{-1}
1. C—H	
(a) C_{sp^3}—H	2800–3000
(b) C_{sp^2}—H	3000–3100
(c) C_{sp}—H	3300
2. C—C	
(c) C—C	1150–1250
(b) C=C	1600–1670
(c) C≡C	2100–2260
3. C—N	
(a) C—N	1030–1230
(b) C=N	1640–1690
(c) C≡N	2210–2260
4. C—O	
(a) C—O	1020–1275
(b) C=O	1650–1800 (see also Table C)
5. C—X	
(a) C—F	1000–1350
(b) C—Cl	800–850
(c) C—Br	500–680
(d) C—I	200–500
6. N—H	
(a) RNH_2, R_2NH	3400–3500 (two)
(b) $R\overset{+}{N}H_3$, $R_2\overset{+}{N}H_2$, $R_3\overset{+}{N}H$	2250–3000
(c) $RCONH_2$, RCONHR′	3400–3500
7. O—H	
(a) ROH	3610–3640 (free)
	3200–3400 (H-bonded)
(b) RCO_2H	2500–3000
8. N—O	
(a) RNO_2	1350, 1560
(b) $RONO_2$	1620–1640, 1270–1285
(c) RN=O	1500–1600
(d) RO—N=O	1610–1680 (two), 750–815
(e) C=N—OH	930–960
(f) $R_3\overset{+}{N}$—O$^-$	950–970

TABLE A (continued)

Bond	$\tilde{\nu}$, cm^{-1}
9. S—O	
(a) $R_2\overset{+}{S}$—O$^-$	1040–1060
(b) $R_2\overset{\text{O}}{\underset{}{S}}$=O	1310–1350, 1120–1160
(c) R—$\overset{\text{O}}{\underset{\text{O}}{S}}$—OR$'$	1330–1420, 1145–1200
10. Cumulated systems	
(a) C=C=C	1950
(b) C=C=O	2150
(c) R_2C=$\overset{+}{N}$=N$^-$	2090–3100
(d) RN=C=O	2250–2275
(e) RN=$\overset{+}{N}$=N$^-$	2120–2160

TABLE B
Useful C—H Out-of-Plane Bending Vibrations

Bond	$\tilde{\nu}$, cm^{-1}
1. Alkynes, C≡C—H	600–700
2. Alkenes	
(a) RCH=CH	910, 990
(b) R_2C=CH$_2$	890
(c) *trans*-RCH=CHR	970
(d) *cis*-RCH=CHR	725–675
(e) R_2C=CHR	790–840
3. Aromatic	
(a) mono-	730–770, 690–710 (two)
(b) *o-*	735–770
(c) *m-*	750–810, 690–710 (two)
(d) *p-*	810–840
(e) 1,2,3-	760–780, 705–745 (two)
(f) 1,3,5-	810–865, 675–730 (two)
(g) 1,2,4-	805–825, 870,885 (two)
(h) 1,2,3,4-	800–810
(i) 1,2,4,5-	855–870
(j) 1,2,3,5	840–850
(k) penta-	870

TABLE C
Summary of Carbonyl Stretching Frequencies

Compound Type	$\tilde{\nu}$, cm^{-1}
1. Aldehydes	
(a) RCHO	1725
(b) C=CCHO	1685
(c) ArCHO	1700
2. Ketones	
(a) $R_2C=O$	1715
(b) C=C—C=O	1675
(c) Ar—C=O	1690
(d) four-membered cyclic	1780
(e) five-membered cyclic	1745
(f) six-membered cyclic	1715
3. Carboxylic acids	
(a) RCOOH	1760 (monomer)
	1710 (dimer)
(b) C=C—COOH	1720 (monomer)
	1690 (dimer)
(c) RCO_2^-	1550–1610, 1400 (two)
4. Esters	
(a) RCOOR	1735
(b) C=C—COOR	1720
(c) ArCOOR	1720
(d) γ-Lactone	1770
(e) δ-Lactone	1735
5. Amides	
(a) $RCONH_2$	1690 (free)
	1650 (associated)
(b) RCONHR′	1680 (free)
	1655 (associated)
(c) $RCONR_2'$	1650
(d) β-lactam	1745
(e) γ-lactam	1700
(f) δ-lactam	1640
6. Acid anhydrides	1820, 1760 (two)
7. Acyl halides	1800

Appendix VII
Symbols and Abbreviations

Å	Ångstrom unit (10^{-8} cm)
Ac	acetyl group, CH_3CO—
Ar	aryl radical
$[\alpha]$	specific optical activity
aq.	aqueous
Boc	t-butoxycarbonyl group, $(CH_3)_3COCO$—
n-Bu	n-butyl group, $CH_3CH_2CH_2CH_2$—
t-Bu	t-butyl group, $(CH_3)_3C$—
cmr	^{13}C magnetic resonance
Cbz	benzyloxycarbonyl group, $C_6H_5CH_2OCO$—
D	Debye (10^{-18} esu cm); measure of dipole moment
DCC	dicyclohexylcarbodiimide, $C_6H_{11}N{=}C{=}NC_6H_{11}$
δ	chemical shift downfield from TMS, given as ppm
Δ	symbol for heat supplied to a reaction
$\Delta G°$	standard Gibbs free energy of reaction
$\Delta G^{\ddagger}$	Gibbs free energy of activation
$\Delta H°$	standard enthalpy of reaction
$\Delta H_f°$	enthalpy of formation from standard states
$\Delta H^{\ddagger}$	enthalpy of activation
$\Delta S°$	standard entropy of reaction
$\Delta S^{\ddagger}$	entropy of activation
$DH°$	bond dissociation energy
DIBAL	diisobutylaluminum hydride, $[(CH_3)_2CHCH_2]_2AlH$
diglyme	di-2-methoxyethyl ether, $(CH_3OCH_2CH_2)_2O$
DMF	dimethylformamide, $(CH_3)_2NCHO$
DMSO	dimethyl sulfoxide, $(CH_3)_2SO$
DNP	2,4-dinitrophenyl group, $2,4\text{-}(O_2N)_2C_6H_3$— or $2,4\text{-dinitrophenylhydrazone}, 2,4\text{-}(O_2N)_2C_6H_3NHN{=}$
E	*entgegen*, opposite sides in (E,Z) nomenclature of alkenes
E1	unimolecular elimination reaction mechanism
E2	bimolecular elimination reaction mechanism
EA	electron affinity
Et	ethyl group, CH_3CH_2—
eu	entropy units, cal deg^{-1} mole^{-1}
f_i	partial rate factor at position i
glyme	1,2-dimethoxyethane, $CH_3OCH_2CH_2OCH_3$
H	magnetic field
HMPT	hexamethylphosphoric triamide, $[(CH_3)_2N]_3PO$
$h\nu$	symbol for light
Hz	Hertz (sec^{-1} or cycles per second)
IP	ionization potential
ir	infrared
J	coupling constant, usually in Hz
k	rate constant for reaction
K	equilibrium constant for reaction
K_a	acid dissociation constant
LDA	lithium diisopropylamide, $LiN[CH(CH_3)_2]_2$
Me	methyl group, CH_3—
m/e	mass-to-charge ratio in mass spectrometry

MHz	megaHertz $\equiv 10^6$ Hz
μ	dipole moment
nmr	nuclear magnetic resonance
NR	no reaction; also indicated by $-/\!/\!\rightarrow$
Ph	phenyl radical, C_6H_5-
pH	measure of acidity $\equiv -\log [H^+]$
pK_a	measure of acid strength $\equiv -\log K_a$
pmr	proton magnetic resonance
PPA	polyphosphoric acid
ψ	wave function or orbital
R	alkyl or cycloalkyl group
(R,S)	designation of stereochemical configuration
S_N1	unimolecular nucleophilic substitution mechanism
S_N2	bimolecular nucleophilic substitution mechanism
τ	chemical shift $\equiv 10 - \delta$ ppm
THF	tetrahydrofuran, $\overline{CH_2CH_2CH_2CH_2O}$
TMS	tetramethylsilane, $(CH_3)_4Si$
Ts	tosyl or p-toluenesulfonyl group, $p\text{-}CH_3C_6H_4SO_2-$
uv	ultraviolet
X	halogen group
xs	excess
Z	*zusammen*, same side in (E,Z) nomenclature of alkenes
$\curvearrowright$	symbol for flow of electron pair

Index

A

Abietic acid, 1167
ABS (plastic polymer), 610
Absolute configuration, 127
Absolute stereostructure, 127
Acenes, 1053
Acetaldehyde
  acidity, 558
  cmr chemical shift, 365 t
  nmr spectrum, 364
  physical properties, 361 t
  pK_a, 559
  polymerization, 387
  preparation, 348
  stereoscopic figure, 357
Acetals, 382–87
  from diols, 385–86, 850
  glycosides, 899
  from monosaccharides, 905–907
  as protecting groups, 386, 904
  pyrolysis, 386
  thio-, 813, 832
Acetamide
  hydrogen bonding, 532
  physical properties, 532 t
p-Acetamidobenzenesulfonyl chloride, 820
Acetamidomalonic ester, 944–45
Acetanilide
  biological oxidation, 1027
  o-methoxy-, nitration, 716
  nitration, 762–63
  properties, 754
Acetate ion, 160
  electronic structure, 503
  resonance stabilization, 502–503
Acetic acid
  in biosynthesis, 1165
  glacial, 521
  ionization, 501–502
  natural occurrence, 521
  physical properties, 501 t
  pK_a, 62 t, 245, 502
  substituted, acidity, 503–504, 1127 t
Acetic anydydride, 526
  in cyclic anhydride formation, 867
  in dehydration of amides, 563
  industrial synthesis, 612
  in Perkin reaction, 609–10
Acetic formic anhydride, 551
Acetoacetic ester
  condensation (*see also* Claisen condensation), 560
  in Michael addition, 883
  pK_a, 561, 875
  synthesis, 560, 877–80
Acetoacetyl coenzyme A, 75
Acetone
  acidity, 370–72, 558
  basicity, 358
  cmr chemical shifts, 365 t
  condensation, 598–99
  from cumene oxidation, 997–98
  dipole moment, 358

Acetone [*cont.*]
  electronic transitions, 628
  physical properties, 362 t
  pK_a, 559
  pyrolysis, 611
  reaction with galactose, 905
  resonance structures, 379
  solvent properties, 164
  ultraviolet spectroscopy, 628
Acetone anion, 371
Acetonitrile
  pK_a, 559
  reaction with organometallic compounds, 553
  structure, 528
  use as solvent, 164, 532
Acetonitrile oxide, 14
Acetophenone, 405
  nitration, 718
  preparation, 698
Acetophenones, as triplet sensitizer, 1159
o-Acetoxybenzoic acid (aspirin), 1001
Acetylacetone, *see* 2,4-Pentanedione
4-Acetylbiphenyl, 1034
Acetyl chloride, 519
  in cyclic anhydride formation, 867
  hydrolysis, 539, 544
  pK_a, 559
  structure, 528
Acetylcholine, 741
Acetyl coenzyme A (acetyl CoA), 521, 562, 1175, 1182
Acetylene
  handling and storage, 340–41
  hydration, 348
  industrial source, 340
  ionization potential, 489 t
  Lewis structure, 8
  physical properties, 336–37, 340
  pK_a, 538
  preparation, 340
  in preparation of chloroprene, 594
  reactions, 344–51
  structure, 334–35
  in synthesis of acrylonitrile, 610
Acetylenedicarboxylic acid, diethyl ester, 615
Acetylenes (*see also* Alkynes)
  cyclic polymers, 351
  cycloaddition to diazomethane, 1087
  cycloaddition to nitrile oxides, 1086
  ultraviolet spectroscopy, 632
2-Acetylfuran, 1076
Acetylide ions, 339, 342–43
  addition to aldehydes and ketones, 390–91
  heavy metal salts, 340
N-Acetyl-DL-leucine, resolution, 948–49
2-Acetylnaphthalene, 1043–44
Acetyl nitrate, 1075
2-Acetylpyrrole, 1076
Acetylsalicylic acid (aspirin), 1001
2-Acetylthiophene, 1076
Achirality, 122
Acid, conjugate, 62–63
Acid anhydrides, 526
  cyclic, 867

1205

is why I'm here.

~~If and that clicks never~~

If this doesn't make
sense, don't worry.
because no one
understands me.

Mary.

Tom—
give Julie Dubelak
a call if you
are not too busy
to confirm the
fact that I
was here but
you weren't

Mary Deb

P.S. she bet me
that I wouldn't
"go to the lab" that

O

Bond-Dissociation Energies $DH°$ for A—B bond, kcal mole^{-1}

A	B:	(52.1) H	(29.0) Cl	(26.7) Br	(25.5) I	(9.3) OH	(40.1) NH_2	(34) CH_3	(78) C_6H_5	(99) CN
(34) CH_3—		104	84	70	56	91	80	88	100	112
(26) C_2H_5—		98	81	68	53	91	77	85	97	113
(18) $i\text{-}C_3H_7$—		95	80	68	53	92	78	84	95	111
(7) $t\text{-}C_4H_9$—		91	79	65	50	91	76	81	90	
(78) C_6H_5—		110	95	80	64	110	97	100	113	125
(47) $C_6H_5CH_2$—		87	72	57	47	80		74	88	
(40) $CH_2{=}CHCH_2$—		87	69	55	42	79		74		
(—5) CH_3CO—		87	82	67	50	108		81	93	
(—5) C_2H_5O—		103				43		81	99	
(68) $CH_2{=}CH$—		108	88	76				97	111	123
(52.1) H—		104.2	103.2	87.5	71.3	119.2	103.2	104	110	120

Numbers in parentheses: $(\Delta H_f°)$ for A· or B·.